# PYRIDINE AND ITS DERIVATIVES

## SUPPLEMENT
## PART THREE

*Edited by*

**R. A. Abramovitch**

*University of Alabama*

AN INTERSCIENCE® PUBLICATION

JOHN WILEY & SONS

NEW YORK • LONDON • SYDNEY • TORONTO

An Interscience ® Publication



***Library of Congress Cataloging in Publication Data:***

Abramovitch, R. A. 1930–
Pyridine supplement.

(The Chemistry of heterocyclic compounds, v. 14)
"An Interscience publication."
Supplement to E. Klingsberg's Pyridine and its derivatives.
Includes bibliographical references.
1. Pyridine. I. Klingsberg, Erwin, ed. Pyridine and its derivatives. II. Title.

QD401.A22 547'.593 73–9800

ISBN 0–471–37915–8

Printed in the United States of America

10 9 8 7 6 5 4 3 2 1

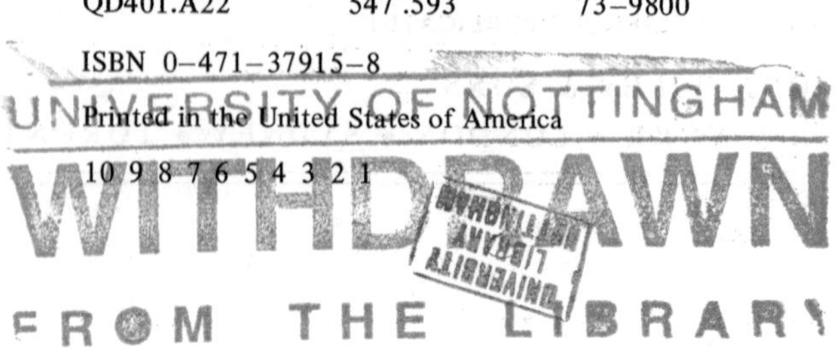

# Contributors

C. S. GIAM, *Chemistry Department, Texas A&M University, College Station, Texas*

RENAT H. MIZZONI, *Ciba Pharmaceutical Company, Division of Ciba-Geigy Corporation, Summit, New Jersey*

M. E. NEUBERT, *Department of Chemistry, Kent State University, Kent, Ohio*

PETER I. POLLAK (deceased)

HOWARD TIECKELMANN, *State University of New York, Buffalo, New York*

MARTHA WINDHOLZ, *Merck Sharp & Dohme Research Laboratories, Rahway, New Jersey*

TO THE MEMORY OF

Michael

## The Chemistry of Heterocyclic Compounds

The chemistry of heterocyclic compounds is one of the most complex branches of organic chemistry. It is equally interesting for its theoretical implications, for the diversity of its synthetic procedures, and for the physiological and industrial significance of heterocyclic compounds.

A field of such importance and intrinsic difficulty should be made as readily accessible as possible, and the lack of a modern detailed and comprehensive presentation of heterocyclic chemistry is therefore keenly felt. It is the intention of the present series to fill this gap by expert presentations of the various branches of heterocyclic chemistry. The subdivisions have been designed to cover the field in its entirety by monographs which reflect the importance and the interrelations of the various compounds, and accommodate the specific interests of the authors.

In order to continue to make heterocyclic chemistry as readily accessible as possible new editions are planned for those areas where the respective volumes in the first edition have become obsolete by overwhelming progress. If, however, the changes are not too great so that the first editions can be brought up-to-date by supplementary volumes, supplements to the respective volumes will be published in the first edition.

Arnold Weissberger

*Research Laboratories*
*Eastman Kodak Company*
*Rochester, New York*

Edward C. Taylor

*Princeton University*
*Princeton, New Jersey*

# Preface

Four volumes covering the pyridines were originally published under the editorship of Dr. Erwin Klingsberg over a period of four years, Part I appearing in 1960 and Part IV in 1964. The large growth of research in this specialty is attested to by the fact that a supplement is needed so soon and that the four supplementary volumes are larger than the original ones. Pyridine chemistry is coming of age. The tremendous variations from the properties of benzene achieved by the replacement of an annular carbon atom by a nitrogen atom are being appreciated, understood, and utilized.

Progress has been made in all aspects of the field. New instrumental methods have been applied to the pyridine system at an accelerating pace, and the mechanisms of many of the substitution reactions of pyridine and its derivatives have been studied extensively. This has led to many new reactions being developed and, in particular, to an emphasis on the direct substitution of hydrogen in the parent ring system. Moreover, many new and important pharmaceutical and agricultural chemicals are pyridine derivatives (these are usually ecologically acceptable, whereas benzene derivatives usually are not). The modifications of the properties of heteroaromatic systems by *N*-oxide formation are being exploited extensively.

For the convenience of practitioners in this area of chemistry and of the users of these volumes, essentially the same format and the same order of the supplementary chapters are maintained as in the original. Only a few changes have been made. Chapter I is now divided into two parts, Part A on pyridine derivatives and Part B on reduced pyridine derivatives. A new chapter has been added on pharmacologically active pyridine derivatives. It had been hoped to have a chapter on complexes of pyridine and its derivatives. This chapter was never received and it was felt that Volume IV could not be held back any longer.

The decision to publish these chapters in the original order has required sacrifices on the part of the authors, for while some submitted their chapters on time, others were less prompt. I thank the authors who finished their chapters early for their forebearance and understanding. Coverage of the literature starts as of 1959, though in many cases earlier references are also given to present sufficient background and make the articles more readable. The literature is covered until 1970 and in many cases includes material up to 1972.

I express my gratitude to my co-workers for their patience during the course of this undertaking, and to my family, who saw and talked to me even less than usual during this time. In particular, I acknowledge the inspiration given me by the strength and smiling courage of my son, Michael, who will never know how much the time spent away from him cost me. I hope he understood.

R. A. ABRAMOVITCH

*University, Alabama*
*June 1973*

# Contents

## Part Three

## Part One



## Part Two

## Part Four

# PYRIDINE AND ITS DERIVATIVES

SUPPLEMENT IN FOUR PARTS

PART THREE

*This is the fourteenth volume in the series*

THE CHEMISTRY OF HETEROCYCLIC COMPOUNDS

CHAPTER VIII

# Nitropyridines and Reduction Products (Except Amines)

RENAT H. MIZZONI

*Ciba Pharmaceutical Co.*
*Division, Ciba-Geigy Corp.*
*Summit, New Jersey*

# I. Nitropyridines

## 1. Preparation

### A. *Synthesis from Aliphatic Intermediates*

Gundermann and Alles[1] have studied the stepwise reaction of potassium 2,2-dinitroethanol with formaldehyde, dinitrogen tetroxide, and dilute acid **(VIII-1)**. They concluded that the reaction product was 2,4,6-trinitropyridine-1-oxide on the basis of spectral evidence and mode of formation. The reaction is analogous to one employing potassium nitroacetonitrile to give 2,4,6-tricyano-pyridine-1-oxide.

$$\left[HOCH_2\overset{NO_2}{\overset{|}{C}}=\underset{O}{\underset{\downarrow}{N}}-O\right]K^+ \xrightarrow[2N-H_2SO_4]{HCHO,\ N_2O_4} \text{2,4,6-trinitropyridine-1-oxide } (O_2N,\ NO_2,\ NO_2;\ N^+-O^-) \qquad \textbf{(VIII-1)}$$

### B. *By Nitration of Substituted Pyridines*

The nitration of 2-dimethylaminopyridine-1-oxide under mild conditions gives 2-dimethylamino-5-nitropyridine-1-oxide; significantly, none of the 4-nitro isomer is formed in the reaction.[2]

DeSelms[3] has reinvestigated the nitration of 2-methyl- and 2-chloro-3-pyridinol. The entering nitro group is directed to the 4- and 6- positions in a 4 to 1 ratio. Electrophilic nitration of 3-pyridinol to give 2-nitro-3-pyridinol[4] and 2,6-dinitro-3-pyridinol[3] was confirmed. This seems to be the only example of 4-nitration except for the case of the pyridine-1-oxides.

### C. *By Oxidation of Aminopyridines*

The preparation of 3-fluoro-4-nitropyridine can be effected by oxidation of the aminofluoro compound with persulfuric acid.[5] A similar reaction yields 4-nitrotetrafluoropyridine from the corresponding amino precursor.[6,7]

4-Nitrotetrafluoropyridine is a liquid whose boiling point (152 to 154°) is appreciably lower than that of 2-nitropyridine (256°), or of 3-nitropyridine (216°).

In futher examples, 2-aminopyridine and 2-amino-5-bromopyridine give 2-nitropyridine-1-oxide and 5-bromo-2-nitropyridine-1-oxide directly, in low yields, on oxidation with peroxytrifluoroacetic acid.[8]

## D. *From Nitropyridine-1-oxides*

Kroehnke and Schaefer[9] have studied the deoxygenation of 4-nitropyridine-1-oxides by various reagents. Nitrosylsulfuric acid and "nitration acid" give yields of deoxygenated products in excess of 90%; the conventional reagent ($PCl_3-CHCl_3$) is somewhat less effective and gives up to 71% of products.

Simultaneous nitration-deoxygenation has also been observed by these workers. For example, pyridine-1-oxide undergoes nitration and deoxygenation with concentrated sulfuric acid and fuming nitric acid at 130 to 165° to give 4-nitropyridine in 71% yield. As additional examples, 3-picoline-1-oxide gives 4-nitro-3-picoline (81%), and 3-bromopyridine-1-oxide affords 3-bromo-4-nitropyridine (75%) on treatment with nitric oxide and sulfuric acid at 150 to 200°.

The *N*-oxide function is retained on treatment with nitric and sulfuric acids at somewhat lower temperatures. Thus Talik and Talik[10] prepared 3-chloro-4-nitropyridine-1-oxide (84.5%) and 3-iodo-4-nitropyridine-1-oxide (56.4%) with this reagent at steam-bath temperature.

## E. *Side-Chain Nitro Compounds*

Rubinstein, Hazen, and Zerfing[11] noted the occurrence of appreciable side-chain nitration during the oxidation of 5-ethyl-2-picoline with nitric acid **(VIII-2)**. The product of this reaction gives methyl 2-methyl-5-pyridyl ketoxime on reduction with tin and hydrochloric acid.

Et, Me, N $\xrightarrow{HNO_3}$ $C(NO_2)_2$–Me, Me, N $\xrightarrow{Sn,\ HCl}$ C(=NOH)–Me, Me, N

**(VIII-2)**

# 2. Reactions of Nitropyridines

## A. *Reduction*

Yamada and Kikugawa[12] reported that 2- and 4-nitropyridines give the hydrazo- and azo- compounds, respectively, on reduction with sodium borohydride in boiling ethanol; nitrobenzene, however, does not react under these conditions. The reduction of picolinonitrile and isonicotinonitrile with this reagent further exemplifies the enhanced reactivity of 2- and 4- substitutents on the pyridine ring.

## B. *Reactivity of Nitropyridines and Halonitropyridines*

The relative reactivity of substituents in nitropyridines, halonitropyridines, and halonitropyridine-1-oxides has been studied extensively during recent years.

Johnson[13] investigated the reactivities of 2- and 4-halo- and 2- and 4-nitropyridine-1-oxides toward sodium methoxide and found that the energies of activation were lower for the nitropyridine-1-oxides than for the corresponding halo compounds.

Talik[14,15] studied the behavior of 3-chloro-4-nitropyridine-1-oxide with various reagents, and showed that sodium methoxide causes replacement of the nitro group, while amines, on the other hand, effect displacement of the halogen.

2-Halo-4-nitropyridine-1-oxides react with two equivalents of sodium methoxide at room temperature to effect replacement of both halogen and nitro groups. One equivalent of sodium methoxide at that temperature, however, causes replacement of the nitro group alone to give 2-chloro-4-methoxypyridine-1-oxide in 84% yield.[16,17] The use of two equivalents of the base in boiling methanol gives 2,4-dimethoxypyridine-1-oxide.

Boiling aqueous potassium hydroxide converts 2-chloro-4-nitropyridine into 2-chloro-4-pyridone.[18] In general, the reactivity pattern of 2-halo-4-nitropyridines parallels that of the corresponding 1-oxides.[17,19]

3-Fluoro-4-nitropyridine-1-oxide undergoes facile displacement of the halogen under mild conditions. Alkoxides in general lead to replacement of fluorine at room temperature, and of both substituents at higher temperatures.[20,22]

Abramovitch and his co-workers[21] have studied the reaction kinetics of variously substituted halopyridines with methoxide ion in methanol. As part of this study, energies of activation were determined for 2-chloro-3-nitro- and 2-chloro-5-nitropyridines; they were found to be 18.7 and 18.1 kcal/mole, respectively.

4-Nitro-3-chloropyridine-1-oxide is reduced with hydrazine to 4-amino-3-chloropyridine-1-oxide.[17] 2-Chloro-4-nitropyridine gives 1,2-bis-(2-chloro-4-pyridyl)hydrazine on treatment with hydrosulfide.[18]

### C. *Reactions of Nitroaminopyridines*

2-, 3-, And 4-nitroaminopyridines react with halogens and red phosphorus in boiling chloroform or carbon tetrachloride to give chloro-, bromo- and iodopyridines.[23]

## II. Nitrosopyridines and Hydroxylaminopyridines

4-Nitropyridine-1-oxide undergoes reduction with phenylhydrazine to give 4-hydroxylaminopyridine-1-oxide in nearly quantitative yield.[24] This product is very reactive; it undergoes oxidation in aqueous ammonia to form 4,4′-azopyridine-1,1′-dioxide, and with potassium permanganate in acid solution to give 4-nitrosopyridine-1-oxide.

Photolysis of 4-nitropyridine in ethanol yields 4-hydroxylaminopyridine.[25]

Yates and his co-workers[26] have studied the reactions of 2,6-dialkyl-4-pyrones with hydroxylamine. Thus, 2,6-dimethyl-4-pyrone and 2,6-diethyl-4-pyrone give the corresponding 2,6-dialkyl-4-hydroxylaminopyridine-1-oxides in 17 to 20% yields **(VIII-3)**.

$NH_2OH \cdot HCl$
$C_5H_5N$

NHOH

*hν*

$H_2O$, air
pH 7–9

$H_2O$, air
pH 12–13

**(VIII-3)**

The hydroxylaminopyridine-1-oxides are reactive compounds that are oxidized by air in strongly alkaline solutions to azopyridines, and that undergo photochemical conversion to azoxy derivatives. Mixtures of azo and azoxy compounds are produced by atmospheric oxidation under less alkaline conditions **(VIII-3)**.

## III. Azopyridines and Azoxypyridines

Brown and his collaborators[27] found that 5-amino-2-dimethylaminopyridine does not react with nitrosobenzene to give the expected 2-dimethylamino-5-phenylazopyridine. Instead, the desired compound is obtained by reaction of 2-chloro-5-phenylazopyridine with dimethylamine. In contrast to the biological action of 3-(*p*-dimethylaminophenyl)azopyridine, this substance is not carcinogenic.

Elslager and his co-workers[28] have prepared a variety of pyridylazo compounds for testing as chemotherapeutic agents.

Czuba[29] has investigated the behavior of a large number of substituted 3-nitraminopyridines on treatment with sulfuric acid. The products of the reaction are substituted 3-azopyridines, 3-azoxypyridines, and 3-pyridinol.

The oxidation of 2-(*p*-nitrophenylazo)pyridine with perbenzoic acid gives a mixture of the 1-oxide and the $\alpha$-azoxy-1-oxide.[4]

Gladstone and Norman[30] have subjected benzoylpyridine phenylhydrazones to lead tetraacetate oxidation. The intermediate side-chain azo compounds thus formed undergo conversion to 3-pyridylindazoles with Lewis acids **(VIII-4)**.

$$\text{PhC(Py)=NNHPh} \xrightarrow{\text{Pb(OAc)}_4} \text{PhC(Py)(OAc)-N=NPh} \xrightarrow{\text{Lewis acids}} \text{1-phenyl-3-pyridylindazole}$$

**(VIII-4)**

2-Substituted-5-aminopyridines react with nitrosobenzene under basic conditions to afford 2-substituted-5-phenylazopyridines in yields of 53 to 84%.[31]

In a significant reaction, pyridine couples with phenyldiazonium salts in the presence of sodium bisulfite to give 3-phenylazopyridine. The pyridine-sodium bisulfite adduct is thought to be the reactive heterocyclic moiety[32] **(VIII-6)**.

$$\text{pyridine} \xrightarrow{3\text{NaHSO}_3} \text{2,4,6-tris(NaO}_3\text{S)piperidine (2-H, 4-H, 6-H)} \xrightarrow[\text{NaHCO}_3]{\text{PhN}_2^+}$$

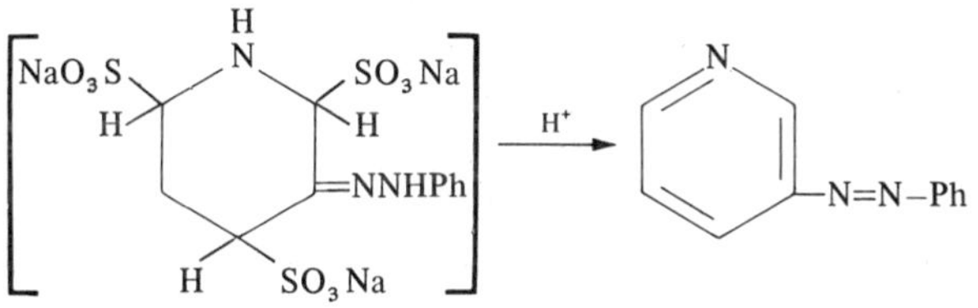

(VIII-6)

## IV. Hydrazinopyridines

A number of reactive fluoropyridines have been used to synthesize hydrazinopyridines. Thus 3-fluoropyridine-1-oxide reacts readily with hydrazine to give 3-hydrazinopyridine-1-oxide.[33] Similarly, pentafluoropyridine gives 4-hydrazinotetrafluoropyridine.[34] In like manner, 3,5-difluoro-4-hydrazino-pyridine is readily prepared.[35]

Pyridine and some of its homologs have been subjected to direct hydrazination with substituted hydrazines. Reaction occurs almost exclusively at the 2-position, although in one case a trihydrazino compound forms as a by-product **(VIII-5)**.[36]

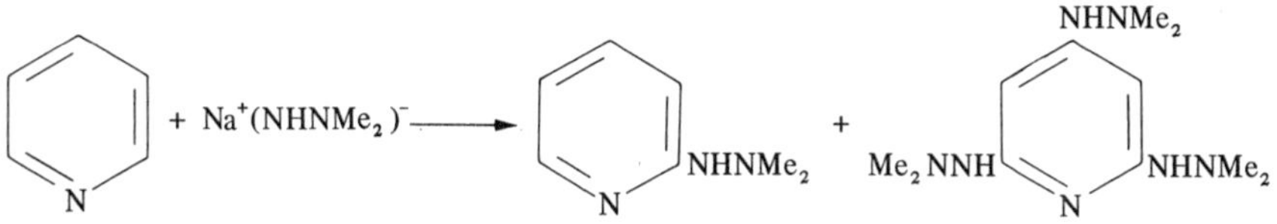

**(VIII-5)**

The reaction of 2-chloropyridine with monosubstituted hydrazines in the presence of sodium hydride gives 1,1-disubstituted hydrazines.[37]

## V. Pyridyl Azides

A number of pyridyl azides have been prepared by conventional methods. The reaction of 4-hydrazino-2-picoline with nitrous acid, for example, gives 4-azido-2-picoline. 4-Azidopyridine-1-oxide is obtained in a similar manner. The reaction of 4-chloropyridine with sodium azide is less satisfactory, and gives the product in low yield.[38]

3-Pyridylazide is formed by reaction of 3-pyridyldiazonium chloride with sodium azide.[39] 2-Aminopyridine-1-oxides can be diazotized, and treatment of the salt with azide ion gives rise to the 2-azidopyridine-1-oxide in good yields.[39a]

4-Azido-2-picoline is oxidized with hydrogen peroxide to 4,4′-azoxy-2,2′-dimethylpyridine,[38] which reacts with propargyl alcohol to give the pyridyl-(hydroxymethyl)-1,2,3-triazole. 4-Azidopyridine-1-oxide yields 4,4′-azoxypyridine-1,1′-dioxide on photolysis in acetone.[40]

Huisgen and his co-workers have investigated the reaction of 2-pyridylazide with various acetylenes. Although the equilibrium is largely toward the tetrazole,[40a, 40b] the substance reacts to give 1-(2-pyridyl)-1,2,3-triazoles **(VIII-7)**.[40a] (See Ch. IA for some reactions of 2-azidopyridine-1-oxides.)

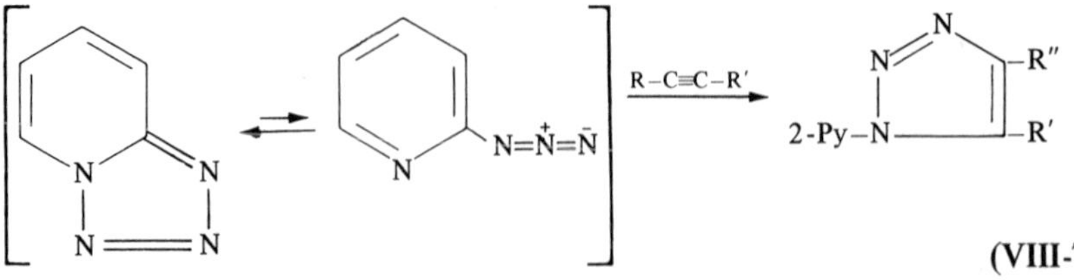


**(VIII-7)**

R′ = H , R″ = $CO_2Me$
R′, R″ = $CO_2Me$
R′, R″ = Ph

TABLE VIII-1. Nitropyridines

| Compound | Method of preparation | Yield (%) | Properties | Ref. |
|---|---|---|---|---|
| 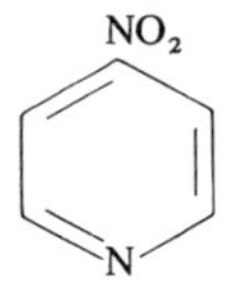 $NO_2$, N | 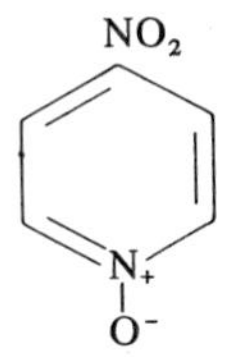 $NO_2$, $N^+$–$O^-$ + (a) Nitrosylsulfuric acid<br>(b) NO, conc. $H_2SO_4$<br>(c) $PCl_3$–$CHCl_3$ | 93<br>91.5<br>71 | | 9 |
| 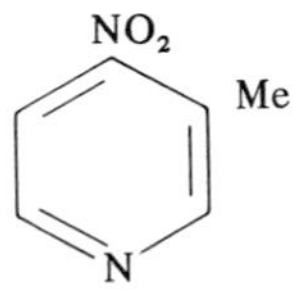 $NO_2$, Me, N | 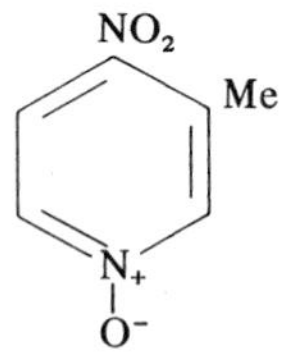 $NO_2$, Me, $N^+$–$O^-$ , NO, $H_2SO_4$ | 81 | m.p. 28° | 9 |
| 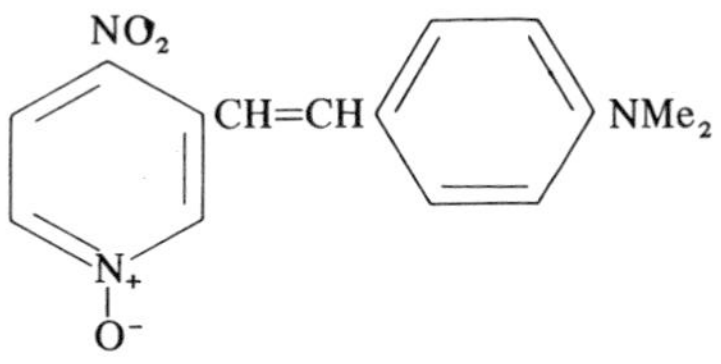 $NO_2$, CH=CH, $NMe_2$, $N^+$–$O^-$ | 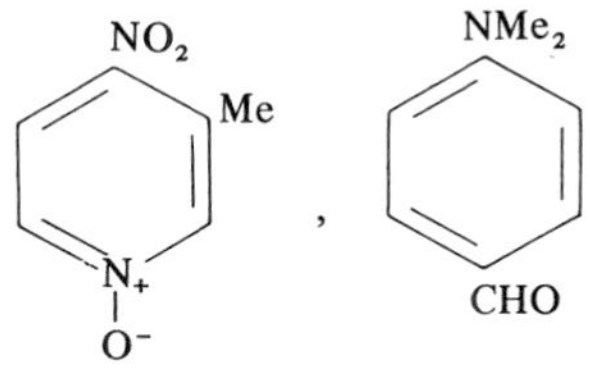 $NO_2$, Me, $N^+$–$O^-$ , $NMe_2$, CHO , (piperidine) | | m.p. 208° | 41 |
| 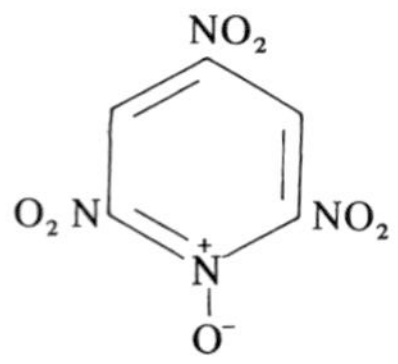 $NO_2$, $O_2N$, $NO_2$, $N^+$–$O^-$ | 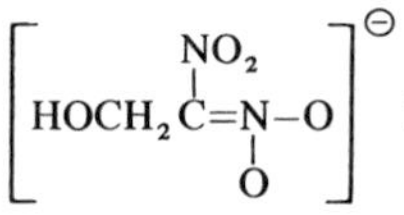 $[HOCH_2C(NO_2)=N(O)–O]^{\ominus}$ $K^{\oplus}$, $H_2SO_4$, HCHO | | m.p. 190° (dec.) | 1 |

TABLE VIII-2. Preparation and Properties of Halonitropyridines and 1-Oxides

| Compound | Method of preparation | Yield (%) | Properties | Ref. |
|---|---|---|---|---|
| 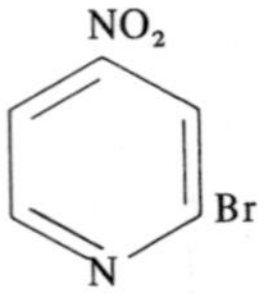 $NO_2$, Br, N | 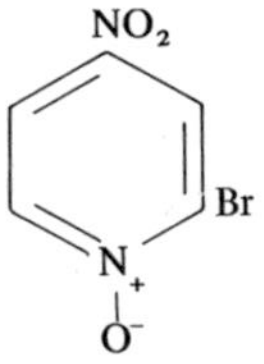 $NO_2$, Br, $N^+$, $O^-$, $PCl_3-CHCl_3$ | | m.p. 64° | 42 |
| 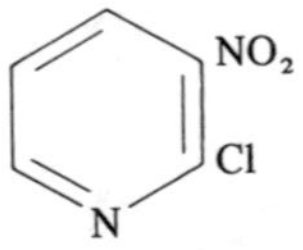 $NO_2$, Cl, N | 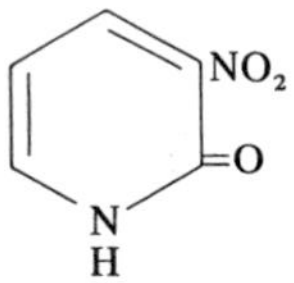 $NO_2$, =O, N, H, $POCl_3-HCONMe_2$ | 97 | m.p. 99° | 43 |
| 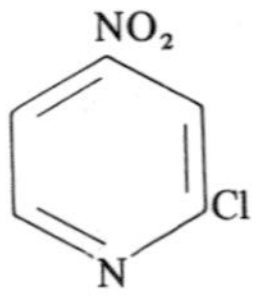 $NO_2$, Cl, N | 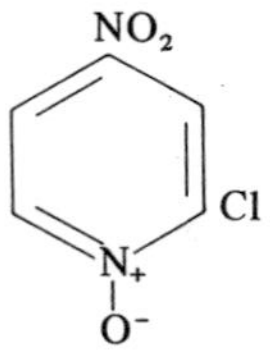 $NO_2$, Cl, $N^+$, $O^-$ | | m.p. 53° | 41 |
| 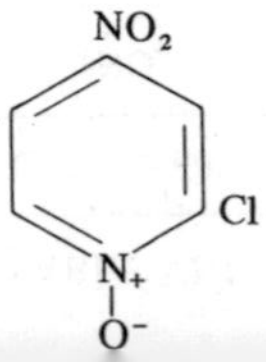 $NO_2$, Cl, $N^+$, $O^-$ | 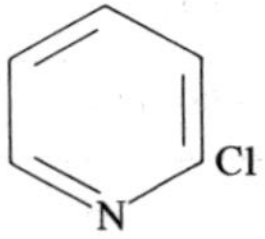 Cl, N, $HNO_3$ | | m.p. 151–152° | 44 |

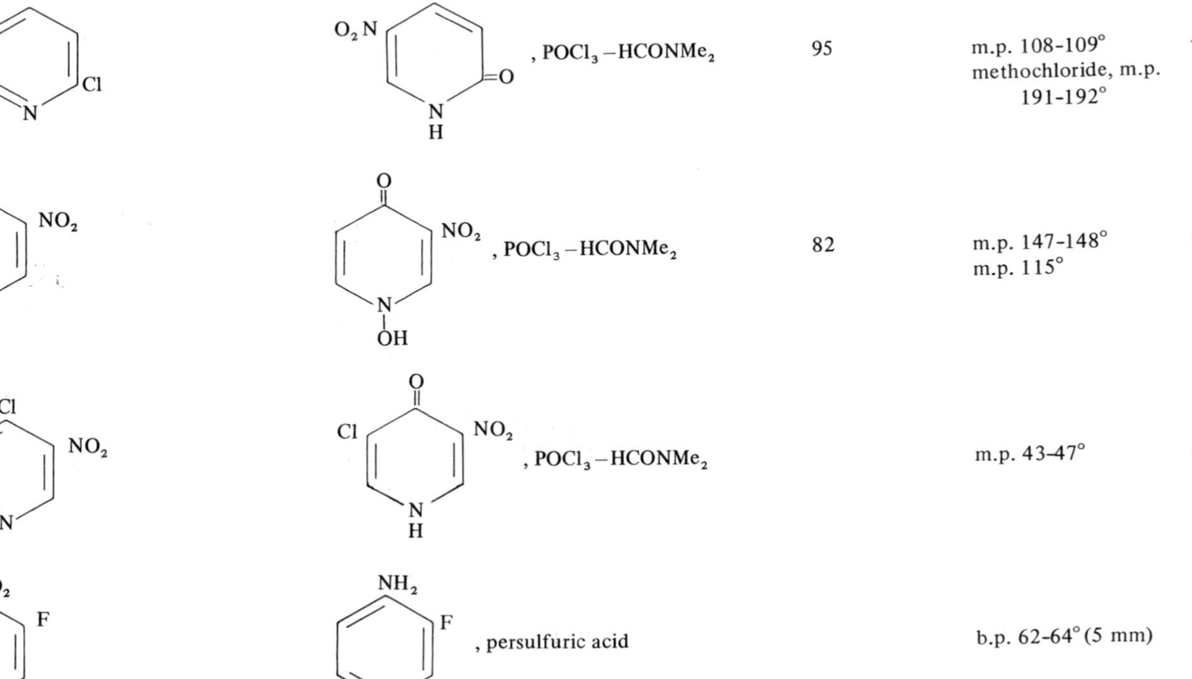

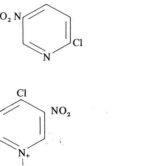

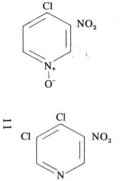

| | | | | |
|---|---|---|---|---|
| , $POCl_3-HCONMe_2$ | 95 | m.p. 108-109°<br>methochloride, m.p. 191-192° | 43 |
| , $POCl_3-HCONMe_2$ | 82 | m.p. 147-148°<br>m.p. 115° | 10,43 |
| , $POCl_3-HCONMe_2$ | | m.p. 43-47° | 45 |
| , persulfuric acid | | b.p. 62-64° (5 mm) | 5 |

TABLE VIII-2. Preparation and Properties of Halonitropyridines and 1-Oxides (Continued)

| Compound | Method of preparation | Yield (%) | Properties | Ref. |
|---|---|---|---|---|
| 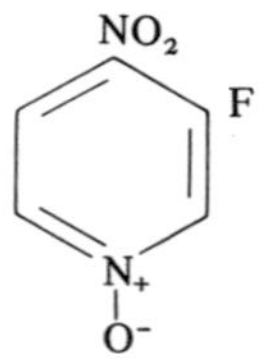 $NO_2$, F, $N^+$–$O^-$ | 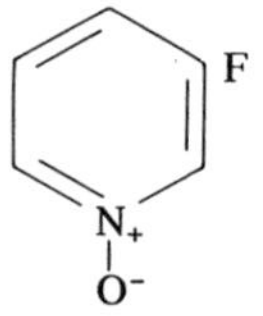 F, $N^+$–$O^-$, $HNO_3-H_2SO_4$ | | m.p. 128° | 20 |
| 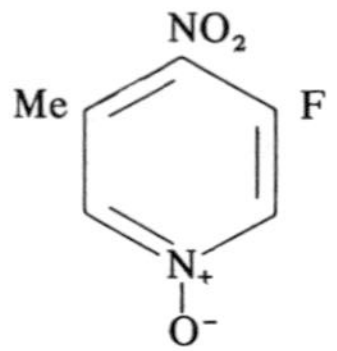 $NO_2$, Me, F, $N^+$–$O^-$ | 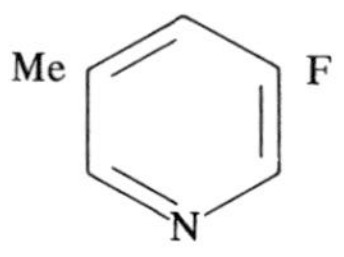 Me, F, N, (i) $H_2O_2-Ac_2O$ (ii) $HNO_3-H_2SO_4$ | | m.p. 119° | 46 |
| 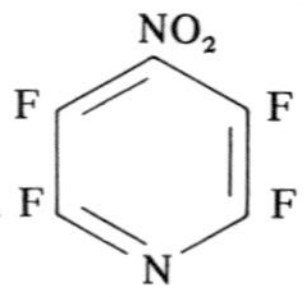 $NO_2$, F, F, F, F, N | 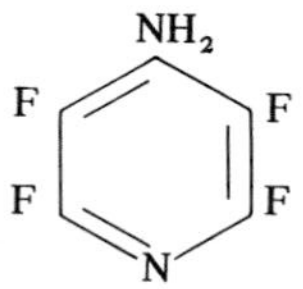 $NH_2$, F, F, F, F, N, $H_2O_2-(CF_3CO_2)O$ | 56 | b.p. 152–154° $n_D^{20} = 1.4459$ | 6,7 |

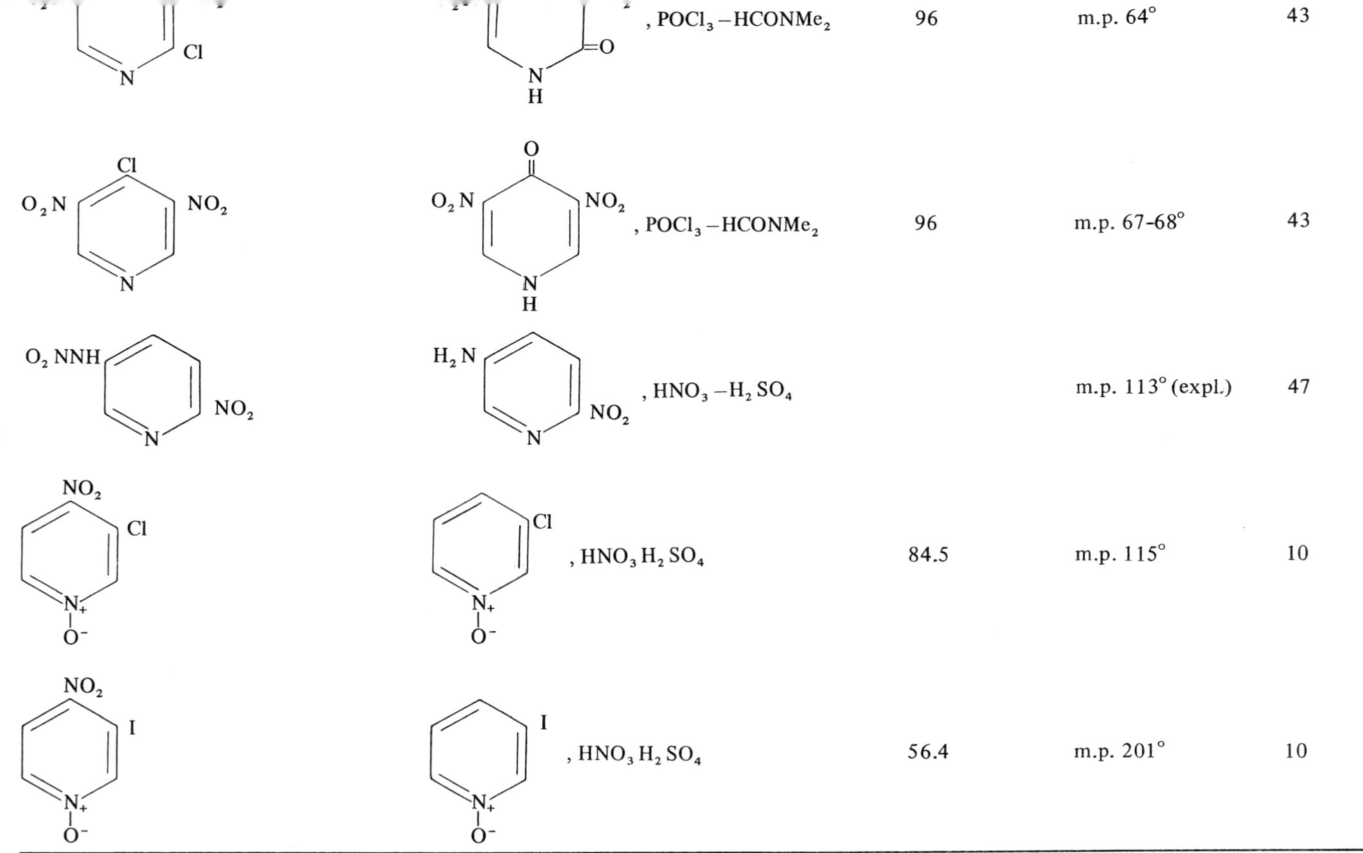

| Compound | Starting material, reagents | Yield (%) | Properties | Ref. |
|---|---|---|---|---|
| [structure] | [structure], $POCl_3-HCONMe_2$ | 96 | m.p. 64° | 43 |
| [structure] | [structure], $POCl_3-HCONMe_2$ | 96 | m.p. 67-68° | 43 |
| [structure] | [structure], $HNO_3-H_2SO_4$ | | m.p. 113° (expl.) | 47 |
| [structure] | [structure], $HNO_3\,H_2SO_4$ | 84.5 | m.p. 115° | 10 |
| [structure] | [structure], $HNO_3\,H_2SO_4$ | 56.4 | m.p. 201° | 10 |

TABLE VIII-3. Side-Chain Nitro Compounds

| Compound | Method of preparation | Properties | Ref. |
|---|---|---|---|
| 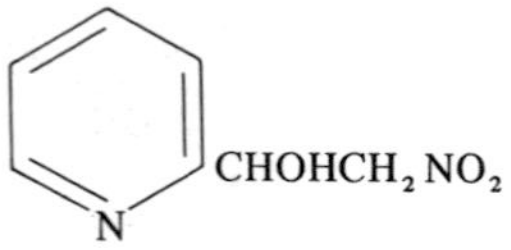 2-pyridyl–$CHOHCH_2NO_2$ | 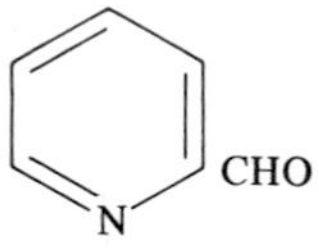 2-pyridyl–CHO, $MeNO_2-K_2CO_2$ | m.p. 68°<br>HCl salt, m.p. 136-137° | 48 |
| 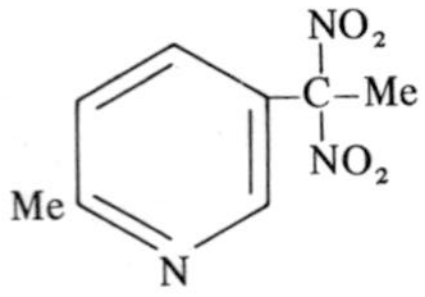 6-Me-3-pyridyl–$C(NO_2)_2$–Me | 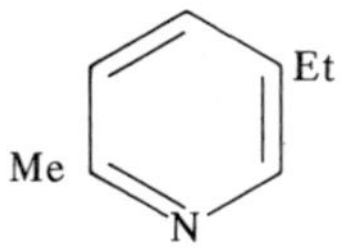 6-Me-3-Et-pyridine, $HNO_3$ | b.p. 112°(1 mm);<br>138°(4 mm) | 11 |

TABLE VIII-4. Pyridylhydrazines

| Compound | Method of preparation | Yield | Properties | Ref. |
|---|---|---|---|---|
| 2-Py$NHNH_2$ | 2-Py$SO_3H$, $N_2H_4 \cdot H_2O$, $ZnCl_2$ | | picrate, m.p. 187–189° | 49 |
| | Pyridine, $NaNHNH_2$, Heat | | m.p. 46–47°<br>mono-HCl salt, m.p. 183°<br>di-HCl salt, m.p. 214–215° | 36 |
| Me, N, $NHNH_2$ | 4-Picoline, $NaNHNH_2$, Heat | | m.p. 74-75° | 36 |
| Me, N, $NHNH_2$ | 2-Picoline, $NaNHNH_2$, Heat | | m.p. 58-59° | 36 |
| Me, Me, N, $NHNH_2$ | 2, 4-Lutidine, $NaNHNH_2$, Heat | | m.p. 67-68° | 36 |
| 2-PyNHNHMe | Pyridine, NaNHNHMe, Heat | | m.p. 46°<br>picrate, m.p. 145–146° | 36 |
| 2-Py$NHNMe_2$ | Pyridine, $NaNHNMe_2$, Heat | | m.p. 95°<br>picrate, m.p. 185° | 26 |
| 2-PyNHNHPy-2 | 2-Py$NHNH_2$, Pyridine, $NaNH_2$ | | | 36 |

TABLE VIII-4. Pyridylhydrazines (Continued)

| Compound | Method of preparation | Yield | Properties | Ref. |
|---|---|---|---|---|
| 2-pyridyl-N(Me)$NH_2$ | 2-PyCl, NaNHNHMe | | b.p. 68° (5 mm) | 37 |
| 2-pyridyl-N(Bu)$NH_2$ | 2-PyCl, NaNHNHBu | | b.p. 120° (0.2 mm) | 37 |
| 2-pyridyl-N(Ph)$NH_2$ | 2-PyCl, NaNHNHPh | | b.p. 140° (0.2 mm) | 37 |
| 2-pyridyl-N($CH_2Ph$)$NH_2$ | 2-PyCl, $NaNHNHCH_2Ph$ | | b.p. 142° (0.3 mm) | 37 |
| 2-pyridyl-NHNH-C$_6$H$_4$-$NO_2$ (p) | 2-PyBr, $[p\text{-}NO_2C_6H_4NHNH]^{\ominus}Na^{\oplus}$, 140° | | m.p. 158-159°<br>HBr salt, dec. 227-228° | 4 |

| | | | |
|---|---|---|---|
| 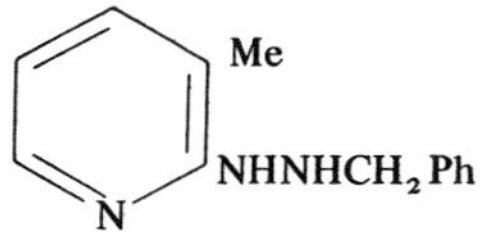 Me, $NHNHCH_2Ph$, N | | m.p. 198-200° | 50 |
| 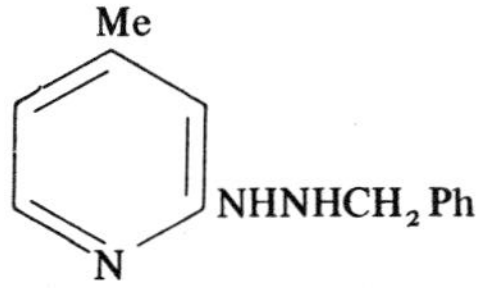 Me, $NHNHCH_2Ph$, N | | m.p. 175-178° | 50 |
| 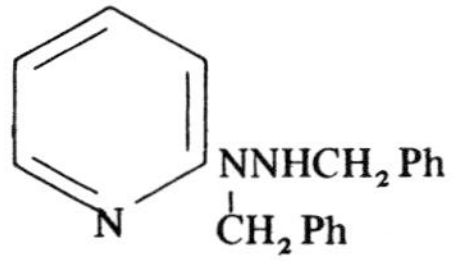 $NNHCH_2Ph$, $CH_2Ph$, N | | m.p. 165° | 50 |
| 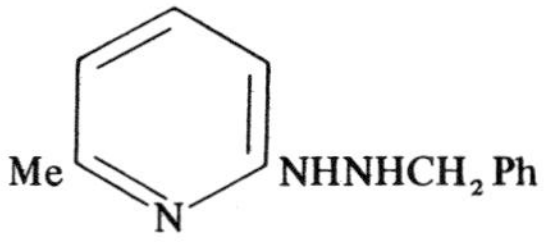 Me, $NHNHCH_2Ph$, N | | m.p. 147–148° | 50 |
| Me, Me, $NHNHCH_2Ph$, N | | m.p. 172–173° | 50 |
| $NHNH_2$, $N^+$, $O^-$ | 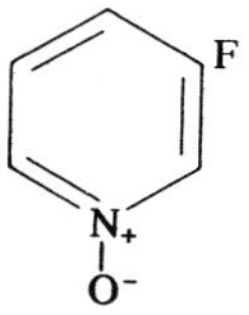 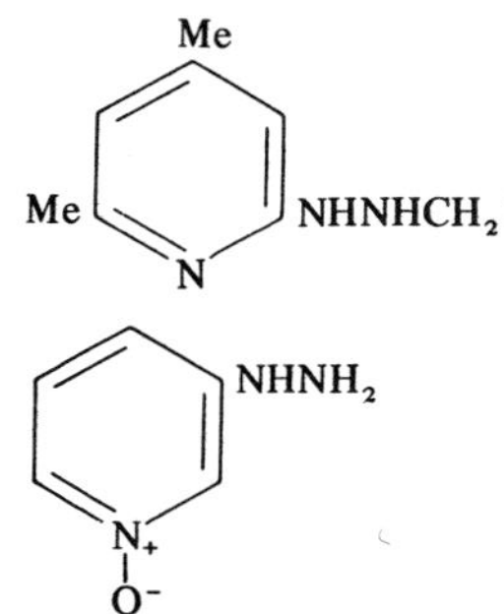 | m.p. 148° | 33 |

TABLE VIII-4. Pyridylhydrazines (Continued)

| Compound | Method of preparation | Yield | Properties | Ref. |
|---|---|---|---|---|
| $NO_2$, $NHNH_2$, N (pyridine ring) | $NO_2$, Br, $N^+$–$O^-$ (pyridine N-oxide ring), $N_2H_4 \cdot H_2O$(50°) | | m.p. 192° | 14 |
| $NO_2$, Me, $NHNH_2$, $N^+$–$O^-$ (pyridine N-oxide ring) | $NO_2$, Me, F, $N^+$–$O^-$ (pyridine N-oxide ring), $N_2H_4 \cdot H_2O$ | 84% | m.p. 192° | 46 |
| 4–$PyNHNH_2$ | 4-$PySO_3H$, $N_2H_4 \cdot H_2O$, $ZnCl_2$ | | HCl salt, m.p. 242–244°<br>dibenzoyl deriv.<br>m.p. 234 to 250°<br>deriv. with $MeCOCO_2Et$,<br>m.p. 128 to 130° | 49 |
| $NHNMe_2$, $Me_2NHN$, $NHNMe_2$, N (pyridine ring) | Minor product in reaction of pyridine with $NaNHNMe_2$ | | m.p. 154°<br>Tri-HCl salt, m.p. 214° | 36 |

TABLE VIII-5. Miscellaneous Pyridylhydrazines and Derivatives

| Compound | Method of preparation | Yield | Properties | Ref. |
|---|---|---|---|---|
| 2-PyN(Me)NHCONH$_2$ | | | m.p. 220° | 37 |
| 2-PyN(Bu)NHCO$_2$Et | 2-PyN(CH$_2$Ph)NH$_2$, ClCO$_2$Et | | b.p. 130° (0.2 mm) | 37 |
| 2-PyN(Bu)NHCONH$_2$ | | | m.p. 190° (dec.) | 37 |
| 2-PyN(CH$_2$Ph)NHCO$_2$Et | 2-PyN(CH$_2$Ph)NH$_2$, ClCO$_2$Et | | m.p. 50° | 37 |
| 2-PyN(CH$_2$Ph)NHCONH$_2$ | | | m.p. 220° (dec.) | 37 |
| 2-PyN(Ph)NHCO$_2$Et | 2-PyN(Ph)NH$_2$, ClCO$_2$Et | | m.p. 103–104° | 37 |
| 2-PyN(Ph)NHCONH$_2$ | | | m.p. 227° | 37 |
| 5-$O_2N$-2-pyridyl–NHN (1-hydantoinyl: 2,4-dioxoimidazolidin-1-yl) | 5-$O_2N$-2-chloropyridine, 1-Aminohydantoin | | m.p. 247-249° | 51 |

TABLE VIII-6. Halopyridylhydrazines

| Compound | Method of preparation | Properties | Ref. |
|---|---|---|---|
| ring: Cl, N, $NHNH_2$ | ring: Cl, N, Cl, $N_2H_4 \cdot H_2O$, Heat | m.p. 118–120° | 52 |
| ring: $NHNH_2$; F, F; N | ring: F; F, F; N, $N_2H_4 \cdot H_2O$ | m.p. 134–135° sublimes *in vacuo* | 35 |
| ring: $NHNH_2$; Cl, Cl; F, F; N | ring: F; Cl, Cl; F, F; N, $N_2H_4$, Heat | | 7 |
| ring: $NHNH_2$; F, Cl; F, F; N | ring: F; F, Cl; F, F; N, $N_2H_4$, Heat | m.p. 101–102° | 7 |

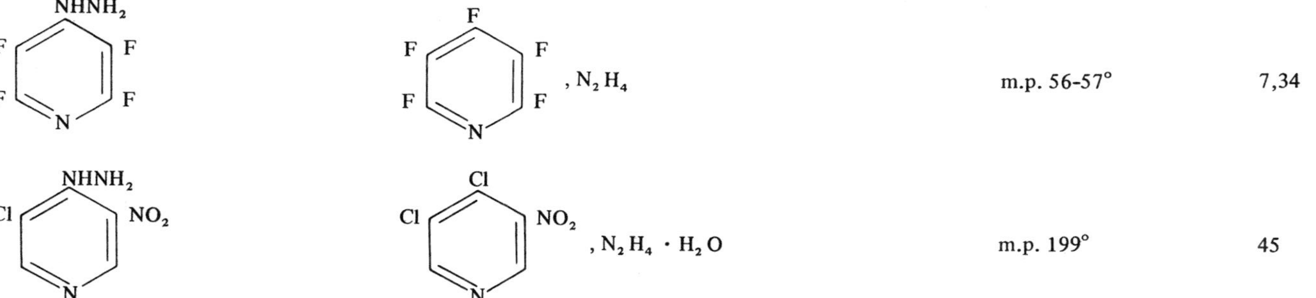

| Compound | Reactants | Properties | Ref. |
| --- | --- | --- | --- |
| 4-hydrazino-2,3,5,6-tetrafluoropyridine | pentafluoropyridine, $N_2H_4$ | m.p. 56-57° | 7,34 |
| 3-chloro-4-hydrazino-5-nitropyridine | 3,4-dichloro-5-nitropyridine, $N_2H_4 \cdot H_2O$ | m.p. 199° | 45 |

TABLE VIII-7. Preparation and Properties of Nitroso- and Hydroxylaminopyridines and Derivatives

| Compound | Method of preparation | Yield (%) | Properties | Ref. |
|---|---|---|---|---|
| 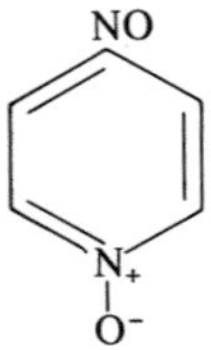 NO, $N^+$–$O^-$ | 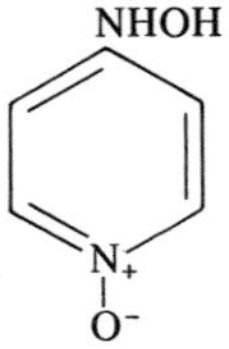 NHOH, $N^+$–$O^-$, $KMnO_4$, dil. $H_2SO_4$ | | m.p. 139° | 24 |
| 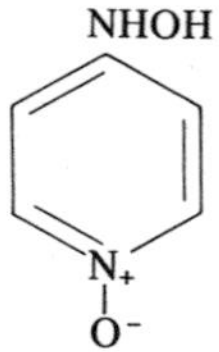 NHOH, $N^+$–$O^-$ | 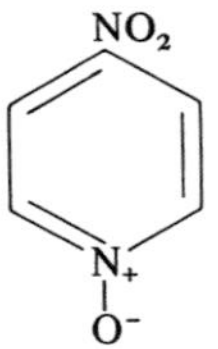 $NO_2$, $N^+$–$O^-$, $PhNHNH_2$ | 100 | m.p. 237° | 24 |
| 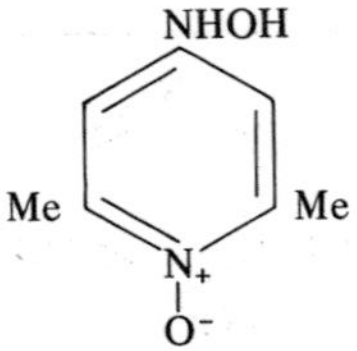 NHOH, Me, Me, $N^+$–$O^-$ | 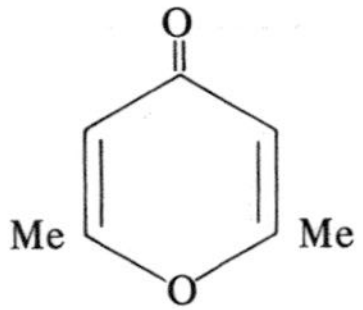 O, Me, Me, O, $NH_2OH \cdot HCl$–$C_5H_5N$ | 17 | m.p. not sharp | 26 |
| 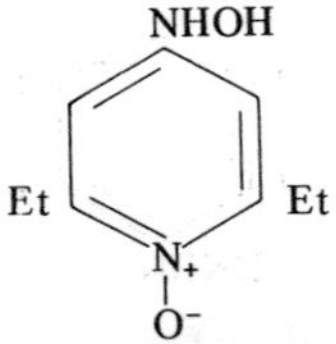 NHOH, Et, Et, $N^+$–$O^-$ | 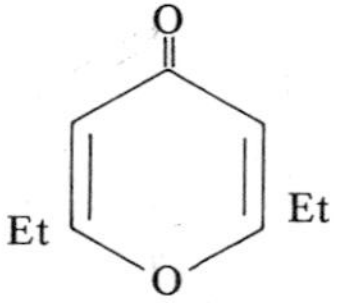 O, Et, Et, O, $NH_2OH \cdot HCl$–$C_5H_5N$ | 20 | m.p. 170° (dec.) | 26 |

TABLE VIII-8. Preparation and Properties of Azidopyridines and Derivatives

| Compound | Method of preparation | Yield (%) | Properties | Ref. |
|---|---|---|---|---|
| 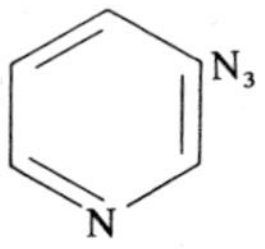 [structure: pyridine, $N_3$; N] | 3-Py$N_2^+$, $Cl^-$, $NaN_3$ | | b.p. 75°(2 mm)<br>$n$ 1.5752<br>$d_{19}$ 1.196 | 39 |
| 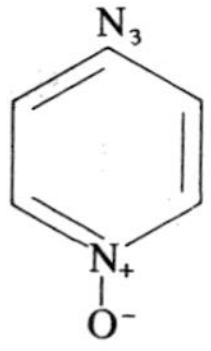 [structure: $N_3$; $N^+$–$O^-$] | 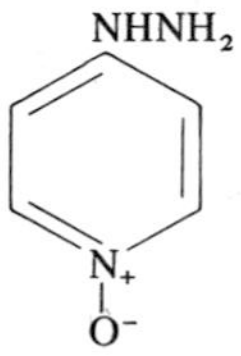 [structure: $NHNH_2$; $N^+$–$O^-$], $HNO_2$ | 55 | m.p. 139-140°(dec.)[a] | 38 |
| | 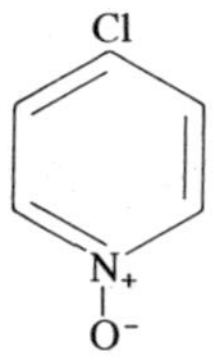 [structure: Cl; $N^+$–$O^-$], $NaN_3$ | 10 | m.p. 142-143°[a] | 38 |
| 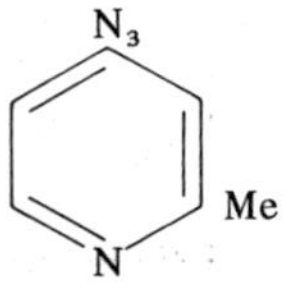 [structure: $N_3$; Me; N] | 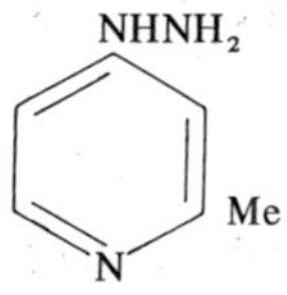 [structure: $NHNH_2$; Me; N], $HNO_2$ | 69 | b.p. 74-78°(6 mm)<br>(bath temp.) | 38 |

TABLE VIII-8. Preparation and Properties of Azidopyridines and Derivatives (Continued)

| Compound | Method of preparation | Yield (%) | Properties | Ref. |
|---|---|---|---|---|
| $N_3$; $\overset{+}{N}$–$O^-$ | $NH_2$; $\overset{+}{N}$–$O^-$ (i)$HNO_2$ (ii)$NaN_3$ | | m.p. 84.5-85.5° | 39a |
| $CH_3$, $N_3$; $\overset{+}{N}$–$O^-$ | $CH_3$, $NH_2$; $\overset{+}{N}$–$O^-$ (i)$HNO_2$ (ii)$NaN_3$ | | m.p. 89-90° | 39a |
| $CH_3$, $N_3$; $\overset{+}{N}$–$O^-$ | $CH_3$, $NH_2$; $\overset{+}{N}$–$O^-$ (i)$HNO_2$ (ii)$NaN_3$ | | m.p. 43–46° | 39a |
| $NO_2$, $N_3$; $\overset{+}{N}$–$O^-$ | $NO_2$, Cl; $\overset{+}{N}$–$O^-$ , $NaN_3$ | | m.p. 86-88° | 39a |

[a]Identical IR spectra

TABLE VIII-9. Preparation and Properties of Azopyridines

| Compound | Method of preparation | Properties | Ref. |
|---|---|---|---|
| 2-pyridyl–N=NPh | $2\text{-PyNH}_2$, PhNO, aq. NaOH | m.p. 32°<br>picrate, m.p. 133–135°<br>1-oxide, m.p. 137–138°<br>1-oxide picrate, m.p. 126–128° | 53 |
| 2-pyridyl–N=N–C$_6$H$_4$–$NO_2$ (4) | $O_2N$–C$_6$H$_4$–NHNHPy-2, $NaNO_2$, aq. AcOH | m.p. 160–162°<br>1-oxide, m.p. 214–215° | 4 |
| pyridine: $H_2N$, MeO, $NH_2$, N=N·Ph | pyridine: $H_2N$, MeO, $NH_2$, $PhN_2^{\oplus}Cl^{\ominus}$ | m.p. 141° | 27 |
| pyridine: $H_2N$, EtO, $NH_2$, N=NPh | pyridine: $H_2N$, EtO, $NH_2$, $PhN_2^{\oplus}Cl^{\ominus}$ | m.p. 119° | 27 |

TABLE VIII-9. Preparation and Properties of Azopyridines (Continued)

| Compound | Method of preparation | Properties | Ref. |
|---|---|---|---|
| 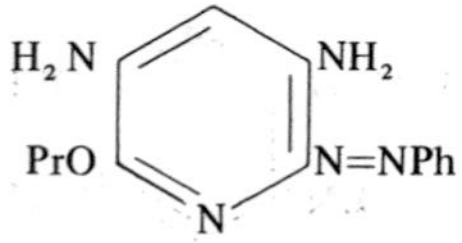 $H_2N$, $NH_2$, PrO, N=NPh, N (pyridine ring) | 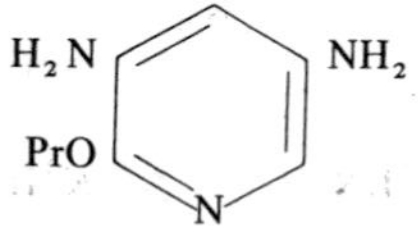 $H_2N$, $NH_2$, PrO, N (pyridine ring), $PhN_2^{\oplus}Cl^{\ominus}$ | m.p. 122° | 27 |
| 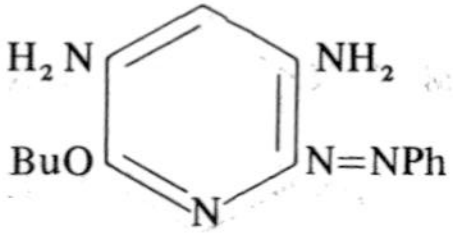 $H_2N$, $NH_2$, BuO, N=NPh, N (pyridine ring) | 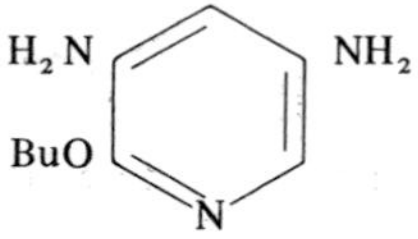 $H_2N$, $NH_2$, BuO, N (pyridine ring), $PhN_2^{\oplus}Cl^{\ominus}$ | m.p. 126° | 27 |
| 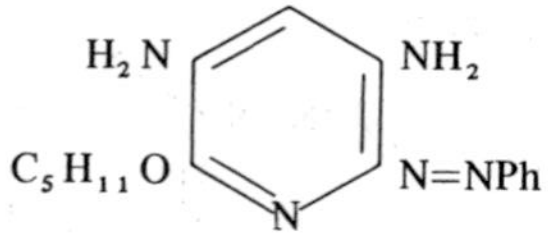 $H_2N$, $NH_2$, $C_5H_{11}O$, N=NPh, N (pyridine ring) | 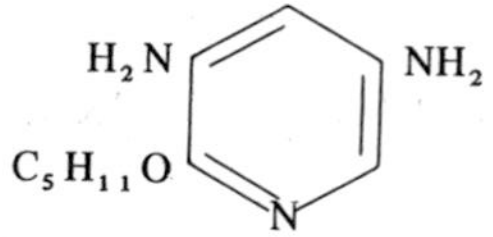 $H_2N$, $NH_2$, $C_5H_{11}O$, N (pyridine ring), $PhN_2^{\oplus}Cl^{\ominus}$ | m.p. 96° | 27 |
| 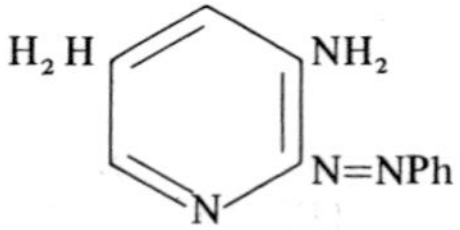 $H_2H$, $NH_2$, N=NPh, N (pyridine ring) | 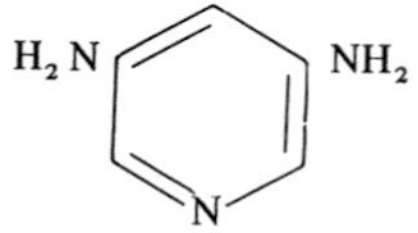 $H_2N$, $NH_2$, N (pyridine ring), $PhN_2^{\oplus}Cl^{\ominus}$ | m.p. 192° | 27 |
| 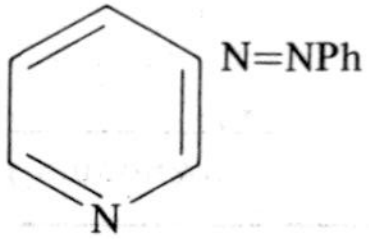 N=NPh, N (pyridine ring) | 3-$PyNH_2$, PhNO, aq. NaOH | m.p. 53–54°<br>picrate, m.p. 163–165°<br>1-oxide, m.p. 85–87°<br>1-oxide picrate, m.p. 159–161° | 53 |

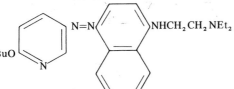

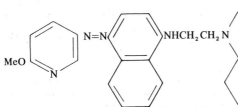

| | | | |
|---|---|---|---|
| $Me_2N$, N, N=NPh | Cl, N, N=NPh, $Me_2NH$ | m.p. 114.5–115.2° | 27 |
| BuO, N, N=N, Me, $NCH_2CH_2NEt_2$ | | m.p. 90–95° | 28 |
| 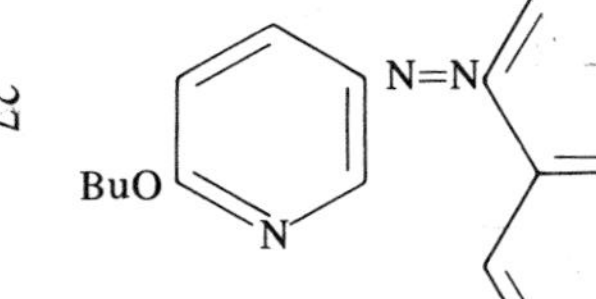 BuO, N, N=N, $NHCH_2CH_2NEt_2$ | | m.p. 92–94° | 28 |
| MeO, N, N=N, $NHCH_2CH_2N$ | | m.p. 214–216° (dec.) | 28 |

TABLE VIII-9. Preparation and Properties of Azopyridines (Continued)

| Compound | Method of preparation | Properties | Ref. |
|---|---|---|---|
| 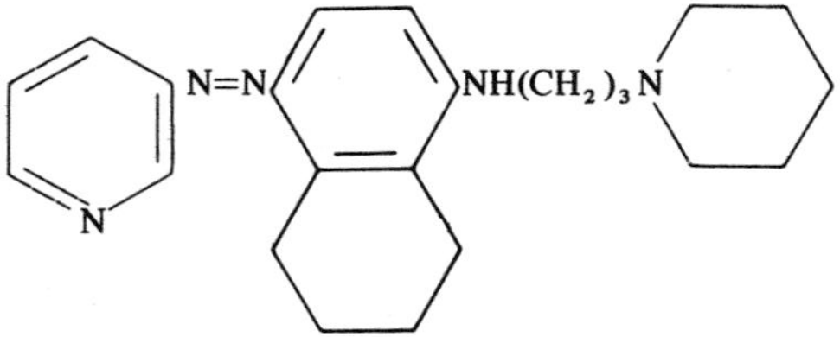 N=N NH$(CH_2)_3$N N | 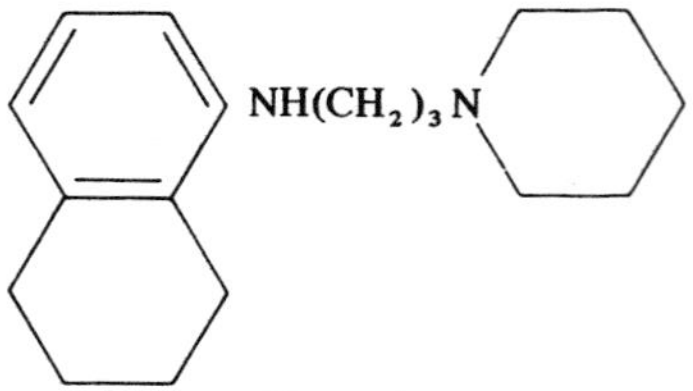 NH$(CH_2)_3$N + 3-Py$N_2^{\oplus}X^{\ominus}$ | m.p. 117–119° | 28 |
| 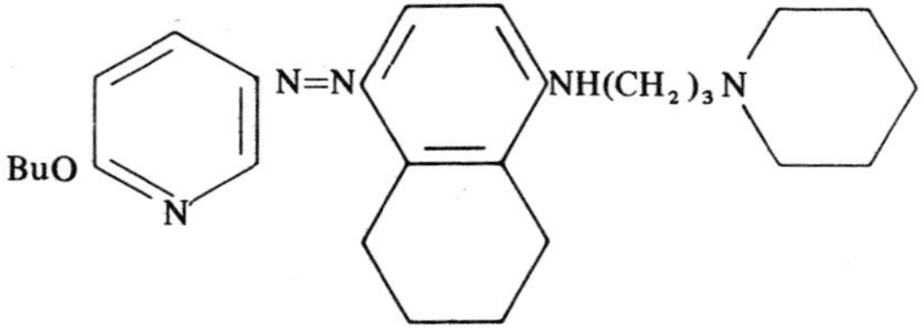 BuO N N=N NH$(CH_2)_3$N | 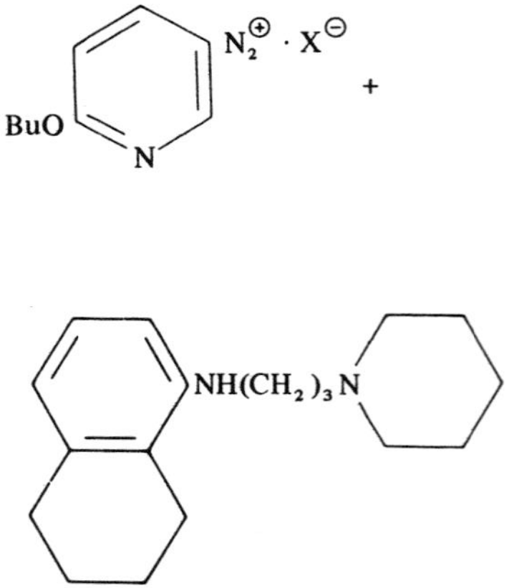 BuO N $N_2^{\oplus} \cdot X^{\ominus}$ + NH$(CH_2)_3$N | m.p. 123–124° | 54 |

| | | |
|---|---|---|
|  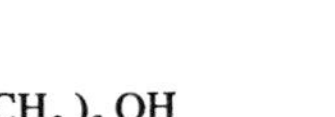  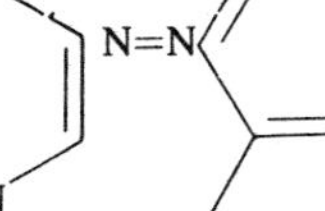 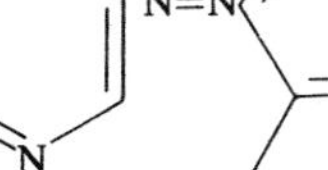  | m.p. 166–167° | 28 |
|  | m.p. 171–172° | 28 |
|  | m.p. 139–141° | 28 |
|  | m.p. 234° (dec.) | 28 |

TABLE VIII-9. Preparation and Properties of Azopyridines (Continued)

| Compound | Method of preparation | Properties | Ref. |
|---|---|---|---|
| 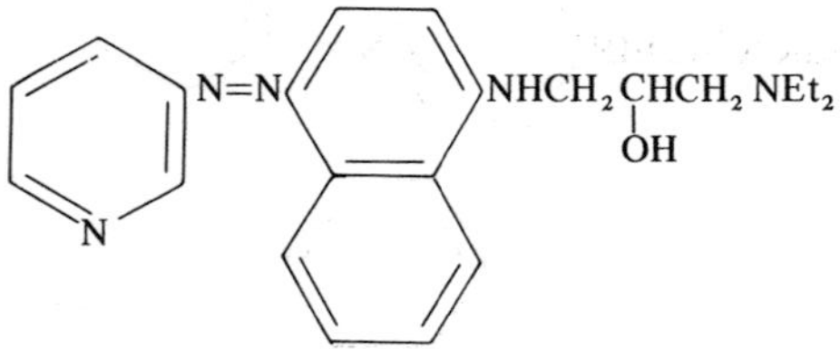 N=N; $NHCH_2CHCH_2NEt_2$; OH; N |  | m.p. 154–157° | 28 |
| 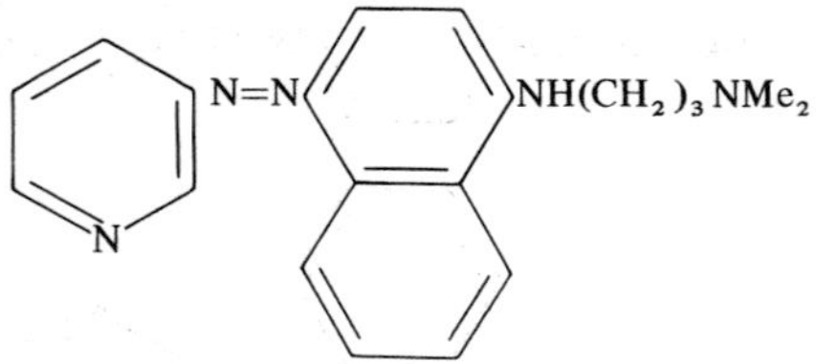 N=N; $NH(CH_2)_3NMe_2$; N |  | m.p. 130–132° | 28 |
| 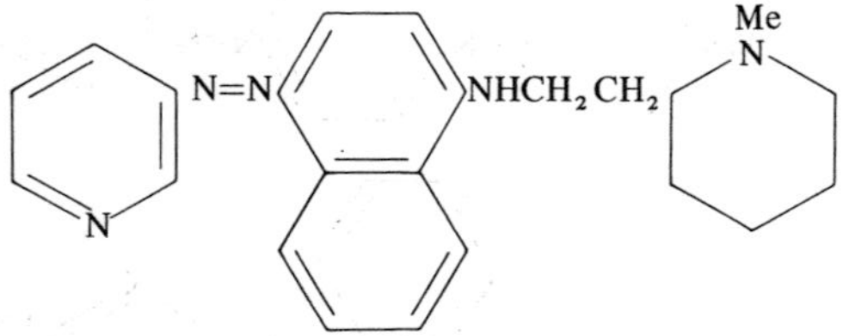 N=N; $NHCH_2CH_2$; Me; N; N |  | m.p. 166° (dec.) | 28 |
| 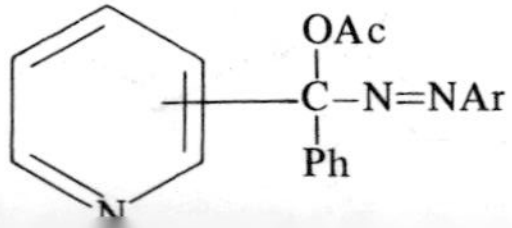 OAc; C–N=NAr; Ph; N | 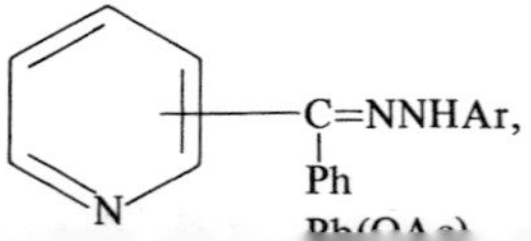 C=NNHAr,; Ph; N; [illegible] |  | 30 |

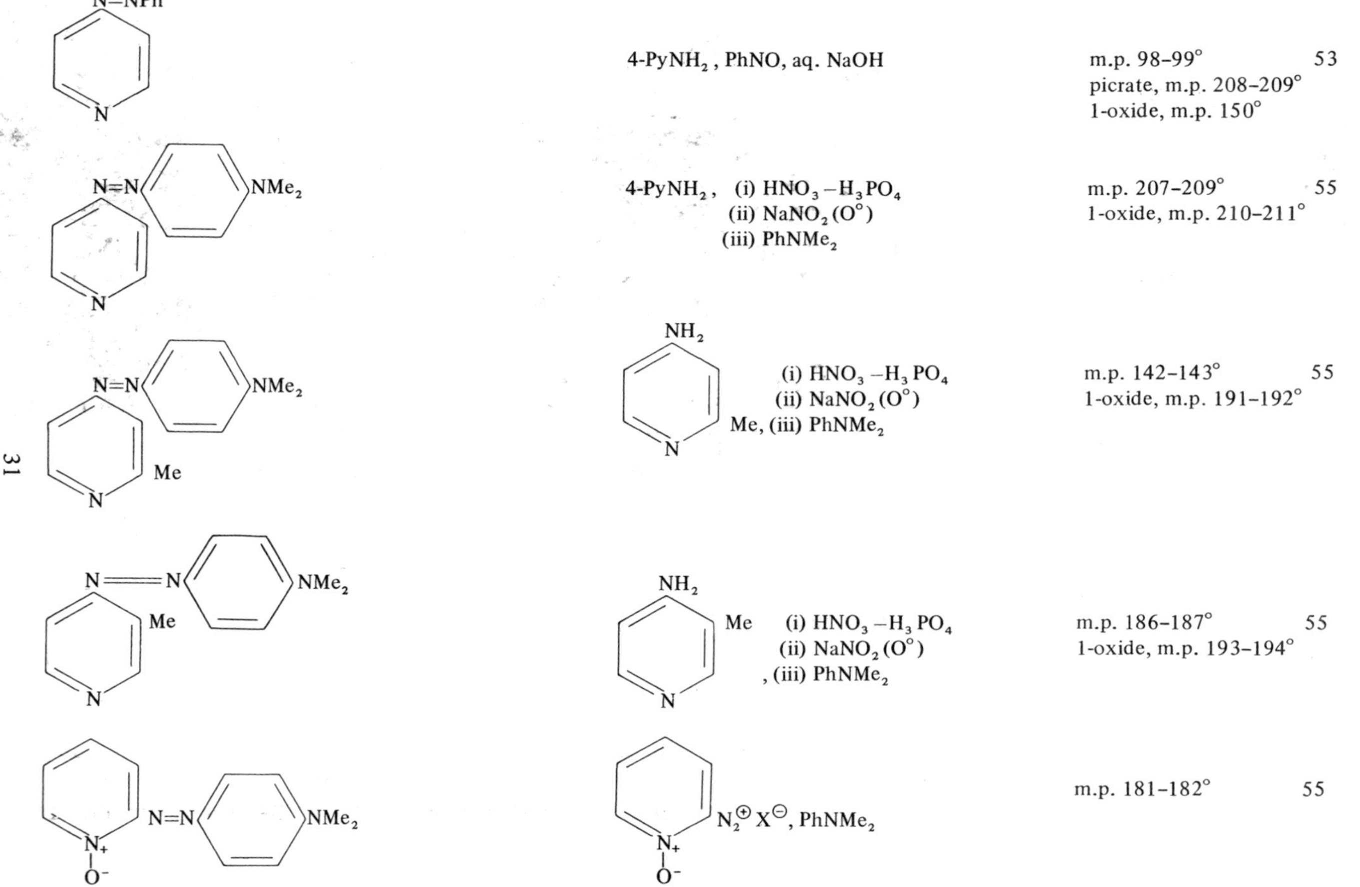

| Reagents | Properties | Ref. |
|---|---|---|
| 4-PyNH$_2$, PhNO, aq. NaOH | m.p. 98–99°<br>picrate, m.p. 208–209°<br>1-oxide, m.p. 150° | 53 |
| 4-PyNH$_2$, (i) HNO$_3$–H$_3$PO$_4$ (ii) NaNO$_2$ (O°) (iii) PhNMe$_2$ | m.p. 207–209°<br>1-oxide, m.p. 210–211° | 55 |
| (i) HNO$_3$–H$_3$PO$_4$ (ii) NaNO$_2$ (O°) (iii) PhNMe$_2$ | m.p. 142–143°<br>1-oxide, m.p. 191–192° | 55 |
| (i) HNO$_3$–H$_3$PO$_4$ (ii) NaNO$_2$ (O°) (iii) PhNMe$_2$ | m.p. 186–187°<br>1-oxide, m.p. 193–194° | 55 |
| $N_2^{\oplus}X^{\ominus}$, PhNMe$_2$ | m.p. 181–182° | 55 |

TABLE VIII-9. Preparation and Properties of Azopyridines (Continued)

| Compound | Method of preparation | Properties | Ref. |
|---|---|---|---|
| 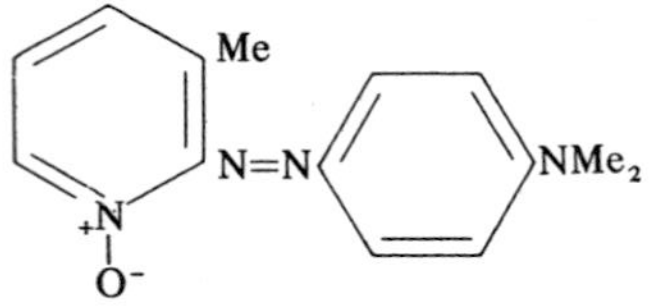 Me; $N{=}N$; $NMe_2$; $\overset{+}{N}$–$O^-$ | 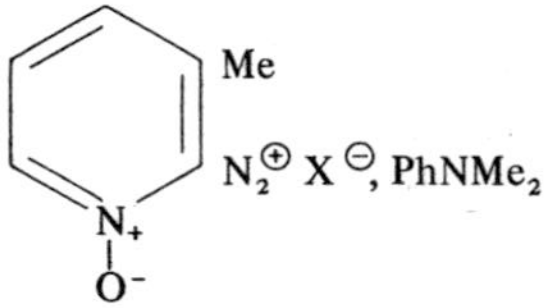 Me; $N_2^{\oplus}X^{\ominus}$, $PhNMe_2$; $N^+$–$O^-$ | m.p. 193–194° | 55 |
| 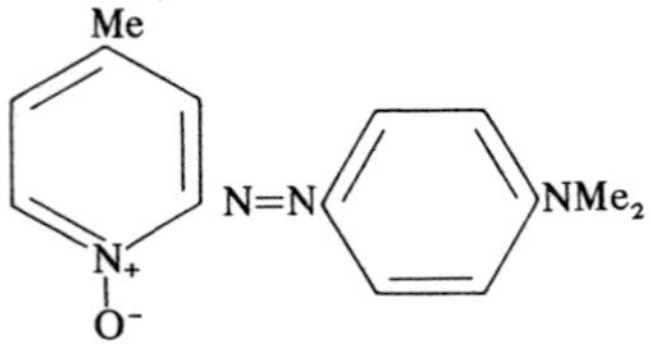 Me; $N{=}N$; $NMe_2$; $N^+$–$O^-$ | 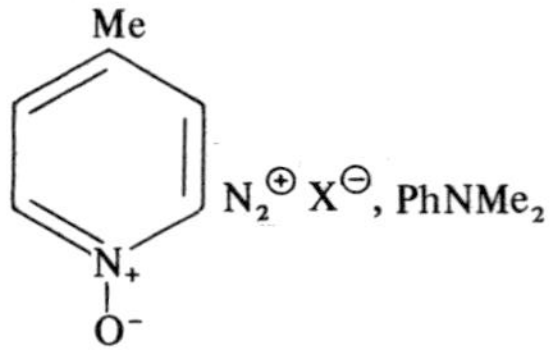 Me; $N_2^{\oplus}X^{\ominus}$, $PhNMe_2$; $N^+$–$O^-$ | m.p. 184–185° | 55 |
| 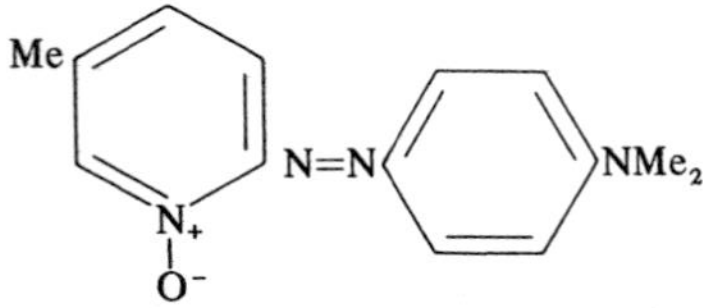 Me; $N{=}N$; $NMe_2$; $N^+$–$O^-$ | 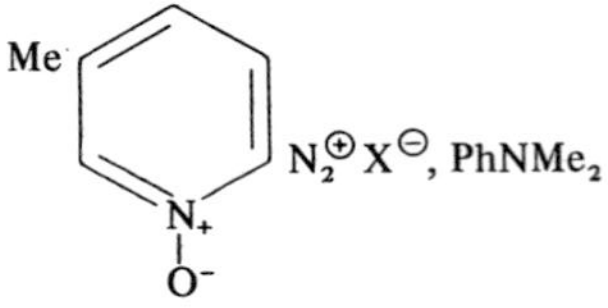 Me; $N_2^{\oplus}X^{\ominus}$, $PhNMe_2$; $N^+$–$O^-$ | m.p. 201–202° | 55 |
| 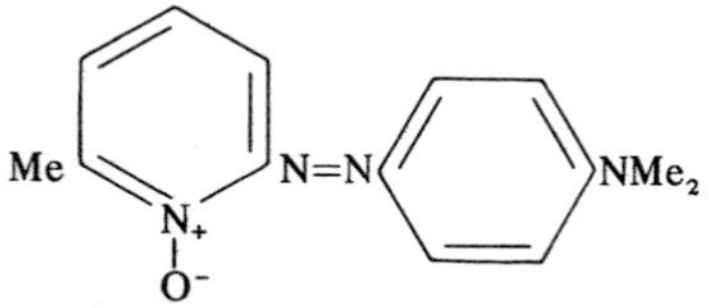 Me; $N{=}N$; $NMe_2$; $N^+$–$O^-$ | 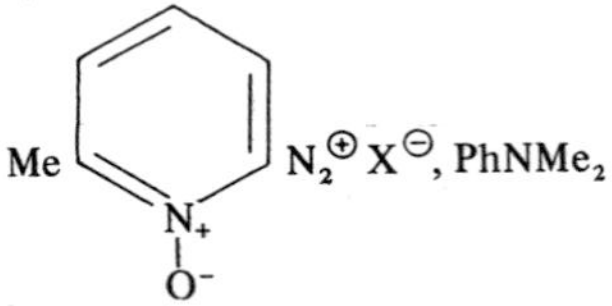 Me; $N_2^{\oplus}X^{\ominus}$, $PhNMe_2$; $N^+$–$O^-$ | m.p. 170–171° | 55 |

| Compound | Preparation | Properties | Ref. |
|---|---|---|---|
| 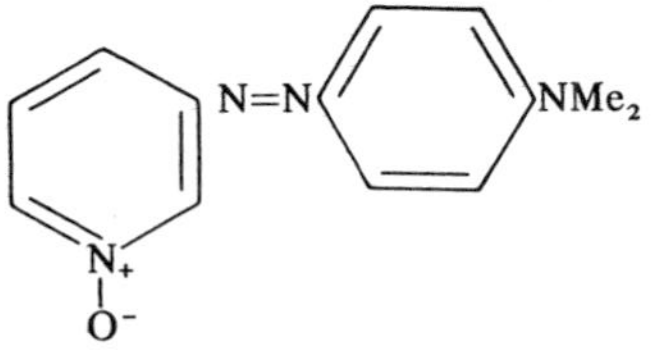 N=N, $NMe_2$, $N^+$, $O^-$ | 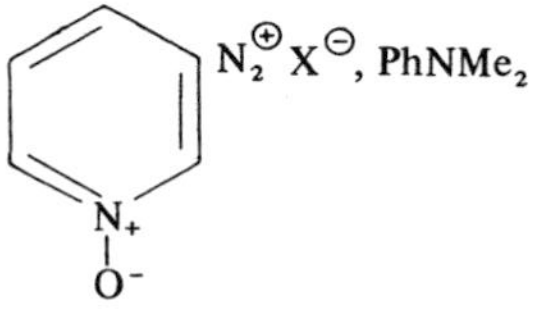 $N_2^{\oplus}X^{\ominus}$, $PhNMe_2$; $N^+$, $O^-$ | m.p. 189–191° | 55 |
| 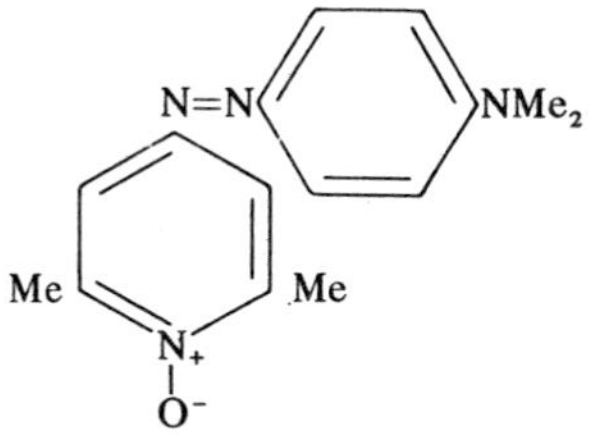 N=N, $NMe_2$, Me, Me, $N^+$, $O^-$ | 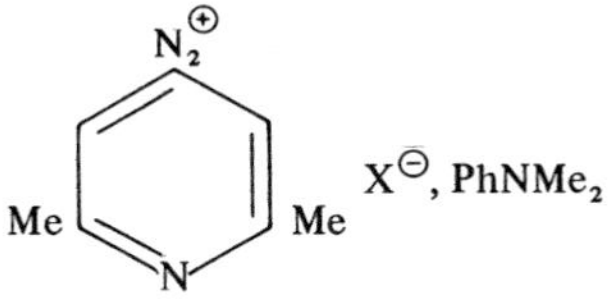 $N_2^{\oplus}$, Me, Me, N, $X^{\ominus}$, $PhNMe_2$ | m.p. 197–198° | 55 |
| 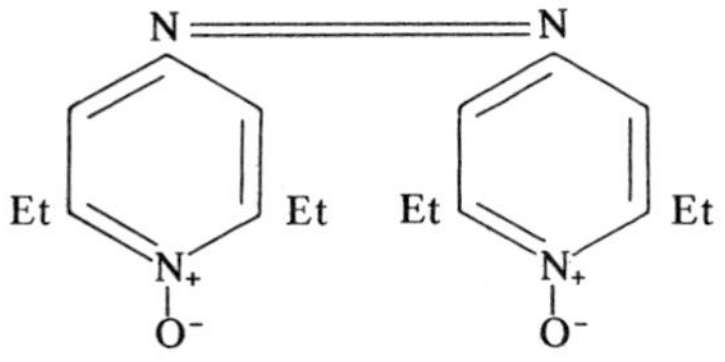 N=N, Et, Et, $N^+$, $O^-$; Et, Et, $N^+$, $O^-$ | 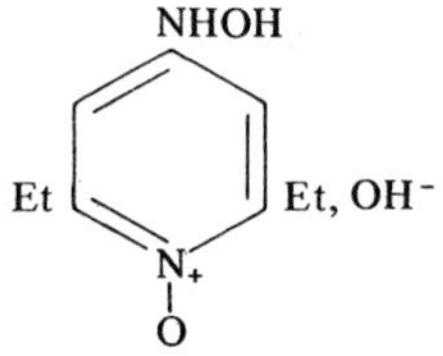 NHOH, Et, Et, $N^+$, O, $OH^-$ | m.p. 176–177° | 26 |
| 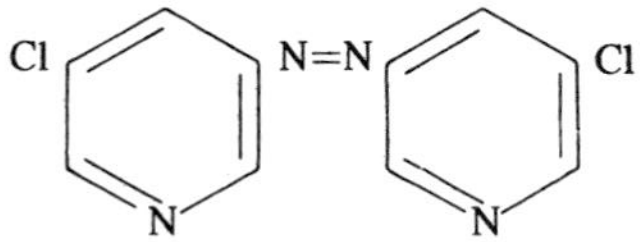 Cl, N=N, Cl, N, N | 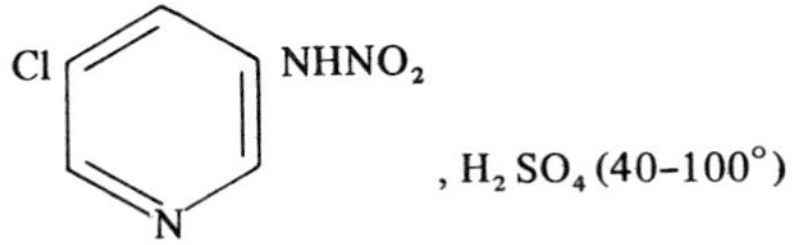 Cl, $NHNO_2$, N, $H_2SO_4$ (40–100°) | m.p. 183° | 8 |
| 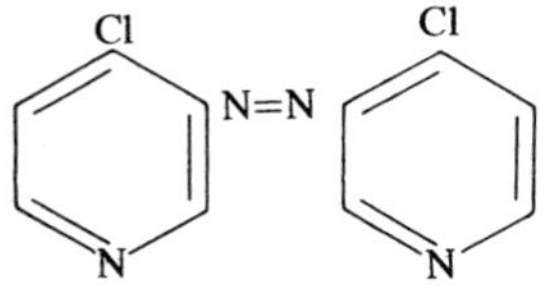 Cl, N=N, Cl, N, N | 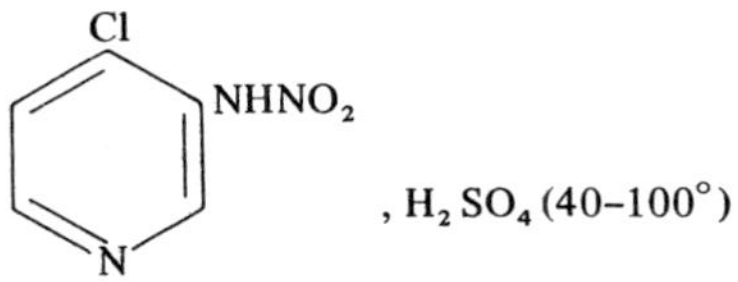 Cl, $NHNO_2$, N, $H_2SO_4$ (40–100°) | m.p. 164° | 8 |

TABLE VIII-9. Preparation and Properties of Azopyridines (Continued)

| Compound | Method of preparation | Properties | Ref. |
|---|---|---|---|
| 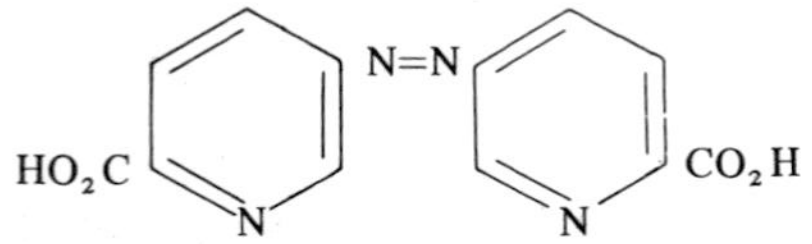 $HO_2C$, N=N, $CO_2H$, N, N | 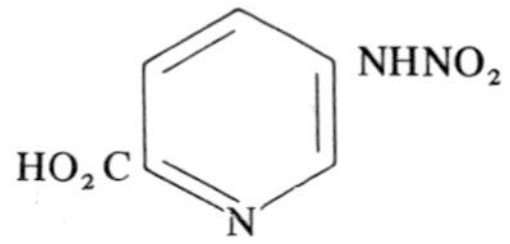 $HO_2C$, $NHNO_2$, N, $H_2SO_4$ (40–100°) | m.p. 246° | 8 |
| 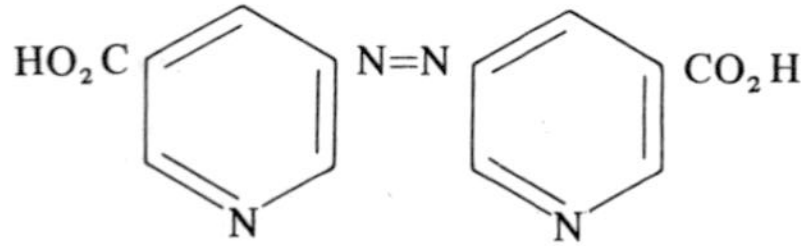 $HO_2C$, N=N, $CO_2H$, N, N | 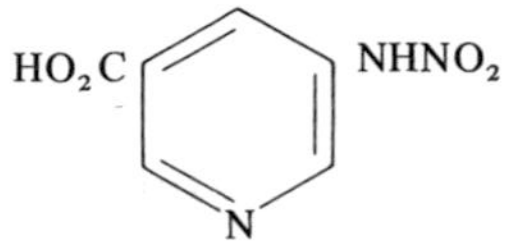 $HO_2C$, $NHNO_2$, N, $H_2SO_4$ (40–100°) | m.p. 298° | 8 |
| 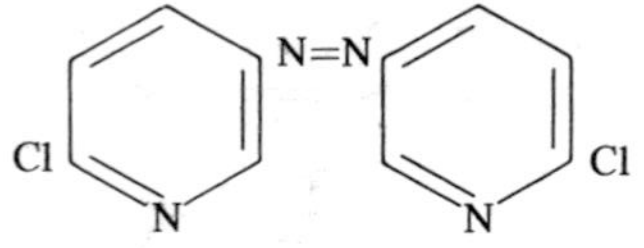 Cl, N=N, Cl, N, N | 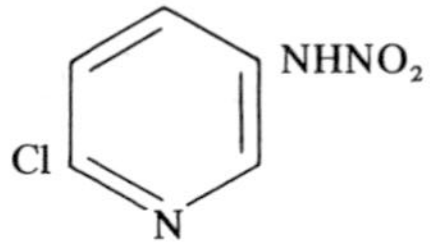 Cl, $NHNO_2$, N, $H_2SO_4$ (40–100°) | m.p. 215–217° | 8 |
| 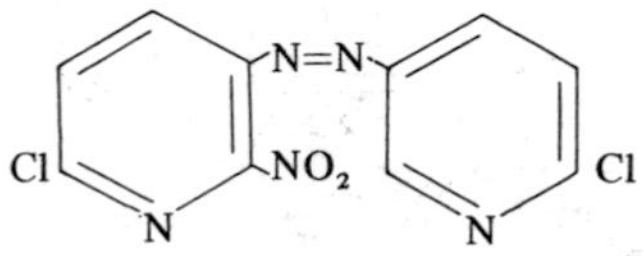 Cl, N=N, $NO_2$, Cl, N, N (minor product) | 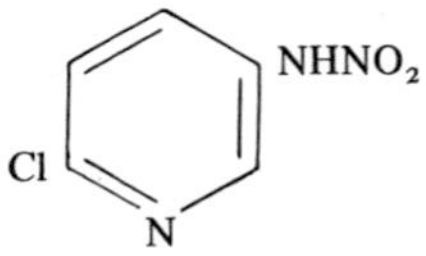 Cl, $NHNO_2$, N, $H_2SO_4$ (40–100°) | m.p. 165° | 8 |
| 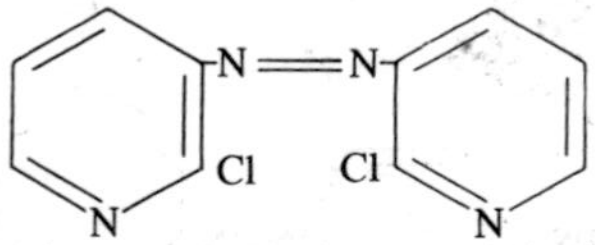 N=N, Cl, Cl, N, N | 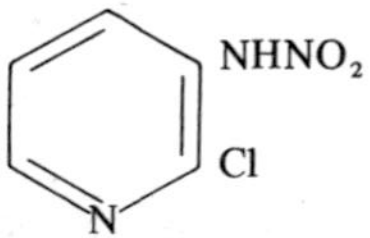 $NHNO_2$, Cl, N, $H_2SO_4$ (40–100°) | m.p. 237–239° | 8 |

TABLE VIII-10. Preparation and Properties of Azoxypyridines and Derivatives

| Compound | Method of preparation | Properties | Ref. |
|---|---|---|---|
| 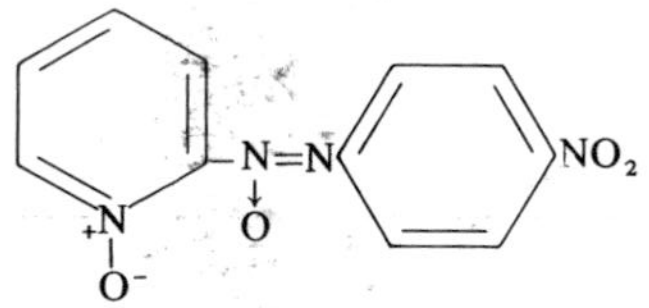 N=N, ↓O, $NO_2$, $\overset{+}{N}$–$O^-$ | 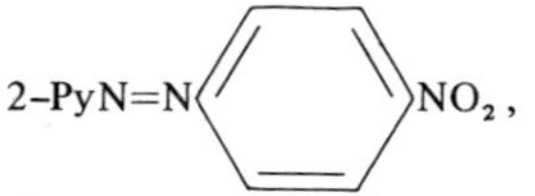 2-PyN=N–$NO_2$, $PhCO_3H$ | m.p. 192–193° | 4 |
| 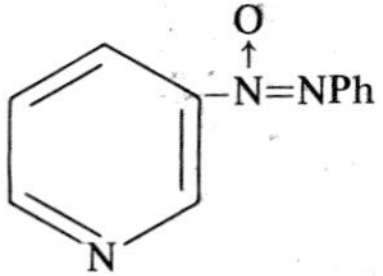 O↑, N=NPh, N | $3\text{-}PyNH_2$, PhNO, NaOH; 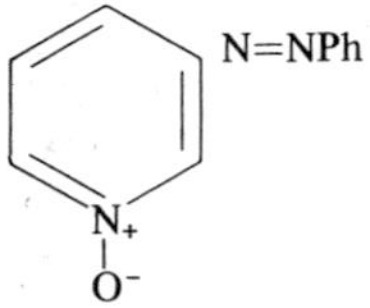 N=NPh, $N^+$–$O^-$, $PCl_3$–$CHCl_3$ | m.p. 61–62°<br>picrate, m.p. 180° | 53 |
| 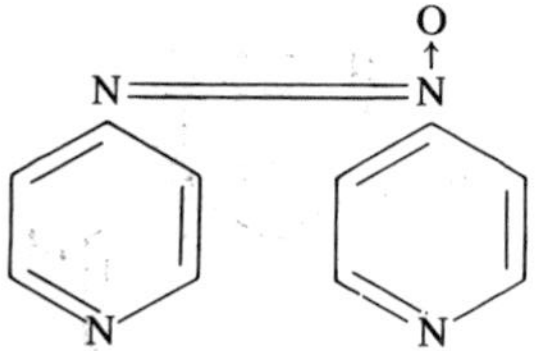 N=N→O, N, N | $4\text{-}PyN_3$, photolysis in $Me_2CO$ | m.p. 235° (dec.) | 40 |
| 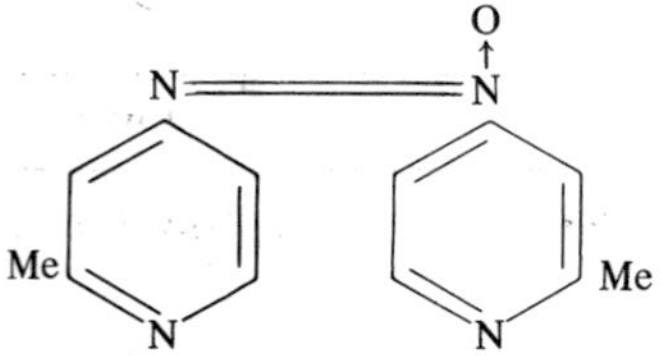 N=N→O, Me, N, N, Me | 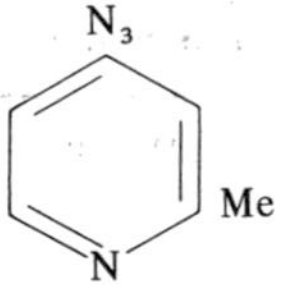 $N_3$, Me, N, $H_2O_2$ | m.p. 218° (dec.) | 38 |

TABLE VIII-11. Preparation and Properties of Pyridotriazoles[56]

| Compound | Method of Preparation | | Yield (%) | Properties |
|---|---|---|---|---|
| 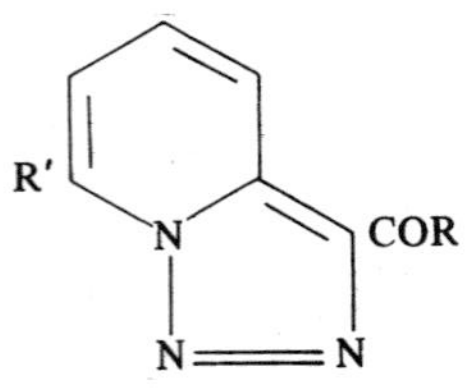 R′ N COR N═N | 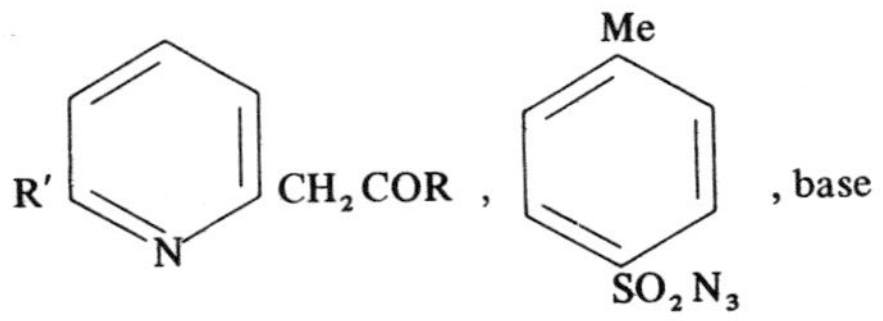 R′ N $CH_2COR$ , | Me $SO_2N_3$ , base | | |
| | *R* | *R′* | | |
| | Me | H | 50 | m.p. 157–158° |
| | Pr | H | 79 | m.p. 98° |
| | *t*-Bu | H | 90 | m.p. 95–96° |
| | 2-Furyl | H | 94 | m.p. 215° |
| | Ph | H | 88 | m.p. 111–112° |
| | 2-Py | H | 72 | m.p. 153° |
| | 3-Py | H | 57 | m.p. 134–135° |
| | 2-Furyl | Me | 77 | m.p. 183° |
| | 2-Thienyl | Me | 88 | m.p. 150° |
| | 3-Py | Me | 75 | m.p. 160° |

# VII. Supplement

The following table contains references to the spectral data on the material covered in this chapter. The studies cited are of a more comprehensive nature. In most cases the original literature was consulted; chemical abstract citations are included for all references for the purpose of convenience, however.

| Substance | Type of Study | Reference |
|---|---|---|
| 2-Nitropyridine | UV | 61 |
| 3-Nitropyridine | ESR | 59, 67 |
| | UV | 61 |
| 4-Nitropyridine | UV | 61 |
| | NMR | 72, 73, 81 |
| | IR | 63 |
| | ESR | 65 |
| 4-Nitropyridine-1-oxide | UV | 71, 77 |
| | NMR | 70, 73 |
| | Mass Spec. | 58 |
| | ESR | 66, 75 |
| 2-Nitro-3-picoline | Mass Spec. | 74 |
| 2-Nitro-4-picoline | Mass Spec. | 74 |
| 2-Nitro-4-picoline | Mass Spec. | 74 |
| 2-Nitro-6-picoline | Mass Spec. | 74 |
| 4-Nitro-2-picoline | IR | 79 |
| 4-Nitro-3-picoline | IR | 60, 69, 79 |
| | UV | 60 |
| 3-Ethyl-4-nitropyridine | UV, IR | 60 |
| 4-Nitro-3-propylpyridine | UV, IR | 60 |
| 2-Chloro-3-nitropyridine | UV | 57 |
| 3-Bromo-4-nitropyridine-1-oxide | IR | 69 |
| 3-Methoxy-2-nitropyridine | UV | 62 |
| 2-Methoxy-3-nitropyridine | UV | 62 |
| 4-Ethoxy-3-nitropyridine | UV | 57 |
| 2-Methoxy-5-nitropyridine | UV | 62 |
| 3,5-Dinitropyridine | ESR | 59 |
| 4-Nitro-2,6-lutidine | IR | 79 |
| 4-Nitro-2,6-lutidine-1-oxide | IR | 79 |
| 4-Nitro-3,5-lutidine | IR, UV | 60 |
| 2,3,5,6-Tetramethyl-4-nitropyridine | IR, UV | 60 |
| 4-Nitrosopyridine-1-oxide | ESR | 68 |
| Dinitro-(2-pyridyl)methane | IR, UV | 78 |
| 4-Hydrazinopyridine-1-oxide | IR study of hydrazones from | 80 |
| 4-Hydrazino-2,3,5,6-tetrafluoro-pyridine | NMR | 76 |
| 2,2′-Azopyridine | NMR | 64 |
| 3,3′-Azopyridine | NMR | 64 |

| Substance | Type of Study | Reference |
|---|---|---|
| 4,4′-Azopyridine | NMR | 64 |
| 3,3′-Dimethyl-4,4′-azopyridine-1,1′-dioxide | IR | 69 |
| 2,2′,6,6′-Tetramethyl-4,4′-azopyridine-1,1′-dioxide | IR | 79 |

## References


1. K. D. Gundermann and H. U. Alles, *Angew. Chem. Intern. Ed. Engl.*, **5,** 846 (1966).
2. J. S. Wieczorek and E. Plazek, *Rec. Trav. Chim. Pays-Bas,* **83,** 249 (1964); *Chem. Abstr.,* **60,** 15822b (1964).
3. R. C. De Selms, *J. Org. Chem.,* **33,** 478 (1968).
4. L. Pentimalli, *Gazz. Chim. Ital.,* **93,** 404 (1963); *Chem. Abstr.,* **59,** 5127b (1963).
5. T. Talik and Z. Talik, *Rocz. Chem.,* **40,** 1187 (1966); *Chem. Abstr.,* **66,** 11165u (1967).
6. R. D. Chambers, J. Hutchinson, and W. K. R. Musgrave, *J. Chem. Soc.,* 5040 (1965).
7. R. D. Chambers, J. Hutchinson, and W. K. R. Musgrave, Belg. patent 660,873 (1965).
8. M. D. Coburn, *J. Heterocycl. Chem.,* **7,** 455 (1970).
9. F. Kroehnke and H. Schaefer, *Chem. Ber.,* **95,** 1098 (1962).
10. T. Talik and Z. Talik, *Rocz. Chem.,* **36,** 539 (1962); *Chem. Abstr.,* **57,** 12421a (1962).
11. H. Rubinstein, G. Hazen, and R. Zerfing, *J. Chem. Eng. Data,* **12,** 149 (1967); *Chem. Abstr.,* **66,** 55347m (1967).
12. S. Yamada and Y. Kikugawa, *Chem. Ind.* (London), 1325 (1967).
13. R. M. Johnson, *J. Chem. Soc., B,* 1058 (1966).
14. T. Talik, *Rocz. Chem.,* **36,** 1465 (1962); *Chem. Abstr.,* **59,** 6360b (1963).
15. T. Talik, *Rocz. Chem.,* **36,** 1563 (1962); *Chem. Abstr.,* **59,** 6231d (1963).
16. Z. Talik, *Rocz. Chem.,* **35,** 475 (1961); *Chem. Abstr.,* **57,** 15065h (1962).
17. Z. Talik, *Bull. Acad. Polon. Sci., Ser. Sci. Chim.,* **9,** 561 (1961); *Chem. Abstr.,* **60,** 2884d (1964).
18. Z. Talik, *Bull. Acad. Polon. Sci., Ser. Sci. Chim.,* **9,** 567 (1961); *Chem. Abstr.,* **60,** 2883h (1964).
19. Z. Talik, *Bull. Acad. Polon. Sci., Ser. Sci. Chim.,* **9,** 571 (1961); *Chem. Abstr.,* **60,** 2885b (1964).
20. T. Talik and Z. Talik, *Rocz. Chem.,* **38,** 777 (1964); *Chem. Abstr.,* **61,** 10653e (1964).
21. R. A. Abramovitch, F. Helmer, and M. Liveris, *J. Chem. Soc., B,* 492 (1968).
22. T. Talik and Z. Talik, *Rocz. Chem.,* **40,** 1675 (1966); *Chem. Abstr.,* **66,** 94889j (1967).
23. A. Puszynski and T. Talik, *Rocz. Chem.,* **43,** 1771 (1969).
24. E. Ochiai and H. Mitarashi, *Chem. Pharm. Bull.* (Tokyo), **11,** 1084 (1963); *Chem. Abstr.,* **59,** 12755a (1963).
25. C. Kaniko and S. Yamada, *Tetrahedron Lett.,* 4729 (1966).
26. P. Yates, M. J. Jorgensen, and S. K. Roy, *Can. J. Chem.,* **40,** 2146 (1962).
27. E. V. Brown, A. F. Smetana, and A. A. Hambden, *J. Med. Chem.,* **8,** 252 (1965).
28. E. W. Elslager, D. B. Capps, D. H. Kurtz, L. M. Werbel, and D. R. Worth, *J. Med. Chem.,* **6,** 646 (1963).
29. W. Czuba, *Bull. Acad. Polon. Sci., Ser. Sci. Chem.,* **8,** 281 (1960); *Chem. Abstr.,* **60,** 2883g (1964).
30. W. A. F. Gladstone and R. O. C. Norman, *J. Chem. Soc., C,* 1527 (1966).

31. P. Tomasik, *Rocz. Chem.*, **44**, 509 (1970).
32. Z. J. Allan, J. Podstata, and Z. Vrba, *Tetrahedron Lett.*, 4855 (1969).
33. M. Bellas and H. Suschitsky, *J. Chem. Soc.*, 4007 (1963).
34. R. D. Chambers, J. Hutchinson, and W. K. R. Musgrave, *Proc. Chem. Soc.*, 83 (1964).
35. M. T. Chaudhry, G. A. Powers, R. Stephens, and J. C. Tatlow, *J. Chem. Soc.*, 874 (1964).
36. T. Kauffman, J. Hansen, C. Hansen, C. Kosel, and W. Schoeneck, *Ann. Chem.*, **656**, 103 (1962).
37. G. Palazzo and L. Baiocci, *Ann. Chim.* (Rome), **55**, 935 (1965); *Chem. Abstr.*, **63**, 16335h (1965).
38. T. Itai and S. Kamiya, *Chem. Pharm. Bull.* (Tokyo), **9**, 87 (1961).
39. V. Ya. Pochinok and L. F. Ayromenko, *Ukr. Chim. Zh.*, **28**, 511 (1962); *Chem. Abstr.*, **58**, 2348g (1963).
39a. R. A. Abramovitch and B. W. Cue, *J. Org. Chem.*, **38**, 173 (1973).
40. S. Kamiya, *Chem. Pharm. Bull.* (Tokyo), **10**, 471 (1962).
40a. R. Huisgen and K. von Frauenberg, *Tetrahedron Lett.*, 2595 (1969).
40b. T. Sasaki, K. Kanematsu, and M. Murata, *Tetrahedron*, **27**, 5121 (1971).
41. L. Pentimalli, *Ann. Chim.* (Rome), **53**, 1123 (1963); *Chem. Abstr.*, **60**, 10645d (1964).
42. Z. Talik and T. Talik, *Rocz. Chem.*, **36**, 417 (1962); *Chem. Abstr.*, **58**, 5627b (1963).
43. A. Signor, E. Scoffone, L. Biondi, and S. Bezzi, *Gazz. Chim. Ital.*, **93**, 65 (1963); *Chem. Abstr.*, **59**, 2811g (1963).
44. P. C. Jain, S. K. Chatterjee, and N. Anand, *Ind. J. Chem.*, **4**, 403 (1966); *Chem. Abstr.*, **66**, 46552x (1967).
45. P. Nantka-Namirski, *Acta Polon. Pharm.*, **18**, 391 (1961); *Chem. Abstr.*, **57**, 16554b (1962).
46. T. Talik and Z. Talik, *Rocz. Chem.*, **40**, 1457 (1966); *Chem. Abstr.*, **66**, 9488h (1967).
47. K. Lewicka and E. Plazek, *Rocz. Chem.*, **39**, 643 (1945); *Chem. Abstr.*, **63**, 8311a (1965).
48. K. W. Merz and H. J. Janssen, *Arch. Pharm.* (Weinheim), **297**, 10 (1964); *Chem. Abstr.*, **60**, 10643f (1964).
49. Y. Suzuki, *Yakugaku Zasshi*, **81**, 1146 (1961); *Chem. Abstr.*, **56**, 3450f (1962).
50. K. T. Potts and H. R. Burton, *J. Org. Chem.*, **31**, 251 (1966).
51. K. M. Ghoneim, M. Khalifa, and Y. M. Abou-Zeid, *J. Pharm. Sci.*, **55**, 349 (1966); *Chem. Abstr.*, **64**, 12637b (1966).
52. J. Buchi, P. Fabiani, H. U. Frey, A. Hofstetter, and A. Schorno, *Helv. Chim. Acta*, **49**, 272 (1966).
53. L. Pentimalli, *Ann. Chim.* (Rome), **55**, 435 (1965); *Chem. Abstr.*, **63**, 6963d (1965).
54. L. M. Werbel, E. W. Elslager, M. W. Fisher, Z. B. Gavrilis, and A. A. Phillips, *J. Med. Chem.*, **11**, 411 (1968).
55. R. W. Faessinger and E. V. Brown, *Trans. Kentucky Acad. Sci.*, **24**, 349 (1966); *Chem. Abstr.*, **60**, 14465a (1964).
56. M. Regitz and A. Liedhegener, *Chem. Ber.*, **99**, 2918 (1966).
57. G. B. Barlin, *J. Chem. Soc.*, 2150 (1964); *Chem. Abstr.*, **61**, 7763d (1964).
58. N. Bild and M. Hesse, *Helv. Chim. Acta*, **50**, 1885 (1967); *Chem. Abstr.*, **67**, 120957x (1967).
59. P. T. Cottrell and P. H. Reger, *Mol. Phys.*, **12**, 149 (1967); *Chem. Abstr.*, **67**, 69314f (1967).
60. J. M. Essery and K. Schofield, *J. Chem. Soc.*, 225 (1963); *Chem. Abstr.*, **58**, 12389f (1963).

61. G. Favini, A. Gamba, and I. R. Bellobono, *Spectrochim. Acta,* **23A,** 89 (1967); *Chem. Abstr.,* **66,** 50516f (1967).
62. G. Favini, M. Raimondi, and G. Gandolfo, *Spectrochim. Acta,* **24A,** 207 (1968); *Chem. Abstr.,* **68,** 118201y (1968).
63. C. R. Frank and L. B. Rogers, *Inorg. Chem.,* **5,** 615 (1966); *Chem. Abstr.,* **64,** 15186g (1966).
64. G. Giacometti and G. Rigatti, *Nucl. Mag. Res. Chem. Proc. Symp. Cagliari, Italy* 173 (1964); *Chem. Abstr.,* **66,** 15277g (1967).
65. A. Ishitani and S. Nagakura, *Bull. Chem. Soc. Jap.,* **38,** 367 (1965); *Chem. Abstr.,* **63,** 5144c (1965).
66. M. Itô, T. Okamoto, and S. Nagakura, *Bull. Chem. Soc. Jap.,* **36,** 1665 (1963); *Chem. Abstr.,* **60,** 7595d (1964).
67. M. Itô and S. Nagakura, *Bull. Chem. Soc. Jap.,* **38,** 825 (1965); *Chem. Abstr.,* **63,** 5497b (1965).
68. M. Itô and T. Okamoto, *Chem. Pharm. Bull.* (Tokyo), **15,** 435 (1967); *Chem. Abstr.,* **67,** 43237h (1967).
69. R. A. Jones and R. P. Rao, *Aust. J. Chem.,* **18,** 583 (1965); *Chem. Abstr.,* **62,** 15593e (1965).
70. C. R. Kanekar and H. V. Venkatasetty, *Current Sci.* (India), **34,** 555 (1965); *Chem. Abstr.,* **64,** 2886a (1966).
71. C. Kaneko, S. Yamada and I. Yokoe, *Tetrahedron Lett.,* 4729 (1966); *Chem. Abstr.,* **66,** 2043v (1967).
72. P. D. Kaplan and M. Orchin, *Inorg. Chem.,* **4,** 1393 (1965); *Chem. Abstr.,* **64,** 14238f (1965).
73. A. R. Katritzky and J. M. Lagowski, *J. Chem. Soc.,* 43 (1961); *Chem. Abstr.,* **56,** 4727a (1962).
74. G. H. Keller, L. Bauer, and C. L. Bell, *J. Heterocycl. Chem.,* **5,** 647 (1968); *Chem. Abstr.,* **69,** 111306y (1968).
75. T. Kubota, K. Nishikida, H. Miyazaki, K. Iwatani, and Y. Oishi, *J. Amer. Chem. Soc.,* **90,** 5080 (1968); *Chem. Abstr.,* **69,** 95819k (1968).
76. J. Lee and K. G. Orrell, *J. Chem. Soc.,* 582 (1965); *Chem. Abstr.,* **62,** 6455h (1965).
77. T. Okano, K. Uekama, Y. Isawa, and K. Tsukuda, *Yakugaku Zasshi,* **87,** 1309 (1967); *Chem. Abstr.,* **68,** 95185e (1968).
78. V. I. Slovetskii, L. I. Khmal'nitskii, O. V. Lebedev, T. S. Novikova, and S. S. Novikov, *Khim. Geterotsikl. Svedin., Akad. Nauk Latv. SSR,* 835 (1965); *Chem. Abstr.,* **64,** 15705f (1966).
79. J. Suszko and M. Szafran, *Bull. Acad. Polon. Sci., Ser. Sci. Chem.,* **10,** 233 (1962); *Chem. Abstr.,* **58,** 7519c (1963).
80. G. Tacconi and S. Pietra, *Ann. Chim.* (Rome), **55,** 810 (1965); *Chem. Abstr.,* **64,** 686f (1966).
81. T. K. Wu and B. P. Dailey, *J. Chem. Phys.,* **41,** 3307 (1964); *Chem. Abstr.,* **62,** 1228d (1965).

# CHAPTER IX

# Aminopyridines

C. S. GIAM

*Chemistry Department,*
*Texas A & M University*
*College Station, Texas*

# I. Nuclear Amines

## 1. Preparation of Primary Amines

### A. *From Nonpyridine Starting Materials*

Aminopyridines were detected in the complex mixture of products resulting from the condensation of paraldehyde with excess ammonia at 210° at 100 atm pressure.[1]

Aminohalopyridines are obtained by the action of anhydrous halogen acids (hydrogen bromide or hydrogen iodide) on 3-hydroxyglutaronitriles, glutaconitriles or 1,3-dicyano-2-propanol; thus, a mixture of 1,3-dicyano-2-propanol and anhydrous hydrogen bromide gives (after neutralization of the product) 2-amino-6-bromopyridine.[2,3]

2,6-Diamino-3-picoline **(IX-1)** is obtained in 30% yield by the dehydrogenation of α-methylglutarimidine with a Pd catalyst. 2,6-Diamino-3,5-dimethylpyridine was prepared similarly from α,α-dimethylglutarimidine.[4]

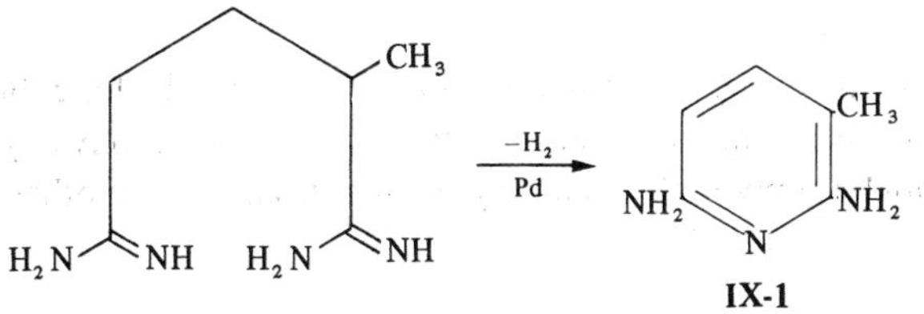

$$\xrightarrow[\text{Pd}]{-H_2}$$

**IX-1**

The reaction of 3-hydroxyglutaronitrile with aniline hydrobromide yields 2,6-dianilinopyridine;[5] using thiophenol, hydrogen bromide, and acetic acid instead of aniline hydrobromide, 2-amino-6-phenylthiopyridine results.[6]

Interesting cases of ring formation have been reported, for example, the synthesis of 2-amino-3,5-dicyano-6-ethoxy-4-phenylpyridine **(IX-4)** from benzalmalonitrile **(IX-3)** and ethanolic potassium hydroxide.[7] The presence of

$$\underset{\textbf{IX-3}}{C_6H_5CH{=}C(CN)_2} + KOH \xrightarrow{EtOH} C_6H_5CHO + \underset{\textbf{IX-2}}{K^+CH(CN)_2^-}$$

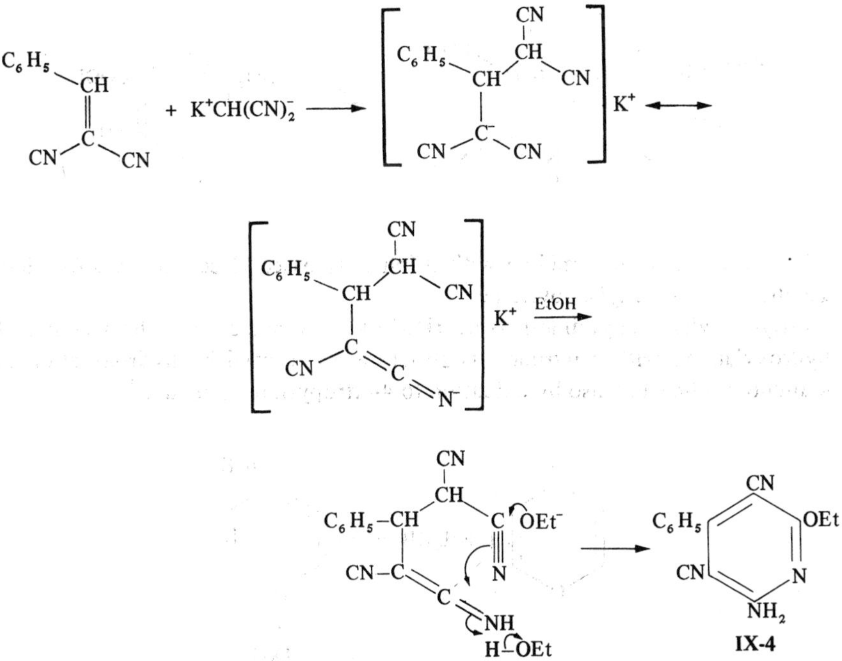

$$+\ K^+CH(CN)_2^- \longrightarrow \quad K^+ \longleftrightarrow$$

$$K^+ \xrightarrow{EtOH}$$

$$\longrightarrow$$

**IX-4**

benzaldehyde and the absence of malonitrile in the reaction products suggested a mechanism involving a reverse aldol condensation to form benzaldehyde and the potassium salt of malonitrile (**IX-2**). **IX-2** then condenses with **IX-3** to yield the substituted pyridine **IX-4**. If methanol is used instead of ethanol, the 6-methoxy derivative is formed.

Cyclizations involving activated carbonyl compounds have been reported. 1-Ethoxy-2,4-dioxopentane condenses with β-amino-β-ethoxyacrylate to yield ethyl 2-amino-4-ethoxymethyl-6-methylnicotinate (**IX-5**).[8] Another condensation

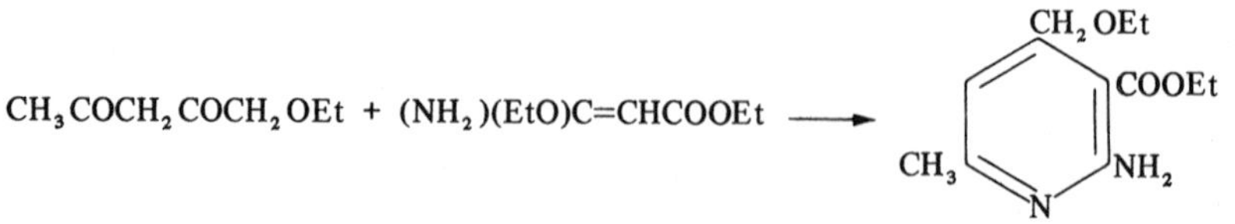

$CH_3COCH_2COCH_2OEt + (NH_2)(EtO)C{=}CHCOOEt \longrightarrow$

**IX-5**

reaction involves nitromalondialdehyde and ethyl amidinoacetate; this results in the formation of ethyl 2-amino-5-nitronicotinate (**IX-6**):[9]

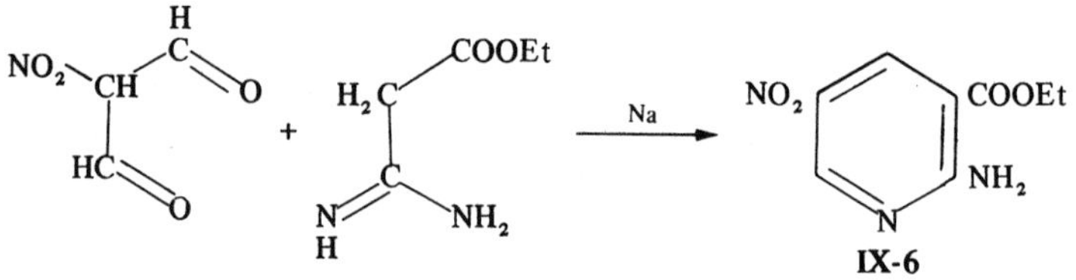

**IX-6**

The reaction of acetamidine with 3-(phenylazo)acetylacetone gives 6-amino-2,4-dimethyl-3-phenylazopyridine.[10]

4-Hydroxylaminopyridine-1-oxide (**IX-7**) was obtained from the reaction of hydroxylamine with γ-pyrone. Its structure was proved by hydrogenation to 4-aminopyridine and also by oxidation to 4-nitropyridine-1-oxide.[11]

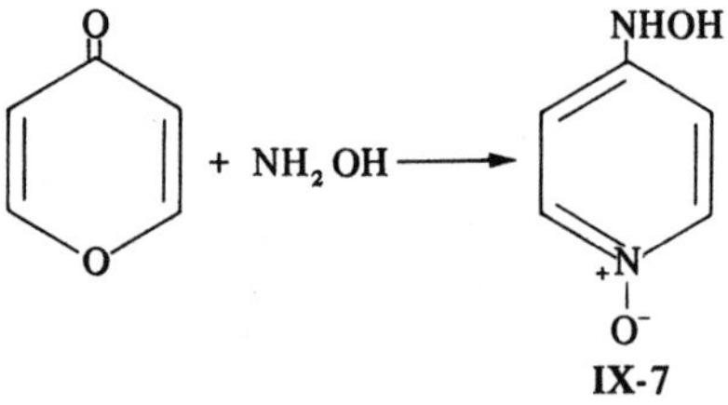

$+ NH_2OH \longrightarrow$

**IX-7**

## B. *Amination of Pyridines with Alkali Amides (The Tschitschibabin Reaction)*

Direct amination of alkylpyridines with sodium amide generally results in the formation of the 2- or 6-amino-derivatives. 2-Amino-3-picoline and 2-amino-4-picoline were obtained in good yields (70–80%) from the direct amination of the crude picoline fraction of coal tar with sodamide.[12] 4-Alkylpyridines having more than 5 carbon atoms in the alkyl group can be aminated with sodamide in inert solvents, or with sodium in liquid ammonia, to give 2-amino-4-alkylpyridine;[13, 14] for example, 4-isopentylpyridine yields 2-amino-4-isopentylpyridine when boiled under reflux with sodamide in xylene at 130 to 140° for 8 hours.[13] Even with 2-methyl-4-pyridone, direct amination results in the formation of 6-amino-2-methyl-4-pyridone.[15] Amination of 3-substituted pyridines by sodamide gives predominantly the 2-amino-3-substituted derivative rather than the 6-isomer;[16] for example, 3-picoline gives a mixture of 2-amino-3-picoline and 6-amino-3-picoline in the ratio of 10.5:1.[17, 18]

When 2,5-dimethyl-4-phenylpyridine is heated with sodium amide at 180° for 5 hours, 2-amino-3,6-dimethyl-4-phenylpyridine **(IX-8)** is obtained.[19] If the 2- and 6-positions are not available, direct amination takes place at the 4-position; thus, the reaction of 2,6-lutidine with sodium in liquid ammonia yields 4-amino-2,6-lutidine **(IX-9)**.[20] The 4-chloro substituent in perchloropyridine is

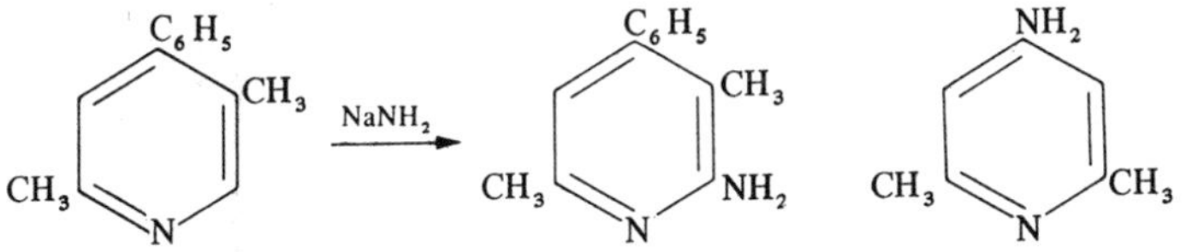

$C_6H_5$, $CH_3$, $CH_3$, N $\xrightarrow{NaNH_2}$ $C_6H_5$, $CH_3$, $CH_3$, $NH_2$, N    $NH_2$, $CH_3$, $CH_3$, N

**IX-8**    **IX-9**

easily replaced by an amino group; for example, perchloropyridine readily reacts with sodium amide to give 4-amino-2,3,5,6-tetrachloropyridine[21] (see Chapter VI).

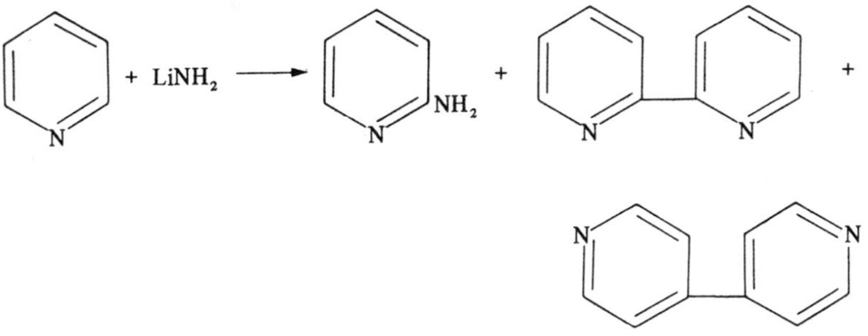

N + $LiNH_2$ ⟶ $NH_2$, N + N, N + N, N

**IX-10**

The reaction of lithium amide with pyridine has been studied; 2,2′- and 4,4′-dipyridyl, rather than 2-aminopyridine, were the predominant products **(IX-10)**.[22] Whereas the reaction of sodium amide with pyridine in various solvents gave yields of 2-aminopyridine of at least 43%, lithium amide under similar reaction conditions produced less than 1% of 2-aminopyridine.[22] The mechanism of the Tschitschibabin reaction in which a ring hydrogen atom was replaced by an amino group has been the subject of much recent discussion. The "pyridyne" mechanism **(IX-11)** proposed by Levitt and Levitt[23] and supplemented by molecular orbital calculations[24] was proved unacceptable by Abramovitch and by other workers.[17,18,25-27] Instead, convincing evidence indicated that the overall mechanism was an $S_NAr2$ type addition-elimination pathway **(IX-12)**. The subject has been reviewed by Abramovitch and Saha.[16]

The mechanism of ammonodehalogenation reactions of halopyridines by alkali amides probably involves two mechanistic pathways. The reactions of 3- or 4-halopyridines with potassium amide in liquid ammonia proceed mainly, if not exclusively, *via* 3,4-pyridines to give mixtures of 3- and 4-aminopyridines; in contrast, 2-halopyridines react *via* an addition-elimination pathway, giving only 2-aminopyridines.[28-31]

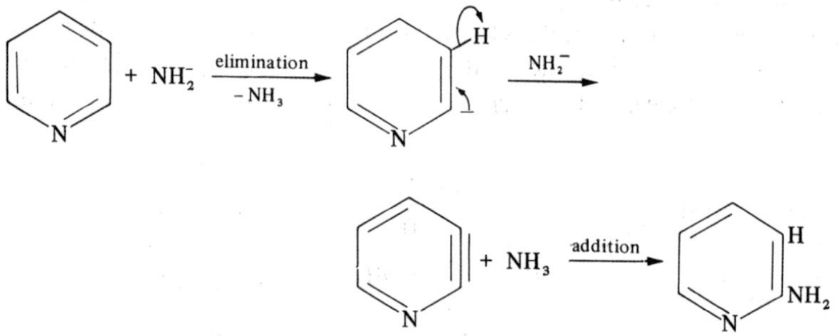


**(IX-11)**

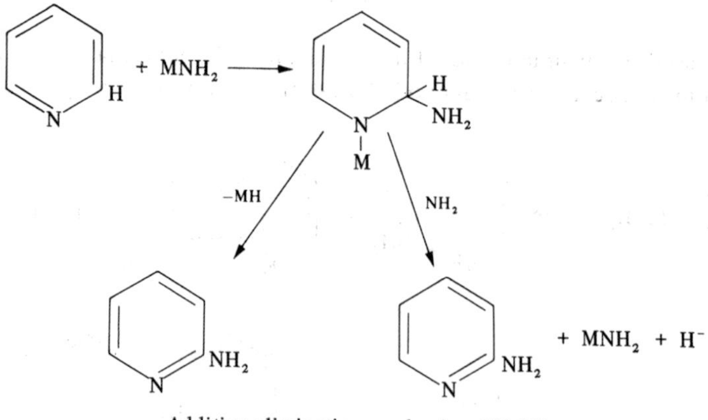


Addition-elimination mechanism **(IX-12)**

The mechanistic pathways of the ammonodehalogenation reactions of aminohalopyridine-1-oxides have been studied[30-32] and are discussed in Chapter IV.6.

The reactions of dibromopyridines and various substituted bromopyridines with alkali amides have been reported;[33-36] aminopyridines were the products in most but not all cases. For example, 2,6-dibromopyridine gives 4-amino-2-methylpyridine when treated with potassium amide in liquid ammonia;[37] under similar conditions but using 2-amino-3-bromopyridine, 3-cyanopyrrole is formed[38] (see Chapter I.A.).

## C. *Ammonolysis of Halopyridines*

Halopyridines are converted to aminopyridines under a variety of conditions. The conversion of 3-bromopyridine to 3-aminopyridine requires a catalyst (copper sulfate), ammonium hydroxide, and 20 atm pressure.[39-40] When 2-bromo-5-chloro-3-methylaminopyridine was heated with aqueous ammonium hydroxide and copper sulfate in a sealed tube, the bromine atom was preferentially replaced to yield 2-amino-5-chloro-3-methylaminopyridine.[41] Concentrated ammonium hydroxide is used to ammonodehalogenate 4-chloronicotinic acid-1-oxide[42] and 4-chloro-3,5-dimethylpyridine-1-oxide[43] to the corresponding 4-amino derivatives. Nitrohalopyridines undergo amination under milder conditions;[44] for example, 2-chloro-3,5-dinitropyridine[45] and 2,4-dichloro-3-nitropyridine[42] react with ammonia at ambient temperatures to give the corresponding 2-amino derivatives. The reactions of some chloro-, bromo-, or iodonitropyridines with ammonia do not necessarily result in the displacement of the halogen atom. Thus, 2-halo- or 3-halo-4-nitropyridines react with

$NO_2$ / Cl / N $\xrightarrow{NH_4OH}$ $NH_2$ / Cl / N **IX-13**

ammonium hydroxide to give 2-halo- and 3-halo-4-aminopyridines, respectively[47, 48] (e.g., **IX-13**).

The fluorine atom in 2-fluoro-4-nitropyridine is more labile than the nitro group, and it is readily replaced by ammonium hydroxide to give 2-amino-4-nitropyridine **(IX-14)** rather than 2-fluoro-4-aminopyridine.[49] Similarly, 3-fluoro-4-nitropyridine and its *N*-oxide give the corresponding 3-amino-4-nitro derivatives.[50-52]

The ammonolysis reactions of various polyfluoropyridines have been studied.[53-57] The 4-fluoro substituent in pentafluoropyridine is easily

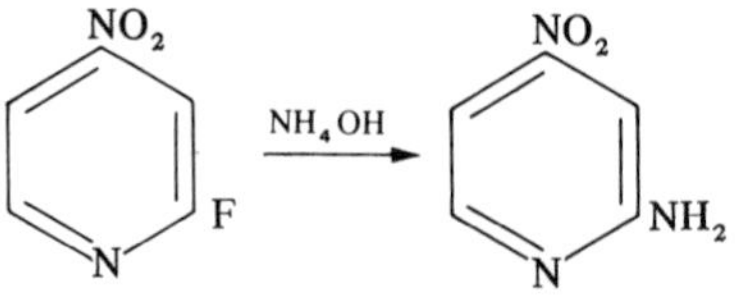

$NO_2$ $NO_2$

$NH_4OH$

F $NH_2$

N N

**IX-14**

replaced by ammonia (or other amines) to give 4-amino-2,3,5,6-tetrafluoropyridine;[53,54] similarly, 3-chloro-2,4,5,6-tetrafluoropyridine and 3,5-dichloro-2,4,6-trifluoropyridine yield the corresponding 4-amino derivatives.[56,58] If the 4-position does not bear a fluorine atom, as in 4-iodo-2,3,5,6-tetrafluoropyridine, the 2-fluoro substituent undergoes ammonolysis preferentially to give 2-amino-4-iodo-3,5,6-trifluoropyridine[59] (see also Chapter VI).

## D. *Hofmann, Schmidt, and Curtius Reactions*

Hofmann degradations were carried out on 5-chloro- and 5-bromonicotinamide and 5-ethyl-2-picolinamide to yield 3-amino-5-chloropyridine,[60] 3-amino-5-bromopyridine,[61] and 2-amino-5-ethylpyridine,[62] respectively. When pyridine carboxylic acids are treated with sodium azide in an oleum medium, a good yield (69%) of 3-aminopyridine and poorer yields (<30%) of 2- and 4-aminopyridine are realized.[63] 3-Amino-5-nitropyridine is similarly prepared by the Schmidt reaction using 5-nitronicotinic and hydrazoic acid.[64,65]

Several aminofluoropyridines are prepared either by the Hofmann reaction or the Curtius degradation. Thus, 2-amino-6-fluoropyridine is easily obtained from either 6-fluoropicolinamide or 6-fluoropicolinic hydrazide.[66] 3-Amino-2-fluoropyridine and 5-amino-2-fluoropyridine are prepared best from the appropriate 2-fluoropyridine carboxamide rather than from the hydrazide because the α-fluorine atom is easily replaced by hydrazine.

## E. *Reduction of Nitro Compounds*

The preparation of aminopyridines through the reduction of nitropyridines and nitropyridine-1-oxides, may be subdivided into catalytic hydrogenation and noncatalytic reducing systems. 3-Ethyl-4-nitropyridine-1-oxide and 4-amino-2,6-lutidine-1-oxide are hydrogenated over platinum oxide to give 3-ethyl-4-aminopyridine and 4-amino-2,6-lutidine,[68] respectively. Palladium on various inert supports[69-77] has been used instead of platinum oxide. If a halogen substituent is present, not only is the nitro group reduced but the halogen atom is removed as well; for example, 6-chloro-5-cyano-4-ethoxymethyl-3-nitro-2-picoline gives

3-amino-5-cyano-4-ethoxymethyl-2-picoline **(IX-15)** when hydrogenated in absolute ethanol in the presence of palladium on calcium carbonate.[77] Often the

$$\text{6-Cl-5-CN-4-}CH_2OEt\text{-3-}NO_2\text{-2-}CH_3\text{-pyridine} \xrightarrow{H_2/Pd-CaCO_3} \text{5-CN-4-}CH_2OEt\text{-3-}NH_2\text{-2-}CH_3\text{-pyridine}$$

**IX-15**

halogen substituent may be more effectively removed by treatment of the halopyridine with hydrazine prior to hydrogenation.[69]

Catalytic hydrogenation of 6-chloro-5-cyano-4-ethoxymethyl-3-nitro-2-picoline over nickel formate selectively reduces the nitro function but spares the chloro and cyano groups, giving 3-amino-6-chloro-5-cyano-4-ethoxymethyl-2-picoline.[78] Both nitropyridines and nitropyridine-1-oxides are converted to aminopyridines by catalytic hydrogenation over Raney nickel.[79-88] For example, 4-nitropyridine-1-oxide and 4-nitro-3-picoline-1-oxide are hydrogenated over Raney nickel to give 4-aminopyridine and 4-amino-3-picoline, respectively.[84-86] Similarly, halonitropyridine-1-oxides may be converted to the corresponding aminohalopyridines;[87] the halogen atom is not removed.

Metal-acid solutions have been employed in the reduction of nitropyridines to aminopyridines. Zinc dust and acetic acid have been used in the preparation of 3-amino-1-benzyl-4-pyridone or 3-amino-1-benzyl-5-iodo-4-pyridone from the corresponding 3-nitro derivatives.[89] Tin or iron in acids is more popular.[74, 90-97] For example, 3-amino-2-(*o*-bromobenzyloxy)-pyridine is obtained by the reduction of 2-(*o*-bromobenzyloxy)-3-nitropyridine using iron and hydrochloric acid.[94] Various ferrous salts have also been effective reducing agents. Several derivatives of phenyl 5-amino-2-pyridyl sulfide **(IX-16)** (or the sulfone) are

$$\text{5-}NO_2\text{-2-pyridyl-S-}C_6H_5 \xrightarrow{Fe^{2+}} \text{5-}NH_2\text{-2-pyridyl-S-}C_6H_5$$

**IX-16**

obtained by the reduction of phenyl 5-nitro-2-pyridyl sulfide (or the sulfone) with ferrous salts in ammonium chloride solution.[98] These amino derivatives are given in Tables IX-19 and IX-20. Ferrous sulfate successfully reduces 3-nitro-2-picoline-4-carboxaldehyde to the unstable 3-amino-2-picoline-4-carboxaldehyde;[99] ferrous hydroxide converts 3-iodo-4-nitropyridine-1-oxide to 3-iodo-4-aminopyridine.[100]

4-Nitropyridine is reduced to 4-aminopyridine by sulfur dioxide dissolved in dilute sulfuric acid containing hydrogen iodide as a catalyst.[101] Aqueous solutions of sodium sulfide or ammonium sulfide reduce only the 3-nitro substituent in 3,5-dinitropyridine and 3,5-dinitro-2-pyridone.[102-104] 4-Nitro-2,6-lutidine-1-oxide **(IX-17)** reacts with aqueous hydrosulfite to give 4-amino-2,6-lutidine **(IX-18)**;[105] however, **IX-17** reacts with sodium borohydride to give 4,4′-azo-2,6-lutidine-1,1′-dioxide **(IX-19)**.[106]

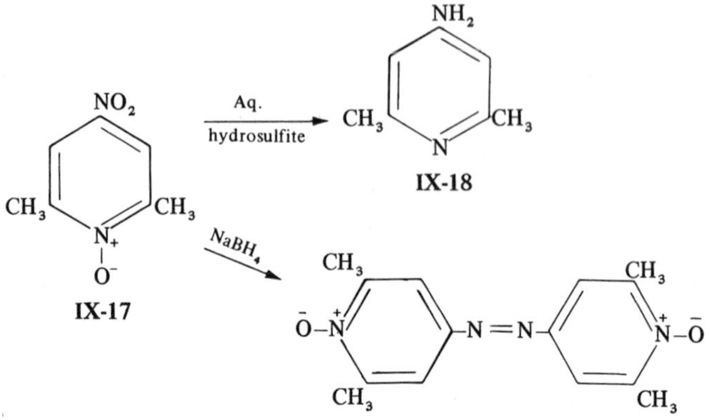

$NO_2$ $CH_3$ $CH_3$ $N^+$ $O^-$ **IX-17** —Aq. hydrosulfite→ $NH_2$ $CH_3$ $CH_3$ N **IX-18**

**IX-17** —$NaBH_4$→ **IX-19**

When 2-halo-4-nitropyridine-1-oxides are treated with hydrazine hydrate in ethanol, 4-amino-2-halopyridine-1-oxides result.[107] The reaction of 4-nitropyridine-1-oxide (**IX-20**, R = H) and of 4-nitro-2-picoline-1-oxide (**IX-20**, R = $CH_3$) with phenylhydrazine gives 4-hydroxylaminopyridine-1-oxide (**IX-21**, R = H) and 4-hydroxylamino-2-picoline-1-oxide (**IX-21**, R = $CH_3$), respectively, rather than the corresponding 4-amino derivatives[108, 109] (See also Chapter VIII).

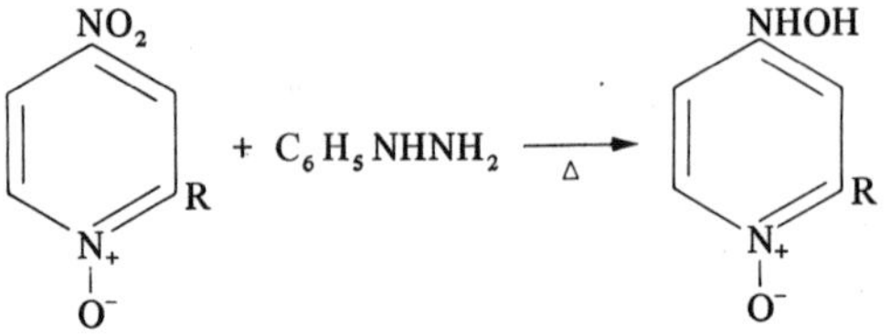

**IX-20** ($NO_2$, R, $N^+$, $O^-$) + $C_6H_5NHNH_2$ —Δ→ **IX-21** (NHOH, R, $N^+$, $O^-$)

## F. *From Pyridylpyridinium Halides and N-(Pyridyl)-2-pyridones*

When *N*-(2-methyl-3-pyridyl)pyridinium perchlorate is dissolved in piperidine and ethanol and heated, 3-amino-2-picoline **(IX-22)** is obtained.[110] Similarly, other *N*-(2- and 4-pyridyl)pyridinium salts give the corresponding 2- and 4-aminopyridines.[111]

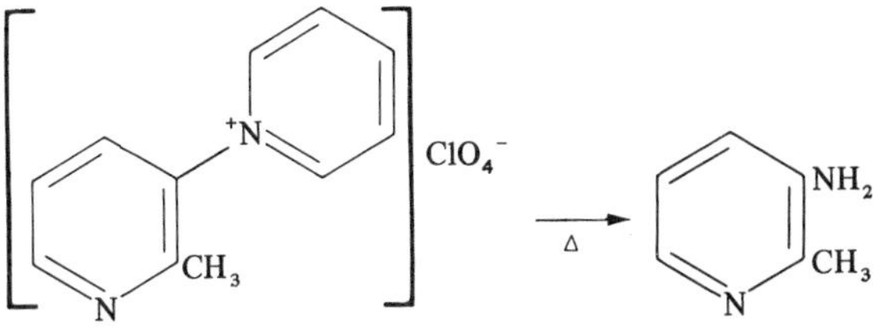

$ClO_4^-$ Δ $NH_2$ $CH_3$

**IX-22**

Heating 1-(4-hydroxy-2-pyridyl)pyridinium chloride in aniline and acetic anhydride gives a product that, on hydrogenation over palladium-charcoal, proved to be 2-aminopyridine.[112]

Contrary to a previous report,[113] the reaction of pyridine-1-oxide with acetic anhydride does not give exclusively 2-pyridone; 2-aminopyridine and *N*-(2-pyridyl)-2-pyridone **(IX-23)** are also formed.[114] Similarly, 6-amino-3-picoline and 3-methyl-*N*-(5-methyl-2-pyridyl)-2-pyridone were isolated in the reaction of 3-picoline-1-oxide with acetic anhydride[115] (see also Chapter IV).

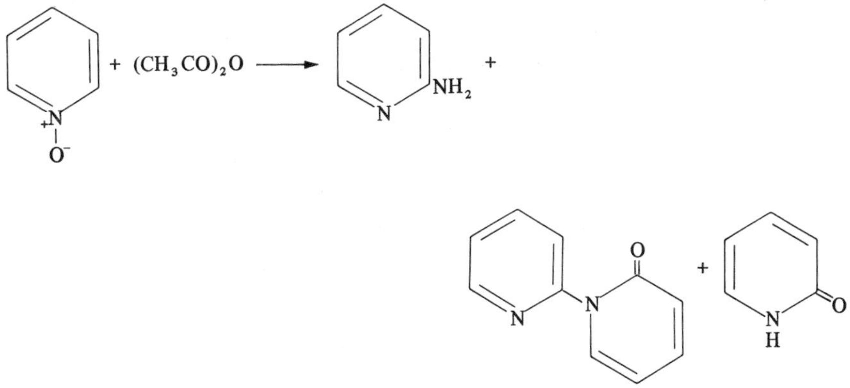

+ $(CH_3CO)_2O$ → $NH_2$ +

**IX-23**

When 3-methoxypyridine-1-oxide is treated with 2-bromopyridine, 1-(5-methoxy-2-pyridyl)-2-pyridone **(IX-24)**, 1-(3-methoxy-2-pyridyl)-2-pyridone **(IX-25)**, and some other unidentified products are obtained. Heating **IX-24** and **IX-25** with aqueous sodium hydroxide gives the corresponding aminomethoxypyridines **IX-26** and **IX-27**. Several other aminomethoxypyridines are prepared similarly.[116]

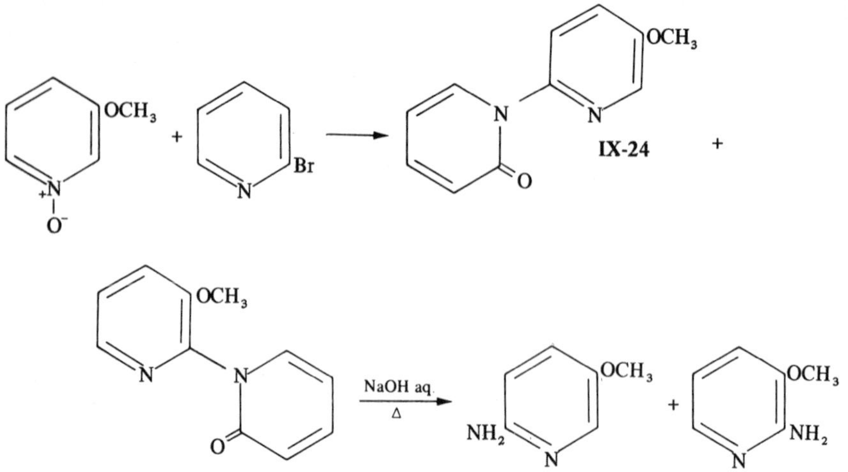

## G. *Miscellaneous Methods*

Hydroxylamine reacts with 2,6-dimethylpyrone or with the barium salt of diacetyl acetone to give 4-hydroxylamino-2,6-lutidine-1-oxide, which can be reduced catalytically to 4-amino-2,6-lutidine using hydrogen and platinum oxide.[117]

When pyridine-4-sulfonic acid is heated with ammonium hydroxide or with an alkylamine in the presence of a small amount of zinc chloride it gives 4-aminopyridine and 4-alkylaminopyridine, respectively.[118] 4-Amino-2,6-di-*t*-butylpyridine is obtained on heating 2,6-di-*t*-butylpyridine-4-sulfonic acid with aqueous ammonia in a small sealed tube.[119] The cyano group in the 4-position is also labile; thus 4-cyanopyridinium methiodide reacts with ammonium hydroxide to give 4-aminopyridine methiodide **(IX-28)**.[120]

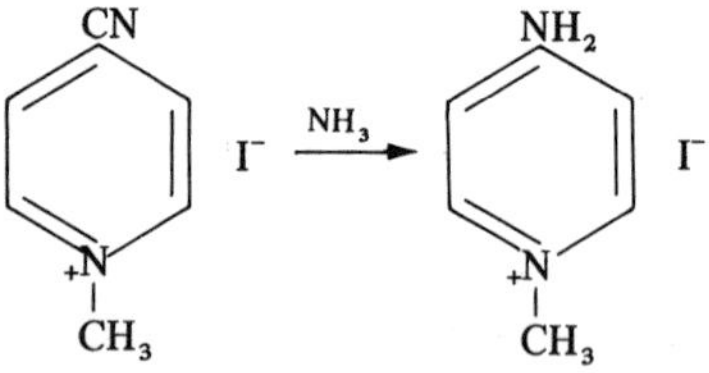

2-Amino-6-fluoropyridine is prepared by the hydrolysis of 6-fluoropyridine-2-isocyanate.[121] *N*-(4-Methyl-5-phenylnicotinyl)-*N'*-(4-methyl-5-phenyl-3-pyridyl)

urea is hydrolyzed with hydrochloric acid to give 3-amino-4-methyl-5-phenylpyridine **(IX-29)**.[12]

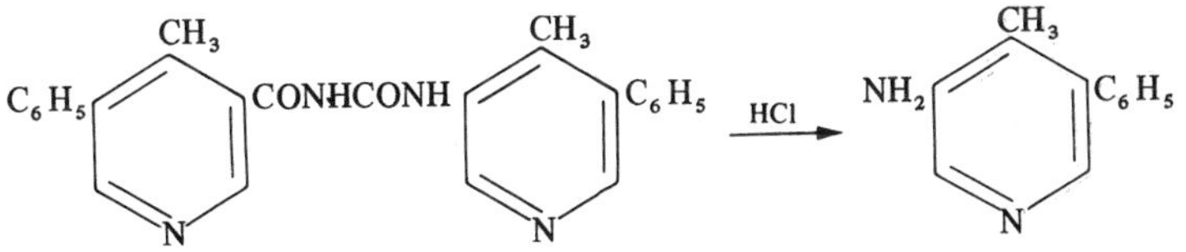

**IX-29**

When the thiazoline **IX-30** was sublimed at 400 to 500°, small amounts of 2-amino-3,4,5,6-tetramethylpyridine **(IX-31)** were isolated.[122] 4,4-Dimethyl-1-(2-pyridylamino)glutamide is obtained from the pyrolysis of the mono-2-pyridylhydrazide of 3,3-dimethylglutaric acid.[123]

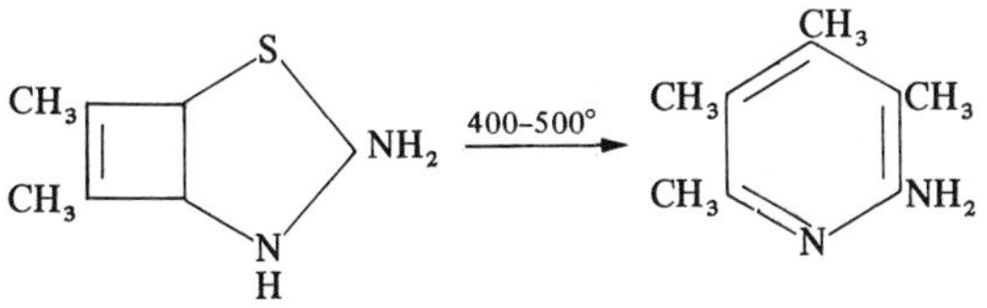

**IX-30** **IX-31**

4-Aminonicotinamide is obtained by the hydrogenation of 4-benzyloxynicotinamide over a Raney nickel catalyst.[124]

2-Aminopyridine-1-oxide can now be converted to 2-aminopyridine by heating it with granulated lead and ferrous oxalate; this procedure is valuable for aminopyridine-1-oxides that cannot be deoxygenated by phosphorous trichloride.[125]

The thermal decomposition of oximes of $\gamma$-nitro or $\gamma$-cyano ketones results in a complex mixture of products, the major component of which is a substituted 2-aminopyridine **(IX-32)**.[126]

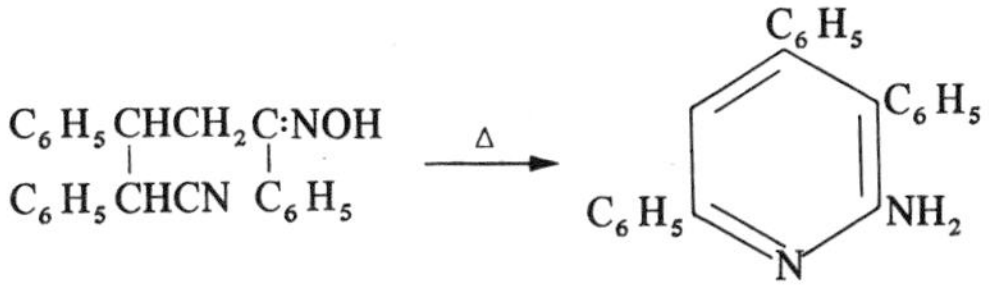

**IX-32**

A mixture of chloramine and ammonia with pyridine, 4-picoline, 2,4-lutidine, or nicotinic acid is reported to give the corresponding 2-amino derivative. 1-Aminopyridine hydrochloride as well as 2-aminopyridine hydrochloride are detected in aqueous solutions of chloramine and pyridine.[127]

## 2. Preparation of Secondary and Tertiary Amines

Vajda and Kovacs[128, 129] described three methods labeled A, B, and C, for the direct amination of the pyridine with alkyl amines in the presence of sodium amide or finely divided potassium or sodium metal. An example of method A was the preparation of 2-*n*-butylaminopyridine from a mixture of *n*-butylamine, pyridine, and sodium amide in boiling toluene (60 hours). In method B, powdered sodium (or potassium) metal was used instead of sodium amide and the reflux time was 20 hours. Method C employed sodium metal, a bath temperature of 120°, boiling under reflux with stirring for 3 hours and an additional 7 hours of boiling without stirring. The yields of 2-*n*-butylaminopyridine in the above three methods were 38%, 50%, and 33%, respectively. These authors claimed that the reaction proceeded via a radical rather than an ionic (nucleophilic) pathway because dipyridyls were also formed. When a pyridine is heated with a three- to fourfold excess of a primary aliphatic amine in the presence of finely divided sodium metal, 70 to 80% yields of 2-alkylaminopyridines are obtained.[130, 131]

Another general method of direct alkyl- and aryl- amination of the pyridine ring has recently been reported by Abramovitch and Singer.[132] Pyridine-1-oxide **(IX-33)** and *N*-phenylbenzimidoyl chloride **(IX-34)** in boiling ethylene chloride give *N*-benzoyl-2-anilinopyridine **(IX-36)**, which, on hydrolysis, gives 2-anilinopyridine **(IX-37)**; the reaction is believed to proceed *via* an intramolecular nucleophilic substitution. Various *N*-oxides (except 4-nitropyridine-1-oxide) and imidoyl chlorides have been successfully employed in this synthesis (see Tables IX-1 and IX-2). Reduction of **IX-36** gives the corresponding tertiary amine.[133]

$+ C_6H_5C(Cl){=}NC_6H_5 \xrightarrow{\Delta}$

**IX-33** **IX-34** **IX-35**

$\xrightarrow{-HCl}$ $\xrightarrow{H_2O}$

**IX-36** **IX-37**

Pyridine-1-oxide also reacts with phenyl isocyanate to give good yields of 2-anilinopyridine.[134]

Secondary amines are also prepared from aminopyridines and alkyl halides.[135-138] For example, sodium amide and 2-aminopyridine are boiled under reflux in ether for 1 hour and then with allyl chloride for 3 hours to give 61% of 2-allylaminopyridine; similarly 2-allylamino-5-bromopyridine and 2-allylamino-3,5-dibromopyridine were prepared.[140, 141]

2-Aminopyridine condenses with alcohols to yield 2-alkylaminopyridines.[142-144] 2-Aminopyridine and cyclohexanol are dissolved in 80% sulfuric acid and heated at 60 to 70° for 6 hours; the reaction mixture is then poured over ice and neutralized with ammonia to give 2-(cyclohexylamino)pyridine (70% yield). 2-Isopropylaminopyridine is similarly prepared from isopropyl alcohol.

When 2-aminopyridine is treated with sodium amide and styrene oxide, a mixture of 2-(β-hydroxy-β-phenethylamino)pyridine **(IX-38)** and 2-(β-hydroxy-α-phenethylamino)pyridine **(IX-39)** in the ratio 3:1 is obtained.[145]

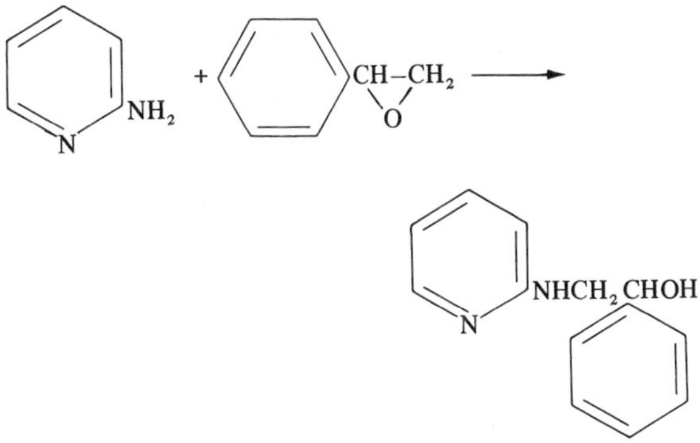

$NH_2$ + CH–$CH_2$ (O) →

$NHCH_2CHOH$ + $NHCHCH_2OH$

**IX-38** **IX-39**

2-(β-Hydroxy-β-phenethyl)aminopyridine **(IX-40)** can be prepared by the reduction of 2-mandelaminopyridine with lithium hydride.[146]

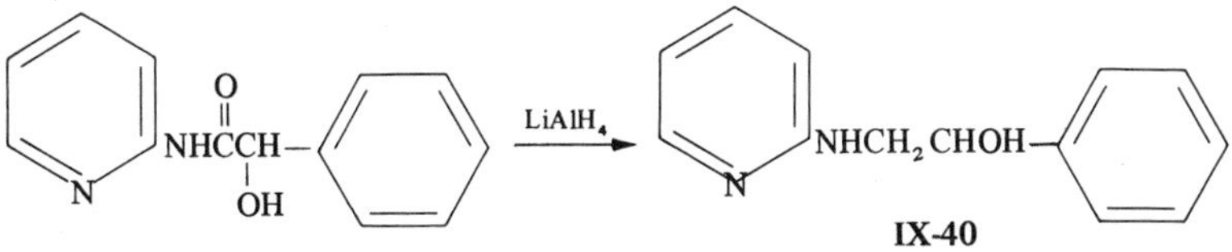

NHC(O)CH(OH)– $\xrightarrow{LiAlH_4}$ $NHCH_2CHOH$–

**IX-40**

Wibaut and Broekman reported that 4-chloropyridine reacted slowly with primary and secondary amines.[147] Tertiary amines do not react with

4-chloropyridine. A mixture of 4-chloropyridine and methylamine in benzene was heated in a sealed tube at 160° for 6 hours to yield 34% of 4-methylaminopyridine. The formation of various secondary aminopyridines from a halopyridine and an alkyl amine include the preparations of 4-isopropylaminopyridine and 4-dodecylaminopyridine.[147-151]

4-Chloropyridine-1-oxide when heated with 30% aqueous methylamine at 140° for 18 hours gives 4-methylaminopyridine-1-oxide. Several substituted 4-methylaminopyridines are similarly prepared.[43] 2-Propylaminopyridine and 2-phenethylaminopyridine are obtained by heating 2-bromopyridine with propylamine[152] and phenethylamine,[153] respectively. 2-Bromopyridine also reacts with 1-aminoindane to give 2-(1-indanylamino)pyridine.[154,155] 2,6-Dichloropyridine reacts with butylamine to give 2-butylamino-6-chloropyridine.[156] 2-Benzylamino-3-nitropyridine and 2-(chlorobenzylamino)-3-nitropyridine are readily prepared from 2-chloro-3-nitropyridine with benzyl amine and *p*-chlorobenzyl amine, respectively.[157]

Anilinopyridines are prepared by heating aniline and a halopyridine, generally in the presence of an inorganic base.[158-161] For example, 2-nitro-*N*-(3-pyridyl)-aniline is obtained by boiling a mixture of 3-bromopyridine, *o*-nitroaniline, and potassium carbonate in nitrobenzene for 24 hours under reflux.[158] 2-Anilinopyridine is also isolated (14% yield) from a reaction mixture containing aniline, iodobenzene, 2-chloropyridine, methyl salicyate and potassium carbonate, though the main product is diphenylamine.[159] When 2,6-dichloropyridine is heated with aniline, both substituents are replaced and 2,6-dianilinopyridine is obtained.[160] 2,6-Dianilinopyridine **(IX-41)** has also been obtained by the cyclization of 3-hydroxyglutaronitrile with aniline hydrobromide.[6] 2-Amino-

$$(CNCH_2)_2CHOH + C_6H_5NH_2 \longrightarrow$$ 2,6-$(C_6H_5NH)_2$-pyridine

**IX-41**

3,5-dinitropyridine can be diazotized and undergoes the Schiemann reaction to give 2-fluoro-3,5-dinitropyridine,[1] which, with aniline, gives the 2-anilino derivative **(IX-42)**.[161]

2-$NH_2$-3,5-$(NO_2)_2$-pyridine $\xrightarrow[H_3BO_3,\ 0°]{(i)\ HNO_2\ (ii)\ HF,}$ 2-F-3,5-$(NO_2)_2$-pyridine $\xrightarrow{C_6H_5NH_2}$ 2-$NHC_6H_5$-3,5-$(NO_2)_2$-pyridine

**IX-42**

4-Chloropyridine reacts with *p*-(2-diethylaminoethoxy)aniline to give 4-[*p*-(2-diethylaminoethoxy)anilino] pyridine; other substituted *p*-dialkylamino-alkoxyanilinopyridines are prepared similarly.[162, 163]

Alkoxy groups in the 2- or 4-positions of alkoxynitropyridines are easily replaced by ammonia, alkylamines, or dialkylamines. Thus, 4-methylamino-3-nitropyridine **(IX-43)** is obtained in near quantitative yield by heating 4-ethoxy-3-nitropyridine hydrochloride in an autoclave with aqueous methyl-amine.[152]

$OC_2H_5$, $NO_2$ (pyridine) + $CH_3NH_2$ ⟶ $NHCH_3$, $NO_2$ (pyridine)

**IX-43**

Other methods of preparing secondary aminopyridines have also been reported.

4-(*m*-Hydroxyphenyl)aminopyridine is obtained by heating the sodium salt of *m*-aminophenol with 4-chloropyridine.[164] A number of 4-arylaminopyridines have been prepared similarly (see Table IX-41).

A preparation of the secondary aminopyridine **IX-45** involves the reaction of 2-aminopyridine with the salt **IX-44**.[165]

$(CH_3)_2NCH{=}CHCHO + (CH_3)_2SO_4 \longrightarrow [(CH_3)_2N{\cdots}CH{\cdots}CH{\cdots}CHOCH_3]^+CH_3SO_4^-$

**IX-44**

$\xrightarrow{\text{2-aminopyridine } (NH_2)}$ $[\text{pyridyl-}NH{\cdots}CH{\cdots}CH{\cdots}CH{\cdots}N(CH_3)_2]^+ \; CH_3SO_4^-$

**IX-45**

2-Chloro-5-nitropyridine condenses with 2-aminobenzenethiol to give 2-(5-nitropyridylthio)aniline, which, when heated in ethanolic solutions of hydrogen chloride, gives **IX-46**.[166] This Smiles rearrangement has been investigated[166-170] (see also Chapter XV).

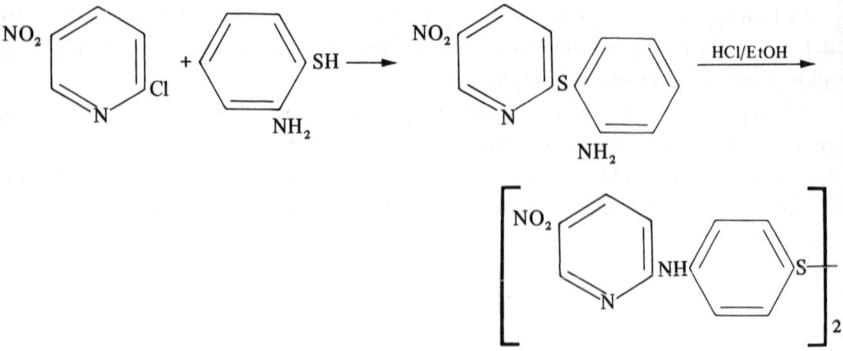

IX-46

Aralkylaminopyridines are prepared by reductive alkylation; for example, when 2-aminopyridine is heated with 4-alkoxybenzaldehyde and formic acid, 2-(4-alkoxybenzylamino)pyridines are formed. The latter compounds, when treated with sodium amide and dialkylaminoethyl chloride, give *N*-4-alkoxybenzyl-*N*-2-pyridyl-*N*′,*N*′-dialkylethylenediamine **(IX-47)**.[171]

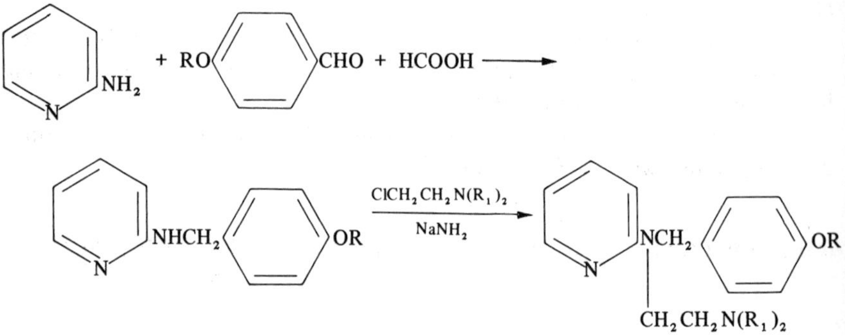

$ClCH_2CH_2N(R_1)_2$, $NaNH_2$

$CH_2CH_2N(R_1)_2$

IX-47

When 4-pyridinesulfonic acid is heated with an alkyl amine in the presence of zinc chloride, 4-alkylaminopyridines **(IX-48)** result.[172]

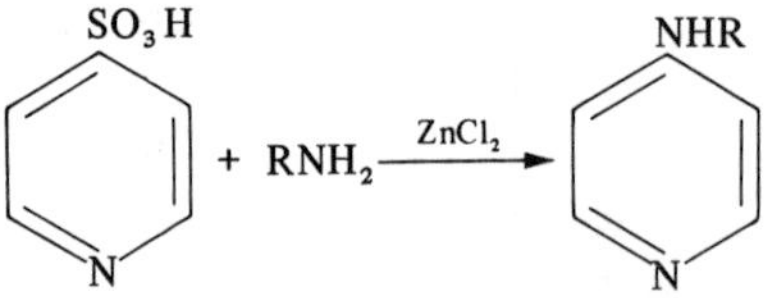

$SO_3H$ + $RNH_2$ $\xrightarrow{ZnCl_2}$ NHR

IX-48

Tertiary aminopyridines are generally prepared either by the amination of halopyridines with secondary alkyl or aryl amines or by the amination of a secondary pyridylamine with an alkyl or aryl halide.[162-167] The reactions of 2-halopyridines and their *N*-oxides with an excess of dimethylamine have been studied. Based on the yields of diethylaminopyridines and dimethylaminopyridine-1-oxides produced, the following is the order of decreasing reactivities of the halopyridines:[174]

2-bromopyridine-1-oxide > 2-chloropyridine-1-oxide >

(2-iodopyridine-1-oxide, 2-iodopyridine, 2-chloropyridine)

2-Dimethylamino-3,5-dinitropyridine is obtained from 2-chloro-3,5-dinitropyridine with 40% aqueous dimethylamine. 2-Diethylamino-3,5-dinitropyridine is formed from anhydrous diethylamine and 2-(β-chloroethoxy)-3,5-dinitropyridine.[181]

*N*-(2-Pyridyl)-*N*-bromobenzyl-*N'*,*N'*-dialkylethylenediamines are prepared from 2-(β-dialkylaminoethylamino)pyridine, alkali amide and an alkyl bromide (preferably bromobenzyl bromide; see Table IX-35). Thus, *N*-(2-pyridyl)-*N*-(*m*-bromobenzyl)-*N'*,*N'*-diethylethylene diamine **(IX-49)** is obtained from the condensation of 2-(β-diethylaminethylamino)pyridine with *m*-bromobenzyl bromide in the presence of sodium amide.[181]

$NHCH_2CH_2N(C_2H_5)_2$ + $CH_2Br$, Br → $NCH_2$, Br, $CH_2CH_2N(C_2H_5)_2$

**IX-49**

2-(2-Dimethylaminoethylamino)pyridine reacts with 1-phenethyl chloride in the presence of sodamide to give *N*-(1-phenethyl)-*N*-(2-pyridyl)-*N'*,*N'*-dimethylaminoethylene diamine **(IX-50)**;[182] other aryl derivatives have been prepared[183] (see Table IX-36).

$NHCH_2CH_2N(CH_3)_2$ + $CH_3CH(Cl)C_6H_5$ → $NCH_2CH_2N(CH_3)_2$, $CH(CH_3)C_6H_5$

**IX-50**

Heating *N,N*-diethyl-*N'*-2-pyridylethylenediamine with sodium amide in toluene followed by treatment with 2-chloromethylselenophene gives *N,N*-diethyl-*N'*-2-pyridyl-*N'*-2-selenylethylene diamine **(IX-51)**. Other pyridyl derivatives of selenylethylene diamine were prepared.[184, 185]

$$\text{2-}C_5H_4N\text{-}NHCH_2CH_2N(C_2H_5)_2 + \text{Cl-}C_4H_3Se \xrightarrow{NaNH_2} \text{2-}C_5H_4N\text{-}N(C_4H_3Se)\text{-}CH_2CH_2N(C_2H_5)_2$$

**IX-51**

An unusual preparation is that of 2-(*N*-cyano-*N*-phenyl)aminopyridine **(IX-52)**, which is obtained by heating pyridine with *N*-anilino-*t*-butoxycarbonyl imidoyl chloride at 170° for 15 minutes.[186]

$$C_5H_5N + C_6H_5NHN{=}C(Cl){-}COOBu^t \longrightarrow \text{2-}C_5H_4N\text{-}N(CN)\text{-}C_6H_5$$

**IX-52**

When an aqueous solution of ammonium chloride and a polymethinium perchlorate salt **(IX-53**, $R = H$ or $C_6H_5$) is heated, it gives 4-dimethylaminopyridine (78%) and 4-dimethylamino-2-phenylpyridine (49%), respectively[187]

$$[(CH_3)_2N{=}CHCH{=}C(N(CH_3)_2)C{=}C(R){-}N(CH_3)_2]^+ClO_4^- \longrightarrow \text{4-}N(CH_3)_2\text{-2-R-}C_5H_3N$$

**IX-53**

4-Aminopyridine reacts with acrylonitrile to give a quantitative yield of 4-[*N,N*-bis(2-cyanoethyl)] aminopyridine.[188]

Aminopyridines also react with aldehydes, ketones, and other carbonyl derivatives to give products that can be converted to secondary or tertiary amines; these reactions are discussed in Section I.4.C.

### 3. Structures and Properties

Many excellent books, reviews, and reports[189-196] are now available on the investigations of the structure and properties of aminopyridines.

The dipole moments of many aminopyridines in several solvents have been measured.[189, 197, 198] These studies contain many interesting observations. One amino-hydrogen in 2-aminopyridine is reported to be in the same plane as the heterocyclic ring.[199] While the resonance interactions of the methyl and chloro substituents in methyl- and chloropyridine are not very very marked, those of the amino group in 4-aminopyridine are significant.[191] Dioxane forms complexes with 2-, 3-, and 4-aminopyridine through hydrogen bonds to the amino-nitrogen; intermolecular associations by hydrogen bonding with 3- and 4-aminopyridine in benzene have also been reported.[200]

The ultraviolet spectra of various aminopyridines,[190, 201-206] their methoxy derivatives[207] and dipyridyl amines[208] were recorded. The investigation of amino-imino tautomerism continues;[192, 200, 209-212] the amino form generally predominates. 2-, 3-, And 4-acetylamidopyridine exist predominantly in the acylamino form.[213] The ultraviolet spectra of 3-monosubstituted tertiary 4-aminopyridines or 3,5-disubstituted secondary 4-aminopyridines show changes in intensities of bands that were attributed to steric inhibition of resonance. No evidence for such inhibition in the case of the primary 3- and 4-aminopyridines or the 3-monosubstituted 4-methylaminopyridines was observed. The spectra of the monocations of all the above aminopyridines indicated that the ring nitrogen atom was the basic center.[214] Theoretical calculations of transition energies, oscillator strengths, and dipole moments of $\pi \rightarrow \pi^*$ bands agree well with experimental results; these calculations were based on the localized-orbital model,[215] LCAO-MO,[216-218] SCF[219] and CNDO[220] methods.

The basicities ($pK_a$) of a large number of pyridines have been determined.[189, 190, 193, 221-227] The variation of $pK_a$ values with temperature for monovalent and divalent organic cations may be calculated from the following expressions (the entropy change of $-4 \pm 6$ cal/degree was assumed to be a constant):[228]

$$\left( \begin{array}{l} \dfrac{-\mathrm{d}(\mathrm{p}K_a)}{\mathrm{dT}} = (\mathrm{p}K_a - 0.9)_\mathrm{T} \pm 0.004 \\[2ex] \dfrac{-\mathrm{d}(\mathrm{p}K_a)}{\mathrm{dT}} = \dfrac{\mathrm{p}K_a}{\mathrm{T}} \end{array} \right)$$

Protonation first occurs at the 2- or 4-amino group in 2,3- and 3,4-diaminopyridines; a second protonation then takes place at the 3-amino group. Replacement of a hydrogen atom by a methyl group on an extranuclear $NH_2$ gives rise to a small bathochromic shift for the cation of 3,4-diaminopyridine or 4-amino-3-nitropyridine, but not with the neutral species.[225]

The fluorescence properties of 2-, 3-, and 4-aminopyridines in various solvents have been investigated.[229] The 2- and 3-isomers are efficient fluorescers with quantum yields approaching unity in $0.1N$ $H_2SO_4$. 4-Aminopyridine, however, fluoresces weakly due to an n → π* (rather than π → π*) lowest excited singlet and a second excited state that has large charge-transfer character. The $pK_a$ values of the excited states of 2- and 3-aminopyridine indicate that the excited states are weaker conjugate acids than the ground states. The fluorescence of biacetyl sensitized by 2-aminopyridine (but not by the 3- or 4-isomer) has been observed.[230] The europium complex of 2-aminopyridine, but not that of the 3-isomer, fluoresces weakly red.[231]

Considerable activity in the study of the infrared spectroscopy of aminopyridines,[189, 195, 196, 205, 232-240] methylaminopyridines,[239, 241] and aminomethyl pyridines[194, 242, 243] is noted. In addition, the Raman spectra of aminopicolines were reported.[244, 245] In the crystalline state, 2-amino-, 2-amino-5-halo-, and 2-amino-5-nitropyridines were reported to exist in the tautomeric amino form, and the $NH_2$ group absorbed at 3280, 3400, and 1640 $cm^{-1}$.[205] Whereas the vibrational spectrum of 3-aminopyridine hardly changes on protonation, the vibrational spectra of the 2- and 4-isomer change distinctly. The hydrochloride of 2-(and of 4-)aminopyridine is best represented as the aminium ion **IX-54**. The fundamental symmetrical and asymmetrical N–H

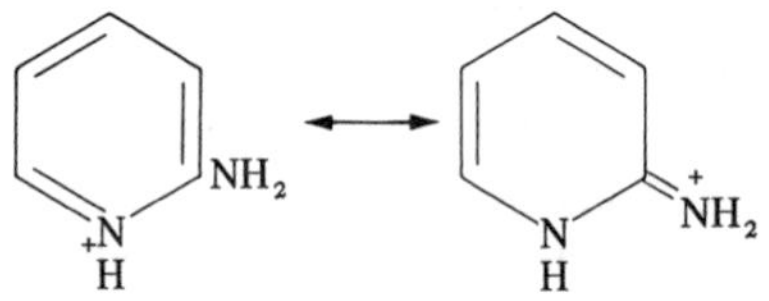

**IX-54**

stretching vibrations of 2-, 3-, and 4-aminopyridines were studied in eleven solvents. It was concluded that the excess basicities were due to excess π-charges on the nuclear nitrogen. The hydrogen bonding strength was governed by the charges on the extranuclear nitrogen.[246]

Magnetic resonance spectroscopy employing the nuclei $^{1}H$,[190, 247-251] $^{19}F$[252, 253] and $^{14}N$[254-256] has been used to study the π-electron densities and the various structural and electronic properties of the amino group in aminopyridines. The PMR spectra of various aminopyridines in various solvents give good linear relationships between chemical shifts and electron densities derived from molecular orbital calculations.[248, 255, 257, 258] Inter-, intra-, and solvent hydrogen-bonding studies have also been reported.[259, 260] The preferred configuration of amides is related to the magnitudes of the downfield shifts.[261] Proton resonance spectra confirm that aminopyridines protonate at the ring nitrogen.[262]

The crystal structures of 2-amino-6-pyridone,[263] 2-amino-5-chloropyridine hydrochloride[264,265] and 3-pyridylmethylamine dihydrochloride[266] were examined. Electron-impact induced fragmentations of aminopicolines showed unexpected loss of ammonia from both 2-amino-3-picoline and 2-amino-6-picoline.[267]

Miscellaneous studies include the use of 2-aminopyridine as a standard for low wave-length spectrofluorimetry,[268] the change-transfer complexes of iodine and aminopyridine,[269, 270] the linear isotherm free-energies of absorption of 2- and 3-aminopyridine on alumina,[271] rotation mobility of an amino group by dielectric relaxation[272] and the free energies, enthalpies and entropies of protonation.[271, 273]

Various rearrangement reactions involving aminopyridines have been studied; the rearrangement of nitraminopyridines will be discussed later. Reversible hydrolytic ring opening of 2-amino-$^{15}N$-pyridine has been reported, but 3-amino-$^{15}N$-pyridine yielded no rearrangment of $^{15}N$.[274]

Hydrogen-deuterium exchange at the $\alpha$- and $\beta$-positions of 4-aminopyridine and 4-dimethylaminopyridine in $DClO_4$ and NaOD solutions have been studied by NMR spectroscopy.[275] Base-catalyzed H–D exchange occurred at both positions, but acid-catalyzed exchange took place preferentially at the $\beta$-position. Mechanisms of exchange were postulated. Hydrogen-deuterium exchange is believed to occur between the $-ND_2$ group and the ring-hydrogen in 2- or 3-aminopyridine.[276]

Photolysis of aminopyridines has been studied by Taylor and his co-workers.[277, 278] Ultraviolet irradiation of 2-aminopyridine (and several substituted 2-aminopyridines) in hydrochloric acid solution results in the formation of the 1,4-dimer having the *anti-trans* configuration **IX-55**.

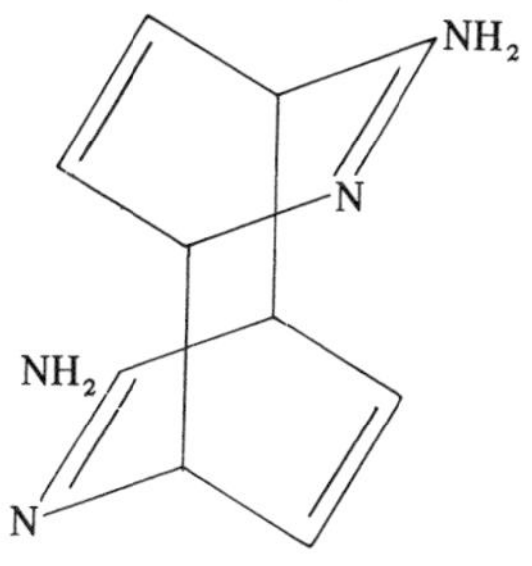

NH2

N

NH2

N

**IX-55**

## 4. Reactions

### A. *Oxidation*

Various studies on the oxidation of aminopyridines have been reported. Taylor and Driscoll[279] suggested that the best way to convert 3-aminopyridine to 3-nitropyridine is by oxidation with persulfuric acid followed by peracetic acid.

The oxidation of 2-aminopyridine, 2-methylaminopyridine, and 2-dimethylaminopyridine with perbenzoic acid at 10° and room temperature gives the pyridine-1-oxide of 2-amino- and 2-methylaminopyridine but the *amine-*N′*-oxide* of 2-dimethylaminopyridine. The difference in behavior is attributed to steric factors.[280] The perbenzoic acid oxidation of 4-methylaminopyridine and 4-dimethylaminopyridine gives the corresponding pyridine-1-oxides;[281] however, the reaction of the cyclic 2-amino-tetrachloropyridines with performic or trifluoroperacetic acid gives the corresponding hydroxylamines **(IX-56)** rather than the 1-oxides.[282]

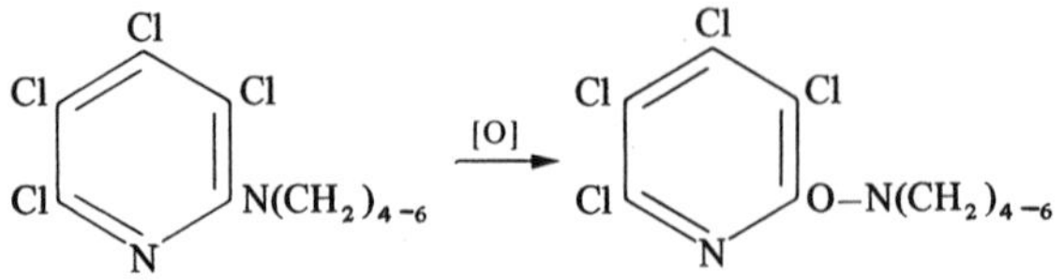

**IX-56**

Chloro- and bromoaminopyridines are oxidized by persulfuric acid at 0° to their nitro derivatives;[283, 284] thus, 3-chloro-, and 3-bromo-4-aminopyridine are converted to the respective 3-halo-4-nitropyridines. However, 4-amino-3-iodopyridine is *not* oxidized under these conditions.[283] 4-Amino-2,3,5,6-tetrafluoropyridine is difficult to oxidize and requires refluxing peroxytrifluoroacetic acid for 22 hours in order to yield the 4-nitro derivative. Potassium bromate has been used to oxidize 5-amino-3-methyl-2-pyridone, but the product was not a nitro compound; instead 3-hydroxy-6-methyl-2-aza-1,4-benzoquinone-4-(2,6-dihydroxy-5-methyl-3-pyridyl)imine **(IX-57)** was obtained.[284]

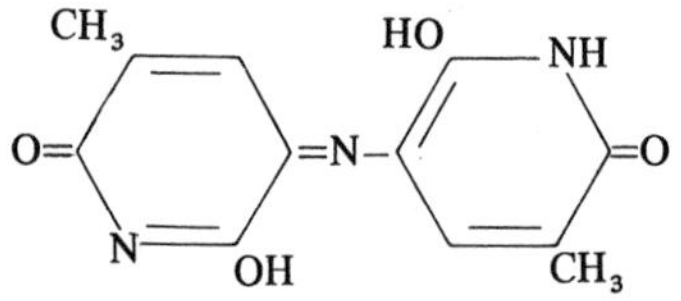

**IX-57**

In some cases in which other functional groups (not the amino group) in aminopyridine have to be oxidized, protection of the amino groups and the use of mild oxidizing agents are resorted to. Thus, 6-amino-2-picoline is converted to 6-acetamido-2-picoline, which is then oxidized with aqueous potassium permanganate to 6-acetaminopicolinic acid. The latter, on hydrolysis, gives 6-aminopicolinic acid.[285] In the preparation of intermediates of nor-vitamin $B_6$, 3-amino-6-chloro-5-cyano-4-ethoxycarbonyl-2-picoline is oxidized by selenium dioxide to 3-amino-6-chloro-5-cyano-4-ethoxycarbonyl-2-formylpyridine **(IX-58)**.[286]

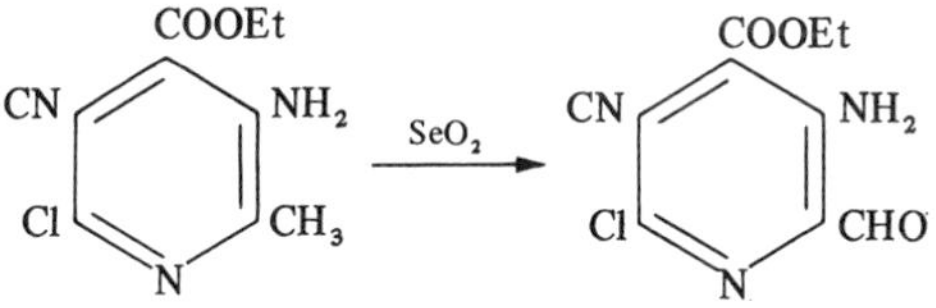

IX-58

## B. *Reactions with Aldehydes and Ketones*

When a solution of 2- or 3-aminopyridine in 48% formalin is allowed to stand for 2 days at room temperature, crystals of *N,N'*-di-2- **(IX-59)** and *N,N'*-di-3-pyridylmethylenediamine **(IX-59)**, respectively, are isolated. Under similar reaction conditions, 4-aminopyridine requires 7 days to produce *N,N'*-di-4-pyridylmethylenediamine **(IX-59)**.[287]

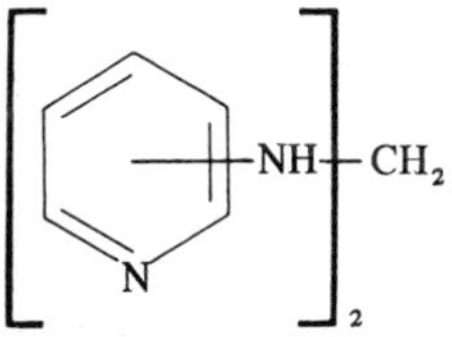

IX-59 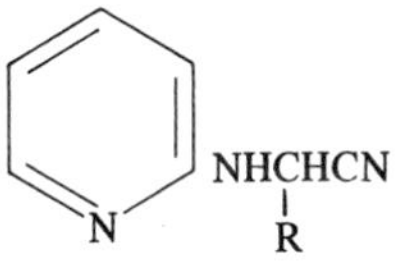 IX-60

With 2-aminopyridine-1-oxide, formaldehyde gives 2-methylolaminopyridine-1-oxide; similarly, 5-bromo-2-methylolaminopyridine-1-oxide is obtained from 5-bromo-2-aminopyridine-1-oxide.[288] When 2-aminopyridine is treated with a hot aqueous solution of sodium bisulfite and formalin followed by sodium cyanide, 2-pyridylaminoacetonitrile (**IX-60**, R = H) is obtained; when benzaldehyde instead of formalin is used, α-(2-pyridylamino)phenylacetonitrile (**IX-60**, R = $C_6H_5$) is formed.[289]

It has been shown that two moles of 2-aminopyridine condense with one mole of an aliphatic aldehyde ($C_1-C_{10}$) to give a 2,2'-(alkylidenedimino)dipyridine **(IX-61)**; these are suitable crystalline derivatives for the identification of aldehydes.[290] The reaction of 2-aminopyridine with trichloroacetaldehyde had

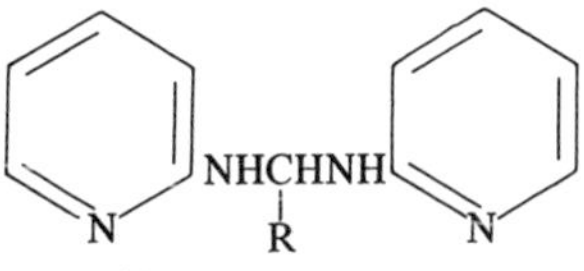

IX-61 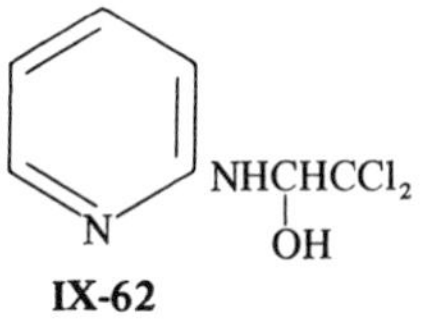 IX-62

been carried out previously,[291-293] but the structure of the product was not established. A patent now claims the structure of the product to be **IX-62**.[294]

Substituted benzylidene-2-aminopyridines **(IX-63)** are obtained by boiling methanolic solutions of equimolar quantities of the appropriate *o*-hydroxybenzaldehyde and 2-aminopyridine for 1 hour.[295] Other azomethines are also

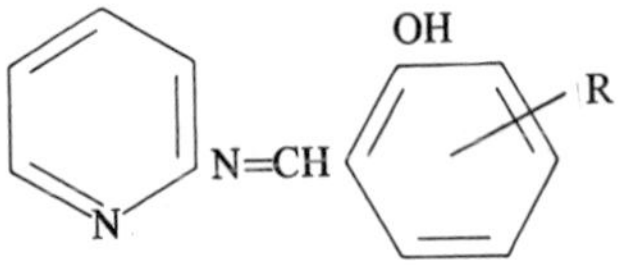

OH R N=CH N

**IX-63**

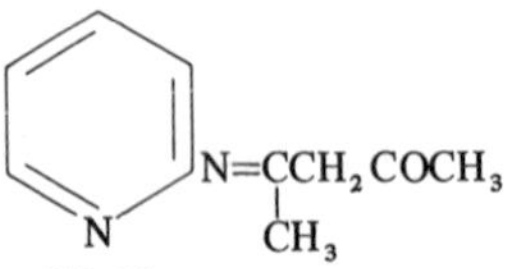

$N{=}C(CH_3)CH_2COCH_3$ N

**IX-64**

formed from aminopyridines with carbonyl compounds. Thus **IX-64** is obtained by boiling 2-aminopyridine with acetyl acetone.[296, 297] 2-Aminopyridine reacts with 1-pyrenecarboxaldehyde to give *N*-(2-pyridyl)-1-pyrenyl methenimine.[298]

The imines formed in the reactions of aminopyridine with aldehydes and ketones may be reduced to the amines either *in situ* in the presence of formic acid, or with sodium borohydride. Thus, 2-(4-methylthiobenzylamino)pyridine **(IX-65)** is obtained from 2-aminopyridine, 4-methylthiobenzaldehyde and formic acid.[299]

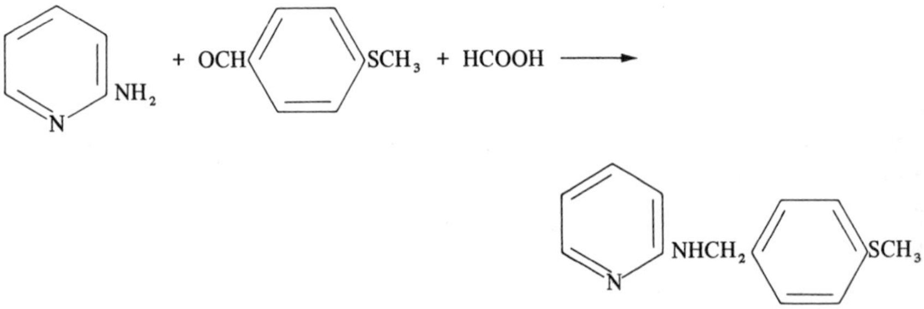

$NH_2$ N + OCH $SCH_3$ + HCOOH ⟶

N $NHCH_2$ $SCH_3$

**IX-65**

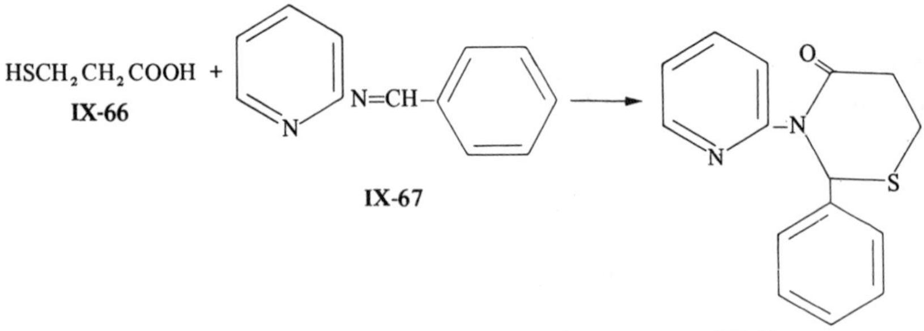

$HSCH_2CH_2COOH$ +
**IX-66**

N N=CH–

**IX-67**

⟶ O N –N S

**IX-68**

When 3-mercaptoproponic acid **(IX-66)** is added to the Schiff's base **IX-67** (from 2-aminopyridine and benzaldehyde), 2-phenyl-3-(2-pyridyl)-1,3-thiazan-4-one **(IX-68)** is obtained. Similarly, Schiff's bases from 3- and 4-aminopyridine give the corresponding phenylpyridyl-1,3-thiazan-4-one.[299-301]

2-Aminopyridine condenses with cinnamaldehyde in the presence of anhydrous zinc chloride to give a product that, on treatment with sodium borohydride, gives 2-cinnamylaminopyridine; similarly, various 2-(substituted-cinnamyl)aminopyridines are prepared from 2-aminopyridine and substituted cinnamaldehydes.[302, 303] When 2-aminopyridine and formylferrocene are boiled under reflux in toluene containing phosphorus oxychloride, there is isolated a product which, when reduced with sodium borohydride, gives 2-pyridylamino-methyl ferrocene.[304]

2-Aminopyridine condenses with heteroaromatic aldehydes; with furfural or 3-chloro-2-formylbenzofuran it gives the 2-(furfurylideneamino)pyridine and 2-[(3-chloro-2-benzofuranyl)-methylene] aminopyridine **(IX-69)**, respectively.[305-308] With ninhydrin and 2-aminopyridine, 2-hydroxy-2-pyridylamino-1,3-indandione **(IX-70)** results.[309]

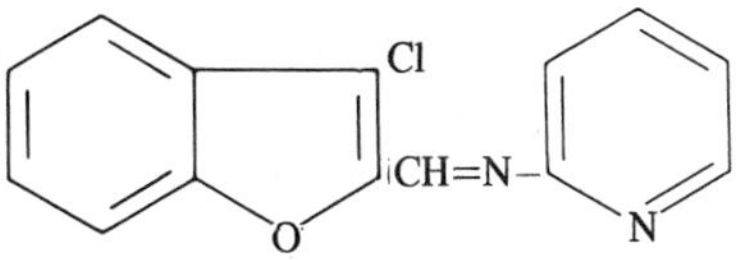

Cl CH=N– O N

**IX-69**

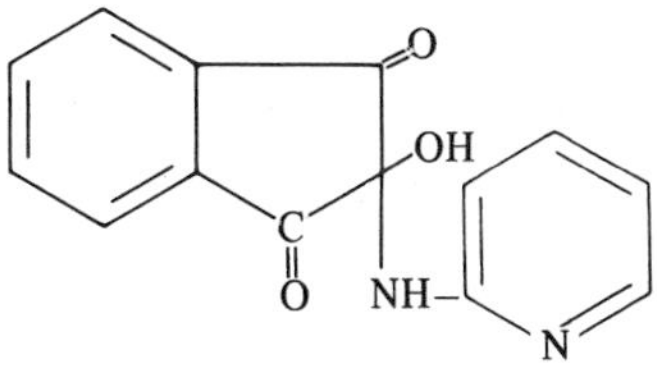

O OH C O NH– N

**IX-70**

2,3-Dichlorothiophene-5-carboxaldehyde condenses with 2-aminopyridine to yield 2-(4,5-dichloro-2-thienylidene)aminopyridine **(IX-71)**.[310]

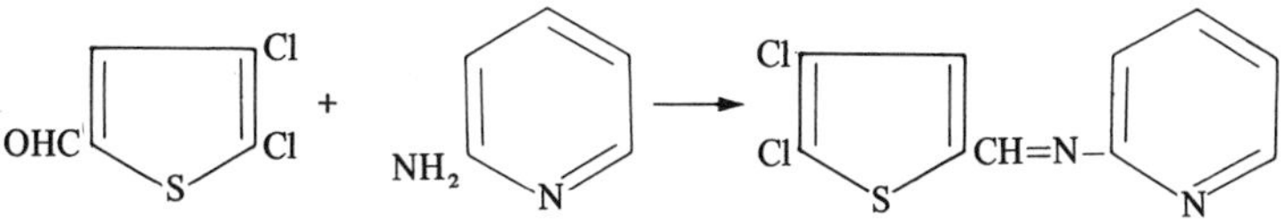

OHC S Cl Cl + $NH_2$ N ⟶ Cl Cl S CH=N– N

**IX-71**

When diketene is added to a benzene solution of 2-aminopyridine and the reaction mixture is heated for 5 minutes, 2-acetoacetamidopyridine (61% yield) is obtained. Diketene reacts with 2-aminopyridine-1-oxide to yield 2-acetoacetamidopyridine-1-oxide (80%). A similar reaction is observed with 3-aminopyridine and with 3-aminopyridine-1-oxide.[311]

The complex reaction of 2-aminopyridine, isobutyraldehyde, cyclohexyl isocyanide, and hydrazoic acid gives 2-[1-(1-cyclohexyl-1*H*-tetrazolyl-5-yl)-2-methylpropyl]aminopyridine **(IX-72)**.[312]

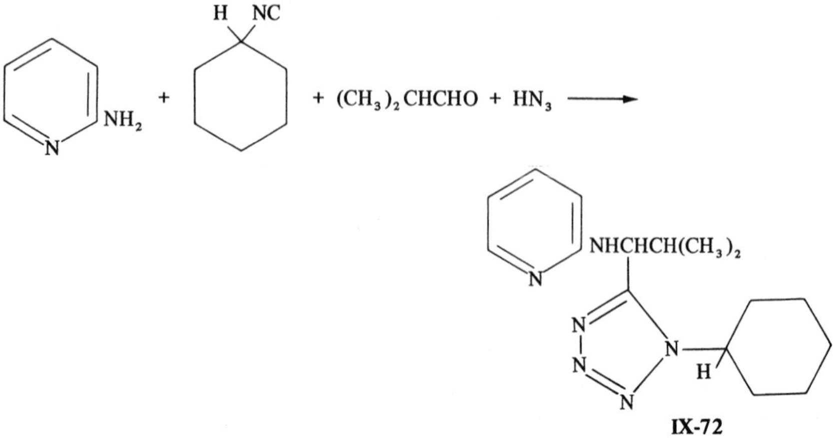

$$\text{2-aminopyridine} + \text{cyclohexyl isocyanide} + (CH_3)_2CHCHO + HN_3 \longrightarrow$$

**IX-72**

The reaction of 2,6-diaminopyridine with acetonylacetone produces 2-amino-6-(2,5-dimethyl-1-pyrrolyl)pyridine **(IX-73)**, which reacts with another mole of acetonylacetone to yield 2,6-bis-(2,5-dimethyl-1-pyrrolyl)-pyridine **(IX-74)**.[313]

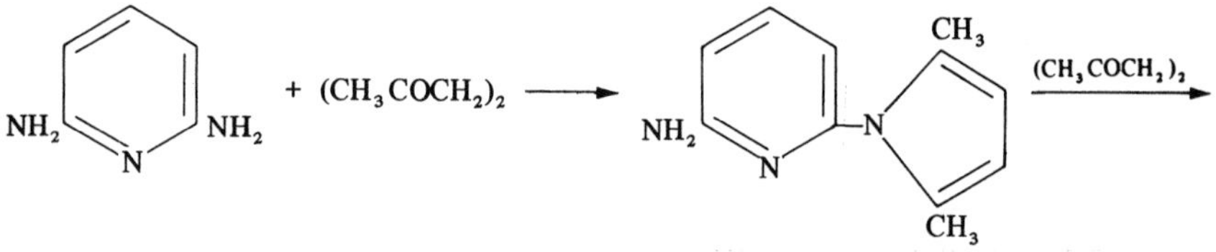

$$\text{2,6-diaminopyridine} + (CH_3COCH_2)_2 \longrightarrow \textbf{IX-73} \xrightarrow{(CH_3COCH_2)_2} \textbf{IX-74}$$

**IX-73**

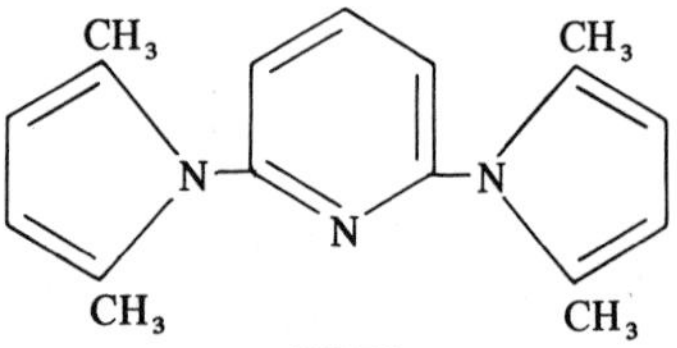

**IX-74**

## C. *Acylation*

a. CARBONYL DERIVATIVES 4-Aminopyridine is formylated with a mixture of formic acid and acetic anhydride. Similarly, 4-amino-3-picoline is converted to 4-formamido-3-picoline.[314] 2-, 3-, And 4-acetamidopyridines are obtained by heating the appropriate aminopyridine either with acetic anhydride or with acetic acid in tetrahydrofuran.[315-317] The benzamido-, phenylacetamido-, and chloroacetamidopyridines are similarly prepared from benzoic, phenyacetic, and monochloracetic acids, respectively.[315] Various 2-(substituted benzamido)pyridines are also prepared from substituted benzyl chlorides and 2-aminopyridine;[318] 2-(2-aminoethyl)pyridine gives 2-(2-benzamidoethyl)pyridine when treated with benzoyl chloride.[319] A quantitative yield of 2-[2-chloro-2,2-bis-(*p*-chlorophenyl)acetamido] pyridine (**IX-75**) was obtained from the reaction of 2-aminopyridine with bis(*p*-chlorophenyl)chloroacetyl chloride.[320] The reaction of 2- and 4-aminopyridines with chloroacetyl chloride or chloropropionyl

$NH_2$ + $(p\text{-}ClC_6H_4)_2CClCOCl$ ⟶ $NHCOCCl(p\text{-}ClC_6H_4)_2$

N N

**IX-75**

chloride followed by various amines ($R_2NH$) gives products having the general structure **IX-76**.[321, 322]

$NH_2$ + $Cl(CH_2)_nCOCl$ ⟶ $\xrightarrow{R_2NH}$ $NHCO(CH_2)_nNR_2$

N N

**IX-76**

Aminopyridyl derivatives of carbazole, phenoxazine, iminodibenzyl, and phenothiazine have been reported.[323] Thus 3-(10-phenothiazinyl)propionic acid gives the corresponding 2-, 3-, and 4-pyridylpropionamides when boiled with 2-, 3-, or 4-aminopyridine in triethylamine. These pyridylamides are reduced with lithium aluminum hydride to yield the corresponding 2-, 3-, or 4-[3-(10-phenothiazinyl)propylamino] pyridines.

3-Aminopyridine reacts with 2-benzo-1,5-dioxepanylformyl chloride to form *N*-3-pyridyl-2-benzo-1,5-dioxepanylformamide, which, on reduction with lithium aluminum hydride, yields 3-[3,4-dihydro-2*H*-1,5-benzodioxepin-2-yl)-methyl] aminopyridine.[324]

Aminopyridines react with acyl or aroyl chlorides to give the corresponding amides (**IX-77**);[325-329] for example, 4-aminopyridine reacts with 10-undecenoyl chloride to give the amide **IX-78**.[326]

$$\text{X-C}_5\text{H}_3\text{N-NH}_2 + \text{RCOCl} \longrightarrow \text{X-C}_5\text{H}_3\text{N-NHCOR}$$

**IX-77**

$$\text{4-NH}_2\text{C}_5\text{H}_4\text{N} + \text{CH}_2\text{=CH(CH}_2)_8\text{COCl} \longrightarrow \text{4-[NHCO(CH}_2)_8\text{CH=CH}_2]\text{C}_5\text{H}_4\text{N}$$

**IX-78**

Attempts to prepare 2,6-diacyldiamino-3-aminopyridine by means of the reductive cleavage of 2,6-diacyldiamino-2′-butoxy-3,5′-azopyridine did not give the desired product. Thus, heating 2,6-diphthaloydiamino-2′-butoxy-3,5′-azopyridine **(IX-79)** with iron and hydrochloric acid leads to 5-amino-2-butoxypyridine **(IX-80)** and 2,6-diamino-3-hydroxypyridine **(IX-81)**.[330]

$$\text{2,6-(RNH)}_2\text{-3-(N=N-[2-OC}_4\text{H}_9\text{-5-pyridyl])pyridine} \longrightarrow \text{5-NH}_2\text{-2-OC}_4\text{H}_9\text{C}_5\text{H}_3\text{N} + \text{2,6-(NH}_2)_2\text{-3-OH-C}_5\text{H}_2\text{N}$$

(R = phthaloyl)

**IX-79** **IX-80** **IX-81**

When 2-aminopyridine and mandelic acid are azeotroped for 20 hours in toluene, 2-mandelamidopyridine **(IX-82)** is obtained; several substituted 2-mandelaminopyridines are similarly prepared.[331-333] Generally, acids react

$$\text{2-NH}_2\text{C}_5\text{H}_4\text{N} + \text{C}_6\text{H}_5\text{CHOHCOOH} \longrightarrow \text{2-(NHCOCHOHC}_6\text{H}_5)\text{C}_5\text{H}_4\text{N}$$

**IX-82**

with aminopyridines to give salts[334-337] that are often crystalline and are used as derivatives for identification or purification purposes. Thus, the reactions 2-, 3-, or 4-aminopyridines with fatty acids,[334, 338] uric acid,[336] dinitrophthalic acid,[339] and antipyrine-4-sulfonic acid[340] all yield crystalline salts. Phosphoric, perchloric, and nitric acids have also been used.[341-344]

Unsaturated acids react with aminopyridines to form condensation products rather than salts. Thus, 2-aminopyridine reacts with propiolic acid to form *trans*-2-imino-1(2*H*)-pyridineacrylic acid **(IX-83)** and 2*H*-pyrido[1,2-*a*]pyrimidin-2-one **(IX-84)**.[345] The reactions of aminopyridines with acrylonitrile and

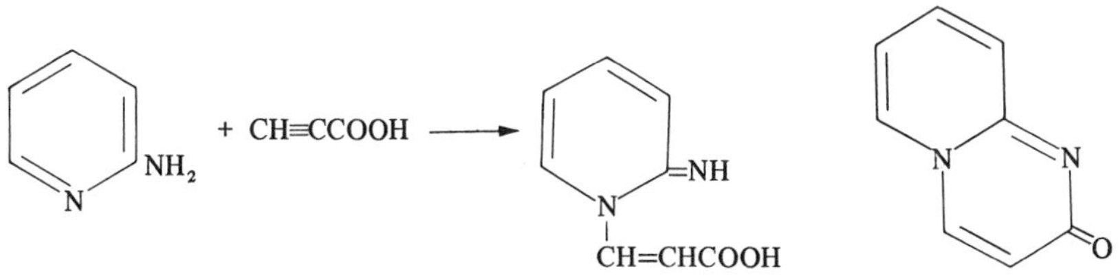

+ CH≡CCOOH ⟶

**IX-83** **IX-84**

acrylic acid and its derivatives have been investigated.[346, 347, 188] Heating 2-amino-4-picoline with acrylic acid, its nitrile, amide, methyl or ethyl ester in water at 100° gives *β*-[1-(4-methyl-2-aminopyridyl)] propionic acid **(IX-85)**.[346] The latter, on heating with concentrated hydrochloric acid, cyclizes to **IX-86**.

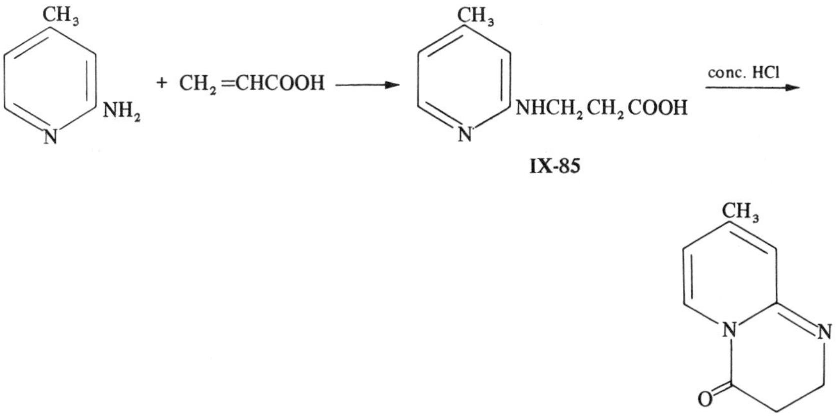

+ $CH_2$=CHCOOH ⟶ conc. HCl ⟶

**IX-85**

**IX-86**

2-Aminopyridine reacts with ethyl α-cyano-α-(*o*-nitrobenzoyl)acetate to form a 1:1 adduct **(IX-87)**, the structure of which has not been established.[348]

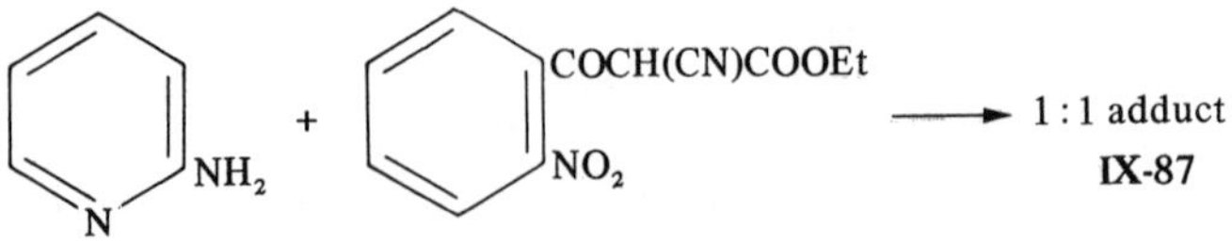

NH$_2$ + COCH(CN)COOEt NO$_2$ ⟶ 1:1 adduct

**IX-87**

Metathiazanones **(IX-88)** are prepared by heating a benzene solution of thiosalicylic acid, 2-(or 3-)-aminopyridine, and *o*-chloro-(or *m*-nitro-)benzaldehyde.[357]

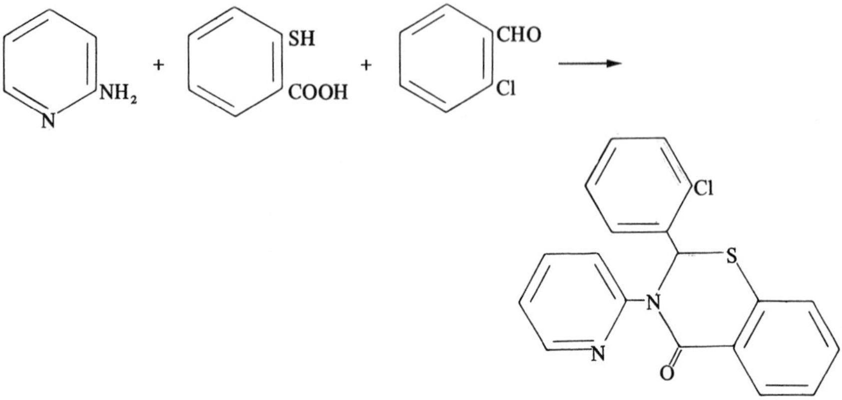

NH$_2$ + SH COOH + CHO Cl ⟶

Cl S N N O

**IX-88**

b. SULFONYL DERIVATIVES Aryl sulfonamidopyridines are formed by the direct reaction of arylsulfonyl chlorides with aminopyridines.[92, 349-352] Thus, 2,6-dimethyl-3-aminopyridine reacts with benzenesulfonyl chloride to give *N*-(2,6-dimethyl-3-pyridyl)benzenesulfonamide.[351]

Methyl 2-aminoisonicotinate in anhydrous pyridine reacts with *p*-nitrobenzenesulfonyl chloride to give methyl 2-(*p*-nitrobenzenesulfonamido)isonicotinate **(IX-89)**.[353]

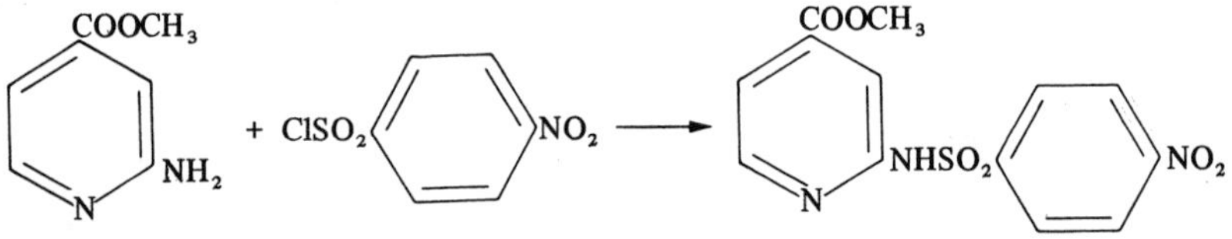

COOCH$_3$ NH$_2$ + ClSO$_2$ NO$_2$ ⟶ COOCH$_3$ NHSO$_2$ NO$_2$

**IX-89**

Chloromethanesulfonyl chloride reacts with 2-aminopyridine to give 2-(chloromethylsulfonamido)pyridine.[354] *N*-2-Pyridylalkanesulfonamides are formed from 2-aminopyridine and an alkane sulfonic acid; these sulfonamides are amphoteric.[355]

Miscellaneous reactions of aminopyridines with sulfur derivatives have been observed. *p*-Chlorophenyl hydroxymethylsulfone reacts with 2-aminopyridine to give *p*-chlorophenyl-2-pyridylaminomethyl sulfone **(IX-90)**.[356]

$$\text{2-aminopyridine} + \text{Cl-C}_6\text{H}_4\text{-SO}_2\text{CH}_2\text{OH} \longrightarrow \text{2-pyridyl-NHCH}_2\text{SO}_2\text{-C}_6\text{H}_4\text{-Cl}$$

**IX-90**

c. UREAS, AMIDINES 2-Amino-5-chloropyridine reacts with methyl isocyanate to give 1-methyl-3-(5-chloro-2-pyridyl)urea; 1-methyl-3-(5-iodo-2-pyridyl)urea and 1-methyl-3-(2-methoxy-5-pyridyl)urea are prepared similarly.[358] 3-Amino-4-pyridone condenses with urea to give *N,N'*-bis-(4-hydroxy-3-pyridyl)urea **(IX-91)**;[358] with ethyl chloroformate, it gives 1,3-diethoxycarbonylamino-4-pyridone **(IX-92)**.

$$\text{3-amino-4-pyridone} \xrightarrow{CO(NH_2)_2} \text{(4-oxo-1H-pyridin-3-yl)-NHCONH-(4-oxo-1H-pyridin-3-yl)}$$

**IX-91**

$$\text{3-amino-4-pyridone} \xrightarrow{ClCOOEt} \text{1-COOEt-3-NHCOOEt-4-pyridone}$$

**IX-92**

When 2-amino-6-ethoxypyridine in ethanol is boiled under reflux with carbon disulfide, *N,N'*-bis-(6-ethoxy-2-pyridyl)thiourea **(IX-93)** is obtained. Similarly, treatment with an aryl isothiocyanate gives the corresponding thiourea.[359]

$$C_2H_5O\text{-C}_5H_3N\text{-NH}_2 + CS_2 \longrightarrow [EtO\text{-C}_5H_3N\text{-NH}]_2CS$$

**IX-93**

When a mixture of 2,6-lutidine, ethyl iodide, and *N,N'*-di-(5-bromo-2-pyridyl)-formamidine is heated with ethyl orthoformate, **IX-94** results.[360]

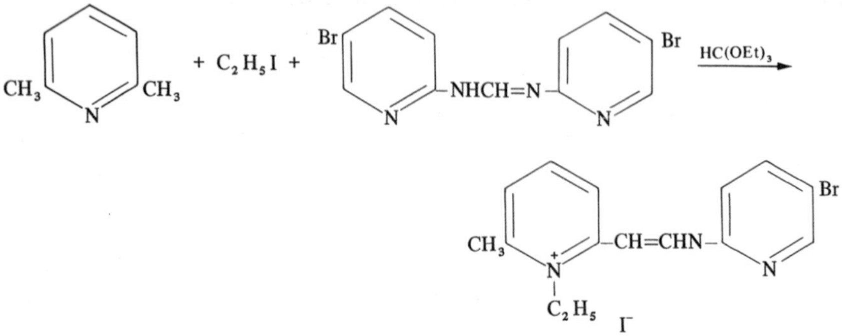

**IX-94**

## D. *Diazotization Reactions*

If 2- and 4-aminopyridines are diazotized in dilute mineral acids, diazonium salts are formed; however, the diazonium ions are not stable and they hydrolyze rapidly to the corresponding pyridones.[361-364] For example, the diazotization of 2-amino-4-nitropyridine in sulfuric acid yields 85% of 4-nitro-2-pyridone. Diazonium salts of 2-substituted-4-aminopyridines with electron-withdrawing substituents such as COOH, $CONH_2$ and CN are particularly unstable.[365]

The kinetics of the diazotization of 4-aminopyridine in 0.0025–5.0 *M* perchloric acid were studied,[366] and the reaction was believed to proceed by only one mechanism whose kinetic form is first order with respect to the amine as well as to nitrous acid. The rate of the reaction showed an exponential catalytic dependence on the concentration of added perchloric acid and sodium perchlorate. In solutions of constant ionic strength the rate was directly related to the acidity function ($H_0$). The 4-diazonium ion was characterized as 4-(2-hydroxy-1-naphthylazo)pyridine.[361] When 4-aminopyridine is diazotized and treated with *N,N*-dimethylaniline, 4-(*p*-dimethylaminophenylazo)pyridine **(IX-95)** is obtained.[367] 2-(*p*-Dimethylaminophenyl)pyridine is isolated when 2-aminopyridine is used instead of 4-aminopyridine.[368]

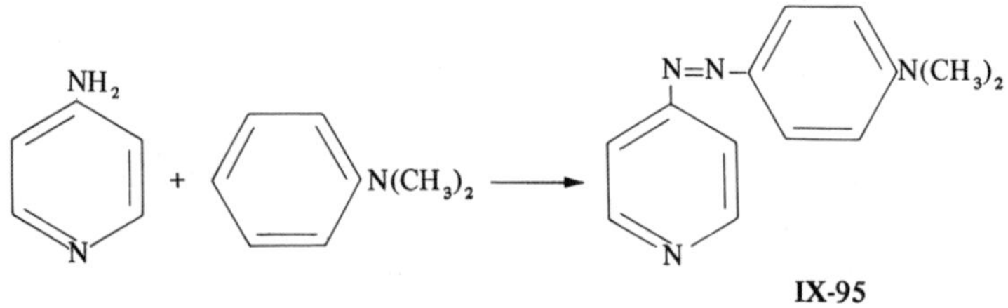

**IX-95**

When 2-aminopyridine and isoamylnitrite are added to sodium ethoxide and the mixture boiled for 8 hours, good yields of the diazonium salt are obtained. When β-naphthol is added to the diazonium salt, azo coupling to **IX-96** occurs.[369]

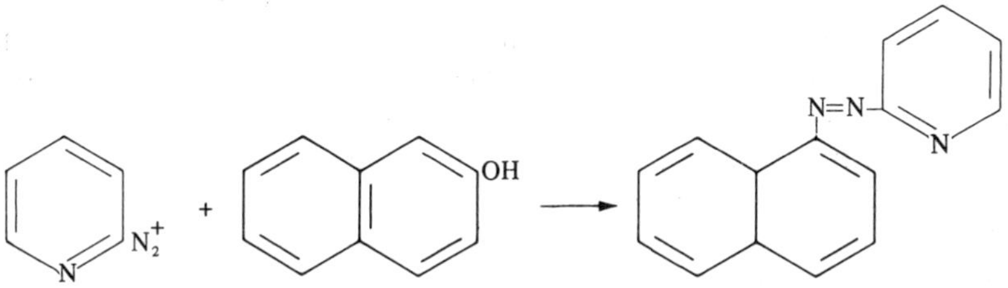

**IX-96**

2-Aminopyridine-1-oxide and its derivatives are diazotized and coupled with anilines to give azo dyes for acrylic fibers.[370-372]

The amino function in 4-amino-3-halopyridines behaves unexceptionally. Thus 4-amino-3-chloropyridine gives the 3-chloro-4-pyridinediazonium salt when treated with nitrous acid; the diazonium salt decomposes in the presence of potassium iodide to yield 3-chloro-4-iodopyridine.[373]

4-Amino-2,3,5,6-tetrafluoropyridine is a very weak base but it can be diazotized in 80% hydrofluoric acid. The diazonium salt is converted to 4-bromo-2,3,5,6-tetrabromopyridine with cuprous bromide, but its reaction with water or with *N,N*-dimethylaniline are complex.[374]

4-Amino-2-chloro-3-nitropyridine is *not* converted to 2-chloro-3-nitro-4-pyridone **(IX-97)** on diazotization with mineral acids and sodium nitrite or with isoamyl nitrite in glacial acetic acid; with the latter reagent 2-chloro-3-nitropyridine **(IX-98)** is formed.[46] Nitrous acid reacts with 4-amino-2-pyridone to give 4-amino-3-nitroso-2-pyridone instead of the diazonium salt.[375]

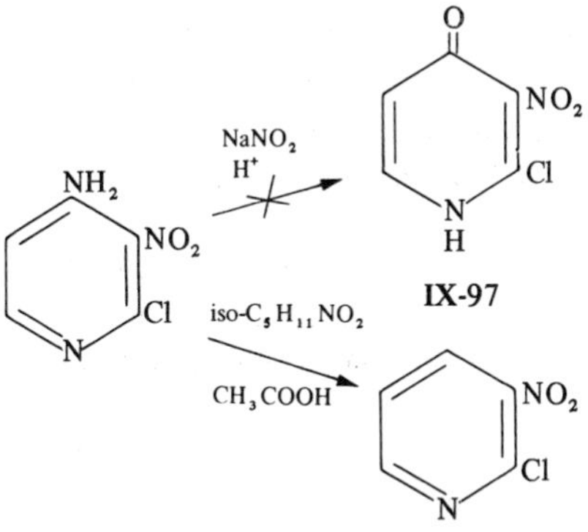

$NaNO_2$, $H^+$ (no reaction); iso-$C_5H_{11}NO_2$, $CH_3COOH$

**IX-97**

**IX-98**

When 3-aminopyridine is diazotized with hydrochloric acid and sodium nitrite and then treated with acetic acid containing sulfur dioxide and cupric chloride, 3-pyridinesulfonic acid can be isolated.[376] Similarly, 4-aminopyridine-1-oxide gives 4-pyridinesulfonic acid-1-oxide, but the 2-isomer gives very poor yields of the corresponding sulfonic acid. The preparations and reactions of various 3-pyridinediazonium salts have been reported.[52,377,378] 3-Amino-2,6-lutidine is diazotized with hydrochloric acid and sodium nitrite and then decomposed in the presence of copper powder to give 3-chloro-2,6-lutidine.[68] The diazonium salt reacts with potassium cyanide and with β-naphthol to give 3-cyano-2,6-lutidine and (2,6-dimethyl-3-pyridyl)-1-azo-2-hydroxynaphthalene, respectively.[379]

3-Amino-5-hydroxypyridine and nitrous acid give 3,5-dihydroxypyridine,[380] but diazotization of 3-amino-5-nitro-2-pyridone yields 3-diazo-5-nitro-2-pyridone.[379]

The Sandmeyer reaction on 1-phenethyl-4(or 5)-amino-2-pyridone has been reported.[381] Thus, 5-chloro-1-phenethyl-2-pyridone is obtained when 5-amino-1-phenethyl-2-pyridone is diazotized and then treated with cuprous chloride.

## E. *Nuclear Substitution Reactions*

5-Chloro- and 5-bromo-2-aminopyridine are prepared by the action of hydrogen peroxide on a mixture of 2-aminopyridine and the appropriate hydrogen halide.[382,383] 2-Amino-5-chloropyridine is also obtained by bubbling chlorine gas into a solution of 2-aminopyridine in sulfuric acid.[384]

When a solution of 2-aminopyridine is treated with bromine, 2-amino-5-bromopyridine is obtained.[385] Bromination of 2-amino-3-nitropyridine and 2-amino-5-nitropyridine gives 2-amino-5-bromo-3-nitropyridine and 2-amino-3-bromo-5-nitropyridine, respectively.[386]

6-Amino-2-bromo-3,4,5-tricyanopyridine reacts with the sodium salt of malonitrile to form 6-amino-α,α,3,4,5-pentacyano-2-picoline.[387]

When 2-amino-4-chloropyridine is added to sodium in absolute methanol containing a little copper powder, and the mixture is heated for 12 hours at 150° in a sealed tube, 2-amino-4-methoxypyridine is obtained. This reaction has been used to prepare a large number of aminomethoxypyridines (see Tables IX-16 to IX-18).[388]

4-Amino-2-methoxypyridine is acetylated to give the acetyl derivative that, when treated with perbenzoic acid in chloroform, gives 4-amino-2-methoxypyridine-1-oxide.[389] Similarly, 4-amino-3-chloropyridine-1-oxide is obtained from 4-amino-2-chloropyridine.

Glutazine **(IX-99)** reacts with *p*-chlorobenzenediazonium chloride to give **IX-100**.[390] 4-Amino-2,6-dichloropyridine is prepared in 60% yield from the reaction of glutazine with phosphorus oxychloride at 150° for 5 hours.[390]

5-Hydroxy-6-methylpyridine-3-carboxylic acid **(IX-102)**, a metabolite of pyridoxamine in *Pseudomonas MA*, has now been synthesized from 5-amino-2-chloro-6-methylpyridine-3-carboxylic acid **(IX-101)** by hydrogenation

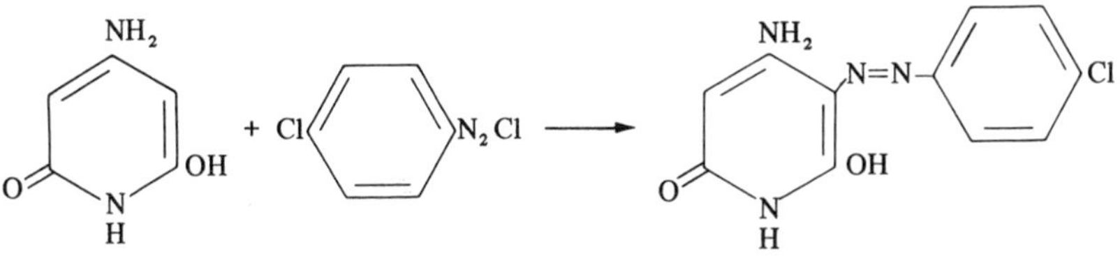

**IX-99**

**IX-100**

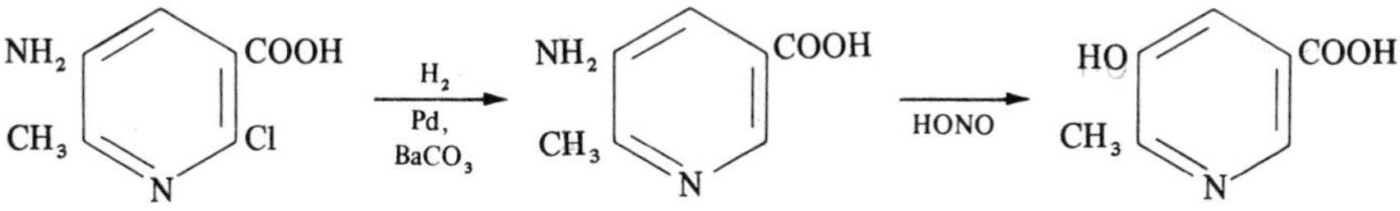

$H_2$ / Pd, $BaCO_3$ ; HONO

**IX-101**

**IX-102**

of an aqueous suspension using palladium on barium carbonate followed by diazotization with nitrous acid.[378, 391, 392]

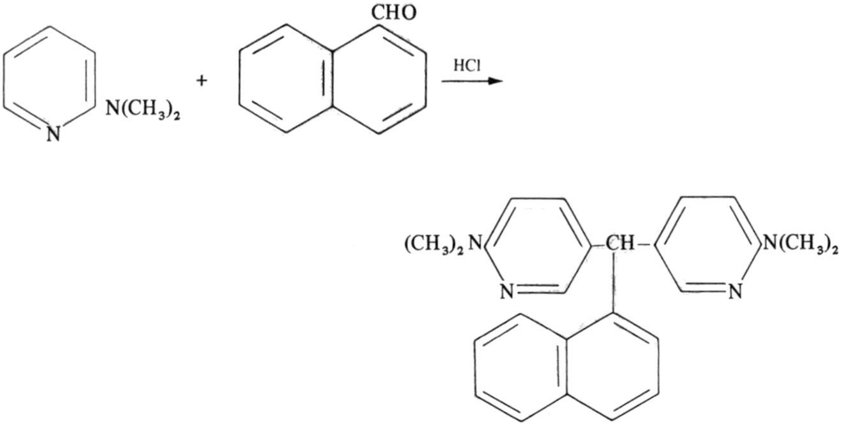

HCl

**IX-103**

2-Dimethylaminopyridine condenses with α-naphthaldehyde in the presence of concentrated hydrochloric acid to give bis[5-(2-dimethylamino)pyridyl]-1-naphthylmethane **(IX-103)**.[393]

## F. *Synthesis of Polycyclic Systems*

The present discussion on the synthesis of condensed-ring systems is limited to aminopyridines in which only one amino group is involved; syntheses involving more than one amino group are discussed in Section I.7., on "Diamino- and Triaminopyridines."

a. NAPHTHYRIDINES An excellent review of the chemistry of naphthyridines has recently appeared.[394] Aminopyridines have been used to prepare 1,5-, 1,6-, 1,7-, and 1,8-naphthyridines.

Various substituted 3-aminopyridines have been cyclized to 1,5-naphthyridines.[394, 395] Three products, **IX-104** to **IX-106**, have been isolated from the Skraup reaction with 3,5-diaminopyridine; Czuba has suggested that 3-amino-1,5-naphthyridine **(IX-104)** was the logical precursor to **IX-105** and

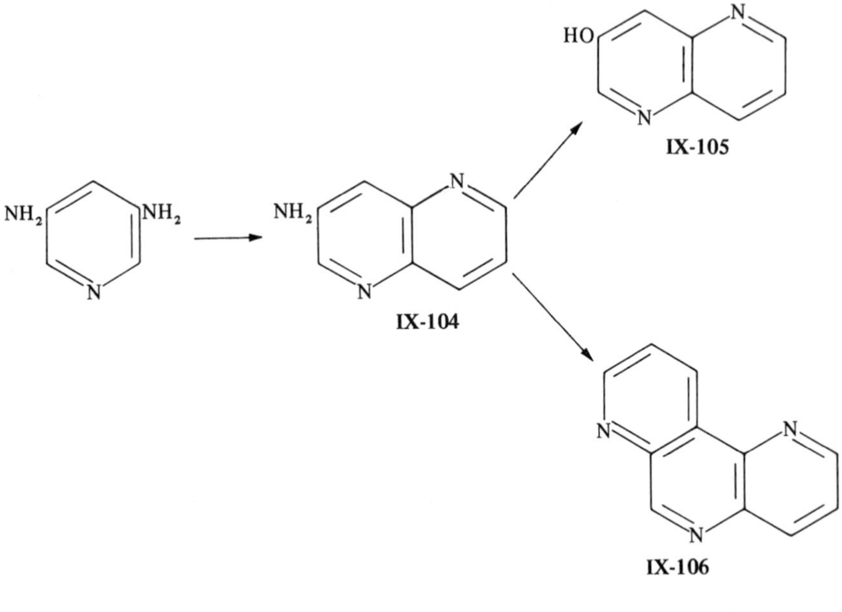

HO N N

**IX-105**

NH$_2$ NH$_2$ N NH$_2$ N N

**IX-104**

N N N

**IX-106**

**IX-106**.[396] In contrast, 2,5-diaminopyridine afforded only 2-amino-1,5-naphthyridine.[61, 397] When 3-amino-5-bromopyridine was used, 3-bromo-1,5-naphthyridine was produced.[61] Besides glycerol, other reagents were employed (see Table IX-3).

After several failures[398, 399] to prepare 1,6-naphthyridine by means of the Skraup reaction with 4-aminopyridine, it can now be obtained using glycerol and "sulfomix".[400, 401] The Skraup reaction yields 1,6-naphthyridine-6-oxide

**(IX-107)** and its derivatives when it is applied to 4-aminopyridine-1-oxide and some of its derivatives.[402, 403]

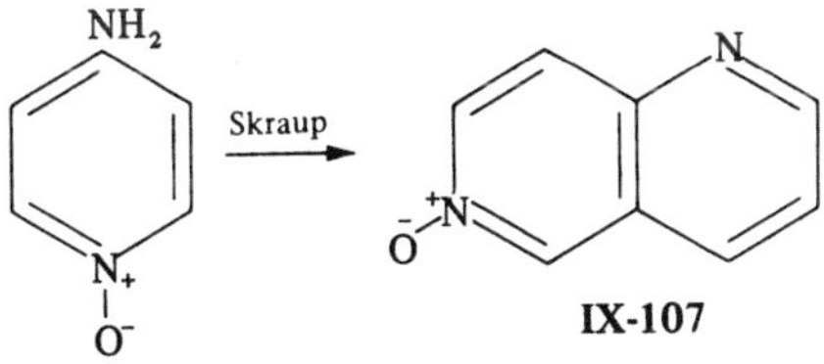

NH2
Skraup
N
N+
O-
IX-107

The reagents for cyclization are not limited to glycerol; they include crotonaldehyde, methacrolein, and methyl vinyl ketone, which afford 2-methyl-,[404] 3-methyl[405] and 4-methyl-1,6-naphthyridines,[404] respectively.

The cyclization of the condensation product of 3-aminopyridine-1-oxide with diethyl ethoxymethylenemalonate (EMME) yields ethyl 4-hydroxy-1,7-naphthyridine-7-oxide-3-carboxylate **(IX-108)**;[406] the *N*-oxide function need not be present if the α-positions of 3-aminopyridines are blocked.[407] The Skraup reaction with 3-amino-2-pyridone yields 1,7-naphthyridin-8-one.[408] Several

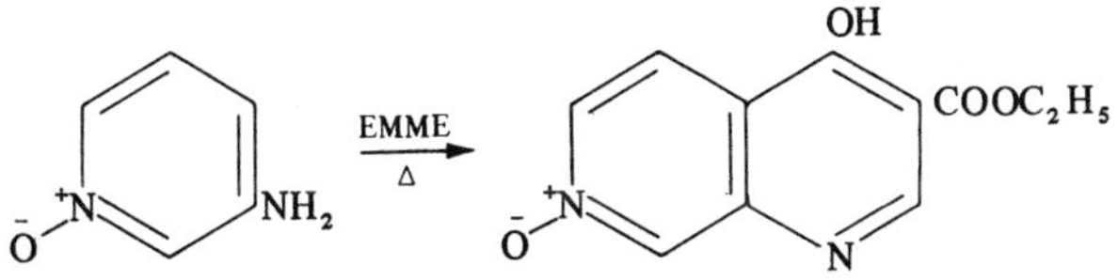

OH
$COOC_2H_5$
EMME
Δ
NH2
N
IX-108

4-substituted-3-aminopyridines are also cyclized to 1,7-naphthyridine derivatives **(IX-109)**;[409, 410] for example, 2-aminopyridine and some of its derivatives are cyclized under Skraup conditions to 1,8-naphthyridines. Table IX-4 summarizes some of these cyclizations.[405, 411] 2-Aminopyridine and diethyl ethoxymethylenemalonate also gives 1,8-naphthyridine in boiling "Dowtherm A".[412]

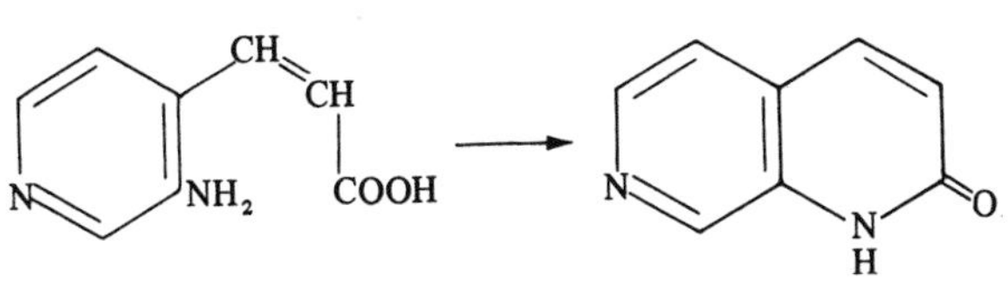

CH
CH
COOH
NH2
N
O
H
IX-109

The condensation of 2,6-diaminopyridine with ethyl acetoacetate under Conrad-Limpach conditions is now believed to give 7-amino-4-methyl-1,8-naphthyridin-2-one **(IX-110)**[412-414] and not 7-amino-2-methyl-1,8-naphthyridin-4-one as previously claimed.[415] Other 7-amino-1,8-naphthyridines are prepared similarly.[413-421]

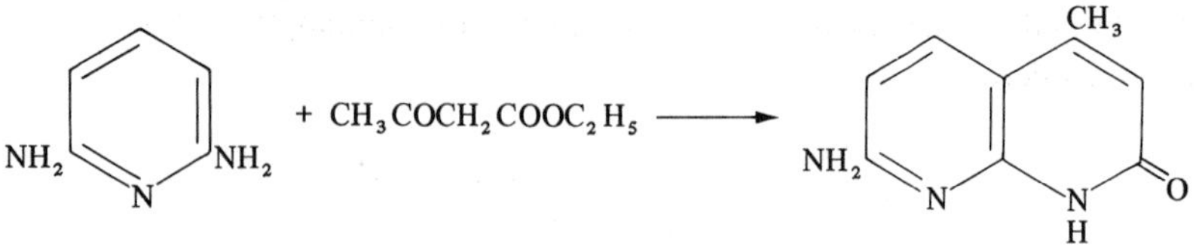

$$\text{2,6-diaminopyridine} + CH_3COCH_2COOC_2H_5 \longrightarrow \text{IX-110}$$

**IX-110**

1,8-Naphthyridines have been prepared by means of a Niementowski synthesis from ethyl 2-amino-6-phenylnicotinate and by means of Friedlander synthesis from 2-aminonicotinaldehydes.[422, 423] Ethyl 2-amino-6-phenylnicotinate **(IX-111)**, which is readily prepared from ethyl α-ethoxycarbonyl acetimidate and benzoylacetaldehyde, condenses with the simple esters **(IX-122**; R = H, $CH_3$, $C_6H_5$) in the presence of sodium to give good yields of 1,8-naphthyridinones **(IX-113)**.

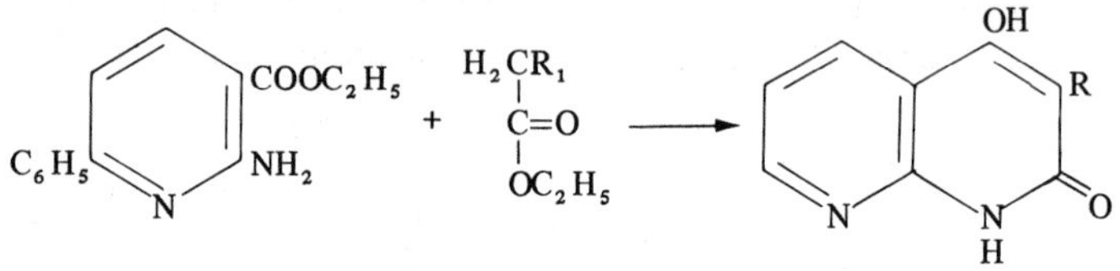

$$\text{IX-111} + H_2CR_1-C(=O)-OC_2H_5 \ (\text{IX-112}) \longrightarrow \text{IX-113}$$

**IX-111** **IX-112** **IX-113**

2-Aminonicotinaldehydes **(IX-114)** are potentially more reactive than 2-aminonicotinates. Thus, **IX-114** condenses readily with various ketones or aldehydes **(IX-115)** in the presence of piperidine to give 2,3-disubstituted-1,8-naphthyridines **(IX-116)**.[422] Malononitrile and cyanoacetamide may be used in place of **IX–115**.

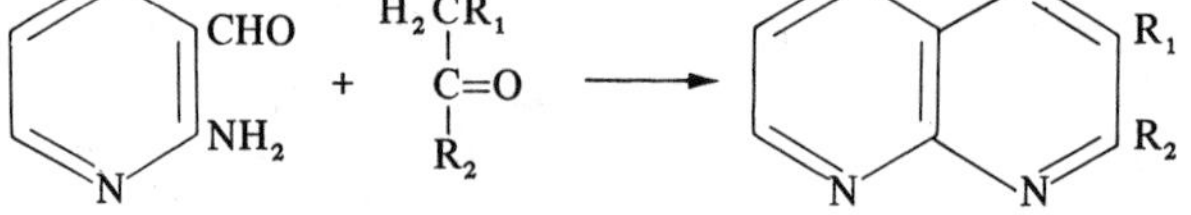

$$\text{IX-114} + H_2CR_1-C(=O)-R_2 \ (\text{IX-115}) \longrightarrow \text{IX-116}$$

**IX-114** **IX-115** **IX-116**

A mixture of 3-amino-2,6-dimethylpyridine **(IX-117)** and diethyl ethoxymethylenemalonate when heated on a steam bath under reduced pressure gives diethyl (2,6-dimethyl-3-pyridyl) methylenemalonate **(IX-118)**. The latter yields ethyl 4-hydroxy-6,8-dimethyl-1,7-naphthyridine-3-carboxylate **(IX-119)** when boiled for 4 minutes in diethyl phthalate.

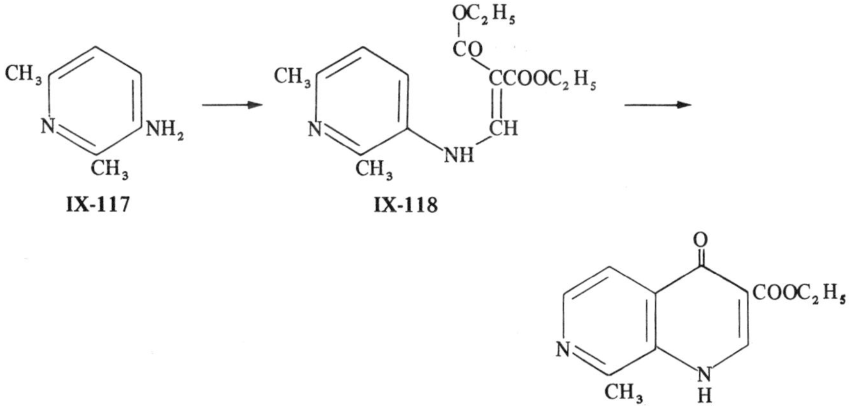

IX-117 → IX-118 → IX-119

A new heterocyclic ring system, pyrrolo[3,2-3,4]-1,8-naphthyridine **(IX-121)** is obtained from the cyclization of mono- and di-*N*-acyl derivatives of 2-aminonicotyrine **(IX-120)** with phosphorus oxychloride.[424] Thus, 2-dibenzoylaminonicotyrine is dissolved in toluene and phosphorus oxychloride and heated

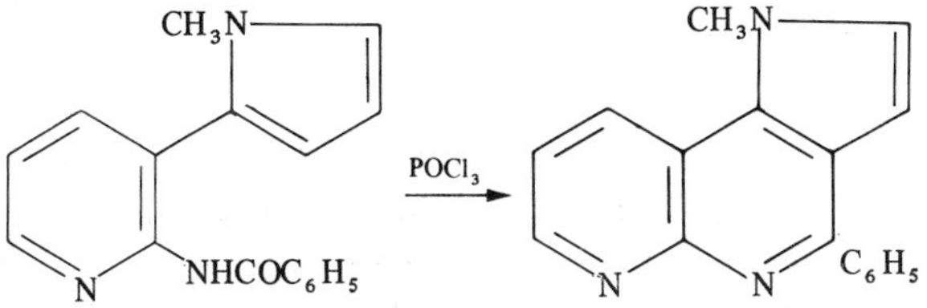

IX-120 —$POCl_3$→ IX-121

under reflux for 3 hours to yield 1-methyl-2-phenylpyrrolo[3,2-3,4]-1,8-naphthyridine **(IX-121)**.

b. PYRIDOPYRIMIDINE The reaction of 2-aminopyridine **(IX-124)** with methyl acrylate **(IX-122)** in the presence of an acid to give 2*H*-pyrido[1,2-*a*]-pyrimidin-2-one **(IX-126)** has been studied. It was suggested that **IX-122** was "activated" by protonation of the carboxyl group **IX-123**, followed by a nucleophilic attack at the β-carbon atom by the ring-nitrogen to form the intermediate **IX-125**, which cyclized to **IX-126**.[425, 426] When **IX-124** is heated in water with 1–3 moles of a derivative of acrylic acid (the nitrile, amide, methyl or ethyl ester), good yields (30–83%) of β-[1-(2-aminopyridyl)] propionic acid are obtained. This, on heating with concentrated hydrochloric acid, can be cyclized to the hydrochloride of **IX-126**.[427]

Lappin[428, 429] claimed that a cold ethereal solution of 2-aminopyridine and methyl propiolate give **IX-126** as well as a di-adduct, methyl 2-(2-carbomethoxyvinylimino)-1(2*H*)-pyridineacrylate **(IX-127)**. Subsequently,

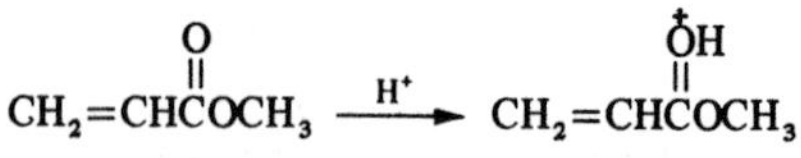

$$CH_2{=}CHCOCH_3 \xrightarrow{H^+} CH_2{=}CH\overset{+}{C}(OH)OCH_3$$

IX-122 IX-123

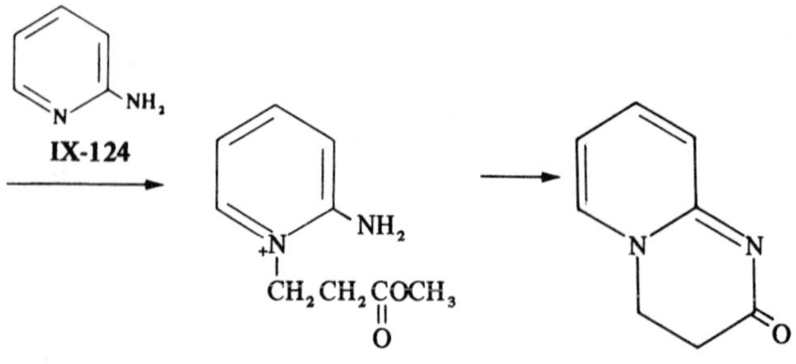

IX-124

IX-125 IX-126

the structure of the di-adduct was proved by Wilson and Bottomley to be **IX-128** and not **IX-127**.[430]

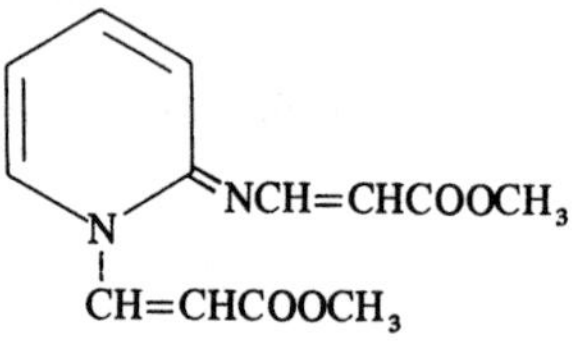

IX-127 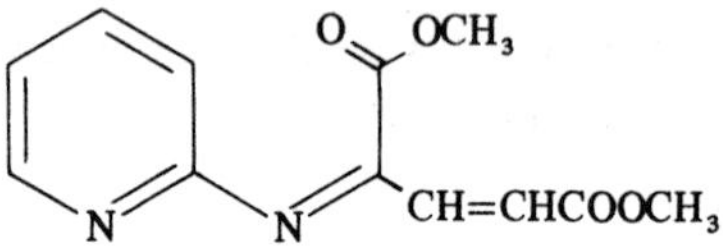 IX-128

High yields (80% and better) of 4*H*-pyrido[1,2-*a*]pyrimidin-4-ones **(IX-131)** are obtained in a one-step synthesis involving the condensation of 1 mole of a 2-aminopyridine **(IX-129)** with 1.5 mole of β-keto ester **(IX-130)** at 100° for 1 hour in the presence of polyphosphoric acid.[431]

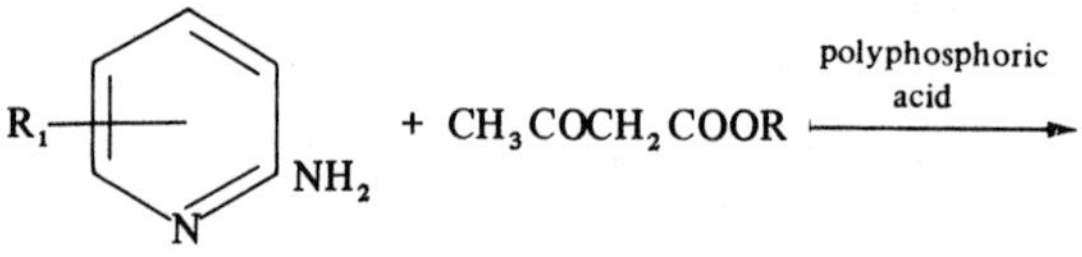

$$+ \; CH_3COCH_2COOR \xrightarrow{\text{polyphosphoric acid}}$$

IX-129 IX-130

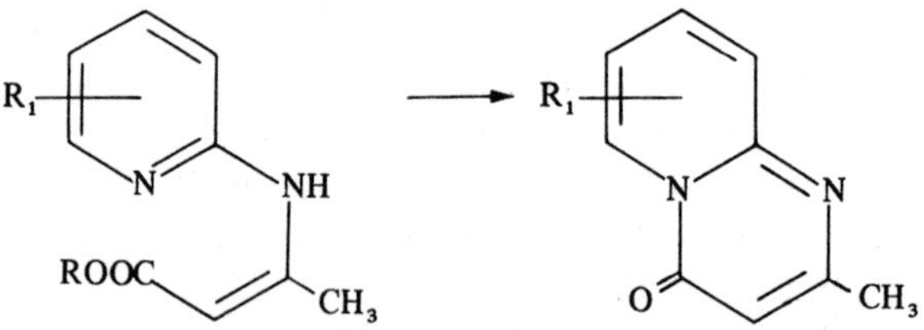

IX-131

$R = CH_3, C_2H_5$

$R_1$ = 3-$CH_3$, 5-Cl, 5-Br, 3,5-$Br_2$

If 2-aminopyridine is boiled with ethyl malonate for 5 minutes, it gives *N,N*-di-(2-pyridyl)malonamide; when the reaction time is extended to 6 hours, the product is pyrido[1,2-*a*]pyrimidine-4,6-dione (**IX-132**, R = H). When

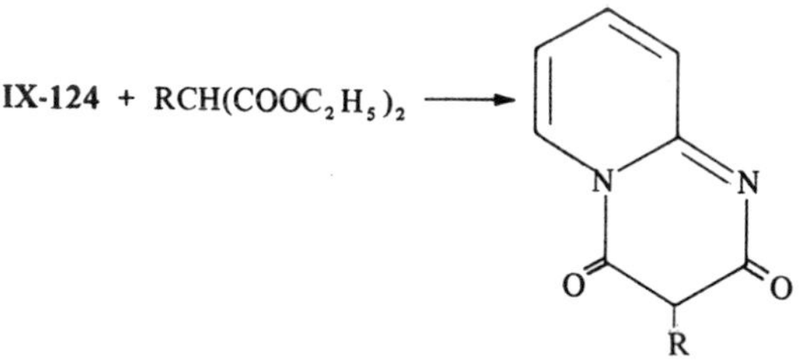

IX-124 + $RCH(COOC_2H_5)_2$ ⟶

**IX-132**

alkylmalonyl chloride is used instead of ethyl malonate, the 5-alkyl derivative **IX-132** is obtained in good yield.[432, 433] The chemistry of these cyclized "malonyl-α-aminopyridines" is being investigated.[434, 435]

The reaction of 2-aminopyridine with epichlorohydrin yields the hydrochloride of 3-hydroxy-3,6-dihydro-2*H*-pyrido[1,2-*a*]pyrimidine (**IX-133**).[436] If chlorohydrin is used instead of epichlorohydrin, no pyridopyrimidine is formed.[437]

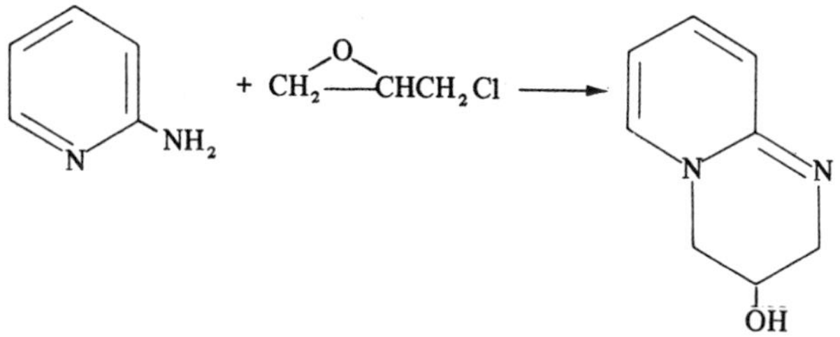

+ $CH_2$—(O)—$CHCH_2Cl$ ⟶

**IX-133**

c. DIAZAINDENES 4-, 5-, And 6-methyl-1,7-diazaindene are prepared by the Madelung ring closure of the corresponding 2-formamidolutidines, which are obtained by formylation of aminolutidines; for example, 2-formamido-3,4-lutidine (**IX-134**) cyclizes to 4-methyl-1,7-diazaindene (**IX-135**) in the presence of potassium formate and sodium anilide.[438] 3-Formamido-4-picoline gives only 3-amino-4-picoline and no cyclized product when treated with various bases.[439]

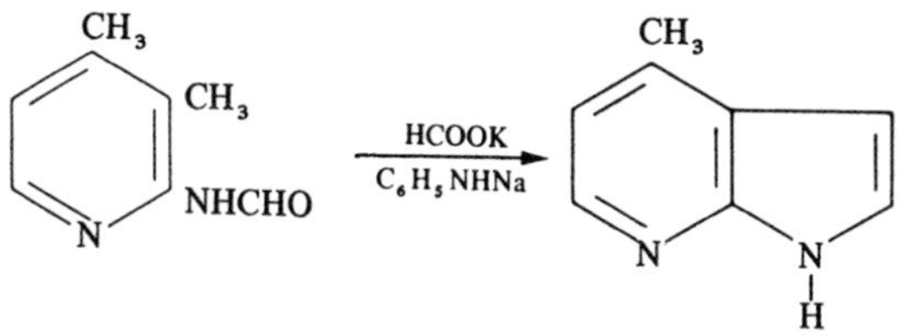

$\xrightarrow[C_6H_5NHNa]{HCOOK}$

**IX-134** **IX-135**

d. CARBOLINES The Pschorr-type ring closure for the synthesis of the carboline ring system was first investigated by Abramovitch, Hey, and Mulley.[440] When an aqueous solution of the diazonium salt of 2-amino-*N*-methyl-*N*-2-pyridylaniline **(IX-136)** is heated, a small amount (7%) of *ind*-*N*-methyl-α-carboline **(IX-137)** is formed, though the main product is 5-methylpyrido[1,2-*a*]benzimidazolium chloride **(IX-138)**

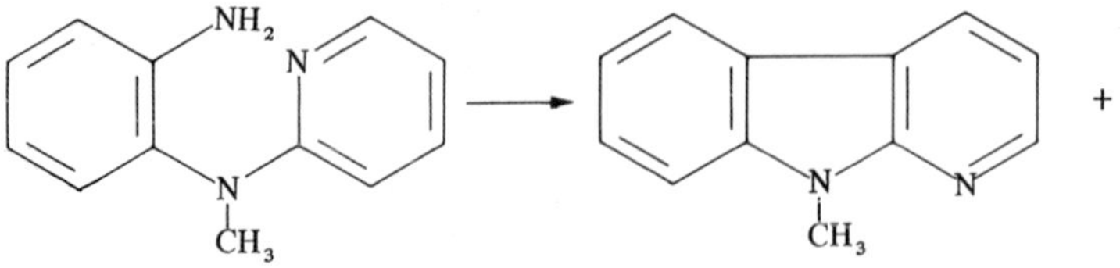

NH$_2$ N N CH$_3$

**IX-136**

N N CH$_3$ +

**IX-137**

$\overset{+}{N}$ N CH$_3$ Cl$^-$

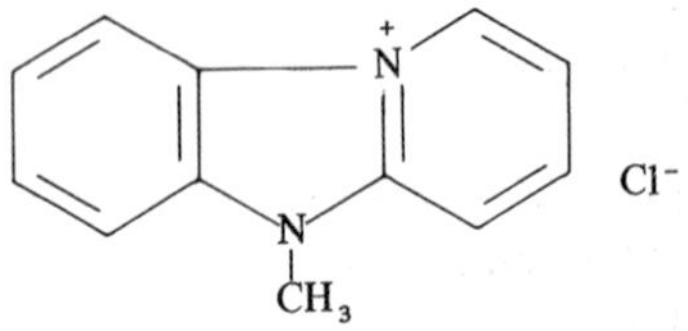

**IX-138**

The first synthesis of δ-carboline **(IX-141)** was achieved by the thermal decomposition of 3-azidopyridine **(IX-140)**, prepared from 3-amino-2-phenylpyridine **(IX-139)**.[441] When the diazonium salt from 2-amino-*N*-methyl-*N*,3′-

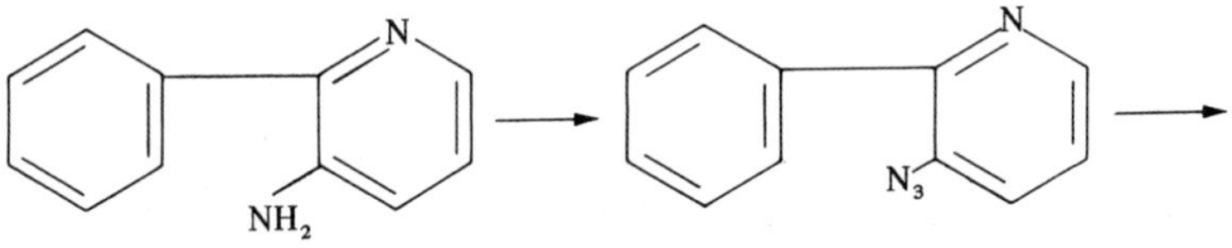

N NH$_2$

**IX-139**

N N$_3$

**IX-140**

N N H

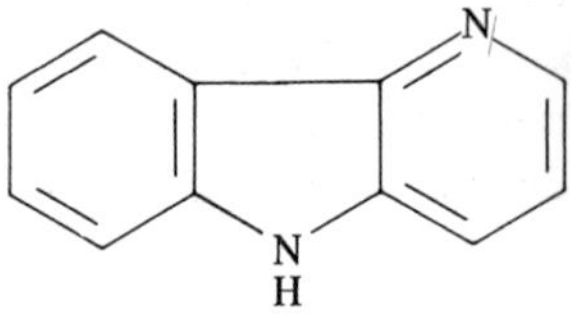

**IX-141**

pyridylaniline **(IX-142)** in aqueous acid solution is decomposed in the presence of copper powder, a mixture of *ind*-*N*-methyl-δ-carboline **(IX-143)** and *ind*-*N*-methyl-β-carboline **(IX-144)** is obtained.[442, 443]

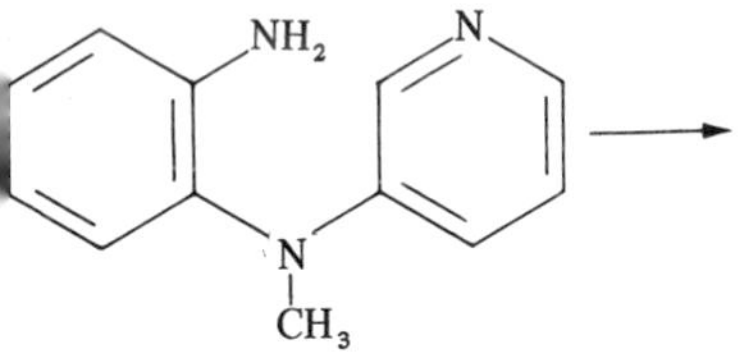

IX-142

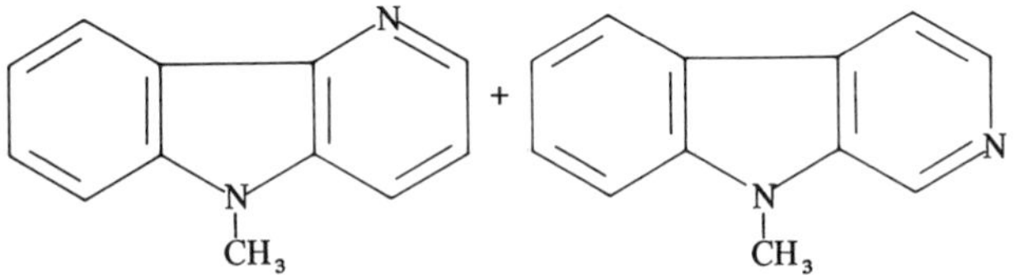

IX-143

IX-144

Nantka-Namirski[44] has employed the Graebe-Ullmann method to prepare γ-carbolines **(IX-147)** from phenyl-substituted triazolopyridines **(IX-145)** or

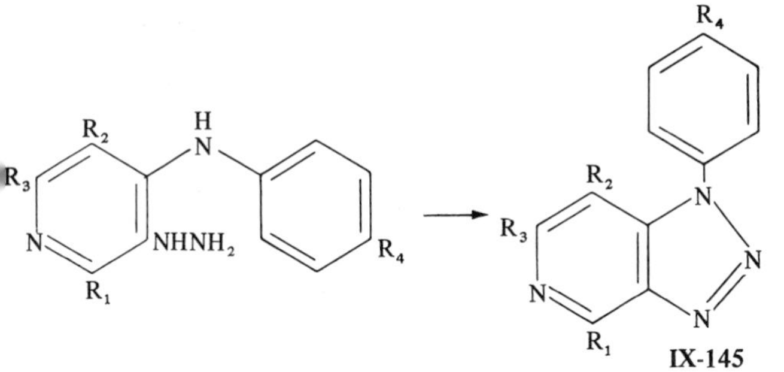

IX-145

pyridyl-substituted benzotriazoles **(IX-146)** in pyrophosphoric acid. Various aspects of carboline synthesis have been discussed.[16, 444]

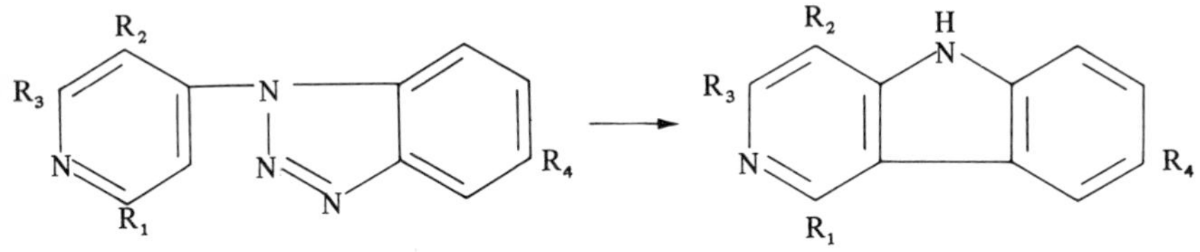

IX-146

IX-147

e. IMIDAZOPYRIDINES Improved laboratory procedures for the synthesis of imidazopyridines have been reported.[331, 445] 7-Chloro-, 5-chloro-, 5-amino-, 5-amino-2-methyl and 5-amino-2,3-dimethyl-imidazo[1,2-*a*]pyridines were prepared by condensation of haloaminopyridines with haloaldehydes. Thus bromoacetaldehyde and 2-amino-4-chloropyridine give 7-chloroimidazo[1,2-*a*]-pyridine **(IX-148)**.[445]

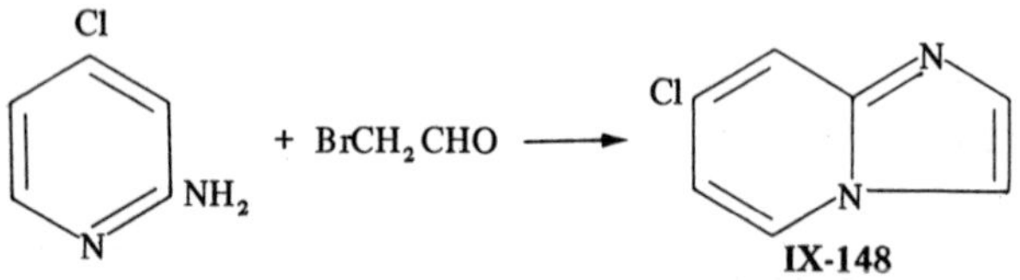

$$\text{2-amino-4-chloropyridine} + BrCH_2CHO \longrightarrow \textbf{IX-148}$$

Various imidazopyridinium salts were also prepared. 2-Alkylamino- or arylaminopyridines react with α-bromoketones to give 1-alkyl- or 1-arylimidazo-[1,2-*a*]pyridinium salts substituted at position 2. 2-Anilinopyridine **(IX-149)** and bromoacetone in boiling acetone give the 2-methyl-1-phenylimidazo[1,2-*a*]-pyridinium bromide **IX-151** rather than the 1-acetonyl salt **IX-150**.[446] When

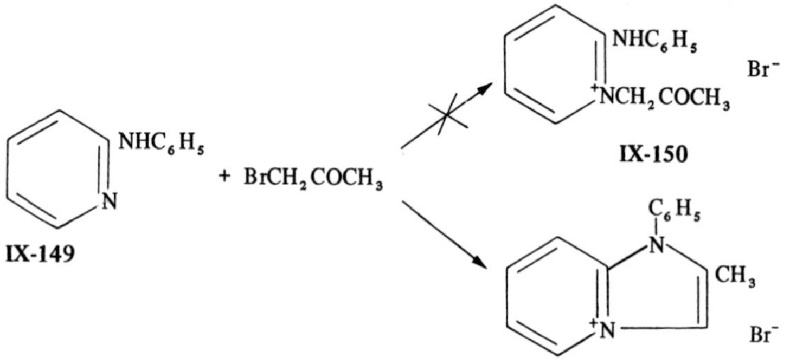

$$\textbf{IX-149} + BrCH_2COCH_3 \longrightarrow \textbf{IX-151} \quad (\text{not } \textbf{IX-150})$$

2-(*N*-methylanilino)pyridine was used instead of 2-anilinopyridine, cyclization was not possible. The bromination in benzene of 2-allylaminopyridine gives **IX-152**.[447]

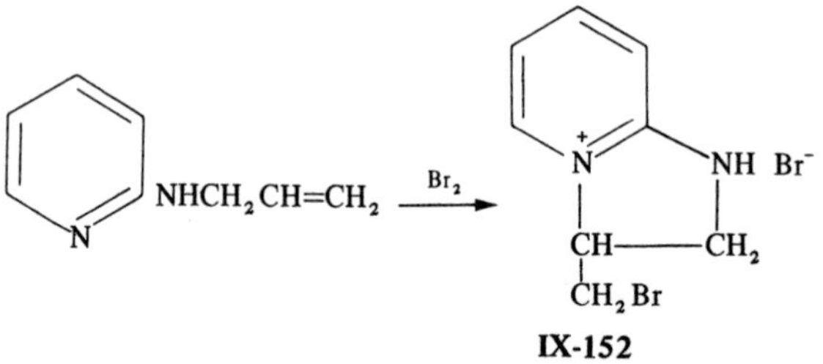

$$\text{2-}(NHCH_2CH{=}CH_2)\text{pyridine} \xrightarrow{Br_2} \textbf{IX-152}$$

In the presence of dry hydrochloric acid, 3-amino-2-(β-diethylaminoethyl-amino)pyridine **(IX-153)** condenses with *p*-ethoxyphenylacetonitrile to give

3-(β-diethylaminoethyl)-2-(*p*-ethoxylbenzyl)imidazo[4,5-*c*]pyridine **(IX-154)**. Using other arylacetonitriles, various imidazopyridines are prepared.[80]

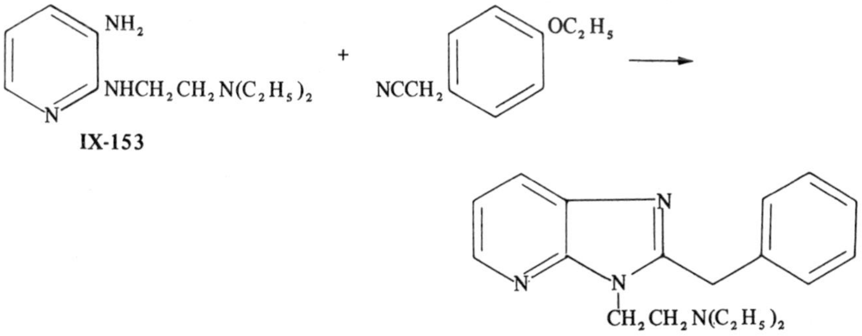

$NH_2$ $NHCH_2CH_2N(C_2H_5)_2$ + $NCCH_2$ $OC_2H_5$ →

**IX-153**

$CH_2CH_2N(C_2H_5)_2$

**IX-154**

f. MISCELLANEOUS CYCLIZATIONS 2-Aminopyridine condenses with 3-chloro-2-ethoxy-1,4-naphthoquinone **(IX-155)** to yield 6*H*,11*H*-benzo(*f*)pyrido-[1,2-*a*]benzimidazole-6,11-dione **(IX-156)**;[448] with 2,3-chloro-1,4-naphthoquinone, it yields 5*H*,6*H*-benzo[*e*]pyrido[1,2-*a*]benzimidazole-5,6-dione **(IX-158)**.[449, 450]

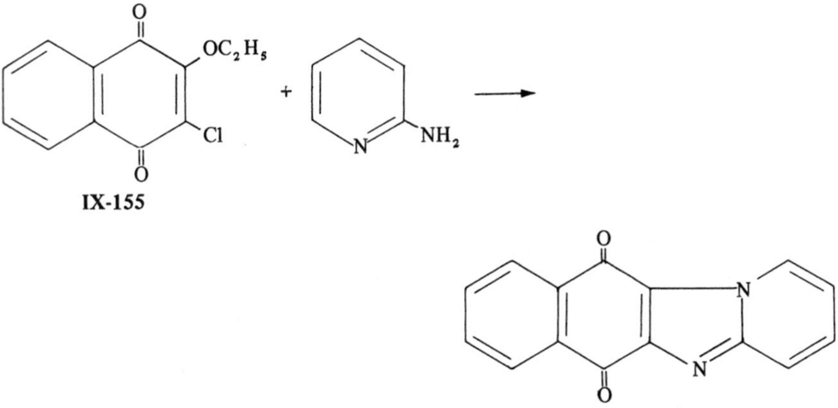

$OC_2H_5$ Cl + $NH_2$ →

**IX-155**

**IX-156**

Cl Cl + $NH_2$ →

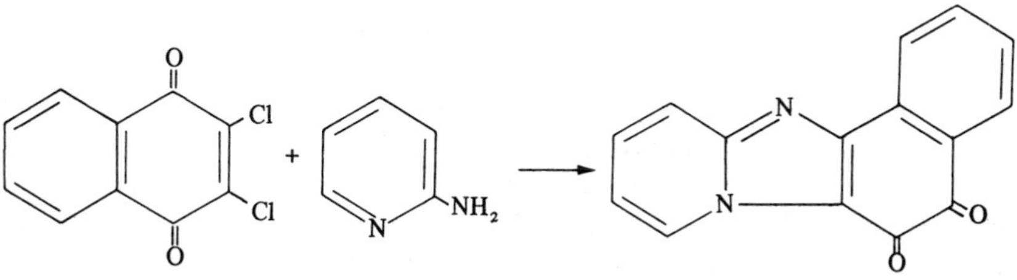

**IX-157**

**IX-158**

10-Substituted aminoalkylazaphenothiazines are obtained by intramolecular cyclization of aminoalkylaminohalophenylthiopyridine. Thus, 10-(3-dimethylaminopropyl)-2-azaphenothiazine **(IX-160)** is obtained by heating 4-(2-bromophenylthio)-3-(3-dimethylaminopropyl)aminopyridine **(IX-159)** in dimethylformamide with sodium and copper powder.[449] Similarly, 3-amino-4-(2-bromo-4-chlorophenythio)pyridine gives 8-chloro-2-azaphenothiazine **(IX-161)**.[451]

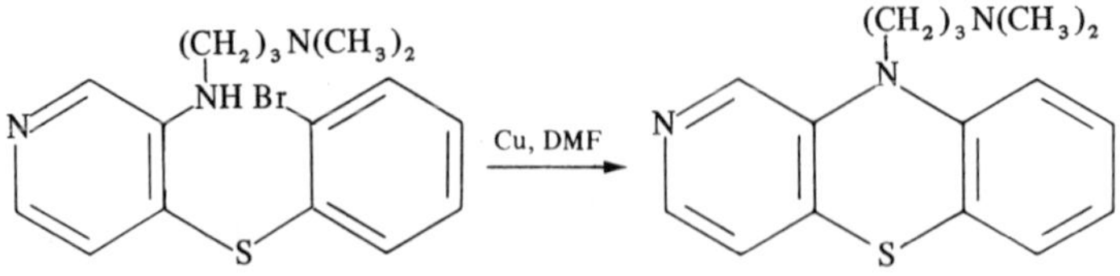

$(CH_2)_3N(CH_3)_2$ NH Br N S — Cu, DMF → — $(CH_2)_3N(CH_3)_2$ N N S

**IX-159** **IX-160**

2-Chloronicotinic acid, when heated gradually to 145° with methyl *o*-aminobenzoate and potassium iodide, gives **IX-162**; with 3-amino-*o*-xylene, however, the 2-anilinonicotinic acid derivative **(IX-163)** is obtained.[452] When a

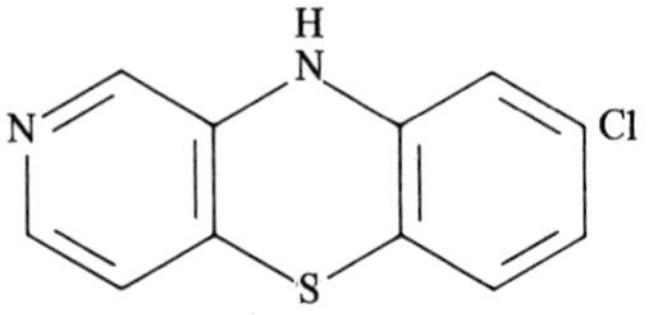

H N N Cl S

**IX-161**

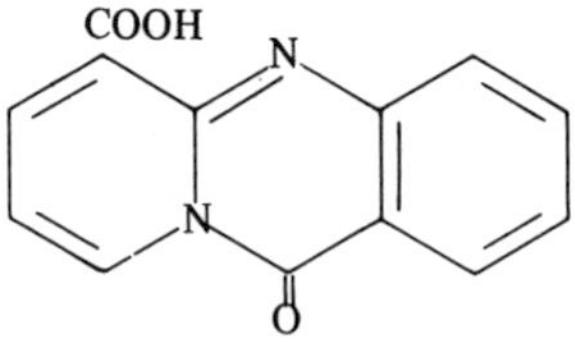

COOH N N O

**IX-162**

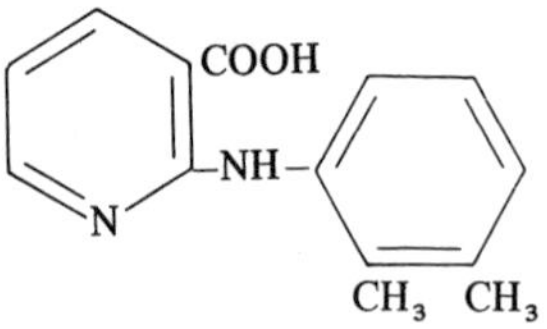

COOH NH– N $CH_3$ $CH_3$

**IX-163**

mixture of 2-aminopyridine and *N*-acetylanthranilic acid is heated to 120° for 4 hours and the product is then hydrolyzed, 2-methyl-3-(2-pyridyl)-3,4-dihydro-4-oxoquinazoline **(IX-164)** is obtained.[453] The latter can also be prepared by heating *o*-amino-*N*-(2-pyridyl)benzamide **(IX-165)** with acetic anhydride.[453]

3-Amino-4-benzoylpyridine **(IX-166)** reacts with hydroxylamine hydrochloride **(IX-167)** to give 3-amino-4-benzoylpyridine oxime **(IX-168)**. With chloroacetyl chloride **(IX-168)** at room temperature for 72 hours, this gives a

NHCOCH$_3$ COOH NH$_2$ N + N CH$_3$ N N O IX-164 NH$_2$ CONH N Ac$_2$O IX-165

chloro intermediate that, on treatment with methylamine, cyclizes to give 2-methylamino-5-phenyl-3*H*-pyrido[3,4-*e*]-1,4-diazepine **(IX-169)**.[454]

O N NH$_2$ + $NH_2OH \cdot HCl$ → IX-166 IX-167

NOH N NH$_2$ IX-168 $ClCH_2COCl$ → $CH_3NH_2$ → Ph N N N NHCH$_3$ IX-169

## G. *Miscellaneous Reactions*

Aminopyridines undergo a variety of reactions that have not been discussed in the above sections.

Triphenylphosphine dibromide and 2-aminopyridine give the phosphineimine **IX-170**.[455]

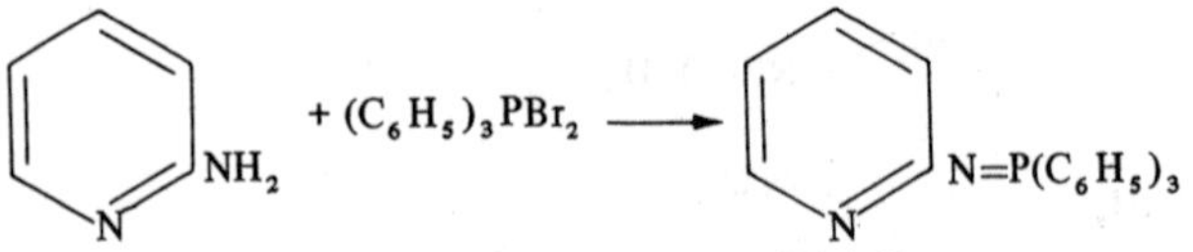

$$\text{2-aminopyridine} + (C_6H_5)_3PBr_2 \longrightarrow \text{2-pyridyl-}N{=}P(C_6H_5)_3$$

**IX-170**

The identification and determination of 2,4-, 2,5-, and 2,6-dinitrophenols by their color reactions with 2-aminopyridine have been reported.[456a]

2-Aminopyridine-1-oxide gives a blue color when it is heated with *N,N*-dimethylaniline and concentrated hydrochloric acid.[455] On diazotization and treatment with azide ion it yields 2-azidopyridine-1-oxide.[456b]

If 2-aminopyridine is boiled with methyl iodide in ethanol and the resultant precipitate (2-imino-1-methyl-1,2-dihydropyridine) is heated with sodium hydroxide it gives 2-methylaminopyridine (Dimroth rearrangement).[457]

2-Aminopyridine condenses with *p*-dimethylaminonitrosobenzene to give 2-(*p*-dimethylaminophenylazo) pyridine **(IX-171)**.[458] 3-Aminopyridine reacts with nitrosobenzene to yield 3-phenylazopyridine.[459]

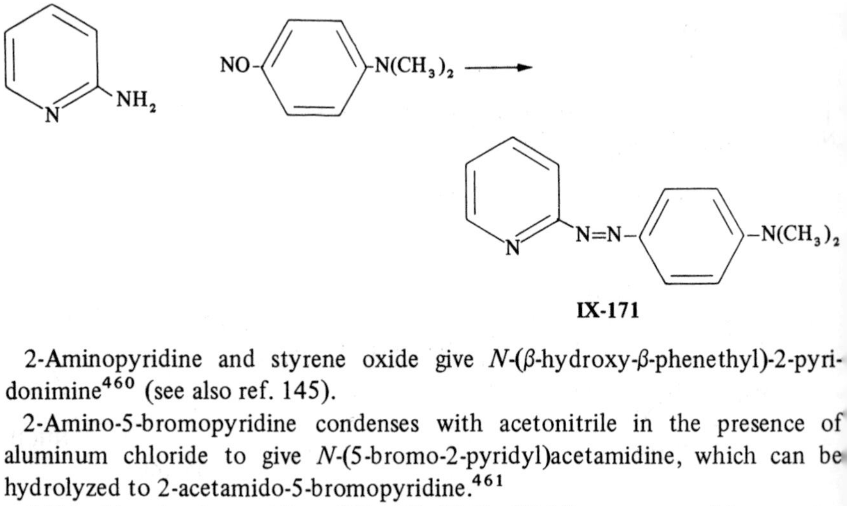

$$\text{2-aminopyridine} + NO\text{-}C_6H_4\text{-}N(CH_3)_2 \longrightarrow \text{2-pyridyl-}N{=}N\text{-}C_6H_4\text{-}N(CH_3)_2$$

**IX-171**

2-Aminopyridine and styrene oxide give *N*-(β-hydroxy-β-phenethyl)-2-pyridonimine[460] (see also ref. 145).

2-Amino-5-bromopyridine condenses with acetonitrile in the presence of aluminum chloride to give *N*-(5-bromo-2-pyridyl)acetamidine, which can be hydrolyzed to 2-acetamido-5-bromopyridine.[461]

5-Halothienylaminopyridines **(IX-172)** (Table IX-26) are prepared by treating thienylaminopyridines with sulfuryl halides.[462]

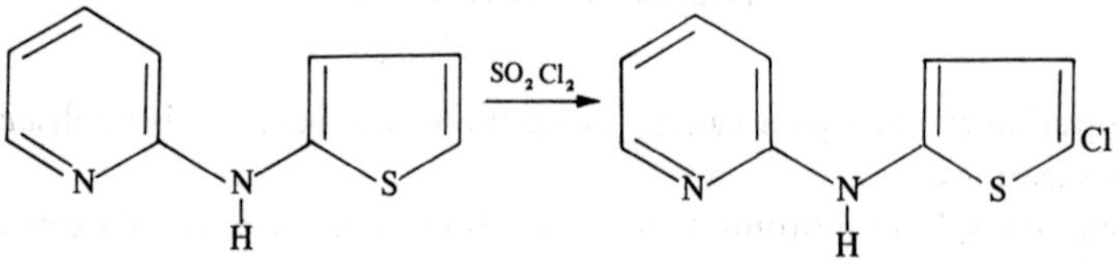

$$\text{2-pyridyl-NH-2-thienyl} \xrightarrow{SO_2Cl_2} \text{2-pyridyl-NH-(5-chloro-2-thienyl)}$$

**IX-172**

## 5. Nitraminopyridines

Geller and Samosvat[463] found that the rearrangement of 2-nitraminopyridine in sulfuric acid in the presence of $Na^{15}NO_3$ resulted in incorporation of $^{15}N$ into 3- **(IX-173)** and 5-nitro-2-aminopyridines **(IX-174)** to a smaller extent than remained in $NaNO_3$. They claimed that intramolecular rearrangement had occurred.

NHNO₂ —$Na^{15}NO_3$→ **IX-173** ($^{15}NO_2$, $NH_2$) + **IX-174** ($^{15}NO_2$, $NH_2$)

An improved preparation of 3-nitraminopyridine has been reported.[464] The reaction of nitraminopyridines with phosphorus halides gives the corresponding halopyridine.[464] Thus, 2-, 3-, and 4-nitraminopyridine react with phosphorous trichloride to yield 2-chloro- (38%), 3-chloro (45%), and 4-chloropyridine (65%), respectively.

In the presence of an *ortho*- or a *para*-nitro substituent, a nitramino function in the 2- or 4-position is easily replaced by a hydrazino group. Thus, a solution of 2-nitramino-3-nitropyridine in methanol gives 2-hydrazino-3-nitropyridine when treated with hydrazine hydrate. Various hydrazinonitropyridines are prepared similarly;[465] these are listed in Table IX-51.

Hydrolysis of nitraminopyridines generally gives hydroxypyridines; 5-nitramino-3-picoline and 3-nitramino-4-picoline are hydrolyzed with sulfuric acid to 5-hydroxy-3-picoline and 3-hydroxy-4-picoline, respectively.[466]

3-Nitramino-2-pyridinesulfonic acid gives 3-hydroxy-2-pyridinesulfonic acid when heated in concentrated sulfuric acid at 60 to 80°; but 3-nitramino-6-pyridinesulfonic acid gives only 3-hydroxypyridine.[467]

Aminopyridinecarboxylic acids are converted into nitramino derivatives by treating a cold solution of the amino compounds in sulfuric acid with nitric acid. Thus 5-aminonicotinic acid and 5-aminopicolinic acid give 5-nitraminonicotinic and 5-nitraminopicolinic acid, respectively.[468]

The nitration of 4-amino-3-picoline yields 3-methyl-4-nitraminopyridine, which, on heating, gives 4-amino-5-nitro-3-picoline. Reduction of this with iron and acetic acid gives 4,5-diamino-3-picoline.[469]

5-Amino-2-nitropyridine gives 2-nitro-5-nitraminopyridine when treated with a mixture of concentrated nitric acid and sulfuric acid below 0°; similarly, 4-amino-2-iodopyridine is nitrated to 2-iodo-4-nitraminopyridine. When the latter is dissolved in concentrated sulfuric acid, it rearranges to 4-amino-2-iodo-3-nitropyridine.[470]

When a mixture of concentrated nitric acid and sulfuric acid is added to a cold solution of 5,5′-diamino-2,2′-dipyridyl and the mixture is allowed to come to room temperature 5,5′-dinitramino-2,2′-dipyridyl **(IX-175)** is formed. At room temperature, **IX-175** decomposes to give 5,5′-diamino-4,4′-dinitro-2,2′-dipyridyl **(IX-176)**.[476]

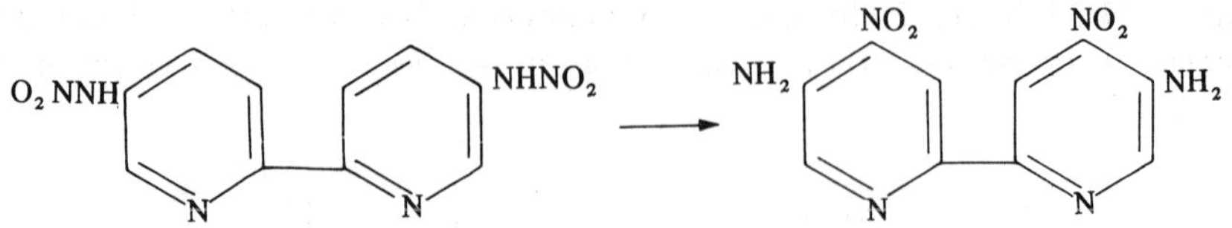

$O_2NNH$ $NHNO_2$ → $NH_2$ $NO_2$ $NO_2$ $NH_2$

**IX-175** **IX-176**

## 6. Pyridonimines

Pyridonimines are usually prepared by the reaction of aminopyridines with alkyl halides. Thus, 2-aminopyridine and allyl bromide form, predominantly, 1-allyl-2-pyridonimine **(IX-177)** together with *N*-allylaminopyridine **(IX-178)**.[472]

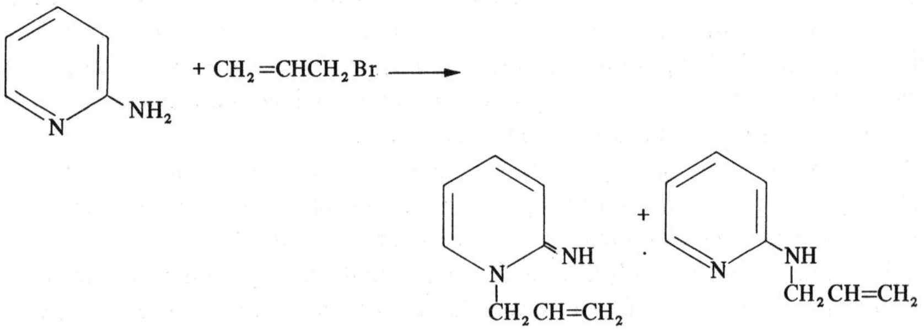

$NH_2$ + $CH_2{=}CHCH_2Br$ →

NH, $CH_2CH{=}CH_2$ + NH, $CH_2CH{=}CH_2$

**IX-177** **IX-178**

Other methods of preparing pyridonimines have been reported.[473, 474] 1-Cyano-2-methylpenta-3-yne-1-ene **(IX-179)** reacts with alkylamines to yield *N*-alkyl-2-pyridonimines; for example, methylamine yields 1,4,6-trimethyl-2-pyridonimine (**IX-180**, R = $CH_3$).[473]

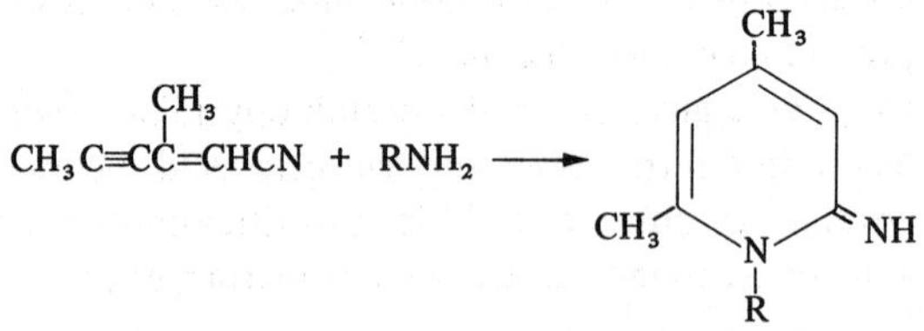

$CH_3C{\equiv}CC(CH_3){=}CHCN$ + $RNH_2$ → $CH_3$, $CH_3$, NH, R

**IX-179** **IX-180**

When furfural is treated with sulfamic acid (preferably in the presence of chlorine and chlorination catalysts) 3-hydroxy-2-imino-1(2*H*)-pyridinesulfonic acid monohydrate **(IX-181)** is isolated.[475]

$$\text{furfural (CHO)} + HSO_3NH_2 \xrightarrow[\text{Cl}_2]{\text{Catalyzed by}} \text{3-hydroxy-2-imino-1(2H)-pyridine-}N\text{-}SO_3H \cdot H_2O$$

**IX-181**

## 7. Diamino- and Triaminopyridines

### A. *Preparation*

Most investigations of the synthesis of diamino- and triaminopyridines are directed toward the substituted rather than the unsubstituted polyaminopyridines. Generally, the procedures are similar to those used for the preparation of monoaminopyridines.

a. REDUCTION OF NITRO GROUPS This may be effected chemically or catalytically. Stannous chloride and hydrochloric acid introduce a chloro substituent into the 2-position of the pyridine ring, in addition to reducing nitro groups. Thus, 4,5-diamino-3-picoline is obtained by the reduction of 4-amino-5-nitro-3-picoline with iron and acetic acid; if stannous chloride and hydrochloric acid are used, the reduction product is 2-chloro-4,5-diamino-3-picoline.[469]

The preparation and reactions of 3-bromo-4,5-diaminopyridine have been studied. The nitration of 4-amino-3-bromopyridine **(IX-182)** and the subsequent rearrangement of the nitramino derivative **IX-183** give 4-amino-3-bromo-5-nitropyridine **(IX-184)**. The reduction of **IX-184** with iron and acetic acid produces 5-bromo-3,4-diaminopyridine **(IX-185)**, but treatment of **IX-184** with stannous chloride again gives a chloro derivative, 5-bromo-2-chloro-3,4-diaminopyridine **(IX-186)**.[476]

Sodium dithionite has been used as a reducing reagent for 2-amino-5-chloro-3-nitropyridine to yield 5-chloro-2,3-diaminopyridine.[477]

2-Amino-3,5-dinitropyridine is selectively reduced with freshly prepared ammonium sulfide to 2,3-diamino-5-nitropyridine; catalytic reduction with palladium and hydrochloric acid gives 2,3,5-triaminopyridine trihydrochloride.[478]

Treatment of 2-bromo-3,5-dinitropyridine with hydrogen bromide and granulated tin gives 3,5-diaminopyridine.[175]

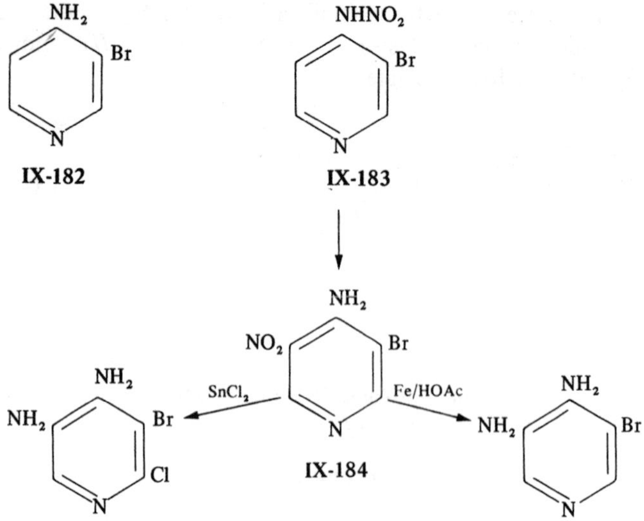

IX-182

IX-183

IX-184

IX-186

IX-185

2-Chloro-3,5-dinitropyridines give 2-alkoxy-3,5-dinitropyridines when heated with various alcohols in a sealed tube at 140° for 3 to 5 hours. Reduction of these compounds over $PtO_2$ gives the corresponding diamino derivatives

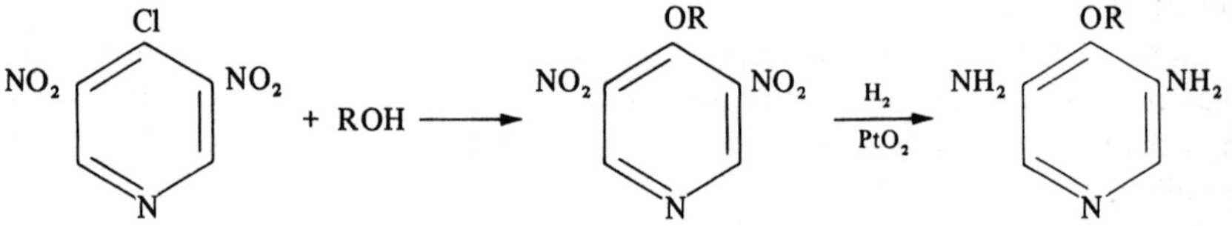

IX-187

(**IX-187**) (see also Table IX-61).[479]

Nitration of 2,6-dichloropyridine-1-oxide (**IX-188**) gives 2,6-dichloro-4-nitropyridine-1-oxide (**IX-189**), which, on catalytic reduction using Raney nickel as the catalyst, gives 4-amino-2,6-dichloropyridine (**IX-190**). From the amino compound, the 2,6-dichloro-4-nitraminopyridine (**IX-191**) is readily formed and rearranges to **IX-192**. Catalytic reduction of **IX-192** gives an almost quantitative yield of 2,6-dichloro-3,4-diaminopyridine (**IX-193**).[87]

2,4-Dichloro-3-nitro-6-*n*-propylpyridine reacts with alcoholic ammonia to give 2,4-diamino-3-nitro-6-propylpyridine, which can be reduced further to 6-propyl-2,3,4-triaminopyridine on catalytic hydrogenation over palladium chloride.[480]

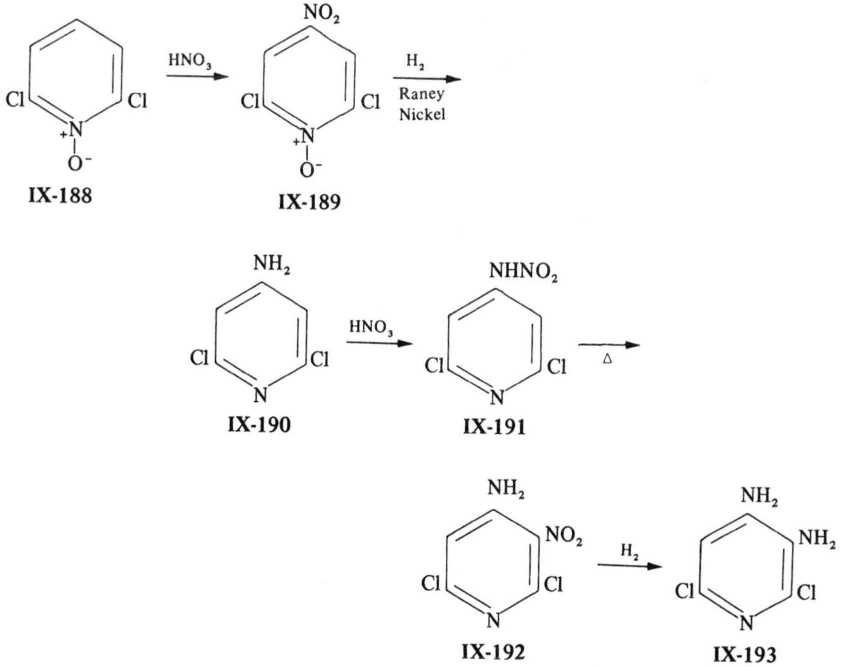

IX-188 IX-189

IX-190 IX-191

IX-192 IX-193

A series of 4-alkoxy-, 4-aralkoxy-, and 4-aryloxy-2,6-diaminopyridines and some of their thio and seleno analogues are prepared by the Curtius reaction, employing the appropriately substituted pyridinecarboxylic acid azides.[481, 482] For example, 4-butoxypyridine-2,6-dicarboxylic acid diazide is heated to give 4-butoxy-2,6-diaminopyridine.[481] Table IX-58 lists the various 4-substituted 2,6-diaminopyridines prepared this way.

## B. *Properties and Reactions*

Tables IX-52 to IX-61 list the physical properties of the diaminopyridines. Miscellaneous reactions, including the oxidation and diazotizations of diaminopyridines, have been reported. 2,6-Diaminopyridine reacts with 2-bromopyridine to give 2-amino-6-(2-pyridylamino)pyridine **(IX-194)**.[116]

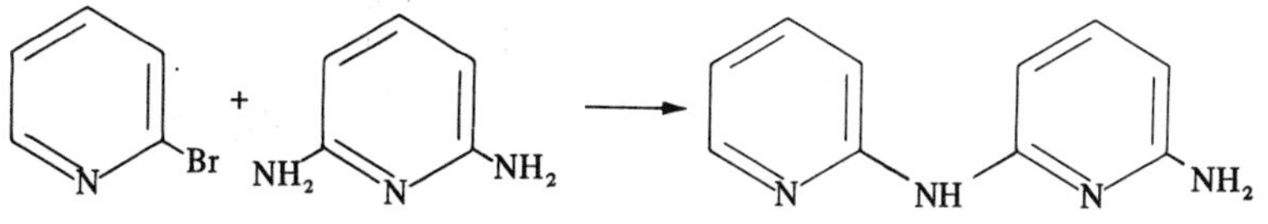

IX-194

When a solution of 4,5-diamino-3-picoline (**IX-195**, R = H) in sulfuric acid is diazotized with sodium nitrite and then neutralized with potassium bicarbonate, 3-methyl-4,5-pyridotriazole (**IX-196**, R = H) is obtained. When **IX-195** is treated with freshly distilled formic acid, 3-methyl-4,5-pyridoimidazole (**IX-197**, R = H) results.

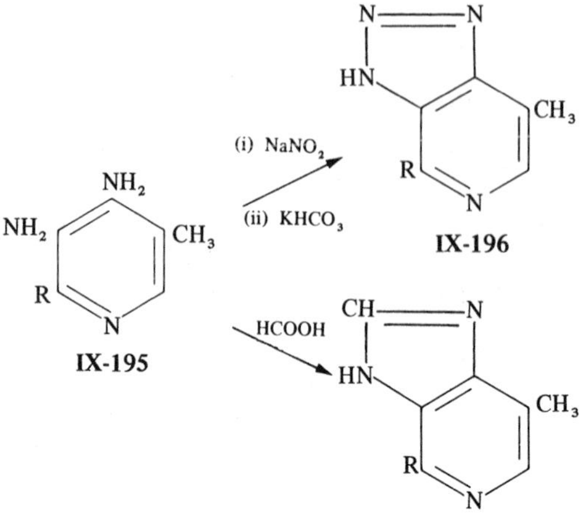

(i) $NaNO_2$; (ii) $KHCO_3$; HCOOH

**IX-195** **IX-196** **IX-197**

A solution of 6-chloro-4,5-diamino-3-picoline (**IX-195**, R = Cl) and hydrazine hydrate in absolute alcohol is boiled for 3 hours to yield 6-hydrazino-3-methyl-4,5-pyridotriazole (**IX-196**, R = $NHNH_2$). Oxidation of 2,5-diaminopyridines produces dyes of various hues, provided the 5-amino group is unsubstituted.[483]

2,6-Diamino-3-nitrosopyridine is easily oxidized to the 3-nitro compound by a solution of 30% hydrogen peroxide in trifluoroacetic acid—the same procedure used for oxidizing 5-nitrosopyrimidines to 5-nitropyrimidines.[484]

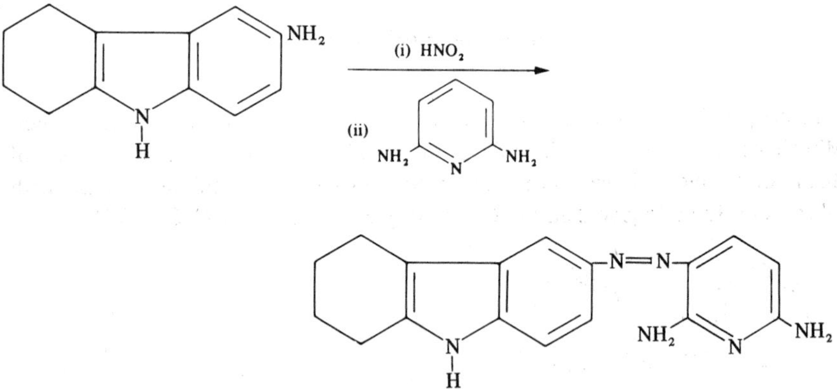

(i) $HNO_2$; (ii) 2,6-diaminopyridine

**IX-198**

3-Amino-2,4,6-triiodobenzoic acid is diazotized and allowed to react with 2,6-diaminopyridine to give 3-(2,6-diamino-3-pyridylazo)-2,4,6-triiodobenzoic acid.[485] Diazotization of 6-amino-1,2,3,4-tetrahydrocarbazole followed by treatment with 2,6-diaminopyridine yields **IX-198**.[486]

## C. *Synthesis of Condensed Heterocyclic Systems*

a. PYRIDOPYRAZINES (AZAQUINOXALINES) *o*-Diaminopyridines condense with α-carbonyl compounds to form pyridopyrazines. The reaction of 2,3,4,6-tetraaminopyridine with pyruvaldehyde under acidic conditions gives mainly 6,8-diamino-3-methylpyrido[2,3-*b*]pyrazine **(IX-199)**.[487]

$NH_2$ $NH_2$ $NH_2$ $NH_2$ N + $CH_3COCHO$ ⟶ $NH_2$ N $NH_2$ N N $CH_3$

**IX-199**

The condensation of benzil with ethyl 4,5,6-triamino-2-pyridylcarbamate **(IX-200)** gives a mixture of 8-amino-2,3-diphenylpyrido[2,3-*b*]pyrazines **(IX-201)** and **(IX-202)** under neutral conditions, and mainly **IX-202** under acidic conditions.[487]

$NH_2$ $NH_2$ EtOOCNH N $NH_2$ $\xrightarrow{(C_6H_5CO)_2}$

**IX-200**

$NH_2$ N Ph EtOOCNH N N Ph + $NH_2$ N N Ph EtOOCNH N Ph

**IX-201** **IX-202**

Benzil also condenses with 2,3-diamino-4,6-lutidine to yield 6,8-dimethyl-2,3-diphenylpyrido[2,3-*b*]pyrazine **(IX-203)**.[488] The reaction of diacetyl with 3,4-diaminopyridine yields 2,3-dimethylpyrido-[3,4-*b*]pyrazine **(IX-204)**.[489]

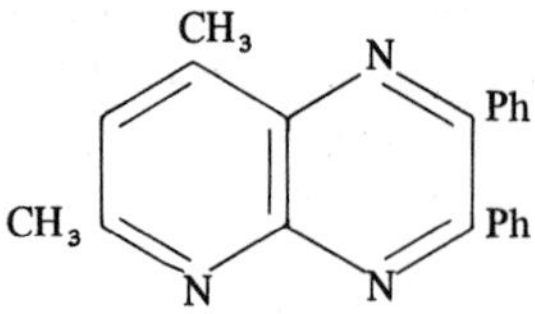

IX-203 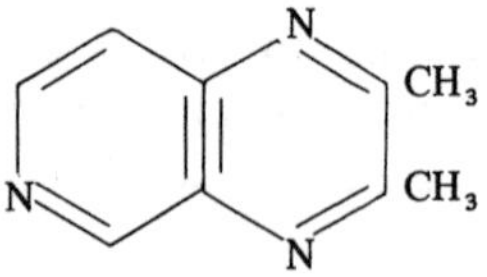 IX-204

Other azaquinoxalines are prepared from 2,3-diaminopyridine and α-dicarbonyl compounds. For example, 2,3-diaminopyridine and **IX-205** in hot acetic acid gives 5-aza-2-styryl-3-quinoxalone **(IX-206)**.[490]

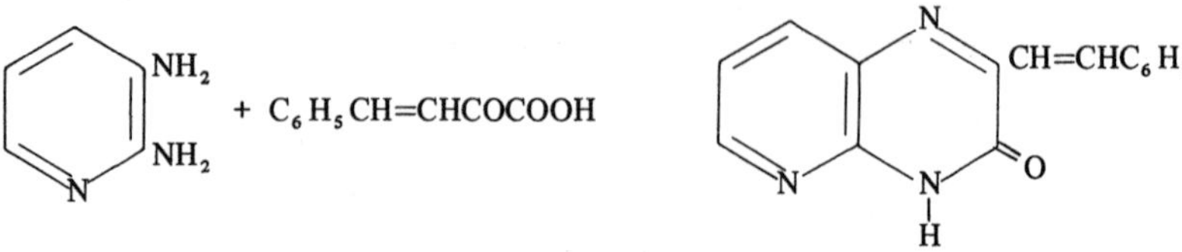

$+ C_6H_5CH{=}CHCOCOOH$

IX-205 IX-206

b. IMIDAZOPYRIDINES 2,3- and 3,4-diaminopyridines cyclize with a variety of reagents to give imidazopyridines. When benzyl alcohol and 2,3-diaminopyridine are stirred with polyphosphoric acid, 2-phenylimidazo-[4,5-*b*] pyridine **(IX-207)** is formed. Using this procedure, and the appropriate alcohols, 2-thienyl- and 2-pyridylvinylimidazo [4,5-*b*] pyridine can be prepared.[491]

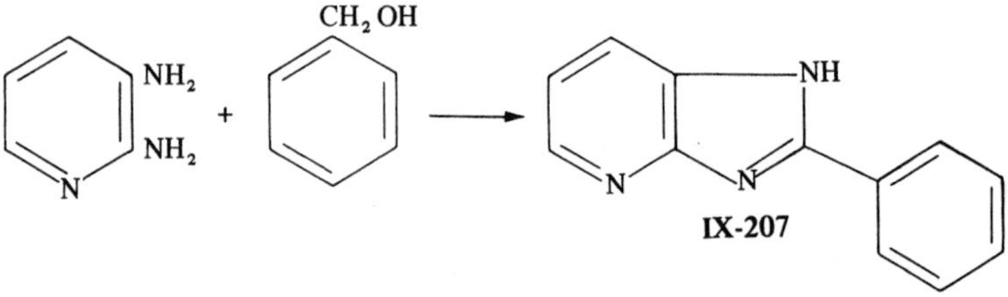

IX-207

2,3- And 3,4-diaminopyridine condense with phenylacetaldehyde to give 2-benzylimidazo [4,5-*b*] pyridine (**IX-208**, R = H) and 2-benzylimidazo [4,5-*c*]-

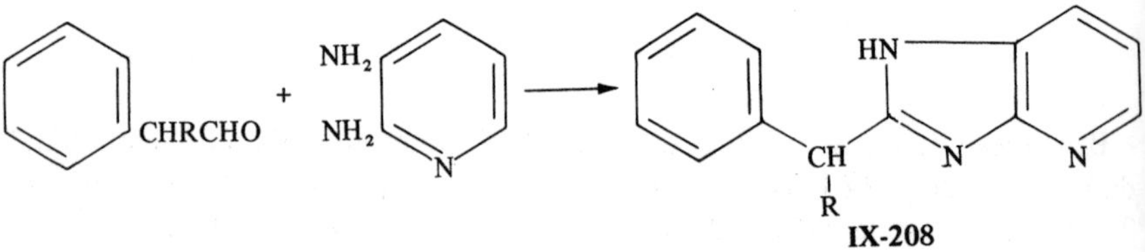

IX-208

pyridine (**IX-209**, R = H), respectively;[492] the iminoether hydrochloride of benzyl cyanide may be used instead of phenyl acetaldehyde. If mandelic acid is

CHRCHO + $NH_2$ $NH_2$ N →

HN N N CH R

**IX-209**

used, the products are 2-(α-hydroxybenzyl)imidazo[4,5-*b*]pyridine (**IX-208**, R = OH) and 2-[α-hydroxybenzyl]imidazo[4,5-*c*]pyridine (**IX-209**, R = OH), respectively.[493]

A simple procedure for preparing 6-nitroimidazo[4,5-*b*]pyridine involves the action of formic acid on 2,3-diamino-5-nitropyridine.[493] Similarly, when 3-amino-2-(β-diethylaminoethylamino)-5-nitropyridine (**IX-210**) is heated with formic acid (**IX-211**, R = H), the product is 3-(β-diethylaminoethyl)-6-nitroimidazo[4,5-*b*]pyridine (**IX-212**, R = H). Using propionic acid (**IX-211**, R = $C_2H_5$), **IX-212** (R = $C_2H_5$) is obtained.[494]

$NO_2$ $NH_2$ NH N $C_2H_4N(Et)_2$ + RCOOH → $NO_2$ N R N N $C_2H_4N(Et)_2$

**IX-210** **IX-211** **IX-212**

When 6-*n*-propyl-2,3,4-triaminopyridine hydrochloride is heated with formalin in the presence of cupric acetate, 7-amino-5-*n*-propylimidazo[4,5-*b*]pyridine

$NH_2$ $NH_2$ $NH_2$ n–$C_3H_7$ N · HCl + $CH_2O$ $\xrightarrow{Cu(OAc)_2}$ $NH_2$ N n–$C_3H_7$ N N H

**IX-213**

(**IX-213**) is formed.[480] 3,4-Diamino-6-propyl-2-pyridone reacts with formic acid to yield 5-propylimidazo[5,4-*c*]-4-pyridone (**IX-214**).[480]

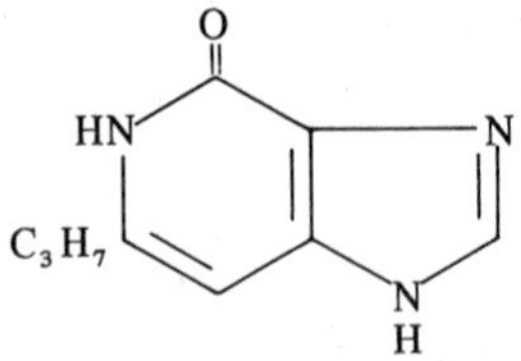

IX-214

The preparation of 6-chloroimidazo[4,5-*c*]pyridine (**IX-216**, $R_1$ = H, $R_2$ = Cl) and 4,6-dichloroimidazo[4,5-*c*]pyridine (**IX-216**, $R_1$ = $R_2$ = Cl) has been accomplished by the ring closure of the appropriate chloro-3,4-diaminopyridine (**IX-215**) with ethyl orthoformate and acetic anhydride.[87] 6-Aminoimidazo-[4,5-*c*]pyridine (**IX-215**, $R_1$ = H, $R_2$ = $NH_2$) was obtained similarly.

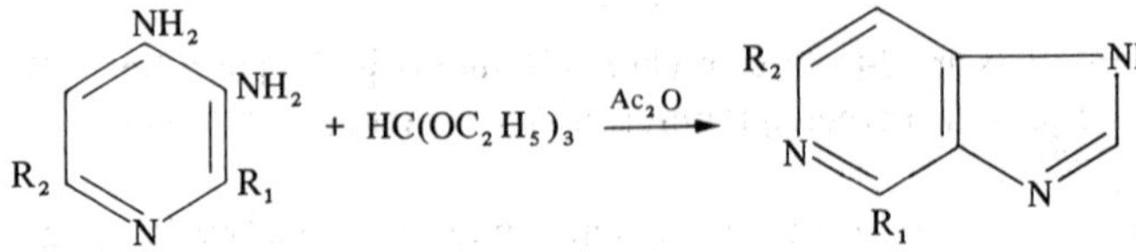

$$\text{IX-215} + HC(OC_2H_5)_3 \xrightarrow{Ac_2O} \text{IX-216}$$

Heating 3,4-diaminopyridine with chloroacetic acid gives 6-aza-2-quinazolone (**IX-217**). However, if glycollic acid or acetic anhydride is used instead of chloroacetic acid, 2-hydroxymethylimidazo[4,5-*c*]pyridine and 2-methylimidazo[4,5-*c*]pyridine are obtained, respectively.[495]

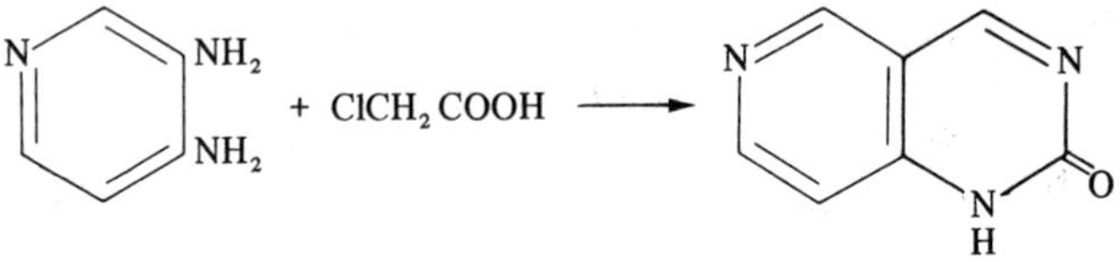

$$+ ClCH_2COOH \longrightarrow \text{IX-217}$$

c. MISCELLANEOUS CYCLIZATIONS Miscellaneous cyclizations include the reactions of diaminopyridines with phenanthraquinone or with triphenyl borate. Thus, 2,4-dimethyl-5,6-diaminopyridine condenses with 9,10-phenanthraquinone to yield **IX-218**.[488]

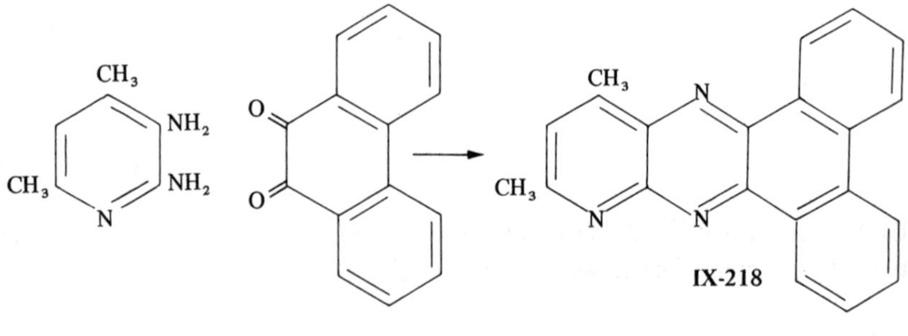

IX-218

2,3-Diaminopyridine is reported to react with triphenyl borate to give phenol and tris[1*H* (or 3*H*)-pyridino[2,3-*d*]-1,3,2-diazaborolo]borazine **(IX-219)**.[496]

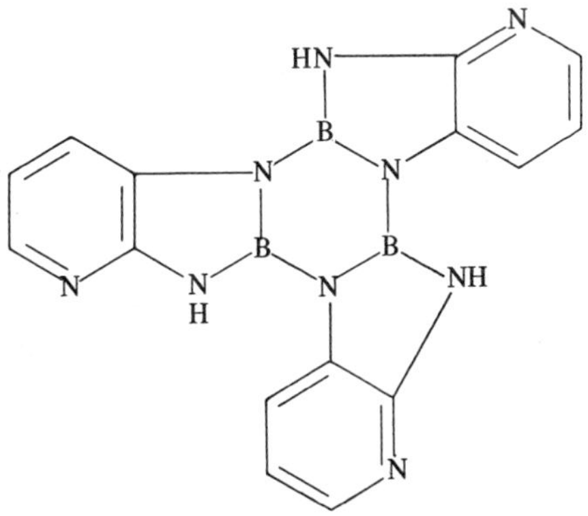

**IX-219**

## II. Side-Chain Amines

### 1. Preparation

### A. *Aminolysis of Side-Chain Halides*

Side-chain aminopyridines are easily prepared by the action of ammonia or of an amine on haloalkylpyridines.[497]

2-Chloromethylpyridine reacts with aniline in the presence of potassium carbonate to yield 2-(phenylaminomethyl)pyridine. When the latter is treated with sodium amide followed by 2-chloroethyl-*N,N*-diethylamine, 2-[*N*-phenyl-*N*-(2-*N,N*-diethylaminoethyl)]picolylamine **(IX-220)** is formed.[498] Other 2-, 3-, and 4-(pyridylmethyl)-*N,N'*-diethylethylenediamines (Table IX-62) have been prepared similarly.[499, 500]

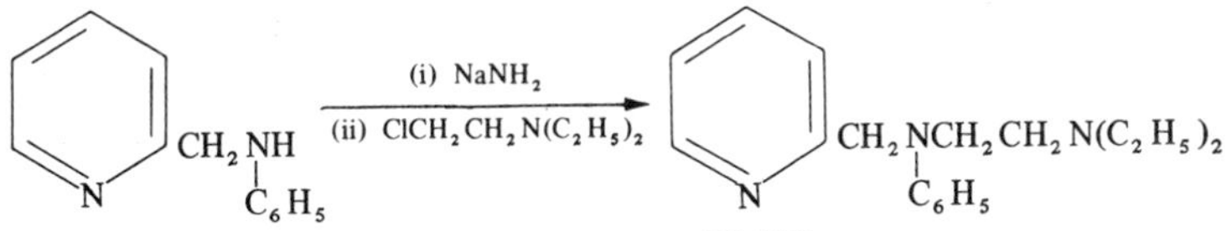

$$\text{C}_5\text{H}_4\text{N}\text{-CH}_2\text{NH(C}_6\text{H}_5) \xrightarrow[\text{(ii) ClCH}_2\text{CH}_2\text{N(C}_2\text{H}_5)_2]{\text{(i) NaNH}_2} \text{C}_5\text{H}_4\text{N}\text{-CH}_2\text{N(C}_6\text{H}_5)\text{CH}_2\text{CH}_2\text{N(C}_2\text{H}_5)_2$$

**IX-220**

When 2-chloromethylpyridine is treated with ethanolamine, *N*-2-hydroxyethyl-*N*-2-picolylamine **(IX-221)** is isolated.[501] A mixture of 2-chloromethylpyridine and *N*-(4-hydroxybutyl)aminoethanol is heated for 2 hours to yield *N*-(4-hydroxybutyl)-*N*-(2-hydroxyethyl)-2-pyridylmethylamine.[502]

$$\text{2-Pyridylmethyl chloride} + NH_2CH_2CH_2OH \longrightarrow \text{2-}(CH_2NHCH_2CH_2OH)\text{pyridine}$$

**IX-221**

2-Pyridylmethyl chloride and potassium phthalimide give a solid that is decomposed by hydrogen bromide to give 2-pyridylmethylamine hydrobromide.[503]

*o*-(4-Pyridylmethylamino)phenol **(IX-222)** is obtained from the reaction of 4-chloromethylpyridine with the hydrochloride of *o*-aminophenol, diethylamine, and methanol.[504]

$$\text{4-}(CH_2Cl)\text{pyridine} + o\text{-}HOC_6H_4NH_2 \cdot HCl \xrightarrow[MeOH]{HN(C_2H_5)_2} \text{IX-222}$$

**IX-222**

2-, 3-, And 4-pyridylmethyl chloride condense with 1,2-diaminoethane to give the corresponding *N*-(pyridylmethyl)-1,2-diaminoethanes. For example, 4-pyridylmethyl chloride and 1,2-diaminoethane give *N*-(4-pyridylmethyl)-1,2-diaminoethane **(IX-223)**.[505]

$$\text{4-}(CH_2Cl)\text{pyridine} + H_2NCH_2CH_2NH_2 \longrightarrow \text{4-}(CH_2NHCH_2CH_2NH_2)\text{pyridine}$$

**IX-223**

Substituted side-chain amines may also be prepared by aminolysis of substituted haloalkyl pyridines.[506-508] Thus, 2-bromomethyl-3-nitropyridine yields the expected 2-arylaminomethyl-3-nitropyridines when treated with aromatic amines at room temperature. At higher reaction temperatures, however, 2*H*-pyrazolo[4,3-*b*]pyridines **(IX-224)** result.[506]

$$NC_6H_4R\text{-}p, \quad CNHC_6H_4R\text{-}p$$

**IX-224**

When a mixture of aniline hydrochloride and 2-chloromethylpyridine is heated at 135 to 140° for 2 hours and then made alkaline with sodium carbonate, *N*-(2-picolyl)aniline (**IX-225**) results. If the reaction temperature is raised to 225 to 230°, rearrangement occurs and 4-(2-picolyl)aniline (**IX-226**) is obtained instead.[506]

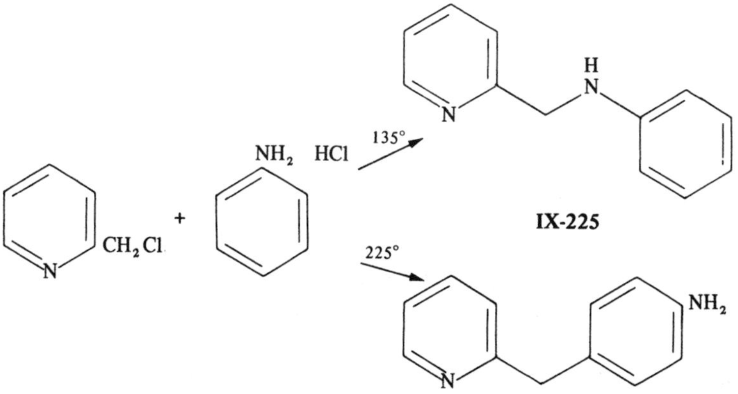

$CH_2Cl$ + $NH_2$ HCl 135° 225°

**IX-225**

**IX-226**

## B. *Reduction of Nitriles*

Cyanopyridines have been reduced to aminoethylpyridines by a variety of methods, 2- And 4-cyanopyridine are reduced with sodium borohydride to 2- and 4-aminomethylpyridine, respectively.[509] In acidic solutions, the polarographic reduction of 4-cyanopyridine yields 4-aminomethylpyridine.[510]

Hydrogenation of 6-chloro-5-cyano-4-methoxymethyl-3-nitro-2-picoline over palladium-charcoal in aqueous hydrochloric acid gives 3-amino-5-aminomethyl-4-methoxymethyl-2-picoline hydrochloride.[511, 512]

3-Amino-5-aminomethyl-2,4-lutidine is prepared in good yield by the catalytic reduction (Pd–C in methanol) of 6-chloro-3-cyano-2,4-dimethyl-3-nitropyridine.[513] The hydrogenation of 3-(β-pyridylmethoxy)propionitrile over Raney Nickel yields 3-[(3-aminopropoxy)methyl] pyridine.[514]

3-Aminomethyl-4-picoline is obtained by the catalytic reduction of 3-cyano-2,6-dichloro-4-picoline using 30% Pd–$CaCO_3$ as the catalyst.[515]

Pyridoxamine dihydrochloride may be prepared by the reduction of 4-cyano-3-hydroxy-5-hydroxymethy-2-picoline hydrochloride in methanol over a palladium-on-carbon catalyst.[516]

## C. *Reduction of Amides*

The use of lithium aluminum hydride for the reduction of pyridinecarboxylic acid amides **(IX-227)** to side-chain aminopyridines **(IX-228)** has been reported; thus, *N,N*-dibenzylnicotinamide **(IX-227**, $R_1 = R_2 = C_6H_5CH_2$) is reduced to

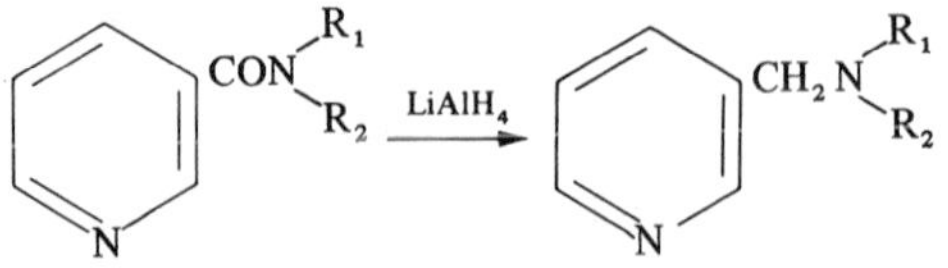

CON(R₁)(R₂) —LiAlH₄→ CH₂N(R₁)(R₂)

**IX-227** **IX-228**

*N*-(3-pyridylmethyl)dibenzylamine **(IX-228**, $R_1 = R_2 = C_6H_5CH_2$).[517] Diisobutyl aluminum hydride has also been used, as in the preparation of 3-diethylaminomethylpyridine **(IX-227**, $R_1 = R_2 = C_2H_5$) from *N,N*-diethylnicotinamide **(IX-228**, $R_1 = R_2 = C_2H_5$).[518] 7-Methoxy-1-naphthylacetyl chloride reacts with 2-aminopyridine to give the corresponding amide, which is reduced by lithium aluminum hydride to **IX-229**.[519]

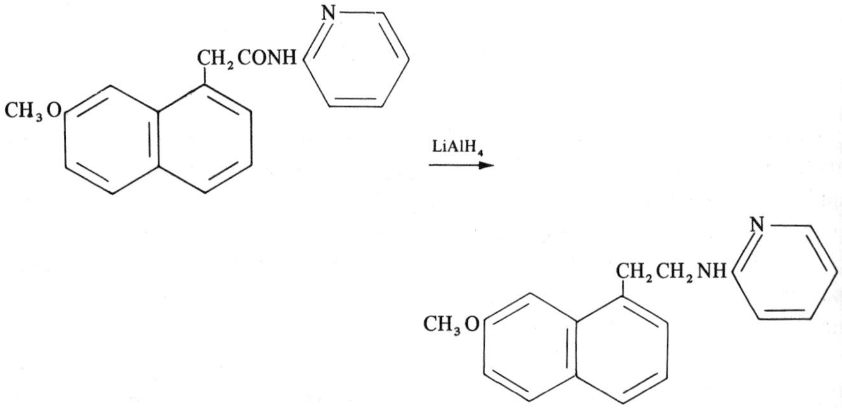

$CH_3O$ — $CH_2CONH$ —LiAlH₄→ $CH_3O$ — $CH_2CH_2NH$

**IX-229**

Aminoalkylpyridines have been prepared by heating thioamides with an alkyl amine and Raney Nickel in an autoclave under 10 atm of hydrogen at 70° for several hours. Under these conditions, β-(2-pyridyl)propionic thioamide reacts with diethylamine to give 2-(3-diethylaminopropyl)pyridine **(IX-230)**.[520]

2-Pyridyl-$CH_2CH_2C(=S)NH_2$ $\xrightarrow[\text{Raney Ni/H}_2]{(C_2H_5)_2NH}$ 2-Pyridyl-$CH_2CH_2CH_2N(C_2H_5)_2$

**IX-230**

## D. *Reduction of Oximes*

Aldoximes and ketoximes have been reduced to aminoalkyl pyridines; reduction methods include polarographic, catalytic and chemical means. 2-, 3-, And 4-aminomethylpyridine have been prepared by the polarographic reduction of the corresponding pyridinealdoximes.[521]

The oximes of 1-(2-pyridylcarbonyl)-, 1-(3-pyridylcarbonyl)-, and 1-(4-pyridylcarbonyl)-1-phenylacetamide **(IX-231)** are hydrogenated over Raney Nickel at 50° and 55 atm to give **IX-232.**[522]

Pyridyl-$C(=NOH)CH(CONH_2)C_6H_5$ $\xrightarrow[\text{Raney Ni}]{H_2}$ Pyridyl-$CH(NH_2)CH(CONH_2)C_6H_5$

**IX-231** **IX-232**

4-Amino-3-hydroxy-5-hydroxymethylpyridine dihydrochloride is prepared by hydrogenation of the oxime of 4-formyl-3-hydroxyl-5-hydroxymethylpyridine in aqueous hydrochloric acid over palladium-charcoal.[523]

2-(α-Aminobenzyl)pyridine is obtained by the reduction of phenyl 2-pyridyl ketone oxime with zinc dust and acetic acid.[524]

Lithium aluminum hydride is used to reduce the oxime of (3-methyl-2-pyridyl)acetone **(IX-233)** to 2-amino-1-(3-methyl-2-pyridyl)propane **(IX-234)** Oximes of other 2-pyridylacetones can be used similarly.

3-$CH_3$-2-Pyridyl-$CH_2C(=NOH)CH_3$ $\xrightarrow{LiAlH_4}$ 3-$CH_3$-2-Pyridyl-$CH_2CH(NH_2)CH_3$

**IX-233** **IX-234**

If a mixture of 4-pyridinecarboxaldehyde and its oxime are hydrogenated in methanol in the presence of Raney Nickel, bis(4-pyridylmethyl)amine **(IX-235)** is formed.[525]

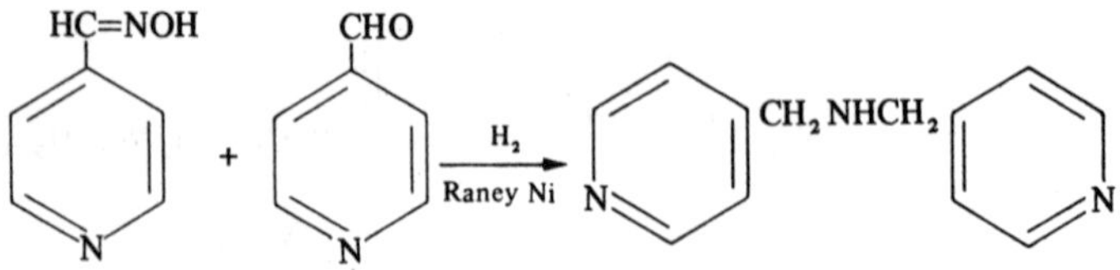

HC=NOH CHO

$\xrightarrow[\text{Raney Ni}]{H_2}$ $CH_2NHCH_2$

**IX-235**

## E. *Reduction of Schiff Bases*

Schiff bases that are obtained from pyridine aldehydes and amines are reduced to aminoalkylpyridines by various reagents.

The Schiff bases obtained by condensing 2,6-pyridinedicarboxylaldehyde with aliphatic amines are reduced by sodium borohydride to bis(alkylaminomethyl)-pyridines. For example, the product obtained from 2,6-pyridinedicarboxaldehyde and *n*-butylamine gives 2,6-bis(*n*-butylaminomethyl)pyridine **(IX-236)** on reduction.[526]

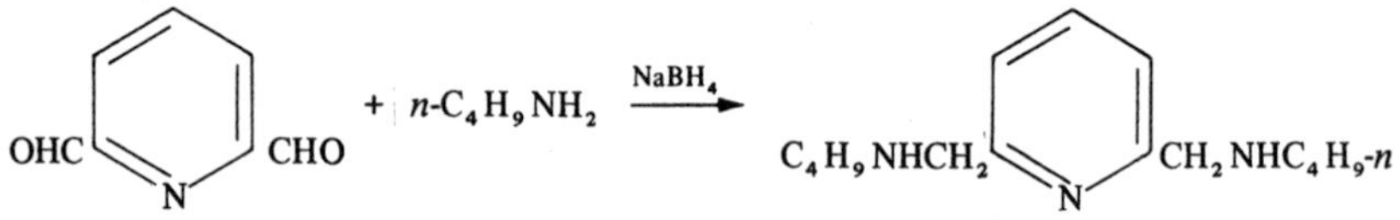

OHC CHO + $n\text{-}C_4H_9NH_2$ $\xrightarrow{NaBH_4}$ $C_4H_9NHCH_2$ $CH_2NHC_4H_9\text{-}n$

**IX-236**

Various secondary pyridylamines have been prepared by the sodium borohydride reduction of the corresponding arylidene-amines.[527, 528] For example, pyridine-3-aldehyde condenses with *p*-nitroaniline to give **IX-237**, which, on reduction, gives the secondary amine **IX-238**.[527]

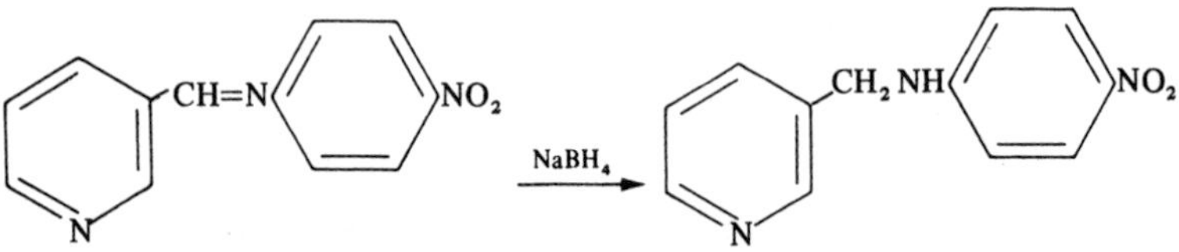

CH=N $NO_2$ $\xrightarrow{NaBH_4}$ $CH_2NH$ $NO_2$

**IX-237** **IX-238**

Catalytic reduction of Schiff bases has also been used to produce side-chain aminopyridines.[529-532] 2-Acetylpyridine condenses with ethylene diamine to give an imine which is reduced with hydrogen in the presence of platinum oxide to give *N,N*-bis[α-(2-pyridyl)ethyl]ethylenediamine **(IX-239)**. Table IX-65 lists

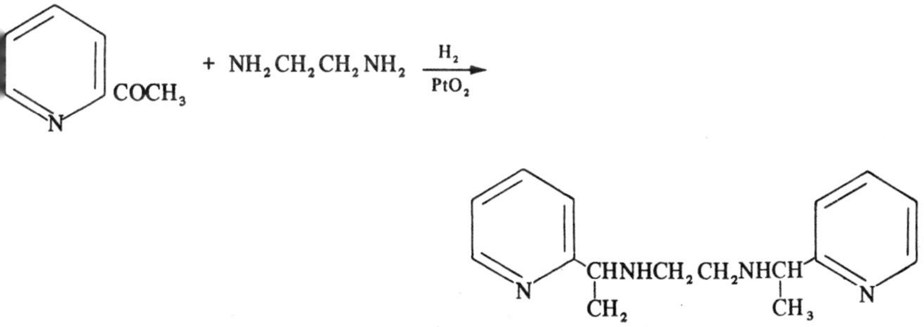

$+\ NH_2CH_2CH_2NH_2 \xrightarrow[PtO_2]{H_2}$ $CHNHCH_2CH_2NHCH$ $CH_3$ $CH_3$

**IX-239**

compounds so obtained.

Compounds of the general structure **IX-241** are prepared from the appropriate ketone **(IX-240)** and amines ($R_4R_5NH$) by reductive amination; thus 1-(5-methyl-2-pyridyl)-2-aminopropane **(IX-243)** is obtained by autoclaving (5-methyl-2-pyridyl)acetone **(IX-242)** in the presence of ammonia, hydrogen, and Raney Nickel.[533]

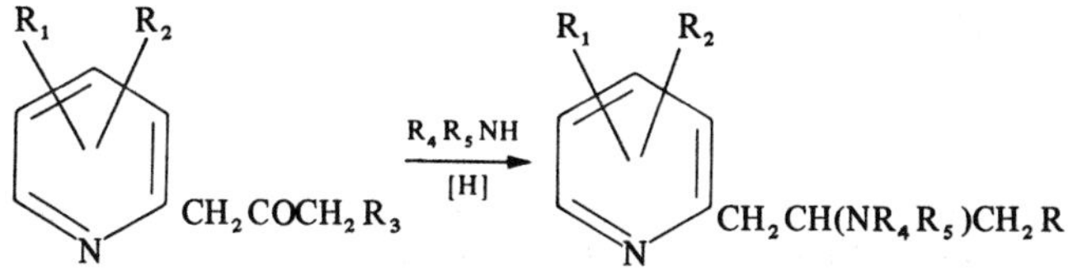

$R_1$ $R_2$ $CH_2COCH_2R_3 \xrightarrow[[H]]{R_4R_5NH}$ $R_1$ $R_2$ $CH_2CH(NR_4R_5)CH_2R$

**IX-240** **IX-241**

$CH_3$ $CH_2\overset{O}{\overset{\|}{C}}CH_3 \xrightarrow[H_2/Ni]{NH_3}$ $CH_3$ $CH_2CH(NH_2)CH_3$

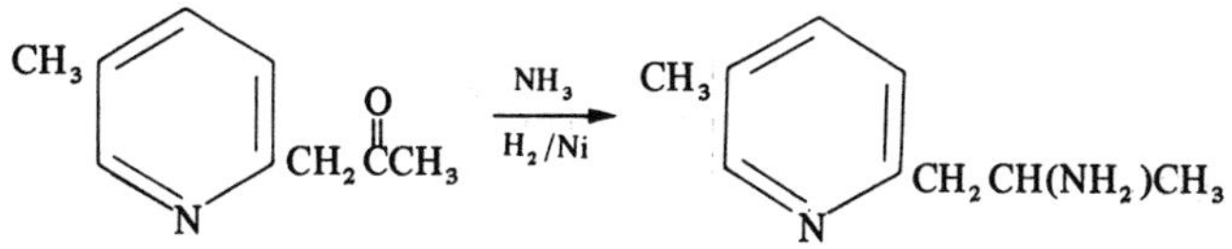

**IX-242** **IX-243**

A good source of *N*-(2-pyridylmethyl)anilines and *N*-(4-pyridylmethyl)anilines is the reaction of 2- and 4-pyridinemethanol with anilines in the presence of potassium hydroxide at high temperature. Thus, a mixture of 4-pyridinemethanol, *p*-ethoxyaniline, potassium hydroxide, and a trace of amyl nitrite gives *N*-(4-pyridylmethyl)-*p*-ethoxyaniline **(IX-244)**.[534] Amyl nitrite is not a necessary reagent.[535]

$$\text{4-(CH}_2\text{OH)pyridine} + NH_2C_6H_4OCH_2CH_3 \xrightarrow{KOH} \text{4-(CH}_2NHC_6H_4OCH_2CH_3\text{)pyridine}$$

**IX-244**

## F. *Side-Chain Alkylation with Aminoalkyl Halides*

In the presence of sodamide, 2-picoline or 2-benzylpyridine condenses wit 2-dimethylaminoethyl chloride to give 1-(2-pyridyl)-3-dimethylaminopropan (**IX-245**, R = H) and 1-phenyl-1-(2-pyridyl)-3-dimethylaminopropane (**IX-245**, R = $C_6H_5$), respectively.[536, 537]

$$\text{2-(CH}_2\text{R)pyridine} + (CH_3)_2NCH_2CH_2Cl \xrightarrow{NaNH_2} \text{2-[CH(R)CH}_2CH_2N(CH_3)_2\text{]pyridine}$$

**IX-245**

2-Benzylpyridine is boiled with sodium amide and then treated wit *N*-(3-bromopropyl)-3-phenyl-2-propylamine to yield 3-phenyl-*N*-[4-(phenyl) 4-(2-pyridyl)] butyl-2-propylamine (**IX-246**).[538]

Other 3-substituted-2-propylamines have been prepared in an analogou manner.

$$\text{2-(CH}_2C_6H_5\text{)pyridine} + NaNH_2 + C_6H_5CH_2CH(CH_3)NH(CH_2)_3Br \longrightarrow \text{2-[CH(C}_6H_5)(CH_2)_3NHCH(CH_3)CH_2C_6H_5\text{]pyridine}$$

**IX-246**

The reaction of di-(2-pyridyl)methane with 2-(2-bromoethyl)-1-phenylaminc propane in the presence of sodamide produces 2-amino-2-benzyl-1,1-(2,2-d pyridyl)pentane (**IX-247**).[539]

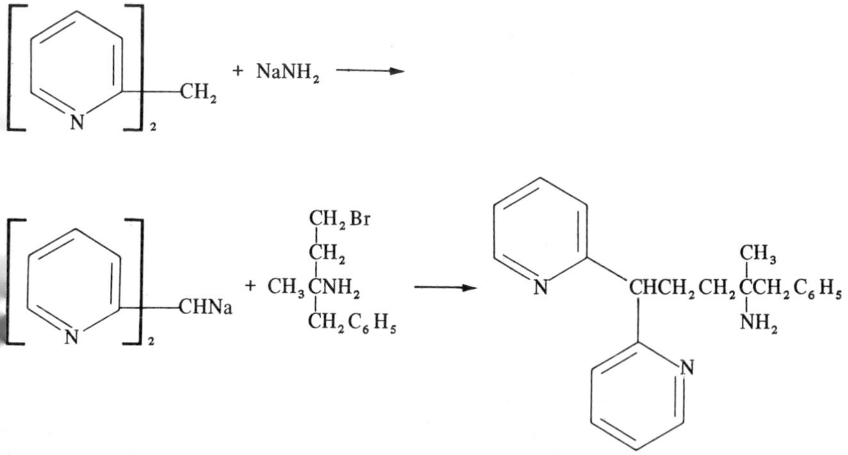

$$\left[\text{2-pyridyl}\right]_2CH_2 + NaNH_2 \longrightarrow$$

$$\left[\text{2-pyridyl}\right]_2CHNa + CH_3C(NH_2)(CH_2CH_2Br)(CH_2C_6H_5) \longrightarrow (\text{2-pyridyl})_2CHCH_2CH_2C(CH_3)(NH_2)CH_2C_6H_5$$

**IX-247**

## G. *Mannich Reactions*

When 2-benzylpyridine, paraformaldehyde, and dimethylamine hydrochloride are heated in water for 24 hours, 2-(2-dimethylaminoethyl-1-phenyl)pyridine dihydrochloride **(IX-248)** results.[540]

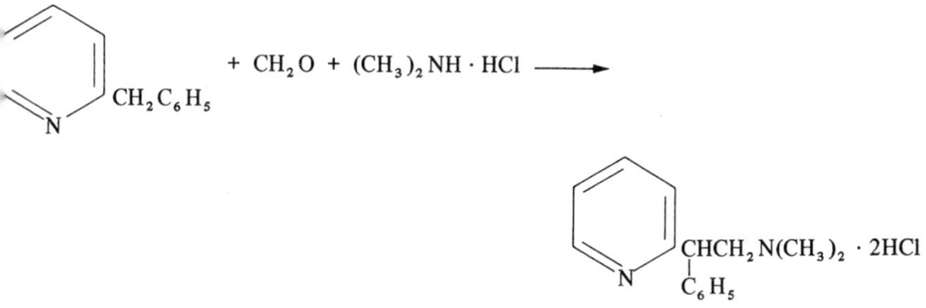

$$\text{2-pyridyl-}CH_2C_6H_5 + CH_2O + (CH_3)_2NH \cdot HCl \longrightarrow \text{2-pyridyl-}CH(C_6H_5)CH_2N(CH_3)_2 \cdot 2HCl$$

**IX-248**

Heating 3-hydroxypyridine in water with secondary amines and formaldehyde gives various 2-aminoalkyl-3-hydroxypyridines; for example, with dimethylaminoethylamine, 2-(dimethylaminomethyl)-3-hydroxypyridine is obtained in 73% yield.[541]

A mixture of 3-hydroxy-6-nitro-2-picoline, dimethylamine, and 30% formalin gives 2-[2-(dimethylamino)ethyl]-3-hydroxy-6-nitropyridine.[542] 2-Aminomethyl-5-hydroxy-6-nitropyridine is prepared similarly from 3-hydroxy-2-nitropyridine.[543]

The 4-position can also be aminomethylated. Thus, 4-dimethylaminomethyl-6-ethyl-3-hydroxy-2-picoline **(IX-249)** is obtained from condensation of formalin, dimethylamine, and 6-ethyl-3-hydroxy-2-picoline.[544]

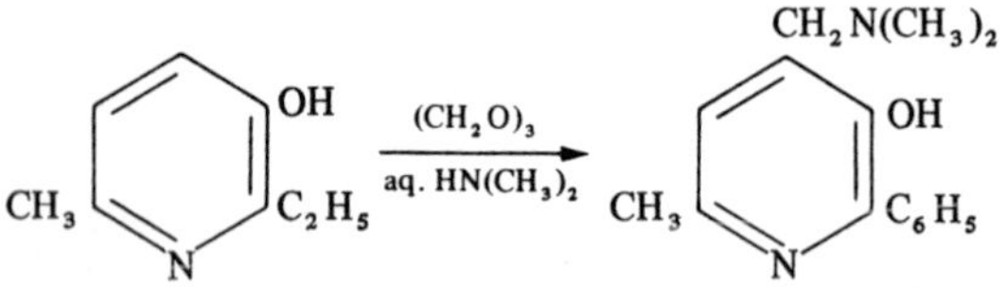

$CH_2N(CH_3)_2$

OH, $CH_3$, $C_2H_5$, N $\xrightarrow[\text{aq. } HN(CH_3)_2]{(CH_2O)_3}$ OH, $CH_3$, $C_6H_5$, N

**IX-249**

A study by NMR and chemical methods shows that aminomethylation of 2-alkyl-3-hydroxypyridines under the above conditions produces 6-aminomethyl and 4,6-diaminomethyl-2-alkyl-3-hydroxypyridines; the 6-position is the site of initial aminomethylation.[545, 546]

## H. *From Alkynes*

Side-chain pyridineamines, particularly those containing an alkyne function, are prepared from pyridylalkynes or by the action of alkynes on pyridyl ketones.

Miocque has prepared various *N,N*-dimethylaminopyridylalkynes of the type **IX-251** by aminomethylation of **IX-250** with dimethylamine and trioxane;[546, 548] for example, 1-dimethylamino-5-(2-pyridyl)-2-pentyne (**IX-251**, $n$ = 1) is obtained by heating 4-(2-pyridyl)-1-butyne (**IX-250**, $n$ = 1), trioxane, and dimethyl amine. Furthermore, this and other aminoalkynylpyridines may be reduced to the corresponding aminoalkylpyridines by catalytic hydrogenation.[547]

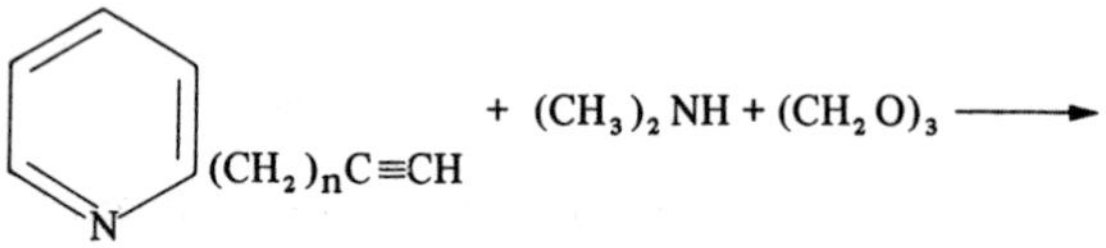

$(CH_2)_nC{\equiv}CH$ + $(CH_3)_2NH$ + $(CH_2O)_3$ ⟶

**IX-250**

$(CH_2)_nC{\equiv}CCH_2N(CH_3)_2$

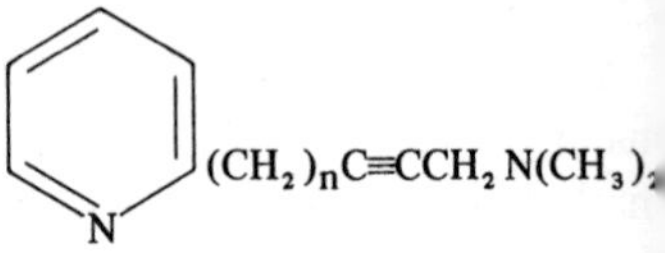

**IX-251**

$N$-substituted derivatives of aminoalkynyl pyridines are readily prepared when propargyl bromide is heated with a 2-, 3-, or 4-picolylamine in acetone to give the corresponding $N$-methyl-$N$-picolyl-2- propynylamine **(IX-252)**.[549, 550]

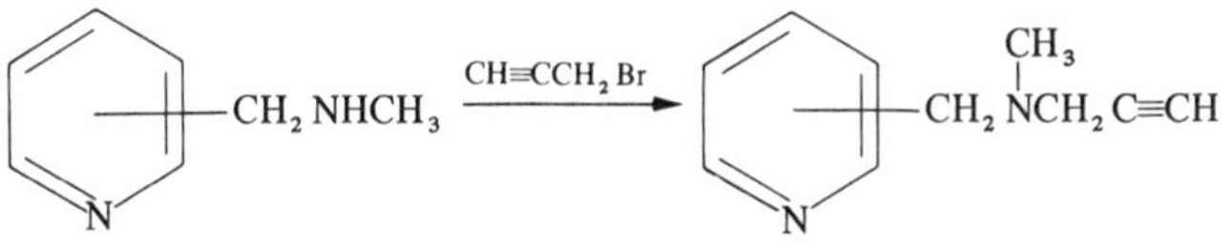

**IX-252**

5-Ethynyl-2-methylpyridine heated with piperidine gives a 70% yield of 2-methyl-5-(2-piperidinoethenyl)pyridine **(IX-253)**. Similarly, 2-methyl-5-(2-morpholinoethenyl)pyridine is obtained with morpholine.[551]

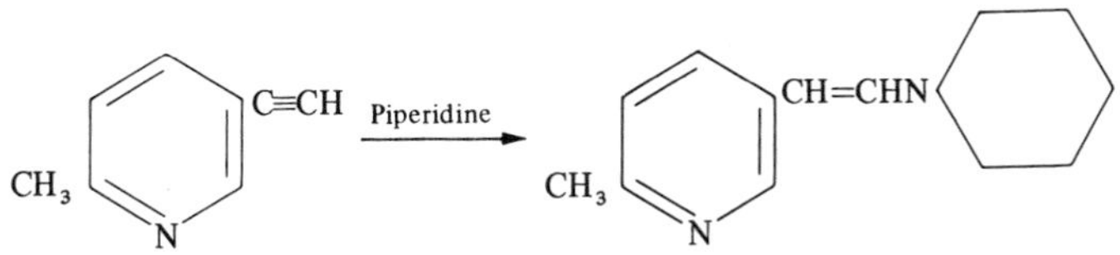

**IX-253**

Another method of preparing aminoalkynylpyridines involves the reaction of lithium alkynes with pyridyl ketones. Thus, 4-diethylamino-1-phenyl-1-(2-pyridyl)-2-butyn-1-ol **(IX-254)** is obtained by the reaction of lithium amide with diethylamino-2-propyne followed by 2-benzoylpyridine. Similarly, the use of 3-

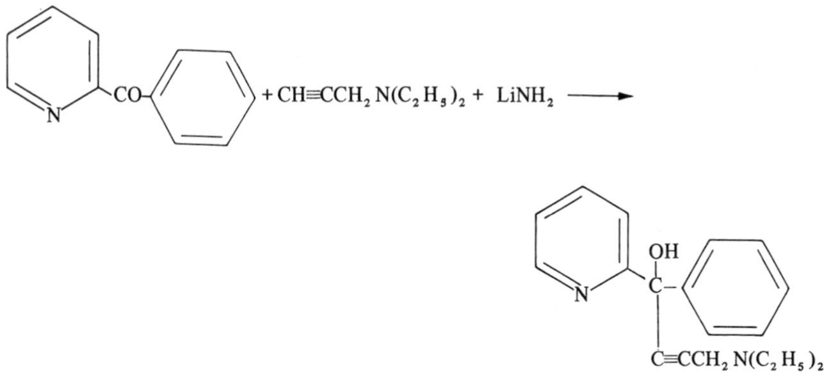

**IX-254**

and 4-benzoylpyridine yields **IX-255** and **IX-256**, respectively.[552]

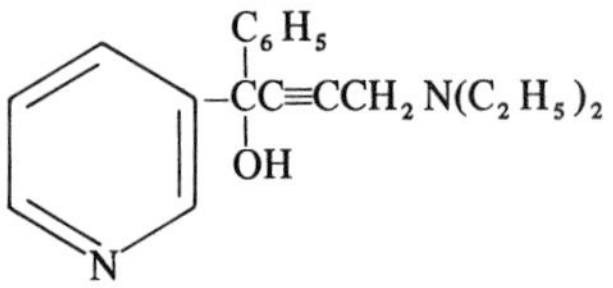

IX-255 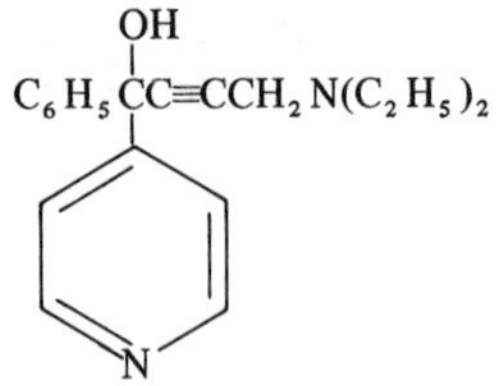 IX-256

## I. *Leuckart, Clarke-Eschweiler, and Hofmann Reactions*

The Leuckart reaction has been used for the preparation of 4-(α-aminobenzyl)-pyridine, which was obtained by heating a mixture of concentrated ammonia, formic acid, and 4-benzoylpyridine.[553]

The related Clarke-Eschweiler reaction in which the reductive alkylation of an amine is carried out with formaldehyde and formic acid was used to prepare 3-dimethylamino-1-phenyl-1-(2-pyridyl)propane **(IX-258)** from 3-phenyl-3-(2-pyridyl)propylamine **(IX-257)**.[554]

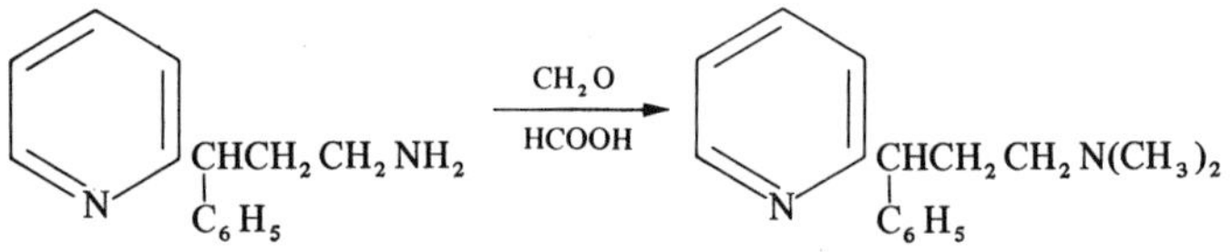

$$\text{IX-257: 2-Py-CH(C}_6\text{H}_5\text{)CH}_2\text{CH}_2\text{NH}_2 \xrightarrow[\text{HCOOH}]{\text{CH}_2\text{O}} \text{IX-258: 2-Py-CH(C}_6\text{H}_5\text{)CH}_2\text{CH}_2\text{N(CH}_3\text{)}_2$$

The Hofmann rearrangement has not been used often for the preparation of aminoalkylpyridines; it has been reported in the synthesis of 5-aminomethyl-2-picoline **(IX-260)**, which is isolated from the reaction of sodium hypobromide with **IX-259**.[555]

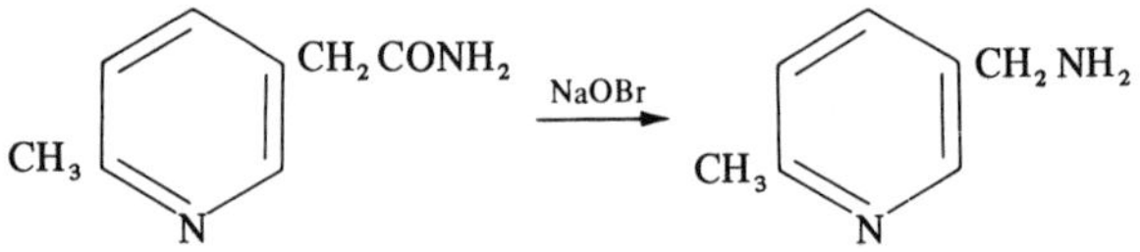

$$\text{IX-259: } CH_3\text{-Py-}CH_2CONH_2 \xrightarrow{\text{NaOBr}} \text{IX-260: } CH_3\text{-Py-}CH_2NH_2$$

## J. *Addition of Amines to Vinylpyridines*

The addition of amines to vinylpyridines has been the source of a large variety of side-chain aminopyridines. For example, the reactions of 2-vinylpyridine with four aliphatic diamines ethylene-, tetramethylene-, hexamethylene-, and octamethylenediamine—and the mono-, di-, tri-, and tetra-addition compounds obtained were studied.[556, 557] Mono- and di-addition compounds are best prepared in benzene with acetic or propionic acid as a catalyst; tri- and tetra-addition compounds require the same catalyst but higher temperatures and longer reaction times. Some examples are given below.

Pyridylethylation resulting from the addition of monoamines to 2-vinylpyridine was shown to be strongly catalyzed by organic and inorganic acids and by ammonium salts of inorganic acids, for example, the alkanoic ($C_1-C_5$) acids, stearic acid, phenol, and piperidine hydrochloride. Protic solvents affect the reactions slightly; aprotic solvents have no influence. Thus, heating 2-vinylpyridine in piperidine for 8 hours gives 63% 2-(β-piperidinoethyl)pyridine **(IX-261)**; in the presence of the above catalysts yields of up to 92% are obtained.[558]

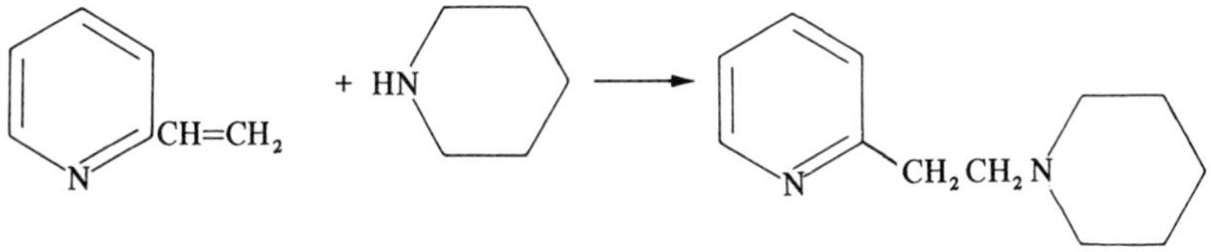

$CH=CH_2$ + HN → $CH_2CH_2N$

**IX-261**

The reactions of 2- and 4-vinylpyridines with various amines include the following: 2-vinylpyridine reacts with ammonium chloride to yield 2-(2-aminoethyl)pyridine (35%);[559] when treated with methylamine it yields 2-(2-methylaminoethylpyridine).[540]

2- And 4-vinylpyridine when heated for 5 hours with dimethylamine give 2- and 4-(2-dimethylaminoethyl)pyridine **(IX-262)**, respectively.[540, 560]

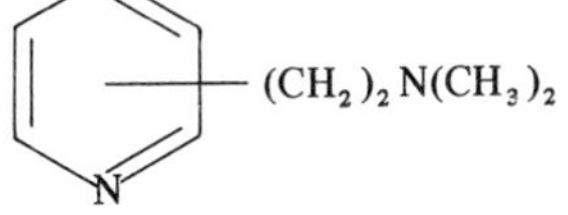

$(CH_2)_2N(CH_3)_2$

**IX-262**

Secondary aromatic amines such as **IX-263** add to 2-vinylpyridine to give *N*-dimethylaminoalkyl-*N*-(pyridylethyl)anilines **(IX-264)**[561-563] (see Table IX-76).

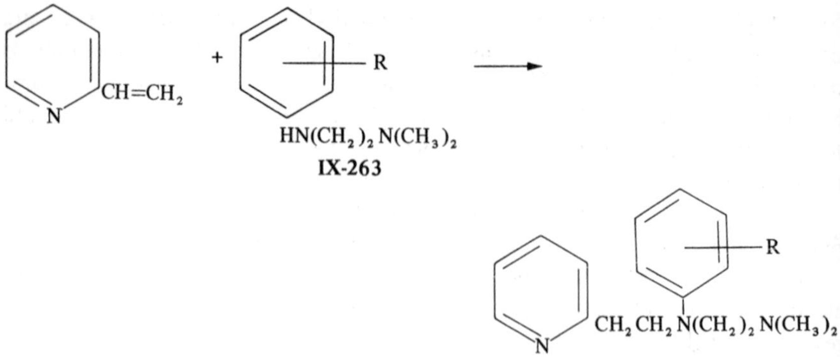

IX-263

IX-264

Heterocyclic amines also add to vinylpyridines. 2-Aminopyridine and 2-aminoquinoline react with 2-vinylpyridine in the presence of sodium metal to give **IX-265** and **IX-266**, respectively.[564, 565]

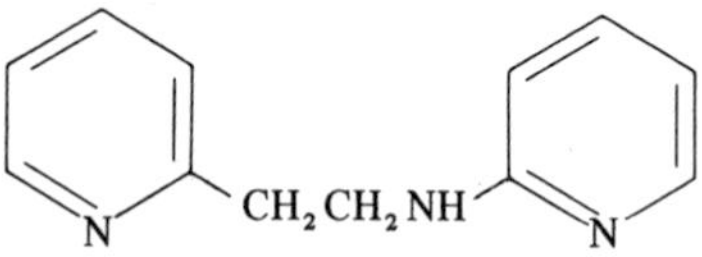

IX-265

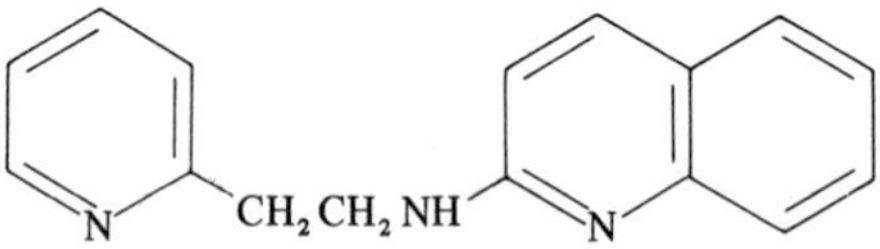

IX-266

The reaction of 2-vinyl- or 4-vinylpyridine with *N*-hydroxypyrrolidine give the oxygen-free product **IX-268** rather than the expected **IX-267**. A variety of such

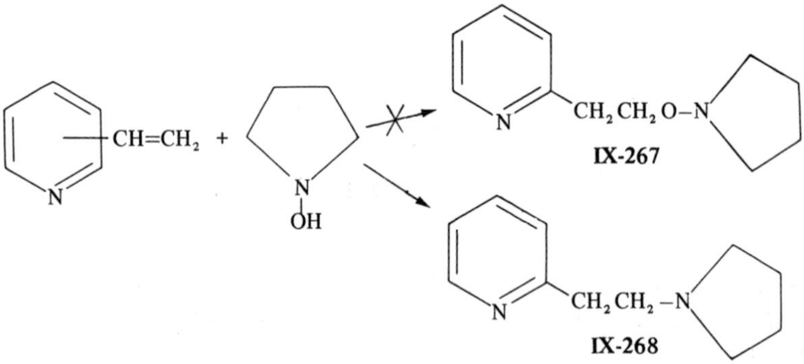

IX-267

IX-268

pyridylethylamines rather than the oxygenated products have been obtained in this way.[566]

Ethylenimine condenses with 2-vinylpyridine in the presence of metallic sodium to give 2-ethyleniminoethylpyridine.[567]

The reactions of substituted vinylpyridines have not been intensively studied. *N*-Methylaniline and 5-vinyl-2-picoline condense in the presence of sodium metal to give 5-(*N*-methyl-2-phenylaminoethyl)pyridine **(IX-269)**.[568]

$CH{=}CH_2$, $CH_3$, N + Na + $C_6H_5NHCH_3$ ⟶

$CH_2CH_2NHC_6H_5$, $CH_3$, N

**IX-269**

When *d*-α-methylphenethylamine **(IX-270)** and 5-ethyl-2-vinylpyridine are heated in acetic acid and methanol, *N*-(*d*-α-methylphenethyl)-2-(5-ethyl-2-pyridyl)ethylamine **(IX-271)** is formed.[569, 570]

$C_2H_5$, $CH{=}CH_2$, N + $C_6H_5CH_2CH(CH_3)NH_2$ $\xrightarrow{\text{Acetic acid}}$

**IX-270**

$C_2H_5$, N, $CH_2CH_2NHCH(CH_3)CH_2C_6H_5$

**IX-271**

2-[β-Alkyl- or arylaminoethyl]-5-vinylpyridines were prepared by treating 2,5-divinylpyridine with an amine in protic solvents in the presence of an acid catalyst at 90 to 120°.[571]

## K. *Miscellaneous Reactions*

The reduction of pyridine derivatives also provides a source of side-chain amines. Nicotine and its analogues are reduced to side-chain aminopyridines;

thus, hydrogenation of nicotine over palladium-charcoal at 54° until one mole of hydrogen is consumed gives 3-(4-methylaminobutyl)pyridine **(IX-272)**.[572] If the latter is treated with formic acid and formaldehyde, 3-(4-dimethylaminobutyl)-

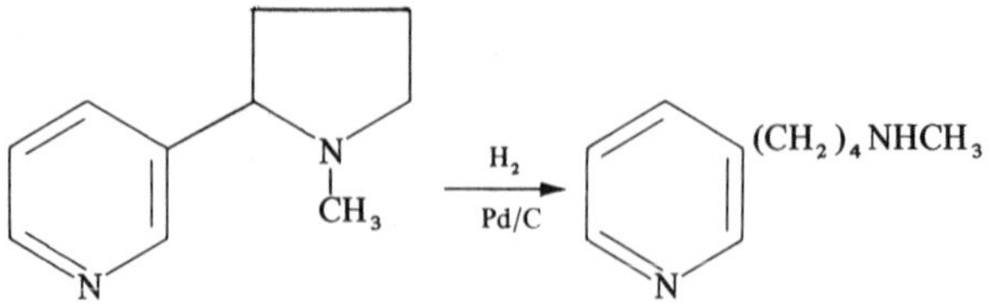

$H_2$, Pd/C → $(CH_2)_4NHCH_3$

**IX-272**

pyridine results.

At a cathode potential of −0.75 volts, isonicotinanilide is reduced to 4-anilinomethylpyridine and 4-pyridylcarbinol.[573]

1-Nitro-2-(3-pyridyl)ethylene is reduced by copper coated with zinc dust to β-(3-pyridyl)ethylamine **(IX-273)**, which can also be obtained from the reduction of 3-pyridylacetaldoxime.[574]

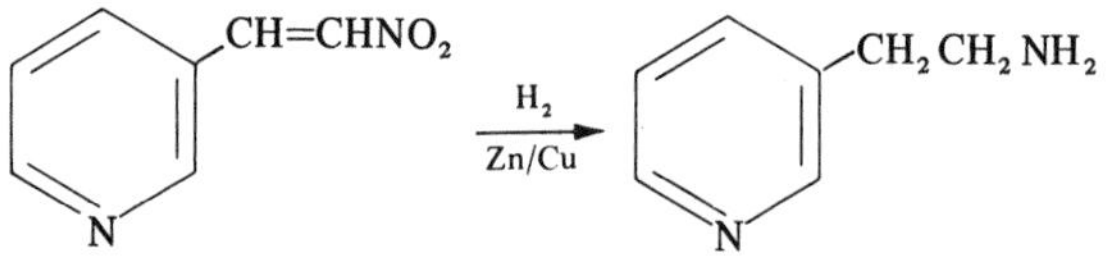

$CH{=}CHNO_2$ — $H_2$, Zn/Cu → $CH_2CH_2NH_2$

**IX-267**

## L. *Miscellaneous Preparations*

The preparations discussed here did not fit in previous sections. They include side-chain amines obtained from hydrolysis or condensation reactions. The preparation of aminoethers and aminothioethers have also been included.

Several aminocyclohexylpyridines are prepared by the hydrolysis of pyridylcyclohexyl carbamates. Thus 2-(1-aminocyclohexyl)pyridine **(IX-274)** is obtained by the alkaline hydrolysis of methyl *N*-[1-(2-pyridyl)cyclohexyl] carbamate.[575] Other aminocyclohexyl pyridines have been reported.[576]

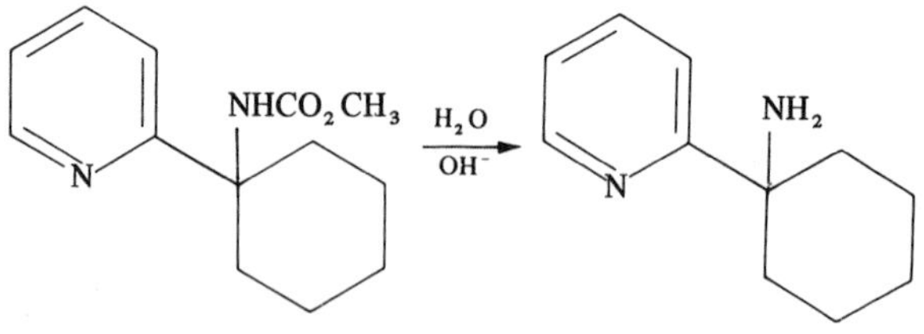

$NHCO_2CH_3$ — $H_2O$, $OH^-$ → $NH_2$

**IX-274**

It has been reported[577] that 3-(4-aminobutyl)piperidine is obtained by the hydrogenation of 1,8-diamino-4-aminomethyloctane in the presence of ammonia. Acetylation of this piperidine, followed by dehydrogenation at 200° over palladium-charcoal gives 3-(4-acetylaminobutyl)pyridine **(IX-275)**.

$$NH_2(CH_2)_3CH(CH_2NH_2)(CH_2)_4NH_2 + NH_3 \xrightarrow{[H]}$$

$$\text{3-}[(CH_2)_4NH_2]\text{piperidine} \xrightarrow[\text{(ii)}-[H]]{\text{(i) } CH_3COCl} \text{3-}[(CH_2)_4NHCOCH_3]\text{pyridine}$$

**IX-275**

When 2-picolyllithium is added to an azomethine linkage, various substituted 2-(2-pyridyl)ethylamines **(IX-276)** are formed in good yield.[578, 579]

$$\text{2-}(CH_2Li)\text{pyridine} + R_2C{=}NR_1 \longrightarrow \text{2-}[CH_2C(R)_2NHR_1]\text{pyridine}$$

**IX-276**

Heating *N*-benzylidine aniline with a mixture of magnesium turnings (or aluminum), mercuric chloride, and pyridine, gives 1,*N*-diphenyl-1-(2-pyridyl)-methylamine **(IX-277)**. Several such picolylamines are prepared by this modification of the Emmert-Asendorf reaction.[580] When **IX-277** is heated with

$$C_6H_5CH{=}NC_6H_5 + \text{pyridine} \xrightarrow{Mg(Hg)} \text{2-}[CH(C_6H_5)NHC_6H_5]\text{pyridine}$$

**IX-277**

$$\xrightarrow[NaNH_2]{(CH_3)_2NCH_2CH_2Cl} \text{2-}[CH(C_6H_5)N(C_6H_5)CH_2CH_2N(CH_3)_2]\text{pyridine}$$

**IX-278**

sodium amide and $\beta$-dimethylaminoethyl chloride, it yields 2-[$\alpha$-(*N*-2-dimethylaminoethyl-*N*-phenylamino)benzyl] pyridine **(IX-278)**.

2-Chloromethyl-1,4-benzodioxan reacts with picolylamine to give 2-picolylaminomethyl-1,4-benzodioxan **(IX-279)**.[581] Table IX-64 lists the 1,4-benzodioxans thus prepared.

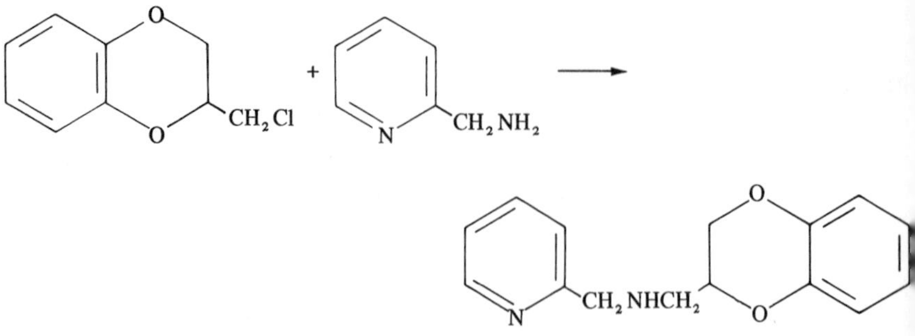

$CH_2Cl$ + $CH_2NH_2$ →

$CH_2NHCH_2$

**IX-279**

When 2-pyridinecarboxaldehyde is added to a cooled solution of nitroethane and diethylamine in ethanol, 2-nitro-1-(2-pyridyl)propanol **(IX-280)** is obtained.[582] Dehydration of **IX-280** followed by hydrogenation gives

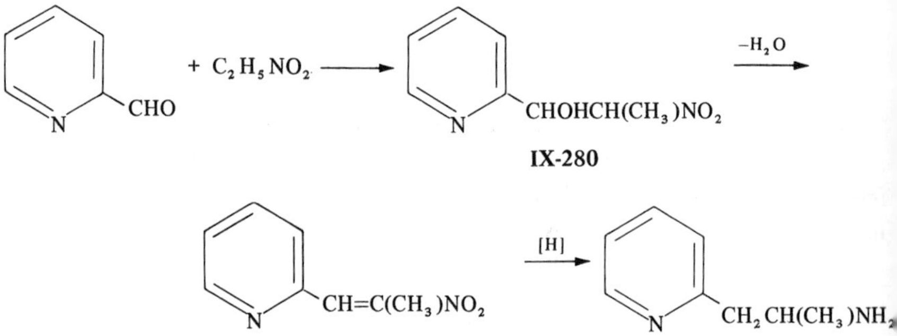

CHO + $C_2H_5NO_2$ → $CHOHCH(CH_3)NO_2$ $\xrightarrow{-H_2O}$

**IX-280**

$CH{=}C(CH_3)NO_2$ $\xrightarrow{[H]}$ $CH_2CH(CH_3)NH_2$

**IX-281** **IX-282**

3-amino-1-(2-pyridyl)propane **(IX-282)**.[582, 583] The 3- and 4-pyridyl analogs are prepared similarly.

The preparation of aminoethers and aminothioethers of pyridine have been reported. 4-Chloropyridine-1-oxide reacts with ethanolamine to give 4-(β-aminoethoxy)pyridine-1-oxide.[584]

β-Dialkylaminoethyl chlorides have been used to prepare β-aminoalkyl 2-pyridyl sulfides **(IX-283)**.[585, 586] The preparation involves heating 2-pyridinethiones and β-dialkylaminoethyl chlorides in ethanol. Compounds **IX-283** are also obtained from the reaction of chloropyridines with β-dialkylaminoethyl mercaptans (see also Chapter XV).

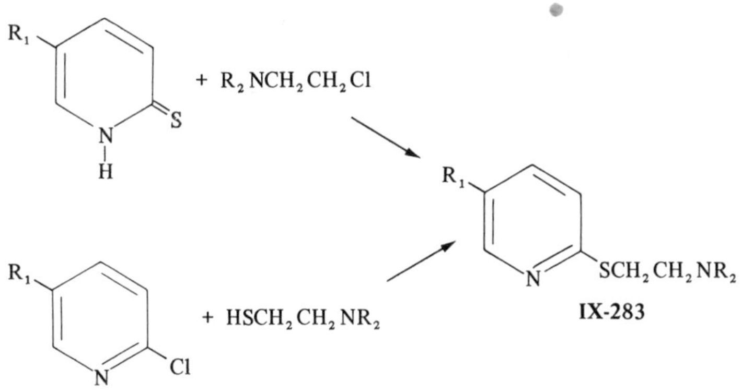

+ $R_2NCH_2CH_2Cl$

+ $HSCH_2CH_2NR_2$

$SCH_2CH_2NR_2$

**IX-283**

Dimethylaminoethyl ethers **(IX-284)** are obtained when aryl 2-pyridyl carbinols are treated with (β-dimethylamino)ethyl chloride. For example,

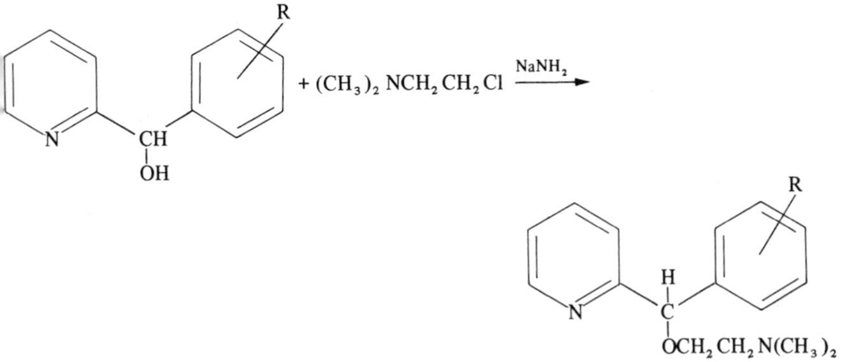

+ $(CH_3)_2NCH_2CH_2Cl \xrightarrow{NaNH_2}$

$OCH_2CH_2N(CH_3)_2$

**IX-284**

pyridine-2-aldehyde reacts with phenylmagnesium bromide to give phenyl 2-pyridyl carbinol, which, when treated with sodium amide and 2-dimethylaminoethyl chloride hydrochloride, gives dimethylaminoethyl α-2-pyridylbenzyl

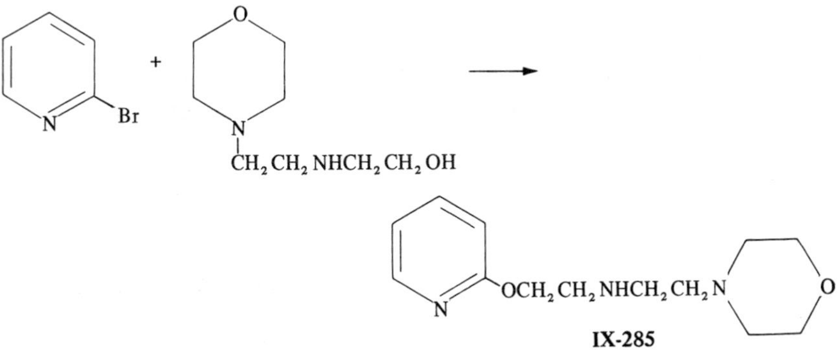

$CH_2CH_2NHCH_2CH_2OH$

$OCH_2CH_2NHCH_2CH_2N$

**IX-285**

ether (**IX-284**, R = H).[587, 588] When 2-bromopyridine is added to a suspension of 2-(2-*N*-morpholinoethylamino)ethanol and sodamide in boiling toluene, 2-[2-(2-*N*-morpholinoethylamino)]ethoxypyridine (**IX-285**) results.[589]

2-Pyridylalkylguanidines (**IX-286**) are obtained by the reaction of 2-pyridylalkylamine with 2-methyl-2-thiopseudourea.[590]

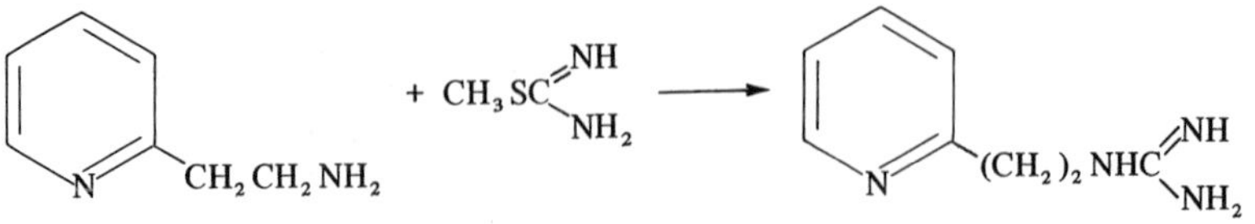

**IX-286**

Various 5-aminoethyl-2-picolines are prepared by nitrosation of the corresponding 5-(2-phenylaminoethyl)-2-picolines, with subsequent cleavage of the resulting nitrosyl derivatives by heating them with alkali hydroxide; no further details are reported in this patent.[591]

Nitrile derivatives of dialkylaminoalkylpyridines have been synthesized.[592, 593] Thus, in the presence of sodamide, *m*-tolylacetonitrile condenses with 2-bromopyridine and β-diisopropylaminoethyl chloride to give 4-diisopropylamino-2-(2-pyridyl)-2-(3-tolyl)butyronitrile (**IX-287**). 3-[(Dimethylamino)methyl]-2,6-diphenylpyridine (**IX-288**) is obtained by heating 1,3-dibenzoyl-2-dimethylaminomethylpropane hydrochloride with hydroxylamine hydrochloride.[594]

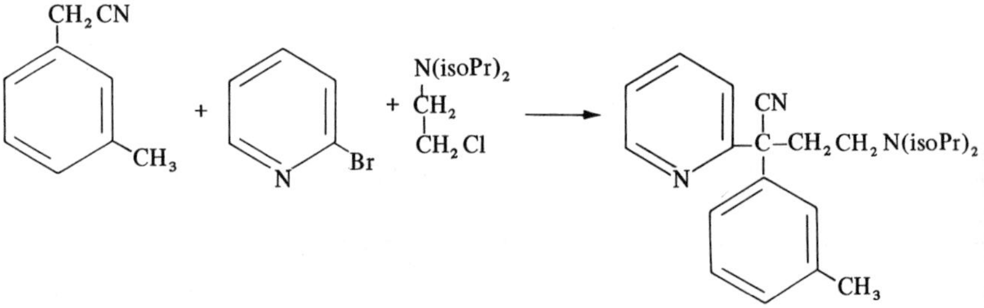

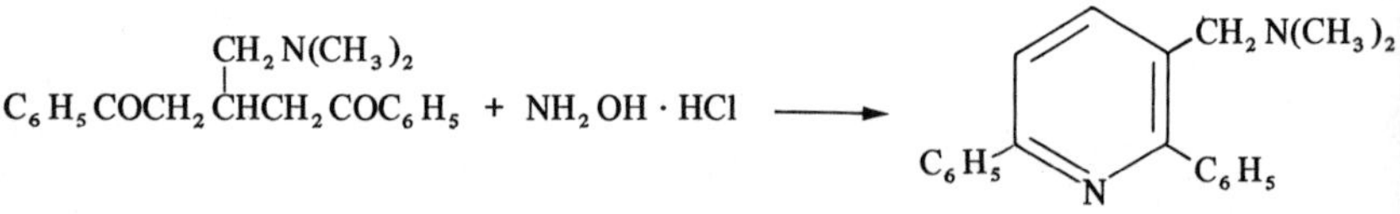

**IX-287**

**IX-288**

## 2. Properties and Reactions

The physical properties of side-chain aminopyridines are listed in Tables IX-63 to IX-80. Their chemical reactions are reported below.

The acylation of 3-dimethylamino-1-(2-pyridyl)propane **(IX-289)** and 3-dimethylamino-1-phenyl-1-(2-pyridyl)propane **(IX-290)** with various esters using phenyllithium or phenylsodium as the base to produce the carbanion yields the ketones **IX-291**. For example, when the ester is ethyl benzoate, the ketone **IX-291**, ($R_1$ = H, $R_2$ = $C_6H_5$) is obtained.[537]

$CH_2CH_2CH_2N(CH_3)_2$

**IX-289**

$C_6H_5$
$CHCH_2CH_2N(CH_3)_2$

**IX-290**

$R_1$
$CCH_2CH_2N(CH_3)_2$
CO
$R_2$

**IX-291**

When treated with hydrochloric acid and an aqueous solution of sodium nitrite, *N*-methyl-2-picolylamine gives 2-(*N*-methyl-*N*-nitroso)picolylamine.[594]

1-(*p*-Chlorophenyl)-3-(4-picolyl)urea **(IX-292)** results from the action of 4-aminomethylpyridine on *p*-chlorophenylisocyanate.[596] Other ureas are similarly prepared (Table IX-88).

$CH_2NH_2$ + Cl–$C_6H_4$–NCO ⟶

Cl–$C_6H_4$–NHCNHCH$_2$–$C_5H_4$N (with O above C)

**IX-292**

3-Picolylamine is alkylated with 7-β-bromoethyltheophylline **(IX-293)** to give **IX-294**.[596]

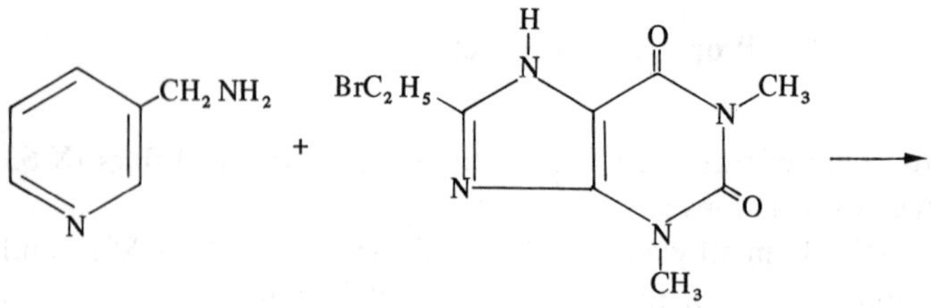

**IX-293**

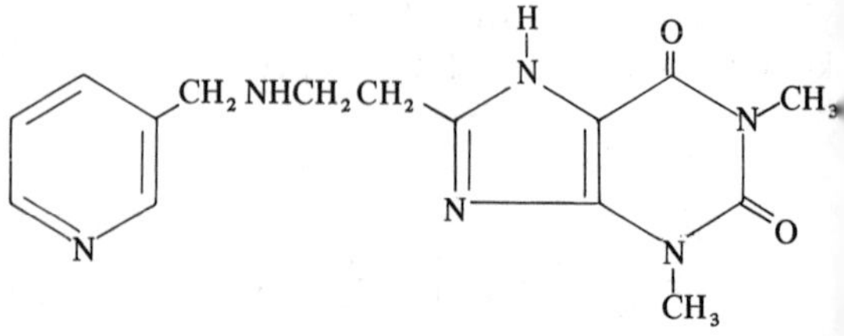

**IX-294**

## III. List of Tables

## IV. Acknowledgments

For their help and cooperation, I wish to thank the secretaries of the Chemistry Department and several of my colleagues, especially, Mr. Emil Krochmal, Mr. Kirby Lowery, Dr. Klaus Adam, Dr. James L. Lyle, and Dr. A. E. Martell. I am especially grateful to the Robert A. Welch Foundation whose grant permitted me to continue my research projects during the writing of this chapter, and to the many helpful comments of our untiring editor.

## V. Guide to Locating Compounds in Tables

The reader should first consult the list of tables (Section III), which summarizes the general classes of aminopyridines compiled; generally, derivatives are listed in the same tables as the parent compounds.

Compounds in a particular table are arranged in alphabetical order except when abbreviations (e.g., Me, Et, *n*-Pr, Ph, etc.) for alkyl or aryl substituents are used. In this case, groups with fewer carbon atoms have priority over those with more carbon atoms (thus Me before Et). When several columns of substituents are present in a table, the column on the left is first filled, then the next column, and the last to be filled is that on the extreme right. Monosubstituted compounds are listed before di- or trisubstituted aminopyridines.

## VI. TABLES

TABLE IX-1. Reaction of 3-Substituted Pyridine-1-oxides with *N*-Phenylbenzimidoylchlorides (See Chapter IV)

TABLE IX-2. Reaction of Pyridine-1-oxides with *N*-Arylbenzimidoyl Chlorides (See Chapter IV)

TABLE IX-3. Preparation of 1,5-Naphthyridines from 3-Aminopyridines

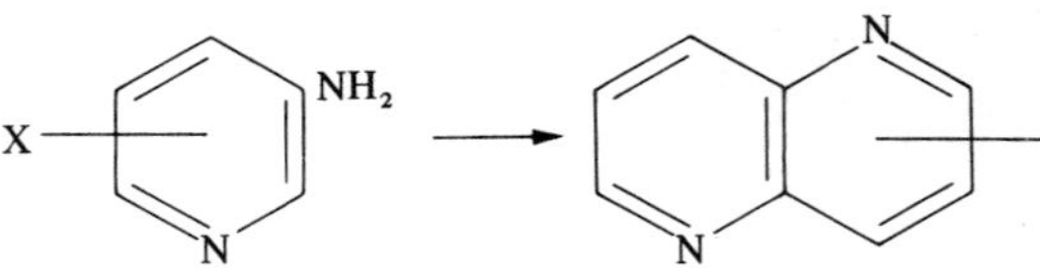


| X | Reagent | R 2 | R 3 | R 4 | R 6 | % Yield | Ref. |
|---|---|---|---|---|---|---|---|
| H | Glycerol | H | H | H | H | 31 | 394, 598 |
| H | Crotonaldehyde | Me | H | H | H | 8 | 394, 598 |
| H | Methacrolein | H | Me | H | H | 30 | 394, 598 |
| H | Methylvinylketone | H | H | Me | H | 11 | 394, 598 |
| H | Ethylacrolein | H | Et | H | H | 4 | 394, 598 |
| H | Acetaldehyde and acetone | Me | H | Me | H | – | 394, 398 |
| 2-OH | Glycerol | OH | H | H | H | 15 | 394, 599 |
| 6-OH | Acetaldehyde | OH | H | H | Me | 27 | 394, 599 |
| 6-Cl | Acetaldehyde | Me | H | H | Cl | – | 394, 600 |
| 4-OH | Glycerol | H | H | OH | H | 37 | 394, 601 |
| 1-Me-5-$NH_2$-2-pyridone | Acetaldehyde | 1,6-$Me_2$-1,5-naphthyridine-2-one | | | | 30 | 394, 599 |

TABLE IX-4. Preparation of 1,8-Naphthyridines from 2-Aminopyridines

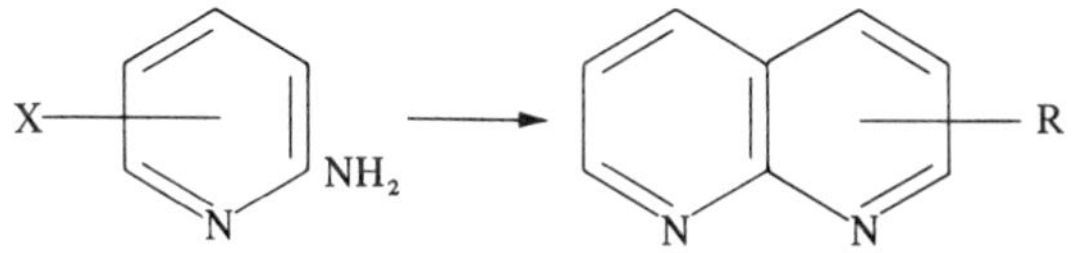

| X | Reagent | R 2 | R 3 | R 4 | R 5 | R 6 | R 7 | % Yield | Ref. |
|---|---|---|---|---|---|---|---|---|---|
| H | Glycerol | H | H | H | H | H | H | 30 | 411 |
| 4-Me | Glycerol | H | H | Me | H | H | H | 17 | 411 |
| 4-Me | Crotonaldehyde | Me | H | H | Me | H | H | 17 | 405 |
| 4-Me | Methylvinylketone | H | H | Me | Me | H | H | 3 | 405 |
| 5-Me | Glycerol | H | Me | H | H | H | H | 18 | 405 |
| 5-Me | Crotonaldehyde | Me | H | H | H | Me | H | 15 | 405 |
| 5-Me | Methylvinylketone | H | Me | H | Me | H | H | 3 | 405 |
| 5-Me | Methacrolein | H | Me | H | H | Me | H | 1 | 405 |
| 6-Me | Glycerol | Me | H | H | H | H | H | 10 | 411 |
| 6-Me | Crotonaldehyde | Me | H | H | H | H | Me | 15 | 405 |
| 4,6-$Me_2$ | Glycerol | Me | H | Me | H | H | H | 10 | 411 |
| 4,6-$Me_2$ | Crotonaldehyde | Me | H | Me | H | H | Me | 16 | 405 |

TABLE IX-5. Preparation of 7-Amino-1,8-naphthyridines from 2,6-Diaminopyridine

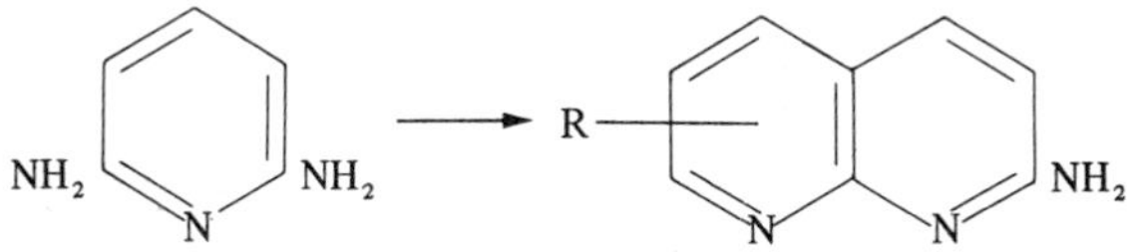

| Reagent | R 2 | R 3 | R 4 | R 7 | Ref. |
|---|---|---|---|---|---|
| Ethyl 2-oxalylpropionate | OH | Me | COOH | $NH_2$ | 413 |
| Ethyl oxalacetate | OH | COOEt | – | $NH_2$ | 416 |
| Malic acid | OH | – | – | $NH_2$ | 417 |
| Ethyl β-methylmalate | OH | Me | – | $NH_2$ | 417 |
| Citric acid | OH | – | $CH_2COOH$ | $NH_2$ | 417 |
| Acetone dicarboxylic acid | OH | – | $CH_2COOH$ | $NH_2$ | 417 |
| Ethyl oxalate | OH | – | COOH | $NH_2$ | 421 |
| Ethyl acetoacetate | OH | – | Me | $NH_2$ | 414 |

TABLE IX-6. Alkyl- and Aryl-2-aminopyridines

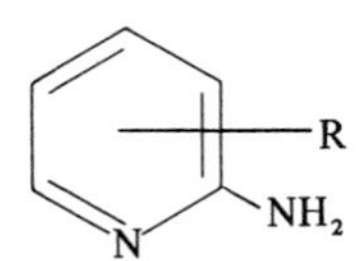


| Substituents | | | | | |
|---|---|---|---|---|---|
| 3 | 4 | 5 | 6 | Physical properties | Refs. |
| H | H | H | H | m.p. 57–8°; | 112 |
| | | | | picrate, m.p. 223–5°; | 111 |
| | | | | 1-oxide, HCl, m.p. 154–6° | 280 |
| Me | H | H | H | acetyl, m.p. 204–5°; | 367 |
| | | | | *N,N*-dibenzoyl, m.p. 124–5° | 367 |
| *t*-Bu | H | H | H | m.p. 128–9°; | 41 |
| | | | | picrate, m.p. 242° | 41 |
| H | 1,2,4-Tri-methyl-pentyl | H | H | b.p. 152–4°/4 mm; | 13 |
| | | | | picrate, m.p. 134° | 13 |
| (1-Me-2-pyrrolidinyl) | H | H | H | | 602, 609 |
| (1-Me-pyrrol-2yl | H | H | H | m.p. 76–8° | 424 |
| H | H | (1′-Me-3′-nitro-2′-pyrrolidinyl) | H | m.p. 171–2° | 602 |
| H | Me | H | H | m.p. 182° (dec.); picrate, m.p. 162.5–3.5°; HCl, m.p. 194.5–5.5°; perchlorate, m.p. 143.5–4.5° | 436 |

| | | | | | |
|---|---|---|---|---|---|
| H | Et | H | H | | 606 |
| H | *n*-Pentyl | H | H | b.p. 161°/20 mm; | 13 |
| | | | | b.p. 150–60°/6 mm; | 13, 14 |
| | | | | m.p. 54–5°; | 13 |
| | | | | m.p. 50–8.5° | 14 |
| H | 2-Pentyl | H | H | b.p. 145–50°/10 mm; m.p. 31–2° picrate, m.p. 150° | 13 |
| H | Isopentyl | H | H | b.p. 150–60°/10 mm | 13 |
| H | *n*-Hexyl | H | H | b.p. 180–200°/20 mm; m.p. 58–60° | 13 |
| H | Isohexyl | H | H | b.p. 175–80°/10 mm; m.p. 74°; picrate, m.p. 161–2° | 13 |
| H | *n*-Heptyl | H | H | b.p. 155–65°/7 mm; m.p. 59°; picrate, m.p. 136° | 13 |
| H | *n*-Octyl | H | H | b.p. 230–40°/35 mm; m.p. 64° | 13 |
| H | 2-Octyl | H | H | b.p. 152–5°; m.p. 54°; picrate, m.p. 134–6° | 13 |
| H | 5-Nonyl | H | H | b.p. 170–80°/20 mm; m.p. 56°; picrate, m.p. 112° | 13 |
| H | Benzyl | H | H | m.p. 103.5–4° | 14 |
| H | H | Me | H | picrate, m.p. 247.5–9.5° | 115, 436 |
| H | H | Et | H | b.p. 90–2°/3 mm | 27, 62 |
| H | H | (1-Me-2-pyrrolidinyl) | H | | 602, 609 |
| H | H | (1-Me-2-pyrrol-2-yl) | H | m.p. 90–2° | 424 |
| H | H | H | Et | picrate, m.p. 191–3° | 603 |

TABLE IX-6. Alkyl- and Aryl-2-aminopyridines (Continued)

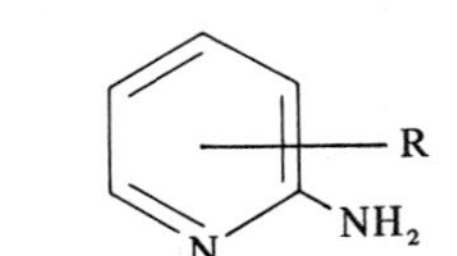


| Substituents | | | | | |
|---|---|---|---|---|---|
| 3 | 4 | 5 | 6 | Physical properties | Refs. |
| H | H | H | *n*-Heptyl | | 604 |
| H | H | H | (2,5-$Me_2$-pyrrol-l-yl) | b.p. 142°/0.15 mm; m.p. 121.5° | 313 |
| Me | Me | H | H | b.p. 124/10 mm; HCl, m.p. 239–40° | 438 |
| Me | H | Me | H | HCl, m.p. 203–5° | 438 |
| Me | H | H | Me | m.p. 50–2° | 605 |
| | | | | HCl, m.p. 243–4° | 438 |
| Et | H | H | Me | | 26 |
| H | Ph | H | Ph | picrate, m.p. 198–200° | 607 |
| H | *p*-Nitrophenyl | H | Ph | m.p. 245–50° | 607 |
| H | H | Et | Pr | b.p. 67°/0.15 mm | 480 |
| H | H | Me | Me | m.p. 54–5° | 603, 608 |
| Ph | Ph | H | Ph | m.p. 156–7° | 126 |
| Me | Me | Me | Me | m.p. 113–14° | 122, 214 |

TABLE IX-7. Alkyl- and Aryl-3-aminopyridines

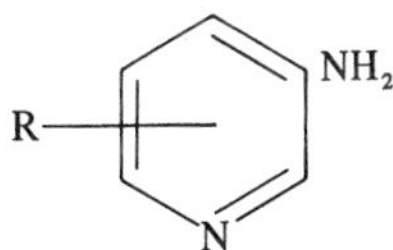


| Substituents | | | | | |
|---|---|---|---|---|---|
| 2 | 4 | 5 | 6 | Physical properties | Ref. |
| Me | H | H | H | m.p. 114–15° | 40, 110 |
| Ph | H | H | H | b.p. 119–21°/0.35 mm | 110, 392 |
| | | | | m.p. 65–7° | 441 |
| | | | | m.p. 62–4°; picrate, m.p. 204–6° | 110, 392, 441 |
| *p*-$BrC_6H_4$ | H | H | H | m.p. 99–101° | 110 |
| | | | | m.p. 134.5–7°; picrate, m.p. 177–9° | 610 |
| H | Me | H | H | m.p. 106–7°, 104–6° | 40, 468 |
| | | | | picrate, m.p. 177–8° | 95 |
| H | H | Me | H | m.p. 61–3°, 58°; picrate, m.p. 225° (dec.) | 40, 468 |
| H | H | H | Me | m.p. 95–6°; acetyl, m.p. 158–60° | 40, 555 |
| Me | Me | H | H | b.p. 246–56° m.p. 70–4° | 91, 92, 352 611 |
| | | | | picrate, m.p. 227–8° | 91 |
| | | | | sulfonyl, m.p. 54° | 92 |
| Me | H | Me | H | | 379 |
| Me | H | H | Me | m.p. 123°; picrate, m.p. 180–1°; 1-oxide, m.p. 155–8° | 68 |
| H | Me | Ph | H | m.p. 136-6.5° | 612 |
| H | Me | H | Me | b.p. 255–7°; m.p. 62–3°, m.p. 68–71°; picrate, m.p. 186–7°; sulfonyl, m.p. 62–3° | 91, 92, 352 |
| Me | Me | H | Me | 1-oxide HCl, m.p. 211° | 351 |

TABLE IX-8. Alkyl-4-aminopyridines

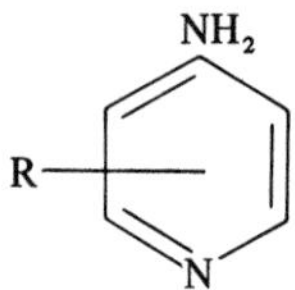

| Substituents | | | | | |
|---|---|---|---|---|---|
| 2 | 3 | 5 | 6 | Physical properties | Ref. |
| H | H | H | H | m.p. 152–5°; picrate, m.p. 215–17°; 1-oxide, HCl, m.p. 181–3° | 119 |
| Me | H | H | H | m.p. 94–5°; | 86, 118 |
| | | | | 1-oxide, m.p. 190° | 71 |
| | | | | 1-oxide-HCl, m.p. 191–2° | 75 |
| Et | H | H | H | b.p. 128–30°/4–5 mm; 117–20°/2.5 mm | 88, 614 |
| H | Me | H | H | m.p. 108–9°; | 86, 118 |
| | | | | 1-oxide, m.p. 120–5°; | 71 |
| | | | | 1-oxide HCl, m.p. 219–20° | 75 |
| H | Et | H | H | m.p. 42–3°; picrate, m.p. 202–3° | 41, 67, 615 |
| | | | | HCl, m.p. 209–10° | 62 |
| | | | | hemihydrate, m.p. 52–3° | 214 |
| | | | | hemihydrate, picrate, m.p. 196–7° | 41, 214 |
| H | isoPr | H | H | hemihydrate, m.p. 69–70° hemihydrate, picrate, m.p. 156–7° | 41, 214 |
| H | Styryl | H | H | 1-oxide, m.p. 179° | 616 |
| H | H | Et | H | m.p. 91–2° | 97, 617, 618 |
| Me | H | $COOCH_3$ | H | m.p. 125–6° | 97 |
| Me | H | $CH_2OH$ | H | m.p. 132–4° | 97 |
| Me | H | Et | H | 1-oxide HCl, m.p. 181–3°; picrate, m.p. 181–2° | 75 |
| Me | H | H | Me | m.p. 192–3°; 190–1° | 74, 118 |
| | | | | m.p. 188–90°, 192° | 20, 105 |
| | | | | 1-oxide, m.p. 265° | 74, 619 |
| | | | | 1-oxide HCl, m.p. 257° | 74 |
| *t*-Bu | H | H | *t*-Bu | chloroaurate, m.p. 193–5°; | 119 |
| | | | | picrate, m.p. 146–50° | 119, 620 |
| H | Me | Me | H | 1-oxide, m.p. 227–9°; 1-oxide, picrate, m.p. 221–3° | 41, 214 |
| Me | Me | Me | Me | hemihydrate, m.p. 196–7°; picrate, m.p. 225–6° | 41 |

ABLE IX-9. Halo-2-aminopyridines

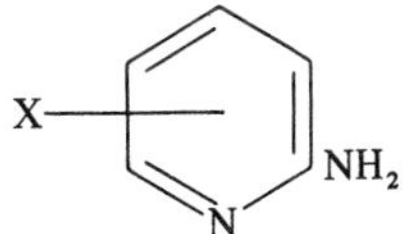


| | Substituents | | | | |
|---|---|---|---|---|---|
| | 4 | 5 | 6 | Physical properties | Ref. |
| | Cl | H | H | | 445 |
| | H | Br | H | | 382, 385 |
| | H | Cl | H | m.p. 135–6° | 278, 382, 384 |
| | H | H | Br | m.p. 89–90° | 2 |
| | H | H | F | m.p. 53–4.5°, 58–9° | 66, 121 |
| | H | H | I | m.p. 109–10° | 2, 3 |
| r | Br | H | H | | 34 |
| | Br | H | Br | | 34 |
| | H | F | F | m.p. 94.5–5.5° | 53, 374 |
| | Br | F | F | m.p. 116–17° | 53, 374 |
| | I | F | F | m.p. 114–5° | 59 |

TABLE IX-10. Halo-3-aminopyridines

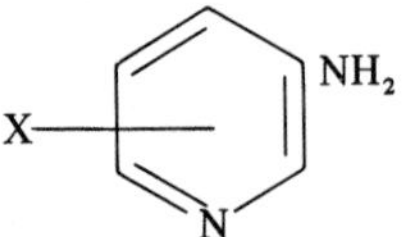


| Substituents | | | | | |
|---|---|---|---|---|---|
| 2 | 4 | 5 | 6 | Physical properties | Ref. |
| Cl | H | H | H | 2-nitrobenzoyl, m.p. 158–61.5°<br>2-aminobenzoyl, m.p. 168–72°<br>4-Cl-2-$NO_2$-benzoyl, m.p. 205°<br>2-$NH_2$-4-Cl-benzoyl, m.p. 202°<br>5-Me-2-$NO_2$-benzoyl, m.p. 154–5°<br>2-$NH_2$-5-Me-benzoyl, m.p. 186–8° | 621 |
| F | H | H | H | b.p. 116–17°/24 mm; HCl, m.p. 187–8° | 66 |
| H | I | H | H | m.p. 75–6°; picrate, m.p. 253° (dec.) | 373 |
| H | H | Br | H | | 386 |
| H | H | H | F | m.p. 87–7.5° | 66 |
| SH | | Cl | | m.p. 204–5° | 168 |
| H | Cl | Br | H | m.p. 108.5°; nitrate, m.p. 176° | 44 |
| H | Cl | Cl | H | m.p. 108°; nitrate, m.p. 169° | 44, 622 |
| H | Cl | I | H | m.p. 102° | 623 |
| H | Cl | H | Cl | m.p. 84–5° | 223, 225, 624 |
| H | $NHNH_2$ | H | Cl | m.p. 167–9° | 624 |
| Me | $CH_2Br$ | $CH_2Br$ | H | HBr, m.p. 220° | 625 |

TABLE IX-11. Halo-4-aminopyridines

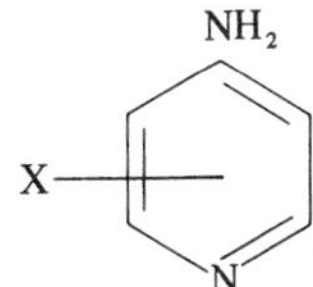


| Substituents | | | | | |
|---|---|---|---|---|---|
| 2 | 3 | 5 | 6 | Physical properties | Ref. |
| Br | H | H | H | m.p. 95–6° | 47 |
| Cl | H | H | H | m.p. 92°; 1-oxide HCl, m.p. 152–3.5° | 47, 87, 96, 389 |
| [illegible] | H | H | H | m.p. 99°, 1-oxide m.p. 110° (dec.) | 47, 107, 177, 470 |
| H | Br | H | H | m.p. 110° | 373 |
| | | | | 1-oxide, m.p. 67° | 47, 107, 177 |
| H | Cl | H | H | m.p. 59°; 60° | 48, 373 |
| H | F | H | H | m.p. 77° | 51 |
| H | I | H | H | m.p. 71°; 75–6°; | 48, 100 |
| | | | | picrate, m.p. 253° (dec.) | 47, 177, 107 |
| | | | | 1-oxide, picrate, m.p. 189° | |
| H | H | H | F | m.p. 89°; picrate, m.p. 223° | 49 |
| Cl | H | H | Cl | m.p. 172–3°, 170° | 87, 390 |
| $CCl_3$ | Cl | Cl | H | m.p. 136–8° | 56 |
| Cl | Cl | H | $CCl_3$ | m.p. 81–4° | 56 |
| Cl | Cl | Cl | Cl | m.p. 220–1° | 21 |
| Cl | Cl | Cl | $CCl_3$ | m.p. 113–18° | 56 |
| F | Cl | Cl | F | m.p. 112–13° | 53, 58 |
| F | Cl | F | F | m.p. 117–18° | 53, 58, 374 |
| F | F | F | F | m.p. 85–6° | 53, 54, 57, 252, 626 |
| F | F | F | MeO | m.p. 92.5–3° | 252 |

TABLE IX-12. Alkylhalo-(2,3, or 4)-aminopyridines

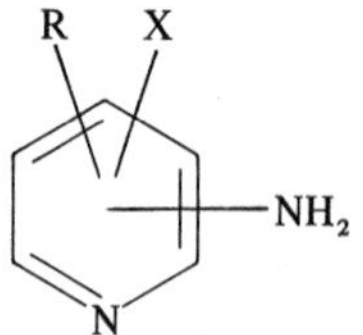

| Substituents | | | | | | |
|---|---|---|---|---|---|---|
| 2 | 3 | 4 | 5 | 6 | Physical properties | Ref. |
| $NH_2$ | Me | H | Br | H | m.p. 90–2° | 40 |
| $NH_2$ | Me | H | H | Br | m.p. 114–14.5° | 3 |
| $NH_2$ | H | Me | H | Br | m.p. 115–16° | 2, 3 |
| $NH_2$ | H | Et | H | Br | m.p. 113–13.5° | 3 |
| $NH_2$ | H | Et | H | I | m.p. 112–13° | 3 |
| $NH_2$ | H | Cl | H | Me | m.p. 108–9° | 388 |
| $NH_2$ | H | Ph | H | Br | m.p. 141–3° | 3 |
| $NH_2$ | H | Ph | H | I | m.p. 154–5.5° | 3 |
| $NH_2$ | H | H | Me | Br | m.p. 97.5–8° | 3 |
| Me | $NH_2$ | H | Cl | H | | 627 |
| $NH_2$ | Br | Me | Br | H | m.p. 123–4° | 40 |
| Me | $NH_2$ | H | H | COOH | m.p. 262–3° | 391 |
| Me | $NH_2$ | COOH | H | Cl | m.p. 218–19° (dec.) | 391 |
| Me | COOH | H | $NH_2$ | Cl | m.p. 227–8° | 391 |
| Me | $NH_2$ | $CH_2Br$ | $CH_2Br$ | H | HBr, m.p. 220° | 625 |
| Me | $NH_2$ | $EtOCH_2$ | CN | Cl | m.p. 146° | 78 |
| Me | H | $NH_2$ | H | Cl | m.p. 155–7° | 388 |
| Me | F | $NH_2$ | H | Me | m.p. 92°; picrate m.p. 235° | 52 |
| F | F | $NH_2$ | F | MeO | m.p. 92.5–3° | 252 |

TABLE IX-13. Aminonitropyridines

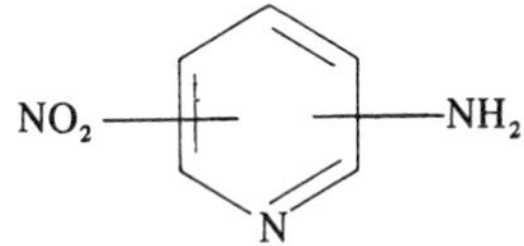


| Substituents | | | | | | |
|---|---|---|---|---|---|---|
| 2 | 3 | 4 | 5 | 6 | Physical properties | Ref. |
| $NH_2$ | $NO_2$ | H | H | H | | 386 |
| $NH_2$ | H | $NO_2$ | H | H | m.p. 96°; *N*-acetyl, m.p. 211° | 628 |
| | | | | | 1-oxide, m.p. 245–6°; acetyl, m.p. 206° | 371, 628 |
| $NH_2$ | H | H | $NO_2$ | H | m.p. 188° | 9 |
| | | | | | *N*-acetyl, m.p. 194–6°; 1-oxide, m.p. 196–7° | 371, 386 |
| $NH_2$ | $NO_2$ | H | $NO_2$ | H | m.p. 188° (190–2°), 188° | 45, 176, 367, 478 |
| $NH_2$ | H | $NO_2$ | $NO_2$ | H | m.p. 217° | 628 |
| OH | $NH_2$ | H | $NO_2$ | H | m.p. 199–201°; *N*-acetyl, m.p. 281° | 102 |
| H | $NH_2$ | $NO_2$ | H | H | b.p. 62–4°/0.5 mm | 50 |
| | | | | | m.p. 138° | 51 |
| | | | | | 1-oxide, m.p. 237° | 178 |
| H | $NH_2$ | H | $NO_2$ | H | m.p. 141°; acetyl, m.p. 146–8° | 64, 68 |
| | | | | | benzoyl, m.p. 170–2° | 103 |
| H | $NO_2$ | $NH_2$ | H | H | m.p. 204°; HCl, m.p. 260–1°; picrate, m.p. 197–8° | 152 |

TABLE IX-14. Alkylnitro-(2,3, or 4)-aminopyridines

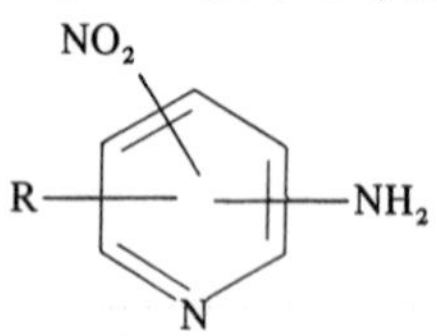


| Substituents | | | | | | |
|---|---|---|---|---|---|---|
| 2 | 3 | 4 | 5 | 6 | Physical properties | Ref. |
| $NH_2$ | $NO_2$ | H | H | Pr | m.p. 150°; picrate, m.p. 165–6° | 480 |
| $NH_2$ | $NO_2$ | Me | H | Me | m.p. 164° | 488 |
| $NH_2$ | $NO_2$ | H | Me | Me | m.p. 165–6° | 608 |
| $NH_2$ | H | Me | $NO_2$ | Me | m.p. 188–9°, 184° | 10, 488 |
| Me | $NH_2$ | $NO_2$ | H | Me | m.p. 112°; 1-oxide, m.p. 189–90° | 52 |
| Me | $NO_2$ | $NH_2$ | H | Me | m.p. 126° | 469 |
| Pr | H | $NH_2$ | $NO_2$ | H | m.p. 193° | 480 |
| H | Me | $NH_2$ | $NO_2$ | H | m.p. 193° | 469 |

TABLE IX-15. Haloaminonitropyridines

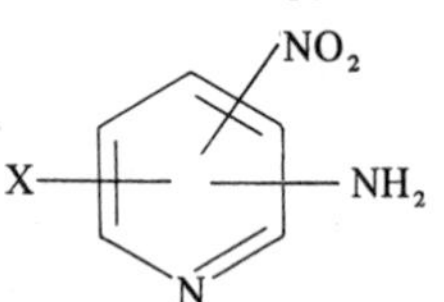


| Substituents | | | | | | |
|---|---|---|---|---|---|---|
| 2 | 3 | 4 | 5 | 6 | Physical properties | Ref. |
| $NH_2$ | $NO_2$ | Cl | H | H | m.p. 176° | 46 |
| $NH_2$ | Br | H | $NO_2$ | H | m.p. 213°, 222° | 9, 386 |
| $NH_2$ | $NO_2$ | H | Br | H | m.p. 205° | 386 |
| $NH_2$ | $NO_2$ | H | Cl | H | | 477 |
| $NH_2$ | $NO_2$ | H | $NO_2$ | H | m.p. 205° | 386 |
| $NH_2$ | $NO_2$ | Me | Br | H | | 477 |
| $NH_2$ | F | $NO_2$ | F | F | | 252 |
| $NH_2$ | $NO_2$ | Me | Br | Me | m.p. 169–70° | 630 |
| F | $NH_2$ | $NO_2$ | F | F | m.p. 84–4.5° | 252 |
| Br | $NO_2$ | $NH_2$ | H | H | m.p. 197° | 375 |
| Cl | $NO_2$ | $NH_2$ | H | H | m.p. 209–10° | 87, 96, 375, 629 |
| I | $NO_2$ | $NH_2$ | H | H | m.p. 269–71° (dec.) | 470 |
| Cl | H | $NH_2$ | $NO_2$ | H | m.p. 190–1°, 206° | 87, 96 |
| H | Cl | $NH_2$ | $NO_2$ | H | m.p. 181° | 44 |
| Cl | $NO_2$ | $NH_2$ | H | Cl | m.p. 142–3° | 87 |

TABLE IX-16. Alkoxy-2-aminopyridines

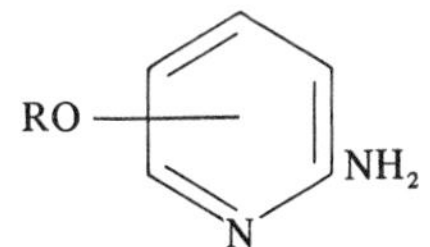

| Substituents | | | | | |
|---|---|---|---|---|---|
| 3 | 4 | 5 | 6 | Physical properties | Ref. |
| $CH_3O$ | H | H | H | | 116 |
| $C_2H_5O$ | H | H | H | m.p. 79–82° | 33, 35, 38 |
| H | $CH_3O$ | H | H | m.p. 115–16° | 116, 388 |
| H | $C_2H_5O$ | H | H | | 34, 35 |
| H | $n$–$C_3H_7O$ | H | H | m.p. 143–5° | 482 |
| H | iso$C_3H_7O$ | H | H | m.p. 150–2° | 482 |
| H | $n$–$C_5H_{11}O$ | H | H | m.p. 135–7° | 482 |
| H | iso$C_5H_{11}O$ | H | H | m.p. 129–31° | 482 |
| H | $n$–$C_6H_{13}O$ | H | H | m.p. 132–4° | 482 |
| H | H | H | $C_2H_5O$ | b.p. 120–2°/ 16 mm | 359 |
| H | H | H | iso$C_5H_{11}O$ | b.p. 162–4°/23 mm | 359 |
| Br | $C_2H_5O$ | H | H | m.p. 147–8° | 34 |
| H | $CH_3O$ | H | $CH_3$ | m.p. 141–2° | 388 |
| H | $C_2H_5O$ | Br | H | m.p. 149–50° | 34, 35 |
| H | $C_2H_5O$ | H | $C_2H_5O$ | m.p. 35–6°; picrate, m.p. 176–7° | 34 |
| Br | $C_2H_5O$ | Br | H | m.p. 100–1° | 34 |

TABLE IX-17. Alkoxy- and Aryloxy-3-aminopyridines

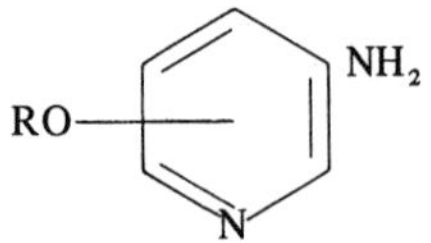


| Substituents | | | | | |
|---|---|---|---|---|---|
| 2 | 4 | 5 | 6 | Physical properties | Ref. |
| *o*-$BrC_6H_4O$ | H | H | H | – | 94 |
| Ethoxy | H | H | H | b.p. 105–6°/10 mm; m.p. 31–2° | 468 |
| 1-*β*-*D*-Gluco-pyranosyloxy | H | H | H | m.p. 147–53° | 81 |
| tetra-*O*-acetyl-1-*β*-*D*-gluco-pyranosyloxy | H | H | H | m.p. 147–8°; acetyl, m.p. 129–31° | 81 |
| H | $C_2H_5O$ | H | H | m.p. 83° HCl, m.p. 216° | 83 152 35, 30, 3 |
| H | H | $CH_3O$ | H | b.p. 166–8°/15 mm; m.p. 64–5° | 388 |
| H | H | $C_2H_5O$ | H | picrate, m.p. 189–91° | 28, 35, 4 |
| H | H | H | $CH_3O$ | b.p. 141°/25 mm | 313 |
| H | H | H | $C_2H_5O$ | | 28 |
| H | H | H | $C_4H_9O$ | | 330 |
| H | H | H | 4-$NH_2$-2-$ClC_6H_3O$ | m.p. 174–5° | 98 |
| H | H | H | 4-$NH_2$-2-$ClC_6H_3O$ | m.p. 145–6° di-*N*-acetyl, m.p. 195–6° | 98 631 |

ABLE IX-17. Alkoxy- and Aryloxy-3-aminopyridines (Continued)

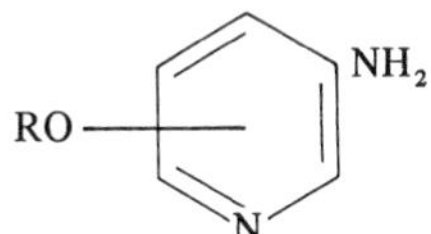


| | Substituents | | | | |
|---|---|---|---|---|---|
| | 4 | 5 | 6 | Physical properties | Ref. |
| I | H | H | 4-$NH_2$-2,6-$Cl_2C_6H_2O$ | m.p. 147–8°<br>di-*N*-acetyl, m.p. 226–7° | 98<br>631 |
| ( | H | H | tetra-*O*-acetyl-1-β-*D*-glucopyranosyloxy | m.p. 59–63°; acetyl, m.p. 176–7° benzoyl, m.p. 253–5° | 82 |
| le | MeO | H | H | m.p. 102–4.5° | 632 |
| leO | H | H | Br | m.p. 78–9° | 388 |
| leO | H | H | $CH_3$ | m.p. 74–5° | 388 |
| leO | H | H | $CH_3O$ | b.p. 116–17°/10 mm; m.p. 43–5°; acetyl, m.p. 85–7° | 388 |
| ( | H | Br | $C_2H_5O$ | m.p. 142–4° | 61 |
| ( | H | Cl | 4-$NH_2$-2-$ClC_6H_3O$ | m.p. 164–5°<br>di-*N*-acetyl, m.p. 171–2° | 98<br>631 |
| I | H | Cl | 4-$NH_2$-2,6-$Cl_2C_6H_2O$ | m.p. 215–16° di-*N*-acetyl, m.p. 252–3° | 98 |
| le | EtO | H | Me | m.p. 62–3° | 105 |
| I | MeO | MeO | Me | b.p. 140–1°/15 mm; HCl, m.p. 163–4° | 388 |

TABLE IX-18. Alkoxy- and Aryloxy-4-aminopyridines

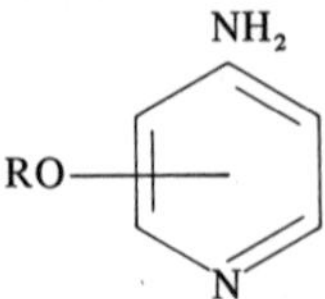

| Substituents | | | | | |
|---|---|---|---|---|---|
| 2 | 3 | 5 | 6 | Physical properties | Ref. |
| MeO | H | H | H | m.p. 88–9°; 1-oxide | 388, 389, 633 |
| | | | | acetyl, m.p. 93° | 389 |
| EtO | H | H | H | m.p. 87–9° | 28, 33, 35 |
| H | EtO | H | H | – | 35 |
| MeO | H | H | Me | m.p. 98–9° | 388 |
| MeO | H | H | MeO | m.p. 82–3° | 388 |

TABLE IX-19. 5-Amino-2-arylthiopyridines and Other Thioaminopyridines

| $R_1$ | $R_2$ | $R_3$ | $R_4$ | $R_5$ | m.p. | Acetate m.p. | Ref. |
|---|---|---|---|---|---|---|---|
| H | H | H | OH | H | 169–70° | 168–9° | 98 |
| Cl | H | H | OH | H | 217–18° | 192–3° | 98 |
| H | Cl | H | Cl | H | 84–5° | 125–6° | 98 |
| H | Cl | H | OH | H | 205–8° | 159–60° | 98 |
| H | H | Cl | OH | H | 148–9° | 169–70° | 98 |
| H | H | Cl | H | Cl | 87–8° | – | 98 |
| Cl | Cl | H | Cl | H | 114–15° | 121–2° | 98 |
| Cl | Cl | H | OH | H | 172–4° | 217–18° | 98 |
| Cl | H | Cl | Cl | H | 103–4° | 118–19° | 98 |
| Cl | H | Cl | OH | H | 171–2° | 137–8° | 98 |
| Cl | H | Cl | H | Cl | 129–30° | 148–9° | 98 |
| H | Cl | H | OH | Cl | 213–14° | – | 98 |
| H | H | Cl | OH | Cl | 222–3° | 172–3° | 98 |
| Cl | Cl | H | OH | Cl | 178–9° | 137–8° | 98 |
| Cl | H | Cl | OH | Cl | 223–4° | 167–8° | 634 |
| 4-(2′-Bromophenylthio) | | | | | – | – | 449 |
| 2-Methylthio-3-(2,4-dinitrophenyl)amino-5-chloropyridine | | | | | 223–24° | – | 168 |

TABLE IX-20. 5-Amino-2-pyridyl aryl sulfones

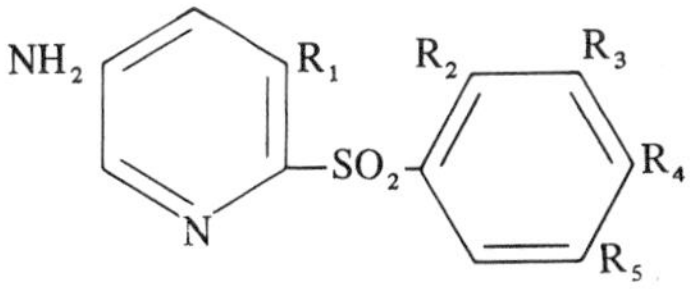

| $R_1$ | $R_2$ | $R_3$ | $R_4$ | $R_5$ | m.p. | Acetate m.p. | Ref. |
|---|---|---|---|---|---|---|---|
| H | H | H | H | H | – | – | 98 |
| H | H | H | OH | H | 219–20° | 97–8° | 98 |
| Cl | H | H | OH | H | 214–15° | – | 98 |
| H | Cl | H | Cl | H | 166–7° | 195–6° | 98 |
| H | Cl | H | OH | H | 216–7° | – | 98 |
| H | H | Cl | Cl | H | 242–3° | 153–4° | 98 |
| H | H | Cl | OH | H | 232–3° | 107–8° | 98 |
| H | H | Cl | H | Cl | 104–5° | 138–9° | 98 |
| Cl | Cl | H | Cl | H | 185–6° | – | 98 |
| Cl | Cl | H | OH | H | 276–7° | – | 98 |
| Cl | H | Cl | Cl | H | 212–13° | 183–4° | 98 |
| Cl | H | Cl | OH | H | 211–13° | 263–4° | 98 |
| Cl | H | Cl | H | Cl | 184–5° | – | 98 |
| H | Cl | H | OH | Cl | 230–1° | 157–8° | 98 |
| H | H | Cl | OH | Cl | 232–3° | 227–8° | 98 |
| Cl | H | Cl | OH | Cl | 238–9° | 213–14° | 98 |

TABLE IX-21. Miscellaneous 2-Aminopyridines

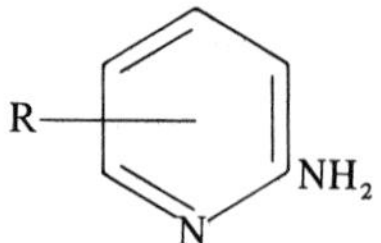

| Substituents | | | | | |
|---|---|---|---|---|---|
| 3 | 4 | 5 | 6 | Physical properties | Ref. |
| CHO | H | H | H | m.p. 86–8° | 422 |
| CHO | H | H | Ph | m.p. 124–5° | 422 |
| H | H | H | COOH | m.p. 317–19°<br>acetyl, m.p. 227–9° | 285 |
| H | H | H | COOMe | m.p. 88°<br>acetyl, m.p. 180° | 285 |
| H | H | H | COOEt | m.p. 56° | 285 |
| H | H | H | $CONH_2$ | – | 285 |
| H | H | H | $(-N{=}C(Me)-CH_2-COOEt)$ | m.p. 68–70° | 414 |
| H | H | H | $C_6H_5S$ | m.p. 117–7.5° | 6 |
| H | H | H | $\beta$-($o$–$NH_2C_6H_4Et$) | m.p. 96–7° | 635 |
| H | H | H | (2-ethoxalyl-propylamido) | m.p. 245° | 415 |
| $NO_2$ | –S– | H | H | m.p. 98° | 38 |
| COOEt | H | 2-naphthylazo | H | m.p. 25° | 9 |
| H | OH | H | Me | nitrate, m.p. 250–1°;<br>picrate, m.p. 182–3° | 15 |

TABLE IX-22. 3-Aminopyridinecarboxylic Acids and Esters

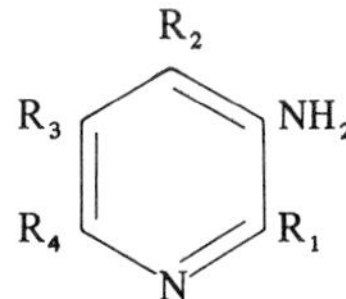


| $R_1$ | $R_2$ | $R_3$ | $R_4$ | Physical properties | Ref. |
|---|---|---|---|---|---|
| COOH | H | H | H | m.p. 210° | 62 |
| Me | H | COOH | H | m.p. 217° | 95 |
| | | | | 0.5 $H_2SO_4$, | 378, 391 |
| | | | | m.p. 262–3° | 392 |
| H | COOH | COOH | H | $H_2O$, m.p. 240–5° | 286 |
| Me | H | COOH | OH | m.p. 319–20° | 391 |
| Me | H | COOH | Cl | m.p. 227–8° | 378, 391, 392 |
| COOH | COOEt | CN | H | m.p. 160–2° | 286 |
| Me | COOH | COOH | H | $H_2O$, m.p. 233–5° (dec.) | 641 |
| COOH | COOH | COOH | H | $H_2O$, m.p. 205–10° | 286 |
| Me | COOH | COOH | Cl | m.p. 218–20° | 636 |
| H | COOEt | CN | H | m.p. 129–31° | 286 |
| H | COOMe | COOMe | H | m.p. 74–6° | 286 |
| Me | COOEt | CN | H | m.p. 130–1° | 77 |
| CHO | COOEt | CN | H | m.p. 85–8° | 286 |
| OH | COOEt | CN | H | m.p. 235–40° | 286 |
| Me | $CO_2Me$ | CN | Cl | | 391 |
| CHO | COOEt | CN | Cl | m.p. 131–2° | 286 |
| Me | COOMe | COOMe | Cl | m.p. 121–2° | 636 |

TABLE IX-23. Miscellaneous 3-Aminopyridines

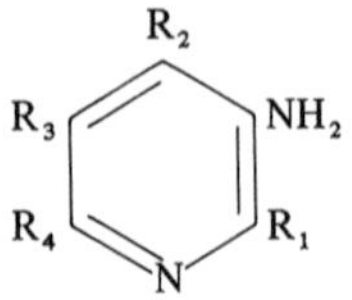

| $R_1$ | $R_2$ | $R_3$ | $R_4$ | Physical properties | Ref. |
|---|---|---|---|---|---|
| H | OH | H | H | *N*-benzyl, m.p. 161–2° | 89 |
| OH | H | H | Pr | m.p. 167–8° | 480 |
| H | OH | I | H | *N*-benzyl, m.p. 232–3° | 89 |
| Me | $CH_2CHO$ | H | H | phenylhydrazone | 99 |
| Me | $CH_2OH$ | $CH_2OH$ | H | m.p. 150–60°; picrate, m.p. 214.5–16.5°; HCl, m.p. 160–70° | 636 |
| Me | Me | $NH_2CH_2$ | H | di-HCl, m.p. 310° dipicrate, m.p. 225° | 513 78, |
| Me | $CF_3$ | $NH_2CH_2$ | H | tri-HCl, m.p. 276–8° | 639 |
| Me | $CH_2OMe$ | $NH_2CH_2$ | H | di-HCl, m.p. 230–1°; dipicrate, m.p. 214–15° | 78 |
| Me | $CH_2OEt$ | $NH_2CH_2$ | H | di-HCl, m.p. 127° | 78 |
| Me | $CH_2OEt$ | CN | H | m.p. 80° | 77 |
| Me | $CH_2OEt$ | CN | Cl | m.p. 146° | 78 |
| 1-(2-Pyridone) | H | H | H | acetyl, m.p. 243–4° | 116 |
| H | H | H | 1-(2-Pyridone) | acetyl, m.p. 210–11° | 116 |
| 2′-Pyridyl | H | H | H | m.p. 105–8° | 169 |
| (3′-Nitro-2′-pyridyl) | H | H | H | m.p. 167–8° | 169, |
| (5′-Nitro-2′-pyridyl) | H | H | H | m.p. 176–7° | 169 |
| (5′-Methyl-3′-nitro-2′-pyridyl) | H | H | H | m.p. 145–6° | 169 |
| (3′-Methyl-5′-nitro-2′-pyridyl) | H | H | H | m.p. 159–60° | 169 |
| (1-Pyrrolidinyl) | H | H | H | m.p. 70°; picrate, m.p. 184° | 637, |
| H | (1-Pyrrolidinyl) | H | H | b.p. 150°/1.1 mm; picrate, m.p. 170° | 637, |
| 2,4-Dinitrophenyl | H | Cl | H | m.p. 183–4°; picrate, m.p. 170°; acetyl, m.p. 162–4° | 168 |

TABLE IX-24. Miscellaneous 4-Aminopyridines

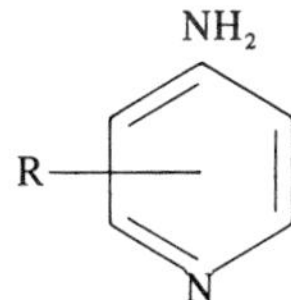

| Substituents | | | | | |
|---|---|---|---|---|---|
| 2 | 3 | 5 | 6 | Physical properties | Ref. |
| COOH | H | H | H | m.p. 319°; amide, m.p. 169° | 365 |
| H | COOH | H | H | 1-oxide, m.p. 270.5° | 42 |
| CN | H | H | H | m.p. 145° | 365 |
| Me | $SO_3H$ | H | Me | m.p. 293–5° | 106 |
| OH | H | H | OH | – | 390 |
| OH | *p*-$ClC_6H_4N{=}N-$ | H | OH | m.p. 315° (dec.) | 390 |
| OH | NO | H | H | m.p. 330° (dec.) | 375 |
| 2-[(5-Nitro-2-furyl)vinyl] | H | H | H | m.p. 203°; acetyl, m.p. 223–5° | 316 |
| 1-(2-Pyridone) | H | H | H | m.p. 169.5–70°; picrate, m.p. 256–8° (dec.) | 116 |
| H | HgCl | H | H | m.p. 293–5° | 642 |
| H | HgSCN | H | H | m.p. 195° | 642 |
| H | HgCl | Cl | H | m.p. 283–5° | 642 |

TABLE IX-25. 2-Alkylaminopyridines

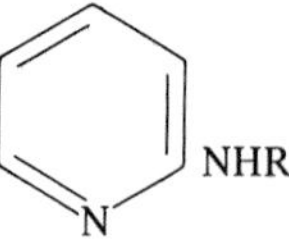

| R | Physical properties | Ref. |
|---|---|---|
| β-Acetoxy-β-phenylethyl | | 146 |
| Allyl | b.p. 108°/12 mm | 447 |
| | b.p. 56–8°/1 mm; picrate, m.p. 151–2° | 141 |
| 2-Aminoethyl | | 643 |
| 1-(3-Amino)propyl | | 643 |
| *n*-Butyl | b.p. 100–5°/0.4 mm; b.p. 90–100°/0.6 mm | 128 |
| | dipicrate, m.p. 181–3° | 60 |
| β-(*N*-Butylcarbamoyloxy)-β-phenylethyl | m.p. 95–8° | 146 |
| 3-(9-Carbazoyl)propyl | m.p. 98.9° | 323 |
| Carboxymethyl | HCl, m.p. 206°; picrate, m.p. 199° | 289 |
| β-[β-Chloro)phenethyl] | picrate, m.p. 170° | 137 |
| (5-Chloro-1-indanyl) | b.p. 160–70°/0.5 mm; m.p. 117° | 155 |
| 1-Cyanoethyl | m.p. 133–5° | 289 |
| Cyanomethyl | m.p. 126° | 289 |
| Cyclohexyl | m.p. 124–5°; picrate, m.p. 184–5° | 60, 142 |
| 2-(Diethylamino)ethyl | b.p. 110–15°/0.6 mm; dipicrate, m.p. 197–8° $n_D^{20}$ 1.5270 | 60 |
| Dihydro-β-ionyl | b.p. 180°/0.5 mm | 644, 645 |
| 2-(Dimethylamino)ethyl | b.p. 96–102°/0.4 mm $n_D^{20}$ 1.5412 di-HCl, m.p. 221–3° | 60 |
| Dimethylaminovinyl | m.p. 113–15° | 165 |
| Dodecyl | b.p. 160–85°/0.6 mm m.p. 57–9° dipicrate, m.p. 134–5° | 60, 129 |

TABLE IX-25. 2-Alkylaminopyridines (Continued)

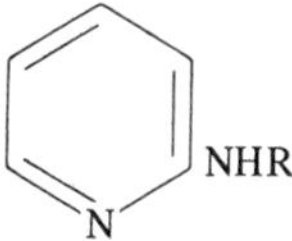


| R | Physical properties | Ref. |
|---|---|---|
| Ethyl | HCl, m.p. 138–40° | 60 |
| β-Hydroxy-β-cyclohexylethyl | m.p. 86–8°;<br>HCl, m.p. 143–4° | 146 |
| β-Hydroxy-β, β-diphenylethyl | m.p. 167–9°<br>HCl, m.p. 201–2° | 146 |
| 6-β-Hydroxy-β-phenylethyl | m.p. 83–5°;<br>HCl, m.p. 140–2° | 146 |
| β-[(β-Hydroxyl)phenethyl] | m.p. 84°; picrate,<br>m.p. 151–2° | 137 |
| 2-Hydroxypropyl | HCl, m.p. 164–5° | 331 |
| 1-Indanyl | b.p. 140–50°/0.2 mm;<br>m.p. 126° | 155 |
| Isopropyl | b.p. 84–6°/4–5 mm | 142, 144 |
| (2-Isopropyl-2-propynylamino)ethyl | tri-HCl, m.p. 209–10° | 135 |
| 2-(7-Methoxy-1-naphthyl)ethyl | HCl, m.p. 168.5–9.5° | 519 |
| Methyl | b.p. 84–5°/15 mm; | 509 |
| | 1-oxide, m.p. 67–8°;<br>picrate, m.p. 156–8° | 280, 434 |
| Methylcyclohexyl | m.p. 163° | 176 |
| 1-Methyl-3-(2,6,6-trimethyl-1-cyclohexen-1-yl)propyl | | 644 |
| 1-Methyl-3-(2,6,6-trimethyl-1-cyclohexyl)propyl | | 644 |
| 3-(10-Phenothiazinyl)propyl | m.p. 98–9°; HCl,<br>m.p. 172° | 323 |
| Propyl | b.p. 110–11°/15 mm;<br>picrate, m.p. 149–50.5° | 434 |
| | b.p. 66–7°/1.5 mm;<br>picrate, m.p. 148.5–9.5° | 141 |
| Tetrahydro-β-ionyl | b.p. 132–4°/0.03 mm | 644 |

TABLE IX-26. 2-Arylaminopyridines

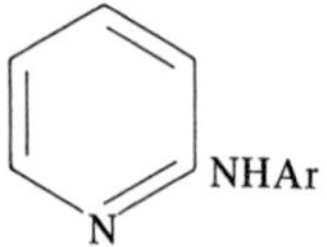


| R | Physical properties | Ref. |
|---|---|---|
| 5-Chloroindol-1-yl | m.p. 117° | 154 |
| | b.p. 170°/0.5 mm; | |
| 4-(7-Chloroquinolyl) | m.p. 171° | 138 |
| 5-Chloro-2-thienyl | b.p. 163–8°/1.9 mm | 462 |
| | m.p. 87–90°; | |
| 3-Cyano-2-pyridyl | m.p. 350° | 151 |
| 1-(1′-Cyclohexyl-5′-tetrazolyl)-2-methyl-2-propyl | m.p. 166–8° | 312 |
| 5-(5′-Nitro-2′-furyl)-1,3,4-thiadiazol-2-yl | m.p. 330–40° | 646–649 |
| 1-Indolyl | b.p. 140–50°/0.2 mm; m.p. 126° | 154 |

TABLE IX-27. 2-Aralkylaminopyridines

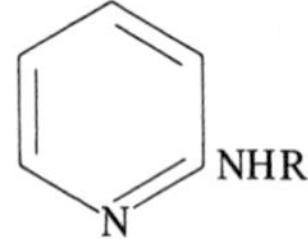


| R | Physical properties | Ref. |
|---|---|---|
| 1,4-Benzodioxan-2-ylmethyl | HCl, m.p. 265–7° | 581 |
| Benzyl | b.p. 116–31°/0.6 mm; m.p. 94–5° | 131 |
| α-Cyanobenzyl | m.p. 210–11° | 289 |
| α-[*p*-(β-Diethylaminoethoxy)-phenyl]-β-phenylethyl | m.p. 75–6° | 579 |
| 2-Diphenylmethyl | m.p. 100–1°; HCl, 197–8° | 331 |
| α-Hydroxybenzyl | m.p. 123–4° | 146 |
| (2-Ethyl-3-benzofuranyl)methyl | b.p. 118–25°/0.001 mm | 650 |
| (4-Methylthiobenzyl) | m.p. 128° | 299 |
| *m*-Nitrobenzylidene | | 300 |
| *p*-Nitrobenzylidene | 0.5 $H_2O$, m.p. 147° | 652 |
| Phenethyl | m.p. 112–13° | 137 |
| (2-Phenyl-1,3-dioxolan-2-yl-methyl) | b.p. 140–65°/0.3 mm; m.p. 29°; picrate, m.p. 143–4° | 153, 651 |
| 1-Pyrenylmethylene | m.p. 158–60° | 298, 653 |

TABLE IX-28. 2,2'-(Alkylidenediamino)dipyridines[290]

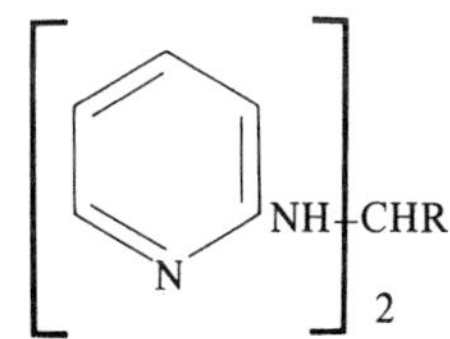

| R | m.p. | R | m.p. |
|---|---|---|---|
| Amyl | 100° | Hexyl | 92° |
| Butyl | 122° | Methyl | 122° |
| Isobutyl | 108° | Nonyl | 72° |
| Decyl | 66° | Octyl | 76° |
| Ethyl | 154° | Propyl | 110° |
| Heptyl | 81° | Isopropyl | 146° |

TABLE IX-29. 2-(Alkoxybenzyl)aminopyridines

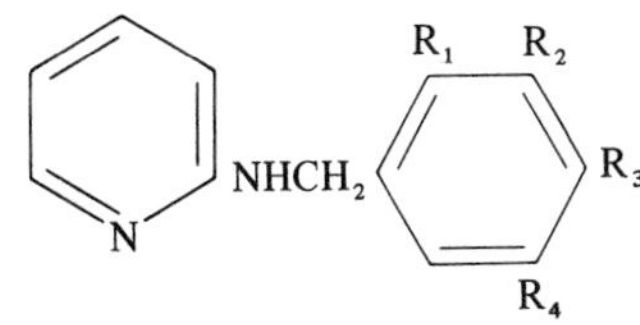

| $R_1$ | $R_2$ | $R_3$ | $R_4$ | b.p./10 mm | m.p. | Ref. |
|---|---|---|---|---|---|---|
| MeO | H | H | H | 181–4° | 120–1° | 171 |
| EtO | H | H | H | 184–7° | 90–1° | 171 |
| *n*-PrO | H | H | H | 190–3° | 87–8° | 171 |
| IsoPrO | H | H | H | 186–9° | 105–6° | 171 |
| *n*-BuO | H | H | H | 202–5° | 86–7° | 171 |
| IsoBuO | H | H | H | 188–91° | 109–10° | 171 |
| *n*-Pentyloxy | H | H | H | 206–9° | 80–1° | 171 |
| isoPentyloxy | H | H | H | 207–10° | 87–8° | 171 |
| MeO | H | H | F | – | 77° | 654 |
| EtO | H | H | F | – | 77.5° | 654 |
| *n*-PrO | H | H | F | – | 70° | 654 |
| *n*-BuO | H | H | F | – | 58° | 654 |
| H | F | MeO | H | – | 113° | 654 |
| H | F | EtO | H | – | 88.5° | 654 |
| H | F | *n*-PrO | H | – | 75° | 654 |
| H | MeO | MeO | H | – | – | 72 |
| H | MeO | MeO | MeO | – | 167–8° | 72 |

TABLE IX-30. Halo-2-alkylaminopyridines

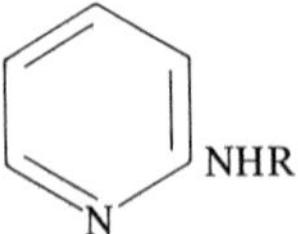


| Substituents | | | | | | |
|---|---|---|---|---|---|---|
| R | 3 | 4 | 5 | 6 | Physical properties | Ref. |
| Pr | H | H | Br | H | m.p. 40.3–41°; picrate, m.p. 146–7° | 141 |
| Pr | H | H | Cl | H | m.p. 31.5–32°; picrate, m.p. 138–9° | 140 |
| Pr | Br | H | Br | H | b.p. 95–6°/2 mm; picrate, m.p. 119.5–20.5° | 141 |
| Pr | Cl | H | Cl | H | b.p. 100–2°/2 mm; picrate, m.p. 199–201° (dec.) | 140 |

TABLE IX-31. Nitro Secondary 2-Aminopyridines

$R_3$ $R_2$ $NHR_1$ N

| $R_1$ | $R_2$ | $R_3$ | m.p. | Ref. |
|---|---|---|---|---|
| Benzyl | $NO_2$ | H | 80° | 157 |
| *p*-Chlorobenzyl | $NO_2$ | H | 105° | 157 |
| 3,4-Dimethoxy-β-phenethyl | $NO_2$ | H | 110° | 655 |
| 4-Dimethylamino-β-phenethyl | $NO_2$ | H | 110° | 655 |
| 3,4-Dimethyl-β-phenethyl | $NH_2$ | H | 98–9° | 655 |
| Dodecyl | $NO_2$ | H | 102° | 655 |
| β-Hydroxy-β-phenethyl | $NO_2$ | H | 90–2° | 655 |
| 2-(3-Indolyl)ethyl | $NO_2$ | H | 161° | 655 |
| Phenethyl | $NO_2$ | H | 85–6° | 655 |
| Benzyl | H | $NO_2$ | 131–4° | 656 |
| 4-Dimethylaminophenyl | H | $NO_2$ | 187–9° | 656 |
| β-Hydroxy-β-phenethyl | H | $NO_2$ | 137–8° | 655 |
| Morpholino | H | $NO_2$ | 142–3° | 656 |
| Methyl | H | $NO_2$ | 180–1° | 457, 656 |

TABLE IX-31. Nitro Secondary 2-Aminopyridines (Continued)

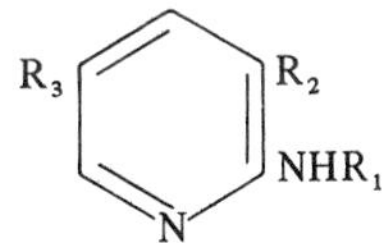


| $R_1$ | $R_2$ | $R_3$ | m.p. | Ref. |
|---|---|---|---|---|
| 2,2′-(*p,p*′-Methylenedianilino)-bis | H | $NO_2$ | 203–7° | 656 |
| Phenyl | H | $NO_2$ | 136° | 723 |
| | | | 110° | 166 |
| Piperazino | H | $NO_2$ | 84–5° | 656 |
| *p*-Tolyl | H | $NO_2$ | 136–7° | 656 |
| *p*-Acetamidophenyl | $NO_2$ | $NO_2$ | 276° | 176 |
| *p*-Acetophenyl | $NO_2$ | $NO_2$ | 186° | 180 |
| *o*-Anisyl | $NO_2$ | $NO_2$ | 200° | 176 |
| Benzyl | $NO_2$ | $NO_2$ | 112° | 176 |
| *p*-Chlorophenyl | $NO_2$ | $NO_2$ | 178° | 176 |
| Cyclohexyl | $NO_2$ | $NO_2$ | 117°, 116–7° | 176, 180 |
| Ethyl | | | 103–4° | 180 |
| β-Hydroxyethyl | $NO_2$ | $NO_2$ | 104° | 176 |
| Methyl | $NO_2$ | $NO_2$ | 163° | 176 |
| | | | 148° | 493 |
| | | | 142–3° | 180 |
| α-Naphthyl | $NO_2$ | $NO_2$ | 234° | 176 |
| β-Naphthyl | $NO_2$ | $NO_2$ | 229° | 176 |
| *o*-Nitrophenyl | $NO_2$ | $NO_2$ | 167° | 176 |
| *m*-Nitrophenyl | $NO_2$ | $NO_2$ | 158° | 176 |
| *p*-Nitrophenyl | $NO_2$ | $NO_2$ | 200° | 176 |
| 1-Piperidinyl | $NO_2$ | $NO_2$ | 138–9° | 180 |
| Phenyl | $NO_2$ | $NO_2$ | 149° | 176 |
| *p*-$NH_2SO_2C_6H_4$ | $NO_2$ | $NO_2$ | 249° | 176 |
| *p*-$H_2O_3AsC_6H_4$ | $NO_2$ | $NO_2$ | >300° | 176 |
| *o*-$HO_2CC_6H_4$ | $NO_2$ | $NO_2$ | 239° | 176 |
| *p*-$HO_2CC_6H_4$ | $NO_2$ | $NO_2$ | 298° | 176 |
| *p*-$CH_3O_2CC_6H_4$ | $NO_2$ | $NO_2$ | 163° | 176 |
| *p*-$C_2H_5O_2CC_6H_4$ | $NO_2$ | $NO_2$ | 160° | 176 |
| [(2-$C_5H_4N$)CONH–] | $NO_2$ | $NO_2$ | 229° | 176 |
| *o*-Tolyl | $NO_2$ | $NO_2$ | 170° | 176, 180 |
| *p*-Tolyl | $NO_2$ | $NO_2$ | 162° | 176, 180 |

TABLE IX-32. 2-Anilinonicotinic Acids

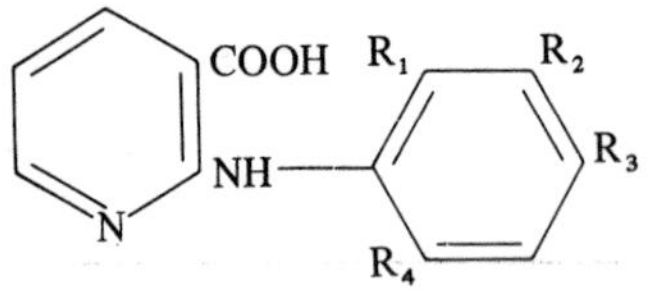

| $R_1$ | $R_2$ | $R_3$ | $R_4$ | m.p. | Ref. |
|---|---|---|---|---|---|
| Me | H | H | H | 107–8° | 151 |
| OH | H | H | H | 234–5° | 452 |
| H | $CF_3$ | H | H | 204° | 452 |
| H | H | MeO | H | 208° | 452 |
| H | H | EtO | H | 202° | 452 |
| Me | Me | H | H | 248° | 452 |
| | | | | acetyl deriv. | 452 |
| | | | | m.p. 195° | 452 |
| Me | H | Cl | H | 208–9° | 452 |
| OH | H | Cl | H | 276° | 452 |
| Me | H | H | Me | 219–20° | 452 |
| H | OH | $CO_2Me$ | H | 218° | 452 |
| Me | H | Me | H | 245–50° | 151 |

TABLE IX-33. 2-Dialkylaminoalkylaminopyridines

$R_3$, $R_4$, $R_2$, $R_5$, N, $NH(CH_2)_nN(R_1)_2$

| n | $R_1$ | $R_2$ | $R_3$ | $R_4$ | $R_5$ | Physical properties | Ref. |
|---|---|---|---|---|---|---|---|
| 2 | Me | H | H | H | Me | b.p. 105–15°/0.6 mm<br>di-HCl, m.p. 226–8° | 131<br>128 |
| 2 | Me | H | H | H | Cl | b.p. 124–6°/0.16 mm; picrate, m.p. 151–2° | 156 |
| 2 | Me | H | H | H | BuO | b.p. 127–9°/.05 mm; HCl, m.p. 104–5° | 156 |
| 2 | Et | $NH_2$ | H | H | H | b.p. 140°/0.05 mm | 80 |
| 2 | Et | H | H | H | Cl | b.p. 150–2°/0.18 mm; picrate, m.p. 118–19° | 156 |
| 2 | Et | H | H | H | BuO | b.p. 163–5°/0.2 mm; HCl, m.p. 76–7° | 156 |
| 2 | Et | $NO_2$ | H | H | H | b.p. 120°/0.05 mm<br>m.p. 58–9° | 80, 149<br>494 |
| 2 | Me | $NH_2$ | $NO_2$ | H | H | m.p. 137–8° | 494 |
| 2 | Me | $NO_2$ | $NO_2$ | H | H | HCl, m.p. 268–70° | 494 |
| 2 | Et | $NH_2$ | $NO_2$ | H | H | m.p. 83°<br>HCl, m.p. 200–5° | 80<br>494 |
| 2 | Et | $NO_2$ | $NO_2$ | H | H | m.p. 66°; HCl, m.p. 179–80° | 80, 149, 494 |
| 2 | Et | $NO_2$ | H | $NO_2$ | H | m.p. 66° | 181 |
| 3 | Me | H | H | H | Cl | b.p. 176–82°/11 mm | 657 |
| 3 | Me | H | H | $NO_2$ | H | m.p. 64° | 143 |
| 3 | Et | H | H | $NO_2$ | H | m.p. 78–80° | 143 |
| 3 | Morpholino | H | H | $NO_2$ | H | m.p. 102°; picrate, m.p. 112° | 143 |
| 3 | Piperidino | H | H | $NO_2$ | H | m.p. 80–2°; picrate, m.p. 195° | 143 |
| 3 | $CH_2CH_2Cl$ | H | H | H | H | m.p. 144–6° | 658 |

TABLE IX-34. Nuclear Substituted Secondary 2-Aminopyridines

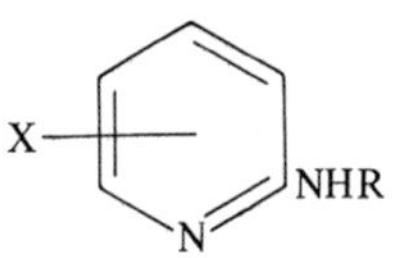


| R | Substituents | | | | Physical properties | Ref. |
|---|---|---|---|---|---|---|
| | 3 | 4 | 5 | 6 | | |
| Allyl | H | H | H | H | b.p. 56–8°/1 mm; picrate, m.p. 151–2° | 141 |
| Allyl | H | H | Cl | H | b.p. 98–9°/2 mm; picrate, m.p. 150–1° | 140 |
| Allyl | H | H | Br | H | m.p. 50°–50.7°; picrate, m.p. 170–70.5° | 141 |
| Allyl | Cl | H | Cl | H | b.p. 111–13°/6 mm; picrate, m.p. 201–3° | 140 |
| Allyl | Br | H | Br | H | b.p. 108–10°/1.5 mm picrate, m.p. 125.5–6.5° | 141 |
| Pr | H | H | Br | H | m.p. 40.3–41°; picrate, m.p. 146–7° | 141 |
| Bu | H | H | H | Cl | b.p. 99–105°/0.02 mm; benzoate, m.p. 91–2° | 156 |
| Bu | $NO_2$ | H | Me | Me | b.p. 26° | 608 |
| Benzyl | H | Isopentyl | H | H | b.p. 24°–50°/20 mm | 13 |
| Benzyl | H | Pentyl | H | H | b.p. 25°/23 mm | 13 |
| Benzyl | H | 5-Nonyl | H | H | b.p. 220–30°/10 mm | 13 |
| Benzyl | $NO_2$ | H | Me | Me | m.p. 104–5° | 608 |
| 5-Chloro-2-pyridyl | Me | H | H | H | | 170 |
| 6-(β-Hydroxy-β-phenylethyl-) | H | H | Me | H | m.p. 97–9°; HCl, m.p. 116–17° | 146 |
| 2-(β-Hydroxy-β-phenylethyl-) | H | H | Cl | H | m.p. 102–3°; HCl, m.p. 177–8° | 146 |
| 2-(β-Hydroxyl-β-phenylethyl-) | H | H | Br | H | m.p. 110–11°; HCl, m.p. 187–9° | 146 |
| 6-(β-Hydroxy-β-phenylethyl-) | Me | H | Me | H | m.p. 77–8°; HCl, m.p. 132–3° | 146 |
| 2-MeS-3-pyridyl | Me | H | $NO_2$ | H | | 170 |
| 2-MeS-3-pyridyl | $NO_2$ | H | Me | H | | 170 |
| *p*-Methoxybenzyl | H | Heptyl | H | H | b.p. 225–30°/2 mm | 13 |
| 3-Me-2-pyridyl | H | H | Cl | H | m.p. 68–9° | 723 |
| Phenethyl | $NO_2$ | H | Me | Me | m.p. 93° | 608 |
| Phenyl | Cl | H | H | H | m.p. 49–50° | 723 |
| 4-Pyridylmethyl | H | Me | H | H | | 659 |
| $CH_2CH_2N(CH_3)_2$ $C_6H_4OCH_3$-*p* | H | H | H | BuO | b.p. 195–200°/0.4 mm HCl, m.p. 129–30° | 156 |

TABLE IX-35. 2-[(*N*-Benzyl-*N*-(β-dialkylaminoethyl)] aminopyridines

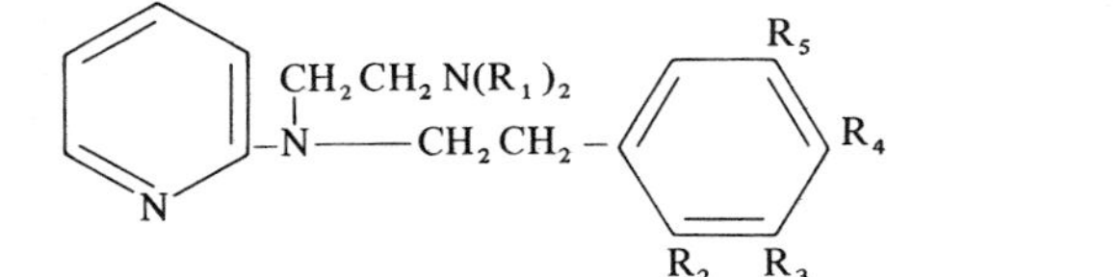


| $R_1$ | $R_2$ | $R_3$ | $R_4$ | $R_5$ | b.p./mm | HCl m.p. | di-HCl m.p. | Ref. |
|---|---|---|---|---|---|---|---|---|
| Et | Br | H | H | H | 210°/1 | 145° | – | 181, 660 |
| Et | EtO | H | H | F | 169°/0.5 | 118° | – | 654 |
| Et | MeO | H | H | F | 166°/0.1 | 138.5° | 121° | 654 |
| Et | H | Br | H | H | 210°/1 | – | – | 181, 660 |
| | | | | | perchlorate, m.p. 88° | | | 181 |
| | | | | | maleate, m.p. 110° | | | |
| Et | H | F | EtO | H | 180°/0.1 | 127° | 115° | 654 |
| Et | H | F | MeO | H | 178°/0.09 | 128° | 110° | 654 |
| Et | H | H | Br | H | 210°/1 | – | – | 181, 660 |
| | | | | | perchlorate, m.p. 123° | | | 181 |
| | | | | | maleate, m.p. 115° | | | |
| | | | | | citrate, m.p. 127° | | | |
| Et | H | H | BuO | H | 217–20°/1 | – | – | 171 |
| Et | H | H | isoBuO | H | 214–17°/1 | – | – | 171 |
| Et | H | H | EtO | H | 200–3°/1 | – | – | 171 |
| Et | H | H | MeO | H | 195–8°/1 | – | – | 171 |
| Et | H | H | $C_5H_{11}O$ | H | 221–4°/1 | – | – | 171 |
| Et | H | H | iso$C_5H_{11}O$ | H | 217–20°/1 | – | – | 171 |
| Et | H | H | PrO | H | 208–11°/1 | – | – | 171 |
| Et | H | H | isoPrO | H | 202–5°/1 | – | – | 171 |

TABLE IX-35. 2-[*N*-Benzyl-*N*-(β-dialkylaminoethyl)]aminopyridines (Continued)

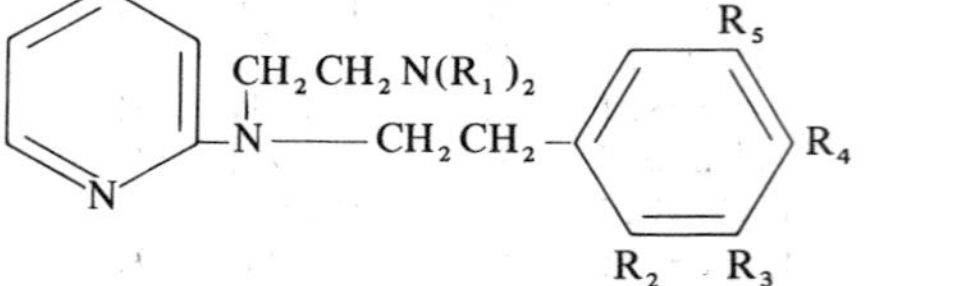

| $R_1$ | $R_2$ | $R_3$ | $R_4$ | $R_5$ | b.p./mm | HCl m.p. | di-HCl m.p. | Ref. |
|---|---|---|---|---|---|---|---|---|
| Me | BuO | H | H | F | 171°/0.5 | 105° | – | 654 |
| Me | EtO | H | H | F | 159°/0.6 | 152° | 116° | 654 |
| MeO | MeO | H | H | F | 163°/0.1 | 174° | 143° | 654 |
| Me | PrO | H | H | F | 164°/0.6 | 114° | – | 654 |
| Me | H | F | BuO | H | 182°/0.5 | 108° | – | 654 |
| Me | H | F | EtO | H | 161°/0.4 | 144° | 108° | 654 |
| Me | H | F | PrO | H | 170°/0.5 | 116.5° | 76° | 654 |
| Me | H | H | n-BuO | H | 221–4°/1 | – | – | 171 |
| Me | H | H | isoBuO | H | 218–21°/1 | – | – | 171 |
| Me | H | H | EtO | H | 202–5°/1 | – | – | 171 |
| Me | H | H | MeO | H | 198–210°/1 | – | – | 171 |
| Me | H | H | n-$C_5H_{11}O$ | H | 226–9°/1 | – | – | 171 |
| Me | H | H | iso$C_5H_{11}$ | H | 223–6°/1 | – | – | 171 |
| Me | H | H | PrO | H | 208–11°/1 | – | – | 171 |
| Me | H | H | isoPr | H | 204–7°/1 | – | – | 171 |
| Me | Me | H | Me | H | 160°/0.5 | – | 222° | 154 |
| Morpholino | MeO | H | H | F | 200°/0.6 | 171° | 138° | 654 |
| Morpholino | H | F | MeO | H | 208°/0.2 | 158.5° | 139° | 654 |
| Piperidino | MeO | H | H | F | 182°/0.4 | 179° | 126.5° | 654 |
| Piperidino | H | F | MeO | H | 202°/0.2 | 165° | 124° | 654 |

| | | | | | | | | |
|---|---|---|---|---|---|---|---|---|
| isoPr | MeO | H | H | F | 171°/0.5 | 143° | 122° | 654 |
| isoPr | H | F | MeO | H | 188°/0.2 | 140° | 117° | 654 |
| 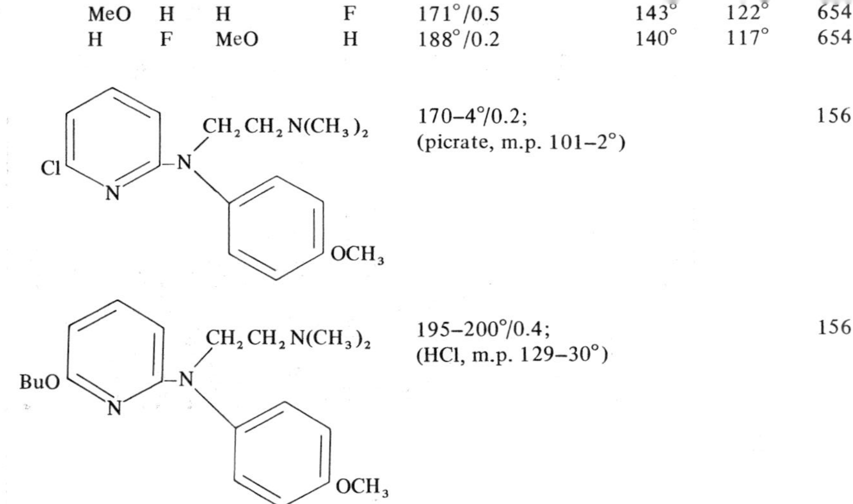 | | | | | 170–4°/0.2; (picrate, m.p. 101–2°) | | | 156 |
| | | | | | 195–200°/0.4; (HCl, m.p. 129–30°) | | | 156 |

TABLE IX-36. Other 2-*N*-Substituted-*N*-(β-dialkylaminoethyl)aminopyridines

$$\text{2-Pyridyl}-N(R_2)-CH_2CH_2N(R_1)_2$$

| $R_1$ | $R_2$ | Physical properties | Ref. |
|---|---|---|---|
| Et | *p*-Anisyl$COCH_2$ | b.p. 189–92° | 661 |
| Et | β-Hydroxy-β-Phenethyl | – | 183 |
| Me | 4-Bromophenethyl | b.p. 160–5°/0.1 mm;<br>HCl, m.p. 178° | 154 |
| Me | 5-Bromo-2-selenophenyl | b.p. 194°/4 mm; | 185 |
| Me | 4-Chlorophenethyl | b.p. 145–50°/0.1 mm;<br>HCl, m.p. 171° | 154, 182 |
| Me | 5-Chloro-l-indanyl | b.p. 160–70°/0.2 mm;<br>HCl, m.p. 191° | 154, 155 |
| Me | 5-Chloro-2-selenophenyl | b.p. 199–200°/4.5 mm | 185 |
| Me | 5-Chloro-2-thienyl | b.p. 162–6°/0.75 mm;<br>$n_D^{25}$ 1.5860;<br>citrate, m.p. 114–16° | 462 |
| Me | 4-Fluorophenethyl | b.p. 190–200°/10 mm,<br>b.p. 122°/0.1 mm;<br>di-HCl, m.p. 156° | 154<br>182 |
| Me | 1-Indanyl | b.p. 156°/0.2 mm;<br>HCl, m.p. 186° | 154, 155 |
| Me | α-Phenethyl | b.p. 140°/0.15 mm;<br>HCl, m.p. 160° | 182 |
| Me | 2-Thienyl | – | 462 |
| Me | α-(*p*-Tolyl)ethyl | b.p. 160°/0.5 mm;<br>di-HCl, m.p. 222° | 182 |

Other Tertiary 2-Aminopyridines

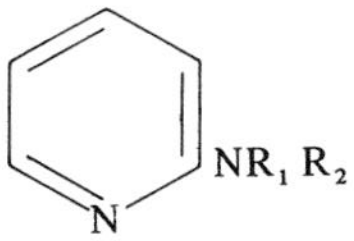

| $R_1$ | $R_2$ | Substituents 3 | 4 | 5 | 6 | Physical properties | Ref. |
|---|---|---|---|---|---|---|---|
| Me | Me | H | $NO_2$ | H | H | m.p. 99°, 98–9°; 1-oxide, m.p. 126° | 107, 177 |
| Me | Me | H | H | $NO_2$ | H | m.p. 154–5° | 656, 179 |
| | | | | | | m.p. 150–1° | 180 |
| Me | Me | H | H | H | 2-Amino-propyl | b.p. 94–6°/1.3 mm | 662 |
| Me | Me | H | H | H | Me | – | 662 |
| Me | Me | Me | H | $NO_2$ | H | m.p. 181–91° | 656 |
| Me | Me | $NO_2$ | H | $NO_2$ | H | m.p. 124–5°, 119°, | 175, 176 |
| | | | | | | m.p. 118–19° | 180 |
| Me | Me | Me | H | $NO_2$ | H | m.p. 181–91° | 656 |
| Me | Me | H | H | H | H | 2-*N*-oxide; m.p. 59–60°; picrate, m.p. 132–4° | 280 |
| Me | Ph | $NO_2$ | H | $NO_2$ | H | m.p. 134–35° | 180 |
| Me | Ph | H | H | $NO_2$ | H | m.p. 102–4° | 656 |
| Me | 2-Amino-3,5-dibromobenzyl | H | H | H | H | HCl, m.p. 198.5–201° HBr, m.p. 208–11° | 664 |
| Me | β-Hydroxyl-β-phenethyl | H | H | H | H | b.p. 147–57° | 146 |
| Me | 1-Pyrenylmethyl | H | H | H | H | m.p. 165–7°; m.p. 158–60° | 298, 653 |
| Et | Et | $NH_2$ | H | H | H | m.p. 260°; diazonium salt, m.p. 165° | 656 |

TABLE IX-37. Other Tertiary 2-Aminopyridines (Continued)

2-($NR_1R_2$)pyridine

| $R_1$ | $R_2$ | Substituents 3 | 4 | 5 | 6 | Physical properties | Ref. |
|---|---|---|---|---|---|---|---|
| Et | Et | $NO_2$ | H | H | H | m.p. 154–5° | 656 |
| Et | Et | H | $NO_2$ | H | H | m.p. 33°; 1-oxide, m.p. 90° | 107 |
| Et | Et | $NO_2$ | H | $NO_2$ | H | m.p. 62°, 59° | 181, 176 |
| Et | Ph | $NO_2$ | H | $NO_2$ | H | m.p. 135° | 176 |
| Bu | Bu | 2-Chloroethyl | Me | H | Cl | b.p. 175° | 663 |
| Bu | Bu | Et | Me | H | H | b.p. 132–4°/4 mm; picrate, m.p. 139–40° | 150 |
| Bu | Bu | H | Me | 2-Chloro-ethyl | H | b.p. 138–40°/13 mm; HCl, m.p. 133–4° | 150 |
| Bu | Bu | H | Me | Et | H | picrate, m.p. 143–4° | 150 |
| Bu | Bu | H | Me | Vinyl | Cl | b.p. 168°/6 mm | 150 |
| Ph | $(CH_2)_2$–N[CH($CH_3$)–$CH_2$Ph] | H | H | H | H | maleate, m.p. 136–8° | 322 |
| Ph | Ph | Vinyl | Me | H | Cl | b.p. 220–1°/0.5 mm m.p. 182–3° | 663 |
| Me | Ph | $NO_2$ | H | $NO_2$ | H | m.p. 134–5° | 180 |
| CN | Ph | H | Ph | H | H | m.p. 92° | 186 |
| CN | Ph | H | Me | H | Me | b.p. 120°/0.1 mm | 186 |
| CN | Ph | H | H | H | H | m.p. 52° | 186 |
| CN | o-$ClC_6H_4$ | H | H | H | H | m.p. 116° | 186 |
| CN | p-$ClC_6H_4$ | H | H | H | H | m.p. 105° | 186 |
| CN | 2,4-$Me_2C_6H_3$ | H | H | H | H | m.p. 58° | [illegible] |

ABLE IX-38. Secondary 3-Aminopyridines

$R_2$, $R_3$, $R_4$, $R_5$, $NHR_1$, N

| | $R_2$ | $R_3$ | $R_4$ | $R_5$ | Physical properties | Ref. |
|---|---|---|---|---|---|---|
| nzyl | $NO_2$ | H | H | H | 1-oxide, m.p. 189° | 178 |
| -(1,4-Benzodioxan-2-yl)ethyl] – | H | H | H | H | HCl, m.p. 104–5° | 666 |
| tyloxypropyl | H | H | H | H | b.p. 210–13°/10 mm; m.p. 30° | 514 |
| clohexyl | $NO_2$ | H | H | H | 1-oxide, m.p. 197–9° | 178 |
| Dimethylaminopropyl | 2-bromo-4-chlorophenylthio | H | H | H | b.p. 208–11°/0.3 mm | 449 |
| Dimethylaminopropyl | 2-bromophenylthio | H | H | H | b.p. 198–202°/0.3 mm | 448 |
| Dimethylaminopropyl | 2-bromo-4-chlorophenylthio | H | H | H | b.p. 210–15°/0.3 mm | 449 |
| 4-Dinitrophenyl | H | Cl | H | H | methyl sulfonyl, m.p. 211–13° | 168 |
| | H | $NO_2$ | H | H | m.p. 96° | 50 |
| | H | H | H | Me | b.p. 104–5°; picrate, m.p. 207–8° | 555 |
| Hydroxyethyl | H | $NO_2$ | H | H | m.p. 129° | 50 |
| e | Me | H | H | H | b.p. 116°/10 mm; picrate, 230°; HCl, m.p. 278–9° | 515, 665 |
| e | $NO_2$ | H | H | H | 1-oxide m.p. 227° | 152 |
| e | H | H | H | Me | m.p. 61–2°; picrate, m.p. 209° | 555 |
| e | H | Cl | H | Br | m.p. 43–4° | 41 |
| | H | $NO_2$ | Me | Me | 1-oxide, m.p. 146° | 52 |
| | H | H | Me | H | m.p. 126–6.5° | 158 |
| ı | H | H | H | Me | m.p. 102–3° | 158 |
| ı | H | H | H | H | m.p. 125.5–6° | 158 |
| 4,5-Trimethoxybenzyl | H | H | H | H | m.p. 109–10° | 72 |
| eratryl | H | H | H | H | | 72 |

TABLE IX-39. Diethyl (2,6-Disubstituted-3-pyridylamino)-methylene Malonates[407]

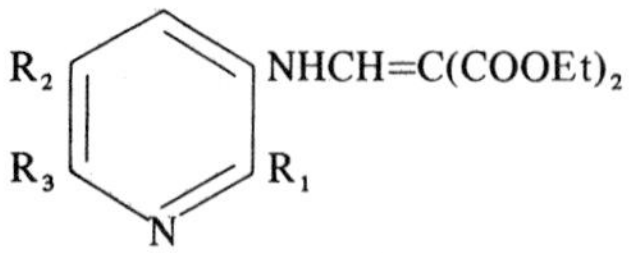

(Structure: pyridine ring with $R_1$ at C-2, $NHCH{=}C(COOEt)_2$ at C-3, $R_2$ at C-5, $R_3$ at C-6)

| $R_1$ | $R_2$ | $R_3$ | m.p. |
|---|---|---|---|
| Me | H | H | 100–100.8° |
| MeO | H | H | 81–3° |
| H | H | Me | 83.6–7.6° |
| Me | H | Et | 85.8–6.2° |
| Me | H | $CH_3CONH$ | 151.6–2.8° |
| Et | H | Et | 81.0–3.6° |
| $CH_3CONH$ | H | $CH_3CONH$ | 217–18° |
| Me | Me | Me | 108–9° |

TABLE IX-40. Tertiary 3-Aminopyridines[178]

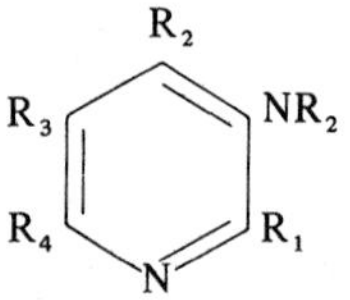

(Structure: pyridine ring with $R_1$ at C-2, $NR_2$ at C-3, $R_2$ at C-4, $R_3$ at C-5, $R_4$ at C-6)

| R | $R_1$ | $R_2$ | $R_3$ | $R_4$ | Physical properties |
|---|---|---|---|---|---|
| Me | H | $NO_2$ | H | H | 1-oxide, m.p. 146° |
| Et | H | $NO_2$ | H | H | 1-oxide, m.p. 73–4° |

TABLE IX-41. Secondary 4-Aminopyridines

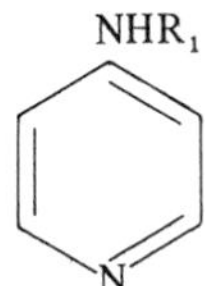


| $R_1$ | Physical properties | Ref. |
|---|---|---|
| *n*-Butyl | 1-oxide, picrate, m.p. 189–90° | 43 |
| 4-Carbethoxyphenyl | m.p. 197–8° | 164 |
| 4-Carboxyphenyl | m.p. above 250° | 164 |
| 4-Chlorophenyl | m.p. 245–7° | 164 |
| 2-(3,4-Dihydroxyphenyl)ethyl | HBr, m.p. 140° | 655 |
| 2-(3,4-Dimethoxyphenyl)ethyl | oil | 655 |
| Dodecyl | m.p. 83–4°; picrate, m.p. 94.2–5.4° | 147 |
| Ethyl | 1-oxide, m.p. 117–18°; picrate, m.p. 182–3° | 43 |
| Hydroxy | 1-oxide, m.p. 237° | 43 |
| 3-Hydroxyphenyl | m.p. 210–12° | 164 |
| 4-Methoxyphenyl | m.p. 178–9° | 164 |
| Methyl | b.p. 87–90°/6 mm; di-HCl, m.p. 271–2°; | 509, 510, 525 |
| | picrate, m.p. 122–4° | 118 |
| | 1-oxide, m.p. 192–4° | 118, 509, 510, 525 |
| β-Phenethyl | m.p. 39° | 655 |
| isoPr | m.p. 82–2.5°; picrate, m.p. 118–19° | 147 |
| *p*-$NH_2SO_2C_6H_4$ | m.p. 210–212° | 164 |
| $C_6H_5CH_2CO$ | oil | 655 |
| $C_6H_3$-3,4-$(OCH_3)_2CH_2CO$ | HCl, m.p. 203–5° | 655 |

TABLE IX-42. Nuclear Substituted Secondary 4-Aminopyridines

(Structure: pyridine ring with $NHR_1$ at position 4, substituents $R_2$, $R_3$, $R_4$, $R_5$)

| $R_1$ | $R_2$ | $R_3$ | $R_4$ | $R_5$ | Physical properties | Ref. |
|---|---|---|---|---|---|---|
| | Substituents | | | | | |
| Allyl | $CCl_3$ | Cl | Cl | H | m.p. 93–4° | 56 |
| 2-Aminoethyl | $CCl_3$ | Cl | Cl | H | m.p. 119–20° | 56 |
| 2-(2-Aminoethyl)-aminoethyl | $CCl_3$ | Cl | Cl | H | HCl, m.p. 170–200° | 56 |
| 6-Aminohexyl | $CCl_3$ | H | Cl | Cl | – | 56 |
| *o*-Aminophenyl | Me | H | H | Me | m.p. 180–1° | 105 |
| *o*-Aminophenyl | H | $NO_2$ | H | H | m.p. 169–74° | 667 |
| 3-(3-Aminopropyl)-aminopropyl | $CCl_3$ | Cl | Cl | H | – | 56 |
| Amyl | Cl | Cl | H | $CCl_3$ | – | 56 |
| *p*-Aminophenyl | Me | $NO_2$ | H | Me | m.p. 109–10° | 105 |
| Benzyl | $CCl_3$ | Cl | Cl | H | m.p. 63.5–4.5° | 56 |
| Bis(2-hydroxypropyl) | Cl | Cl | H | $CCl_3$ | – | 56 |
| Bu | $CCl_3$ | Cl | Cl | H | m.p. 36–8° | 56 |
| 3-Carboxyphenyl | H | $NO_2$ | H | H | m.p. 266° | 669 |
| Cyclohexyl | $CCl_3$ | Cl | Cl | H | m.p. 72–3° | 56 |
| 2-Diethylaminoethyl | H | $NH_2$ | H | H | b.p. 155–60°/0.07 mm | 80 |
| 2-Diethylaminoethyl | H | $NO_2$ | H | H | b.p. 141–3°/0.05 mm | 80, 494 |

| | | | | | | |
|---|---|---|---|---|---|---|
| 2-Diethylaminoethyl | H | Phenyl-acetamide | H | H | m.p. 135–7° | 80 |
| 2-Diethylaminoethyl | Cl | Cl | H | $CCl_3$ | | 56 |
| 2-Dimethylaminoethyl | H | $NH_2$ | H | H | b.p. 144–8°/0.05 mm | 494 |
| 2-Dimethylaminoethyl | H | $NO_2$ | H | H | m.p. 97–8.5° | 494 |
| 2-Dimethylaminoethyl | H | $NH_2$ | $NO_2$ | H | m.p. 98–100° | 494 |
| Dodecyl | Cl | Cl | H | $CCl_3$ | | 56 |
| Et | $CCl_3$ | Cl | Cl | H | | 56 |
| OH | Me | H | H | H | 1-oxide, m.p. 212–14° | 109 |
| OH | H | COOH | H | H | | 42 |
| OH | Me | H | H | Me | | 73 |
| OH | $C_6H_5$ | H | H | $C_6H_5$ | 1-oxide, m.p. 152° | 670 |
| OH | $CH_3$ | H | H | $NH_2$ | nitrate, m.p. 250–1°; picrate, m.p.182–3° | 43 |
| OH | H | H | H | H | 1-oxide, m.p. 237° | 108 |
| 2-Hydroxycyclo-hexyl | $CCl_3$ | Cl | Cl | H | | 56 |
| Hydroxyethyl | $CCl_3$ | Cl | Cl | H | m.p. 88–90° | 56 |
| 2-Hydroxypropyl | Cl | Cl | H | $CCl_3$ | | 56 |
| β-Hydroxy-β-phenethyl | H | $NO_2$ | H | H | m.p. 174° | 655 |
| Hexadecyl | Cl | Cl | Cl | $CCl_3$ | – | 56 |
| *p*-Iodophenyl | H | $NO_2$ | I | H | m.p. 149° | 669 |
| Me | H | Br | H | H | m.p. 92° | 43 |
| Me | H | $NO_2$ | H | H | m.p. 160° | 118, 152, 509, 510, 525 |
| Me | H | $CH_3$ | H | H | m.p. 125–6°; picrate, m.p. 199–200°; 1-oxide, m.p. 94.5–5.5° | 43 |
| Me | H | Et | H | H | m.p. 117–18° | 214 |
| Me | H | isoPr | H | H | m.p. 95–6° | 43 |

TABLE IX-42. Nuclear Substituted Secondary 4-Aminopyridines (Continued)

| Substituents | | | | | | |
|---|---|---|---|---|---|---|
| $R_1$ | $R_2$ | $R_3$ | $R_4$ | $R_5$ | Physical properties | Ref. |
| Me | H | Me | Me | H | m.p. 94.5–5.5°; picrate, m.p. 172–3° | 43, 158 |
| Me | $CCl_3$ | Cl | Cl | H | m.p. 140–2° | 56 |
| Me | Me | Me | Me | Me | m.p. 118–19°; picrate, m.p. 160–1°; 1-oxide picrate, m.p. 140–1° | 43 |
| Me | $CCl_3$ | Cl | Cl | Cl | 81–7° | 56 |
| 6-Me-2-pyridyl | $CCl_3$ | Cl | Cl | H | | 56 |
| Morpholino | $CCl_3$ | Cl | Cl | H | m.p. 108.5–12° | 56 |
| Octadecyl | $CCl_3$ | Cl | Cl | H | m.p. 30–2° | 56 |
| Octyl | $CCl_3$ | Cl | Cl | H | | 56 |
| Octyl | $CCl_3$ | Cl | H | Cl | | 56 |
| Ph | H | Br | $NO_2$ | H | m.p. 153.5–4° | 44 |
| Ph | H | Cl | $NO_2$ | H | m.p. 148–8.5° | 44 |
| Ph | H | $NO_2$ | I | H | m.p. 148.5–49° | 669 |
| Ph | Me | $NO_2$ | H | Me | m.p. 109–10° | 105 |
| Ph | Me | $NO_2$ | Cl | Me | m.p. 114–15° | 105 |
| 3-Phenylpropyl | H | $NO_2$ | H | H | HCl, m.p. 170–2° | 655 |
| Phenylpropylamino | $CCl_3$ | Cl | Cl | H | | 56 |
| Piperazino | $CCl_3$ | Cl | Cl | H | HCl, m.p. 192–5° (dec.) | 56 |

| | | | | | | |
|---|---|---|---|---|---|---|
| Propargyl | $CCl_3$ | Cl | Cl | H | m.p. 137–8° | 56 |
| 2-Pyridyl | Cl | H | Cl | $CCl_3$ | | 56 |
| 2-Pyrimidinyl | Cl | Cl | H | $CCl_3$ | | 56 |
| Pyrrolidino | Cl | Cl | Cl | $CCl_3$ | m.p. 98–103° | 56 |
| $CH_2-CH_2R$ | | | | | | |
| $R$ = | | | | | | |
| *m*-Anisyl | H | $NO_2$ | H | H | HCl, m.p. 223° (dec.) | 655 |
| *p*-Anisyl | H | $NO_2$ | H | H | m.p. 99° | 655 |
| 3-Benzyloxy-4-methoxyphenyl | H | $NO_2$ | H | H | m.p. 123° | 655 |
| 3,4-Dichlorophenyl | H | $NO_2$ | H | H | m.p. 189–90° | 655 |
| 3-Indolyl | H | $NO_2$ | H | H | m.p. 193° | 655 |
| α-Naphthyl | H | $NO_2$ | H | H | m.p. 180° | 655 |
| β-Naphthyl | H | $NO_2$ | H | H | m.p. 114° | 655 |
| *N*-Phenyl-*N*-piperazinyl | H | $NO_2$ | H | H | m.p. 115–17° | 655 |

TABLE IX-43. Tertiary 4-Aminopyridines

$NRR_1$

| R | $R_1$ | 2 | 3 | 5 | 6 | Physical properties | Ref. |
|---|---|---|---|---|---|---|---|
| *n*-Bu | *n*-Bu | H | H | H | H | b.p. 124–32°/1.3 mm;<br>picrate, m.p. 124° | 147 |
| $CH_2CN$ | $CH_2CN$ | H | H | H | H | m.p. 208–9° | 188 |
| Me | Me | Ph | H | H | H | m.p. 83–4° | 187 |
| Me | Me | H | Br | H | H | b.p. 82–4°/0.5 mm;<br>picrate, m.p. 182–3°;<br>1-oxide picrate, m.p. 160–1° | 43 |
| Me | Me | H | Et | H | H | b.p. 82–3°/0.8 mm;<br>picrate, m.p. 118–19°;<br>1-oxide, b.p. 178–80°/1 mm;<br>m.p. 139–40° | 43 |
| Me | Me | H | isoPr | H | H | b.p. 79–80°/0.45 mm;<br>picrate, m.p. 138–9°;<br>1-oxide picrate, m.p. 151–2° | 43 |
| Me | Me | H | Me | H | H | b.p. 73–5°/1 mm;<br>picrate, m.p. 172–3°;<br>1-oxide, b.p. 142–4°/0.15 mm;<br>1-oxide picrate, m.p. 130–1° | 43, 214<br><br><br>43 |
| Me | Me | H | Me | Me | H | b.p. 69–70°;<br>picrate, m.p. 172–73° | 43, 214 |
| Me | Me | $CCl_3$ | Cl | Cl | H | m.p. 72–3° | 56 |

| | | | | | | | |
|---|---|---|---|---|---|---|---|
| Me | Me | Cl | Cl | Cl | $CCl_3$ | m.p. 52.5–4.5° | 56 |
| Me | Me | H | H | H | H | m.p. 110–12° | 147 |
| | | | | | | m.p. 112–13° | 118 |
| | | | | | | m.p. 114° | 187 |
| | | | | | | 1-oxide, m.p. 97–9° | 281 |
| | | | | | | picrate, m.p. 204° | 118 |
| | | | | | | picrate, m.p. 206–8° | 147 |

TABLE IX-44. 2-Nitraminopyridines

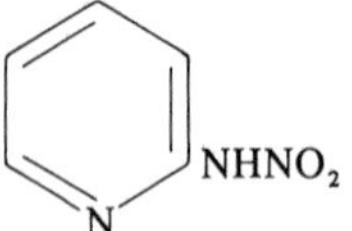

| Substituents | | | | | |
|---|---|---|---|---|---|
| 3 | 4 | 5 | 6 | Physical properties | Ref. |
| Br | H | H | H | m.p. 181° | 46[illegible] |
| $NO_2$ | H | H | H | m.p. 137° | 46[illegible] |
| H | $NO_2$ | H | H | m.p. 154° | 46[illegible] |
| H | H | $NO_2$ | H | m.p. 161° | 46[illegible] |
| H | H | H | Pr | m.p. 97° | 48[illegible] |
| Me | H | $NO_2$ | H | m.p. 165° | 46[illegible] |
| H | Me | H | Me | m.p. 154° (dec.) | 48[illegible] |
| H | H | Me | Me | m.p. 141–2° | 60[illegible] |
| 1-Me-2-pyrrolidinyl | H | 1-Me-2-pyrrolydinyl | H | | 60[illegible] |
| $NO_2$ | H | Me | H | m.p. 62° | 46[illegible] |
| $NO_2$ | Me | H | H | m.p. 162° | 46[illegible] |
| H | Me | $NO_2$ | H | m.p. 163° | 46[illegible] |
| H | $NO_2$ | $NO_2$ | H | m.p. 105° | 46[illegible] |
| H | H | $NO_2$ | Me | m.p. 116° | 46[illegible] |
| H | Me | Br | Me | m.p. 159–60° | 63[illegible] |

TABLE IX-45. 3-Nitraminopyridines

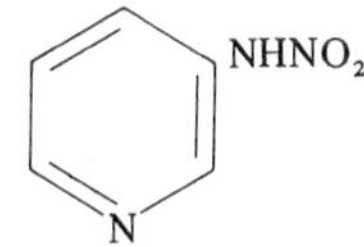


| Substituents | | | | | |
|---|---|---|---|---|---|
| 2 | 4 | 5 | 6 | Physical properties | Ref. |
| H | Me | H | H | m.p. 184° | 466, 468 |
| H | H | Me | H | m.p. 172° | 466 |
| H | $NO_2$ | H | H | m.p. 104°; 1-oxide, m.p. 108° | 465 |
| H | H | H | $NO_2$ | m.p. 113° (explosive dec.) | 470 |
| H | H | H | Cl | m.p. 179° | 60 |
| H | H | EtO | H | m.p. 160° | 468 |
| COOH | H | H | H | m.p. 186° | 468 |
| H | H | COOH | H | m.p. 201° | 468 |
| $SO_3H$ | H | H | H | m.p. 184° (dec.) | 467 |
| H | H | H | $SO_3H$ | m.p. 217–20° (dec.) | 467 |
| Me | H | H | Me | m.p. 162° | 351 |
| Me | Me | H | Me | m.p. 162° | 351 |

TABLE IX-46. 4-Nitraminopyridines

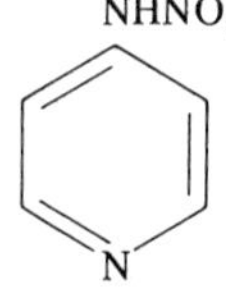


| Substituents | | | | | |
|---|---|---|---|---|---|
| 2 | 3 | 5 | 6 | m.p. | Ref. |
| Br | H | H | H | 181° | 375 |
| Cl | H | H | H | 148° | 465 |
| I | H | H | H | 195° | 465 |
| H | Br | H | H | 203° | 465 |
| H | Cl | H | H | 206° | 465 |
| H | I | H | H | 204° | 465 |
| H | Me | H | H | 212° | 469 |
| H | $NO_2$ | H | H | 204° | 465 |
| Me | $NO_2$ | H | H | 187° | 465 |
| Cl | H | H | Cl | | 87 |

TABLE IX-47. 2-Azopyridines

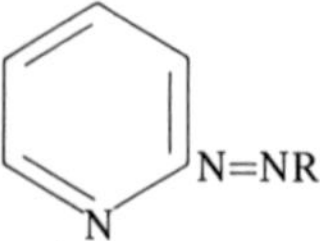

| Substituents | | | | | |
|---|---|---|---|---|---|
| R | 3 | 5 | 6 | Physical properties | Ref. |
| 4-Amino-1-naphthyl | H | H | H | 1-oxide, m.p. 192–3° | 673 |
| *p*-Anilino | H | H | H | 1-oxide | 370, 671, 672 |
| *p*-Dimethyl-aminophenyl | H | H | H | m.p. 142–3° | 367 |
| 2-Hydroxy-1-naphthyl | H | H | H | | 369 |
| Ph | $NH_2$ | $NH_2$ | MeO | di-HCl, m.p. 141° | 674 |
| Ph | $NH_2$ | $NH_2$ | EtO | di-HCl, m.p. 119° | 674 |
| Ph | $NH_2$ | $NH_2$ | PrO | di-HCl, m.p. 122° | 674 |
| Ph | $NH_2$ | $NH_2$ | BuO | di-HCl, m.p. 126° | 674 |
| Ph | $NH_2$ | $NH_2$ | $C_5H_{11}O$ | di-HCl, m.p. 96° | 674 |

ABLE IX-48. 3-Azopyridines

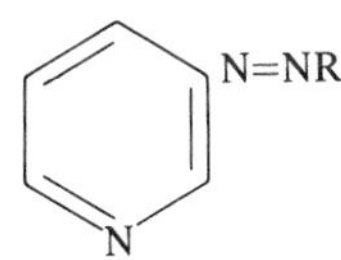


| | Substituents | | | | | |
|---|---|---|---|---|---|---|
| | 2 | 4 | 5 | 6 | Physical properties | Ref. |
| $_1$ = *Naphthyl* | | | | | | |
| ·$NH_2$-1-$R_1$ | H | H | H | H | m.p. 185–7° | 675 |
| ·$NH_2$-1-$R_1$ ·$NH_2$-8-Cl | H | H | H | Cl | – | 673 |
| 2-Me-1-$R_1$ | H | H | H | H | – | 673 |
| *ienyl* | | | | | | |
| ·Chlorophenyl | $NH_2$ | H | H | H | – | 676, 677 |
| ·Chlorophenyl | $NH_2$ | H | H | $NH_2$ | – | 676, 677 |
| (2-Diethylaminoethyl-amino)phenyl | OH | $NH_2$ | H | OH | m.p. 315° (dec.) | 390 |
| | $NH_2$ | H | H | $NH_2$ | m.p. 234° | 675 |
| ·(2-Diethylamino-ethoxy)phenyl | $NH_2$ | H | H | $NH_2$ | HCl, m.p. 243° | 679 |
| ·Iodophenyl | $NH_2$ | H | H | $NH_2$ | m.p. 204.5–5° | 623 485 |
| ,4,6-Tri-iodophenyl | $NH_2$ | H | H | $NH_2$ | – | 485 |
| h | $NH_2$ | H | H | $NH_2$ | – | 678 |
| h | Me | Me | H | $NH_2$ | m.p. 131–2° | 10 |
| $_2$=*Pyridyl* | | | | | | |
| ′-BuO-3′-$R_2$ | $NH_2$ | H | H | $NH_2$ | – | 680, 681 |
| ′-BuO-3′-$R_2$ | $NH_2$ | H | H | $NH_2$ | – | 330 |
| ′-Cl-5′-I-3′-$R_2$ | $NH_2$ | H | H | $NH_2$ | m.p. 270° | 623 |
| ′-Cl-3′-$R_2$ | H | Cl | H | H | – | 60 |
| ′-Cl-3′-$R_2$ | H | H | Cl | H | m.p. 183° | 60 |

TABLE IX-49. 4-Azopyridines

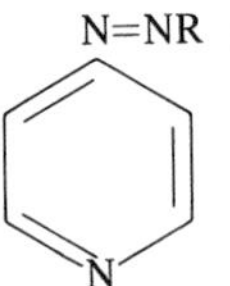


| R | Substituents 2 | 3 | 5 | 6 | Physical properties | Ref. |
|---|---|---|---|---|---|---|
| 4-Amino-1-naphthyl | H | H | H | H | m.p. 164–6° | 673 |
| 4-Dimethylaminophenyl | H | H | H | H | m.p. 207–9° | 367 |
| 2′,6′-$Me_2$-4′-pyridyl | Me | H | H | H | m.p. 124–6° | 74 |
| 3′-Me-4′-pyridyl | H | Me | H | H | m.p. 150° | 86 |
| 4′-Pyridyl | H | H | H | H | m.p. 106–7.5° | 509 |

TABLE IX-50. 2-, 3-, and 4-Arylazoaminopyridines

| Substituted pyridine | $R_1$ | $R_2$ | $R_3$ | Physical properties | Ref. |
|---|---|---|---|---|---|
| 2-Pyridyl | H | H | H | – | 652 |
| 2-Pyridyl | $NO_2$ | H | H | m.p. 185–6° | 652 |
| 2-Pyridyl | H | $NO_2$ | H | m.p. 212° | 652 |
| 2-Pyridyl | H | H | $NO_2$ | m.p. 246°, 249° | 652 |
| 3-Me-2-pyridyl | H | H | $(CH_3)_2N$ | m.p. 193–4° | 367 |
| 4-Me-2-pyridyl | H | H | $(CH_3)_2N$ | m.p. 184–5° | 367 |
| 5-Me-2-pyridyl | H | H | $(CH_3)_2N$ | m.p. 201–2° | 367 |
| 6-Me-2-pyridyl | H | H | $(CH_3)_2N$ | m.p. 170–1° | 367 |
| 3-Pyridyl | $NO_2$ | H | H | m.p. 128° | 652 |
| 3-Pyridyl | H | $NO_2$ | H | m.p. 205°, 212° | 652 |
| 3-Pyridyl | H | H | $NO_2$ | m.p. 226° | 652 |
| 3-Pyridyl | H | H | $(CH_3)_2N$ | m.p. 189–91° | 367 |
| 4-Pyridyl | H | H | $(CH_3)_2N$ | m.p. 210–11° | 367 |
| 2-Me-4-pyridyl | H | H | $(CH_3)_2N$ | m.p. 191–2° | 367 |
| 3-Me-4-pyridyl | H | H | $(CH_3)_2N$ | m.p. 193–4° | 367 |
| 2,6-$Me_2$-4-pyridyl | H | H | $(CH_3)_2N$ | m.p. 197–8° | 367 |

TABLE IX-51. 2- and 4-Hydrazinopyridines

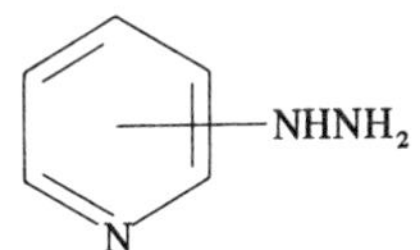

| Hydrazino | Substituents 2 | 3 | 4 | 5 | 6 | Physical properties | Ref. |
|---|---|---|---|---|---|---|---|
| 2 | – | H | H | $NO_2$ | H | m.p. 204° | 465 |
| 2 | – | Br | H | $NO_2$ | H | m.p. 170° | 386 |
| 2 | – | $NO_2$ | H | Br | H | m.p. 138° | 386 |
| | | | | | | acetyl deriv., m.p. 172° | 386 |
| 2 | – | $NO_2$ | H | Me | H | m.p. 166° | 465 |
| 2 | – | Me | H | $NO_2$ | H | m.p. 177° | 465 |
| 2 | – | H | Me | $NO_2$ | H | m.p. 192° | 465 |
| 2 | – | H | $NHNO_2$ | $NO_2$ | H | m.p. 218° | 465 |
| 2 | – | H | H | $NO_2$ | Me | m.p. 122° | 465 |
| 2 | – | $NO_2$ | Me | H | Me | m.p. 153° | 488 |
| 2 | – | H | Me | $NO_2$ | Me | m.p. 145–6° | 488 |
| 2 | – | H | H | H | H | picrate, m.p. 162–3° | 118 |
| 4 | H | $NO_2$ | – | H | H | m.p. 207° | 465 |
| 4 | Br | H | – | $NO_2$ | H | m.p. 152° (dec.) | 628 |
| 4 | H | Cl | – | $NO_2$ | H | m.p. 199° | 44 |
| 4 | $CCl_3$ | Cl | – | Cl | H | m.p. 171–2° | 56 |
| 4 | Me | $NO_2$ | – | H | Me | m.p. 117–8° | 105 |
| 4 | F | F | – | F | F | m.p. 56–7° | 53 |
| 4 | Me | $NO_2$ | – | Cl | Me | m.p. 171–3° | 105 |
| 4 | H | H | – | H | H | HCl, m.p. 242–4° | 118 |

TABLE IX-52. Alkyl 2,3-Diaminopyridines

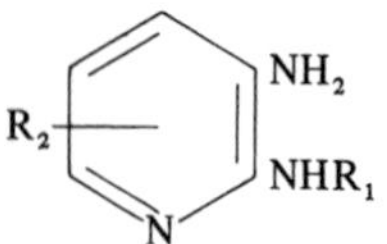


| Substituents | | | | | |
|---|---|---|---|---|---|
| | $R_2$ | | | | |
| $R_1$ | 4 | 5 | 6 | Physical properties | Ref. |
| Benzyl | H | Me | Me | m.p. 58–9° | 608 |
| *n*-Bu | H | Me | Me | m.p. 183–4° | 608 |
| 2-(Diethylamino)-ethyl | H | $NO_2$ | H | m.p. 83°; HCl, m.p. 200–5° | 149 |
| 2-(Diethylamino)-ethyl | H | H | H | b.p. 140°/0.05 mm | 149 |
| 3,4-Difluorophenyl | H | Me | Me | m.p. 123–4° | 608 |
| *p*-Fluorophenyl | H | Me | Me | m.p. 90–1° | 608 |
| β-Hydroxy-β-phenethyl | H | H | H | di-HCl, m.p. 157–8° | 655 |
| Phenethyl | H | Me | Me | m.p. 96–7° | 608 |
| Phenethyl | H | H | H | HCl, m.p. 144° | 682 |
| Ph | H | H | MeO | m.p. 210° (dec.) | 723 |
| Ph | H | Me | Me | m.p. 69–70° | 608 |
| H | H | H | Pr | picrate, m.p. 193.5–195° | 480 |
| H | Me | H | Me | m.p. 183° | 488 |
| H | H | Me | Me | m.p. 179–81° | 608 |
| $CH_2CH_2R_3$ $R_3$= | | | | | |
| 3,4-Dihydroxyphenyl | H | H | H | di-HBr, 174° | 655 |
| 3,4-Dimethoxyphenyl | H | H | H | $HCl.H_2O$, m.p. 162–4° | 655 |
| 4-Dimethylamino-phenyl | H | carboxy-methyl | H | m.p. 134–5° | 655 |
| 3-Indolyl | H | H | H | di-HCl, m.p. 240° | 655 |
| α-Naphthyl | H | H | H | HCl, m.p. 175° | 655 |
| 3,4-Xylyl | H | H | H | HCl, m.p. 168° | 655 |

TABLE IX-53. Halo 2,3-Diaminopyridines

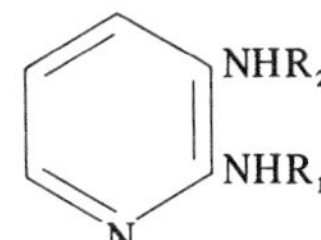


| Substituents | | | | | | |
|---|---|---|---|---|---|---|
| $R_1$ | $R_2$ | 4 | 5 | 6 | Physical properties | Ref. |
| Benzyl | H | H | Cl | H | m.p. 113° | 683 |
| Diethylaminoethyl | H | H | Br | H | b.p. 70°/1 mm | 684 |
| 2-Dimethylamino-ethyl | H | H | Cl | H | picrate, m.p. 202° | 683 |
| Et | H | H | Cl | H | m.p. 107–8° | 683 |
| Me | H | H | Br | H | m.p. 126° | 684 |
| Me | H | H | Cl | H | m.p. 132° | 683 |
| Ph | H | H | Cl | H | m.p. 144–5° | 723 |
| Ph | H | H | H | Cl | m.p. 232–3° | 723 |
| H | Me | H | Cl | H | m.p. 122–4° | 41 |
| H | H | H | Br | H | m.p. 158–60° | 41 |
| H | H | H | Cl | H | – | 477 |
| H | H | H | Cl | Cl | m.p. 167° | 477 |
| H | H | Me | Br | Me | m.p. 186°; picrate, m.p. 217° | 630 |

TABLE IX-54. Nitro 2,3-Diaminopyridines

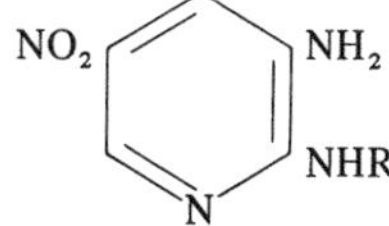


| R | Physical properties | Ref. |
|---|---|---|
| H | m.p. 260° | 45, 478, 685 |
| | m.p. 256° | 493 |
| 2-Diethylaminoethyl | m.p. 83°; HCl, m.p. 200–5° | 80 |
| 3,4-Dimethoxyphenethyl | m.p. 174–6° | 685 |
| 2-Dimethylaminoethyl | m.p. 137–8° | 494 |
| Me | m.p. 222–4° | 493 |
| 2-Methylphenethyl | HCl, m.p. 208–11° | 685 |
| Phenethyl | m.p. 136° | 682 |

TABLE IX-55. 2,4-Diaminopyridines

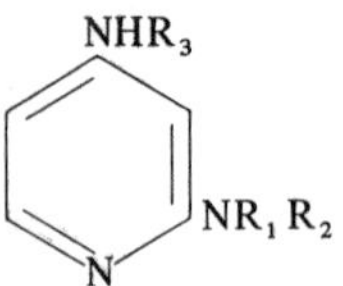


| | | | Substituents | | | | |
|---|---|---|---|---|---|---|---|
| $R_1$ | $R_2$ | $R_3$ | 3 | 5 | 6 | Physical properties | Ref. |
| H | H | H | H | H | $NO_2$ | m.p. 197.5–98° | 87, 628 |
| H | H | H | H | H | OH | m.p. 178° | 390 |
| H | H | H | $NO_2$ | H | H | m.p. 212° | 375 |
| Me | Me | H | H | H | H | m.p. 151°; picrate, m.p. 216° | 107, 177 |
| Et | Et | H | H | H | H | m.p. 117°; picrate, m.p. 172° | 107, 177 |
| H | H | Me | H | Cl | H | m.p. 122–4° | 41 |
| H | H | H | $NO_2$ | H | Pr | m.p. 145.5° | 480 |
| H | H | H | F | F | F | m.p. 111–12° | 252 |

TABLE IX-56. 5-Amino-2-anilinopyridines

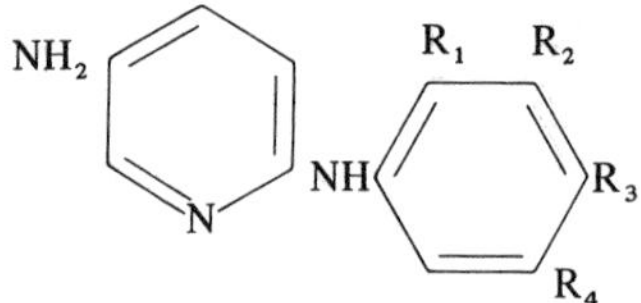


| $R_1$ | $R_2$ | $R_3$ | $R_4$ | Physical properties | Ref. |
|---|---|---|---|---|---|
| H | H | H | H | HCl, m.p. 219–23° | 656, 686, 723 |
| H | H | F | H | m.p. 141° | 723 |
| H | H | Me | H | HCl, m.p. 237–42° | 656 |
| H | H | $NMe_2$ | H | HCl, m.p. 245° | 656 |
| H | H | $C_5H_{11}O$ | H | b.p. 225–35°/ 0.5 mm | 723 |
| H | $CF_3$ | H | H | m.p. 115° | 723 |
| Me | H | H | H | b.p. 178–85°/ 0.2 mm | 723 |
| MeO | H | H | Cl | b.p. 190–5° | 723 |
| Me | Me | H | H | b.p. 200–5°/0.7 mm | 723 |
| | | | | m.p. 105° | 723 |

TABLE IX-57. Other 2,5-Diaminopyridines

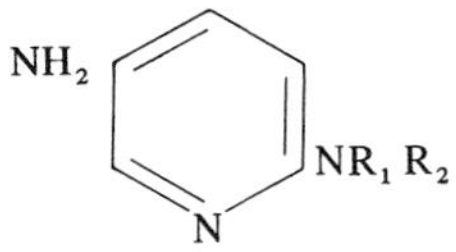

| Substituents | | | | | | |
|---|---|---|---|---|---|---|
| $R_1$ | $R_2$ | 3 | 4 | 6 | Physical properties | Ref. |
| H | H | Carboxy | H | H | m.p. 302° | 9 |
| H | H | H | Me | H | – | 483 |
| H | H | H | H | Me | – | 483 |
| H | H | H | Me | Me | m.p. 175–7° | 10 |
| H | Benzoyl | H | H | H | m.p. 141–2° | 687 |
| H | Benzoyl | H | H | H | HCl, m.p. 210–20° | 656 |
| H | Cyclohexyl | H | H | H | – | 483 |
| H | 3-Dimethyl-aminopropyl | H | H | H | – | 483 |
| H | β-Hydroxy-β-phenethyl | H | H | H | m.p. 179–80° | 655 |
| H | Me | H | H | H | HCl, m.p. 239–41° | 656 |
| H | MeO | H | H | Me | m.p. 126–8° | 669 |
| H | 2-Methoxy-propyl | H | H | H | – | 483 |
| H | 3-Methoxy-propyl | H | H | H | – | 483 |
| H | Morpholino | H | H | H | HCl, m.p. 250° | 656 |
| H | Piperazino | H | H | H | HCl, m.p. 240–5° | 656 |

TABLE IX-57. Other 2,5-Diaminopyridines (Continued)

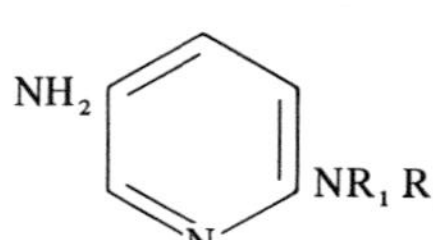


| Substituents | | | | | | |
|---|---|---|---|---|---|---|
| $R_1$ | $R_2$ | 3 | 4 | 6 | Physical properties | Ref. |
| H | 2,2'-(*p,p'*-Methylenedianilino)bis | H | H | H | HCl, m.p. 235° | 656 |
| H | 2-[(2-nitro-2-furyl)vinyl]- | H | H | H | m.p. 255–8°; HCl, m.p. 235–45° | 317 |
| Me | Me | H | H | H | di-HCl, m.p. 244–7°; picrate, m.p. 246–8°; HCl, m.p. 260° | 179, 656 483, 686 179, 656, 686 |
| Me | Me | Me | H | H | HCl, m.p. 221–4° | 656 |

TABLE IX-58. 2,6-Diaminopyridines

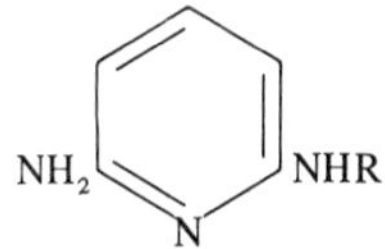


| Substituents | | | | | |
|---|---|---|---|---|---|
| R | 3 | 4 | 5 | Physical properties | Ref. |
| H | Me | H | H | m.p. 156–7° | 4 |
| H | $NO_2$ | H | H | – | 484 |
| H | Ph | H | H | – | 688 |
| H | PhN=N | H | H | acetyl, m.p. 208° | 181 |
| H | H | EtO | H | – | 482, 689, 690 |
| H | H | MeO | H | – | 482, 689, 690 |
| H | H | PrO | H | – | 482 |
| H | H | IsoPro | H | m.p. 150–2° | 482 |
| H | H | BuO | H | m.p. 146.5–7° | 481, 482 |
| H | H | Hexyloxy | H | m.p. 132–4° | 482 |
| H | H | $C_5H_{11}O$ | H | m.p. 135–7° | 482 |
| H | H | iso$C_5H_{11}O$ | H | m.p. 129–31° | 482 |
| H | H | $CH_2$=$CHCH_2O$ | H | m.p. 114.5–16° | 691 |
| H | H | $CH_3OCH_2CH_2O$ | H | m.p. 97–9° | 691 |
| H | H | $C_6H_5O$ | H | m.p. 195° | 691 |
| H | H | $C_6H_5CH_2O$ | H | m.p. 166–68.5° | 691 |
| H | H | $C_6H_5CH_2CH_2O$ | H | m.p. 133–5° | 691 |
| H | H | $C_6H_5CH_2CH_2CH_2O$ | H | m.p. 138° | 691 |
| H | H | $C_6H_5CH$=$CHCH_2O$ | H | m.p. 146–7° | 691 |
| H | H | EtS | H | m.p. 165° | 691 |
| H | H | *n*-$C_4H_2S$ | H | m.p. 157–9° | 691 |
| H | H | PhS | H | m.p. 108–10° | 691 |
| H | H | *p*-$ClC_6H_4S$ | H | m.p. 133–5° | 691 |
| H | H | $C_6H_5CH_2S$ | H | m.p. 202–3° | 691 |
| H | H | $C_6H_5Se$ | H | m.p. 99–100° | 691 |
| H | H | $EtSO_2$ | H | m.p. 170–71.5° | 691 |
| H | H | *n*-$BuSO_2$ | H | m.p. 161–3° | 691 |
| H | H | $PhSO_2$ | H | m.p. 285° | 691 |
| H | I | H | I | – | 485 |
| H | Me | H | Me | m.p. 186–7° | 4 |
| H | $NO_2$ | $Ph_2CH$ | H | m.p. 260° | 692 |
| Ph | Phenyl-aminoethyl | $Bu_2N$ | H | b.p. 202–4°/0.5 mm | 150 |

TABLE IX-59. Alkyl and Aryl 3,4-Diaminopyridines

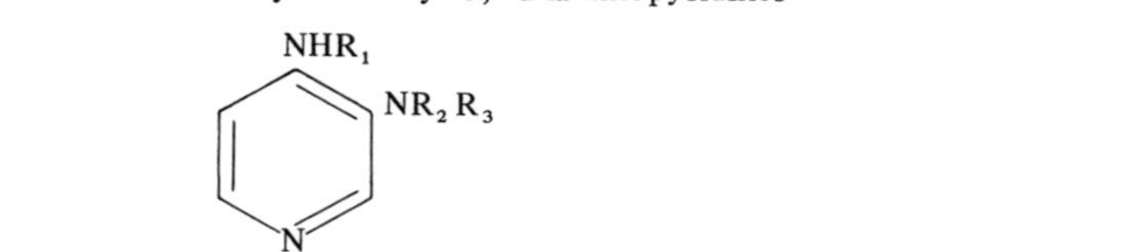


| $R_1$ | $R_2$ | $R_3$ | Substituents 2 | 5 | 6 | Physical properties | Ref. |
|---|---|---|---|---|---|---|---|
| *o*-Aminophenyl | H | H | H | H | H | m.p. 141–2° | 667 |
| *p*-Anisyl | H | H | H | H | H | m.p. 175–6° | 667 |
| *p*-Chlorophenyl | H | H | H | H | H | m.p. 172–3° | 667 |
| 3,4-Dimethoxyphen-ethyl | H | H | H | H | H | di-HCl, m.p. 202° | 682, 693 |
| 2-Dimethylaminoethyl | H | H | H | H | H | b.p. 144–8°/0.05 mm | 494 |
| Methyl | H | H | H | H | H | m.p. 169–70°; picrate, m.p. 185°; di-HCl, m.p. 227–8° | 152 |
| α-Methylphenethyl | H | H | H | H | H | b.p. 210–20°/0.001 mm | 682, 693 |
| Phenethyl | H | H | H | H | H | HCl, m.p. 174–5° | 682, 693 |
| Ph | H | H | Me | H | Me | m.p. 174° | 44 |
| Ph | H | H | Me | Cl | Me | m.p. 199–200° | 44 |
| 3-Phenylpropyl | H | H | H | H | H | di-HCl, m.p. 189–92° | 655 |
| H | H | Cyclohexyl | H | H | H | m.p. 164°; picrate, m.p. 199° | 178 |
| H | H | 2-Diethylaminoethyl | H | H | H | m.p. 103° | 682 |
| H | H | Me | H | H | H | m.p. 114°; picrate, m.p. 233° | 152, 225, 238 |
| H | H | Phenethyl | H | H | H | HCl, m.p. 174–5° | 682, 693 |

| | | | | | | | |
|---|---|---|---|---|---|---|---|
| 2-Diethylaminoethyl | H | Phenylacetamido | H | H | H | m.p. 135–7° | 80 |
| H | Me | Me | H | H | H | m.p. 79–80°; picrate, m.p. 185° | 178 |
| H | H | H | H | Me | H | m.p. 149°; picrate, m.p. 198° | 469 |
| H | H | H | Me | H | Me | m.p. 181°; picrate, m.p. 215° | 52, 109, **469** |
| H | H | H | Cl | Me | H | – | 469 |
| *$CH_2CH_2R$* | | | | | | | |
| *R* = | | | | | | | |
| *m*-Anisyl | H | H | H | H | H | di-HCl, m.p. 185° | 655 |
| *p*-Anisyl | H | H | H | H | H | di-HCl, m.p. 188–9° | 655 |
| 3,4-Dichlorophenyl | H | H | H | H | H | m.p. 135–6°; di-HCl, m.p. 274° | 655 |
| *p*-Hydroxyphenyl | H | H | H | H | H | di-HBr, m.p. 217–8° | 655 |
| 3-Hydroxy-4-methoxyphenyl | H | H | H | H | H | $HCl \cdot H_2O$, m.p. 170° | 655 |
| *β*-Hydroxy-*β*-phenethyl | H | H | H | H | H | di-HCl, m.p. 178–9° (dec.) | 655 |
| 3-Indolyl | H | H | H | H | H | di-HCl, m.p. 198° | 655 |
| *α*-Naphthyl | H | H | H | H | H | di-HCl, m.p. 218° | 655 |
| *β*-Naphthyl | H | H | H | H | H | $HCl \cdot H_2O$, m.p. 243° | 655 |
| *N*-Phenyl-*N*-piperazinyl | H | H | H | H | H | tri-HCl, m.p. 250–2° | 655 |
| $CH_2CH_2$– ; $OCH_2Ph$ ; OMe | H | H | H | H | H | di-HCl, m.p. 177° | 655 |
| H | H | 3,4-Dimethoxyphenyl | H | H | H | oil | 655 |
| H | H | 3,4-Dihydroxyphenyl | H | H | H | | 655 |

TABLE IX-60. Other 3,4-Diaminopyridines

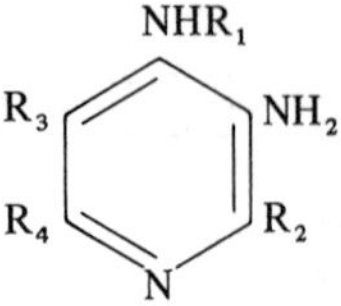


| $R_1$ | $R_2$ | $R_3$ | $R_4$ | Physical properties | Ref. |
|---|---|---|---|---|---|
| H | Cl | H | H | HCl, m.p. 209–10°; di-HCl, m.p. 175–7° | 87 |
| H | H | Br | H | m.p. 140–1°; picrate, m.p. 220–1° | 476 |
| H | H | H | Cl | m.p. 146° | 624 |
| | | | | HCl, m.p. 238–9° | 87 |
| H | H | $NO_2$ | H | – | 682 |
| H | Cl | Br | H | m.p. 206–8° | 476 |
| H | Cl | Me | H | m.p. 157° | 469 |
| H | Cl | H | Cl | HCl, m.p. 205–7° | 87 |
| H | OH | H | Pr | m.p. 198° | 480 |
| H | Me | H | Me | 1-oxide HCl, m.p. 211° | 351 |
| H | H | Br | Cl | m.p. 206–8° | 476 |
| H | F | F | F | m.p. 117.5–18.5° | 252 |
| H | H | H | H | m.p. 220°; picrate, m.p. 234–6° | 152 |
| *p*-Anisyl | H | Br | H | m.p. 134–5° | 667 |
| *p*-Chlorophenyl | H | Br | H | m.p. 157–8° | 667 |
| 2-Diethylamino-ethyl | H | Br | H | HCl, m.p. 208–9° | 682 |
| 2-Diethylamino-ethyl | H | $NO_2$ | H | m.p. 82–3° | 682 |
| 2-Dimethylamino-ethyl | H | $NO_2$ | H | – | 494 |
| *p*-Iodophenyl | H | I | H | m.p. 202–5.5° | 624 |
| Me | H | H | H | m.p. 169–70°; picrate, m.p. 185°; di-HCl, m.p. 227–8° | 152 |
| Phenethyl | H | Br | H | HCl, m.p. 203–4° | 682 |
| Phenethyl | H | $NO_2$ | H | m.p. 112° | 682 |
| Ph | H | Br | H | m.p. 144.5° | 44 |
| Ph | H | Cl | H | $H_2O$, m.p. 149.5° | 44 |
| Ph | H | I | H | m.p. 187.5–8.5° | 624 |
| Ph | Me | H | Me | m.p. 174° | 105 |
| Ph | Me | Br | Me | m.p. 177–9° | 44 |
| Ph | Me | Cl | Me | m.p. 199–200° | 105 |

TABLE IX-61. 3,5-Diaminopyridines

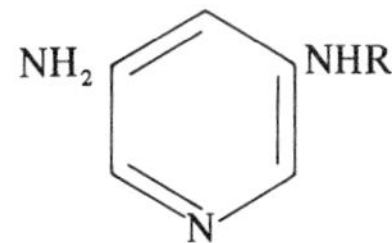

| | Substituents | | | | |
|---|---|---|---|---|---|
| R | 2 | 4 | 6 | Physical properties | Ref. |
| H | H | H | H | m.p. 107–8°; acetyl, m.p. 246–8° | 102, 175 75, 102 |
| H | H | H | H | 1-oxide, m.p. 229–30° | 694 |
| H | H | H | PhN=N– | m.p. 192° | 674 |
| H | Cl | H | H | m.p. 101–3°; picrate, m.p. 213–4° diacetyl, m.p. 210–12° | 102 |
| H | $Me_2N$ | H | H | – | 695 |
| H | OH | H | H | di-HCl, m.p. 205°, acetyl, m.p. 267–9° | 102 |
| H | MeO | H | H | di-HCl, m.p. 190° | 674 |
| H | EtO | H | H | di-HCl, m.p. 168° | 674 |
| H | PrO | H | H | di-HCl, m.p. 192° | 674 |
| H | BuO | H | H | di-HCl, m.p. 175° | 674 |
| H | $C_5H_{11}O$ | H | H | di-HCl, m.p. 150° | 674 |
| H | 2-Hydroxyethoxy | H | H | di-HCl, m.p. 210° | 181 |
| H | 2-Acetoxyethoxy | H | H | acetyl, m.p. 203° | 181 |
| H | MeO | H | PhN=N– | m.p. 141° | 674 |
| H | EtO | H | PhN=N– | m.p. 119° | 674 |
| H | EtO | H | PhN=N– | acetyl, m.p. 173° | 181 |
| H | PrO | H | PhN=N– | m.p. 122° | 674 |
| H | BuO | H | PhN=N– | m.p. 126° | 674 |
| H | AmO | H | PhN=N– | m.p. 96° | 674 |
| H | Me | Cl | Me | m.p. 283–5° | 696, 697 |
| OH | Me | Cl | Me | m.p. 97.5–8° | 696 |

TABLE IX-62. Triaminopyridines

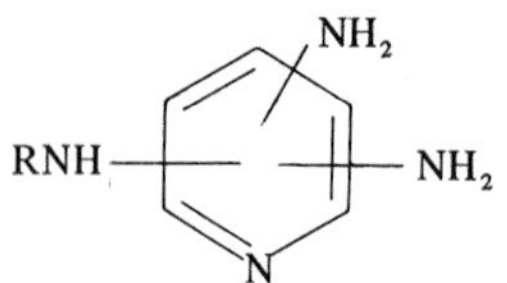


| R | Position of NHR | Position of $NH_2$ | Position of $NH_2$ | Physical properties | Ref. |
|---|---|---|---|---|---|
| $NH_2$ | 2 | 3 | 5 | tetra-HCl, m.p. 140° | 478 |
| 2-Diethylaminoethyl | 4 | 3 | 5 | picrate, m.p. 164–7° | 682 |
| Ph | 6 | 2 | 3 | m.p. 144° | 723 |
| Ph | 4 | 3 | 5 | – | 619 |
| $m$-$CF_3C_6H_4$ | 6 | 2 | 3 | m.p. 300° | 723 |
| H | 2 | 3 | 4 | m.p. 298° (dec.); di-HCl, m.p. 260° | 87 |
| H | 2 | 3 | 5 | tri-HCl, m.p. 160° (dec.) | 478 |
| (2,3,4-Triamino-6-propylpyridine) | | | | picrate, m.p. 155–6° | 478 |

TABLE IX-63. *N*-Aryl-2-picolylamines

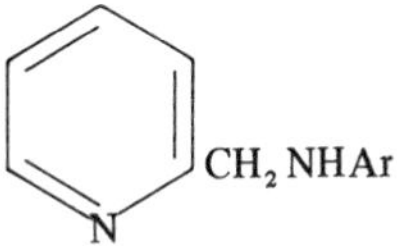


| Ar | Physical properties | Ref. |
|---|---|---|
| *o*-Anisyl | b.p. 173–6°/7 mm | 534 |
| | picrate, m.p. 144° (dec.) | 534 |
| *N,N'*-Bis(2-pyridylmethyl)-*O*-phenylenediamine | b.p. 235–43°/7 mm | 534 |
| *p*-Carboxyphenyl | m.p. 207–9° | 527 |
| 1-(3,4-Methylenedioxyphenyl)-2-propyl | m.p. 210° | 527 |
| *o*-Ethoxyphenyl | m.p. 102–4° | 534 |
| β-Hydroxy-β-(3,4-dimethoxyphenethyl) | m.p. 145° | 527 |
| β-Hydroxy-β-(4-methoxyphenethyl) | m.p. 78° | 527 |
| β-Hydroxy-β-(4-methylphenethyl) | m.p. 115° | 527 |
| β-Hydroxy-β-phenethyl | m.p. 166° | 527 |
| *p*-Hydroxyphenyl | m.p. 164–6° | 527 |
| 1-(4-Hydroxyphenyl)-2-propyl | m.p. 221° | 527 |
| 1-Hydroxy-1-phenyl-2-propyl | m.p. 218° | 527 |
| 1-(4-Methoxyphenyl)-2-propyl | m.p. 197° (dec.) | 527 |
| α-Naphthyl | b.p. 211–13°/6 mm; picrate, m.p. 172° (dec.) | 534 |
| α-Phenethyl | m.p. 224° | 527 |
| β-Phenethyl | m.p. 170° | 527 |
| Ph | m.p. 50–3° | 698 |
| | m.p. 50–4° | 534 |
| 1-Phenyl-2-propyl | m.p. 203° | 527 |
| *o*-Tolyl | b.p. 153–8°/4 mm; picrate, m.p. 147° | 534 |

TABLE IX-64. Simple 2-Aminoalkylpyridines

(Structure: pyridine ring with $R_3$, $R_4$, $R_5$, $R_6$ substituents and $(CH_2)_nNR_1R_2$ at position 2)

| n | $R_1$ | $R_2$ | $R_3$ | $R_4$ | $R_5$ | $R_6$ | Physical properties | Ref. |
|---|---|---|---|---|---|---|---|---|
| 1 | 2-Aminoethyl | H | H | H | H | H | b.p. 77°/0.01 mm.<br>tri-HCl, m.p. 174–5° | 224, 556, 699<br>505 |
| 1 | *p*-Anisyl | H | H | H | H | H | m.p. 74–5° | 535 |
| 1 | 1,4-Benzodioxan-2-ylmethyl | H | H | H | H | Me | di-HCl, m.p. 142–5° | 581 |
| 1 | 1,4-Benzodioxan-2-ylmethyl | H | H | H | H | H | di-HCl, m.p. 165–70° | 581 |
| 1 | Benzyl | H | H | H | H | H | b.p. 156–9°/3 mm;<br>picrate, m.p. 162–2.5° | 524 |
| 1 | Benzyl | H | H | H | Me | H | b.p. 143–52°/9.3 mm;<br>di-HCl, m.p. 178–9° | 533 |
| 1 | 2-Diethylaminoethyl | H | H | H | H | H | b.p. 109–10°/0.9 mm;<br>tri-picrate,<br>m.p. 185–6° | 499 |
| 1 | Diphenylmethyl | Me | H | H | H | H | m.p. 56–8° | 528 |
| 1 | *p*-Ethoxyphenyl | H | H | H | H | H | m.p. 72–3° | 535 |
| 1 | Et | Et | H | H | H | H | b.p. 109–10°/0.9 mm | 499 |
| 1 | Et | Et | H | H | Acetoxy | Me | b.p. 110–5°/1–2 mm | 545 |
| 1 | Et | Et | H | H | OH | Me | m.p. 137.5–8.5°,<br>di-HCl, m.p. 205–6°<br>tripicrate, m.p.<br>m.p. 185–6° | 545<br>545<br>499 |
| 1 | 2-Hydroxyethyl | H | H | H | H | H | b.p. 150–5°/6.5 mm; | 501 |

| | | | | | | | | |
|---|---|---|---|---|---|---|---|---|
| 1 | 2-Hydroxyethyl | 2-Hydroxyethyl | H | H | H | H | b.p. 187–90°/3.5 mm; di-HCl, m.p. 164–6° | 501 |
| 1 | 4-Hydroxybutyl | 2-Hydroxyethyl | H | H | H | H | b.p. 195–205°/3 mm | 501 |
| 1 | Me | 2-$NH_2$-5-Br-benzyl | H | H | H | H | – | 664 |
| 1 | Me | 2-$NH_2$-3,5-$Br_2$-benzyl | H | H | H | H | HCl, m.p. 198.5–201° | 664 |
| 1 | Me | 4-$NH_2$-3,5-$Br_2$-benzyl | H | H | H | H | HCl, m.p. 208–11° | 664, 700, 702 |
| 1 | Me | Me | H | H | OH | $NO_2$ | m.p. 181–1.5° | 542 |
| 1 | Me | Me | OH | H | H | H | m.p. 59–60° | 541 |
| 1 | Me | Me | OH | H | H | Me | b.p. 100–2°/1 mm | 541 |
| 1 | Me | Me | H | H | OH | Me | m.p. 159–60°; di-HCl, m.p. 210–11° | 545 |
| 1 | Me | Me | OH | *n*-Bu | H | IsoAm | m.p. 113–14° | 546 |
| 1 | Me | Me | OH | IsoBu | H | IsoAm | m.p. 118–19° | 546 |
| 1 | Me | H | H | H | Et | H | b.p. 135–7°/12 mm; di-HCl, m.p. 178–9° | 533 |
| 1 | Me | H | H | H | H | Me | *N*-nitroso, m.p. 49° | 594 |
| 1 | Me | H | H | H | H | H | *N*-nitroso, b.p. 100–1°/0.2 mm | 594 |
| 1 | α-[3,4-Methylenedioxy]-phenethyl | H | H | H | H | H | HCl, b.p. 150–2° | 577 |
| 1 | β-Naphthyl | H | H | H | H | H | m.p. 82–3° (sealed capillary) | 535 |
| 1 | β-Phenethyl | H | H | H | H | H | HCl, m.p. 183–5° di-HCl, m.p. 186–8° | 530 |
| 1 | Piperidino | – | OH | *n*-Pr | H | IsoAm | m.p. 165–6° | 546 |
| 1 | Piperidino | – | OH | IsoPr | H | IsoAm | m.p. 155–6° | 546 |
| 1 | Piperidino | – | OH | *n*-Bu | H | IsoAm | m.p. 168–9° | 546 |
| 1 | Piperidino | – | OH | IsoBu | H | IsoAm | m.p. 192–3° | 546 |
| 1 | Piperidino | – | OH | IsoAm | H | IsoAm | m.p. 170–1° | 546 |

TABLE IX-64. Simple 2-Aminoalkylpyridines (Continued)

| n | $R_1$ | $R_2$ | $R_3$ | $R_4$ | $R_5$ | $R_6$ | Physical properties | Ref. |
|---|---|---|---|---|---|---|---|---|
| 1 | Ph | H | H | H | H | Me | m.p. 61–2° | 698 |
| 1 | Ph | H | H | H | H | H | m.p. 54–5°, 47–9°; | 498, 535 |
| | | | | | | | b.p. 105–10°/0.04 mm | 498 |
| | | | | | | | b.p. 137–40°/3 mm; picrate, m.p. 154–5° (dec.); dipicrate, m.p. 151–2°; di-HCl, m.p. 190–2° | 535 |
| 1 | 2-Pyridyl | H | H | H | H | H | b.p. 152–3°/3 mm; m.p. 53–4°; picrate, m.p. 202–4° | 535 |
| 1 | *p*-Tolyl | H | H | H | H | H | b.p. 150–2°/3 mm; m.p. 41–3° | 535 |
| 1 | H | H | H | H | Me | H | b.p. 121–4°/13 mm; di-HBr, m.p. 200–3° | 533 |
| 1 | H | H | H | H | Et | H | b.p. 130–2°/13 mm; di-HCl, m.p. 208–9° | 533 |
| 1 | H | H | H | H | OH | $NO_2$ | m.p. 128–9° | 542 |
| 1 | Benzoyl | Benzoyl | H | H | H | H | b.p. 183°/0.25 mm; methiodide, m.p. 140–1° | 499 |

| | | | | | | | | |
|---|---|---|---|---|---|---|---|---|
| 1 | 2-Benzyloxyethyl | 2-Benzyloxy-ethyl | H | H | H | H | di-HCl, m.p. 163.5°; dipicrate, m.p. 150–1° | 501 |
| 1 | 4-Chlorobutyl | 2-Chloroethyl | H | H | H | H | picrate, m.p. 122–3° | 501 |
| 1 | Diphenylacetyl | Diphenylacetyl | H | H | H | H | b.p. 190°/0.15 mm; m.p. 65–6° | 499 |
| 1 | Et | Et | OH | H | H | Me | b.p. 79–80°/1 mm | 541 |
| 1 | Et | Et | OH | H | H | H | b.p. 85–6°/1 mm; di-HCl, m.p. 201–2° | 541 |
| 1 | Me | 2-Propynyl | H | H | H | H | b.p. 97°/4 mm; HCl, m.p. 177° | 549, 550 |
| 1 | Ph | Diethylamino-ethyl | H | H | H | H | b.p. 128–30°/0.04 mm; dipicrate, m.p. 164–6° | 498 |
| 1 | 2-Picolyl | Ph | H | H | H | Me | m.p. 81–83° | 698 |
| 1 | 2-Picolyl | Ph | H | H | H | H | m.p. 152–3° | 535 |
| 2 | 4-Aminobutyl | H | H | H | H | H | b.p. 69°/0.001 mm | 556 |
| 2 | 2-Aminoethyl | H | H | H | H | H | b.p. 77°/0.01 mm | 556 |
| 2 | 6-Aminohexyl | H | H | H | H | H | b.p. 91–3°/0.002 mm | 556 |
| 2 | 8-Aminooctyl | H | H | H | H | H | b.p. 113°/0.002 mm | 556 |
| 2 | Butyl | H | H | H | H | H | b.p. 117–18°/8 mm picrate, m.p. 144.5–5.5°; HCl, m.p. 117–18° | 558 |
| 2 | Hexamethylenimino | H | H | H | H | H | b.p. 150–2°/10 mm; MeI, m.p. 113–14.5°; picrate, m.p. 120–20.5° | 558 |
| | | | | | | | m.p. 75–6° | 578 |

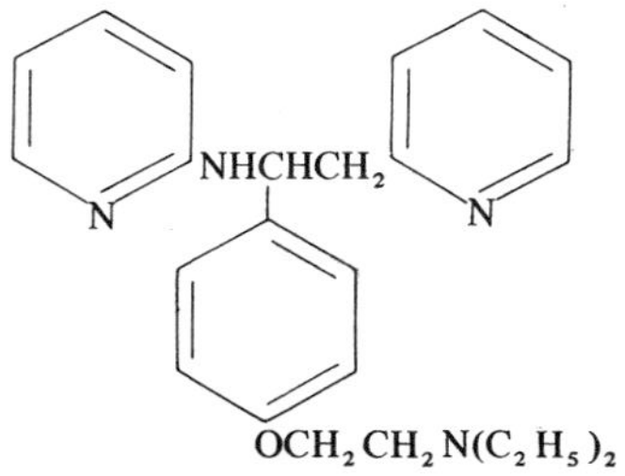

NHCHCH$_2$

OCH$_2$CH$_2$N(C$_2$H$_5$)$_2$

TABLE IX-64. Simple 2-Aminoalkylpyridines (Continued)

$$\text{pyridine ring: } R_3, R_4, R_5, R_6 \text{ substituted; 2-position } (CH_2)_nNR_1R_2$$

| n | $R_1$ | $R_2$ | $R_3$ | $R_4$ | $R_5$ | $R_6$ | Physical properties | Ref. |
|---|---|---|---|---|---|---|---|---|
| 2 | Diphenylmethyl | H | H | H | H | H | m.p. 69–71° | 528 |
| 2 | Benzyl | Benzyl | H | H | H | H | b.p. 179°/0.1 mm | 559 |
| 2 | Et | Et | $NO_2$ | H | H | H | b.p. 170°/0.05 mm | 80 |
| 2 | Et | Et | OH | H | H | $NO_2$ | m.p. 113–14° | 542 |
| 2 | Et | Me | H | H | H | H | HCl, m.p. 178–90° | 540 |
| 2 | Me | Me | OH | H | H | $NO_2$ | m.p. 143–4° | 542 |
| 2 | Me | Me | H | H | H | H | b.p. 120–3°/0.4 mm; 51–3°/0.1 mm, di-HCl, m.p. 195–7° | 128, 540 |
| 2 | Me | 2-Propynyl | H | H | H | H | b.p. 150°/ 30 mm | 550 |
| 2 | IsoPr | IsoPr | H | H | H | H | b.p. 91–2°/7 mm | 559 |
| 2 | H | H | H | H | Et | H | b.p. 69–71°/0.7 mm; di-HCl, m.p. 205°; | 532 |
| | | | | | | | b.p. 105°/10 mm; picrate, m.p. 194°; chloroplatinate, m.p. 211° | 547, 703 |
| 2 | H | H | OH | H | H | $NO_2$ | di-HCl, m.p. 125–7° | 542 |
| 2 | H | H | H | H | H | H | b.p. 87°/10 mm; b.p. 88–90°/8 mm | 559, 703 |
| | | | | | | | picrate, m.p. 164°; chloroplatinate, m.p. 230°; | 703 |

| | | | | | | | | |
|---|---|---|---|---|---|---|---|---|
| 2 | α-Methylphenethyl | H | H | H | Et | H | b.p. 148–50°/0.4 mm | 570 |
| 2 | Piperidino | – | OH | H | H | $NO_2$ | m.p. 210° | 542 |
| 2 | Piperidino | – | H | H | OH | $NO_2$ | m.p. 160°; m.p. 172–3° | 542, 543 |
| 2 | β-Piperidino | | H | H | H | H | b.p. 136–6.5°/10 mm; picrate, m.p. 126.5°; HCl, m.p. 173–4° | 563 |
| 2 | 1-Ph | H | H | H | H | H | acetyl, b.p. 170–90°/ /0.16 mm; methiodide, m.p. 151–2° | 569 |
| 2 | 1-Phenyl-2-propyl | H | H | H | H | H | 4-methoxyphenylcar-boxamide, b.p. 238–4°/0.03 mm | 569 |
| 2 | 1-Phenyl-2-propyl | H | H | H | H | H | b.p. 120°/0.05 mm | 569 |
| 2 | *p*-Anisyl | 2-Dimethylamino-ethyl | H | H | H | H | b.p. 152–4°/0.07 mm; dipicrate, m.p. 179–80° | 561, 562 |
| 2 | *m*-Chlorophenyl | 2-Dimethylamino-ethyl | H | H | H | H | b.p. 140–8°/0.03 mm; dipicrate, m.p. 180–3° | 561, 562 |
| 2 | *p*-Chlorophenyl | 2-Dimethylamino-ethyl | H | H | H | H | b.p. 152–4°/0.09 mm; dipicrate, m.p. 142–5° | 561, 562 |
| 2 | *m*-Chlorophenyl | 3-Dimethylamino-propyl | H | H | H | H | b.p. 138°/0.02 mm | 561, 562 |
| 2 | *p*-Chlorophenyl | 3-Dimethylamino-propyl | H | H | H | H | b.p. 166–70°/0.05 mm; dipicrate, m.p. 160–3° | 561, 562 |
| 2 | α-Methylphenethyl | H | H | H | Et | H | b.p. 148–50°/0.4 mm | 570 |
| 3 | Me | Me | OH | H | H | $NO_2$ | – | 542 |
| 5 | Me | Et | H | H | H | H | b.p. 100–8°/2 mm; methobromide m.p. 199–200° | 508 |
| 7 | H | H | H | H | H | H | dipicrate, m.p. 103° | 547 |
| 8 | H | H | H | H | H | H | b.p. 183°/15 mm | 547 |
| 13 | H | H | H | H | H | H | b.p. 202°/1.5 mm | 547 |

TABLE IX-65. 2-(α-Alkyl-branched)aminoalkylpyridines

[Structure: 2-pyridyl–C($R_3$)($R_4$)–$(CH_2)_nNR_1R_2$]

| n | $R_1$ | $R_2$ | $R_3$ | $R_4$ | Physical properties | Ref. |
|---|---|---|---|---|---|---|
| | Substituents | | | | | |
| 0 | EtNH | H | Me | H | b.p. 74–5°/0.07 mm; fumarate, m.p. 159–61° | 529 |
| 0 | *t*-Bu | H | Me | H | b.p. 63–9°/0.03 mm; maleate, m.p. 124–5° | 529 |
| 0 | Me | Me | Me | H | methobromide, m.p. 189–91° | 529 |
| 0 | Ph | H | Me | IsoBu | b.p. 145°/0.08 mm; picrate, m.p. 145–6° | 580 |
| 0 | Ph | H | Me | Hexyl | b.p. 141°/0.03 mm | 580 |
| 2 | Me | Me | H | H | b.p. 67–9°/0.6 mm | 537 |
| 2 | Me | Me | H | Acetyl | b.p. 86–7°/0.33 mm; dipicrate, m.p. 179.7–80.4° | 537 |
| 2 | Me | Me | H | Benzoyl | b.p. 132–3°/0.28 mm; dipicrate, m.p. 179.2–9.8° | 537 |
| 2 | Me | Me | H | Butyroyl | b.p. 95–6°/0.26 mm; dipicrate, m.p. 149.6–50.6° | 537 |
| 2 | Me | Me | H | Isobutyroyl | b.p. 89–90°/0.28 mm; dipicrate, m.p. 155.8–6.7° | 537 |
| 2 | Me | Me | H | Propionyl | b.p. 90–1°/0.29 mm; dipicrate, m.p. 169–70° | 537 |

TABLE IX-65. 2-(α-Alkyl-branched)aminoalkylpyridines (Continued)

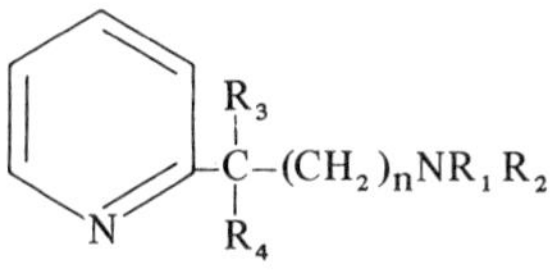

$$\text{2-Pyridyl}-\underset{R_4}{\overset{R_3}{C}}-(CH_2)_nNR_1R_2$$

| n | Substituents $R_1$ | $R_2$ | $R_3$ | $R_4$ | Physical properties | Ref. |
|---|---|---|---|---|---|---|
| 2 | Me | Me | H | Trimethyl-acetyl | b.p. 90–1°/0.29 mm; dipicrate, m.p. 171.3–2.5° | 537 |
| 2 | Me | Me | Ph | Acetyl | b.p. 151–2°/0.62 mm; dipicrate, m.p. 159.2–60.4° | 537 |
| 2 | Me | Me | Ph | Benzoyl | b.p. 200–5°/1 mm; m.p. 86–7.4°; tripicrate, m.p. 199–200.5° | 537 |
| 2 | Me | Me | Ph | Butyroyl | b.p. 152–3°/0.29 mm; dipicrate, m.p. 150.5–1.4° | 537 |
| 2 | Me | Me | Ph | Isobutyroyl | b.p. 145–6°/0.29 mm; dipicrate, m.p. 159–60° | 537 |
| 2 | Me | Me | Ph | 2-Ethyl-hexanoyl | b.p. 176–7°/9.28 mm | 537 |
| 2 | Me | Me | Ph | Propionyl | b.p. 144.5°/0.27 mm; maleate, m.p. 144.2–5.3° | 537 |
| 2 | Me | Me | Ph | Trimethyl-acetyl | b.p. 156–8°/0.25 mm; m.p. 65–70°; dipicrate, m.p. 156.6–7.4° | 537 |

TABLE IX-66. 2-(α-Aryl-branched)aminoalkylpyridines

| *n* | $R_1$ | $R_2$ | $R_3$ | $R_4$ | $R_5$ | b.p./mm | m.p. | Picrate, m.p. | Ref. |
|---|---|---|---|---|---|---|---|---|---|
| 0 | *o*-Anisyl | H | H | H | H | 176°/0.02 | – | 146° | 580 |
| 0 | *p*-Anisyl | H | H | H | H | 185°/0.03 | 91° | 131° | 580 |
| 0 | Benzyl | H | H | H | H | 184°/0.03 | – | 131° | 580 |
| 0 | *o*-$ClC_6H_4$ | H | H | H | H | 168°/0.05 | 100° | 154° | 580 |
| 0 | *p*-$ClC_6H_4$ | H | H | H | H | 171°/0.08 | 113° | 158° | 580 |
| 0 | Et | H | H | H | H | 106°/0.50 | – | 150° | 580 |
| 0 | *o*-$MeOC_6H_4$ | H | H | H | H | 136–41°/1 | – | 130–1° | 580 |
| 0 | Me | H | H | H | H | 122°/0.50 | – | – | 580 |
| 0 | Ph | H | H | H | H | 158°/0.01<br>155–62°/0.1 | 78° | 175° | 580 |
| 0 | Ph | H | H | 2-Cl | H | – | – | 165–6° | 580 |
| 0 | Ph | H | H | 3-Cl | H | 183°/0.09 | 96° | 163° | 580 |
| 0 | Ph | H | H | 4-Cl | H | 163°/0.03 | 86° | 169° | 580 |
| 0 | Ph | H | H | 4-Me | H | 167°/0.02 | – | 163° | 580 |
| 0 | *o*-Tolyl | H | H | H | H | 161°/0.01 | – | 156° | 580 |
| 0 | *p*-Tolyl | H | H | H | H | 166°/0.03 | 81° | 128° | 580 |
| 0 | Ph | H | Ph | H | H | 156°/0.08 | 53° | 195° | 580 |
| 1 | H | H | H | 4-Br | H | 160–3°/1 | – | – | 497 |
| 1 | H | H | H | 4-Cl | H | 160–4°/1 | – | – | 497 |
| 1 | H | H | H | 4-Me | H | 148–50°/1 | – | – | 497 |
| 1 | H | H | H | H | H | 130–5°/2 | – | – | 497 |

| | | | | | | | | |
|---|---|---|---|---|---|---|---|---|
| 2 | 1-(4'-Anisyl)-2-propyl | H | H | H | H | 200°/0.01 | (1:1 maleate, m.p. 126°) | | 538 |
| 2 | Bu | H | H | 4-Br | IsoBu | 175–6°/0.01 | – | – | 704 |
| 2 | Et | H | H | 4-Br | IsoBu | 167–70°/0.05 | – | dipicrate 167–70° | 704 |
| 2 | Me | H | H | 4-Br | IsoBu | 160–1°/0.05 | – | dipicrate 140–2° | 704 |
| 2 | 2-[2-Methyl-3-(4-chlorophenyl)]-propyl | H | H | H | H | 210–23°/0.3 | (1:1 maleate, m.p. 137–8°) | | 538 |
| 2 | 1-Phenyl-3-butyl | H | H | H | H | 210–12°/0.4 | (1:5 fumaric m.p. 157–8°) | | 538 |
| 2 | IsoPr | H | H | 4-Br | IsoBu | 165–7°/0.01 | – | 84–9° | 704 |
| 2 | Cl | Me | H | 4-Cl | IsoBu | 148–50°/0.1 | – | 185–7° | 704 |
| 2 | Et | Et | H | 4-Br | IsoBu | 158–62°/0.1 | – | 149–50° | 704 |
| 2 | Et | Et | H | 4-Cl | IsoBu | 155–8°/0.01 | – | 157–60° | 704 |
| 2 | Me | Me | H | 4-Cl | H | 136–8°/0.3 | – | – | 554 |
| | | | | | | 155–60°/3.0 | (maleate, m.p. 132–3°) dipicrate, m.p. 194–6° | | 705 |
| 2 | Me | Me | H | 4-Br | IsoBu | 137–40°/0.01 | – | – | 704 |
| 2 | Me | Me | H | H | H | 132–6°/0.4 | – | – | 554 |
| | | | | | | 128–9°/0.85 | – | 201–2° | 537 |
| 2 | Me | IsoPr | H | 4-Br | IsoBu | – | – | 150–6° | 704 |
| 2 | *N*-Pyrrolidino | | H | H | H | 136–8° | – | dipicrate 170–1° | 537 |
| 2 | Pr | Pr | Amido | H | H | – | 63–4° | – | 592 |
| 2 | IsoPr | IsoPr | Amido | 4-Cl | H | – | – | – | 592 |
| 2 | IsoPr | IsoPr | Amido | 3,4-Dimethoxy | H | – | 102–3° | – | 592 |
| 2 | IsoPr | IsoPr | Amido | 3,5–$Me_2$ | H | – | 152–5° | – | 592 |
| 2 | IsoPr | IsoPr | Amido | 2-F | H | – | 77–8° | – | 592 |

TABLE IX-66. 2-(α-Aryl-branched)aminoalkylpyridines (Continued)



| n | $R_1$ | $R_2$ | $R_3$ | $R_4$ | $R_5$ | b.p./mm | m.p. | Picrate, m.p. | Ref. |
|---|---|---|---|---|---|---|---|---|---|
| 2 | IsoPr | IsoPr | Amido | 4-MeO | H | – | – | – | 592 |
| 2 | IsoPr | IsoPr | Amido | 3-Me | H | – | 113–14° | – | 592 |
| 2 | IsoPr | IsoPr | Amido | H | Cl | – | – | – | 592 |
| 2 | IsoPr | IsoPr | Amido | H | H | – | 94.5–5° | – | 592 |
| 2 | IsoPr | IsoPr | Amido | 2-Naphthyl | H | – | 152–5° | – | 592 |
| 2 | Pr | Pr | CN | H | H | 164.5°/0.1 | – | – | 592 |
| 2 | IsoPr | IsoPr | CN | 4-Cl | H | 176–9°/0.4 | – | – | 592 |
| 2 | IsoPr | IsoPr | CN | 3,4–$(MeO)_2$ | H | 185–90°/0.1 | – | – | 592 |
| 2 | IsoPr | IsoPr | CN | 3,5-$Me_2$ | H | – | – | – | 592 |
| 2 | IsoPr | IsoPr | CN | 2-F | H | – | – | – | 592 |
| 2 | IsoPr | IsoPr | CN | 4-F | H | 154°/0.4 | – | – | 592 |
| 2 | IsoPr | IsoPr | CN | 4-MeO | H | – | – | – | 592 |
| 2 | IsoPr | IsoPr | CN | 3-Me | H | 164–8°/0.2 | – | – | 592 |
| 2 | IsoPr | IsoPr | CN | H | Cl | 165–72°/0.25 | – | – | 592 |
| 2 | IsoPr | IsoPr | CN | H | H | 145–60°/0.3 | – | – | 592 |
| 2 | H | H | H | 4-Cl | H | 138–40°/0.13 | – | – | 554 |
| 2 | H | H | H | H | H | 128–32°/0.25 | – | | 554 |

| | | | | | | | | |
|---|---|---|---|---|---|---|---|---|
| 3 | 2-[1-Methyl-3-(4-chlorophenyl)]-propyl | H | H | H | H | 216–20°/0.2 | (1:1 maleate, m.p. 129–30°) | 538 |
| 3 | 1-Phenyl-3-butyl | H | H | H | H | 207–12°/0.1 | (1:1 maleate, m.p. 129–30°) | 538 |
| 3 | 1-Phenyl-2-propyl | H | H | H | H | 215–17°/0.8 | (1:1 maleate, m.p. 127–8°) | 538 |

TABLE IX-67. 2-(β-Alkyl-branched)aminoalkylpyridines[533]

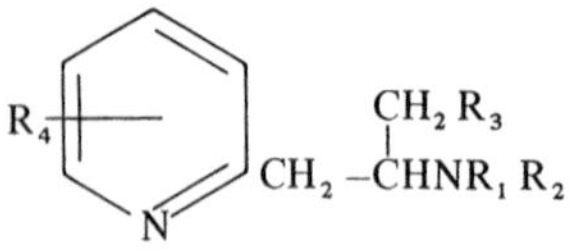

$$R_4\text{-}C_5H_3N\text{-}2\text{-}CH_2\text{-}CH(CH_2R_3)NR_1R_2$$

| $R_1$ | $R_2$ | $R_3$ | $R_4$ 4 | $R_4$ 5 | $R_4$ 6 | Physical properties |
|---|---|---|---|---|---|---|
| Benzyl | H | H | H | Et | H | b.p. 143–52°/0.3 mm; di-HCl, m.p. 178–9° |
| 2-Chloroethyl | H | H | H | Me | H | m.p. 179–80° |
| 2-Hydroxyethyl | H | H | H | Me | H | b.p. 162–5°/12 mm |
| 2-Hydroxyethyl | H | H | Me | H | Me | b.p. 161–4° |
| Me | H | H | H | Me | H | b.p. 112–16°/12 mm; di-HCl, m.p. 175–6° |
| Me | H | H | Me | H | Me | b.p. 109–10° |
| H | H | Me | H | Et | H | b.p. 130–2°/13 mm; di-HCl, m.p. 208–9° |
| 2-Chloroethyl | Me | H | H | Me | H | di-HCl, m.p. 171–2° |
| 2-Chloroethyl | Me | H | Me | H | Me | di-HCl, m.p. 175–6° |
| 2-Hydroxyethyl | Me | H | H | Me | H | b.p. 163–6°/13 mm |
| 2-Hydroxyethyl | Me | H | Me | H | Me | b.p. 152–5° |
| Me | Me | H | H | Me | H | b.p. 123–4°/15 mm; di-HCl, m.p. 181–2° |
| Me | Me | H | H | H | Me | b.p. 101–3°/14 mm |
| Me | Me | H | Me | H | Me | b.p. 123–6°/17 mm; dipicrate, m.p. 155–60°; methiodide, m.p. 170° |
| 2-Chloroethyl | H | Me | H | Et | H | – |
| 2-Hydroxyethyl | H | Me | H | Et | H | b.p. 175–80°/10 mm; di-HCl, m.p. 129–30° |
| Me | H | Me | H | Et | H | b.p. 135–7°/12 mm; di-HCl, m.p. 173–5° |
| 2-Chloroethyl | Me | Me | H | Et | H | di-HCl, m.p. 152–3° |
| 2-Hydroxyethyl | Me | Me | H | Et | H | b.p. 148–51°/0.18 mm |
| H | H | H | Me | H | H | b.p. 127–32°/15 mm |
| H | H | H | H | Et | H | b.p. 121–4°/13 mm |
| H | H | H | Me | H | Me | b.p. 111–40°; di-HCl, m.p. 209–10° |

TABLE IX-68. 2-(β-Aryl-branched)aminoalkylpyridines

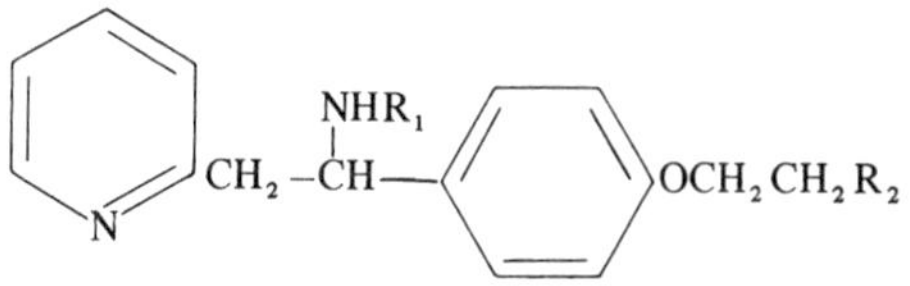

| $R_1$ | $R_2$ | m.p. | Ref. |
|---|---|---|---|
| p-Anisyl | Diethylamino | 66–8° | 579 |
| m-$ClC_6H_4$ | Diethylamino | 62–4° | 579 |
| p-$ClC_6H_4$ | Diethylamino | 76–7° | 579 |
| p-$FC_6H_4$ | Diethylamino | – | 579 |
| p-$MeOC_6H_4$ | Pyrrolidino | 93–5° | 579 |
| p-Tolyl | Morpholino | 103–5° | 579 |
| p-Tolyl | Piperidino | 93–5° | 579 |
| Ph | Diethylamino | 61–3° | 579 |
| Ph | Piperidinomethyl | 88–91° | 579 |
| p-Tolyl | Diethylamino | 77–9° | 579 |
| 2-Pyridyl | (Phenyl[a]) | 84–5° (b.p. 182–7°/0.1 mm) | 578 |

[a]This is a β-phenyl substituent and not an alkoxyphenyl group.

TABLE IX-69. Longer Chain 2-Aminoalkylpyridines

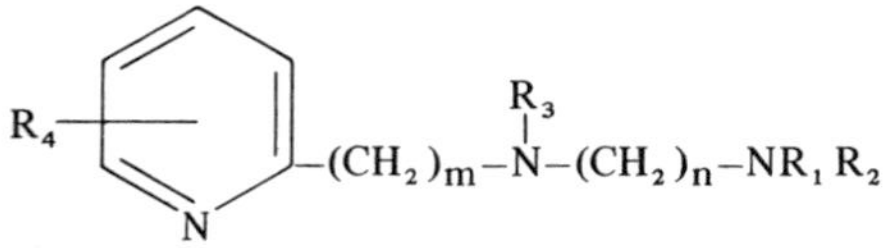

| m | n | $R_1$ | $R_2$ | $R_3$ | $R_4$ | Physical properties | Ref. |
|---|---|---|---|---|---|---|---|
| 2 | 2 | Acetyl | H | Acetyl | H | b.p. 136°/0.003 mm | 556 |
| 2 | 2 | Acetyl | H | 2-(2-Pyridyl)ethyl | H | b.p. 138–40°/0.003 mm | 556 |
| 2 | 2 | 2-Benzoyloxyethyl | H | 2-(2-Pyridyl)ethyl | H | m.p. 172° | 556 |
| 2 | 2 | Bu | H | H | H | b.p. 69°/0.001 mm | 556 |
| 2 | 2 | Carbanilido | H | Carbanilido | H | m.p. 138° | 556 |
| 2 | 2 | Ethoxymethyl | H | 2-(2-Pyridyl)-3-hydroxypropyl | H | b.p. 168°/0.001 mm | 556 |
| 2 | 2 | 2-Hydroxyethyl | H | H | H | b.p. 110°/0.003 mm | 556 |
| 2 | 2 | Octylsulfonyl | H | Octylsulfonyl | H | m.p. 77° | 556 |
| 2 | 2 | 2-(2-Pyridyl)ethyl | H | 2-(2-Pyridyl)ethyl | H | b.p. 148°/0.1 mm | 556 |
| 2 | 2 | Acetyl | Acetyl | 2-(2-Pyridyl)ethyl | H | b.p. 156°/0.003 mm | 556 |
| 2 | 2 | Acetyl | 2-(Pyridyl)ethyl | 2-(2-Pyridyl)ethyl | H | b.p. 180°/0.01 mm | 556 |
| 2 | 2 | Benzyl | *p*-Chlorophenyl | Ethyl | H | dimaleate, m.p. 103–5° | 563 |
| 2 | 2 | Benzyl | *p*-Chlorophenyl | Me | H | methiodide, m.p. 130–1° | 563 |
| 2 | 2 | Benzyl | Ph | Me | H | methiodide, m.p. 125°; dimaleate, m.p. 114–16° | 563 |
| 2 | 2 | Benzyl | Ph | H | H | tri-HCl, m.p. 151–2° | 563 |
| 2 | 2 | Butylsulfonyl | Butylsulfonyl | Butylsulfonyl | H | m.p. 270° (dec.) | 556 |
| 2 | 2 | 4-Chlorobenzyl | Ph | Me | H | dimaleate, m.p. 106.5–7° | 563 |
| 2 | 2 | Dichloroacetoxyethyl | Dichloroacetoxyethyl | Dichloroacetyl | H | m.p. 170° | 556 |
| 2 | 2 | Ethoxymethyl | 2-(2-Pyridyl)allyl | 2-(2-Pyridyl)allyl | H | – | 556 |
| 2 | 2 | Ethoxymethyl | 2-(2-Pyridyl)ethyl | 2-(2-Pyridyl)ethyl | H | b.p. 90–5°/0.02 mm | 556 |

| | | | | | | | |
|---|---|---|---|---|---|---|---|
| 2 | 2 | 2-Hydroxyethyl | 2-Hydroxyethyl | H | H | b.p. 15[illegible]°/0.01 mm | 556 |
| 2 | 2 | Me | Me | *m*-Chlorophenyl | 5-Et | b.p. 172–4°/0.02 mm | 561, 562 |
| 2 | 2 | Me | Me | *m*-Chlorophenyl | H | b.p. 140–8°/0.03 mm | 561, 562 |
| 2 | 2 | Me | Me | *p*-Chlorophenyl | H | b.p. 152–4°/0.09 mm; dipicrate, m.p. 142–5° | 561, 562 |
| 2 | 2 | Me | Me | *p*-Methoxyphenyl | H | b.p. 152–4°/0.07 mm; dipicrate, m.p. 179–80° | 561, 562 |
| 2 | 2 | Morpholino | H | H | H | b.p. 75–80°/15 mm | 556 |
| 2 | 2 | 2-(2-Pyridyl)ethyl | 2-(2-Pyridyl)ethyl | 2-(2-Pyridyl)ethyl | H | b.p. 183°/0.01 mm | 556 |
| 2 | 2 | H | H | Acetyl | H | b.p. 108°/0.003 mm | 556 |
| 2 | 2 | H | H | Benzoyloxy | H | m.p. 152° | 556 |
| 2 | 2 | H | H | 2-(2-Pyridyl)ethyl | H | b.p. 120°/0.004 mm | 556 |
| 2 | 2 | H | H | H | H | b.p. 77°/0.01 mm | 556 |
| 2 | 3 | Benzyl | Ph | H | H | maleate, m.p. 117.5–18.5° | 563 |
| 2 | 3 | Benzyl | Ph | Me | H | di-HCl, m.p. 133.5–4.5° | 563 |
| 2 | 3 | Me | Me | *m*-Chlorophenyl | 5-Et | b.p. 158–60°/0.03 mm | 561, 562 |
| 2 | 3 | Me | Me | *p*-Chlorophenyl | 5-Et | b.p. 170–2°/0.02 mm | 561, 562 |
| 2 | 3 | Me | Me | *m*-Chlorophenyl | H | b.p. 138°/0.02 mm | 561, 562 |
| 2 | 3 | Me | Me | *p*-Chlorophenyl | H | b.p. 166–70°/0.05 mm dipicrate, m.p. 160–3° | 561, 562 |
| 2 | 4 | Acetyl | H | H | H | b.p. 108°/0.001 mm | 556 |
| 2 | 4 | Acetyl | H | Acetyl | H | b.p. 138°/0.001 mm | 556 |
| 2 | 4 | Carbanilido | H | Carbanilido | H | m.p. 156° | 556 |
| 2 | 4 | 2-Hydroxyethyl | H | H | H | b.p. 120°/0.001 mm | 556 |
| 2 | 4 | 2-Hydroxyethyl | H | 2-Hydroxyethyl | H | b.p. 150°/0.01 mm | 556 |
| 2 | 4 | 2-(2-Pyridyl)ethyl | H | 2-(2-Pyridyl)ethyl | H | b.p. 156°/0.001 mm | 556 |
| 2 | 4 | Acetyl | 2-(2-Pyridyl)ethyl | 2-(2-Pyridyl)ethyl | H | b.p. 175°/0.005 mm | 556 |
| 2 | 4 | 2-Cyanoethyl | 2-(2-Pyridyl)ethyl | 2-(2-Pyridyl)ethyl | H | b.p. 158°/0.02 mm | 556 |
| 2 | 4 | 2-Hydroxyethyl | 2-Hydroxyethyl | 2-Hydroxyethyl | H | b.p. 175°/ [a] | 556 |
| 2 | 4 | | (Morpholino) | H | H | b.p. 100°/1 mm | 556 |
| 2 | 4 | 2-(2-Pyridyl)ethyl | 2-(2-Pyridyl)ethyl | 2-(2-Pyridyl)ethyl | H | b.p. 183–5°/0.001 mm | 556 |
| 2 | 4 | H | H | 2-(2-Pyridyl)ethyl | H | b.p. 128°/0.001 mm | 556 |

TABLE IX-69. Longer Chain 2-Aminoalkylpyridines (Continued)

$R_4$–(pyridin-2-yl)$-(CH_2)_m-N(R_3)-(CH_2)_n-NR_1R_2$

| m | n | $R_1$ | $R_2$ | $R_3$ | $R_4$ | Physical properties | Ref. |
|---|---|---|---|---|---|---|---|
| 2 | 4 | H | H | H | H | b.p. 83°/0.01 mm | 556 |
| 2 | 6 | Acetyl | H | H | H | b.p. 118°/0.002 mm | 556 |
| 2 | 6 | Acetyl | H | Acetyl | H | b.p. 143°/0.002 mm | 556 |
| 2 | 6 | Acetyl | H | 2-(2-Pyridyl)ethyl | H | b.p. 143°/0.001 mm | 556 |
| 2 | 6 | 2-Hydroxyethyl | H | H | H | b.p. 125°/0.01 mm | 556 |
| 2 | 6 | 2-(2-Pyridyl)ethyl | H | 2-(2-Pyridyl)ethyl | H | b.p. 163°/0.001 mm | 556 |
| 2 | 6 | Acetyl | Acetyl | Acetyl | H | b.p. 205°/0.02 mm | 556 |
| 2 | 6 | Acetyl | Acetyl | 2-(2-Pyridyl)ethyl | H | b.p. 163°/0.002 mm | 556 |
| 2 | 6 | Acetyl | 2-(2-Pyridyl)ethyl | 2-(2-Pyridyl)ethyl | H | b.p. 175°/0.005 mm | 556 |
| 2 | 6 | 3-Aminopropyl | 2-(2-Pyridyl)ethyl | 2-(2-Pyridyl)ethyl | H | b.p. 152°/0.002 mm | 556 |
| 2 | 6 | 2-Carbamoylethyl | 2-(2-Pyridyl)ethyl | 2-(2-Pyridyl)ethyl | H | m.p. 342° | 556 |
| 2 | 6 | Carbanilido | 2-(2-Pyridyl)ethyl | 2-(2-Pyridyl)ethyl | H | m.p. 134° | 556 |
| 2 | 6 | Cyanoethyl | 2-(2-Pyridyl)ethyl | 2-(2-Pyridyl)ethyl | H | b.p. 193°/0.08 mm | 556 |
| 2 | 6 | 2-Hydroxyethyl | 2-Hydroxyethyl | H | H | b.p. 166–8°/0.004 mm | 556 |
| 2 | 6 | 2-Hydroxyethyl | 2-Hydroxyethyl | 2-Hydroxyethyl | H | b.p. 186°/0.01 mm | 556 |
| 2 | 6 | 2-Hydroxyethyl | 2-Hydroxyethyl | 2-(2-Pyridyl)ethyl | H | b.p. 160°/0.005 mm | 556 |
| 2 | 6 | 2-Hydroxyethyl | 2-(2-Pyridyl)ethyl | 2-(2-Pyridyl)ethyl | H | b.p. 139°/0.001 mm | 556 |
| 2 | 6 | (Morpholino) | | H | H | b.p. 110°/1 mm | 556 |
| 2 | 6 | Propen-2-yl | 2-(2-Pyridyl)ethyl | 2-(2-Pyridyl)ethyl | H | b.p. 95–100°/0.001 mm | 556 |
| 2 | 6 | 2-(2-Pyridyl)ethyl | 2-(2-Pyridyl)ethyl | 2-(2-Pyridyl)ethyl | H | b.p. 220°/0.02 mm | 556 |
| 2 | 6 | Thiocarbanilido | 2-(2-Pyridyl)ethyl | 2-(2-Pyridyl)ethyl | H | m.p. 85° | 556 |
| 2 | 6 | H | H | 2-(2-Pyridyl)ethyl | H | b.p. 136°/0.002 mm | 556 |

| | | | | | | | |
|---|---|---|---|---|---|---|---|
| 2 | 8 | H | H | H | H | b.p. 91–3°/0.002 mm | 556 |
| 2 | 8 | Acetyl | H | H | H | b.p. 125°/0.001 mm | 556 |
| 2 | 8 | Acetyl | H | Acetyl | H | b.p. 165°/0.002 mm | 556 |
| 2 | 8 | 2-Hydroxyethyl | H | H | H | b.p. 126°/0.001 mm | 556 |
| 2 | 8 | 2-Hydroxyethyl | H | 2-Hydroxyethyl | H | b.p. 172°/0.02 mm | 556 |
| 2 | 8 | 2-(2-Pyridyl)ethyl | H | 2-(2-Pyridyl)ethyl | H | b.p. 185°/0.001 mm | 556 |
| 2 | 8 | Acetyl | Acetyl | Acetyl | H | b.p. 208°/0.001 mm | 556 |
| 2 | 8 | Acetyl | 2-(2-Pyridyl)ethyl | 2-(2-Pyridyl)ethyl | H | b.p. 193°/0.001 mm | 556 |
| 2 | 8 | 2-Cyanoethyl | 2-(2-Pyridyl)ethyl | 2-(2-Pyridyl)ethyl | H | b.p. 162°/0.02 mm | 556 |
| 2 | 8 | 2-(2-Pyridyl)ethyl | 2-(2-Pyridyl)ethyl | 2-(2-Pyridyl)ethyl | H | b.p. 225°/0.001 mm | 556 |
| 2 | 8 | H | H | 2-(2-Pyridyl)ethyl | H | b.p. 148°/0.001 mm | 556 |
| 2 | 8 | H | H | H | H | b.p. 113°/0.002 mm | 556 |
| 3 | 2 | Benzyl | *p*-Bromophenyl | Me | H | maleate, m.p. 133–4°; dipicrate, m.p. 155–60° | 563 |
| 3 | 2 | Benzyl | *p*-Chlorophenyl | H | H | di-HCl, m.p. 171–1.5° | 563 |

[a]Pressure not reported.

TABLE IX-70. 2,2′-{[(Aminoalkyl)imino] diethylene}dipyridines[556]

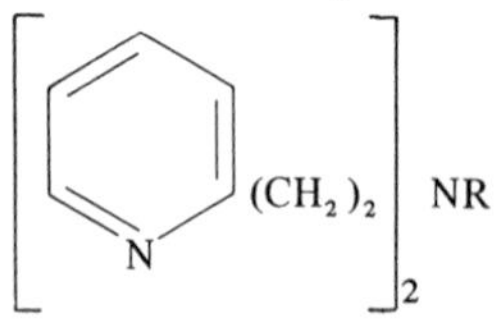

| R | Physical properties |
|---|---|
| 2-Aminoethyl | |
| 4-Aminobutyl | b.p. 128°/0.001 mm |
| 6-Aminohexyl | b.p. 136°/0.002 mm |
| 8-Aminooctyl | b.p. 148°/0.001 mm |

TABLE IX-71. *N,N*-Bis[2-, 3-, 4-(pyridylalkyl)amino] alkylenediamine

| Pyridine ring position | $R_1$ | $R_2$ | n | Physical properties | Ref. |
|---|---|---|---|---|---|
| 2 | H | H | 2 | b.p. 151–8°/0.8 mm | 530 |
| | | | | dimaleate, m.p. 174–5° | 529 |
| 2 | Me | H | 2 | b.p. 154–6°/0.25 mm; dimaleate, m.p. 162–3° | 530 |
| 2 | Me | H | 2 | b.p. 153–6°/0.25 mm; difumarate, m.p. 197–8° | 529 |
| 2 | Me | CHO | 2 | b.p. 165–8°/0.01–.03 mm; | 530 |
| | | | | dimaleate, m.p. 145–7° | 529 |
| 2 | Me | Me | 2 | b.p. 168–72°/0.08 mm tetra-HCl, salt, m.p. 170° | 530 |
| 2 | Me | H | $-CH_2CH(CH_3)-$ | b.p. 140–5°/0.04 mm; | 530 |
| | | | | difumarate, m.p. 151–3° | 529, 5 |
| 2 | Me | H | $(CH_3CH)_2$ | b.p. 160–3°/0.75 mm; difumarate, m.p. 172–4° | 529 |
| 2 | Me | H | 3 | b.p. 148–52°/0.2 mm | 530 |
| 2 | Me | H | 5 | b.p. 167–8°/0.04 mm; dimaleate, m.p. 148–50° | 530 |
| 2 | $CH_3$ | H | (*trans*-1,2-Cyclohexenyl) | b.p. 153–5°/0.005 mm; tetra-HCl, m.p. 244–6° | 529 |

ABLE IX-71. *N,N*-Bis[2-, 3-, 4-(pyridylalkyl)amino]alkylenediamine (Continued)

| 'yridine ing osition | $R_1$ | $R_2$ | n | Physical properties | Ref. |
|---|---|---|---|---|---|
| | Me | H | 2 | b.p. 176–80°/0.1 mm; dimaleate, m.p. 151–3° | 529 |
| | Me | H | 2 | b.p. 170–4° [a] dimaleate, m.p. 145–7° | 529 |

[a]Pressure not reported.

TABLE IX-72. *N*-Aryl-3-picolylamines

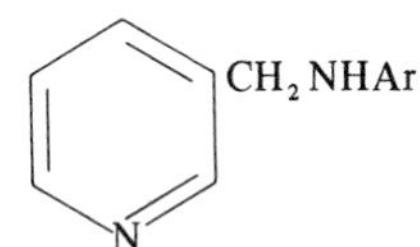


| Ar | m.p. | Ref. |
|---|---|---|
| 3is(3-chloro-2-methyl-4-picolyl)amine | 114–14.5°; tri-HCl, m.p. 292–5° (dec.) | 640 |
| *n*-Carboxyphenyl | 190–2° | 527 |
| )-Carboxyphenyl | 220–2° | 527 |
| -(3,4-Dimethoxyphenyl-2-propyl) | 202° | 527 |
| -(3,4-Dimethoxyphenethyl) | 226° | 527 |
| )-Hydroxyphenyl | 177–8° | 527 |
| -Hydroxy-β-(3,4-dimethoxyphenethyl) | 197° | 527 |
| -Hydroxy-β-(4-methylphenethyl) | 198° | 527 |
| -Hydroxy-β-(4-methoxyphenethyl) | 153° | 527 |
| -Hydroxy-1-phenyl-2-propyl | 227° | 527 |
| -(4-Hydroxyphenyl)-2-propyl | 228° (dec.) | 527 |
| -Hydroxy-β-phenethyl | 179° | 527 |
| -Hydroxyphenyl | 145–6° | 527 |
| -(3,4-Methylenedioxyphenyl)-2-propyl | 232° | 527 |
| -(4-Methoxyphenyl)-2-phenyl | 216° | 527 |
| -Nitrophenyl | 114–15° | 527 |
| -Phenethyl | 199° | 527 |
| -Phenethyl | 205° | 527 |
| -Phenyl-2-propyl | 203° (dec.) | 527 |

TABLE IX-73. 3-Aminoalkylpyridines

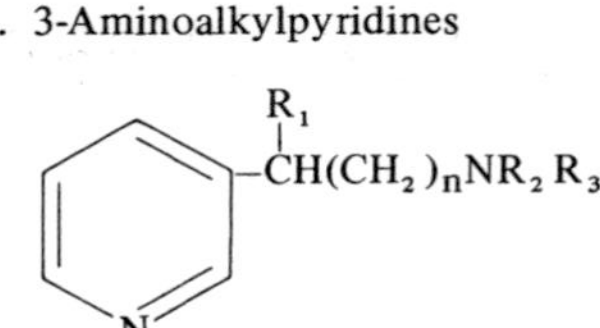


| n | $R_1$ | $R_2$ | $R_3$ | Substituents 2 | 4 | 5 | 6 | Physical properties | Ref. |
|---|---|---|---|---|---|---|---|---|---|
| 0 | Benzyl | H | H | H | H | H | H | – | 422 |
| 0 | Et | H | H | H | H | H | H | b.p. 105–8°/12 mm; phenylthiourea, m.p. 164–5° | 572 |
| 0 | Me | H | H | H | H | H | H | b.p. 100–2°/12 mm; dipicrate, m.p. 204–5° | 572 |
| 0 | *n*-Pr | H | H | H | H | H | H | b.p. 109–11°/12 mm; dipicrate, m.p. 237–8° | 572 |
| 0 | H | 6-Chloro-1,4-benzodioxan-2-ylmethyl | H | H | H | H | H | di-HCl, m.p. 144–50° | 581 |
| 0 | H | Diethylaminoethyl | H | H | H | H | H | b.p. 120–2°/1 mm; tripicrate, m.p. 185–6° | 499 |
| 0 | H | Me | H | H | H | H | H | *N*-nitroso, b.p. 115–6°/0.2 mm | 594 |
| 0 | H | Methylaminoethyl | H | H | H | H | H | – | 224 |
| 0 | H | Methylaminoethyl | H | H | H | H | H | – | 224 |
| 0 | H | Methylaminopropyl | H | H | H | H | H | – | 707 |
| 0 | H | 6-Methyl-1,4-benzodioxan-2-ylmethyl | H | H | H | H | H | di-HCl, m.p. 137–42° | 581 |

| | | | | | | | | | |
|---|---|---|---|---|---|---|---|---|---|
| 0 | H | Methyleneamino | H | H | H | H | H | b.p. 100.5°/0.04 mm; | 505 |
| | | | | | | | | $n_D$ 1.5450; | |
| | | | | | | | | $d_{20}$ 1.0604; | |
| | | | | | | | | tri-HCl, | 706 |
| | | | | | | | | m.p. 184–6° | |
| 0 | H | 3,4-Methylenedioxyphen-ethyl | H | H | H | H | H | HCl, m.p. 229–31° | 583 |
| 0 | H | 5-Methyl-8-isopropyl-1,4-benzodioxan-2-ylmethyl | H | H | H | H | H | di-HCl, m.p. 142–51° | 581 |
| 0 | H | Ph | H | H | H | H | H | m.p. 92–3° | 498 |
| | | | | | | | | b.p. 130–5°/0.06 mm; | |
| | | | | | | | | di-HCl, m.p. 194–5°; | |
| | | | | | | | | dipicrate, m.p. 137–9° | |
| 0 | H | Ph | H | H | H | H | H | m.p. 92–3°; | 498 |
| | | | | | | | | b.p. 130–5°/0.06 mm; | |
| | | | | | | | | di-HCl, m.p. 194–5°; | |
| | | | | | | | | dipicrate, m.p. 137–9° | |
| 0 | H | Phenethyl | H | H | H | H | H | HCl, m.p. 216–18° | 531 |
| 0 | H | $n$-Pr | H | H | H | H | H | b.p. 100–100.5°/0.04 mm | 505 |
| 0 | H | 3,4,5-Trimethoxybenzyl | H | H | H | H | H | m.p. 205–7° | 72 |
| 0 | Ph | 2-Dimethylaminoethyl | H | H | H | H | H | b.p. 160–3°/0.03 mm | 580 |
| 0 | $n$-Pr | Me | H | H | H | H | H | b.p. 112°/10 mm; | 572 |
| | | | | | | | | dipicrate, | |
| 0 | H | Benzoyl | 2-Diethylaminoethyl | H | H | H | H | m.p. 234–5° | |
| | | | | | | | | b.p. 190°/0.5 mm; | 499, 505 |
| 0 | H | Benzyl | Benzyl | H | H | H | H | dipicrate, | |
| 0 | H | Cinnamoyl | 2-Diethylaminoethyl | H | H | H | H | m.p. 138–9° | |
| | | | | | | | | m.p. 62–3° | 517 |
| | | | | | | | | b.p. 220°/0.5 mm; | 499, 505 |
| | | | | | | | | $HCl \cdot H_2O$, | |
| | | | | | | | | 102–4° | |
| 0 | H | α-Cyanobenzyl | Benzyl | H | H | H | H | m.p. 80–1° | 517 |

TABLE IX-73. 3-Aminoalkylpyridines (Continued)

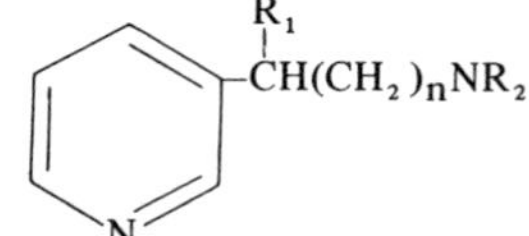

| n | $R_1$ | $R_2$ | $R_3$ | Substituents 2 | 4 | 5 | 6 | Physical properties | Ref. |
|---|---|---|---|---|---|---|---|---|---|
| 0 | H | Diphenylacetyl | 2-Diethylaminoethyl | H | H | H | H | b.p. 235°/0.75 mm; m.p. 66–7°, dipicrate, m.p. 159–61° | 499, 505 |
| 0 | H | Et | Et | H | H | H | H | b.p. 106–8°/12 mm; b.p. 96–7°/13 mm | 572 708 |
| 0 | H | Me | 2-Propynyl | H | H | H | H | b.p. 100°/3 mm | 550 |
| 0 | H | 3-Methylpyridyl | Benzyl | H | H | H | H | b.p. 225°/5 mm | 517 |
| 0 | H | 4-Methoxybenzyl | Benzyl | H | H | H | H | b.p. 230°/3–4 mm | 517 |
| 0 | H | Me | Et | H | H | H | H | b.p. 96–8°/12 mm; dipicrate, m.p. 172–3° | 572 |
| 0 | H | Me | Me | H | H | H | H | b.p. 72–5°/3 mm | 708 |
| 0 | H | Me | Me | Ph | H | H | Ph | b.p. 164–8°/0.1 mm | 593 |
| 0 | H | Ph | 2-Diethylaminoethyl | H | H | H | H | b.p. 129–30°/0.03 mm | 498 |
| 0 | H | H | H | H | Me | Ph | H | b.p. 120–3°/5 mm; dipicrate, m.p. 218–21° | 612 |
| 1 | Benzyl | H | H | H | H | H | H | – | 709 |
| 1 | Et | H | H | H | H | H | Me | b.p. 132–6°/10 mm | 568 |
| 1 | Me | H | H | H | H | H | Me | b.p. 109–10°/7 mm | 568 |
| 1 | OH | Benzyl | H | H | H | H | H | b.p. 189°/0.2 mm; m.p. 81°; | 538 |

| | | | | | | | | | |
|---|---|---|---|---|---|---|---|---|---|
| | | | | | | | | di-HCl, m.p. 190–5° | |
| 1 | OH | Me | H | H | H | H | H | b.p. 115–18°/0.4 mm; dibenzyl, m.p. 139–41° | 538 |
| 1 | OH | Pr | H | H | H | H | H | b.p. 126–8°/0.2 mm; di-HCl, m.p. 136° | 538 |
| 1 | Ph | Bu | H | H | H | H | Me | b.p. 180–3°/3 mm; picrate, m.p. 205–7° | 568 |
| 1 | Ph | Me | H | H | H | H | Me | b.p. 175–7°/3 mm | 568 |
| 1 | H | Et | Et | H | Me | H | H | b.p. 104°/4 mm; dipicrate, m.p. 202–3°; HCl, m.p. 153–4° | 507 |
| 1 | H | H | H | H | H | H | Me | b.p. 118°/8 mm; dipicrate, m.p. 210–11° | 568 |
| 2 | Ph | H | H | H | H | H | H | b.p. 180–90°/0.3 mm | 592 |
| 4 | H | *n*-BuO | H | H | H | H | H | – | |
| 3 | H | Et | Et | H | H | H | H | b.p. 133°/15 mm | 710 |

TABLE IX-74. 3-Amino-α-hydroxyalkylpyridines[532]

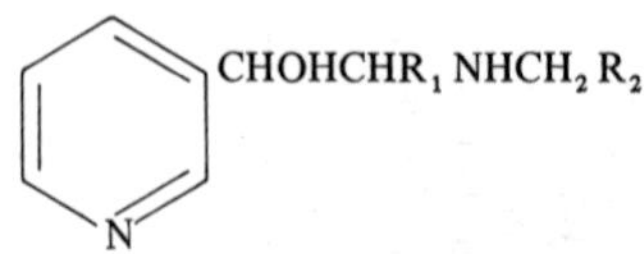


| $R_1$ | $R_2$ | Properties |
|---|---|---|
| Me | H | b.p. 115–18°/0.2 mm; di-HCl, m.p. 222–4° |
| H | $CH_2OH$ | b.p. 180–5°/0.54 mm; m.p. 90° dipicrate, m.p. 172° |
| Me | Me | b.p. 107–10°/0.14 mm; di-HCl, m.p. 232–4° |
| Me | Et | b.p. 119–23°/0.34 mm; di-HCl, m.p. 208–12° |
| Me | Ph | b.p. 163°/0.19 mm; di-HCl, m.p. 210–11° |
| Et | $CH_2OH$ | b.p. 184–8°/0.01 mm |

TABLE IX-75. 4-Picolylamines

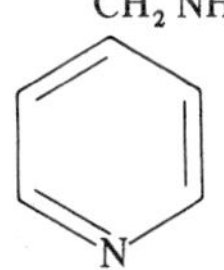


| R | m.p. | Ref. |
|---|---|---|
| Anilino | tri-HCl, 228–30° | 525 |
| *p*-Anisyl | 78–80°; b.p. 184–6°/4 mm | 534 |
| *p*-Carboxyphenyl | 244–7° | 527 |
| β-(3,4-Dimethoxyphenethyl) | 210° | 527 |
| 1-(3,4-Dimethoxyphenyl)-2-propyl | 153° | 527 |
| β-Hydroxy-β-(3,4-dimethoxyphenethyl) | 203° | 527 |
| β-Hydroxy-β-(4-methylphenethyl) | 201° | 527 |
| β-Hydroxy-β-(4-methoxyphenethyl) | 161° | 527 |
| 1-(4-Hydroxyphenyl)-2-propyl | 230° (dec.) | 527 |
| 1-Hydroxy-1-phenyl-2-propyl | 228° | 527 |
| β-Hydroxy-β-phenethyl | 143° | 527 |
| *o*-Hydroxyphenyl | 176–8° | 504, 527 |
| 1-(3,4-Methylenedioxyphenyl)-2-propyl | 168° | 527 |
| 1-(4-Methoxyphenyl)-2-phenyl | 190° | 527 |
| α-Phenethyl | 232° | 527 |
| β-Phenethyl | 198° (dec.) | 527 |
| Ph | 101–3°, 103–4°; | 534, 698 |
| | b.p. 125–50°/0.04 mm | 498 |
| | di-HCl, m.p. 197–9°; | |
| | picrate, m.p. 147–8° | |
| 1-Phenyl-2-propyl | 185° | 527 |
| 2-Pyridyl | 109–11°; b.p. 166–7°/4 mm | 534 |
| 3,4,5-Trimethoxyphenethyl | 210° (dec.) | 527 |
| *o*-Tolyl | 75–6°; b.p. 165–7°/3 mm | 534 |
| *p*-Tolyl | 74°; b.p. 162–6°/4 mm | 534 |

TABLE IX-76. 4-(Aminoalkyl)pyridines

| n | $R_1$ | $R_2$ | $R_3$ | $R_4$ | $R_5$ 2 | $R_5$ 3 | $R_5$ 5 | $R_5$ 6 | Physical properties | Ref. |
|---|---|---|---|---|---|---|---|---|---|---|
| 0 | Diethylaminoethyl | H | H | H | H | H | H | H | b.p. 110–12°/0.4 mm; tripicrate, m.p. 176–7° | 499 |
| 0 | Ethylamino | H | H | H | H | H | H | H | b.p. 108.5°/0.1 mm; $n_D^{20}$ 1.5454 $d_{20}$ 1.0580 tri-HCl, m.p. 247° | 505 |
| 0 | 3,4-(Methylenedioxy)phenethyl | H | H | H | H | H | H | H | HCl, m.p. 216–18° | 583 |
| 0 | Piperidino | – | H | H | IsoBu | OH | H | Me | HCl, m.p. 209–10° | 546 |
| 0 | Aminomethyl | H | H | Ph | H | H | H | H | m.p. 178–81° | 591 |
| 0 | Me | H | H | Ph | H | H | H | H | m.p. 178–81° | 591 |
| 0 | H | H | H | Ph | H | H | H | H | b.p. 143–7° | 595 |
| 0 | H | H | H | H | H | OH | Hydroxymethylene | H | di-HCl, m.p. 165–9° (dec.) | 523 |
| 0 | Benzyl | Benzyl | H | H | H | H | H | H | m.p. 85–6° | 517 |
| 0 | Benzoyl | Diethylamino- | H | H | H | H | H | H | b.p. 170°/0.04 mm; | 499 |

| | | | | | | | | | |
|---|---|---|---|---|---|---|---|---|---|
| 0 | Cinnamoyl | Diethylamino-ethyl | H | H | H | H | H | H | b.p. 195°/0.04 mm; dipicrate, m.p. 151–3° | 499 |
| 0 | Diethylamino-ethyl | Diphenylacetyl | H | H | H | H | H | H | b.p. 215°/0.09 mm; m.p. 93–4°; dipicrate, m.p. 165–6° | 499 |
| 0 | Diphenylmethyl | Me | H | H | H | H | H | H | m.p. 135–7° | 528 |
| 0 | Diethylamino-ethyl | Ph | H | H | H | H | H | H | b.p. 135–40°/0.04 mm; tripicrate, m.p. 292–3° | 498 |
| 0 | Et | Et | H | H | H | H | H | H | b.p. 101–3°/5 mm | 708 |
| 0 | Me | 2-Propynyl | H | H | H | H | H | H | b.p. 79°/1.2 mm; HCl, m.p. 183–5° | 549, 550 |
| 0 | 2-Phenylpropyl | 3,4,5-Trimethoxy-phenyl | H | H | H | H | H | H | m.p. 214–16° | 72 |
| 0 | Me | Me | H | H | Me | OH | H | Me | b.p. 110–12°/2–3 mm; di-HCl, m.p. 255–6° | 544 |
| 0 | Me | Me | H | H | Et | OH | H | Me | b.p. 115–20°/2–3 mm; di-HCl, m.p. 190.5–1.5° | 544 544 |
| 1 | H | H | H | H | H | H | H | H | b.p. 104°/9 mm; picrate, m.p. 152° | 564 |
| 1 | Benzoyl | H | H | H | H | H | H | H | 1-oxide, di-HCl, m.p. 177–8° | 574 |
| 1 | H | H | OH | H | H | H | H | H | b.p. 138°/0.7 mm | 574 |
| 1 | *o*-Chlorophenyl | Dimethylamino-ethyl | H | H | H | H | H | H | m.p. 99–100°; dipicrate, m.p. 138–42° | 561, 562 |
| 1 | *o*-Chlorophenyl | Dimethylamino-ethyl | H | H | H | H | H | H | b.p. 156°/0.09 mm | 561, 562 |
| 1 | *m*-Chlorophenyl | Dimethylamino-ethyl | H | H | H | H | H | H | b.p. 164–8°/0.02 mm | 561, 562 |
| 1 | *m*-Chlorophenyl | Dimethylamino-propyl | H | H | H | H | H | H | b.p. 178–80°/0.02 mm | 561, 562 |
| 1 | *p*-Chlorophenyl | Dimethylamino-ethyl | H | H | H | H | H | H | b.p. 168–70°/0.02 mm; dipicrate, m.p. 165–6° | 561, 562 |

TABLE IX-76. 4-(Aminoalkyl)pyridines (Continued)

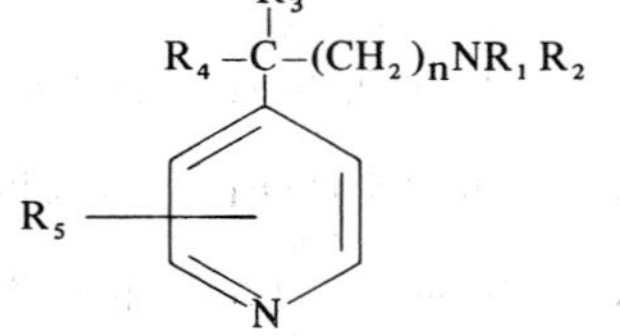


| n | $R_1$ | $R_2$ | $R_3$ | $R_4$ | $R_5$: 2 | 3 | 5 | 6 | Physical properties | Ref. |
|---|---|---|---|---|---|---|---|---|---|---|
| 1 | *p*-Chlorophenyl | Dimethylamino-propyl | H | H | H | H | H | H | b.p. 168–70°/0.05 mm | 561, 562 |
| 1 | Dimethylamino-ethyl | *m*-Methylphenyl | H | H | H | H | H | H | b.p. 158–60°/0.05 mm; dipicrate, m.p. 178–9° | 561, 562 |
| 1 | Dimethylamino-ethyl | Ph | H | H | H | H | H | H | b.p. 154–6°/0.02 mm; dipicrate, m.p. 176–7° | 561, 562 |
| 1 | 2-Ethoxybenzoyl | 2-Phenylpropyl | H | H | H | H | H | H | m.p. 103–6° | 569 |
| 1 | Me | Me | H | H | H | H | H | H | b.p. 70°/2 mm | 508 |
| 1 | Me | 2-Propynyl | H | H | H | H | H | H | b.p. 153°/20 mm | 550 |
| 1 | 2-Phenylpropyl | H | H | H | H | H | H | H | b.p. 138–40°/0.08 mm | 569 |
| 2 | H | H | H | H | H | H | H | H | b.p. 90–2°/3 mm | 508 |
| 2 | 3-Pyrrolidino | H | H | H | H | H | H | H | b.p. 132–5°/5 mm; methobromide, m.p. 191–3° | 508<br>508 |
| 2 | Me | Me | H | H | H | H | H | H | b.p. 90–2°/3 mm; methobromide, m.p. 253–4° | 508<br>508 |
| 2 | Me | Me | H | Ph | H | H | H | H | b.p. 133–5°/5 mm | 508 |

| | | | | | | | | | | |
|---|---|---|---|---|---|---|---|---|---|---|
| 2 | IsoPr | IsoPr | Amido | Ph | H | H | H | H | m.p. 138.5–9° | 592 |
| 2 | IsoPr | IsoPr | CN | Ph | H | H | H | H | b.p. 164–8°/0.7 mm | 592 |
| 3 | Me | Me | H | H | H | H | H | H | b.p. 95–105°/2 mm; methobromide, m.p. 209–10° | 508 |
| 4 | H | H | H | H | H | H | H | H | dipicrate, m.p. 134–5° | 547 |
| 7 | H | H | H | H | H | H | H | H | dipicrate, m.p. 108° | 547 |
| 12 | H | H | H | H | H | H | H | H | dipicrate, m.p. 110° | 547 |

TABLE IX-77. 6-Substituted-2,4-bis(alkylaminomethyl)-pyridines[546]

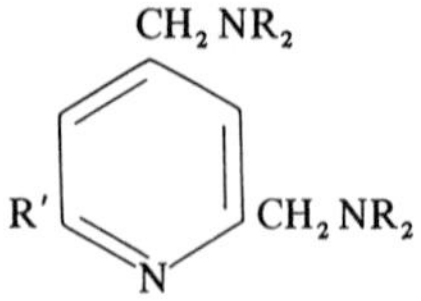

$$\text{2,4-bis}(CH_2NR_2)\text{-6-R}'\text{-pyridine}$$

| $R_2N$ | $R_1$ | Physical properties |
|---|---|---|
| $Me_2N$ | *n*-Bu | b.p. 180–2°/1 mm |
| $Me_2N$ | IsoAm | b.p. 150–1°/1 mm |
| Piperidino | *n*-Bu | tri-HCl, m.p. 224–5.5° |
| Piperidino | *n*-Pr | b.p. 98–100°/1–2 mm; tri-HCl, m.p. 213–4° |
| Piperidino | IsoPr | m.p. 76–7°; tri-HCl, m.p. 198–9° |

TABLE IX-78. Miscellaneous Aminopyridines

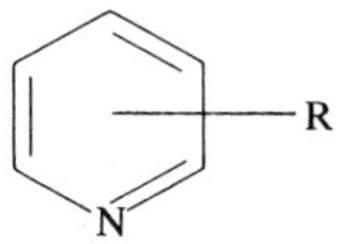

| | Physical properties | Ref. |
|---|---|---|
| 3-Amino-5-aminomethyl-4-methoxymethyl-2-methylpyridine | di-HCl, m.p. 230–1°; dipicrate, m.p. 214–15° | 511 |
| 3-Amino-5-aminomethyl-2-methylpyridine | di-HCl, m.p. 295–7° | 711 |
| 3-Amino-4,5-bis(aminomethyl)-2-methylpyridine | tri-HCl | 378 |
| 3-Amino-4,5-bis(carbethoxyaminomethyl)-2-methylpyridine | m.p. 195–6° | 378 |
| 2-[1-(2-Aminoethyl)amino] ethylpyridine | b.p. 74–5°/0.07 mm | 529 |
| 4,5-Bis(aminomethyl)-3-chloro-2-methylpyridine | tri-HCl | 378 |
| 2,2′-{3-[(*p*-Chloro-α-methyl)amino] propylidene}dipyridine | b.p. 201–10°/0.5 mm | 539 |
| 1-Methylamino-2-pyridylcyclohexane | b.p. 72–4°/0.08 mm | 575 |
| 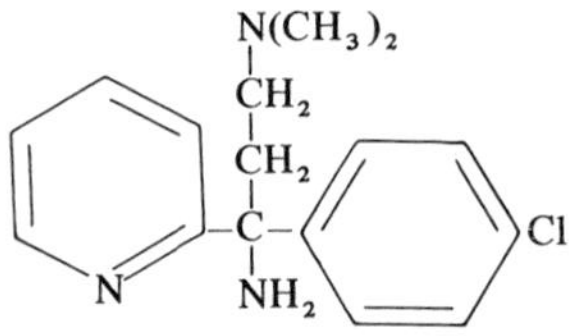 | b.p. 155–60°/3 mm; maleate, m.p. 132–3°; dipicrate, m.p. 194–6° | 705 |
| 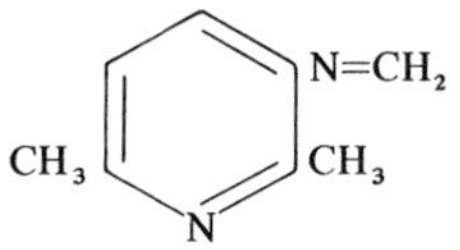 | m.p. 146° | 351 |

TABLE IX-78. Miscellaneous Aminopyridines (Continued)

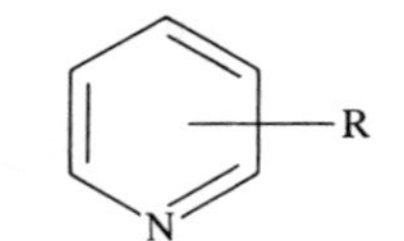


| | | Physical properties | Ref. |
|---|---|---|---|
| R-(2-pyridyl)–CH(2-pyridyl)$CH_2CH_2NHCH(CH_3)CH_2$–$C_6H_5$ | R = H | b.p. 210–18°/1.5 mm<br>m.p. 69–70°<br>HCl, m.p. 130–1° | 539 |
| | R = $CH_3$ | b.p. 204–8°/0.5 mm;<br>HCl, m.p. 146–7° | 539 |
| 1-(2-Pyridyl)-1-aminocyclohexane | | di-HCl, m.p. 241–2°; acetamide, m.p. 141–2° | 575 |
| 1-(2-Pyridyl)-1-ethylaminocyclohexane | | m.p. 63.5–64° | 575 |
| 1-(2-Pyridyl)-1-ethylmethlaminocyclohexane | | b.p. 90–100°/0.1 mm | 575 |
| 1-(2-Pyridyl)-1-*N*-piperidinocyclohexane | | b.p. 105–8°/0.03 mm | 575 |
| 2-(2-Pyridylthio)-1-aminoethane | | di-HCl, m.p. 201° | |

TABLE IX-79. 2-, 3-, and 4-Dimethylamino-2-alkynylpyridines

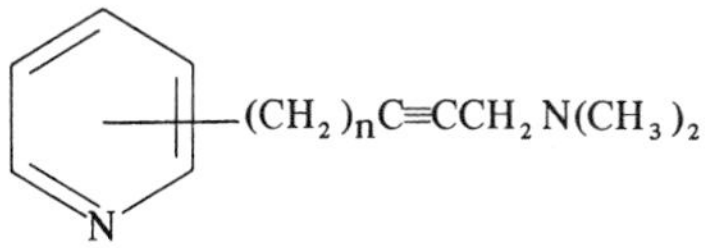

(Structure: pyridine ring–$(CH_2)_nC{\equiv}CCH_2N(CH_3)_2$)

| Ring position | n | b.p./mm | Dipicrate, m.p. | Ref. |
|---|---|---|---|---|
| 2 | 2 | 142–3°/13 | 159° | 547 |
| 2 | 3 | 159–60°/14 | – | 547 |
| 2 | 4 | 177°/17 | 157–8° | 547 |
| 2 | 5 | 187°/14 | 134° | 547 |
| 2 | 10 | 182°/0.7 | 70° monopicrate, m.p. 88° | 547 |
| 3 | 3 | 151–2°/2 | – | 548 |
| 3 | 4 | 147–8°/0.8 | – | 548 |
| 4 | 3 | 142–3°/1.4 | – | 548 |
| | | 162°/14 | 135° | 547 |
| 4 | 5 | 166°/1.5 | 158.5° | 547 |
| 4 | 10 | 213–14°/1.6 | 75° | 547 |

TABLE IX-80. Other 2-, 3-, or 4-Dialkylamino-2-butynylpyridines[552]

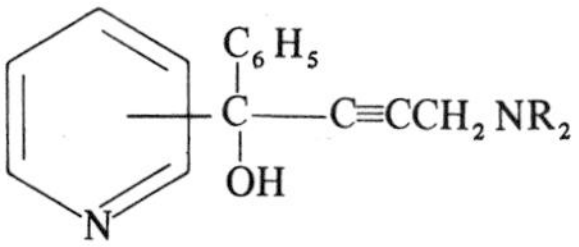

(Structure: pyridine ring–$C(C_6H_5)(OH)$–$C{\equiv}CCH_2NR_2$)

| Ring position | R | Properties |
|---|---|---|
| 2 | Et | m.p. 70–2°; methobromide, m.p. 151–2° |
| 2 | Me | m.p. 116–7°; methobromide, m.p. 154–5° |
| 3 | Et | m.p. 81.2° |
| 4 | Et | m.p. 72–4° |

TABLE IX-81. β-Dialkylaminoethyl Pyridyl Sulfides

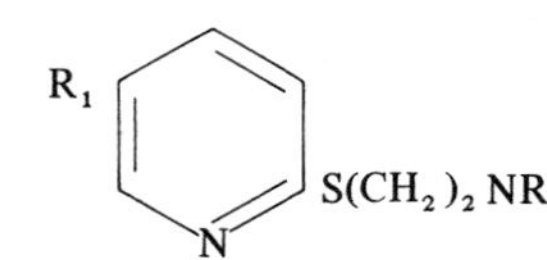

| Ring position | $R_1$ | R | Physical properties | m.p. of derivatives | | | | Ref. |
|---|---|---|---|---|---|---|---|---|
| | | | | Picrate | HCl | MeI | EtI | |
| 2 | $NO_2$ | Me | m.p. 40–2° | 167–70° | 182–3° | 190–2° | 218–20° | 585 |
| 2 | $NO_2$ | Et | m.p. 50–2° | – | 173–4° | 172–4° | 206–8° | 585 |
| 2 | H | Me | b.p. 91–3°/3 mm | 30–2° | 194–6° | 183–4° | 150–2° | 585 |
| 2 | H | Et | b.p. 130–2°/4 mm | 103–5° | – | 130–2° | 182–4° | 585 |
| 2 | H | H | (di-HCl, m.p. 201°; (acetyl, m.p. 70°) (diacetamide, m.p. 102°) (diacetamide sulfone, m.p. 120°) | | | | | 586 |
| 4 | H | H | (di-HCl, m.p. 236°) (acetyl, m.p. 75°) (diacetamide sulfone, m.p. 124°) | | | | | 586 |

TABLE IX-82. 2-Aminoethoxypyridines[589]

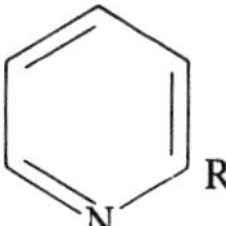

| R | Physical properties |
|---|---|
| [2-(2-Morpholinoethylamino)ethoxy]- | b.p. 170–8°/1 mm |
| [2-(3-Morpholinopropylamino)ethoxy]- | di-HCl, m.p. 189–90° |
| [2-(*N*-methyl-2-morpholinoethylamino)-ethoxy]- | di-HCl, m.p. 180–2° |
| [2-(*N*-methyl-2-piperidinoethylamino)-ethoxy]- | di-HCl, m.p. 208–10° |

TABLE IX-83. Mandelamidopyridines

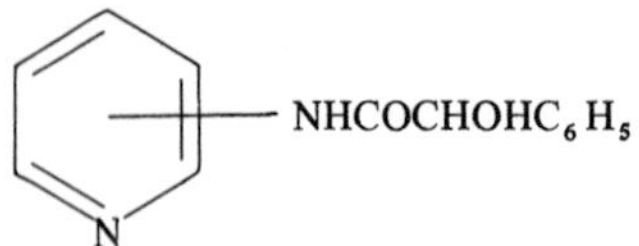


| Compound | m.p. | HCl, m.p. | Ref |
|---|---|---|---|
| 2-(*O*-Acetylmandelamido)pyridine | – | – | 331 |
| 4-(*O*-Acetylmandelamido)pyridine | | | 331 |
| 2-(4-Bromomandelamido)pyridine | 146–7° | 197–8° | 331 |
| 5-Bromo-2-mandelamidopyridine | 155–6° | 175° (dec.) | 331 |
| 2-[β-(*N*-Butylcarbamyloxy)phenethylamino] pyridine | 95–8° | | 332 |
| 5-Chloro-2-mandelamidopyridine | 146–8° | 169° | 332 |
| 2-[β-(*N*-Ethylcarbamoyloxy)phenethylamino] pyridine | 132–4° | 75–8° | 332 |
| 2-(Hexahydromandelamido)pyridine | 118–20° | – | 331 |
| 2-(β-Hydroxy-4-bromophenethylamino)-pyridine | 105–6° | 135–6° | 332 |
| 2-(β-Hydroxy-4-chlorophenethylamino)-pyridine | 91–8° | 112–23° | 332 |
| 2-(β-Hydroxy-β-cyclohexylethylamino)-pyridine | 86–8° | 143–4° | 322 |
| 2-(β-Hydroxyphenethylamino)-5-bromo-pyridine | 110–11° | 187–9° | 322 |
| 2-(β-Hydroxyphenethylamino)-5-chloro-pyridine | 102–3° | 177–8° | 332 |
| 6-(β-Hydroxyphenethylamino)-2,4-lutidine | 77–8° | 132–3° | 322 |
| 2-[*N*-(β-Hydroxyphenethyl)methyl-amino] pyridine | b.p. 147–57°/0.3 mm | 170–3° | 332 |
| 6-(β-Hydroxyphenethylamino)-3-picoline | 97–9° | 116–17° | 332 |
| 2-(β-Hydroxy-β-phenethylamino)-4-picoline | 90–1° | 135–6° | 332 |
| 2-(β-Hydroxyphenethylamino)pyridine | 102–5° | 124–6° | 332 |
| 3-(β-Hydroxyphenethylamino)pyridine | b.p. 195–202°/0. 3 mm | – | 322 |
| 2-(γ-Hydroxy-γ-phenylpropylamino)-pyridine | 123–4° | 138–60° | 332 |
| 2-Mandelamido-4-picoline | 143–6° | 188–9° | 332 |
| 6-Mandelamido-2,4-lutidine | 167–9° | 195–6° | 332 |
| 6-Mandelamido-3-picoline | 141–2° | 203° | 332 |

TABLE IX-84. 2-Sulfanilamidopyridines[388]

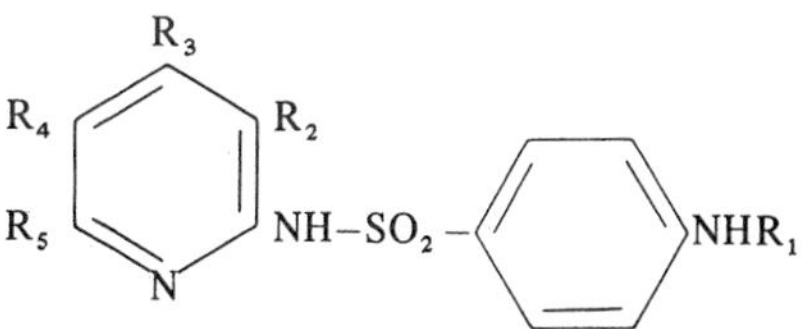


| $R_1$ | $R_2$ | $R_3$ | $R_4$ | $R_5$ | m.p. |
|---|---|---|---|---|---|
| Acetyl | OMe | H | H | H | 212° |
| Acetyl | H | OMe | H | H | 262–3° |
| Acetyl | H | H | OMe | H | 228–30° |
| Acetyl | H | OMe | H | OMe | 201–3° |
| Acetyl | H | OMe | H | Me | 236–8° |
| H | OMe | H | H | H | 214–15° |
| H | H | OMe | H | H | 237–9° |
| H | H | H | OMe | H | 200–1° |
| H | H | OMe | H | OMe | 158–9° |
| H | H | OMe | H | Me | 184–5° |

TABLE IX-85. 3-Sulfanilamidopyridines[388]

| $R_1$ | $R_2$ | $R_3$ | $R_4$ | $R_5$ | m.p. |
|---|---|---|---|---|---|
| Acetyl | H | H | Br | H | 250–1° |
| Acetyl | H | H | Cl | H | 255–6° |
| Acetyl | OMe | H | H | H | 194–5° |
| Acetyl | H | OMe | H | H | 257–8° |
| Acetyl | H | H | OMe | H | 263–4° |
| Acetyl | OMe | H | H | OMe | 193–5° |
| Acetyl | OMe | H | H | Br | 210–11° |
| Acetyl | H | OMe | H | Me | 228–9° |
| Acetyl | OMe | OMe | H | Me | 236–7° |
| H | H | H | Br | H | 208–10° |
| H | H | H | Cl | H | 208–9° |
| H | OMe | H | H | H | 134–6° |
| H | H | OMe | H | H | 222–3° |
| H | H | H | OMe | H | 220–1° |
| H | OMe | H | H | Br | 172–3° |
| H | OMe | H | H | OMe | 165–7° |
| H | H | OMe | H | Me | 210–11° |
| H | OMe | OMe | H | Me | 225–7° |

TABLE IX-86. 4-Sulfanilamidopyridines[388]

| $R_1$ | $R_2$ | $R_3$ | $R_4$ | $R_5$ | m.p. |
|---|---|---|---|---|---|
| Acetyl | OMe | H | H | H | 218–19° |
| Acetyl | H | OMe | H | H | 232–3° |
| Acetyl | OMe | H | OMe | H | 230–1° |
| Acetyl | OMe | H | Me | H | 204–5° |
| H | OMe | H | H | H | 151–2° |
| H | H | OMe | H | H | 173–4° |
| H | OMe | H | OMe | H | 181–2° |
| H | OMe | H | Me | H | 69–71° |

TABLE IX-87. 1-(2-Pyridylamino)glutarimides[123]

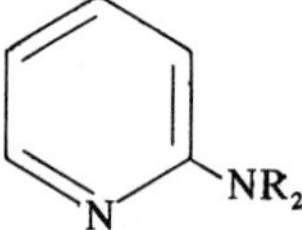


| $R_2$ | m.p. |
|---|---|
| 4,4-Dimethylglutarimide | 196° |
| 4-Ethyl-4-methylglutarimide | 137–8° |
| 4,4-Tetramethyleneglutarimide | 172–3° |
| 4-Phenylglutarimide | 201–3° |

TABLE IX-88. *N*-Aryl-*N'*-(Pyridylalkyl)ureas[595]

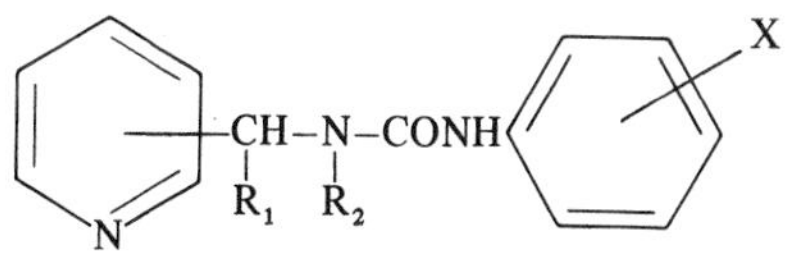

| Pyridine ring position | $R_1$ | $R_2$ | X | Physical properties |
|---|---|---|---|---|
| 2 | Me | H | 2-Cl | HBr, m.p. 134–5° |
| 2 | Me | H | 4-OEt | – |
| 2 | H | H | 2-Cl | m.p. 148–50°; HCl, m.p. 173–5° |
| 3 | Me | H | 2-Cl | – |
| 3 | Me | H | 4-OEt | m.p. 118–19° |
| 3 | H | H | 4-Cl | m.p. 171–3° |
| 3 | H | H | 3,4,5-$(OMe)_3$ | m.p. 164–6°; HCl, m.p. 169–70° |
| 4 | Me | H | 4-Br | m.p. 159–60° |
| 4 | Me | H | 2-Cl | HCl, m.p. 145–6° |
| 4 | Me | H | 3-Cl | HCl, m.p. 143–7° |
| 4 | Me | H | 4-Cl | m.p. 136–7° |
| 4 | Me | H | 2,5-$Cl_2$ | m.p. 95–7° |
| 4 | Me | H | 3-F | – |
| 4 | Me | H | 4-I | – |
| 4 | Ph | H | 2-Cl | m.p. 192–5° |
| 4 | H | H | 4-Cl | HCl, m.p. 235–6° |
| 4 | H | H | 2-OEt | – |
| 4 | H | H | 3,4,5-$(OMe)_3$ | m.p. 170.5–2.5°; HCl, m.p. 222–4° |

TABLE IX-89. Arylmethylene-2-aminopyridines

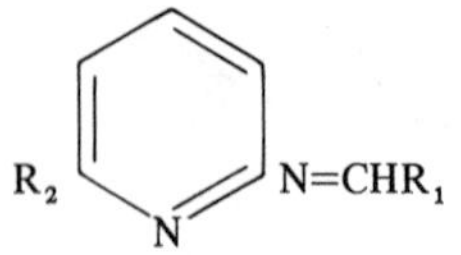

[Structure: pyridine ring with $R_2$ and N=CHR$_1$ substituents]

| $R_1$ | $R_2$ | Physical properties | Ref. |
|---|---|---|---|
| 2-Furanyl | Me | b.p. 138–9°/0.3 mm | 305 |
| *o*-Aminobenzylidene | H | m.p. 125° | 297 |
| *o*-Anilino | H | m.p. 125° | 296, 297 |
| Benzyl | H | b.p. 123–5°/1 mm | 578 |
| (3-Chloro-2-benzyfuranyl)methyl | H | m.p. 107–9° | 307 |
| Dibenzoylmethylene | H | m.p. 70° | 297 |
| 4,5-Dichloro-2-thienyl | H | m.p. 98.5° | 310 |
| *p*-(2-Diethylaminoethoxy)phenyl | H | b.p. 160–5°/0.1 mm | 578 |
| 2-Et-3-benzofuranyl | H | b.p. 105–15°/0.001 mm | 650 |
| 2-Furanyl | H | m.p. 88–9° | 305 |
| Isopropylidene | H | Acetyl, m.p. 36° | 297 |
| *p*-Nitrobenzyl | H | m.p. 147–8° | 712 |

TABLE IX-90. Aldimines

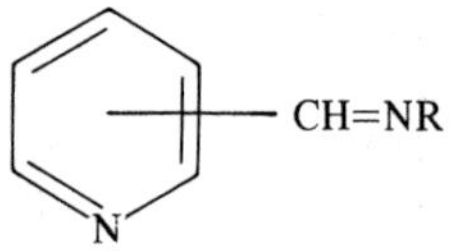

[Structure: pyridine ring with CH=NR substituent]

| Ring position | R | m.p. | Ref. |
|---|---|---|---|
| 2 | *p*-Carboxyphenyl | 235–7° | 527 |
| 2 | *p*-Hydroxyphenyl | 186° | 527 |
| 2 | *β*-Phenethyl | 39–41°; b.p. 147–8° | 531 |
| 3 | *p*-Aminophenyl | 144° | 527 |
| 3 | *o*-Carboxyphenyl | 87–8° | 527 |
| 3 | *m*-Carboxyphenyl | 219–21° | 527 |
| 3 | *p*-Carboxyphenyl | 241–3° | 527 |
| ? | Diphenylmethyl | 105–7° | 528 |
| 3 | *p*-Hydroxyphenyl | 212–13° | 527 |
| 3 | *m*-Nitrophenyl | 109–11° | 527 |
| 3 | *β*-Phenethyl | 46–7° | 531 |
| 4 | *p*-Carboxyphenyl | 288–90° | 527 |
| 4 | *o*-Hydroxyphenyl | 166–8° | 527 |
| 4 | *p*-Hydroxyphenyl | 201–2° | 527 |
| 4 | *β*-Phenethyl | b.p. 150–1° | 531 |

TABLE IX-91. Amino-2-imino-1,2-dihydropyridines

| $R_1$ | $R_2$ | $R_3$ | $R_4$ | Physical properties | Ref. |
|---|---|---|---|---|---|
| $NH_2$ | H | H | 2-Pyridyl | | 656 |
| H | $NH_2$ | H | 2-Pyridyl | m.p. 132.5–4° | 116 |
| $NH_2$ | $NH_2$ | H | 2-Pyridyl | m.p. > 300° | 723 |
| 2-Amino-3,3,4,4,5,5-hexafluoro-6-imino-piperidine | | | | m.p. 162–73° | 713 |

TABLE IX-92. 1,2,3,6-Tetrahydroaminopyridines

| $R_1$ | $R_2$ | $R_3$ | Physical properties | Ref. |
|---|---|---|---|---|
| Me | *p*-Aminobenzamido | Me | m.p. 168–70° | 718 |
| Me | *p*-Aminobenzylamino | Me | m.p. 126–7° | 718 |
| -Aminoethyl | Et | H | b.p. 102–4°/10 mm | 717 |
| -Aminoethyl | Ph | H | – | 329 |
| -Aminobutyl | Ph | H | b.p. 135–6°/0.2 mm | 329 |
| Me | $NH_2C_6H_4CH_2CH_2CONH$ | Me | m.p. 153–4° | 719, 720 |

TABLE IX-93. 1,2,3,4-Tetrahydroaminopyridines

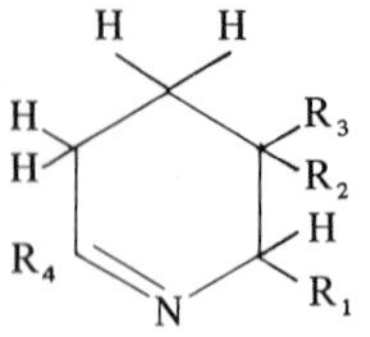

| $R_1$ | $R_2$ | $R_3$ | $R_4$ | Physical properties | Ref. |
|---|---|---|---|---|---|
| 2-Dimethyl-amino | H | H | H | – | 383 |
| 2-Diethyl-amino | H | H | H | – | 383 |
| Me | 3-Amino-propyl | Me | H | b.p. 92°;[a] picrate, m.p. 162–3° | 722 |
| H | H | H | 4-Amino-butyl | – | 721 |

[a]Pressure not reported.

# References


1. M. I. Farberov, A. M. Kut'in, B. F. Ustaushchikov, and N. K. Shemyakina, *Zh. Prikl. Khim.*, **37,** 661 (1964); *Chem. Abstr.*, **60,** 14467 (1964).
2. F. Johnson, U.S. patent, 3,096,377 (1963); *Chem. Abstr.*, **59,** 12768 (1963).
3. F. Johnson, J. P. Panella, A. A. Carlson, and D. H. Hummeman, *J. Org. Chem.*, **27,** 2473 (1962).
4. T. Takata, *Bull. Chem. Soc. Jap.*, **35,** 1438 (1962); *Chem. Abstr.*, **58,** 2451 (1963).
5. F. Johnson, U.S. patent, 3,255,041 (1965); *Chem. Abstr.*, **64,** 6623 (1966).
6. F. Johnson, U.S. patent, 3,247,214 (1966); *Chem. Abstr.*, **64,** 19570 (1966).
7. M. R. S. Weir, K. E. Helmer, and J. B. Hyne, *Can. J. Chem.*, **41,** 1042 (1963).
8. Daiichi Seiyaku Co., Ltd., Jap. patent, 26,386 (1963); *Chem. Abstr.*, **60,** 1199 (1964).
9. D. J. Collins, *J. Chem. Soc.*, 1337 (1963).
10. R. Urban, L. H. Chopard-dit-Jean, and O. Schnider, *Gazz. Chim. Ital.*, **93,** 163 (1963); *Chem. Abstr.*, **59,** 1582 (1963).
11. F. Parisi, P. Bovina, and A. Quilico, *Gazz. Chim. Ital.*, **90,** 903 (1969); *Chem. Abstr.*, **55,** 21116 (1961).
12. P. Nantka-Namirski, Pol. patent, **43,**117 (1960); *Chem. Abstr.*, **55,** 13451 (1961).
13. W. Mathes and A. Wolf, Ger. patent, 1,119,274 (1961); *Chem. Abstr.*, **56,** 14247 (1962).
14. F. H. Case and W. A. Butte, *J. Org. Chem.*, **26,** 4415 (1961).
15. H. Bojarska-Dahlig and I. Gruda, *Rocz. Chem.*, **33,** 505 (1959); *Chem. Abstr.*, **54,** 522 (1960).
16. R. A. Abramovitch and J. G. Saha, in "Advances in Heterocyclic Chemistry," A. R. Katritzky and A. J. Boulton, Eds., Vol. 6, Academic Press, New York, 1966, p. 295.
17. R. A. Abramovitch, F. Helmer, and J. G. Saha, *Chem. Ind.*, (London) 659 (1964).
18. R. A. Abramovitch, F. Helmer, and J. G. Saha, *Can. J. Chem.*, **43,** 725 (1965).

19. N. S. Prostakov, N. N. Mikheeva, and D. Pkhal'gumani, *Khim. Geterotsikl. Soedin.*, 671 (1967); *Chem. Abstr.*, **68**, 104915X (1968).
20. R. F. Evans and H. C. Brown, *J. Org. Chem.*, **27**, 1329 (1962).
21. A. Roedig and K. Grohe, *Chem. Ber.*, **98**, 923 (1965).
22. Presented by C. S. Giam, First International Congress of Heterocyclic Chemistry, June 15, 1967, Albuquerque, N. Mex. Abstr. No. 92.
23. L. S. Levitt and B. W. Levitt, *Chem. Ind.* (London), 1621 (1963).
24. H. L. Jones and D. L. Beveridge, *Tetrahedron Lett.*, 1577 (1964).
25. G. C. Barrett and K. Schofield, *Chem. Ind.* (London), 1980 (1963).
26. R. F. Childs and A. W. Johnson, *Chem. Ind.* (London), 542 (1964).
27. Y. Ban and T. Wakamatsu, *Chem. Ind.* (London), 710 (1964).
28. M. J. Pieterse and H. J. den Hertog, *Rec. Trav. Chim. Pays-Bas*, **80**, 1376 (1961); *Chem. Abstr.*, **58**, 5625 (1963).
29. R. J. Martens, H. J. den Hertog, and M. Van Ammers, *Tetrahedron Lett.*, 3207 (1964).
30. R. J. Martens and H. J. den Hertog, *Rec. Trav. Chim. Pays-Bas*, **83**, 621 (1964); *Chem. Abstr.*, **61**, 9460 (1964).
31. T. Kato, T. Niitsuma, and N. Kusaka, *Yakugaku Zasshi*, **84**, 432 (1964); *Chem. Abstr.*, **61**, 4171 (1964).
32. R. J. Martens and H. J. den Hertog, *Rec. Trav. Chim. Pays-Bas*, **86**, 655 (1967).
33. M. J. Pieterse and H. J. den Hertog, *J. S. African Chem. Inst.*, **17**, 41 (1964); *Chem. Abstr.*, **61**, 5604 (1964).
34. M. J. Pieterse and J. H. den Hertog, *Rec. Trav. Chim. Pays-Bas*, **81**, 855 (1962); *Chem. Abstr.*, **58**, 11325 (1963).
35. H. J. den Hertog, M. J. Pieterse, and D. J. Buurman, *Rec. Trav. Chim. Pays-Bas*, **82**, 1173 (1963).
36. J. W. Streef and H. J. den Hertog, *Rec. Trav. Chim. Pays-Bas*, **85**, 803 (1966); *Chem. Abstr.*, **66**, 75975t (1967).
37. H. J. den Hertog, H. C. Van der Plas, M. J. Pieterse, and J. W. Streef, *Rec. Trav. Chim. Pays-Bas*, **84**, 1569 (1965); *Chem. Abstr.*, **64**, 12670 (1966).
38. H. J. den Hertog, R. J. Martens, H. C. Van der Plas, and J. Bon, *Tetrahedron Lett.*, 4325 (1966).
39. Yu. I. Chumakov, *Metody Polucheniya Khim. Reaktivov i Preparatov, Gos. Kom. Sov. Min. SSSR po Khim.*, **11**, 77, (1964); *Chem. Abstr.*, **64**, 15832 (1966).
40. L. van der Does and H. J. den Hertog, *Rec. Trav. Chim. Pays-Bas*, **84**, 951 (1965); *Chem. Abstr.*, **63**, 14805 (1965).
41. T. Takahasi, K. Kanematsu, R. Ohishi, and T. Mizutani, *Chem. Pharm. Bull.* (Tokyo), **8**, 539 (1960); *Chem. Abstr.*, **55**, 13429 (1961).
42. E. C. Taylor and J. S. Driscoll, *J. Amer. Chem. Soc.*, **82**, 3141 (1960).
43. J. M. Essery and K. Schofield, *J. Chem. Soc.*, 4953 (1960).
44. P. Nantka-Namirski, *Acta Polon. Pharm.*, **18**, 391 (1961); *Chem. Abstr.*, **57**, 16554 (1962).
45. A. Hunger, H. Keberle, A. Rossi, and K. Hoffman, Swiss patent 386,442 (1965); *Chem. Abstr.*, **63**, 14876 (1965).
46. J. A. Montgomery and K. Hewson, *J. Med. Chem.*, **9**, 354 (1966).
47. Z. Talik, *Rocz. Chem.*, **36**, 1313 (1962); *Chem. Abstr.*, **59**, 6358 (1963).
48. T. Talik, *Rocz. Chem.*, **37**, 69 (1963); *Chem. Abstr.*, **59**, 8698 (1963).
49. T. Talik and Z. Talik, *Rocz. Chem.*, **41**, 1721 (1967); *Chem. Abstr.*, **69**, 2823u (1968).
50. T. Talik and Z. Talik, *Rocz. Chem.*, **40**, 1187 (1966); *Chem. Abstr.*, **66**, 11165u (1967).

51. T. Talik and Z. Talik, *Rocz. Chem.*, **38**, 777 (1964); *Chem. Abstr.*, **61**, 10653 (196
52. Z. Talik, T. Talik, and A. Puszynski, *Rocz. Chem.*, **39**, 601 (1965); *Chem. Abstr.*, **6** 16298 (1965).
53. R. D. Chambers, J. Hutchinson, W. K. R. Musgrave, Belg. patent, 660,873 (196 *Chem. Abstr.*, **65**, 7152 (1966).
54. R. E. Banks, J. E. Burgess, W. M. Cheng, and R. N. Hazeldine, *J. Chem. Soc.*, 5 (1965).
55. R. D. Chambers, J. Hutchinson, and W. K. R. Musgrave, *J. Chem. Soc.*, 3573 (196
56. C. T. Redemann, Belg. patent, 628,486 (1963); *Chem. Abstr.*, **60**, 15840 (1964).
57. R. D. Chambers, J. Hutchinson, and W. K. R. Musgrave, *Proc. Chem. Soc.*, **83**, (196
58. R. D. Chambers, J. Hutchinson, and W. K. R. Musgrave, *J. Chem. Soc., Suppl.*, **1**, 56 (1964).
59. R. E. Banks, R. N. Haszeldine, E. Phillips, and I. M. Young, *J. Chem. Soc., C*, 20 (1967).
60. W. Czuba, *Rocz. Chem.*, **34**, 905 (1960).
61. W. Czuba, *Rec. Trav. Chim. Pays-Bas*, **82**, 988 (1963); *Chem. Abstr.*, **60**, 1723 (196
62. L. N. Yakhontov, E. I. Lapan, and M. V. Rubtsov, *Khim. Geterotsikl. Soedin.*, 10 (1967); *Chem. Abstr.*, **69**, 86786v (1968).
63. V. L. Zbarskii, G. M. Shutov, V. F. Zhilin, and E. Yu. Orlova, *Khim. Geterotsik Soedin.*, **1**, 178 (1967); *Chem. Abstr.*, **67**, 73500f (1967).
64. M. Nakadate, Y. Takano, T. Hirayama, S. Sakizawa, T. Hirano, K. Okamoto, K. Hira T. Kawamura, and M. Kimura, *Chem. Pharm. Bull.* (Tokyo), **13**, 113 (1965); *Che Abstr.*, **62**, 14620 (1965).
65. V. L. Zbarskii, L. M. Glazunova, G. M. Shutov, V. F. Zhilin, and E. Yu. Orlov U.S.S.R. patent, 195,455 (1967); *Chem. Abstr.*, **68**, 104990t (1968).
66. G. C. Finger, L. D. Starr, A. Roe, and W. J. Link, *J. Org. Chem.*, **27**, 3965 (1962).
67. B. M. Ferrier and N. Campbell, *Proc. Roy. Soc. Edinburgh*, **A65**, 231 (1959).
68. T. Kato and T. Niitsuma, *Chem. Pharm. Bull.* (Tokyo), **13**, 963 (1965); *Chem. Abst* **63**, 14804 (1965).
69. W. L. Mosby, *Chem. Ind.* (London), 1348 (1959).
70. S. Pietra and G. Tacconi, *Farmaco, Ed. Sci.*, **19**, 741 (1964); *Chem. Abstr.*, **62**, 16 (1965).
71. J. DeLarge, *Farmaco., Ed. Sci.*, **22**, 99 (1967); *Chem. Abstr.*, **66**, 94891d (1967).
72. G. N. Walker, M. A. Moore, and B. N. Weaver, *J. Org. Chem.*, **26**, 2740 (1961).
73. P. Yates, M. J. Jorgenson, and S. K. Roy, *Can. J. Chem.*, **40**, 2146 (1962).
74. J. Suszko and M. Szafran, *Rocz. Chem.*, **39**, 709 (1965); *Chem. Abstr.*, **64**, 346 (1966).
75. G. Tocconi and S. Pietra, *Ann. Chim.* (Rome), **55**, 810 (1965); *Chem. Abstr.*, **64**, 6 (1966).
76. E. V. Brown, A. F. Smetana, and A. A. Hamdan, *J. Med. Chem.*, **8**, 252 (1965); *Che Abstr.*, **62**, 12279 (1965).
77. B. Benke, S. Joger, L. Szabo, I. Koczka, G. Losonczi, and I. Hoffmann, Hung. pate 150,277, (1963); *Chem. Abstr.*, **60**, 2909 (1964).
78. V. A. Kon'kova and L. A. Petrova, *Trudy Vsesoyuz, Nauch.-Issledovatel, Vitam Inst.*, **6**, 10 (1959); *Chem. Abstr.*, **55**, 12399 (1961).
79. P. Krumholz, *Inorg. Chem.*, **4**, 609 (1965).
80. K. Hoffmann, A. Hunger, J. Kerbrle, and A. Ross, U.S. patent, 2,987,518 (1961 *Chem. Abstr.*, **55**, 24795 (1961).
81. G. Wagner and E. Fickweiler, *Arch. Pharm.* (Weinheim), **298**, 297 (1965); *Che Abstr.*, **63**, 5728 (1965).

2. G. Wagner and E. Fickweiler, *Arch. Pharm.* (Weinheim), **298**, 62 (1965); *Chem. Abstr.*, **63**, 1847 (1965).
3. T. Wieland, C. Fest, and G. Pfleiderer, *Ann. Chem.*, **642**, 163 (1961)
4. E. Hayashi and H. Yamanaka, Jap. patent, 15,616 (1961); *Chem. Abstr.*, **56**, 12863 (1962).
5. H. J. Knackmuss, *Chem. Ber.*, **101**, 1148 (1968).
6. R. P. Rao, *Austr. J. Chem.*, **17**, 1434 (1964).
7. R. J. Rousseau and R. K. Robins, *J. Heterocycl. Chem.*, **2**, 196 (1965).
8. J. Dadrowski, W. Jasiobedzki, T. Jaworski, M. Midon, and J. Terpinski, Pol. patent, 45,711, (1962); *Chem. Abstr.*, **58**, 9033 (1963).
9. A. Raczka, A. Swirska, and H. Bojarska-Dahling, *Acta Polon. Pharm.*, **20**, 155 (1963); *Chem. Abstr.*, **62**, 1630 (1965).
0. Z. Skrowaczewska and H. Ban, *Bull. Acad. Polon. Sci., Ser. Chem.*, **9**, 213 (1961); *Chem. Abstr.*, **59**, 7474 (1963).
1. L. Archremowicz, T. Batkowski, and Z. Skrowaczewska, *Rocz. Chem.*, **38**, 1317 (1964); *Chem. Abstr.*, **62**, 1630 (1965).
2. W. Sliwa and E. Plazek, *Acta Polon. Pharm.*, **23**, 105 (1966); *Chem. Abstr.*, **65**, 10559 (1966).
3. L. Achremowicz and Z. Skrowaczewska, *Rocz. Chem.*, **39**, 1417 (1965); *Chem. Abstr.*, **64**, 17530 (1966).
4. H. L. Yale, F. A. Sowinski, and J. Bernstein, U.S. patent, 3,123,614 (1964); *Chem. Abstr.*, **61**, 4385 (1964).
5. E. V. Brown and R. H. Neil, *J. Org. Chem.*, **26**, 3546 (1961).
6. P. C. Jain, S. K. Chatterjee, and N. Anand, *Indian J. Chem.*, **4**, 403 (1966); *Chem. Abstr.*, **66**, 46522x (1967).
7. T. Nakashima and M. Endo, *Nippon Kagaku Zasshi*, **81**, 816 (1960); *Chem. Abstr.*, **56**, 5923 (1962).
8. S. P. Acharya and K. S. Nargund, *J. Sci. Res.*, **21B**, 483 (1962); *Chem. Abstr.*, **58**, 5623 (1963).
9. L. Achremowicz and Z. Skrowaczewska, *Rocz. Chem.*, **41**, 1555 (1967); *Chem. Abstr.*, **69**, 10335b (1969).
0. Z. Talik and T. Talik, *Rocz. Chem.*, **36**, 417 (1962); *Chem. Abstr.*, **58**, 5627 (1963).
1. C. Koenig, H. Pelster, and H. Puetter, Ger. patent, 1,270,563 (1968); *Chem. Abstr.*, **69**, 96341k (1968).
2. P. Tomasik and E. Plazek, *Rocz. Chem.*, **39**, 1671 (1965); *Chem. Abstr.*, **64**, 19549 (1966).
3. P. Tomasik and E. Plazek, *Rocz. Chem.*, **38**, 709 (1964); *Chem. Abstr.*, **62**, 4003 (1965).
4. B. Glowiak and R. Kulik, *Rocz. Chem.*, **36**, 959 (1962); *Chem. Abstr.*, **58**, 5631 (1963).
5. P. Nantka-Namirski, *Acta Polon. Pharm.*, **18**, 391, 449 (1961); *Chem. Abstr.*, **58**, 3424 (1963).
6. R. F. Evans and H. C. Brown, *J. Org. Chem.*, **27**, 1665 (1962).
7. Z. Talik, *Bull. Acad. Polon. Sci., Ser. Sci. Chim.*, **9**, 561 (1961); *Chem. Abstr.*, **60**, 2884 (1964).
8. E. Ochiai and H. Miarashi, *Chem. Pharm. Bull.* (Tokyo), **11**, 1586 (1963); *Chem. Abstr.*, **60**, 5446 (1964).
9. P. Nantka-Namirski, C. Kaczmarczyk, and L. Toba, *Acta Pol. Pharm.*, **24**, 231 (1967); *Chem. Abstr.*, **69**, 2827y (1968).
0. K. Dickorè and R. Kröhnke, *Chem. Ber.*, **93**, 2479 (1960)

111. M. Hamana and K. Funakoshi, *Yakugaku Zasshi*, **82**, 518 (1962); *Chem. Abstr.*, **5**
4513 (1963).
112. M. Hamana and K. Funakoshi, *Yakugaku Zasshi*, **84**, 23 (1964); *Chem. Abstr.*, **6**
3068 (1964).
113. M. Katada, *J. Pharm. Soc. Jap.*, **67**, 51 (1947); *Chem. Abstr.*, **45**, 9536 (1951).
114. D. M. Pretorius and P. A. de Villiers, *J. S. African Chem. Inst.*, **18**, 48 (1965); *Cher*
*Abstr.*, **63**, 14807 (1965).
115. G. F. van Rooyen, C. V. D. M. Brink, and P. A. de Villiers, *Tydskr. Naturwetenskapp*
**4**, 182 (1964); *Chem. Abstr.*, **63**, 13202 (1965).
116. S. Kajihara, *Nippon Kagaku Zasshi*, **86**, 839 (1965); *Chem. Abstr.*, **65**, 16935 (1966
117. F. Parisi, P. Bovina, and A. Quilico, *Gazz. Chim. Ital.*, **92**, 1138 (1962); *Chem. Abstr*
**58**, 13908 (1963).
118. Y. Suzuki, *Yakugaku Zasshi*, **81**, 1146 (1961); *Chem. Abstr.*, **56**, 3450 (1962).
119. H. C. van der Plas and H. J. den Hertog, *Rec. Trav. Chim. Pays-Bas*, **81**, 841 (1962
*Chem. Abstr.*, **59**, 1579 (1963).
120. E. J. Poziomek, *J. Org. Chem.*, **28**, 590 (1963).
121. G. C. Finger and L. D. Starr, *Nature*, **191**, 595 (1961).
122. R. Criegee and M. Krieger, *Chem. Ber.*, **98**, 387 (1965).
123. J. Beuchi, H. Muehle, H. Braunschweiger, and P. Fabiani, *Helv. Chim. Acta*, **45**, 44
(1962); *Chem. Abstr.*, **57**, 776 (1962).
124. T. Weiland and H. Biener, *Chem. Ber.*, **96**, 266 (1963).
125. R. A. Abramovitch and K. A. H. Adams, *Can. J. Chem.*, **39**, 2134 (1961).
126. C. F. H. Allen, *Can. J. Chem.*, **43**, 2486 (1965).
127. B. Rudner, U.S. patent, 2,982,841 (1959); *Chem. Abstr.*, **54**, 1558 (1960).
128. T. Vajda and K. Kovacs, *Rec. Trav. Chim. Pays-Bas*, **80**, 47 (1961); *Chem. Abstr.*, **5**
16547 (1961).
129. K. Kovacs and T. Vajda, *Acta Chim. Acad. Sci. Hung.*, **21**, 445 (1959); *Chem. Abst*
**55**, 1608 (1961).
130. K. Kovacs and T. Vajda, *Acta Pharm. Hung.*, **31**, Suppl. 72 (1961); *Chem. Abstr.*, **5**
5922 (1962).
131. K. Kovacs and T. Vajda, *Acta Chim. Acad. Sci. Hung.*, **29**, 245 (1961); *Chem. Abstr*
**57**, 5892 (1962).
132. R. A. Abramovitch and G. M. Singer, *J. Amer. Chem. Soc.*, **91**, 5672 (1969).
133. R. A. Abramovitch, Private Communication.
134. R. Huisgen, *Angew. Chem.*, **75**, 604 (1963).
135. N. D. Dawson, U.S. patent, 3,106,553 (1963); *Chem. Abstr.*, **60**, 2959 (1964).
136. C. E. Frost and Co., Nether. patent, 6,401,902 (1965); *Chem. Abstr.*, 8200 (1966).
137. R. Delaby, J. Mandereau, and P. Reynaud, *Bull. Soc. Chim. Fr.*, 2065 (1961).
138. Industrial Iberica Quimico-Farmaceutica, S. A., Span. patent, 299,510 (1964); *Cher*
*Abstr.*, **61**, 14646 (1964).
139. S. M. Roberts and H. Suschitzky, *Chem. Commun.*, 893 (1967).
140. B. I. Mikhant'ev and E. I. Fedorov, *Zh. Obshch. Khim.*, **33**, 865 (1963); *Chem. Abstr*
**54**, 24714 (1960).
141. B. I. Mikhant'ev and E. I. Fedorov, *Zh. Obshch. Khim.*, **30**, 568 (1960); *Chem. Abstr*
**54**, 24714 (1960).
142. S. I. Burnistov and V. A. Krasovskii, *Khim. Geterotsikl. Soedin*, **1**, 173 (1967); *Chem*
*Abstr.*, **67**, 64214p (1967).
143. A. B. Sen and S. K. Gupta, *J. Indian Chem. Soc.*, **39**, 129 (1962); *Chem. Abstr.*, **5**
4662 (1962).
144. S. I. Burmistrov and Y. A. Krasovskii, USSR patent, 159,843 (1964); *Chem. Abstr*
**60**, 11991 (1964).

45. L. Halski, Pol. patent, 48,154 (1963); *Chem. Abstr.*, **62**, 1633 (1965).
46. Irwin, Neisler and Co., Brit. patent, 901,311 (1962); *Chem. Abstr.*, **58**, 7915 (1963).
47. J. P. Wibaut and F. W. Broekman, *Rec. Trav. Chim. Pays-Bas,* **80**, 309 (1961); *Chem. Abstr.*, **56**, 24745 (1961).
48. W. G. Duncan and D. W. Henry, *J. Med. Chem.*, **11**, 909 (1968); *Chem. Abstr.*, **69**, 51953v (1968).
49. K. Hoffmann, A. Hunger, J. Kebrle, and A. Rossi, Ger. patent 1,120,454 (1959); *Chem. Abstr.*, **57**, 836 (1962).
50. L. N. Yakhontov and M. V. Rubtsov, *Zh. Obshch. Khim.*, **34**, 1129 (1964); *Chem. Abstr.*, **61**, 3066 (1964).
51. E. Kretzschmar, H. Barth, H. Goldhahn, and E. Carstens, Ger. (East) patent, 52,138 (1966); *Chem. Abstr.*, **68**, 49457x (1968).
52. J. W. Clark-Lewis and R. P. Singh, *J. Chem. Soc.*, 2379 (1962).
53. S. Biniecki and W. Modrzejewska, *Acta Polon. Pharm.*, **19**, 103 (1962); *Chem. Abstr.*, **59**, 1613 (1963).
54. L. Novak and M. Protiva, Czech. patent, 106,406 (1963); *Chem. Abstr.*, **60**, 1716 (1964).
55. L. Novak and M. Protiva, *Collect. Czech. Chem. Commun.*, **27**, 2413 (1962); *Chem. Abstr.*, **58**, 12477 (1963).
56. J. Beuchi, S. Allisson, and W. Vetsch, *Helv. Chim. Acta,* **48**, 1216 (1965); *Chem. Abstr.*, **63**, 11488 (1965).
57. F. Yoneda and T. Kato, Jap. patent, 15,194 (1965); *Chem. Abstr.*, **63**, 14826 (1965).
58. T. V. Tsaranova, *Khim. Geterotsikl. Soedin., Akad. Nauk. Latv. SSSR,* 909 (1965); *Chem. Abstr.*, **64**, 12637 (1966).
59. W. T. Dent, Brit. patent, 850,870 (1960); *Chem. Abstr.*, **55**, 22237 (1961).
60. M. Hamana and M. Yamasaki, *Yakugaku Zasshi,* **81**, 612 (1961); *Chem. Abstr.*, **55**, 24742 (1961).
61. T. Talik and Z. Talik, *Rocz. Chem.*, **41**, 1507 (1967); *Chem. Abstr.*, **69**, 10337d (1969).
62. J. P. English, F. L. Bach, Jr., and S. Gordon, Brit. patent, 1,034,538 (1966); *Chem. Abstr.*, **65**, 15397 (1966).
63. J. P. English, F. L. Bach, Jr., and S. Gordon, U.S. patent, 3,330,832 (1967); *Chem. Abstr.*, **68**, 49453t (1968).
64. N. N. Vereshchagina and I. Ya. Postovski, *Trudy Ural. Politekh. Inst. im. S. M. Kirova,* 24 (1960); *Chem. Abstr.*, **56**, 8681 (1962).
65. H. Bredereck, F. Effenberger, and D. Zeyfang, *Angew. Chem.*, 77, 219 (1965); *Chem. Abstr.*, **62**, 14494 (1965).
66. Y. Maki, K. Kawasaki, and K. Sato, *Gitu Yakka Daigaku Kiyo,* **13**, 34 (1963); *Chem. Abstr.*, **60**, 13220 (1964).
67. Y. Maki, K. Yamane, and M. Sato, *Yakugaku Zasshi,* **86**, 50 (1966); *Chem. Abstr.*, **64**, 11165 (1966).
68. Y. Maki, Y. Okada, Y. Yoshida, and K. Obata, *Gifu Yakka Daigaku Kiyo,* (12), 54 (1962); *Chem. Abstr.*, **59**, 11479 (1963).
69. O. R. Rodig, R. E. Collier, and R. K. Schlatzer, *J. Org. Chem.*, **29**, 2652 (1964).
70. O. R. Rodig, R. E. Collier, and R. K. Schlatzer, *J. Med. Chem.*, **9**, 116 (1966).
71. A. A. Aroyan and M. A. Iradyan, *Arm. Khim. Zh.*, **20**, 915 (1967); *Chem. Abstr.*, **69**, 77083m (1969).
72. Y. Sazuki, *Yakugaku Zasshi,* **81**, 1146 (1961); *Chem. Abstr.*, **56**, 3450 (1962).
73. J. B. Petersen, J. Lei, N. Clauson-Kaas, and K. Norris, *Kg. Dan. Vidensk, Selsk., Mat.-Fys. Medd.*, **36**, 23 (1967); *Chem. Abstr.*, **69**, 59063a (1968).

174. Z. Talik, *Bull. Acad. Polon. Sci., Ser. Sci. Chim.*, **9**, 571 (1961); *Chem. Abstr.*, **6**
2885 (1964).
175. P. Tomasik and Z. Skrowaczewska, *Rocz. Chem.*, **41**, 275 (1967); *Chem. Abstr.*, **6**
11404u (1967).
176. Z. Talik and E. Plazek, *Bull. Acad. Polon. Sci., Ser. Sci. Chim.*, **8**, 219 (1960); *Che*
*Abstr.*, **60**, 9241 (1964).
177. Z. Talik, *Rocz. Chem.*, **35**, 475 (1961); *Chem. Abstr.*, **57**, 15065 (1962).
178. T. Talik, *Rocz. Chem.*, **36**, 1465 (1962); *Chem. Abstr.*, **59**, 6360 (1963).
179. J. S. Wieczorek and E. Plazek, *Rec. Trav. Chim. Pays-Bas*, **83**, 249 (1964); *Che*
*Abstr.*, **60**, 15822 (1964).
180. Z. Talik, *Rocz. Chem.*, **34**, 465 (1960); *Chem. Abstr.*, **54**, 24705a (1960).
181. J. Barycki and E. Plazek, *Rocz. Chem.*, **39**, 1811 (1965); *Chem. Abstr.*, **64**, 158
(1966).
182. M. Protiva, L. Novak, and Z. Sedivy, *Collect. Czech. Chem. Commun.*, **27**, 21
(1962); *Chem. Abstr.*, **59**, 483 (1963).
183. H. Zellner, Austrian patent 229,312 (1963); *Chem. Abstr.*, **60**, 507 (1964).
184. Y. K. Yur'ev and M. A. Gal'bershtam, *Zh. Obshch. Khim.*, **32**, 1301 (1962); *Che*
*Abstr.*, **58**, 13909 (1963).
185. Y. K. Yur'ev and M. A. Gal'bershtam, *Zh. Obshch. Khim.*, **33**, 1789 (1963); *Che*
*Abstr.*, **59**, 9956 (1963).
186. R. Fusco, C. P. Dalla, and A. Salvi, *Gazz. Chim. Ital.*, **98**, 511 (1968); *Chem. Abst*
**69**, 96598z (1968).
187. Z. Arnold and A. Holy, *Collect. Czech. Chem. Commun.*, **28**, 2040 (1963); *Che*
*Abstr.*, **59**, 13802 (1963).
188. P. Buckus and A. Buckiene, *Zh. Obshch. Khim.*, **33**, 3112 (1963); *Chem. Abstr.*, **6**
1687 (1964).
189. A. Albert, in "Physical Methods in Heterocyclic Chemistry," (A. R. Katritzky, ed
Vol. I, Academic Press, N. Y., 1963, p. 1.
190. F. Mason, in "Physical Methods in Heterocyclic Chemistry," (A. R. Katritzky, ed
Vol. II, Academic Press, New York, 1963, p. 1.
191. K. Schofield, "Hetero-aromatic Nitrogen Compounds," Plenum Press, New Yo
1967, p. 120.
192. A. R. Katritzky and J. M. Lagowski, in "Advances in Heterocyclic Chemistry," Vol.
Academic Press, New York, 1963, p. 341.
193. A. Albert, "Heterocyclic Chemistry," Oxford University Press, New York, 1968,
67.
194. E. D. Schmid and R. Goeckle, *Spectrochim. Acta*, **22**, 1645 (1966); *Chem. Abstr.*, **6**
16247 (1966).
195. R. Joeckle, E. D. Schmid, and R. Mecke, *Z. Naturforsch.*, **21**, 1906 (1966); *Che*
*Abstr.*, **67**, 16291q (1967).
196. R. Isaac, F. F. Bentley, H. Sternglanz, W. C. Coburn, Jr., C. V. Stephenson, and W.
Wilcox, *Appl. Spectr.*, **17**, 90 (1963).
197. C. W. N. Cumper, R. F. A. Ginman, D. G. Redford, and A. I. Vogel, *J. Chem. So*
1731 (1963).
198. I. R. Bellobono and A. Gamba, *Gazz. Chim. Ital.*, **96**, 935 (1966); *Chem. Abstr.*, **6**
16192 (1966).
199. H. Lumbroso and J. Barassin, *Bull. Soc. Chim. Fr.*, 3143 (1965); *Chem. Abstr.*, **6**
6466 (1966).
200. J. Barassin and H. Lumbroso, *Bull. Soc. Chim. Fr.*, 492 (1961); *Chem. Abstr.*, **5**
4204 (1962).

1. C. W. N. Cumper and A. Singleton, *J. Chem. Soc., B,* 649 (1968).
2. N. Mataga and S. Mataga, *Bull. Soc. Chem. Jap.,* **32,** 600 (1969); *Chem. Abstr.,* **54,** 6303 (1960).
3. V. I. Bliznyukou and V. A. Grin, *Trudy Kar'kov. Farm. Inst.,* 26 (1957); *Chem. Abstr.,* **55,** 12031 (1961).
4. T. N. Misra, *Indian J. Phys.,* **35,** 420 (1961); *Chem. Abstr.,* **56,** 5540 (1962).
5. S. G. Bogomolov, M. G. Bystritskaya, and M. M. Kirillova, *Izvest. Akad. Nauk S.S.S.R., Ser. Fiz.,* **23,** 1199 (1959); *Chem. Abstr.,* **54,** 7337 (1960).
6. S. F. Mason, *J. Chem. Soc.,* 219 (1960).
7. G. Favini, M. Raimondi, and C. Gandolfo, *Spectrochim. Acta,* 207 (1968).
8. L. Sobczyk and A. Koll, *Bull. Acad. Polon. Sci., Ser. Sci. Chim.,* **13,** 97 (1965); *Chem. Abstr.,* **63,** 6483 (1965).
9. Y. N. Sheinker and E. M. Peresleni, *Zh. Fiz. Khim.,* **35,** 2623 (1961); *Chem. Abstr.,* **59,** 1454 (1963).
0. E. M. Peresleni, L. N. Yakhontov, D. M. Krasnolutskaya, and Yu. N. Sheinker, *Dokl. Akad. Nauk SSSR,* **177,** 592 (1967); *Chem. Abstr.,* **68,** 95634a (1968).
1. Yu. N. Sheinker and E. M. Peresleni, *Stroenie Veschestva i Spektroskopiya, Akad. Nauk. S.S.S.R.,* 28 (1960); *Chem. Abstr.,* **55,** 10067 (1961).
2. H. Sterk and H. Junek, *Monatsh.,* **98,** 1763 (1967); *Chem. Abstr.,* **67,** 116432e (1967).
3. R. A. Y. Jones and A. R. Katritzky, *J. Chem. Soc.,* 1317 (1959).
4. J. M. Essery and K. Schofield, *J. Chem. Soc.,* 3939 (1961).
5. G. Favini, A. Gamba, and I. R. Bellobono, *Spectrochim. Acta,* **23A,** 89 (1967); *Chem. Abstr.,* **66,** 50516f (1967).
6. J. Sandstrom, *Acta Chem. Scand.,* **18,** 871 (1964).
7. O. E. Polansky and M. A. Grassberger, *Monatsh.,* **94,** 662 (1963); *Chem. Abstr.,* **59,** 13367 (1963).
8. G. Tsoucaris, *J. Chim. Phys.,* **58,** 613 (1961).
9. J. S. Kwiatowski, *Acta Phys. Pol.,* **30,** 963 (1966); *Chem. Abstr.,* **68,** 29164w (1968).
0. J. Del Bene and H. H. Jaffe, *J. Chem. Phys.,* **49,** 1221 (1968).
1. H. Deelstra, W. Vanderleen, and F. Verbeek, *Bull. Soc. Chim. Belg.,* **72,** 632 (1963); *Chem. Abstr.,* **60,** 54 (1964).
2. F. Peradejordi, *C. R. Acad. Sci., Paris, Ser. C,* **258,** 1241 (1964); *Chem. Abstr.,* **60,** 10518 (1964).
3. G. B. Barlin, *J. Chem. Soc., B,* 285 (1966).
4. D. Wagler and E. Hoyer, *Z. Anorg. Allgem. Chem.,* **337,** 169 (1965); *Chem. Abstr.,* **63,** 9421 (1965).
5. G. B. Barlin, *J. Chem. Soc.,* 2150 (1964).
6. M. Paabo, R. A. Robinson, and R. C. Bates, *Anal. Chem.,* **38,** 1573 (1966).
7. R. Nasanen, I. Virtamo, and P. Merilainen, *Suomen Kemist,* **37B,** 67 (1964); *Chem. Abstr.,* **63,** 12399 (1965).
8. D. D. Perrin, *Austr. J. Chem.,* **17,** 484 (1964).
9. A. Weisstuch and A. C. Testa, *J. Phys. Chem.,* **72,** 1982 (1968).
0. J. Hennessy and A. C. Testa, *J. Chem. Phys.,* **49,** 956 (1968).
1. G. Kallistratos, A. Pfau, and B. Ossowski, *Naturwissenshaften,* **47,** 468 (1960); *Chem. Abstr.,* **55,** 7036 (1961).
2. M. R. Yagudaev and Y. N. Sheinker, *Dokl. Akad. Nauk. SSSR,* **144,** 177 (1962); *Chem. Abstr.,* **57,** 6759 (1962).
3. B. Glowiak. *Bull. Acad. Polon. Sci., Ser. Sci. Chim.,* **10,** 9 (1962); *Chem. Abstr.,* **58,** 501 (1963).

234. J. Suszko and M. Szafran, *Bull. Acad. Polon. Sci., Ser. Sci. Chim.*, **10,** 233 (196 *Chem. Abstr.*, **58,** 7519 (1963).
235. M. R. Yagudaev, E. M. Popov, I. P. Yakovlev, and Yu. N. Sheinker, *Izv. Akad. Na SSSR, Ser. Khim.*, 1189 (1964); *Chem. Abstr.*, **64,** 18706 (1966).
236. W. E. Thompson, R. J. Warren, I. B. Eisdorfer, and J. E. Zarembo, *J. Pharm. Sci.*, 1819 (1965); *Chem. Abstr.*, **64,** 4906 (1966).
237. M. L. Josien, *Proc. Int. Conf. Spectrosc.* (Bombay), **2,** 282 (1967); *Chem. Abstr.*, 47819n (1968).
238. R. A. Jones and R. P. Rao, *Austr. J. Chem.*, **18,** 583 (1965).
239. A. R. Katritzky and R. A. Jones, *J. Chem. Soc.*, 3674 (1959).
240. M. R. Yagudaev and Y. N. Sheinker, *Uzbeksk. Khim. Zh.*, **8,** 86 (1964); *Chem. Abs* **62,** 11303 (1965).
241. G. C. Kulasingam, W. R. McWhinnie, and R. R. Thomas, *Spectrochim. Acta*, **22,** 1 (1966); *Chem. Abstr.*, **65,** 4849 (1966).
242. P. G. Puranik and K. V. Ramiah, *Proc. Indian Acad. Sci.*, **54A,** 121 (1961); *Ch Abstr.*, **56,** 8183 (1962).
243. P. S. Puranik and K. V. Ramiah, *Proc. Indian Acad. Sci.*, **54A,** 146 (1964).
244. K. V. Ramiah and V. V. Chalapathi, *Current Sci.*, **34,** 687 (1965).
245. E. Spinner, *J. Chem. Soc.*, 3119 (1962).
246. K. C. Medhi and G. S. Kastha, *Indian J. Phys.*, **38,** 483 (1964); *Chem. Abstr.*, **62,** 63 (1965).
247. W. Bruegel, *Z. Elektrochem.*, **66,** 159 (1962).
248. T. K. Wu and B. P. Dailey, *J. Chem. Phys.*, **41,** 3307 (1964).
249. N. Das, N. Chatterjee, and M. Bose, *Proc. Int. Conf. Spectros.* (Bombay), **2,** 4 (1967); *Chem. Abstr.*, **69,** 23498g (1968).
250. W. W. Paudler and H. L. Blewitt, *J. Org. Chem.*, **31,** 1295 (1966).
251. G. R. Bedford, H. Dorn, G. Hilgetag, and A. R. Katritzky, *Rec. Trav. Chim. Pays-B* **83,** 189 (1964).
252. R. D. Chambers, J. Hutchinson, and W. K. R. Musgrave, *J. Chem. Soc., C,* 220 (196
253. C. S. Giam and J. L. Lyle, *J. Chem. Soc., B,* 1516 (1970).
254. D. H. Evans and R. E. Richards, *Mol. Phys.*, **8,** 19 (1964).
255. M. L. Martin, J. P. Dorie, and F. Peradejordi, *J. Chim. Phys.*, **64,** 1193 (1967).
256. L. Guibe, *Ann. Phys.*, **7,** 177 (1962); *Chem. Abstr.*, **58,** 6352 (1963).
257. B. M. Lynch, B. C. Macdonald, and J. G. K. Webb, *Tetrahedron*, **24,** 3595 (1968).
258. A. Veilland and B. Pullman, *C. R. Acad. Sci., Paris, Ser. C,* **253,** 2418 (1961).
259. F. Genet, *Bull. Soc. Fr., Mineral Crist.*, **88,** 463 (1965); *Chem. Abstr.*, **64,** 4 (1966).
260. C. Giessner-Prettre, *Ann. Phys.*, **9,** 557 (1964).
261. R. F. C. Brown, L. Radom, S. Sternhell, and I. D. Rae, *Can. J. Chem.*, **46,** 2 (1968).
262. A. R. Katritzky and R. E. Reavill, *J. Chem. Soc.*, 3825 (1965).
263. B. D. Sharma, *Acta Cryst.*, **20,** 921 (1966).
264. B. T. Gorres, Univ. Microfilms (Ann Arbor, Mich.), Order No. 64-12,125; **25**(7), 3 (1965); *Chem. Abstr.*, **62,** 13963 (1965).
265. B. T. Gorres and R. A. Jocobson, *Acta Cryst.*, **17,** 1599 (1964).
266. F. Genet, *C. R. Acad. Sci., Paris, Ser. C,* **251,** 1397 (1960); *Chem. Abstr.*, **55,** 5 (1961).
267. G. H. Keller, L. Bauer, and C. L. Bell, *J. Heterocycl. Chem.*, **5,** 647 (1968).
268. R. Rusakowicz and A. C. Testa, *J. Phys. Chem.*, **72,** 2680, (1968).
269. G. Aloisi, G. Cauzzo, and U. Mazzucato, *Trans. Faraday Soc.*, **63,** 1858 (1967).

70. K. R. Bhaskar and S. Singh, *Spectrochim. Acta,* **33A,** 1155 (1967).
71. C. D. Ritchie and P. D. Heffley, *J. Amer. Chem. Soc.,* **87,** 5402 (1965).
72. P. Knobloch, *Z. Naturforsch.,* **20a,** 854 (1965); *Chem. Abstr.,* **63,** 10810 (1965).
73. O. Chalvet, R. Daudel and F. Peradejordi, *C. R. Acad. Sci., Paris, Ser. C,* **254,** 1283 (1962).
74. M. Wahren, *Z. Chem.,* **6,** 181 (1966); *Chem. Abstr.,* **65,** 7005 (1966).
75. J. A. Zoltewicz and J. D. Meyer, *Tetrahedron Lett.,* 421 (1968).
76. I. F. Tupitsyn and V. I. Komarov, *Tr. Gos. Inst. Prikl. Khim.,* **52,** 160 (1964); *Chem. Abstr.,* **63,** 13005 (1965).
77. E. C. Taylor, R. O. Kan, and W. W. Paudler, *J. Amer. Chem. Soc.,* **83,** 4484 (1961).
78. E. C. Taylor and R. O. Kan, *J. Amer. Chem. Soc.,* **85,** 776 (1963).
79. E. C. Taylor and J. S. Driscoll, *J. Org. Chem.,* **25,** 1716 (1960).
80. L. Pentimalli, *Gazz. Chim. Ital.,* **94,** 458 (1964); *Chem. Abstr.,* **61,** 11963 (1964).
81. L. Pentimalli, *Gazz. Chim. Ital.,* **94,** 902 (1964); *Chem. Abstr.,* **62,** 1628 (1965).
82. S. M. Roberts and H. Suschitzky, *J. Chem. Soc., C,* 1537 (1968).
83. Z. Talik and T. Talik, *Rocz. Chem.,* **36,** 545 (1962); *Chem. Abstr.,* **57,** 12420 (1962).
84. T. Talik and Z. Talik, *Rocz. Chem.,* **36,** 539 (1962); *Chem. Abstr.,* **57,** 12421 (1962).
85. G. Ferrari and E. Marcon, *Farmaco, Ed. Sci.,* **14,** 594 (1959); *Chem. Abstr.,* **54,** 6709 (1959).
86. B. van der Wal, Th. J. de Boer, and H. O. Huisman, *Rec. Trav. Chim. Pays-Bas,* **80,** 203 (1961); *Chem. Abstr.,* **55,** 21116 (1961).
87. M. Marzona and R. Carpignano, *Ann. Chim.,* **55,** 1007 (1965); *Chem. Abstr.,* **64,** 6607 (1966).
88. D. V. Maier, U.S. patent, 3,163,655 (1964); *Chem. Abstr.,* **62,** 10419 (1962).
89. M. Ohta and M. Masaki, *Bull. Chem. Soc. Jap.,* **33,** 1392 (1960); *Chem. Abstr.,* **55,** 14457 (1961).
90. R. Tiollais, G. Bouget, and H. Bouget, *C. R. Acad. Sci., Paris, Ser. C,* **254,** 2597 (1962); *Chem. Abstr.,* **57,** 651 (1962).
91. L. Schmidt and B. Becker, *Monatsh.,* **46,** 674 (1926); *Chem. Abstr.,* **21,** 94 (1927).
92. K. Feist, W. Awe, J. Schultz, and F. Klatt, *Arch. Pharm.* (Weinheim), **272,** 100 (1934); *Chem. Abstr.,* **28,** 3407 (1934).
93. E. Steinhäuser and E. Diepolder, *J. Prakt. Chem.,* **93,** 387 (1916).
94. CIBA Ltd., Fr. patent, 1,363,599 (1964); *Chem. Abstr.,* **61,** 1388 (1965).
95. O. A. Osipov, V. I. Minkin, D. Sh. Verkhovodova, and M. I. Knyazhanskii, *Zh. Neorg. Khim.,* **12,** 1549 (1967); *Chem. Abstr.,* **68,** 87107r (1968).
96. A. S. Kudryavtsev and I. A. Savich, *Zh. Obshch. Khim.,* **33,** 1351 (1963); *Chem. Abstr.,* **59,** 12952 (1963).
97. A. S. Kudryavtsev, and I. A. Savich, *Vestn. Mosk. Univ., Ser. II, Khim.,* **17,** no. 2, 57 (1962); *Chem. Abstr.,* **58,** 4512 (1963).
98. E. Marcus and J. T. Fitzpatrick, *J. Org. Chem.,* **25,** 199 (1960).
99. L. Novak, Czech. patent, 110,210 (1964); *Chem. Abstr.,* **61,** 9475 (1964).
00. G. Fenech and M. Basile, *Gazz. Chim. Ital.,* **91,** 145 (1961); *Chem. Abstr.,* **56,** 12898 (1962).
01. G. Fenech and M. Basile, *Gazz. Chim. Ital.,* **91,** 163 (1961); *Chem. Abstr.,* **56,** 12898 (1962).
02. M. Protiva and K. Petz, Czech. patent, 111,880 (1964); *Chem. Abstr.,* **62,** 1463 (1965).
03. K. Pelz and M. Protiva, *Collect. Czech. Chem. Commun.,* **30,** 2231 (1965); *Chem. Abstr.,* **63,** 8235 (1965).
04. I. K. Barben, *J. Chem. Soc.,* 1827 (1961).

305. I. P. Tsukervanik and S. A. Israilova, *Dokl. Akad. Nauk. Uz. SSSR*, **20**, 25 (196. *Chem. Abstr.*, **59**, 11392 (1963).
306. Z. M. Rodionova, *Tr. po Khim. i Khim. Tekhnol.*, **2**, 262 (1964); *Chem. Abstr.*, 6 9899 (1965).
307. Taisho Pharmaceutical Co. Ltd., Jap. patent 1430 (1964); *Chem. Abstr.*, **60**, 119 (1964).
308. Y. Anmo, Y. Tsuruta, S. Ito and K. Noda, *Yakugaku Zasshi*, **83**, 807 (1963); *Che Abstr.*, **59**, 15239 (1963).
309. M. Friedman, *Can. J. Chem.*, **45**, 2275 (1967).
310. E. Profft and H. Mitternacht, *J. Prakt. Chem.*, **16**, 13 (1962); *Chem. Abstr.*, **57**, 165 (1962).
311. T. Kato, H. Yamanaka, T. Niitsuma, K. Wagatsuma, and M. Oizumi, *Chem. Phar. Bull.* (Tokyo), **12**, 910 (1964); *Chem. Abstr.*, **63**, 2949 (1965).
312. I. Ugi, *Agnew. Chem.*, **74**, 9 (1962).
313. N. P. Buu-Hoi, R. Rips, and C. Derappe, *Bull. Soc. Chim. Fr.*, 3456 (1965).
314. S. Okuda and M. M. Robison, *J. Org. Chem.*, **24**, 1008 (1959).
315. A. Buzas, F. Canac, C. Egnell, and P. Freon, *C. R. Acad. Sci., Paris, Ser. C*, **262**, 6 (1966); *Chem. Abstr.*, **64**, 15831 (1966).
316. E. Haach, H. Berger, and W. Vomel, Brit. patent, 1,014,050 (1965); *Chem. Abstr.*, 6 6624 (1966).
317. C. F. Boehringer and G. Soehne, m.b.H., Neth. Appl., 6,402,960 (1964); *Che. Abstr.*, **62**, 7732 (1965).
318. R. F. Coles and R. A. Miller, U.S. patent, 3,149,990 (1964); *Chem. Abstr.*, **61**,162 (1964).
319. H. O. Hankovszky and K. Hideg, *J. Med. Chem.*, **9**, 151 (1966).
320. C. R. McArthur, Belg. patent, 665,984 (1965); *Chem. Abstr.*, **65**, 659 (1966).
321. K. N. Gaind and S. D. Popli, *Indian J. Chem.*, **5**, 524 (1967); *Chem. Abstr.*, 6 10336c (1969).
322. Deutsch Gold-und Silber-Scheideanstalt vorm. Roessler, Belg. patent, 626,079 (196. *Chem. Abstr.*, **59**, 1000 (1963).
323. M. Thiel and K. Stach, *Monatsh.*, **93**, 1080 (1962); *Chem. Abstr.*, **59**, 6389 (1963).
324. F. Leonard and J. Koo, Belg. patent, 613,212 (1962); *Chem. Abstr.*, **57**, 166 (1962).
325. J. R. Geigy A.-G., Neth. patent, 6,509,930 (1966); *Chem. Abstr.*, **64**, 19571 (196
326. G. N. Pershin, N. S. Bogdanova, S. N. Milovanova, E. G. Popova, and M. G. Karase *Farmakol. i Toksikol.*, **27**, 716 (1964); *Chem. Abstr.*, **63**, 2950 (1965).
327. M. Hooper, D. A. Patterson, and D. G. Wibberley, *J. Pharm. Pharmacol.*, **17**, 7 (1965); *Chem. Abstr.*, **64**, 17535 (1966).
328. J. Mirek, *Rocz. Chem.*, **40**, 205 (1966); *Chem. Abstr.*, **65**, 7137 (1966).
329. G. E. Bonvicino, U.S. patent, 3,072,649 (1963); *Chem. Abstr.*, **58**, 13924 (1963).
330. M. Melandri, G. Vittorina, and A. Buttini, *Ann. Chim.* (Rome), **50**, 125 (1960); *Che. Abstr.*, **55**, 27306 (1961).
331. A. P. Gray and D. E. Heitmeier, *J. Amer. Chem. Soc.*, **81**, 4347 (1959).
332. A. P. Gray, U.S. patent, 3,165,527 (1965); *Chem. Abstr.*, **62**, 11787 (1965).
333. Irwin, Neisler and Co., Brit. patent, 901,312 (1962); *Chem. Abstr.*, **58**, 7914 (196
334. E. L. Skau, R. R. Mod, and F. C. Magne, U.S. patent, 2,951,849 (1958); *Chem. Abst* **57**, 3418 (1962).
335. P. Papini and G. Auzzi, *Gazz. Chim. Ital.*, **96**, 125 (1966); *Chem. Abstr.*, **64**, 195 (1966).
336. L. P. Kulev and A. V. Boldyreva, *Izv. Tomsk. Politekhm. Inst.*, **102**, 41 (1959); *Che Abstr.*, **59**, 2812 (1963).

7. E. V. Vladzimirskaya, *Biol. Aktivn. Soedin., Akad. Nauk. SSSR*, 262 (1965); *Chem. Abstr.*, **63**, 18019 (1965).
8. R. R. Mod, F. C. Magne, and E. L. Shau, *J. Amer. Oil Chemists Soc.*, **36**, 616 (1959).
9. T. Momose and M. Nakamura, *Chem. Pharm. Bull.* (Tokyo), **10**, 544 (1962); *Chem. Abstr.*, **59**, 5242 (1963).
0. B. Janik, B. Lucka, and I. Zagala, *Dissertationes Pharm.*, **14**, 165 (1962); *Chem. Abstr.*, **59**, 1618 (1963).
1. B. A. Arbuzov, V. M. Zoroastrova, and S. P. Achesbak, *Zh. Obshch. Khim.*, **1**, 2190 (1966); *Chem. Abstr.*, **66**, 94896j (1967).
2. B. A. Arbuzov, V. M. Zorastrova, and M. P. Osipova, *Izv. Akad. Nauk SSSR, Otd. Khim. Nauk.*, 2163 (1961); *Chem. Abstr.*, **57**, 8536 (1962).
3. C. J. Barnes and A. J. Matuszko, *U.S. Dept. Com., Office Tech. Sew., A. D.* 415, 706, 11 pp. (1963); *Chem. Abstr.*, **60**, 10642 (1964).
4. C. J. Barnes and A. J. Matuszko, *J. Org. Chem.*, **27**, 2239 (1962).
5. I. R. Pachter, *J. Org. Chem.*, **26**, 4157 (1961).
6. P. Buckus, N. Raguotiene, and B. Buckiene, *Zh. Obshch. Khim.*, **34**, 3847 (1964); *Chem. Abstr.*, **62**, 9128 (1965).
7. H. Zimmer and J. P. Bercz, *Ann. Chem.*, **686**, 107 (1965); *Chem. Abstr.*, **63**, 14731 (1965).
8. R. T. Coutts and D. G. Wibberley, *J. Chem. Soc.*, 2518 (1962).
9. H. Dorn, *Monatsber. Deut. Akad. Wiss. Berlin*, **3**, No. 1, 51 (1961); *Chem. Abstr.*, **58**, 2439d (1963).
0. A. Puszyniski and T. Talik, *Rocz. Chem.*, **41**, 1625 (1967); *Chem. Abstr.*, **68**, 114379q (1968).
1. T. Batkowski and E. Plazek, *Rocz. Chem.*, **37**, 273 (1963); *Chem. Abstr.*, **59**, 74786 (1963).
2. W. Sliwa and E. Plazek, *Acta Polon. Pharm.*, **20**, 253 (1963); *Chem. Abstr.*, **62**, 514 (1965).
3. C. Casagrande, M. Canova, and G. Ferrari, *Boll. Chim. Farm.*, **104**, 424 (1965); *Chem. Abstr.*, **64**, 5037 (1966).
4. A. G. Kostsova, *Trudy Voronezh. Gosudarst. Univ.*, **49**, 15 (1958); *Chem. Abstr.*, **56**, 2358b (1962).
5. A. G. Kostsova, E. P. Kosheleva, *Zh. Obshch. Khim.*, **32**, 1009 (1962); *Chem. Abstr.*, **58**, 5630 (1963).
6. I. K. Fel'dman, Z. E. Ovsyanaya, and E. S. Kozlov, *Tr. Leninyr. Khim. Farmatsevt. Inst.*, (11), 34 (1960); *Chem. Abstr.*, **59**, 11311 (1963).
7. G. Fenech, M. Basile, and G. Vigorita, *Ann. Chim.* (Rome), **54**, 607 (1964); *Chem. Abstr.*, **62**, 2772 (1965).
8. Mobil Oil Corp., Brit. patent, 1,078,288 (1967); *Chem. Abstr.*, **68**, 78139a (1968).
9. N. P. Buu-Hoï, M. Gauthier, and N. Dat Xuong, *Bull. Soc. Chim. Fr.*, 52 (1965).
0. T. Takahashi, Jap. patent, 4,216 (1965); *Chem. Abstr.*, **62**, 16202 (1965).
1. E. Kalatzis, *J. Chem. Soc., B*, 273 (1967).
2. E. Meuller and H. Huber-Enden, *Ann. Chem.*, **649**, 70 (1961).
3. C. R. Kolder and H. G. den Hertog, *Rec. Trav. Chim., Pays-Bas*, **79**, 474 (1960).
4. J. Kozlowska and E. Plazek, *Rocz. Chem.*, **33**, 831 (1959).
5. T. Talik and E. Plazek, *Rocz. Chem.*, **35**, 463 (1961); *Chem. Abstr.*, **53**, 18954 (1959).
6. E. Kalatzis, *J. Chem. Soc., B*, 277 (1967).
7. R. W. Faessinger and E. V. Brown, *Trans. Kentucky Acad. Sci.*, **24** (3–4) 106 (1963); *Chem. Abstr.*, **60**, 14465 (1964).
8. L. Pentimali, *Gazz. Chim. Ital.*, **89**, 1843 (1959); *Chem. Abstr.*, **55**, 4505 (1961).

369. G. S. Petrova and V. A. Platonova, *Metody Poluch. Khim. Reaktivov Prep.*, No. 120 (1967); *Chem. Abstr.*, **68**, 14381j (1968).
370. C. E. Lewis, A. P. Paul, Sien-Moo Tsang, and J. J. Leavitt, U.S. patent, 3,051,6 (1962); *Chem. Abstr.*, **59**, 5291 (1964).
371. J. W. Dehn, Jr. and A. J. Salina, U.S. patent, 3,249,597; *Chem. Abstr.*, **65**, 90 (1966).
372. Sien-Moo Tsang, C. E. Lewis, and A. P. Paul, U.S. patent, 2,893,816 (1959); *Che Abstr.*, **54**, 1868 (1961).
373. T. Talik, *Rocz. Chem.*, **36**, 1049 (1962); *Chem. Abstr.*, **59**, 7480 (1964).
374. R. D. Chambers, J. Hutchinson, and W. K. R. Musgrave, *J. Chem. Soc.*, 5040 (196
375. T. Talik and Z. Talik, *Rocz. Chem.*, **37**, 75 (1963); *Chem. Abstr.*, **59**, 8698 (1963).
376. L. Thunus and J. Delange, *J. Pharm. Belg.*, **21**, 485 (1966); *Chem. Abstr.*, **66**, 377 (1966).
377. L. O. Shnaidman, M. I. Siling, I. N. Kushchinskava, T. N. Eremiva, O. N. Shevyre V. P. Shishkov, N. A. Kosolapova, L. G. Kazakevich, and T. P. Timofeeva, *Tr. In Ekspevim. Klinich. Med. Akad. Nauk. Latv. SSSR*, **27**, 1 (1962); *Chem. Abstr.*, 4508 (1963).
378. M. Gadekar, J. L. Frederick, and E. C. DeRenzo, *J. Med. Pharm. Chem.*, **5**, 531 (196
379. T. Batkowski and E. Plazek, *Rocz. Chem.*, **36**, 51 (1962); *Chem. Abstr.*, **57**, 150 (1962).
380. P. Tomasik and T. Plazek, *Rocz. Chem.*, **39**, 365 (1965); *Chem. Abstr.*, **63**, 162 (1965).
381. H. Tomisawa, *Yakugaku Zasshi*, **79**, 1167 (1959); *Chem. Abstr.*, **54**, 3416 (1960).
382. F. Friedrich and R. Pohloudek-Fabini, *Pharmazie*, **19**, 677 (1964); *Chem. Abstr.*, 7720 (1965).
383. F. Freifelder, R. W. Matton, and Y. H. Ng, *J. Org. Chem.*, **9**, 3730 (1964).
384. P. A. Van Zwieten, J. A. Valthuijsen, and H. O. Huisman, *Rec. Trav. Chim. Pays-B* **80**, 1066 (1961); *Chem. Abstr.*, **56**, 3448 (1962).
385. S. N. Godovikova, USSR patent, 1,611,756 (1964); *Chem. Abstr.*, **61**, 3078 (1964).
386. T. Batkowski, *Rocz. Chem.*, **41**, 729 (1967); *Chem. Abstr.*, **67**, 82061r (1967).
387. W. J. Middleton, U.S. Patent, 2,914,534 (1959); *Chem. Abstr.*, **54**, 9962 (1960).
388. R. Urban and O. Schnider, *Helv. Chim. Acta*, **47**, 363 (1964).
389. C. S. Mizukami, E. Hirai, and M. Morimoto, *Schionogi Kenkyusho Nempo*, **16**, (1966); *Chem. Abstr.*, **66**, 10827g (1967).
390. H. N. Rydon and K. Undheim, *J. Chem. Soc.*, 4676 (1962).
391. C. J. Argoudelis and F. A. Kummerow, *J. Org. Chem.*, **26**, 3420 (1961).
392. R. A. Abramovitch and A. D. Notation, *Can. J. Chem.*, **38**, 1445 (1960).
393. B. N. Dashkevich and Yu. Yu. Tsmur, *Invest. Vysshikh, Ucheb. Zavedenil, Khim Khim. Teknol.*, **3**, 754 (1960); *Chem. Abstr.*, **55**, 2636 (1961).
394. W. W. Paudler and T. J. Kress, in "Advances in Heterocyclic Chem.," (A. Katritzky, ed.), Vol. II, 1961, p. 123
395. E. Ziegler and E. Noelken, *Monatsh.*, **92**, 1184 (1961).
396. W. Czuba, *Rocz. Chem.*, **41**, 289 (1967).
397. K. Miyaki, *J. Pharm. Soc. Jap.*, **62**, 257 (1942); *Chem. Abstr.*, **45**, 2950 (1951).
398. B. Bobranski and E. Sacharda, *Chem. Ber.*, **60**, 1081 (1926).
399. B. Bobranski and E. Sucharda, *Rocz. Chem.*, **7**, 192 (1927).
400. T. J. Kress and W. W. Paudler, *Chem. Commun.*, **3**, (1967).
401. W. P. Utermohlen, Jr., *J. Org. Chem.*, **8**, 544 (1943).
402. T. Kato, F. Hamaguchi, and T. Oiwa, *Chem. Pharm. Bull.* (Tokyo), **4**, 178 (195 *Chem. Abstr.*, **51**, 7367 (1957).

3. S. Tamura, T. Kudo, and Y. Yanagihara, *Yakugaku Zasshi,* **80,** 562 (1960); *Chem. Abstr.,* **54,** 22650 (1960).
4. W. W. Paudler and T. J. Kress, *J. Org. Chem.,* **31,** 3055 (1966).
5. W. W. Paudler and T. J. Kress, *J. Heterocycl. Chem.,* **4,** 284 (1967).
6. J. G. Murray and C. R. Hauser, *J. Org. Chem.,* **19,** 2008 (1954).
7. Sterling Drug Inc., Brit. patent, 1,022,214 (1966); *Chem. Abstr.,* **64,** 19618 (1966).
8. W. W. Paudler and T. J. Kress, *J. Org. Chem.,* **33,** 1384 (1968).
9. H. E. Baumgarten and K. C. Cook, *J. Org. Chem.,* **22,** 138 (1957).
10. H. E. Baumgarten and A. L. Krieger, *J. Amer. Chem. Soc.,* 77, 2438 (1955).
11. W. W. Paudler and T. J. Kress, *J. Org. Chem.,* **32,** 832 (1967).
12. E. V. Brown, *J. Org. Chem.,* **30,** 1607 (1965).
13. S. Carboni, A. DaSettimo, and G. Pirisino, *Ann. Chim.* (Rome), **54,** 677 (1964); *Chem. Abstr.,* **61,** 11980 (1964).
14. S. Carboni, S. DaSettimo, G. Pirisino, and D. Segnini, *Gazz. Chim. Ital.,* **96,** 103 (1966).
15. S. Carboni and G. Pirisino, *Ann. Chim.* (Rome), **52,** 340 (1962); *Chem. Abstr.,* **57,** 9825 (1962).
16. S. Carboni and G. Pirisino, *Ann. Chim.* (Rome), **52,** 279 (1962); *Chem. Abstr.,* **57,** 791 (1962).
17. S. Carboni, A. DaSettimo, and G. Pirisino, *Ann. Chim.* (Rome), **54,** 883 (1964); *Chem. Abstr.,* **63,** 5620 (1965).
18. S. Carboni, A. DaSettimo, D. Segnini, and I. Tennetti, *Gazz. Chim. Ital.,* **96,** 1443 (1966).
19. S. Carboni, A. DeSettimo, and P. L. Ferrarini, *Gazz. Chim. Ital.,* **97,** 1961 (1967).
20. S. Carboni, A. DaSettimo, P. I. Ferrarini, and G. Pirisino, *Gazz. Chim. Ital.,* **96,** 1456 (1966); *Chem. Abstr.,* **67,** 100084g (1967).
21. V. Burckhardt, H. Suter, and W. Kundig, Ger. patent, 829,894 (1954); *Chem. Abstr.,* **52,** 11,127 (1958).
22. E. M. Hawes and D. B. Wibberley, *J. Chem. Soc., C,* 315 (1966).
23. E. M. Hawes and D. G. Wibberley, *J. Chem. Soc., C,* 1564 (1966).
24. A. W. Johnson, T. J. King, and J. R. Turner, *J. Chem. Soc.,* 1509 (1960).
25. R. Baltrusis and A. Maciulus, *Lietuvos TSR Mokslu Akad, Darbai Ser.,* **B,** 163 (1964); *Chem. Abstr.,* **61,** 4165 (1964).
26. R. Baltrusis, A. Maciulis, and A. Purenas, *Lietuvos TSR Mokslu Akad. Darbai, Ser. B.,* No. 2, 125 (1962); *Chem. Abstr.,* **58,** 6827 (1963).
27. P. Buckus, G. Denis, and N. Raquotiene, *Zh. Obshch. Khim.,* **33,** 1236 (1963); *Chem. Abstr.,* **59,** 10036 (1963).
28. G. R. Lappin, *J. Org. Chem.,* **26,** 2350 (1961).
29. G. R. Lappin, *Org. Chem. Bull.,* **33,** No. 1, 6 (1961); *Chem. Abstr.,* **55,** 14453 (1961).
30. J. G. Wilson and W. Bottomley, *J. Heterocycl. Chem.,* **4,** 360 (1967).
31. M. Shur and S. S. Israelstam, *J. Org. Chem.,* **33,** 3015 (1968).
32. M. Khalifa, *Bull. Fac. Pharm.,* **1,** 149 (1961); *Chem. Abstr.,* **61,** 5643 (1964).
33. M. Khalifa and Y. M. Abou-Zeid, *Bull. Fac. Pharm.,* **1,** 159 (1961); *Chem. Abstr.,* **61,** 5643 (1964).
34. A. R. Katritzky and A. J. Waring, *J. Chem. Soc.,* 1540 (1962).
35. E. A. Ingalls and F. D. Popp, *J. Heterocycl. Chem.,* **4,** 523 (1967).
36. V. Klusis and S. Kutkevicius, *Lietuvos TSR Aukstuju Mokykly Mokslo Dorhai, Chem. ir Chem. Technol.,* **6,** 51 (1965); *Chem. Abstr.,* **64,** 19607 (1966).
37. V. Klusis and S. Kutkevicius, *Lietuvos TSR Audstuju Mokylu Mokslo Darbai, Chem. ir Chem. Technol.,* 7, 67 (1965); *Chem. Abstr.,* **64,** 19607 (1966).

438. A. Albert and R. E. Willette, *J. Chem. Soc.*, 4063 (1964).
439. W. Herz and D. R. K. Murty, *J. Org. Chem.*, **25**, 2242 (1960).
440. R. A. Abramovitch, D. H. Hey, and R. D. Mulley, *J. Chem. Soc.*, 4623 (1954).
441. R. A. Abramovitch, K. A. H. Adams, and A. D. Notation, *Can. J. Chem.*, **38**, 21 (1960).
442. R. A. Abramovitch, *Can. J. Chem.*, **38**, 2273 (1960).
443. T. V. Tsaranova and Tr. Kishinev, *Politekh. Inst.*, No. 5, 65 (1966); *Chem. Abstr.*, 32612a (1967).
444. R. A. Abramovitch and I. D. Spenser, in "Advances in Heterocylic Chemistry," ( R. Katritzky, ed.), Vol. 3, Academic Press, New York, 1964, p. 79.
445. J. P. Paolini and R. K. Robins, *J. Heterocycl. Chem.*, **2**, 53 (1965).
446. C. K. Bradsher, E. F. Litzinger, and M. F. Zinn, *J. Heterocycl. Chem.*, **2**, 331 (196
447. V. I. Staninets and E. A. Shilov, *Ukr. Khim. Zh.*, **31**, 1286 (1965); *Chem. Abstr.*, 12626 (1966).
448. W. L. Mosby, *J. Org. Chem.*, **26**, 1316 (1961).
449. Sociéte des usines chimiques Rhône-Poulenc, Fr. patent, 1,167,657 (1958); *Che Abstr.*, **55**, 9436 (1961).
450. W. L. Mosby and R. J. Boyle, *J. Org. Chem.*, **24**, 374 (1959).
451. Sociéte des usines chimiques Rhône-Poulenc, Fr. patent, 1,170,119 (1959); *Che Abstr.*, **55**, 9437 (1961).
452. Laboratoires U.P.S.A., Neth. patent, 6,414,717 (1965); *Chem. Abstr.*, **64**, 712 (196
453. W. Kaupmann and S. Funke, Ger. patent, 1,168,435 (1964); *Chem. Abstr.*, **61**, 31 (1965).
454. R. Littell and D. S. Allen, U.S. patent, 3,314,941 (1967); *Chem. Abstr.*, **67**, 6445 (1967).
455. L. Homer and H. Oediger, *Ann. Chem.*, **627**, 142 (1959); *Chem. Abstr.*, **55**, 14 (1961).
456. (a) I. M. Bortovoi, *Dokl. 2-oi Vtoroi Mezhvuz. Konf. po Khim. Organ. Komplek Soedin.*, 94 (1963); *Chem. Abstr.*, **61**, 13876 (1964).
(b) R. A. Abramovitch and B. W. Cue, Jr., *J. Org. Chem.*, **38**, 173 (1973).
457. T. Goerdeler and W. Roth, *Chem. Ber.*, **96**, 534 (1963); *Chem. Abstr.*, **58**, 125 (1963).
458. A. Spiliadis, M. Hilsenrath, V. Cornea-Ivan, and E. Balta, *Bull. Soc. Chim. Fr.*, 9 (1966); *Chem. Abstr.*, **65**, 10700 (1966).
459. L. Pentimalli, *Ann. Chim.* (Rome), **55**, 435 (1965); *Chem. Abstr.*, **63**, 6963 (1965).
460. J. Klosa, *J. Prakt. Chem.*, **8**, 168 (1959); *Chem. Abstr.*, **54**, 5640 (1961).
461. T. Okamoto, M. Hirobe, and E. Yabe, *Chem. Pharm. Bull.* (Tokyo), **14**, 523 (196 *Chem. Abstr.*, **65**, 8897 (1966).
462. F. C. Meyer and H. C. Godt, U.S. patent Appl., 2,983,729 (1958); *Chem. Abstr.*, 1430 (1962).
463. B. A. Geller and L. S. Samosvat, *Zh. Obshch. Khim.*, **34**, 613 (1964); *Chem. Abs* **60**, 13112 (1964).
464. A. Puszynski and T. Talik, *Rocz. Chem.*, **41**, 917 (1967); *Chem. Abstr.*, **67**, 8206 (1967).
465. T. Talik and Z. Talik, *Rocz. Chem.*, **41**, 483 (1967); *Chem. Abstr.*, **67**, 64192e (196
466. W. Czuba, *Rocz. Chem.*, **56**, 1647 (1962); *Chem. Abstr.*, **56**, 3445 (1962).
467. W. Czuba, *Rocz. Chem.*, **35**, 1347 (1961); *Chem. Abstr.*, **57**, 8543 (1962).
468. W. Czuba, *Rocz. Chem.*, **34**, 1639 (1960); *Chem. Abstr.*, **56**, 3445 (1962).
469. B. Brekiesz-Lewandowska and Z. Talik, *Rocz. Chem.*, **41**, 1887 (1967); *Chem. Abs* **69**, 2199w (1968).

70. K. Lewicka and E. Plazek, *Rocz. Chem.*, **39**, 643 (1965); *Chem. Abstr.*, **63**, 8310 (1965).

71. M. Balme and M. Gruffaz, Fr. patent, 1,477,734 (1967); *Chem. Abstr.*, **68**, 68889t (1968).

72. E. I. Fedorov and B. I. Mikhant'ev, *Tr. Pobl. Lab. Khim. Vysokoml. Soedin., Voronezh Gos. Univ.*, (4), 48 (1966); *Chem. Abstr.*, **69**, 18986k (1968).

73. F. Ya. Perveev and N. V. Koshmina, *Zh. Org. Khim.*, **4**, 177 (1968); *Chem. Abstr.*, **68**, 78086f (1968).

74. T. Okamoto, M. Hirobe, Y. Tamai, and E. Yabe, *Chem. Pharm. Bull.* (Tokyo), **14**, 506 (1966); *Chem. Abstr.*, **65**,8896 (1966).

75. J. R. Geigy, Brit. patent, 1,108,975 (1968); *Chem. Abstr.*, **69**, 67226w (1968).

76. J. S. Wieczorek and T. Talik, *Rocz. Chem.*, **36**, 967 (1962); *Chem. Abstr.*, **58**, 5675 (1963).

77. M. Israel and A. R. Day, *J. Org. Chem.*, **24**, 1455 (1959).

78. P. Tomasik and Z. Skrowaczewska, *Rocz. Chem.*, **40**, 637 (1966); *Chem. Abstr.*, **56**, 3826 (1966).

79. J. Barycki and E. Plazek, *Rocz. Chem.*, **37**, 1443 (1963); *Chem. Abstr.*, **60**, 7987 (1964).

80. C. A. Salemink, *Rec. Trav. Chim. Pays-Bas*, **80**, 545 (1961).

81. Smith, Kline & French Laboratories, Brit. patent, 1,020,060 (1966); *Chem. Abstr.*, **64**, 11181 (1966).

82. D. G. Markees, V. C. Dewey, and X. Y. Kidder, *Arch. Biochem. Biophys.*, **86**, 179 (1960).

83. F. W. Lange, *Seifen-Oele-Fette-Wachse*, **91**, 593 (1965); *Chem. Abstr.*, **64**, 11349 (1966).

84. E. C. Taylor and A. McKillop, *J. Org. Chem.*, **30**, 3153 (1965).

85. H. Bojarska-Dahlig and P. Nantka-Namirski, *Pharm. Acta Helv.*, **35**, 423 (1960); *Chem. Abstr.*, **55**, 1607 (1961).

86. J. Kotler-Brajtburg, *Acta Polon. Pharm.*, **20**, 169 (1963); *Chem. Abstr.*, **62**, 1623 (1965).

87. R. D. Elliot, C. Temple, Jr., and J. A. Montgomery, *J. Org. Chem.*, **33**, 2393 (1968).

88. T. Batkowski, *Rocz. Chem.*, **37**, 385 (1963); *Chem. Abstr.*, **59**, 11486 (1963).

89. R. C. De Selms and H. S. Mosher, *J. Amer. Chem. Soc.*, **82**, 3762 (1960).

90. S. Bodforss, *Ann. Chem.*, **676**, 136 (1964); *Chem. Abstr.*, **62**, 560 (1965).

91. D. L. Garmaise and J. Kombossy, *J. Org. Chem.*, **29**, 3404 (1964).

92. S. K. Chatterjee, P. C. Jain, and N. Anand, *Indian J. Chem.*, **3**, 138 (1965); *Chem. Abstr.*, **63**, 5628 (1965).

93. P. C. Jain and N. Anand, *Indian J. Chem.*, **6**, 123 (1968); *Chem. Abstr.*, **69**, 67294s (1968).

94. A. Hunger, J. Kabrle, A. Rossi, and K. N. Hoffman, U.S. patent, 3,004,978 (1960); *Chem. Abstr.*, **56**, 4771 (1962).

95. W. Knoblock and H. Kuehne, *J. Prakt. Chem.*, **17**, 199 (1962); *Chem. Abstr.*, **58**, 3431 (1963).

96. K. Pilgrim and F. Korte, *Tetrahedron*, **19**, 137 (1963).

97. F. J. Villani, U.S. patent, 2,898,338 (1959); *Chem. Abstr.*, **54**, 580 (1961).

98. V. Carelli, M. Cardellini, and F. Liberatore, *Ann. Chim.* (Rome), **48**, 1342 (1958); *Chem. Abstr.*, **58**, 1428 (1963).

99. V. Carelli, M. Cardellini, and F. Liberatore, *Farmaco, Ed. Sci.*, **15**, 803 (1960); *Chem. Abstr.*, **55**, 21109 (1961).

00. V. Carelli, M. Cardellini, and F. Liberatore, *Farmaco, Ed. Sci.*, **16**, 375 (1961); *Chem. Abstr.*, **56**, 5921 (1962).

501. B. K. Varma and A. B. Lal, *J. Indian Chem. Soc.*, **43**, 416 (1966); *Chem. Abstr.*, **65** 18556 (1966).
502. Yoshitomi Pharmaceutical Industries Ltd., Jap. patent, 26,852 (1964); *Chem. Abstr.* **62**, 10418 (1965).
503. S. Kuwata, *Bull. Chem. Soc. Jap.*, **33**, 1668 (1960).
504. H. Kakimoto, I. Sekigawa, and K. Yamamoto, Jap. patent, 20,716 (1965); *Chem. Abstr.*, **64**, 2065 (1966).
505. E. Hoyer, *Chem. Ber.*, **93**, 2475 (1960).
506. J. Hurst and D. G. Wibberly, *J. Chem. Soc., C,* 1487 (1968).
507. L. N. Yakhontov and M. V. Rubtsov, *Zhur. Obshch. Khim.*, **30**, 1507 (1960); *Chem Abstr.*, **55**, 1606 (1961).
508. J. Krapcho and W. A. Lott, U.S. patent, 2,918,470 (1959); *Chem. Abstr.*, **54**, 676 (1960).
509. S. Yamada and Y. Kikugawa, *Chem. Ind.* (London), 1325 (1967).
510. J. Volke and J. Holubek, *Collect. Czech. Chem. Commun.*, **28**, 1597 (1963); *Chem Abstr.*, **59**, 8361 (1963).
511. B. V. Balyakina, E. S. Zhanovich, and N. A. Preobrazhenskii, *Zhur. Obshch. Khim.* **31**, 542 (1961); *Chem. Abstr.*, **55**, 22308 (1961).
512. J. Rus, V. Loeffelmann, V. Prokes, and A. Kana, Czech. patent 101,790 (1961) *Chem. Abstr.*, **59**, 1013 (1963).
513. J. P. Wibaut, J. H. Uhlenbroek, E. C. Kooijman, and D. K. Kettenes, *Rec. Trav. Chim Pays-Bas,* **79**, 481 (1960); *Chem. Abstr.*, **55**, 1609 (1961).
514. G. Erlemann, W. Guea, and O. Schnider, Fr. patent, M3398 (1965); *Chem. Abstr.*, **64** 716 (1966).
515. Y. Sawa and R. Maeda, Jap. patent, 2147 (1965); *Chem. Abstr.*, **62**, 1463 (1965).
516. T. Maruyama, M. Yasumatsu, and K. Kurisono, Jap. patent, 4074 (1965); *Chem Abstr.*, **62**, 16211 (1965).
517. T. S. Gardner, E. Wenis, and J. Lee, *J. Med. Pharm. Chem.*, **3**, 461 (1961).
518. L. I. Zakharkin and I. M. Khorlina, *Izvest. Akad. Nauk. SSSR, Otdel. Khim. Nauk.* 2146 (1959); *Chem. Abstr.*, **54**, 10932 (1960).
519. J. Ankrieux, D. Anker, and C. Mentzer, *Chim. Therap.*, **2**, 57 (1966); *Chem. Abstr.* **65**, 8837 (1966).
520. M. Nakanishi and T. Numakata, Jap. patent, 14,366 (1960); *Chem. Abstr.*, **55**, 1551 (1961).
521. J. Volke, R. Kubicek, and F. Santavy, *Collect. Czech. Chem. Commun.*, **25**, 87 (1960); *Chem. Abstr.*, **58**, 13448 (1963).
522. L. Karzynski, J. Nowak, and H. Witek, *Diss. Pharm. Pharmacol.*, **20**, 169 (1968) *Chem. Abstr.*, **69**, 43648s (1968).
523. V. L. Florent'ev, N. A. Drobinskaya, L. V. Ionova, M. Y. Karpeiskii, and K. F Turchin, *Dokl. Akad. Nauk. SSSR,* **177**, 617 (1967); *Chem. Abstr.*, **69**, 27198 (1968).
524. S. O. Winthrop and G. Gavin, *J. Org. Chem.*, **24**, 1936 (1959).
525. S. Binieck, and B. Kabzinska, Pol. patent, 51,864 (1966); *Chem. Abstr.*, **68**, 95700 (1968).
526. R. H. Mizzoni, U.S. patent, 2,963,486 (1960); *Chem. Abstr.*, **55**, 10475 (1961).
527. G. N. Walker and M. A. Klett, *J. Med. Chem.*, **9**, 624 (1966); *Chem. Abstr.*, **65**, 700 (1966).
528. R. I. Meltzer, U.S. patent, 3,313,822 (1967); *Chem. Abstr.*, **67**, 5403w (1967).
529. J. E. Robertson, J. H. Biel, T. F. Mitchell, Jr., W. K. Hoya, and H. A. Leiser, *J. Med Chem.*, **6**, 805 (1963); *Chem. Abstr.*, **60**, 1692 (1964).

530. J. H. Biel and W. K. Hoya, U.S. patent, 3,205,133 (1965); *Chem. Abstr.*, **65**, 3496 (1966).

531. S. Biniecki and W. Zlakowska, *Acta Polon. Pharm.*, **23**, 89 (1966); *Chem. Abstr.*, **65**, 3813 (1966).

532. F. Zymalkowski and F. Koppe, *Arch. Pharm.* (Weinheim), **294**, 453 (1961); *Chem. Abstr.*, **56**, 2415 (1962).

533. F. Hoffman-La Roche & Co., Belg. patent, 615,478 (1962); *Chem. Abstr.*, 7496 (1963).

534. S. Miyano and N. Abe, *Chem. Pharm. Bull.* (Tokyo), **15**, 511 (1967); *Chem. Abstr.*, **67**, 73494g (1967).

535. S. Miyano, *Chem. Pharm. Bull.* (Tokyo), **13**, 1135 (1965); *Chem. Abstr.*, **64**, 688 (1966).

536. W. A. Shuler and A. Gross, U.S. patent, 2,953,562 (1960); *Chem. Abstr.*, **55**, 4552 (1961).

537. S. Raynolds and R. Levine, *J. Amer. Chem. Soc.*, **82**, 1152 (1960).

538. Deutsche Gold-und Silber-Scheideanstalt vorm. Roessler, Fr. patent, 1,380,771 (1964); *Chem. Abstr.*, **63**, 584 (1965).

539. K. Thiele, U.S. patent, 3,042,680 (1962); *Chem. Abstr.*, **57**, 13736 (1962).

540. F. F. Blicke and J. L. Hughes, *J. Org. Chem.*, **26**, 3257 (1961).

541. L. D. Smirnov, V. P. Lezina, V. F. Bystrov, and K. M. Dyumaer, *Izv. Akad, Nauk. SSSR, Ser. Khim,* (1), 198 (1965); *Chem. Abstr.*, **62**, 11774 (1965).

542. L. D. Smirnov, V. P. Lezina, T. P. Kartasheva, and K. M. Dyumaev, *Isv. Akad. Nauk. SSSR, Ser. Khim.*, 199 (1968); *Chem. Abstr.*, **69**, 77086g (1968).

543. L. D. Smirnov, T. P. Kartasheva, V. P. Lezina, and K. M. Dyumaev, *Izv. Akad. Nauk. SSSR, Ser. Khim.*, 2742 (1967); *Chem. Abstr.*, **69**, 67191f (1968).

544. K. M. Dyumaev, L. D. Smirnov, and V. F. Bystrov, *Izv. Akad. Nauk. SSSR. Otd. Khim. Nauk.*, 883, (1962); *Chem. Abstr.*, **57**, 12424 (1962).

545. L. D. Smirnov, V. P. Lezina, V. F. Bystrov, and K. M. Dyumaev, *Izv. Akad. Nauk. SSSR, Ser. Khim.*, **10**, 1836 (1965); *Chem. Abstr.*, **64**, 5038 (1966).

546. L. D. Smirnov, S. L. Orlova, V. P. Lezina and K. M. Dymaev, *Izv. Akad. Nauk. SSSR, Ser. Khim.*, 1816 (1967); *Chem. Abstr.*, **68**, 87111n (1968).

547. M. Miocque, *Bull. Soc. Chim. Fr.*, **2**, 330 (1960).

548. M. Mioque and J. A. Gautier, *C. R. Acad. Sci., Paris, Ser. C*, **252**, 2416 (1961); *Chem. Abstr.*, **56**, 14232 (1962).

549. Abbott Laboratories, Fr. patent, M2474 (1964); *Chem. Abstr.*, **61**, 13290 (1964).

550. Abbott Laboratories, Span. patent, 281,171 (1962); *Chem. Abstr.*, **60**, 2904 (1964).

551. A. N. Kost, P. B. Terent'ev, and T. Zavada, *Dokl. Akad. Nauk. SSSR*, **130**, 326 (1960); *Chem. Abstr.*, **54**, 11015 (1961).

552. R. Dahlbom, B. Karlen, S. Ramsey, I. Kraft, and R. Mollberg, *Acta Pharm. Suecica*, **1**, 237 (1964); *Chem. Abstr.*, **63**, 5595 (1965).

553. C. Lovell and K. J. Rorig, U.S. patent, 3,025,302 (1962); *Chem. Abstr.*, **57**, 11173 (1962).

554. G. Ehrhart, H. Ruschig, and K. Schmitt, Ger. patent, 1,116,226 (1961); *Chem. Abstr.*, **57**, 2199 (1962).

555. M. Masamichi, K. Isagawa, T. Kuki, and Y. Fushizaki, *Nippon Kagaku Zasshi*, **83**, 212 (1962); *Chem. Abstr.*, **59**, 3879 (1963).

556. E. Profft and S. Lojack, *Rev. Chim., Acad. Rep. Populaire Roumaine*, **7**, 405 (1962); *Chem. Abstr.*, **59**, 8696 (1963).

557. D. Wagler and E. Hoyer, *Chem. Ber.*, **98**, 1073 (1965); *Chem. Abstr.*, **63**, 14352 (1965).

558. S. I. Suminov and A. N. Kost, *Zh. Obshch. Khim.*, **34,** 2421 (1964); *Chem. Abstr.*, 6 11966 (1964).
559. L. E. Brady, M. Freifelder, and G. H. Stone, *J. Org. Chem.*, **26,** 4757 (1961).
560. N. F. Kazarinova, N. V. Dzhigirei, and N. A. Shabaeva, *Metody Poluch. Khi Reaktivov Prep.*, No. 14, 38 (1966); *Chem. Abstr.*, **67,** 64187g (1967).
561. S. L. Shapiro, L. Freedman, H. Soloway, U.S. patent, 2,993,905 (1959); *Che Abstr.*, **56,** 2429 (1962).
562. S. L. Shapiro, H. Soloway, E. Chodos, and L. Freedman, *J. Pharm. Sci.*, **50,** 10 (1961).
563. N. V. Organon, Belg. patent, 627,240 (1964); *Chem. Abstr.*, **60,** 9284 (1964).
564. G. N. Shibanova and R. G. Kulikova, USSR patent, 195,454 (1967); *Chem. Abstr.*, 6 104989z (1968).
565. G. N. Shibanova and R. G. Kulikova, USSR patent 195,456 (1967); *Chem. Abstr.*, 6 104991u (1968).
566. L. A. Paquette, *J. Org. Chem.*, **27,** 2870 (1962).
567. N. F. Kazarinova and A. O. Grinberg, USSR patent, 178,378 (1966); *Chem. Abst* **64,** 19570 (1966).
568. A. N. Kost, E. V. Vinogradova, and V. Kozler, *Zh. Obshch. Khim.*, **33,** 3602 (196 *Chem. Abstr.*, **60,** 9239 (1963).
569. S. L. Shapiro, I. M. Rose, F. C. Testa, and L. Freedman, *J. Org. Chem.*, **26,** 13 (1961).
570. S. L. Shapiro and L. Freedman, U.S. patent, 3,055,906 (1962); *Chem. Abstr.*, 5 6808 (1963).
571. M. I. Druzin, A. K. Val'kova, A. B. Pashkov, I. V. Zaitseva, N. V. Bogdanova, L. Pertsov, S. F. Kalinkin, and B. M. Kuindzhi, USSR patent, 215,997 (1968); *Che Abstr.*, **69,** 52016d (1968).
572. H. Erdtman, F. Haglid, I. Wellings, and U.S. v. Euler, *Acta Chem. Scand.*, **17,** (6) 17 (1963).
573. H. Lund, *Acta Chem. Scand.*, **17,** 2325 (1963).
574. K. W. Merz and H. Stolte, *Arch. Pharm.* (Weinheim), **292,** 496 (1959); *Chem. Abs* **54,** 8809 (1961).
575. Parke, Davis & Co., Brit. patent, 851,577 (1960); *Chem. Abstr.*, **55,** 12428 (1961).
576. CIBA, Neth. patent, 6,513,784 (1966); *Chem. Abstr.*, **65,** 15342 (1966).
577. M. M. Baizer, U.S. patent, 3,426,000 (1966); *Chem. Abstr.*, **64,** 19568 (1966).
578. R. F. Shuman and E. D. Amstutz, *Rec. Trav. Chim. Pays-Bas*, **84,** 441 (1965); *Che Abstr.*, **63,** 5596 (1965).
579. E. D. Amstutz, F. P. Papopoli, C. H. Tilford, and R. F. Shyman, Brit. patent, 990,3 (1965); *Chem. Abstr.*, **63,** 6978 (1965).
580. G. B. Bachman and M. Karickhoff, U.S. patent, 3,217,012 (1965); *Chem. Abstr.*, 6 2062 (1966).
581. C. E. Frosst & Co., Belg. patent, 645,217 (1964); *Chem. Abstr.*, **63,** 11576 (1965).
582. S. Biniecki and S. Emiljan, *Acta Polon. Pharm.*, **24,** 345 (1967); *Chem. Abstr.*, 6 114387r (1968).
583. S. Biniecki and W. Zladkowska, *Acta Polon. Pharm.*, **21,** 521 (1964); *Chem. Abst* **63,** 2948 (1965).
584. E. Profft and G. Shulz, *Arch. Pharm.* (Weinheim), **294,** 292 (1961); *Chem. Abstr.*, 5 18725 (1961).
585. A. V. Voropaeva and I. Kh. Feldman, *Khim. Ceterotsikl. Soedin., Akad. Nauk. La SSSR*, 271 (1965); *Chem. Abstr.*, **63,** 6967 (1965).
586. Kh. Feldman and A. V. Voropaeva, *Tr. Leningr. Khim.-Farmatsuet. Inst.*, (16), (1962); *Chem. Abstr.*, **61,** 639 (1964).

7. P. A. Roukema, *Sci. Commun.*, **10**, No. 162, 1 (1960–61); *Chem. Abstr.*, **58**, 4513 (1963).
8. N. V. Koninklijke Pharmaceutische Fabrieken Voorheen Bracades-Sfeeman & Pharmacia, Fr. patent, M1801 (1963); *Chem. Abstr.*, **59**, 12769 (1963).
9. N. V. Nederlands Conbinatie Voor Chemische Industrie, Belg. patent, 624,975 (1963); *Chem. Abstr.*, **60**, 13257 (1964).
0. CIBA Ltd., Brit. patent, 890,602 (1962); *Chem. Abstr.*, **56**, 15489 (1962).
1. A. N. Kost, E. V. Vinogradova, and W. Kozler, USSR patent, 161,758 (1964); *Chem. Abstr.*, **61**, 3075 (1964).
2. Smith, Kline & French Laboratories, Neth. patent, 6,512,616 (1966); *Chem. Abstr.*, **65**, 7095 (1966).
3. J. W. Cusic and H. W. Sause, U.S. patent, 3,255,054 (1965); *Chem. Abstr.*, **64**, 6625 (1966).
4. M. N. Tilichenko and G. V. Pavel, *Zh. Organ. Khim.*, **1**, 1992 (1965); *Chem. Abstr.*, **64**, 8132 (1966).
5. F. Bergel, S. S. Brown, C. L. Leese, C. M. Timmis, and R. Wade, *J. Chem. Soc.*, 846 (1963).
6. K. J. Rorig, U.S. patent, 3,128,280 (1964); *Chem. Abstr.*, **60**, 14481 (1964).
7. Eprova Ltd., Neth. patent, 6,600,250 (1966); *Chem. Abstr.*, **65**, 18600 (1966).
8. H. Rapoport and A. D. Batcho, *J. Org. Chem.*, **28**, 1753 (1963).
9. V. Petrow and B. Sturgeon, *J. Chem. Soc.*, 1157 (1949).
0. T. Takahashi, T. Yatsuka, and S. Senda, *J. Pharm. Soc. Jap.*, **64**, 9 (1944); *Chem. Abstr.*, **46**, 110 (1952).
1. E. P. Hart, *J. Chem. Soc.*, 212 (1956).
2. Ya. L. Gol'dfarb, L. V. Antik, and V. A. Petukhov, *Izvest. Akad. Nauk. SSSR., Otdel Khim. Nauk.*, 887 (1961); *Chem. Abstr.*, **55**, 22307 (1961).
3. G. Y. Lesher and D. Gruett, Belg. patent, 612,258 (1962); *Chem. Abstr.*, **58**, 7954 (1963).
4. D. E. Orr and K. Weisner, *Chem. Ind.* (London), 672 (1959).
5. A. M. Roe, *J. Chem. Soc.*, 2195 (1963).
6. E. A. Mistryukov, N. I. Aronova, and V. F. Kucherov, *Izv. Akad. Nauk. SSSR, Ser. Khim.*, 512 (1964).
7. W. Zecher and F. Krohnke, *Chem. Ber.*, **94**, 698 (1961).
8. A. Dornow and E. H. Rohe, *Chem. Ber.*, **93**, 1093 (1960); *Chem. Abstr.*, **54**, 18519 (1960).
9. A. N. Hambly and B. V. O'Grady, *Austr. J. Chem.*, **17**, 860 (1964).
0. W. Herz and D. R. K. Murty, *J. Org. Chem.*, **26**, 418 (1961).
1. B. M. Bain and J. E. Saxton, *J. Chem. Soc.*, 5216 (1961).
2. J. A. Moore and H. H. Puschner, *J. Amer. Chem. Soc.*, **81**, 6041 (1959).
3. E. Profft and F. Melichar, *J. Prakt. Chem.*, **2**, 87 (1955); *Chem. Abstr.*, **54**, 1514 (1960).
4. N. F. Kucherova, R. M. Khomutov, E. I. Budovskii, V. P. Evdakov, and N. K. Kotchekov, *Zhur. Obshch. Khim.*, **29**, 915 (1959); *Chem. Abstr.*, **54**, 1515 (1959).
5. V. A. Zasosov, E. I. Metel'kova, and V. S. Onoprienko, *Med. Prom. SSSR*, **15**, No. 3, 35 (1961); *Chem. Abstr.*, **55**, 24748 (1961).
6. L. Pentimalli, *Ann. Chim.* (Rome), **53**, 1123 (1963); *Chem. Abstr.*, **60**, 10645 (1964).
7. T. Sakuragi, C. Argoudelis, and F. A. Kummerow, *Arch. Biochem. Biophys.*, **89**, 160 (1960); *Chem. Abstr.*, **54**, 25201 (1960).
8. A. N. Kost, L. A. Golovleva, P. B. Terent'ev, and M. Islam, *Zh. Analit. Khim.*, **21**, 859 (1966); *Chem. Abstr.*, **65**, 19290 (1966).

619. L. D. Goodhue, A. J. Reinert, and R. P. Williams, U.S. patent, 3,150,041 (1964 *Chem. Abstr.*, **61**, 15289 (1964).
620. H. C. Van der Plas and H. J. den Hertog, *Tetrahedron Lett.*, 13 (1960).
621. G. Schmidt, Ger. patent, 1,179,943 (1964); *Chem. Abstr.*, **62**, 1677 (1965).
622. P. Nantka-Namirski, S. Kurzepa, J. Kazimierczyh, H. Kierylowicz, and M. Kobylinsk *Acta Physiol. Polon*, **17**, 145 (1966); *Chem. Abstr.*, **64**, 17936 (1966).
623. H. Bojarska-Dhalig and P. Nantka-Namirski, *Rocz. Chem.*, **34**, 189 (1960); *Chem Abstr.*, **54**, 19674 (1960).
624. A. Albert and G. B. Barlin, *J. Chem. Soc.*, 5156 (1963).
625. G. E. McCasland, L. K. Gottwald, and A. Furst, *J. Org. Chem.*, **26**, 3541 (1961).
626. R. C. Chambers, J. Hutchinson, and W. K. R. Musgrave, *J. Chem. Soc.*, 3736 (1964
627. C. A. I. Goring, U.S. patent, 3,050,382 (1962); *Chem. Abstr.*, **57**, 14218 (1962).
628. T. Talik and Z. Talik, *Nitro Compds., Proc. Intern. Symp. Warsaw*, 81 (1963); *Chem Abstr.*, **64**, 2046 (1966).
629. J. A. Montgomery and K. Hewson, *J. Med. Chem.*, **9**, 105 (1966).
630. T. Batkowski and M. Tuszynska, *Rocz. Chem.*, **38**, 585 (1964); *Chem. Abstr.*, **61** 10654 (1964).
631. S. P. Acharya and K. S. Nargund, *J. Sci. Ind. Res.*, **21B**, 452 (1962); *Chem. Abstr.*, **5** 2429 (1963).
632. B. E. Fisher and J. E. Hodge, *J. Org. Chem.*, **29**, 776 (1964).
633. T. Talik and E. Plazek, *Rocz. Chem.*, **33**, 1343 (1959); *Chem. Abstr.*, **54**, 1312 (1960).
634. I. Kh. Fel'dman and A. V. Voropaeva, *Tr. Leningr. Khim.-Farmatsevt. Inst.*, 1 (1962); *Chem. Abstr.*, **61**, 638 (1964).
635. F. J. Villani and T. A. Mann, *J. Med. Chem.*, **11**, 894 (1968); *Chem. Abstr.*, **6** 42647w (1968).
636. B. Van der Wal, Th. J. de Boer, and H. O. Huisman, *Rec. Trav. Chim. Pays-Bas*, **8** 228 (1961).
637. O. Meth-Cohn, R. K. Smalley, and H. Suschitzky, *J. Chem. Soc.*, 1666 (1963).
638. R. K. Smalley, *J. Chem. Soc., C*, 82 (1966).
639. J. L. Greene and J. A. Montgomery, *J. Med. Chem.*, **6**, 294 (1963); *Chem. Abstr.*, **5** 12507 (1963).
640. S. M. Gadehar, U.S. patent, 3,206,464 (1965); *Chem. Abstr.*, **64**, 711 (1966).
641. C. H. Huang and G. Ya Kondrat'eva, *Izv. Akad. Nauk. SSSR, Otd. Khim. Nauk.*, 52 (1962); *Chem. Abstr.*, **57**, 15064 (1962).
642. E. Profft and K. H. Otto, *J. Prakt. Chem.*, **8**, 156 (1959).
643. E. Neuzil, J. C. Breton, and H. Plagnol, *Bull. Mem. Ecole Natl. Med. Phar. Dakar*, **7** 157, 159 (1959); *Chem. Abstr.*, **57**, 4993 (1962).
644. T. Kralt and J. Van Dijk, U.S. patent, 3,046,280 (1963); *Chem. Abstr.*, **58**, 460 (1963).
645. N. V. Philips' Gloeilampenfobrieken, Brit. patent, 848,690 (1960); *Chem. Abstr.*, **57** 875 (1962).
646. Aktiebolaget Pharmacia, Brit. patent, 852,795 (1960); *Chem. Abstr.*, **55**, 1344 (1961).
647. E. N. Ifversen, K. Rubinstein, and K. T. J. Skagius, Ger. patent, 1,077,220 (1960) *Chem. Abstr.*, **55**, 22343 (1961).
648. K. T. J. Skagius, Swed. patent, 174,845 (1959); *Chem. Abstr.*, **56**, 7339 (1962).
649. K. Shagius, K. Rubinstein, and E. Ifversen, *Acta Chem. Scand.*, **14**, 1054 (1960) *Chem. Abstr.*, **56**, 12904 (1962).
650. R. Landi-Vittory, F. Gatta, F. Toffler, S. Chiavarelli, and G. L. Gatta, *Farmaco, Ed Sci.*, **18**, 465 (1963); *Chem. Abstr.*, **59**, 9941 (1963).

551. A. P. Gray, D. E. Heitmeier, and E. E. Spinner, *J. Amer. Chem. Soc.*, **81**, 4351 (1959).
552. P. Grammaticakis, *Bull. Soc. Chim. Fr.*, 480 (1959).
553. E. Marcus, J. T. Fitzpatrick and F. C. Frostick, Jr., U.S. patent, 2,993,894 (1961); *Chem. Abstr.*, **56**, 1416 (1962).
554. B. Maziere and N. Dat-Xuong, *Chim. Ther.*, **3** (1), 1 (1968); *Chem. Abstr.*, **69**, 43880d (1968).
555. P. C. Jain, V. Kapoor, N. Anand, G. K. Patnaik, A. Ahmad, and M. M. Vohra, *J. Med. Chem.*, **11**, 87 (1968).
556. R. G. D. Moore and R. J. Cox, Brit. patent, 870,027 (1961); *Chem. Abstr.*, **55**, 23134 (1961).
557. W. A. Schuler and H. Beschke, Belg. patent, 630,125 (1963); *Chem. Abstr.*, **60**, 14515 (1964).
558. R. M. Peck, R. K. Preston, and H. J. Creech, *J. Org. Chem.*, **26**, 3409 (1961).
559. A. Wolf and E. V. Haxthausen, *Arzneimittel-Forsch.*, **10**, 50 (1969); *Chem. Abstr.*, **54**, 11289 (1960).
560. K. Credner, G. Renwanz, and H. Engelhard, Ger. patent, 1,049,863 (1959); *Chem. Abstr.*, **55**, 4541 (1961).
561. H. Zellner, Austrian patent, 226,703 (1963); *Chem. Abstr.*, **59**, 10,000 (1963).
562. A. Burger and H. Hu Ong, *J. Med. Chem.*, **6**, 205 (1963).
563. L. N. Yakhontov and M. V. Rubtsov, *Zhur. Obshch. Khim.*, **30**, 3300 (1960); *Chem. Abstr.*, **55**, 18721 (1961).
564. J. Keck, *Ann. Chem.*, **662**, 171 (1963).
565. E. T. Holmes and H. R. Synder, *J. Org. Chem.*, **29**, 2155 (1964).
566. F. Leonard and J. Koo, Belg. patent, 613,213 (1962); *Chem. Abstr.*, **57**, 16629 (1962).
567. P. Nantka-Namirski, *Acta Polon. Pharm.*, **19**, 299 (1962); *Chem. Abstr.*, **59**, 15260 (1963).
568. N. Kinoshita, M. Hamana, and T. Kawasahi, *Chem. Pharm. Bull.* (Tokyo), **10**, 753 (1962); *Chem. Abstr.*, **58**, 6785 (1963).
569. H. Bojarska-Dahlig and P. Nantka-Namirski, *Congr. Sci. Farm., Conf. Comun., Pisa*, **21**, 203 (1961); *Chem. Abstr.*, **59**, 6408 (1963).
570. I. E. El-Kholy, F. K. Rafla, and G. Soliman, *J. Chem. Soc.*, 1857 (1962).
571. F. Brody and W. J. Sydor, U.S. patent, 3,118,871 (1964); *Chem. Abstr.*, **60**, 16021 (1964).
572. Sien-Moo Tsang and C. E. Lewis, U.S. patent, 3,151,106 (1964); *Chem. Abstr.*, **61**, 1619 (1964).
573. E. F. Elslager and D. F. Worth, *J. Med. Chem.*, **6**, 444 (1963).
574. J. Barycki and E. Plazek, *Rocz. Chem.*, **38**, 553 (1964); *Chem. Abstr.*, **61**, 10654 (1964).
575. E. F. Elslager, D. F. Worth, D. B. Copps and L. M. Werbel, U.S. patent, 3,139,421 (1964); *Chem. Abstr.*, **61**, 6973 (1964).
576. E. Jeney and T. Zsolnai, *Zentr. Bakteriol, Parasitenk*, **180**, 84 (1960); *Chem. Abstr.*, **55**, 5657 (1961).
577. T. Zsolnai, *Biochem. Pharmacol.*, **11**, 995 (1962); *Chem. Abstr.*, **58**, 837 (1963).
578. American Cyanamid Co., Brit. patent, 871,624 (1961); *Chem. Abstr.*, **56**, 14437 (1962).
579. H. Martin and E. Habicht, Swiss patent, 337,098 (1959); *Chem. Abstr.*, **54**, 19720 (1960).
580. G. Nabert-Bock, *Arzneimittel-Forsch.*, **10**, 125 (1960); *Chem. Abstr.*, **54**, 13264 (1960).
581. J. Matzke, Austr. patent, 242,701 (1965); *Chem. Abstr.*, **63**, 18112 (1965).

682. M. Vohra, S. Pradhan, P. Jain, S. Chatterjee, and N. Anand, *J. Med. Chem.*, **8**, 29 (1965).
683. T. Takahashi, F. Yoneda, and R. Oishi, *Chem. Pharm. Bull.* (Tokyo), 7, 602 (1959 *Chem. Abstr.*, **54**, 14246 (1960).
684. T. Kurihara and T. Chiba, *Tohoku Uakka Diagaku Kenyu Nempo,* **10**, 65 (1963 *Chem. Abstr.*, **61**, 651 (1964).
685. L. A. Elson, *Proc. Intern. Congr. Intem. Soc. Hematol.*, (7th Rome), **3**, 647 (1958 *Chem. Abstr.*, **55**, 4772 (1961).
686. I. Tanaka, *Kagaku To Kogyo,* **17**, 961 (1964); *Chem. Abstr.*, **61**, 11349 (1964).
687. J. A. Moore and F. J. Marascia, *J. Amer. Chem. Soc.*, **81**, 6049 (1959).
688. M. Balassa, *Orvosi Hetilap,* **101**, No. 8, 259 (1960); *Chem. Abstr.*, **55**, 1767 (1961).
689. H. Schwarzkopf, Belg. patent, 615,394 (1962); *Chem. Abstr.*, **58**, 8850 (1963).
690. F. W. Lange, Ger. patent, 1,142,045 (1963); *Chem. Abstr.*, **58**, 13708 (1963).
691. D. G. Markees, V. C. Dewey, and W. George, *J. Med. Chem.*, **11**, 126 (1968).
692. R. D. Elliot, C. Temple, Jr., and J. A. Montgomery, *J. Org. Chem.*, **31**, 1890 (1966
693. A. Ahmad, C. K. Patnaik, and M. M. Vohra, *Indian J. Exptl. Biol.*, **4**, 154 (1966 *Chem. Abstr.*, **65**, 15969 (1966).
694. W. Sliwa, *Acta Polon. Pharm.*, **24**, 359 (1967); *Chem. Abstr.*, **68**, 39420r (1968).
695. Firma Hans Schwarzkopf, Fr. patent, 1,397,551 (1965); *Chem. Abstr.*, **63**, 1124 (1965).
696. M. Fujimoto, Jap. patent, 3370 (1959); *Chem. Abstr.*, **54**, 14275 (1960).
697. M. Fujimoto, Jap. patent, 2169 (1959); *Chem. Abstr.*, **54**, 11055 (1960).
698. M. Hamana and K. Funakoshi, *Yakugaku Zasshi,* **82**, 523 (1962); *Chem. Abstr.*, **5** 3385 (1963).
699. R. G. Lacoste and A. E. Martell, *Inorg. Chem.*, **3**, 881 (1964); *Chem. Abstr.*, **61**, 25 (1964).
700. G. J. Sutton, *Austr. J. Chem.*, **16**, 371 (1963).
701. G. J. Sutton, *Austr. J. Chem.*, **16**, 1137 (1963).
702. G. J. Sutton, *Austr. J. Chem.*, **16**, 1134 (1963).
703. K. Nakajima, *Nippon Kagaku Zasshi,* **81**, 1746 (1960); *Chem. Abstr.*, **56**, 46 (1962).
704. K. Schulte, U.S. patent, 2,991,289 (1961); *Chem. Abstr.*, **56**, 3465 (1962).
705. M. Susai and S. Miyamoto, Jap. patent, 11,829 (1962); *Chem. Abstr.*, **59**, 100 (1963).
706. R. P. Mull and W. E. Barrett, U.S. patent, 3,252,860 (1966); *Chem. Abstr.*, **65**, 53 (1966).
707. F. Kuffner and T. Kirchenmayer, *Monatsh.*, **92**, 701 (1961); *Chem. Abstr.*, **56**, 142 (1962).
708. H. Kamimura, A. Matsumoto, Y. Miyazaki, and I. Yamamoto, *Agr. Biol. Chem* (Tokyo), **27**, 684 (1963); *Chem. Abstr.*, **60**, 6160 (1964).
709. C. H. Krauch and W. Metzner, *Chem. Ber.*, **99**, 88 (1966).
710. T. Kisaki and E. Tamaki, *Nippon Nogei Kagaku Kaishi,* **38**, 549 (1964); *Chem. Abstr* **63**, 7059 (1965).
711. T. Naito, T. Yoshikawa, F. Ishikawa, S. Isoda, Y. Omura, and I. Takamura, *Chem Pharm. Bull.* (Tokyo), **13**, 869 (1965); *Chem. Abstr.*, **63**, 11488 (1965).
712. Yu. A. Zhdanov, I. D. Sadekov, A. D. Garnovskii, and V. I. Minkin, *Izv. Vysshik Uchebn. Zavedenii, Khim. i Khim. Teckhnol.*, **8**, 954 (1965); *Chem. Abstr.*, **64**, 1757 (1966).
713. E. H. Kober, R. F. W. Raetz, and H. Ulrich, U.S. patent, 3,041,346 (1962); *Chem Abstr.*, **57**, 13731 (1962).

14. H. Ubrich, E. Kober, H. Schroeder, R. Raetz, and C. Grundmann, *J. Org. Chem.*, **27,** 2585 (1962).
15. H. Brown and P. Schuman, *J. Org. Chem.*, **28,** 1122 (1963).
16. H. Brown, *J. Polymer Sci.*, **44,** 9 (1960); *Chem. Abstr.*, **55,** 4039 (1961).
17. Farbenfabriken Bayer A.-G., Brit. patent, 985,354 (1965); *Chem. Abstr.*, **63,** 586 (1965).
18. F. Hoffmann-La Roche & Co., A.-G., Neth. Appl. 6,407,413 (1965); *Chem. Abstr.*, **63,** 1774 (1965).
19. L. Berger, A. J. Corraz, and J. Lee, Fr. patent, M3502 (1965); *Chem. Abstr.*, **64,** 2104 (1966).
20. F. Hoffmann-La Roche & Co., Neth. Appl., 6,407,463 (1965); *Chem. Abstr.*, **62,** 16207 (1965).
21. Kanegafuchi Spinning Co., Ltd., Belg. patent, 652,893 (1964); *Chem. Abstr.*, **64,** 11084 (1966).
22. T. Takata, K. Sayama, and M. Takigawa, *Nippon Kagaku Zasshi,* **85,** 237 (1964); *Chem. Abstr.*, **63,** 2950 (1965).
23. Deutshe Gold-und Silber-Scheideanstalt vorm. Roessler, Neth. patent, 6,511,104 (1966); *Chem. Abstr.*, **65,** P2231 (1966).

# CHAPTER X

# Pyridinecarboxylic Acids

PETER I. POLLAK* AND MARTHA WINDHOLZ

*Merck Sharp & Dohme Research Laboratories*
*Rahway, New Jersey*



## I. Preparation

### 1. From Nonpyridine Starting Materials

Additional syntheses of pyridine ring systems from 1,5-dicarbonyl precursors ve been described (see also Chapter II).

Deceased, 1971.

Citric acid **(X-1)** when treated with ammonia or urea at 130 to 200°[1] or wi $p$-toluenesulfonic acid under autoclaving conditions[2] affords citrazinic acid **(X-** X = OH) or its amide **(X-2**, X = $NH_2$) in yields exceeding 60% of theory

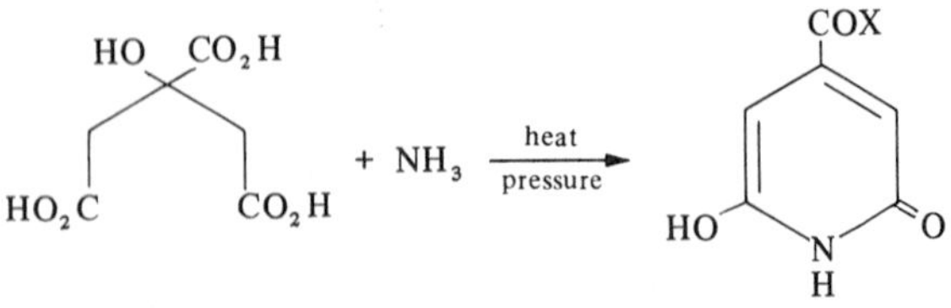

**X-1** **X-2**

Citrazinic acid can be converted in high yield to 2,6-dichloroisonicotinic ac **(X-3**, X = Cl) with $POCl_3$. The latter is easily reduced catalytically isonicotinic acid **(X-3**, X = H).[2]

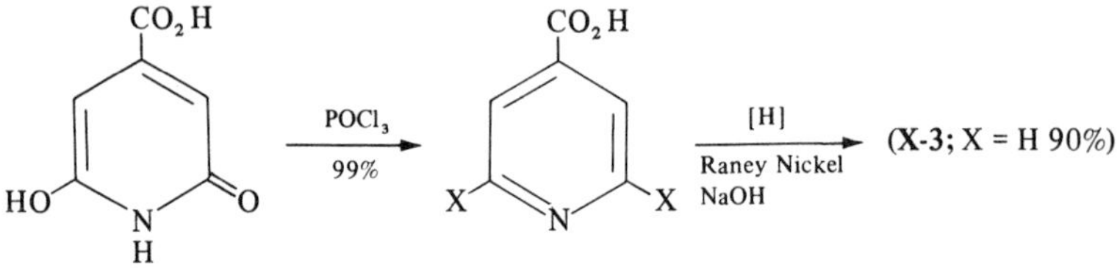

**X-2** **X-3** (X = Cl) (**X-3**; X = H 90%)

2-Pyrone **(X-4**, R = R′ = H) and its 5- **(X-4**; R = $CO_2H$, R′ = H) a 6-carboxylic acid **(X-4**; R = H, R′ = $CO_2H$) or its 5,6-dicarboxylic acid **(X-4**, R R′ = $CO_2H$) were converted to the corresponding pyridine carboxylic acids **(X-** by treatment with $LiAlH_4$ followed by aminolysis.[3] The intermedia

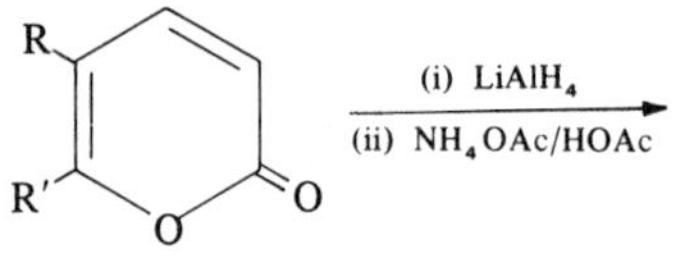

**X-4** **X-5**

*cis,trans*-muconic acid semialdehydes could be isolated.

The sodium enolate of acetoacetaldehyde **(X-6)** could be condensed wi formaldehyde and ammonolyzed to yield the dihydropyridine **(X-7)**. Subseque nitric acid oxidation gave pyridine-3,5-dicarboxylic acid **(X-8)**.[4] When th

†In this chapter all 2- or 4-oxipyridine derivatives will be formulated as pyridones rath than as hydroxypyridines in keeping with modern structural evidence (A. R. Katritzk *Principles of Heterocyclic Chemistry*, Academic Press, New York, 1968).

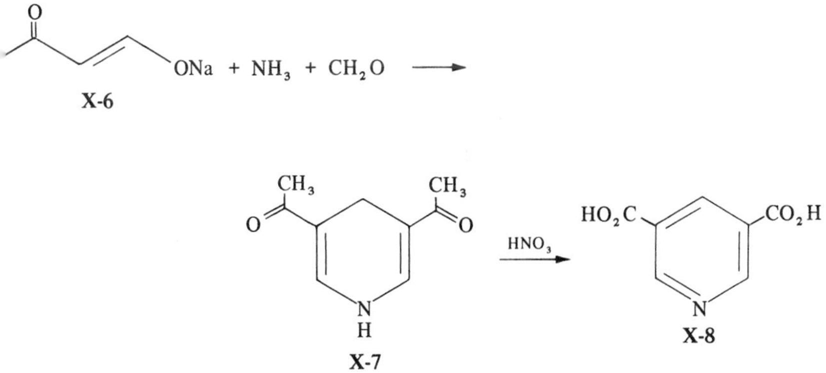

ONa + $NH_3$ + $CH_2O$ →

**X-6**

$HNO_3$

**X-7** **X-8**

action was carried out with acetaldehyde and ammonium carbonate, the roduct was 4-methyl-3,5-diacetyl-1,4-dihydropyridine **(X-9)**, which is readily xidized with nitrous acid to the pyridine **(X-10)**. Sodium hypobromite

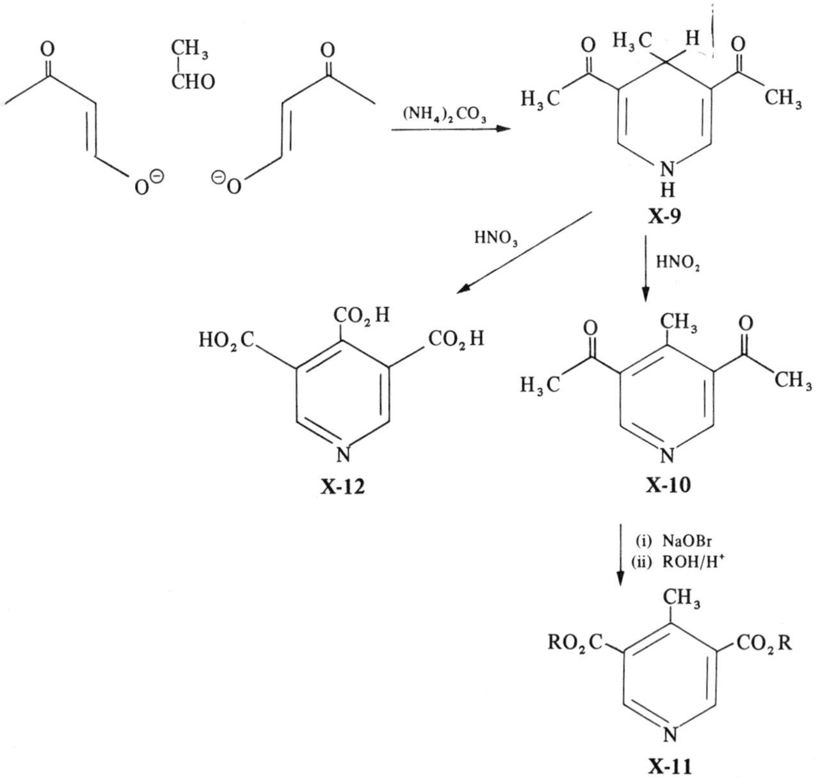

$(NH_4)_2CO_3$

**X-9**

$HNO_3$

$HNO_2$

**X-12** **X-10**

(i) NaOBr
(ii) $ROH/H^+$

**X-11**

oxidation followed by esterification in methanol or ethanol gave t corresponding dialkyl 4-methylpyridine-3,5-dicarboxylates (**X-11**, R = $CH_3$ $C_2H_5$). Oxidation of the dihydropyridine (**X-9**) with nitric acid afford pyridine-3,4,5-tricarboxylic acid (**X-12**), which could be esterified w diazomethane.[5]

Related bimolecular condensations of enamines derived from β-diketones w an aldehyde have also been described.[6]

Similar reaction systems in which the aldehyde has been replaced by a suita malonic acid or malononitrile precursor, undergo similar condensations pyridones.[7–14] With appropriately substituted precursors, the synthesis is suitab

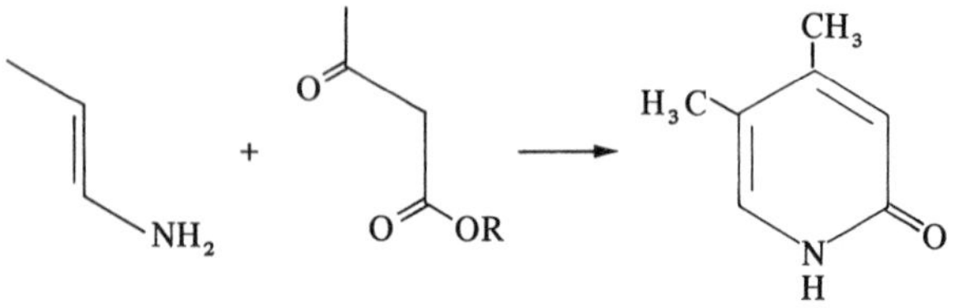

$NH_2$ + O OR → $CH_3$ $H_3C$ N H O

for the preparation of pyridonecarboxylic acids.

A related reaction is the bimolecular self-condensation of ethyl β-meth aminoacrylate (**X-13**) to *N*-methyl-5-carbethoxy-2-pyridone (**IX-14**).[15]

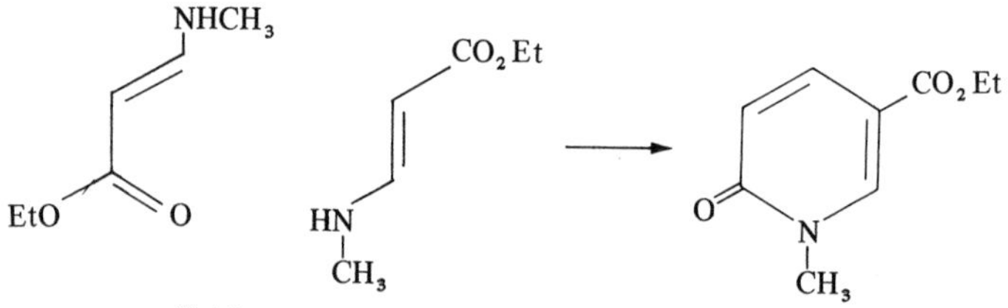

$NHCH_3$ EtO O $CO_2Et$ HN $CH_3$ → $CO_2Et$ O N $CH_3$

**X-13** **X-14**

Nicotinic acid (**X-17**) can be obtained by an interesting electroly trimerisation of acrylonitrile to 1,3,6-tricyanohexane (**X-15**). The latter reduced with Raney Cobalt to the corresponding triamine, which, in turn, catalytically cyclized to 3-(ω-amino-*n*-butyl)piperidine (**X-16**, R = H). T acetylated piperidine (**X-16**, R = OAc) is dehydrogenated and oxidized to t vitamin (**X-17**).[16]

Huisgen and Herbig[17] condensed *N*-benzylphenylazomethine (**X-18**) with t moles of dimethyl acetylenedicarboxylate to obtain the correspondi 1,2-dihydropyridinetetracarboxylic acid ester (**X-19**). Oxidation with bromi afforded tetramethyl 2-phenylpyridine 3,4,5,6-tetracarboxylate (**X-20**).

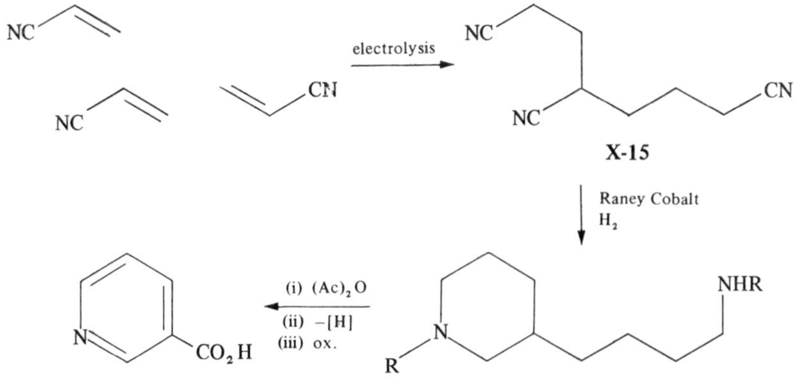

The acid-catalyzed trimerization of potassium $\beta$-hydroxy-$\alpha$-nitropropionitrile -21) yields 2,4,6-tricyanopyridine-1-oxide (**X-22**, R = CN). The latter could be

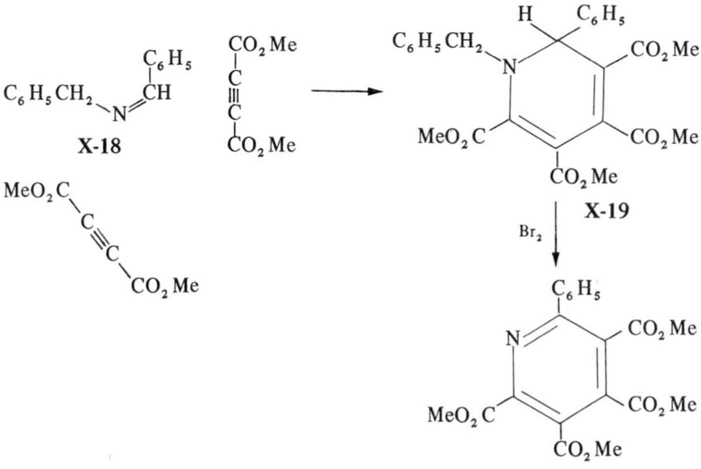

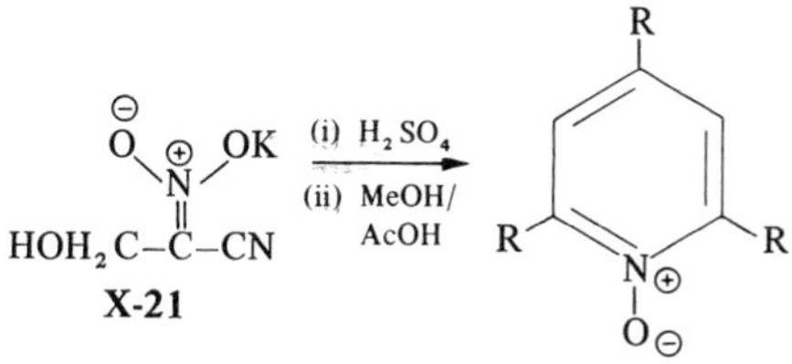

converted to the trimethyl ester (**X-22**, R = $CO_2Me$) by methanolysis in acet
acid.[18]

Isoxazoles (**X-23**) have been used to protect β-iminocarbonyl systems in th
manipulation of complex starting materials. Thus they yield dihydro-β-acety
pyridine-β′-carboxylic acids (**X-24**).[19]

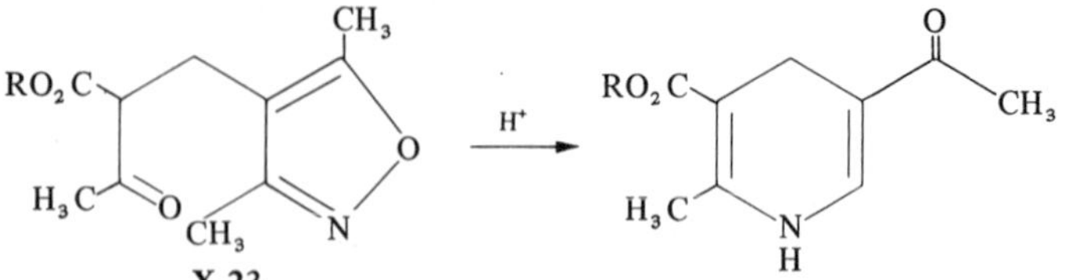

**X-23** $\xrightarrow{H^+}$ **X-24**

A number of miscellaneous condensations have been recorded in the last te
years. Thus, the Hantzsch synthesis has been reexamined and a 53% yield o
**X-25** was obtained.[20]

**X-25**

Treatment of aminopyranodioxins (**X-26**) and aminopyranooxazines (**X-27**
with alkoxides or hydroxides afforded 4,6-dihydroxy-2-oxopyridine-3-carboxy
ates (**X-28**) or carboxanilides (**X-29**) *via* the following mechanism:[21]

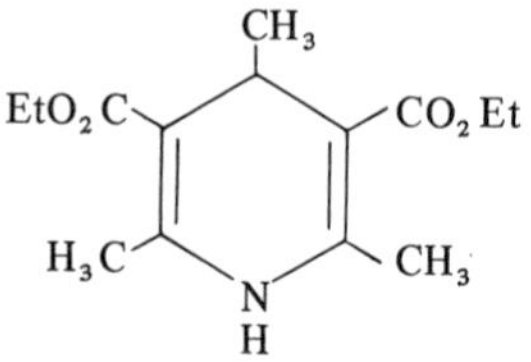

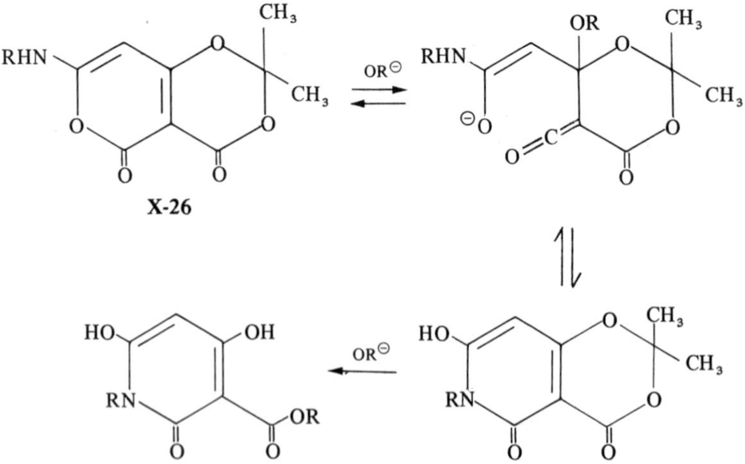

**X-26**

**X-28**

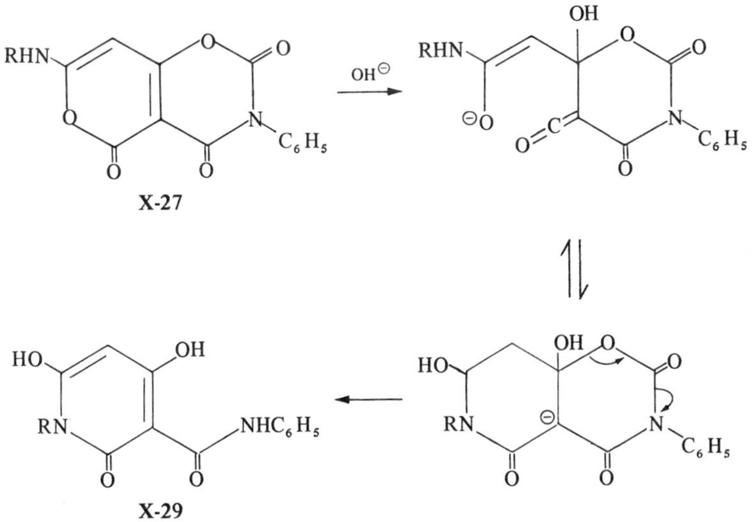

X-27

X-29

[M]aleic acid, when treated with sulfuric acid containing oleum and methanol, [ga]ve a 50% yield of methyl coumalate (**X-30**). Ammonolysis afforded a 75% [y]ield of 2-pyridone-5-carboxylic acid (**X-31**).[22] Cinnamoylacetaldehyde (**X-32**)

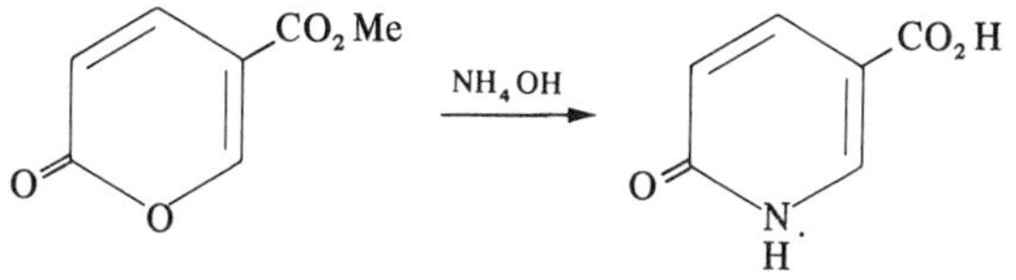

X-30 X-31

[an]d cyanoacetamide were condensed to 3-cyano-6-styryl-2-pyridone (**X-33**, X = [C]N), which was hydrolyzed to the carboxylic acid **X-34** (X = $CO_2H$).

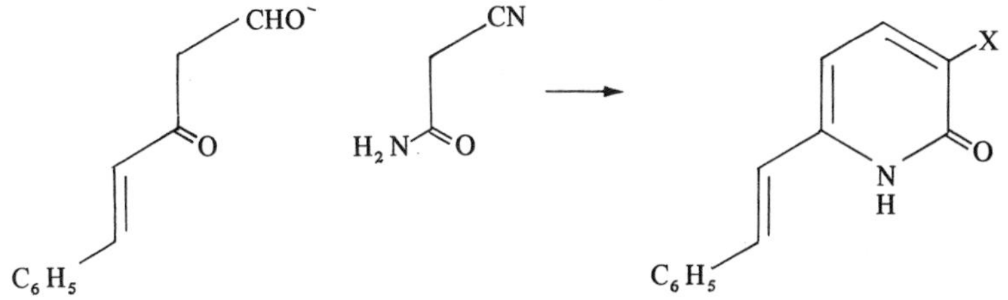

X-32

**X-33** (X = CN)
**X-34** (X = $CO_2H$)

## 2. By Electrocyclic Reactions

A new class of pyridine syntheses is based on the proclivity of som heteroazoles to add dienophiles with the intermediate formation hetero-2,2,1-bicycloazaheptenes. Thus, substituted oxazoles **(X-35)** react with typical dienophile such as maleimide **(X-36)** at elevated temperatures to yie substituted pyridines **(X-37)**.[24] The same type of reaction utilizing substitut

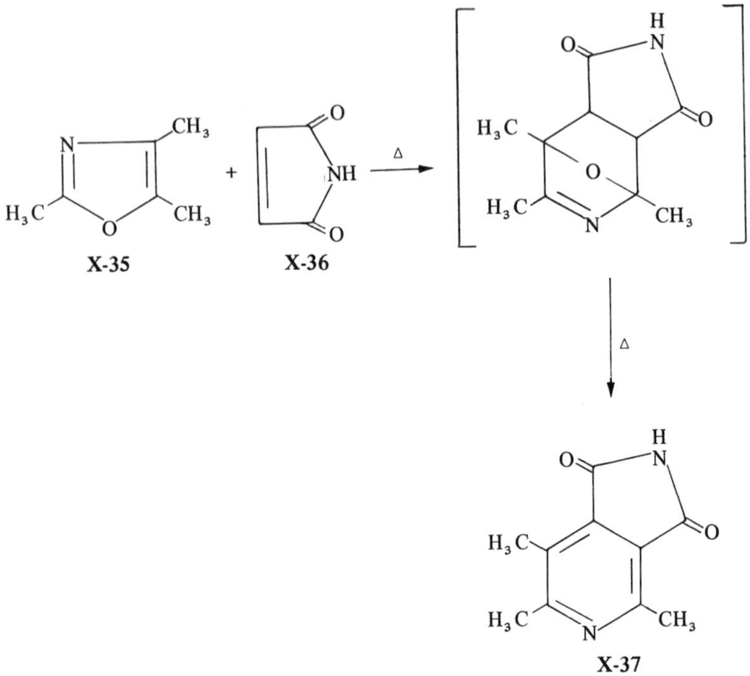

**X-35** **X-36**

**X-37**

5-alkoxyoxazoles **(X-38)** and maleic anhydride **(X-39)** affords 4,5-dicarboxy pyridinols **(X-40)**.[25] (See also Chapter XII.)

It was immediately recognized that this system offers a new route to pyrido (vitamin $B_6$): 4-methyl-5-ethoxyoxazole **(X-41)**, on treatment with dietl maleate **(X-42**, X = $CO_2Et$) or maleonitrile **(X-42**, X = CN), gave the expect bicyclic adducts **(X-43**, X = $CO_2Et$ or CN) which were readily solvolyzed 2-methyl-4,5-dicarbethoxy-3-pyridinol **(X-44**, X = $CO_2Et$) or the correspondi nitrile **(X-44**, X = CN). The latter are easily converted to pyridoxol **(X-44**, X $CH_2OH$).[26–28]

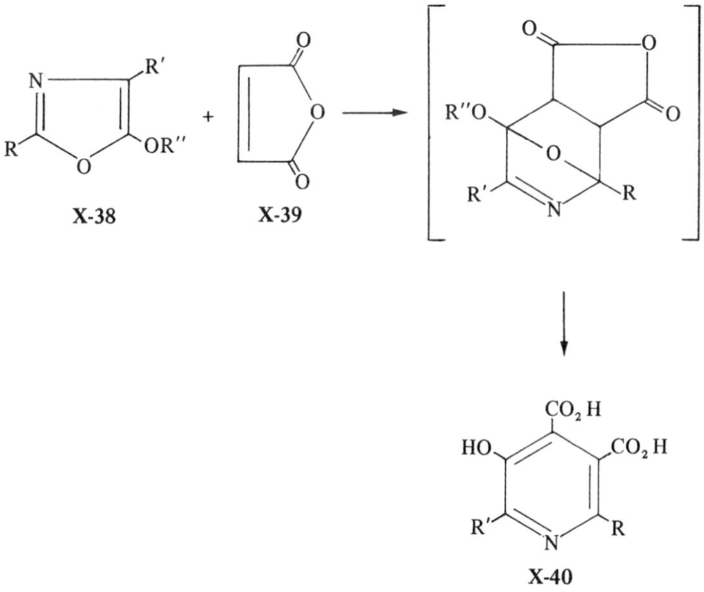

X-38 + X-39 → [ ] → X-40

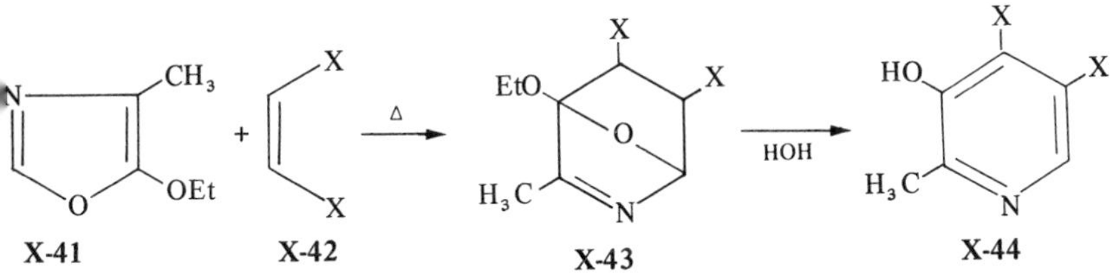

X-41 + X-42 —Δ→ X-43 —HOH→ X-44

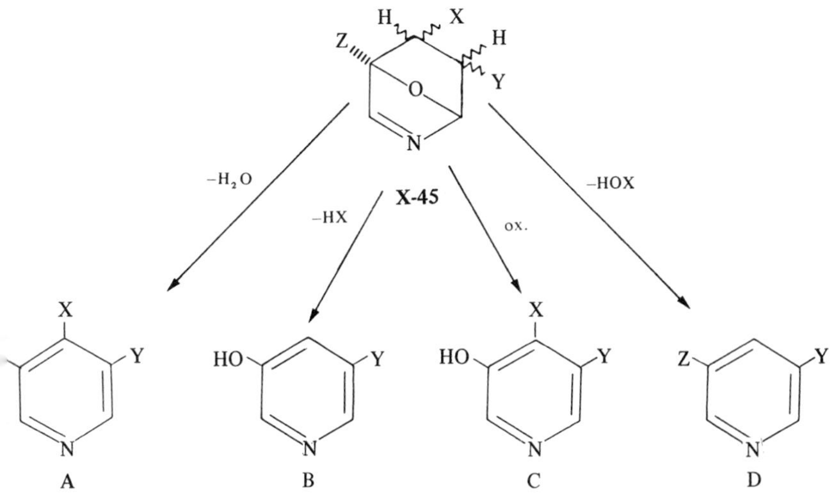

X-45

$-H_2O$ → A

$-HX$ → B

ox. → C

$-HOX$ → D

Investigation of the mechanism and stereochemistry of this Diels-Alde reaction showed that the intermediate **X-45** could lead to four different produc types depending on the electronic nature of X, Y, and Z, on the stereochemica relationship of the 4-proton to the oxide bridge, and on whether or not th medium contains an oxidant.[29]

The use of unsymmetrical dienophiles such as acrylonitrile fits well into th scheme in that the two possible intermediates (**X-45**; X = CN, Y = Z = H) an (**X-45**; X = Z = H, Y = CN), formed from 4-methyloxazole, are transformed t the expected product mixture.[30, 31]

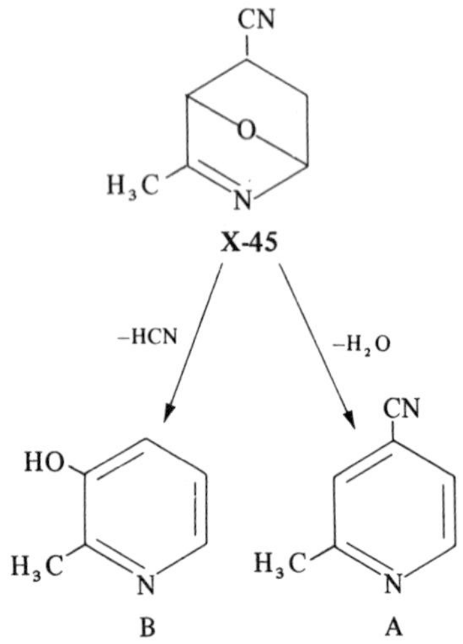

CN
O
$H_3C$
N
**X-45**

–HCN
$-H_2O$

HO
$H_3C$
N
B

CN
$H_3C$
N
A

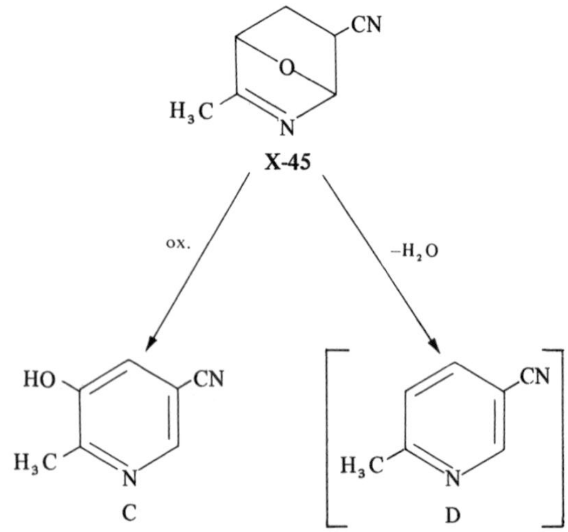

CN
O
$H_3C$
N
**X-45**

ox.
$-H_2O$

HO
CN
$H_3C$
N
C

CN
$H_3C$
N
D

Results similar, but less complete, than those reported for the above dienophile ›re reported for acrylic acid. The products isolated correspond to only one -46) of the two possible isomeric adducts that could be formed. Thermolysis pyridinecarboxylic acid occurred as expected in about 50% overall yield.[32]

$CO_2H$

O

$H_3C$

N

**X-46**

Tetraphenylcyclopentadienone **(X-47)** added phenyl cyanoformate **(X-48)** at 6° to yield phenyl 3,4,5,6-tetraphenylpyridine-2-carboxylate **(X-50)**. An :ermediate, [2.2.1]-azabicycloheptadienone **(X-49)**, is postulated.[33] The

**X-47** **X-48** **X-49**

–CO

**X-50**

›nyl ester was hydrolyzed to give the free acid. The same reaction was run h tetrachlorocyclopentadienone. Continuous extraction with methanol ›rded methyl phenyl 3,4,5-trichloropyridine-2,6-dicarboxylate **(X-52)**, which y have arisen from the solvolytic ring-opening of the intermediate **X-51**.

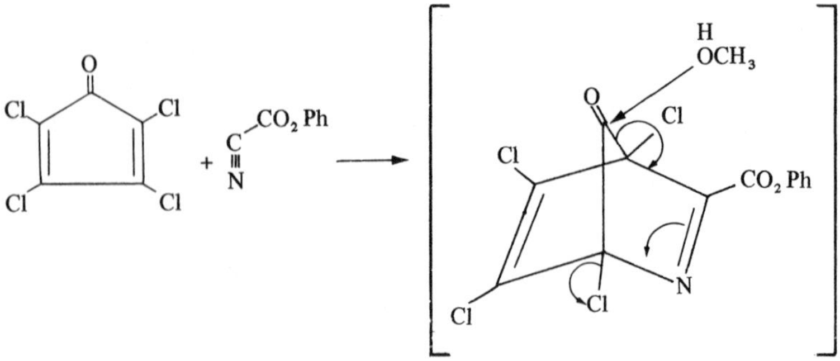

X-51

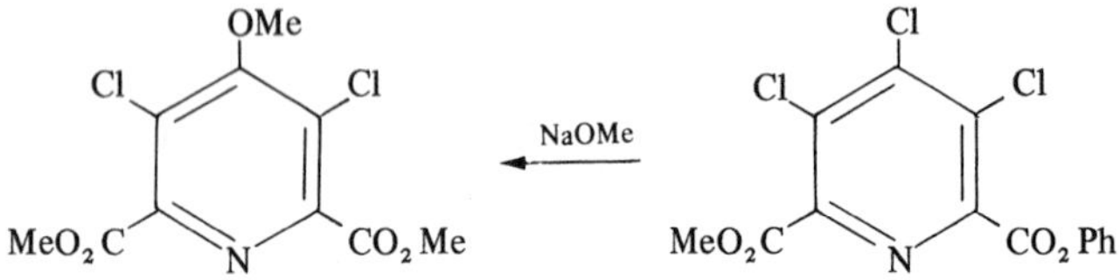

NaOMe

X-53 X-52

Treatment of the mixed ester (**X-50**) with sodium methoxide led to dimet 3,5-dichloro-4-methoxypyridine-2,6-dicarboxylate (**X-53**).

### 3. By Oxidation of Alkylpyridines, Quinolines, and Isoquinolines

The oxidation of appropriate alkylpyridines and quinolines continues represent the most important process for the manufacture of nicotinic acid. aspects of the field have seen continuing development, and a Czech review available.[34] A general developmental study comparing seven oxidizing agents a quantitative basis has also appeared.[35]

The oxidation of alkylpyridines in general, and 5-ethyl-2-methylpyrid (**X-54**) in particular, by dilute nitric acid at elevated pressures and temperatu has been the subject of many publications.[36-47]

The use of cupric nitrate as a catalyst in these systems has also b described.[48,49] The reaction usually proceeds to isocinchomeronic acid (**X-** R = $CO_2H$), which is decarboxylated to nicotinic acid (**X-56**) either *in situ* o a subsequent operation. Under specific conditions of temperature and pressu the reaction can be stopped at the 6-methylnicotinic acid (**X-55**, R = CH stage.[50] It thus seems clear that under those conditions the 5-alkyl group is m prone to oxidative attack. The same nitric acid reaction system, with or with

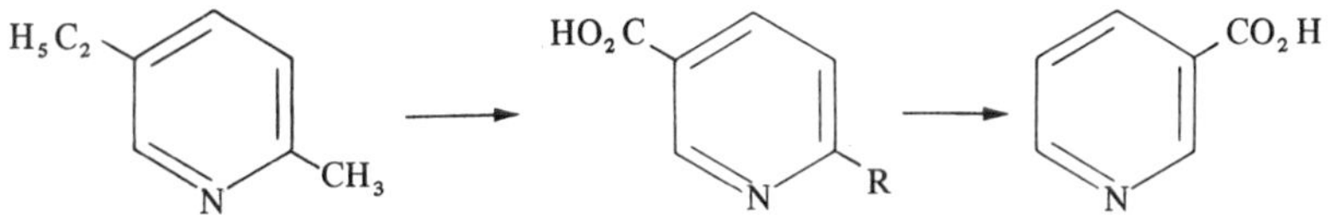

X-54 X-55 X-56

u$^{+2}$ or $MoO_4^{-2}$ catalysis, converts quinoline **(X-57)** to quinolinic acid
X-58).[51-54] If quinoline is first converted to the 8-sulfonic acid with 65%

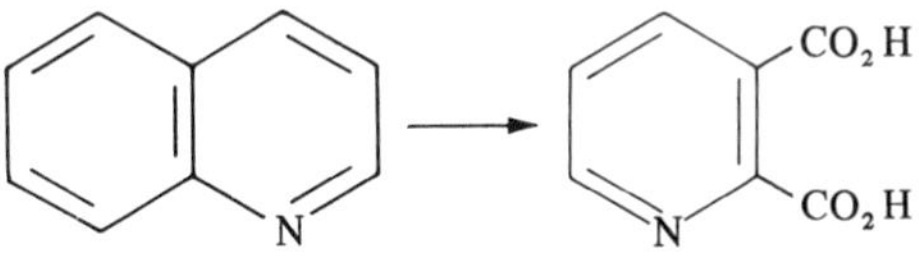

X-57 X-58

eum, oxidation with nitric acid yields an improved direct yield of nicotinic id.[55]

Picolines and 5-ethyl-2-methylpyridine **(X-54)**[1] can also be converted to the orresponding acids in basic media containing lead dioxide[56] or iron-aluminum xide catalysts.[57] Again elevated pressures and temperatures are required.

A number of publications also describe the air oxidation of the same pyridine ibstrates in the presence of cupric nitrate.[58-62] Extensive work has also been ported on the fixed or fluidized bed vapor phase of alkylpyridines using nadium catalysts.[63-67] The intermediate aldehydes can be isolated,[68] but sually the conditions are set to minimize this side reaction. Vanadates, olybdates, and tin salts of these two oxyacids are also described as effective talysts in the high temperature oxidation of quinolines and isoquinolines.[69-71]

In the laboratory, selenium dioxide in organic solvents is a good reagent for e oxidation of alkylpyridines to carboxylic acids.[72-74] The intermediate dehydes can be obtained depending on the position of the alkyl side-chain and e substrate/$SeO_2$ ratio. The reaction is selective in the case of ethynyl-2-methylpyridine **(X-59)** yielding 5-ethynyl-2-picolinic acid in 40% eld.[75]

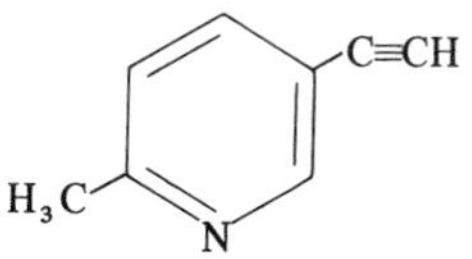

X-59

A number of industrially important systems have been described that use $SeO_2$ rmed *in situ* from Se metal and nitric or sulfuric acid containing $SO_3$.[76-78]

The same oxidants can be used with quinoline or isoquinoline as substrates.[79-] In most of these reactions Se metal is used in catalytic amounts and continuously regenerated by the other oxidants present.

Ozone has been used to convert 8-hydroxyquinoline to nicotinic acid.[83, 84] the case of alkylquinolines, acridine, and phenanthridine, ozonolysis leads alkylpyridine carboxylic acids, their *N*-oxides, and more extensive degradati products derived from the subsequent C–N bond cleavage in the *N*-oxides. 5-Fluoronicotinic acid was obtained from 3-fluoroquinoline using a Cu(I acetate/hydrogen peroxide couple.[86] Manganese dioxide[87, 88] and permang nate[89, 90] are also effective, even in the case of acetamidopicolines.[91] The use chromic acid seems an obvious extension.[92-94] An interesting mechanistic stu concerning the air oxidation of alkylpyridines in strongly basic media shows th the reaction proceeds through direct electrophilic attack of oxygen on t alkylpyridine carbanions concerned.[95]

$CH_3$, $CH_3$, N
**X-60**

HCHO

$CH_2CH_2OH$, $CH_3$, N
**X-61**

$C(CH_2OH)_3$, $CH_3$, N
**X-62**

$CH_3$, $CH_2CH_2O$, N
**X-63**

$HNO_3$

$CO_2H$, $CH_3$, N
**X-64**

$CH_3$, $CO_2H$, N
**X-65**

An interesting variant on this scheme to form alkylpyridinecarboxylic acids by itric acid oxidation has been described by Czech workers: starting with 2,4- or ,6-lutidine these bases are condensed with formaldehyde to yield nsymmetrical β-(hydroxyethyl)methylpyridines. Nitric acid oxidation now ccurs preferentially at the hydroxylated side chain yielding the corresponding ethylpicolinic acids.[96-99] Thus, 2,4-lutidine **(X-60)** is converted to a separable ixture of 4-(β-hydroxyethyl)-2-methylpyridine **(X-61)**, 4-[tris(hydroxy-ethyl)methyl]-2-methylpyridine **(X-62)**, and 2-(β-hydroxymethyl)-4-methyl-yridine **(X-63)**. Nitric acid oxidation yields 2-methylpyridine-4-carboxylic acid **-64)** from **X-61** and **X-62**, and 4-methylpyridine-2-carboxylic acid **(X-65)** om **X-63**. A similar reaction sequence carried out with 2,6-lutidine **(X-66)** fords ultimately 6-methylpyridine-2-carboxylic acid **(X-67)**.

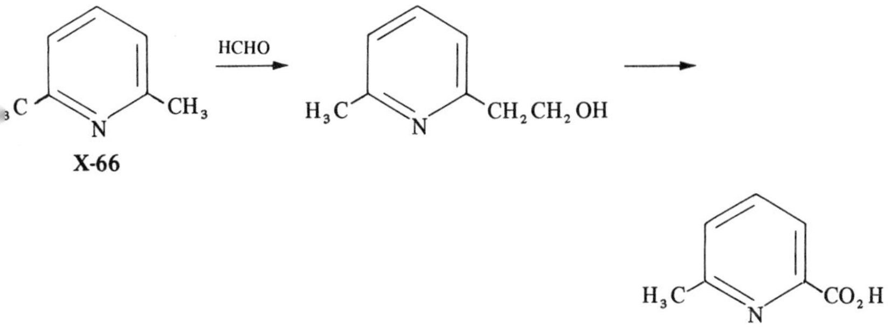

HCHO

$_3$C CH$_3$ N **X-66** → H$_3$C N CH$_2$CH$_2$OH →

H$_3$C N CO$_2$H **X-67**

Vigorous oxidation conditions of course vitiate any selectivity adduced by nctionalizing one of the side-chains: complete oxidation of all alkyl bstituents is the result.[100]

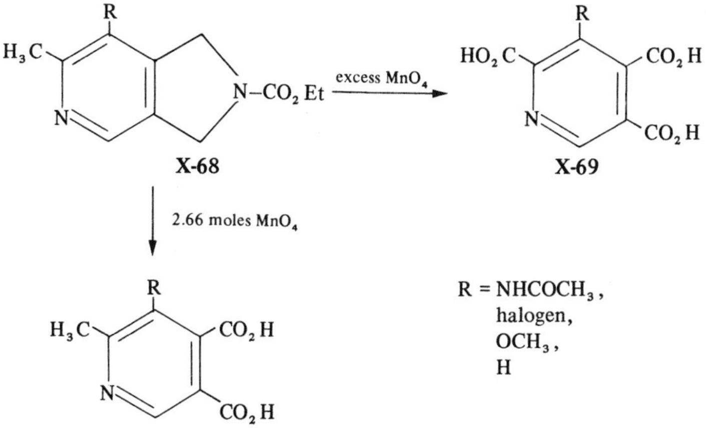

R, H$_3$C, N, N–CO$_2$Et **X-68** —excess MnO$_4$→ R, HO$_2$C, CO$_2$H, N, CO$_2$H **X-69**

↓ 2.66 moles MnO$_4$

R, H$_3$C, CO$_2$H, N, CO$_2$H **X-70**

R = NHCOCH$_3$,
halogen,
OCH$_3$,
H

Substituted merimines **(X-68)** yield tricarboxylic acids **(X-69)** on vigor oxidation with excess permanganate. With exactly 2.66 moles of permangana selective oxidation can be accomplished to a number of substitu 6-methylcinchomeronic acids **(X-70)**.[101]

A nice example of a selective oxidation with manganese dioxide has b reported in the pyridoxine series:[102] in ethanolic KOH pyridoxine **(X-** yields 3-hydroxy-5-hydroxymethyl-2-methylpyridine-4-carboxylic acid **(X-7** The same reagent in dilute sulfuric acid yields the hemiacetal **X-73**.

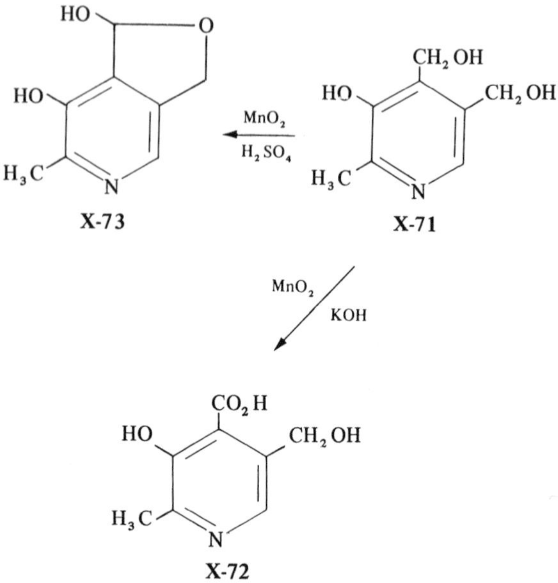

HO, HO, $H_3C$, O, N — **X-73** ← $MnO_2$ / $H_2SO_4$ — $CH_2OH$, HO, $CH_2OH$, $H_3C$, N — **X-71**

$MnO_2$ / KOH ↙

$CO_2H$, HO, $CH_2OH$, $H_3C$, N — **X-72**

## 4. Electrolytic Oxidation

In efforts to develop industrial processes for the manufacture of nicotinic ac Russian workers have studied the electrolytic oxidation of alkylpyridines a quinolines **(X-74)** using neutral or acidic media and platinum and le anodes.[103-110] Reactions occurred at 40 to 100° with high current efficienci but reported yields are low (<50%). Desired products appeared, of course, in t anolyte compartment. In the catholyte compartment, on the other har quinoline substrate was, *inter alia*, reduced to 1,2,3,4-tetrahydroquinoli **(X-75)**, isolated as the *N*-acetyl derivative.

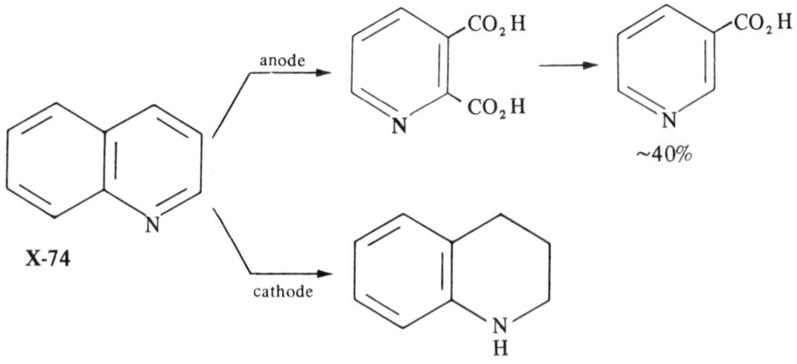

anode

cathode

~40%

X-74

X-75

## 5. Synthesis via Nitriles

The solvolysis of pyridine nitriles **(X-76)** is an important route to yridinecarboxylic acids (**X-77**, X = OH) or their derivatives **(X-77)**.[111] In this ection, only those new syntheses that proceed through nitriles that are not erived in the first place from precursor carboxylic acids[112-118] will be described.

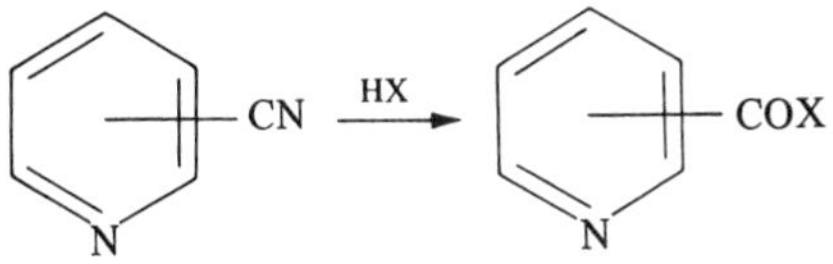

CN —HX→ COX

X-76

X-77

X = OH, $NH_2$, OR

Much progress has been made in the catalytic air ammoxidation of kylpyridines to cyanopyridines **(X-76)**. Yields are generally high. The catalysts mployed are oxides of B, Al, P, Bi, Mo and V.[119-128] In the case of -ethyl-2-methylpyridine **(X-78)** the principal products observed were icotinonitrile **(X-79)**, 6-cyanonicotinamide **(X-80)**, and 5-ethylpicolinonitrile **(X-81)**. These products indicate that initial oxidative attack in **X-78** may occur the 2-methyl position, and that initial oxidative attack to the alcohol quivalent is the slow step in the reaction sequence.[129]

Additional support for this interpretation comes from work on isomeric tidines: 2,3-lutidine **(X-82)** is converted to 2-cyano-3-methylpyridine **(X-83)** nd 3,5-lutidine **(X-84)** affords 5-methylnicotinonitrile **(X-85)**.[130]

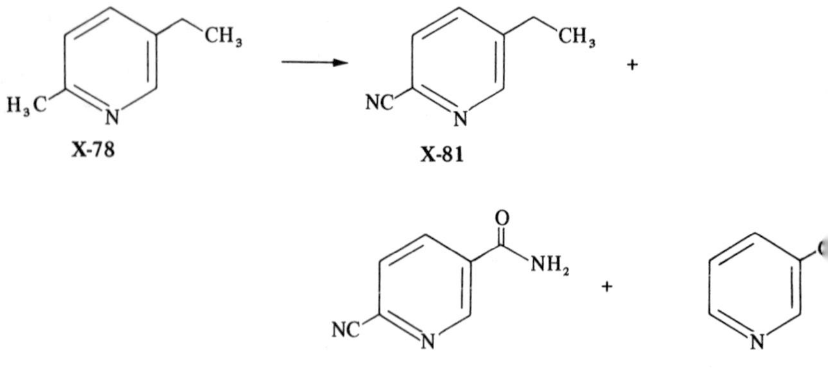

A related reaction is the oxidative ammonolysis of 1,1,1-trihydroxymethyl-4-picoline **(X-86)** to isonicotinonitrile **(X-87)**.[131]

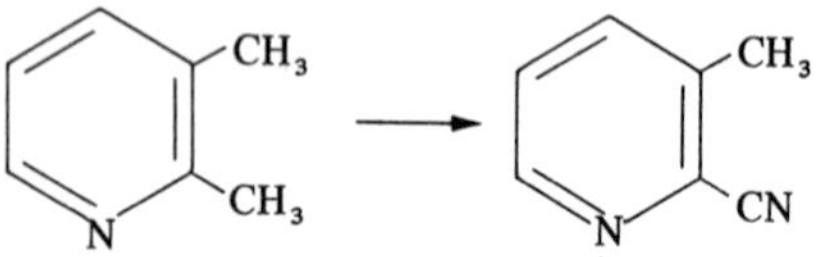

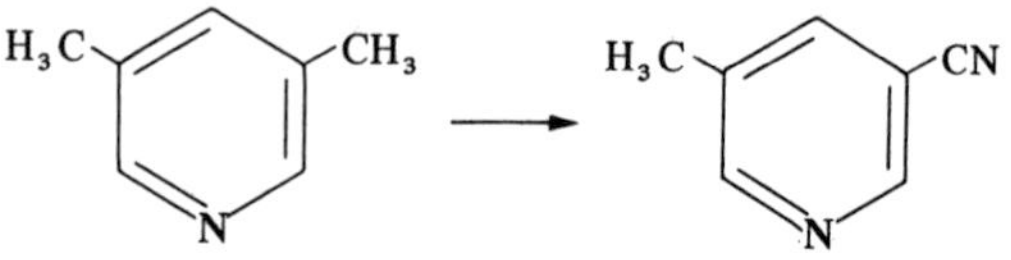

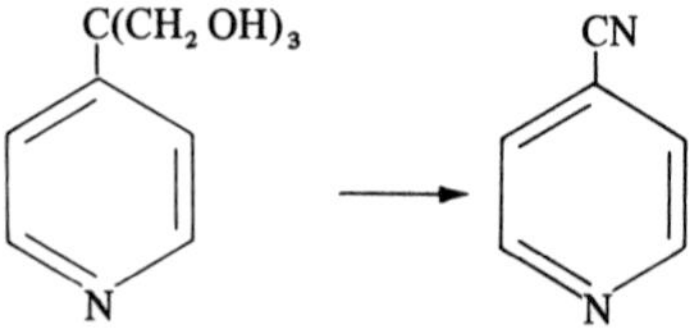

The cyanation of alkylated pyridine-1-oxides **(X-88)** leads to 2- or 4-cyanopyridines[132-134] (see also Chapter IV). The intermediate quaternary salt **(X-88)** can be isolated. Related acylated pyridinum-1-imines **(X-89)** undergo a similar reaction.[135]

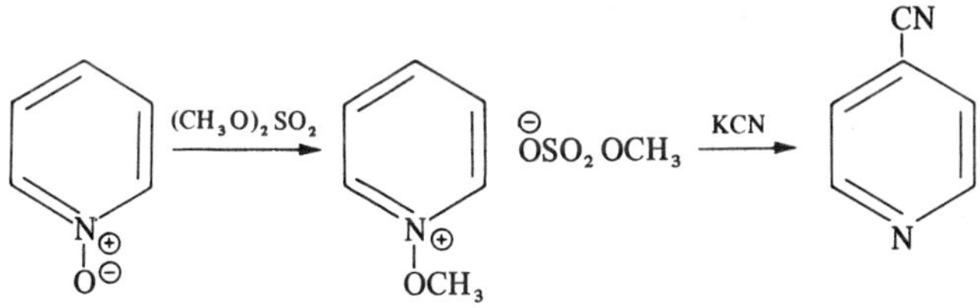

An internal redox reaction of 2-picoline-1-oxide (**X-90**, X = H) and its alkyl rivatives with nitrite ion and an acylating agent yields 2-cyanopyridines (**X-91**, = H).[136] The following mechanism is proposed:

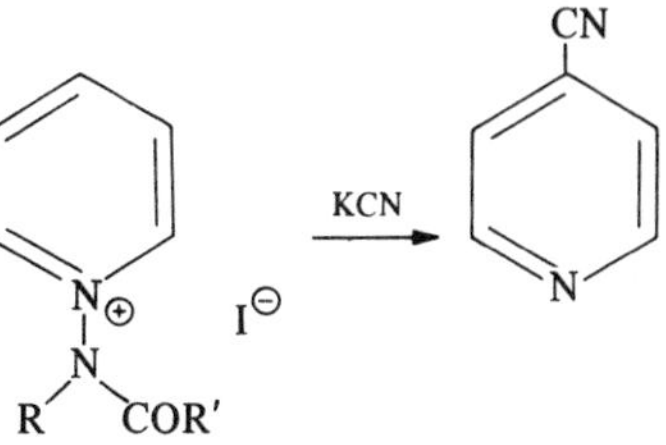

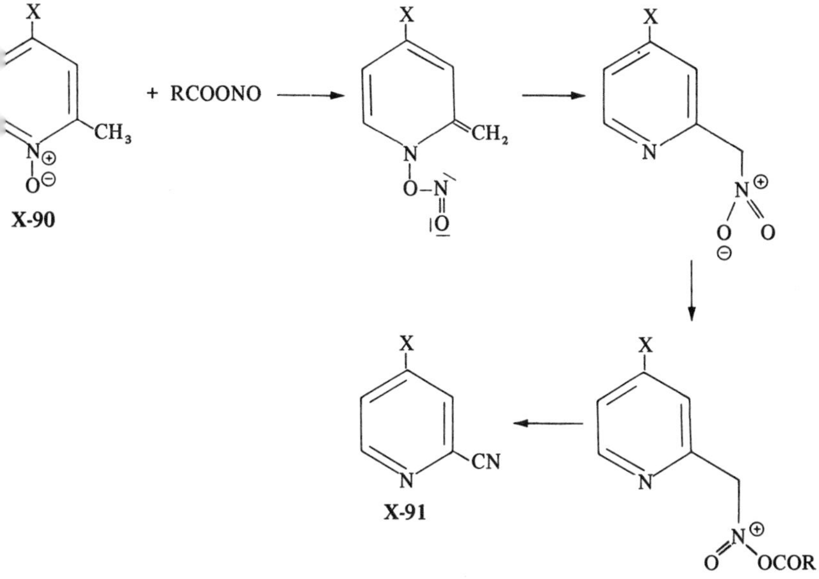

is reaction can also be achieved with an intramolecular source of nitrite ion: nitro-2-picoline-1-oxide (**X-90**, X = $NO_2$), on treatment with acetyl chloride

under cooling, yields a 68% yield of 4-chloro-2-picoline-1-oxide (**X-90**, X = C
If this reaction is carried out with warming, some 4-chloro-2-cyanopyridi
(**X-91**, X = Cl), and a large yield (57%) of 4-chloro-2-cyanopyridine-1-oxide
claimed.[137]

2- or 4-Methylcyanopyridines, convertible to 2- or 4-methylpyridinecarboxy
acids, can be obtained by the action of methylmagnesium iodide
2,5-dicyanopyridines (**X-92**, X = H or $CH_3$). The intermediate 1,2-
1,4-dihydropyridines (**X-93**, X = H or $CH_3$) are isolable yellow solids. They c
be dehydrogenated catalytically (Pd/C) or with silver oxide.[138, 139]

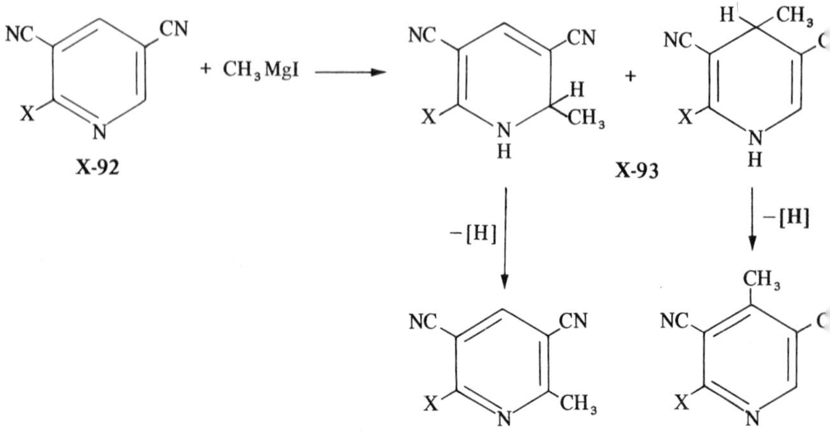

X-92 + $CH_3MgI$ → X-93 —(−[H])→

X-94 —($H_3O^+$)→ X-95 —(Δ)→ X-96 —($H_3O^+$)→ X-97

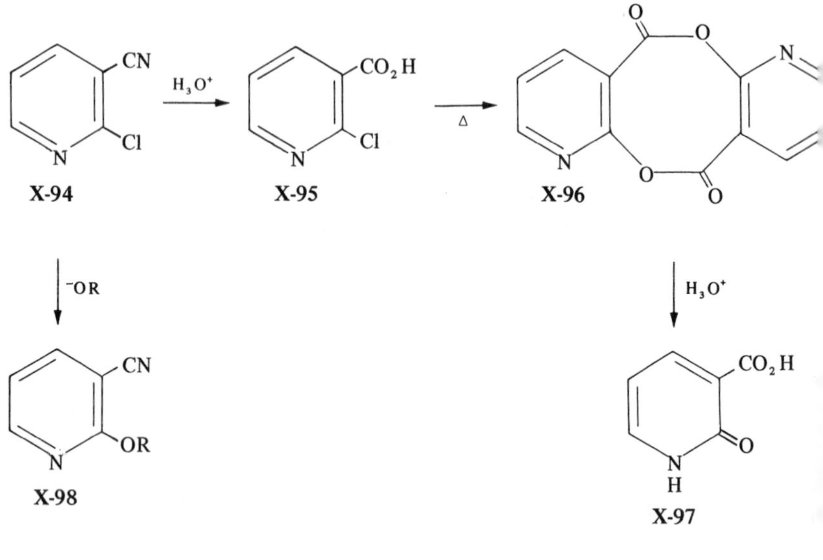

X-94 —($^-OR$)→ X-98

In the acid-catalyzed hydrolysis of halocyanopyridines such as 2-chloro-3-anopyridine **(X-94)**, an intermediate 2-chloronicotinic acid **(X-95)** is formed ıich, on heating, undergoes a bimolecular condensation to the bis-ester **(X-96)**. ırther hydrolysis leads to 2-pyridone-3-carboxylic acid **(X-97)**. Treatment of **94** with alkoxides leads to the 2-alkoxy-3-cyanopyridine **(X-98)**.[140]

$HO_3S$ $SO_3H$ N N $HO_3S$ $SO_3H$ **X-99** $\xrightarrow[350°]{K_3Fe(CN)_6}$ N N CN **X-100**

$KMnO_4$ → $CO_2H$ $CO_2H$ N **X-102**

HCl → N N $CO_2H$ **X-101**

CN $N^{\oplus}$ $O^{\ominus}$ **X-103** + N Br → $N^{\oplus}$ H O N CN **X-104** → N O N CN **X-105** → $NH_2$ N $CO_2H$ **X-106**

Treatment of (4,4′-bipyridyl)-3,3′,5,5′-tetrasulfonic acid **(X-99)** with pot sium ferricyanide at 350° affords 3-cyano-4,4′-bipyridyl (**X-100**) as the distilla The latter could be hydrolyzed with concentrated hydrochloric acid (4,4′-bipyridyl)-3-carboxylic acid **(X-101)**. Oxidation of **X-100** with neut permanganate yielded cinchomeronic acid **(X-102)**.[141]

Nicotinonitrile-1-oxide **(X-103)** on treatment with 2-bromopyridine afford 1-(3-cyano-6-pyridyl)-2-pyridone **(X-105)**, which could be hydrolyzed 6-aminonicotinic acid **(X-106)**. It is proposed that the reaction proceeds throu the intermediate 2,4-oxadiazoline **(X-104)**.[142] (See also Chapter IV.)

## 6. Synthesis of Pyridinecarboxylic Acids Containing Sulfur Substituents

5-Amino-2-chloropyridine (**X-107**, X = $NH_2$), obtained by reduction of t 5-nitro precursor (**X-107**, X = $NO_2$), was converted by a Sandmeyer reaction the 5-cyano-compound (**X-107**, X = CN). Saponification afforded 5-carboxy acid (**X-107**, X = $CO_2H$), which, on treatment with potassium hydrogen sulfic yielded 2-mercaptopyridine-5-carboxylic acid (**X-108**, Y = SH). The latter w oxidized with alkaline permanganate to the 2-sulfonic acid (**X-108**, Y = $SO_3$I

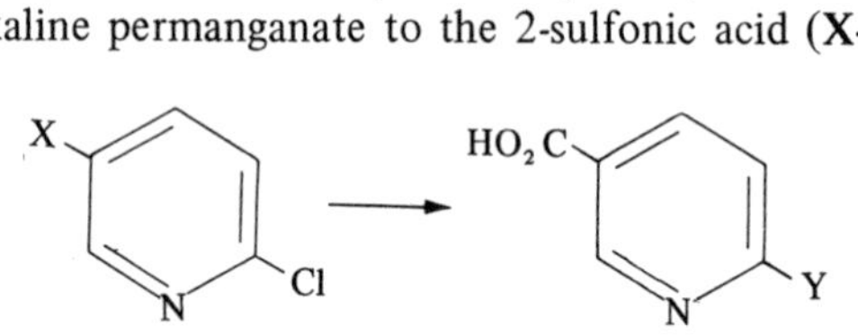

**X-107**

X = $NO_2$, $NH_2$, CN, $CO_2H$

**X-108**

Y = SH, $SO_3H$

A second Sandmeyer reaction on 5-amino-2-chloropyridine (**X-107**, X = NH and copper-catalyzed coupling with $SO_2$-saturated acetic acid gave t 2-chloropyridine-5-sulfonic acid (**X-109**, X = Cl). This substance was convert

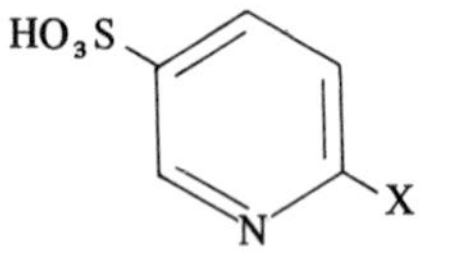

**X-109**

X = Cl, SH, $SO_3H$

to the pyridine-2,5-disulfonic acid (**X-109**, X = $SO_3H$) by treatment with KI and permanganate.[143, 144]

A second excellent method of introducing a mercapto substituent consists treating 6-methyl-2-pyridone **(X-110)** with $P_2S_5$. Oxidation with concentrat itric acid leads directly to the 6-carboxypyridine-2-sulfonic acid X-111).[144, 145] The same sequence can be carried out with 4-methyl-2-yridone.

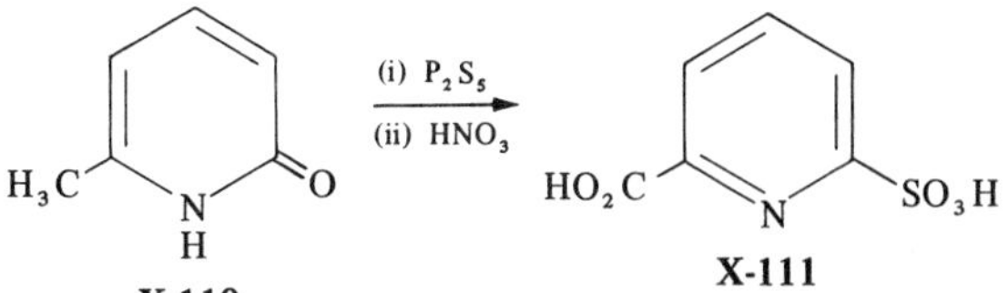

X-110 → X-111

## 7. Miscellaneous Syntheses

Photochlorination of 2-picoline **(X-112)** at 50 to 150° produces chloro-2-trichloromethylpyridine **(X-113)** and 3,6-dichloro-2-trichloromethyl-ridine **(X-114)**. Hydrolysis with concentrated nitric acid gave a 95% yield the corresponding 2-carboxylic acids.[146] Continued photochlorination affords perchlorinated trichloromethylpyridine **(X-115)**, which, after ammonolysis and iling in sulfuric acid, afforded picloram **(X-116)**.[147, 148]

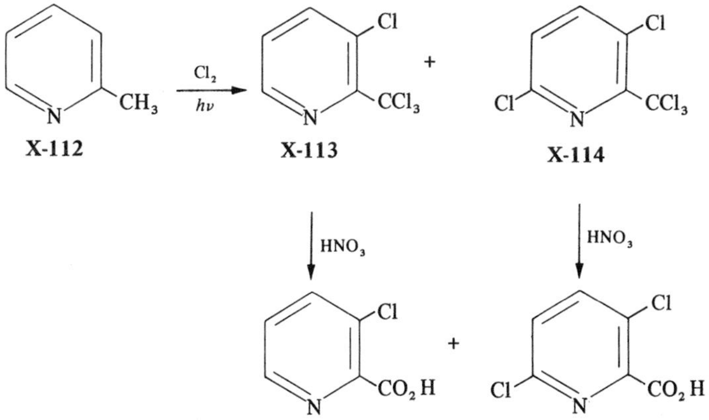

X-112 → X-113 + X-114

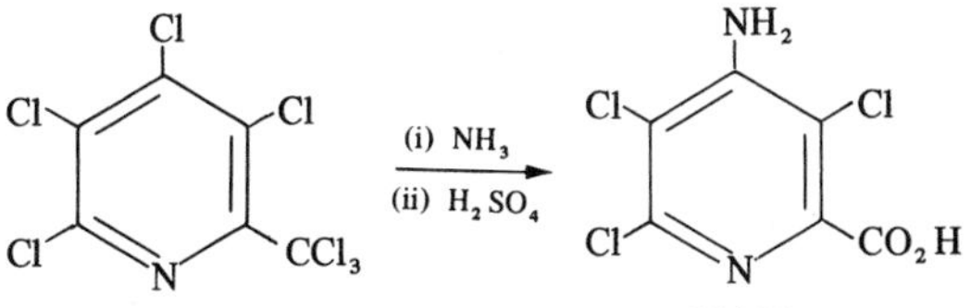

X-115 → X-116

Potassium nicotinate **(X-117)** when autoclaved with $CdF_2$ under $CO_2$ pressu yields isocinchomeronic acid **(X-118)**. The same product is obtained when pota

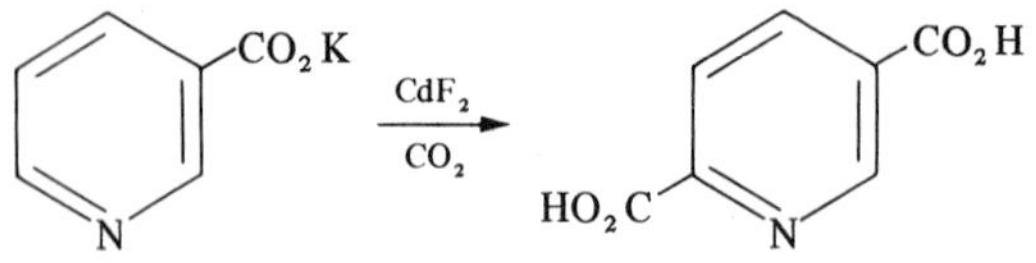

X-117 X-118

sium picolinate **(X-119)** or potassium quinolinate **(X-120)** is exposed to the san reaction conditions. Potassium isonicotinate **(X-121)** affords **X-118** ar pyridine-2,4,6-tricarboxylic acid **(X-122)**.[149]

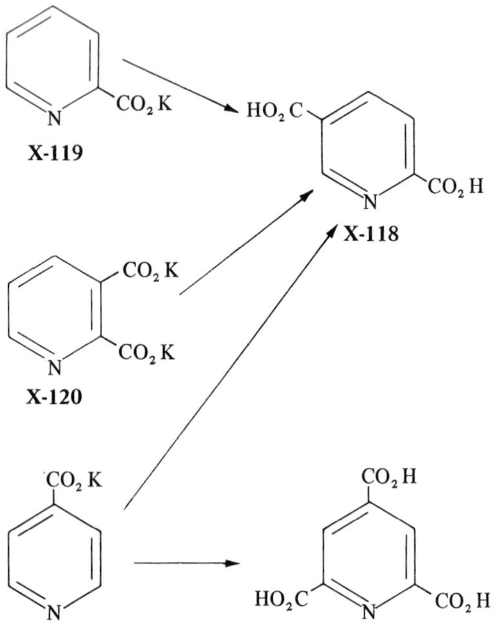

X-119 X-118 X-120 X-121 X-122

Preparation of substituted pyridinecarboxylic acid 1-oxides by means H-abstraction is reported by Abramovitch and others.[150] Thus, treatment of t lithium derivative of 4-chloropyridine-1-oxide or 4-chloro-3-methylpyridine- oxide with $CO_2$ gave the corresponding 4-chloro- and 4-chloro-5-methylpicolin acid-1-oxides. Under the same conditions, 4-picoline-1-oxide gave the 2,6-c carboxylic acid.

X = H or $CH_3$

## II. Physical Properties

The apparent ionization constants and ultraviolet absorption spectra of the ree isomeric pyridine monocarboxylic acids were determined at 25° in queous KCl. As expected, it was confirmed that these acids exist in aqueous oelectric solutions mainly in the dipolar ion form.[151] Similar data for onocarboxylic and monosulfonic acids[152] and polyvalent acids[153] were btained more recently. The dissociation constants and electrophoretic obilities of fifteen substituted alkyl-, nitro-, and chloropyridine carboxylic cids were also determined.[154] The ultraviolet absorption spectra of a series of ihydropyridine-3,5-dicarboxylic acids were correlated with electronic and steric ctors.[155]

Intermolecular hydrogen bonding has been observed by infrared spectroscopy r nicotinic and isonicotinic acid, while the expected intramolecular henomenon was observed with picolinic acid.[156] Spectral abnormalities of the arboxylate ion absorption at 1640 $cm^{-1}$ and 1380 $cm^{-1}$ have been corded.[157] The infrared spectra of isomeric pyridinecarboxylic acid-1-oxides ave been reported and Hammett $\sigma$ values determined for the 4- ($\sigma = 0.25$) and - ($\sigma = 1.18$) positions.[158]

The NMR spectra of 154 pyridine derivatives were examined neat or in DMSO lution. With the exception of strong electron withdrawing substituents in the -position, the coupling constants are normal.[159]

The acid dissociation constants of chelidamic acid chelates [2,6-dicarboxy-4-ydroxypyridine **(X-123)**] as a function of the metal M were determined.[160, 161] The p$K_a$ of the hydroxyl group in these chelates increased in the rder: Cu(II) < Co(II) < Zn(II) < Ni(II) < Mn(II).

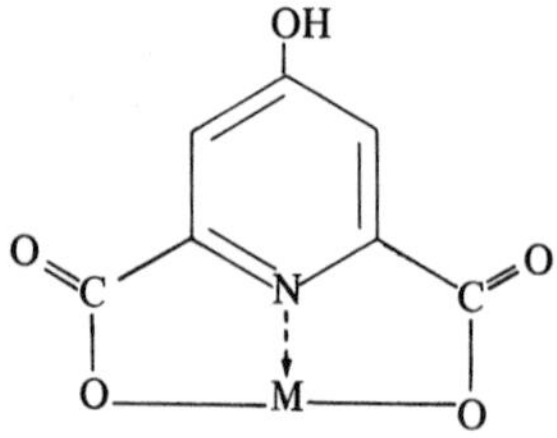

X-123

Separation schemes for pyridinecarboxylic acids[162] and nicotinic an isonicotinic acids[163] are reported. Paper chromatography[164] and ga chromatography[165] of pyridinecarboxylic acids have been discussed, and th crystal structures of picolinic acid hydrochloride[166] and picolinamide[167] ha been described.

## III. Reactions

### 1. Chemical Reduction

Reduction of pyridinecarboxylic acid methyl esters with a large excess o sodium borohydride in methanol is reported to give the correspondin alcohol.[168] Thus, using a twentyfold excess of hydride, nicotinic acid gave hig yields of 3-hydroxymethylpyridine. Similar reductions of unsaturate pyridinecarboxylic acid esters such as **X-124** gave predominantly the saturate alcohols **(X-125)**.

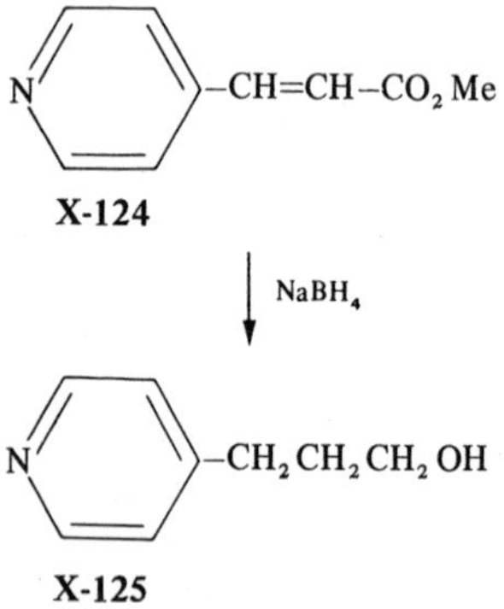

$-CH{=}CH{-}CO_2Me$

X-124

$NaBH_4$

$-CH_2CH_2CH_2OH$

X-125

Similar reductions in ethanol gave the same result for methyl nicotinate, bu diethyl 2,6-dimethyl-3,5-pyridine dicarboxylate **(X-126)** gave the dihydro yridine **(X-127)** and the partially reduced ethyl 5-hydroxymethyl-2,6-dimethyl-icotinate **(X-128)**.[169]

Reduction of isomeric methyl pyridine dicarboxylates (2,5-; 2,3-; 3,4-) with thium aluminum hydride at $-80°$ afforded good yields of the corresponding ialdehydes.[170]

tO$_2$C CO$_2$Et H$_3$C N CH$_3$

**X-126**

EtO$_2$C CO$_2$Et H$_3$C N H CH$_3$ + EtO$_2$C CH$_2$OH H$_3$C N CH$_3$

**X-127** **X-128**

Electrolytic reductions of isonicotinic acid and isonicotinamide in acid olutions were investigated.[171-172] The main reduction product was the dehyde. Its stability to further reduction is explained by hydrate formation. In n acetate buffer the reduction of the amide proceeds to the carbinol. If the mide is *N*-phenylsubstituted the *N*-phenylaminomethyl compound is obtained.

Reduction of 2,6-dichloroisonicotinic acid with hydrazine hydrate in presence f palladium on charcoal is claimed to yield isonicotinic acid.[173]

Intermediates in the sodium dithionite reduction of pyridinium salts have een isolated.[174] Thus 1-(*p*-chlorobenzyl)nicotinamide **(X-129)** yielded the ihydrosulfinate **X-130**, which is desulfinated in alkaline media to -(*p*-chlorobenzyl)-1,4-dihydronicotinamide **(X-131)**. The desulfination of **-130** was confirmed by deuterium exchange experiments.

CONH$_2$ N+ CH$_2$C$_6$H$_4$Cl → H SO$_2$ CONH$_2$ N CH$_2$C$_6$H$_4$Cl → CONH$_2$ N CH$_2$C$_6$H$_4$Cl

**X-129** **X-130** **X-131**

## 2. Catalytic Reduction

Rhodium (5%) on carbon is a better catalyst for the reduction of the isomeri pyridinecarboxylic acids, their esters, and amides to the correspondin piperidine analogues than is rhodium on alumina.[175] Reductions are generall slow, but yields are very satisfactory. Catalytic reductions of pyridylalkan

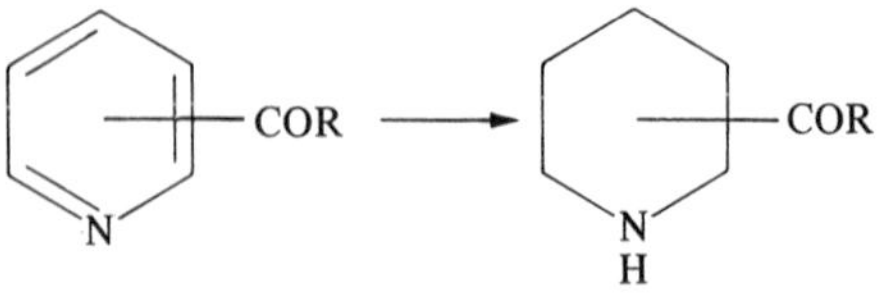

R = OH, OMe, OEt, $NH_2$, $NEt_2$

carboxylic acids using rhodium catalysts and ammonia are also successful.[17 Good results have also been claimed with a nickel, copper, chromic oxide, an alumina catalyst system.[177]

Catalytic reductions of substituted pyridinecarboxylic acids continue t play an important role in some syntheses of pyridoxol. Thus 4-carbethoxy 6-chloro-5-cyano-2-methyl-3-nitropyridine **(X-132)** is reduced over Raney Nicke in water to 3-amino-5-aminomethyl-4-carboxy-2-methylpyridine-4,5-lactar **(X-133)**.[178-180]

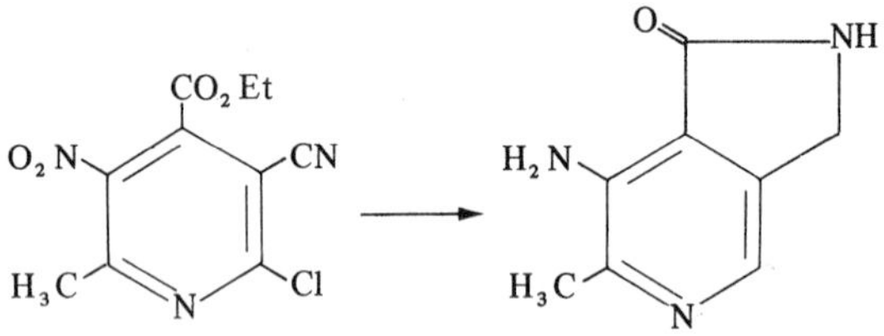

**X-132** **X-133**

Raney Nickel catalyzed reductions of pyridinium salts **(X-134)** afforded t corresponding *N*-(aminoalkyl)piperidinecarboxylic acid **(X-135)**.[181]

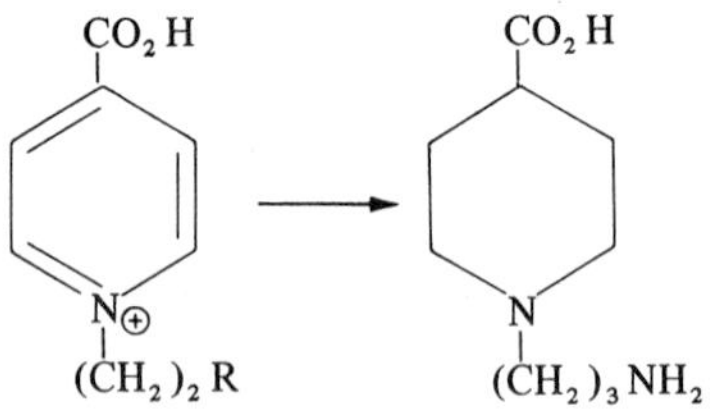

**X-134** **X-135**

R = CN, $CH_2NH_2$

### 3. Esterification

The selective esterification of isocinchomeronic acid continues to be of obvious interest in the large-scale preparation of nicotinic acid. Selectivity is attempted by hydrolysis of the diesters as well as by selective monoesterification.[182-184]

Alcoholic KOH hydrolyzed the diester **X-137** predominantly to the -monoester **X-138**. The 2-monoester **X-139** was isolated from the mother liquors. Esterification of **X-136** with large amounts of $H_2SO_4$ and alcohol afforded the diester **X-137**, but reduced amounts of solvent and acid led to the -monoester **X-139**.

$CO_2H$ / $HO_2C$ / N

**X-136**

$CO_2H$ / $RO_2C$ / N

**X-139**

$CO_2R$ / $RO_2C$ / N

**X-137**

$CO_2R$ / $HO_2C$ / N

**X-138**

The second order rate constants $k$ of the esterification of the three isomeric pyridinecarboxylic acids with diphenyldiazomethane in absolute ethanol were determined to be 0.108 for picolinic acid, 0.096 for nicotinic acid, and 0.182 for isonicotinic acid (temperature not specified in *Chemical Abstracts*). The reactivity ratios agree with theoretical expectations.[185]

Diazomethane reacted with 4-carboxy-3-cyano-6-methyl-2-pyridone **(X-140)** in the expected manner to form the methyl ester **(X-141)**. Further treatment of **X-141** with diazomethane led to a mixture of *O*- **(X-142)** and *N*-methyl **(X-143)** derivatives. Treatment of **X-141** with methyl iodide in sodium methoxid yielded **X-143** exclusively.[186]

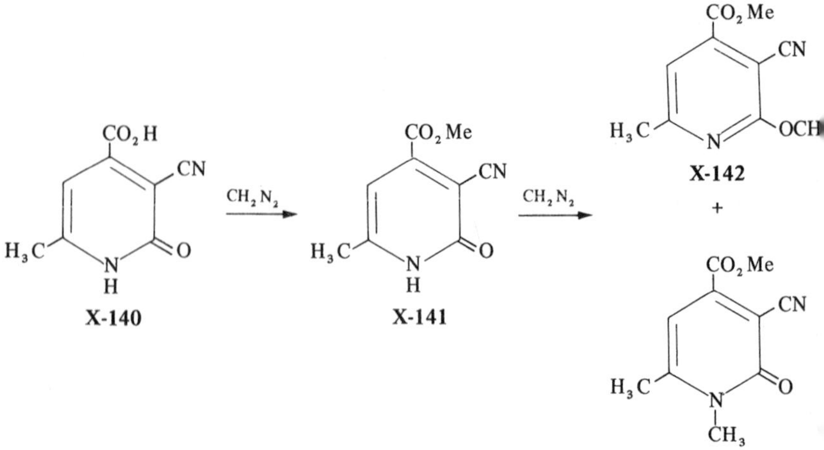

$CH_2N_2$ $CH_2N_2$

X-140 X-141 X-142 + X-143

Vinyl esters of pyridine-2,5- and 2,6-dicarboxylic acids were prepared from th diacid chlorides and mercuriacetaldehyde. The allyl and propargylic esters o pyridine 2,4- and 2,6-diacids were prepared similarly from the alcohols.[187]

## 4. Decarboxylation

An extensive study of the decarboxylation of picolinic acid in twelve pol solvents has been concluded.[188] Thirty-two sets of activation parameters we obtained. The data favor the uncharged molecule **(X-144)** over the zwitteric **(X-145)** as the entity involved in the formation of the transition state.

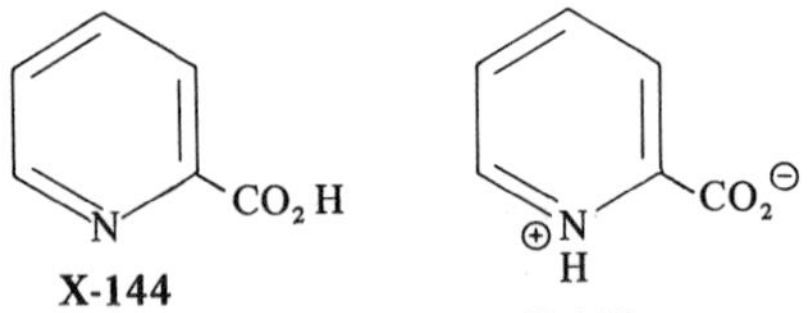

X-144 X-145

The decarboxylation of 2-substituted pyridinecarboxylic acids was studied i ethylene glycol at constant pressure[189] and in sulfuric acid and in ammoniu bisulfate.[190, 191] The reactions obeyed apparent first order kinetics. Th determined parameters of activation are given in Table X-1. The data were aga

ABLE X-1. Activation Energies (E), Activation Enthalpies (ΔH*), and Activation Entropies (ΔS*) in the Decarboxylation Process

| ridine rboxylic id | ethylene glycol[189] | | | $H_2SO_4$ | | | $NH_4HSO_4$ [190, 191] | | |
|---|---|---|---|---|---|---|---|---|---|
| | E kcal/mole | ΔH* kcal/mole | ΔS* e.u. | E | ΔH* | ΔS* | E | ΔH* | ΔS* |
| | 44.03 | 43.1 | 20.84 | | | | | | |
| 6- | 38.97 | 38.1 | 8.71 | | | | | | |
| 5- | 47.64 | 46.1 | 30.01 | | | | 53.6 | 52.6 | 31.3 |
| 3- | 31.63 | 30.9 | 6.04 | 36.0 | 35.1 | -5.1 | 31.8 | 31.0 | -2.5 |
| 4- | | | | | | | 55.8 | 54.9 | 36.6 |

terpreted as favoring the neutral molecule **(X-144)** as the reactive species. The carboxyl group was not lost over the temperature ranges studied (up to 164° $NH_4HSO_4$ and 233° in $H_2SO_4$). Its effect as a neighboring group in the carboxylation of the 2,3-dicarboxylic acid (quinolinic acid) is apparent.

The $^{14}C$ kinetic isotope effect in the decarboxylation of picolinic acid **(X-144)** as measured in a variety of solvents. The observed effects are related to drogen bonding.[192]

The deutero-decarboxylation of isomeric pyridinecarboxylic acids is a good ay of introducing deuterium into the pyridine nucleus.[193] The 2- and

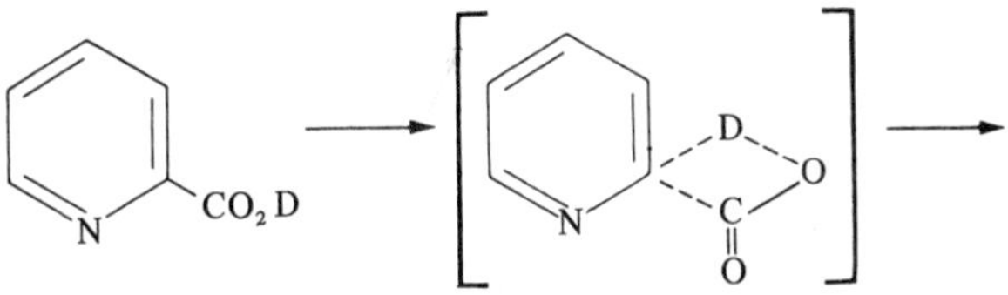

**X-146**

positions are especially favorable. The role of the hypothetical intermediate **(X-146)** is reviewed.

The relative rates of decarboxylation of picolinic acid **(X-144)**, its *N*-methyl omologue, homarine **(X-147)**, and its *N*-oxide **(X-148)** in ethylene glycol 134° are 1:720:160. Homarine decarboxylates $10^3$ times faster than e 3-position isomer, trigonelline, and *N*-methylisonicotinic acid.[194] The

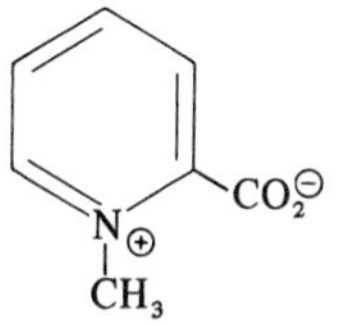

**X-147** 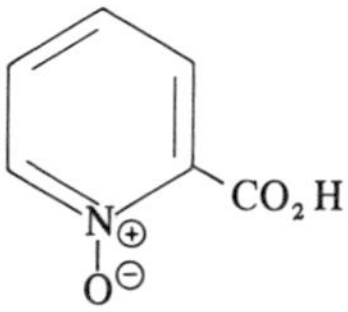 **X-148**

decarboxylations of **X-144** and **X-148** are inhibited by divalent metal ions such Cu(II), Mg(II), and Mn(II).[194] The rapid decarboxylation of betaine **X-147** h biological significance.

The decarboxylation of 2,3- or 3,4-substituted pyridinedicarboxylic acids still of importance in technical syntheses of nicotinic acid[195] and pyridoxol.[1

The photodecarboxylation of isomeric pyridinedicarboxylic acids has be studied.[197] Monocarboxylic acids are photostable in aqueous solutio Dicarboxylic acids with 1,3-related carboxyl groups decarboxylate. Oth dicarboxylic acids are photostable. Dissolved oxygen has no effect, whi suggests that the photodecarboxylations are related to a $\pi \rightarrow \pi^*$ singlet excit state. In contrast to its thermal stability the photolability of the 3-carbox group is remarkable.

## 5. Pyridinecarboxylic Acid Amides

The conversion of nicotinic acid **(X-149)** to its amide **(X-150)** continues to examined.

$CO_2H$ → $CONH_2$

**X-149** **X-150**

The use of urea at 150 to 250° is described for this reaction and for t analogous conversion of isonicotinic acid.[198] Urea has also been used to amida quinolinic acid and substituted pyridine-3,4-dicarboxylic acids.[199] Sulfamic ac and its ammonium salt at 165° in the presence of ammonia also converts **X-1** to **X-150** and amidates isonicotinic acid.[200]

More exotic procedures are the Raney Nickel-catalyzed hydrogenation of t nicotinic hydroxamic acid to **X-150**,[201] and a sequence of reactions starting wi 4-amino-3-cyano-1,2,3,6-tetrahydropyridine **(X-151)**. This compound is hydr lyzed to 1-acetyl-4-oxohexahydronicotinamide **(X-152)**, which is reduced a

$NH_2$, CN, N, H → O, $CONH_2$, N, $COCH_3$ → $CONH_2$, N

**X-151** **X-152** **X-150**

en hydrogenated, deacetylated and dehydrated to **X-150** in 60 to 75% overall ld.[202-204]

Some new work in very classical procedures is also reported such as the n-exchange catalyzed hydration of nicotinonitrile,[205,206] the ammonolysis of ters of nicotinic acid,[207,208] as well as of the free acid.[209]

Substituted nicotinamides have been prepared by the reaction of the acid -149) with phosphorus oxychloride and diethylamine.[210] The Bodroux action using a dibutylaminomagnesium salt has been applied to ethyl colinate.[211]

The diamide of pyridine-3,5-dicarboxylic acid **(X-154)** was prepared by idizing 3,5-diacetyl-1,4-dihydropyridine **(X-153)** with dilute nitric acid llowed by treatment with thionyl chloride and dimethylformamide.[212]

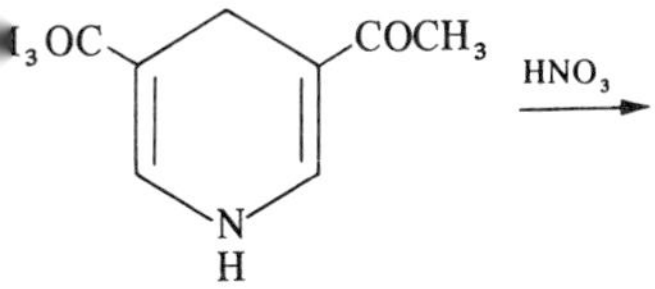

$_3$OC COCH$_3$ $\xrightarrow{HNO_3}$

N
H

**X-153**

HO$_2$C CO$_2$H ⟶ H$_2$NOC CONH$_2$

N N

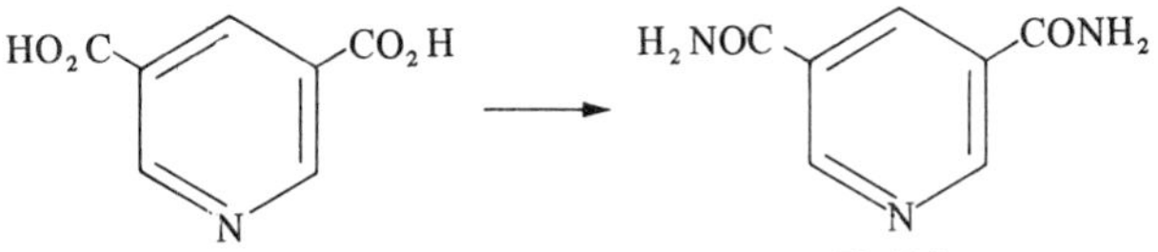

**X-154**

The alkali- and acid-catalyzed hydrolysis of the isomeric pyridine onocarboxamides was examined. In the second order alkaline process, the lculated Hammett $\sigma$-values agreed well with constants obtained by molecular bital calculations. The methods were extended to the acid-catalyzed process here Hammett constants are not available for comparison.[213] The te of acid-catalyzed hydrolysis of picolinamide and *N*-methylpicolinamide is mewhat decreased by Cu(II) ions.[214]

Picolinamide and isonicotinamide can be $N^1$-alkylated by treatment with alkyl omides or iodides. These can be converted to the chlorides by treatment with eshly precipitated silver chloride.[215]

Pyridine carboxamides are readily converted to the corresponding aldehydes treatment with lithium aluminum hydride in tetrahydrofuran.[216]

Nicotinamide and quinolinamide are readily converted to nitriles by treatment ith phosphorus pentoxide[217] or thionyl chloride in dimethylformamide.[218] ne methylamide of nicotinic acids is converted to the *N*-methylimidochloride -155) by treatment with phosphorus pentachloride. Reaction of **X-155** with drazoic acid yields the tetrazole **(X-156)**.[219]

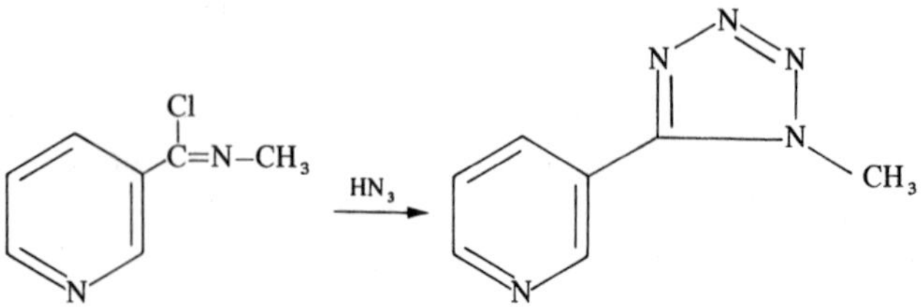

X-155 X-156

An interesting amidation procedure utilizes the condensation of nicotinic ac hydrazide **(X-157)** with chloral. The resulting Schiff's base **(X-158)** is smooth converted to the *N*-cyclohexylamide **(X-159)** on treatment with cyclohex amine.[220] The reaction has general utility.

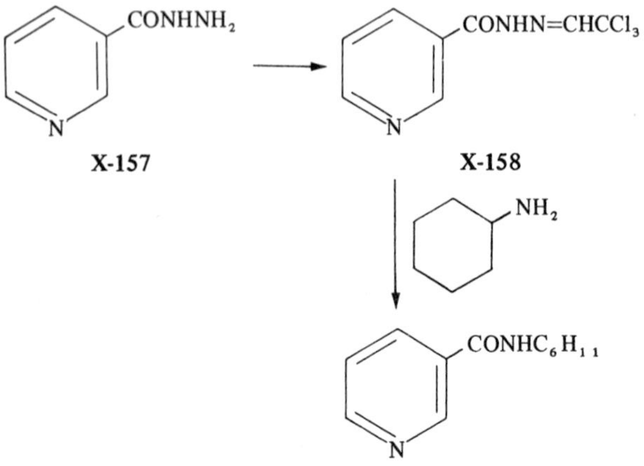

X-157 X-158

X-159

Hydrolysis of the isomeric pyridinecarboxylic acid azides has been e amined.[221] The rate-determining step involves hydroxide ion attack carbonyl carbon. The rates fall in the order: isonicotinic > picolinic > nicotin Azides are relatively easily hydrolyzed.

## 6. Pyridinethiocarboxylic Acids

Mild alkaline hydrolysis of 2-ethylpyridine-4-thioamide **(X-160)** leads to t corresponding thiocarboxylic acid **(X-163)** through the iminothiol **(X-161)** a the carbonylthiol **(X-162)**.[222] The carboxamide and the free acid are si products.

S ‖ C–NH$_2$ / H$_3$C / N — **X-160** ⇌ SH | C=NH / H$_3$C / N — **X-161** ⇌ SH | C=O / H$_3$C / N — **X-162** ⇌ S ‖ C–OH / H$_3$C / N — **X-163**

4-(3-Hydroxypropyl)picolinonitrile **(X-164)** is converted to the thioamide **(X-165)** by treatment with alcoholic ammonia and sulfur.[223]

$CH_2CH_2CH_2OH$ / CN / N — **X-164** $\xrightarrow[S]{NH_3}$ $CH_2CH_2CH_2OH$ / S ‖ C–NH$_2$ / N — **X-165**

2-*n*-Propylnicotinothioamide **(X-166)** yields the *S*-oxide derivative **(X-167)** on action with the hydrogen peroxide-sodium perborate couple.[224]

S ‖ C–NH$_2$ / $CH_2CH_2CH_3$ / N — **X-166** $\xrightarrow[NaBO_3]{H_2O_2}$ $O^{\ominus}$–$S^{\oplus}$ ‖ C–NH$_2$ / $CH_2CH_2CH_3$ / N — **X-167**

Treatment of 2-picoline with methylformamide and sulfur leads to iopicolinic acid methylamide **(X-168)**.[225]

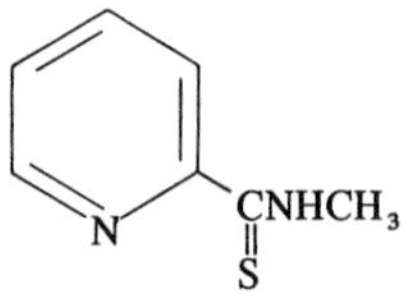

$CNHCH_3$ (C=S)

X-168

## 7. Miscellaneous Reactions

Treatment of nicotinic acid and isonicotinic acid, or their correspond amides, with sodamide followed by oxidation with concentrated nitric a affords 2,6-diaminopyridine.[226]

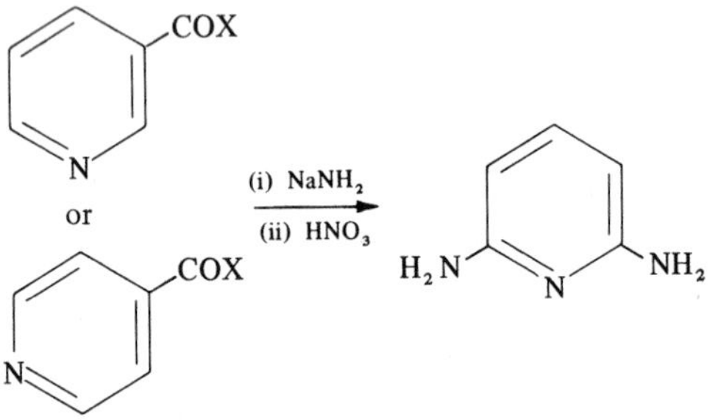

COX, N, or, COX, N; (i) $NaNH_2$, (ii) $HNO_3$; $H_2N$, N, $NH_2$

Cyanopyridines are prepared by passage of nicotinic or isonicotinic acid, th ammonium salts, or amides over a dehydration catalyst at 350° in the prese of ammonia. Boron phosphate and aluminum phosphate are mentio specifically as catalysts.[227]

2- And 4-cyano-3-methoxypyridines **(X-170)** are prepared from the c responding 3-nitropyridines **(X-169)** by the action of methanol and cyan ion.[228] The mechanism might involve a series of addition-elimination steps indicated. The final dehydrogenation of the hypothetical dihydropyrid intermediate **(X-171)** must be accomplished by the expelled nitrite ion. T reaction is general and yields are reported to exceed 50% of theory.

Isocinchomeronic acid **(X-172,** X = OH) when treated with $PCl_5$ at 1( afforded the corresponding diacid chloride **(X-172,** X = Cl). Ammonolysis of latter followed by dehydration with $POCl_3$ gave 2,5-dicyanopyridine **(X-173).**

Nicotinic or isonicotinic acids when heated to 200° with *p*-toluenesulfonam yield the corresponding nitriles. Carboxyl groups in the 2-position are elimina in this reaction.[230]

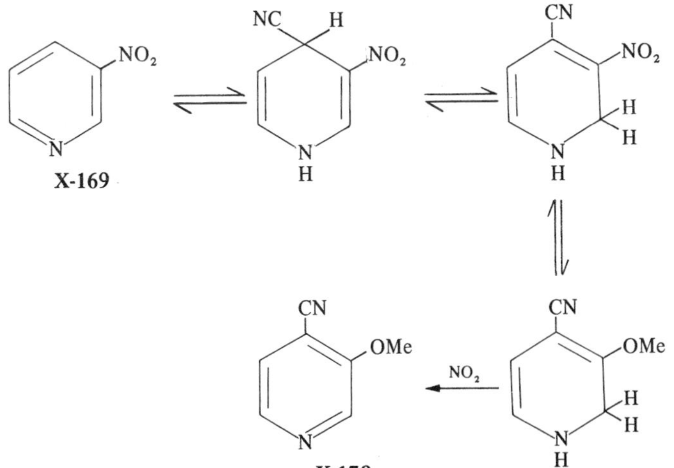

X-169 X-170 X-171

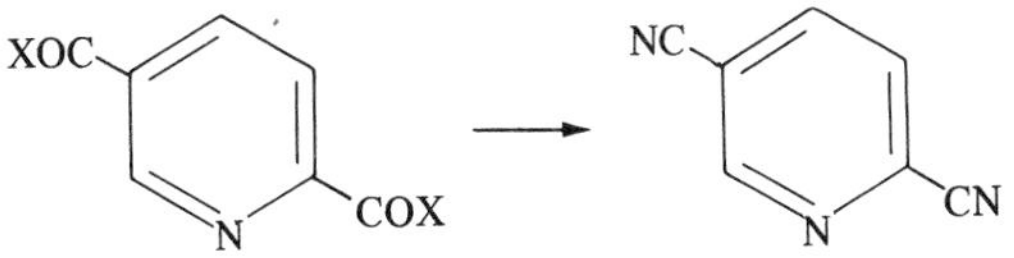

X-172 X-173

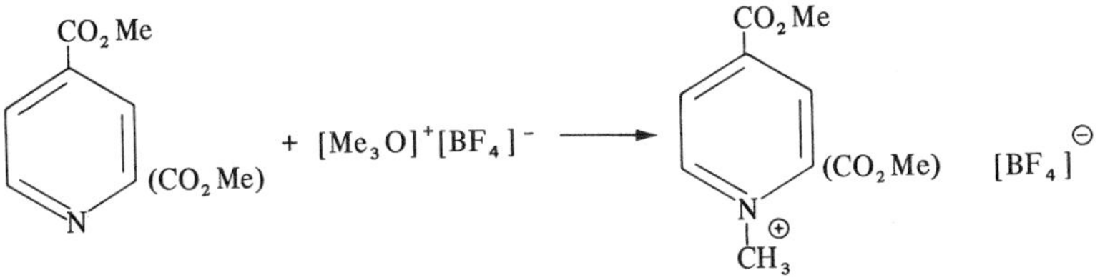

$+ [Me_3O]^+[BF_4]^-$

X-174

Methyl 2- or 4-pyridinecarboxylates gave high yields of *N*-methylated ridinium salts **(X-174)** when treated with the Meerwein reagent. Mild drolysis of **X-174** gave the corresponding betaines **(X-175)**, which were adily decarboxylated under mild conditions in dipolar aprotic solvents. The termediate betaines **(X-176)** could be trapped with electrophiles such as nzaldehyde or diazonium salts.[232]

Hydrogenation of dimethyl pyridine-3,4-dicarboxylate **(X-177)** over lladium-on-charcoal affords the 1,4,5,6-tetrahydro derivative **(X-178)**, which, n further reduction over platinum oxide, yields the piperidine derivative **-179)**. On the other hand, dimethyl pyridine-2,3-dicarboxylate **(X-180)** is duced over palladium-on-charcoal directly to the piperidine **(X-181)**.[233]

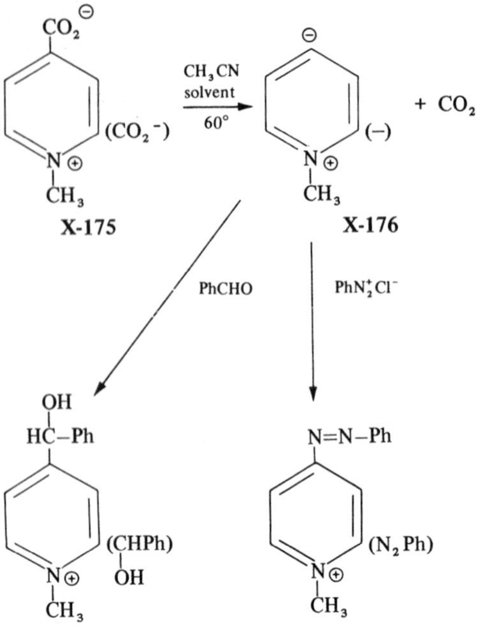

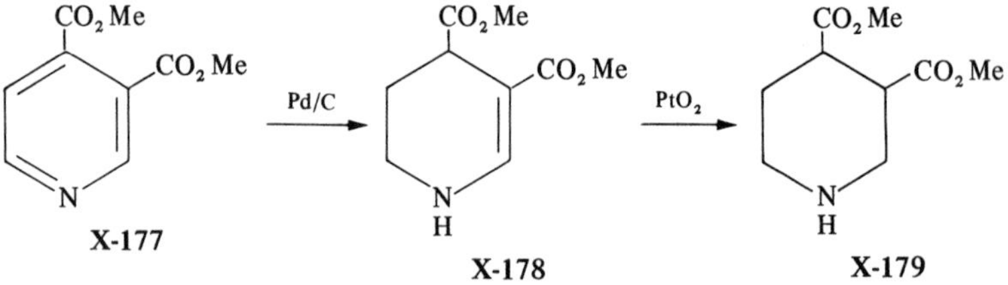

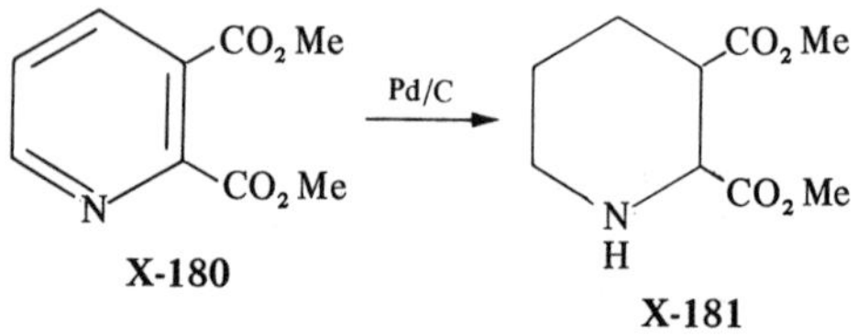

X-175 X-176

X-177 X-178 X-179

X-180 X-181

Reductive alkylation of methyl isonicotinate with benzyl chloride yi
4-pyridyl benzyl ketone **(X-182)**.[234]

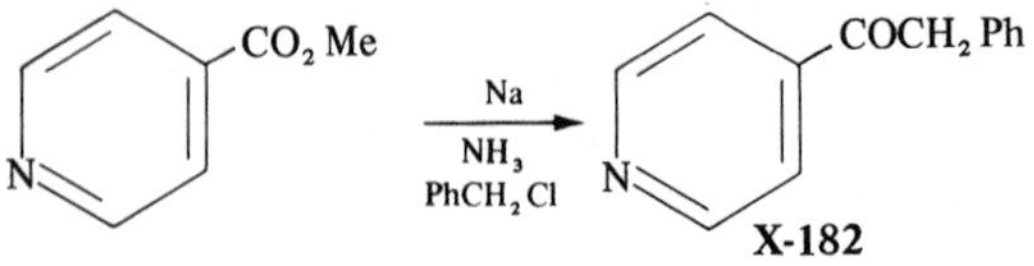

X-182

Mercapto derivatives derived from 5-amino-2-picolinic acid **(X-183)** and aminonicotinic acid **(X-184)** are obtained by diazotization in the presence of lfur and sodium sulfide. The thiols can be converted to the corresponding lfonic acids with permanganate.[235]

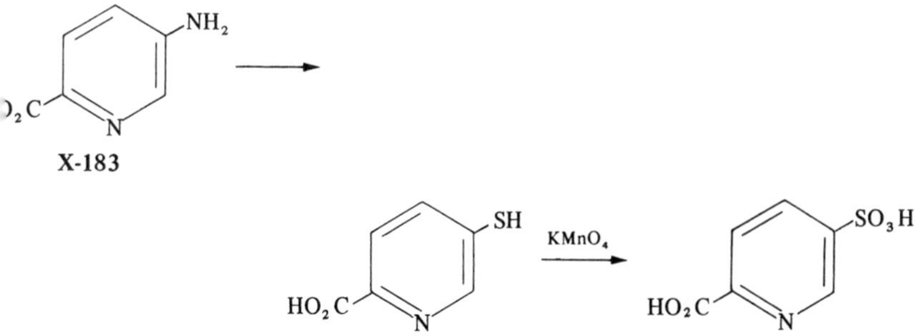

X-183

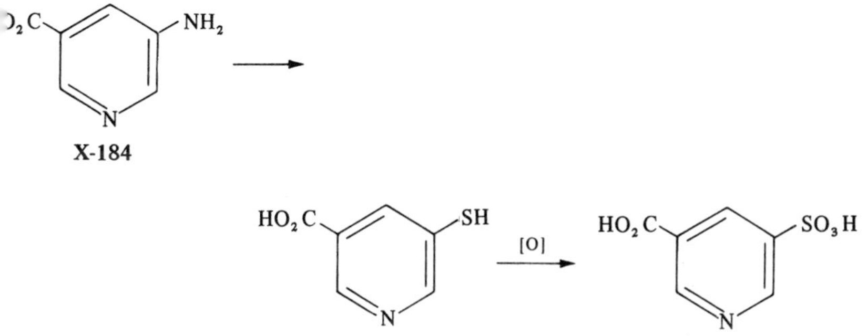

$KMnO_4$

X-184

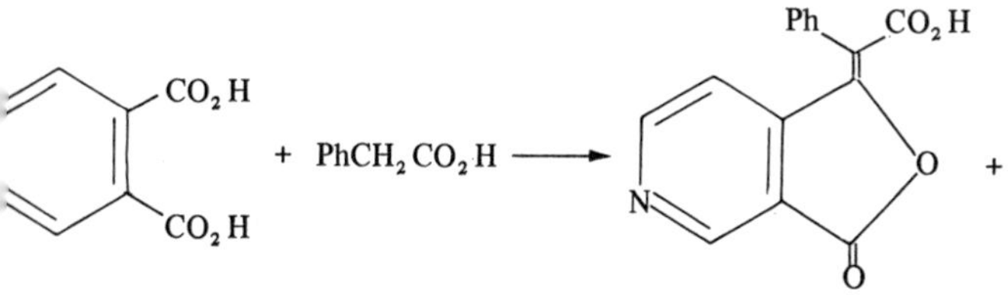

[O]

+ $PhCH_2CO_2H$

X-185

X-186

NaOMe

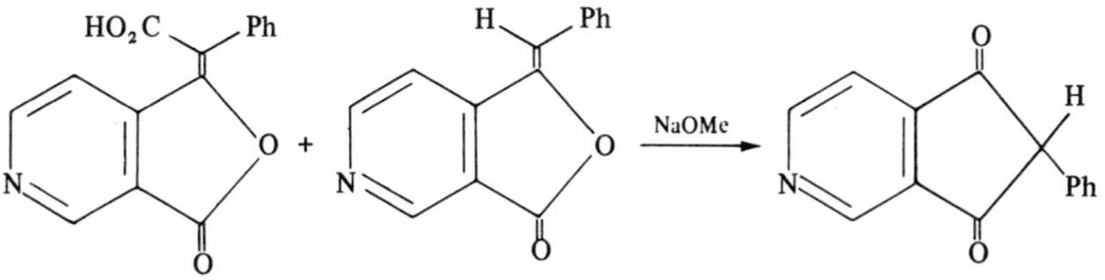

X-187

X-188

X-189

Cinchomeronic acid **(X-185)** condenses with phenylacetic acid in the prese of acetic anhydride and triethylamine to yield 1-benzylidene-3-oxo-1,5-dihyd furo[3,4-*c*]pyridine **(X-188)** and a mixture of the isomeric acids **(X-186)** a **(X-187)**.[236] The lactone **(X-188)** was converted to the dione **(X-189)** w sodium methoxide.

Dimethyl 4-chloropyridine-2-6-dicarboxylate **(X-190)** on treatment w sodium salts of cyanoacetic esters, or malononitrile, yields dihydropyrid derivatives **(X-191)**.[237]

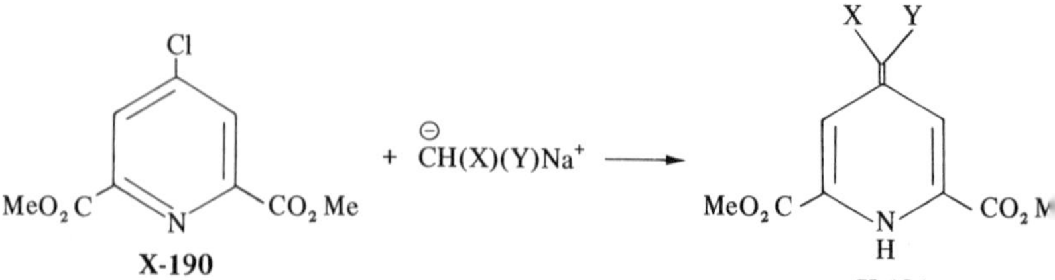

Cl; $MeO_2C$; N; $CO_2Me$ **X-190** + $\overset{\ominus}{C}H(X)(Y)Na^+$ → X Y; $MeO_2C$; N H; $CO_2M$ **X-191**

X = CN Y = $CO_2Bu$-*t*
X = CN Y = $CO_2Et$
X = Y = CN

Ethyl 2-chloronicotinate reacts with ethyl acetate and sodium ethoxide form ethyl α-(2-ethoxynicotinoyl)acetate **(X-192)**. Reduction of the latter to diol **(X-193)** followed by treatment with hydrobromic acid affo 2*H*-pyrano[2,3-*b*]pyridine · HBr **(X-194)** (R.I. 1701).[238]

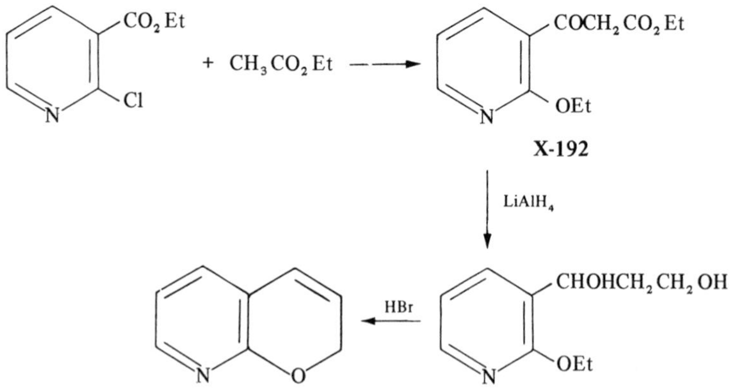

$CO_2Et$; Cl; N + $CH_3CO_2Et$ → $COCH_2CO_2Et$; OEt; N **X-192**

$LiAlH_4$ ↓

$CHOHCH_2CH_2OH$; OEt; N **X-193** —HBr→ N; O **X-194**

A similar reaction sequence with acetophenone but omitting the reducti step yields azaflavone (2-phenyl-4*H*-pyrano[2,3-*b*]pyridin-4-one) **(X-195)**.

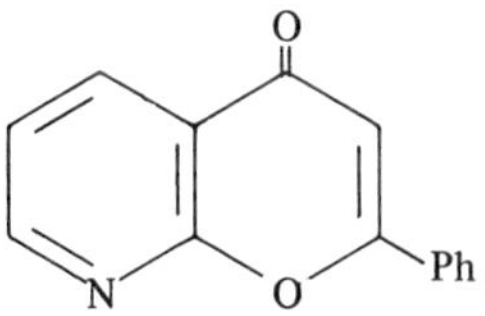

X-195

Cinchomeronic acid, on treatment with ammonium molybdate, urea, and pric chloride at elevated temperatures affords copper tetra-3,4-pyridinopor-yrazine, the nitrogen analogue of copper phthalocyanine.[239]

Quinolinic anhydride **(X-196)** reacts with *o*-aminothiophenol in dimethylform-ide to yield 2-(2-benzothiazolyl)pyridine-3-carboxylic acid **(X-197)**, which uld be decarboxylated to the known 2-(2-pyridyl)benzothiazole.[240]

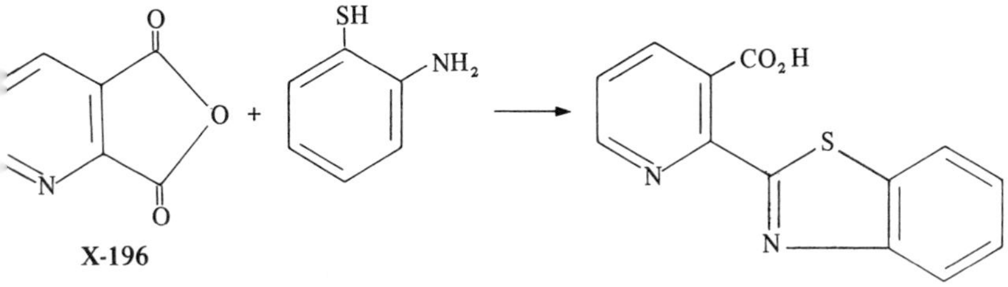

X-196 X-197

ıchomeronic anhydride yielded a mixture of analogous 4-(2-benzothiazolyl)-ridine-3- and 3-(2-benzothiazolyl)pyridine-4-carboxylic acids.

Another example of the Schmidt reaction was reported: isomeric ridinecarboxylic acids were treated with sodium azide in sulfuric acid at vated temperatures to give 3-aminopyridine (69% yield) and 2- and minopyridines in less than 30% yield.[241]

Wynberg described the photolysis of some pyridinecarboxylic acid ivatives.[242] 3,5-Dicarbethoxy-2,4,6-trimethylpyridine **(X-198)** gave 3,5-di-bethoxy-2,4,6-trimethyl-1,4-dihydropyridine **(X-199)** and 3-carbethoxy-,5-trimethylpyrrole **(X-200)** upon irradiation in ethanol. Irradiation of the lesmethyl derivative of **X-198** in methanol afforded only reduction products those of addition of solvent at the 2- or 4-positions.

The characteristic color reaction of nicotinic acid with cyanogen bromide and aphthylamine has been studied. The product is believed to be **X-201**.[243]

Argentic picolinate is a powerful oxidizing agent.[244] Thus, toluene is oxidized benzyl alcohol, benzaldehyde, and benzoic acid in a stepwise, controllable nner. Primary alcohols were converted to aldehydes, secondary alcohols to tones. Amines were oxidized to aldehydes or ketones in dimethylsulfoxide.

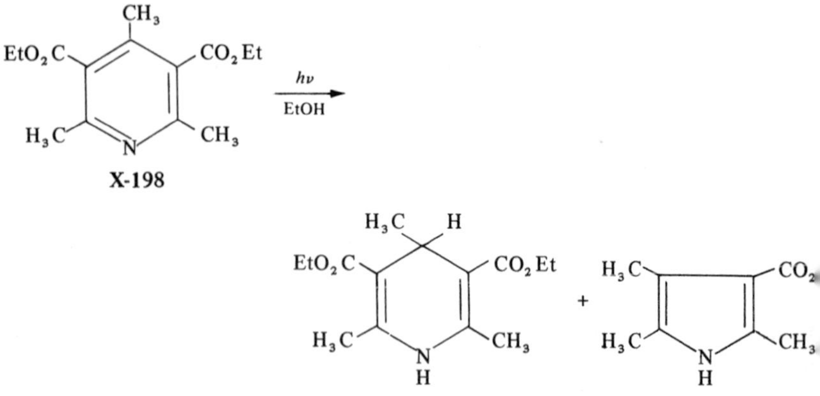

X-198

X-199

X-200

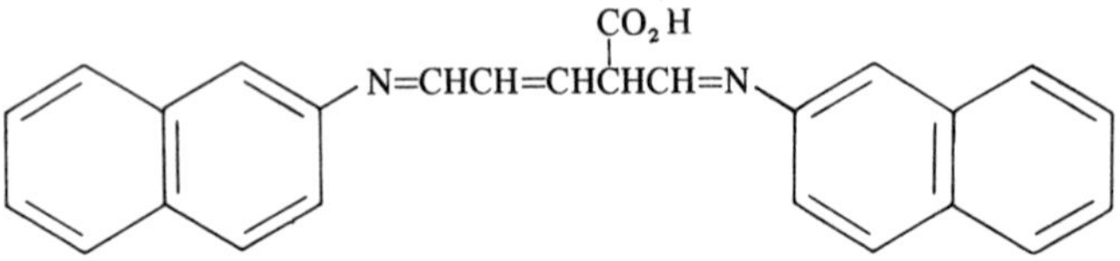

X-201

Cyclohexanol was converted to cyclohexanone, cyclohexanediol and the vici diketone. Activated methylenes reacted readily to yield mixtures of produc The method has potential utility in carbohydrate chemistry.

Quinolinic acid anhydride undergoes fragmentation under electron imp analogously to phthalic anhydride.[245]

Pyridine-2,6-dicarboxylic acid and pyridine-2,4,6-tricarboxylic acid valuable reagents for the precise photometric determination of iron.[246, 247]

Pyrolysis of the 1,4-dihydropyridine (**X-202**) gave mainly the lactone (**X-20** but some ring contraction to **X-204** was observed.[248]

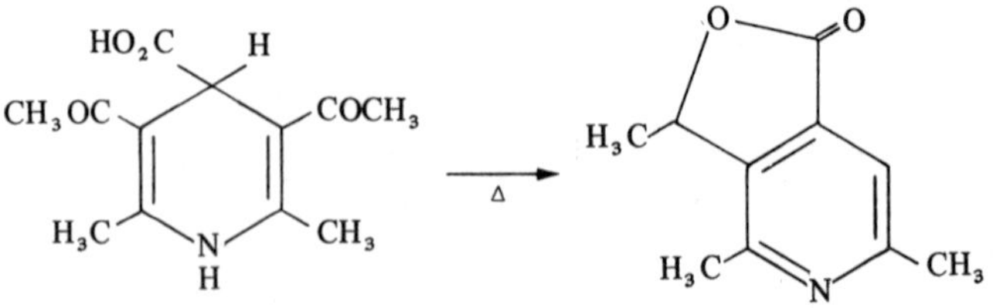

X-202

X-203

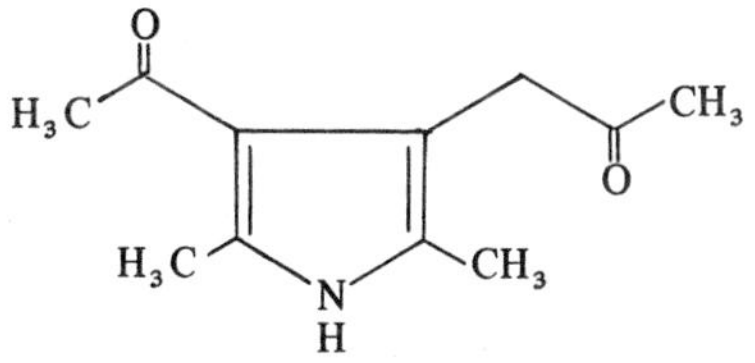

X-204

Treatment of alkyl isonicotinates or 4-cyanopyridine with alkyl free radicals enerated from diacyl peroxides) afforded 2-alkyl-substituted molecules.[249]

An improved synthesis of pyridoxol has appeared[250] and 5-hydroxy-6-methyl-otinic acid (**X-205**), a metabolite of pyridoxal in *Pseudomonas* MA has been epared.[251]

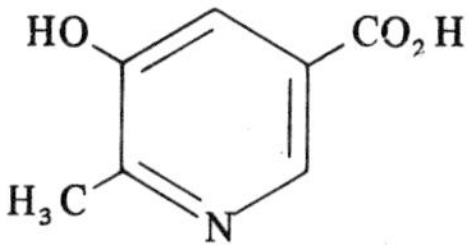

X-205

## IV. Pyridinecarboxylic Acid-1-oxides

*N*-Oxides of pyridinecarboxylic acids are prepared by oxidation of their tassium salts in hydrogen peroxide/acetic acid.[252] (See also Chapter IV.) The oxides, particularly those derived from isocinchomeronic acid, are reputed to hibit the decomposition of peracids.[253]

The reaction has been applied to substituted pyridinecarboxylic acids such as methoxyisonicotinic acid (**X-206**),[254] or to the parent 2-pyridone. These acid oxides form amides and nitriles *via* the esters.

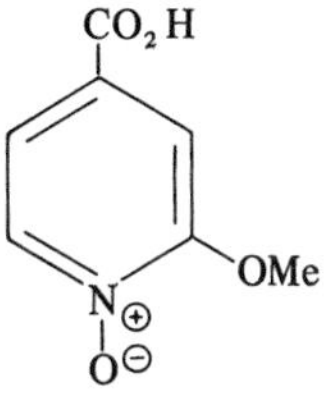

X-206

icolinic acid amide undergoes oxidation to the *N*-oxide with the $H_2O_2$/HOAc. e latter undergoes a normal Hoffman degradation with potassium pobromite to form 2-aminopyridine-1-oxide.[255] The amine *N*-oxide can also be obtained directly by the oxidation of 2-aminopyridine with or without t intermediate formation of the acetylated *N*-oxide.

4-Picoline-1-oxide was converted to the *N*-methoxy-4-picolinium iod (**X-207**) by treatment with methyl iodide. The latter was treated with cyanide yield 2-cyano-4-picoline (**X-208**, X = CN) which was hydrolyzed to the a (**X-208**, X = $CO_2H$) with dilute hydrochloric acid. The nitrile (**X-208**, X = C was converted to the 1-oxide (**X-209**, X = CN), which could also be hydroly to the corresponding acid (**X-209**, X = $CO_2H$). The same *N*-oxide acid (**X-2** X = $CO_2H$) could be obtained directly by *N*-oxidation of the acid (**X-208**, X $CO_2H$).[256]

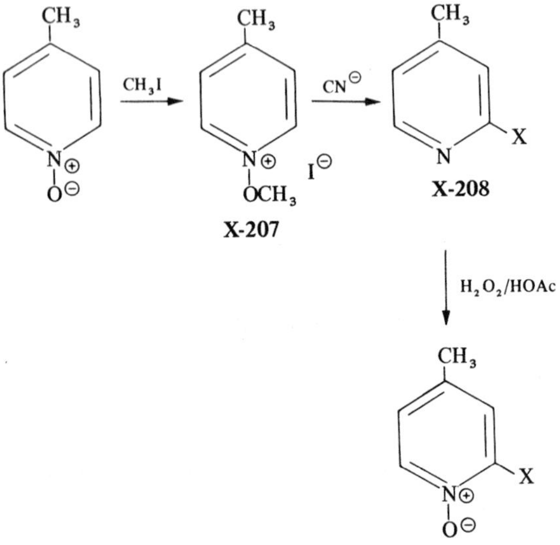

$CH_3I$ → **X-207** ($I^{\ominus}$) ; $CN^{\ominus}$ → **X-208** ; $H_2O_2$/HOAc → **X-209**

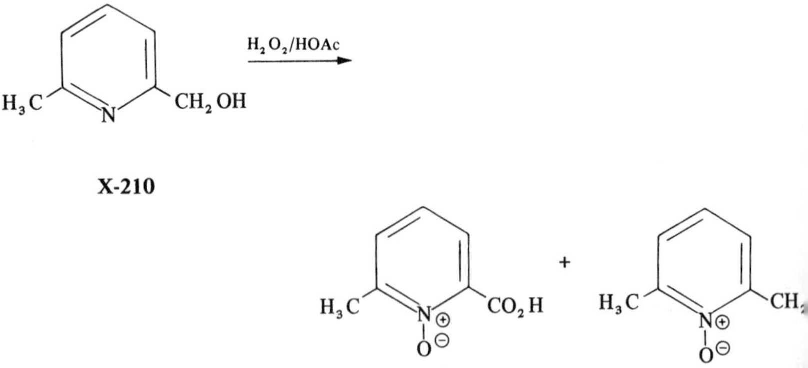

**X-210** — $H_2O_2$/HOAc → **X-211** + **X-212**

Oxidation of hydroxymethyl alkylpyridines with the $H_2O_2$/HOAc affords *N*- d *C*-oxidation products: 2-hydroxymethyl-6-methylpyridine **(X-210)** yielded nethylpicolinic acid-1-oxide **(X-211)** and 2-hydroxymethyl-6-methylpyri- ne-1-oxide **(X-212)**. Similarly, 2,6-bis(hydroxymethyl)pyridine **(X-213)** yielded e expected pyridine-1-oxide **(X-214)** and a small amount of 6-hydroxymethyl- colinic acid-1-oxide **(X-215)**. On the other hand, permanganate oxidations lead

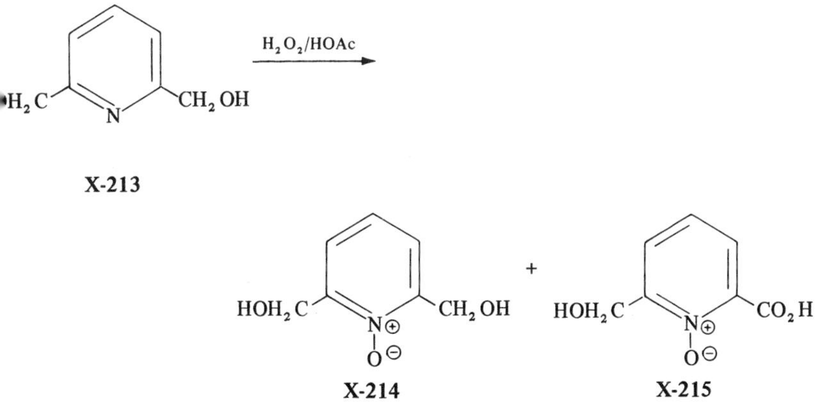

$H_2O_2$/HOAc

**X-213**

**X-214** + **X-215**

clusively to *C*-oxidation even with appropriate *N*-oxides: thus lutidine **(X-216)** orded pyridine-2,6-dicarboxylic acid **(X-218)** and the intermediate 6-methyl-

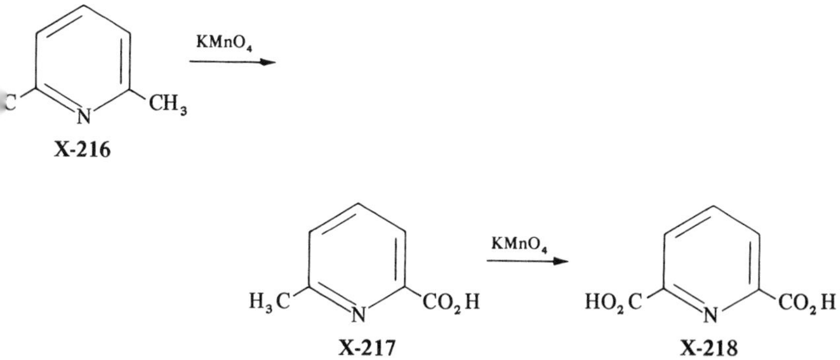

$KMnO_4$

**X-216**

$KMnO_4$

**X-217** **X-218**

olinic acid **(X-217)**. Lutidine-1-oxide **(X-219)** was oxidized to **X-220** and 21 with permanganate.[257]

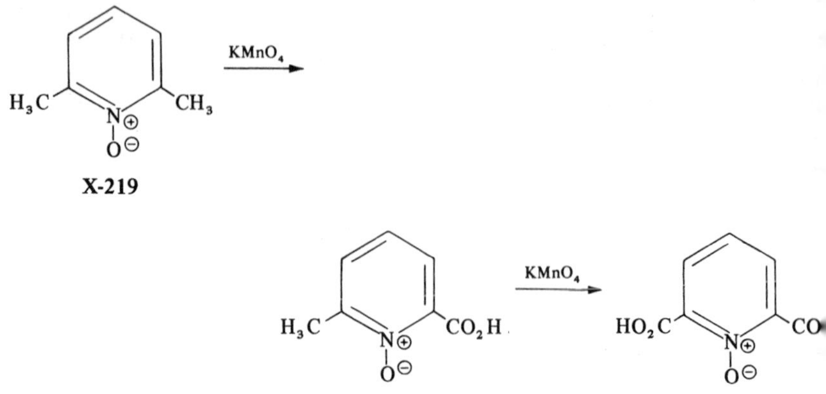

$\xrightarrow{KMnO_4}$

X-219

$\xrightarrow{KMnO_4}$

X-220 X-221

Nicotinic acid-1-oxide (**X-222**) on treatment with acetic anhydride converted to 2-acetylnicotinic acid-1-oxide (**X-223**) and 3-carboxy- (**X-224**) a 5-carboxy-2-pyridone (**X-225**).[258] The structure of **X-223** was proved by deri

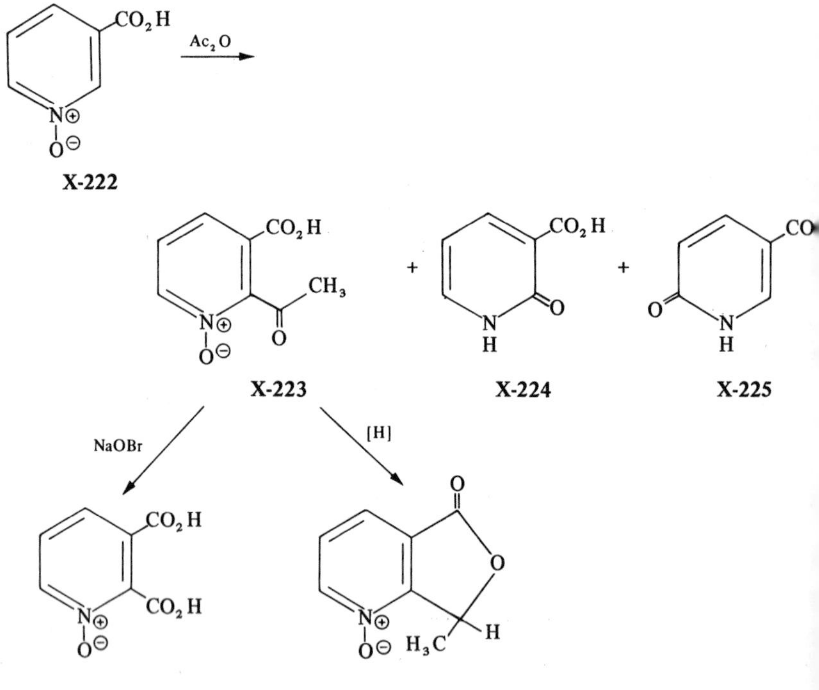

$\xrightarrow{Ac_2O}$

X-222

X-223 + X-224 + X-225

NaOBr [H]

tization, oxidation, and reduction. Reaction of **X-222** with propionic hydride yielded **X-224**, **X-225**, and the diketone-1-oxide (**X-226**). Treatment

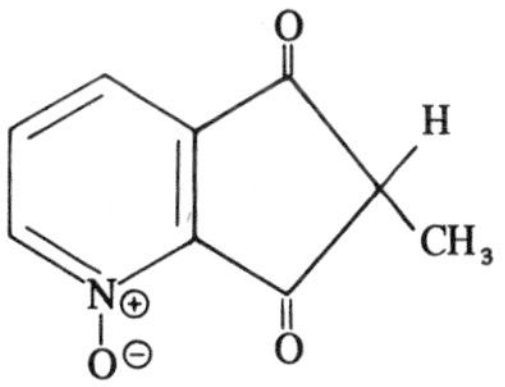

X-226

›nicotinic acid-1-oxide with acetic anhydride afforded 4-carboxy-2-pyridone d an unidentified material. Cinchomeronic acid-1-oxide **(X-227)** is claimed to ›ld cinchomeronic acid **(X-228)** in boiling acetic acid.[259]

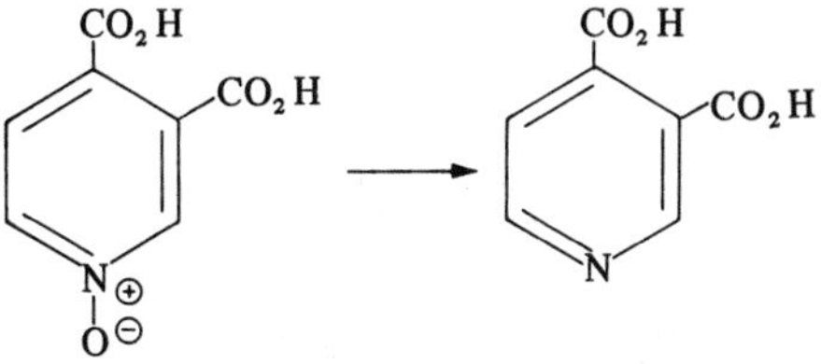

X-227 X-228

A rearrangement of alkylpyridine-1-oxides that is related was also ›vestigated.[259] Ethyl 2-methylnicotinate-1-oxide **(X-229)** afforded an oil on ›atment with acetic anhydride, which, after acid hydrolysis, yielded the

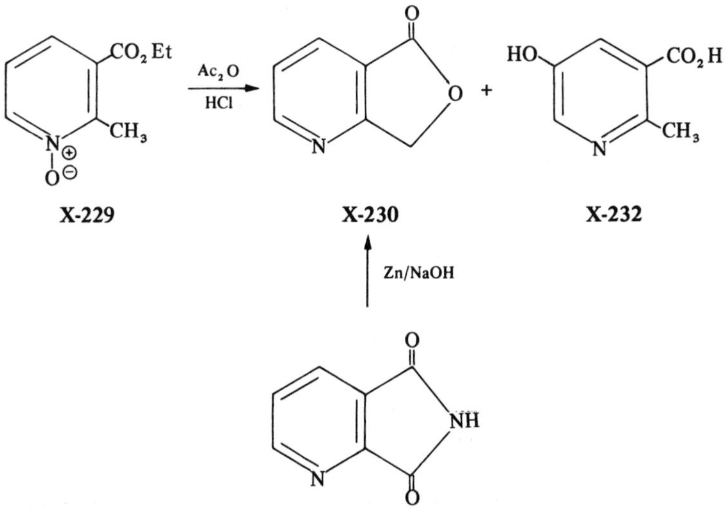

$Ac_2O$ / HCl

X-229 X-230 X-232

Zn/NaOH

X-231

lactone **(X-230)** [synthesized independently from quinolinimide **(X-231)** t reduction with zinc and alkali], and 5-hydroxy-2-methylnicotinic acid **(X-232** Under the same conditions ethyl 2,6-dimethylnicotinate-1-oxide **(X-233)** yield ethyl 2-acetoxymethyl- **(X-234)** and 6-acetoxymethylnicotinate **(X-235** Hydrolysis of crude **X-234** afforded 2,6-dimethyl-5-hydroxynicotinic ac **(X-236)**, 5-methyl-4-azaphthalide **(X-237)**, and 6-hydroxymethyl-2-methylnic tinic acid **(X-238)**.

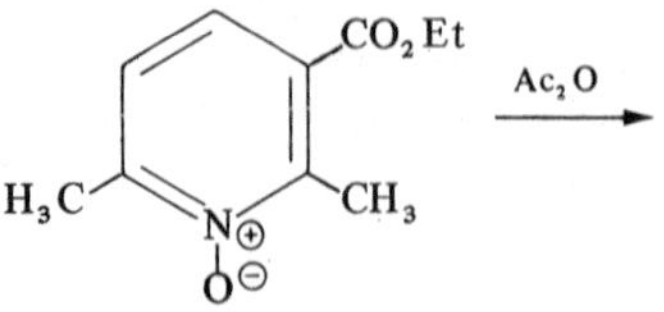

$CO_2Et$ $H_3C$ $CH_3$ $N^{\oplus}$ $O^{\ominus}$ $\xrightarrow{Ac_2O}$

**X-233**

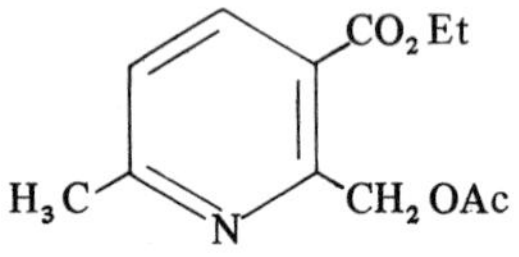

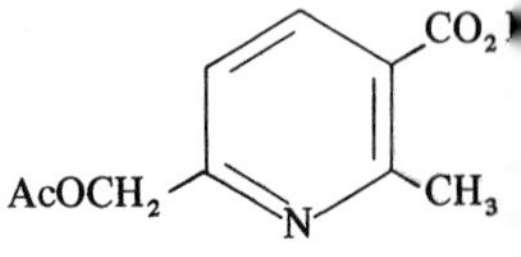

$CO_2Et$ $H_3C$ N $CH_2OAc$ + AcOCH$_2$ N $CO_2$ $CH_3$

**X-234** **X-235**

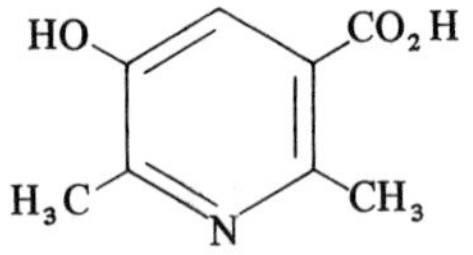

HO $CO_2H$ $H_3C$ N $CH_3$ O O $H_3C$ N

**X-236** 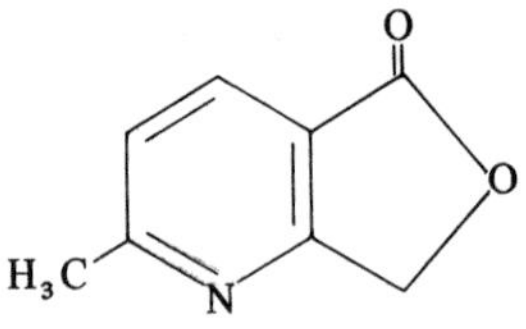 **X-237**

$CO_2H$ $HOH_2C$ N $CH_3$

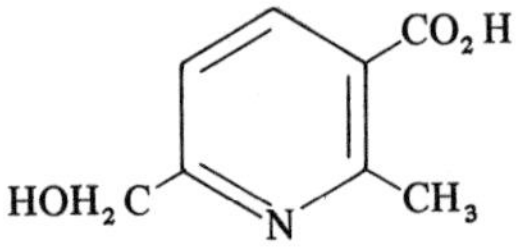

**X-238**

A reaction deals with the preparation of 4-halogenated pyridinecarboxyl acids **(X-240)** from 4-nitronicotinic acid-1-oxide **(X-239)** by explosi interaction with phosphorous trichloride or tribromide.[260] The requisi formation of nitrosyl chloride is not recorded. The 4-iodoacid (**X-240**, X = was prepared from the chloro-acid with hydrogen iodide. The authors also repo

$NO_2$ $CO_2H$ $\xrightarrow{POX_3}$ X $CO_2H$

X-239 X-240

e formation of 5-iodonicotinic acid from 3-iodo-5-methylpyridine by oxida-
on with permanganate.

Picolinic acid-1-oxide is readily converted to the 4-nitro derivative. The nitro oup is easily replaced by nucleophiles such as alkoxide, hydroxide, hydrogen lfide, ammonia, chloride, and phenoxide.[261,262] Derivatization of the acid nction and other reactions remain as expected.

The rate constants for the esterification of the *N*-oxides of picolinic, nicotinic, onicotinic, and dipicolinic acids with diphenyldiazomethane were determined. he rates reflect the decrease in electron availability at the various nuclear ositions as $2 < 4 < 3$. The 2,6-diacid had the highest rate constant indicating at the slow rate for picolinic acid is not due to steric effects.[263]

## V. Tables

TABLE X-2. Pyridine Monocarboxylic Acids and Their Derivatives

| Substituent and position | | | | | | | |
|---|---|---|---|---|---|---|---|
| 2 | 3 | 4 | 5 | 6 | m.p. | Derivatives | Ref. |
| $CH_3$ | $CH_3$ | $CO_2H$ | | | 224–6° | | 32 |
| $CH_3$ | $CH_3$ | $CO_2H$ | | $CH_3$ | 270–1° | | 32 |
| $CH_3$ | OH | $CO_2H$ | | $CH_3$ | 290–1° | | 32 |
| $C_6H_5$ | $CH_3$ | $CO_2H$ | | | 272–3° | | 32 |
| $CO_2H$ | Ph | Ph | Ph | Ph | 106–8° | Ph ester, m.p. 206–8° | 33 |
| $CO_2H$ | | | C≡CH | | 160° | Et ester, m.p. 110-12° | 75 |
| $CO_2H$ | | | $C{\equiv}CC(OH)(CH_3)_2$ | | 171–2° | | 75 |
| $CO_2H$ | | | C≡CC(OH)(Me)(*t*-Bu) | | 153–4° | | 75 |
| $CO_2H$ | | | $C{\equiv}CC(OH)(CH_2)_5$ | | 166–8° | | 75 |
| $CH_3$ | | $CO_2H$ | | | 263° | thioamide, m.p. 187° | 115 |
| $CH_3$ | | $CO_2H$ | | $CH_3$ | 283–4° | Et ester, m.p. 164° | |
| | | | | | | thioamide, m.p. 205–7° | 115 |
| | $CH_3$ | $CO_2H$ | | | | thioamide, m.p. 185–7° | 115 |
| | $CO_2H$ | 4′-Pyridyl | | | 65° | | 141 |
| $CH_3$ | $CO_2H$ | | $CH_2OH$ | $CH_3$ | | Et ester, m.p. 104° | 169 |
| $CO_2H$ | | | | | | *N*-cyclohexyl amide, m.p. 57–8° | 220 |
| | | | | | | anilide, m.p. 75–6° | 220 |
| | $CO_2H$ | | | | | *N*-butyl amide, b.p. 186–8°/11 mm | 220 |
| | | | | | | *N*-cyclohexyl amide, m.p. 138° | 220 |
| | | | | | | anilide, m.p. 123° | 220 |
| | | | $CO_2H$ | | | *N*-butyl amide, b.p. 140°/15 mm | 220 |
| | | | | | | *N*-cyclohexyl amide, m.p. 138–9° | 220 |
| | | | | | | anilide, m.p. 170° | 220 |
| 2-Benzothiazolyl | | $CO_2H$ | | | 206–7° | Me ester, m.p. 61–2° | 240 |

| | | | | | | | |
|---|---|---|---|---|---|---|---|
| 2-Benzothiazolyl | $CO_2H$ | | | | | 117° | 240 |
| | 2-Benzothiazolyl | $CO_2H$ | | | 245° | Me ester, m.p. 124° | 240 |
| $-\underset{\text{OH}}{\underset{\vert}{\text{CH}}}CH_3$ | $CO_2H$ | | | | | lactone, m.p. 77–9° | 258 |
| $CO_2H$ | | | | | | *N*-methylthioamide, m.p. 74–8° | 225 |
| | | | | | | *N,N*-dimethylthioamide, b.p. 129–33°/0.55 mm | 225 |
| | | $CO_2H$ | | | | *N*-methylthioamide, m.p. 94–7° | 225 |
| | | | | | | *N,N*-dimethylthioamide, m.p. 64–5° | 225 |
| | | | | | | *N*-butylthioamide, m.p. 75–8° | 225 |
| $C_2H_2$ | | CSOH | | | 146° | | 222 |
| | Ac | $CO_2H$ | Ac | | 205° | Et ester, m.p. 86° | 6 |
| $n$-$C_3H_7$ | | $CO_2H$ | | | | thioamide *S*-oxide, m.p. 117–18° | 224 |
| $n$-$C_4H_9$ | | $CO_2H$ | | | | thioamide, *S*-oxide, m.p. 92–3° | 224 |
| $C_2H_5$ | | $CO_2H$ | | | | Me ester, b.p. 91°/3 mm | 249 |
| $C_3H_7$ | | $CO_2H$ | | | | Et ester, b.p. 108–10°/4 mm | 249 |
| $AcOCH_2$ | $CO_2H$ | | | $CH_3$ | | Et ester, m.p. 52–4° | 259 |
| $CH_3$ | $CO_2H$ | | | $CH_2OH$ | | Et ester, m.p. 58–62° | 259 |

TABLE X-3. Pyridine Polycarboxylic Acids

| Substituent and position | | | | | Physical Properties | Derivatives | Ref. |
|---|---|---|---|---|---|---|---|
| 2 | 3 | 4 | 5 | 6 | | | |
| | $CO_2H$ | | $CO_2H$ | | | bis(ethylamide), m.p. 161–2.5° | 4 |
| | | | | | | bis(diethylamide), m.p. 81–2.5° | 4 |
| | | | | | | bis(piperidide), m.p. 125–6° | 4 |
| | | | | | | diamide .½$H_2O$, m.p. 318–19° | 4 |
| | $CO_2H$ | $CO_2H$ | $CO_2H$ | | | tri-Me ester, m.p. 87–8° | 5 |
| $CH_3$ | $CO_2H$ | $CO_2H$ | OH | $CH_3$ | | di-Et ester HC1, m.p. 145–7° | 29 |
| $CH_3$ | $CO_2H$ | | $CO_2H$ | | | di-Et ester, m.p. 60–4° | 29 |
| C1 | $CO_2H$ | | $CO_2H$ | $CH_3$ | | di-Et ester, m.p. 52–5° | 29 |
| $CO_2H$ | C1 | C1 | C1 | $CO_2H$ | | di-Me ester, m.p. 73–5° | 33 |
| $CO_2H$ | C1 | $CH_3O$ | C1 | $CO_2H$ | | di-Me ester, m.p. 141–3° | 33 |
| $CO_2H$ | $CO_2H$ | | F | | m.p. 147° | | 86 |
| $CO_2H$ | | | $CO_2H$ | | | divinyl ester, m.p. 68° | 187 |
| | | | | | | dipropynyl, m.p. 92° | 187 |
| $CO_2H$ | | | | $CO_2H$ | | divinyl ester, m.p. 68° | 187 |
| | | | | | | diallyl ester, b.p. 164–6°/3 mm | 187 |
| | | | | | | dipropynyl ester, m.p. 124° | 187 |
| $CO_2H$ | | $CO_2H$ | | | b.p. 130°/3 mm | diallyl ester, b.p. 171–3°/4.5 mm | 187 |
| | | | | | | dipropynyl, m.p. 78–9° | 187 |

| | | | | | | |
|---|---|---|---|---|---|---|
| $CH_3$ | $CO_2H$ | $CO_2H$ | $CH_3$ | | imide, m.p. 271–2° | 199 |
| | $CO_2H$ | $CO_2H$ | $CH_3$ | $CH_3$ | imide, m.p. 273–4° | 199 |
| $CH_3$ | $CO_2H$ | $CO_2H$ | $CH_3$ | $CH_3$ | imide, m.p. 220–1° | 199 |
| $CH_3$ | $CO_2H$ | $CO_2H$ | $-(CH_2)_4-$ | | imide, m.p. 246–7° | 199 |
| $CO_2H$ | | $CO_2H$ | | | bis(dimethylthioamide), m.p. 169–71° | 225 |
| | | | | | bis(methylthioamide), m.p. 158–62° | 225 |
| $CO_2H$ | | $CH_3$ | | | *N*-methylthioamide, b.p. 174–6°/13 mm | 225 |
| $CO_2H$ | | | | $CO_2H$ | bis(dimethylthioamide), m.p. 194–7° | 225 |
| | | | | | bis(methylthioamide), m.p. 168-70° | 225 |
| $CO_2H$ | | $CO_2H$ | | $CO_2H$ | tris(dimethylthioamide), m.p. 239–42° | 225 |
| $CO_2H$ | | | $CO_2H$ | | bis(*N*-methylthioamide), m.p. 163–6° | 225 |

TABLE X-4. Polyalkyl or Aryl Polycarboxylic Acids

| Substituent and position | | | | | | | |
|---|---|---|---|---|---|---|---|
| 2 | 3 | 4 | 5 | 6 | m.p. | Derivatives | Ref. |
| | $CO_2H$ | $CO_2H$ | $CH_3$ | $CH_3$ | | imide, m.p. 268–70° | 24 |
| $CH_3$ | $CO_2H$ | $CO_2H$ | $CH_3$ | | | imide, m.p. 271–2° | 24 |
| $CH_3$ | $CO_2H$ | $CO_2H$ | $CH_3$ | $CH_3$ | | imide, m.p. 220–1° | 24 |
| | | | | | | 3-monoamide, m.p. 183–5.5° | |
| | | | $-(CH_2)_4-$ | | | 3-monoamide, m.p. 246–7° | 24 |
| | $CO_2H$ | $CO_2H$ | $t$-$C_4H_9$ | | | 3-monoamide, m.p. 176° | 24 |
| $CH_3$ | $CO_2H$ | $CO_2H$ | OH | | 239° | imide, m.p. 308° | 24, 25 |
| $C_2H_5$ | $CO_2H$ | $CO_2H$ | OH | | 240° | imide, m.p. 258° | 24, 25 |
| $C_3H_7$ | $CO_2H$ | $CO_2H$ | OH | | 216–17° | imide, m.p. 251° | 24, 25 |
| $C_4H_9$ | $CO_2H$ | $CO_2H$ | OH | | 198–99° | imide, m.p. 234–5° | 24, 25 |
| $C_5H_{11}$ | $CO_2H$ | $CO_2H$ | OH | | 181–3° | imide, m.p. 234° | 24, 25 |
| $C_6H_5$ | $CO_2H$ | $CO_2H$ | OH | | 286–7° | | 24 |
| $CH_3$ | $CO_2H$ | $CO_2H$ | OH | $CH_3$ | 269–70° | imide, m.p. 275–7° | 24, 25 |
| $CH_3$ | $CO_2H$ | $CO_2H$ | OH | $C_2H_5$ | 255–7° | | 24 |
| $C_2H_5$ | $CO_2H$ | $CO_2H$ | OH | $CH_3$ | 233–4° | | 25 |
| $C_5H_{11}$ | $CO_2H$ | $CO_2H$ | OH | $CH_3$ | 213–14° | | 25 |
| | $CO_2H$ | $CO_2H$ | OH | $C_2H_5OCOCH_2$ | | diethyl ester, m.p. 124–5° | 28 |
| | $CO_2H$ | $CO_2H$ | | $CH_3$ | 252° | imide, m.p. 266–8° | 30 |
| | | $CO_2H$ | $CO_2H$ | $CH_3$ | | 5-Et ester, m.p. 201° | 32 |

| Substituent and position | | | | | | | | |
|---|---|---|---|---|---|---|---|---|
| 1 | 2 | 3 | 4 | 5 | 6 | m.p. | Derivatives | Ref. |
| | OH | | $CONH_2$ | | OH | > 270° with charring | | 1 |
| | OH | | $CO_2H$ | | OH | | | 1 |
| Ph | OH | | OH | $CO_2H$ | Me | | Et ester, m.p. 175° | 7 |
| 2,4-$(Me)_2C_6H_3$ | OH | | OH | $CO_2H$ | Me | | Et ester, m.p. 198–9° | 7 |
| *p*-$CH_3OC_6H_4$ | OH | | OH | $CO_2H$ | Me | | Et ester, m.p. 207–8° | 7 |
| | OH | CN | $CO_2H$ | | Et | | Me ester, m.p. 205° | 8 |
| | OH | CN | $CO_2H$ | | Me | | Me ester, m.p. 228° | 8 |
| | OH | CN | $CO_2H$ | | Me | | Et ester, m.p. 217° | 8 |
| | OH | CN | $CO_2H$ | | Et | | Et ester, m.p. 211° | 8 |
| | OH | CN | $CO_2H$ | | Me | | amide, m.p. 300° | 8 |
| | OH | $CO_2H$ | | CN | OH | no data given | | 9 |
| | OH | CN | $CO_2H$ | | Et | 316–18° | Et ester, m.p. 218° | 12 |
| | OH | CN | $CO_2H$ | | Pr | | Et ester, m.p. 150–1° | 12 |
| | OH | CN | $CO_2H$ | | $PhCH_2$ | | Et ester, m.p. 168–72° | 12 |
| | OH | CN | $CO_2H$ | | $C_6H_{11}CH_2$ | | Et ester, m.p. 168–70° | 12 |
| | OH | | $CO_2H$ | | Et | 308–10° | Et ester, m.p. 103–5° | 12 |
| | OH | | $CO_2H$ | | Pr | 276–8° | | 12 |
| | OH | | $CO_2H$ | | $C_6H_{11}CH_2$ | | Me ester, m.p. 162–4° | 12 |
| Ph | OH | $CO_2H$ | OH | | OH | | PhNHCO, m.p. 234° | 21 |
| | | $CO_2H$ | | OH | $CH_3$ | 201° | amide, m.p. 305°, hydrazide, m.p. 330°, thioamide, m.p. 259-60°, diethyl ester HCl, m.p. 143–4° | 30 |
| | OH | | $CO_2H$ | | | 270° (decomp.) | Me ester, m.p. 208–10° | 115 |
| | OH | | $CO_2H$ | | $CO_2H$ | 318° | | 23 |
| | OH | $CO_2H$ | | | | | benzyl ether, m.p. 135° | 140 |
| | OH | | $CO_2H$ | | $CH_3$ | 300° | Me ester, m.p. 224–5° | 186 |

TABLE X-5. Hydroxypyridine (or Pyridone) Carboxylic Acids (continued)

| Substituent and position | | | | | | | | |
|---|---|---|---|---|---|---|---|---|
| 1 | 2 | 3 | 4 | 5 | 6 | m.p. | Derivatives | Ref. |
| $CH_3$ | OH | CN | $CO_2H$ | | $CH_3$ | | Et ester, m.p. 134° | 186 |
| | $CH_3$ | OH | $CO_2H$ | $CO_2H$ | | | acetate, m.p. 207–11°<br>acetate di-Me ester, m.p. 60–1° | 250 |
| | $CH_3$ | $CH_3O$ | $CO_2H$ | $CO_2H$ | | | di-Me ester, b.p. 145–50°/1–3 mm | 250 |
| | $CH_3$ | $C_6H_5CH_2O$ | $CO_2H$ | $CO_2H$ | | | bisbenzyl ester HC1, m.p. 110° | 250 |
| | | $CO_2H$ | | OH | $CH_3$ | | Me ester, m.p. 243–4° | 251 |
| | $CH_3$ | $CO_2H$ | | OH | $CH_3$ | 300° | Et ester, m.p. 179–80° | 259 |
| | | $CO_2H$ | OH | | | | amide, benzyl ether, m.p. 178° | 262 |
| | OH | CN | | | Styryl | 304° | 2-Me ether, m.p. 122° | 23 |
| | OH | CN | | | α-methylstyryl | 238° | *N*-Me, m.p. 124° | 23 |
| | OH | CN | | | Ac | 213° | | 23 |
| | OH | CN | | | α-Phenylstyryl | 257° | | 23 |
| | OH | CN | | | Benzoyl | 253° | | 23 |
| | OH | $CO_2H$ | $CO_2H$ | | $CO_2H$ | 262° | tri-Me ester, Me ether, m.p. 132° | 23 |
| | OH | | $CO_2H$ | | α-Methylstyryl | 314° | Me ester, Me ether, m.p. 128° | 23 |
| | OH | $CO_2H$ | $CO_2H$ | | Ac | 254° | oxime, m.p. 350°, di-Me ester, Me ether, m.p. 80° | 23 |
| | $CO_2H$ | OH | $CO_2H$ | $CO_2H$ | | | acid, dihydrate, m.p. 198–9° | 101 |
| | OH | $CO_2H$ | | | Styryl | 256° | methyl ether, m.p. 63° | 23 |
| | OH | CN | | | $CO_2H$ | 275° | Me ether, Me ester, m.p. 124° | 23 |
| | OH | $CO_2H$ | | | $CO_2H$ | 304° | | 23 |
| | OH | $CO_2H$ | | | α-Methylstyryl | | Me ether, Me ester, m.p. 73° | 23 |
| | OH | $CO_2H$ | | | Ac | 271° | Me ester, Me ether, m.p. 80° | 23 |
| | OH | $CO_2H$ | | | Benzoyl | 257° | | 23 |
| | OH | | $CO_2H$ | | Styryl | 343° | Me ether, Me ester, m.p. 98° | 23 |

TABLE X-6. Halopyridinecarboxylic Acids

| Substituent and position | | | | | | | |
|---|---|---|---|---|---|---|---|
| 2 | 3 | 4 | 5 | 6 | m.p. | Derivatives | Ref. |
| Cl | | $CO_2H$ | | Cl | 209–10° | | 2 |
| | $CO_2H$ | | F | | | | 86 |
| Cl | | $CO_2H$ | | | 251° | | |
| Br | | $CO_2H$ | | | 229° | | 115 |
| | $CO_2H$ | $CO_2H$ | Br | $CH_3$ | 247–8° (decomp.) | monoamide, m.p. 230–1° | 101 |
| | $CO_2H$ | $CO_2H$ | Cl | $CH_3$ | 248–50° (decomp.) | monoamide, m.p. 275–7° (decomp.) | 101 |
| | $CO_2H$ | $CO_2H$ | I | $CH_3$ | | acid, monohydrate, m.p. 241–6° (decomp.) | 101 |
| $CO_2H$ | | Cl | | | 183–4° | | 137 |
| | $CO_2H$ | | Br | | | Et ester, HBr, m.p. 147–8° | 138 |
| Cl | | | $CO_2H$ | | 199° | | 143 |
| SH | | | $CO_2H$ | | 273° | | 143 |
| $SO_3H$ | | | $CO_2H$ | | 310° | | 143 |
| Cl | | | | $CO_2H$ | 190° | | 144 |
| Cl | | $CO_2H$ | | | 245° | | 144 |
| SH | | $CO_2H$ | | | 304° | | 144 |
| $SO_3H$ | | | | $CO_2H$ | 260–2° | | 144 |
| $SO_3H$ | | $CO_2H$ | | | 296–7° | | 144 |
| | $CO_2H$ | SH | | | 234–7° | | 145 |
| $CO_2H$ | | SH | | | 188–90° | | 145 |
| | SH | $CO_2H$ | | | 259–60° | | 145 |
| $CO_2H$ | Cl | | | Cl | 150–2° | Me ester, m.p. 53°; amide m.p. 188° | 146 |
| $CO_2H$ | Cl | $NH_2$ | Cl | Cl | 218–19° | Me ester, m.p. 116–18° | 148 |
| $CO_2H$ | Cl | $NHCH_3$ | Cl | Cl | 134–7° | | 148 |
| $CO_2H$ | Cl | NH–(pyrrolidin-2-yl, N–H) | Cl | Cl | 146–8° | | 148 |

TABLE X-6. Halopyridinecarboxylic Acids (continued)[a]

| Substituent and positions | | | | | | | |
|---|---|---|---|---|---|---|---|
| 2 | 3 | 4 | 5 | 6 | m.p. | Derivatives | Ref. |
| $CO_2H$ | Cl | $N(CH_3)_2$ | Cl | Cl | 122–4° | | 148 |
| | $CO_2H$ | | SH | | 162–5° | | 235 |
| $CO_2H$ | | | SH | | 205–8° | | 235 |
| | $CO_2H$ | | $SO_3H$ | | 333–5° | | 235 |
| $CO_2H$ | | | $SO_3H$ | | 285° | | 235 |
| $CO_2H$ | | Cl | | $CO_2H$ | | dimethyl ester, m.p. 142° | 237 |
| $CH_3$ | $NH_2$ | $CO_2H$ | CN | Cl | | Et ester, m.p. 171–2° | 250 |
| Cl | $CO_2H$ | $CO_2H$ | $NH_2$ | $CH_3$ | 218–19° | | 251 |
| | $CO_2H$ | Br | | | 165–6° | | 260 |
| | $CO_2H$ | I | | | 138–40° | | 260 |
| | $CO_2H$ | | I | | 224–5° | | 260 |
| | $CO_2H$ | Cl | | | | amide, m.p. 100° | 262 |
| | $CO_2H$ | Br | | | | amide, m.p. 140° | 262 |
| | $CO_2H$ | SH | | | | amide, m.p. 250° | 262 |

[a]See also Table X-8

TABLE X-7. Nitropyridinecarboxylic Acids

| Substituent and position | | | | | | | |
|---|---|---|---|---|---|---|---|
| 2 | 3 | 4 | 5 | 6 | m.p. | Derivatives | Ref. |
| Cl | CN | $CO_2H$ | $NO_2$ | Et | | Et ester, m.p. 70–1.5° | 11 |
| $CH_3$ | $NO_2$ | | | $CO_2H$ | 126° | | 90 |
| $CO_2H$ | $NO_2$ | | | | 122–3° | | 90 |
| $CH_3$ | $NO_2$ | $CH_3$ | | $CO_2H$ | 138° | | 90 |
| $CH_3$ | $NO_2$ | $CO_2H$ | | $CH_3$ | 238° | | 90 |
| $CO_2H$ | $NO_2$ | $CH_3$ | | | 138° | | 90 |
| $CH_3$ | $NO_2$ | $CO_2H$ | | | 250–2° | | 90 |
| $CH_3$ | $NO_2$ | $CO_2H$ | CN | =O | | Et ester, m.p. 196–8° | 250 |
| $CH_3$ | $NO_2$ | $CO_2H$ | CN | Cl | | Et ester, m.p. 56–8° | 250 |

TABLE X-8. Aminopyridinecarboxylic Acids[a]

| Substituent and position | | | | | | | |
|---|---|---|---|---|---|---|---|
| 2 | 3 | 4 | 5 | 6 | m.p. | Derivatives | Ref. |
| $CO_2H$ | $NH_2$ | $CO_2H$ | $CO_2H$ | | 215–17° | | 11 |
| $CO_2H$ | $NH_2$ | | | | 217° | | 90 |
| $NH_2$ | | $CO_2H$ | | | > 360° | Me ester, m.p. 148°<br>Et ester, m.p. 118°<br>amide, m.p. 258° | 91 |
| $NHCOCH_3$ | | $CO_2H$ | | | 286° | Me ester, m.p. 198°<br>amide, m.p. 258–60° | 91 |
| $NH_2$ | | $CO_2H$ | | | | *N*-methylthioamide, m.p. 134–7° | 225 |
| $CH_3$ | $NH_2$ | $CO_2H$ | CN | Cl | | Et ester, m.p. 171–2° | 250 |
| $CH_3$ | $NH_2$ | $CO_2H$ | CN | | | Et ester, m.p. 127° | 250 |
| $CH_3$ | $NH_2$ | $CO_2H$ | $CO_2H$ | | | monohydrate, m.p. 240–5° | 250 |
| | $CO_2H$ | | $NH_2$ | $CH_3$ | 262–3° | | 251 |
| Cl | $CO_2H$ | $CO_2H$ | $NH_2$ | $CH_3$ | 218–19° | | 251 |
| | $CO_2H$ | $NH_2$ | | | | amide, m.p. 233-4° | 262 |

[a]See also table X-6.

TABLE X-9. Cyanopyridines and Cyanopyridinecarboxylic Acids

| Substituent and position | | | | | | | | |
|---|---|---|---|---|---|---|---|---|
| 1 | 2 | 3 | 4 | 5 | 6 | m.p. | Derivatives | Ref. |
| | $CH_3$ | OH | CN | CN | | 189–91° | | 26 |
| | $CH_3$ | | CN | | | 43° | picrate, m.p. 163–4° | 30 |
| | $CH_3$ | | CN | $NH_2$ | | 174° | | 30 |
| | $CH_3$ | OH | | CN | | 190–1° | | 30 |
| | $CH_3$ | CN | | OH | $CH_3$ | 249–51° | | 29 |
| | Cl | | CN | | | 67–8° | | 115 |
| | CN | OH | | | | 211–12° | | 118 |
| | CN | $CH_3$ | | | | 85° | | 130 |
| | CN | | | | $CH_3$ | 71° | | 130 |
| | | $CO_2H$ | | CN | | | Et ester, m.p. 89–90°; | 130 |
| | | | | | | | amide, m.p. 220–1° | 138 |
| | | CN | | CN | | 113–14° | | 138 |
| | Cl | CN | | CN | $CH_3$ | 143–4° | | 138 |
| | $CH_3$ | CN | | CN | | 76° | | 138 |
| | | CN | $CH_3$ | CN | | 84–5° | | 138 |
| | $CH_3$ | CN | | CN | $CH_3$ | 118–19° | | 138 |
| | $CH_3$ | CN | $CH_3$ | CN | | 115° | | 138 |
| | $CH_3$ | CN | $C_2H_5$ | CN | | 68° | | 139 |
| | $C_2H_5O$ | CN | | | | 35° | | 140 |
| | $C_6H_5O$ | CN | | | | 109° | | 140 |
| | | CN | 4′-Pyridyl | | | 152° | | 141 |
| | N O Br | | | CN | | 264° | | 142 |

TABLE X-9. Cyanopyridines and Cyanopyridinecarboxylic Acids (Continued)

| Substituent and position | | | | | | | | |
|---|---|---|---|---|---|---|---|---|
| 1 | 2 | 3 | 4 | 5 | 6 | m.p. | Derivatives | Ref. |
| | | | | CN | | 204–5° | | 142 |
| | CN | Cl | $NH_2$ | Cl | Cl | 240–1° | | 148 |
| | $OCH_3$ | CN | $CO_2H$ | | $CH_3$ | | Et ester, m.p. 70° | 186 |
| | | | | | | | Me ester, m.p. 114° | 186 |
| $CH_3$ | $OCH_3$ | CN | $CO_2H$ | | $CH_3$ | | Et ester, m.p. 174° | 186 |
| | | $CH_3O$ | CN | | | 61–2° | | 228 |
| | $CH_3$ | $CH_3O$ | CN | $C_2H_5$ | | 87–9° | | 228 |
| | CN | | | CN | | 112–13° | | 229 |
| | $C_2H_2$ | | CN | | | b.p. 82–5° | | 249 |
| | CN | | $CH_3$ | | | 89° | | 256 |

| Substituent and position | | | | | | | | |
|---|---|---|---|---|---|---|---|---|
| 1 | 2 | 3 | 4 | 5 | 6 | m.p. | Derivatives | Ref. |
| H | | Ac | $H_2$ | Ac | | 195° | | 6 |
| | OH | $H_2$ | $H_2$ | $CO_2H$ | Me | | amide | 13 |
| H | Me | $CO_2H$ | $H_2$ | Et | | | Et ester | 14 |
| $CH_3$ | | $CO_2H$ | | | =O | 241–3° | Et ester, m.p. 78–80°<br>amide, m.p. 211° | 15 |
| $CH_3$ | | $CO_2H$ | =O | | | 245–7° | amide, m.p. 183–4° | 15 |
| $CH_3$ | $CO_2H$ | $CO_2H$ | $CO_2H$ | $CO_2H$ | Ph; H | | tetra-Me ester, m.p. 165–7° | 17 |
| $CH_2Ph$ | $CO_2H$ | $CO_2H$ | $CO_2H$ | $CO_2H$ | Ph; H | | tetra-Me ester, m.p. 128–30° | |
| $CH_2CH_2Ph$ | $CO_2H$ | $CO_2H$ | $CO_2H$ | $CO_2H$ | Ph; H | 150–2° | | |
| H | Me | $CO_2H$ | $H_2$ | Ac | | | Et ester | 19 |
| | OH | CN; H | H; OH | | Styryl | 315° | 4-acetate, m.p. 289 | 23 |
| $CH_3NAc$ | | | H; CN | | | 98–9° | | 135 |
| H | $CH_3$; H | CN | | CN | | 114–15° | | 138 |
| H | $CH_3$ | CN | $H_2$ | CN | | 180–1° | | 138 |
| H | $CH_3$; H | CN | | CN | $CH_3$ | 152–3° | | 138 |
| H | $CH_3$ | CN | H; $CH_3$ | CN | | 129–30° | | 138 |
| H | | CN | $H_2$ | CN | | 205–6° | | 139 |
| H | $H_2$ | CN | | CN | | 188–90° | | 139 |
| H | $H_2$ | CN | $CH_3$ | CN | | 214–15° | | 139 |
| H | H; $CH_3$ | CN | $C_2H_5$ | CN | $CH_3$ | 101–2° | | 139 |
| H | $CH_3$ | CN | $H_2$ | CN | $CH_3$ | 215–20° | | 169 |
| H | $CH_3$ | $CO_2H$ | $H_2$ | $CO_2H$ | $CH_3$ | | di-Et ester, m.p. 175–8° | 169 |
| Ac | H | $CO_2H$ | | | | | amide, m.p. 175–80° | 203 |
| H | $CO_2H$ | | $(CN)_2C=$ | | $CO_2H$ | | di-Me ester, m.p. 218–19° | 237 |
| H | $CO_2H$ | | (*t*-Butyl-OOC)(CN)C= | | $CO_2H$ | | di-Me ester, m.p. 175° | 237 |
| H | $CO_2H$ | | $(C_2H_5OOC)(CN)C=$ | | $CO_2H$ | | di-Me ester, m.p. 149° | |
| H | H, $CH(OH)CH_3$ | $CO_2H$ | | $CO_2H$ | | | di-Me ester, m.p. 160–6° (2 isomers)<br>(2 isomers) 150–8° | 242 |
| H | | $CO_2H$ | H, $CH(OH)CH_3$ | $CO_2H$ | | | di-Me ester, m.p. 135–6° | 242 |

TABLE X-11. Cyano- and Carboxypyridine-1-oxides

| Substituent and position | | | | | | | |
|---|---|---|---|---|---|---|---|
| 2 | 3 | 4 | 5 | 6 | m.p. | Derivatives | Ref. |
| CN | | CN | | CN | 222° | | 18 |
| $CO_2H$ | | $CO_2H$ | | $CO_2H$ | | tri-Me ester | 18 |
| | | $CO_2H$ | | | | amide | 115 |
| $CH_3$ | | $CO_2H$ | | | | thioamide, m.p. 188° | 115 |
| | $CH_3$ | $CO_2H$ | | | | thioamide, m.p. 191–3° | 115 |
| $CH_3$ | | $CO_2H$ | | $CH_3$ | | thioamide, m.p. 215–16° | |
| $CH_3$ | $CO_2H$ | $CO_2H$ | | | 235–40° (decomp.) | | 85 |
| $CH_3$ | | $CO_2H$ | $CO_2H$ | | 205–10° (decomp.) | | |
| CN | | Cl | | | 130.5–31° | | 137 |
| CN | | $NO_2$ | | | 178–9° | | 137 |
| $CH_3$ | $CH_3O$ | CN | | | 175–6° | | 229 |
| $CO_2H$ | | | | | 162° | | 253 |
| $CO_2H$ | | | | $CO_2H$ | 155–7° | | 253 |
| $CO_2H$ | | | $CO_2H$ | | 241–4° | | 253 |
| $CH_3O$ | | $CO_2H$ | | | 186–8° | Me ester, m.p. 129–30° | 254 |
| OH | | $CO_2H$ | | | | Et ester, m.p. 147–8°; amide, m.p. 270–2° | 254 |
| OH | | $CO_2H$ | $NO_2$ | | 185° | | 254 |
| $CO_2H$ | | | | | | amide, m.p. 161–2° | |
| $CO_2H$ | | $CH_3$ | | | 169° | | 256 |
| $CO_2H$ | | $CH_3$ | | $CO_2H$ | 160° | | 150 |
| $CO_2H$ | | $CO_2H$ | | | 262° | | 256 |
| $CO_2H$ | | | | $CH_2OH$ | 195° | | 257 |
| $C_2H_5CO$ | $CO_2H$ | | | | 195–6° | | 258 |
| $CH_3$ | $CO_2H$ | | | | 50–2° | | 259 |
| $CH_3$ | $CO_2H$ | | | $CH_3$ | | di-Et ester, m.p. 35–45° | 259 |
| $CO_2H$ | | | | | 162–5° | | 261 |

| | | | | | | |
|---|---|---|---|---|---|---|
| $CO_2H$ | | $C_2H_5O$ | | 144° | amide, m.p. 169° | 261 |
| $CO_2H$ | | $C_3H_7O$ | | 107° | amide, m.p. 126° | 261 |
| $CO_2H$ | | $IsoC_3H_7O$ | | 147° | | 261 |
| $CO_2H$ | | OH | | 213°, 243° | | 261 |
| $CO_2H$ | | $OCH_2CH_2OH$ | | 152° | | 261 |
| $CO_2H$ | | *p*-Cresyloxy | | 145° | | 261 |
| $CO_2H$ | | Cl | | 144° | Me ester, m.p. 102° | 261 |
| | | | | 135–6° | | 150 |
| | $CH_3$ | Cl | $CO_2H$ | 160° | | 150 |
| $CO_2H$ | | $NH_2$ | | 217° | Me ether, m.p. 139° | 261 |
| $CO_2H$ | | $SO_3H$ | | | Na salt monohydrate, m.p. 317° | 261 |
| $CO_2H$ | | SH | | 155° | benzyl ether, m.p. 157° | 261 |
| | $CO_2H$ | $NO_2$ | | | amide, m.p. 204° | 262 |
| | $CO_2H$ | $C_6H_5CH_2O$ | | 147° | | 262 |
| | $CO_2H$ | $CH_3O$ | | 200° | | 262 |
| | $CO_2H$ | Cl | | 173–5° | | 262 |
| | $CO_2H$ | Br | | 155° | | 262 |
| | $CO_2H$ | $NH_2$ | | 300° | | 262 |
| | $CO_2H$ | SH | | 165–7° | | 262 |

## References


1. M. DeMaldè and E. Alneri, *Chim. Ind.* (Milan) **38,** 473 (1956).
2. M. M. Baizer, M. Dub, S. Gister, and N. G. Steinberg, *J. Amer. Pharm. Assoc.*, **45,** 4 (1956).
3. L. R. Morgan, Jr., *J. Org. Chem.*, **27,** 343 (1962).
4. J. Kuthan and J. Palecek, *Collect. Czech. Chem. Commun.*, **31,** 2618 (1966).
5. J. Palecek and J. Kuthan, *Collect. Czech. Chem. Commun.*, **34,** 3336 (1969).
6. F. Micheel and H. Dralle, *Ann. Chem.*, **670,** 57 (1963).
7. E. Ziegler, F. Hradetzky, and K. Belegratis, *Monatsh. Chem.*, **96,** 1347 (1965).
8. F. Cuiban, S. Cilianu-Bibian, S. Popescu, and I. Rogozea, Fr. patent, 1,366,064 (19 to Ministry of Petroleum and Chemical Industry, Romania; *Chem. Abstr.*, **61,** 146 (1964).
9. H. Hellmann and K. Seegmuller, *Angew. Chem.*, **70,** 271 (1958).
10. H. Junek, *Monatsh. Chem.*, **96,** 2046 (1965).
11. S. M. Gadekas, J. L. Frederick, J. Semb, and J. R. Vaughan, *J. Org. Chem.*, **26,** 4 (1961).
12. O. Fuchs, V. Senkariuk, A. Nemes, G. Zolyomi, T. Somogyi, and A. Lazar, Hu patent, 157,008 (1970); *Chem. Abstr.*, **72,** 100533f (1970).
13. T. Kato, H. Yamanaka, and J. Kawamata, *Chem. Pharm. Bull.* (Tokyo), **17,** 24 (1969).
14. H. D. Eilhauer and I. Kaempfer, *Z. Chem.*, **9,** 188 (1969).
15. S. Tanaka and J. M. Price, *J. Org. Chem.*, **32,** 2351 (1967).
16. M. Baizer, U.S. patent, 3,246,000 (1966); *Chem. Abstr.*, **64,** 19568h (1966).
17. R. Huisgen and K. Herbig, *Ann. Chem.*, **688,** 98 (1965).
18. K. D. Gundermann and H. U. Alles, *Angew. Chem.*, **78,** 906 (1966); *Angew. Che Intern. Ed. Engl.*, **5,** 846 (1966).
19. M. Ohashi, H. Kamachi, H. Kakisawa, and G. Stork, *J. Amer. Chem. Soc.*, **89,** 54 (1967).
20. I. Ehsan and Karimullah, *Pakistan J. Sci. Ind. Res.*, **11,** 5 (1968).
21. A. Butt, I. A. Akhter, and M. Akhter, *Tetrahedron*, **23,** 199 (1967).
22. H. Gault, J. Gilbert, and D. Briancourt, *C. R. Acad. Sci., Paris, Ser. C*, **266, 1** (1968).
23. L. Rateb, G. A. Mina, and G. Soliman, *J. Chem. Soc., C*, 2140 (1968).
24. G. Ya. Kondrat'eva and C. -H. Huang, *Dokl. Akad. Nauk SSSR*, **141,** 628 (1961).
25. G. Ya. Kondrat'eva and C. -H. Huang, *Dokl. Akad. Nauk SSSR*, **141,** 961 (1961).
26. E. E. Harris, R. A. Firestone, K. Pfister, III, R. R. Boettcher, F. J. Cross, R. B. Cur M. Monaco, E. R. Peterson, and W. Reuter, *J. Org. Chem.*, **27,** 2705 (1962).
27. K. Pfister, III, E. E. Harris, and R. A. Firestone, U.S. patent, 3,227,721 (1966); *Ch Abstr.*, **64,** 9690d (1966).
28. M. Kawazu, Jap. patent, 18627 (1967); *Chem. Abstr.*, **69,** 10366n (1968).
29. T. Naito and T. Yoshikawa, *Chem. Pharm. Bull.* (Tokyo), **14,** 918 (1966).
30. T. Naito, T. Yoshikawa, F. Ishikawa, S. Isoda, Y. Omura, and I. Takamura, *Ch Pharm. Bull.* (Tokyo), **13,** 869 (1965).
31. P. Colin, Fr. patent, 1,550,352, (1968); *Chem. Abstr.*, **72,** 31629c (1970).
32. G. Ya. Kondrat'eva and C. -H. Huang, *Dokl. Akad. Nauk SSSR*, **164,** 816 (1965).
33. T. Jaworski and B. Korybut-Daszkiewicz, *Rocz. Chem.*, **41,** 1521 (1967).
34. N. Kucharczyk, *Chem. Listy*, **55,** 1199 (1961).

5. L. O. Shnaidman and M. I. Siling, *Tr. Vses. Nauchn.-Issled. Vitamin. Inst.*, **7,** 18 (1961); *Chem. Abstr.*, **60,** 7988h (1964).

6. Swiss patent, 339,625 (1955), to Lonza Elektrizitaetswerke & Chemische Fabriken A.-G.; *Chem. Abstr.*, **56,** 2430e (1962).

7. E. Bartholome, H. Nienburg, and K. Scherf, Ger. patent, 1,119,842 (1961); *Chem. Abstr.*, **57,** 786b (1962).

8. Ya. N. Ivashchenko, *Pererabotka Vydelenie, i Analizy Koksokhim. Produktov Sb.*, 45 (1961); *Chem. Abstr.*, **60,** 4102d (1964).

9. Yu. I. Chumakov, L. A. Rusakova, A. I. Mednikov, and R. I Virnik, *Metody Polucheniya Khim. Reaktivov i Preparatov, Gos. Kom Sov. Min. SSSR po Khim.*, **7,** 79 (1963); *Chem. Abstr.*, **61,** 5604f (1964).

0. V. Kudlacek and V. Stverka, *Sb. Ved. Praci, Vysoka Skola Chem.-Technol., Pardubice* (1), 33 (1965); *Chem. Abstr.*, **65,** 681b (1966).

1. B. F. Ustavshchikov, T. S. Titova, E. V. Degtyarev, and M. I. Farberov, *Zh. Prikl. Khim.*, **39,** 1388 (1966); *Chem. Abstr.*, **65,** 10557g (1966).

2. El Mutafchieva, N. Gospodinov, and Ya. Todorova, *Farmatsiya* (Sofia), **17,** 20 (1967); *Chem. Abstr.*, **68,** 87110m (1968).

3. W. Swietoslawski, J. Bialek, and A. Bylicki, Polish patent, 47,460 (1963); *Chem. Abstr.*, **61,** 5614d (1964).

4. J. Zundel, Fr. patent, 1,509,120 (1968); *Chem. Abstr.*, **70,** 68170f (1969).

5. R. Aries, Fr. patent, 1,509,049 (1968); *Chem. Abstr.*, **70,** 37657r (1969).

6. T. S. Titova, B. F. Ustavshchikov, M. Farberov, and E. V. Degtyarev, *Zh. Prikl. Khim.*, **42,** 910 (1969); *Chem. Abstr.*, **71,** 38750y (1969).

7. W. Hoefling, H. D. Eilhauer, G. Krautschik, and R. Mohrhauer, East Ger. patent, 58,090 (1967); *Chem. Abstr.*, **70,** 3842g (1969).

8. B. F. Ustavshchikov, M. I. Farberov, A. M. Kut'in, and G. S. Levskaya, *Uch. Zap. Yaroslavsk. Tekhnol. Inst.*, **5,** 71 (1960); *Chem. Abstr.*, **57,** 11153h (1962).

9. M. I. Farberov, B. F. Ustavshchikov, A. M. Kut'in, and I. T. Baranova, *Metody Polucheniya Khim. Reaktivov i Preparatov, Gos. Kom. Sov. Min. SSSR po Khim.* (11), 60 (1964); *Chem. Abstr.*, **65,** 758b (1966).

0. J. E. Mahan and R. P. Williams, U.S. patent, 2,993,904 (1957); *Chem. Abstr.*, **56,** 1434a (1962).

1. E. B. Bengtsson, Ger. patent, 1,161,563 (1964); *Chem. Abstr.*, **60,** 10654g (1964).

2. E. S. Zhdanovich, I. B. Chekmareva, L. I. Kaplina, and N. A. Preobrazhenskii, USSR patent, 148,411 (1962); *Chem. Abstr.*, **58,** 9031e (1963).

3. W. L. F. Armarego and R. F. Evans, *J. Appl. Chem.* (London), **12,** 45 (1962).

4. I. B. Chekmareva, E. S. Zhdanovich, and N. A. Preobrazhenskii, *Zh. Prikl. Khim.*, **38,** 220 (1965); *Chem. Abstr.*, **62,** 13119g (1965).

5. J. O'Brochta, U.S. patent, 2,999,094 (1959); *Chem. Abstr.*, **56,** 2432b (1962).

6. W. Schwarze, Ger. patent, 1,071,085 (1959); *Chem. Abstr.*, **57,** 11172g (1962).

7. A. Matsuura, Y. Suzuki, and K. Matsuki, Jap. patent, 17,340 (1963); *Chem. Abstr.*, **60,** 2904c (1964).

8. T. Kato, *Bull. Chem. Soc. Jap.*, **34,** 636 (1961).

9. H. Sobue, A. Tomita, Y. Sumita, T. Kato, and S. Hasegawa, Jap. patent 13,081 (1963); *Chem. Abstr.*, **60,** 507f (1964).

0. T. I. Baranova, A. M. Kut'in, M. I. Farberov, and B. F. Ustavshchikov, USSR patent, 161,755 (1964); *Chem. Abstr.*, **61,** 4325d (1964).

1. K. Uda, A. Sakurai, and K. Sakakibara, Jap. patent, 23,792 (1965); *Chem. Abstr.*, **64,** 3502g (1966).

62. T. I. Baranova, L. F. Titova, and A. M. Kut'in, *Khim. Prom.*, **43,** 204 (1967); *Che Abstr.*, **67,** 21792h (1967).
63. S. K. Bhattacharyya, V. Shankar, and A. K. Kar, *Ind. Eng. Chem., Prod. R Develop.*, **5,** 65 (1966).
64. S. K. Bhattacharyya and A. K. Kar, *Indian J. Appl. Chem.*, **30,** 35 (1967).
65. S. K. Bhattacharyya and A. K. Kar, *Indian J. Appl. Chem.*, **30,** 42 (1967).
66. T. A. Afanas'eva, A. D. Kagarlitskii, and B. V. Surorov, *Khim. Geterotsikl. Soed* (1), 142 (1968); *Chem. Abstr.*, **70,** 3791q (1969).
67. B. V. Suvorov, A. D. Kagarlitskii, T. A. Afanas'eva, and I. I. Kan, *Khim. Geterots Soedin.* (6), 1129 (1969); *Chem. Abstr.*, **72,** 121320z (1970).
68. L. Leitis and M. V. Shimanskaya, *Khim. Geterotsikl. Soedin.* (3), 507 (1967); *Che Abstr.*, **68,** 29553x (1968).
69. F. Komatsu, *Muroran Kogyô Daigaku Kenkyû Hôkoku*, **3,** 61 (1958); *Chem. Abs* **53,** 13149e (1959).
70. F. Komatsu, Y. Ozono, K. Sakurei, and H. Komori, *Koru Taru* **12,** 420 (1960); *Che Abstr.*, **61,** 4309h (1964).
71. A. B. Sanz, Span. patent, 272,411 (1962); *Chem. Abstr.*, **60,** 2905h (1964).
72. (a) D. Jerchel, J. Heider, and H. Wagner, *Ann. Chem.*, **613,** 153 (1958); (b) *Ibid.*, 1
73. B. Marcot and R. Palland, *C. R. Acad. Sci., Paris, Ser. C.*, **248,** 252 (1959).
74. T. Slebodzinski, H. Kielczewka, and W. Biernacki, *Przem. Chem.*, **48,** 90 (196 *Chem. Abstr.*, **71,** 38751z (1969).
75. A. N. Kost, P. B. Terent'ev, and L. V. Moshentseva, *Khim.-Farm. Zh.*, **1,** 12 (196 *Chem. Abstr.*, **68,** 68846q (1968).
76. S. de Groot and J. Strating, *Rec. Trav. Chim. Pays-Bas,* **80,** 944 (1961).
77. C. R. Adams, *J. Heterocycl. Chem.*, **4,** 137 (1967).
78. K.-M. Wu, *Hua Hsueh* (4), 147 (1966); *Chem. Abstr.*, **67,** 116802a (1967).
79. Brit. patent, 979,761 (1965) to Instytut Chemii Ogolnej; *Chem. Abstr.*, **62,** 162 (1965).
80. A. Kotarski, *Chim. Ind.* (Paris) **94,** 366 (1965).
81. W. Swietoslawski, J. Bialek, A. Bylicki, and A. Kotarski, U.S. patent, 3,261,8 (1966); *Chem. Abstr.*, **65,** 12177b (1966).
82. J. Dialek and J. Malczynski, *Chim. Ind.* (Paris) **95,** 69 (1966).
83. I. B. Chekmareva, E. S. Zhdanovich, A. I. Reznik, and N. A. Preobrashenskii, *Prikl. Khim.*, **38,** 707 (1965); *Chem. Abstr.*, **62,** 16186a (1965).
84. N. D. Rus'yanova, N. V. Malysheva, and L. P. Yurkina, *Khim. Prod. Koksovan Uglei Vostoka SSSR, Poluch., Obrab., Ispol'z, Metody Anal., Vost. Nauch. -Iss Uglekhim. Inst., Sb. Statei*, **4,** 231 (1967); *Chem. Abstr.*, **69,** 2825w (1968).
85. E. J. Moriconi and F. A. Spano, *J. Amer. Chem. Soc.*, **86,** 38 (1964).
86. N. J. Leonard and L. R. Peters, U.S. patent, 3,027,380 (1962); *Chem. Abstr.*, 5896e (1962).
87. M. Levi and Sh. Levi, *Farmatsiya* (Sofia), **8,** No. 4, 20 (1958); *Chem. Abstr.*, 18951h (1959).
88. M. Levi and Ch. Ivanov, *Farmatsiya* (Sofia), **15**(2), 85 (1965); *Chem. Abstr.*, 11492d (1965).
89. M. I. Farberov, B. F. Ustavshchikov, and T. T. Titov, *Metody Polucheniya Kh Reaktivov i Preparatov, Gos. Kom. Sov. Min. SSSR po Khim.* (11), 58 (1964); *Ch Abstr.*, **64,** 15832d (1966).
90. E. V. Brown and R. H. Neil, *J. Org. Chem.*, **26,** 3546 (1961).
91. G. Ferrari and E. Marcon, *Farmaco, Ed. sci.*, **13,** 485 (1958); *Chem. Abstr.*, **53,** 71 (1959).

R. D. Lekberg, R. A. Jensen, and W. Buiter, U.S. patent, 3,313,821 (1967); *Chem. Abstr.*, **67**, 90685n (1967).

L. H. Beck, U.S. patent, 3,154,549 (1964); *Chem. Abstr.*, **62**, 1673e (1965).

J. Kuthan and J. Palecek, Czech. patent, 124,047 (1967); *Chem. Abstr.*, **69**, 35953q (1968).

W. Bartok, D. D. Rosenfeld, and A. Schriesheim, *J. Org. Chem.*, **28**, 410 (1963).

R. Lukes, V. Galik, and J. Jizba, *Collect. Czech. Chem. Commun.*, **26**, 2727 (1961).

R. Lukes, J. Jizba, and V. Galik, *Collect. Czech. Chem. Commun.*, **26**, 3044 (1961).

R. Bodalski, J. Michalski, and K. Studniarski, *Rocz. Chem.* **38**, 1337 (1964).

T. A. Afanas'eva, A. D. Kagarlitskii, and B. V. Surorov, *Khim. Geterotsikl. Soedin.* (1), 142 (1968); *Chem. Abstr.*, **70**, 3791q (1969).

E. Kobayashi and K. Matsumoto, Jap. patent, 7481 (1962); *Chem. Abstr.*, **59**, 1604a (1963).

S. M. Gadekas and J. L. Frederick, *J. Heterocycl. Chem.*, **5**, 125 (1968).

H. Ahrens and W. Korytnyl, *J. Heterocycl. Chem.*, **4**, 625 (1967).

T. Mutavchiev and A. Marinov, *Godishnik Khim.-Tekhnol. Inst.*, **2**, 193 (1956); *Chem. Abstr.*, **52**, 11631e (1958).

V. G. Khomyakov, S. S. Kruglikov, and L. I. Kazakova, *Tr. Mosk. Khim.-Tekhnol. Inst.*, No. 32, 189 (1961); *Chem. Abstr.*, **57**, 15065e (1962).

S. S. Kruglikov and V. G. Khomyakov, *Tr. Mosk. Khim.-Tekhnol. Inst.*, No. 32, 194 (1961); *Chem. Abstr.*, **57**, 16542i (1962).

L. D. Borkhi and V. G. Khomyakov, USSR patent, 187,024 (1966); *Chem. Abstr.*, **67**, 17395p (1967).

V. G. Khomyakov, N. A. Dzbanovskii, and L. D. Borkhi, *Tr., Vses. Nauch. Issled. Inst. Khim. Reaktivov Osobo Chist. Khim. Veshchestv.*, **29**, 304 (1966); *Chem. Abstr.*, **67**, 116796b (1967).

V. V. Tsodikov, L. D. Borkhi, V. G. Brudz, N. E. Khomutov, and V. G. Khomyakov, *Khim. Geterotsikl. Soedin.* (1), 112 (1967); *Chem. Abstr.*, **67**, 17256u (1967).

L. D. Borkhi and V. G. Khomyakov, *Khim. Geterotsikl. Soedin.* (1), 167 (1967); *Chem. Abstr.*, **67**, 64209r (1967).

N. A. Dzbanovskii and L. D. Borkhi, USSR patent, 166,654 (1964); *Chem. Abstr.*, **62**, 11789h (1965).

B. Lipka and E. Treszczanowicz, Polish patent, 50,077 (1965); *Chem. Abstr.*, **65**, 5446a (1966).

I. B. Chekmareva, E. S. Zhdanovich, G. I. Sazonova, and N. A. Preobrazhenskii, USSR patent, 164,601 (1964); *Chem. Abstr.*, **62**, 9115d (1965).

I. B. Chekmareva, E. S. Zhdanovich, and N. A. Preobrazhenskii, *Med. Prom. SSSR*, **19**(7), 11 (1965); *Chem. Abstr.*, **63**, 11488a (1965).

I. B. Chekmareva, E. S. Zhdanovich, and N. A. Preobrazhenskii, *Zh. Prikl. Khim.*, **38**, 2387 (1965); *Chem. Abstr.*, **64**, 689d (1966).

Y. Suzuki, *Yakugaku Zasshi*, **81**, 1204 (1961); *Chem. Abstr.*, **56**, 3445d (1962).

I. A. Arkhipova, S. R. Rafikov, and B. V. Suvorov, *Zh. Prikl. Khim.*, **35**, 389 (1962); *Chem. Abstr.*, **57**, 9808d (1962).

J. Pennington and B. Yeomans, Fr. patent, 1,327,679 (1963); *Chem. Abstr.*, **59**, 12768g (1963).

F. W. Broekman, A. van Veldhuizen, and H. Jansen, *Rec. Trav. Chim. Pays-Bas*, **81**, 792 (1962).

B. Scherhag, S. Hartung, A. Hausweiler, and H. Gruenewald, Belg. patent, 636,800 (1963); *Chem. Abstr.*, **61**, 14643f (1964).

R. S. Aries, U.S. patent, 3,029,245 (1962); *Chem. Abstr.*, **57**, 11115i (1962).

121. Japan Catalytic Chemical Industry Co., Ltd., Fr. patent 1,368,494 (1964); *Che Abstr.*, **62,** 1606b (1965).
122. E. S. Zhdanovich, I. B. Chekmareva, and N. A. Preobrazhenskii, *Zh. Obshch. Khi* **31,** 3272 (1961).
123. V. I. Trubnikov, E. S. Zhdanovich, and N. A. Preobrazhenskii, *Khim.-Farm. Zh.*, **3,** (1969); *Chem. Abstr.*, **72,** 111241k (1970).
124. B. Lipka, B. Buszynska, and E. Treszczanowicz, *Przem. Chem.*, **49,** 34 (1970); *Che Abstr.*, **72,** 111231g (1970).
125. T. A. Afanas'eva, A. D. Kagarlitskii, V. A. Serazetdinova, L. S. Saltybaeva, and B. Suvorov, *Khim. Geterotsikl. Soedin.* (4), 672 (1969); *Chem. Abstr.*, **72,** 4336 (1970).
126. T. A. Afanas'eva, A. D. Kagarlitskii, I. I. Kan, and B. V. Suvorov, *Khim. Geterots Soedin.* (4), 675 (1969); *Chem. Abstr.*, **72,** 43366c (1970).
127. B. V. Suvorov, A. D. Kagarlitskii, T. A. Afanas'eva, O. B. Lebedeva, A. I. Loika, a V. A. Serazetdinova, *Khim. Geterotsikl. Soedin.* (6), 1024 (1969); *Chem. Abstr.*, 132456x (1970).
128. V. I. Trubnikov, V. V. Petrov, E. S. Zhdanovich, and N. A. Preobrazhens *Khim.-Farm. Zh.*, **3**(9), 49 (1969); *Chem. Abstr.*, **72,** 12498t (1970).
129. E. A. Pavlov and B. V. Suvorov, *Tr. Kaz. Sel'skokhoz. Inst.*, **10,** 229 (1965); *Che Abstr.*, **66,** 85676k (1967).
130. N. Kucharczyk and A. Zvakova, *Collect. Czech. Chem. Commun.*, **28,** 55 (1963).
131. B. V. Suvorov, A. D. Kagarlitskii, and T. A. Afanas'eva, USSR patent, 197,591 (196 *Chem. Abstr.*, **69,** 27267d (1968).
132. W. E. Feely, U.S. patent, 2,991,285 (1961); *Chem. Abstr.*, **56,** 7282 (1962).
133. W. E. Feely, G. Evanega, and E. M. Beavers, *Org. Synth.*, **42,** 30 (1962).
134. T. Okamoto and H. Tani, Jap. patent, 9089 (1962); *Chem. Abstr.*, **59,** 5140g (196
135. T. Okamoto, M. Hirobe, C. Mizushima, and A. Osawa, *Chem. Pharm. Bull.* (Toky **11,** 780 (1963).
136. F. E. Cislak, U.S. patent, 2,989,534 (1961); *Chem. Abstr.*, **56,** 3466f (1962).
137. T. Kato and H. Hayashi, *Yakugaku Zasshi,* **83,** 352 (1963); *Chem. Abstr.*, **59,** 747 (1963).
138. J. Kuthan, E. Janeckova, and M. Havel, *Collect. Czech. Chem. Commun.*, **29,** 1 (1964).
139. J. Kuthan and E. Janeckova, *Collect. Czech. Chem. Commun.*, **29,** 1654 (1964).
140. P. Nantka-Namirski, *Acta Pol. Pharm.*, **23,** 403 (1966); *Chem. Abstr.*, **66,** 8567 (1967).
141. A. A. Ziyaev, O. S. Otroshchenko, and A. S. Sadykov, *Zh. Obshch. Khim.*, **34,** 3 (1964).
142. S. Kajihari, *Nippon Kagaku Zasshi,* **86,** 839 (1965); *Chem. Abstr.*, **65,** 16935b (196
143. J. Delarge, D. Fernandez, and C. L. Lapiere, *J. Pharm. Belg.*, **22,** 213 (1967); *Che Abstr.*, **68,** 59406u (1968).
144. J. Delarge, *Farmaco, Ed. Sci.*, **22,** 1069 (1967; *Chem. Abstr.*, **69,** 2830u (1968).
145. J. Delarge, *J. Pharm. Belg.*, **22,** 257 (1967); *Chem. Abstr.*, **68,** 104913v (1968).
146. H. Johnston, U.S. patent, 3,317,549 (1967); *Chem. Abstr.*, **68,** 29611q (1968).
147. Yu. V. Shcheglov, M. S. Sokolov, A. N. Kasikhin, N. P. Zhukov, Yu. V. Boronin, M. Kirmalova, and V. P. Litvinov, *Agrokhimiya* (5), 105 (1967); *Chem. Abstr.*, **68,** 2179 (1968).
148. H. Johnston and M. S. Tomita, Belg. patent, 628,487 (1963); *Chem. Abstr.*, **61,** 18 (1964).

9. B. Raecke, B. Blaser, W. Stein, H. Shirp, and H. Schuett, Ger. patent, 1,095,281 (1955); *Chem. Abstr.*, **56,** 2425e (1962).

0. R. A. Abramovitch, M. Saha, E. M. Smith, and R. T. Coutts, *J. Amer. Chem. Soc.*, **89,** 1537 (1967).

0a. R. A. Abramovitch, E. M. Smith, and R. T. Coutts, *J. Org. Chem.*, **37,** 3584 (1972).

1. P. O. Lumme, *Suomen Kemistilehti,* **30B,** 168 (1957); *Chem. Abstr.*, **52,** 3514c (1958).

2. V. M. Reznikov and V. I. Bliznyukov, *Nekotorye Vopr. Emission. i Molekulyarh. Spektroskopii, Krasnoyarsk, Sb.*, 193 (1960); *Chem. Abstr,* **58,** 9768c (1963). L. Thunus, *J. Pharm. Belg.*, **21,** 491 (1966); *Chem. Abstr.*, **66,** 32415t (1967).

3. K. C. Ong, B. Douglas, and R. A. Robinson, *J. Chem. Eng. Data,* **11,** 574 (1966).

4. A. N. Kost, P. B. Terent'ev, L. A. Golovleva, and A. A. Stolyarchuk, *Khim.-Farm. Zh.*, **1**(5), 3 (1967); *Chem. Abstr.*, **68,** 29556a (1968).

5. P. J. Brignell, U. Eisner, and P. G. Farrell, *J. Chem. Soc., B,* 1083 (1966).

6. R. F. Evans and W. Kynaston, *J. Chem. Soc.*, 1005 (1961). (1962)

7. L. D. Taylor, *J. Org. Chem.*, **27,** 4064 (1962).

8. H. Shindo, *Chem. Pharm. Bull.* (Tokyo) **6,** 117 (1958); *Chem. Abstr.*, **53,** 19457c (1958).

9. W. Bruegel, *Z. Elektrochem.*, **66,** 159 (1962).

0. S. P. Bag, Q. Fernando, and H. Freiser, *Inorg. Chem.*, **1,** 887 (1962).

1. G. Anderegg and E. Bottari, *Helv. Chim. Acta,* **48,** 887 (1965).

2. J. Bialek, A. Bylicki, A. Chwistek, A. Galazka, and J. Szaton, Polish patent, 50,079 (1965); *Chem Abstr.*, **65,** 540a (1966).

3. E. Katscher and W. Moroz, U.S. patent, 3,147,269 (1964); *Chem. Abstr.*, **62,** 4013h (1965).

4. F. Schindler and F. Kuffner, *Monatsh. Chem.*, **94,** 252 (1963).

5. H. Liliedahl, *Acta Chem. Scand.*, **20,** 95 (1966).

6. A. Laurent, *C. R. Acad. Sci., Paris, Ser. C,* **256,** 916 (1963).

7. T. Takano, Y. Sasada, and M. Kakudo, *Acta Cryst.*, **21,** 514 (1966).

8. M. S. Brown and H. Rapoport, *J. Org. Chem.*, **28,** 3261 (1963).

9. S. Yamada and Y. Kikugawa, *Chem. Ind.* (London), 2169 (1966).

0. G. Queguiner and P. Pastour, *C. R. Acad. Sci., Paris, Ser. C,* **258,** 5903 (1964).

1. H. Lund, *Acta Chem. Scand.*, **17,** 972 (1963).

2. H. Lund, *Acta Chem. Scand.*, **17,** 2325 (1963).

3. S. L. Mukherjee and G. M. Shah, Indian patent, 72,814 (1963); *Chem. Abstr.*, **60,** 2890d (1964).

4. J. F. Bjellmann and H. Callot, *Bull. Soc. Chim. Fr.*, 1154 (1968).

5. M. Freifelder, R. M. Robinson, and G. R. Stone, *J. Org. Chem.*, **27,** 284 (1962).

6. M. Freifelder, U.S. patent, 3,159,639 (1964); *Chem. Abstr.*, **62,** 7732c (1965).

7. G. von Schuckmann and O. Weissel, Ger. patent, 1,233,870 (1967); *Chem. Abstr.*, **67,** 43682z (1967).

8. H. Fukawa and H. Kurihara, Jap. patent, 9091 (1962); *Chem. Abstr.*, **59,** 5140b (1963).

9. B. Benke, S. Jager, L. Szabo, I. Koczka, G. Losonczi, and I. Hoffmann, Hung. patent, 150,277 (1963); *Chem. Abstr.*, **60,** 2909c (1964).

0. R. Schliessel, L. Szabo, S. Jager, and M. Pogany, Hung. patent, 150,349 (1963); *Chem. Abstr.*, **60,** 2909b (1964).

1. F. E. Cislak and W. H. Rieger, Fr. patent, 1,390,118, (1965); *Chem. Abstr.*, **62,** 16208b (1965).

2. Z. D. Tadic and M. D. Muskatirovic, *Glasnik Hem. Drustva, Beograd,* **25–26,** 491 (1960–1961); *Chem. Abstr.*, **59,** 6358b (1963).

183. K. Isagawa, M. Kawai, and Y. Fushizaki, *Nippon Kagaku Zasshi,* **88,** 553 (196 *Chem. Abstr.,* **68,** 68840h (1968).
184. L. Thunus and M. Dejardin-Duchene, *J. Pharm. Belg.,* **24,** 3 (1969); *Chem. Abstr.,* 81100y (1969).
185. D. M. Dimitrijevic, Z. D. Tadic, and M. D. Muskatirovic, *Glasnik Hem. Drust Beograd,* **28,** 83 (1963); *Chem. Abstr.,* **60,** 14346g (1964).
186. C. Musante and S. Fatutta, *Ann. Chim.* (Rome), **47,** 385 (1957).
187. V. E. Blokhin, Z. Yu. Kokoshko, L. V. Kireeva, and S. Kirova, *Khim. Geterots Soedin.* (4), 744 (1969); *Chem. Abstr.,* **72,** 90227t (1970).
188. L. W. Clark, *J. Phys. Chem.,* **69,** 2277 (1965).
189. A. Kaneda and T. Hara, *Doshisha Daigaku Rikogaku Kenkyu Hokoku,* **7,** 161 (196 *Chem. Abstr.,* **67,** 90244t (1967).
190. J. Biakel, *Bull. Acad. Polon. Sci., Ser. Sci. Chim.,* **10,** 621 (1962).
191. J. Bialek, *Bull. Acad. Polon. Sci., Ser. Sci. Chim.,* **10,** 625 (1962).
192. M. Zelinskii, *Tr. po. Khim. i Khim. Tekhnol.,* No. 4, 707 (1961); *Chem. Abstr.,* 9670d (1963).
193. J. A. Zoltewicz, C. L. Smith, and J. D. Meyer, *Tetrahedron,* **24,** 2269 (1968).
194. P. Haake and J. Mantecon, *J. Amer. Chem. Soc.,* **86,** 5230 (1964).
195. K. Uda, A. Sakurai, and K. Sakabibara, Jap. patent, 20,555 (1965); *Chem. Abstr.,* 2069c (1966).
196. D. Palm, A. A. Smucker, and E. E. Snell, *J. Org. Chem.,* **32,** 826 (1967).
197. C. Azuma and A. Sugimori, *Kogyo Kagaku Zasshi,* **72,** 239 (1969); *Chem. Abstr.,* 96575k (1969).
198. W. Hoefling, D. Eilhauer, and G. Reckling, Ger. patent, 1,189,995 (1965); *Che Abstr.,* **63,** 583b (1965).
199. G. Ya. Kondrat'eva and C. -H. Huang, *Zhur. Priklad. Khim.,* **35,** 199 (1962); *Che Abstr.,* **56,** 14230b (1962).
200. P. Mueller and R. Trefzer, U.S. patent, 3,026,324 (1962); *Chem. Abstr.,* **57,** 111 (1962).
201. R. M. Gipson, F. H. Pettit, C. G. Skinner, and W. Shive, *J. Org. Chem.,* **28,** 14 (1963).
202. D. Taub, C. H. Kuo, and N. L. Wendler, *J. Chem. Soc., C,* 1558 (1967).
203. N. L. Wendler, D. Taub, and C. H. Kuo, U.S. patent, 3,441,568 (1969); *Chem. Abs* **71,** 81191d (1969).
204. N. L. Wendler, D. Taub, and C. H. Kuo, U.S. patent, 3,435,044 (1969); *Chem. Abs* **71,** 38820w (1969).
205. Z. Blaszkowska and H. Grochowska, *Przemysl. Chem.,* **45,** 145 (1966); *Chem. Abs* **64,** 19547f (1966).
206. E. F. Kozlova, M. I. Kustanovich, M. M. Yanina, and I. B. Chekmareva, *Khim. Fa Zh.,* **2**(7), 28 (1968); *Chem. Abstr.,* **70,** 3772j (1969).
207. I. B. Chekmareva, E. S. Zhdanovich, T. S. Novopokrovskaya, and N. A. Preobrazl skii, *Zh. Prikl. Khim.,* **35,** 1157 (1962); *Chem. Abstr.,* **57,** 8546c (1962).
208. N. V. Dormidontova, B. F. Ustavshchikov, M. I. Farberov, and L. M. Malinnikova, *Prikl. Khim.,* **42,** 666 (1969); *Chem. Abstr.,* **71,** 38746b (1969).
209. E. S. Zhdanovich, E. B. Chekmareva, T. S. Novopokrovskaya, and N. A. Preobrazl skii, *Zh. Obshch. Khim.,* **32,** 2828 (1962).
210. A. Pelka, L. Grabowski, S. Poradowski, and T. Kosinski, Polish patent, 43,588 (19 *Chem. Abstr.,* **57,** 13736g (1962).
211. R. P. Houghton and C. S. Williams, *Tetrahedron Lett.,* 3929 (1967).

(a) J. Palecek and J. Kuthan, Czech. patent, 125,165 (1967); *Chem. Abstr.*, **69,** 96476h (1968); (b) *Ibid.*, Czech. patent, 125,164 (1967); *Chem. Abstr.*, **69,** 96475g (1968).

G. Favini, *Rend. Ist. Lombardo Sci.*, Pt. I, **91,** 162 (1957); *Chem. Abstr.*, **52,** 11539e (1958).

A. Agren, G. Ekenved, S. O. Nilsson, and E. Svensjo, *Acta Pharm. Suecica,* **2,** 421 (1965); *Chem. Abstr.*, **64,** 9532c (1966).

H. Hjedo, *Acta Chem. Scand.*, **17,** 2351 (1963).

W. Ried and G. Neidhardt, *Ann. Chem.*, **666,** 148 (1963).

W. Kirsten and H. Schulz, East Ger. patent, 30,872 (1965); *Chem. Abstr.*, **64,** 8152g (1966).

M. G. Gal'pern and E. A. Luk'yanets, *Zh. Vses. Khim. Obshchest.*, **12,** 474 (1967); *Chem. Abstr.*, **68,** 2789s (1968).

G. F. Holland and J. N. Pereira, *J. Med. Chem.*, **10,** 149 (1967).

T. Kametani, S. Takano, O. Umezawa, H. Agui, K. Kanno, Y. Konno, F. Sato, H. Nemoto, K. Yamaki, and H. Ueno, *Yakugaku Zasshi,* **86,** 823 (1966); *Chem. Abstr.*, **65,** 20093h (1966).

R. Camain-Giabicani and A. Broche, *Bull. Soc. Chim. Fr.*, 1254 (1964).

J. Syedel, *Tetrahedron Lett.*, 1145 (1966).

F. E. Cislak, U.S. patent, 3,045,024 (1962); *Chem. Abstr.*, **57,** 15080e (1962).

Rhône-Poulenc S.A., Fr. patent, M1861 (1963); *Chem. Abstr.*, **60,** 2904h (1964).

W. Schaeffer and R. Wegler, Ger. patent, 1,149,356 (1963); *Chem. Abstr.*, **59,** 11441f (1963).

H. Bojarska-Dahlig and P. Nantka-Namirski, *Rocz. Chem.*, **30,** 621 (1956).

W. Hoefling, D. Eilhauer, and H. Reckling, East Ger. patent, 33,621 (1964); *Chem. Abstr.*, **63,** 11518g (1965).

T. Okamoto and H. Takahashi, Jap. patent, 70 03,380 (1970); *Chem. Abstr.*, **72,** 111308n (1970).

E. A. Pavlov, V. A. Serazetdinova, A. D. Kagarlitskii, and B. V. Surorov, *Khim. Geterotsikl. Soedin.*, **4,** 665 (1968); *Chem. Abstr.*, **70,** 28778d (1969).

T. Hirakata, S. Kubota, and T. Akita, *Yakugaku Zasshi,* **77,** 219 (1957); *Chem. Abstr.*, **51,** 11341i (1957).

H. Quast and E. Schmitt, *Ann. Chem.*, **732,** 64 (1970).

H. Quast and E. Schmitt, *Ann. Chem.*, **732,** 43 (1970).

E. Wenkert and G. D. Reynolds, *Aust. J. Chem.*, **22,** 1325 (1969).

M. Winn, D. A. Dunningan, and H. E. Zaugg, *J. Org. Chem.*, **33,** 2388 (1968).

J. Delarge, *Pharm. Acta Helv.*, **44,** 637 (1969); *Chem. Abstr.*, **71,** 112761w (1969).

L. Neilands and G. Vanags, *Latvijas PSR Zinatnu Akad. Vestis, Kim. Ser.* (2), 203 (1964); *Chem. Abstr.*, **61,** 6999d (1964).

Y. Omote, K. -T. Kuo, and N. Sugiyama, *Bull. Chem. Soc. Jap.*, **40,** 1695 (1967).

H. Sliwa, *C. R. Acad. Sci., Paris, Ser. C,* **264,** 1893 (1967).

M. Yokote, F. Shibamiya, and S. Tokairin, *Kogyo Kagaku Zasshi* **67,** 166 (1964); *Chem. Abstr.*, **61,** 3235f (1964).

F. S. Babichev, L. A. Kirpianova, and T. A. Dashevskaya, *Ukr. Khim. Zh.*, **32,** 706 (1966); *Chem. Abstr.*, **65,** 13681f (1966).

V. L. Sbarskii, G. M. Shutov, V. Zhilin, and E. Yu. Orlova, *Khim. Geterotsikl. Soedin.* (1), 178 (1967); *Chem. Abstr.*, **67,** 73500f (1967).

R. M. Kellogg, T. J. Van Bergen, and H. Wynberg, *Tetrahedron Lett.*, 5211 (1969).

A. M. Aliev and M. A. Salimov, *Aptechn. Delo,* **13,** 36 (1964).

J. B. Lee and T. G. Clarke, *Tetrahedron Lett.*, 415 (1967).

245. M. P. Cava, M. J. Mitchell, D. C. DeJongh, and R. Y. Van Fossen, *Tetrahedron Le*
2947 (1966).
246. H. Hartkamp, *Z. Anal. Chem.*, **190**, 66 (1962).
247. I. Morimoto and S. Tanaka, *Nippon Kagaku Zasshi*, **83**, 357 (1962); *Chem Abstr.*,
5033f (1963).
248. J. F. Biellmann, R. J. Highet, and M. P. Goeldner, *J. Chem. Soc., D*, 295 (1970).
249. Fr. patent, 1,393,092 (1965); *Chem. Abstr.*, **63**, 1774e (1965).
250. H. M. Wuest, J. A. Bigot, Th. J. de Boer, B. van der Wal, and J. P. Wibaut, *Rec. Tr*
*Chim. Pays-Bas*, **78**, 226 (1959).
251. C. J. Argoudelis and F. A. Kummerow, *J. Org. Chem.*, **26**, 3420 (1961).
252. W. Steinke, East Ger. patent, 23,754 (1962); *Chem. Abstr.*, **59**, 8713b (1963).
253. J. T. Dunn and D. L. Heywood, U.S. patent, 3,048,624 (1962); *Chem. Abstr.*,
6369f (1963).
254. S. Mizukami, E. Hirai, and M. Morimoto, *Shionogi Kenkyusho Nempo*, **16**,
(1966); *Chem. Abstr.*, **66**, 10827q (1967).
255. J. Delarge and L. Thunus, *Farmaco, Ed. Sci.*, **21**, 846 (1966); *Chem. Abstr.*,
75889t (1967).
256. M. Szafran and Z. Sarbak, *Rocz. Chem.*, **43**, 309 (1969).
257. M. Szafran and B. Brezinski, *Rocz, Chem.*, **43**, 653 (1969).
258. B. M. Bain and J. E. Saxton, *J. Chem. Soc.*, 5216 (1961).
259. H. J. Rimek, *Ann. Chem.*, **670**, 69 (1963).
260. M. Celadnik, L. Novacek, and K. Palat, *Chem. Zvesti*, **21**, 109 (1967); *Chem. Abs*
**67**, 64196j (1967).
261. E. Profft and W. Steinke, *J. Prakt. Chem.*, **13**, 58 (1961).
262. T. Wieland and H. Biener, *Chem. Ber.*, **96**, 266 (1963).
263. Z. D. Tadic, M. M. Misic, and D. M. Dimitrijevic, *Glasnik Hem. Drustva, Beograd*,
407 (1962); *Chem. Abstr.*, **60**, 15698f (1964).

# CHAPTER XI

# Pyridine Side-Chain Carboxylic Acids

MARY E. NEUBERT

*Liquid Crystal Institute*
*Kent State University*
*Kent, Ohio*

The area of pyridine side-chain carboxylic acids continues to be a rapi expanding segment of pyridine chemistry. In addition to the general chem interests in these compounds, which are discussed throughout this chapter, m have been investigated for a wide variety of practical applications. They h been employed in the synthesis of natural products such as hydroxycotinir nicotine;[2] a variety of isoquinoline,[3-8] lupine,[9] and indole[10-13] alkaloids; *DL*-castoramins;[14] porphobilinogen;[15] and flavonoids.[16] Structural determina studies of dioscorine,[17] retamine,[18, 19] and evonine[20] involved the use pyridine side-chain acids. Wilfordic and hydroxywilfordic acids have been sh to be pyridine side-chain acids.[21] Pyridine side-chain acids are also formed in metabolism of nicotine[22, 23] and have been employed in metabolic stu performed with cotinine.[24] Many have been investigated as therapeutic ag such as antibacterial and antifungal,[25-35] antiviral,[36] anti-inflammatory,[3 antifertility,[41] hypoglycemic,[42] hypocholesterolemic,[41-43] and eurh mic[44, 45] agents. Additionally, they have been examined as CNS dep sants[46-49] and stimulants,[50] analgesics,[51] antispasmodics,[52-55] analeptics hypotensors,[56, 57] monoamine oxidase inhibitors,[58] and corticosuprar inhibitors.[59] A few have been studied as bactericides and fungicides,[6 pesticides,[64] insecticides,[65] solubilizing agents,[66] and sensitizers[67] desensitizers in photographic emulsions.[68]

# I. Preparations

## 1. From Nonpyridine Starting Materials (Table XI-1)

everal esters (**XI-3**) were obtained from a Diels-Alder cycloaddition between oxazole **XI-1** and the activated olefins **XI-2**.[69]

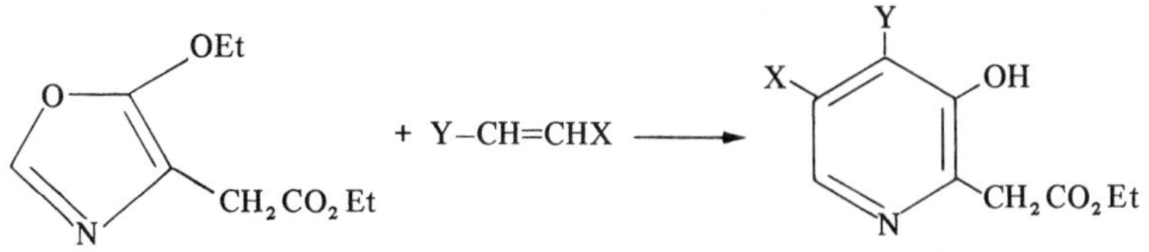

+ Y–CH=CHX

**XI-1** **XI-2** **XI-3**

X = Y = $CO_2Et$
X = Y = CN
X = Y = $CH_2OMe$
X = CN, Y = $CH_2OAc$

hen the proper conditions were employed, the 3*H*-azepine (**XI-4**) was verted to the pyridine ester **XI-5** in good yields.[70]

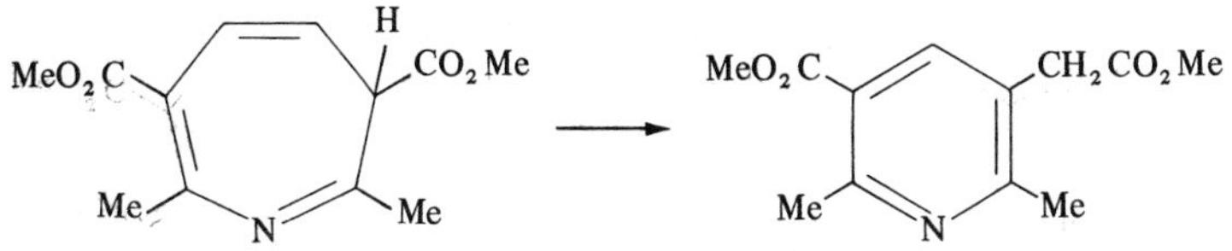

**XI-4** **XI-5**

he 2-pyridylalanine (**XI-7**) was synthesized from the pyrone (**XI-6**)[71] and

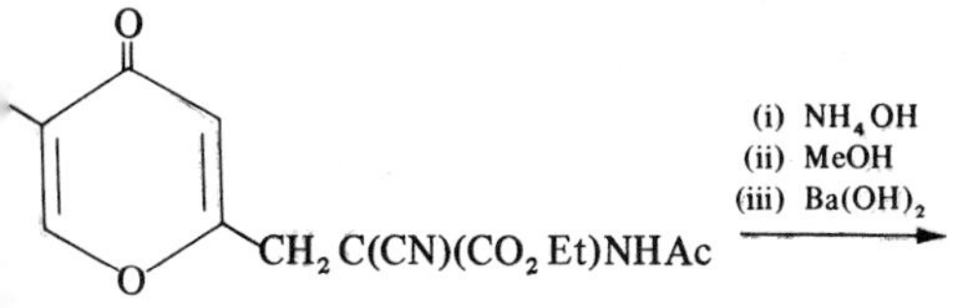

(i) $NH_4OH$
(ii) MeOH
(iii) $Ba(OH)_2$

**XI-6**

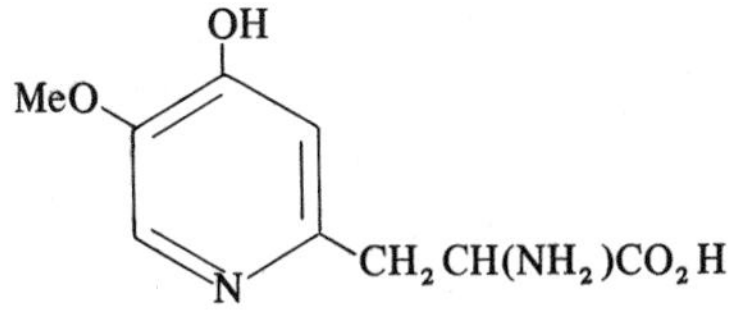

**XI-7**

treatment of the isoquinoline **XI-8** with benzenesulfonyl chloride produced t 4-pyridylacrylic acid **XI-9**.[72]

$$\textbf{XI-8} + PhSO_2Cl \longrightarrow \textbf{XI-9}$$

The remaining syntheses of pyridine side-chain acids from nonpyridi precursors involves typical condensation reactions. The preparations of ac containing a partially reduced pyridine nucleus were performed either throu condensation reactions or by the treatment of pyrones with amines.

## 2. From Pyridine Starting Materials

### A. *Side-Chain Oxidation* (Table XI-2)

Side-chain acids have been obtained by employing the following oxidizi agents on a variety of functional groups:

| | | | |
|---|---|---|---|
| (i) Chromic Acid | $-CH_2OH$ | $\longrightarrow$ | $-CO_2R$ |
| (ii) $AgNO_3$ | $-CHO$ | $\longrightarrow$ | $-CO_2H$ |
| (iii) $SeO_2$ | $-COMe$ | $\longrightarrow$ | $-COCO_2H$ |
| (iv) $KMnO_4$ | $-CH{=}CH_2$ | $\longrightarrow$ | $-CO_2H$ |
| (v) $MnO_2$ | $-COCO_2Et$ | $\longrightarrow$ | $-CO_2H$ |

In three instances, enzymes were used as the oxidizing agents.[73, 74] Oxidation *l*-anabasine **(XI-10)** with 10% hydrogen peroxide under mild conditions caus cleavage of the piperidine ring to form the acid **XI-11**.[75] When 34% hydrog peroxide in acetic acid was employed at a higher temperature, howev oxidation of the pyridine nitrogen atom occurred to form the *N*-oxide a **(XI-12)**. This latter acid yielded the *dl*-lactam **(XI-13)** on reduction with zinc acetic acid. The *dl*-lactam was then converted to *dl*-anabasine by reduction w

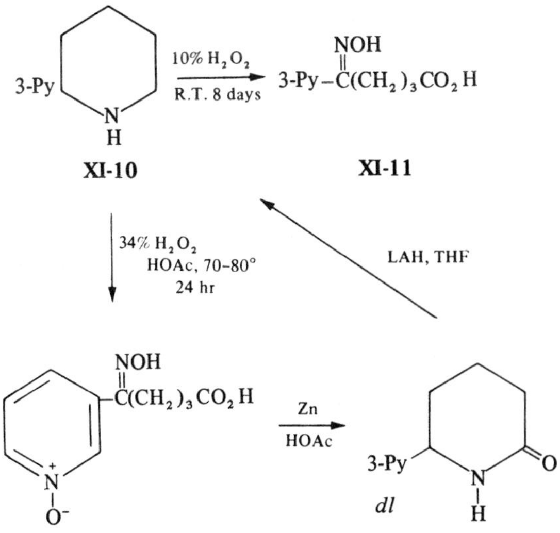

3-Py-piperidine (XI-10) —[$10\%\ H_2O_2$, R.T. 8 days]→ $3\text{-Py}-C(=NOH)(CH_2)_3CO_2H$ (XI-11)

XI-10 —[$34\%\ H_2O_2$, HOAc, 70–80°, 24 hr]→ XI-12

XI-12: pyridine N-oxide with $-C(=NOH)(CH_2)_3CO_2H$ —[Zn, HOAc]→ XI-13 (3-Py lactam, *dl*)

XI-13 —[LAH, THF]→ XI-11

**XI-10** **XI-11** **XI-12** **XI-13**

ium aluminum hydride in tetrahydrofuran. The *l*-isomer of the lactam (-13) was synthesized by using the following reaction scheme:

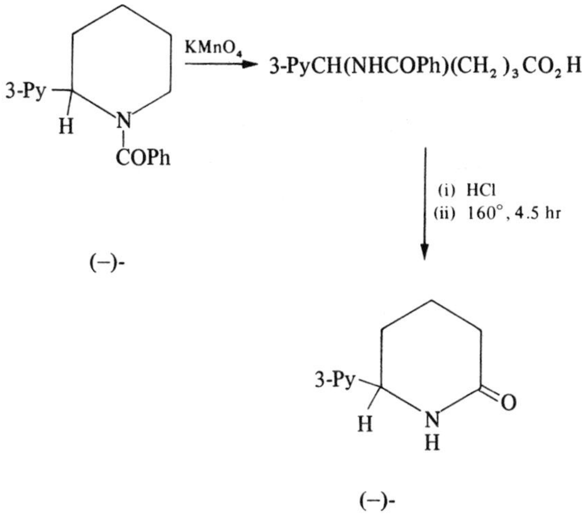

(−)-3-Py-N-COPh-piperidine —[$KMnO_4$]→ $3\text{-PyCH(NHCOPh)}(CH_2)_3CO_2H$ —[(i) HCl (ii) 160°, 4.5 hr]→ (−)-lactam

(−)- (−)-

one instance,[76] a dihydropyridine acetic acid was oxidized by oxygen to the ridylacetic acid.

## B. *Carbonation of Organometallic Compounds* (Table XI-3)

The carbonation of side-chain metallated lutidines has been investigated.[77,
As expected, the lithium derivatives of 2,3- and 2,5-lutidine were carbonat
only at the C-2 methyl group. The addition of carbon dioxide to the side-ch
lithium or sodium derivatives of 2,4- and 2,6-lutidine yielded primarily the (
acetic acids. In one instance,[77] the 2,6-diacetic acid was also isolated. T
sodium salts of pyridylacetylenes were converted to the corresponding acids
good yields when treated with carbon dioxide.[79-81]

## C. *Increase of Chain Length by the Arndt-Eistert Method* (Table XI-4)

Only two examples employing this method were found.

## D. *Condensations of Halogenated Pyridines with Active Methylene Compounds* (Table XI-5)

Although many phenylacetamides were successfully alkylated with 2-brom
pyridine, the corresponding thioamides were not.[82] The pyrrolidone **XI-14** w
employed as the active methylene compound in a condensation wi
2-bromopyridines.[50] Treatment of quaternary salts of halo- or alkoxy-pyridin

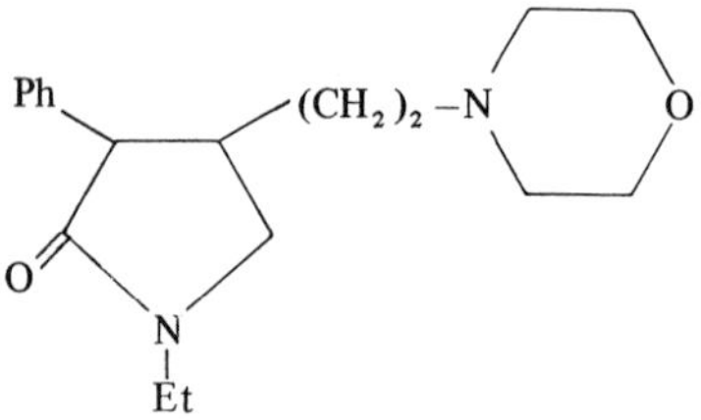

Ph
$(CH_2)_2$–N
O
O
N
Et

**XI-14**

(**XI-15** and **XI-17**) with an active methylene compound gave the anhydro bas
**XI-16** and **XI-18**. The electron-attracting substituents stabilized the anhyd
bases.

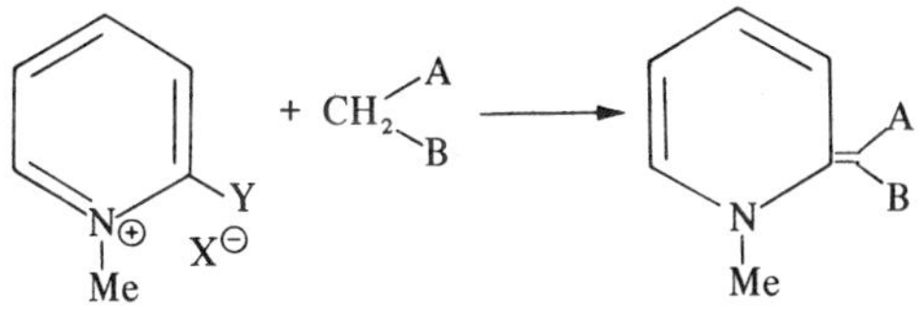

XI-15
Y = Br, Cl, or OPh

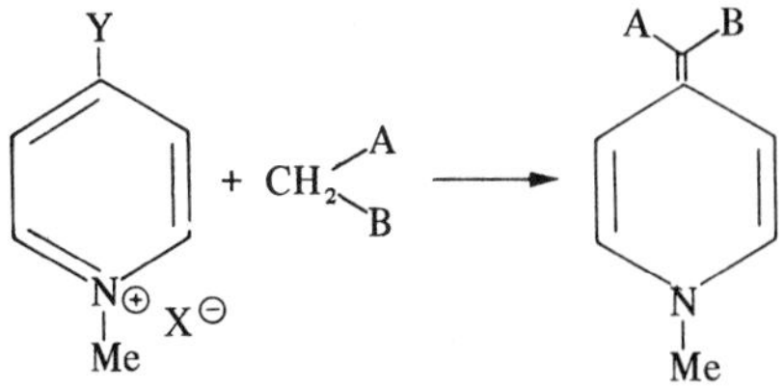

XI-16

XI-17
Y = OMe, OPh

XI-18

Attempts to hydrolyze the ester group in the *o*-nitro-2- and 4-pyridyl acetates I-19, synthesized by treating the corresponding pyridyl halide with a malonic cid derivative, were unsuccessful using either basic or acidic conditions.[84] The

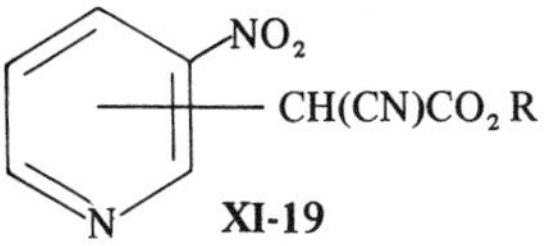

XI-19

enzyl esters (**XI-19**, R = $CH_2Ph$) also did not undergo hydrogenolysis when 0% Pd-C, 30% Pd-C, Ra-Ni, or $PtO_2$ were employed as catalysts. Only reduction f the nitro group to the amine occurred.

## E. *Condensations of Picolines and Related Compounds*

a. *Condensation of Chloral with Picolines (Table XI-6) and Hydrolysis of the Products Table XI-7)*

b. *Condensations with Picolines to Give Side-Chain Acids and Esters (Table XI-8) (see lso Chapter IV, Section V.1.)* The previously unknown ethyl 2- and 4-pyridyl yruvates (**XI-20**) have been synthesized, although only in 10% yields.[85, 86] he C-2 isomer was prepared by treating the lithium derivative of 2-picoline vith diethyl oxalate in the presence of a mixture of the cadmium salts, 2-$PyCH_2)_2Cd$ and $CdCl_2$ (in Chapter XI of the 1964 edition, attempts to

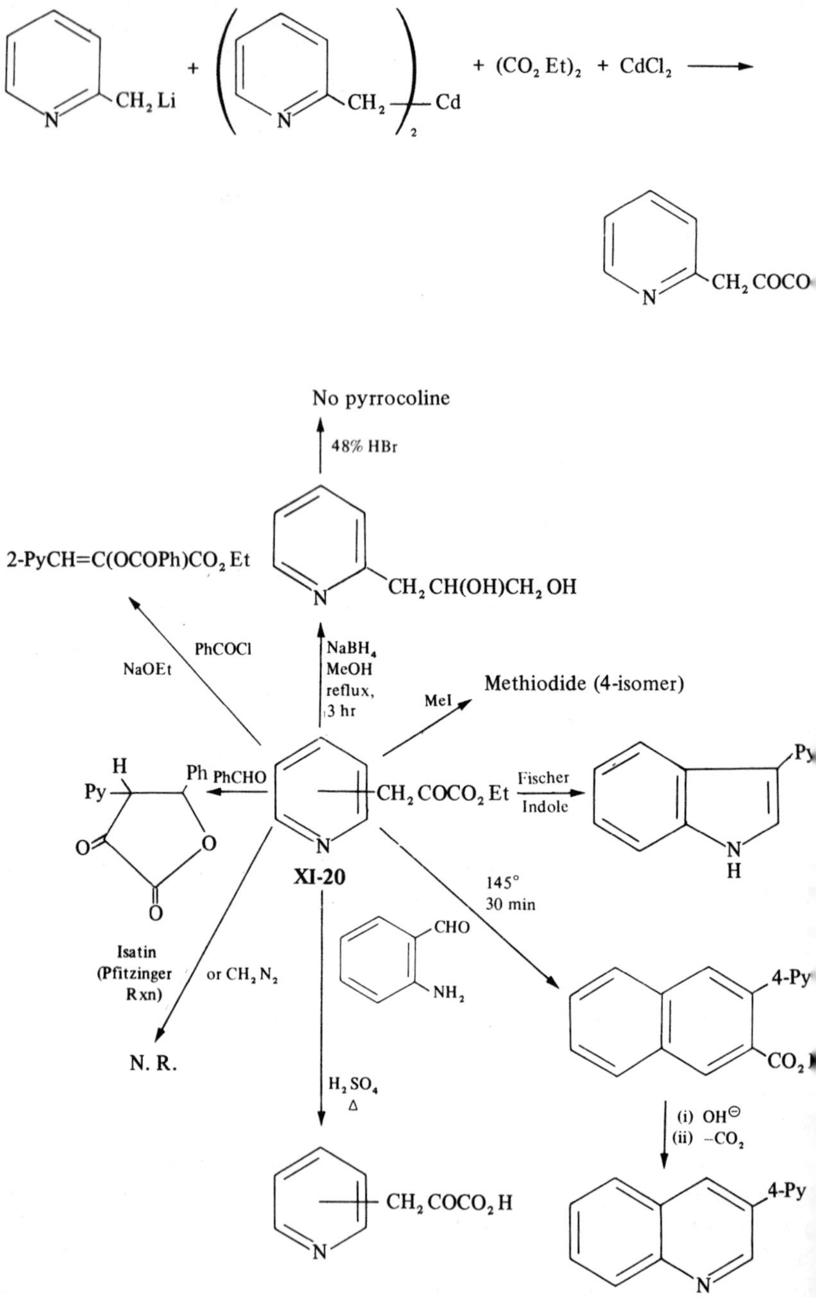

$CH_2Li$ + ( $CH_2$ )$_2$Cd + $(CO_2Et)_2$ + $CdCl_2$ ⟶

$CH_2COCO$

No pyrrocoline

48% HBr

2-PyCH=C(OCOPh)$CO_2$Et

$CH_2CH(OH)CH_2OH$

PhCOCl

NaOEt

$NaBH_4$
MeOH
reflux,
3 hr

MeI

Methiodide (4-isomer)

H

Py

Ph PhCHO

O

O

O

$CH_2COCO_2Et$

Fischer
Indole

Py

N
H

XI-20

145°
30 min

CHO

$NH_2$

Isatin
(Pfitzinger
Rxn)

or $CH_2N_2$

4-Py

$CO_2$

N. R.

$H_2SO_4$
Δ

(i) $OH^{\ominus}$
(ii) $-CO_2$

4-Py

$CH_2COCO_2H$

Scheme XI-1

nthesize these pyruvates using diethyl oxalate and base were reported to give ubstitution). In the synthesis of the 4-isomer, the mercury salts, $(PyCH_2)_2Hg$ and $HgCl_2$ had to be employed. These pyruvates were reported undergo the reactions illustrated in Scheme XI-1.

Pyridylpyruvic acids have been synthesized in good yields by treating several olines containing electron withdrawing substituents with oxalyl chloride in e presence of phosphoryl chloride.[87] These pyruvates exist primarily in the ol form with hydrogen bonding between the enol hydroxyl group and the ridine nitrogen atom **(XI-21)**. This probably explains the reluctance of the ruvate **XI-21** to undergo decarbonylation to the acetate.[17] Pyruvates also have

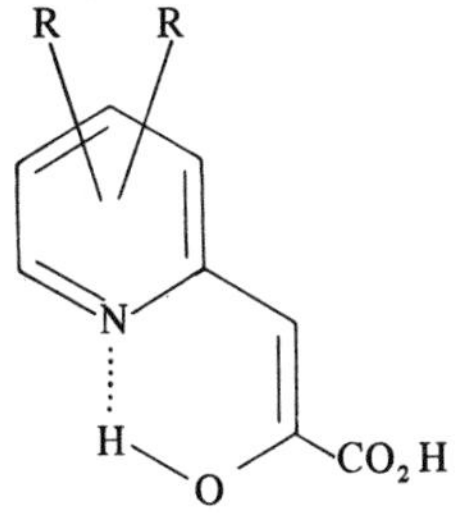

**XI-21**

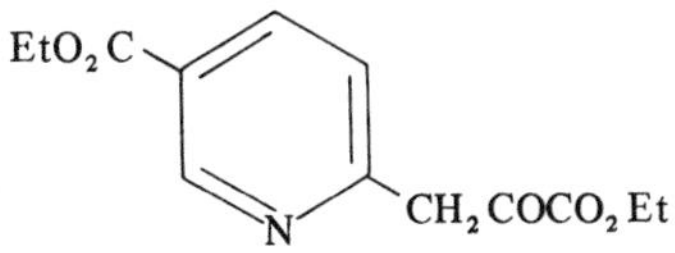

**XI-22**

en reported to be formed by treatment of 2-picoline-1-oxides with diethyl alate in the presence of base.[135, 139]

Although a picoline was usually converted to a pyridine acetate when treated th ethyl carbonate under basic conditions, ethyl 6-methylnicotinate **(XI-23)** ve the ketone **XI-24** instead.[17]

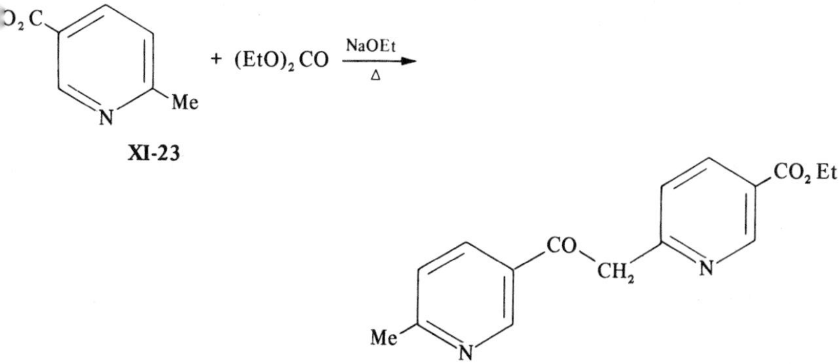

**XI-23**

**XI-24**

## F. *Condensations of Vinylpyridines with Esters* (Table XI-9)

Two syntheses of α-amino-γ-(2- and 4-pyridyl)butyric acids employing t[h]
Michael addition have been reported. In the one instance, 2-vinylpyridine w[as]
allowed to react with the activated ester **XI-25** and the resulting product w[as]
hydrolyzed to give the amino acid ester **(XI-26)**.[89] The second synthesis involv[es]

$$\underset{\textbf{XI-25}}{\text{2-PyCH=CH}_2 + \text{AcNHCH(CN)CO}_2\text{Et}} \xrightarrow[\text{(ii) HCl}]{\text{(i) NaOEt}} \underset{\textbf{X-26}}{\text{2-Py(CH}_2)_2\underset{\text{NH}_2}{\text{CH}}\text{CO}_2\text{Et}}$$

the condensation of 4-vinylpyridine with the malonic ester **(XI-27)** on a bas[ic]
ion-exchange resin. Acid treatment of the product yielded the amino ac[id]
**XI-28** isolated as the hydrochloride.[30]

$$\underset{\textbf{XI-27}}{\text{4-PyCH=CH}_2 + \text{AcNHCH(CO}_2\text{Et)}_2} \xrightarrow[\text{(ii) HCl}]{\text{(i) ion exchange OH}^{\ominus}} \underset{\textbf{XI-28}}{\text{4-Py(CH}_2)_2\underset{\text{NH}_2}{\text{CH}}\text{CO}_2\text{H}}$$

An electrolytic condensation of 2-vinylpyridine with the activated olef[in]
diethyl fumarate **(XI-29)** produced the diester **XI-30**.[90] The Michael addition [of]

$$\text{2-PyCH=CH}_2 + \underset{\textbf{XI-29}}{(\text{H})(\text{EtO}_2\text{C})\text{C=C}(\text{CO}_2\text{Et})(\text{H})} \xrightarrow{e^{\ominus}} \underset{\textbf{XI-30}}{\text{2-Py(CH}_2)_2\underset{\text{CO}_2\text{Et}}{\text{CH}}\text{CH}_2\text{CO}_2\text{Et}}$$

diethylmalonate to compound **XI-31** occurred in the normal manner, except th[at]
the α-acetate group was lost to yield the olefin **XI-32**, probably by the mechanis[m]
shown.[14] Hydrolysis of the acetate function also occurred.

Barbituric acids have been employed as the active methylene compounds i[n]
the Michael addition to 4-vinylpyridine to yield the barbiturates **XI-33**.[91] [A]
variety of 3-vinylpyridyl ketones of the type **XI-34** were treated with este[rs]
containing active methylene groups to form the δ-keto esters **(XI-35)**.[208]

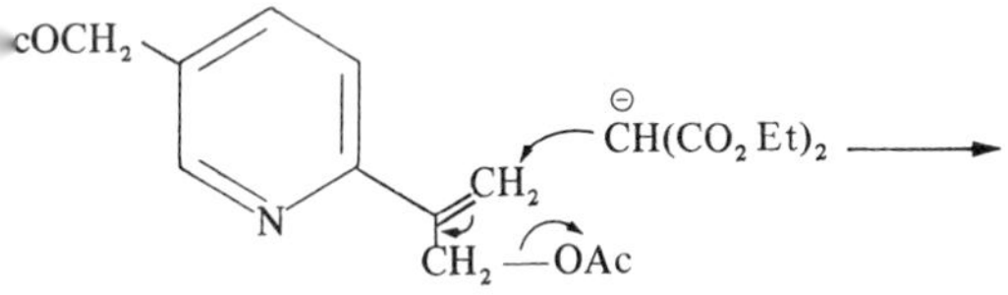

$AcOCH_2$ … $\overset{\ominus}{C}H(CO_2Et)_2$ →

$CH_2$ $CH_2$—OAc

**XI-31**

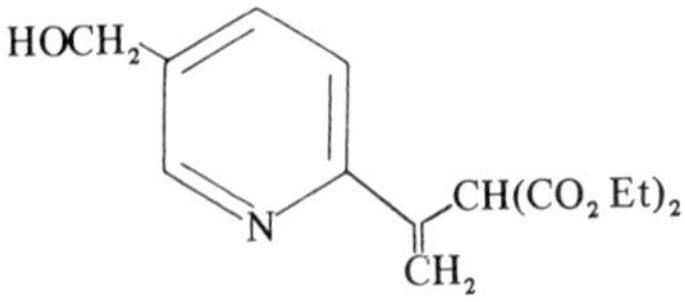

$HOCH_2$ … $CH(CO_2Et)_2$ $CH_2$

36%

**XI-32**

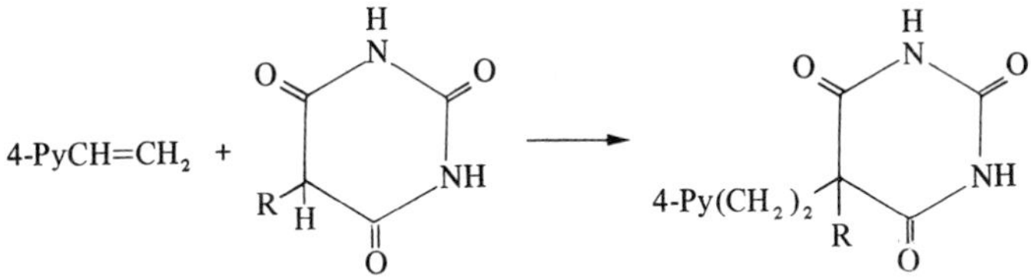

4-PyCH=CH$_2$ + … →  4-Py(CH$_2$)$_2$ …

R = Et, Ph

**XI-33**

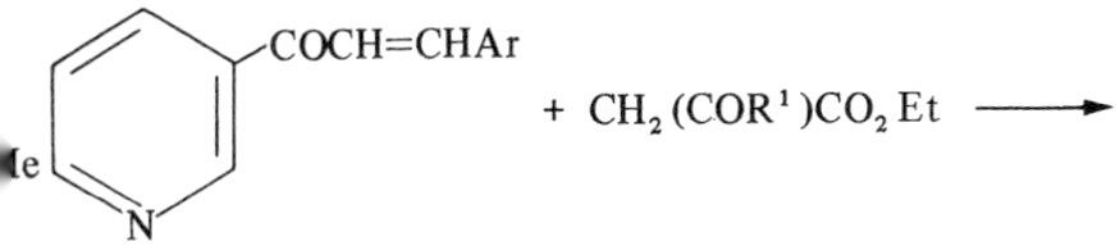

COCH=CHAr + $CH_2(COR^1)CO_2Et$ →

**XI-34**

Me … $COCH_2CHArCH(COR^1)CO_2Et$

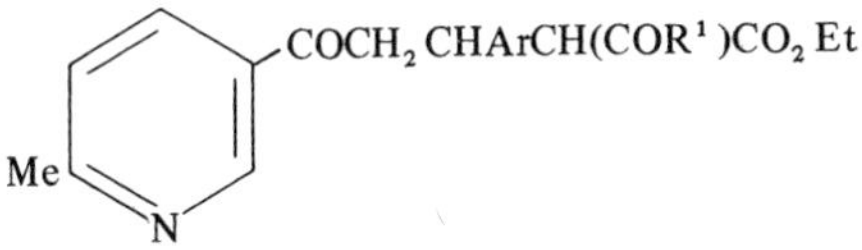

**XI-35**

## G. *Condensations of Pyridinealdehydes and Ketones* (Table XI-10)

The use of acylaminoacetates as the active methylene compounds in condensations with pyridinealdehydes and -ketones has produced a variety of pyridineacetic acids. When monoethylacetamidomalonate was employed in t[he] presence of base, *N*-acetyl-β-pyridylserines **(XI-36)** were formed.[93] Hippu[ric] acid, when allowed to react with these aldehydes in the presence of sulf[ur] trioxide yielded azlactones **(XI-37)** that were converted either to α-aminopr[o]pionic acids **(XI-38)** or to acrylic acids **(XI-39)**.[94] In these instances, the aceta[ls] of the pyridinealdehydes had to be used to avoid the formation of ta[rs]. α-Amino-β-(2-pyridyl)propionic acid (**XI-38**, 2-isomer) was also synthesized fro[m] the hydantoin **(XI-40)**.[95] Nitrosoamino esters[96] and alkynes[79, 97] have al[so]



been employed as the active methylene compounds.

The stereochemistry of the β-carboxystilbazoles formed in a Perki[n] condensation of pyridinealdehydes with arylacetic acids has been shown to b[e]

(**XI-41**)[98-100] rather than *trans* as reported earlier.[10] The *cis* stereochemistry ·s shown to be correct by the following transformations:

3-PyCHO + $PhCH_2CO_2H$ —(i) NaOH/EtOH (ii) $Ac_2O$ reflux 2 hr→ 3-Py(H)C=C(Ph)($CO_2H$) **XI-41** —copper chromite, quinoline 250°→ 3-Py(H)C=C(Ph)(H) **XI-42**

**XI-42** —$I_2$, $PhNO_2$ reflux, 20 min→ 3-Py(H)C=C(H)(Ph)

3-Py($HO_2C$)C=C(Ph)(H) —copper chromite, quinoline 230°→ **XI-42**

ıis stereochemistry is the opposite of that found in the β-cyanostibazoles **I-43)** obtained from a Knoevenagel condensation of arylacetonitriles with ridinealdehydes (see Table XI-25). Unfortunately, these results do not provide ntheses for both the *cis-* (**XI-41**) and the *trans-* (**XI-45**) carboxystilbazoles cause, although the nitriles can be hydrolyzed to the *trans* amides (**XI-44**), ese amides isomerize when hydrolyzed further to yield the *cis* acids.[98]

3-PyCHO + $PhCH_2CN$ —NaOEt, EtOH 50°→ 3-Py(H)C=C(CN)(Ph) **XI-43** —85% $H_2SO_4$, steam-bath, 3 hr→ 3-Py(H)C=C($CONH_2$)(Ph) **XI-44**

**XI-44** —60% $H_2SO_4$, 100°, 3 hr→ 3-Py(H)C=C(Ph)($CO_2H$) **XI-41**

**XI-44** ↛ 3-Py(H)C=C($CO_2H$)(Ph) **XI-45**

## H. *Condensations of Pyridinecarboxylic Esters* (Table XI-11)

Some of the keto esters, obtained from condensation reactions with ridinecarboxylic esters as listed in Table XI-11, have been hydrolyzed and carboxylated,[11, 23, 75, 101-104] condensed with aldehydes in Aldol condensa- ·ns,[105, 106] and converted to barbiturates,[91] azomethine dyes,[107, 108] aryl- ·drazones,[176] and acylpyridines.[109]

Ethyl picolinate failed to undergo condensation with diethyl oxaloacetate **I-46)** (using Claisen conditions) to give the desired product **XI-47**.[110]

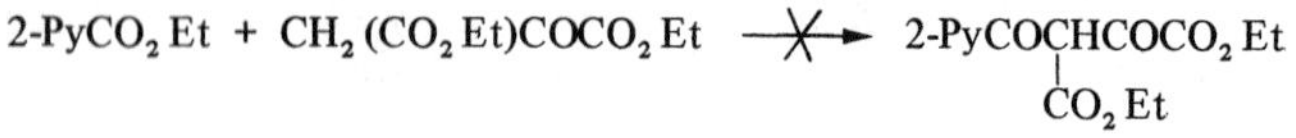

2-PyCO$_2$Et + CH$_2$(CO$_2$Et)COCO$_2$Et ↛ 2-PyCOCH(CO$_2$Et)COCO$_2$Et

XI-46 XI-47

## I. *Condensations of Pyridineacetic Esters* (Table XI-12)

Usually the condensation of ethyl pyridylacetates with aromatic aldehyd requires the presence of a base to form the required anion. When, however, t methiodide of the pyridine ester is employed no added base is needed. Quaternization of the pyridine nitrogen atom facilitates proton removal. T stereochemistry of the cinnamate isolated from the condensation of eth 2-pyridylacetate methiodide **(XI-48)** with benzaldehyde has been shown to *cis* **(XI-49)** whereas the product obtained by using the free base has the *tra* configuration **(XI-50)**.[111] On the other hand, the 4-isomer yielded only t *trans*-cinnamate in both instances. Apparently the steric crowding between t phenyl group and the quaternary nitrogen atom in the 2-pyridyl-*trans*-cinnama methiodide **(XI-51)** is too great to allow its formation, whereas in the 4-isom this crowding is absent.

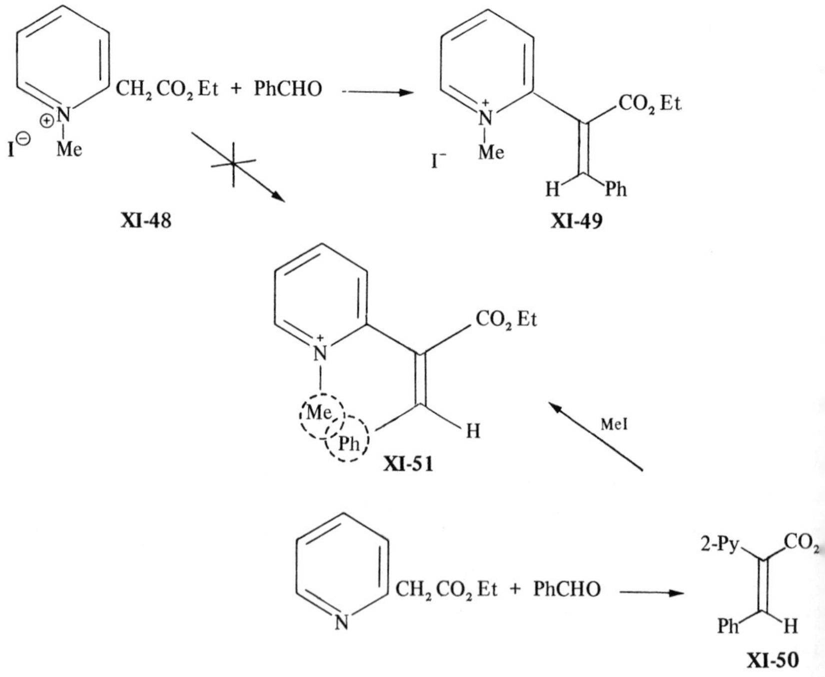

$I^{\ominus}$ N-Me pyridinium-2-CH$_2$CO$_2$Et + PhCHO → XI-49 ($I^-$; N$^+$-Me; C(CO$_2$Et)=CH(Ph), H and Ph)

XI-48 ↛ XI-51

XI-51 (N$^+$-Me; CO$_2$Et; H; Me / Ph crowding) ← MeI

Pyridine-2-CH$_2$CO$_2$Et + PhCHO → 2-Py(CO$_2$)C=C(Ph)H

XI-50

When the aromatic aldehyde contained an *ortho* hydroxyl or amino group, yclization to the coumarins or quinolines, respectively, occurred.[111] Attempts o condense ethyl 2- and 4-pyridylacetate (either as the free base or as the nethiodide) with phenyl methyl ketone or with *o*-hydroxyphenyl methyl etone were unsuccessful. However, the methiodide of the 4-isomer yielded a uinoline when treated with *o*-aminophenyl methyl ketone.

The condensation of ethyl 2-pyridylacetate with benzoyl chloride in the resence of base gave only the disubstituted product **(XI-52)** rather than the nonosubstituted derivative.[112] Acetyl chloride did not produce the correspondng diacetyl ester.

$$\text{2-PyC(PhCO)}_2\text{CO}_2\text{Et}$$

**XI-52**

everal Mannich-type reactions have been performed using 3- and 4-pyridylcetates to form the expected products in fair yields.[113]

## J. *Reduction of Side-Chain Functions* (Table XI-13)

Reduction of the double bond of β-(2-pyridyl)acrylates **(XI-53)** without educing the ester group in order to obtain 2-pyridylpropionates **(XI-54)** was ccomplished in good yields by using either a mixture of phosphorus with 57% ydriodic acid in acetic acid[114] or by hydrogenation over platinum oxide[114] or 0% palladium-on-carbon[34] in acetic acid. Treatment of the betaine of either the 2- or 4-pyridineacrylate **(XI-55)** with fused potassium formate in formic acid also

Py–CH=CHCO$_2$Et (**XI-53**) —[P, HOAc, 57% HI; or $H_2/PtO_2$, HOAc]→ Py–CH$_2$CH$_2$CO$_2$Et (**XI-54**)

**XI-53** —[MeI, acetone]→ N-Me-pyridinium–CH=CHCO$_2$Et $I^{\ominus}$ —[$Ag_2O$, $H_2O$]→ N-Me-pyridinium–CH=CHCO$_2^-$ (**XI-55**)

**XI-55** —[(i) 95% HCO$_2$H, HCO$_2$K, reflux, 10–13 hr; (ii) EtOH, HCl]→ **XI-54**

gave the propionates, but in poor yields, and accompanied by the correspondi piperidine propionates.[114]

The use of sodium borohydride in the reduction of the acrylate **XI-56** cause the formation of a mixture of products resulting from the reduction of both t ester and the double bond functions.[115] As expected, the composition of th mixture was depéndent on the amount of sodium borohydride employed.

$$\underset{\textbf{XI-56}}{4\text{-PyCH=CHCO}_2\text{Me}} + \text{NaBH}_4 \xrightarrow[\text{reflux, 1-2 hr}]{\text{MeOH}} \underset{\textbf{XI-57}}{4\text{-Py(CH}_2)_2\text{CO}_2\text{Me}} +$$

$$\underset{\textbf{XI-58}}{4\text{-PyCH=CHCO}_2\text{Me}} + \underset{\textbf{XI-59}}{4\text{-Py(CH}_2)_3\text{OH}} + \underset{\textbf{XI-60}}{4\text{-PyCH=CHCH}_2\text{OH}}$$

| Molar Ratio | | % Product in Mixture | | | |
|---|---|---|---|---|---|
| **XI-56** | $NaBH_4$ | **XI-57** | **XI-58** | **XI-59** | **XI-60** |
| 1 | 2 | 34 | 36 | 8 | 22 |
| 1 | 5 | 10 | 0 | 54 | 36 |
| 1 | 10 | 1 | 0 | 91 | 8 |

Sodium borohydride has also been used to reduce the carbonyl group to hydroxyl group in some pyridine keto acids[23] and esters[19, 116] without reducin the acid function. Reduction of the oxime of ethyl 4-oxo-4-(3-pyridyl)butyra in the presence of a lead catalyst yielded the amino acid **XI-61**.[2] Attempts reduce the carbonyl groups of the diketo ester **XI-62** failed in the presence Raney Nickel.[110]

$$3\text{-PyC(=NOH)CH}_2\text{CH}_2\text{CO}_2\text{Et} \longrightarrow \underset{\textbf{XI-61}}{3\text{-PyCH(NH}_2)\text{CH}_2\text{CH}_2\text{CO}_2\text{Et}} \qquad \underset{\textbf{XI-62}}{2\text{-PyCOCH}_2\text{COCO}_2\text{E}}$$

## K. *Willgerodt Reaction* (Table XI-14)

Only a few examples of the use of this reaction to synthesize pyridin side-chain acids were found. 4-Pyridyl propyl ketone was converted to th methyl ester of 4-pyridylbutyric acid.[117]

### 3. Miscellaneous Methods (Table XI-14-1)

Several methods that employ the opening of a variety of side-chain or fused yridine ring systems have been developed for the synthesis of side-chain acids d derivatives.[95, 102, 117–122] An interesting example is the ring opening of clohexanones **XI-63** and **XI-64**.[117, 119]

3- or 4-PyCO–(2-oxocyclohexyl, H) $\xrightarrow[\text{(ii) EtOH, HCl}]{\text{(i) reflux, HCl}}$ 3- or 4-PyCO$(CH_2)_5CO_2$Et

**XI-63**

3-nitro-4-(2-oxocyclohexyl)pyridine $\xrightarrow[\Delta]{\text{acid}}$ 4-$(CH_2)_5CO_2H$-3-$NO_2$-pyridine

**XI-64**

The α-keto ester **XI-65** was synthesized in good yields by treating 3-nicotinoyl loride with a Wittig reagent.[123, 124] Treatment of the oxime of

$$\text{3-PyCOCl} + Ph_3\overset{\oplus}{P}CH_2OMe\ Br^{\ominus} \text{ or } Ph_3P{=}CHOMe \xrightarrow[Et_2O]{PhLi} \text{3-PyCOC(OMe)=}PPh_3 \xrightarrow{PhI(OAc)_2} \text{3-PyCO}CO_2\text{Me}$$

**XI-65**

oxo-4-(3-pyridyl)butyric acid **(XI-66)** with phosphorus oxychloride yielded pyridylacetic acid.[22]

$$\text{3-Py}CH_2C(=NOH)CH_2CO_2H \xrightarrow{POCl_3} \text{3-Py}CH_2CO_2H$$

**XI-66**

## II. Properties and Reactions

### 1. Esterification and Ester Hydrolysis

Several methyl esters were synthesized by the treatment of pyridine aceti acids with diazomethane.[3, 102, 114, 125] A variety of C-21 steroidal alco hols[37-40, 126] and the following alcohols have been employed in esterificatio reactions with pyridine side-chain acids.[53, 127]

$HOCH_2CH_2NEt_2$

$HOCH_2CH_2\overset{\oplus}{N}Me_3\ Br^{\ominus}$

$HOCH_2CH_2Br$

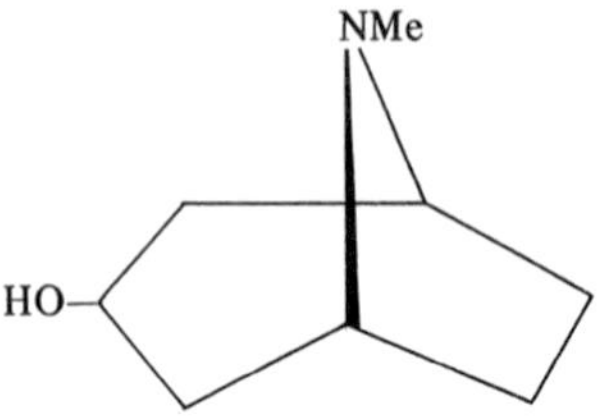

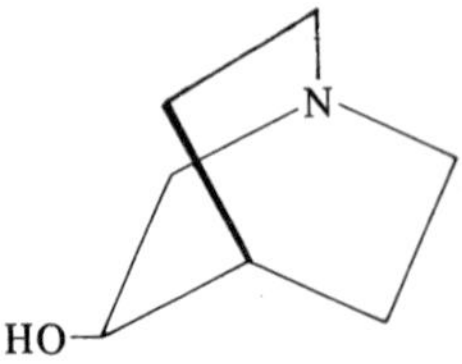

HO– NMe

N HO–

In one instance, a pyridinecarboxylic acid **XI-67** was converted to a pyridin acetic acid **XI-68** through the Arndt-Eistert reaction.[128]

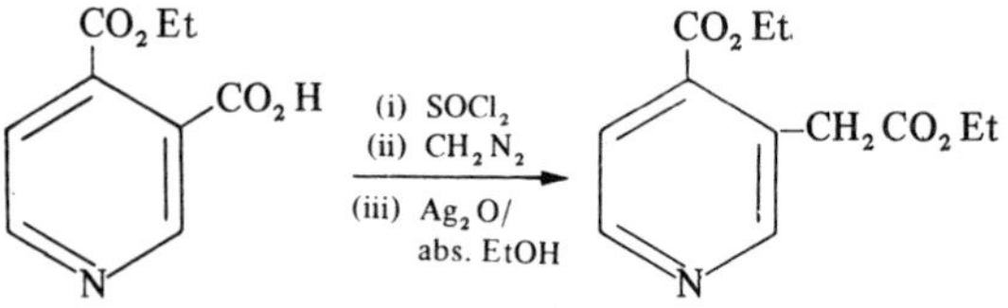

$CO_2Et$ $CO_2H$ N **XI-67** (i) $SOCl_2$ (ii) $CH_2N_2$ (iii) $Ag_2O$/abs. EtOH → $CO_2Et$ $-CH_2CO_2Et$ N **XI-68**

### 2. Decarboxylation (Tables XI-16 and XI-17)

Heating the pyrrolidones **(XI-69)** caused cleavage of the lactam bond, whic was then followed by decarboxylation and enamine formation to yield th pyrrolines **(XI-70)**.[129]

The decarboxylation of α-pyridylcinnamic acids yielded a variety o stilbazoles. Beard and Katritzky reported that they obtained the *trans*-stilbazol **(XI-72)** (as shown by i.r.) by the decarboxylation of the α-3-pyridylcinnami acid **(XI-71)** which itself was obtained by the condensation of benzaldehyd with 3-pyridylacetic acid.[101] Clarke and co-workers found, however, that th yCO— (ring) NR $\xrightarrow{\Delta}$

O

**XI-69**

PyCO— NHR $CO_2H$ ⟶ Py— O NHR

↓

2-Py— N R + $CO_2$

**XI-70**

$$\text{3-PyCH}_2\text{CO}_2\text{H} + \text{PhCHO} \longrightarrow \text{3-PyC(CO}_2\text{H)=CHPh} \xrightarrow{\Delta} \text{3-Py(H)C=C(H)Ph}$$

**XI-71** **XI-72**

is-stilbazole **(XI-74)** was formed when they repeated this work as well as when he *trans*-cinnamic acid **(XI-73)** (formed by the condensation of 3-pyridinealde-yde with sodium phenylacetate) was decarboxylated.[98] This latter reaction has

$$\text{3-PyCHO} + \text{PhCH}_2\text{CO}_2\text{Na} \longrightarrow \text{3-Py(H)C=C(Ph)CO}_2\text{H} \xrightarrow{\Delta} \text{3-Py(H)C=C(H)Ph}$$

**XI-73** **XI-74**

een discussed earlier under pyridinealdehyde condensations. Although the rans-stilbazoles are apparently not formed by the condensation of a yridylacetic acid with an aldehyde, they have been synthesized by irradiation f the *cis*-isomer with a tungsten lamp.[100] Although the base **XI-73** underwent ecarboxylation readily on heating, the corresponding pyridine-1-oxide did ot.[101]

Hydrolysis, decarboxylation followed by dehydration of the nitrile **XI-75** gave he olefin **XI-76**, which was converted to the di-(pyridine-2,6-dimethylene) **XI-77)** (RIS–8671).[130]

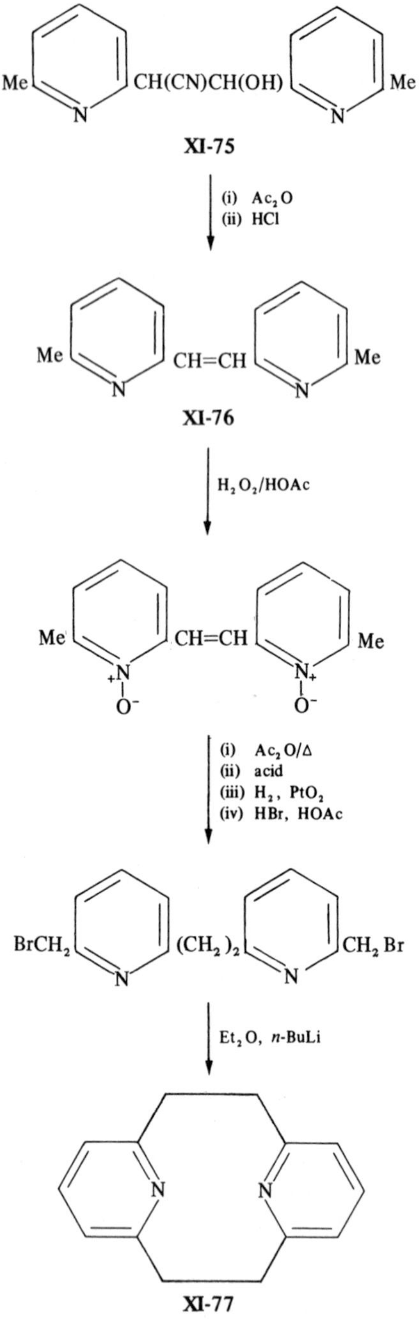

Me CH(CN)CH(OH) Me

**XI-75**

(i) $Ac_2O$
(ii) HCl

Me CH=CH Me

**XI-76**

$H_2O_2$/HOAc

Me CH=CH Me

(i) $Ac_2O/\Delta$
(ii) acid
(iii) $H_2$, $PtO_2$
(iv) HBr, HOAc

$BrCH_2$ $(CH_2)_2$ $CH_2Br$

$Et_2O$, *n*-BuLi

**XI-77**

An attempt to perform a Hammick reaction on 2-pyridylacetic acid to obtain a riety of alcohols gave only 2-picoline.[131] However, the sodium salt of this acid as successfully converted to a variety of alcohols (**XI-78**). Apparently, the free id provides the proton necessary to form 2-picoline, whereas the sodium salt es not.

2-PyCH$_2$CO$_2$H $\xrightarrow[\Delta]{-CO_2}$ (ylide ⟷ methylene-dihydropyridine) → 2-PyMe

2-PyCH$_2$CO$_2$Na $\xrightarrow{\Delta}$ (2-pyridylmethyl anion ⟷ resonance form) $\xrightarrow{RR^1C{=}O}$ 2-PyCH$_2$CRR$^1$OH

**XI-78**

| R | R$^1$ |
|---|---|
| Me | Ph |
| H | Ph |
| –(CH$_2$)$_5$– | |

## 3. Active Methylene Reactions

No new examples of this type of reaction were found.

## 4. Reduction (Table XI-18)

In general the reduction of a pyridine side-chain acid or ester using platinum ide, Raney Nickel, rhodium-on-carbon, rhodium-on-alumina, or ruthenium ide as the catalyst gives the piperidine acid or ester. Partial reduction of the ridine ring to a tetrahydropyridine usually occurred when palladium-on-rbon was employed as the catalyst, although two exceptions were ported.[132,133] Either a mixture of the piperidine and the tetrahydropyridine ter[133] or the tetrahydropyridine ester alone[134] was formed when sodium rohydride was used at room temperature in the reduction of pyridine le-chain ester salts. When the free bases were employed, reduction of the ester oup occurred instead of nuclear reduction.[46,47,115,177] The use of lithium ıminum hydride gave the same results (see Table XI-18). Many acetamides were reduced to tetrahydropyridylacetamides with sodium borohydride at roc temperature.[367]

The diol **XI-80**, obtained by the side-chain reduction of the keto est **(XI-79)** with sodium borohydride, served as a precursor to 5-methylindolizi **(XI-82)**.[135] However, the yield was low in this synthesis and the indolizine cou be prepared by a more favorable route from the aldehyde **XI-81**.

Me, $CH_2COCO_2Et$, $\overset{+}{N}$, $O^-$ **XI-79** —(i) Na; (ii) $NaBH_4$, MeOH, reflux, 4 hr→ Me, $CH_2CH(OH)CH_2OH$, N **XI-80**

(i) LAH, THF, reflux, 3 h; (ii) NaOAc, $Ac_2O$, overni; (iii) 450°

N, Me **XI-82**

(i) $Ac_2O$; (ii) 450°

Me, CHO, N **XI-81** + $CH_2{=}CHMgBr$ —THF→ Me, $CH(OH)CH{=}CH_2$, N

Catalytic reduction of the α-(*o*-nitroaryl)acrylonitriles **(XI-84)** over palladium catalyst at room temperature gave the amine **(XI-85)**.[10] Reducti with iron in boiling acetic acid yielded the indole **XI-83**, however.[10] In an earl work by Walker,[136] catalytic (palladium, 80°) reduction of the correspondi β-phenylacrylonitriles **(XI-86)** caused reductive cyclization of the nitrile to gi the indole **XI-87**. An attempt was made to apply Walker's method to t 2-pyridyl isomer of **XI-84** but the expected indole was not isolated.[137] T mechanism proposed for the cyclization of **XI-84** to the indole **(XI-83**; Py 2-Pyridyl) involves an intramolecular Michael addition of a hydroxylami group to the double bond rather than attack by an amino group. Support s mechanism was shown by the lack of cyclization of the amino compound **88.**

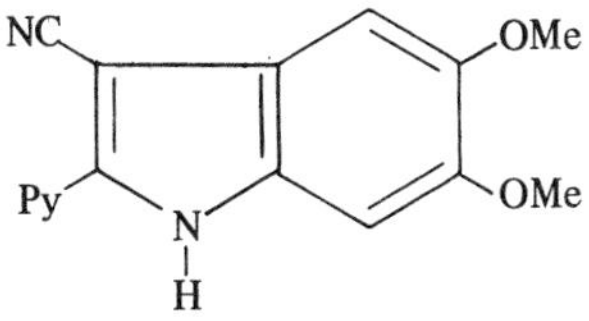

XI-83

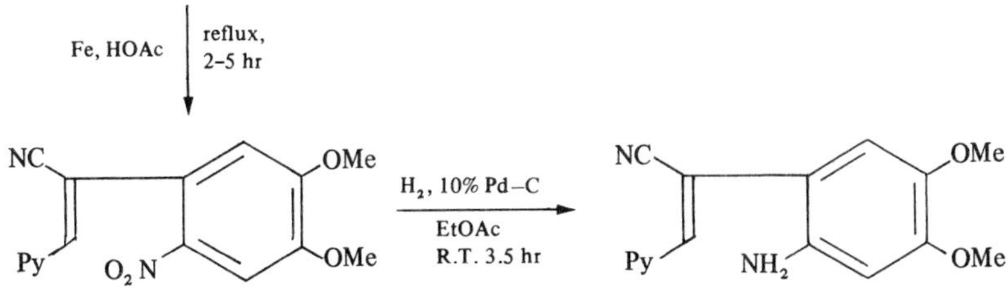

Fe, HOAc

reflux, 2–5 hr

XI-84

$H_2$, 10% Pd–C

EtOAc

R.T. 3.5 hr

XI-85

Py = 3- or Pyridyl

$H_2$

10% Pd–C

EtOAc

80°, 1.3 hr

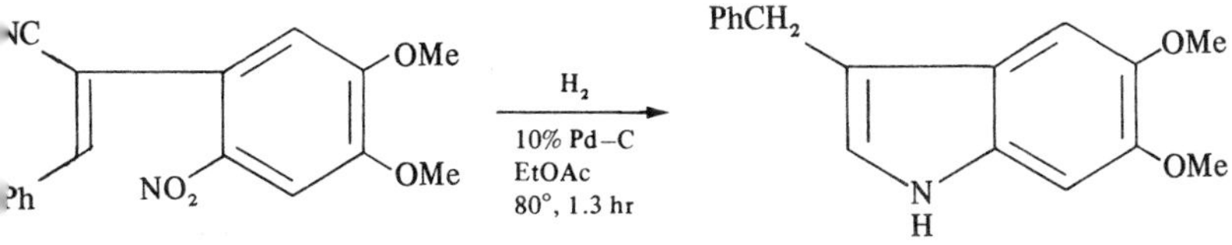

XI-86

XI-87

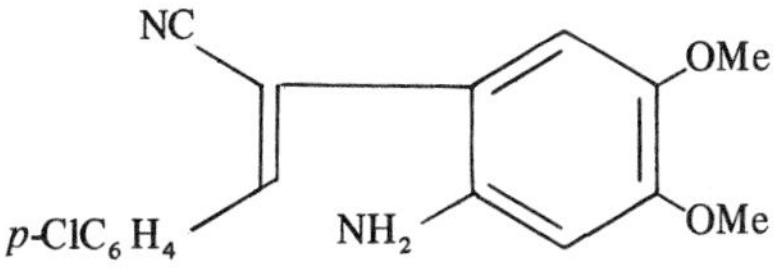

XI-88

### 5. Synthesis of Condensed Heterocycles–Quinolizines and Quinolizidines (Table XI-19)

The quinolizidine **XI-90** obtained by reduction of the diester **XI-89** was nverted to the azaphenalene **XI-91.**[138] The diester **XI-93** (R = H) failed to

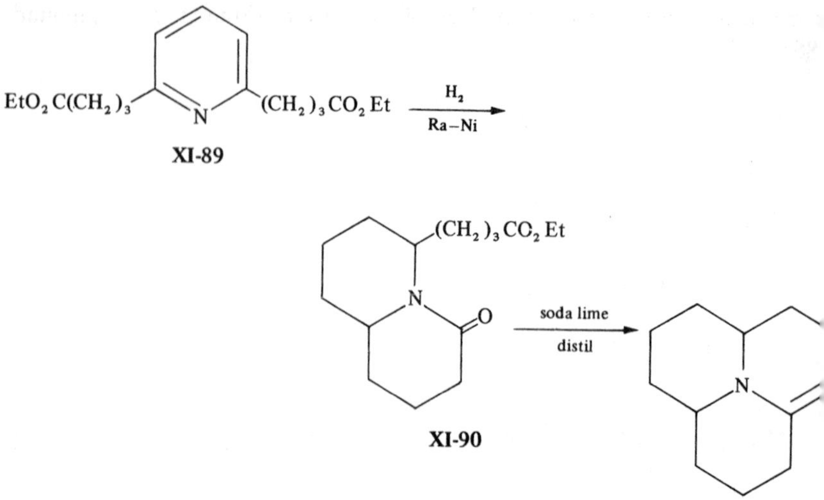

$EtO_2C(CH_2)_3$ … $(CH_2)_3CO_2Et$ $\xrightarrow[\text{Ra-Ni}]{H_2}$

XI-89

$(CH_2)_3CO_2Et$ … N … O $\xrightarrow[\text{distil}]{\text{soda lime}}$

XI-90

XI-91

condense with either compound **XI-92** or **XI-95** to yield the correspondi quinolizines **XI-94** and **XI-96**, respectively.[139] However, when R = OE condensation did take place.

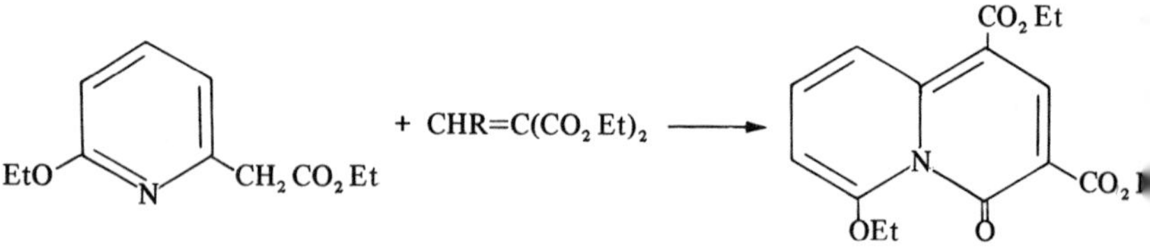

EtO … $CH_2CO_2Et$ + $CHR{=}C(CO_2Et)_2$ ⟶ $CO_2Et$, $CO_2$, OEt, O

XI-92 XI-93 XI-94

O … N H … $CH_2CO_2Et$ + $CH_2{=}C(CO_2Et)_2$ ⟶ (crossed) $CO_2Et$, $CO_2Et$, O, H, O

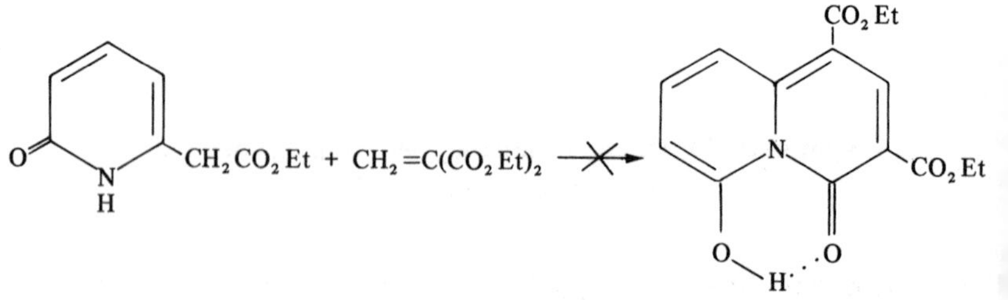

XI-95 XI-96

Several quinolizines were synthesized by heating 2-pyridyl 3-butyrolactonyl tones (**XI-97**) in acid.[140]



**XI-97**

### 6. Synthesis of Condensed Heterocycles Other Than Quinolizines (Table XI-20)

A great deal of research has been performed in this area since the earlier view, especially in the synthesis of indolizines, chromones, and naphthyridines.

#### A. *Indolizines*

The most common procedure employed for the synthesis of indolizines is the ndensation of a 2-pyridylacetate or acetonitrile with an α-halo activated ethylene compound.[141-145] The yields were often good but decreased when R

$$2\text{-PyCH}_2\text{Y} + \text{XCHRCOR}^1 \xrightarrow[\Delta]{\text{solvent}}$$



$Y{=}CO_2Et$, CN X=Br, Cl

as a large alkyl residue or when X was a chlorine, instead of a bromine, atom. nce the indolizine was formed without employing a base, the earlier belief of epanov and Grineva that a base is required for cyclization to occur[136] is not stified. A mechanism for this cyclization has been proposed as shown in heme XI-2.[143] In a few instances, the 1-carbethoxyindolizines (**XI-98**) that casionally formed were hydrolyzed and decarboxylated to yield the C-1 substituted indolizines (**XI-99**).[142]

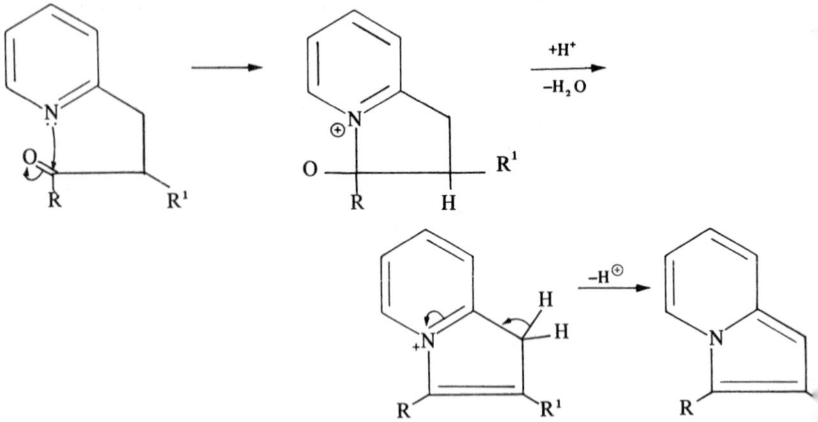

$+H^+$

$-H_2O$

$-H^{\oplus}$

Scheme XI-2

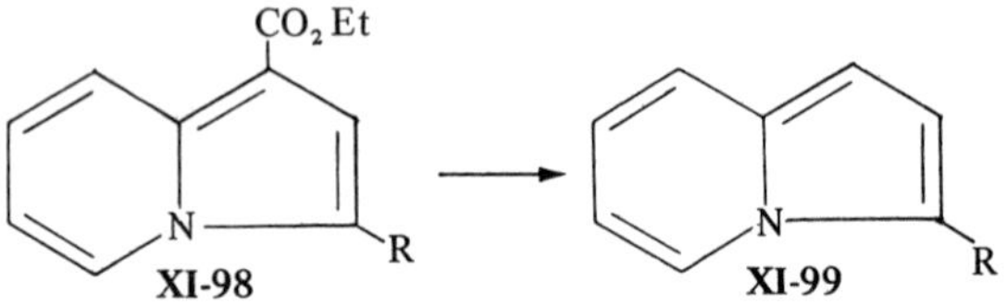

$CO_2Et$

XI-98

XI-99

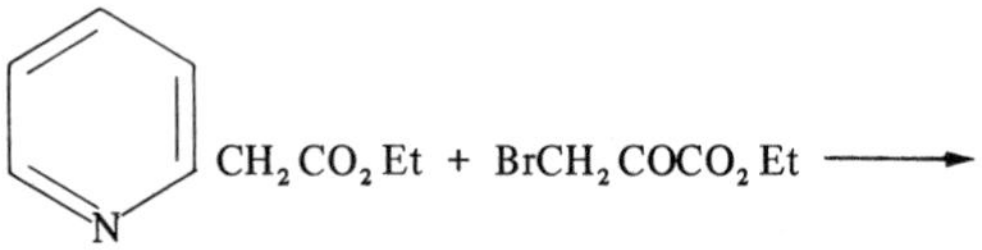

$CH_2CO_2Et$ + $BrCH_2COCO_2Et$ →

XI-100

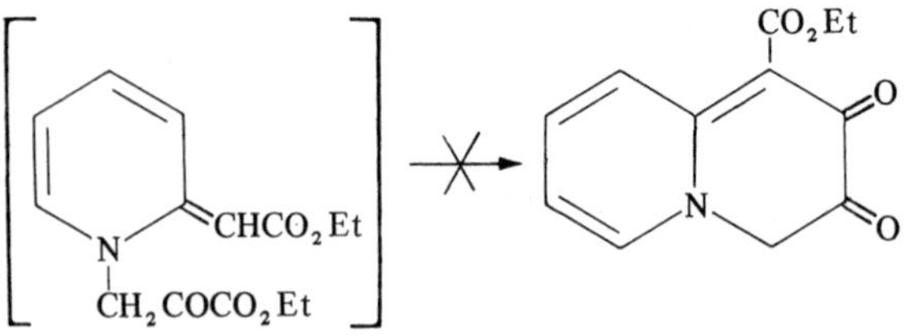

$CHCO_2Et$

$CH_2COCO_2Et$

$CO_2Et$

XI-101

$CO_2Et$ $CO_2Et$ + $CO_2Et$ $CO_2H$

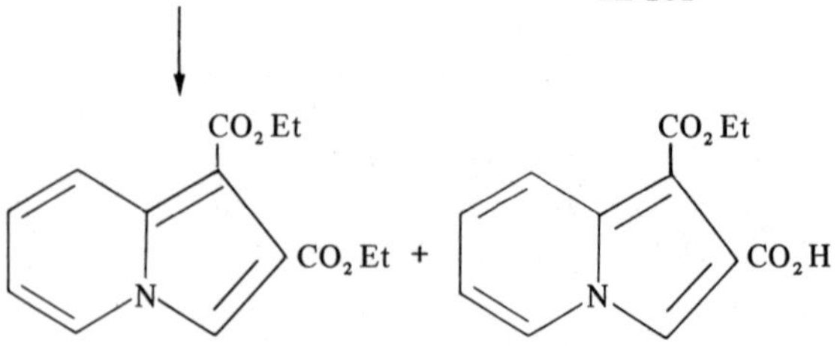

XI-102

XI-103

nitially the quinolizine **XI-101** was thought to be formed in the reaction of ıyl 2-pyridylacetate **(XI-100)** with ethyl bromopyruvate,[146] but it was later ›wn that indolizines **XI-102** and **XI-103** were actually produced.[144, 147]

Several indolizines were synthesized by cyclizing pyridine side-chain acid rivatives or nitriles in boiling acetic anhydride.[143, 145] Only two indolizidines ve been prepared from pyridine side-chain acids.[132, 148]

## B. *Coumarins and Chromones*

Coumarins and chromones containing pyridine substituents were synthesized ›m pyridine side-chain acids by employing the Knoevenagel,[48, 49, 149] chmann,[16, 149] and Simonis[149] reactions. The chromones **XI-106** were tained from a Simonis condensation of ethyl nicotinoylacetate **(XI-105)** with variety of phenols **(XI-104)**, whereas a Pechmann condensation yielded the umarins **(XI-107)**.

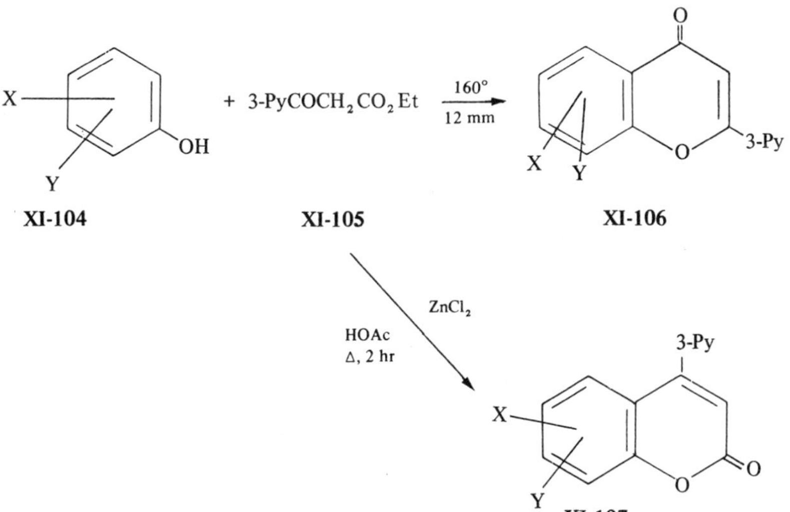


XI-104 XI-105 XI-106 XI-107

## C. *Naphthyridines*

2,6-Naphthyridine **(XI-113)** has been synthesized from pyridine side-chain ids *via* two different routes. Treatment of the dinitrile **XI-108** with anhydrous drogen bromide yielded the naphthyridine **(XI-107)**. Boiling **XI-109** with drazine gave the dihydrazino naphthyridine **XI-110**. The hydrazino groups were removed by treating **XI-110** with copper sulfate in boiling acetic acid yield 2,6-naphthyridine **(XI-113)**.[150] The yields were good and the dinitr **XI-108** could be obtained by a four-step synthesis from ethyl 3-pyridylacetat The other route involved cyclizing the diamide **XI-111** to the imide **(XI-11** which was then reduced to 2,6-naphthyridine **(XI-113)**.[128] No yields we reported in this synthesis. The diamide **XI-111** was synthesized in four ste

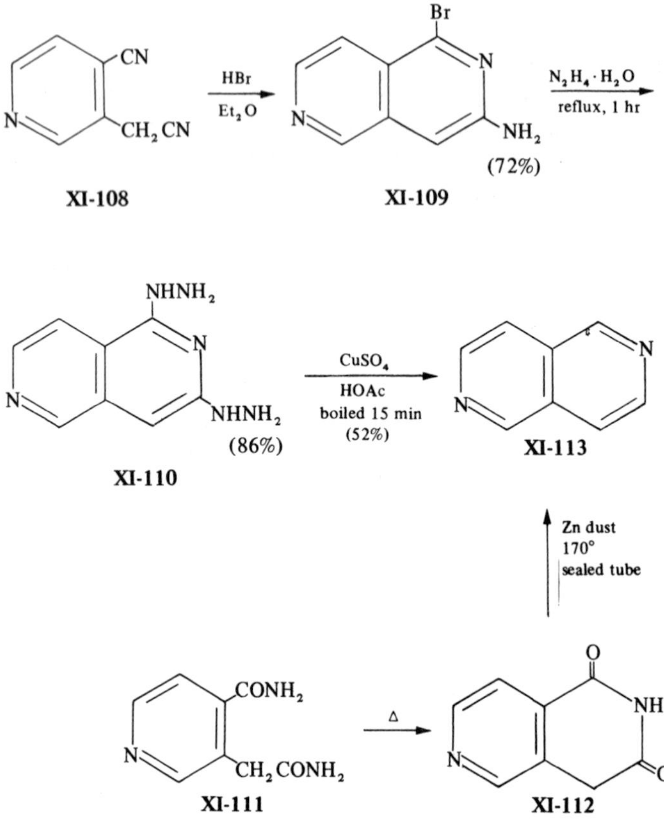

CN
$CH_2CN$
HBr
$Et_2O$
Br
$NH_2$
(72%)
$N_2H_4 \cdot H_2O$
reflux, 1 hr
**XI-108**
**XI-109**

$NHNH_2$
$NHNH_2$
(86%)
**XI-110**
$CuSO_4$
HOAc
boiled 15 min
(52%)
**XI-113**

Zn dust
170°
sealed tube

$CONH_2$
$CH_2CONH_2$
**XI-111**
Δ
O
NH
O
**XI-112**

from 4-carbethoxynicotinic acid. Similarly, 1,7-naphthyridine **(XI-115)** w prepared from the dinitrile **(XI-114)**.[92]

Syntheses of 1,5-[151] and 1,6-[152, 153] naphthyridine from pyridine side-chai acids have also been reported.

$CH_2CN$ CN N HBr $Et_2O$

I-114

$NH_2$ N N Br (72%) $H_2$ alc. KOH 10% Pd–C $NH_2$ N N (81%)

$NH_2NH_2$, dioxane, reflux, 1 hr

NHNH$_2$ N N NHNH$_2$ (64.5%) 30% HOAc hot $CuSO_4$ boil 15 min N N (18%)

XI-115

# III. Functional Derivatives

## 1. Esters

The esterification of side-chain acids has already been discussed under :erification (Section II.1.).

## 2. Acid Chlorides and Anhydrides

Several acid chlorides have been prepared, usually as intermediates (Table -21). Two of these have been characterized, one through its boiling point[128] d one through its hydrochloride salt.[298,299] As in the earlier review, no hydrides have been reported.

## 3. Amides

Amides prepared by conventional methods are listed in Tables XI-14 and -27.

The pyridine-1-oxide **XI-117** was synthesized by treating compound **XI-1** with ammonium hydroxide.[154] Reaction of the sulfonamide **XI-118** w

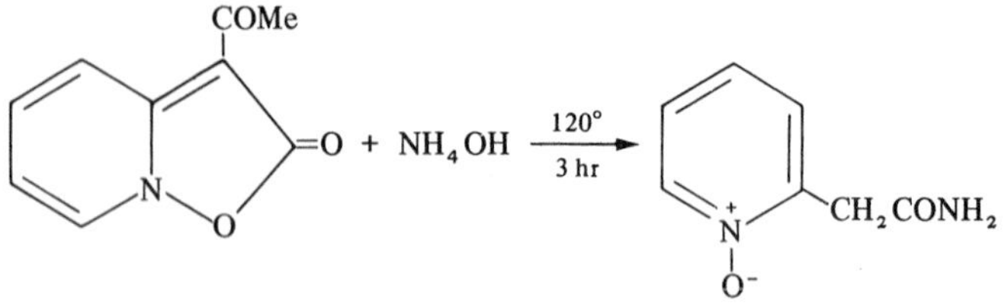

COMe

=O + $NH_4OH$ $\xrightarrow[3\ hr]{120°}$ $CH_2CONH_2$

N O N O⁻

**XI-116** **XI-117**

sodium hydroxide gave the amide **XI-119**, whereas the 2-isomer **(XI-120)** form a mixture of the amide **(XI-121)** and **XI-122**.[154] The proposed mechanism

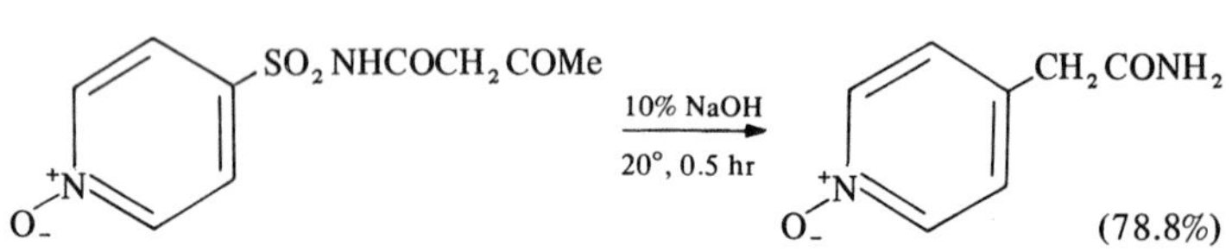

$SO_2NHCOCH_2COMe$ $\xrightarrow[20°,\ 0.5\ hr]{10\%\ NaOH}$ $CH_2CONH_2$

(78.8%)

**XI-118** **XI-119**

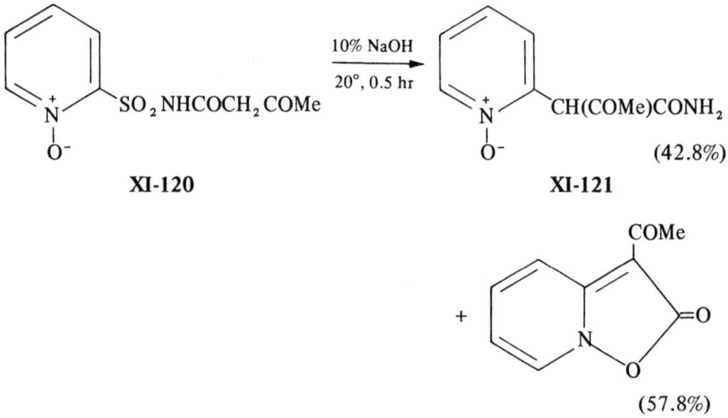

$SO_2NHCOCH_2COMe$ $\xrightarrow[20°,\ 0.5\ hr]{10\%\ NaOH}$ $CH(COMe)CONH_2$

(42.8%)

**XI-120** **XI-121**

COMe

+ =O

N O

(57.8%)

**XI-122**

this reaction is the same as that proposed for the rearrangement of nitrophe sulfonamides.[155] A few acids have also been prepared by this method.

6-Methyl-3-pyridylacetamide **(XI-123)** was converted to the 3-aminometh pyridine **(XI-124)** by the Hofmann reaction.[156] Alkylation of the amides **XI-1** with aminoalkyl halides yielded amides containing substituted amino gro **(XI-126)**[157, 158]

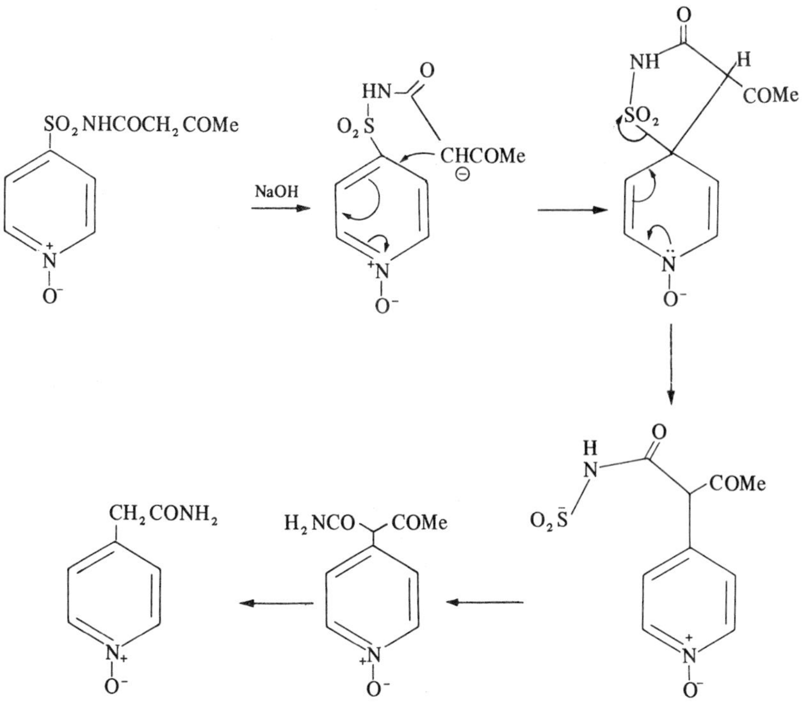

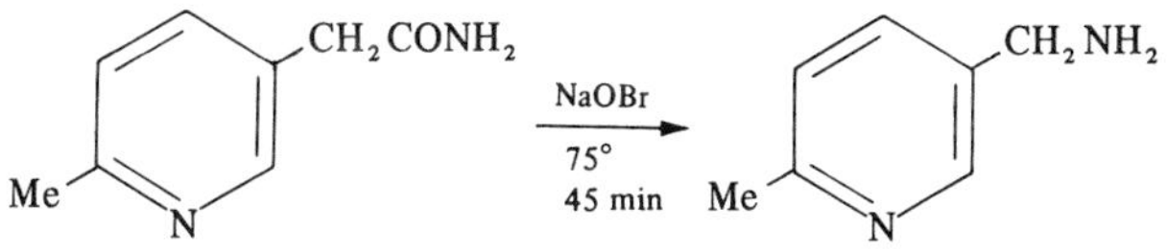

$SO_2NHCOCH_2COMe$ ... NaOH → ... $CHCOMe^{\ominus}$ ... → ...

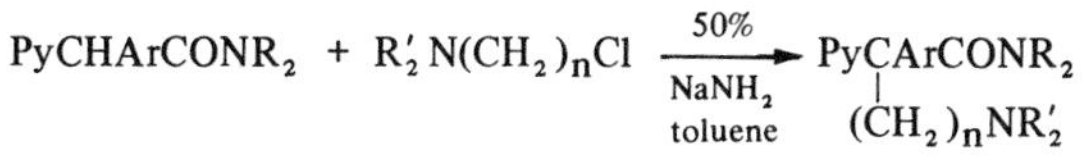

$H_2NCO \diagdown\!\!\diagup COMe$ ← ... ← ... $CH_2CONH_2$

$$\text{XI-123 (}CH_2CONH_2\text{, Me)} \xrightarrow[\text{75°, 45 min}]{\text{NaOBr}} \text{XI-124 (}CH_2NH_2\text{, Me)}$$

$$\underset{\textbf{XI-125}}{PyCHArCONR_2} + R'_2N(CH_2)_nCl \xrightarrow[\text{NaNH}_2\text{, toluene}]{50\%} \underset{\textbf{XI-126}}{PyC(ArCONR_2)(CH_2)_nNR'_2}$$

## 4. Hydrazides, Hydroxamic Acids, and Amidines

A few hydrazides have been treated with ketones and the resulting lhydrazones **XI-127** were reduced to yield substituted hydrazides -128).[58,159,160] The properties of these hydrazides are listed in Table XI-33.

$$Py(CH_2)_nCONHNH_2 + RR'CO \longrightarrow Py(CH_2)_nCONHN{=}CRR' \; (\textbf{XI-127}) \xrightarrow{H_2} Py(CH_2)_nCONHNHCHRR' \; (\textbf{XI-128})$$

Two examples of the conversion of a hydrazide to an amide through the Curti reaction have been reported.[8, 291]

The guanidine **XI-130** and not 2-amino-4-hydroxy-6-(3-pyridyl)pyrimidi (**XI-131**) was reportedly formed when the ester **XI-129** was treated wi guanidine carbonate at 140°.[161] Treatment of the ester **XI-132** with guanidi

$$\underset{\textbf{XI-129}}{3\text{-}PyCOCH_2CO_2Et} + NH_2C({=}NH)NH_2 \xrightarrow[1\ hr]{140°} \underset{\textbf{XI-130}}{3\text{-}PyCOCH_2CONHC({=}NH)NH_2}$$

(not formed: **XI-131**, 2-amino-4-hydroxy-6-(3-pyridyl)pyrimidine; structure labels: OH, N, N, 3-Py, $NH_2$)

gave the substituted guanidine **XI-133** in 85% yield.[162]

$$\underset{\textbf{XI-132}}{3\text{-}PyCOC({=}NH)CO_2Et} + NH_2C({=}NH)NH_2 \longrightarrow \underset{\textbf{XI-133}}{3\ PyCOC({=}NH)CONHC({=}NH)NH_2}$$

## 5. Nitriles

### A. *Synthesis*

As reported in the earlier edition, side-chain nitriles have been synthesized pyridylation of nitriles (Table XI-22), alkylation of pyridylacetonitriles (Ta XI-23), employing the Michael addition (Table XI-24), and the Knoevena (Table XI-25) and Strecker (Table XI-26) reactions, by the dehydration ides (Table XI-26), and the nucleophilic displacement of halogens or alkoxyl ups by cyanide (Table XI-26).

n the Michael addition of acrylonitrile to the methiodides of a variety of olines **(XI-134)** and **(XI-135)**, two molecules of the nitrile added to a C-2 thyl group whereas three molecules added to a C-4 methyl group.[312]

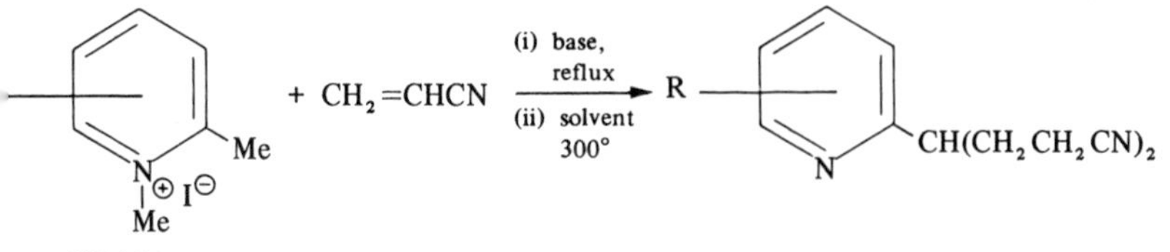

$$\text{XI-134} + CH_2{=}CHCN \xrightarrow[\text{(ii) solvent } 300^\circ]{\text{(i) base, reflux}} R\text{-}C_5H_3N\text{-}2\text{-}CH(CH_2CH_2CN)_2$$

**XI-134**

$$\text{XI-135} + CH_2{=}CHCN \xrightarrow[\text{(ii) } 300^\circ]{\text{(i) Base, reflux, solvent}} R\text{-}C_5H_3N\text{-}4\text{-}C(CH_2CH_2CN)_3$$

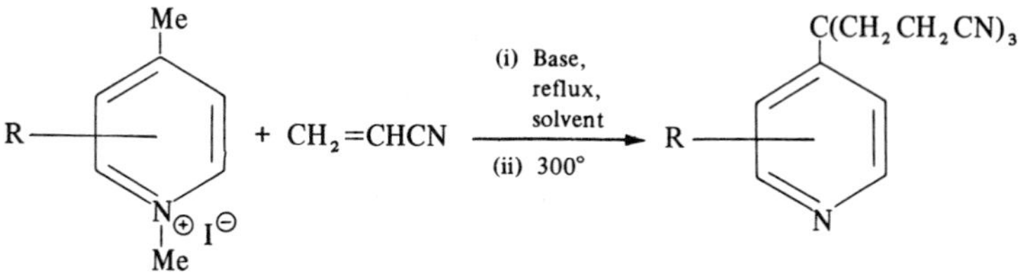

**XI-135**

parently steric hindrance between the C-1 methyl and the cyanoethyl groups the anhydro base intermediate of the C-2 isomer **XI-136** prevented its mation and therefore the addition of a third molecule of acrylonitrile. No h steric hindrance is present in the C-4 isomer. The same reasoning was

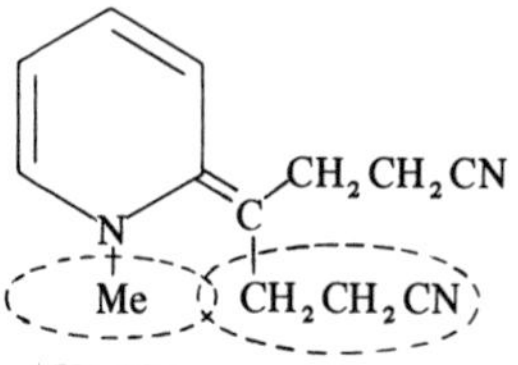

**XI-136**

ployed to explain the addition of only one molecule of acrylonitrile to a C-2 yl group in compound **XI-137**.

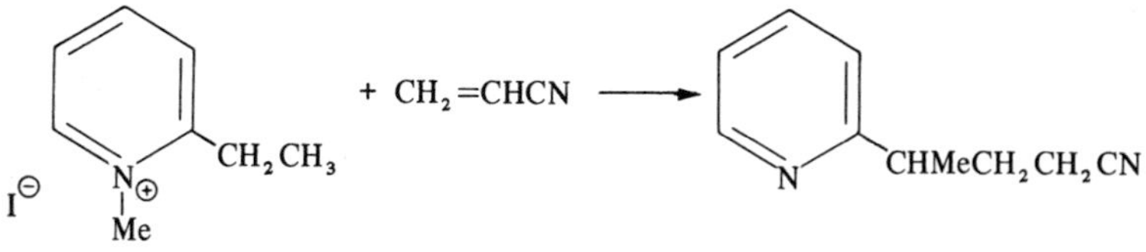

$$\text{XI-137} + CH_2{=}CHCN \longrightarrow C_5H_4N\text{-}2\text{-}CHMeCH_2CH_2CN$$

**XI-137**

Investigation of the reaction of 2-pyridylacetonitrile with acid anhydri showed that the α-acylpyridyl acetonitriles formed exist in two pH depend tautomeric forms **XI-138** and **XI-139**, as determined by uv and spectroscopy.[163] When 4-pyridylacetonitrile was treated with acetic anhyd

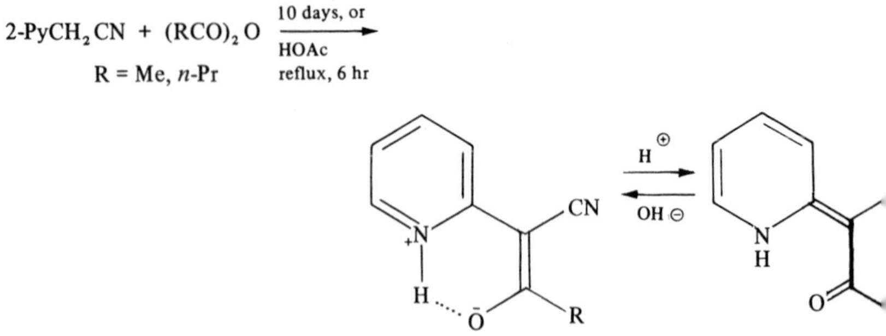

$$\text{2-PyCH}_2\text{CN} + (\text{RCO})_2\text{O} \xrightarrow[\text{HOAc reflux, 6 hr}]{\text{10 days, or}}$$

R = Me, *n*-Pr

**XI-138** ⇌ ($H^{\oplus}$ / $OH^{\ominus}$) **XI-139**

at room temperature, acylation occurred on the nitrogen rather than on α-carbon atom to give a compound that appeared to be a resonance hybri

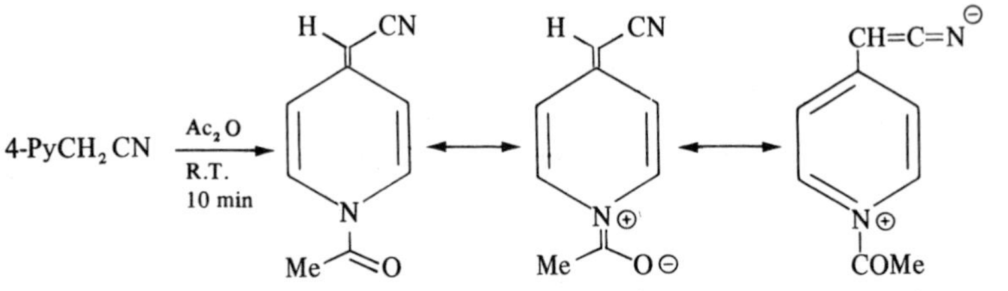

$$\text{4-PyCH}_2\text{CN} \xrightarrow[\text{R.T. 10 min}]{\text{Ac}_2\text{O}}$$

**XI-140** ⟷ **XI-141** ⟷ **XI-142**

$\downarrow$ $Ac_2O$, Δ

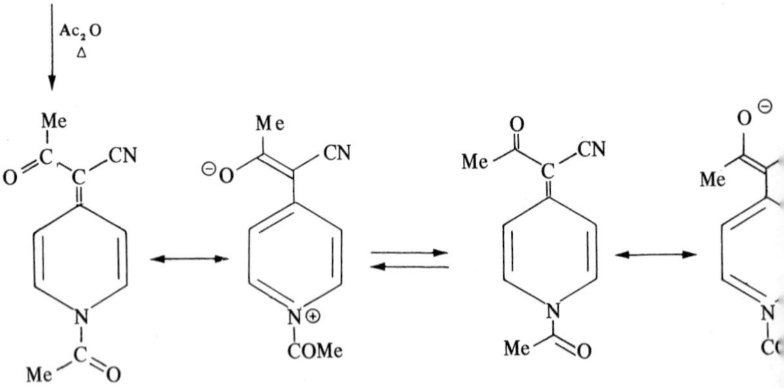

**XI-143** ⇌ **XI-144**

ıctures **XI-140 – XI-142**. Acylation occurred on both the nitrogen and the arbon atoms to give a mixture of tautomers **XI-143** and **XI-144** when the tion mixture was heated.

## B. *Solvolysis and Aminolysis* (Table XI-27)

'he α-(*p*-aminophenyl)-β-pyridylpropionitriles **(XI-145)** proved difficult to drolyze to the corresponding acids.[164] This lack of reactivity is probably ısed by the effect of the polarizability of the amino group on the methine drogen atom alpha to the nitrile group, which would allow charged complex mation with polar reagents. Also, there is the tendency for the *p*-amino group localize partial negative charges at the C-1, C-3, and C-5 positions of the matic ring (see **XI-146**), which would reduce the usual charge distribution r the arylacetonitrile group. The C-3 and C-4 isomers of **XI-145** were

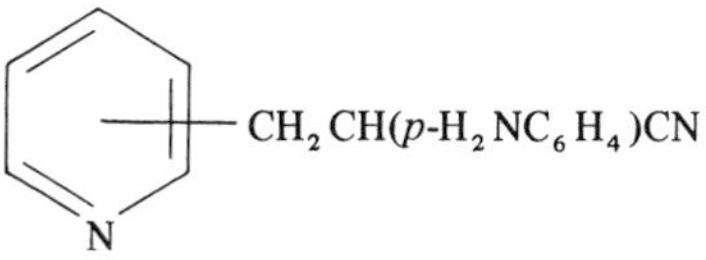

$CH_2CH(p\text{-}H_2NC_6H_4)CN$ (on pyridine ring, N)

**XI-145**

$PyCH_2CH(CN)$, $\delta^{\ominus}$, $\delta^{\ominus}$, $NH_2$ $\delta^{\oplus}$, $\delta^{\ominus}$

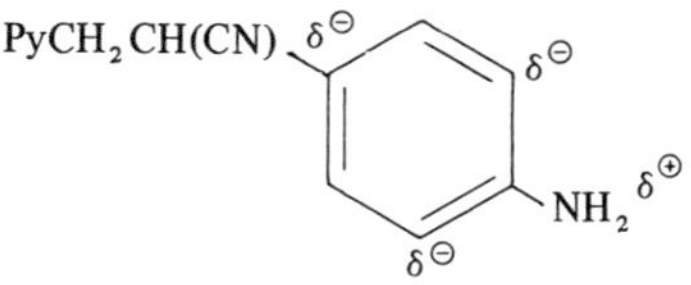

**XI-146**

cessfully hydrolyzed by first forming the esters in the presence of hydrogen oride (through the imino ether dihydrochloride) and then hydrolyzing the r with base. An attempt to hydrolyze nitrile **XI-147** failed. Nitrile **XI-148**

$$4\text{-PyCH(CN)CO(CH}_2)_3CO_2Et$$

**XI-147**

ıld be hydrolyzed and dehydrated by acid to give the amide **XI-149** but this ıld not be further hydrolyzed to the acid **XI-150**.[99]

$$4\text{-PyCH(CN)CH}(o\text{-}O_2NPh)OH \xrightarrow[H_2SO_4]{75\%} 4\text{-PyC(CONH}_2)\text{=CH}(o\text{-}O_2NPh)$$

**XI-148** **XI-149**

$$\not\downarrow$$

$$4\text{-PyC(CO}_2H)\text{=CH}(o\text{-}O_2NPh)$$

**XI-150**

cid hydrolysis of 1,3-dicyano-3-(2-pyridyl)hexane **(XI-151)** resulted in ization to the imide **XI-152**.[178] Lactam **XI-154** was isolated from the rolysis of nitrile **XI-153** followed by treatment with thionyl chloride.[57] The

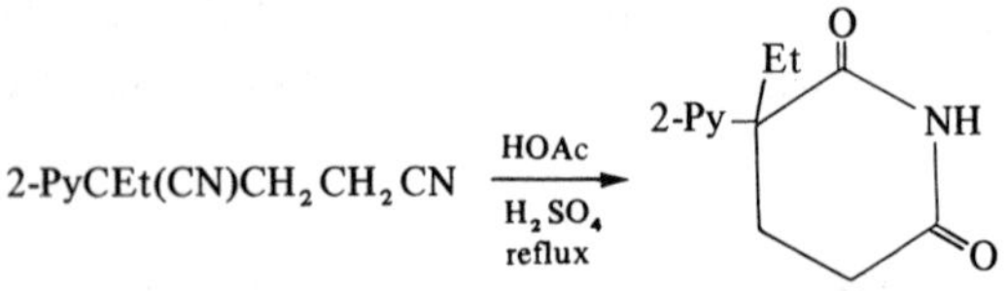

$$\text{2-PyCEt(CN)CH}_2\text{CH}_2\text{CN} \xrightarrow[\text{H}_2\text{SO}_4\text{, reflux}]{\text{HOAc}}$$

2-Py, Et, O, NH, O

**XI-151** **XI-152**

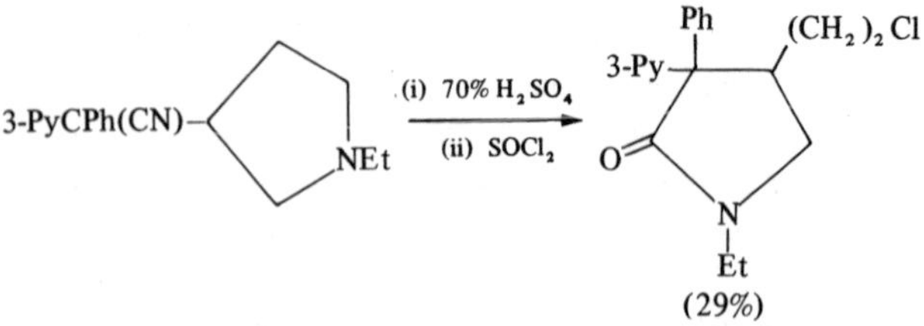

3-PyCPh(CN)–, NEt

(i) 70% $H_2SO_4$
(ii) $SOCl_2$

Ph, $(CH_2)_2Cl$, 3-Py, O, N, Et

(29%)

**XI-153** **XI-154**

acids obtained from nitriles **XI-155** were found to decarboxylate readily duri the hydrolyses.[50]

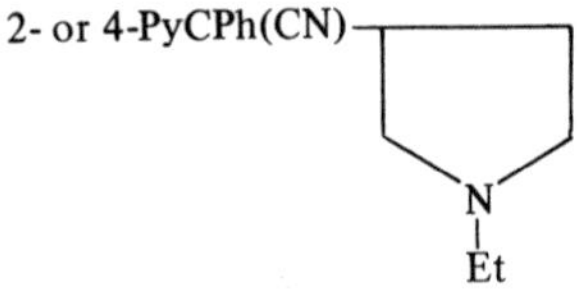

2- or 4-PyCPh(CN)–, N, Et

**XI-155**

Several thioamides have been prepared by treating nitriles with hydrog sulfide.[166-168] A few $\Delta^2$-thiazoline derivatives **(XI-156)** have been isolated fr the reaction of pyridylacetonitriles with 2-aminoethyl thiol.[168] 3-Pyridylaceto

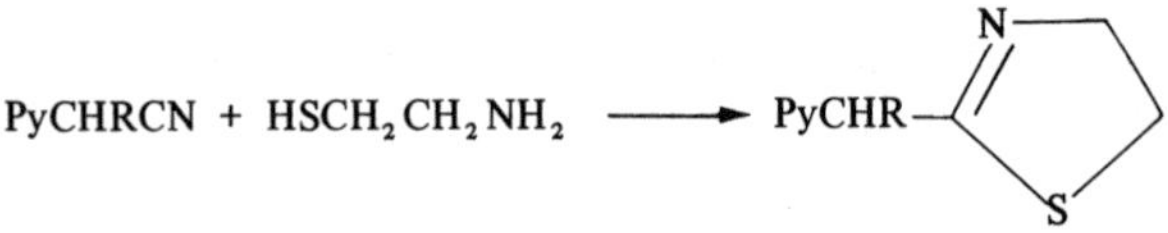

$$\text{PyCHRCN} + \text{HSCH}_2\text{CH}_2\text{NH}_2 \longrightarrow \text{PyCHR–}$$

N, S

**XI-156**

trile was converted to the tetrazole **XI-157** by treatment with boil butanol-acetic acid followed by sodium azide.[169]

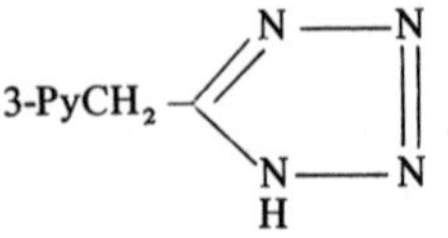

$3\text{-PyCH}_2$–, N, N, N, N, H

**XI-157**

### C. *Reduction* (Table XI-28)

The reduction of 3-pyridylacetonitrile and 3-pyridylacrylonitrile using -*p*-menthene **(XI-158)** as the hydrogen source and 10% palladium-on-carbon as e catalyst yielded 3-ethylpyridine and 3-propylpyridine, respectively (yields 5–95%).[170]

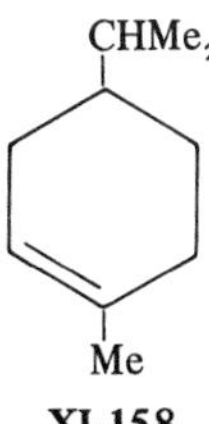


**XI-158**

### D. *Reactions with Organometallic Compounds*

These reactions and the products obtained are summarized in Table XI-29.

## IV. Derivatives with Side-Chains of Mixed Function

### 1. Carbonyl Derivatives (Table XI-30)

Treatment of the keto ester **XI-159** with hydroxylamine yielded (4-pyridyl)isoxazol-5-one **(XI-160)** rather than the expected oxime.[105]

$(4\text{-PyCOCHCO}_2\text{Et})_2\text{CHPh} + \text{NH}_2\text{OH}$

**XI-159**

4-Py
N
O
O

**XI-160**

(γ-Pyridyl)-5-aminopyrazoles **(XI-162)** were synthesized by treating the pyridyl-β-ketonitriles **(XI-161)** with substituted hydrazines.[171] No yields were ported.

Keto esters have also been employed in the synthesis of a wide variety of con-nsed heterocycles. These reactions have already been discussed under Section 6. and recorded in Table XI-20.

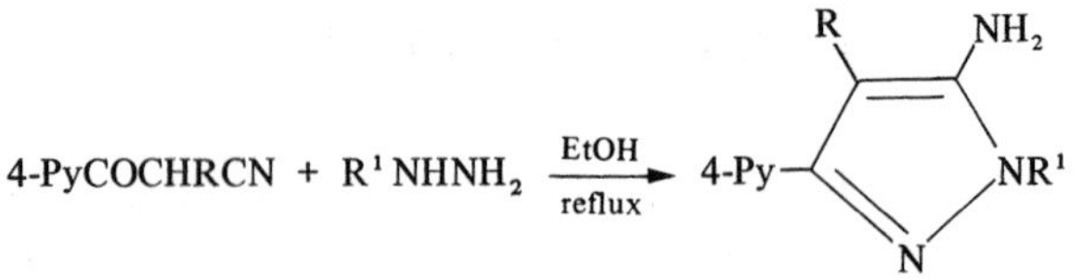

4-PyCOCHRCN + $R^1NHNH_2$ $\xrightarrow[\text{reflux}]{\text{EtOH}}$ 4-Py (R, $NH_2$, $NR^1$, N)

**XI-161** **XI-162**

## 2. Hydroxyl Derivatives

Hydroxyl derivatives have been employed to synthesize indenones[172] a quinolines[87, 173] (Table XI-20). The lactone **XI-164**, obtained from es **XI-163**, was converted to a variety of *N*-substituted lactams **(XI-165)**, whi were then reduced to the tetrahydronaphthyridines **(XI-166)**[174] (Table XI-2

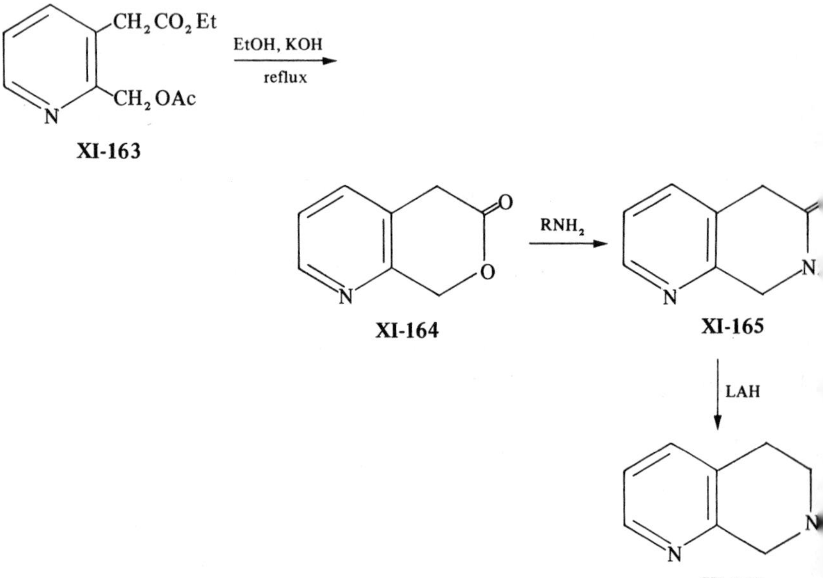

$CH_2CO_2Et$, $CH_2OAc$, N $\xrightarrow[\text{reflux}]{\text{EtOH, KOH}}$

**XI-163**

O, O, N $\xrightarrow{\text{RNH}_2}$ N, N

**XI-164** **XI-165**

LAH

N, N

**XI-166**

## 3. Ethylenic Derivatives

### A. *Syntheses*

These are summarized in Table XI-31.

## B. *Reactions* (Table XI-32)

The cyclopropanes **XI-168** were obtained when the unsaturated side-chain ketones **XI-167** were treated either with diazomethane or with the dimsyl ion.[94]

-, 3-, or 4-PyCH, O, N, O, Ph, **XI-167** + $CH_2N_2$ or $\overset{\ominus}{C}H_2SOMe$ ⟶ 2-, 3-, or 4-Py, O, N, O, Ph, **XI-168**

### 4. Displacement of Side-Chain Substituents

The chlorine atoms of a mixture of compounds **XI-169** and **XI-170** were displaced by amines to yield the enamines **XI-171**.[175]

$$\underset{\textbf{XI-169}}{\text{3-PyCCl=CHCO}_2\text{Et}} + \underset{\textbf{XI-170}}{\text{3-PyCCl}_2\text{CH}_2\text{CO}_2\text{Et}} \xrightarrow{R_2NH} \underset{\textbf{XI-171}}{\text{3-PyC(NR}_2\text{)=CHCO}_2\text{Et}}$$

## V. Tables

TABLE XI-1. Preparation of Side-Chain Acids From Nonpyridine Starting Materials

| Starting materials | Conditions | Product | Yield | Properties | Ref. |
|---|---|---|---|---|---|
| $EtO_2CCH{=}CHCO_2Et$ 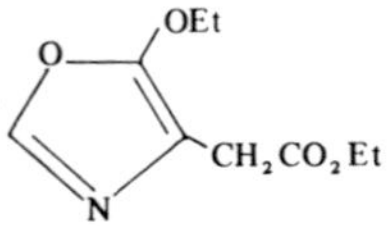 + O OEt $CH_2CO_2Et$ N | 3 hr, 30°, $N_2$; anhyd HCl, EtOH | 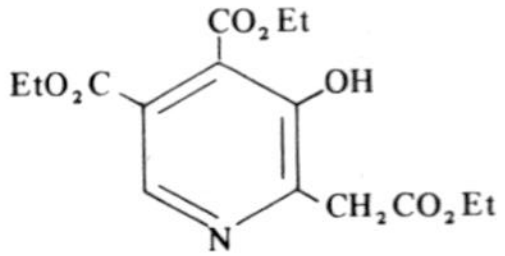 $CO_2Et$ $EtO_2C$ OH $CH_2CO_2Et$ N | | hydrochloride, m.p. 124–125° | 69 |
| NCCH=CHCN 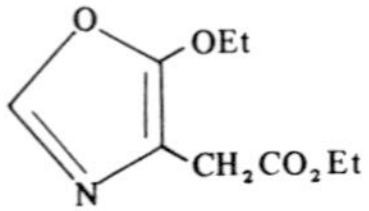 + O OEt $CH_2CO_2Et$ N | same as above | 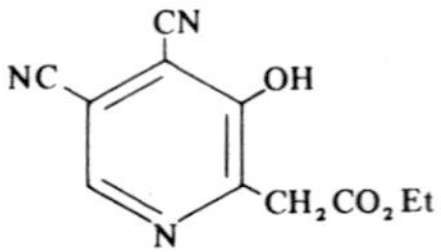 CN NC OH $CH_2CO_2Et$ N | | m.p. 164–166° | 69 |
| $AcOCH_2CH{=}CHCN$ 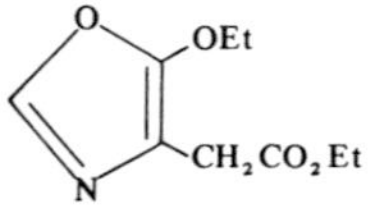 + O OEt $CH_2CO_2Et$ N | same as above | 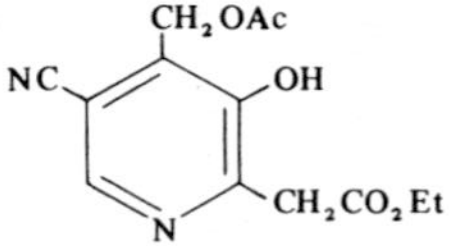 $CH_2OAc$ NC OH $CH_2CO_2Et$ N | | m.p. 118–120° | 69 |
| $MeOCH_2CH{=}CHCH_2OMe$ 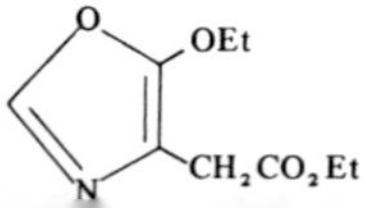 + O OEt $CH_2CO_2Et$ N | same as above | 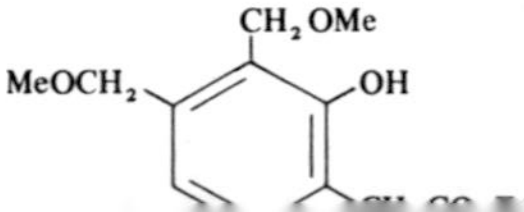 $CH_2OMe$ $MeOCH_2$ OH | | m.p. 139–142° | 69 |

| Starting material | Conditions | Product | Yield | Properties | Ref. |
| --- | --- | --- | --- | --- | --- |
| 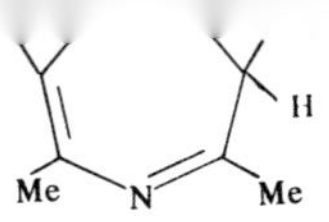 Me, N, Me, H | Conc $H_2SO_4$, 15 min, steam-bath *or* NaOH, MeOH, reflux 2 hr | 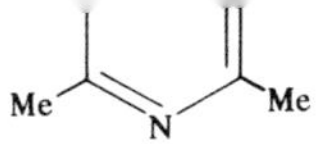 Me, N, Me | 72% 78% | m.p. 57.5–58.5°; ir, uv, NMR | 70 |
| 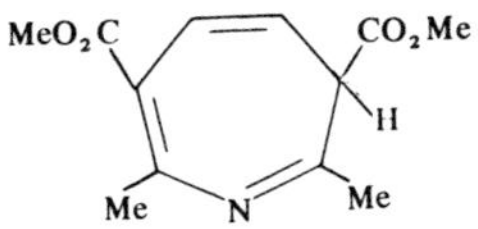 $MeO_2C$, $CO_2Me$, H, Me, N, Me | MeOH, HCl, R.T., 19 hr | 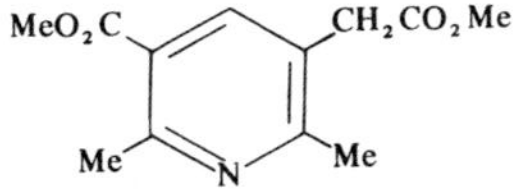 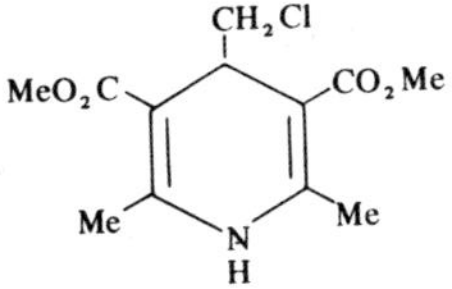 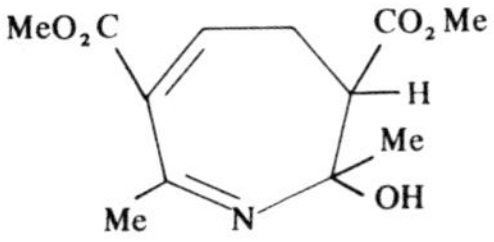 $MeO_2C$, $CH_2CO_2Me$, Me, N, Me + $CH_2Cl$, $MeO_2C$, $CO_2Me$, Me, N, H, Me + $MeO_2C$, $CO_2Me$, H, Me, OH, Me, N | | | 70 |
| 2 $EtO_2CC(Me)_2Br$ + $NC(CH_2)_3CO_2Et$ or $NC(CH_2)_3CO_2Me$ | Zn | 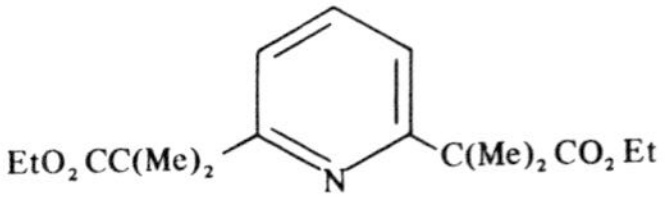 $EtO_2CC(Me)_2$, N, $C(Me)_2CO_2Et$ | | b.p. 174–175°/15 mm; m.p. 58–60° | 179, 180 |
| 2 $MeO_2CC(Me)_2Br$ + $NC(CH_2)_3CO_2Et$ | Zn | 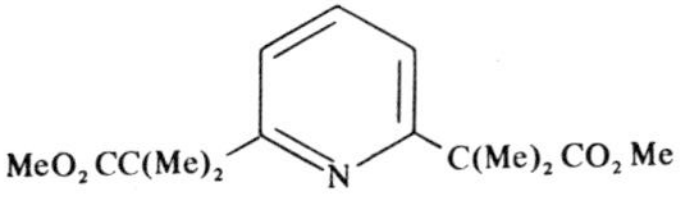 $MeO_2CC(Me)_2$, N, $C(Me)_2CO_2Me$ | | b.p. 171–172°/15 mm; m.p. 63–65° | 179 |

TABLE XI-1. Preparation of Side-Chain Acids From Nonpyridine Starting Materials (Continued)

| Starting materials | Conditions | Product | Yield | Properties | Ref. |
|---|---|---|---|---|---|
| $EtO_2CCH_2COCH_2CO_2Et + NCCH_2CO_2Et$ | $EtNH_2$, EtOH, 7 days; overnight. conc. $H_2SO_4$ | [pyridine ring: 4-$CH_2CO_2Et$, 3-$CO_2Et$, 2-OH, 6-HO] | 2.5 g (nitrile) ↓ 1.4 g | | 181 |
| $NCCH_2C(OH)(CH_2CN)_2$ | 30% HBr, HOAc, 1.5 hr; $NaHCO_3$; MeOH, R.T., 1.5 hr; reflux 1 hr | [pyridine ring: 4-$CH_2CO_2Me$, 2,6-Br] | | | 182 |
| [4-pyrone: 5-HO, 2-$CH_2C(CN)(CO_2Et)NHAc$] | $NH_4OH$, sealed bomb, 85–90°, 2.5 hr; $Ba(OH)_2$, $H_2O$, reflux several hr | [4-pyridone (N-H): 5-MeO, 2-$CH_2CH(NH_2)CO_2H$] | 29 g ↓ 14 g | m.p. 252–255° (decomp.) uv | 71 |
| [isoquinoline: 7-OH, 8-NO] + $PhSO_2Cl$ | NaOH, acetone-$H_2O$, 55-60°, 10 min | [pyridine: 4-CH=$CHCO_2H$, 3-CN] | 40% | m.p. 232.4–232.8° (decomp.) | 72 |
| [ring: $CH_2CH(NH_2)CO_2H$] | conc. HCl, reflux | [ring: H, $CH_2CO_2H$] | | uv | 76 |

| Reactants | Conditions | Product | Yield | Properties | Ref. |
| --- | --- | --- | --- | --- | --- |
| $MeCH(Br)CO_2$ *s*-Bu + $NC(CH_2)_3CO_2Me$ | Zn | 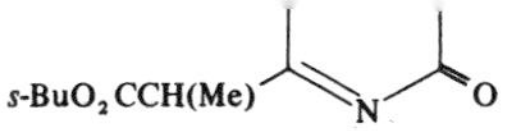 *s*-$BuO_2CCH(Me)$, N, O | | | 180 |
| $MeNHCH{=}CHAc + NCCH_2C(NH_2){=}C(CN)_2$ | boil 15 min, EtOH, pyridine | 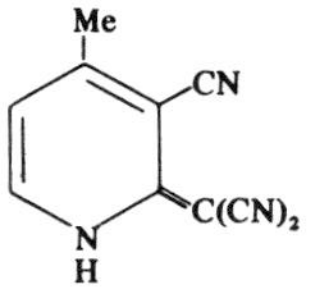 Me, CN, $C(CN)_2$, N H | 1 g (ketone) → 1.3 g | m.p. 225° (decomp.) | 183 |
| $NH_2CH{=}C(Me)Ac + NCCH_2C(NH_2){=}C(CN)_2$ | boil 15 min, HOAc | 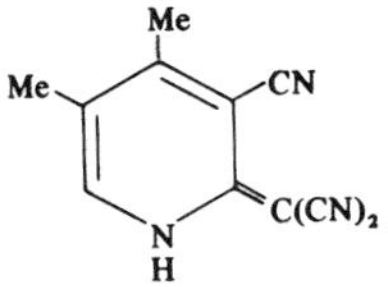 Me, Me, CN, $C(CN)_2$, N H | 0.7 g (ketone) → 0.3 g | m.p. 238° (decomp.) | 183 |
| $MeNHCH{=}C(Me)Ac + NCCH_2C(NH_2){=}C(CN)_2$ | HOAc | 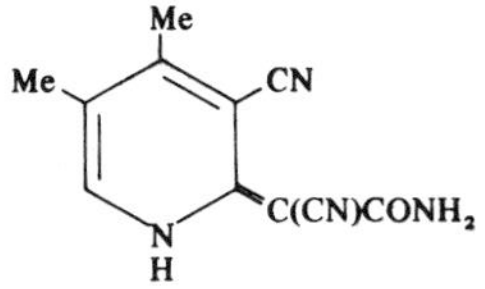 Me, Me, CN, $C(CN)CONH_2$, N H | | | 153 |
| $MeCOCH_2COMe + NCCH_2C(NH_2){=}C(CN)_2$ | 10% aq. NaOH, 10 min or Na, EtOH, reflux 2 min | 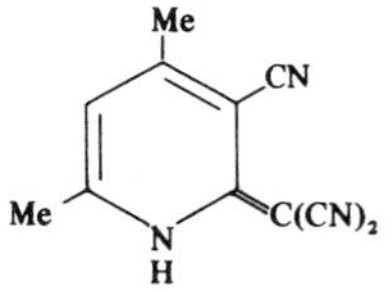 Me, Me, CN, $C(CN)_2$, N H | 0.5 g (nitrile) → 1 g | | 184 |

TABLE XI-1. Preparation of Side-Chain Acids From Nonpyridine Starting Materials (Continued)

| Starting materials | Conditions | Product | Yield | Properties | Ref. |
|---|---|---|---|---|---|
| $NH_2COC(Me){=}CHCOMe + NCCH_2C(NH_2){=}C(CN)_2$ | 24 hr, $H_2O$, or reflux 10 min, HOAc | 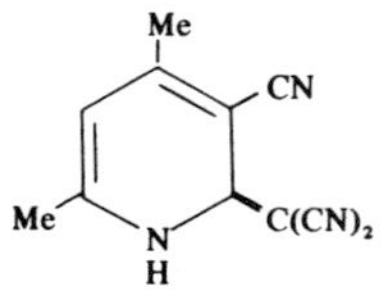 Me, CN, Me, N, H, $C(CN)_2$ | 0.6 g (nitrile) ↓ 0.5 g | m.p. 255° | 183 |
| $NH_2COC(Me){=}C(Me)COMe$ + $NCCH_2C(NH_2){=}C(CN)_2$ | same as above | 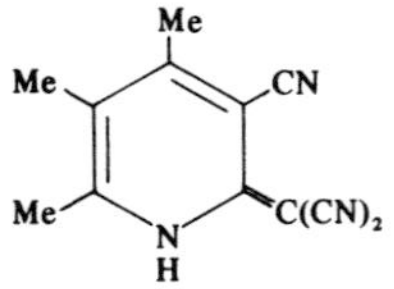 Me, Me, CN, Me, N, H, $C(CN)_2$ | | m.p. 246° (decomp) | 184 |
| $NH_2COC(Et){=}CHCOMe + NCCH_2C(NH_2){=}C(CN)_2$ | same as above | 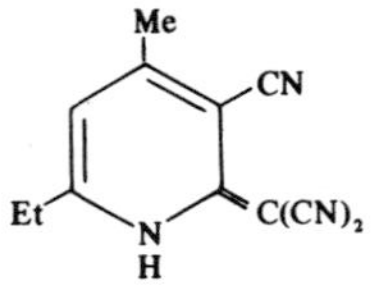 Me, CN, Et, N, H, $C(CN)_2$ | | m.p. 224–225° | 184 |
| $NH_2COC(n\text{-}Bu){=}CHCOMe + NCCH_2C(NH_2){=}C(CN)_2$ | same as above | 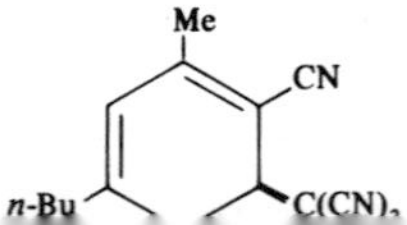 Me, CN, n-Bu, $C(CN)_2$ | | m.p. 224° | 184 |

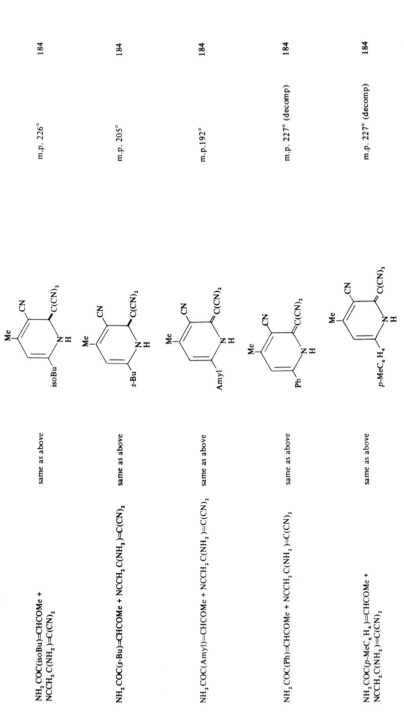

| | | | | |
|---|---|---|---|---|
| $NH_2COC(isoBu){=}CHCOMe$ + $NCCH_2C(NH_2){=}C(CN)_2$ | same as above | 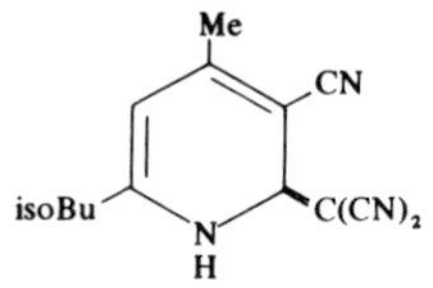 Me, CN, isoBu, $C(CN)_2$, N H | m.p. 226° | 184 |
| $NH_2COC(s\text{-}Bu){=}CHCOMe$ + $NCCH_2C(NH_2){=}C(CN)_2$ | same as above | 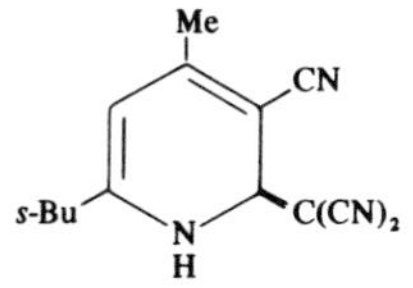 Me, CN, *s*-Bu, $C(CN)_2$, N H | m.p. 205° | 184 |
| $NH_2COC(Amyl){=}CHCOMe$ + $NCCH_2C(NH_2){=}C(CN)_2$ | same as above | 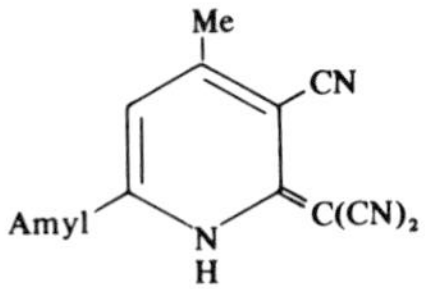 Me, CN, Amyl, $C(CN)_2$, N H | m.p.192° | 184 |
| $NH_2COC(Ph){=}CHCOMe$ + $NCCH_2C(NH_2){=}C(CN)_2$ | same as above | 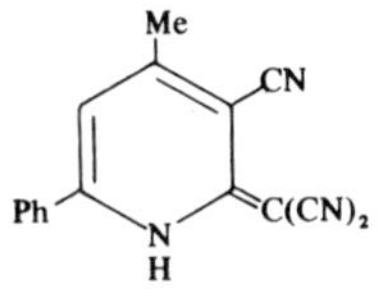 Me, CN, Ph, $C(CN)_2$, N H | m.p. 227° (decomp) | 184 |
| $NH_2COC(p\text{-}MeC_6H_4){=}CHCOMe$ + $NCCH_2C(NH_2){=}C(CN)_2$ | same as above | 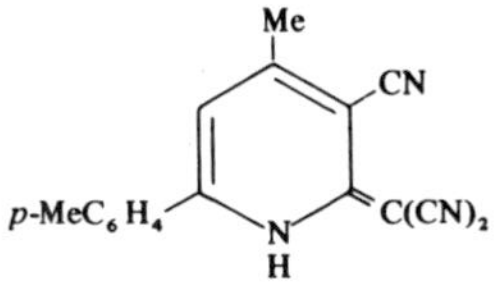 Me, CN, *p*-$MeC_6H_4$, $C(CN)_2$, N H | m.p. 227° (decomp) | 184 |

TABLE XI-1. Preparation of Side-Chain Acids From Nonpyridine Starting Materials (Continued)

| Starting materials | Conditions | Product | Yield | Properties | Ref. |
|---|---|---|---|---|---|
| MeNHCH=CHCOPh + NCCH$_2$C(NH$_2$)=C(CN)$_2$ | HOAc | 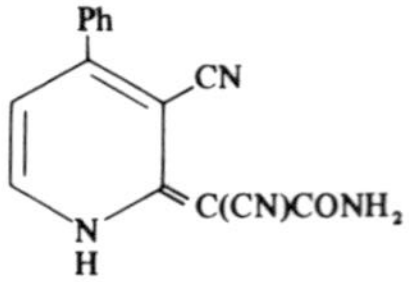 Ph; CN; =C(CN)CONH$_2$; NH | | m.p. 330° (decomp) | 153 |
| MeNHCH=CHCOPh + NCCH$_2$C(NH$_2$)=C(CN)$_2$ | boil 15 min, EtOH, pyridine | 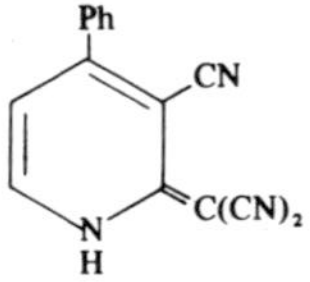 Ph; CN; =C(CN)$_2$; NH | 1 g (ketone) ↓ 0.4 g | m.p. 238° | 183 |
| MeNHC(Me)=CHCOPh + NCCH$_2$C(NH$_2$)=C(CN)$_2$ | same as above | 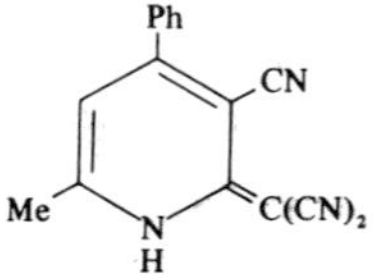 Ph; CN; Me; =C(CN)$_2$; NH | 0.9 g (ketone) ↓ 0.3 g | m.p. 230° (decomp) | 183 |
| PhCH$_2$NHC(Me)=CHAc + NCCH$_2$C(NH$_2$)=C(CN)$_2$ | same as above | 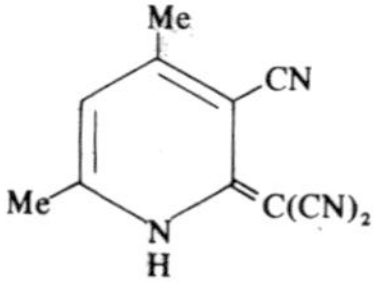 Me; CN; Me; =C(CN)$_2$; NH | | m.p. 255° | 183 |
| Me$_2$CO + (CO$_2$Et)$_2$ + NH$_2$COCH$_2$CN | Na, MeOH, PhMe; | 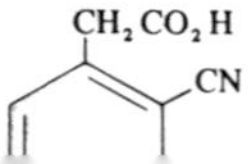 CH$_2$CO$_2$H; CN | | | |

| | | | | | |
|---|---|---|---|---|---|
| EtCOMe + $(CO_2Et)_2$ + $NH_2COCH_2CN$ | same as above | Et, N, H, O | 58% | m.p. 205° | 185 |
| $Me_2CO$ + $CH_2(CO_2Et)_2$ + $NH_2COCH_2CN$ | same as above | $(CH_2)_2CO_2H$, CN, Me, N, H, O | | m.p. 217° | 185 |
| EtCOMe + $CH_2(CO_2Et)_2$ + $NH_2COCH_2CN$ | same as above | $(CH_2)_2CO_2H$, CN, Et, N, H, O | | m.p. 211° | 185 |
| $C(CN)_2$, Me, O, Me + $HCONH_2$ | 1 hr, 150° | $C(CN)_2$, Me, N, H, Me | 5 g → 1.7 g | m.p. 330–331° | 186 |
| $C(CN)_2$, Me, O, Me + $MeNH_2$ | EtOH, reflux 30 min | $C(CN)_2$, Me, N, Me, Me | | m.p. 225–228° | 361 |

TABLE XI-1. Preparation of Side-Chain Acids From Nonpyridine Starting Materials (Continued)

| Starting materials | Conditions | Product | Yield | Properties | Ref. |
|---|---|---|---|---|---|
| 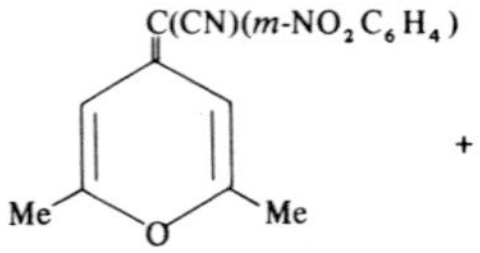 $C(CN)(m\text{-}NO_2C_6H_4)$, Me, O, Me + $MeNH_2$ | same as above | 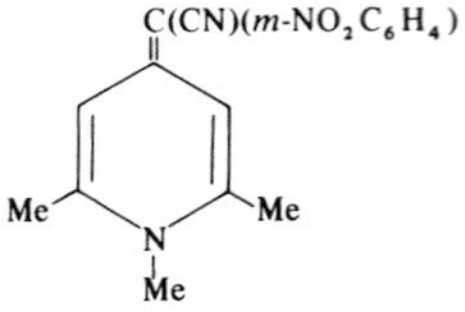 $C(CN)(m\text{-}NO_2C_6H_4)$, Me, N, Me, Me | | | 361 |
| 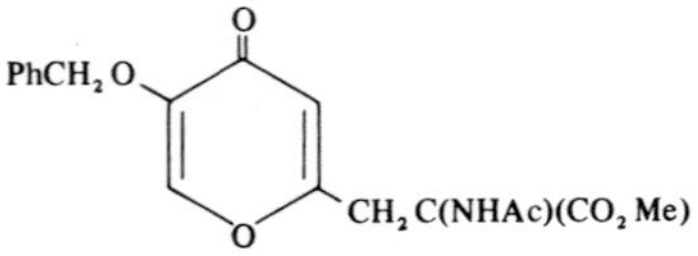 $PhCH_2O$, O, O, $CH_2C(NHAc)(CO_2Me)$ | $NH_3$, 100° sealed tube, 2 hr; 4 *N* $H_2SO_4$, reflux 8 hr | 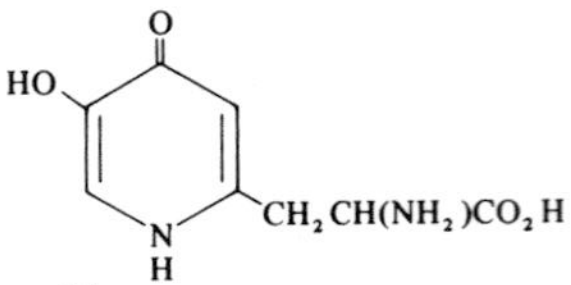 HO, O, N, H, $CH_2CH(NH_2)CO_2H$ *DL*– | 41.6% | | 187 |
| 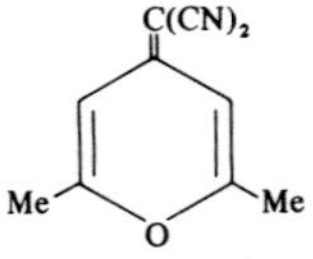 $C(CN)_2$, Me, O, Me + $PhNH_2$ | | 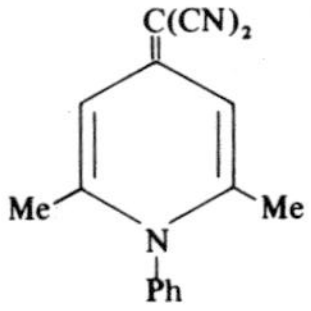 $C(CN)_2$, Me, N, Me, Ph | 20% | m.p. 314–315° | 186, 188 |
| 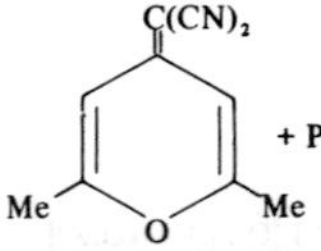 $C(CN)_2$, Me, O, Me + $PhCH_2NH_2$ | | 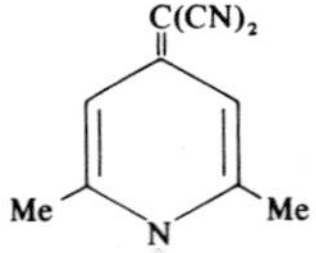 $C(CN)_2$, Me, N, Me | 34% | m.p. 242–245° | 186, 188 |

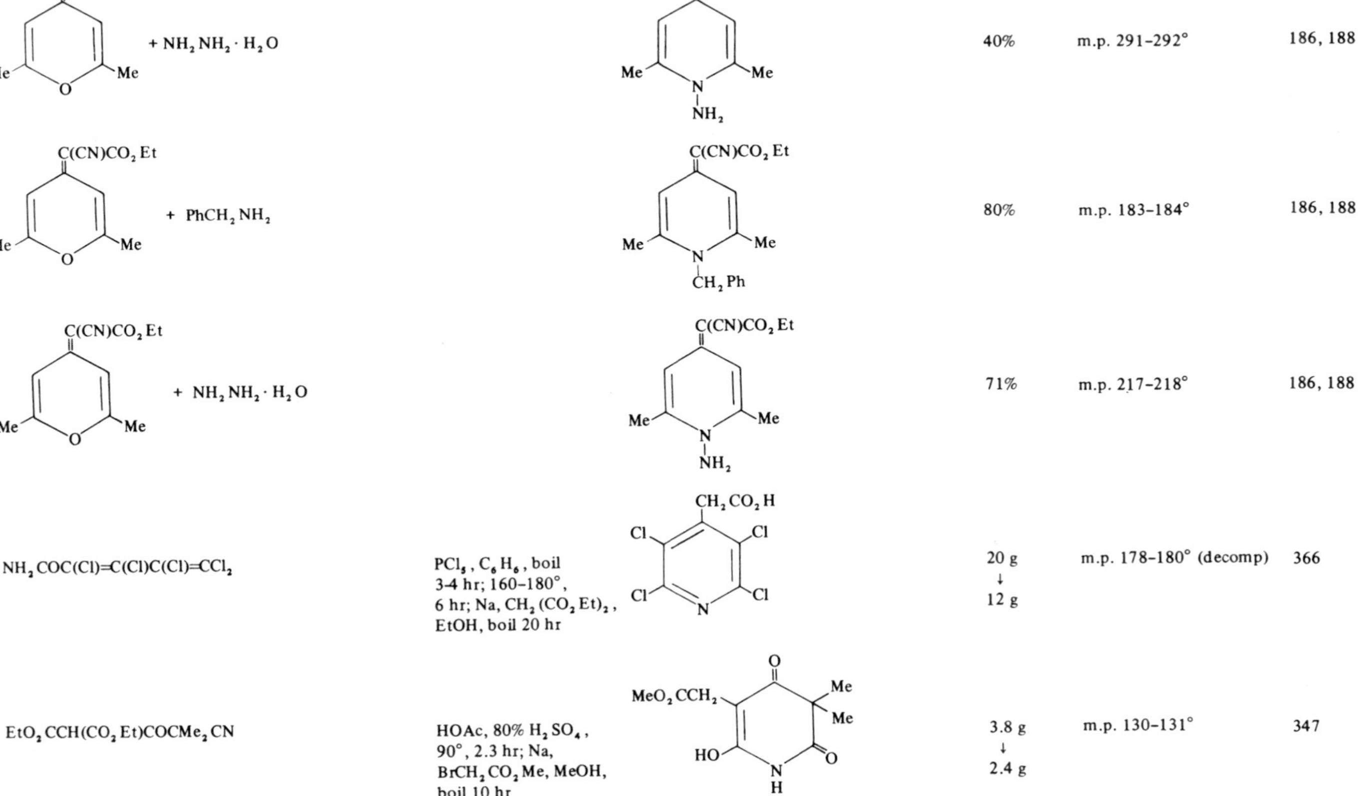

| Starting materials | Conditions | Product | Yield | Properties | Ref. |
|---|---|---|---|---|---|
| + $NH_2NH_2 \cdot H_2O$ | | | 40% | m.p. 291–292° | 186, 188 |
| + $PhCH_2NH_2$ | | | 80% | m.p. 183–184° | 186, 188 |
| + $NH_2NH_2 \cdot H_2O$ | | | 71% | m.p. 217–218° | 186, 188 |
| $NH_2COC(Cl){=}C(Cl)C(Cl){=}CCl_2$ | $PCl_5$, $C_6H_6$, boil 3-4 hr; 160–180°, 6 hr; Na, $CH_2(CO_2Et)_2$, EtOH, boil 20 hr | | 20 g → 12 g | m.p. 178–180° (decomp) | 366 |
| $EtO_2CCH(CO_2Et)COCMe_2CN$ | HOAc, 80% $H_2SO_4$, 90°, 2.3 hr; Na, $BrCH_2CO_2Me$, MeOH, boil 10 hr | | 3.8 g → 2.4 g | m.p. 130–131° | 347 |

TABLE XI-2. Preparation of Side-Chain Acids by Oxidation

| Starting material | Oxidant | Conditions | Product | Yield | Properties | Ref. |
|---|---|---|---|---|---|---|
| 2-Py$(CH_2)_2CH{=}CH_2$ | $KMnO_4$ | acetone, < 35° | 2-Py$(CH_2)_2CO_2H$ | 52% | m.p. 140–141° | 189 |
| 2-Py$(CH_2)_4CH{=}CH_2$ | $KMnO_4$ | acetone, < 35° | 2-Py$(CH_2)_4CO_2H$ | 53.5% | m.p. 96.5° | 189 |
| $CH_2OH$; 2-Py; HO; N | Chromic acid | 3 hr, 100°; MeOH, HCl, reflux 1 hr | $CO_2Me$; 2-Py; HO; N | 50% | b.p. 140–190°/0.05 mm | 19 |
| $CH_2OH$; 2-Py; HO; N | Chromic acid | 3 hr, 100°; MeOH, HCl, reflux 4 hr | $CO_2Me$; 2-Py; HO; N | 0.5 g → 0.4 g | b.p. 160–165°/0.02 mm | 19 |
| 3-Py–; –CHO; N H | $AgNO_3$ | NaOH, 50% EtOH, 60–65°, $H_2O$, 4 hr | 3-Py–; –$CO_2H$; N H | 60% | m.p. 200–201° (decomp) | 102 |
| 2-Py$(CH_2)_2CHO$ | | | 2-Py$(CH_2)_2CO_2H$ | | | 189 |
| 2-Py$(CH_2)_4CHO$ | | | 2-Py$(CH_2)_4CO_2H$ | | | 189 |
| COMe; $\overset{+}{N}$ | $SeO_2$ | Py, 80°, 105 min | 3-Py$COCO_2H$ | 44% | m.p. 178–179°; phenylhydrazone, m.p. 173–174°; 2,4-DNP, | 218 |

| | | | | | | |
|---|---|---|---|---|---|---|
| 3-PyCOMe | $SeO_2$ | Py, 80°, 120 min | 3-PyCOCO$_2$H | [illegible] | | [illegible] |
| 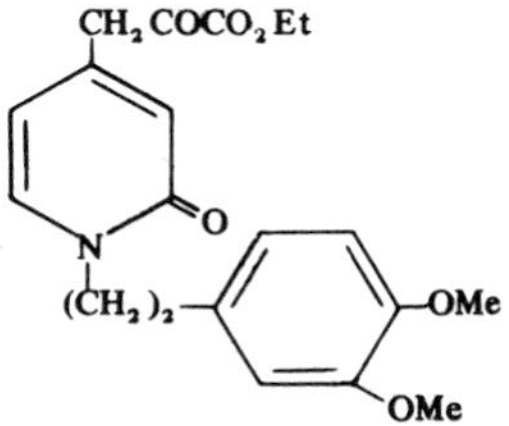 | $MnO_2$, $H_2O_2$ | dil. alkali, 0°; oxidant, 16 hr, 0° | 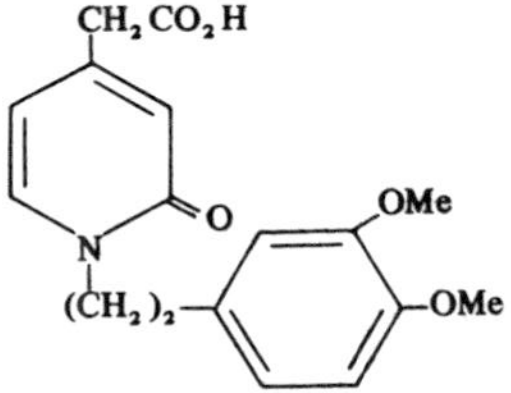 | 78% | m.p. 156.5–157° (decomp.) | 3 |
| 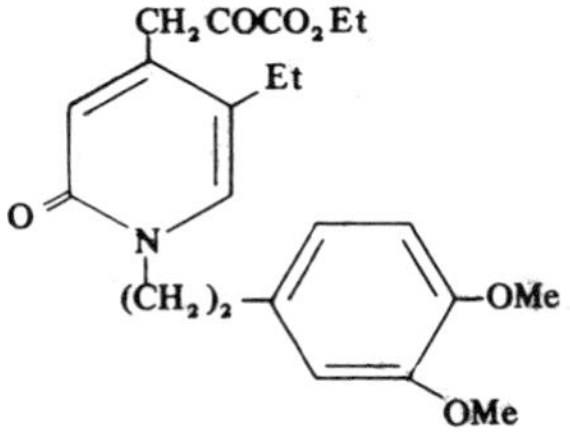 | $MnO_2$ | 10% aq. NaOH | 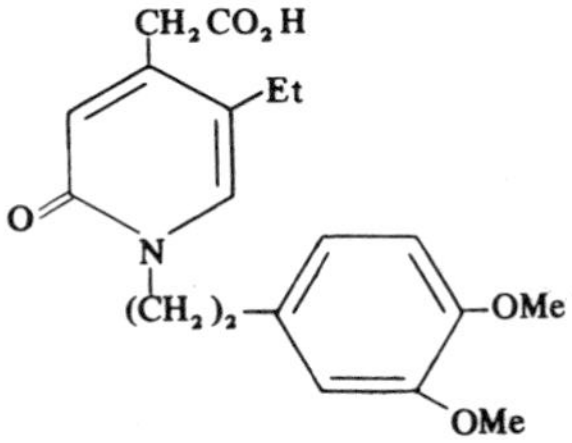 | 76.5% | hydrochloride, m.p. 160.5–162.5° | 4 |
| 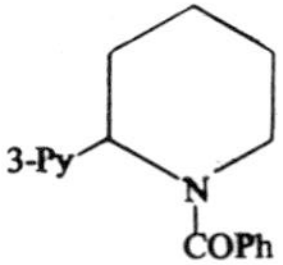 | $KMnO_4$ | | 3-PyCHNHCOPh<br>$(CH_2)_3CO_2H$ | 22% | m.p. 145–146°; $[\alpha]_D^{21}$ 9.5° | 75 |
| 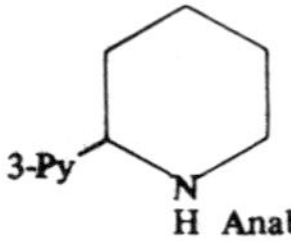 Anabasine | 34% $H_2O_2$ | HOAc, 70–80°, 24 hr | 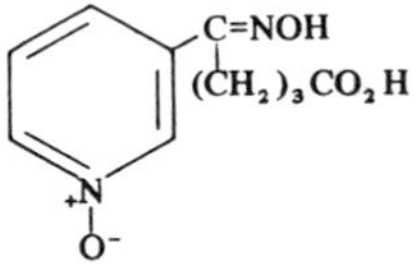 | | m.p. 229–231° (decomp.) | 75 |

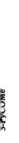

TABLE XI-2. Preparation of Side-Chain Acids by Oxidation (Continued)

| Starting material | Oxidant | Conditions | Product | Yield | Properties | Ref. |
|---|---|---|---|---|---|---|
| 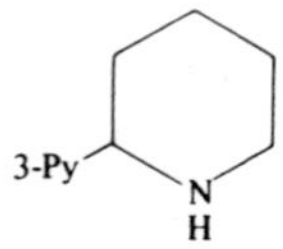 3-Py (2-piperidyl, N–H) | 10% $H_2O_2$ | R.T., 8 days | 3-PyC=NOH with $(CH_2)_3CO_2H$ |  | m.p. 161–163° | 75 |
| 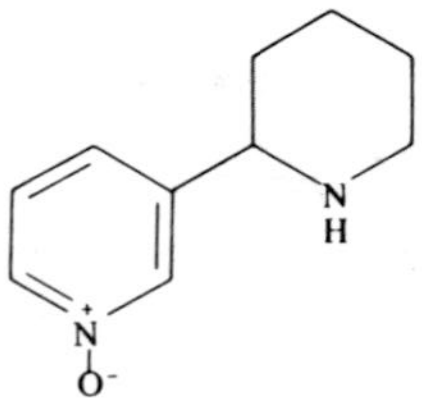 3-(2-piperidyl)pyridine N-oxide (N–H) | $H_2O_2$ | HOAc, 65–80°, 11 hr | 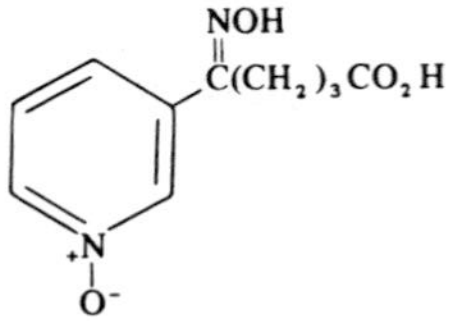 pyridine N-oxide, 3-C(=NOH)$(CH_2)_3CO_2H$ |  | m.p. 229–231° (decomp) | 190 |
| 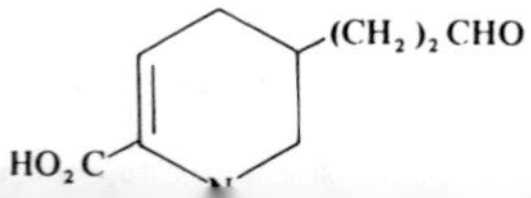 ring with $HO_2C$, N, $(CH_2)_2CHO$ | *Pseudomonas fluorescens* |  | 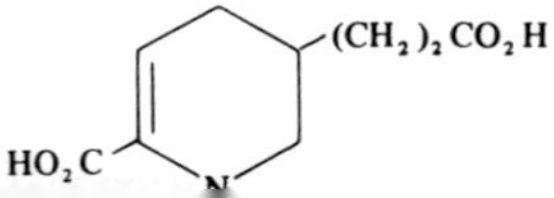 ring with $HO_2C$, N, $(CH_2)_2CO_2H$ |  |  | 73 |

| | | | | | |
|---|---|---|---|---|---|
| HO, OH, N, $CO_2H$ | oxygenase | | N, H, $CO_2H$ | | 73 |
| 3-Py, N, Me — nicotine | mature resting cell enzyme in *Anicotinophagum* | | O, N, H, $CO(CH_2)_2CO_2H$ | | 74 |
| 4-$PyCH_2CO_2Me$ | $SeO_2$ | HOAc, benzene, reflux 30 min | 4-$PyC(OH)_2CO_2Me$ | b.p. 150°/3 mm, m.p. 114–118° | 31 32 |
| H, $CH_2CO_2H$, N, H, $CO_2H$ | $O_2$ | acid, boil | $CH_2CO_2H$, N, $CO_2H$ | | 76 |

TABLE XI-3. Carbonation of Organometallic Compounds

| Starting materials | Conditions | Product | Yield | Properties | Ref. |
|---|---|---|---|---|---|
| 2-PyMe | PhNa, PhMe, 1 hr, 35°; $CO_2$, $Et_2O$; EtOH, $H_2SO_4$, reflux 3 hr | $2\text{-PyCH}_2CO_2Et$ | 46.5 g ↓ 41.5 g | b.p. 119–122°/12 mm; picrate, m.p. 139° | 77 |
| | or PhBr, Li, reflux $Et_2O$, 30 min, dry ice; EtOH, HCl | | 139.7 g ↓ 83.1 g | b.p. 109–112°/6 mm | 66 |
| | PhLi, dry ice; $Br(CH_2)_2OH$, R.T., 4.5 days | $2\text{-PyCH}_2CO_2(CH_2)_2Br$ | | picrate, m.p. 126–128° | 191 |
| 2,3-$Me_2$Py | PhLi, $Et_2O$, 30 min; $CO_2$; EtOH, HCl, 24 hr, R.T. | 3-Me-2-Py$CH_2CO_2Et$ | 33–34% | b.p. 141–145°/18 mm; picrate, m.p. 125.5–126.5° | 78 |
| 2,4-$Me_2$Py | same as above | 4-Me-2-Py$CH_2CO_2Et$ | | b.p. 142–146°/17 mm; m.p. 130–131°; picrolonate, m.p. 165–166° | 78 |
| 2,5-$Me_2$Py | same as above | 5-Me-2-Py$CH_2CO_2Et$ + 2,5-$Me_2$-biphenyl (Me, Ph, Me benzene) | 56% | b.p. 100–105°/2 mm; picrate, m.p. 136–137° | 78 |

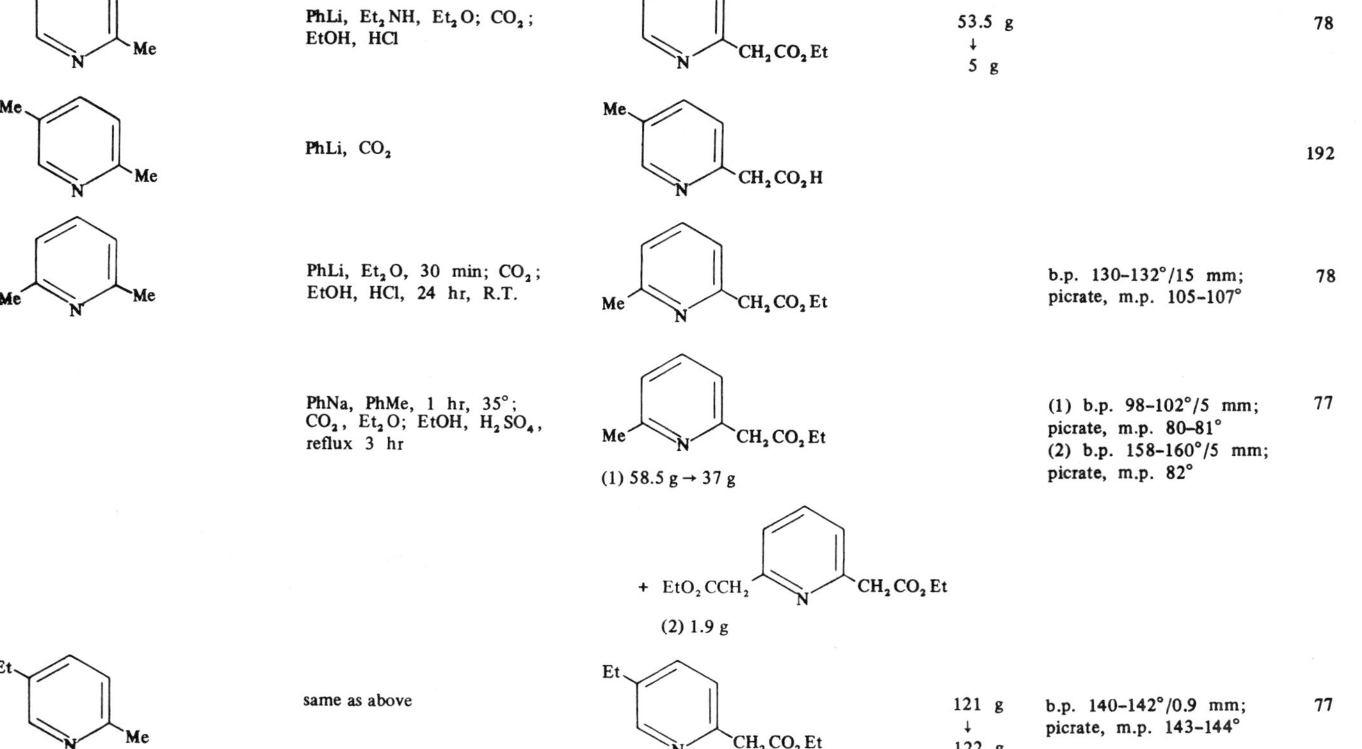

| | | | | |
|---|---|---|---|---|
| PhLi, $Et_2NH$, $Et_2O$; $CO_2$; EtOH, HCl | $CH_2CO_2Et$ | 53.5 g → 5 g | | 78 |
| PhLi, $CO_2$ | $CH_2CO_2H$ | | | 192 |
| PhLi, $Et_2O$, 30 min; $CO_2$; EtOH, HCl, 24 hr, R.T. | $CH_2CO_2Et$ | | b.p. 130–132°/15 mm; picrate, m.p. 105–107° | 78 |
| PhNa, PhMe, 1 hr, 35°; $CO_2$, $Et_2O$; EtOH, $H_2SO_4$, reflux 3 hr | $CH_2CO_2Et$ (1) 58.5 g → 37 g + $EtO_2CCH_2$ … $CH_2CO_2Et$ (2) 1.9 g | | (1) b.p. 98–102°/5 mm; picrate, m.p. 80–81° (2) b.p. 158–160°/5 mm; picrate, m.p. 82° | 77 |
| same as above | $CH_2CO_2Et$ | 121 g → 122 g | b.p. 140–142°/0.9 mm; picrate, m.p. 143–144° | 77 |

TABLE XI-3. Carbonation of Organometallic Compounds (Continued)

| Starting material | Conditions | Product | Yield | Properties | Ref. |
|---|---|---|---|---|---|
| OMe, N, Me (4-methoxy-2-methylpyridine structure) | PhLi, PhBr, $Et_2O$, dry ice, overnight; EtOH; HCl, 10 hr, R.T. | OMe, N, $CH_2CO_2Et$ (structure) | 44 g ↓ 20.8 g | b.p. 94.0–94.5°/0.5 mm; $n_D^{21}$ 1.5038; picrate, m.p. 156–158°; ir | 193 |
| 2-PyEt | PhLi, $Et_2O$, 30 min; $CO_2$; EtOH, HCl, 24 hr, R.T. | 2-PyCH(Me)$CO_2$Et | | b.p. 124-127°/13 mm | 78 |
| 2-Py$(CH_2)_4$C≡CH | $NaNH_2$, $Et_2O$, $N_2$, overnight; $CO_2$, 3 hr, R.T. | 2-Py$(CH_2)_4$C≡C$CO_2$H | 55.6% | m.p. 95°, 130°; hydrochloride, m.p. 109–110°, 107–109°; picrate, m.p. 97°; ir | 80, 81 |
| 2-Py$(CH_2)_5$C≡CH | same as above | 2-Py$(CH_2)_5$C≡C$CO_2$H | 72% | m.p. 95°, hydrochloride, m.p. 109–110°; ir | 80 |
| 2-Py$(CH_2)_{10}$C≡CH | $NaNH_2$, $NH_3$, $Et_2O$, reflux 7 hr; $CO_2$, R.T., overnight; $Et_2O$, reflux several hr | 2-Py$(CH_2)_{10}$C≡C$CO_2$H | 80% | m.p. 87°; hydrochloride, m.p. 97–98°; picrate, m.p. 68°; ir | 80, 81 |

| | | | | | |
|---|---|---|---|---|---|
| 2-pyridyl–C(Me)(OH)C≡CH (partial structure) | same as above | 2-pyridyl–C(Me)(OH)C≡$CCO_2$Me (partial structure) | | m.p. 115–117° | 79 |
| 3-PyCH$(Ph)_2$ | $[Ph_2Li]Na$, $CO_2$, $Et_2O$, $N_2$ | 3-PyC($CO_2$H)$(Ph)_2$ | | m.p. 113–114° (decomp) | 194 |
| $(3\text{-Py})_2$CHPh | same as above | $(3\text{-Py})_2$C(Ph)$CO_2$H | 40% | unstable | 194 |
| 4-PyMe | PhNa, PhMe, 1 hr, 35°; $CO_2$, $Et_2O$; EtOH, $H_2SO_4$, reflux 3 hr | 4-Py$CH_2CO_2$Et | 46.5 g ↓ 33.5 g | b.p. 129–132°/12 mm; picrate, m.p. 122° | 77 |
| 4-Py$(CH_2)_4$C≡CH | $NaNH_2$, $NH_3$, $Et_2O$; reflux 7 hr; $CO_2$, R.T. overnight; $Et_2O$, reflux several hr | 4-Py$(CH_2)_4$C≡$CCO_2$H | 70% | m.p. 235°; hydrochloride, m.p. 187°; picrate, m.p. 100–101°; ir | 80, 81 |
| 4-Py$(CH_2)_5$C≡CH | same as above | 4-Py$(CH_2)_5$C≡$CCO_2$H | 71% | m.p. 139–140°; picrate, m.p. 103–104°; ir | 80 |

TABLE XI-4. Arndt-Eistert Synthesis

| Starting materials | Conditions | Product | Yield | Properties | Ref. |
|---|---|---|---|---|---|
| 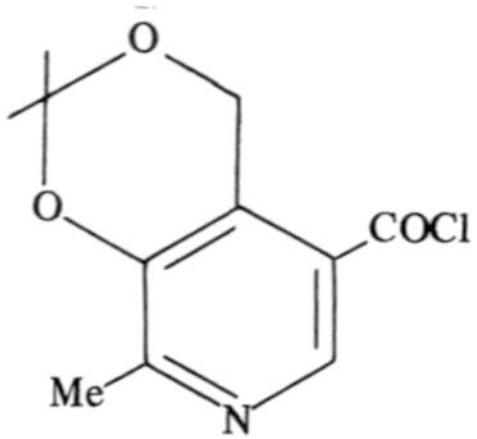 O, O, COCl, Me, N | $CH_2N_2$, conc. HCl, $Et_2O$, R.T., overnight; $Ag_2O$, $Na_2CO_3$, 50–60°, aq. sodium thiosulfate | 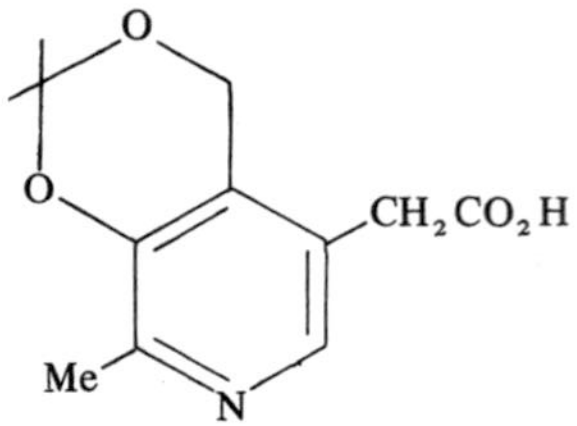 O, O, $CH_2CO_2H$, Me, N | 22% | m.p. 186–187° (decomp.); ir, nmr | 195 |
| 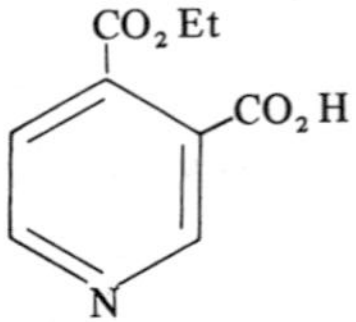 $CO_2Et$, $CO_2H$, N | $SOCl_2$ ; $CH_2N_2$ ; $Ag_2O$ abs. EtOH | 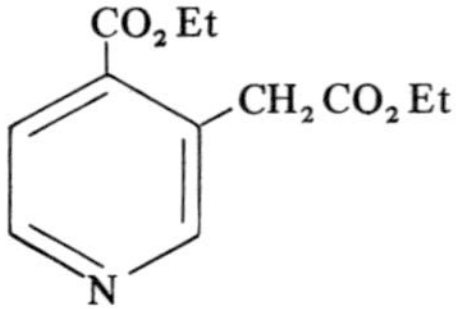 $CO_2Et$, $CH_2CO_2Et$, N | | b.p. 116–117°/0.05 mm; picrate, m.p. 104–106° | 128 |

| Halide | Active methylene compound | Conditions | Product | Yield | Properties | Ref. |
|---|---|---|---|---|---|---|
| 2-PyBr | $PhCH_2CONMe_2$ | $NaNH_2$, PhMe, reflux 2 hr | 2-PyCH(Ph)$CONMe_2$ | 57% | m.p. 92–93.5° | 82, 158 |
| 2-PyBr | 2-Py$CH_2$CON | same as above | $(2\text{-Py})_2$CHCON | | b.p. 220–225°/3 mm; m.p. 103–104° | 158 |
| 2-PyBr | $PhCH_2CON$ | same as above | 2-PyCH(Ph)CON | | m.p. 107–109° | 158 |
| 2-PyBr | $PhCH_2CONEt_2$ | same as above | 2-PyCH(Ph)$CONEt_2$ | 50% | b.p. 185–190°/1.0 mm; m.p. 63–64.5° | 82 |
| 2-PyBr | $PhCH_2CON(isoPr)_2$ | same as above | 2-PyCH(Ph)$CON(isoPr)_2$ | 45% | m.p. 104–106° | 82 |
| 2-PyBr | $PhCH_2CON$ | same as above | 2-PyCH(Ph)CON | 55% | m.p. 108.5–110° | 82 |
| 2-PyBr | $PhCH_2CON$ | same as above | 2-PyCH(Ph)CON | 71% | b.p. 195–197°/0.4 mm; m.p. 72.5–73.5° | 82 |
| 2-PyCl | $PhCH_2CON$ | same as above | 2-PyCH(Ph)CON | 66% | | 82 |
| 2-PyBr | $PhCH_2CON$ (O) | same as above | 2-PyCH(Ph)CON (O) | 51% | b.p. 195–198°/0.4 mm; m.p. 84–85.5° | 82 |

TABLE XI-5. Condensation of Halopyridines with Active Methylene Compounds (Continued)

| Halide | Active methylene compound | Conditions | Product | Yield | Properties | Ref. |
|---|---|---|---|---|---|---|
| 2-PyBr | $PhCH_2CON(Me)Ph$ | same as above | 2-PyCH(Ph)CON(Me)Ph | 41% | m.p. 97–98.5° | 82 |
| 2-PyBr | $p\text{-}ClC_6H_4CH_2CONMe_2$ | same as above | $2\text{-}PyCH(p\text{-}ClC_6H_4)CONMe_2$ | 80% | b.p. 180–205°/0.2 mm; m.p. 75–76.5° | 82 |
| 2-PyBr | $p\text{-}ClC_6H_4CH_2CONEt_2$ | same as above | $2\text{-}PyCH(p\text{-}ClC_6H_4)CONEt_2$ | 70% | m.p. 69.5–70° | 82 |
| 2-PyBr | $p\text{-}BrC_6H_4CH_2CONEt_2$ | same as above | $2\text{-}PyCH(p\text{-}BrC_6H_4)CONEt_2$ | | b.p. 190–200°/1.0 mm | 82 |
| 2-PyBr | $p\text{-}BrC_6H_4CH_2CON$ 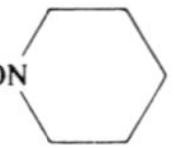 (ring) | same as above | $2\text{-}PyCH(p\text{-}BrC_6H_4)CON$  (ring) | | b.p. 210–220°/1.0 mm | 82 |
| 2-PyBr | $p\text{-}MeC_6H_4CH_2CONMe_2$ | same as above | $2\text{-}PyCH(p\text{-}MeC_6H_4)CONMe_2$ | | b.p. 180–185°/1.0 mm | 82 |
| 2-PyBr | $(\alpha\text{-naphthyl})CH_2CONEt_2$ | same as above | $2\text{-}PyCH(\alpha\text{-naphthyl})CONEt_2$ | 59% | m.p. 165.5–166° | 82 |
| 2-PyBr | 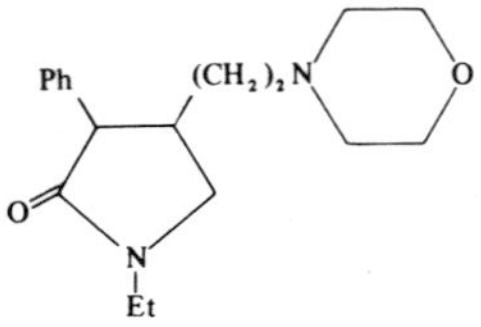 Ph; $(CH_2)_2N$ (ring) O; O; N–Et | $NaNH_2$, PhMe | 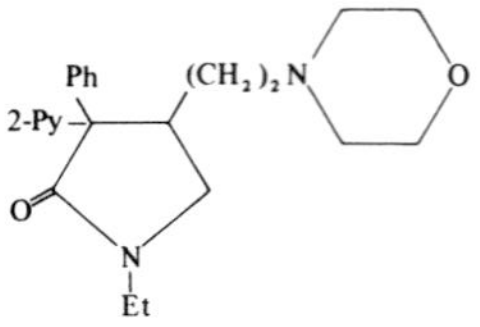 Ph; 2-Py; $(CH_2)_2N$ (ring) O; O; N–Et | 27% | m.p. 91–92°; hydrochloride hydrate, m.p. 230° | 50 |
| $2\text{-}Cl\text{-}3\text{-}NO_2Py$ | $NCCH_2CO_2Me$ | $Me_3CO^{\ominus}K^{\oplus}$, $Me_3COH$, reflux 5–12 hr | 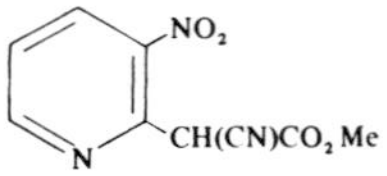 $NO_2$; N; $CH(CN)CO_2Me$ | 82% | m.p. 186–188° | 84 |
| $2\text{-}Cl\text{-}3\text{-}NO_2Py$ | $NCCH_2CO_2Et$ | same as above | 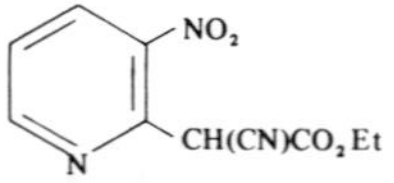 $NO_2$; N; $CH(CN)CO_2Et$ | 87% | m.p. 136–137° | 84 |

| | | | | | | |
|---|---|---|---|---|---|---|
| 2-Cl-3-$NO_2$Py | $NCCH_2CO_2CH_2Ph$ | same as above | 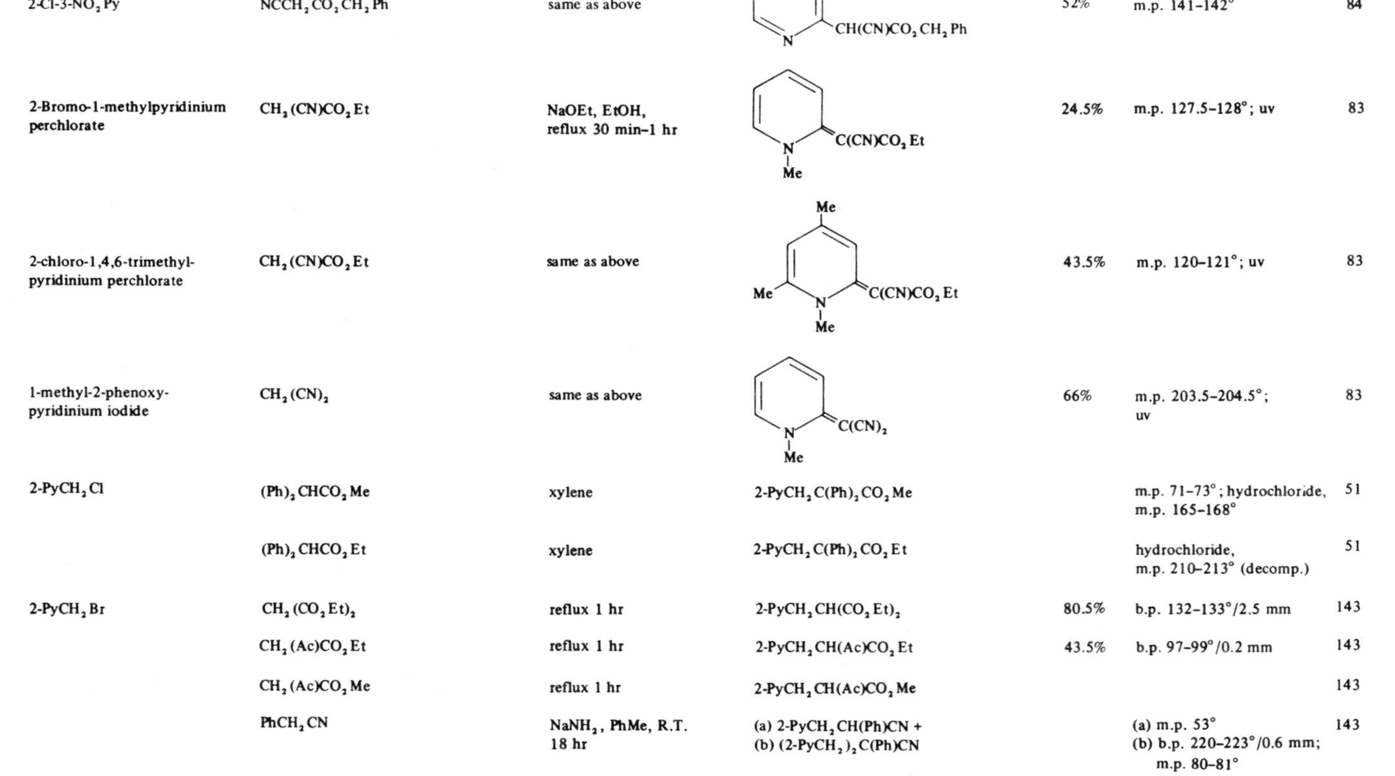 $CH(CN)CO_2CH_2Ph$ | 52% | m.p. 141–142° | 84 |
| 2-Bromo-1-methylpyridinium perchlorate | $CH_2(CN)CO_2Et$ | NaOEt, EtOH, reflux 30 min–1 hr | N–Me, $C(CN)CO_2Et$ | 24.5% | m.p. 127.5–128°; uv | 83 |
| 2-chloro-1,4,6-trimethyl-pyridinium perchlorate | $CH_2(CN)CO_2Et$ | same as above | Me, Me, N–Me, $C(CN)CO_2Et$ | 43.5% | m.p. 120–121°; uv | 83 |
| 1-methyl-2-phenoxy-pyridinium iodide | $CH_2(CN)_2$ | same as above | N–Me, $C(CN)_2$ | 66% | m.p. 203.5–204.5°; uv | 83 |
| 2-$PyCH_2Cl$ | $(Ph)_2CHCO_2Me$ | xylene | 2-$PyCH_2C(Ph)_2CO_2Me$ | | m.p. 71–73°; hydrochloride, m.p. 165–168° | 51 |
| | $(Ph)_2CHCO_2Et$ | xylene | 2-$PyCH_2C(Ph)_2CO_2Et$ | | hydrochloride, m.p. 210–213° (decomp.) | 51 |
| 2-$PyCH_2Br$ | $CH_2(CO_2Et)_2$ | reflux 1 hr | 2-$PyCH_2CH(CO_2Et)_2$ | 80.5% | b.p. 132–133°/2.5 mm | 143 |
| | $CH_2(Ac)CO_2Et$ | reflux 1 hr | 2-$PyCH_2CH(Ac)CO_2Et$ | 43.5% | b.p. 97–99°/0.2 mm | 143 |
| | $CH_2(Ac)CO_2Me$ | reflux 1 hr | 2-$PyCH_2CH(Ac)CO_2Me$ | | | 143 |
| | $PhCH_2CN$ | $NaNH_2$, PhMe, R.T. 18 hr | (a) 2-$PyCH_2CH(Ph)CN$ + (b) (2-$PyCH_2)_2C(Ph)CN$ | | (a) m.p. 53° (b) b.p. 220–223°/0.6 mm; m.p. 80–81° | 143 |

TABLE XI-5. Condensation of Halopyridines with Active Methylene Compounds (Continued)

| Halide | Active methylene compound | Conditions | Product | Yield | Properties | Ref. |
|---|---|---|---|---|---|---|
| 2-chloromethyl-6-methylpyridine | $(Ph)_2CHCO_2Me$ | xylene | 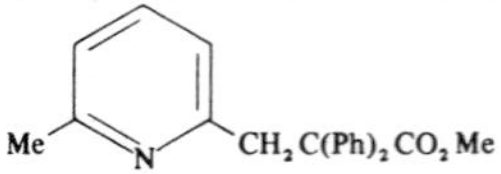 Me–(pyridine, N)–$CH_2C(Ph)_2CO_2Me$ | | m.p. 88–90°; hydrochloride, m.p. 170–172° (decomp.) | 51 |
| 2-chloromethyl-6-methylpyridine | $(Ph)_2CHCO_2Et$ | xylene | 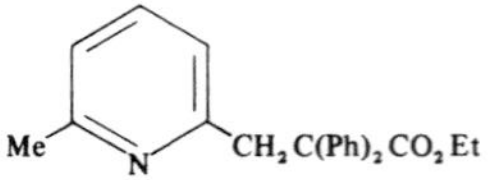 Me–(pyridine, N)–$CH_2C(Ph)_2CO_2Et$ | | hydrochloride, m.p. 194–197° (decomp) | 51 |
| 2-chloromethyl-6-methylpyridine | $(Ph)_2CHCO_2(CH_2)_2NEt_2$ | xylene | 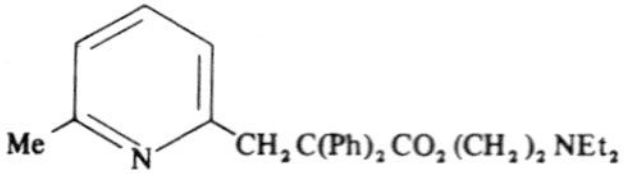 Me–(pyridine, N)–$CH_2C(Ph)_2CO_2(CH_2)_2NEt_2$ | | dihydrochloride, m.p. 144–147° | 51 |
| 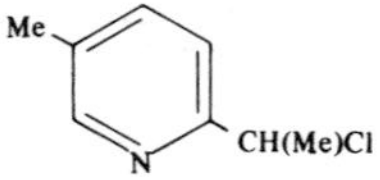 Me–(pyridine, N)–CH(Me)Cl | $CH_2(CO_2Et)_2$ | PhMe, Na, reflux 15 hr | 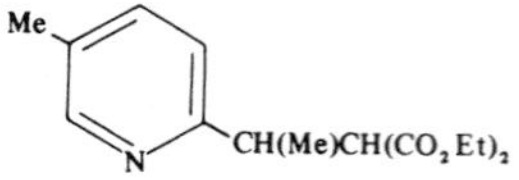 Me–(pyridine, N)–$CH(Me)CH(CO_2Et)_2$ | 4.3 g ↓ 4 g | b.p. 163–164° | 196 |
| 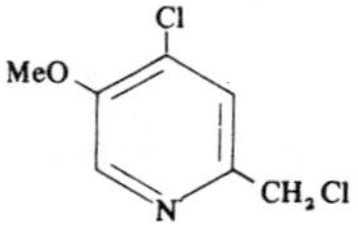 Cl, MeO–(pyridine, N)–$CH_2Cl$ | $AcNHCH(CO_2Et)_2$ | Na, EtOH, reflux 24 hr | 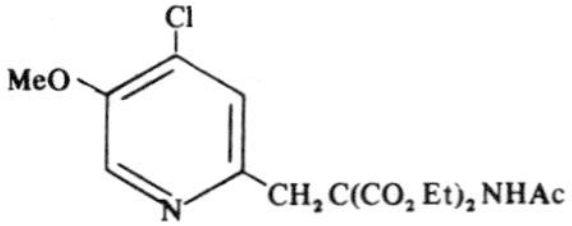 Cl, MeO–(pyridine, N)–$CH_2C(CO_2Et)_2NHAc$ | 32 g ↓ 38.4 g | m.p. 150–151° | 95 |
| 3-PyCl |  $PhCH_2CON$(ring) | $NaNH_2$, PhMe reflux 2 hr |  3-PyCH(Ph)CON(ring) | | m.p. 149°; hydrochloride | 158 |

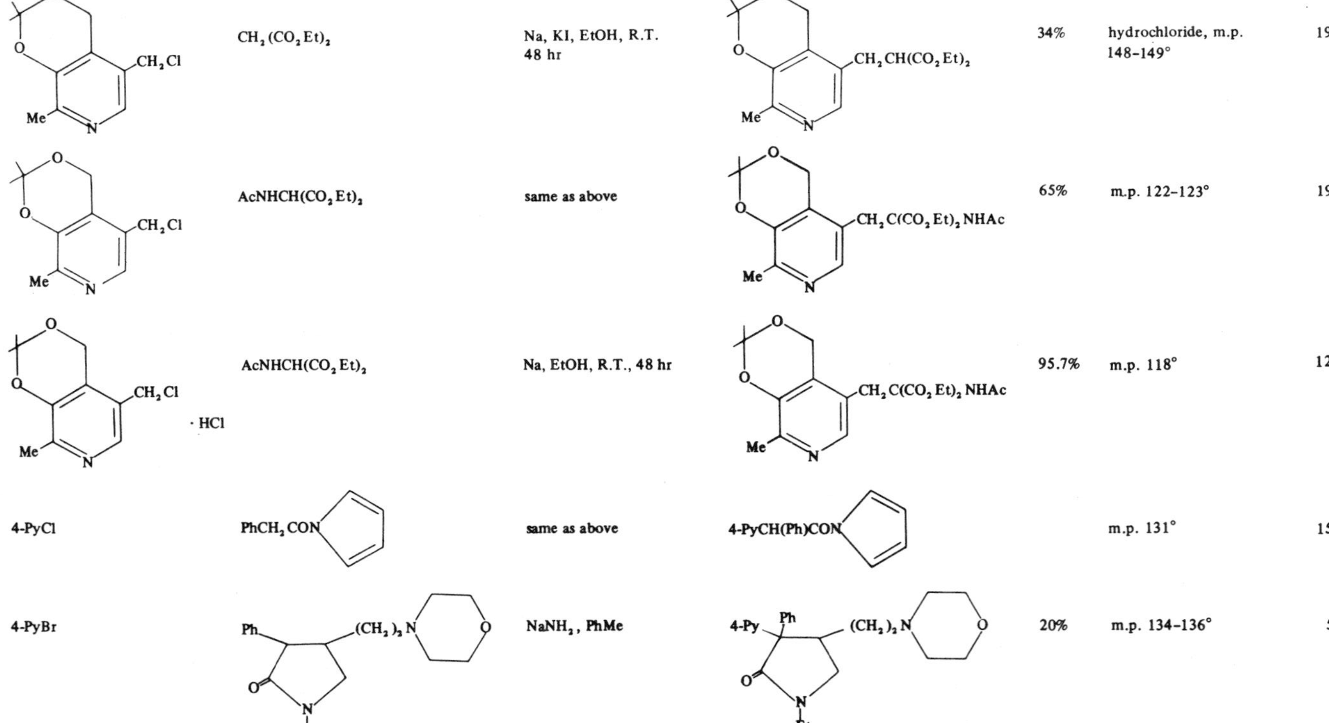

| | | | | | | |
|---|---|---|---|---|---|---|
| | $CH_2(CO_2Et)_2$ | Na, KI, EtOH, R.T. 48 hr | | 34% | hydrochloride, m.p. 148–149° | 195 |
| | $AcNHCH(CO_2Et)_2$ | same as above | | 65% | m.p. 122–123° | 195 |
| · HCl | $AcNHCH(CO_2Et)_2$ | Na, EtOH, R.T., 48 hr | | 95.7% | m.p. 118° | 125 |
| 4-PyCl | | same as above | | | m.p. 131° | 158 |
| 4-PyBr | | $NaNH_2$, PhMe | | 20% | m.p. 134–136° | 50 |

TABLE XI-5. Condensation of Halopyridines with Active Methylene Compounds (Continued)

| Halide | Active methylene compound | Conditions | Product | Yield | Properties | Ref. |
|---|---|---|---|---|---|---|
| 4-chloro-3-nitropyridine | 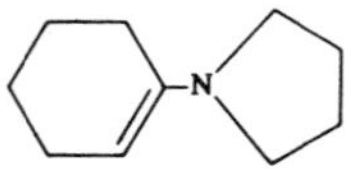 1-(1-cyclohexenyl)pyrrolidine [structure] | $CH_2Cl_2$, $Et_3N$, $N_2$, 4 days | 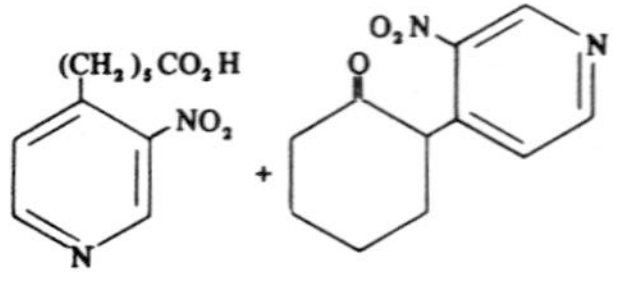 $(CH_2)_5CO_2H$, $NO_2$, N (a) + O, $O_2N$, N [structures] | | (a) m.p. 112–114° | 119 |
| 4-methoxy-3-nitropyridine | $NCCH_2CO_2Et$ | $Me_3CO^{\ominus}K^{\oplus}$, $Me_3COH$, reflux 5–12 hr | 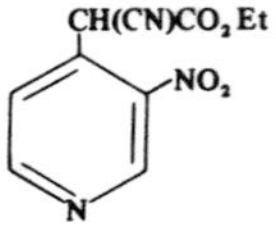 $CH(CN)CO_2Et$, $NO_2$, N [structure] | 52% | m.p. 177–178° | 84 |
| 4-methoxy-3-nitropyridine | $NCCH_2CO_2CH_2Ph$ | same as above | 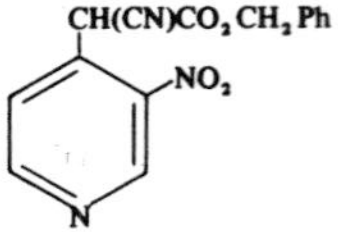 $CH(CN)CO_2CH_2Ph$, $NO_2$, N [structure] | 60% | m.p. 204° (decomp) | 84 |
| 1-methyl-4-phenoxy-pyridinium iodine | $CH_2(CN)_2$ | same as above | 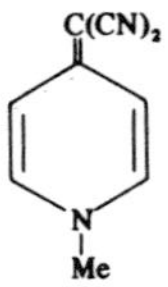 $C(CN)_2$, N–Me [structure] | 63% | m.p. 237–239° | 83 |
| 1-methyl-4-phenoxy-pyridinium iodide | $CH_2(CN)CO_2Et$ | same as above | 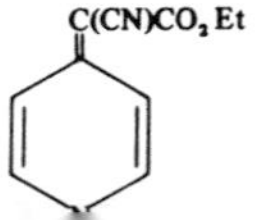 $C(CN)CO_2Et$ [structure] | 37% | m.p. 179–179.5° ; uv | 83 |

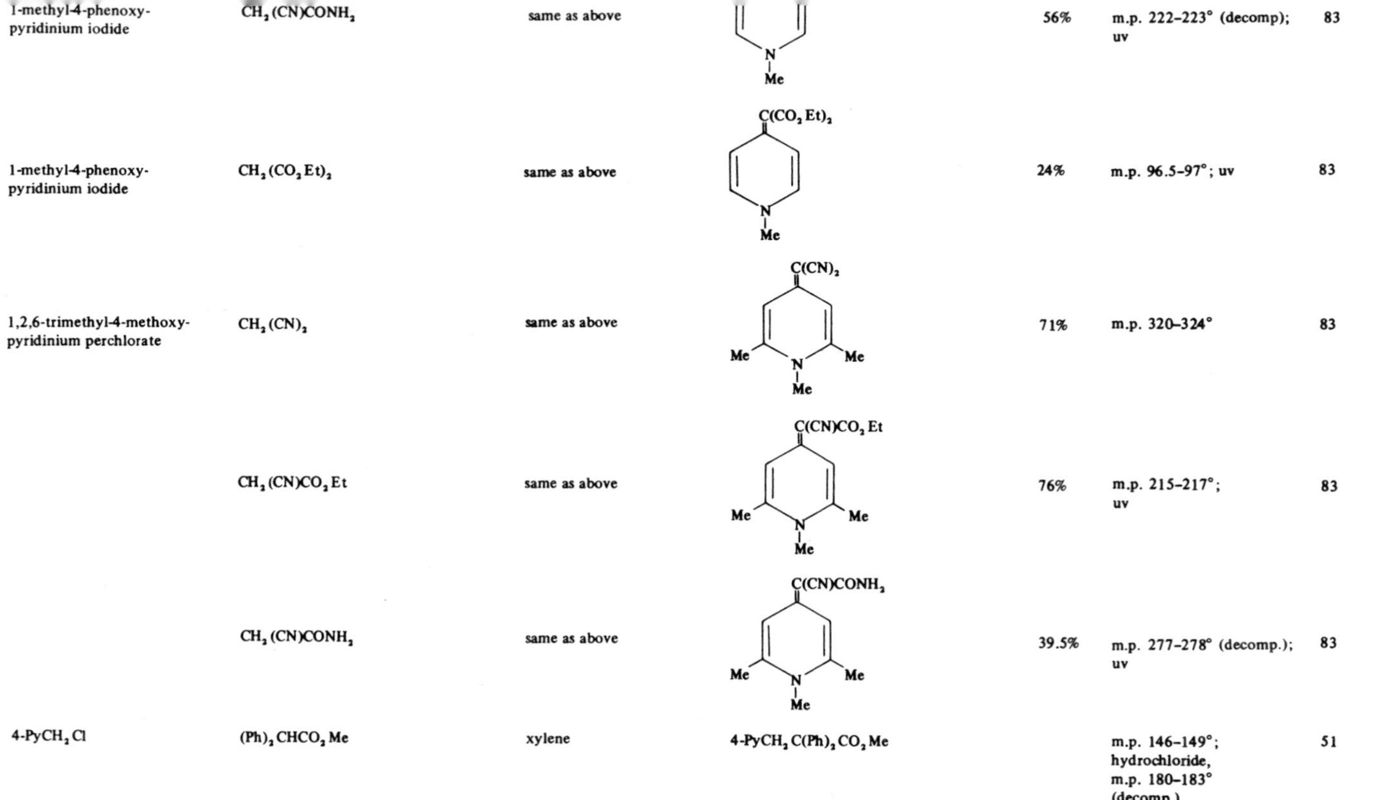

| | | | | | | |
|---|---|---|---|---|---|---|
| 1-methyl-4-phenoxy-pyridinium iodide | $CH_2(CN)CONH_2$ | same as above | (structure) N–Me | 56% | m.p. 222–223° (decomp); uv | 83 |
| 1-methyl-4-phenoxy-pyridinium iodide | $CH_2(CO_2Et)_2$ | same as above | $C(CO_2Et)_2$ (structure) N–Me | 24% | m.p. 96.5–97°; uv | 83 |
| 1,2,6-trimethyl-4-methoxy-pyridinium perchlorate | $CH_2(CN)_2$ | same as above | $C(CN)_2$ (structure) Me, N–Me, Me | 71% | m.p. 320–324° | 83 |
| | $CH_2(CN)CO_2Et$ | same as above | $C(CN)CO_2Et$ (structure) Me, N–Me, Me | 76% | m.p. 215–217°; uv | 83 |
| | $CH_2(CN)CONH_2$ | same as above | $C(CN)CONH_2$ (structure) Me, N–Me, Me | 39.5% | m.p. 277–278° (decomp.); uv | 83 |
| 4-$PyCH_2Cl$ | $(Ph)_2CHCO_2Me$ | xylene | 4-$PyCH_2C(Ph)_2CO_2Me$ | | m.p. 146–149°; hydrochloride, m.p. 180–183° (decomp.) | 51 |

TABLE XI-6. Condensation of Chloral with Picolines

| Picoline | Conditions | Product | Yield | Ref. |
|---|---|---|---|---|
| 2-PyMe | HOAc, piperidine, xylene, 12 hr, 150° | 2-$PyCH_2CH(OH)CCl_3$ | 28.5% | 114 |
| 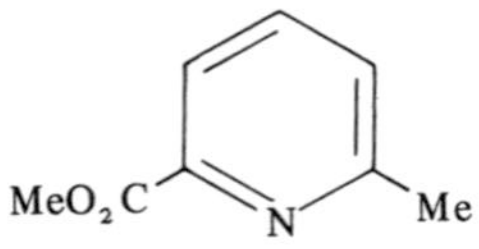 $MeO_2C$, N, Me | 100°, 100 hr | 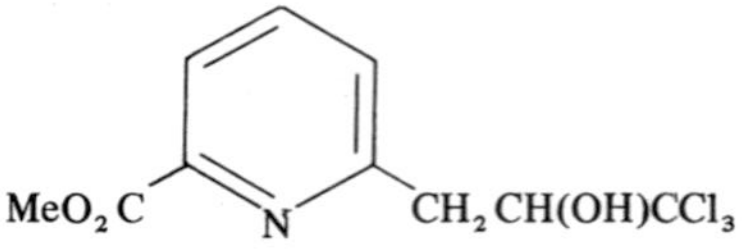 $MeO_2C$, N, $CH_2CH(OH)CCl_3$ | 45 g → 35 g | 197 |
| 4-PyMe | HOAc, piperidine, xylene, 12 hr, 150° | 4-$PyCH_2CH(OH)CCl_3$ | | 114 |
| | AmOAc | 4-$PyCH_2CH(OH)CCl_3$ | | 29 |
| 4-$PyCOCH_2CO_2Et$ | 3 hr, HOAc, steam-bath | 4-$PyCOCH(CO_2Et)CH(OH)CCl_3$ | | 105 |

| Py isomer | Conditions | Product | Yield | Properties | Ref. |
|---|---|---|---|---|---|
| 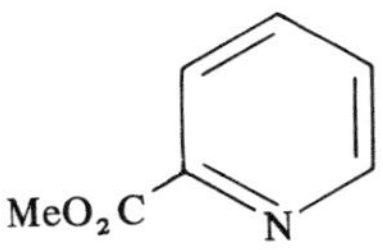 6-(methoxycarbonyl)pyridin-2-yl ($MeO_2C$-substituted pyridine ring, structure drawn) | KOH, EtOH, R.T., reflux 2 hr; dry HCl, 0° | 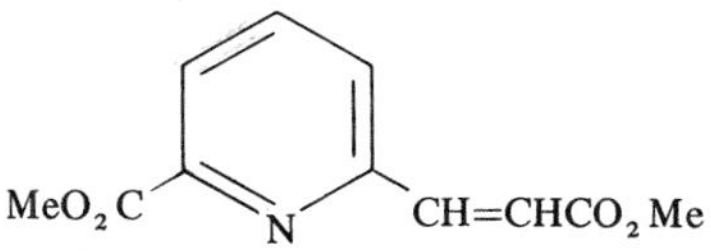 $MeO_2C$-substituted pyridine ring with $CH{=}CHCO_2Me$ (structure drawn) | 30 g → 15 g | m.p. 107–109°; ir | 197 |
| 2-Py | KOH, EtOH; EtOH, $H_2SO_4$, 4 hr, steam-bath | $2\text{-PyCH}{=}\text{CHCO}_2\text{Et}$ | 85% | b.p. 118–120°/6 mm; methiodide, m.p. 157–159° | 114 |
| | | $2\text{-PyCH}{=}\text{CHCO}_2\text{H}$ | | m.p. 200.1–200.6° | 198 |
| 4-Py | KOH, EtOH | $4\text{-PyCH}{=}\text{CHCO}_2\text{H}$ | | | 29 |
| | KOH, EtOH | $4\text{-PyCH}{=}\text{CHCO}_2\text{H}$ | 36% | ethyl ester, m.p. 64°; methiodide, m.p. 170–171° | 114 |

TABLE XI-8. Condensation with Picolines to Give Side-Chain Acids and Esters

| Picoline | Condensed with | Conditions | Product | Yield | Properties | Ref. |
|---|---|---|---|---|---|---|
| 2-PyMe | $Br(CH_2)_{10}CONH_2$ | $NaNH_2$, THF | $2\text{-Py}(CH_2)_{11}CONH_2$ | | m.p. 92–94° | 199 |
| $2\text{-PyCH}_2Li$ | $(CO_2Et)_2$ | $(2\text{-PyCH}_2)_2Cd$, $CdCl_2$, $Et_2O$, −70° | $2\text{-PyCH}_2COCO_2Et$ | 10% | m.p. 82.5–83.5°; picrate, m.p. 150–151°; 2,4-DNP, m.p. 141–142°; 2,4-DNP·$H_2SO_4$, m.p. 168–169°; oxime, m.p. 120–121° | 85, 8 |
| 3-Methoxymethyl-2-picoline | $(EtO)_2CO$ | $KNH_2$, $Et_2O$, reflux 2 hr | $CH_2OMe$, N, $CH_2CO_2Et$ | 6 g → 2 g | b.p. 150–152° | 9 |
| 5-Carbethoxy-2-picoline | $(CO_2Et)_2$ | NaOEt, $Et_2O$, 16 hr | $EtO_2C$, N, CH=C(OH)$CO_2Et$ | 3.3 g ↓ 4.4 g | m.p. 102°; copper complex, m.p. 220° | 17 |
| $EtO_2C$, Ph, N, Me | $(CO_2Et)_2$ | NaOMe, MeOH, reflux 0.5 hr | $EtO_2C$, Ph, N, CH=C(OMe)$CO_2Me$ | | m.p. 173–174°; ir; uv | 87 |
| $EtO_2C$, Ph, O, N H, Me | $(COCl)_2$ | $POCl_3$, reflux 40 min; EtOH | $EtO_2C$, Ph, O, N H, CH=C(OH)$CO_2Et$ | good | m.p. 168–170°; ir; uv | 87 |
| $HO_2C$, Ph, O, N H, Me | $(COCl)_2$ | $POCl_3$, reflux 40 min; $H_2O$ | $HO_2C$, Ph, O, N H, CH=C(OH)$CO_2H$ | 85% | m.p. 190° (decomp); ir; uv | 87 173 |

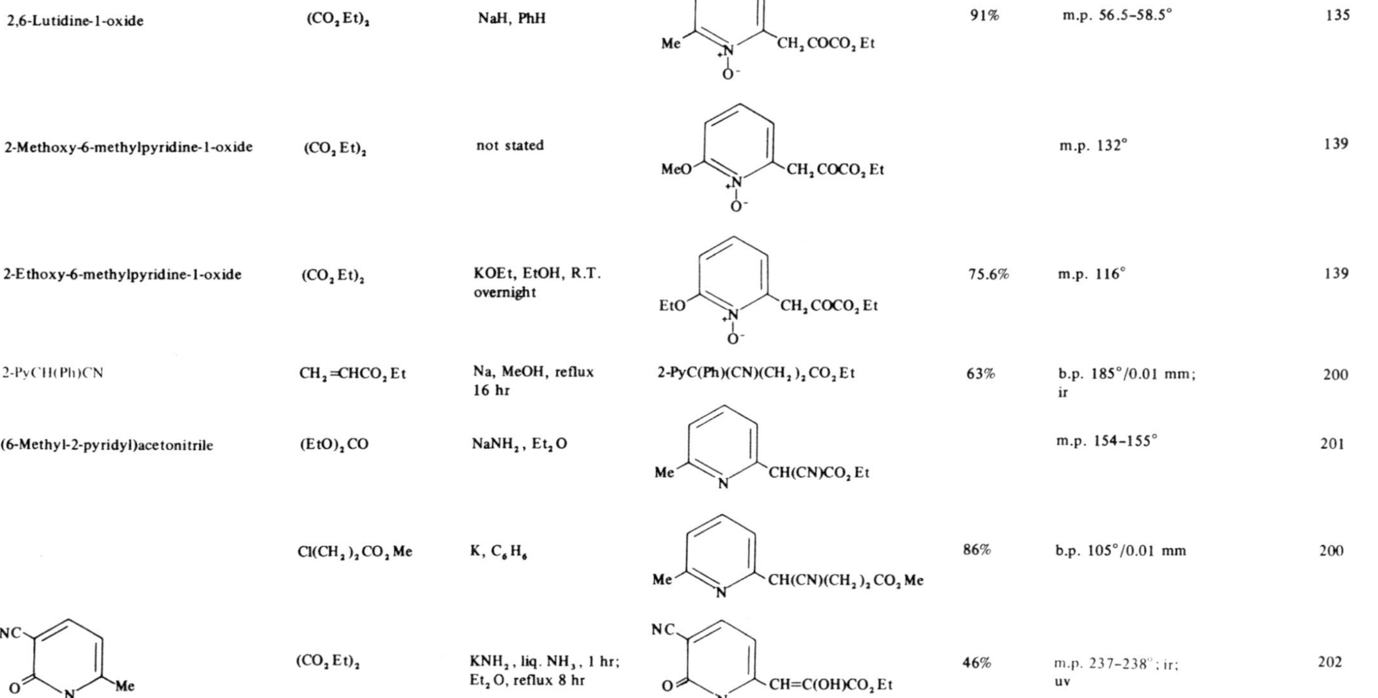

| | | | | | | |
|---|---|---|---|---|---|---|
| 2,6-Lutidine-1-oxide | $(CO_2Et)_2$ | NaH, PhH | | 91% | m.p. 56.5–58.5° | 135 |
| 2-Methoxy-6-methylpyridine-1-oxide | $(CO_2Et)_2$ | not stated | | | m.p. 132° | 139 |
| 2-Ethoxy-6-methylpyridine-1-oxide | $(CO_2Et)_2$ | KOEt, EtOH, R.T. overnight | | 75.6% | m.p. 116° | 139 |
| 2-PyCH(Ph)CN | $CH_2{=}CHCO_2Et$ | Na, MeOH, reflux 16 hr | 2-PyC(Ph)(CN)$(CH_2)_2CO_2Et$ | 63% | b.p. 185°/0.01 mm; ir | 200 |
| (6-Methyl-2-pyridyl)acetonitrile | $(EtO)_2CO$ | $NaNH_2$, $Et_2O$ | | | m.p. 154–155° | 201 |
| | $Cl(CH_2)_2CO_2Me$ | K, $C_6H_6$ | | 86% | b.p. 105°/0.01 mm | 200 |
| | $(CO_2Et)_2$ | $KNH_2$, liq. $NH_3$, 1 hr; $Et_2O$, reflux 8 hr | | 46% | m.p. 237–238°; ir; uv | 202 |

TABLE XI-8. Condensation with Picolines to Give Side-Chain Acids and Esters (Continued)

| Picoline | Condensed with | Conditions | Product | Yield | Properties | Ref. |
|---|---|---|---|---|---|---|
| 1-Methoxy-6-methyl-2-pyridone | $(CO_2Et)_2$ | KOEt, EtOH | O, N–OMe, $CH_2COCO_2Et$ |  | m.p. 153° | 139 |
| 1-Benzyl-6-methyl-2-pyridone | $(CO_2Et)_2$ | KOEt; 6% $H_2SO_4$, reflux | O, N–$CH_2Ph$, $CH_2COCO_2H$ | 99.5% | m.p. 211° | 139 |
| 3-PyMe | $Br(CH_2)_{10}CONH_2$ | $NaNH_2$, THF | 3-Py$(CH_2)_{11}CONH_2$ |  | m.p. 93–96° | 199 |
| 3-PyCOCH$_2$Br | NaC(NHAc)$(CO_2Et)_2$ | $C_6H_6$, reflux 48 hr | 3-PyCOCH$_2$C(NHAc)$(CO_2Et)_2$ | 145 g → 105 g | m.p. 139.5–141.5° | 1 |
| 3-PyCH$_2$CN | $(EtO)_2CO$ | $NaNH_2$, $Et_2O$ | 2-PyCH(CN)$CO_2Et$ |  | m.p. 107–108° | 201 |
| 4-PyMe | $Br(CH_2)_{10}CONH_2$ | $NaNH_2$, THF | 4-Py$(CH_2)_{11}CONH_2$ | 10 g → 4.1 g | m.p. 135.5° | 199 |
| 4-PyCH$_2$Li | $(CO_2Et)_2$ | (4-PyCH$_2$)$_2$Hg, $Et_2O$, -70°, $HgCl_2$ | 4-PyCH$_2$COCO$_2$Et | 10% | m.p. 138–139°; picrate, m.p. 164–165° (decomp); 2, 4-DNP, m.p. 139–140° (decomp); methiodide, m.p. 199–200° | 85, 86 |
| 4-PyCH$_2$Li | ClCO$(CH_2)_4CO_2Et$ | $Et_2O$, 25 min; 10% HCl, reflux 5 hr | 4-PyCH$_2$CO$(CH_2)_4CO_2H$ | 76 g → 29 g | m.p. 221–222° | 117 |
| 2-Methoxy-4-methyl-5-nitropyridine | $(CO_2Et)_2$ | KOEt; pH 3 | $CH_2COCO_2Et$, $NO_2$, MeO, N | 93% | m.p. 97–98°; ir; uv nmr | 15 |

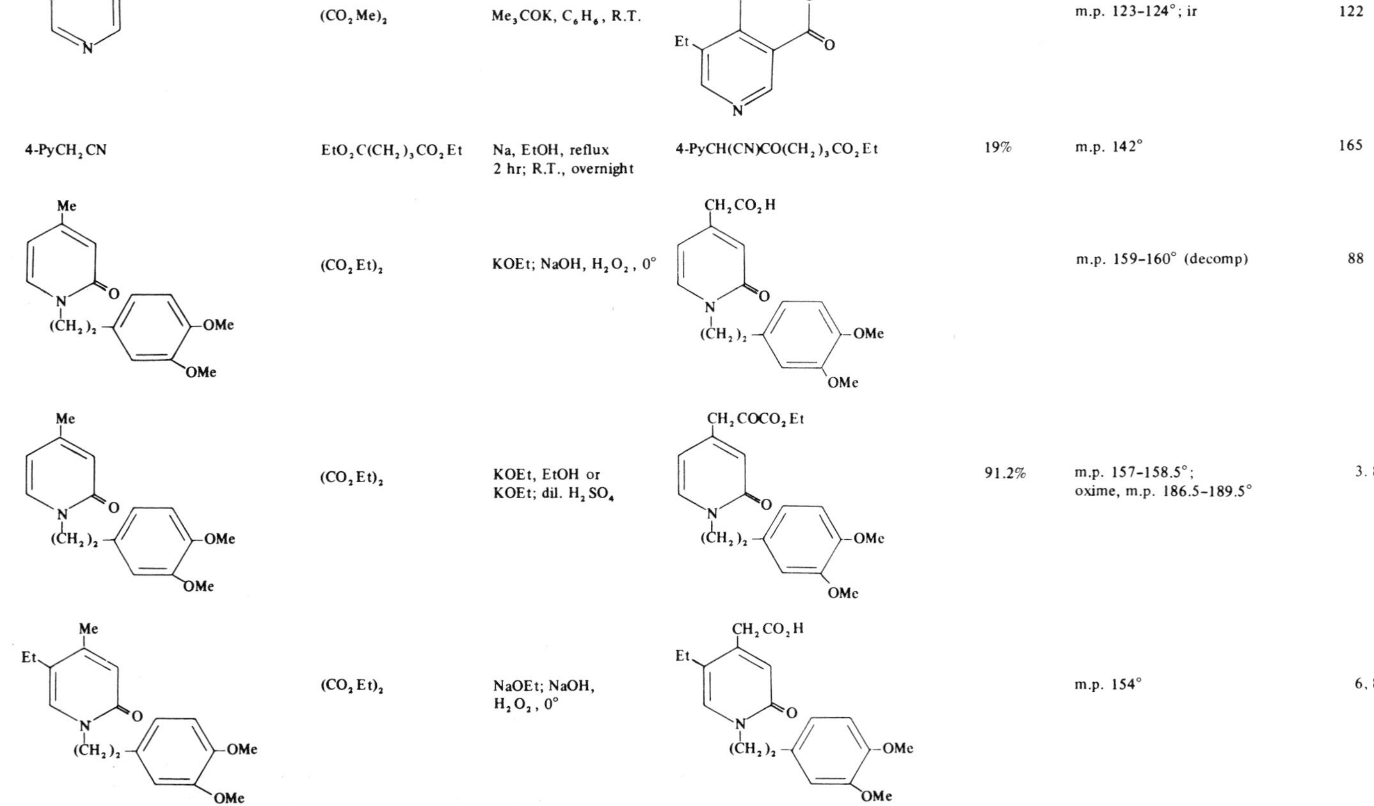

| Starting material | Reagent | Conditions | Product | Yield | Properties | Ref. |
|---|---|---|---|---|---|---|
| (structure) | $(CO_2Me)_2$ | $Me_3COK$, $C_6H_6$, R.T. | (structure, Et) | | m.p. 123–124°; ir | 122 |
| 4-$PyCH_2CN$ | $EtO_2C(CH_2)_3CO_2Et$ | Na, EtOH, reflux 2 hr; R.T., overnight | 4-$PyCH(CN)CO(CH_2)_3CO_2Et$ | 19% | m.p. 142° | 165 |
| (structure: Me, O, N, $(CH_2)_2$, OMe, OMe) | $(CO_2Et)_2$ | KOEt; NaOH, $H_2O_2$, 0° | (structure: $CH_2CO_2H$, O, N, $(CH_2)_2$, OMe, OMe) | | m.p. 159–160° (decomp) | 88 |
| (structure: Me, O, N, $(CH_2)_2$, OMe, OMe) | $(CO_2Et)_2$ | KOEt, EtOH or KOEt; dil. $H_2SO_4$ | (structure: $CH_2COCO_2Et$, O, N, $(CH_2)_2$, OMe, OMe) | 91.2% | m.p. 157–158.5°; oxime, m.p. 186.5–189.5° | 3, 88 |
| (structure: Me, Et, O, N, $(CH_2)_2$, OMe, OMe) | $(CO_2Et)_2$ | NaOEt; NaOH, $H_2O_2$, 0° | (structure: $CH_2CO_2H$, Et, O, N, $(CH_2)_2$, OMe, OMe) | | m.p. 154° | 6, 88 |

TABLE XI-8. Condensation with Picolines to Give Side-Chain Acids and Esters (Continued)

| Picoline | Condensed with | Conditions | Product | Yield | Properties | Ref. |
|---|---|---|---|---|---|---|
| 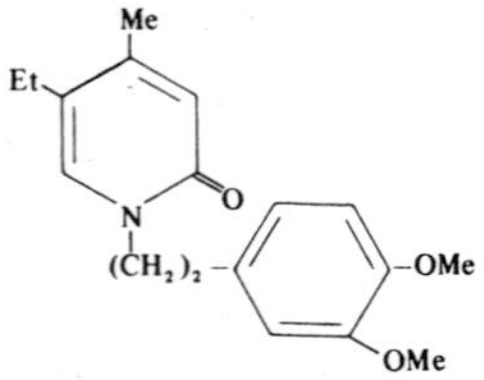 Me, Et, N, O, $(CH_2)_2$, OMe, OMe | $(CO_2Et)_2$ | KOEt; dil. $H_2SO_4$ or K, abs. EtOH, 0°; reflux 24 hr; 2*N* $H_2SO_4$ | 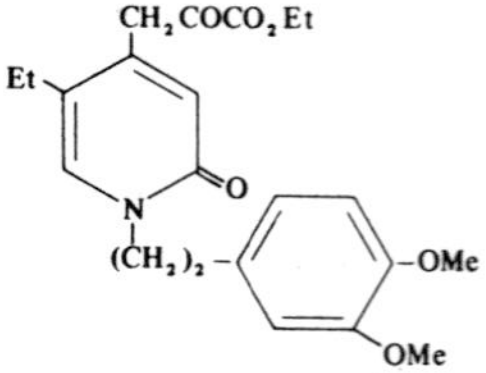 $CH_2COCO_2Et$, Et, N, O, $(CH_2)_2$, OMe, OMe | | m.p. 141.5–142.5° | 6, 4, 88 |

| Vinylpyridine | Condensed with | Conditions | Product | Yield | Properties | Ref. |
|---|---|---|---|---|---|---|
| 2-PyCH=CH$_2$ | $CH_2(CO_2Et)_2$ | hydroquinone, reflux 6 hr | 2-Py$(CH_2)_2CH(CO_2Et)_2$ | 26% | b.p. 130–134°/0.1 mm | 91 |
| | $MeCH(CO_2Et)_2$ | Na, EtOH, reflux 2 hr | 2-Py$(CH_2)_2CMe(CO_2Et)_2$ | 75% | b.p. 140°; hydrochloride, m.p. 124–126° | 200 |
| | $EtCH(CO_2Et)_2$ | hydroquinone, reflux 6 hr | 2-Py$(CH_2)_2CEt(CO_2Et)_2$ | 65% | b.p. 150–160°/0.15 mm | 91 |
| | $n$-$BuCH(CO_2Et)_2$ | Na, EtOH, reflux 7 hr | 2-Py$(CH_2)_2CBu(CO_2Et)_2$ | | b.p. 150°/0.05 mm; picrate, m.p., 102° | 203 |
| | $PhCH(CO_2Et)_2$ | hydroquinone, reflux 6 hr | 2-Py$(CH_2)_2CPh(CO_2Et)_2$ | 16% | b.p 150–152°/0.6 mm | 91 |
| | $CH_2(CN)CO_2Et$ | Na, 100°; 110°, 5 hr | 2-Py$(CH_2)_2CH(CN)CO_2Et$ | | b.p. 148–151°/0.8 mm | 204 |
| | $AcNHCH(CN)CO_2Et$ | NaOEt, $C_6H_6$, reflux 7 hr; conc. HCl, 7 hr, 120–130° | 2-Py$(CH_2)_2CH(NH_2)CO_2Et$ | 3.1 g ↓ 2.1 g | b.p. 138–140°/4 mm; picrolonate, m.p. 222° (decomp); picrate, m.p. 186° | 89 |
| | 2-Py$(CH_2)_2CH(CO_2Et)_2$ | Na, hydroquinone | (2-Py$CH_2CH_2)_2C(CO_2Et)_2$ | | dipicrate, m.p. 150–152° | 205 |
| | $EtO_2CCH{=}CHCO_2Et$ | $MeSO_3^{\ominus}$ $NEt_4^{\oplus}$, $Me_2NNO$, electrolysis at -1.3 to -1.35V reference calomel electrode for 1.88 amp. hr, Mercury Cathode | 2-Py$(CH_2)_2CH(CO_2Et)CH_2CO_2Et$ | | b.p. 141–143°/25 mm | 90 |
| 3-Me-2-PyCH=CH$_2$ | $(CH_2CO_2Et)_2$ | Na, abs. EtOH | 3-Me-2-Py$(CH_2)_2CH(CO_2Et)_2$ | 34 g ↓ 31 g | b.p. 160°/0.001 mm; picrate, m.p. 104–105.5° | 206 |
| 5-Et-2-PyCH=CH$_2$ | $EtCH(CO_2Et)_2$ | hydroquinone, reflux 6 hr | 5-Et-2-Py$(CH_2)_2CEt(CO_2Et)_2$ | 41% | b.p. 160–168°/0.3 mm | 91 |
| $CH_2{=}CH$-2,6-Py-$CH{=}CH_2$ | $CH_2(CO_2Et)_2$ | $Me_3COK$, DMF, 8 hr | $(EtO_2C)_2CH(CH_2)_2$-2,6-Py-$(CH_2)_2CH(CO_2Et)_2$ | 32-33% | b.p. 186–192°/0.03 mm | 138 |
| 5-$AcOCH_2$-2-Py$C(CH_2OAc){=}CH_2$ | | Na, EtOH | 5-$HOCH_2$-2-Py$C({=}CH_2)CH_2CH(CO_2Et)_2$ | 32% | b.p. 200°/0.001 mm; ir | 14 |

TABLE XI-9. Condensations of Vinylpyridines with Esters (Continued)

| Vinylpyridine | Condensed with | Conditions | Product | Yield | Properties | Ref. |
|---|---|---|---|---|---|---|
| 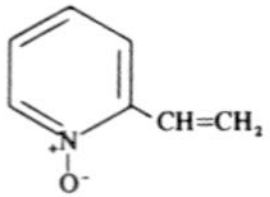 $CH{=}CH_2$, $\overset{+}{N}$–$O^-$ | $CH_2(CO_2Et)_2$ | Na, reflux 6 hr | 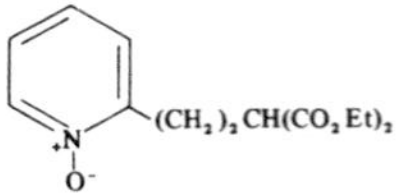 $(CH_2)_3CH(CO_2Et)_2$, $\overset{+}{N}$–$O^-$ | 42% | b.p. 140–170°/0.1 mm $n_D^{25}$ 1.5040 | 207 |
| | $CH_2(COMe)CO_2Et$ | Na, reflux 6 hr | 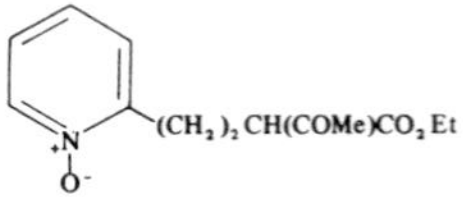 $(CH_2)_3CH(COMe)CO_2Et$, $\overset{+}{N}$–$O^-$ | 80% | b.p. 180–200°/0.1 mm $n_D^{25}$ 1.5120 | 207 |
| 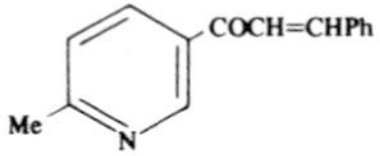 $COCH{=}CHPh$, Me, N | $CH_2(CO_2Me)_2$ | $Et_2NH$, NaOMe *or* Triton B, solvent | 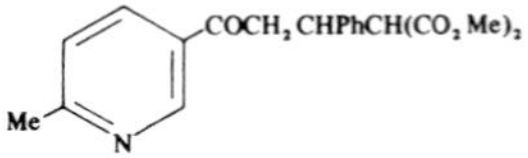 $COCH_2CHPhCH(CO_2Me)_2$, Me, N | 88% | m.p. 94°; picrate, m.p. 134° | 208 |
| | $CH_2(CO_2Et)_2$ | same as above | 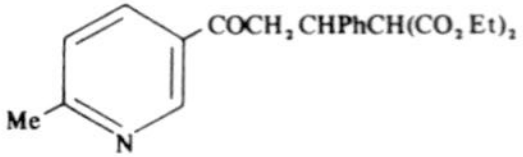 $COCH_2CHPhCH(CO_2Et)_2$, Me, N | 59% | m.p. 96–97°; picrate, m.p. 111° | 208 |
| | $MeCOCH_2CO_2Et$ | same as above | 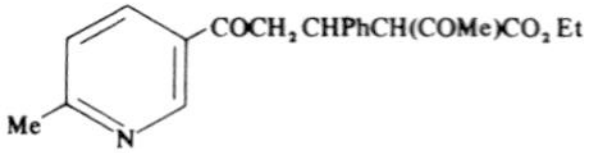 $COCH_2CHPhCH(COMe)CO_2Et$, Me, N | 82.3% | m.p. 105–106°; picrate, m.p. 166–167° | 208 |
| 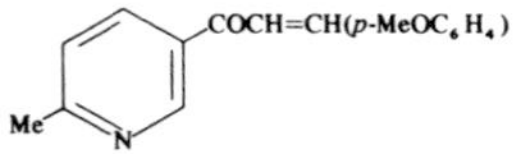 $COCH{=}CH(p\text{-}MeOC_6H_4)$, Me, N | $CH_2(CO_2Et)_2$ | same as above | 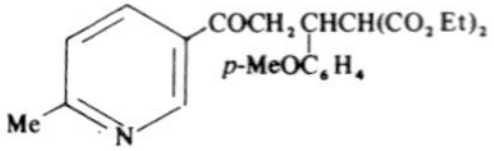 $COCH_2CHCH(CO_2Et)_2$, $p\text{-}MeOC_6H_4$, Me, N | 79.8% | m.p. 75–75.5°; picrate, m.p. 169° | 208 |
| 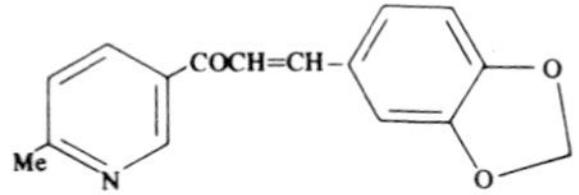 $COCH{=}CH$–, O, O, Me, N | $CH_2(CO_2Et)_2$ | same as above | 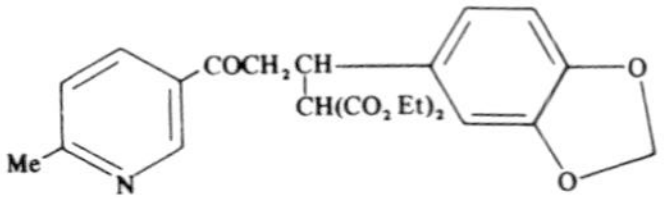 $COCH_2CH$–, $CH(CO_2Et)_2$, O, O, Me, N | 92% | m.p. 150°; picrate, m.p. 203–205° | 208 |

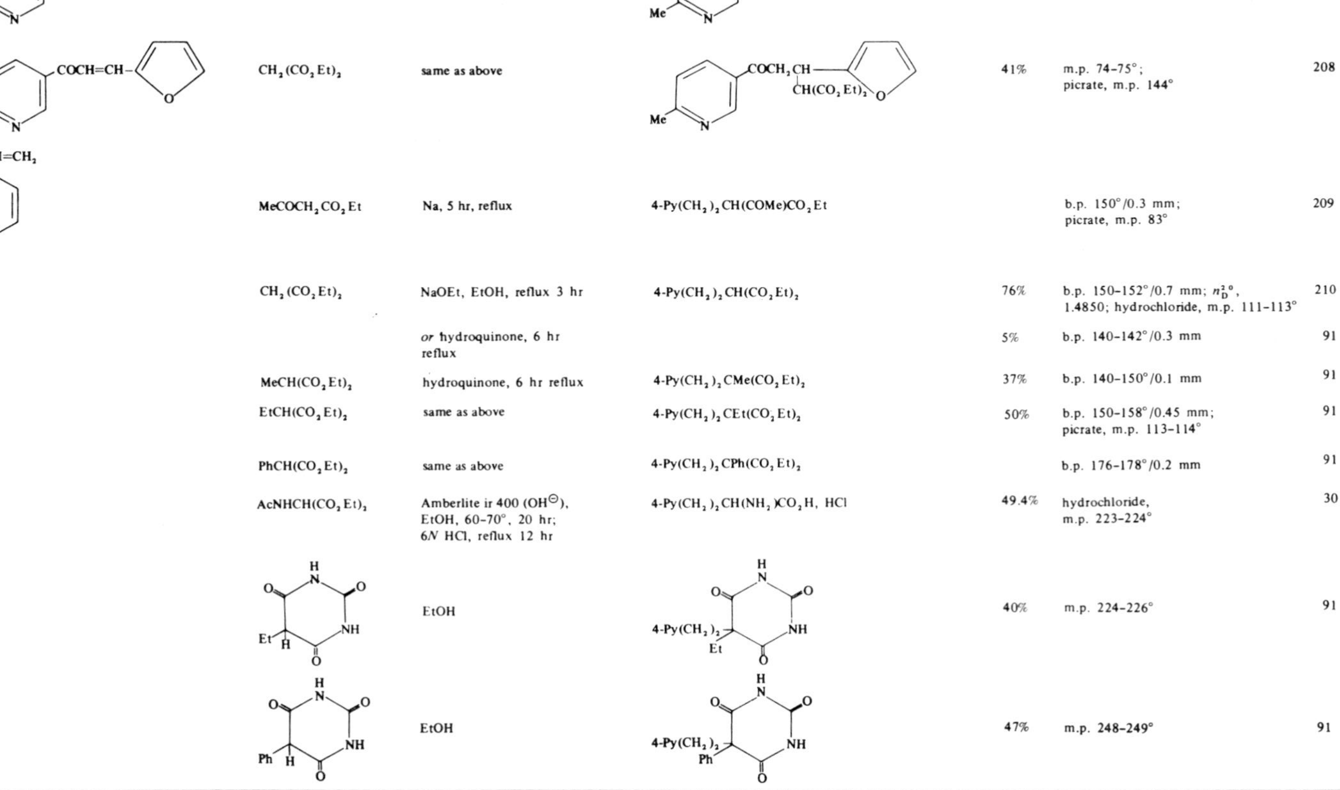

| | | | | | | |
|---|---|---|---|---|---|---|
| | | | $CH(CO_2Et)_2$ | | | |
| | $CH_2(CO_2Et)_2$ | same as above | | 41% | m.p. 74–75°; picrate, m.p. 144° | 208 |
| | $MeCOCH_2CO_2Et$ | Na, 5 hr, reflux | 4-Py$(CH_2)_2CH(COMe)CO_2Et$ | | b.p. 150°/0.3 mm; picrate, m.p. 83° | 209 |
| | $CH_2(CO_2Et)_2$ | NaOEt, EtOH, reflux 3 hr | 4-Py$(CH_2)_2CH(CO_2Et)_2$ | 76% | b.p. 150–152°/0.7 mm; $n_D^{20}$, 1.4850; hydrochloride, m.p. 111–113° | 210 |
| | | *or* hydroquinone, 6 hr reflux | | 5% | b.p. 140–142°/0.3 mm | 91 |
| | $MeCH(CO_2Et)_2$ | hydroquinone, 6 hr reflux | 4-Py$(CH_2)_2CMe(CO_2Et)_2$ | 37% | b.p. 140–150°/0.1 mm | 91 |
| | $EtCH(CO_2Et)_2$ | same as above | 4-Py$(CH_2)_2CEt(CO_2Et)_2$ | 50% | b.p. 150–158°/0.45 mm; picrate, m.p. 113–114° | 91 |
| | $PhCH(CO_2Et)_2$ | same as above | 4-Py$(CH_2)_2CPh(CO_2Et)_2$ | | b.p. 176–178°/0.2 mm | 91 |
| | $AcNHCH(CO_2Et)_2$ | Amberlite ir 400 ($OH^{\ominus}$), EtOH, 60–70°, 20 hr; 6*N* HCl, reflux 12 hr | 4-Py$(CH_2)_2CH(NH_2)CO_2H$, HCl | 49.4% | hydrochloride, m.p. 223–224° | 30 |
| | | EtOH | | 40% | m.p. 224–226° | 91 |
| | | EtOH | | 47% | m.p. 248–249° | 91 |

TABLE XI-10. Condensation of Pyridinealdehydes and Ketones Yielding Side-Chain Acid Derivatives

| Aldehyde | Condensed with | Conditions | Product | Yield | Properties | Ref. |
|---|---|---|---|---|---|---|
| 2-PyCHO | $CH_2(CN)CO_2Et$ | EtOH<br>or | 2-PyCH=C(CN)$CO_2Et$ | 87.6% | m.p. 94–95° | 211 |
| | | EtOH, piperidine, reflux 1–5 hr | | 64% | m.p. 94.5–96° | 137 |
| | $CH_2(COMe)CO_2Et$ | $Et_2NH$ | 2-PyCH=C(COMe)$CO_2Et$ | 80% | m.p. 85–87° | 211 |
| | $PhCOCH_2CO_2Et$ | piperidine, overnight | 2-PyCH=C(COPh)$CO_2Et$ | 58% | m.p. 98–99° | 211 |
| | $EtO_2CCOCHFCO_2Et$ | reflux 45 min | 2-PyCH=CF$CO_2Et$ | 35% | b.p. 118–120°/4 mm | 212 |
| | $p$-$ClC_6H_4CH_2CO_2H$ | $Et_3N$, $Ac_2O$, stir 0.5 hr; 100° within 0.5 hr; 100°, 5 hr | 2-PyCH=C($p$-$ClC_6H_4$)$CO_2H$ | | | 53 |
| | $CH_2(NHAc)CO_2Et$ | $Et_3N$, 14 hr | 2-PyCH(OH)CH(NHAc)$CO_2Et$ | 56% | m.p. 124–126° | 93 |
| 2-PyCH(OEt)$_2$ | $PhCONHCH_2CO_2H$ | $SO_3$, DMF, heat, 1 hr, R.T. overnight | 2-PhCH= (ring: O, O, N=, Ph) | 22% | | 94 |
| | $PhCONHCH_2CO_2H$ | $SO_3$, DMF, heat, 1 hr, R.T. overnight; red P, $Ac_2O$, HI, reflux 4 hr | 2-PyCh$_2$CH($NH_2$)$CO_2H$ | 62% | m.p. 209–212°; ir | 94 |
| | $PhCONHCH_2CO_2H$ | $SO_3$, DMF, heat, 1 hr, R.T., overnight; 10% NaOH, steam bath, 4 hr; HCl-HOAc (1:1), reflux 20 hr | 2-PyCH=CH$CO_2H$ | 83% | m.p. 206° | 94 |
| 2-PyCHO | ONNH$(CH_2)_2CO_2Me$ | | 2-PyCO$(CH_2)_2CO_2Me$ | 65% | b.p. 104–106°/0.3 mm; m.p. 46–47°; 2,4-DNP, m.p. 174° | 96 |
| | ONNH($CH_2$)[illegible]$CO_2Me$ | | 2-PyCO($CH_2$)[illegible]$CO_2Me$ | 63% | b.p. 114°/0.3 mm; 2,4-DNP, | 96 |

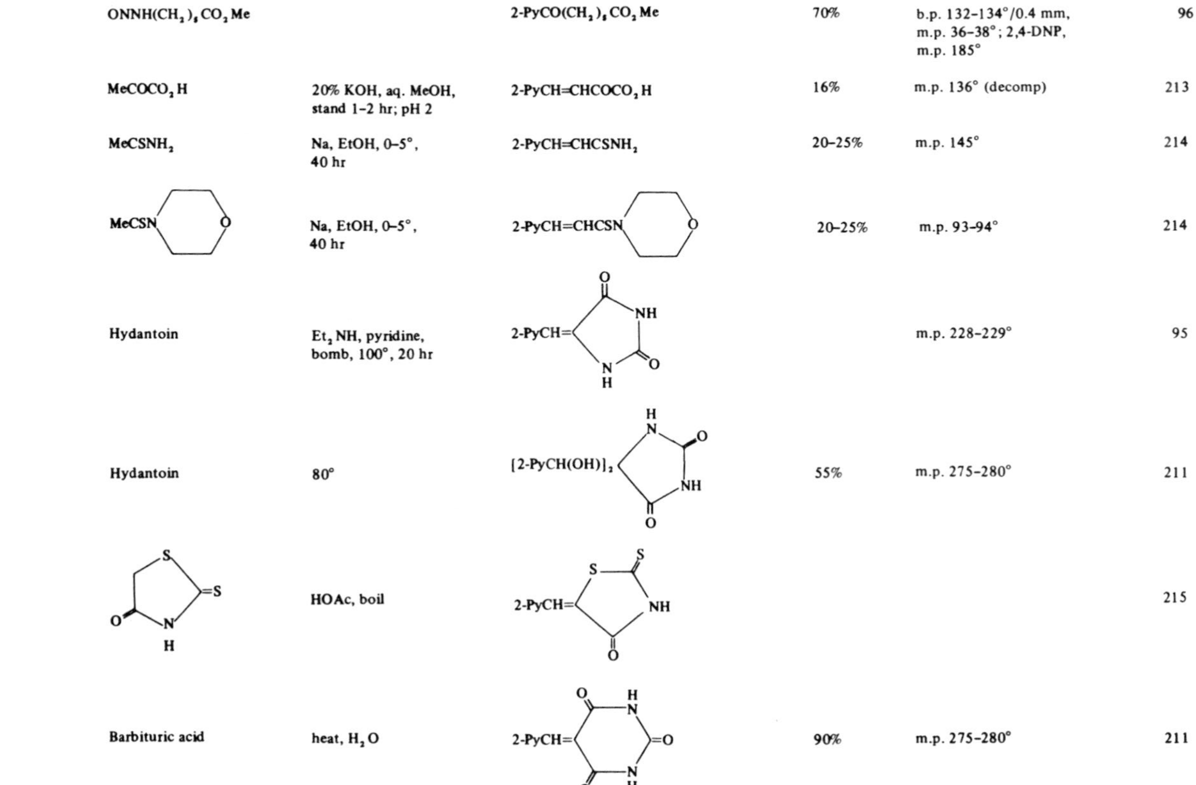

| | | | | | |
|---|---|---|---|---|---|
| | $ONNH(CH_2)_5CO_2Me$ | | $2\text{-}PyCO(CH_2)_5CO_2Me$ | 70% | b.p. 132–134°/0.4 mm, m.p. 36–38°; 2,4-DNP, m.p. 185° | 96 |
| | $MeCOCO_2H$ | 20% KOH, aq. MeOH, stand 1–2 hr; pH 2 | $2\text{-}PyCH{=}CHCOCO_2H$ | 16% | m.p. 136° (decomp) | 213 |
| | $MeCSNH_2$ | Na, EtOH, 0–5°, 40 hr | $2\text{-}PyCH{=}CHCSNH_2$ | 20–25% | m.p. 145° | 214 |
| | MeCSN (O) | Na, EtOH, 0–5°, 40 hr | 2-PyCH=CHCSN (O) | 20–25% | m.p. 93–94° | 214 |
| | Hydantoin | $Et_2NH$, pyridine, bomb, 100°, 20 hr | 2-PyCH= (O, NH, N H, O) | | m.p. 228–229° | 95 |
| 2-PyCHO (2 equiv.) | Hydantoin | 80° | $[2\text{-}PyCH(OH)]_2$ (H N, O, NH, O) | 55% | m.p. 275–280° | 211 |
| 2-PyCHO | (S, S, O, N H) | HOAc, boil | 2-PyCH= (S, S, NH, O) | | | 215 |
| | Barbituric acid | heat, $H_2O$ | 2-PyCH= (O, H N, O, N H, O) | 90% | m.p. 275–280° | 211 |

TABLE XI-10. Condensation of Pyridinealdehydes and Ketones Yielding Side-Chain Acid Derivatives (Continued)

| Aldehyde | Condensed with | Conditions | Product | Yield | Properties | Ref. |
|---|---|---|---|---|---|---|
| | 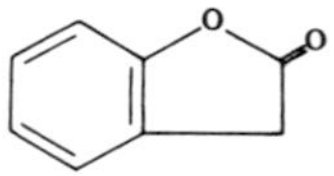 | $Et_3N$ | 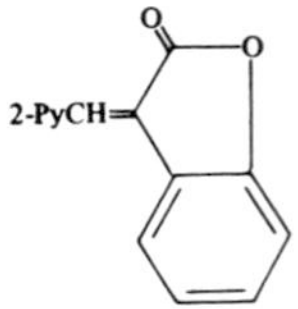 2-PyCH= | 69% | | 216 |
| | KCN, $(NH_4)_2CO_3$ | $H_2O$, EtOH, 50–55°, 4 hr; HOAc, 24 hr | 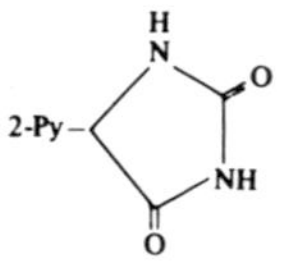 2-Py– | | m.p. 301–302° (decomp); ir | 217 |
| | KCN,$(NH_4)_2CO_3$ | $H_2O$, EtOH, 50–55°, 4 hr; $Ba(OH)_2$, 40 hr, reflux | 2-PyCH($NH_2$)$CO_2H$ | | m.p. 130–132° | 217 |
| 3-$NO_2$-2-PyCHO | $CH_2(CO_2H)_2$ | pyridine, piperidine, steam-bath, 2 hr | 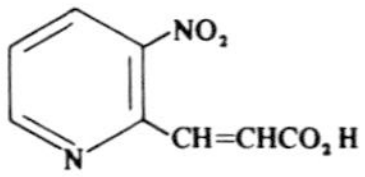 $NO_2$ CH=CH$CO_2H$ | | m.p. 192.5–194° | 151 |
| 2-PyCHO-1-oxide | $CH_2(CO_2H)_2$ | pyridine, 60 min, steam-bath | 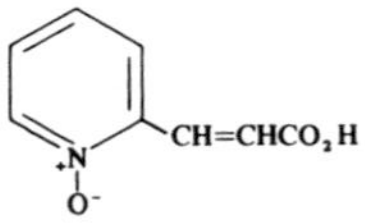 CH=CH$CO_2H$ | 70% | m.p. 257–258° | 218 |
| 3-Me-2-PyCHO-1-oxide | $CH_2(CO_2H)_2$ | pyridine, 60 min, steam-bath | 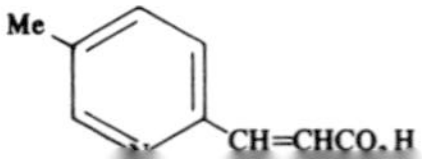 Me CH=CH$CO_2H$ | 55% | m.p. 245° (decomp) | 218 |

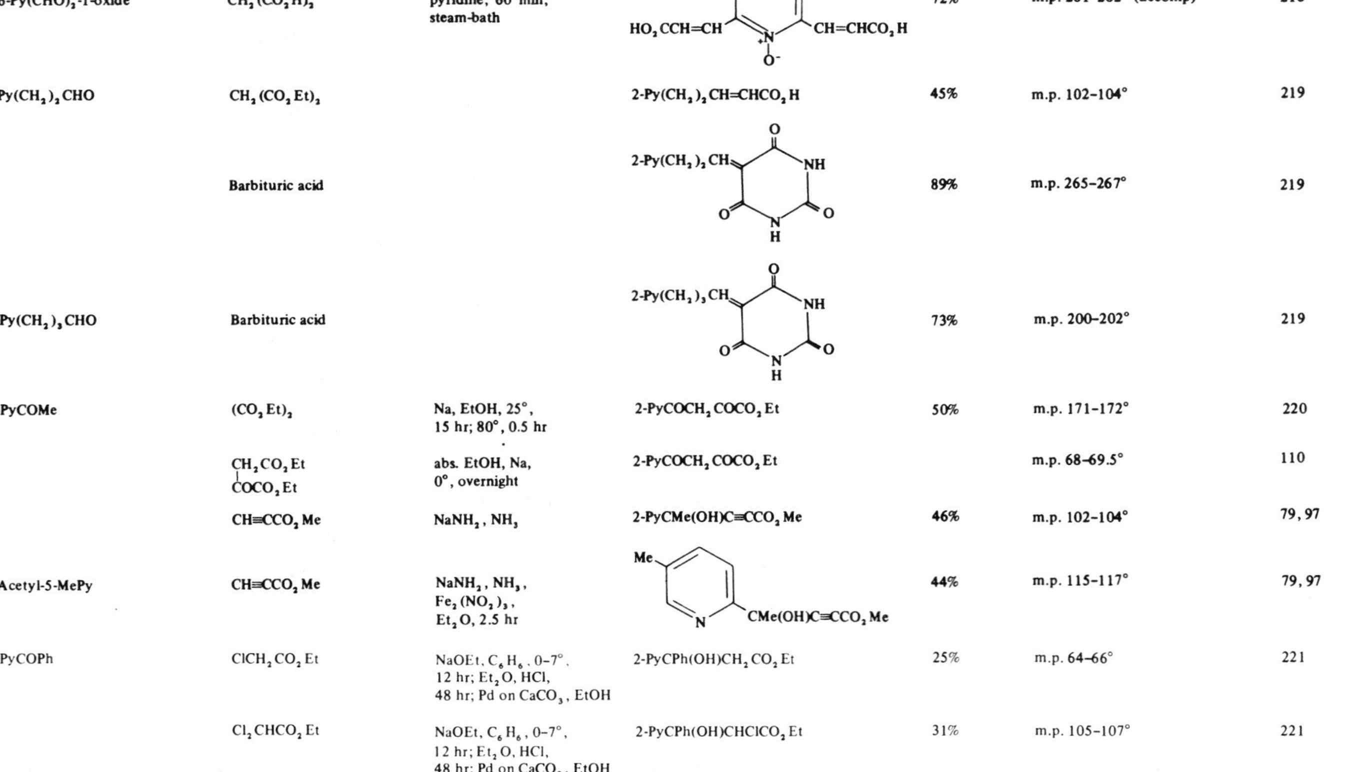

| | | | | | | |
|---|---|---|---|---|---|---|
| 2,6-Py(CHO)$_2$-1-oxide | $CH_2(CO_2H)_2$ | pyridine, 60 min, steam-bath | $HO_2CCH{=}CH$–(pyridine N-oxide)–$CH{=}CHCO_2H$ | 72% | m.p. 281–282° (decomp) | 218 |
| 2-Py$(CH_2)_2$CHO | $CH_2(CO_2Et)_2$ | | 2-Py$(CH_2)_2CH{=}CHCO_2H$ | 45% | m.p. 102–104° | 219 |
| | Barbituric acid | | 2-Py$(CH_2)_2CH{=}$(barbituric acid ring) | 89% | m.p. 265–267° | 219 |
| 2-Py$(CH_2)_3$CHO | Barbituric acid | | 2-Py$(CH_2)_3CH{=}$(barbituric acid ring) | 73% | m.p. 200–202° | 219 |
| 2-PyCOMe | $(CO_2Et)_2$ | Na, EtOH, 25°, 15 hr; 80°, 0.5 hr | 2-PyCOCH$_2$COCO$_2$Et | 50% | m.p. 171–172° | 220 |
| | $CH_2CO_2Et$–$COCO_2Et$ | abs. EtOH, Na, 0°, overnight | 2-PyCOCH$_2$COCO$_2$Et | | m.p. 68–69.5° | 110 |
| | $CH{\equiv}CCO_2Me$ | $NaNH_2$, $NH_3$ | 2-PyCMe(OH)C$\equiv$CCO$_2$Me | 46% | m.p. 102–104° | 79, 97 |
| 2-Acetyl-5-MePy | $CH{\equiv}CCO_2Me$ | $NaNH_2$, $NH_3$, $Fe_2(NO_2)_3$, $Et_2O$, 2.5 hr | 5-Me-2-Py–CMe(OH)C$\equiv$CCO$_2$Me | 44% | m.p. 115–117° | 79, 97 |
| 2-PyCOPh | $ClCH_2CO_2Et$ | NaOEt, $C_6H_6$, 0–7°, 12 hr; $Et_2O$, HCl, 48 hr; Pd on $CaCO_3$, EtOH | 2-PyCPh(OH)CH$_2$CO$_2$Et | 25% | m.p. 64–66° | 221 |
| | $Cl_2CHCO_2Et$ | NaOEt, $C_6H_6$, 0–7°, 12 hr; $Et_2O$, HCl, 48 hr; Pd on $CaCO_3$, EtOH | 2-PyCPh(OH)CHClCO$_2$Et | 31% | m.p. 105–107° | 221 |

TABLE XI-10. Condensation of Pyridinealdehydes and Ketones Yielding Side-Chain Acid Derivatives (Continued)

| Aldehyde | Condensed with | Conditions | Product | Yield | Properties | Ref. |
|---|---|---|---|---|---|---|
| | $BrCH_2CO_2Et$ | Zn | 2-PyCPh(OH)$CH_2CO_2Et$ | 55% | m.p. 59–61° | 132 |
| | | | | | m.p. 63–64°; b.p. 160–170°/2 mm | 54 |
| | | | | | m.p. 65–67° | 222 |
| | $BrCHMeCO_2Et$ | Zn | 2-PyCPh(OH)$CHMeCO_2Et$ | | m.p. 51–53° | 222 |
| | $PhCH(MgBr)CO_2H$ | $C_6H_6$, reflux 4 hr | 2-PyC(OH)$PhCHPhCO_2H$ | | m.p. 162.5° (decomp) | 172 |
| | $MeCONMe_2$ | $NaNH_2$, $Et_2O$, 1.5 hr | 2-PyC(OH)$PhCH_2CONMe_2$ | | m.p. 105° | 223 |
| 3-PyCHO | $CH_2(CO_2H)_2$ | pyridine, piperidine | 3-PyCH=$CHCO_2H$ | 78% | m.p. 235–235.5° | 224 |
| | $CH_2(CN)CO_2Et$ | piperidine, EtOH, reflux 1–5 hr | 3-PyCH=C(CN)$CO_2Et$ | 83% | m.p. 87.5–88.5° | 137 |
| | $PhCH_2CO_2H$ *or* $PhCH_2CO_2Na$ | NaOH, EtOH; $Ac_2O$, reflux 2 hr *or* $Ac_2O$, reflux 2 hr | (3-Py)(H)C=C(Ph)($CO_2H$) | 12 g → 35 g<br>55% | m.p. 197–200° | 98, 100 |
| | $p$-$MeC_6H_4CH_2CO_2H$ | NaOH, EtOH; $Ac_2O$ reflux 2 hr | (3-Py)(H)C=C($p$-$MeC_6H_4$)($CO_2H$) | | m.p. 198° | 100 |
| | $p$-$BrC_6H_4CH_2CO_2H$ | same as above | (3-Py)(H)C=C($p$-$BrC_6H_4$)($CO_2H$) | | m.p. 183° | 100 |
| | $p$-$ClC_6H_4CH_2CO_2H$ | same as above | (3-Py)(H)C=C($p$-$ClC_6H_4$)($CO_2H$) | | m.p. 220° | 100 |
| | $p$-$IC_6H_4CH_2CO_2H$ | same as above | (3-Py)(H)C=C($p$-$IC_6H_4$)($CO_2H$) | | m.p. 189–195° | 100 |
| | | | 3-Py, ($p$-$MeOC_6H_4$) | | | |

| | | | | | | |
|---|---|---|---|---|---|---|
| | Sodium homoveratric acid | $Ac_2O$, reflux 18 hr | 3-PyCH=C($CO_2H$)–[ring]–OMe | | | 225 |
| | $o$-$NO_2C_6H_4CH_2CO_2Na$ | $Ac_2O$, fused $ZnCl_2$, steam bath, 20 hr | 3-Py(H)C=C($o$-$O_2NC_6H_4$)($CO_2H$) | 43% | m.p. 203–204° | 99 |
| | $p$-$O_2NC_6H_4CH_2CO_2H$ | NaOH, EtOH; $Ac_2O$, reflux 2 hr | 3-Py(H)C=C($p$-$O_2NC_6H_4$)($CO_2H$) | | m.p. 155° | 100 |
| | $EtO_2CCOCHFCO_2Et$ | reflux 45 min | 3-PyCH=CF$CO_2Et$ | 40% | b.p. 120–122°/4 mm | 212 |
| | $Me_2NCH_2CO_2Et$ | Na, EtOH | 3-PyCH=C($NMe_2$)$CO_2Et$ | 71% | b.p. 97–98°/0.05 mm; ir | 226 |
| | $CH_2(NHAc)CO_2Et$ | $Et_3N$, 10 hr | 3-PyCH(OH)CH(NHAc)$CO_2Et$ | 65% | m.p. 154–157° | 93 |
| 3-PyCH$(OMe)_2$ | $PhCONHCH_2CO_2H$ | $SO_3$, DMF, heat, 1 hr; R.T., overnight; red P, $Ac_2O$, HI, reflux 4 hr | 3-Py$CH_2$CH($NH_2$)$CO_2H$ | 72% | m.p. 278–280°; ir | 94 |
| | $PhCONHCH_2CO_2H$ | $SO_3$, DMF, heat, 1 hr; R.T., overnight; 10% NaOH, steam-bath, 4 hr; HCl-HOAc(1:1), reflux 20 hr | 3-PyCH=CH$CO_2H$ | 76% | m.p. 236–241° | 94 |
| 3-PyCHO | $MeCOCO_2H$ | 20% KOH, aq. MeOH, 1-2 hr; pH 2 | 3-PyCH=CHCO$CO_2H$ | 51% | m.p. 169° (decomp); oxime, m.p. 176° (decomp) | 213 |
| | $MeCSNH_2$ | Na, EtOH, 0–5°, 40 hr | 3-PyCH=CHCS$NH_2$ | 20–25% | m.p. 184° | 214 |
| | MeCSN$(CH_2CH_2)_2$O | same as above | 3-PyCH=CHCSN$(CH_2CH_2)_2$O | 20–25% | m.p. 115° | 214 |

## TABLE XI-10. Condensation of Pyridinealdehydes and Ketones Yielding Side-Chain Acid Derivatives (Continued)

| Aldehyde | Condensed with | Conditions | Product | Yield | Properties | Ref. |
|---|---|---|---|---|---|---|
| | $p\text{-ClC}_6\text{H}_4\text{CO(CH}_2)_2\text{CO}_2\text{H}$ | $Ac_2O$, NaOAc | 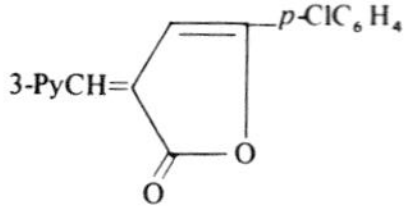 | 73% | m.p. 242–243″ | 227 |
| | 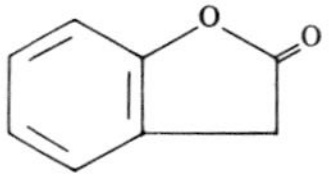 | $Et_3N$ | 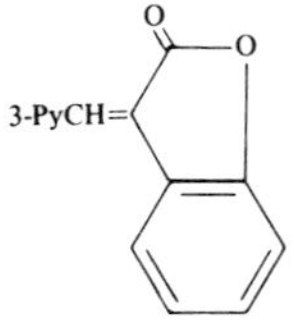 | 34.5% | m.p. 147° | 216 |
| 4-Me-3-PyCHO | $CH_2(CO_2Et)_2$ | piperidine, $C_6H_6$, reflux | 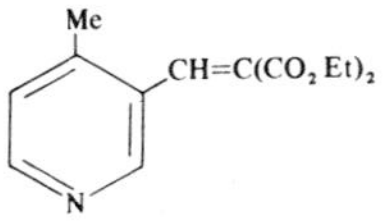 | 69% | b.p. 139–143°/0.5 mm; picrate, m.p. 107–110.5° | 5 |
| 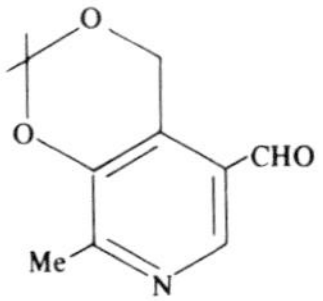 | $CH_2(CO_2H)_2$ | pyridine, 95% EtOH, reflux 45 min | 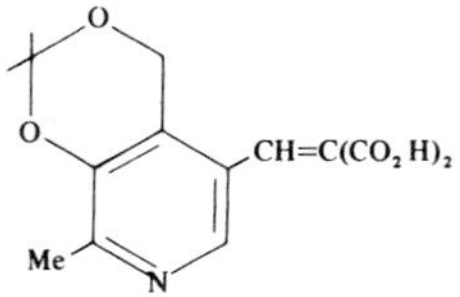 | 56% | m.p. 200–201° (decomp) | 228 |
| | $CH_2(CO_2H)_2$ | pyridine, piperidine, 2 hr, steam-bath; 5°, overnight | 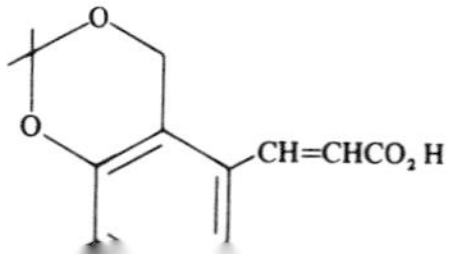 | 82% | m.p. 220–221° | 228 |

| | | | | | | |
|---|---|---|---|---|---|---|
| [structure: O, N, $(CH_2)_2$, OMe, OMe] | $CH_2(CO_2H)_2$ | piperidine, pyridine, steam-bath, 2 hr | [structure: O, N, $(CH_2)$, OMe, OMe] | 81% | two forms, m.p. 155–157°, m.p. 178–181° | 7 |
| 3-PyCOMe | $(CO_2Et)_2$ | Na, EtOH, 25°, 15 hr; 80°, 0.5 hr | 3-PyCOCH$_2$COCO$_2$Et | 10% | m.p. 185–187° | 220 |
| | NaCH$(CO_2Et)_2$ | 32% HBr, HOAc, Py. HBr; $CH_2(CO_2Et)_2$, 12 hr, ice; 25°, 12 hr | 3-PyCO$(CH_2)_2CO_2H$ | 60% | | 229 |
| | BrCHMeCO$_2$Me | Zn, HgCl$_2$, $C_6H_6$, reflux 1 hr | 3-PyC(OH)MeCHMeCO$_2$Me 37% *threo*, 63% *erythro* | 50% | NMR | 230 |
| [structure: Me, COMe, Me, N, H, O] | $(CO_2Et)_2$ | Na, EtOH, 90–100% 3 hr | [structure: Me, COCH$_2$COCO$_2$Et, Me, N, H, O] | 2 g → 2.1 g | m.p. 175–176°; ir | 231 |
| 3-PyCOPh | BrCH$_2$CO$_2$Et | Zn | 3-PyC(OH)PhCH$_2$CO$_2$Et | 56% | m.p. 77–78° | 54 |
| 4-PyCHO | $o$-$O_2NC_6H_4CH_2CO_2Na$ | Ac$_2$O, fused ZnCl$_2$, steam-bath, 20 hr | [structure: H, CO$_2$H, 4-Py, ($o$-$O_2NC_6H_4$)] | 11.5 g → 5.5 g | picrate, m.p. 203–204° | 99 |
| | $CH_2$(NHAc)CO$_2$Et | Et$_3$N, 10 days | 4-PyCH(OH)CH(NHAc)CO$_2$Et | 60% | m.p. 120–122° | 93 |
| | AcNHCH(CO$_2$H)CO$_2$Me | Et$_3$N, EtOH, R.T., 6 days | 4-PyCH(OH)CH(NHAc)CO$_2$Me | 75% | m.p. 130° | 232 |
| 4-PyCH(OMe)$_2$ | PhCONHCH$_2$CO$_2$H | SO$_3$, DMF, heat, 1 hr; R.T., overnight; red P, Ac$_2$O, HI, reflux 4 hr | 4-PyCH$_2$CH(NH$_2$)CO$_2$H | 65% | m.p. 265–270°; ir | 94 |

TABLE XI-10. Condensation of Pyridinealdehydes and Ketones Yielding Side-Chain Acid Derivatives (Continued)

| Aldehyde | Condensed with | Conditions | Product | Yield | Properties | Ref. |
|---|---|---|---|---|---|---|
| | $PhCONHCH_2CO_2H$ | $SO_3$, DMF, heat, 1 hr; R.T., overnight; 10% NaOH, steam-bath, 4 hr; HCl-HOAc (1:1), reflux 20 hr | 4-PyCH=$CHCO_2H$ | 80% | m.p. 230° | 94 |
| 4-PyCHO | $MeCOCO_2H$ | 20% KOH, MeOH, stand 1-2 hr; pH 2 | 4-PyCH=$CHCOCO_2H$ | 24% | m.p. 290–291° (decomp) | 213 |
| | $MeCSNH_2$ | Na, EtOH, 0–5°, 40 hr | 4-PyCH=$CHCSNH_2$ | 20–25% | m.p. 163–164° | 214 |
| | MeCSN O | Na, EtOH, 0–5°, 40 hr | 4-PyCH=CHCSN O | 20–25% | m.p. 98° | 214 |
| | O O | $Et_3N$ | 4-PyCH= | 100% | m.p. 148° | 216 |
| 4-PyCOMe | $(CO_2Et)_2$ | NaOAc, | 4-$PyCOCH_2COCO_2Et$ | | m.p. 96° | 233 |
| | | *or* Na, EtOH, 25°, 15 hr; 80°, 0.5 hr | | 5% | m.p. 98–100° | 220 |
| | | *or* Na, $Et_2O$, 0°, 2hr; several hr, R.T.; reflux, | | | m.p. 75° | 234 |

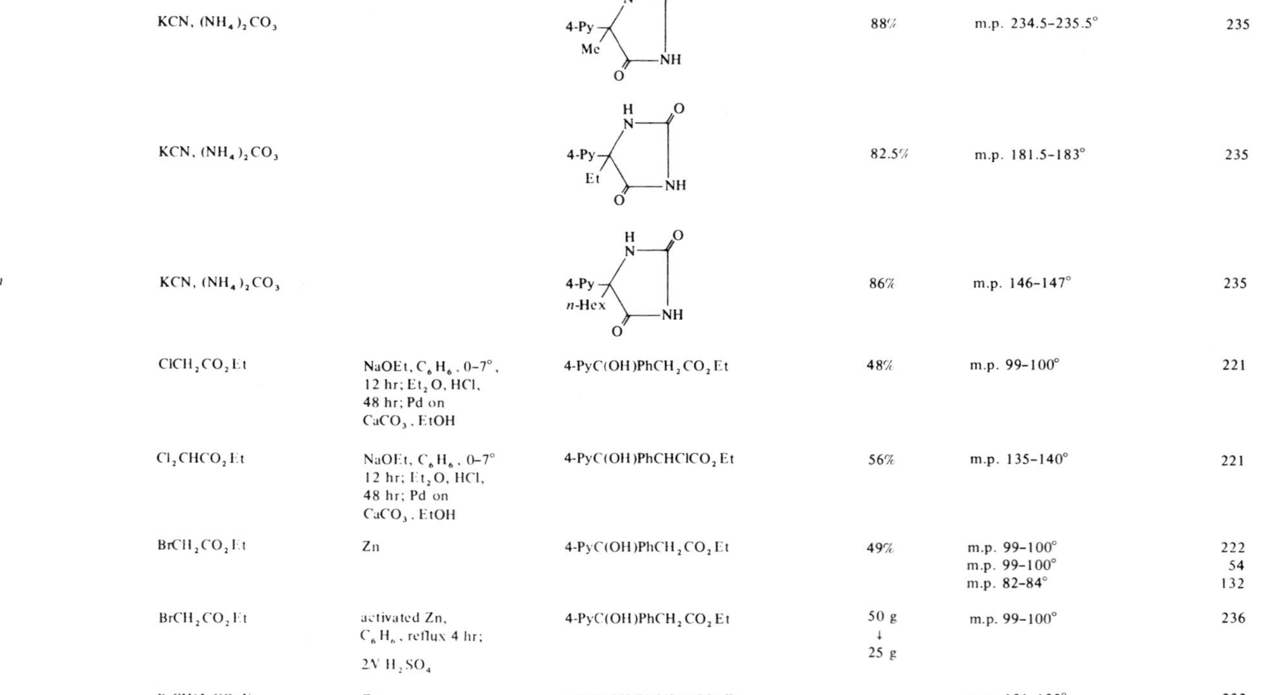

| | | | | | | |
|---|---|---|---|---|---|---|
| | KCN, $(NH_4)_2CO_3$ | | 4-Py, Me hydantoin (structure) | 88% | m.p. 234.5–235.5° | 235 |
| 4-PyCOEt | KCN, $(NH_4)_2CO_3$ | | 4-Py, Et hydantoin (structure) | 82.5% | m.p. 181.5–183° | 235 |
| 4-PyCOHex-*n* | KCN, $(NH_4)_2CO_3$ | | 4-Py, *n*-Hex hydantoin (structure) | 86% | m.p. 146–147° | 235 |
| 4-PyCOPh | $ClCH_2CO_2Et$ | NaOEt, $C_6H_6$, 0–7°, 12 hr; $Et_2O$, HCl, 48 hr; Pd on $CaCO_3$, EtOH | $4\text{-}PyC(OH)PhCH_2CO_2Et$ | 48% | m.p. 99–100° | 221 |
| | $Cl_2CHCO_2Et$ | NaOEt, $C_6H_6$, 0–7° 12 hr; $Et_2O$, HCl, 48 hr; Pd on $CaCO_3$, EtOH | $4\text{-}PyC(OH)PhCHClCO_2Et$ | 56% | m.p. 135–140° | 221 |
| | $BrCH_2CO_2Et$ | Zn | $4\text{-}PyC(OH)PhCH_2CO_2Et$ | 49% | m.p. 99–100°<br>m.p. 99–100°<br>m.p. 82–84° | 222<br>54<br>132 |
| | $BrCH_2CO_2Et$ | activated Zn, $C_6H_6$, reflux 4 hr; 2*N* $H_2SO_4$ | $4\text{-}PyC(OH)PhCH_2CO_2Et$ | 50 g → 25 g | m.p. 99–100° | 236 |
| | $BrCHMeCO_2Et$ | Zn | $4\text{-}PyC(OH)PhCHMeCO_2Et$ | | m.p. 121–122° | 222 |

TABLE XI-10. Condensation of Pyridinealdehydes and Ketones Yielding Side-Chain Acid Derivatives (Continued)

| Aldehyde | Condensed with | Conditions | Product | Yield | Properties | Ref. |
|---|---|---|---|---|---|---|
| | $PhCH(MgBr)CO_2H$ | $C_6H_6$, reflux<br>4 hr | $4\text{-}PyC(OH)PhCHPhCO_2H$ | | | 172 |
| $2,4\text{-}Py(CHO)_2$ | $CH_2(CN)CO_2Et$ | pyridine, piperidine,<br>steam-bath 2 hr | 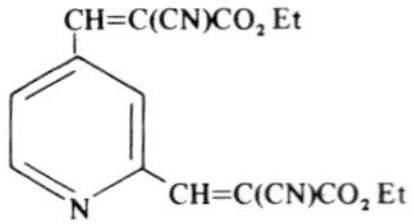 $CH{=}C(CN)CO_2Et$<br>N<br>$CH{=}C(CN)CO_2Et$ | | | 237 |

| Pyridine ester | Condensed with | Conditions | Product | Yield | Properties | Ref. |
|---|---|---|---|---|---|---|
| 2-$PyCO_2Et$ | $MeCO_2Et$ | 6.5 hr 115°, NaOEt | 2-$PyCOCH_2CO_2Et$ | 84.3% | b.p. 92–97°/0.1 mm | 104 |
| | $EtO_2C(CH_2)_3CO_2Et$ | Na, $NaNH_2$, $Et_2O$, $N_2$, 45°, 13 hr | 2-$PyCOCH(CO_2Et)(CH_2)_2CO_2Et$ | 42% | b.p. 160–170°/0.7 mm; $n_D^{20}$ 1.5011 | 102 |
| | 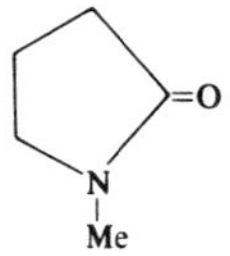 | K, $C_6H_6$ | 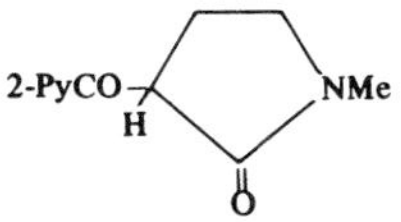 2-PyCO– (H) NMe | 41% | b.p. 123°/0.05 mm; m.p. 92° | 129 |
| | 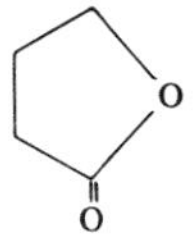 | NaH, PhMe, reflux 8 hr; HOAc | 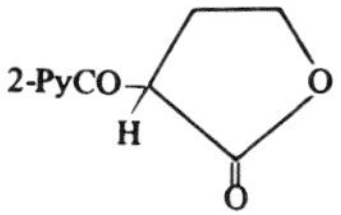 2-PyCO– (H) | 45 g → 39.7 g | b.p. 130°/3 × $10^{-6}$ mm; m.p. 53.5–54.5°; ir; uv | 140 |
| | 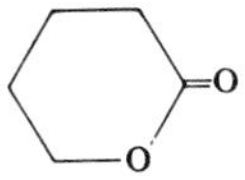 | K, $C_6H_6$, reflux 5 hr | 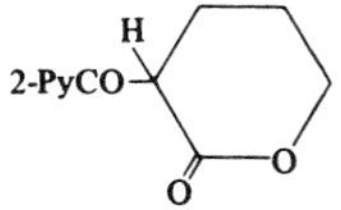 2-PyCO– (H) | | b.p. 122–124°/3 × $10^{-6}$ mm; ir; uv | 140 |
| 2-PyCOCl · HCl | $EtCH(CO_2Et)_2$ | Na, EtOH, reflux 2.5 hr | 2-$PyCOCEt(CO_2Et)_2$ | 18% | b.p. 122–128°/0.05 mm | 91 |
| 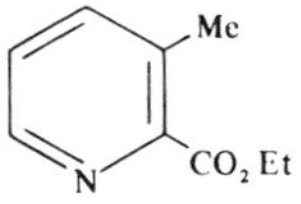 (3-Me-pyridine-2-$CO_2Et$) | 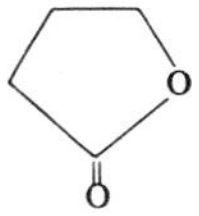 | NaH, PhMe, reflux 8 hr; HOAc | 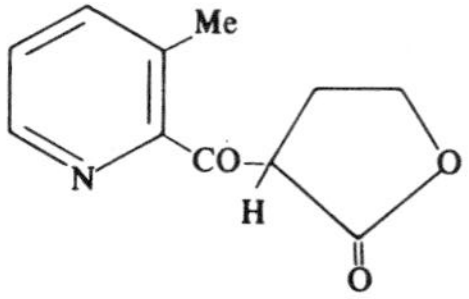 | 17.8 g → 6.6 g | b.p. 145°/5× $10^{-4}$ mm; ir; uv | 140 |

TABLE XI-11. Condensations of Pyridinecarboxylic Esters Yielding Side-Chain Acid Derivatives (Continued)

| Pyridine ester | Condensed with | Conditions | Product | Yield | Properties | Ref. |
|---|---|---|---|---|---|---|
| 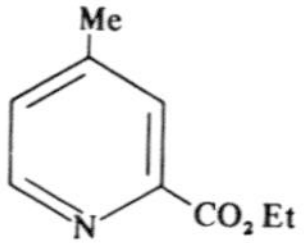 Me, N, $CO_2Et$ | $MeCO_2Et$ | Na, $Et_2O$, reflux 3 hr; kept over night | 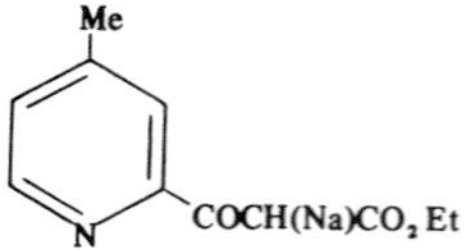 Me, N, $COCH(Na)CO_2Et$ | 10 g → 13 g | | 109 |
| 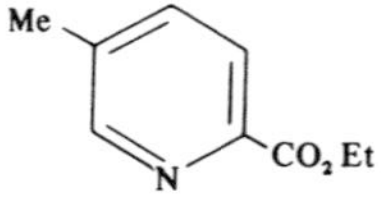 Me, N, $CO_2Et$ | 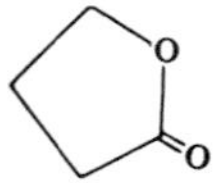 O, O | NaH, PhMe, reflux 8 hr; HOAc | 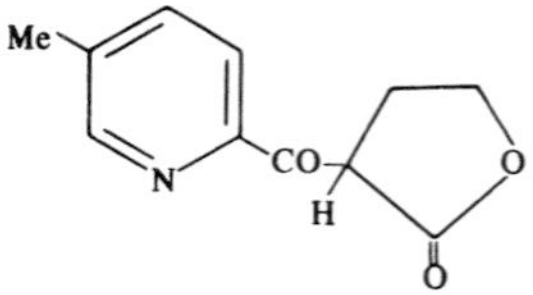 Me, N, CO, H, O, O | | b.p. 136–137°/5 × $10^{-5}$ mm; m.p. 73–74°; ir; uv | 140 |
| 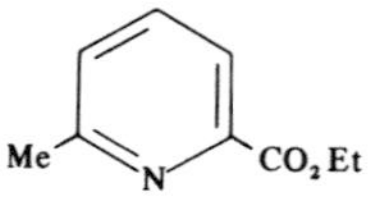 Me, N, $CO_2Et$ | 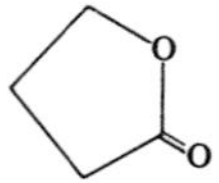 O, O | NaH, PhMe, reflux 8 hr; HOAc | 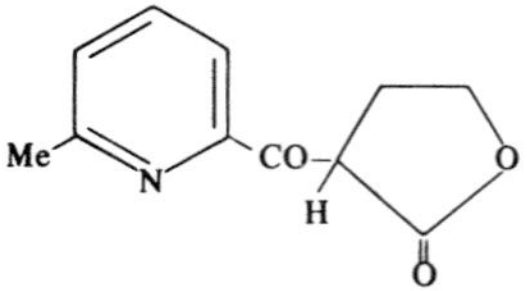 Me, N, CO, H, O, O | | b.p. 127–130°/3 × $10^{-5}$ mm; m.p. 91–93°; ir; uv | 140 |
| | | | | | b.p. 144–149°/3 mm | 238 |
| 3-$PyCO_2Et$ | $MeCO_2Et$ | NaOEt, 115°, 6.5 hr *or* | 3-$PyCOCH_2CO_2Et$ | 84% | b.p. 122–132°/0.35 mm | 104 |
| | | NaOEt, $C_6H_6$, 10 hr 80°; aq. HCl | | 65% | b.p. 165–168°/5 mm | 107 |
| | $MeCO_2Et$ | Na, PhMe | 3-$PyC(ONa){=}CHCO_2Et$ | 95% | | 176 |
| | $(CH_2CO_2Et)_2$ | NaH, reflux in $C_6H_6$, EtOH; 5*N* HCl | 3-$PyCOCH(CO_2Et)CH_2CO_2Et$ | 58.8% | | 23 |

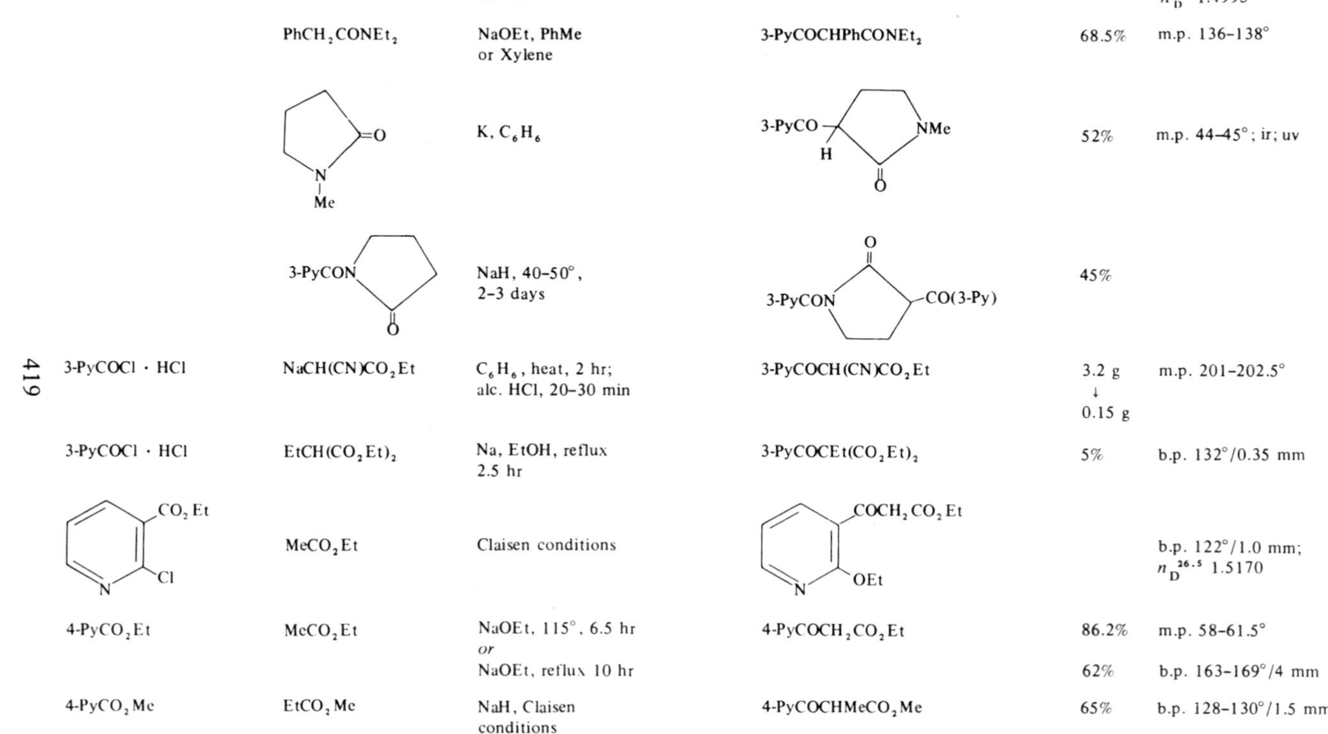

| | | | | | | |
|---|---|---|---|---|---|---|
| | | 1.5 hr | | | $n_D^{20}$ 1.4995 | |
| | $PhCH_2CONEt_2$ | NaOEt, PhMe or Xylene | $3\text{-}PyCOCHPhCONEt_2$ | 68.5% | m.p. 136–138° | 82 |
| | N-Me, =O (ring) | K, $C_6H_6$ | 3-PyCO, H, NMe, O (ring) | 52% | m.p. 44–45°; ir; uv | 129 |
| | 3-PyCON, O (ring) | NaH, 40–50°, 2–3 days | O, 3-PyCON, CO(3-Py) (ring) | 45% | | 129 |
| 3-PyCOCl · HCl | $NaCH(CN)CO_2Et$ | $C_6H_6$, heat, 2 hr; alc. HCl, 20–30 min | $3\text{-}PyCOCH(CN)CO_2Et$ | 3.2 g ↓ 0.15 g | m.p. 201–202.5° | 103 |
| 3-PyCOCl · HCl | $EtCH(CO_2Et)_2$ | Na, EtOH, reflux 2.5 hr | $3\text{-}PyCOCEt(CO_2Et)_2$ | 5% | b.p. 132°/0.35 mm | 91 |
| $CO_2Et$, Cl, N (ring) | $MeCO_2Et$ | Claisen conditions | $COCH_2CO_2Et$, OEt, N (ring) | | b.p. 122°/1.0 mm; $n_D^{26.5}$ 1.5170 | 239 |
| $4\text{-}PyCO_2Et$ | $MeCO_2Et$ | NaOEt, 115°, 6.5 hr *or* | $4\text{-}PyCOCH_2CO_2Et$ | 86.2% | m.p. 58–61.5° | 104 |
| | | NaOEt, reflux 10 hr | | 62% | b.p. 163–169°/4 mm | 108 |
| $4\text{-}PyCO_2Me$ | $EtCO_2Me$ | NaH, Claisen conditions | $4\text{-}PyCOCHMeCO_2Me$ | 65% | b.p. 128–130°/1.5 mm | 11 |

TABLE XI-11. Condensations of Pyridinecarboxylic Esters Yielding Side-Chain Acid Derivatives (Continued)

| Pyridine ester | Condensed with | Conditions | Product | Yield | Properties | Ref. |
|---|---|---|---|---|---|---|
| 4-PyCOCl · HCl | $EtCH(CO_2Et)_2$ | Na, EtOH, reflux 2.5 hr | $4\text{-PyCOCEt}(CO_2Et)_2$ | 14% | b.p. 138–142°/0.3 mm | 91 |
| $4\text{-PyCO}_2Et$ (1 equiv) | $EtO_2C(CH_2)_6CO_2Et$ | NaH, 130–140°, 4 hr; 20% HCl; $Na_2CO_3$; 20% $H_2SO_4$, reflux 8 hr, $-CO_2$ | $4\text{-PyCO}(CH_2)_6CO_2H$ | 15.6% | m.p. 163–164° | 117 |
| $4\text{-PyCO}_2Et$ (2 equiv.) | $EtO_2C(CH_2)_6CO_2Et$ | NaOEt, 12 hr, 130–140°; 20% HCl | $4\text{-PyCOCH}(CO_2Et)(CH_2)_4CH(CO_2Et)$ CO(4-Py) | | m.p. 75° | 117 |
| | $EtO_2C(CH_2)_8CO_2Et$ | NaOEt, 12 hr, 130–140°; 20% HCl | $4\text{-PyCOCH}(CO_2Et)(CH_2)_6CH(CO_2Et)$ CO(4-Py) | 14.5% | m.p. 94–95° | 117 |
| $4\text{-PyCO}_2Et$ | 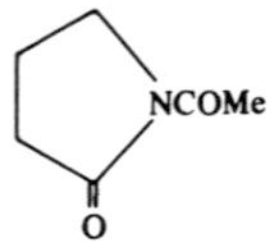 NCOMe, O | K, $C_6H_6$; hydrolysis | 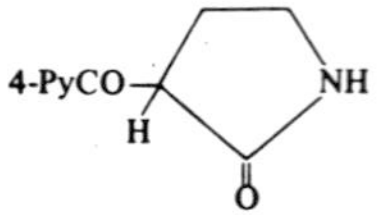 4-PyCO, H, NH, O | 32% | m.p. 115°; ir | 129 |
| | 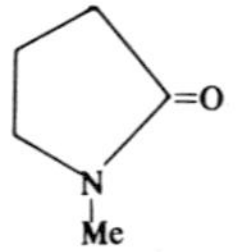 =O, N, Me | K, $C_6H_6$ | 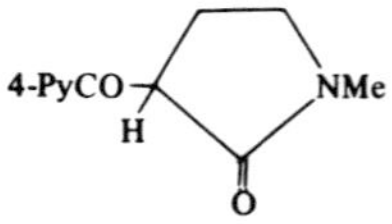 4-PyCO, H, NMe, O | 42% | m.p. 85°; ir | 129 |
| | 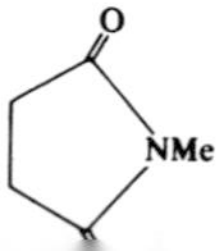 O, NMe | Na, $C_6H_6$; 120–130°, 12 hr | 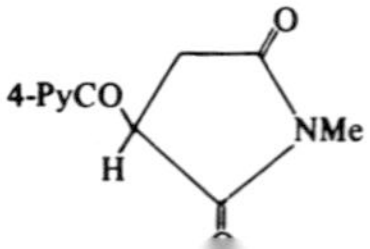 4-PyCO, H, O, NMe | | m.p. 136–137° | 117 |

| Reactant | Conditions | Product | Yield | Properties | Ref. |
|---|---|---|---|---|---|
| 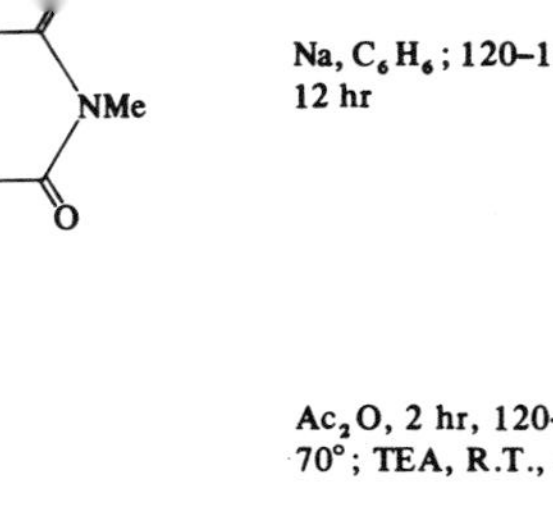  | Na, $C_6H_6$; 120–130°, 12 hr | 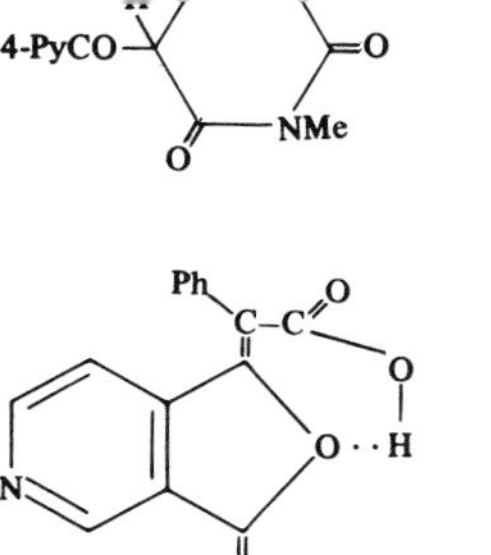  | 87% | m.p. 109–111° | 117 |
| 3, 4-$Py(CO_2H)_2$ | $Ac_2O$, 2 hr, 120–125°; $PhCH_2CO_2H$, 70°; TEA, R.T., 48 hr; HCl, R.T., 36 hr | | 0.7 g → 1.0 g | m.p. 221–222° | 240 |

TABLE XI-12. Condensations of Pyridineacetic Acids, Esters, and Amides to give Longer-Chain Acid Derivatives

| Pyridine | Condensed with | Conditions | Product | Yield | Properties | Ref. |
|---|---|---|---|---|---|---|
| 2-$PyCH_2CO_2H$ | 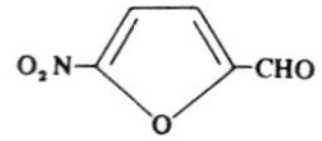 $O_2N$–[furan, O]–CHO | $Ac_2O$, 20–30°, 4 hr | 2-$PyC(CO_2H)$=CH–[furan, O]–$NO_2$ 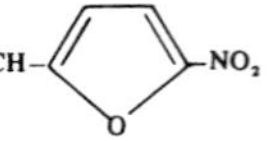 | | m.p. 126–128° | 28 |
| 2-$PyCH_2CO_2K$ | 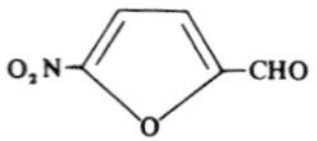 $O_2N$–[furan, O]–CHO | same as above | 2-$PyC(CO_2H)$=CH–[furan, O]–$NO_2$ 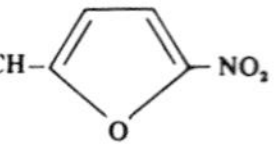 | | m.p. 240–242° (decomp.) | 61 |
| 2-$PyCH_2CO_2Me$ | $BrCH_2CO_2Me$ | (i) K, $C_6H_6$, 4 hr; (ii) overnight with halide | 2-$PyCH(CO_2Me)CH_2CO_2Me$ | 89.3 g ↓ 52.7 g | b.p. 130–140°/0.5 mm | 19 |
| | 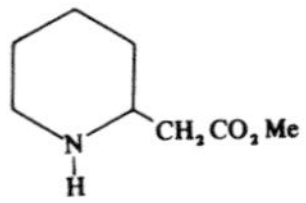 [piperidine, N–H]–$CH_2CO_2Me$ | 40% HCHO, $H_2O$, 2 hr, steam-bath | 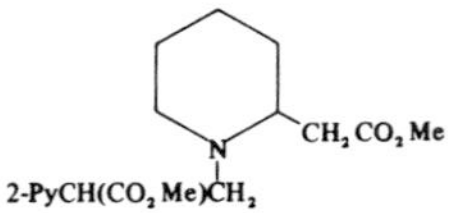 [piperidine]–$CH_2CO_2Me$, N–$CH_2(CO_2Me)CH$-2-Py (2-$PyCH(CO_2Me)CH_2$) | 42% | b.p. 130–150°/0.05 mm; ir | 19 |
| | *n*-BuBr | (i) NaOEt, EtOH or Na, $C_6H_6$, 50°, 6 hr; (ii) reflux 12 hr with halide | 2-PyCH(*n*-Bu)$CO_2Et$ | 50% | b.p. 117–121°/25 mm; $n_D^{25}$ 1.4906 | 112 |
| | *n*-HexBr | same as above | 2-PyCH(*n*-Hex)$CO_2Et$ | 50% | b.p. 181–184°/21 mm; $n_D^{25}$ 1.4855 | 112 |
| | $PhCH_2Cl$ | same as above | 2-$PyCH(CH_2Ph)CO_2Et$ | 52% | b.p. 204–209°/18 mm; $n_D^{20}$ 1.5451 | 112 |
| | $CH_2$=$CHCH_2Br$ | same as above | 2-$PyCH(CH_2CH$=$CH_2)CO_2Et$ | 27% | b.p. 163–168°/18 mm | 112 |
| | *n*-HexCHO | Piperidine, $C_6H_6$, HOAc | 2-PyC(=CH-Hex)$CO_2Et$ | 35% | b.p. 194–196°/25 mm | 112 |
| | PhCHO | Piperidine, $C_6H_6$, HOAc | 2-PyC(=CHPh)$CO_2Et$ | 41% | b.p. 145°/0.3 mm; $n_D^{25}$ 1.5965 | 112 |
| | PhCHO | piperidine, |  2-Py, Ph (C=C) | 46% | b.p. 160–161°/1.0 mm; | 111 |

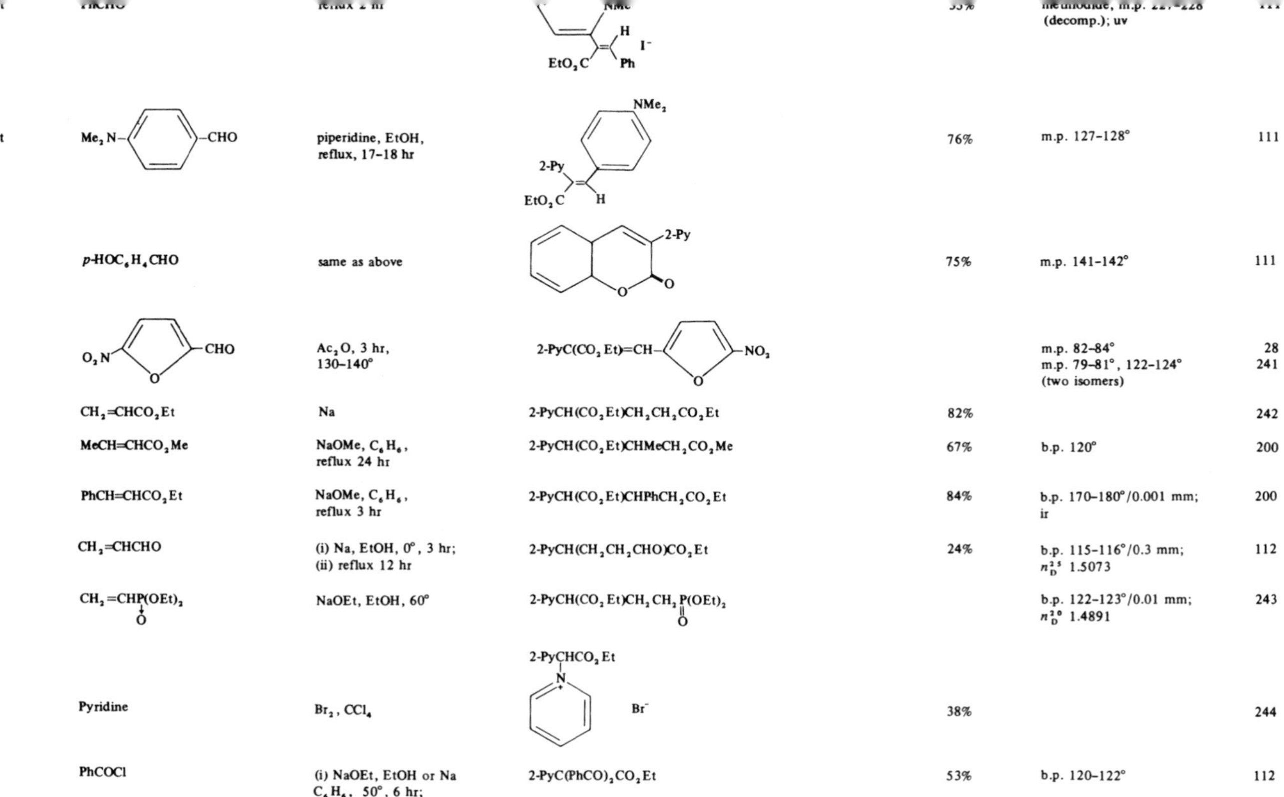

| | | | | | | |
|---|---|---|---|---|---|---|
| $2\text{-PyCH}_2CO_2Et$ MeI | PhCHO | reflux 2 hr | | | methiodide, m.p. 227–228 (decomp.); uv | |
| $2\text{-PyCH}_2CO_2Et$ | $Me_2N$–$C_6H_4$–CHO | piperidine, EtOH, reflux, 17–18 hr | | 76% | m.p. 127–128° | 111 |
| | $p\text{-HOC}_6H_4CHO$ | same as above | | 75% | m.p. 141–142° | 111 |
| | $O_2N$–furyl–CHO | $Ac_2O$, 3 hr, 130–140° | $2\text{-PyC}(CO_2Et)\text{=CH}$–furyl–$NO_2$ | | m.p. 82–84°<br>m.p. 79–81°, 122–124° (two isomers) | 28<br>241 |
| | $CH_2\text{=CHCO}_2Et$ | Na | $2\text{-PyCH}(CO_2Et)CH_2CH_2CO_2Et$ | 82% | | 242 |
| | $MeCH\text{=CHCO}_2Me$ | NaOMe, $C_6H_6$, reflux 24 hr | $2\text{-PyCH}(CO_2Et)CHMeCH_2CO_2Me$ | 67% | b.p. 120° | 200 |
| | $PhCH\text{=CHCO}_2Et$ | NaOMe, $C_6H_6$, reflux 3 hr | $2\text{-PyCH}(CO_2Et)CHPhCH_2CO_2Et$ | 84% | b.p. 170–180°/0.001 mm; ir | 200 |
| | $CH_2\text{=CHCHO}$ | (i) Na, EtOH, 0°, 3 hr; (ii) reflux 12 hr | $2\text{-PyCH}(CH_2CH_2CHO)CO_2Et$ | 24% | b.p. 115–116°/0.3 mm; $n_D^{25}$ 1.5073 | 112 |
| | $CH_2\text{=CHP(O)(OEt)}_2$ | NaOEt, EtOH, 60° | $2\text{-PyCH}(CO_2Et)CH_2CH_2P(O)(OEt)_2$ | | b.p. 122–123°/0.01 mm; $n_D^{20}$ 1.4891 | 243 |
| | Pyridine | $Br_2$, $CCl_4$ | $2\text{-PyCHCO}_2Et$ (N-pyridinium) $Br^-$ | 38% | | 244 |
| | PhCOCl | (i) NaOEt, EtOH or Na $C_6H_6$, 50°, 6 hr; (ii) reflux 12 hr with halide | $2\text{-PyC(PhCO)}_2CO_2Et$ | 53% | b.p. 120–122° | 112 |

TABLE XI-12. Condensations of Pyridineacetic Acids, Esters, and Amides to Give Longer-Chain Acid Derivatives (Continued)

| Pyridine | Condensed with | Conditions | Product | Yield | Properties | Ref. |
|---|---|---|---|---|---|---|
| $2\text{-}PyCH_2CO_2Et$ or $2\text{-}PyCH(CO_2Et)\text{-}N^+C_5H_5$ Br | $ON\text{-}C_6H_4\text{-}N(CH_2CH_2Cl)_2$ | Py·HBr, KOH, EtOH | $2\text{-}PyC(CO_2Et){=}N(\rightarrow O)\text{-}C_6H_4\text{-}N(CH_2CH_2Cl)_2$ | | m.p. 139–140° | 244 |
| $2\text{-}PyCH_2CO_2Et \cdot MeI$ | $ON\text{-}C_6H_4\text{-}N(CH_2CH_2Cl)_2$ | | 1-Me-pyridinium-2-$C(CO_2Et){=}N\text{-}C_6H_4\text{-}N(CH_2CH_2Cl)_2$ $I^-$ | | m.p. 151–152° | 244 |
| $[C_5H_5N^+CH(2\text{-}Py)(CO_2Et)]\ Br^{\ominus}$ | $ON\text{-}C_6H_4\text{-}N(CH_2CH_2OH)_2$ | $K_2CO_3$, EtOH, 0.5–1 hr | $2\text{-}PyC(CO_2Et){=}N(\rightarrow O)\text{-}C_6H_4\text{-}N(CH_2CH_2OH)_2$ | | m.p. 150–153° (decomp.) | 244 |
| $2\text{-}PyCH_2CO_2Et \cdot MeI$ | $ON\text{-}C_6H_4\text{-}N(CH_2CH_2OH)_2$ | | 1-Me-pyridinium-2-$C(CO_2Et){=}N\text{-}C_6H_4\text{-}N(CH_2CH_2OH)_2$ $I^-$ | | m.p. 161–163° (decomp.) | 244 |
| $2\text{-}PyCH_2CONH_2$ | 5-$O_2N$-furan-2-CHO | $Ac_2O$, 3 hr, 130–140° | $2\text{-}PyC(CONH_2){=}CH$-2-(5-$NO_2$-furyl) | | m.p. 180–182°, 180–181.5° (two isomers)<br>m.p. 180–182° | 241<br>28 |
| 5-Me-2-$PyCH_2CO_2Et$ | MeI | (i) K, $C_6H_6$, reflux 15 hr; (ii) halide 2–6 hr | 5-Me-2-$PyCH(Me)CO_2Et$ | 62% | b.p. 132–136°/14 mm;<br>picrate, m.p. 170° | 206<br>192 |
| | $ICH_2CH_2CO_2Et$ | (i) K, $C_6H_6$, reflux, 15 hr; (ii) halide, reflux 2 hr | 5-Me-2-$PyCH(CH_2)_2CO_2Et$ [illegible] | 50% | b.p. 140–150°/0.01 mm | 206 |

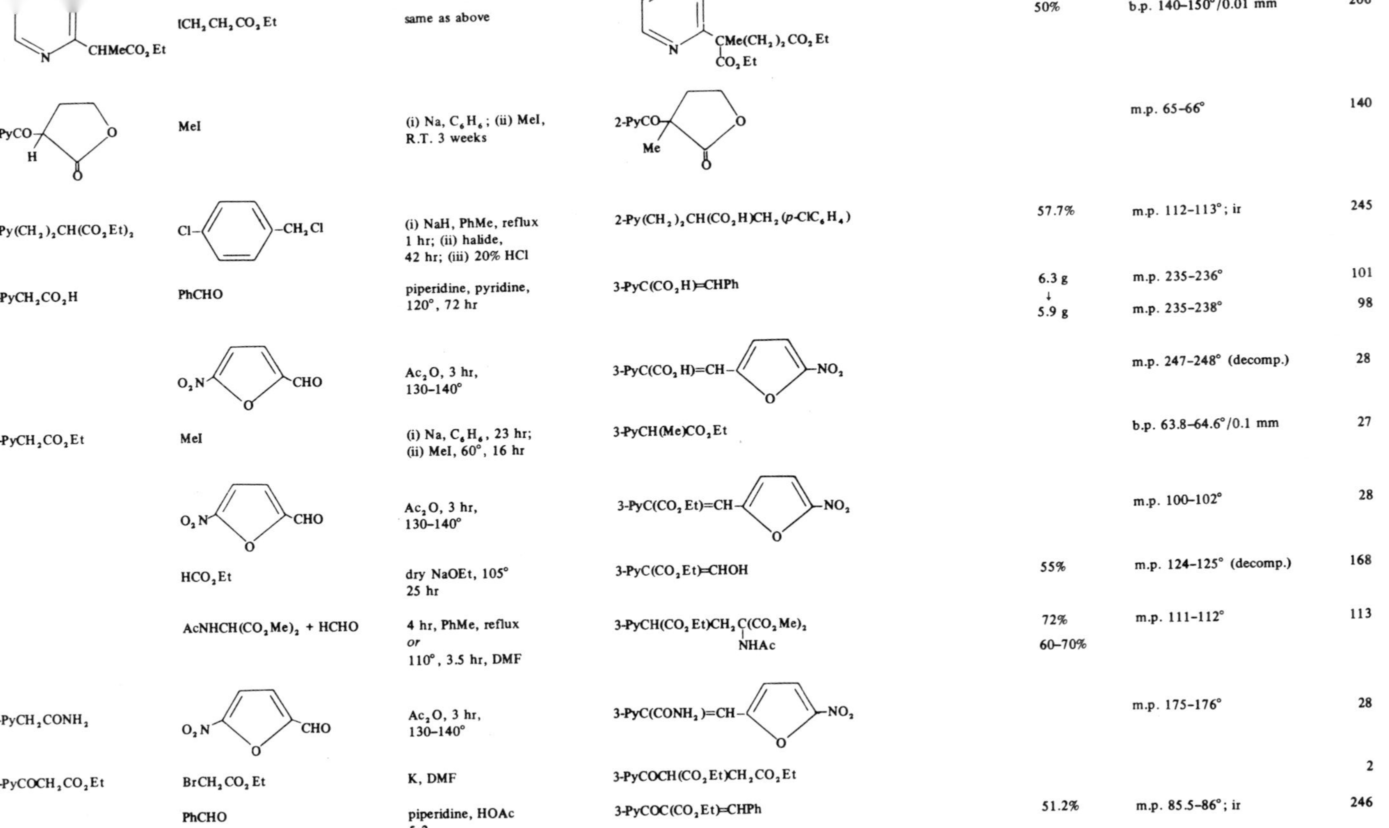

| | | | | | | |
|---|---|---|---|---|---|---|
| N, $CHMeCO_2Et$ | $ICH_2CH_2CO_2Et$ | same as above | N, $CMe(CH_2)_2CO_2Et$, $CO_2Et$ | 50% | b.p. 140–150°/0.01 mm | 206 |
| 2-PyCO, H, O, O | MeI | (i) Na, $C_6H_6$; (ii) MeI, R.T. 3 weeks | 2-PyCO, Me, O, O | | m.p. 65–66° | 140 |
| $2\text{-Py}(CH_2)_2CH(CO_2Et)_2$ | Cl, $CH_2Cl$ | (i) NaH, PhMe, reflux 1 hr; (ii) halide, 42 hr; (iii) 20% HCl | $2\text{-Py}(CH_2)_2CH(CO_2H)CH_2(p\text{-}ClC_6H_4)$ | 57.7% | m.p. 112–113°; ir | 245 |
| $3\text{-PyCH}_2CO_2H$ | PhCHO | piperidine, pyridine, 120°, 72 hr | $3\text{-PyC}(CO_2H){=}CHPh$ | 6.3 g ↓ 5.9 g | m.p. 235–236° m.p. 235–238° | 101 98 |
| | $O_2N$, O, CHO | $Ac_2O$, 3 hr, 130–140° | $3\text{-PyC}(CO_2H){=}CH$–, O, $NO_2$ | | m.p. 247–248° (decomp.) | 28 |
| $3\text{-PyCH}_2CO_2Et$ | MeI | (i) Na, $C_6H_6$, 23 hr; (ii) MeI, 60°, 16 hr | $3\text{-PyCH(Me)}CO_2Et$ | | b.p. 63.8–64.6°/0.1 mm | 27 |
| | $O_2N$, O, CHO | $Ac_2O$, 3 hr, 130–140° | $3\text{-PyC}(CO_2Et){=}CH$–, O, $NO_2$ | | m.p. 100–102° | 28 |
| | $HCO_2Et$ | dry NaOEt, 105° 25 hr | $3\text{-PyC}(CO_2Et){=}CHOH$ | 55% | m.p. 124–125° (decomp.) | 168 |
| | $AcNHCH(CO_2Me)_2$ + HCHO | 4 hr, PhMe, reflux *or* 110°, 3.5 hr, DMF | $3\text{-PyCH}(CO_2Et)CH_2C(CO_2Me)_2$, NHAc | 72% 60–70% | m.p. 111–112° | 113 |
| $3\text{-PyCH}_2CONH_2$ | $O_2N$, O, CHO | $Ac_2O$, 3 hr, 130–140° | $3\text{-PyC}(CONH_2){=}CH$–, O, $NO_2$ | | m.p. 175–176° | 28 |
| $3\text{-PyCOCH}_2CO_2Et$ | $BrCH_2CO_2Et$ | K, DMF | $3\text{-PyCOCH}(CO_2Et)CH_2CO_2Et$ | | | 2 |
| | PhCHO | piperidine, HOAc 5:2 | $3\text{-PyCOC}(CO_2Et){=}CHPh$ | 51.2% | m.p. 85.5–86°; ir | 246 |

TABLE XI-12. Condensations of Pyridineacetic Acids, Esters, and Amides to Give Longer-Chain Acid Derivatives (Continued)

| Pyridine | Condensed with | Conditions | Product | Yield | Properties | Ref. |
|---|---|---|---|---|---|---|
| | 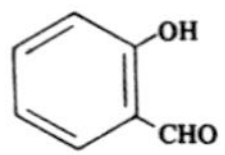 OH<br>CHO | same as above | 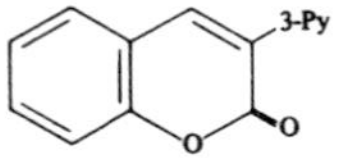 3-Py<br>O<br>O | 38.5% | m.p. 158–159°; hydrochloride, m.p. 236–237° | 246 |
| | 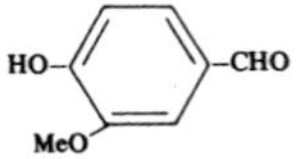 HO– –CHO<br>MeO | same as above | $3\text{-PyCOC(CO}_2\text{Et)=CH-}$ 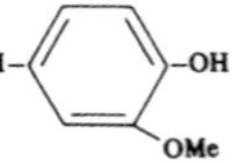 –OH<br>OMe | 61.3% | m.p. 145–146° | 246 |
| | $p\text{-Me}_2\text{NC}_6\text{H}_4\text{CHO}$ | same as above | $3\text{-PyCOC(CO}_2\text{Et)=CH(}p\text{-Me}_2\text{NC}_6\text{H}_4\text{)}$ | 46% | m.p. 121° | 246 |
| | $o\text{-O}_2\text{NC}_6\text{H}_4\text{CHO}$ | same as above | $3\text{-PyCOC(CO}_2\text{Et)=CH(}o\text{-O}_2\text{NC}_6\text{H}_4\text{)}$ | 40.4% | m.p. 103–104° | 246 |
| | $p\text{-O}_2\text{NC}_6\text{H}_4\text{CHO}$ | same as above | $3\text{-PyCOC(CO}_2\text{Et)=CH(}p\text{-O}_2\text{NC}_6\text{H}_4\text{)}$ | 86.5% | m.p. 110–111° | 246 |
| | $p\text{-C}_6\text{H}_4\text{(CHO)}_2$ | same as above | $3\text{-PyCOC(CO}_2\text{Et)=CH-}$ 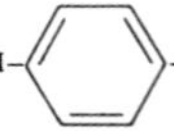 $\text{-CH=C(CO}_2\text{Et)CO(3-Py)}$ | 50% | m.p. 179; ir | 246 |
| | 2-FurylCHO | same as above | $3\text{-PyCOC(CO}_2\text{Et)=CH(2-furyl)}$ | 77.9% | m.p. 114–115° | 246 |
| | 2-PyCHO | same as above | $3\text{-PyCOC(CO}_2\text{Et)=CH(2-Py)}$ | 85% | m.p. 119–120° | 246 |
| | 3-PyCHO | same as above | $3\text{-PyCOC(CO}_2\text{Et)=CH(3-Py)}$ | 34% | m.p. 84°; ir | 246 |
| | 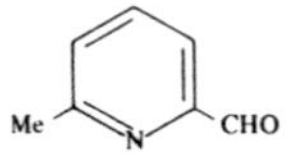 Me N CHO | same as above | $3\text{-PyCOC(CO}_2\text{Et)=CH-}$ 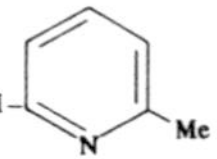 N Me | 87.8% | m.p. 123.5–124.5° | 246 |
| | 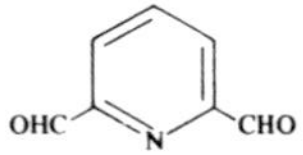 OHC N CHO | same as above | $3\text{-PyCOC(CO}_2\text{Et)=CH-}$ 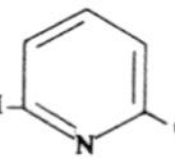 N $\text{CH=C(CO}_2\text{Et)CO(3-Py)}$ | 89% | m.p. 135° | 246 |
| $4\text{-PyCH}_2\text{CO}_2\text{Et}$ | PhCHO | $\text{Ac}_2\text{O}$, 5 hr, 150–160°; 2*N* MeOH, KOH | $4\text{-PyC(=CHPh)CO}_2\text{H}$ | 1.8 g → 1 g | m.p. 203° (decomp.) | 29 |
| | PhCHO | piperidine, EtOH, |  4-Py Ph | 75% | b.p. 175–176°/2.3 mm; | 111 |

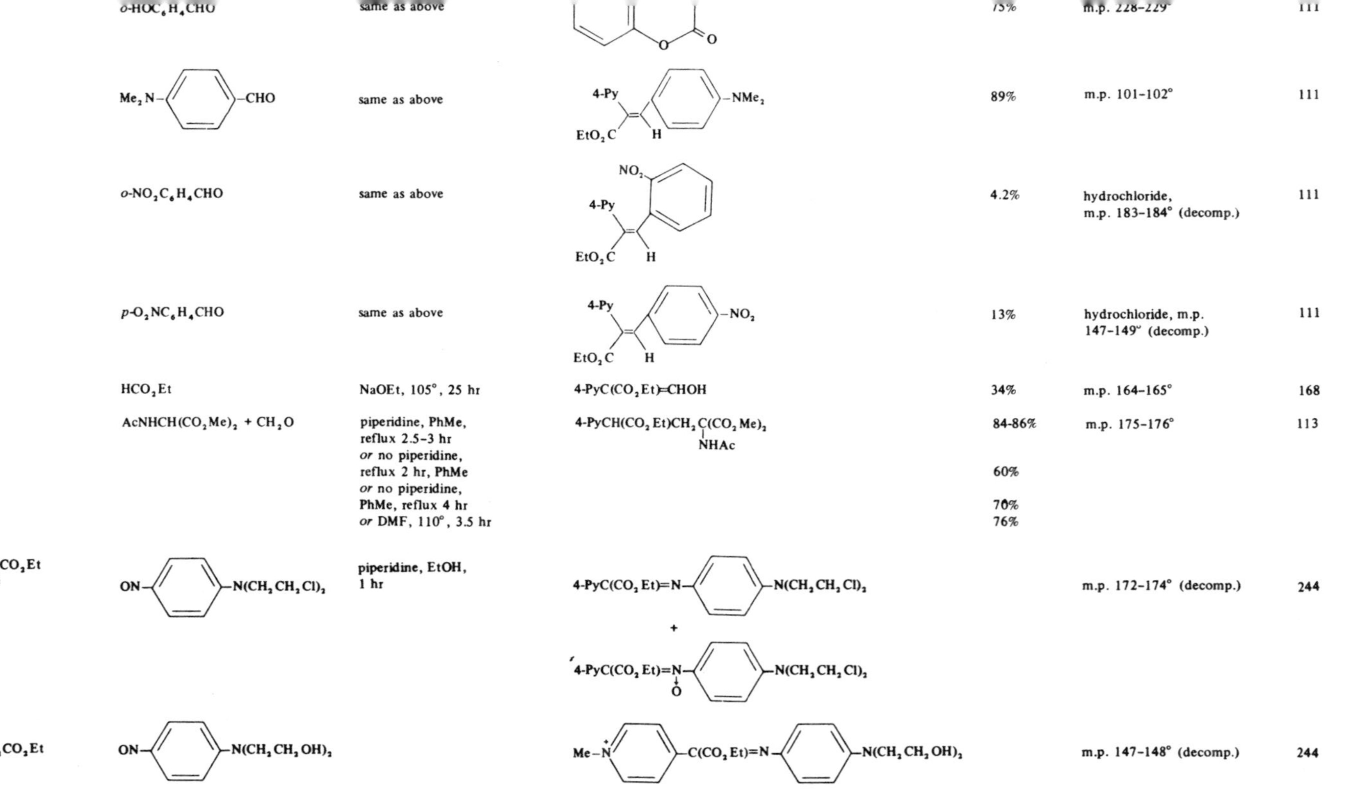

| | | | | | | |
|---|---|---|---|---|---|---|
| | $o\text{-}HOC_6H_4CHO$ | same as above | O, O | 75% | m.p. 228–229° | 111 |
| | $Me_2N$–, –CHO | same as above | 4-Py, $EtO_2C$, H, $-NMe_2$ | 89% | m.p. 101–102° | 111 |
| | $o\text{-}NO_2C_6H_4CHO$ | same as above | $NO_2$, 4-Py, $EtO_2C$, H | 4.2% | hydrochloride, m.p. 183–184° (decomp.) | 111 |
| | $p\text{-}O_2NC_6H_4CHO$ | same as above | 4-Py, $EtO_2C$, H, $-NO_2$ | 13% | hydrochloride, m.p. 147–149° (decomp.) | 111 |
| | $HCO_2Et$ | NaOEt, 105°, 25 hr | $4\text{-}PyC(CO_2Et){=}CHOH$ | 34% | m.p. 164–165° | 168 |
| | $AcNHCH(CO_2Me)_2 + CH_2O$ | piperidine, PhMe, reflux 2.5–3 hr | $4\text{-}PyCH(CO_2Et)CH_2C(CO_2Me)_2$ (NHAc) | 84-86% | m.p. 175–176° | 113 |
| | | *or* no piperidine, reflux 2 hr, PhMe | | 60% | | |
| | | *or* no piperidine, PhMe, reflux 4 hr | | 70% | | |
| | | *or* DMF, 110°, 3.5 hr | | 76% | | |
| $4\text{-}PyCH_2CO_2Et$ free base | ON–, $-N(CH_2CH_2Cl)_2$ | piperidine, EtOH, 1 hr | $4\text{-}PyC(CO_2Et){=}N$–, $-N(CH_2CH_2Cl)_2$ + $4\text{-}PyC(CO_2Et){=}N(\rightarrow O)$–, $-N(CH_2CH_2Cl)_2$ | | m.p. 172–174° (decomp.) | 244 |
| $4\text{-}PyCH_2CO_2Et$ ·MeI | ON–, $-N(CH_2CH_2OH)_2$ | | Me–$\overset{+}{N}$, $-C(CO_2Et){=}N$–, $-N(CH_2CH_2OH)_2$, $I^-$ | | m.p. 147–148° (decomp.) | 244 |

TABLE XI-12. Condensations of Pyridineacetic Acids, Esters, and Amides to Give Longer-Chain Derivatives (Continued)

| Pyridine | Condensed with | Conditions | Product | Yield | Properties | Ref. |
|---|---|---|---|---|---|---|
| 4-$PyCOCH_2CO_2Et$ | $BrCH_2CH_2CO_2Et$ | Na, EtOH; 70°, 2 hr | 4-$PyCOCH(CO_2Et)CH_2CH_2CO_2Et$ | | b.p. 155–160°/0.1 mm | 117 |
| | PhCHO | piperidine, 3 hr, steam-bath | (4-$PyCOCHCO_2Et)_2CHPh$ | | m.p. 102–103° | 105 |
| | $CH_2(COPh)_2$ + $CH_2O$ | PhMe, reflux 4 hr | 4-$PyCOCH(CO_2Et)CH_2CH(COPh)_2$ | 50% | m.p. 112–113° | 113 |
| | $AcNHCH(CO_2Et)_2$ + $CH_2O$ | same as above | 4-$PyCOCH(CO_2Et)CH_2C(NHAc)(CO_2Et)_2$ | 52% | m.p. 160–161° | 113 |
| | $AcNHCH(Ac)CO_2Me$ + $CH_2O$ | same as above | 4-$PyCOCH(CO_2Et)CH_2C(NHAc)(Ac)CO_2Me$ | 47% | m.p. 122–123° | 113 |
| 4-$PyC(CO_2Et)$=$S(O)Me_2$ | MeI | $CHCl_3$ | 4-$PyC(I)(Me)CO_2Et$ | | m.p. 63° | 247 |

| Starting material | Conditions | Product | Yield | Properties | Ref. |
|---|---|---|---|---|---|
| 2-PyCH=CHCO$_2$H | P, HOAc, 57% HI, 14 hr; $Na_2HPO_4$, heat, *n*-BuOH | 2-Py(CH$_2$)$_2$CO$_2$H | | m.p. 143–144° | 114 |
| 2-PyCH=CHCO$_2$Et | MeI, acetone; $H_2O$, $Ag_2O$; $HCO_2H$ | 2-Py(CH$_2$)$_2$CO$_2$H | | | 114 |
| | $H_2$, $PtO_2$ HOAc | 2-Py(CH$_2$)$_2$CO$_2$Et | 50% | m.p. 96–97°/3 mm; methiode, m.p. 81–82° | 114 |
| | *or* $H_2$, 10% Pd-C, HOAc | | | | 34 |
| | *or* MeI, acetone; $Ag_2O$, $H_2O$; 95% $HCO_2H$, $HCO_2K$, 13 hr, 143–148°; EtOH, HCl, reflux 4 hr | | 9.5% | | 114 |
| $MeO_2C$ N $CH{=}CHCO_2Me$ | $H_2$, $PtO_2$ | $MeO_2C$ N $CH{=}CHCO_2Me$ | 1.0 g → 0.5 g | m.p. 53–54°; ir | 197 |
| $MeO_2C$ N $CH{=}CHCO_2Me$ | $H_2$, $PtO_2$, 5 atm, 6 hr, MeOH, 5 *N* HCl; PhCOCl, Pyridine; NaH, THF, 2 days, EtOH, $N_2$, 36 hr, R.T.; reflux 4 hr | $PhCH_2N$ OH $CO_2Me$ | | | 197 |
| 2-PyC(=CH$_2$)CO$_2$H | $H_2$, 5% Pd-C, aq. NaOH, 2 atm, < 2 hr | 2-PyCHMeCO$_2$H | 67.4% | m.p. 144.5–145° | 248 |
| 2-Py(CH$_2$)$_4$C≡CCO$_2$H | $H_2$, 0.5% Pd on alumina R.T., pressure | 2-Py(CH$_2$)$_4$CH=CHCO$_2$H | | m.p. 93–94°; ir | 80 |

TABLE XI-13. Preparation of Acids and Derivatives by Reduction of Side-Chain Functions (Continued)

| Starting material | Conditions | Product | Yield | Properties | Ref. |
|---|---|---|---|---|---|
| | $H_2$, Ni or $PtO_2$, R.T., aq. EtOH | 2-Py$(CH_2)_6CO_2H$ | | m.p. 60°; ir | 80 |
| 2-Py$(CH_2)_5C{\equiv}CCO_2H$ | $H_2$, 0.5% Pd on alumina R.T., pressure | 2-Py$(CH_2)_5CH{=}CHCO_2H$ | | m.p. 43–44° | 80<br>81 |
| | $H_2$, Ni or $PtO_2$, R.T., aq. EtOH | 2-Py$(CH_2)_7CO_2H$ | | m.p. 45° | 80 |
| 2-Py$(CH_2)_{10}C{\equiv}CCO_2H$ | $H_2$, 0.5% Pd on alumina, R.T., pressure | 2-Py$(CH_2)_{10}CH{=}CHCO_2H$ | | m.p. 75° | 80 |
| | $H_2$, Ni or $PtO_2$, aq. EtOH, R.T., pressure | 2-Py$(CH_2)_{12}CO_2H$ | | m.p. 77° | 80<br>81 |
| 2-PyCOCH$_2$CO$_2$Et | $H_2$, $PtO_2$, 2 hr, 3 atm, EtOH | 2-PyCH(OH)CH$_2$CO$_2$Et | | b.p. 89–90°/0.1 mm; picrate, m.p. 99–100°; hydrochloride, m.p. 129–130° | 249 |
| 2-PyCOCHPhCO$_2$Me | $NaBH_4$, MeOH | 2-PyCH(OH)CHPhCO$_2$Me | 75% | m.p. 105–106°; hydrochloride, m.p. 108–110° | 116 |
| $CO_2Me$, O, 2-Py, H, N | $NaBH_4$, MeOH, 20°, 2 hr | H, O, N, $CO_2Me$, 2-Py | 82% | m.p. 135–137°; ir | 19 |
| | | + $CO_2Me$, OH | | m.p. 149° | |

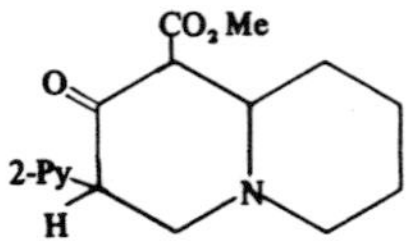

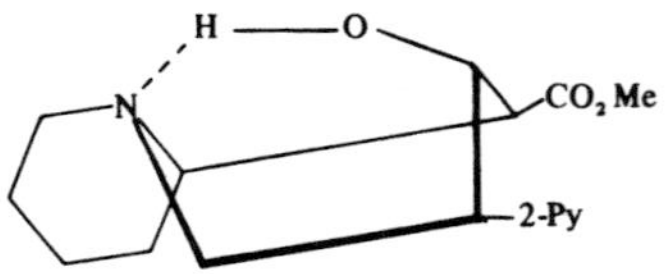

| | | | | | |
|---|---|---|---|---|---|
| | and then repeat | | | | |
| 3-PyCCl=CHCO$_2$Et + 3-PyCCl$_2$CH$_2$CO$_2$Et | $H_2$, 5% Pd-C, EtOH, Py, R.T. | 3-PyCH$_2$CH$_2$CO$_2$H | | $n_D^{20}$ 1.5140; picrate, m.p. 103° | 175 |
| 3-PyCPh=CHCO$_2$Na | $H_2$, Ni | 3-PyCHPhCH$_2$CO$_2$H | 90% | m.p. 171–172° | 54 |
| 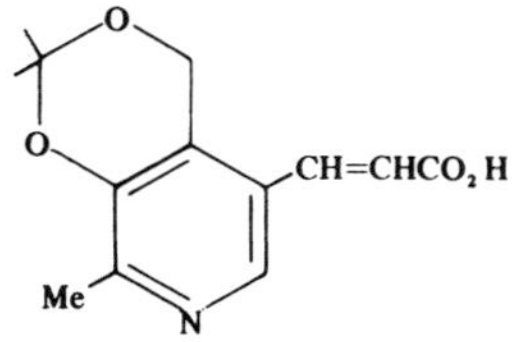 O, O, Me, N, CH=CHCO$_2$H | $H_2$, Pd, EtOH, 4 hr | 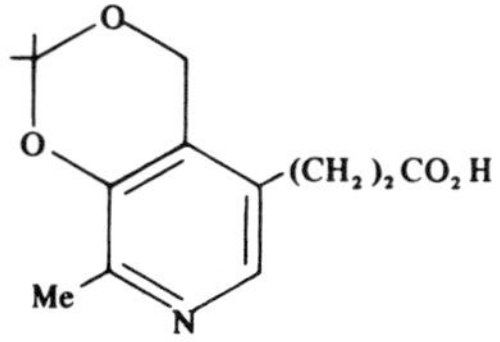 O, O, Me, N, (CH$_2$)$_2$CO$_2$H | 97.5% | m.p. 188–190°; ir; uv | 228 |
| 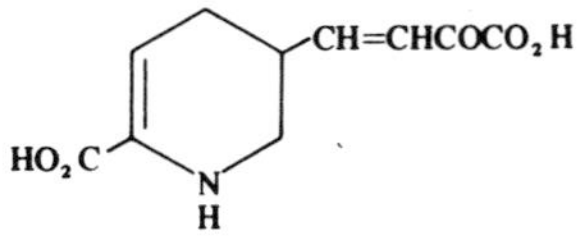 HO$_2$C, N, H, CH=CHCOCO$_2$H | enzyme reduction | 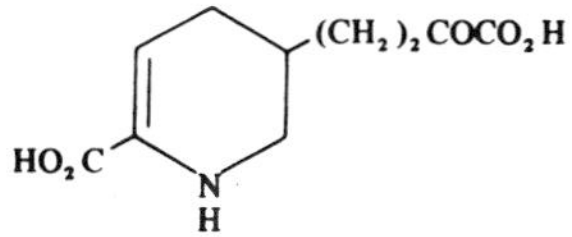 HO$_2$C, N, H, (CH$_2$)$_2$COCO$_2$H | | | 73 |
| 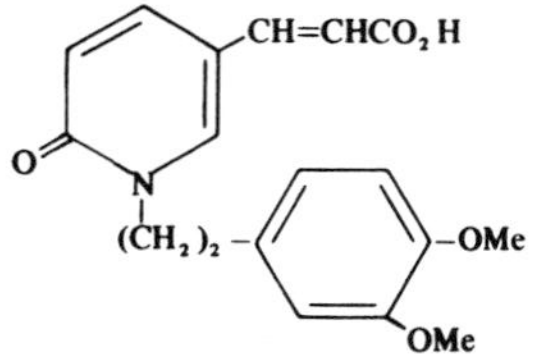 O, N, (CH$_2$)$_2$–, CH=CHCO$_2$H, OMe, OMe | $H_2$, PtO$_2$, EtOH, 40–45° | 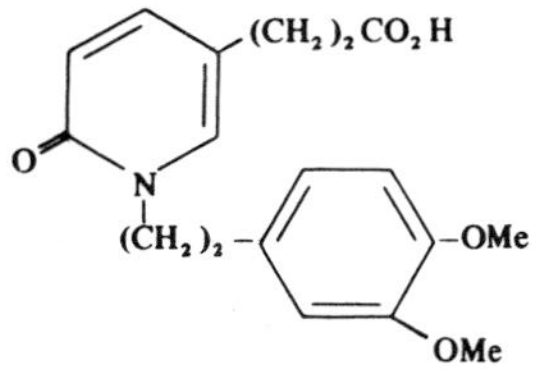 O, N, (CH$_2$)$_2$–, (CH$_2$)$_2$CO$_2$H, OMe, OMe | 3 g → 3 g | m.p. 58–59° | 7 |
| 3-PyC(=CH$_2$)CO$_2$H | $H_2$, 5% Pd-C, aq. NaOH, 2 atm, < 2 hr | 3-PyCHMeCO$_2$H | 63% | m.p. 163–165° | 248 |
| 3-PyCOCHPhCONH$_2$ | NaBH$_4$, MeOH–H$_2$O | 3-PyCH(OH)CHPhCONH$_2$ | 80% | m.p. 193–194° | 116 |
| 3-PyCH$_2$COCH$_2$CONHMe | Wolff-Kishner; KOH, 4 hr, 195° | 3-Py(CH$_2$)$_3$CO$_2$H | | m.p. 121–122° | 22 |

TABLE XI-13. Preparation of Acids and Derivatives by Reduction of Side-Chain Functions (Continued)

| Starting material | Conditions | Product | Yield | | Ref. |
|---|---|---|---|---|---|
| 3-PyCO$(CH_2)_2CO_2H$ | $NaBH_4$, KOH, $H_2O$, 2 hr | 3-PyCH(OH)$(CH_2)_2CO_2H$ | | | 23 |
| 3-PyCO$(CH_2)_5CO_2Et$ | Wolff-Kishner | 3-Py$(CH_2)_6CO_2H$ | 2.5 → 0.7 g | m.p. 92–93° | 117 |
| 3-PyC(=NOH)$(CH_2)_2CO_2Et$ | Pb-$BaSO_4$, $H_2$ 1 atm | 3-PyCH($NH_2$)$(CH_2)_2CO_2Et$ *d,l* | | | 2 |
| 4-PyCH=CH$CO_2H$ | $H_2$ | 4-Py$(CH_2)_2CO_2H$ | | | 115 |
| 4-PyCH=CH$CO_2Me$ | $NaBH_4$, MeOH, reflux 1-2 hr | 4-Py$(CH_2)_2CO_2Me$ | | | 115 |
| 4-PyCH=CH$CO_2Et$ | MeI, acetone; $Ag_2O$, $H_2O$; $HCO_2H$ | 4-Py$(CH_2)_2CO_2H$ | | | 114 |
| | MeI, acetone; $Ag_2O$, $H_2O$; 95% $HCO_2H$, $HCO_2K$, 10 hr, 156–157°; HCl, EtOH, reflux 4 hr | 4-Py$(CH_2)_2CO_2Et$ | 30 → 49 g | b.p. 113°/4 mm; picrate, m.p. 119–120°; methiodide, m.p. 88–89° | 114 |
| 4-PyCH=CHCONHN$H_2$ | $H_2$, $PtO_2$, EtOH, R.T., 1 atm | 4-Py$(CH_2)_2$CONHN$H_2$ | 0.3 → 0.15 g | m.p. 84°; hydrate, m.p. 64° | 166 |
| 4-PyC(=$CH_2$)$CO_2H$ | $H_2$, 5% Pd-C, aq. NaOH, 2 atm, < 2 hr | 4-PyCHMe$CO_2H$ | 80% | m.p. 225–227° | 248 |
| 4-PyC(=CHPh)CONHN$H_2$ | $H_2$, $PtO_2$, EtOH, R.T., 1 atm | 4-PyCH($CH_2$Ph)CONHN$H_2$ | | m.p. 138° | 166 |
| 4-Py$(CH_2)_4$C≡C$CO_2H$ | $H_2$, 0.5% Pd on alumina, R.T., pressure | 4-Py$(CH_2)_4$CH=CH$CO_2H$ | | m.p. 120° | 80 |
| | $H_2$, Ni or $PtO_2$, aq. EtOH, R.T., pressure | 4-Py$(CH_2)_6CO_2H$ | | m.p. 165–166° | 80 81 |
| 4-Py$(CH_2)_5$C≡C$CO_2H$ | $H_2$, 0.5% Pd on alumina, | 4-Py$(CH_2)_5$CH=CH$CO_2H$ | | m.p. 98° | 80 |

| | | | | | |
|---|---|---|---|---|---|
| | EtOH, R.T., pressure | | | | |
| 4-PyCOCHPhCONH$_2$ | NaBH$_4$, MeOH-H$_2$O | 4-PyCH(OH)CHPhCONH$_2$ | 82.8% | m.p. 144–145° | 116 |
| 4-PyCO(CH$_2$)$_5$CO$_2$Et | Na, 65% NH$_2$NH$_2$, O(CH$_2$CH$_2$OH)$_2$; EtOH, 2 hr, 115°; 200°, 3 hr | 4-Py(CH$_2$)$_6$CO$_2$H | 39% | m.p. 163–164° | 117 |

TABLE XI-14. The Willgerodt Reaction

| Starting material | Conditions | Product | Yield | Properties | Ref. |
|---|---|---|---|---|---|
| 2-PyCH=$CH_2$ | S, DMF, 30 hr, 140–160° | 2-Py$CH_2$CSN$Me_2$ | | b.p. 153–157°/1.2 mm; | 250 |
| 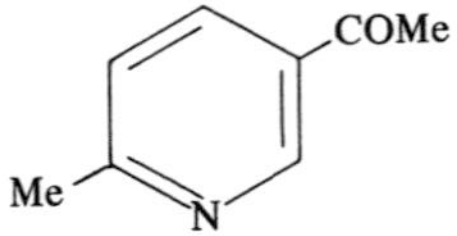 COMe, Me, N | S, morpholine, reflux 6 hr; 95% EtOH, 60% NaOH, reflux 16 hr | 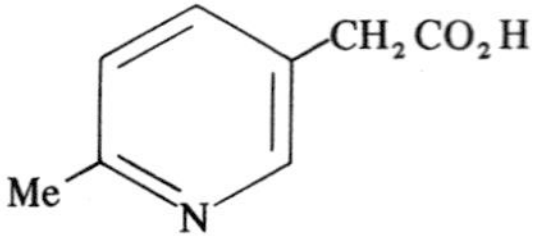 $CH_2CO_2H$, Me, N | | hydrochloride, m.p. 170–172° | 251 |
| | $NH_4SH$, dioxane, heat 6 hr | 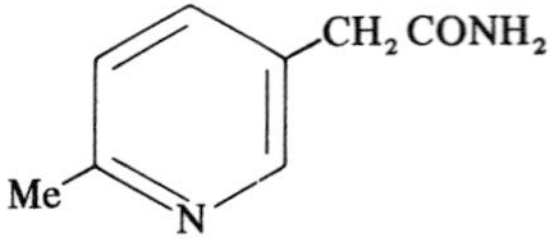 $CH_2CONH_2$, Me, N | 27 g → 26 g | m.p. 158–160°; picrate, m.p. 168–169° | 156 |
| 4-PyEt | S, DMF, 30 hr, 140–160° | 4-Py$CH_2$CSN$Me_2$ | | m.p. 108–112° | 250 |
| | S, MeNHCHO, 30 hr, 140–160° | 4-Py$CH_2$CSNHMe | | m.p. 136–141° | 250 |
| 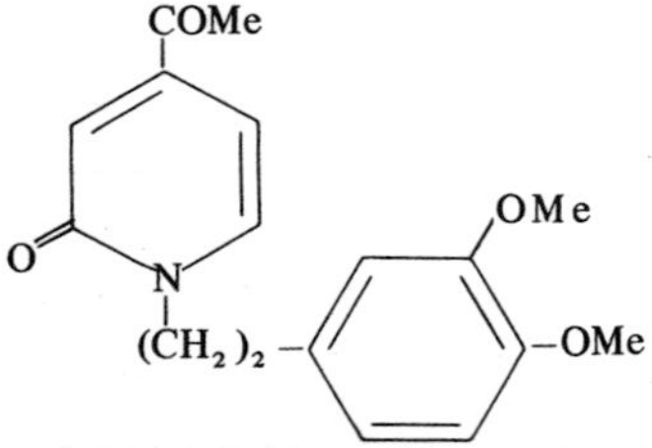 COMe, O, N, $(CH_2)_2$, OMe, OMe | S, morpholine, 150–160°, reflux 5 hr | 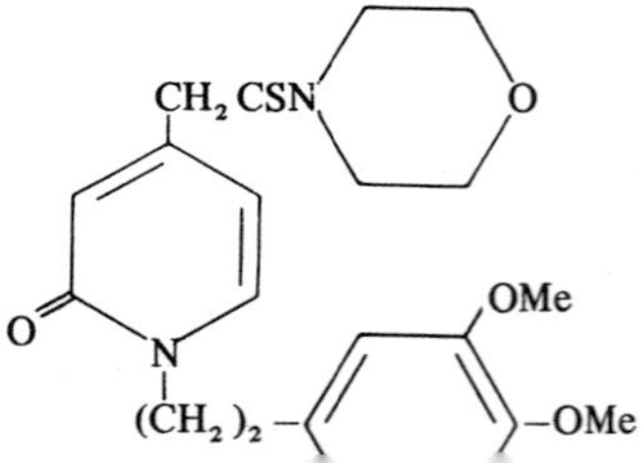 $CH_2$CSN, O, O, N, $(CH_2)_2$, OMe, OMe | 52% | | 8 |

| | | | | | |
|---|---|---|---|---|---|
| [illegible] | [illegible] reflux 11 hr, 50% NaOH; HCl, 5 hr, boiled in MeOH | [illegible] | [illegible] ↓ 1.4 g | [illegible] | [illegible] |
| 4-PyCH=$CH_2$ | S, morpholine, reflux 16 hr | 4-Py$CH_2$CSN(morpholino: $-N(CH_2CH_2)_2O$) | 53% | m.p. 105–107° | 3 |

TABLE XI-14-1. Preparation of Acids and Derivatives by Miscellaneous Methods

| Reactants | Conditions | Product | Yield | Properties | Ref. |
|---|---|---|---|---|---|
| 2-PyMgCl + $BrCH_2CO_2Et$ | | $2\text{-}PyCH_2CO_2Et$ | | | 252 |
| 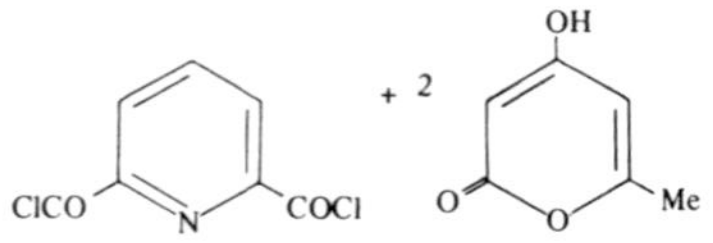 ClCO, COCl (pyridine-2,6-dicarbonyl chloride) + 2 (OH, O, O, Me pyranone) | Py, $C_6H_6$, R.T., 4–5 hr | 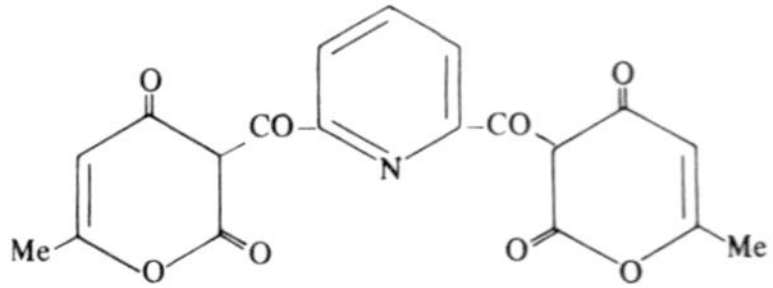 (pyridine-2,6-diyl bis-CO- pyranone; O, Me labels) | 79% | m.p. 196° | 253 |
| $2\text{-}PyCH_2CO_2Et$ | $NaNO_2$, HOAc, 15–25° 30 min | $2\text{-}PyC(=NOH)CO_2Et$ | | m.p. 149–150°; ir | 112 |
| $2\text{-}PyCH_2CO_2Et$ | MeI, EtOH; 1*N* NaOH | 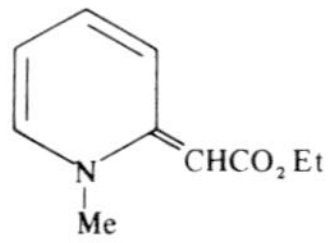 (N-Me pyridylidene) $CHCO_2Et$ | | m.p. 52–54° | 254 |
| $2\text{-}PyCH_2CN$ + $EtCH_2I$ | 6 hr steam-bath; NaOH | 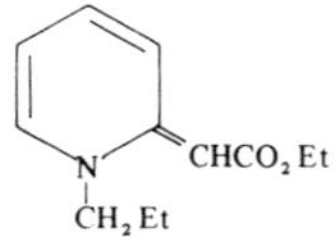 (N-$CH_2Et$ pyridylidene) $CHCO_2Et$ | | | 145 |
| $2\text{-}PyCH_2CO_2Et$ + $PhCH_2Br$ | 3 hr, steam-bath; NaOH | 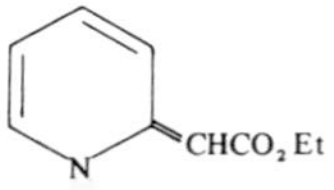 (N pyridylidene) $CHCO_2Et$ | | m.p. 94–95°; ir | 145 |

| Reactants | Conditions | Product(s) | Yield | Properties | Refs. |
|---|---|---|---|---|---|
| 2-$PyCH_2CN$ + $PhCH_2Br$ | 1 hr, steam-bath; NaOH | N—$CH_2Ph$, CHCN | | m.p. 128–129°; ir | 145 |
| 2-$PyCH_2CONH_2$ + $PhCH_2Br$ | 6 hr, steam-bath; NaOH | N—$CH_2Ph$, $CHCONH_2$ | | m.p. 153–154° (decomp.); ir | 145 |
| 2-$PyCH_2$—(H-N, O, O, NH) | $Ba(OH)_2$, $H_2O$, heat | 2-$PyCH_2CH(NH_2)CO_2H$ | | m.p. 209–210° | 95 |
| $\overset{+}{N}$—$O^-$, $SO_2NHCOCH_2Ph$ | 10% NaOH, 24°, 1 hr | $\overset{+}{N}$—$O^-$, $CHPhCO_2H$ (77.8%) + Ph, N, O, =O (6.7%) | | m.p. 100–102° (decomp.) | 154 |
| | 10% NaOH, 95°, 96 min | $\overset{+}{N}$—$O^-$, $CHPhCO_2H$ | 80% | m.p. 100–102° (decomp.) | 255, 256, 257 |
| $\overset{+}{N}$—$O^-$, $SO_2NHCH_2CN$ | 10% NaOH, 90–95°, 1 hr | $\overset{+}{N}$—$O^-$, $CH_2CO_2H$ | | m.p. 126–127° (decomp.) | 154 |

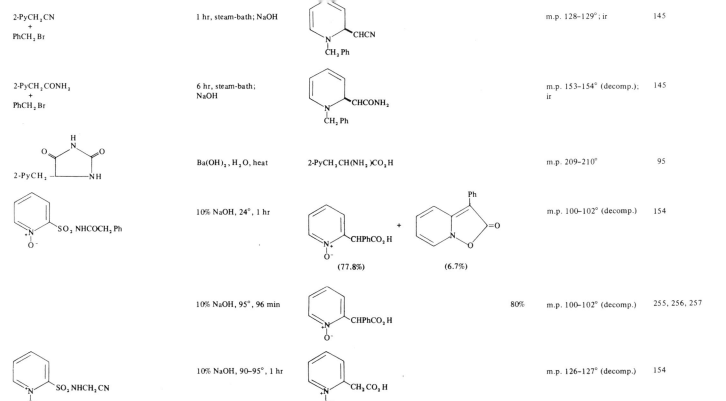

TABLE XI-14-1. Preparation of Acids and Derivatives by Miscellaneous Methods (Continued)

| Reactants | Conditions | Product | Yield | Properties | Ref. |
|---|---|---|---|---|---|
| 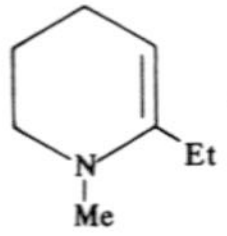 (N-Me, 2-Et ring) + $CH_2$=$CHCO_2Et$ | dioxane | 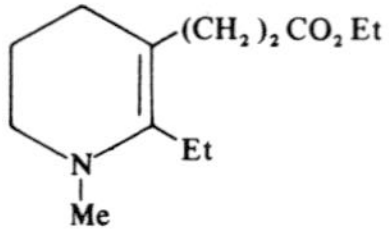 (N-Me, 2-Et ring) $(CH_2)_2CO_2Et$ | 42% | b.p. 112–114°/2 mm | 258 |
| 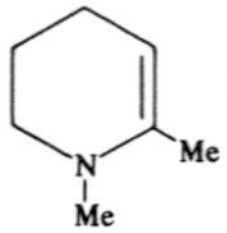 (N-Me, 2-Me ring) + $CH_2$=$CHCO_2Et$ | anhyd. dioxane, R.T., 4 days | 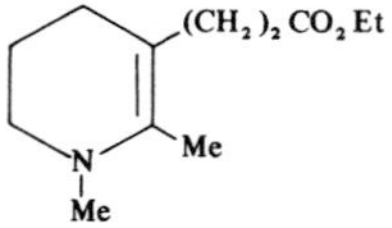 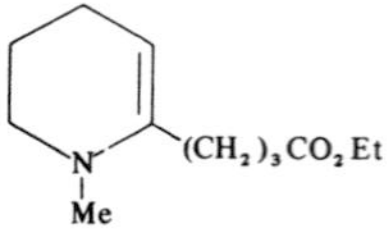 (N-Me, 2-Me ring) $(CH_2)_2CO_2Et$ + (N-Me ring) $(CH_2)_3CO_2Et$ | | | 259 |
| 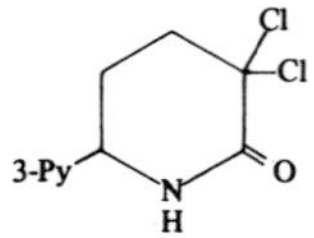 3-Py, Cl, Cl, O, N-H (ring) | HCl, reflux 4 hr; dry HCl | 3-PyCH($NH_2$)$CH_2CH_2C(CO_2H)Cl_2$ | 90% | m.p. 181–182° (decomp.) | 102 |
| 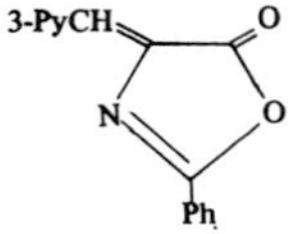 3-PyCH=, O, N, O, Ph (ring) + PhSH | steam-bath, 2 hr | 3-PyCH(SPh)CH(NHCOPh)COSPh | 82% | m.p. 139° | 120 |
| 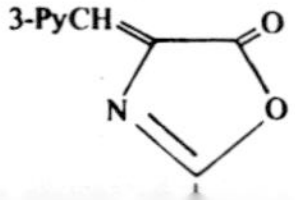 3-PyCH=, O, N, O (ring) + *p*-$MeC_6H_4SH$ | same as above | 3-PyCHCH(NHCOPh)COSPh<br>S(*p*-$MeC_6H_4$) | 83% | m.p. 165° | 120 |

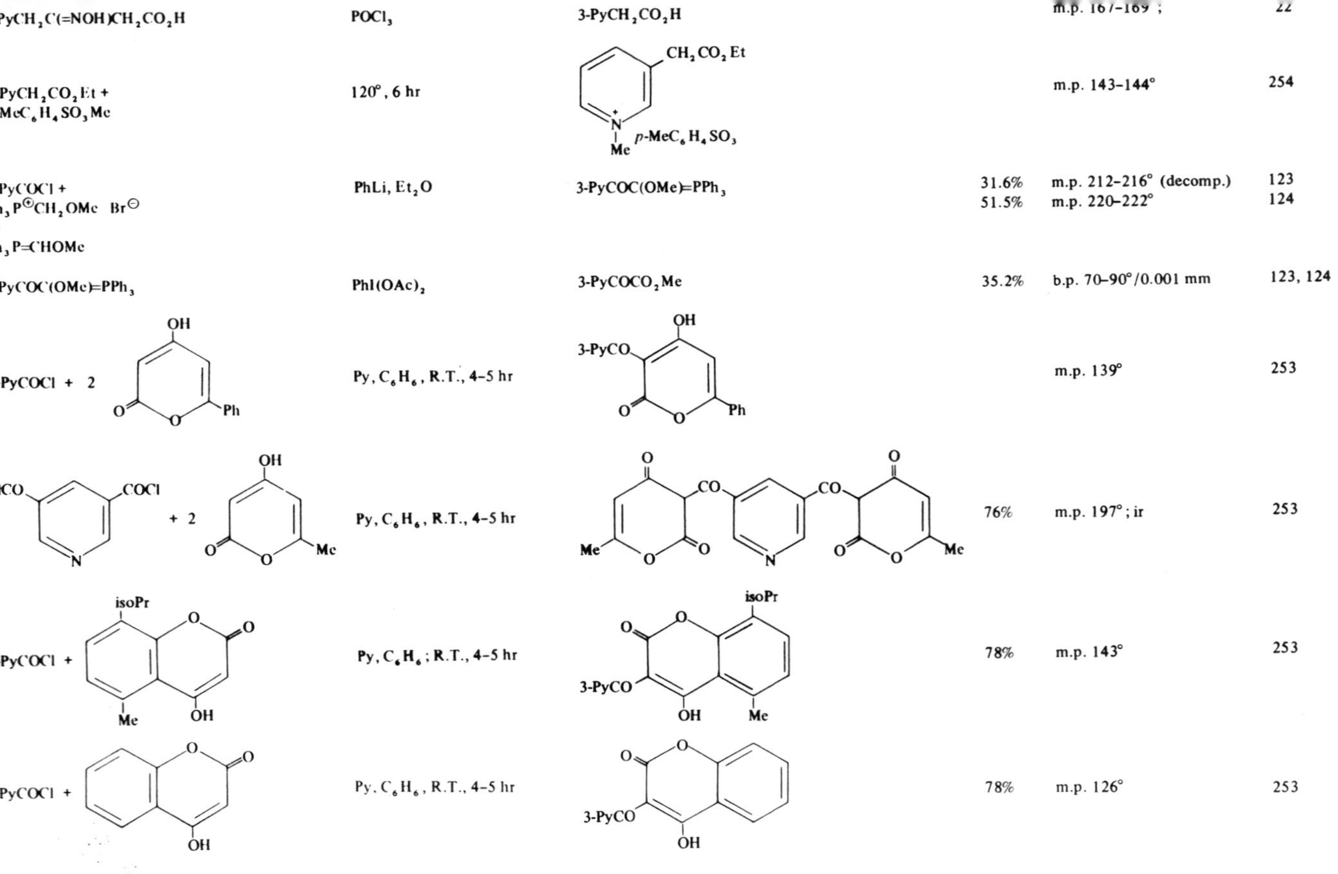

| | | | | | |
|---|---|---|---|---|---|
| 3-$PyCH_2C(=NOH)CH_2CO_2H$ | $POCl_3$ | 3-$PyCH_2CO_2H$ | | m.p. 167–169°; | 22 |
| 3-$PyCH_2CO_2Et$ + *p*-$MeC_6H_4SO_3Me$ | 120°, 6 hr | | | m.p. 143–144° | 254 |
| 3-PyCOCl + $Ph_3P^{\oplus}CH_2OMe\ Br^{\ominus}$ *or* $Ph_3P{=}CHOMe$ | PhLi, $Et_2O$ | 3-$PyCOC(OMe){=}PPh_3$ | 31.6%<br>51.5% | m.p. 212–216° (decomp.)<br>m.p. 220–222° | 123<br>124 |
| 3-$PyCOC(OMe){=}PPh_3$ | $PhI(OAc)_2$ | 3-$PyCOCO_2Me$ | 35.2% | b.p. 70–90°/0.001 mm | 123, 124 |
| 3-PyCOCl + 2 | Py, $C_6H_6$, R.T., 4–5 hr | | | m.p. 139° | 253 |
| + 2 | Py, $C_6H_6$, R.T., 4–5 hr | | 76% | m.p. 197°; ir | 253 |
| 3-PyCOCl + | Py, $C_6H_6$; R.T., 4–5 hr | | 78% | m.p. 143° | 253 |
| 3-PyCOCl + | Py, $C_6H_6$, R.T., 4–5 hr | | 78% | m.p. 126° | 253 |

TABLE XI-14-1. Preparation of Acids and Derivatives by Miscellaneous Methods (Continued)

| Reactants | Conditions | Product | Yield | Properties | Ref. |
|---|---|---|---|---|---|
| 3-PyCOCl 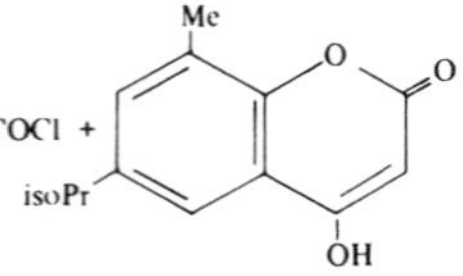 + (Me, isoPr, OH, O) | Py, $C_6H_6$, R.T., 4–5 hr | 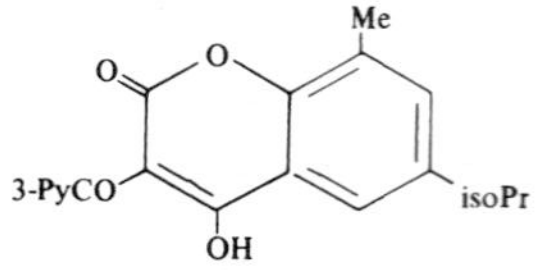 3-PyCO (Me, isoPr, OH, O) | 76% | m.p. 85° | 253 |
| 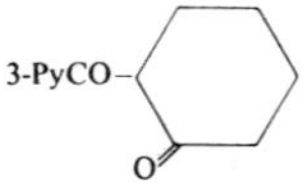 3-PyCO– (O) | HCl, reflux 1.5 hr; EtOH, reflux; HCl | 3-$PyCO(CH_2)_5CO_2Et$ | 27.4 g → 13.3 g | b.p. 155–158°/0.4 mm | 117 |
| 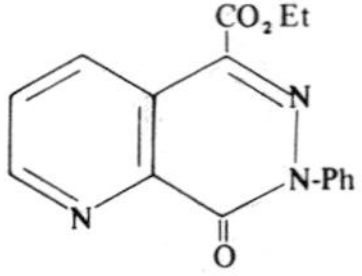 ($CO_2Et$, N, N-Ph, O) | aq. NaOH, cold | 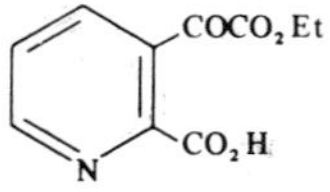 ($COCO_2Et$, $CO_2H$, N) |  | m.p. 223° (decomp.) | 260 |
|  | aq. NaOH, warm | 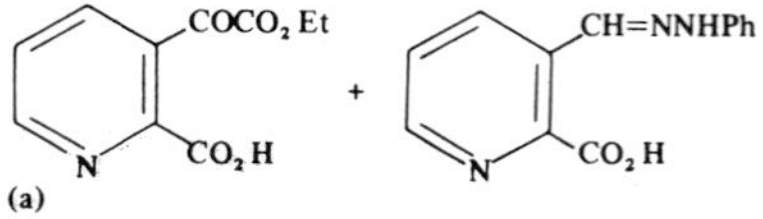 ($COCO_2Et$, $CO_2H$, N) (a) + (CH=NNHPh, $CO_2H$, N) |  | (a) m.p. 223° (decomp.) | 260 |
| 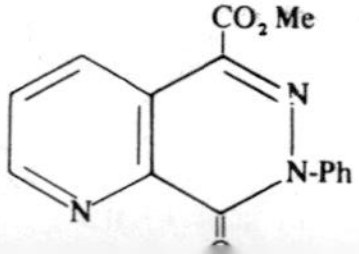 ($CO_2Me$, N, N-Ph, O) | aq. NaOH, warm | 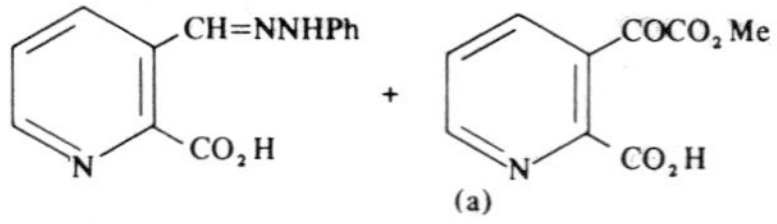 (CH=NNHPh, $CO_2H$, N) + ($COCO_2Me$, $CO_2H$, N) (a) |  | (a) m.p. 227–229° (decomp.) | 260 |

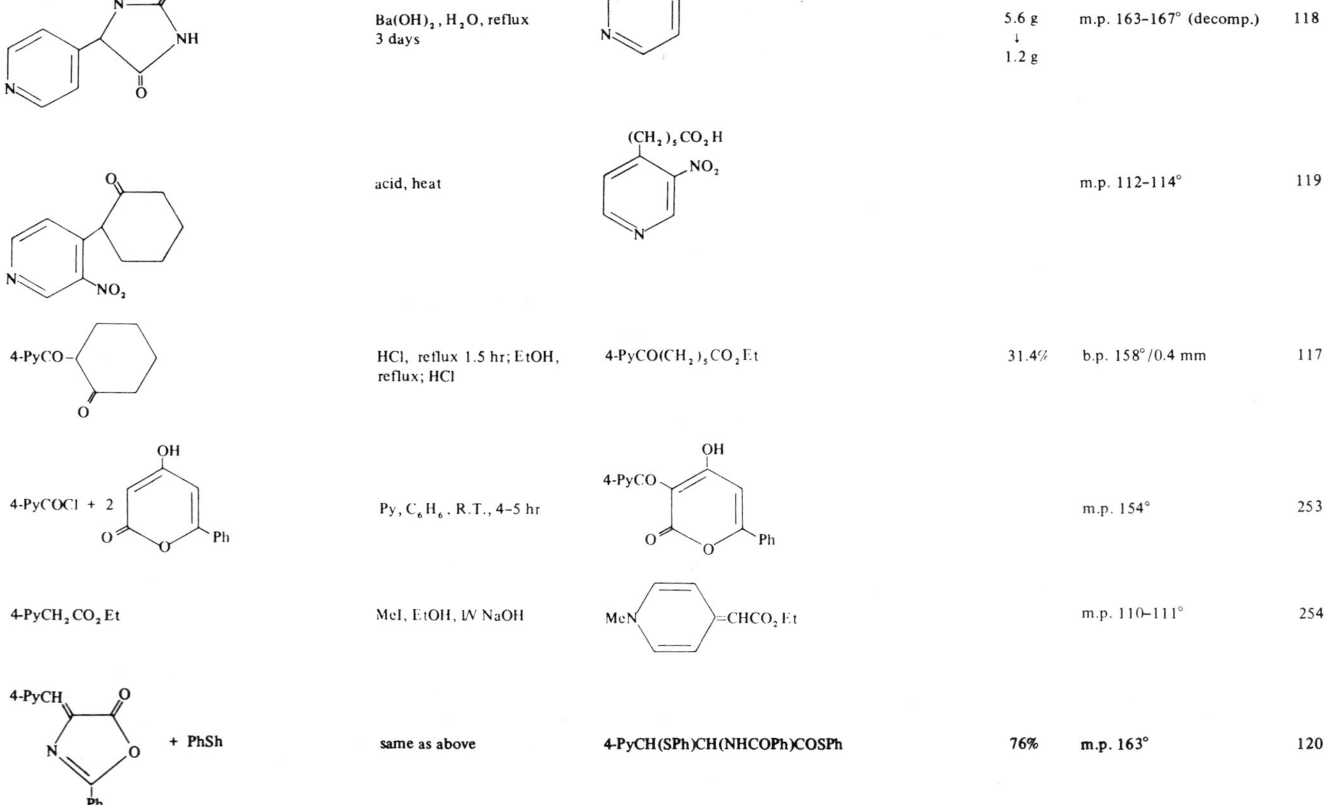

| | | | | | |
|---|---|---|---|---|---|
| | $Ba(OH)_2$, $H_2O$, reflux 3 days | | 5.6 g ↓ 1.2 g | m.p. 163–167° (decomp.) | 118 |
| | acid, heat | | | m.p. 112–114° | 119 |
| 4-PyCO– | HCl, reflux 1.5 hr; EtOH, reflux; HCl | 4-PyCO$(CH_2)_5CO_2Et$ | 31.4% | b.p. 158°/0.4 mm | 117 |
| 4-PyCOCl + 2 | Py, $C_6H_6$, R.T., 4–5 hr | | | m.p. 154° | 253 |
| 4-Py$CH_2CO_2Et$ | MeI, EtOH, 1*N* NaOH | | | m.p. 110–111° | 254 |
| + PhSh | same as above | **4-PyCH(SPh)CH(NHCOPh)COSPh** | **76%** | **m.p. 163°** | 120 |

TABLE XI-14-1. Preparation of Acids and Derivatives by Miscellaneous Methods (Continued)

| Reactants | Conditions | Product | Yield | Properties | Ref. |
|---|---|---|---|---|---|
| 4-PyCH=(2-Ph-oxazol-5-one) + $o$-$MeC_6H_4SH$ | same as above | 4-PyCH(S($o$-$MeC_6H_4$))CH(NHCOPh)COS($o$-$MeC_6H_4$) | 78% | m.p. 149° | 120 |
| 4-PyCH=(2-Ph-oxazol-5-one) + $p$-$MeC_6H_4SH$ | same as above | 4-PyCH(S($p$-$MeC_6H_4$))CH(NHCOPh)COS($p$-$MeC_6H_4$) | 72% | m.p. 146° | 120 |
| 5-$n$-Pr-4-$CH_2CO_2Et$-3,3-di-Cl-piperidin-2-one (N–H) | $Ba(OH)_2 \cdot 8H_2O$, $H_2O$, reflux 3 hr | 5-$n$-Pr-4-$CH_2CO_2H$-3-Cl-5,6-dihydropyridin-2-one (N–H) | | m.p. 200° | 261 |
| 5-isoPr-4-$CH_2CO_2Et$-3,3-di-Cl-piperidin-2-one (N–H) | same as above | 5-isoPr-4-$CH_2CO_2H$-3-Cl-5,6-dihydropyridin-2-one (N–H) | | m.p. 203° | 261 |
| 4-($SO_2NHCOCH_2Ph$)pyridinium ($^+N$) | 10% NaOH, 27°, 17 hr *or* | 4-($CHPhCO_2H$)pyridine N-oxide ($^-O$–$^+N$) | 84.4% | m.p. 144–145° (decomp.) | 154 |

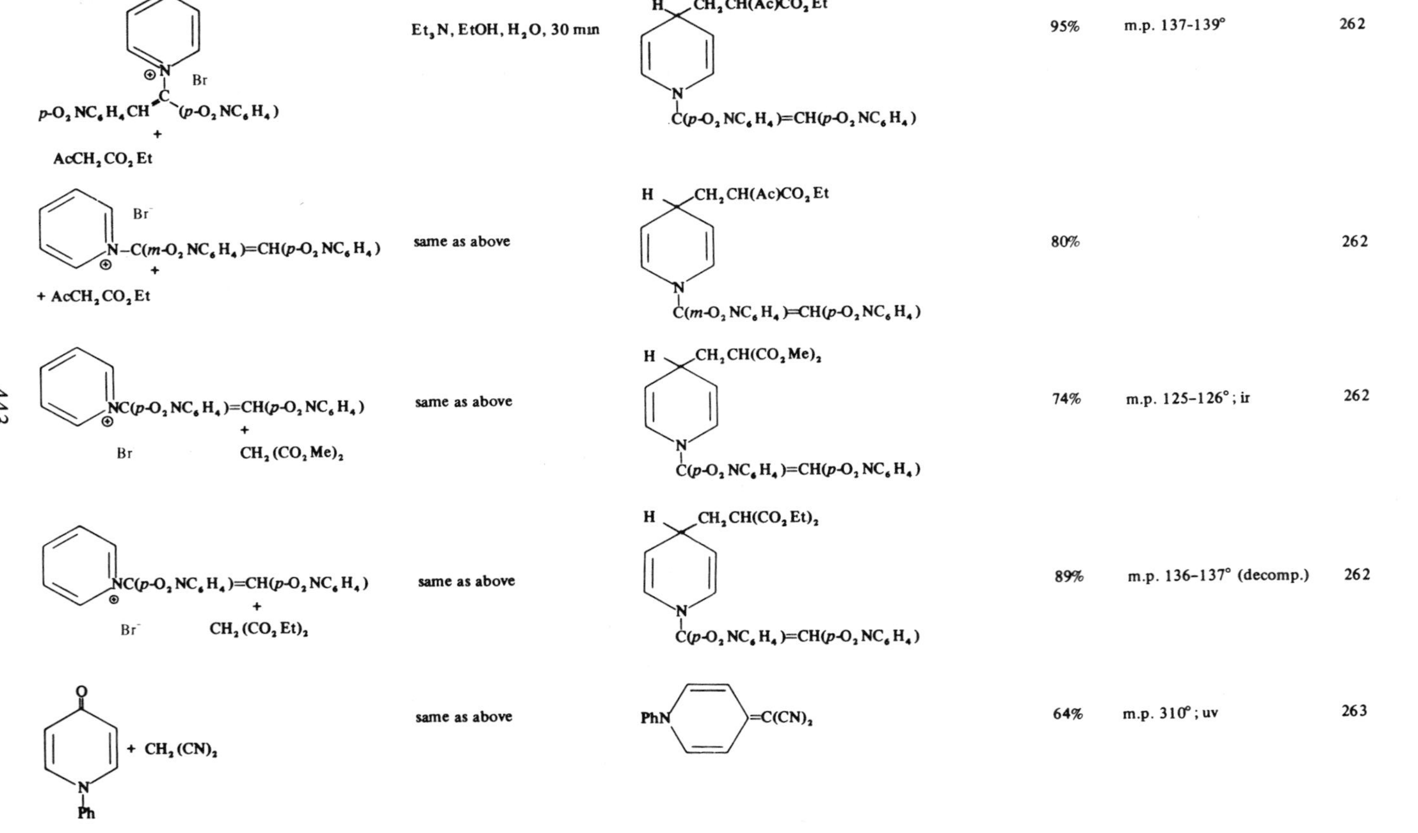


| Conditions | Yield | Properties | Ref. |
|---|---|---|---|
| $Et_3N$, EtOH, $H_2O$, 30 min | 95% | m.p. 137-139° | 262 |
| same as above | 80% | | 262 |
| same as above | 74% | m.p. 125–126°; ir | 262 |
| same as above | 89% | m.p. 136–137° (decomp.) | 262 |
| same as above | 64% | m.p. 310°; uv | 263 |

TABLE XI-14-1. Preparation of Acids and Derivatives by Miscellaneous Methods (Continued)

| Reactants | Conditions | Product | Yield | Properties | Ref. |
|---|---|---|---|---|---|
| 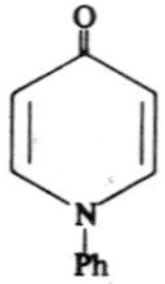 1-phenyl-4-pyridone + $CH_2(CN)CO_2Et$ | same as above | 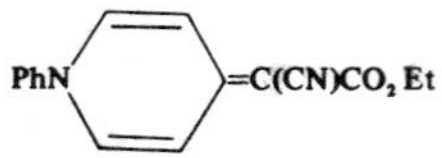 PhN(pyridylidene)=$C(CN)CO_2Et$ | | m.p. 182°; uv | 263 |
| 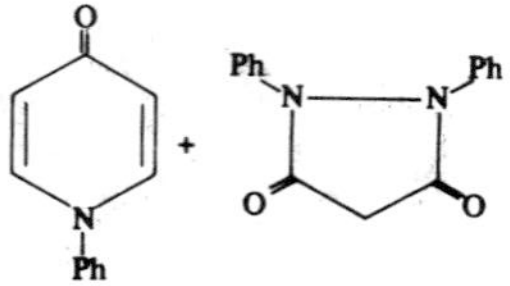 1-phenyl-4-pyridone + 1,2-diphenylpyrazolidine-3,5-dione | same as above | 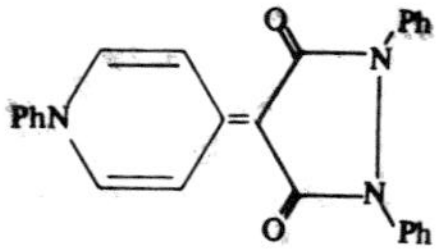 4-(1-phenylpyridin-4-ylidene)-1,2-diphenylpyrazolidine-3,5-dione | 79% | m.p. 279°; uv | 263 |
| 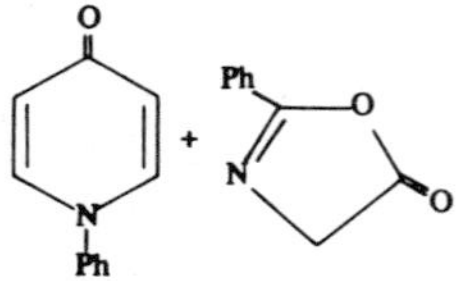 1-phenyl-4-pyridone + 2-phenyl-oxazol-5(4H)-one | same as above | 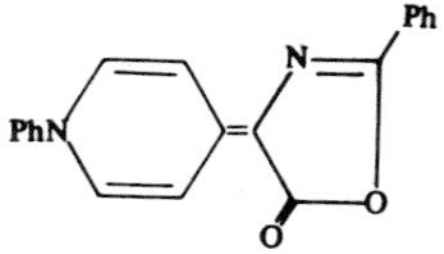 4-(1-phenylpyridin-4-ylidene)-2-phenyloxazol-5-one | 63% | m.p. 195°; uv | 263 |
| 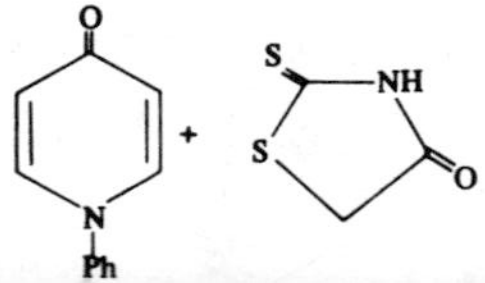 1-phenyl-4-pyridone + rhodanine | same as above | 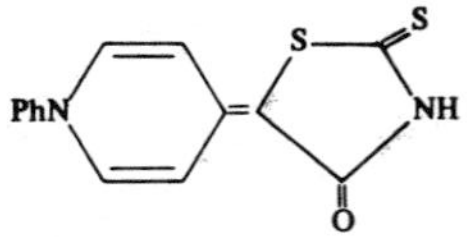 5-(1-phenylpyridin-4-ylidene)rhodanine | 49% | m.p. 294°; uv | 263 |

| | | | | | |
|---|---|---|---|---|---|
| 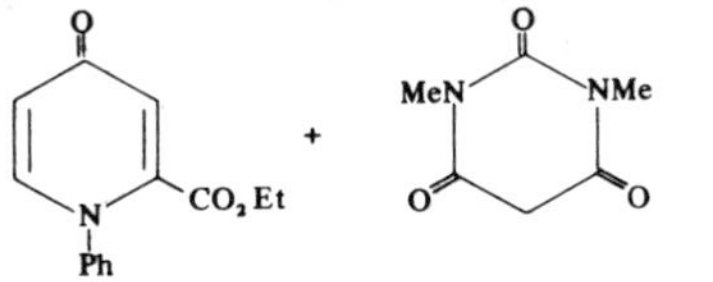 | same as above | 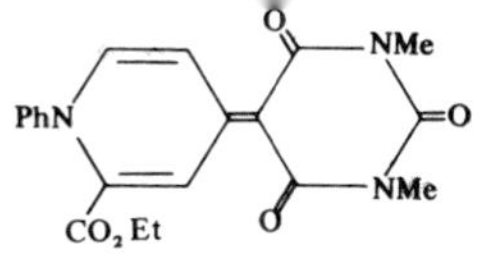 | | m.p. 257°; uv | 263 |
| 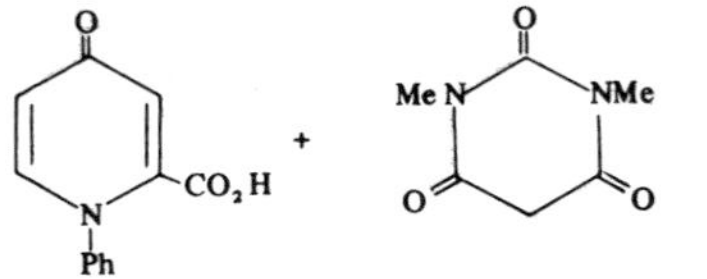 | same as above | 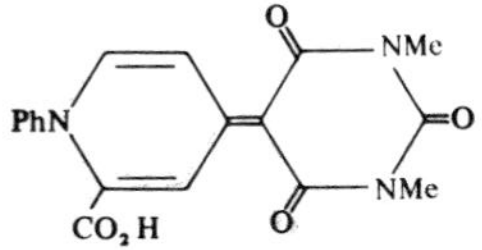 | | m.p. 252°; uv | 263 |
| 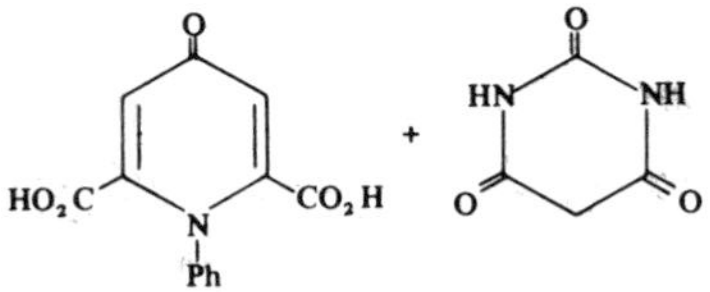 | $Ac_2O$, HOAc, heat | 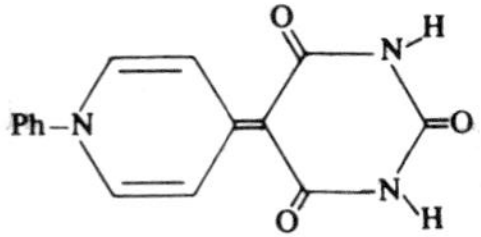 | 85% | m.p. 325°; uv | 263 |
| 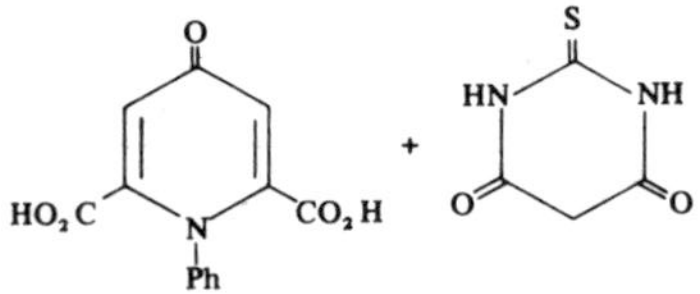 | same as above | 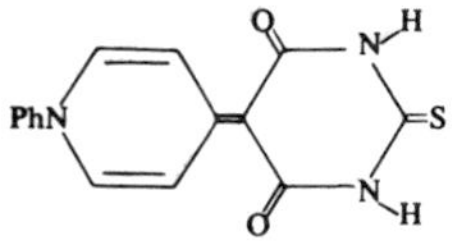 | 50% | m.p. 330°; uv | 263 |
| 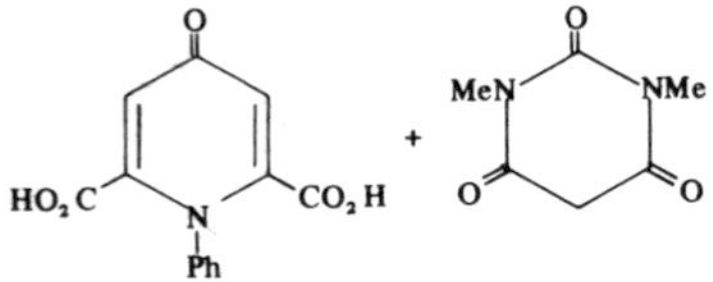 | same as above | 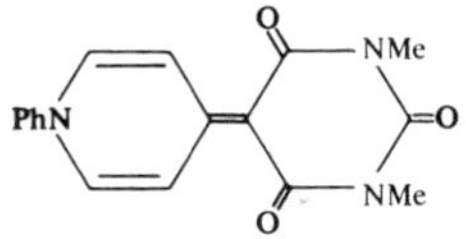 | 80% | m.p. 310°; uv | 263 |

TABLE XI-14-1. Preparation of Acids and Derivatives by Miscellaneous Methods (Continued)

| Reactants | Conditions | Product | Yield | Properties | Ref. |
|---|---|---|---|---|---|
| 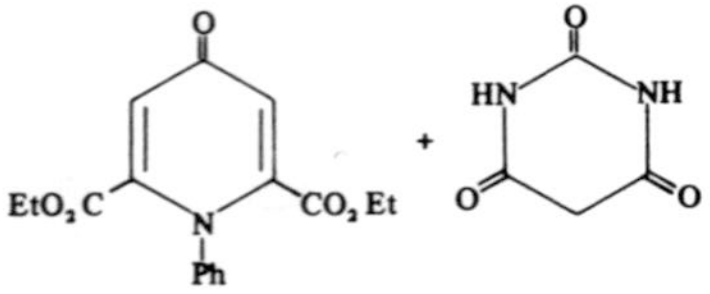 | same as above | 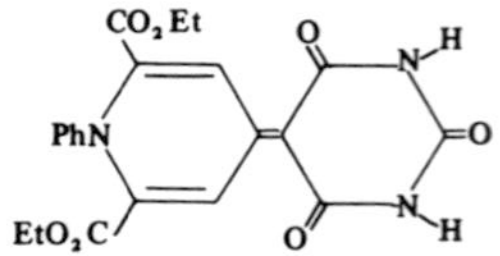 | | m.p. 247°; uv | 263 |
| 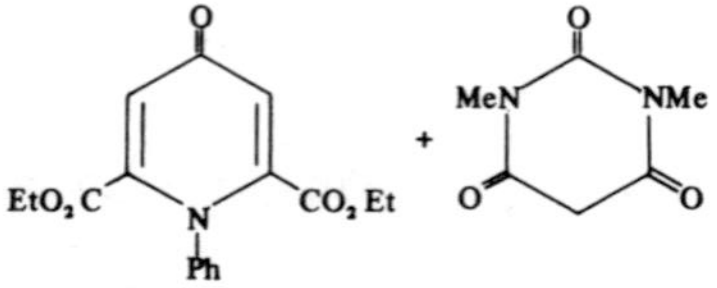 | | 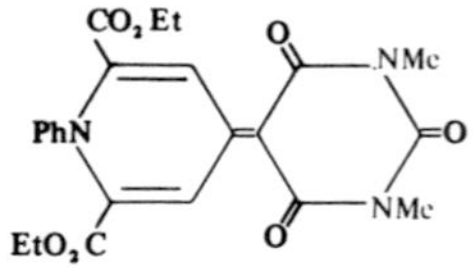 | | m.p. 157–158°; uv | 263 |
| 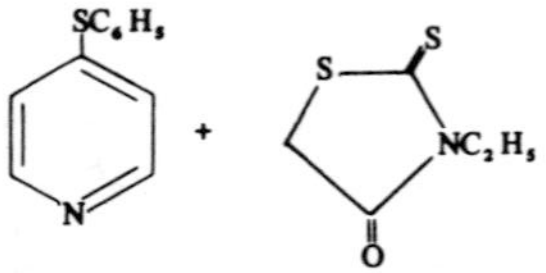 | EtOH, $Et_3N$, boil, 20 min | 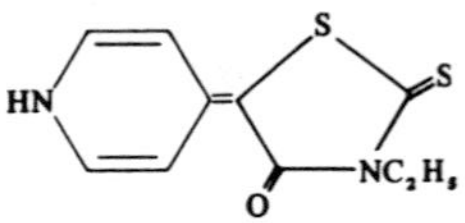 | 23% | m.p. 270–271° (decomp.) | 67 |
| 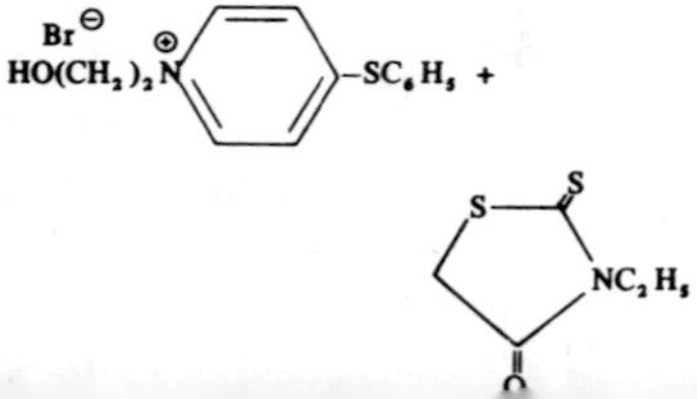 | same as above | 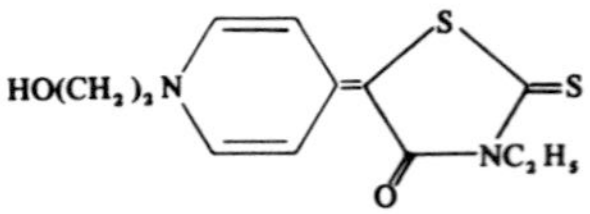 | 48% | m.p. 249–251° (decomp.) | 67 |

| Reactants | Conditions | Product | Properties | Ref. |
|---|---|---|---|---|
| $Br^{\ominus}$ $HO(CH_2)_3\overset{\oplus}{N}$ (pyridinium)–$SC_6H_5$ + rhodanine ($NC_2H_5$) | same as above | $HO(CH_2)_3N$ (pyridylidene)= rhodanine ($NC_2H_5$) | m.p. 189–191° (decomp.) | 67 |
| $Br^{\ominus}$ $EtO_2CCH_2\overset{\oplus}{N}$ (pyridinium)–$SC_6H_5$ + rhodanine ($NC_2H_5$) | same as above | $EtO_2CCH_2N$ (pyridylidene)= rhodanine ($NC_2H_5$) | m.p. 237–238° (decomp.) | 67 |
| $C_6H_5CH_2\overset{+}{N}$ (pyridinium)–$SC_6H_5$ $Br^-$ + rhodanine ($NC_2H_5$) | same as above | $C_6H_5CH_2N$ (pyridylidene)= rhodanine ($NC_2H_5$) | m.p. 246–248° (decomp.) | 67 |
| $Br^{\ominus}$ $C_6H_5(CH_2)_2\overset{\oplus}{N}$ (pyridinium)–$SC_6H_5$ + rhodanine ($NC_2H_5$) | same as above | $C_6H_5(CH_2)_2N$ (pyridylidene)= rhodanine ($NC_2H_5$) | m.p. 232–234° (decomp.) | 67 |

TABLE XI-14-1. Preparation of Acids and Derivatives by Miscellaneous Methods (Continued)

| Reactants | Conditions | Product | Yield | Properties | Ref. |
|---|---|---|---|---|---|
| 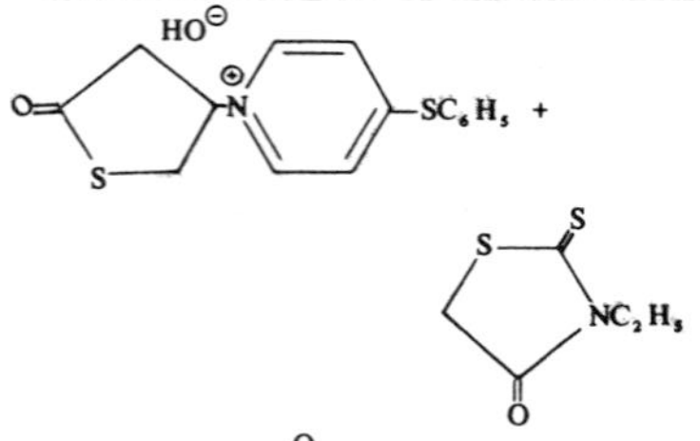 | same as above | 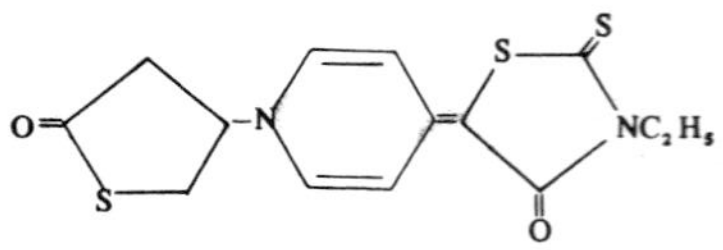 | 69% | sodium salt, m.p. > 300° | 67 |
| 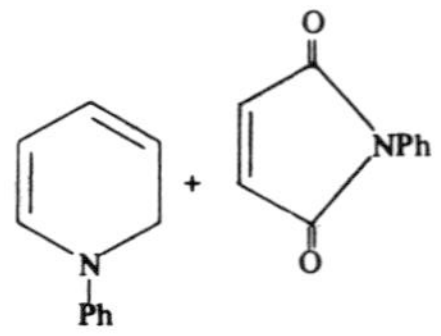 | $Et_2O$, vac., 6 hr | 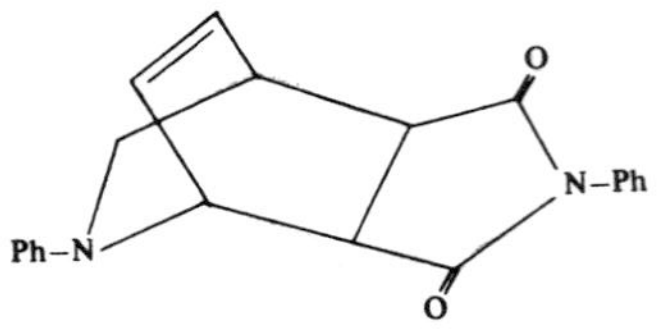 | 71% | m.p. 215–216°; ir; nmr | 353 |
| 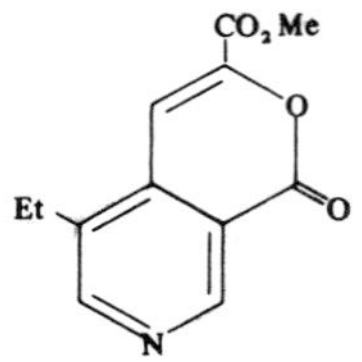 | aq. KOH, R.T., 0.5 hr; 30% $H_2O_2$, 5°, 4 hr | 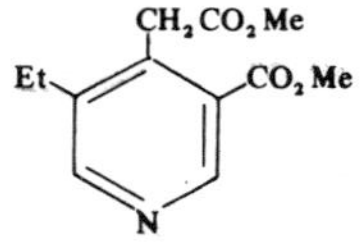 | | b.p. 135–145°/0.5 mm; m.p. 39–40° | 122 |
| 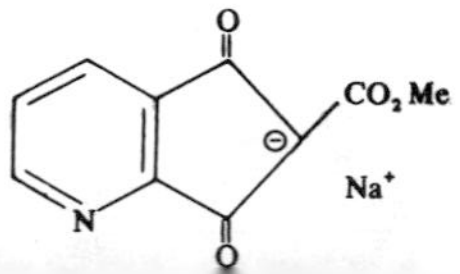 | $Cl_2$, $H_2O$ | 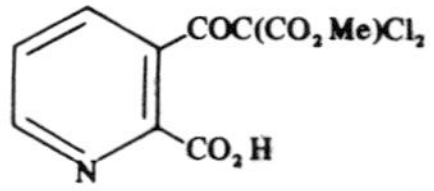 | | m.p. 127° (decomp.) | 121 |

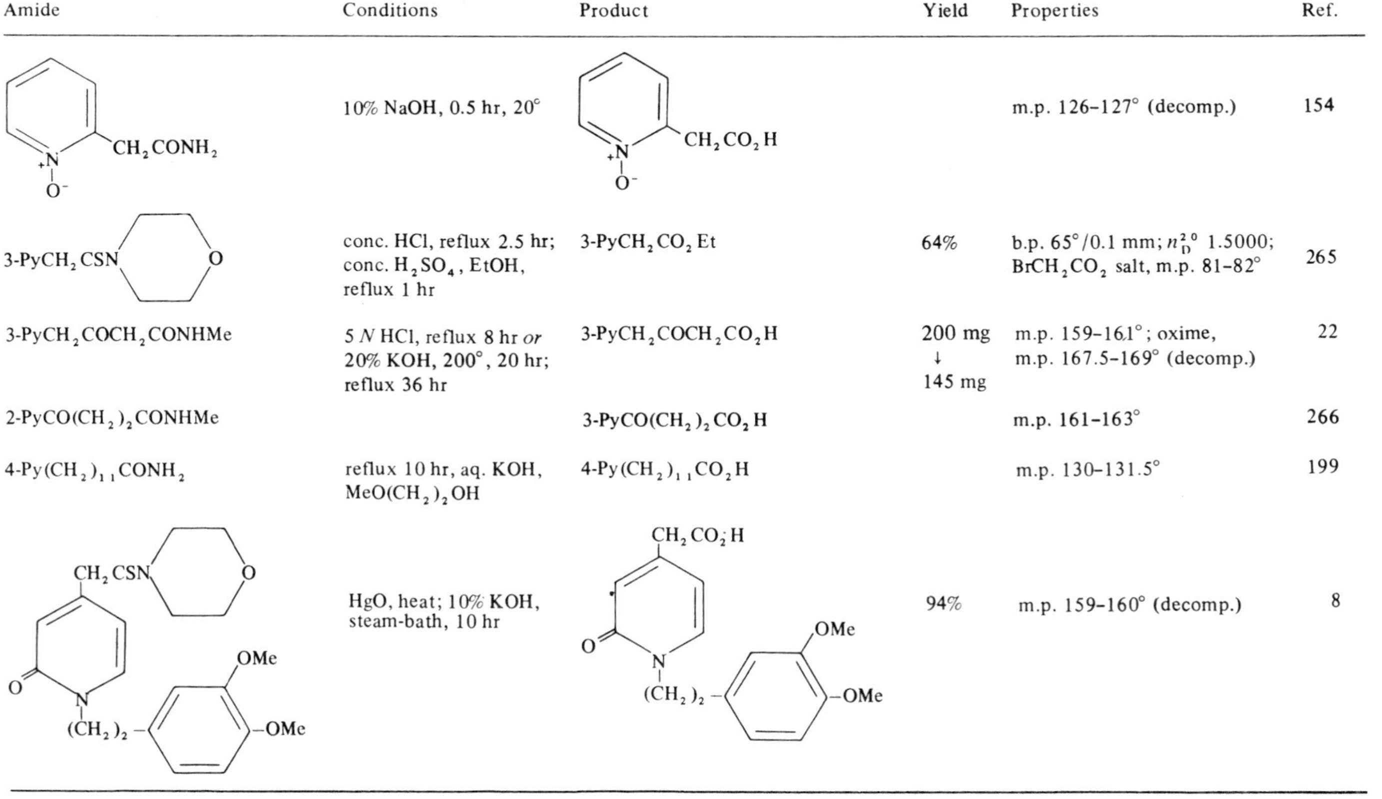

| Amide | Conditions | Product | Yield | Properties | Ref. |
|---|---|---|---|---|---|
| $CH_2CONH_2$; $\overset{+}{N}$–$O^-$ | 10% NaOH, 0.5 hr, 20° | $CH_2CO_2H$; $\overset{+}{N}$–$O^-$ | | m.p. 126–127° (decomp.) | 154 |
| 3-$PyCH_2CSN$ (morpholine ring, O) | conc. HCl, reflux 2.5 hr; conc. $H_2SO_4$, EtOH, reflux 1 hr | 3-$PyCH_2CO_2Et$ | 64% | b.p. 65°/0.1 mm; $n_D^{20}$ 1.5000; $BrCH_2CO_2$ salt, m.p. 81–82° | 265 |
| 3-$PyCH_2COCH_2CONHMe$ | 5 *N* HCl, reflux 8 hr *or* 20% KOH, 200°, 20 hr; reflux 36 hr | 3-$PyCH_2COCH_2CO_2H$ | 200 mg ↓ 145 mg | m.p. 159–161°; oxime, m.p. 167.5–169° (decomp.) | 22 |
| 2-$PyCO(CH_2)_2CONHMe$ | | 3-$PyCO(CH_2)_2CO_2H$ | | m.p. 161–163° | 266 |
| 4-$Py(CH_2)_{11}CONH_2$ | reflux 10 hr, aq. KOH, $MeO(CH_2)_2OH$ | 4-$Py(CH_2)_{11}CO_2H$ | | m.p. 130–131.5° | 199 |
| $CH_2CSN$ (morpholine ring, O); O; N; $(CH_2)_2$–; OMe; –OMe | HgO, heat; 10% KOH, steam-bath, 10 hr | $CH_2CO_2H$; O; N; $(CH_2)_2$–; OMe; –OMe | 94% | m.p. 159–160° (decomp.) | 8 |

TABLE XI-16. Decarboxylation Reactions Yielding Acidic Products

| Starting material | Conditions | Product | Yield | Properties | Ref. |
|---|---|---|---|---|---|
| 2-PyCH($CO_2$Et)CHMeCH$_2$CO$_2$Me | MeOH, $H_2O$, KOH, reflux 14 hr | 2-PyCH$_2$CHMeCH$_2$CO$_2$Me | 65% | b.p. 120°/0.01 mm; picrate, m.p. 118° | 200 |
| 2-PyCH($CO_2$Et)CHPhCH$_2$CO$_2$Et | 10% NaOH, 3 hr, 80°; HCl; residue, 140° | 2-PyCH$_2$CHPhCH$_2$CO$_2$Et | | b.p. 180–190°/0.05 mm; picrolonate, m.p. 168–169° | 200 |
| 2-Py(CH$_2$)$_2$CBu(CO$_2$H)$_2$ | 140°, reduced pressure | 2-Py(CH$_2$)$_2$CHBuCO$_2$H | 86% | | 203 |
| (2-PyCH$_2$CH$_2$)$_2$C(CO$_2$Et)$_2$ | 20% HCl, 120–123°, 3 hr | (2-PyCH$_2$CH$_2$)$_2$CHCO$_2$Et | | dipicrate, m.p. 171–172° | 205 |
| HO, CH$_2$OH, CH$_2$CH(CO$_2$Et)$_2$, Me, N | conc. HCl, reflux 4.5 hr | HO, CH$_2$OH, (CH$_2$)$_2$CO$_2$H, Me, N | 68% | m.p. 213–215° (decomp.); ir; nmr; hydrochloride, m.p. 214–215° (decomp.) | 195 |
| O, O, CH$_2$CH(CO$_2$Et)$_2$, Me, N | | HO, CH$_2$OH, (CH$_2$)$_2$CO$_2$H, Me, N | | | 195 |
| O, O, CH=C(CO$_2$H)$_2$ | 95% EtOH, pyridine, reflux, 7 hr | O, O, CH=CHCO$_2$H | | m.p. 220–221°; ir; uv | 228 |

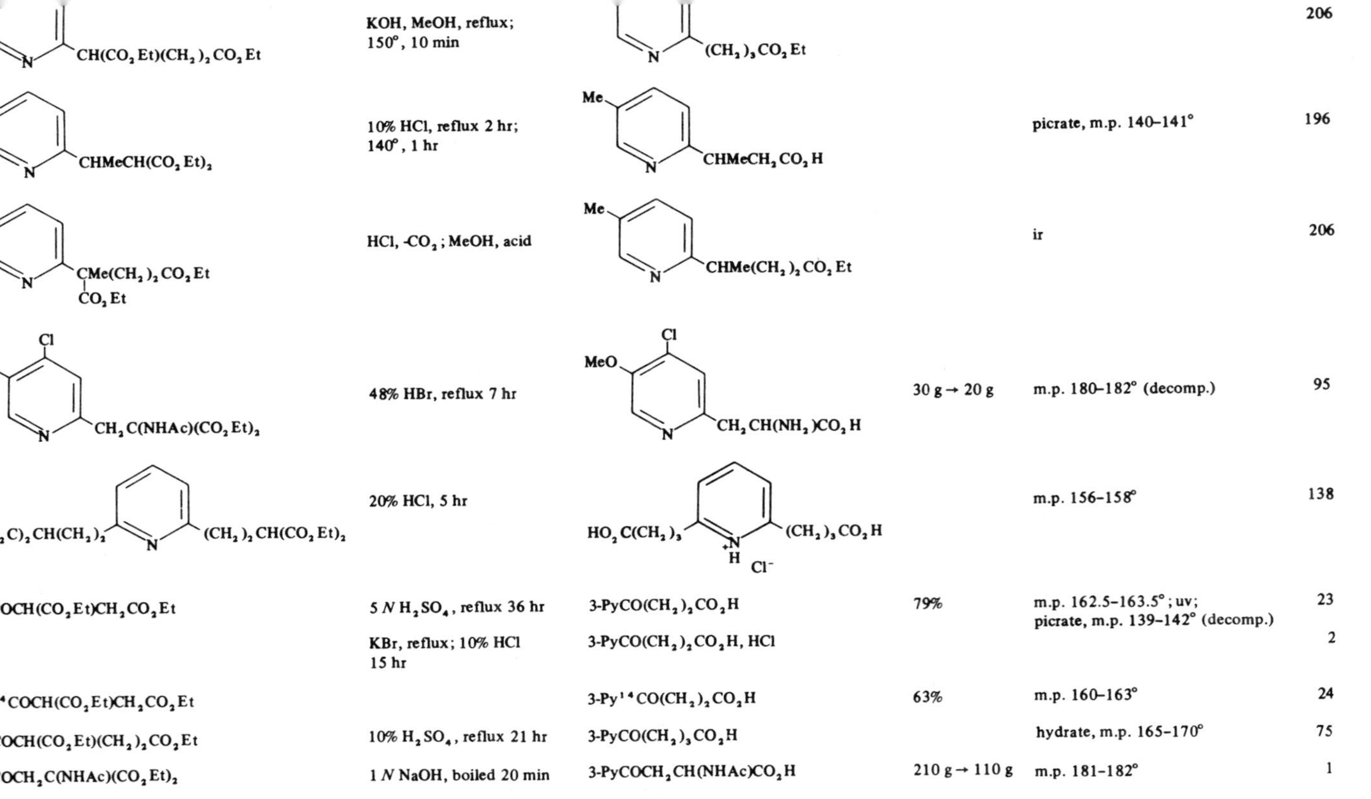

| | | | | | |
|---|---|---|---|---|---|
| $CH(CO_2Et)(CH_2)_2CO_2Et$ (N) | KOH, MeOH, reflux; 150°, 10 min | $(CH_2)_3CO_2Et$ (N) | | | 206 |
| Me; $CHMeCH(CO_2Et)_2$ (N) | 10% HCl, reflux 2 hr; 140°, 1 hr | Me; $CHMeCH_2CO_2H$ (N) | | picrate, m.p. 140–141° | 196 |
| Me; $CMe(CH_2)_2CO_2Et$ / $CO_2Et$ (N) | HCl, $-CO_2$; MeOH, acid | Me; $CHMe(CH_2)_2CO_2Et$ (N) | | ir | 206 |
| Cl; MeO; $CH_2C(NHAc)(CO_2Et)_2$ (N) | 48% HBr, reflux 7 hr | Cl; MeO; $CH_2CH(NH_2)CO_2H$ (N) | 30 g → 20 g | m.p. 180–182° (decomp.) | 95 |
| $(EtO_2C)_2CH(CH_2)_2$; $(CH_2)_2CH(CO_2Et)_2$ (N) | 20% HCl, 5 hr | $HO_2C(CH_2)_3$; $(CH_2)_3CO_2H$ ($\overset{+}{N}$H, $Cl^-$) | | m.p. 156–158° | 138 |
| 3-$PyCOCH(CO_2Et)CH_2CO_2Et$ | 5 *N* $H_2SO_4$, reflux 36 hr | 3-$PyCO(CH_2)_2CO_2H$ | 79% | m.p. 162.5–163.5°; uv; picrate, m.p. 139–142° (decomp.) | 23 |
| | KBr, reflux; 10% HCl 15 hr | 3-$PyCO(CH_2)_2CO_2H$, HCl | | | 2 |
| 3-$Py^{14}COCH(CO_2Et)CH_2CO_2Et$ | | 3-$Py^{14}CO(CH_2)_2CO_2H$ | 63% | m.p. 160–163° | 24 |
| 3-$PyCOCH(CO_2Et)(CH_2)_2CO_2Et$ | 10% $H_2SO_4$, reflux 21 hr | 3-$PyCO(CH_2)_3CO_2H$ | | hydrate, m.p. 165–170° | 75 |
| 3-$PyCOCH_2C(NHAc)(CO_2Et)_2$ | 1 *N* NaOH, boiled 20 min | 3-$PyCOCH_2CH(NHAc)CO_2H$ | 210 g → 110 g | m.p. 181–182° | 1 |
| 3-$PyCOCH(CO_2Et)(CH_2)_2CO_2Et$ | 1 *N* $H_2SO_4$, reflux 24 hr | 3-$PyCO(CH_2)_3CO_2H$ | 95% | m.p. 126–127° | 102 |

TABLE XI-16. Decarboxylation Reactions Yielding Acidic Products (Continued)

| Starting material | Conditions | Product | Yield | Properties | Ref. |
|---|---|---|---|---|---|
| 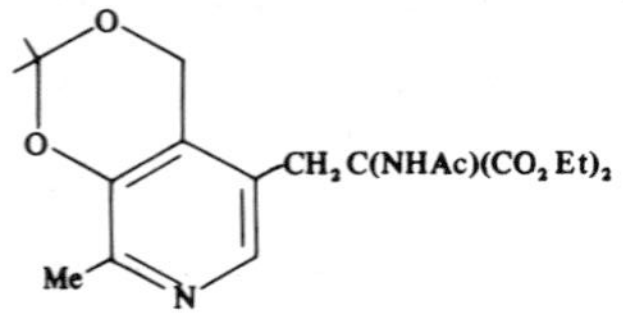 O, O, Me, N, $CH_2C(NHAc)(CO_2Et)_2$ | conc. HCl, 2 hr, R.T.; reflux 2 hr | 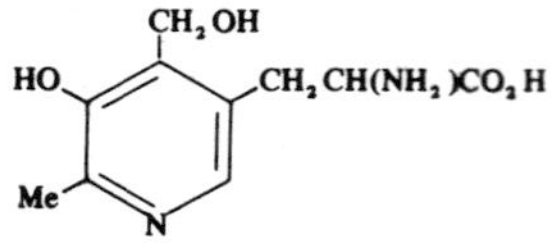 $CH_2OH$, HO, $CH_2CH(NH_2)CO_2H$, Me, N | 61.5% | m.p. 248–249° (decomp.) | 125 |
| 4-Py$(CH_2)_2CH(CO_2Et)_2$ | conc. HCl, reflux 18 hr | 4-Py$(CH_2)_3CO_2H \cdot HCl$ | | m.p. 199–202° | 210 |
| 4-Py$(CH_2)_2CPh(CO_2Et)_2$ | aq. NaOH, reflux 2 hr | 4-Py$(CH_2)_2CHPhCO_2H$ | 7% | m.p. 125–127° | 91 |

| Starting material | Conditions | Product | Yield | Properties | Ref. |
|---|---|---|---|---|---|
| | | *A. Side-chain acid derivatives* | | | |
| 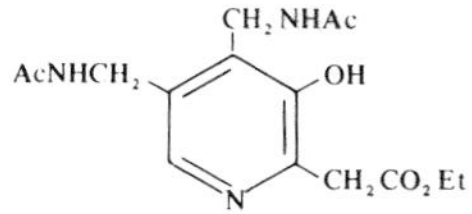 $CH_2NHAc$; $AcNHCH_2$; OH; N; $CH_2CO_2Et$ | 20% HCl, $N_2$, reflux 3 hr | 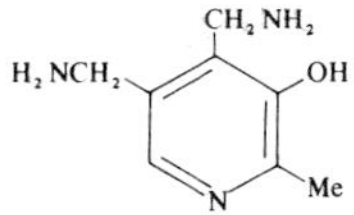 $CH_2NH_2$; $H_2NCH_2$; OH; N; Me | | | 69 |
| 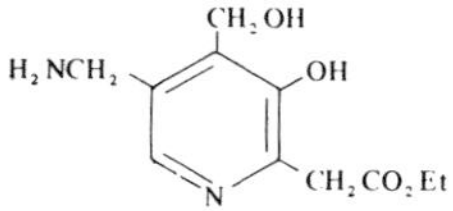 $CH_2OH$; $H_2NCH_2$; OH; N; $CH_2CO_2Et$ | HCl, reflux; $NaNO_2$ | 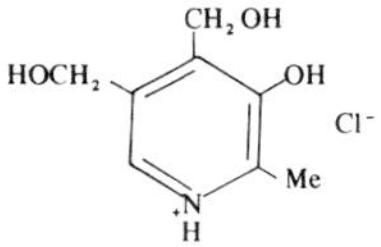 $CH_2OH$; $HOCH_2$; OH; $\overset{+}{N}$H; Me; $Cl^-$ | | | 69 |
| 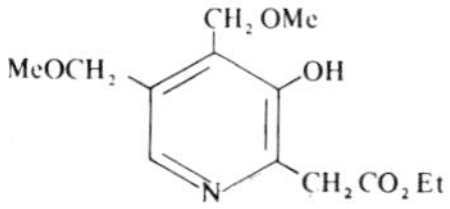 $CH_2OMe$; $MeOCH_2$; OH; N; $CH_2CO_2Et$ | conc. HCl, 3 hr; sealed tube, 130° | 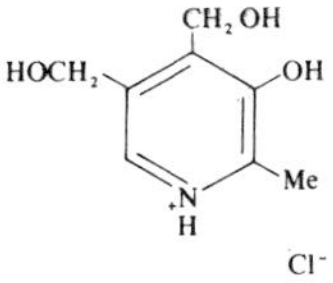 $CH_2OH$; $HOCH_2$; OH; $\overset{+}{N}$H; Me; $Cl^-$ | | | 69 |
| 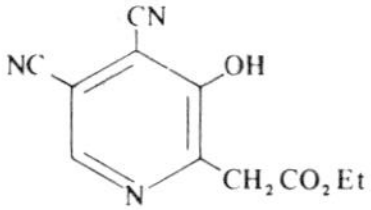 CN; NC; OH; N; $CH_2CO_2Et$ | 10% KOH, MeOH, reflux 1 hr | 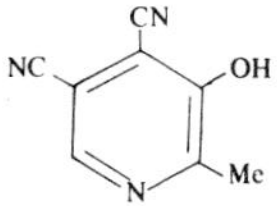 CN; NC; OH; N; Me | | | 267 |
| 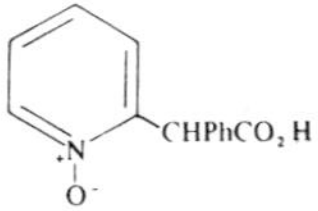 $\overset{+}{N}$–$O^-$; $CHPhCO_2H$ | EtOH, 1 hr, 70–80° | 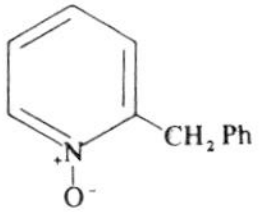 $\overset{+}{N}$–$O^-$; $CH_2Ph$ | 93.4% | | 257 |

## TABLE XI-17. Decarboxylation Reactions Yielding Nonacidic Products (Continued)

| Starting material | Conditions | Product | Yield | Properties | Ref. |
|---|---|---|---|---|---|
| 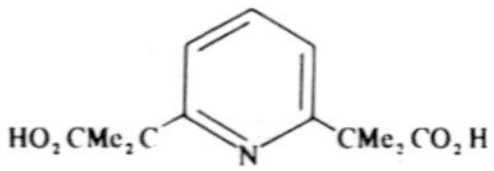 $HO_2CMe_2C$, N, $CMe_2CO_2H$ | 102–103° | 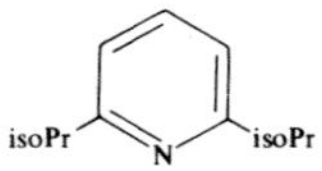 isoPr, N, isoPr | | | 179 |
| 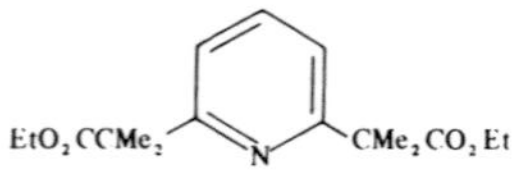 $EtO_2CCMe_2$, N, $CMe_2CO_2Et$ | base, fusion | 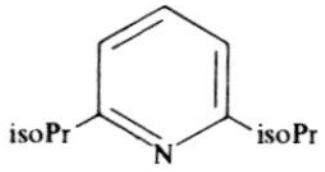 isoPr, N, isoPr | | | 180 |
| 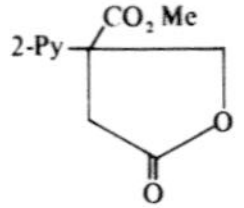 2-Py, $CO_2Me$, O, O | conc. HCl, reflux 13 hr; $K_2CO_3$, $CHCl_3$, $H_2O$, 2 hr | 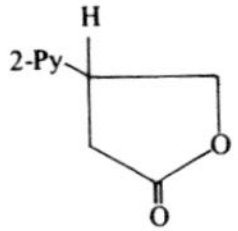 H, 2-Py, O, O | 80% | | 19 |
| $2\text{-Py}(CH_2)_2CH(CN)CO_2Et$ | base; heat | $2\text{-Py}(CH_2)_3CN$ | | b.p. 95–97°/1.0 mm | 204 |
| $2\text{-PyCH}(CO_2Et)(CH_2)_2P(O)(OEt)_2$ | 30% HCl, reflux 6 hr; base | $2\text{-Py}(CH_2)_3P(O)(OH)_2$ | 50% | | 243 |
| $2\text{-PyCOCH}_2CO_2Et$ | 0°, HBr; $Br_2$, 5–10°; 35°, $-CO_2$; 1.5 hr, 40–45° | $2\text{-PyCOCH}_2Br \cdot HBr$ | 89% | | 104 |
| 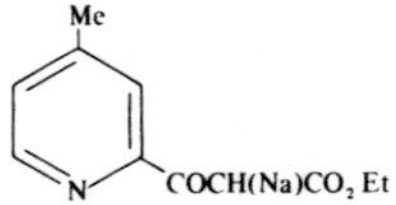 Me, N, $COCH(Na)CO_2Et$ | 20% HCl, reflux 4 hr | 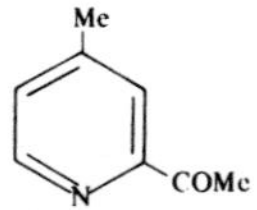 Me, N, COMe | 42% | | 109 |
|  2-PyCO, NMe | heat, sealed tube | 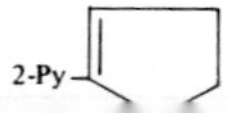 2-Py | | | 129 |

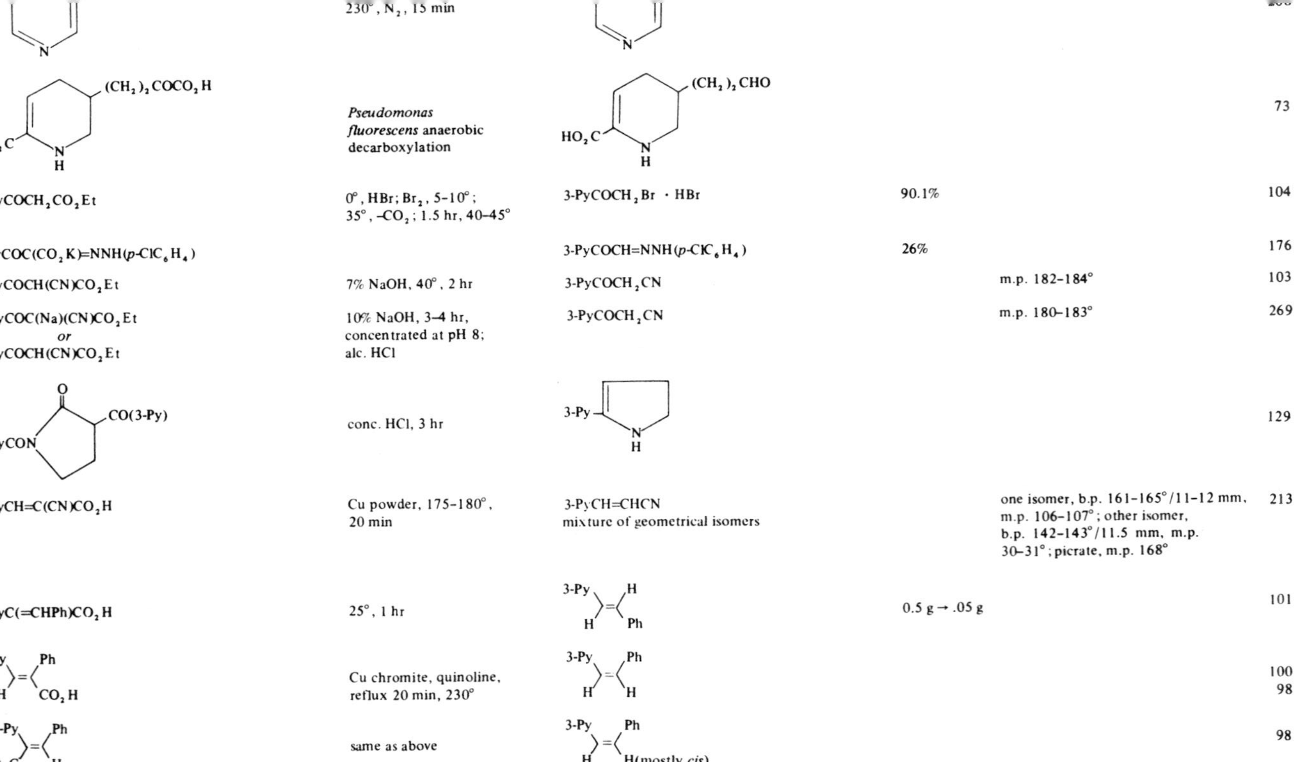

| Starting material | Conditions | Product | Yield | Remarks | Ref. |
|---|---|---|---|---|---|
| | $230^\circ$, $N_2$, 15 min | | | | |
| | *Pseudomonas fluorescens* anaerobic decarboxylation | | | | 73 |
| 3-$PyCOCH_2CO_2Et$ | $0^\circ$, HBr; $Br_2$, $5-10^\circ$; $35^\circ$, $-CO_2$; 1.5 hr, $40-45^\circ$ | 3-$PyCOCH_2Br \cdot HBr$ | 90.1% | | 104 |
| 3-$PyCOC(CO_2K)$=NNH($p$-$ClC_6H_4$) | | 3-PyCOCH=NNH($p$-$ClC_6H_4$) | 26% | | 176 |
| 3-PyCOCH(CN)$CO_2Et$ | 7% NaOH, $40^\circ$, 2 hr | 3-$PyCOCH_2CN$ | | m.p. $182-184^\circ$ | 103 |
| 3-PyCOC(Na)(CN)$CO_2Et$ *or* 3-PyCOCH(CN)$CO_2Et$ | 10% NaOH, 3–4 hr, concentrated at pH 8; alc. HCl | 3-$PyCOCH_2CN$ | | m.p. $180-183^\circ$ | 269 |
| | conc. HCl, 3 hr | | | | 129 |
| 3-PyCH=C(CN)$CO_2H$ | Cu powder, $175-180^\circ$, 20 min | 3-PyCH=CHCN mixture of geometrical isomers | | one isomer, b.p. $161-165^\circ$/11–12 mm, m.p. $106-107^\circ$; other isomer, b.p. $142-143^\circ$/11.5 mm, m.p. $30-31^\circ$; picrate, m.p. $168^\circ$ | 213 |
| 3-PyC(=CHPh)$CO_2H$ | $25^\circ$, 1 hr | | 0.5 g → .05 g | | 101 |
| | Cu chromite, quinoline, reflux 20 min, $230^\circ$ | | | | 100, 98 |
| | same as above | (mostly *cis*) | | | 98 |

TABLE XI-17. Decarboxylation Reactions Yielding Nonacidic Products (Continued)

| Starting material | Conditions | Product | Yield | Properties | Ref. |
|---|---|---|---|---|---|
| 3-Py(H)C=C($p$-MeC$_6$H$_4$)($CO_2H$) | same as above | 3-Py(H)C=C($p$-MeC$_6$H$_4$)(H) | | | 100 |
| 3-Py(H)C=C($p$-BrC$_6$H$_4$)($CO_2H$) | same as above | 3-Py(H)C=C($p$-BrC$_6$H$_4$)(H) | | | 100 |
| 3-Py(H)C=C($p$-ClC$_6$H$_4$)($CO_2H$) | same as above | 3-Py(H)C=C($p$-ClC$_6$H$_4$)(H) | | | 100 |
| 3-Py(H)C=C($p$-IC$_6$H$_4$)($CO_2H$) | same as above | 3-Py(H)C=C($p$-IC$_6$H$_4$)(H) | 59% | | 100 |
| 3-Py(H)C=C($p$-MeOC$_6$H$_4$)($CO_2H$) | same as above | 3-Py(H)C=C($p$-MeOC$_6$H$_4$)(H) | | | 100 |
| 3-Py(H)C=C($p$-$O_2$NC$_6$H$_4$)($CO_2H$) | same as above | 3-Py(H)C=C($p$-$O_2$NC$_6$H$_4$)(H) | | | 100 |
| 3-PyCH=C($CO_2$H)-(3,4-di-OMe-C$_6$H$_3$) | 210–220°, quinoline, copper chromite, 4 hr 220–230° | 3-PyCH=CH-(3,4-di-OMe-C$_6$H$_3$) | | | 225 |
| [quinolizine-2,3-dione bearing $CO_2$Et] | 25% HCl, reflux 5 hr | [quinolizine-2,3-dione] | 15% | | 148 |
| [aryl ring bearing $CH_2CO_2H$] | | [aryl ring bearing Me] | | | 76 |

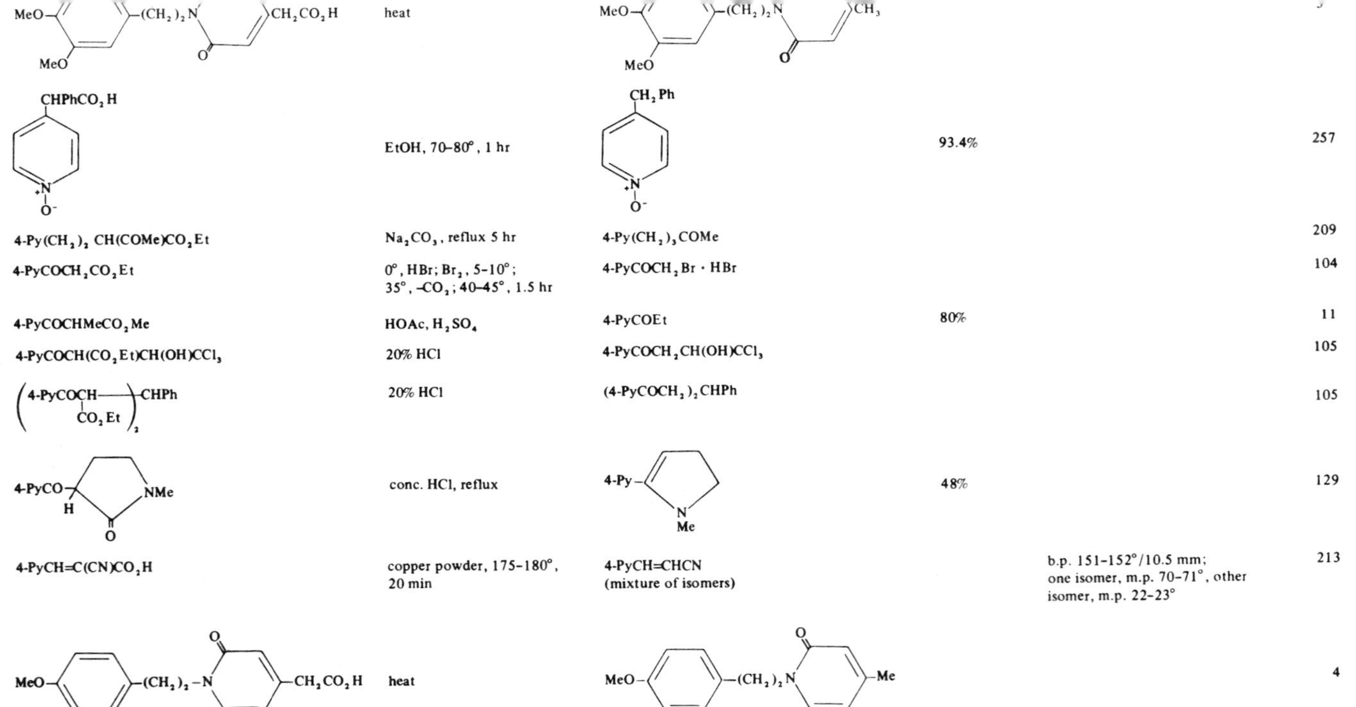

| | | | | | |
|---|---|---|---|---|---|
| MeO-/MeO-substituted phenyl-$(CH_2)_2N$ pyridone with $CH_2CO_2H$ [structure] | heat | MeO-/MeO-substituted phenyl-$(CH_2)_2N$ pyridone with $CH_3$ [structure] | | | |
| 4-$CHPhCO_2H$-pyridine N-oxide [structure] | EtOH, 70–80°, 1 hr | 4-$CH_2Ph$-pyridine N-oxide [structure] | 93.4% | | 257 |
| 4-Py$(CH_2)_2$ CH(COMe)$CO_2$Et | $Na_2CO_3$, reflux 5 hr | 4-Py$(CH_2)_3$COMe | | | 209 |
| 4-PyCO$CH_2CO_2$Et | 0°, HBr; $Br_2$, 5–10°; 35°, $-CO_2$; 40–45°, 1.5 hr | 4-PyCO$CH_2$Br · HBr | | | 104 |
| 4-PyCOCHMe$CO_2$Me | HOAc, $H_2SO_4$ | 4-PyCOEt | 80% | | 11 |
| 4-PyCOCH($CO_2$Et)CH(OH)$CCl_3$ | 20% HCl | 4-PyCO$CH_2$CH(OH)$CCl_3$ | | | 105 |
| (4-PyCOCH($CO_2$Et)$)_2$CHPh | 20% HCl | (4-PyCO$CH_2)_2$CHPh | | | 105 |
| 4-PyCO-(H)-pyrrolidinone, NMe [structure] | conc. HCl, reflux | 4-Py-pyrroline, N-Me [structure] | 48% | | 129 |
| 4-PyCH=C(CN)$CO_2$H | copper powder, 175–180°, 20 min | 4-PyCH=CHCN (mixture of isomers) | | b.p. 151–152°/10.5 mm; one isomer, m.p. 70–71°, other isomer, m.p. 22–23° | 213 |
| MeO-/MeO-phenyl-$(CH_2)_2$–N pyridone, $CH_2CO_2H$, Et ·HCl [structure] | heat | MeO-/MeO-phenyl-$(CH_2)_2$N pyridone, Me, Et [structure] | | | 4 |

TABLE XI-17. Decarboxylation Reactions Yielding Nonacidic Products (Continued)

| Starting material | Conditions | Product | Yield | Properties | Ref. |
|---|---|---|---|---|---|
| | | B. *Hydrolysis and decarboxylation of pyridine acetonitriles* | | | |
| 2-Py$(CH_2)_2C(CN)(NHAc)CO_2Et$ | conc. HCl, 120–130°, 7 hr; EtOH | 2-Py$(CH_2)_2CH(NH_2)CO_2Et$ | 5 g → 2.1 g | b.p. 138–140°/4 mm; picrolonate, m.p.222° (decomp), picrate, m.p. 186° | 89 |
| 2-PyCPh(CN)$CH_2CO_2Et$ | 48% HBr, HOAc, 175°, 24 hr; EtOH, HCl | 2-PyCHPh$(CH_2)_2CO_2Et$ | | b.p. 164°/0.01 mm | 200 |
| 2-PyCH(CN)CH(2-Py)$CH_2$COPh | 50% $H_2SO_4$, reflux 1 hr | 2-Py$CH_2$CH(2-Py)$CH_2$COPh | 1.7 g → 9 g | | 270 |
| 2-Py(p-Cl$C_6H_4$)$(CH_2CH_2NMe_3)$CN | KOH, xylene, reflux 20-24 hr | 2-PyCH(p-Cl$C_6H_4$)$CH_2CH_2NMe_2$ | | | 271 |
| 2-PyC(p-MeO$C_6H_4)_2$CN *or* 2-PyC(p-HO$C_6H_4)_2$CN | 48% HBr. reflux; aq. KOH. reflux | 2-PyCH(p-HO$C_6H_4)_2$ | | | 272, 273, 274 |
| 2-PyCOCHPhCN | 75% $H_2SO_4$ | 2-PyCO$CH_2$Ph | 74.5% | | 275 |
| 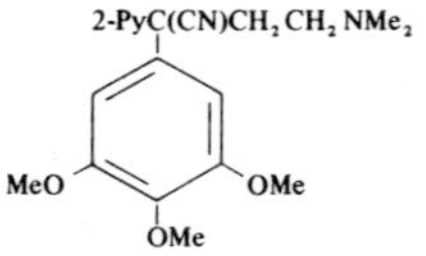 2-PyC(CN)$CH_2CH_2NMe_2$ (ring: MeO, OMe, OMe) | $LiNH_2$, xylene, reflux 32 hr | 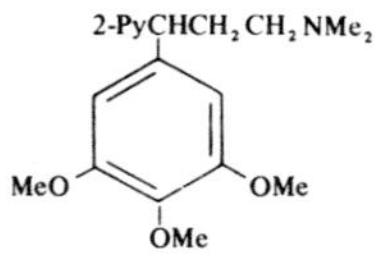 2-PyCH$CH_2CH_2NMe_2$ (ring: MeO, OMe, OMe) | 89.9% | | 276 |
| 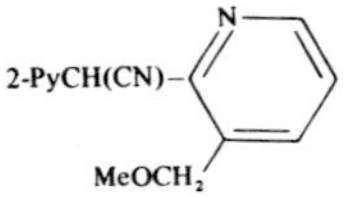 2-PyCH(CN)– (pyridine ring, N; $MeOCH_2$) | 70% $H_2SO_4$, boiled 4 hr | 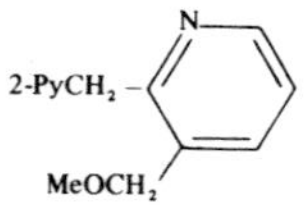 2-Py$CH_2$ – (pyridine ring, N; $MeOCH_2$) | | | 9 |
|  | heat, 14 hr, HCl; MeOH, $H_2SO_4$ |  | 91% | b.p. 115°/0.7 mm | 200 |

| | | | | | |
|---|---|---|---|---|---|
| Me, N, CH(CN)CH(OH), N, Me | $Ac_2O$; HCl | Me, N, CH=CH, N, Me | | | [illegible] |
| 3-PyC($p$-$FC_6H_4$)PhCN | $H_2SO_4$ | 3-PyCH($p$-$FC_6H_4$)Ph | | | 277 |
| 3-PyCOCHPhCN | 75% $H_2SO_4$ | 3-PyCOCH$_2$Ph | 87.7% | | 275 |
| | *or* conc. HBr, 16 hr | | | | 225 |
| 3-PyCOCH(CN)$CO_2$Et | $H_2O$, boiled | 3-PyCOCH$_2$$CO_2$H | | | 269 |
| 4-PyCH(CN)CH($o$-$O_2NC_6H_4$)OH | 75% $H_2SO_4$ | 4-PyC($CONH_2$)=CH($o$-$O_2NC_6H_4$) | | m.p. 208–209° | 99 |
| [4-PyCOCH($CO_2$Et)$CH_2CH_2$]$_2$ | 20% HCl, reflux 3 hr | [4-PyCO($CH_2$)$_3$]$_2$ | 14.8% | | 117 |
| 3-PyCPh(CN), H, NEt | 70% $H_2SO_4$, 130–140°, 48 hr; HCl, $SOCl_2$, 3 hr | Ph, 3-Py, ($CH_2$)$_2$Cl, O, N, Et | | | 56 |
| 3-PyCO$_2$CPh(CN), H, isoPr, NPh | same as above | Ph, ($CH_2$)$_2$Cl, 3-PyCO$_2$, NEt | | | 56 |

TABLE XI-18. Reduction of Side-Chain Acids

| Starting material | Conditions | Product (pip = piperidine) | Yield | Properties | Ref. |
|---|---|---|---|---|---|
| | | *Nuclear reduction* | | | |
| 2-Py$CH_2CO_2Me$ | TsOMe, $N_2$, 65°, 24 hr; NaOMe, 10% Pd-C, $H_2$, MeOH, 45 psi | 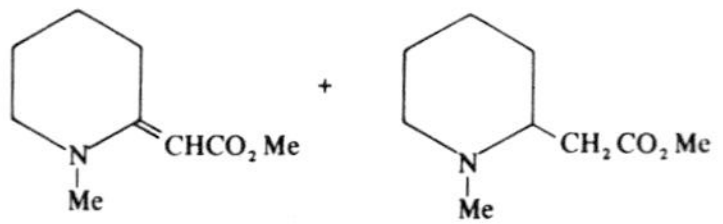 (N-Me piperidine)=$CHCO_2Me$ + (N-Me piperidine)–$CH_2CO_2Me$ | | | 133 |
| 2-PyCHMe$CO_2H$ | $H_2$, 5% Rh-C, $H_2O$, 29% $NH_4OH$, 3 hr | 2-PipCHMe$CO_2H$ | 96.4% | | 248 |
| 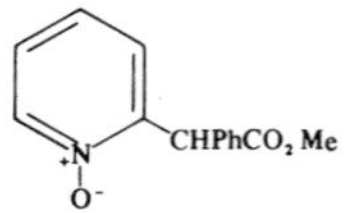 (pyridine $N^+$–$O^-$)–CHPh$CO_2Me$ | $H_2$, $PtO_2$, 10% HCl, EtOH, 1 atm | 2-PipCHPh$CO_2Me$ · HCl | 75.6% | | 257<br>256 |
| 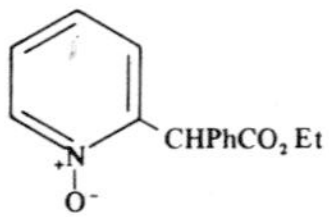 (pyridine $N^+$–$O^-$)–CHPh$CO_2Et$ | same as above | 2-PipCHPh$CO_2Et$ · HCl | | | 257 |
| 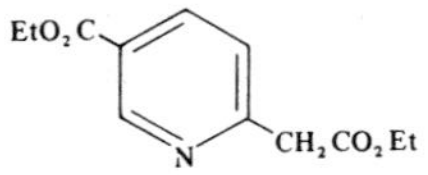 $EtO_2C$–(pyridine)–$CH_2CO_2Et$ | $H_2$, Rh-$Al_2O_3$, HOAc, R.T., 10 hr | 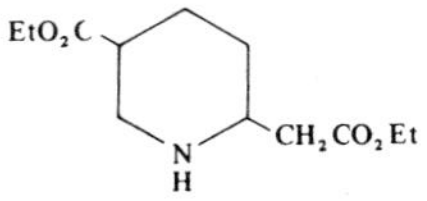 $EtO_2C$–(piperidine, NH)–$CH_2CO_2Et$<br>*trans* (predominates) + *cis* | | | 17 |
| 2-Py$(CH_2)_2CO_2Et$ | $H_2$, $PtO_2$, HOAc, 24 hr | 2-Pip$(CH_2)_2CO_2Et$ | 82% | | 148 |
| 2-PyCH(OH)$CH_2CO_2Et$ | $H_2$ | 2-PipCH(OH)$CH_2CO_2Et$ | | | 249 |

| | | | | |
|---|---|---|---|---|
| 2-PyCPh(OH)CH$_2$CO$_2$Et | H$_2$, PtO$_2$, HCl, 3 hr |   + 2-PipC(OH)PhCH$_2$CO$_2$H<br>16.5% 78.1% | | 54 |
| 3-Me-2-Py(CH$_2$)$_2$CH(CO$_2$Et)$_2$ | H$_2$, PtO$_2$, HOAc-EtOH(1:1) | 3-Me-2-Pip(CH$_2$)$_2$CH(CO$_2$Et)$_2$ | | 206 |
| 2-Py-(γ-butyrolactone) | H$_2$, PtO$_2$, HOAc, 70% HClO$_4$ | 2-Pip-(γ-butyrolactone) | 52.5 g<br>↓<br>51.2 g | 19 |
| 2-PyCH$_2$CHMeCH$_2$CO$_2$Me | H$_2$, PtO$_2$, HOAc; PhCOCl | 1-COPh-2-Pip-CH$_2$CHMeCH$_2$CO$_2$Me | 95% | 200 |
| 2-PyCH$_2$CHPhCH$_2$CO$_2$Et | same as above | 1-COPh-2-Pip-CH$_2$CHPhCH$_2$CO$_2$Et | 83% | 200 |
| 2-Py(CH$_2$)$_2$CMe(CO$_2$Et)$_2$ | HCl, heat; H$_2$, PtO$_2$, HOAc | 2-Pip(CH$_2$)$_2$CHMeCO$_2$H | | 200 |
| 2-Py(CH$_2$)$_2$CHBuCO$_2$Et | H$_2$, PtO$_2$, HCl, EtOH | 2-Pip(CH$_2$)$_2$CHBuCO$_2$Et | | 203 |
| 2-PyCH(CO$_2$Et)CH$_2$CH$_2$CN | H$_2$, PtO$_2$, HCl, EtOH, 5 atm | 2-PipCH(CO$_2$Et)CH$_2$CH$_2$CN | | 242 |
| 2-PyCOCH$_2$CO$_2$Et | H$_2$, HOAc, 6 hr | 2-PipCH(OH)CH$_2$CO$_2$Et | 10 g<br>↓<br>9.5 g | 249 |

TABLE XI-18. Reduction of Side-Chain Acids (Continued)

| Starting material | Conditions | Product (pip = piperidine) | Yield | Properties | Ref. |
|---|---|---|---|---|---|
| 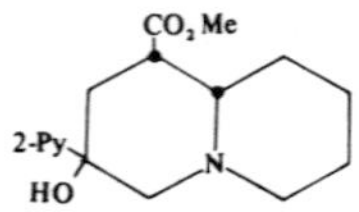 [structure: $CO_2Me$, 2-Py, HO] | $H_2$, $PtO_2$, HOAc | 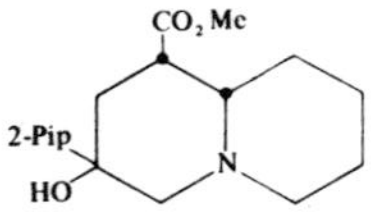 [structure: $CO_2Me$, 2-Pip, HO] | | | 19 |
| 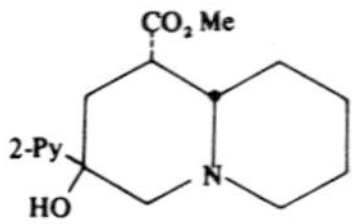 [structure: $CO_2Me$, 2-Py, HO] | $H_2$, $PtO_2$, HOAc | 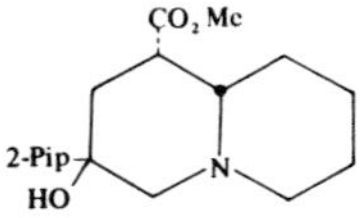 [structure: $CO_2Me$, 2-Pip, HO] | | | 19 |
| 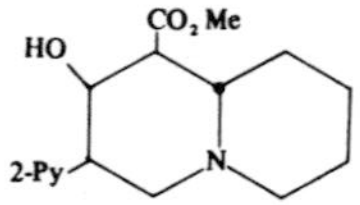 [structure: $CO_2Me$, HO, 2-Py] (mixture of isomers) | $H_2$, $PtO_2$, HOAc | 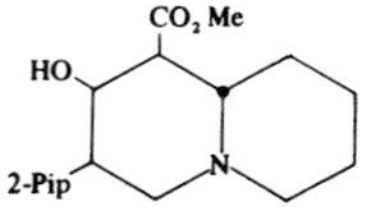 [structure: $CO_2Me$, HO, 2-Pip] | 92% | | 19 |
| 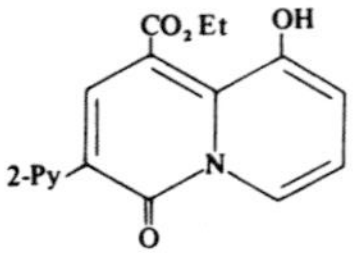 [structure: $CO_2Et$, OH, 2-Py, O] | $H_2$, Ra-Ni, dioxane, 170–180°; 200°, vac., 4 hr | 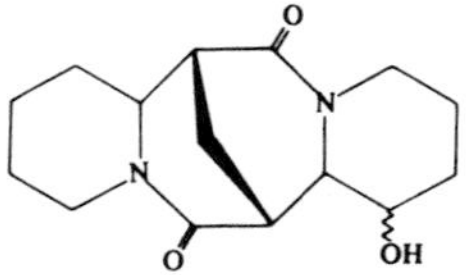 [structure: O, O, OH] | | | 18 |
| 3-$PyCH_2CO_2Me$ | MeI, $N_2$, 15 hr; $NaBH_4$, MeOH, $N_2$, R.T., 15 hr | 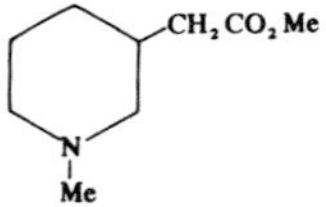 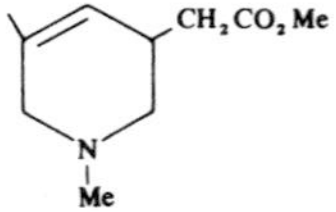 [structure: $CH_2CO_2Me$, N–Me] + [structure: $CH_2CO_2Me$, N–Me] | | nmr; picrate, m.p. 114–117.5°; ir,  [structure: $(CH_2)_2Br$] | 133 |

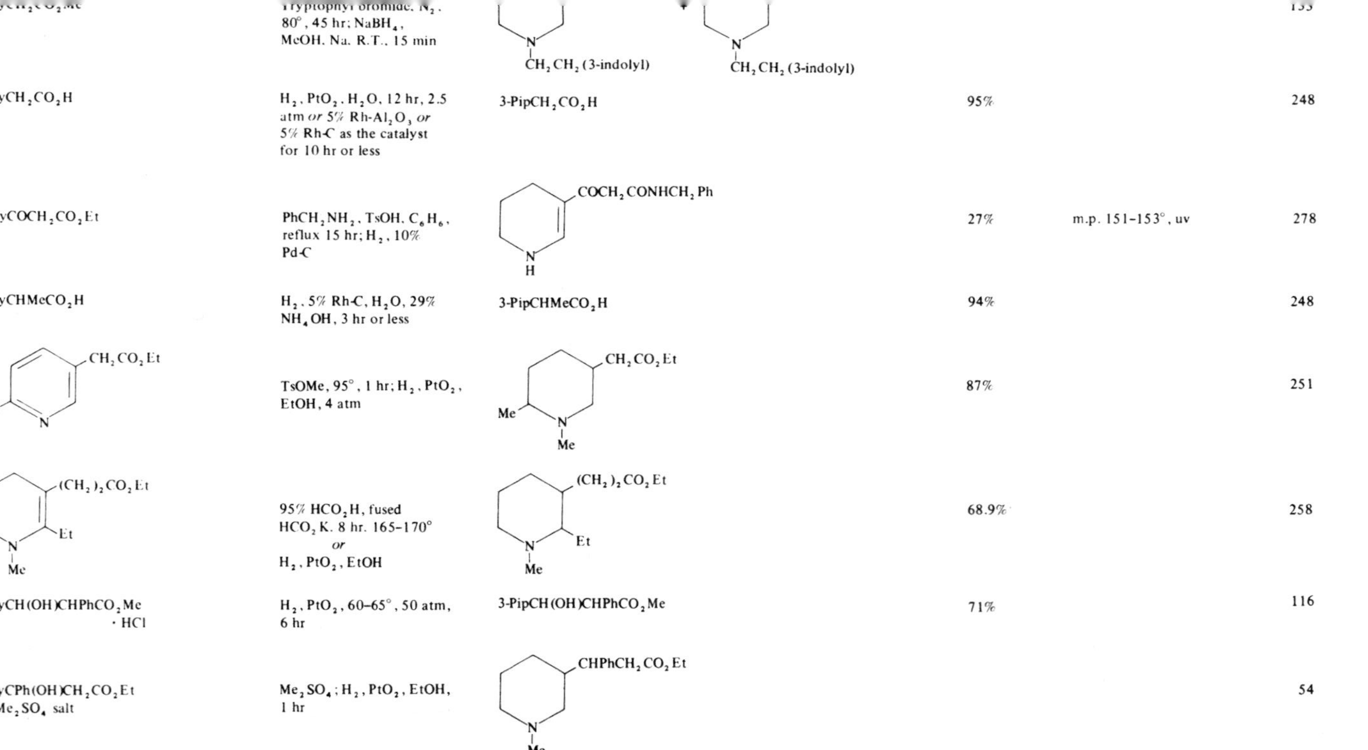

| | | | | | |
|---|---|---|---|---|---|
| 3-Py$CH_2CO_2$Me | Tryptophyl bromide, $N_2$, 80°, 45 hr; $NaBH_4$, MeOH, Na, R.T., 15 min | [structure] N–$CH_2CH_2$(3-indolyl) + [structure] N–$CH_2CH_2$(3-indolyl) | | | 135 |
| 3-Py$CH_2CO_2H$ | $H_2$, $PtO_2$, $H_2O$, 12 hr, 2.5 atm *or* 5% Rh-$Al_2O_3$ *or* 5% Rh-C as the catalyst for 10 hr or less | 3-Pip$CH_2CO_2H$ | 95% | | 248 |
| 3-PyCO$CH_2CO_2$Et | Ph$CH_2NH_2$, TsOH, $C_6H_6$, reflux 15 hr; $H_2$, 10% Pd-C | [structure] CO$CH_2$CONH$CH_2$Ph | 27% | m.p. 151–153°, uv | 278 |
| 3-PyCHMe$CO_2H$ | $H_2$, 5% Rh-C, $H_2O$, 29% $NH_4OH$, 3 hr or less | 3-PipCHMe$CO_2H$ | 94% | | 248 |
| [structure] $CH_2CO_2$Et, Me | TsOMe, 95°, 1 hr; $H_2$, $PtO_2$, EtOH, 4 atm | [structure] $CH_2CO_2$Et, Me, N–Me | 87% | | 251 |
| [structure] $(CH_2)_2CO_2$Et, Et, N–Me | 95% $HCO_2H$, fused $HCO_2K$, 8 hr, 165–170° *or* $H_2$, $PtO_2$, EtOH | [structure] $(CH_2)_2CO_2$Et, Et, N–Me | 68.9% | | 258 |
| 3-PyCH(OH)CHPh$CO_2$Me · HCl | $H_2$, $PtO_2$, 60–65°, 50 atm, 6 hr | 3-PipCH(OH)CHPh$CO_2$Me | 71% | | 116 |
| 3-PyCPh(OH)$CH_2CO_2$Et *or* $Me_2SO_4$ salt | $Me_2SO_4$; $H_2$, $PtO_2$, EtOH, 1 hr | [structure] CHPh$CH_2CO_2$Et, N–Me | | | 54 |

## TABLE XI-18. Reduction of Side-Chain Acids (Continued)

| Starting material | Conditions | Product (pip = piperidine) | Yield | Properties | Ref. |
|---|---|---|---|---|---|
| 3-PyCPh(OH)$CH_2CO_2Et$ | $H_2$, $PtO_2$, HCl, EtOH, 24 hr | 3-PipCHPh$CH_2CO_2Et$ | | | 54 |
| 3-Py$COCH_2CO_2Me$ | TsOMe $C_6H_6$, reflux 3 hr; $H_2$, 10% Pd-C, $Et_3N$, MeOH | 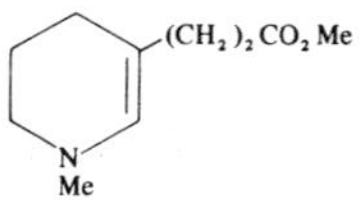 $(CH_2)_2CO_2Me$; N; Me | 2.2 g ↓ 0.9 g | ir; nmr | 133 |
| 3-PyCO$(CH_2)_2CO_2Et$ | $H_2$, Pd-C, 95% EtOH, 7 hr | 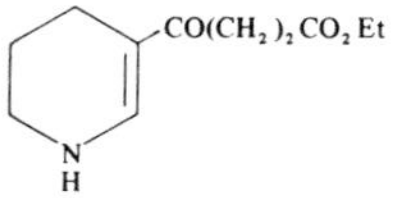 CO$(CH_2)_2CO_2Et$; N; H | 50% | m.p. 81–83° | 278 |
| 3-PyCH=$CHCO_2H$ | $H_2$, $RuO_2$, $H_2O$ *or* $H_2$, $PtO_2$, $H_2O$ | 3-Pip$(CH_2)_2CO_2H$ | 82.6% | | 224<br>152 |
| 4-Py$CH_2CO_2H$ · HCl | 5% Rh-C, $NH_4OH$ | 4-Pip$CH_2CO_2H$ | | | 248 |
| 4-Py$CH_2CONH_2$ | MeI, MeOH, 3 hr reflux; $NaBH_4$, NaOH, $H_2O$, R.T., 3.5 hr | 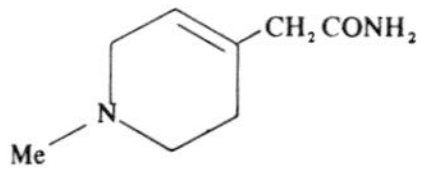 $CH_2CONH_2$; N; Me | | citrate, m.p. 152–153° | 367 |
| 4-Py$CH_2$CONHBu | same as above | 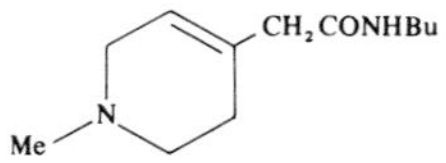 $CH_2$CONHBu; N; Me | | maleate, m.p. 120° | 367 |
| 4-Py$CH_2CONHC_6H_{13}$ | same as above | 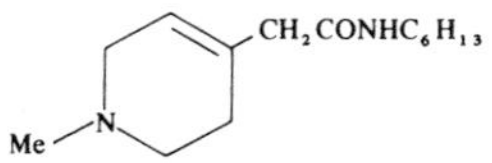 $CH_2CONHC_6H_{13}$; N; Me | | b.p. 180–190°/0.01 mm | 367 |
| 4-Py$CH_2$CONHPh | same as above | 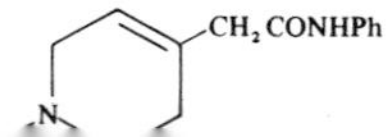 $CH_2$CONHPh; N | | citrate, m.p. 90–92° | 367 |

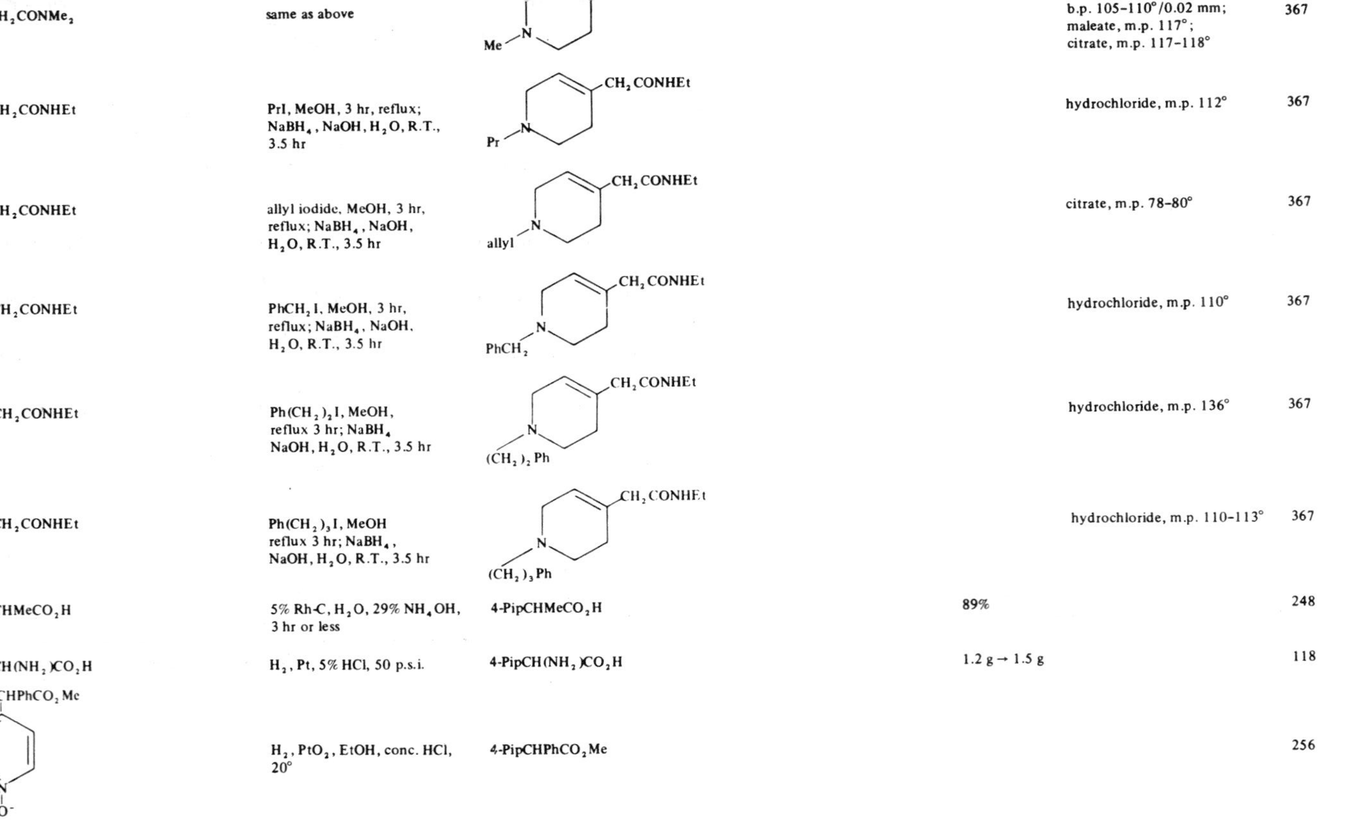

| Starting material | Conditions | Product | Yield | Properties | Ref. |
|---|---|---|---|---|---|
| 4-$PyCH_2CONMe_2$ | same as above | Me–N (ring) | | b.p. 105–110°/0.02 mm; maleate, m.p. 117°; citrate, m.p. 117–118° | 367 |
| 4-$PyCH_2CONHEt$ | PrI, MeOH, 3 hr, reflux; $NaBH_4$, NaOH, $H_2O$, R.T., 3.5 hr | Pr–N (ring)–$CH_2CONHEt$ | | hydrochloride, m.p. 112° | 367 |
| 4-$PyCH_2CONHEt$ | allyl iodide, MeOH, 3 hr, reflux; $NaBH_4$, NaOH, $H_2O$, R.T., 3.5 hr | allyl–N (ring)–$CH_2CONHEt$ | | citrate, m.p. 78–80° | 367 |
| 4-$PyCH_2CONHEt$ | $PhCH_2I$, MeOH, 3 hr, reflux; $NaBH_4$, NaOH, $H_2O$, R.T., 3.5 hr | $PhCH_2$–N (ring)–$CH_2CONHEt$ | | hydrochloride, m.p. 110° | 367 |
| 4-$PyCH_2CONHEt$ | $Ph(CH_2)_2I$, MeOH, reflux 3 hr; $NaBH_4$ NaOH, $H_2O$, R.T., 3.5 hr | $(CH_2)_2Ph$–N (ring)–$CH_2CONHEt$ | | hydrochloride, m.p. 136° | 367 |
| 4-$PyCH_2CONHEt$ | $Ph(CH_2)_3I$, MeOH reflux 3 hr; $NaBH_4$, NaOH, $H_2O$, R.T., 3.5 hr | $(CH_2)_3Ph$–N (ring)–$CH_2CONHEt$ | | hydrochloride, m.p. 110–113° | 367 |
| 4-$PyCHMeCO_2H$ | 5% Rh-C, $H_2O$, 29% $NH_4OH$, 3 hr or less | 4-$PipCHMeCO_2H$ | 89% | | 248 |
| 4-$PyCH(NH_2)CO_2H$ | $H_2$, Pt, 5% HCl, 50 p.s.i. | 4-$PipCH(NH_2)CO_2H$ | 1.2 g → 1.5 g | | 118 |
| $CHPhCO_2Me$ (pyridine N-oxide) | $H_2$, $PtO_2$, EtOH, conc. HCl, 20° | 4-$PipCHPhCO_2Me$ | | | 256 |

TABLE XI-18. Reduction of Side-Chain Acids (Continued)

| Starting material | Conditions | Product (pip = piperidine) | Yield | Properties | Ref. |
|---|---|---|---|---|---|
| 4-Py$(CH_2)_2CO_2Me \cdot MeI$ | $H_2$, $PtO_2$, MeOH | 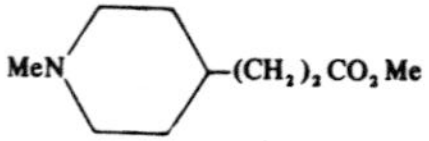 MeN(piperidin-4-yl)–$(CH_2)_2CO_2Me$ | | | 114 |
| | $NaBH_4$, MeOH | 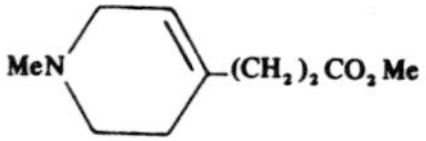 MeN(1,2,3,6-tetrahydropyridin-4-yl)–$(CH_2)_2CO_2Me$ | | | 114 |
| 4-PyCPh(OH)$CH_2CO_2Et$ | $H_2$, 10% Pd-C, HOAc, 25%, 3-4 atm | 4-PipCPh(OH)$CH_2CO_2Et$ | | | 132 |
| 4-PyCH(OH)CHPh$CO_2Me$ ·HCl | $H_2$, $PtO_2$, 60–65°, 50 atm, 6 hr | 4-PipCH(OH)CHPh$CO_2Me$ | | | 116 |
| 4-PyCPh(OH)$CH_2CO_2Et$ | $H_2$, $PtO_2$, HCl, EtOH, 1.5 hr | 4-PipCPh(OH)$CH_2CO_2Et$ | | | 54 |
| 4-Py$(CH_2)_2CH(CO_2Et)_2$ | $H_2$, Pt, EtOH | 4-Pip$(CH_2)_2CH(CO_2Et)_2$ ·HCl | | | 210 |
| 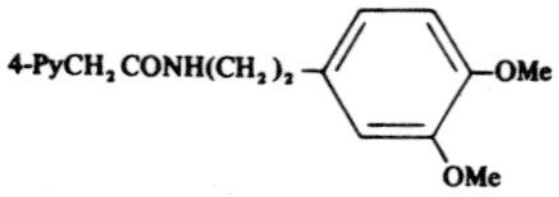 4-Py$CH_2CONH(CH_2)_2$–(3,4-di-OMe-phenyl) | $H_2$, Ra-Ni, EtOH, 1.5 hr; $H_2$, $PtO_2$, HOAc | 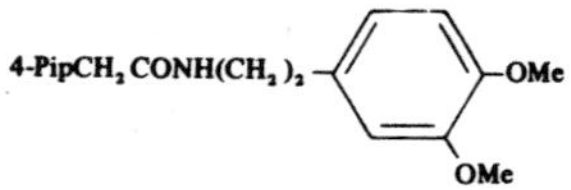 4-Pip$CH_2CONH(CH_2)_2$–(3,4-di-OMe-phenyl) | 3 g → 1.7 g | | 3 |
| 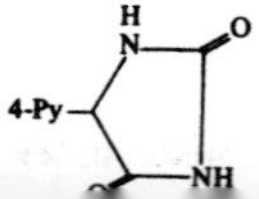 4-Py–(imidazolidinone ring: HN–C(=O)–NH) | $H_2$, Pt, 5% HCl, 50 p.s.i., 3 hr | 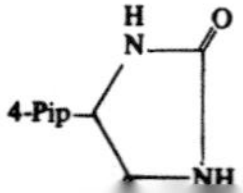 4-Pip–(imidazolidinone ring: HN–C(=O)–NH) | | | 118 |

| Substrate | Conditions | Product | Yield | Remarks | Ref. |
|---|---|---|---|---|---|
| 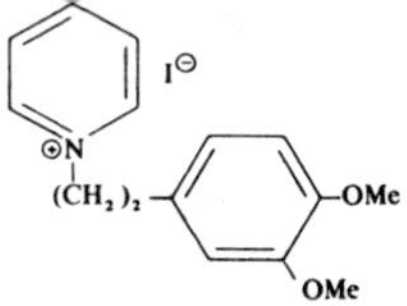 $I^{\ominus}$ ⊕N, $(CH_2)_2$, OMe, OMe | $H_2$, Ra-Ni, 144–145°, $Et_3N$, EtOH, 3 hr | 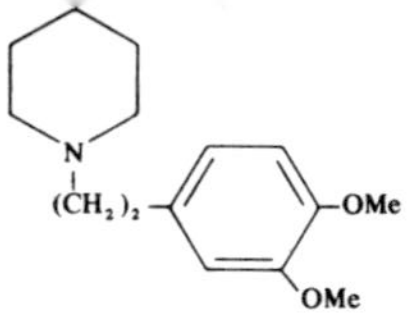 N, $(CH_2)_2$, OMe, OMe | 1.6 g ↓ 1.2 g | | 3 |
| 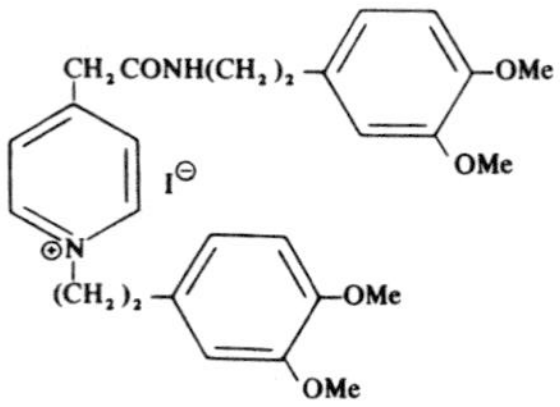 $CH_2CONH(CH_2)_2$, OMe, OMe, $I^{\ominus}$ ⊕N, $(CH_2)_2$, OMe, OMe | $H_2$, $PtO_2$, R.T. | 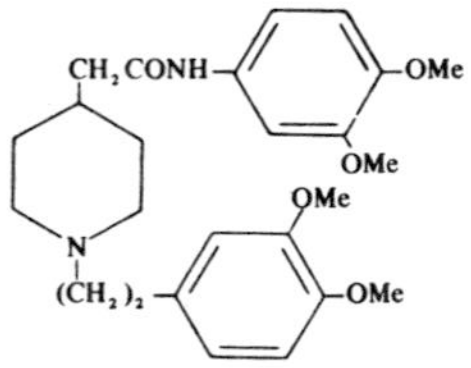 $CH_2CONH$, OMe, OMe, N, $(CH_2)_2$, OMe, OMe | 88% | | 3 |
| 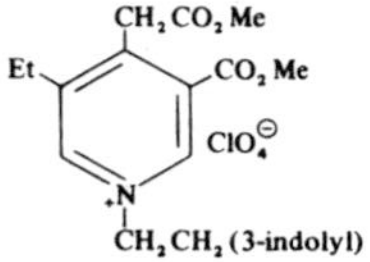 $CH_2CO_2Me$, Et, $CO_2Me$, $ClO_4^{\ominus}$, $\overset{+}{N}$, $CH_2CH_2$ (3-indolyl) | $H_2$, Pd-C, $Et_3N$, abs. MeOH | 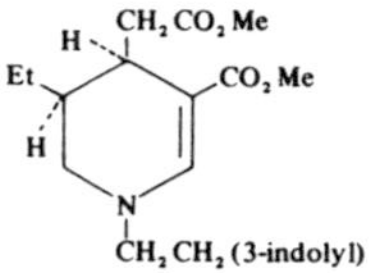 H, $CH_2CO_2Me$, Et, $CO_2Me$, H, N, $CH_2CH_2$ (3-indolyl) | | unstable in air | 12 |
| 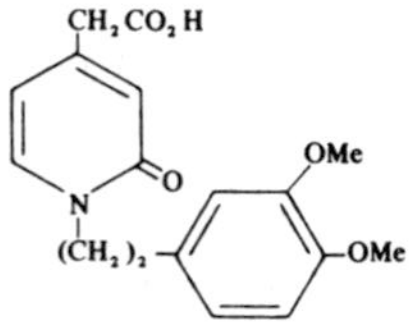 $CH_2CO_2H$, N, O, $(CH_2)_2$, OMe, OMe | $H_2$ | 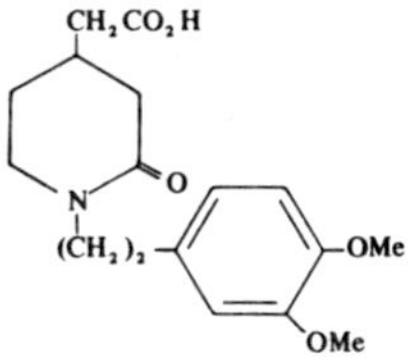 $CH_2CO_2H$, N, O, $(CH_2)_2$, OMe, OMe | | | 88 |
| | $H_2$, $PtO_2$, R.T., MeOH | same as above | 86% | | 3 |

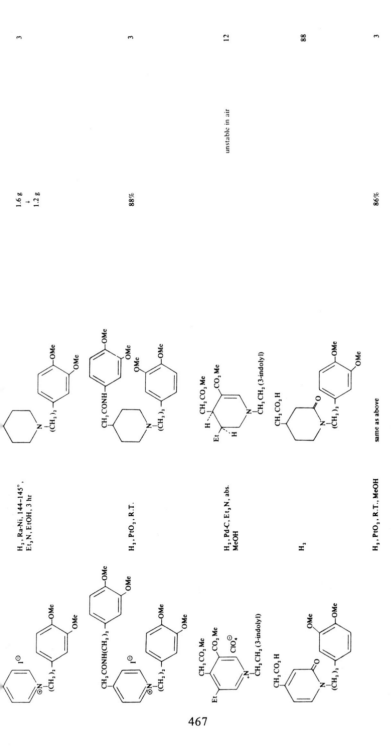

TABLE XI-18. Reduction of Side-Chain Acids (Continued)

| Starting material | Conditions | Product (pip = piperidine) | Yield | Properties | Ref. |
|---|---|---|---|---|---|
| 1-[2-(3,4-dimethoxyphenyl)ethyl]-4-[$CH_2CONH(CH_2)_2$-(3,4-dimethoxyphenyl)]-2-pyridone | $H_2$, $PtO_2$, EtOH 40–50°, 3 hr | 1-[2-(3,4-dimethoxyphenyl)ethyl]-4-[$CH_2CONH(CH_2)_2$-(3,4-dimethoxyphenyl)]-2-piperidone | | | 8 |
| 1-[2-(3,4-dimethoxyphenyl)ethyl]-5-Et-4-$CH_2CO_2H$-2-pyridone | no details given | 1-[2-(3,4-dimethoxyphenyl)ethyl]-5-Et-4-$CH_2CO_2H$-2-piperidone | 84% | | 6 |
| | $H_2$, $PtO_2$, MeOH, R.T., 1 atm, 2–3 days | same as above | *trans*-(70%) *cis*-(15%) | | 4 |
| 4-($CHPhCO_2Me$)pyridine N-oxide | $H_2$, $PtO_2$, 10% HCl, EtOH, 1 atm | 4-Pip$CHPhCO_2Me$ · HCl | 45% | | 257 |
| 4-PyCH=$CHCO_2Me$ · MeI | $NaBH_4$, MeOH, 90 min | 1-Me-4-(CH=$CHCO_2Me$)-1,2,3,6-tetrahydropyridine | | b.p. 188.5°/2.5 mm; picrate, m.p. 175–176° | 134 |
| 1-Me-pyridinium-4-CH=$CHCO_2^{\ominus}$ | $H_2$, $PtO_2$, $H_2O$, 26 hr; | 1-Me-4-[$(CH_2)_2CO_2Et$]piperidine | | | 134 |

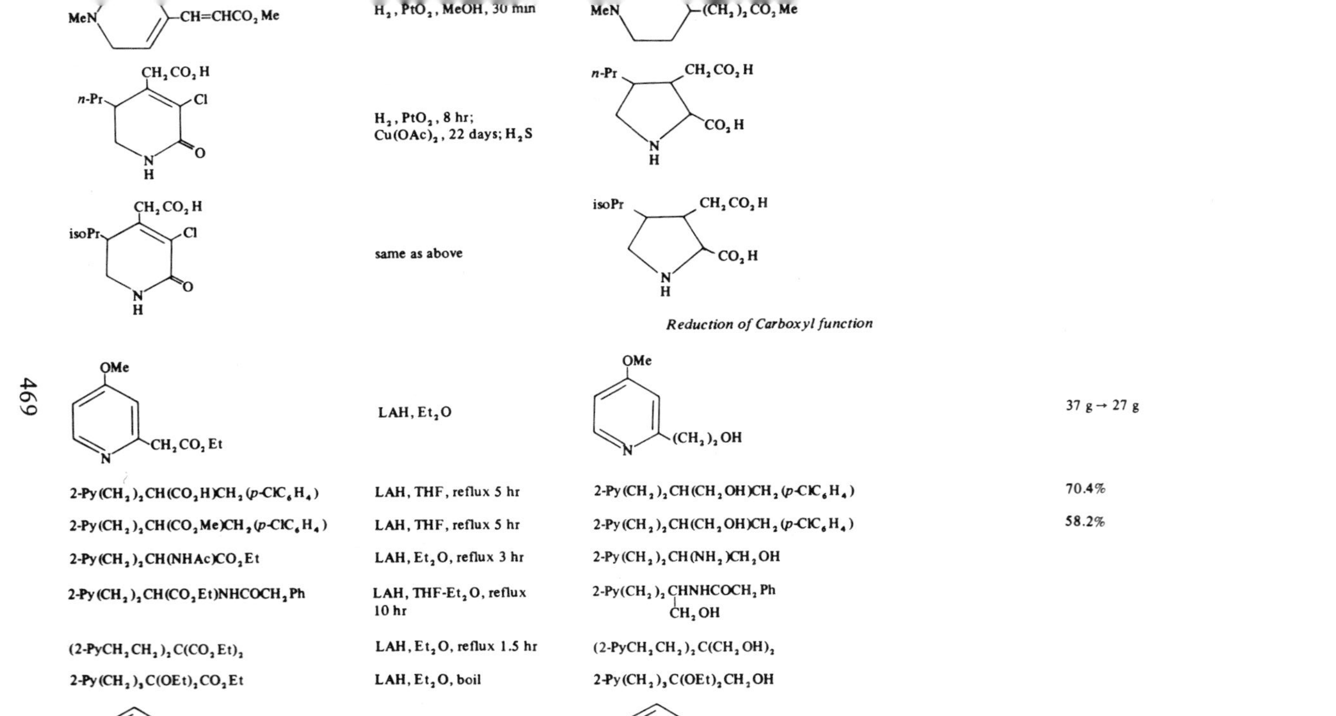

| Starting material | Reagents and conditions | Product | Yield | Ref. |
|---|---|---|---|---|
| | $H_2$, $PtO_2$, MeOH, 30 min | | | |
| | $H_2$, $PtO_2$, 8 hr; $Cu(OAc)_2$, 22 days; $H_2S$ | | | 261 |
| | same as above | | | 261 |

*Reduction of Carboxyl function*

| Starting material | Reagents and conditions | Product | Yield | Ref. |
|---|---|---|---|---|
| | LAH, $Et_2O$ | | 37 g → 27 g | 193 |
| $2\text{-}Py(CH_2)_2CH(CO_2H)CH_2(p\text{-}ClC_6H_4)$ | LAH, THF, reflux 5 hr | $2\text{-}Py(CH_2)_2CH(CH_2OH)CH_2(p\text{-}ClC_6H_4)$ | 70.4% | 245 |
| $2\text{-}Py(CH_2)_2CH(CO_2Me)CH_2(p\text{-}ClC_6H_4)$ | LAH, THF, reflux 5 hr | $2\text{-}Py(CH_2)_2CH(CH_2OH)CH_2(p\text{-}ClC_6H_4)$ | 58.2% | 245 |
| $2\text{-}Py(CH_2)_2CH(NHAc)CO_2Et$ | LAH, $Et_2O$, reflux 3 hr | $2\text{-}Py(CH_2)_2CH(NH_2)CH_2OH$ | | 89 |
| $2\text{-}Py(CH_2)_2CH(CO_2Et)NHCOCH_2Ph$ | LAH, THF-$Et_2O$, reflux 10 hr | $2\text{-}Py(CH_2)_2CH(CH_2OH)NHCOCH_2Ph$ | | 89 |
| $(2\text{-}PyCH_2CH_2)_2C(CO_2Et)_2$ | LAH, $Et_2O$, reflux 1.5 hr | $(2\text{-}PyCH_2CH_2)_2C(CH_2OH)_2$ | | 205 |
| $2\text{-}Py(CH_2)_3C(OEt)_2CO_2Et$ | LAH, $Et_2O$, boil | $2\text{-}Py(CH_2)_3C(OEt)_2CH_2OH$ | | 280 |
| | Na; $NaBH_4$, MeOH, reflux 4 hr | | | 135 |

TABLE XI-18. Reduction of Side-Chain Acids (Continued)

| Starting material | Conditions | Product (pip = piperidine) | Yield | Properties | Ref. |
|---|---|---|---|---|---|
| 2-PyCH=C(CN)$CO_2$Et | $NaBH_4$, isoPrOH, 8 hr | 2-Py$CH_2$CH(CN)$CH_2$OH | | | 46<br>47 |
| 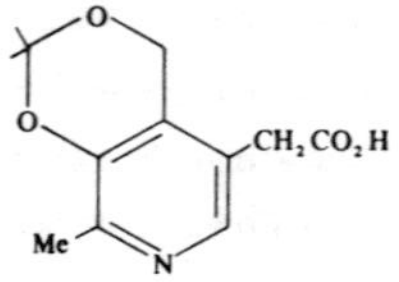 O, O, Me, N, $CH_2CO_2H$ | LAH, THF, $N_2$, 2.5 hr | 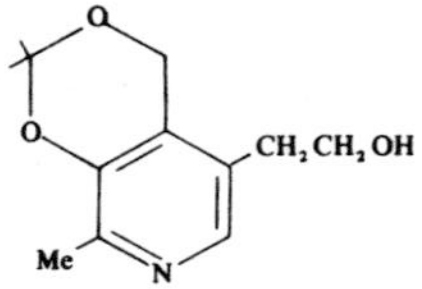 O, O, Me, N, $CH_2CH_2OH$ | | | 125 |
| 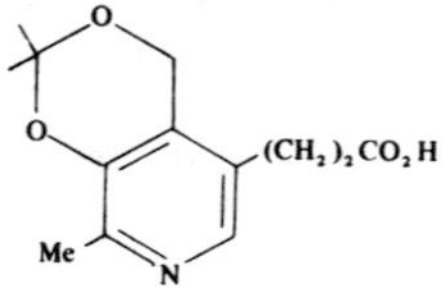 O, O, Me, N, $(CH_2)_2CO_2H$ | LAH, THF, R.T., 1 hr | 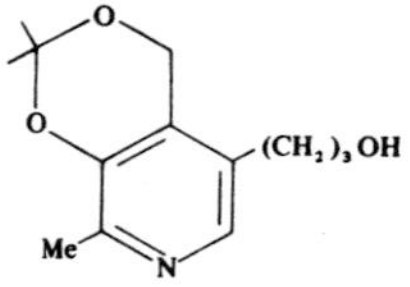 O, O, Me, N, $(CH_2)_3OH$ | | | 228 |
| 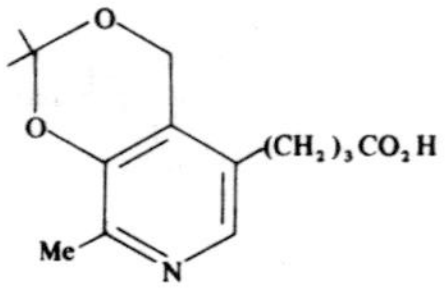 O, O, Me, N, $(CH_2)_3CO_2H$ | LAH, THF, R.T., $N_2$, 2 hr | 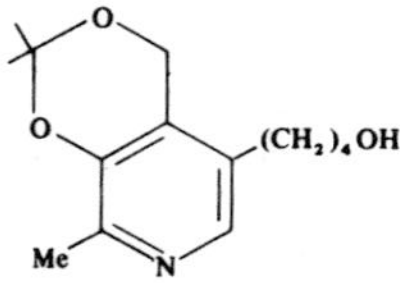 O, O, Me, N, $(CH_2)_4OH$ | | | 125 |
| 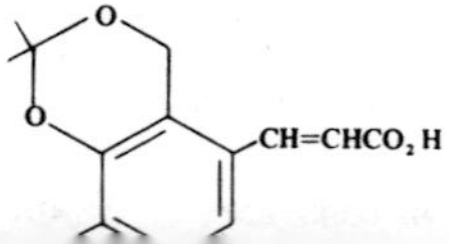 O, O, CH=$CHCO_2H$ | LAH, THF | 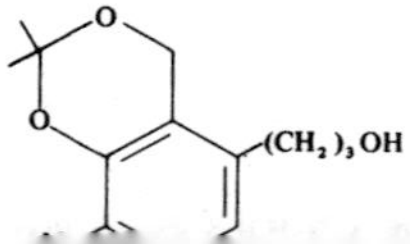 O, O, $(CH_2)_3OH$ | 71% | | 228 |

| | | | | | |
|---|---|---|---|---|---|
| 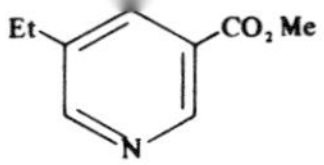 Et / $CO_2Me$ / N | LAH, $Et_2O$, reflux 4 hr | 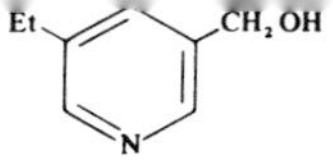 Et / $CH_2OH$ / N | 2.2 g → 1.9 g | | 122 |
| 4-PyCH=$CHCO_2Me$ *or* 4-PyCH=$CHCO_2Et$ | $NaBH_4$, MeOH, reflux 1–2 hr | 4-Py$(CH_2)_3OH$ | | | 115 |
| 4-PyCH=C(CN)$CO_2Et$ | $NaBH_4$, isoPrOH, 8 hr *or* in diglyme, R.T., 3 hr | 4-$PyCH_2CH(CN)CH_2OH$ | 68% | m.p. 68–69° | 46, 47, 177 |
| | | *Reduction of Other Substituents* | | | |
| 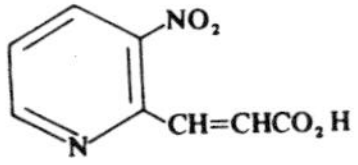 $NO_2$ / CH=$CHCO_2H$ / N | $FeSO_4$, $NH_3$ | 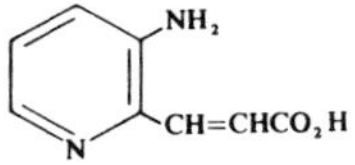 $NH_2$ / CH=$CHCO_2H$ / N | 57% | m.p. 207–208° | 151 |
| 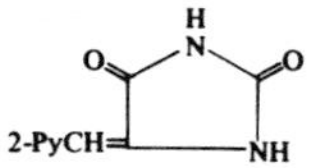 H / N / O / O / 2-PyCH= / NH | $H_2$, Pd-C, 95% EtOH, 3 atm, 12 hr | 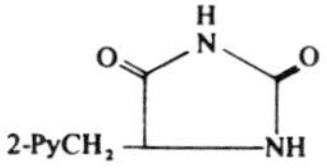 H / N / O / O / 2-$PyCH_2$ / NH | 5 g → 4 g | m.p. 171–172° | 95 |
| 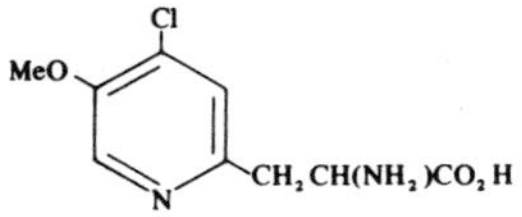 Cl / MeO / N / $CH_2CH(NH_2)CO_2H$ | $H_2$, Pd, 3 atm, 3 hr | 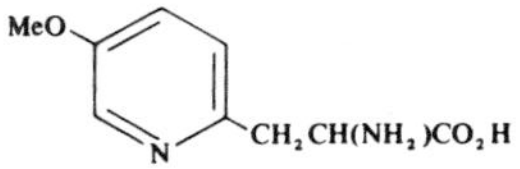 MeO / N / $CH_2CH(NH_2)CO_2H$ | 1.4 g → 0.7 g | m.p. 224–227° (decomp.) | 95 |
| 2-PyC(=NOH)$CO_2Et$ | 5% Pd-C, abs. EtOH | 2-PyCH$(NH_2)CO_2Et$ | 75% | b.p. 103°/0.3 mm; $n_D^{25}$ 1.5163 | 112 |
| 2-PyC(CN)=CHMe | $H_2$, 1 atm, 10% Pd-C, EtOH, R.T. | 2-PyCHEtCN | 88% | b.p. 73–76°/0.5 mm; ir; nmr | 163 |
| 3-PyCH=C($p$-$ClC_6H_4$)CN | $H_2$, 3–4 atm, 10% Pd-C, EtOAc, 70° | 3-$PyCH_2CH$($p$-$ClC_6H_4$)CN | 15% | m.p. 76–80°; ir | 164 |
| 3-PyCH=C($p$-$O_2NC_6H_4$)CN | $H_2$, 3–4 atm, 10% Pd-C, EtOAc, heat | 3-$PyCH_2CH$($p$-$H_2NC_6H_4$)CN | | m.p. 99–100°; ir | 164 |

TABLE XI-18. Reduction of Side-Chain Acids (Continued)

| Starting material | Conditions | Product (pip = piperidine) | Yield | Properties | Ref. |
|---|---|---|---|---|---|
| 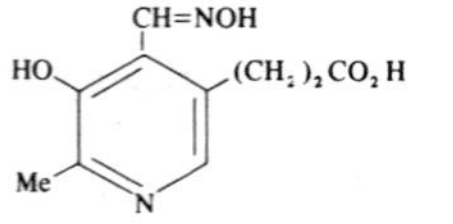 HO, CH=NOH, $(CH_2)_2CO_2H$, Me, N | $H_2$, 10% Pd-C, conc. HCl, R.T., 1 atm, 1 hr | 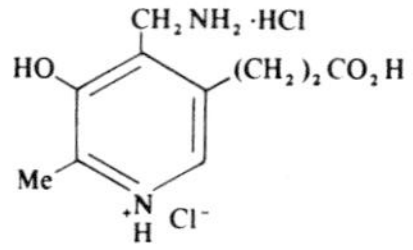 HO, $CH_2NH_2\cdot HCl$, $(CH_2)_2CO_2H$, Me, $\overset{+}{N}$H $Cl^-$ | 59% | ir; uv | 279 |
| 4-PyCH=C($p$-$O_2NC_6H_4$)CN | $H_2$, 3–4 atm, 10% Pd-C, EtOAc, heat | 4-$PyCH_2CH$($p$-$H_2NC_6H_4$)CN | | m.p. 80–84°; ir | 164 |
| | | *Miscellaneous* | | | |
| 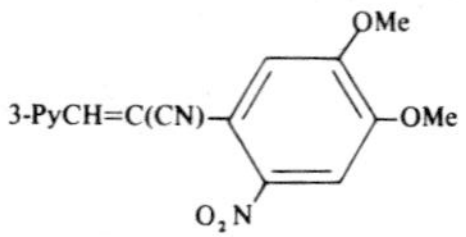 3-PyCH=C(CN)–, OMe, OMe, $O_2N$ | Fe, HOAc, reflux 2–5 hr | 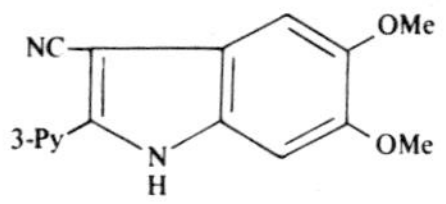 NC, 3-Py, NH, OMe, OMe | 39% | m.p. 238–239° | 10 |
| 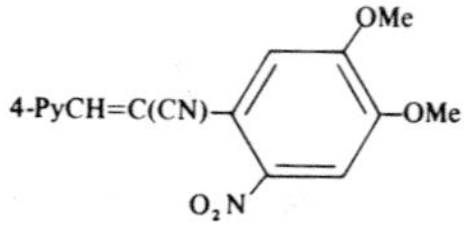 4-PyCH=C(CN)–, OMe, OMe, $O_2N$ | Fe, HOAc, reflux 2–5 hr | 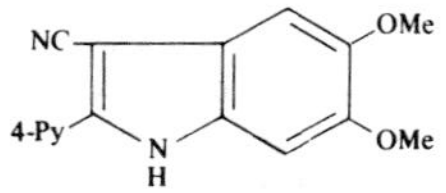 NC, 4-Py, NH, OMe, OMe | 72.5% | m.p. 321°; ir; uv; nmr | 10 |
| 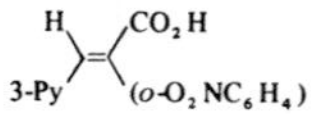 H, $CO_2H$, 3-Py, ($o$-$O_2NC_6H_4$) | $FeSO_4$, $NH_4OH$, $H_2O$, 80–90°, 20 min | 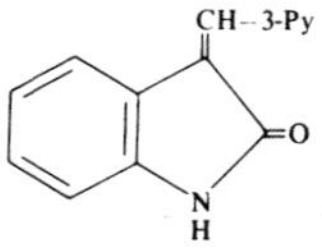 CH–3-Py, =O, NH *trans* | | m.p. 187–188° | 99 |
| 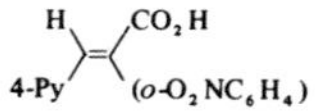 H, $CO_2H$, 4-Py, ($o$-$O_2NC_6H_4$) | $FeSO_4$, $NH_4OH$, $H_2O$, 80–90°, 20 min | 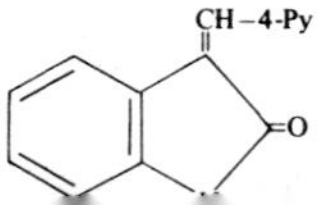 CH–4-Py, =O | | m.p. 225° | 99 |

| Reactants | Conditions | Products | Properties | Ref. |
|---|---|---|---|---|
| 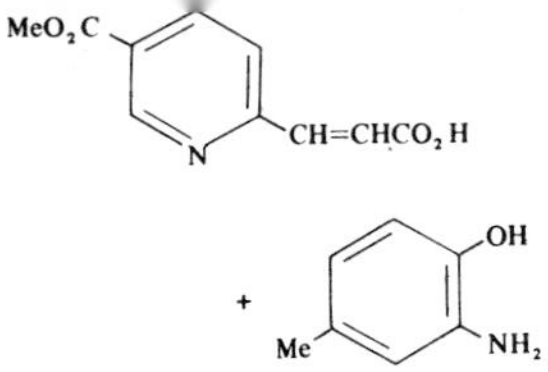 $MeO_2C$, N, CH=CHCO$_2$H + Me, OH, $NH_2$ | 1, 2-dichlorobenzene, $H_3BO_3$, 120°, 8 hr, $N_2$ | 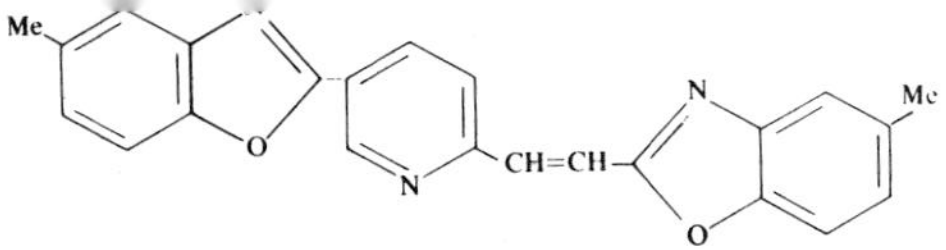 Me, O, N, CH=CH, N, O, Me | | 281 |
| 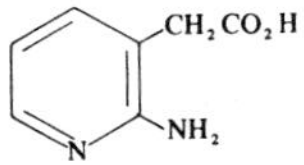 $CH_2CO_2H$, N, $NH_2$ | 225° | 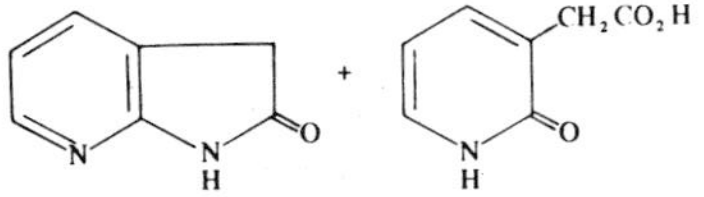 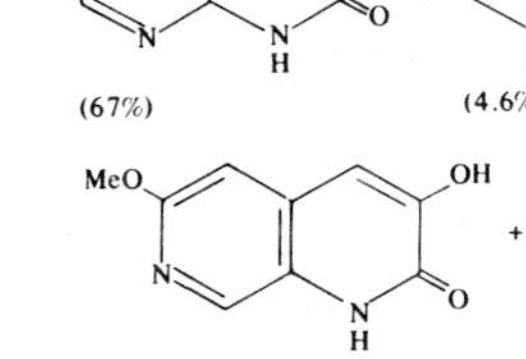  | m.p. 240–241° (decomp.) | 268 |
| 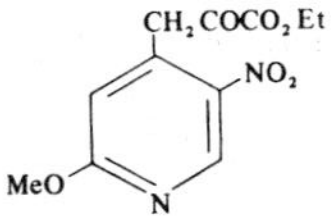 $CH_2COCO_2Et$, $NO_2$, MeO, N | $H_2$, $PtO_2$, EtOH | 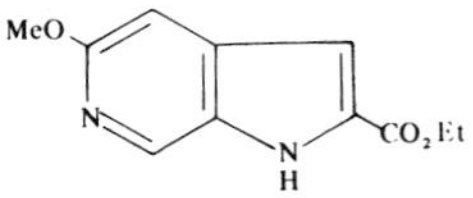 MeO, N, OH, N H, O + MeO, N, N H, $CO_2Et$ | m.p. 103–106°; sublimes 90°/0.1 mm; uv; nmr | 15 |

TABLE XI-19. Synthesis of Quinolizidines and Quinolizines

| Starting materials | Conditions | Product | Yield | Properties | Ref. |
|---|---|---|---|---|---|
| 2-$PyCH_2CO_2Et$ + $BrCH_2COCO_2Et$ | acetone, 2 hr | 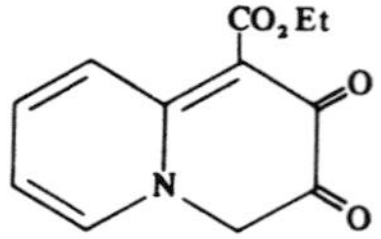 $CO_2Et$, O, O, N | | | 148 |
| 2-$PyCH(CO_2Me)CH_2$—N, $CH_2CO_2Me$ 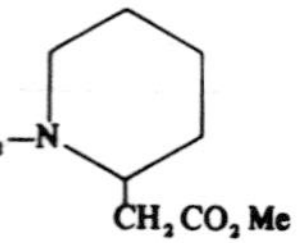 | | 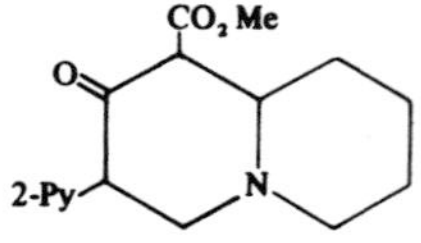 $CO_2Me$, O, 2-Py, N | 28% | m.p. 119–120° | 19 |
| 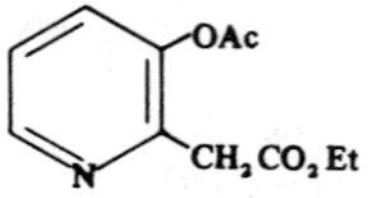 OAc, $CH_2CO_2Et$, N + 2-$PyC(CO_2Et)$=CHOH | $Ac_2O$, reflux 3 hr | 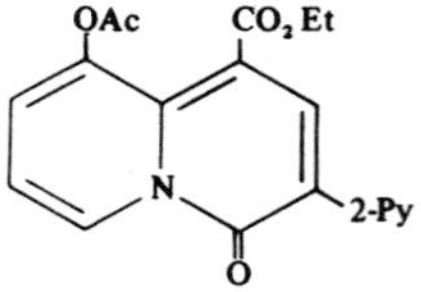 OAc, $CO_2Et$, N, 2-Py, O | 10 g → 4 g | b.p. 176°; ir | 18 |
| 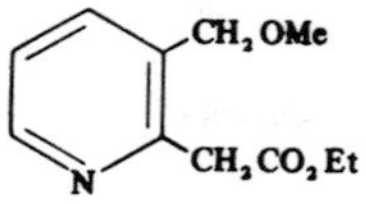 $CH_2OMe$, $CH_2CO_2Et$, N + $EtOCH=C(CO_2Et)_2$ | reflux 8 hr | 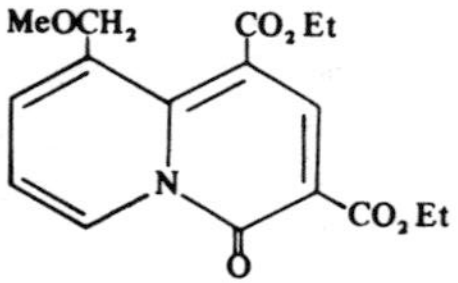 $MeOCH_2$, $CO_2Et$, N, $CO_2Et$, O | 2 → 0.4 g | m.p. 75–77° | 9 |
| 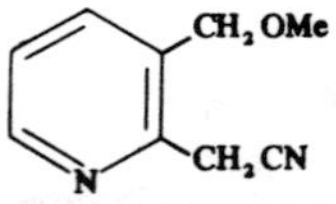 $CH_2OMe$, $CH_2CN$, N + $EtOCH=C(CO_2Et)_2$ | reflux 2 hr | 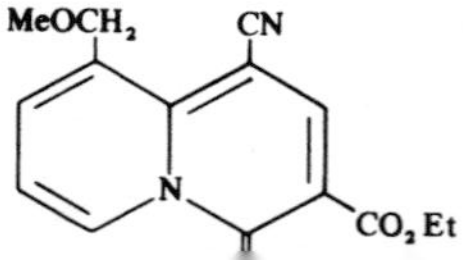 $MeOCH_2$, CN, N, $CO_2Et$ | 40% | | 9 |

| | | | | |
|---|---|---|---|---|
| 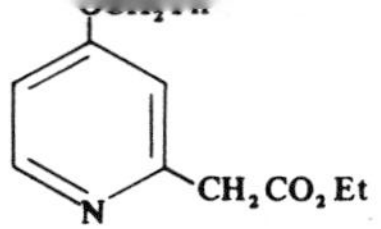 $CH_2CO_2Et$ (pyridine, N) + 2-PyC($CO_2$Et)=CHCO$_2$H | $Ac_2O$, reflux 2 hr | 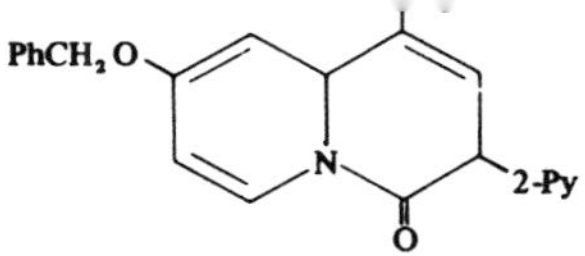 $PhCH_2O$, N, O, 2-Py | 25 g → 10 g | 282 |
| 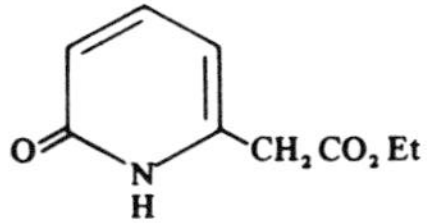 O, N, H, $CH_2CO_2Et$ + EtOCH=C($CO_2$Et)$_2$ | Na, EtOH | 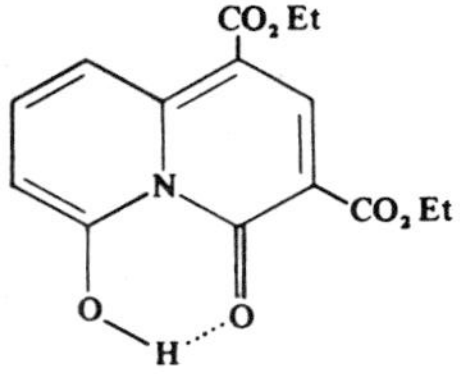 $CO_2Et$, N, $CO_2Et$, O, H, O | 72.1% | 139 |
| 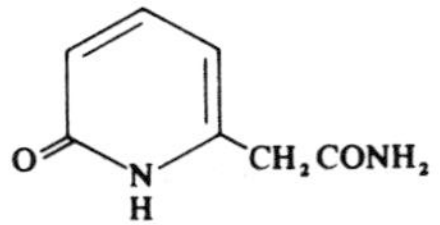 O, N, H, $CH_2CONH_2$ + EtOCH=C($CO_2$Et)$_2$ | Na, EtOH | 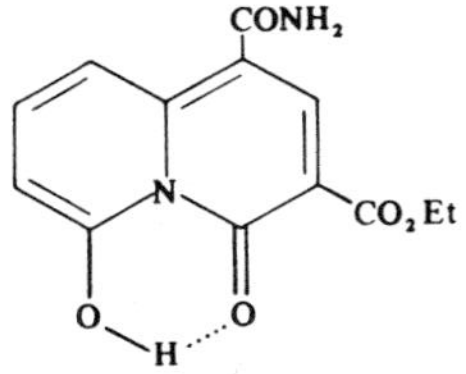 $CONH_2$, N, $CO_2Et$, O, H, O | 80.8% | 139 |
| 2-PyCHMe($CH_2$)$_2CO_2$Et | $H_2$, $PtO_2$, HCl, EtOH; 200°, 2 hr | 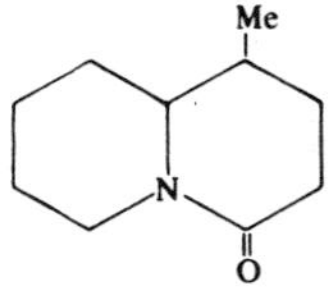 Me, N, O | 4.6 g → 2.2 g | 283 |

TABLE XI-19. Synthesis of Quinolizidines and Quinolizines (Continued)

| Starting materials | Conditions | Product | Yield | Properties | Ref. |
|---|---|---|---|---|---|
| 2-PyCHPh$(CH_2)_2CO_2Et$ | $H_2$, $PtO_2$, HOAc, 200°, 4 hr | 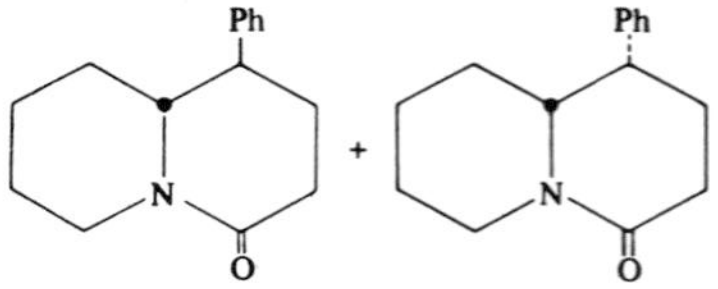 Ph, N, O + Ph, N, O | | | 200 |
| 2-PyC$(=CH_2)CH_2CH(CO_2Et)_2$ | $H_2$, $PtO_2$, HOAc | 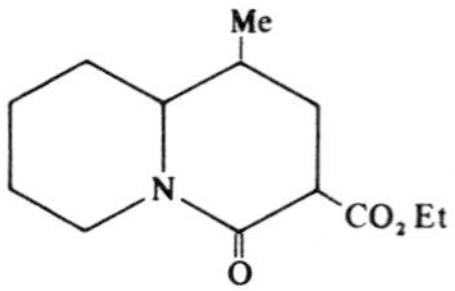 Me, N, O, $CO_2Et$ | 80% | | 206 |
| 2-PyCH$(CO_2Et)CH_2CH_2CO_2Et$ | $H_2$, Ra-Ni, 175°, 100 atm | 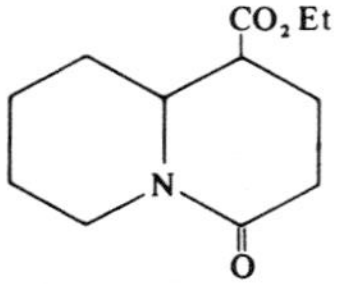 $CO_2Et$, N, O | 95% | | 242 |
| 2-PyCMe(OH)C≡C $CO_2Me$ | $H_2$, $PtO_2$, HCl; $SOCl_2$; $H_2$, $PtO_2$; 200°, 2 hr | 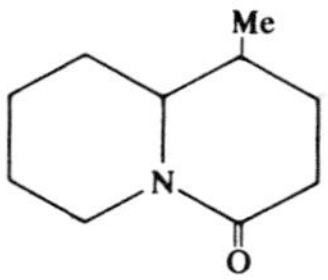 Me, N, O | | | 79, 97 |
| 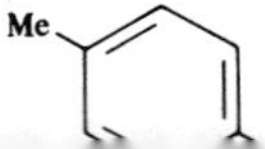 Me | $H_2$, $PtO_2$, HCl, EtOH; 200°, 2 hr | 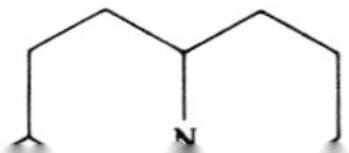 N | | | 283 |

| Reactant | Conditions | Product | Yield | Ref. |
|---|---|---|---|---|
| 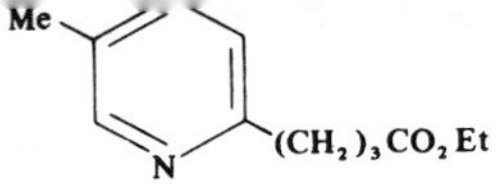 Me; N; $(CH_2)_3CO_2Et$ | $H_2$, $PtO_2$, HOAc; 200°, 1 hr | 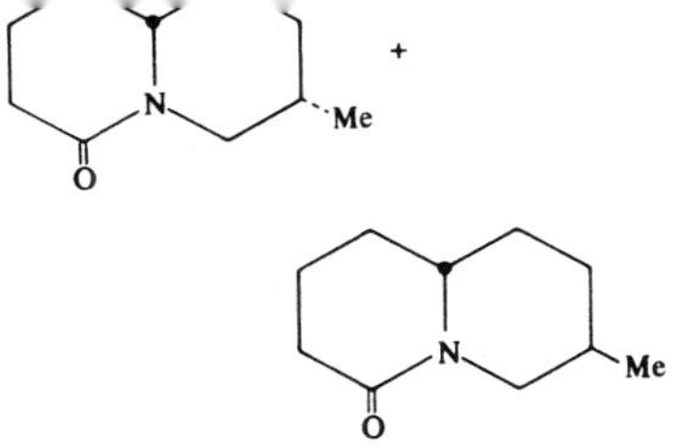 N; O; Me + N; O; Me | | 206 |
| 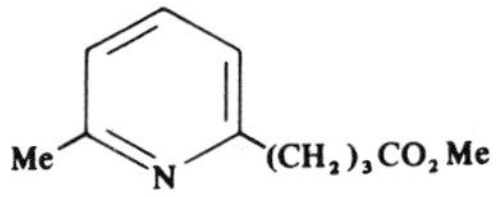 Me; N; $(CH_2)_3CO_2Me$ | $H_2$, $PtO_2$, HOAc, 200°, 5 hr | 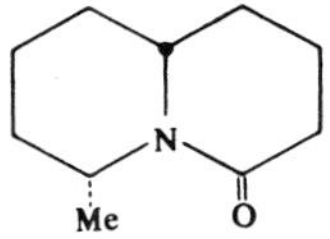 N; Me; O | | 200 |
| 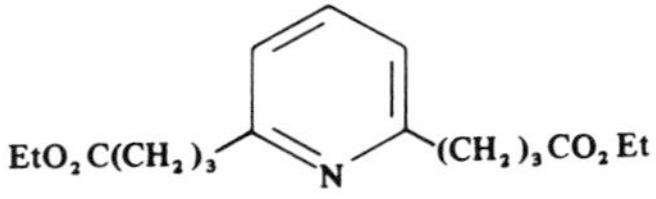 $EtO_2C(CH_2)_3$; N; $(CH_2)_3CO_2Et$ | $H_2$, Ra-Ni, 150–160°, 150 atm | 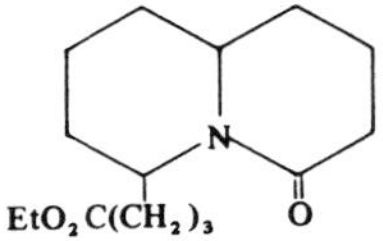 N; $EtO_2C(CH_2)_3$; O | 85–90% | 138 |
| 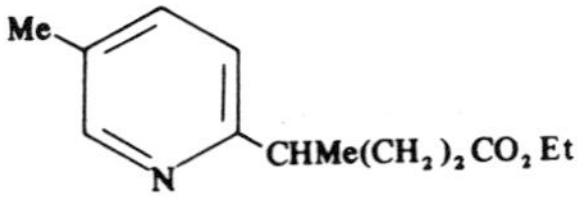 Me; N; $CHMe(CH_2)_2CO_2Et$ | $H_2$, $PtO_2$, HCl, EtOH; 200°, 2 hr | 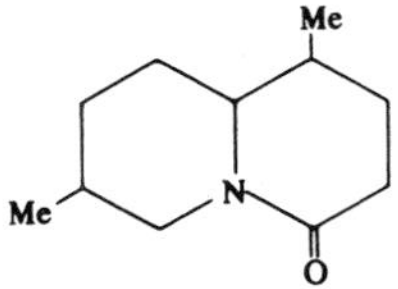 Me; N; Me; O | | 283 |

TABLE XI-19. Synthesis of Quinolizidines and Quinolizines (Continued)

| Starting materials | Conditions | Product | Yield | Properties | Ref. |
|---|---|---|---|---|---|
| 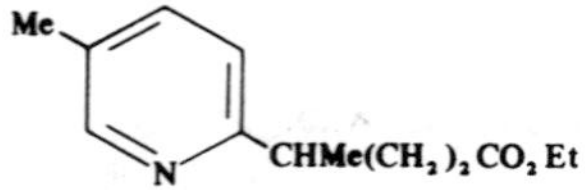 Me, $CHMe(CH_2)_2CO_2Et$ | $H_2$ $PtO_2$, HOAc; 200°, 1 hr | 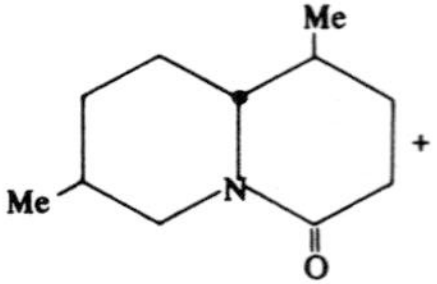 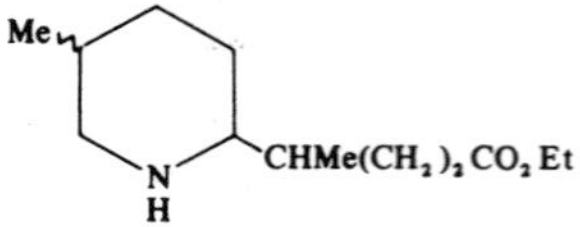 Me, Me, N, O + Me, N H, $CHMe(CH_2)_2CO_2Et$ | | | 206 |
| 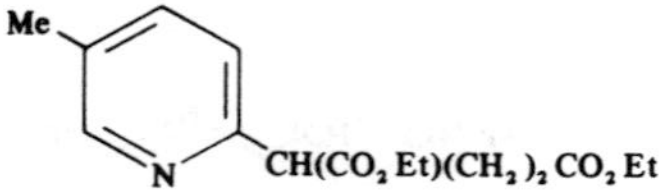 Me, N, $CH(CO_2Et)(CH_2)_2CO_2Et$ | $H_2$, $PtO_2$, HOAc; 200°, 1 hr | 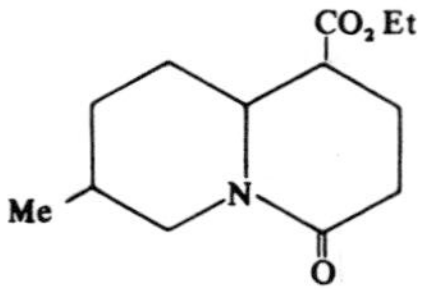 $CO_2Et$, Me, N, O | | | 206 |
| 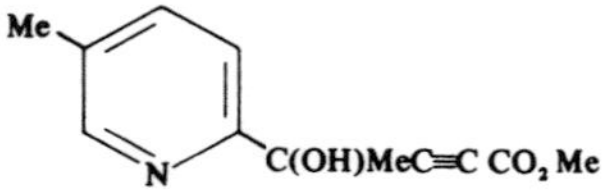 Me, N, $C(OH)MeC{\equiv}C\ CO_2Me$ | $H_2$, $PtO_2$, HCl; $SOCl_2$; $H_2$, $PtO_2$; 200°, 2 hr | 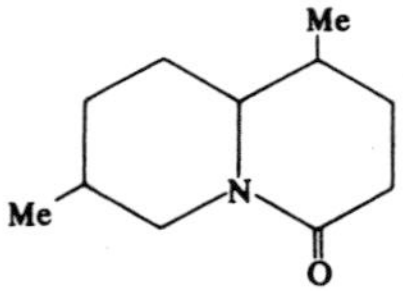 Me, Me, N, O | | | 79, 97 |
| 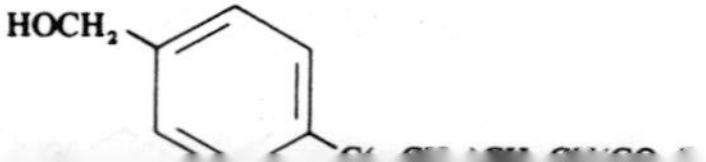 $HOCH_2$ | $H_2$, $PtO_2$, HOAc, 24 hr; HCl, reflux; EtOH, HCl | 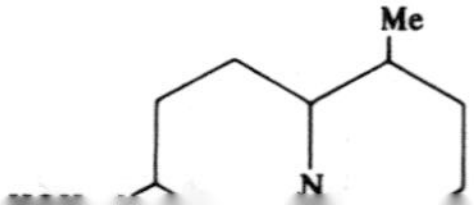 Me, N | 77% | | 14 |

| | | | | |
|---|---|---|---|---|
| 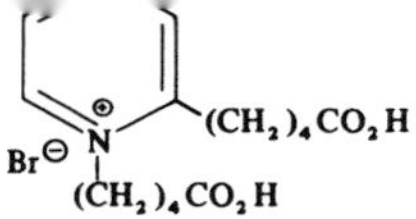 $Br^{\ominus}$ $\overset{\oplus}{N}$ $(CH_2)_4CO_2H$ $(CH_2)_4CO_2H$ | $H_2$, $PtO_2$ | 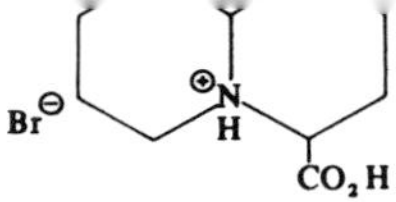 $Br^{\ominus}$ $\overset{\oplus}{N}$ H $CO_2H$ | 21 g → 20 g | 284 |
| 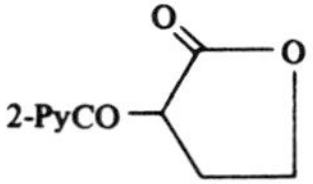 2-PyCO | 48% HBr, 110°, 1.5 hr | 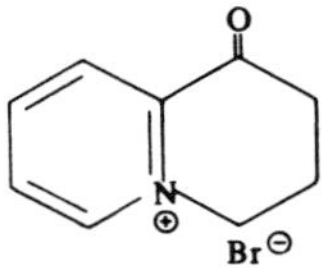 $Br^{\ominus}$ | 82.5% | 140 |
| 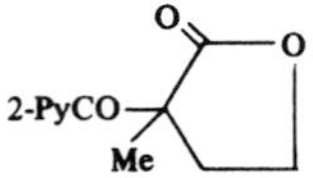 2-PyCO Me | HBr, heat; $Ac_2O$, heat | 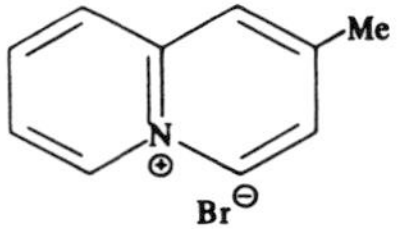 Me $Br^{\ominus}$ | | 140 |
| 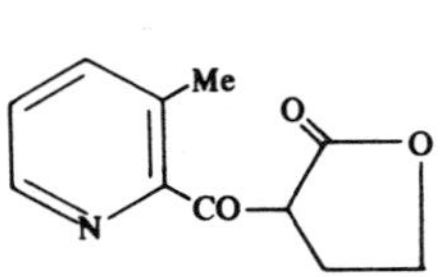 Me CO | 48% HBr, 110°, reflux 1.5 hr; heat 2.5 hr | 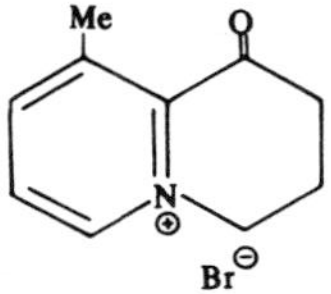 Me $Br^{\ominus}$ | 8 g → 6.8 g | 140 |
| 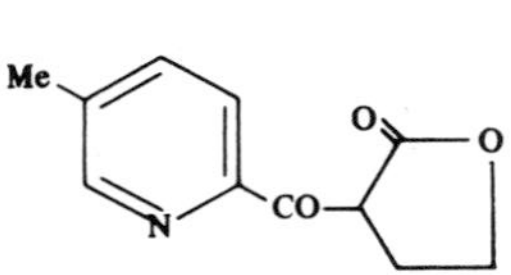 Me CO | same as above | 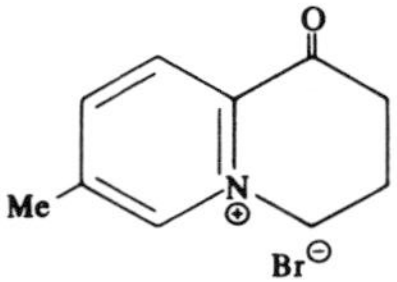 Me $Br^{\ominus}$ | 19 g → 21 g | 140 |

TABLE XI-19. Synthesis of Quinolizidines and Quinolizines (Continued)

| Starting materials | Conditions | Product | Yield | Properties | Ref. |
|---|---|---|---|---|---|
| 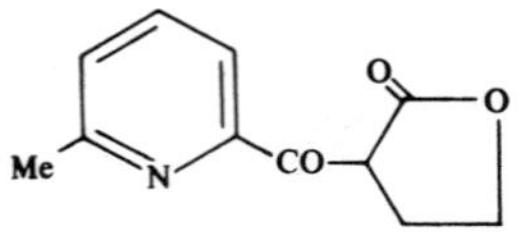 Me, N, CO, O, O | same as above | 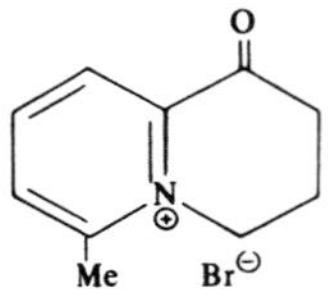 O, N, ⊕, Me, $Br^{\ominus}$ | 49 g → 49 g | | 140 |
| 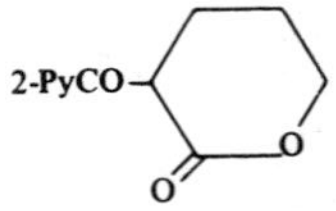 2-PyCO, O, O | 6*N*, $H_2SO_4$, $-CO_2$; $PBr_3$, steam-bath; $Ac_2O$, heat | 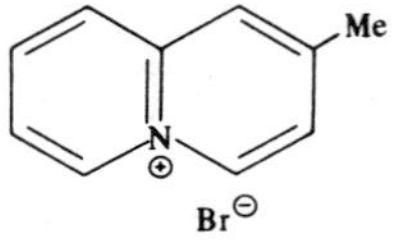 Me, N, ⊕, $Br^{\ominus}$ | | | 140 |

| Starting materials | Conditions | Product | Yield | Properties | Ref. |
|---|---|---|---|---|---|
| | | *Indolizines* | | | |
| 2-$PyCH_2CHPhCN$ | $Ac_2O$, reflux 2 hr | 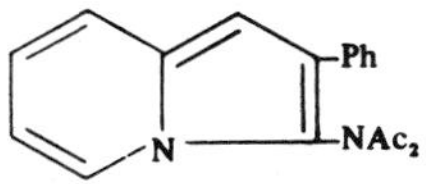 Ph; N; $NAc_2$ | 18% | ir | 143 |
| 2-$PyCH_2CH(COMe)CO_2Me$ | $Ac_2O$, reflux 1–2 hr | 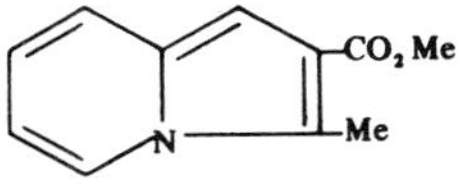 $CO_2Me$; N; Me | 9% | m.p. 87–88° | 143 |
| 2-$PyCH_2CH(COMe)CO_2Et$ | same as above | 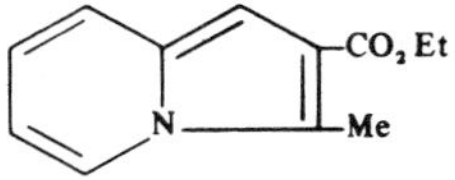 $CO_2Et$; N; Me | 19% | m.p. 49–50° | 143 |
| 2-$PyCH_2CH(CO_2Et)_2$ | same as above | 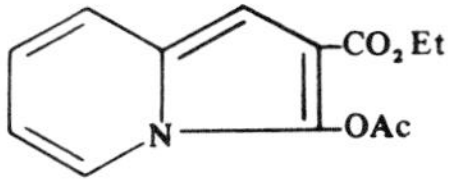 $CO_2Et$; N; OAc | 36% | m.p. 73–74° | 143 |
| 2-$PyCH_2Br$ + $NCCH_2CO_2Et$ | R.T. 1 hr | 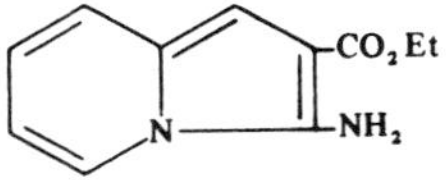 $CO_2Et$; N; $NH_2$ | | m.p. 71–72°; b.p. 170–200°/2 mm; ir | 143 |
| 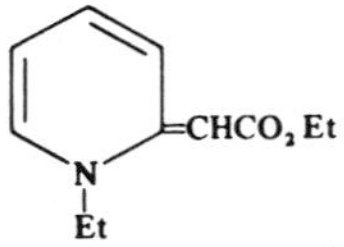 $=CHCO_2Et$; N; Et | $Ac_2O$, reflux 4 hr | 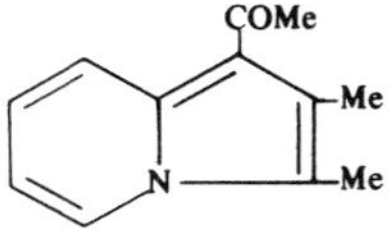 COMe; Me; N; Me | 25% | | 145 |

TABLE XI-20. Synthesis of Condensed Heterocycles Other Than Quinolizidines (Continued)

| Starting materials | Conditions | Product | Yield | Properties | Ref. |
|---|---|---|---|---|---|
| 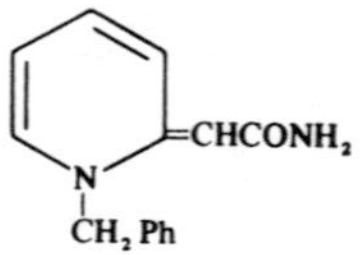 =CHCONH$_2$; N; $CH_2Ph$ | $Ac_2O$, reflux 1 hr | 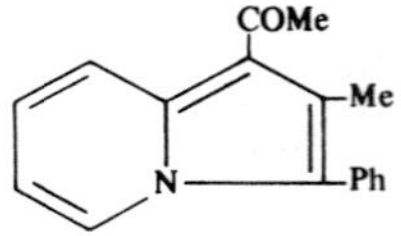 COMe; Me; Ph; N | 96% | | 145 |
| 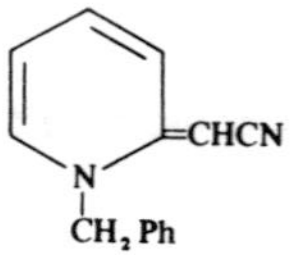 =CHCN; N; $CH_2Ph$ | $Ac_2O$, reflux 12 hr | 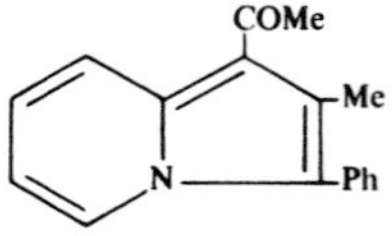 COMe; Me; Ph; N | 11% | | 145 |
| 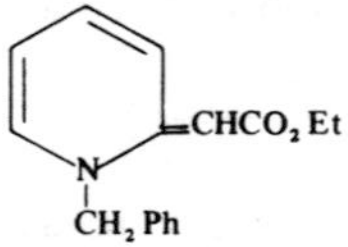 =CHCO$_2$Et; N; $CH_2Ph$ | $Ac_2O$, reflux 2 hr | 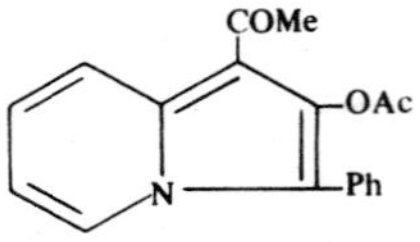 COMe; OAc; Ph; N | 83% | | 145 |
| 2-PyCH$_2$CO$_2$Et + BrCH$_2$CHO | heat, steam-bath, 2 hr | 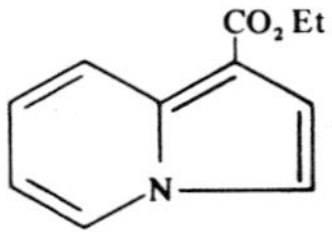 CO$_2$Et; N | 18% | b.p. 124°/3 mm | 142 |
| 2-PyCH$_2$CO$_2$Et + ClCH$_2$COMe | EtOH, reflux 14 hr | 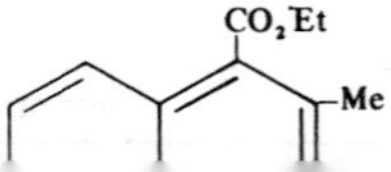 CO$_2$Et; Me | 36% | m.p. 43–44° | 142 |

| Reactants | Conditions | Product | Yield | Properties | Ref. |
| --- | --- | --- | --- | --- | --- |
| 2-$PyCH_2CO_2Et$ + $BrCH_2COMe$ | acetone, reflux 17 hr | 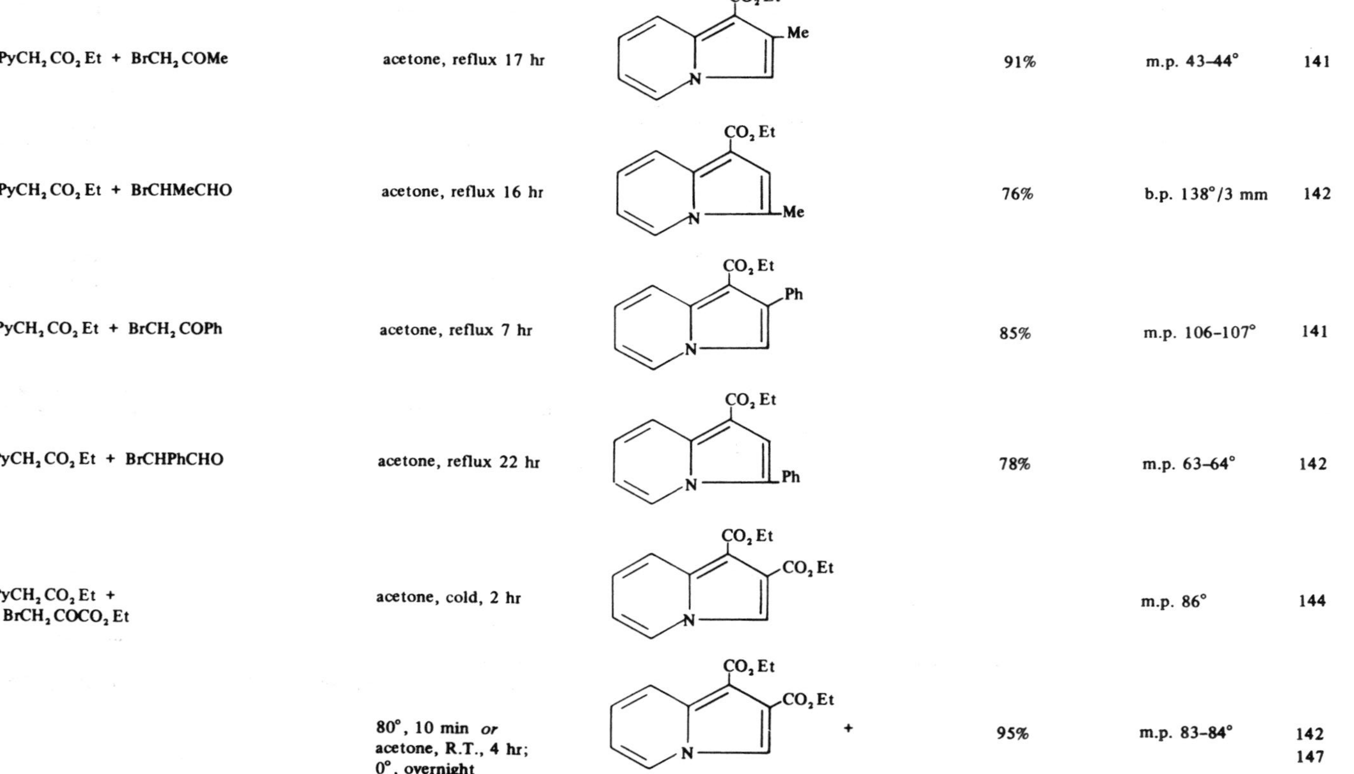 | 91% | m.p. 43–44° | 141 |
| 2-$PyCH_2CO_2Et$ + BrCHMeCHO | acetone, reflux 16 hr | | 76% | b.p. 138°/3 mm | 142 |
| 2-$PyCH_2CO_2Et$ + $BrCH_2COPh$ | acetone, reflux 7 hr | | 85% | m.p. 106–107° | 141 |
| 2-$PyCH_2CO_2Et$ + BrCHPhCHO | acetone, reflux 22 hr | | 78% | m.p. 63–64° | 142 |
| 2-$PyCH_2CO_2Et$ + $BrCH_2COCO_2Et$ | acetone, cold, 2 hr | | | m.p. 86° | 144 |
| | 80°, 10 min *or* acetone, R.T., 4 hr; 0°, overnight | + | 95% | m.p. 83–84° | 142<br>147 |

TABLE XI-20. Synthesis of Condensed Heterocycles Other Than Quinolizidines (Continued)

| Starting materials | Conditions | Product | Yield | Properties | Ref. |
|---|---|---|---|---|---|
| | | Indolizine: 1-$CO_2Et$, 2-$CO_2H$ | 2% | m.p. 228–229° | |
| 2-$PyCH_2CO_2Et$ + $BrCHMeCOMe$ | acetone, reflux 17 hr | Indolizine: 1-$CO_2Et$, 2-Me, 3-Me | 45% | m.p. 62–63° | 141 |
| 2-$PyCH_2CO_2Et$ + $BrCHMeCOPh$ | same as above | Indolizine: 1-$CO_2Et$, 2-Ph, 3-Me | 38% | m.p. 101–102° | 141 |
| 2-$PyCH_2CO_2Et$ + $BrCHPhCOPh$ | same as above | Indolizine: 1-$CO_2Et$, 2-Ph, 3-Ph | 24% | m.p. 151–152° | 141 |
| 2-$PyCH_2CO_2Et$ + | acetone, heat on | $CO_2Et$, OH | 48% | m.p. 51–52°; | 142 |

| | | | | | |
|---|---|---|---|---|---|
| 2-Py$CH_2CO_2$Et + BrCHPh$CO_2$Et | acetone, steam-bath 17 hr | 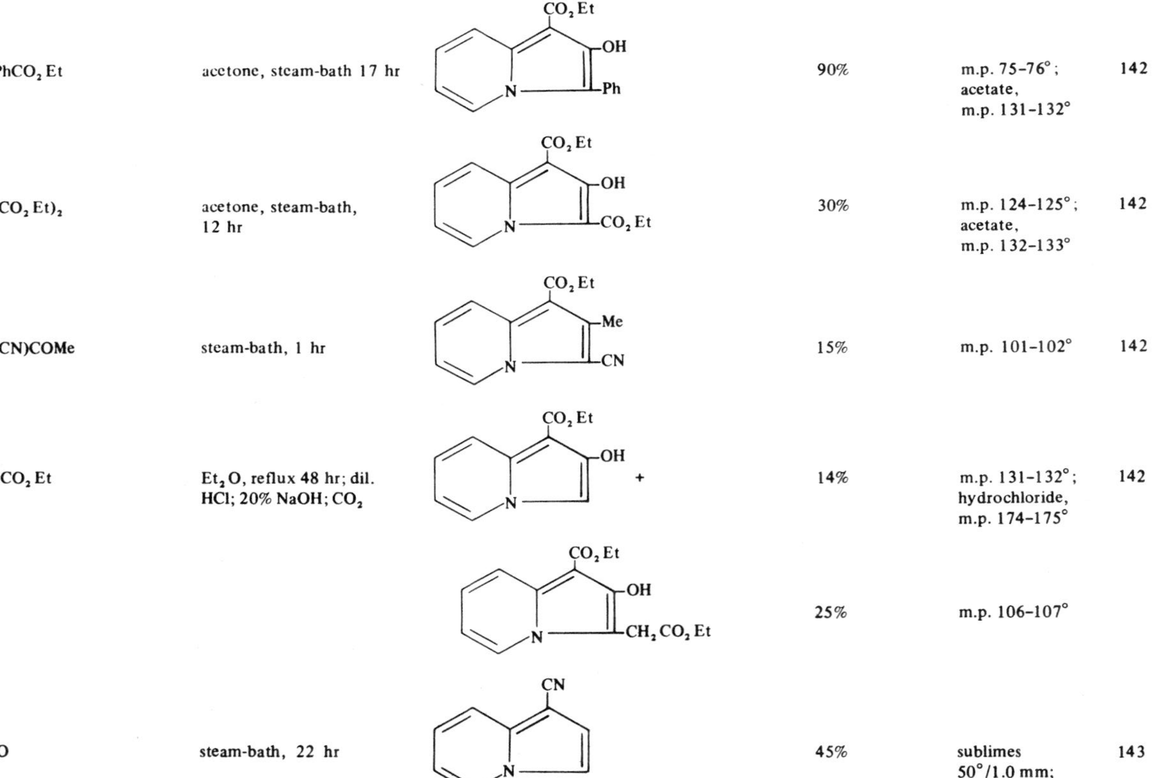 | 90% | m.p. 75–76°; acetate, m.p. 131–132° | 142 |
| 2-Py$CH_2CO_2$Et + BrCH($CO_2$Et)$_2$ | acetone, steam-bath, 12 hr | | 30% | m.p. 124–125°; acetate, m.p. 132–133° | 142 |
| 2-Py$CH_2CO_2$Et + BrCH(CN)COMe | steam-bath, 1 hr | | 15% | m.p. 101–102° | 142 |
| 2-Py$CH_2CO_2$Et + Br$CH_2CO_2$Et | $Et_2O$, reflux 48 hr; dil. HCl; 20% NaOH; $CO_2$ | + | 14% | m.p. 131–132°; hydrochloride, m.p. 174–175° | 142 |
| | | | 25% | m.p. 106–107° | |
| 2-Py$CH_2$CN + Br$CH_2$CHO | steam-bath, 22 hr | | 45% | sublimes 50°/1.0 mm; m.p. 52–53°; ir | 143 |

TABLE XI-20. Synthesis of Condensed Heterocycles Other Than Quinolizidines (Continued)

| Starting materials | Conditions | Product | Yield | Properties | Ref. |
|---|---|---|---|---|---|
| | | *Indolizines* | | | |
| | KOH, EtOH, reflux 6 hr | 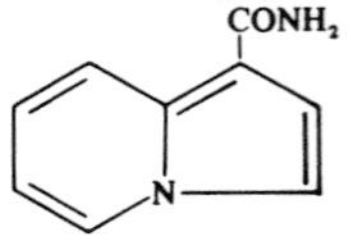 $CONH_2$ | 82% | sublimes 130°/0.5 mm; m.p. 148–149°; ir | 145 |
| 2-$PyCH_2CN$ + $BrCH_2COMe$ | reflux 17 hr | 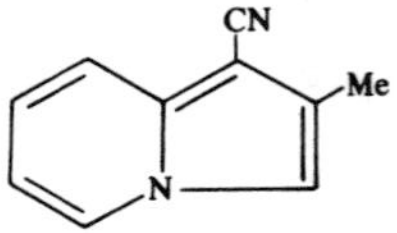 CN, Me | 69% | sublimes 90°/1.0 mm; m.p. 100–101°; ir | 143 |
| 2-$PyCH_2CN$ + $BrCH_2COPh$ | $Me_2CO$, reflux 14 hr | 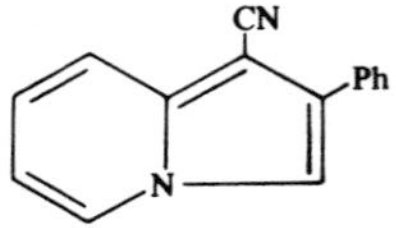 CN, Ph | 48% | m.p. 101–102° | 142 |
| 2-$PyCH_2CN$ + BrCHPhCHO | reflux 6 hr | 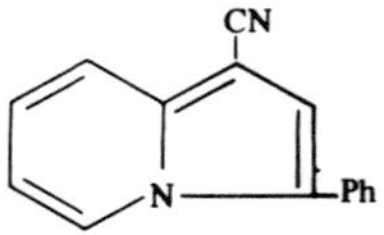 CN, Ph | 50% | sublimes 90°/1.0 mm; m.p. 95–96°; ir | 143 |
| 2-$PyCH_2CN$ + $BrCH_2COCO_2Et$ | reflux 1 hr | 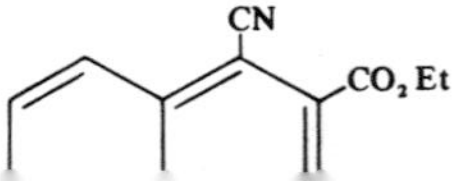 CN, $CO_2Et$ | 77% | m.p. 125–126°; | 143 |

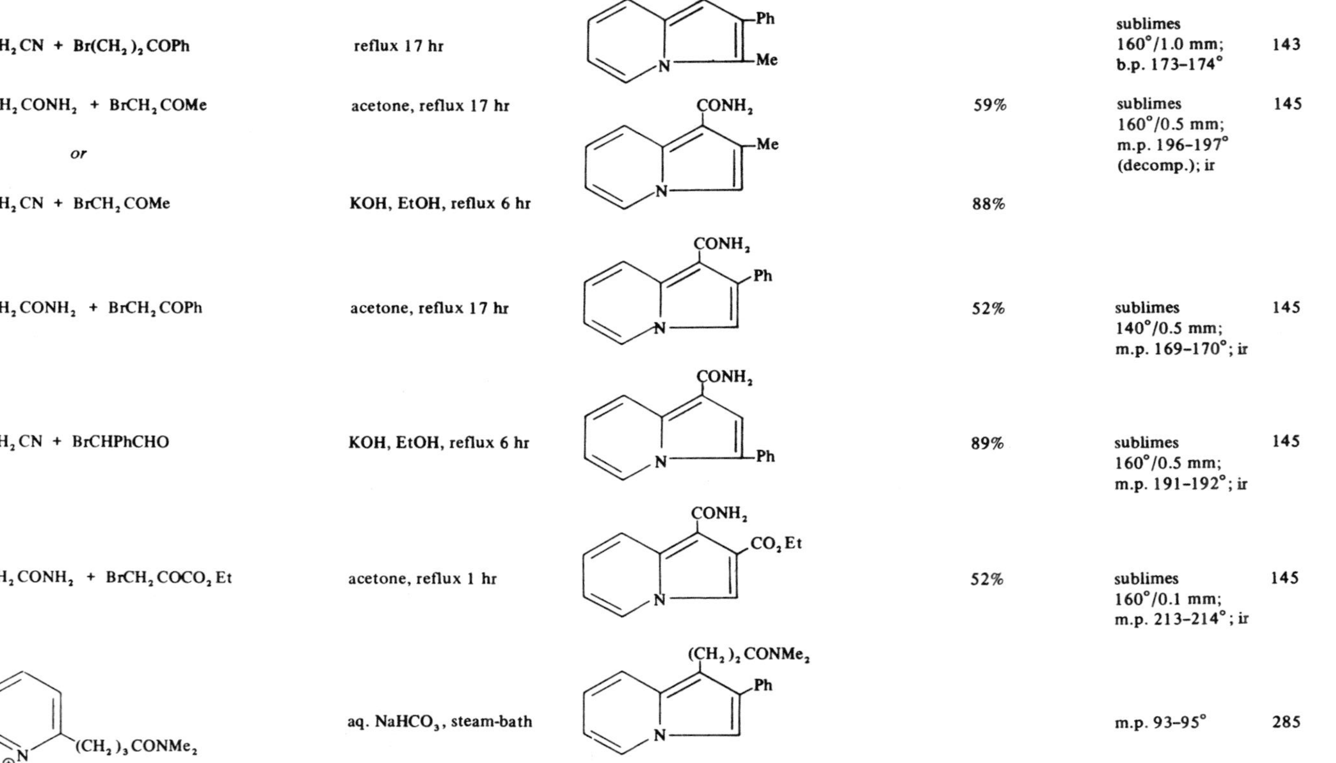

| Starting materials | Conditions | Product | Yield | Properties | Ref. |
|---|---|---|---|---|---|
| 2-$PyCH_2CN$ + $Br(CH_2)_2COPh$ | reflux 17 hr | | | sublimes 160°/1.0 mm; b.p. 173–174° | 143 |
| 2-$PyCH_2CONH_2$ + $BrCH_2COMe$ | acetone, reflux 17 hr | | 59% | sublimes 160°/0.5 mm; m.p. 196–197° (decomp.); ir | 145 |
| *or* | | | | | |
| 2-$PyCH_2CN$ + $BrCH_2COMe$ | KOH, EtOH, reflux 6 hr | | 88% | | |
| 2-$PyCH_2CONH_2$ + $BrCH_2COPh$ | acetone, reflux 17 hr | | 52% | sublimes 140°/0.5 mm; m.p. 169–170°; ir | 145 |
| 2-$PyCH_2CN$ + BrCHPhCHO | KOH, EtOH, reflux 6 hr | | 89% | sublimes 160°/0.5 mm; m.p. 191–192°; ir | 145 |
| 2-$PyCH_2CONH_2$ + $BrCH_2COCO_2Et$ | acetone, reflux 1 hr | | 52% | sublimes 160°/0.1 mm; m.p. 213–214°; ir | 145 |
| | aq. $NaHCO_3$, steam-bath | | | m.p. 93–95° | 285 |

TABLE XI-20. Synthesis of Condensed Heterocycles Other Than Quinolizidines (Continued)

| Starting materials | Conditions | Product | Yield | Properties | Ref. |
|---|---|---|---|---|---|
| 2-Py$CH_2NO_2$ + Br$CH_2COCO_2$Et | reflux 1 hr | 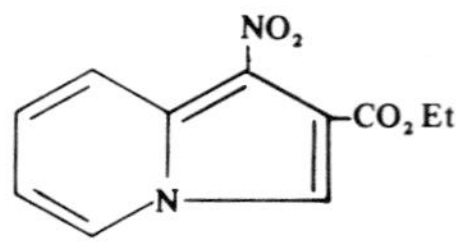 $NO_2$; $CO_2$Et; N | 78% | m.p. 117–118° | 143 |
| 2-Py$(CH_2)_2CO_2$Et + $(CO_2Et)_2$ | Na, EtOH, $Et_2O$ | 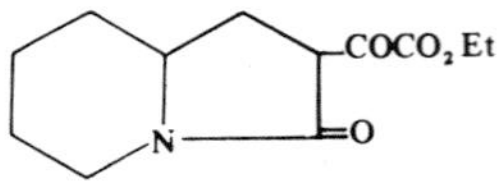 $COCO_2$Et; N; O | | | 148 |
| 2-PyCPh(OH)$CH_2CO_2$Et | $H_2$, 10% Pd-C, HOAc; 25°, 3–4 atm | 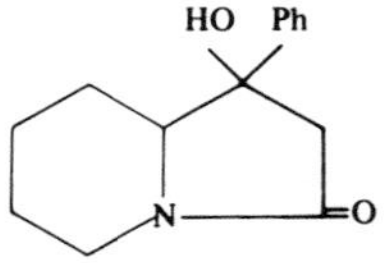 HO Ph; N; O | | | 132 |
| | | *Indenes and naphthalenes* | | | |
| 2-PyCPh(OH)$CHCO_2$Et | conc. $H_2SO_4$, 100° briefly | 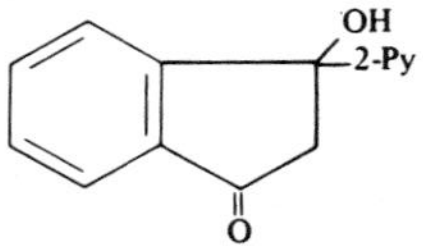 OH; 2-Py; O | | | 172 |
| 2-PyCPh(OH)$CMeCO_2$Et | same as above | 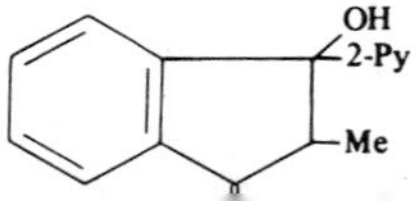 OH; 2-Py; Me | | | 172 |

| | | | | |
|---|---|---|---|---|
| 2-PyCPh(OH)CHPhCO$_2$Et | conc. $H_2SO_4$, R.T. | | | 172 |
| 2-PyCPh(OH)CHPhCO$_2$H | 98% $H_2SO_4$, 1 hr | 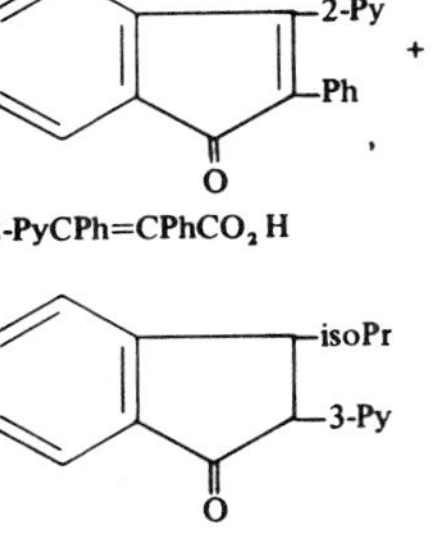 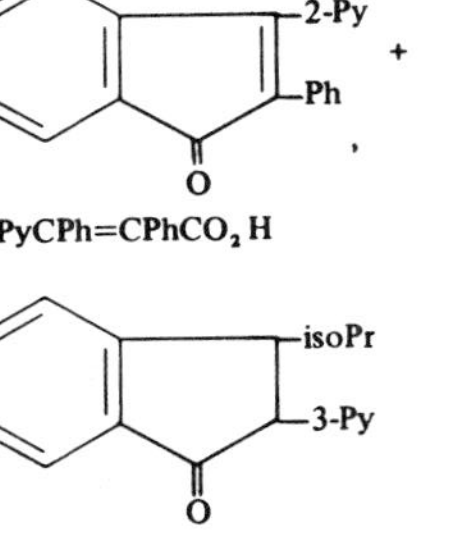 +, 2-PyCPh=CPhCO$_2$H | | 172 |
| 3-PyCH(CN)CHPh-isoPr | PPA, 105–110° | 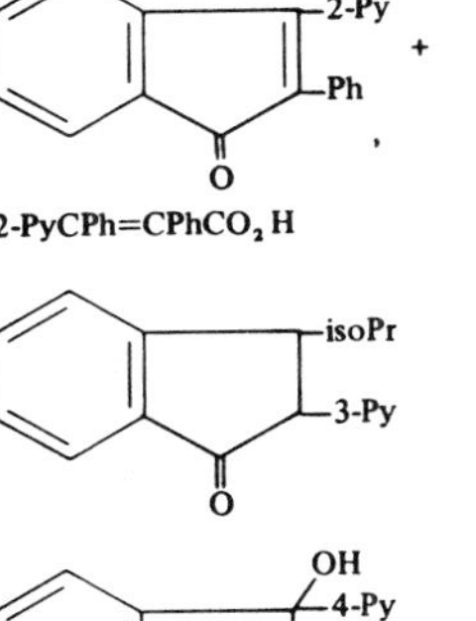 | 73% | 286 |
| 4-PyCPh(OH)CHCO$_2$Et | conc. $H_2SO_4$, 100° briefly | 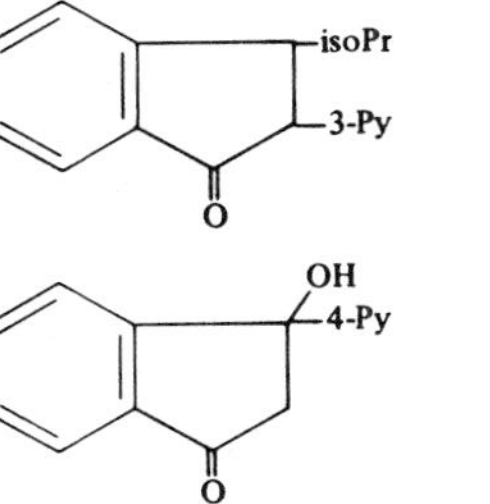 | | 172 |
| 4-PyCPh(OH)CHMeCO$_2$Et | conc. $H_2SO_4$, 100° briefly | | | 172 |
| 4-PyCPh(OH)CHPhCO$_2$Et | conc. $H_2SO_4$, R.T. | | | 172 |

TABLE XI-20. Synthesis of Condensed Heterocycles Other Than Quinolizidines (Continued)

| Starting materials | Conditions | Product | Yield | Properties | Ref. |
|---|---|---|---|---|---|
| | | *Indenes and naphthalenes* | | | |
| 4-PyCPh(OH)CHPhCO$_2$H | 98% $H_2SO_4$, 1 hr | 4-Py, Ph, O + | | | 172 |
| | | 4-PyCPh=CPhCO$_2$H | | m.p. 248–255° | |
| 3-PyCH(CO$_2$H)(CH$_2$)$_2$Ph | PPA, 105–110° | 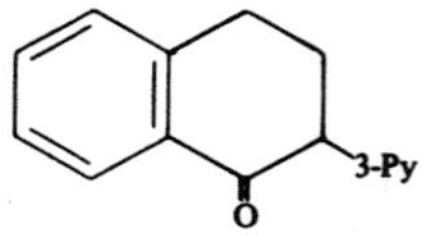 3-Py, O | 87% | | 286–288 |
| 3-PyCH(CO$_2$H)CH$_2$CH$_2$(*m*-MeOC$_6$H$_4$) | PPA, 105–110° | 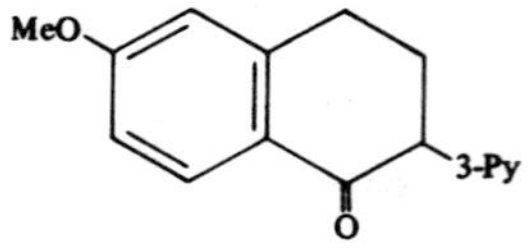 MeO, 3-Py, O | 55% | | 286, 289 |
| 3-PyCH(CONH$_2$)CH$_2$CH$_2$(*m*-MeOC$_6$H$_4$) | PPA, 125–130° | 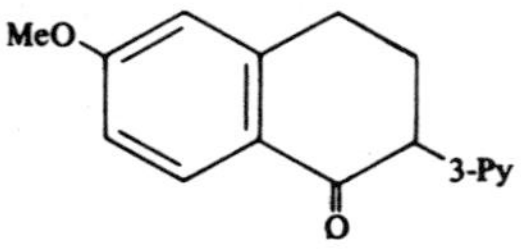 MeO, 3-Py, O | | | 290 |
| 3-PyCH(CO$_2$H)CH$_2$CH$_2$(*p*-tolyl) | PPA, 90–110° | 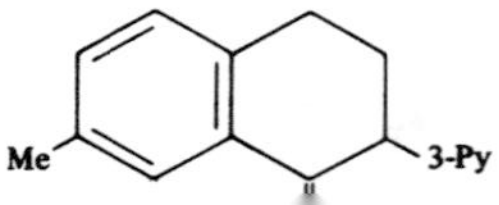 Me, 3-Py | | | 288 |

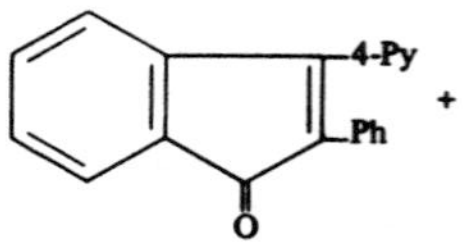

| | | | | | |
|---|---|---|---|---|---|
| 3-PyCH($CO_2$H)$CH_2CH_2$(p-Cl$C_6H_4$) | PPA, 105–110° | [structure: Cl, O, 3-Py] | 86% | | 286–289 |
| 3-PyCH($CO_2$H)$CH_2CH_2$(m-Cl$C_6H_4$) | same as above | [structure: Cl, O, 3-Py] + | 51% | | 288, 289 |
| | | [structure: Cl, O, 3-Py] | 65% | | 286 |
| 3-PyCH($CO_2$H)$(CH_2)_3$Ph | same as above | [structure: O, 3-Py] | | | 288 |
| | | *Coumarins and chromones* | | | |
| 2-Py$CH_2CO_2$H + [structure: Br, CHO, OH] | $Ac_2O$, $Et_3N$ | [structure: Br, 2-Py, O, O] | 24% | m.p. 187.5–188.5° | 48<br>49 |
| 2-Py$CH_2CO_2$Me + [structure: $O_2N$, CHO, OH] | piperidine | [structure: $O_2N$, 2-Py, O, O] | 64% | m.p. 215–216.5° | 48<br>49 |

TABLE XI-20. Synthesis of Condensed Heterocycles Other Than Quinolizidines (Continued)

| Starting materials | Conditions | Product | Yield | Properties | Ref. |
|---|---|---|---|---|---|
| | | *Coumarins and chromones* | | | |
| $3\text{-PyCH}_2CO_2H$ + 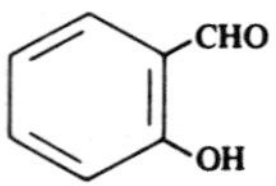CHO<br>OH | $Ac_2O, Et_3N$ | 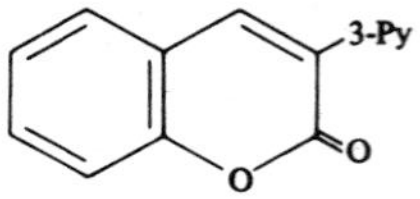3-Py<br>O<br>O | 74% | m.p. 167.5–169° | 48, 49 |
| $3\text{-PyCH}_2CO_2H$ + 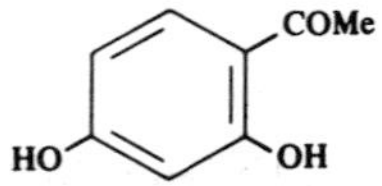COMe<br>HO<br>OH | $Ac_2O, Et_3N$ | 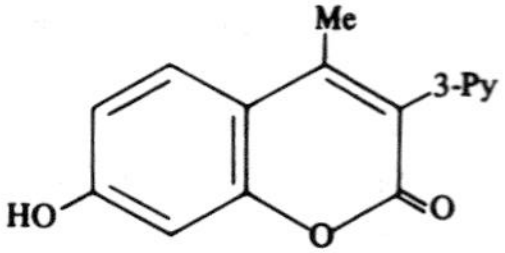Me<br>3-Py<br>HO<br>O<br>O | 28% | m.p. 257–258.5°; hydrochloride, m.p. 313–315° (decomp.) | 48, 49 |
| $3\text{-PyCH}_2CO_2H$ + 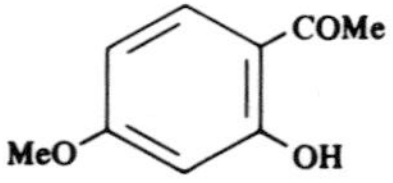COMe<br>MeO<br>OH | $Ac_2O, Et_3N$ | 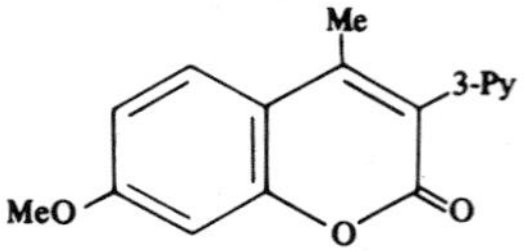Me<br>3-Py<br>MeO<br>O<br>O | 51% | m.p. 138–141° | 48, 49 |
| $3\text{-PyCH}_2CO_2H$ + 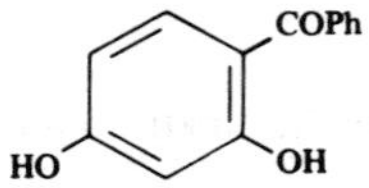COPh<br>HO<br>OH | $Ac_2O, Et_3N$ | 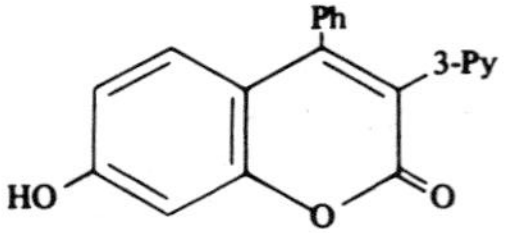Ph<br>3-Py<br>HO<br>O<br>O | 40% | m.p. 288–290.5° | 48, 49 |
| $3\text{-PyCH}_2CO_2H$ + 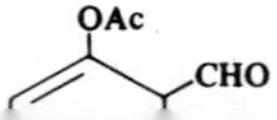OAc<br>CHO | | 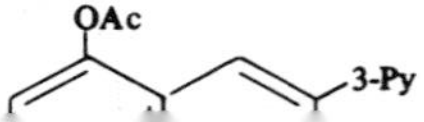OAc<br>3-Py | | | |

| Reactants | | Conditions | Product | Yield | Properties | Ref. |
|---|---|---|---|---|---|---|
| 3-$PyCH_2CO_2H$ + | 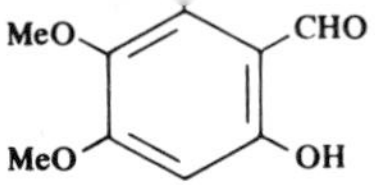 MeO, MeO, CHO, OH | $Ac_2O$, $Et_3N$ | 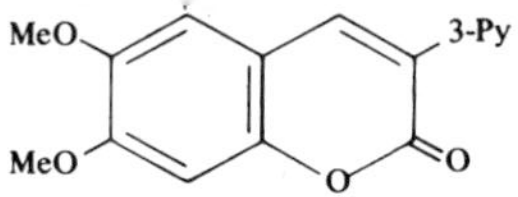 MeO, MeO, 3-Py, O, O | 10% | m.p. 150.5–153.5° | 48, 49 |
| 3-$PyCH_2CO_2Me$ + | 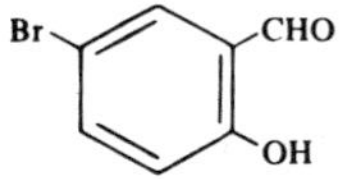 Br, CHO, OH | piperidine | 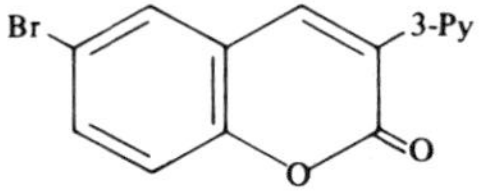 Br, 3-Py, O, O | 89% | m.p. 230–231° | 48, 49 |
| 3-$PyCH_2CO_2Et$ + | 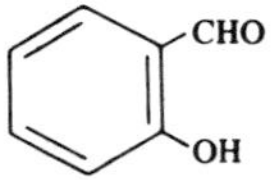 CHO, OH | piperidine, R.T., 24 hr | 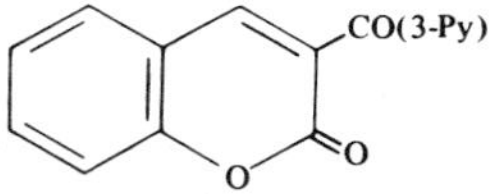 CO(3-Py), O, O | | m.p. 158° | 149 |
| 3-$PyCOCH_2CO_2Et$ + | 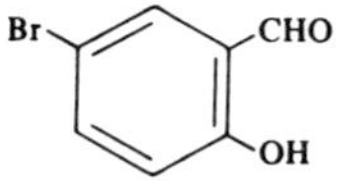 Br, CHO, OH | piperidine, R.T., 24 hr | 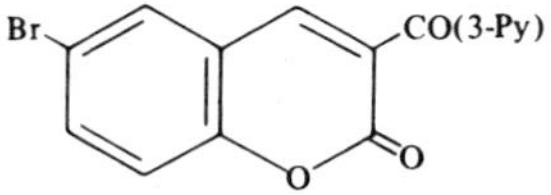 Br, CO(3-Py), O, O | | m.p. 188° | 149 |
| 3-$PyCOCH_2CO_2Et$ + | 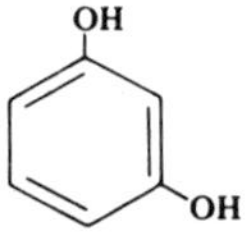 OH, OH | 160°, 12 mm | 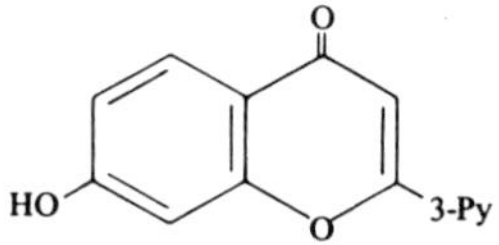 O, HO, O, 3-Py | | m.p. 248° | 149 |
| 3-$PyCOCH_2CO_2Et$ + | 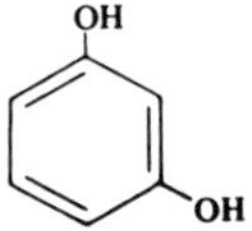 OH, OH | $ZnCl_2$, HOAc, steam-bath, 2 hr | 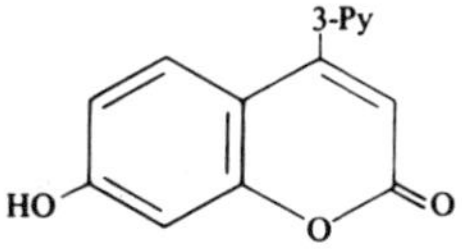 3-Py, HO, O, O | | m.p. 265° | 149 |

TABLE XI-20. Synthesis of Condensed Heterocycles Other Than Quinolizidines (Continued)

| Starting materials | Conditions | Product | Yield | Properties | Ref. |
|---|---|---|---|---|---|
| | | *Coumarins and chromones* | | | |
| 3-PyCOCH$_2$CO$_2$Et 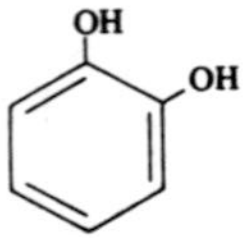 + OH, OH | 160°, 12 mm | 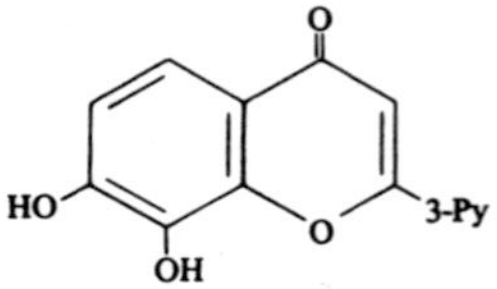 O; HO; OH; O; 3-Py | | m.p. 313° | 149 |
| 3-PyCOCH$_2$CO$_2$Et 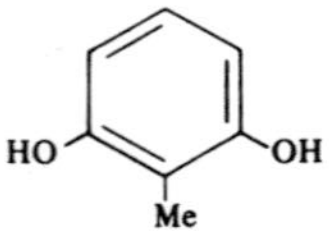 + HO, OH, Me | 160°, 12 mm | 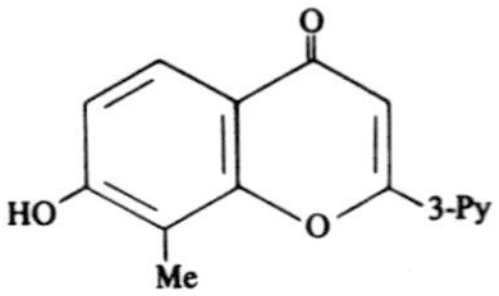 O; HO; Me; O; 3-Py | | m.p. 232° | 149 |
| 3-PyCOCH$_2$CO$_2$Et 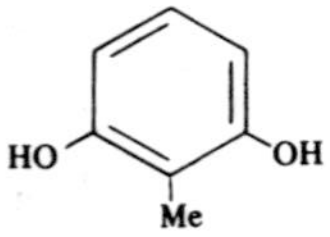 + HO, OH, Me | $ZnCl_2$, HOAc, steam-bath, 2 hr | 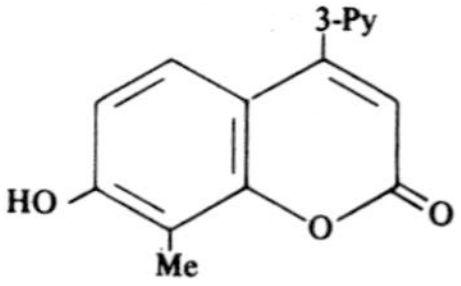 3-Py; HO; Me; O; O | | m.p. 250° | 149 |
| 3-PyCOCH$_2$CO$_2$Et 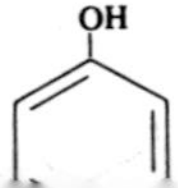 + OH | 160°, 12 mm | 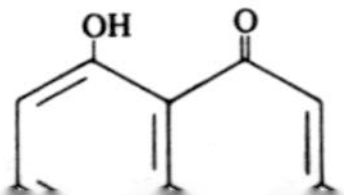 OH; O | | m.p. 323° | 149 |

| Reactants | Conditions | Product | Yield | Properties | Ref. |
|---|---|---|---|---|---|
| 3-PyCOCH$_2$CO$_2$Et + 1,2,3-trihydroxybenzene (HO, OH, OH) | $ZnCl_2$, HOAc, steam-bath, 2 hr | 7,8-dihydroxy-4-(3-Py)coumarin (HO, OH) | | m.p. 300° | 149 |
| 3-PyCOCH$_2$CO$_2$Et + 1,3,5-trihydroxybenzene (OH, HO, OH) | $ZnCl_2$, HOAc steam-bath, 2 hr | 5,7-dihydroxy-4-(3-Py)coumarin (OH, 3-Py, HO) | | m.p. 360° | 149 |
| 3-PyCOCH$_2$CO$_2$Et + 3,5-dibromo-2-hydroxybenzaldehyde (Br, Br, CHO, OH) | piperidine, R.T., 24 hr | 6,8-dibromo-3-(CO–3-Py)coumarin (Br, Br, CO–3-Py) | | m.p. 210° | 149 |
| 3-PyCOCH($p$-MeOC$_6$H$_4$)CO$_2$Et + resorcinol (HO, OH) | PPA, 80–85°, 0.5 hr, R.T. | 7-hydroxy-4-(3-Py)-3-($p$-MeOC$_6$H$_4$)coumarin (3-Py, $p$-MeOC$_6$H$_4$, HO) | 81% | m.p. 201–202° | 16 |
| 4,6-dimethyl-3-(COCH$_2$COCO$_2$Et)-2-pyridone (Me, Me, N–H, O) | HCl, 10% EtOH, reflux 1 hr | pyrano[2,3-b]pyridine (Me, O, Me, N, O, CO$_2$Et) | 0.5 g → 0.3 g | m.p. 170–171°; ir; uv; nmr | 231 |

TABLE XI-20. Synthesis of Condensed Heterocycles Other Than Quinolizidines (Continued)

| Starting materials | Conditions | Product | Yield | Properties | Ref. |
|---|---|---|---|---|---|
| $4\text{-PyCH}_2CO_2Me$ 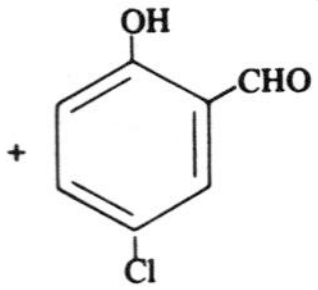 + (OH, CHO, Cl-substituted benzene) | piperidine | 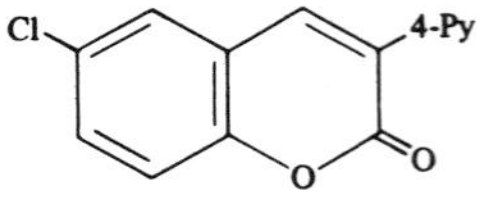 Cl, 4-Py, O, O | 91% | m.p. 271–272.5° | 48<br>49 |
| $4\text{-PyCH}_2CO_2Me$ 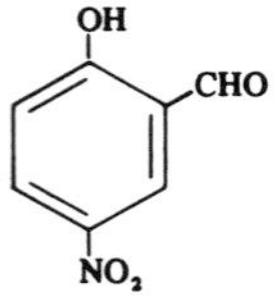 + (OH, CHO, $NO_2$-substituted benzene) | piperidine | 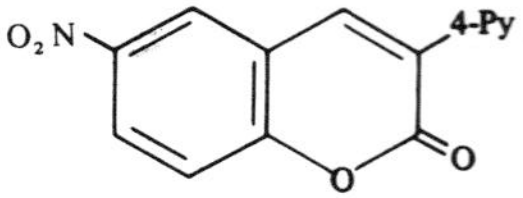 $O_2N$, 4-Py, O, O | 75% | m.p. 270–271° | 48, 49 |
| $4\text{-PyCH}_2CO_2Me$ 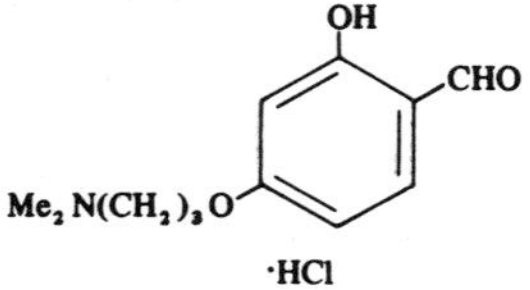 + (OH, CHO, $Me_2N(CH_2)_3O$-substituted benzene) ·HCl | piperidine | 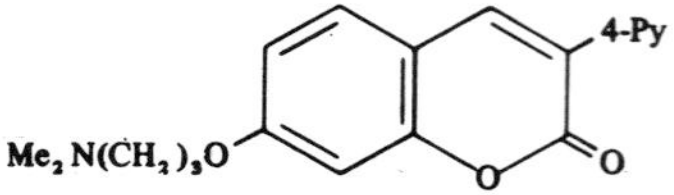 $Me_2N(CH_2)_3O$, 4-Py, O, O | 35% | hydrochloride, m.p. 136-138° | 48, 49 |
| $4\text{-PyCH}_2COCO_2Me$ 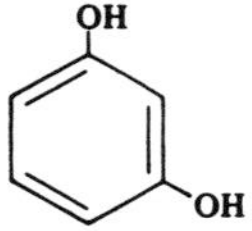 + (OH, OH-substituted benzene) | PPA, 8 hr, 2 days | 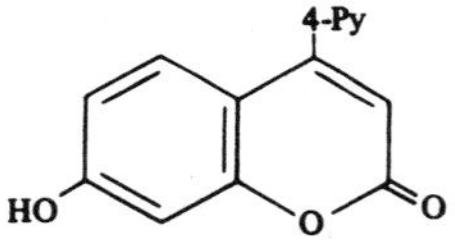 HO, 4-Py, O, O | 57% | m.p. 307–312° (decomp.) | 48, 49 |
| $4\text{-PyCH}_2COCO_2Me$ 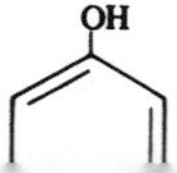 + (OH-substituted benzene) | same as above | 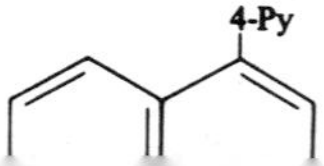 4-Py | 31% | m.p. 212–214.5° | 48, 49 |

| | | | | | |
|---|---|---|---|---|---|
| $4\text{-}PyCH_2COCO_2Me$ + 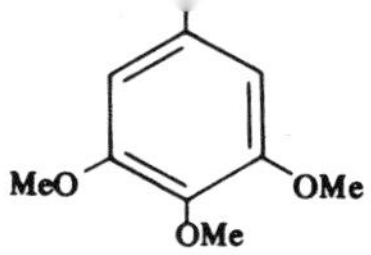 | $H_2SO_4$, 8 hr, 2 days | 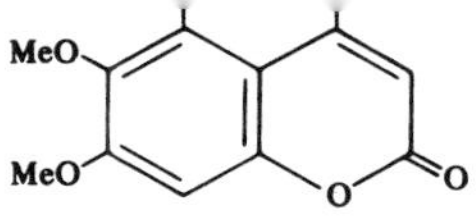 | 80% | m.p. 183–186° | 48, 49 |
| | | *Quinolines* | | | |
| $2\text{-}PyCH_2CH_2CH(NHAc)CO_2Et$ | $H_2$, $PtO_2$, 5 atm, EtOH, HCl; 20% HCl, 110–120°, 3 hr | 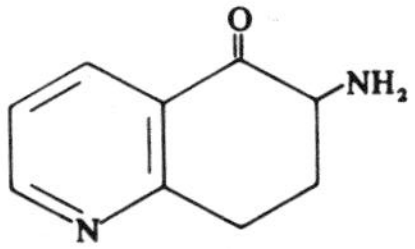 | | | 89 |
| $2\text{-}PyCH_2CONHCHMeCH(OMe)$– 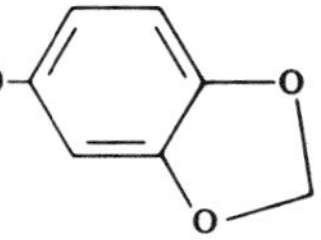 | $POCl_3$, reflux 40 min | 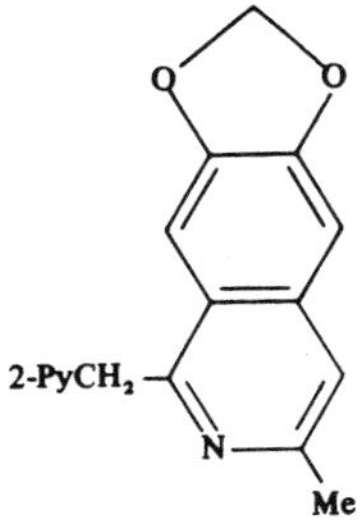 $2\text{-}PyCH_2$– | | | 291 |
| 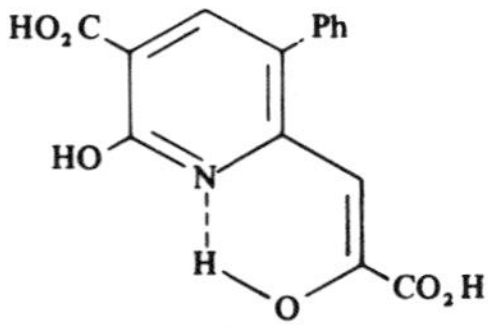 | conc $H_2SO_4$, 3–4 hr; R.T., 2 days | 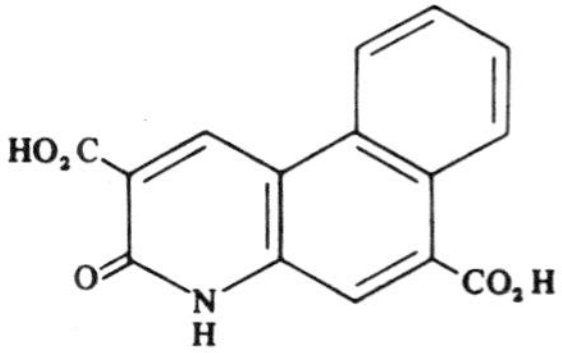 | 74% | green $FeCl_3$ test; m.p. > 360°; ir; uv; diethyl ester, m.p. 211° | 87, 173 |

TABLE XI-20. Synthesis of Condensed Heterocycles Other Than Quinolizidines (Continued)

| Starting materials | Conditions | Product | Yield | Properties | Ref. |
|---|---|---|---|---|---|
| 3-PyC(NHPh)=CHCONHPh | PPA, 120–130°, 20 min | 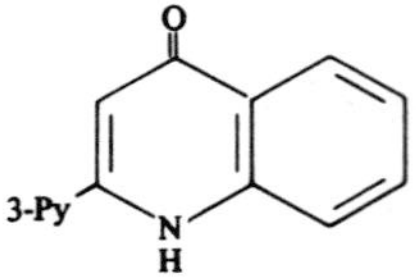 3-Py, O, N, H | 72% | | 292 |
| $3\text{-Py}(CH_2)_3CONH(CH_2)_2$–, 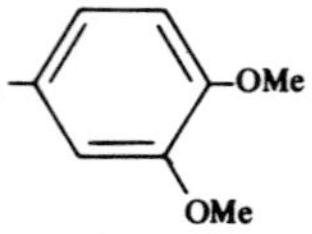 OMe, OMe | $POCl_3$, reflux 2 hr, $N_2$ | $3\text{-Py}(CH_2)_2$–, 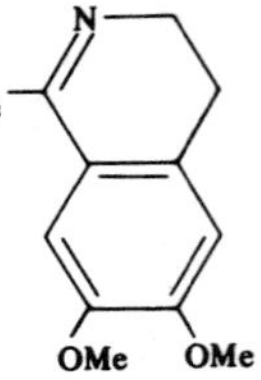 N, OMe, OMe | | | 293 |
| 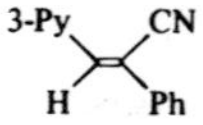 3-Py, CN, H, Ph | cyclohexane, light, (40 w Hanovia medium pressure mercury vapor lamp) 3.5 hr | 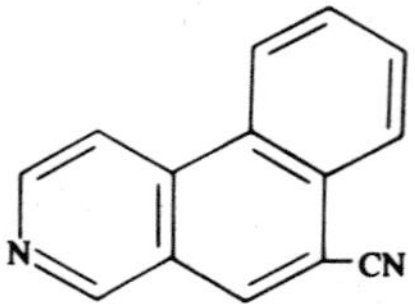 N, CN | 41% | b.p. 180°/10 mm; m.p. 196–196.5° | 294 |
| $4\text{-Py}(CH_2)_3CONH(CH_2)_2$–, 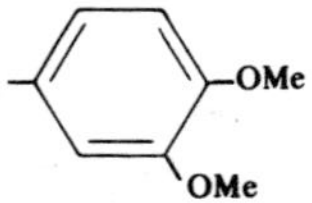 OMe, OMe | $POCl_3$, $N_2$, reflux 2 hr | $4\text{-Py}(CH_2)_2$–, 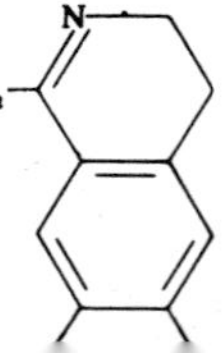 N | | | 293 |

| Starting material | Reagent/conditions | Product | Yield | Properties | Ref. |
|---|---|---|---|---|---|
| 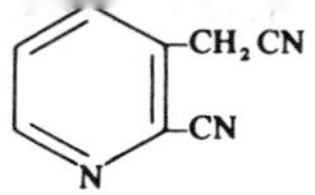 [2-cyano-3-(cyanomethyl)pyridine: $CH_2CN$, CN] | anhyd. HBr, $Et_2O$ | 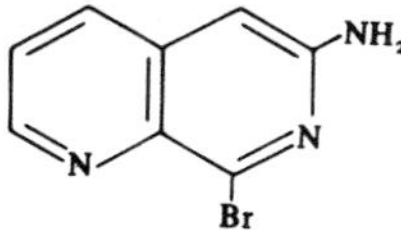 [6-amino-8-bromo-1,7-naphthyridine: $NH_2$, Br] | 72% | | 92 |
| 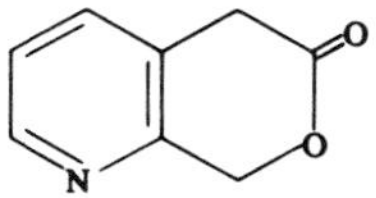 [pyrano[3,4-b]pyridin-one] + $2Me_2N(CH_2)_2NH_2$ | 200° | 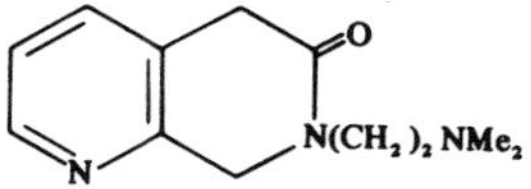 [lactam: $N(CH_2)_2NMe_2$] | 68% | m.p. 140–150° | 174 |
| 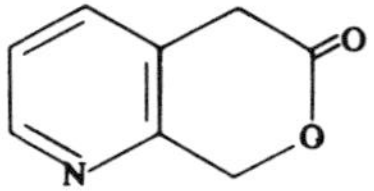 [pyrano[3,4-b]pyridin-one] + $2Et_2N(CH_2)_2NH_2$ | 200° | 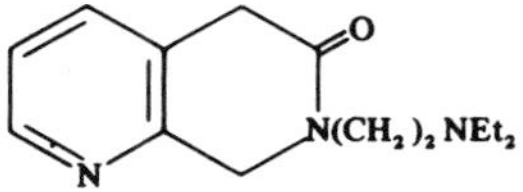 [lactam: $N(CH_2)_2NEt_2$] | 62% | m.p. 145–155° | 174 |
| 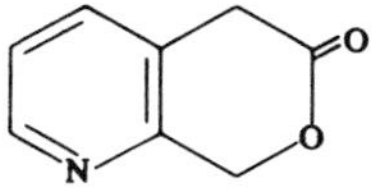 [pyrano[3,4-b]pyridin-one] + $2(isoPr)_2N(CH_2)_2NH_2$ | 200° | 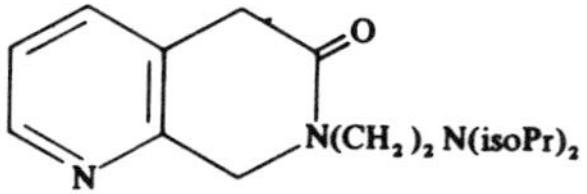 [lactam: $N(CH_2)_2N(isoPr)_2$] | 50% | m.p. 165–170° | 174 |
| 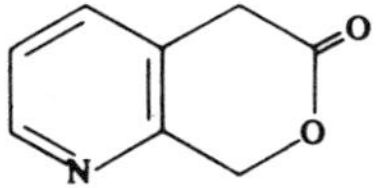 [pyrano[3,4-b]pyridin-one] + $2PhNH_2$ | 200° | 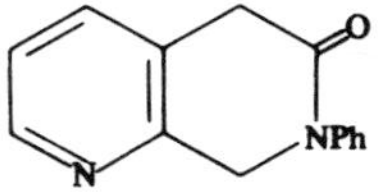 [lactam: NPh] | 53% | m.p. 131–132° | 174 |
| 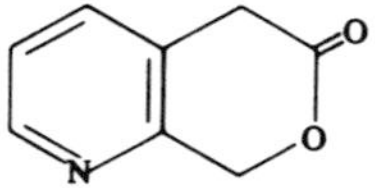 [pyrano[3,4-b]pyridin-one] + $2PhCH_2NH_2$ | 200° | 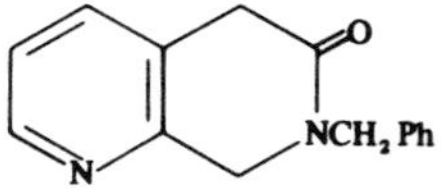 [lactam: $NCH_2Ph$] | 61% | m.p. 117–118° | 174 |
| 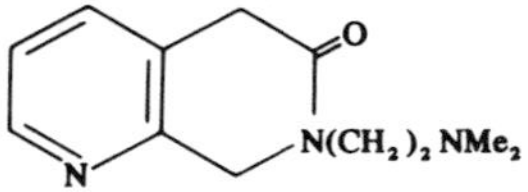 [lactam: $N(CH_2)_2NMe_2$] | LAH, $Et_2O$ | 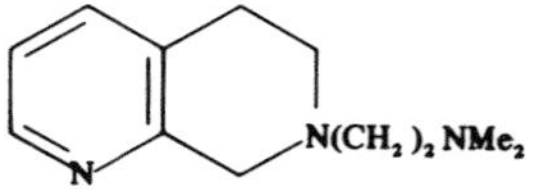 [tetrahydro-1,7-naphthyridine: $N(CH_2)_2NMe_2$] | 60% | | 174 |

TABLE XI-20. Synthesis of Condensed Heterocycles Other Than Quinolizidines (Continued)

| Starting materials | Conditions | Product | Yield | Properties | Ref. |
|---|---|---|---|---|---|
| | | *Naphthyridines* | | | |
| 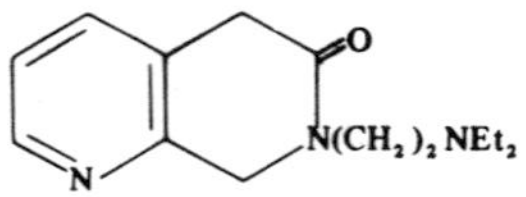 O; N; $N(CH_2)_2NEt_2$ | LAH, $Et_2O$ | 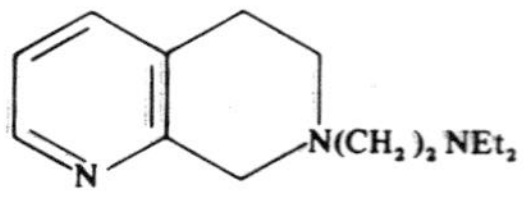 N; $N(CH_2)_2NEt_2$ | 63% | | 174 |
| 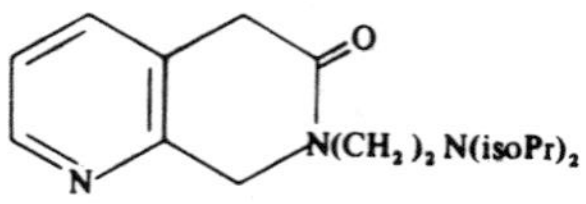 O; N; $N(CH_2)_2N(isoPr)_2$ | LAH, $Et_2O$ | 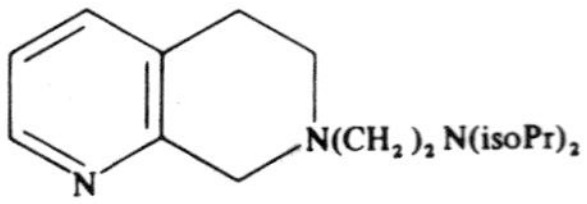 N; $N(CH_2)_2N(isoPr)_2$ | 45% | | 174 |
| 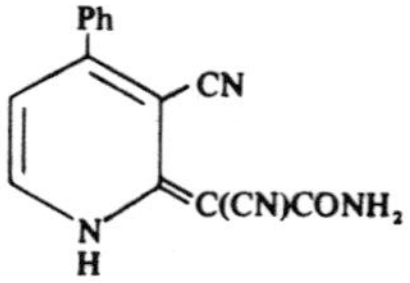 Ph; CN; $C(CN)CONH_2$; N H | 50% $H_2SO_4$ | 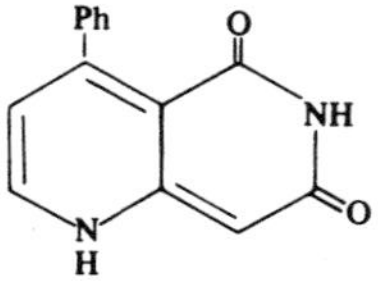 Ph; O; NH; O; N H | | m.p. 227° | 153 |
| 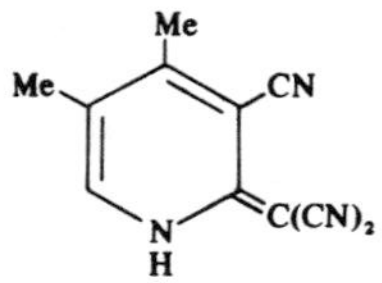 Me; Me; CN; $C(CN)_2$; N H | 50% $H_2SO_4$ | 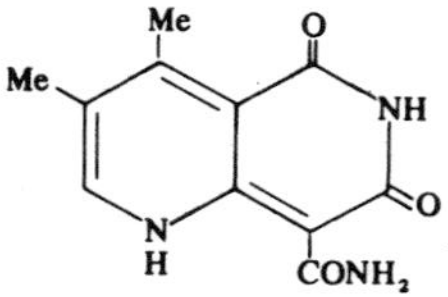 Me; Me; O; NH; O; $CONH_2$; N H | | m.p. >300° (decomp.) | 153 |
| 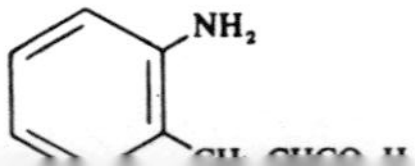 $NH_2$; … | NaOMe, $NH_2OH$ | 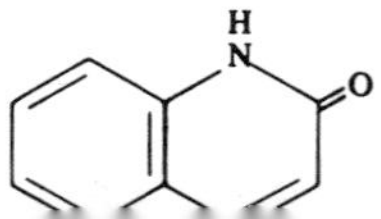 H N; O | | m.p. 260–261° | 151 |

| | | | | | |
|---|---|---|---|---|---|
| 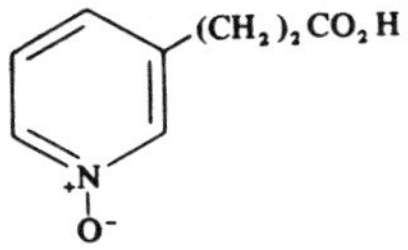 $(CH_2)_2CO_2H$, $\overset{+}{N}$–$O^-$ | fuming $HNO_3$, $H_2SO_4$; reduced Fe powder, HOAc | 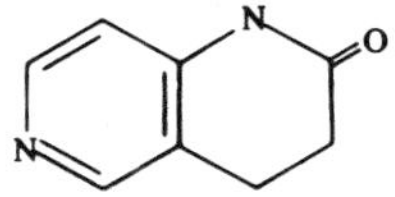 N, O | | m.p. 208° | 152 |
| 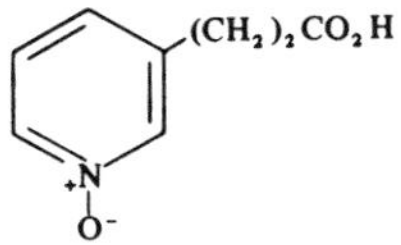 $(CH_2)_2CO_2H$, $\overset{+}{N}$–$O^-$ | Zn, HCl, $NH_4OH$ | 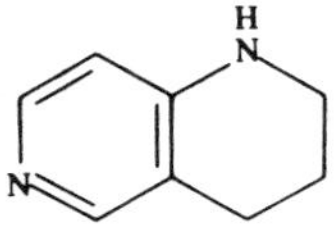 H, N | | | 152 |
| 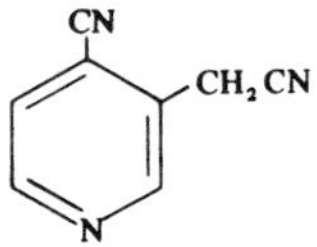 CN, $CH_2CN$, N | anhyd. HBr, $Et_2O$; aq. $NaHCO_3$ | 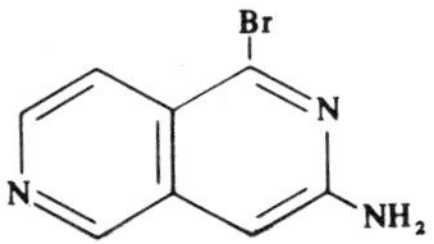 Br, N, $NH_2$ | 80.3% | | 150 |
| 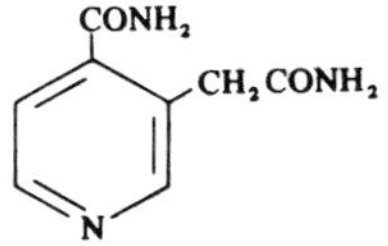 $CONH_2$, $CH_2CONH_2$, N | heat | 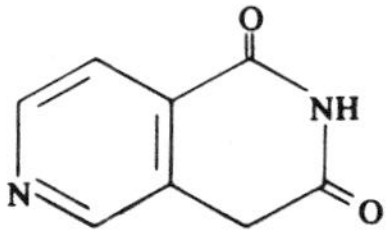 O, NH, O, N | | m.p. 229–230° | 128 |
| 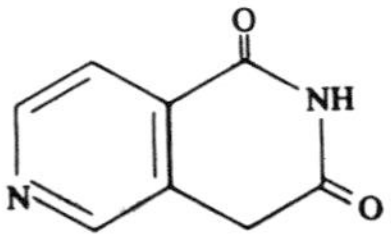 O, NH, O, N | Zn dust, 170°, sealed tube | 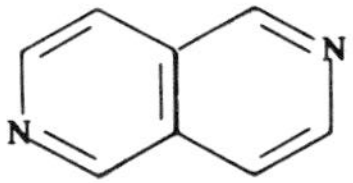 N, N | | | 128 |
| 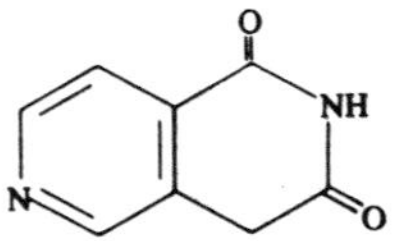 O, NH, O, N | $POCl_3$, 120°, sealed tube, several days | 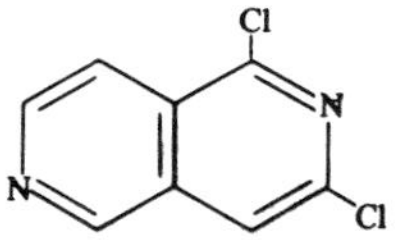 Cl, N, Cl, N | | | 128 |

TABLE XI-20. Synthesis of Condensed Heterocycles Other Than Quinolizidines (Continued)

| Starting materials | Conditions | Product | Yield | Properties | Ref. |
|---|---|---|---|---|---|
| | | *Miscellaneous* | | | |
| 3-PyCH=C(CN)– 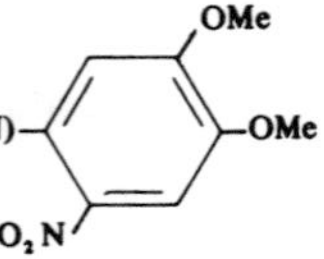 (2-$O_2N$-4,5-(OMe)$_2$-phenyl) | Fe, HOAc, reflux 2–5 hr | 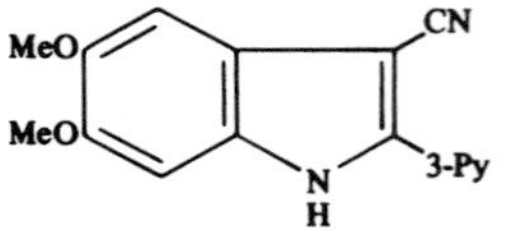 5,6-(MeO)$_2$-3-CN-2-(3-Py)indole (N–H) | 39% | m.p. 238–239°; ir; uv; nmr | 10 |
| 4-PyCH=C(CN)– 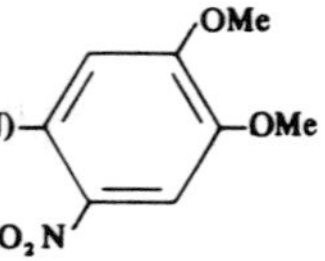 (2-$O_2N$-4,5-(OMe)$_2$-phenyl) | same as above | 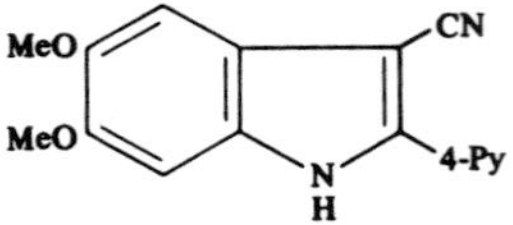 5,6-(MeO)$_2$-3-CN-2-(4-Py)indole (N–H) | 72.5% | m.p. 321°; ir; uv; nmr | 10 |
| 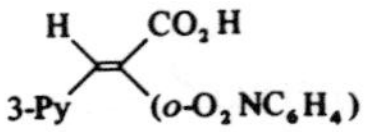 H(3-Py)C=C($CO_2H$)(o-$O_2NC_6H_4$) | $FeSO_4$, $NH_4OH$, $H_2O$, 80–90°, 20 min | 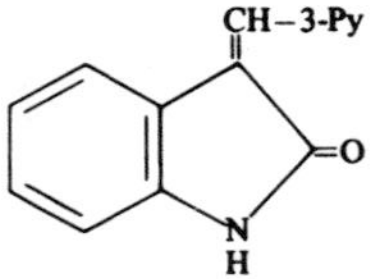 3-(CH–3-Py)oxindole (N–H, C=O) | | m.p. 187–188° | 99 |
| 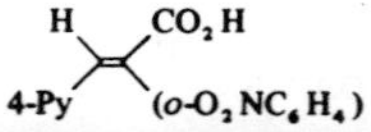 H(4-Py)C=C($CO_2H$)(o-$O_2NC_6H_4$) | same as above | 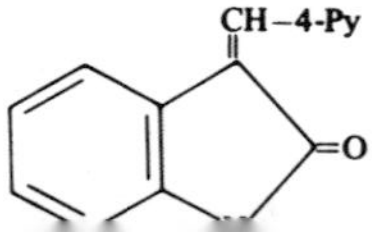 3-(CH–4-Py)oxindole (C=O) | | m.p. 225° | 99 |

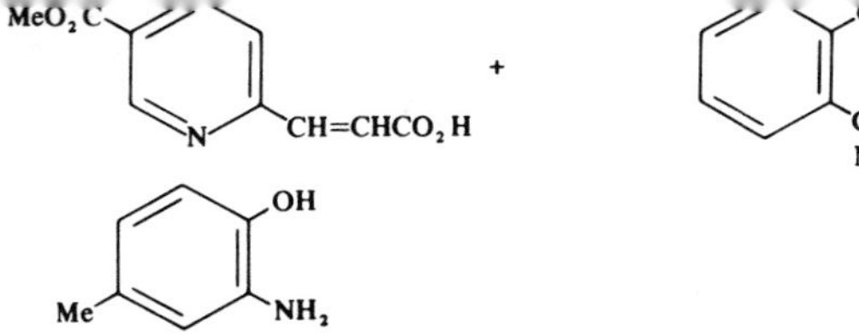

| Reactants | Conditions | Product | Properties | Ref. |
|---|---|---|---|---|
| $MeO_2C$ … $CH{=}CHCO_2H$ + OH, $NH_2$, Me | Cl, $H_3BO_3$, 120°, $N_2$, 8 hr | Me … O … N, CH=CH, O, N, Me | | 281 |
| $CH_2CO_2H$, N, $NH_2$ | 225° | N, N H, O (67%) (a) + $CH_2CO_2H$, N H, O (4.6%) (b) | (a) m.p. 175°<br>(b) m.p. 240–241° (decomp.) | 268 |

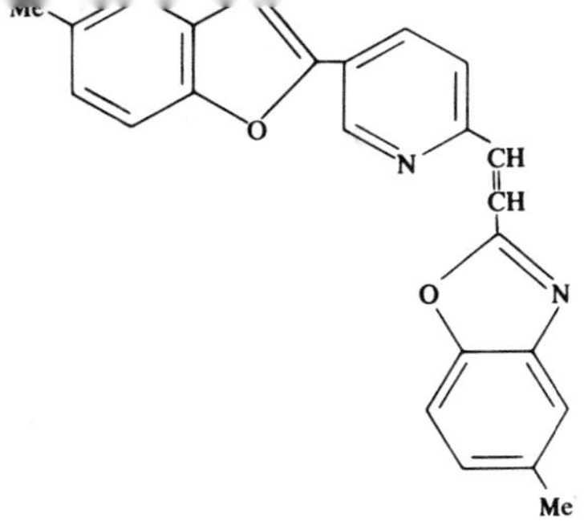

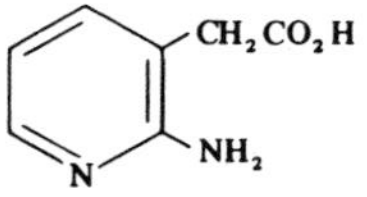

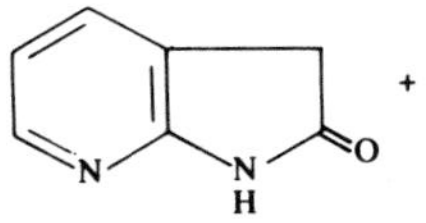

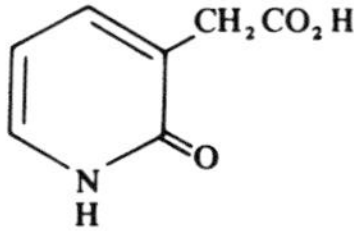

TABLE XI-20. Synthesis of Condensed Heterocycles Other Than Quinolizidines (Continued)

| Starting materials | Conditions | Product | Yield | Properties | Ref. |
|---|---|---|---|---|---|
| | | *Miscellaneous* | | | |
| 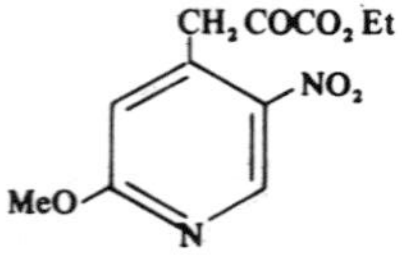 $CH_2COCO_2Et$, $NO_2$, MeO, N | $H_2$, $PtO_2$, EtOH | 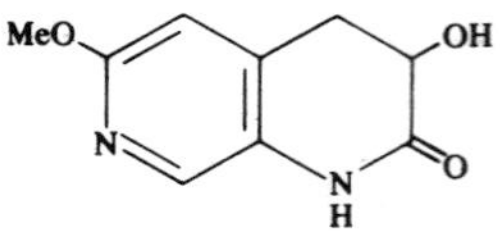 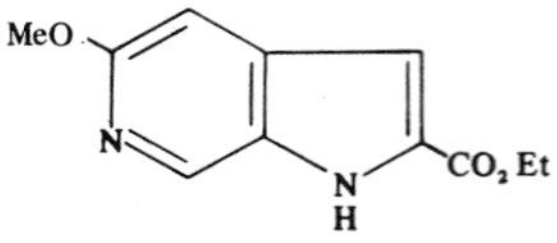 MeO, OH, N, N H, O + MeO, N, N H, $CO_2Et$ | | | 15 |
| 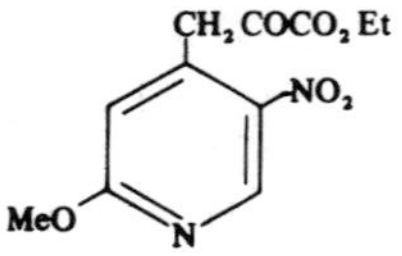 $CH_2COCO_2Et$, $NO_2$, MeO, N | $H_2$, Pd-C, EtOH | 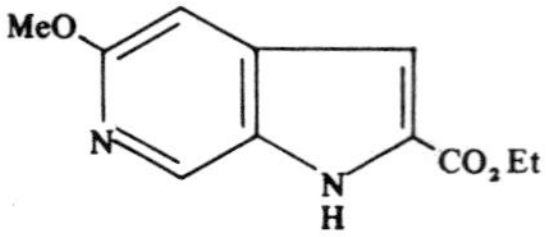 MeO, N, N H, $CO_2Et$ | 85% | sublimes 90%/0.1 mm; m.p. 103–106°; u.v; nmr | 15 |
| 2-PyCH($CH_2$Ph)$CO_2$Me | $N_2H_4 \cdot H_2O$, MeOH, heat, 6 hr; $NaNO_2$, 3% HCl | 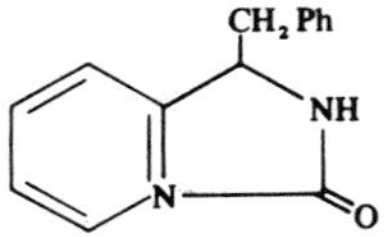 $CH_2$Ph, NH, N, O | | m.p. 161–163° | 305 |
| 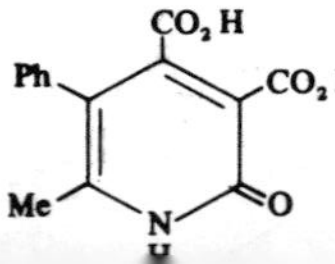 $CO_2$H, Ph, $CO_2$H, Me, N H, O + $(COCl_2)_2$ | $POCl_3$ | 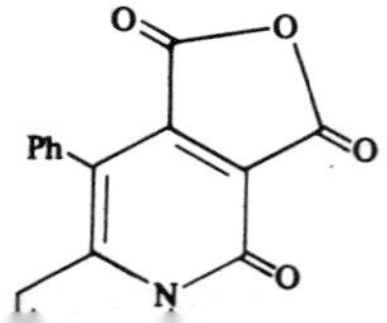 O, O, O, Ph, N, O | 13 g → 10 g | | 87 |

| | | | | | |
|---|---|---|---|---|---|
| 4-PyC(OH)$_2$CO$_2$Me 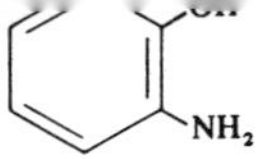 + (OH, NH$_2$) | MeOH, 35–50°, 5 hr; steam-bath 20 min | 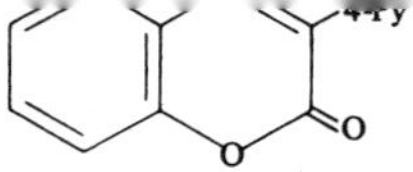 (4-Py, O, O) | 38% | sublimes 162°/0.002 mm; m.p. 168–169° | 31 |
| | | 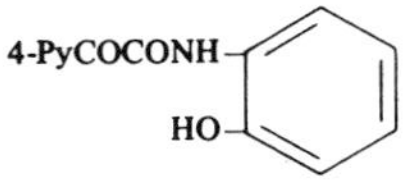 + 4-PyCOCONH– (HO–) | 11.4% | m.p. 221.5–222.5° (decomp.) | 32 |
| 3-PyCOC(CN)=N($p$-Me$_2$NC$_6$H$_4$) 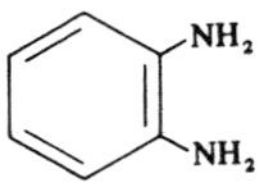 + (NH$_2$, NH$_2$) | | 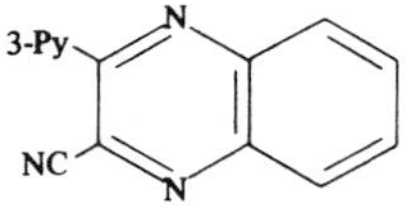 (3-Py, NC, N, N) | 86% | m.p. 193–194° | 295 |
| 4-PyCOC(CN)=N($p$-Me$_2$NC$_6$H$_4$) 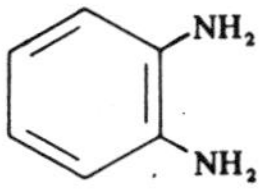 + (NH$_2$, NH$_2$) | | 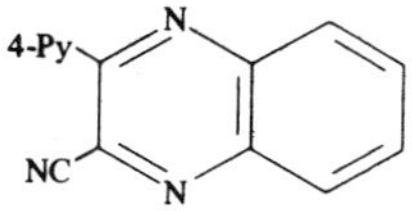 (4-Py, NC, N, N) | 90% | m.p. 228–229° | 295 |
| 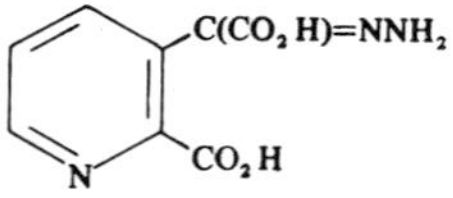 (N, C(CO$_2$H)=NNH$_2$, CO$_2$H) | | 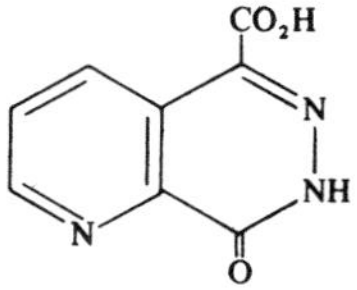 (CO$_2$H, N, NH, O, N) | | m.p. 300–301° (decomp.); methyl ester, m.p. 227–229° (decomp.) | 260 |

TABLE XI-20. Synthesis of Condensed Heterocycles Other Than Quinolizidines (Continued)

| Starting materials | Conditions | Product | Yield | Properties | Ref. |
|---|---|---|---|---|---|
| | | *Miscellaneous* | | | |
| 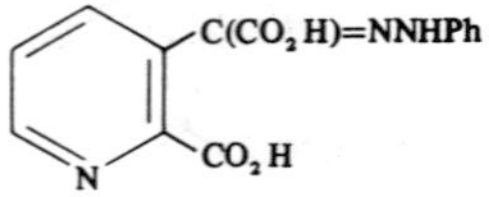 $C(CO_2H){=}NNHPh$, $CO_2H$, N | DMF | 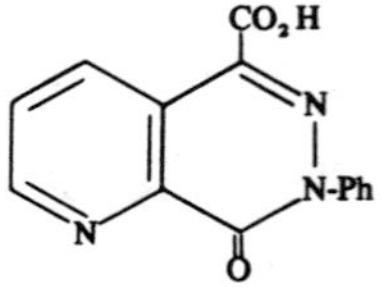 $CO_2H$, N, N-Ph, O, N | | m.p. 257° (decomp.); methyl ester, m.p. 178–179°; ethyl ester, m.p. 138–139° | 260 |
| 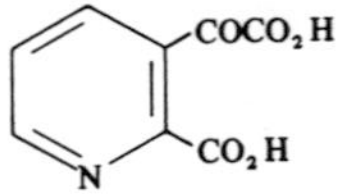 $COCO_2H$, $CO_2H$, N | $N_2H_4 \cdot H_2SO_4$, $H_2O$, 100°, 1 hr | 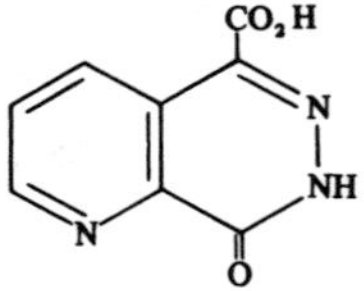 $CO_2H$, N, NH, O, N | 80% | m.p. 300–301° (decomp.) | 260 |
| 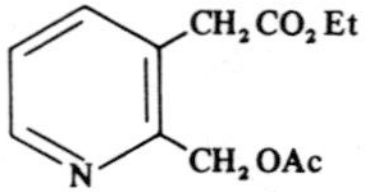 $CH_2CO_2Et$, $CH_2OAc$, N | KOH, EtOH, reflux 9 hr | 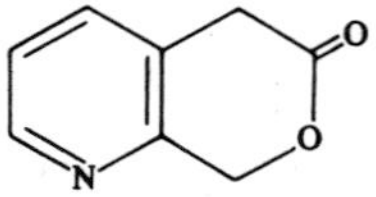 O, O, N | 5.9 g ↓ 1.3 g | m.p. 118–119° | 174 |
| 2-$PyCH_2CN$ 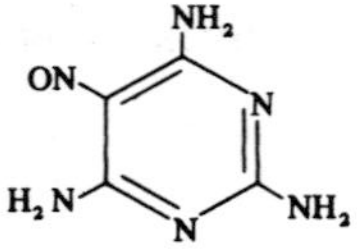 + $NH_2$, ON, N, $H_2N$, N, $NH_2$ | Na, $HOCH_2CH_2OEt$ | 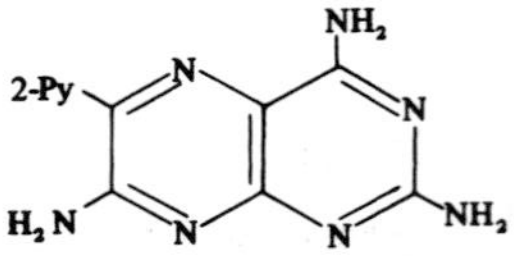 $NH_2$, 2-Py, N, N, $H_2N$, N, N, $NH_2$ | | | 296 |
| 3-$PyCH_2CN$ 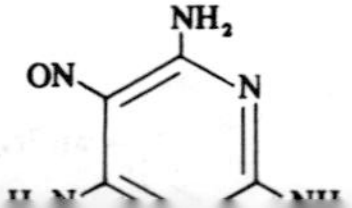 + $NH_2$, ON, N | Na, $HOCH_2CH_2OEt$ | 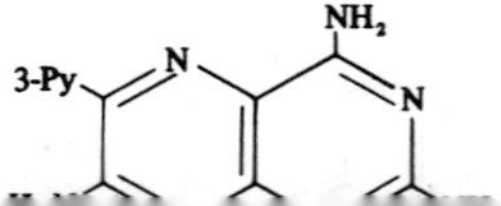 $NH_2$, 3-Py, N, N | | | 296 |

| | | | | |
|---|---|---|---|---|
| 4-$PyCH_2CN$ 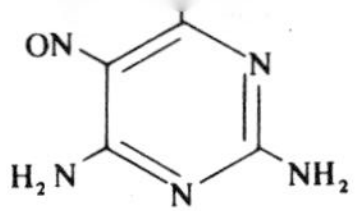 + | Na, $HOCH_2CH_2OEt$ | 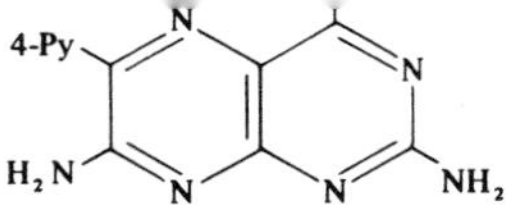 | | 296 |
| 3-$PyCOCH_2CO_2H$ 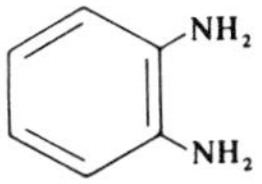 + | inert solvent, 180° *or* 120° without solvent | 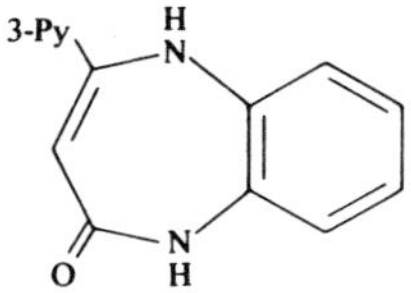 | m.p. 219–220°; ir; uv | 297 |
| 3-$PyCOCH_2CO_2H$ 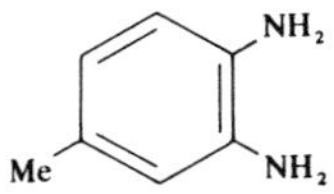 + | same as above | 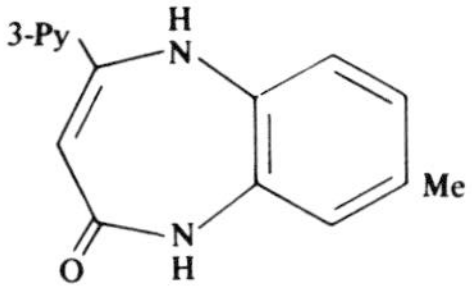 | m.p. 213–217°; ir; uv | 297 |
| 3-$PyCOCH_2CO_2H$ 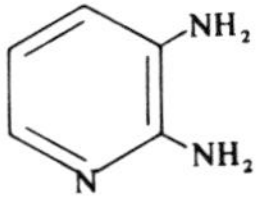 + | same as above | 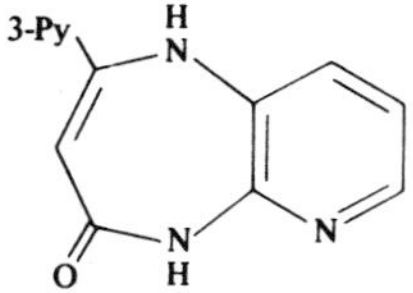 | m.p. 249–252° | 297 |
| 3-$PyCOCH_2CO_2H$ 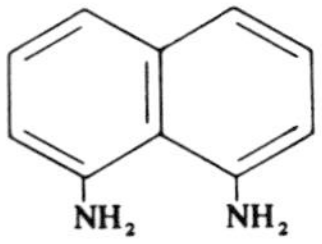 + | same as above | 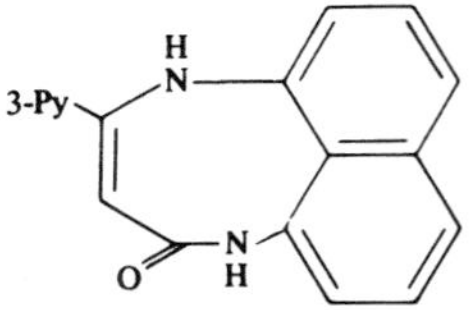 | m.p. 271–272° | 297 |

TABLE XI-20. Synthesis of Condensed Heterocycles Other Than Quinolizidines (Continued)

| Starting materials | Conditions | Product | Yield | Properties | Ref. |
|---|---|---|---|---|---|
| | | ***Miscellaneous*** | | | |
| 3-$PyCOCH_2CO_2H$ 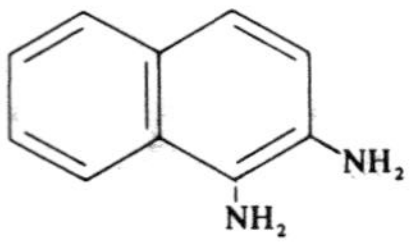 + [structure: 1,2-diaminonaphthalene, $NH_2$, $NH_2$] | same as above | 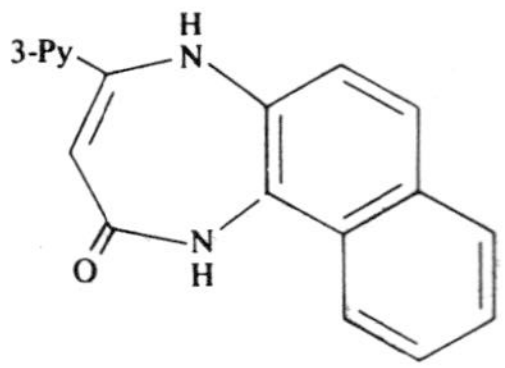 [structure: 3-Py, NH, NH, O] | | m.p. 209–230°; ir; uv | 297 |

ABLE XI-21. Side-Chain Acid Chlorides

| cid chloride | Properties | Ref. |
|---|---|---|
| PyCMe(OH)C=CCOCl | | 79, 97 |
| 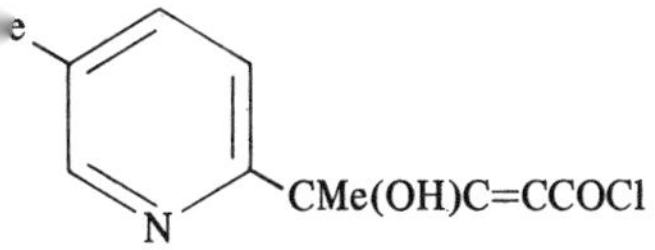 e, N, CMe(OH)C=CCOCl | | 79, 97 |
| $PyCH_2COCl$ | | 37–40 |
| PyCOCOCl | | 37 |
| PyCH=CHCOCl·HCl | m.p. 196° (decomp.) | 298, 299, 343 |
| 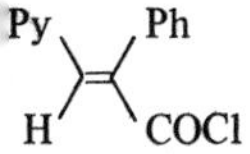 Py, Ph, H, COCl | | 98 |
| 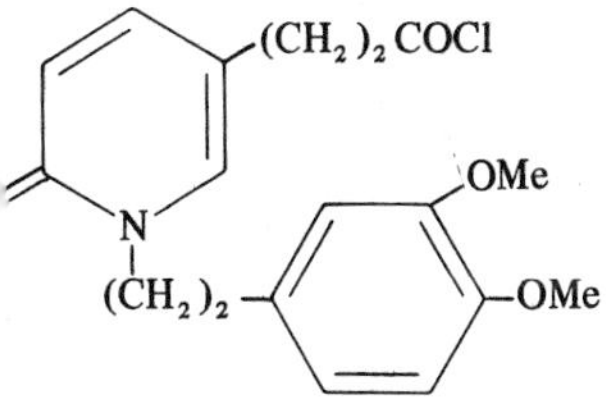 $(CH_2)_2COCl$, N, $(CH_2)_2$, OMe, OMe | | 7 |
| 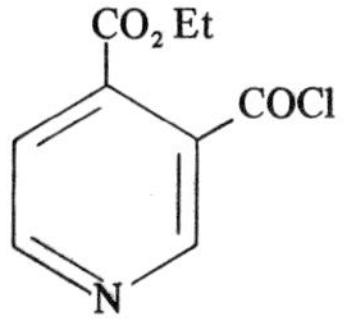 $CO_2Et$, COCl, N | b.p. 102°/0.1 mm | 128 |
| PyCPh=CHCOCl | | 172 |

Table XI-22. Pyridylation of Nitriles

| Halide | Condensed with | Conditions | Product | Yield | Properties | Ref. |
|---|---|---|---|---|---|---|
| 2-PyBr | $PhCH_2CN$ | $NaNH_2$, PhMe, 45 min; 100°, 2 hr | 2-PyCHPhCN | | m.p. 87–88° | 45, 200, 339 |
| | $p$-$FC_6H_4CH_2CN$ | same as above | 2-PyCH($p$-$FC_6H_4$)CN | | b.p. 125–129°/0.5 mm | 45 |
| | $p$-$ClC_6H_4CH_2CN$ | same as above | 2-PyCH($p$-$ClC_6H_4$)CN | | m.p. 67.5–68° | 45 |
| | MeO, MeO, $CH_2CN$ | same as above | 2-PyCH(CN)–, OMe, OMe | | b.p. 175–178°/0.1 mm | 45 |
| | MeO, MeO, MeO, $CH_2CN$ | Na, $NH_3$, $Et_2O$, stir 1 hr; reflux 1 or 15 hr | 2-PyCH(CN)–, OMe, OMe, OMe | 38.4% | | 45, 276 |
| | $(p$-$MeOC_6H_4)_2CHCN$ *or* $p$-$MeOC_6H_4CH_2CN$ + $p$-$MeOC_6H_4Br$ | $NaNH_2$, PhMe, reflux until $NH_3$ evolution ceases | 2-PyC($p$-MeOPh$)_2$CN | | | 272–274, 300 |
| | N, $CH_2OMe$, $CH_2CN$ | $NaNH_2$ | 2-PyCH(CN)–, N, $MeOCH_2$ | 15.9% | m.p. 79–80°; picrate, m.p. 132–134° | 9 |
| | $PhCH_2CN$ + $Pr_2N(CH_2)_2Cl$ | $NaNH_2$, PhMe, 105°, 1 hr; amine added; 108–110°, 3.5 hr; R.T., 12 hr | 2-PyCPh(CN)$CH_2CH_2NPr_2$ | | b.p. 164–165°/0.1 mm | 45 |
| | $PhCH_2CN$ + iso$Pr_2N(CH_2)_2Cl$ | same as above | 2-PyCPh(CN)$CH_2CH_2N$iso$Pr_2$ | | b.p. 145–160°/0.3 mm | 45 |

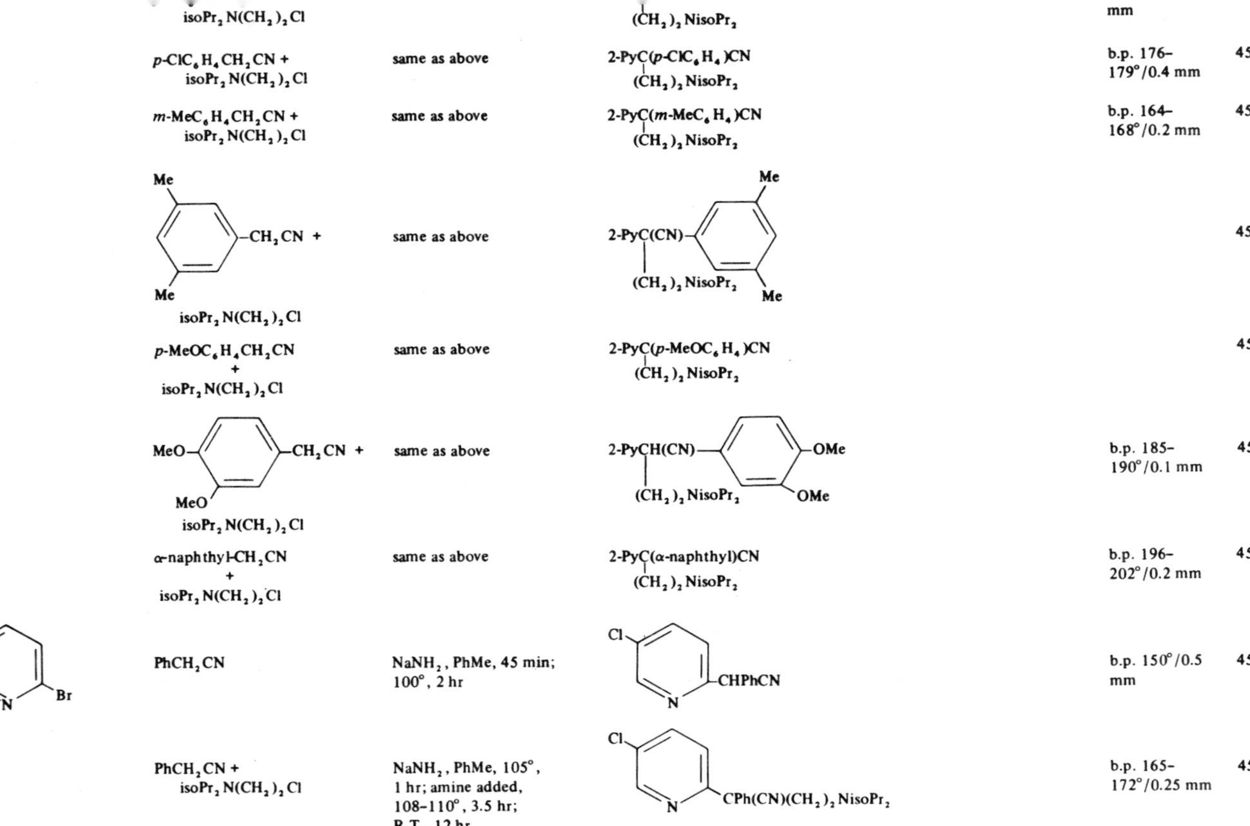

| | | | | | |
|---|---|---|---|---|---|
| | $isoPr_2N(CH_2)_2Cl$ | | $(CH_2)_2NisoPr_2$ | mm | |
| | $p\text{-}ClC_6H_4CH_2CN$ + $isoPr_2N(CH_2)_2Cl$ | same as above | 2-PyC($p\text{-}ClC_6H_4$)CN, $(CH_2)_2NisoPr_2$ | b.p. 176–179°/0.4 mm | 45 |
| | $m\text{-}MeC_6H_4CH_2CN$ + $isoPr_2N(CH_2)_2Cl$ | same as above | 2-PyC($m\text{-}MeC_6H_4$)CN, $(CH_2)_2NisoPr_2$ | b.p. 164–168°/0.2 mm | 45 |
| | Me, Me –$CH_2CN$ + $isoPr_2N(CH_2)_2Cl$ | same as above | 2-PyC(CN)– Me, Me, $(CH_2)_2NisoPr_2$ | | 45 |
| | $p\text{-}MeOC_6H_4CH_2CN$ + $isoPr_2N(CH_2)_2Cl$ | same as above | 2-PyC($p\text{-}MeOC_6H_4$)CN, $(CH_2)_2NisoPr_2$ | | 45 |
| | MeO, MeO –$CH_2CN$ + $isoPr_2N(CH_2)_2Cl$ | same as above | 2-PyCH(CN)– OMe, OMe, $(CH_2)_2NisoPr_2$ | b.p. 185–190°/0.1 mm | 45 |
| | α-naphthyl-$CH_2CN$ + $isoPr_2N(CH_2)_2Cl$ | same as above | 2-PyC(α-naphthyl)CN, $(CH_2)_2NisoPr_2$ | b.p. 196–202°/0.2 mm | 45 |
| Cl, N, Br | $PhCH_2CN$ | $NaNH_2$, PhMe, 45 min; 100°, 2 hr | Cl, N, –CHPhCN | b.p. 150°/0.5 mm | 45 |
| | $PhCH_2CN$ + $isoPr_2N(CH_2)_2Cl$ | $NaNH_2$, PhMe, 105°, 1 hr; amine added, 108–110°, 3.5 hr; R.T., 12 hr | Cl, N, –$CPh(CN)(CH_2)_2NisoPr_2$ | b.p. 165–172°/0.25 mm | 45 |

Table XI-22. Pyridylation of Nitriles (Continued)

| Halide | Condensed with | Conditions | Product | Yield | Properties | Ref. |
|---|---|---|---|---|---|---|
| 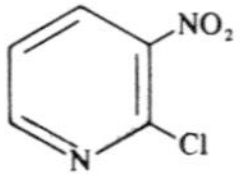 $NO_2$, Cl, N | $NCCH_2CO_2Me$ | $Me_3CO^{\ominus}K^{\oplus}$, $Me_3COH$, reflux 5–12 hr | 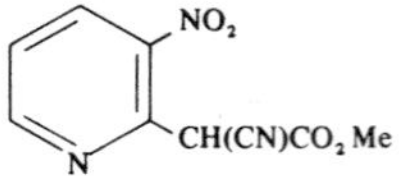 $NO_2$, $CH(CN)CO_2Me$, N | 82% | m.p. 186–188° | 84 |
|  | $NCCH_2CO_2Et$ | same as above | 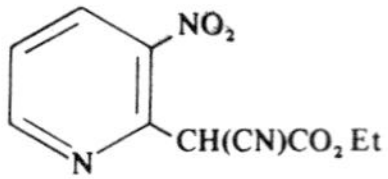 $NO_2$, $CH(CN)CO_2Et$, N | 87% | m.p. 136–137° | 84 |
|  | $NCCH_2CO_2CH_2Ph$ | same as above | 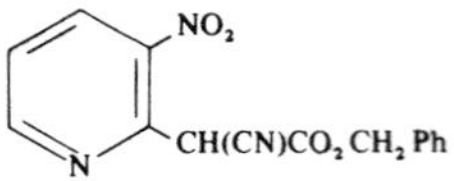 $NO_2$, $CH(CN)CO_2CH_2Ph$, N | 52% | m.p. 139–140° | 84 |
| $2\text{-}PyCH_2Cl$ | $(Ph)_2CH_2CN$ | xylene | $2\text{-}PyCH_2C(Ph)_2CN$ |  | m.p. 116–118°; hydrochloride, m.p. 208–212° (decomp) | 51 |
| 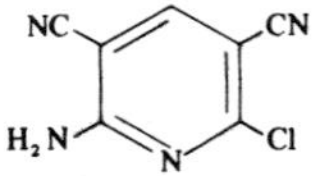 NC, CN, $H_2N$, N, Cl | $NaC(CN)_2CH{=}C(CN)_2$ | R.T., 1 hr; ion exchange resin (sulfonic acid $H^+$) | 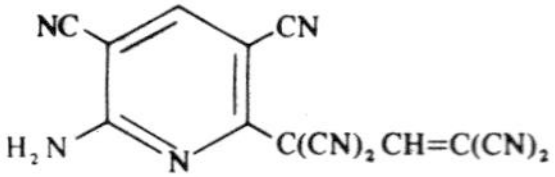 NC, CN, $H_2N$, N, $C(CN)_2CH{=}C(CN)_2$ |  | sublimes 200° | 301 |
| 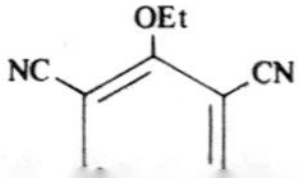 OEt, NC, CN | $NaC(CN)_2C(OEt){=}C(CN)_2$ | same as above | 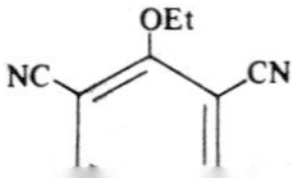 OEt, NC, CN |  | m.p. 264–265° | 301 |

| | | | | | | |
|---|---|---|---|---|---|---|
| 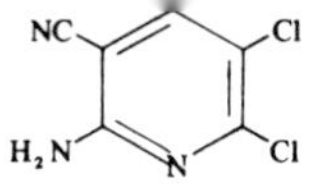 NC, Cl, $H_2N$, N, Cl | $NaCH(CN)_2$ | R.T., 1 hr | 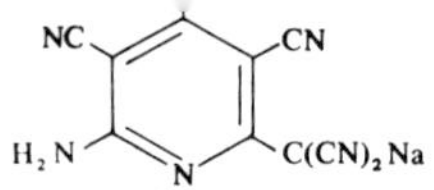 NC, CN, $H_2N$, N, $C(CN)_2Na$ | | $Me_4N^+$ salt, m.p. > 300° | 301 |
| 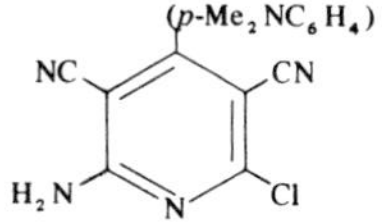 ($p$-$Me_2NC_6H_4$), NC, CN, $H_2N$, N, Cl | $NaC(CN)_2C(=C(CN)_2)(p\text{-}NMe_2C_6H_4)$ | R.T., 1 hr; ion exchange resin (sulfonic acid $H^+$) | 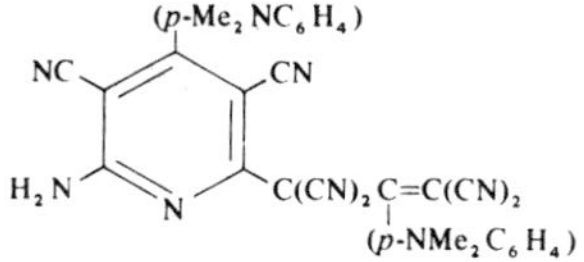 ($p$-$Me_2NC_6H_4$), NC, CN, $H_2N$, N, $C(CN)_2C(=C(CN)_2)(p\text{-}NMe_2C_6H_4)$ | | m.p. >320° | 301 |
| 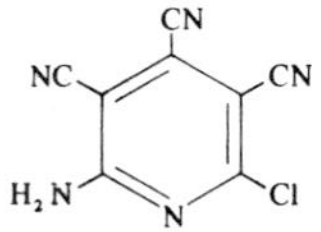 CN, NC, CN, $H_2N$, N, Cl | $NaCH(CN)C(NH_2)=C(CN)_2$ | same as above | 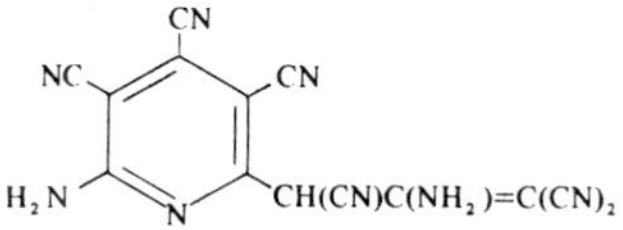 CN, NC, CN, $H_2N$, N, $CH(CN)C(NH_2)=C(CN)_2$ | | m.p. 170–173°; sodium salt, m.p. > 300° | 301 |
| 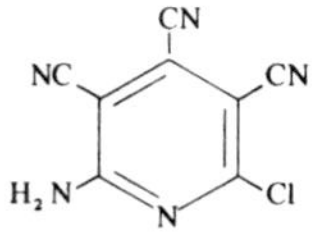 CN, NC, CN, $H_2N$, N, Cl | $NaC(CN)_2C(CN)=C(CN)_2$ | same as above | 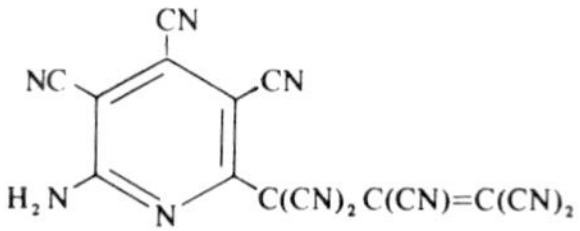 CN, NC, CN, $H_2N$, N, $C(CN)_2C(CN)=C(CN)_2$ | | m.p. 228–229° | 301 |
| 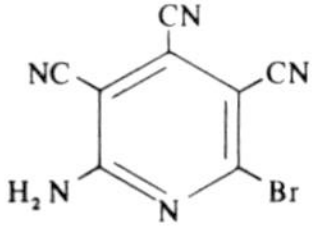 CN, NC, CN, $H_2N$, N, Br | $NaCH(CN)_2$ | same as above | 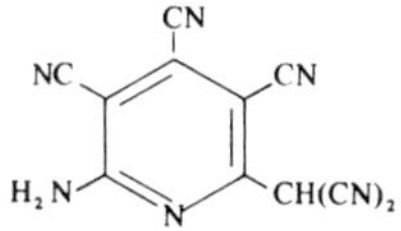 CN, NC, CN, $H_2N$, N, $CH(CN)_2$ | | $Me_4\overset{+}{N}$ salt, m.p. >300°; $Et_4\overset{+}{N}$ salt, m.p. 190–191°; $Me_3\overset{+}{S}$ salt, m.p. >260° | 301 |
| 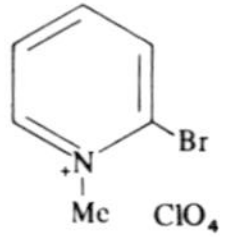 Br, $\overset{+}{N}$–Me, $ClO_4$ | $CH_2(CN)CO_2Et$ | NaOEt, EtOH, reflux 30 min–1 hr | 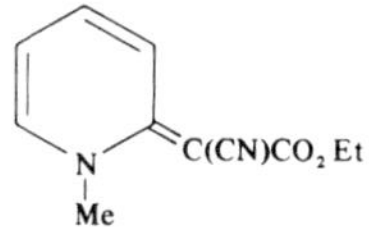 N–Me, $=C(CN)CO_2Et$ | 24.5% | m.p. 127.5–128° | 83 |

Table XI-22. Pyridylation of Nitriles (Continued)

| Halide | Condensed with | Conditions | Product | Yield | Properties | Ref. |
|---|---|---|---|---|---|---|
| (pyridinium ring: Me, Me, Cl; $\overset{+}{N}$–Me; $ClO_4^-$) | $CH_2(CN)_2$ | NaOEt, EtOH, reflux 30 min–1 hr | (dihydropyridine ring: Me, Me; N–Me; $=C(CN)_2$) | 34% | m.p. 201–202°; uv | 83 |
| | $CH_2(CN)CO_2Et$ | same as above | (dihydropyridine ring: Me, Me; N–Me; $=C(CN)CO_2Et$) | 43.5% | m.p. 120–121° | 83 |
| (pyridinium ring: Me, Me, Cl; $\overset{+}{N}$–Me; $ClO_4^-$) | $CH_2(CN)CONH_2$ | NaOEt, EtOH, reflux 30 min–1 hr | (dihydropyridine ring: Me, Me; N–Me; $=C(CN)CONH_2$) | 44% | m.p. 162–164 | 83 |
| 3-PyBr | $PhCH_2CN$ | $NaNH_2$, PhMe, 45 min; 100°, 2 hr | 3-PyCHPhCN | | b.p. 150°/0.5 mm | 45 |
| | | | | | b.p. 155–170°/0.5 mm | 277 |
| | $PhCH(p\text{-}FC_6H_4)CN$ | $Me_3CO^{\ominus}K^{\oplus}$ | $3\text{-}PyCPh(p\text{-}FC_6H_4)CN$ | | b.p. 170–182°/0.4 mm | 277 |
| | $PhCH_2CN$ + $isoPr_2N(CH_2)_2Cl$ | $NaNH_2$, PhMe, 105° 1 hr; amine added, [illegible] | $3\text{-}PyCPh(CN)(CH_2)_2NisoPr_2$ | | b.p. 180–190°/0.3 mm | 45 |

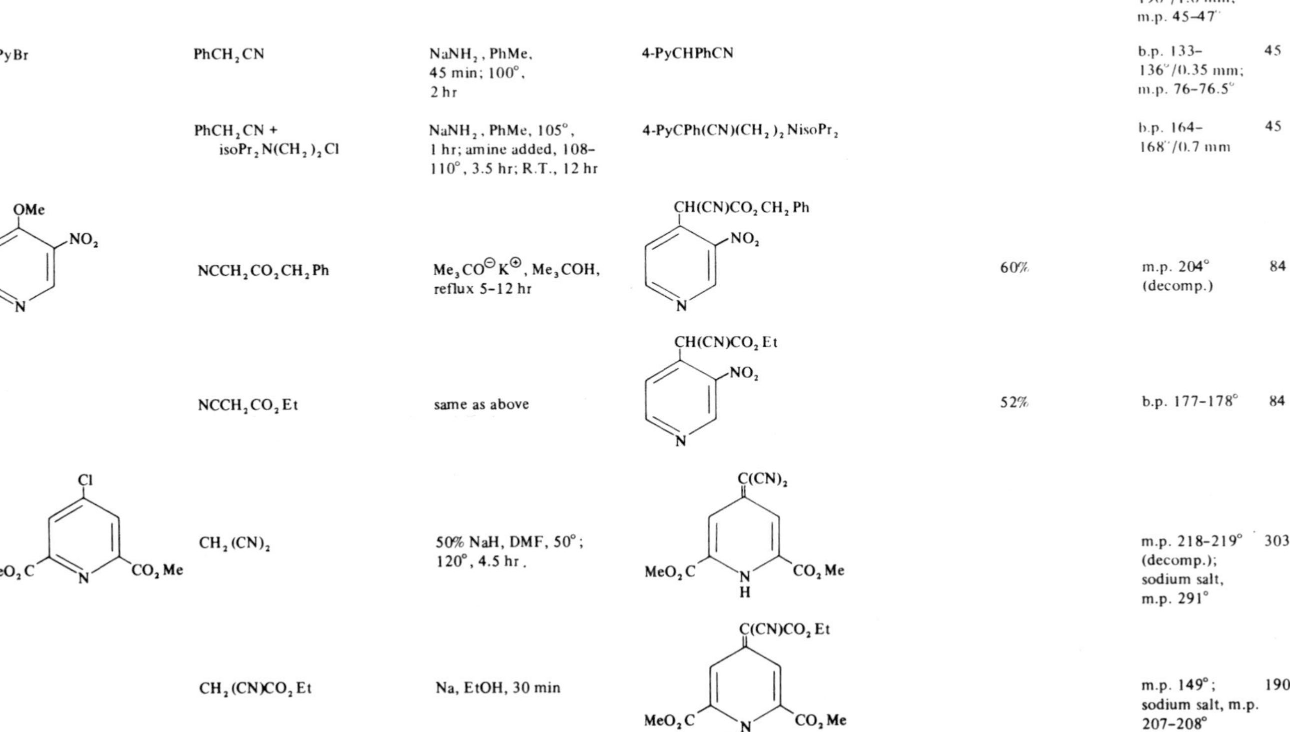

| | | | | | | |
|---|---|---|---|---|---|---|
| | | | | | 190°/1.0 mm; m.p. 45–47° | |
| 4-PyBr | $PhCH_2CN$ | $NaNH_2$, PhMe, 45 min; 100°, 2 hr | 4-PyCHPhCN | | b.p. 133–136°/0.35 mm; m.p. 76–76.5° | 45 |
| | $PhCH_2CN$ + $isoPr_2N(CH_2)_2Cl$ | $NaNH_2$, PhMe, 105°, 1 hr; amine added, 108–110°, 3.5 hr; R.T., 12 hr | $4\text{-}PyCPh(CN)(CH_2)_2NisoPr_2$ | | b.p. 164–168°/0.7 mm | 45 |
| | $NCCH_2CO_2CH_2Ph$ | $Me_3CO^{\ominus}K^{\oplus}$, $Me_3COH$, reflux 5–12 hr | | 60% | m.p. 204° (decomp.) | 84 |
| | $NCCH_2CO_2Et$ | same as above | | 52% | b.p. 177–178° | 84 |
| | $CH_2(CN)_2$ | 50% NaH, DMF, 50°; 120°, 4.5 hr. | | | m.p. 218–219° (decomp.); sodium salt, m.p. 291° | 303 |
| | $CH_2(CN)CO_2Et$ | Na, EtOH, 30 min | | | m.p. 149°; sodium salt, m.p. 207–208° (decomp.) | 190, 303 |

TABLE XI-22. Pyridylation of Nitriles (Continued)

| Halide | Condensed with | Conditions | Product | Yield | Properties | Ref. |
|---|---|---|---|---|---|---|
| | $CH_2(CN)CO_2(t\text{-Bu})$ | 50% NaH, DMF, 50°; 120°, 4.5 hr; HCl, EtOH | 4-[$CH(CO_2\,t\text{-Bu})CN$]-ylidene-1,4-dihydropyridine with $MeO_2C$ at C-2 and $CO_2Me$ at C-6 (N–H) | | m.p. 175° (decomp.); sodium salt, m.p. 247° (decomp.) | 303 |
| 4-Py$CH_2Cl$ | $p\text{-MeC}_6H_4CHCN$, $p$-substituent $O(CH_2)_2NEt_2 \cdot HCl$ | Na, liq. $NH_3$ | 4-Py$CH_2C(p\text{-MeC}_6H_4)CN$, $p$-substituent $O(CH_2)_2NEt_2$ | 57% | p$Ka$ = 8.02 | 43 |

| PyCHRCN | $R^1X$ | Conditions | Product | Yield | Properties | Ref. |
|---|---|---|---|---|---|---|
| 2-$PyCH_2CN$ | EtBr | $NaNH_2$, PhMe, 1 hr; halide reflux 2 hr | 2-PyCHEtCN | 10 g → 8.5 g | b.p. 75–77°/0.05 mm; picrate, m.p. 132–133° | 178 |
| | PhCl | $NaNH_2$ | 2-PyCHPhCN | | m.p. 87–88° | 304 |
| | *p*-$MeOC_6H_4Cl$ | $NaNH_2$ | 2-PyCH(*p*-$MeOC_6H_4$)CN | | | 304 |
| | $PhCH_2Cl$ | $NaNH_2$, $Et_2O$ | 2-PyCH(CN)$CH_2$Ph<br>+ | 6 g → 4.5 g | m.p. 67.5–68.5°; b.p. 153–154°/3.0 mm; picrate, m.p. 161–162° | 305 |
| | | | 2-PyC(CN)$(CH_2Ph)_2$ | 6 g → 3.5 g | m.p. 93–97.5° | |
| | $PhCH_2OAc$ | Na, $PhCH_2OH$, reflux 2 hr, 170–180° | 2-PyCH($CH_2$Ph)CN | 84% | m.p. 66–67° | 306 |
| | 2-$PyCH_2CN$ | NaOEt, EtOH, reflux 2 hr | 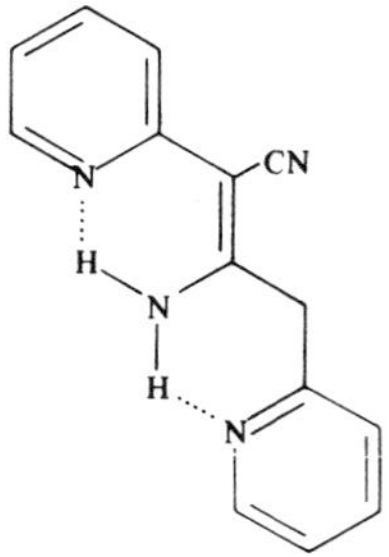 [structure: CN, N, H, N–H, N] | 96% | m.p. 124–126°; ir; uv; nmr | 163 |
| 2-PyCHPhCN | 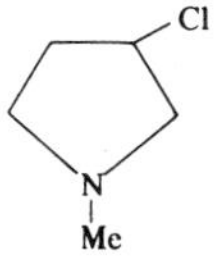 [structure: Cl, N–Me] | $NaNH_2$ | 2-PyCPhCN 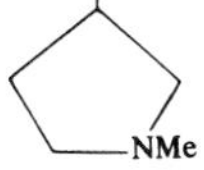 [structure: NMe] | | m.p. 148–151°/0.07 mm | 304 |

Table XI-23. Alkylation of Pyridylacetonitriles (Continued)

| PyCHRCN | R'X | Conditions | Product | Yield | Properties | Ref. |
|---|---|---|---|---|---|---|
| | 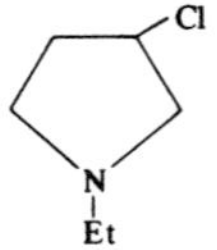 Cl, N–Et | $NaNH_2$ | 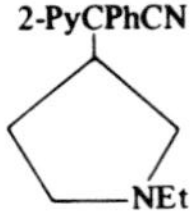 2-PyCPhCN, NEt | | m.p. 110–119° | 304 |
| | 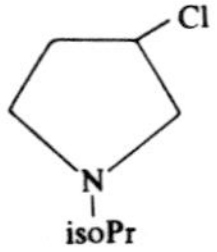 Cl, N–isoPr | $NaNH_2$ | 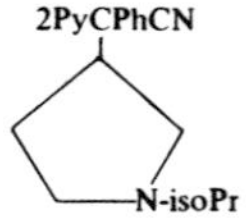 2PyCPhCN, N-isoPr | | m.p. 107–109° | 304 |
| | 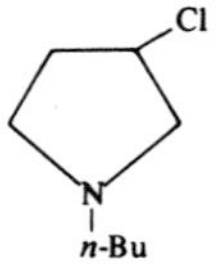 Cl, N–*n*-Bu | $NaNH_2$ | 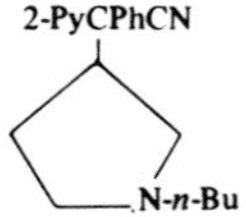 2-PyCPhCN, N-*n*-Bu | | b.p. 170–175°/0.08 mm | 304 |
| | 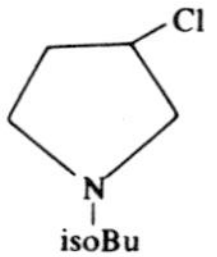 Cl, N–isoBu | $NaNH_2$ | 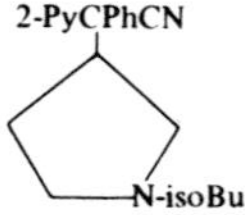 2-PyCPhCN, N-isoBu | | b.p. 161–165°/0.07 mm | 304 |
| | 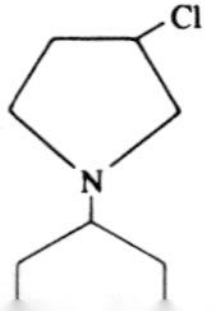 Cl, N | $NaNH_2$ | 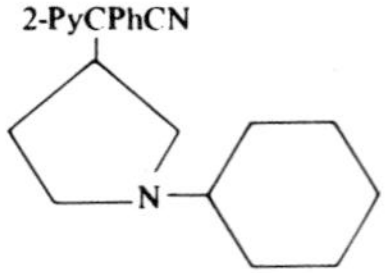 2-PyCPhCN, N | | b.p. 200–208°/0.05 mm | 304 |

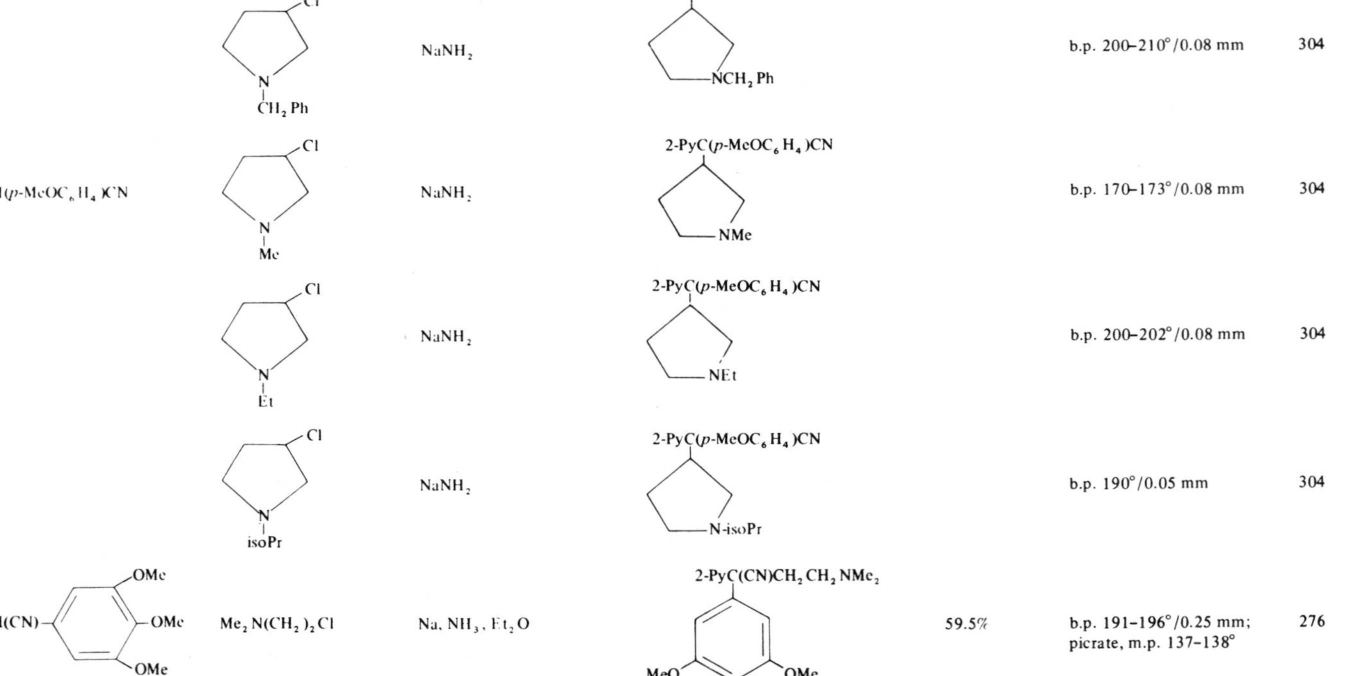

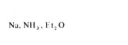

| Reactant | Alkylating agent | Conditions | Product | Yield | Properties | Ref. |
|---|---|---|---|---|---|---|
| | 3-Chloro-1-benzylpyrrolidine ($CH_2Ph$) | $NaNH_2$ | (pyrrolidine)–$NCH_2Ph$ | | b.p. 200–210°/0.08 mm | 304 |
| 2-PyCH($p$-$MeOC_6H_4$)CN | 3-Chloro-1-methylpyrrolidine (N–Me) | $NaNH_2$ | 2-PyC($p$-$MeOC_6H_4$)CN (3-pyrrolidinyl, NMe) | | b.p. 170–173°/0.08 mm | 304 |
| | 3-Chloro-1-ethylpyrrolidine (N–Et) | $NaNH_2$ | 2-PyC($p$-$MeOC_6H_4$)CN (3-pyrrolidinyl, NEt) | | b.p. 200–202°/0.08 mm | 304 |
| | 3-Chloro-1-isopropylpyrrolidine (N–isoPr) | $NaNH_2$ | 2-PyC($p$-$MeOC_6H_4$)CN (3-pyrrolidinyl, N-isoPr) | | b.p. 190°/0.05 mm | 304 |
| 2-PyCH(CN)–(3,4,5-trimethoxyphenyl; OMe, OMe, OMe) | $Me_2N(CH_2)_2Cl$ | Na, $NH_3$, $Et_2O$ | 2-PyC(CN)$CH_2CH_2NMe_2$ (3,4,5-trimethoxyphenyl; MeO, OMe, OMe) | 59.5% | b.p. 191–196°/0.25 mm; picrate, m.p. 137–138° | 276 |
| 2-PyCH($CH_2Ph$)CN | $Me_2N(CH_2)_2Cl$ | $NaNH_2$, PhMe heat, 2.5 hr | 2-PyC(CN)($CH_2Ph$)$CH_2CH_2NMe_2$ | | b.p. 155–167°/2 mm; methiodide, m.p. 167–167.5° | 305 |

TABLE XI-23. Alkylation of Pyridylacetonitriles (Continued)

| PyCHRCN | R'X | Conditions | Product | Yield | Properties | Ref. |
|---|---|---|---|---|---|---|
| 3-$PyCH_2CN$ | EtCl | $PhCH_2\overset{\oplus}{N}(CH_2CH{=}CH_2)_3\overset{\ominus}{Cl}$ 50% NaOH, < 25° | 3-PyCHEtCN | | b.p. 73–75°/0.1 mm | 307 |
| | $o$-$ClC_6H_4CH_2Cl$ | NaH, PhH, DMF, R.T., 15 hr | 3-$PyCH(CN)CH_2(o$-$ClC_6H_4)$ | | b.p. 125–135°/0.08 mm, m.p. 64–65° | 59 |
| | 2-($CH_2CH{=}CH_2$)$C_6H_4CH_2Cl$ | NaH, DMF; 25°, overnight | 3-$PyCH(CN)CH_2$-$C_6H_4$-2-($CH_2$–$CH{=}CH_2$) | | | 290 |
| | $Ph(CH_2)_2Br$ | NaH, DMF, R.T.; PhMe, R.T., 3 hr; R.T. overnight | 3-$PyCH(CN)CH_2CH_2Ph$ | 60% | b.p. 143–150°/0.01 mm<br>b.p. 147–150°/0.01 mm | 308<br>286–289, 309, 310 |
| | $m$-$ClC_6H_4(CH_2)_2Br$ | same as above | 3-$PyCH(CN)CH_2CH_2(m$-$ClC_6H_4)$ | 61% | b.p. 160–162°/0.05 mm | 287–289, 308–310 |
| | $p$-$ClC_6H_4(CH_2)_2Br$ | same as above | 3-$PyCH(CN)CH_2CH_2(p$-$ClC_6H_4)$ | 58% | b.p. 168–169°/0.05 mm<br>b.p. 168–175°/0.05 mm | 286–289<br>308–310 |
| | $p$-$MeC_6H_4(CH_2)_2Br$ | same as above | 3-$PyCH(CN)CH_2CH_2(p$-$MeC_6H_4)$ | | b.p. 170–173°/0.04 mm | 288 |
| | isoPrCHPhCl | same as above | 3-PyCH(CN)CHPhisoPr | | m.p. 96–97° | 286 |
| | $m$-$MeOC_6H_4(CH_2)_2Br$ | same as above *or* | 3-$PyCH(CN)CH_2CH_2(m$-$MeOC_6H_4)$ | 60% | b.p. 160–163°/0.04 mm | 286 |

| | | | | | | |
|---|---|---|---|---|---|---|
| | | PhMe, R.T., 3 hr, R.T., overnight | | | m.p. 96–97 | |
| 3-PyCHPhCN | 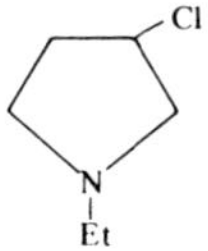 Cl, N, Et | PhMe, reflux 4 hr; halide added, reflux 3 hr | 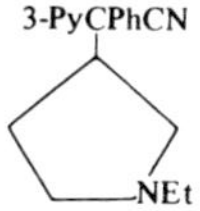 3-PyCPhCN, NEt | 57% | b.p. 168–171°/0.005 mm | 57 |

TABLE XI-24. Synthesis of Side-Chain Nitriles by Michael Addition

| Vinyl compound | Addend | Conditions | Product | Yield | Properties | Ref. |
|---|---|---|---|---|---|---|
| $CH_2$=CHCN | 2-PyMe | Na, 100–110°, reflux 3–5 hr | 2-Py$(CH_2)_3$CN | | b.p. 95–97°/1.0 mm | 311 |
| $CH_2$=CHCN | 2-PyMe · MeI | $Et_3N$, reflux 2 hr, $H_2O$, EtOH; 300° | 2-PyCH$(CH_2CH_2CN)_2$ | 78% | b.p. 182°/1.2 mm; methiodide, m.p. 141–142° | 312 |
| $CH_2$=CHCN | Et, N, Me | Na, 100–110°; reflux 3–5 hr | Et, N, $(CH_2)_3$CN | | | 311 |
| $CH_2$=CHCN | Me, Me, $\overset{+}{N}$, Me, $I^-$ | $Et_3N$, $H_2O$, EtOH, reflux 48 hr | C$(CH_2CH_2CN)_3$, CH$(CH_2CH_2CN)_2$, $\overset{+}{N}$, Me, $I^-$ | 44% | methiodide, m.p. 192–193° | 312 |
| $CH_2$=CHCN | Me, Me, $\overset{+}{N}$, Me, $I^-$ | $Et_3N$, $H_2O$, EtOH, reflux 24 hr | Me, $(CH_2)_3$CN, $\overset{+}{N}$, Me, $I^-$ | 44% | m.p. 192–193° | 312 |
| $CH_2$=CHCN | Et, N, $I^-$ | $Et_3N$, reflux 24 hr, EtOH, $H_2O$; 300° | 2-PyCHMe$CH_2CH_2$CN | 65% | b.p. 170°/45 mm; methiodide, m.p. 97° | 312 |

| | | | | | | |
|---|---|---|---|---|---|---|
| | | Na, $N_2$ | | 71% | | 242 |
| $CH_2{=}CHCN$ | 6-Me-2-Py$CH_2CO_2Et$ | Na, 180° | 6-Me-2-Py$CH(CO_2Et)CH_2CH_2CN$ | | $n_D^{20}$ 1.5030 | 283 |
| $CH_2{=}CHCN$ | 6-Me-2-Py$CHMeCO_2Et$ | Na, 180° | 6-Me-2-Py$CMe(CO_2Et)CH_2CH_2CN$ | | b.p. 160°/1.0 mm | 283 |
| 2-Py$CH{=}CH_2$ | $NCCH_2CO_2Et$ | Na, 100°; 100–110°, 5 hr | 2-Py$(CH_2)_2CH(CN)CO_2Et$ | | | 204 |
| $CH_2{=}CHCO_2Et$ | 2-Py$CH_2CN$ | Na, heat, 2 hr | 2-Py$CH(CN)CH_2CH_2CO_2Et$ | 25 g → 13 g | b.p. 143–148°/0.2 mm | 178 |
| $CH_2{=}CHCN$ | 2-PyCHEtCN | Na, 150°, 2 hr | 2-PyCEt(CN)$CH_2CH_2CN$ | 8 g → 16 g | b.p. 108–110°/0.05 mm | 178 |
| $CH_2{=}CHP(O)(OEt)_2$ | 2-Py$CH_2CN$ | NaOEt, EtOH, 60° | 2-PyCH(CN)$CH_2CH_2P(O)(OEt)_2$ | 61% | $n_D^{20}$ 1.4935; picrolonate, m.p. 102–103° | 243 |
| PhCH=CHCOMe | 2-Py$CH_2CN$ | 30% KOH, MeOH, $Et_2O$, 20–25°, 2 hr | 2-PyCH(CN)CHPh$CH_2$COMe | 34 g → 27 g | m.p. 109°; phenylhydrazone, m.p. 150–151° | 270 |
| PhCH=CHCOPh (0.75 equiv.) | 2-Py$CH_2CN$ | 30% KOH, MeOH, $Et_2O$, 20–25°, 2 hr | 2-PyCH(CN)CHPh$CH_2$COPh | 20 g → 35 g | m.p. 119°; phenylhydrazone, m.p. 148° | 270 |
| PhCH=CHCOPh (1 equiv.) | 2-Py$CH_2CN$ | 30% KOH, MeOH, $Et_2O$, 20–25°, 2 hr | 2-PyCH(CN)CHPh$CH_2$COPh + | 20 g → 8 g | | 270 |
| | | | 2-PyC(CN)[CHPh$CH_2$COPh$]_2$ | 16 g | m.p. 212–213° | |
| 2-PyCH=CHCOPh | 2-Py$CH_2CN$ | 30% KOH, MeOH, $Et_2O$, 20–25°, 2 hr | 2-PyCH(CN)CH(2-Py)$CH_2$COPh | 34 g → 38 g | m.p. 158° | 270 |
| $CH_2{=}CHCN$ | 2-PyCHMe$CO_2Et$ | Na, 180° | 2-PyCMe$(CO_2Et)CH_2CH_2CN$ | 68% | yellow oil, $n_D^{20}$ 1.5046 | 283 |

TABLE XI-24. Synthesis of Side-Chain Nitriles by Michael Addition (Continued)

| Vinyl compound | Addend | Conditions | Product | Yield | Properties | Ref. |
|---|---|---|---|---|---|---|
| $CH_2$=CHCN | 2-Py$CH_2$P(O)$Ph_2$ | Triton B, MeCN, overnight | 2-PyCHP(O)$Ph_2$ with $(CH_2)_2CN$ on the CH | 1 g → 0.85 g | | 314 |
| $CH_2$=CHCN | 2-PyCH=$CH_2$ |  $Me_2N$–$C_6H_4$–NO, $H_2O$, $e^{\ominus}$, $Et_4N^{\oplus}Tos^{\ominus}$, | 2-Py$(CH_2)_4$CH | | b.p. 136–140°/0.5 mm | 90 |
| $CH_2$=CHCN | 2-Py$(CH_2)_2$CN | heat | 2-PyCH($CH_2$CN)$CH_2CH_2$CN | | | 311 |
| $CH_2$=CHCN | 2-Py$(CH_2)_2$CHPhCOMe | *t*-BuOH, 25–45°, 3 hr | 2-Py$(CH_2)_2$CPh$(CH_2)_2$CN with COMe on the CPh | 69% | b.p. 195–200°/0.1 mm | 36 |
| 2-PyCH=$CH_2$ | Ph$CH_2$CN | Triton B, MeOH, reflux 3 hr | 2-Py$(CH_2)_2$CHPhCN + (2-Py$CH_2CH_2)_2$CPhCN | | b.p. 72°/0.8 mm; b.p. 194°/0.5 mm | 167 |
| 2-PyCH=$CH_2$ | MeCN | same as above | 2-Py$(CH_2)_3$CN | | b.p. 111°/1.5 mm | 167 |
| $CH_2$=CHCN | 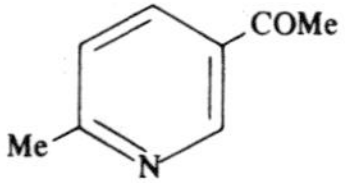 pyridine ring with Me and COMe | 40% Triton B, dioxane, 10–20° | 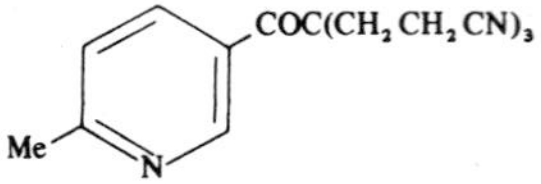 pyridine ring with Me and COC$(CH_2CH_2CN)_3$ | 34 g → 10 g | m.p. 144.5–146°; picrate, m.p. 168–170° | 156 |
| | 4-PyMe | Na, 100–110°; reflux 3–5 hr | 4-Py$(CH_2)_3$CN | | b.p. 122–125°/1.0 mm | 311 |
| $CH_2$=CHCN (2 equiv.) | 4-PyMe | same as above | 4-PyCH$(CH_2CH_2CN)_2$ | | b.p. 140–145° | 311 |

| | | | | | | |
|---|---|---|---|---|---|---|
| | 4-*n*-PrPy · MeI | $Et_3N$, $H_2O$, EtOH, reflux 24 hr; 300° | 4-PyCHEt$(CH_2CH_2CN)_2$ | 85% | b.p. 230–235°/1.4 mm; m.p. 74.5°; methiodide, m.p. 232.5–235° | 312 |
| | 4-$PyCH_2P(O)Ph_2$ | Triton B, MeCN, overnight | 4-PyCHP(O)$Ph_2$ with $(CH_2)_2CN$ on the CH | 58% | m.p. 199–200° | 314 |
| | 4-$Py(CH_2)_3CN$ | heat | 4-PyCH$(CH_2CH_2CN)_2$ | | b.p. 140–145° | 311 |
| | 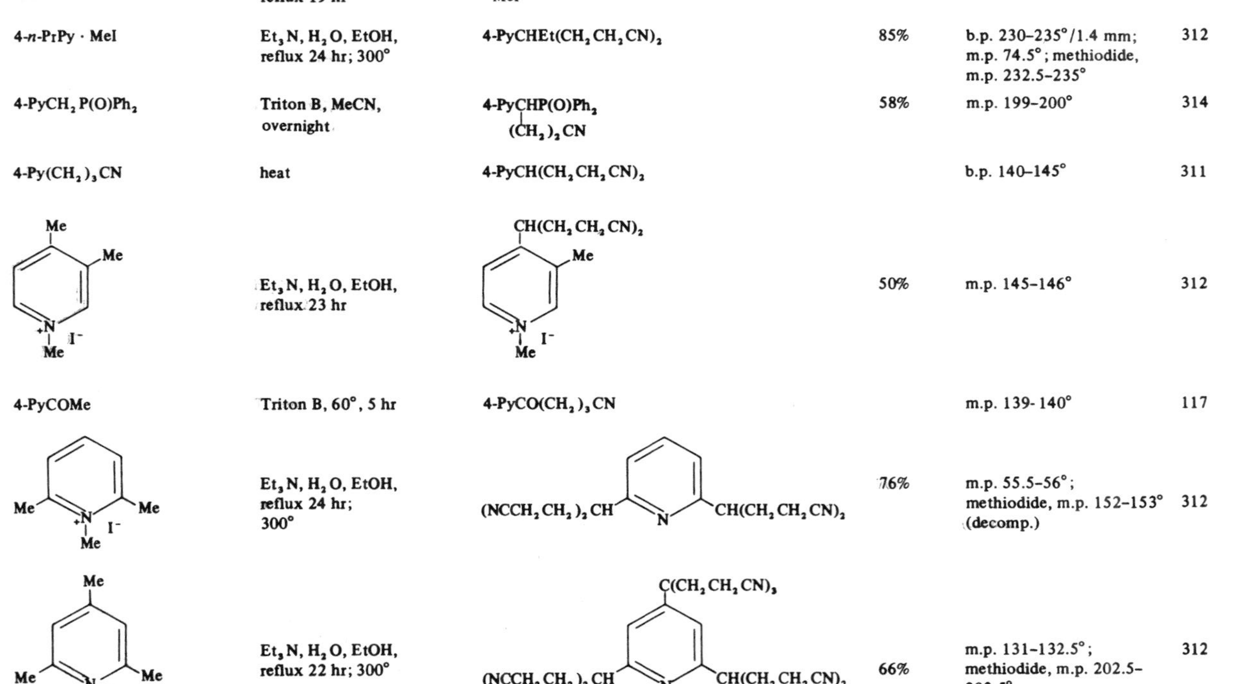 | $Et_3N$, $H_2O$, EtOH, reflux 23 hr | | 50% | m.p. 145–146° | 312 |
| | 4-PyCOMe | Triton B, 60°, 5 hr | 4-PyCO$(CH_2)_3CN$ | | m.p. 139-140° | 117 |
| $CH_2$=CHCN | | $Et_3N$, $H_2O$, EtOH, reflux 24 hr; 300° | | 76% | m.p. 55.5–56°; methiodide, m.p. 152–153° (decomp.) | 312 |
| | | $Et_3N$, $H_2O$, EtOH, reflux 22 hr; 300° | | 66% | m.p. 131–132.5°; methiodide, m.p. 202.5–203.5° | 312 |

TABLE XI-24. Synthesis of Side-Chain Nitriles by Michael Addition (Continued)

| Vinyl compound | Addend | Conditions | Product | Yield | Properties. | Ref. |
|---|---|---|---|---|---|---|
| | 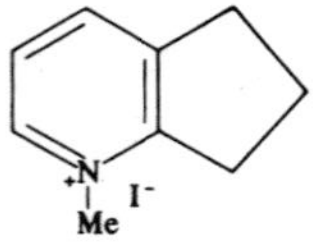 N⁺, Me, $I^-$ | $Et_3N$, $H_2O$, EtOH, reflux 17 hr | 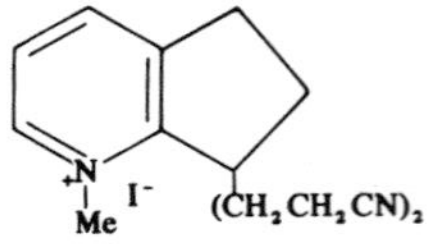 N⁺, Me, $I^-$, $(CH_2CH_2CN)_2$ | 71% | methiodide, m.p. 195.5–196° | 312 |
| 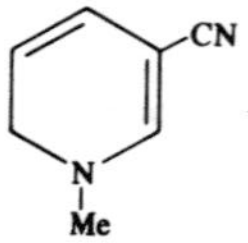 CN, N, Me | $PhCH_2CN$ | Triton B, 100°, $N_2$, 12 hr | 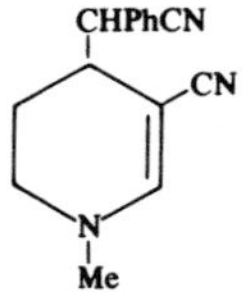 CHPhCN, CN, N, Me | 17.5 g → 11.6 g | b.p. 165–180°/0.1 mm; ir | 315 |

| Carbonyl component | Methylene component | Conditions | Product | Yield | Properties | Ref. |
|---|---|---|---|---|---|---|
| 2-PyCHO | $NCCH_2CO_2Et$ | EtOH, piperidine; reflux 1-5 hr | 2-PyCH=C(CN)$CO_2$Et | 64% | m.p. 94.5-96° | 137 |
| | $NCCH_2CO_2Et$ | EtOH | 2-PyCH=C(CN)$CO_2$Et | 88% | m.p. 94-95° | 211 |
| | $PhCH_2CN$ | NaOH, EtOH, heat | 2-PyCH$(CHPhCN)_2$ | 75% | m.p. 171-173° | 211 |
| | $p$-$ClC_6H_4CH_2CN$ | EtOH, piperidine, reflux 1-5 hr | 2-PyCH=C(CN)($p$-$ClC_6H_4$) | 67% | m.p. 165-167° | 137 |
| | $p$-$Me_2NC_6H_4CH_2CN$ | NaOEt, EtOH, 50°; 1 hr with no heat | (2-Py)(H)C=C(CN)($p$-$Me_2NC_6H_4$) | >70% | m.p. 136-138°; uv | 98 |
| | $p$-$O_2NC_6H_4CH_2CN$ | EtOH, piperidine, reflux 1-5 hr | 2-PyCH=C(CN)($p$-$O_2NC_6H_4$) | 96% | m.p. 199-201° | 137 |
| | | MeOH, piperidine, heat | (2-Py)(H)C=C(CN)($p$-$O_2NC_6H_4$) | 99% | m.p. 200-201.5°; ir; uv | 164 |
| | 2-$NO_2$-4,5-$(MeO)_2C_6H_2CH_2CN$ | EtOH, piperidine reflux 1-5 hr | 2-PyCH=C(CN)-[2-$O_2N$-4,5-$(OMe)_2C_6H_2$] | 96% | m.p. 192.5-193.5° | 137 |
| 2-PyCOMe | $PhNH_2$ | $CN^{\ominus}$, HOAc | 2-PyCMe(CN)NHPh | | m.p. 127.5-128.5° | 316 |
| | MeCN | $NaNH_2$, $Et_2O$, 1.5 hr | 2-PyCPh(OH)$CH_2CN$ | | m.p. 90-92° | 223 |

TABLE XI-25. Synthesis of Side-Chain Nitriles by Knoevenagel Condensation (Continued)

| Carbonyl component | Methylene component | Conditions | Product | Yield | Properties | Ref. |
|---|---|---|---|---|---|---|
| 2-PyCOPh | EtCN | $NaNH_2$, $Et_2O$, 1.5 hr | 2-PyCPh(OH)CHMeCN | | m.p. 100–102° | 223 |
| | $PhCH_2CN$ | $NaNH_2$, $C_6H_6$, reflux 1 hr | 2-PyCPh=CPhCN | | m.p. 150–151°; uv | 41 |
| | $p$-$ClC_6H_4CH_2CN$ | $NaNH_2$, $C_6H_6$, reflux 1 hr | 2-PyCPh=C($p$-$ClC_6H_4$)CN | | m.p. 165–167°; uv | 41 |
| | $p$-$MeOC_6H_4CH_2CN$ | $NaNH_2$, $C_6H_6$, reflux 1 hr | 2-PyCPh=C($p$-$MeOC_6H_4$)CN | | m.p. 138–140°; uv | 41 |
| MeCHO | $2$-$PyCH_2CN$ | piperidine, HOAc $C_6H_6$, 45 min | 2-PyC(CN)=CHMe | 74% | b.p. 70–71°/0.25 mm | 163 |
| $p$-$O_2NC_6H_4CHO$ | $2$-$PyCH_2CN$ | piperidine, EtOH, reflux 1 hr | 2-PyC(CN)=CH($p$-$O_2NC_6H_4$) | | | 68 |
| 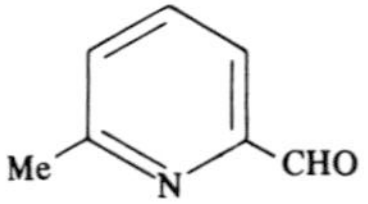 Me N CHO | 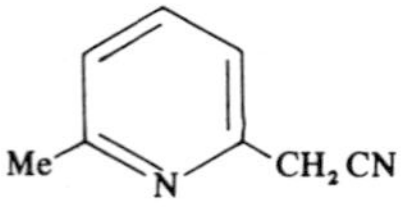 Me N $CH_2CN$ | *N*-ethylpiperidine *or* heat only | 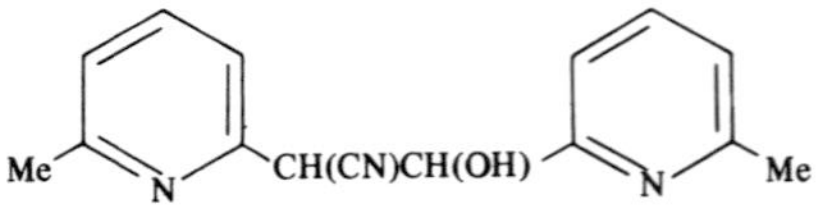 Me N CH(CN)CH(OH) N Me | | m.p. 51–52° | 130 |
|  $O_2N$ O CHO | $2$-$PyCH_2CN$ | $Ac_2O$, overnight | 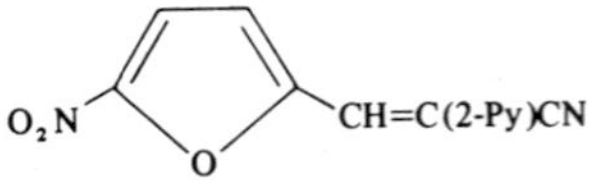 $O_2N$ O CH=C(2-Py)CN | | m.p. 197–200° | 241 |
| 3-PyCHO | $NCCH_2C_6H_5$ | NaOEt, EtOH, 50°; [illegible] |  3-Py CN | 98% | m.p. 92–93°; uv | 98 |

| | | | | | |
|---|---|---|---|---|---|
| | | H (p-ClC$_6$H$_4$) | | | |
| | piperidine, MeOH, heat | 3-PyCH=C(p-ClC$_6$H$_4$)CN | 76% | m.p. 137–138.5°; ir, uv | 164 |
| NCCH$_2$(p-O$_2$NC$_6$H$_4$) | same as above | 3-PyCH=C(p-O$_2$NC$_6$H$_4$)CN | 88% | m.p. 156–157° | 164 |
| | NaOEt, EtOH, 50°; 1 hr with no heat | 3-Py, CN / H, (p-O$_2$NC$_6$H$_4$) | >70% | m.p. 156–157° | 98 |
| NCCH$_2$(p-H$_2$NC$_6$H$_4$) | same as above | 3-Py, CN / H, (p-H$_2$NC$_6$H$_4$) | >70% | m.p. 169–171° | 98 |
| NCCH$_2$(p-NMe$_2$C$_6$H$_4$) | same as above | 3-Py, CN / H, (p-Me$_2$NC$_6$H$_4$) | >70% | m.p. 122–124° | 98 |
| NCCH$_2$(p-AcNHC$_6$H$_4$) | same as above | 3-Py, CN / H, (p-AcNHC$_6$H$_4$) | >70% | m.p. 223–224° | 98 |
| NCCH$_2$(p-MeOC$_6$H$_4$) | same as above | 3-Py, CN / H, (p-MeOC$_6$H$_4$) | >70% | m.p. 112–114° | 98 |
| 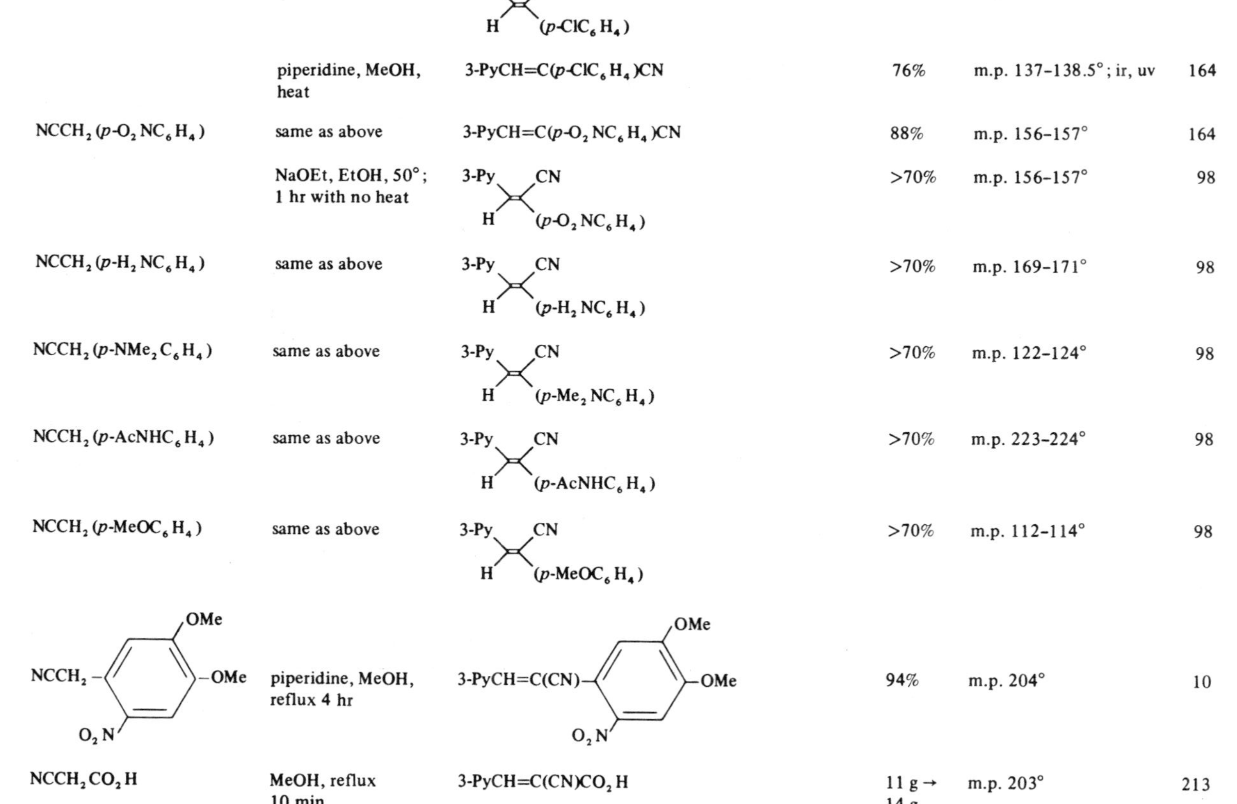 NCCH$_2$–(OMe, OMe, O$_2$N) | piperidine, MeOH, reflux 4 hr | 3-PyCH=C(CN)–(OMe, OMe, O$_2$N) | 94% | m.p. 204° | 10 |
| NCCH$_2$CO$_2$H | MeOH, reflux 10 min | 3-PyCH=C(CN)CO$_2$H | 11 g → 14 g | m.p. 203° | 213 |

TABLE XI-25. Synthesis of Side-Chain Nitriles by Knoevenagel Condensation (Continued)

| Carbonyl component | Methylene component | Conditions | Product | Yield | Properties | Ref. |
|---|---|---|---|---|---|---|
| | $NCCH_2CO_2Et$ | piperidine, EtOH reflux 1–5 hr | 3-PyCH=C(CN)$CO_2$Et | 83% | m.p. 87.5–88.5° | 137 |
| 3-PyCOPh | $PhCH_2CN$ | $NaNH_2$, $C_6H_6$, reflux 1 hr | 3-PyCPh=CPhCN | | m.p. 154.5–157° | 41 |
| | *p*-$ClC_6H_4CH_2CN$ | same as above | 3-PyCPh=C(*p*-$ClC_6H_4$)CN | | m.p. 185–187° | 41 |
| | *p*-$MeOC_6H_4CH_2CN$ | same as above | 3-PyCPh=C(*p*-$MeOC_6H_4$)CN | | m.p. 164–165° | 41 |
| PhCHO | 3-$PyCH_2CN$ | NaOEt, EtOH, 50° | 3-Py, Ph<br>C=C<br>NC, H | | m.p. 107–109°; uv | 98 |
| *o*-$ClC_6H_4CHO$ | 3-$PyCH_2CN$ | NaOEt, 95% EtOH, R.T. | 3-PyC(CN)=CH(*o*-$ClC_6H_4$) | | m.p. 117–118° | 59 |
| *p*-$Me_2NC_6H_4CHO$ | 3-$PyCH_2CN$ | MeOH, $H_2O$ | 3-PyC(CN)=CH(*p*-$Me_2NC_6H_4$) | | m.p. 142–143° | 317 |
| | | NaOEt, EtOH, 50° | 3-Py, p-$Me_2NC_6H_4$<br>C=C<br>NC, H | | m.p. 139–141° | 98 |
| $O_2N$–(furan, O)–CHO | 3-$PyCH_2CN$ | $Ac_2O$, overnight | $O_2N$–(furan, O)–CH=C(3-Py)CN | | m.p. 167–169° | 241 |
| 4-PyCHO | $NCCH_2CO_2H$ | MeOH, reflux | 4-PyCH=C(CN)$CO_2H$ | 68% | hemihydrate, | 213 |

| | | | | | | |
|---|---|---|---|---|---|---|
| | | heat | H, $(p\text{-}O_2NC_6H_4)$ | | | |
| | $NCCH_2(p\text{-}Me_2NC_6H_4)$ | NaOEt, EtOH, 50°; 1 hr with no heat | 4-Py(H)C=C(CN)$(p\text{-}Me_2NC_6H_4)$ | >70% | m.p. 154–157°; uv | 98 |
| | 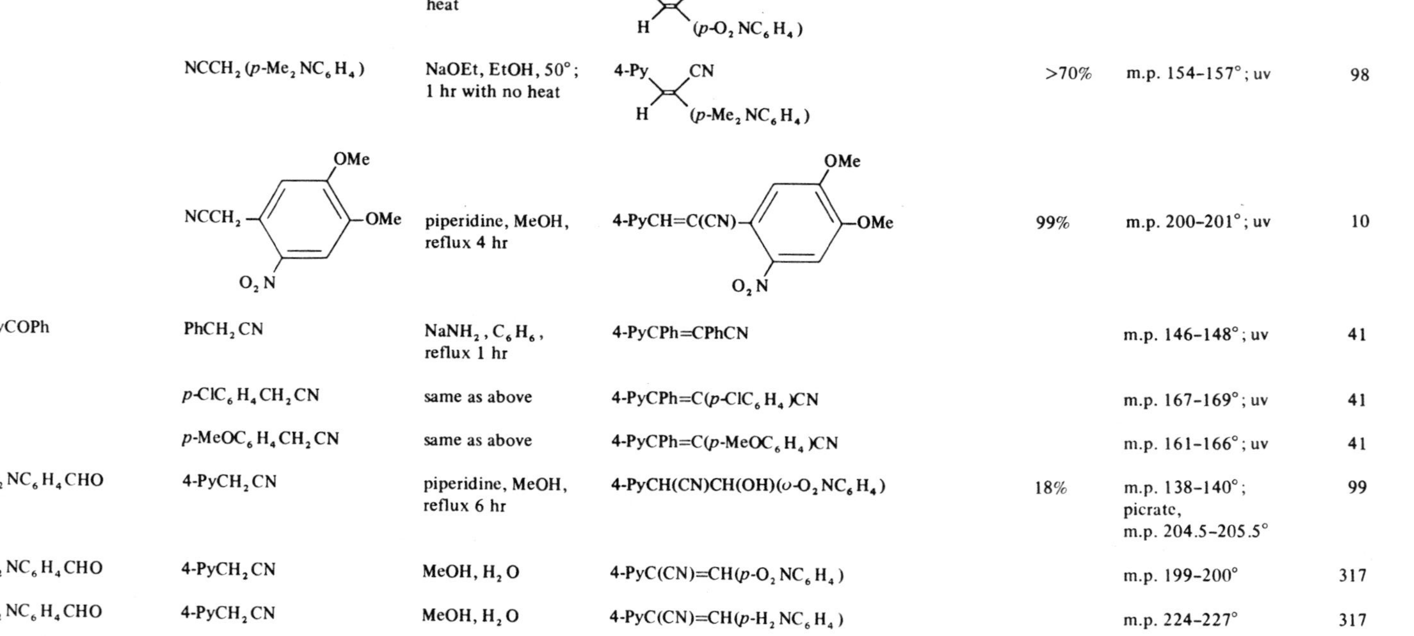 $NCCH_2$-(2,4-dimethoxy-5-nitrophenyl) | piperidine, MeOH, reflux 4 hr | 4-PyCH=C(CN)-(2,4-dimethoxy-5-nitrophenyl) | 99% | m.p. 200–201°; uv | 10 |
| 4-PyCOPh | $PhCH_2CN$ | $NaNH_2$, $C_6H_6$, reflux 1 hr | 4-PyCPh=CPhCN | | m.p. 146–148°; uv | 41 |
| | $p\text{-}ClC_6H_4CH_2CN$ | same as above | 4-PyCPh=C$(p\text{-}ClC_6H_4)$CN | | m.p. 167–169°; uv | 41 |
| | $p\text{-}MeOC_6H_4CH_2CN$ | same as above | 4-PyCPh=C$(p\text{-}MeOC_6H_4)$CN | | m.p. 161–166°; uv | 41 |
| $o\text{-}O_2NC_6H_4CHO$ | $4\text{-}PyCH_2CN$ | piperidine, MeOH, reflux 6 hr | 4-PyCH(CN)CH(OH)$(o\text{-}O_2NC_6H_4)$ | 18% | m.p. 138–140°; picrate, m.p. 204.5–205.5° | 99 |
| $p\text{-}O_2NC_6H_4CHO$ | $4\text{-}PyCH_2CN$ | MeOH, $H_2O$ | 4-PyC(CN)=CH$(p\text{-}O_2NC_6H_4)$ | | m.p. 199–200° | 317 |
| $p\text{-}H_2NC_6H_4CHO$ | $4\text{-}PyCH_2CN$ | MeOH, $H_2O$ | 4-PyC(CN)=CH$(p\text{-}H_2NC_6H_4)$ | | m.p. 224–227° | 317 |
| $p\text{-}Me_2NC_6H_4CHO$ | $4\text{-}PyCH_2CN$ | MeOH, $H_2O$ | 4-PyC(CN)=CH$(p\text{-}Me_2NC_6H_4)$ | | m.p. 181–182° | 317 |

TABLE XI-26. Side-Chain Nitrile Synthesis: Miscellaneous Methods

| Starting material | Conditions | Product | Yield | Properties | Ref. |
|---|---|---|---|---|---|
| | | *Dehydration* | | | |
| 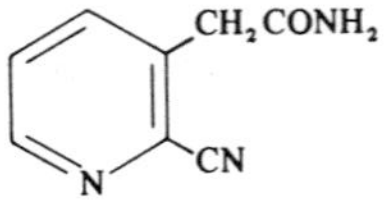 $CH_2CONH_2$, CN, N | $POCl_3$, pyridine, 65° 2.5 hr | 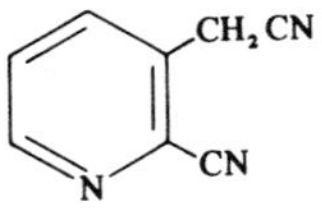 $CH_2CN$, CN, N | | m.p. 64–65.5° | 92 |
| 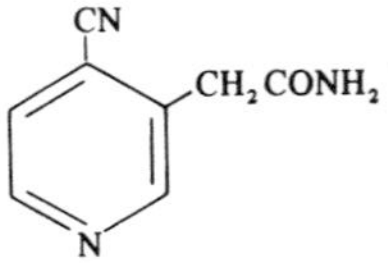 CN, $CH_2CONH_2$, N | same as above | 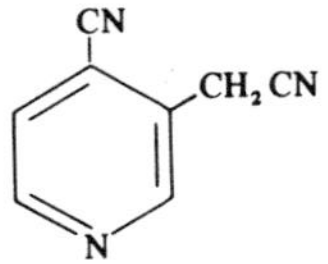 CN, $CH_2CN$, N | 87% | m.p. 78.5–79.5° | 150 |
| 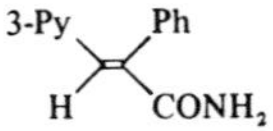 3-Py, Ph, H, $CONH_2$ | TsCl, pyridine, R.T., 1 hr | 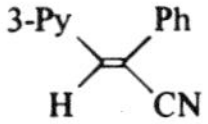 3-Py, Ph, H, CN | 72% | m.p. 64–66° | 98 |
| 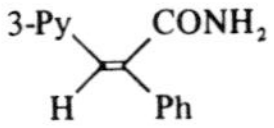 3-Py, $CONH_2$, H, Ph | 60% $H_2SO_4$, 100°, 3 hr; R.T., overnight | 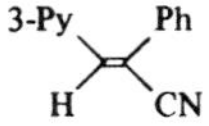 3-Py, Ph, H, CN | | | 98 |
| 4-$PyCH_2CONH_2$ | $POCl_3$, heat, 6 hr | 4-$PyCH_2CN$ | 58.5% | m.p. 43–44° | 168 |
| | | *Metathesis* | | | |
| 2-$PyCH_2Br$ | NaCN, $Et_2SO$ or MeEtSO, 92–94°, 5 hr | 2-$PyCH_2CN$ | | | 318 |

| Starting material | Conditions | Product | Yield | Properties | Ref. |
|---|---|---|---|---|---|
| 2-$PyCH_2Cl \cdot HCl$ | $NaHCO_3$, $H_2O$; NaCN, MeOH, 10 hr *or* KCN, KI, 60% EtOH, 45–50°, 9 hr *or* NaCN, DMSO | 2-$PyCH_2CN$ | 100 g → 50 g<br>87% | b.p. 85–90°/1–2 mm<br>b.p. 118–120°/0.3 mm | 270<br>296<br>319 |
| 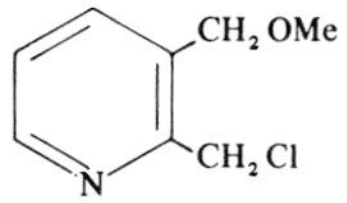 $CH_2OMe$, $CH_2Cl$, N | NaCN, EtOH, reflux 10 hr | 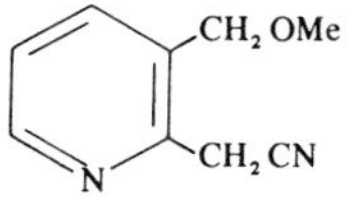 $CH_2OMe$, $CH_2CN$, N | 83.5% | b.p. 110–115°/0.05 mm; picrate, m.p. 146–148° | 9 |
| 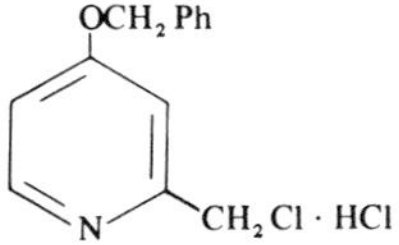 $OCH_2Ph$, $CH_2Cl \cdot HCl$, N | KCN, KI, EtOH, heat 3 hr | 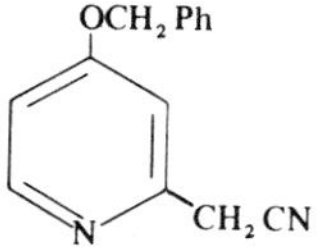 $OCH_2Ph$, $CH_2CN$, N | 81% | picrate, m.p. 175° | 282 |
| 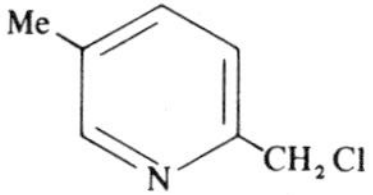 Me, $CH_2Cl$, N | KCN, EtOH, reflux 8 hr | 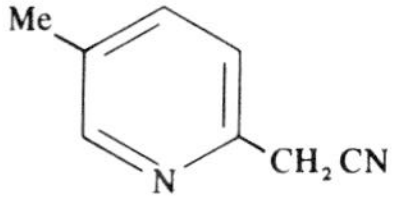 Me, $CH_2CN$, N | 7 g → 5 g | b.p. 105–106°/3.0 mm | 192 |
| 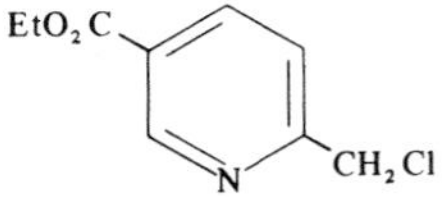 $EtO_2C$, $CH_2Cl$, N | NaCN, EtOH, $H_2O$, reflux 23 min | 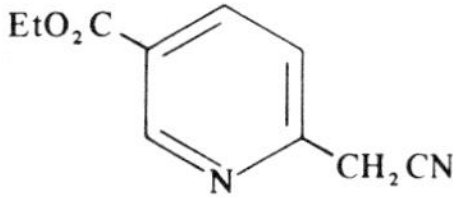 $EtO_2C$, $CH_2CN$, N | 1.75 g → 1.2 g | | 17 |
| 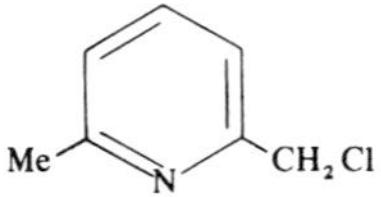 Me, $CH_2Cl$, N | NaCN, MeOH, 13 hr *or* KCN, MeOH, steam-bath 2 hr | 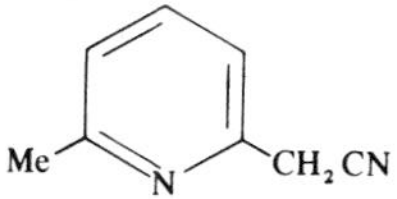 Me, $CH_2CN$, N | 65%<br>5 g → 2.5 g | b.p. 105°/12 mm; picrate, m.p. 151°<br>m.p. 40–41° | 200<br>130 |

TABLE XI-26. Side-Chain Nitrile Synthesis: Miscellaneous Methods (Continued)

| Starting material | Conditions | Product | Yield | Properties | Ref. |
|---|---|---|---|---|---|
| 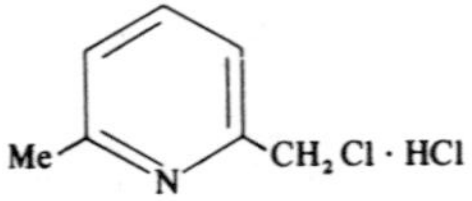 Me, N, $CH_2Cl \cdot HCl$ | KCN, KI, 60% EtOH, 45–50°, 9 hr | 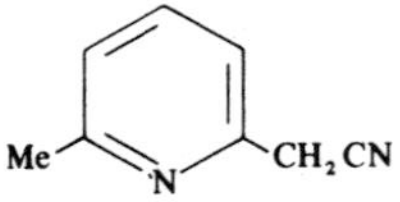 Me, N, $CH_2CN$ | | | 296 |
| 2-Py$(CH_2)_2CH(OH)SO_3H$ | aq. KCN | 2-Py$(CH_2)_2CH(OH)CN$ | 63% | m.p. 81–82° | 219 |
| 2-Py$(CH_2)_4CH(OH)SO_3H$ | aq. KCN | 2-Py$(CH_2)_4CH(OH)CN$ | 78.5% | m.p. 62° | 219 |
| 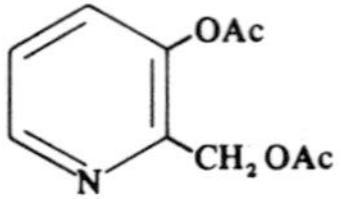 OAc, N, $CH_2OAc$ | KCN, MeOH, 60–65°, 2 hr | 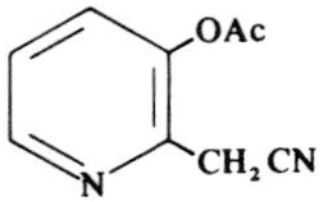 OAc, N, $CH_2CN$ | | | 18 |
| 2-PyC(Cl)=NOH or as a HCl salt | KCN, MeOH, 60°, 1 hr | 2-PyC(=NOH)CN | | m.p. 218–221° | 320 |
| 3-Py$CH_2Cl \cdot HCl$ | NaCN, 95% EtOH, $H_2O$, reflux 1 hr *or* NaCN, DMSO | 3-Py$CH_2CN$ | 76.5–82.5% | b.p. 126°/7 mm; $n_D^{20}$ 1.5279<br>b.p. 143–145°/15 mm; methiodide, m.p. 120–121° | 268<br>319 |
| O, O, $CH_2Cl$, Me | KCN, acetone, $H_2O$, reflux 16–20 hr | O, O, $CH_2CN$, Me | 86% | m.p. 90–91° | 195 |
| | *or* NaCN, DMSO, 140°, 15 min | | 91% | m.p. 89–90°; nmr | 125 |

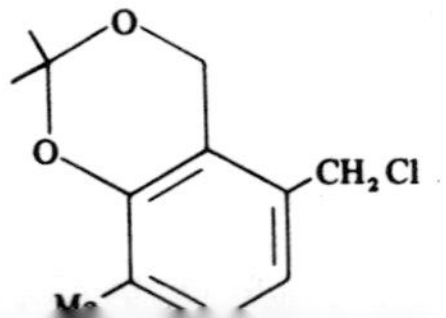

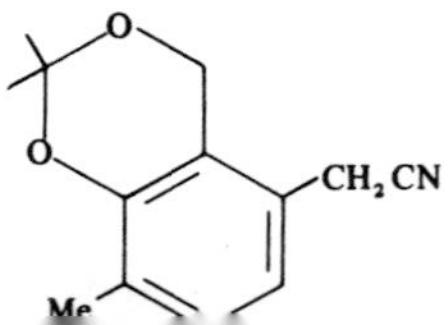

| Starting material | Reagents / conditions | Product | Yield | Properties | Ref. |
|---|---|---|---|---|---|
| 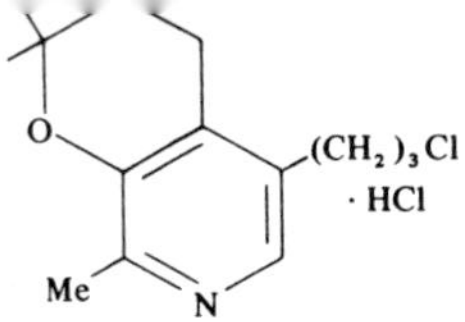 O, Me, N, $(CH_2)_3Cl \cdot HCl$ | NaCN, DMSO, 140° | 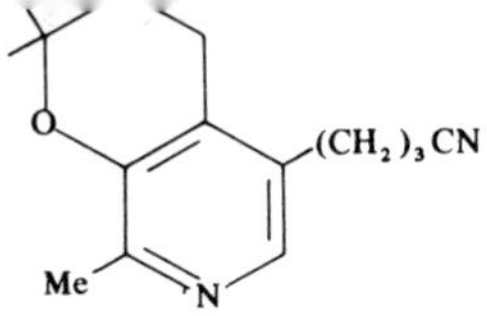 O, Me, N, $(CH_2)_3CN$ | | | 125 |
| 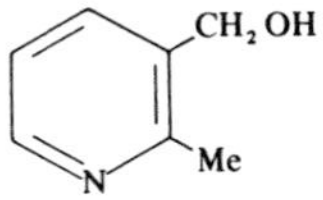 $CH_2OH$, Me, N | $SOCl_2$, reflux 2.5 hr; KI, KCN, EtOH, reflux 5 hr | 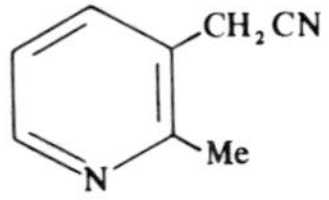 $CH_2CN$, Me, N | 119 g → 89 g | b.p. 136–137°; $n_D^{25}$ 1.5255; picrate, m.p. 149–149.5° | 174 |
| 4-Py$CH_2Cl$·HCl | KCN, MeOH, $H_2O$, reflux 2 hr | 4-Py$CH_2CN$ | 45 g → 36 g | b.p. 92°/0.5 mm; m.p. 36°; hydrochloride, m.p. 270° (decomp.) | 165 |
| | *or* KCN, KI, 60% EtOH, 45–50°, 9 hr | | | m.p. 43° | 296 |
| | *or* NaCN, DMSO | | | b.p. 145°/18 mm; methiodide, m.p. 156–158° | 319 |
| 4-PyC(Cl)=NOH·HCl | KCN, 60°, 1 hr, MeOH | 4-PyC(=NOH)CN | | m.p. 275–278° (decomp.) | 320 |
| | | *Strecker syntheses* | | | |
| 2-PyCHO | KCN, dil. HCl | 2-PyCH(OH)CN | | | 321, 217 |
| | KCN, $Me_2NH$, $H_2O$, overnight | 2-PyCH(CN)$NMe_2$ | | b.p. 107°/0.8 mm | 52 |
| | KCN, $Et_2NH$, $H_2O$, overnight | 2-PyCH(CN)$NEt_2$ | | b.p. 140–145°/0.4 mm | 52 |
| | KCN, pyrrolidine, $H_2O$, overnight | 2-PyCH(1-pyrrolidinyl)CN | | b.p. 145°/0.8 mm | 52 |

TABLE XI-26. Side-Chain Nitrile Synthesis: Miscellaneous Methods (Continued)

| Starting material | Conditions | Product | Yield | Properties | Ref. |
|---|---|---|---|---|---|
| | KCN, piperidine, $H_2O$, overnight | 2-PyCH(piperidino)CN | | b.p. 150–152°/0.5 mm; m.p. 66–67° | 52 |
| | KCN, morpholine, $H_2O$, overnight | 2-PyCH(morpholino)CN | | m.p. 90–92° | 52 |
| | KCN, $COCl_2$, $H_2O$, PhMe, 5°, 1 hr | $[2\text{-PyCH(CN)CO}]_2CO$ | 60% | | 322 |
| 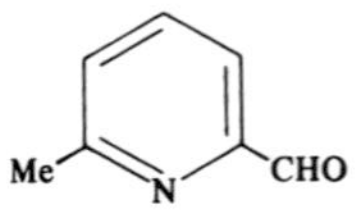 Me–N–CHO | KCN, dil. HCl | 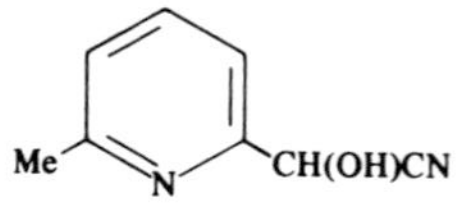 Me–N–CH(OH)CN | | m.p. 134° | 321 |
| | KCN, $Me_2NH$, $H_2O$, overnight | 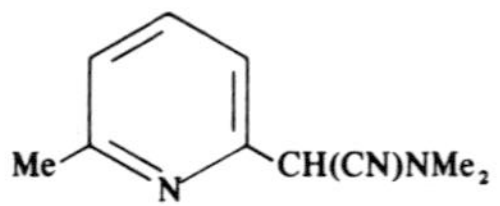 Me–N–$CH(CN)NMe_2$ | | b.p. 112–113°/0.6 mm | 52 |
| | KCN, $Et_2NH$, $H_2O$, overnight | 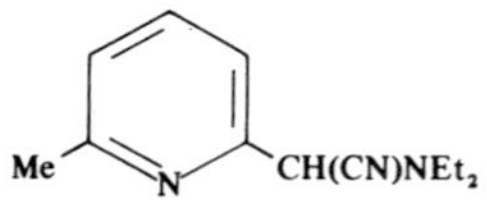 Me–N–$CH(CN)NEt_2$ | | b.p. 120–125°/0.7 mm | 52 |
| | KCN, pyrrolidine, $H_2O$, overnight | 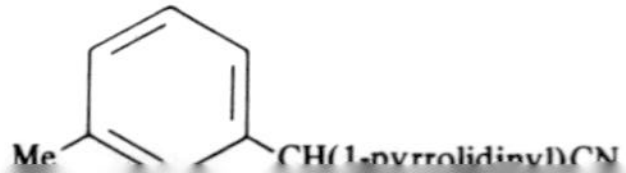 Me–N–CH(1-pyrrolidinyl)CN | | b.p. 136–138°/0.07 mm | 52 |

| | | | | | |
|---|---|---|---|---|---|
| | KCN, piperidine, $H_2O$, overnight | 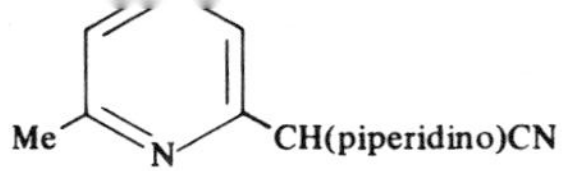 Me … N … CH(piperidino)CN | | b.p. 160°/1.0 mm | 52 |
| | KCN, morpholine, $H_2O$, overnight | 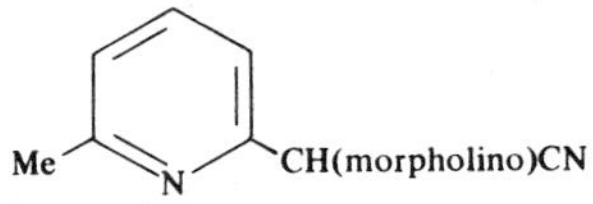 Me … N … CH(morpholino)CN | | b.p. 164–165°/0.25 mm | 52 |
| 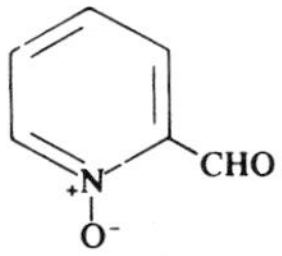 $\overset{+}{N}$–$O^-$ … CHO | KCN, R. T., 1 hr | 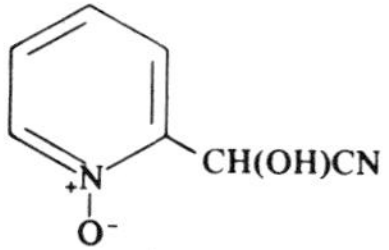 $\overset{+}{N}$–$O^-$ … CH(OH)CN | 61% | m.p. 125° | 323 |
| 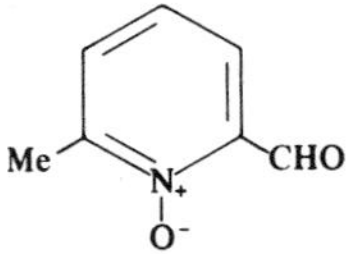 Me … $N^+$–$O^-$ … CHO | KCN, R.T., 1 hr | 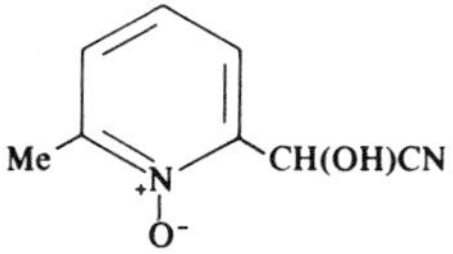 Me … $\overset{+}{N}$–$O^-$ … CH(OH)CN | 49% | m.p. 125° | 323 |
| 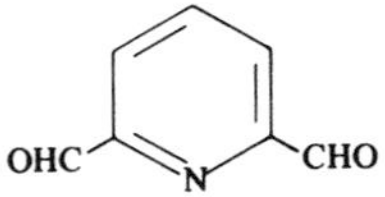 OHC … N … CHO | KCN, dil. HCl | 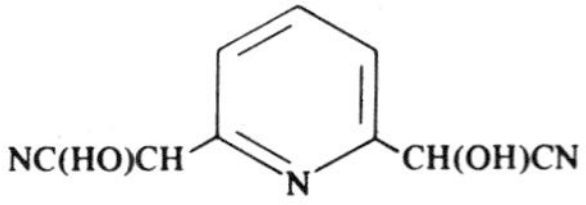 NC(HO)CH … N … CH(OH)CN | | m.p. 105° (decomp.) | 321 |
| 2-PyCOMe | KCN, dil. HCl | 2-PyCMe(OH)CN | | m.p. 50–51° | 321 |
| | | *Strecker syntheses* | | | |
| 3-PyCHO | same as above | 3-PyCH(OH)CN | | | 321 |
| | KCN, $NH_4Cl$ | 3-PyCH($NH_2$)CN | | | 27 |

TABLE XI-26. Side-Chain Nitrile Synthesis: Miscellaneous Methods (Continued)

| Starting material | Conditions | Product | Yield | Properties | Ref. |
|---|---|---|---|---|---|
|  | PhCOCl, $C_6H_6$, –5°; aq. KCN, 4 hr | 3-PyC(OH)(CN)COPh | 31% | m.p. 95–96° | 324 |
| 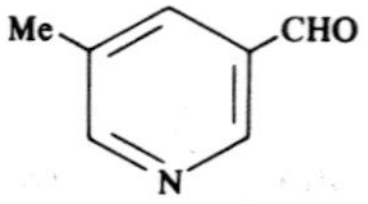 Me, CHO, N | 2*N* HCl, –4 to –5° 20% aq. KCN | 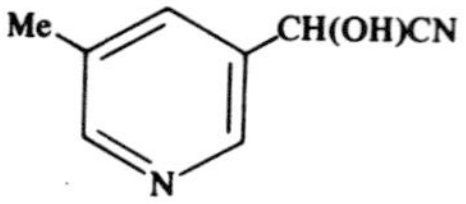 Me, CH(OH)CN, N |  | m.p. 99.5° | 325 |
| 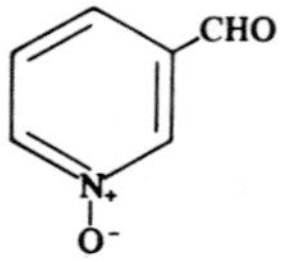 CHO, $N^+$, $O^-$ | KCN, R.T., 1 hr | 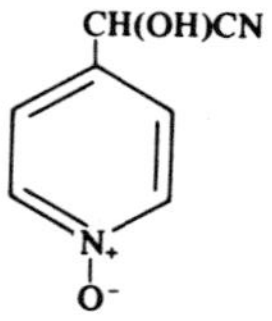 CH(OH)CN, $N^+$, $O^-$ | 36% | m.p. 207° | 323 |
| 4-PyCHO | KCN, $NH_4Cl$ | 4-PyCH($NH_2$)CN |  |  | 27 |
|  | $MeNH_2$, $NaHSO_3$, EtOH; NaCN, $H_2O$, 15 hr | 4-PyCH(NHMe)CN·2HCl |  | m.p. 165–166° | 63 |
| 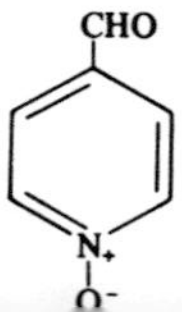 CHO, $N^+$, $O^-$ | KCN, R.T., 1 hr | 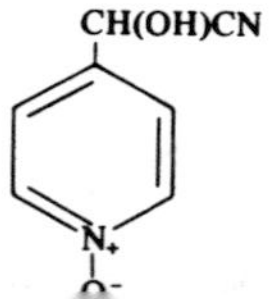 CH(OH)CN, $N^+$, $O^-$ | 98% | m.p. 118° | 323 |

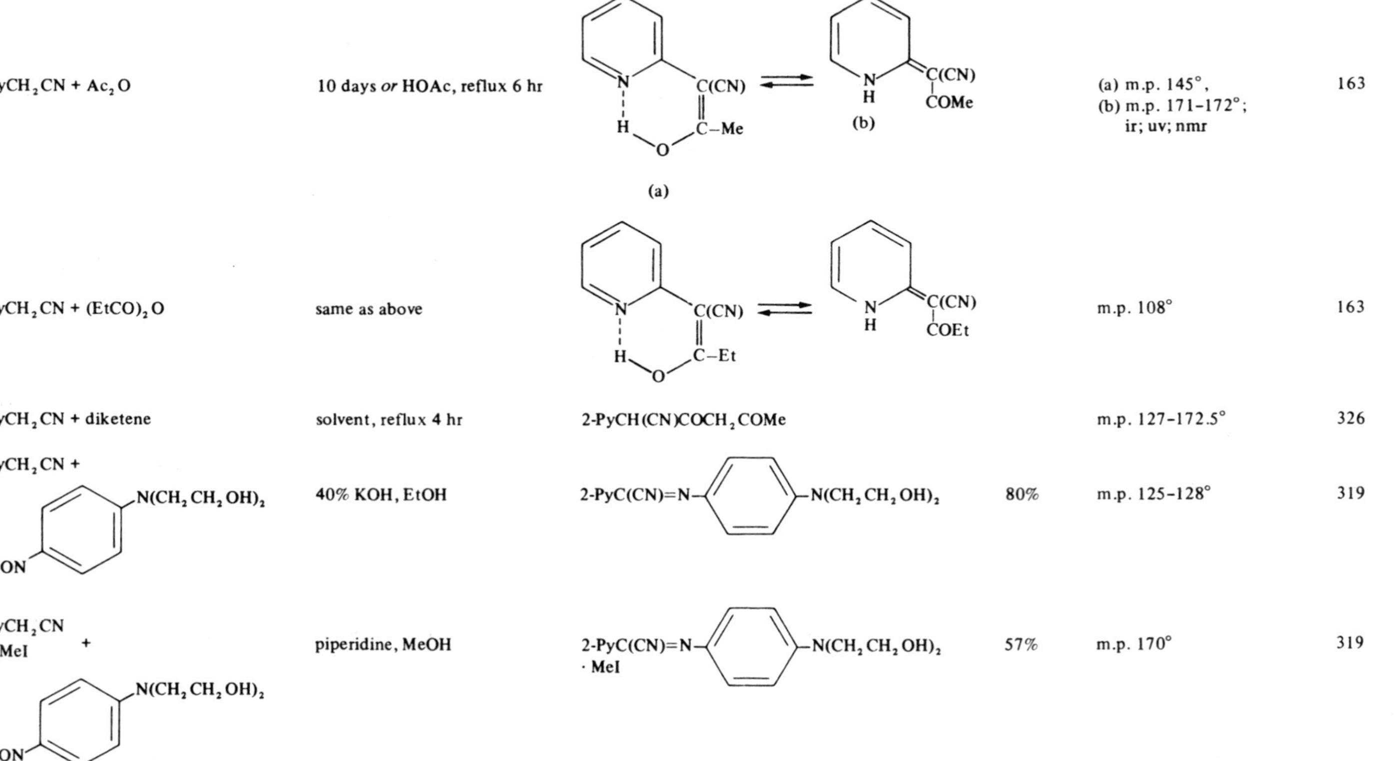

| | | | | | |
|---|---|---|---|---|---|
| $2\text{-PyCH}_2\text{CN} + \text{Ac}_2\text{O}$ | 10 days *or* HOAc, reflux 6 hr | (a) ⇌ (b) | | (a) m.p. 145°, (b) m.p. 171–172°; ir; uv; nmr | 163 |
| $2\text{-PyCH}_2\text{CN} + (\text{EtCO})_2\text{O}$ | same as above | ⇌ | | m.p. 108° | 163 |
| $2\text{-PyCH}_2\text{CN}$ + diketene | solvent, reflux 4 hr | $2\text{-PyCH(CN)COCH}_2\text{COMe}$ | | m.p. 127–172.5° | 326 |
| $2\text{-PyCH}_2\text{CN}$ + ON–C₆H₄–$\text{N(CH}_2\text{CH}_2\text{OH)}_2$ | 40% KOH, EtOH | $2\text{-PyC(CN)=N}$–C₆H₄–$\text{N(CH}_2\text{CH}_2\text{OH)}_2$ | 80% | m.p. 125–128° | 319 |
| $2\text{-PyCH}_2\text{CN}$ · MeI + ON–C₆H₄–$\text{N(CH}_2\text{CH}_2\text{OH)}_2$ | piperidine, MeOH | $2\text{-PyC(CN)=N}$–C₆H₄–$\text{N(CH}_2\text{CH}_2\text{OH)}_2$ · MeI | 57% | m.p. 170° | 319 |

TABLE XI-26. Side-Chain Nitrile Synthesis: Miscellaneous Methods (Continued)

| Starting material | Conditions | Product | Yield | Properties | Ref. |
|---|---|---|---|---|---|
| | | *Others* | | | |
| 2-Py$CH_2CN$ + ON–$C_6H_4$–N$(CH_2CH_2Cl)_2$ | 40% KOH, EtOH | 2-PyC(CN)=N–$C_6H_4$–N$(CH_2CH_2Cl)_2$ | 72% | m.p. 96–97° | 319 |
| 2-Py$CH_2CN$ · MeI + ON–$C_6H_4$–N$(CH_2CH_2Cl)_2$ | piperidine, MeOH | 2-PyC(CN)=N–$C_6H_4$–N$(CH_2CH_2Cl)_2$ · MeI | 67% | m.p. 154–156° (decomp.) | 319 |
| 2-Py$CH_2$CH(CN)$CO_2$Me | $(MeO)_2$P(=S)SH, $Et_3$N, steam bath, 1 hr | 2-PyCH(SP(=S)$(OMe)_2$)CH(CN)$CO_2$Me | | | 65 |
| 2-PyC(CN)=C(OH)Me | EtI, aq. NaOH, $Ag^{\oplus}$ | 2-PyC(CN)=C(OEt)Me | | m.p. 81–82°; ir; uv; nmr | 163 |
| 2-Py$CO_2$Et | NaOH, PhMe; MeCN; reflux 4 hr; R.T. overnight | 2-PyCO$CH_2$CN | 18% | m.p. 93–94° | 360 |
| 2-Py$CO_2$Et + Ph$CH_2$CN | NaOEt, reflux | 2-PyCOCHPhCN | 94% | m.p. 122–123° | 275 |
| 2,6-Me$_2$-1-OMe-pyridinium $I^-$ | KCN, R.T., 1 hr | 2-NC$CH_2$-6-Me-pyridine (33%) + 4-CN-2,6-Me$_2$-pyridine (13%) | | b.p. 125–133°/22 mm; picrate, m.p. 176–179° | 327 |

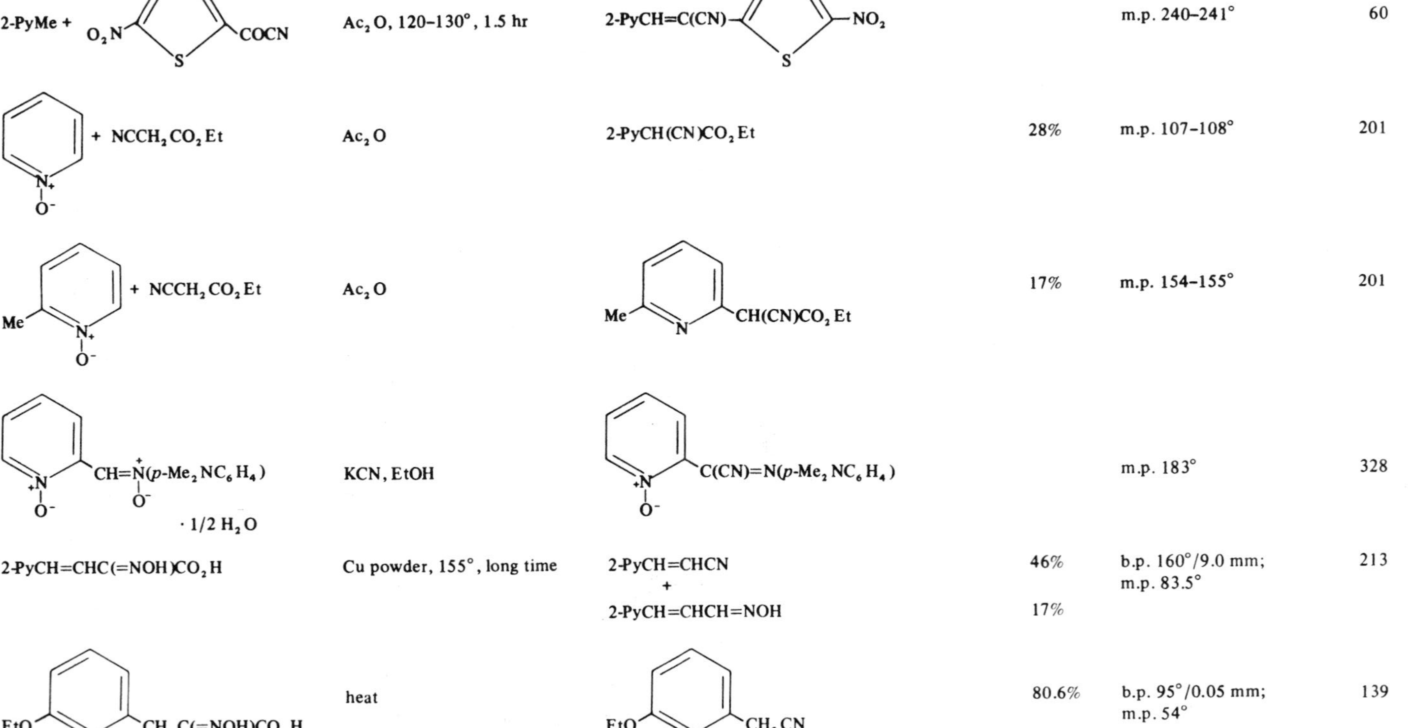

| | | | | | |
|---|---|---|---|---|---|
| 2-PyMe + $O_2N$–(thiophene)–COCN | $Ac_2O$, 120–130°, 1.5 hr | 2-PyCH=C(CN)–(thiophene)–$NO_2$ | | m.p. 240–241° | 60 |
| Pyridine N-oxide + $NCCH_2CO_2Et$ | $Ac_2O$ | 2-PyCH(CN)$CO_2$Et | 28% | m.p. 107–108° | 201 |
| 2-Me-pyridine N-oxide + $NCCH_2CO_2Et$ | $Ac_2O$ | 6-Me-2-Py–CH(CN)$CO_2$Et | 17% | m.p. 154–155° | 201 |
| Pyridine N-oxide–CH=$\overset{+}{N}$($O^-$)(p-$Me_2NC_6H_4$) · 1/2 $H_2O$ | KCN, EtOH | Pyridine N-oxide–C(CN)=N(p-$Me_2NC_6H_4$) | | m.p. 183° | 328 |
| 2-PyCH=CHC(=NOH)$CO_2$H | Cu powder, 155°, long time | 2-PyCH=CHCN<br>+<br>2-PyCH=CHCH=NOH | 46%<br><br>17% | b.p. 160°/9.0 mm;<br>m.p. 83.5° | 213 |
| EtO–(pyridine)–$CH_2$C(=NOH)$CO_2$H | heat | EtO–(pyridine)–$CH_2$CN | 80.6% | b.p. 95°/0.05 mm;<br>m.p. 54° | 139 |

TABLE XI-26. Side-Chain Nitrile Synthesis: Miscellaneous Methods (Continued)

| Starting material | Conditions | Product | Yield | Properties | Ref. |
|---|---|---|---|---|---|
| | | *Others* | | | |
| 2-PyC(Cl)=NOH · HCl | KCN, 60°, 1 hr | 2-PyC(=NO⊖K⊕)CN | 11% | m.p. 218–221° | 329 |
| 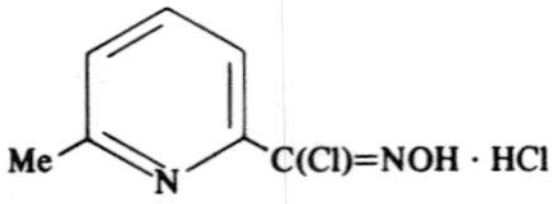 Me, N, C(Cl)=NOH · HCl | KCN, 60°, 1 hr | 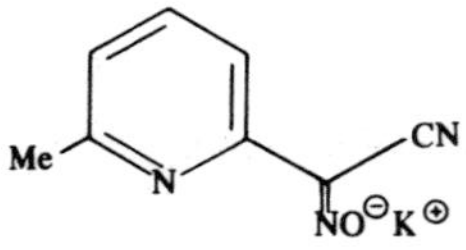 Me, N, C(=NO⊖K⊕)CN | 47% | m.p. 209–210° | 329 |
| 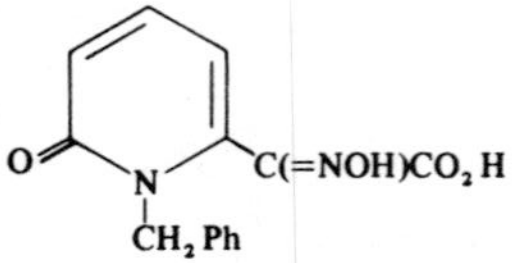 O, N, $CH_2Ph$, C(=NOH)$CO_2H$ | heat | 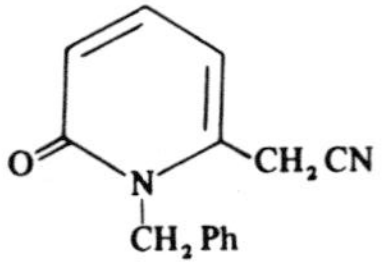 O, N, $CH_2Ph$, $CH_2CN$ | 88.5% | m.p. 92° | 139 |
| 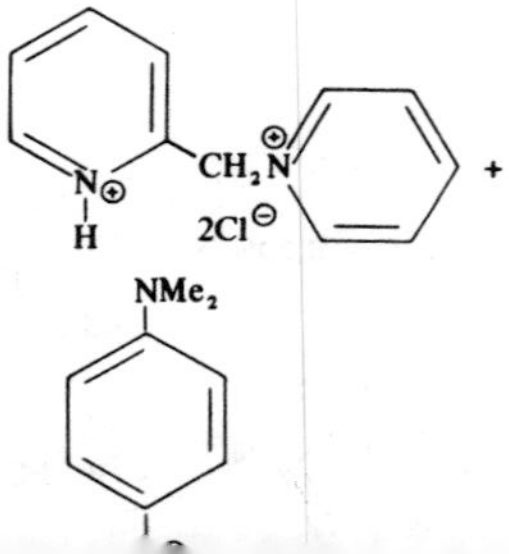 N⊕–H, $CH_2N$⊕, 2Cl⊖ + $NMe_2$ | NaCN, EtOH, 0°, 3 hr | 2-PyC(CN)=N(*p*-$Me_2NC_6H_4$) | 68% | m.p. 116–118° | 295 |

| Reactants | Conditions | Product | Yield | Properties | Ref. |
|---|---|---|---|---|---|
| 1-Methyl-2-(pyridiniomethyl)pyridinium $2I^{\ominus}$ + p-$Me_2NC_6H_4NO$ | NaCN | 2-PyC(CN)=N(p-$Me_2NC_6H_4$) · MeI | 81% | m.p. 189–191° | 295 |
| 1-(1-Oxido-2-pyridylmethyl)pyridinium $I^{\ominus}$ + p-$Me_2NC_6H_4NO$ | KCN, 95% EtOH, reflux 2 hr | 1-Oxido-2-pyridyl-C(CN)=N(p-$Me_2NC_6H_4$) | 64% | m.p. 183° | 328 |
| 2-PyCOCH$_2$-pyridinium $ClO_4^{\ominus}$ + p-$Me_2NC_6H_4NO$ | NaCN | 2-PyCOC(CN)=N(p-$Me_2NC_6H_4$) | 97% | m.p. 160–161° | 295 |

TABLE XI-26. Side-Chain Nitrile Synthesis: Miscellaneous Methods (Continued)

| Starting materials | Conditions | Product | Yield | Properties | Ref. |
|---|---|---|---|---|---|
| | | *Others* | | | |
| Me, Me, N, Me, $CH_2$ | ClCN, $C_6H_6$, 65°, 5 hr | Me, $NCCH_2$, N, Me, CHCN | | m.p. 196–197° | 330 |
| 3-$PyCH_2CN$ | $Et_2CO_3$, NaOEt, EtOH, reflux 4 hr | 3-$PyCH(CO_2Et)CN$ | 80% | m.p. 103–104° | 168 |
| 3-$PyCH_2CN$ + $MeCO_2Et$ | NaOEt | 3-PyCH(CN)COMe | | | 331 |
| 3-$PyCH_2CN$ + $PhCO_2Et$ | NaOMe, EtOH, reflux 2.5 hr | 3-PyC(CN)=CPhOH | | m.p. 245–247° | 42 |
| | | + 3-$PyCH_2$(H$_2$N)C=C(3-Py)CN | | m.p. 117–119.5° | |
| 3-$PyCH_2CN$ + $p$-$ClC_6H_4CO_2Et$ | same as above | 3-PyC(CN)=C($p$-$ClC_6H_4$)OH | | m.p. 248–250° | 42 |
| 3-$PyCH_2CN$ + $p$-$Me_2NC_6H_4CO_2Et$ | same as above | 3-PyC(CN)=C($p$-$Me_2NC_6H_4$)OH | | m.p. 165–166° | 332 |
| 3-$PyCH_2CN$ + $Me_2C$($p$-$ClC_6H_4$)–$C_6H_4$–$CO_2Et$ | same as above | 3-PyC(CN)=C(OH)–$C_6H_4$–($p$-$ClC_6H_4$)$Me_2$ | | m.p. 131–135° | 332 |

| Reactants | Conditions | Product | Yield | Properties | Ref. |
|---|---|---|---|---|---|
| $3\text{-PyCH}_2\text{CN}$ + $p\text{-ON-C}_6\text{H}_4\text{-N(CH}_2\text{CH}_2\text{OH)}_2$ | 40% KOH, EtOH | $3\text{-PyC(CN)=N-C}_6\text{H}_4\text{-}p\text{-N(CH}_2\text{CH}_2\text{OH)}_2$ | 80% | m.p. 131° | 319 |
| $3\text{-PyCH}_2\text{CN}\cdot\text{MeI}$ + $p\text{-ON-C}_6\text{H}_4\text{-N(CH}_2\text{CH}_2\text{OH)}_2$ | piperidine, MeOH | $3\text{-PyC(CN)=N-C}_6\text{H}_4\text{-}p\text{-N(CH}_2\text{CH}_2\text{OH)}_2\cdot\text{MeI}$ | 80% | m.p. 225° (decomp.) | 319 |
| $3\text{-PyCH}_2\text{CN}$ + $p\text{-ON-C}_6\text{H}_4\text{-N(CH}_2\text{CH}_2\text{Cl)}_2$ | 40% KOH, EtOH | $3\text{-PyC(CN)=N-C}_6\text{H}_4\text{-}p\text{-N(CH}_2\text{CH}_2\text{Cl)}_2$ | 72% | m.p. 120° | 319 |
| $3\text{-PyCH}_2\text{CN}\cdot\text{MeI}$ + $p\text{-ON-C}_6\text{H}_4\text{-N(CH}_2\text{CH}_2\text{Cl)}_2$ | piperidine, MeOH | $3\text{-PyC(CN)=N-C}_6\text{H}_4\text{-}p\text{-N(CH}_2\text{CH}_2\text{Cl)}_2\cdot\text{MeI}$ | 98% | m.p. 203–204° (decomp.) | 319 |

TABLE XI-26. Side-Chain Nitrile Synthesis: Miscellaneous Methods (Continued)

| Starting material | Conditions | Product | Yield | Properties | Ref. |
|---|---|---|---|---|---|
| | | *Others* | | | |
| 3-PyCOCH$_2$–$\overset{+}{N}$C$_5$H$_5$ + $p$-Me$_2$NC$_6$H$_4$NO | NaCN | 3-PyCOC(CN)=N($p$-Me$_2$NC$_6$H$_4$) | 93% | m.p. 156–157° | 295 |
| 3-PyCO$_2$Et + MeCN | NaOEt *or* boiled 9 hr; alc. HCl | 3-PyCOCH$_2$CN | | m.p. 180–183°<br>m.p. 182–184° | 269<br>103 |
| 3-PyCO$_2$Et + PhCH$_2$CN | Na, EtOH, reflux | 3-PyCOCHPhCN | | m.p. 137–141° | 225 |
| 3-PyCO$_2$Et + $p$-FC$_6$H$_4$CH$_2$CN | NaOMe, EtOH, reflux 2.5 hr | 3-PyC(OH)=C($p$-FC$_6$H$_4$)CN | | m.p. 220–223° | 42 |
| 3-PyCO$_2$Et + $m$-ClC$_6$H$_4$CH$_2$CN | same as above | 3-PyC(OH)=C($m$-ClC$_6$H$_4$)CN | | m.p. 188–190° | 42 |
| 3-PyCO$_2$Et + $p$-ClC$_6$H$_4$CH$_2$CN | NaOMe, EtOH, reflux 2.5 hr | 3-PyC(OH)=C($p$-ClC$_6$H$_4$)CN | | m.p. 217–219° | 42 |
| 3-PyCO$_2$Et + $p$-CH$_3$C$_6$H$_4$CH$_2$CN | NaOMe, EtOH, reflux 2.5 hr | 3-PyC(OH)=C($p$-CH$_3$C$_6$H$_4$)CN | | m.p. 189–191° | 42 |
| 3-PyCO$_2$Et + $p$-MeOC$_6$H$_4$CH$_2$CN | Na$^{\oplus}$ $\overset{\ominus}{C}$H$_2$SOMe | 3-PyCOCH($p$-MeOC$_6$H$_4$)CN | 72% | m.p. 155–156° | 16 |
| 3-PyCO$_2$Et + 3-PyCH$_2$CN | NaOMe, EtOH, reflux 2.5 hr | 3PyC(OH)=C(3-Py)CN | | m.p. 251–252° (decomp.); uv | 333 |
| 3-PyCH=CHC(=NOH)CO$_2$H | 155° | 3-PyCH=CHCN | 79% | b.p. 161–165°/11–12 mm | 213 |

| $MeCOCH(CH_2CH_2CN)_2$ | Na, EtOH, reflux 6 hr | 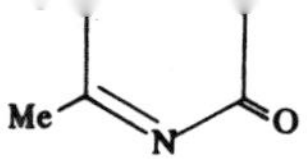 Me, N, O (ring) | 53% | m.p. 166° | 334 |
|---|---|---|---|---|---|
| $MeCOCMe(CH_2CH_2CN)_2$ | Na, EtOH, reflux 6 hr | 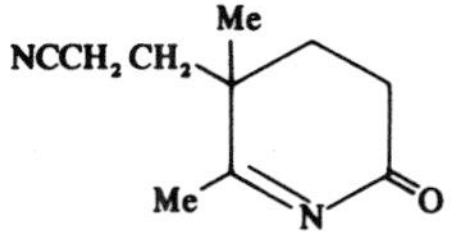 $NCCH_2CH_2$, Me, Me, N, O (ring) | 79.5% | m.p. 157–158° | 334 |
| $MeCOC(CH_2CH_2CN)_3$ | Na, EtOH, reflux 6 hr | 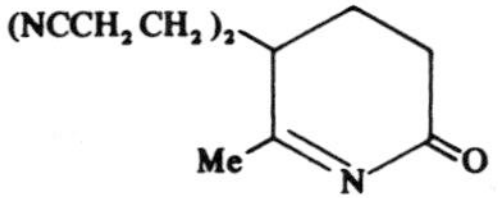 $(NCCH_2CH_2)_2$, Me, N, O (ring) | 67% | m.p. 222–223° | 334 |
| 4-PyCN + $TsCH_2CN$ | NaOMe; $H_2SO_4$ | $4\text{-}PyC(NH_2)=C(CN)Ts$ | 57% | m.p. 246–249° | 335 |
| $4\text{-}PyCH_2CN$ | $Ac_2O$, heat | 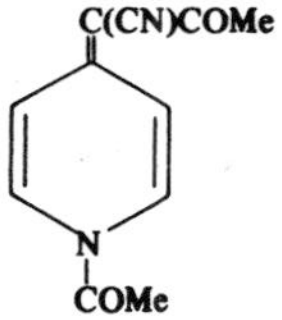 C(CN)COMe, N, COMe (ring) | 62% | m.p. 216–217°; ir; nmr | 163 |
| $4\text{-}PyCH_2CN$ | $Ac_2O$, R.T., 10 min | 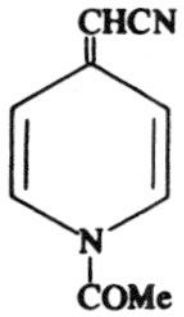 CHCN, N, COMe (ring) | 87% | m.p. 149–150°; ir; uv; nmr | 163 |
| $4\text{-}PyCH_2CN$ | $Et_2CO_3$, NaOEt, EtOH, reflux 4 hr | $4\text{-}PyCH(CN)CO_2Et$ | 64% | m.p. 210–212° | 168 |

TABLE XI-26. Side-Chain Nitrile Synthesis: Miscellaneous Methods (Continued)

| Starting materials | Conditions | Product | Yield | Properties | Ref. |
|---|---|---|---|---|---|
| | | *Others* | | | |
| 4-PyCH$_2$CN + ON–C$_6$H$_4$–N(CH$_2$CH$_2$OH)$_2$ | 40% KOH, EtOH | 4-PyC(CN)=N–C$_6$H$_4$–N(CH$_2$CH$_2$OH)$_2$ | 80% | m.p. 167–168° | 319 |
| 4-PyCH$_2$CN · MeI + ON–C$_6$H$_4$–N(CH$_2$CH$_2$OH)$_2$ | piperidine, MeOH | 4-PyC(CN)=N–C$_6$H$_4$–N(CH$_2$CH$_2$OH)$_2$ · MeI | 78% | methiodide, m.p. 218–219° (decomp.) | 319 |
| 4-PyCH$_2$CN + ON–C$_6$H$_4$–N(CH$_2$CH$_2$Cl)$_2$ | 40% KOH, EtOH | 4-PyC(CN)=N–C$_6$H$_4$–N(CH$_2$CH$_2$Cl)$_2$ | 85% | m.p. 149–150° | 319 |
| 4-PyCH$_2$CN · MeI + ON–C$_6$H$_4$–N(CH$_2$CH$_2$Cl)$_2$ | piperidine, MeOH | 4-PyC(CN)=N–C$_6$H$_4$–N(CH$_2$CH$_2$Cl)$_2$ | 50% | m.p. 195–196° (decomp.) | 319 |

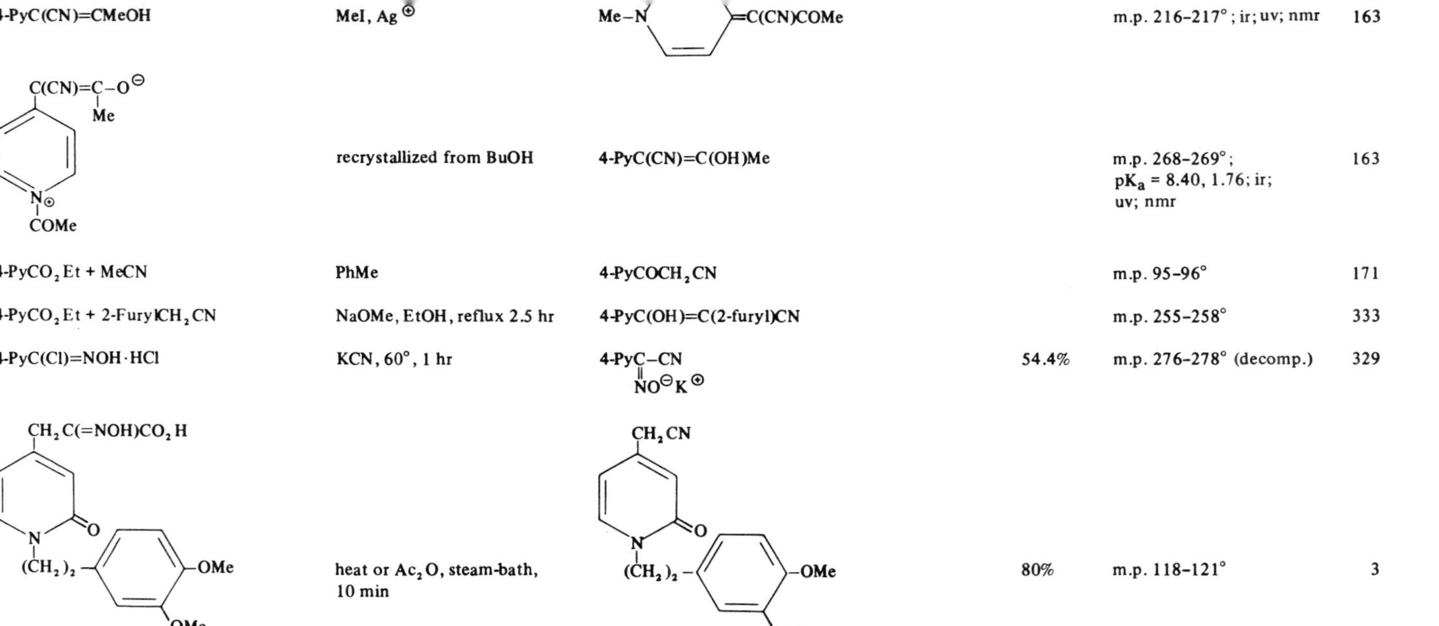

| Starting material | Conditions | Product | Yield | Properties | Ref. |
|---|---|---|---|---|---|
| 4-PyC(CN)=CMeOH | MeI, $Ag^{\oplus}$ | Me–N(pyridylidene)=C(CN)COMe | | m.p. 216–217°; ir; uv; nmr | 163 |
| 4-[C(CN)=C(Me)–$O^{\ominus}$]-1-COMe-pyridinium | recrystallized from BuOH | 4-PyC(CN)=C(OH)Me | | m.p. 268–269°; $pK_a$ = 8.40, 1.76; ir; uv; nmr | 163 |
| 4-$PyCO_2Et$ + MeCN | PhMe | 4-$PyCOCH_2CN$ | | m.p. 95–96° | 171 |
| 4-$PyCO_2Et$ + 2-Furyl$CH_2CN$ | NaOMe, EtOH, reflux 2.5 hr | 4-PyC(OH)=C(2-furyl)CN | | m.p. 255–258° | 333 |
| 4-PyC(Cl)=NOH·HCl | KCN, 60°, 1 hr | 4-PyC(–CN)=$NO^{\ominus}K^{\oplus}$ | 54.4% | m.p. 276–278° (decomp.) | 329 |
| 4-[$CH_2C(=NOH)CO_2H$]-1-[$(CH_2)_2$-3,4-(OMe)$_2$-phenyl]-2-pyridone | heat or $Ac_2O$, steam-bath, 10 min | 4-$CH_2CN$-1-[$(CH_2)_2$-3,4-(OMe)$_2$-phenyl]-2-pyridone | 80% | m.p. 118–121° | 3 |

TABLE XI-26. Side-Chain Nitrile Synthesis: Miscellaneous Methods (Continued)

| Starting materials | Conditions | Product | Yield | Properties | Ref. |
|---|---|---|---|---|---|
| | | *Others* | | | |
| 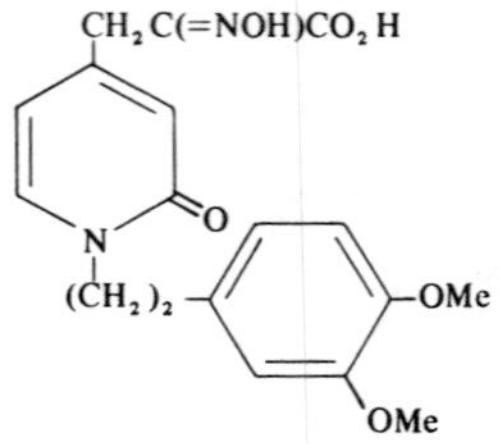 $CH_2C(=NOH)CO_2H$; N–$(CH_2)_2$–(OMe, OMe); =O | $Ac_2O$ | 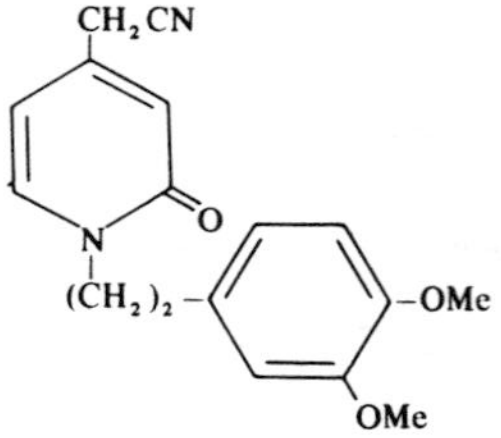 $CH_2CN$; N–$(CH_2)_2$–(OMe, OMe); =O | | m.p. 118–120° | 88 |
| 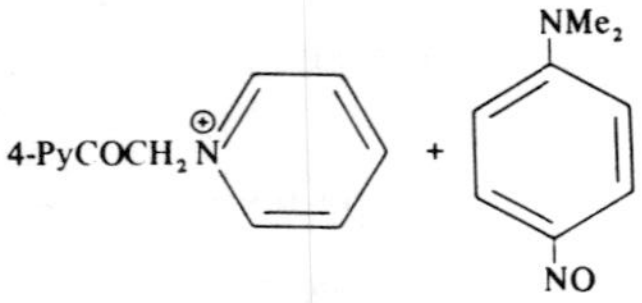 4-PyCOCH$_2$N$^{\oplus}$ + $NMe_2$, NO | NaCN | 4-PyCOC(CN)=N($p$-$Me_2NC_6H_4$) | 93% | m.p. 188–189° | 295 |
| 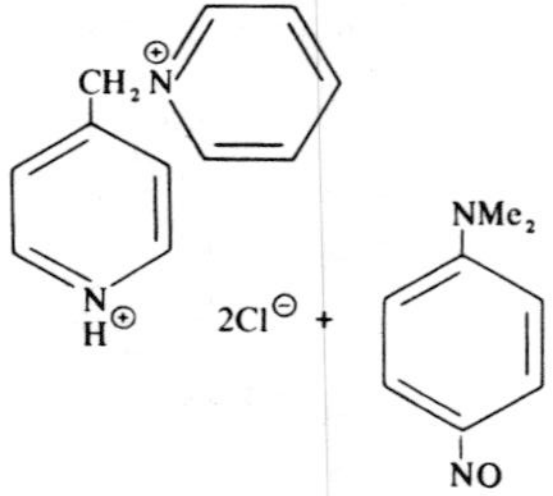 $CH_2N^{\oplus}$; $N^{\oplus}$H $2Cl^{\ominus}$ + $NMe_2$, NO | NaCN, EtOH, 0°, 3 hr | 4-PyC(CN)=N($p$-$Me_2NC_6H_4$) | 64% | m.p. 145–146° | 295 |

| Nitrile | Conditions | Product | Yield | Properties | Ref. |
|---|---|---|---|---|---|
| 2-PyCH$_2$CN | Amberlite IRA-400 (base form), reflux 2 hr | 2-PyCH$_2$CONH$_2$ | | m.p. 119–120° | 163 |
| 2-PyCHPhCN | 96% H$_2$SO$_4$, 24 hr; MeOH, reflux 45 min | 2-PyCHPhCO$_2$Me | 95.2% | m.p. 74–75° | 336 |
| | 96% H$_2$SO$_4$, 24 hr; EtOH, reflux 45 min | 2-PyCHPhCO$_2$Et | 90.6% | b.p. 158°/0.5 mm | 336 |
| | H$_2$S, Et$_3$N, pyridine, R. T., 17 hr *or* P$_2$S$_5$, xylene, heat, 4.5 hr | 2-PyCHPhCSNH$_2$ | | m.p. 137.5–138° | 166 |
| 2-PyCH(*p*-ClC$_6$H$_4$)CN | H$_2$S, Et$_3$N, pyridine, R.T., 6 hr | 2-PyCH(*p*-ClC$_6$H$_4$)CSNH$_2$ | | m.p. 160–160.5° | 166 |
| 2-PyCH(CH$_2$Ph)CN | 98% H$_2$SO$_4$, 2 days | 2-PyCH(CH$_2$Ph)CONH$_2$ | 0.5 g → 0.2 g | m.p. 123–125° | 305 |
| | MeOH, HCl, reflux | 2-PyCH(CH$_2$Ph)CO$_2$Me | 3 g → 2.7 g | | 305 |
| 2-PyC(CH$_2$Ph)$_2$CN | 98% H$_2$SO$_4$, 2 days | 2-PyC(CH$_2$Ph)$_2$CONH$_2$ | | | 305 |
| 2-PyCEt(CN)CH$_2$CH$_2$CN | HOAc, H$_2$SO$_4$, reflux | 3-Et-3-(2-Py)piperidine-2,6-dione (structure: 2-Py, Et, O, NH, O) | | m.p. 84–86°; picrate, m.p. 152–154° | 178 |
| 2-PyCPh(CN)CH$_2$CH$_2$CN | 75% H$_2$SO$_4$, 130°, 15 hr; alc. NH$_3$; HOAc, 0.5 hr steam-bath | 2-PyCHPhCH$_2$CH$_2$CO$_2$H | 124 g → 105 g | m.p. 113–114° | 337 |
| 2-PyC(p-ClC$_6$H$_4$)(CN)CH$_2$CH$_2$CN | same as above | 2-PyCH(*p*-ClC$_6$H$_4$)CH$_2$CH$_2$CO$_2$H | | m.p. 100–101° | 337 |

TABLE XI-27. Solvolysis of Side-Chain Nitriles (Continued)

| Nitrile | Conditions | Product | Yield | Properties | Ref. |
|---|---|---|---|---|---|
| 2-PyCPh(CN)– NMe | 70 % $H_2SO_4$, 147°, 48 hr | 2-PyCPh($CONH_2$)– NMe | | m.p. 150–153° | 304 |
| 2-PyCPh(CN)– NEt | same as above | 2-PyCPh($CONH_2$)– NEt | | m.p. 160–161° | 304 |
| 2-PyCPh(CN)– N-isoPr | same as above | 2-PyCPh($CONH_2$)– N-isoPr | | m.p. 127.5–133° | 304 |
| 2-PyCPh(CN)– N-*n*-Bu | same as above | 2-PyCPh($CONH_2$)– N-*n*-Bu | | m.p. 108–111° | 304 |
| 2-PyCPh(CN)$CH_2CH_2NMe_2$ | conc. $H_2SO_4$, steam-bath 4 hr | 2-PyCPh($CONH_2$)$CH_2CH_2NMe_2$ | | Sulfuric acid salt, m.p. 206–207° | 44 |
| 2-PyCPh(CN)$CH_2CH_2NEt_2$ | same as above | 2-PyCPh($CONH_2$)$CH_2CH_2NEt_2$ | | m.p. 63.5–64.5° | 44 |
| 2-PyCPh(CN)$CH_2CH_2NPr_2$ | same as above | 2-PyCPh($CONH_2$)$CH_2CH_2NPr_2$ | | m.p. 63–64° | 44, 45 |
| 2-PyCPh(CN)$CH_2CH_2N(isoPr)_2$ | same as above | 2-PyCPh($CONH_2$)$CH_2CH_2N(isoPr)_2$ | | m.p. 94.5–95° | 44, 45 |
| 2-PyCPh(CN)$CH_2CH_2N$ | same as above | 2-PyCPh($CONH_2$)$CH_2CH_2N$ | | m.p. 109.5–110.5° | 44 |

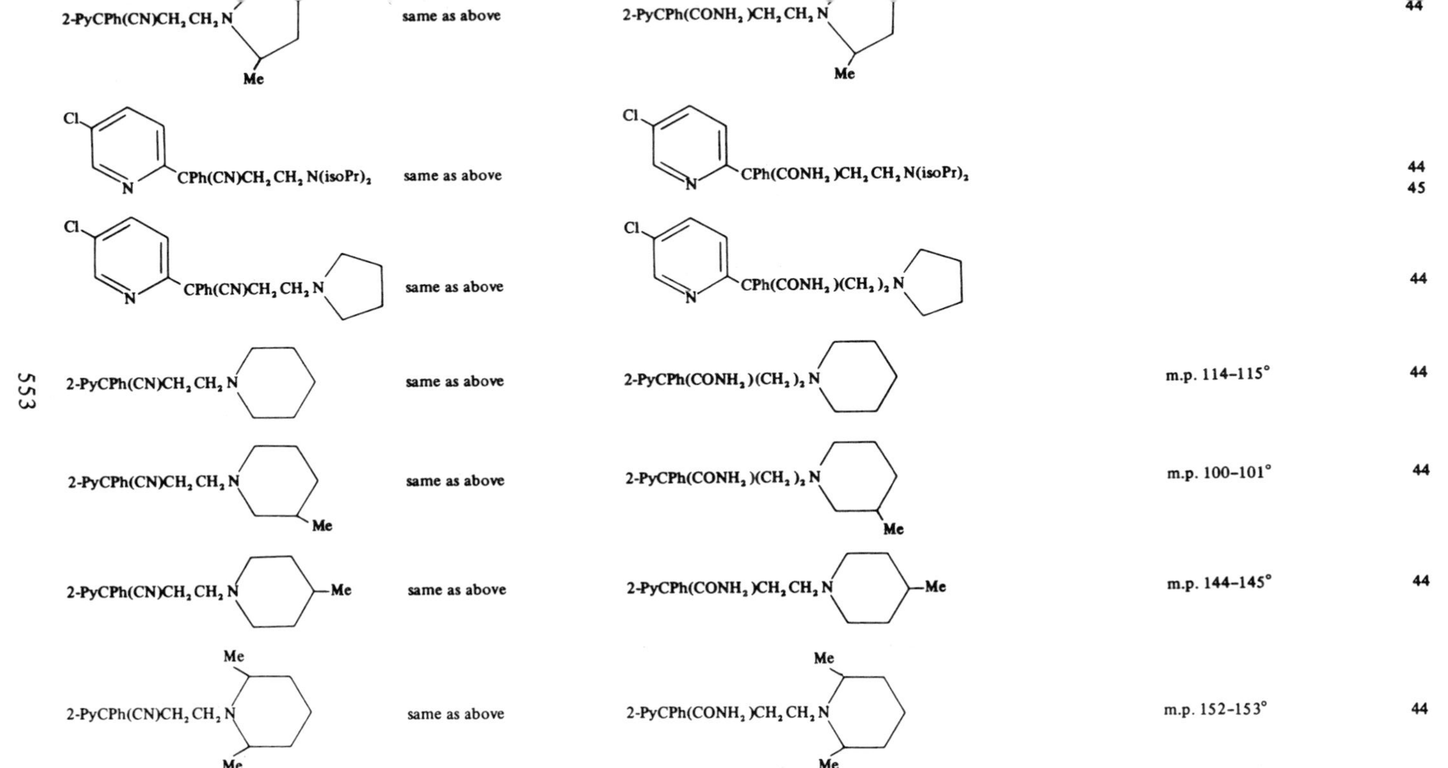

| | | | | |
|---|---|---|---|---|
| 2-PyCPh(CN)$CH_2CH_2N$ (2-methylpyrrolidinyl) | same as above | 2-PyCPh($CONH_2$)$CH_2CH_2N$ (2-methylpyrrolidinyl) | | 44 |
| 5-Cl-2-pyridyl-CPh(CN)$CH_2CH_2N(isoPr)_2$ | same as above | 5-Cl-2-pyridyl-CPh($CONH_2$)$CH_2CH_2N(isoPr)_2$ | | 44<br>45 |
| 5-Cl-2-pyridyl-CPh(CN)$CH_2CH_2N$ (pyrrolidinyl) | same as above | 5-Cl-2-pyridyl-CPh($CONH_2$)$(CH_2)_2N$ (pyrrolidinyl) | | 44 |
| 2-PyCPh(CN)$CH_2CH_2N$ (piperidinyl) | same as above | 2-PyCPh($CONH_2$)$(CH_2)_2N$ (piperidinyl) | m.p. 114–115° | 44 |
| 2-PyCPh(CN)$CH_2CH_2N$ (3-methylpiperidinyl) | same as above | 2-PyCPh($CONH_2$)$(CH_2)_2N$ (3-methylpiperidinyl) | m.p. 100–101° | 44 |
| 2-PyCPh(CN)$CH_2CH_2N$ (4-methylpiperidinyl) | same as above | 2-PyCPh($CONH_2$)$CH_2CH_2N$ (4-methylpiperidinyl) | m.p. 144–145° | 44 |
| 2-PyCPh(CN)$CH_2CH_2N$ (2,6-dimethylpiperidinyl) | same as above | 2-PyCPh($CONH_2$)$CH_2CH_2N$ (2,6-dimethylpiperidinyl) | m.p. 152–153° | 44 |

TABLE XI-27. Solvolysis of Side-Chain Nitriles (Continued)

| Nitrile | Conditions | Product | Yield | Properties | Ref. |
|---|---|---|---|---|---|
| 2-PyCPh(CN)$CH_2CH_2N$ 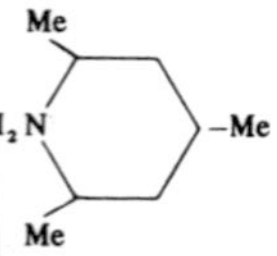 (2,4,6-trimethylpiperidino; Me, Me, –Me) | same as above | 2-PyCPh($CONH_2$)$CH_2CH_2N$ 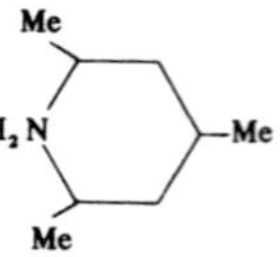 (2,4,6-trimethylpiperidino; Me, Me, –Me) | | m.p. 117–119° | 44 |
| 2-PyCPh(CN)$CH_2CH_2N$ 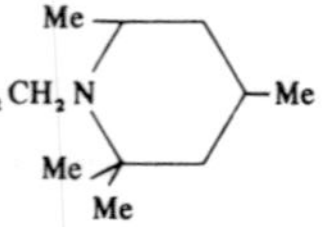 (2,2,4,6-tetramethylpiperidino; Me, Me, Me, Me) | same as above | 2-PyCPh($CONH_2$)$CH_2CH_2N$ 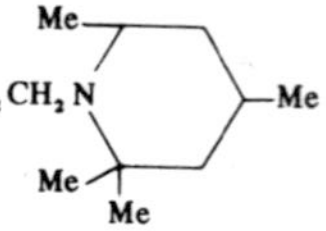 (2,2,4,6-tetramethylpiperidino; Me, Me, Me, Me) | | m.p. 100–103° | 44 |
| 2-PyCH(CN)$CH_2CH_2N$  (hexamethyleneimino) | same as above | 2-PyCH($CONH_2$)$CH_2CH_2N$ 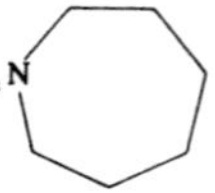 (hexamethyleneimino) | | m.p. 123° | 44 |
| 2-PyCPh(CN)$CH_2CH_2N$ 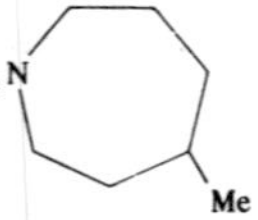 (Me-substituted hexamethyleneimino) | same as above | 2-PyCPh($CONH_2$)$CH_2CH_2N$ 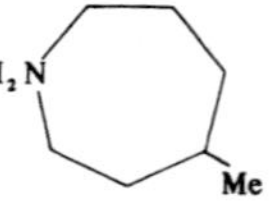 (Me-substituted hexamethyleneimino) | | | 44 |
| 2-PyC($m$-$MeC_6H_4$)(CN)$CH_2CH_2NisoPr_2$ | KOH, EtOH, $H_2O$, reflux 2 hr, *or* $H_2SO_4$, steam-bath 4 hr | 2-PyC($m$-$MeC_6H_4$)($CONH_2$)$CH_2CH_2NisoPr_2$ | | m.p. 113–114° | 44, 45 |
| 2-PyC(CN)$CH_2CH_2NisoPr_2$ (on 3,5-dimethylphenyl; Me, Me) 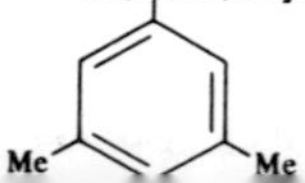 | $H_2SO_4$, 4 hr, steam-bath *or* KOH, EtOH, $H_2O$, [illegible] | 2-PyC($CONH_2$)$CH_2CH_2NisoPr_2$ (on 3,5-dimethylphenyl; Me, Me) 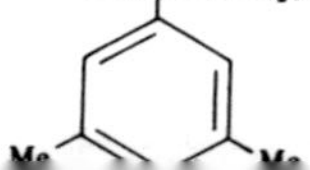 | | | 44, 45 |

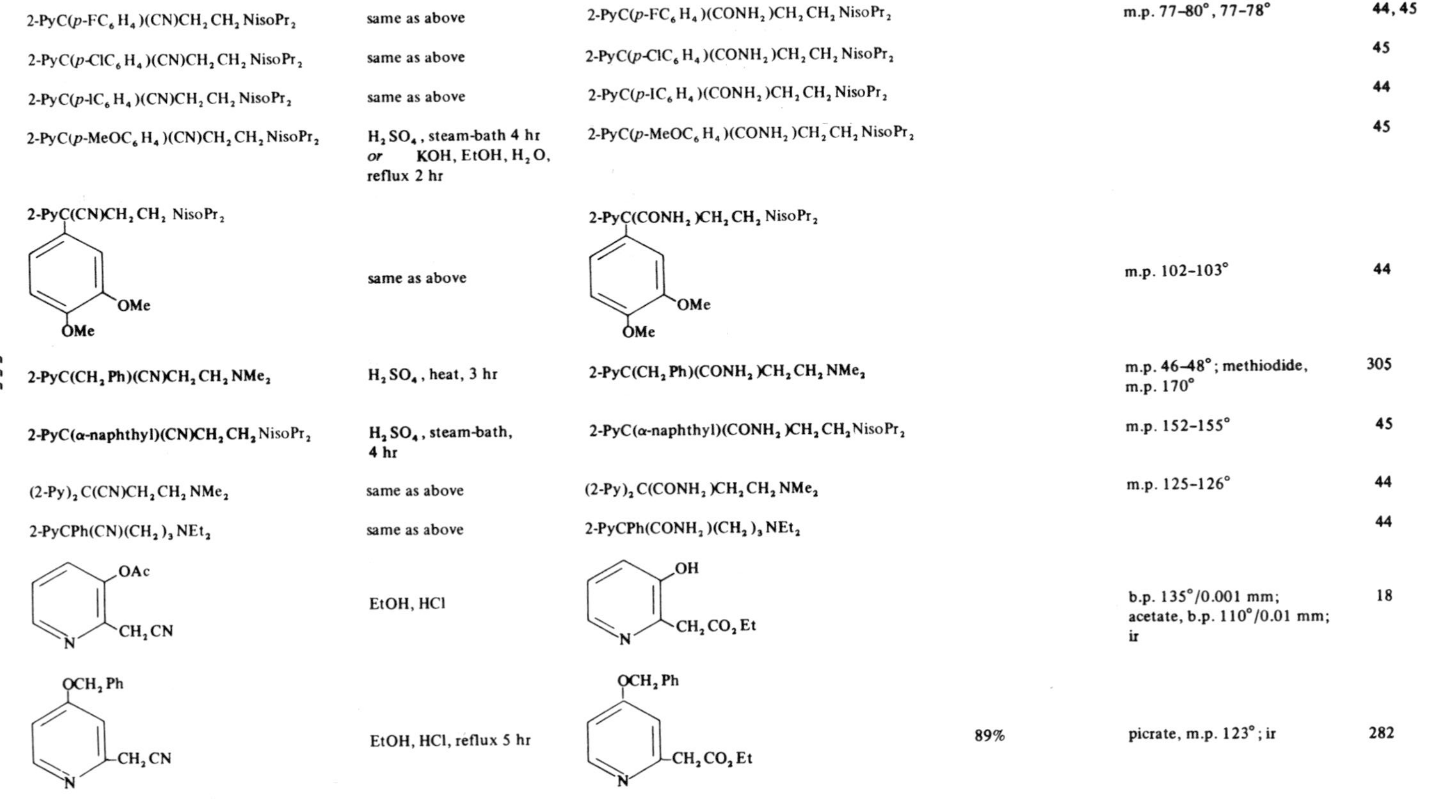

| | | | | | |
|---|---|---|---|---|---|
| | steam-bath | | | | |
| 2-PyC($p$-$FC_6H_4$)(CN)$CH_2CH_2$NisoPr$_2$ | same as above | 2-PyC($p$-$FC_6H_4$)($CONH_2$)$CH_2CH_2$NisoPr$_2$ | | m.p. 77–80°, 77–78° | 44, 45 |
| 2-PyC($p$-$ClC_6H_4$)(CN)$CH_2CH_2$NisoPr$_2$ | same as above | 2-PyC($p$-$ClC_6H_4$)($CONH_2$)$CH_2CH_2$NisoPr$_2$ | | | 45 |
| 2-PyC($p$-$IC_6H_4$)(CN)$CH_2CH_2$NisoPr$_2$ | same as above | 2-PyC($p$-$IC_6H_4$)($CONH_2$)$CH_2CH_2$NisoPr$_2$ | | | 44 |
| 2-PyC($p$-$MeOC_6H_4$)(CN)$CH_2CH_2$NisoPr$_2$ | $H_2SO_4$, steam-bath 4 hr *or* KOH, EtOH, $H_2O$, reflux 2 hr | 2-PyC($p$-$MeOC_6H_4$)($CONH_2$)$CH_2CH_2$NisoPr$_2$ | | | 45 |
| 2-PyC(CN)$CH_2CH_2$ NisoPr$_2$ (OMe, OMe) | same as above | 2-PyC($CONH_2$)$CH_2CH_2$NisoPr$_2$ (OMe, OMe) | | m.p. 102–103° | 44 |
| 2-PyC($CH_2$Ph)(CN)$CH_2CH_2NMe_2$ | $H_2SO_4$, heat, 3 hr | 2-PyC($CH_2$Ph)($CONH_2$)$CH_2CH_2NMe_2$ | | m.p. 46–48°; methiodide, m.p. 170° | 305 |
| 2-PyC(α-naphthyl)(CN)$CH_2CH_2$NisoPr$_2$ | $H_2SO_4$, steam-bath, 4 hr | 2-PyC(α-naphthyl)($CONH_2$)$CH_2CH_2$NisoPr$_2$ | | m.p. 152–155° | 45 |
| (2-Py)$_2$C(CN)$CH_2CH_2NMe_2$ | same as above | (2-Py)$_2$C($CONH_2$)$CH_2CH_2NMe_2$ | | m.p. 125–126° | 44 |
| 2-PyCPh(CN)$(CH_2)_3NEt_2$ | same as above | 2-PyCPh($CONH_2$)$(CH_2)_3NEt_2$ | | | 44 |
| OAc, $CH_2CN$, N | EtOH, HCl | OH, $CH_2CO_2Et$, N | | b.p. 135°/0.001 mm; acetate, b.p. 110°/0.01 mm; ir | 18 |
| $OCH_2Ph$, $CH_2CN$, N | EtOH, HCl, reflux 5 hr | $OCH_2Ph$, $CH_2CO_2Et$, N | 89% | picrate, m.p. 123°; ir | 282 |

TABLE XI-27. Solvolysis of Side-Chain Nitriles (Continued)

| Nitrile | Conditions | Product | Yield | Properties | Ref. |
|---|---|---|---|---|---|
| 5-Me-2-pyridyl-$CH_2CN$ | EtOH, HCl, 50°, 2 hr; overnight; 80°, 8 hr *or* conc. $H_2SO_4$, EtOH, reflux 8 hr | 5-Me-2-pyridyl-$CH_2CO_2Et$ | 5 g → 1.5 g | b.p. 110–113°/4.5 mm | 192 |
| 5-$EtO_2C$-2-pyridyl-$CH_2CN$ | EtOH, HCl, R.T. overnight | 5-$EtO_2C$-2-pyridyl-$CH_2CO_2Et$ | 1.3 g → 1.4 g | b.p. 106–108°/0.1 mm; picrate, m.p. 106–107° | 17 |
| 1-$CH_2Ph$-2-oxo-6-($CH_2CN$)-pyridine | 95% EtOH, HCl, reflux 2 hr; R.T., 12 hr | 1-$CH_2Ph$-2-oxo-6-($CH_2CO_2Et$)-pyridine | 98.5% | m.p. 79° | 139 |
| 2-Py$(CH_2)_2CN$ | EtOH, HCl, $Et_2O$, 10 hr; 30–40°, overnight | 2-Py$(CH_2)_2CO_2Et$ | 89% | b.p. 76–79°/0.3 mm | 148 |
| 2-PyCPh(OH)$CH_2CN$ | EtOH, $H_2SO_4$, reflux 1.5 hr | 2-PyCPh(OH)$CH_2CO_2Et$ | | m.p. 62° | 223 |
| 2-PyCPh(OH)CHMeCN | EtOH, $H_2SO_4$, reflux 1.5 hr | 2-PyCPh(OH)CHMe$CO_2Et$ | | m.p. 63° | 223 |
| 2-Py$CH_2$CH(α-naphthyl)CN | KOH, $PhCH_2OH$, reflux 15 hr | 2-Py$CH_2$CH(α-naphthyl)$CO_2K$ | 76% | m.p. 137°; free acid, m.p. 177–178° (decomp.) | 338 |
| 2-PyCOCHPhCN | 90% $H_2SO_4$, 50°, 3 hr | 2-PyCOCHPh$CONH_2$ | 84% | m.p. 184–186° | 116 |

| | | | | | |
|---|---|---|---|---|---|
| | 50°, 1 hr | | | | |
| | pyridine, $Et_3N$; $H_2S$, several hr | 2-Py$(CH_2)_3CSNH_2$ | 3 g → 2 g | m.p. 86° | 167 |
| 2-Py$(CH_2)_2$CHPhCN | KOH, MeOH, 3% $H_2O_2$, 50°, 1 hr | 2-Py$(CH_2)_2$CHPhCON$H_2$ | 4.6 g → 4.3 g | m.p. 129° | 167 |
| | pyridine, $Et_3N$, $H_2S$, several hr | 2-Py$(CH_2)_2$CHPhCS$NH_2$ | 4.6 g → 2.6 g | m.p. 159° | 167 |
| | 18% HCl, reflux 3 hr; NaOH | 2-Py$(CH_2)_2$CHPh$CO_2$Na | 4.6 g → 5.2 g | m.p. 161–162°; sodium salt, m.p. 207° | 167 |
| | dry HCl, MeOH | 2-Py$(CH_2)_2$CHPh$CO_2$Me | 10 g → 7.3 g | b.p. 178°/2 mm | 167 |
| 2-Py$(CH_2)_2$CH(OH)CN | | 2-Py$(CH_2)_2$CH(OH)$CO_2$H | 41% | m.p. 93–94°; hydrochloride, m.p. 162° | 219 |
| 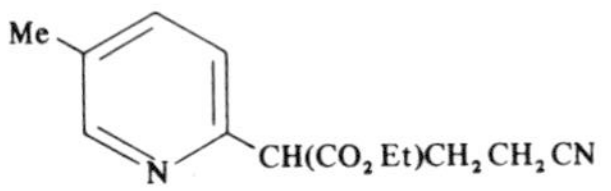 Me, N, CH($CO_2$Et)$CH_2CH_2$CN | 16% aq. HCl, reflux 10 hr; 150°, 15 min, -$CO_2$; HCl, EtOH, 3 hr | 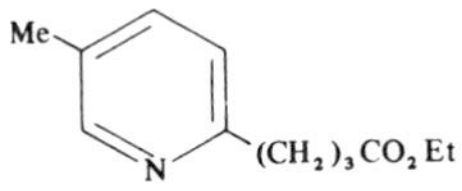 Me, N, $(CH_2)_3CO_2$Et | | b.p. 102°/0.2 mm | 283 |
| 2-PyCMe($CO_2$Et)$CH_2CH_2$CN | same as above | 2-PyCHMe$CH_2CH_2CO_2$Et | 4.6 g → 2.2 g | b.p. 110°/1.0 mm | 283 |
| 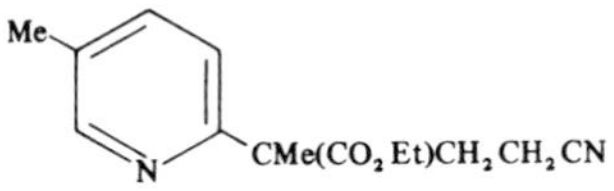 Me, N, CMe($CO_2$Et)$CH_2CH_2$CN | same as above | 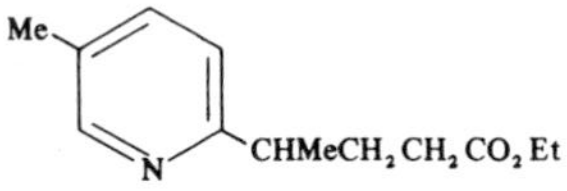 Me, N, CHMe$CH_2CH_2CO_2$Et | | b.p. 105°/0.3 mm | 283 |
| 2-Py$(CH_2)_3$CH(OH)CN | | 2-Py$(CH_2)_3$CH(OH)$CO_2$H | 38% | m.p. 143° | 219 |
| 2-Py$(CH_2)_2$CPh(COMe)$CH_2CH_2$CN | 46% $H_2SO_4$, reflux 3 hr | 2-Py$(CH_2)_2$CPh(COMe)$CH_2CH_2CO_2$H | 88% | m.p. 171.5–173° | 36 |
| 3-Py$CH_2$CN | $H_2S$ | 3-Py$CH_2CSNH_2$ | | m.p. 135–136.5° | 168 |

TABLE XI-27. Solvolysis of Side-Chain Nitriles (Continued)

| Nitrile | Conditions | Product | Yield | Properties | Ref. |
|---|---|---|---|---|---|
| | $HSCH_2CH_2NH_2$ | 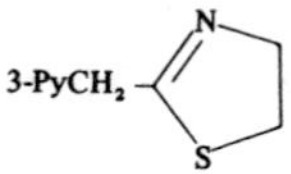 3-$PyCH_2$– (N, S) | 65% | | 168 |
| 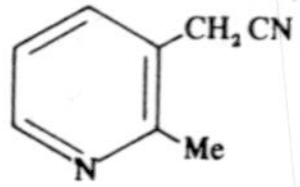 $CH_2CN$, Me, N | HCl, EtOH, cold for 1.5 hr *or* reflux 5 hr | 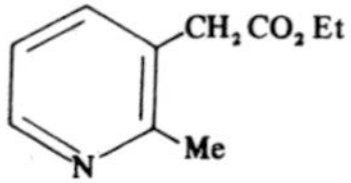 $CH_2CO_2Et$, Me, N | 45 g → 43 g | b.p. 124–125°/7 mm, $n_D^{25}$ 1.4982; picrate, m.p. 154–155.5° | 174 |
| 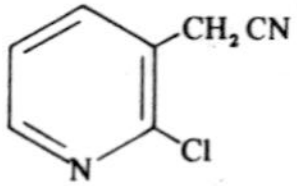 $CH_2CN$, Cl, N | conc. HCl, reflux 5.5 hr | 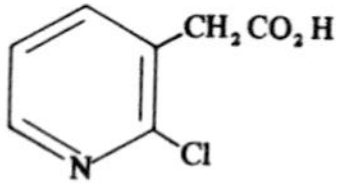 $CH_2CO_2H$, Cl, N | 76% | m.p. 203–204° | 268 |
| 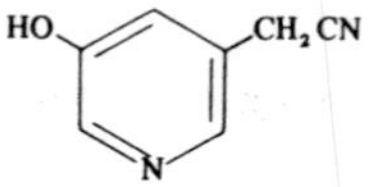 HO, $CH_2CN$, N | conc. HCl, reflux 5 hr | 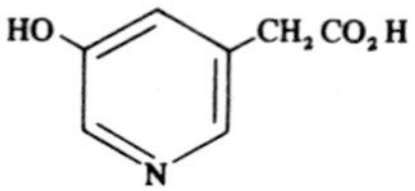 HO, $CH_2CO_2H$, N | 78% | m.p. 197° | 268 |
| 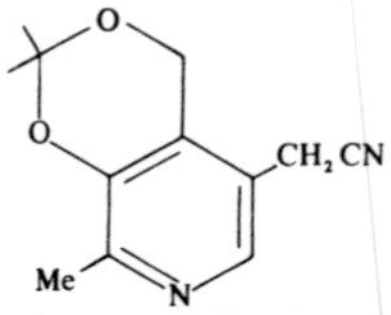 O, O, $CH_2CN$, Me, N | 10% alc. KOH, reflux 3 hr<br>*or* NaOH, 70% EtOH, reflux 3 hr | 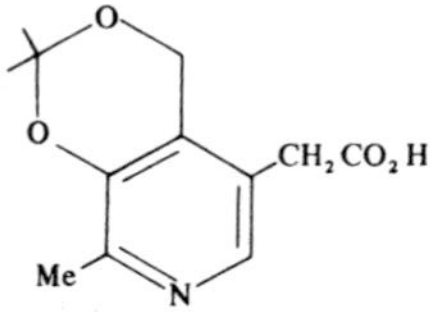 O, O, $CH_2CO_2H$, Me, N | 0.7 g → 0.4 g<br>83% | m.p. 194°<br>m.p. 186–187° | 125<br>195 |
| | conc. HCl, 40°, 45 min | 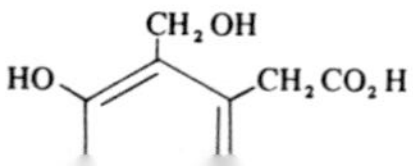 $CH_2OH$, HO, $CH_2CO_2H$ | 43% | m.p. 217–218° (decomp.); | 195 |

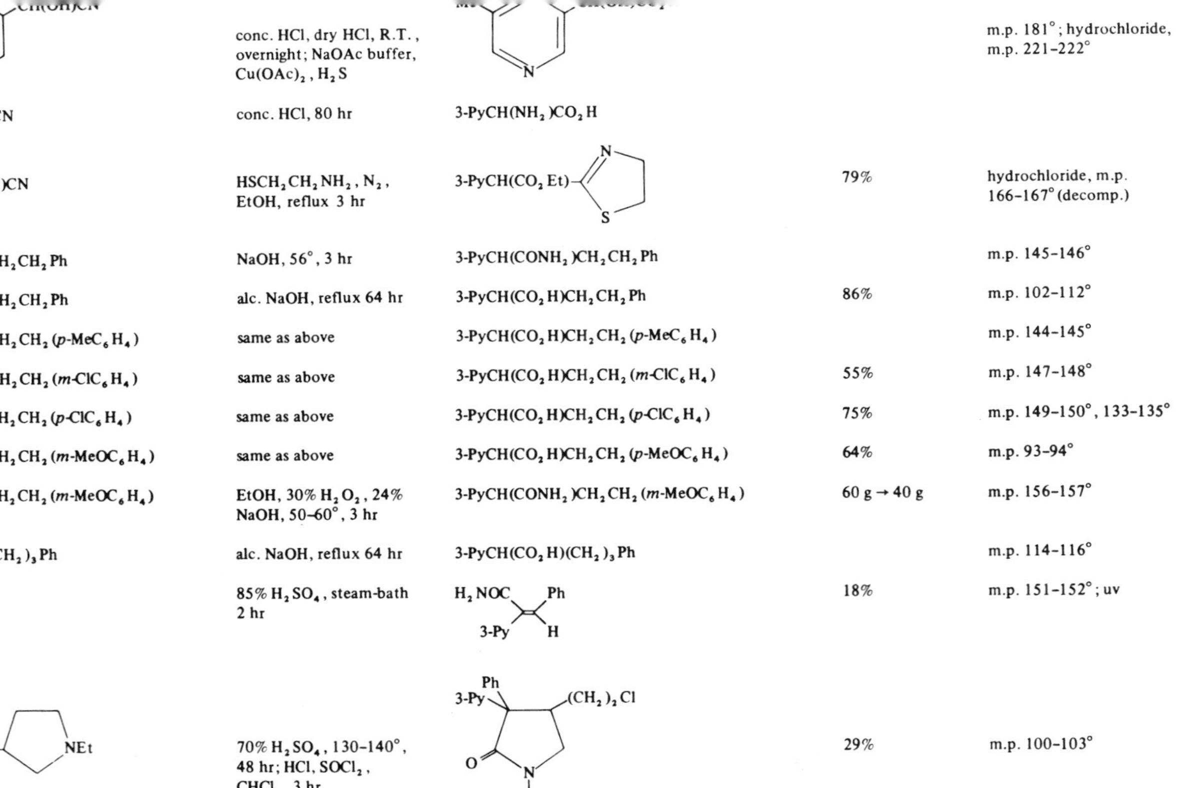

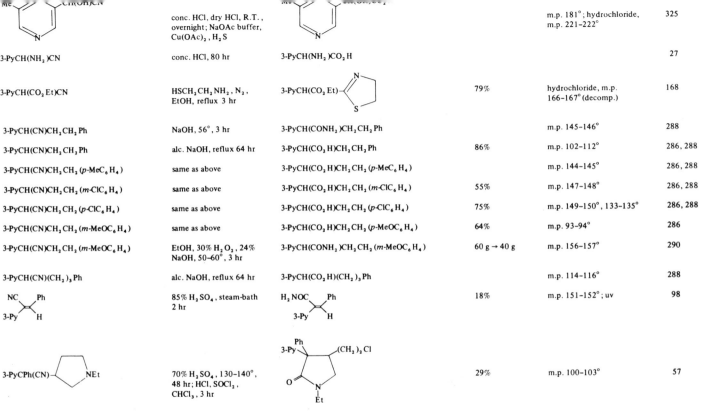

| | | | | | |
|---|---|---|---|---|---|
| (Me-pyridyl-CH(OH)CN) | conc. HCl, dry HCl, R.T., overnight; NaOAc buffer, $Cu(OAc)_2$, $H_2S$ | (Me-pyridyl-CH(OH)CO$_2$...) | | m.p. 181°; hydrochloride, m.p. 221–222° | 325 |
| 3-PyCH($NH_2$)CN | conc. HCl, 80 hr | 3-PyCH($NH_2$)$CO_2H$ | | | 27 |
| 3-PyCH($CO_2$Et)CN | $HSCH_2CH_2NH_2$, $N_2$, EtOH, reflux 3 hr | 3-PyCH($CO_2$Et)-(2-thiazolinyl) | 79% | hydrochloride, m.p. 166–167° (decomp.) | 168 |
| 3-PyCH(CN)$CH_2CH_2$Ph | NaOH, 56°, 3 hr | 3-PyCH($CONH_2$)$CH_2CH_2$Ph | | m.p. 145–146° | 288 |
| 3-PyCH(CN)$CH_2CH_2$Ph | alc. NaOH, reflux 64 hr | 3-PyCH($CO_2$H)$CH_2CH_2$Ph | 86% | m.p. 102–112° | 286, 288 |
| 3-PyCH(CN)$CH_2CH_2$(*p*-Me$C_6H_4$) | same as above | 3-PyCH($CO_2$H)$CH_2CH_2$(*p*-Me$C_6H_4$) | | m.p. 144–145° | 286, 288 |
| 3-PyCH(CN)$CH_2CH_2$(*m*-Cl$C_6H_4$) | same as above | 3-PyCH($CO_2$H)$CH_2CH_2$(*m*-Cl$C_6H_4$) | 55% | m.p. 147–148° | 286, 288 |
| 3-PyCH(CN)$CH_2CH_2$(*p*-Cl$C_6H_4$) | same as above | 3-PyCH($CO_2$H)$CH_2CH_2$(*p*-Cl$C_6H_4$) | 75% | m.p. 149–150°, 133–135° | 286, 288 |
| 3-PyCH(CN)$CH_2CH_2$(*m*-MeO$C_6H_4$) | same as above | 3-PyCH($CO_2$H)$CH_2CH_2$(*p*-MeO$C_6H_4$) | 64% | m.p. 93–94° | 286 |
| 3-PyCH(CN)$CH_2CH_2$(*m*-MeO$C_6H_4$) | EtOH, 30% $H_2O_2$, 24% NaOH, 50–60°, 3 hr | 3-PyCH($CONH_2$)$CH_2CH_2$(*m*-MeO$C_6H_4$) | 60 g → 40 g | m.p. 156–157° | 290 |
| 3-PyCH(CN)($CH_2$)$_3$Ph | alc. NaOH, reflux 64 hr | 3-PyCH($CO_2$H)($CH_2$)$_3$Ph | | m.p. 114–116° | 288 |
| (NC)(3-Py)C=CH(Ph) | 85% $H_2SO_4$, steam-bath 2 hr | ($H_2$NOC)(3-Py)C=CH(Ph) | 18% | m.p. 151–152°; uv | 98 |
| 3-PyCPh(CN)-(1-ethyl-3-pyrrolidinyl) | 70% $H_2SO_4$, 130–140°, 48 hr; HCl, $SOCl_2$, $CHCl_3$, 3 hr | 3-Ph-3-(3-Py)-4-($CH_2$)$_2$Cl-1-Et-2-pyrrolidinone | 29% | m.p. 100–103° | 57 |

TABLE XI-27. Solvolysis of Side-Chain Nitriles (Continued)

| Nitrile | Conditions | Product | Yield | Properties | Ref. |
|---|---|---|---|---|---|
| 3-PyCPh(CN)$CH_2CH_2$NisoPr$_2$ | conc. $H_2SO_4$, steam-bath 4 hr | 3-PyCPh($CONH_2$)$CH_2CH_2$NisoPr$_2$ | | | 44 |
| 3-PyCH$_2$CH(p-NH$_2$C$_6$H$_4$)CN | MeOH, dry HCl, 3–4 hr; MeOH, 10–11% NaOH, reflux 3 hr | 3-PyCH$_2$CH(p-NH$_2$C$_6$H$_4$)CO$_2$H | 63% | m.p. 181–183° | 164 |
| 3-PyCH$_2$CH(α-naphthyl)CN | KOH, PhCH$_2$OH, reflux 15 hr | 3-PyCH$_2$CH(α-naphthyl)CO$_2$K | 84% | m.p. 222–224° (decomp.); free acid, m.p. 192.5–193.5° | 338 302 |
| 3-PyCOCHPhCN | 90% $H_2SO_4$, 50°, 3 hr | 3-PyCOCHPhCONH$_2$ | 71% | m.p. 136–137° | 116 |
| 3-PyCOCH(p-MeOC$_6$H$_4$)CN | HCl, EtOH, overnight | 3-PyCOCH(p-MeOC$_6$H$_4$)CO$_2$Et | 15% | m.p. 70–71° | 16 |
| 3-PyCH=CHCN | aq. KOH, reflux 4 hr; pH 4 with HCl | 3-PyCH=CHCO$_2$H | | m.p. 234.5°; | 213 |
| (H)(3-Py)C=C(Ph)(CN) | 85% $H_2SO_4$, steam-bath 2 hr | (H)(3-Py)C=C(Ph)(CONH$_2$) | 65% | m.p. 134–137° (PrOH-2-PrOAc), m.p. 129–130° (Me$_2$CO-pentane); uv | 98 |
| (H)(3-Py)C=C(p-ClC$_6$H$_4$)(CN) | 85% $H_2SO_4$, steam-bath 2 hr | (H)(3-Py)C=C(p-ClC$_6$H$_4$)(CONH$_2$) | 39% | m.p. 155–159°; uv | 98 |
| (H)(3-Py)C=C(p-Me$_2$NC$_6$H$_4$)(CN) | 50% $H_2SO_4$, steam-bath 2 hr | (H)(3-Py)C=C(p-Me$_2$NC$_6$H$_4$)(CONH$_2$) | 95% | m.p. 201–203°; uv | 98 |
| 3-PyC(OH)=C(p-MeC$_6$H$_4$)CN | conc. $H_2SO_4$, R.T., 1 hr | 3-PyC(OH)=C(p-MeC$_6$H$_4$)CONH$_2$ | | m.p. 136–138° | 42 |
| 3-PyC(OH)=C(3-Py)CN | conc. fuming $H_2SO_4$, 25%, 3 hr | 3-PyC(OH)=C(3-Py)CONH$_2$ | | | 333 |

| | | | | | |
|---|---|---|---|---|---|
| | $HSCH_2CH_2NH_2$ | $4\text{-PyCH}_2$– S | 57% | | 168 |
| $CH_2CN$ N O $(CH_2)_2$ OMe OMe | NaOH, EtOH, reflux overnight | $CH_2CO_2H$ N O $(CH)_2$ OMe OMe | | m.p. 156.5–157° (decomp.) | 3, 88 |
| $4\text{-PyCH}(NH_2)CN$ | conc. HCl, R.T., 5 days | $4\text{-PyCH}(NH_2)CO_2H\cdot HCl$ | 172 g → 132 g | | 27 |
| $4\text{-PyCH}(CO_2Et)CN$ | $HSCH_2CH_2NH_2$, $N_2$, EtOH, reflux 3 hr | $4\text{-PyCH}(CO_2Et)$– N S | 64% | m.p. 100–101° | 168 |
| **$4\text{-PyCH(CN)CH(OH)}(o\text{-}O_2NC_6H_4)$** | 75% $H_2SO_4$; base | $4\text{-PyC}(CONH_2){=}CH(o\text{-}O_2NC_6H_4)$ | 0.7 g → 0.2 g | m.p. 208–209° | 99 |
| $4\text{-PyCPh(CN)}CH_2CH_2N\text{isoPr}_2$ | $H_2SO_4$, steam-bath, 4 hr | $4\text{-PyCPh}(CONH_2)CH_2CH_2N\text{isoPr}_2$ | | m.p. 138.5–139° | 44, 45 |
| **$4\text{-PyCPh(CN)}(CH_2)_3NC_5H_{10}$** | conc. $H_2SO_4$, steam-bath, 4 hr | $4\text{-PyCPh}(CONH_2)(CH_2)_3$–N | | | 44 |
| **$4\text{-PyCH}_2CH(p\text{-}H_2NC_6H_4)CN$** | HCl, MeOH, 3–4 hr; MeOH, 10–11% NaOH, reflux 3 hr. | $4\text{-PyCH}_2CH(p\text{-}H_2NC_6H_4)CO_2H$ | 27% | m.p. 196–198° | 164 |
| **$4\text{-PyCH}_2CH(\alpha\text{-naphthyl})CN$** | KOH, $PhCH_2OH$, reflux 15 hr | $4\text{-PyCH}_2CH(\alpha\text{-naphthyl})CO_2K$ | 89% | m.p. 211–212.5° | 338 |
| **4-PyCOCHPhCN** | 90% $H_2SO_4$, 50°, 3 hr | $4\text{-PyCOCHPhCONH}_2$ | 97% | m.p. 161–162° | 116 |
| **4-PyCH=CHCN** | aq. KOH, reflux 4 hr | $4\text{-PyCH{=}CHCO}_2H$ | 74% | m.p. 290–291° (decomp.) | 213 |

TABLE XI-27. Solvolysis of Side-Chain Nitriles (Continued)

| Nitrile | Conditions | Product | Yield | Properties | Ref. |
|---|---|---|---|---|---|
| 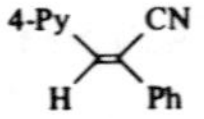 4-Py(H)C=C(CN)Ph | 85% $H_2SO_4$, steam-bath 2 hr | 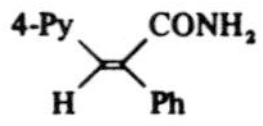 4-Py(H)C=C($CONH_2$)Ph | 54% | m.p. 178–180° | 98 |
| 4-PyCHP(O)$Ph_2$ ($CH_2$)$_2$CN | HCl | 4-PyCHP(O)$Ph_2$ ($CH_2$)$_2CO_2H$ | 57% | m.p. 200° | 314 |
| 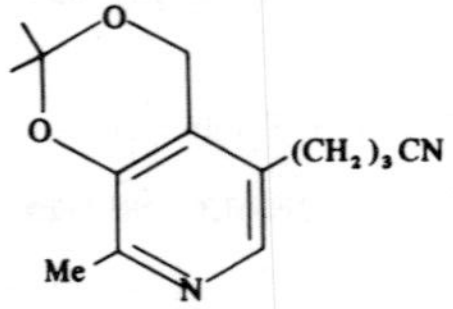 O, O, Me, N, ($CH_2$)$_3$CN | 10% alc. KOH, 3 hr | 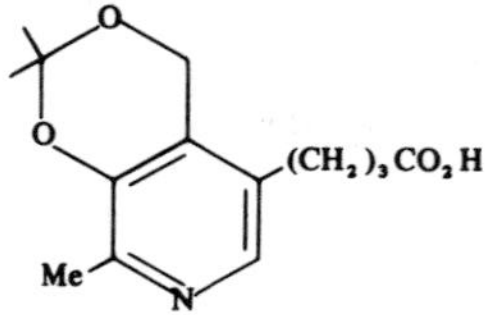 O, O, Me, N, ($CH_2$)$_3CO_2H$ | | m.p. 160–161° | 125 |

| Nitrile | Conditions | Product | Yield | Properties | Ref. |
|---|---|---|---|---|---|
| 2-Py$(CH_2)_2CN$ | $H_2$, Ra-Ni, $NH_3$, EtOH, 70–75° | 2-Py$(CH_2)_3NH_2$ | | | 285 |
| 2-Py$(CH_2)_3CN$ | $H_2$, Ra-Ni, $Et_2NH$, 2000 atm, 75–100° | 2-Py$(CH_2)_4NEt_2$ | | | 236 |
| | $H_2$, Ra-Ni, pyrrolidine, 2000 atm, 75–100° | 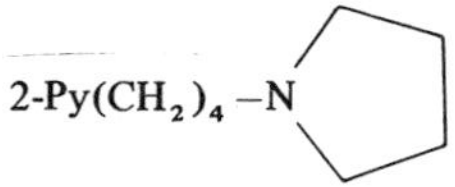 2-Py$(CH_2)_4$–N | | | 236 |
| | $H_2$, Ra-Ni, piperidine, 2000 atm, 75–100° | 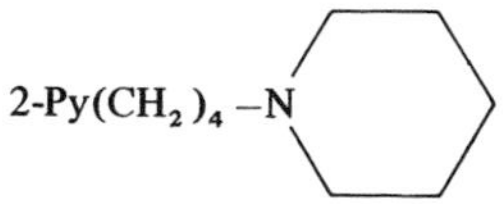 2-Py$(CH_2)_4$–N | | | 236 |
| 2-PyCH(CN)$(CH_2)_2CO_2Et$ | $H_2$, $PtO_2$, EtOH, 2 atm | 2-PyCH$(CH_2NH_2)CH_2CH_2CO_2Et$ | | picrate, m.p. 128–129° | 178 |
| 2-Py$(CH_2)_2$CHPhCN | LAH, $Et_2O$ | 2-Py$(CH_2)_2$CHPh$CH_2NH_2$ | 13.8 g → 8 g | | 167 |
| 3-Py$CH_2CN$ | $\Delta^1$-*p*-menthene, 10% Pd-C, 200 min | 3-PyEt | | | 170 |
| 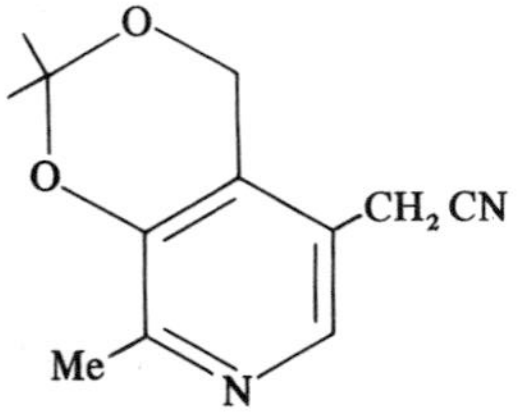 O, O, $CH_2CN$, Me, N | LAH, $Et_2O$, 0°, 1 hr | 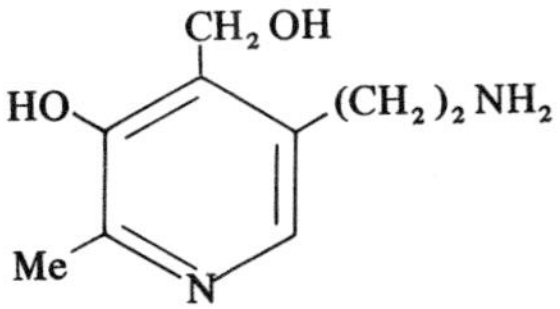 $CH_2OH$, HO, $(CH_2)_2NH_2$, Me, N | 45% | | 195 |

TABLE XI-28. Reduction of Side-Chain Nitriles (Continued)

| Nitrile | Conditions | Product | Yield | Properties | Ref. |
|---|---|---|---|---|---|
| 3-PyCH=CHCN | $\Delta^1$-*p*-menthene, 10% Pd-C, 210 min | 3-*n*-PrPy | | | 170 |
| 4-PyCHPhCN | $H_2$, Ra-Ni, MeOH | 4-PyCHPhCH$_2$NH$_2$ | | | 339 |

TABLE XI-29. Reactions of 2-Pyridineacetonitriles with Organometallic Reagents

| Starting material | Reagent | Conditions | Product | Yield | Ref. |
|---|---|---|---|---|---|
| 2-PyC(CH$_2$Ph)(CN)CH$_2$CH$_2$NMe$_2$ | EtMgBr | Et$_2$O, reflux 4 hr | 2-PyC(CH$_2$Ph)(COEt)CH$_2$CH$_2$NMe$_2$ | 0.93 g → 0.73 g | 305 |

| Starting material | Reagent and conditions | Product | Yield | Properties | Ref. |
|---|---|---|---|---|---|
| 2-PyCOCO$_2$Et | PhOH, 87% H$_2$SO$_4$, 4–5 hr overnight | 2-PyCH($p$-HOC$_6$H$_4$)$_2$ | | | 340 |
| 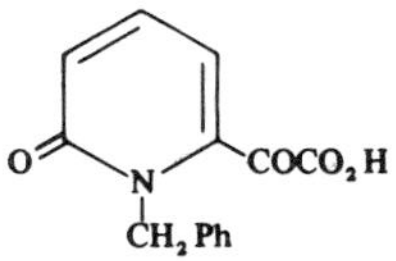 O, N, COCO$_2$H, CH$_2$Ph | NH$_2$OH · HCl, NaOH, 36 hr | 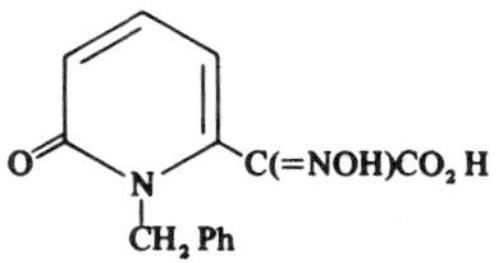 O, N, C(=NOH)CO$_2$H, CH$_2$Ph | 91.4% | m.p. 191° | 139 |
| 2-PyCOCHPhCO$_2$Me | PhNHNH$_2$ · HCl, MeOH-H$_2$O, reflux, Py, 1 hr | 2-PyC(=NNHPh)CHPhCO$_2$Me | 89% | m.p. 142–144° | 341 |
| 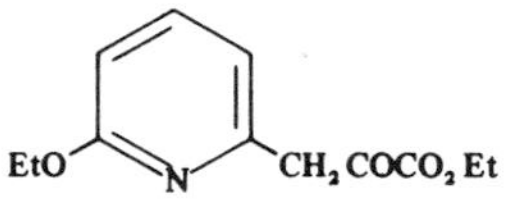 EtO, N, CH$_2$COCO$_2$Et | NH$_2$OH · HCl, Py, EtOH, steam-bath, 3 hr | 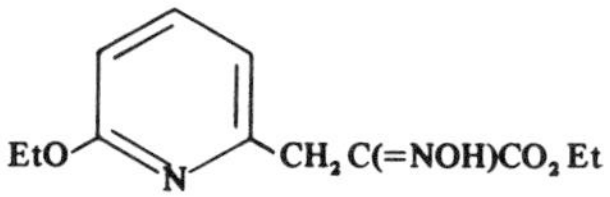 EtO, N, CH$_2$C(=NOH)CO$_2$Et | 92.4% | m.p. 73° | 139 |
| 2-PyCH=CHCOCO$_2$H | NH$_2$OH · HCl, aq. KOAc | 2-PyCH=CHC(=NOH)CO$_2$H | 90% | m.p. 129° | 213 |
| 2-PyCH$_2$COCO$_2$Et | PhCHO | 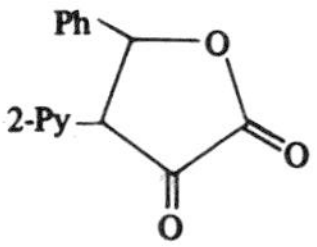 Ph, O, 2-Py, O, O | 0.5 g → 0.3 g | m.p. 225–226° | 86 |
| 2-PyCOCHPhCO$_2$Me | NH$_2$CONHNH$_2$ · HCl | 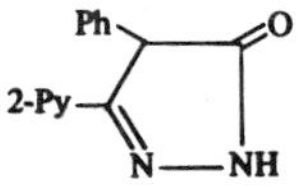 Ph, O, 2-Py, N—NH | 78.7% | m.p. 228–230°; picrate, m.p. 210° | 341 |

TABLE XI-30. Carbonyl Reactions of Keto Acids (Continued)

| Starting material | Reagent and conditions | Product | Yield | Properties | Ref. |
|---|---|---|---|---|---|
| | $PhNHNH_2$, isoBuOH, reflux 3 hr | 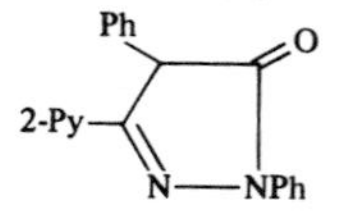 Ph, O, 2-Py, N—NPh | | m.p. 188–189° | 341 |
| $2\text{-}PyCOCHPhCONH_2$ | $NH_2CSNHNH_2$ | 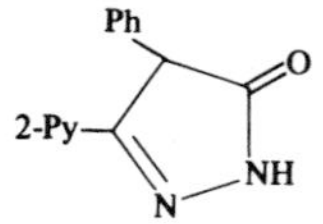 Ph, O, 2-Py, N, NH | | | 341 |
| | $PhNHNH_2$, 100–110°, 2 hr | 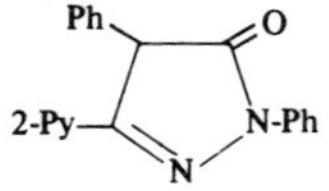 Ph, O, 2-Py, N, N-Ph | | | 341 |
| $2\text{-}PyC(=NNHPh)CHPhCO_2Me$ | PhMe *or* isoBuOH, reflux 4 hr | 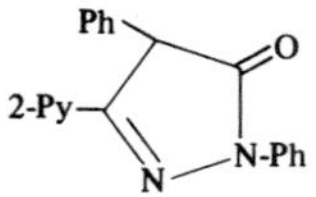 Ph, O, 2-Py, N, N-Ph | 82% | | 341 |
| $3\text{-}PyCOCO_2H$ | $CH_2(CO_2H)_2$, Py, steam-bath, 1 hr | $3\text{-}PyC(OH)(CO_2H)CH_2CO_2H$ | 66% | m.p. 217–218° | 218 |
| 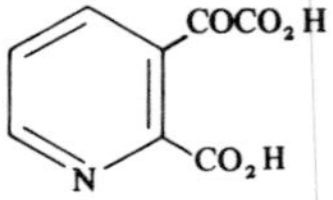 $COCO_2H$, $CO_2H$, N | $NH_2NH_2 \cdot H_2SO_4$, NaOAc, $H_2O$, R.T., overnight | 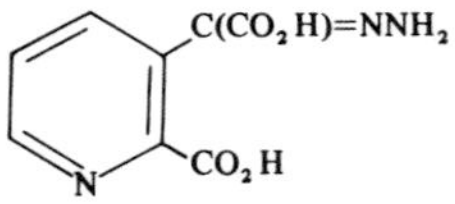 $C(CO_2H){=}NNH_2$, $CO_2H$, N | 74% | m.p. 302° (decomp.); ir | 260 |
| | $NH_2NHPh \cdot HCl$, NaOAc, | 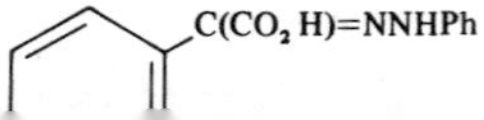 $C(CO_2H){=}NNHPh$ | 5 g → 6.2 g | m.p. 218–220° (decomp.); | 260 |

| | | | | | |
|---|---|---|---|---|---|
| | heat, 2–3 min | | | | |
| 3-PyCOCH$_2$CO$_2$Et | H$_2$NNHCSNH$_2$, boiling alcohol, 5–6 min | 3-PyC(CH$_2$CO$_2$Et)=NNHCSNH$_2$ | | m.p. 153° | 25 |
| | dry HCl, PCl$_5$, C$_6$H$_6$, 2 hr, 0°; R.T., 2 hr; 80°, 20 min | 3-PyC(Cl)=CHCO$_2$Et + 3-PyC(Cl)$_2$CH$_2$CO$_2$Et + 3-PyC(Cl$_2$)Me + 3-PyC(Cl)=CH$_2$ | | b.p. 85°/0.01 mm; $n_D^{20}$ 1.5578 | 175 |
| | CH$_2$N$_2$, Et$_2$O, 3 days | 3-PyC(OMe)=CHCO$_2$Et | 11% | b.p. 86°/0.1 mm; $n_D^{20}$ 1.5479; picrate, m.p. 139–140°; ir | 175 |
| 3-PyCOCHPhCO$_2$Me | PhNHNH$_2$, R.T., 24 hr | 3-PyC(=NNHPh)CHPhCO$_2$Me | | m.p. 207–209° | 341 |
| 3-PyCOCH$_2$CN | NH$_2$–C$_6$H$_4$–NEt$_2$, AgCl, Na$_2$CO$_3$, 1 hr, EtOH-H$_2$O | 3-PyC(=N–C$_6$H$_4$–NEt$_2$)CH$_2$CN | 0.36 → 0.06 g | m.p. 99–101° | 103 |
| 3-PyCO(CH$_2$)$_2$CO$_2$Me | NH$_2$OH · HCl, Py | 3-PyC(=NOH)CH$_2$CH$_2$CO$_2$Me | | m.p. 70° | 22 |
| 3-PyCO(CH$_2$)$_2$CO$_2$Et | NH$_2$OH · HCl | 3-PyC(=NOH)CH$_2$CH$_2$CO$_2$Et | | m.p. 132–133° | 2 |
| 3-PyCOCH$_2$CONHPh | PhNH$_2$, PhNH$_2$ · HCl, benzene, reflux | 3-PyC(NHPh)=CHCONHPh | | | 292 |
| 3-Py$^{14}$COCH$_2$CH$_2$CO$_2$H | MeNH$_2$, EtOH, overnight, R.T.; H$_2$, 5% Pt-BaSO$_4$, 6 hr | (±)Py$^{14}$CH(NHMe)CH$_2$CH$_2$CO$_2$H | 53% | m.p. 120–123° | 24 |
| 3-PyCO(CH$_2$)$_3$CO$_2$H | NH$_2$OH · HCl, aq. KOH, 7 hr | 3-PyC(=NOH)CH$_2$CH$_2$CH$_2$CO$_2$H | | m.p. 161–163° | 75 |
| 3-PyCH$_2$COCH$_2$CONHMe | NH$_2$OH · HCl, 10% NaOH, overnight, R.T. | 3-PyCH$_2$C(=NOH)CH$_2$CONHMe | | m.p. 160–165° | 22 |

TABLE XI-30. Carbonyl Reactions of Keto Acids (Continued)

| Starting material | Reagent and conditions | Product | Yield | Properties | Ref. |
|---|---|---|---|---|---|
| 3-PyCH=CHCOCO$_2$H | NH$_2$OH · HCl, aq. KOAc | 3-PyCH=CHC(=NOH)CO$_2$H | quantitative | m.p. 176° (decomp.) | 213 |
| 3-PyCOCH$_2$CH(NHAc)CO$_2$H | MeNH$_2$, Ra-Ni, H$_2$, 4 hr, 85°; 5*N* HCl, reflux 5 hr | 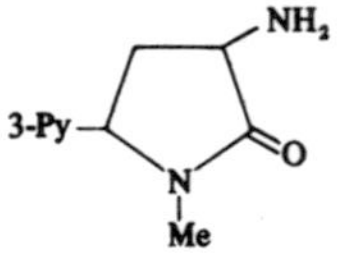 NH$_2$, 3-Py, O, N, Me | | | 1 |
| 3-PyCOCHPhCONH$_2$ | NH$_2$CSNHNH$_2$ | 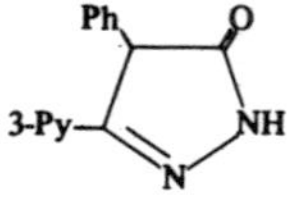 Ph, O, 3-Py, NH, N | | m.p. 255–257°; picrate, m.p. 231–235° | 341 |
| 3-PyCOCHPhCO$_2$Me | NH$_2$CONHNH$_2$ · HCl | 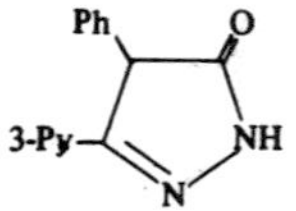 Ph, O, 3-Py, NH, N | 89.2% | | 341 |
| 3-PyCOCHPhCO$_2$Me | PhNHNH$_2$, N$_2$, isoBuOH, reflux | 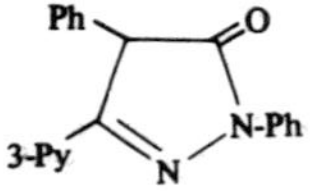 Ph, O, 3-Py, N-Ph, N | 90.2% | m.p. 221–223°; picrate m.p. 208–211° | 341 |
| 3-PyC(=NNHPh)CHPhCO$_2$Me | isoBuOH or PhMe, reflux | 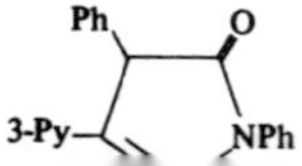 Ph, O, 3-Py, NPh | | | 341 |

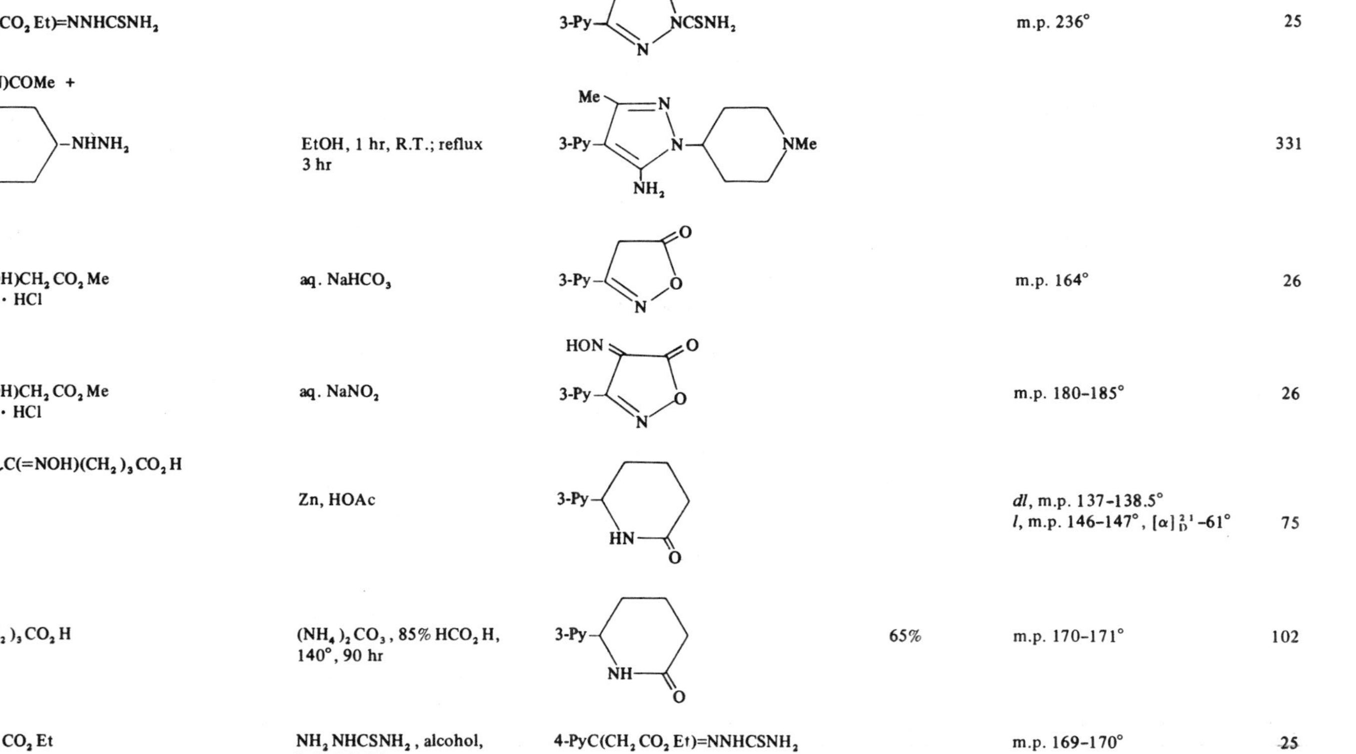

| Starting material | Conditions | Product | Yield | Properties | Ref. |
|---|---|---|---|---|---|
| 3-PyC($CH_2CO_2Et$)=NNHCSNH$_2$ | | | | m.p. 236° | 25 |
| 3-PyCH(CN)COMe + | EtOH, 1 hr, R.T.; reflux 3 hr | | | | 331 |
| 3-PyC(=NOH)$CH_2CO_2Me$ · HCl | aq. $NaHCO_3$ | | | m.p. 164° | 26 |
| 3-PyC(=NOH)$CH_2CO_2Me$ · HCl | aq. $NaNO_2$ | | | m.p. 180–185° | 26 |
| | Zn, HOAc | | | *dl*, m.p. 137–138.5°<br>*l*, m.p. 146–147°, $[\alpha]_D^{21}$ –61° | 75 |
| 3-PyCO$(CH_2)_3CO_2H$ | $(NH_4)_2CO_3$, 85% $HCO_2H$, 140°, 90 hr | | 65% | m.p. 170–171° | 102 |
| 4-PyCO$CH_2CO_2Et$ | $NH_2NHCSNH_2$, alcohol, boiled 5–6 hr | 4-PyC($CH_2CO_2Et$)=NNHCSNH$_2$ | | m.p. 169–170° | 25 |

TABLE XI-30. Carbonyl Reactions of Keto Acids (Continued)

| Starting material | Reagent and conditions | Product | Yield | Properties | Ref. |
|---|---|---|---|---|---|
| 4-$PyCOCHPhCO_2Me$ | $PhNHNH_2$, R.T., 24 hr | 4-$PyC(=NNHPh)CHPhCO_2Me$ | | m.p. 241–243° | 341 |
| 4-$PyCOCH_2CONHPh$ | $PhNH_2$, $PhNH_2 \cdot HCl$<br>$C_6H_6$, reflux | 4-$PyC(NHPh)=CHCONHPh$ | | m.p. 192–193° | 292 |
| 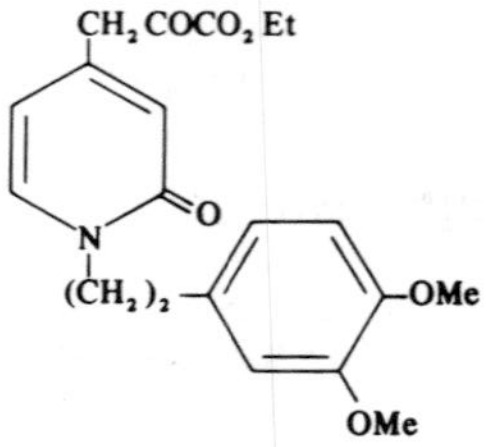 $CH_2COCO_2Et$<br>O<br>N<br>$(CH_2)_2$<br>OMe<br>OMe | $NH_2OH \cdot HCl$, NaOAc, EtOH, reflux 30 min | 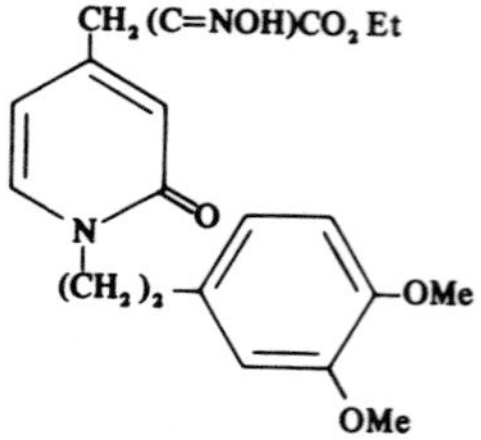 $CH_2(C=NOH)CO_2Et$<br>O<br>N<br>$(CH_2)_2$<br>OMe<br>OMe | 82% | m.p. 186.5–189.5° | 3, 88 |
| 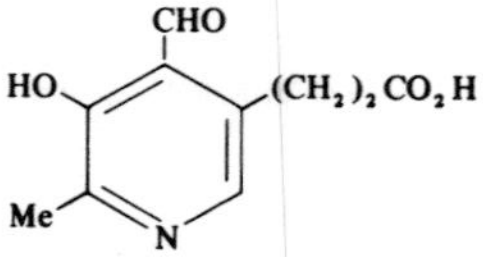 CHO<br>HO<br>$(CH_2)_2CO_2H$<br>Me<br>N | $NH_2OH \cdot HCl$, NaAc, EtOH, reflux 3 hr | 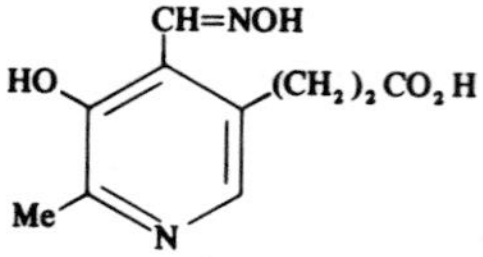 CH=NOH<br>HO<br>$(CH_2)_2CO_2H$<br>Me<br>N | 76% | m.p. 211–212° | 279 |
| 4-$PyCOCHPhCO_2Me$ | $NH_2CONHNH_2 \cdot HCl$ | 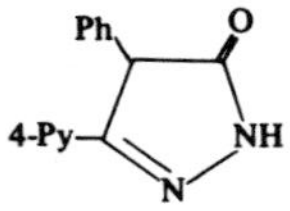 Ph<br>O<br>4-Py<br>NH<br>N | 90.6% | m.p. 269–271°; picrate, m.p. 229–230° | 341 |
| 4-$PyCOCHPhCO_2Me$ | $PhNHNH_2$, $N_2$, isoBuOH, | 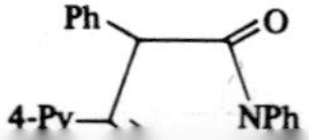 Ph<br>O<br>4-Py<br>NPh | 87.9% | m.p. 235–237°; picrate | 341 |

| | | | |
|---|---|---|---|
| 4-PyCOCHPhCONH$_2$ | NH$_2$CSNHNH$_2$ | 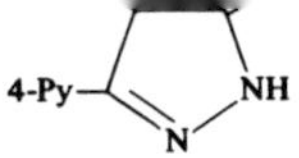 4-Py, NH, N | 341 |
| 4-PyCOCH$_2$CN + N$_2$H$_4$ | abs. EtOH, reflux | 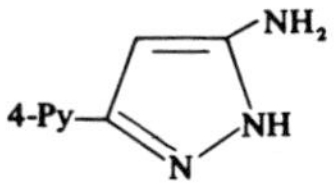 4-Py, NH$_2$, NH, N | 171 |
| 4-PyCOCH$_2$CN + MeNHNH$_2$ | abs. EtOH, reflux | 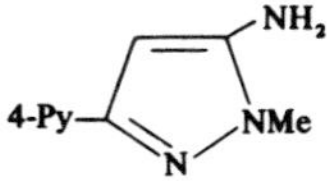 4-Py, NH$_2$, NMe, N | 171 |
| 4-PyCOCH$_2$CN + isoPrNHNH$_2$ | abs. EtOH, reflux | 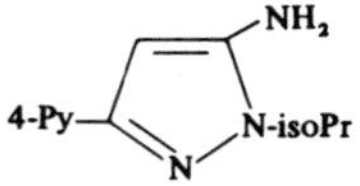 4-Py, NH$_2$, N-isoPr, N | 171 |
| 4-PyCOCH$_2$CN + 2-BuNHNH$_2$ | abs. EtOH, reflux | 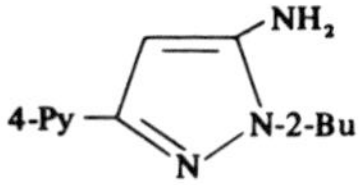 4-Py, NH$_2$, N-2-Bu, N | 171 |
| 4-PyCOCH$_2$CN + 2-AmNHNH$_2$ | same as above | 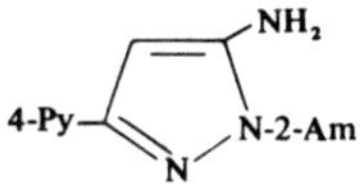 4-Py, NH$_2$, N-2-Am, N | 171 |
| 4-PyCOCH$_2$CN 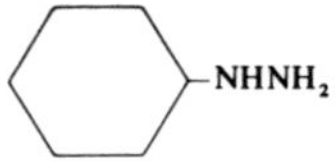 +     -NHNH$_2$ | Na, 98% EtOH, reflux 10 hr | 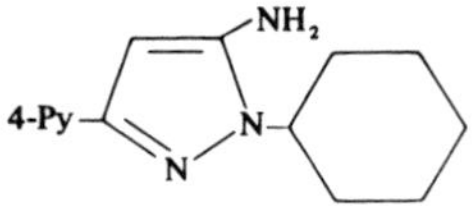 4-Py, NH$_2$, N, N | 171 |

TABLE XI-30. Carbonyl Reactions of Keto Acids (Continued)

| Starting material | Reagent and conditions | Product | Yield | Properties | Ref. |
|---|---|---|---|---|---|
| 4-$PyCOCH_2CN$ + $HOCH_2CH_2NHNH_2$ | abs. EtOH, reflux | 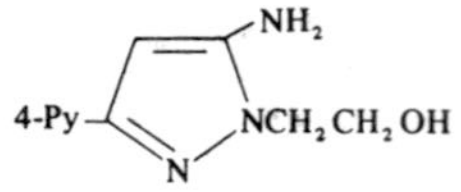 4-Py, $NH_2$, $NCH_2CH_2OH$, N | | | 171 |
| 4-$PyCOCH_2CN$ + $Et_2NCH_2CH_2NHNH_2$ | same as above | 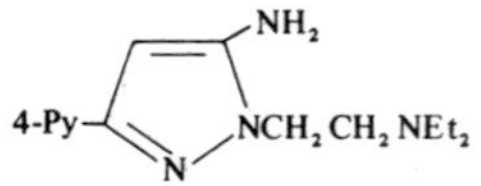 4-Py, $NH_2$, $NCH_2CH_2NEt_2$, N | | | 171 |
| 4-$PyCOCH_2CN$ + MeN–NHNH$_2$ 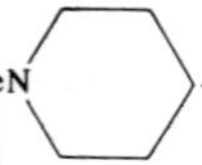 | Na, 98% EtOH, reflux 10 hr | 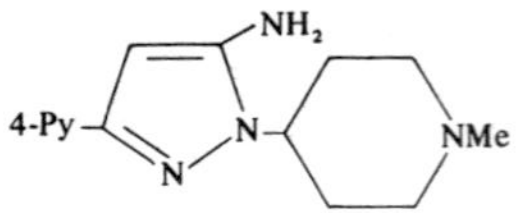 4-Py, $NH_2$, N, N, NMe | | | 171 |
| 4-$PyCOCH_2CN$ + $PhCH_2NHNH_2$ | abs. EtOH, reflux | 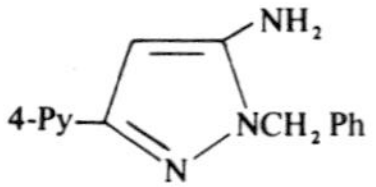 4-Py, $NH_2$, $NCH_2Ph$, N | | | 171 |
| 4-$PyCOCH_2CN$ + $PhNHNH_2$ | same as above | 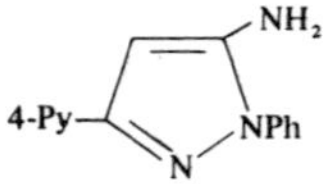 4-Py, $NH_2$, NPh, N | | | 171 |
| 4-$PyCOCH_2CN$ + $m$-$ClC_6H_4NHNH_2$ | same as above | 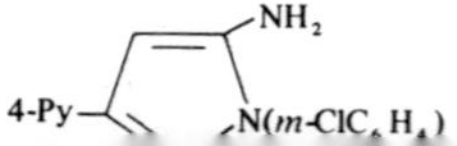 4-Py, $NH_2$, N($m$-$ClC_6H_4$) | | | 171 |

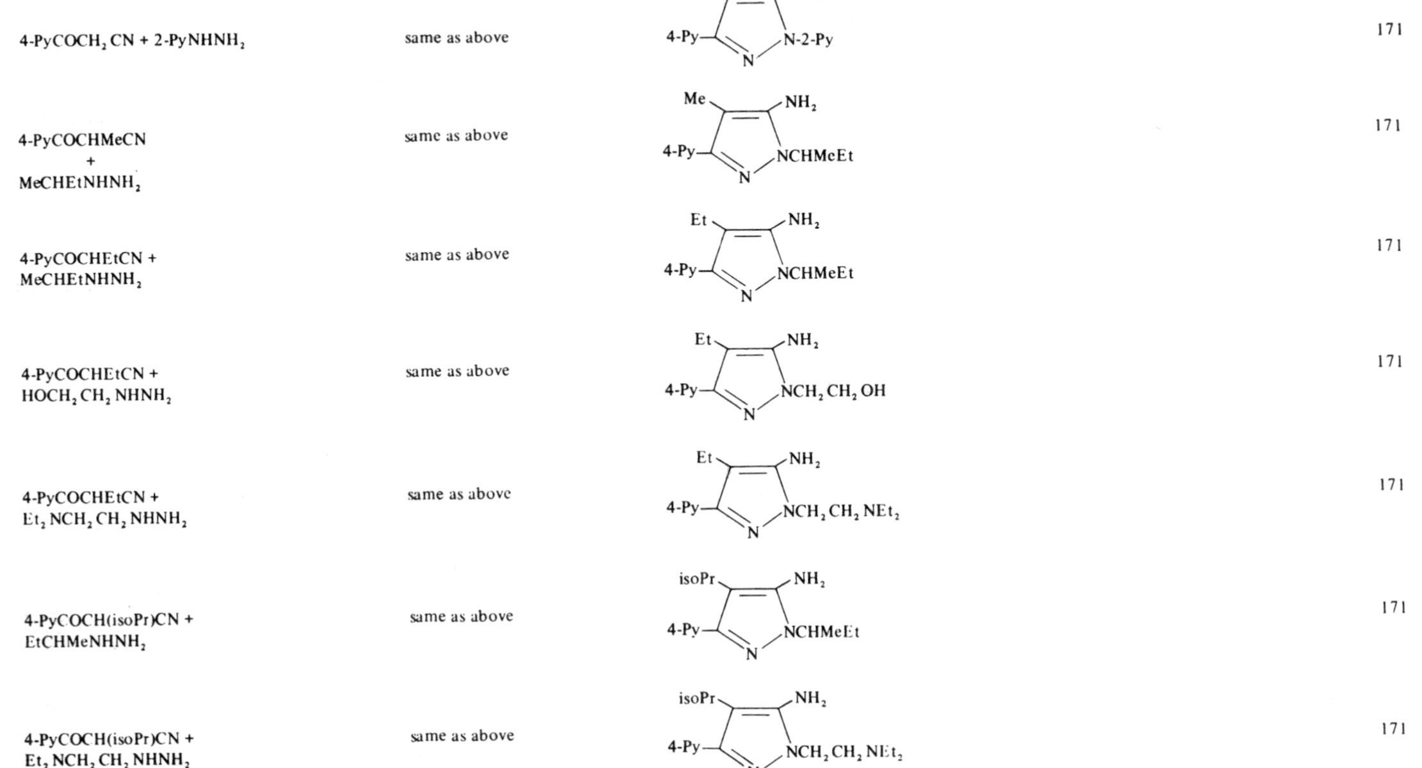

| | | | |
|---|---|---|---|
| 4-PyCOCH$_2$CN + 2-PyNHNH$_2$ | same as above | | 171 |
| 4-PyCOCHMeCN + MeCHEtNHNH$_2$ | same as above | | 171 |
| 4-PyCOCHEtCN + MeCHEtNHNH$_2$ | same as above | | 171 |
| 4-PyCOCHEtCN + HOCH$_2$CH$_2$NHNH$_2$ | same as above | | 171 |
| 4-PyCOCHEtCN + Et$_2$NCH$_2$CH$_2$NHNH$_2$ | same as above | | 171 |
| 4-PyCOCH(isoPr)CN + EtCHMeNHNH$_2$ | same as above | | 171 |
| 4-PyCOCH(isoPr)CN + Et$_2$NCH$_2$CH$_2$NHNH$_2$ | same as above | | 171 |

TABLE XI-30. Carbonyl Reactions of Keto Acids (Continued)

| Starting material | Reagent and conditions | Product | Yield | Properties | Ref. |
|---|---|---|---|---|---|
| 4-PyCOCH(isoPr)CN 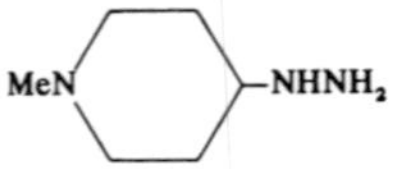 + MeN⟨piperidine⟩–$NHNH_2$ | same as above | 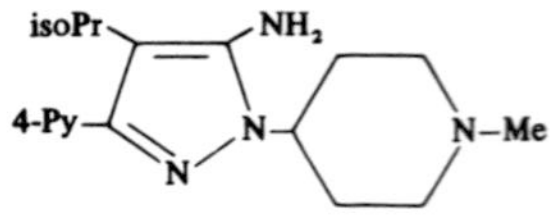 isoPr, $NH_2$, 4-Py, N, N, N–Me | | | 171 |
| 4-PyC(=NNPh)CHPh$CO_2$Me | isoBuOH or PhMe, reflux 4 hr | 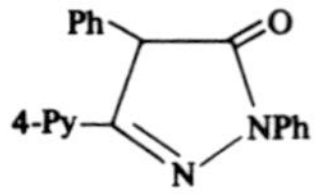 Ph, O, 4-Py, NPh, N | | m.p. 235–237°; picrate, m.p. 246–249° | 341 |
| 4-PyC($CH_2CO_2$Et)=NNHCSN$H_2$ | | 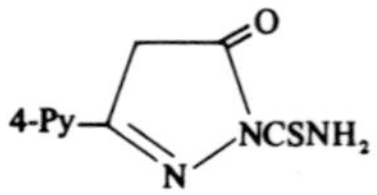 O, 4-Py, $NCSNH_2$, N | | m.p. 222–223° | 25 |
| 4-Py$CH_2$CO$CO_2$Et | PhCHO | 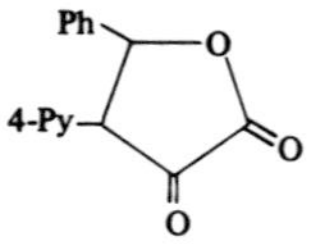 Ph, O, 4-Py, O, O | 0.5 g → 0.2 g | m.p. 266–267° | 86 |
| (4-PyCOCH$CO_2$Et)$_2$CHPh | $NH_2OH \cdot HCl$ | 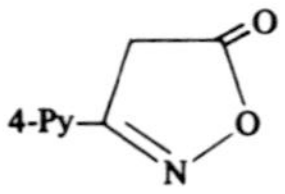 O, 4-Py, O, N | | m.p. 194–195° (decomp.) | 105 |

| Starting material | Conditions | Product | Yield | Properties | Ref. |
|---|---|---|---|---|---|
| 2-PyCPh(OH)$CH_2CO_2H$ | $SOCl_2$, reflux 2 hr;<br>10% NaOH, 1 hr | 2-PyCPh=$CHCO_2H$ | 53% | m.p. 203–204° | 54 |
| 2-PyCPh(OH)$CH_2CO_2Et$ | $H_2SO_4$, 90 min | 2-PyCPh=$CHCO_2Et$ | | m.p. 47–48.5° | 222 |
| 2-PyCPh(OH)$CHMeCO_2Et$ | $H_2SO_4$, 60 min | 2-PyCPh=$CMeCO_2Et$ | | m.p. 96–98° | 222 |
| 2-PyCPh(OH)$CHPhCO_2H$ | 98% $H_2SO_4$, 1 hr | 2-Py, Ph, O +<br>2-PyCPh=$CPhCO_2H$ | | <br>m.p. 190–195° (decomp.) | 172 |
| 2-PyC$(SMe)_2CH_2CO_2Et$ | DMSO, 40–50°<br>*or*<br>$Me_2SO_4$, 1–2 hr, 90–100°;<br>boiled 10–15 min,<br>$Me_3S^{\oplus}$ $MeSO_4^{\ominus}$ | 2-PyC(SMe)=$CHCO_2Et$ | | m.p. 157–158° | 342 |
| 3-PyCPh(OH)$CH_2CO_2H$ | $SOCl_2$, reflux 2 hr; 10%<br>NaOH, 1 hr | 3-PyCPh=$CHCO_2H$ | 56% | m.p. 201–202° | 54 |
| 4-PyCPh(OH)$CH_2CO_2H$ | same as above | 4-PyCPh=$CHCO_2H$ | 78% | m.p. 241–242° | 54 |
| 4-PyCPh(OH)$CH_2CO_2Et$ | $P_2O_5$, $C_6H_6$, reflux 8 hr<br>*or*<br>98% $H_2SO_4$, 20°, 45 min | 4-PyCPh=$CHCO_2Et$ | | m.p. 104–105° | 222 |

TABLE XI-31. Introduction of a Double Bond (Continued)

| Starting material | Conditions | Product | Yield | Properties | Ref. |
|---|---|---|---|---|---|
| 4-PyCPh(OH)CHMeCO$_2$Et | $P_2O_5$ or $H_2SO_4$, 15 min | 4-PyCPh=CMeCO$_2$Et | | m.p. 53–55° | 222 |
| 4-PyCPh(OH)CHPhCO$_2$H | 98% $H_2SO_4$, 1 hr | 4-PyCPh=CPhCO$_2$H 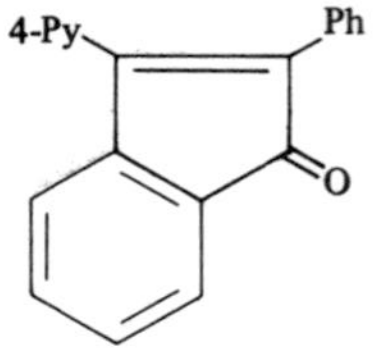 + 3-(4-Py)-2-phenylinden-1-one [structure] | | m.p. 248–255° | 172 |
| 4-PyC(SMe)$_2$CH$_2$CO$_2$Et | DMSO, 40–50° *or* Me$_2$SO$_4$, 1–2 hr, 90–100°; boiled 10–15 min, $Me_3S^{\oplus}MeSO_4^{\ominus}$ | 4-PyC(SMe)=CHCO$_2$Et | | m.p. 67–68° | 342 |

TABLE XI-32. Addition Reactions of Unsaturated Side Chains

| Starting material | Reagent and conditions | Product | Yield | Properties | Ref. |
|---|---|---|---|---|---|
| $(2\text{-PyCH=NCH}_2)_2$ | HCN, 5–10° | $[2\text{-PyCH(CN)NHCH}_2]_2$ | 65.7% | m.p. 96–102° (decomp.) | 343 |
| 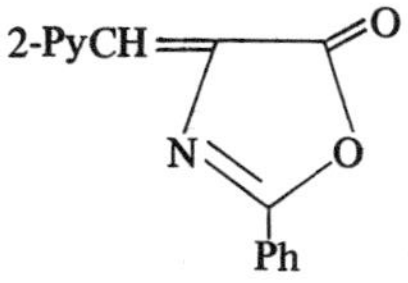 2-PyCH= (O, N, O, Ph) | $CH_2N_2$ or DMSO | 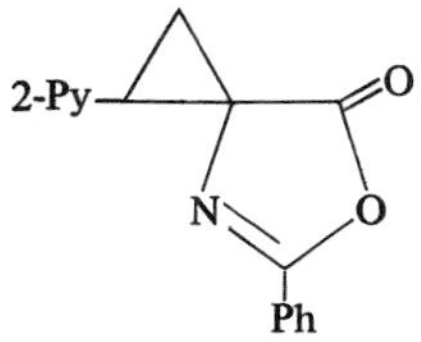 2-Py– (O, N, O, Ph) | 12%<br>19% | m.p. 141° | 94 |
| 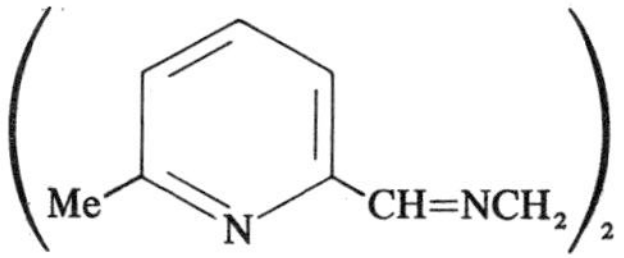 (Me, N, $CH=NCH_2$)$_2$ | HCN, 5–10° | 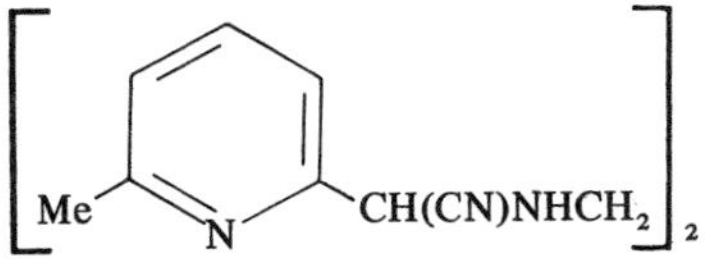 [Me, N, $CH(CN)NHCH_2$]$_2$ | 72.2% | | 343 |
| 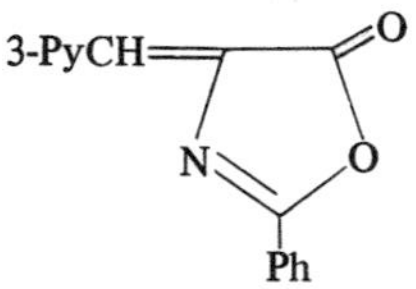 3-PyCH= (O, N, O, Ph) | $CH_2N_2$ or DMSO | 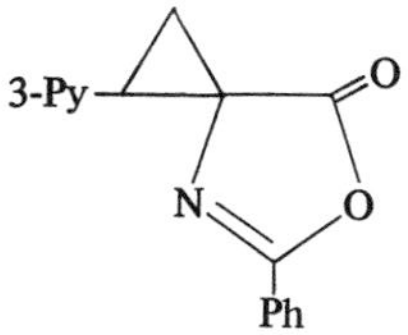 3-Py– (O, N, O, Ph) | 15%<br>25% | m.p. 145° | 94 |
| $3\text{-PyC(Cl)=CHCO}_2\text{Et}$<br>+<br>$3\text{-PyCCl}_2\text{CHCO}_2\text{Et}$<br>(mixture) | $Et_2NH$, 80°, 24 hr | $3\text{-PyC(NEt}_2\text{)=CHCO}_2\text{Et}$ | 44.7% | b.p. 101°/0.01 mm; $n_D^{20}$ 1.5582; picrate, m.p. 110° | 175 |

TABLE XI-32. Addition Reactions of Unsaturated Side-Chains (Continued)

| Starting Material | Reagent and Conditions | Product | Yield | Properties | Ref. |
|---|---|---|---|---|---|
| | morpholine, $C_6H_6$, reflux 4 hr | 3-PyC(morpholino)=$CHCO_2Et$ | 57.5% | b.p. 129–131°/0.01 mm; $n_D^{20}$ 1.5728; picrate, m.p. 125° | 175 |
| | piperidine, $C_6H_6$, reflux 4 hr | 3-PyC(Piperidino)=$CHCO_2Et$ | 68% | b.p. 128°/0.01 mm; $n_D^{20}$ 1.5788; picrate m.p. 111° | 175 |
| | pyrrolidine, $C_6H_6$, reflux 4 hr; $H_2$; EtOH, Py, 5% Pd-C | 3-PyC(1-pyrrolidinyl)=$CHCO_2Et$ | 71% | b.p. 130–131°/0.01 mm; $n_D^{20}$ 1.5869; picrate, m.p. 127° | 175 |
| 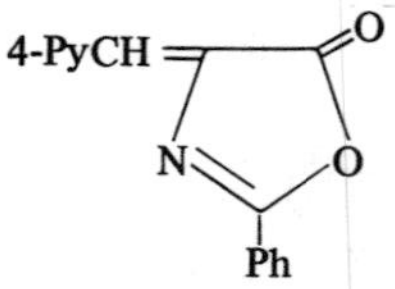 4-PyCH= (oxazolone ring: C=O, O, N, Ph) | $CH_2N_2$ or DMSO | 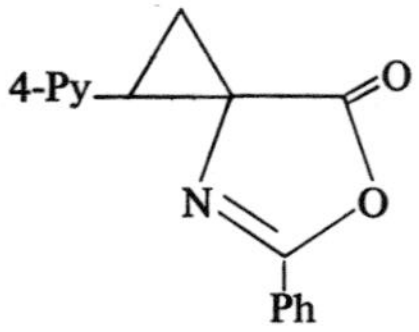 4-Py (cyclopropane spiro-fused to oxazolone ring: C=O, O, N, Ph) | 17%<br>22% | m.p. 152° | 94 |

| Compound | b.p./mm | m.p. | Derivative | Ref. |
|---|---|---|---|---|
| 2-$PyCH_2CONHNH_2$ | | 120° | hydrochloride, m.p. 175–177° | 351 |
| | | 117–119° | hydrochloride, m.p. 195–199° | 362 |
| | | 117–118° | picrate, m.p. 138–139° | 291 |
| 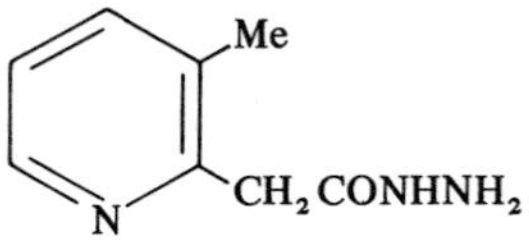  | | 136–138° | | 34 |
| 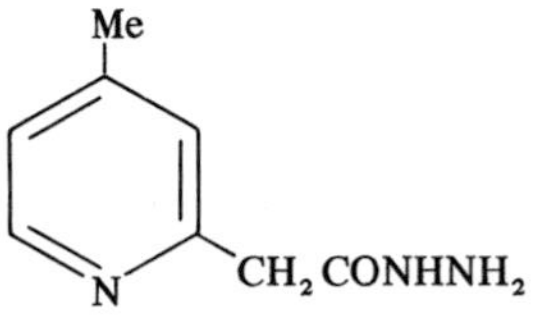  | | 105.5° | | 34 |
| | | 106.5° | | |
| 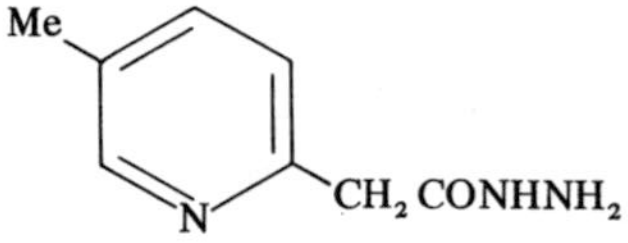  | | 122–123° | | 34 |
| 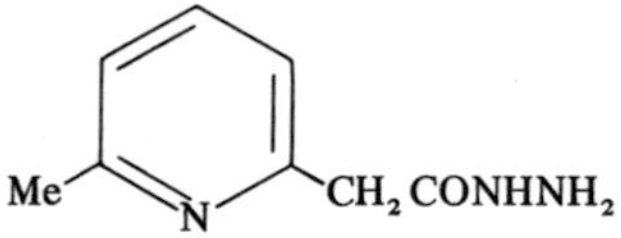  | | 143–144.5° | | 34 |

TABLE XI-33. Pyridine Side-Chain Hydrazides, Hydroxamic Acids, Amidines, and Imides (Continued)

| Compound | b.p./mm | m.p. | Derivative | Ref. |
|---|---|---|---|---|
| Me; O; N H; $CH_2CONHNH_2$ | | 189–190° | | 363 |
| Me; O; N H; $CH_2CONHN{=}CMeEt$ | | 171–172° | | 363 |
| $2\text{-}PyCH_2CONHNH{=}$(cyclohexylidene) | | 99–102° | | 362 |
| Me; O; N H; $CH_2CONHN{=}$(cyclohexylidene) | | 229–230° | | 363 |

| | | |
|---|---|---|
| 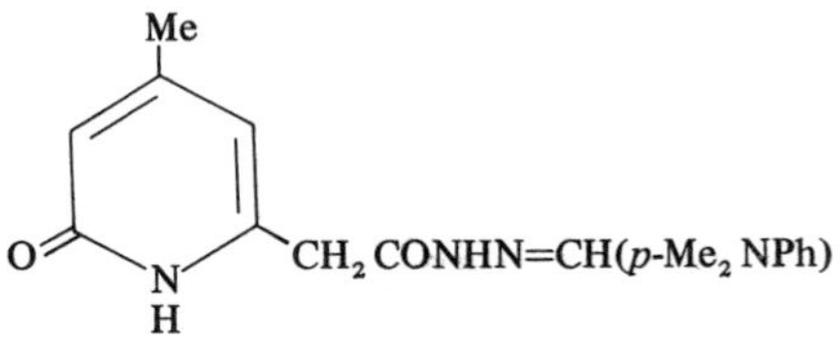 | 220–221° | 363 |
| 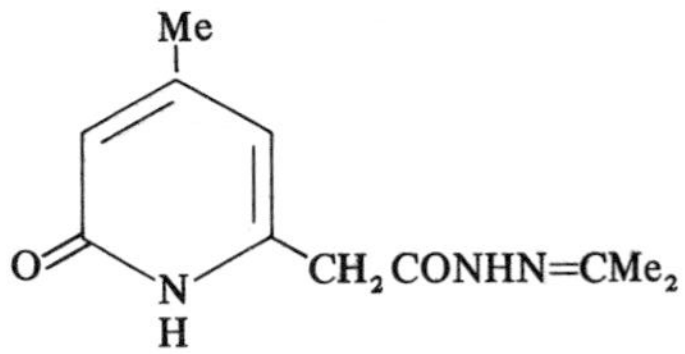 | 229–239° (decomp.) | 363 |
| 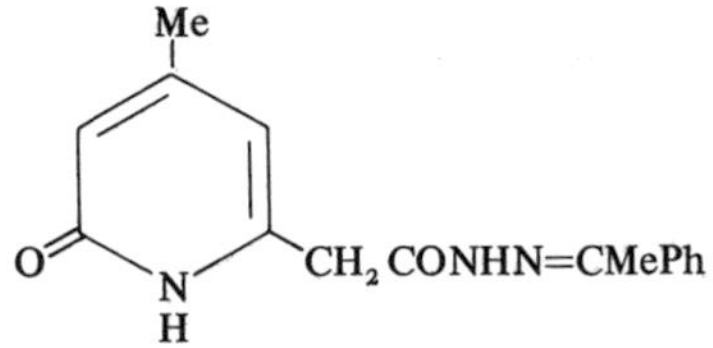 | 206–207° | 363 |
| 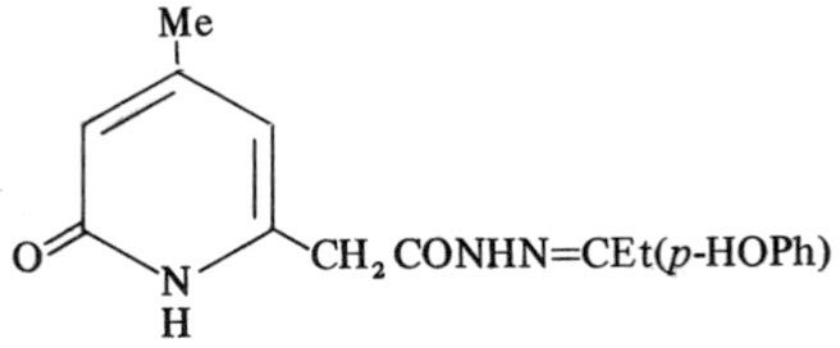 | 206–207° | 363 |

TABLE XI-33. Pyridine Side-Chain Hydrazides, Hydroxamic Acids, Amidines, and Imides (Continued)

| Compound | b.p./mm | m.p. | Derivative | Ref. |
|---|---|---|---|---|
| 2-PyCHMeCONHNH$_2$ | | 77-78° | | 34 |
| 2-PyCH(OH)CONHNH$_2$ | | 102° | | 35 |
| 2-Py(CH$_2$)$_2$ CONHNH$_2$ | 143–149°/0.002 | | $p$-Me$_2$ NPhCHO derivative, m.p. 142–145° | 34 |
| 3-PyCH$_2$ CONHNH$_2$ | | 108–109° | hydrochloride, m.p. 182–185° | 351 |
| 3-PyCH$_2$ CONHNHCHMe(3-Py) | | 122–124° | | 159 |
| 3-PyCH$_2$ CONHN=CMe(3-Py) | | 156–158° (decomp.) | | 159 |
| 3-PyCH(OH)CONHNH$_2$ | | 141° | | 35 |
| 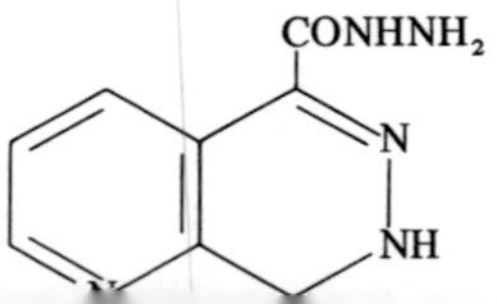 [structure: CONHNH$_2$, N, NH] | | 296° (decomp.) | | 260 |

| | | | | |
|---|---|---|---|---|
|  | | 198° | | 260 |
| 3-PyCOC(=NH)NHC(=NH)$NH_2$ | | 150–151° | | 162 |
| 3-Py$COCH_2$ CONHC(=NH)$NH_2$ | | 283–288° (decomp.) | | 161 |
| 3-PyCOCONHC(=NH)$NH_2$ | | 283–288° (decomp.) | | 161 |
| 4-Py$CH_2$ CONH$NH_2$ | | 90–91° | | 364 |
| | | | hydrochloride, m.p. 185–188° | 351 |
| 4-Py$CH_2$ CONH$NH_2$ | | 94–95° | | 58 |
| 4-Py$CH_2$ CONHNHisoPr | no data | | | 58 |
| 4-Py$CH_2$ CONHNH$CH_2$ Ph | | 119–120° | | 159 |
| 4-Py$CH_2$ CONHN=CHPh | | 69–71° (hydrate)<br>116–117° | | 159<br>364 |
| 4-Py$CH_2$ CONHN=CH(*p*-HOPh) | | 217–218° | | 364 |
| 4-Py$CH_2$ CONHN=CH(*p*-AcNHPh) | | | hydrate, m.p. 226–227° | 364 |

TABLE XI-33. Pyridine Side-Chain Hydrazides, Hydroxamic Acids, Amidines, and Imides (Continued)

| Compound | b.p./mm | m.p. | Derivative | Ref. |
|---|---|---|---|---|
| 4-$PyCH_2CONHN{=}CH$– –OH, OMe | | 208–209° | | 364 |
| 4-$PyCH_2CONHN{=}CMe_2$ | no data | | | 58 |
| $CH_2CONHNH_2$, O, N, $(CH_2)_2$– –OMe, OMe | | 110–112° | | 8 |
| 4-$PyCH(CH_2Ph)CONHNH_2$ | | 138° | | 166 |
| 4-$Py(CH_2)_2CONHNH_2$ | | 61–62° | | 364 |
| | | 84° | hydrate, m.p. 64° | 166 |
| 4-$Py(CH_2)_2CONHNHCHMe_2$ | | 74° | | 160 |

| | | | |
|---|---|---|---|
| $4\text{-}Py(CH_2)_2CONHN{=}CH\text{-}C_6H_3(4\text{-}OH)(3\text{-}OMe)$ | | 212–214° | 364 |
| $4\text{-}Py(CH_2)_2CONHN{=}CMe_2$ | no data | | 160 |
| $4\text{-}PyCH{=}CHCONHNH_2$ | | 109–110° | 29 |
| $4\text{-}PyCH{=}CHCONHN{=}CH(p\text{-}AcNHPh)$ | | 169–171° | 364 |
| $4\text{-}PyCH{=}CHCONHN{=}CH\text{-}C_6H_3(4\text{-}OH)(3\text{-}OMe)$ | | 124–126° | 364 |

## References


1. H. McKennis, Jr., L. B. Turnbull, E. R. Bowmann, and E. Tamaki, *J. Org. Chem.*, 383 (1963).
2. F. Zymalkowski and B. Trenktrog, *Arch. Pharm.* (Weinheim), **292,** 9 (1959).
3. M. Barash and J. M. Osbond, *J. Chem. Soc.,* 2157 (1959).
4. M. Barash, J. M. Osbond, and J. C. Wickens, *J. Chem. Soc.,* 3530 (1959).
5. J. M. Bobbitt and D. A. Scola, *J. Org. Chem.,* **25,** 560 (1960).
6. A. Cohen and J. M. Osbond, U.S. patent, 2,877,227 (1959); *Chem. Abstr.,* **55,** 74 (1961).
7. M. Kirisawa, *Chem. Pharm. Bull.* (Tokyo), **7,** 35 (1959).
8. M. Kirisawa, *Chem. Pharm. Bull.* (Tokyo), **7,** 38 (1959).
9. Y. Sato, *Chem. Pharm. Bull.* (Tokyo), **7,** 241 (1959).
10. J. T. Suh and B. M. Puma, *J. Org. Chem.,* **30,** 2253 (1965).
11. A. Jackson and J. A. Joule, *Chem. Commun.,* 459 (1967).
12. E. Wenkert, K. G. Dave, R. G. Lewis, and P. W. Sprague, *J. Amer. Chem. Soc.,* 6741 (1967).
13. J. A. Weisbach, J. L. Kirkpatrick, K. R. Williams, E. L. Anderson, N. C. Yim, anc Douglas, *Tetrahedron Lett.,* 3457 (1965).
14. F. Bohlmann, E. Winterfeldt, H. Laurent, and W. Ude. *Tetrahedron,* **19,** 195 (196
15. B. Frydman, M. E. Despuy, and H. Rapoport, *J. Amer Chem. Soc.,* **87,** 3530 (196
16. C. E. Cook, R. C. Corley, and M. E. Wall, *J. Org. Chem.,* **30,** 4114 (1965).
17. I. G. Morris and A. R. Pinder, *J. Chem. Soc.,* 1841 (1963).
18. F. Bohlmann, E. Winterfeldt, D. Schumann, and B. Gatscheff, *Chem. Ber.,* **98,** (1965).
19. F. Bohlmann, E. Winterfeldt, D. Schumann, U. Zarnack, and P. Wandrey, *Chem. B* **95,** 2365 (1962).
20. M. Pailer and R. Libiseller, *Monatsh. Chem.,* **93,** 511 (1962); *Chem. Abstr.* **57,** 47 (1962).
21. M. Beroza, *J. Org. Chem.,* **28,** 3562 (1963).
22. H. McKennis, Jr., E. R. Bowman, and L. B. Turnbull, *J. Amer. Chem. Soc.,* **82,** 3 (1960).
23. H. McKennis, Jr., S. L. Schwartz, L. B. Turnbull, E. Tamaki, and E. R. Bowman *Biol. Chem.,* **239,** 3981 (1964).
24. P. L. Morselli, H. H. Ong, E. R. Bowman, and H. McKennis, Jr., *J. Med. Chem.,* 1033 (1967).
25. C. Belzecki and T. Urbański, *Rocz. Chem.,* **32,** 779 (1958); *Chem. Abstr.* **53,** 101 (1959).
26. C. Caradonna, M. L. Stein, and M. Ikram, *Ann. Chim.* (Rome), **49,** 2083 (19 *Chem. Abstr.,* **54,** 19646c (1960).
27. L. C. Cheney and J. C. Godfrey, U.S. patent, 3,202,653 (1965); *Chem Abstr.,* 18094c (1965).
28. A. Fujita, T. Yamamoto, J. Matsumoto, S. Minami, and H. Takamatsu, *Yakug Zasshi,* **85,** 565 (1965); *Chem. Abstr.,* **63,** 9909f (1965).
29. S. Kakimoto, J. Nishie, and K. Yamamoto, *Japan J. Tuberc.,* **7,** 76 (1959); *Ch Abstr.,* **54,** 21079h (1960).
30. F. Leonard and W. Tschannen, *J. Med. Chem.,* **9,** 140 (1966).
31. R. B. Moffett, *J. Med. Chem.,* **9,** 475 (1966).
32. R. B. Moffett, U.S. patent, 3,255,186 (1966); *Chem. Abstr.,* **65,** 8928c (1966).

Y. G. Perron, L. C. Cheney, and J. C. Godfrey, Belg. patent, 631,631 (1963); *Chem. Abstr.*, **61**, 7021h (1964).

J. Izdebski, *Rocz. Chem.*, **39**, 717 (1965); *Chem. Abstr.*, **64**, 3465e (1966).

E. G. Brain and J. H. C. Nayler, Brit. patent, 986,544 (1965); *Chem. Abstr.*, **62**, 16255c (1965).

E. J. Cragoe, Jr., A. M. Pietruszkiewicz, and C. M. Robb, *J. Org. Chem.*, **23**, 971 (1959).

J. Heider and D. Jerchel, Ger. patent 1,166,190 (1964); *Chem. Abstr.*, **60**, 15947d (1964).

J. Heider and D. Jerchel, Ger. patent 1,180,367 (1964); *Chem. Abstr.*, **62**, 1725f (1965).

J. Heider and D. Jerchel, U.S. patent, 3,167,544 (1965); *Chem. Abstr.*, **62**, 11887c (1965).

J. Heider and D. Jerchel, U.S. patent, 3,314,854 (1967); *Chem. Abstr.*, **67**, 108849f (1967).

G. M. K. Hughes, U.S. patent, 3,094,533 (1963); *Chem. Abstr.*, **60**, 2912e (1964).

W. L. Bencze and G. N. Walker, U.S. patent, 3,337,568 (1967); *Chem. Abstr.*, **68**, 87179r (1968).

M. Perelman and S. Mizsak, *J. Med. Chem.*, **6**, 533 (1963).

J. W. Cusic and H. W. Sause, Belg. patent, 617,730 (1962); *Chem. Abstr.*, **58**, 12522d (1963).

J. W. Cusic and H. W. Sause, U.S. patent, 3,225,054 (1965); *Chem. Abstr.*, **64**, 6625f (1966).

J. A. Meschino, U.S. patent, 3,185,707 (1965); *Chem. Abstr.*, **63**, 2934f (1965).

J. A. Meschino, U.S. patent, 3,254,092 (1966); *Chem. Abstr.*, **65**, 12179d (1966).

R. B. Moffett, *J. Med. Chem.*, **7**, 446 (1964).

R. B. Moffett, U.S. patent, 3,156,697 (1964); *Chem. Abstr.*, **62**, 5257 (1965).

A. D. Cale, Jr., H. Jenkins, B. V. Franko, J. W. Ward, and C. D. Lunsford, *J. Med. Chem.*, **10**, 214 (1967).

R. W. Temple and L. F. Wiggins, Brit. patent, 930,459 (1963); *Chem. Abstr.*, **59**, 13956b (1963).

E. Seeger and A. Kottler, Ger. patent 1,026,318 (1958); *Chem. Abstr.*, **54**, 11058a (1960).

Smith, Kline, and French Laboratories, Nether. patent appl., 6,507,155 (1965); *Chem. Abstr.*, **65**, 761c (1966).

F. J. Villani, M. S. King, and F. J. Villani, *J. Med. Chem.*, **6**, 142 (1963).

Laboratori Panacea S.r.1., Belg. patent, 628,483 (1963); *Chem. Abstr.*, **60**, 14509a (1964).

C. D. Lunsford and A. D. Cale, Jr., Belg. patent, 613,734 (1962); *Chem. Abstr.*, **58**, 507g (1963).

C. D. Lunsford and A. D. Cale, Jr., U.S. patent, 3,192,207 (1965); *Chem. Abstr.*, **63**, 11503a (1965).

T. S. Gardner, E. Wenis, and J. Lee, *J. Med. Pharm. Chem.*, **5**, 503 (1962).

W. L. Bencze, Belg. patent 620,631 (1963); *Chem. Abstr.*, **59**, 11449b (1963).

S. Minami, A. Fujita, J. Matsumoto, K. Fujimoto, M. Shimizu, and Y. Takase, Jap. patent, 23,662 (1965); *Chem. Abstr.*, **64**, 3504b (1966).

S. Minami, A. Fujita, K. Yamamoto, K. Fujimoto, M. Shimizu, and Y. Takase, Jap. patent, 6,908 (1966); *Chem. Abstr.*, **65**, 7154b (1966).

S. Minami, A. Fujita, K. Yamamoto, K. Fujimoto, and Y. Takase, Jap. patent, 1,178 (1967); *Chem. Abstr.*, **66**, 94912m (1967).

63. H. M. Taylor, U.S. patent, 3,313,683; *Chem. Abstr.*, **67**, 64073s (1967).
64. Farbenfabriken Bayer AG, Nether. appl., 6,605,907 (1966); *Chem. Abstr.*, **67**, 2201 (1967).
65. M. A. McCall and R. L. McConnell, U.S. patent, 3,106,566 (1963); *Chem. Abstr.*, 6 4115g (1964).
66. M. Samejima, *Yakugaku Zasshi,* **80,** 1706 (1960); *Chem. Abstr.*, **55**, 10439b (196
67. Kodak, Soc. anon., Belg. patent, 588,862 (1960); *Chem. Abstr.*, **55**, 1251i (1961).
68. N. J. Doorenbos, U.S. patent, 2,953,561 (1960); *Chem. Abstr.*, **55**, 1250d (1961).
69. Tanabe Seiyaku Co. Ltd., Belg. patent, 671,385 (1966); *Chem. Abstr.*, **65**, 1534 (1966).
70. M. Anderson and A. W. Johnson, *J. Chem. Soc., C,* 1075 (1966).
71. S. J. Norton and E. Sanders, *J. Med. Chem.*, **10**, 961 (1967).
72. C. H. Kao, *K'o Hsüch T'ung Pao* No. 14, 434 (1957); *Chem. Abstr.*, **55**, 235 (1961).
73. K. Horibata, H. Taniuchi, M. Tashiro, S. Kuno, O. Hayaishi, T. Sakan, T. Tokuyan and S. Seno, *Koso Kagaku Shinpojiumu,* **15**, 117 (1961); *Chem. Abstr.*, **56**, 511 (1962).
74. J. W. Hylin, *Arch. Biochem. Biophys.*, **83**, 528 (1959); *Chem. Abstr.*, **53**, 2224 (1959).
75. Y. L. Gol'dfarb, F. D. Alashev, and V. K. Zvorykina, *Izv. Akad. Nauk SSR, O Khim. Nauk,* 2209 (1962); *Chem. Abstr.*, **58**, 14014d (1963).
76. S. Senoh, S. Imamoto, and Y. Maeno, *Tetrahedron Lett.*, 3431 (1964).
77. R. Hiltmann and H. Wollweber, Ger. patent, 1,073,497 (1960); *Chem. Abstr.*, 5 19956a (1961).
78. J. Izdebski, *Rocz. Chem.*, **39**, 1625 (1965); *Chem. Abstr.*, **64**, 17536b (1966).
79. Z. Dabrowski and J. T. Wrobel, *Chem. Ind.* (London), 1758 (1964).
80. J. A. Gautier, I. Marszak, M. Olomucki, and M. Miocque, *Bull. Soc. Chim. Fr.*, 25 (1965).
81. J. A. Gautier, I. M. Olomycki, and M. Miocque, *C. R. Acad. Sci., Paris, Ser. C,* 25 562 (1960); *Chem. Abstr.*, **55**, 9397e, 11407g (1961).
82. O. Martensson and E. Nilsson, *Acta Chem. Scand.*, **14**, 1129 (1960).
83. G. V. Boyd and A. D. Ezekiel, *J. Chem. Soc., C,* 1866 (1967).
84. R. E. Willette, *J. Chem. Soc.*, 5874 (1965).
85. E. D. Amstutz and M. M. Besso, *J. Org. Chem.*, **25**, 1687 (1960).
86. M. M. Besso, PhD thesis, Lehigh University, Bethlehem, Pa. (1959).
87. G. N. Walker and B. N. Weaver, *J. Org. Chem.*, **26**, 4441 (1961).
88. J. M. Osbond, Brit. patent, 808,046 (1959); *Chem. Abstr.*, **53**, 18063 (1959).
89. Y. Noike, *Yakugaku Zasshi,* **79**, 1514 (1959); *Chem. Abstr.*, **54**, 11021b (1960).
90. M. M. Baizer, U.S. patent, 3,218,245 (1965); *Chem. Abstr.*, **64**, 17554f (1966).
91. S. L. Shapiro, V. Bandurco, and L. Freedman, *J. Org. Chem.*, **27**, 174 (1962).
92. R. Tan and A. Taurins, *Tetrahedron Lett.*, 1233 (1966).
93. H. Hellmann and H. Piechota, *Ann. Chem.*, **631**, 175 (1960).
94. G. Slater and A. W. Somerville, *Tetrahedron,* **13**, 2823 (1967).
95. S. J. Norton, C. G. Skinner, and W. Shive, *J. Org. Chem.*, **26**, 1495 (1961).
96. S. Hauptmann and K. Hirshberg, *J. Prakt. Chem.*, **34**, 272 (1966); *Chem. Abstr.*, 6 55360k (1967).
97. J. Wrobel and Z. Dabrowski, *Rocz. Chem.*, **39**, 1239 (1965); *Chem. Abstr.*, 6 15936g (1966).
98. F. H. Clarke, G. A. Felock, G. B. Silverman, and C. M. Watnick, *J. Org. Chem.*, 2 533 (1962).

W. Herz and D. R. K. Murty, *J. Org. Chem.*, **26**, 418 (1961).

G. Galiazzo, *Gazz. Chim. Ital.*, **95**, 1322 (1965).

J. A. T. Beard and A. R. Katritzky, *Rec. Trav. Chim. Pays-Bas,* **78**, 592 (1959).

G. B. D. de Graaff, W. Ch. Melger, J. Van Bragt, and S. Schukking, *Rec. Trav. Chim. Pays-Bas,* **83**, 910 (1964).

N. S. Vul'fson and L. I. Lukashina, *Sbornik Statei, Nauch.-Issledovatel. Inst. Org. Poluprod. i Krasitelei No. 2,* 137 (1961); *Chem. Abstr.*, **56**, 10094e (1962).

W. Wunderlich, *J. Prakt. Chem.*, **2**, 302 (1955); *Chem. Abstr.*, **54**, 4579i (1960).

O. Y. Magidson, *Zh. Obshch. Khim.*, **29**, 165 (1959); *Chem. Abstr.*, **54**, 530d (1960).

K. W. Merz and R. Barchet, *Arch. Pharm.* (Weinheim), **297**, 423 (1964).

N. S. Vul'fson, V. E. Kolchin, and L. K. Artemchick, *Zh. Obshch. Khim.*, **32**, 3382 (1962); *Chem. Abstr.*, **59**, 1784e (1963).

N. S. Vul'fson and V. E. Kolchin, *Zh. Obshch. Khim.*, **34**, 2387 (1964); *Chem. Abstr.*, **61**, 14818c (1964).

Nishimoto and T. Nakashima, *Yakugaku Zasshi,* **81**, 88 (1961); *Chem. Abstr.*, **55**, 13420h (1961).

K. Winterfeld and H. Buschbeck, *Arch. Pharm.* (Weinheim), **294**, 468 (1961).

D. R. Bragg and D. G. Wiberley, *J. Chem. Soc., C,* 5074 (1961).

G. VanZyl, D. L. DeVries, R. H. Decker, and E. T. Niles, *J. Org. Chem.*, **26**, 3373 (1961).

H. Hellmann and D. Dieterich, *Ann. Chem.*, **656**, 53 (1962).

R. Lukeš, J. N. Zvonkova, A. F. Mironov, and M. Ferles, *Collect. Czech. Chem. Commun.*, **25**, 2668 (1960); *Chem. Abstr.*, **55**, 4501d (1961).

M. S. Brown and H. Rapoport, *J. Org. Chem.*, **28**, 3261 (1963).

L. Kuczynski, Z. Machon, and L. Wykret, *Dissertationes Pharm.*, **16**, 479 (1964); *Chem. Abstr.*, **63**, 11491f (1965).

J. Čejka, M. Fereles, S. Chládek, J. Labsky, and M. Zelinka, *Collect Czech. Chem. Commun.*, **26**, 1429 (1961); *Chem. Abstr.*, **55**, 27309b (1961).

A. L. Davis, C. G. Skinner, and W. Shive, *Arch. Biochem. Biophys.*, **87**, 88 (1960); *Chem. Abstr.*, **55**, 2513i (1961).

M. E. Kuehne, *J. Amer. Chem. Soc.*, **84**, 837 (1962).

A. Mustafa and M. M. M. Sallam, *J. Org. Chem.*, **27**, 2406 (1962).

L. Neilande, A. Karklins, V. Veiss, and G. Vanags, *Latvijas PSR Zinatnu Akad. Vestis, Kim Series* 7 (1964); *Chem. Abstr.*, **61**, 3699g (1964).

E. Wenkert, K. G. Dave, and F. Haglid, *J. Amer. Chem. Soc.*, **87**, 5461 (1965).

E. Zbiral and E. Werner, *Tetrahedron Lett.*, 2001 (1966).

E. Zbiral and E. Werner, *Monatsh. Chem.*, **97**, 1797 (1966); *Chem. Abstr.*, **66**, 85824g (1967).

W. Kornytnyk, B. Paul, A. Bloch, and C. A. Nichol, *J. Med. Chem.*, **10**, 345 (1967).

Dr. Karl Thomae GmbH, Belg. patent, 621,453 (1963); *Chem. Abstr.*, **59**, 10185f (1963).

A. Stempel, Belg. patent, 635,615 (1964); *Chem. Abstr.*, **61**, 10661e (1964).

G. Giacomello, F. Gualtieri, F. M. Riceieri, and M. L. Stein, *Tetrahedron Lett.*, 1117 (1965).

F. Korte and H. J. Schulze-Steinen, *Chem. Ber.*, **95**, 2444 (1963).

W. Baker, K. M. Buggle, J. F. W. McOmie, and D. A. M. Watkins, *J. Chem. Soc.*, 3594 (1958).

M. J. Betts and B. R. Brown, *J. Chem. Soc., C,* 1730 (1967).

G. N. Walker, *J. Org. Chem.*, **27**, 2966 (1962).

133. E. Wenkert, K. G. Dave, F. Haglid, R. G. Lewis, T. Oishi, R. V. Stevens, and Terashima, *J. Org. Chem.*, **33,** 747 (1968).
134. A. Mironov, M. Fereles, and M. Pergal, *Sb. Vysoke Skoly Chem.-Technol. Praze, C Technol,* 5, 83 (1961); *Chem. Abstr.*, **62,** 13017b (1965).
135. V. Boekelheide and R. J. Windgasse, Jr., *J. Amer. Chem. Soc.*, **81,** 1456 (1959).
136. G. N. Walker, *J. Amer. Chem. Soc.*, 77, 3844 (1955).
137. J. Sam, *J. Pharm. Sci.*, **56,** 1360 (1967).
138. I. Murakoski, A. Kubo, J. Saito, and J. Haginiwa, *Chem. Pharm. Bull.* (Tokyo), 747 (1964).
139. R. Adams and W. Reifschneider, *J. Amer. Chem. Soc.*, **81,** 2537 (1959).
140. T. Miyadera and I. Iwai, *Chem. Pharm. Bull.* (Tokyo), **12,** 1338 (1964).
141. D. R. Bragg and D. G. Wibberley, *J. Chem. Soc.*, 2627 (1962).
142. D. R. Bragg and D. G. Wibberley, *J. Chem. Soc.*, 3277 (1963).
143. J. Hurst, T. Melton, and D. G. Wibberley, *J. Chem. Soc.*, 2948 (1965).
144. T. Kappe, *Monatsh. Chem.*, **98,** 1858 (1967); *Chem. Abstr.*, **68,** 39447e (1968).
145. T. Melton and D. G. Wibberley, *J. Chem. Soc., C,* 983 (1967).
146. K. Winterfeld and W. Erning, *Arch. Pharm.* (Weinheim), **298,** 220 (1965).
147. D. R. Bragg and D. G. Wibberley, *J. Chem. Soc., C,* 2120 (1966).
148. K. Winterfeld and W. Erning, *Arch. Pharm.* (Weinheim), **298,** 220 (1965); *Che Abstr.*, **63,** 14809h (1965).
149. H. Pacheco and R. Gatto, *Bull. Soc. Chim. Fr.*, 95 (1960).
150. R. Tan and A. Taurins, *Tetrahedron Lett.*, 2737 (1965).
151. H. E. Baumgarten, H. C. F. Su, and R. P. Barkley, *J. Heterocycl. Chem.*, **3,** 3 (1966).
152. B. M. Ferrier and N. Campbell, *Proc. Roy. Soc. Edinburgh Sect. A,* **65,** 2 (1959–60); *Chem. Abstr.*, **55,** 24748e (1961).
153. H. Junek, *Monatsh. Chem.*, **96,** 2046 (1965); *Chem. Abstr.*, **64,** 11167h (1966).
154. R. Dohmori, *Chem. Pharm. Bull.* (Tokyo), **12,** 595 (1964).
155. T. Naito, R. Dohmori, and M. Shimoda, *Pharm. Bull.* (Japan), **3,** 34 (1955).
156. M. Maruoka, K. Isagawa, T. Kuki, and Y. Fushizaki, *Nippon Kaguku Zasshi,* **83,** 2 (1962); *Chem. Abstr.*, **59,** 3879d (1963).
157. O. Martensson and E. Nilsson, *Acta Chem. Scand.*, **15,** 1026 (1961).
158. W. A. Schuler and A. Gross, U.S. patent, 2,953,562 (1960); *Chem. Abstr.*, **55,** 45 (1961).
159. H. Bojarska-Dahlig, Polish patent 47,469 (1963); *Chem. Abstr.*, **61,** 10661g (1964).
160. Roussel UCLAF, Fr. patent, M3268 (1965); *Chem. Abstr.*, **63,** 16314c (1965).
161. R. P. Mariella and J. J. Zelko, *J. Org. Chem.*, **25,** 647 (1960).
162. K. Odo and E. Ichikawa, Jap. patent 9565 (1967); *Chem. Abstr.*, **67,** 64250x (196
163. C. D. Gutsche and H. W. Voges, *J. Org. Chem.*, **32,** 2685 (1967).
164. G. N. Walker, *J. Med. Chem.*, **8,** 583 (1965).
165. B. E. Betts and W. Davey, *J. Chem. Soc.*, 3333 (1961).
166. H. W. Sause, Belg. patent, 669,165 (1966); *Chem. Abstr.*, **65,** 13666c (1966).
167. E. Profft and W. Steinke, *Chem. Ber.*, **94,** 2267 (1961).
168. A. Banashek and M. N. Shckukina, *Zh. Obshch. Khim.*, **31,** 1479 (1961); *Che Abstr.*, **55,** 24739e (1961).
169. G. F. Holland and J. N. Pereira, *J. Med. Chem.*, **10,** 149 (1967).
170. K. Kindler and K. Lührs, *Chem. Ber.*, **99,** 227 (1966).
171. CIBA Ltd., Belg. patent 612,971 (1962); *Chem. Abstr.*, **58,** 3438f (1969).
172. R. DeFazi and A. Marsili, *Gazz. Chim. Ital.*, **89,** 1709 (1959); *Chem. Abstr.*, **55,** 45 (1961).

3. G. N. Walker and B. N. Weaver, *J. Org. Chem.*, **25**, 484 (1960).
4. Y. Sato, T. Iwashige, and T. Miyadera, *Chem. Pharm. Bull.* (Tokyo), **8**, 427 (1960).
5. K. M. Naef and H. Schaltegger, *Helv. Chim. Acta*, **45**, 1018 (1962).
6. W. E. Hahn and T. Zieliński, *Łódź. Towarz. Nauk., Wydzial III*, No. 6, 23 (1960); *Chem. Abstr.*, **55**, 5525d (1960).
7. J. A. Meschino and C. H. Bond, *J. Org. Chem.*, **28**, 3129 (1963).
8. V. Carelli, *Ann. Chim.* (Rome), **51**, 713 (1961); *Chem. Abstr.*, **56**, 4720i (1962).
9. V. C. Arsenijevic, H. Lapin, and A. Horeau, *C. R. Acad. Sci., Paris, Ser. C*, **248**, 3309 (1959).
0. H. Lapin and A. Horeau, *Chimia*, **15**, 551 (1961); *Chem. Abstr.*, **59**, 560d (1963).
1. B. M. Ferrier and N. Campbell, *J. Chem. Soc.*, 3513 (1960).
2. F. Johnson, J. P. Panella, A. A. Carlson, and D. H. Hunneman, *J. Org. Chem.*, **27**, 2473 (1962).
3. H. Junek, *Monatsh. Chem.*, **95**, 1473 (1964); *Chem. Abstr.*, **62**, 13124b (1965).
4. H. Junek, *Monatsh. Chem.*, **95**, 1201 (1964); *Chem. Abstr.*, **62**, 1629g (1965).
5. F. Cuiban, S. Cilianu-Bibian, S. Popescu, and I. Rogozea, Fr. patent 1,366,064 (1964); *Chem. Abstr.*, **61**, 14643c (1964).
6. H. Kato, T. Ogawa, and M. Ohta, *Bull. Chem. Soc. Jap.*, **33**, 1468 (1960); *Chem. Abstr.*, **55**, 27301b (1961).
7. H. W. R. Williams, Brit. patent, 980,417 (1965); *Chem. Abstr.*, **62**, 7864d (1965).
8. H. Kato, T. Ogawa, and M. Ohta, *Chem. Ind.* (London), 1300 (1960).
9. E. Profft and R. Stumpf, *J. Prakt. Chem.*, **19**, 266 (1963); *Chem. Abstr.*, **59**, 15252b (1963).
0. Y. L. Gol'dfarb, F. D. Alashev, and V. K. Zyorykina, *Izv. Akad. Nauk SSSR, Ser. Khim*, 2241 (1964); *Chem. Abstr.*, **62**, 7819g (1965).
1. F. H. Clarke and C. M. Watnick, *J. Org. Chem.*, **24**, 1574 (1959).
2. Y. Arata and K. Achiwa, *Yakugaku Zasshi*, **79**, 108 (1959); *Chem. Abstr.*, **53**, 10211d (1959).
3. A. L. Logothetis, *J. Org. Chem.*, **29**, 1834 (1964).
4. G. W. J. M. Groenendaal, *Rec. Trav. Chim. Pays-Bas*, **78**, 446 (1959).
5. I. Tomita, H. G. Brooks, and D. E. Metzler, *J. Heterocycl. Chem.*, **3**, 178 (1966).
6. Y. Arata, *Yakugaku Zasshi*, **80**, 709 (1960); *Chem. Abstr.*, **54**, 21084d (1960).
7. B. J. Calvert and J. D. Hobson, *J. Chem. Soc.*, 5378 (1964).
8. A. Jart, A. J. Bigler, and V. Bitsch, *Anal. Chim. Acta*, **31**, 472 (1964); *Chem. Abstr.*, **62**, 1082d (1965).
9. D. E. Ames and B. T. Warren, *J. Chem. Soc.*, 5518 (1965).
0. F. Bohlmann, E. Winterfeldt, G. Boroschewski, R. Mayer-Mader, and B. Gatscheff, *Chem. Ber.*, **96**, 1792 (1963).
1. M. Hamama and M. Yamazaki, *Chem. Pharm. Bull.* (Tokyo), **11**, 415 (1963).
2. S. Boatman, T. M. Harris, and C. R. Hauser, *J. Amer. Chem. Soc.*, **87**, 5198 (1965).
3. G. A. Taylor, *J. Chem. Soc.*, 3332 (1965).
4. J. Mills, U.S. patent, 2,903,459 (1959); *Chem. Abstr.*, **54**, 5702a (1960).
5. S. Ohki, Y. Noike, T. Ohishi, K. Nara, and T. Kato, *Yakugaku Zasshi*, **79**, 1522 (1959); *Chem. Abstr.*, **54**, 11022h (1960).
6. F. Bohlmann, E. Winterfeldt, P. Studt, H. Laurent, G. Borosehewski, and K. M. Kleine, *Chem. Ber.*, **94**, 3151 (1961).
7. V. Boekelheide and R. Scharrer, *J. Org. Chem.*, **26**, 3802 (1961).
8. M. Maruoka, K. Isagawa, and Y. Fushizaki, *Nippon Kagaku Zasshi*, **82**, 1279 (1961); *Chem. Abstr.*, **59**, 563b (1963).
9. A. M. Parsons, Brit. patent, 847,051 (1960); *Chem. Abstr.*, **55**, 7441c (1961).

210. E. E. Mikhlina and M. V. Rubtsov, *Zh. Obshch. Khim.*, **32**, 2177 (1962); *Che Abstr.*, **58**, 9024g (1963).
211. E. Profft, F. Schneider and H. Beyer, *J. Prakt. Chem.*, **2**, 147 (1955).
212. E. D. Bergmann and I. Shahak, *J. Chem. Soc.*, 4033 (1961).
213. M. Strell and E. Kopp, *Chem. Ber.*, **91**, 1621 (1958).
214. G. Pappalardo, B. Tornetta, and G. Seapini, *Farmaco, Ed. Sci.*, **21**, 740 (1966); *Che Abstr.*, **66**, 46363m (1967).
215. B. M. Ferrier and N. Campbell, *Chem. Ind.* (London), 1089 (1958).
216. R. Walter and H. Zimmer, *J. Heterocycl. Chem.*, **1**, 205 (1964).
217. M. Viscontini and H. Raschig, *Helv. Chim. Acta*, **42**, 570 (1959).
218. D. Jerchel, J. Heider, and H. Wagner, *Ann. Chem.*, **613**, 153 (1958).
219. E. Profft and R. Stumpf, *Arch. Pharm.* (Weinheim), **296**, 79 (1963).
220. T. S. Gardner, E. Wenis, and J. Lee, *J. Org. Chem.*, **26**, 1514 (1961).
221. Finanz and Kompensationanstalt, Fr. patent, 1,427,135 (1966); *Chem. Abstr.*, 7151c (1966).
222. R. DeFazi, S. Carboni, and A. Marsili, *Gazz. Chim. Ital.*, **89**, 1701 (1959); *Che Abstr.*, **55**, 4497c (1961).
223. C. Runti and L. Sindellari, *Boll. Chim. Farm.*, **99**, 499 (1960); *Chem. Abstr.*, 10468a (1961).
224. H. K. Hall, Jr., *J. Amer. Chem. Soc.*, **82**, 1209 (1960).
225. F. J. Villani, U.S. patent, 3,301,863 (1967); *Chem. Abstr.*, **67**, 21833x (1967).
226. L. Horner and E. O. Reuth, *Ann. Chem.*, **703**, 37 (1967).
227. P. Truitt and S. G. Truitt, *J. Med. Chem.*, **9**, 637 (1966).
228. W. Korytnyk, *J. Med. Chem.*, **8**, 112 (1965).
229. H. N. Wingfield, Jr., *J. Org. Chem.*, **24**, 872 (1959).
230. J. Canceill, J. J. Basselier, and J. Jacques, *Bull. Soc. Chim. Fr.*, 1024 (1967).
231. C. Bonsall and J. Hill, *J. Chem. Soc., C,* 1836 (1967).
232. H. Hellmann, H. Piechota, H. Henecka, and H. Timmler, Ger. patent, 1,081,8 (1960); *Chem. Abstr.*, **55**, 19813a (1961).
233. J. Jančulev and M. Jančevska, *Bull. Sci., Conseil Acad. RPF Yougoslavie*, **6**, 1 (196 *Chem. Abstr.*, **55**, 27305h (1961).
234. S. Fatutta and A. Stener, *Gazz. Chim. Ital.*, **88**, 89 (1958); *Chem. Abstr.*, **53**, 224 (1959).
235. C. C. Chu and P. C. Teague, *J. Org. Chem.*, **23**, 1578 (1958).
236. Finanz and Kompensationsanstalt, Fr. patent, M3143 (1965); *Chem. Abstr.*, 2933d (1965).
237. G. Queguiner and P. Pastour, *C. R. Acad. Sci., Paris, Ser. C*, **258**, 5903 (1964); *Che Abstr.*, **61**, 6987c (1964).
238. R. P. Eustigneeva and N. A. Preobrazhenskii, *Zh. Obshch. Khim.*, **28**, 3085 (195 *Chem. Abstr.*, **53**, 10215a (1959).
239. H. Sliwa and P. Maitte, *C. R. Acad. Sci. Paris, Ser. C*, **259**, 2255 (1964); *Chem. Abs* **62**, 2760d (1965).
240. L. Neilande and G. Vanags, *Latvijas PSR Zinatnu Akad. Vestis, Kim. Ser.*, 203 (196 *Chem. Abstr.*, **61**, 6999d (1964).
241. H. Takamatsu, S. Minami, A. Fujita, K. Yamamoto, K. Fujimoto, M. Shimizu, Takase, and I. Nakajima, Jap. patent 6233 (1965); *Chem. Abstr.*, **63**, 5609e (1965).
242. K. Winterfield and R. Knieps, *Arch. Pharm.* (Weinheim), **293**, 478 (1960).
243. E. Maruszewska-Wieczorkowska and J. Michalski, *Rocz. Chem.*, **37**, 1315 (196 *Chem. Abstr.*, **60**, 5546a (1964).

4. W. Schulze and H. Willitzer, *J. Prakt. Chem.*, **23**, 20 (1964); *Chem. Abstr.*, **60**, 13219h (1964).

5. I. Matsuo, K. Sugimoto, and S. Ohki, *Chem. Pharm. Bull.* (Tokyo), **14**, 691 (1966).

6. K. W. Merz and R. Barchet, *Arch. Pharm.* (Weinheim), **297**, 412 (1964).

7. Badische Anilin und Soda-Fabrik AG, Nether. appl. 6,514,088 (1966); *Chem. Abstr.*, **65**, 13605g (1966).

8. M. Freifelder, *J. Org. Chem.*, **28**, 602 (1963).

9. V. Carrelli, F. Liberatore, and F. Morlacchi, *Ann. Chim.* (Rome), **51**, 467 (1961); *Chem. Abstr.*, **56**, 7273i (1962).

50. W. Shaeffer and R. Wegler, Ger. patent, 1,149,356 (1963); *Chem. Abstr.*, **59**, 11441c (1963).

51. N. Sperber, M. Sherlock, D. Papa, and D. Kender, *J. Amer. Chem. Soc.*, **81**, 704 (1959).

52. H. E. Ramsden, Brit. patent, 820,083 (1959); *Chem. Abstr.*, **54**, 24818c (1960).

53. F. Dallacker, P. Kratzer, and M. Lipp, *Ann. Chem.*, **643**, 97 (1961).

54. R. A. Jones and A. R. Katritzky, *Aust. J. Chem.*, **17**, 455 (1964); *Chem. Abstr.*, **60**, 15824e (1964).

55. T. Naito and T. Kotake, Jap. patent, 428 (1959); *Chem. Abstr.*, **54**, 6763e (1960).

56. T. Naito and T. Kotake, Jap. patent, 2929 (1960); *Chem. Abstr.*, **54**, 24812i (1960).

57. T. Naito, R. Dohmori, and T. Kotake, *Chem. Pharm. Bull.* (Tokyo), **12**, 588 (1964).

58. O. Červinka, *Collect. Czech. Chem. Commun.*, **25**, 2675 (1960); *Chem. Abstr.*, **55**, 4507b (1961).

59. O. Červinka, *Collect. Czech. Chem. Commun.*, **25**, 1174 (1960); *Chem. Abstr.*, **54**, 13126b (1960).

60. I. Matsuura, F. Yoneda, and Y. Nitta, *Chem. Pharm. Bull.* (Tokyo), **14**, 1010 (1966).

61. K. Osugi, *Yakugaku Zasshi*, **78**, 1353 (1958); *Chem. Abstr.*, **53**, 8112f (1959).

62. H. Ahlbrecht and F. Kroehnke, *Ann. Chem.*, **704**, 133 (1967).

63. F. Eiden and P. Peter, *Arch. Pharm.* (Weinheim), **297**, 1 (1964).

64. F. E. Cislak, U.S. patent, 3,069,427 (1962); *Chem. Abstr.*, **58**, 125220c (1963).

65. M. J. Martell, Jr., and T. O. Soine, *J. Pharm. Sci.*, **52**, 331 (1963).

66. H. McKennis, Jr., L. B. Turnbull, E. R. Bowman, and S. L. Schwartz, *J. Amer. Chem. Soc.*, **84**, 4598 (1962).

67. Daiichi Seiyaku Co. Ltd., Nether. appl. 6,607,005 (1966); *Chem. Abstr.*, **67**, 32593v (1967).

68. S. Okuda and M. M. Robison, *J. Amer. Chem. Soc.*, **81**, 740 (1959).

69. N. S. Vul'fson, L. I. Lukashina, and S. L. Davydova, *Khim. Tekhnol. i Primenenie Proizv. Piridina i Khinolina, Materialy Soveshch., Inst. Khim. Akad. Nauk Latv. SSR, Riga*, 243 (1957); *Chem. Abstr.*, **57**, 16604a (1962).

70. H. Beyer and K. Leverenz, *Chem. Ber.*, **94**, 263 (1961).

71. T. J. Kasper and D. E. Rivard, U.S. patent, 3,116,297 (1963); *Chem. Abstr.*, **61**, 645h (1964).

72. A. Buzas, C. Dufour, and J. Roy, U.S. patent, 2,827,465 (1958); *Chem. Abstr.*, **53**, 413d (1959).

73. A. Buzas, C. Dufour, J. Roy, and Société d'Etudes et d'Applications Chimiques, Fr. patent, 1,214,807 (1960); *Chem. Abstr.*, **55**, 19955 (1961).

74. Laboratories Fher SA, Span. patent, 273,594 (1962); *Chem. Abstr.*, **60**, 508b (1964).

75. L. Kuczynski, Z. Machon, and L. Wykret, *Dissertationes Pharm.*, **13**, 299 (1961); *Chem. Abstr.*, **57**, 8540h (1962).

76. L. T. DiFazio and P. F. Smith, *J. Med. Chem.*, **9**, 631 (1966).

277. R. H. Mizzone and R. P. Mull, Belg. patent, 654,416 (1965); *Chem. Abstr.*, 14171e (1966).
278. P. M. Quan and L. D. Quin, *J. Org. Chem.*, **31**, 2487 (1966).
279. C. Iwata and D. E. Metzler, *J. Heterocycl. Chem.*, **4**, 319 (1967).
280. J. Pĭtha and I. Ernest, *Chem. Listy*, **52**, 1937 (1958); *Chem. Abstr.*, **53**, 3258a (195
281. T. Maruyama, N. Kuraki, and K. Konishi, *Kogyo Kagaku Zasshi*, **68**, 2428 (196 *Chem. Abstr.*, **65**, 13849f (1966).
282. F. Bohlmann, E. Winterfeldt, and H. Brackel, *Chem. Ber.*, **91**, 2194 (1958).
283. J. Wrobel and Z. Dabrowski, *Rocz. Chem.*, **40**, 317 (1966); *Chem. Abstr.*, **65**, 6 (1966).
284. F. Mareš and M. Hudlický, *Chem. Listy*, **52**, 1933 (1958); *Chem. Abstr.*, **53**, 32 (1959).
285. L. W. Walter and P. Margolis, *J. Med. Chem.*, **10**, 498 (1967).
286. W. L. Bencze and L. I. Barsky, *J. Med. Pharm. Chem.*, **5**, 1298 (1962).
287. W. L. Bencze, Belg. patent, 626,022 (1963); *Chem. Abstr.*, **60**, 14481h (1960).
288. W. L. Bencze, U.S. patent, 3,238,218 (1966); *Chem. Abstr.*, **65**, 694c (1966).
289. W. L. Bencze, Belg. patent, 611,278 (1962); *Chem. Abstr.*, **57**, 15080f (1962).
290. Upjohn Co., Brit. patent, 1,053,312 (1966); *Chem. Abstr.*, **66**, 85701q (1967).
291. Y. Morimoto, *Yakugaku Zasshi*, **82**, 386 (1962); *Chem. Abstr.*, **58**, 3421g (1963).
292. J. Moszev and B. Fisher, *Zesz. Nauk. Univ. Jagiellon, Pr. Chem.*, **11**, 103 (196 *Chem. Abstr.*, **68**, 114382k (1968).
293. CIBA Ltd., Brit. patent 974,137 (1964); *Chem. Abstr.*, **62**, 7737f (1965).
294. C. E. Loader and C. J. Timmons, *J. Chem. Soc.*, 1078 (1966).
295. F. Kröhnke and K. F. Gross, *Chem. Ber.*, **92**, 22 (1959).
296. R. Tschesche and G. Sturm, U.S. patent, 3,290,312 (1966); *Chem. Abstr.*, **66**, 3795 (1967).
297. R. Barchet and K. W. Merz. *Tetrahedron Lett.*, 2239 (1964).
298. H. Burford, A. R. Gault, S. T. Coker, and W. L. Nobles, *J. Amer. Pharm. Assoc. S Ed.*, **48**, 669 (1959); *Chem. Abstr.*, **54**, 3859d (1960).
299. N. P. Buu-Hoi and P. Jacquignon, Fr. patent, M2379 (1964); *Chem. Abstr.*, **61**, 184 (1964).
300. J. B. Marco, Span. patent 275,507 (1962); *Chem. Abstr.*, **60**, 1713d (1964).
301. W. J. Middleton, U.S. patent, 2,914,534 (1959); *Chem. Abstr.*, **54**, 9962e (1960).
302. E. Szarvasi, Fr. patent Addn., 83,531 (1964); *Chem. Abstr.*, **62**, 535a (1965).
303. Y. Ormote, K. T. Kuo, and N. Sugiyama, *Bull. Chem. Soc. Jap.*, **40**, 1695 (196 *Chem. Abstr.*, **68**, 21803h (1968).
304. C. D. Lunsford and A. D. Cale, Jr., U.S. patents 3,192,210, 3,192,221, and 3,192,2 (1965); *Chem. Abstr.*, **63**, 11504b (1965).
305. T. Kato, T. Atsumi, and H. Sasaki, *Yakugaku Zasshi*, **85**, 812 (1965); *Chem. Abs* **63**, 18018g (1965).
306. S. Miyano and N. Abe, *Tetrahedron Lett.*, 1509 (1966).
307. T. Urbanski, B. Serafinowa, C. Belzecki, J. Lange, H. Makurukowa, and M. Makos Polish patent, 47,902 (1964); *Chem. Abstr.*, **61**, 4279g (1964).
308. W. L. Bencze, Belg. patent, 622,295 (1963); *Chem. Abstr.*, **59**, 10001h (1963).
309. C. F. Huebner and W. L. Bencze, Belg. patent, 626,200 (1963); *Chem. Abstr.*, 10620e (1964).
310. C. F. Huebner and W. L. Bencze, Belg. patent, 626,201 (1963); *Chem. Abstr.*, 9221b (1964).
311. F. E. Cislak, U.S. patent, 2,868,794 (1959); *Chem. Abstr.*, **53**, 10255d (1959).
312. J. A. Adamick and R. J. Flores, *J. Org. Chem.*, **29**, 572 (1964).

3. R. K. Hill, C. E. Glassick, and L. J. Fliedner, *J. Amer. Chem. Soc.*, **81,** 737 (1959).
4. E. Maruszewska-Wieczorkowska and J. Michalski, *Rocz. Chem.*, **38,** 625 (1964); *Chem. Abstr.*, **61,** 10702h (1964).
5. K. Schenker and J. Druey, *Helv. Chim. Acta,* **42,** 2571 (1959).
6. J. S. Walia, L. Heindl, H. Lader, and P. S. Walia, *Chem. Ind.* (London), 155 (1968).
7. W. L. Bencze, Belg. patent 626,962 (1963); *Chem. Abstr.*, **60,** 6828f (1964).
8. K. Fukui, M. Kondo, and H. Kitano, Jap. patents, 12,418 and 12,419 (1961); *Chem. Abstr.*, **56,** 8536d (1962).
9. W. Schulze, *J. Prakt. Chem.*, **19,** 91 (1962); *Chem. Abstr.*, **58,** 12509f (1963).
0. E. J. Poziomek and A. R. Melvin, U.S. patent, 3,209,008 (1965); *Chem. Abstr.*, **63,** 18041g (1965).
1. W. Sauermilch and A. Wolf, *Arch. Pharm.* (Weinheim), **292,** 38 (1959).
2. F. Gude, Ger. patent, 1,217,384 (1966); *Chem. Abstr.*, **65,** 5374g (1966).
3. W. Mathes and W. Sauermilch, *Ann. Chem.*, **618,** 152 (1958).
4. A. Banashek and M. N. Shchukina, *Zh. Obshch. Khim.*, **32,** 205 (1962); *Chem. Abstr.*, **57,** 16586b (1962).
5. W. Mathes and W. Sauermilch, *Chem. Ber.*, **93,** 286 (1960).
6. T. Kato and T. Atsumi, *Yakugaku Zasshi,* **87,** 961 (1967); *Chem. Abstr.*, **68,** 49422g (1968).
8. H. Hamana, B. Umezawa, Y. Goto, and K. Noda, *Chem. Pharm. Bull.* (Tokyo), **8,** 692 (1960); *Chem. Abstr.*, **55,** 18723d (1961).
9. E. J. Poziomek and A. R. Melvin, *J. Org. Chem.*, **26,** 3769 (1961).
0. O. Weissel, Ger. patent, 1,118,784 (1961); *Chem. Abstr.*, **57,** 4639c (1962).
1. E. Jucker and A. J. Lindemann, U.S. patent, 3,041,343 (1962); *Chem. Abstr.*, **57,** 13764c (1962).
2. W. L. Bencze and G. N. Walker, U. S. patent, 3,337,565 (1967); *Chem. Abstr.*, **68,** 114446j (1968).
3. G. N. Walker, W. L. Bencze, and J. B. Ziegler, U.S. patent, 3,337,566 (1967); *Chem. Abstr.*, **68,** 29608u (1968).
4. Dynamit-Nobel, Belg. patent, 635,427 (1963); *Chem. Abstr.*, **61,** 11977f (1964).
5. H. J. Kabbe, *Ann. Chem.*, **704,** 144 (1967).
6. J. Kreidl, S. Jager, and B. Benke, Hung. patent, 149,888 (1962); *Chem. Abstr.*, **60,** 9253e (1964).
7. G. Ehrhart, H. Ruschig, and K. Schmitt, Ger. patent, 1,116,226 (1961); *Chem. Abstr.*, **57,** 2199c (1962).
8. E. Szarvasi, *C. R. Acad. Sci. Paris, Ser. C,* **259,** 166 (1964); *Chem. Abstr.*, **61,** 8266e (1964).
9. Smith, Kline, and French Labs., Nether. patent appl., 6,512,616 (1966); *Chem. Abstr.*, **65,** 7095c (1966).
0. R. Fusco and G. Bianchetti, Brit. patent, 879,247 (1961); *Chem. Abstr.*, **56,** 12859f (1962).
1. L. Kuczynski and L. Wykret, *Dissertations Pharm.*, **16,** 485 (1964); *Chem. Abstr.*, 11537b (1965).
2. J. Grosselck, L. Beress, H. Schenk, and G. Schmidt, *Angew. Chem. Intern. Ed.*, **4,** 1080 (1965).
3. A. E. Frost and H. H. Freedman, *J. Org. Chem.*, **24,** 1905 (1959).
4. R. Selleri, G. Orzalesi, and O. Caldini, *Boll. Chim. Farm.*, **105,** 117 (1966); *Chem. Abstr.*, **65,** 3821 (1966)..
5. S. Tatsuoka, Y. Yeno, K. Tanaka, and T. Ueyanagi, Jap. patent, 11,831 (1962); *Chem. Abstr.*, **59,** 10006c (1963).

346. K. Patzelt and J. Ctvrtnik, Czech. patent, 117,825 (1966); *Chem. Abstr.*, **66**, 24
(1967).
347. Lepetit S.p.A., Belg. patent, 624,836 (1963); *Chem. Abstr.*, **59**, 11445g (1963).
348. G. Jommi, F. Pelizzoni, and A. Fiecchi, *Ann. Chim.* (Rome), **51**, 1340 (1961); *Che*
*Abstr.*, **56**, 15478d (1962).
349. A. R. Katritzky, J. A. T. Beard, and N. A. Coats, *J. Chem. Soc.*, 3680 (1959).
350. J. A. Weisbach, K. R. Williams, N. Yim, J. L. Kirkpatrick, E. L. Anderson, and Douglas, *Chem. Ind.* (London), 662 (1966).
351. F. Zymalkowski and B. Trentrog. *Arch. Pharm.* (Weinheim), **293**, 47 (1960); *Che*
*Abstr.*, **55**, 21108e (1961).
352. K. H. Büchel and F. Korte, *Z. Anal. Chem.*, **190**, 243 (1962); *Chem. Abstr.*, **58**, 633
(1962).
353. M. Saunders and E. H. Gold, *J. Org. Chem.*, **27**, 1439 (1962).
354. W. Ciusa and G. Barbiroli, *Ann. Chim.* (Rome), **56**, 3 (1966); *Chem. Abstr.*, **65**, 382
(1966).
355. I. A. Calo, V. Evdokimoff, and C. Cardini, *Rend. Ist. Super. Sanita*, **25**, 529 (196
*Chem. Abstr.*, **59**, 2808d (1963).
356. V. Quercia, *Boll. Chim. Farm.*, **100**, 834 (1961); *Chem. Abstr.*, **57**, 5883i (1962).
357. G. A. Robison and F. W. Schueler, *J. Pharm. Sci.*, **50**, 602 (1961).
358. W. Bruegel, *Z. Elektrochem.*, **66**, 159 (1962); *Chem. Abstr.*, **57**, 307b (1962).
359. K. Hoffmann, J. Heer, E. Sury, and E. Urech, U.S. patent, 2,957,879 (1960); *Che*
*Abstr.*, **55**, 10478f (1961).
360. F. H. Case and W. A. Butte, *J. Org. Chem.*, **26**, 4415 (1961).
361. J. Kelemen and R. Wizinger, *Helv. Chim. Acta*, **45**, 1908 (1962).
362. J. Swiderski and J. Izdebski, *Rocz. Chem.*, **36**, 963 (1962); *Chem. Abstr.*, **58**, 562
(1963).
363. A. Zaitsev, M. M. Shestaeva, and V. A. Zagorevskii, *Khim Geterotsikl Soedin*, 1
(1967); *Chem. Abstr.*, **67**, 73498m (1967).
364. E. S. Nikitsikaya, E. E. Mikhlina, L. N. Yakontov, and V. Ya Furshtatova, *Z*
*Obschch. Khim.*, **28**, 2786 (1958); *Chem. Abstr.*, **53**, 9207d (1959).
365. M. Freymann, R. Freymann, and C. Geissner-Prettre, *Arch. Sci.* (Geneva) **13**, Spec. N
506 (1960); *Chem. Abstr.*, **58**, 3289d (1963).
366. A. Roedig and K. Grohe, *Chem. Ber.*, **98**, 923 (1965).
367. J. R. Geigy A. G., Nether. appl. 6,600,522 (1966); *Chem. Abstr.*, **65**, 16948b (1966)

# CHAPTER XII

# Pyridinols and Pyridones

**HOWARD TIECKELMANN**

*State University of New York*
*Buffalo, New York*

## I. Preparation

A large variety of ring closure reactions have been applied to the direct rmation of pyridinols and pyridones. Judicious choices of open-chain ermediates often can provide excellent routes to a variety of substituted ridines and fused ring systems.

The ring closures presented in this section are classified according to the mber of ring carbons furnished by each reagent: (1) one reactant provides all e ring carbons; (2) one reactant provides four ring carbons; and (3) one actant provides three ring carbons.

This classification of ring closure is arbitrary. The isolation of an acyclic ermediate often is the choice of the investigator. For example, 6-phenyl-2-ridone **(XII-4)** has been directly prepared from **XII-1** and **XII-2** or from the lable intermediate **XII-3**.[1]

The further classification used here and earlier[2] into (1) ring closures and (2) eparations from other rings is also arbitrary and often appears inappropriate in at a useful relationship such as that between furfural and the acyclic oxoglutaconaldehyde **(XII-5)** may not be apparent.[3]

$N(CH_3)_2$ — $CH_2$ — $CH_2$ — $C(=O)C_6H_5$

**XII-1**

+

**XII-2** ($H_2NC(=O)CH_2-N^+C_5H_5\ Cl^-$) → **XII-3**

**XII-4**

$\xrightarrow{RNH_2}$

$Cl_2$, $H_3O^+$

**XII-5**

$\xrightarrow{RNH_2}$

## 1. Ring Closures of Acyclic Compounds That Provide Five Ring Carbons

### A. *Glutaconic Acid and Its Derivatives*

Glutaconic acid derivatives have been used extensively for the synthesis of 5-pyridinediols;[4] however, there has been relatively little interest in these mpounds during the past ten years. β-(*p*-Methoxyphenyl)glutaconic acid **II-6)** and methylamine or benzylamine give *N*-methyl- or *N*-benzyl-6-hydroxy-(*p*-methoxyphenyl)-2-pyridone **(XII-7)**.[5] β-Methylglutaconic anhydride and

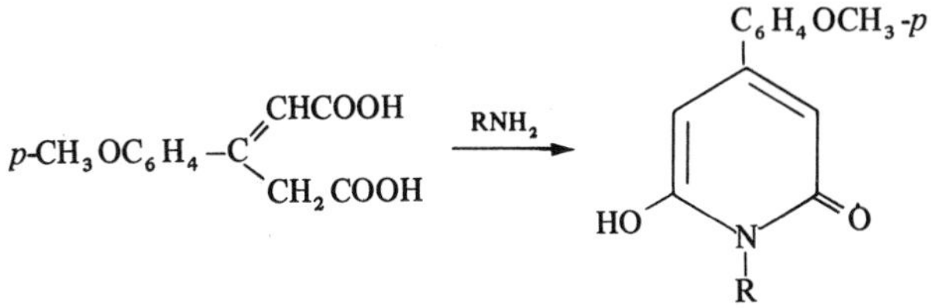

$p$-$CH_3OC_6H_4$–C(=CHCOOH)$CH_2COOH$ $\xrightarrow{RNH_2}$ (4-$C_6H_4OCH_3$-$p$, 6-HO, N-R 2-pyridone)

**XII-6** **XII-7**

examethylenediamine give the dipyridone **(XII-8)** when boiled in xylene.[6]

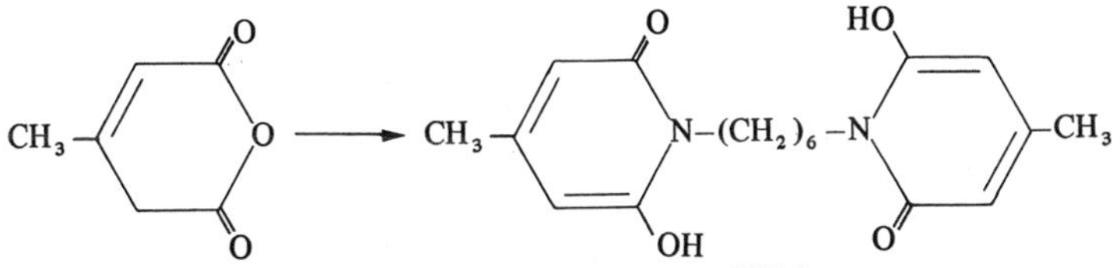

**XII-8**

In addition to cyclizations where glutaconic acid derivatives provide five ring arbons, cyclizations are known where a glutaconate provides four ring carbons nd a 2-pyridone is formed (Section I.2.B., p. 616).

Alkylations of diethyl α-alkylmalonates with trichloroacrylonitrile give the -cyanovinylmalonates **(XII-9)**, which cyclize in sulfuric acid to 3-alkyl-4,5-ichloro-2,6-pyridinediones **(XII-10)** in good yield.[7]

β-Ketoesters and cyanoacetic acid in benzene containing ammonium acetate nd acetic acid give ethyl γ-cyano-2-butenoates **(XII-11**, X = $OC_2H_5$), which are aponified to the corresponding acids **(XII-11**, X = OH) and then are converted o the acyl chlorides **(XII-11**, X = Cl) with phosphorus pentachloride. The acyl hlorides cyclize to 6-chloro-2-pyridones **(XII-12)** in *n*-butyl ether containing ydrogen chloride.[8]

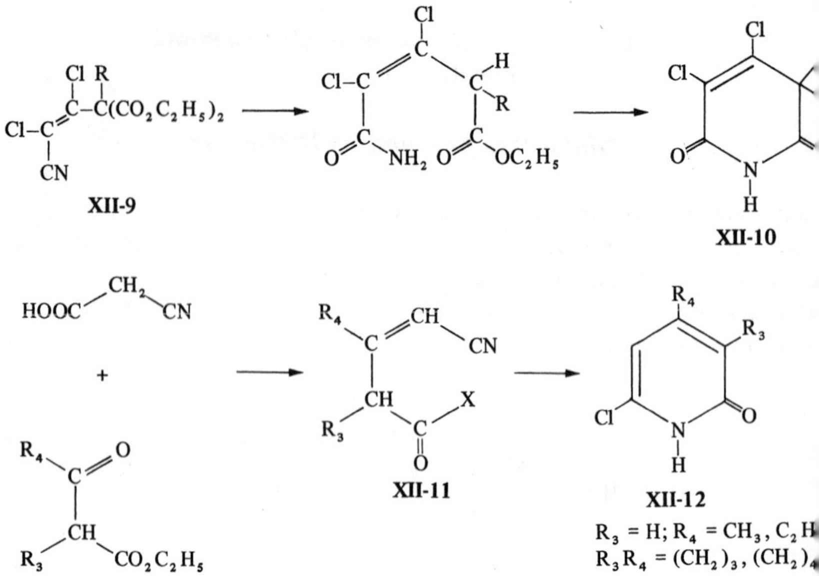

XII-9

XII-10

XII-11

XII-12

$R_3 = H; R_4 = CH_3, C_2H$

$R_3R_4 = (CH_2)_3, (CH_2)_4$

When the dimer of malononitrile, 2-amino-1,1,3-tricyanopropene (**XII-13**) heated with hydrochloric acid, it gives glutazinamide (**XII-14**; R = $CONH_2$, 7 yield). Cyanoacetamide and sodium ethoxide in ethanol form glutazinonitr (**XII-14**, R = CN).[9]

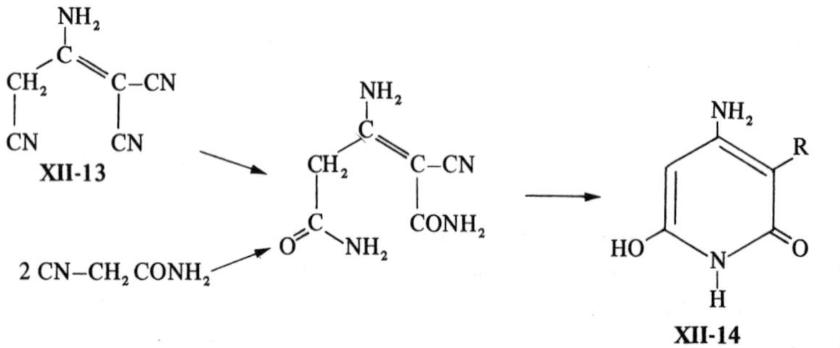

XII-13

XII-14

The potassium salt of tetracyanopropene (**XII-15**, M = K) gives 2-amino- cyano-6-hydroxy-3-pyridinecarboxamide (**XII-16**, R = H) when heated aqueous alkali.[10] In aqueous alcohols, **XII-15** forms 6-alkoxy-2-amino-3,5- cyanopyridines (**XII-17**; R = H, R′ = $CH_3, C_2H_5$) or 6-alkoxy-2-amino-5-cyan 3-pyridinecarboxamide (**XII-18**, R = H) depending on conditions.[10] Bromofor and chloroform react with malononitrile in the presence of sodium alkoxides give **XII-17** (R = H).[11]

Ethoxymethylenemalononitrile **(XII-19)**, a useful intermediate for the synthesis of 4-amino-5-cyanopyrimidines **(XII-20)**, gives **XII-17** (R = H, R′ = $C_2H_5$) in aqueous alcohol. A retrocondensation of **XII-19** forms some malononitrile, which reacts with more **XII-19** to form **XII-15** (R = H) and then **XII-17** (R = H).[10] Benzylidenemalononitrile (**XII-21**, R = $C_6H_5$) reacts similarly. However, a Michael addition of the malononitrile anion to **XII-21** gives the anion of 2-phenyltetracyanopropane (**XII-22**, R = $C_6H_5$), which, through autooxidation or hydride transfer, forms **XII-15** (R = $C_6H_5$).[12] Aldehydes and malononitrile have been used to form **XII-17** (R = alkyl, aryl; R′ = $C_2H_5$) in one step.[13]

$$\text{XII-20 (2-RS-4-}NH_2\text{-5-CN-pyrimidine)} \xleftarrow{H_2N-C(SR)=NH} C_2H_5OCH{=}C(CN)_2\ (\text{XII-19}) \rightleftharpoons CH_2(CN)_2 + HCO_2C_2H_5$$

$$\text{RCH}{=}C(CN)_2\ (\text{XII-21}) \xrightarrow{\bar{C}H(CN)_2}$$

$$\underset{\text{XII-22}}{(CN)_2C^{-}\text{–CH(R)–CH(CN)}_2} \xrightarrow{(O)} \underset{\text{XII-15}}{M^+\ (CN)_2\bar{C}\text{–C(R)}{=}C(CN)_2} \xleftarrow[C_2H_5O^-]{CHX_3} \bar{C}H(CN)_2$$

XII-15 → XII-18 (2-RO-3-CN-4-R-5-CONH₂-6-NH₂-pyridine), XII-17 (2-R′O-3-CN-4-R-5-CN-6-NH₂-pyridine), XII-16 (2-HO-3-CN-4-R-5-CONH₂-6-NH₂-pyridine)

## B. *Pentadienoic Acid Derivatives*

Condensation of perchloropropene with trichloroethylene gives 4*H*-nonachloro-1-pentene, which can be converted to perchlorocyclopentene-3-one **(XII-23)**[14] or converted to perchloro-2,4-pentadieneamides **(XII-24)** *via* perchloro-2,4-pentadienoic acid.[15] When **XII-23** is treated with ammonia, ring opening occurs to give **XII-24**. Cyclization of **XII-24** with $PCl_5$ gives perchloro-2 pyridinol **(XII-25)** and pentachloropyridine **(XII-26**, 67%).[16] *N*-Substitute amides **(XII-24**; R = $CH_3$,$C_6H_5$) can be cyclized to pyridone by heating ther alone or with $PCl_5$.[15]

Cl Cl Cl Cl Cl O XII-23 → Cl Cl C C C–Cl Cl C Cl CONHR XII-24

$PCl_5$, R = H → XII-25 (Cl, Cl, Cl, Cl, N, OH) + XII-26

$PCl_5$, R = $CH_3$, $C_6H_5$ → Cl, Cl, Cl, Cl, N–R, O

Lithium aluminum hydride and methyl perchloro-2,4-pentadienoate **(XII-27)**[1] at −78° form dienes **XII-28** and **XII-29**.[15] The mixture of isomers **XII-28** an **XII-29**, or pure **XII-28**, is converted to **XII-29** by heating in $CCl_4$.[17] The mixtur reacts with aryl hydrazines *via* hydrazones **(XII-30**, R = arylamino) to forn *N*-arylamino-3,4,5-trichloro-2-pyridones **(XII-31)**, with semicarbazide to forn *N*-ureido-3,4,5-trichloro-2-pyridone (**XII-31**, R = ureido) (Section IV.3., p. 839 and with ammonia or primary amines to form amides **(XII-32)**, which are readil cyclized to pyridones **(XII-31)**.[15]

Compounds **XII-28** and **XII-29** react with 2,4-dinitrophenylhydrazine t form isomeric products **XII-33** and **XII-34**, which cyclize to 3,4,5-trichlor 1-(2,4-dinitroanilino)-2-pyridone **(XII-35)**.[17]

Perchloropentadienoyl chloride[18] and 2,3,4,5-tetrachloro-5-bromo-2,4-pent dienoyl chloride[19] react with aniline to form *N*-phenylperchloro-2-pyridon **(XII-36)** *via* the corresponding *N*-phenylamide. The 2-pyridones **XII-37** an **XII-38** are prepared similarly from the appropriate bromo analogs.[19]

XII-28 → XII-30 → XII-31

XII-27 → XII-28 ; XII-27 → XII-29

XII-29 → XII-32 → XII-31

XII-28 ⟶ $Cl_2C{=}CCl{-}CCl{=}CCl{-}C(=NNHAr)H$ (XII-33) $\xrightarrow{AcOH}$ XII-35

XII-29 ⟶ $CHCl{=}CCl{-}CCl{=}CCl{-}C(=O)NHNHAr$ (XII-34) $\xrightarrow{Al_2O_3}$ XII-35

2,3,4,5-Tetrachloro-5-phenyl-2,4-pentadienamide **(XII-39)**, prepared from erchloropentadienoyl chloride *via* a Friedel Crafts phenylation, is cyclized ɔ 3,4,5-trichloro-6-phenyl-2-pyridone **(XII-40)** by pyrolysis or by heating vith sodium hydroxide in methanol.[20]

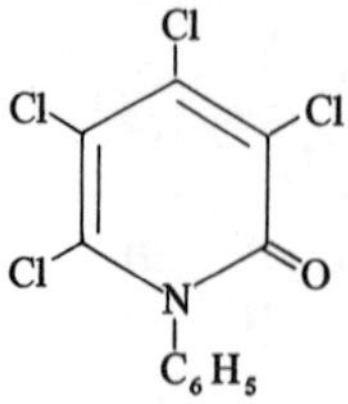

**XII-36** 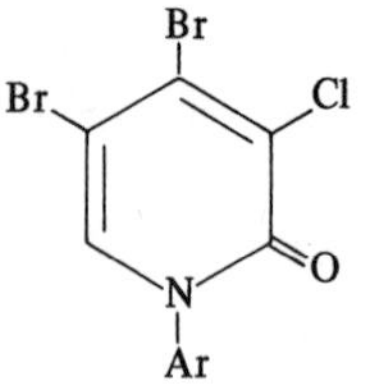 **XII-37** 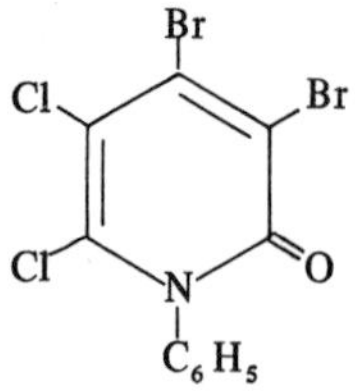 **XII-38**

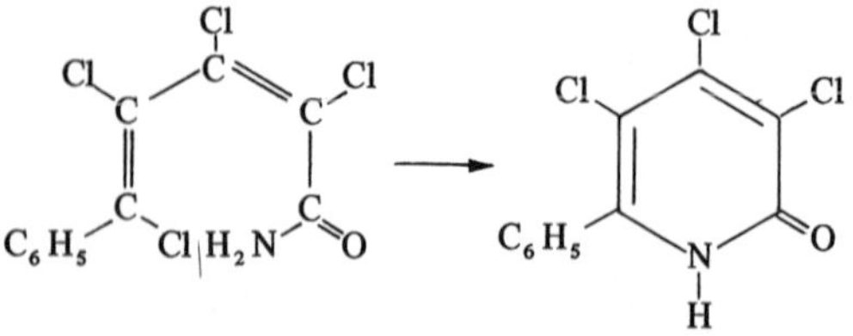

**XII-39** → **XII-40**

1,1,3,3-Tetraethoxypropane (malonaldehyde tetraethylacetal) **(XII-41)** a diethyl malonate in acetic anhydride give diethyl 3-ethoxyallylidenemalona **(XII-42)**, which cyclizes in polyphosphoric acid or formic acid to th pyrone **XII-43**. Reaction of **XII-42** with ammonia or benzylamine forr 3-aminoallylidenemalonates **(XII-44**; R = H, $CH_2C_6H_5$), which are cycliz to the ethyl α-pyridone-3-carboxylates **(XII-45)** in alcoholic sodium ethoxi or piperidine.[21]

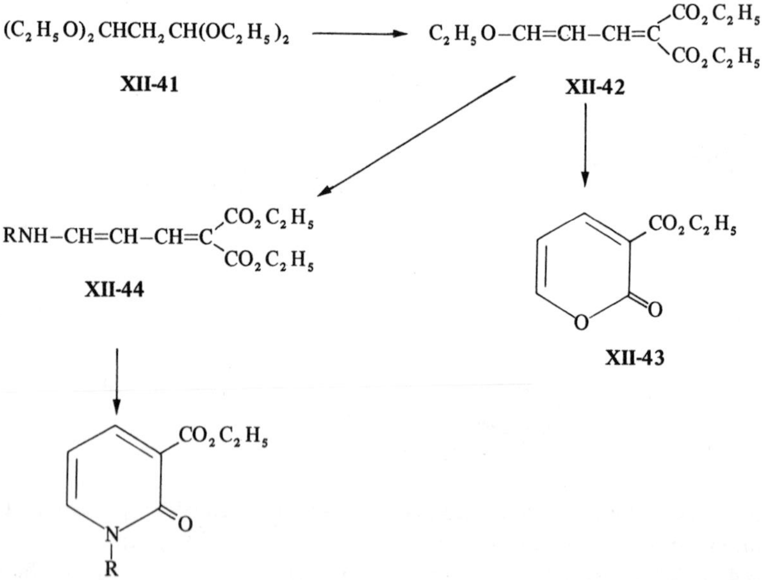

$$(C_2H_5O)_2CHCH_2CH(OC_2H_5)_2 \longrightarrow C_2H_5O{-}CH{=}CH{-}CH{=}C(CO_2C_2H_5)_2$$

**XII-41** **XII-42**

$$RNH{-}CH{=}CH{-}CH{=}C(CO_2C_2H_5)_2$$

**XII-44** **XII-43**

**XII-45**

Nitroso compounds and carbomethoxydienes form Diels-Alder adducts, the ,6-dihydro-1,2-oxazines **XII-46** and **XII-47**,[22,23] which are reduced by zinc a acetic acid to the 3,6-dihydro-2-pyridones **(XII-48)** and the 2-pyrones XII-49) respectively. Because of the directive effect of the carbomethoxy roup, **XII-47**, if not formed exclusively, is the major product. Dehydration f **XII-48** to 1-aryl-6-substituted-2-pyridones **(XII-50)** occurs on heating with oncentrated hydrochloric acid.[24]

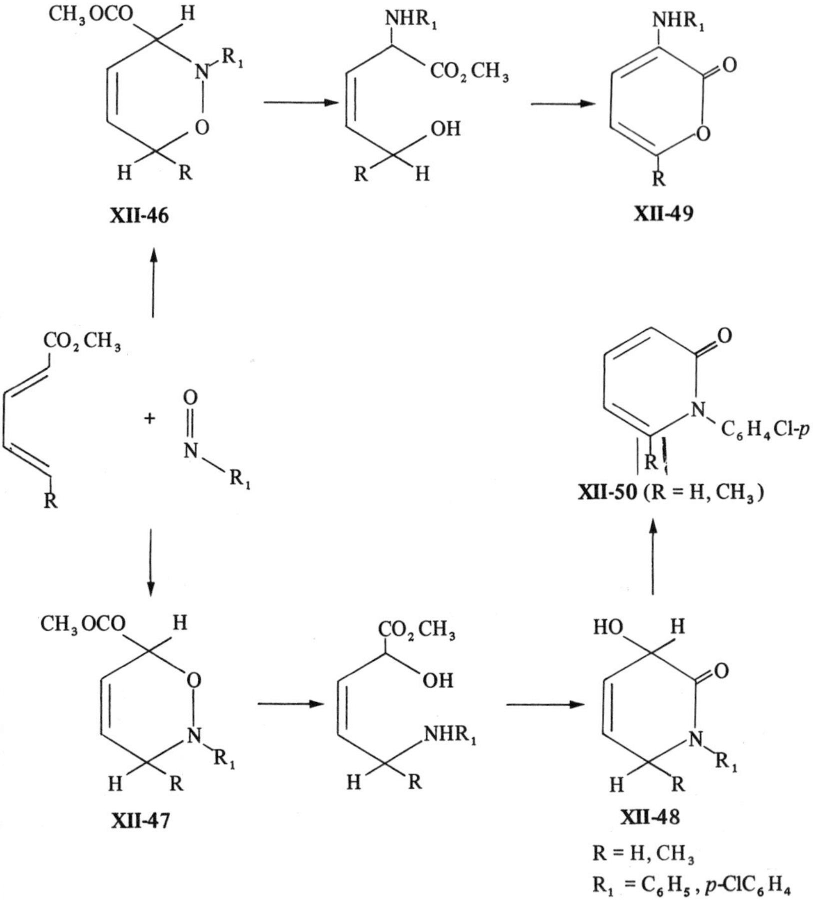

**XII-46**

**XII-49**

**XII-50** (R = H, $CH_3$)

**XII-47**

**XII-48**

R = H, $CH_3$
$R_1 = C_6H_5, p\text{-}ClC_6H_4$

The oxazines of structure **XII-47** form pyrroles when they are treated ith potassium hydroxide in methanol. However, under appropriate condi- ons **XII-51** in methanolic KOH forms 63% of 3-hydroxy-6-methyl-*N*- -chlorophenyl)-2-pyridone **(XII-52)** and 33% of 1-(*p*-chlorophenyl)-5- ethyl-2-pyrrolecarboxylic acid.[24]

XII-51

$\xrightarrow{KOH/CH_3OH}$

XII-52

## C. *Triketones, Ketodialdehydes (Oxoglutaconaldehydes), and Ketodicarboxylic Acid Derivatives*

Hauser and co-workers have prepared a number of triketones by (1) acylati of benzoylacetone with aliphatic esters employing lithium amide,[25] (2) twofold aroylation of acetone with methyl esters using sodium hydride,[26] (3) aroylation and acylation of aliphatic diketones with potassium amide in liqu ammonia[27] and (4) by aroylation of sodioacetoacetaldehyde[28] in the presence potassium amide.[27] Disodio- and dipotassiobenzoylacetone are not acylated ethyl acetate or by phenyl propionate.[25, 29] However, dilithiobenzoylacetone a dilithioacetylacetone with an excess of lithium amide are acylated by alipha esters. It is usually more convenient to synthesize 4-pyridones directly fro these triketones by cyclization with ethanolic ammonia rather than by way the intermediate 4-pyrone.[25]

5-Phenyl-3,5-diketopentanal **(XII-53)** is converted to 2-phenyl-4-pyrido **(XII-54)** in good yield with ethanolic ammonia.[27]

$$C_6H_5CO{-}CH_2{-}CO{-}CH_2{-}CHO \xrightarrow{NH_3} \text{2-}C_6H_5\text{-4-pyridone}$$

XI-53 → XII-54

[The]he γ-benzyl derivative of **XII-53 (XII-55)** prepared from the sodium salt of [p]henyl-1,3-pentanedione, potassium amide, and methyl benzoate, was not [iso]lated but was converted to 5-benzyl-6-phenyl-4-pyridone with ammonia in [meth]anol.[27]

$$[C_6]H_5CH_2CH_2CO{-}\overset{Na^+}{C}H{-}CHO \xrightarrow[NH_3]{KNH_2} C_6H_5CH_2CHCO{-}CH{-}CHO \xrightarrow{C_6H_5CO_2CH_3}$$

$$C_6H_5CO{-}CH(CH_2C_6H_5){-}CO{-}CH_2{-}CHO \xrightarrow{NH_3} \text{5-}C_6H_5CH_2\text{-6-}C_6H_5\text{-4-pyridone}$$

XII-55

[C]ondensation at the terminal methyl group of 1-phenyl and 1-(*p*-methoxy-[ph]enyl)-1,3,5-hexanetrione **(XII-56)** with benzophenone or anisaldehyde in the [pr]esence of three equivalents of sodamide in liquid ammonia gives the [hy]droxytriketones **(XII-57)**, which cyclize with ammonia to give the 4-pyridones **[(X]II-58a–c)**.[30]

$$[R_1]COCH_2COCH_2COCH_3 \longrightarrow$$

XII-56

$$R_1COCH_2COCH_2COCH_2CR_2R_3OH \longrightarrow \text{2-}R_1\text{-6-}(CH_2CR_2R_3OH)\text{-4-pyridone}$$

XII-57 → XII-58a-c

| | $R_1$ | $R_2$ | $R_3$ |
|---|---|---|---|
| a: | $C_6H_5$ | $C_6H_5$ | $C_6H_5$ |
| b: | $C_6H_5$ | $p$-$CH_3OC_6H_4$ | H |
| c: | $p$-$CH_3OC_6H_4$ | $C_6H_5$ | $C_6H_5$ |

The trisodio salts of **XII-56** are also alkylated with $\alpha,\omega$-dibromoalkanes to g the corresponding bis(triketones), which can be converted to the polymethyle bis-2-(6-aryl-4-pyridones) (**XII-59**, $n$ = 5,6) with ammonia.[30a]

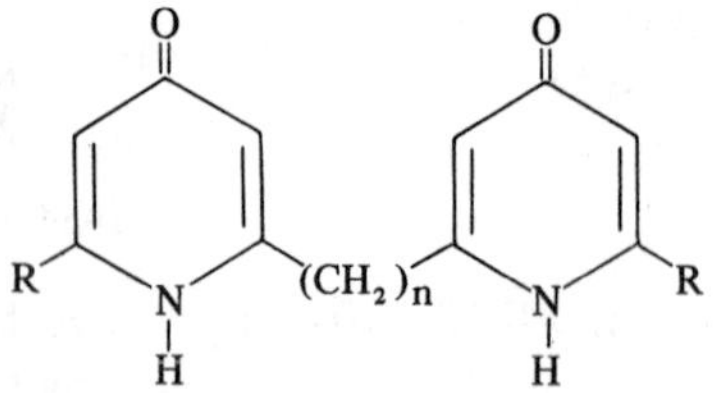

R (CH$_2$)$_n$ R

**XII-59**

Although three different carbonyl groups are available, 6-phenyl-2,4,6-hexa trione (**XII-60**) reacts with aromatic amines to form the ketoenamines (**XII-** R = H, $o$-Cl, $p$-Cl, $p$-$OCH_3$), which cyclize to $N$-aryl-2-methyl-6-phenyl-4-py dones (**XII-62**) in hot polyphosphoric acid. $n$-Butylamine and **XII-60** give corresponding ketoenamines, but these could not be cyclized to the pyridone

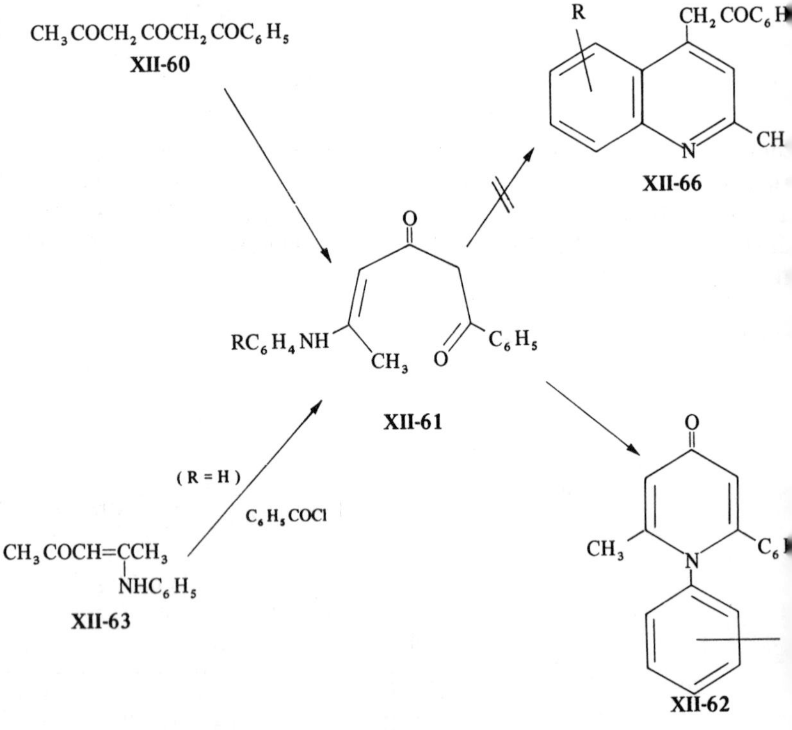

$CH_3COCH_2COCH_2COC_6H_5$
**XII-60**

R $CH_2COC_6H$
CH
**XII-66**

$RC_6H_4NH$ $CH_3$ $C_6H_5$
**XII-61**

(R = H)
$C_6H_5COCl$

$CH_3COCH{=}CCH_3$
$NHC_6H_5$
**XII-63**

$CH_3$ $C_6$
**XII-62**

s method. Instead they gave the pyrone.[31] The enamine **(XII-63)** forms a otassio salt with potassium amide in liquid ammonia, which reacts with nzoyl chloride to give the diketoenamine (**XII-61**, R = H).[32] Although the toenamine **(XII-64)** is cyclized to the quinoline **(XII-65)** by sulfuric acid,[33] inolines **(XII-66)** were not detected in these studies.[31]

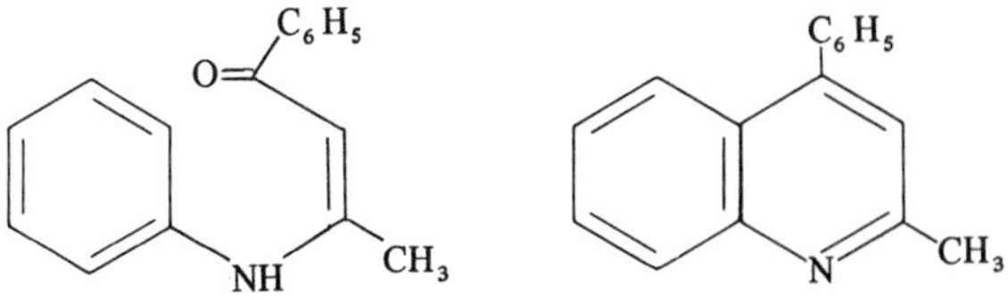

**XII-64** **XII-65**

Vhen furfural is oxidized with chlorine or bromine in an aqueous acidic dium, a solution of hydrates of oxoglutaconaldehyde **(XII-67)**, to which the vial name endialone has been given, is formed in high yield.[3] Endialone and monia or aliphatic amines give 3-hydroxy-2-pyridones **(XII-68)**, usually in low ld.[3] Endialone **(XII-67)** and sulfamic acid give **XII-69** (43%) and **XII-70** %), which are slowly transformed to **XII-71** and **XII-68** (R = H), respectively,

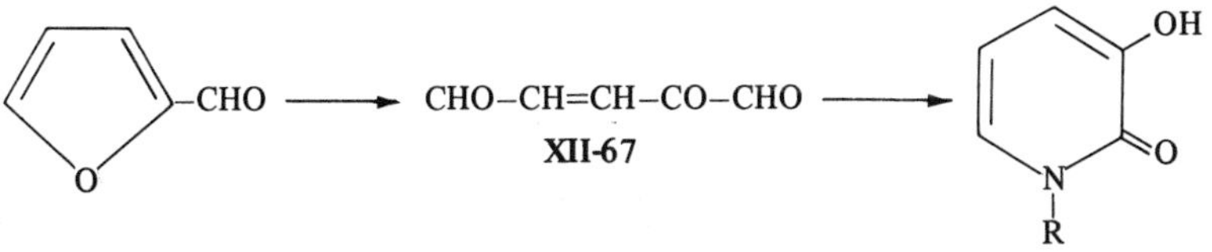

CHO–CH=CH–CO–CHO
**XII-67**

**XII-68**

R = H(69%), $CH_3$(19%), $(CH_2)_3N(CH_3)_2$(24%), $CH_2COOH$(8%)

standing. Chlorine adds to **XII-67** to form 2,3-dichloro-4-oxoglutaraldehyde **I-72)**, which reacts with ammonia at pH 7, with methylamine at pH 7.5, to e 4-chloro-3-hydroxy-2-pyridone (**XII-73**; R = H, $CH_3$), with anilines in ongly acidic solution to give 5-chloro-3-hydroxy-2-pyridones (**XII-74**, **R** = l), and with sulfamic acid to give **XII-74** (R = H).[3]

Diethyl acetone-1,3-dicarboxylate and a mixture of cold aqueous ammonia l liquid ammonia (1:1) are converted to 4,6-diamino-2-pyridone **(XII-75)**.[34]

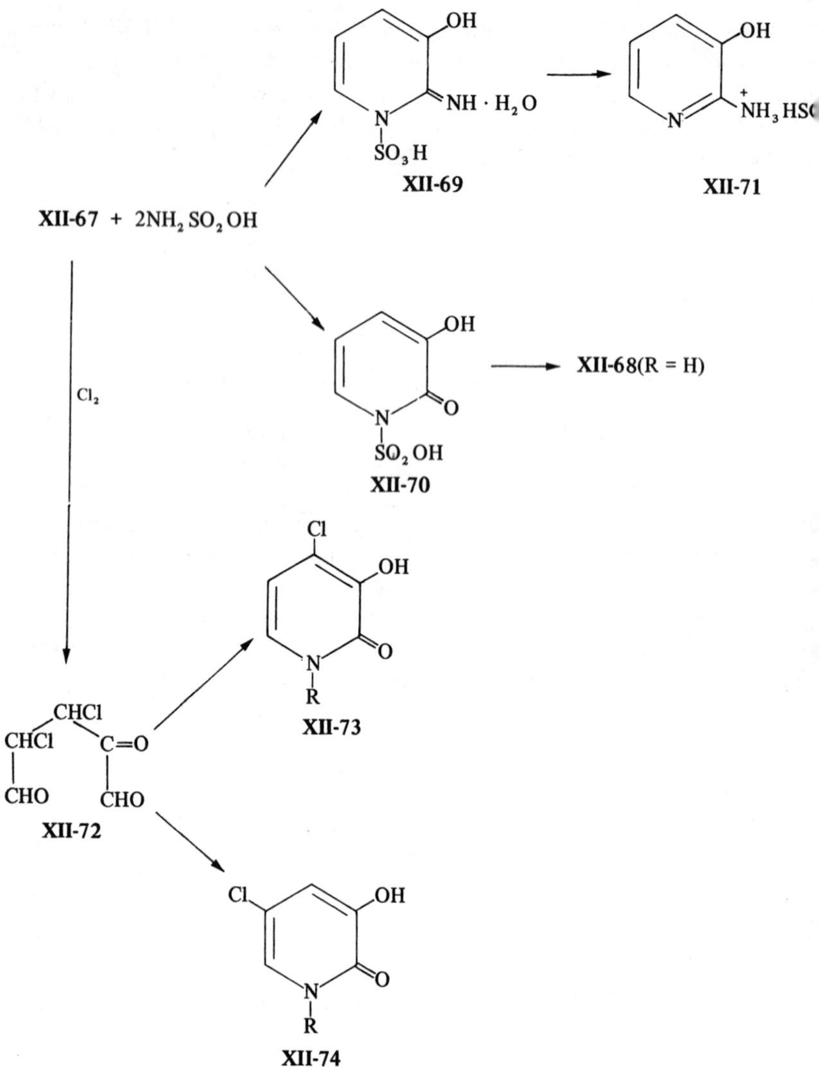

XII-67 + $2NH_2SO_2OH$

$Cl_2$

OH

$NH \cdot H_2O$

N

$SO_3H$

XII-69

OH

$\overset{+}{N}H_3HSO$

N

XII-71

OH

O

N

$SO_2OH$

XII-70

XII-68(R = H)

CHCl

CHCl

C=O

CHO

CHO

XII-72

Cl

OH

O

N

R

XII-73

Cl

OH

O

N

R

XII-74

O

C

$CH_2$

$CH_2$

$C_2H_5OCO$

$CO_2C_2H_5$

$NH_2$

$H_2N$

O

N

H

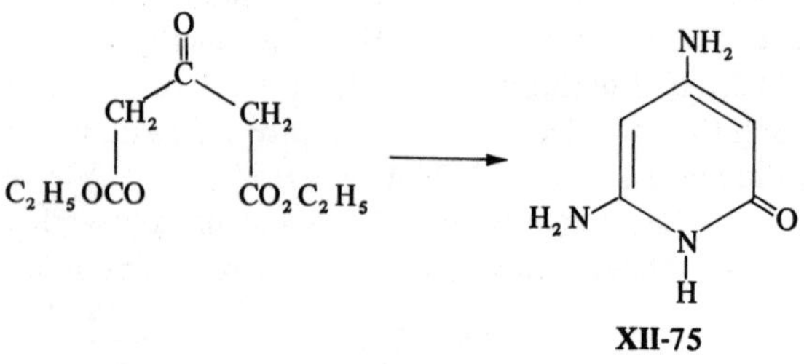

XII-75

## D. δ-*Ketonitriles (3,4-Dihydro-2-pyridones)*

Monocyanoethylation of deoxybenzoin and phenylacetone in *t*-butyl alcohol es γ-cyanoketones (**XII-76**; $R_5 = C_6H_5$, $R_6 = CH_3, C_6H_5$), which are cyclized potassium hydroxide in *t*-butyl alcohol/methanol to 3,4-dihydro-5-phenyl-6-ostituted-2-pyridones (**XII-77**), which can be dehydrogenated to 5-phenyl-6-ostituted-2-pyridones (**XII-78**).[35] When γ-benzoylbutyronitrile is passed over $_2O_3$ at 400° or over $Cr_2O_3/Al_2O_3$ at 300 to 310°, 6-phenyl-2-pyridone **II-79**) is formed.[36] Mixtures of products can be formed in these reactions; for

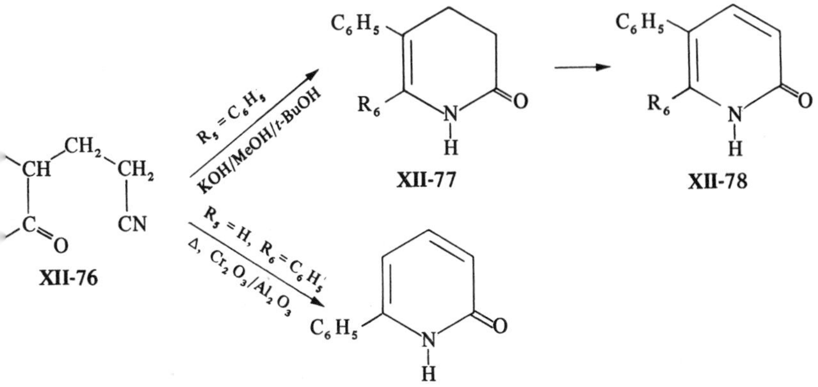

XII-76 → ($R_5 = C_6H_5$; KOH/MeOH/*t*-BuOH) XII-77 → XII-78

XII-76 → ($R_5$ = H, $R_6 = C_6H_5$; Δ, $Cr_2O_3/Al_2O_3$) XII-79

ample, γ-acetylvaleronitrile (**XII-76**, $R_5 = R_6 = CH_3$) is converted to 3-dimethylpyridine (6.6%), 5,6-dimethyl-1,6-dehydropiperidone (23%), and 6-dimethyl-3,4-dihydro-2-pyridone (6%) at 340 to 360°. At 400 to 410° some 6-dimethyl-2-pyridone (6.6%) is formed.[36]

Trichloroacrylonitrile and ethyl α-alkylacetoacetates (**XII-80**) give **XII-81**, hich cyclizes with sulfuric acid to 5-alkyl-3,4-dichloro-6-methyl-2-pyridones **XII-82**).[37]

$$CH_3CO{-}CH(R){-}CO_2C_2H_5 \xrightarrow[t\text{-BuOK}]{Cl_2C{=}CClCN} CH_3CO{-}C(R)(CO_2C_2H_5){-}CCl{=}CCl{-}CN \xrightarrow{H_2SO_4} \text{XII-82}$$

XII-80 XII-81 XII-82

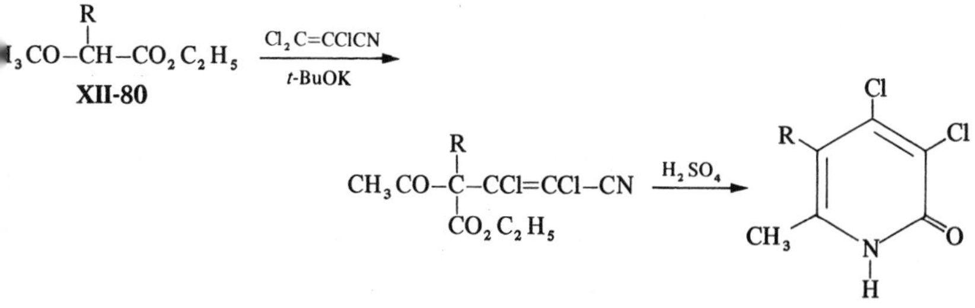

$R = H, CH_3, C_2H_5, n\text{-}C_4H_9, C_6H_5CH_2$

4-Carbethoxy-4-cyano-1,4-diphenyl-2-butene-1-one **(XII-83)**, from et α-cyano-α-phenylacetate and phenyl β-chlorovinyl ketone, is cyclized in 8 sulfuric acid to 3,6-diphenyl-2-pyridone.[38]

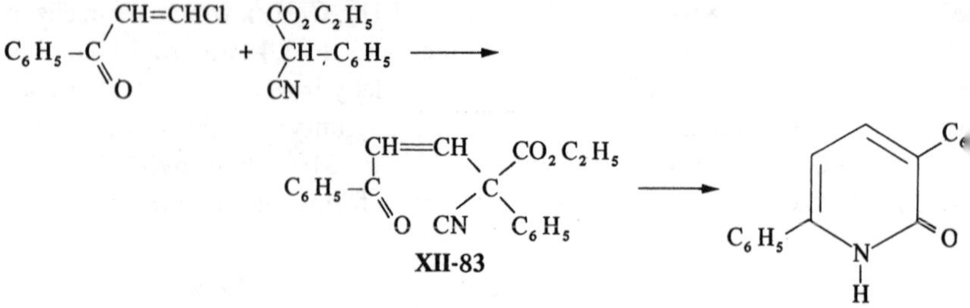

**XII-83**

## E. *Miscellaneous*

β-Diketones (**XII-84**; R = $CH_3$, $C_6H_5$) condense with benzylideneaniline, b not with aliphatic Schiff bases, in the presence of potassium amide to g 2,3-dihydro-4-pyridones **(XII-86)** either directly (R = $CH_3$) or by treatment **XII-85** (R = $C_6H_5$) with sulfuric acid. Under these conditions 6-phenyl-2 hexanedione (**XII-84**, R = $CH_2CH_2C_6H_5$) gives 1-anilino-1,7-diphenyl-3 heptanedione (**XII-85**, R = $CH_2CH_2C_6H_5$), which forms **XII-86** (R $CH_2CH_2C_6H_5$) on treatment with sulfuric acid. The isomeric dihydropyrid

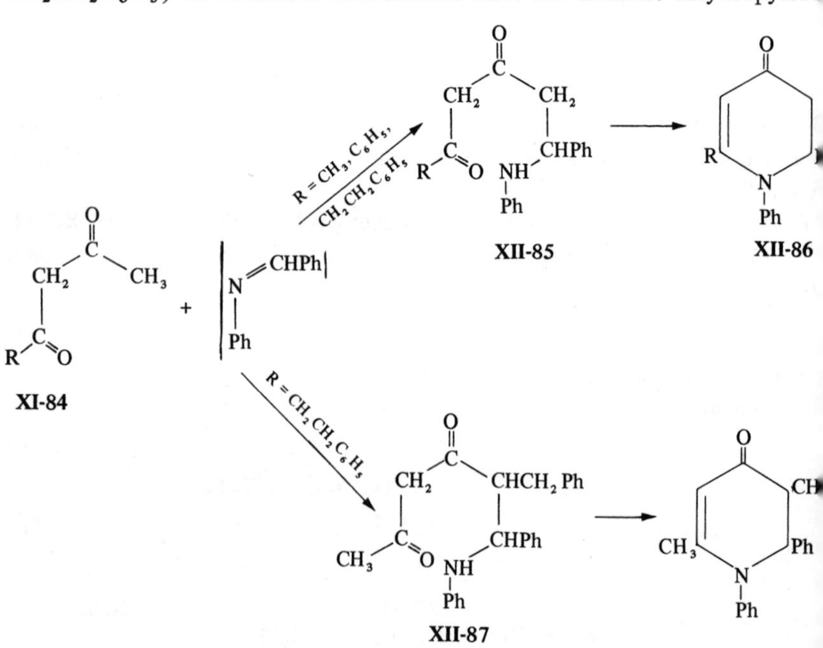

R = $CH_3$, $C_6H_5$, $CH_2CH_2C_6H_5$

R = $CH_2CH_2C_6H_5$

**XI-84**

**XII-85**

**XII-86**

**XII-87**

a by-product in the formation of **XII-86** (R = $C_6H_5CH_2CH_2$) presumably *via* **I-87**.[39]

The β-diketones (**XII-84**; R = $CH_3$, $C_6H_5$, $C_6H_5CH_2CH_2$) are condensed with nzonitrile and potassium amide to give the 6-substituted-2-phenyl-4-pyridones **II-88**) directly.[40] The possible products **XII-89**, isomeric with **XII-88** (R =

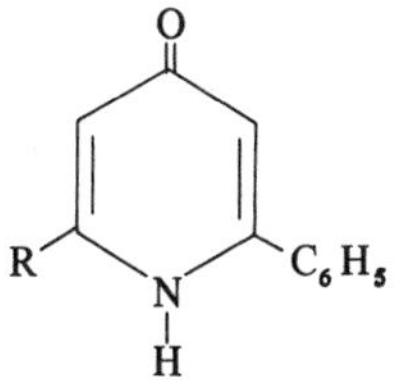

**XII-88**

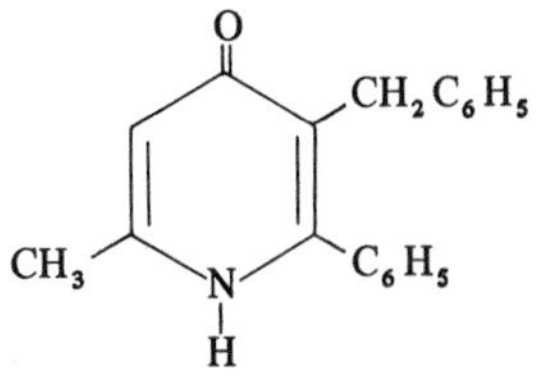

**XII-89**

$H_5CH_2CH_2$), which could arise from the condensation of benzonitrile at C-5, re not isolated.[40]

Di-1-propynyl ketone (**XII-90**) and primary amines (ethylamine, isopropyl-ine) or amino acids that are unsubstituted in the α-position (glycine, lanine, ε-aminocaproic acid, glycylglycine, glycylleucine, alanylglycine) give amines (**XII-91**), which form *N*-substituted lutidones when their sodium salts heated in water or their esters are heated in xylene. When 2-substituents are esent, cyclization occurs when their calcium salts or amides are heated in ueous alkali. When heated in the presence of Triton B and DMF, α-alanyl-ide and di-1-propynyl ketone give the lutidone.[41]

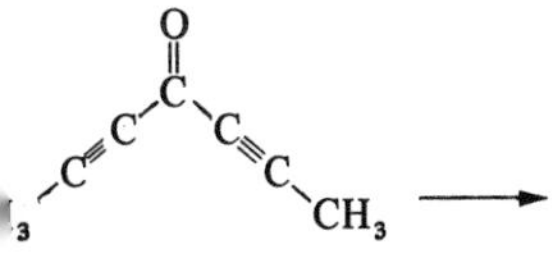

**XII-90**

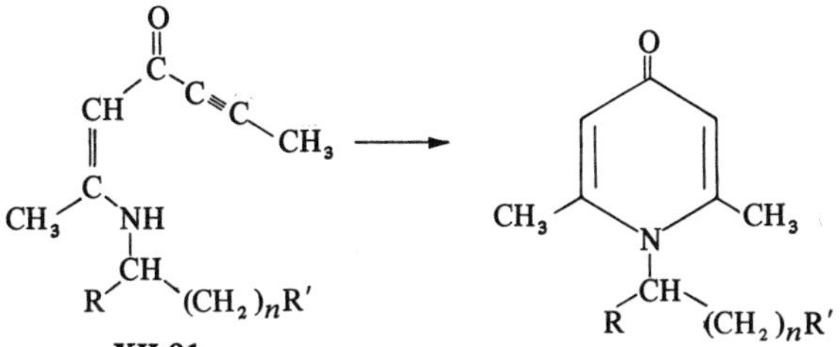

**XII-91**

-Nitrophenylpropiolyl chloride and sodium phenylacetylene give the diynone **II-92**), which reacts with aniline in alcohol to give **XII-93**, which cyclizes in iling xylene to 1,6-diphenyl-2-(*p*-nitrophenyl)-4-pyridone.[42]

$p\text{-}NO_2C_6H_4{-}C{\equiv}C{-}COCl \xrightarrow{C_6H_5C\equiv CNa} p\text{-}NO_2C_6H_4{-}C{\equiv}C{-}CO{-}C{\equiv}C{-}C_6H_5 \longrightarrow$

**XII-92**

**XII-93**

## 2. Ring Closures in Which One Acyclic Compound Provides Four Ring Carbons

### A. *Ketoamides*

α,α-Diethylacetoacetamide reacts with methyl formate in the presence sodium ethoxide in toluene to form 3,3-diethyl-2,4-pyridinedione.[43] Using t equivalents of sodium, α-ethylacetoacetanilide, and methyl formate in toluel methanol gives 3-ethyl-4-hydroxy-1-phenyl-2-pyridone **(XII-94)**.[44]

$\xrightarrow{HCO_2CH_3}$

**XII-94**

### B. *Unsaturated Acid Derivatives*

Diethyl glutaconate and two moles of *N*-benzylidenemethylamine **(XII-95,** R = $CH_3$) or benzylideneaniline **(XII-95**, R = $C_6H_5$) in xylene condense to g compounds characterized as 1-substituted-3-benzylidene-5-carbethoxy-2-oxo phenyl-$\Delta^4$-piperidenes (interpretation of NMR), which form 1-substituted benzyl-5-carboxy-6-phenyl-2-pyridones after heating in methanolic potassi hydroxide. The carboxypyridones have been decarboxylated to **XII-96** α-Cyano-β-methylcinnamide and ethyl formate are cyclized in the presence sodium hydride to 3-cyano-4-phenyl-2-pyridone **(XII-97)**.[46]

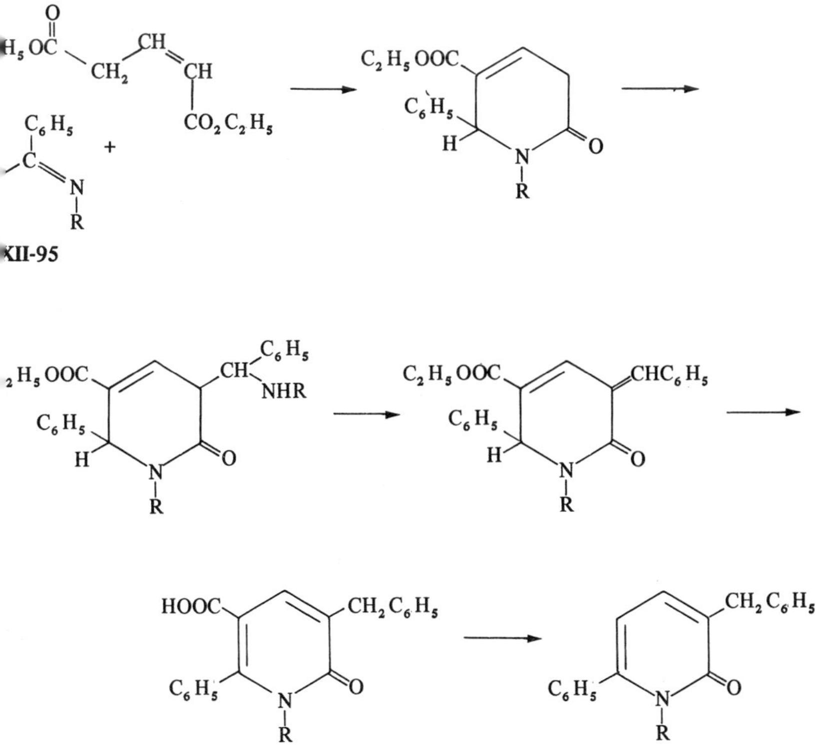

XII-95

XII-96
$R = CH_3, C_6H_5$

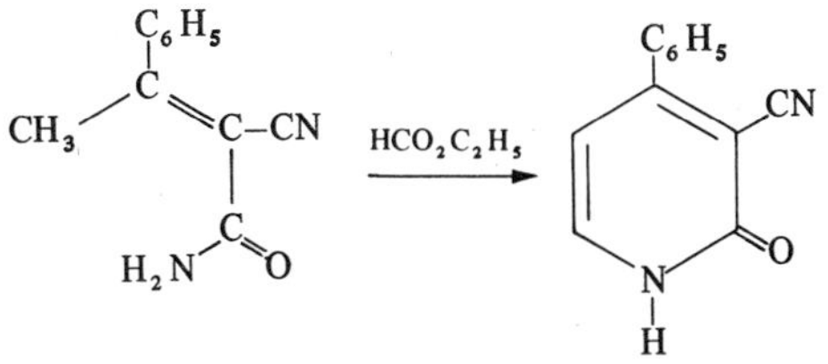

$HCO_2C_2H_5$

XII-97

## C. *Dienes*

Chlorosulfonyl isocyanate and isoprene in ether give 5-hydroxy-3-methyl-2-pentenoic acid lactone (4-methyl-5,6-dihydro-2-pyrone, **XII-98**) and 5-amino-3-methyl-3-pentenoic acid lactam (4-methyl-3,6-dihydro-2-pyridone, **XII-99**) by hydrolysis of intermediate *N*-sulfochlorides that were not isolated.[47]

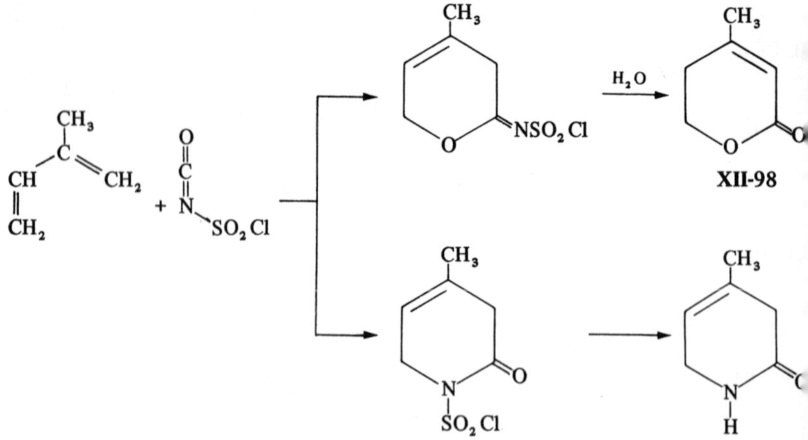

XII-98

XII-99

## D. *Diketones*

$\beta$-Diketones **(XII-100)** and benzaldazine give 1-benzylideneimino-2,3-dihyd 4-pyridones **(XII-101)**, which can be converted to the corresponding 2,3-hydro-4-pyridones **(XII-102)** by hydrogenolysis. Dehydrogenation with chlora gives the 4-pyridone.

$RCOCH_2COCH_3$ $\xrightarrow[C_6H_5CH=N-N=CHC_6H_5]{KNH_2}$

XII-100

XII-101

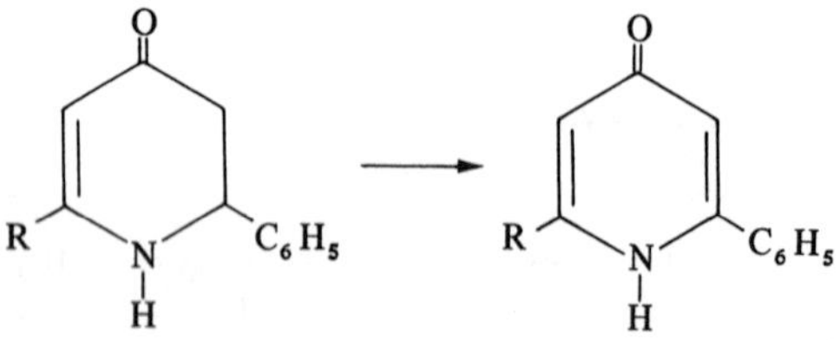

XII-102

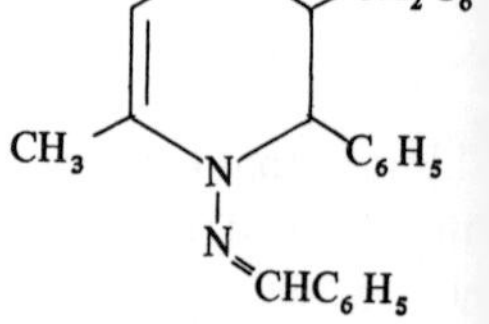

XII-103

Vhen 6-phenyl-2,4-hexanedione (**XII-100**, R = $C_6H_5CH_2CH_2$) was used, **I-101** (R = $CH_2CH_2C_6H_5$, 52%) and the isomeric pyridone (**XII-103**, 21%) re formed.[48] The β-diketones (**XII-100**) and benzylideneaniline give *N*-phenyl ivatives of **XII-102**.[39]

## E. β-*Ketoesters*

'he sodium salt of diethyl acetonedicarboxylate reacts with aryl isocyanates give 1-aryl-4,6-dihydroxy-2-pyridones (**XII-104**) and with phenyl isothiocya- :e to give 3-carbethoxy-4-hydroxy-2-mercapto-1-phenyl-6-pyridone.[49] The npound formed on treatment of **XII-104** (R = H) with acetic anhydride has :n proposed to be 3,5-diacetyl-4,6-dihydroxy-1-phenyl-2-pyridone. Character- tion was based on the infrared spectrum.[49]

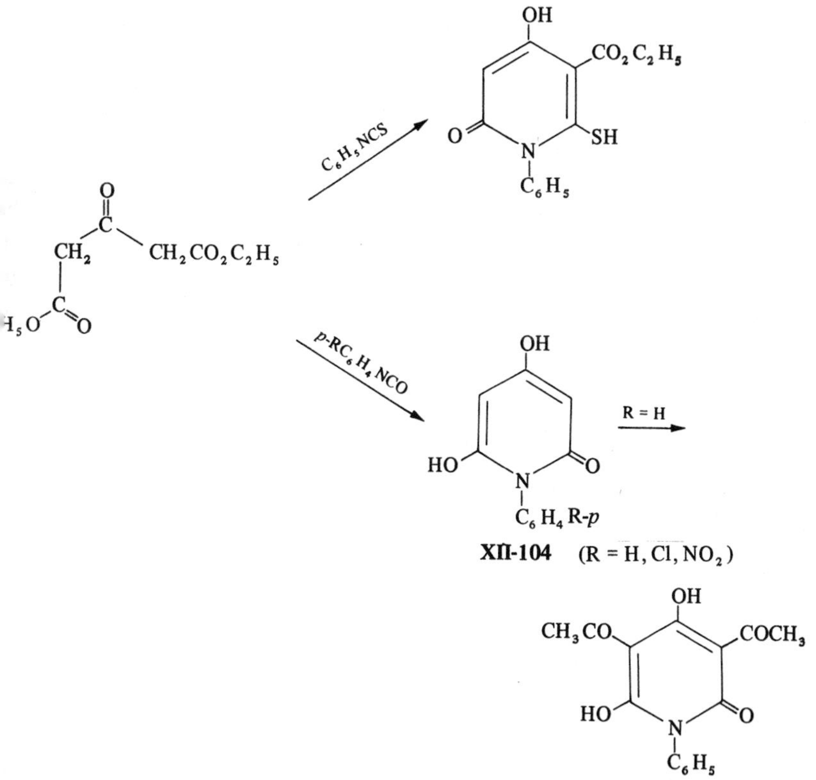

**XII-104** (R = H, Cl, $NO_2$)

'hioamides react with diketene in acetic acid or xylene to form ubstituted-4-hydroxy-2-pyridones (**XII-105**).[50]

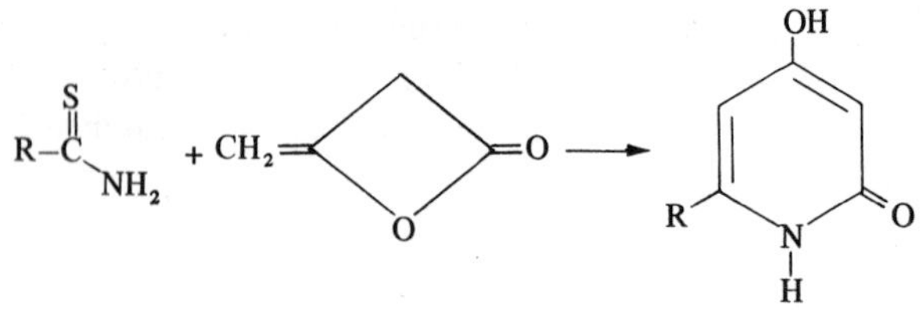

**XII-105**
R = alkyl, aryl

## 3. Ring Closures in Which One Acyclic Compound Provides Three Ring Carbons

### A. *β-Diketones, β-Ketoaldehydes, and Malonaldehyde*

There are two possible orientations for ring closure between reacta furnishing two and three carbons, respectively, to the ring. When the carbo reagent is an unsymmetrical diketone, the electrophilicity of the two carbo groups often is similar and both possible products are detected. The methyl group of cyanoacetamide prefers to react with the more electrophilic or the hindered carbonyl.[51] Reaction of 1-methoxy-3-phenylhydrazono-2,4-penta dione (**XII-106**) and cyanoacetamide employing sodium ethoxide gives b possible products (**XII-107** and **XII-108**), with **XII-107** predominating.[52] sumably the inductive effect of the 1-methoxy group makes the 2-carbonyl m electrophilic, and it is preferentially attacked by the conjugate base of cya acetamide [for leading references see (53)].

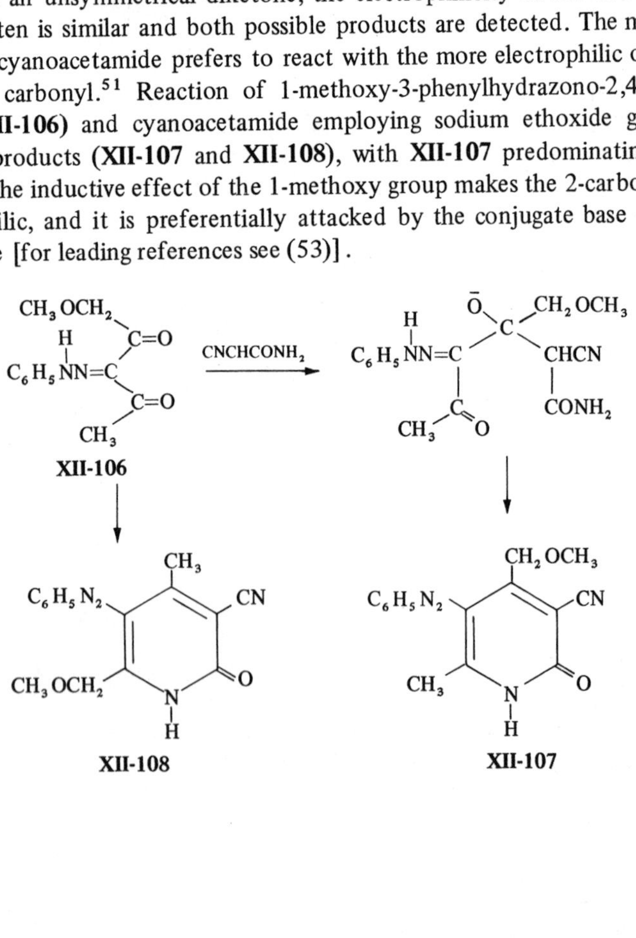

$CNCHCONH_2$

**XII-106**

**XII-108**

**XII-107**

-Hydroxymethylene-1-phenyl-2-propanone (**XII-109**) condenses with cyano-tamide in piperidine to give 3-cyano-6-methyl-5-phenyl-2-pyridone I-110a) in low yield. Cyanoacetamide and the sodium salt of **XII-109** give yano-4-methyl-5-phenyl-2-pyridone (**XII-110b**). With ethyl cyanoacetate in thanol containing piperidine, 3-carbomethoxy-6-methyl-5-phenyl-2-pyridone I-110c) is formed. It has been suggested that the formation of the sodium salt the hydroxymethylene group of **XII-109** blocks attack by the methylene of noacetamide at that carbon.[53] 2,5-Diphenyl-1,3-pentanedione and cyano-tamide in aqueous sodium carbonate give 3-cyano-5-phenyl-6-(β-phenylethyl)-yridone (**XII-111**).[27]

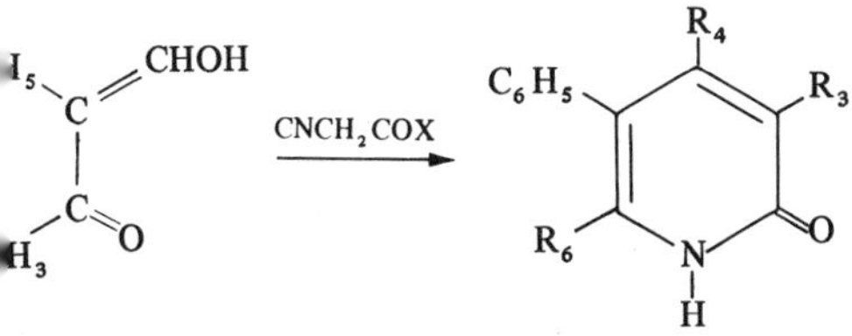

**II-109** $\xrightarrow{CNCH_2COX}$ **XII-110**

a: $R_3 = CN, R_4 = H, R_6 = CH_3$
b: $R_3 = CN, R_4 = CH_3, R_6 = H$
c: $R_3 = CO_2CH_3, R_4 = H, R_6 = CH_3$

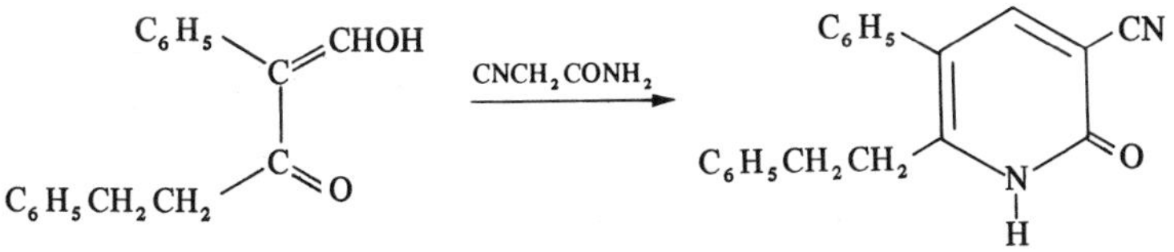

$C_6H_5C(=CHOH)C(=O)CH_2CH_2C_6H_5 \xrightarrow{CNCH_2CONH_2}$ **XII-111**

he dicarbanion of acetoacetaldehyde is alkylated in the γ-position to give s and copper chelates of β-ketoaldehydes (**XII-112**; R = $C_6H_5CH_2$, $CH_3$, $_4H_9$ and $n$-$C_8H_{17}$), which are converted to 3-cyano-2-pyridones (**XII-113**; = $CH_3$, $C_6H_5CH_2$, $n$-$C_4H_9$) by reaction with cyanoacetamide. $n$-Butylation the dianion of α-benzylacetoacetaldehyde followed by cyclization with eous cyanoacetamide gives 5-benzyl-3-carboxamido-6-$n$-pentyl-2-pyridone I-114).[54]

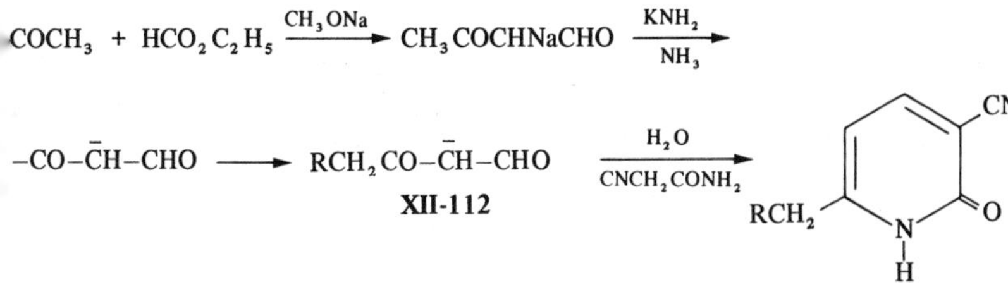

$COCH_3 + HCO_2C_2H_5 \xrightarrow{CH_3ONa} CH_3COCHNaCHO \xrightarrow[NH_3]{KNH_2}$

$-CO-\bar{C}H-CHO \longrightarrow RCH_2CO-\bar{C}H-CHO$ (**XII-112**) $\xrightarrow[CNCH_2CONH_2]{H_2O}$ **XII-113**

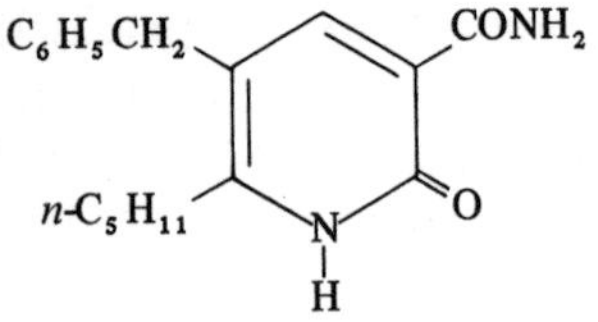

**XII-114**

Ethyl 2-keto-3-phenyllevulinate and cyanoacetamide in methanol contain piperidine give a mixture of methyl and ethyl 3-cyano-6-methyl-5-phenyl-2-p done-4-carboxylates **(XII-115)**.[53]

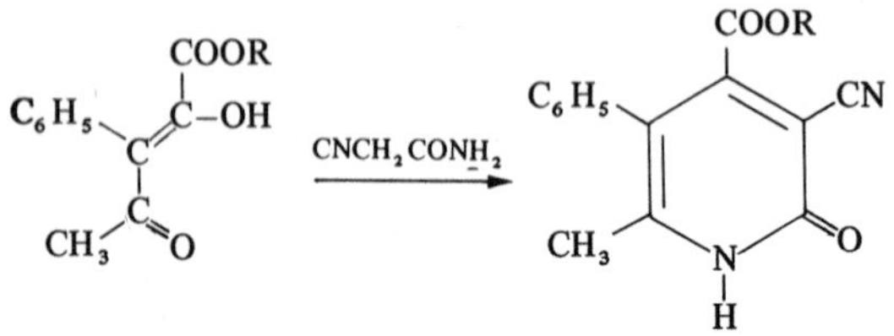

$\xrightarrow{CNCH_2CONH_2}$

**XII-115**

$R = CH_3, C_2H_5$

A series of 3-cyano-4-trifluoromethyl-2-pyridones **(XII-117)** has been prepa employing trifluoromethyldiketones **(XII-116)** or ethyl 4,4,4-trifluoroace acetate (Section I.3.B., p. 625) and cyanoacetamide.[55] The stron

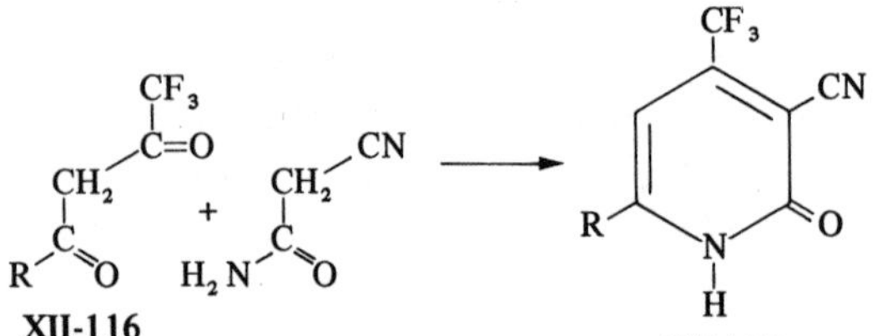

**XII-116**

**XII-117**

$R = CH_3, CF_3, C_6H_5$, 2-thienyl

electron-withdrawing trifluoromethyl group in the diketones activates carbonyl group to give products **XII-117** rather than the isom 6-trifluoromethyl-4-alkylpyridones.[55] The conversion of **XII-117** ($R = CH_3$ the vitamin $B_6$ analog, 5-hydroxy-6-methyl-4-trifluoromethyl-3-pyrid methanol **(XII-118)**, illustrates a utility of this compound.[56] Reactions cinnamoylacetaldehydes with cyanoacetamides give 3-cyano-6-styryl-2-pyrid **(XII-119)**.[57]

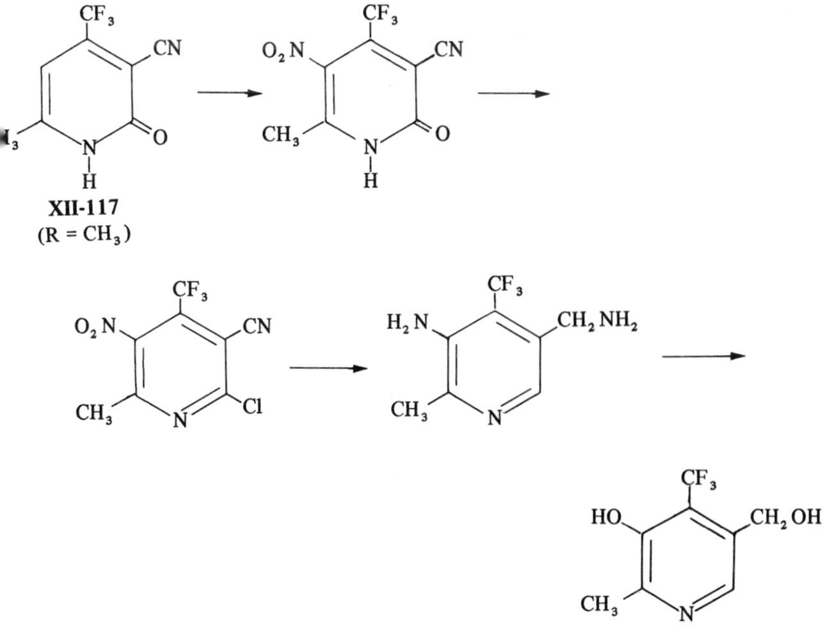

XII-117
($R = CH_3$)

XII-118

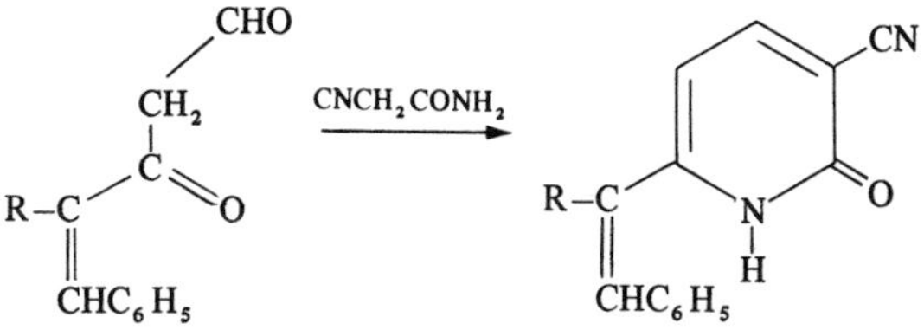

$CNCH_2CONH_2$

XII-119

Ethyl cyanoacetate and 1-amino-2-benzoylethylene cyclize to 3-cyano-4-enyl-2-pyridone **(XII-120)** in the presence of sodium ethoxide.[46]

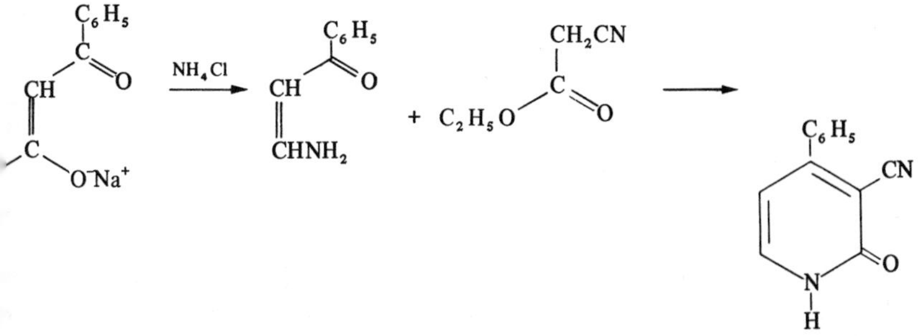

$NH_4Cl$

XII-120

1,1,3,3-Tetraethoxypropane (malonaldehyde diethyl acetal) and cyanoac mide in aqueous triethylamine give 3-cyano-2-pyridone (**XII-123**, 57%).[58]

$$(C_2H_5O)_2CH-CH_2-CH(OC_2H_5)_2 \xrightarrow[(C_2H_5)_3N]{CNCH_2CONH_2} \text{3-cyano-2-pyridone}$$

**XII-123**

The reaction between β-dicarbonyl compounds and cyanoacetamide to fo the corresponding 3-cyano-2-pyridones, followed by conversion to the 2-h derivatives and reduction in the presence of 5% Pd-C, is a useful route 3-cyano-4- and 3-cyano-6-substituted pyridines (**XII-121**).[59] Acetoacetamide

$$R_4COCH_2COR_6 \xrightarrow{CNCH_2CONH_2} \text{4-}R_4\text{-6-}R_6\text{-3-cyano-2-pyridone} \longrightarrow \text{4-}R_4\text{-6-}R_6\text{-2-chloro-3-cyanopyridine} \longrightarrow$$

**XII-121**

acetylacetone in 10% ethanolic hydrogen chloride give 3-acetyl-4,6-dimethy pyridone (**XII-122**).[60]

$$CH_3COCH_2COCH_3 + CH_3COCH_2CONH_2 \longrightarrow \text{3-COCH}_3\text{-4,6-(CH}_3)_2\text{-2-pyridone}$$

**XII-122**

An enamine from 1-ethoxy-2,4-pentanedione, previously described **XII-124**[61, 62] has been shown to be **XII-125**, which has been prepared also fr ethoxyacetonitrile and acetone.[63] The pyridone **XII-126** is formed by treatm of **XII-125** with malononitrile followed by ammonia and cyclization.[63]

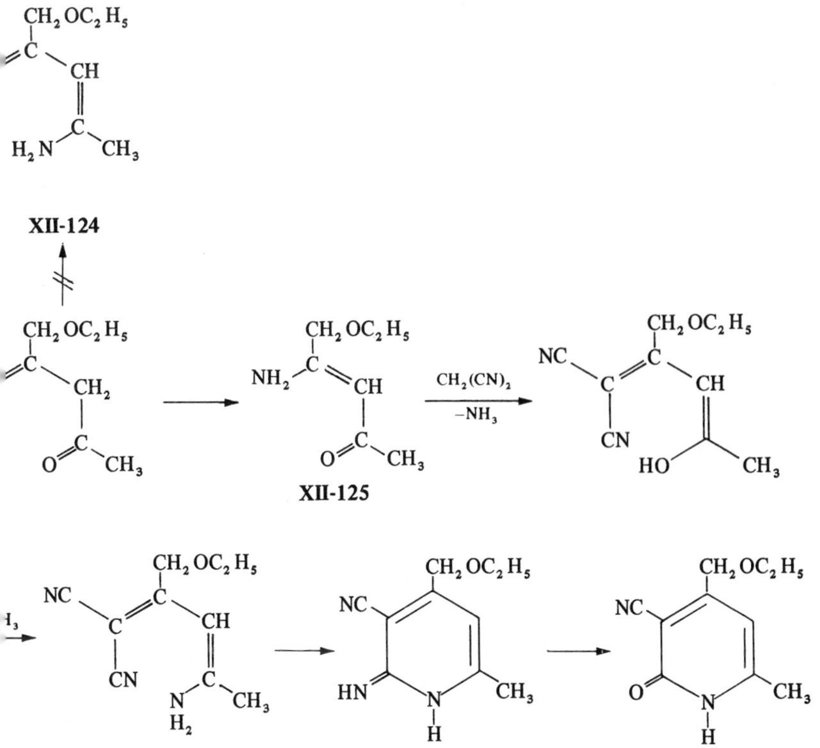

XII-124

XII-125

XII-126

## B. β-Ketoacid Derivatives

Cyclizations of β-ketoesters with cyanoacetamide or with ethyl cyanoacetate d ammonia give 5-cyano-6-hydroxy-2-pyridones.[64] Ethyl 4,4,4-trifluoroaceto-tate and cyanoacetamide in the presence of sodium methoxide form 3-cyano-6-droxy-4-trifluoromethyl-2-pyridone (**XII-127**) after crystallization from 15% drochloric acid. This pyridone appears to tautomerize in the solid state to yano-6-hydroxy-4-trifluoromethyl-2-pyridone.[55]

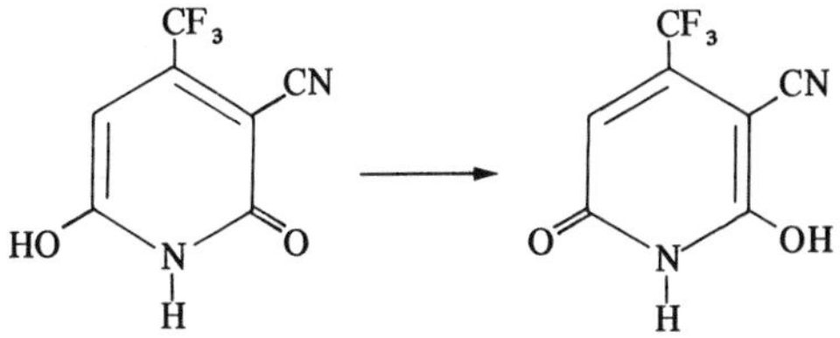

XII-127

Only small amounts of 6-aryl-4-methyl-2-pyridones (**XII-128**; R = $C_6H_5$, 6 $\alpha$-$C_{10}H_7$, 8.5%) are formed from ethyl acetoacetate, aryl methyl ketones a ammonium acetate.[65]

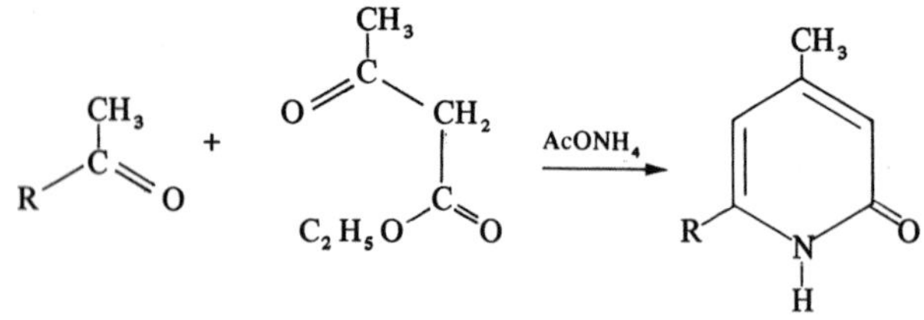

$$\text{R}-\text{CO}-\text{CH}_3 + \text{CH}_3\text{CO}\text{CH}_2\text{CO}_2\text{C}_2\text{H}_5 \xrightarrow{\text{AcONH}_4}$$

**XII-128**

The procedure of Hauser and Eby[66] has been used to prepare seve substituted 4-phenyl-2-pyridones related to streptonigrin (**XII-129**). In mo experiments, aroylacetonitriles and ketones in polyphosphoric acid g 4-aryl-2-pyridones (**XII-130**). 5,6-Dimethyl-4-phenyl-2-pyridone (**XII-130a**) v

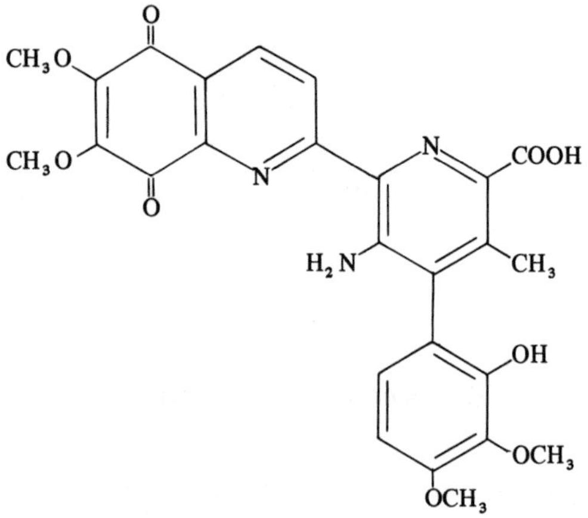

**XII-129**

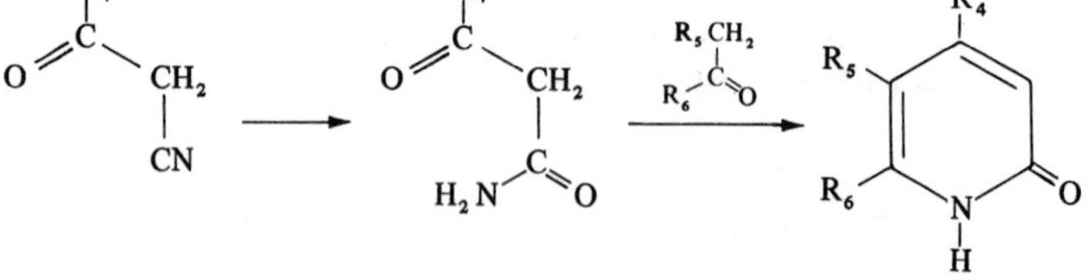

$$\text{R}_4\text{CO}\text{CH}_2\text{CN} \longrightarrow \text{R}_4\text{CO}\text{CH}_2\text{CONH}_2 \xrightarrow{\text{R}_5\text{CH}_2\text{COR}_6}$$

**XII-130 a, b**

| | $R_4$ | $R_5$ | $R_6$ |
|---|---|---|---|
| a: | $C_6H_5$ | $CH_3$ | $CH_3$ |
| b: | 2,3,4-$(CH_3O)_3C_6H_2$ | H | $CH_3$ |

only pyridone isolated when ethyl methyl ketone and benzoylacetonitrile re used.[67] Dimethylaminoacetonitrile can be aroylated to give α-aroyl-α-*N*,*N*-nethylaminoacetonitriles (**XII-131**; R = $C_6H_5$, *p*-$CH_3OC_6H_4$, *p*-$ClC_6H_4$), ich are cyclized with acetophenone to 4-aryl-3-dimethylamino-6-phenyl-2-idones.[68]

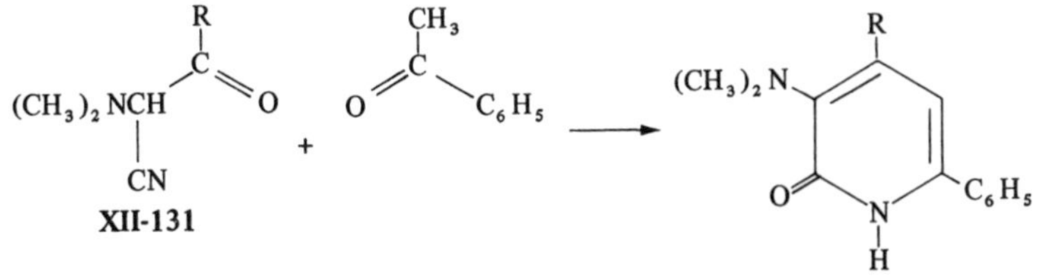

$(CH_3)_2NCH(CN)C(=O)R$ + $O{=}C(CH_3)C_6H_5$ → 3-$(CH_3)_2N$-4-R-6-$C_6H_5$-2(1H)-pyridone

**XII-131**

rimary and secondary enamines prepared from ethyl acetoacetate or iketones react with diketene to form 4-pyridones (**XII-132**). The arbethoxy-4-pyridones (**XII-132**, $R_3$ = $CO_2Et$) are readily saponified and

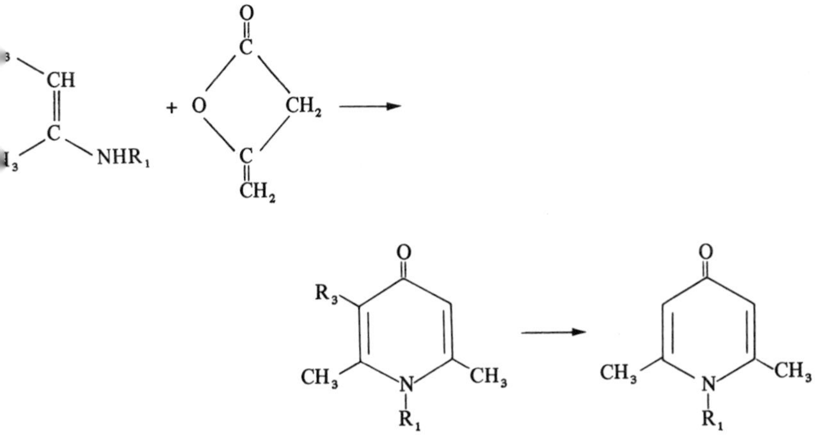

**XII-132** **XII-133**

$R_1$ = H, $C_6H_5$, $CH_2C_6H_5$; $R_3$ = $CO_2C_2H_5$.
$R_1$ = H; $R_3$ = $COCH_3$, $COC_6H_5$.

carboxylated to *N*-substituted-4-lutidones (**XII-133**, $R_1$ = H, $C_6H_5$).[69] Acylated acetoacetic esters are cleaved with methanolic potassium hydroxide give **XII-134**, which can be converted to enamines **XII-135** with ammonia in presence of a small amount of ammonium nitrate. Methyl β-alanate and I-135 in ethanol give the enamines **XII-136**, which are cyclized to the -dihydro-4-pyridones using sodium methoxide in methanol.[70] When the

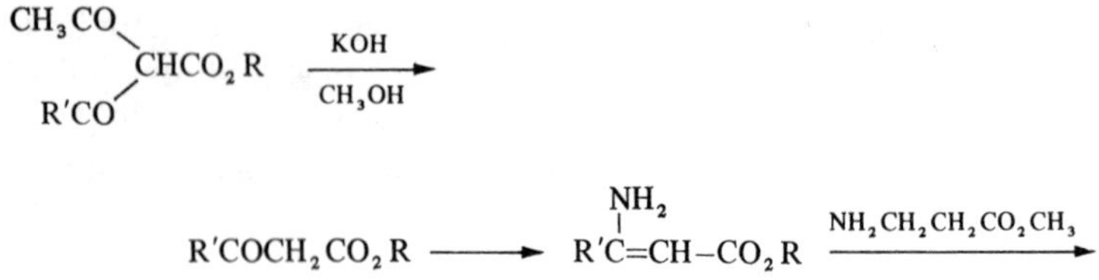

$$\begin{matrix} CH_3CO \\ R'CO \end{matrix} \!\!> CHCO_2R \xrightarrow[CH_3OH]{KOH}$$

$$R'COCH_2CO_2R \longrightarrow R'C(NH_2){=}CH{-}CO_2R \xrightarrow{NH_2CH_2CH_2CO_2CH_3}$$

**XII-134** **XII-135**

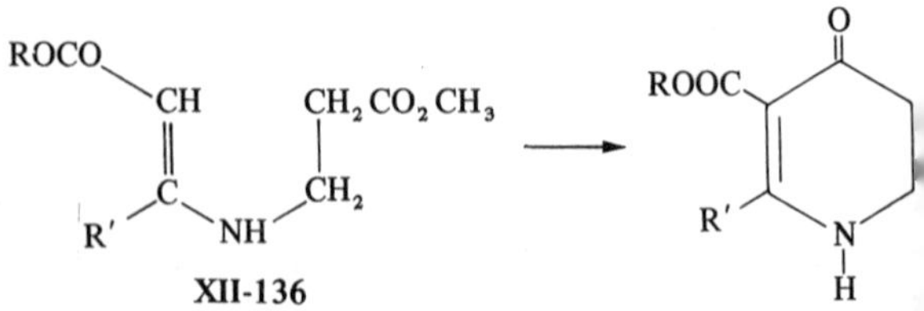

$$ROCO{-}CH{=}C(R'){-}NH{-}CH_2{-}CH_2CO_2CH_3 \longrightarrow$$

**XII-136**

| R | R′ |
|---|---|
| $CH_3$ | $C_6H_5CH_2-$ |
| *t*-Bu | $C_6H_5CH_2-$ |
| $CH_3$ | *p*-$CH_3OC_6H_4CH_2-$ |
| $CH_3$ | 3,4-$(CH_3O)_2C_6H_3CH$ |
| $CH_3$ | $CH_3$ |

enamines **XII-135** are boiled under reflux with acrylic esters, 3,4-dihydro pyridones **(XII-137)** are formed.[70]

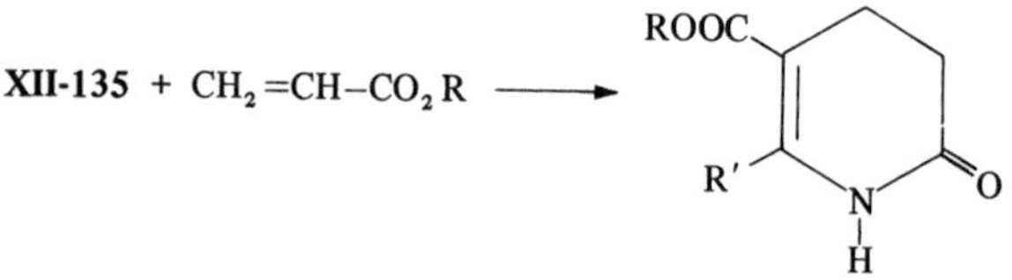

$$\textbf{XII-135} + CH_2{=}CH{-}CO_2R \longrightarrow$$

**XII-137**

| R | R′ |
|---|---|
| $C_2H_5$ | $CH_3$ |
| $CH_3$ | $CH_2C_6H_5$ |

## C. β-*Substituted Ketones*

Both dialkylamine and pyridine can function as leaving groups in this type pyridone synthesis. For example, 6-phenyl-2-pyridone can be formed fro ω-dimethylaminopropiophenone and *N*-(carbamylmethyl)pyridinium chlori **(XII-138)**[1] (Section I.1., p. 599).

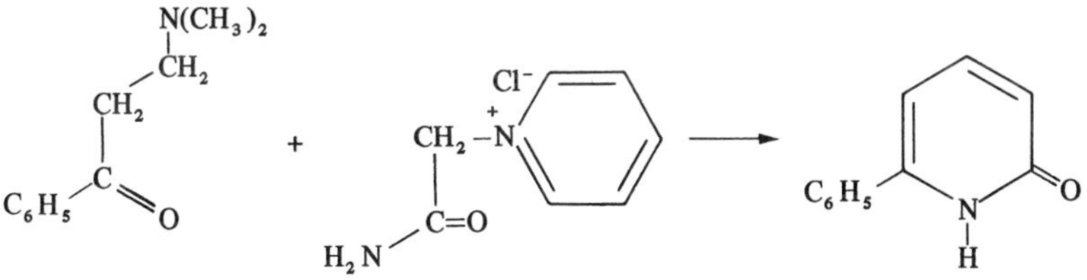

**XII-138**

## D. *Unsaturated Carbonyl Compounds*

Compound **XII-138** or its *N*-methyl derivative in methanol contain- dimethylamine or sodium hydroxide, or in acetic acid containing nmonium acetate, reacts with benzylideneacetophenone, benzylideneacetone, nnamaldehydes, or ω-4-(picolinylidenyl)acetophenone probably by paths tiated by a Michael addition to give **XII-140**.[1, 71-74] *N*-Carbethoxymethyl- ridinium bromide **(XII-139)** and ammonium acetate in acetic acid has

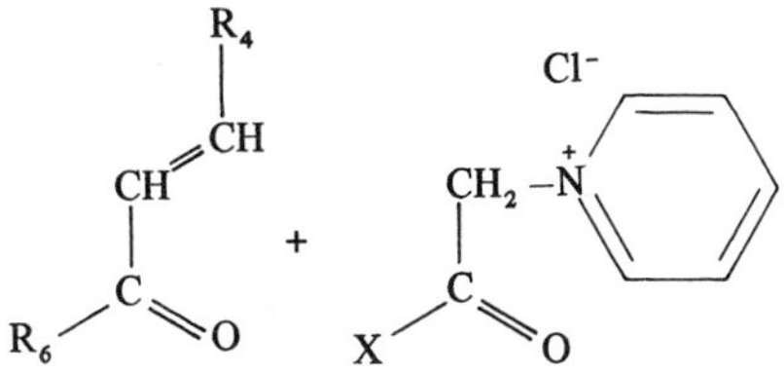

X = $NH_2$ **(XII-138)**
X = $OC_2H_5$ **(XII-139)**

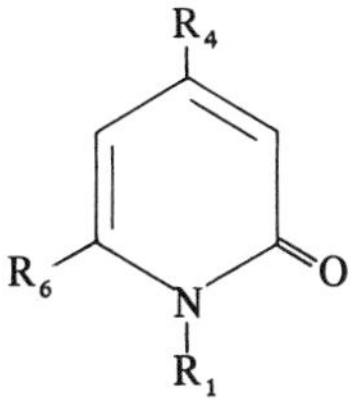

**XII-140**

| | $R_1$ | $R_4$ | $R_6$ |
|---|---|---|---|
| a–c: | H | $C_6H_5$ | H, $CH_3$, $C_6H_5$ |
| d: | $CH_3$ | $C_6H_5$ | $C_6H_5$ |
| e: | H | 4-pyridyl | $C_6H_5$ |
| f: | H | $C_6H_5$ | 3-pyridyl |
| g, h: | H | *p*-$NO_2C_6R_4$ | $C_6H_5$, 2-pyridyl |
| i: | H | *o*-$NO_2C_6H_4$ | H |

vantages over **XII-138** for synthesis of **XII-140e-g.**[72] Benzylideneaceto- enone and **XII-138** or **XII-139** with ammonium acetate in acetic acid at 140° e both **XII-140a** and 2,4,6-triphenylpyridine.[74] A retrocondensation of benzyl- neacetophenone occurs to form benzaldehyde and acetophenone, which react th benzylideneacetophenone and ammonium acetate to form triphenylpyridine a typical Chichibabin reaction,[75] which includes a dehydrogenation.

Nicotellin **(XII-141)**, a tobacco alkaloid, has been synthesized from **XII-** and the Claisen condensation product of nicotinaldehyde and 3-acetylpyridi

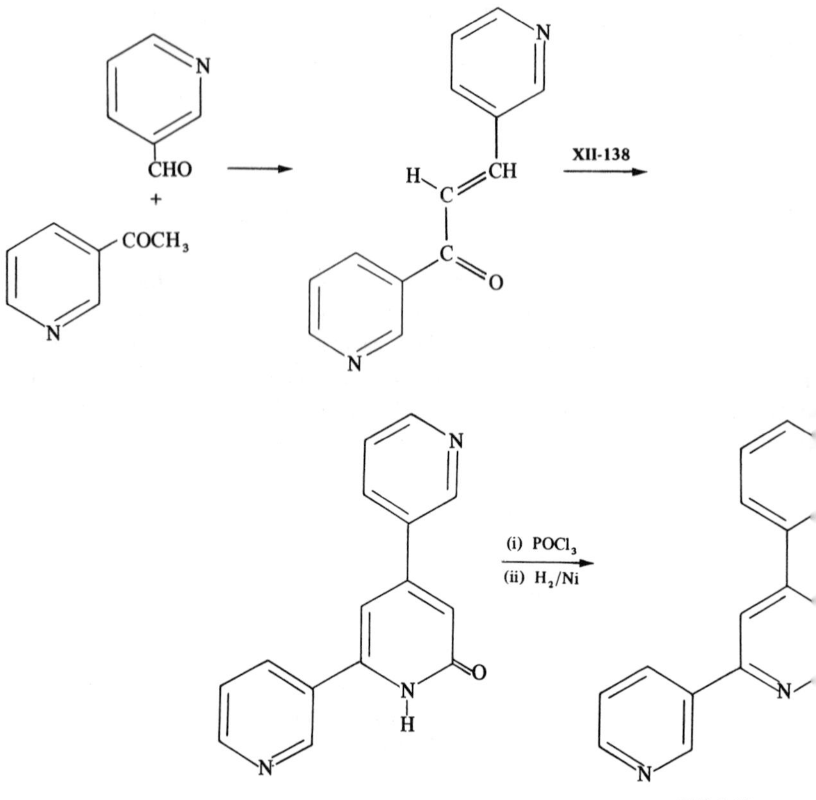

XII-138

(i) $POCl_3$

(ii) $H_2/Ni$

**XII-141**

Vinyl ketones, ethyl cyanoacetate and ammonium acetate give 4,6-disub tuted 3-cyano-2-pyridones **(XII-142)** by a reaction path that includes a Mich addition and a dehydrogenation. The intermediate 3,4-dihydro-2-pyridones w not detected. When benzylideneacetophenone was used in benzene, the reacti proceeded by a Knoevenagel condensation and it was possible to isolate intermediate 3-cyano-5,6-dihydro-4,6-diphenyl-2-pyridone, which was dehydr enated to **XII-142** by boiling in acetone.[76]

Condensation of 2-substituted veratrals with ethyl methyl ketones gives sub tuted vinyl ketones that react with ethyl cyanoacetate in the presence ammonium acetate to give 4-aryl-3-cyano-5,6-dimethyl-2-pyridones **(XII-143)**

Cyanoacetamide and 1-dimethylamino-3-phenyl-1-propene-3-one in acetic a give 3-carboxamido-4-phenyl-2-pyridone **(XII-144)**.[78]

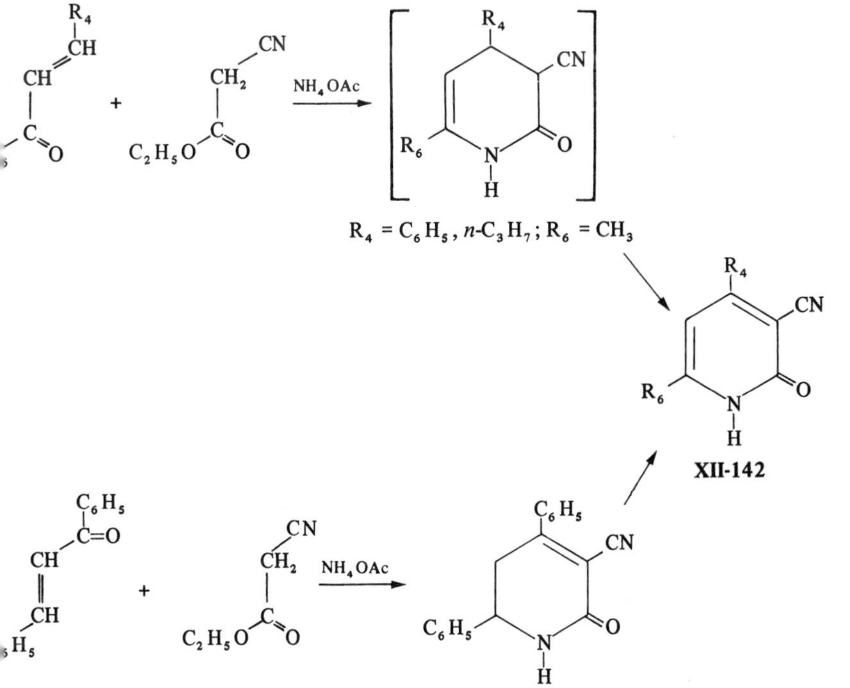

$R_4 = C_6H_5, n\text{-}C_3H_7; R_6 = CH_3$

NH₄OAc

**XII-142**

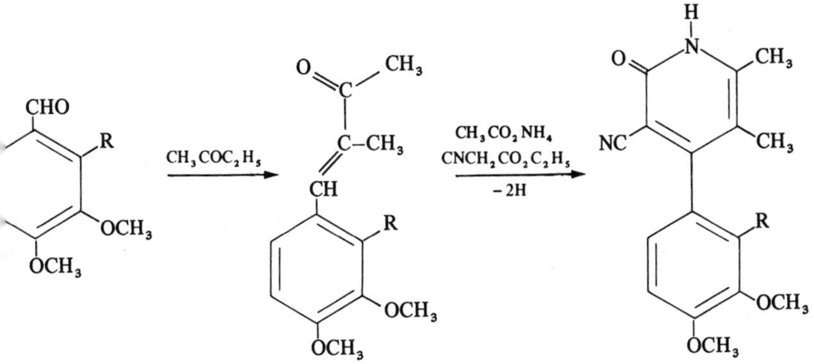

NH₄OAc

$CH_3COC_2H_5$

$CH_3CO_2NH_4$

$CNCH_2CO_2C_2H_5$

$-2H$

**XII-143**

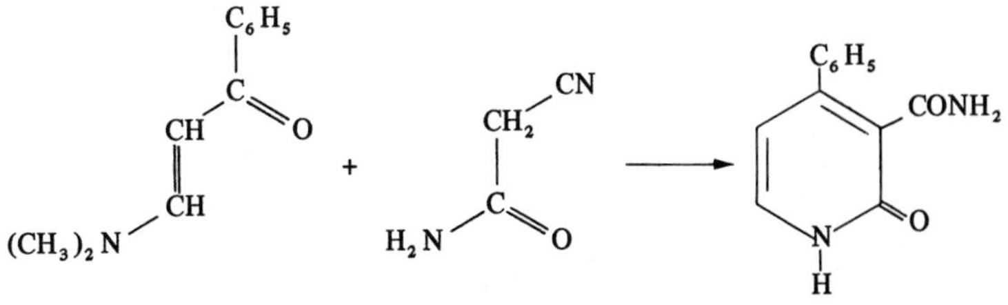

**XII-144**

1,3,3-Tricyano-2-aminopropene(dimeric malononitrile) **(XII-145)** and enami ketones **(XII-146**; R = $NHCH_3$, $N(CH_3)_2$, $R_4$ = $C_6H_5$, $p\text{-}CH_3C_6H_4$, 2,4-$(CH_3)_2C_6H_3$, $p\text{-}CH_3OC_6H_4$) in acetic acid give bipyridonyls **(XII-147)**, as well as smaller amounts of **XII-148** and the 5,6-dihydro-2-pyridone **(XII-149)**, a reaction that is not influenced by radical inhibitors or initiators or by the exclusion of oxygen.[78-80]

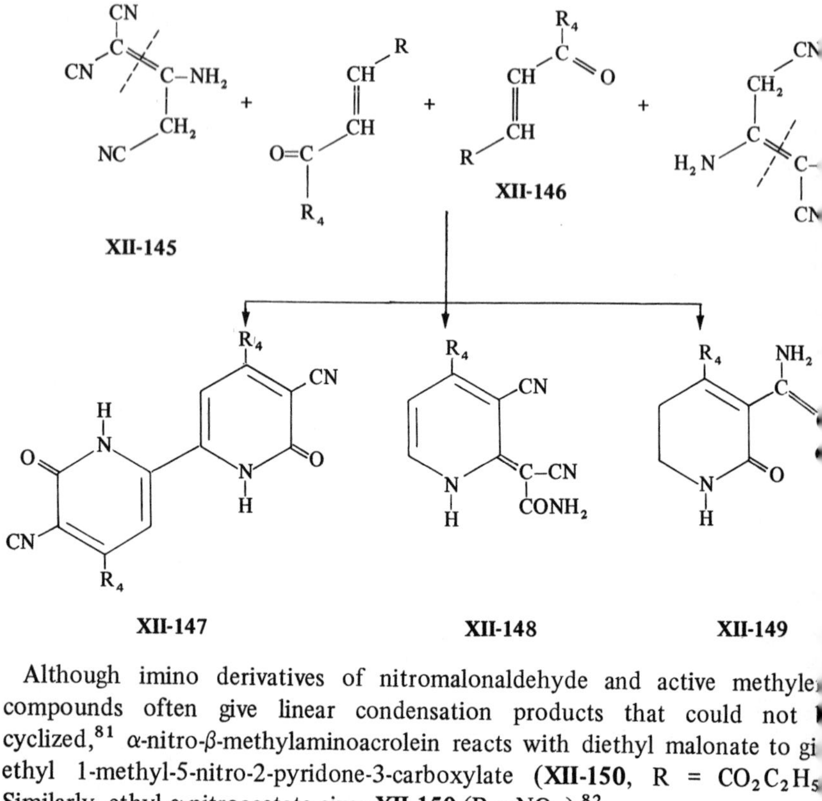

XII-145 XII-146

XII-147 XII-148 XII-149

Although imino derivatives of nitromalonaldehyde and active methylene compounds often give linear condensation products that could not be cyclized,[81] α-nitro-β-methylaminoacrolein reacts with diethyl malonate to give ethyl 1-methyl-5-nitro-2-pyridone-3-carboxylate (**XII-150**, R = $CO_2C_2H_5$). Similarly, ethyl α-nitroacetate gives **XII-150** (R = $NO_2$).[82]

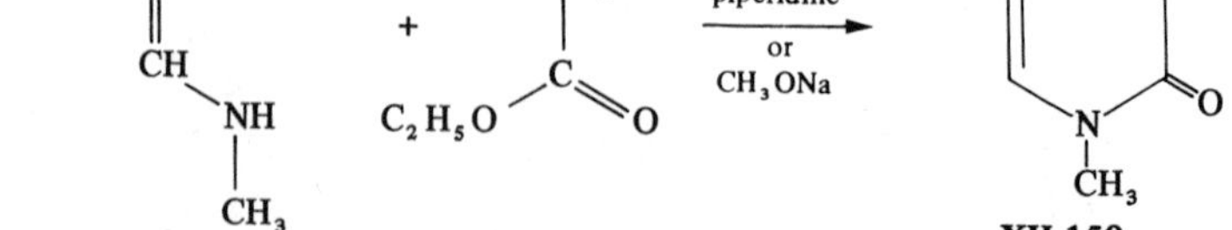

piperidine or $CH_3ONa$

XII-150

Dehydrochlorination of dichloroacetyl chloride by triethylamine in the presence of a Schiff's base with a cinnamaldehyde structure through

loaddition gives dihydro-2-pyridones (**XII-151**), which are converted to
ıloro-2-pyridones (**XII-152**) by an excess of triethylamine.[83]

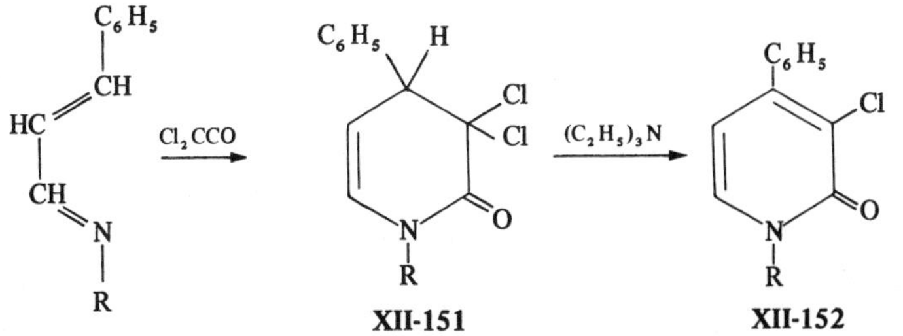

**XII-151** **XII-152**

R = $C_6H_5$, $CH_3C_6H_4$, $C_6H_5CH{=}CH{-}CH{=}N{-}$

ıamines **XII-153**, **XII-154**, and **XII-155** and propiolactone or acrylic acids
n dihydropyridones, **XII-156**, **XII-157**, and **XII-158**, when boiled in
ɔrobenzene.[84]

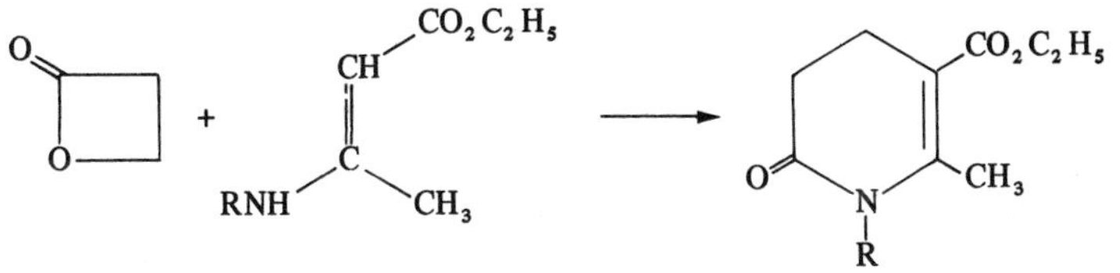

**XII-153** **XII-156**

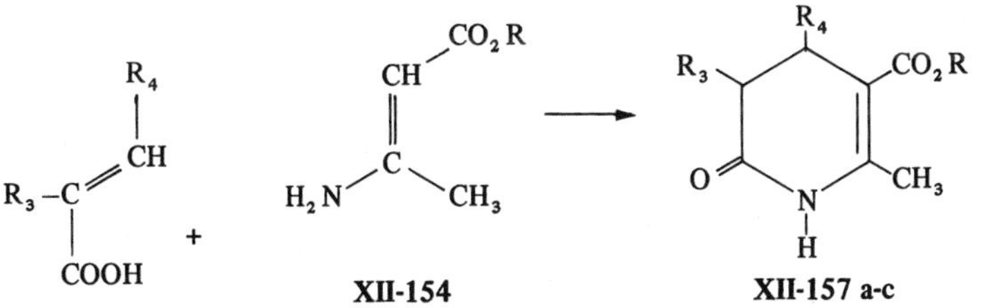

**XII-154** **XII-157 a-c**

| | $R_3$ | $R_4$ | R |
|---|---|---|---|
| a: | H | H | *t*-Bu |
| b: | $CH_3$ | H | $C_2H_5$ |
| c: | H | $CH_3$ | $C_2H_5$ |

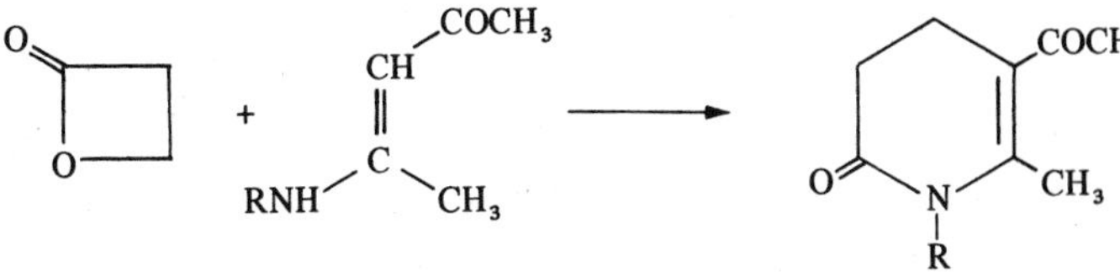

**XII-155** **XII-158**

## E. *Malonic Acid Derivatives*

In the presence of alkoxides, diethyl malonate reacts with acetylacetone im to give the 4,6-dimethyl-2-pyridone **(XII-159)** and with ethyl β-aminocroton to give the 4-hydroxy-2-pyridone **(XII-160)**.[85] In these reactions, et malonate, like ethyl cyanoacetate and cyanoacetamide, provides two of the r carbons. Ethyl nitromalonate, which cannot form a 2-pyridone by this pa reacts with enamines **(XII-161**; R = $C_6H_5$, $OC_2H_5$) through its two carbetho groups to give the 5-substituted-4-hydroxy-3-nitro-2-pyridones **(XII-162)**. T 5-carbethoxy-2-pyridone **(XII-162**, R = $OC_2H_5$) is readily saponified a decarboxylated.[86] In contrast, it had been suggested earlier that when the rela ethyl alkylmalonates are used, intermediates are formed which undergo arom. zation through loss of a carbethoxy group from the malonate moiety.[85]



Primary and secondary enamines and anils **(XII-163)** react with carb suboxide **(XII-164)** to form 3-unsubstituted-4-hydroxy-2-pyrido **(XII-165)**,[87,88] with carbethoxy ethyl ketene **(XII-166)** to form 3-ethyl hydroxy-2-pyridones **(XII-167)**,[89] and with malonyl chloride **(XII-168)** malonic acids and acetic anhydride **(XII-169)** to give 3-substituted-4-hydroxy pyridones **(XII-170)**[90-92] (Table XII-1). Anils of cyclic ketones **(XII-163**, $R_5$ = $(CH_2)_n$, where $n$ = 3,4,5,6; $R_1$ = aryl) and substituted malonyl chlori ($R_3$ = alkyl) give 5,6-polymethylene-4-hydroxy-2-pyridones.[93] A comparis

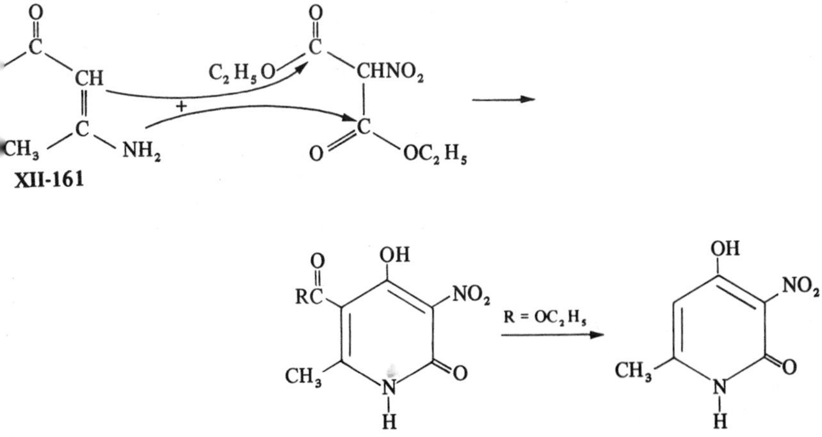

XII-161

XII-162

ith known isomeric pyridones[94] showed that the product formed from enzoylacetoneamine (**XII-163**; $R_5$ = $COCH_3$, $R_6$ = $C_6H_5$, $R_1$ = H) and arbon suboxide (**XII-164**) is 5-acetyl-4-hydroxy-6-phenyl-2-pyridone XII-165; $R_1$ = H, $R_5$ = $COCH_3$, $R_6$ = $C_6H_5$) and not the isomeric benzoyl-4-hydroxy-6-methyl-2-pyridone[87, 95] (Section I.5.C., p. 658).

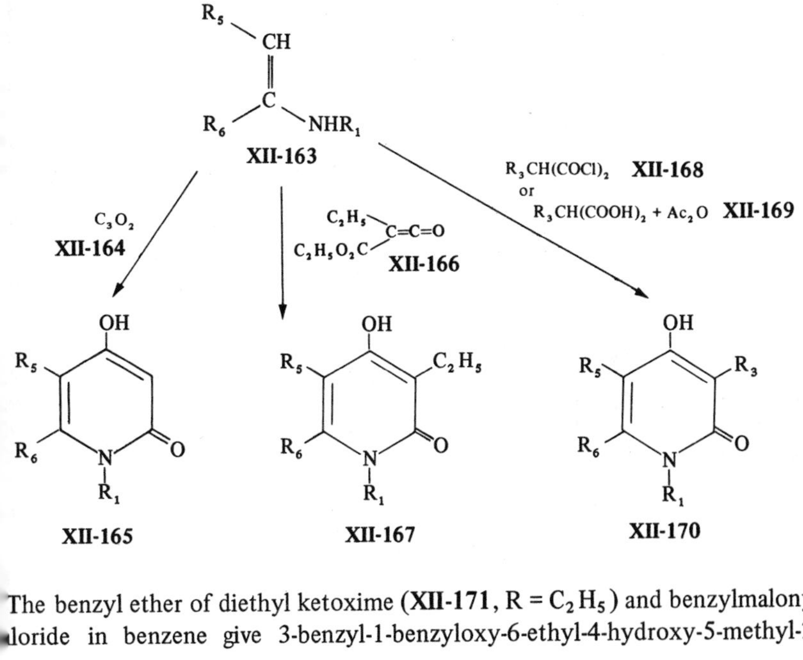

XII-163

$C_3O_2$ XII-164

$R_3CH(COCl)_2$ XII-168 or $R_3CH(COOH)_2$ + $Ac_2O$ XII-169

XII-166

XII-165 XII-167 XII-170

The benzyl ether of diethyl ketoxime (**XII-171**, R = $C_2H_5$) and benzylmalonyl hloride in benzene give 3-benzyl-1-benzyloxy-6-ethyl-4-hydroxy-5-methyl-2-

TABLE XII–1. 4-Hydroxy-2-pyridones from Derivatives of Malonic Acid

| $R_1$ | $R_3$ | $R_5$ | $R_6$ | Reagent | % Yield | F |
|---|---|---|---|---|---|---|
| H | H | $CO_2C_2H_5$ | $CH_3$ | XII-164 | 56 | 8 |
| H | H | $CO_2C_2H_5$ | $C_2H_5$ | XII-164 | 31 | 8 |
| H | H | $CO_2C_2H_5$ | $n$-$C_3H_7$ | XII-164 | 47 | 8 |
| $C_6H_5$ | H | $CO_2C_2H_5$ | $CH_3$ | XII-164 | 38 | 8 |
| $(CH_3)_2C_6H_3$ | H | $CO_2C_2H_5$ | $CH_3$ | XII-164 | 46 | 8 |
| | | | | XII-168 | 100 | 9 |
| $p$-$CH_3OC_6H_4$ | H | $CO_2C_2H_5$ | $CH_3$ | XII-164 | 47 | 8 |
| | | | | XII-168 | 100 | 9 |
| H | H | $COCH_3$ | $CH_3$ | XII-164 | 80 | 8 |
| | | | | XII-169 | 42 | 9 |
| $C_6H_5$ | H | $COCH_3$ | $CH_3$ | XII-164 | 46 | 8 |
| $C_6H_5CH_2$ | H | $COCH_3$ | $CH_3$ | XII-164 | 92 | 8 |
| H | H | $COCH_3$ | $C_6H_5$ | XII-164 | 74 | 8 |
| | | | | XII-169 | 33 | 9 |
| $C_6H_5$ | H | $COCH_3$ | $C_6H_5$ | XII-164 | 31 | 8 |
| | | | | XII-168 | 61 | 9 |
| $CH_3$ | H | $COCH_3$ | $C_6H_5$ | XII-164 | 38 | 8 |
| $C_6H_5$ | H | $(CH_2)_3$ | | XII-164 | 46 | 8 |
| $C_6H_5$ | H | $(CH_2)_4$ | | XII-164 | 48 | 8 |
| $C_6H_5$ | H | $(CH_2)_5$ | | XII-164 | 46 | 8 |
| $C_6H_5$ | H | $(CH_2)_6$ | | XII-164 | 19 | 8 |
| $C_6H_5$ | $C_2H_5$ | $(CH_2)_4$ | | XII-166 | 26 | 8 |
| H | $C_2H_5$ | $CH_3CO$ | $CH_3$ | XII-166 | 39 | 8 |
| | | | | XII-168 | 33 | 9 |
| $C_6H_5$ | H | $CO_2C_2H_5$ | H | XII-168 | 68 | 9 |
| H | $C_6H_5CH_2$ | $COCH_3$ | $C_6H_5$ | XII-168 | 92 | 9 |
| H | $C_6H_5CH_2$ | $COCH_3$ | $CH_3$ | XII-168 | 54 | 9 |
| | | | | XII-169 | 58 | 9 |
| $C_6H_5$ | $C_6H_5CH_2$ | $COCH_3$ | $CH_3$ | XII-168 | 29 | 9 |
| H | $C_2H_5$ | $CH_3$ | $C_6H_5$ | XII-168 | 65 | 9 |
| H | $CH_2C_6H_5$ | $CH_3$ | $C_6H_5$ | XII-168 | 68 | 9 |
| H | $C_6H_5CH_2$ | $CH_2$<br>- CH　　CH -<br>$C(CH_3)_2$ | | XII-168 | 30 | 9 |
| $C_6H_5$ | $C_6H_5CH_2$ | H | $C_6H_5$ | XII-168 | 42 | 9 |
| $C_6H_5$ | $CH(CH_3)_2$ | H | $C_6H_5$ | XII-168 | 56 | 9 |
| $p$-$CH_3C_6H_4$ | $C_6H_5CH_2$ | H | $C_6H_5$ | XII-168 | 82 | 9 |
| $p$-$CH_3C_6H_4$ | $n$-$C_4H_9$ | H | $C_6H_5$ | XII-168 | 42 | 9 |
| $p$-$CH_3C_6H_4$ | $CH(CH_3)_2$ | H | $C_6H_5$ | XII-168 | 66 | 9 |
| $C_6H_5$ | $C_6H_5CH_2$ | $CH_3$ | $C_6H_5$ | XII-168 | 38 | 9 |
| $C_6H_5$ | $C_6H_5CH_2$ | $C_2H_5$ | $C_6H_5$ | XII-168 | 39 | 9 |

ridone, which on hydrogenolysis gives 3-benzyl-6-ethyl-1,4-dihydroxy-5-thyl-2-pyridone (**XII-172**, R = $C_2H_5$). The pyridone (**XII-173**) is formed by ating **XII-172** at reduced pressure. When the benzyl ether of propiophenone ime (**XII-171**, R = $C_6H_5$) is heated at 250 to 280° with diethyl nzylmalonate, the pyridone (**XII-173**, R = $C_6H_5$) is obtained directly.[96]

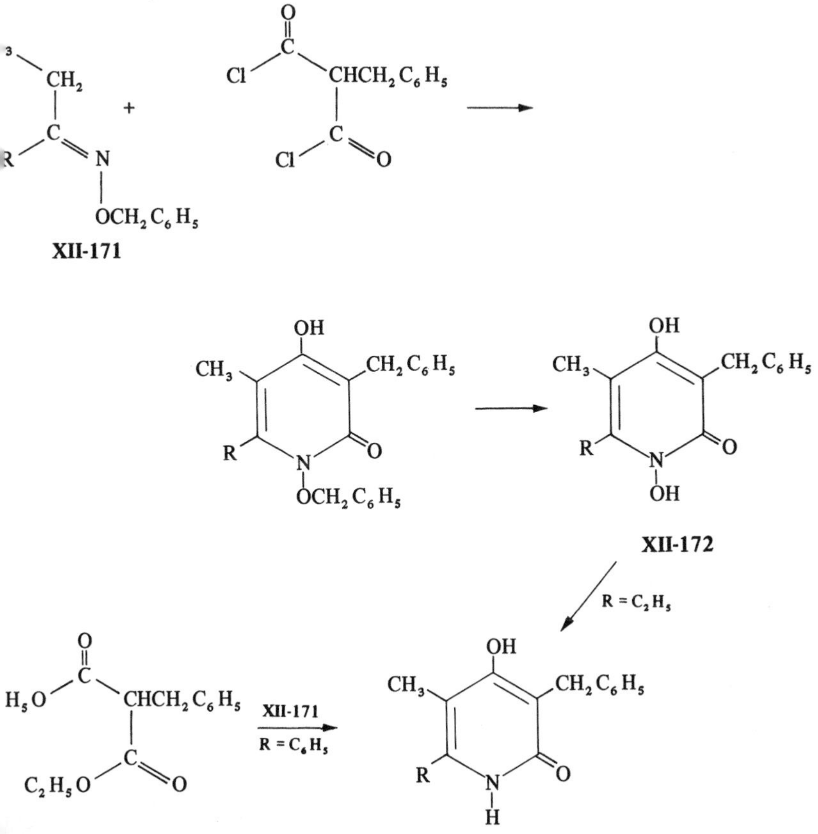

**XII-171**

**XII-172**

R = $C_2H_5$

XII-171

R = $C_6H_5$

**XII-173**

Nitriles condense with malonyl chloride to give 2-chloro-4,6-dihydroxypyrines (**XII-174**) in 23 to 63% yield[97] and/or the pyrimidones (**XII-175**)[98] or loropyranooxazines (Section I.5.C.) depending on conditions.[97,99]

5-Bromo-3-carbethoxy-2-chloro-4,6-dihydroxypyridine (**XII-176**, R = $O_2C_2H_5$) is prepared either by bromination of **XII-174** (R = $CO_2Et$) or by clization of bromomalonyl chloride and ethyl cyanoacetate. Methylmalonyl loride did not react with propionitrile.[97] The product from acetonitrile and alonyl chloride, first described as 2-chloro-4,6-dihydroxypyridine (**XII-174**, R H),[97] is 6-chloro-2-methyl-4-pyrimidone (**XII-175**, R = H).[98]

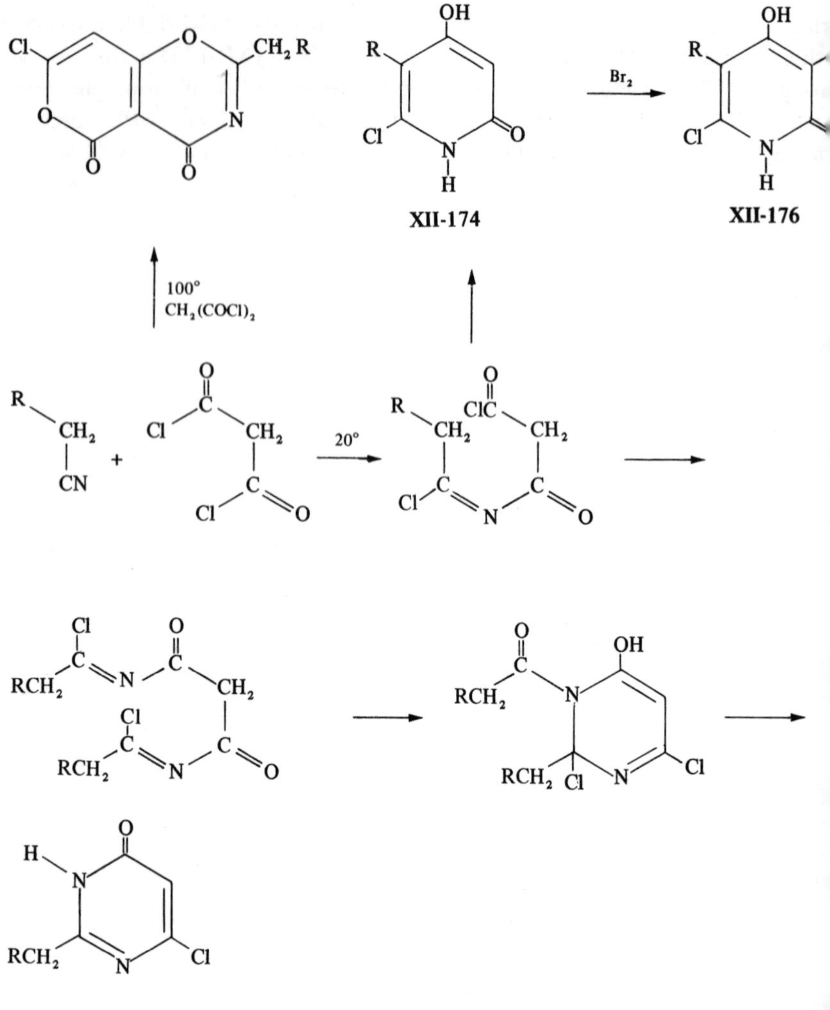

XII-174

XII-176

100°
$CH_2(COCl)_2$

20°

XII-175

## F. *Acetone Derivatives*

Dimethyl acetonedicarboxylate or ethyl acetoacetate can provide the 3-, and 5-carbons to give 3,5-disubstituted-4-pyridones (**XII-177**, $R_5 = CH_3C$ $CO_2CH_3$). *s*-Triazine has been used to provide the 2- and 6-carbons.[100] compound, previously described as 4-hydroxy-3-pyridinecarboxylic ac (**XII-177**, R = $R_5$ = H)[101] was shown to be nicotinic acid-1-oxide (**XII-178**).[10

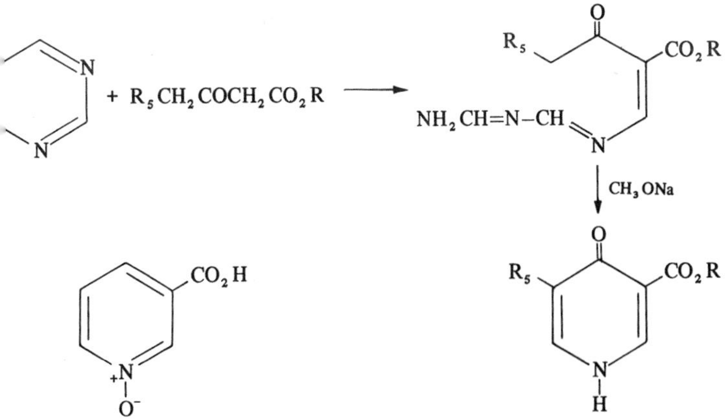

XII-178

XII-177

### G. *Isocyanates*

In isocyanates the incipient ring carbons are not in a continuous carbon chain. :yryl isocyanates (**XII-179**) react with ynamines in acetonitrile to give amino-5-substituted-2-pyridones by 1,4-cycloaddition. Further reaction occurs ıder these conditions to give 1-(*N*-substituted-carbamoyl)-2-pyridones **(XII-180)**, which are readily decomposed thermally.[102]

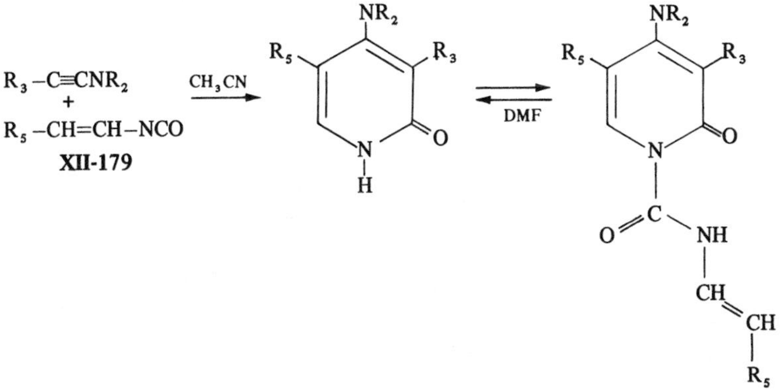

XII-179

XII-180

## 4. Miscellaneous Ring Closures

### A. *Autocondensations*

There is some confusion in the literature that describes autocondensations of :etoacetamides. In 1902, Claisen and Meyer[103] proposed that the pyridone product formed by heating acetoacetamide was 5-carboxamido-4,6-dimethyl-pyridone **(XII-181)** and its hydrolysis product was **XII-182**.[103] A reinvestigati of this reaction led to assignment of **XII-183** as the structure because it gives t

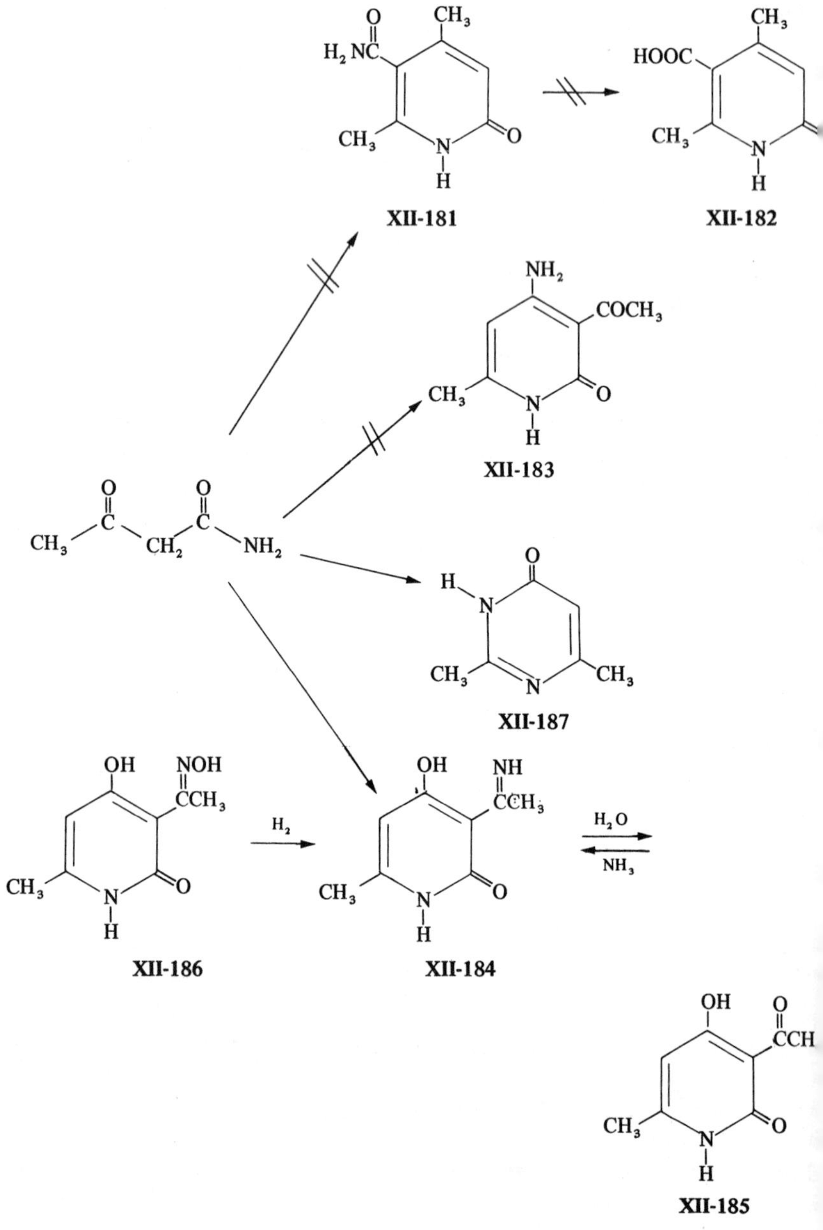

XII-181

XII-182

XII-183

XII-187

XII-186

XII-184

XII-185

nown 3-acetyl-4-hydroxy-6-methyl-2-pyridone (**XII-185**) on hydrolysis.[104] owever, easy conversion of **XII-185** to the original cyclization product with nmonia suggested the structure **XII-184** for the latter.[105] Product **XII-184** was so prepared by reduction of the corresponding oxime (**XII-186**). The main roduct of cyclization of acetoacetamide is 2,6-dimethyl-4-pyrimidone (**XII-187**, 2%). When acetoacetamide was heated under reduced pressure, the yield of **II-184** could be increased from 3% to 30–45%.[105] A small amount of **XII-185** formed during the formation of acetoacetamide from diketene and nmonia.[106] *N*-Alkylacetoacetamides also undergo auto-condensations when eated, but the yields of pyridones are low once more. Water formed appears to ydrolyze the reactant to acetone, carbon dioxide, and amine. The products ave been characterized as 5-(*N*-alkylcarbamoyl)-2-pyridones (**XII-188**; R = $CH_3$, $_2H_5$) because the 1-methyl homolog (**XII-188**, R = $CH_3$) was hydrolyzed and ecarboxylated to 1,4,6-trimethyl-2-pyridone.[107]

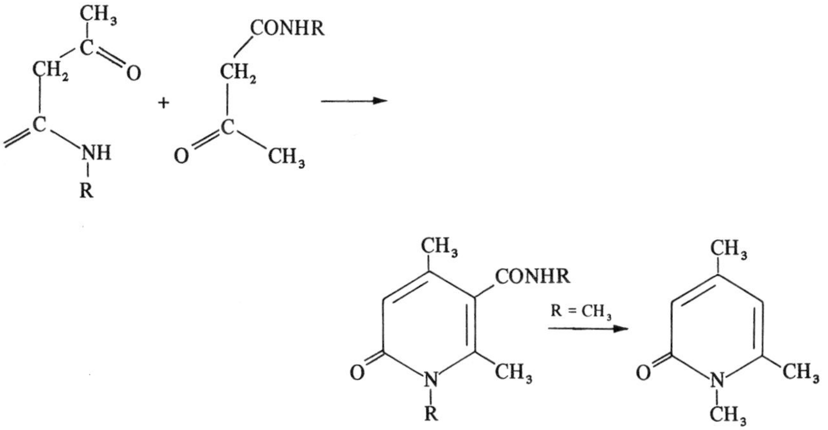

XII-188

*β*-Ketoamides substituted in the *α*-position cannot give products analogous to **II-188** but react to give 5,6-dialkyluracils.[107]

Diketene and glycine in a basic solution give 3-acetyl-1-carboxymethylene-4-droxy-6-methyl-2-pyridone (**XII-189**) by a reaction path in which hydroacetic acid is not an intermediate. However, *N*-acetoacetylglycine was t detected.[154] Deacylation of **XII-189** to **XII-190** occurs in sulfuric acid.

5-Aminotropolone and diketene give 3-acetyl-4-hydroxy-6-methyl-1-(5-tropo-nyl)-2-pyridone (**XII-192**) directly, in the presence of triethylamine. In the sence of triethylamine, 5-acetoacetamidotropolone (**XII-191**) could be olated.[109]

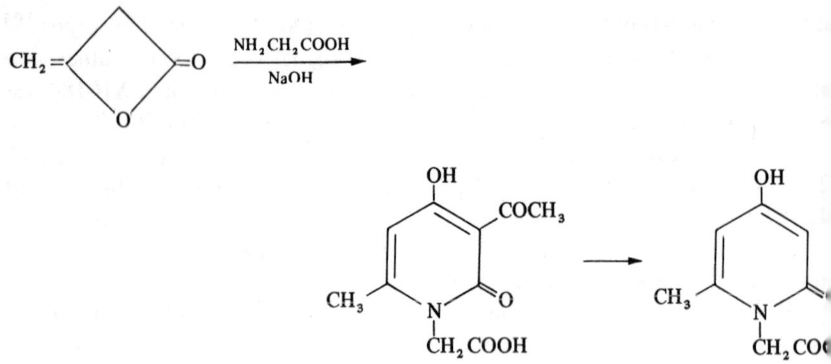

$CH_2$= (diketene) =O $\xrightarrow[NaOH]{NH_2CH_2COOH}$

XII-189

XII-190

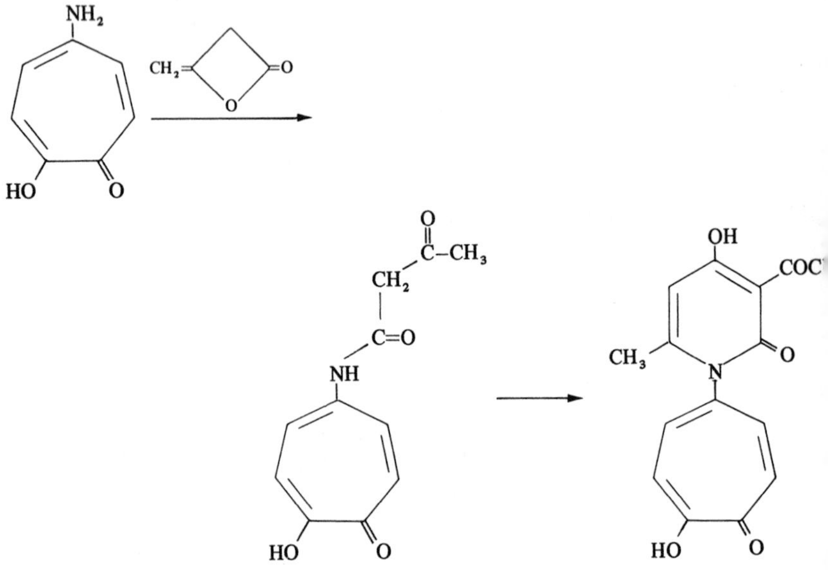

XII-191

XII-192

4-Aminotropolone and diketene form 3-acetyl-4-hydroxy-6-methyl-1-(trop lon-4-yl)-2-pyridone **(XII-193)** and N-(tropolon-4-yl)-2,6-dimethyl-4-oxo-4 pyran-3-carboxamide **(XII-194)**.[110]

A preparation of glutazine **(XII-196)** from ethyl cyanoacetate has be improved.[34] 3-Carbethoxyglutazine is readily prepared from ethyl α-cyano iminoglutarate **(XII-195)** and can be decarboxylated by heating with ammon to glutazine **(XII-196)** in an overall yield of 35%.[34]

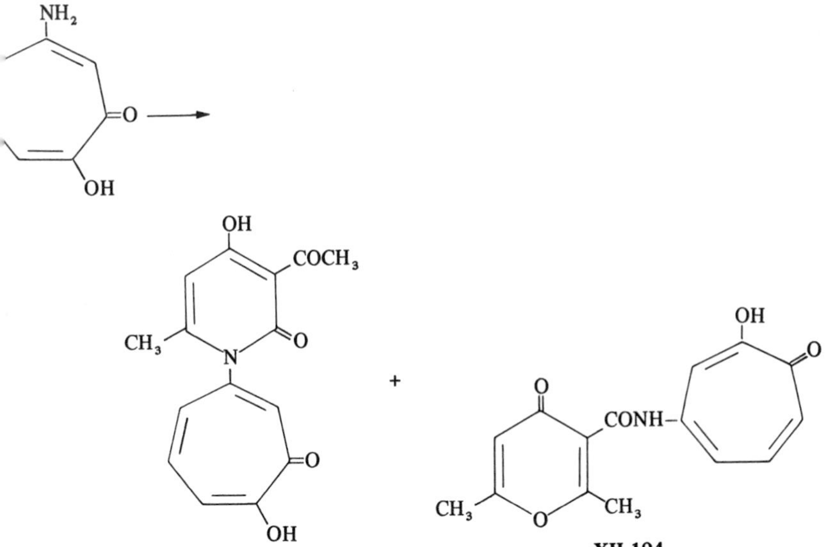

XII-193 + XII-194

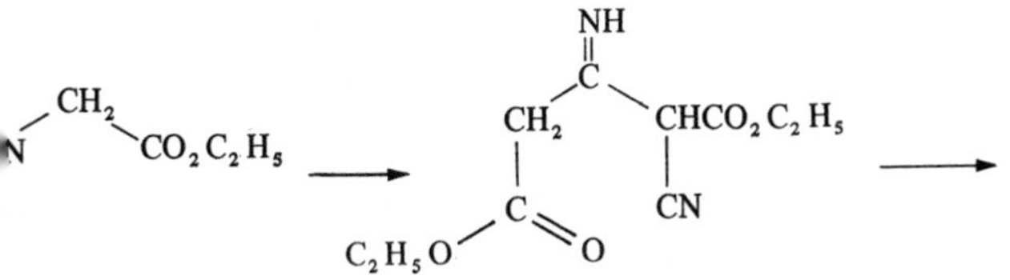

XII-195

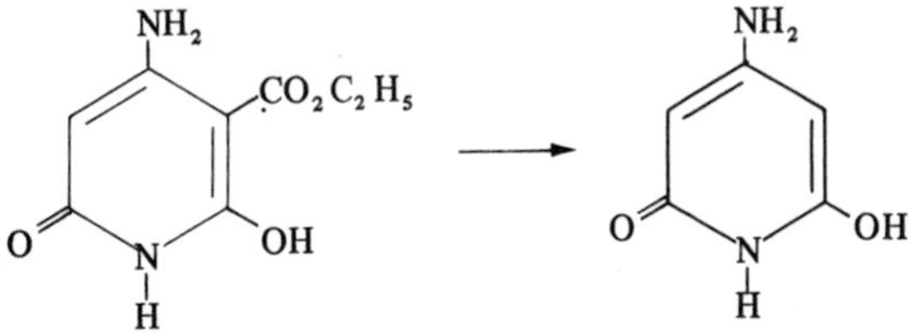

XII-196

## B. *Cyclizations of Three or More Acyclic Molecules*

The condensation of an aldehyde with ethyl cyanoacetate and ammonia to ive 4-alkyl-3,5-dicyano-6-hydroxy-2-pyridone **(XII-198)** probably proceeds rough the oxidation of an intermediate glutaric acid imide **(XII-197)**. Only nall amounts of the reduction products **(XII-199)** could be isolated. Additional athways for the oxidation of **XII-197** to **XII-198** were not excluded.[111]

s-Triazine and ethyl α-cyanoacetimidate give 3,5-dicyano-2,6-diethoxypyridi **(XII-200)**.[100]

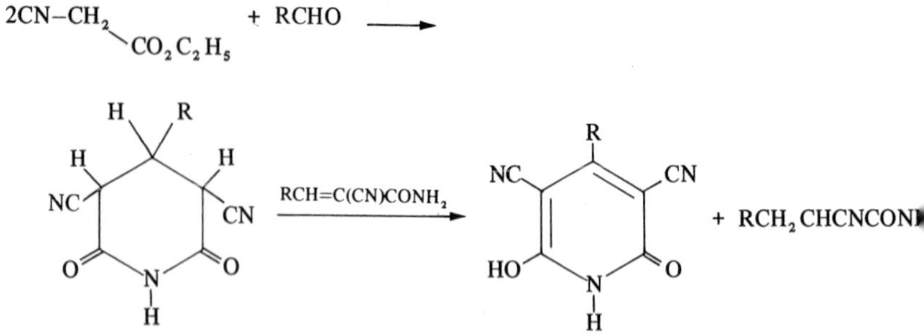

$2CN{-}CH_2{-}CO_2C_2H_5$ + RCHO ⟶

XII-197 $\xrightarrow{RCH=C(CN)CONH_2}$ XII-198 + $RCH_2CHCNCONH$ XII-199

s-Triazine + $2CNCH_2C(=NH)OC_2H_5$ ⟶ XII-200

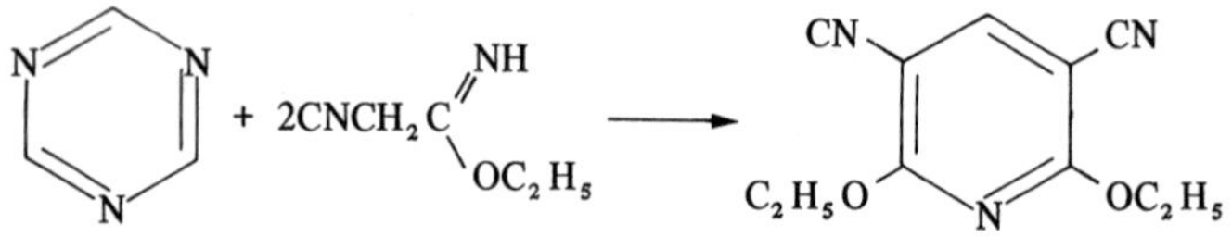

Treatment of α-unsubstituted-*N*-(2-pyridyl)acetoacetamides with trieth orthoformate and zinc chloride in ethanol gives 3-acetyl-6-methyl-1-(2-pyridyl 5-(*N*-2-pyridylcarboxamido)-2-pyridone **(XII-201)**.[112]

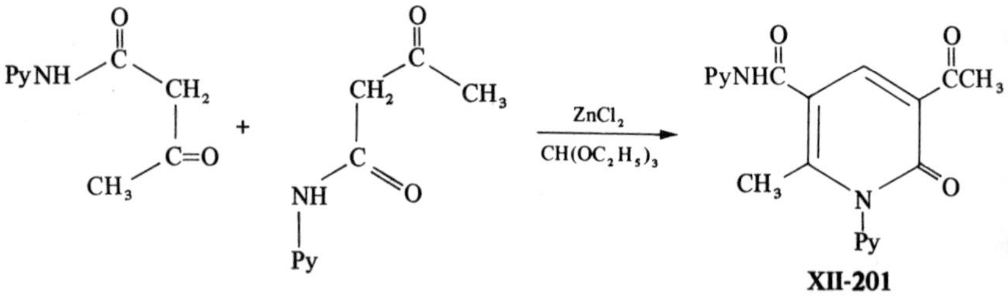

$PyNHCOCH_2COCH_3$ + $CH_3COCH_2CONHPy$ $\xrightarrow[CH(OC_2H_5)_3]{ZnCl_2}$ XII-201

Py = ; R = H, $CH_3$

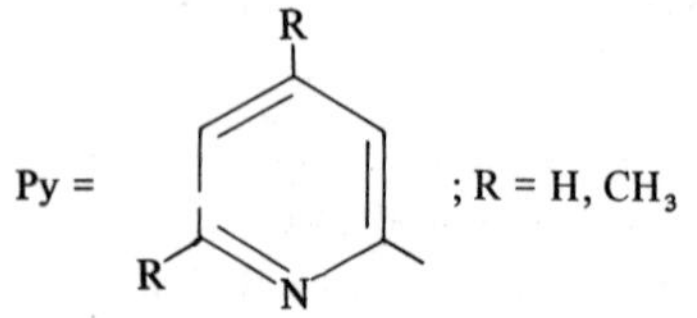

## 5. From Other Ring Compounds

### A. *2-Pyrones*

ımmonia, primary amines, and related acyclic nitrogen nucleophiles react :h a large variety of pyrones to form pyridones. These reactions often have ·n used to characterize the pyrones or to remove them from mixtures. Under ·se circumstances, yields are often not reported and experimental conditions not optimal, and, therefore, in many instances these reactions are difficult to ·luate, particularly as alternative routes to the somewhat more carefully died direct ring closures to form pyridones.

ılthough pyrones and pyridones are potentially interconvertible, the ıversion of pyridones to pyrones has been reported only occasionally. For ımple, 2,3,6-trimethyl-4-pyrone in aqueous methylamine gives 1,2,3,6-tetra- thyl-4-pyridone **(XII-202)** (59%, isolated). The pyridone is converted to the rone (31%, isolated) in dilute sulfuric acid containing mercuric sulfate.[113]

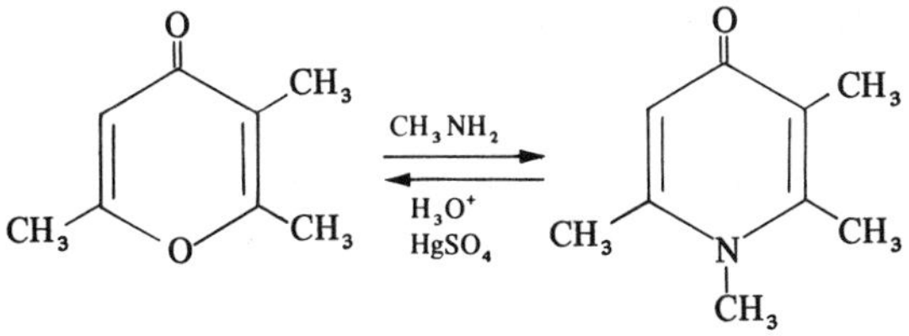

**XII-202**

ı wide range of experimental conditions has been used to effect these nsformations. For example, scillaridin A, a 17-β-(5-pyronyl)-steroid **(XII-203)** es not react with ammonia in methanol or with ammonium acetate in boiling ·tic acid and gives only a low yield of pyridone with ammonia in aqueous ·thanol at 120°. It gives the pyridone **(XII-204)** in 58% yield with ammonium ·tate and acetic acid in dimethylformamide at 175° in a sealed tube.[114]

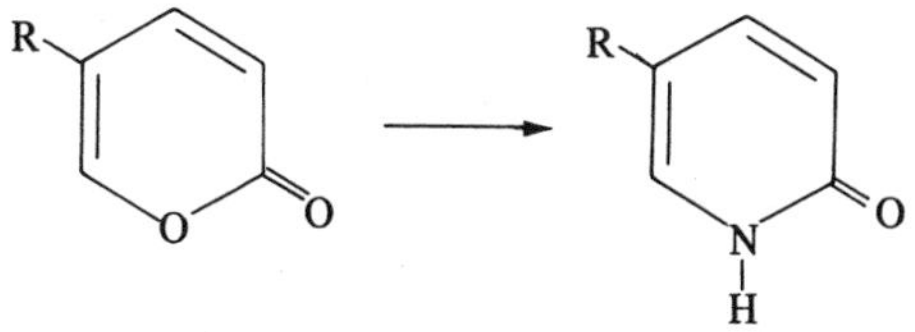

**XII-203** **XII-204**

Methyl coumalate reacts with amines to give Schiff's bases that can be cycli in aqueous sodium carbonate to *N*-substituted-2-pyridones [**XII-205**, R 2-thiazolyl, 2-pyridyl, 4-methyl-2-pyridyl, 2-(1,3,4-thiadiazolylmethyl)].[115] A

$CH_3OOC$ RNH$_2$ →

R–N=CH–C(=CH–CH$_2$COOH)(CO$_2$CH$_3$) $\xrightarrow[Na_2CO_3]{H_2O}$ COOH … N–R

**XII-205**

boiling under reflux for 4 hours, 2-pyrone-6-carboxylic acid and ammonium a tate in acetic acid give a 51% yield of 2-pyridone-6-carboxylic acid. 4-Methyl-6-phenyl-2-pyridone is prepared in 98% yield by heating the co sponding α-pyrone in methanolic ammonia at 100°.[117]

Butylamine and 2,6-dimethyl-4-pyrone (**XII-206**, X = O) and 2,6-dimethy thiopyrone (**XII-206**, X = S) give the corresponding 4-pyridone thiopyridone, respectively. Benzylamine, a weaker base and weaker nucleoph reacts with the thiopyrone but not with the pyrone.[118]

X, $CH_3$, O, $CH_3$ RNH$_2$ → X, $CH_3$, N–R, $CH_3$

**XII-206**

3,4,5-Trichloro-6-phenyl-2-pyridone (**XII-208**, R = H) can be prepared fr the corresponding α-pyrone by boiling it under reflux for 36 hours w ammonium acetate in acetic anhydride and acetic acid. It is, however, m conveniently prepared by cyclization of the amide **XII-207**.[20] 3,5-Dichlor pyridone (**XII-209**, R = H) is formed in low yield from the pyrone. pyridones **XII-209** (R = H, $CH_3$, $C_6H_5$) are, however, formed in good yield reduction of **XII-210** (see Section I.1.B., p. 604).[18]

3,5-Dichloro-6-phenyl-2-pyridone can be prepared from 6-phenyl-2-pyrone chlorination with sulfuryl chloride to give **XII-211**, which is then treated w ammonium acetate, or by chlorination of 6-phenyl-2-pyridone.[119] Perchloro-2 pentadienoyl chloride and the corresponding methyl ester (**XII-212**) are redu

$C_6H_5(CCl{=}CCl)_2CONHR$

**XII-207**

**XII-208**

$C_6H_5(CCl{=}CCl)_2COCl$

$C_6H_5(CCl{=}CCl)_2COOH$

**XII-209**

**XII-210**

**XII-211**

$SO_2Cl_2$

$NH_4OAc$

$SO_2Cl_2$

by lithium aluminum hydride at $-10°$ to perchloro-2,4-pentadien-1-Cyclization in the presence of manganese dioxide or chromic acid in *t*-bu alcohol gives 3,4,5-trichloro-2-pyrone, which is converted to the 2-pyrido **(XII-213)** with ammonium acetate in acetic acid[15] (see Section I.1.B., p. 60 3,4,5-Trichloro-2-pyridone **(XII-213)** can be formed also from perchloro-2 pentadienal by heating in acetic acid.[17]

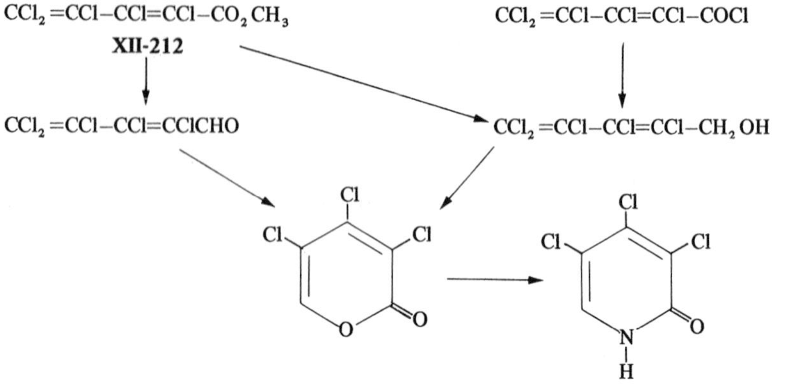

$CCl_2{=}CCl{-}CCl{=}CCl{-}CO_2CH_3$ **XII-212**

$CCl_2{=}CCl{-}CCl{=}CCl{-}COCl$

$CCl_2{=}CCl{-}CCl{=}CClCHO$

$CCl_2{=}CCl{-}CCl{=}CCl{-}CH_2OH$

**XII-213**

ω-Methoxy- and ω-aryloxyacetophenones and benzyl phenyl keto condense with ethyl phenylpropiolate to give 5-methoxy-[120] and 5-aryloxy-4 diphenyl-2-pyrones,[121] and 4,5,6-triaryl-2-pyrones.[122] Ammonia and meth amine give the corresponding 2-pyridones **(XII-214)**. *N*-Methyl-2-thiopyrido **(XII-215)** are formed from 2-thiopyrones and methylamine.[122]

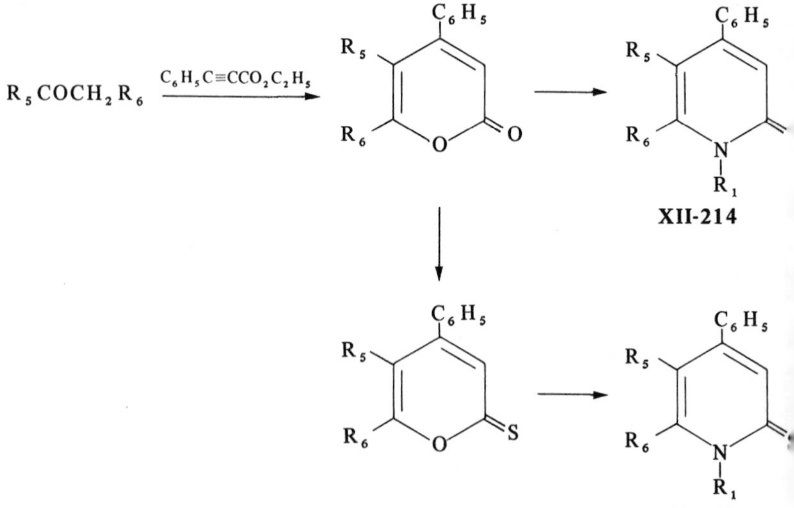

$R_5COCH_2R_6 \xrightarrow{C_6H_5C\equiv CCO_2C_2H_5}$

**XII-214**

**XII-215**

## B. *4-Pyrones*

laltol **(XII-216)**, allomaltol **(XII-217)**, kojic acid **(XII-218)**, chelidonic acid **I-219)**, meconic acid **(XII-220)**, and comenic acid **(XII-221**, R = H) are mples of well-known natural products and their derivatives that have been

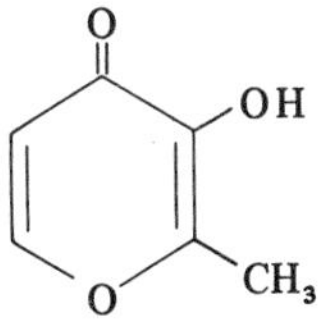

**XII-216** 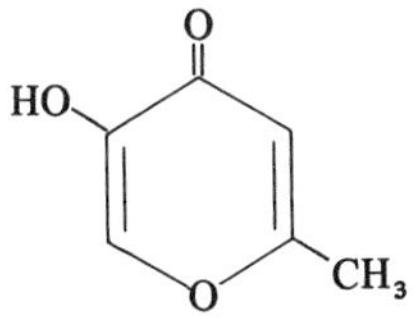 **XII-217** 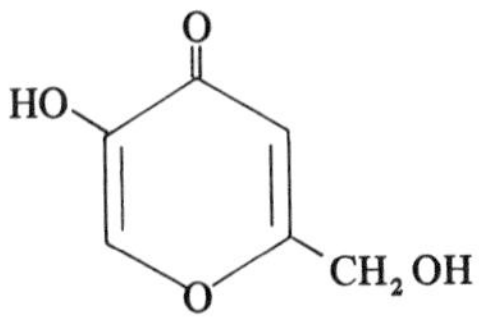 **XII-218**

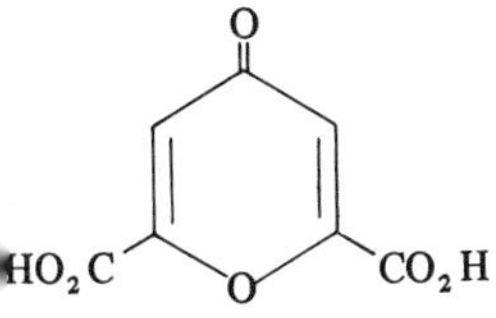

**XII-219** 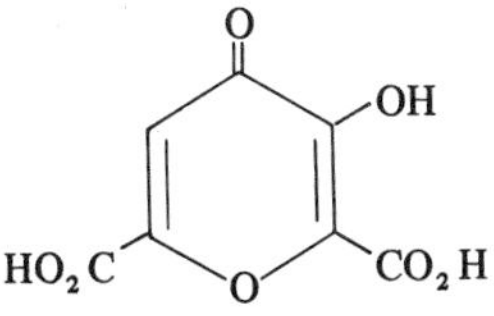 **XII-220** 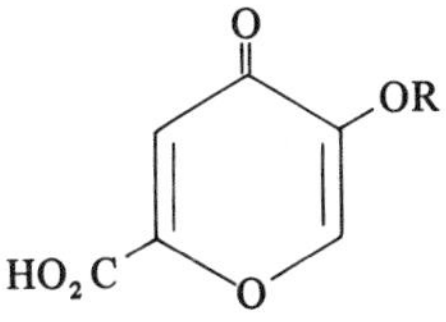 **XII-221**

verted to 4-pyridones. Many of these conversions have been instrumental in cidating structures of naturally occurring furanols and pyrones.[123]

somaltol, first described as a pyrone, has been shown to be 3-hydroxy-2-furyl thyl ketone **(XII-222)**.[124, 125] This structure has been confirmed by its version to 3-hydroxy-2-methyl-4-pyridone,[125] which has been synthesized m maltol **(XII-216)**.[125, 126]

ojic acid **(XII-218)** is acylated in the 2-position by butyrolactone in luoroacetic acid. The product and aqueous ammonia give 3-hydroxy-2-(γ-lroxybutyryl)-6-hydroxymethyl-4-pyridone.[127]

reatment of **XII-220** or **XII-221** (R = H, $CH_3$) with aromatic amines in water es 1-aryl-5-hydroxy(or methoxy)-4-pyridone-2-carboxylic acids, which are dily decarboxylated to **XII-223**, (R = H, $CH_3$) or converted to ethyl ers.[128, 129] 2-Hydroxymethyl-5-methoxy-4-pyrone, a monomethyl ether of ic acid, and 2-dialkylaminoethylamines in water give 1-(2-dialkylaminoethyl)-ydroxymethyl-5-methoxy-4-pyridones (**XII-224**; R = $CH_3$, $C_2H_5$).[130] Natural ltol glucoside and ammonia or primary amines give the corresponding idones **(XII-225)** when boiled in methanol.[131]

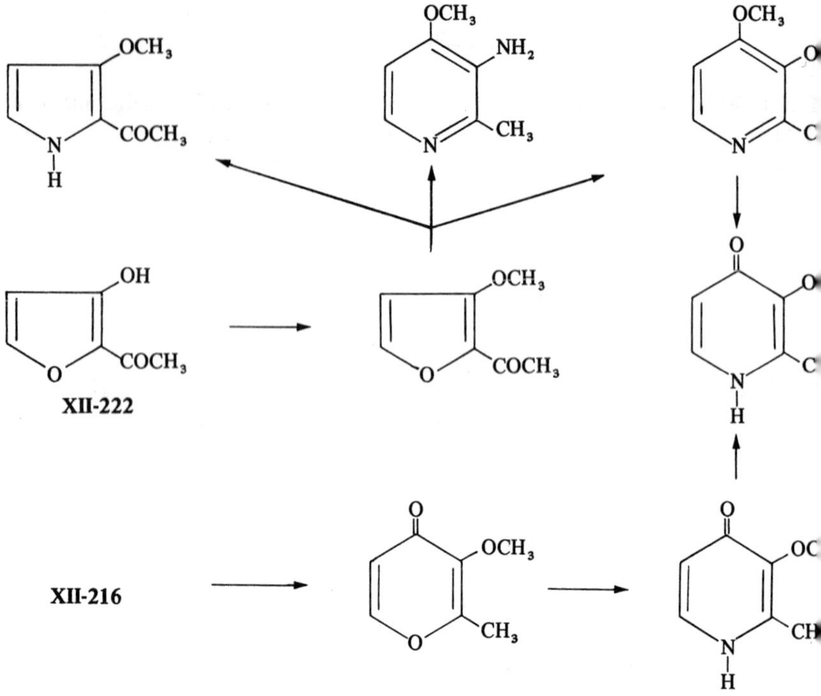

XII-222

XII-216

XII-218

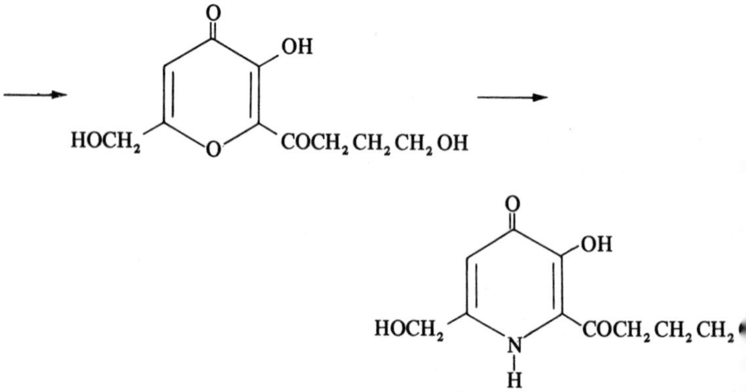

XII-220

XII-221

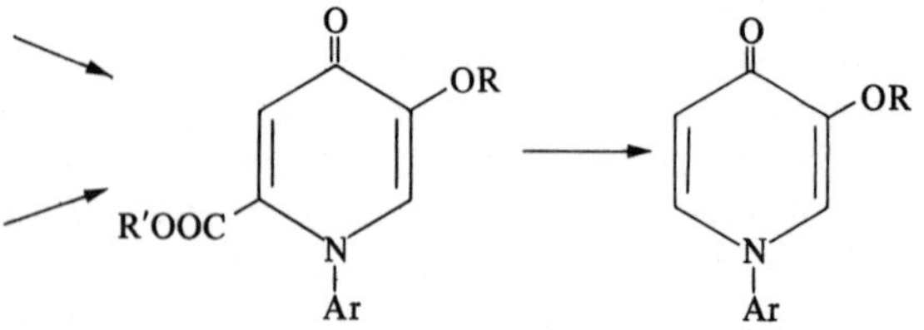

XII-223

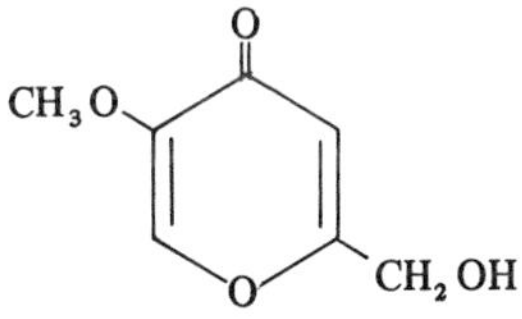

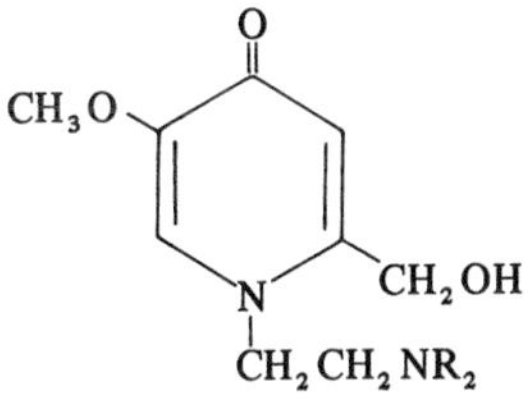

$CH_3O$ $CH_2OH$ $CH_3O$ $CH_2OH$ $CH_2CH_2NR_2$

**XII-224**

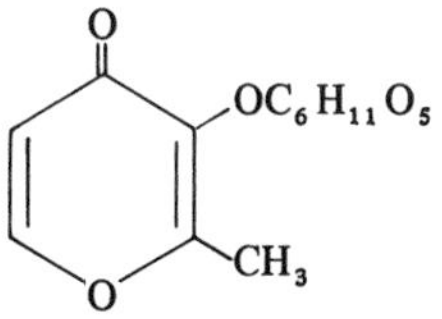

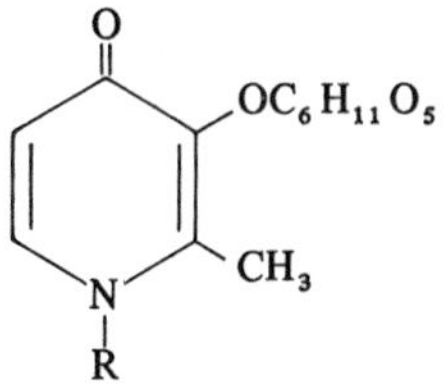

$OC_6H_{11}O_5$ $CH_3$ $OC_6H_{11}O_5$ $CH_3$ R

**XII-225**

γ-Pyrones and cyanamide in aqueous ethanol give *N*-cyano-4-pyridones (XII-226); however, reactions of **XII-226** with acids or bases have not given characterizable products.[132]

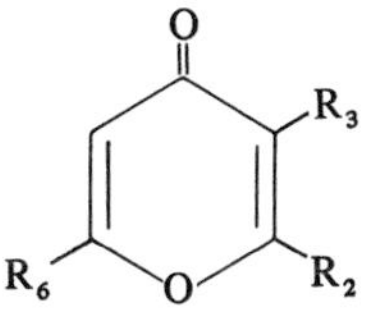

$R_3$ $R_6$ $R_2$

$R_2$ = H, $CH_3$; $R_3$ = H, OH, $OCH_3$; $R_6$ = $CH_2OH$, $CH_2Cl$, $CH_3$

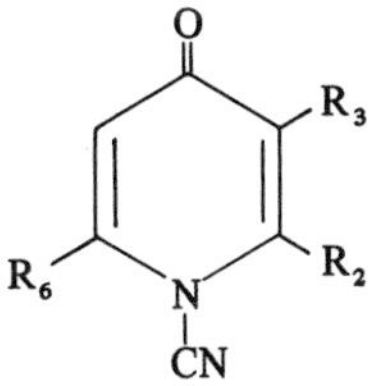

$R_3$ $R_6$ $R_2$ CN

**XII-226**

The naturally occurring amino acid mimosine (leucaenine; **XII-229**) has been synthesized from meconic acid **(XII-220)**, which is decarboxylated and then converted to the benzyl ether **(XII-227)**. Mimosine is obtained *via* **XII-228.**[133] 3-Hydroxy-4-pyridone was formed by treatment of mimosine with zinc dust or by hydrogenolysis of 3-benzyloxy-4-pyridone in a procedure designed to locate the label from aspartic acid incorporation into 4-pyridone in *Mimosa pudica*.[134] Although the 4-pyridones **XII-230** can be prepared from the pyrones, it is more convenient to obtain them directly from triketones by treatment with ammonia in ethanol[25] (see Section I.1.C., p. 608). 1-Phenyl-1,3,5-hexanetrione is acetylated and cyclized by boron trifluoride and acetic anhydride to the pyrone **(XII-231)**. Conversion to 3-acetyl-2-methyl-6-phenyl-4-pyridone in low yield is accomplished in anhydrous ammonia.[135] 1,5-Diphenyl-1,3,5-pentanetrione gives 3-benzoyl-2-methyl-6-phenyl-4-pyrone **(XII-232)**, which reacts with

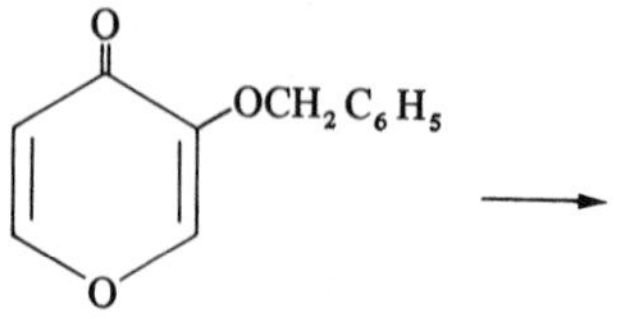

XII-227

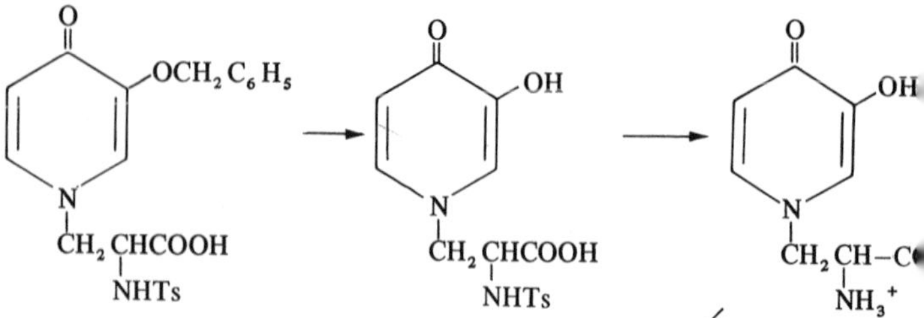

XII-228

XII-229

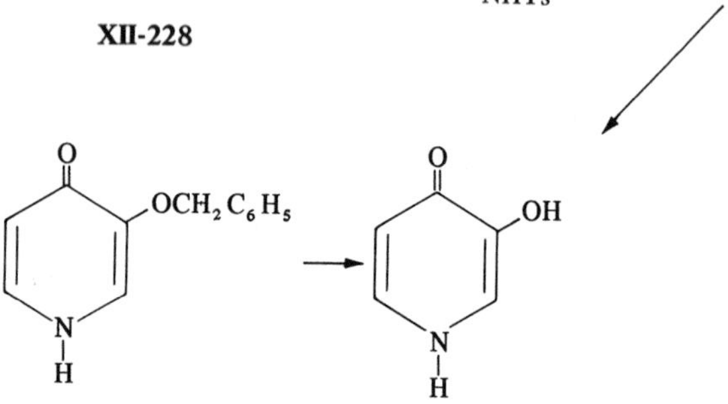

liquid ammonia to form the pyridone **XII-233** in 86% yield. Products isome with **XII-232** and **XII-233**, such as **XII-234** (X = O, NH), were not found.[1] β-Ketoesters react with acyl and aroyl phosphorus ylids to give the sa 4-pyrones. For example, ethyl α-acetylacetoacetate and benzoylmethylenet phenylphosphorane give **XII-231**.[136] Additional conversions of 4-pyrones 4-pyridones are summarized in Table XII-2.

Lutidones rather than 4-hydroxy-2-pyridones are formed from dehydroace acid **(XII-235)** and ammonia or primary amines.[142] With one equivalent aqueous or ethanolic primary amine at ambient temperatures, dehydroace acid gives 3-(α-alkylaminoethylidene)-6-methylpyran-2,4-diones **(XII-236)**. excess of amine (R = Me, Et, *n*-Pr, *n*-Bu, $CH_2C_6H_5$, $CH_2CH_2C_6H_5$) a dehydroacetic acid give 2,6-bisalkylaminohepta-2,5-dien-4-ones **(XII-237)** or t lutidones **(XII-238)**, depending on conditions.[108, 143-148] Aqueous ammo and dehydroacetic acid give some 3-carboxylutidone **XII-239**, which is n decarboxylated under the reaction conditions and therefore is not an intermedi in the formation of lutidone under these conditions.[147, 148] These dieno (l-237) in which the two alkyl groups are different were not isolated on tment of **XII-236** (R = Et) with methylamine.[108] Aqueous dimethylformde has been used as the solvent to give **XII-236**.[145] Compounds **XII-236** ld not be converted directly to the lutidones **XII-238**,[143] which indicates the *N*-methyllutidone is not formed by rearrangement of **XII-236** through intramolecular nucleophilic attack by the nitrogen on C-6.[143] However, -236 (R = $CH_2CH_2C_6H_5$, $CH(CH_3)CH_2C_6H_5$) in dilute mineral acid does give -238.[145] Under appropriate conditions diacetylacetone and amines give the

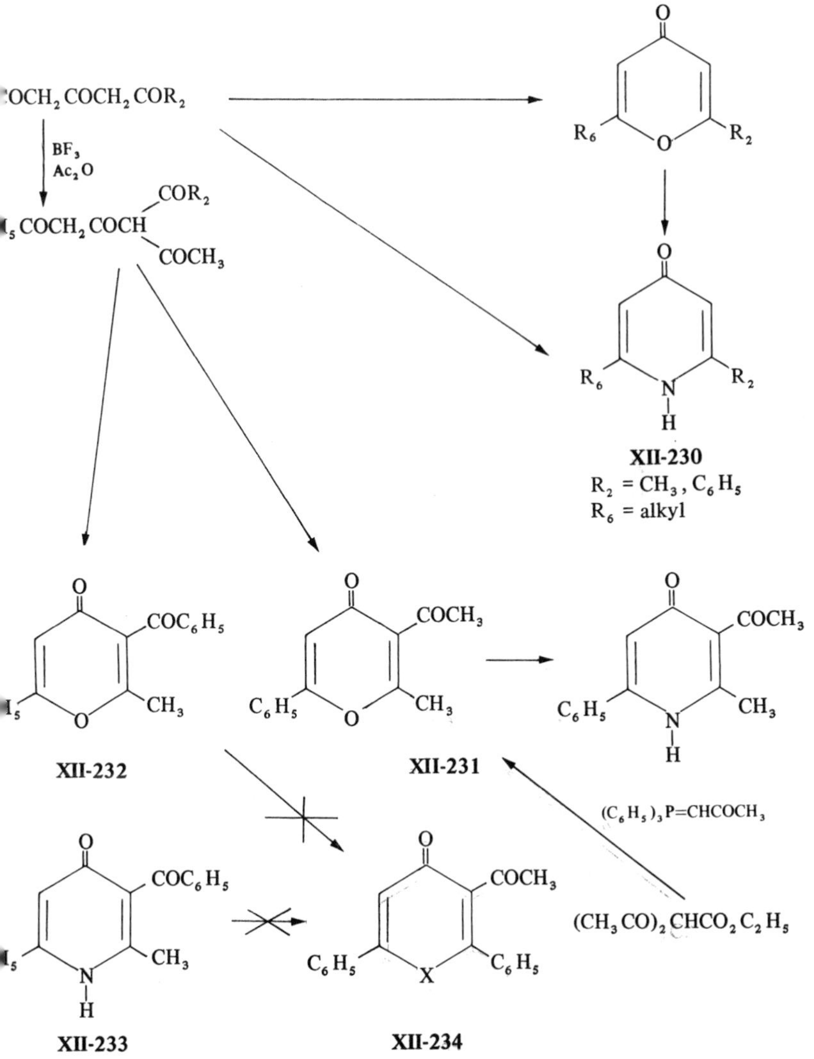

TABLE XII-2. Conversion of 4-Pyrones to 4-Pyridones

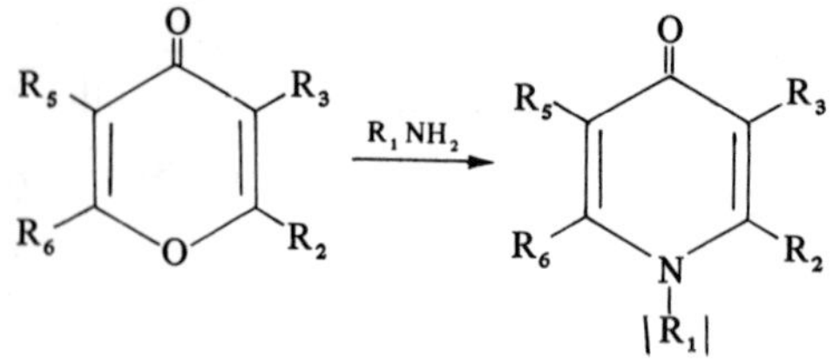

$R_1NH_2$

| $R_1$ | $R_2$ | $R_3$ | $R_5$ | $R_6$ | % Yield |
|---|---|---|---|---|---|
| H | $CH_3$ | H | H | $CH_3$ | 65 |
| H | $CH_3$ | H | H | 2-quinolylmethyl | 80 |
| $CH_3$ | H | $CH_3$ | $CH_3$ | H | 74 |
| $CH_3$ | $CH_3$ | H | H | $CH_3$ | ~100 |
| H | $CH_3$ | $COCH_3$ | H | $CH_3$ | – |

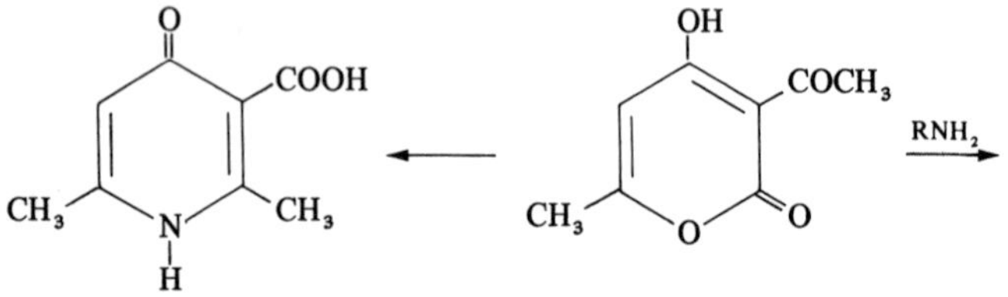

**XII-239** **XII-235** ($RNH_2$)

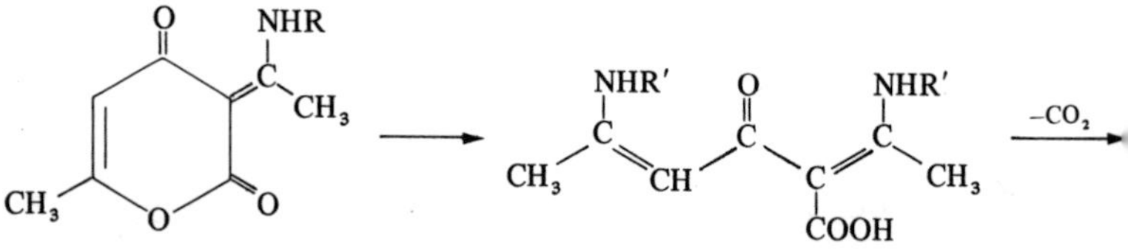

**XII-236** ($-CO_2$)

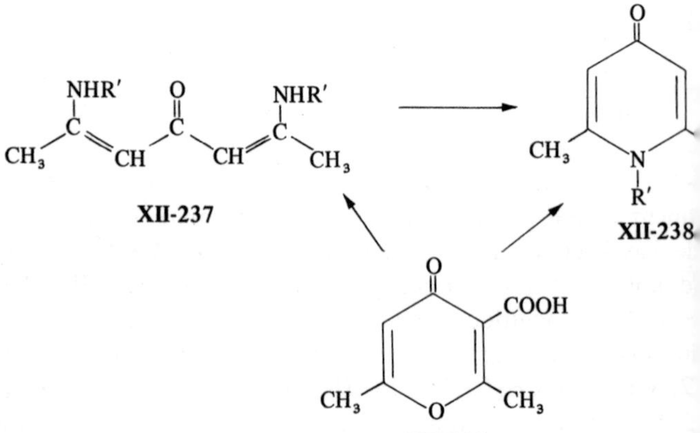

**XII-237** **XII-238**

**XII-240**

enone **XII-237** or the lutidone **XII-238**.[143, 144] 3-Carboxy-2,6-dimethyl-4-rone (**XII-240**) and amines also give **XII-237** or **XII-238**.[149] Ethanolamine and hydroacetic acid give **XII-236** and **XII-238** (R′ = $CH_2CH_2OH$), but not I-237.[147, 148] *N*-Phenyllutidone (**XII-238**, R = $C_6H_5$) is prepared from hydroacetic acid and aniline by boiling in concentrated hydrochloric acid for hours.[147]

## C. *4-Hydroxy-2-pyridones from Pyrones*

Ethyl *n*-butyroacetate, on treatment with sodium bicarbonate, gives the hydroacetic acid homolog **XII-241**, which, on treatment with 90% $H_2SO_4$, is nverted to 4-hydroxy-6-propyl-2-pyrone, which gives 4-hydroxy-6-*n*-propyl-2-ridone in low yield when treated with ammonia.[150]

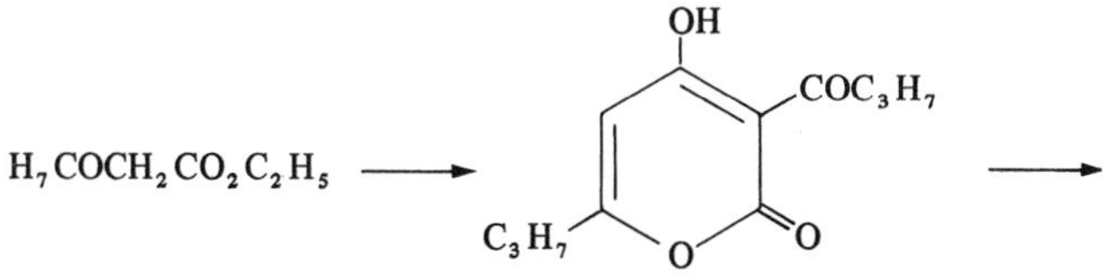

$H_7COCH_2CO_2C_2H_5$ ⟶ **XII-241** ⟶

**XII-241**

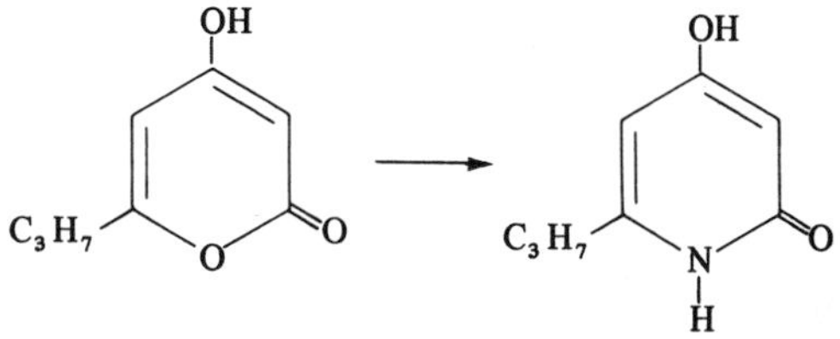

Dehydroacetic acid oxime rearranges in polyphosphoric acid to 3-acetamido-4-droxy-6-methyl-(2*H*)-2-pyrone and 2,6-dimethyl-(4*H*)-pyrano[3,4-*d*] oxazol-4-e (**XII-242**). Treatment of **XII-242** with ammonia gives 2,6-dimethyl-(5*H*)-azolo[4,5-*c*]-4-pyridone, which, in turn, gives 3-acetamido-4-hydroxy-6-ethyl-2-pyridone with hydrochloric acid at room temperature and amino-4-hydroxy-6-methyl-2-pyridone on heating.[151]

Reactions between triacetic acid lactone (**XII-243**) and ammonia or amines to e 4-hydroxy-6-methyl-2-pyridones are well known.[152, 153] For example, cine and **XII-243** give 1-carboxymethylene-4-hydroxy-6-methyl-2-pyri-ne.[154] This latter 4-hydroxy-2-pyridone has also been formed from diketene d glycine in aqueous base *via* **XII-244**, which can be deacetylated in ncentrated sulfuric acid. Dehydroacetic acid does not appear to be an ermediate in the formation of **XII-244**. Under these conditions it reacted with cine to form an isomeric product, which, however, was not characterized.[154]

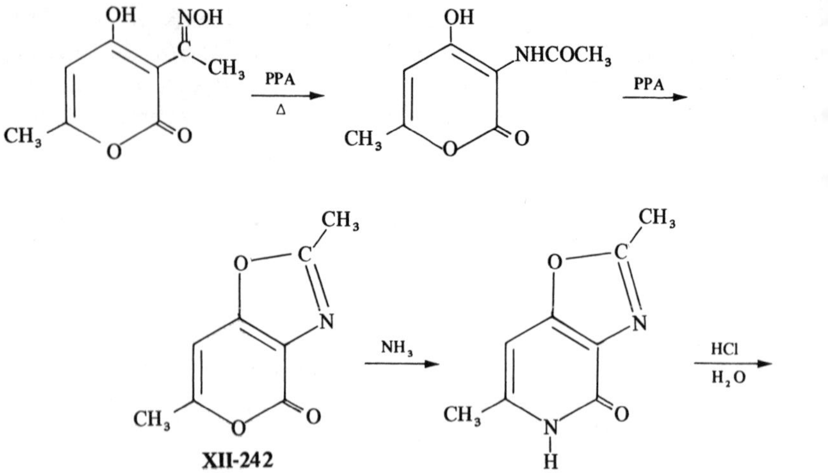

XII-242

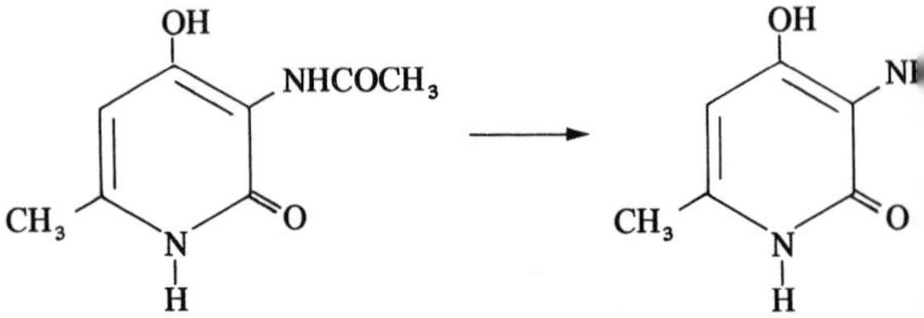

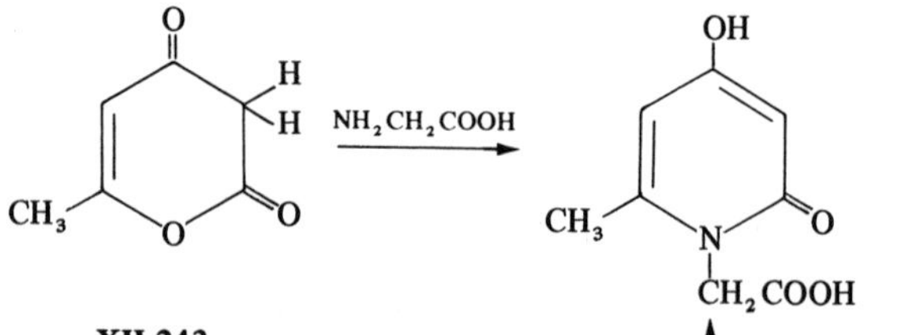

XII-243

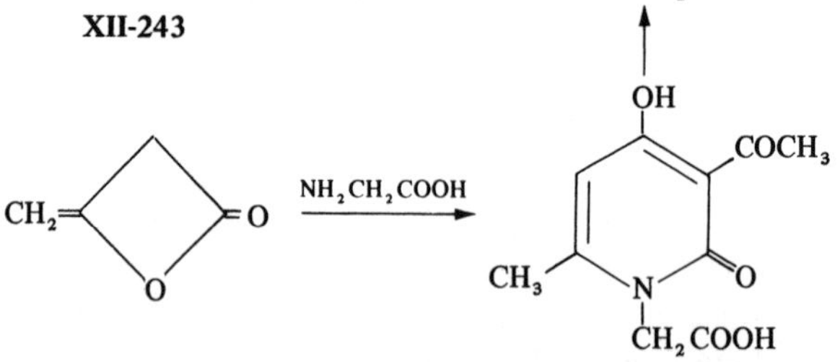

XII-244

On heating with polyphosphoric acid *o*-haloacetoacetanilides give mixtures 4-hydroxyquinaldines and 3-arylcarbamyl-2,6-dimethyl-4-pyrones (**XII-245**; R 2-$ClC_6H_4$, 2,4-$Cl_2C_6H_3$, 2,5-$Cl_2C_6H_3$, 2-$BrC_6H_4$). The 4-pyrones (**XII-245**) a drolyzed in 70% sulfuric acid to give mixtures of the 4-pyrone-3-carboxylic ds **(XII-246)** and the *N*-aryl-4-hydroxy-6-methyl-2-pyridones **(XII-247)**.[149]

A compound previously described as 3-benzoyl-6-phenyl-2,4-(1*H*,3*H*) pyridine-one **(XII-248)**[155] and earlier as 2,6-diphenyl-4-pyridone,[156] which is formed

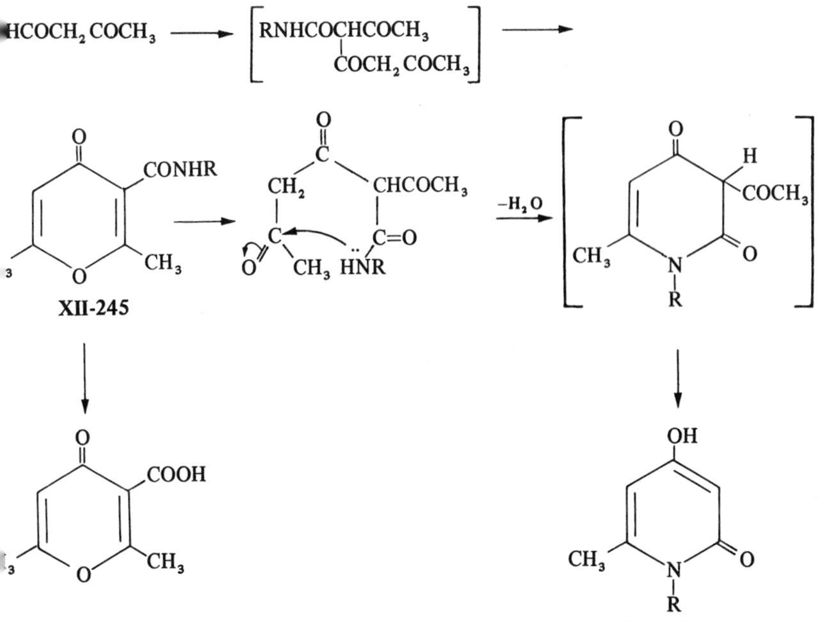

HCOCH$_2$COCH$_3$ → [RNHCOCHCOCH$_3$ / COCH$_2$COCH$_3$] →

CONHR, CH$_3$, **XII-245** → CH$_2$, CHCOCH$_3$, C=O, CH$_3$ HNR → $-H_2O$ → [H, COCH$_3$, CH$_3$, N–R]

**XII-246** (COOH, CH$_3$) **XII-247** (OH, CH$_3$, N–R)

om 3-benzoyl-6-phenylpyran-2,4-(3*H*)-dione and ammonia, has been shown to 3-α-aminobenzylidene-6-phenyl-2*H*-pyran-2,4-(3*H*)dione **(XII-249)**[143] (see o **XII-236**, p. 653).

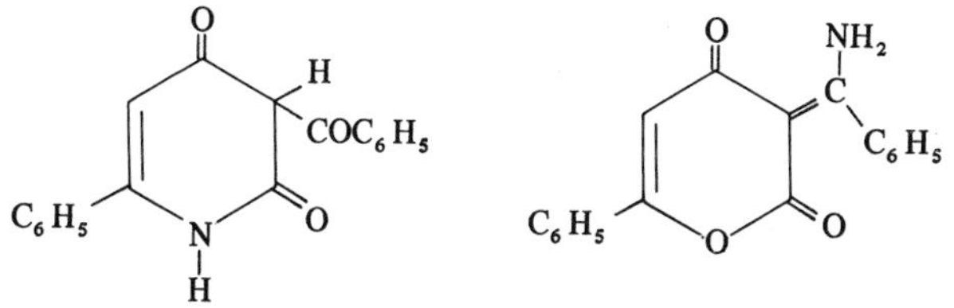

**XII-248** (H, COC$_6$H$_5$, C$_6$H$_5$, N–H) **XII-249** (NH$_2$, C, C$_6$H$_5$, C$_6$H$_5$)

Although ketones, including ethyl acetoacetate and several related compounds, act with malonyl chloride to form chloropyranodioxins (this section, p. 661), -diketones give pyrones. It was proposed originally that benzoylacetone ms the 4-hydroxy-2-pyrone that tautomerizes to the 2-hydroxy-4-pyrone

**(XII-250).**[94] It was shown later, however, that the reaction gives a mixture the isomers **XII-251** (60%) and **XII-252** (40%) and that both g 5-acetyl-4-hydroxy-6-phenyl-2-pyridone on treatment with aqueous ammonia

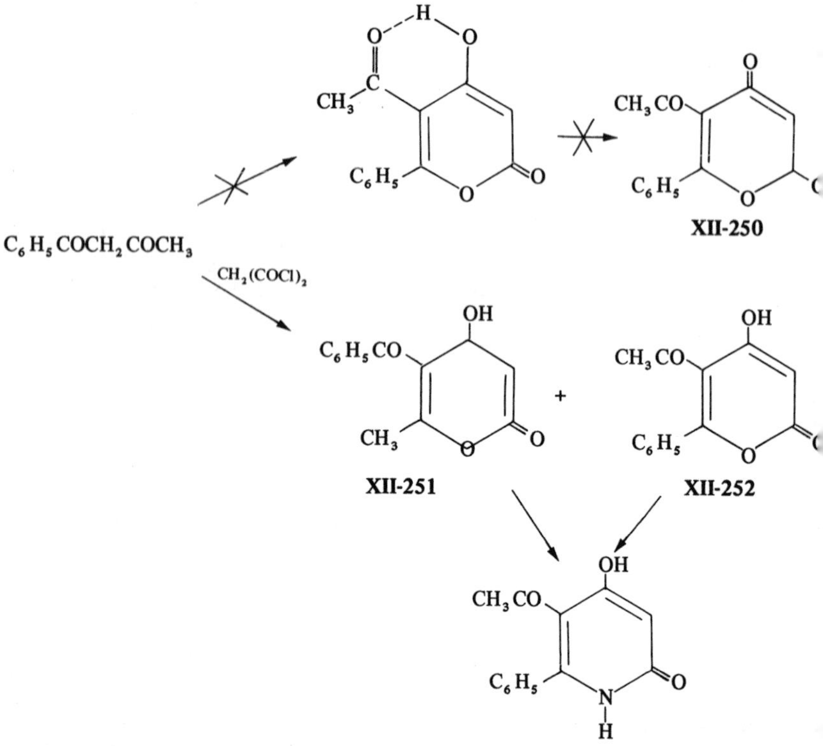

$C_6H_5COCH_2COCH_3$ $\xrightarrow{CH_2(COCl)_2}$

**XII-250**

**XII-251** + **XII-252**

Acetylacetone and malonyl chloride give 5-acetyl-4-hydroxy-6-methyl pyrone **(XII-253)**, a positional isomer of dehydroacetic acid, which reacts wi aqueous ammonia or aqueous methylamine to form **XII-254** (R = H, $CH_3$), which can also be prepared from the enamine of acetylacetone and carb suboxide[87] (see Section I.3.E., p. 635). The *N*-methylpyridone **XII-254** (R $CH_3$) is deacetylated to 4-hydroxy-1,6-dimethyl-2-pyridone with sulfuric ac however, **XII-254** (R = H) was not deacetylated under these conditions. Acetylacetone and carbon suboxide give 8-acetyl-4-hydroxy-7-methylpyra [4,3-*b*] pyrane-2,5-dione **(XII-255)** with catalytic amounts of sulfuric acid. T is also formed from acetylacetone and malonyl chloride or from carb suboxide and the intermediate 5-acetyl-4-hydroxy-6-methyl-2-pyro **(XII-253)**.[157] 3,5-Diacetyl-4-hydroxy-6-methyl-2-pyrone **(XII-256)** can formed by acetylation of **XII-253** with acetic acid and phosphorus oxychlori or by degradation of **XII-255**. 3,5-Diacetyl-4-hydroxy-6-methyl-2-pyridone c prepared quantitatively from **XII-256** and ammonia or by acetylation of **I-254.**[157]

A product of the tropolone-producing mold *Penicillium stipitatum* was entified as 3,6-dimethyl-4-hydroxy-2-pyrone by its conversion to 3,6-dimethyl-hydroxy-2-pyridone **(XII-257)** with ammonia in a sealed tube at 120° and to e *N*-methyl-2-pyridone with aqueous methylamine.[158, 159]

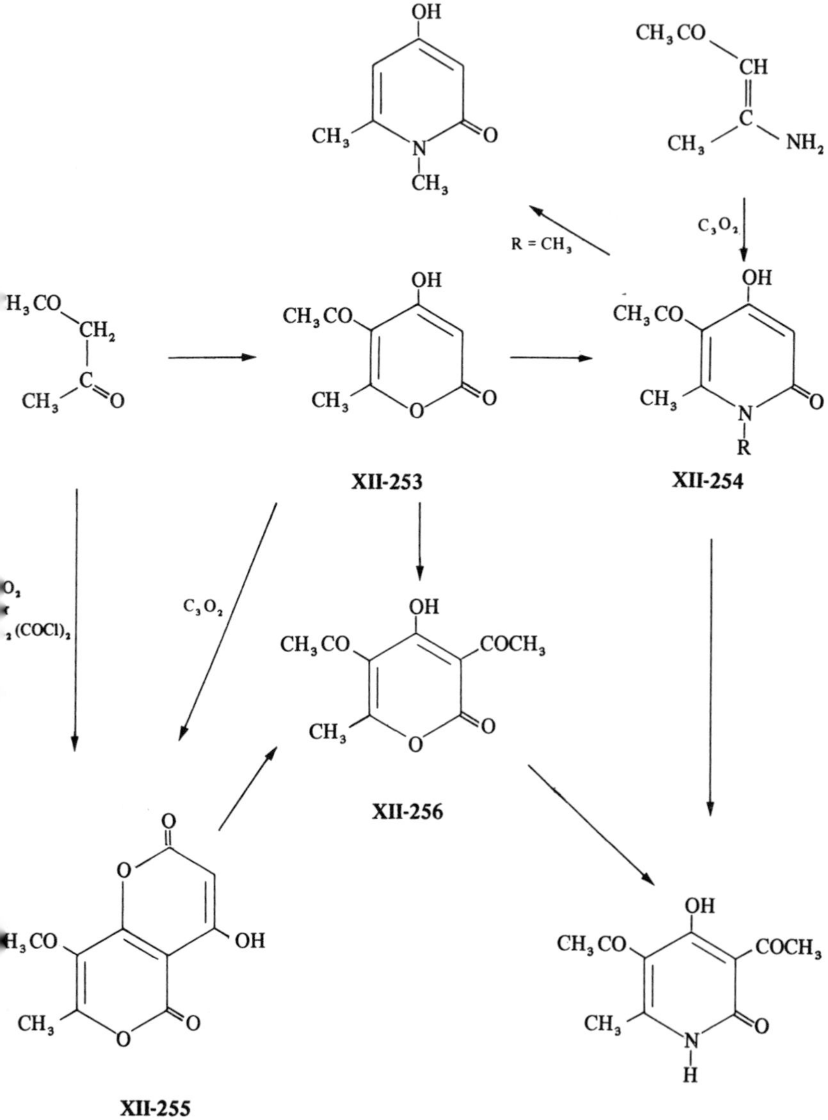

$CH_3CO$ $CH$ $C$ $CH_3$ $NH_2$

$C_3O_2$

R = $CH_3$

$H_3CO$ $CH_2$ $C$ $CH_3$ O

**XII-253**

**XII-254**

$O_2$

$_2(COCl)_2$

$C_3O_2$

**XII-256**

**XII-255**

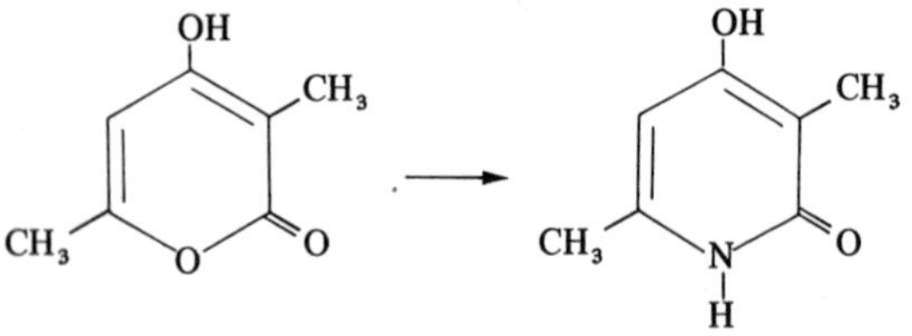

XII-257

Treatment of 2-bromo-6-hydroxymethyl-3-methoxy-4-pyrone with aque ammonia or aqueous methylamine gives 4-hydroxy-2-pyridones.[160] 6-Chloro hydroxy-2-pyrone-3-carboxylic acid and *m*-anisidine give *N,N*-di-*m*-metho phenylacetonedicarboxamide, which can be converted to 4-hydroxy-*N*-( methoxyphenyl)-6-(*m*-methoxyanilino)-2-pyridone **(XII-258)** by heating w phosphorus oxychloride.[161]

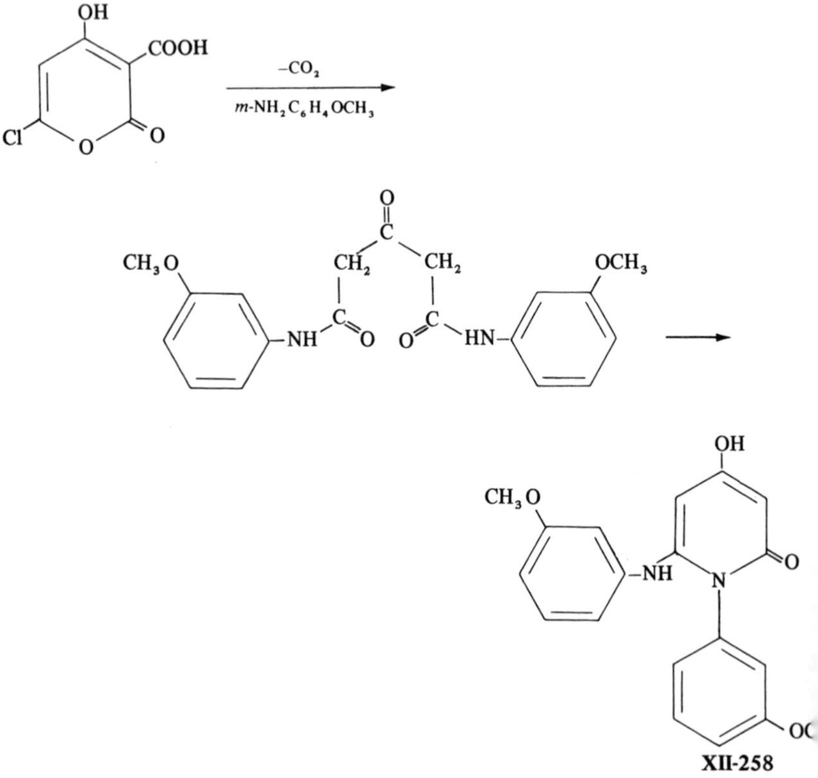

$\xrightarrow[m\text{-}NH_2C_6H_4OCH_3]{-CO_2}$

XII-258

Ammonia or ammonium acetate in acetic acid and aureothin, desmethyli aureothin (**XII-259**, R = H), and isoaureothin (**XII-259**, R = $CH_3$) g trogen-containing products. Only the product from **XII-259** (R = H) could, owever, be identified as the corresponding pyridone.[162]

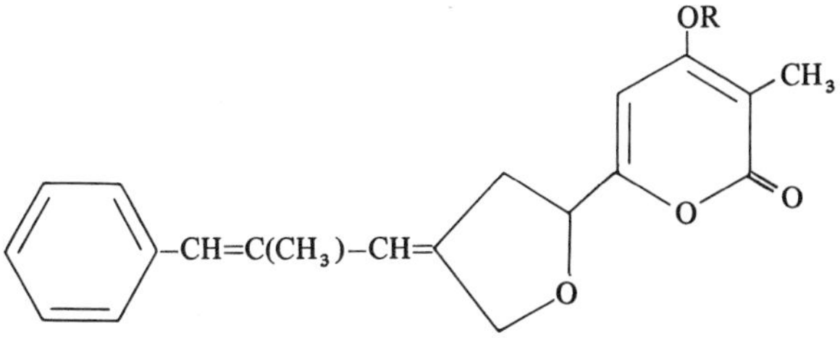

**XII-259**

1-Thiopyran-2-ones **(XII-260, XII-261)**, as well as the corresponding pyrones, ve pyridones when treated with ammonia or amines. Under the conditions used 8 hr at 120°) decarboxylation at the 5-position occurs.[163]

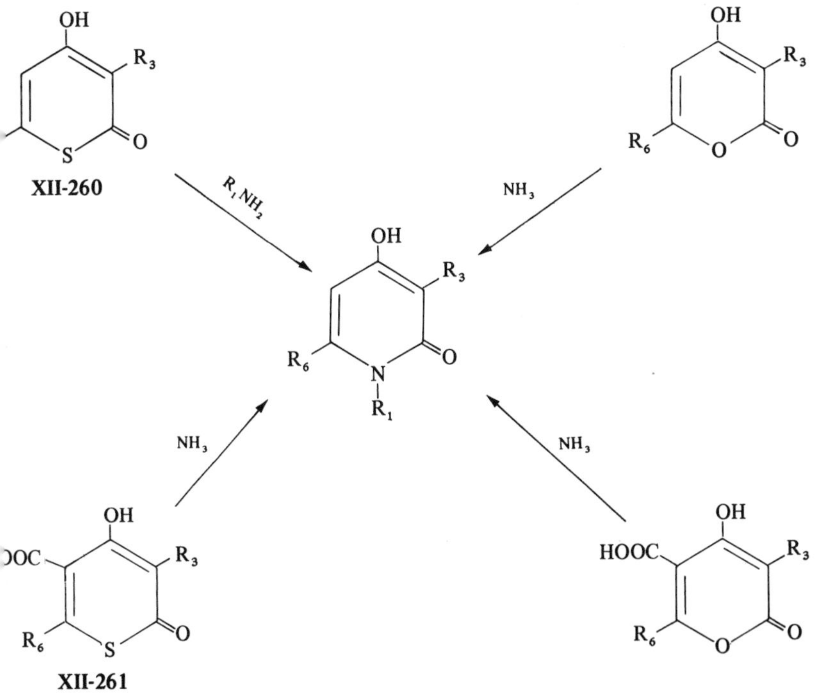

**XII-260**

**XII-261**

7-Chlorodioxopyrano-1,2-dioxins **(XII-262)**, prepared from malonyl chloride d ketones,[164–166] are converted to 7-substituted aminodioxopyrano-1,3-di- ns when treated with amines.[164, 167] The 7-amino derivatives isomerize in the presence of alkoxides to 6-aryl-2,2-dimethyl-7-hydroxy-4,5-dioxopyrido[4,3-*d*] 1,3-dioxins **(XII-263)**, which are converted to 3-carbomethoxy-4,6-dihydroxy- pyridones **(XII-264)**.[167-169] The intermediate pyridodioxins **(XII-263)** a isolated when sodium phenoxide is used.[170] The 7-aminopyranodioxin (R′ $C_6H_5$) and phosphorus oxychloride form the 7-chloropyridinodioxin, whi gives 4,6-dialkoxy-3-carboxy-*N*-phenyl-2-pyridones (**XII-265**; R = $CH_3$, $C_2H$ when treated with alkoxide ion.[167]

$CH_2(COCl)_2$ → ; $R_2CO$ → **XII-262** ; $R'NH_2$ → ; $C_6H_5O^-$ → **XII-263** ; $CH_3O^-$ → **XII-264** ; $POCl_3$ ($R' = C_6H_5$) → → **XII-265**

2,2-Dimethyl-7-phenylamino-4,5-dioxopyrano[4,3-*d*]-1,3-dioxin reacts w aniline to form a 1,3-dioxindiamide, which gives an acetone-tricarboxylic a derivative when treated with sodium methoxide.[164] Cyclization to 4,6-dil

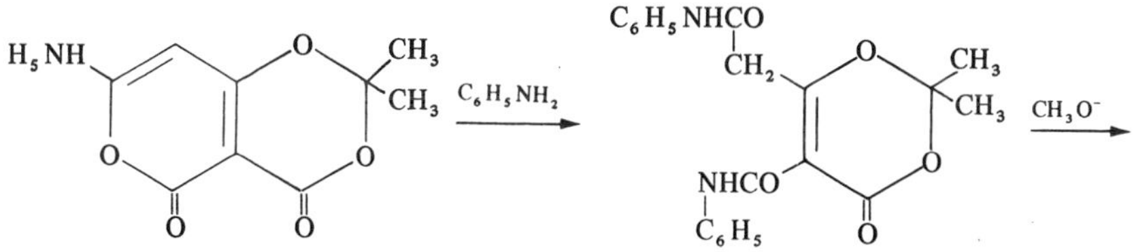

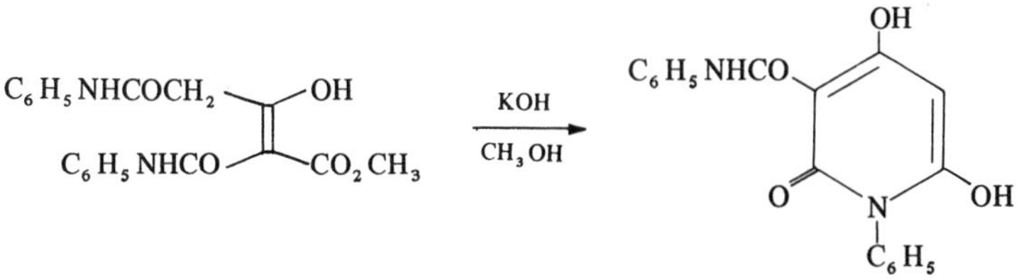

**XII-266**

oxy-1-phenyl-3-(*N*-phenylcarbamoyl)-2-pyridone **(XII-266)** occurs on boiling ith methanol containing potassium hydroxide.[171, 172]

Isocyanates and malonyl chloride or isocyanates and 6-chloro-4-hydroxy-2-opyran-3-carbonyl chloride give 7-chloro-2,4,5-trioxopyrano[3,4-*e*]-1,3-azines **(XII-267)**, which can be converted to 2,4-dioxo-1,3-oxazines and ethyl 2-carbamoyl-3-ethoxy-2-pentenedioates **(XII-268**, R = Et) when heated th ethanol. Isothiocyanates react similarly to form thio analogs. When **XII-268** = Et) is heated in ethanolic potassium hydroxide, 3-carbethoxy-4-ethoxy-6-droxy-2-pyridones are formed that can be saponified and decarboxylated to **I-269**. Hydrolysis of the enol ethers **(XII-268)** gives the enol **(XII-268**, R = H), hich can be cyclized in ethanolic potassium hydroxide to the *N*-substituted-3-rbethoxy-4,6-dihydroxy-2-pyridone. The dioxo-oxazines in ethanolic potas-m hydroxide give 3,5-dicarbethoxy-4,6-dihydroxy-2-pyridones **(XII-270)**.[171]

7-Amino-2,4,5-trioxo-3-phenylpyrano[3,4-*e*]-1,3-oxazines[172] and sodium phen-ide in phenol as solvent give 6-hydroxy-2-oxo-4-phenoxy-1-substituted ridine-3-carboxanilides **(XII-271**; $R_1$ = aryl, R = phenyl). It has been suggested at the aryloxy ion attacks the ring with concomitant loss of $CO_2$ followed by merization to the product. Intermediates could not be isolated.[172, 173] coholic potassium hydroxide and the oxazines give **XII-271** (R = H).[174]

$R_1$ = $C_6H_5$; *o*-, *m*-, and *p*-$CH_3C_6H_4$; *o*-, *m*-, and *p*-$CH_3OC_6H_4$, *m*-$OHC_6H_4$, $CH_2$=CH–$CH_2$, iso$C_4H_9$.

## D. *3-Pyridinols from Furans*

Acylfurans, convenient intermediates for the synthesis of 2-substituted-3-·droxypyridines,[175] are synthesized by acylation of furans or by acylation of omatic substrates with 2-furoyl halides. 3-Pyridinols **(XII-272)**, pyrrolyl ·tones, and pyrroyl ketimines are formed on treatment with ammonia.[176-182]

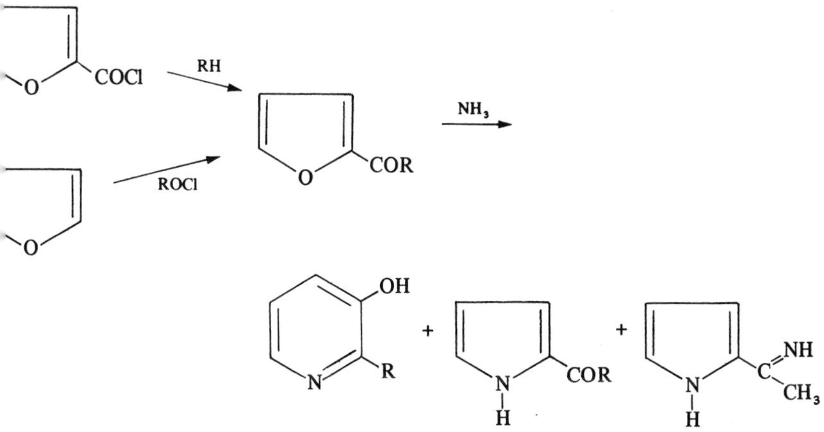

COCl, RH, ROCl, COR, $NH_3$, OH, R, N, +, COR, N, H, NH, C, $CH_3$, N, H

**XII-272**

·e products can be separated conveniently by "sublimatography".[179-181]

6-Methyl-3-pyridinol **(XII-273)** can be prepared directly from furfurylamine d formalin in hydrochloric acid[183] or by way of 2-aminomethyl-5-hydroxy-ethylfuran.[184]

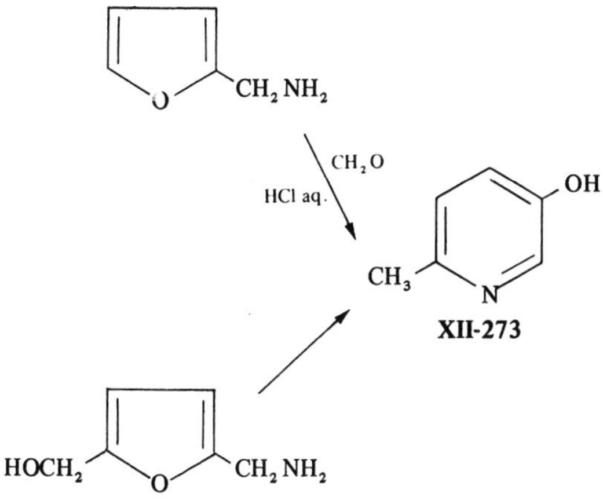

$CH_2NH_2$, $CH_2O$, HCl aq., OH, $CH_3$, N, **XII-273**, $HOCH_2$, $CH_2NH_2$

The reaction of 2-acylfurans to form 3-pyridinols is not prevented by sm alkyl groups at C-3. 3-Methyl- and 3-ethyl-5-methyl-2-acetylfuran (**XII-274**; R = $CH_3$, $R_4$ = $CH_3$, $CH_3CH_2$)[185] and 2-aroyl-3-methylfurans (**XII-274**; R = ar $R_4$ = $CH_3$)[186] form 3-pyridinols without difficulty.

**XII-274**

Oxidation of furfural with chlorine water or sodium hypochlorite followed treatment of the product (see Section I.1.C.) with sulfamic acid gi 3-hydroxy-2-imino-1(2*H*)-pyridinesulfonic acid, which is converted to 2-ami 3-pyridinol (**XII-275**) on hydrolysis.[187, 188]

**XII-275**

Oxidation of *N*-monosubstituted-2-(α-aminoalkyl)furans (**XII-276**) w chlorine water gives *N*-alkyl-3-hydroxypyridinium chlorides (**XII-277**; $R_1$ alkyl, $R_2$ = H, CN, $CONH_2$). The 2-cyanomethylfurans (**XII-276**; $R_1$ = alk $R_2$ = CN) are prepared from furfural by a Strecker reaction and hydrolyzed the amides (**XII-277**, $R_2$ = $CONH_2$). Pyridinium salts of the amides (**XII-2** $R_1$ = $CH_3$, $C_2H_5$, $CH_2C_6H_5$; $R_2$ = $CONH_2$) give 3-hydroxypicolinami (**XII-278**, $R_2$ = $CONH_2$) on heating.[189]

**XII-276** **XII-277** **XII-278**

A variety of 2-substituted-3-pyridinols can be prepared by well-known ethods.[190] For example, diarylmethyl 2-furyl ketones are prepared in good eld from ethyl furoate and diarylmethanes using $KNH_2$ or $NaNH_2$ in liquid nmonia and are conveniently converted to **XII-279**. *p*-Chlorobenzyl 2-furyl tones, which can be converted to **XII-280**, can be prepared from chlorophenylacetonitrile and ethyl furoate by condensation in the presence of koxide followed by hydrolysis and decarboxylation. 2-Pyrryl ketones are also rmed in the last step. Several 2-furyl ketones are cleaved under these reaction nditions and do not give 3-pyridinols.[191]

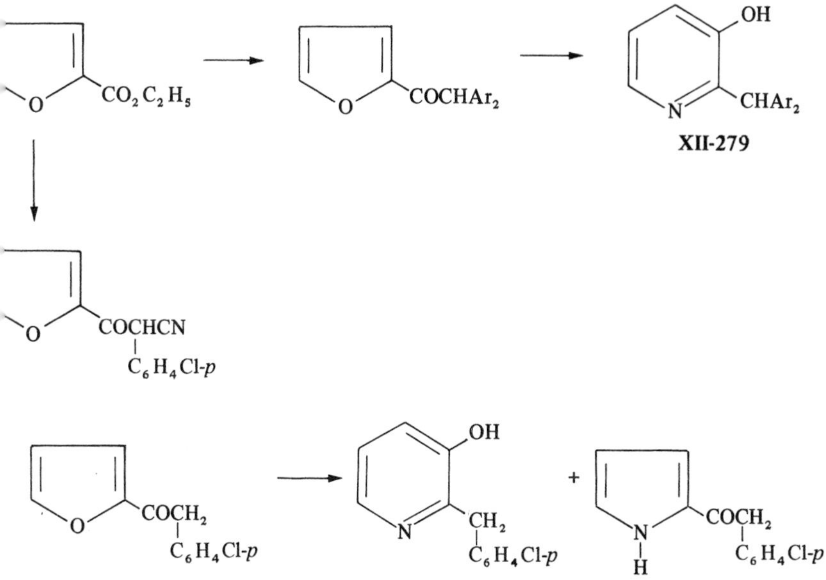

$CO_2C_2H_5$ → $COCHAr_2$ → OH, $CHAr_2$, N

**XII-279**

COCHCN, $C_6H_4Cl$-*p*

$COCH_2$, $C_6H_4Cl$-*p* → OH, $CH_2$, $C_6H_4Cl$-*p*, N + $COCH_2$, $C_6H_4Cl$-*p*, N, H

**XII-280**

Derivatives of 5-aminoaldoses have been partially characterized by conversion 3-pyridinols. For example, 6-amino- and 6-nitro-5-acetamido-1,2-*O*-cyclo- exylidene-5,6-dideoxy-L-iodofuranose and the corresponding D-glucofuranoses hen boiled under reflux with hydrochloric acid form 6-aminomethyl- and nitromethyl-3-pyridinol (**XII-281**, R = $NH_2$, $NO_2$).[192, 193]

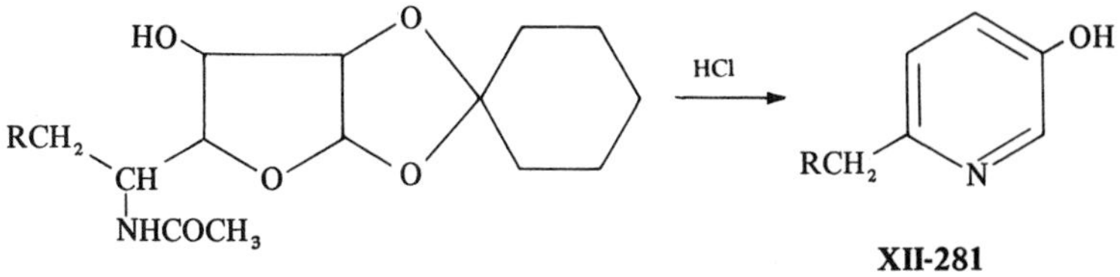

HO, $RCH_2$, CH, $NHCOCH_3$, O, O, O → (HCl) OH, $RCH_2$, N

**XII-281**

## E. *3-Pyridinols from Pyrones*

3-Pyridinols have been prepared indirectly from kojic acid (**XII-218**, p. 64[illegible] which reacts with dimethyl sulfate to give 2-hydroxymethyl-5-methoxy-4[illegible] pyran-4-one, which can be converted to 2-hydroxymethyl-5-methoxy-4-pyrido[illegible] (**XII-282**) with aqueous ammonia. Treatment of **XII-282** with phosphor[illegible]

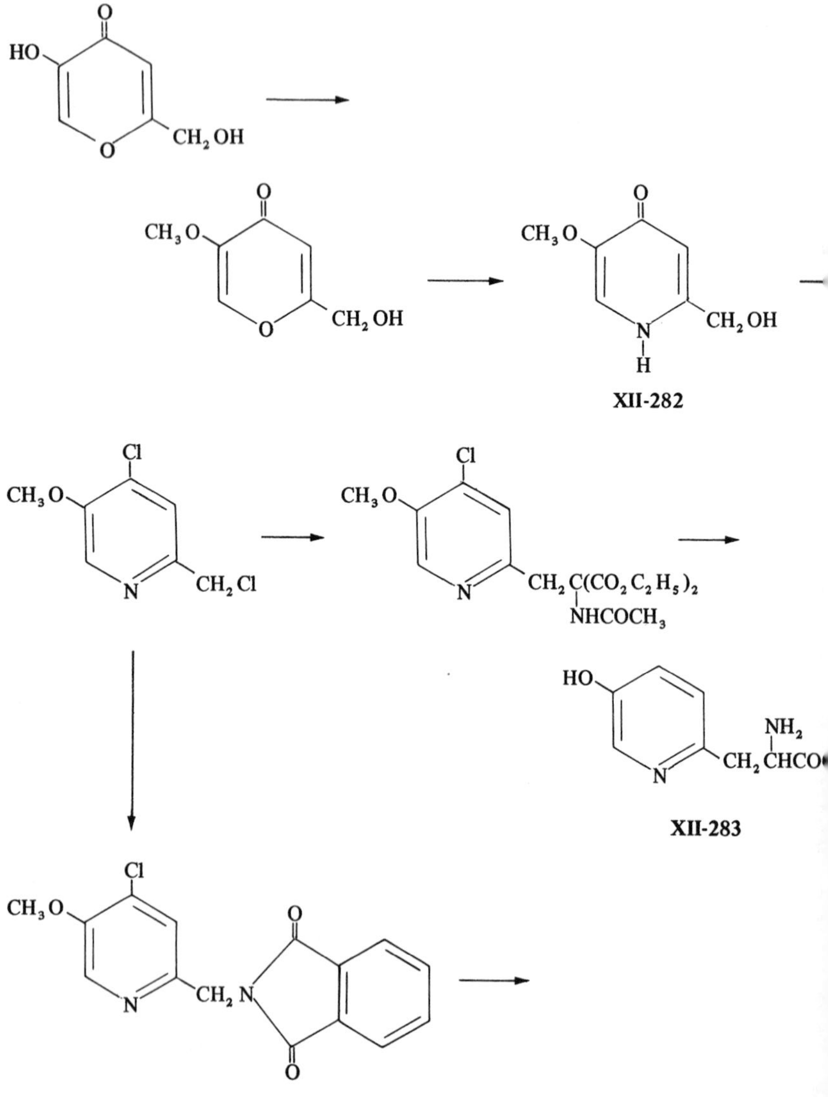

HO O CH$_2$OH

CH$_3$O O O CH$_2$OH

CH$_3$O O N H CH$_2$OH

**XII-282**

CH$_3$O Cl N CH$_2$Cl

CH$_3$O Cl N CH$_2$C(CO$_2$C$_2$H$_5$)$_2$ NHCOCH$_3$

HO N NH$_2$ CH$_2$CHCO[illegible]

**XII-283**

CH$_3$O Cl N CH$_2$N O O

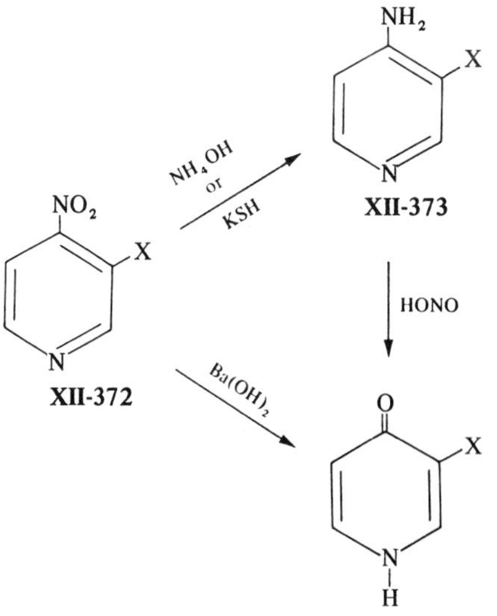

NH₄OH or KSH; HONO; Ba(OH)₂

XII-372 XII-373 XII-374

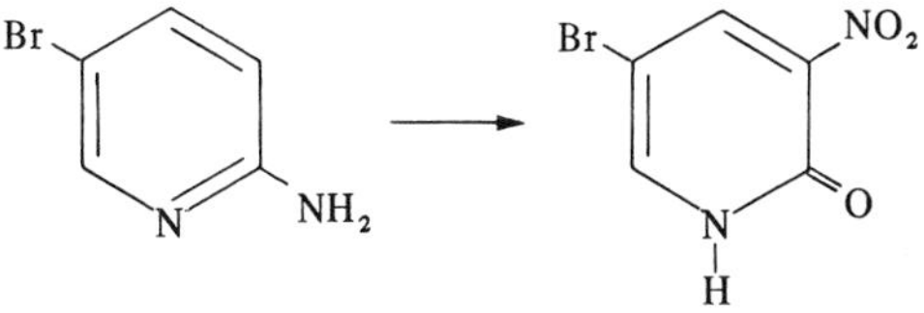

XII-375

e nitramine can be converted to 4-amino-3-nitro-2-pyridone **(XII-376)** with tassium hydroxide containing hydrogen peroxide. 3-Nitro-4-hydroxy-pyridone has been formed from **XII-376** by treatment with nitrous acid.[305]

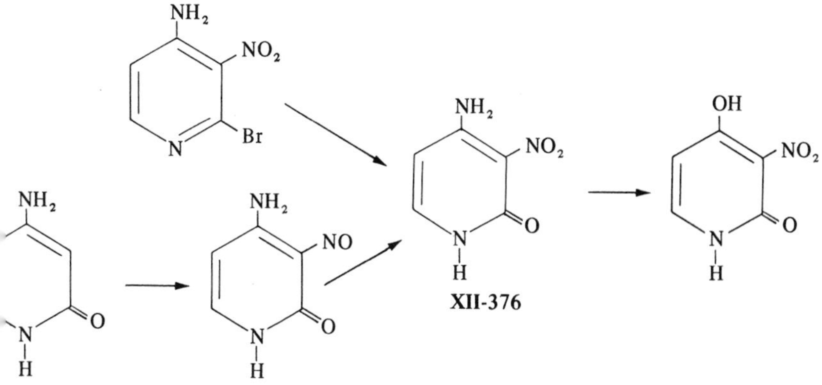

XII-376

XII-377

5-Ethoxy-4-methyloxazole has been used to form pyridoxine (**XII-286**, R H). With maleic anhydride it gives an adduct that, on treatment with ethanc hydrogen chloride, forms the products **XII-287** and **XII-288** (R = R′ = Et) an monoester (**XII-288**; R = Et, R′ = H or R = H, R′ = Et).[199] The diethyl es **XII-288** is also formed from ethyl maleate or ethyl fumarate a 5-ethoxy-4-methyloxazole.[200] Fumaronitrile and 5-ethoxy-4-methyloxazole gi 4,5-dicyano-2-methyl-3-pyridinol,[199, 201] also a known precursor to pyridoxi 5-Ethoxy-4-methyloxazoles and 2-butene-1,4-diol give pyridoxine (**XII-286**, R H), which is difficult to purify when prepared in this way (assay, 23%).

TABLE XII-3. Reactions of 5-Alkoxyoxazoles with Maleimide and Maleic Anhydride

| $R_2$ | $R_6$ | R | % **XII-284** Ref. 196 | % **XII-285** Ref. 197 |
|---|---|---|---|---|
| $CH_3$ | H | $C_2H_5$ | 43 | 54 |
| $C_2H_5$ | H | $C_2H_5$ | 80 | 47 |
| $n$-$C_3H_7$ | H | $C_2H_5$ | 77 | 31 |
| $n$-$C_4H_9$ | H | $C_2H_5$ | 83 | 42 |
| $n$-$C_5H_{11}$ | H | $C_2H_5$ | 81 | 40 |
| $C_6H_5$ | H | $C_2H_5$ | 43 | 0 |
| $CH_3$ | $CH_3$ | $CH_3$ | 36 | 36 |
| $C_2H_5$ | $CH_3$ | $CH_3$ | 73 | 25 |
| $n$-$C_5H_{11}$ | $CH_3$ | $CH_3$ | 68 | 16 |
| $CH_3$ | $C_2H_5$ | $C_2H_5$ | 85 | 0 |

5-Ethoxy-4-methyloxazole reacts similarly with a variety of ethers and est of 2-butene-1,4-diol.[202] 1,4-Dimethoxy-2-butene forms **XII-286** (R = CH 2,5-dihydrofuran forms the pyridoxine cyclic ether (**XII-289**)[199, 201] a 1,4-diacetoxy-2-butene forms pyridoxine.[200]

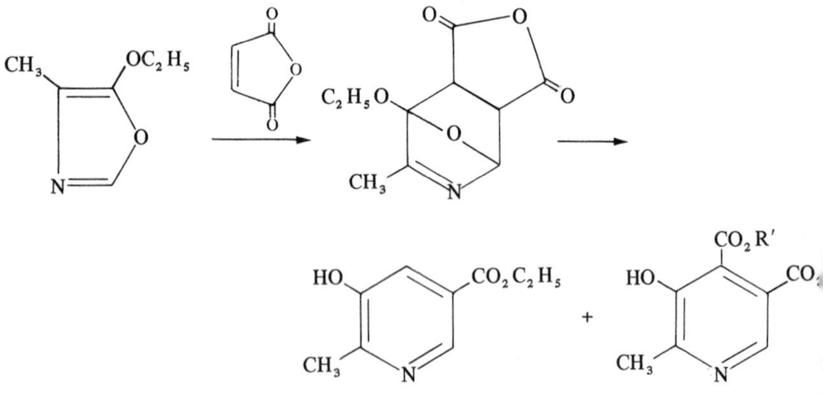

$CH_3$ $OC_2H_5$ O N · O O O → $C_2H_5O$ O O O O $CH_3$ N →

HO $CO_2C_2H_5$ $CH_3$ N + $CO_2R'$ HO $CO_2$ $CH_3$ N

**XII-287** **XII-288**

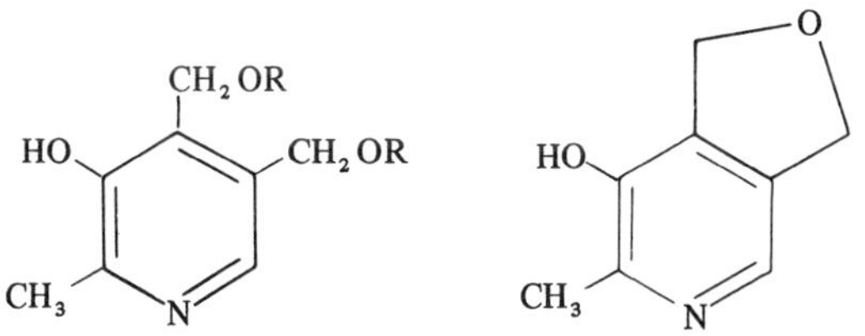

XII-286 XII-289

-Ethoxy-4-methyloxazole reacts with *cis*- and *trans*-2,5-dimethoxy-2,5-di-
lrofuran to form both *endo*- **(XII-290)** and *exo*- **(XII-291)** adducts. All four
lucts have been characterized. The adducts give *cis*- or *trans*- **XII-292** when
ated with base. Both of the latter isomers give 3-hydroxy-2-methylpyridine-
-dicarboxaldehyde on treatment with hydrochloric acid.[203]

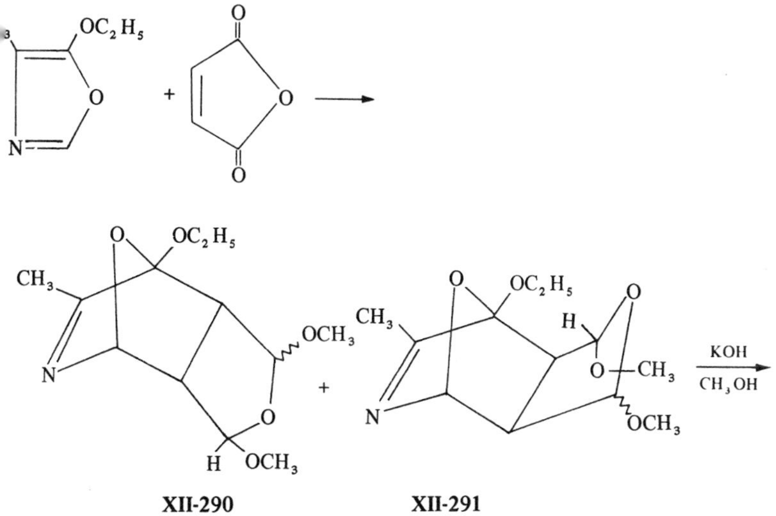

XII-290 XII-291

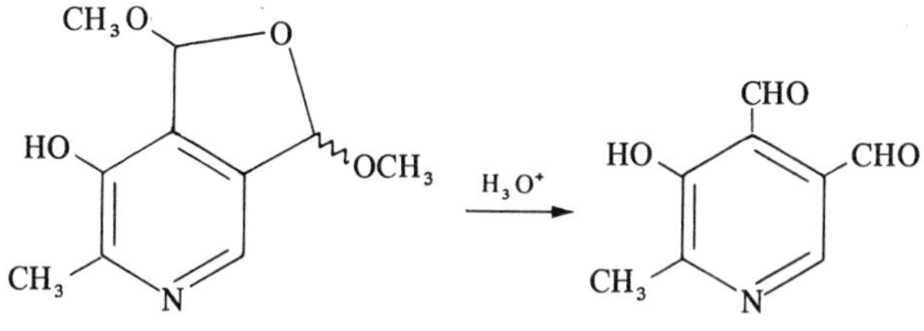

XII-292

n unstable anhydride has been isolated from the reaction between maleic
ydride and ethyl 5-ethoxy-4-oxazolylacetate and has been characterized as
I-293.[204]

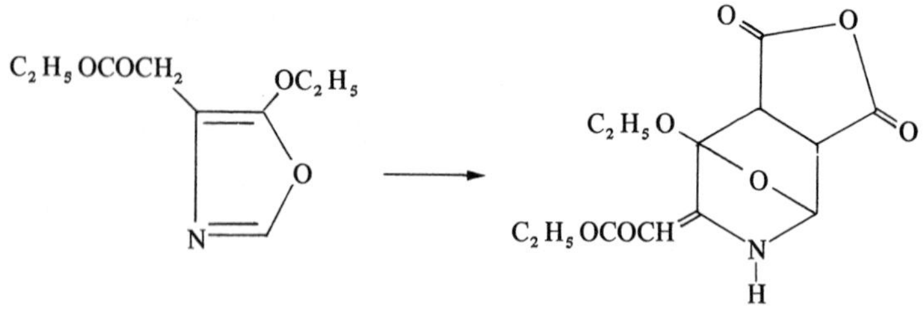

XII-293

4-Carboxymethyl-5-ethoxyoxazole reacts with fumaronitrile to give 4,5-cyano-2-methyl-3-pyridinol.[205] It reacts with 4,7-dihydro-1,3-dioxepine to g **XII-294**, which is hydrolyzed with dilute hydrochloric acid to pyridoxine. T ethyl ester of 4-carboxymethyl-5-ethoxyoxazole reacts with 1,3-dioxepine form **XII-294** and **XII-295**, which can also be hydrolyzed to pyridoxi 5-Cyano-4-methyloxazole and 4,7-dihydro-1,3-oxepines have been used

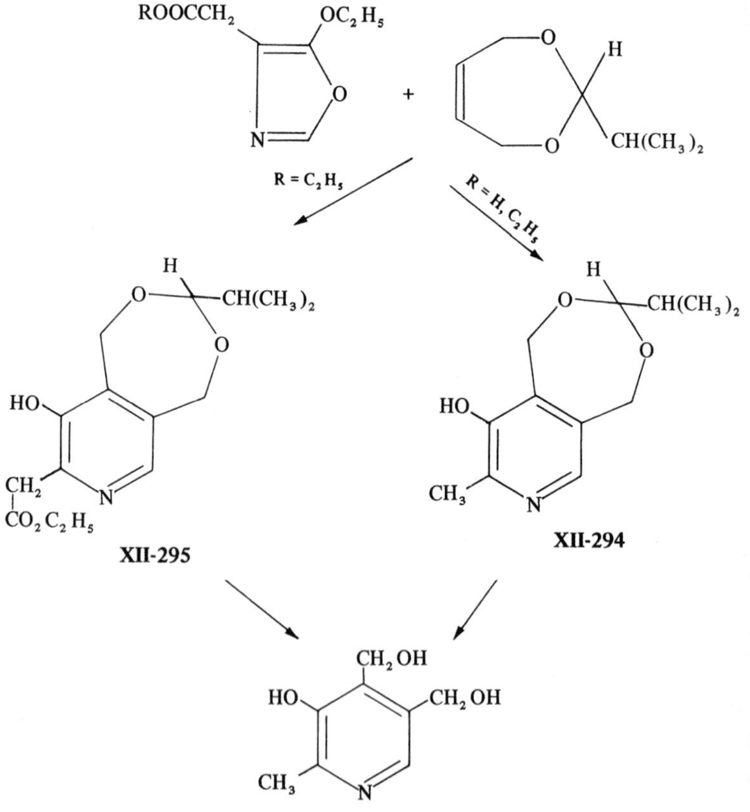

R = $C_2H_5$ R = H, $C_2H_5$

**XII-295** **XII-294**

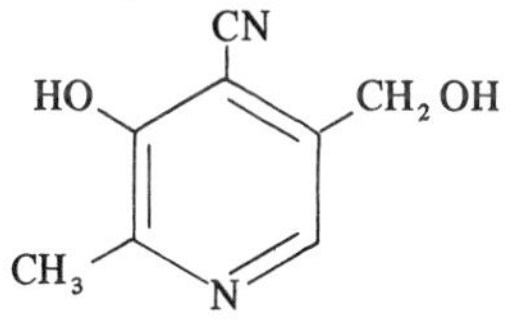

XII-296

pare pyridoxine by a route in which cyano is the leaving group during the ;radation of the adduct.[206]

oth 4-methyl-[207] and 4-carboxymethyl-5-ethoxyoxazole[208] and γ-hydroxy-tonitrile give 4-cyano-5-hydroxymethyl-2-methyl-3-pyridinol **(XII-296)**.

roducts from the four reaction paths (*a*-*d*) are observed in Diels-Alder additions oxazoles **(XII-297)**, depending on the nature of the groups on the 4- and ositions of the incipient pyridine ring.[209] A 5-unsubstituted oxazole can give -pyridinol even though the dienophile is an alkene, if the dienophile can vide to the adduct a leaving group (X) alternate to the 3-hydroxyl that is duced by cleavage of the oxygen bridge (path *b*). 3-Pyridinols can also be med by dehydrogenation (path *c*, $R_5$ = H). The yields of 3-pyridinols are ally low, however, when 5-unsubstituted oxazoles are used because of npeting reactions. 4-Methyloxazole reacts with acrylonitrile in toluene to give yano-2-methylpyridine (path *a*), 2-methyl-3-pyridinol **(XII-298)** (path *b*) and mino-6-methylnicotinonitrile (by dehydrogenation, path *c*, $R_5$ = H), all in yields. A similar reaction in aqueous acetic acid gave only **XII-298** (28%) l hydrogen cyanide. 4-Phenyloxazole and 2,4-dimethyloxazole in acetic acid

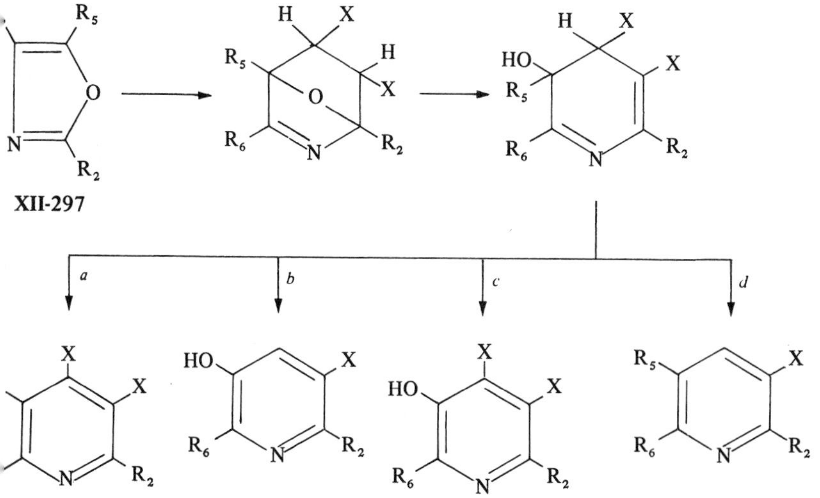

XII-297

$R_5$ = H, OR, CN, and so on

react with acrylonitrile to give 2-phenyl-3-pyridinol and 2,6-dimethyl-3-p dinol, respectively.[210]

Fumaronitrile and 4-methyloxazole give 5-cyano-2-methyl-3-pyridi (**XII-299**) in unusually high yield for this type of reaction (71%). Both diet fumarate and diethyl maleate give low yields of **XII-300** and of **XII-301** ( $CO_2C_2H_5$), which is formed by dehydrogenation (path *c*).[211] Et

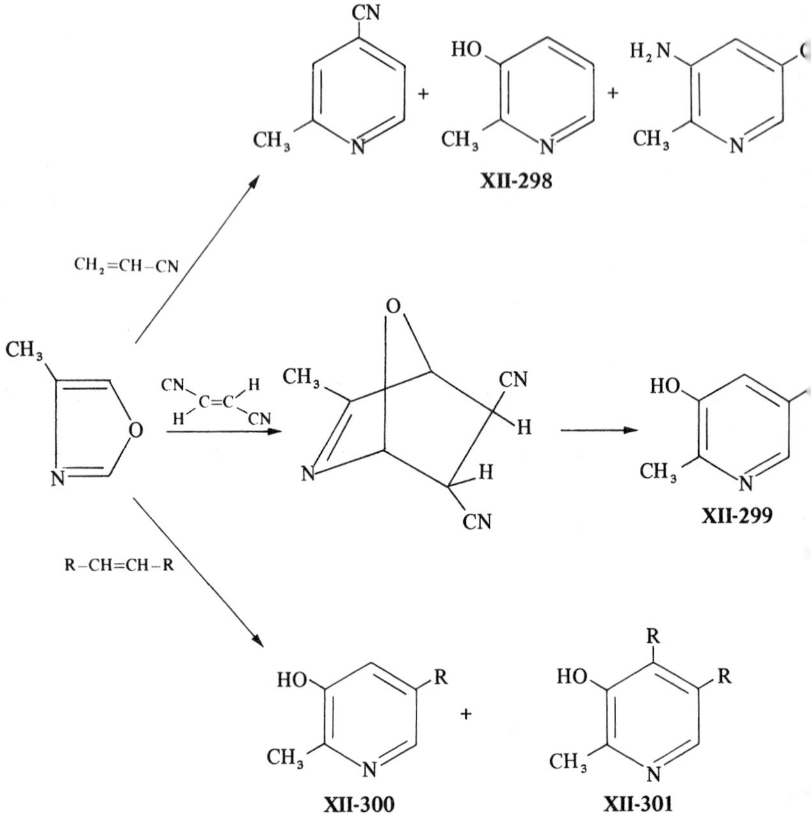

*trans*-3-cyanoacrylate gives both possible products of structure **XII-300** ( $COOC_2H_5$, CN). Ethyl acrylate gives **XII-300** (R = H) and 5-carbethoxy-2-met 3-pyridinol (**XII-300**, R = $CO_2C_2H_5$).[211] 5-Carbethoxy-2,4-dimethyl-3-pyridi a precursor for 4-deoxypyridoxine, and **XII-302**, a pyridoxine intermediate, can be formed from 4-methyloxazole.[212] Depending on conditions, 4-met oxazole and *N*-phenylmaleimide in ethyl acetate form either of two add (**XII-303** or **XII-304**) that were not completely characterized and that 6-methyl-*N*′-phenyl-3,4-pyridinedicarboximide, (path *a*) and 5-hydrox methyl-*N*′-phenyl-3,4-pyridinedicarboximide (path *c*), respectively.[209] Sim

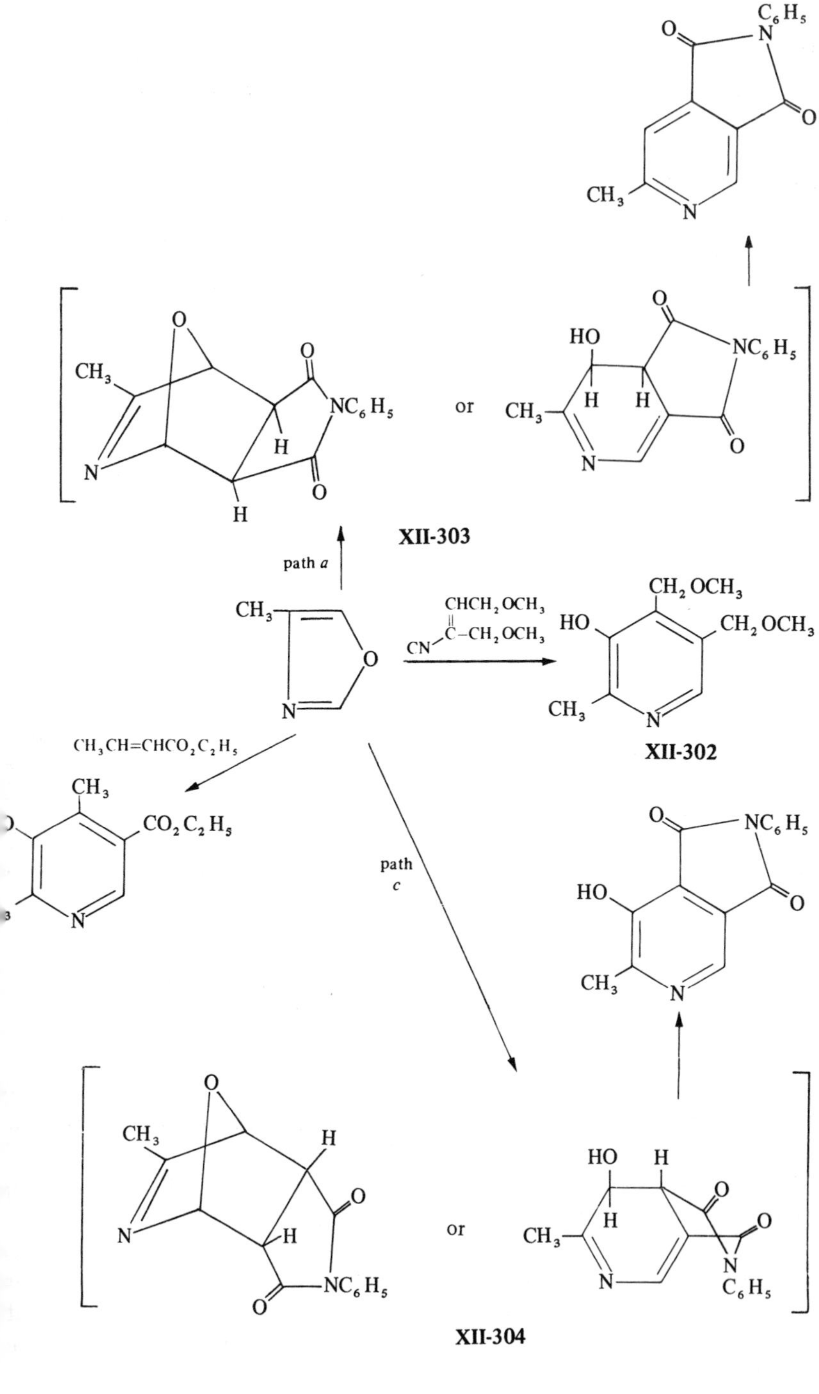

path $a$

XII-303

$CHCH_2OCH_3$

$CN-\overset{\|}{C}-CH_2OCH_3$

XII-302

$CH_3CH{=}CHCO_2C_2H_5$

path $c$

XII-304

Diels-Alder reactions have been reported for 2,4-dimethyloxazole, eth 4-methyl-5-oxazolocarboxylate, 4-methyl-5-oxazolocarbonitrile, 4-methyl-5-oxazolocarboxylic acid and 2-amino-4,5-dimethyloxazole.[209] The formation of trace of diethyl 2-methyl-3,5-pyridinedicarboxylate from ethyl 4-methyl-5-oxazolecarboxylate and diethyl fumarate is the only example cited for path *d*.[209]

It has been suggested that in paths *b* and *c* reaction of the adduct occurs $C_1-O$ cleavage of the oxygen bridge followed by nucleophilic attack at $C_5$ displace X and to give **XII-305**, or at $C_4$ to displace hydride and give **XII-306**.[1]

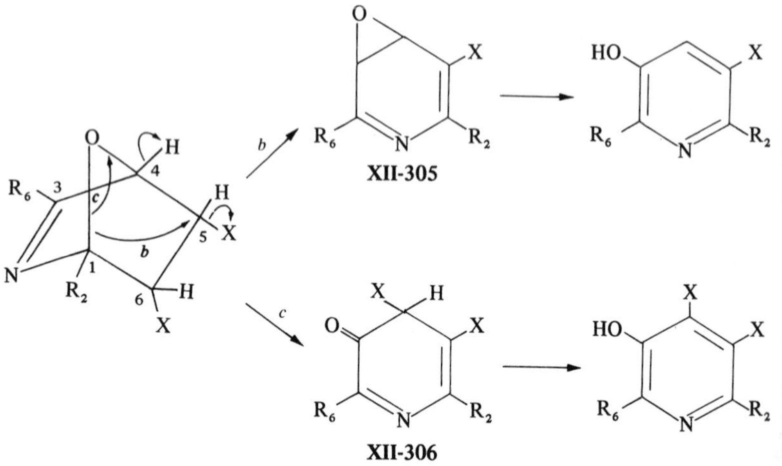

XII-305

XII-306

The dehydrogenation product is usually formed in small amounts; however, the presence of hydride acceptors such as hydrogen peroxide or nitrobenze the amount of dehydrogenation is increased.[195]

## G. *3-Pyridinols and Pyridones from Other Nitrogen-Containing Heterocycles*

Diels-Alder addition of dimethyl acetylenedicarboxylate to 3-benzyl-6-meth 2,5-dihydroxypyrazine gives an adduct that decomposes on heating to give eq amounts of the 2-pyridones **XII-307** and **XII-308**.[213]

4,6-Dihydroxy-2-methylpyrimidine and dimethyl acetylenedicarboxylate fo an unstable adduct that decomposes to give the pyridone **XII-309**.[213]

Acetophenone-anil and benzylmalonic acid react in acetic anhydride to fo 5-benzyl-2-methyl-4,6-dioxo-2,3-diphenylhexahydro-1,3-oxazine, which re ranges to 3-benzyl-4-hydroxy-1,6-diphenyl-2-pyridone **(XII-310)** when heat with phosphorus pentoxide. A number of tetrahydrocarbostyrils have be prepared similarly from cyclohexanone-anil *via* **XII-311**.[214]

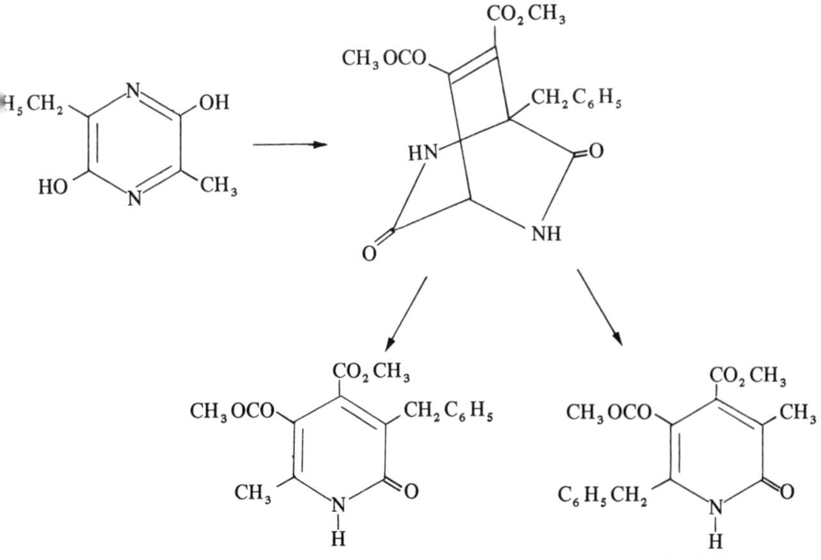

XII-307

XII-308

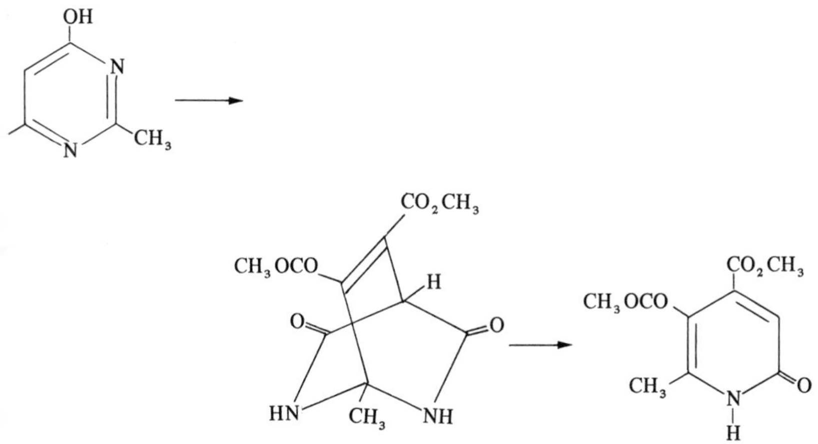

XII-309

,3-Dihydro-1,2-diazabicyclo[3.2.0]-3-hepten-6-ones **XII-312** and **XII-313** and zepinones **XII-314** are readily interconvertible ring systems[215-217] and all ee can give 1-amino-3-pyridinol salts (see Section IV.3., p. 839) or yridinols.[218, 219] When heated in methanol, the 2-acetyl- or 2-benzoyl- ivative of **XII-313** rearranges to several products including 6-acetamido- or enzamido-3-hydroxy-4-methyl-5-phenylpyridine **(XII-317)** in 17% and 68% ld, respectively, presumably through intermediates such as **XII-315** and

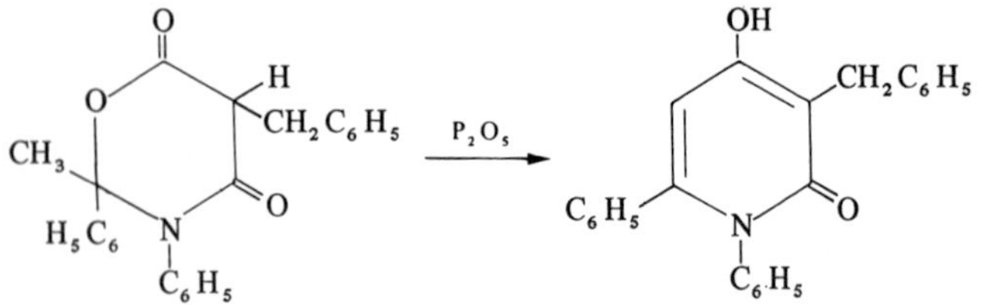

XII-310

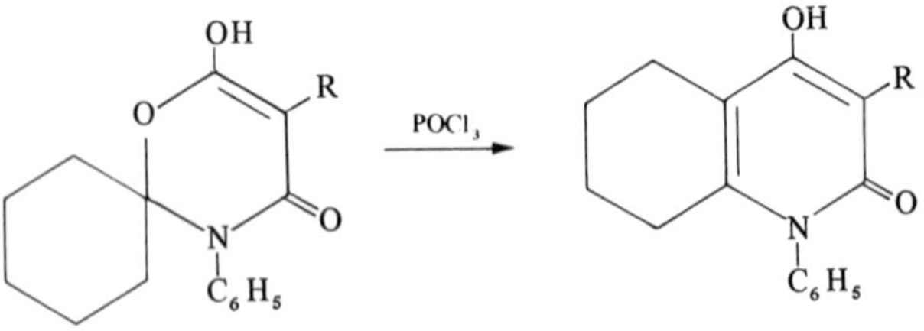

XII-311

**XII-316.** In basic aqueous solution **XII-313** (R = $CH_3$) is converted 4-amino-3-hydroxy-4-methyl-5-phenyl-5-piperidein-2-one **(XII-320)** and 3-droxy-4-methyl-5-phenyl-2-pyridone **(XII-321)** by a process beginning w **XII-318**, involving hydration, azetinone ring opening, disproportionation **XII-319**, and recyclization.[220, 221]

Although earlier reports indicated that **XII-314** did not react w aqueous base, later work has shown that it rearranges to 6- 2-amino-4-methyl-5-phenyl-3-pyridinol, presumably through intermedia **XII-322** and **XII-323**.[222, 223]

Dimethyl sulfate and **XII-314** give equal amounts of the two *N*-met derivatives **XII-324** and **XII-325**.[216] Tautomerization of **XII-325** to **XII-** occurs on standing. The betaine **XII-325** rearranges to 3-hydroxy-4-methy methylamino-5-phenylpyridinium chloride when treated with hydrochl acid.[216] The *N*-methyldihydroazepinones **XII-324** and **XII-326** rearrange 3-pyridinols when warmed with methanolic alkali.[222, 224, 225]

The alkaloid dendrobine gives 4-isopropyl-2-pyridone **(XII-327)** when hea with selenium at 300°.[226]

2-Ethyl-5-methyl-3-pyridinol **(XII-328)** has been characterized as a degradat product of the steroid alkaloid leptinidine.[227]

The acid azide **XII-329**, freshly prepared from (2,3-diphenylcycloprop-2-en acetyl chloride, gives 3,4-diphenyl-2-pyridone **(XII-332)** (60–70%) and a sm amount of the urethane **(XII-333)** on treatment with boiling ethanol. W heated, **XII-329** rearranges quantitatively to the isocyanate **(XII-330)**.[228, 22 has been suggested that **XII-332** is formed *via* **XII-331** through an intramolec cycloaddition.[228] Photolysis of **XII-329**, followed by treatment with ethano

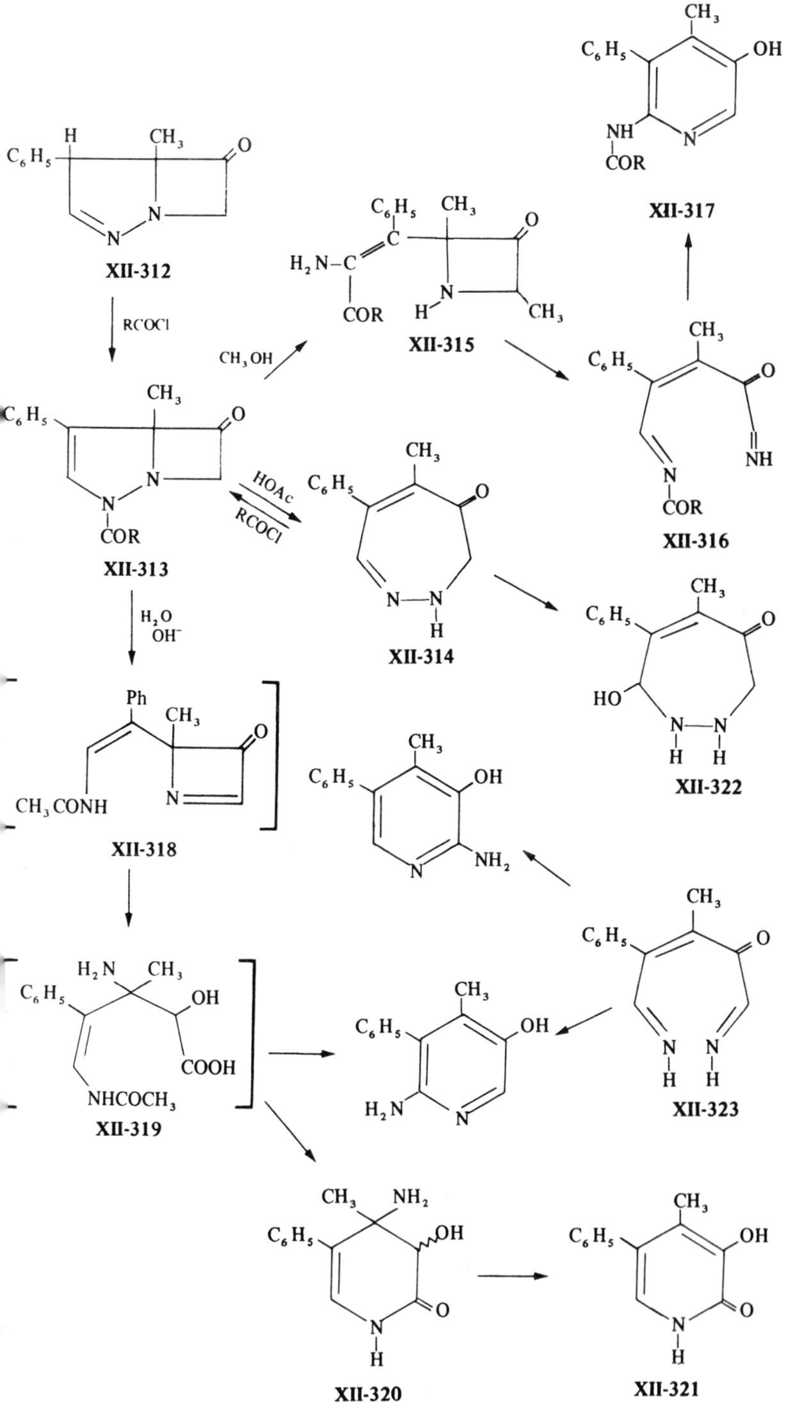

$C_6H_5$ H $CH_3$ O N N

**XII-312**

RCOCl

$CH_3OH$

$C_6H_5$ $CH_3$ O N N COR

**XII-313**

$H_2O$ $OH^-$

$C_6H_5$ $CH_3$ $H_2N-C$ C O COR N H $CH_3$

**XII-315**

HOAc RCOCl

$C_6H_5$ $CH_3$ O N N H

**XII-314**

$CH_3$ $C_6H_5$ OH NH COR N

**XII-317**

$CH_3$ $C_6H_5$ O N COR NH

**XII-316**

$CH_3$ $C_6H_5$ O HO N N H H

**XII-322**

Ph $CH_3$ O $CH_3CONH$ N

**XII-318**

$CH_3$ $C_6H_5$ OH $NH_2$ N

$H_2N$ $CH_3$ $C_6H_5$ OH COOH $NHCOCH_3$

**XII-319**

$CH_3$ $C_6H_5$ OH $H_2N$ N

$CH_3$ $C_6H_5$ O N N H H

**XII-323**

$CH_3$ $NH_2$ $C_6H_5$ OH O N H

**XII-320**

$CH_3$ $C_6H_5$ OH O N H

**XII-321**

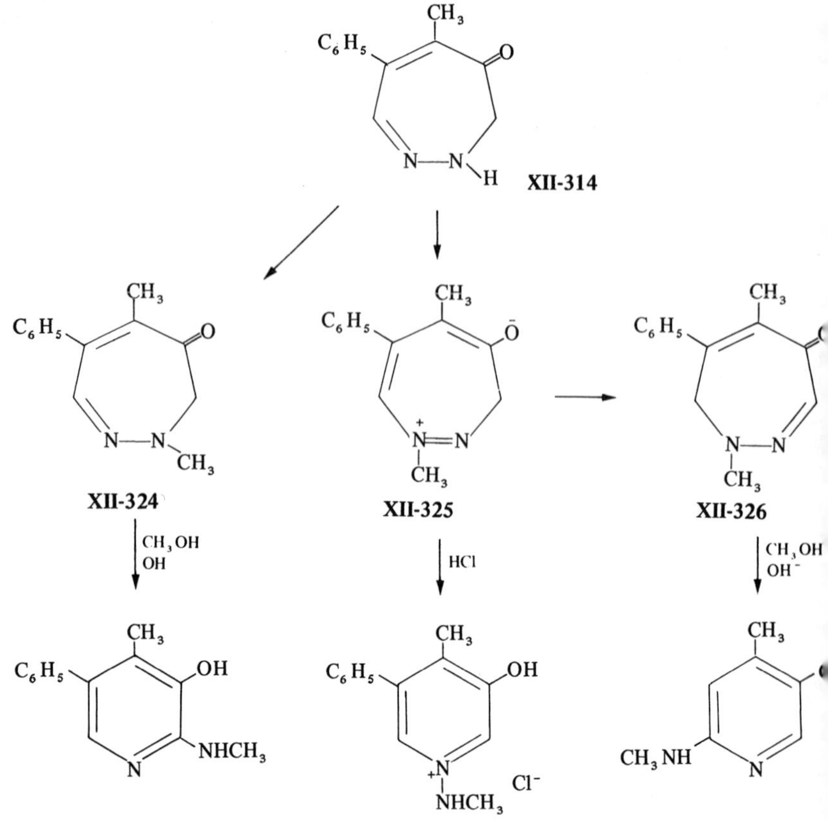

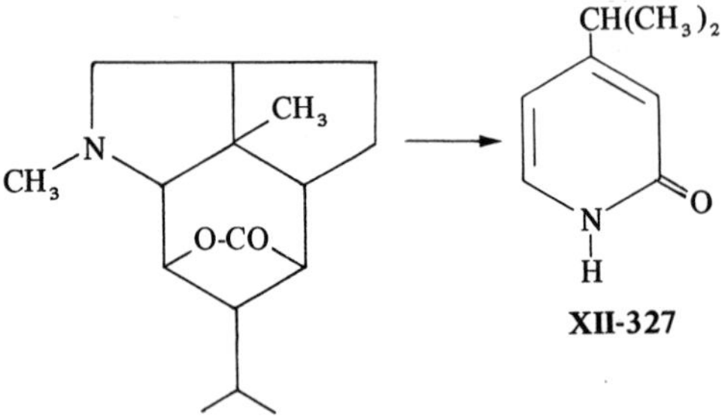

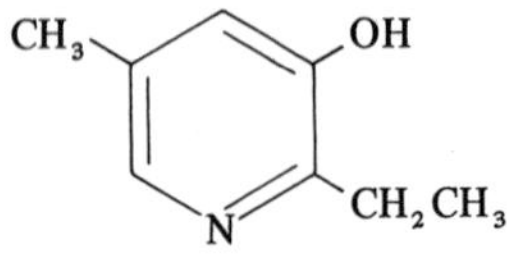

XII-314

XII-324

$CH_3OH$
OH

XII-325

HCl

XII-326

$CH_3OH$
$OH^-$

XII-327

XII-328

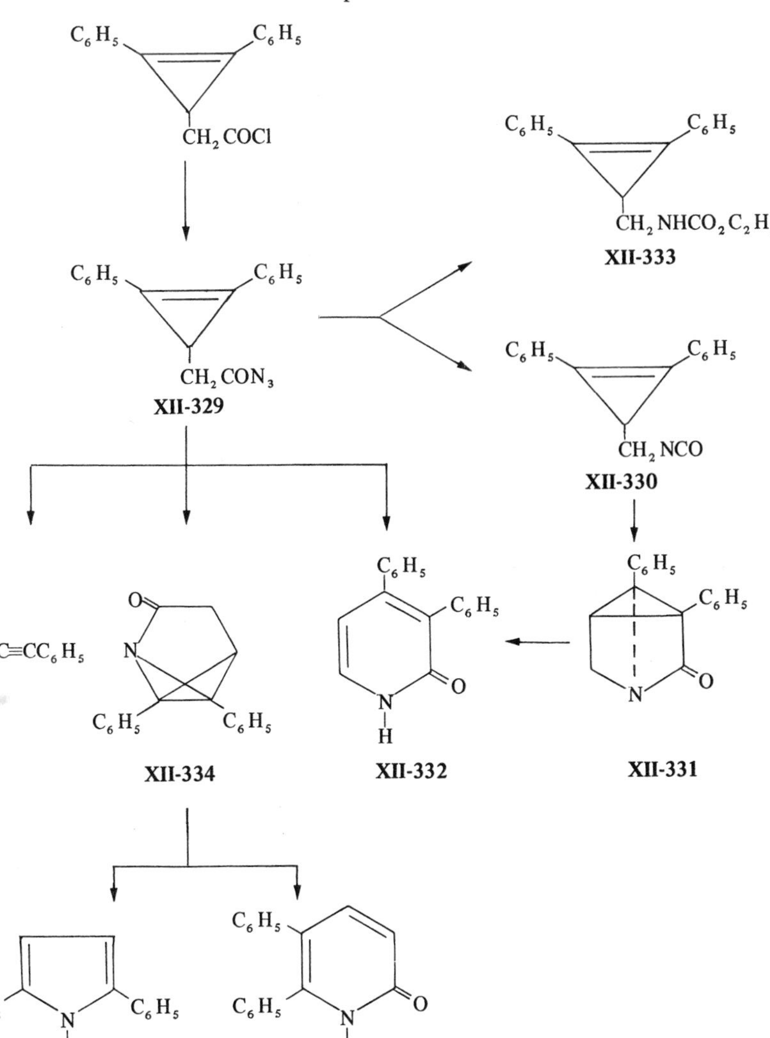


', gives 2,5-diphenylpyrrole (6%), 5,6-diphenyl-2-pyridone (5%), diphenyl-tylene, and **XII-332**. The nitrene adduct **XII-334** is postulated as a common ermediate in the formation of 2,5-diphenylpyrrole and 5,6-diphenyl-2-pyri-ie.[229] (1,2,3-Triphenylcycloprop-2-enyl)acetyl azide gives, on similar itment with ethanol, the corresponding carbamate and 3,4,5-triphenyl-2-idone (6–8%).[230] Photolysis gives 4,5,6-triphenyl-2-pyridone.[230]

## H. *Pyridinols and Pyridones From Carbocyclics*

Phenylcyclobutenediones and enamines give 1:1 adducts that rearrange w
bases to 3-hydroxy-2-pyridones (**XII-335**).[231]

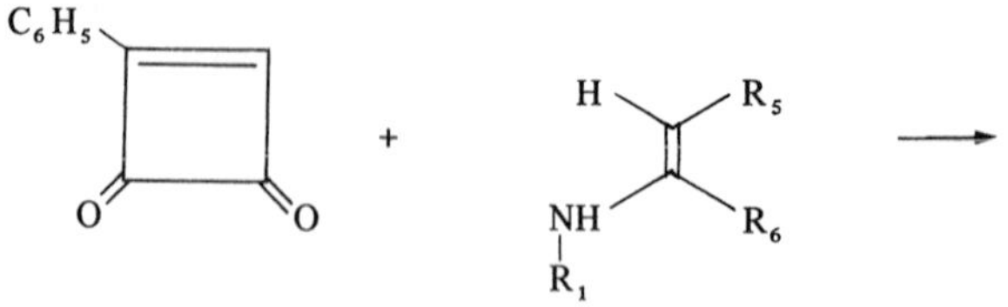

$R_1$ and $R_6$ = H, alkyl, aryl
$R_5$ = $CO_2R$, COR, CN

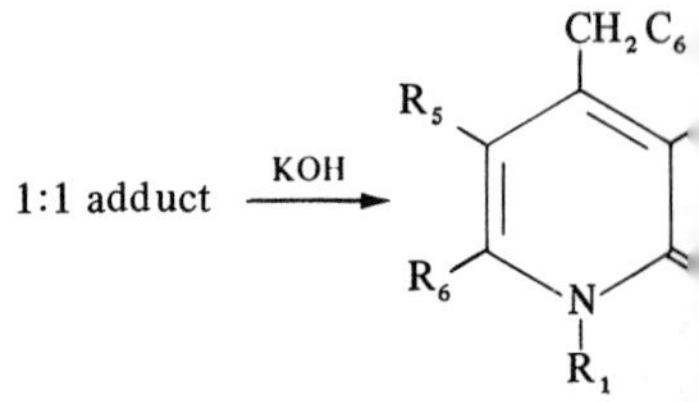

1:1 adduct $\xrightarrow{KOH}$

**XII-335**

The secondary amine **XII-336**, which can be prepared conveniently f
chloranil, reacts with aqueous sodium hydroxide to give the 4-pyridone **XII-**
which is decarboxylated and deacetylated by hydrochloric acid in isopro
alcohol to *N*-(*p*-aminophenyl)-2,3,5-trichloro-4-pyridone.[232]

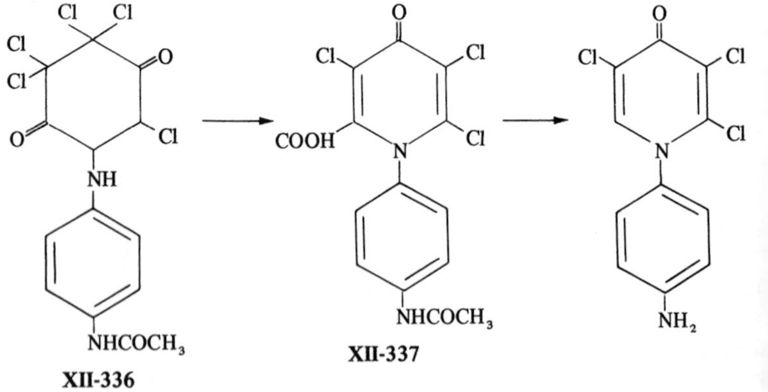

**XII-336**

**XII-337**

## 6. From Other Pyridine Derivatives

ıe activating influence of the ring nitrogen provides an important versatility the synthesis of pyridones, pyridinols, and pyridine ethers, by nucleophilic ;titution at the 2- and 4- and even the 3-position. Further, the potential of ›mplishing these substitutions *via* pyridine *N*-oxides contributes to the scope his method.

### A. *Halopyridines*

ıe preparation of pyridinols from halopyridines has generally not been ›rtant. More often pyridinols have been the source for the halopyridines. 'ever, the recent development of new routes to several halopyridines has e them available as pyridinol precursors (see also Chapter VI). For example, 'halopyridines, formed by direct ring closure[233] or through treatment of ırimide with phosphorus pentachloride,[234] give polychloro 2- and ridones on hydrolysis[235] and 4-alkoxypolychloropyridines when treated alkoxides.[236] Glutarimide and phosphorus pentachloride form a mixture **-338)** of 2,6-dichloro- (10%), 2,3,6-trichloro- (75%) and 2,3,5,6-tetrachloro- dine (15%)[234, 235] that is hydrolyzed to a mixture of 2-pyridinols by sodium roxide in aqueous *t*-butyl alcohol at 150 to 160°.[235] Further chlorination of acidified hydrolysate gives 3,5,6-trichloro-2-pyridinol **(XII-339)**.[235] 2,6-Di- ropyridine is converted to 6-chloro-2-pyridinol by sodium hydroxide in tyl alcohol-water.[237] Photochlorination of **XII-338** gives pentachloropyri- , which forms 2,3,5,6-tetrachloropyridine by reductive dehalogenation.

$Cl_n$-pyridine (**XII-338**) $\xrightarrow[H_2O/t\text{-BuOH}]{NaOH}$ $Cl_{n-1}$-2-pyridinol $\xrightarrow{Cl_2}$ 3,5,6-trichloro-2-pyridinol (**XII-339**)

**XII-338** $\xrightarrow{Cl_2}$ pentachloropyridine $\xrightarrow[HCl]{Zn}$ 2,3,5,6-tetrachloropyridine $\rightarrow$ **XII-339**

Hydrolysis gives **XII-339** in high yields, but the product by this route is n difficult to purify.[235] Pentachloropyridine can also be prepared f perchlorocyclopentenone[16] (see Section I.1.B.) or from 2,6-diaminopyridin chlorination in hydrochloric acid followed by treatment with phosph oxychloride and phosphorous trichloride.[238] Pentachloropyridine and nuc philes give mixtures of 2- but mainly 4-substituted products, with bul reagents giving increased amounts of substitution at the less sterically hind 2-position. The ratios of monosubstituted products (**XII-340**:**XII-341**) for by treatment of pentachloropyridine with alkoxides are 85:15 ($R = CH_3$), 6 ($R = C_2H_5$), and 57:43 ($R = n\text{-}C_4H_9$).[238] 4-Methoxyperchloropyridin converted to the 4-pyridinol in concentrated hydrochloric acid at 170°.[239]

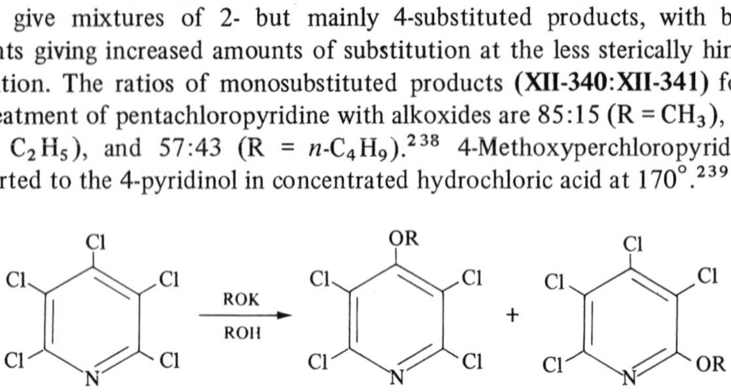

ROK / ROH

**XII-340** **XII-341**

Perchloropyridine also reacts with secondary amines to form mainly 2- some 4-substituted aminotetrachloropyridines. Oxidation with peroxyaci followed by intramolecular rearrangement to give *O*-pyridylhydroxylami Perchloro-2-pyridinol (**XII-343**) has been formed in good yields f perchloropyridine through formation and oxidation of 2-dialkylaminot chloropyridines followed by pyrolysis.[240] Electron withdrawal by the chlo atoms and the ring nitrogen favors the rearrangement of the intermed *N*-oxide (**XII-342**) to the hydroxylamine. Although 4-methoxy-2-piperid perchloropyridine gives the hydroxylamine on oxidation, the correspon 4-pyridinol (**XII-344**) gives the stable piperidine *N*-oxide, presumably becaus

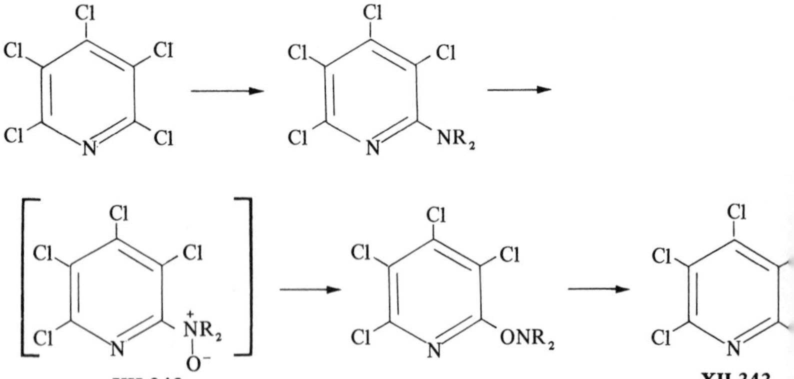

**XII-342** **XII-343**

tron release by the 4-oxygen that exists in the hydroxyl form.[240, 241] ormic acid oxidations of the less highly substituted 2-chloro-6-piperidino-dine also gives the *N*-oxide and no rearrangement.[240]

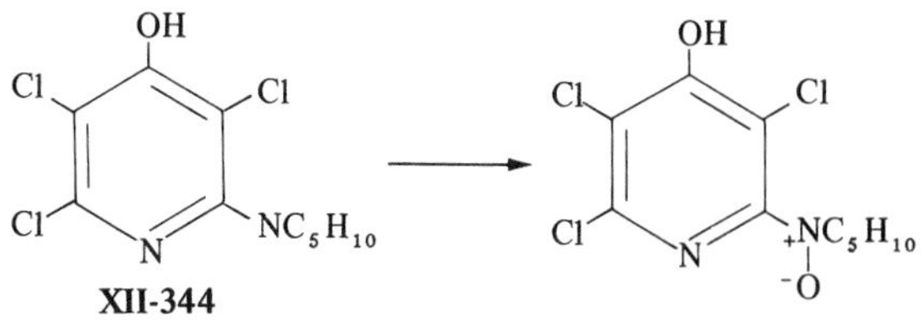

XII-344

3-Diphenylglutarimide and phosphorus pentachloride give 2,5,6-trichloro-diphenylpyridine **(XII-345)** and 5,6-dichloro-3,4-diphenyl-2-pyridinol. Treat-t of **XII-345** with dimethyl sulfate in water and then with aqueous sodium roxide gives the pyridones **XII-346** and **XII-347**, which are hydrogenated to and 3,4-diphenyl-1-methyl-2-pyridone, respectively.[242]

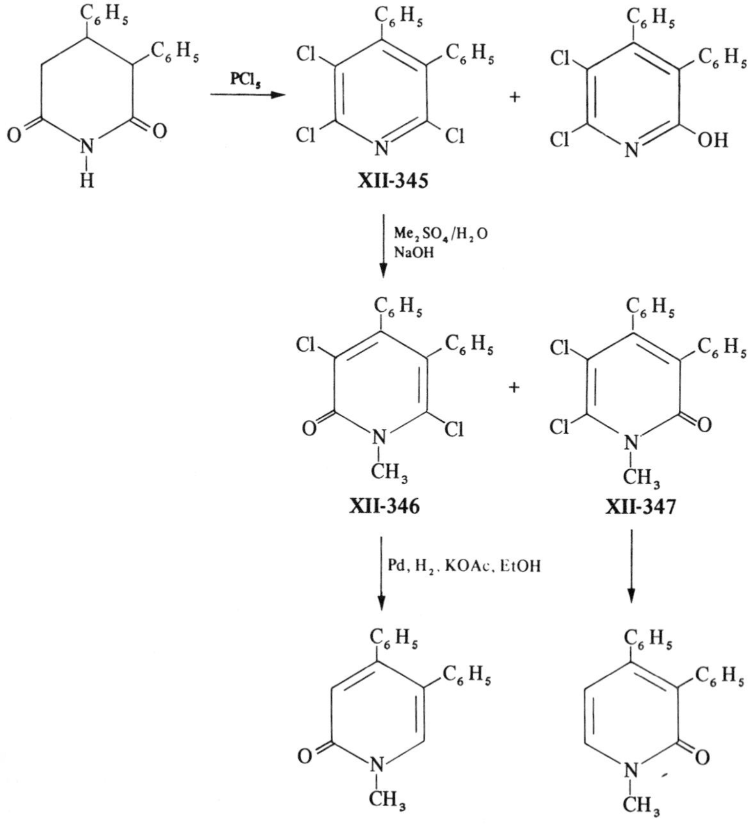

$PCl_5$

XII-345

$Me_2SO_4/H_2O$
NaOH

XII-346 XII-347

Pd, $H_2$, KOAc, EtOH

Pentafluoropyridine, 3-chlorotetrafluoropyridine, and 3,5-dichlorotriflu pyridine, can be obtained from pentachloropyridine[243, 244] and are conve compounds for the preparation of chlorofluoropyridinols[244-247] (*vide in* Pentafluoropyridine can also be prepared from pyridine *via* undecafluoropi dine, although less conveniently.[244, 248] It usually reacts with nucleophil give the 4-substituted product initially. With sodium methoxide it 4-methoxytetrafluoropyridine **(XII-348)**, 2,4-dimethoxytrifluoropyridine, 2,4,6-trimethoxydifluoropyridine.[245, 248] Under very mild conditions possible to isolate **XII-348** in good yield.[245] When *t*-butyl alcohol is used i hydrolysis, a steric participation by solvent is indicated. Dilute aqueous so or potassium hydroxide and perfluoropyridine give only 4-hydroxytetrafl pyridine[246, 248] but with potassium hydroxide in *t*-butyl alcohol, 4- 2-hydroxyperfluoropyridine are formed in the ratio of 9:1.[246] 3,5,6-Trifl

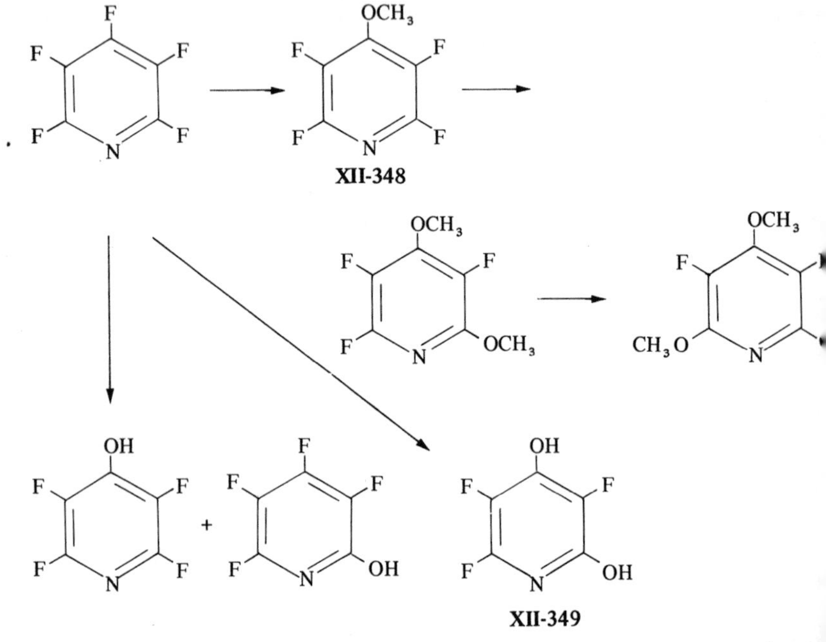

2,4-dihydroxypyridine **(XII-349)** is formed from pentafluoropyridi treatment with 40% aqueous sodium hydroxide.[248] 3-Chlorotetrafluoropyr and aqueous potassium hydroxide form 3-chloro-2,5,6-trifluoro-4-hydroxy dine **(XII-350)** and 5-chloro-3,4,6-trifluoro-2-hydroxypyridine **(XII-351)** ratio of 9:1, but in *t*-butyl alcohol **XII-350**, **XII-351**, and 3-chloro-4.5, fluoro-2-hydroxypyridine **(XII-352)** are formed in the ratio of 5.5:3.5: Aqueous potassium hydroxide and 3,5-dichlorotrifluoropyridine give an yield of the two hydroxypyridines **XII-353** and **XII-354** also in a 9:1 rati

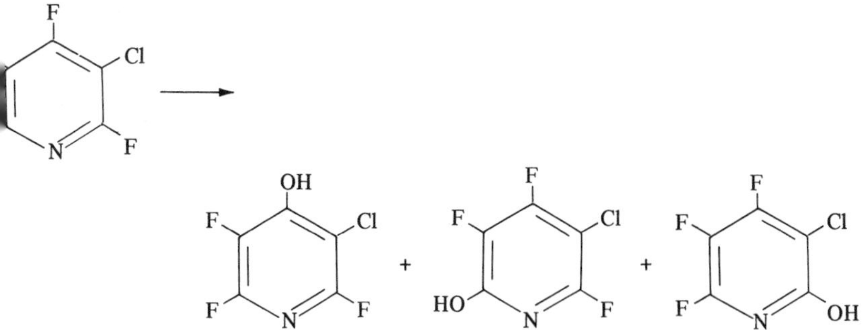

**XII-350** **XII-351** **XII-352**

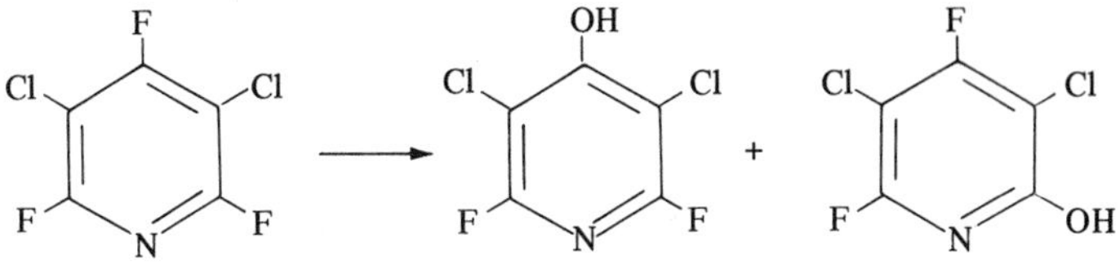

**XII-353** **XII-354**

utyl alcohol, the ratio is 7:3.[246] 4-Bromotetrafluoropyridine and potassium droxide in *t*-butyl alcohol form 4-bromo-3,5,6-trifluoro-2-hydroxypyridine in ;h yield.[247]

\n increase in susceptibility to nucleophilic substitution of halogen at all ring sitions by an *N*-oxide function has been observed. Relative reactivities toward lium methoxide have been reported for 4-chloropyridine, 4-chloro-2,6-luti-ıe, 4-chloro-3-nitro-2,6-lutidine, and their *N*-oxides,[249] and for 2-bromo-, romo-3-methyl- and 2-bromo-5-methylpyridine, and their *N*-oxides.[250]

n the reactions of 2-halopyridines and their *N*-oxides with methanolic tassium methoxide, the order of reactivities is 2-X > 2-X-3-$CH_3$ > 2-X-5-$CH_3$ = Cl, Br), which is not predictable on the basis of steric and inductive effects the methyl group. Arrhenius parameters have been obtained for these ctions and have been used to rationalize observations. For example, it has en suggested that ion-dipole attraction between the 3-methyl and attacking thoxide is in part responsible for the greater reactivity of 2-bromo-3-methyl-ridine over 2-bromo-5-methylpyridine. However, in 2-chloropyridine it is ;gested that bond breaking is more important in the rate-determining step.[250] ward thiophenoxide ion in methanol, the order 2-Br-3-$CH_3$ > 2-Br- > 3r-5-$CH_3$ is observed.[251] Further extensions of these reactions to include 2,3-d 2,5-dibromopyridine indicate that more than one *ortho* effect appears to be erative.[251]

The syntheses and nucleophilic reactivities of a number of chloro and oronitropyridines have been studied extensively by Talik and Talik. In studies with 2-halo-4-nitropyridine *N*-oxides, it was demonstrated that the 4-nitro gro is more reactive than the 2-halo function toward aqueous alkali and alkoxid The ethers **XII-355** and **XII-356** have been formed by reduction.[252, 253] Tow diethylamine the 2-halo is more reactive than the 4-nitro group (Br > Cl > I).

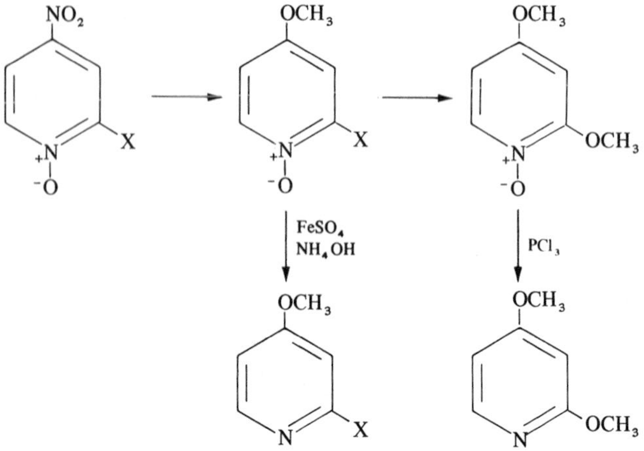

**XII-355** **XII-356**

3-Halo-4-nitropyridines and their *N*-oxides react at the *C*-nitro group wh treated with bases or alkoxides to give **XII-357** or **XII-358** (X = Cl, Br, However, 3-fluoro-4-nitropyridine and its 1-oxide form 4-nitro-3-pyridinols a 4-nitro-3-alkoxypyridines, respectively.[254-256a] The 3-alkoxy-4-nitropyridine oxides have been converted to 3,4-dialkoxypyridine-1-oxides.[256] Because of t marked reactivity of the 3-fluoro substituent, these studies have been extend to 3-fluoro-5-methyl-4-nitropyridine-1-oxide,[257] 3-fluoro-2-methyl-4-nitropy dine-1-oxide,[258] and 2,6-dimethyl-3-fluoro-4-nitropyridine-1-oxide.[258] Seve of these fluoronitropyridines have been extensively studied as potential reage for formation of amino acid derivatives.[256a-260] 2-Fluoro-3,5-dinitropyridine typical example, is hydrolyzed by hot water and reacts with hot alcohols form 2-alkoxy-3,5-dinitropyridines and reacts with amino acids and th derivatives to give well-defined products.[260] The reactions of a number fluoronitropyridines and their *N*-oxides have been summarized by Talik a Talik[261] and the relative reactivities toward simple nucleophiles have be observed,[258,262] as shown on p. 689.

2-Fluoro-3-nitropyridines are prepared from the corresponding 2-aminopy dines in only fair yield but are hydrolyzed in good yield to the 2-pyridinols. 2-Fluoro-6-methyl-5-nitropyridine,[263] 2-fluoro-3,5-dinitro-4-methylpyridine, 2-fluoro-3,5-dinitro-6-methylpyridine[264] and 2-chloro-3,5-dinitropyridine

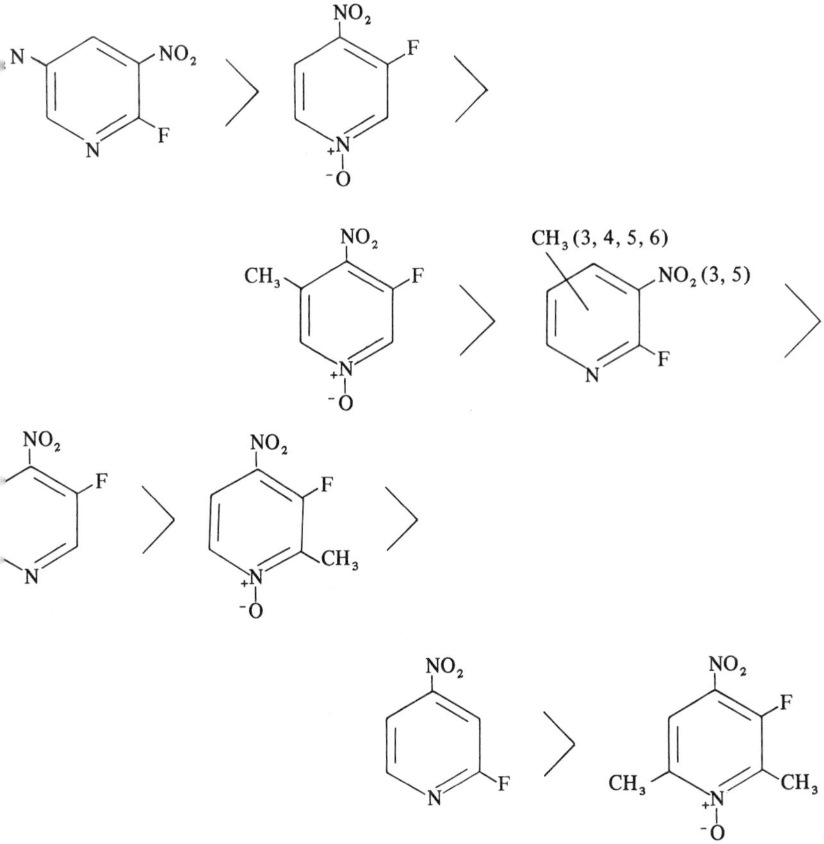

.ve also been converted to the corresponding ethers and 2-pyridones by lvolyses.

4-Cyano-2-fluoropyridine is hydrolyzed by aqueous acid to 2-hydroxy-4-ridinecarboxylic acid.[266]

The rate of displacement of the halogen in 2-chloro- and 2-iodo-5-nitropyri-ne by water decreases with increasing acid concentration and is proportional to e 4th power of $a_{H_2O}$ and to the fraction of the halopyridine protonated in etic acid-water and dioxane-water. Rates in $D_2O$ with $D_2SO_4$ give an isotope fect of 2.36. The following mechanism has been suggested:[267]

2-, 3- and 4-Methoxypyridine are cleaved to the pyridone by sodium ethoxide in methanol. 3-Chloro- and 3-bromopyridine react with methoxide n to give 3-methoxypyridine **(XII-359)**, which is then cleaved by methoxide in

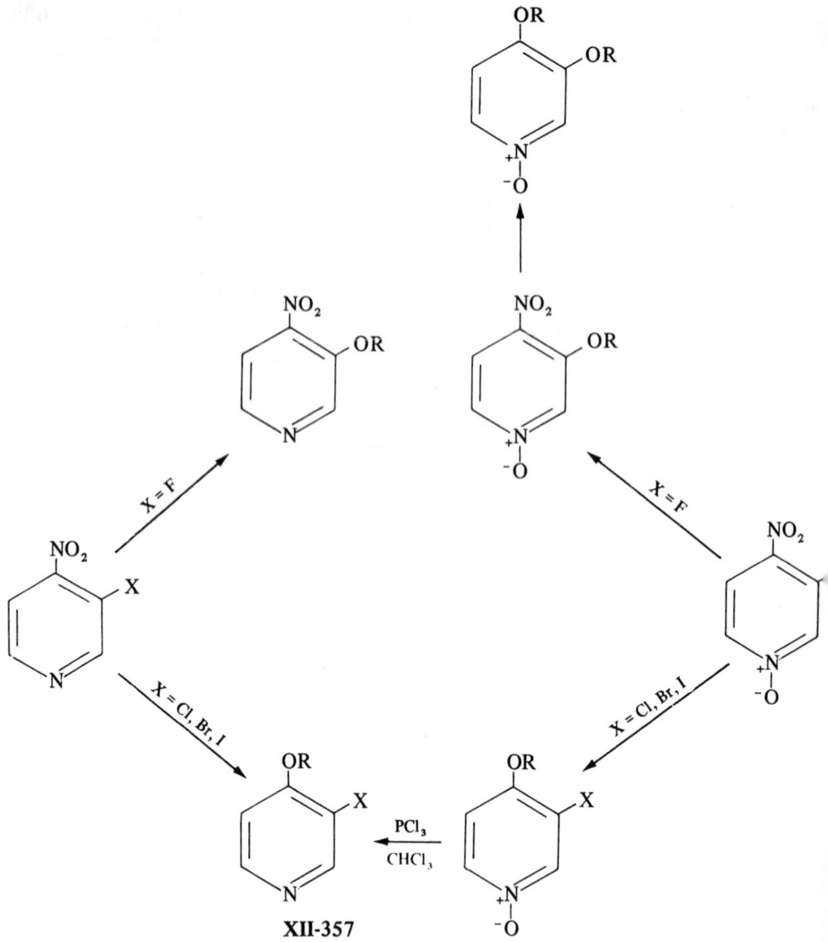

$X = F$

$X = Cl, Br, I$

$PCl_3$ / $CHCl_3$

**XII-357**

**XII-358**

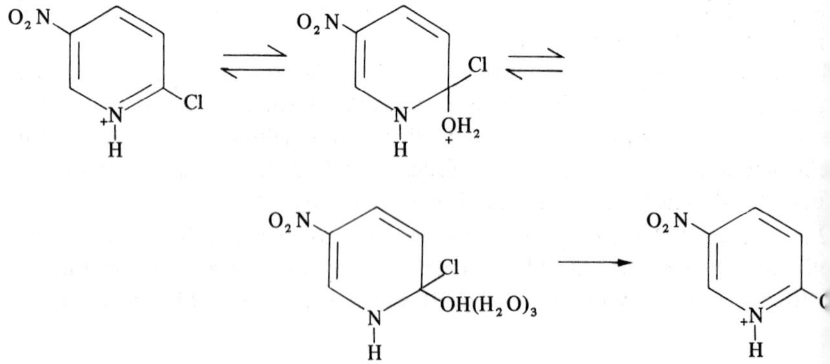

$S_N2$ mechanism.[268] Treatment of 3-bromopyridine with methanolic tassium hydroxide has been reported to give an 87% yield of nethoxypyridine.[269] However, a reinvestigation of this reaction has shown at the yield of 3-methoxypyridine rises to 36% after 9 hours and then creases.[268] The ratios of second-order rate constants at 218° for formation of nethoxypyridine[268, 270] and its subsequent cleavage are 0.53 for 3-chloropyri-ne and 0.75 for 3-bromopyridine.[268] 3-Chloropyridine and sodium methoxide methanol at 230° give 3-pyridinol in 70% yield, a reaction that appears to ve preparative value.[268] Reactions of 2- and 4-chloropyridines with methoxide relatively fast and ether cleavage is not an important side reaction.[268, 270] Methoxypyridine with methoxide in aqueous methanol gives both ether and cohol by cleavage and undergoes a more rapid methoxyl exchange as shown by periments with methanol-$d_3$ and deuterium oxide.[268]

$CH_3O^-$ ; $^-OCH_3$ ; $+ CH_3OCH_3$

XII-359

Treatment of 2-, 3-, and 4-substituted pyridine-1-oxides **(XII-360)** with bromopyridine gives 1-(substituted-2-pyridyl)-2-pyridones, usually in low elds, 2,3′-dipyridyl ethers, and 2-pyridones (see also Chapter IV). In several stances, the *N*-(2-pyridyl)-2-pyridones have been hydrolyzed to aminopyridine th aqueous sodium hydroxide.[271] Electron donating substituents, particularly the 3-position, favor, and electron withdrawing groups on the 2- or 4-position event, pyridyl-2-pyridone formation.

NaOH

**XII-360**

= $CH_3$, $CH_3O$, AcNH, $NO_2$

Observations of the reaction between pyridine-1-oxide, quinoline-1-oxide, and oquinoline-2-oxide with 2-bromopyridine, 2-bromoquinoline, or 1-bromoiso-

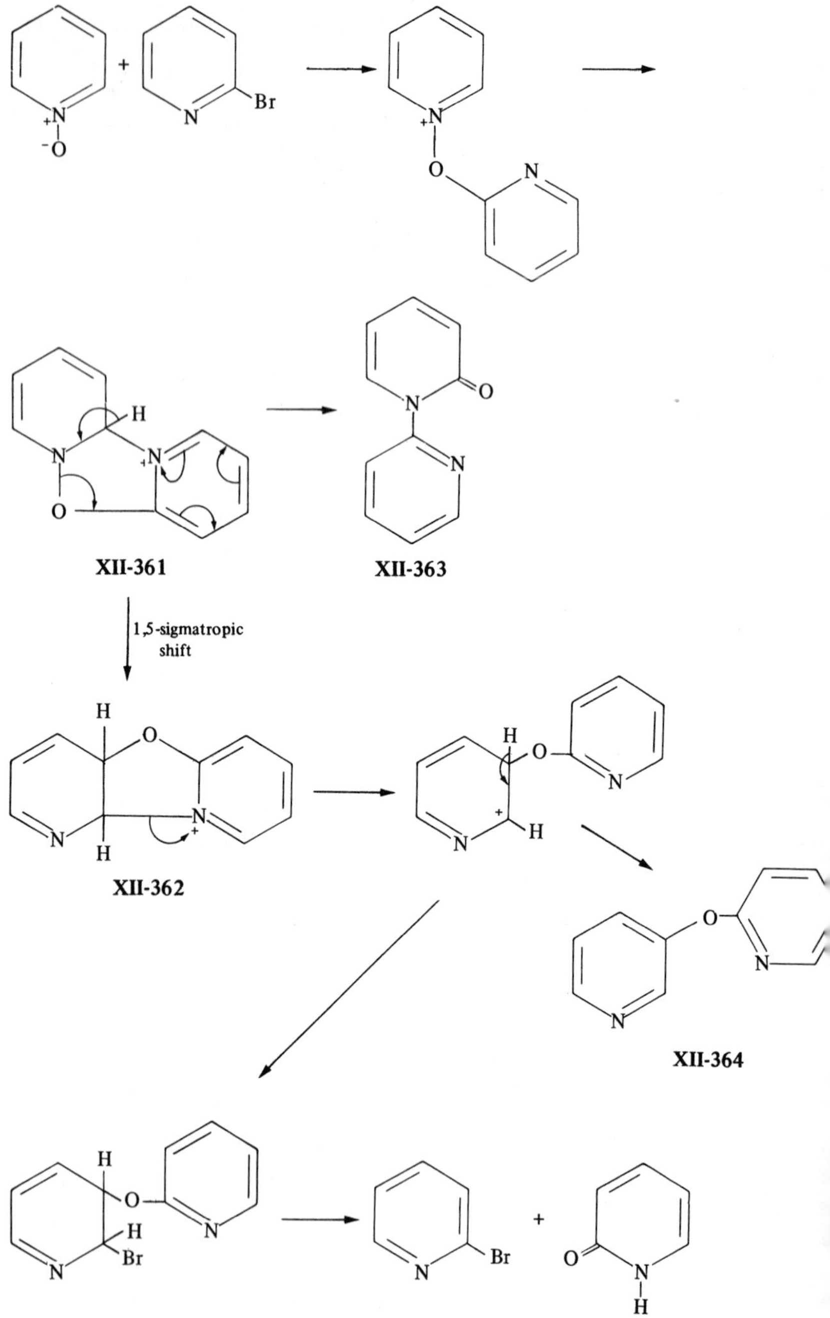

XII-361

XII-363

1,5-sigmatropic shift

XII-362

XII-364

XII-365

inoline are not consistent with previously proposed mechanisms[272] and have
to the suggestions that nucleophilic attack by pyridine-1-oxide on the carbon of 2-bromopyridine followed by ring closure gives the intermediate **I-361** which rearranges to **XII-362** through cleavage of the N–O bond. bsequent cleavages of the C–N or C–O bond give the observed products **I-363** to **XII-365**.[273]

The intermediate **XII-366**, formed by cyclization at the 1-position of oquinoline-2-oxide, cannot rearrange to a cyclic intermediate comparable to **II-362**, and an ether product analogous to **XII-364** is not formed.[274]

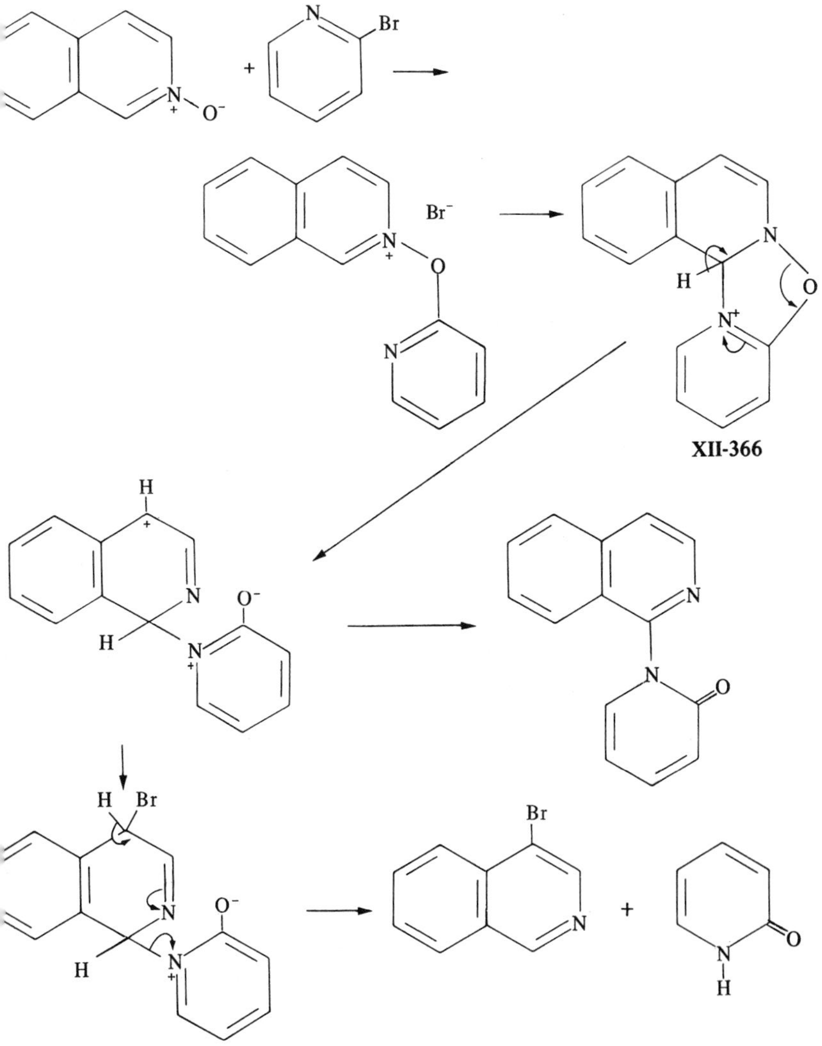

Br

$Br^-$

H

XII-366

H

H

H Br

H

Br

4-Alkoxypyridines rearrange at elevated temperatures to *N*-alkyl-4-pyridon Isomerizations of this kind are catalyzed by acids and alkyl halides. Spinner a White, however, have prepared 4-methoxypyridine from 4-chloropyridine a sodium methoxide in 86% yield through temperature control and neutralization of the reaction mixture with solid carbon dioxide.[275]

2-(3,5-Dinitro-2-pyridyl)pyridinium chloride **(XII-367)**, formed from 3,5- nitro-2-chloropyridine and pyridine in benzene or ether, is readily hydrolyzed an alkaline medium to 3,5-dinitro-2-pyridone.[276]

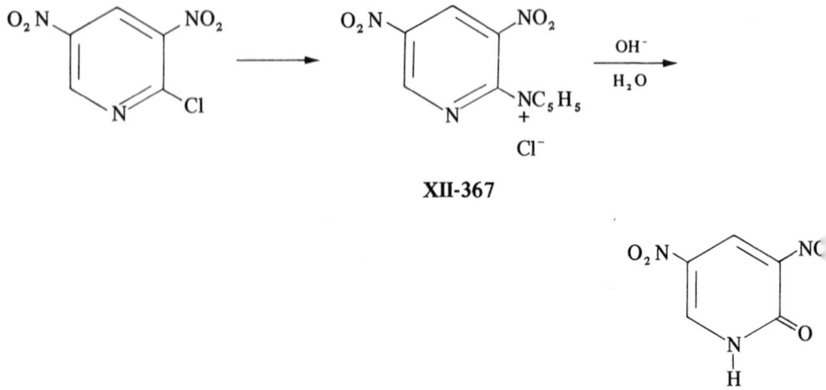

**XII-367**

## B. *Nitropyridines*

A nitro group in the 2- or 4-position of the pyridine ring is susceptible nucleophilic substitution under relatively mild conditions, and in these reactio it competes with a 3- or 5-halogen.[277] 3-Halo-4-nitropyridines (halo = Cl, Br, react with aqueous barium or potassium hydroxide to form 3-halo-4-pyridon and with alkoxides to form 4-alkoxy-3-halopyridines.[254] However, 3-fluoro- nitropyridine is converted to 4-nitro-3-pyridinol or 3-alkoxy-4-nitropy dines.[256a] The corresponding *N*-oxides react similarly.[256] (See Section I.6.A., 688).

4-Nitro-3-picoline **(XII-368)** is hydrolyzed on standing to 1-(3′-methyl- pyridyl)-3-methyl-4-pyridone and 3-methyl-4-pyridone.[278] 4-Nitropyridine a benzyl bromide give *N*-benzyl-3,5-dibromo-4-pyridone **(XII-369)** in 33% yie The quaternary salt first formed is hydrolyzed to the 4-pyridone. Oxidation hydrogen bromide by the nitrous acid product provides the bromine and wat The yield is increased to 71% by the addition of bromine. 3,5-Dibromo-4-chlor pyridine and benzyl bromide in base also give **XII-369**.[279] Similar 3-methyl-4-nitropyridine and benzyl bromide give *N*-benzyl-3-bromo-5-methyl-

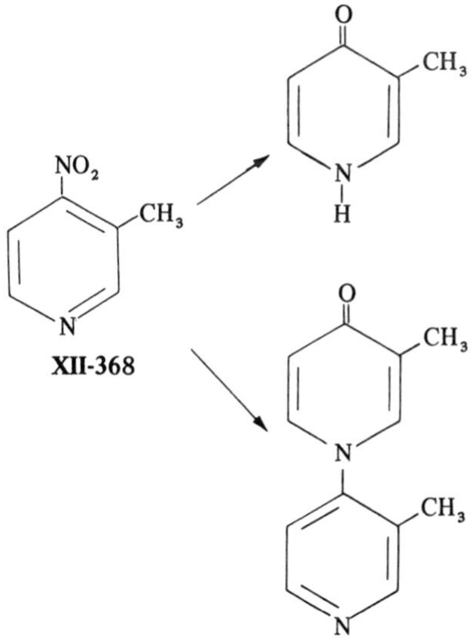

XII-368

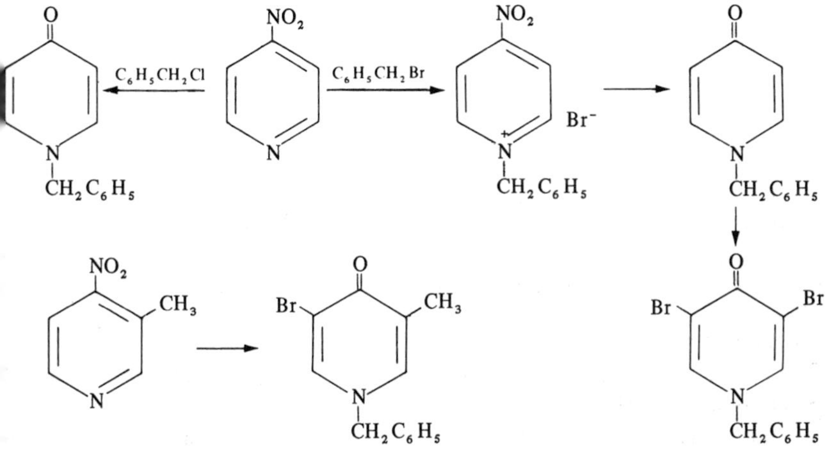

XII-370

XII-369

ridone (**XII-370**). *N*-Benzyl-4-pyridone is prepared in 74% yield from nitropyridine and benzyl chloride.

4-Nitropyridine and 4-hydroxypyridine-1-oxide in boiling methanol react give *N*-(4′-pyridyloxy)-4(1*H*)pyridone (**XII-371**), which rearranges to 3′-(4,4′-dihydroxy)bipyridyl on heating at 100° in benzene or dioxane. vidence supports a homolytic fission of the N–O bond in **XII-371** to give a radical intermediate.[280]

**XII-371**

## C. *Aminopyridines*

Although many 3-aminopyridines can be diazotized and coupled, the ri nitrogen activates the diazonium ions from 2- and 4-aminopyridines, and th are rapidly converted to pyridones. A relatively large number of pyridinols ar pyridones have been formed by this route. Examples are listed in Tables XII-4 XII-6.

Although aminopyridines can be prepared by the reduction of nitropyridin 2- or 4-nitro groups are readily displaced by nucleophiles. For examp 3-halo-4-nitropyridines **(XII-372)** can be converted to the intermedia 4-amino-3-halopyridines **(XII-373)** by reduction with potassium hydrosulfide by displacement with ammonia, or they can be hydrolyzed directly 3-halo-4-pyridones **(XII-374)**.[254] In those cases where both methods have be used to prepare pyridones the route **XII-372** → **XII-373** → **XII-374** usually giv lower yields.[299] Nitration of 2-amino-5-bromopyridine with warm sulfu acid-nitric acid gives 5-bromo-3-nitro-2-pyridone **(XII-375)**.[304] Howev 4-amino-2-bromo-3-nitropyridine prepared from 4-amino-2-bromopyridine

BLE XII-4. 2-Pyridones from 2-Aminopyridines

$$R_3,R_4,R_5,R_6\text{-2-}NH_2\text{-pyridine} \longrightarrow R_3,R_4,R_5,R_6\text{-2(1}H\text{)-pyridone}$$

| $R_n$ | | | | | |
|---|---|---|---|---|---|
| | 4 | 5 | 6 | % Yield | Ref. |
| $_3$ | | | | 73 | 281 |
| | | $CH_3$ | | 58 | 281 |
| $H_5$ | $CH_3$ | | | mixture, 75 | 281 |
| | $CH_3$ | $C_2H_5$ | | same as above | 281 |
| | | $CH_3$ | | 75 | 282 |
| | $C_2H_5$ | | | 69 | 283 |
| | EtO | Br | | 85 | 284 |
| $)_2$ | | | | mixture, 88 | 285 |
| | | $NO_2$ | | same as above | 285 |
| | $NO_2$ | | | 85 | 286 |
| | $NO_2$ | $NO_2$ | | 80 | 286 |
| $)_2$ | | I | | 75 | 287 |
| | | $NO_2$ | | 80 | 287 |
| $)_2$ | $CH_3$ | | $CH_3$ | 97 | 288 |
| | $CH_3$ | $NO_2$ | $CH_3$ | – | 288 |
| | $CH_3$ | Br | $CH_3$ | 50 | 289 |
| $)_2$ | $CH_3$ | Br | $CH_3$ | – | 289 |
| | | $NO_2$ | | 67 | 290 |
| $)_2$ | | Br | | 83 | 290 |
| | | $NO_2$ | | 72 | 291 |
| $)_2$ | | Cl | | 70 | 291 |
| | | OH | | – | 292 |
| $H_5$ | $CH_3$ | OH | | 60 | 292 |
| | $p$-$C_6H_4NO_2$ | | $C_6H_5$ | 61 | 74 |
| | $CH(CH_3)_2$ | | | 73 | 226 |

TABLE XII-5. 4-Pyridones from 4-Aminopyridines

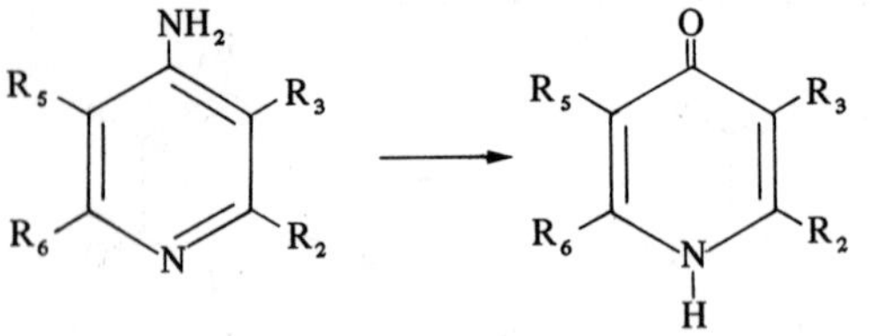

| 2 | 3 | 5 | 6 | % Yield |
|---|---|---|---|---|
| $CH_3$ | | | | 80 |
| $CH_3$ | | | $CH_3$ | 85 |
| $CH_3$ | | | $CH_3$ | 33 |
| Cl | | | Cl | 65 |
| F | | | | 79 |
| | F | | | 58 |
| $CH_3$ | | $CH_3$ | | 60 |
| | Br | | | 45 |
| | I | | | – |

TABLE XII-6. 3-Pyridinols from 3-Aminopyridines

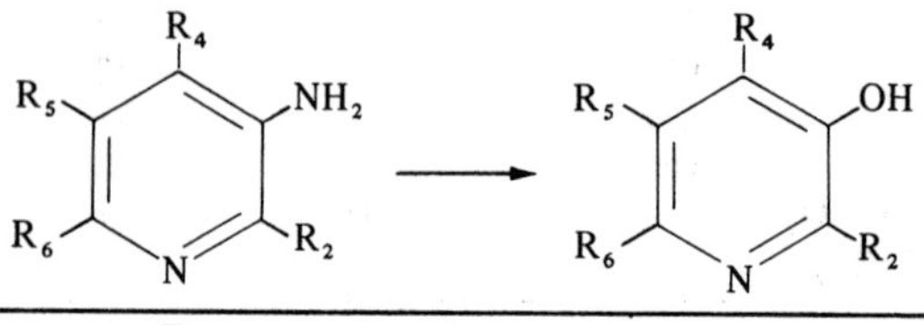

| 2 | 4 | 5 | 6 | % Yield |
|---|---|---|---|---|
| $CH_3$ | | | $CH_3$ | 62 |
| $CH_3$ | $CH_3$ | | $CH_3$ | 52 |
| | | | $C_6H_5CONH$ | 54 |
| | | OH | | 20.7 |
| | $COCH_3$ | | | 76 |
| $COCH_3$ | | | | 52 |
| I | | | I | 14 |
| $CH_3$ | COOH | COOH | | 80 |

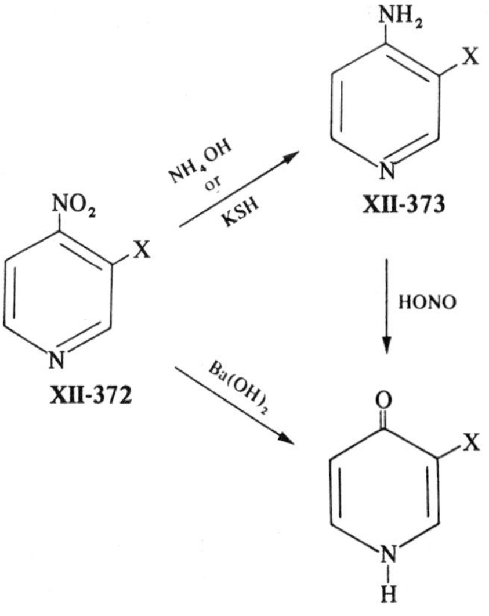

NH₂ X N XII-373

$NH_4OH$ or KSH

$NO_2$ X N XII-372

HONO

$Ba(OH)_2$

O X N H XII-374

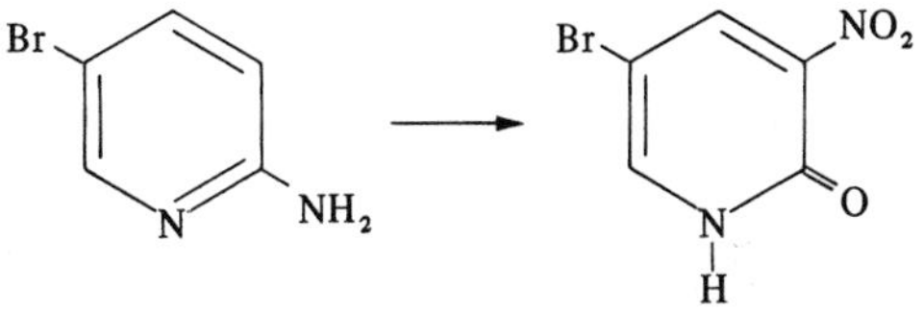

Br N $NH_2$ → Br $NO_2$ O N H XII-375

nitramine can be converted to 4-amino-3-nitro-2-pyridone **(XII-376)** with
:assium hydroxide containing hydrogen peroxide. 3-Nitro-4-hydroxy-
yridone has been formed from **XII-376** by treatment with nitrous acid.[305]

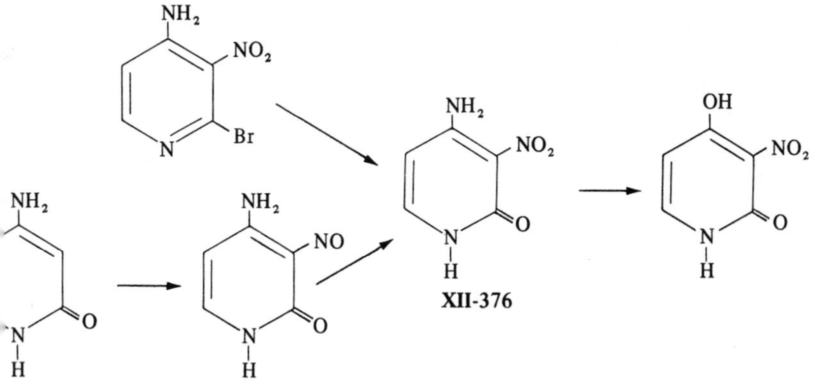

$NH_2$ $NO_2$ N Br

$NH_2$ O N H XII-377 → $NH_2$ NO O N H → $NH_2$ $NO_2$ O N H XII-376 → OH $NO_2$ O N H

Both 4-amino-2-pyridone (**XII-377**)[305] and 4-amino-1,6-dimethyl-2-pyridone[3 form 3-nitroso derivatives when treated with nitrous acid. Oxidation 4-amino-3-nitroso-2-pyridone also gives **XII-376.**[305] 2,6-Diaminopyridine is nit sated and diazotized with nitrous acid in aqueous sulfuric acid. Ultraviol infrared and nuclear magnetic resonance spectra are consistent with t 1,2,3,6-tetrahydro-3-hydroxyimino-2,6-pyridinedione structure **XII-378.**[3 Under mild hydrolytic conditions the nitrogens of the ring and of the ami

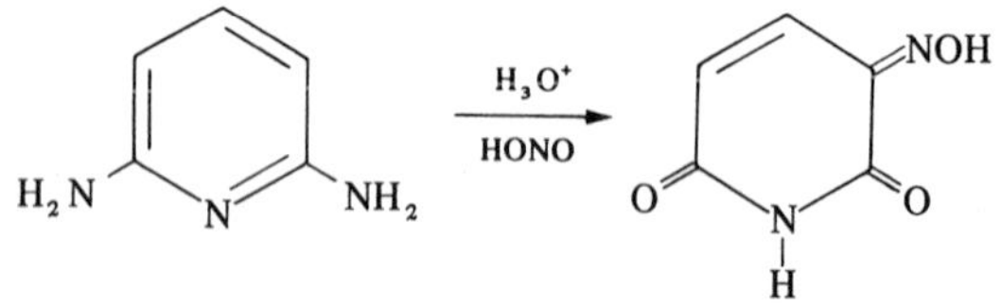

$H_3O^+$ / HONO

**XII-378**

group in 3- and 5-nitro-2-aminopyridine **XII-379** exchange positions (a Dimro rearrangement). In reactions with nitrous acid, kinetic studies with **XII-3** indicate that the intact pyridines and not the ring-opened species rea preferentially, (**XII-379a → XII-380a, XII-379b → XII-380b**).[308, 309]

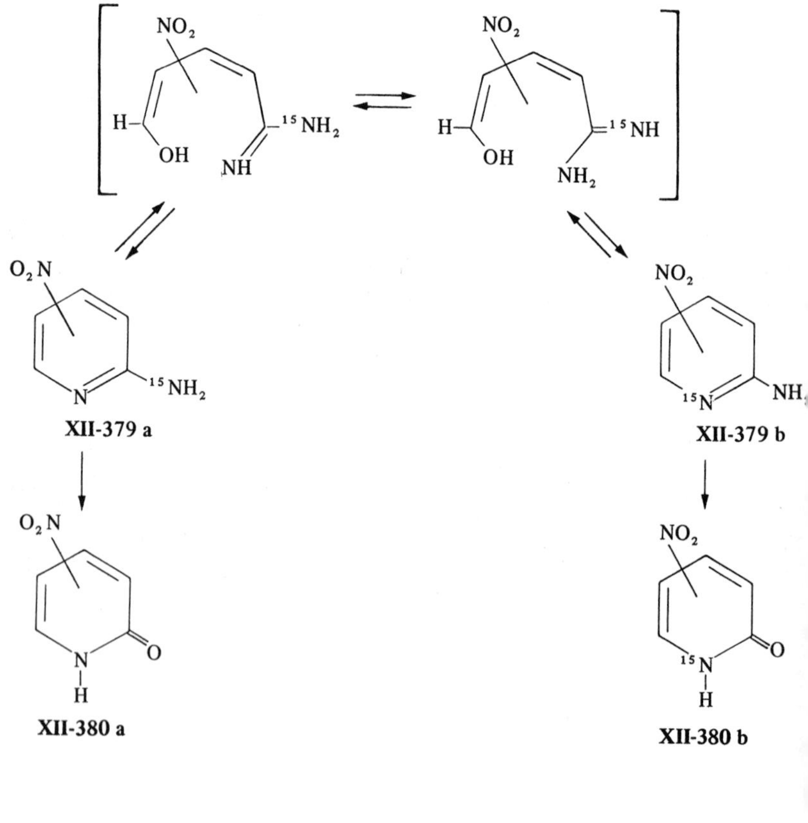

**XII-379 a**

**XII-379 b**

**XII-380 a**

**XII-380 b**

4-Amino-3-(β-phenylethyl)pyridine and nitrous acid give 4-hydroxy-3-phenylethyl)pyridine (**XII-381**) with or without Gattermann copper powder, t 3-amino-4-(β-phenylethyl)pyridine undergoes a Pschorr cyclization to **I-382** in the presence of copper.[310]

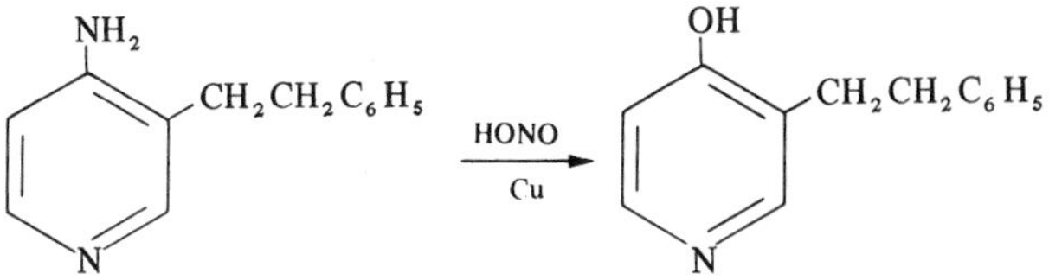

HONO / Cu

**XII-381**

HONO / Cu

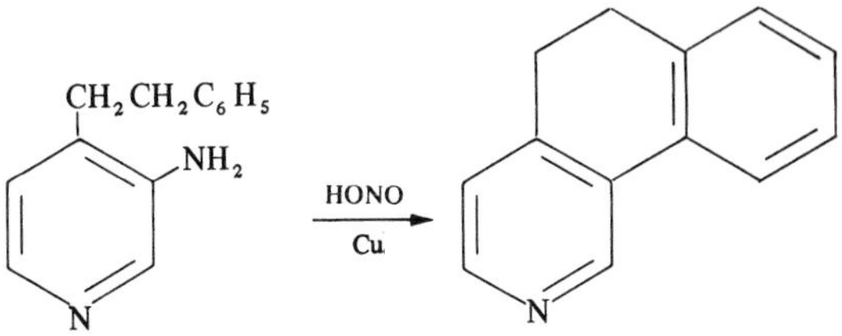

**XII-382**

2-Amino-5-pyridinol (**XII-383**) and 5-hydroxy-2-pyridone can be prepared om 2-amino-5-nitropyridine. After protecting the 2-amino group by nzoylation, hydrogenation followed by treatment with nitrous acid and acid ydrolysis gives **XII-383**. 5-Hydroxy-2-pyridone can be isolated as the onobenzoate from the reaction between **XII-383** and nitrous acid after

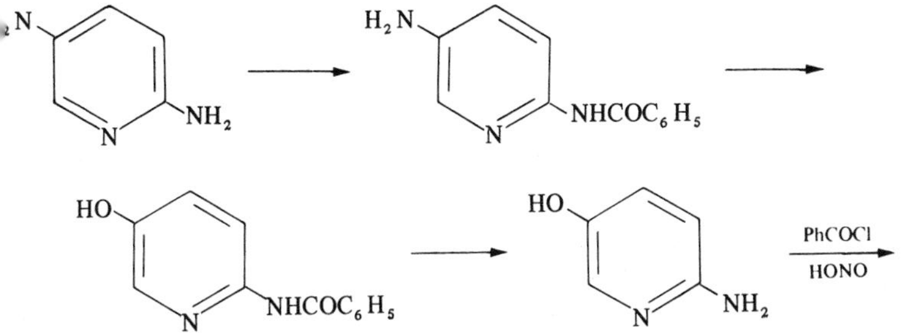

PhCOCl / HONO

**XII-383**

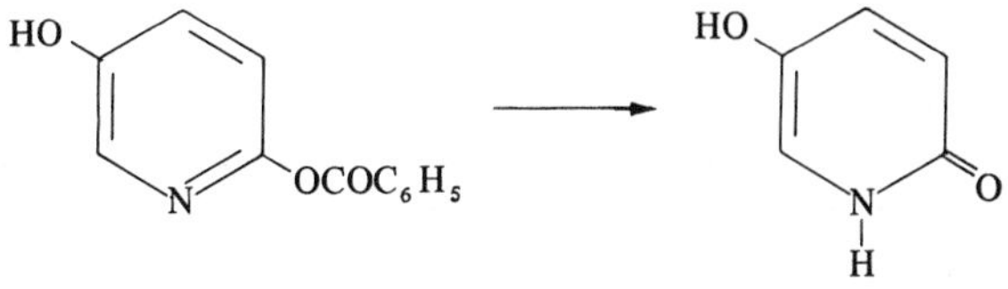

treatment with benzoyl chloride.[292] It has been prepared in low yield fro 2-aminopyridine[311] or from 3-ethoxypyridine[312] or more conveniently and improved yields by the Elbs peroxydisulfate oxidation of 2-pyridone.[313]

2-(3-Amino-2-pyridyl)propene **(XII-384)**, on treatment with nitrous acid hydrochloric acid, gives 3-chloro-2-isopropenylpyridine and 2-acetyl-3-pyridin Bicyclic products were not detected. Nitrous acid oxidation of 2-isopropenyl- pyridinol or treatment of 2-acetyl-3-aminopyridine with nitrous acid also giv 2-acetyl-3-pyridinol.[302]

$NH_2$ $C{=}CH_2$ $CH_3$ N **XII-384** → Cl $C{=}CH_2$ $CH_3$ N + OH $COCH_3$ N

OH $C{=}CH_2$ $CH_3$ N ; $NH_2$ $COCH_3$ N

The composition of a mixture of 3- and 5-ethyl-2-aminopyridines from t direct amination of 3-ethylpyridine has been estimated by conversion *N*-methyl-2-pyridones.[314]

## D. *Nitramines*

Nitration of aminopyridines with mixed acid gives nitramines. 2- a 4-Nitraminopyridines when heated with sulfuric acid rearrange to aminonit pyridines. For example, 2-nitraminopyridine **(XII-385)** gives a mixture 3-nitro- and 5-nitro-2-aminopyridine and only a trace of 2-pyridone. The yield 2-pyridone is increased significantly by using acetic acid-acetic anhydride place of sulfuric acid. 3-Nitraminopyridine when treated with sulfuric acid acetic acid-acetic anhydride gives 3-pyridinol as the major product.[3 2-Aminopyridine has been used to prepare 3,5-dinitro-2-pyridone in an over yield of 37% *via* the nitramine **(XII-385)** and the aminonitropyridines.[285]

2-Amino-5-bromo-3-nitropyridine and nitric acid in sulfuric acid give t 2-nitramine, which reacts further with sulfuric acid-nitric acid to g

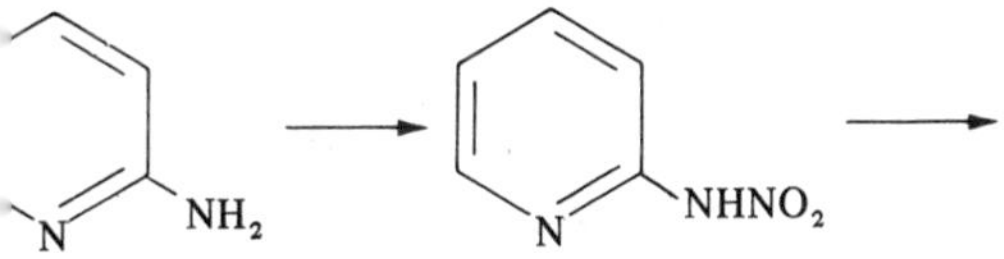

XII-385

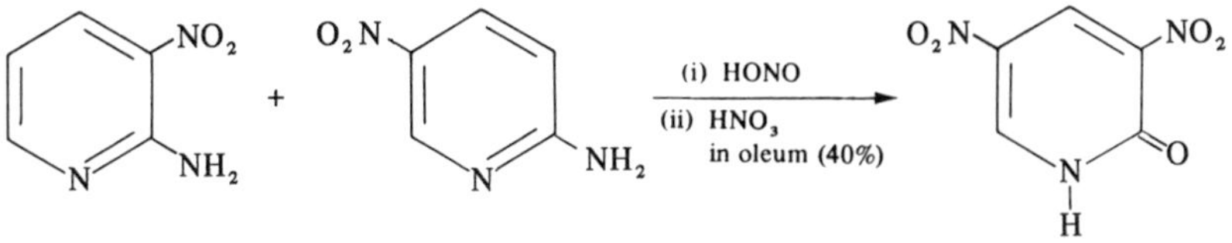

ɔromo-3-nitro-2-pyridone (**XII-386**),[304] a behavior characteristic of 3,5-disub-tuted-2-nitramines.[315]

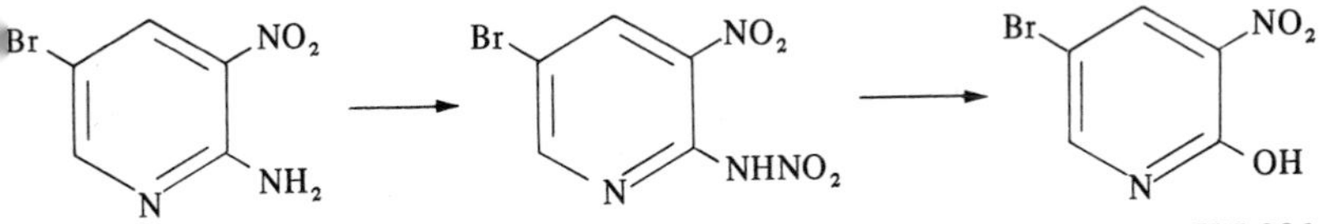

XII-386

2-Nitraminonicotine and 6-nitraminonicotine and sulfuric acid give nitrous ɛide and hydroxynicotines (**XII-387**, 54%; **XII-388**, 74%) and amino-5-nitro-cotines. The nitraminonicotines and acetic anhydride in acetic acid give some

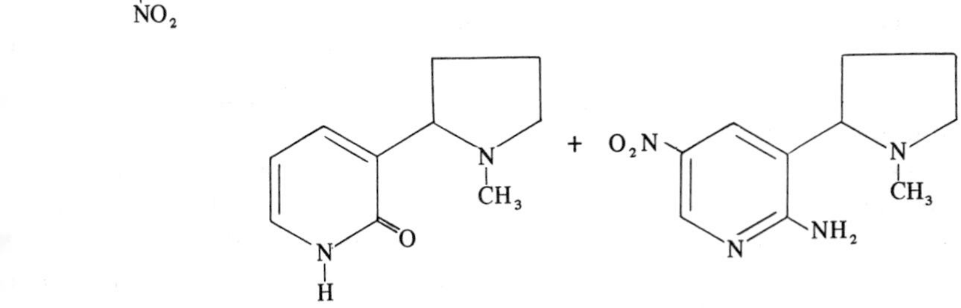

XII-387

hydroxynicotine, tars, and nitrogen. Also with an equimolecular amount triethylamine in acetic anhydride, 2-nitraminopyridine reacts similarly to give good yield of nitrogen, 2-pyridone and some diethylacetamide. It has be suggested that an intermediate oxidation of the *N*-methylpyrrolidine ring or triethylamine by the nitramine is responsible for nitrogen formation (**XII-38 XII-390**).[316]

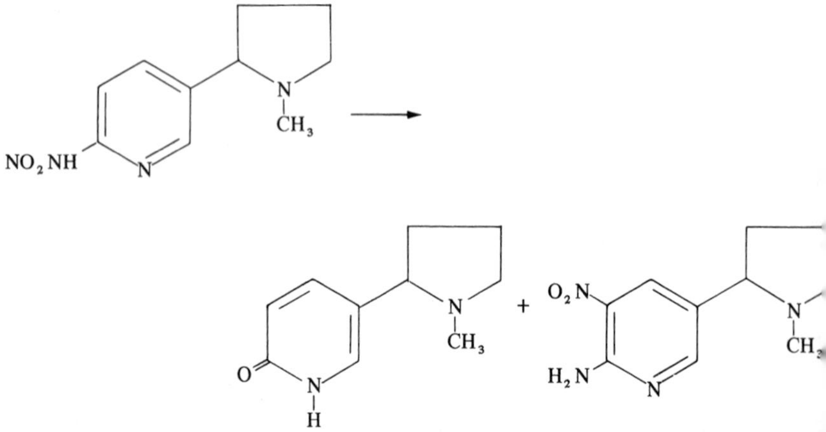

**XII-388**

$$RNHNO_2 + Et_3N \longrightarrow ROH + N_2 + Et_3NO \quad \textbf{(XII-38}$$

$$Et_3NO + Ac_2O \longrightarrow AcNEt_2 + CH_3CHO + AcOH \quad \textbf{(XII-39}$$

Czuba has studied 3-nitramines in some detail.[317] Nitramines of the fo chloro-3-aminopyridines,[318] 3-amino-5-ethoxypyridine, 5-aminonicotinic ac 5-aminopicolinic acid,[319] 3-amino- and 5-amino-2-pyridinesulfonic acids,[3 3-amino-4-picoline and 5-amino-3-picoline[321] have been prepared. Attempts prepare nitramines from the four 3-aminopyridinols were unsuccessful.[3 However, 3-amino-5-ethoxypyridine forms the nitramine that reacts in sulfu acid to give 5-ethoxy-3-pyridinol.[319]

3-Nitraminopyridines that normally do not rearrange when heated wi sulfuric acid give azopyridines (**XII-391**; R = 6-Cl, 2-Cl, 5-Cl, 4-Cl, 5-COOH) azoxypyridines (**XII-392**, R = 6-COOH) in addition to 3-pyridinols (**XII-393**; R 5-Cl, 5-$OC_2H_5$, 4-$CH_3$, 5-$CH_3$, 2-$SO_2OH$).[317] 5-Nitramino-2-pyridinesulfo acid gives 3-pyridinol.[320]

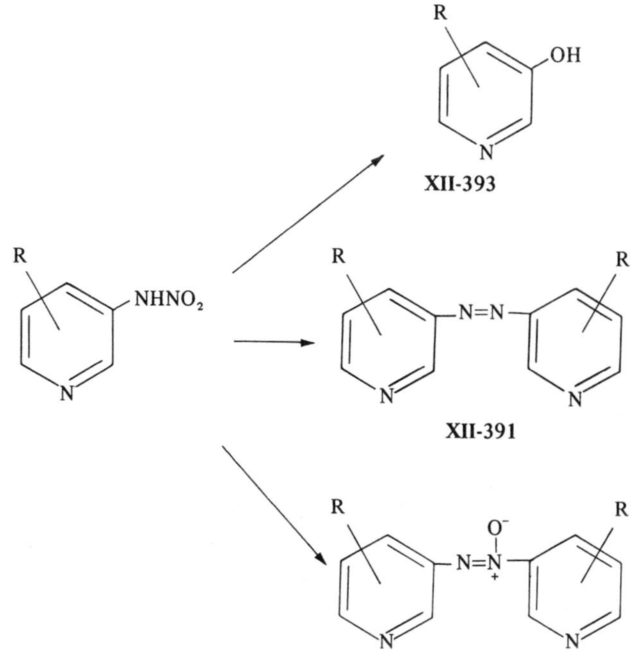

XII-393

XII-391

XII-392

## E. *Pyridonimines*

- or 4-Pyridonimines can be prepared by treatment of aminopyridines with yl halides and then with silver oxide.[322] 2-Aminopyridines and styrene oxides aqueous ethanol form *N*-(β-hydroxy-β-phenylethyl)-2-pyridonimines (**XII-394**; = H, $CH_3$).[323] Treatment of **XII-394** with aqueous alkali gives β-hydroxy-β-phenylethyl)-2-pyridone.[323, 324] 2-Aminopyridine and 3-(*o*-yloxy)propylene oxide give **XII-395** after hydrolysis.[325] In the presence of amide in liquid ammonia *N*-alkylation of 2-aminopyridine occurs to give -396.[324]

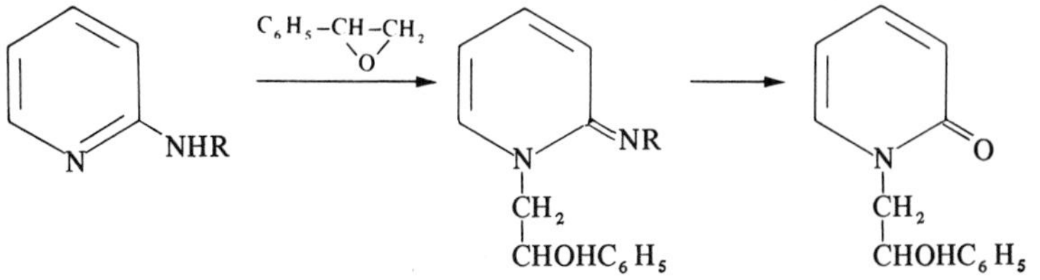

$C_6H_5-CH-CH_2$ (O)

XII-394

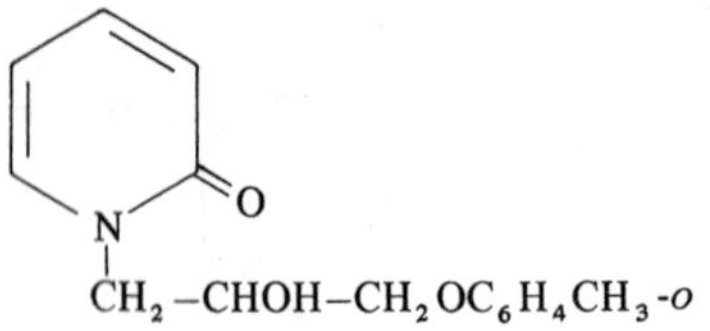

$CH_2-CHOH-CH_2OC_6H_4CH_3$-*o*

**XII-395**

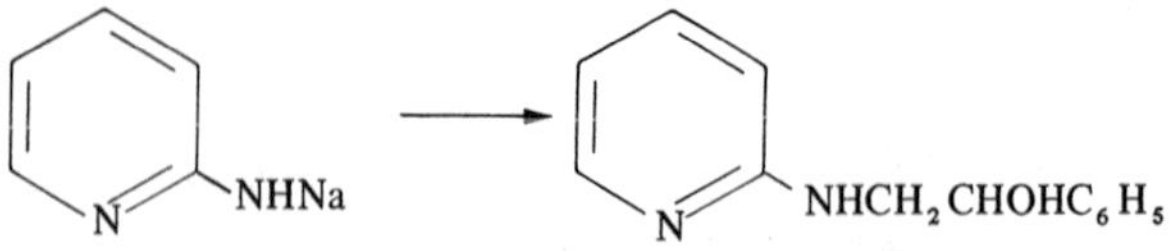

NHNa → $NHCH_2CHOHC_6H_5$

**XII-396**

1-(β-Hydroxyalkyl)-2-pyridones can be synthesized by the direct alkylation 2-pyridone with alkylene oxides.[326]

Although 3- and 5-nitro-1-methyl-2-pyridonimines undergo the Dimrc rearrangement rapidly to give **XII-397** when treated with base, analogs simila substituted with less powerful electron withdrawing groups ($R_3 = R_5 =$ H; $R_3$ $R_5$ = Cl; $R_5$ = Cl; $R_3$ = H) undergo hydrolysis to give the *N*-methyl-2-pyrido The corresponding 5-cyano-2-imino-1-methylpyridine did not react under conditions used.[327]

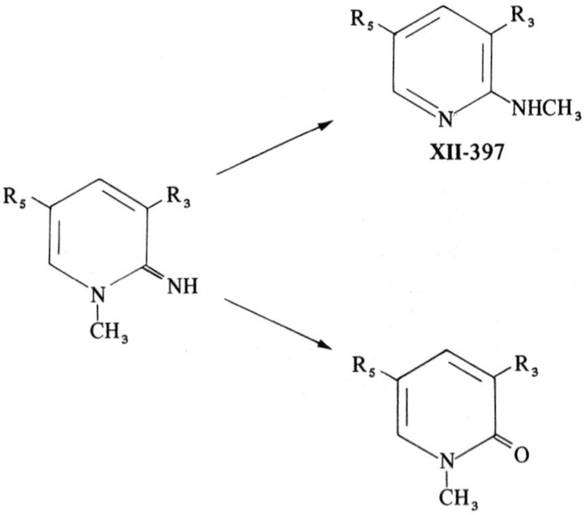

$R_5$ $R_3$ NH $CH_3$ → $R_5$ $R_3$ $NHCH_3$ **XII-397**; → $R_5$ $R_3$ O $CH_3$

## F. *Quaternary Salts*

Alkaline potassium ferricyanide oxidation of *N*-substituted pyridinium salts ten is a method of choice for the preparation of *N*-alkyl-2-pyridones. For examp methyl-2-pyridone is prepared conveniently from pyridine by conversion to e quaternary methosulfate followed by oxidation with potassium ferricyanide.[328] Certain quinones with high redox potentials are reduced by methylpyridonium methosulfate, but in these reactions *N*-methyl-2-pyridone uld not be isolated.[329] However, *N*-methyl-2-pyridone is isolated when ueous methylpyridinium hydroxide **(XII-398)** is treated with *p*-benzoinone.[329, 330]

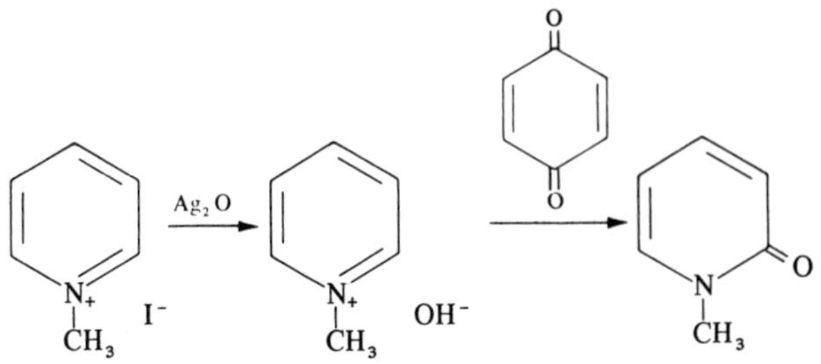

$Ag_2O$

**XII-398**

Dipole moments have indicated that the oxidation of 1-phenethyl-3-phenoxyridinium salts by alkaline ferricyanide gives the 6-pyridone **(XII-399)** and that e corresponding 3-bromo- and 3-cyanopyridinium salts (R = Br, CN) form the pyridones **(XII-400)**.[331] Subsequently these structures have been conmed.[332] The acetal of *N*-(3,4-dimethoxyphenethyl)-3-ethyl-4-formylridinium bromide gives about equal amounts of the 2- and 6-pyridones.[333] Substituents on the 3-position can contribute both electronic and steric fects.[333-335] It has been shown that a 3-methyl group is slightly activating

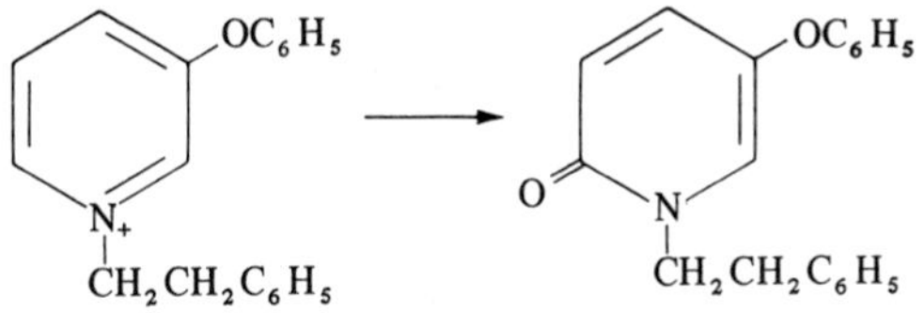

**XII-399**

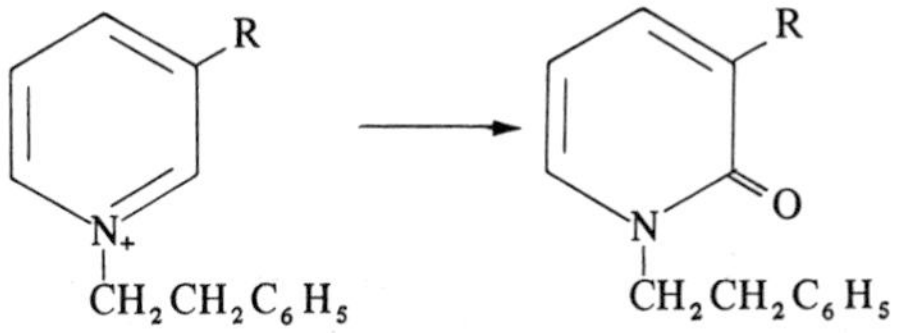

**XII-400**

d directs oxidation to the 2-position (**XII-401**:**XII-402**–96.6:3.4). The cyano group is appreciably activating and also directs oxidation to the

2-position but 6- and 4-pyridones (**XII-402** and **XII-403**) are also form (83.1:15.9:1). The carbomethoxy group is deactivating and directs only to t 6-position.[334] Neither C–H bond breaking nor the addition of hydroxide ion rate-determining. It has been suggested that the rate-determining step is t formation of a complex (**XII-404**) that then reacts with more ferricyanide form a second complex that gives the pyridone.[334]

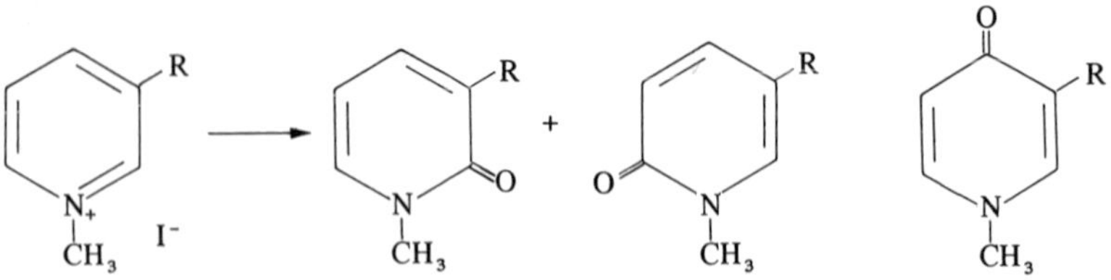

**XII-401** **XII-402** **XII-403**

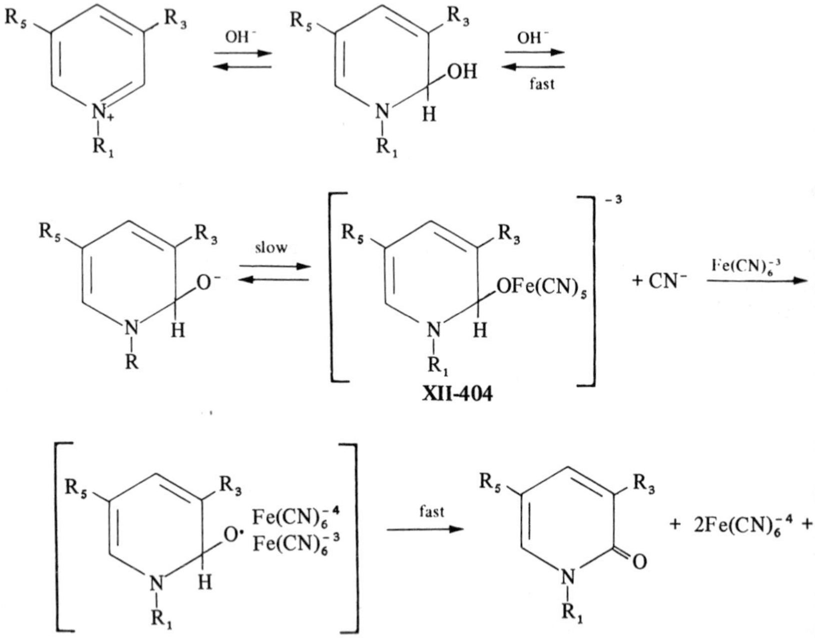

**XII-404**

The alkylation of the alkaloid anabasin (**XII-405**) with methyl iodide gives t hydroiodide of *N*-methylanabasin iodomethylate which gives *N,N'*-dimeth anabasone (**XII-406**) when treated with potassium ferricyanide.[336]

The oxidation of **XII-406** to 2(S)(−)-*N*-methylpipecolinic acid has establish the absolute configuration of **XII-405**.[336]

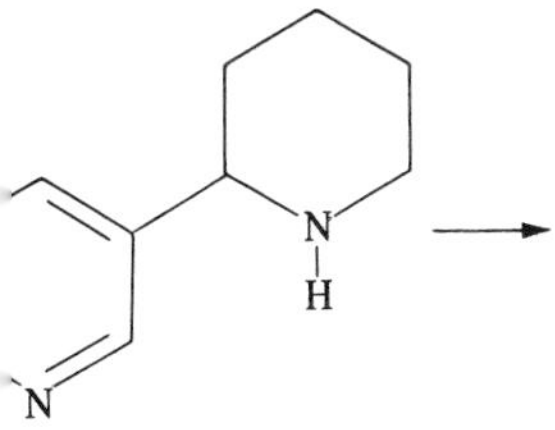

N H N

XII-405

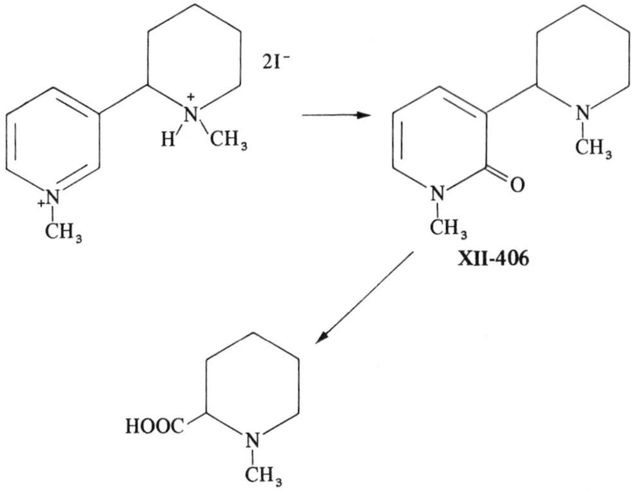

2I⁻ N H CH₃ N CH₃ N CH₃ N CH₃ O

XII-406

HOOC N CH₃

-(1-Methyl-2-phenylethyl)-2-pyridone,[337] 1-β-(3,4-dimethoxyphenethyl)-2-idone,[338] 1-(3,4-dimethoxyphenethyl)-5-(1-methyl-2-pyrrolidinyl)-2-pyri-ıe,[339] 1,1′-hexamethylene-di-2-pyridone,[6] and 1-[3-(3,4-methylenedioxy-enyl)propyl]-2-pyridone[340] have been formed by oxidation of pyridinium :s by alkaline potassium ferricyanide. However, attempts to oxidize β-6-methoxy-1-naphthylethyl)pyridinium bromide[341] or 1-methyl-3-pyri-ol[342] to the pyridone with alkaline potassium ferricyanide have been uccessful.

'yano-, halo-, and amino- groups on the 2- and 4-positions of pyridinium ions susceptible to nucleophilic substitution. For example, 2- and 4-cyano-1-thylpyridinium perchlorate (**XII-407** and **XII-408**) and aqueous sodium lroxide give both amides and pyridones. The ratio of rates of reaction of the :ile group are 50:5.7:1 (2-CN > 4-CN > 3-CN) and at the ring carbon are )0:43:1 (2-CN > 4-CN > 6-position of 3-CN).[343] 1-Methyl-3-cyanopyri-ium perchlorate (**XII-409**, X = $ClO_4$) in aqueous sodium hydroxide has been orted to give 3-carbamidopyridinium perchlorate (87%) and smaller amounts

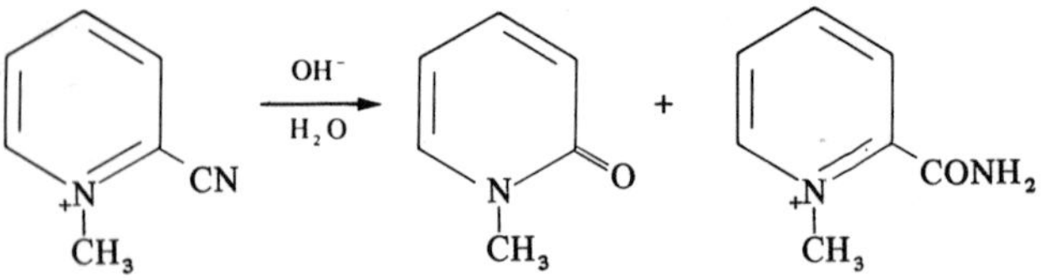

XII-407

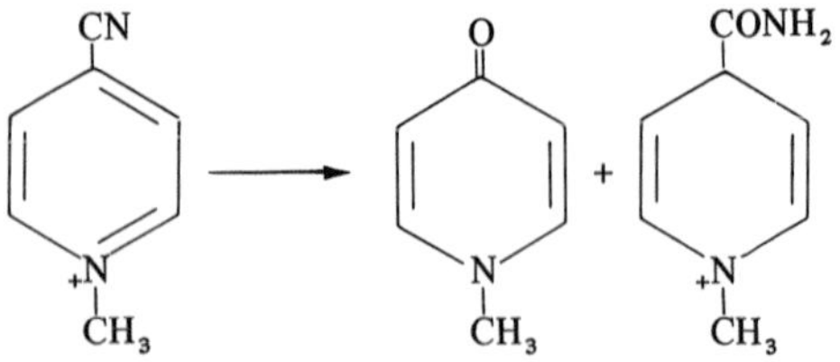

XII-408

of 4-cyano-5-methylamino-2,4-pentadienal (6%) and 1-methyl-5-cyano-2-p done (~1%).[343] However, oxidation of the corresponding iodide (**XII-409**, I) with alkaline aqueous potassium ferricyanide has been shown to give only 3-cyano-*N*-methylpyridones and no hydrolysis products.[334]

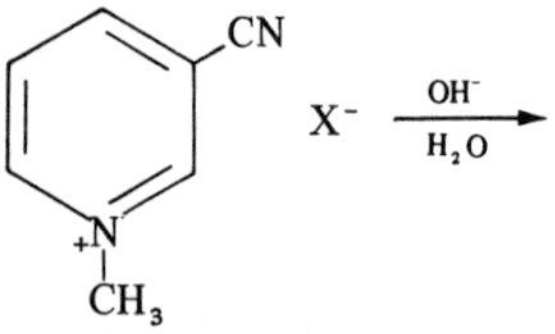

XII-409

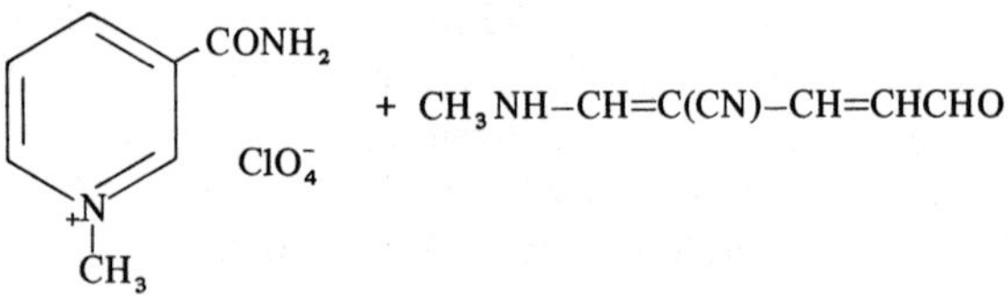

$+ CH_3NH–CH{=}C(CN)–CH{=}CHCHO$

4- And 2-bromo-1-alkylpyridinium fluoroborates (**XII-410**) and *para*-s stituted phenylhydrazines form phenylhydrazino derivatives that are oxidized nitrous acid to the arylazopyridinium fluoroborates (**XII-411**)[344] that can attacked by nucleophiles at positions *a*, *b*, or *c*.[345, 346] Reaction between al and **XII-411** (X = H) gives *N*-ethyl-2-pyridone and benzene.[344, 345] When X i or Cl only 1% of **XII-412** is formed (attack at *c*). Sulfinates and sodium bisul react at *b* to give adducts.

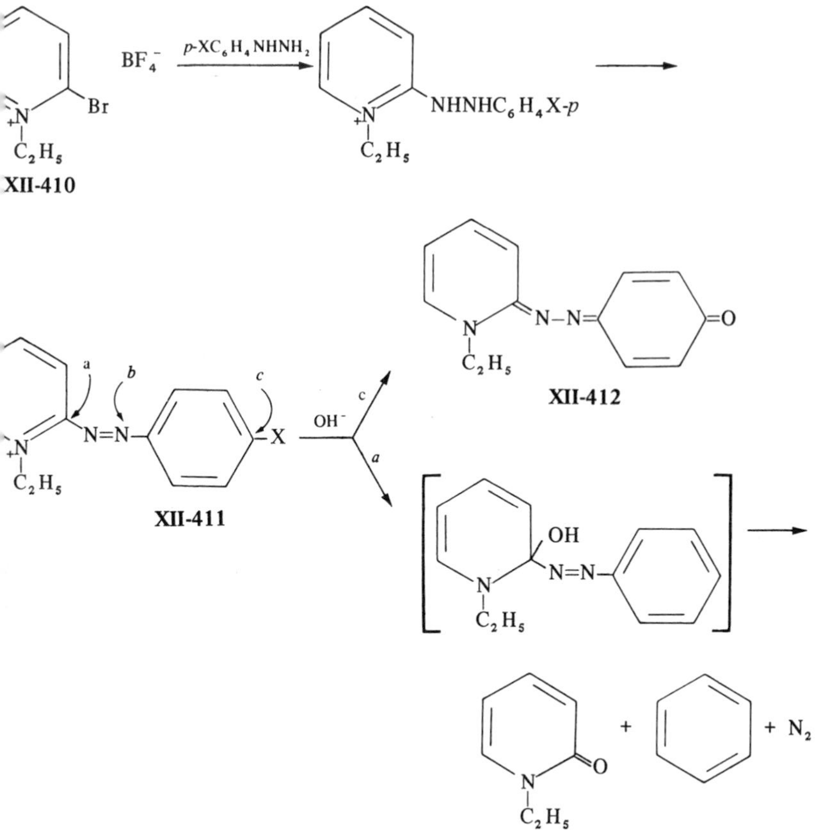

XII-410

XII-411

XII-412

'he 2-cyano-*N*-methylpyridinium ion, the 2-carboxy-1-methylpyridinium ion 1 *N*-methyl-2-pyridone have been identified as metabolites of *N*-methylpyri-ium-2-aldoxime iodide (2-PAM) **(XII-413)**[347, 348] in rats.

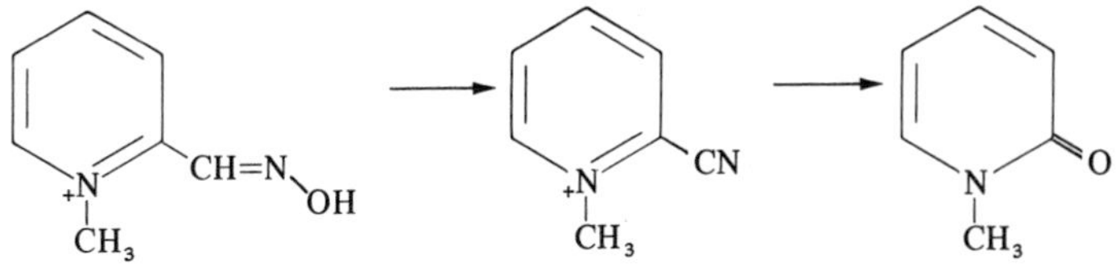

XII-413

.s part of a program to synthesize ipecac alkaloids, dipyridinium diiodides **[-414)**, which are formed from 1,5-disubstituted-2-methylpyridinium salts [ iodine in pyridine, were converted to 2-pyridones **(XII-415)** with aqueous

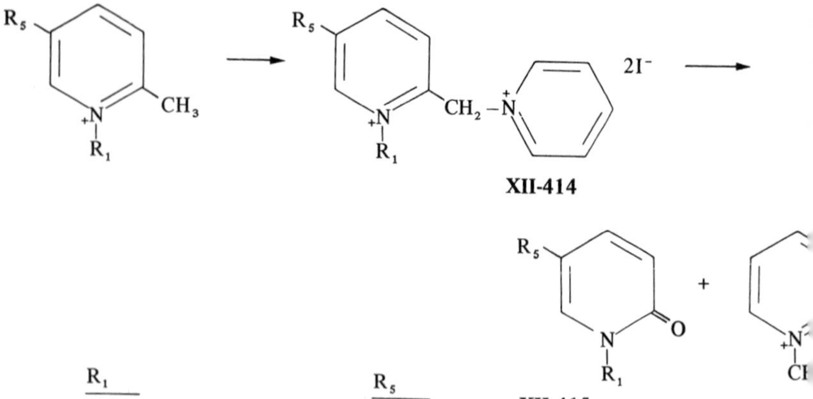

XII-414

XII-415

| $R_1$ | $R_5$ |
|---|---|
| $CH_3$ | H |
| $(CH_2)_2C_6H_5$ | H, $C_2H_5$ |
| 3, 4-$(CH_3O)_2C_6H_3(CH_2)_2$ | H, $C_2H_5$ |

alkali.[349-351] An attempt to use this method with **XII-416**, which has electron withdrawing carboxyl or carbomethoxy group in the 4-position, unsuccessful. Cleavage gives pyridine and not the pyridone.[349, 350]

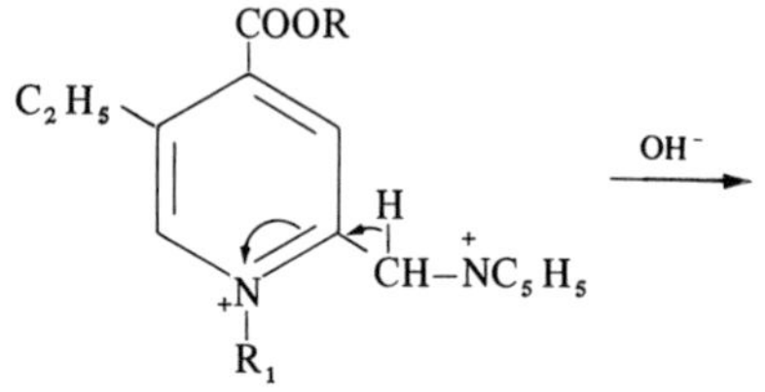

XII-416

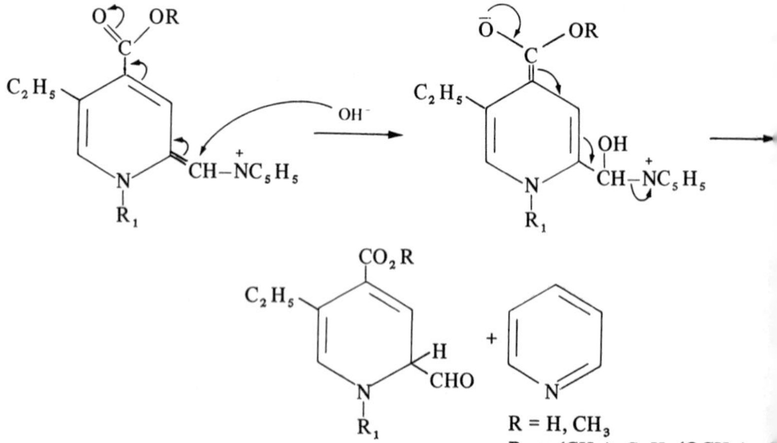

R = H, $CH_3$
$R_1 = (CH_2)_2C_6H_3(OCH_3)_2$-

plication of this route to the synthesis of **XII-415** [$R_1 = (CH_2)_2C_6H_5$, 4-$(CH_3O)_2C_6H_3(CH_2)_2$; $R_5 = C_2H_5$] has also been unsuccessful.[352] (Carboxaldehydediethylacetal)-1-(3,4-dimethoxyphenethyl)-5-ethyl-2-pyridone (**XII-417**) has been prepared in low yield by this method and was used to characterize the products of alkaline ferricyanide oxidation of **XII-418**.[333]

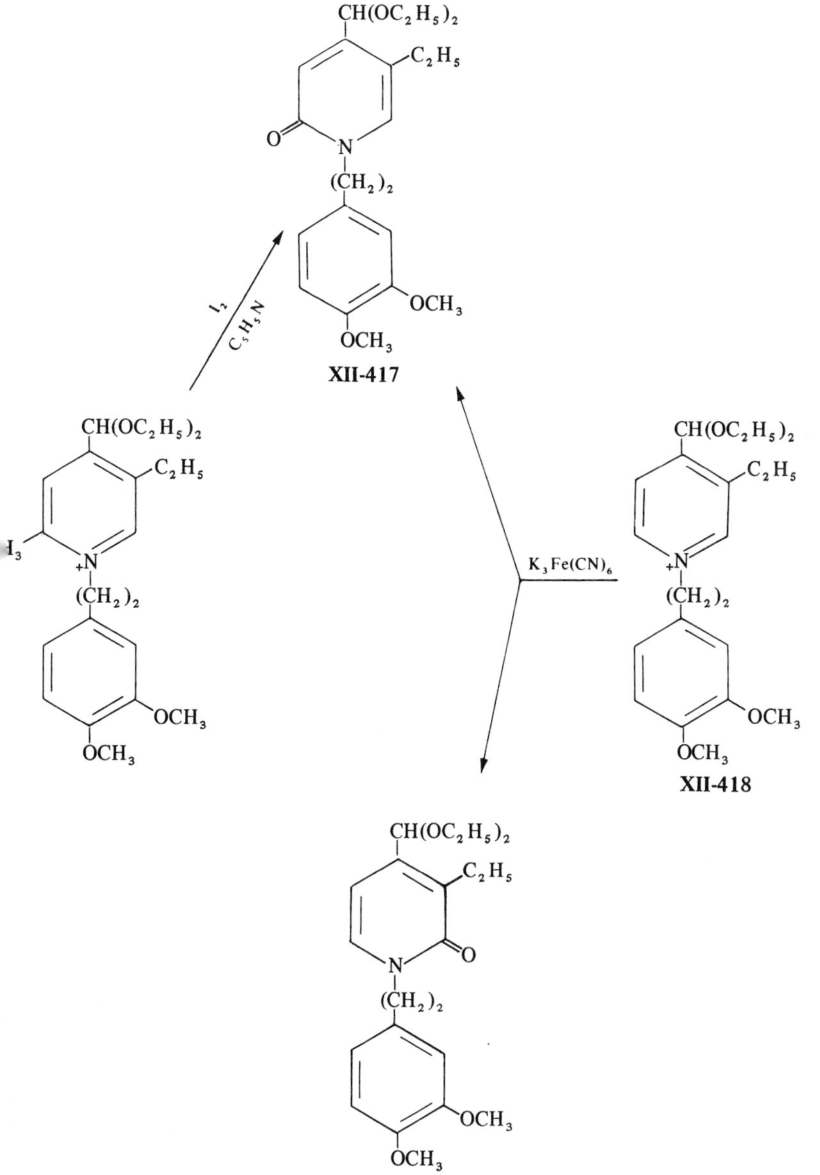

$CH(OC_2H_5)_2$ $C_2H_5$ O N $(CH_2)_2$ $OCH_3$ $OCH_3$

**XII-417**

$I_2$ $C_5H_5N$

$CH(OC_2H_5)_2$ $C_2H_5$ $I_3^-$ $^+N$ $(CH_2)_2$ $OCH_3$ $OCH_3$

$K_3Fe(CN)_6$

$CH(OC_2H_5)_2$ $C_2H_5$ $^+N$ $(CH_2)_2$ $OCH_3$ $OCH_3$

**XII-418**

$CH(OC_2H_5)_2$ $C_2H_5$ O N $(CH_2)_2$ $OCH_3$ $OCH_3$

2-Methylpyridine-1-oxides react with tosyl chloride to give 2-chlorometh derivatives (**XII-419**), which react with pyridine to form 1-(2′-pyridylmethy pyridinium chlorides. These, on methylation, give **XII-420** which form t 2-pyridone[353] by the procedure previously described.[349]

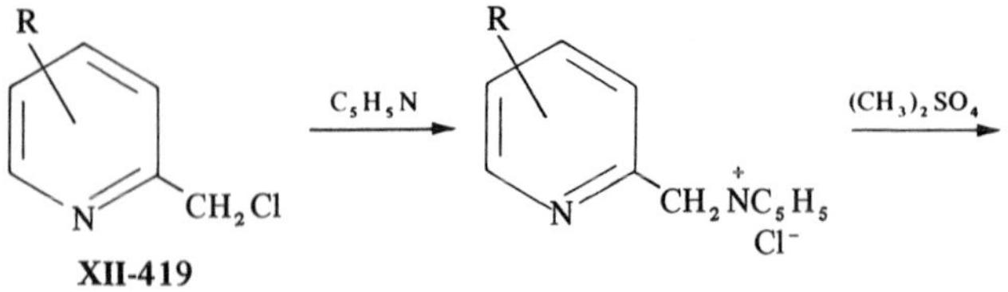

R $CH_2Cl$ N **XII-419** $\xrightarrow{C_5H_5N}$ R N $CH_2\overset{+}{N}C_5H_5$ $Cl^-$ $\xrightarrow{(CH_3)_2SO_4}$

R $^+N$ $CH_3$ $CH_2\overset{+}{N}C_5H_5$ **XII-420** $\xrightarrow{KOH}$ R N H O

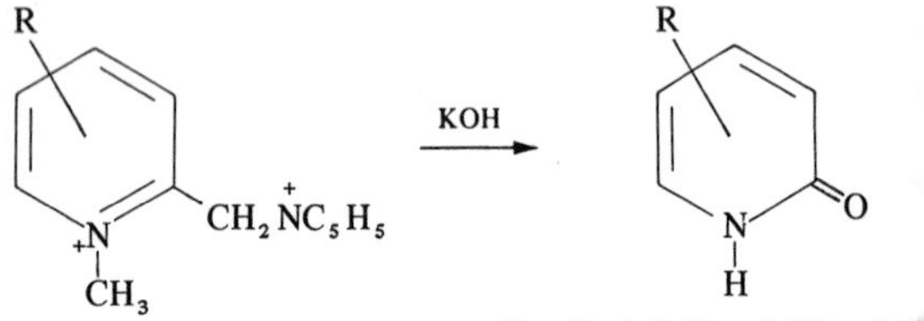

$R = H, 5\text{-}C_2H_5, 6\text{-}CH_3, 4\text{-}C$

5-Acetyl-2-methyl-1-phenylethylpyridinium perchlorate when treated w aqueous sodium hydroxide gives the anhydro base that forms 5-acetyl-1-ph ethyl-2-pyridone (**XII-421**) with potassium permanganate in acetone.[354]

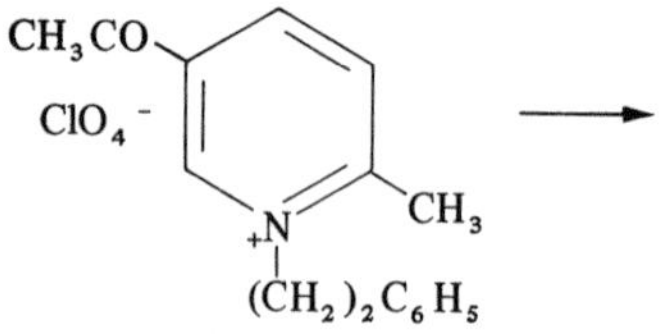

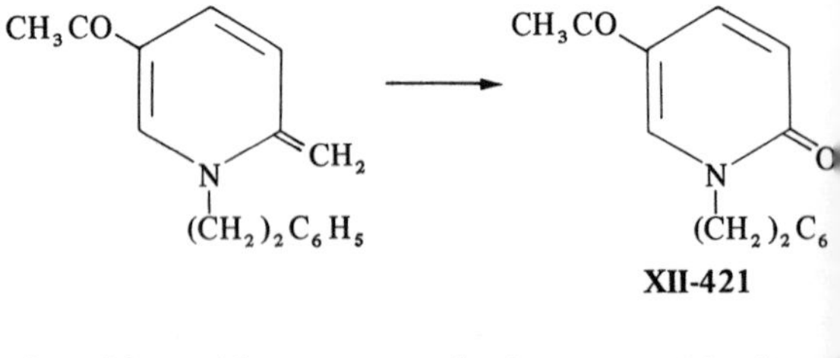

$CH_3CO$ $ClO_4^-$ $^+N$ $CH_3$ $(CH_2)_2C_6H_5$ $\longrightarrow$

$CH_3CO$ N $CH_2$ $(CH_2)_2C_6H_5$ $\longrightarrow$ $CH_3CO$ N O $(CH_2)_2C_6$ **XII-421**

A single *N*-methylnicotinamide oxidase apparently is responsible for t oxidation of *N*-methylnicotinamide to *N*-methyl-2-pyridone-5-carboxam (**XII-422**) and *N*-methyl-4-pyridone-3-carboxamide (**XII-423**) and the oxidati of pyridoxal to pyridoxic acid in man. A xanthine oxidase appears to oxid *N*-methylnicotinamide ion to **XII-422** but not to **XII-423**.[355]

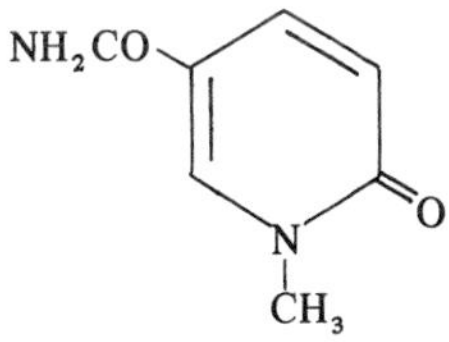

XII-422 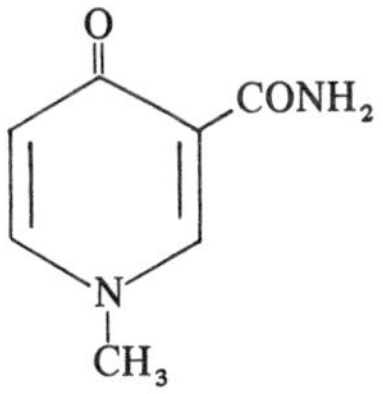 XII-423

## G. N-*Oxides and Anhydrides*

/hen pyridine-1-oxides and related heterocyclic *N*-oxides in which there are alkyl groups at the 2- or 4-positions are heated with acetic anhydride, .cetylation occurs. This is followed by rearrangement to the acetoxypyridine I-424), which is hydrolyzed readily to the pyridone.[356-360] Acetoxylation ally occurs at the 2-position[361-364] (see also Chapter IV).

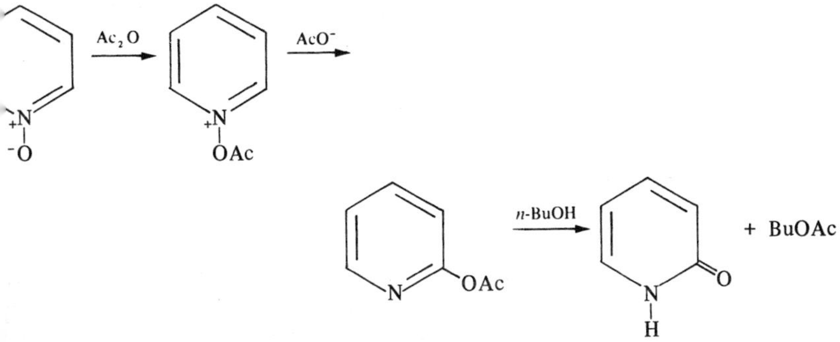

$Ac_2O$ → AcO⁻ →

*n*-BuOH → + BuOAc

XII-424

\ method of preparation of 2-pyridone from pyridine in 50% overall yield the *N*-oxide and **XII-424** is recommended by Chumakov.[365] The 2-pyridone, e of water and acetic acid, is prepared from **XII-424** by heating with utyl alcohol.[365] When an alkyl group occupies the 2- or 4-position (e.g., **I-425** to **XII-427**) substitution at the α-position of the side-chain dominates and is usually accompanied by some reaction at the

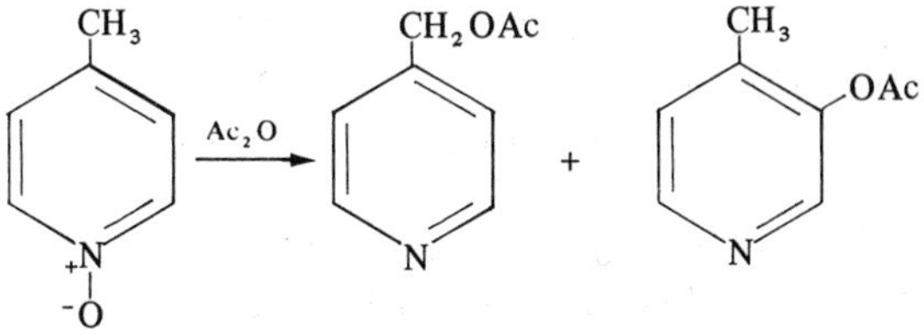

$Ac_2O$ → +

XII-425 (53%) (7%)

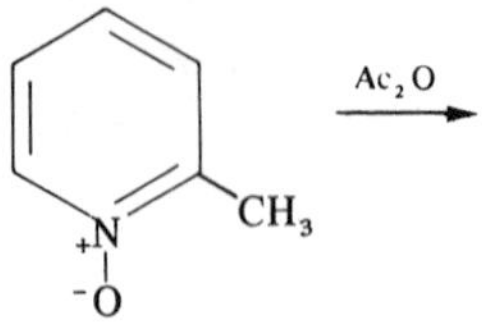

$\xrightarrow{Ac_2O}$

XII-426

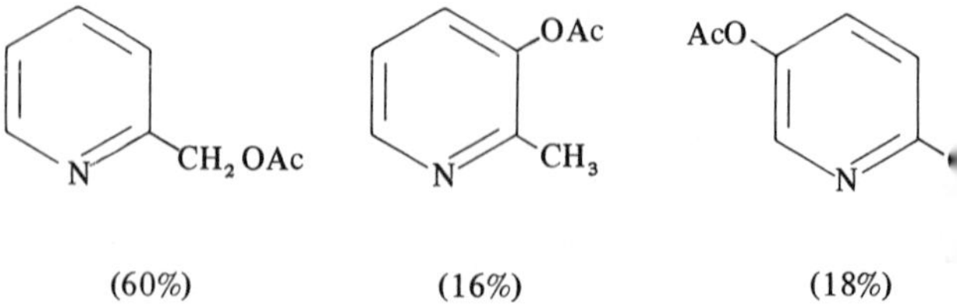

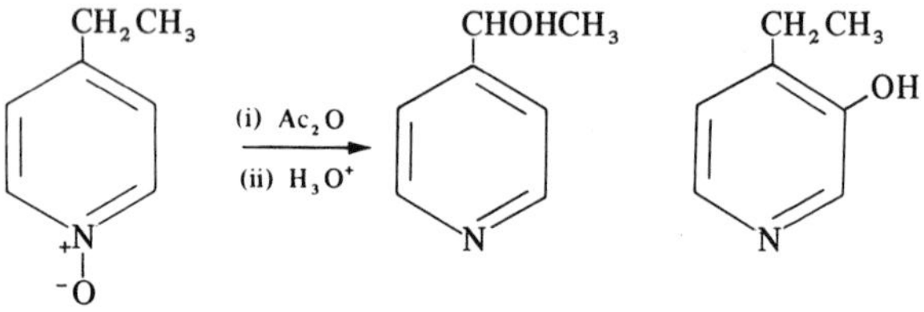

$CH_2OAc$ (60%)  OAc, $CH_3$ (16%)  AcO (18%)

$\xrightarrow{(i)\ Ac_2O;\ (ii)\ H_3O^+}$

XII-427

CHOHCH$_3$ (27%)  $CH_2CH_3$, OH (6%)

3-position.[366, 367] Further treatment of *N*-oxides of rearrangement produ with acetic anhydride gives diols and aldehydes after hydrolysis. For examp 2,6-lutidine-1-oxide has been converted to the aldehyde and the diols (**XII-** to **XII-430**).[368, 369] All possible pyridinols and pyridinylcarbinols are forr from 2,4-lutidine-1-oxide and from 2,4,6-collidine-1-oxide when treated w acetic anhydride.[370]

The product of rearrangement of 6-methylpicolinic acid-1-oxide (**XII-431**) acetic anhydride, previously characterized as 2-acetoxymethylpyridine,[371] been shown to be 6-acetoxy-2-picoline (**XII-432**).[372] The observation t 2-picoline-1-oxide (**XII-426**), the decarboxylation product of **XII-431**, gi mainly 2-acetoxymethylpyridine indicates that decarboxylation, N–O cleava and acetoxylation of **XII-431** at C-2 are concerted.[372]

The preparation of 6-methyl-2-pyridone from 2,6-lutidine *via* **XII-431** appe to be preferred[372] over diazotization and hydrolysis of 6-methyl-2-ami pyridine.[373]

2-Carboxypyridine-1-oxide is decarboxylated quantitatively with ac anhydride.[361, 362, 374] However, conversion to the methyl ester blo decarboxylation and methyl 2-pyridone-6-carboxylate is formed after hyd

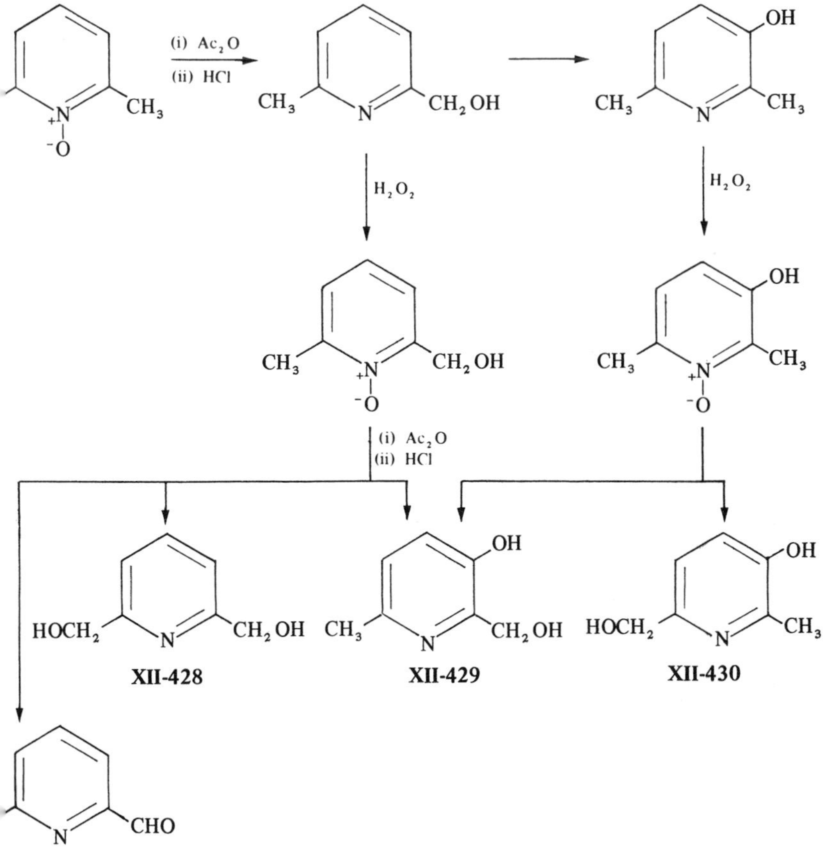

(i) $Ac_2O$
(ii) HCl

$H_2O_2$

$H_2O_2$

(i) $Ac_2O$
(ii) HCl

$HOCH_2$ N $CH_2OH$

**XII-428**

$CH_3$ N OH $CH_2OH$

**XII-429**

$HOCH_2$ N OH $CH_3$

**XII-430**

N CHO

$Ac_2O$

$H_2O$

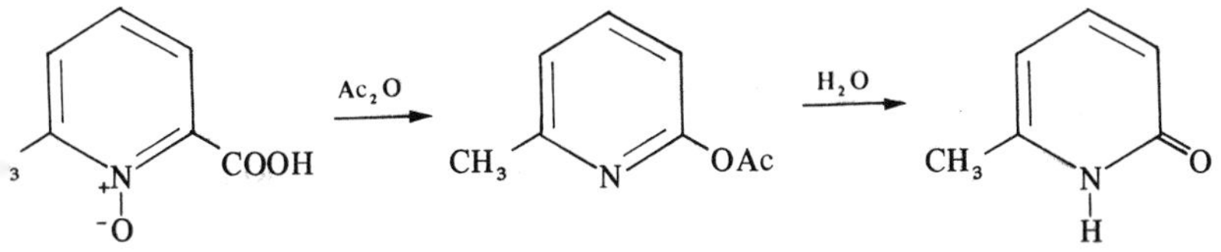

**XII-431**

**XII-432**

,[360] Methyl isonicotinate-1-oxide gives methyl 2-pyridone-4-carboxylate in r yield after hydrolysis.[283]

Trifluoromethylpyridine-1-oxide and acetic anhydride give a poor yield of ifluoromethyl-2-pyridone. 4-Trifluoromethylpyridine-1-oxide does not react ı acetic anhydride. It has been suggested that in the latter case the *N*-oxygen is not acetylated because of a strong hyperconjugative electron withdrawa the 4-trifluoromethyl group.[375]

Studies with 3-substituted pyridines indicate that the product distribu depends on the substituent. When the group in the 3-position is methyl, the isomeric 2-pyridones are formed in about equal amounts. However, when group is inductively electron withdrawing, the 3-substituted-2-pyridone ei predominates or is the only isolated product. 3-Picoline-1-oxide and ac anhydride give 3-methyl-2-pyridone (35–40%), 5-methyl-2-pyridone (35–4 and 3-methyl-1-(5-methyl-2-pyridyl)-2-pyridone (4%) after hydrolysis water.[361] In a later study, much less 3-methyl-2-pyridone was detected.[376]

3-(Trimethylsilyl)pyridine-1-oxide and acetic anhydride give 5-(trimethylsi and 3-(trimethylsilyl)-2-pyridone in a ratio of approximately 1.5:1.[377]

3-Acetylamino-2,6-lutidine rearranges to the acetate of 3-acetylamin methyl-2-pyridinemethanol.[378] However, 3-hydroxypyridine-1-oxide and ac anhydride give 2,3-dihydroxypyridine only,[361] and 3-halopyridine-1-oxides the 2-acetoxy-3-halopyridines.[363] 3-Nitropyridine-1-oxide gives 3-nitro-2- done in 50% yield.[364] Although only 3- and 5-carbomethoxy-2-pyridone isolated from a reaction between methyl nicotinate-1-oxide and a anhydride,[362] nicotinic acid-1-oxide (**XII-433**) forms 2-acetylnicotinic acid smaller amounts of 2- and 6-hydroxynicotinic acid after hydrolysis

COOH (i) $Ac_2O$ (ii) $H_2O$

**XII-433**

COOH, $COCH_3$ + COOH + C

(25–30%) (10%) (3%)

3-Trifluoromethylpyridine-1-oxide and acetic anhydride give an 83% yiel 3-trifluoromethyl-2-pyridone.[375]

A nitro group at position 3 or 5 of 2,4-[379] or 2,6-dimethylpyridine-1 ide [295, 379] or of *sym*-collidine-1-oxide has no effect on the ease rearrangement but has some effect on the product distribution. Decreases in yields of 4-hydroxymethyl containing products and of pyridinols observed.[379] The 4-nitro group of methylpyridine-1-oxides is a reactive lea oup in nucleophilic substitutions.[380] 4-Nitro-2-picoline-1-oxide (**XII-434**, R = ) and 4-nitro-2,6-lutidine-1-oxide (**XII-434**, R = $CH_3$) and acetic anhydride give e 4-nitro-2-hydroxymethylpyridine and the 2-hydroxymethyl-4-pyridone.[381] Nitro-2,5-lutidine-1-oxide forms 2-hydroxymethyl-5-methyl-4-nitropyridine d 2,5-dimethyl-4-pyridone.[297]

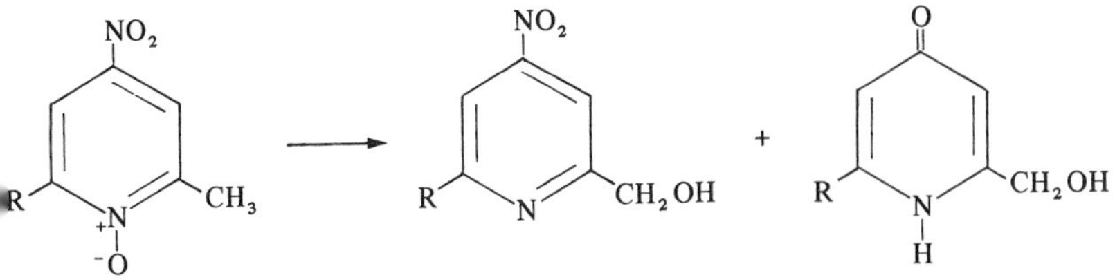

**XII-434**

2-Chloro-, 2-ethoxy-, and 2-phenoxypyridine-1-oxides (**XII-435**) do not arrange but form 1-hydroxy-2-pyridones after hydrolysis. However, 2-chloro-4-ethylpyridine-1-oxide (**XII-436**, R = 4-$CH_3$) and 2-chloro-6-methylpyridine-1-xide (**XII-436**, R = 6-$CH_3$) give the corresponding acetoxymethyl-2-chloropyri-ines.[374]

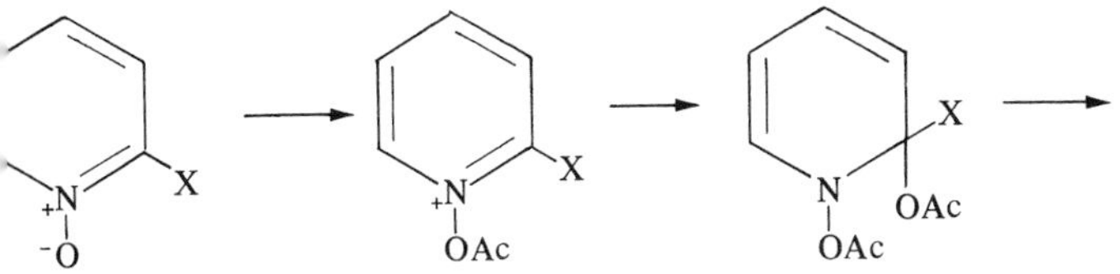

**XII-435**

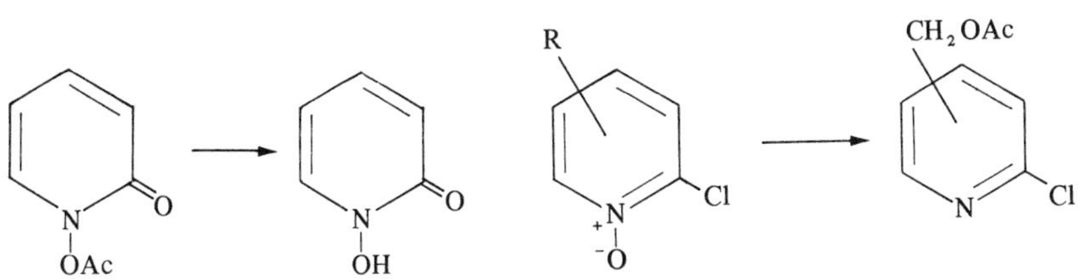

**XII-436**

Considerable attention has been devoted to the mechanisms of these reactions luring the past decade and with only partial success. It should be emphasized hat rearrangement to the nucleus at the 2- or 3-position or to the α-carbons of he side-chain on the 2- or 4-positions are different kinds of reactions. Studies vith $^{18}O$ labeled acetic anhydride show that rearrangement of 2-picoline-1-oxide s intramolecular and that equilibration of the two oxygens from the V-acetyloxy group takes place.[382, 383] Although this suggests a "radical pair" intermediate and many investigators earlier favored a radical mechanis (**XII-438**), later work has shown that 2- and 4-picoline-1-oxides react throu anhydro bases (e.g., **XII-437**) and has suggested that 2-picoline-1-oxi reacts by an intramolecular path *via* an ion pair intermediate (**XI 439**).[360-362, 366, 383-398] (See also Ch. IV).

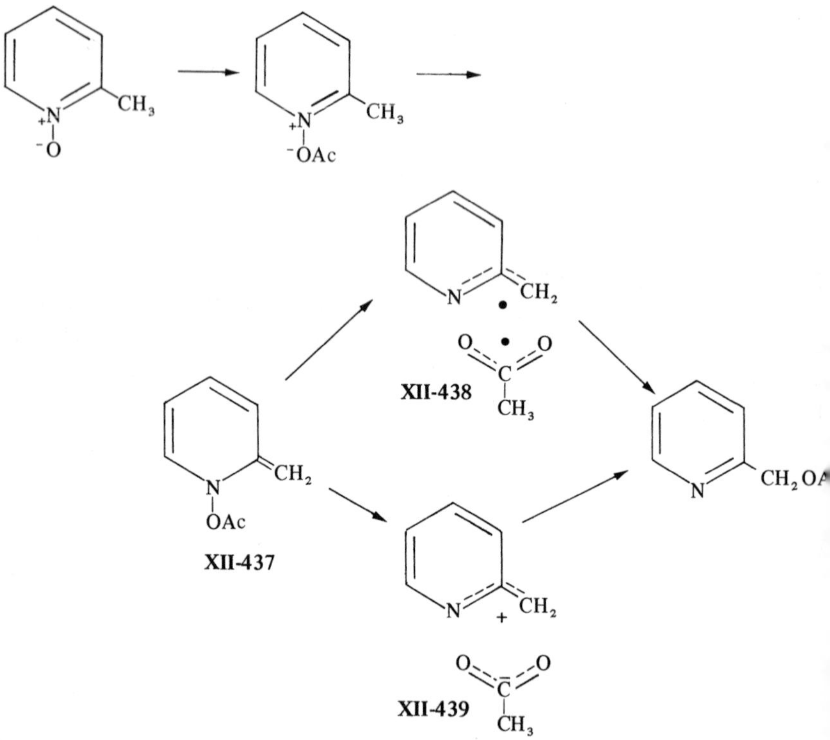

Studies using 1-acetoxy-2-(α,α-dideuterobenzyl)pyridinium perchlorat sodium acetate, and acetic acid show that proton transfer to form the anhydr base is irreversible.[389] Relatively large isotope effects ($k_H/k_D$) for reactions c *N*-oxides of 2-benzylpyridine, 2-picoline, and 4-picoline also show that proto removal to form the anhydro base is the rate-determining step.[396]

In the intramolecular rearrangement of 2-picoline-1-oxides it appears that afte N–O cleavage, the rearrangement process is completed so quickly that th oxygen atoms of the acetoxy are not completely scrambled. The extent o scrambling appears to be related to the structure and conformational preferenc of the anhydro base.[399]

It has been suggested that in 2-picoline-1-oxide either a (1,3)-sigmatropic shi of the nitrogen atom between the two oxygen atoms of the acetoxy grou efore the N → $C_\alpha$ migration or competitive (1,3)- and (3,3)-sigmatropic arrangements of the anhydro base could be operative.[397]

The rearrangement of the anhydro base from 4-picoline-1-oxide may be tramolecular (through ion-pairs or radical pairs) or intermolecular depending n solvent and concentration.[387, 390, 391] Experiments with $^{18}O$-acetic nhydride indicate that, in the absence of solvent or in acetic acid, 4-acetoxy-ethylpyridine, and 3-acetoxy-4-methylpyridine are formed by intermolecular ucleophilic attack of the acetate on the anhydro base.[392] Recently, chemically duced nuclear spin polarization during the reaction of 4-picoline-1-oxide with cetic anhydride has provided direct evidence for the intermediate formation of radical pair and strengthens the suggestion that a dual mechanism is operative volving cleavage of the anhydro base to both radical and ion pairs.[400] MO onsideration of data from the action of acetic anhydride on 2- and -methoxypyridine-1-oxides has led to the conclusion that this rearrangement is oncerted and ionic.[369]

The observation that rearrangements to the 4-methyl group do not occur with -nitro-2,4- and 5-nitro-2,4-lutidine-1-oxide and 3-nitro-2,4,6-collidine-1-oxide dicates a greater stabilization of the 2- rather than the 4-methylene anhydro ase by the 3-nitro group. The 3-nitro group also increases the reactivity of the nethylene group relative to that of the pyridine nucleus.[379]

The reaction of pyridine-1-oxide with acetic anhydride to form 2-acetoxy-yridine, which cannot proceed through an analogous anhydro base, exhibits *seudo*-first-order kinetics and a secondary isotope effect (2,6-$d_2$-; $k_H/k_D$ = .92). An ionic process is consistent with the kinetic data and with the effects of -substituents on product distribution. The pathway involves attack of acetate on within the ion pair **(XII-440)** or attack of acetate at C-2 of the cation, but oes not involve intramolecular rearrangement of the free cation.[360]

$$\text{C}_5\text{H}_5\overset{+}{\text{N}}\text{–O}^- + \text{Ac}_2\text{O} \rightleftharpoons \left[\text{C}_5\text{H}_5\overset{+}{\text{N}}\text{–OAc} \quad \text{AcO}^-\right] \rightleftharpoons \text{C}_5\text{H}_5\overset{+}{\text{N}}\text{–OAc} + \text{AcO}^-$$

**XII-440**

A study with $^{18}O$ labeled acetic anhydride suggests that the rearrangement of -picoline-1-oxide to give a mixture of 3- and 5-methyl-2-acetoxypyridines roceeds by an intermolecular ionic process.[394]

In addition to rearrangement, pyridine-1-oxides undergo redox reactions on reatment with acid anhydrides to give carbon dioxide, an aldehyde or ketone, he parent acid, and the pyridine as well as pyridinols and alkylated yridines.[387, 398, 401, 402] When 2- and 4-picoline-1-oxides are treated with phenylacetic anhydride in benzene, at least 46% and 79% respectively of *N*-oxides are consumed by redox reactions[401] (e.g., equations **XII-441** a **XII-442**). Additional products from 2-picoline-1-oxide are 2-(β-phenyleth

$$2C_6H_7NO + (C_6H_5CH_2CO)_2O \longrightarrow C_6H_5COOH + C_6H_5CHO + CO_2 + 2\,C_6H_7N \quad \textbf{(XII-4}$$

$$2C_6H_7NO + 3(C_6H_5CH_2CO)_2O \longrightarrow \text{(2,3-diphenylmaleic anhydride)} + 4C_6H_5CH_2COOH + 2\,C_6H_7N \quad \textbf{(XII-4}$$

pyridine, benzylpicolines, hydroxypicolines and toluene.[401] Pyridine-1-oxi reacts with acetic anhydride at a relatively slow rate to give 2-acetoxypyridine the principal product[403, 404] but reacts with higher normal and branched ch anhydrides and phenylacetic anhydride to give considerable amounts of red products (Table XII-7).[405, 406] Higher carboxylic acid anhydrides can replaced by mixtures of acetic anhydride and the corresponding carboxylic a with similar results.[405, 406] Although radicals are present in these mixtures, io mechanisms are available and appear to be more reasonable to ma investigators.[402, 405-407]

TABLE XII-7. Redox Reaction Products from Pyridine-1-oxide and Anhydrides

| Anhydride | Solvent | Product (% yield) | % Yield $CO_2$ | Ref. |
|---|---|---|---|---|
| Phenylacetic | Benzene | Benzaldehyde (69) | 76 | 406 |
| Diphenylacetic | Benzene | Benzophenone (68) | 83 | 406 |
| Isobutyric | Toluene | Acetone (39) | 69 | 406 |
| Butyric | Xylene | Propionaldehyde (12) | 19 | 406 |
| Acetic | Xylene, neat | 2-Pyridone (65) | nil | 403, 4 |
| *p*-Nitrophenylacetic | Neat | *p*-Nitrobenzaldehyde (20) | 21 | 407 |

3-(Trimethylsilyl)pyridine-1-oxide and phosphorus oxychloride give 2- a 4-chloro-3-(trimethylsilyl)pyridine, which are difficult to separate but wh ve been converted to the pyridones by conversion to the benzyl ethers fol- wed by hydrogenolysis.[377]

Acetyl or benzoyl chloride and 2-picoline-1-oxide in nonaqueous solvents give e acetate or benzoate of 2-pyridinemethanol and a small amount of pyridylmethyl chloride.[408] Pyridine-1-oxide and tosyl chloride in benzene give 3-dipyridyl ether, *N*-(2′-pyridyl)-2-pyridone, *N*-(2′-pyridyl)-5-chloro-2-pyri- ›ne, *N*-(2′-pyridyl)-3-chloro-2-pyridone, and 3-tosyloxypyridine.[409] Tracer ıdies have suggested that the 3-tosyloxypyridine is formed *via* an intimate ion ir **(XII-443)**.[410] (For alternate suggestion, see Ch. I).

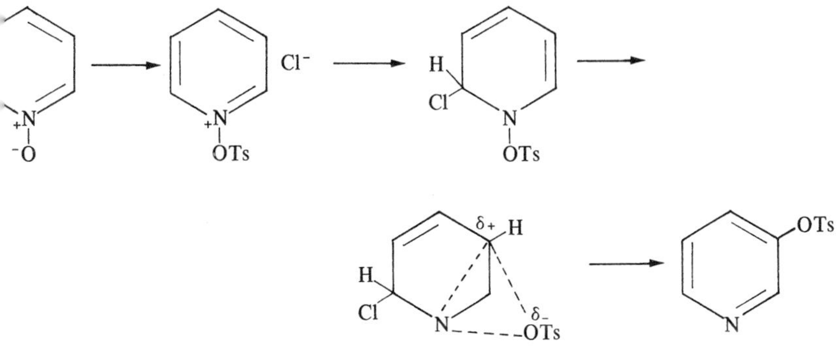

**XII-443**

4-Picoline-1-oxide, 2,6-lutidine-1-oxide, and 4-nitro-2,6-lutidine-1-oxide react th ketene to form the corresponding hydroxymethylpyridine and the rresponding 3-pyridinol in relatively low yields. *N*-oxides of 3-picoline, nitropyridine, and 4-nitro-2-picoline did not give comparable products when eated with ketene.[411]

*N*-Oxides undergo deoxidative substitution in the presence of acetic anhydride d mercaptans to give mainly 2- and 3-alkylthiopyridines, **(XII-444** and **II-445**). 4-Alkylthiopyridines were not detected but acetates are formed by arrangement of the intermediate *N*-acetoxypyridinium salt. The mechanism, hich is consistent with observations,[412-444] has been proposed, as shown in echanisms **XII-44** and **XII-45**.

Pyridine *N*-oxides and butyllithium give lithiopyridine *N*-oxides that react with ygen to form hydroxamic acids in poor yield. They react with sulfur to give hydroxy-2-pyridinethiones (**XII-446** to **XII-448**) in better yields; however, no tempt was made to optimize yields.[415] (See also Chs. VII and XV).

3-Hydroxy-1-methoxypyridinium methyl sulfate and the sodium salt of butyl mercaptan give 2-butylthio-3-hydroxypyridine **(XII-449)**.[416]

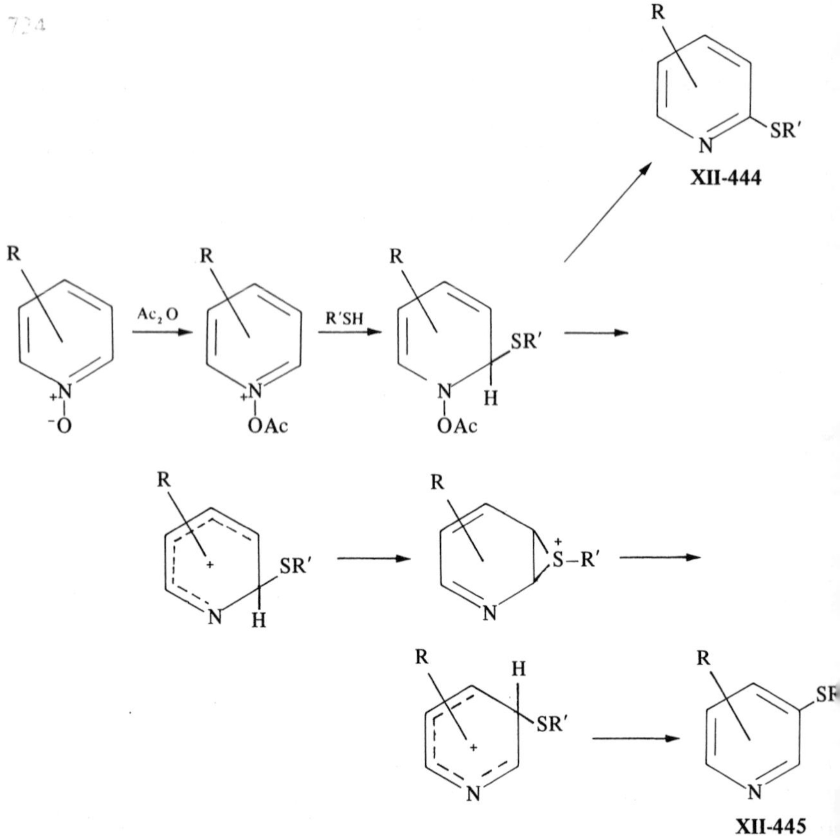

724

$Ac_2O$ ; R′SH

XII-444

XII-445

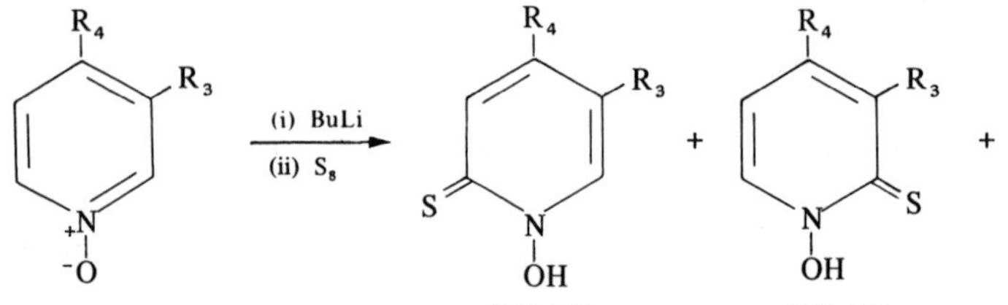

(i) BuLi (ii) $S_8$

XII-446 + XII-447 +

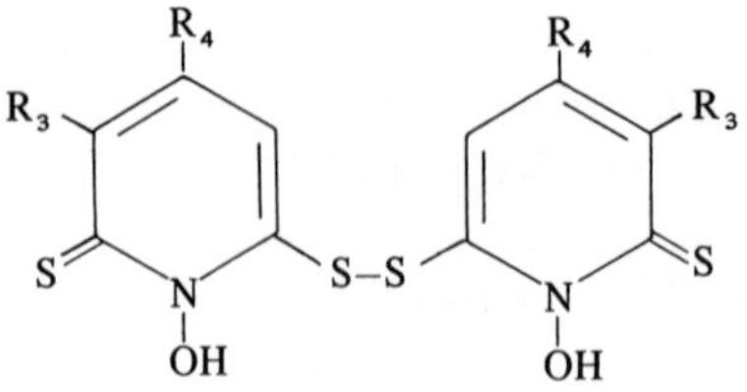

XII-448

| $R_3$ | $R_4$ | % Yield XII-446 | XII-447 | XII-448 |
|---|---|---|---|---|
| H | H | 8 | – | – |
| H | $CH_3$ | 39 | – | – |
| $CH_3$ | Cl | 11.5 | – | – |
| $CH_3$ | $CH_3$ | 24 | 12.5 | 37.4 |

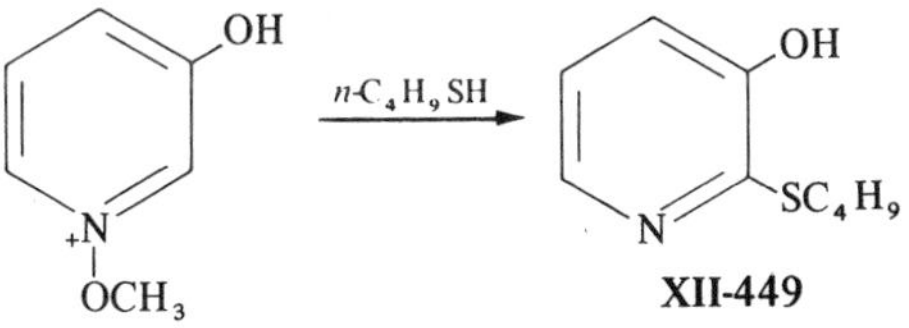

$n$-$C_4H_9SH$

OH, $\overset{+}{N}$, $OCH_3$ → OH, $SC_4H_9$, N

**XII-449**

## H. N-*Oxides (by Reduction)*

Pyridine-1-oxides are comparatively resistant to reduction because of ·sonance stabilization by the aromatic system. Typical reagents that have been sed for the formation of pyridones and pyridinols are Raney Nickel in ıethanol, palladium-on-charcoal, phosphorous trichloride, or phosphorus xychloride in ethyl acetate.[380] The *N*-oxides of pyridoxine, pyridoxal, and yridoxamine have been deoxygenated catalytically.[417] 4-Alkoxy-3-halopyri-ine-1-oxides are *N*-deoxygenated by phosphorous trichloride in chloroform.[255] -Amino-3-pyridinol can be prepared from 2-nitro-3-pyridinol-1-oxide **(XII-450)** ı acetic acid by treatment with iron and mercuric chloride and then with inc.[418] 2-Halo-3-pyridinols can be prepared from **XII-450** by treatment with hosphorous trihalides in chloroform.[418] 2,4-Diiodo- and 2,4,6-triiodo-3-

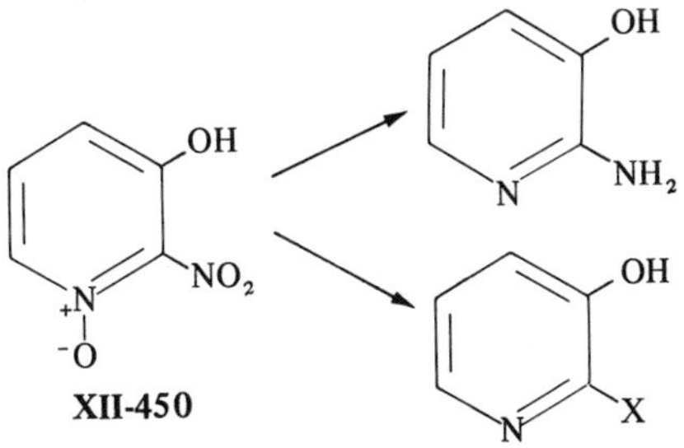

OH, $NO_2$, $\overset{+}{N}$, $^{-}O$ → OH, $NH_2$, N ; OH, X, N

**XII-450**

yridinol can be formed from their *N*-oxides by deoxygenation with iron and ›etic acid.[298]

Irradiation of pyridine-1-oxides in the gas phase with ultraviolet light can lead › deoxygenation. In benzene the pyridine and some phenol are often formed. ›ther products include 3-pyridinols, 2-pyridones, and 2-formyl- or acylpyr-›les.[419, 420] 2-Picoline-1-oxide gives small amounts of both 2- and ·methyl-3-pyridinol, in addition to other products.[421, 422] Irradiation of ,6-lutidine-1-oxide in benzene gives phenol, 2,6-lutidine, 2,6-dimethyl-3-pyri-ınol, and 2-acetyl-5-methylpyrrole.[420, 421] In ether, 2,5-dimethyl-3-formylpyr-›le and 3-acetyl-2-methylpyrrole are also formed.[420, 422] Both 2,3,6- and ,4,6-collidine-1-oxide give some of the corresponding 3-pyridinol.[420]

The 1-benzyloxy- group has been used as a protecting group in pyridone ·ntheses. 2-Benzyloxypyridine-1-oxides undergo facile rearrangement to

*N*-benzyloxy-2-pyridone even at ambient temperature.[423] For example, compound previously described as 2-benzyloxy-6-methylpyridine-1-oxi (**XII-451**)[424] was later shown to be 1-benzyloxy-6-methyl-2-pyridone.[425] T intermediate **XII-451** can be isolated under mild conditions.[423]

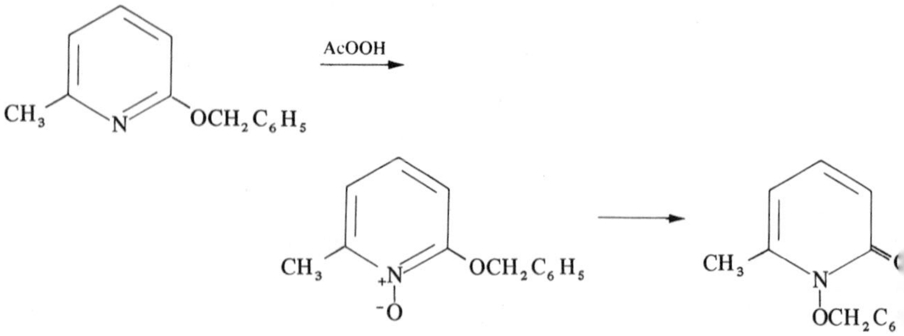

**XII-451**

Although *m*-chloroperbenzoic acid oxidation of 2-benzyloxy-4-methylpyridi gives the *N*-oxide,[426] this oxidizing agent and 1-benzyloxy-6-methyl-2-pyrido give only the rearranged product.[425] 1-Benzyloxy-6-methyl-2-pyridone co denses smoothly with ethyl oxalate to give the β-(6-pyridonyl)pyruva (**XII-452**), which can be converted to 6-(β-alanyl)-2-pyridone *via* the oxin (**XII-453** to **XII-455**) and the 1-hydroxy-6-(β-alanyl)-2-pyridone.[425]

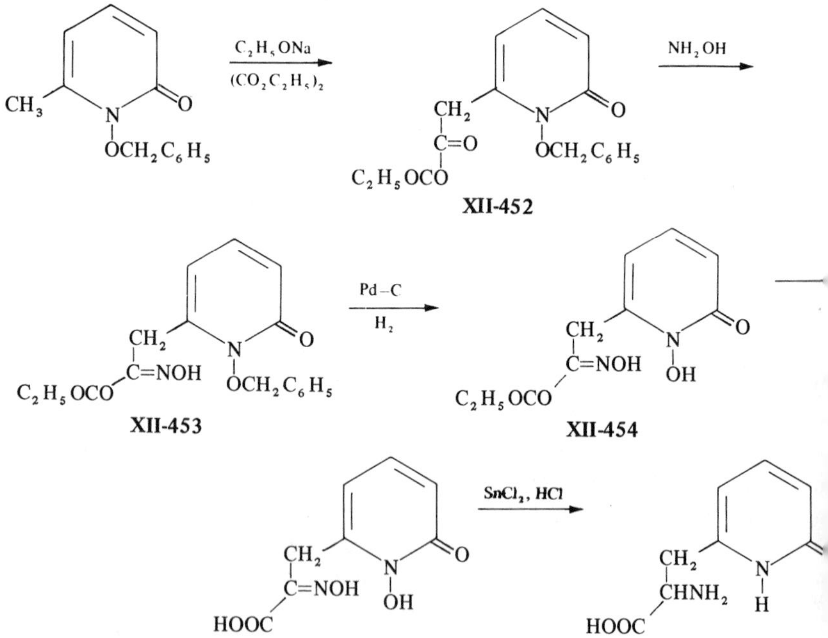

**XII-452**

**XII-453**

**XII-454**

**XII-455**

## I. *Direct Hydroxylation*

2-Pyridone has been oxidized to 5-hydroxy-2-pyridone by the Elbs method th potassium peroxidisulfate.[342] 2-Pyridone has been formed by $^{60}$cobalt adiation of an aqueous solution of pyridine.[427] The mechanism of attack and activity of OH radicals with pyridine have been studied.[428] Rates of reactions pyridines with the electrophilic hydroxyl radicals are predictably somewhat wer than analogous rates of reactions of benzene derivatives, which are ry fast and approach diffusion control rates. For β-substituted pyridines, good rrelation was obtained between rate data and $\sigma_m$. Rates for pyridines ntaining α-, or γ-substituents that are affected by the ring nitrogen, did not rrelate with $\sigma_o$ and $\sigma_p$. Hydroxylations occurred at both carbon and trogen.[429] 2-Pyridone has been isolated as a minor product from the reaction nitrobenzenes with pyridine at 600°.[430] Ruthenium tetroxide and pyridine in rbon tetrachloride give a 1:2 complex that yields ruthenium and 2-pyridone hen heated in a stream of hydrogen (equation **XII-456**).[431]

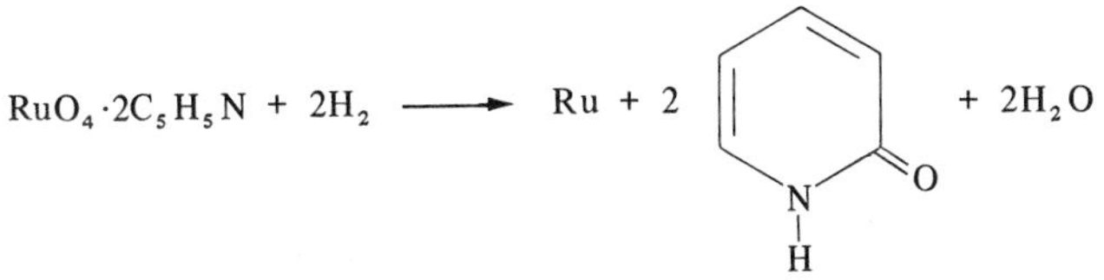

$$RuO_4 \cdot 2C_5H_5N + 2H_2 \longrightarrow Ru + 2 \; [\text{N–H, =O ring}] + 2H_2O \qquad \textbf{(XII-456)}$$

3-Hydroxy-4-methyl-5-phenyl-2-pyridone and sodium nitrite in sulfuric acid ve 2-hydroxy-4-methyl-5-phenyl-1-azaquinone (**XII-457**).[221]

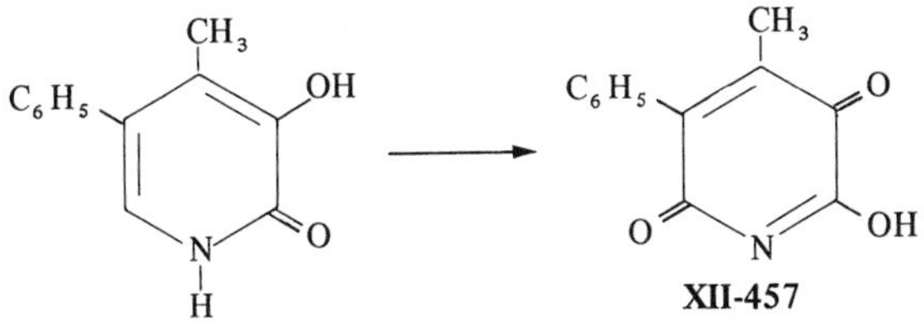

[Structures: CH$_3$, C$_6$H$_5$, OH, O, N, H $\longrightarrow$ CH$_3$, C$_6$H$_5$, O, O, N, OH]

**XII-457**

The three monohydroxypyridines and 4-hydroxy-2-pyridone are not oxidized y hydrogen peroxide and peroxidase. However, 5-hydroxy-2-pyridone is xidized to **XII-458** and **XII-459**.[432]

A relatively large number of metabolic products of 3-acetylpyridine, an ntimetabolite of nicotinamide, have been isolated and identified. They include -(α-hydroxyethyl)pyridine,[433] its glucuronide,[434] and its *N*-oxide,[435] -(α-hydroxyethyl)-6-pyridone,[435] 5-acetyl-2-pyridone,[436-438] 5-acetyl-1-ethyl-2-pyridone,[437] 3-carboxamido-1-methyl-4-pyridone[435] and 3-carbox-

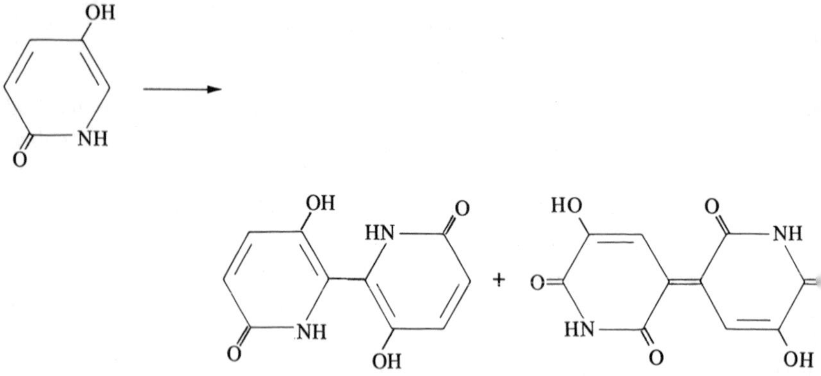

XII-458 XII-459

amido-1-methyl-6-pyridone.[435] 3-Hydroxy-4-pyridone is an intermediate in t metabolism of 4-pyridone by *Agrobacterium sp.*[439,440]

## J. *Hydropyridinols*

Pyridinols and pyridones are sometimes prepared by dehydrogenation hydropyridinols. For example, the available 1-phenyl- and 1-methyl-5,6-dihydr 2-pyridones (**XII-460**; R = $CH_3$, $C_6H_5$)[441] are readily dehydrogenated to th pyridones with palladium-on-carbon.[442] γ-Acetyl- and γ-benzoyl-γ-phenylbutyr

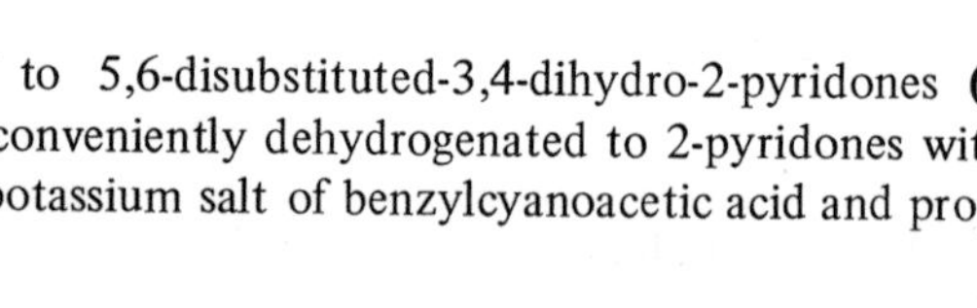

XII-460

nitriles cyclize to 5,6-disubstituted-3,4-dihydro-2-pyridones (**XII-461**),[35,44 which are also conveniently dehydrogenated to 2-pyridones with palladium-o carbon.[35] The potassium salt of benzylcyanoacetic acid and propiophenone giv

XII-461

2-cyano-3,4-diphenyl-4-acetylbutanoic acid, which cyclizes and decarboxylates on standing to 6-methyl-4,5-diphenyl-3,4-dihydro-2-pyridone.[444]

Cycloalkano-2-pyridones (**XII-463**) are formed from 2-(2-cyanoethyl)cycloalkanones by cyclization to **XII-462** with cold concentrated sulfuric acid, which is reduced to sulfur dioxide in a subsequent aromatization.[445]

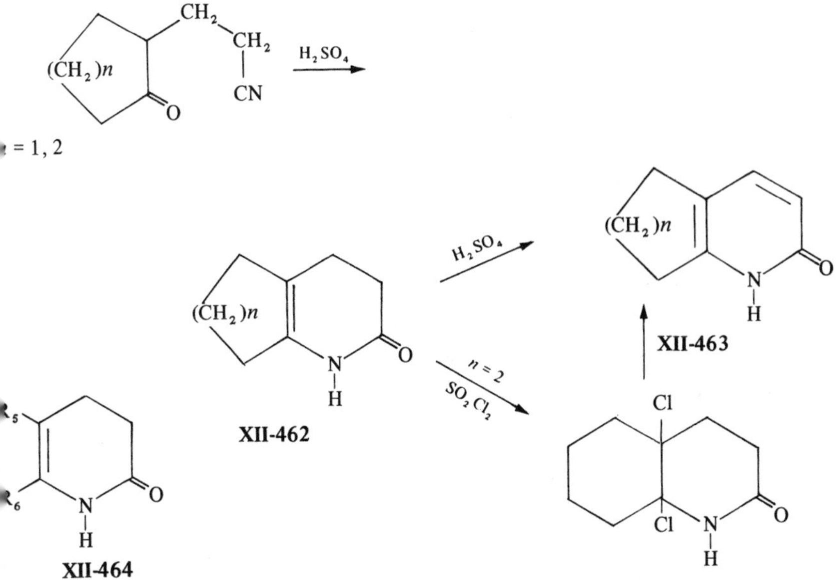

n = 1, 2

**XII-462**

**XII-463**

**XII-464**

Treatment of 5,6-dialkyl-3,4-dihydro-2-pyridones (**XII-464**; $R_5 = R_6 = CH_3$, $R_5 = C_3H_7$, $R_6 = CH_3$) with bromine[446] or sulfuryl chloride[447] results in addition of halogen, which is followed by dehydrohalogenation to the 2-pyridone. The tetrahydroquinolone (**XII-463**, $n = 2$) was also prepared by this route from **XII-462**.[447]

Diethyl 2,6-diphenyl-4-piperidone-3,5-dicarboxylates (**XII-465**) are dehydrogenated by chromic oxide or ceric sulfate to the 4-pyridone (**XII-466**; R = H, $CH_3$, $C_6H_5CH_2$, $CH_2{=}CH{-}CH_2$). Some dealkylation of the *N*-substituted piperidones occurs to give the pyridone (**XII-466**, R = H) when chromic oxide is used.[448]

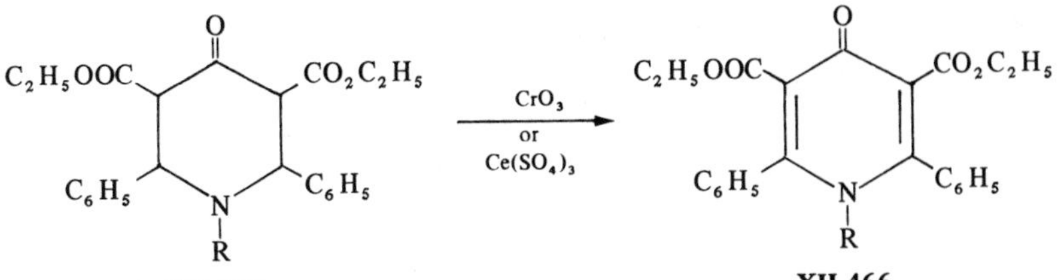

**XII-465**

**XII-466**

1,2-Diphenyl-2,3-dihydro-6-methyl-4-pyridone (**XII-467**) and chloranil tetrahydrofuran or mercuric acetate in water form 1,2-diphenyl-6-methyl-pyridone.[449]

CH3 N C6H5 C6H5 O XII-467 → (chloranil, or $Hg(OAc)_2$) → CH3 N C6H5 C6H5 O

Acrolein (**XII-468**, $R_4$ = H) and crotonaldehyde (**XII-468**, $R_4$ = $CH_3$) rea with 1,2-dialkoxyethylenes to form 2,3-dialkoxy-3,4-dihydro-1,2-pyrans, whi have been converted to 3-alkoxypyridines with ammonia in the presence Pt/$Al_2O_3$ at 180 to 300°.[450]

XII-468 + HC(OR)=HC(OR) → → ($NH_3$, Pt/$Al_2O_3$)

$R_4$ = H; R = $C_2H_5$, $C_3H_7$.
$R_4$ = $CH_3$; R = $C_2H_5$.

Cassine (**XII-469**), a crystalline alkaloid from the legume *Cassia excelsa* Shrad has been dehydrogenated to 2-methyl-6-(10-acetylundecyl)-3-pyridinol wit palladium-on-carbon at 220°.[451]

HO CH3 N $(CH_2)_9CH(CH_3)COCH_3$ XII-469 → (Pd-C)

HO CH3 N $(CH_2)_9CH(CH_3)COCH$

## II. Properties and Structure

### 1. Physical Methods

No attempt will be made here to review and correlate the large number of traviolet, infrared, nuclear magnetic resonance, and mass spectra that have en recorded, interpreted, and used in structure determination of pyridinols d pyridones. Articles that contain significant spectral data and/or discussions e noted in the Tables. The role of spectroscopic techniques, ionization nstants, and dipole moments in the studies of structure and tautomerism of terocyclic compounds has been reviewed recently by Albert.[452] Although nization constants have been used to estimate pyridinol-pyridone equilibrium nstants with some success,[241, 453-457] caution must be exercised about awing conclusions from this type of data.[458]

The infrared literature was reviewed by Katritzky and Ambler in 1963.[459] terpretation of infrared bands of pyridinols and pyridones has, however, been e subject of several investigations during recent years and merits a brief scussion. The infrared spectrum of 4-pyridone (**XII-470a**, R = H) in the solid ate contains three significant bands in the double bond stretching region. A nd near 1590 $cm^{-1}$ (I = 0.25) that disappears on both *N*-deuteration and -methylation is assigned to NH in-plane bending.[460] A high frequency band at 535 $cm^{-1}$ at one time was assigned to carbonyl stretching. The low frequency nd at 1534 $cm^{-1}$ was assigned to C=C bond stretching. Subsequently, because an inability to correlate these bands with those of analogous oxygen terocycles it was suggested that these assignments be reversed.[461] In solution, e low frequency band (solid, 1550 $cm^{-1}$) in *N*-methyl-4-pyridone (**XII-470,** = $CH_3$) was more sensitive than the high frequency band to changes in solvent $HCl_3$, 1581 $cm^{-1}$; $C_6H_6$, 1598 $cm^{-1}$), a characteristic of carbonyl group sorptions, which also supported a reversal of the assignment.[461] The low rbonyl stretching frequency was presumably due to the polar nature of that ond attributed to contributions to **XII-470a** from **XII-470b**.

**XII-470 a** ⟷ **XII-470 b**

A comparison of the infrared spectra of 4-pyridone and 1-methyl-4-pyridon in the solid states shows that the bands at 1635 $cm^{-1}$ are not much different bu that the 1534 $cm^{-1}$ band of 4-pyridone is shifted to near 1550 $cm^{-1}$ an sharpens in 1-methyl-4-pyridone, consistent with the expected shift from hydrogen bonded to a nonhydrogen bonded carbonyl.[461] However, comparison of the spectra of 3,5-dihalo-, 3,5-dihalo-1-methyl-, 2,3,5,6-$d_4$, 1-$d$ and 1-methyl-4-pyridones has led to the conclusion that both C=O and C= stretchings are extensively mixed to give composite vibrations at bot frequencies.[460, 462] These difficulties have not negated the earlier structur assignments of 4-pyridones that have been based on infrared interpretations.

In 1-methyl-2-pyridone the higher frequency bands have been assigned to carbonyl stretching vibration.[461] The inductive effect of the nitrogen alpha t the carbonyl operates in a direction opposite to that of the resonance effect an this less polar carbonyl gives a higher frequency band than that observed fo 4-pyridones.[461] However, 2-methoxypyridinium hexachloroantimonate, whic does not have a carbonyl group, shows a strong ring mode absorption at 163 $cm^{-1}$.[463, 464] A comparison of the spectra of 2-pyridone and 2-pyridone-$^{18}$( also shows extensive mixing of vibrations.[463, 465]

The assignments of the 1477–1443 $cm^{-1}$ bands of 2-pyridone to skeleta stretching[466] has been questioned[463, 465, 467] since these bands are sensitive t phase, solvent, and isotopic substitution.[465]

## 2. Association of 2- and 4-Pyridones

2-Pyridone and its thio and seleno analogs all form strong dimers i chloroform[466] and in benzene.[468] Hydrogen bonding in sulfur and seleniur analogs is lower than in 2-pyridone but nevertheless all three are classified a powerful hydrogen bonders.[469] 2-Pyridone forms an exceptionally stron hydrogen bonded dimer in nonpolar solvents.[469-471] This is supported by NM studies where an unusually low field resonance for the enolizable proton i observed, indicative of strong hydrogen bonding.[465, 472] NMR studies indicat that 4-pyridone is more strongly associated in solution than is 2-pyridone. I dilute solution in chloroform, it has a molecular weight that correspond approximately to a trimer.[462]

In solid 2-pyridone, hydrogen bonding is of the NH···O type.[47 6-Chloro-2-pyridinol exists as the pyridinol tautomer in the solid state and form hydrogen bonded dimers of the OH···N type.[474] Molecular weigh determinations in chloroform show that concentrated solutions of 4-methyl- 4-phenyl-, 3,4-trimethylene-, and 3,4-tetramethylene-6-chloro-2-pyridone con tain mainly hydrogen bonded dimers.[457]

Rate and equilibrium constants have been measured for the hydrogen bon dimerization of 2-pyridone in chloroform-dimethyl sulfoxide and in carbor etrachloride-dimethyl sulfoxide[475] and of 2-pyridone and 2-thiopyridine in chloroform,[471] by ultrasonic attenuation, ultraviolet, infrared, and NMR methods. In the first solvent system, association appeared to be diffusion controlled. In the second solvent it was suggested that solvation-desolvation of -pyridone by dimethyl sulfoxide is rate limiting.[475]

### 3. The Pyridone Structure

2- And 4-pyridones are aromatic and do not have the properties of simple nsaturated lactams. It has been estimated that 2-pyridones have about 35% of he aromaticity of benzene as defined by their ability to sustain an induced ring urrent[476] and as calculated by the SCF treatment.[477]

The infrared spectra of 2- and 4-pyridones are consistent with the pyridone tructure.[478] A comparison of the infrared spectra of 2-pyridone and -pyridone-$^{18}O$ indicates that the lactam description is better than the dipolar ne.[463] A comparison of infrared and Raman spectra of 4-pyridone (**XII-470**, R = I) with those of 4-pyridinium ions and a rationalization of the dipole moment as led to the estimate that the upper limit of the contribution to 4-pyridone by he dipolar form (**XII-470b**, R = H) is 10 to 15%.[460]

The chemical shifts of the $\gamma$-protons of several 2-substituted pyridines ncluding 2-pyridone, the 2-pyridone anion, and 2-phenoxy- and 2-methoxy-yridine have been correlated with the electron donating and withdrawing roperties of the substituents.[479]

Charge distributions and dipole moments have been calculated by a modified Hückel method for *N*-methyl-2-pyridone ($\mu$ = 4.0 D) and *N*-methyl-4-pyridone $\mu$ = 6.7, 6.4 D).[480, 481] Dipole moments have been measured in benzene for *N*-methyl-2-pyridone ($\mu$ = 4.15 D;[453] 4.0 D;[482] 4.04 D[469]) and for *N*-methyl-4-pyridone ($\mu$ = 6.9 D[453]).

### 4. Tautomerism

In water, the ultraviolet spectra of 2-pyridone, 2-thiopyridone, and 2-selenopyridone resemble the spectra of their respective *N*-methyl derivatives, and all exist as the amide tautomer (>99%).[454, 468, 483, 484] 2-Pyridinethiones **XII-471**; X = S, R = $CH_3$, $\mu$ = 5.26 D) and 2-pyridineselenones (**XII-471**; X = Se, R = $CH_3$, $\mu$ = 5.73 D) in benzene have progressively higher dipole moments than 2-pyridones[469] because of increasing contributions of structure **XII-471a** X = O, S, Se) to **XII-471**.[485] Acid strengths are in the expected order: **XII-471**: R = H, X = O, p$K_a$ 11.62, S, 9.97; Se, 9.36.[468]

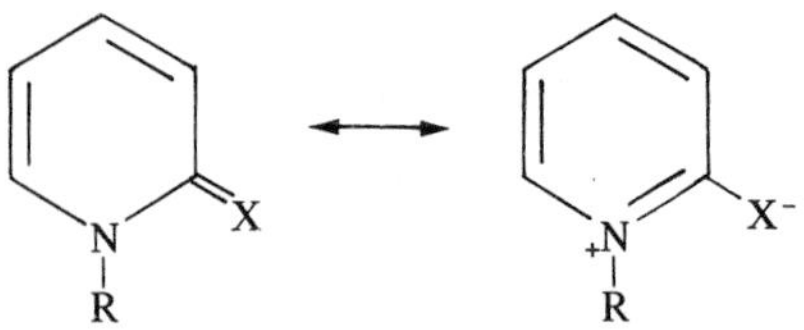

XII-471 a XII-471 b

The NMR spectrum of 4-pyridone in aqueous solution is consistent with pyridone structure.[486] In deuterochloroform, the NMR spectra of 2- a 4-pyridone-$^{15}$N show a very rapid intermolecular exchange of the enolizab proton.[462,465] The observed $^{15}$N–H coupling of 90 Hz at low temperatu indicates that the relative amount of the 2-pyridinol tautomer is less than 2%.[4] A study of $^{15}$nitrogen-hydrogen coupling in 4-pyridone was thwarted by its po solubility at lower temperatures.[462]

It has been demonstrated that appreciable pyridinol tautomer is present in t vapor of 2-pyridone[487] and 6-phenyl-2-pyridone.[488]

Unlike the corresponding pyridones, 2- and 4-aminopyridines prefer the ami forms **(XII-472)**.

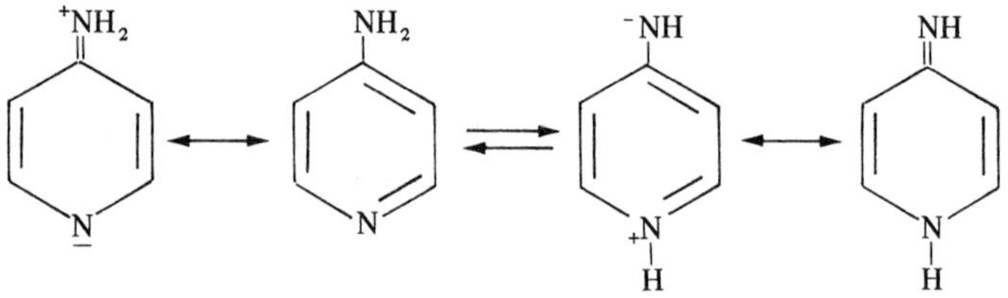

XII-472 a XII-472 b XII-473 a XII-473 b

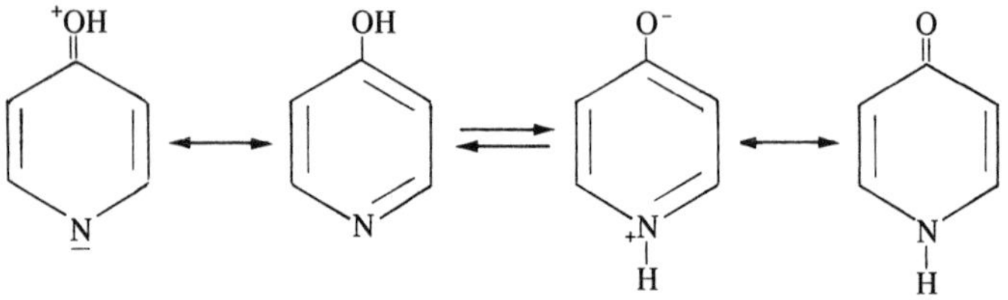

XII-474 a XII-474 b XII-475 a XII-475 b

In 4-aminopyridine, stabilization of **XII-472** and **XII-473** by charge separate forms **(XII-472a** and **XII-473a)** are similar and moderate and the aromatic natu of **XII-472b** is responsible for the predominance of **XII-472**. However, in 2- an 4-pyridone, stabilization of **XII-474** by the charge separated form **XII-474** where oxygen bears a positive charge is much less than stabilization of **XII-47** by **XII-475a** where oxygen is negatively charged, and **XII-475** predom nates.[453,489]

lthough 2-pyridone, 4-pyridone, and many of their derivatives appears to st predominantly in the pyridone form under all conditions, electron hdrawing groups alpha to the nitrogen shift the equilibrium in favor of the ridinol tautomer.[7, 241, 242, 246, 456, 457, 490-497] For example, tetrafluoro-4-ridinol ($pK_a$ = 3.21; 20°, $H_2O$), which is a stronger acid than ntafluorophenol ($pK_a$ = 5.53, 25°) exists mainly (>95%) in the pyridinol m.[248] Electron withdrawing substituents increase the acidity of NH of the ridone tautomer and decrease the basicity of the pyridinol nitrogen. Both ects lead to a relative stabilization of the pyridinol form.[241] Groups in the osition usually produce smaller effects. For example, 3-nitro-4-pyridone, in ich the nitro group has an inductive base-weakening effect on both forms 4.0 p$K$ units), exists as the pyridone by a factor of $10^{3.4}$:1, presumably by bilization of the pyridone by canonical form **XII-476**.[497] Comparable ntributing structures to the pyridinol, requiring the positive charge to be on ygen, are not important.[497]

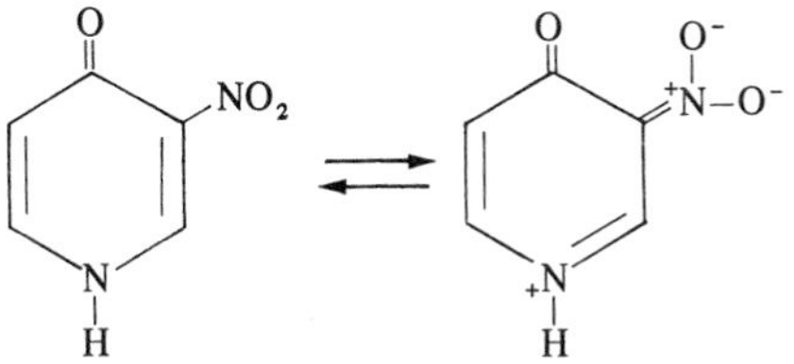

**XII-476**

'automeric equilibrium constants $K_t$ = [pyridone]/[pyridinol] have been asured for a number of 2-pyridones (Table XII-8). The constant $K_t$ has been imated by a Hückel approximation for 2- and 4-pyridone.[498] The $K_t$ values tained from ultraviolet spectra and those obtained from acidity constants are poor agreement for 6-chloro-4-substituted-2-pyridones (**XII-477**; R = $CH_3$, $H_5$) and 6-chloro-3,4-polymethylene-2-pyridones (**XII-478**; $n$ = 3,4) in ter.[457] For example, in **XII-477** where R = $CH_3$, values of 6% and 44% respec-ly of pyridinol tautomer (**XII-477b**) are obtained from ultraviolet data and m acidity constants. However, values obtained from ultraviolet studies ear to be more reliable.[457, 458]

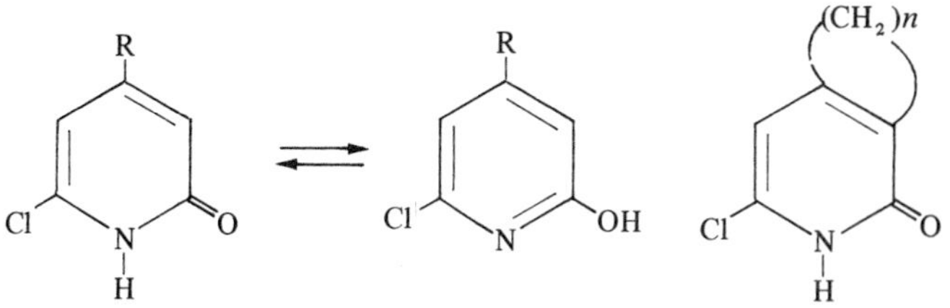

**XII-477 a** **XII-477 b** **XII-478**

TABLE XII-8. Tautomeric Equilibrium Constants: $K_t$ = [pyridone]/[pyridinol]

| $R_4$ | $R_6$ | Solvent | $K_t$ | Method | |
|---|---|---|---|---|---|
| H | Cl | 98% $H_2O$, 2% EtOH | 1.66 | uv | 4 |
| | | 50% $H_2O$, 50% EtOH | 1.16 | | |
| | | EtOH | 0.14 | | |
| | | 50% EtOH, 50% Dioxane | 0.09 | | |
| | | Dioxane | 0.07 | | |
| H | $NH_2$ | Ethanol | 99 | uv | 4 |
| | | 50% EtOH, 50% Dioxane | 46.6 | | |
| | | Dioxane | 0.64 | | |
| $CH_3$ | Cl | $H_2O$ (20°) | 15 | uv | 4 |
| $CH_3$ | Cl | $H_2O$ (20°) | 1.3 | p$K$ | |
| $CH_3$ | Cl | Methanol | 0.48 | uv | |
| | | Chloroform | 0.79 | uv | |
| | | Dimethyl sulfoxide | 0.04 | uv | |
| | | Cyclohexane | 0.05 | uv | |

The equilibrium shifts toward the pyridinol as the medium becomes polar[241, 458, 489, 490, 494–496] (Table XII-8) and as the solvent becomes less a to solvate the carbonyl group.[457] Protic solvents favor the pyridone tauto (Table XII-8) by hydrogen bonding to the carbonyl.[457] For example, in b dimethyl sulfoxide ($\epsilon$ = 45.0) and cyclohexane ($\epsilon$ = 2.0) the pyridinol f predominates (85–96%) for **XII-477** and **XII-478**. The pyridinol content is (28–69%) in protic solvents and in chloroform. In the solid state, th compounds exist in the pyridone form.[457] The temperature dependence of tautomeric equilibrium for 6-amino-, 6-bromo-, and 6-chloro-2-pyridone been studied in several solvents.[496] Heteroaromatic tautomer equilibria chan with solvent for 6-chloro-2-pyridone, 2,3,5-trichloro-4-pyridone, 2-chlor pyridone, 3-pyridinol and 2,6-di-(methoxycarbonyl)-4-pyridone have b correlated by equation **XII-478a**,[490] where $K_{T1}$ is the equilibrium constan one solvent and $P_1$ is the polarity of the solvent (ethanol-water, methanol-wa ethylene glycol, acetonitrile, chloroform, and isooctane) and Kosower $Z$ va are used as a measure of solvent polarity.[499] An abnormality has been obser for 2,6-di-(methoxycarbonyl)-4-pyridone, which exists as the pyridone chloroform and carbon tetrachloride but not in isooctane.[490]

$$\ln \frac{K_{T1}}{K_{T2}} = a(P_1 - P_2) \qquad \text{XII-4}$$

ıe ultraviolet and infrared spectra of the dimethyl and diethyl esters of dicarboxy-4-pyridones (**XII-479**; $R_2$ = $R_6$ = phenyl, α-pyridinyl, ıinolinyl) indicate they exist as conjugate pyridinol chelates in chloroform, on tetrachloride and ethanol.[448, 500]

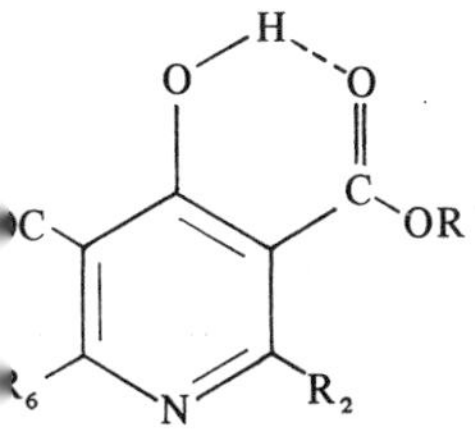

**XII-479** 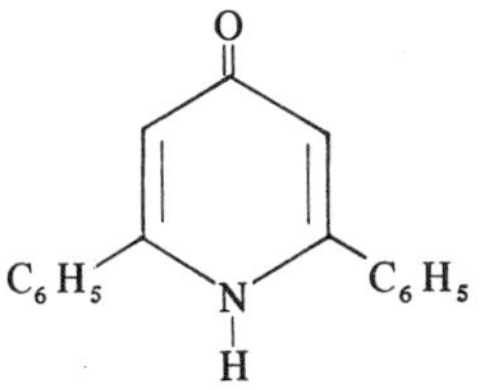 **XII-480** 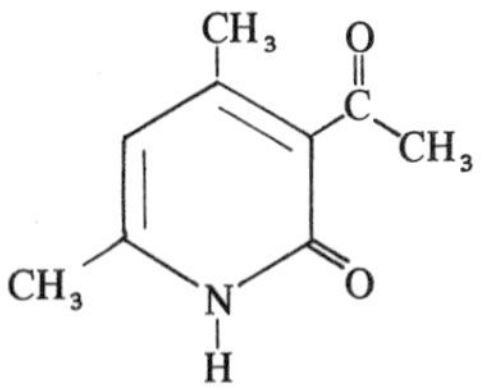 **XII-481**

ıe dimethyl and diethyl esters of 2,6-di-α-pyridinyl- and 2,6-di-α-quinolinyl- ⁄droxy-3,5-pyridinedicarboxylic acids (**XII-479**) also exist in the solid state as ugate chelates.[500] 2,6-Diphenyl-4-pyridone, which does not possess the o-hydroxy ester structural feature of **XII-479**, exists as the pyridone **-480)** in the solid state and in solution.[448] 3-Acetyl-4,6-dimethyl-2-pyridone **-481)** exists as the pyridone tautomer in methanol. Stabilization of the idinol form by hydrogen bonding does not occur.[60] The infrared (KBr) and aviolet spectra (ethanol) of 3-hydroxy-2-pyridone are consistent with the idone structure (**XII-482**).[501]

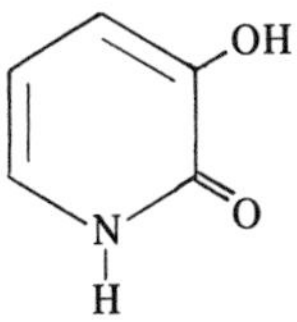

**XII-482** 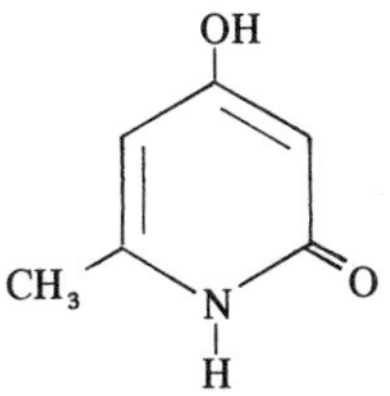 **XII-483**

ıe 4-hydroxy-6-methyl-2-pyridone structure (**XII-483**) is consistent with its R, ir, and uv spectra[502] and is similar to that of 4-hydroxy-2-pyridone.[503] traviolet spectra in water and ethanol and infrared spectra (KBr) show that azine and several of its derivatives (**XII-484**; R = H, CN, $CONH_2$, $CO_2Et$) t as an equilibrium mixture of the 4-amino-6-hydroxy-2-pyridone (**XII-484a**) the 4-amino-2,6-pyridinedione (**XII-484b**).[504]

utaconimide, which exists predominately in lactam forms,[467] is mainly droxy-2-pyridone (**XII-485a:b:c** = 25:60:15) in water, as shown by a parison of its ultraviolet spectra with those of model compounds.[455] The

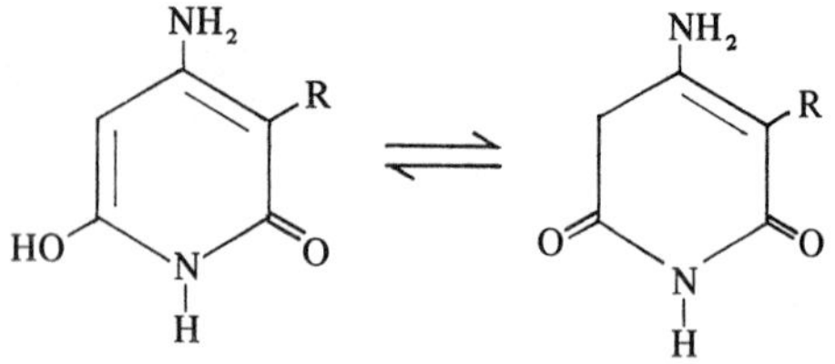

XII-484 a ⇌ XII-484 b

relatively large amounts of the pyridinediol (**XII-485a**) at equilibriun considered to be related to the reduced basicity of the glutaconimide nitrc compared to that of the nitrogen of 2-pyridones, which have only

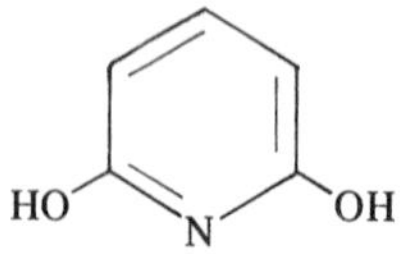

XII-485 a 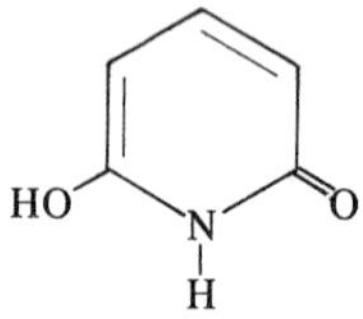 XII-485 b 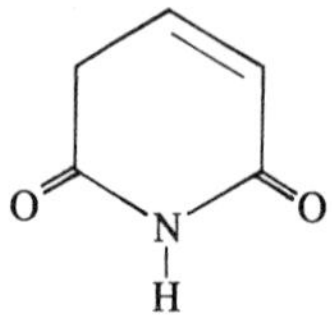 XII-485 c

α-oxygen.[455] The *O*-methyl derivative appears to exist mainly as struct **XII-486** and the *N*-methyl derivative as approximately equal amounts **XII-487a** and **XII-487b** in aqueous solution; only the diketo tauto (**XII-487a**) is observed in chloroform solution, however.[505]

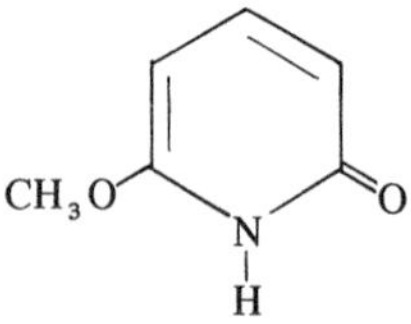

XII-486 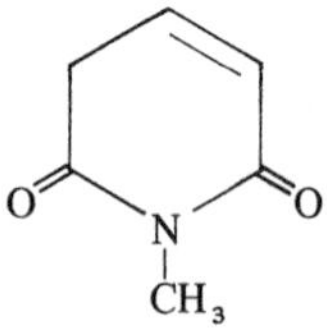 XII-487a 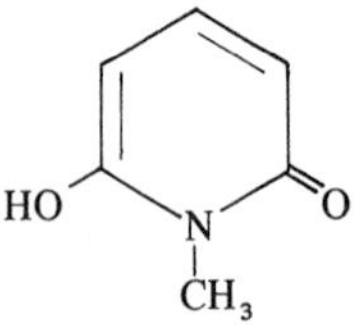 XII-487b

The infrared spectra of **XII-488** (R = $CH_3$, $C_2H_5$, $n$-$C_4H_9$, $n$-$C_5H_{11}$) in solid state exclude tautomers containing OH groups; in carbon tetrachlor solution, however, absorptions due to OH containing tautomers are observe

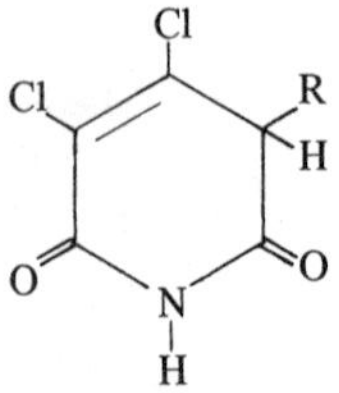

XII-488

TABLE XII-9. Tautomerism Constants ($K_T$ = [CH form]/[OH form]) for 3-Hydroxy-2,6-pyridinediones (XII-489)

| $R_4$ | $R_5$ | $K_T$(DMSO-$d_6$) | $K_T$($CF_3COOH$) |
|---|---|---|---|
| H | $CH_3$ | 0.4 | 0 |
| $CH_3$ | $CH_3$ | 2.0 | 0 |
| $CH_3$ | H | 4.2 | 0.3 |

-Hydroxy-2,6-pyridinediones (**XII-489**) have been obtained in both the CH OH forms. Tautomeric constants have been calculated from NMR data. In ISO, the CH form predominates when there is a methyl group at the osition. In trifluoroacetic acid, the OH form either exists exclusively or dominates[506] (Table XII-9).

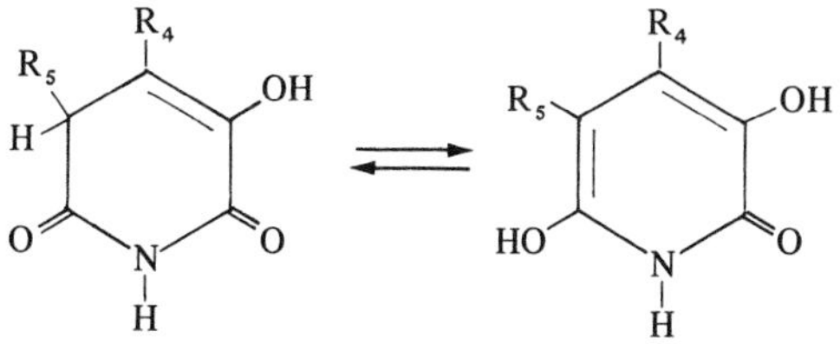

**XII-489**

nfrared studies have ruled out **XII-490a** and **XII-490b** as structures for stalline citrazinic acid. Esters of citrazinic acid exist in form **XII-490c** in

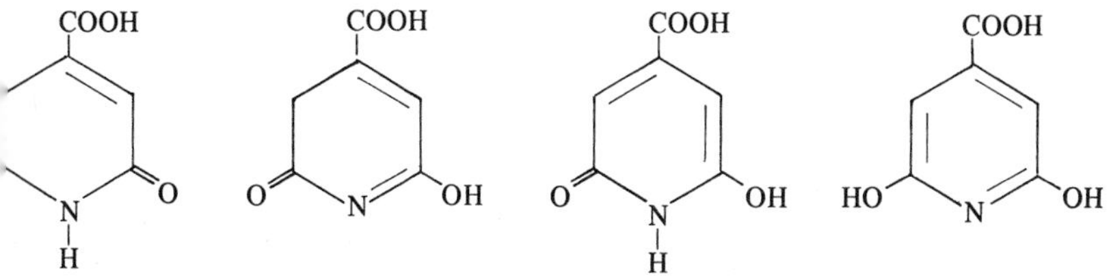

**XII-490 a** **XII-490 b** **XII-490 c** **XII-490 d**

thanol and in the imide form (**XII-490a**) in chloroform, dioxane, and tonitrile.[507]

he reaction of either the cyanoglutaconimide **XII-491** or the cyanofuropyri-ne **XII-492** with hydrochloric acid gives two isomeric products that were first sidered to be the furopyridone (**XII-493a**) and the furopyridinol I-493b).[508] Subsequently it has been demonstrated that only one tautomer be isolated[458, 509] and that the compound earlier identified as **XII-493a** is bably 4-methyl-3-spirocyclopropane-2,6-pyridinedione (**XII-494**),[510] which is med from **XII-495** during alkaline workup of the reaction mixture.[510]

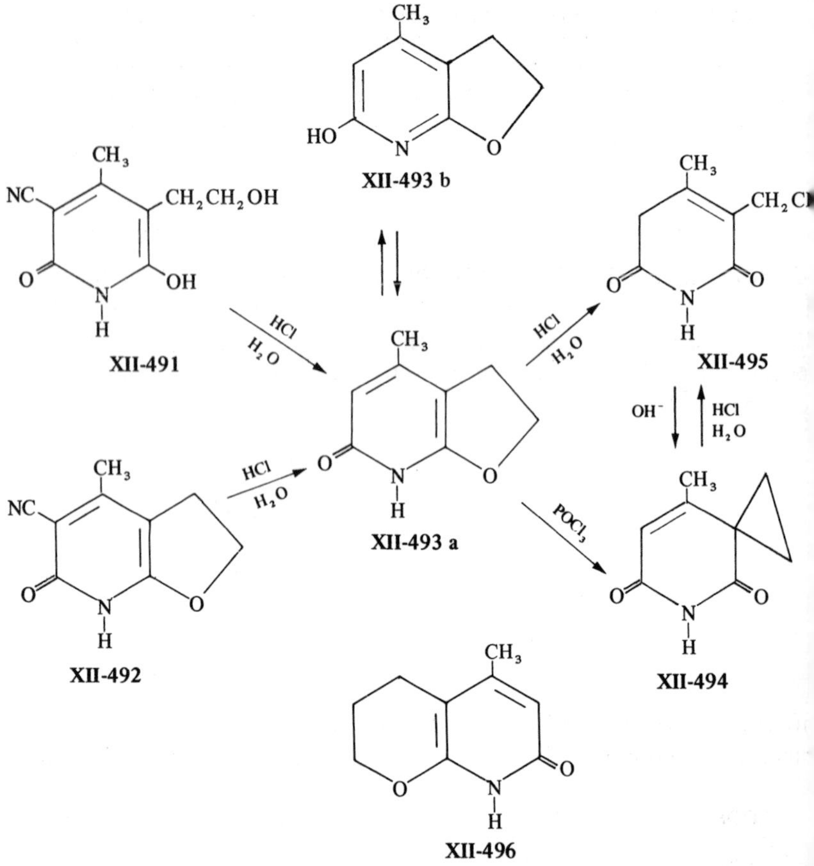

In ethanol, 2,3-dihydro-6-hydroxy-4-methylfuro[2,3-*b*]pyridine exists ma as the pyridinol **(XII-493b)**, while the corresponding 2,3-dihydropyranopyri **XII-496** exists mainly as the pyridone. Under the same condit 6-methoxy-2-pyridone exists as equal amounts of the two tautomers. preference, on energetic grounds, by a five membered ring for exo double bo (a possible Kékulé localization for **XII-493b**) and the preference by dihydropyrano ring for endo double bonds can account for t observations.[458]

## 5. 3-Pyridinols

Spectroscopic data and p*K* measurements have indicated that 3-pyridin aqueous solution is present in a tautomeric equilibrium between approxima al amounts of **XII-497a** and **XII-497b**[511] among other ionic forms.[511, 512] infrared spectrum of 3-pyridinol in the solid state[513] and its NMR spectrum onpolar solvents[514] show the presence of intermolecular hydrogen bonding is compatible with a dipolar structure. A comparison of the NMR spectra of ethyl-3-pyridinol and 1,2-dimethyl-3-hydroxypyridinium iodide has provided itional evidence for the existence of the zwitterion form in aqueous tion.[515] Calculated singlet-singlet transition energies of the zwitterionic form 3-pyridinol by the Pariser-Parr-Pople method are in good agreement with erimental data, but the discrepancies are higher than those realized for the lic and anionic forms.[516]

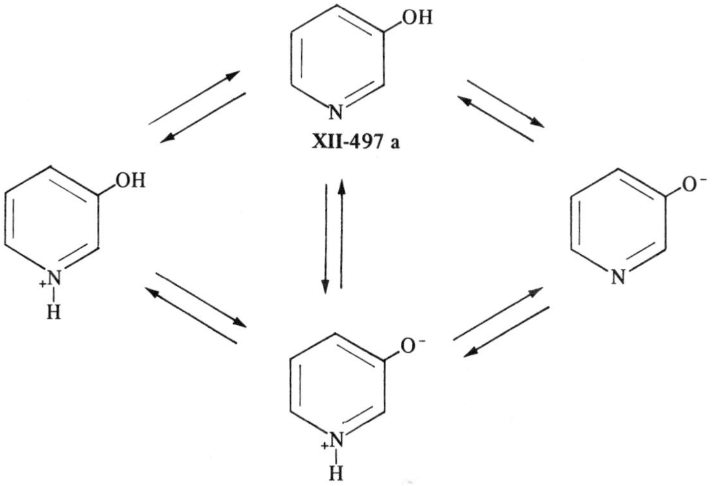

XII-497 a

XII-497 b

nhydro 3-hydroxy-*N*-methylpyridinium hydroxide, "*N*-methyl-3-pyridone," **I-499)** has been prepared and characterized and has been classified a mesionic compound.[517] The similarity of spectral properties of the e *N*-methylpyridones **(XII-498, XII-499, XII-500)** and the observations that olar structures make only a small contribution to **XII-498** and **XII-500** has gested that there is also only a small contribution by the polar structure **-499a** to "1-methyl-3-pyridone." A delocalization of the negative charge into ring that is accommodated through canonical structures **XII-499b–d** can ount for the carbonyl properties of **XII-499**. It has been suggested that nulas such as **XII-499e** be employed for mesionics;[517] however, undesirable notations such as an implication of *d* orbital participation could result from notation. A PPP ASMO calculation also predicts that 2-, 3-, and 4-pyridone uld all have similar properties.[517, 518]

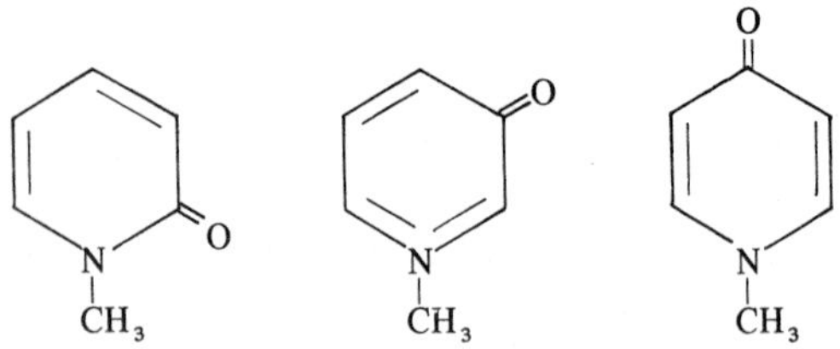

XII-498 XII-499 e XII-500

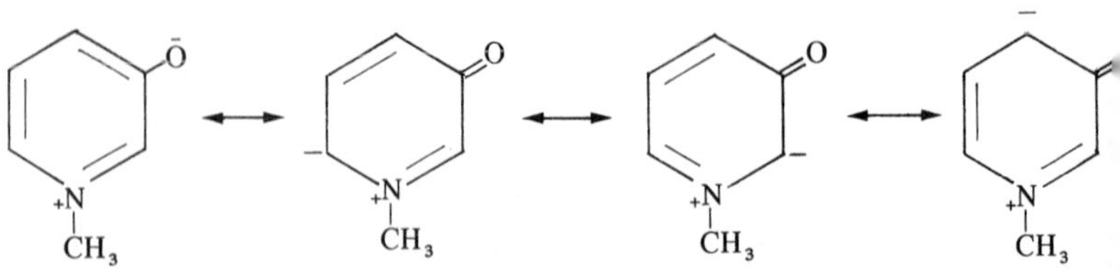

XII-499 a XII-499 b XII-499 c XII-499 d

The ultraviolet spectra of bis(3-hydroxy-2-pyridylmethylene)ethylenedian **(XII-501)** indicate that in ethanol the keto form predominates and protonation occurs on the annular nitrogen atoms.[519]

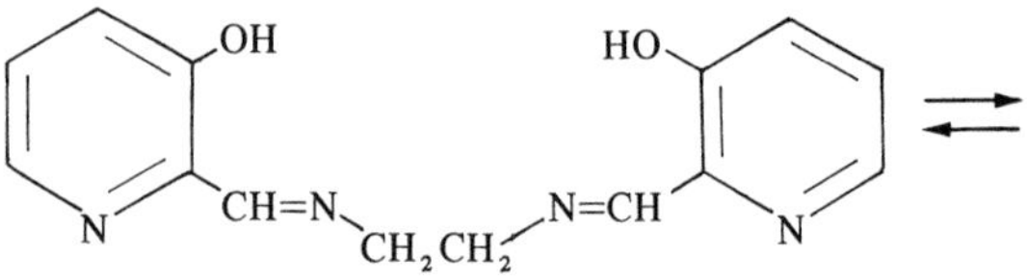

XII-501

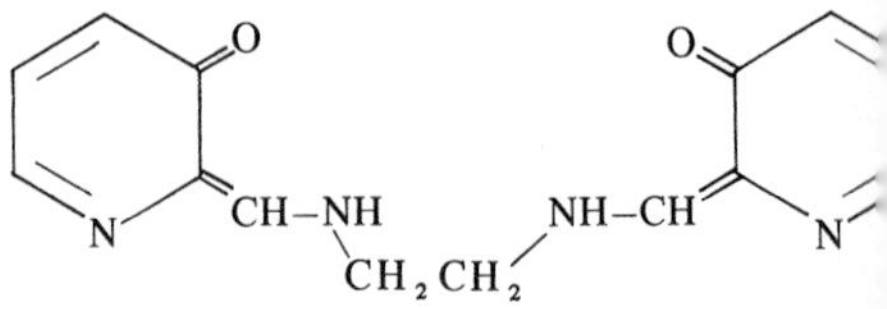

Chemical shifts have been used to evaluate pi electron densities for severa 4-, and 6-alkyl- and 2-, and 6-aminomethyl-3-pyridinols. Reasonable agreeme observed in a comparison of these charge distributions and those obtaine MO LCAO calculations.[520]

Earlier data have indicated that 3-pyridinol[453, 521] is a weaker base pyridine.[522] A reexamination of these basicities suggests that the reverse is and that the 3-hydroxyl group is electron releasing here and its introduc increases the basicity of the nitrogen of pyridine and alkylpyridines.[523] values for the individual acid dissociation steps for pyridoxamine, pyridc 3-hydroxypyridine-4-carboxaldehyde, and 3-hydroxypyridine-2-carboxaldel have been obtained by NMR measurements in $D_2O$ solution.[524]

### 6. Fluorescence

luorescence studies have been reported for several 3-pyridinols including idinols of the vitamin $B_6$ series, and 2- and 4-pyridones.[512, 525-527] yridone and 4-methoxypyridine are not fluorescent at any p*H* values.[525] The ions, anions, and dipolar ions of 3-pyridinol derivatives are fluorescent, but ıtral forms are not.[525, 526] All forms of 2-pyridone and its derivatives are ›rescent.[525]

xcited state ionization (ionization of acids and bases in the excited e at p*H* ranges where unexcited molecules do not ionize)[528] has been lied for cations of 3-pyridinols,[525, 528] pyridoxine,[525, 527, 528] pyridoxine,[525, 527, 528] pyridoxal,[525, 527, 528] 2-pyridone,[525] and *N*-methyl-2-pyrie and *N*-methyl-3-pyridinol.[525] Excited state p$K_a$ values for the 3-pyridinol on and cation have been studied using photopotentiometry, the measurement the potential developed between one illuminated electrode and one dark :trode in solution.[529]

### 7. Relative Stabilities of *N*-Alkylpyridones and Alkoxypyridines

ı equilibration studies in the liquid state by using the corresponding nethyl-2-(or)4-methoxypyridinium tetrafluoroborate as the catalyst, it has n determined that the *N*-methyl isomer is more stable than the *O*-methyl ner for both pairs: 2-methoxypyridine ⇌ *N*-methyl-2-pyridone ($\Delta G° > -9.3$ l/mole) (**XII-502** ⇌ **XII-503**) and 4-methoxypyridine ⇌ *N*-methyl-4-pyridone **I-504** ⇌ **XII-505**) ($\Delta G° > -7.4$ kcal/mole) at 130°.[530, 531] On the other hand, sulfur analog of **XII-502** is more stable than the *N*-methylpyridinethione.[532]

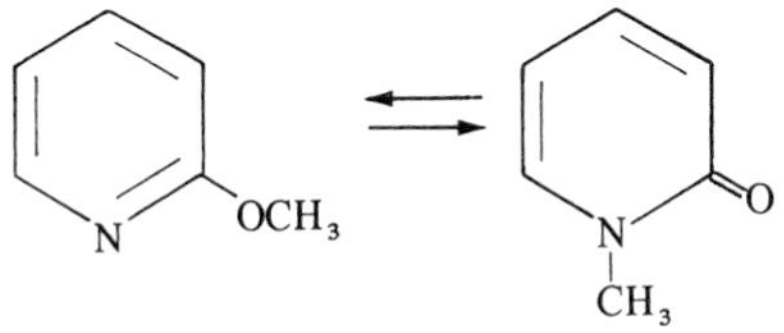

**XII-502** ⇌ **XII-503**

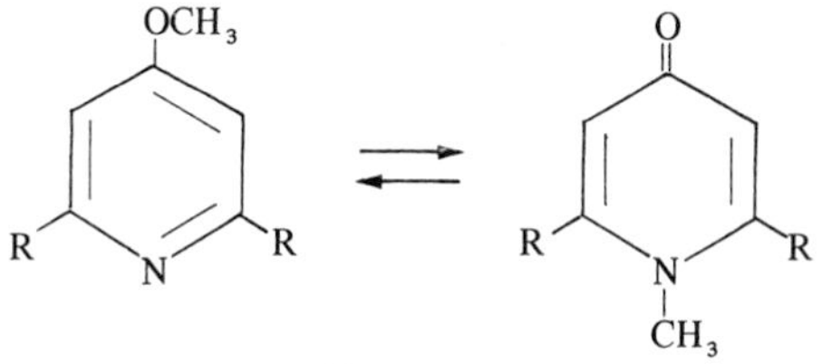

**XII-504** ⇌ **XII-505**

The relative stabilities of these two sulfur analogs should be compared with tautomeric equilibrium of 2-thiopyridine where the pyridinethione tauto predominates in aqueous solution. These observations demonstrate tautomeric equilibria in protic systems cannot be used reliably to predict relative stabilities of the methylated isomers.[532] It has been shown recently *O*-mesyloxypyridines **(XII-506)** are more stable than *N*-mesyl-2-pyridones.[533]

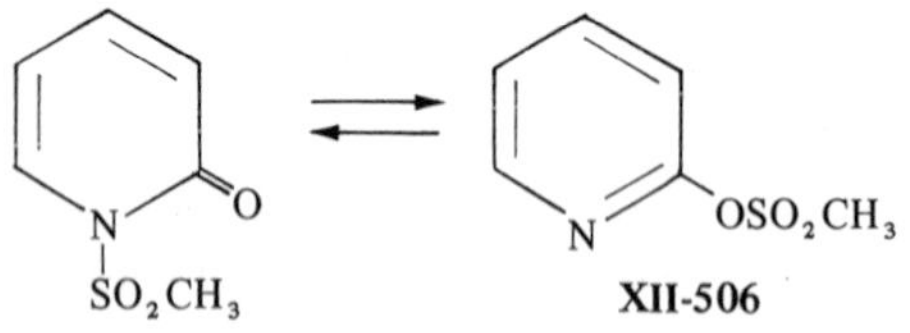

$SO_2CH_3$ ⇌ $OSO_2CH_3$

**XII-506**

4-Methoxy-2,6-diphenylpyridine (**XII-504**, R = $C_6H_5$) is more stable t *N*-methyl-2,6-diphenyl-4-pyridone due to 1,2,6-steric interactions.[530, 531]

## 8. Protonation of 2- and 4-Pyridones

Although it had been suggested on the basis of Raman and infrared evide that 2- and 4-pyridone are protonated on nitrogen,[534] subsequent informa has established that protonation occurs on oxygen.[464, 472, 486, 535-542]

Protonated 6-methoxy-2-pyridone, 6-methoxy-1-methyl-2-pyridone, 6 droxy-1-methyl-2-pyridone, and 2,6-dimethoxypyridine all exist as struc **XII-507**.[455]

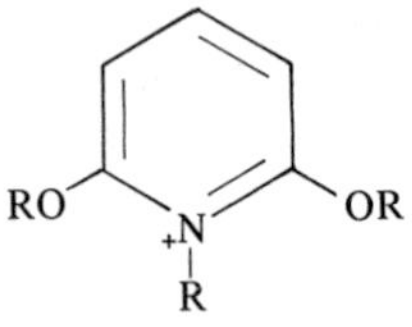

RO OR +N R

**XII-507**

The crystal structure of 2-hydroxypyridinium chloride monohydrate 2,6-dihydroxypyridinium chloride have been determined. The cations essentially planar and are in the mono- and dipyridinol forms respectively.[543]

## 9. The Pyridone Anion

Although at one time there was some disagreement as to the location of charge on the 2-pyridone anion in its ground state,[544, 545] later work demonstrated that the charge lies mainly on the oxygen atom.[472, 546]

# III. Reactions

## 1. Involving *O* and *N*

### A. *O- and N-Alkylation and Arylation*

GENERAL Alkylations of 2-pyridones have been studied extensively and lly are accomplished by reactions between the pyridone anion and an alkyl de or dialkyl sulfate or between the pyridone and a diazoalkane. Tosylates, nes, and epoxides have been used as alkylating agents, but less extensively. ough there are fewer reports of 3-pyridinol and 4-pyridone alkylations, rest in 3-pyridinols has increased over the past decade, but the course of r alkylations and the structures of products have been less clearly elucidated. nd/or *N*-Alkylations usually are observed, although side-chain $\alpha$-*C*-alkylation be accomplished. It has been shown in alkylations of 2-pyridone salts, that product distribution in these reactions results from kinetic control and ends on the nature of the pyridone, the cation, the alkylating agent, and the ent.

2-PYRIDONES The course of alkylations of 2-pyridone salts is far more itive to changes in cation and solvent than are comparable alkylations of rimidone salts.[547, 548] The pi-electron deficiency in pyrimidones due to two nitrogens imposes more severe restrictions on polarizability than are rved in alkylations of pyridones. In alkylations of alkali metal and silver ridone salts with halides, important solvent effects are observed. Solvent has greatest effect on silver salt alkylations, where alkoxypyridine formation is red in poor ion-solvating media.[547, 549] Under heterogeneous conditions y alkylations give only 2-alkoxypyridines.[549] For example, the silver salt of ridone and methyl iodide give *N*-methyl-2-pyridone as the major product in ethylformamide but give 2-methoxypyridine (97%) when benzene is used.[547] lations of alkali metal salts are less solvent sensitive but an increase in rate more *O*-alkylation are observed in dimethylformamide[547] and dimethyl xide[550] compared with protic and nonpolar solvents. Although alkali metal of 2-pyridone and methyl iodide in dimethylformamide give *N*-methyl-2-done as the major product (>90%), *O*-alkylation increases as the halide is d from methyl to ethyl and predominates with isopropyl iodide due to a c effect (Table XII-10).[547] The potassium salt of 6-acetamino-2-pyridone thanol-water is *N*-alkylated by isoamyl bromide.[551] The "traditional" nts employed in 2-pyridone salt alkylations, methanol and ethanol, ar to be relatively poor for ethylations and isopropylations of 2-pyridone; iderable elimination occurs as is evidenced by 2-pyridone formation.[547]

In alkylations of silver salts in DMF, considerable amounts of 2-pyrid are also regenerated. An $S_N2$ alkyl-oxygen cleavage of 2-methoxypyri may be responsible for the pyridone that is formed (see Section IV). ( elimination is observed with the silver salt of 2-pyridone and isopropyl bror in DMF (Table XII-11).[547]

Although the product distributions listed in Tables XII-10 and XII-11 calculated from gas chromatographic data, simple procedures have b developed for bench-scale preparations by direct alkylation to f chromatographically pure 2-isopropyloxypyridine (76%), 2-benzyloxypyri (78%), and 2-methoxypyridine (57%). These procedures avoid the m circuitous route where a 2-halopyridine is treated with an alkoxide to form ether.[547]

In view of the relatively large solvent effects on the alkylation site of the si salt of 2-pyridone, generalizations concerning the site of alkylation ring-substituted 2-pyridone salts need to be reconsidered. These generalizati sometimes have been made by comparing reactions run in different solvents

TABLE XII-10. Alkylation of Salts of 2-Pyridone in DMF[547]

| Alkylating agent | Cation | % Yield | Product Composition, % | | |
|---|---|---|---|---|---|
| | | | *N*-Alkyl | *O*-Alkyl | 2-Pyric |
| MeI | Na | 93 | 95 | 5 | |
| EtBr | Na | 94 | 77 | 23 | |
| IsoPrBr | Na | 84 | 29 | 68 | 3 |
| $C_6H_5CH_2Br$ | Na | 95 | 97 | 3 | |
| MeI | Ag | 81 | 74 | 12 | 21 |
| EtBr | Ag | 80 | 20 | 38 | 42 |
| IsoPrBr | Ag | No alkylation, only 2-pyridone was formed | | | |
| $C_6H_5CH_2Br$ | Ag | 85 | 54 | 46 | |

TABLE XII-11. Solvent Effects on Alkylations of Salts of 2-Pyridone[547]

| Alkyl halide | Solvent | Cation | % Yield | Product Composition, % | | |
|---|---|---|---|---|---|---|
| | | | | *N*-Alkyl | *O*-Alkyl | 2-Pyri |
| EtBr | DMF | Na | 94 | 77 | 23 | |
| EtBr | MeOH | Na | 80 | 66 | 5 | 29 |
| EtBr | $(CH_3O)_2(CH_2)_2$ | Na | 88 | 87 | 6 | 7 |
| EtBr | DMF | Ag | 80 | 20 | 38 | 42 |
| EtI | $(CH_3O)_2(CH_2)_2$ | Ag | 90 | 27 | 54 | 19 |
| EtI | EtOH | Ag | 91 | 1 | 80 | 19 |
| EtI | $C_6H_6$ | Ag | 100 | | 100 | |
| $C_6H_5CH_2Br$ | $C_6H_6$ | Ag | 100 | | 100 | |
| IsoPrI | $C_6H_6$ | Ag | 100 | | 100 | |

omparing reactions where only a part of the total reaction products were tified.[547, 552]

has been verified by comparing alkylations of sodium salts and silver salts of ridone, 5-nitro-2-pyridone, and 5-carbethoxy-2-pyridone that electron drawing groups at the 5-position cause increased *N*-alkylation. This tituent effect, like the solvent effect, is more pronounced in alkylations of salts.[549] When electron withdrawing groups occupy both the 3- and sitions, *N*-alkylation is strongly favored. For example, silver salts of o-5-nitro-2-pyridones and methyl iodide in ethanol give mainly the ethyl-2-pyridones and only small amounts of the ethers (3–6%).[287, 290, 291] omo-1-methyl-5-nitro-2-pyridone has also been prepared from the sodium and dimethyl sulfate.[290]

melting a mixture of 2-pyridone and picrylpyridinium chloride, the *O*- and kylated products (**XII-508** and **XII-509**) are formed in 17% and 82% yield, ectively. Rearrangement of **XII-508** to **XII-509** occurs at its melting point.[553] silver salt of 2-pyridone and picryl chloride in benzene gives 97% of **XII-508** 3% of **XII-509**.[553]

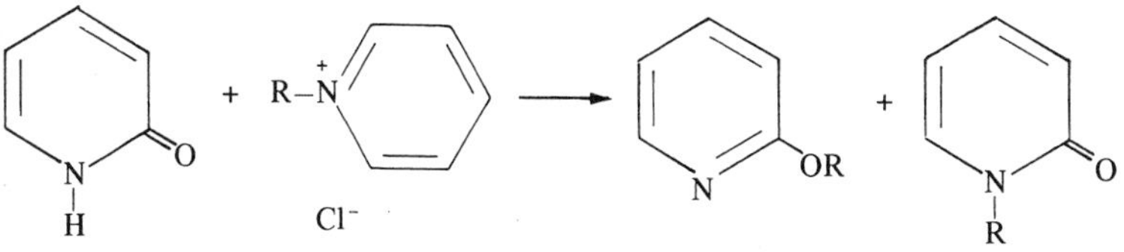

+ R–N⁺ Cl⁻ ⟶ **XII-508** (R = Picryl) + **XII-509**

e silver salt of 2-pyridone and trimethylene iodide in dioxane gives lihydropyrido[2,1-*b*]oxinium iodide (**XII-510**).[554] Potassium salts of

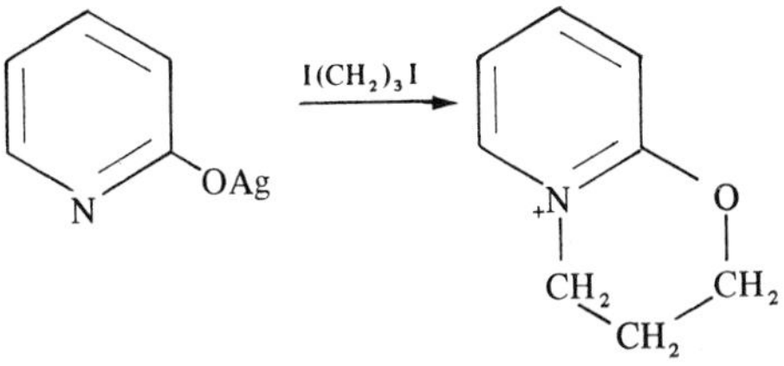

$I(CH_2)_3I$

**XII-510**

5,6-tetrafluoro-2-pyridinol and 2,3,5,6-tetrafluoro-4-pyridinol are methyl- on oxygen by methyl iodide in methanol and by dimethyl sulfate.[246]

ver salts of 4,5,6-triphenyl-2-pyridone and 4,6-diphenyl-5-phenoxy-2- lone are mainly *O*-alkylated by methyl iodide in methanol or benzyl ide in benzene. The potassium salt of 4,5,6-triphenyl-2-pyridone in an excess of methyl iodide gives more *N*-alkylation product, but the sodium and benzyl chloride in ethanol gives more *O*-alkylation product.[121] This app to be due to a steric effect, since sodium salts of 2-pyridones that are unsu tuted in the 3- and 6-position give similar product distributions in reactions benzyl halides and with methyl iodide under similar conditions.[547, 549] steric effect has been observed in the benzylation of 6-methyl-2-pyridon

Butylation of the sodium salt of 2-pyridone in dimethyl sulfoxide 2-*n*-butoxypyridine and *N*-*n*-butyl-2-pyridone, in the same product ratio pressure of 1 atm and at 1360 atm. Similarly, the product distributio benzylation of the potassium salt in ethanol does not vary with pressure, therefore the transition states leading to *O*- and *N*-alkylation have n identical molar volumes.[550] It has been concluded from these data that branching of the reaction pathway occurs at or beyond the transition state. model for the transition state **(XII-511)** implies that the products are determ by minor differences in conformation and energy distribution.[550]

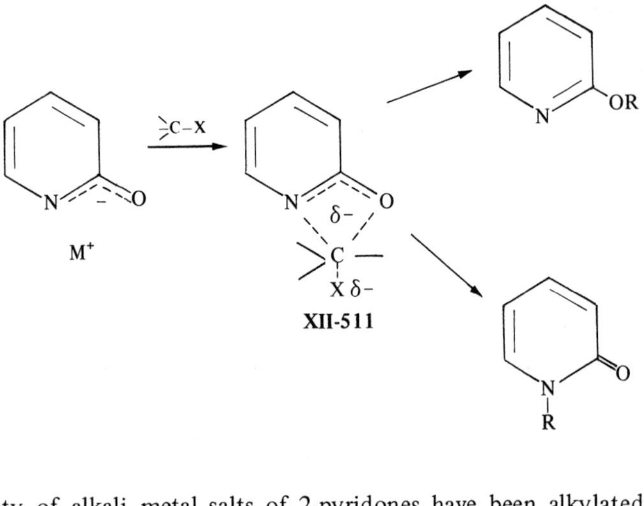

**XII-511**

A variety of alkali metal salts of 2-pyridones have been alkylated in solvents to give *N*-alkyl-2-pyridones (Table XII-12).

The sodium salts of 2-pyridone and 4-pyridone react with 1-bromoisoquin and copper-bronze to give mixtures of products. For example, under Ull conditions the sodium salt of 2-pyridone gave 39% of *N*-(2-isoquinol pyridone **(XII-512)** and minor amounts of 2-(2-pyridyl)isocarbos 2-(1-isoquinolyl)isocarbostyril, 1-(2-pyridyl)-2-pyridone and isocarbostyr (See also Section I.6.A., p. 691.) Under these conditions, 3-acetamido-2-br pyridine forms 1-(3-acetamido-2-pyridyl)-2-pyridone, which can be hydro to the aminopyridyl-2-pyridone and then cyclized in the presenc polyphosphoric acid to dipyrido[1,2-*a*:3,2-*d*]imidazole **(XII-513)**.[271]

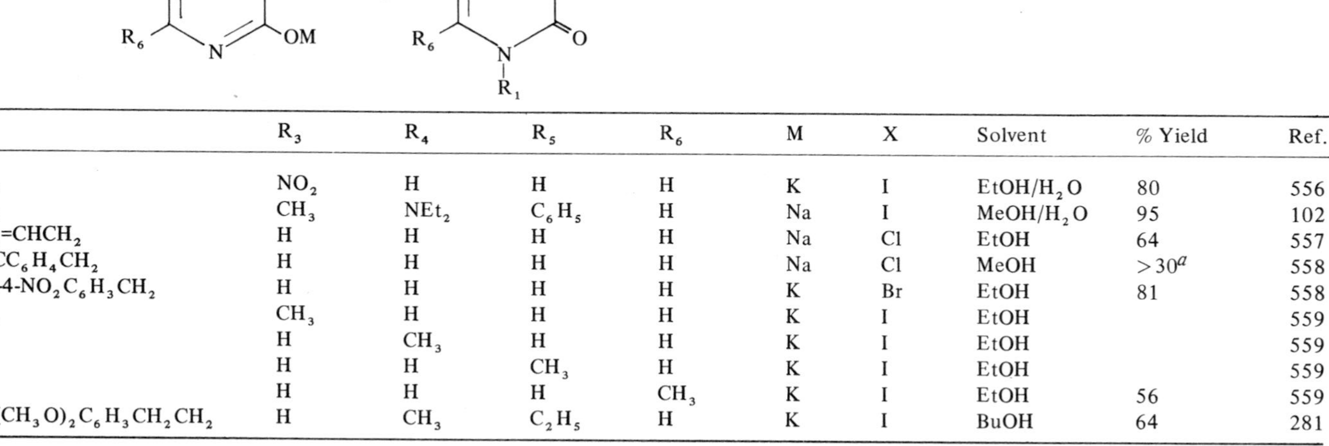


| $R_1$ | $R_3$ | $R_4$ | $R_5$ | $R_6$ | M | X | Solvent | % Yield | Ref. |
|---|---|---|---|---|---|---|---|---|---|
| $CH_3$ | $NO_2$ | H | H | H | K | I | $EtOH/H_2O$ | 80 | 556 |
| $CH_3$ | $CH_3$ | $NEt_2$ | $C_6H_5$ | H | Na | I | $MeOH/H_2O$ | 95 | 102 |
| $CH_2$=$CHCH_2$ | H | H | H | H | Na | Cl | EtOH | 64 | 557 |
| *p*-$NCC_6H_4CH_2$ | H | H | H | H | Na | Cl | MeOH | >30[a] | 558 |
| 2-Cl-4-$NO_2C_6H_3CH_2$ | H | H | H | H | K | Br | EtOH | 81 | 558 |
| $CH_3$ | $CH_3$ | H | H | H | K | I | EtOH | | 559 |
| $CH_3$ | H | $CH_3$ | H | H | K | I | EtOH | | 559 |
| $CH_3$ | H | H | $CH_3$ | H | K | I | EtOH | | 559 |
| $CH_3$ | H | H | H | $CH_3$ | K | I | EtOH | 56 | 559 |
| 3,4-$(CH_3O)_2C_6H_3CH_2CH_2$ | H | $CH_3$ | $C_2H_5$ | H | K | I | BuOH | 64 | 281 |

[a]Yield reported is for the product of hydrolysis, *N*-(*p*-carboxybenzyl)-2-pyridone.

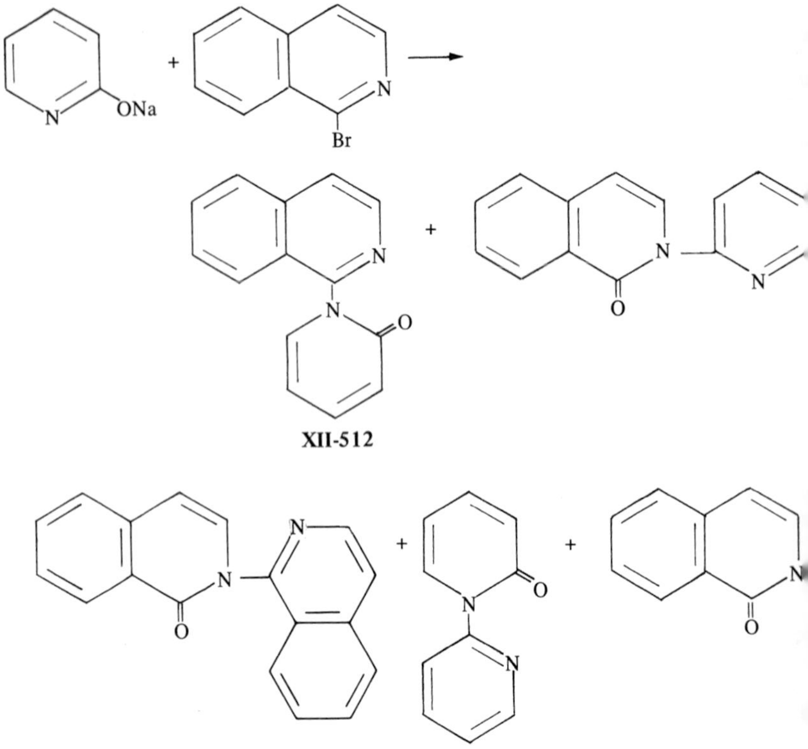

ONa
Br
N
N
N
O
N
O
XII-512 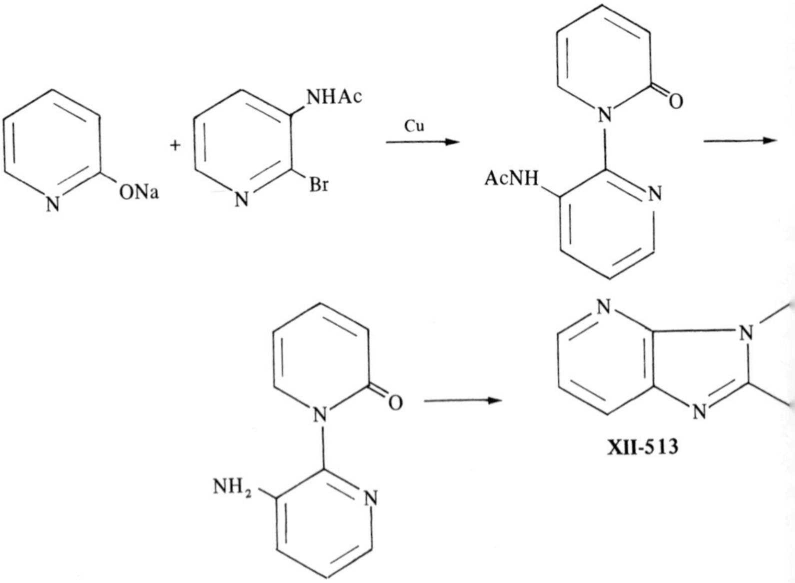
N
N
O
N
O
N
N
O
NHAc
Cu
N
ONa
N
Br
N
O
AcNH
N
N
N
O
NH$_2$
N
N
N
XII-513

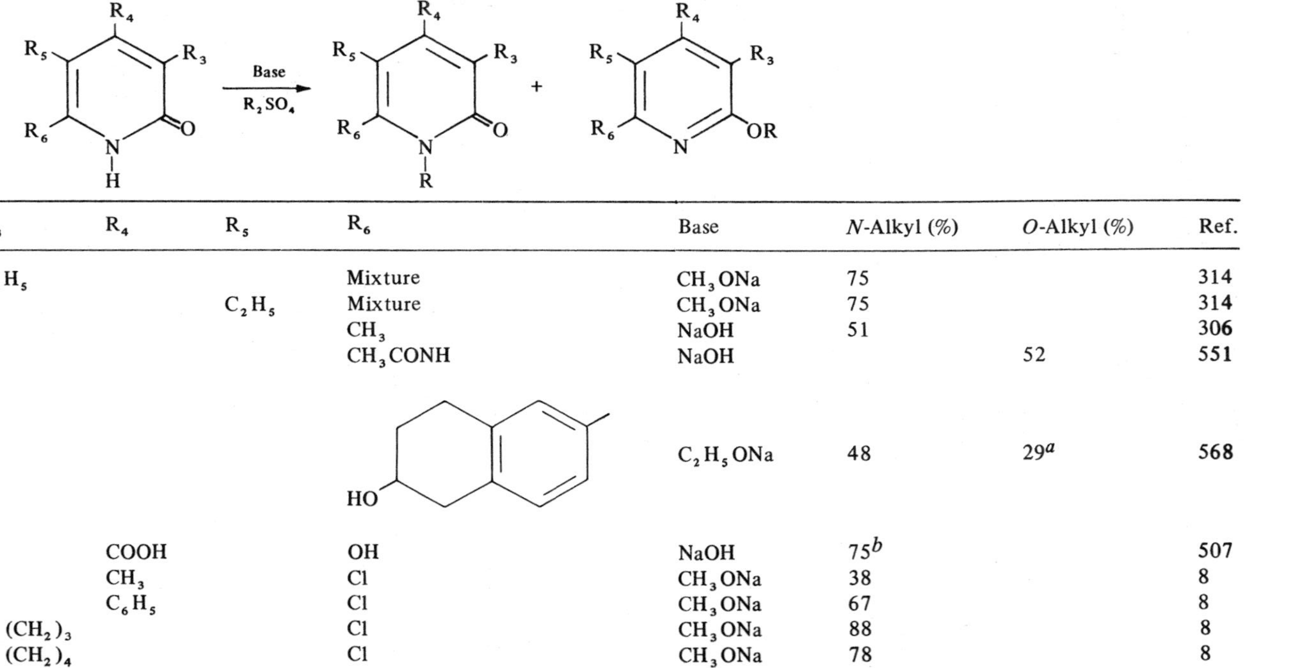


| R | $R_3$ | $R_4$ | $R_5$ | $R_6$ | Base | *N*-Alkyl (%) | *O*-Alkyl (%) | Ref. |
|---|---|---|---|---|---|---|---|---|
| $CH_3$ | $C_2H_5$ | | | Mixture | $CH_3ONa$ | 75 | | 314 |
| $CH_3$ | | | $C_2H_5$ | Mixture | $CH_3ONa$ | 75 | | 314 |
| $CH_3$ | | | | $CH_3$ | NaOH | 51 | | 306 |
| $C_2H_5$ | | | | $CH_3CONH$ | NaOH | | 52 | 551 |
| $CH_3$ | | | | HO | $C_2H_5ONa$ | 48 | 29[a] | 568 |
| $CH_3$ | | COOH | | OH | NaOH | 75[b] | | 507 |
| $CH_3$ | | $CH_3$ | | Cl | $CH_3ONa$ | 38 | | 8 |
| $CH_3$ | | $C_6H_5$ | | Cl | $CH_3ONa$ | 67 | | 8 |
| $CH_3$ | $(CH_2)_3$ | | | Cl | $CH_3ONa$ | 88 | | 8 |
| $CH_3$ | $(CH_2)_4$ | | | Cl | $CH_3ONa$ | 78 | | 8 |

[a] 6-(6-Methoxy-1,2,3,4-tetrahydro-2-naphthyl)-2-pyridone was also formed.
[b] Methyl *N*-methylcitrazinate.

A large number of *N*-substituted-2-pyridones have been synthesized potential pharmaceuticals, principally by Gogolimska[561–563] Raczka[337, 564, 565] and Sacha.[566, 567] These compounds have been prepared alkylation of alkali metal salts with halides,[337, 561–564] by modification of *N*-substituent,[561–567] and by oxidation of pyridinium salts with ferricyanide. Careful analyses of product distribution from alkylations of 2-pyridones w dialkyl sulfates are not generally reported. When dimethyl sulfate is used v salts of 2-pyridones that are unsubstituted or have small groups or chlorine the 6-position, *N*-alkylation appears to predominate. For example, 3- 5-ethyl-2-pyridone,[314] 6-chloro-2-pyridones[8] and 6-methyl-2-pyridone[306] *N*-alkylated by dimethyl sulfate and base (Table XII-13). On the other hand, sodium salt of 6-acetamido-2-pyridone and diethyl sulfate give 2-acetamido ethoxypyridine.[551] 6-(6-Hydroxy-1,2,3,4-tetrahydro-2-naphthyl)-2-pyridone dimethyl sulfate give the products of *O*- and *N*-alkylation[568] (Table XII-13).

Although it had been assumed earlier that diazomethane and 2-pyridone g exclusive *O*-alkylation,[569] it has been shown that both *N*-methyl-2-pyridone 2-methoxypyridine (Table XII-14) are formed and that the product ratio is same in methanol-ethyl ether as in methylene chloride-ethyl ether. An incre in *O*-alkylation of 2-pyridone by diazoethane as compared with alkylation diazomethane (Table XII-14) has been rationalized on steric grounds or resulting from a decrease in the $S_N2$ character of a graded $S_N1$–$S_N2$ factor.

4,5,6-Triphenyl-2-pyridone and 5-phenoxy-4,6-diphenyl-2-pyridone are ported to be *O*-alkylated by diazomethane.[121] Citrazinic acid and its met ester **(XII-514)** give mixtures of the products of *O*- and *N*-methylati 4-Carbomethoxy-6-methoxy-1-methyl-2-pyridone **(XII-515)** can be prepa from citrazinic acid by dimethyl sulfate to give methyl *N*-methylcitrazin (Table XII-13), which is then treated with diazomethane.

**XII-514** $\xrightarrow{CH_2N_2}$ 2,6-dimethoxy-4-carbomethoxypyridine + **XII-515**

Citrazinic acid $\xrightarrow[NaOH/H_2O]{(CH_3)_2SO_4}$ methyl *N*-methylcitrazinate $\xrightarrow[C_4H_8O]{CH_2N_2}$ **XII-515**

-Hydroxy-2-pyridone **(XII-516)** and a large excess of diazomethane give 1ethoxy-2-pyridone (2%), 6-methoxy-1-methyl-2-pyridone (4%), 2,6-dimeth- pyridine (8%), and the 3-(β-methylhydrazone) of 1-methyl-2,3,6-pyridine- one **(XII-517)** (14%).[505] 6-Methoxy-2-pyridone and diazomethane give a ture of *O*- and *N*-methylation products. 6-Hydroxy-1-methyl-2-pyridone es 6-methoxy-1-methyl-2-pyridone and **XII-517**.[505]

HO, N, H, O
**XII-516**
$CH_2N_2$ $CH_2N_2$
N, $CH_3$, O; $CH_3O$, N, $CH_3$, O; $CH_3O$, N, H, O
$CH_2N_2$ $CH_2N_2$
$CH_3O$, N, $OCH_3$
N, $CH_3$, $O^-$ + $N{\equiv}\overset{+}{N}{-}CH_3$ ⟶ N=N, $CH_3$, H, O, N, $CH_3$, O ⟶
N–NH$CH_3$, O, N, $CH_3$, O
**XII-517**

TABLE XII-14. Alkylations of 2-Pyridones and 2-Pyridinols with Diazoalkanes

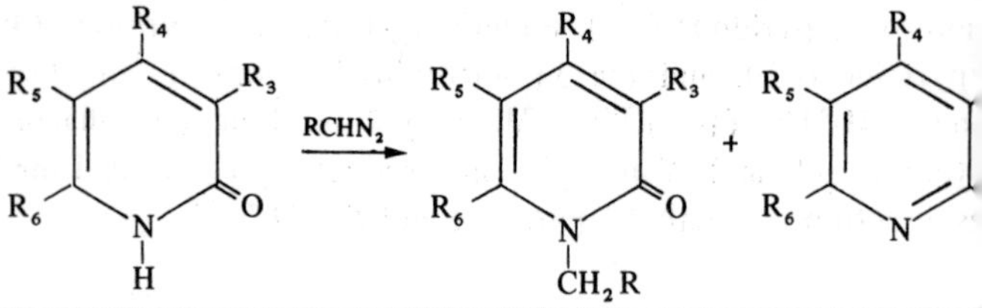

| R | $R_3$ | $R_4$ | $R_5$ | $R_6$ | *N*-Alkyl (%) | *O*-Alkyl (%) |
|---|---|---|---|---|---|---|
| H | H | H | H | H | 55 | 35 |
| $CH_3$ | H | H | H | H | 66 | 24 |
| H | | | | $NH_2$ | 37 | 48 |
| H | | | | Cl | 57 | 17 |
| $CH_3$ | | $C_2H_5O$ | Br | | | 62 |

4-Methyl-3-(spirocyclopropane)-2,6-pyridinedione **(XII-518)** and dia methane in methanol-ether give the *N*-methyl derivative.

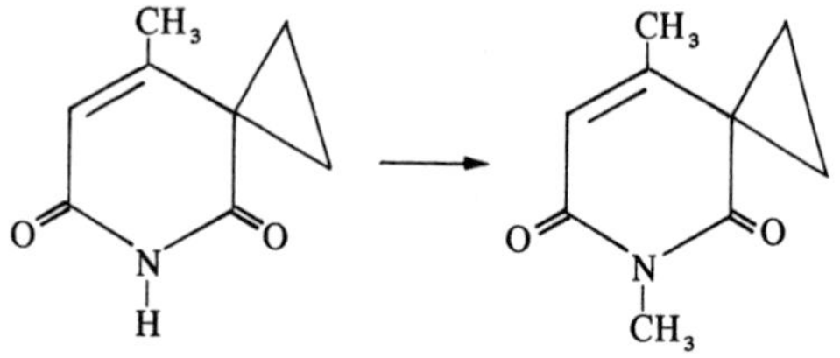

**XII-518**

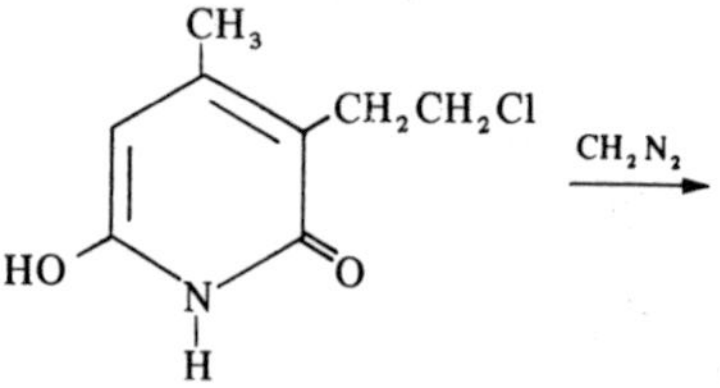

**XII-519**

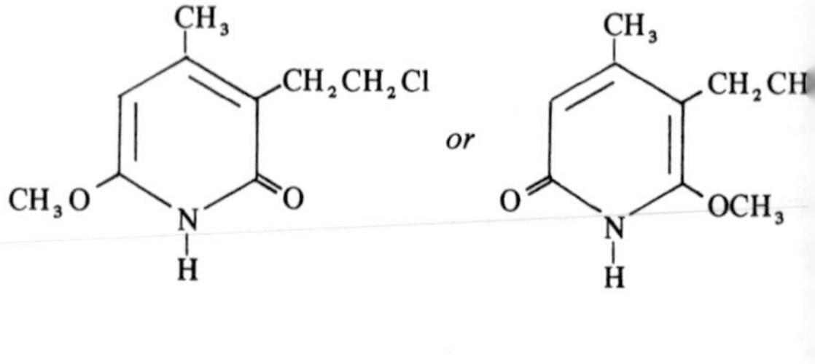

*or*

the other hand, 3-(β-chloroethyl)-4-methyl-2,6-pyridinediol **(XII-519)** gives roduct of *O*-alkylation that was not characterized.[510] Additional alkylations 2-pyridones by diazomethane and diazoethane are summarized in Table -14.

'he sodium salt of 2-pyridone and triethyloxonium tetrafluoroborate in thylene chloride gives 2-ethoxypyridine (21%), *N*-ethyl-2-pyridone (40%), l *N*-methyl-2-ethoxypyridinium tetrafluoroborate **(XII-520)**, 29%.[571] A wer alkylation of the sodium salt by **XII-520** produces 2-ethoxypyridine (9%)

$(C_2H_5)_3\overset{+}{O}BF_4^-$

$OC_2H_5$

ONa

$C_2H_5$

$BF_4^-$

**XII-520**

d *N*-ethyl-2-pyridone (79%).[571] The product distribution in this alkylation anges with time.[571] In ethanol and with an excess of triethyloxonium rafluoroborate, the monoethylation products are formed in 19% and 77% ld, respectively.[570] Product **XII-520** can be formed in 97% yield from thoxypyridine and is hydrolyzed with aqueous sodium hydroxide to *N*-ethyl-yridone (89%).[571]

-Ethyl-2-methoxy-5-nitropyridinium tetrafluoroborate **(XII-521)** is a strong thylating agent. Sodium benzoate, sodium iodide, and lithium bromide are thylated by this reagent.[572]

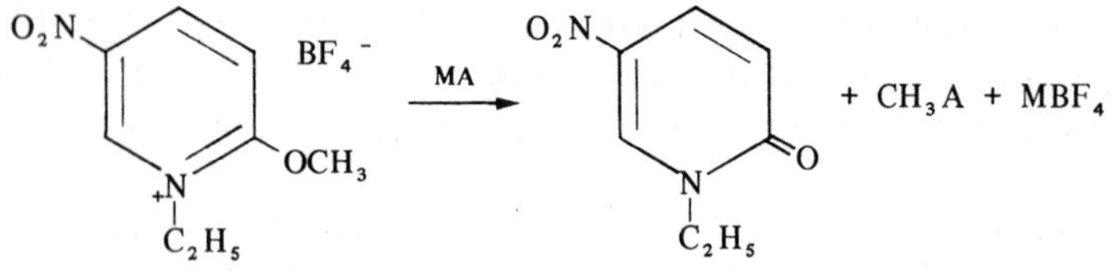

**XII-521**

$A = C_6H_5CO_2, Br, I$

The sodium salt of 2-pyridone and *O*-(2-hydroxyethyl)acetoneoxi *p*-toluenesulfonate gives the *N*-alkylation product **XII-522**.[573]

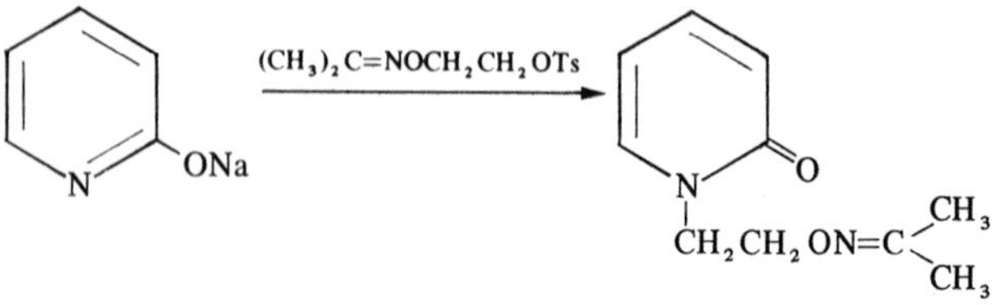

**XII-522**

Treatment of 2-pyridones (**XII-523**, X = O) or 2-pyridinethiones (**XII-523**, X = S) with dihydropyran in the presence of *p*-toluenesulfonic acid gi 1-tetrahydropyranyl-2-pyridones or 1-tetrahydropyranyl-2-pyridinethi (**XII-524**).[574]

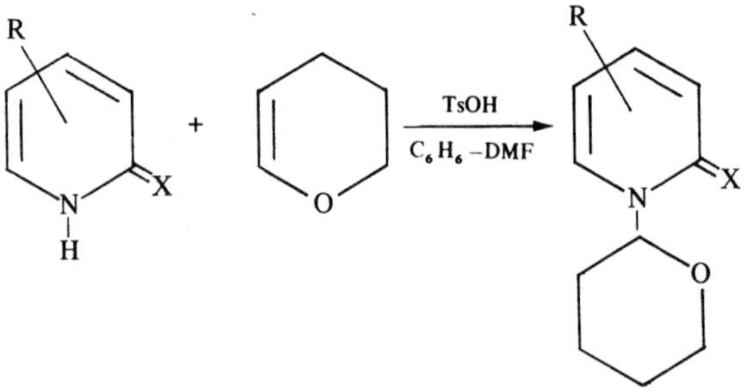

**XII-523** **XII-524**

2-Pyridone and *p*-nitrostyrene oxide give 1-(β-hydroxy-β-*p*-nitrophenethyl) pyridone (**XII-525**), which was also synthesized by alkylation of 2-pyridone w ω-bromo-*p*-nitroacetophenone to give the ketone **XII-526**, which was reduc with sodium borohydride.[326]

Dianions of 4- and 6-alkyl-3-cyano-2-pyridones can be formed from pyridone by treatment with two equivalents of potassium amide in liq ammonia and can be alkylated, aroylated, or acylated selectively on

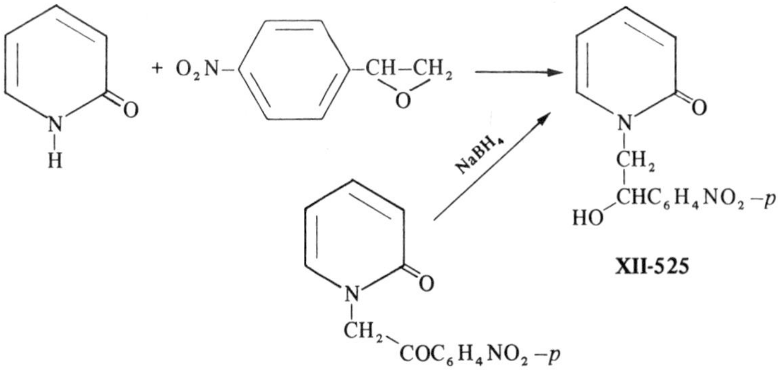

XII-525

XII-526

e-chain in good yields. By this method 6-methyl-2-pyridones **(XII-527)** are thylated to give the 6-ethylpyridone **(XII-528)** and benzylated to give the henethyl-2-pyridones **(XII-529)**, which can be alkylated further. 3-Cyano-4-thyl-6-phenyl-2-pyridone **(XII-530)** gives the 4-phenethyl derivative in 90% ld.[575-577]

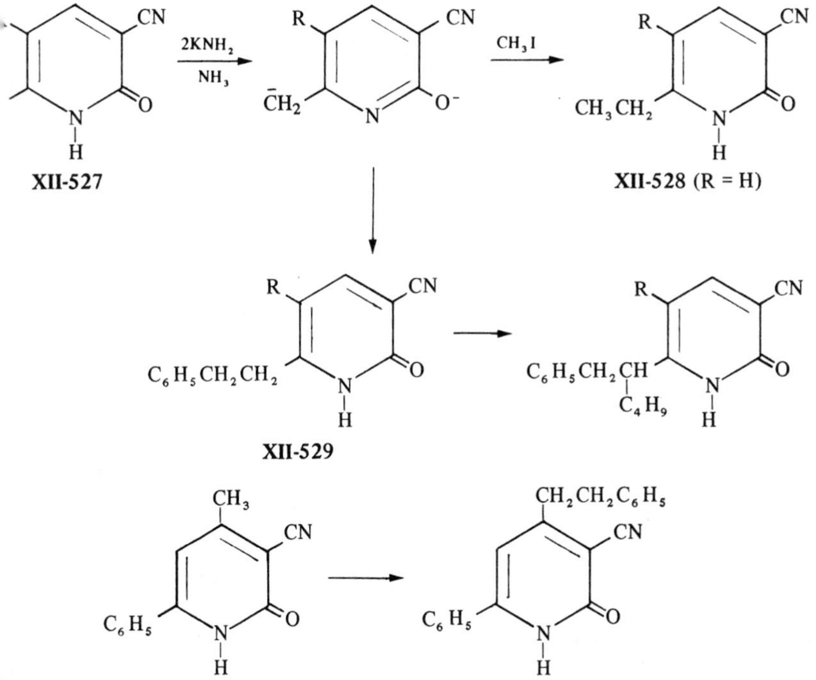

XII-527

XII-528 (R = H)

XII-529

XII-530

The dianion of 3-cyano-4,6-dimethyl-2-pyridone **(XII-531)** and benzyl chlo[illegible] give only one product, which was not characterized. However, **XII-531** [illegible] three equivalents of potassium amide and an excess of benzyl chloride gi[illegible] 30% yield of stilbene and 35% each of a monobenzyl- (**XII-532a** or **XII-53**[illegible] and the dibenzyl- derivatives.[575] This method gives better yields of

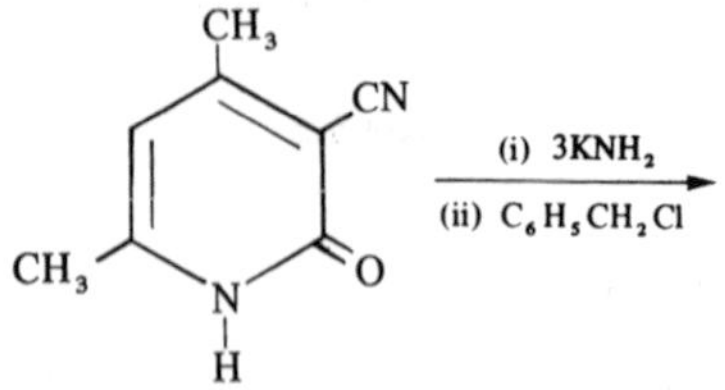

**XII-531** —(i) $3KNH_2$; (ii) $C_6H_5CH_2Cl$→

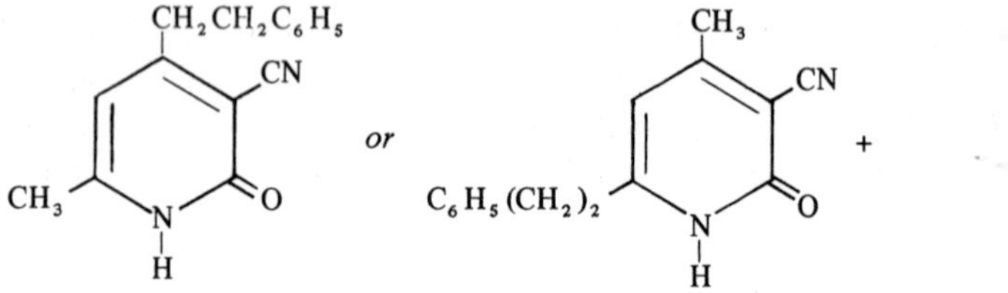

**XII-532 a** *or* **XII-532 b** +

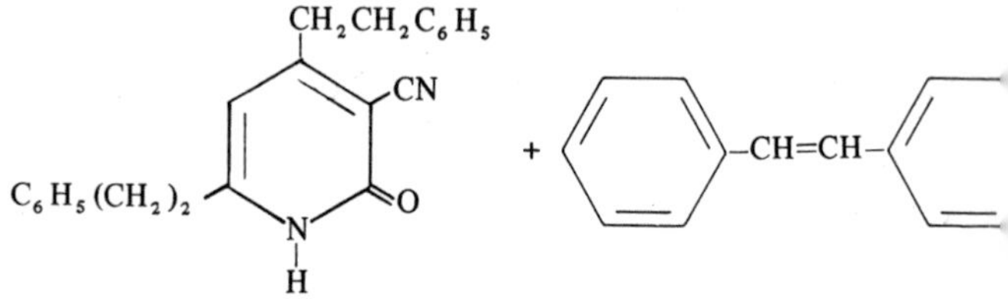

6-alkyl-3-cyano-2-pyridones than the route where dicarbanions of β-k[illegible] aldehydes or β-diketones **XII-533** are alkylated and then cyclized w[illegible] cyanoacetamide.[54] In the latter method, mixtures of products are form[illegible]

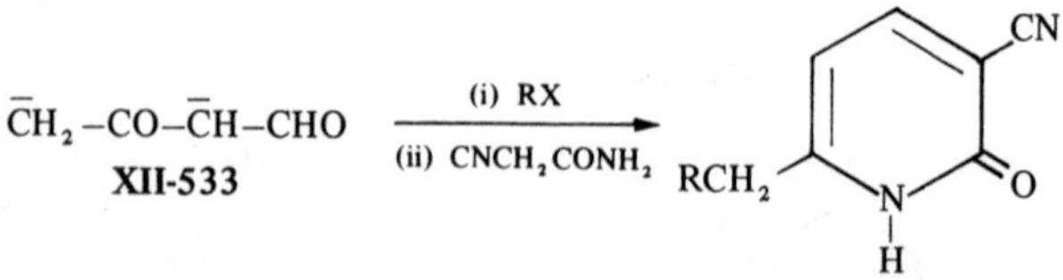

$\bar{C}H_2-CO-\bar{C}H-CHO$ **XII-533** —(i) RX; (ii) $CNCH_2CONH_2$→

$R = C_6H_5CH_2, CH_3$

although the use of an imine of a β-diketone has overcome this disadvantage.

The dianions from **XII-527** ($R = H, C_6H_5$) and **XII-530** are aroylated w[illegible] methyl benzoate to give the ketones **XII-534** and **XII-535**. They react w[illegible] nzaldehyde and benzophenone to give the alcohols **XII-536**. Benzophenone d the dianion **XII-530** give **XII-537**.[576]

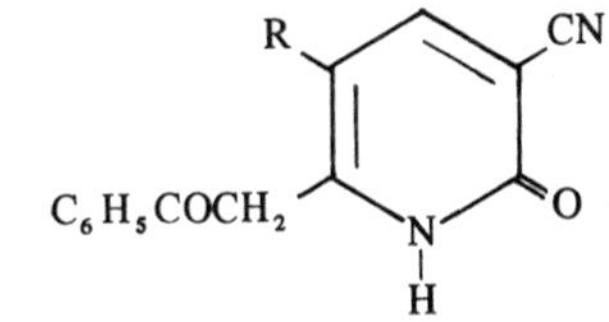

**XII-534** (R = H, $C_6H_5$)

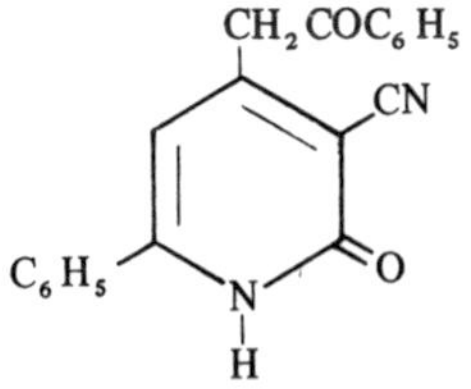

**XII-535**

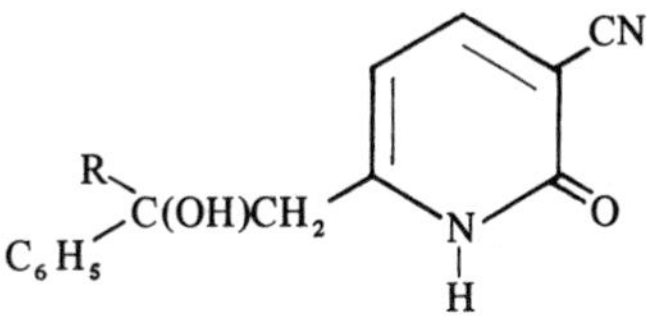

**XII-536**

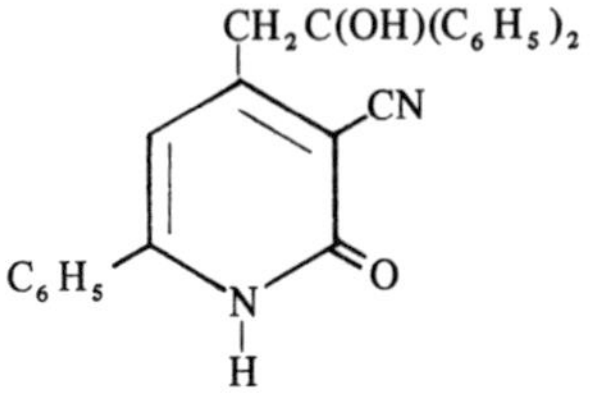

**XII-537**

The monoanion of 6-methyl-2-pyridone is not sufficiently acidic to form a isfactory dipotassium salt when treated with two equivalents of potassium ide. It does, however, react with *n*-butyllithium in tetrahydrofuran or ether to e the dilithio salt, which condenses readily with carbonyl compounds[579, 580] ., to give the tertiary alcohol **XII-538** in 94% yield. Less conveniently, the hydration product of **XII-538** was prepared from the dipotassium salt of **I-527** (R = H) *via* **XII-536** (R = $C_6H_5$). 4-Methyl- and 3-methyl-2-pyridones as

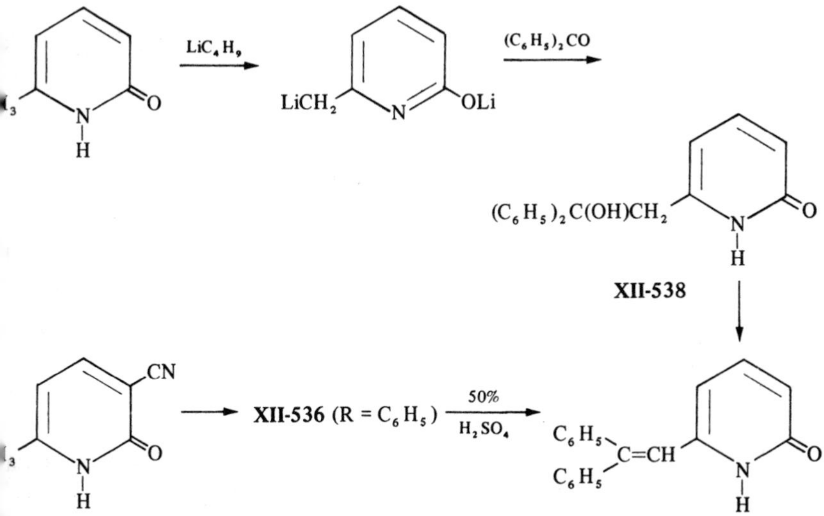

**XII-538**

**II-527** (R = H) → **XII-536** (R = $C_6H_5$) $\xrightarrow[H_2SO_4]{50\%}$

well as 3-methyl- and 6-methyl-2-pyridinethiones form dilithio salts that rea similarly.[579] Alkylation of the side-chain of methylpyridines has been extend to 2-benzyloxymethylpyridines (**XII-539, XII-540, XII-541**). The products c be converted to alkyl-2-pyridones by hydrogenolysis.[6, 581]

Attempted alkylations of 3-benzyloxy-6- and 2-benzyloxy-5-methylpyridi with tetramethylene dibromide have been unsuccessful.[6]

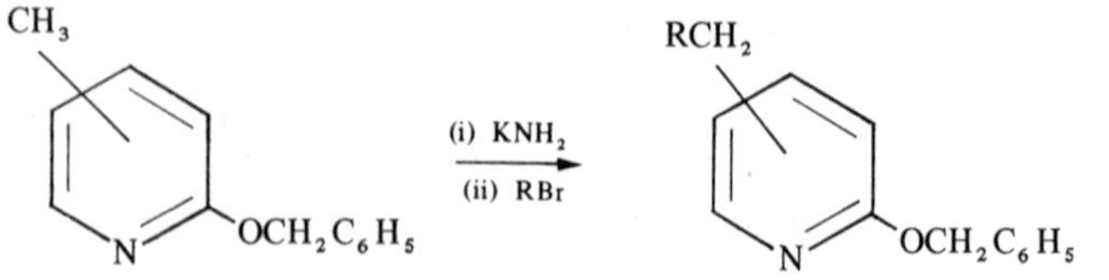

$CH_3$ $OCH_2C_6H_5$ **XII-539** (i) $KNH_2$ (ii) RBr $RCH_2$ $OCH_2C_6H_5$

R = 3-, 4-, 5- and 6-hexyl; 4-, and 5-undecyl; 4-benzyl.

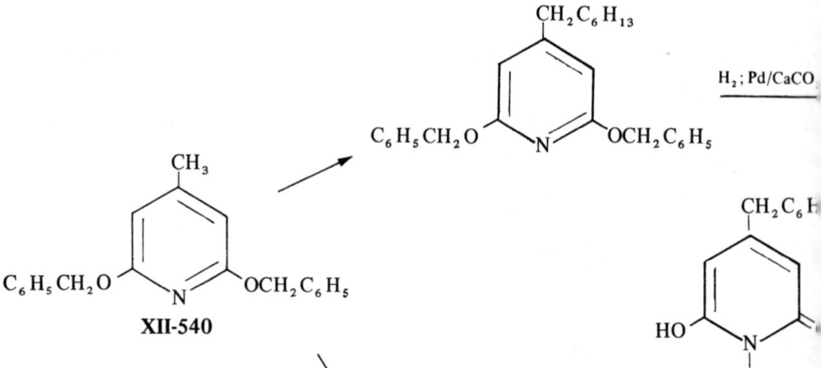

$CH_3$ $C_6H_5CH_2O$ $OCH_2C_6H_5$ **XII-540** → $CH_2C_6H_{13}$ $C_6H_5CH_2O$ $OCH_2C_6H_5$ $H_2$; Pd/CaCO $CH_2C_6H$ HO N H

HO H–N O $-(CH_2)_6-$ O N– OH

2-Pyridinethione and 1-bromo-2-chloroethane in dimethylformamide conta ing potassium carbonate give 2,3-dihydrothiazolo[3,2-*a*]pyridinium bromide

I. 1145). 3-Hydroxy-2-pyridinethione (**XII-542**, R = H) and 3-hydroxy-6-thyl-2-pyridinethione (**XII-542**, R = $CH_3$) and ethylene bromide in methanol taining sodium methoxide react to give dihydrothiazolo[3,2-*a*]pyridinium xides **(XII-543)**.[583] Methyl 2,3-dibromopropionate and **XII-542** form

$C_6H_5CH_2O$ N $CH_3$ **XII-541** → O NH $-(CH_2)_6-$ HN O

ydroxy-dihydrothiazolo[3,2-*a*]pyridinium 2-carboxylates **(XII-544)** after ting in benzene, followed by saponification. These reactants, in the presence sodium methoxide in methanol, give the 3-carboxylate **(XII-545)** after lrolysis, presumably *via* β-elimination of the ester to give ethyl romoacrylate, which then reacts with **XII-542**.[583] 8-Hydroxy-5-methyldi-lrothiazolo[3,2-*a*]pyridinium 3-carboxylate (**XII-545**, $R_6$ = $CH_3$) is a blue orescent material that has been isolated from bovine liver hydrolysates.[584] imilarly, **XII-542** reacts with α-bromoacrylonitrile to give the 3-cyano ivatives and with α-bromo-α-unsaturated acids by *trans* addition to form stituted 3-carboxyldihydrothiazolo[3,2-*a*]pyridinium compounds that are dily decarboxylated.[585] Under these conditions **XII-542** does not react with romostyrene.

he *trans* product **XII-546** from bromomaleic anhydride and **XII-542** requires t the anhydride ring of the intermediate adduct be opened to assume the formation necessary for *trans* displacement of Br by the annular nitrogen.[585]

4-PYRIDONES A combination of *ortho* and electronic effects appears to ermine the course of alkylations of 4-pyridones. *N*-Alkylation is more sitive to steric effects than is *O*-alkylation. For example, the potassium salt of itro-4-pyridone is *N*-alkylated by methyl iodide,[586] and the sodium salt of -diiodo-4-pyridone is *N*-alkylated by ω-bromoalkanoates in ethanol.[587] On other hand, the dimethyl and diethyl esters of 2,6-di-(α-pyridyl)-4-hydroxy--pyridinedicarboxylic acid (**XII-547**; R = $CH_3$, $C_2H_5$) undergo *O*-alkylation h methyl iodide in alcoholic potassium hydroxide,[500] and the potassium salt tetrafluoro-4-pyridone and either methyl iodide or dimethyl sulfate in thanol give tetrafluoro-4-methoxypyridine. The product of *N*-methylation ; not detected.[246] Alkylation of the sodium salt of 4-pyridone with henoxybutyl bromide followed by treatment with hydriodic acid gives 4-iodobutyl)-4-pyridone **(XII-548)**, which forms a polymer in water rather n the *N*-spiro derivative.[588] Dehydrohalogenation of 2-chloro-3-(β-chloro-

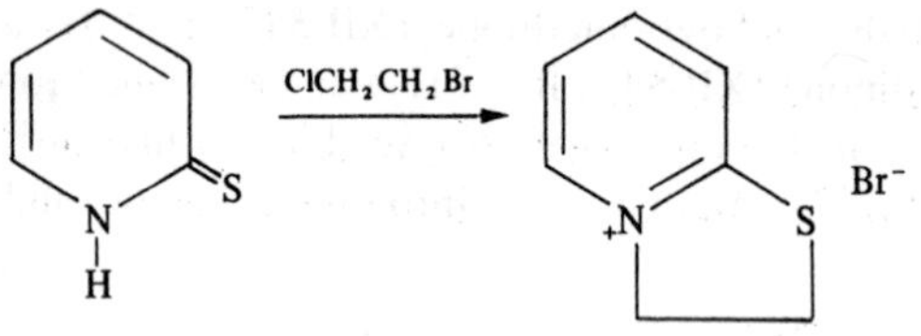

$ClCH_2CH_2Br$

$Br^-$

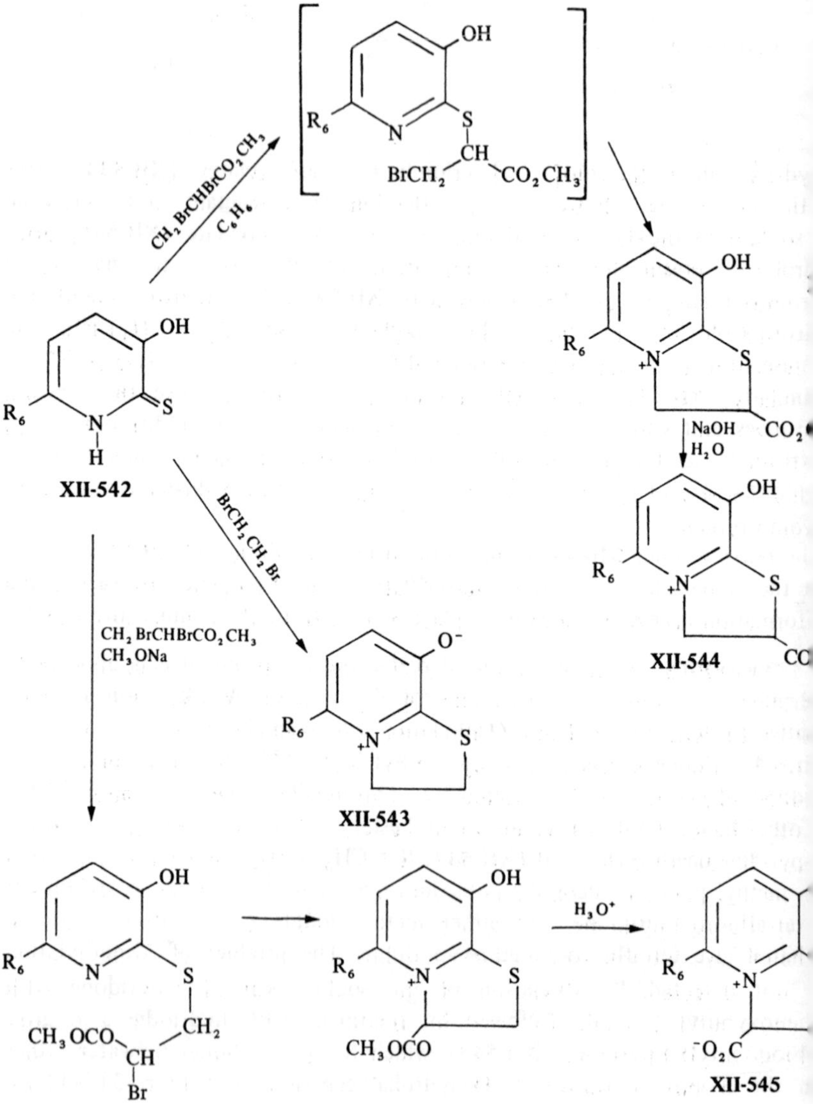

$CH_2BrCHBrCO_2CH_3$
$C_6H_6$

OH

$R_6$

$BrCH_2$ $CO_2CH_3$

NaOH
$H_2O$

**XII-542**

$BrCH_2CH_2Br$

$CH_2BrCHBrCO_2CH_3$
$CH_3ONa$

$O^-$

**XII-543**

**XII-544**

$CH_3OCO$

$H_3O^+$

$^-O_2C$

**XII-545**

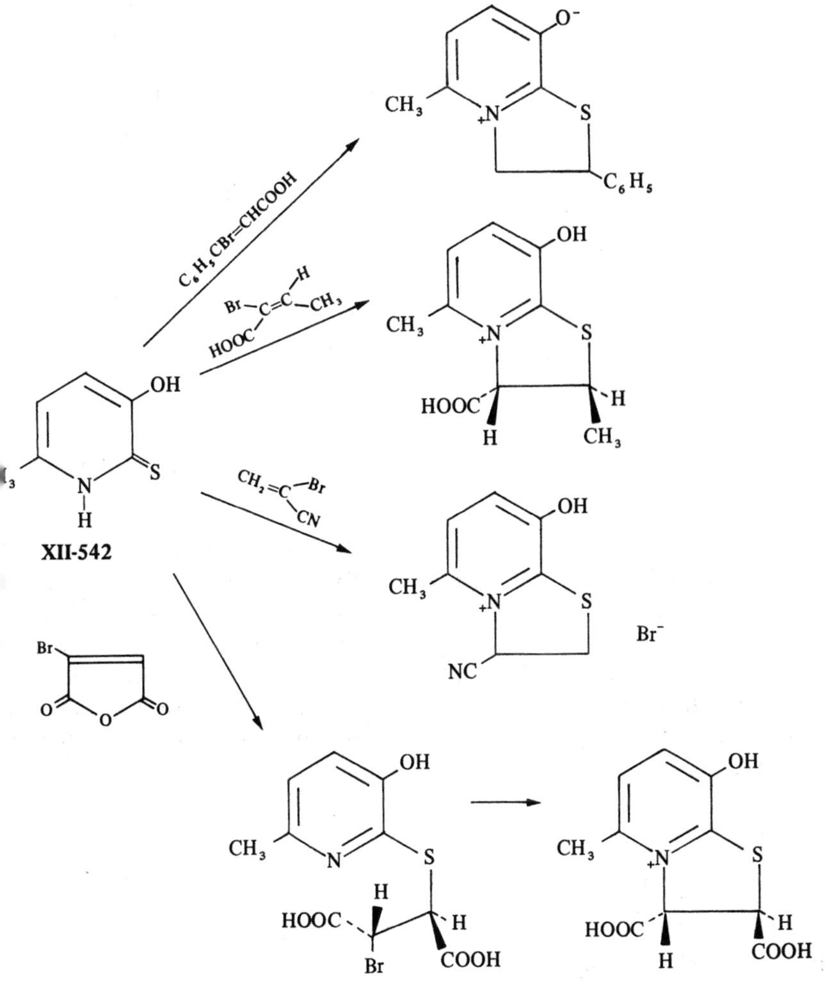

$C_6H_5CBr{=}CHCOOH$

XII-542

XII-546

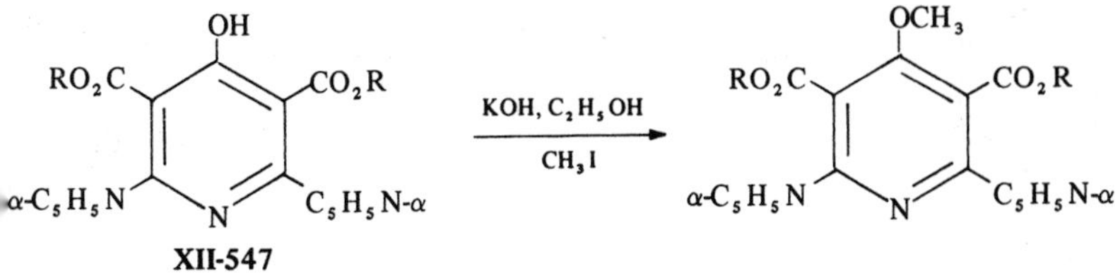

KOH, $C_2H_5OH$

$CH_3I$

XII-547

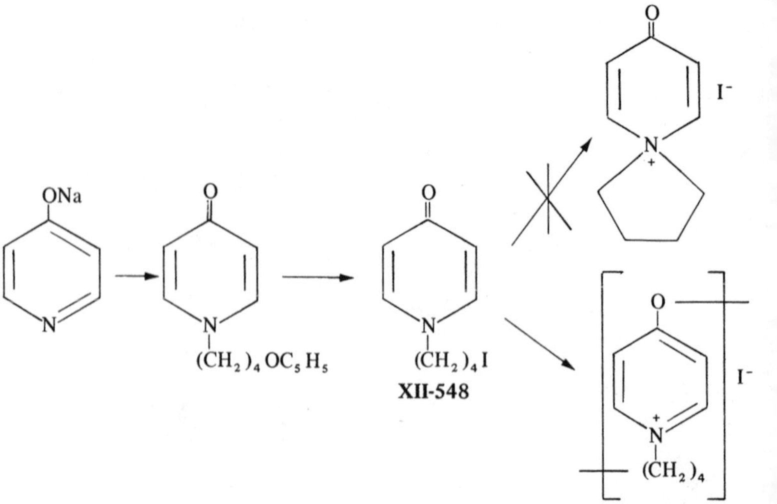

ethyl)-4,6-dihydroxypyridine (**XII-549**) in ethanolic ammonia gives 4-chloro-2 dihydro-6-oxofuro[3,2-*c*]pyridine (**XII-550**)[589] and not 2-chloro-3-vinyl-4 pyridinediol as previously reported.[590] Treatment of **XII-549** with sodi carbonate and methyl or ethyl iodide produces 5-alkyl-4-chloro-2,3-dihydro oxofuro[3,2-*c*]pyridine (**XII-551**) (R.I. 1306). The *N*-alkylfuro[3,2-*c*]pyridi (**XII-551**; R = $CH_3$,$C_2H_5$) and phosphorus oxychloride give 2,4,6-trichloro $\beta$-chloroethylpyridine,[589] which can be prepared directly from **XII-549** **XII-550**.[590]

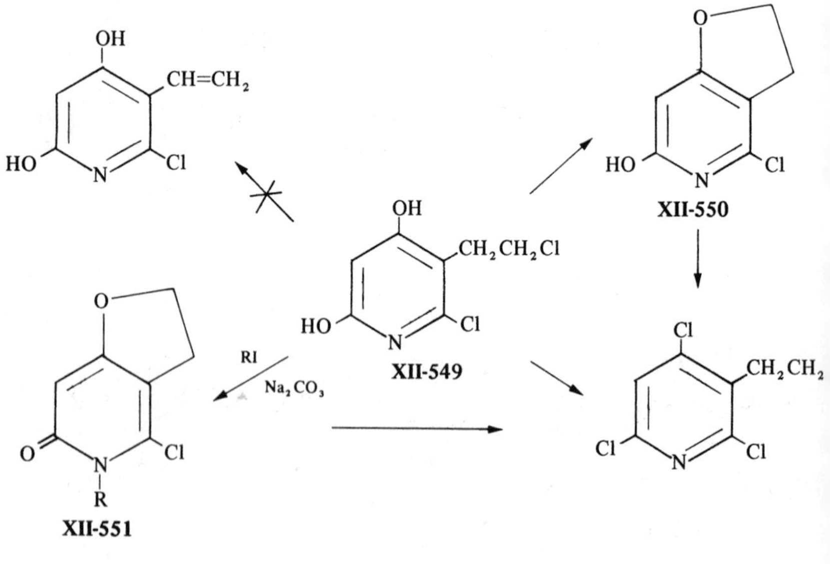

-Acetyl-4-hydroxy-6-methyl-2-pyridone and dimethyl sulfate in sodium thoxide give the *N*-methyl-2-pyridone (40%) but diazomethane causes nethylation (52%).[94]

Alkylation of 4-hydroxy-6-methyl-3-nitro-2-pyridone (**XII-552**, R = H) and its arbethoxy derivative (**XII-552**, $R = CO_2C_2H_5$) with diazomethane in ether or nzene gives a mixture of the 2,4-dimethoxypyridine and the 4-methoxy-methyl-2-pyridone (**XII-553**).[86] Alkylation of **XII-552** (R = H) with an cess of dimethyl sulfate gives a good yield of the 4-hydroxy-*N*-methyl-2-ridone, which gives **XII-553** on treatment with diazomethane.[86]

4-Hydroxy-6-hydroxymethyl-3-methoxy-2-pyridone (**XII-554**) and diazo-thane in ether-ethanol give both 6-hydroxymethyl-2,3,4-trimethoxypyridine d 6-hydroxymethyl-3,4-dimethoxy-2-pyridone.[591] Acetylation of **XII-554**

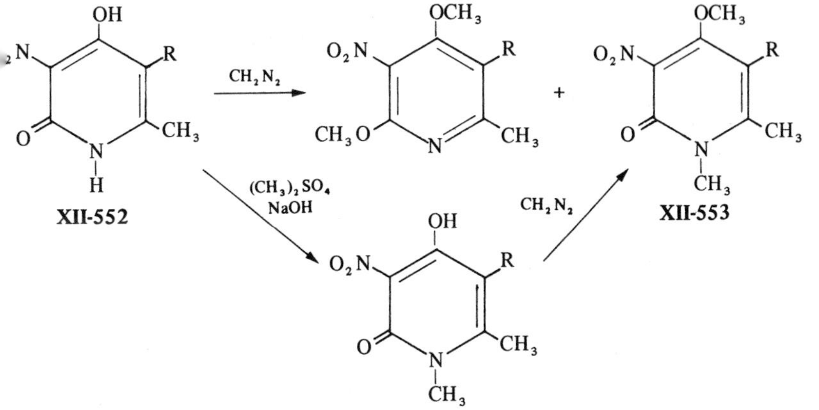

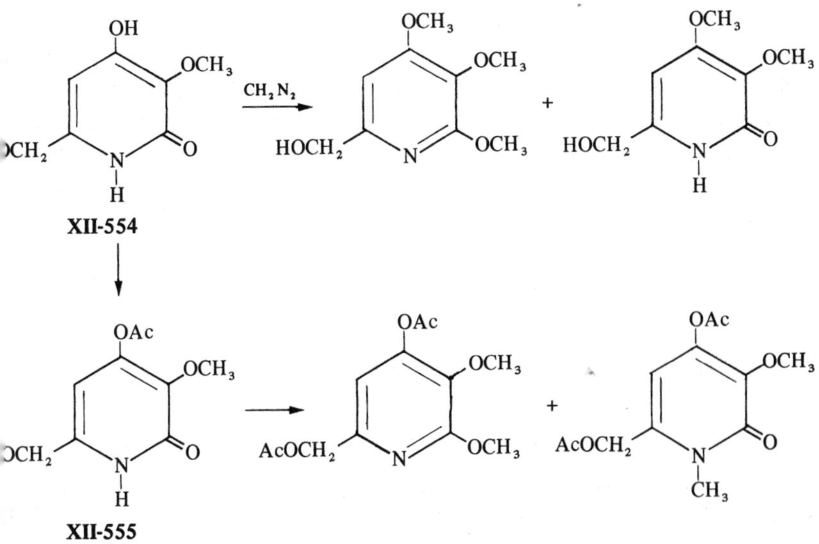

$CH_2N_2$

$(CH_3)_2SO_4$, NaOH

$CH_2N_2$

**XII-552**

**XII-553**

$CH_2N_2$

**XII-554**

**XII-555**

gives **XII-555**, which reacts with diazomethane to give both 4-acetoxy-6-aceto: methyl-2,3-dimethoxypyridine and 4-acetoxy-6-acetoxymethyl-3-methoxy- methyl-2-pyridone.[591]

*N*-Methyl-4-pyridone and methyl iodide in ethylene chloride containing sil tetrafluoroborate give 4-methoxy-*N*-methylpyridinium tetrafluoroborate (**X 556**).[139]

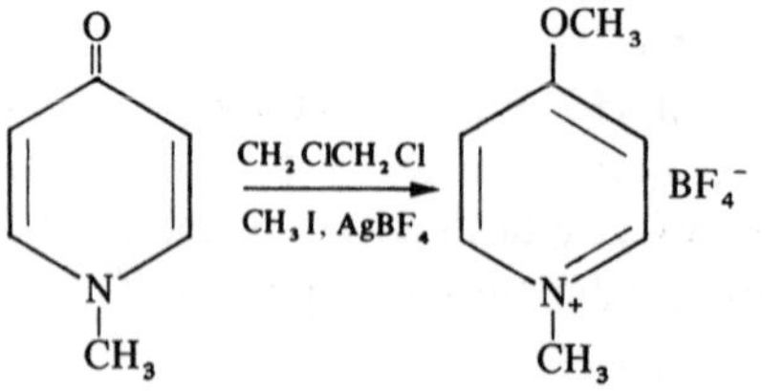

$CH_2ClCH_2Cl$, $CH_3I$, $AgBF_4$ → $BF_4^-$

**XII-556**

4-Pyridone and cyanogen bromide in chloroform give *N*-cyano-4-pyrid( (**XII-557**).[460]

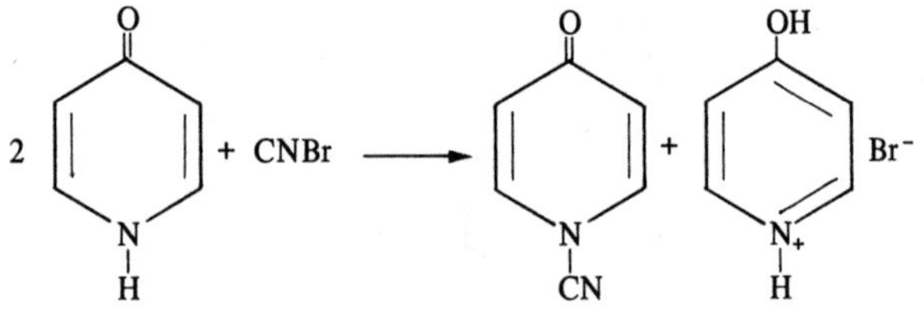

2 + CNBr → + $Br^-$

**XII-557**

d. 3-PYRIDINOLS Alkylations of 3-pyridinols have been reported to give f types of products: 3-alkoxypyridines (**XII-558**), 3-hydroxypyridinium s (**XII-559**, R = H), betaines (**XII-560**), and betaine complexes (**XII-561**).

Shapiro, Weinberg, and Freedman[592] have assigned the betaine hydrohal structure to a number of *N*-alkylation products, several of which are conside

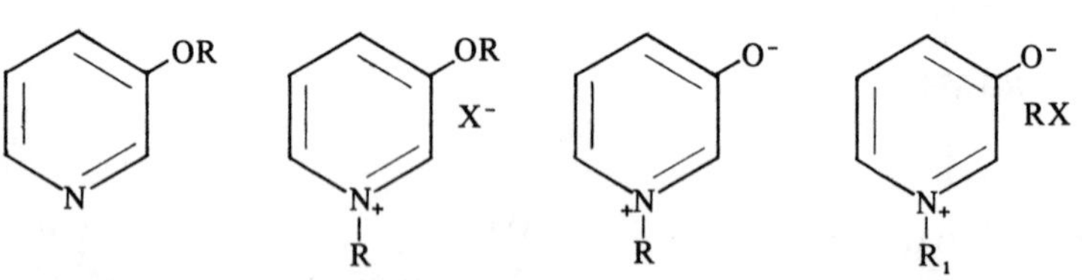

**XII-558** **XII-559** **XII-560** **XII-561**

by others to be pyridinium salts.[189, 342, 593] For example, the *N*-methylati product from 3-pyridinol and methyl iodide has been described *N*-methyl-3-oxypyridyl betaine hydroiodide (**XII-561**; $R_1 = CH_3$, R = H, X

[592] and more often has been assigned the *N*-methyl-3-hydroxypyridinium dide structure (**XII-559**).[342, 593, 594] Alkylation of 3-methoxypyridine **II-558**, R = $CH_3$) with benzyl halides gives *N*-benzyl-3-methoxypyridinium lides (**XII-559**; R = $CH_2C_6H_5$, X = Br; R = $CH_2C_6H_4Cl$-*o* and *p*).[594] Salts of ructure **XII-559** (R = H, $R_1$ = alkyl, aryl) have been prepared from furfural and nines[190] (see Section I.5.D.) Alkylations of 3-pyridinols have often given low elds of 3-alkoxypyridines. Reactions usually take place at nitrogen to give the ridinium salt[595, 596] or the betaine. The synthesis of 3-methoxypyridine from pyridinol or its sodium salt and dimethyl sulfate appears to be satisfactory.[594] Alkylation with diazomethane at ambient temperature gives ly a low yield of 3-methoxypyridine and considerable amounts of -alkylation; the yield, however, can be increased to 75% by carrying out the actions in *t*-butyl alcohol at −15 to −20°.[594] Following this procedure, bromo-3-methoxy-6-methylpyridine has been prepared in 90% yield.[597] dium salts of 3-pyridinols are preferentially *O*-alkylated by alkyl halides in methyl sulfoxide or dimethylformamide. By this method, thirty-five new alkoxypyridines have been prepared in good yield in dimethyl sulfoxide.[596] Pyridinol is *O*-alkylated by ethyl α-bromopropionate in 37% yield in sodium hoxide in ethanol and in 21% yield in ethanolic potassium hydroxide. Using dium hydride in dimethyl sulfoxide, a 67% yield is realized.[595] Ethyl hydroxyisonicotinate is alkylated similarly by ethyl bromoacetate and ethyl bromopropionate. 3-Alloxypyridine is prepared in only 5% yield from allyl romide and the sodium salt of 3-pyridinol in dimethylformamide.[598] The yield increased to 45% when sodium hydride in dimethyl sulfoxide is used.[595]

The course of alkylation of 2- and 6-substituted-3-pyridinols is subject to a eric effect. Sodium salts of 3-pyridinol in ethanol are *N*-alkylated by methyl romo- or iodoacetate. However, 2-bromo-3-pyridinol is *N*-alkylated by methyl nd ethyl iodide in dimethylformamide but is *O*-alkylated by haloacetates to **II-563** ($R_2$ = Br, $R_6$ = H). 2-Bromo-6-methyl-3-pyridinol (**XII-562**, $R_2$ = Br, $_6$ = $CH_3$) is *N*-alkylated by methyl and ethyl iodide and *O*-alkylated by iazomethane, but reacts with bromoacetic acid in chlorobenzene to give -hydroxy-6-methyl-2-pyridone (**XII-565**), possibly *via* 2-(α-bromoacetoxy)-6-ethyl-3-pyridinol (**XII-564**).[597] With 6-methyl-2-methylthio-3-pyridinol **XII-562**; $R_2$ = $CH_3S$, $R_6$ = $CH_3$), *N*-alkylation should be favored electronically, articularly by electron release by the 2-methylthio group. However, uaternization is difficult even with the simple alkyl halides and only -alkylation is observed when methyl iodoacetate is used.[597]

2,4,6-Triiodo-3-pyridinol, potassium carbonate, and ethyl chloroacetate in ethyl ethyl ketone give the *O*-alkylation product (**XII-566**) in 80% yield.[599]

2-Diphenylmethyl-3-pyridinol and dimethyl sulfate in alkaline water-dioxane ive both 2-diphenylmethyl-3-methoxypyridine and 2-diphenylmethyl-3-ydroxy-1-methylpyridinium hydroxide.[191]

$R_2$ = Br, $R_6$ = H; $R_1$ = $CH_3$, $C_2H_5$

$R_2$ = Br, $R_6$ = H; $XCH_2COOH$

$R_2$ = Br, $R_6$ = $CH_3$; $BrCH_2COOH$

XII-562 XII-563

XII-564 XII-565

$ClCH_2CO_2C_2H_5$

XII-566

The alkaloid Syphilobin F, a 2-substituted-3-pyridinol (see Section VI., p. 86] is *O*-methylated by diazomethane in diethyl ether-benzene in 43% yield.[600]

Anhydro-3-hydroxy-*N*-methylpyridinium hydroxide **(XII-567)** can be forme by the reaction between 3-pyridinol and diazomethane, by the reaction *N*-methyl-3-hydroxypyridinium chloride with anhydrous sodium carbonate an by thermal decomposition of 3-methoxypyridine iodomethylate. However, i preparation from the dimer complex of *N*-methyl-3-hydroxypyridinium iodid **(XII-568)**[593] by treatment with silver oxide was the most convenier method.[517] It has also been prepared from the quaternary iodide.[755]

Paoloni notes that the ground state properties of **XII-567** are in line with thos of its isomers *N*-methyl-2-pyridone and *N*-methyl-4-pyridone and suggests tha this is a mesionic compound for which the betaine structure unsatisfactory[517, 518] (see, however Section II.5.).

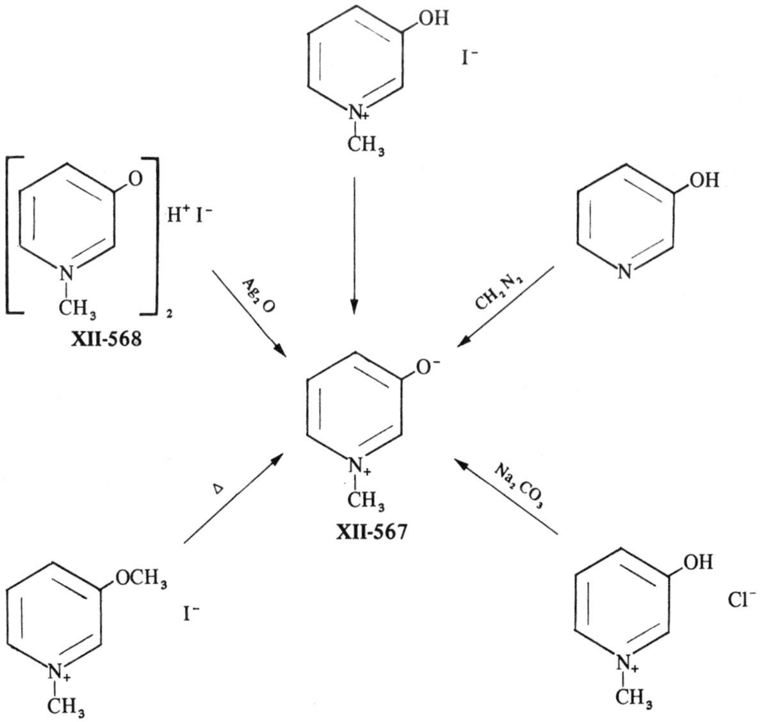

The two diols, 3- and 5-hydroxy-2-pyridone (**XII-569**, 3-OH, 5-OH) are converted to the corresponding *N*-methyl-2-pyridones by treatment with excess methyl iodide.[342]

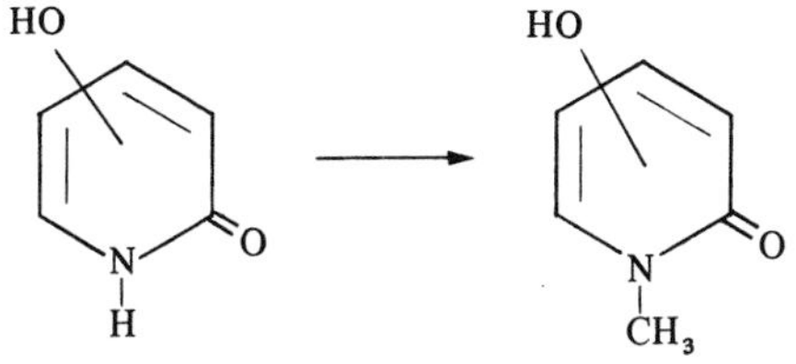

## B. *Pyridine Nucleosides*

The chemistry of pyridine nucleoside analogs is similar to that of the more widely studied pyrimidine nucleosides. The nucleosides are prepared from pyridone salts or alkoxypyridines or trimethylsilyloxypyridines and halogenose The products of direct alkylation of salts appear to be the *O*-glycosides, b rearrangements can occur *in situ* or by subsequent treatment with mercur bromide.

The silver salt of 2-pyridone gives the *O*-glycoside **(XII-570)** when treated wi 1-bromo-2,3,4-tri-*O*-benzoyl-D-ribopyranose. *O*-Ribofuranosides **(XII-571** ar **XII-574)** are formed from silver salts of 2- and 4-pyridone a 1-chloro-2,3,5-tri-*O*-benzoyl-β-D-ribofuranose in boiling toluene. Rearrangemen of the *O*-glycosides to *N*-glycosides (**XII-572**, **XII-573**, **XII-575**; R = $C_6H_5C($ are accomplished by heating with mercuric bromide in toluene.[601, 6 Deblocking to the nucleoside analogs [**XII-572** and **XII-573** (R = H)] accomplished by trans-esterification with sodium methoxide in methanol.[602]

The blocked nucleosides **XII-572** and **XII-573** can also be prepared fro 2-ethoxypyridine and the halogenoses. 4-Ethoxypyridine and 1-chloro-2,3,5-t *O*-benzoyl-β-D-ribofuranose gives both the β- **(XII-575)** and α-anome **(XII-576)**.[602] *N*-(D-Glucopyranosyl)-2-pyridone is prepared from the silver sa

HgBr$_2$

N O

O

RO OR OR

**XII-570**

N O

O

RO OR OR

**XII-572**

N OAg

N $OC_2$

HgBr$_2$

N O

RO O

OR OR

**XII-571**

N O

RO O

OR OR

**XII-573**

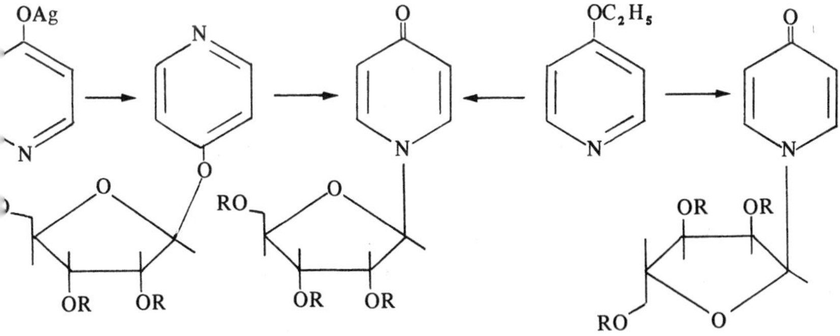

XII-574 XII-575 XII-576

2-pyridone and tetra-*O*-acetyl-α-D-glucopyranosyl bromide (**XII-577**) by milar procedures. It can also be prepared from 2-ethoxypyridine or the ercuri-chloride salt of 2-pyridone with the halogenose.[603]

Silver salts of substituted 2-pyridones (**XII-578**; R = 3-$CO_2CH_3$, 4-$CO_2CH_3$, $CO_2CH_3$, 4-$C_6H_5$, 6-$C_6H_5$, 5-$NO_2$, 5-CN, 5-Cl, 5-Br, 5-I) and **XII-577** give β-glucosides. The 6-substituted-*O*-β-glucosides (6-$CO_2CH_3$, 6-$C_6H_5$) do not, wever, rearrange when heated with mercuric bromide in toluene.[604] Some β → anomerization occurs[605, 606] when the silver salts of 5-halopyridones are arranged. When R is the electron withdrawing 5-$NO_2$ or 5-CN, only omerization is observed.[605]

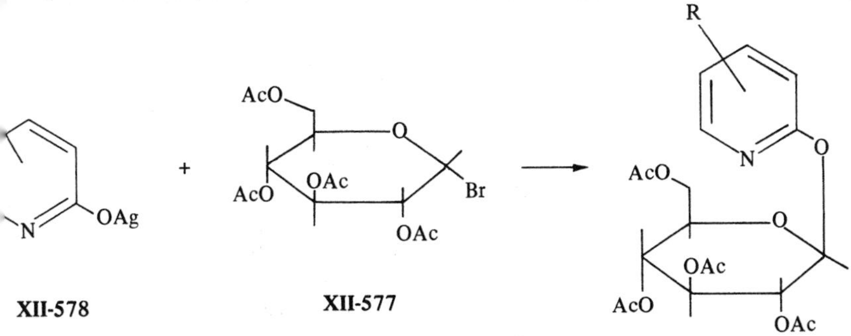

XII-578 XII-577

In the preferred conformation of the *N*-β-glucosides, the aglycone moiety is rpendicular to the axis of the sugar group and the C=O or C=S group is igned with the vicinal proton (**XII-579**).[607]

3-Substituted and 3,5-disubstituted 2-(tetra-*O*-acetyl-β-D-glucopyranosyloxy)-ridones also undergo *O* → *N* rearrangement when heated with mercuric romide[608] and often are accompanied by some formation of the -α-anomer.[609, 610]

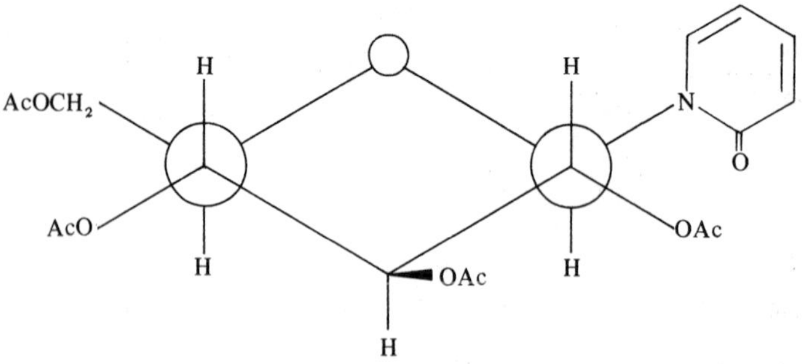

**XII-579**

The silver salt of 2-pyridone and 2-deoxy-3,5-di-*O*-(*p*-tolyl)-α-D-ribofuranos chloride give both the α- and β-*O*-glycosides. Rearrangement of either anomer the presence of mercuric bromide gives the *N*-deoxyribosides with an α:β ratio 1:2.[611, 612]

The silver salt of 3-carbamoyl-6-pyridone and 1-chloro-2,3,5-tri-*O*-benzoyl-β-I ribofuranose give 3-carbamoyl-*N*-(β-D-ribofuranosyl)-6-pyridone after rearrang ment and deblocking with sodium methoxide. This nucleoside has bee converted to 3-carbamoyl-6-pyridone adenine dinucleotide.[613]

The silver salt of 2-pyridone and 1-bromo-2,3-di-*O*-benzoyl-5-*O*-dipheny phosphoryl-D-ribofuranose in toluene gives the *O*-riboside **(XII-580)**, which

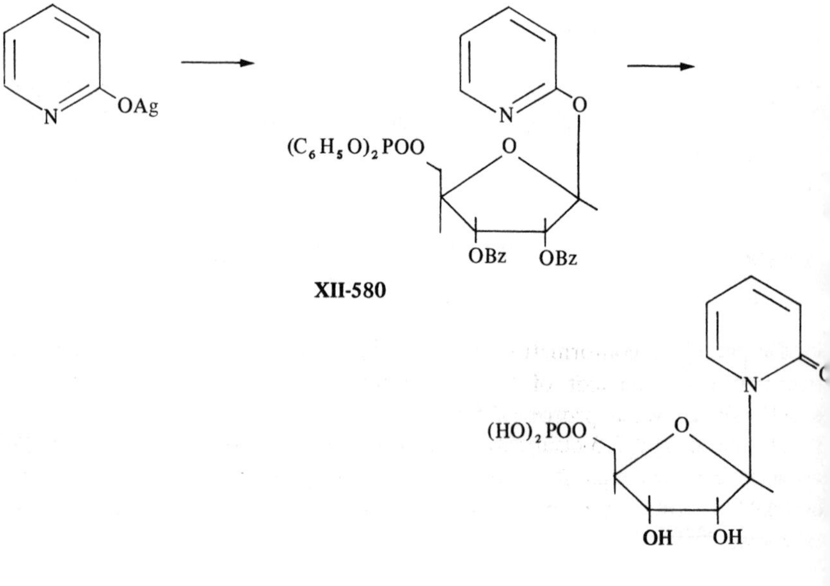

**XII-580**

rn gives the *N*-riboside 5′-phosphate after rearrangement and then hydrolysis st with alkali and then with phosphordiesterase.[614]

The nucleotide can also be prepared from *N*-(2′,3′-*O*-isopropylidene-β-D-ofuranosyl)-2-pyridone through phosphorylation with β-cyanoethyl phosphate the presence of *N,N*′-dicyclohexylcarbodiimide followed by deblocking.[615]

2-Pyridone and trimethylchlorosilane in toluene containing triethylamine form trimethylsilyloxy)pyridine, which reacts with tetra-*O*-acetyl-α-D-gluco-ranosyl bromide to give *N*-(tetra-*O*-acetyl-D-glucopyranosyl)-2-pyridone **II-581**).[616] 2,4-Bis(trimethylsilyloxy)pyridine and 2,3,5-tri-*O*-benzoyl-D-

$(CH_3)_3SiCl$, $Et_3N$; $OSi(CH_3)_3$; $AcOCH_2$; OAc; AcO; OAc

**XII-581**

ofuranosyl bromide in acetonitrile give the *N*-riboside after boiling in ethanol. Hydroxy-*N*-(β-D-ribofuranosyl)-2-pyridone (3-deazauridine) **(XII-582)** is rmed by deblocking with alcoholic ammonia.[616]

Reaction of **XII-582** with diphenyl carbonate gives 2,2′-anhydro-2-hydroxy-*N*-) -arabinosyl-4-pyridone **(XII-583)**.[617]

The "1-deazauridine" **(XII-587)** has been prepared from 3-bromo-2,6-di-nzyloxypyridine by means of conversion to the 3-pyridyllithium and then to e di-3-pyridylcadmium **(XII-584)**. Treatment of **XII-584** with 1-chloro-2,3,5--*O*-benzoyl-β-D-ribofuranose gives a 2,6-dibenzyloxy-3-(D-ribofuranosyl)-ridine **(XII-585)** and the benzylidene derivative **(XII-586)**. Hydrogenolysis of **I-585** gives a 3-ribosyl-2,6-dihydroxypyridine **(XII-587)**.

The cadmium derivative **XII-584** is converted to a 2′-deoxyribofuranoside **II-588)** by a similar procedure beginning with 3,5-di-*p*-toluyl-2′-deoxy-D-ofuranosyl chloride.[618]

The configurations of anomers of *N*-(5′-toluyl-2′-deoxy-D-ribofuranosyl)-2-ridone and the corresponding *O*-deoxyribofuranosides have been determined m their chemical shifts and coupling constants.[619]

The silver salt of 2-pyridone has also been used to prepare *N*-(β-D-gluco-ranosyl)-2-pyridone[620] and *N*-(5′-deoxy-β-D-ribofuranosyl)-2-pyridone.[621] ridyl ethers have been used to prepare *N*-(β-2′-deoxyribofuranosyl)-2-pyri-ne,[622] *N*-(β-D-glucopyranosyl)-4-pyridone,[623] and 5-nitro-1-(β-D-gluco-

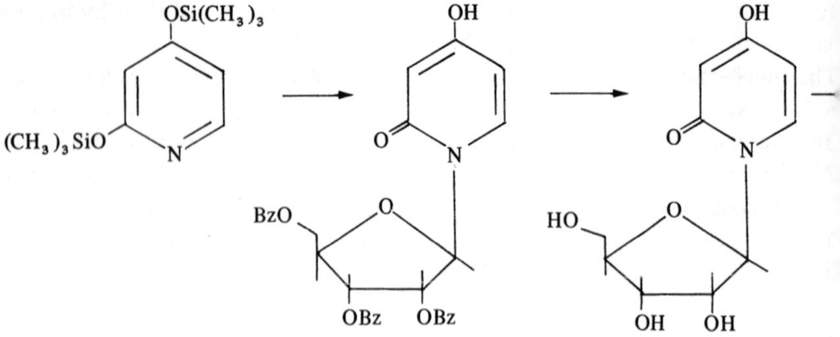

**XII-582**

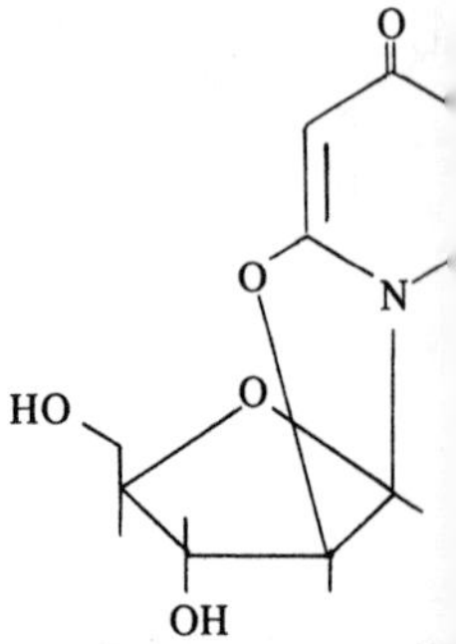

**XII-583**

pyranosyl)-2-pyridone.[624] The mercuric salt method has been used to prep*N*-(β-glucopyranosyl)-2-pyridone,[620] *N*-(β-D-ribofuranosyl)-2-pyridone,[625] a *N*-(β-2-deoxyribofuranosyl)-2-pyridone.[622]

Sodium salts of 2-pyridone and of a number of 3- and 5-substitut 2-pyridones and 3,5-disubstituted-2-pyridones when treated with tetra-*O*-acet α-D-glucopyranosyl bromide in acetone give both the *O*- and *N*-β-glu sides.[624, 626]

Blocked pyridine nucleosides have been phosphorylated with β-cyanoet phosphate and dicyclohexylcarbodiimide,[615, 625, 627] diphenyl phosphorochl date,[615, 628] and with triethyl phosphate followed by oxidation.[627]

*N*-(β-D-Ribofuranosyl)-2-pyridone 5′-diphosphate **(XII-590)** has been prepa from *N*-(2′,3′-*O*-ethoxymethylene-β-D-ribofuranosyl)-2-pyridone by phosphory tion with triimidazoyl phosphate to give **XII-589**, which is treated w bis(tributylammonium) orthophosphate and deblocked with hydrochlo acid.[628] The synthesis of *N*-(3′-*O*-phosphoryl-β-D-ribofuranosyl)-2-pyrid (3′ → 5′) uridine has been reported.[627]

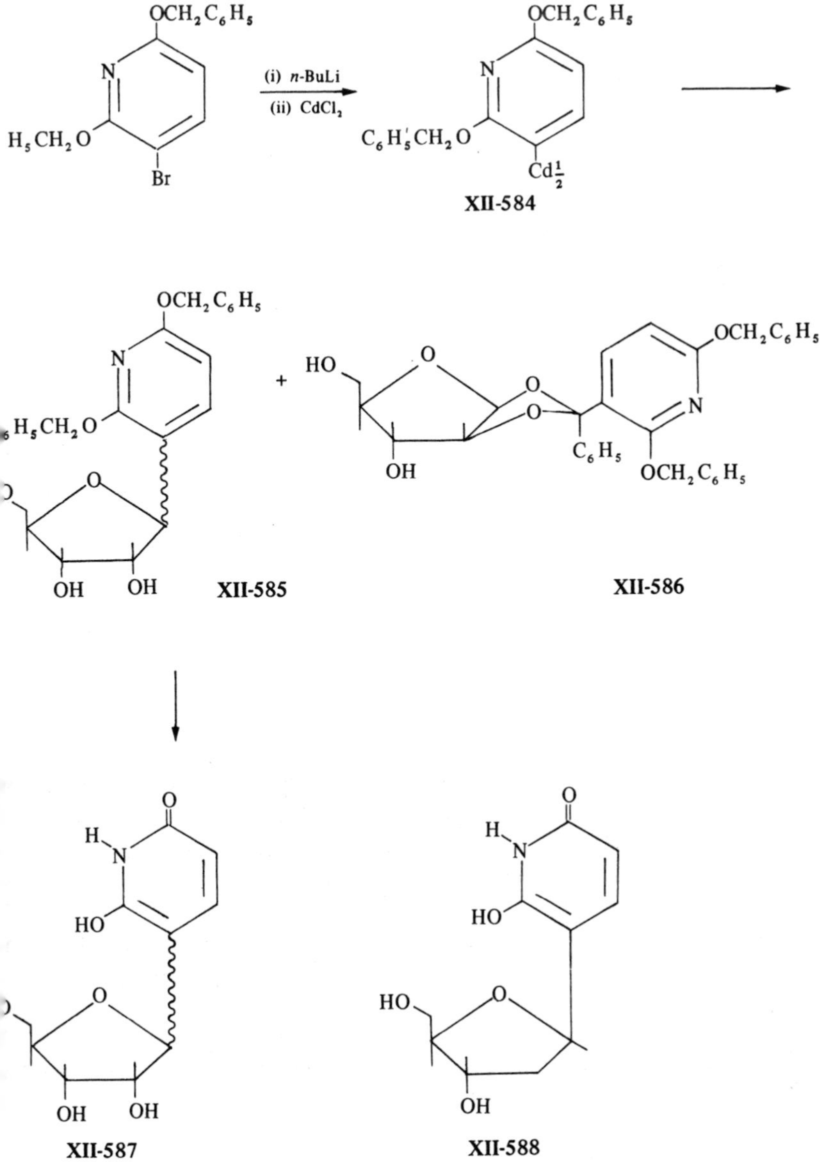

(i) n-BuLi
(ii) $CdCl_2$

$OCH_2C_6H_5$
N
$H_5CH_2O$
Br

$OCH_2C_6H_5$
N
$C_6H_5CH_2O$
$Cd\frac{1}{2}$
**XII-584**

$OCH_2C_6H_5$
N
$H_5CH_2O$
O
OH OH
**XII-585**

+

HO
O
O
O
OH
$C_6H_5$
$OCH_2C_6H_5$
N
$OCH_2C_6H_5$
**XII-586**

O
H
N
HO
O
OH OH
**XII-587**

O
H
N
HO
HO
O
OH
**XII-588**

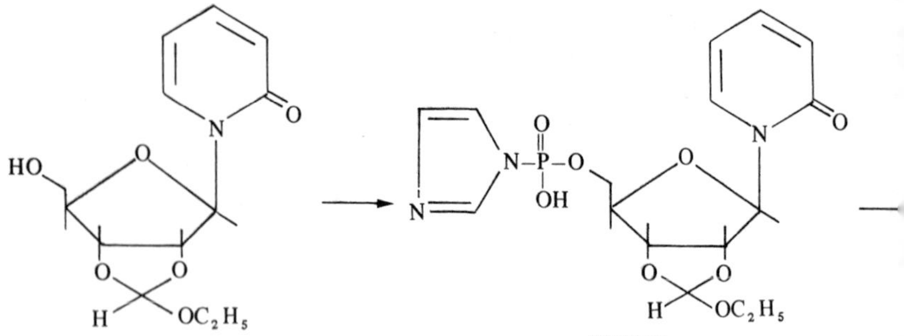

**XII-589**

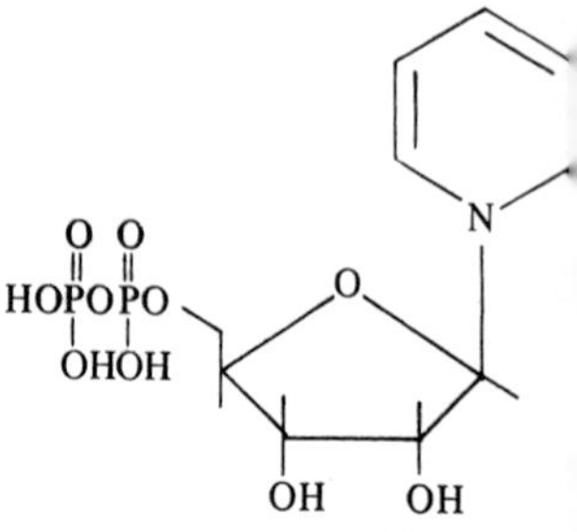

**XII-590**

## C. O- *and* N-*Acylation*

Esters of pyridinols and pyridones are readily prepared by conventior procedures, although early investigators experienced some difficulty in t preparation of esters of 2- and 4-pyridones because of their reactivit particularly their susceptibility to hydrolysis.[629] *N*-Acylated pyridones ha been elusive until recently.

It has been suggested that *N*-acylations have often been overlooked and th *N*-acylations of 2-pyridones may be general but that rapid rearrangement of t *N*- to the *O*-acylpyridine occurs.[630]

*N*-Acetyl-2-pyridone **(XII-591)** was first prepared by acetylation of t thallium(I) salt of 2-pyridone that gives both **XII-591** and 2-acetoxypyridin

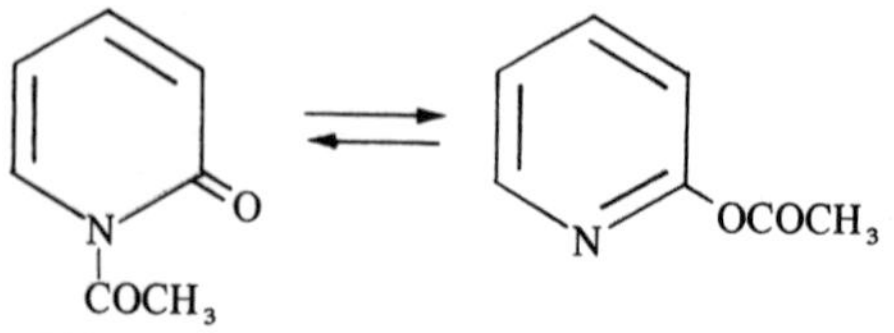

**XII-591**

e reaction product contains approximately 40% of **XII-591** at $-40°$; at ıbient temperatures the mixture is almost entirely 2-acetoxypyridine ›proximately 90%).[631] Treatment of 2-pyridone with acetyl chloride in the :sence of a tertiary amine catalyst gives a 46:54 mixture of 2-acetoxypyridine d *N*-acetyl-2-pyridone **(XII-591)**. After 24 hours in dimethyl sulfoxide, the xture contains 92% of 2-acetoxypyridine. This $N \rightarrow O$ migration is accelerated 2-pyridone.[632]

\cetylation of 4-pyridone or its thallium salt with acetic anhydride gives acetyl-4-pyridone **(XII-592)**.[633] Earlier this product had been described as ıcetoxypyridine.[634, 635] In methylene chloride, **XII-592** equilibrates with

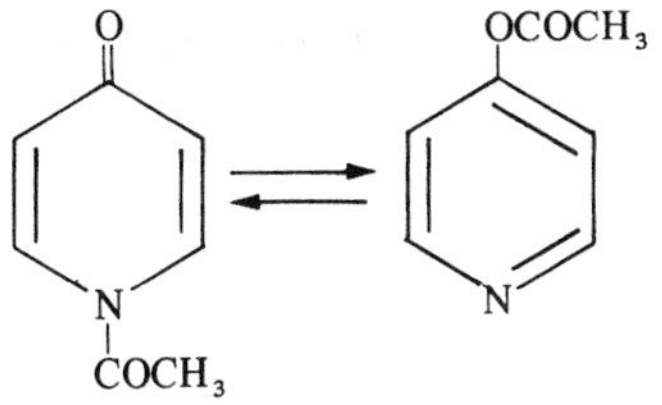

O N $COCH_3$ ⇌ $OCOCH_3$ N

**XII-592**

ıcetoxypyridine in the ratio 53:47.[633] 4-Pyridyl benzoate, however, exists tirely in this form in methylene chloride.[633] The isomeric pyridyl benzoates : prepared conveniently from benzoyl chloride and the pyridinol by boiling in loroform.[636]

2- And 3-acetoxypyridine have been examined as potential acetylating ents.[637] Generally 2-acetoxypyridine gives better yields with the amines and enols surveyed. *n*-Butyl alcohol is acetylated in 92% yield after 3 hours with ıcetoxypyridine in tetrahydrofuran at ambient temperatures, but requires 7 urs under reflux (93%) with 3-acetoxypyridine in tetrahydrofuran.

3-Acetoxypyridine and benzoic acid in boiling xylene give 3-benzoyloxypyri- ıe in 85% yield.[637] 2-Butyroxypyridine is prepared in 80% yield from ›utyric acid and 2-acetoxypyridine in xylene by distilling off the acetic acid as is formed. 2-Benzoyloxypyridine is formed in 63% yield by this method. The rresponding acylthiopyridines are prepared from 2-acetylthiopyridine.[638] ıese reactions are subject to considerable steric effects.

2-Pyridyl methacrylate has been prepared from the sodium salt of 2-pyridone d methacryl chloride and has been polymerized by radicals but not by ions.[639] 2-Pyridone and phosgene (1:4) in dry tetrahydrofuran or benzene th or without pyridine form di-($\alpha$-pyridyl)carbonate, which is a useful reagent : the formation of the benzoate and the monophenyl phosphate of ›yridone.[640]

2-Amino-6-pyridone and acetic anhydride in water give the diacetylated ıterial **(XII-593)**, which can be hydrolyzed to give 2-acetamido-6-pyridone; s can be *O*-alkylated conveniently.[641]

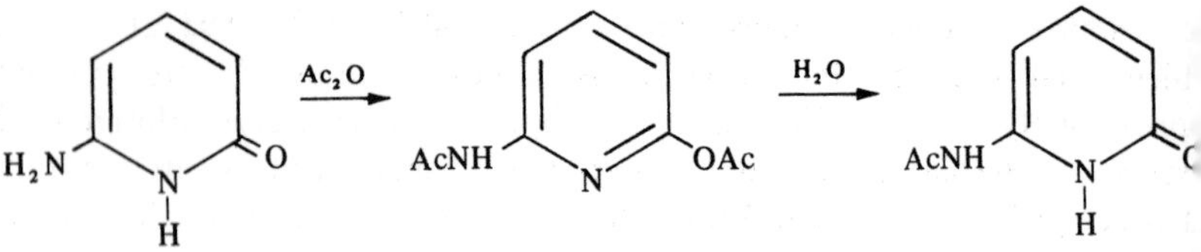

XII-593

2-Alkyl(or aryl)oxazolo[5,4-*b*-]pyridines (**XII-594**) are formed by acylation 3-amino-2-pyridones followed by distillation from $P_2O_5$.[556, 642, 643] 6-Brom phenyloxazolo[5,4-*b*]pyridine (**XII-594**; R = 6-Br, R′ = $C_6H_5$) is prepa directly from 3-amino-5-bromo-2-pyridone and benzoic anhydride.[643]

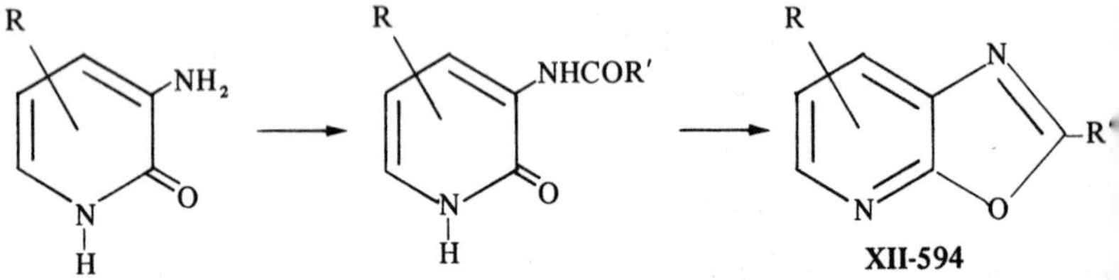

XII-594

Treatment of 5-cyano-4-methoxymethyl-2-methyl-3-phenylazo-6-pyrid (**XII-595**) with tosyl chloride in acetone-pyridine gives the tosylate; this can hydrogenated in two steps to 3-amino-5-aminomethyl-4-methoxymethy methylpyridine (**XII-596**) or to 3-amino-5-aminomethyl-4-methoxymethy methyl-6-tosyloxypyridine in the presence of Raney Nickel, depending conditions.[52]

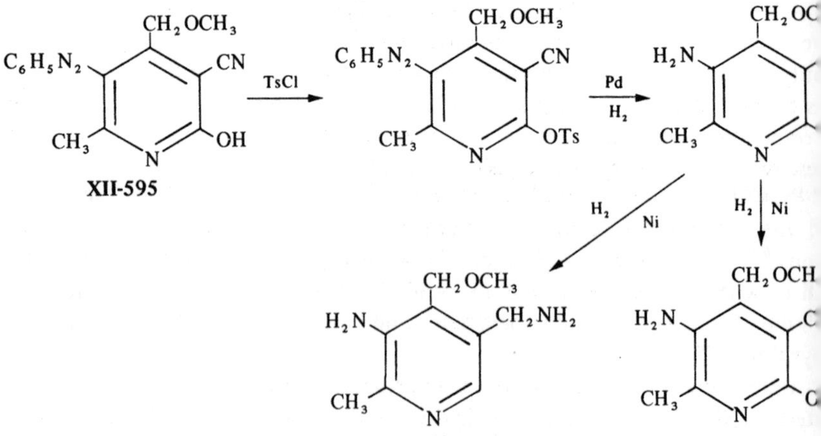

XII-595

XII-596

-Hydroxy-2-pyridones form 4-acetoxy-2-pyridones readily. 4-Acetoxy-1,6-di-thyl-2-pyridone[644] and 4-acetoxy-5-acetyl-6-methyl-2-pyridone[87] have been pared employing acetic anhydride. 4-Acetoxy-6-methyl-1-phenyl-2-pyridone been formed with phosphoryl chloride and acetic acid.[645] Methyl citrazinate **(I-597)** and acetic anhydride give methyl 2,6-diacetoxyisonicotinate.[507]

$CO_2CH_3$ HO N H O $\xrightarrow{Ac_2O}$ $CO_2CH_3$ AcO N OAc

**XII-597**

-Nitro-4-pyridone and 3-amino-4-pyridone are *N*-acylated by ethyl chloro-mate in aqueous sodium carbonate.[586]

Reactions of 3-pyridinols with organic and inorganic acid derivatives to give ers are well known.[646] Recently 2-nitro-3-pyridinol was converted to cetoxy-2-nitropyridine with acetyl chloride in benzene-pyridine. 6-Nitro-3-ridinol and acetyl chloride in acetone containing sodium carbonate gave the tate quantitatively.[647]

3-Pyridinol or 2-nitro-3-pyridinol and chlorosulfonic acid in chloroform ntaining dimethylaniline give the 3-pyridyl bisulfate. 2-Pyridone gives the stable *N*-sulfonate and 4-pyridone gives both the *N*-sulfonate and the yridyl bisulfate, all isolated as their potassium salts.[648] 3-Methanesulfonoxy-ridine is prepared from 3-pyridinol and methanesulfonyl chloride in loroform containing 2,6-lutidine.[649]

Pyridoxylidenebenzylamine **(XII-598)** and benzoyl chloride in ether-pyridine rm 7-benzoyloxy-1-(*N*-benzylbenzamido)-6-methylfuro[3,4-*c*] pyridine **II-599)**. However, treatment of **XII-598** with acetic anhydride in pyridine ves *N*-($\alpha^4$-acetoxy-3,$0^5$-diacetylpyridoxyl)-*N*-acetylbenzylamine.[650]

3,5-Pyridinediol and acetic acid containing sodium acetate form the diacetate. nzoylation is accomplished with benzoyl chloride in pyridine.[301]

3-Pyridinol and benzyl chloroformate in tetrahydrofuran give benzyl-3-pyridyl rbonate **(XII-600)**, which has been used to introduce the *N*-benzyloxycarbonyl oup into amino acids. It reacts with L-lysine in water-dimethylformamide to ve $\epsilon$-(*N*-benzyloxycarbonyl)-L-lysine (33%). Somewhat higher yields are tained from benzyl 8-quinolyl carbonate and 4-nitrobenzyl-8-quinolyl rbonate employing 3-pyridinol as a catalyst.[651]

*N*-Benzyloxycarbonylamino acids and 3-pyridinol form 3-pyridyl esters in the esence of dicyclohexylcarbodiimide. The esters in ethyl acetate are treated th an amino acid ethyl ester hydrochloride and triethylamine to give peptides

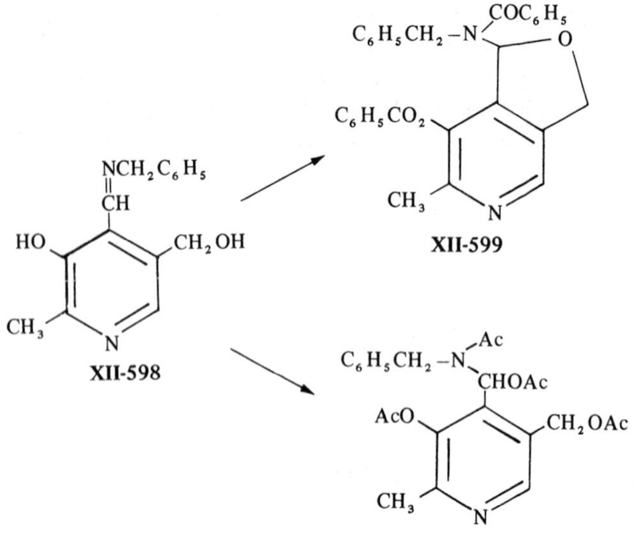

$C_6H_5CH_2-N(COC_6H_5)$ … O

$C_6H_5CO_2$

$CH_3$ N

**XII-599**

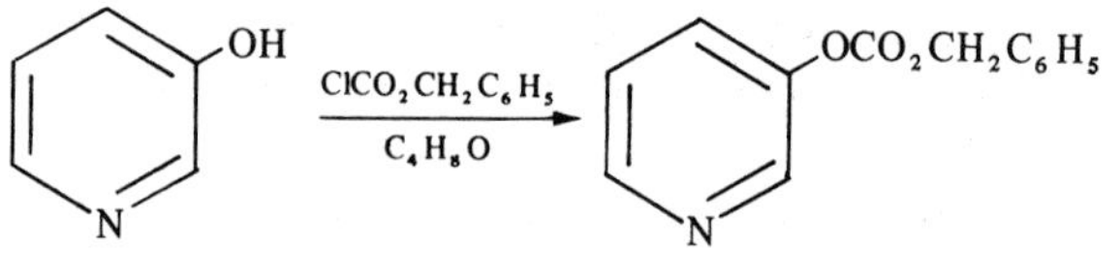

$NCH_2C_6H_5$
‖
CH

HO $CH_2OH$

$CH_3$ N

**XII-598**

$C_6H_5CH_2-N(Ac)$ CHOAc

AcO $CH_2OAc$

$CH_3$ N

OH

$ClCO_2CH_2C_6H_5$

$C_4H_8O$

N

$OCO_2CH_2C_6H_5$

N

**XII-600**

OH

N

R

$C_6H_5CH_2OCONHCHCOOH$

DCC

R

$C_6H_5CH_2OCONHCHCO_2$

N

$R'CHNH_2CO_2C_2H_5$

R R₁

$C_6H_5CH_2OCONHCHNHCHCO_2C_2$

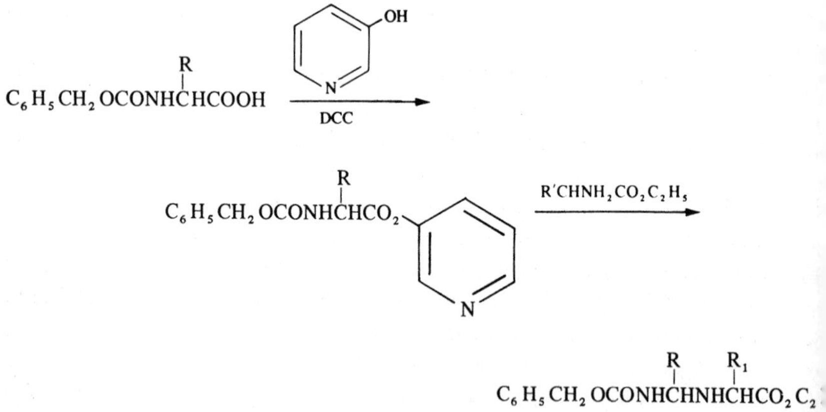

**XII-601**

**(XII-601).**[652,653] Similarly, 2-methyl-4-nitro-3-pyridinol, 2,6-dimethyl-4-nitr 3-pyridinol, 2-isobutyl-6-methyl-4-nitro-3-pyridinol, and 2-methyl-6-(*N*-meth eridyl)-3-pyridinol have been converted to *N*-benzyloxycarbonylamino acid ers for use in peptide synthesis.[654]

-(*N*-Benzyloxycarbonyl)-α-*N*-toluene-*p*-sulfonyl L-lysine **(XII-602)**, 3-pyri-ol, and *N,N'*-dicyclohexylcarbodiimide in tetrahydrofuran give the pyridyl er, which can be converted to the hydrobromide by hydrogen bromide in romethane.[655] Esterification of the *N*-tosyl derivatives of phenylalanine,

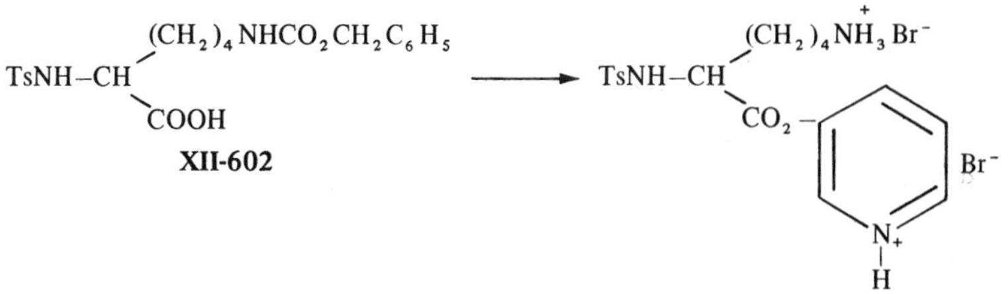

$$\text{TsNH-CH}(\text{(CH}_2)_4\text{NHCO}_2\text{CH}_2\text{C}_6\text{H}_5)(\text{COOH}) \longrightarrow \text{TsNH-CH}((\text{CH}_2)_4\overset{+}{\text{N}}\text{H}_3\,\text{Br}^-)(\text{CO}_2\text{-3-C}_5\text{H}_4\overset{+}{\text{N}}\text{H})\ \text{Br}^-$$

**XII-602**

nethylphenylalanine, and L-valine with 3-pyridinol in the presence of V'-dicyclohexylcarbodiimide yields 3-pyridyl esters that can be isolated as ir hydrochlorides.[656]

'he 3-pyridyl ester of *N*-benzoyl-DL-alanine **(XII-603)** has been prepared from iixed anhydride and 3-pyridinol.[657]

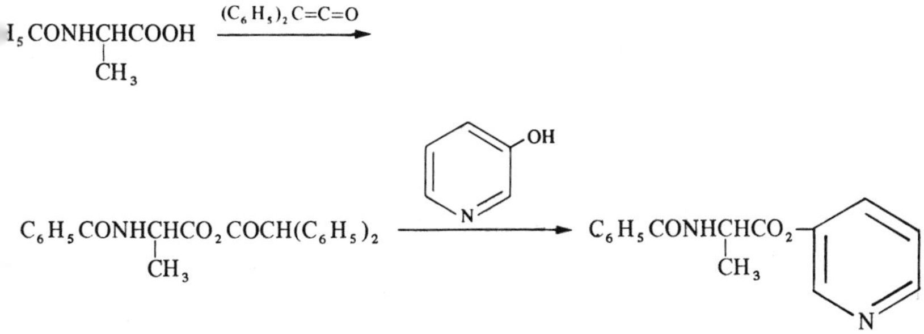

$$\text{I}_5\text{CONHCH(CH}_3\text{)COOH} \xrightarrow{(\text{C}_6\text{H}_5)_2\text{C=C=O}}$$

$$\text{C}_6\text{H}_5\text{CONHCH(CH}_3\text{)CO}_2\text{COCH(C}_6\text{H}_5)_2 \xrightarrow{\text{3-pyridinol (OH)}} \text{C}_6\text{H}_5\text{CONHCH(CH}_3\text{)CO}_2\text{-3-pyridyl}$$

**XII-603**

-Pyridone and phosgene in tetrahydrofuran give di-(2-pyridyl) carbonate **I-604)**.[658] Ethyl pyrocarbonate and pyridinols in aqueous ethanol give the responding pyridyl ethyl carbonates.[659]

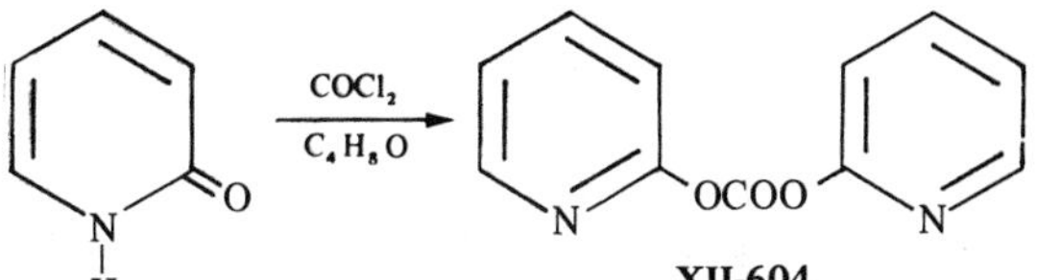

$$\text{2-pyridone (N-H)} \xrightarrow[\text{C}_4\text{H}_8\text{O}]{\text{COCl}_2} \text{2-pyridyl-OCOO-2-pyridyl}$$

**XII-604**

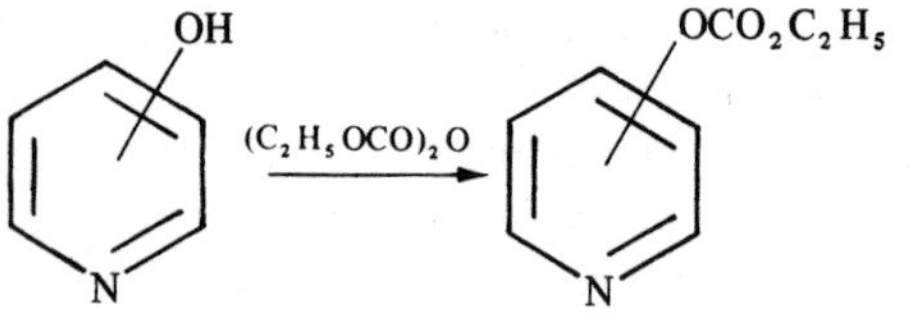

OH

$(C_2H_5OCO)_2O$

N

$OCO_2C_2H_5$

N

## D. *Pyridyl Phosphates*

α-Pyridyl phosphates (**XII-605**; R = thymyl, benzyl, β-naphthyl, phe
*p*-chlorophenyl) can be prepared from di-(2-pyridyl) carbonate (**XII-604**) .
monoesters of phosphoric acid and isolated as their cyclohexylammonium
sodium salts. Some of these mixed phosphates (**XII-605**; R = $C_6H_5$, *p*-$ClC_6$
$C_2H_4CN$) can also be prepared from 2-pyridone and aryl phosphorodichlorid
or from 2-pyridone and monoesters and *N,N'*-dicyclohexylcarbodiimide (DC
2-Pyridone and phosphorus oxychloride and pyridine give the tripyri
phosphate, which is readily hydrolyzed to the dipyridyl phosphate (**XII-605**,
-$C_5H_4N$).[658]

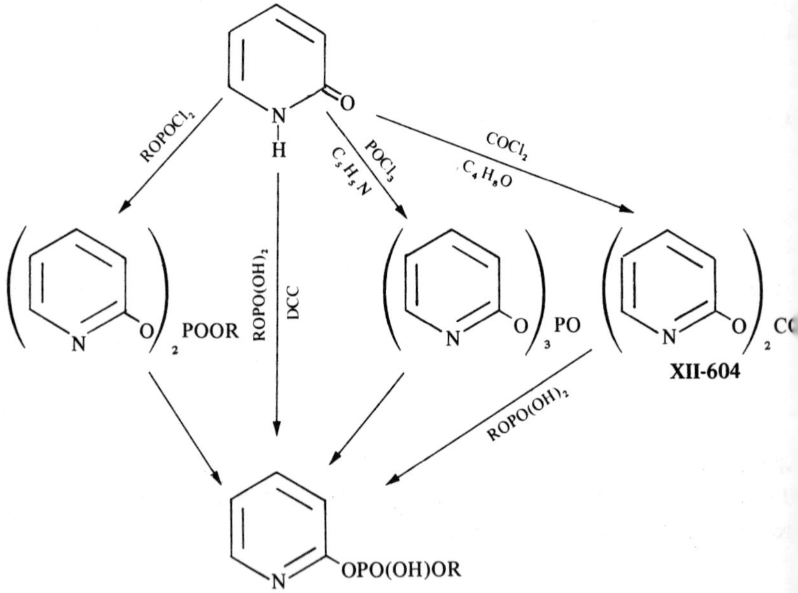

N H O

$ROPOCl_2$ $POCl_3$ $C_5H_5N$ $COCl_2$ $C_4H_8O$

$ROPO(OH)_2$ DCC

POOR 2 PO 3 CO 2

**XII-604**

$ROPO(OH)_2$

OPO(OH)OR

**XII-605**

Because of the reactivity of the α-pyridyl moiety, phosphates of struct
**XII-605** are convenient reagents for the formation of derivatives of phosph
acid. They react with monoesters of phosphoric acid to give pyrophospha

:h alcohols to form dialkyl phosphates **(XII-606)**, and with amines to form noalkyl phosphoramides.[660]

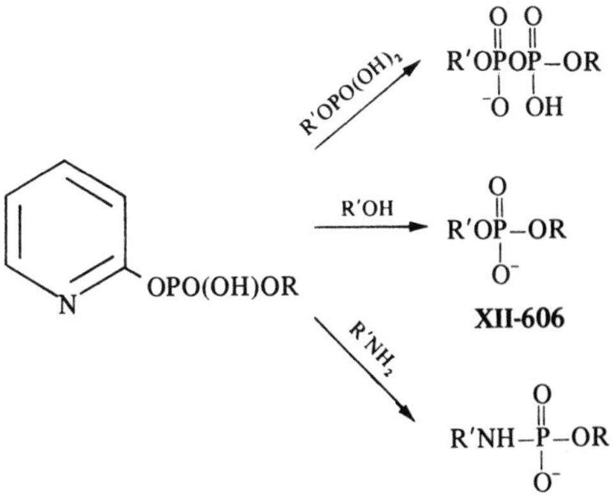

R'OPO(OH)₂ → R'OPOP—OR (O⁻, OH); R'OH → R'OP—OR (O⁻) **XII-606**; R'NH₂ → R'NH—P—OR (O⁻)

α-Pyridyl esters of nucleoside 5′-phosphates **(XII-607)** can be prepared from 5′-phosphate and 2-pyridone using DCC. The α-pyridyl ester **XII-607** and nose phosphate give the unsymmetrical *P,P*′-pyrophosphate diesters **II-608)**. With phosphoric acid salts the pyrophosphate **XII-609** is formed.[661]

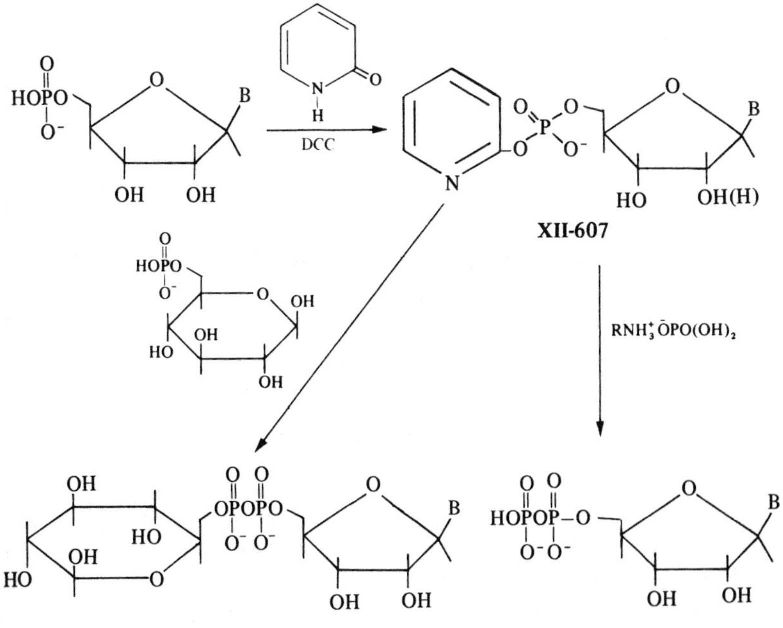

DCC

**XII-607**

$RNH_3^+\bar{O}PO(OH)_2$

**XII-608**

**XII-609**

β-Cyanoethyl 2-pyridyl phosphate and uridine-, cytidine-, and adenosine-phosphates give the *P,P'*-pyrophosphate diesters **(XII-610)** which can converted to the pyrophosphate with base.[662]

$$CNCH_2CH_2OP(O)(O^-)O\text{-}C_5H_4N + HOP(O)(O^-)O\text{-ribose(B)(OH)(OH)} \longrightarrow CNCH_2CH_2OP(O)(O^-)OP(O)(O^-)O\text{-ribose(OH)(OH)}$$

**XII-610**

2-Pyridone and 2-pyridinethione anions react with *O,O'*-diphenylphospho chloridothioate to give *O,O'*-diphenyl *O*-2-pyridyl phosphorothioate (**XII-61** X = O) and *O,O'*-diphenyl *S*-2-pyridyl phosphorodithioate (**XII-611**, X =

$$C_5H_4N\text{-}X^- + (C_6H_5O)_2P(=S)Cl \longrightarrow C_5H_4N\text{-}X\text{-}P(=S)(OC_6H_5)_2 + Cl^-$$

**XII-611**

Phosphates and phosphorothioates are insensitive to anion polarizability in th reaction. 2-Pyridone is a better nucleophile than 2-pyridinethione (relative rat = 28:1). The rates of these reactions have been correlated with acidities f several carboxylic acids, phenols, and thiophenols.[663]

## E. *Displacement by Halogen*

2- And 4-halopyridines are often prepared from the corresponding pyridon (Tables XII-15 and XII-16). 3-Pyridinols, like phenols, are unreactive towa direct displacement, but 3- and 5-halopyridines can be obtained by dire halogenation. Di- and trihalopyridines can be prepared by combinations these procedures.[98, 287, 291, 294, 664, 665] For example, 3,4-dichloro-5-nitı ridine **(XII-612)**, an intermediate used in the preparation of γ-carbolines, has en prepared from 3-nitro-4-pyridone by chlorination to give 3-chloro-5-nitro-pyridone, which is then treated with phosphoryl chloride and phosphorus ntachloride.[664]

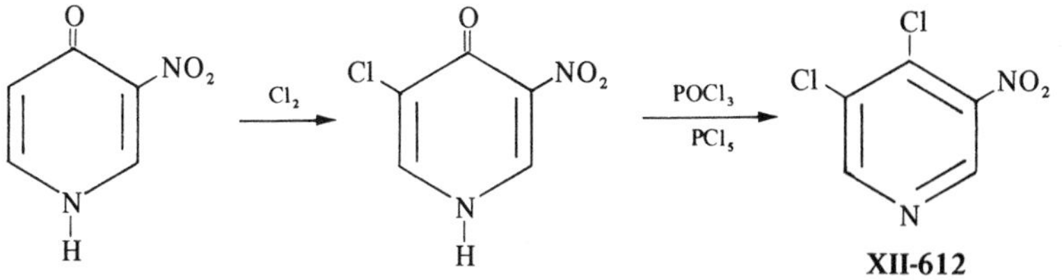

**XII-612**

Direct displacement of hydroxyl by halogen is accomplished with phos-oryl chloride,[34, 97, 98, 291, 305, 338, 502, 510, 666-668] phosphorus pentachlo-le,[37, 293, 669] a mixture of phosphoryl chloride and phosphorus pentachlo-le,[53, 282, 290, 291, 294, 306, 664, 665, 670-675] phosphorus pentachloride and osphorous trichloride,[102] phosphoryl chloride or thionyl chloride in methylformamide,[676] phosphoryl chloride with dimethylaniline,[98] and enylphosphoric dichloride.[672] 2-Pyridones are converted to 2-bromopyridines th phosphorus pentabromide.[287, 290, 291]

In the conversion of 2,6-dimethyl-4-pyridone to 4-chloro-2,6-lutidine a small ount of 4-chloro-2-methyl-6-trichloromethylpyridine **(XII-613)** is isolated.[670]

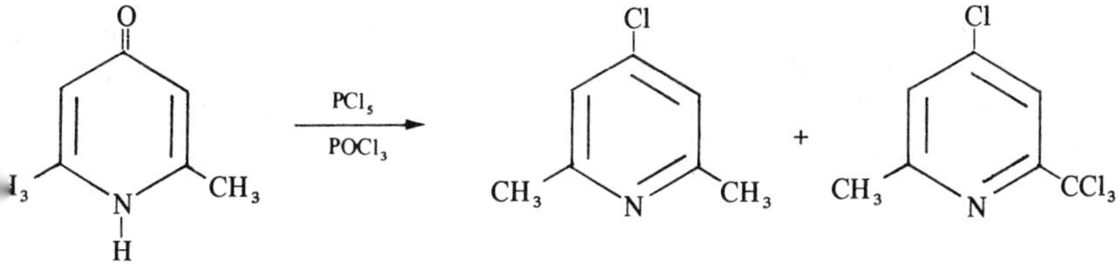

**XII-613**

3-Cyano-6-substituted-4-trifluoromethyl-2-pyridones are not converted to the chloro derivatives under a number of conditions, but several 6-substituted-4-ifluoromethyl-2-pyridones do react to give **XII-614**.[672]

Under conditions where 3-carboxy-6-methyl-5-phenyl-2-pyridone **(XII-615)** rms a 2-chloro-3-chlorocarbonylpyridine, only the carboxy function of **I-616** reacted with phosphorus pentachloride-phosphorus oxychloride to give **I-617**.[53]

Heating ammonium salts of 4-alkyl-3,5-dicyano-6-hydroxy-2-pyridone and osphorus oxychloride under pressure at 220° is required to form 4-alkyl-2,6-chloro-3,5-dicyanopyridines **(XII-618)**. The salts are also converted to **XII-618** heating with phenylphosphonic dichloride at 200°.[111]

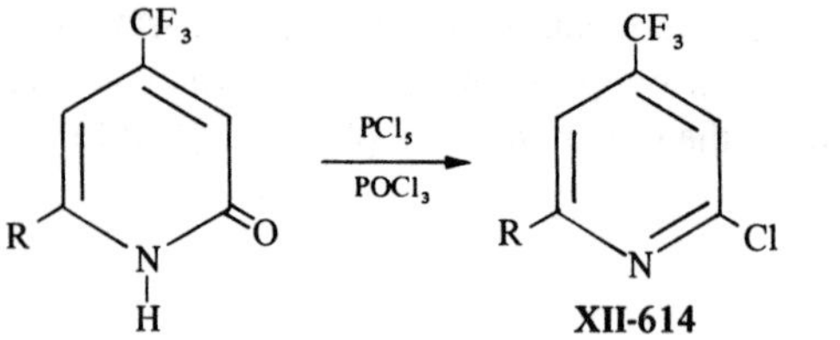

**XII-614**

R = $C_6H_5$, 2-$C_4H_3S$, $CH_3$

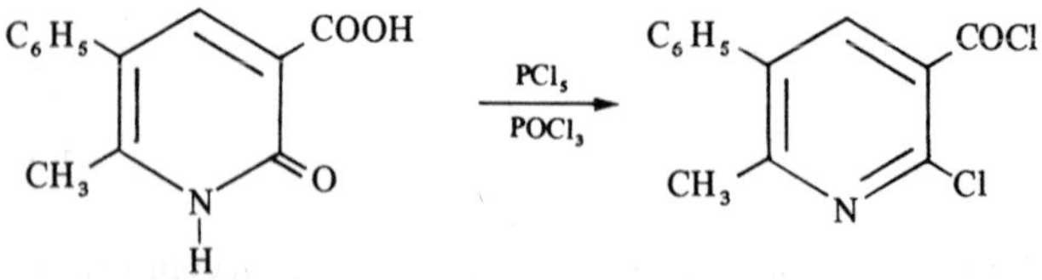

**XII-615**

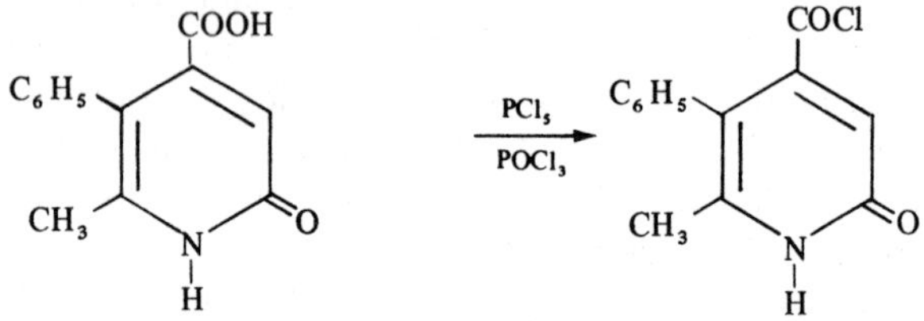

**XII-616** **XII-617**

4-Hydroxy-2-pyridone is converted to 2,4-dichloropyridine[502] and 5,6-chloro-2,4-pyridinediol **(XII-619)** is converted to 2,3,4,6-tetrachloropyridine phosphoryl chloride. In the presence of dimethylaniline, **XII-619** is converted 2,3,6-trichloro-4-pyridinol.[98]

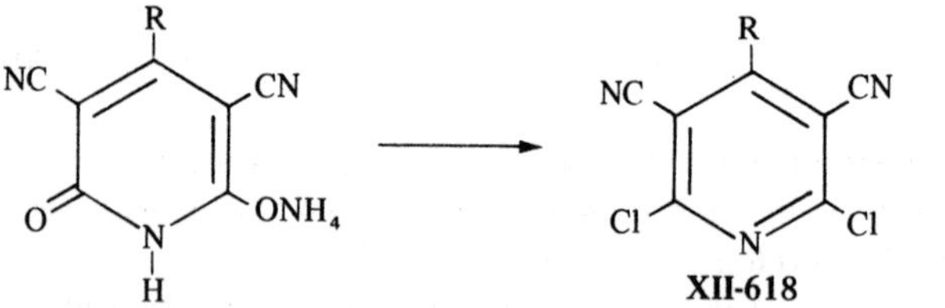

**XII-618**

R = $CH_3$, $C_2H_5$, $CH_2CH_2CH_3$, $C_6H_{13}$

*N*-Alkylpyridones are demethylated when undergoing substitutions of tl kind or are converted to pyridinium salts.[677] For example, *N*-benzyl-3 dibromo-4-pyridone **(XII-620)** and phosphorus pentachloride in toluene foı 3,5-dibromo-4-chloropyridine and a small amount of *N*-benzyl-3,5-dibromo chloropyridinium chloride.[279] *N*-Methyl-5-nitro-2-pyridone and thionyl chlori in dimethylformamide give 2-chloro-1-methyl-5-nitropyridinium chloride.[676]

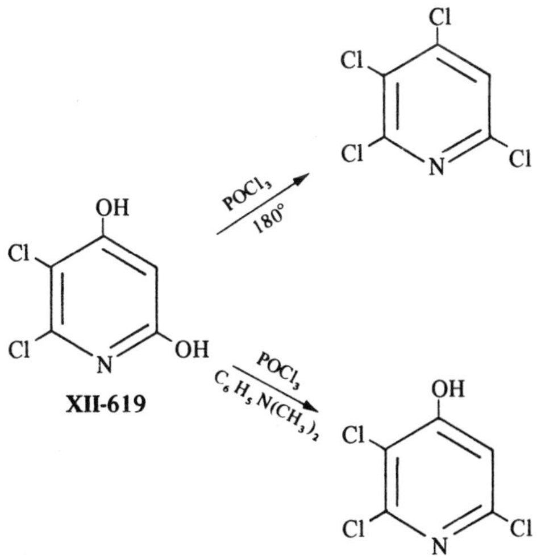

XII-619

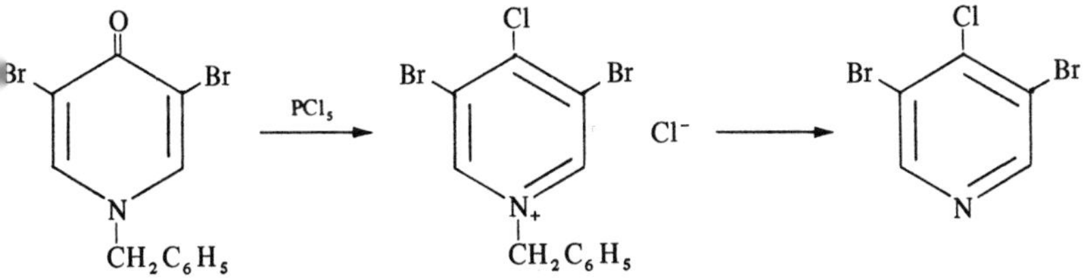

XII-620

-Chloro-1,6-dimethyl-2-pyridone is, however, formed from 4-hydroxy-1,6-nethyl-2-pyridone and phosphoryl chloride-phosphorus pentachloride[306] and arbomethoxy-4,6-dichloro-1-methyl-2-pyridone can be formed from 3-carbo-thoxy-4,6-dihydroxy-1-methyl-2-pyridone.[169]

Treatment of 3-(β-hydroxyethyl)-4-methyl-2,6-pyridinediol (**XII-621**) with onyl chloride gives 3-(β-chloroethyl)-4-methyl-2,6-pyridinediol (**XII-622**), ich cyclizes to the spirane **XII-623** with triethylamine. **XII-623** forms -dichloro-3-(β-chloroethyl)-4-methylpyridine with phosphoryl chloride or is onverted to **XII-622** with hydrochloric acid.[510]

V-Methyl- and *N*-ethyl-4-chloro-2,3-dihydro-6-oxofuro[3,2-*c*]pyridine (**XII-4**) and phosphorus oxychloride give 3-(β-chloroethyl)-2,4,6-trichloropyri-e in 68% and 71% yields, respectively.[589]

V-Substituted-2-pyridones can be used as substrates for ring closure to a mber of polyheterocyclic systems, as an alternate to the Pschorr cycliza-n[678, 679] or to the cyclizations of the corresponding *N*-alkyl-2-halopyridinium ides. 2-Substituted pyridines (**XII-625**; X = Cl, Br, $OC_2H_5$) and 3-(2-bromo-yl)indole give *N*-[2-(3-indolyl)ethyl]-2-substituted pyridinium bromide

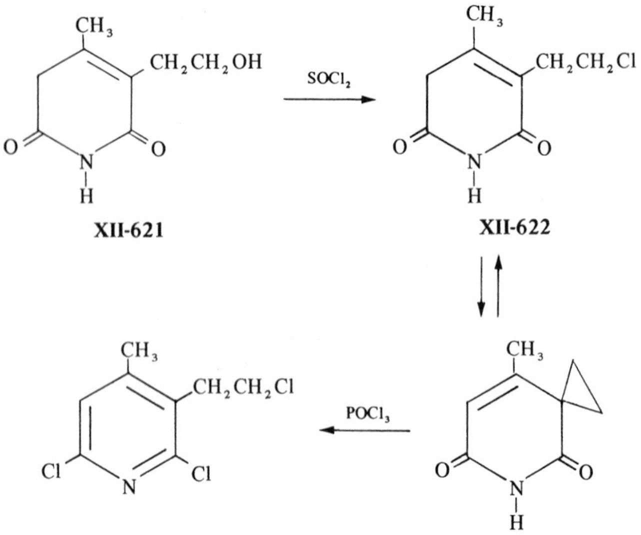

$SOCl_2$

**XII-621** **XII-622**

$POCl_3$

**XII-623**

$POCl_3$

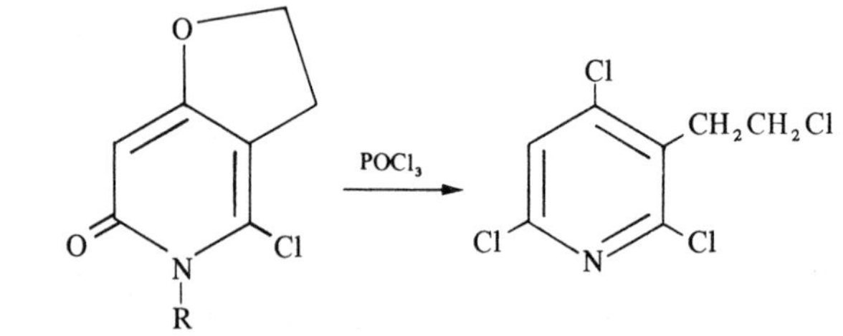

**XII-624**

**(XII-626)**, hydrolysis of which gives the pyridone **(XII-627)** in 80% yield. Ri closure of **XII-626** (X = Br) by fusion in an oil bath or by treatment of **XII-6** with phosphorus oxychloride gives 6,7-dihydro-12*H*-indolo[2,3-*a*]quinoliziniu salts **(XII-628)**. In this series the best yields are obtained in the direct sequen **XII-626** (X = Br) → **XII-628**.[680]

A reaction between *N*-arylethyl-2-pyridones **(XII-629)** and phosphorus ox chloride was originally reported to give quinolizinium salts.[681] Reports that t 5-carbethoxy-[349] and 5-ethyl-[682] derivatives of **XII-629** do not cyclize led reinvestigations of earlier work, which have shown that quinolizinium salts we not formed, but that this reaction gives *N*-arylethyl-2-chloropyridinium sa instead.[341, 683, 684] Additional conversions of pyridones to halopyridines a listed in Tables XII-15 to XII-18.

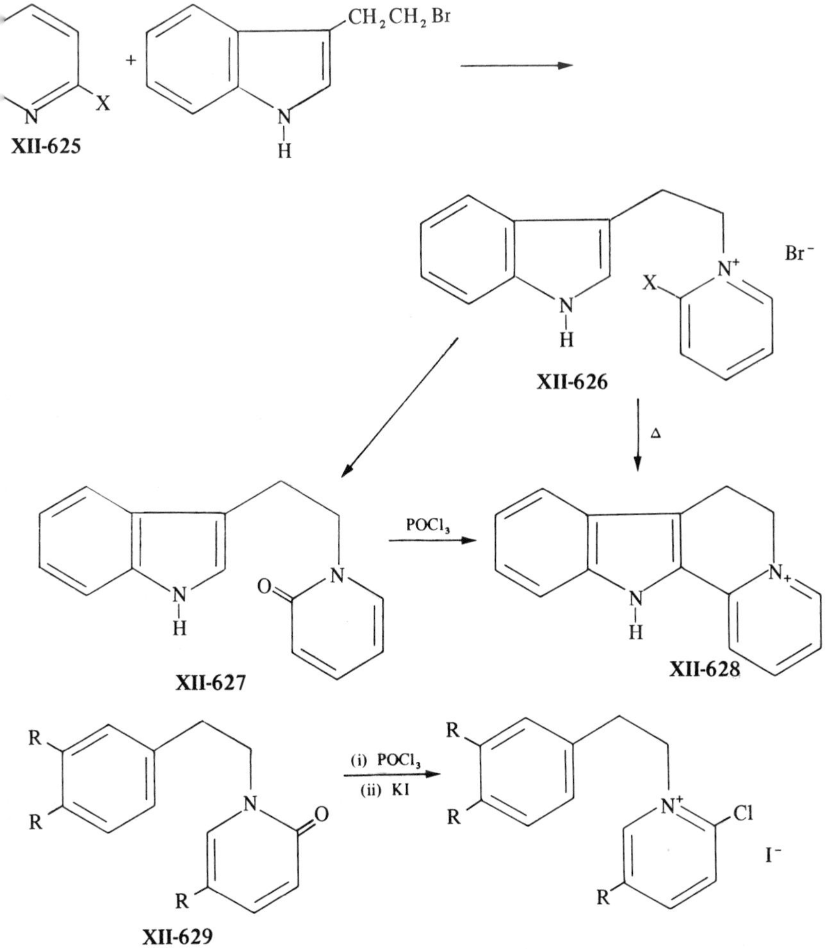

XII-625

$CH_2CH_2Br$

Br⁻

XII-626

Δ

$POCl_3$

XII-627

XII-628

(i) $POCl_3$
(ii) KI

$I^-$

XII-629

TABLE XII-15. Conversion of 2-Pyridones to 2-Chloropyridines

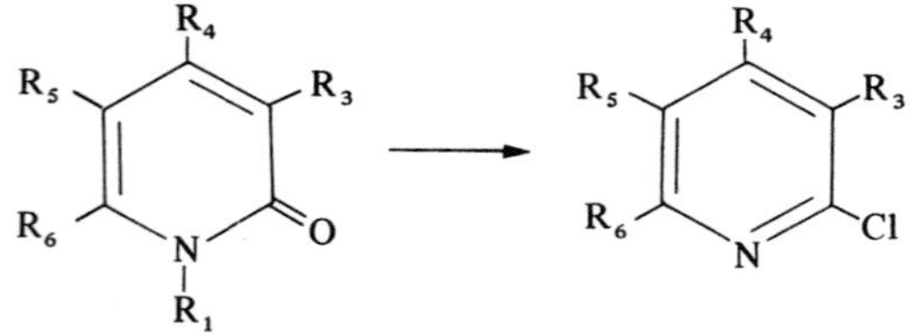

| $R_1$ | $R_3$ | $R_4$ | $R_5$ | $R_6$ | Reagent | % Yield | Ref. |
|---|---|---|---|---|---|---|---|
| | CN | $CH_3$ | $NO_2$ | $CH_3$ | $PCl_5/POCl_3$ | 54 | 674 |
| | | | $CH_3$ | | $PCl_5/POCl_3$ | | 282 |
| | $NO_2$ | $CH_3$ | $NO_2$ | | $PCl_5/POCl_3$ | 27 | 675 |
| | $NO_2$ | | $NO_2$ | $CH_3$ | $PCl_5/POCl_3$ | 83 | 675 |
| | | $CF_3$ | | $C_6H_5$ | $PCl_5/POCl_3$ | 88 | 672 |
| | | $CF_3$ | | $CH_3$ | $PCl_5/POCl_3$ | 66 | 672 |
| | | $CF_3$ | | 2-$C_4H_3S$ | $C_6H_5POCl_2$ | 80 | 672 |
| | | $CF_3$ | | 2-$C_4H_3S$ | $PCl_5/POCl_3$ | *a* | 672 |
| | Cl | Cl | Cl | | $PCl_5$ | 92 | 669 |
| | Cl | Cl | $CH_3$ | $CH_3$ | $PCl_5$ | 96 | 37 |
| | $CH_3$ | $N(C_2H_5)_2$ | $C_6H_5$ | | $PCl_5/PCl_3$ | 75 | 102 |
| | $NO_2$ | | Br | | $PCl_5/PCl_3$ | 95 | 665 |
| | $NO_2$ | | | | $POCl_3$/DMF | 97 | 676 |

| | | | | | | | | |
|---|---|---|---|---|---|---|---|---|
| | Br | | $NO_2$ | | $POCl_3/PCl_5$ | 78 | | 290 |
| | Cl | | $NO_2$ | | $POCl_3/PCl_5$ | 86 | | 291 |
| | $NO_2$ | | Cl | | $POCl_3$ | 60 | | 291 |
| | $NO_2$ | $C_6H_5$ | | $CH_3$ | $POCl_3$ | 58 | | 666 |
| | | | | | | | mixture | |
| | | $C_6H_5$ | $NO_2$ | $CH_3$ | $POCl_3$ | 58 | | 666 |
| | $NO_2$ | | $NO_2$ | | $POCl_3/PCl_5$ | 48 | | 673 |
| | $NO_2$ | $CH_3$ | | $CH_3$ | $PCl_5$ | | | 288 |
| | | $CH_3$ | $NO_2$ | $CH_3$ | $PCl_5$ | | | 288 |
| | $NO_2$ | | $NO_2$ | | $POCl_3$/DMF | 96 | | 676 |
| | I | | $NO_2$ | | $POCl_3/PCl_5$ | 86 | | 287 |
| $C_6H_5CH_2CH_2$ | | | | | $POCl_3$ | | | 338 |
| | $NO_2$ | $NH_2$ | | | $POCl_3$ | 89 | | 305 |
| | COOH | | $C_6H_5$ | $CH_3$ | $POCl_3/PCl_5$ | | | 53 |
| | $NH_2$ | | | $n$-$C_3H_7$ | $POCl_3$ | | | 150 |
| | CN | $CF_3$ | $NO_2$ | $CH_3$ | $PCl_5$ | | | 56 |
| | | $CH_3$ | | Cl | $POCl_3$ | 68 | | 8 |
| | | $C_6H_5$ | | Cl | $POCl_3$ | 88 | | 8 |
| | | 3,4-$(CH_2)_3$- | | Cl | $POCl_3$ | 78 | | 8 |

[a]Treatment of 6-(2-thienyl)-4-trifluoromethyl-2-pyridone with $PCl_5/POCl_3$ gave 6-(5-chloro-2-thienyl)-4-trifluoromethyl-2-pyridone (25%) and 2-chloro-6-(5-chloro-2-thienyl)-4-trifluoromethylpyridine (18%)

TABLE XII-16. Conversion of 4-Pyridones to 4-Chloropyridines

| $R_2$ | $R_3$ | $R_5$ | $R_6$ | Reagent | % Yield | R |
|---|---|---|---|---|---|---|
| $CH_3$ | Cl | $NO_2$ | $CH_3$ | $PCl_5/POCl_3$ | 92 | 2 |
| $CH_3$ | $NO_2$ | | $CH_3$ | $PCl_5/POCl_3$ | 88 | 2 |
| | Cl | $NO_2$ | | $PCl_5/POCl_3$ | 68 | 6 |
| | $NO_2$ | $NO_2$ | | $POCl_3$/DMF | 95 | 6 |
| | $NO_2$ | | | $POCl_3$ | | 6 |
| $CH_3$ | | $CH_3$ | | $POCl_3/PCl_5$ | | 2 |
| COOH | | $CH_3$ | | $SOCl_2$ | | 2 |
| $CH_3$ | | $NO_2$ | | $PCl_5$ | 35 | 2 |
| $C(CH_3)_3$ | | | $C(CH_3)_3$ | $POCl_3/PCl_5$ | | 6 |
| $CH_3$ | $NO_2$ | | $CH_3$ | $POCl_3$ | 61 | 6 |
| $CH_3$ | | | $CH_3$ | $POCl_3/PCl_5$ | 68 | 6 |

TABLE XII-17. Conversion of 2,4-Pyridinediols and 4-Hydroxy-2-pyridones to 2,4-Di chloropyridines

| $R_1$ | $R_3$ | $R_5$ | $R_6$ | Reagent | % Yield | R |
|---|---|---|---|---|---|---|
| | $NO_2$ | | $n$-$C_3H_7$ | $POCl_3$ | 50[a] | 1 |
| | | $CO_2C_2H_5$ | | $POCl_3$ | | 6 |
| | | $CH_3$ | Cl | $POCl_3$ | 58 | |
| | | Cl | Cl | $POCl_3$ | | |
| | | Cl | Cl | $POCl_3/C_6H_5N(CH_3)_2$ | | |
| $CH_3$ | | | $CH_3$ | $POCl_3/PCl_5$ | *b* | 3 |
| | | | $CH_3$ | $POCl_3$ | 65 | 5 |
| | | $CH_2CH_2Cl$ | | $POCl_3$ | 64.5 | 5 |

[a]Forty-five percent of 4-chloro-3-nitro-6-*n*-propyl-2-pyridinol was obtained under mild cc ditions.

[b]Reaction occurred only at the 4-position to give 1,6-dimethyl-4-chloro-2-pyridone.

BLE XII-18. Conversion of 2,6-Pyridinediols and *N*-Substituted Derivatives to 2,6-Di-
loropyridines

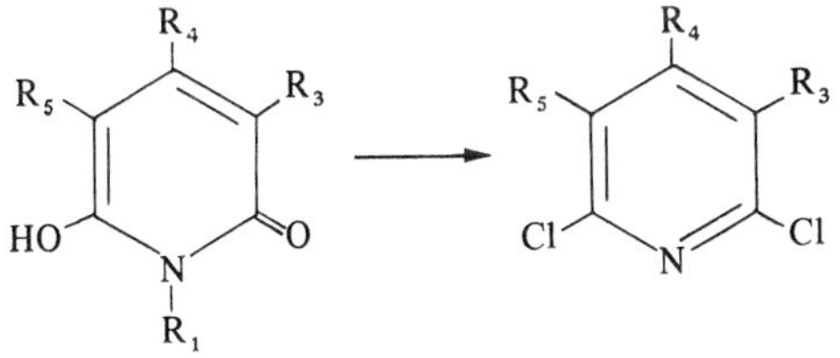

| | $R_3$ | $R_4$ | $R_5$ | Reagent | % Yield | Ref. |
|---|---|---|---|---|---|---|
| | | $NH_2$ | | $POCl_3$ | 60 | 34 |
| $H_5$ | $CONHC_6H_5$ | $OC_6H_5$ | | $POCl_3$ | $90^a$ | 172 |
| $H_5$ | $CONHC_6H_5$ | $OC_6H_4CH_3$-*o* | | $POCl_3$ | $50^b$ | 173 |
| | $CH_3$ | Cl | Cl | $PCl_5$ | 96 | 7 |

roduct was 6-chloro-4-phenoxy-l-phenyl-2-pyridone-3-carboxanilide.
roduct was 6-chloro-l-phenyl-4-(*o*-tolyloxy)-2-pyridone-3-carboxanilide.

## F. *Miscellaneous Replacements*

Pyridinols and pyridones and phosphorus pentasulfide give predictable oducts. 4-Pyridinol and phosphorus pentasulfide in pyridine give 4-pyridine-ione,[685] while 2,6-di-*t*-butyl-4-pyridone gives bis[4-(2,6-di-*t*-butylpyridinyl)] lfide.[671] *N*-Alkyl-3,5-dibromo-2,6-dimethyl- and *N*-alkyl-3,5-dibromo-2,6-di-enyl-4-pyridone are converted to the corresponding 4-pyridinethiones.[118] mmonium salts of 4-alkyl-3,5-dicyano-6-hydroxy-2-pyridones and phosphorus ntasulfide in xylene and pyridine form pyridinium salts of the corresponding mercapto-2-pyridinethiones (**XII-630**).[111]

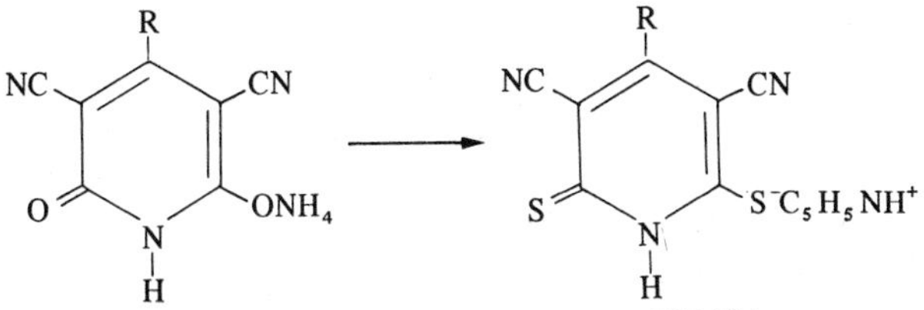

**XII-630**

Pentachloropyridine and phosphorus pentasulfide give 4-mercaptotetrachloro-ridine (**XII-631**), which can be methylated readily with dimethylsulfate in ueous alkali.[686]

A method has been described for the conversion of phenols to thiophenols nploying dimethylthiocarbamoyl chloride. 2- And 4-pyridone and 3-pyridinols ere typical among twenty-nine compounds studied. The carbamyl chloride and

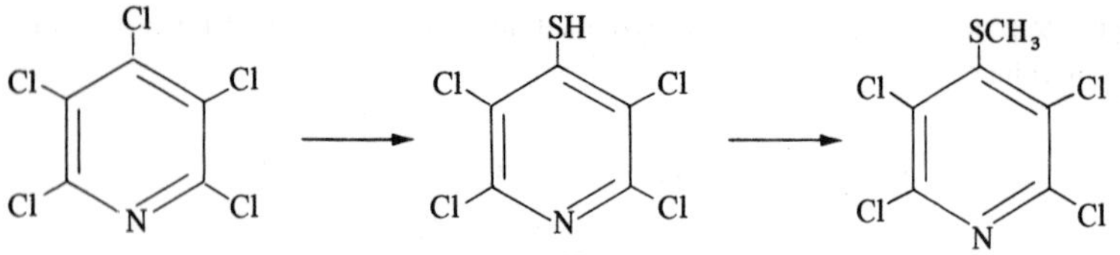

XII-631

the sodium salt of 2-pyridone or 3-pyridinol, or the silver salt of 4-pyridone dimethylformamide gives the *O*-pyridyl dimethylthiocarbamate. Thermal arrangement gives the *S*-pyridyl dimethylthiocarbamates (**XII-632**).[687]

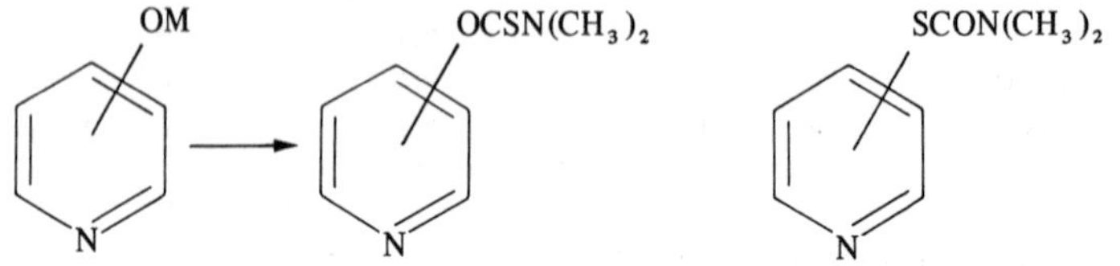

OM    $OCSN(CH_3)_2$    $SCON(CH_3)_2$

**XII-632**

The boron trifluoride and hydrogen chloride salts of 2- and 4-pyrid *N,N'*-dimethylthiocarbamate rearrange at room temperature. The lower tempe ture requirement for phenols substituted with electron withdrawing grou supports a mechanism involving nucleophilic attack by sulfur (**XII-633** **XII-634**).[687] When 4-pyridone is heated with sodium bisulfite a small amount

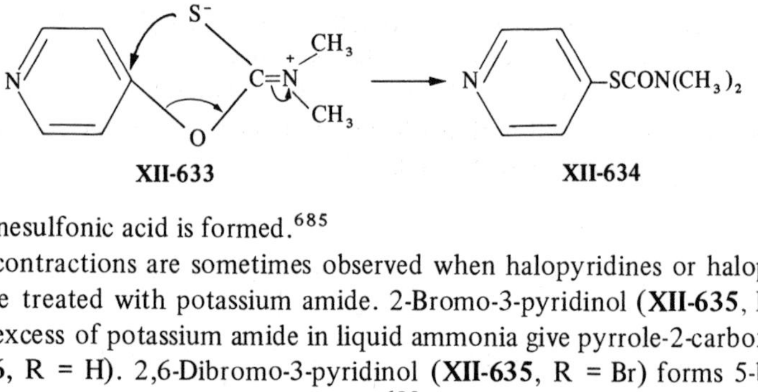

**XII-633**    **XII-634**

4-pyridinesulfonic acid is formed.[685]

Ring contractions are sometimes observed when halopyridines or halopyrin dines are treated with potassium amide. 2-Bromo-3-pyridinol (**XII-635**, R = and an excess of potassium amide in liquid ammonia give pyrrole-2-carboxami (**XII-636**, R = H). 2,6-Dibromo-3-pyridinol (**XII-635**, R = Br) forms 5-brom pyrrole-2-carboxamide (**XII-636**, R = Br).[688]

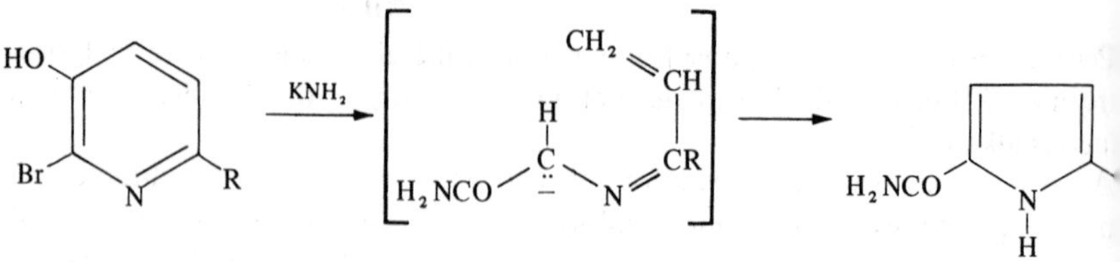

$KNH_2$

**XII-635**    **XII-636**

## 2. Involving the Nucleus

### A. *Reduction*

2-Pyridones are usually hydrogenated employing a Raney Nickel catalyst at vated temperatures and pressures, most often with ethanol as the solvent, hough palladium and platinum catalysts have been used (Table XII-19). At mperatures that are higher than optimum for piperidone formation, piperines are formed.[283] *N*-Benzyl-2-pyridone gives *N*-cyclohexylmethyl-2-piperine (81%) when hydrogenated at 100 atm and at 150° in ethanol employing a ney Nickel catalyst, but gives *N*-benzyl-2-piperidone at 50 atm of hydrogen d 70°.[689] 6-Methyl-5-phenyl-2-pyridone-3-carboxylic acid is not hydrogenated der conditions where the 4-carboxylic acid **(XII-637)** is reduced to 6-methyl-phenyl-2-piperidone-4-carboxylic acid. It was suggested that these properties e a consequence of the pyridinol structure of the 3-carboxylic acid and the ridone structure of **XII-637**.[53] More direct information is needed to verify this

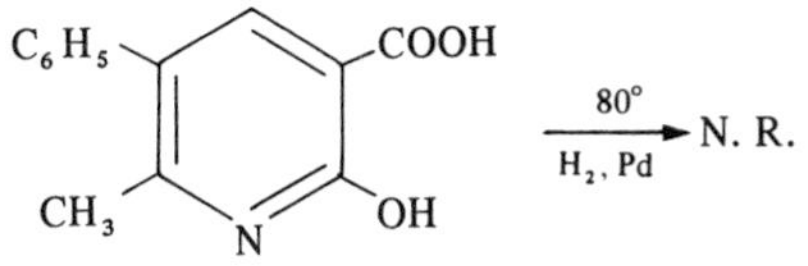

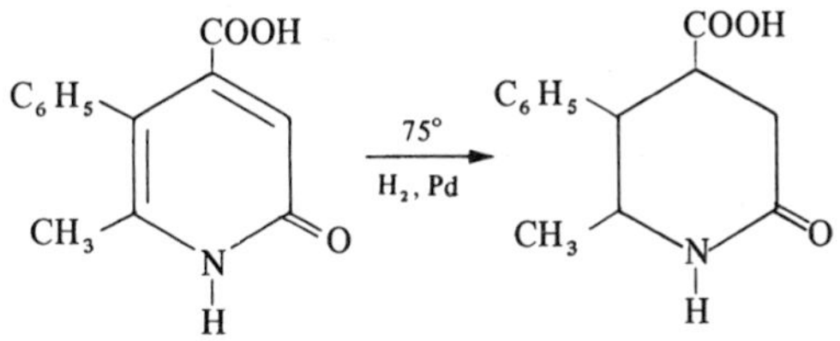

$C_6H_5$, COOH, $CH_3$, N, OH $\xrightarrow[H_2,\ Pd]{80°}$ N. R.

$C_6H_5$, COOH, $CH_3$, N–H, O $\xrightarrow[H_2,\ Pd]{75°}$ $C_6H_5$, COOH, $CH_3$, N–H, O

**XII-637**

nclusion.[489] *N*-Methyl-2-pyridone has been used for the synthesis of 1,7-di-thyl-1,7-diazaspiro[5,5]undecane **(XII-638)** *via* *N*-methyl-2-piperidone in % overall yield.[692]

,3-Dihydro-4*H*-pyrid[2,1-*b*]oxazinium iodide **(XII-639)**, prepared by alkyla-n of the silver salt of 2-pyridone with trimethylene iodide, is hydrogenated in presence of platinum to *N*-(γ-hydroxypropyl)piperidine.[554]

TABLE XII-19. Hydrogenation of 2-Pyridones to 2-Piperidones

| $R_1$ | $R_3$ | $R_4$ | $R_5$ | $R_6$ | Catalyst | Temperature | $P_{atm}$ | % Yield | Ref. |
|---|---|---|---|---|---|---|---|---|---|
| H | H | H | H | H | Raney Ni | 150 | 100 | 72 | 689 |
| $C_6H_5CH_2$ | H | H | H | H | Raney Ni | 70 | 50 | 80 | 689 |
| H | $CH_3$ | H | H | H | Raney Ni | 200–240 | 120 | 80 | 283 |
| H | H | $C_2H_5$ | H | H | Raney Ni | 200–240 | 120 | 80 | 283 |
| H | H | $CO_2CH_3$ | H | H | Raney Ni | 260 | 120 | 50 | 283 |
| H | H | COOH | $C_6H_5$ | $CH_3$ | Pd-C | 75 | 45 p.s.i. | 100 | 53 |
| $CH_3$ | H | H | H | H | Raney Ni | 50 | 150 | 96 | 692 |
| $C_2H_5$ | $CH_3$ | H | H | H | Raney Ni | 150–160 | 135 | 64 | 690 |
| $(CH_2)_2C_6H_3(OCH_3)_2$-3,4 | H | H | (ring) $\underset{\displaystyle CH_3}{\underset{\vert}{N}}$ | H | $Pt_2O$-$PdCl_2$ | 40 | | | 339 |
| $CH_3$ | H | H | $(CH_2)_4\underset{\displaystyle CH_3}{\underset{\vert}{N}}COC_6H_5$ | H | Raney Ni | ~78 | | 85 | 691 |
| $CH_3$ | H | H | $-CH{=}CH(CH_2)_2\underset{\displaystyle CH_3}{\underset{\vert}{N}}COC_6H_5$ | H | Raney Ni | ~78 | | 95 | 691 |
| $CH_2C_6H_4COOH$-*p* | | | | | Raney Ni | 70 | 50 | 60 | 558 |
| $CH_2C_6H_4NH_2$-*p* | | | | | Raney Ni | 70 | 50 | 63 | 558 |
| $CH_2C_6H_4NH_2$-*m* | | | | | Raney Ni | 70 | 50 | 27 | 558 |
| | | | | | Raney Ni | 70 | 50 | 10 | 558 |
| | | | | | Raney Ni | 70 | 50 | 60 | 558 |
| $CH_2C_6H_4Br$-*o* | | | | | Raney Ni | 70 | 50 | 40 | 558 |
| | | | | | Raney Ni | | atm | 83 | 340 |

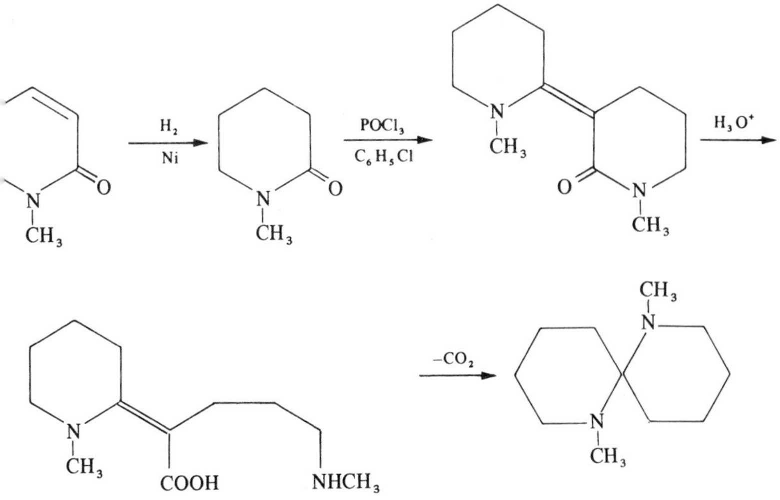

**XII-638**

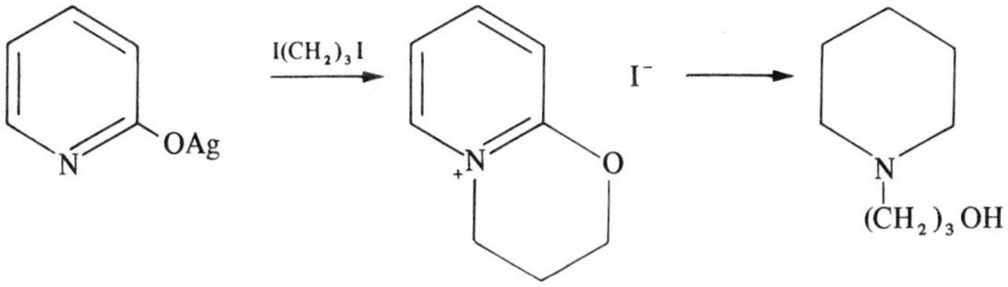

**XII-639**

Hydrogenations of 2-aralkyl-3-pyridinols with platinum catalysts give mixtures products where either or both the pyridyl ring and the phenyl ring are duced. However, the major product of low pressure, controlled hydrogenation 2-diphenylmethyl-3-pyridinol at 50° is 2-(α-cyclohexylbenzyl)-3-pyridinol. ydrogenation of the corresponding methyl ethers, 2-aralkyl-3-methoxypyrines over a platinum catalyst gives cleaner products.[191] 2-Diphenylmethyl-3-ridinol in dioxane containing Raney Nickel is hydrogenated at 130 to 135° d 125 $kg/cm^2$ to 2-diphenylmethyl-3-piperidinol (~70%) and smaller amounts 2-(α-cyclohexylbenzyl)-3-piperidinol and 2-dicyclohexylmethyl-3-piperinol.[191]

Although early attempts were unsuccessful, it is possible to reduce *N*-methyl-2-ridones with lithium aluminum hydride and aluminum chloride in tetrahydroran or ether, or without aluminum chloride in tetrahydrofuran.[693] *N*-Methyl-pyridone (**XII-640**, R = H) gives 1-methyl-3-piperideine (**XII-641**, R = H) as e major product and 1,4-dimethyl-2-pyridone (**XII-640**, R = $CH_3$) gives

1,4-dimethyl-3-piperideine (**XII-641**, R = $CH_3$). However, 1,3-, 1,5-, a 1,6-dimethyl-2-pyridone and 1,4,6-trimethyl-2-pyridone give mixtures of t corresponding piperidines and piperideines.[693, 694] For example, reduction 1,4,6-trimethyl-2-pyridone with lithium aluminum hydride gives **XII-642** as t major product along with **XII-643**, **XII-644**, and **XII-645**. In the presence aluminum chloride, the less stable of the piperidines, the *trans* isomer (**XII-64** is not detected[694] and **XII-645** is the major product. The product distributi

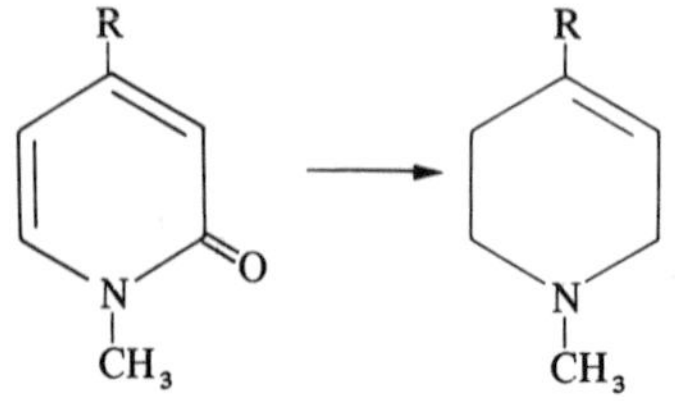

**XII-640** **XII-641**

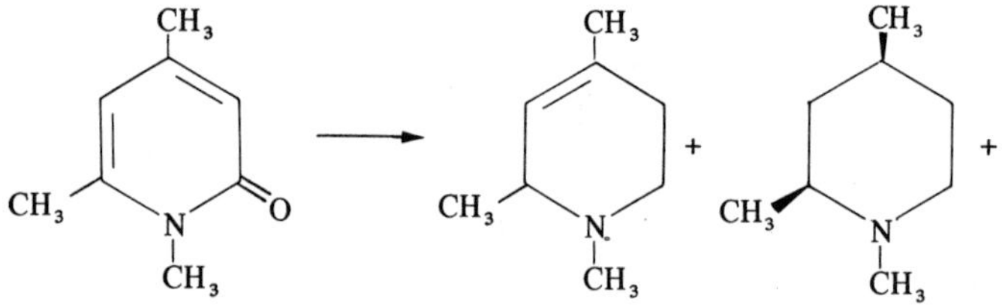

**XII-642** **XII-643**

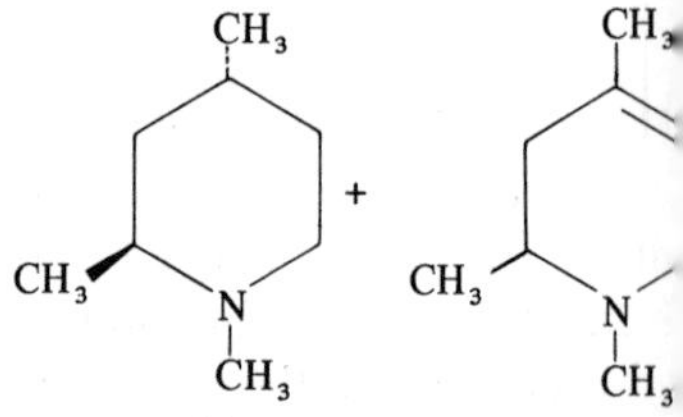

**XII-644** **XII-645**

has been rationalized through steps initiated by 1,6-, 1,4-, or 1,2-additi involving the carbonyl group.[695] 1,6-Dimethyl-2-pyridone and two moles lithium aluminum hydride per mole of pyridone gives mainly 1,2-dimethyl- piperideine (**XII-646**) and 1,2-dimethyl-2-piperideine (**XII-647**) and a sm amount of 1,6-dimethyl-3-piperideine. As an illustration, 1,6-addition of t aluminum hydride ion to 1,6-dimethyl-2-pyridone gives **XII-648**, which followed by reductive cleavage to the enamine **XII-651**; this reacts with mo aluminum hydride ion to give **XII-646** after hydrolysis.[695] With an excess of t

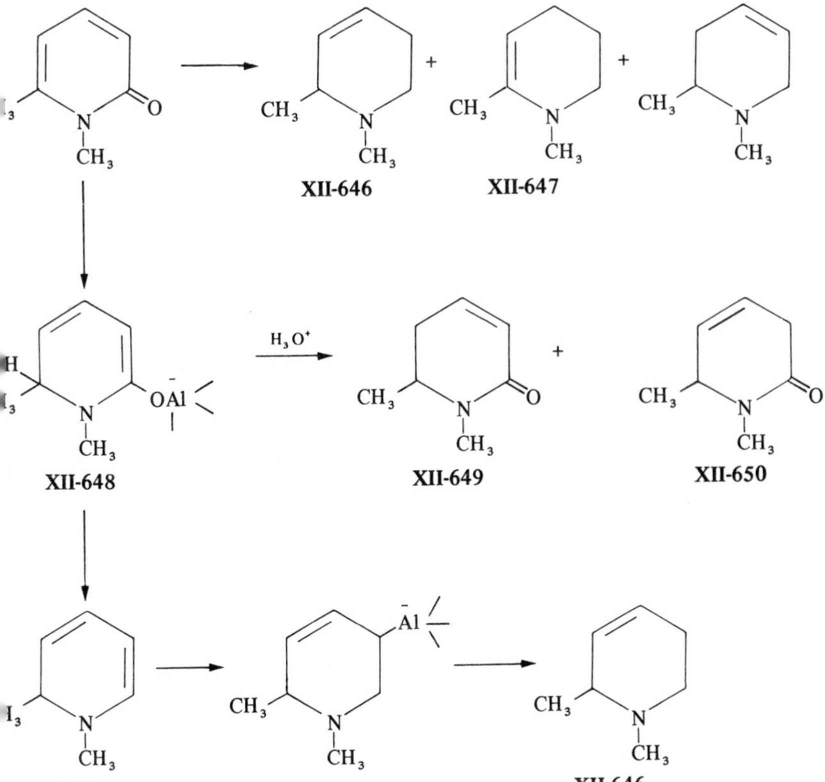

CH3 N CH3 O → CH3 N CH3 + CH3 N CH3 + CH3 N CH3

XII-646 XII-647

H CH3 N CH3 OAl XII-648 $\xrightarrow{H_3O^+}$ CH3 N CH3 O + CH3 N CH3 O

XII-649 XII-650

CH3 N CH3 XII-651 → CH3 N CH3 Al → CH3 N CH3 XII-646

pyridone the dihydropyridones **XII-649** and **XII-650** are isolated after ydrolysis. The product of 1,4-addition **XII-647** was isolated in good yield from 6-dimethyl-2-pyridone, but products of reactions initiated by 1,4-addition are ot detected when other *N*-methyl-2-pyridones were reduced.[695]

Electrolytic reductions of *N*-methyl- and *N*-ethyl-2-pyridone have given mainly alkylpiperidine and some 1-alkyl-3-piperideine. Both 1,3-dimethyl- **(XII-652)** d 1,6-dimethyl-2-pyridone give mainly the piperidine (e.g., **XII-653**) and equal t smaller amounts of the two isomeric piperideines (e.g., **XII-654**, **II-655**).[696]

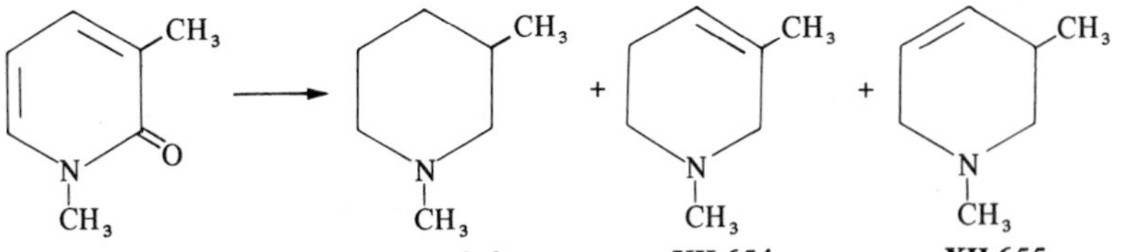

CH3 N CH3 O → CH3 N CH3 + CH3 N CH3 + CH3 N CH3

XII-652 XII-653 XII-654 XII-655

*N*-Benzyl-5,6-dihydro-6-methyl-2-pyridone[697] and 3,4-dihydro-5,6-diphenyl pyridone[443] have been hydrogenated to the corresponding 2-piperidones in t presence of platinum. Piperidones can be converted to piperidines with lithi aluminum hydride.[697]

## B. *Electrophilic Substitution—General*

Electrophilic substitutions of pyridinols and pyridinol ethers have be reviewed recently.[698] Although pyridine reacts with difficulty, aminopyridin pyridinols, and pyridones, undergo electrophilic substitution readily un moderate conditions. The annular NH—CO group of 2-pyridone, which acts as electron releasing group in these reactions, has an acid strengthening effect o second substituent. This group, therefore, appears to be electron withdrawing its ground state and electron releasing (+E effect) in electrophilic substituti transition states through contributions from structures such as **XII-656** at t demand of cationic reagents.[699] A study of electrophilic substitution pyridones is complicated because of the variety of forms that could rea Several studies have, however, been reported for nitration in acid solutic where the reacting species have been determined.[700, 701]

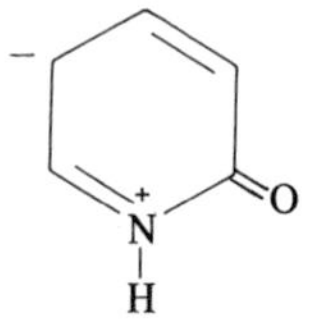

**XII-656**

## C. *Halogenation*

Halogenation of 2- and 4-pyridones occurs at the 3- and 5-positions and, ev under mild conditions, the monohalogenated product from 2-pyridones uns stituted in the 3- and 5-positions is usually not isolated.[698, 702] Rece contributions to the literature describe product distributions that are consiste with earlier chemistry[702] (Tables XII-20 to XII-22).

Although treatment of 4-methyl-, 1-methyl-, 1,4-dimethyl-, and 1,6-dimeth 2-pyridone with *N*-bromosuccinimide (NBS) gives 3,5-dibromo-2-pyridon side-chain bromination occurs with 1,3- and 1,5-dimethyl-2-pyridones to give and 5-bromomethyl-*N*-methyl-2-pyridone, respectively. 3- And 5-methyl-2-py done give oils that decompose on distillation.[559]

TABLE XII-20. 3-Halo-2-pyridones by Halogenation

| $R_4$ | $R_5$ | $R_6$ | Reagent | % Yield | Ref. |
|---|---|---|---|---|---|
| | $NO_2$ | | HCl, $KClO_3$ | 60 | 291 |
| | $NO_2$ | | $Br_2$ | 91 | 290 |
| | $NO_2$ | | KI, $KIO_3$, $H^+$ | 80 | 287 |
| OH | $CH_3$ | Cl | $Br_2$, $C_2H_5OH$ | 95 | 97 |

.BLE XII-21. 5-Halo-2-pyridones by Halogenation

| | $R_3$ | $R_4$ | $R_6$ | Reagent | % Yield | Ref. |
|---|---|---|---|---|---|---|
| | $NO_2$ | | | HCl, $KClO_3$ | 45 | 291 |
| | $NO_2$ | | | KI, $KIO_3$, $H^+$ | 60 | 287 |
| | $CH_3$ | | | $Br_2$, HOAc | | 559 |
| $H_5$ | $CONHC_6H_5$ | $OC_6H_5$ | OH | $Br_2$, $CHCl_3$ | 67 | 172 |
| $H_5$ | $CONHC_6H_5$ | $OC_6H_4CH_3$-*m* | OH | $Br_2$, $CCl_4$ | 50 | 173 |
| $H_5$ | $CONHC_6H_5$ | $OC_6H_4CH_3$-*o* | OH | $Br_2$, $CCl_4$ | 50 | 173 |
| $CH_3OC_6H_4$ | $CO_2CH_3$ | OH | OH | $Br_2$, $CHCl_3$ | 80 | 169 |
| | $NO_2$ | OH | $CH_3$ | $Br_2$, HOAc | 74 | 502 |

odination of 2-pyridone with aqueous iodine-potassium iodide gives 5-iodo-2-ridone and 3,5-diiodo-2-pyridone and iodination of 4-pyridone gives 3-iodo-4-ridone (50%) and 3,5-diiodo-4-pyridone (21%).[704] 3,5-Dibromo-4-methoxy-ridine **(XII-657)** has been prepared from chelidamic acid by bromination, lowed by decarboxylation, treatment with phosphorus oxychloride, and then h potassium methoxide.[705]

TABLE XII-22. 3, 5-Dihalo-2-pyridones by Halogenation

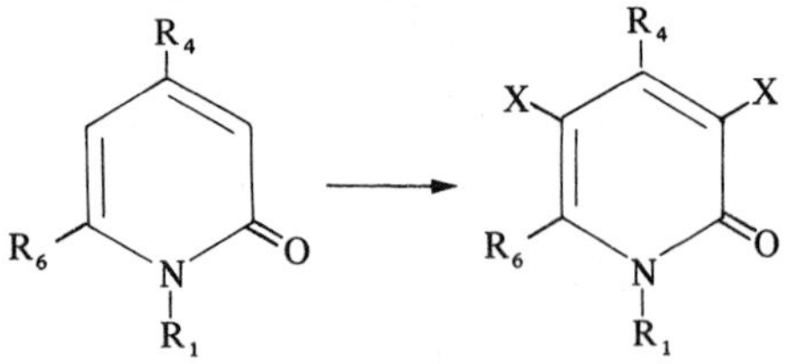

| $R_1$ | $R_4$ | $R_6$ | Reagent | % Yield | Ref. |
|---|---|---|---|---|---|
| H | $CH_3$ | H | | | |
| $CH_3$ | $CH_3$ | H | $Br_2$, HOAc | 75–90 | 559 |
| $CH_3$ | H | $CH_3$ | | | |
| $CH_3$ | H | H | | | |
| H | $CH_3$ | H | NBS, benzoyl peroxide | 75 | 559 |
| $CH_3$ | $CH_3$ | H | NBS, benzoyl peroxide | 45 | 559 |
| $CH_3$ | H | $CH_3$ | NBS, benzoyl peroxide | 57 | 559 |
| $CH_3$ | H | H | NBS, $AlCl_3$ | 79 | 559 |
| H | H | Cl | HCl, $H_2O_2$, HOAc | 80 | 703 |
| H | H | Cl | $Br_2$, 33% $H_2SO_4$ | 93 | 703 |
| | | | KI, $I_2$ | 15 | 704 |

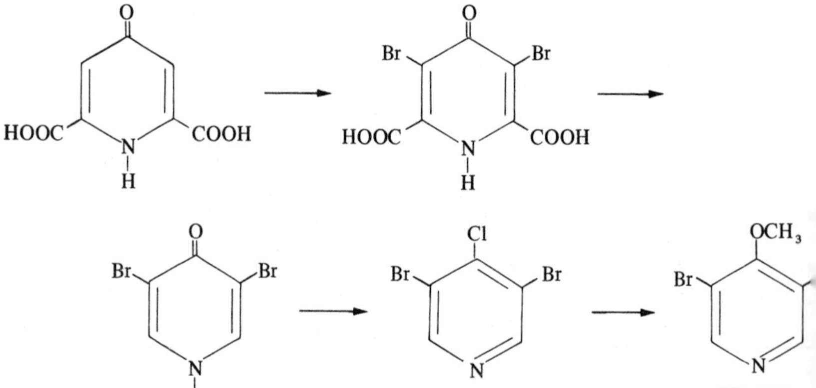

**XII-657**

3,5-Dibromo-1-methyl-4-pyridone has been prepared from *N*-methyl pyridone by modifications[139] of an earlier procedure.[706] *N*-(*n*-Butyl)-2,6-methyl-4-pyridone is also brominated in acetic acid to the 3,5-derivative. 3-Nitro-4-pyridone[664] and 2,6-dimethyl-3-nitro-4-pyridone[294] are halogenated the 5-position to give 5-bromo- and 5-chloro-2,6-dimethyl-3-nitro-4-pyrido which have been converted to the corresponding 4-chloropyridines for use γ-carboline synthesis.

-Acetyl-2,6-dimethyl-4-pyridone and sodium hypobromite form 3-acetyl-5-mo-2,6-dimethyl-4-pyridone (**XII-658**), which then gives 3,5-dibromo-2,6-dithyl-4-pyridone (**XII-659**)[141]

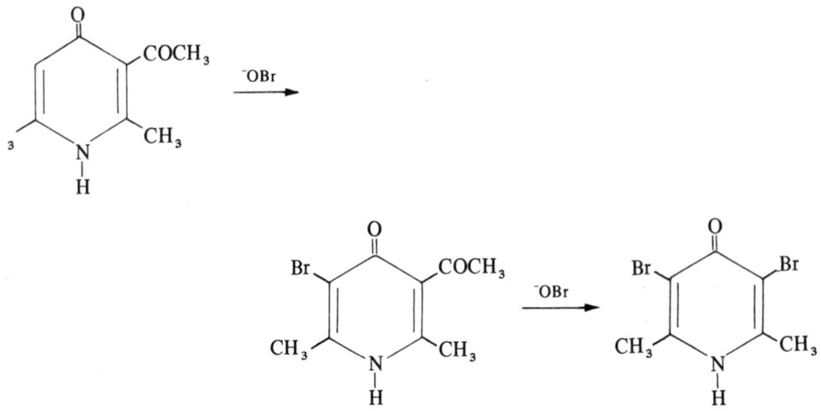

Chlorination of 4-hydroxy-6-methyl-2-pyridone (**XII-660**) in water at 40° gives chloro- and 3,5-dichloro-6-methyl-2,4-pyridinediol and 6-chloromethyl-3,5-chloro-4-hydroxy-2-pyridone in a ratio of 4:4:1.[502]

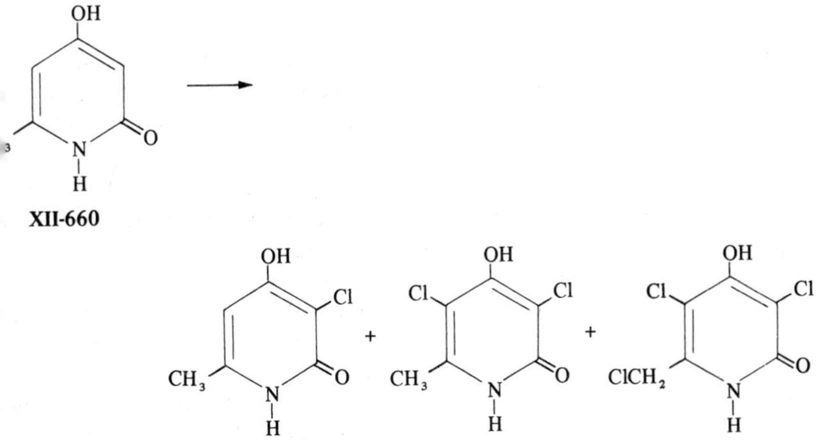

Chlorination of 5-carbethoxy-4-hydroxy-6-methyl-2-pyridone with $SO_2Cl_2$ es **XII-661**, **XII-662**, or **XII-663**, depending on conditions. Products **XII-662** d **XII-663** are converted to **XII-661** with zinc and acetic acid.[707]

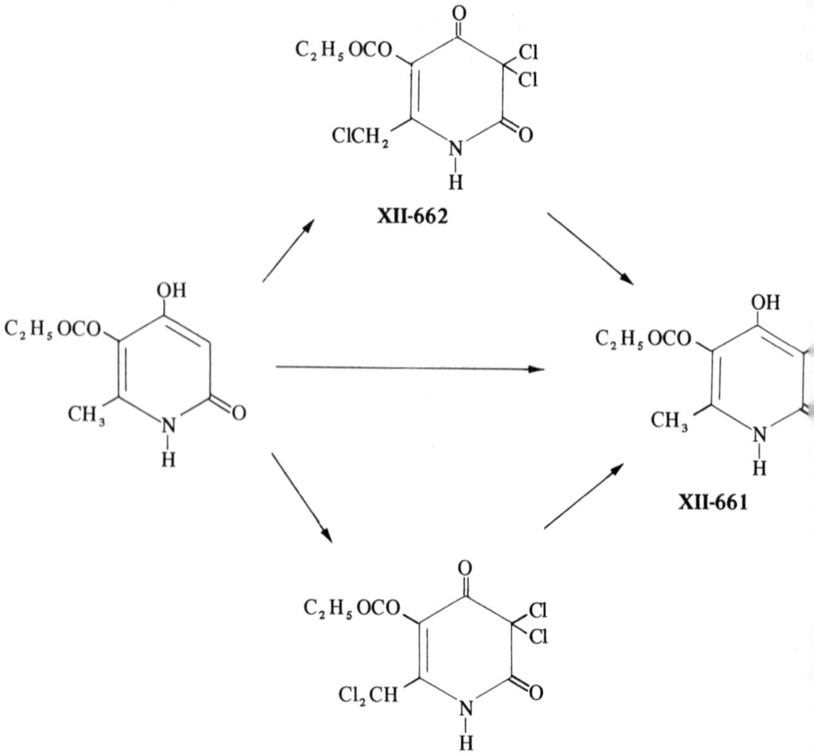

XII-662

XII-661

XII-663

3-Pyridinol is converted to 2-bromo-3-pyridinol by an alkaline solution bromine. Additional activation of the ring can be provided by conversion to *N*-oxide. 3-Hydroxypyridine-1-oxide **(XII-664)** can give 2-bromo- (47%) 2,4,6-tribromo-3-hydroxypyridine-1-oxide (78%), depending on conditio Further bromination of 2-bromo-3-hydroxypyridine-1-oxide can give 2,6-bromo-3-hydroxypyridine-1-oxide (36%). The *N*-oxides are reduced to corresponding 3-pyridinols by iron and acetic acid.[708]

6-Methyl-3-pyridinol is brominated in pyridine to give 2-bromo- and th 2,4-dibromo-6-methyl-3-pyridinol. Iodination with sodium iododichloride gi 2-iodo-6-methyl-3-pyridinol.[583] Iodination of 2-(*p*-substituted phenyl)-3-py dinols (**XII-665**; R = H, $CH_3$, $C_2H_5$, $CH(CH_3)_2$, $C(CH_3)_3$, Cl, $OCH_3$) gi 2-aryl-6-iodo-3-pyridinols.[709] This should be contrasted with nitration of the pyridinols, which occurs first on the phenyl group[710, 711] (Section III.2. p. 810).

Bromination in acetic acid of 3-bromo-5-ethoxypyridine **(XII-666)** gi 2,3-dibromo-5-ethoxypyridine. Similar treatment of 2-bromo-6-ethoxypyrid **(XII-667)** gives 2,3-dibromo-6-ethoxypyridine.[712]

XII-664

XII-665

XII-666

$C_2H_5O$ N Br $\xrightarrow[HOAc]{Br_2}$ $C_2H_5O$ N Br Br

**XII-667**

## D. *Nitration*

The preparative nitration of pyridones has been reviewed recently.[698] nitrations of several 2- and 4-pyridones have been studied kinetically.[700,] Pyridones with $pK_a < 1.5$ are nitrated as their free bases under all conditi 4-Pyridone ($pK_a$ 3.27) is nitrated as the free base in 65 to 85% sulf acid but as the conjugate acid in 85 to 98% $H_2SO_4$.[700] 6-Hydrox pyridone and its *N*- and *O*-methyl derivatives are rapidly converted to 3-nitro-derivative *via* the free bases in 70 to 77% sulfuric acid. At higher acid the reactions were too fast to be followed kinetically.[701]

Since 2-pyridones are nitrated mainly as their free bases and are probabl the pyridone form,[700] it is not likely that a transition state invol hydrogen-bond participation by OH as has been suggested for *ortho*-nitratio phenol[713] and for nitration of 2-substituted-3-pyridinols at the 4-position (this section, p. 809) is responsible for the formation of 3-nitro-2-pyridone, major product.[714]

3,5-Dinitro-2-pyridone is usually prepared by direct nitration of 2-p done.[673, 676, 715] It can be formed in 50% yield employing fuming nitric aci oleum[676] or in 37% yield from 2-aminopyridine **(XII-668)** *via* its nitramine *situ* rearrangement to a mixture of 3- and 5-nitro-2-aminopyridines, is follo by conversion to the pyridones and nitration (Table XII-23).[285]

Nitration of 6-methyl-4-phenyl-2-pyridone **(XII-669)** gives a mixture of th and 5-mononitro derivatives that could not be separated because of insolubility in organic solvents and that did not react with triethyl phosph Conversion to the 2-chloro derivatives gave a mixture that could be converte the $\beta$-carbolines **XII-670** and **XII-671** when treated with triethyl phosph presumably through nitrene intermediates.[666]

Mixtures of 3- and 5-nitro-4-methyl-2-pyridone and of 3- and 5-nitro-6-meth 2-pyridone have been converted to the corresponding 3,5-dinitro-2-pyrido with nitric acid in oleum.[675]

A procedure has been described for the preparation of 3-nitro-4-pyridone fr 4-pyridone and mixed acids that avoids the oleum used by previous investiga (Table XII-24).[667] Both 4-hydroxy-2-picoline and 4-ethoxy-2-picoline nitrated at the 3- and 5-positions but give more 3-nitration.[293] 4-Hydrox

XII-668

XII-670

XII-669

XII-671

TABLE XII-23. Nitration of 2-Pyridones

| $R_3$ | $R_4$ | $R_5$ | $R_6$ | Reagent | % Yield 3-$NO_2$ | % Yield 5-$NO_2$ | Ref. |
|---|---|---|---|---|---|---|---|
| $NO_2$ | | | | 40% oleum $HNO_3$ | | 50 | 285 |
| | | $NO_2$ | | | | | 285 |
| CN | $CF_3$ | | $CH_3$ | $HNO_3$ (fuming, *d* 1.5)<br>$H_2SO_4$ (conc.) | | 72 | 56 |
| | OH | | $CH_3$ | 70% $HNO_3$ | only 3-$NO_2$<br>82 | | 502 |
| | OH | | $n$-$C_3H_7$ | 83% $HNO_3$ | 50 | | 150 |
| | OH | Cl | Cl | 50% $HNO_3$ | 77 | | 98 |
| | OH | $CH_3$ | Cl | 50% $HNO_3$ | 94 | | 97 |

idones are usually nitrated at the 3-position (Table XII-23). Additional ations of 2-pyridones and 4-pyridones are summarized in Tables XII-23 and -24.

LE XII-24. 3-Nitro-4-pyridones

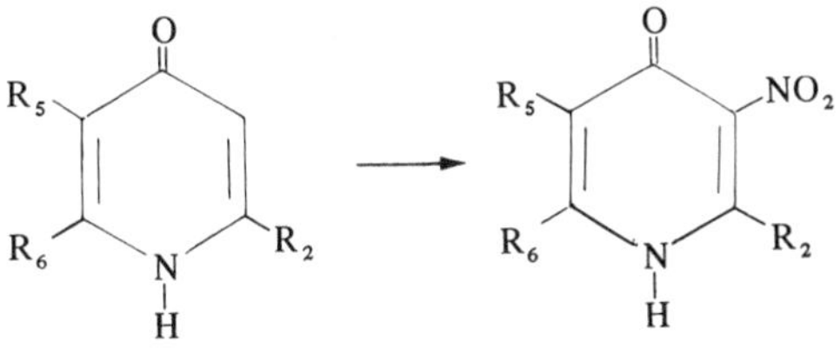

| $R_5$ | $R_6$ | Reagent | % Yield | Ref. |
|---|---|---|---|---|
| H | H | $HNO_3$ (*d* 1.5), $H_2SO_4$ (*d* 1.84) | 61 | 667 |
| H | $CH_3$ | $HNO_3$ (*d* 1.5), $H_2SO_4$ (*d* 1.84) | | 670 |
| H | $CH_3$ | $HNO_3$ (*d* 1.5), $H_2SO_4$ (*d* 1.84) | 83 | 294 |
| H | $CH_3$ | $HNO_3$ (*d* 1.5), $H_2SO_4$ | 70 | 295 |
| $NO_2$ | H | $HNO_3$ (fuming), 30% oleum | 92 | 716 |

-Pyridinol and 3-methoxypyridine undergo nitration with nitric acid in uric acid as their conjugate acids.[701] 3-Pyridinol with sulfuric and nitric 1[647] or with fuming acids[701] gives 2-nitro-3-pyridinol in good yield along h a small amount of 6-nitro-3-pyridinol.[647] Further nitration in acetic l-acetic anhydride gives 2,6-dinitro- and then 2,4,6-trinitro-3-pyridinol.[717] erally, in electrophilic substitutions of 3-pyridinols, the 4-position is less tive than the 2- or the 6-position. However, in the nitrations of 2-alkyl- or alo-3-pyridinol **(XII-672)** at lower temperatures (0–5°) formation of the itro-2-substituted-3-pyridinol predominates.[647, 718] In fact, several 4-nitro-3-idinols were first reported to be 6-nitro-3-pyridinols[719] but were later rectly described.[718] It has been suggested that in these series, hydrogen-ded 6-membered transition states involving the 3-hydroxy group are rgetically favored and are responsible for the preferred reactivity of the osition over the 6-position when the 2-position is occupied.[714] The weakening absence of this hydrogen-bonding should influence orientation. Thus, in the ation of 2-methyl-3-pyridinol (**XII-672**, R = $CH_3$) at higher temperatures °), the 6-nitro product predominates, and the nitration of 3-methoxy-2-hylpyridine **(XII-673)** gives only the 6-nitro derivative.[714] 2-Nitro-3-idinol **(XII-674)**, which is a hydrogen-bonded substrate, is also nitrated erentially at the 6-position to give 2,6-dinitro-3-pyridinol **(XII-675)**.[714] How-, the rate profiles for the nitration of 3-pyridinol and 3-methoxypyridine to 2-nitro derivatives are similar.[701]

2,6-Dialkyl-3-pyridinols are nitrated in good yield at the 4-position using ni acid and sulfuric acid.[719] Nitration of 2,6-dimethyl-3-pyridinol with sod nitrate in oleum gives 2,6-dimethyl-4-nitro-3-pyridinol in only 10% yield.[295]

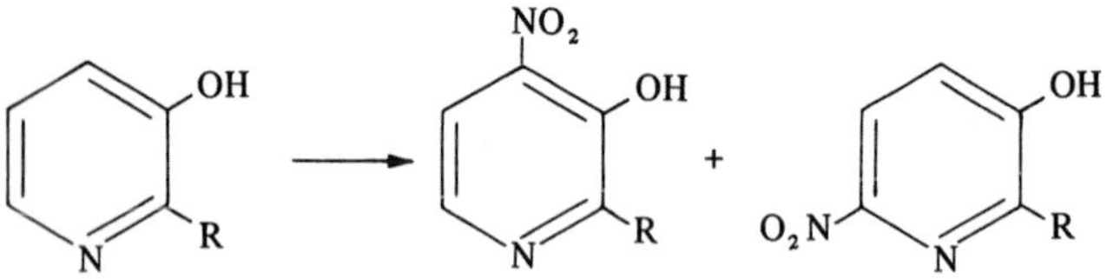

**XII-672**

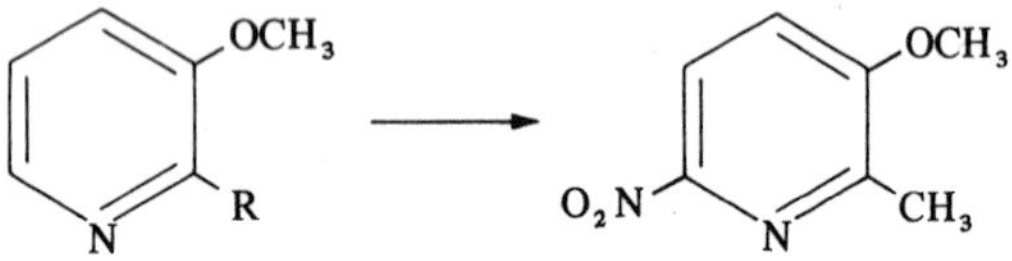

**XII-673**

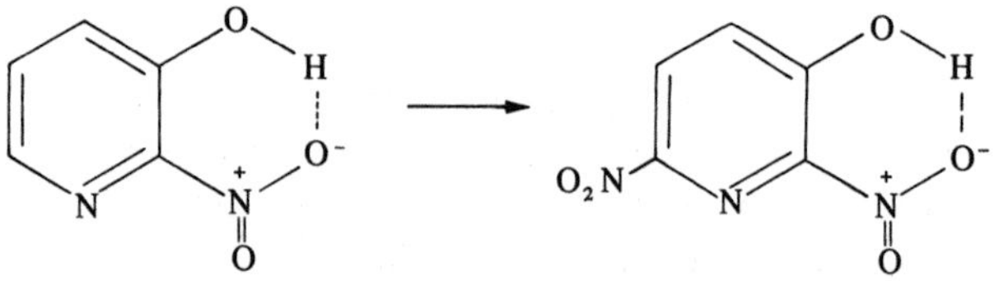

**XII-674** **XII-675**

With concentrated sulfuric acid and nitric acid (*d* 1.34) 2-phenyl-3-pyridi (**XII-676**, R = H) is nitrated first at the phenyl group (88% in the *para* positi rather than at the cationic pyridinol ring, to give mainly 2-(*p*-nitrophenyl pyridinol, which reacts further at the 6-position of the pyridine ring to g 6-nitro-2-(*p*-nitrophenyl)-3-pyridinol.[710] When the *para* position of the phe ring contains *ortho-para-* directing substituents, nitration occurs at the posit *ortho* to the substituent to give 2-(3-nitro-4-substituted phenyl)-3-pyridi (**XII-677**; R = $CH_3$, $C_2H_5$, $CH(CH_3)_2$, $C(CH_3)_3$, $OCH_3$, Cl). Further nitratio **XII-677** (R = $CH_3$) again occurs at the 6-position of the pyridinol ring. Iodination of **XII-676** in the presence of sodium carbonate occurs only at 6-position of the pyridinol ring.[709]

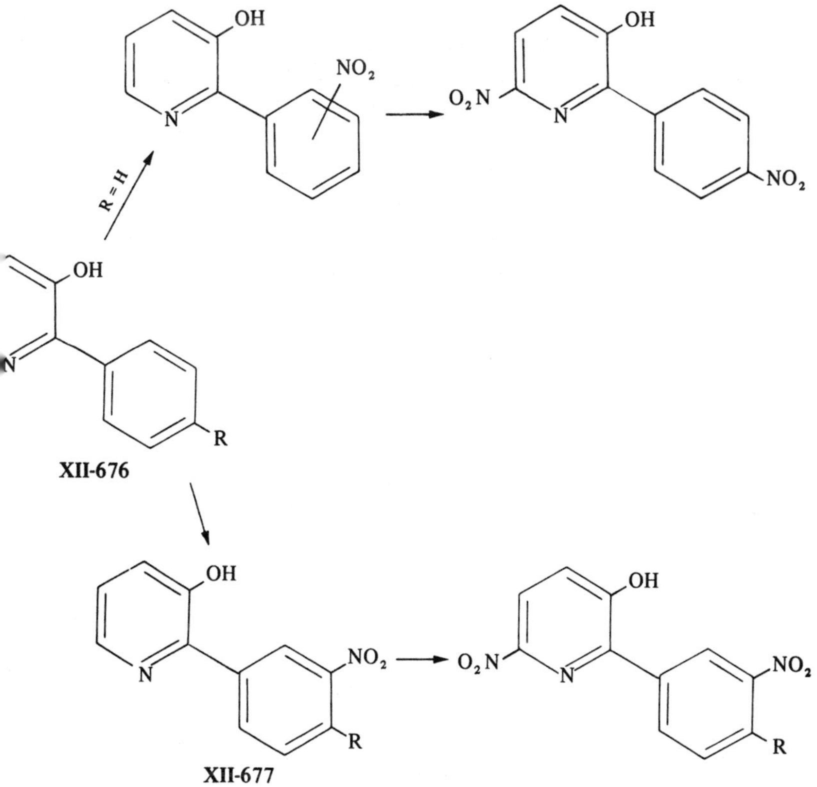

OH NO$_2$ O$_2$N N OH N NO$_2$

R = H

OH N R

**XII-676**

OH N NO$_2$ R

**XII-677**

O$_2$N N OH NO$_2$ R

## E. *The Mannich Reaction*

lthough 3-pyridinol is not sufficiently reactive to undergo chloromethylation Friedel Crafts alkylation or arylation it is readily aminomethylated. The of reaction appears to be in the order of 2- > 6- > 4-positions in absence of steric influences. 3-Pyridinol reacts with secondary amines and naldehyde at the 2- and then the 6-positions and 6-methyl-3-pyridinol reacts he 2-position. However, reaction at the 4-position is not observed, even under stic conditions[720] (unless the α-positions are blocked—see below). Although ethylamine and piperidine react with formaldehyde and 3-pyridinol to give h the 2-mono- and 2,6-bis-products, diethylamine and 3-pyridinol give only ethylaminomethyl-3-pyridinol, a behavior that is similar to that observed in roxymethylation.[720]

Methyl-3-pyridinol reacts with secondary amines and formaldehyde to form 6-dialkylaminomethyl- **(XII-678)** and 4,6-bis(dialkylaminomethyl)-2-methyl-

3-pyridinol (**XII-679**). Hydroxymethyl-3-pyridinols are formed by conver the Mannich products (**XII-678** and **XII-679**, R = $CH_3$) to the acetate follo by hydrolysis. Hydrogenation of the Mannich products gives the expe 2,6-dimethyl- and 2,4,6-trimethyl-3-pyridinol (**XII-680**), respectively.[721] Hi 2-alkyl-3-pyridinols also react stepwise with formaldehyde and piperidin dimethylamine to form 4,6-bisdialkylamino products. When the group at 2-position is bulky (R = isobutyl, isoamyl), the formation of the bis pro requires more severe conditions because of steric influence of the alkyl grou the activating effects of the 3-hydroxyl group.[722]

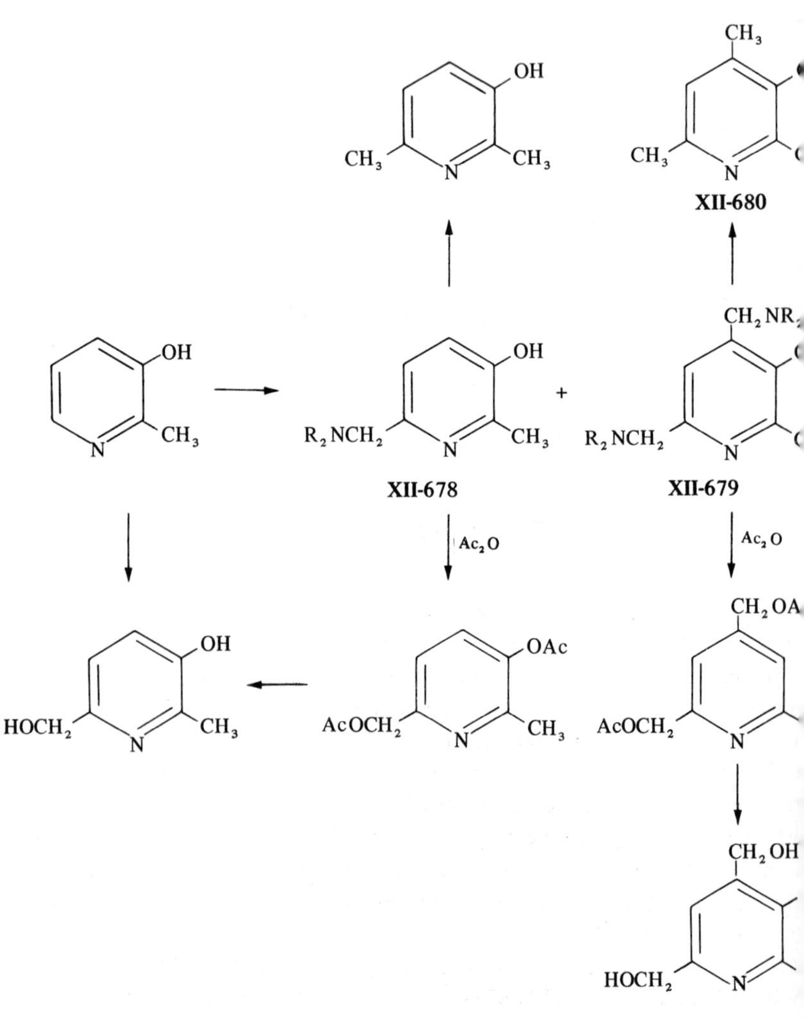

OH CH$_3$ N CH$_3$ CH$_3$ N OH CH$_3$ CH$_3$ **XII-680**

OH N CH$_3$ → $R_2NCH_2$ N OH CH$_3$ **XII-678** + $R_2NCH_2$ N $CH_2NR_2$ **XII-679**

$Ac_2O$ $Ac_2O$

$HOCH_2$ N OH CH$_3$ ← $AcOCH_2$ N OAc CH$_3$ $AcOCH_2$ N $CH_2OA$

$HOCH_2$ N $CH_2OH$

-Ethyl-6-methyl- and 2,6-dimethyl-3-pyridinol react with piperidine, di-hylamine, and morpholine in aqueous formaldehyde to form 4-dialkylamino-hylated products that have been used to prepare vitamin $B_6$ analogs **XII-681** **XII-682**.[723] Reaction at the side-chain is not observed.[724]

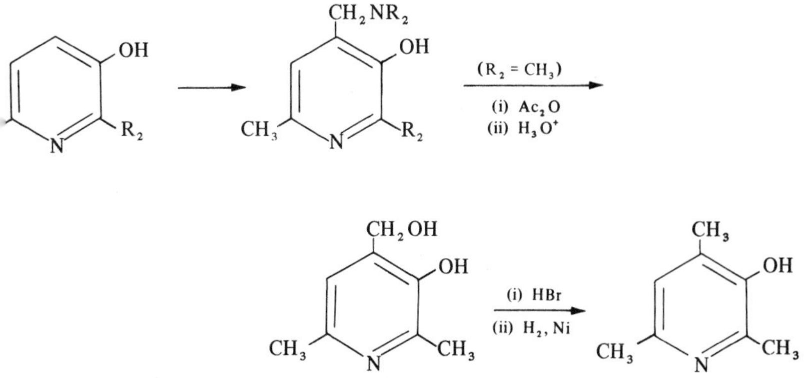

$(R_2 = CH_3)$; (i) $Ac_2O$; (ii) $H_3O^+$

(i) HBr; (ii) $H_2$, Ni

**XII-681** **XII-682**

Nitro-3-pyridinol forms 6-dialkylaminomethyl-2-nitro-3-pyridinols that do react further even under forcing conditions.[725] A compound previously -ribed as 2-methyl-6-nitro-3-pyridinol[726] and later identified as 2-methyl-4-o-3-pyridinol (**XII-683**)[726a] reacts with formalin and dialkylamines to give 1 6-dialkylaminomethyl- and 2-(β-dialkylaminoethyl)-3-pyridinols.

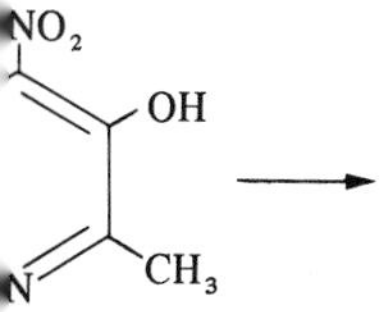

**XII-683**

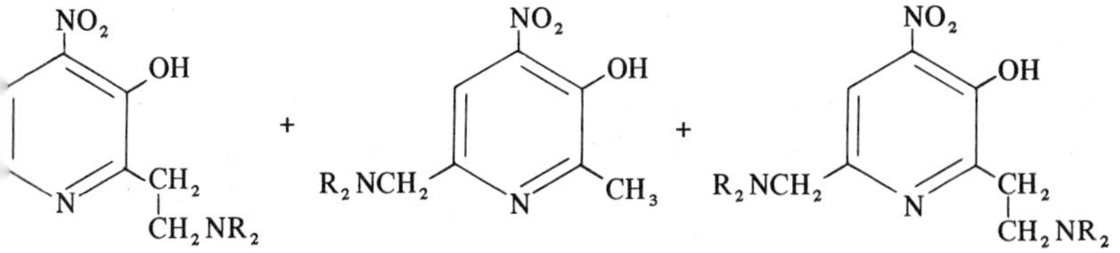

Pyridone and 3-methoxy-2-pyridone (**XII-684**; $R_3$ = H, $CH_3O$) give alkylaminomethylpyridones, but 3-hydroxy-2-pyridone (**XII-684**, $R_3$ = OH) as the 6- and 4,6- Mannich products.[501] The 6-morpholino- Mannich product

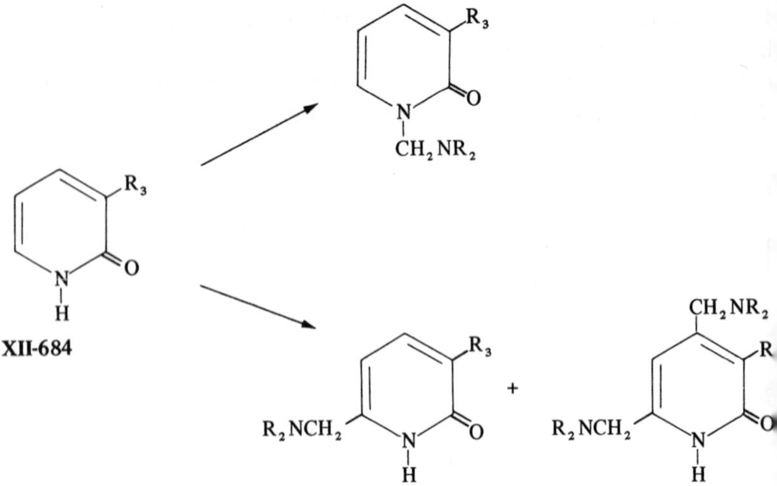

XII-684

reacts with cyclohexanone pyrrolidine enamine to give the dihydroxyhexahy-benzo[b]indolizinone **XII-685**.[501]

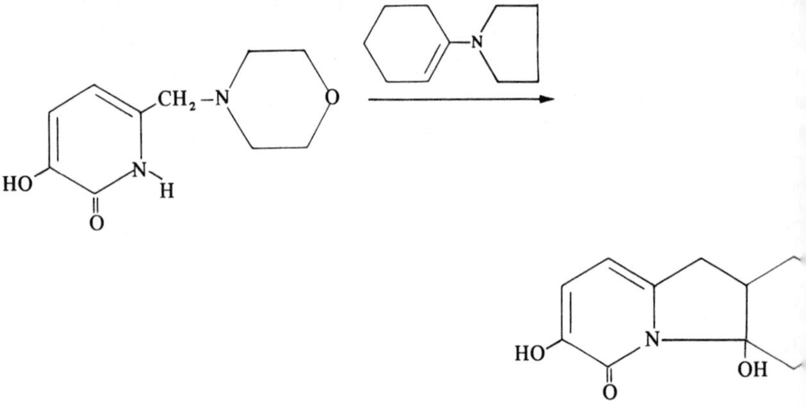

**XII-685**

Aminomethylation of 2-(*p*-anisyl)-3-pyridinol gives 6-mono- and 4,6-bis-methylaminomethyl)-2-(*p*-anisyl)-3-pyridinol. Aminomethylation at the ar group is not observed. Aminomethylation of 2-(*p*-hydroxyphenyl)-3-pyrid occurs at both the 3- and 5-position of the *p*-hydroxyphenyl group, followe aminomethylation at the 6- and then the 4-position of the pyridinol ring.[727]

Proton chemical shifts and LCAO calculated charge distributions have b correlated with reactivity and substitution sequences in the Mannich reaction 3-pyridinols.[520, 728]

## F. *Hydroxymethylations*

1-Methyl-2-pyridone, formaldehyde, and hydrochloric acid give 5-hydroxy-ethyl-1-methyl-2-pyridone. Addition of acetic anhydride and sodium acetate the reaction mixture appears to be necessary to obtain a good yield. A similar action using 5-ethyl-1-methyl-2-pyridone gives 5-ethyl-3-hydroxymethyl-1-ethyl-2-pyridone.[729] 2-Methyl-3-pyridinol undergoes only 6-hydroxymethyla-n when treated with formaldehyde in aqueous base;[721] with formaldehyde d secondary amines, however, 4,6-bis products can be isolated (see Section .2.E.).

## G. *Hydrogen-Deuterium Exchange*

It has been suggested that deuteration of *N*-methyl-2-pyridone occurs at the ethyl group.[730] Base-catalyzed hydrogen-deuterium exchange, however, takes ace at the 2- and 6-positions of 4-pyridones (**XII-686**, R = H, $CH_3$, Br) and nethoxy-1-methylpyridinium tetrafluoroborate (**XII-687**) and at the 6-posi-n of *N*-methyl-2-pyridone in deuterium oxide *via* an ylide mecha-sm.[139, 731, 732] The rates of exchange relative to *N*-methyl-4-pyridone are: methyl-2-pyridone ($D_2O$ at 100°), $10^{-0.8}$; 4-methoxy-1-methylpyridinium trafluoroborate (**XII-687**) ($CH_3OD$ at 40°), $10^{5.3}$.[732] The high rate of corporation into **XII-687** is due to the positive charge that lowers the tivation energy for hydrogen-abstraction.[732] *N*-methyl-2,6-dimethyl-4-

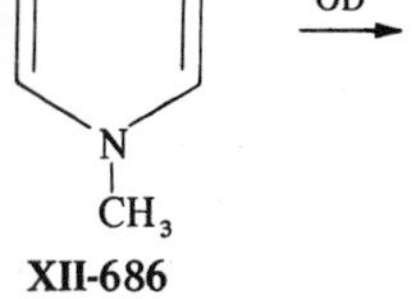

**XII-686**

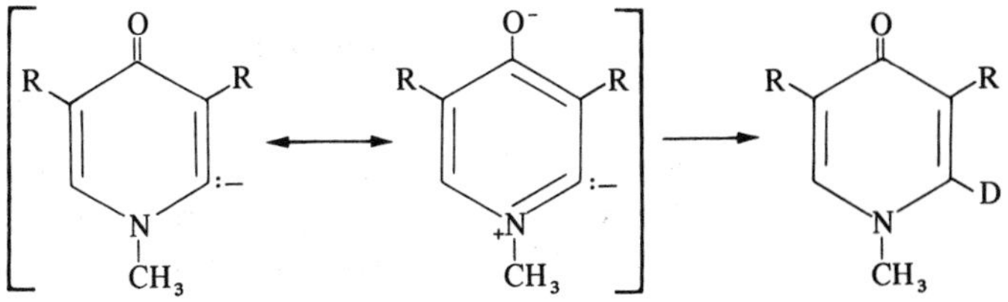

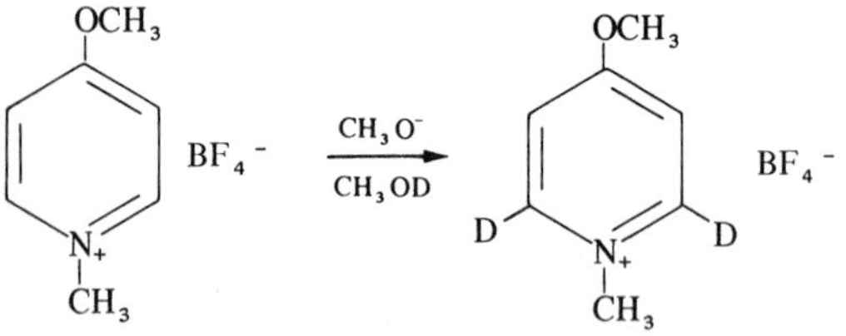

$OCH_3$ $BF_4^-$ $\xrightarrow[CH_3OD]{CH_3O^-}$ D D $BF_4^-$ $N^+$ $CH_3$

**XII-687**

pyridone **(XII-688)** undergoes exchange at the α-side-chains.[139]

Acid-catalyzed hydrogen-deuterium exchange occurs on the 3- and 5-positic of pyridones.[137, 537, 733-736] Over a wide range of acidities (pH 4 - $H_o$ −10), t

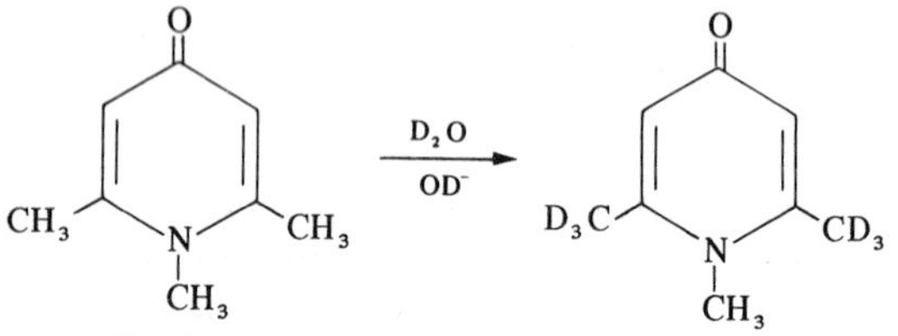

O $CH_3$ N $CH_3$ $CH_3$ $\xrightarrow[OD^-]{D_2O}$ O $D_3C$ N $CD_3$ $CH_3$

**XII-688**

rates of deuterium exchange vary by relatively small increments for 4-pyrido 1-methyl-4-pyridone, and 3- and 5-methyl-2-pyridone, consistent with t suggestion that these compounds react as the free bases.[734] Methyl substituti at the 2- and 6-positions activates the much less reactive conjugate aci 2,6-Dimethyl-4-pyridone exchanges as the free base at lower acidities but as t conjugate acid at acidities above $H_0$ −3.5. A similar changeover in mechani occurs at $H_0$ −2.7 for 1,2,6-trimethyl-4-pyridone.[137, 733] A comparison of r data from 1,2,6-trimethyl-4-pyridone, 4-methoxy-2,6-dimethylpyridine a 2,6-dimethyl-4-pyridone, indicates that exchange occurs on the pyridone fo of the free base.[137] The deactivating effect of the positively charged nitrogen the conjugate acids has been compared with effects of heterocyclic analo containing $=\overset{+}{N}OH-$, $=\overset{+}{N}(O^-)-$, $=\overset{+}{O}-$ and $=\overset{+}{S}-$.[137] 4-Methoxypyridine does n exchange under these conditions.[734, 735] The 3- and 5-positions of 2-pyrido appear to be deuterated at approximately the same rate in 23% deuterosulfu acid.[736]

## H. *Acetylation*

5-Acetyl-4-hydroxy-6-methyl-2-pyridone and phosphoryl chloride in ace acid give 3,5-diacetyl-4-hydroxy-6-methyl-2-pyridone **(XII-689)**.[157] 4-Hydro 6-methyl-2-pyridone **(XII-690**, R = H), 1,6-dimethyl-4-hydroxy-2-pyrido

II-690, R = $CH_3$), and 4-hydroxy-6-methyl-1-phenyl-2-pyridone (**XII-690**, R = $H_5$) and acetic acid in phosphoryl chloride give the 4-acetoxy products.[644, 645] In polyphosphoric acid (PPA), 3-acylation can occur to give I-691 (R = R′ = $CH_3$; R = H, R′ = $CH_3$, $C_2H_5$, $n$-$C_3H_7$, $n$-$C_4H_9$, iso$C_4H_9$, $C_5H_{11}$).[644, 645, 737] Structure assignments have been based on infrared

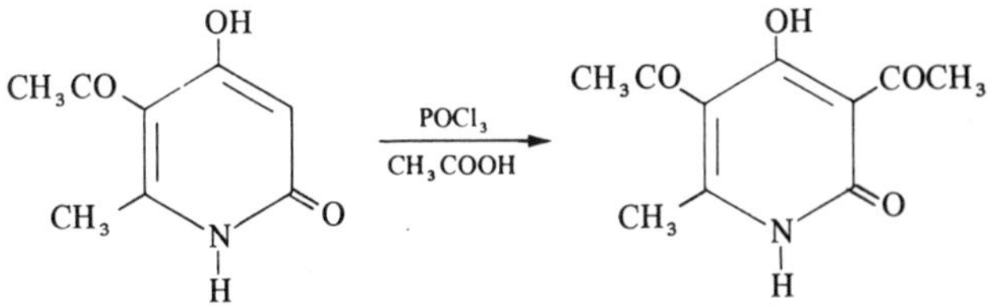

**XII-689**

ctroscopy and on the synthesis of 3,6-dimethyl-1-phenylpyrazolo[4,3-*c*]pyri-4-one (**XII-692**) (R.I. 1186), which was also prepared from the known 6-dimethyl-1-phenylpyrano[4,3-*c*]pyrazol-4-one (**XII-693**).

Both 4-acetoxy-6-methyl-2-pyridone (**XII-694**, R = H) and 4-acetoxy-6-thyl-1-phenyl-2-pyridone (**XII-694**, R = $C_6H_5$) undergo the Fries rearrange-

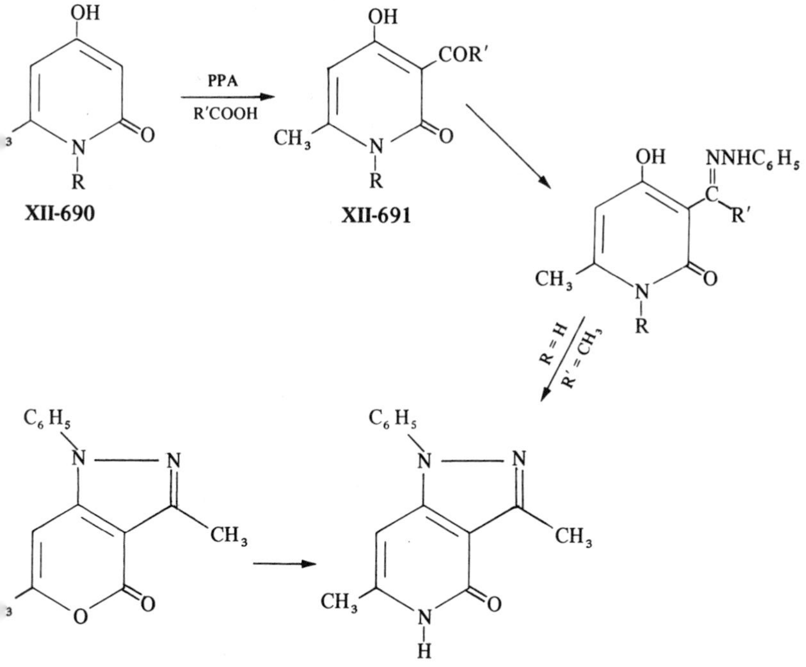

**XII-690** **XII-691**

**XII-693** **XII-692**

ment in the presence of aluminum chloride to give the correspond 3-acetyl-4-hydroxy-2-pyridone.[644, 645]

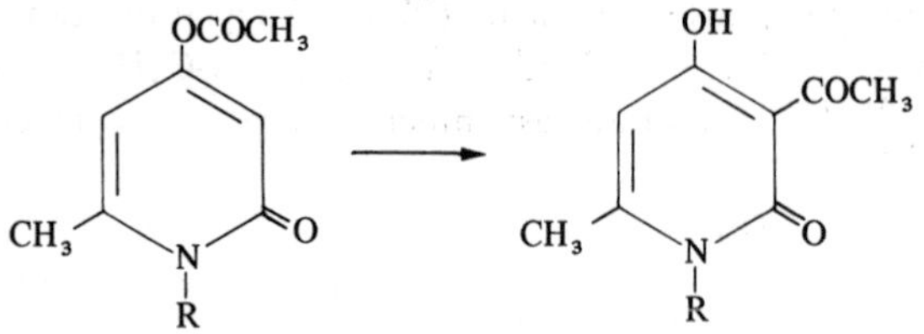

XII-694

## I. *Coupling with Diazonium Salts*

Pyridinols, pyridones, and alkoxypyridines that are sufficiently activated be coupled with diazonium salts. 2-Alkoxy-3,5-diaminopyridine gives 2-alko 3,5-diamino-6-phenylazopyridine **(XII-695)** with benzenediazonium chloride.

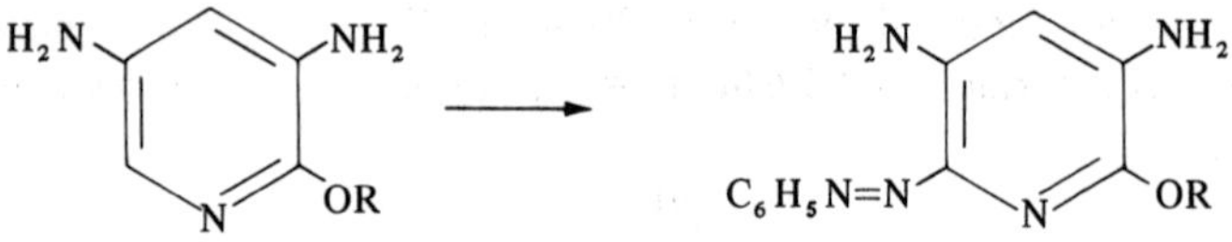

**XII-695**
($R = H, CH_3, C_2H_5, C_3H_7, C_4H_9, C_5H_{11}$)

2-Hydroxy-4-(*p*-methoxyphenyl)-1-phenyl-6-pyridone couples with diazon salts to give colored products **XII-696**.[739]

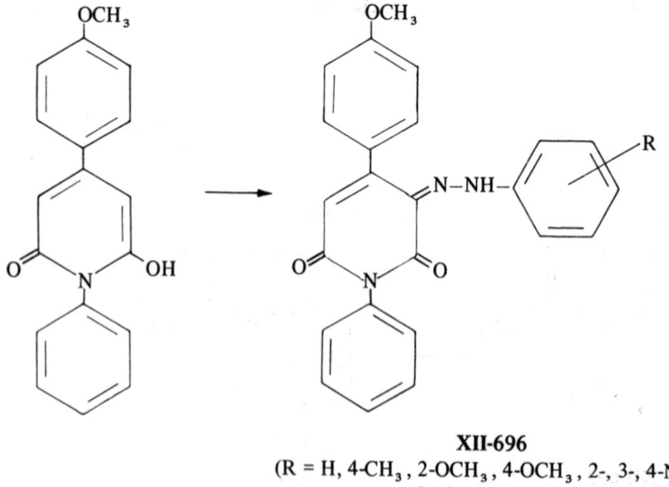

**XII-696**
($R = H$, 4-$CH_3$, 2-$OCH_3$, 4-$OCH_3$, 2-, 3-, 4-$NO_2$, 2-, 3-, 4-Cl, 4-Br)

-Pyridinol and *p*-nitrobenzenediazonium chloride form 6-(*p*-nitrophenylazo)-yridinol as the major product. A low yield of the 2-isomer can be increased to % in a neutral or slightly acidic medium.[292] Glutazine (**XII-697**, R = H) and arbethoxyglutazine (**XII-697**, R = $CO_2C_2H_5$) react with *p*-chlorobenzene-zonium chloride to give 3-*p*-chlorophenylazo derivatives.[34] 6-Hydroxy-2-

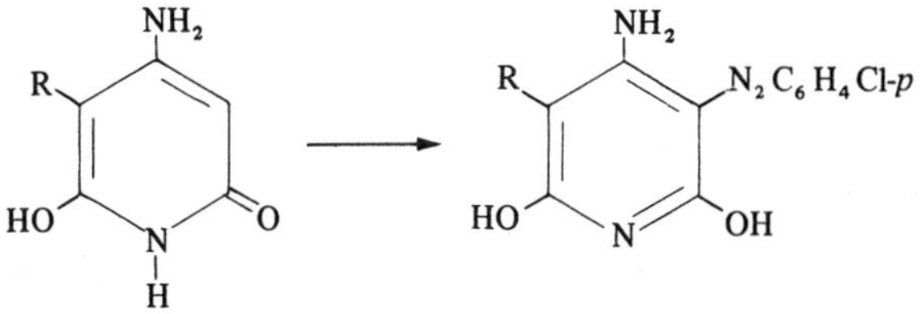

**XII-697**

idone and benzenediazonium chloride give both 3-phenylazo- and 3,5-bis-enylazo)-6-hydroxy-2-pyridone.[740]

ı contrast to nitration[295, 719] and aminomethylation,[724] 2,6-dimethyl-3-idinol does not couple with benzenediazonium or *p*-nitrobenzenediazonium oride in weakly acid or alkaline medium. Only 6-substitution occurs when ıethyl-3-pyridinol is coupled with diazonium salts.[741]

## J. *Nitrosation*

ıoth 4-amino-2-pyridone (**XII-698**, $R_1 = R_6 = H$)[305] and 4-amino-1,6-ıethyl-2-pyridone (**XII-698**, $R_1 = R_6 = CH_3$)[306] react with nitrous acid to e a 4-amino-3-nitroso-2-pyridone rather than a diazonium salt.

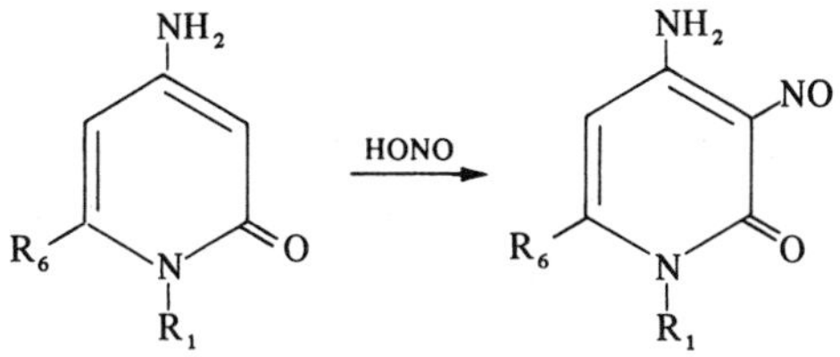

**XII-698**

## K. *Nucleophilic Substitution*

here are a few instances in which direct amination or arylation of pyridones been reported.[742] 2-Methyl-4-pyridone and sodamide give 6-amino-2-methyl-yridone in 79% yield.[743] 4-Ethoxypyridine and *t*-butyllithium give 2-*t*-butyl-

and 2,6-di-*t*-butyl-4-ethoxypyridine. Hydrolysis of the ether to 2,6-di-*t*-butyl pyridone **(XII-700)** is accomplished with hot hydrochloric acid. Simi treatment of the 3-sulfonic acid derivatives **XII-699** with hydrochloric acid lea to displacement of the sulfonic acid group to give **XII-700**; on the other har **XII-699** and aluminum chloride in tetrachloroethane give the 4-hydrox pyridinesulfonic acid (75%).[671] 3-Methoxypyridine and phenyllithium react

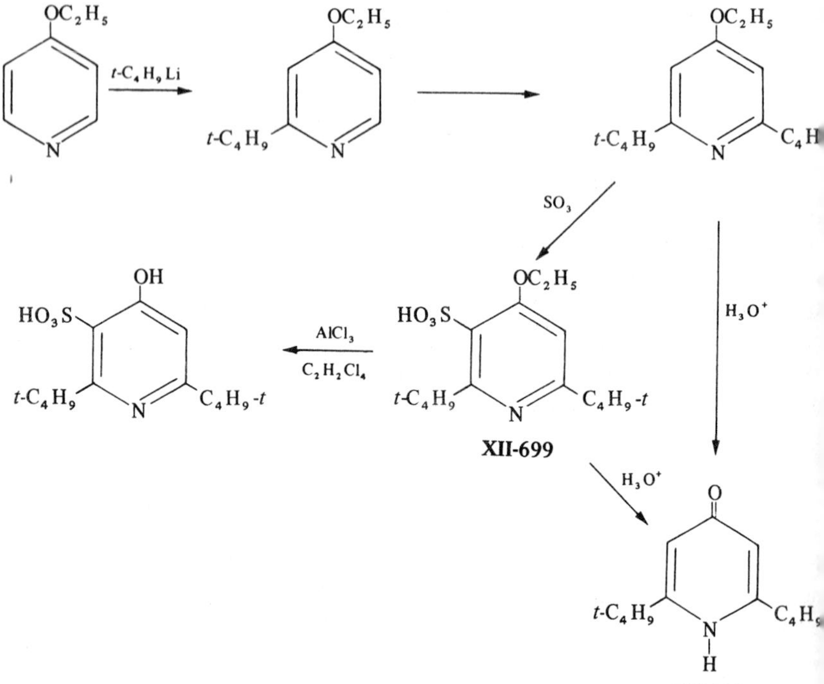

give 3-methoxy-2-phenylpyridine exclusively. It has been suggested that reaction path involving coordination of the oxygen electrons to the lithium operative.[744]

## L. *Diels-Alder Dienophiles*

The structures of 2-pyridone and its *N*-alkyl derivatives suggest that they cou perform as dienes in Diels-Alder reactions. When maleic anhydride a *N*-methyl-2-pyridone are heated alone or in toluene at 100° a 2:1 adduct formed (24%) that dissociates into the pyridone and maleic anhydride in po solvents or when heated to its melting point or when sublimed. The structu

-701 has been proposed for this adduct, and its formation has been onalized according to the following scheme:[730]

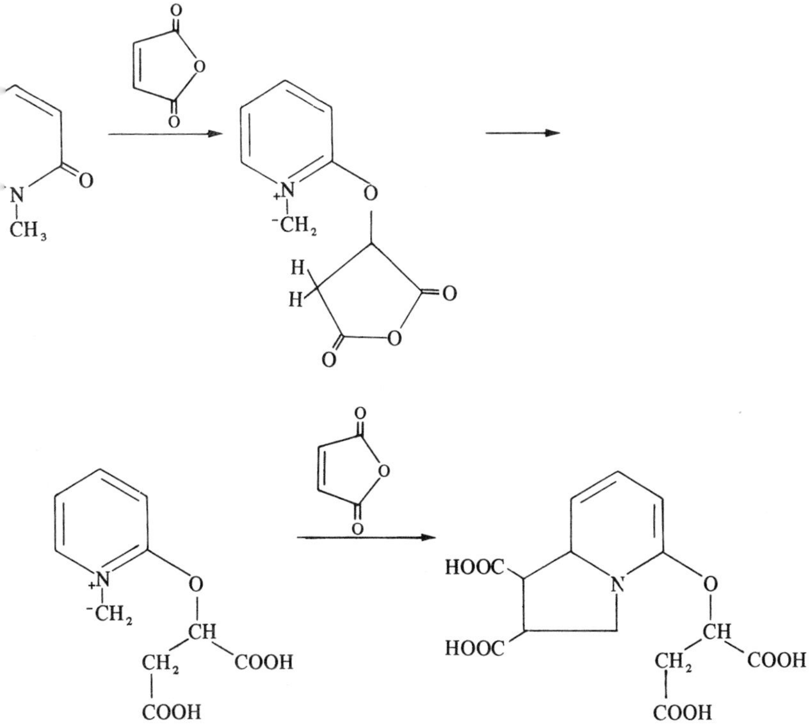

N
CH₃

XII-701

Iethyl-2-pyridone and tetracyanoethylene in diisopropyl ether give a 2:1 tacyanopropene salt **(XII-702)**.[745]

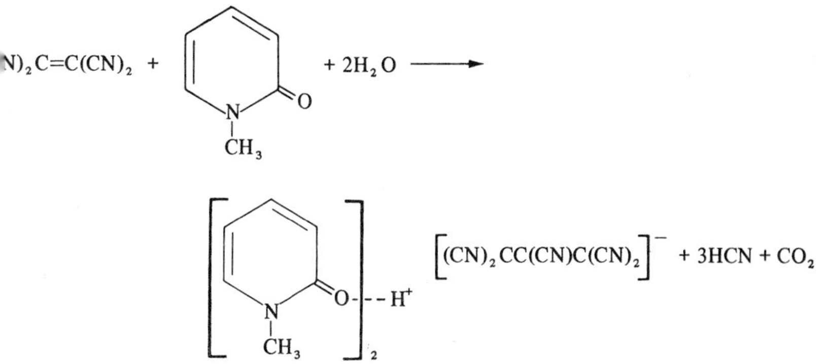

$N)_2C{=}C(CN)_2$ + [N-methyl-2-pyridone] + $2H_2O$ ⟶

$[\text{N-methyl-2-pyridone}\;O\text{---}]_2H^+$ $[(CN)_2CC(CN)C(CN)_2]^-$ + 3HCN + $CO_2$

XII-702

When heated with perfluoro-2-butyne at 175° for 12 hours, 2-pyridone gi 1-[3,3,3-trifluoro-1-(trifluoromethyl)propenyl]-2-pyridone **(XII-703)**. *N*-Meth 2-pyridone, a compound that cannot undergo this addition, gives no charact izable products with perfluoro-2-butyne.[746]

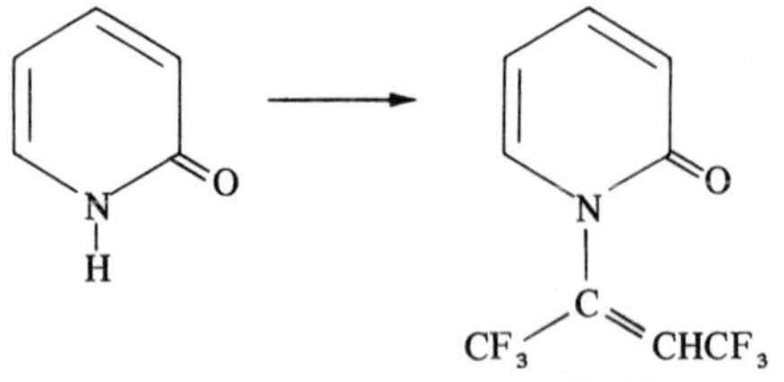

$CF_3$ $CHCF_3$

**XII-703**

2-Pyridone and ethoxyacetylene in chloroform-hexane were heated for days to give a mixture of 2-acetoxypyridine **(XII-705)** (67%) and *N*-(1-etho vinyl)-2-pyridone **(XII-704)** (33%). Similarly, 4-pyridone and ethoxyacetyle gave 4-acetoxypyridine (41%) and *N*-(1-ethoxyvinyl)-4-pyridone (17%) afte days. It has been suggested that **XII-705** is formed by rearrangement of **XII-7** *via* *N*-acetyl-2-pyridone, or through a ketene ketal intermediate by hydr ysis.[747]

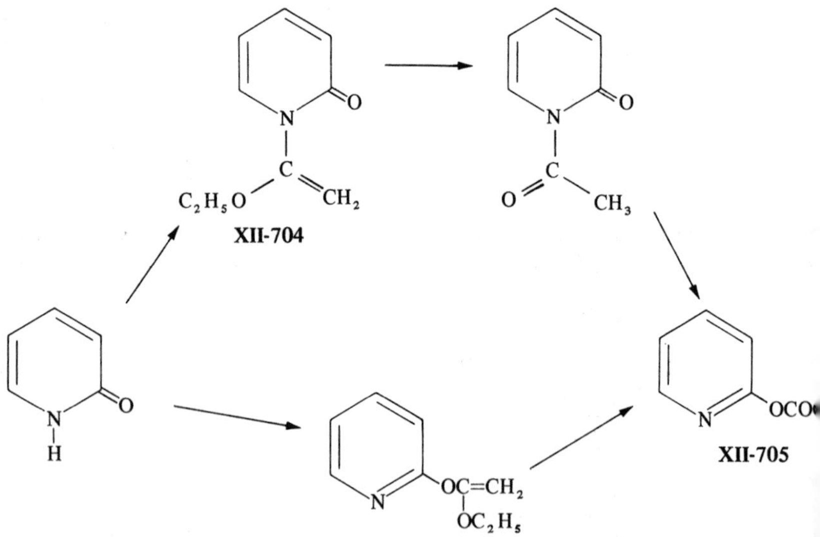

$C_2H_5O$ $CH_2$

**XII-704**

O $CH_3$

OCO

**XII-705**

$OC{=}CH_2$

$OC_2H_5$

-Pyridone, 6-methyl-2-pyridone,[748] 5,6-tetramethylene-2-pyridone, and -dimethyl-2-pyridone[749] give the products of 1,2-addition to dimethyl acetyl-dicarboxylate. *N*-Methyl-2-pyridone gives *N*-methylphthalimide, which is bably formed by decomposition of the Diels-Alder adduct **XII-706**.[748]

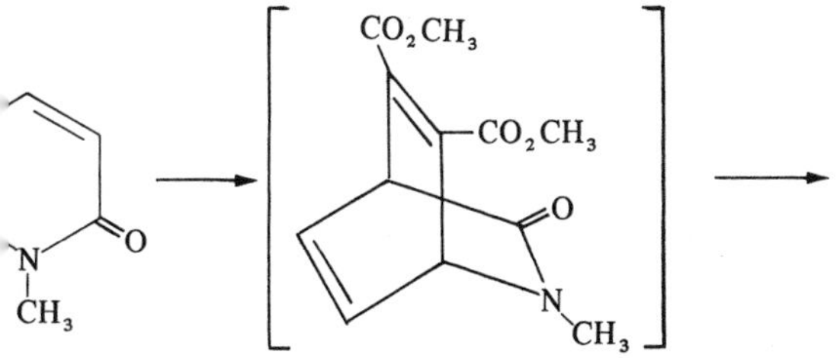

**XII-706**

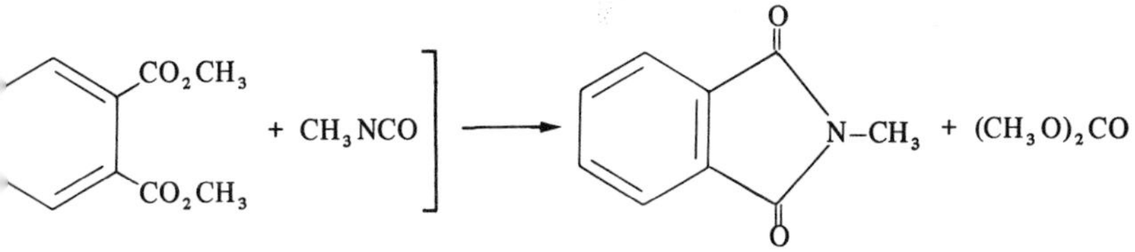

$+ CH_3NCO$ → N–$CH_3$ + $(CH_3O)_2CO$

-Pyridon-1-yl fumarates **(XII-707)**, products of addition of 5,6-dimethyl-2-ridone to dimethylacetylene dicarboxylates are rapidly saponified in cold ieous alkali to the acids **(XII-708)**, presumably by an intramolecular catalysis olving the pyridone oxygen.[749]

Benzyne, from chlorobenzene and sodamide, and *N*-methyl-2-pyridone give henyl-1-methyl-2-pyridone in 5.4% yield, together with a small amount of ′-dimethyl-2,2′-bipyridyl-6,6′-dione **(XII-709)**.[750] 2-Pyridone undergoes *N*-d *O*-arylation in low yield when treated with benzyne prepared from zotized anthranilic acid.[750,751] A small amount of acridone is formed also. wever, *N*-methyl-2-pyridone and the methyl-1-methyl-2-pyridones react with nzyne from anthranilic acid to form the Diels-Alder adducts **(XII-710)**.[750, 751]

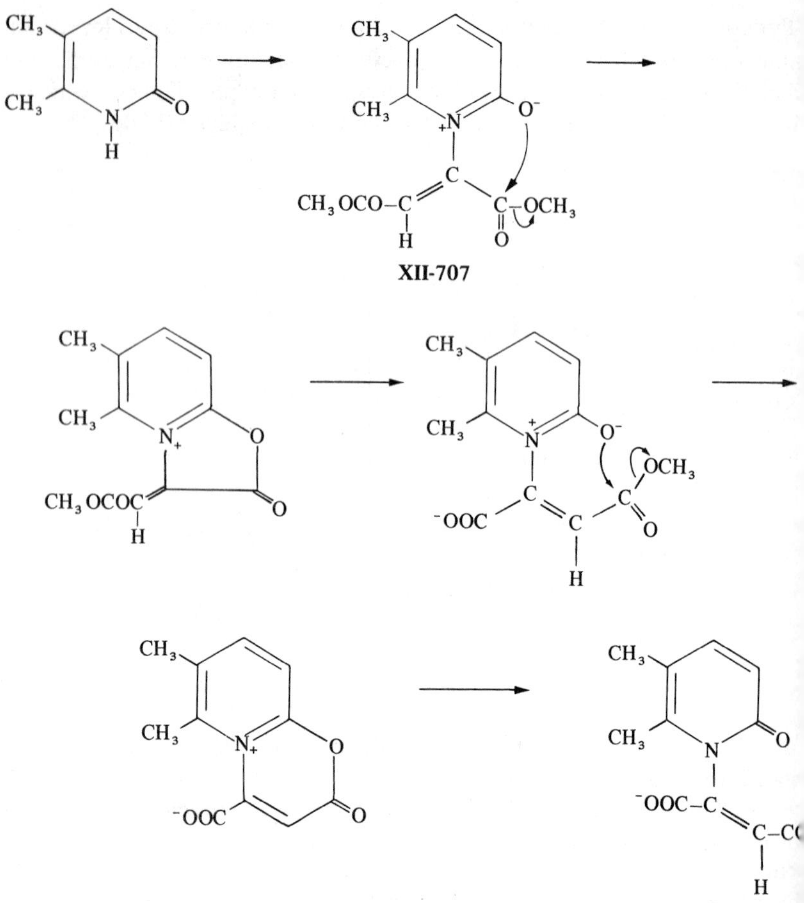

XII-707

XII-708

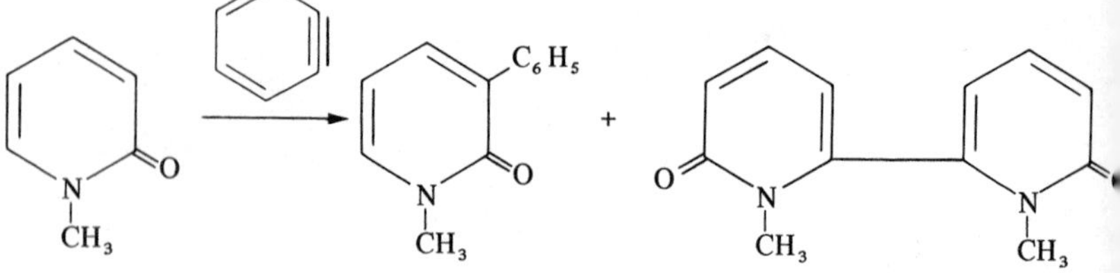

XII-709

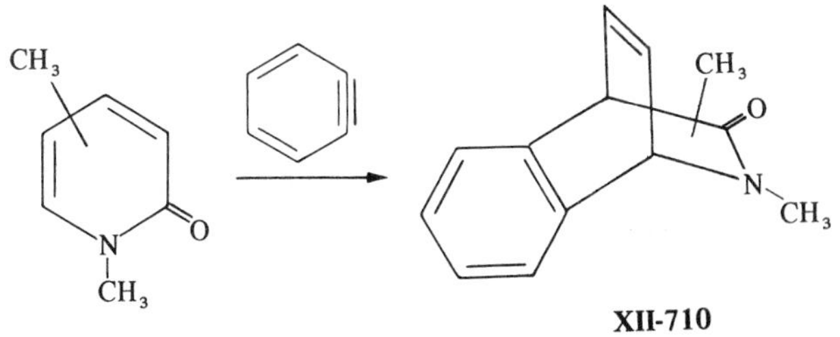

XII-710

ecently a successful Diels-Alder addition of maleic anhydride to 1-methyl-2-idone was accomplished by heating the reactants in toluene for 72 hours, ding the *endo* product **(XII-711)**.[752, 753] 1-Methyl-2-pyridone and fumaroni-

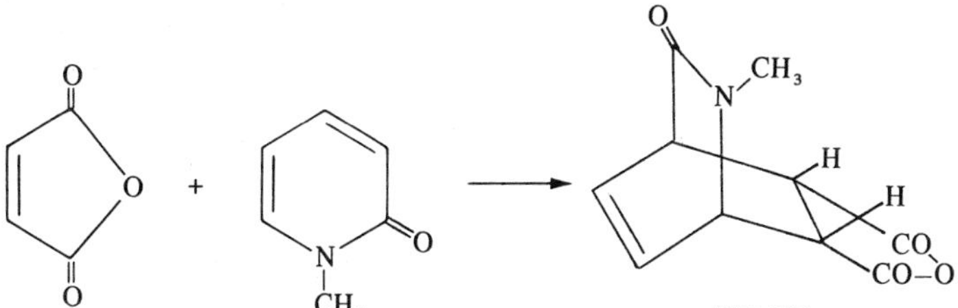

XII-711

, when boiled for 111 hours in toluene formed the adduct **XII-712**, but only % yield.[754]

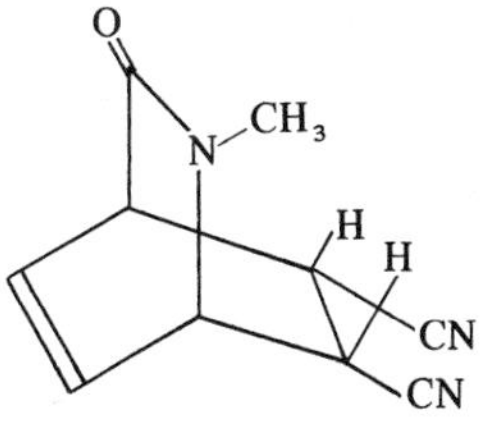

XII-712

ne *endo* dicarboxylic acid can be converted to the dimethyl ester, which nerizes to **XII-713** in good yield on treatment with potassium *t*-butoxide.[754] nhydro-3-hydroxy-1-methylpyridinium hydroxide **(XII-714)** and *N*-phenyl-eimide, acrylonitrile, or methyl acrylate give adducts that are conveniently verted to tropones and tropolones. Acrylonitrile and methyl acrylate and **714** give **XII-715** (R = *exo*-CN; R = $CO_2CH_3$, *endo*:*exo*, 1:1), which, after tment with methyl iodide, gives **XII-716**. This forms 6-dimethylamino-5-oxo-6-cycloheptatriene-1-carbonitrile (**XII-717**, R = CN) and the 1-carbomethoxy vative (**XII-717**, R = $CO_2CH_3$) on reaction with silver oxide.[755] Treatment **XII-716** with sodium bicarbonate gives **XII-717** and **XII-718** (R = CN, $CH_3$).

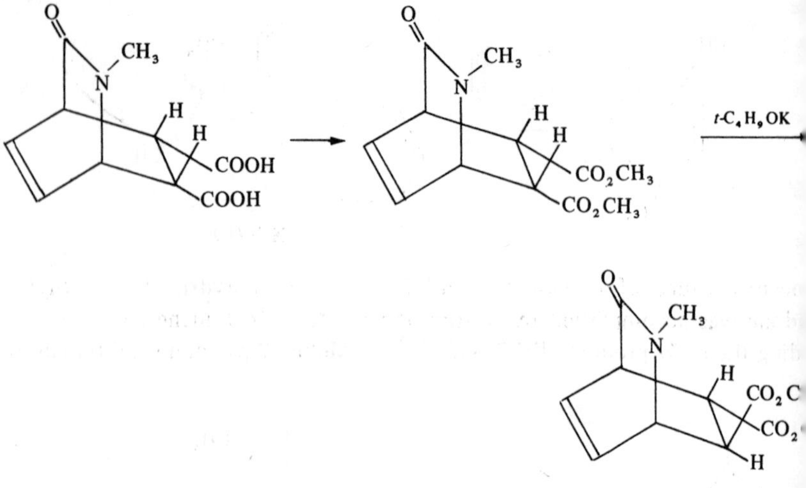

XII-713

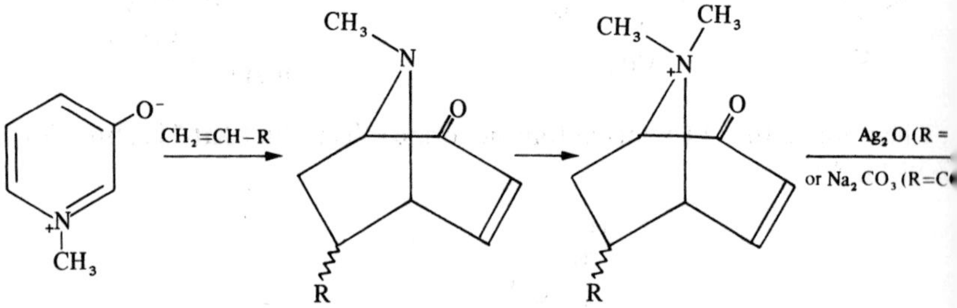

XII-714 XII-715 XII-716

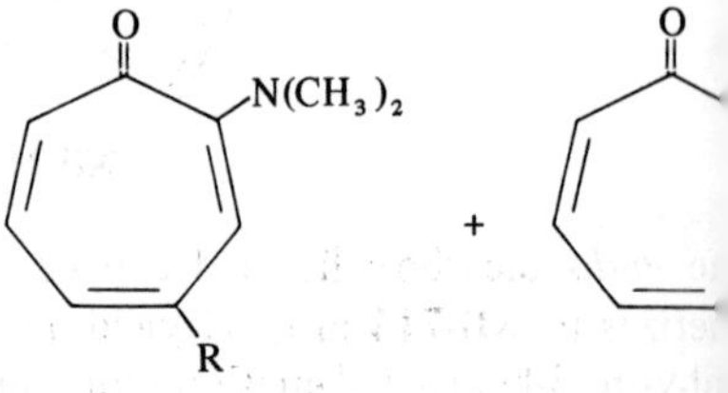

XII-717 XII-718

## M. *Photochemistry*

Ultraviolet irradiation of aqueous solutions of 2-pyridone and *N*-meth
pyridone gives dimers to which the structure **XII-719** (R = H, $CH_3$, respecti

first assigned,[756] but which subsequently have been shown to be s-3,7-diazatricyclo[4.2.2.2$^{2,5}$] dodeca-9,11-dien-4,8-diones **(XII-720)**.[757-760] NMR spectrum of the photodimer **(XII-721)** from 1,4-dimethyl-2-pyridone vs that the methyl group is at the double bond rather than at the gehead,[757] as would be required by the ring system in **XII-719**.

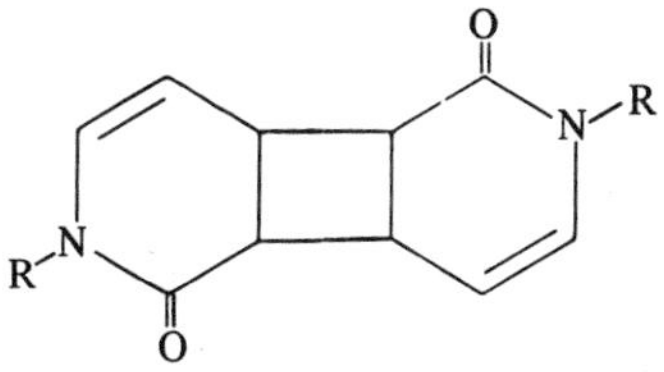

**XII-719**

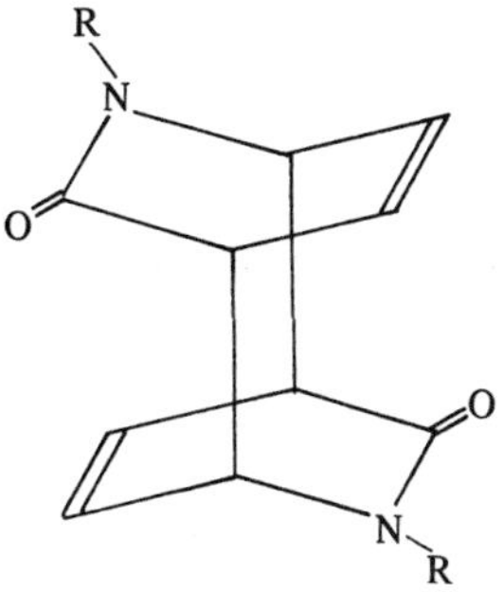

**XII-720**

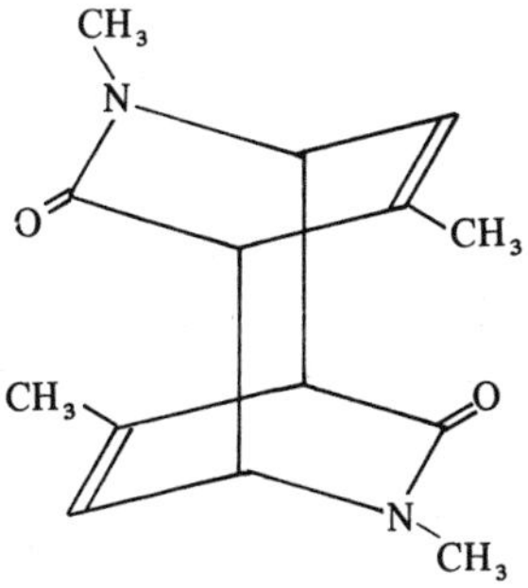

**XII-721**

everal substituted-2-pyridones have been dimerized to **XII-720**.[758, 761, 762] A toinduced diradical intermediate was suggested as the species leading to erization.[762] When irradiated in water *N*-benzyl-2-fluoropyridinium bromide s the same dimer as is formed from *N*-benzyl-2-pyridone in ethanol.[763] toisomerization of *N*-methyl-2-pyridone in ether, followed by careful kup of products, gives **XII-722**.[764]

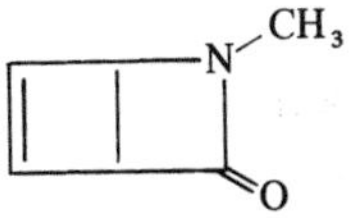

**XII-722**

ltraviolet irradiation of 2,3-dihydro-1,2-diphenyl-6-methyl-4-pyridone **I-723)** in methanol gives 2,5-diphenyl-6-methyl- and 2,3-dihydro-2,6-di- nyl-5-methyl-4-pyridone and 5-anilino-1-phenyl-1,4-hexadiene-3-one.[765]

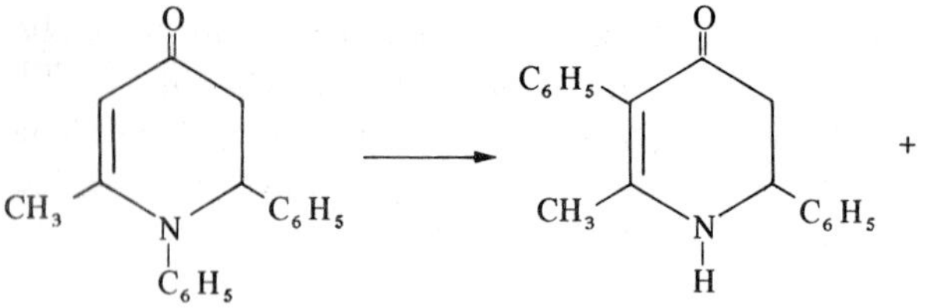

XII-723

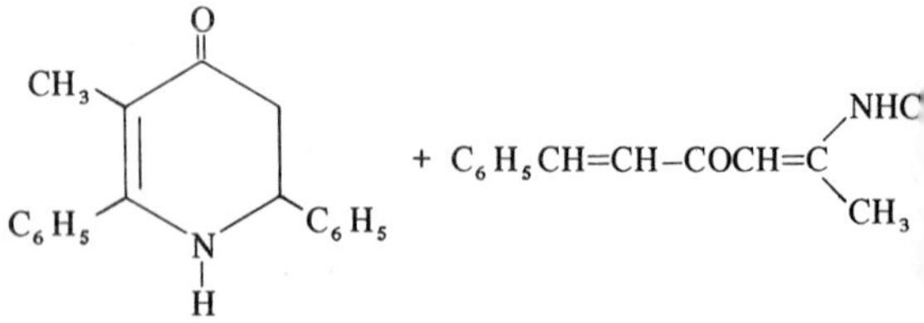

+ $C_6H_5CH{=}CH{-}COCH{=}C(NHC...)(CH_3)$

## N. *Miscellaneous*

2-Pyridone and sulfur dichloride in benzene give 2,2′-dihydroxy-5,5′-dipyri sulfide (**XII-724**). 3-Pyridinol and $SCl_2$ form a poorly defined product 4-pyridone did not react.[766]

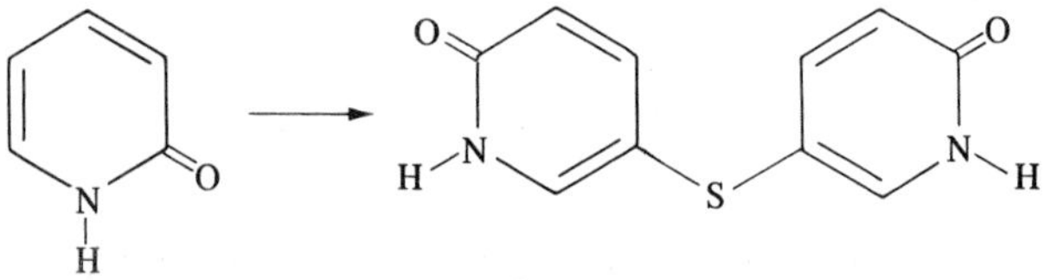

XII-724

4-Methoxy-1-methylpyridinium iodide (**XII-725**) and benzylmagnesium io in ether give 2-benzyl-1,2-dihydro-4-methoxy-1-methylpyridine, an air-sensi red oil, which forms 2-benzyl-2,3-dihydro-1-methyl-4-pyridone when hea with sodium hydroxide in aqueous methanol.[767]

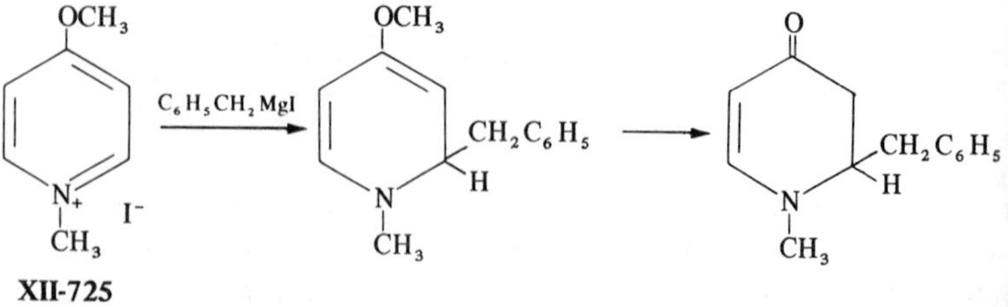

$C_6H_5CH_2MgI$

XII-725

# IV. *O*- and *N*-Substitution Products

## 1. Reactions of Ethers (See also Section I.6.A)

he annular nitrogen of pyridine ethers, like a nitro group, performs not only an activating group in nucleophilic substitutions but also increases the ceptibility to alkyl-oxygen cleavage of the alkoxy group by an $S_N2$ chanism. The relative rates of cleavage at 164.7° of 2-, 3-, and 4-methoxypyri-e **(XII-726)** by sodium methoxide in methanol to give the anion and dimethyl er are 1.0:1.1:2.8.[268] 2,4-Dichloropyridine and sodium methoxide in thanol not only give 2,4-dimethoxypyridine **(XII-727)** as the major product also 2-methoxy-4-pyridone **(XII-728)** and minor amounts of 4-methoxy-2-idone. 2-Chloro-4-methoxypyridine gives **XII-727** and **XII-728**.[768] This

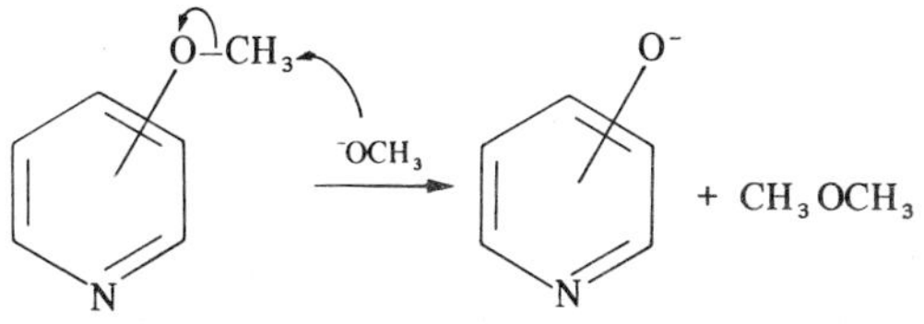

$^{-}OCH_3$ + $CH_3OCH_3$

**XII-726**

avage reaction is observed also as a side reaction during the formation of a isenheimer adduct from 2-methoxy-3,5-dinitropyridine and methanol.[769]

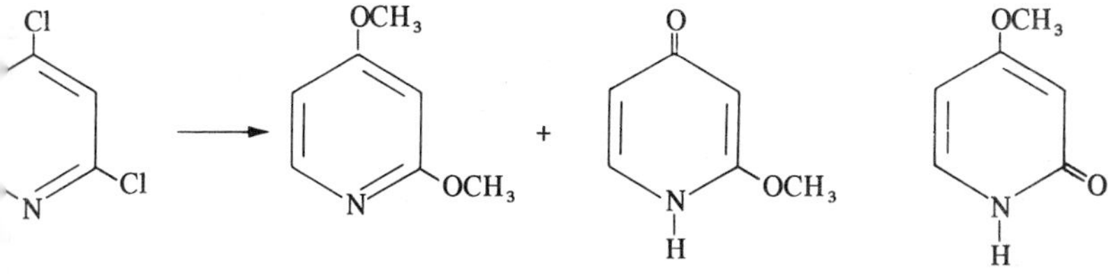

+

**XII-727** **XII-728**

Alkoxy-1-methylpyridinium salts **(XII-729)** have been used to study the avior of carbonium ions in solvolyses. They are easy to prepare and purify, in solvolyses the leaving group does not undergo internal return. The olysis of 4-(cyclopropylcarbinyloxy)-1-methylpyridinium iodide **(XII-729**, $CH_2$—◁) in 80% ethanol gives some cyclopropylcarbinol, cyclobutanol, and lcarbinol but mainly the corresponding ethyl ethers **(XII-730**; R = cyclo-pylcarbinyl, cyclobutyl, allylcarbinyl; R′ = $C_2H_5$) in a ratio of 3.6:1.5:1.0.

Hydrolysis of optically active 4-(cyclopropylmethylcarbinyloxy)-1-met pyridinium iodide (**XII-729**, R = ▷–CH(CH$_3$)–) in the presence of lithium carbo gives cyclopropylmethylcarbinol that is 95.6% racemic.[770]

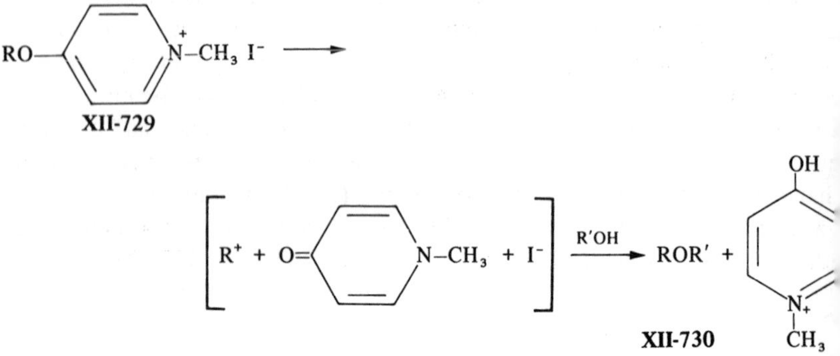

RO–C$_5$H$_4$N$^+$–CH$_3$ I$^-$ ⟶

**XII-729**

[R$^+$ + O=C$_5$H$_4$N–CH$_3$ + I$^-$] $\xrightarrow{R'OH}$ ROR′ + 4-hydroxy-1-methylpyridinium

**XII-730**

A methoxy group *meta* to a reaction site is normally activating in arom nucleophilic substitutions ($\sigma_m$ is positive). However, the methoxy group 2-chloro-4-methoxypyridine retards the reaction with sodium methoxide rela to that with 2-chloropyridine ($k_{MeO}/k_H$ = 0.71 at 15°). It has been prop that a conjugative interaction between the electron releasing methoxyl gr and the ring nitrogen reduces the activation of the 2-halo by nitrogen.[768] Si complexes, formed from sodium methoxide and polynitroanisoles, have b studied extensively to shed light on the nature of nucleophilic arom substitution.[771] These studies have been extended recently to heterocy systems including 3,5-dinitropyridine,[772,773] 4-methoxy-3,5-dinitrop dine,[774–777] and 2-methoxy-3,5-dinitropyridine[769,774,777] whose properties similar to those of the nitroanisoles.

Addition of sodium methoxide to a solution of 3,5-dinitropyridine dimethyl sulfoxide gave the anion **XII-731**, which rapidly rearranged to **XII-** ($t_{0.5}$ 10 min).[773] This was the only anion observed in methanol.

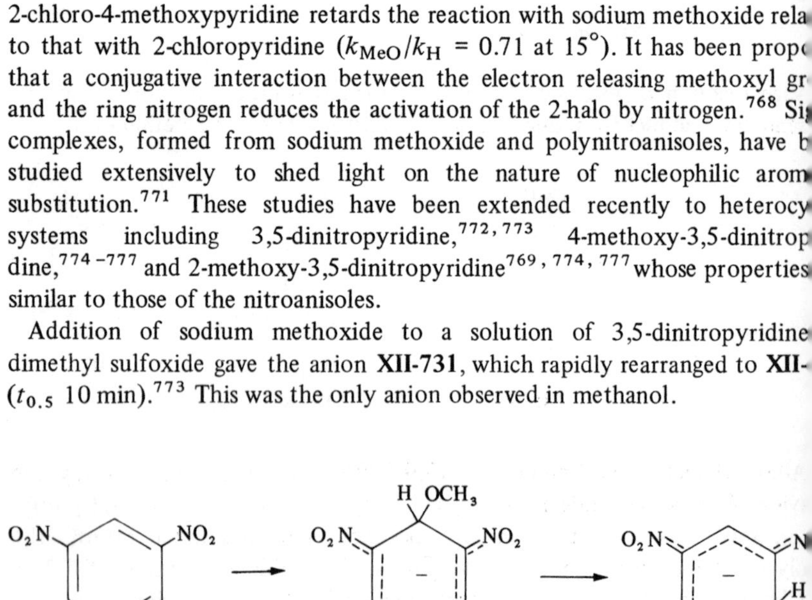

3,5-dinitropyridine ⟶ **XII-731** ⟶ **XII-732**

4-Chloro-3,5-dinitropyridine[775] or 4-methoxy-3,5-dinitropyridine[776] and odium methoxide in methanol give sodium 4-aza-1,1-dimethoxy-2,6-dinitro-yclohexadienate **(XII-733)**. In dimethyl sulfoxide the adduct **XII-734** is ormed,[774] which may be an intermediate in the formation of **XII-733**.[776] The roduct of rearrangement, *N*-methyl-3,5-dinitro-4-pyridone, is formed by treat-nent of a solution containing **XII-733** with acetic acid or by heating -methoxy-3,5-dinitropyridine.[776, 777] The conversion of 2,6-dinitrochloro-

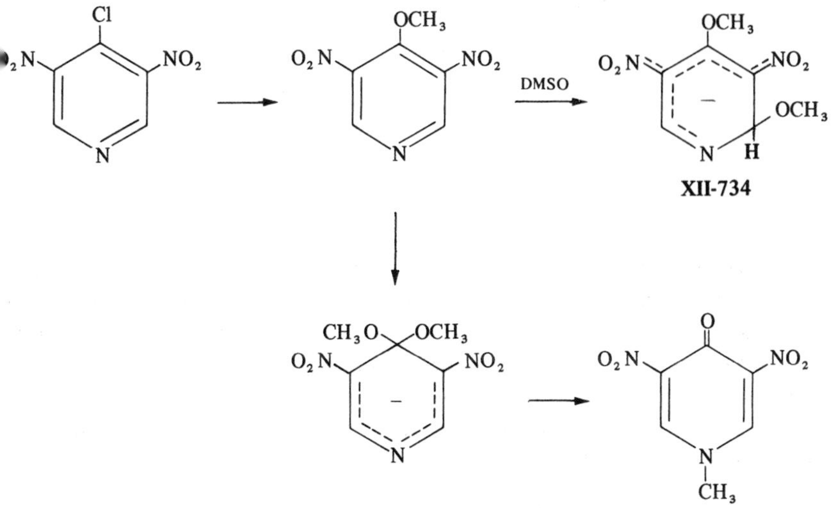

**XII-734**

**XII-733**

enzene to the methyl ether is faster in dimethyl sulfoxide than in methanol, which solvates the methoxide ion. When 3,5-dinitro-2-methoxypyridine is reated with sodium methoxide in methanol, excess solvent causes irreversible ormation of the sodium salt of 3,5-dinitro-2-pyridone by a competitive rather han a consecutive reaction.[769] The only Meisenheimer adduct observed is the -aza-1,3-dimethoxy-4,6-dinitrocyclohexadienate anion **(XII-735)**, which is more onveniently formed in dimethyl sulfoxide.[769, 777] However, the behavior of ,5-dinitro-2-methoxypyridine in dimethyl sulfoxide could not be related irectly to nucleophilic heteroaromatic substitution because of the failure of **XII-735** to rearrange to the 1,1-adduct.[769, 777]

The preferred structure of the sigma complex presumably depends on steric ffects. The methine complex **XII-734** experiences adverse steric interactions etween the methoxyl group bound to the $sp^2$ carbon and the other nitro roups. This compression is enhanced in protic solvents where **XII-733** redominates. In 3,5-dinitro-2-methoxypyridine, the annular nitrogen provides nly a small steric effect and the methine complex **XII-735** is observed.[777]

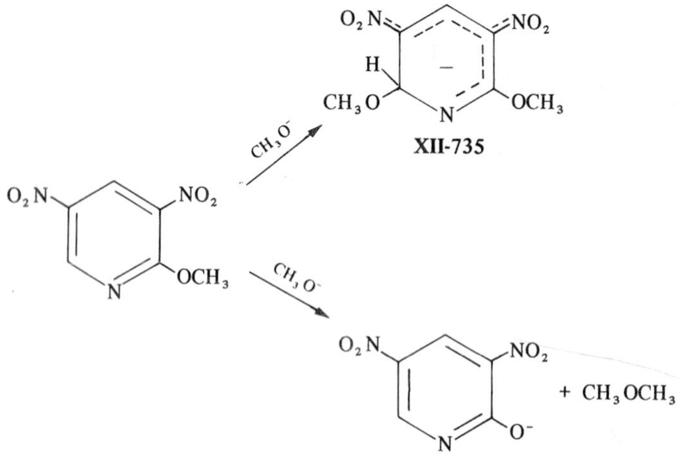

**XII-735**

Equilibrium constants[775, 776] and rates of formation[776] for the adduc **XII-733** have been measured and compared with those of 2,4,6-trinitroanisole An annular nitrogen in the 4-position to a reactive center results in a relativ decrease in stability of the Meisenheimer compound and an increase in its rate o formation when compared to the effect of a *p*-nitro group.[776]

The 4-methoxypyridone **(XII-736)** has been used to prepare a number o 4-hydrazino- and substituted-4-amino-1-methyl-2-pyridones.[86]

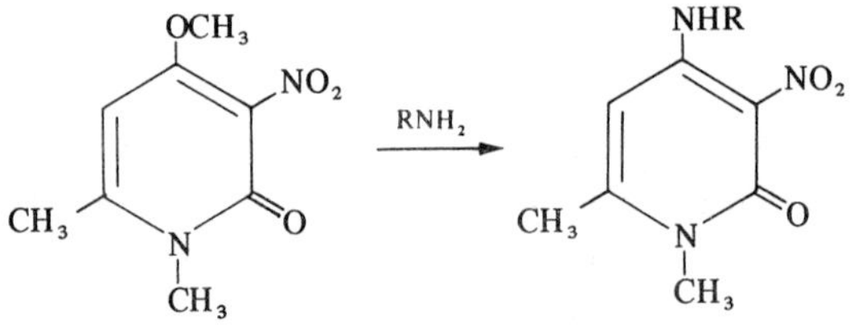

**XII-736**

Grignard[778] and organolithium reagents[778, 779] react with tetrachloro-4 methoxypyridine to form 4-alkyl- and 4-aryltetrachloropyridines. Phenyllithiun and tetrachloro-4-methoxypyridine form tetrachloro-4-phenyl- **(XII-737**, Ar = $C_6H_5$)[778, 779] trichloro-4,6-diphenyl- and dichloro-2,4,6-triphenylpyridine.[77] Under mild conditions a series of aryllithium compounds form 4-aryltetra chloropyridines **(XII-737)** that react with butyllithium to give 4-aryl-trichloro-3 pyridyllithium **(XII-738)**.[778, 779]

3-Phenyl-4-pyridone has been prepared from 4-methoxy-3-nitropyridine by a diazotization of 3-amino-4-methoxypyridine **(XII-739)** with pentyl nitrite in the presence of benzene and in the absence of added acid.[780]

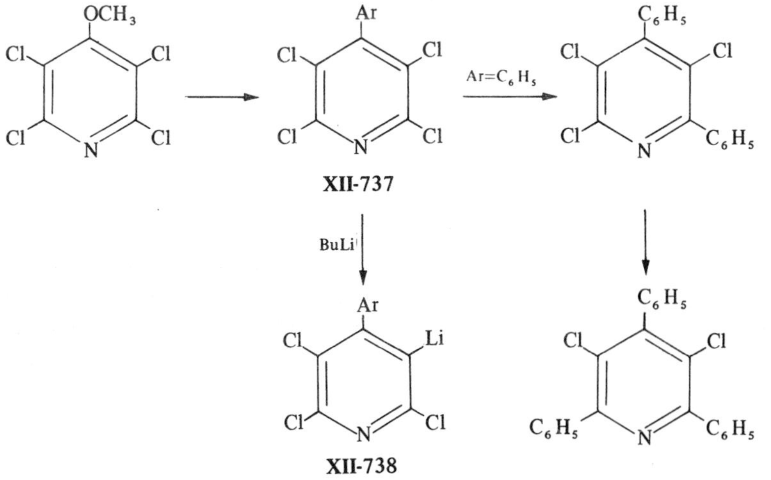

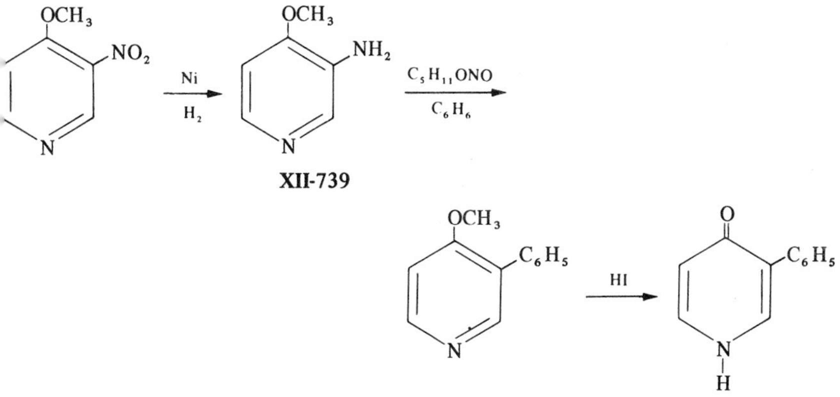

**XII-737**

**XII-738**

**XII-739**

## 2. *O* → *N*-Rearrangement

Although it has been shown that the thermal rearrangement of 2-methoxypyridine to *N*-methyl-2-pyridone **(XII-740)** is intermolecular and is catalyzed by benzoyl peroxide,[781] these features of *O* → *N* rearrangements are not universal.[782, 783] Four centered mechanisms have been suggested for rearrangements of methoxypyrroline[782] or 2-alkoxypyrimidines.[783] However, intramolecularity in rearrangements of 2-alkoxypyrimidines has not been demonstrated.[784] Ion-pair intermediates have been suggested for the rearrangement of 4-methoxypyridine to *N*-methyl-4-pyridone and for the rearrangements of 2-alkoxypyrimidines to *N*-alkyl-2-pyrimidones.[783]

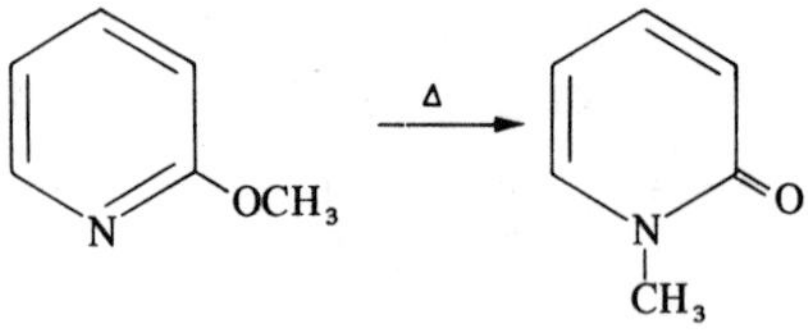

**XII-740**

The $O \rightarrow N$ rearrangement of 4- and 2-alkoxypyridines are catalyzed by alky halides through quaternary nitrogen derivatives (e.g., **XII-741**). Catalyti amounts of methyl iodide have been used to convert the methoxypyridines t *N*-methylpyridones under relatively mild conditions.[139, 530]

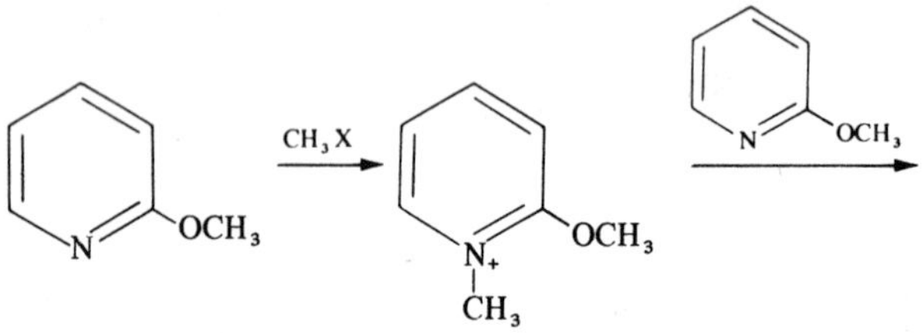

**XII-741**

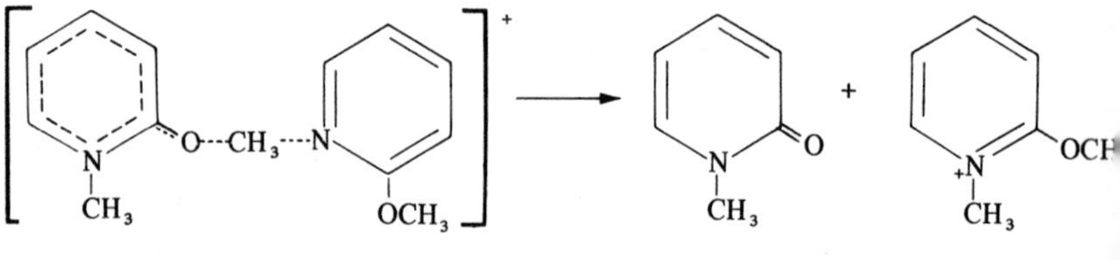

**XII-742**

Acid and alkyl halide-catalyzed rearrangements have been considered t involve an intermolecular ion-pair intermediate.[783]

2-Methoxypyridine, 4-methoxypyridine, and 2,6-diphenyl-4-methoxypyridin have been equilibrated in the liquid phase (130–250°) with their respectiv pyridone isomers in the presence of the corresponding methoxy-1-methyl pyridinium tetrafluoroborate presumably *via* a species isomeric with **XII-742** Although the pyridone (e.g., **XII-743**) is more stable in the above cases 4-methoxy-2,6-diphenylpyridine is more stable than 2,6-diphenyl-1-methyl-4 pyridone **(XII-744)** because of 1,2,6-steric interactions.[530] Experiments usin 2-methoxy- and 4-methoxypyridine in the presence of isotopically labele catalysts showed that oxygen to oxygen transfer occurs in addition to oxygen t nitrogen transfer, but that nitrogen to oxygen transfer does not occur.[530]

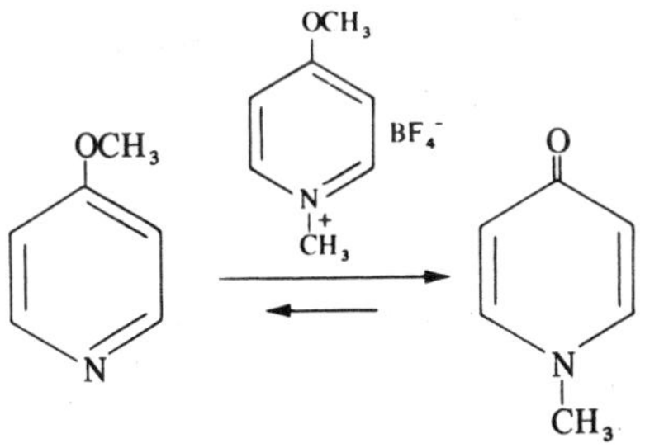

**XII-743**

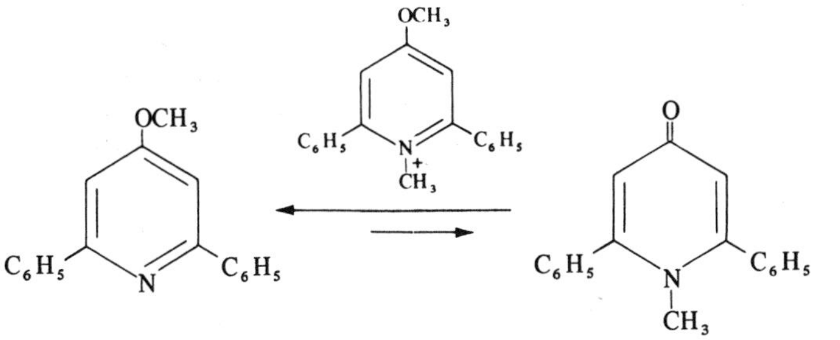

**XII-744**

Equilibration of *N*-methyl-2-pyridinethione **(XII-745)** with 2-methylthiopyridine in the presence of *N*-methyl-2-methylthiopyridinium tetrafluoroborate showed that the thioether is the more stable in the liquid phase [$K_{eq}$ = 14.4 (188°), 10.7 (145°)].[532]

The alkyl halide-catalyzed rearrangement can be extended to reactions where R ≠ R′ and provides a convenient route to *N*-alkylpyridones as an alternative to direct alkylation. For example, although alkylations of 2-pyridone salts with 2-(6-methoxy-1-naphthyl)ethyl halides (**XII-746**, R′ = 6-methoxy-1-naphthyl) appear to be unsuccessful, 2-ethoxypyridine and **XII-746** [X = Br, R′ = 1-(6-$CH_3OC_{10}H_6$), $C_6H_5$] give the *N*-arylethyl derivative in good yield.[341]

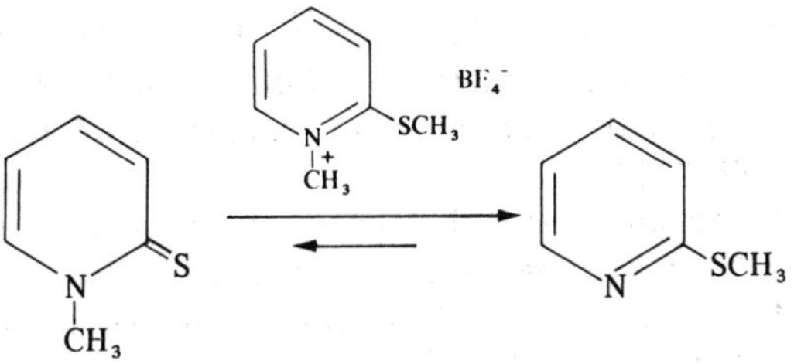

**XII-745**

*p*-Anisyl chloride and an excess of 2-ethoxypyridine give a quantitative yield of 1-(*p*-methoxybenzyl)-2-pyridone **(XII-747)** when heated without a solvent.[762]

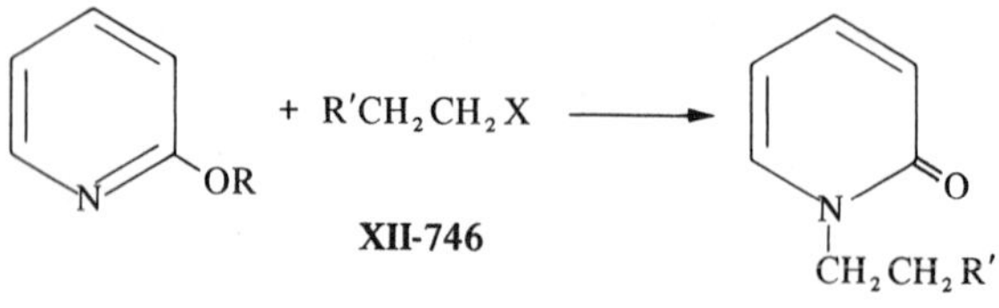

+ $R'CH_2CH_2X$ ⟶

OR

**XII-746**

N, O, $CH_2CH_2R'$

$OC_2H_5$ ⟶ N, O, $CH_2$, $OCH_3$

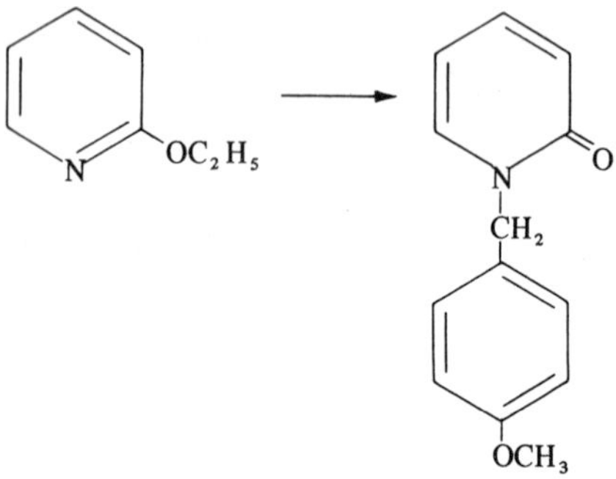

**XII-747**

The product of direct alkylation of a pyridone or pyrimidone salt is the alkoxy derivative if the reagent is bulky. For example, *O*-glycosides are formed from halogenoses and these heterocycles, but can be rearranged to *N*-glycosides by heating with mercuric bromide in a solvent such as toluene. (See section on alkylation, III.1.A.). Recently, catalysis of rearrangement by other Lewis acids has been reported.[609] For example, the main product obtained from the silver salt of 5-nitro-2-pyridone and acetobromoglucose is the *β*-*O*-acetoglucoside **(XII-748)**, which rearranges to the *a*-*O*-glucoside **(XII-749)** and the *β*-*N*-glucoside **(XII-750)** when treated with mercuric bromide in boiling toluene.[605, 609] Stannic chloride, titanium chloride, or antimony chloride and **XII-748** in benzene at room temperature or zinc chloride or cadmium chloride in boiling benzene also give **XII-750**. The *α*-anomer **(XII-749)** and mercuric chloride or stannic chloride in benzene also gives **XII-750**.[609]

Treatment of 2-bromopyridine with *N*-phenylethanolamine gives *N*-(*β*-anilinoethyl)-2-pyridone.[785] It has been suggested that 2-(*N*-*β*-hydroxyethylanilino)-pyridine **(XII-751)** is first formed and that subsequent rearrangement occurs *via* **XII-752** or *via* **XII-753** and **XII-754**.[785]

AcOCH$_2$ O NO$_2$ AcO OAc O N OAc **XII-749**

O$_2$N O N AcOCH$_2$ O AcO OAc OAc **XII-748**

O$_2$N O N AcOCH$_2$ O AcO OAc OAc **XII-750**

N Br $\xrightarrow{C_6H_5NHCH_2CH_2OH}$ N N C$_6$H$_5$ CH$_2$CH$_2$OH **XII-751**

+N N–C$_6$H$_5$ **XII-752**

C$_6$H$_5$ N N H O **XII-753**

O N CH$_2$CH$_2$NHC$_6$H$_5$

N OCH$_2$CH$_2$NHC$_6$H$_5$ **XII-754**

Alkylation of 2-methoxypyridine with a phenylacyl bromide or bromoaceto[ne] gives *N*-phenacyl- or *N*-acetonyl-2-pyridone, which can be cyclized in sulfu[ric] acid or perchloric acid to oxazolo[3,2-*a*]pyridinium salts (R.I. 1115), whi[ch] have been isolated as their perchlorates (**XII-755**).[786, 787]

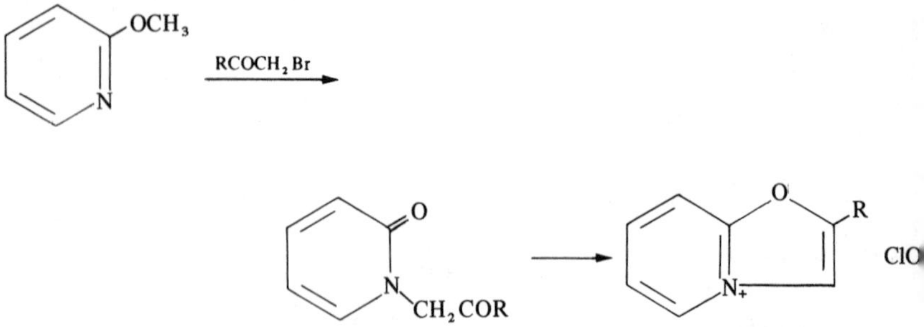

$RCOCH_2Br$

$CH_2COR$

$ClO$

**XII-755**

4-Methoxypyridine and β-(3-indolyl)ethyl bromides (**XII-756**; R = H, $CH_3$) [in] ethanol give 1-[β-(3-indolyl)ethyl]-4-methoxypyridinium bromide, which [is] converted to indolylethylpyridones (**XII-757**) by treatment with sodiu[m] hydroxide in ethanol.[788] 4-Methoxy-3-pyridinecarboxamide and methyl iodi[de]

$OCH_3$ + $CH_2CH_2Br$

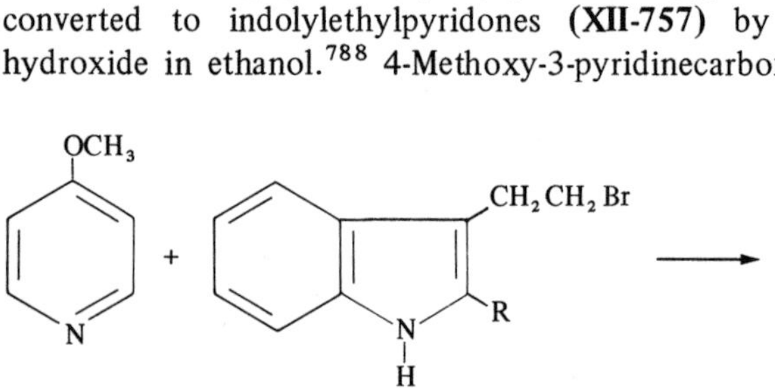

**XII-756**

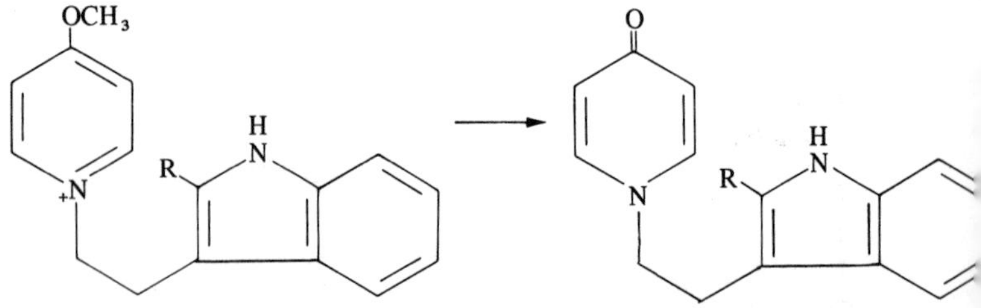

**XII-757**

give 3-carboxamido-1-methyl-4-pyridone.[789]

2-Allyloxy- (**XII-758**, R = R′ = H),[598, 790] 2-methallyloxy- (**XII-758**; R = CH[$_3$], R′ = H)[790] and 2-crotoxypyridine (**XII-758**; R = H, R′ = $CH_3$)[790] give *N*- a[nd] ıbstituted pyridones by a normal Claisen rearrangement when heated in ıethyl- or diethylaniline.[790] When 2-crotoxypyridine is rearranged neat, a

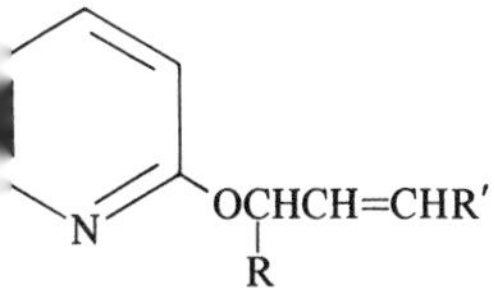

XII-758

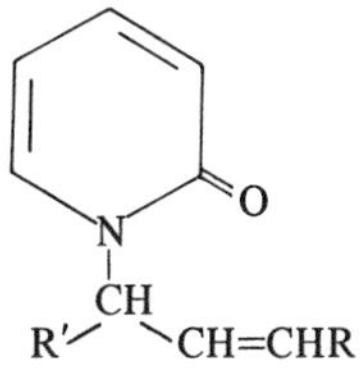

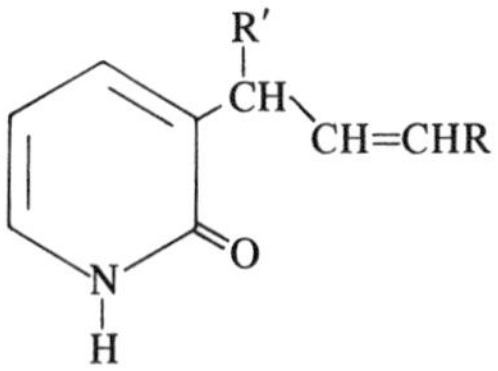

mal Claisen rearrangement gives 3-methallyl-2-pyridone, but the rearrange-nt to nitrogen provides both the Claisen product, *N*-methylallyl-2-pyridone, ı *N*-crotyl-2-pyridone, an abnormal product.[790] Possibly, the abnormal duct could be formed by a [1,3]-sigmatropic process that competes with the isen rearrangement, a [3,3]-sigmatropic rearrangement. Attempted rearrange-nt of 4-allyloxypyridine was unsuccessful.[598] 2-Allyloxypyridines, like alkyl-/pyridines, rearrange in good yield to *N*-allyl-2-pyridones, in the presence of wis acids. 2-Crotoxypyridine rearranges to *N*-methallyl-2-pyridone in the sence of chloroplatinic acid but gives *N*-crotoxy-2-pyridone and some rotyl-2-pyridone in the presence of boron trifluoride etherate, demonstrating different roles of the two catalysts.[791]

### 3. *N*-Amino Derivatives

Kojic acid (**XII-759**, R = H) and hydrazine give the corresponding pyrazole ivative (**XII-760**) and 3,6-bis(hydroxymethyl)-4-pyridazinone (**XII-1**).[792, 793] However, 2-hydroxymethyl-5-methoxy-4-pyrone (Kojic acid nonomethyl ether) (**XII-759**, R = $CH_3$), which contains a blocked 5-position, es 1-amino-2-hydroxymethyl-5-methoxy-4-pyridone (**XII-762**) (20%) and 3-hydroxymethyl-5-pyrazolyl)-α-methoxyacetaldehyde hydrazone (**XII-763**) 5%). These results suggest a nucleophilic attack by hydrazine at the α-position the pyrone, followed by ring opening and nucleophilic attack by nitrogen at 4-, 5-, or 6-carbon.[794, 795] Allomaltol (**XII-764**, R = $CH_3$) and pyromeconic d (**XII-764**, R = H) also give pyridazine and pyrazole derivatives.[794] 3-Pyrazolyl)acetaldehyde hydrazone was the only product isolated from yrone and hydrazine hydrate.[796, 797]

Products formed from 1,3,5-triketones (**XII-765**) or γ-pyrones and *p*-nitro-enylhydrazine had originally been assigned the structures **XII-766** and **I-767**.[798, 799] However, it has been shown that 4-pyrone and *p*-nitrophenyl-drazine give the *p*-nitrophenylhydrazone of the pyrazole (**XII-768**, R = H).[797]

XII-761

XII-760

R = H

XII-759

$R = CH_3$

XII-762

XII-763

XII-764

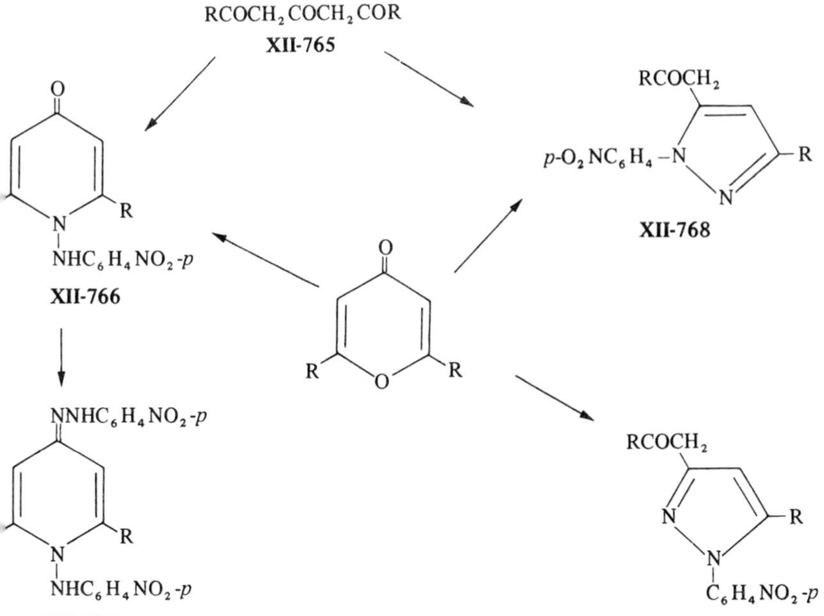


**XII-765** **XII-766** **XII-767** **XII-768** **XII-769**

t has been proposed subsequently that diacetylacetone (**XII-765**, R = $CH_3$) d *p*-nitrophenylhydrazine form **XII-768** or **XII-769** (R = $CH_3$) and that a ond compound that is described as the 4-pyridone **XII-766** (R = $CH_3$) is med from 2,6-dimethyl-4-pyrone. Reaction of **XII-766** with additional itrophenylhydrazine gives **XII-767** (R = $CH_3$).[140] It is proposed that tonedicarboxylic anhydride (**XII-770**) and phenylhydrazines give the pyrine derivatives.[140, 800, 801] For example, **XII-770** and *p*-nitrophenylhydrazine e the *N*-(*p*-nitroanilino)-4-pyridone and its *p*-nitrophenylhydrazone **II-771**).[800]

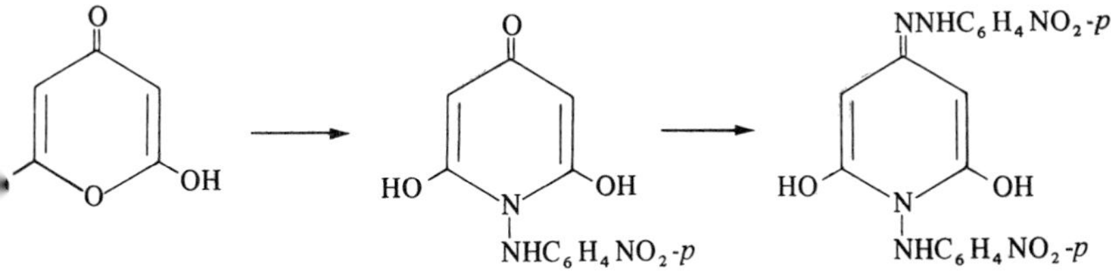


**XII-770** **XII-771**

Hydrazine and 2,6-dimethyl-4-pyrone appear to give the bipyridone **I-772**.[140] Triaryl-2-pyrones react with hydrazine and phenylhydrazine to give

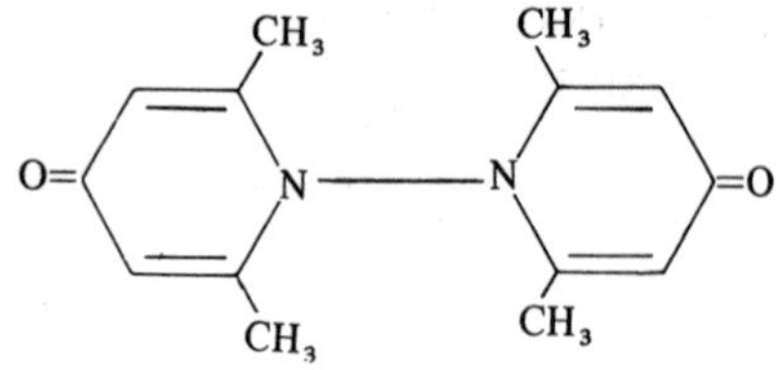

**XII-772**

the *N*-amino- and *N*-phenylamino-2-pyridones **(XII-773)**, resp tively.[120-122,802]

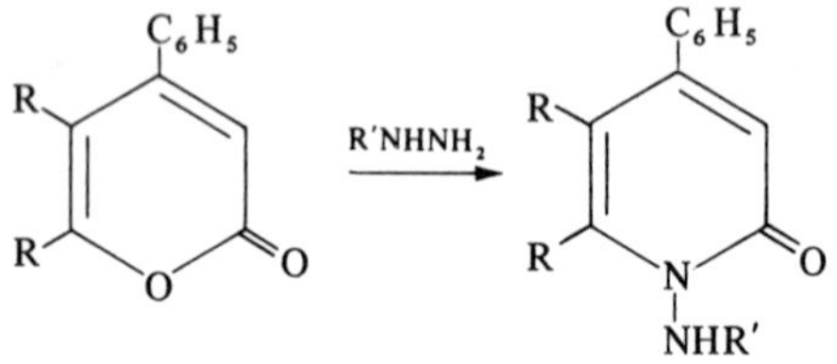

$R'NHNH_2$

**XII-773**

$R' = C_6H_5$, H; $R = C_6H_5$, *p*-$BrC_6H_4$-, *p*-$CH_3C_6H_4$-

Reactions of the 1-amino group of 1-amino-4,5,6-triaryl-2-pyridones have be studied.[802] The ethyl esters of 3-benzoyloxy-4-oxo-4*H*-pyran-6-carboxylic a and hydrazine in ethyl alcohol are reported to give the hydrazide 1-amino-3-benzoyloxy-4-pyridone-6-carboxylic acid (**XII-774**, $R = C_6H_5CO$) only in 8% yield. Similarly, 6-carbethoxy-3-methoxy-4-pyrone and hydraz give **XII-774** ($R = CH_3$).[803]

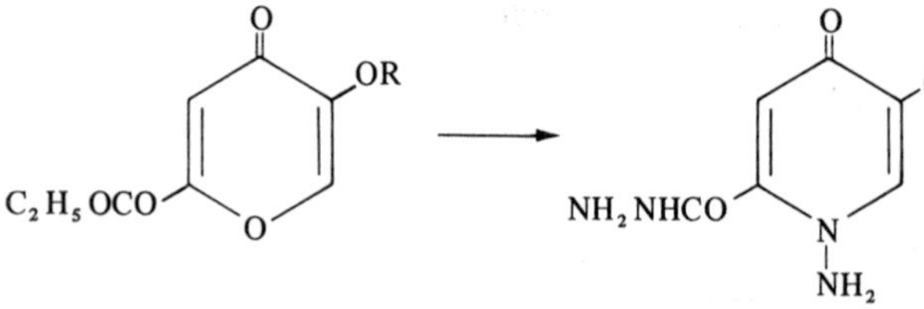

**XII-774**

*cis,cis*-2,4,6-Perchloroheptatriene-6-al phenylhydrazone **(XII-775)** gives 1-a lino-3,5-dichloro-2-trichlorovinyl-4-pyridone when heated in aqueous ethan The corresponding 2,4-dinitrophenylhydrazone required heating in acetic ac hydrochloric acid to effect cyclization. The bromo analog was prepar similarly.[233]

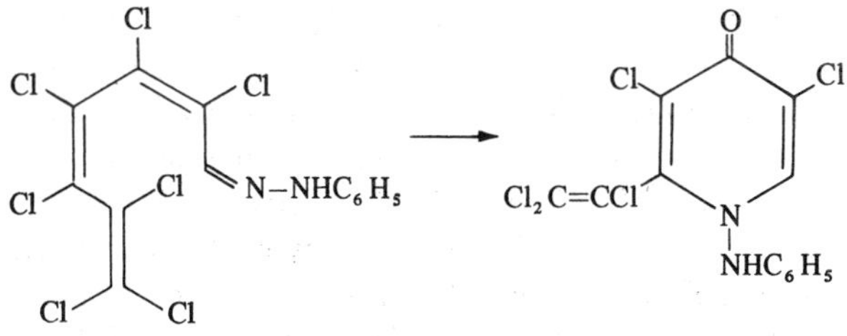

**XII-775**

-Amino- and 1-methylamino-3-hydroxy-4-methyl-5-phenylpyridinium chlo- e react with alkali or carbonate to form the corresponding pyridinium aines **(XII-776)**. The amino group of **XII-776** (R = H) is not basic but has the

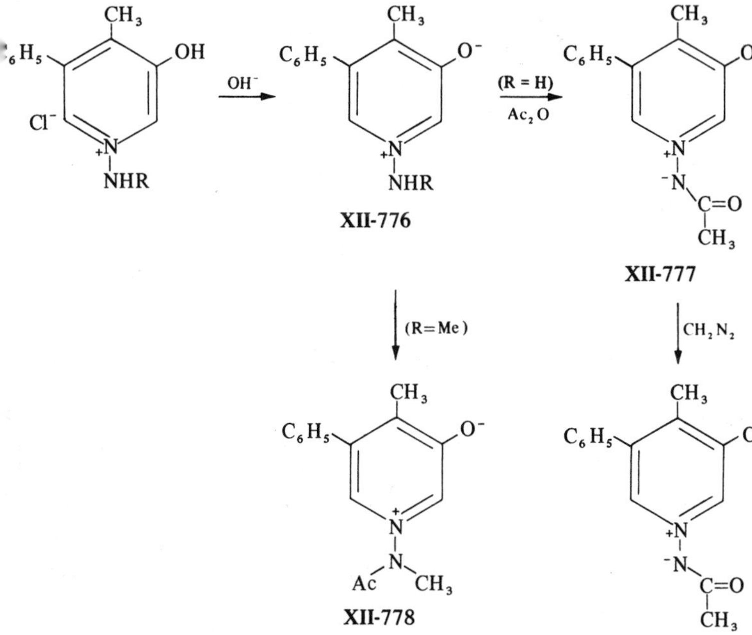

**XII-776**

**XII-777**

**XII-778**

ctivity of a primary amine. For example, **XII-776** (R = H) reacts with acetic hydride to form the ylid **(XII-777)**, which is *O*-alkylated with diazomethane. ifluoroacetic anhydride and benzoyl chloride react similarly with **XII-776** (R = . Acetylation of the *N*-methylaminopyridinium betaine (**XII-776**, R = $CH_3$) es the betaine **XII-778**, which, in contrast to **XII-777**, is labile to drolysis.[219]

The interconvertible ring systems **XII-779** and **XII-780** can give amino-3- ridinols when treated with aqueous acids. In hydrochloric acid-methanol,

2-acetyl- and 2-benzoyl-5-methyl-4-phenyl-1,2,-diazabicyclo[3.2.0]-3-hept 6-one (**XII-779**) rearrange to *N*-acetamido- and *N*-benzamido-3-hydroxy-4-meth 5-phenylpyridinium chloride (**XII-781**; R = $CH_3$, $C_6H_5$) by an allylic shift of bridging bond from C-5 to C-3 (see the section on 3-pyridinols for a discussi of this interconversion).[220, 804] The pyridinium salts **XII-781** are converted betaines (**XII-782**) by treatment with pyridine or by neutralization with base. When the diazapine (**XII-780**) is treated with benzoyl chloride and pyridine methanol, the 1-benzoyl-7-methoxytetrahydrodiazepinone (**XII-783**) is isolat and forms **XII-781** on treatment with acid. Treatment of **XII-780** with 1 hydrochloric acid gives 1-amino-3-hydroxy-4-methyl-5-phenylpyridinium ch ride (**XII-784**),[216, 220] which gives **XII-781** on treatment with benzoyl chlori in pyridine.[804]

$C_6H_5$ CH$_3$ O N N C R O — HOAc ⇌ RCOCl, $R_3N$ — $C_6H_5$ CH$_3$ O N—N H

**XII-779** **XII-780**

$C_6H_5$ CH$_3$ O N—N COR

$C_6H_5$ CH$_3$ O CH$_3$O— N—N H COC$_6$H$_5$

**XII-783**

$C_6H_5$ CH$_3$ N$_+$ NH$_2$

**XII-784**

$C_6H_5$ CH$_3$ OH Cl$^-$ $^+$N NHCOR

**XII-781**

$C_6H_5$ CH$_3$ $^+$N NHCO

**XII-782**

imethyl sulfate and **XII-780** give the *N*-methylated products **XII-785** and -**786**. The hydrochloride of **XII-785** rearranges to **XII-787**.[216]

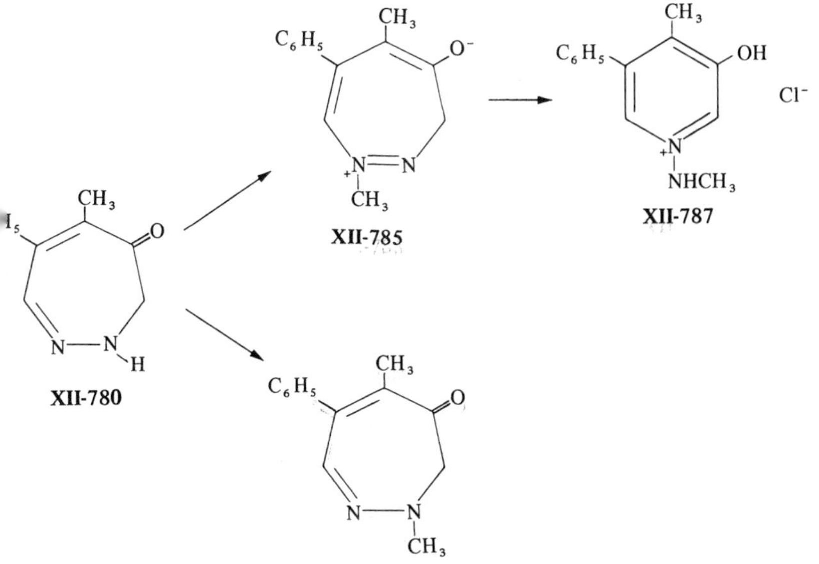

**XII-780** **XII-785** **XII-786** **XII-787**

-Amino-2-pyridone and 2-pyridinecarboxaldehyde in methanol give 1-pyridyl-2-one)-2-pyridinealdimine (**XII-788**, R = H). Although the Schiff's e from 1-amino-2-pyridone and 2-acetylpyridine (**XII-788**, R = $CH_3$) is not ily isolated, it forms complexes with iron(II), nickel(II), and Co(II) chlorates[805] (see Section IV.5.).

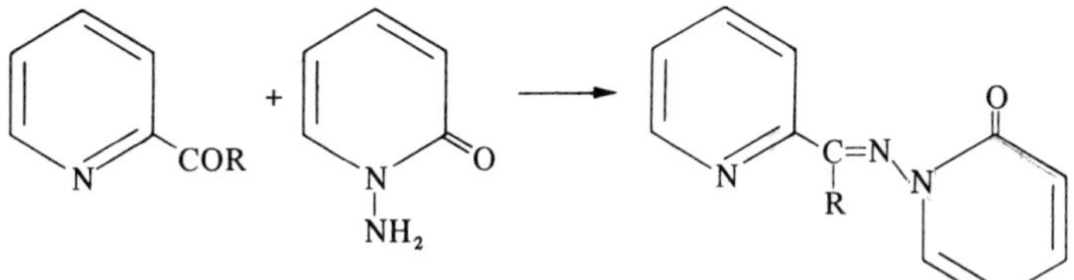

**XII-788**

## 4. Bifunctional Catalysis

lthough early investigators considered that general acid-general base catalysis operative in the mutarotation of tetramethyl-D-glucose in benzene[806] and in the aminolysis of *p*-nitrophenyl acetate,[807] recent studies have demonstra that tautomeric catalysis is consistent with observations. It has been sugges that this mechanism is prevalent particularly in reactions of carbonyl compou where bifunctional catalytic effects have been observed and where 2-pyrid and similar tautomeric catalysts are effective.[807] The existence of a concer general acid-base reaction mechanism has not been substantiated.[808]

A tautomeric catalyst is defined as a "molecule that repeatedly cycles betw two tautomeric states during the course of catalysis in a chemical reaction." A concerted mechanism has been proposed:

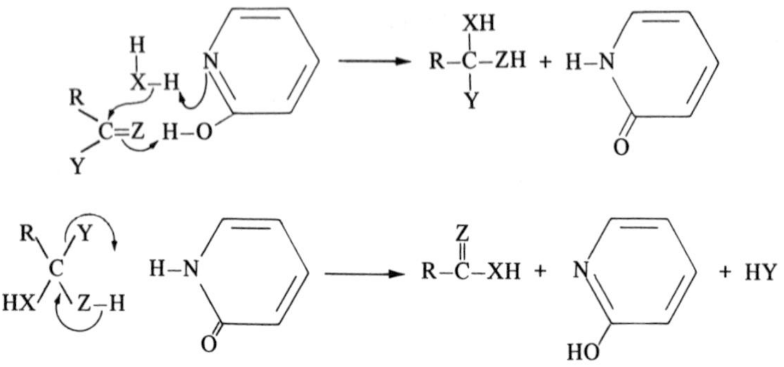

It has, however, been observed that 2-(β-hydroxyethyl)pyridine **(XII-789)** 4-(γ-hydroxypropyl)pyridine **(XII-790)**, which appear to act as bifunctic catalysts, are somewhat more effective than 2-pyridone in the hydrolysis

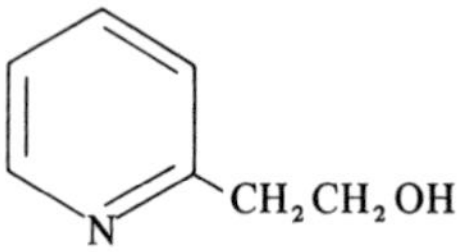

**XII-789**

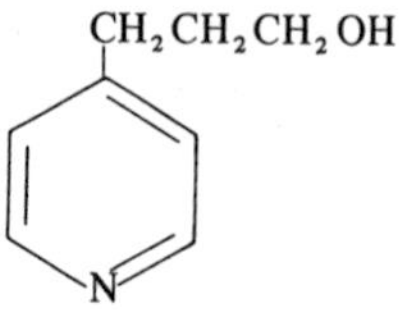

**XII-790**

*p*-nitrophenyl acetate.[809] 2-Pyridone, which acts as a bifunctional catalys peptide synthesis,[810-812] does not cause racemization in the synthesis *N*-benzyloxycarbonyl-L-leucyl-L-phenylalanyl-L-valine *t*-butyl ester from *N*-be yloxycarbonyl-L-leucyl-L-phenylalanine *p*-nitrophenyl ester and L-valine *t*-bu ester in acetonitrile,[811] and in condensations in ethyl acetate solution of car benzoxy-β-cyano-L-alanine with methyl glycinate to give methyl carbobenzo β-cyanoalanyl glycine.[812]

Bifunctional catalysis of the reaction between fluoro-2,4-dinitrobenzene piperidine and chloro-2,4-dinitrobenzene and piperidine by 2-pyridone, but by *N*-methyl-2-pyridone, has been reported.[813] 2-Pyridone has been evaluate

atalyst for the polyurethane forming process.[814] The oxidation of 1-dode-
ethiol by tetramethylene sulfoxide is catalyzed by 2-pyridone and has been
sidered to be general acid and general base catalyzed.[815]

## 5. Complexes and Organometallic Compounds

-Pyridone forms 1:1 adducts with $MnCl_2$, $CoCl_2$, $NiCl_2$, and $CuCl_2$. In
lition, $CuCl_2$ forms a 1:2 adduct, $CuCl_2 \cdot 2(Py\text{-}2\text{-}OH)_2$ and $CoCl_2$ forms a 1:3
luct, $CoCl_2 \cdot 3(Py\text{-}2\text{-}OH)$. Powder diffraction data for the $CuCl_2$ 1:1 adduct
port a chlorine bridged dimeric structure **(XII-791)**. Generally, properties of
other 1:1 adducts are consistent with the dimeric structure.[816]

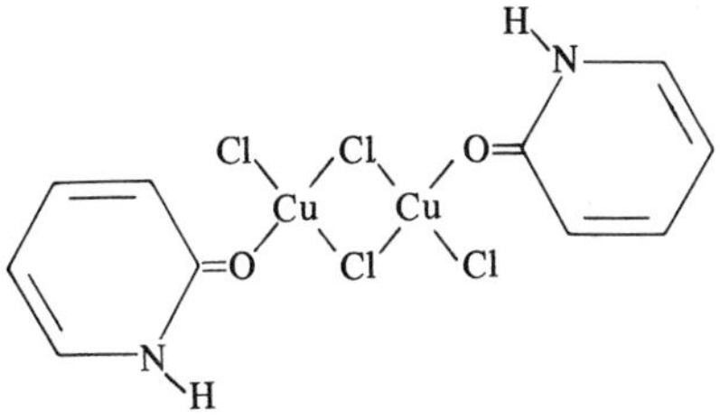

**XII-791**

V-Methyl-2-pyridone forms 1:1 adducts with $HgCl_2$,[817,818] $CdCl_2$,[818] and
$Cl_2$[818] and forms a 1:2 adduct with $SnBr_4$.[818] 1,2,6-Trimethyl-4-pyridone
ms 1:1 complexes with $HgCl_2$, $ZnCl_2$, and $CdCl_2$. *N*-Ethyl-2,6-dimethyl-4-
ridone forms 1:1 adducts with $HgCl_2$, $CdCl_2$, and $CoCl_2$.[819] Infrared spectra
licate that the donor site of these *N*-methylpyridones is the carbonyl and not
nitrogen.[818,819]

- And 4-pyridone coordinate at oxygen to form 1:1 adducts, $Co(NH_3)_5PyOH$-
$O_4)_3$, with pentamminecobalt(III), which are isolated as their perchlorates.
yridinol does not react with aquopentamminecobalt(III) perchlorate.[820]
yridone is reduced by Cr(II) *via* the outer sphere mechanism, while
yridone is reduced *via* ligand transfer.[821]

Complexes of 2-pyridone with the structure $M(Py\text{-}2\text{-}OH)_6(anion)_2$, (M = Mn,
, Co, Ni, Zn, Cd; anion = $ClO_4^-$, $BF_4^-$) have the ligand in the lactam form
ordinated to metal ions also *via* the carbonyl. $Ca(Py\text{-}2\text{-}OH)_8(ClO_4)_2$ and
$(Py\text{-}2\text{-}OH)_6(ClO_4)_3$ have been described also.[822]

4-Pyridone and sodium dinitro-bisacetylacetonato carbonate(III) form a stable
nionic complex **(XII-792)** that appears to be the *trans*-isomer.[823] 4-Pyridone
d hydrated cobalt chloride in alcohol give a blue tetrahedral complex
$(Py\text{-}4\text{-}OH)_2Cl_2$.[824]

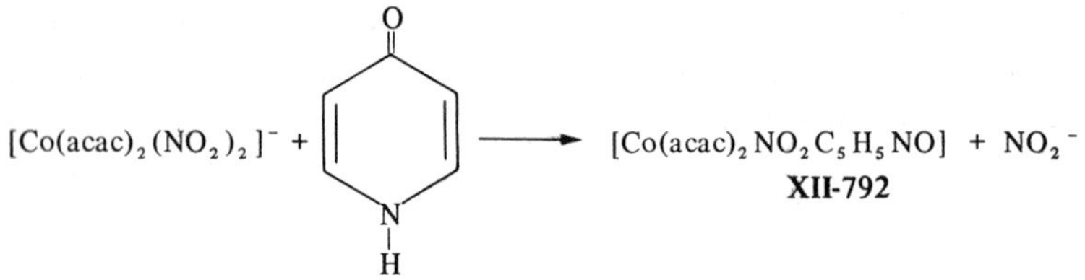

$$[Co(acac)_2(NO_2)_2]^- + \text{(4-pyridone)} \longrightarrow [Co(acac)_2NO_2C_5H_5NO] + NO_2^-$$

**XII-792**

Stability constants have been determined for 1:1 bivalent metal complexes UO$_2$, Cu, Pb, Zn, Be, Ni, Co, Cd, and rare earths with 3-hydroxy-2-pyridone.[8

Stability constants of 1:1 complexes **(XII-793)** formed by chelidamic ac ($H_3L$) and $Ca^{2+}$, $Ba^{2+}$, $Sr^{2+}$, $Mg^{2+}$, and $Mn^{2+}$, and of 1:1 and 1:2 complex $M(HL)_2$, from chelidamic acid and $Co^{2+}$, $Ni^{2+}$, $Cu^{2+}$, and $Zn^{2+}$ have be determined. These complexes are stabilized by loss of the phenolic proton alkaline solution.[826]

Bis(3-hydroxy-2-pyridylmethylene)ethylenediamine **(XII-794)** forms 2 copper(II) and nickel(II) complexes.[827] The 1:1 complex formed from Fe(I

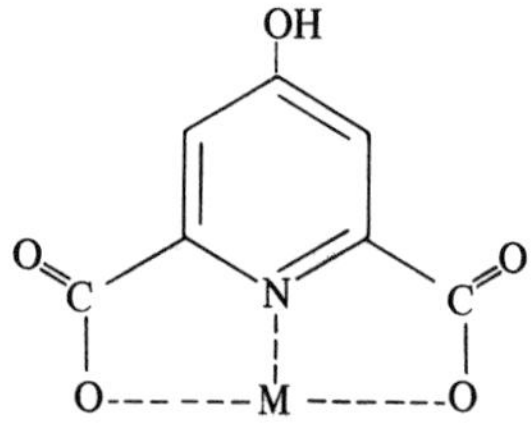

**XII-793**

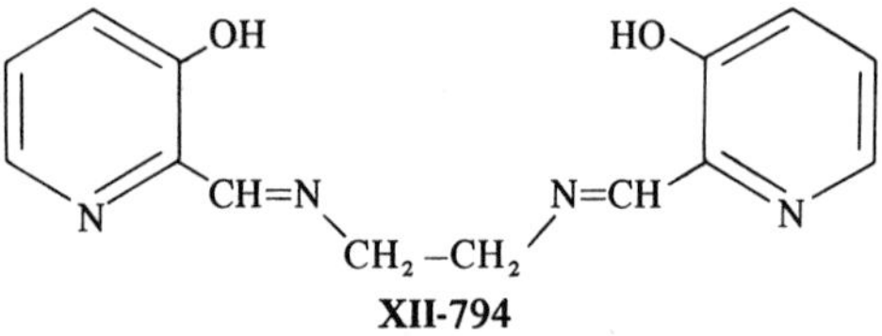

**XII-794**

perchlorate and 2-pyridone has been examined spectroscopically in aqueo solution.[828] The Mossbauer spectrum of several substituted pyridine iron(I derivatives has been measured, including a bis(1,2-cyclohexadienonedioxime iron(II) complex with 3-pyridinol, Fe(II)(Niox)$_2$(Py-3-OH)$_2$.[829]

*N*-Amino-2-pyridone and its Schiff's bases **(XII-795)** (R = H, $CH_3$) fro 2-pyridinecarboxaldehyde or 2-acetylpyridine have been studied as ligands f transition metal ions [Fe(II), Co(II), Ni(II)]. Two series of complexes have bee isolated from 1-amino-2-pyridone: $ML_2Cl_2$ from the metal chlorides a

$L_3(ClO_4)_2$ from the perchlorates. Complexes of the general formula $(L_2)(ClO_4)_2$ were formed from **XII-795**, a tridentate ligand that coordinates rough two nitrogens and one oxygen.[805]

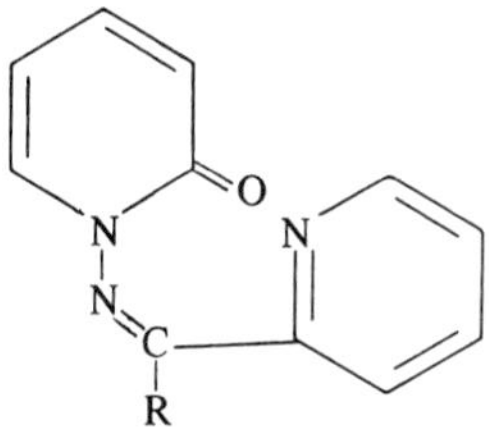

**XII-795**

2-Pyridone reacts with dibutyl dimethyloxytin to form dibutyldi-(2-pyri-onato-1)tin **(XII-796)**, with dimethyl dichlorosilane to form dimethyldi-(2-pyri-nato-1)silane **(XII-797)**, and with aluminum ethoxide to form tri-(2-pyri-nato-1)aluminum **(XII-798)**. Infrared spectra are consistent with an *N*-substi-ted pyridone structure.[830]

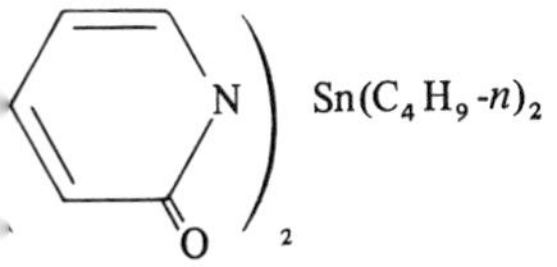

$Sn(C_4H_9\text{-}n)_2$ $Si(CH_3)_2$ Al

**XII-796** 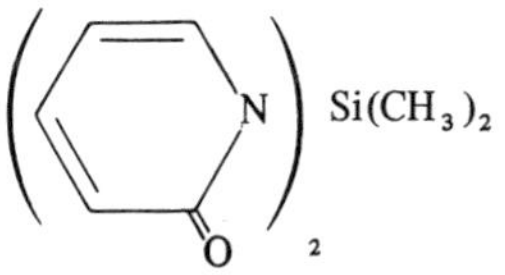 **XII-797** 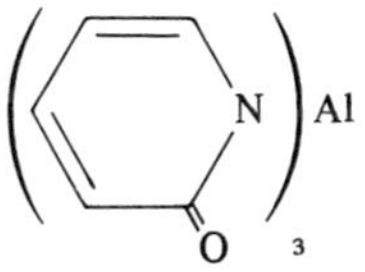 **XII-798**

A comparison of the ultraviolet spectra of arylmetallic derivates of pyridones ith the spectra of their *N*- and *O*-methyl derivatives indicates that phenyl-ercury derivatives of 2-pyridone, 3,5-dichloro-2-pyridone 3,5-dichloro-4-yridone, and 4-pyridone have the lactam structure (e.g., **XII-799**), while the iphenyl tin and triphenyl lead compounds have the lactim structure (e.g., **II-800**).[831]

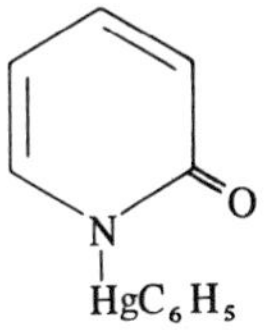

$HgC_6H_5$ $OSn(C_6H_5)_3$

**XII-799** 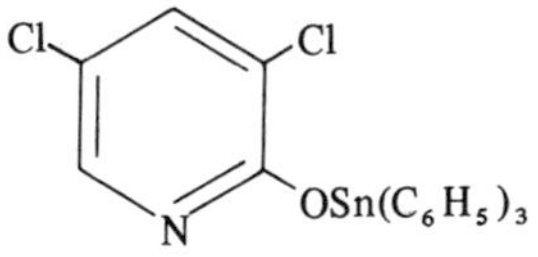 **XII-800**

The rate of substitution of bromide in $[Pt(dien)Br]^+$ by a series of pyridine erivatives including 2-pyridone and 3-pyridinol has been considered to be verned by steric properties of the entering group rather than by its basicity.[832]

Ethyl 4,6-dihydroxy-5-nitrosonicotinate has been studied as an analyti reagent for a photometric determination of iron.[833] 3-Nitroso-2,6-pyridined has been studied as an analytical reagent for the photometric determinations osmium,[834] ruthenium,[833] and palladium.[835] Ruthenium chloride appears form a 3:1 3-nitroso-2,6-pyridinediol to ruthenium anionic complex.[836] A complex is formed from palladium(II) chloride.[835] Methyl and ethyl esters 1-aryl-5-hydroxy-4-pyridone-2-carboxylic acids can be used as chelating age for the extraction and separation of niobium(V) and tantalum(V) from oxal solutions.[837] Triethylborane and 2-pyridone form a 1:1 dimer.[838] *N*-Methy pyridone and cyclomethylenetetranitroamine form a 2:1 complex.[839]

## V. Polyhydroxypyridines

A number of 2,4- and 4,6-dihydroxy-2-pyridones can be prepared fr polyhalopyridines that are available by ring closure and by direct haloge tion.[15,233-235,244-248]

2-Pyridone, 4-pyridone, and 3-pyridinol undergo the Elbs peroxydisulf oxidation, a reaction characteristic of phenols and aromatic amines. A bimolecu ionic reaction in which the 2-pyridyloxy ion attacks the peroxy-bond of persulfate ion with displacement of sulfate ion to give 2-pyridone-5-sulf is consistent with the observations.[840,841]

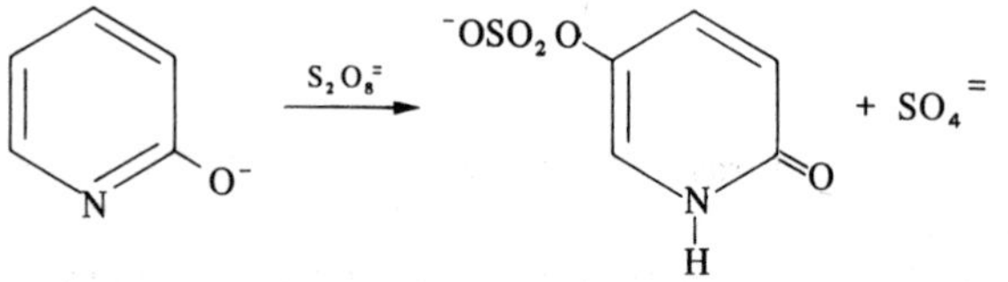

N O$^-$ $\xrightarrow{S_2O_8^{=}}$ $^-OSO_2O$ … O N H + $SO_4^{=}$

4-Methyl-5-phenyl-3-pyridinol couples with *p*-nitrobenzenediazonium chlor to give **XII-801** and **XII-802** (Ar = *p*-$NO_2C_6H_4$), which can be hydrogenated the corresponding aminopyridinols (**XII-803** and **XII-804**). Treatment of **XII-8** with nitrous acid gives **XII-805**, **XII-806**, or **XII-807**, depending on con tions.[292] The autooxidation product **XII-807** can be converted to the triacet of 4-methyl-5-phenyl-2,3,6-pyridinetriol (**XII-808**) by treatment with zinc d in acetic anhydride.[292]

Compounds formed by bromate or chromium trioxide oxidation of 3- a 5-amino-2-pyridone and 3-amino-5-methyl- and 5-amino-3-methyl-2-pyrido were first described as possessing monomeric azaquinone or azaquinhydro type structures.[842] It has, however, been shown that after hydrolysis and oxi tion with potassium bromate, 5-acetamido-3-methyl-2-pyridone forms 3-hydro 6-methyl-2-aza-1,4-benzoquinone-4-(2,6-dihydroxy-5-methylpyridyl)-3-imi

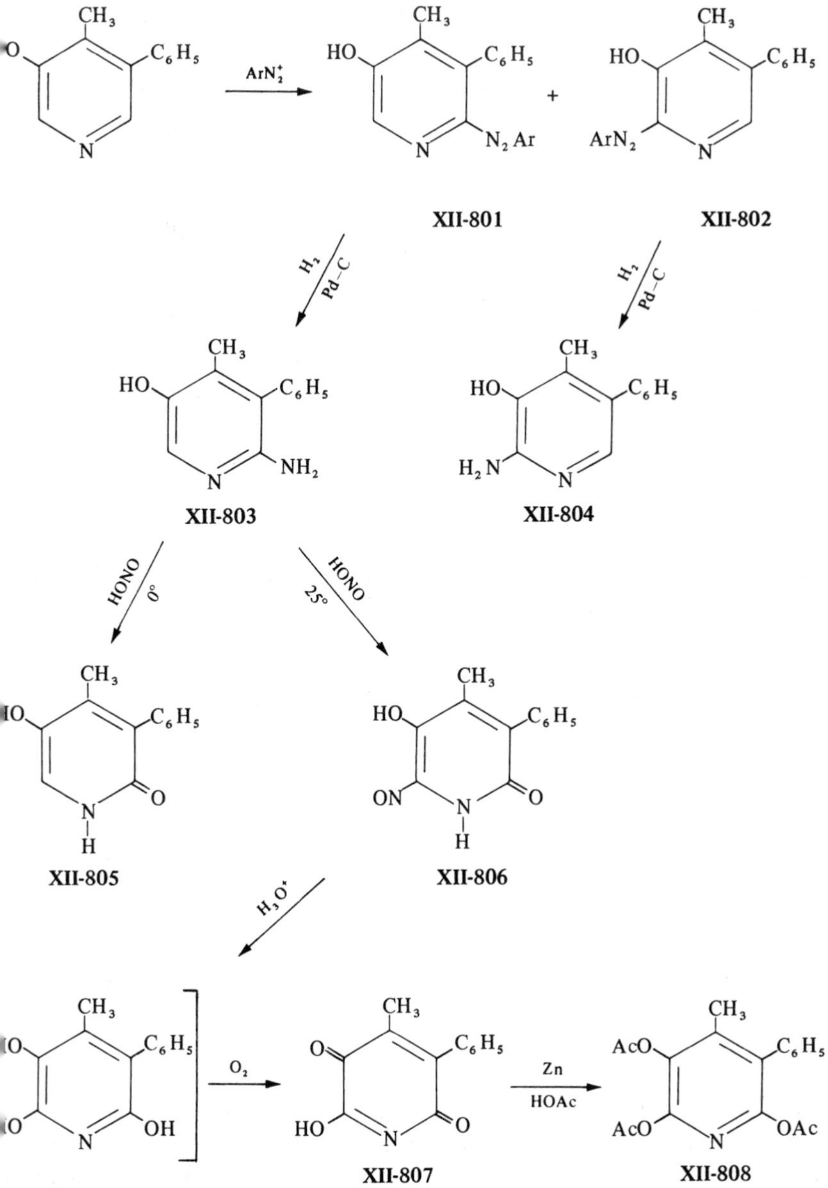

XII-801 XII-802

XII-803 XII-804

XII-805 XII-806

XII-807 XII-808

II-809).[843] Oxidation of **XII-809** with nitric acid produces 5-methyl-2,3,6-ridinetrione **(XII-810)**. Hydrogenation of **(XII-809)** over Pd-$BaSO_4$ gives 4′-dihydroxy-5,5′-dimethyl-3,3′-diaza-2,2′-diphenoquinone **(XII-812)**, through tooxidation of the intermediate 3-amino-5-methyl-2,6-pyridinediol **(XII-**

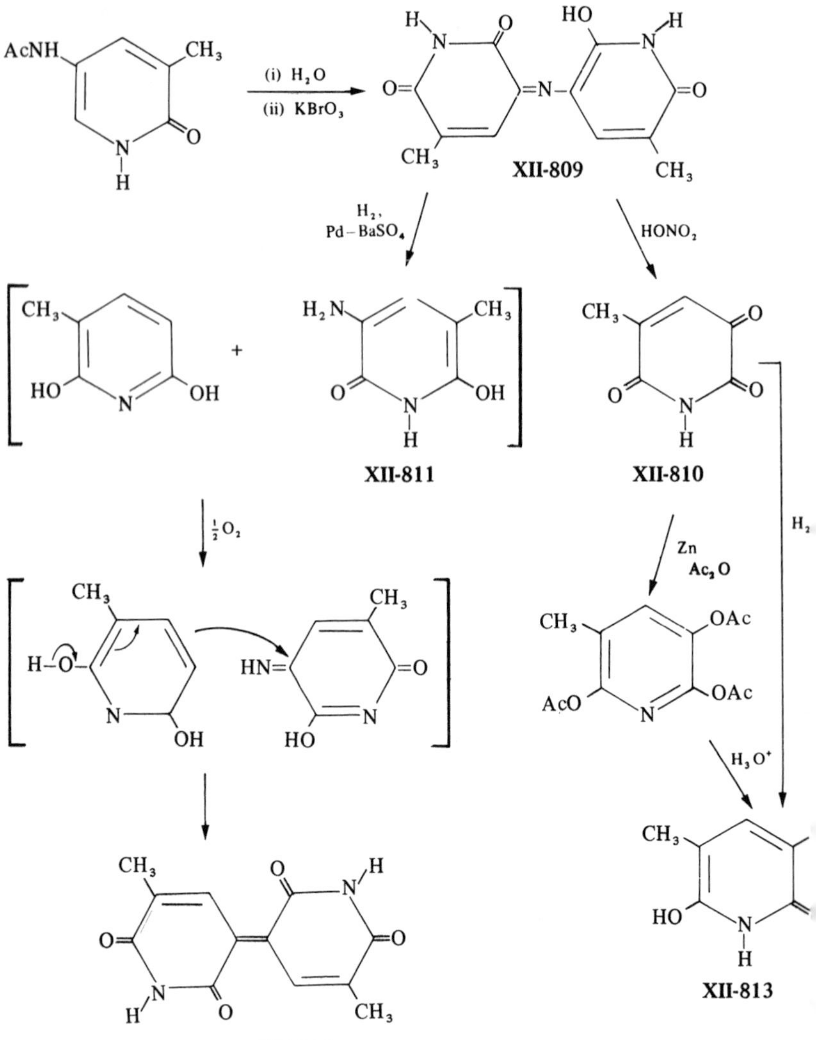

811).[843] Reduction of **XII-810** with zinc in the presence of acetic anhydri gives 2,3,6-triacetoxy-5-methylpyridine, which is hydrolyzed in sulfuric acid **XII-813**, isolated as the hydrochloride hydrate. Hydrogenation of **XII-810** ov palladium also gives **XII-813**.[506]

4-Methyl-3-phenylazo-2,6-pyridinediol (**XII-814**, R = H) and 4,5-dimethyl phenylazo-2,6-pyridinediol (**XII-814**, R = $CH_3$) and zinc in acetic anhydri form 2,6-diacetoxy-3-diacetylaminopyridines that can be converted to t 2,3,6-triacetoxypyridine and then to 3-hydroxy-4-methyl-2,6-pyridinedion

XII-814

XII-815

XII-817

XII-816

XII-818

(XII-815). Autooxidation of **XII-815** (R = H) gives the corresponding bipyridyl (XII-816). Autooxidation of **XII-815** ($R = CH_3$) gives the deep blue azaquinone XII-817. Oxidation of **XII-816** with potassium ferricyanide gives the bisazaquinone (**XII-818**).[506]

Hydrolysis of the azoquinone ketal (**XII-819**) gives a product that had been described earlier as 3-hydroxy-2-azabenzoquinone (**XII-820**),[844] but has been shown to be the diazadiphenoquinone (**XII-821**).[843]

Compounds described as 4-methyl- and 4,5-dimethyltrihydroxypyridines[845] do not have the properties expected from these structures[843] and are probably products of autooxidation.

Although earlier attempts to prepare 3-amino-2,6-pyridinediol were unsuccessful[740] this compound has been prepared from 3-phenylazo-2,6-pyridinediol (XII-822).[846] Hydrogenation of **XII-822** employing Raney Nickel in the presence

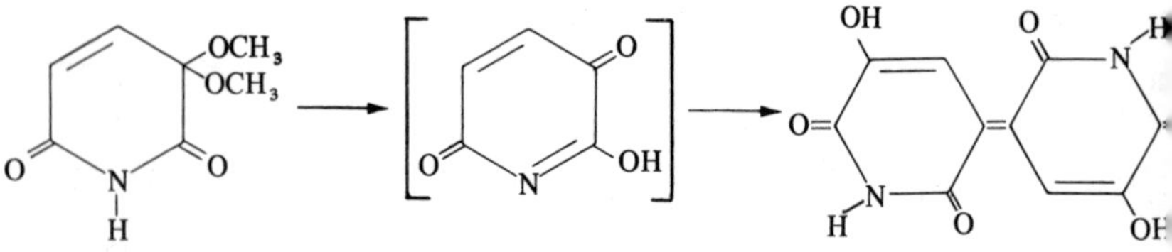

**XII-819** **XII-820** **XII-821**

of acetic anhydride gives the tetraacetate. Reduction of **XII-822** or 3-nitroso-2,6-pyridinediol **(XII-823)** with tin and hydrochloric acid gives t hydrochloride of 3-amino-2,6-pyridinediol **(XII-824)**.[846] Free 3-amino-2,6-py

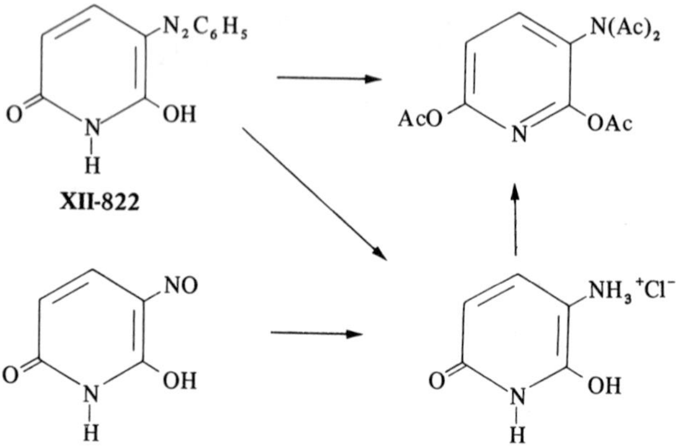

**XII-822**

**XII-823** **XII-824**

dinediol cannot be isolated conveniently because of its autooxidation. When i hydrochloride is dissolved in sodium bicarbonate solution indigoidin **(XII-825)** formed. (See Section VI, p. 860.) Hydrogenation of **XII-825** in acetic anhydri gives 5,5′-bisdiacetylamino-2,6,2′,6′-tetraacetoxy-3,3′-bipyridyl.[846]

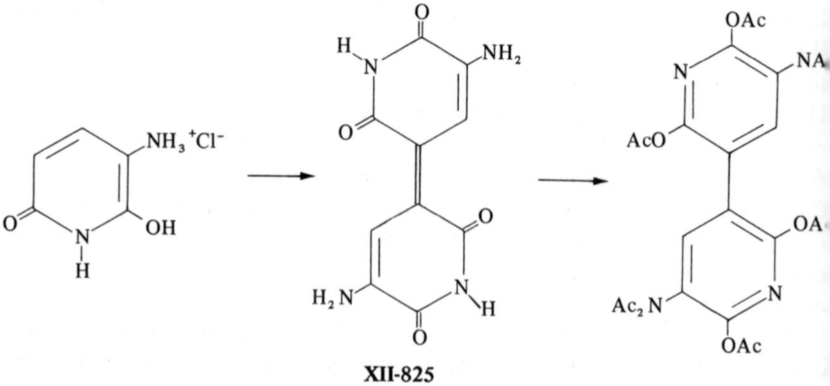

**XII-825**

## VI. Ricinine[847] and Other Pyridine Alkaloids

uccinic acid is incorporated into the 2-, 3- and 7-carbon atoms of ricinine cyano-4-methoxy-1-methyl-2-pyridone). Carbon 1 of succinic acid becomes nitrile carbon of ricinine.[848-850] Nicotinic acid and nicotinamide are highly orporated into ricinine, which suggests that these two compounds are closer ricinine than is succinic acid.[849,851] Ricinine and nicotine show labeling terns that are consistent with a pathway where succinic acid or a related arboxylic acid is a precursor to nicotinic acid, which is an intermediate in the mation of nicotine and ricinine.[849,852,853] The α-carbon of lysine is orporated into carbon 6 of ricinine and the ε-carbon of α-aminoadipic acid is orporated into carbons 2 and 6.[854]

Nudiflorine (5-cyano-1-methyl-2-pyridone) **(XII-826)**, an isomer of ricinidine methoxyricinine), has been isolated from the leaves of *Trewia nudiflora* Linn. has been synthesized from 5-carbomethoxy-2-pyrone by treatment with monia to form 6-oxonicotinic acid, which is alkylated *via* its sodium salt to N-methyl derivative. Esterification, followed by conversion to the amide and hydration, gives nudiflorine. It has been suggested that both ricinidine and

**XII-826**

diflorine arise from nicotinic acid.[855] 4-Acetyl-5-isovaleryl-2,6-pyridinedione **II-827)**, an antibiotic with strong activity against gram-positive and gram-gative bacteria, has been isolated from *Aspergillus flavipes*, a thermophilic ngus.[856]

nnovanamine, 2-methyl-3-β-D-glucopyranosyl-4-pyridone **(XII-829)** has been lated from fallen leaves of *Evodiopanax innovans*,[857] and has been prepared m the natural maltol glucoside **(XII-828)**.[131] Hydrolysis by hydrochloric acid methanol or in the presence of emulsin gives 3-hydroxy-2-methyl-4-pyridone d glucose.[858]

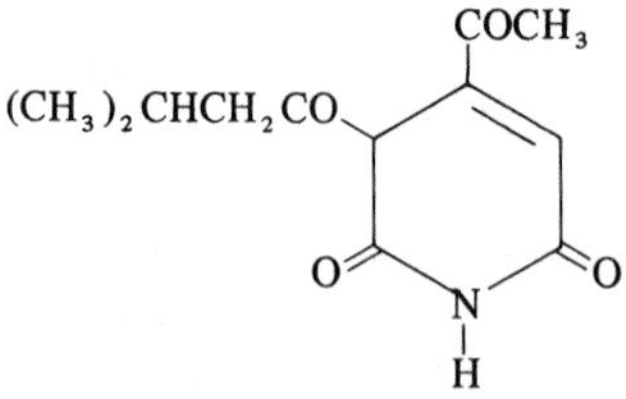

$(CH_3)_2CHCH_2CO$ COCH$_3$ O N H O

**XII-827**

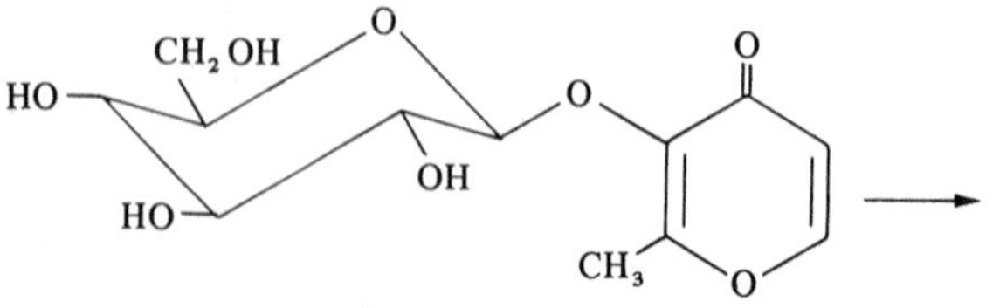

HO CH$_2$OH O O O OH HO CH$_3$ O

**XII-828**

HO CH$_2$OH O O O OH HO CH$_3$ N H HO O CH$_3$ N H

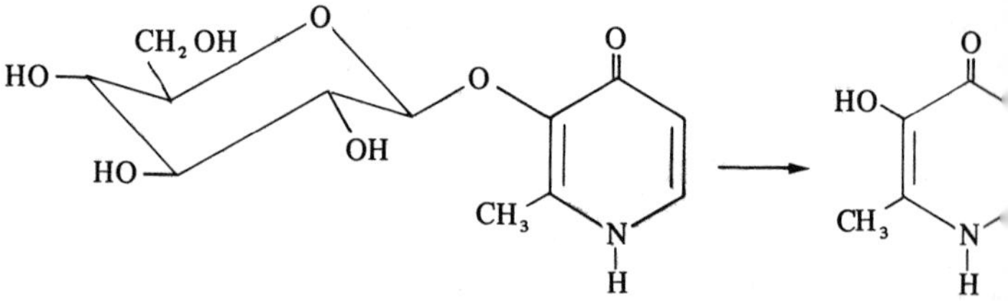

**XII-829**

Perlolidine **(XII-833)**, a minor alkaloid first isolated from the New Zeala perennial rye grass *Lolium perenne* L., has been synthesized from 3-cyano phenyl-2-pyridone (see Section I.2.B.) *via* **XII-830**, **XII-831**, and **XII-832.**

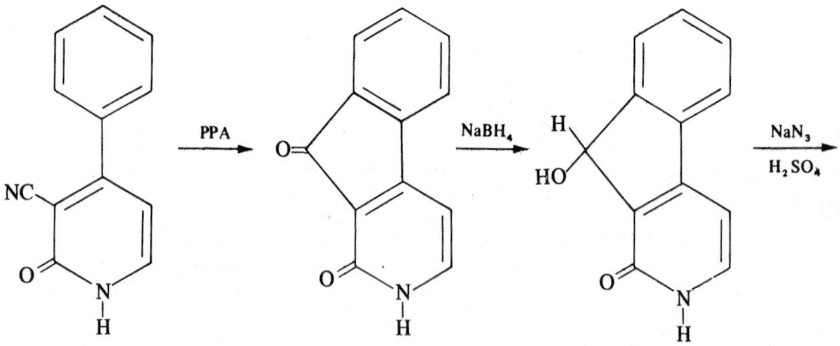

NC O N H PPA O O N H NaBH$_4$ H HO O N H NaN$_3$ H$_2$SO$_4$

**XII-830** **XII-831**

XII-832 XII-833

The alkaloid piplartine **XII-834**, isolated from the roots of *Piper longum* Linn., identical with piperlongumine.[859, 860] A 3,4-dihydro-2-pyridone structure was first proposed,[859] but a detailed NMR study has shown it to be *N*-[β-(3,4,5-trimethoxyphenyl)acrylyl]-5,6-dihydro-2-pyridone **(XII-834)**.[860] Blastidone

XII-834

**XII-835)**, a degradative component of blasticidine S by mild alkaline hydrolysis, has been prepared from methyl isonicotinate through the following sequence:[861, 862]

$CH_3I$ → $K_3Fe(CN)_6$ → $CH_2N_2$ → Ni →

$NH_3$ / $CH_3OH$ → NaOBr → KCNO / HCl →

XII-835

The C-3 of aspartic acid is specifically incorporated into the pyridone nucle of mimosine **(XII-836)**.[134] Lysine is incorporated into the pyridone nucleus

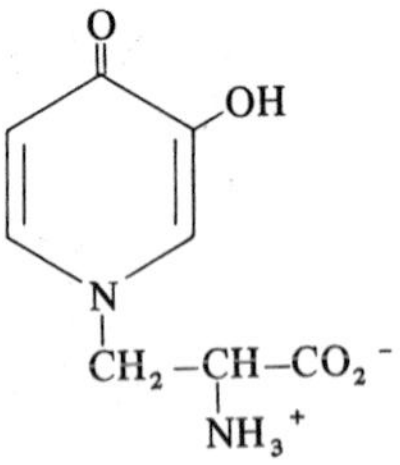

**XII-836**

**XII-836** in *L. Glauca*.[863] The alkaloid tomatillidine has been assigned t structure **XII-837**.[864]

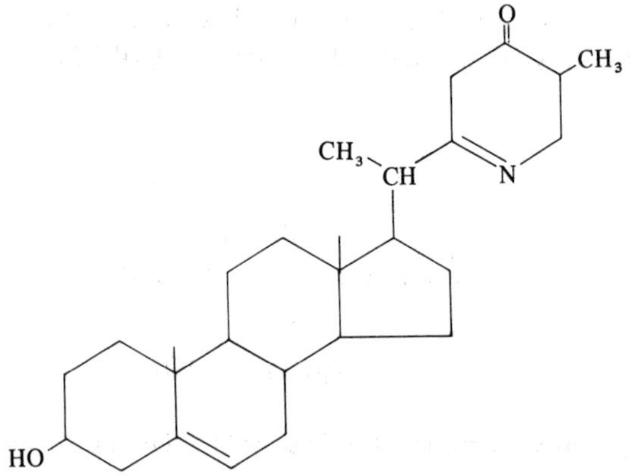

**XII-837**

The structure of piericidin A, a natural insecticide[865] and inhibitor mitochondrial electron transport,[866] has been elucidated **XII-838**.[160, 591, 867–869] Evidence has been presented to assign the S configu tion to C-9 and C-10 of the side chain and a *trans* arrangement at each side-cha double bond.[870] Piericidin B, isolated along with Piericidin A from mycellia

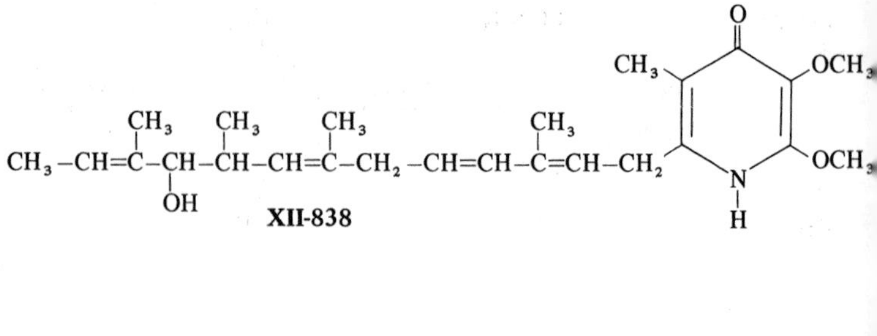

**XII-838**

*treptomyces mobaraensis*, has been identified as the 10-methyl ether of iericidin A.[871, 872] The biosynthesis of Piericidin A and B has been studied.[873]

The metabolism of nicotine by several microorganisms has been studied. ıcluded as products are 6-hydroxynicotine **(XII-839)**, 6-hydroxy- **(XII-841)** and 6-dihydroxy-3-(γ-methylaminobutyryl)pyridine **(XII-842)**, which are obtained om nicotine in a reaction catalyzed by cell-free preparation of a soil acterium.[874-877] and from a *Pseudomonas*.[878] 6-Hydroxynicotine **(XII-839)** ıd **XII-842** are produced from nicotine by oxidation by a cell-free extract of *rthrobacter oxydans*[879, 880] and by a species of *Arthrobacter* present on the ots of tobacco plants.[881]

L-Nicotine oxidation by cell-free extracts of *Arthrobacter oxydans* yields -6-hydroxynicotine exclusively.[882] When *Arthrobacter oxydans* is grown on L-nicotine, L- and D-6-hydroxynicotine oxidases are produced and have been parated.[883] L-6-Hydroxynicotine oxidase converts L-6-hydroxynicotine to (γ-methylaminobutyryl)-2-pyridone **(XII-841)** in the presence of oxygen by a echanism that consists of dehydrogenation followed by hydrolysis of the termediate enamine **(XII-840)**.[884]

2-Pyridone and 4-pyridone are the major persistent metabolites arising from ıe breakdown of pyridine nucleotides following administration of 7-$C^{14}$

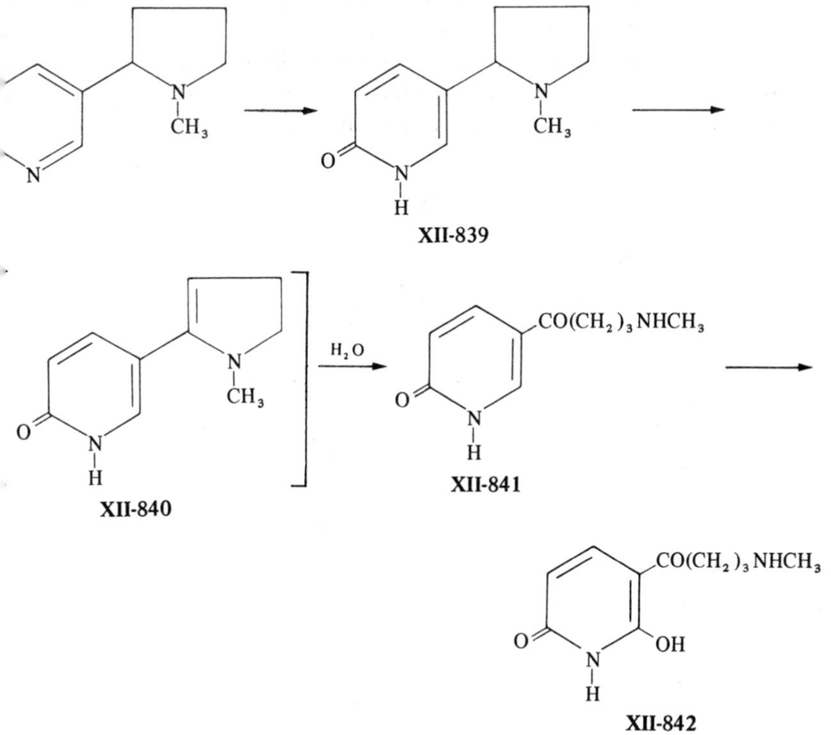

N CH3 → O N H N CH3 →

**XII-839**

O N H N CH3 — $H_2O$ → O N H $CO(CH_2)_3NHCH_3$ →

**XII-840** **XII-841**

O N H OH $CO(CH_2)_3NHCH_3$

**XII-842**

nicotinic acid intramuscularly into a pig.[885] A bacillus isolated on nicotinic aci is shown to oxidize this compound to 6-hydroxynicotinic acid and 2,6-d hydroxynicotinic acid, which is then decarboxylated to 2,6-dihydroxypyr dine.[886] Bacteria utilizing 2- and 3-hydroxypyridine convert these substrates t pyridine-2,5-diol.[439] A number of bacteria produce green or blue pigments b oxidation of 2-pyridone,[887] 5-hydroxy-2-pyridone,[888] nicotine or nicotini acid,[889, 890] 2,6-dihydroxynicotinic acid,[891] isonicotinic acid,[892] and citrazini acid.[892] The blue oxidation products are either indigoidin **(XII-844)** or relate compounds. Indigoidin is a blue compound produced by various bacteria, *I indogofera, C. insidiosum, A. atrocyaneus,* and *A. polychromogenes*.[893-895] Th assignment of structure **XII-844** was based on comparison of its uv, ir, and NM spectra, and acidities with data on structurally related compounds. It has bee prepared directly and in good yield by potassium ferricyanide oxidation c aminocitrazinic acid **(XII-845)**, by decarboxylation of the dicarboxylic aci **XII-846**,[894] by the bromic acid oxidation of 5-amino-2-pyridone,[894] b treatment of the indigoidin hydrolysis product **(XII-847)** with ammoniur acetate, and by the autooxidation of 3-amino-2,6-pyridinol.[846] When th hydrochloride of 3-amino-2,6-pyridinediol is dissolved in sodium bicarbonat solution, a deep blue color is formed that rapidly disappears. On furthe oxidation, the solution turns blue again and deposits indigoidin **(XII-844)** Presumably a blue semiquinoid intermediate is formed that dimerizes to th leuco-indigoidin **(XII-843)**, which is oxidized to **XII-844**. The hydrolysis produc **(XII-847)** is accessible from nitrosocitrazinic acid, from 3,3′-dipyridyl, fror 3-amino-2-pyridone, or from citrazinic acid directly.[894]

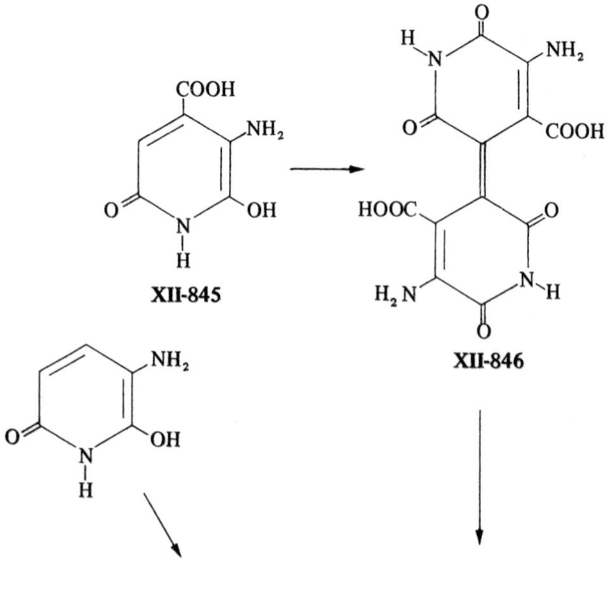

**XII-845**

**XII-846**

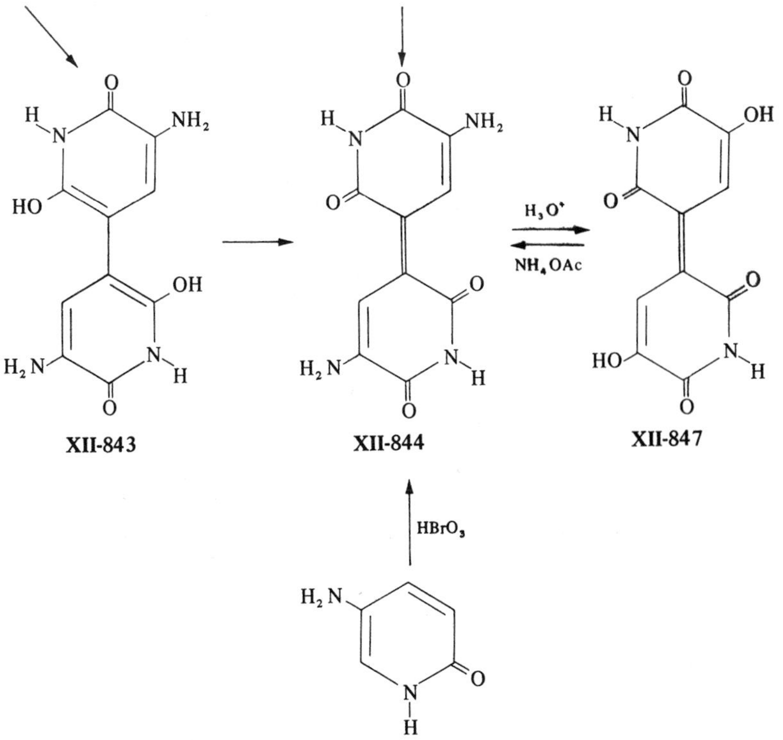

$H_3O^+$ / $NH_4OAc$

HBrO$_3$

**XII-843** **XII-844** **XII-847**

Two alkaloids, syphilobin F **(XII-848)** and syphilobin A, its 3-deoxy erivative, have been isolated from *Lobelia syphilitic* L.[600]

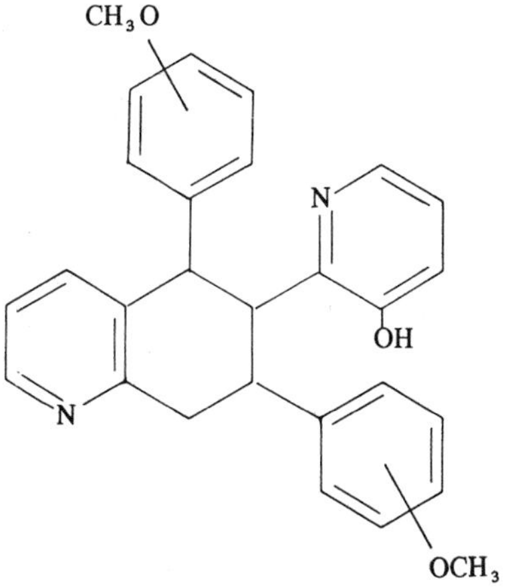

**XII-848**

## VII. Acknowledgments

The author is indebted to Dolores D. Georger who typed the drafts a manuscript, and to Priscilla B. Clarke for many hours of proofreading.

## VIII. Tables of Physical Data

BY PRISCILLA B. CLARKE AND HOWARD TIECKELMANN

Tables XII-25 to XII-145 represent a list of pyridinols and pyridones that ha been described in the literature from 1960 to 1970. A few compounds report during 1958 and 1959 are also included. Melting points, boiling points, and/ analytical data, and references are given for each compound. When the nitrog atom is unsubstituted, the pyridinol structure has been used throughout t tables, which have been divided into the following major categories: 2-, 3-, a 4-pyridinols; 2-, and 4-pyridones; ethers of 2-, 3-, and 4-pyridinols; pyridinedio ethers of pyridinediols, pyridinetriols and ethers of pyridinetriols.

### Format of Tables

To facilitate the location of a specific compound the tables have be subdivided into the following categories:

Alkyl and Aryl; Alkyloxy, Aryloxy, and Hydroxy; Amino, Substituted Amin and Imino; Carboxylic Acids, Aldehydes, Amides and Esters; Nitriles; Hal Nitro and Nitroso; and Sulfur Containing Pyridinols and Pyridones.

If there is more than one substituent on the pyridine ring the compound w be found in the table with the functional group listed last in the above list. F example, 4-chloro-3-nitro-6-*n*-propyl-2-pyridinol is listed in the table with th heading Nitro and Nitroso 2-Pyridinols. In all cases except sulfur containing py dinols and carboxylic acids and their derivatives, the functional group is direct attached to the pyridine ring. In the latter cases the sulfur or carboxylic acid ma appear in any location on the substituent.

# Reference Code

= Molecular orbital calculation
= Crystal data studies
= Column chromatography
= Explosive properties
= Electron spin resonance
= $R_f$ values
= Gas chromatography
= Fluorescence
= Infrared absorption data
= Transition energies
= Optical rotation
= Ligand field spectra
= Mass spectroscopy
= Nuclear magnetic resonance
= Optical rotatory dispersion
= Photochemical decomposition
= Molar conductivity
= Raman spectra
= Spot test
= Thin layer chromatography
= Ultraviolet spectroscopy
= Visible absorption data
= Electronic structure
= X-Ray data
= Stability constants
= Dipole moments

TABLE XII–25. Alkyl and Aryl 2-Pyridinols

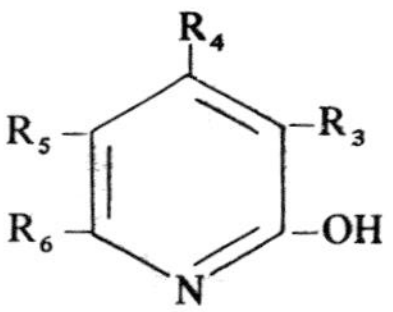

| $R_3$ | $R_4$ | $R_5$ | $R_6$ | m.p. | Derivatives | Ref. |
|---|---|---|---|---|---|---|
| H | H | H | H | 106–108°<br>b.p. 130°/1 mm | | 342 (t); 365 (m.p.);<br>404 (f), (g); 463 (i);<br>465 (i); 472 (n);<br>478 (i), (r); 479 (n);<br>487 (i); 525 (u), (h);<br>546 (i), (r); 550 (b.p.)<br>648 (u); 659 (u);<br>673 (m.p.); 735 (n);<br>896 (i); 897 (m);<br>898 (i); 899 (n);<br>900 (f); 901 (n);<br>902 (i); 903 (n);<br>904 (u); 905 (m.p.), (u);<br>906 (a); 907 (m);<br>908 (n); 909 (u);<br>910 (j); 911 (t);<br>912 (u); 913 (i);<br>914 (m.p.), (u); 915 (m);<br>916 (n); 917 (u);<br>918 (s); 919 (i); |
| | | | | | $^{15}N$ | 465 (i), (n); |
| | | | | | $^{18}O$ | 463 (i); 465 (i);<br>903 (m); |
| | | | | | 2-*O*-acetyl<br>b.p. 74°/0.95 mm | 404 (f), (g);<br>631 (i), (n);<br>747 (b.p.), (i), (n), (u); |
| | | | | | aluminum complex<br>m.p. 186–190° | 920 (m.p.); |
| | | | | | aluminum perchlorate<br>complex, m.p. 200° | 822 (m.p.), (i); |
| | | | | | barium salt | 921 (i); |

TABLE XII-25. Alkyl and Aryl 2-Pyridinols (Continued)

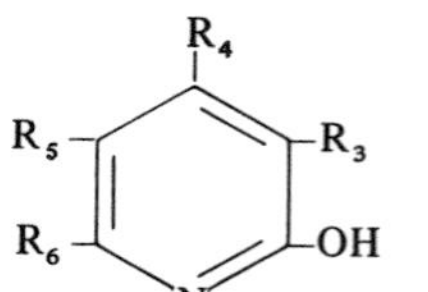

| $R_3$ | $R_4$ | $R_5$ | $R_6$ | m.p. | Derivatives | Ref. |
|---|---|---|---|---|---|---|
| | | | | | dimethyl silicon complex, b.p. 113°/0.8 mm | 920 (b.p.); |
| | | | | | phosphorus complex, m.p. 75° | 920 (m.p.); |
| | | | | | potassium salt | 921 (i); |
| | | | | | silver salt | 921 (i); |
| | | | | | sodium salt | 545 (i), (r); 921 (i); |
| | | | | | trichloroacetyl b.p. 58°/0.7 mm | 922 (b.p.), (i), (n); |
| | | | | | zinc tetrafluoroborate complex, m.p. 118° | 822 (m.p.), (i); |
| | | | | | zinc perchlorate complex, m.p. 139° | 822 (m.p.), (i); |
| $CH_3$ | | | | 139–143° | | 281 (m.p.), (u); 361 (m.p.); 376 (m.p.), (i), (u); 467 (u); 546 (i), (u); 735 (m.p.), (n); 899 (n); 903 (m); 923 (n); 924 (m.p.); |
| | | | | | picrate, m.p. 157–159° | 376 (m.p.); |
| | | | | | $HgCl_2$ m.p. 131–132° | 376 (m.p.); |
| | | | | | $HCl{\cdot}H_2O$ | 546 (i), (u); |
| | $CH_3$ | | | 124–125° | | 467 (i), (u); 546 (i), (u); 903 (m); 907 (m); 923 (n); 924 (m.p.); |
| | | | | | $HCl{\cdot}H_2O$ | 546 (i), (u); |
| | | $CH_3$ | | 184–188° | | 281 (m.p.), (u); 282 (m.p.); 361 (m.p.); 376 (m.p.), (i), (u); 467 (i), (u); |

| | | | | | | |
|---|---|---|---|---|---|---|
| | | | | | | 923 (n), 924 (m.p.), 925 (t), |
| | | | | | picrate, m.p. 146–148° | 376 (m.p.); |
| | | | | | $HgCl_2$, m.p. 206–208° | 376 (m.p.); |
| | | | | | hydrochloride | 546 (i), (u); |
| | | | $CH_3$ | 158–164° | | 372 (m.p.), (i), (n), (u); |
| | | | | | | 420 (m.p.), (i), (m), (n), |
| | | | | | | (u); 422 (i), (n); |
| | | | | | | 467 (i), (u); 903 (m); |
| | | | | | | 907 (m); 923 (n); |
| | | | | | | 924 (m.p.); 925 (t); |
| | | | | | hydrochloride | 546 (i), (u); |
| | | $CH_2OH$ | | 148° | | 880 (m.p.); |
| | | $COCH_3$ | | | | 436 (t); |
| | | | | | | 437 (i), (n), (t), (u); |
| N—N, N—N–H | | | $CH_3$ | 310–312° | | 59 (m.p.); |
| | | –CHCH$_3$ (OH) | | | | 435 (u); |
| | $CH_3$ | | $CH_3$ | 177–186° | | 60 (i), (n); 420 (m.p.), |
| | | | | | | (i), (n), (u); 926 (m.p.); |
| | | $CH_3$ | $CH_3$ | 205–207° | | 927 (m.p.); |
| | $C_2H_5$ | | | 126° | | 283 (m.p.); |
| $–CH_2CH=CH_2$ | | | | 124–128° | | 598 (m.p.), (i), (n), (u); |
| | | | | | | 790 (m.p.); |
| | | $COCH_3$ | $CH_3$ | | | 899 (n); |
| $CH_3$ | $CH_3$ | $CH_3$ | | 76–78° | | 928 (m.p.), (i); |
| $C_2H_5$ | $CH_3$ | | | 169–172° | | 281 (m.p.), (u); |
| | $CH_3$ | $C_2H_5$ | | 160–162° | | 281 (m.p.), (u); |
| $C_2H_5$ | | | $CH_3$ | 102–103° | | 929 (m.p.), (i), (n), (u); |
| | | | | | | 930 (m.p.); |
| | iso$C_3H_7$ | | | oil | | 226 (i); |
| | | | | | picrate, m.p. 146–147° | 226 (m.p.); |
| $CH_2OH$ | | $CH_2OH$ | $CH_3$ | 181–182° | | 931 (m.p.), (i), (n); |
| $Si(CH_3)_3$ | | | | 91–92° | | 377 (m.p.), (i); |
| | $Si(CH_3)_3$ | | | 102–103° | | 377 (m.p.), (i); |
| crotyl | | | | 114–116° | | 790 (m.p.); |
| –CHCH=CH$_2$ (CH$_3$) | | | | oil | | 790 (g); |
| $COCH_3$ | $CH_3$ | | $CH_3$ | 218–219° | | 60 (m.p.), (i), (n), (u); |

TABLE XII-25. Alkyl and Aryl 2-Pyridinols (Continued)

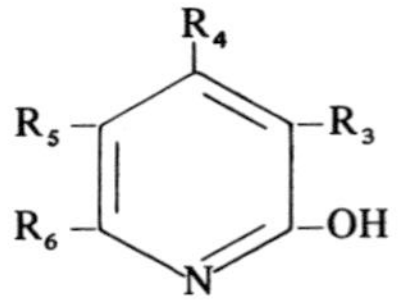

(Structure: pyridine ring with $R_3$, $R_4$, $R_5$, $R_6$ and –OH at C-2; labels: $R_4$, $R_5$, $R_3$, $R_6$, OH, N)

| $R_3$ | $R_4$ | $R_5$ | $R_6$ | m.p. | Derivatives | Ref. |
|---|---|---|---|---|---|---|
| $n$-$C_3H_7$ | | | $CH_3$ | 79–82° | | 929 (m.p.), (i), (n), (u); 930 (m.p.); |
| | | $n$-$C_3H_7$ | $CH_3$ | 127–128° | | 927 (m.p.); 932 (m.p.); |
| | | 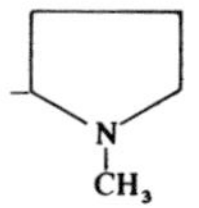 N–$CH_3$ (ring) | | | | 878 (m.p.), (u); |
| | | | | 106° | | 882 (i), (u); |
| | | | | 120–122° | | 881 (m.p.), (u); |
| | | | | | picrate, m.p. 212–215° | 316 (m.p.); |
| | | | | | picrate, m.p. 165–166° | 881 (m.p.), (u); |
| | | $-\underset{\underset{O}{\parallel}}{C}(CH_2)_3NHCH_3$ | | 211–213° | | 878 (u); 880 (m.p.); |
| | | $n$-$C_4H_9$ | $CH_3$ | 93° | | 932 (m.p.); |
| | $C_6H_4NO_2$–$o$ | | | 211–212° | | 73 (m.p.), (i); 933 (m.p.); |
| | | | 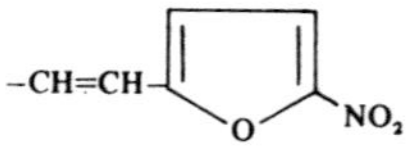 –CH=CH– (O, $NO_2$ ring) | 279° | | 934 (m.p.); 935 (m.p.); 936 (m.p.); 937 (m.p.); |
| | $C_6H_5$ | | | 227–235° | | 1 (m.p.); 46 (m.p.), (i), (m), (u); 71 (m.p.; 907 (m); |
| | | | $C_6H_5$ | 197° | | 71 (m.p.); 488 (i); 907 (m); 938; |
| -N=N$C_6H_4OH$-$p$ | | | | 251° | | 939 (m.p.), (m), (u); |
| | $C_6H_4NH_2$–$o$ | | | 236–237° | | 73 (m.p.), (i), (u); 933 (m.p.); |
| | | | | | $N'$-formyl, m.p. 245–247° | 73 (m.p.), (i); 933 (m.p.); |

O O

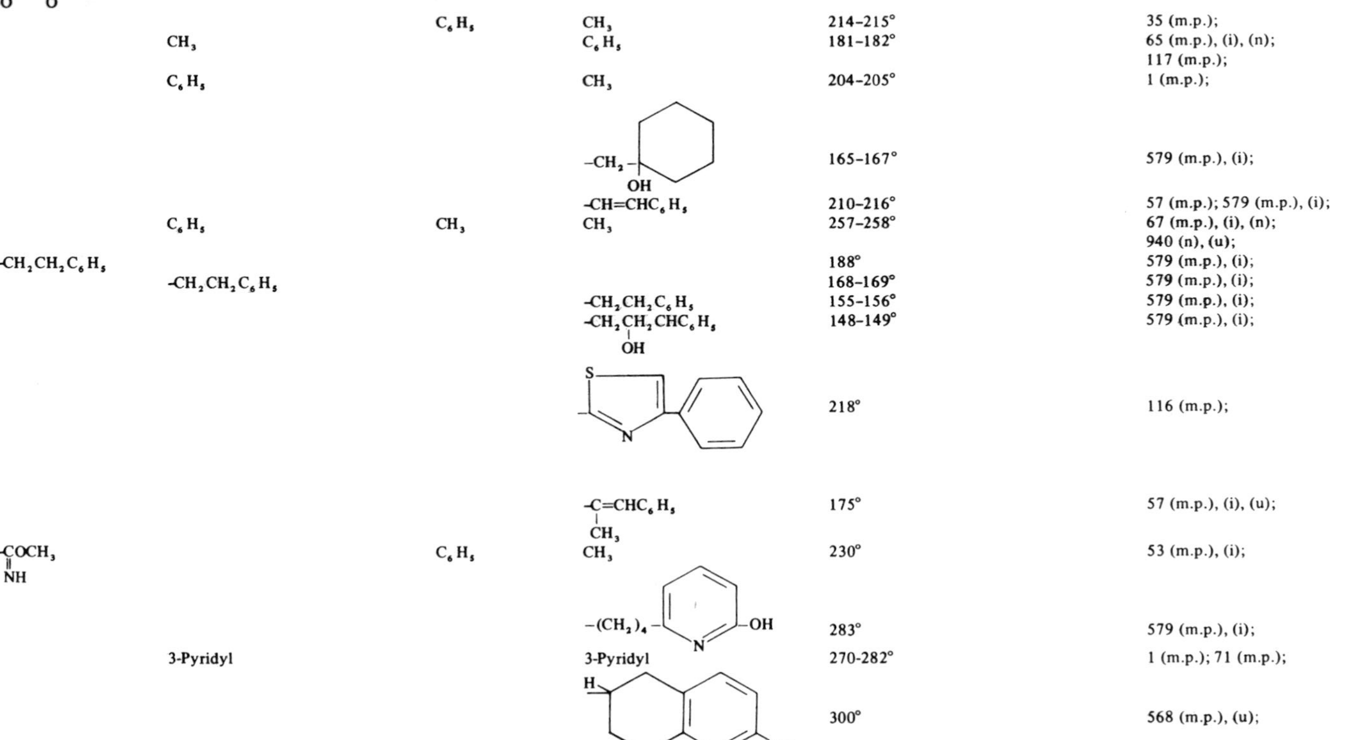

| | | | | | |
|---|---|---|---|---|---|
| | | $C_6H_5$ | $CH_3$ | 214–215° | 35 (m.p.); |
| | $CH_3$ | | $C_6H_5$ | 181–182° | 65 (m.p.), (i), (n); |
| | | | | | 117 (m.p.); |
| | $C_6H_5$ | | $CH_3$ | 204–205° | 1 (m.p.); |
| | | | $-CH_2-$(1-cyclohexyl, OH) | 165–167° | 579 (m.p.), (i); |
| | | | $-CH{=}CHC_6H_5$ | 210–216° | 57 (m.p.); 579 (m.p.), (i); |
| | $C_6H_5$ | $CH_3$ | $CH_3$ | 257–258° | 67 (m.p.), (i), (n); |
| | | | | | 940 (n), (u); |
| $-CH_2CH_2C_6H_5$ | | | | 188° | 579 (m.p.), (i); |
| | $-CH_2CH_2C_6H_5$ | | | 168–169° | 579 (m.p.), (i); |
| | | | $-CH_2CH_2C_6H_5$ | 155–156° | 579 (m.p.), (i); |
| | | | $-CH_2CH_2CH(OH)C_6H_5$ | 148–149° | 579 (m.p.), (i); |
| | | | (S, N thiazolyl structure) | 218° | 116 (m.p.); |
| | | | $-C(CH_3){=}CHC_6H_5$ | 175° | 57 (m.p.), (i), (u); |
| $-C(=NH)OCH_3$ | | $C_6H_5$ | $CH_3$ | 230° | 53 (m.p.), (i); |
| | | | $-(CH_2)_4-$(N, OH structure) | 283° | 579 (m.p.), (i); |
| | 3-Pyridyl | | 3-Pyridyl | 270-282° | 1 (m.p.); 71 (m.p.); |
| | | | (H, OH structure) | 300° | 568 (m.p.), (u); |

TABLE XII-25. Alkyl and Aryl 2-Pyridinols (Continued)

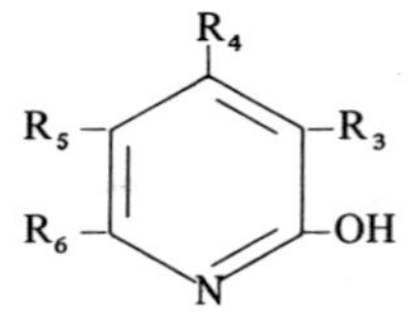

| $R_3$ | $R_4$ | $R_5$ | $R_6$ | m.p. | Derivatives | Ref. |
|---|---|---|---|---|---|---|
| | 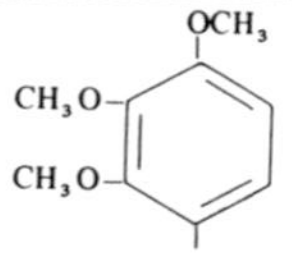 (ring: $OCH_3$, $CH_3O$, $CH_3O$) | | $CH_3$ | 227–229° | | 67 (m.p.), (i); 940 (n); |
| | $C_6H_4NO_2$–*p* | | 2-Pyridyl | 310° | | 72 (m.p.); |
| | 4-Pyridyl | | $C_6H_5$ | 242° | | 72 (m.p.); |
| | $C_6H_5$ | | 3-Pyridyl | 237° | | 72 (m.p.); |
| | $CH_3$ | | α-Naphthyl | 204–206° | | 65 (m.p.), (i), (n); |
| $-\underset{\|\|}{\overset{}{C}}CH_2\underset{\|\|}{C}C_6H_5$ (O, O) | $CH_3$ | | $CH_3$ | 220° | | 60 (m.p.), (i); |
| | | | 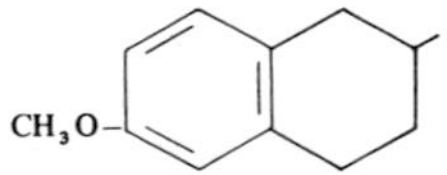 (ring: $CH_3O$–) | 239–240° | | 568 (m.p.), (u); |
| | 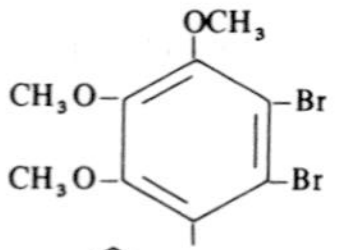 (ring: $OCH_3$, $CH_3O$, $CH_3O$, Br, Br) | $CH_3$ | $CH_3$ | | | 940 (n), (u); |
| | 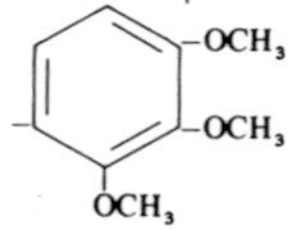 (ring: $OCH_3$, $OCH_3$, $OCH_3$) | $CH_3$ | $CH_3$ | 220° | | 67 (m.p.), (i), (n); 940 (n), (u); |
| | 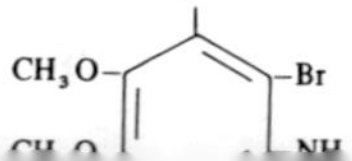 (ring: $CH_3O$, $CH_3O$, Br, $NH$…) | $CH_3$ | $CH_3$ | 233–234 | | 940 (m.p.), (i); |

| | | | | | |
|---|---|---|---|---|---|
| | | | $-(CH_2)_6-$[ring, N]$-OH$ | 239–244° | 6 (m.p.); 579 (m.p.); |
| | $-(CH_2)_6-$[ring, N]$-OH$ | | | 284–285° | 6 (m.p.); |
| | [ring: $OCH_3$, $CH_3O-$, $CH_3O-$, $-NH_2$] | $CH_3$ | $CH_3$ | 280° | 940 (m.p.), (i); |
| | $C_6H_4NO_2-p$ | | $C_6H_5$ | 278–279° | 74 (m.p.); |
| $C_6H_5$ | | | $C_6H_5$ | 260–261° | 38 (m.p.), (i); |
| $C_6H_5$ | $C_6H_5$ | | | 285° | 228 (m.p.), (i), (n), (u); |
| | | | | | 242 (m.p.), (i), (u); |
| | $C_6H_5$ | | $C_6H_5$ | 206–210° | 1 (m.p.); 71 (m.p.); |
| | | | | | 74 (m.p.); |
| | | | | | 941 (m.p.), (i), (u); |
| | | $C_6H_5$ | $C_6H_5$ | 262–263° | 35 (m.p.); |
| | | | | 272–273° | 229 (m.p.); |
| | $(CH_2)_{11}CH_3$ | | | 88–89° | 581 (m.p.); |
| | $C_6H_4OCH_3-p$ | | $C_6H_5$ | 225–226° | 941 (m.p.), (i), (u); |
| $-CH{=}C(C_6H_5)_2$ | | | | 225–226° | 579 (m.p.), (i); |
| | $-CH{=}C(C_6H_5)_2$ | | | 168–170° | 579 (m.p.), (i); |
| | | | $-CH{=}C(C_6H_5)_2$ | 211–213° | 576 (m.p.), (i), (u); |
| | | | | | 580 (m.p.); |
| $-CH_2C(OH)(C_6H_5)_2$ | | | | 294–295° | 579 (m.p.), (i); |
| | $-CH_2C(OH)(C_6H_5)_2$ | | | 259–260° | 579 (m.p.), (i); |
| | | | $-CH_2C(OH)(C_6H_5)_2$ | 271–272° | 580 (m.p.); |
| | $C_6H_5$ | $C_6H_5$ | $C_6H_4Br-p$ | 288° | 122 (m.p.); |
| $C_6H_5$ | $C_6H_5$ | $C_6H_5$ | | 323–325° | 230 (m.p.), (i), (u); |
| $C_6H_5$ | $C_6H_5$ | | $C_6H_5$ | 300–301° | 938 (m.p.), (i); |
| | $C_6H_5$ | $C_6H_5$ | $C_6H_5$ | 277° | 121 (m.p.), (i), (u); |
| | | | | | 122 (m.p.), (i); 230 (m.p.); |

TABLE XII-25. Alkyl and Aryl 2-Pyridinols (Continued)

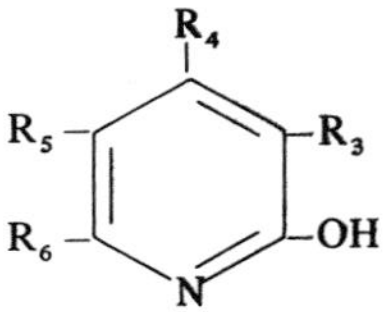

| $R_3$ | $R_4$ | $R_5$ | $R_6$ | m.p. | Derivatives | Ref. |
|---|---|---|---|---|---|---|
| | $C_6H_5$ | $C_6H_5$ | $C_6H_4CH_3$-*p* | 270° | | 122 (m.p.), (i); |
| | $C_6H_5$ | $C_6H_5$ | $C_6H_4OCH_3$-*p* | 245° | | 802 (m.p.), (i); |
| | 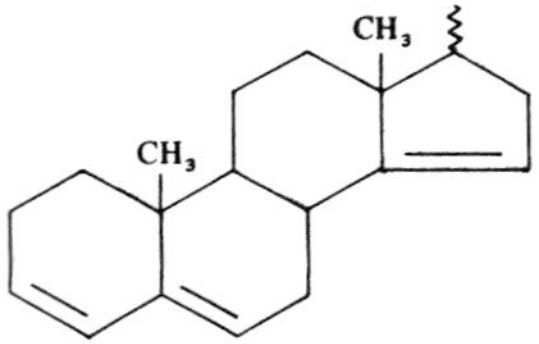 | | | 267–282° | | 114 (m.p.), (i); |
| | 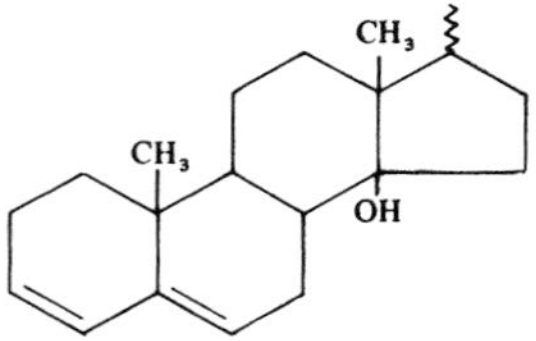 | | | 330–345° | | 114 (m.p.), (i); |
| | 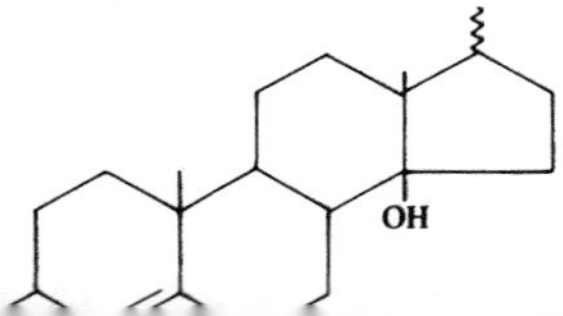 | | | 289–292° | | 114 (m.p.), (i); |

| | m.p. | Derivatives | Ref. |
|---|---|---|---|
| OH<br>O–L-Rhamnose–D-Glucose | 283–284° | | 114 (m.p.), (i); |
| | | hexaacetyl, m.p. 239–270° | 114 (m.p.); |

TABLE XII-26. Alkyloxy and Aryloxy 2-Pyridinols

(Structure: 2-pyridinol ring bearing $R_3$, $R_4$, $R_5$, $R_6$ and OH at C-2)

| $R_3$ | $R_4$ | $R_5$ | $R_6$ | m.p. | Derivatives | Ref. |
|---|---|---|---|---|---|---|
| $OCH_3$ | | | | 114–116° | | 501 (m.p.), (i); |
| | | | $OCH_3$ | 102–104° | | 455 (m.p.), (n), (u); |
| | | | | | | 505 (m.p.), (n), (t); |
| $OCH_3$ | $OCH_3$ | | $CH_2OH$ | 176–177° | | 160 (m.p.), (i), (u); |
| | $OCH_3$ | $COCH_3$ | $CH_3$ | 230° | | 94 (m.p.), (n), (u); |
| | | | $OCH_2C_6H_5$ | 128–129° | | 618 (m.p.), (i), (n); |
| | $C_6H_5$ | $OCH_3$ | $C_6H_5$ | 203° | | 120 (m.p.); 121 (i); |
| | $C_6H_5$ | $OCH_3$ | $C_6H_4CH_3$–*p* | 228° | | 120 (m.p.); |
| | $C_6H_5$ | $OCH_3$ | $C_6H_4OCH_3$–*p* | 206° | | 120 (m.p.); |
| | $C_6H_5$ | $OC_6H_4Cl$–*p* | $C_6H_5$ | 241° | | 121 (m.p.); |
| | $C_6H_5$ | $OC_6H_4Br$–*p* | $C_6H_5$ | 247° | | 121 (m.p.); |
| | $C_6H_5$ | $OC_6H_5$ | $C_6H_5$ | 233° | | 121 (m.p.), (i), (u); |
| | $C_6H_5$ | $OC_6H_4CH_3$–*p* | $C_6H_5$ | 252° | | 121 (m.p.); |

TABLE XII-27. Amino 2-Pyridinols

(Structure: 2-pyridinol ring with substituents $R_3$, $R_4$, $R_5$, $R_6$ and OH)

| $R_3$ | $R_4$ | $R_5$ | $R_6$ | m.p. | Derivatives | Ref. |
|---|---|---|---|---|---|---|
| $NH_2$ | | | | | | 906 (a); |
| | | | | | hydrobromide, m.p. 225–230° | 740 (m.p.), (i); |
| | $NH_2$ | | | | | 906 (a); |
| | | $NH_2$ | | | | 906 (a); |
| | | | $NH_2$ | | | 494 (i), (u); 918 (s); 942 (b); |
| | $NH_2$ | | $NH_2$ | 178° | | 34 (m.p.), (i); |
| $-NHCHO$ | | | | 223–224° | | 642 (m.p.); |
| $-NHCSNH_2$ | | | | 220° | | 642 (m.p.); |
| $-NHCONH_2$ | | | | >310° | | 642 (m.p.), (u); |
| $-NHCOCH_3$ | | | | 215° | | 642 (m.p.), (u); |
| | | | $-NHCOCH_3$ | 212–213° | | 641 (m.p.); |
| $CH_3$ | | $-NHCOCH_3$ | | 246° | | 843 (m.p.), (u); |
| $-NHCOC_2H_5$ | | | | 166° | | 642 (m.p.); |
| $-NHCOOC_2H_5$ | | | | 187° | | 642 (m.p.); |
| $NH_2$ | | | $n$-$C_3H_7$ | | picrate, m.p. 167–168° | 150 (m.p.); |
| $NH_2$ | $NH_2$ | | $n$-$C_3H_7$ | | picrate, m.p. 198° | 150 (m.p.); |
| $-N(COCH_3)_2$ | | | | 133–134° | | 642 (m.p.); |
| $-NHCOCH_3$ | | $-NHCOCH_3$ | | 267–269° | | 943 (m.p.); |
| $CH_3$ | | $-NHCOCH_3$ | | | 2-*O*-acetyl, | |

|  |  |  |  |  | (n); |
|---|---|---|---|---|---|
| $-NHSO_2C_6H_4CH_3$-*p* |  |  |  | 211–212° | 642 (m.p.); |
| $CH_3$ | $-N(C_2H_5)_2$ | $C_6H_5$ |  | 232° | 102 (m.p.), (i), (u); |
| $NH_2$ | 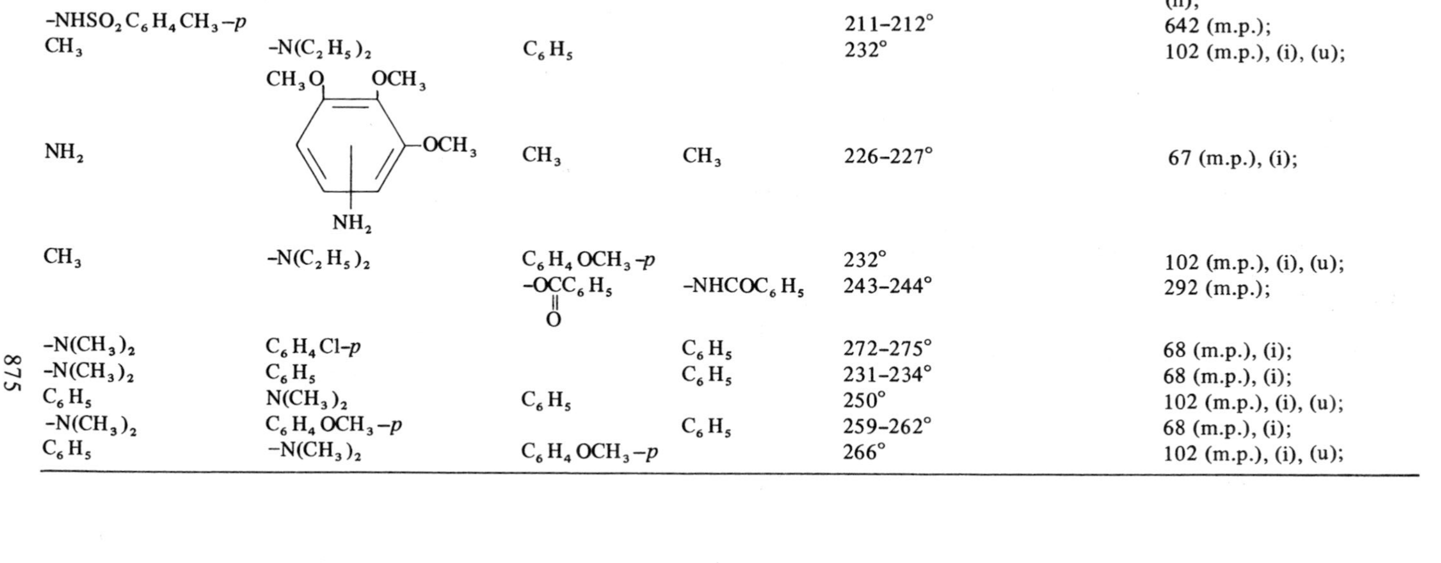 | $CH_3$ | $CH_3$ | 226–227° | 67 (m.p.), (i); |
| $CH_3$ | $-N(C_2H_5)_2$ | $C_6H_4OCH_3$-*p* |  | 232° | 102 (m.p.), (i), (u); |
|  |  | $-OC(=O)C_6H_5$ | $-NHCOC_6H_5$ | 243–244° | 292 (m.p.); |
| $-N(CH_3)_2$ | $C_6H_4Cl$-*p* |  | $C_6H_5$ | 272–275° | 68 (m.p.), (i); |
| $-N(CH_3)_2$ | $C_6H_5$ |  | $C_6H_5$ | 231–234° | 68 (m.p.), (i); |
| $C_6H_5$ | $N(CH_3)_2$ | $C_6H_5$ |  | 250° | 102 (m.p.), (i), (u); |
| $-N(CH_3)_2$ | $C_6H_4OCH_3$-*p* |  | $C_6H_5$ | 259–262° | 68 (m.p.), (i); |
| $C_6H_5$ | $-N(CH_3)_2$ | $C_6H_4OCH_3$-*p* |  | 266° | 102 (m.p.), (i), (u); |

TABLE XII-28. 2-Pyridinol Carboxylic Acids and Derivatives

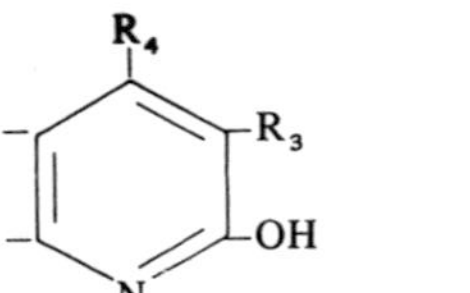


| $R_3$ | $R_4$ | $R_5$ | $R_6$ | m.p. | Derivatives | Ref. |
|---|---|---|---|---|---|---|
| COOH | | CHO | | 219° | | 880 (m.p.); |
| | | | | 246–250° | | 362 (m.p.); |
| | | | | 225° | | 101 (m.p.); |
| | | | | 262° | | 21 (m.p.), (u); |
| | | | | | methyl ester, m.p. 153° | 101 (m.p.); |
| | | | | | ethyl ester, m.p. 139° | 21 (m.p.), (u); |
| | COOH | | | 328° | | 266 (m.p.); |
| | | | | | methyl ester, m.p. 211–212° | 283 (m.p.), 362 (m.p.); |
| | | COOH | | 325° | | 262 (m.p.); 899 (n); |
| | | | | 301–302° | | 101 (m.p.); 855 (m.p.); |
| | | | | | methyl ester, m.p. 164° | 101 (m.p.); 699 (m.p.); |
| | | | | | ethyl ester, m.p. 143–144° | 362 (m.p.); |
| | | | | | m.p. 150° | 880 (m.p.); |
| | | $CONH_2$ | | 325–327° | | 613 (m.p.); |
| | | | COOH | 273–282° | | 57 (m.p.), (i), (u); |
| | | | | | | 101 (m.p.); 362 (m.p.); |
| | | | | | methyl ester, m.p. 109-116° | 116 (m.p.); 362 (m.p.); |
| | | | | | ethyl ester, m.p. 164° | 116 (m.p.); |
| | | | $CONH_2$ | 251° | | 116 (m.p.); |
| | | | $CONHCH_3$ | 176° | | 116 (m.p.); |
| | | | $CONHC_2H_5$ | 176° | | 116 (m.p.); |
| COOH | | | COOH | 304° | | 57 (m.p.), (i), (u); |

| | | | | | | |
|---|---|---|---|---|---|---|
| $CONH_2$ | | | $CH_3$ | 305–306° | | 575 (m.p.), (i), (u); |
| COOH | COOH | | COOH | | hydrate, m.p. 262° | 57 (m.p.), (i), (u); |
| COOH | | | $COCH_3$ | 271° | | 57 (m.p.), (i), (u); |
| COOH | | COOH | $CH_3$ | | diethyl ester, m.p. 196–198° | 931 (m.p.), (i), (n); |
| | COOH | COOH | $CH_3$ | | dimethyl ester, m.p. 169–171° | 213 (m.p.); |
| COOH | $CH_3$ | | $CH_3$ | 260–262° | | 674 (m.p.); |
| | | | | | hydrate m.p. 251–252° | 926 (m.p.); |
| COOH | $CH_3$ | | $CH_3$ | | ethyl ester, m.p. 134–135° | 60 (i), (n), (u); |
| | | | | | | 674 (m.p.); |
| | | | $-CH_2CH(NH_2)COOH$ | 278–280° | | 425 (m.p.), (i); |
| COOH | COOH | | $COCH_3$ | 254° | | 57 (m.p.), (i); |
| | | | | | oxime, m.p. 350° | 57 (m.p.); |
| | | | $-CH_2CH{=}CHCOOH$ | 199–200° | | 944 (m.p.); |
| $CONH_2$ | $C_6H_5$ | | | 285–287° | | 46 (m.p.), (i), (m), (u); |
| $-C(=O)CH_2C(=O)COOH$ | $CH_3$ | | $CH_3$ | | ethyl ester, m.p. 175–176° | 60 (m.p.), (i); |
| COOH | | | $COC_6H_5$ | 257° | | 57 (m.p.), (i), (u); |
| COOH | | $C_6H_5$ | $CH_3$ | 269–271° | | 53 (m.p.), (i), (u); |
| | | | | | methyl ester, m.p. 230–232° | 53 (m.p.), (i), (u); |
| $CONHCH_3$ | | $C_6H_5$ | $CH_3$ | 316–318° | | 53 (m.p.), (i), (u); |
| $CONHNH_2$ | | $C_6H_5$ | $CH_3$ | >350° | | 53 (m.p.), (i); |
| | COOH | $C_6H_5$ | $CH_3$ | 280° | | 53 (m.p.), (i); |
| | | | | | methyl ester, m.p. 183–185° | 53 (m.p.), (i); |
| | | | | | ethyl ester, m.p. 154–155° | 53 (m.p.), (i); |
| $C_6H_5$ | $CH_3$ | $CONH_2$ | | 350–352° | | 218 (m.p.); |
| COOH | | | $-CH{=}CHC_6H_5$ | 256° | | 57 (m.p.), (i), (u); |
| | COOH | | $-CH{=}CHC_6H_5$ | 343° | | 57 (m.p.), (i); |
| COOH | COOH | $C_6H_5$ | $CH_3$ | 225–230° | | 53 (m.p.), (i); |
| $CONH_2$ | | | $-CH_2COC_6H_5$ | 287–289° | | 576 (m.p.); |

TABLE XII-28. 2-Pyridinol Carboxylic Acids and Derivatives (Continued)

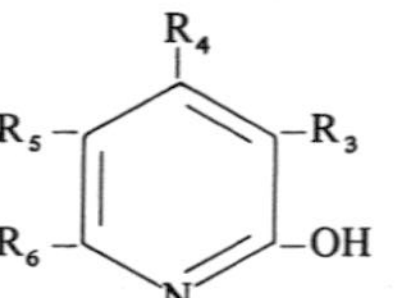


| $R_3$ | $R_4$ | $R_5$ | $R_6$ | m.p. | Derivatives | Ref. |
|---|---|---|---|---|---|---|
| | | | $-CH_2CH(COOH)C_6H_5$ | | methyl ester, m.p. 197-198° | 579 (m.p.), (i); |
| COOH | | $C_6H_5$ | $-CH{=}C(OH)COOH$ | 190° | | 53 (m.p.), (i); |
| COOH | | | $-C(CH_3){=}CHC_6H_5$ | 230° | | 57 (m.p.), (i), (u); |
| | COOH | | $-C(CH_3){=}CHC_6H_5$ | 314° | | 57 (m.p.); |
| $CH_2C_6H_5$ | COOH | COOH | $CH_3$ | | dimethyl ester, m.p. 178–179° | 213 (m.p.); |
| $CH_3$ | COOH | COOH | $CH_2C_6H_5$ | | dimethyl ester, m.p. 191–193° | 213 (m.p.); |
| $CONHC_2H_5$ | | $C_6H_5$ | $CH_3$ | 249–251° | | 53 (m.p.), (i), (u); |
| COOH | | $C_6H_5$ | $-CH{=}C(OH)COOH$ | | diethyl ester, m.p. 168–170° | 53 (m.p.), (i), (u); |
| $-CONHCH_2CH_2N(C_2H_5)_2$ | | $C_6H_5$ | $CH_3$ | 180–182° | | 53 (m.p.), (i); |
| $-CONH(CH_2)_3N(C_2H_5)_2$ | | $C_6H_5$ | $CH_3$ | 181–182° | | 53 (m.p.), (i); |
| $-CONH(CH_2)_2C_6H_5$ | | $C_6H_5$ | $CH_3$ | 248–250° | | 53 (m.p.), (i); |

TABLE XII-29. 2-Pyridinol Nitriles

| $R_3$ | $R_4$ | $R_5$ | $R_6$ | m.p. | Derivatives | Ref. |
|---|---|---|---|---|---|---|
| | | | CN | 214° | | 116 (m.p.); |
| CN | | | COOH | | hydrate, m.p. 275° | 57 (m.p.), (i), (u); |
| CN | | | $CH_3$ | | | 437 (u); 575 (n); |
| CN | | $CONH_2$ | $NH_2$ | | potassium salt hydrate, m.p. 327–332° | 10 (m.p.); |
| CN | $CF_3$ | | $CF_3$ | 119–121° | | 55 (m.p.), (i), (u); |
| | | | | | sodium salt, m.p. 253–255° | 55 (m.p.); |
| CN | $CF_3$ | | $CH_3$ | 232–234° | | 55 (m.p.), (i), (u); |
| CN | | | $COCH_3$ | 213° | | 57 (m.p.), (i), (u); |
| | | | | | oxime, m.p. 284° | 57 (m.p.); |
| CN | $CONH_2$ | | $CH_3$ | 300° | | 945 (m.p.); |
| CN | $CH_3$ | | $CH_3$ | | | 575 (n); |
| CN | | | $CH_2CH_3$ | 246–248° | | 575 (m.p.), (u); |
| CN | CN | $CH_3$ | $CH_3$ | 283° | | 209 (m.p.), (i); |
| CN | $CH_2COOH$ | | $CH_3$ | 228° | | 945 (m.p.); |
| CN | | | iso$C_3H_7$ | 208–210° | | 946 (m.p.); |
| CN | $CH_2OCH_3$ | $NH_2$ | $CH_3$ | 240° | | 52 (m.p.); |
| CN | $CH_3$ | $NH_2$ | $CH_2OCH_3$ | 206° | | 52 (m.p.); |
| CN | $CH_2COOH$ | | $C_2H_5$ | 205° | | 945 (m.p.); |
| CN | $CH_2CH_2COOH$ | | $CH_3$ | 217° | | 945 (m.p.); |
| CN | $CH_2CH_2COOH$ | | $CH_2CH_3$ | 211° | | 945 (m.p.); |
| CN | $C_6H_5$ | | | 233–234° | | 46 (m.p.); |
| CN | | | $COC_6H_5$ | 253° | | 57 (m.p.), (i), (u); |

TABLE XII-29. 2-Pyridinol Nitriles (Continued)

(Structure: 2-pyridinol ring with substituents $R_3$, $R_4$, $R_5$, $R_6$; OH at C-2)

| $R_3$ | $R_4$ | $R_5$ | $R_6$ | m.p. | Derivatives | Ref. |
|---|---|---|---|---|---|---|
| CN | $C_6H_5$ | | $CONH_2$ | 280° | | 79 (m.p.), (i); |
| CN | $CH_3$ | $C_6H_5$ | | 190–192° | | 53 (m.p.), (i), (u); |
| CN | $CH_3$ | | $C_6H_5$ | | | 575 (n); |
| CN | | $C_6H_5$ | $CH_3$ | 294–296° | | 53 (m.p.), (i), (u); 575 (n); |
| CN | $CH_3$ | $C_6H_5$ | $NH_2$ | 346° | | 218 (m.p.); |
| CN | | | $-CH{=}CHC_6H_5$ | 304° | | 57 (m.p.), (i), (u); |
| CN | | | $-CH_2COC_6H_5$ | 199–205° | | 576 (m.p.), (i), (n), (u); |
| CN | COOH | $C_6H_5$ | $CH_3$ | | methyl ester, m.p. 198–199°<br>ethyl ester, m.p. 165–167° | 53 (m.p.), (i), (u);<br>53 (m.p.), (i); |
| CN | $C_6H_5$ | $CH_3$ | $CH_3$ | 302–304° | | 77 (m.p.), (i), (n); |
| CN | | | $CH_2CH_2C_6H_5$ | 204–205° | | 575 (m.p.), (n), (u); |
| CN | $CH_3$ | $C_6H_5$ | $CH_3$ | 354–356° | | 575 (m.p.), (i), (u); |
| CN | | | $-CH_2CH(OH)C_6H_5$ | 205–206° | | 576 (m.p.), (i), (u); |
| CN | | | $-C(CH_3){=}CHC_6H_5$ | 238° | | 57 (m.p.), (i), (u); |
| CN | | | $-CH_2CH(OCOCH_3)C_6H_5$ | 200–308° | | 576 (m.p.), (i), (u); |
| CN | | | (cyanophenyl ring structure, partly cut off) | 269–270° | | 575 (m.p.), (i), (u); |

| | | | | | |
|---|---|---|---|---|---|
| CN | 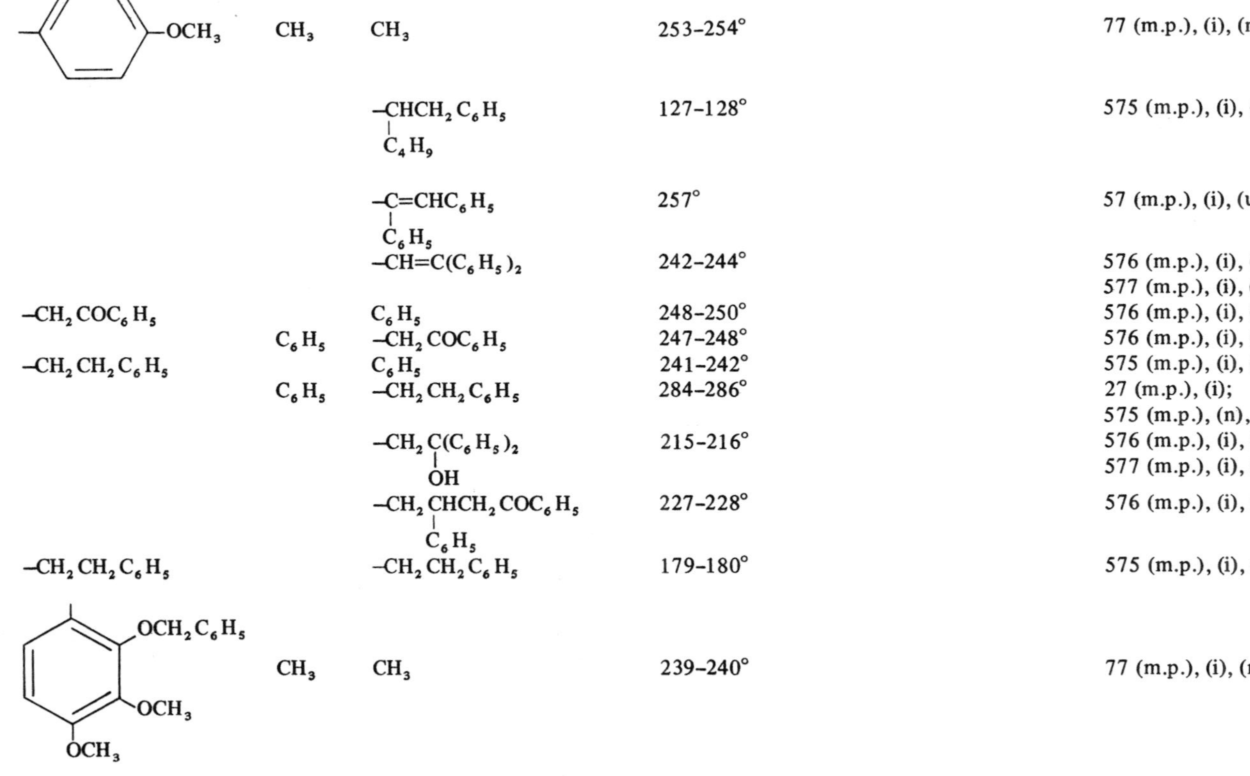 | $CH_3$ | $CH_3$ | 253–254° | 77 (m.p.), (i), (n); |
| CN | | | $-CHCH_2C_6H_5$ (with $C_4H_9$ on CH) | 127–128° | 575 (m.p.), (i), (u); |
| CN | | | $-C(C_6H_5)=CHC_6H_5$ | 257° | 57 (m.p.), (i), (u); |
| CN | | | $-CH=C(C_6H_5)_2$ | 242–244° | 576 (m.p.), (i), (u); 577 (m.p.), (i), (u); |
| CN | $-CH_2COC_6H_5$ | | $C_6H_5$ | 248–250° | 576 (m.p.), (i), (n), (u); |
| CN | | $C_6H_5$ | $-CH_2COC_6H_5$ | 247–248° | 576 (m.p.), (i), (n), (u); |
| CN | $-CH_2CH_2C_6H_5$ | | $C_6H_5$ | 241–242° | 575 (m.p.), (i), (n), (u); |
| CN | | $C_6H_5$ | $-CH_2CH_2C_6H_5$ | 284–286° | 27 (m.p.), (i); 575 (m.p.), (n), (u); |
| CN | | | $-CH_2C(OH)(C_6H_5)_2$ | 215–216° | 576 (m.p.), (i), (u); 577 (m.p.), (i), (u); |
| CN | | | $-CH_2CH(C_6H_5)CH_2COC_6H_5$ | 227–228° | 576 (m.p.), (i), (n), (u); |
| CN | $-CH_2CH_2C_6H_5$ | | $-CH_2CH_2C_6H_5$ | 179–180° | 575 (m.p.), (i), (n), (u); |
| CN | (2-$OCH_2C_6H_5$-3,4-di-$OCH_3$-phenyl) | $CH_3$ | $CH_3$ | 239–240° | 77 (m.p.), (i), (n); |

TABLE XII-29. 2-Pyridinol Nitriles (Continued)

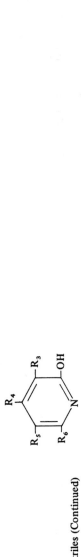

| $R_3$ | $R_4$ | $R_5$ | $R_6$ | m.p. | Derivatives | Ref. |
|---|---|---|---|---|---|---|
| CN | $C_6H_5$ | | [structure: 2-hydroxy-3-carbamoyl-4-phenyl-6-pyridyl; labels N, OH, $CONH_2$, $C_6H_5$] | 360° | | 79 (m.p.); |
| CN | $-CH{=}C(C_6H_5)_2$ | | $C_6H_5$ | 275–276° | | 576 (m.p.), (i), (u); |
| CN | $C_6H_4CH_3$–*p* | | [structure: 2-hydroxy-3-cyano-4-($C_6H_4CH_3$-*p*)-6-pyridyl; labels N, OH, CN, $C_6H_4CH_3$-*p*] | > 300° | | 79 (m.p.); |
| CN | $C_6H_4OCH_3$–*p* | | [structure: 2-hydroxy-3-cyano-4-($C_6H_4OCH_3$-*p*)-6-pyridyl; labels N, OH, CN, $C_6H_4OCH_3$-*p*] | > 300° | | 79 (m.p.); |
| CN | $-CH_2C(C_6H_5)_2$ | | $C_6H_5$ | 250–252° | | 576 (m.p.), (i), (u); |

| | | | |
|---|---|---|---|
| 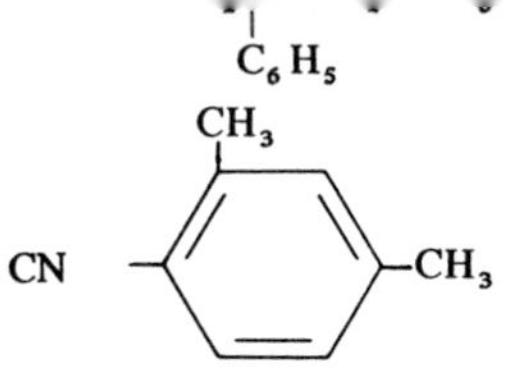 $C_6H_5$; CN, $CH_3$, $CH_3$ (structure) | 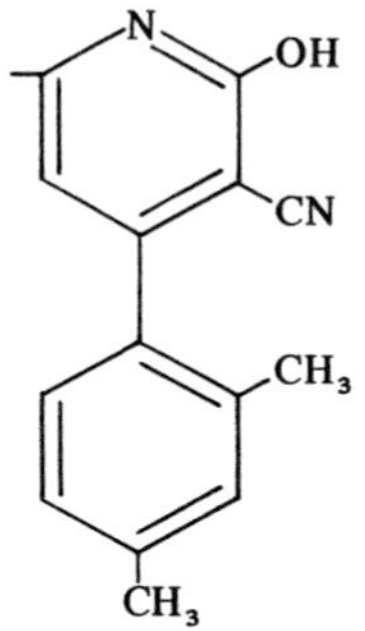 N, OH, CN, $CH_3$, $CH_3$ (structure) | 242° | 79 (m.p.); |

TABLE XII-30. Halo 2-Pyridinols

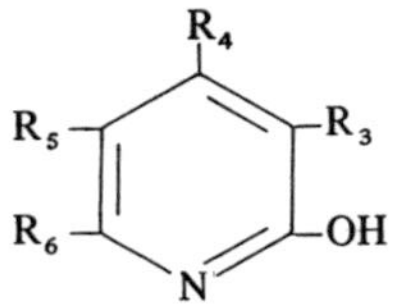

| $R_3$ | $R_4$ | $R_5$ | $R_6$ | m.p. | Derivatives | Ref. |
|---|---|---|---|---|---|---|
| F | | | | 166–167° | | 363 (m.p.), (u); |
| | | | | | 2-*O*-acetyl, b.p. 82°/1 mm | 363 (b.p.); |
| Cl | | | | 181–183° | | 363 (m.p.), (u); 467 (i), (u); 546 (i), (u); |
| | | | | | hexachloro-antimonate·$H_2O$ | 546 (i), (u); |
| | | | | | 2-*O*-acetyl, b.p. 54°/1 mm | 360 (b.p.); |
| | | Cl | | | | 327 (u); 467 (i), (u); 546 (i), (r), (u); |
| | | | | | hexachloro-antimonate | 546 (i), (u); |
| | | | | | hydrochloride | 546 (i), (u); |
| | | | Cl | | | 241 (u); 474 (b); 494 (i); |
| Br | | | | 181–187° | | 363 (m.p.), (u); 476 (i), (u); 546 (i), (u); 947 (m.p.); |
| | | | | | 2-*O*-acetyl b.p. 77–78° | 363 (b.p.); |
| | | Br | | | | 467 (i), (u); 546 (i), |

| | | | | | | |
|---|---|---|---|---|---|---|
| | | I | | 189–191° | | 467 (i), (u); 546 (i), (r), (u); 704 (m.p.), (u); |
| | | | | | hydrochloride | 546 (i), (u); |
| Cl | | Cl | | 178–179° | | 15 (m.p.); 467 (i), (u); 546 (i), (u); 831 (i); |
| | | | | | hexachloro-antimonate | 546 (i), (u); |
| Br | | Br | | | | 467 (i), (u); 546 (i), (u); 831 (i); |
| | | | | | hexachloro-antimonate | 546 (i), (u); |
| I | | I | | 264–265° | | 467 (i), (u); 546 (i), (u); 704 (m.p.), (u); |
| Cl | Cl | Cl | | 228–229° | | 15 (m.p.), (u); |
| Cl | | Cl | Cl | 172–174° | | 703 (m.p.); 948 (p); |
| Br | | Br | Cl | 189–192° | | 703 (m.p.); |
| F | F | Cl | F | | | 246 (n); |
| Cl | F | F | F | | | 246 (n); |
| F | Br | F | F | 130–131° | | 247 (m.p.), (n); 949 (m.p.); |
| Cl | F | Cl | F | 149–153° | | 246 (m.p.), (i), (n); 950 (m.p.); 951 (m.p.); |
| Cl | Cl | Cl | F | 172–176° | | 951 (m.p.); |
| Cl | Cl | Cl | Cl | 221–226° | | 18 (m.p.); 240 (m.p.); 241 (i), (u); |
| Br | Br | Br | Br | 240–241° | | 952 (m.p.), (i); |
| $NH_2$ | | Br | | 182° | | 643 (m.p.); |
| $CF_3$ | | | | 153–154° | | 375 (m.p.), (i), (n); |
| | | | $CF_3$ | 120–121° | | 375 (m.p.), (n); |
| Cl | $OCH_3$ | Cl | F | 179–180° | | 953 (m.p.); |
| Cl | $OCH_3$ | Cl | Cl | 197° | | 240 (m.p.); |
| Br | $OCH_3$ | Br | Br | 210° | | 952 (m.p.), (i); |
| Cl | Cl | | $CH_3$ | 204–206° | | 37 (m.p.); |

TABLE XX-30. Halo 2-Pyridinols (Continued)

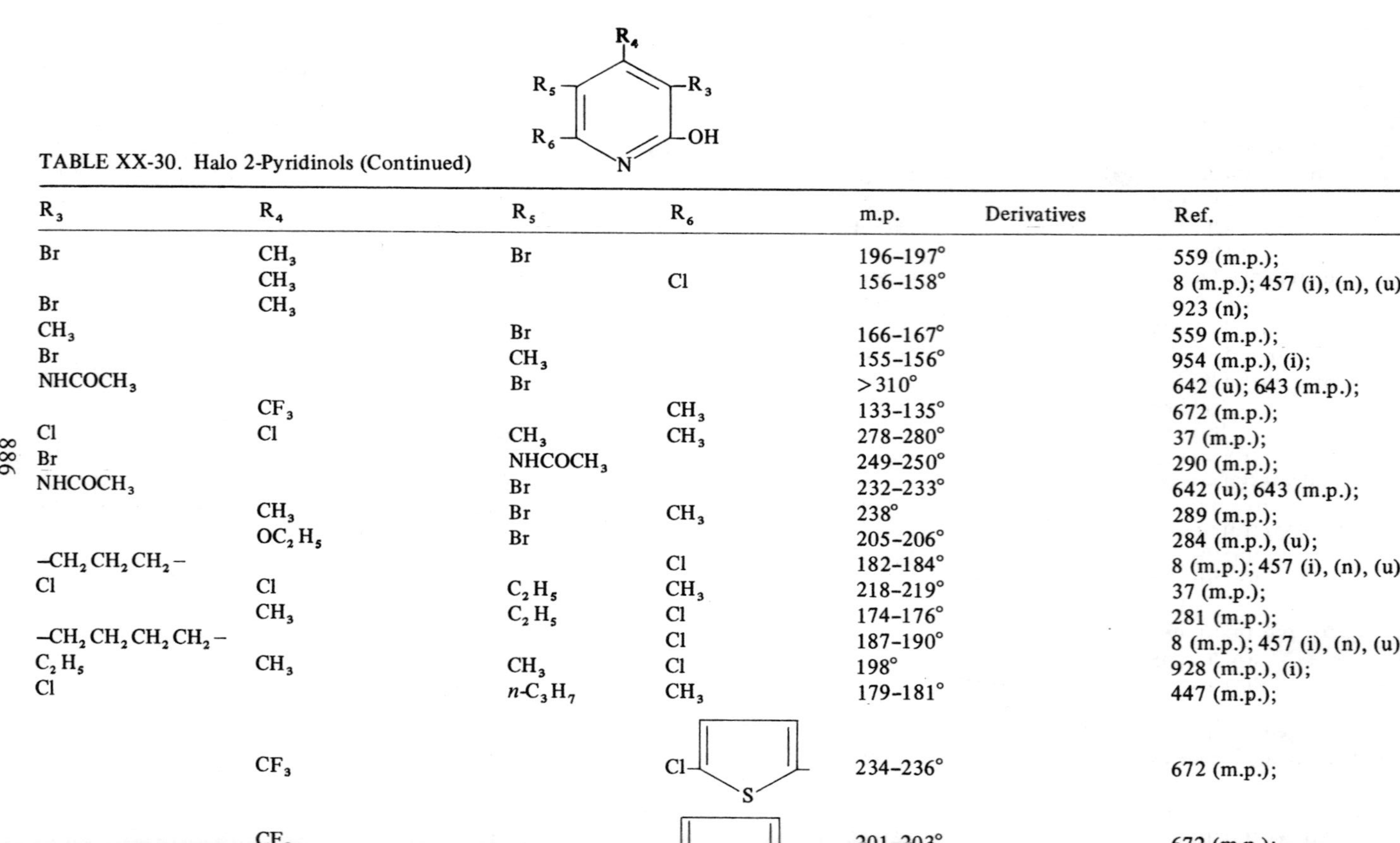


| $R_3$ | $R_4$ | $R_5$ | $R_6$ | m.p. | Derivatives | Ref. |
|---|---|---|---|---|---|---|
| Br | $CH_3$ | Br | | 196–197° | | 559 (m.p.); |
| | $CH_3$ | | Cl | 156–158° | | 8 (m.p.); 457 (i), (n), (u); |
| Br | $CH_3$ | | | | | 923 (n); |
| $CH_3$ | | Br | | 166–167° | | 559 (m.p.); |
| Br | | $CH_3$ | | 155–156° | | 954 (m.p.), (i); |
| $NHCOCH_3$ | | Br | | >310° | | 642 (u); 643 (m.p.); |
| | $CF_3$ | | $CH_3$ | 133–135° | | 672 (m.p.); |
| Cl | Cl | $CH_3$ | $CH_3$ | 278–280° | | 37 (m.p.); |
| Br | | $NHCOCH_3$ | | 249–250° | | 290 (m.p.); |
| $NHCOCH_3$ | | Br | | 232–233° | | 642 (u); 643 (m.p.); |
| | $CH_3$ | Br | $CH_3$ | 238° | | 289 (m.p.); |
| | $OC_2H_5$ | Br | | 205–206° | | 284 (m.p.), (u); |
| $-CH_2CH_2CH_2-$ | | | Cl | 182–184° | | 8 (m.p.); 457 (i), (n), (u); |
| Cl | Cl | $C_2H_5$ | $CH_3$ | 218–219° | | 37 (m.p.); |
| | $CH_3$ | $C_2H_5$ | Cl | 174–176° | | 281 (m.p.); |
| $-CH_2CH_2CH_2CH_2-$ | | | Cl | 187–190° | | 8 (m.p.); 457 (i), (n), (u); |
| $C_2H_5$ | $CH_3$ | $CH_3$ | Cl | 198° | | 928 (m.p.), (i); |
| Cl | | $n$-$C_3H_7$ | $CH_3$ | 179–181° | | 447 (m.p.); |
| | $CF_3$ | | 5-chloro-2-thienyl (Cl–thiophene–S) | 234–236° | | 672 (m.p.); |
| | $CF_3$ | | [thiophene ring] | 201–203° | | 672 (m.p.); |

| | | | | | |
|---|---|---|---|---|---|
| Cl | Cl | Cl | $C_6H_5$ | 282° | 20 (m.p.), (u); |
| Cl | | Cl | $C_6H_5$ | 252° | 119 (m.p.), (i); |
| $C_6H_5$ | | Cl | Cl | 205–208° | 242 (m.p.), (i), (u); |
| | $C_6H_5$ | | Cl | 166–168° | 8 (m.p.); 457 (i), (n), (u); |
| | $CF_3$ | | $C_6H_5$ | 188–189° | 672 (m.p.); |
| $-NHCOC_6H_5$ | | Br | | 252–253° | 643 (m.p.); |
| CN | $CF_3$ | | $C_6H_5$ | 300–301° | 55 (m.p.), (i), (u); |
| $CONH_2$ | $CF_3$ | | $C_6H_5$ | 326–327° | 672 (m.p.); |
| Br | $C_6H_5$ | $CH_3$ | $CH_3$ | 231–232° | 67 (m.p.), (i), (n); 940 (n), (u); |
| Br | 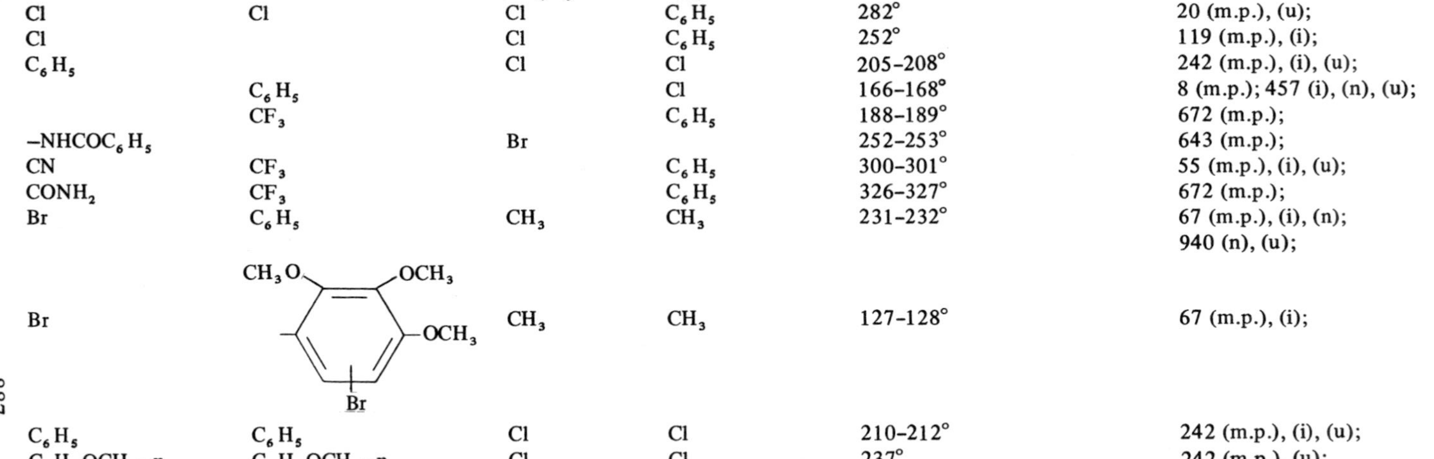  | $CH_3$ | $CH_3$ | 127–128° | 67 (m.p.), (i); |
| $C_6H_5$ | $C_6H_5$ | Cl | Cl | 210–212° | 242 (m.p.), (i), (u); |
| $C_6H_4OCH_3-p$ | $C_6H_4OCH_3-p$ | Cl | Cl | 237° | 242 (m.p.), (u); |

TABLE XII-31. Nitro and Nitroso 2-Pyridinols

| $R_3$ | $R_4$ | $R_5$ | $R_6$ | m.p. | Derivatives | Ref. |
|---|---|---|---|---|---|---|
| $NO_2$ | | | | 224–226° | | 364 (m.p.); 467 (i), (u); 899 (n); 955 (u); 956 (u); |
| | $NO_2$ | | | 242° | | 261 (m.p.); 262 (m.p.); 286 (m.p.); |
| | | $NO_2$ | | 184–187° | | 467 (i), (u); 943 (m.p.); 957 (u); 958 (i); |
| $NO_2$ | | Cl | | 240° | | 291 (m.p.); |
| Cl | | $NO_2$ | | 194–196° | | 291 (m.p.); |
| $NO_2$ | | Br | | 244° | | 290 (m.p.); |
| Br | | $NO_2$ | | 213° | | 290 (m.p.); |
| $NO_2$ | | I | | 248° | | 287 (m.p.); |
| I | | $NO_2$ | | 203° | | 287 (m.p.); |
| NO | $NH_2$ | | | 330° | | 305 (m.p.); |
| $NO_2$ | | $NO_2$ | | 175–178° | | 260 (m.p.); 261 (m.p.); 265 (m.p.); 673 (m.p.), (i); 676 (m.p.); 959 (i); 960 (u); |
| | | | | | silver salt hydrate | 961 (d); |
| | | | | | sodium salt, m.p. 293° | 769 (m.p.), (n); 961 (d); |
| | | | | | copper(II) salt | 961 (d); |
| | | | | | mercury(II) salt | 961 (d); |
| | | | | | lead(II) salt | 961 (d); |

| | | | | | | |
|---|---|---|---|---|---|---|
| | $NO_2$ | $NO_2$ | | 207° | | 286 (m.p.); |
| $NO_2$ | $NH_2$ | | | 314–315° | | 305 (m.p.); |
| $NH_2$ | | $NO_2$ | | 199–201° | | 943 (m.p.); 959 (i); |
| $NO_2$ | $CH_3$ | $NO_2$ | | 233–237° | | 675 (m.p.); |
| | | | | 254° | | 264 (m.p.); |
| $NO_2$ | | $NO_2$ | $CH_3$ | 238–240° | | 264 (m.p.); 675 (m.p.); |
| $NO_2$ | $CH_3$ | | | 229° | | 259 (m.p.); |
| $NO_2$ | | $CH_3$ | | 251–252° | | 259 (m.p.); 700 (m.p.), (u); |
| $CH_3$ | | $NO_2$ | | 229–230° | | 259 (m.p.); 700 (m.p.), (u); |
| | $CH_3$ | $NO_2$ | | 186° | | 259 (m.p.); |
| | | $NO_2$ | $CH_3$ | 230° | | 263 (m.p.); |
| $NO_2$ | | | $OCH_3$ | 172–173° | | 701 (m.p.); |
| $NO_2$ | $CH_3$ | Br | $CH_3$ | 223° | | 289 (m.p.); |
| $NHCOCH_3$ | | $NO_2$ | | 281° | | 943 (m.p.); |
| $NO_2$ | $CH_3$ | | $CH_3$ | 248° | | 288 (m.p.); |
| | $CH_3$ | $NO_2$ | $CH_3$ | 150° | | 288 (m.p.); |
| CN | $CH_3$ | $NO_2$ | $CH_3$ | 271–272° | | 674 (m.p.); |
| COOH | $CH_3$ | $NO_2$ | $CH_3$ | 234–235° | | 674 (m.p.); |
| | | | | | ethyl ester, m.p. 194–195° | 674 (m.p.); |
| $NO_2$ | Cl | | $n\text{-}C_3H_7$ | 182–183° | | 150 (m.p.); |
| $NO_2$ | $NH_2$ | | $n\text{-}C_3H_7$ | 193° | | 150 (m.p.); |
| $NO_2$ | $C_6H_5$ | $CH_3$ | $CH_3$ | 258° | | 67 (m.p.), (i), (n); |
| $NO_2$ | (ring structure: $NO_2$, $-OCH_3$, $CH_3O$, $OCH_3$) | $CH_3$ | $CH_3$ | 263–264° | | 67 (m.p.), (i); |

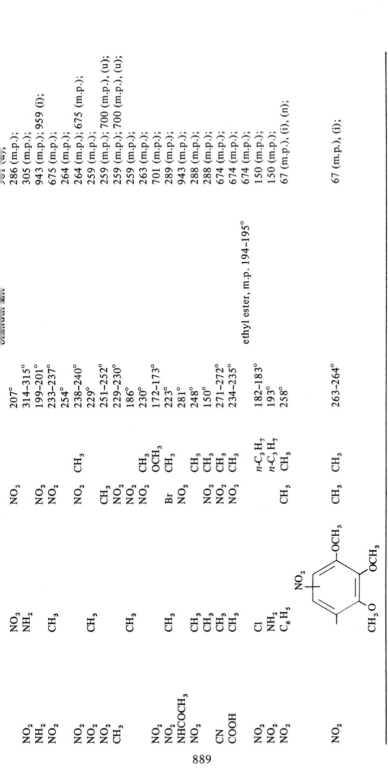

TABLE XII-32. Sulfur containing 2-Pyridinols

| $R_3$ | $R_4$ | $R_5$ | $R_6$ | m.p. | Derivatives | Ref. |
|---|---|---|---|---|---|---|
| | | | $CSNH_2$ | 236° | | 116 (m.p.); |
| | | $SCH_3$ | | 75–76° | | 962 (m.p.); |
| | | $-SOCH_3$ | | 153–155° | | 962 (m.p.); |
| | | $-SO_2CH_3$ | | 247–249° | | 962 (m.p.); |
| $SC_2H_5$ | | Cl | | 158–159° | | 962 (m.p.); |
| $-SOC_2H_5$ | | Cl | | 216–217° | | 962 (m.p.); |
| | | $SC_2H_5$ | | 79–80° | | 962 (m.p.); |
| | | | $SC_2H_5$ | 150–156° | | 962 (m.p.); |
| | | | $-SOC_2H_5$ | 107–110° | | 962 (m.p.); |
| | | | $-SO_2C_2H_5$ | 122–124° | | 962 (m.p.); |
| | | $SCH(CH_3)_2$ | | 53–54° | | 962 (m.p.); |
| | | $-SO_2CH(CH_3)_2$ | | 105–107° | | 962 (m.p.); |
| | | | [S, N, $-CH_3$] | 124° | | 116 (m.p.); |
| | $CH_3$ | $SC_2H_5$ | $CH_3$ | 138–140° | | 962 (m.p.); |
| | | [$-S-$, N, $-OH$] | | 260–263° | | 766 (m.p.), (n), (u); |
| CN | $CF_3$ | | 2-Thienyl | 300–303° | | 55 (m.p.); |
| | | $SC_6H_5$ | | 180–182° | | 962 (m.p.) |
| | $OSO_2C_6H_4Cl$-*p* | | $CH_3$ | 173–175° | | 502 (m.p.); |

| | | | |
|---|---|---|---|
| 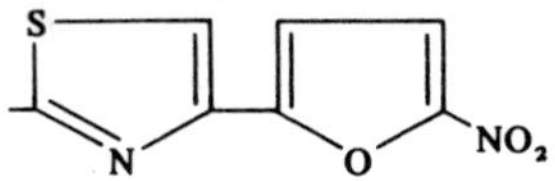 S, N, O, $NO_2$ | $CH_3$ | 336–337° | 963 (m.p.); |
| | 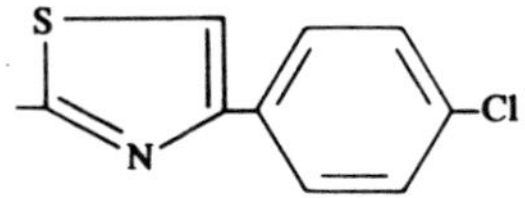 S, N, –Cl | 221° | 116 (m.p.); |
| | 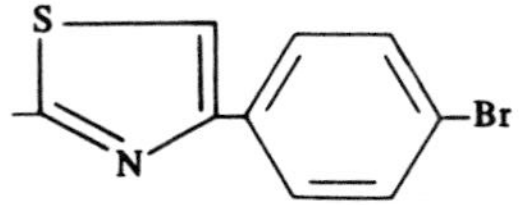 S, N, –Br | 225–226° | 116 (m.p.); |
| | 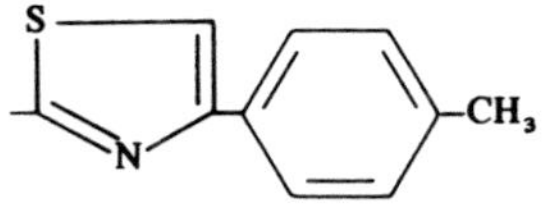 S, N, –$CH_3$ | 184° | 116 (m.p.); |
| | 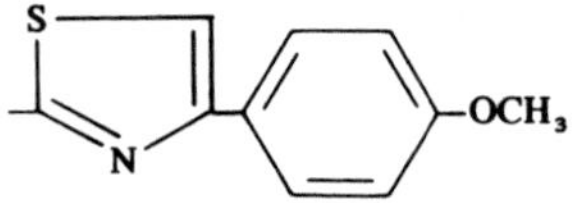 S, N, –$OCH_3$ | 158° | 116 (m.p.); |

TABLE XII-33. Ethers and Esters of Alkyl and Aryl 2-Pyridinols

| R | $R_3$ | $R_4$ | $R_5$ | $R_6$ | m.p. | Derivatives | Ref. |
|---|---|---|---|---|---|---|---|
| $CH_3$ | | | | | b.p. 141–143° | hydrochloride<br>hexachloro-antimonate, m.p. 252–254° | 472 (n); 479 (n); 482 (z); 525 (h), (u); 530 (b.p.), (n); 901 (n); 964 (j); 965 (i); 546 (i); 136 (m.p.), (i); 542 (i); 546 (i); |
| $CH_3$ | 1H-tetrazol-5-yl | | | | 158–160° | | 59 (m.p.); |
| $CH_3$ | | | 1H-tetrazol-5-yl | | 204° | | 59 (m.p.); |
| $CH_3$ | $CH_3$ | | | | b.p. 63°/20 mm | picrate, m.p. 121–122° | 700 (b.p.); 700 (m.p.); |
| $CH_3$ | | | $CH_3$ | | b.p. 163°/722 mm | hydrochloride | 467 (b.p.), (u); 546 (i), (u); |
| $C_2H_5$ | | | | | b.p. 63°/25 mm | | 570 (b.p.); 965 (i); |
| $-CH_2CH{=}CH_2$ | | | | | b.p. 30°/1.5 mm | | 557 (b.p.); |
| $-P(OH)OOCH_2CH_2CN$ | | | | | | picrate, m.p. 105–107°<br>cyclohexyl ammonium salt, [illegible] | 557 (m.p.); [illegible] |

| | | | | |
|---|---|---|---|---|
| | | | hydrochloride, m.p. 148–149° | 966 (m.p.); |
| $n\text{-}C_3H_7$ | | b.p. 43°/7 mm | | 557 (b.p.); |
| | | | picrate, m.p. 116–118° | 557 (m.p.); |
| $iso C_3H_7$ | | b.p. 92°/75 mm | | 547 (b.p.), (g), (u); |
| $Si(CH_3)_3$ | | b.p. 63°/12 mm | | 616 (b.p.); |
| $-\underset{\mid}{COC}=CH_2$ <br> $CH_3$ | | m.p. –15° <br> b.p. 90°/4 mm | | 639 (m.p.), (b.p.), (i), (n); |
| $C_2H_5$ | $CH_2CN$ | 54° | | 944 (m.p.); |
| Crotyl | | b.p. 94°/12 mm | | 790 (b.p.); |
| $-\underset{\mid}{CH}CH=CH_2$ <br> $CH_3$ | | b.p. 86°/20 mm | | 790 (b.p.); |
| $-CO(CH_2)_2CH_3$ | | b.p. 104°/5 mm | | 638 (b.p.), (i); |
| $n\text{-}C_4H_9$ | | b.p. 194°/65 mm | | 550 (b.p.); 899 (n); |
| 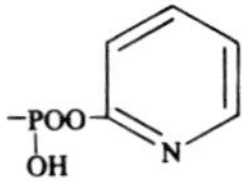 $-\underset{\mid}{P}OO-C_5H_4N$ (2-pyridyl) <br> OH | | | cyclohexyl ammonium salt, m.p. 154–155° | 658 (m.p.); |
| $-\underset{\mid}{P}OO$-Thymyl <br> OH | | 118–120° | | 967 (m.p.); |
| β-D-Xylose | | 121–123° | | 626 (m.p.), (f), (k); |
| 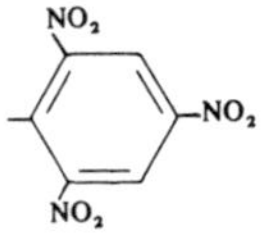 $-C_6H_2(NO_2)_3$ (2,4,6-) | | 148–150° | | 553 (m.p.), (i), (n); |
| $-\underset{\mid}{P}OOC_6H_4Cl\text{-}p$ <br> OH | | 162–164° | | 967 (m.p.); |
| | | | cyclohexyl ammonium salt, m.p. 162–164° | 658 (m.p.), (u); |
| | | | sodium salt | 658; |
| 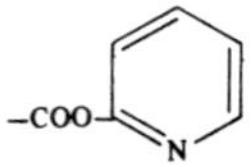 $-COO-C_5H_4N$ (2-pyridyl) | | 110–112° | | 658 (m.p.); 967 (m.p.); |

TABLE XII-33. Ethers and Esters of Alkyl and Aryl 2-Pyridinols (Continued)

| R | $R_3$ | $R_4$ | $R_5$ | $R_6$ | m.p. | Derivatives | Ref. |
|---|---|---|---|---|---|---|---|
| $C_6H_5$ | | | | | 38–42° | | 479 (n); 750 (m.p.); 751 (m.p.); |
| $-POOC_6H_5$ (OH) | | | | | | cyclohexyl ammonium salt, m.p. 166–167° | 658 (m.p.); 967 (m.p.); |
| | | | | | | sodium salt | 658; |
| $-CO(CH_2)_4CH_3$ | | | | | b.p. 122°/4 mm | | 638 (b.p.), (i); |
| β-D-Glucosyl | | | | | 170–173° | | 603 (m.p.), (f); 620 (m.p.); |
| | | | | | | tetraacetyl, m.p. 110–111° | 626 (m.p.); |
| iso$C_3H_7$ | | $CH_3$ | $C_2H_5$ | | b.p. 132°/7.2 mm | | 281 (b.p.); |
| $-COC_6H_4NO_2-p$ | | | | | 116–117° | | 967 (m.p.); |
| $CH_2C_6H_5$ | | | | | b.p. 110°/3 mm m.p. 39–40° | | 550 (b.p.); 965 (i); 967 (m.p.) |
| $-POOCH_2C_6H_5$ (OH) | | | | | | cyclohexyl ammonium salt, m.p. 142° | 658 (m.p.), (u); |
| α-D-Glucosyl | $CH_3$ | | | | 161-163° | | 626 (m.p.), (f), (k); |
| β-D-Glucosyl | $CH_3$ | | | | 95–98° | | 626 (m.p.), (f), (k); |
| β-D-Glucosyl | | $CH_3$ | | | 104–111° | | 626 (m.p.), (f), (k); |
| β-D-Glucosyl | | | | $CH_3$ | 74–76° | | 626 (m.p.), (f), (k); |
| $CH_2C_6H_5$ | $CH_3$ | | | | b.p. 79°/0.1 mm | | 581 (b.p.); |
| $CH_2C_6H_5$ | | | $CH_3$ | | b.p. 118°/0.1 mm m.p.30–31° | | 6 (b.p.), (m.p.); |
| 3-Quinolyl | | | | | 50–51° | picrate, | 273 (m.p.); |

| | | | | | | |
|---|---|---|---|---|---|---|
| OH | | | | | carbonic acid-4-(N,N'-dicyclohexyl-amidinium salt | 661 (f), (u); |
| –POO(5'-Cytidine)<br>OH | | | | | morpholine carbonic acid-4-(N-N'-dicyclohexylamidinium salt | 661 (f), (u); |
| –POO(β-Naphthyl)<br>OH | | | | | cyclohexylammonium salt, m.p. 172–173° | 658 (m.p.), (u); |
| –PO(O-2-Pyridyl)$_2$ | | | | 73–75° | | 658 (m.p.); |
| –POO(5'-Adenosine)<br>OH | | | | | morpholine carbonic acid-4-(*N*,*N*'-dicyclohexyl-amidinium salt | 661 (f), (u); |
| –POO(5'-Guanosine)<br>OH | | | | | morpholine carbonic acid-4-(*N*,*N*'-dicyclohexyl-amidinium salt | 661 (f), (u); |
| –$CH_2C_6H_5$ | $CH_2OH$ | $CH_2OH$ | $CH_3$ | 86–87° | | 931 (m.p.), (i), (n); |
| –POO(Thymyl)<br>OH | | | | | cyclohexyl ammonium salt, m.p. 118–120° | 658 (m.p.), (u); |
| –POO(5'-Thymidine)<br>OH | | | | | ammonium salt | 661 (f), (u); |
| –POO$C_6H_4$Cl-*p*<br>O-2-Pyridyl | | | | 96–97° | | 658 (m.p.); |
| –POO$C_6H_5$<br>O-2-Pyridyl | | | | 56–57° | | 658 (m.p.); |
| –CO$(CH_2)_4$CO 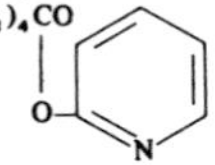 | | | | 98-100° | | 967 (m.p.); |
| α-L-Arabinose | | | | | triacetyl hygroscopic | 626 (k); |
| β-D-Xylosyl | | | | | triacetyl m.p. 114–116° | 626 (m.p.), (k); |
| $C_2H_5$ | | | 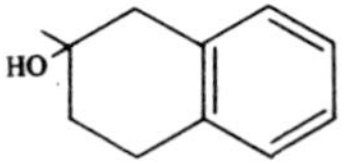 | b.p. 155°/0.1 mm | | 568 (b.p.), (u); |
| | | | | | perchlorate, m.p. 164–165° | 568 (m.p.), (u); |

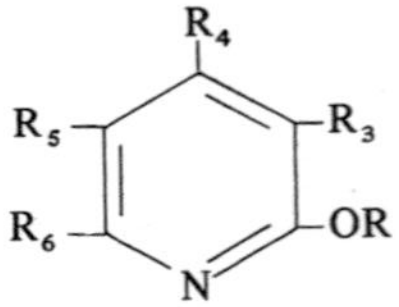

TABLE XII-33. Ethers and Esters of Alkyl and Aryl 2-Pyridinols (Continued)

| R | $R_3$ | $R_4$ | $R_5$ | $R_6$ | m.p. | Derivatives | Ref. |
|---|---|---|---|---|---|---|---|
| β-D-Glucosyl | | $C_6H_5$ | | | 165° | | 604 (m.p.); |
| α-D-Glucosyl | | | | $C_6H_5$ | 186–188° | | 604 (m.p.); |
| β-D-Glucosyl | | | | $C_6H_5$ | 170–173° | | 604 (m.p.); |
| $CH_3$ | | | | 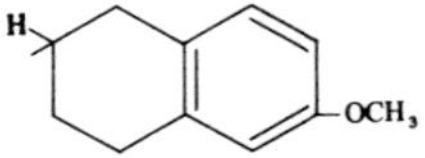 H, $OCH_3$ | | perchlorate, m.p. 168–170° | 568 (m.p.), (u); |
| β-D-Lactosyl | | | | | 182–183° | | 626 (m.p.), (f), (k); |
| $CH_3$ | | $C_6H_5$ | | $C_6H_5$ | 82° | | 941 (m.p.), (i), (u); |
| $C_2H_5$ | | | | 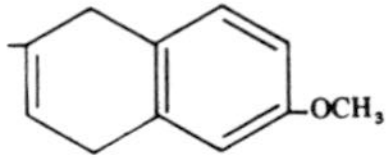 $OCH_3$ | 105-106° | | 568 (m.p.), (u); |
| $C_2H_5$ | | | | 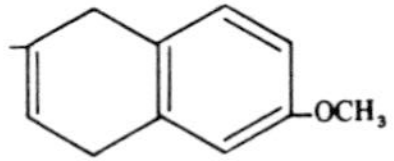 $OCH_3$ | 78° | | 568 (m.p.), (u); |
| $C_2H_5$ | | | | 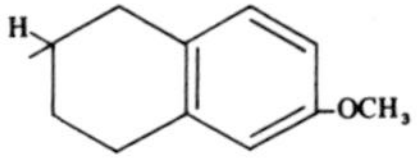 H, $OCH_3$ | b.p. 160°/0.6 mm | perchlorate, m.p. 162–163° | 568 (b.p.), (u); 568 (m.p.); |
| $C_2H_5$ | | | | 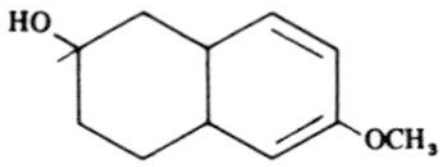 HO, $OCH_3$ | 89–90° | perchlorate, m.p. 169–170° | 568 (m.p.), (u); 568 (m.p.); |
| $CH_3$ | | | | $-(CH_2)_4-$  | b.p. 142°/0.1 mm | | 6 (b.p.); |

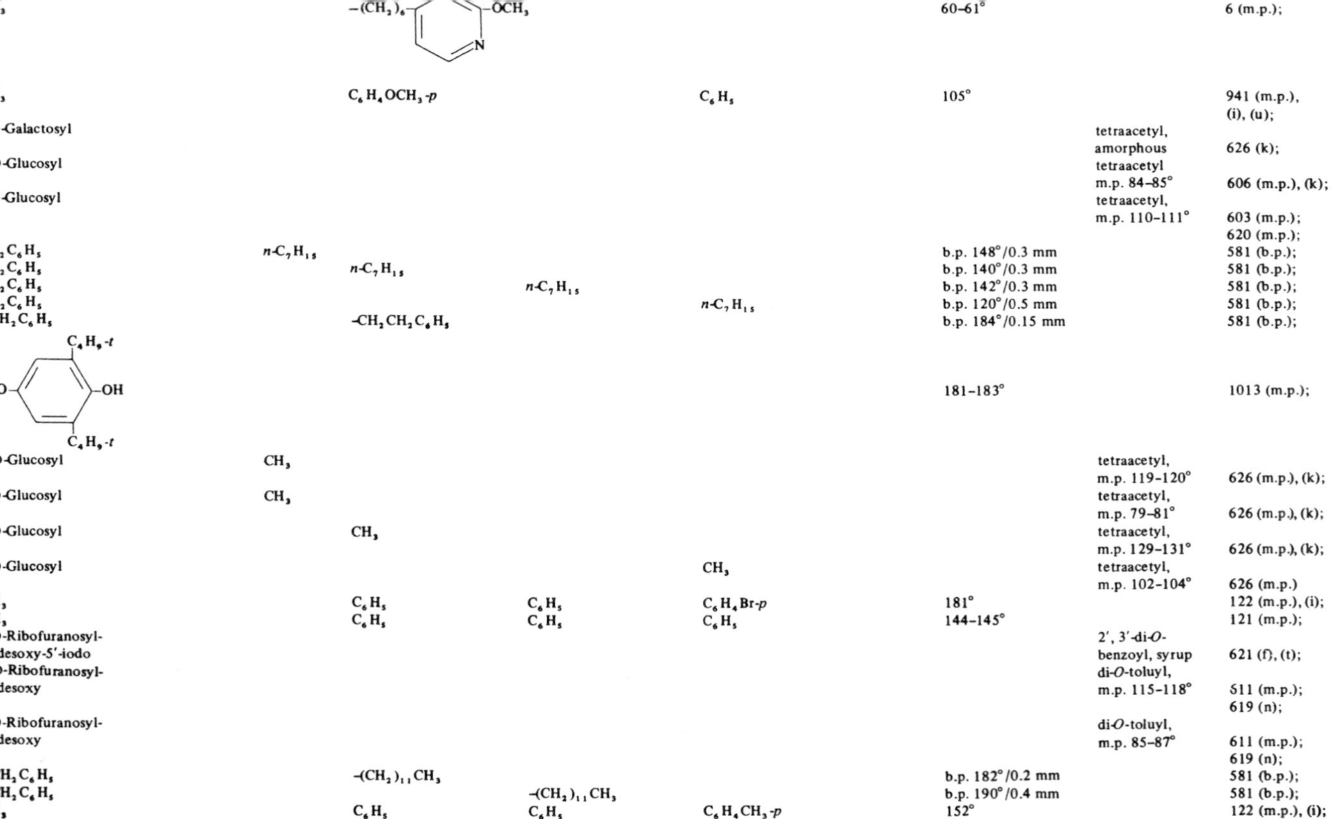

| | | | | | | | |
|---|---|---|---|---|---|---|---|
| $CH_3$ | | $-(CH_2)_6-$(pyridyl)$-OCH_3$ | | | 60–61° | | 6 (m.p.); |
| $CH_3$ | | $C_6H_4OCH_3$-*p* | | $C_6H_5$ | 105° | | 941 (m.p.), (i), (u); |
| β-D-Galactosyl | | | | | | tetraacetyl, amorphous | 626 (k); |
| α-D-Glucosyl | | | | | | tetraacetyl m.p. 84–85° | 606 (m.p.), (k); |
| β-D-Glucosyl | | | | | | tetraacetyl, m.p. 110–111° | 603 (m.p.); 620 (m.p.); |
| $CH_2C_6H_5$ | *n*-$C_7H_{15}$ | | | | b.p. 148°/0.3 mm | | 581 (b.p.); |
| $CH_2C_6H_5$ | | *n*-$C_7H_{15}$ | | | b.p. 140°/0.3 mm | | 581 (b.p.); |
| $CH_2C_6H_5$ | | | *n*-$C_7H_{15}$ | | b.p. 142°/0.3 mm | | 581 (b.p.); |
| $CH_2C_6H_5$ | | | | *n*-$C_7H_{15}$ | b.p. 120°/0.5 mm | | 581 (b.p.); |
| $-CH_2C_6H_5$ | | $-CH_2CH_2C_6H_5$ | | | b.p. 184°/0.15 mm | | 581 (b.p.); |
| $-CO-C_6H_2(C_4H_9\text{-}t)_2-OH$ | | | | | 181–183° | | 1013 (m.p.); |
| α-D-Glucosyl | $CH_3$ | | | | | tetraacetyl, m.p. 119–120° | 626 (m.p.), (k); |
| β-D-Glucosyl | $CH_3$ | | | | | tetraacetyl, m.p. 79–81° | 626 (m.p.), (k); |
| β-D-Glucosyl | | $CH_3$ | | | | tetraacetyl, m.p. 129–131° | 626 (m.p.), (k); |
| β-D-Glucosyl | | | | $CH_3$ | | tetraacetyl, m.p. 102–104° | 626 (m.p.) |
| $CH_3$ | | $C_6H_5$ | $C_6H_5$ | $C_6H_4Br$-*p* | 181° | | 122 (m.p.), (i); |
| $CH_3$ | | $C_6H_5$ | $C_6H_5$ | $C_6H_5$ | 144–145° | | 121 (m.p.); |
| β-D-Ribofuranosyl-5′-desoxy-5′-iodo | | | | | | 2′, 3′-di-*O*-benzoyl, syrup | 621 (f), (t); |
| α-D-Ribofuranosyl-2′-desoxy | | | | | | di-*O*-toluyl, m.p. 115–118° | 511 (m.p.); 619 (n); |
| β-D-Ribofuranosyl-2′-desoxy | | | | | | di-*O*-toluyl, m.p. 85–87° | 611 (m.p.); 619 (n); |
| $-CH_2C_6H_5$ | | $-(CH_2)_{11}CH_3$ | | | b.p. 182°/0.2 mm | | 581 (b.p.); |
| $-CH_2C_6H_5$ | | | $-(CH_2)_{11}CH_3$ | | b.p. 190°/0.4 mm | | 581 (b.p.); |
| $CH_3$ | | $C_6H_5$ | $C_6H_5$ | $C_6H_4CH_3$-*p* | 152° | | 122 (m.p.), (i); |

TABLE XII-33. Ethers and Esters of Alkyl and Aryl 2-Pyridinols (Continued)

(Structure: pyridine ring with $R_3$, $R_4$, $R_5$, $R_6$ substituents and OR at position 2)

| R | $R_3$ | $R_4$ | $R_5$ | $R_6$ | m.p. | Derivatives | Ref. |
|---|---|---|---|---|---|---|---|
| β-D-Glucosyl | | $C_6H_5$ | | | | tetraacetyl,<br>m.p. 129–130° | 604 (m.p.); |
| α-D-Glucosyl | | | | $C_6H_5$ | | tetraacetyl,<br>m.p. 147–149° | 604 (m.p.); |
| β-D-Glucosyl | | | | $C_6H_5$ | | tetraacetyl,<br>m.p. 185–186° | 604 (m.p.); |
| $-(CH_2)_2O(CH_2)_2OC_6H_4C(CH_3)_2CH_2C(CH_3)_3$-*p* | | | 1-methyl-2-pyrrolidinyl | | 78–79° | | 1014 (m.p.); |
| $CH_2C_6H_5$ | | $C_6H_5$ | $C_6H_5$ | $C_6H_5$ | 150° | | 121 (m.p.); |
| β-D-Ribofuranosyl | | | | | | 2,3,5-tri-*O*-benzoyl,<br>m.p. 112–114° | 601 (m.p.);<br>602 (m.p.); |
| | | | | | | 2′,3′-di-*O*-benzoyl-<br>5′-*O*-diphenyl-<br>phosphoryl,<br>m.p. 83–85° | 614 (m.p.), (k);<br>615 (m.p.),<br>(k), (t); |
| β-D-Ribopyranosyl | | | | | | 2,3,5-tri-*O*-benzoyl,<br>amorphous | 601; 602; |
| β-D-Lactosyl | | | | | | heptaacetyl,<br>amorphous | 626 (k); |

TABLE XII-34. Ethers of Alkyloxy and Aryloxy 2-Pyridinols

| R | $R_3$ | $R_4$ | $R_5$ | $R_6$ | m.p. | Derivatives | Ref. |
|---|---|---|---|---|---|---|---|
| β-D-Glucosyl | | $OC_2H_5$ | | | 130–133° | | 604 (m.p.); |
| | | | | | | tetraacetyl m.p. 106–107° | 604 (m.p.); |

TABLE XII-35. Ethers of Amino 2-Pyridinols

| R | $R_3$ | $R_4$ | $R_5$ | $R_6$ | m.p. | Derivatives | Ref. |
|---|---|---|---|---|---|---|---|
| $CH_3$ | | | | $NH_2$ | b.p. 117°/20 mm | | 494 (b.p.), (i), (u); |
| $C_2H_5$ | $NH_2$ | | | | 31–32° | | 319 (m.p.); |
| | | | | | | picrate, m.p. 135–137° | 319 (m.p.); |
| $C_2H_5$ | | | | $NH_2$ | b.p. 120°/16 mm | | 641 (b.p.), (i), (n); |
| | | | | | | N′-acetyl, m.p. 137–138° | 641 (m.p.); |
| $-(CH_2)_2CH(CH_3)_2$ | | | | $NH_2$ | b.p. 162°/23 mm | | 641 (b.p.); |
| | | | | | | N′-acetyl, m.p. 115–116° | 641 (m.p.); |
| $C_6H_5$ | $N_3$ | | | | 40° | | 968 (m.p.); |
| $C_6H_5$ | $NH_2$ | | | | 106° | | 968 (m.p.); |
| $C_6H_5$ | | $NH_2$ | | | 57–58° | | 969 (m.p.); |
| | | | | | | picrate, m.p. 176–177° | 969 (m.p.); |
| $C_6H_5$ | | | | $NH_2$ | 62–63° | | 969 (m.p.); |
| | | | | | | picrate, m.p. 208–209° | 969 (m.p.); |

TABLE XII-35. Ethers of Amino 2-Pyridinols (Continued)

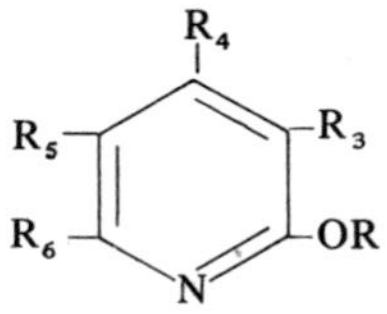

| R | $R_3$ | $R_4$ | $R_5$ | $R_6$ | m.p. | Derivatives | Ref. |
|---|---|---|---|---|---|---|---|
| β-D-Glucosyl | $NH_2$ | | | | 147–153° | tetraacetyl, m.p. 147–148° | 608 (m.p.), (f), (k); 608 (m.p.), (k); |
| β-D-Glucosyl | | | $NH_2$ | | | 5-acetyl, m.p. 210°; tetraacetyl, m.p. 59–63° | 605 (m.p.), (f), (k); 605 (m.p.), (k); |
| $CH_3$ | $NH_2$ | | $NH_2$ | $-N{=}NC_6H_5$ | | dihydrochloride, m.p. 141° | 738 (m.p.); |
| $C_2H_5$ | $NH_2$ | | $NH_2$ | $-N{=}NC_6H_5$ | | dihydrochloride, m.p. 119° | 738 (m.p.); |
| β-D-Glucosyl | $NHCOCH_3$ | | | | | tetraacetyl, m.p. 129–131° | 608 (m.p.), (k); |
| β-D-Glucosyl | | | $NHCOCH_3$ | | | tetraacetyl, m.p. 176–177° | 605 (m.p.), (k); |
| $n$-$C_4H_9$ | | | | $-NHCH_2CH_2N(CH_3)CH_3$ | b.p. 127°/0.05 mm | hydrochloride, m.p. 104–105° | 970 (b.p.); 970 (m.p.); |
| $-(CH_2)_2N(CH_3)_2$ | | | | $-NHC_4H_9$-$n$ | b.p. 120°/0.02 mm | hydrochloride, m.p. 122–124° | 970 (b.p.); 970 (m.p.); |
| $n$-$C_3H_7$ | $NH_2$ | | $NH_2$ | $-N{=}NC_6H_5$ | | dihydrochloride, m.p. 122° | 738 (m.p.); |
| $n$-$C_4H_9$ | $NH_2$ | | $NH_2$ | $-N{=}NC_6H_5$ | | dihydrochloride, m.p. 126° | 738 (m.p.); |
| $n$-$C_4H_9$ | | | | $-NHCH_2CH_2N(C_2H_5)C_2H_5$ | b.p. 163°/0.2 mm | hydrochloride, m.p. 76–77° | 970 (b.p.); 970 (m.p.); |
| $-(CH_2)_2N(C_2H_5)_2$ | | | | $-NHC_4H_9$-$n$ | b.p. 135°/0.02 mm | hydrochloride, m.p. 95–97° | 970 (b.p.); 970 (m.p.); |
| $n$-$C_5N_{11}$ | $NH_2$ | | $NH_2$ | $-N{=}NC_6H_5$ | | dihydrochloride, m.p. 96° | 738 (m.p.) |

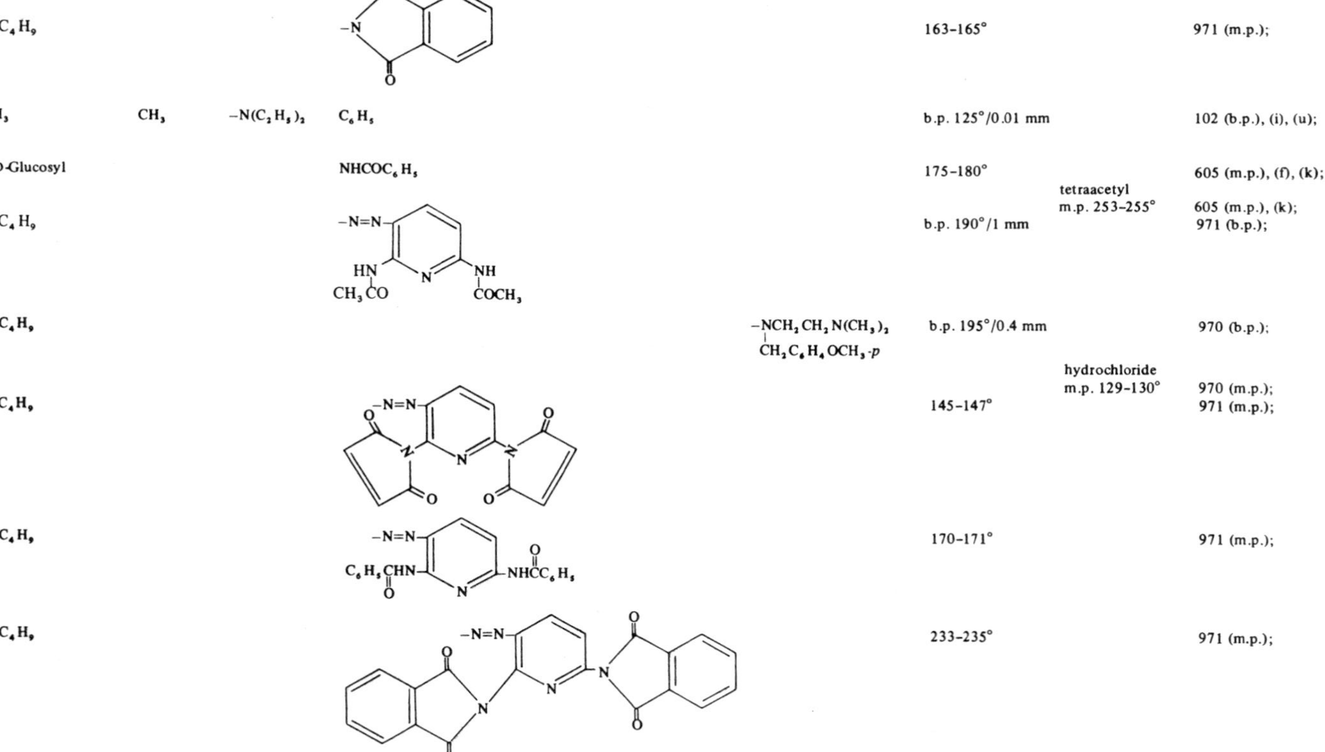

| | | | | | | | |
|---|---|---|---|---|---|---|---|
| $n$-$C_4H_9$ | | | | | 163–165° | | 971 (m.p.); |
| $CH_3$ | $CH_3$ | $-N(C_2H_5)_2$ | $C_6H_5$ | | b.p. 125°/0.01 mm | | 102 (b.p.), (i), (u); |
| β-D-Glucosyl | | | $NHCOC_6H_5$ | | 175–180° | tetraacetyl m.p. 253–255° | 605 (m.p.), (f), (k); |
| $n$-$C_4H_9$ | | | | | b.p. 190°/1 mm | | 605 (m.p.), (k); 971 (b.p.); |
| $n$-$C_4H_9$ | | | | $-N(CH_2C_6H_4OCH_3\text{-}p)CH_2CH_2N(CH_3)_2$ | b.p. 195°/0.4 mm | hydrochloride m.p. 129–130° | 970 (b.p.); 970 (m.p.); |
| $n$-$C_4H_9$ | | | | | 145–147° | | 971 (m.p.); |
| $n$-$C_4H_9$ | | | | | 170–171° | | 971 (m.p.); |
| $n$-$C_4H_9$ | | | | | 233–235° | | 971 (m.p.); |

TABLE XII-36. Ethers of 2-Pyridinol Carboxylic Acids and Derivatives

| R | $R_3$ | $R_4$ | $R_5$ | $R_6$ | m.p. | Derivatives | Ref. |
|---|---|---|---|---|---|---|---|
| $CH_3$ | | | COOH | | | ethyl ester | 549 (g); |
| $C_2H_5$ | | | COOH | | | ethyl ester, b.p. 125°/0.7 mm | 549 (b.p.); |
| $CH_3$ | COOH | COOH | | COOH | | trimethyl ester, m.p. 132° | 57 (m.p.); |
| $CH_3$ | COOH | | | $COCH_3$ | | methyl ester, m.p. 80° | 57 (m.p.), (i), (u); |
| iso$C_3H_7$ | | | COOH | | | ethyl ester | 549; |
| $C_2H_5$ | | | | $CH_2COOH$ | | ethyl ester, m.p. 80° | 944 (m.p.); |
| $CH_3$ | COOH | COOH | | $COCH_3$ | | dimethyl ester, m. p. 80° | 57 (m.p.), (i); |
| $C_2H_5$ | | | | $-CH_2COCOOH$ | | ethyl ester, m.p. 99° | 944 (m.p.); |
| | | | | | | oxime, m.p. 129° | 944 (m.p.); |
| $C_2H_5$ | | | | $CH_2C(=NOH)COOH$ | | ethyl ester, m.p. 73° | 944 (m.p.); |
| β-D-Glucosyl | COOH | | | | | methyl ester, m.p. 80–82° | 604 (m.p.); |
| $CH_2C_6H_5$ | | | COOH | | | ethyl ester, m.p. 50-51° | 549 (m.p.); |

| | | | | | | | |
|---|---|---|---|---|---|---|---|
| $CH_3$ | COOH | | | $-CH{=}CHC_6H_5$ | | methyl ester, m.p. 63° | 57 (m.p.), (i), (u); |
| $CH_3$ | | COOH | | $-CH{=}CHC_6H_5$ | | methyl ester, m.p. 98° | 57 (m.p.), (i); |
| $CH_2C_6H_5$ | COOH | | COOH | $CH_3$ | 186° softens 260° | | 931 (m.p.), (i); |
| $CH_3$ | COOH | | | $-C(CH_3){=}CHC_6H_5$ | | methyl ester, m.p. 73° | 57 (m.p.), (i), (u); |
| $CH_3$ | | COOH | | $-C(CH_3){=}CHC_6H_5$ | | methyl ester, m.p. 128° | 57 (m.p.); |
| β-D-Glucosyl | $COOCH_3$ | | | | | tetraacetyl, m.p. 110–113° | 604 (m.p.); |
| β-D-Glucosyl | | $COOCH_3$ | | | | tetraacetyl, m.p. 93–95° | 604 (m.p.); |
| β-D-Glucosyl | | | | $COOCH_3$ | | tetraacetyl, m.p. 171–173° | 604 (m.p.); |
| β-D-Gluco-pyranosyl | | | $CONH_2$ | | | tetraacetyl, m.p. 179° | 613 (m.p.); |
| β-D-ribo-furanosyl | | | $CONH_2$ | | | 2,3,5-tri-*O*-benzoyl, m.p. 177° | 613 (m.p.); |

TABLE XII-37. Ethers of 2-Pyridinol Nitriles

| R | $R_3$ | $R_4$ | $R_5$ | $R_6$ | m.p. | Derivatives | Ref. |
|---|---|---|---|---|---|---|---|
| $CH_3$ | CN | | | COOH | | methyl ester, m.p. 124° | 57 (m.p.), (i), (u); |
| $CH_3$ | CN | | CN | $NH_2$ | 258–259° | | 10 (m.p.); |
| $C_2H_5$ | CN | | $CONH_2$ | $NH_2$ | 272–273° | | 10 (m.p.); |
| $C_2H_5$ | CN | $CH_3$ | CN | $NH_2$ | 235–237° | | 13 (m.p.); |
| $C_2H_5$ | CN | $C_2H_5$ | CN | $NH_2$ | 203–204° | | 13 (m.p.), (i), (n), (u); |
| β-D-Glucosyl | | | CN | | 180–182° | | 605 (m.p.), (f), (k); |
| $C_2H_5$ | CN | $n$-$C_3H_7$ | CN | $NH_2$ | 156–157° | | 13 (m.p.), (i), (n), (u); |
| $C_2H_5$ | CN | iso$C_3H_7$ | CN | $NH_2$ | 234–235° | | 13 (m.p.), (i), (n), (u); |
| $C_2H_5$ | CN | 2-furyl | CN | $NH_2$ | 203–204° | | 13 (m.p.), (i), (u); |
| $C_2H_5$ | CN | $n$-$C_4H_9$ | CN | $NH_2$ | 142–144° | | 13 (m.p.), (i), (n), (u); |
| $C_2H_5$ | CN | $t$-$C_4H_9$ | CN | $NH_2$ | 230–231° | | 13 (m.p.), (i), (u); |
| $CH_3$ | CN | $C_6H_5$ | CN | $NH_2$ | 259–261° | | 13 (m.p.), (i), (u); |
| $C_2H_5$ | CN | 3-Pyridyl | CN | $NH_2$ | 258° | | 13 (m.p.), (i), (u); |
| $C_2H_5$ | CN | $-CH(C_2H_5)_2$ | CN | $NH_2$ | 123–124° | | 13 (m.p.), (i), (n), (u); |
| $C_2H_5$ | CN | $C_6H_4Cl$-$o$ | CN | $NH_2$ | 263–264° | | 13 (m.p.), (i), (u); |
| $C_2H_5$ | CN | $C_6H_4Cl$-$p$ | CN | $NH_2$ | 180–181° | | 13 (m.p.), (i), (u); |
| $CH_3$ | CN | | | $-CH{=}CH-C_6H_5$ | 122° | | 57 (m.p.), (i), (u); |

| | | | | | | | |
|---|---|---|---|---|---|---|---|
| $C_2H_5$ | CN | $C_6H_5$ | CN | $NH_2$ | 238–239° | [illegible] | [illegible] |
| $C_2H_5$ | CN | $C_6H_{11}$ | CN | $NH_2$ | 159–161° | | 13 (m.p.), (i), (u); |
| $C_2H_5$ | CN | $n$-$C_6H_{13}$ | CN | $NH_2$ | oil | | 13 (i); |
| $C_2H_5$ | CN | (benzene ring with O–$CH_2$–O) | CN | $NH_2$ | 174–175° | | 13 (m.p.), (i), (u); |
| $CH_3$ | CN | | | $-C(CH_3){=}CHC_6H_5$ | 73° | | 57 (m.p.), (i), (u); |
| $C_2H_5$ | CN | $CH_2C_6H_5$ | CN | $NH_2$ | 140–141° | | 13 (m.p.), (i), (n), (u); |
| $C_2H_5$ | CN | $C_6H_4CH_3$-$m$ | CN | $NH_2$ | 230–231° | | 13 (m.p.), (i), (u); |
| $C_2H_5$ | CN | $C_6H_4CH_3$-$p$ | CN | $NH_2$ | 192–194° | | 13 (m.p.), (i), (u); |
| iso$C_3H_7$ | CN | $C_6H_5$ | CN | $NH_2$ | 205–207° | | 13 (m.p.), (i), (u); |
| $n$-$C_3H_7$ | CN | $C_6H_5$ | CN | $NH_2$ | 224–225° | | 13 (m.p.), (i), (u); |
| $C_2H_5$ | CN | $C_6H_4OCH_3$-$p$ | CN | $NH_2$ | 199–200° | | 13 (m.p.), (i), (n), (u); |
| $C_2H_5$ | CN | $C_6H_4N(CH_3)_2$-$p$ | CN | $NH_2$ | 270–271° | | 13 (m.p.), (i), (n), (u); |
| $C_2H_5$ | CN | (benzene ring with $OCH_3$, $OCH_3$, $OCH_3$) | CN | $NH_2$ | 219–221° | | 13 (m.p.), (i), (u); |
| $\beta$-D-Glucosyl | CN | | | | | tetraacetyl, m.p. 197–201° | 608 (m.p.), (k); |
| $\alpha$-D-Glucosyl | | | CN | | | tetraacetyl, m.p. 125–130° | 605 (m.p.), (k); |
| $\beta$-D-Glucosyl | | | CN | | | tetraacetyl, m.p. 153–155° | 605 (m.p.), (k); |

TABLE XII-38. Ethers and Esters of Halo 2-Pyridinols

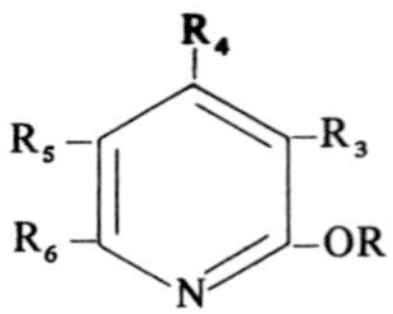

| R | $R_3$ | $R_4$ | $R_5$ | $R_6$ | m.p. | Derivatives | Ref. |
|---|---|---|---|---|---|---|---|
| $CH_3$ | Cl | | | | b.p. 107°/52 mm | | 467 (b.p.), (u); |
| | | | | | | hexachloro-antimonate·HCl | 546 (i), (u); |
| $CH_3$ | | Cl | | | 27–28° | | |
| | | | | | b.p. 71°/17 Torr | | 768 (m.p.), (b.p.); |
| $CH_3$ | | | Cl | | | hydrochloride, m.p. 121° eff. | 467 (m.p.), (u); 546 (i), (u); |
| $CH_3$ | | | | Cl | b.p. 73°/15 mm | | 241 (u); 494 (b.p.); |
| $CH_3$ | Br | | | | b.p. 98°/26 mm | | 467 (b.p.), (u); |
| | | | | | | hydrochloride | 546 (i), (u); |
| $CH_3$ | | | Br | | | hydrochloride, m.p. 125° eff. | 467 (m.p.), (u); 546 (m.p.), (i), (u); |
| $CH_3$ | | | I | | b.p. 106°/30 mm | | 467 (b.p.), (u); |
| | | | | | | hydrochloride, m.p. 146–147° | 467 (m.p.); 546 (i), (u); |
| $CH_3$ | Cl | | Cl | | 38–40° | | 467 (m.p.), (u); 831 (i), (u); |
| | | | | | | hexachloro-antimonate·HCl | 546 (i); |

| | | | | | | | |
|---|---|---|---|---|---|---|---|
| | | | | | | | 831 (i); |
| | | | | | | hexachloro-antimonate·HCl | 546 (i), (u); |
| $CH_3$ | F | | F | F | b.p. 146°/760 mm | | 247 (b.p.); 949 (b.p.); |
| $CH_3$ | F | F | F | F | b.p. 135°/760 mm | | 246 (b.p.), (g), (i), (n); |
| $CH_3$ | F | Br | F | F | b.p. 193°/760 mm | | 247 (b.p.), (n); 949 (b.p.); |
| $CH_3$ | Br | Br | Br | Br | 173° | | 952 (m.p.); |
| $CONHCH_3$ | Cl | Cl | Cl | Cl | 220–223° | | 972 (m.p.); |
| $C_2H_5$ | Cl | Cl | Cl | Cl | | | 241 (u); |
| $CH_2CH_2OH$ | Cl | Cl | Cl | F | 82–83° | | 953 (m.p.); |
| $CONHCH_3$ | Cl | | Cl | Cl | 121–123° | | 972 (m.p.); |
| $CH_3$ | Cl | $CH_3$ | Cl | Cl | 82° | | 778 (m.p.), (n); |
| $N(CH_3)_2$ | Cl | Cl | Cl | Cl | 83° | | 240 (m.p.); |
| $CONHCH_3$ | Cl | | | Cl | 129–132° | | 972 (m.p.); |
| $C_2H_5$ | | | Br | Br | 41–42° | | 712 (m.p.); |
| $PO(OCH_3)_2$ | Cl | | Cl | Cl | 86–88° | | 703 (m.p.); |
| $CH_3$ | | $CH_3$ | | Cl | b.p. 105°/12 mm | | 8 (b.p.); 457 (i), (n), (u); |
| $PO(NHCH_3)_2$ | Cl | | Cl | Cl | 86–89° | | 973 (m.p.); |
| | | | | | 146–148° | | 973 (m.p.); |
| $CON(CH_3)_2$ | Cl | Cl | Cl | Cl | 118–120° | | 972 (m.p.); |
| $-NCH_2CH_2$ / $CH_2-CH_2$ (ring) | Cl | Cl | Cl | Cl | 71° | | 240 (m.p.); |
| $-N(CH_2CH_2)_2O$ (morpholino) | Cl | Cl | Cl | Cl | 130° | | 240 (m.p.); |
| $CH_3$ | $-CH_2CH_2CH_2-$ | | | Cl | b.p. 145°/14 mm | | 8 (b.p.); 457 (i), (n), (u); |
| $-PO(OC_2H_5)_2$ | Cl | | Cl | Cl | 43–44° | | 703 (m.p.); 974 (g), (c); |

TABLE XII-38. Ethers and Esters of Halo 2-Pyridinols (Continued)

| R | $R_3$ | $R_4$ | $R_5$ | $R_6$ | m.p. | Derivatives | Ref. |
|---|---|---|---|---|---|---|---|
| $-\underset{OCH_3}{\underset{\vert}{P}}ONHCH(CH_3)_2$ | Cl | | Cl | Cl | 117–119° | | 973 (m.p.); |
| $-\underset{CH_3NH}{\underset{\vert}{P}}OOCH_2CH_2CH_3$ | Cl | | Cl | Cl | 68–71° | | 973 (m.p.); |
| $-N(CH_2)_5$ | Cl | Cl | Cl | Cl | 87° | | 240 (m.p.); |
| $C_6H_5$ | Cl | Cl | Cl | Cl | 94° | | 240 (m.p.); |
| $-N(CH_2)_6$ | Cl | Cl | Cl | Cl | 54° | | 240 (m.p.); |
| β-D-Glucosyl | Cl | | Cl | | 160–162° | | 608 (m.p.), (f), (k); |
| | | | | | | tetraacetyl, m.p. 164–166° | 608 (m.p.), (k); |
| β-D-Glucosyl | Br | | Br | | 159–160° | | 608 (m.p.), (f), (k); |
| | | | | | | tetraacetyl, m.p. 152–154° | 608 (m.p.), (k); |
| β-D-Glucosyl | Cl | | | | | tetraacetyl, m.p. 147–149° | 608 (m.p.), (k); |
| β-D-Glucosyl | Br | | | | | tetraacetyl, m.p. 135–137° | 608 (m.p.), (k); |
| α-D-Glucosyl | | | Cl | | | tetraacetyl, m.p. 124–125° | 606 (m.p.), (k), (n); 610 (m.p.), (n); |
| β-D-Glucosyl | | | Cl | | 152-155° | | 605 (m.p.), (f), |

| | | | | | | | |
|---|---|---|---|---|---|---|---|
| | | | | | | m.p. 126–129 | 605 (m.p.), (k); 606 (m.p.), (k); |
| β-D-Glucosyl | | | Br | | 164–165° | | 605 (m.p.), (f), (k); |
| | | | | | | tetraacetyl, m.p. 143–144° | 605 (m.p.), (k); |
| α-D-Glucosyl | | | I | | | tetraacetyl, m.p. 139–142° | 605 (m.p.), (k); |
| β-D-Glucosyl | | | I | | 159–162° | | 605 (m.p.), (f), (k); |
| | | | | | | tetraacetyl, m.p. 166–169° | 605 (m.p.), (k); |
| $CH_3$ | | $CON(Et)_2$ | | Cl | b.p. 153°/3 mm | | 975 (b.p.); |
| iso$C_3H_7$ | | $CH_3$ | $C_2H_5$ | Cl | b.p. 78°/0.8 mm | | 281 (b.p.); |
| $Sn(C_2H_5)_3$ | Cl | | Cl | | b.p. 145°/2 mm | | 831 (b.p.), (i), (u); |
| $Sn(C_2H_5)_3$ | Br | | Br | | b.p. 180°/6 mm | | 831 (b.p.); |
| $-CH_2CH_2O-$(2,3,4-trichloro-6-fluorophenyl) | Cl | Cl. | Cl | F | 124–125° | | 953 (m.p.); |
| $CH_3$ | | $C_6H_5$ | | Cl | 65° | | 8 (m.p.); 457 (i), (n), (u); |
| $CH_2C_6H_5$ | | $CH_3$ | $C_2H_5$ | Cl | b.p. 150°/0.6 mm | | 281 (b.p.); |
| $-N(CH_2)_5$ | Cl | –N (piperidino) | Cl | Cl | 82° | | 240 (m.p.); |
| $CH_3$ | $C_6H_5$ | $C_6H_5$ | Cl | Cl | 164–166° | | 242 (m.p.), (u); |
| $CH_3$ | Cl | $C_6H_5$ | $C_6H_5$ | Cl | 152° | | 242 (m.p.), (u); |
| $C_2H_5$ | Cl | $C_6H_5$ | $C_6H_5$ | Cl | 110° | | 242 (m.p.), (m), (u); |
| $CH_2C_6H_5$ | $NHCH_2C_6H_5$ | | | Br | 78–80° | | 740 (m.p.), (i), (n); |

TABLE XII-38. Ethers and Esters of Halo 2-Pyridinols (Continued)

| R | $R_3$ | $R_4$ | $R_5$ | $R_6$ | m.p. | Derivatives | Ref. |
|---|---|---|---|---|---|---|---|
| $CH(C_6H_5)_2$ | | $CH_3$ | $C_2H_5$ | Cl | 104–105° | | 281 (m.p.); |
| $CH(C_6H_5)_2$ | $C_2H_5$ | $CH_3$ | | Cl | 94–95° | | 281 (m.p.); |
| $Pb(C_6H_5)_3$ | Cl | | Cl | | 120° | | 831 (m.p.), (i), (u); |
| $Sn(C_6H_5)_3$ | Cl | | Cl | | 112–113° | | 831 (m.p.), (i), (u); |
| $Sn(C_6H_5)_3$ | Br | | Br | | 127–130° | | 831 (m.p.), (i); |



TABLE XII-39. Ethers of Nitro 2-Pyridinols

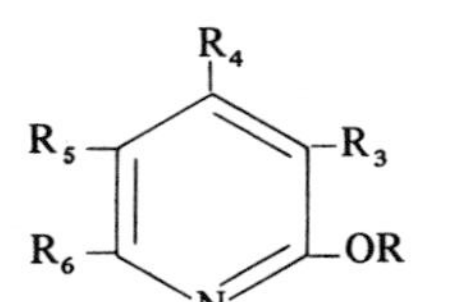


| R | $R_3$ | $R_4$ | $R_5$ | $R_6$ | m.p. | Derivatives | Ref. |
|---|---|---|---|---|---|---|---|
| $CH_3$ | $NO_2$ | | $NO_2$ | D | 89–91° | | 769 (m.p.), (n); |
| $CH_3$ | Cl | | $NO_2$ | | 58° | | 291 (m.p.), (i); |
| $CH_3$ | $NO_2$ | | Cl | | 90° | | 291 (m.p.); |
| $CH_3$ | Br | | $NO_2$ | | 84° | | 290 (m.p.), (i); |

| | | | | | | |
|---|---|---|---|---|---|---|
| | | | | | | 265 (m.p.); 769 (m.p.), |
| | | | | | | (n); 773 (m.p.), (n), |
| | | | | | | (u); 960 (u); |
| $CH_3$ | $NO_2$ | | | | | 956 (u); |
| $CH_3$ | | $NO_2$ | | | 106° | 261 (m.p.); 262 (m.p.); |
| $CH_3$ | | | $NO_2$ | | | 957 (u); 958 (i); |
| $CH_3$ | $NO_2$ | | $NO_2$ | $NHNH_2$ | 180–182° | 769 (m.p.), (n); |
| | | | | | | 265 (m.p.); |
| $C_2H_5$ | $NO_2$ | | $NO_2$ | | 65–71° | 260 (m.p.); 261 (m.p); |
| | | | | | | 265 (m.p.); |
| $CH_3$ | $NO_2$ | $CH_3$ | $NO_2$ | | 129° | 264 (m.p.); |
| $CH_3$ | $NO_2$ | | $NO_2$ | $CH_3$ | 54° | 264 (m.p.); |
| $CH_3$ | $CH_3$ | | $NO_2$ | | 93–95° | 700 (m.p.); |
| $CH_3$ | | | $NO_2$ | $CH_3$ | 68° | 263 (m.p.); |
| $C_2H_5$ | $NO_2$ | $CH_3$ | $NO_2$ | | 87° | 264 (m.p.); |
| $C_2H_5$ | $NO_2$ | | $NO_2$ | $CH_3$ | 63° | 264 (m.p.); |
| iso$C_3H_7$ | | | $NO_2$ | | 51–53° | 549 (m.p.); |
| $n$-$C_4H_9$ | | | $NO_2$ | | | 899 (n); |
| 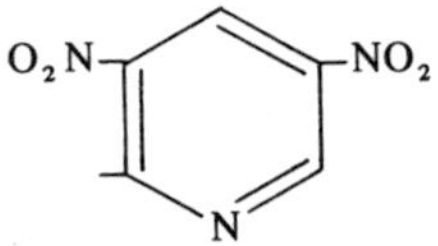 $O_2N$, $NO_2$, N (pyridine ring) | $NO_2$ | | $NO_2$ | | 255–257° | 265 (m.p.); 276 (m.p.); |
| $C_6H_4NO_2$-$p$ | $NO_2$ | | $NO_2$ | | 127° | 265 (m.p.); |
| $C_6H_5$ | Cl | | $NO_2$ | | 92° | 291 (m.p.); |
| $C_6H_5$ | Br | | $NO_2$ | | 121° | 290 (m.p.); |
| $C_6H_5$ | $NO_2$ | | $NO_2$ | | 159° | 265 (m.p.); |
| $C_6H_4OH$-$m$ | $NO_2$ | | | | 168° | 968 (m.p.); |
| $\beta$-D-Glucosyl | | | $NO_2$ | | 165–167° | 605 (m.p.), (f), (k); |

TABLE XII-39. Ethers of Nitro 2-Pyridinols (Continued)

| R | $R_3$ | $R_4$ | $R_5$ | $R_6$ | m.p. | Derivatives | Ref. |
|---|---|---|---|---|---|---|---|
| $-CH_2CH_2O-$(3,5-dinitro-2-pyridyl) | $NO_2$ | | $NO_2$ | | 147–148° | | 265 (m.p.); |
| β-D-Glucosyl | $NO_2$ | | Br | | | tetraacetyl, m.p. 162–165° | 608 (m.p.), (k); |
| β-D-Glucosyl | Br | | $NO_2$ | | | tetraacetyl, m.p. 188–193° sublimes at 160° | 608 (m.p.), (k); |
| β-D-Glucosyl | $NO_2$ | | $NO_2$ | | | tetraacetyl, m.p. 155–161° | 608 (m.p.), (k); |
| α-D-Glucosyl | | | $NO_2$ | | | tetraacetyl, m.p. 166–167° | 605 (m.p.), (k); |
| β-D-Glucosyl | | | $NO_2$ | | | tetraacetyl, m.p. 149-150° | 605 (m.p.), (k); 609 (n); |
| β-D-Glucosyl | $NO_2$ | | | | | tetraacetyl, m.p. 138–140° | 608 (m.p.), (k); |
| β-D-Glucosyl | CN | | $NO_2$ | | | tetraacetyl, m.p. 185–188° | 608 (m.p.), (k); |

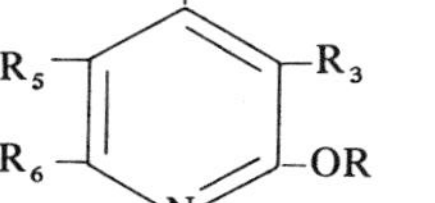


TABLE XII-40. Ethers and Esters of Sulfur Containing 2-Pyridinols

| R | $R_3$ | $R_4$ | $R_5$ | $R_6$ | m.p. | Derivatives | Ref. |
|---|---|---|---|---|---|---|---|
| $COSCH_3$ | Cl | Cl | Cl | Cl | 80–82° | | 976 (m.p.); |
| $COSCH_3$ | | Cl | | Cl | 57–58° | | 976 (m.p.); |
| $PS(OCH_3)_2$ | Cl | | Cl | Cl | 43–44° | | 703 (m.p.); |
| $-PSOCH_3$ with $NHCH_3$ | Cl | | Cl | | 84-86° | | 973 (m.p.); |
| $-PSOCH_3$ with $NHCH_3$ | Br | | Br | | 93–95° | | 973 (m.p.); |
| $-PSOCH_3$ with $NHC_2H_5$ | Cl | | Cl | | 89–91° | | 973 (m.p.); |
| $-PSOC_2H_5$ with $NHCH_3$ | Cl | | Cl | | 71–73° | | 973 (m.p.); |
| $-PSOC_2H_5$ with $NHCH_3$ | Br | | Br | | 88–89° | | 973 (m.p.); |
| $COS(CH_2)_2CH_3$ | Cl | | Cl | Cl | 45–46° | | 976 (m.p.); |
| $PS(OC_2H_5)_2$ | Cl | Cl | Cl | Cl | 47–49° | | 973 (m.p.); |
| $PS(OC_2H_5)_2$ | Cl | | Cl | Cl | 42–43° | | 703 (m.p.); 973 (m.p.); 977 (g); |
| $PS(OC_2H_5)_2$ | Cl | Cl | | Cl | 53–55° | | 973 (m.p.); |
| $PS(OC_2H_5)_2$ | Br | | Br | Cl | 60–61° | | 703 (m.p.); 973 (m.p.); |

TABLE XII-40. Ethers and Esters of Sulfur Containing 2-Pyridinols (Continued)

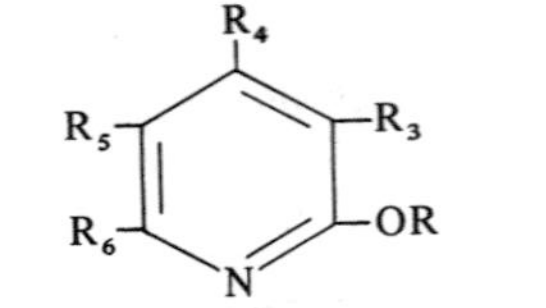

| R | $R_3$ | $R_4$ | $R_5$ | $R_6$ | m.p. | Derivatives | Ref. |
|---|---|---|---|---|---|---|---|
| $-P(S)(OCH_3)NHCH(CH_3)_2$ | Cl | Cl | Cl | Cl | 77–81° | | 973 (m.p.); |
| $-P(S)(OCH_3)NHCH(CH_3)_2$ | Cl | Cl | | Cl | 79–82° | | 973 (m.p.); |
| $-P(S)(OCH_3)NHCH(CH_3)_2$ | Cl | | Cl | Cl | 79–81° | | 973 (m.p.); |
| $-P(S)(NHCH_3)OCH_2CH_2CH_3$ | Cl | | | | 52–55° | | 973 (m.p.); |
| $-P(S)(NHCH_3)OCH(CH_3)_2$ | Cl | | Cl | | 86–89° | | 973 (m.p.); |
| $-P(S)(OCH_3)NHCH(CH_3)_2$ | Cl | | Cl | | 52–56° | | 973 (m.p.); |
| $-P(S)(OCH_3)NHCH(CH_3)_2$ | | Cl | | Cl | 51–54° | | 973 (m.p.); |
| $-P(S)(OC_2H_5)NHC_2H_5$ | Cl | | Cl | | 68–70° | | 973 (m.p.); |
| $-P(S)(OCH_3)NHCH(CH_3)_2$ | Br | | Br | | 63–68° | | 973 (m.p.); |

| | | | | | | |
|---|---|---|---|---|---|---|
| $NHCH_3$ | | | | | | |
| $-PSCH(CH_3)C_2H_5$<br>$NHCH_3$ | Cl | | Cl | | 82–86° | 973 (m.p.); |
| $-PSOC_4H_9$<br>$NHCH_3$ | Cl | | Cl | | 62–64° | 973 (m.p.); |
| $PSOCH_2CH(CH_3)_2$<br>$NHCH_3$ | Cl | | Cl | | 76–78° | 973 (m.p.); |
| $PS(OC_2H_5)_2$ | | | $SO_2CH_3$ | | 56–58° | 978 (m.p.); |
| $PS(OC_2H_5)_2$ | $SOC_2H_5$ | | Cl | | 47–49° | 978 (m.p.); |
| $PSOCH_2CH_2CH_3$<br>$NHCH_2CH_2CH_3$ | Cl | | Cl | | 60–62° | 973 (m.p.); |
| $COSC_6H_5$ | Cl | Cl | Cl | Cl | 103–104° | 976 (m.p.); |
| $COSC_6H_5$ | Cl | | Cl | Cl | 109–110° | 976 (m.p.); |
| $COSC_6H_5$ | | Cl | | Cl | 63° | 976 (m.p.); |
| $COSC_6H_5$ | | | | Cl | 88–89° | 976 (m.p.); |
| $C_2H_5$ | CN | 2-Thienyl | CN | $NH_2$ | 176–178° | 13 (m.p.), (i), (u); |
| $C_2H_5$ | | | | $-NHCSNH$–(4-F-phenyl) | 182–183° | 641 (m.p.); |
| $C_2H_5$ | | | | $-NHCSNH$–(6-$C_2H_5O$-2-pyridyl) | 176–177° | 641 (m.p.); |

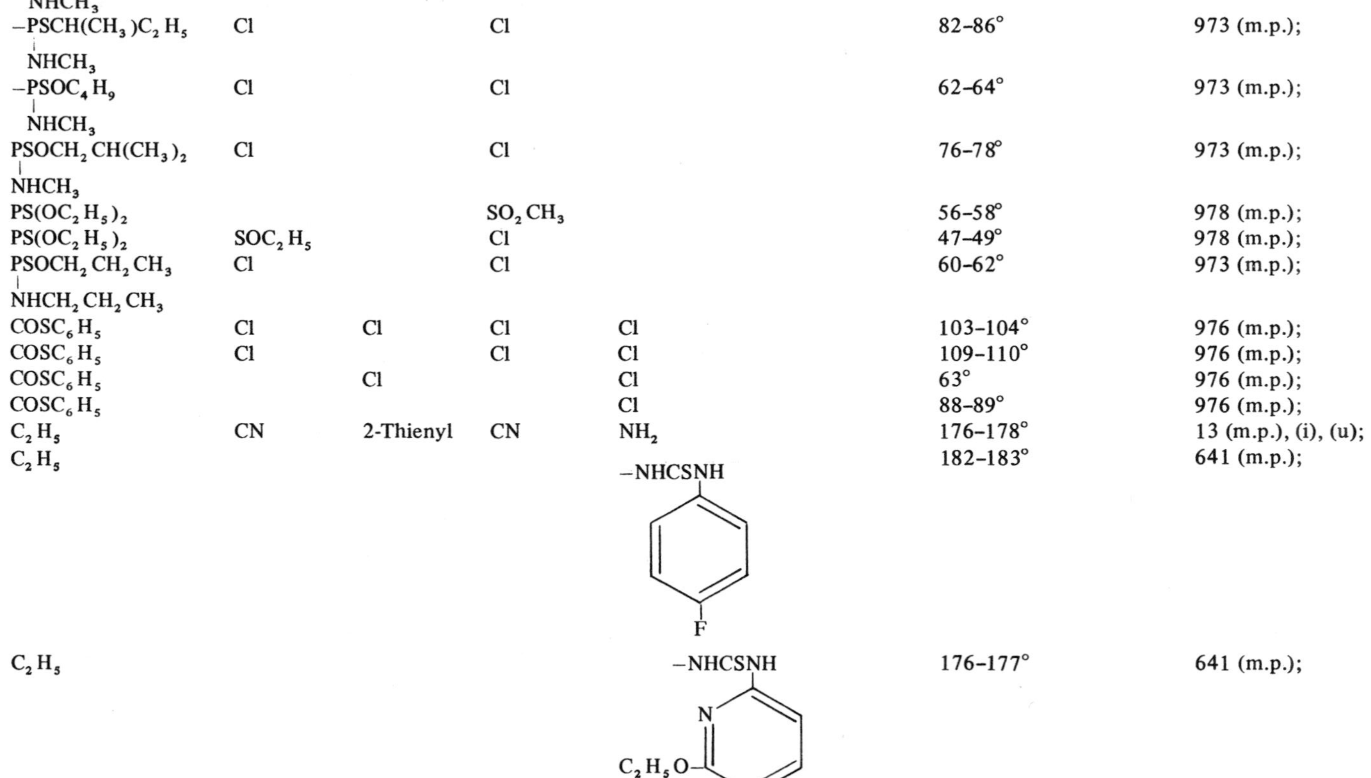

TABLE XII-40. Ethers and Esters of Sulfur Containing 2-Pyridinols (Continued)

| R | $R_3$ | $R_4$ | $R_5$ | $R_6$ | m.p. | Derivatives | Ref. |
|---|---|---|---|---|---|---|---|
| $-SO_2C_6H_4CH_3$-*p* | CN | $CH_2OCH_3$ | $NH_2$ | $CH_3$ | 126° | | 52 (m.p.); |
| $C_2H_5$ | | | | $-NHCSNH-C_6H_4-OC_2H_5$-*p* | 160–161° | | 641 (m.p.); |
| $C_2H_5$ | | | | $-NHCSNH-C_6H_4-OC_2H_5$-*o* | 171–172° | | 641 (m.p.); |
| $C_2H_5$ | | | | $-NHCSNH-C_6H_3(OCH_3)_2$-2,5 | 163–164° | | 641 (m.p.); |
| $-SO_2C_6H_4CH_3$-*p* | $-CH_2NH_2$ | $CH_2OCH_3$ | $NH_2$ | $CH_3$ | | picrate, | |

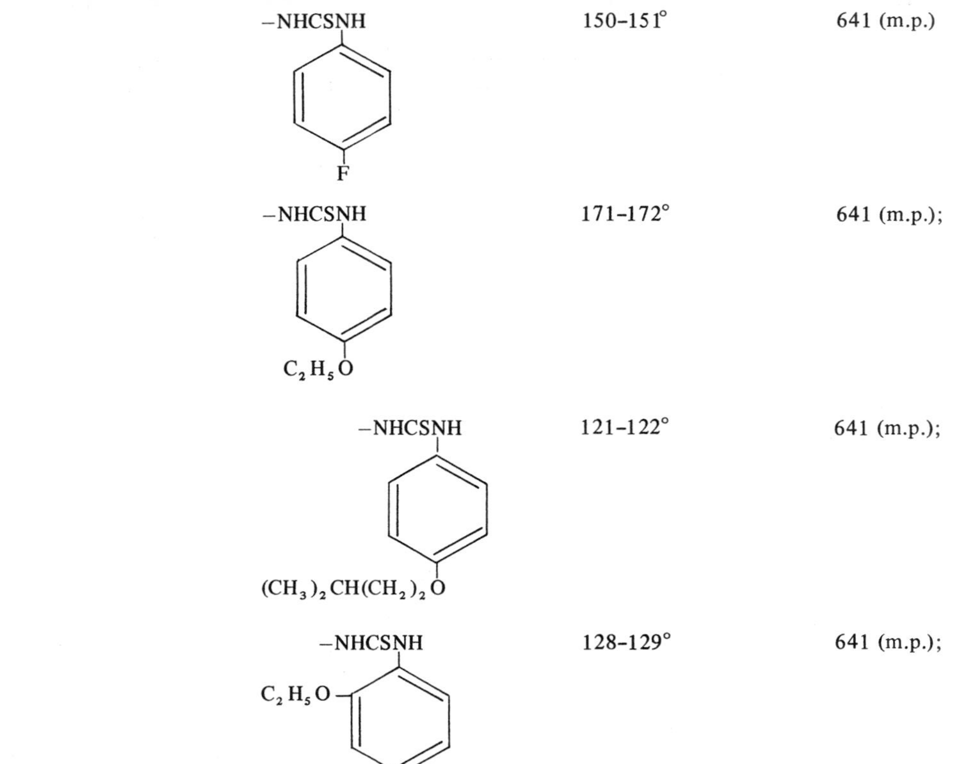

| | | | |
|---|---|---|---|
| iso$C_5H_{11}$ | $-NHCSNH-C_6H_4-F$ (p-F) | 150–151° | 641 (m.p.) |
| iso$C_5H_{11}$ | $-NHCSNH-C_6H_4-OC_2H_5$ (p-$C_2H_5O$) | 171–172° | 641 (m.p.); |
| $C_2H_5$ | $-NHCSNH-C_6H_4-O(CH_2)_2CH(CH_3)_2$ (p-$(CH_3)_2CH(CH_2)_2O$) | 121–122° | 641 (m.p.); |
| iso$C_5H_{11}$ | $-NHCSNH-C_6H_4-OC_2H_5$ (o-$C_2H_5O$) | 128–129° | 641 (m.p.); |

TABLE XII-40. Ethers and Esters of Sulfur containing 2-Pyridinols (Continued)

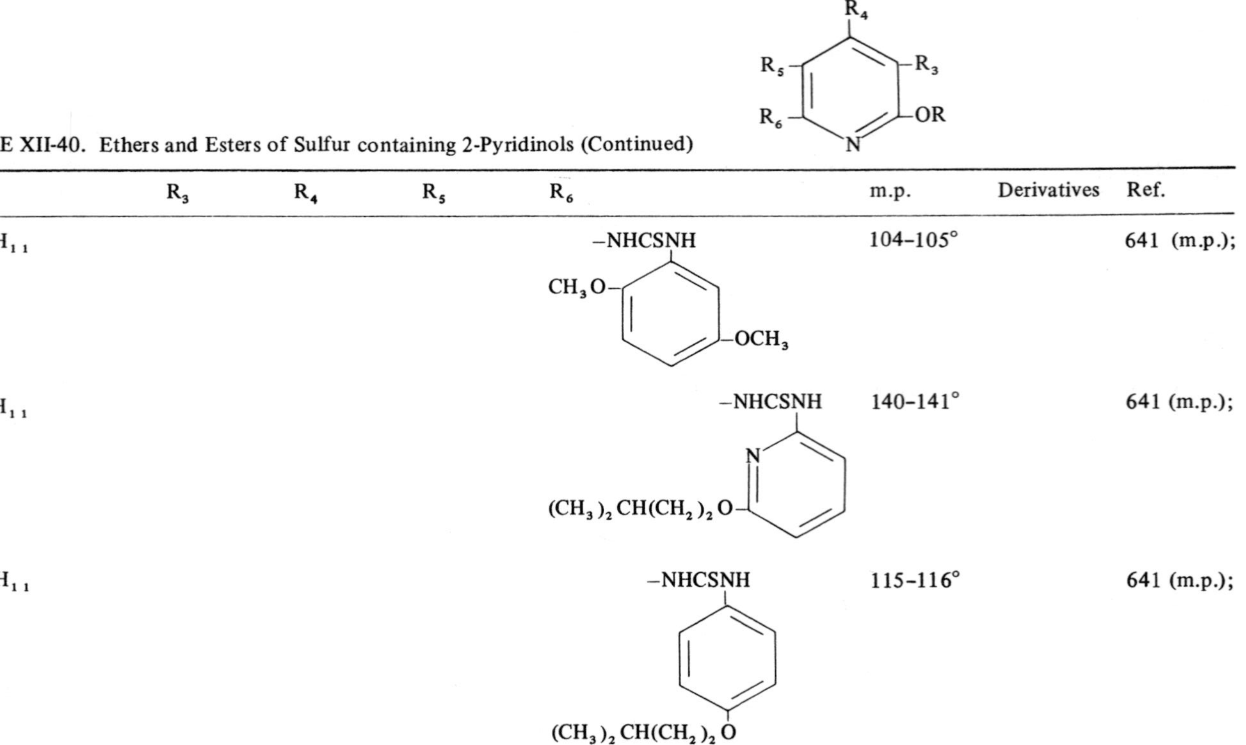


| R | $R_3$ | $R_4$ | $R_5$ | $R_6$ | m.p. | Derivatives | Ref. |
|---|---|---|---|---|---|---|---|
| iso$C_5H_{11}$ | | | | –NHCSNH; $CH_3O$–; –$OCH_3$ | 104–105° | | 641 (m.p.); |
| iso$C_5H_{11}$ | | | | –NHCSNH; N; $(CH_3)_2CH(CH_2)_2O$– | 140–141° | | 641 (m.p.); |
| iso$C_5H_{11}$ | | | | –NHCSNH; $(CH_3)_2CH(CH_2)_2O$ | 115–116° | | 641 (m.p.); |

TABLE XII-41. Alkyl and Aryl 2-Pyridones

(Structure: 2-pyridone ring with $R_1$ on N, =O at C-2, $R_3$, $R_5$, $R_6$ substituents)

| $R_1$ | $R_3$ | $R_4$ | $R_5$ | $R_6$ | m.p. | Derivatives | Ref. |
|---|---|---|---|---|---|---|---|
| D | | | | | | | 465 (i); 487 (i); |
| D | | | | | | $^{15}N$ | 465 (i); |
| D | | | | | | $^{18}O$ | 465 (i); |
| H | H | H | H | H | | $B(C_2H_5)_2$ complex | 838 (n); |
| | | | | | | $C_6H_5Hg$ complex, m.p. 217–218° | 831 (m.p.), (i), (u); |
| | | | | | | $Si(CH_3)_2$ complex, b.p. 116°/1 mm | 830 (b.p.); |
| | | | | | | Al complex m.p. 121° | 830 (m.p.); |
| | | | | | | $Sn(C_4H_9)_2$ complex | 830; |
| $CD_3$ | | | | | | | 485 (n); 907 (m); |
| $CH_3$ | D(H) | | D(H) | D | b.p. 80°/0.4 mm | | 530 (b.p.), (n); |
| $CH_3$ | | | | | b.p. 63°/0.5 mm | | 465 (i); 472 (n); |
| | | | | | b.p. 115°/10 mm | | 476 (b.p.), (n); 481 (z); 525 (h), (u); 530 (b.p.); 541 (i); 616 (b.p.), (i); 692 (b.p.), (i); 764 (m); 923 (n); 979 (c); 342 (t); 348 (f), (g), (h), (u); 353 (b.p.); 466 (i); 480 (i); 482 (z); 485 (n); 745 (u); 907 (m); 919 (i); 980 (w); 981 (b.p.), (i); 982 (a), (u); 983 (i); 984; 985 (n); 1157 (n); |
| | | | | | | $^{15}N$ | 465 (i); |
| | | | | | | $^{18}O$ | 465 (i); |
| | | | | | | picrate, m.p. 142–143° | 351 (m.p.); 353 (m.p.); 745 (m.p.), (u); |
| | | | | | | hydrochloride, m.p. 170–172° | 540 (m.p.), (i); |
| | | | | | | hydrobromide, m.p. 175–176° | 540 (m.p.), (i); |

# TABLE XII-41. Alkyl and Aryl 2-Pyridones (Continued)

| $R_1$ | $R_3$ | $R_4$ | $R_5$ | $R_6$ | m.p. | Derivatives | Ref. |
|---|---|---|---|---|---|---|---|
| | | | | | | hydroiodide · $H_2O$ | |
| | | | | | | m.p. 73–75° | 540 (m.p.), (i); |
| | | | | | | hydroiodide, | |
| | | | | | | m.p. 116–119° | 540 (m.p.), (i); |
| | | | | | | $HSbCl_4$ | |
| | | | | | | m.p. 128–131° | 540 (m.p., (i); |
| | | | | | | $H_2SnBr_6$ | 540 (m.p.), (i); |
| | | | | | | m.p. 132–135° | |
| | | | | | | $H_2SnCl_6$ | 540 (m.p.), (i); |
| | | | | | | m.p. 167–169° | 464 (m.p.), (i); |
| | | | | | | m.p. 250–252° | 464 (m.p.), (i); |
| | | | | | | $HSbCl_6$ | 540 (m.p.), (i); |
| | | | | | | m.p. 186–189° | 542 (i); |
| | | | | | | 2 : 1 complex with | |
| | | | | | | cyclomethylene | |
| | | | | | | tetranitraamine | 839 (i), (x); |
| | | | | | | pentacyanopropene | |
| | | | | | | salt, m.p. 135–136° | 745 (m.p.), (u); |
| | | | | | | 1 : 1 complex $BF_3$ | |
| | | | | | | m.p. 124–126° | 818 (i); 986 (m.p.); |
| | | | | | | $BCl_3$ complex, | |
| | | | | | | m.p. 115–123° | |
| | | | | | | hygroscopic | 818 (m.p.), (i); |
| | | | | | | $BBr_3$ complex, | |
| | | | | | | m.p. 106–109° | 818 (m.p.), (i); |
| | | | | | | $CdCl_2$ complex, | |
| | | | | | | m.p. >290° | 818 (m.p.), (i), |
| | | | | | | $HgCl_2$ | 817 (m.p.); 818 (m.p.), |
| | | | | | | m.p. 126–133° | (i); |
| | | | | | | $SbCl_5$ complex | |
| | | | | | | m.p. 85–87° | |
| | | | | | | hygroscopic | 818 (m.p.), (i); |
| | | | | | | $SnBr_4$ complex, | |
| | | | | | | m.p. 225–227° | 818 (m.p.), (i); |
| | | | | | | $ZnCl_2$ complex | |
| | | | | | | [illegible] | [illegible] |

| | | | | | | | |
|---|---|---|---|---|---|---|---|
| $CO_2CH_3$ | | | | | | | 907 (m); |
| $COCH_3$ | | | | | | | 631 (i), (n); |
| $CH_2CH_2Cl$ | | | | | 62–65°<br>hygroscopic | hydrochloride,<br>m.p. 151–154° | 988 (m.p.);<br>988 (m.p.); 989 (m.p.); |
| $CH_3$ | $CH_2Br$ | | | | 98–99° | | 559 (m.p.); |
| $CH_3$ | | | $CH_2Br$ | | 119° | | 559 (m.p.); |
| $CH_3$ | $CH_3$ | | | | b.p. 78°/0.5 mm | | 476 (b.p.), (i), (n);<br>559 (b.p.); 907 (m);<br>924 (b.p.); 982 (a), (u); |
| $CH_3$ | | $CH_3$ | | | 58–59°<br>b.p. 110°/3 mm | | 353 (b.p.); 559 (m.p.);<br>907 (m); 923 (n);<br>924 (m.p.); 982 (a), (u); |
| $CH_3$ | | | $CH_3$ | | b.p. 88°/0.5 mm<br>b.p. 185°/29 mm<br>m.p. 40–41° | picrate,<br>m.p. 163–164 | 353 (m.p.);<br>476 (m.p.), (i), (n);<br>559 (b.p.); 700 (b.p.);<br>(u); 907 (m);<br>924 (b.p.); 982 (a), (u);<br>984; |
| $CH_3$ | | | | $CH_3$ | 55–58°<br>b.p. 80°/<0.5 mm<br>b.p. 116°/3 mm | picrate<br>m.p. 133–134°<br>hydrochloride,<br>m.p. 198–203° | 353 (b.p.); 442 (b.p.),<br>(m.p.); 476 (m.p.), (n);<br>485 (n); 559 (m.p.);<br>907 (m); 924 (m.p.);<br>982 (a), (u); 984;<br>353 (m.p.);<br>306 (m.p.); 442 (m.p.); |
| $C_2H_5$ | | | | | b.p. 140°/0.5 mm<br>b.p. 118°/7 mm | HgCl$_2$ complex,<br>m.p. 108–109°<br>pentacyanopropene,<br>salt, m.p. 116–117° | 485 (n); 570 (b.p.);<br>907 (m); 987 (b.p.);<br>346 (m.p.); 351 (m.p.);<br>745 (m.p.), (u); |
| $CH_2CH_2OH$ | | | | | 95–97°<br>b.p. 198°/10 mm | | 988 (m.p.), (b.p.); |
| $CH_3$ | | | $CH_2OH$ | | 106–108° | 5-*O*-acetyl,<br>m.p. 45–47° | 729 (m.p.), (i), (n), (u);<br>729 (m.p.), (i), (n); |
| $CH_2CH_2ONH_2$ | | | | | | hydrochloride,<br>m.p. 197–198° | 573 (m.p.); 990 (m.p.); |
| $CH_2COCH_2Cl$ | | | | | 131–132° | | 554 (m.p.); |
| $CH_2CH{=}CH_2$ | | | | | b.p. 88°/1.5 mm | picrate,<br>m.p. 104-106° | 557 (b.p.); 991 (b.p.);<br>557 (m.p.); |
| $CH_3$ | | | $COCH_3$ | | | | 437 (i), (n), (u);<br>438 (n); |
| $CH_3$ | | $CH_3$ | $CH_3$ | | | | 476 (n); 984; |

TABLE XII-41. Alkyl and Aryl 2-Pyridones (Continued)

[structure: 2-pyridone ring, N-$R_1$, =O, $R_3$, $R_4$, $R_5$, $R_6$]

| $R_1$ | $R_3$ | $R_4$ | $R_5$ | $R_6$ | m.p. | Derivatives | Ref. |
|---|---|---|---|---|---|---|---|
| $CH_3$ | | $CH_3$ | | $CH_3$ | 85–89°<br>b.p. 152°/10 mm | | 60 (n); 107 (m.p.);<br>476 (m.p.), (n);<br>694 (b.p.), (m.p.);<br>907 (m);<br>982 (a), (u); |
| $CH_3$ | $C_2H_5$ | | | | picrate,<br>m.p. 59–60° | | 314 (m.p.); |
| $CH_3$ | | | $C_2H_5$ | | b.p. 114°/3 mm | | 353 (b.p.), (i); |
| $C_2H_5$ | | $CH_3$ | | | 32° | picrate,<br>m.p. 122–123° | 314 (m.p.); 353 (m.p.);<br>907 (m.p.), (m); |
| $C_2H_5$ | | | $CH_3$ | | | | 467 (u); |
| $n$-$C_3H_7$ | | | | | b.p. 123°/13 mm | picrate,<br>m.p. 95–97° | 557 (b.p.); 907 (b.p.),<br>(m);<br>557 (m.p.); |
| iso$C_3H_7$ | | | | | | pentacyanopropene<br>salt, m.p. 97° | 745 (m.p.), (u); |
| $CH_3$ | | | CH(OH)$CH_3$ | | | | 485 (n); |
| $CF_3C=CHCF_3$ | | | | | | | 435 (i), (n); |
| $-CH_2CH=CHCH_3$ | | | | | 55–56° | | 746 (m.p.), (i), (u); |
| $CH_2=COC_2H_5$ | | | | | b.p. 94°/0.1 mm | | 790 (b.p.), (g); |
| $n$-$C_4H_9$ | | | | | b.p. 108°/1.2 mm<br>b.p. 172°/33 mm | | 747 (b.p.), (i), (n), (u);<br>550 (b.p.); |
| $CH_3$ | $CH_2OH$ | | $C_2H_5$ | | 78–79° | | 729 (m.p.), (i), (n); |
| [structure: pyridine ring, N, $NO_2$] | | | | | 172–174° | | 271 (m.p.); 1012 (m.p.); |
| [structure: pyridine ring, N] | | | | | 179–180° | | 271 (m.p.); |

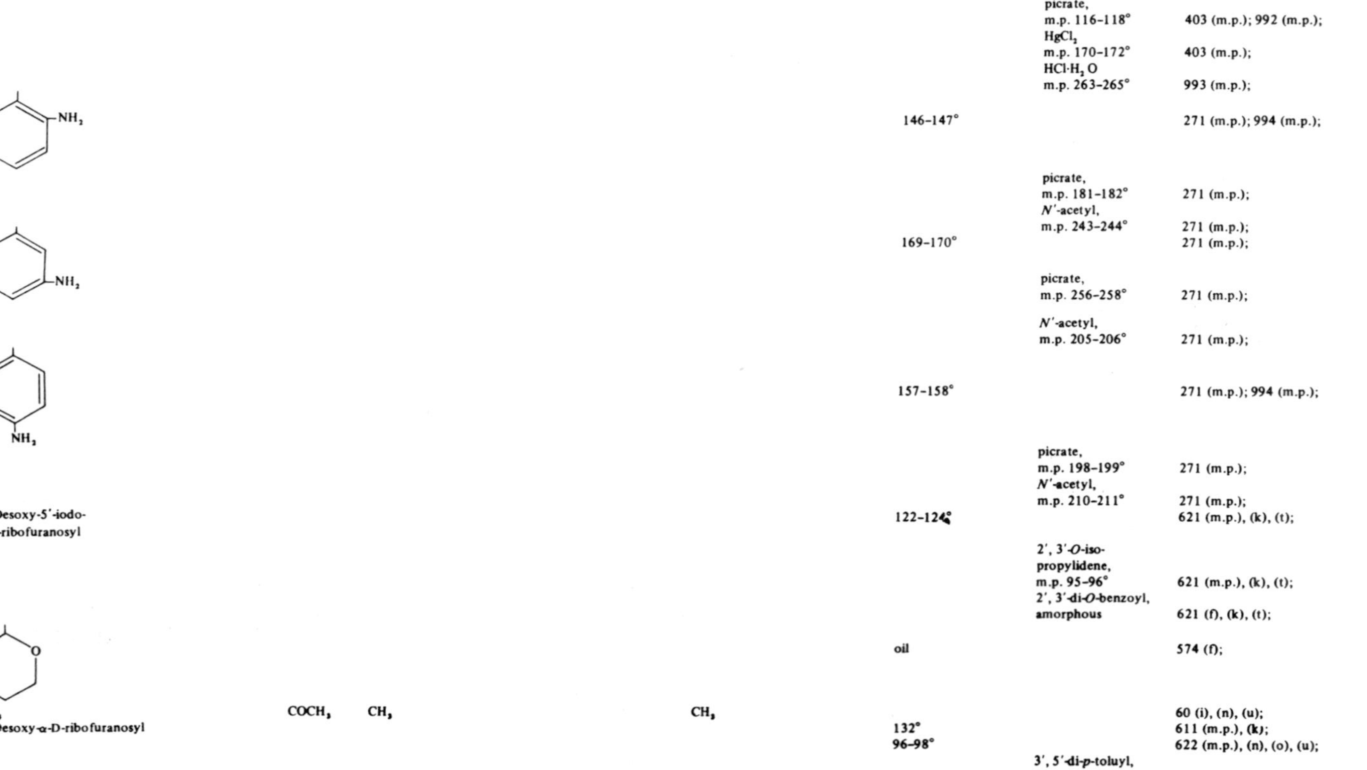

| | | | | | | |
|---|---|---|---|---|---|---|
| | | | | | | (g); 409 (g); |
| | | | | | picrate, m.p. 116–118° | 403 (m.p.); 992 (m.p.); |
| | | | | | $HgCl_2$ m.p. 170–172° | 403 (m.p.); |
| | | | | | $HCl{\cdot}H_2O$ m.p. 263–265° | 993 (m.p.); |
| 2-pyridyl, 3-$NH_2$ | | | | 146–147° | | 271 (m.p.); 994 (m.p.); |
| | | | | | picrate, m.p. 181–182° | 271 (m.p.); |
| | | | | | N′-acetyl, m.p. 243–244° | 271 (m.p.); |
| 2-pyridyl, 4-$NH_2$ | | | | 169–170° | | 271 (m.p.); |
| | | | | | picrate, m.p. 256–258° | 271 (m.p.); |
| | | | | | N′-acetyl, m.p. 205–206° | 271 (m.p.); |
| 2-pyridyl, 5-$NH_2$ | | | | 157–158° | | 271 (m.p.); 994 (m.p.); |
| | | | | | picrate, m.p. 198–199° | 271 (m.p.); |
| | | | | | N′-acetyl, m.p. 210–211° | 271 (m.p.); |
| 5′-Desoxy-5′-iodo-β-D-ribofuranosyl | | | | 122–124° | | 621 (m.p.), (k), (t); |
| | | | | | 2′, 3′-O-isopropylidene, m.p. 95–96° | 621 (m.p.), (k), (t); |
| | | | | | 2′, 3′-di-O-benzoyl, amorphous | 621 (f), (k), (t); |
| tetrahydropyran-2-yl | | | | oil | | 574 (f); |
| $CH_3$ | COCH$_3$ | $CH_3$ | $CH_3$ | | | 60 (i), (n), (u); |
| 2′-Desoxy-α-D-ribofuranosyl | | | | 132° | | 611 (m.p.), (k); |
| | | | | 96–98° | | 622 (m.p.), (n), (o), (u); |
| | | | | | 3′, 5′-di-p-toluyl, m.p. 152–159° | 611 (m.p.), (k); 619 (n); 622 (m.p.); |

TABLE XII-41. Alkyl and Aryl 2-Pyridones (Continued)

| $R_1$ | $R_3$ | $R_4$ | $R_5$ | $R_6$ | m.p. | Derivatives | Ref. |
|---|---|---|---|---|---|---|---|
| 2′-Desoxy-β-D-ribofuranosyl | | | | | 115–116° | | 611 (m.p.), (k); 622 (m.p.), (o), (u); |
| | | | | | | 3′, 5′-di-*O*-toluyl, m.p. 112–115° | 611 (m.p.), (k); 619 (n); |
| 5′-Desoxy-β-D-ribofuranosyl | | | | | syrup | | 621 (f), (k); |
| | | | | | | 2′, 3′-*O*-isopropylidene, m.p. 85–87° | 621 (m.p.), (f), (k), (t); |
| | | | | | | 2′-3′-di-*O*-benzoyl, m.p. 108–111° | 621 (m.p.), (k), (t); |
| α-L-Arabinosyl | | | | | 227–228° | | 626 (m.p.), (f), (k); |
| | | | | | | triacetyl, amorphous | 626 (k); |
| β-D-Ribofuranosyl | | | | | 151–152° | | 601 (m.p.); 602 (m.p.); 615 (u); 625 (m.p.), (i), (u); 628 (u); |
| | | | | | | 2′-phosphate | 627 (u); |
| | | | | | | 3′-phosphate | 627 (u); |
| | | | | | | 5′-phosphate barium salt | 615 (u); |
| | | | | | | 5′-phosphate | 628 (u); |
| | | | | | | 5′-diphosphate | 628 (u); |
| | | | | | | 2′, 3′-cyclophosphate | 627 (c); |
| | | | | | | 2′, 3′-isopropylidene, m.p. 92–93° | 615 (m.p.), (t); |
| | | | | | | 2′, 3′-isopropylidene 5′-phosphate bistriethyl ammonium salt | 615 (u); |
| | | | | | | 2′, 3′-*O*-ethoxymethylene | 628 (t); |
| | | | | | | 5′-phenylphosphate barium salt | 615 (u); |
| | | | | | | 5′-imidazolyl phosphate 2′, 3′-*O*-ethoxymethylene | 628; |
| | | | | | | 2′, 3′-di-*O*-benzoyl m.p. 226–228° | 615 (m.p.), (k), (t); |

| | | | | |
|---|---|---|---|---|
| | | | [illegible] | [illegible] |
| | | | 2, 3, 5-tri-*O*-benzoyl m.p. 139–142° | 601 (m.p.); 602 (m.p.); |
| | | | 2, 3-di-*O*-benzoyl 5′-diphenyl phosphate, m.p. 118–119° | 614 (m.p.), (i), (k); 615 (m.p.), (k), (t); |
| β-D-Ribopyranosyl | | 207–208° | | 601 (m.p.); 602 (m.p.); |
| | | | tri-*O*-benzoyl amorphous | 601; 602; |
| β-D-Xylosyl | | 92–93° | | 626 (m.p.), (f), (k); |
| | | | triacetyl, m.p. 223–225° | 626 (m.p.), (k); |
| $-CH_2CH_2ON{=}C(CH_3)_2$ | | b.p. 125°/0.25 mm | | 573 (b.p.); 990 (b.p.); |
| | | | hydrochloride, m.p. 115–116° | 573 (m.p.); 990 (m.p.); |
| $-CH_2-N$ [morpholine ring, O] | | 79° | | 501 (m.p.); |
| $CH_3$ $-(CH_2)_3N(CH_3)_2$ | *t*-$C_4H_9$ | | | 485 (n); |
| | | | dihydrochloride, m.p. 83–85° | 562 (m.p.); |
| **$C_6H_2(NO_2)_3$-2, 4, 6** | | **206–208°** | | **553 (m.p.), (i), (n);** |
| [pyridine ring: N, CN] | | **204–206°** | | **271 (m.p.);** |
| **$C_6H_4Cl$-*m*** | | **134–136°** | | **565 (m.p.);** |
| **$C_6H_4Cl$-*p*** | | **133°** | | **24 (m.p.);** |
| **$C_6H_4Br$-*m*** | | **145–147°** | | **565 (m.p.);** |
| **$C_6H_4NO_2$-*m*** | | **185–187°** | | **564 (m.p.);** |
| **$C_6H_5$** | | **124–128°** | | **750 (m.p.); 907 (m); 988 (m.p.), (m), (n);** |
| **$C_6H_4OH$-*m*** | | **175–176°** | | **565 (m.p.);** |
| **$C_6H_4NH_2$-*m*** | | **183–185°** | | **564 (m.p.);** |
| [pyridine ring: N, $CH_3$] | | **107–109°** | | **361 (m.p.), (u);** |
| [pyridine ring: N, $CH_3$] | | **93–95°** | | **361 (m.p.), (u);** |
| $-CH_2-$[pyridine ring: N] | | **93°** | | **562 (m.p.);** |

TABLE XII-41. Alkyl and Aryl 2-Pyridones (Continued)

| $R_1$ | $R_3$ | $R_4$ | $R_5$ | $R_6$ | m.p. | Derivatives | Ref. |
|---|---|---|---|---|---|---|---|
| 3-$OCH_3$-2-pyridyl (structure) | | | | | 146–147° | | 271 (m.p.); |
| 4-$OCH_3$-2-pyridyl (structure) | | | | | 106–107° | | 271 (m.p.); |
| 5-$OCH_3$-2-pyridyl (structure) | | | | | 125–127° | | 271 (m.p.); |
| tetrahydropyran-2-yl (structure) | $CH_3$ | | | | oil | | 574 (f); |
| tetrahydropyran-2-yl (structure) | | $CH_3$ | | | oil | | 574 (f); |
| β-D-Galactosyl | | | | | 204–205° | | 626 (m.p.), (f), (k); |
| | | | | | | tetraacetyl, amorphous | 626; |
| β-D-Glucopyranosyl | | | | | 203–205° | | 603 (m.p.), (f); 620 (m.p.); |
| | | | | | | tetraacetyl, m.p. 190° | 603 (m.p.); 607 (n); 616 (m.p.); 620 (m.p.); 626 (m.p.), (k); |

| | | | | | |
|---|---|---|---|---|---|
| $CH_3$ | | $-(CH_2)_4NHCH_3$ | | dihydrochloride, m.p. 171° | 691 (m.p.); |
| | | | | picrate, m.p. 164–165° | 691 (m.p.); |
| $-COCO-N$ (ring, O) | | | 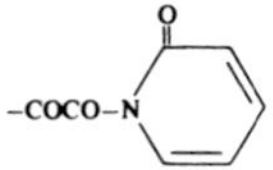 164–174° | | 1016 (m.p.); |
| $-CH_2-$ (ring, Cl, $-NO_2$) | | | 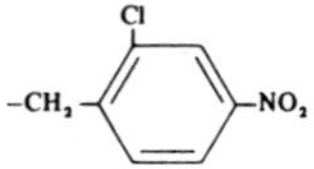 136° | | 558 (m.p.); |
| $-CH_2-$ (ring, Br, $-NO_2$) | | | 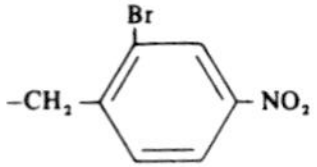 160–161° | | 561 (m.p.); |
| $-CH_2-$ (ring, Br, $NO_2$) | | | 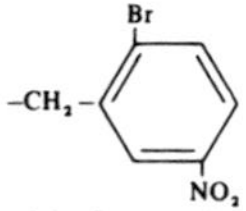 161–163° | | 561 (m.p.); |
| $-CH_2C_6H_4$-F-*o* | | | 76–78° | | 562 (m.p.); |
| $-CH_2C_6H_4$-Cl-*o* | | | 72° | | 561 (m.p.), (u); |
| $-CH_2C_6H_4$-Br-*o* | | | 105–106° | | 1017 (m.p.); |
| $-CH_2C_6H_4$-Br-*p* | | | 97–98° | | 561 (m.p.); |
| $-CH_2C_6H_4$-I-*o* | | | 121–122° | | 561 (m.p.), (u); |
| $-CH_2C_6H_4$-$NO_2$-*o* | | | 135–136° | | 564 (m.p.); 1018 (m.p.); |
| $-CH_2C_6H_4$-$NO_2$-*m* | | | 114–115° | | 564 (m.p.); |
| $-CH_2C_6H_4$-$NO_2$-*p* | | | 147–149° | | 564 (m.p.); |
| $-CH_2C_6H_5$ | | | | | 466 (i); 689 (i); |
| $CH_3$ | $C_6H_5$ | | 130–132° | | 750 (m.p.), (m), (n); |
| $C_6H_5$ | | $CH_3$ | 136–139° | | 24 (m.p.); 442 (m.p.); |
| | | | | hydrochloride, m.p. 201–204° | 442 (m.p.); |
| $C_6H_4CH_3$-*o* | | | | | 907 (m); |
| $-CH_2-$ (ring, Br, $-NH_2$) | | | 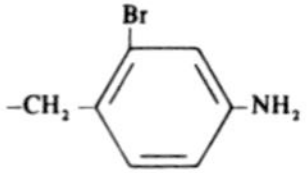 142° | | 561 (m.p.); |

TABLE XII-41. Alkyl and Aryl 2-Pyridones (Continued)

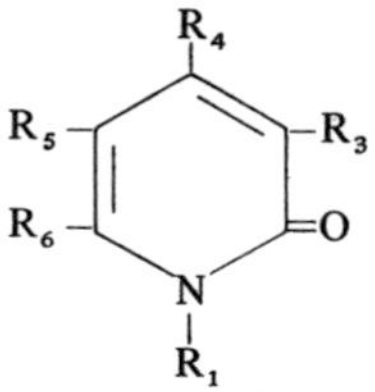

(Structure: 2-pyridone ring with $R_1$ on N, =O at C-2, $R_3$, $R_4$, $R_5$, $R_6$ substituents)

| $R_1$ | $R_3$ | $R_4$ | $R_5$ | $R_6$ | m.p. | Derivatives | Ref. |
|---|---|---|---|---|---|---|---|
| $-CH_2-$[benzene ring with Br, $NH_2$] 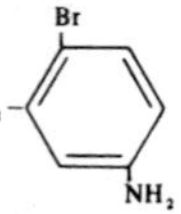 | | | | | 119–120° | | 561 (m.p.); |
| $-CH_2C_6H_4NH_2$-*o* | | | | | 133–134° | | 564 (m.p.); |
| $-CH_2C_6H_4NH_2$-*m* | | | | | 133–134° | | 564 (m.p.); |
| $-CH_2C_6H_4NH_2$-*p* | | | | | 110–112° | | 564 (m.p.); |
| 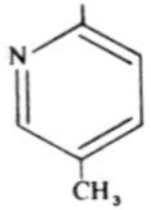 [pyridine ring, N, $CH_3$] | $CH_3$ | | | | 110–111° | | 361 (m.p.), (i), (u); |
| 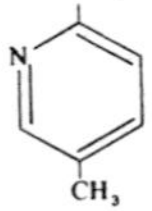 [pyridine ring, N, $CH_3$] | | | $CH_3$ | | 102–103° | | 361 (m.p.), (i), (u); |
| 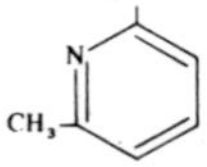 [pyridine ring, N, $CH_3$] | | | | $CH_3$ | 202° | | 993 (m.p.); |
| $CH_3$ | | | | | 206–210° | 6,6′-dimer | 750 (m.p.), (m), (n); |
| β-D-Glucopyranosyl | $CH_3$ | | | | | tetraacetyl, amorphous | 610 (i); |
| β-D-Glucosyl | | $CH_3$ | | | hygroscopic powder | | 626 (f), (k); |
| | | | | | | tetraacetyl, m.p. 163–165° | 626 (m.p.), (k); |
| $CH_3$ |  [ring] | | | | 95–96° | | |

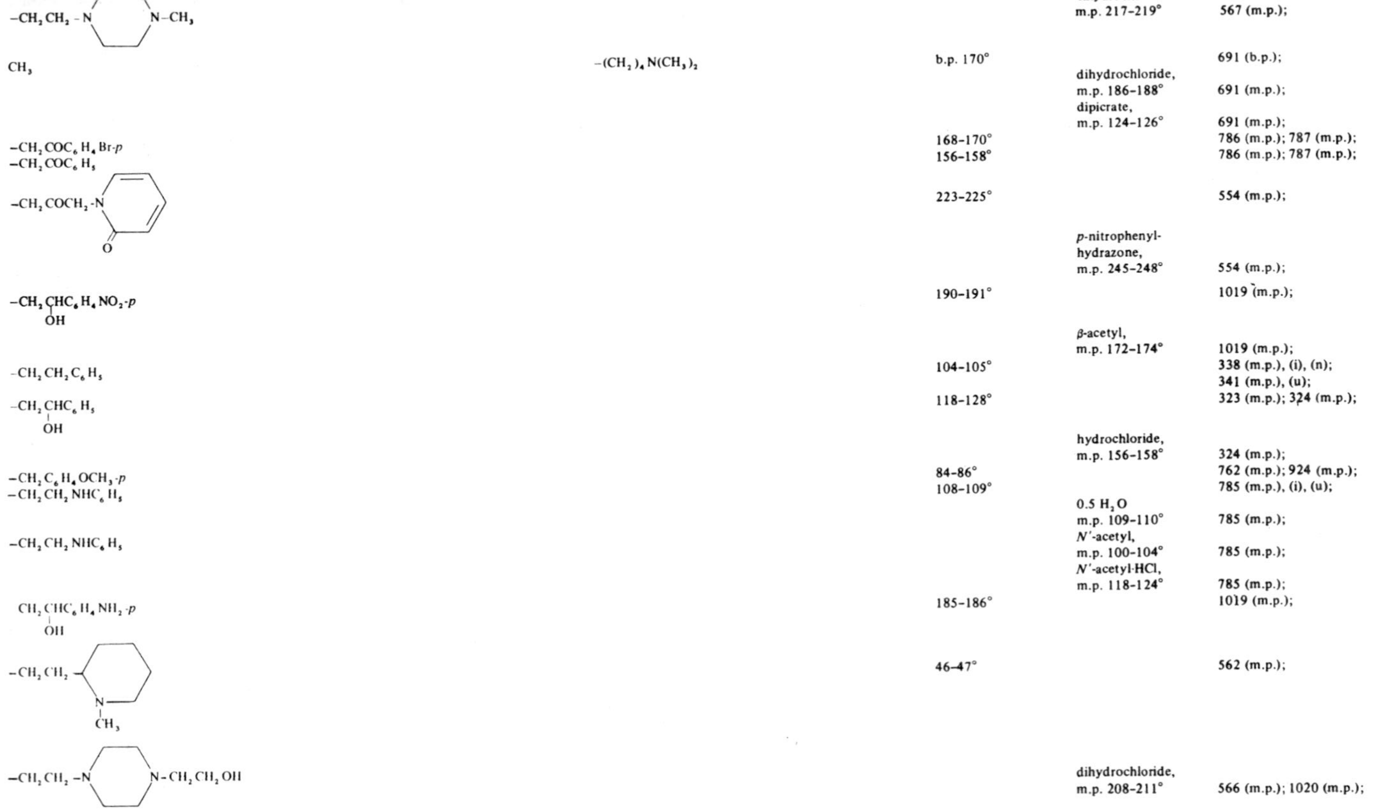

| | | | | |
|---|---|---|---|---|
| $-CH_2CH_2-N\langle\ \rangle N-CH_3$ (piperazine ring) | | | m.p. 217–219° | 567 (m.p.); |
| $CH_3$ | $-(CH_2)_4N(CH_3)_2$ | b.p. 170° | | 691 (b.p.); |
| | | | dihydrochloride, m.p. 186–188° | 691 (m.p.); |
| | | | dipicrate, m.p. 124–126° | 691 (m.p.); |
| $-CH_2COC_6H_4Br$-*p* | | 168–170° | | 786 (m.p.); 787 (m.p.); |
| $-CH_2COC_6H_5$ | | 156–158° | | 786 (m.p.); 787 (m.p.); |
| $-CH_2COCH_2$-N (2-pyridone ring, =O) | | 223–225° | | 554 (m.p.); |
| | | | *p*-nitrophenyl-hydrazone, m.p. 245–248° | 554 (m.p.); |
| $-CH_2CH(OH)C_6H_4NO_2$-*p* | | 190–191° | | 1019 (m.p.); |
| | | | β-acetyl, m.p. 172–174° | 1019 (m.p.); |
| $-CH_2CH_2C_6H_5$ | | 104–105° | | 338 (m.p.), (i), (n); 341 (m.p.), (u); |
| $-CH_2CH(OH)C_6H_5$ | | 118–128° | | 323 (m.p.); 324 (m.p.); |
| | | | hydrochloride, m.p. 156–158° | 324 (m.p.); |
| $-CH_2C_6H_4OCH_3$-*p* | | 84–86° | | 762 (m.p.); 924 (m.p.); |
| $-CH_2CH_2NHC_6H_5$ | | 108–109° | | 785 (m.p.), (i), (u); |
| | | | $0.5\ H_2O$ m.p. 109–110° | 785 (m.p.); |
| $-CH_2CH_2NHC_6H_5$ | | | *N'*-acetyl, m.p. 100–104° | 785 (m.p.); |
| | | | *N'*-acetyl·HCl, m.p. 118–124° | 785 (m.p.); |
| $CH_2CH(OH)C_6H_4NH_2$-*p* | | 185–186° | | 1019 (m.p.); |
| $-CH_2CH_2$- (1-methyl-2-piperidyl, N–$CH_3$) | | 46–47° | | 562 (m.p.); |
| $-CH_2CH_2-N\langle\ \rangle N-CH_2CH_2OH$ (piperazine ring) | | | dihydrochloride, m.p. 208–211° | 566 (m.p.); 1020 (m.p.); |

TABLE XII-41. Alkyl and Aryl 2-Pyridones (Continued)

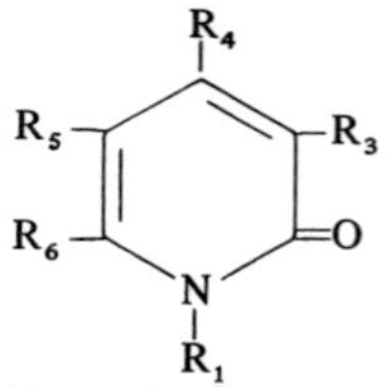

| $R_1$ | $R_3$ | $R_4$ | $R_5$ | $R_6$ | m.p. | Derivatives | Ref. |
|---|---|---|---|---|---|---|---|
| 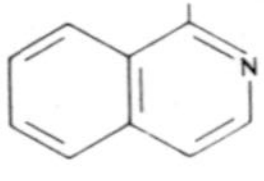 (isoquinoline ring, N) |  |  |  |  | 188–189° |  | 274 (m.p.); 1021 (m.p.); |
| $CH_2C_6H_5$ |  |  |  |  |  | picrate, m.p. 149° | 274 (m.p.); |
| $-CH_2COC_6H_4OCH_3$-*p* |  |  |  | $-CH_2CN$ | 92° |  | 944 (m.p.); |
| $-CH(CH_3)CH_2C_6H_5$ |  |  |  |  | 73° |  | 787; 337 (m.p.); |
| 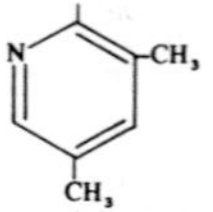 (pyridine ring, N, $-CH_3$, $CH_3$) | $CH_3$ |  | $CH_3$ |  | 104–105° |  | 376 (m.p.), (i); |
|  $-CH_2CH_2$-N(piperazine)N-$CH(CH_3)_2$ |  |  |  |  |  | dihydrochloride, m.p. 237–240° | 566 (m.p.); |
| 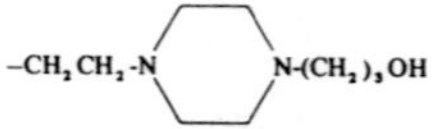 $-CH_2CH_2$-N(piperazine)N-$(CH_2)_3OH$ |  |  |  |  |  | dihydrochloride, m.p. 214–216° | 567 (m.p.); |
| 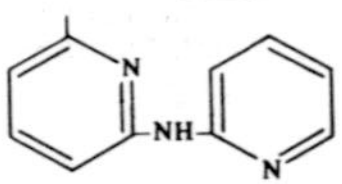 (pyridine ring, N)-NH-(pyridine ring, N) |  |  |  |  | 159–160° | picrate m.p. 224–225° | 271 (m.p.); 271 (m.p.); |
| 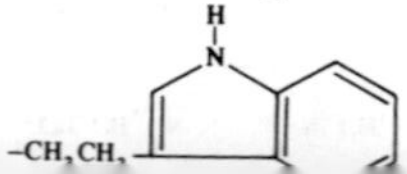 $-CH_2CH_2$-(indole ring, H-N) |  |  |  |  | [illegible] |  | [illegible] |

| | | | | | | |
|---|---|---|---|---|---|---|
| $-CH_2CH_2CH_2-$ (ring with O, $CH_2$) 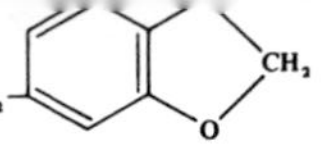 | | | | 102–103° | | 340 (m.p.), (i); |
| $-CH_2CH_2C_6H_5$ | | $-C_2H_5$ | | 56–57° | picrate m.p. 105–106° | 349 (m.p.), (i); 349 (m.p.); |
| $-(CH_2)_4OC_6H_5$ | | | | b.p. 180°/0.8 mm | picrate, m.p. 86–88° | 588 (b.p.); 588 (m.p.); |
| $-CH_2CH_2-$ (ring with $OCH_3$, $OCH_3$) | | | | oily, m.p. 82–83° m.p. 67–68° | chloroplatinate, m.p. 100–103° picrate, m.p. 126–127° | 350 (m.p.), (i), (u); 338 (m.p.), (i), (n); 351 (m.p.); 350 (m.p.); |
| $-CH_2CHCH_2OC_6H_4CH_3$-*o* (OH) | | | | 81–83° | | 325 (m.p.), (i); 1022 (m.p.); |
| 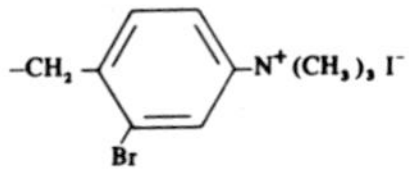 $-CH_2-$ (ring with Br, $N^+(CH_3)_3\ I^-$) | | | | 175–176° | | 561 (m.p.); |
| 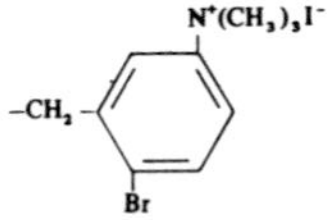 $-CH_2-$ (ring with Br, $N^+(CH_3)_3I^-$) | | | | 174-175° | | 561 (m.p.); |
| | | | 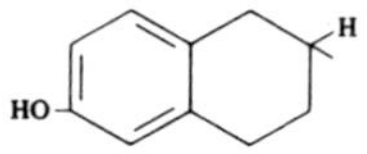 HO (ring) H | 254° | | 568 (m.p.), (u); |
| 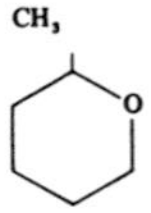 $CH_3$ (ring with O) | $C_6H_5$ | | | 128–130° | | 574 (m.p.), (f); |
| $-(CH_2)_5OC_6H_4NO_2$-*p* | | | | 103° | | 1023 (m.p.); |

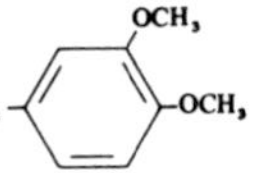

## TABLE XII-41. Alkyl and Aryl 2-Pyridones (Continued)

| $R_1$ | $R_3$ | $R_4$ | $R_5$ | $R_6$ | m.p. | Derivatives | Ref. |
|---|---|---|---|---|---|---|---|
| $-(CH_2)_6$-N(2-pyridon-1-yl) | | | | | 94–95° | | 6 (m.p.); |
| 2-Pyridyl | | | $-COC_6H_5$ | | 168–169° | | 115 (m.p.); |
| $-CH_2CH_2-$(hydroxynaphthyl, OH) | | | | | 202–203° | | 341 (m.p.), (u); |
| | | | | | | 0.5 HBr, m.p. 187–188° | 341 (m.p.); |
| $CH_3$ | | | | $CH_3O$-(tetrahydronaphthyl, H) | 144–146° | | 568 (m.p.), (u); |
| β-D-Glucosyl | | $C_6H_5$ | | | 286–289° | | 604 (m.p.); |
| | | | | | | tetraacetyl, m.p. 135° | 604 (m.p.); |
| $-CH_2CH_2-N$(piperazine)$N-C_6H_4Cl$-*o* | | | | | | hydrochloride, m.p. 236–239° | 566 (m.p.); 1024 (m.p.); |
| $-CH_2CH_2-N$(piperazine)$N-C_6H_4Cl$-*m* | | | | | | hydrochloride, m.p. 240–243° | 566 (m.p.); 1024 (m.p.); |

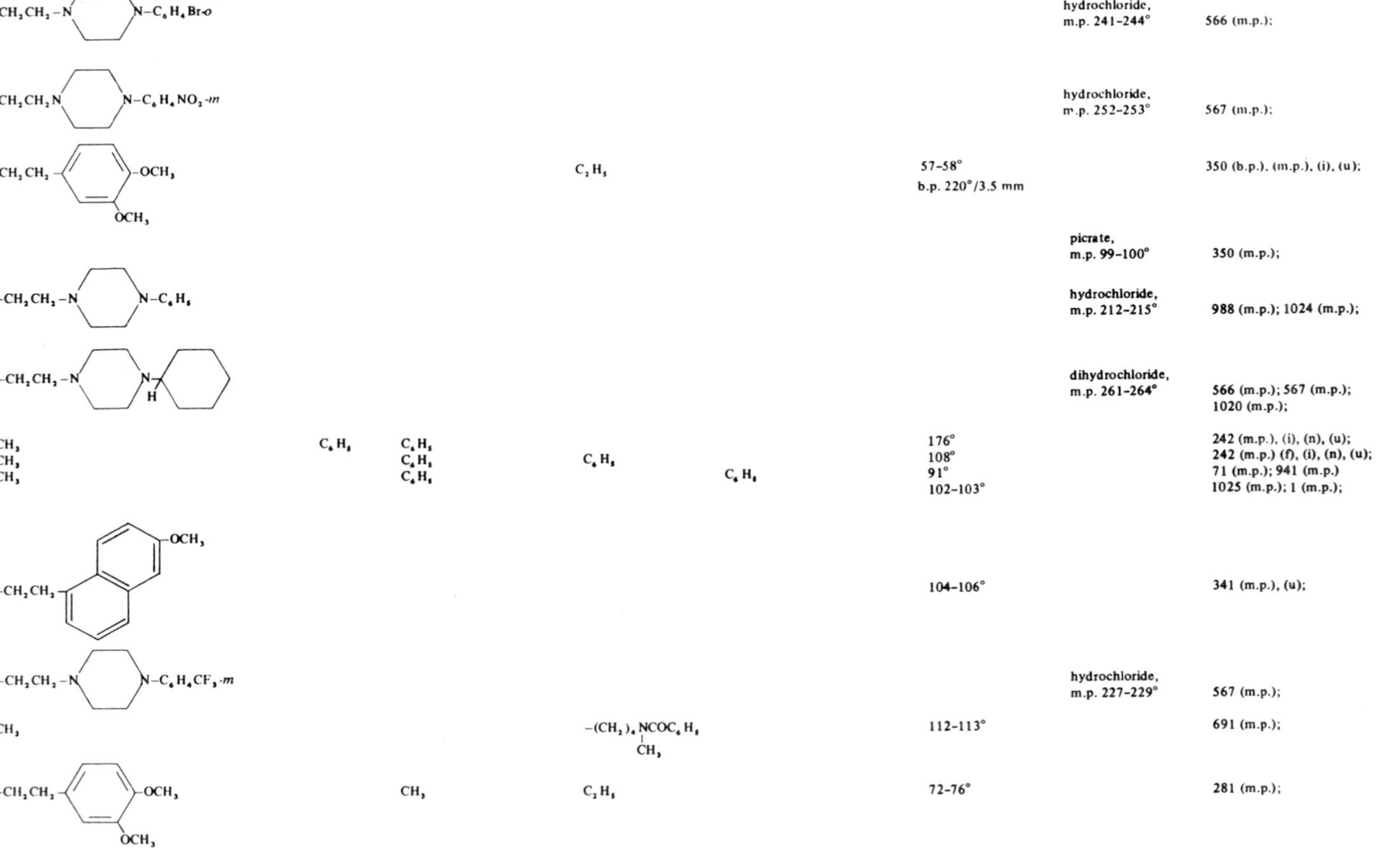

| | | | | | | | |
|---|---|---|---|---|---|---|---|
| $-CH_2CH_2-N$ $N-C_6H_4Br$-*o* | | | | | | hydrochloride, m.p. 241–244° | 566 (m.p.); |
| $-CH_2CH_2N$ $N-C_6H_4NO_2$-*m* | | | | | | hydrochloride, m.p. 252–253° | 567 (m.p.); |
| $-CH_2CH_2-$ $-OCH_3$, $OCH_3$ | | | $C_2H_5$ | | 57–58° b.p. 220°/3.5 mm | | 350 (b.p.), (m.p.), (i), (u); |
| | | | | | | picrate, m.p. 99–100° | 350 (m.p.); |
| $-CH_2CH_2-N$ $N-C_6H_5$ | | | | | | hydrochloride, m.p. 212–215° | 988 (m.p.); 1024 (m.p.); |
| $-CH_2CH_2-N$ $N$ H | | | | | | dihydrochloride, m.p. 261–264° | 566 (m.p.); 567 (m.p.); 1020 (m.p.); |
| $CH_3$ | $C_6H_5$ | $C_6H_5$ | | | 176° | | 242 (m.p.), (i), (n), (u); |
| $CH_3$ | | $C_6H_5$ | $C_6H_5$ | | 108° | | 242 (m.p.) (f), (i), (n), (u); |
| $CH_3$ | | $C_6H_5$ | | $C_6H_5$ | 91° | | 71 (m.p.); 941 (m.p.) |
| | | | | | 102–103° | | 1025 (m.p.); 1 (m.p.); |
| $-CH_2CH_2-$ $-OCH_3$ | | | | | 104–106° | | 341 (m.p.), (u); |
| $-CH_2CH_2-N$ $N-C_6H_4CF_3$-*m* | | | | | | hydrochloride, m.p. 227–229° | 567 (m.p.); |
| $CH_3$ | | | $-(CH_2)_4NCOC_6H_5$ ($CH_3$) | | 112–113° | | 691 (m.p.); |
| $-CH_2CH_2-$ $-OCH_3$, $OCH_3$ | | $CH_3$ | $C_2H_5$ | | 72–76° | | 281 (m.p.); |

TABLE XII-41. Alkyl and Aryl 2-Pyridones (Continued)

| $R_1$ | $R_3$ | $R_4$ | $R_5$ | $R_6$ | m.p. | Derivatives | Ref. |
|---|---|---|---|---|---|---|---|
| $-CH_2CH_2-N(CH_2CH_2)_2N-CH_2C_6H_5$ | | | | | | dihydrochloride, m.p. 252–256° | 988 (m.p.); 1024 (m.p.); |
| $-CH_2CH_2-N(CH_2CH_2)_2N-C_6H_4CH_3$-*o* | | | | | | hydrochloride, m.p. 262–266° | 988 (m.p.); 1024 (m.p.); |
| $-CH_2CH_2N(CH_2CH_2)_2N-C_6H_4CH_3$-*m* | | | | | | dihydrochloride, m.p. 193–201° | 988 (m.p.); 1024 (m.p.); |
| $-CH_2CH_2-N(CH_2CH_2)_2N-C_6H_4CH_3$-*p* | | | | | | hydrochloride, m.p. 227–230° | 988 (m.p.); 1024 (m.p.); |
| $-CH_2CH_2-N(CH_2CH_2)_2N-C_6H_4OCH_3$-*o* | | | | | | hydrochloride, m.p. 220–222° | 567 (m.p.); |
| $-CH_2CH_2-N(CH_2CH_2)_2N-C_6H_4OCH_3$-*p* | | | | | | dihydrochloride, m.p. 210–212° | 567 (m.p.); |
| $CH_3$ | | $-(CH_2)_6-$(1-methyl-2-oxo-1,2-dihydropyridin-4-yl) | | | 141–142° | | 6 (m.p.); |
| $-CH_2CH_2-N(CH_2CH_2)_2N-CH_2CH_2-N$(2-oxo-1(2H)-pyridyl) | | | | | | dihydrochloride, m.p. 272–275° | 566 (m.p.); |

| Structure | Remarks | References |
|---|---|---|
| (nucleotide structure: uridine 3'-phosphate–uridine diester; labels: $CH_2OH$, HO, O, O=P–OH, $OCH_2$, N, O, HO, OH) | | 627 (u); |
| (nucleotide structure; labels: O, N–H, O, N, O, $CH_2OH$, HO, O, $OP$–$OCH_2$, OH, O, HO, OH) | | 627 (u); |
| $-CH_2CH_2-N$(piperazine)$NCH_2C_6H_4CH_3$-*o* | dihydrochloride, m.p. 247–250° | 988 (m.p.); 1024 (m.p.); |
| $-CH_2CH_2-N$(piperazine)$N-CH_2C_6H_4CH_3$-*m* | dihydrochloride, m.p. 246–249° | 988 (m.p.); 1024 (m.p.); |
| $-CH_2CH_2-N$(piperazine)$N-CH_2C_6H_4CH_3$-*p* | dihydrochloride, m.p. 256–259° | 988 (m.p.); 1024 (m.p.); |
| $-CH_2CH_2-N$(piperazine)$N-CH_2CH_2C_6H_5$ | dihydrochloride m.p. 254–258° | 988 (m.p.); 1024 (m.p); |
| $-CH_2CH_2-N$(piperazine)$N-CH(CH_3)C_6H_5$ | dihydrochloride, m.p. 244–247° | 566 (m.p.); |

TABLE XII-41. Alkyl and Aryl 2-Pyridones (Continued)

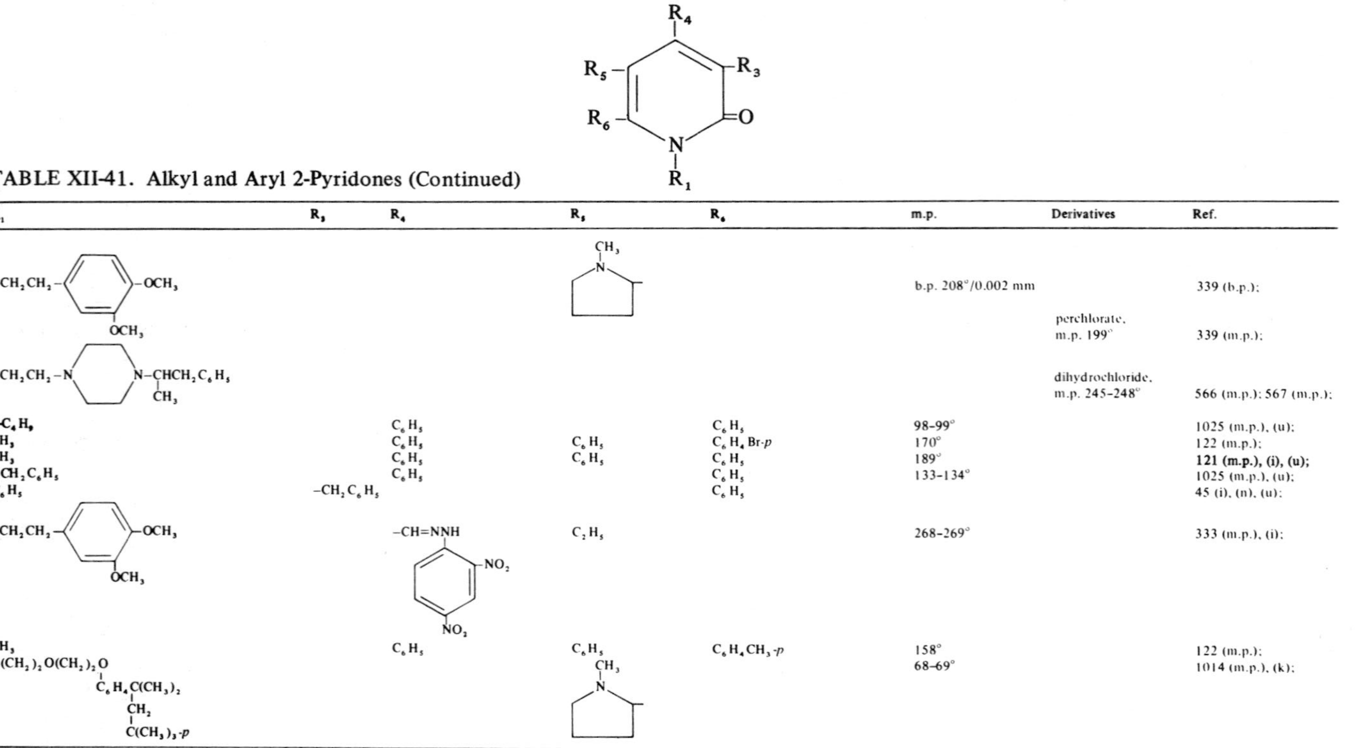

| $R_1$ | $R_3$ | $R_4$ | $R_5$ | $R_6$ | m.p. | Derivatives | Ref. |
|---|---|---|---|---|---|---|---|
| $-CH_2CH_2-C_6H_3(OCH_3)_2$ (3,4-) | | | 1-methyl-2-pyrrolidinyl ($CH_3$–N ring) | | b.p. 208°/0.002 mm | | 339 (b.p.); |
| | | | | | | perchlorate, m.p. 199° | 339 (m.p.); |
| $-CH_2CH_2-N$(piperazine)$N-CHCH_2C_6H_5$ ($CH_3$) | | | | | | dihydrochloride, m.p. 245–248° | 566 (m.p.); 567 (m.p.); |
| ***n*-$C_4H_9$** | | $C_6H_5$ | | $C_6H_5$ | 98–99° | | 1025 (m.p.), (u); |
| $CH_3$ | | $C_6H_5$ | $C_6H_5$ | $C_6H_4Br$-*p* | 170° | | 122 (m.p.); |
| $CH_3$ | | $C_6H_5$ | $C_6H_5$ | $C_6H_5$ | 189° | | **121 (m.p.), (i), (u);** |
| $-CH_2C_6H_5$ | | $C_6H_5$ | | $C_6H_5$ | 133–134° | | 1025 (m.p.), (u); |
| $C_6H_5$ | $-CH_2C_6H_5$ | | | $C_6H_5$ | | | 45 (i), (n), (u); |
| $-CH_2CH_2-C_6H_3(OCH_3)_2$ (2,4-) | | $-CH=NNH-C_6H_3(NO_2)_2$ (2,4-) | $C_2H_5$ | | 268–269° | | 333 (m.p.), (i); |
| $CH_3$ | | $C_6H_5$ | $C_6H_5$ | $C_6H_4CH_3$-*p* | 158° | | 122 (m.p.); |
| $-(CH_2)_2O(CH_2)_2O-C_6H_4C(CH_3)_2CH_2C(CH_3)_3$-*p* | | | 1-methyl-2-pyrrolidinyl ($CH_3$–N ring) | | 68–69° | | 1014 (m.p.), (k); |

R5 R3 R6 =O N R1

TABLE XII-42. Alkyloxy, Aryloxy, and Hydroxy 2-Pyridones

| $R_1$ | $R_3$ | $R_4$ | $R_5$ | $R_6$ | m.p. | Derivatives | Ref. |
|---|---|---|---|---|---|---|---|
| OH | | | | | 148–149° | | 466 (i); 995 (m.p.); |
| | | | | | | thallium salt,<br>m.p. 191–192° | 996 (m.p.); |
| | | | | | | 1-*O*-acetyl,<br>m.p. 93–95° | 995 (m.p.), (u);<br>996 (m.p.); 997 (m.p.), (i); |
| OH | OH | | | | 188-190° | | 3 (m.p.), (n); |
| OH | | | OH | | 220° | | 3 (m.p.), (n); |
| OH | | $CH_3$ | | | 129–135° | | 374 (m.p.); 415 (m.p.); |
| OH | | | | $CH_3$ | 143–145° | | 374 (m.p.); 944 (m.p.); |
| $CH_3$ | OH | | | | 129–132° | | 3 (m.p.); 342 (m.p.), (i),<br>(n), (t), (u);<br>998 (m.p.), (t), (u); |
| | | | | | | 3-*O*-acetyl,<br>m.p. 99–101° | 998 (m.p.), (t), (u); |
| $CH_3$ | | | OH | | 150–153° | | 342 (m.p.), (i), (n), (t), (u); |
| $CH_3$ | | | | OH | 152–154° | | 455 (m.p.), (n), (u); |
| $OCH_3$ | | | | | oil | | 423 (i); 466 (i); |
| OH | | | OH | $CH_3$ | 228–232° | | 999 (m.p.), (n); |
| $OC_2H_5$ | | | | | b.p. 96°/1.1 mm | | 423 (u); 623 (b.p.); |
| OH | $CH_3$ | $CH_3$ | | | 169–170° | | 415 (m.p.); |
| OH | | $CH_3$ | $CH_3$ | | 195° | | 415 (m.p.); |
| $CH_3$ | | OH | | $CH_3$ | 224–230° | | 94 (m.p.); 644 (m.p.); |
| | | | | | | 4-*O*-acetyl,<br>m.p. 144–146° | 644 (m.p.), (i);<br>645 (i), (u); |
| $OCH_3$ | | | | $CH_3$ | 33–34° | | 944 (m.p.); |
| $CH_3$ | | $OCH_3$ | | | 114–116° | | 849 (m.p.); |
| $CH_3$ | | | | $OCH_3$ | 52–54° | | 455 (m.p.), (n), (u); |
| | | | | | | hydrate,<br>m.p. 78–80° | 505 (m.p.), (n), (t); |
| OH | $CH_3$ | $CH_3$ | | OH | 157–158° | | 1000 (m.p.), (i); |
| OH | | | OH | $C_2H_5$ | 167–171° | | 1001 (m.p.), (i), (n), (u); |
| $-OCH_2CH=CH_2$ | | | | | b.p. 91°/0.1 mm | | 423 (g), (i), (u);<br>1002 (b.p.); 1003 (b.p.); |
| $CH_3$ | $CH_3$ | OH | | $CH_3$ | 265° | | 158 (m.p.), (u);<br>159 (m.p. (u); |
| $CH_3$ | $OCH_3$ | OH | | $CH_2OH$ | | diacetyl,<br>m.p. 93–94° | 160 (m.p.), (i), (u); |
| $-CH_2CH=CH_2$ | | OH | | $CH_3$ | 160–163° | | 1004 (m.p.); |

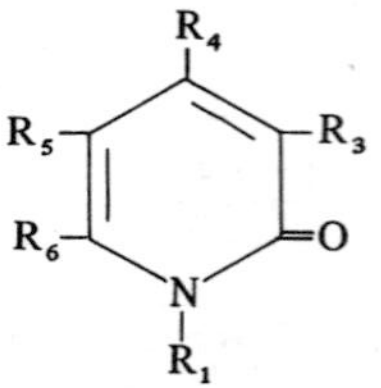

TABLE XII-42. Alkyloxy, Aryloxy, and Hydroxy 2-Pyridones (Continued)

| $R_1$ | $R_3$ | $R_4$ | $R_5$ | $R_6$ | m.p. | Derivatives | Ref. |
|---|---|---|---|---|---|---|---|
| $CH_3$ | $-COCH_3$ | OH | | $CH_3$ | 133–135° | | **644** (m.p.), (i); **645** (i), (u); **1005** (m.p.); |
| | | | | | | phenylhydrazone, m.p. 234–235° | **1005** (m.p.); |
| $CH_3$ | | OH | $-COCH_3$ | $CH_3$ | 280° | | **94** (m.p.), (u); |
| $-O(CH_2)_5CH_3$ | | | | | b.p. 97°/0.1 mm | | **1003** (b.p.), (i), (u); |
| 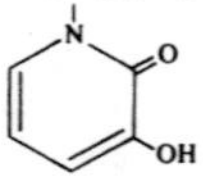 3-hydroxy-2-oxo-1-pyridyl (structure: N, O, OH) | OH | | | | >300° | | **3** (m.p.); |
| 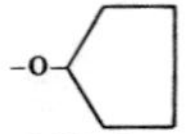 $-O-$cyclopentyl (structure) | | | | | 66–67° | | **1003** (m.p.); **1006** (m.p.), (c); |
| β-D-Ribofuranosyl | | OH | | | 228–229° | | **617** (m.p.), (k), (u); |
| $CH_3$ | $-OP(OC_2H_5)_2$ | | | | 62–63° | | **998** (m.p.), (t), (u); |
| $-(CH_2)_3N(CH_3)_2$ | OH | | | | | phosphate, m.p. 213–214° | **3** (m.p.); |
| $C_6H_4Cl$-*p* | | OH | | OH | 177–179° | | **49** (m.p.); |
| $C_6H_4NO_2$-*p* | | OH | | OH | 185° | | **49** (m.p.); |
| $C_6H_5$ | | OH | | OH | 198–200° | | **49** (m.p.); |
| β-D-Glucosyloxy | | | | | 195–197° | | **623** (m.p.), (f); **1026** (m.p.), (k); |
| | | | | | | tetraacetyl, m.p. 142–143° | **623** (m.p.), (k); |
| $CH_3$ | | $-OP(=O)(OC_2H_5)_2$ | | $CH_3$ | oil | | **1004**; |
| 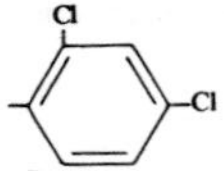 2,4,6-trichlorophenyl (structure: Cl, Cl, Cl) | | OH | | $CH_3$ | | 4-*O*-acetyl, m.p. 143–144° | **149** (m.p.); |
| 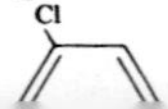 (structure: Cl) | | OH | | $CH_3$ | | 4-*O*-acetyl, | |

| | | | | | | | |
|---|---|---|---|---|---|---|---|
| | | | | | | | 1027 (m.p.); 1028 (i); |
| $C_6H_4Cl$-*p* | OH | | | $CH_3$ | 147° | | 24 (m.p.); |
| $C_6H_4Cl$-*o* | | OH | | $CH_3$ | | 4-*O*-acetyl, m.p. 140–141° | 149 (m.p.); |
| $C_6H_4Br$-*o* | | OH | | $CH_3$ | | 4-*O*-acetyl, m.p. 133–134° | 149 (m.p.); |
| $C_6H_5$ | | OH | | $CH_3$ | | 4-*O*-acetyl, m.p. 148–149° | 645 (m.p.), (i), (u); |
| $-OCH_2C_6H_5$ | | | | | 77–79° | | 423 (m.p.); 466 (i); |
| | | | | | | | 1003 (m.p.); |
| $-CH_2N$ 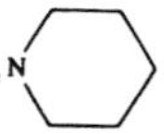 (piperidino) | $OCH_3$ | | | | | 0.5 $H_2O$ m.p. 69–71° | 501 (m.p.); |
| $CH_3$ | $-OCOC_6H_5$ | | | | 136–138° | | 998 (m.p.), (t), (u); |
| 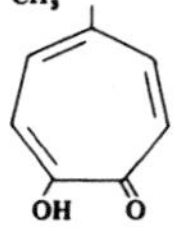 (ring) OH O | | OH | | $CH_3$ | 284–285° | | 109 (m.p.), (u); |
| 4-Pyridyl | $-COCH_3$ | OH | | $CH_3$ | | | 1030 (i), (n); |
| $-OCH_2C_6H_5$ | | $CH_3$ | | | 78–80° | | 425 (m.p.), (i); |
| $-CH_2CH_2C_6H_5$ | | OH | | | 202–204° | | 1008 (m.p.); |
| $-OCH_2CH_2C_6H_5$ | | | | | b.p. 165°/0.25 mm | | 1003 (b.p.); |
| $C_6H_5$ | $C_2H_5$ | OH | | | 306–309° | | 44 (m.p.), (u); |
| $-OCH_2C_6H_4OCH_3$-*p* | | | | | 76–77° | | 1003 (m.p.); |
| OH | | $C_6H_4OCH_3$-*p* | | $-OCH_3$ | 194–196° | | 5 (m.p.), (i), (u); |
| β-D-Glucosyl | | $-OC_2H_5$ | | | amorphous | | 604; |
| | | | | | | tetraacetyl m.p. 78–80° | 604 (m.p.); |
| $C_6H_4Br$-*p* | $-COCH_3$ | OH | | $CH_3$ | 262° | | 1030 (m.p.), (i), (n); |
| $C_6H_5$ | $-COCH_3$ | OH | | $CH_3$ | 217–224° | | 110 (n); 645 (m.p.), (i), (u); |
| | | | | | | | 1005 (m.p.); 1030 (m.p.), (i), (n); |
| | | | | | | phenyl hydrazone, m.p. 239–240° | 1005 (m.p.); |
| $CH_3$ | | OH | $-COCH_3$ | $C_6H_5$ | 270° | | 87 (m.p.); |
| $C_6H_5$ | | OH | $-COCH_3$ | $CH_3$ | 253° | | 87 (m.p.); |
| $-CH_2CH_2C_6H_5$ | $-OCH_3$ | | | | b.p. 140°/0.05 mm | | 1029 (b.p.); |
| $-OCH_3$ | | $C_6H_4OCH_3$-*p* | | $OCH_3$ | 129–130° | | 5 (m.p.), (i), (u); |
| $C_6H_4Cl$-*p* | $-COCH_3$ | | $-COCH_3$ | OH | 238–240° | | 112 (m.p.); |
| $C_6H_5$ | $-COCH_3$ | OH | $-COCH_3$ | OH | 123–125° | | 49 (m.p.); |
| 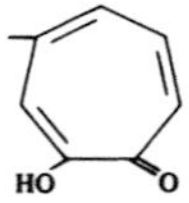 (ring) HO O | $-COCH_3$ | OH | | $CH_3$ | 246° | | 110 (m.p.), (i), (n), (u); 1030 (i), (n); |

## TABLE XII-42. Alkyloxy, Aryloxy, and Hydroxy 2-Pyridones (Continued)

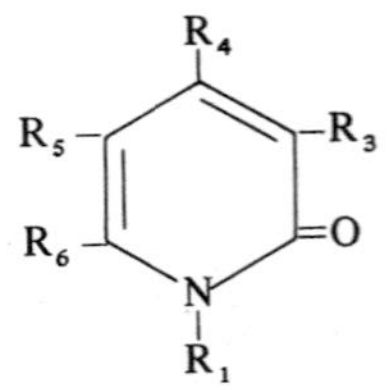

| $R_1$ | $R_3$ | $R_4$ | $R_5$ | $R_6$ | m.p. | Derivatives | Ref. |
|---|---|---|---|---|---|---|---|
| 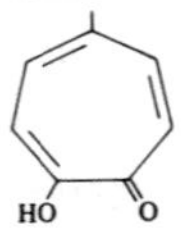 HO … O | $-COCH_3$ | OH | | $CH_3$ | 250°<br>228° | | 109 (m.p.), (i), (n), (u);<br>110 (m.p.), (i), (n), (u);<br>1030 (i), (n); |
| $C_6H_5$ | $CH_3$ | OH | $-CH_2CH_2CH_2-$ | | 263° | | 93 (m.p.); |
| $C_6H_5$ | | OH | $-CH_2CH_2CH_2CH_2-$ | | 325° | | 93 (m.p.); 214 (m.p.); |
| | | | | | | 4-*O*-acetyl m.p. 135° | 214 (m.p.); |
| | | | | | | 4-*O*-benzoyl, m.p. 169° | 214 (m.p.); |
| $CH_3$ | OH | $-CH_2C_6H_5$ | $-COCH_3$ | | 185° | | 231 (m.p.); |
| $-CH_2C_6H_5$ | | OH | $-COCH_3$ | $CH_3$ | 243° | | 87 (m.p.); |
| $C_6H_4CH_3$-*p* | $-COCH_3$ | OH | | $CH_3$ | 209–211° | | 1030 (m.p.), (i), (n); |
| $C_6H_4OCH_3$-*p* | $-COCH_3$ | OH | | $CH_3$ | 192–193° | | 1030 (m.p.), (i), (n); |
| $C_2H_5$ | $-CH_2C_6H_5$ | OH | | $CH_3$ | 262° | | 163 (m.p.); |
| OH | $-CH_2C_6H_5$ | OH | $CH_3$ | $C_2H_5$ | 186° | | 96 (m.p.); |
| $CH_3$ | $-COCH{=}CHC_6H_4Cl$-*p* | OH | | $CH_3$ | 193–194° | | 1031 (m.p.), (u); |
| $CH_3$ | $-COCH{=}CHC_6H_4NO_2$-*m* | OH | | $CH_3$ | 235–236° | | 1031 (m.p.), (u); |
| $CH_3$ | $-COCH{=}CHC_6H_4NO_2$-*p* | OH | | $CH_3$ | 248–250° | | 1031 (m.p.), (u); |
| 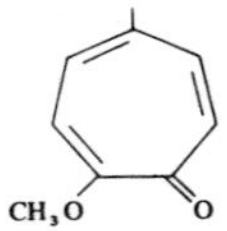 $CH_3O$ … O | $-COCH_3$ | OH | | $CH_3$ | 225° | | 109 (m.p.), (u); |

| | | | | | | |
|---|---|---|---|---|---|---|
| $C_6H_5$ | $CH_3$ | OH | $-CH_2CH_2CH_2CH_2-$ | | 274° | 93 (m.p.); 214 (m.p.); |
| $C_6H_5$ | | $OCH_3$ | $-CH_2CH_2CH_2CH_2-$ | | 198° | 214 (m.p.); |
| $C_6H_5$ | $C_2H_5$ | OH | $-CH_2CH_2CH_2-$ | | 273° | 93 (m.p.); |
| $-OCH_2C_6H_5$ | $CH_3$ | OH | $CH_3$ | $C_2H_5$ | 191° | 96 (m.p.); |
| OH | | $-(CH_2)_6-$ 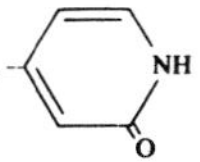 [ring: NH, O] | | | 205–207° | 6 (m.p.); |
| $C_6H_5$ | | OH | | $C_6H_5$ | 308° | 92 (m.p.); |
| OH | | $C_6H_5$ | | $C_6H_5$ | 162° | 941 (m.p.), (i), (u); |
| $C_6H_5$ | | OH | $-CH_2CH_2CH_2CH_2CH_2CH_2-$ | | 287° | 214 (m.p.); |
| $C_6H_5$ | $CH_3$ | OH | $-CH_2CH_2CH_2CH_2CH_2-$ | | 277° | 93 (m.p.); |
| $C_6H_5$ | $C_2H_5$ | OH | $-CH_2CH_2CH_2CH_2-$ | | 282° | 89 (m.p.); 93 (m.p.); 214 (m.p.); |
| $C_6H_5$ | iso$C_3H_7$ | OH | $-CH_2CH_2CH_2-$ | | 272° | 93 (m.p.); |
| $C_6H_4CH_3$-*o* | $CH_3$ | OH | $-CH_2CH_2CH_2CH_2-$ | | 236° | 93 (m.p.); |
| $C_6H_5$ | | OH | $-CO(CH_2)_3CH_3$ | $CH_3$ | 210° | 1032 (m.p.); |
| OH | | $C_6H_5$ | $-OCH_3$ | $C_6H_5$ | 211° | 120 (m.p.); 121 (m.p.), (i); |
| $CH_3$ | $-COCH=CHC_6H_4N(CH_3)_2$-*p* | OH | | $CH_3$ | 236–237° | 1031 (m.p.), (u); |
| $C_6H_5$ | $CH_3$ | OH | $-CH_2CH_2CH_2CH_2CH_2CH_2-$ | | 281° | 93 (m.p.); |
| $C_6H_5$ | $C_2H_5$ | OH | $-CH_2CH_2CH_2CH_2CH_2-$ | | 274° | 93 (m.p.); |
| $C_6H_5$ | iso$C_3H_7$ | OH | $-CH_2CH_2CH_2CH_2-$ | | 280° | 93 (m.p.); |
| $C_6H_4CH_3$-*o* | $C_2H_5$ | OH | $-CH_2CH_2CH_2CH_2-$ | | 240° | 93 (m.p.); |
| $C_6H_5$ | | OH | $-CO(CH_2)_4CH_3$ | $CH_3$ | 193° | 1032 (m.p.); |
| $-(CH_2)_6-N$ [ring: O, OH, $-CH_3$] 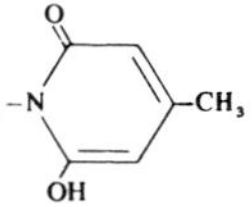 | | $CH_3$ | | OH | 180–182° | 6 (m.p.); |
| $C_6H_5$ | | OH | $-COCH_3$ | $C_6H_4Cl$-*p* | 299° | 1032 (m.p.); |
| $C_6H_5$ | | OH | $-COCH_3$ | $C_6H_5$ | 299–300° | 87 (m.p.); 90 (m.p.); |
| $-CH_2C_6H_5$ | $-OCH_2C_6H_5$ | | | | 115–116° | 596 (m.p.), (i); |
| $-CH_2CH_2C_6H_5$ | 3 (or 5)-Phenoxy | | | | 121–123° | 1011 (m.p.), (z); |
| OH | | $C_6H_5$ | $OCH_3$ | $C_6H_4CH_3$-*p* | 195–196° | 121 (m.p.); |
| $C_6H_5$ | $C_2H_5$ | OH | $-CH_2CH_2CH_2CH_2CH_2CH_2-$ | | 264° | 93 (m.p.); |
| $C_6H_5$ | *n*-$C_4H_9$ | OH | $-CH_2CH_2CH_2CH_2-$ | | 245° | 93 (m.p.); |
| $C_6H_4CH_3$-*o* | iso$C_3H_7$ | OH | $-CH_2CH_2CH_2CH_2-$ | | 230° | 93 (m.p.); |
| $C_6H_5$ | | OH | $-CO(CH_2)_5CH_3$ | $CH_3$ | 176° | 1032 (m.p.); |
| $CH_3$ | OH | $-CH_2C_6H_5$ | $-COC_6H_5$ | | 167–168° | 231 (m.p.); |
| $C_6H_5$ | | OH | $COCH_3$ | $C_6H_4CH_3$-*p* | 310° | 1032 (m.p.); |
| $C_6H_5$ | iso$C_3H_7$ | OH | | $C_6H_5$ | 325° | 92 (m.p.); |
| $OCH_2C_6H_5$ | | $C_6H_4OCH_3$-*p* | | $OCH_3$ | 142–143° | 5 (m.p.), (i), (u); |
| $C_6H_4CH_3$-*o* | *n*-$C_4H_9$ | OH | $-CH_2CH_2CH_2CH_2-$ | | 213° | 93 (m.p.); |
| $C_6H_5$ | $-COCH=CHC_6H_4Cl$-*p* | OH | | $CH_3$ | 219–220° | 1031 (m.p.), (u); |
| $C_6H_5$ | $-COCH=CHC_6H_5$ | OH | | $CH_3$ | 211–213° | 1031 (m.p.), (u); |
| $C_6H_5$ | $-COCH=CHC_6H_4OH$-*p* | OH | | $CH_3$ | 271–272° | 1031 (m.p.), (u); |
| $C_6H_5$ | $N_2C_6H_4NO_2$-*p* | OH | $-CH_2CH_2CH_2CH_2-$ | | 206° | 214 (m.p.); |

TABLE XII-42. Alkyloxy, Aryloxy, and Hydroxy 2-Pyridones (Continued)

| $R_1$ | $R_3$ | $R_4$ | $R_5$ | $R_6$ | m.p. | Derivatives | Ref. |
|---|---|---|---|---|---|---|---|
| $C_6H_5$ | $C_6H_5$ | OH | $-CH_2CH_2CH_2CH_2-$ | | 243° | | 93 (m.p.); |
| $C_6H_5$ | $N_2C_6H_5$ | OH | $-CH_2CH_2CH_2CH_2-$ | | 200° | | 214 (m.p.); |
| $C_6H_5$ | $-CH_2C_6H_5$ | OH | $-COCH_3$ | $CH_3$ | 246–247° | | 90 (m.p.); |
| $C_6H_4CH_3$-*p* | iso$C_3H_7$ | OH | | $C_6H_5$ | 360° | | 92 (m.p.); |
| $C_6H_5$ | $-CH_2C_6H_5$ | OH | $-CH_2CH_2CH_2CH_2-$ | | 260° | | 93 (m.p.); 214 (m.p.); |
| $C_6H_5$ | $C_6H_5$ | OH | $-CH_2CH_2CH_2CH_2CH_2-$ | | 264° | | 93 (m.p.); |
| $C_6H_5$ | $-CH_2C_6H_5$ | OH | | *s*-$C_4H_9$ | 279–281° | | 92 (m.p.); |
| $C_6H_4CH_3$-*p* | *n*-$C_4H_9$ | OH | | $C_6H_5$ | 295–297° | | 92 (m.p.); |
| $-OCH_2C_6H_5$ | $-CH_2C_6H_5$ | OH | $CH_3$ | $C_2H_5$ | 168° | | 96 (m.p.); |
| OH | | $C_6H_5$ | $C_6H_5$ | $C_6H_4Br$-*p* | 216° | | 122 (m.p.), (i); |
| OH | | $C_6H_5$ | $C_6H_5$ | $C_6H_5$ | 232–233° | | 121 (m.p.), (i); |
| OH | | $C_6H_5$ | $OC_6H_5$ | $C_6H_5$ | 224–225° | | 121 (m.p.), (i); |
| $CH_3$ | $-COCH{=}CHC_6H_5$ | OH | | $-CH{=}CHC_6H_5$ | 199–201° | | 1031 (m.p.), (u); |
| $C_6H_5$ | $-COCH{=}CHC_6H_4N(CH_3)_2$-*p* | OH | | $CH_3$ | 272–273° | | 1031 (m.p.), (u); |
| $C_6H_5$ | $C_6H_5$ | OH | $-CH_2CH_2CH_2CH_2CH_2CH_2-$ | | 240° | | 93 (m.p.); |
| $C_6H_5$ | $-CH_2C_6H_5$ | OH | $-CH_2CH_2CH_2CH_2CH_2-$ | | 269° | | 93 (m.p.); |
| $C_6H_4CH_3$-*o* | $-CH_2C_6H_5$ | OH | $-CH_2CH_2CH_2CH_2-$ | | 233° | | 93 (m.p.); |
| $CH_3$ | | $C_6H_5$ | $-OC_6H_4Cl$-*p* | $C_6H_5$ | 200° | | 121 (m.p.); |
| $CH_3$ | | $C_6H_5$ | $-OC_6H_4Br$-*p* | $C_6H_5$ | 209° | | 121 (m.p.); |
| $C_6H_5$ | $-CH_2C_6H_5$ | OH | | $C_6H_5$ | 275–287° | | 92 (m.p.); 214 (m.p.); |
| | | | | | | 4-*O*-acetyl, m.p. 153–154° | 92 (m.p.); |
| $-OCH_3$ | | $C_6H_5$ | $C_6H_5$ | $C_6H_5$ | 163° | | 122 (m.p.), (i); |
| $CH_3$ | | $C_6H_5$ | $-OC_6H_5$ | $C_6H_5$ | 167° | | 121 (m.p.), (i), (u); |
| $-OCH_3$ | | $C_6H_5$ | $-OC_6H_5$ | $C_6H_5$ | 162° | | 122 (m.p.), (i); |
| $C_6H_5$ | $-CH_2C_6H_5$ | OH | $-CH_2CH_2CH_2CH_2CH_2CH_2-$ | | 282° | | 93 (m.p.); 214 (m.p.); |
| $C_6H_5$ | $-CH_2C_6H_5$ | OH | $CH_3$ | $C_6H_5$ | 228–230° | | 92 (m.p.); |
| | | | | | | acetate, m.p. 157–159° | 92 (m.p.); |
| $CH_3$ | | $C_6H_5$ | $-OC_6H_4CH_3$-*p* | $C_6H_5$ | 180° | | 121 (m.p.); |
| $C_6H_4CH_3$-*p* | $-CH_2C_6H_5$ | OH | | $C_6H_5$ | 299–300° | | 92 (m.p.); |
| | | | | | | acetate, m.p. 147–149° | 92 (m.p.); |
| $C_6H_5$ | OH | $-CH_2C_6H_5$ | $-COC_6H_5$ | $CH_3$ | 232–235° | | 231 (m.p.); |
| $C_6H_5$ | $-CH_2C_6H_5$ | OH | $C_2H_5$ | $C_6H_5$ | 283–285° | | 92 (m.p.); |
| $C_6H_5$ | | OH | $-CH_2C_6H_5$ | $-CH_2CH_2C_6H_5$ | 220° | | 1033 (m.p.), (i); |

TABLE XII-43. Amino 2-Pyridones

(Structure: 2-pyridone ring with substituents $R_1$ on N, $R_3$, $R_5$, $R_6$)

| $R_1$ | $R_3$ | $R_4$ | $R_5$ | $R_6$ | m.p. | Derivatives | Ref. |
|---|---|---|---|---|---|---|---|
| $NH_2$ | | | | | | $CoCl_2 \cdot 5H_2O$ complex | 805 (i), (q); |
| | | | | | | $Co(ClO_4)_2 \cdot H_2O$ complex | 805 (i), (q); |
| | | | | | | $FeCl_2$ complex | 805 (i), (q); |
| | | | | | | $Fe(ClO_4)_2 \cdot H_2O$ | 805 (i), (q); |
| OH | | $NH_2$ | | | | 1,4-diacetyl, m.p. 215–217° | 1007 (m.p.); |
| $NH_2$ | OH | | | | 161–163° | | 3 (m.p.); |
| $-NHCONH_2$ | OH | | | | 233–235° | | 3 (m.p.); |
| $CH_3$ | $NH_2$ | | | | | picrate, m.p. 204° | 556 (m.p.); |
| | | | | | | $N'$-acetyl, m.p. 165–166° | 556 (m.p.); |
| $CH_3$ | | | | $NH_2$ | 163–165° | | 494 (m.p.), (i), (u); |
| $CH_3$ | | $NH_2$ | | $CH_3$ | 257–258° | | 306 (m.p.); |
| $-N(CH_3)_2$ | OH | | | | 81–83° | | 3 (m.p.); |
| | | | | | | acetate, m.p. 48–49° | 3 (m.p.); |
| | | | | | | benzoate, m.p. 137–138 ° | 3 (m.p.); |
| $CH_3$ | $NH_2$ | $NH_2$ | | $CH_3$ | 233° | | 86 (m.p.); |

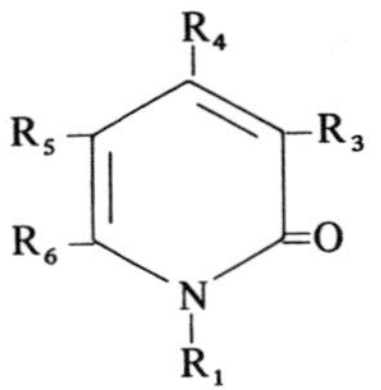

TABLE XII-43. Amino 2-Pyridones (Continued)

| $R_1$ | $R_3$ | $R_4$ | $R_5$ | $R_6$ | m.p. | Derivatives | Ref. |
|---|---|---|---|---|---|---|---|
| 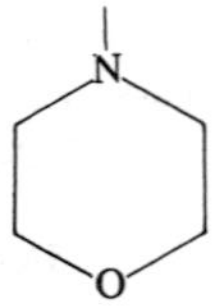 4-morpholinyl | OH | | | | 177–179° | | 3 (m.p.); |
| –N=CH–(2-pyridyl) | | | | | | $Co(ClO_4)_2$ complex | 805 (q); |
| | | | | | | $Cu(ClO_4)_2$ complex | 805 (i), (q); |
| | | | | | | $Fe(ClO_4)_2$ complex | 805 (i), (q); |
| | | | | | | $Ni(ClO_4)_2$ complex $H_2O$ | 805 (q); |
| | | | | | | $Zn(ClO_4)_2$ complex | 805 (i), (q); |
| $CH_3$ | $NH_2$ | $-NHC_4H_9$ | | $CH_3$ | | hydrochloride m.p. 256–257° | 86 (m.p.); |
| $-N{=}CCH_3$ (2-pyridyl) | | | | | | $Co(ClO_4)_2$ complex | 805 (i), (q); |
| | | | | | | $Fe(ClO_4)_2$ complex | 805 (i), (q); |
| | | | | | | $Ni(ClO_4)_2$ complex | 805 (q); |

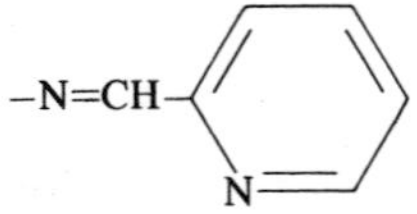

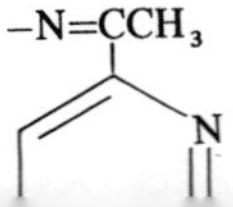

| | | | | | | | |
|---|---|---|---|---|---|---|---|
| $-CH_2CH_2C_6H_5$ | | $NH_2$ | | | | m.p. 92–93° | 332 (m.p.); |
| $-CH_2CH_2C_6H_5$ | | | $NH_2$ | | | hydrochloride, m.p. 166–169° HCl·$H_2O$ m.p. 203–205° | 1008 (m.p.); 1008 (m.p.); |
| $CH_3$ | | $-NHNHC_6H_5$ | | $CH_3$ | 277° | | 306 (m.p.); |
| $NH_2$ | | $C_6H_5$ | | $C_6H_5$ | 166° | diacetate, m.p. 120° | 941 (m.p.), (i), (u); 941 (m.p.); |
| $CH_3$ | $CH_3$ | $-N(C_2H_5)_2$ | $C_6H_5$ | | 96° | | 102 (m.p.), (i), (u); |
| $NH_2$ | | $C_6H_5$ | $-OCH_3$ | $C_6H_5$ | 186° | *N′*-diacetyl, m.p. 128° | 120 (m.p.); 121 (i); 121 (m.p.); |
| $NH_2$ | | $C_6H_4OCH_3$-*p* | | $C_6H_5$ | 163° | | 941 (m.p.) (i); |
| $NH_2$ | | $C_6H_5$ | $-OCH_3$ | $C_6H_4CH_3$-*p* | 174° | | 120 (m.p.); |
| $NH_2$ | | $C_6H_5$ | $-OCH_3$ | $C_6H_4OCH_3$-*p* | 148° | | 120 (m.p.); |
| $C_6H_4OCH_3$-*m* | | OH | | $NHC_6H_4OCH_3$-*m* | 258° | | 161 (m.p.); |
| $-CH_2CH_2C_6H_5$ | | | $-NHCOCH_2C_6H_5$ (C=O) | | 134° | | 1008 (m.p.); |
| $NH_2$ | | $C_6H_5$ | $C_6H_5$ | $C_6H_4Br$-*p* | 197° | *N′*-acetyl, m.p. 252° *N′*-diacetyl, m.p. 195° | 122 (m.p.), (i); 122 (m.p.), (i); 122 (m.p.), (i); |
| $NH_2$ | | $C_6H_5$ | $-OC_6H_4Cl$-*p* | $C_6H_5$ | 203° | *N′*-diacetyl, m.p. 159° | 121 (m.p.); 121 (m.p.); |
| $NH_2$ | | $C_6H_5$ | $-OC_6H_4Br$-*p* | $C_6H_5$ | 222° | *N′*-diacetyl, m.p. 160° | 121 (m.p.); 121 (m.p.); |

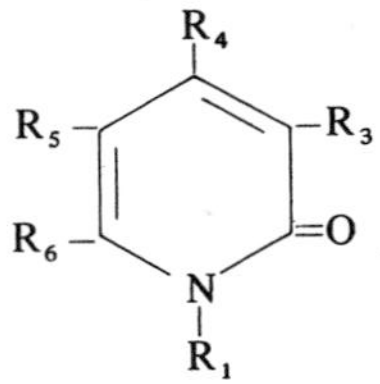

TABLE XII-43. Amino 2-Pyridones (Continued)

| $R_1$ | $R_3$ | $R_4$ | $R_5$ | $R_6$ | m.p. | Derivatives | Ref. |
|---|---|---|---|---|---|---|---|
| $NH_2$ | | $C_6H_5$ | $C_6H_5$ | $C_6H_5$ | 199° | | 121 (m.p.), (i), (u); |
| | | | | | | $N'$-acetyl, m.p. 246° | 121 (m.p.); |
| | | | | | | $N'$-diacetyl, m.p. 167 | 121 (m.p.); |
| | | | | | | $CuCl_2$ complex, m.p. 246° | 802 (m.p.), (i); |
| $NH_2$ | | $C_6H_5$ | $OC_6H_5$ | $C_6H_5$ | 181° | | 121 (m.p.), (i), (u); |
| | | | | | | $N'$-acetyl, m.p. 234° | 121 (m.p.); |
| | | | | | | $N'$-diacetyl, m.p. 158° | 121 (m.p.); |
| $-N{=}CHC_6H_5$ | | $C_6H_5$ | | $C_6H_5$ | 179° | | 941 (m.p.); |
| $NH_2$ | | $C_6H_5$ | $C_6H_5$ | $C_6H_4CH_3$-$p$ | 189° | | 122 (m.p.), (i); |
| $-NHCH_3$ | | $C_6H_5$ | $OC_6H_5$ | $C_6H_5$ | | $N'$-acetyl, m.p. 173° | 121 (m.p.); |
| $NH_2$ | | $C_6H_5$ | $C_6H_5$ | $C_6H_4OCH_3$-$p$ | 157–158° | | 802 (m.p.); |
| | | | | | | $CuCl_2$ complex, m.p. 231–232° | 802 (m.p.), (i); |
| $NH_2$ | | $C_6H_5$ | $-OC_6H_4CH_3$-$p$ | $C_6H_5$ | 202° | | 121 (m.p.); |
| | | | | | | $N'$-acetyl, | |

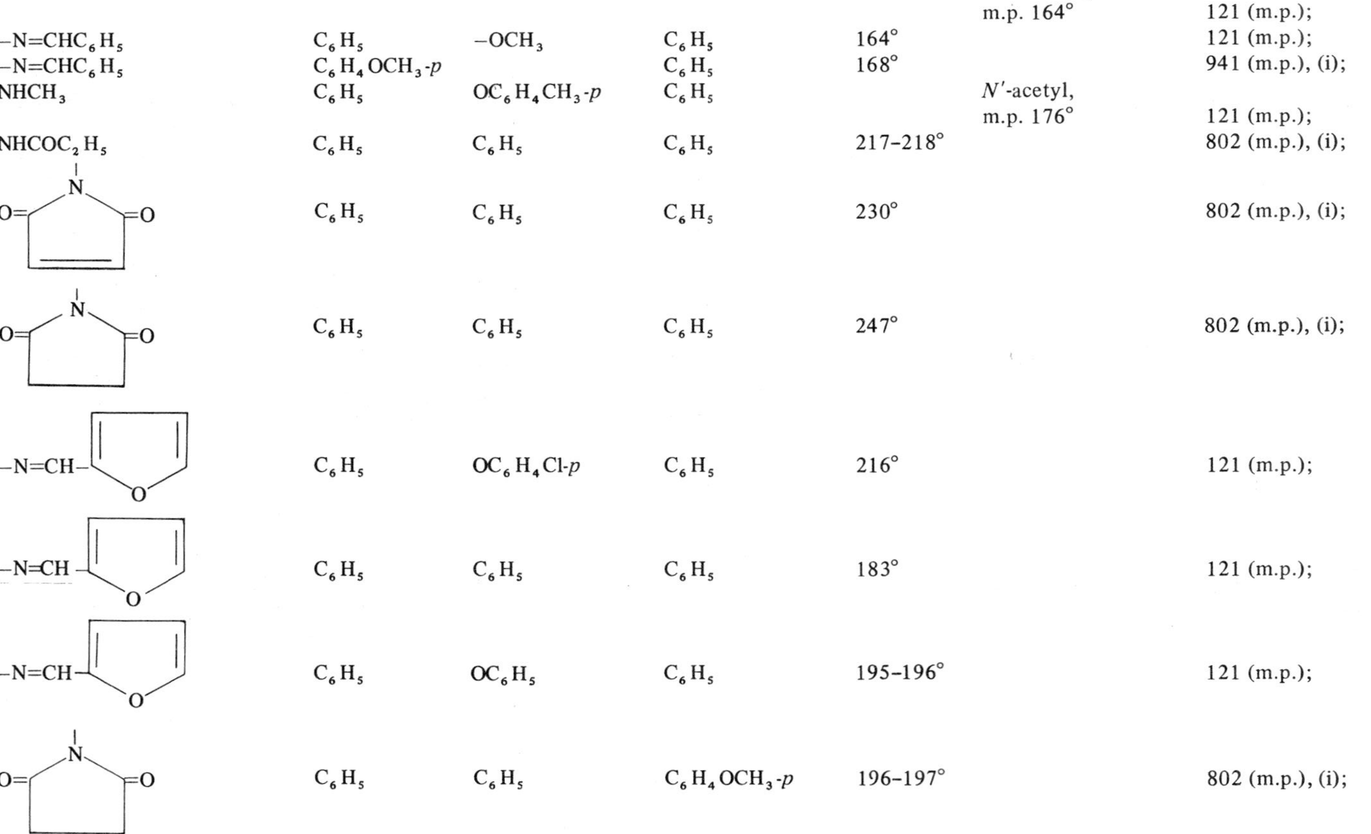

| | | | | | | |
|---|---|---|---|---|---|---|
| | | | | | m.p. 164° | 121 (m.p.); |
| $-N=CHC_6H_5$ | $C_6H_5$ | $-OCH_3$ | $C_6H_5$ | 164° | | 121 (m.p.); |
| $-N=CHC_6H_5$ | $C_6H_4OCH_3$-*p* | | $C_6H_5$ | 168° | | 941 (m.p.), (i); |
| $NHCH_3$ | $C_6H_5$ | $OC_6H_4CH_3$-*p* | $C_6H_5$ | | *N'*-acetyl, m.p. 176° | 121 (m.p.); |
| $NHCOC_2H_5$ | $C_6H_5$ | $C_6H_5$ | $C_6H_5$ | 217–218° | | 802 (m.p.), (i); |
| | $C_6H_5$ | $C_6H_5$ | $C_6H_5$ | 230° | | 802 (m.p.), (i); |
| | $C_6H_5$ | $C_6H_5$ | $C_6H_5$ | 247° | | 802 (m.p.), (i); |
| $-N=CH-$ | $C_6H_5$ | $OC_6H_4Cl$-*p* | $C_6H_5$ | 216° | | 121 (m.p.); |
| $-N=CH-$ | $C_6H_5$ | $C_6H_5$ | $C_6H_5$ | 183° | | 121 (m.p.); |
| $-N=CH-$ | $C_6H_5$ | $OC_6H_5$ | $C_6H_5$ | 195–196° | | 121 (m.p.); |
| | $C_6H_5$ | $C_6H_5$ | $C_6H_4OCH_3$-*p* | 196–197° | | 802 (m.p.), (i); |

TABLE XII-43. Amino 2-Pyridones (Continued)

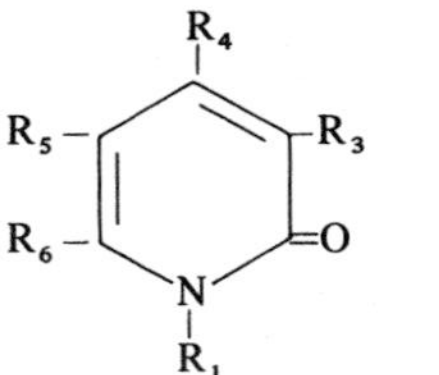


| $R_1$ | $R_3$ | $R_4$ | $R_5$ | $R_6$ | m.p. | Derivatives | Ref. |
|---|---|---|---|---|---|---|---|
| –N=CH– O | | $C_6H_5$ | $OC_6H_4CH_3$-*p* | $C_6H_5$ | 220° | | 121 (m.p.); |
| $N(COC_2H_5)_2$ | | $C_6H_5$ | $C_6H_5$ | $C_6H_5$ | 185° | | 802 (m.p.); |
| –N=$CHC_6H_5$ | | $C_6H_5$ | $C_6H_5$ | $C_6H_4Br$-*p* | 188° | | 122 (m.p.); |
| Salicylideneiminato | | $C_6H_5$ | $C_6H_5$ | $C_6H_5$ | 273° | biscopper (II) complex | 802 (m.p.); |
| –N=$CHC_6H_4NO_2$-*o* | | $C_6H_5$ | $C_6H_5$ | $C_6H_5$ | 193° | | 802 (m.p.); |
| –N=$CHC_6H_4NO_2$-*m* | | $C_6H_5$ | $C_6H_5$ | $C_6H_5$ | 190° | | 802 (m.p.), (i); |
| –N=$CHC_6H_4NO_2$-*p* | | $C_6H_5$ | $C_6H_5$ | $C_6H_5$ | 210° | | 802 (m.p.), (i); |
| –N=$CHC_6H_5$ | | $C_6H_5$ | $C_6H_5$ | $C_6H_5$ | 158° | | 121 (m.p.);<br>802 (m.p.), (i); |
| –N=$CHC_6H_5$ | | $C_6H_5$ | $OC_6H_5$ | $C_6H_5$ | 236° | | 121 (m.p.); |
| –N=$CHC_6H_4OH$-*o* | | $C_6H_5$ | $C_6H_5$ | $C_6H_5$ | 128° | | 802 (m.p.), (i); |
| –N=$CHC_6H_4OH$-*m* | | $C_6H_5$ | $C_6H_5$ | $C_6H_5$ | 262° | | 802 (m.p.), (i); |
| –N=$CHC_6H_4OH$-*p* | | $C_6H_5$ | $C_6H_5$ | $C_6H_5$ | 267° | | 802 (m.p.), (i); |
| –$NHOCH_2C_6H_5$ | | $C_6H_5$ | $C_6H_5$ | $C_6H_5$ | 252° | | 121 (m.p.); |
| –$NHOCH_2C_6H_5$ | | $C_6H_5$ | $OC_6H_5$ | $C_6H_5$ | 242° | | 121 (m.p.); |

| | | | | | | |
|---|---|---|---|---|---|---|
| 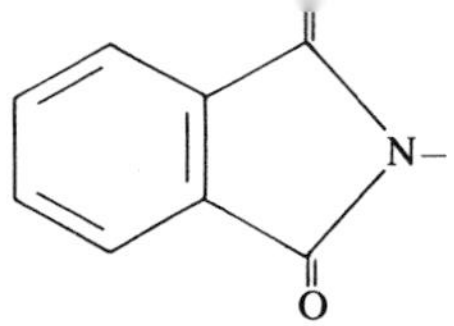 N– | $C_6H_5$ | $C_6H_5$ | $C_6H_5$ | 255–256° | | 802 (m.p.); |
| Salicylideneiminato | $C_6H_5$ | $C_6H_5$ | $C_6H_4OCH_3$-*p* | 247° | biscopper (II) complex | 802 (m.p.), (i); |
| $-N=CHC_6H_5$ | $C_6H_5$ | $C_6H_5$ | $C_6H_4OCH_3$-*p* | 202° | | 802 (m.p.), (i); |
| $-N=CHC_6H_5$ | $C_6H_5$ | $OC_6H_4CH_3$-*p* | $C_6H_5$ | 246° | | 121 (m.p.); |
| $-N=CHC_6H_4OCH_3$-*p* | $C_6H_5$ | $C_6H_5$ | $C_6H_5$ | 209° | | 802 (m.p.), (i); |
| $-N=CHC_6H_4OH$-*o* | $C_6H_5$ | $C_6H_5$ | $C_6H_4OCH_3$-*p* | 230° | | 802 (m.p.), (i); |
| $-N=CHC_6H_4OH$-*m* | $C_6H_5$ | $C_6H_5$ | $C_6H_4OCH_3$-*p* | 259° | | 802 (m.p.), (i); |
| $-N=CHC_6H_4OH$-*p* | $C_6H_5$ | $C_6H_5$ | $C_6H_4OCH_3$-*p* | 275° | | 802 (m.p.), (i); |
| $-NHOCH_2C_6H_5$ | $C_6H_5$ | $OC_6H_4CH_3$-*p* | $C_6H_5$ | 150° | | 121 (m.p.); |
| 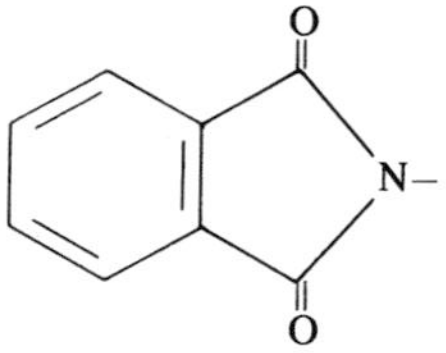 N– | $C_6H_5$ | $C_6H_5$ | $C_6H_4OCH_3$-*p* | 294° | | 802 (m.p.), (i); |
| $-N=CHC_6H_4N(CH_3)_2$-*p* | $C_6H_5$ | $C_6H_5$ | $C_6H_5$ | 242° | | 802 (m.p.), (i); |
| $-N(OCH_2C_6H_5)_2$ | $C_6H_5$ | $C_6H_5$ | $C_6H_5$ | 272° | | 121 (m.p.); |
| $-N(OCH_2C_6H_5)_2$ | $C_6H_5$ | $OC_6H_5$ | $C_6H_5$ | 219° | | 121 (m.p.); |
| $-N(OCH_2C_6H_5)_2$ | $C_6H_5$ | $OC_6H_4CH_3$-*p* | $C_6H_5$ | 198° | | 121 (m.p.); |

## TABLE XII-44. 2-Pyridone Carboxylic Acids and Derivatives

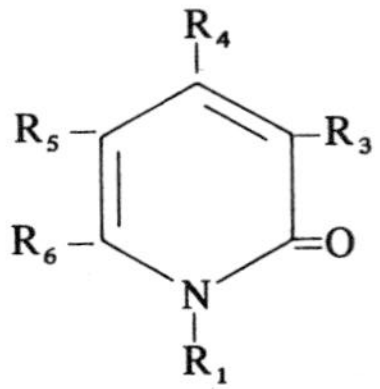

| $R_1$ | $R_3$ | $R_4$ | $R_5$ | $R_6$ | m.p. | Derivatives | Ref. |
|---|---|---|---|---|---|---|---|
| $CH_3$ | COOH | | | | | amide, m.p. 216–217° | 1009 (m.p.); |
| | | | | | | cyanomethyl, m.p. 106–107° | 1009 (m.p.); |
| $CH_3$ | | COOH | | | | methyl ester, m.p. 177–178° | 861 (m.p.), (i); |
| $CH_3$ | | | COOH | | 238–243° | | 101 (m.p.), (u); 699 (m.p.); 855 (m.p.); |
| | | | | | | oxime, m.p. 186–187° | 998 (m.p.), (t), (u); |
| | | | | | | amide, m.p. 209–210° | 855 (m.p.); |
| | | | | | | ethyl ester, m.p. 65–67° | 729 (m.p.), (i), (n); |
| | | | | | | *N'*-benzamido-, m.p. 133–135° | 1009 (m.p.); |
| | | | | | | methyl ester, m.p. 139° | 101 (m.p.); 855 (m.p.); 1009 (i); |
| | | | | | | tropyl ester methiodide, m.p. 330–332° | 998 (m.p.), (t), (u); |
| | | | | | | 1-methyl-2-oxo-3-pyridyl, m.p. 249–251° | 998 (m.p.), (t), (u); |
| $CH_2COOH$ | OH | | | | 217–219° | | 3 (m.p.); |
| $CH_3$ | COOH | OH | | OH | | methyl ester, m.p. 202° | 169 (m.p.), (i), (u); |
| | | | | | | ethyl ester, m.p. 170° | 171 (m.p.), (i), (u); |

| | | | | | | | |
|---|---|---|---|---|---|---|---|
| $CH_3$ | COOH | | $NH_2$ | | | ethyl ester·HCl m.p. 187° | 82 (m.p.); |
| | | | | | | ethyl ester·picrate, m.p. 189° | 82 (m.p.); |
| OH | | | | $CH_2C(=NOH)COOH$ | 178° | | 425 (m.p.), (i); |
| | | | | | | ethyl ester, m.p. 190–191° | 425 (m.p.), (i); |
| $C_2H_5$ | | | COOH | | | ethyl ester, b.p. 130°/0.7 mm | 549 (b.p.); |
| $CH_3$ | | | | $CH_2COOH$ | | ethyl ester, m.p. 100–101° | 944 (m.p.); |
| $CH_3$ | | COOH | | $OCH_3$ | | methyl ester, m.p. 83–84° | 507 (m.p.), (t); |
| $CH_2CH_3$ | COOH | OH | | OH | | methyl ester, m.p. 184° | 168 (m.p.), (i), (u); |
| | | | | | | morpholinium salt, m.p. 142° | 168 (m.p.); |
| | | | | | | ethyl ester, m.p. 185° | 168 (m.p.), (i), (u); |
| $CH_3$ | $OCONHCH_3$ | | | | 135–137° | | 998 (m.p.), (t), (u); |
| OH | | | | $CH_2CH(NH_2)COOH$ | | hydrochloride, m.p. 208–210° | 425 (m.p.), (i); |
| $CH_3$ | $NH_2$ | $NH_2$ | COOH | $CH_3$ | | ethyl ester, m.p. 163° | 86 (m.p.); |
| $-C(COOH)=CHCOOH$ | | | | | | dimethyl ester, m.p. 103–104° | 78 (m.p.), (i), (m), (n), (u); |
| $OCH_3$ | | | | $CH_2COCOOH$ | | ethyl ester, m.p. 153° | 944 (m.p.); |
| $CH_2COOH$ | $OCH_2COOH$ | | | | | dimethyl ester, m.p. 102–103° | 596 (m.p.), (i); |
| iso$C_3H_7$ | | | COOH | | | ethyl ester, m.p. 91–92° | 549 (m.p.); |
| $CH_3$ | $OCON(CH_3)_2$ | | | | 143–144° | | 998 (m.p.), (t), (u); |
| $CH_3$ | $OCONHC_2H_5$ | | | | 142–144° | | 998 (m.p.), (t), (u); |
| $-C(COOH)=CHCOOH$ | | | | $CH_3$ | | dimethyl ester, m.p. 120–121° | 748 (m.p.), (i), (m), (n), (u); |
| $n$-$C_4H_9$ | COOH | OH | | OH | | methyl ester, m.p. 182° | 168 (m.p.), (i), (u); |

TABLE XII-44. 2-Pyridone Carboxylic Acids and Derivatives (Continued)

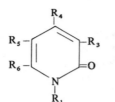

| $R_1$ | $R_3$ | $R_4$ | $R_5$ | $R_6$ | m.p. | Derivatives | Ref. |
|---|---|---|---|---|---|---|---|
|  |  |  |  |  |  | ethyl ester, m.p. 160° | 168 (m.p.), (i); |
| iso$C_4H_9$ | COOH | OH |  | OH |  | methyl ester, m.p. 158° | 168 (m.p.), (i), (u); |
|  |  |  |  |  |  | ethyl ester, m.p. 170° | 168 (m.p.), (i), (u); |
| $CH_3$ |  | $CH_3$ | $-CONHCH_3$ | $CH_3$ | 198–199° |  | 107 (m.p.); |
| 2-Pyridyl |  |  | COOH |  | 277–279° |  | 115 (m.p.); |
|  |  |  |  |  |  | methiodide, m.p. 266–267° | 115 (m.p.); |
|  |  |  |  |  |  | amide | 115; |
|  |  |  |  |  |  | methyl ester, m.p. 163–165° | 115 (m.p.); |
|  |  |  |  |  |  | 1-pyrrolidine amide, m.p. 183–185° | 115 (m.p.); |
|  |  |  |  |  |  | 1-morpholine amide, m.p. 199–200° | 115 (m.p.); |
| 2-Pyridyl |  |  | $-CON(C_2H_5)_2$ |  | 149–150° |  | 115 (m.p.); |
| 2-Pyridyl |  |  | –CONH– (2-pyridyl ring, N) |  | 281–282° |  | 115 (m.p.); |
|  |  |  |  |  |  | dimethiodide, m.p. 233–235° | 115 (m.p.); |
| 2-Pyridyl |  |  | –CONH– (phenyl ring) |  | 266–268° |  | 115 (m.p.); |

| | | | | | | | |
|---|---|---|---|---|---|---|---|
| 2-Pyridyl | | | $-CONHCH_2-$ (ring, O) | | 193–194° | | 115 (m.p.); |
| 2-Pyridyl | | | $-CON$ (ring) $N-CH_3$ | | 152–153° | | 115 (m.p.); |
| 2-Pyridyl | | | $CONH(CH_2)_2N(C_2H_5)_2$ | | 141–143° | | 115 (m.p.); |
| $-C(COOH)=CHCOOH$ | | | $CH_3$ | $CH_3$ | 145–147° | | 749 (m.p.); |
| $CH_3$ | $-COCH_3$ | $CH_3$ | COOH | $CH_3$ | | ethyl ester, m.p. 91–92° | 476 (m.p.), (n); |
| β-D-Ribosyl | | | $CONH_2$ | | 219–220° | | 613 (m.p.), (t); |
| | | | | | | 2′,3′-*O*-isopropylidene m.p. 159–161° | 613 (m.p.); |
| $CH_3$ | $-OCON(C_2H_5)_2$ | | | | 80–81° | | 998 (m.p.), (t), (u); |
| $C_6H_4Cl$-*o* | COOH | OH | | OH | | methyl ester, m.p. 184° | 169 (m.p.), (i), (u); |
| $C_6H_4Cl$-*m* | COOH | OH | | OH | | methyl ester, m.p. 197° | 168 (m.p.), (i), (u); |
| | | | | | | morpholinium salt, m.p. 183° | 168 (m.p.); |
| $C_6H_4Cl$-*p* | COOH | OH | | OH | | methyl ester, m.p. 170° | 168 (m.p.), (i), (u); |
| $C_6H_4Cl$-*o* | COOH | OH | | OH | | ethyl ester, m.p. 192° | 169 (m.p.), (i), (u); |
| $C_6H_4Cl$-*m* | COOH | OH | | OH | | ethyl ester, m.p. 188° | 168 (m.p.), (i), (u); |
| $C_6H_4Cl$-*p* | COOH | OH | | OH | | ethyl ester, m.p. 178° | 168 (m.p.), (i), (u); |
| $C_6H_4Br$-*m* | COOH | OH | | OH | | methyl ester, m.p. 188° | 168 (m.p.), (i), (u); |
| $C_6H_4Br$-*p* | COOH | OH | | OH | | methyl ester, m.p. 196° | 168 (m.p.); |
| $C_6H_4Br$-*m* | COOH | OH | | OH | | ethyl ester, m.p. 192° | 168 (m.p.), (i), (u); |
| $C_6H_4Br$-*p* | COOH | OH | | OH | | ethyl ester, m.p. 203° | 168 (m.p.); |
| $C_6H_4NO_2$ | COOH | OH | | OH | | ethyl ester, m.p. 190° | 169 (m.p.), (i), (u); |

TABLE XII-44. 2-Pyridone Carboxylic Acids and Derivatives (Continued)

| $R_1$ | $R_3$ | $R_4$ | $R_5$ | $R_6$ | m.p. | Derivatives | Ref. |
|---|---|---|---|---|---|---|---|
| $C_6H_5$ | COOH | OH | | OH | | methyl ester, m.p. 210° or m.p. 176° | 168 (m.p.); 169 (m.p.), (i), (u); 171 (m.p.); |
| | | | | | | morpholine salt, m.p. 196° | 171 (m.p.); |
| $C_6H_5$ | COOH | OH | | OH | | ethyl ester, m.p. 204–205° or m.p. 163° | 169 (m.p.), (i), (u); 171 (m.p.), (i), (u); |
| | | | | | | morpholine salt, m.p. 182° | 171 (m.p.), (u); |
| | | | | | | carboxanilide, m.p. 234° | 171 (m.p.), (u); |
| $C_6H_4OH$-*m* | COOH | OH | | OH | | methyl ester, m.p. 193° | 169 (m.p.), (i), (u); |
| (4-methyl-2-pyridyl; drawn as structure) | | | COOH | | 259–261° | | 115 (m.p.); |
| | | | | | | methyl ester, m.p. 212–213° | 115 (m.p.); |
| β-D-Glucosyl | COOH | | | | | methyl ester, amorphous | 604; |
| β-D-Glucosyl | $COOCH_3$ | | | | | tetraacetyl, m.p. 75–76° | 604 (m.p.); |
| β-D-Glucosyl | | COOH | | | | methyl ester, amorphous | 604; |
| β-D-Glucosyl | | $COOCH_3$ | | | | tetraacetyl, m.p. 186–189° | 604 (m.p.); |

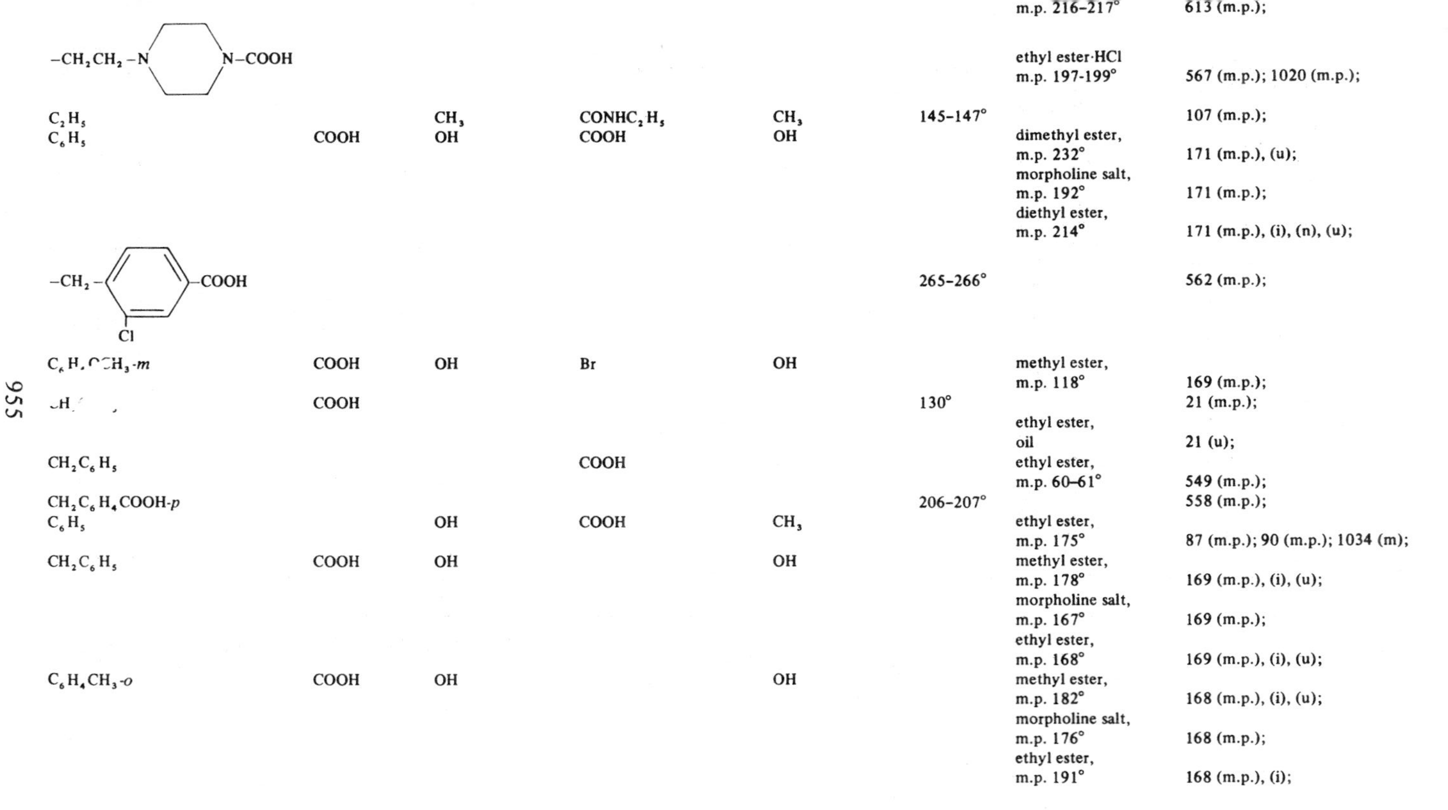

| | | | | | | | |
|---|---|---|---|---|---|---|---|
| | | | | | | m.p. 216–217° | 613 (m.p.); |
| $-CH_2CH_2-N$(piperazine)$N-COOH$ | | | | | | ethyl ester·HCl m.p. 197-199° | 567 (m.p.); 1020 (m.p.); |
| $C_2H_5$ | | $CH_3$ | $CONHC_2H_5$ | $CH_3$ | 145–147° | | 107 (m.p.); |
| $C_6H_5$ | COOH | OH | COOH | OH | | dimethyl ester, m.p. 232° | 171 (m.p.), (u); |
| | | | | | | morpholine salt, m.p. 192° | 171 (m.p.); |
| | | | | | | diethyl ester, m.p. 214° | 171 (m.p.), (i), (n), (u); |
| $-CH_2-C_6H_3(Cl)-COOH$ | | | | | 265–266° | | 562 (m.p.); |
| $C_6H_4OCH_3$-*m* | COOH | OH | Br | OH | | methyl ester, m.p. 118° | 169 (m.p.); |
| [illegible] | COOH | | | | 130° | | 21 (m.p.); |
| | | | | | | ethyl ester, oil | 21 (u); |
| $CH_2C_6H_5$ | | | COOH | | | ethyl ester, m.p. 60–61° | 549 (m.p.); |
| $CH_2C_6H_4COOH$-*p* | | | | | 206–207° | | 558 (m.p.); |
| $C_6H_5$ | | OH | COOH | $CH_3$ | | ethyl ester, m.p. 175° | 87 (m.p.); 90 (m.p.); 1034 (m); |
| $CH_2C_6H_5$ | COOH | OH | | OH | | methyl ester, m.p. 178° | 169 (m.p.), (i), (u); |
| | | | | | | morpholine salt, m.p. 167° | 169 (m.p.); |
| | | | | | | ethyl ester, m.p. 168° | 169 (m.p.), (i), (u); |
| $C_6H_4CH_3$-*o* | COOH | OH | | OH | | methyl ester, m.p. 182° | 168 (m.p.), (i), (u); |
| | | | | | | morpholine salt, m.p. 176° | 168 (m.p.); |
| | | | | | | ethyl ester, m.p. 191° | 168 (m.p.), (i); |

TABLE XII-44. 2-Pyridone Carboxylic Acids and Derivatives (Continued)

| $R_1$ | $R_3$ | $R_4$ | $R_5$ | $R_6$ | m.p. | Derivatives | Ref. |
|---|---|---|---|---|---|---|---|
| $C_6H_4CH_3$-*m* | COOH | OH | | OH | | methyl ester, m.p. 190° | 169 (m.p.); |
| | | | | | | ethyl ester, m.p. 196° | 169 (m.p.), (i), (u); |
| $C_6H_4CH_3$-*p* | COOH | OH | | OH | | methyl ester, m.p. 198° | 168 (m.p.); |
| | | | | | | morpholine salt, m.p. 180° | 168 (m.p.); |
| | | | | | | ethyl ester, m.p. 199° | 168 (m.p.); |
| $C_6H_5$ | COOH | $OCH_3$ | | OH | | methyl ester m.p. 252° | 171 (m.p.), (u); |
| $C_6H_4OCH_3$-o | COOH | OH | | OH | | methyl ester, m.p. 198° | 169 (m.p.), (i), (u); |
| | | | | | | ethyl ester, m.p. 190° | 169 (m.p.), (i), (u); |
| $C_6H_4OCH_3$-*m* | COOH | OH | | OH | | methyl ester, m.p. 190° | 169 (m.p.), (i), (u); |
| | | | | | | ethyl ester, m.p. 180° | 169 (m.p.), (i), (u); |
| $C_6H_4OCH_3$-*p* | COOH | OH | | OH | | methyl ester, m.p. 193° | 167 (m.p.), (i), (u); 168 (m.p.), (i), (u); |
| | | | | | | ethyl ester, m.p. 205° | 168 (m.p.), (i), (u); |
| $-C(COOH)=CHCOOH$ | | | $-CH_2CH_2CH_2CH_2-$ | | 190–191° | | 749 (m.p.); |
| | | | | | | hydrate, m.p. 165–166° | 749 (m.p.); |

| | | | | | | | |
|---|---|---|---|---|---|---|---|
| CHCOOH 1-(2-Pyridone) | | | | | | dimethyl ester, m.p. 189–190° | 748 (m.p.), (i), (m), (n), (u); |
| $CH_2C_6H_5$ | | | | $CH_2COOH$ | 169° | | 944 (m.p.); |
| | | | | | | amide, m.p. 250–252° | 944 (m.p.); |
| | | | | | | ethyl ester, m.p. 79° | 944 (m.p.); |
| $-CH_2CH_2C_6H_5$ | COOH | | | | 161–163° | | 332 (m.p.); |
| | | | | | | ethyl ester, b.p. 191°/0.015 mm | 332 (b.p.); |
| | | | | | | $NHNH_2 \cdot HCl \cdot H_2O$, m.p. 88–90° | 332 (m.p.); |
| $-CH_2CH_2C_6H_5$ | | COOH | | | 260° | | 1008 (m.p.); |
| | | | | | | ethyl ester, m.p. 102–104° | 1008 (m.p.); |
| | | | | | | $NHNH_2$ m.p. 191–193° | 1008 (m.p.); |
| $-CH_2CH_2C_6H_5$ | | | COOH | | | ethyl ester, m.p. 55–56° | 1008 (m.p.); |
| | | | | | | $NHNH_2 \cdot HCl \cdot H_2O$ m.p. 252° | 1008 (m.p.); |
| $C_6H_5$ | COOH | $OCH_3$ | | $OCH_3$ | 195–197° | | 167 (m.p.), (u); |
| $C_6H_5$ | COOH | $OC_2H_5$ | | OH | | ethyl ester, m.p. 160° | 171 (m.p.), (i), (n), (u); |
| $C_6H_4OCH_3$-*p* | | OH | COOH | $CH_3$ | | ethyl ester, m.p. 208° | 87 (m.p.); 90 (m.p.); 1034 (m); |
| $C_6H_4OCH_3$-*o* | COOH | OH | | $OCH_3$ | | methyl ester, m.p. 205° | 169 (m.p.); |
| $C_6H_4Cl$-*p* | $COCH_3$ | | COOH | $CH_3$ | 272–274° | | 112 (m.p.), (n); |
| | | | | | | methyl ester, m.p. 207° | 112 (m.p.); |
| | | | | | | ethyl ester, m.p. 185–187° | 112 (m.p.), (n); |
| | | | | | | *t*-butyl ester, m.p. 228° | 112 (m.p.); |
| $C_6H_4Cl$-*p* | $COCH_3$ | | $-CONH-C_5H_3N-CH_3$ | $CH_3$ | 215–218° | | 112 (m.p.), (n); |
| $CH_2C_6H_5$ | | | | $-CH_2COCOOH$ | 211° | | 944 (m.p.); |
| | | | | | | α-oximino, m.p. 191° | 944 (m.p.); |

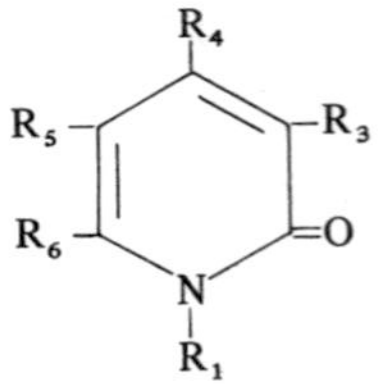

TABLE XII-44. 2-Pyridone Carboxylic Acids and Derivatives (Continued)

| $R_1$ | $R_3$ | $R_4$ | $R_5$ | $R_6$ | m.p. | Derivatives | Ref. |
|---|---|---|---|---|---|---|---|
| $C_6H_4COOH$-*p* | $COCH_3$ | OH | | $CH_3$ | | ethyl ester, m.p. 176–177° | 1030 (m.p.), (i), (n); |
| $CH_3$ | OH | $CH_2C_6H_5$ | COOH | $CH_3$ | | ethyl ester, m.p. 141° | 231 (m.p.); |
| 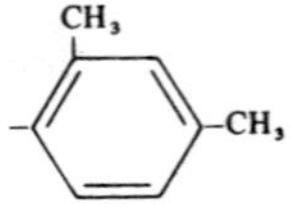 $C_6H_3(CH_3)_2$ (ring structure with two $CH_3$) | | OH | COOH | $CH_3$ | | ethyl ester | 1034 (m); |
| 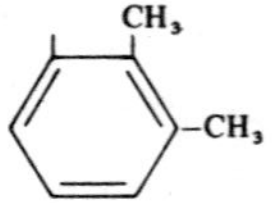 $C_6H_3(CH_3)_2$ (ring structure with two $CH_3$) | | OH | COOH | $CH_3$ | | ethyl ester, m.p. 198–199° | 87 (m.p.); 90 (m.p.); |
| α-$C_{10}H_7$ | COOH | OH | | OH | | methyl ester, m.p. 250° | 171 (m.p.), (u); |
| | | | | | | m.p. 203° | 169 (m.p.), (i), (u); |
| | | | | | | ethyl ester, | 171 (m.p.), (i), (u); |
| | | | | | | m.p. 208–210° | 169 (m.p.), (i), (u); |
| | | | | | | morpholine salt, m.p. 188° | 171 (m.p.), (u); |
| β-$C_{10}H_7$ | COOH | OH | | OH | | methyl ester, m.p. 205° | 168 (m.p.), (i), (u); |
| | | | | | | morpholine salt, m.p. 178° | 168 (m.p.); |
| | | | | | | ethyl ester, m.p. 184° | 168 (m.p.), (i), (u); |

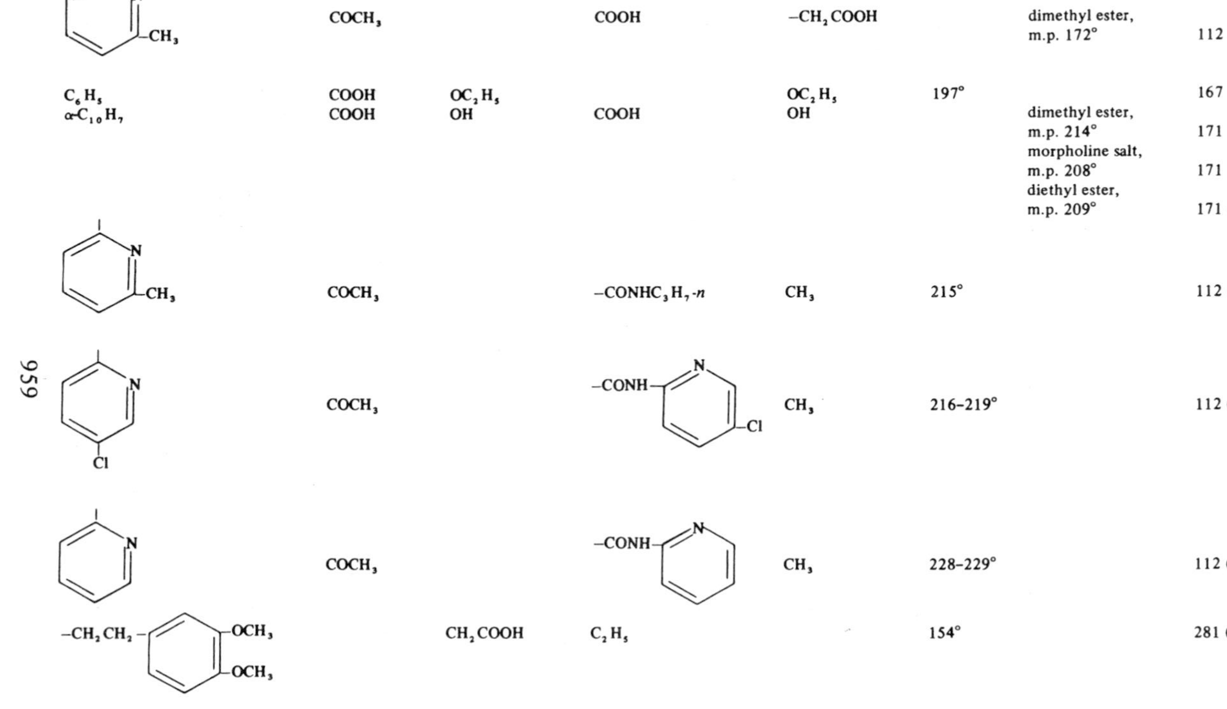

| | | | | | | | |
|---|---|---|---|---|---|---|---|
| | $COCH_3$ | | COOH | $-CH_2COOH$ | | dimethyl ester, m.p. 172° | 112 (m.p.); |
| $C_6H_5$ | COOH | $OC_2H_5$ | | $OC_2H_5$ | 197° | | 167 (m.p.), (u); |
| $\alpha$-$C_{10}H_7$ | COOH | OH | COOH | OH | | dimethyl ester, m.p. 214° | 171 (m.p.), (u); |
| | | | | | | morpholine salt, m.p. 208° | 171 (m.p.); |
| | | | | | | diethyl ester, m.p. 209° | 171 (m.p.), (u); |
| | $COCH_3$ | | $-CONHC_3H_7$-*n* | $CH_3$ | 215° | | 112 (m.p.); |
| | $COCH_3$ | | $-CONH-$ | $CH_3$ | 216–219° | | 112 (m.p.), (n); |
| | $COCH_3$ | | $-CONH-$ | $CH_3$ | 228–229° | | 112 (m.p.), (n); |
| $-CH_2CH_2-$ | | $CH_2COOH$ | $C_2H_5$ | | 154° | | 281 (m.p.), (u); |
| | | | | | | hydrochloride, m.p. 160–163° | 281 (m.p.); |

TABLE XII-44. 2-Pyridone Carboxylic Acids and Derivatives (Continued)

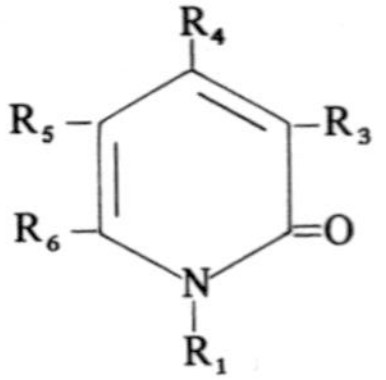

| $R_1$ | $R_3$ | $R_4$ | $R_5$ | $R_6$ | m.p. | Derivatives | Ref. |
|---|---|---|---|---|---|---|---|
| 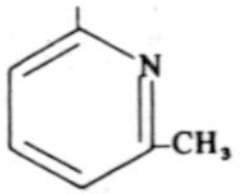 (pyridyl, N, –$CH_3$) | $COCH_3$ | | –$CONHC_4H_9$-*t* | $CH_3$ | 243–246° | | 112 (m.p.); |
| $CH_3$ | –$CH_2C_6H_5$ | | COOH | $C_6H_5$ | 293–294° | | 45 (m.p.), (i), (u); |
| | | | | | | methyl ester, m.p. 119–120° | 45 (m.p.), (n); |
| | | | | | | ethyl ester, m.p. 95–96° | 45 (m.p.), (i), (n), (u); |
| $C_6H_5$ | OH | $CH_2C_6H_5$ | COOH | $CH_3$ | | ethyl ester, m.p. 142° | 231 (m.p.); |
| 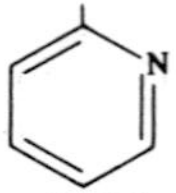 (pyridyl, N) | –$COCH_3$ | | –$CONHC_6H_5$ | $CH_3$ | 230–232° | | 112 (m.p.); |
| 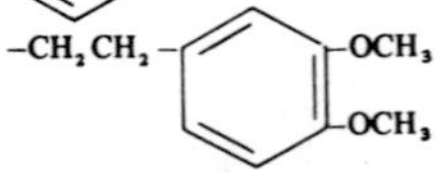 –$CH_2CH_2$–(phenyl, –$OCH_3$, –$OCH_3$) | | –$CH_2COCOOH$ | $C_2H_5$ | | | ethyl ester, m.p. 141–143° | 281 (m.p.); |
| 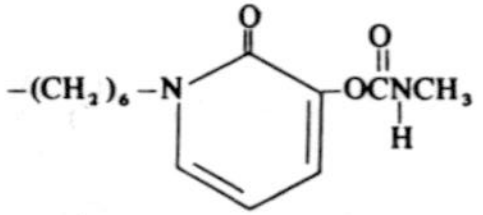 –$(CH_2)_6$–N(2-pyridone, 3-$OC(=O)NHCH_3$) | –$OCONHCH_3$ | | | | 166–168° | | 998 (m.p.), (t), (u); |
| 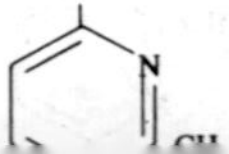 (pyridyl, N, $CH_3$) | $COCH_3$ | | –$CONHC_6H_4Cl$-*p* | $CH_3$ | 231–233° | | 112 (m.p.), (n); |

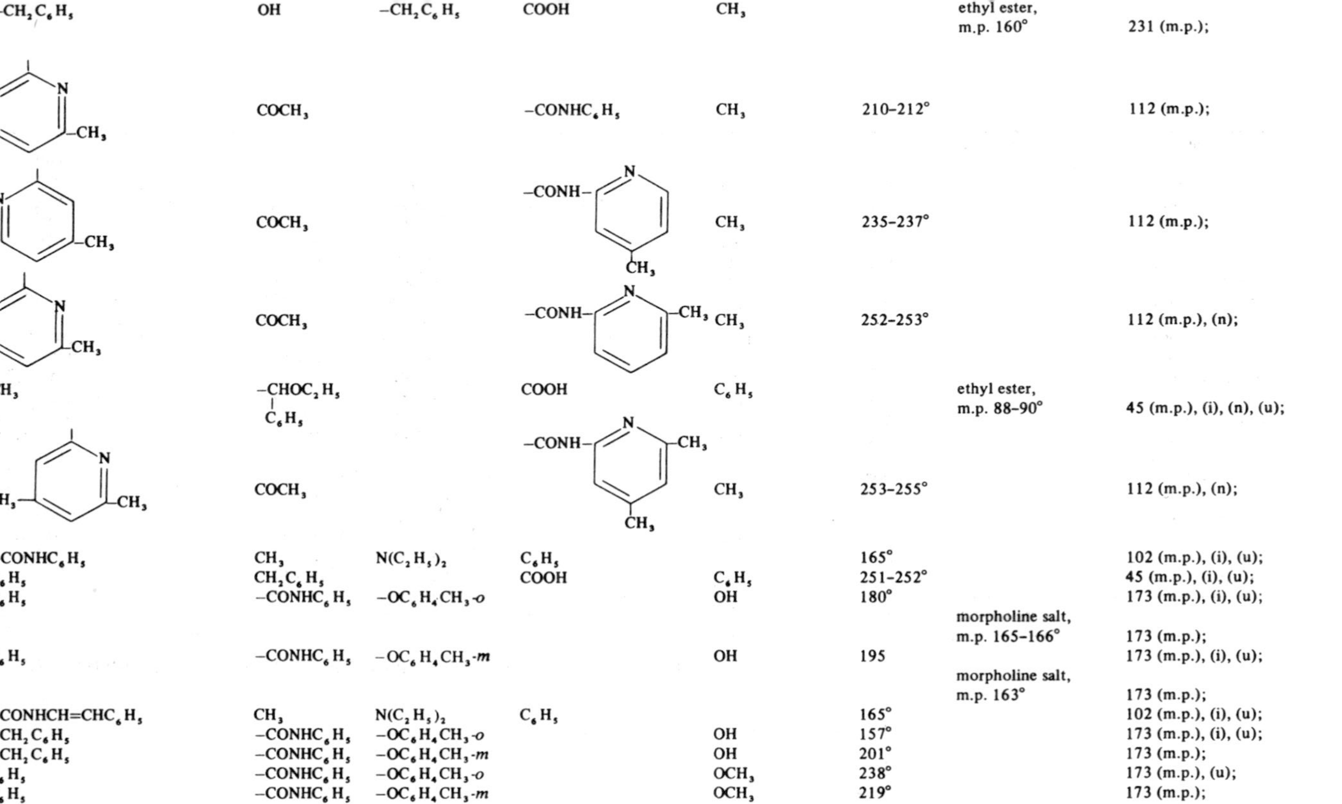

| | | | | | | | |
|---|---|---|---|---|---|---|---|
| $-CH_2C_6H_5$ | OH | $-CH_2C_6H_5$ | COOH | $CH_3$ | | ethyl ester, m.p. 160° | 231 (m.p.); |
| | $COCH_3$ | | $-CONHC_6H_5$ | $CH_3$ | 210–212° | | 112 (m.p.); |
| | $COCH_3$ | | –CONH– (ring) | $CH_3$ | 235–237° | | 112 (m.p.); |
| | $COCH_3$ | | –CONH– (ring) | $CH_3$ | 252–253° | | 112 (m.p.), (n); |
| $CH_3$ | $-CH(OC_2H_5)C_6H_5$ | | COOH | $C_6H_5$ | | ethyl ester, m.p. 88–90° | 45 (m.p.), (i), (n), (u); |
| | $COCH_3$ | | –CONH– (ring) | $CH_3$ | 253–255° | | 112 (m.p.), (n); |
| $-CONHC_6H_5$ | $CH_3$ | $N(C_2H_5)_2$ | $C_6H_5$ | | 165° | | 102 (m.p.), (i), (u); |
| $C_6H_5$ | $CH_2C_6H_5$ | | COOH | $C_6H_5$ | 251–252° | | 45 (m.p.), (i), (u); |
| $C_6H_5$ | $-CONHC_6H_5$ | $-OC_6H_4CH_3$-*o* | | OH | 180° | | 173 (m.p.), (i), (u); |
| | | | | | | morpholine salt, m.p. 165–166° | 173 (m.p.); |
| $C_6H_5$ | $-CONHC_6H_5$ | $-OC_6H_4CH_3$-*m* | | OH | 195 | | 173 (m.p.), (i), (u); |
| | | | | | | morpholine salt, m.p. 163° | 173 (m.p.); |
| $-CONHCH{=}CHC_6H_5$ | $CH_3$ | $N(C_2H_5)_2$ | $C_6H_5$ | | 165° | | 102 (m.p.), (i), (u); |
| $-CH_2C_6H_5$ | $-CONHC_6H_5$ | $-OC_6H_4CH_3$-*o* | | OH | 157° | | 173 (m.p.), (i), (u); |
| $-CH_2C_6H_5$ | $-CONHC_6H_5$ | $-OC_6H_4CH_3$-*m* | | OH | 201° | | 173 (m.p.); |
| $C_6H_5$ | $-CONHC_6H_5$ | $-OC_6H_4CH_3$-*o* | | $OCH_3$ | 238° | | 173 (m.p.), (u); |
| $C_6H_5$ | $-CONHC_6H_5$ | $-OC_6H_4CH_3$-*m* | | $OCH_3$ | 219° | | 173 (m.p.); |

TABLE XII-44. 2-Pyridone Carboxylic Acids and Derivatives (Continued)

| $R_1$ | $R_3$ | $R_4$ | $R_5$ | $R_6$ | m.p. | Derivatives | Ref. |
|---|---|---|---|---|---|---|---|
| $C_6H_4CH_3$-*o* | $-CONHC_6H_5$ | $-OC_6H_4CH_3$-*o* | | OH | 175° | | 173 (m.p.), (i), (u); |
| $C_6H_4CH_3$-*o* | $-CONHC_6H_5$ | $-OC_6H_4CH_3$-*m* | | OH | 211° | | 173 (m.p.), (i), (u); |
| $C_6H_4CH_3$-*o* | $-CONHC_6H_5$ | $-OC_6H_4CH_3$-*o* | | OH | 170° | | 173 (m.p.), (i), (u); |
| $C_6H_4CH_3$-*m* | $-CONHC_6H_5$ | $-OC_6H_4CH_3$-*m* | | OH | 177° | | 173 (m.p.), (i), (u); |
| $C_6H_4CH_3$-*p* | $-CONHC_6H_5$ | $-OC_6H_4CH_3$-*o* | | OH | 178° | | 173 (m.p.), (i), (u); |
| $C_6H_4CH_3$-*p* | $-CONHC_6H_5$ | $-OC_6H_4CH_3$-*m* | | OH | 204° | | 173 (m.p.); |
| $C_6H_4OCH_3$-*o* | $-CONHC_6H_5$ | $-OC_6H_4CH_3$-*o* | | OH | 200° | | 173 (m.p.), (i), (u); |
| $C_6H_4OCH_3$-*o* | $-CONHC_6H_5$ | $-OC_6H_4CH_3$-*m* | | OH | 208° | | 173 (m.p.); |
| $C_6H_4OCH_3$-*m* | $-CONHC_6H_5$ | $-OC_6H_4CH_3$-*o* | | OH | 182° | | 173 (m.p.), (i), (u); |
| $C_6H_4OCH_3$-*m* | $-CONHC_6H_5$ | $-OC_6H_4CH_3$-*m* | | OH | 210° | | 173 (m.p.); |
| $C_6H_4OCH_3$-*p* | $-CONHC_6H_5$ | $-OC_6H_4CH_3$-*o* | | OH | 205° | | 173 (m.p.), (i), (u); |
| $C_6H_4OCH_3$-*p* | $-CONHC_6H_5$ | $-OC_6H_4CH_3$-*m* | | OH | 223° | | 173 (m.p.); |
| $-NHCO(CH_2)_2COOH$ | | $C_6H_5$ | $C_6H_5$ | $C_6H_5$ | 170° | | 802 (m.p.), (i); |
| $-CONHCH{=}CHC_6H_4OCH_3$-*p* | $CH_3$ | $N(C_2H_5)_2$ | $C_6H_4OCH_3$-*p* | | 154° | | 102 (m.p.), (i), (u); |
| $-CONHCH{=}CHC_6H_5$ | $C_6H_5$ | $N(CH_3)_2$ | $C_6H_5$ | | 176° | | 102 (m.p.), (i), (u); |
| α-$C_{10}H_7$ | $-CONHC_6H_5$ | $-OC_6H_4CH_3$-*o* | | OH | 210° | | 173 (m.p.), (i), (u); |
| α-$C_{10}H_7$ | $-CONHC_6H_5$ | $-OC_6H_4CH_3$-*m* | | OH | 230° | | 173 (m.p.); |
| β-$C_{10}H_7$ | $-CONHC_6H_5$ | $-OC_6H_4CH_3$-*o* | | OH | 207° | | 173 (m.p.); (i), (u); |
| β-$C_{10}H_7$ | $-CONHC_6H_5$ | $-OC_6H_4CH_3$-*m* | | OH | 232° | | 173 (m.p.); |
| $-CONHCH{=}CHC_6H_4OCH_3$-*p* | $C_6H_5$ | $N(CH_3)_2$ | $C_6H_4OCH_3$-*p* | | 200° | | 102 (m.p.), (i), (u); |
| $-NHCOC_6H_4COOH$-*o* | | $C_6H_5$ | $C_6H_5$ | $C_6H_5$ | 254–256° | | 802 (m.p.), (i); |
| β-D-Ribofuranosyl | | | $CONH_2$ | | | 2,3,5-tri-*O*-benzoyl m.p. 232° | 613 (m.p.) |

TABLE XII-45. 2-Pyridone Nitriles

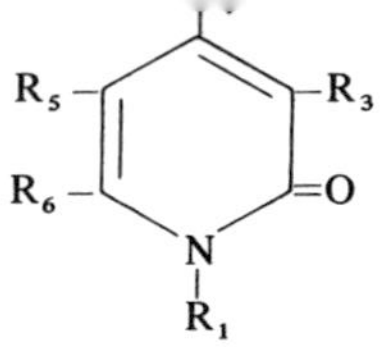

(Structure: 2-pyridone ring with $R_1$ on N, $R_3$, $R_5$, $R_6$ substituents, =O at C-2)

| $R_1$ | $R_3$ | $R_4$ | $R_5$ | $R_6$ | m.p. | Derivatives | Ref. |
|---|---|---|---|---|---|---|---|
| $CH_3$ | | | CN | | 157–161° | | 343 (m.p.); 855 (m.p.), (i), (n), (u); |
| $CH_3$ | CN | $OCH_3$ | | | 201–202° | | 849 (m.p.); 854 (m.p.); |
| 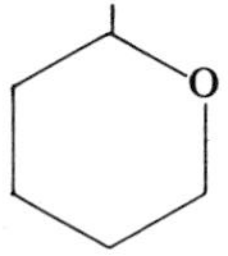 tetrahydropyran-2-yl | | | CN | | 98–100° | | 574 (m.p.), (f); |
| β-D-Glucosyl | | | CN | | 214–216° | | 624 (m.p.), (f); |
| | | | | | | tetraacetyl m.p. 98–100° | 624 (m.p.), (f); |
| $CH_2CH_2C_6H_5$ | CN | | | | 115–116° | | 332 (m.p.); 1011 (m.p.), (z); |
| $CH_2CH_2C_6H_5$ | | | CN | | 144–146° | | 1008 (m.p.); |
| $CH_3$ | CN | | | $-C(CH_3)=CHC_6H_5$ | 124° | | 57 (m.p.), (i), (u); |
| $COCH_3$ | CN | $C_6H_5$ | | 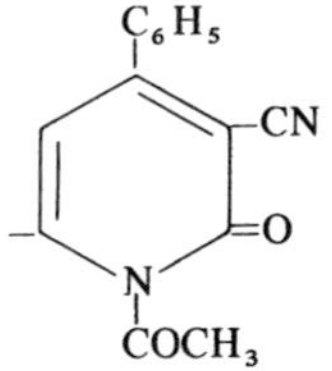 1-acetyl-3-cyano-4-phenyl-2-oxo-1,2-dihydropyridin-6-yl | 291° | | 79 (m.p.), (t); |

TABLE XII-46. Halo 2-Pyridones

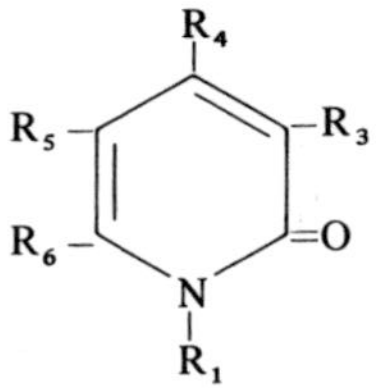

| $R_1$ | $R_3$ | $R_4$ | $R_5$ | $R_6$ | m.p. | Derivatives | Ref. |
|---|---|---|---|---|---|---|---|
| OH | OH | Cl | | | none | | 3 (n); |
| OH | OH | | Cl | | >180° | | 3 (m.p.), (n); |
| OH | Cl | Cl | Cl | | 202–203° | | 17 (m.p.); |
| $CH_3$ | Cl | Cl | Cl | Cl | 148–150° | | 18 (m.p.); 241 (u); |
| $CH_3$ | Cl | Cl | Cl | | 193–194° | | 15 (m.p.), (u); |
| $NHCONH_2$ | Cl | Cl | Cl | | 235° | | 15 (m.p.), (u); |
| $CH_3$ | Cl | | Cl | | 140–141° | | 15 (m.p.); 18 (m.p.); 327 (u); 831 (i), (u); |
| $CH_3$ | Br | | Br | | 175–187° | | 530 (m.p.), (i), (m), (n); 559 (m.p.); 570 (m.p.), (n); 745 (m.p.); 831 (i); |
| $CH_3$ | | | Cl | | | | 327 (u); |
| $CH_3$ | | | | Cl | 61–65° | | 241 (m.p.), (u); 494 (m.p.), (i); |
| $CH_3$ | | | Br | | | | 467 (u); |
| $CH_3$ | OH | Cl | | | 186–187° | | 3 (m.p.), (n); |
| $CH_3$ | Br | | $NH_2$ | | 164° | | 290 (m.p.); |
| | | | | | | $N'$-acetyl, m.p. 244° | 290 (m.p.); |
| $CH_3$ | COOH | Cl | | Cl | | methyl ester, m.p. 126° | 169 (m.p.); |

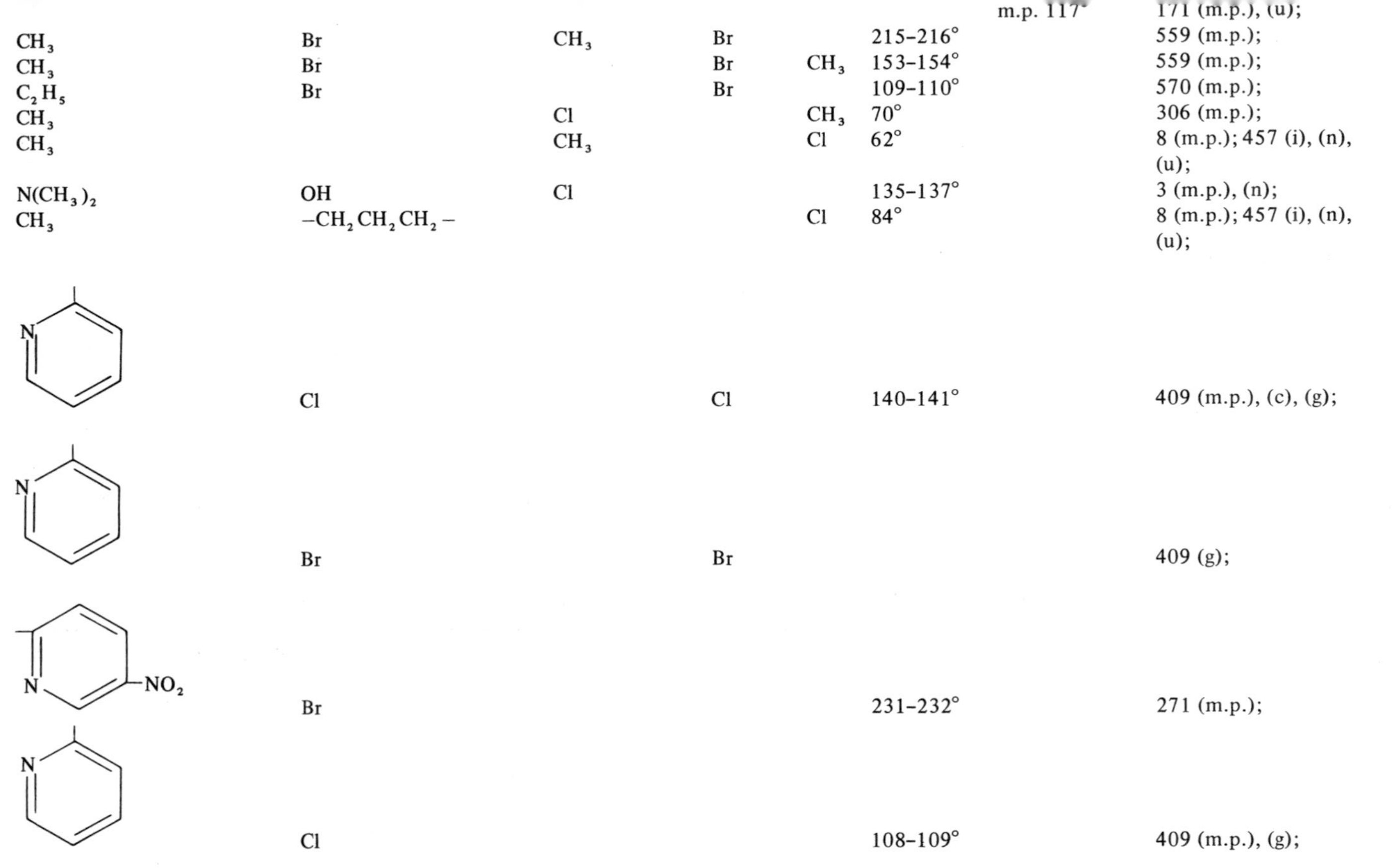

| R1 | R2 | R3 | R4 | R5 | m.p. | Derivative | References |
|---|---|---|---|---|---|---|---|
| | | | | | | m.p. 117° | 171 (m.p.), (u); |
| $CH_3$ | Br | $CH_3$ | Br | | 215–216° | | 559 (m.p.); |
| $CH_3$ | Br | | Br | $CH_3$ | 153–154° | | 559 (m.p.); |
| $C_2H_5$ | Br | | Br | | 109–110° | | 570 (m.p.); |
| $CH_3$ | | Cl | | $CH_3$ | 70° | | 306 (m.p.); |
| $CH_3$ | | $CH_3$ | | Cl | 62° | | 8 (m.p.); 457 (i), (n), (u); |
| $N(CH_3)_2$ | OH | Cl | | | 135–137° | | 3 (m.p.), (n); |
| $CH_3$ | $-CH_2CH_2CH_2-$ | | | Cl | 84° | | 8 (m.p.); 457 (i), (n), (u); |
| | Cl | | Cl | | 140–141° | | 409 (m.p.), (c), (g); |
| | Br | | Br | | | | 409 (g); |
| | Br | | | | 231–232° | | 271 (m.p.); |
| | Cl | | | | 108–109° | | 409 (m.p.), (g); |

TABLE XII-46. Halo 2-Pyridones (Continued)

| $R_1$ | $R_3$ | $R_4$ | $R_5$ | $R_6$ | m.p. | Derivatives | Ref. |
|---|---|---|---|---|---|---|---|
| 2-pyridyl | | | Cl | | | | 409 (g); |
| 2-pyridyl | Br | | | | 111–112° | | 409 (g); 992 (m.p.); |
| 2-pyridyl | | | Br | | | | 409 (g); |

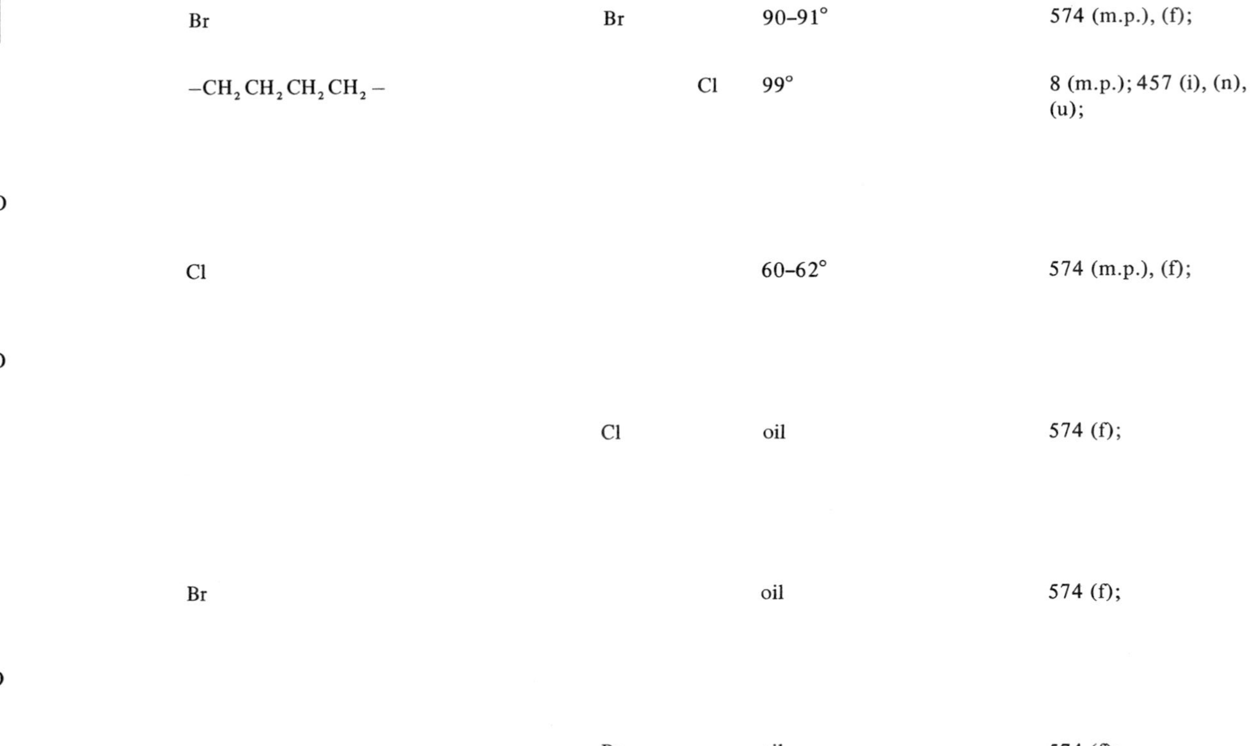

| | | | | |
|---|---|---|---|---|
| | Br | Br | | 90–91° | 574 (m.p.), (f); |
| $CH_3$ | $-CH_2CH_2CH_2CH_2-$ | | Cl | 99° | 8 (m.p.); 457 (i), (n), (u); |
| | Cl | | | 60–62° | 574 (m.p.), (f); |
| | | Cl | | oil | 574 (f); |
| | Br | | | oil | 574 (f); |
| | | Br | | oil | 574 (f); |

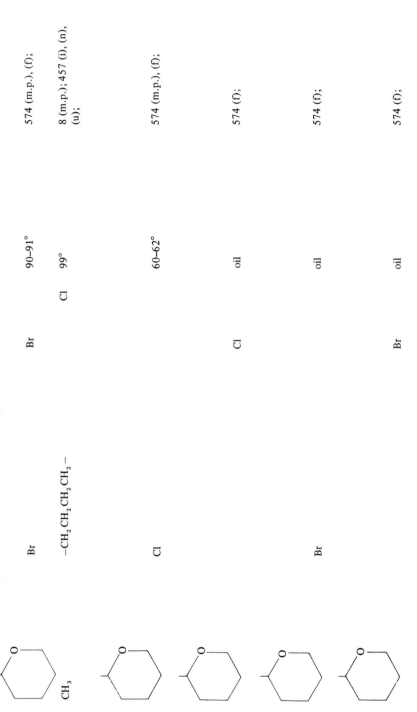

TABLE XII-46. Halo 2-Pyridones (Continued)

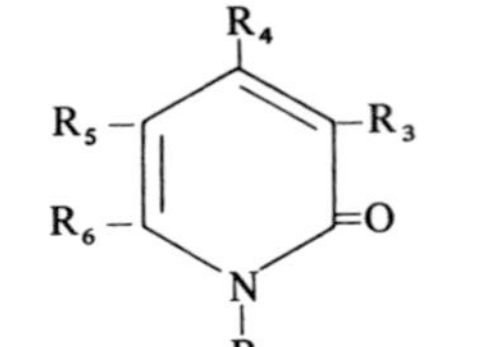

| $R_1$ | $R_3$ | $R_4$ | $R_5$ | $R_6$ | m.p. | Derivatives | Ref. |
|---|---|---|---|---|---|---|---|
| $C_6H_5$ | Cl | Cl | Cl | Cl | 149–156° | | 18 (m.p.); 19 (m.p.), (u); |
| $C_6H_5$ | Br | Br | Cl | Cl | 156–157° | | 19 (m.p.), (u); |
| $-NH-C_6H_3(NO_2)_2$ (2,4-dinitrophenyl) | Cl | Cl | Cl | | 244–252° | | 15 (m.p.); 17 (m.p.); 19 (m.p.); |
| $-NH-C_6H_3(NO_2)_2$ (2,4-dinitrophenyl) | Cl | Br | Br | | 264–266° | | 19 (m.p.), (u); |
| $C_6H_5$ | Cl | Cl | Cl | | 148–149° | | 15 (m.p.), (u); |
| $C_6H_3Cl_2$ (2,3-dichlorophenyl) | OH | | Cl | | 222–223° | | 3 (m.p.), (n); |

| | | | | | | |
|---|---|---|---|---|---|---|
| 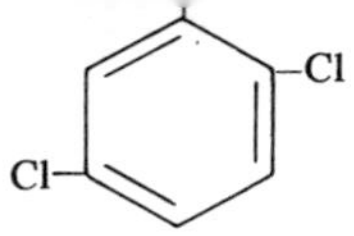 Cl– –Cl | OH | | Cl | 145–146° | | 3 (m.p.), (n); |
| 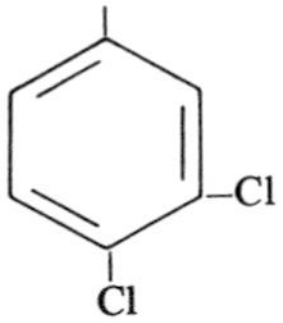 –Cl Cl | OH | | Cl | 193–194° | | 3 (m.p.), (n); |
| 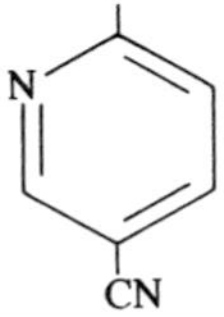 N CN | Br | | | 264–265° | | 271 (m.p.); |
| $NHC_6H_4NO_2$-*p* | Cl | Cl | Cl | 223–225° | | 15 (m.p.); |
| $C_6H_5$ | Cl | | Cl | 126–127° | | 18 (m.p.); |
| $C_6H_5Hg$– | Cl | | Cl | 100–102° | | 831 (m.p.), (i), (u); |
| $C_6H_4Cl$-*o* | OH | | Cl | 164–165° | | 3 (m.p.), (n); |
| $C_6H_4Cl$-*m* | OH | | Cl | 185–186° | | 3 (m.p.), (n); |
| $–NHC_6H_5$ | Cl | Cl | Cl | 173–174° | | 15 (m.p.), (u); |
| $C_6H_4NO_2$-*m* | OH | | Cl | 235–236° | | 3 (m.p.), (n); |
| $C_6H_5$ | OH | | Cl | 147–149° | | 3 (m.p.), (n); |
| $–NHC_6H_5$ | Cl | OH | Cl | 249–250° | | 15 (m.p.); |
| β-D-Glucosyl | Cl | | Cl | 186–188° | | 624 (m.p.), (f); |
| | | | | | tetraacetyl, m.p. 159–161° | 624 (m.p.), (f); |

TABLE XII-46. Halo 2-Pyridones (Continued)

| $R_1$ | $R_3$ | $R_4$ | $R_5$ | $R_6$ | m.p. | Derivatives | Ref. |
|---|---|---|---|---|---|---|---|
| β-D-Glucosyl | Br | | Br | | 219–222° | | 624 (m.p.), (f); |
| | | | | | | tetraacetyl, m.p. 179–180° | 624 (m.p.), (f); |
| β-D-Glucosyl | I | | I | | 232–234° | | 624 (m.p.), (f); |
| | | | | | | tetraacetyl, m.p. 175–176° | 624 (m.p.), (f); |
| β-D-Glucosyl | Cl | | | | | tetraacetyl, m.p. 87–90° | 624 (m.p.), (f); |
| β-D-Glucosyl | | | Cl | | 210–212° | | 605 (m.p.), (f), (k); |
| | | | | | | tetraacetyl, m.p. 229–233° | 605 (m.p.); |
| β-D-Glucosyl | | | Br | | 217–218° | | 605 (m.p.), (f), (k); |
| | | | | | | tetraacetyl, m.p. 224° | 605 (m.p.), (k); |
| β-D-Glucosyl | | | I | | 213–215° | | 605 (m.p.), (f), (k); |
| | | | | | | tetraacetyl, m.p. 199–200° | 605 (m.p.); |
| $-CH_2CH_2N$ (2,3-dichloro-substituted pyridone ring, C=O; structure continues) | Cl | Cl | Cl | | 306–308° | | 15 (m.p.); |

| | | | | | | | |
|---|---|---|---|---|---|---|---|
| | | | | | | m.p. 154° | 171 (m.p.), (u); |
| | | | | | | ethyl ester, m.p. 156° | 171 (m.p.), (i), (n), (u); |
| $-CH_2C_6H_5$ | Cl | | Cl | | 86–87° | | 563 (m.p.); |
| $CH_3$ | | $C_6H_5$ | | Cl | 118° | | 8 (m.p.); 457 (i), (n), (u); |
| $C_6H_4CH_3$-*o* | OH | | Cl | | 172–173° | | 3 (m.p.), (n); |
| $C_6H_4CH_3$-*m* | OH | | Cl | | 158–160° | | 3 (m.p.), (n); |
| $C_6H_4OCH_3$-*p* | OH | | Cl | | 172-174° | | 3 (m.p.), (n); |
| $-CH_2-C_6H_3(Cl)-COOH$ | Cl | | Cl | | 216–217° | | 563 (m.p.); |
| $-CH_2-C_6H_3(Cl)-COOH$ | | | Cl | | 235–237° | | 563 (m.p.); |
| $-CH_2C_6H_4COOH$-*p* | Cl | | Cl | | 238–240° | | 563 (m.p.); |
| $-CH_2C_6H_4COOH$-*p* | | | Cl | | 240–242° | | 563 (m.p.); |
| $CH_2CH_2C_6H_5$ | Cl | | | | 129–130° | | 332 (m.p.); |
| $CH_2CH_2C_6H_5$ | | Cl | | | 101–103° | | 1008 (m.p.); |
| $CH_2CH_2C_6H_5$ | | | Cl | | 122–123° | | 1008 (m.p.); |
| $CH_2CH_2C_6H_5$ | Br | | | | 139–140° | | 332 (m.p.); 1011 (m.p.), (z); |
| $CH_2CH_2C_6H_5$ | | | Br | | 114–115° | | 1008 (m.p.); |
| $CH_2CH_2C_6H_5$ | I | | | | 115–116° | | 332 (m.p.); |
| $CH_2CH_2C_6H_5$ | | | I | | 119–121° | | 1008 (m.p.); |
| $C_4H_9$-*n* | $OC_4H_9$-*n* | | Cl | | 46–47° | | 596 (m.p.), (i); |

TABLE XII-46. Halo 2-Pyridones (Continued)

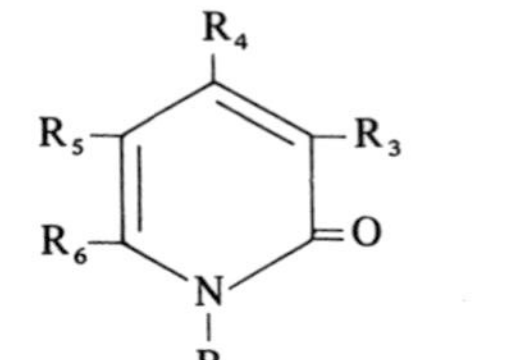


| $R_1$ | $R_3$ | $R_4$ | $R_5$ | $R_6$ | m.p. | Derivatives | Ref. |
|---|---|---|---|---|---|---|---|
| $CH_2C_6H_4COO(CH_2)_2N(CH_3)_2$-p | Cl | | Cl | | | hydrochloride, m.p. 182–184° | 563 (m.p.); |
| $CH_3$ | Cl | $C_6H_5$ | $C_6H_5$ | Cl | 231–233° | | 242 (m.p.), (i), (u); |
| $CH_3$ | $C_6H_5$ | $C_6H_5$ | Cl | Cl | 135–136° | | 242 (m.p.), (i), (u); |
| $CH_2C_6H_4COO(CH_2)_3N(CH_3)_2$-p | Cl | | Cl | | | hydrochloride, m.p. 127–128° | 563 (m.p.); |
| $C_6H_5$ | $CONHC_6H_5$ | $OC_6H_4CH_3$-o | | Cl | 245° | | 173 (m.p.), (u); |
| $C_6H_5$ | $CONHC_6H_5$ | $OC_6H_4CH_3$-o | Br | OH | 165–166° | | 173 (m.p.); |

TABLE XII-47. Nitro and Nitroso 2-Pyridones

(General structure: 2-pyridone ring with $R_1$ on N, $R_3$, $R_5$, $R_6$ substituents, =O at C-2)

| $R_1$ | $R_3$ | $R_4$ | $R_5$ | $R_6$ | m.p. | Derivatives | Ref. |
|---|---|---|---|---|---|---|---|
| $CH_3$ | Cl | | $NO_2$ | | 115° | | 291 (m.p.), (i); |
| $CH_3$ | Br | | $NO_2$ | | 122–124° | | 290 (m.p.), (i); |
| $CH_3$ | I | | $NO_2$ | | 193° | | 287 (m.p.), (i); |
| $CH_3$ | $NO_2$ | | $NO_2$ | | 174–175° | | 82 (m.p.); |
| $CH_3$ | $NO_2$ | | | | 179–180° | | 556 (m.p.); |
| $CH_3$ | | | $NO_2$ | | 172–176° | | 82 (m.p.); 861 (m.p.), (i); |
| $CH_3$ | $NO_2$ | | | OH | 160° | | 701 (m.p.); |
| $CH_3$ | COOH | | $NO_2$ | | 204° | | 82 (m.p.); |
| | | | | | | ethyl ester, m.p. 92–93° | 82 (m.p.); |
| $CH_3$ | $NO_2$ | | $CH_3$ | | 179–180° | | 700 (m.p.); |
| $C_2H_5$ | $NO_2$ | | | | 122° | | 572 (m.p.); |
| $CH_3$ | $NO_2$ | $OCH_3$ | | | 172–173° | | 849 (m.p.), (n); |
| $CH_3$ | $NO_2$ | OH | | $CH_3$ | 218–219° | | 86 (m.p.); |
| $CH_3$ | NO | $NH_2$ | | $CH_3$ | 246° | | 306 (m.p.); |
| $CH_3$ | $NO_2$ | $NH_2$ | | $CH_3$ | 314–315° | | 86 (m.p.); |
| $CH_3$ | $NO_2$ | $NHNH_2$ | | $CH_3$ | 191–192° | | 86 (m.p.); |
| $CH_3$ | $NO_2$ | $NH_2$ | COOH | $CH_3$ | | ethyl ester, m.p. 131–132° | 86 (m.p.); |
| $CH_3$ | $NO_2$ | $OCH_3$ | | $CH_3$ | 188° | | 86 (m.p.); |
| $CH_3$ | $NO_2$ | $-NHCH_2CH_2NH-$ (pyridone ring with $CH_3$, $N{-}CH_3$, $NO_2$, O) | | $CH_3$ | 324–325° | | 86 (m.p.); |
| $CH_3$ | $NO_2$ | $NHNH_2$ | $-CONHNH_2$ | $CH_3$ | 218–220° | | 86 (m.p.); |
| $CH_3$ | $NO_2$ | $OCH_3$ | COOH | $CH_3$ | | ethyl ester, m.p. 131–132° | 86 (m.p.); |

TABLE XII-47. Nitro and Nitroso 2-Pyridones (Continued)

(Structure: 2-pyridone ring with substituents $R_1$ on N, =O, $R_3$, $R_4$, $R_5$, $R_6$)

| $R_1$ | $R_3$ | $R_4$ | $R_5$ | $R_6$ | m.p. | Derivatives | Ref. |
|---|---|---|---|---|---|---|---|
| $CH_3$ | $NO_2$ | $-NHCH_2CH_2Cl$ | | $CH_3$ | 175° | | 86 (m.p.); |
| $CH_3$ | $NO_2$ | $-NHCH_2CH_2OH$ | | $CH_3$ | 229° | | 86 (m.p.); |
| (pyridyl ring, N, $NO_2$, $NO_2$) | $NO_2$ | | $NO_2$ | | 230° | | 276 (m.p.); |
| (tetrahydropyranyl ring, O) | $NO_2$ | | $NO_2$ | | 137–141° | | 574 (m.p.), (f); |
| (tetrahydropyranyl ring, O) | $NO_2$ | | | | 85–87° | | 574 (m.p.), (f); |
| (tetrahydropyranyl ring, O) | | | $NO_2$ | | 88–90° | | 574 (m.p.), (f), |
| $-OCOC_6H_4Br$-*m* | | | $NO_2$ | | 140–143° | | 1028 (m.p.), (i); |
| β-D-Glucosyl | Br | | $NO_2$ | | amorphous | | 608 (f), (k); |
| β-D-Glucosyl | $NO_2$ | | Br | | 228–229° | | 608 (m.p.), (f), (k); |

| | | | | | | | |
|---|---|---|---|---|---|---|---|
| | | | | | | tetraacetyl, m.p. 99–100°, or amorphous | 609 (m.p.), (n), (u); 624 (f), (k); |
| $CH_3$ | $NO_2$ | $NHC_4H_9$ | | $CH_3$ | 113° | | 86 (m.p.); |
| $-OCO-C_6H_3(NO_2)_2$-2,4 | $NO_2$ | | | | 191–193° | | 1028 (m.p.), (i); |
| $-OCO-C_6H_3(NO_2)_2$-3,5 | $NO_2$ | | | | 214° | | 1028 (m.p.), (i); |
| $-OCO-C_6H_3(NO_2)_2$-2,4 | | | $NO_2$ | | 174–175° | | 1028 (m.p.), (i); |
| $-OCOC_6H_4Cl$-*o* | $NO_2$ | | | | 139–141° | | 1028 (m.p.), (i); |
| $-OCOC_6H_4Cl$-*p* | $NO_2$ | | | | 135–137° | | 1028 (m.p.), (i); |
| $-OCOC_6H_4Cl$-*o* | | | $NO_2$ | | 140–142° | | 1028 (m.p.), (i); |
| $-OCOC_6H_4Cl$-*p* | | | $NO_2$ | | 151–153° | | 1028 (m.p.), (i); |
| $-OCOC_6H_4Br$-*m* | $NO_2$ | | | | 135–137° | | 1028 (m.p.), (i); |
| $-OCOC_6H_4NO_2$-*m* | $NO_2$ | | | | 170–172° | | 1028 (m.p.), (i); |
| $-OCOC_6H_4NO_2$-*p* | $NO_2$ | | | | 197–199° | | 1028 (m.p.), (i); |
| $-OCOC_6H_4NO_2$-*m* | | | $NO_2$ | | 186–188° | | 1028 (m.p.), (i); |
| $-OCOC_6H_4NO_2$-*p* | | | $NO_2$ | | 198–201° | | 1028 (m.p.), (i); |
| $-OCOC_6H_5$ | $NO_2$ | | | | 118–120° | | 1028 (m.p.), (i); |
| $-OCOC_6H_5$ | | | $NO_2$ | | 149–150° | | 1028 (m.p.), (i); |
| β-D-Glucosyl | CN | | $NO_2$ | | 144–145° | | 608 (m.p.), (f), (k); |
| | | | | | | tetraacetyl, m.p. 136–140° | 608 (m.p.), (k); |
| $-OCOC_6H_4CN$-*p* | $NO_2$ | | | | 197–199° | | 1028 (m.p.), (i); |

TABLE XII-47. Nitro and Nitroso 2-Pyridones (Continued)

| $R_1$ | $R_3$ | $R_4$ | $R_5$ | $R_6$ | m.p. | Derivatives | Ref. |
|---|---|---|---|---|---|---|---|
| $-OCOC_6H_4CH_3$-*p* | $NO_2$ | | | | 150–152° | | 1028 (m.p.), (i); |
| $-OCOC_6H_4CH_3$-*p* | | | $NO_2$ | | 164–166° | | 1028 (m.p.), (i); |
| $-OCOC_6H_4OCH_3$-*p* | $NO_2$ | | | | 156–158° | | 1028 (m.p.), (i); |
| $-OCOC_6H_4OCH_3$-*p* | | | $NO_2$ | | 139–141° | | 1028 (m.p.), (i); |
| $-CH_2CH_2C_6H_5$ | $NO_2$ | | | | 150–151° | | 332 (m.p.); |
| $-CH_2CH_2C_6H_5$ | | | $NO_2$ | | 155–158° | | 1008 (m.p.); |
| $CH_3$ | $NO_2$ | $-NHNHC_6H_5$ | | $CH_3$ | 169–171° | | 86 (m.p.); |
| $CH_3$ | $NO_2$ | $-NHCH_2C_6H_4NO_2$-*o* | | $CH_3$ | 203–204° | | 86 (m.p.); |
| $CH_3$ | $NO_2$ | $-NHCH_2C_6H_5$ | | $CH_3$ | 143–144° | | 86 (m.p.); |

TABLE XII-48. Sulfur containing 2-Pyridones

| $R_1$ | $R_3$ | $R_4$ | $R_5$ | $R_6$ | m.p. | Derivatives | Ref. |
|---|---|---|---|---|---|---|---|
| $SO_3H$ | | | | | | potassium salt | 648 (i), (u); |
| $OSO_2CH_3$ | | | | | 77–79° | | 995 (m.p.), (i), (u); |
| [2-thiazolyl structure] | | | COOH | | 264–265° | | 115 (m.p.); |
| | | | | | | methiodide, m.p. 244–247° | 115 (m.p.); |
| | | | | | | amide, m.p. 268–270° | 115 (m.p.); |
| | | | | | | methyl ester, m.p. 161–163° | 115 (m.p.); |
| [1,3,4-thiadiazol-2-yl-$CH_2$– structure] | | | COOH | | 254–256° | | 115 (m.p.); |
| | | | | | | methyl ester, m.p. 213–214° | 115 (m.p.); |
| $CH_3$ | | $-OPS(OCH_3)_2$ | | $CH_3$ | oil | | 1004; |
| $OSO_2C_6H_4F$-$p$ | | | | | 130–132° | | 995 (m.p.), (i), (u); |
| $OSO_2C_6H_5$ | | | | | 135–137° | | 995 (m.p.), (i), (u); |

TABLE XII-48. Sulfur containing 2-Pyridones (Continued)

[Structure: 2-pyridone ring with substituents $R_3$, $R_4$, $R_5$, $R_6$, =O and N–$R_1$]

| $R_1$ | $R_3$ | $R_4$ | $R_5$ | $R_6$ | m.p. | Derivatives | Ref. |
|---|---|---|---|---|---|---|---|
| $CH_2CH{=}CH_2$ | | $-OPS(OCH_3)_2$ | | $CH_3$ | 54–57° | | 1004 (m.p.); |
| $CH_3$ | | $-OPS(OC_2H_5)_2$ | | $CH_3$ | 62–63° | | 1004 (m.p.); |
| $C_6H_5$ | | OH | COOH | SH | | ethyl ester, m.p. 126–128° | 49 (m.p.); |
| $-OSO_2C_6H_4CH_3$-*p* | | | | | 101–103° | | 995 (m.p.), (i), (u); |
| $-OSO_2CH_2C_6H_5$ | | | | | 101–103° | | 1035 (m.p.), (i); |
| $CH_3$ | Br | | $-NHCSNHC_6H_5$ | | 232° | | 290 (m.p.); |
| [structure: ring with S, N] | | | –CO–N [ring structure] | | 204–207° | | 115 (m.p.); |
| [structure: ring with S, N] | | | $-CON(C_2H_5)_2$ | | 141° | | 115 (m.p.); |
| $-CH_2-$ [structure: ring with S] | | | $-CON(C_2H_5)_2$ | | 200–201° | | 115 (m.p.); |

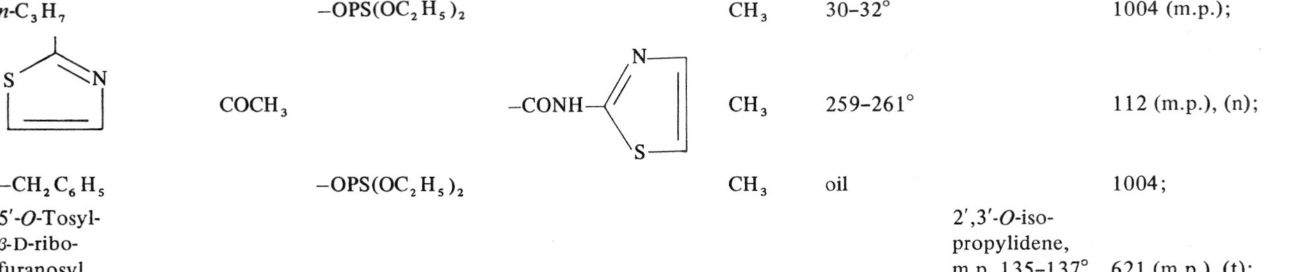

| | | | | | | | |
|---|---|---|---|---|---|---|---|
| $n$-$C_3H_7$ | | $-OPS(OC_2H_5)_2$ | | $CH_3$ | 30–32° | | 1004 (m.p.); |
| | $COCH_3$ | | –CONH– | $CH_3$ | 259–261° | | 112 (m.p.), (n); |
| $-CH_2C_6H_5$ | | $-OPS(OC_2H_5)_2$ | | $CH_3$ | oil | | 1004; |
| 5′-*O*-Tosyl-β-D-ribo-furanosyl | | | | | | 2′,3′-*O*-iso-propylidene, m.p. 135–137° | 621 (m.p.), (t); |

TABLE XII-49. Alkyl and Aryl 3-Pyridinols

(Structure: pyridine ring with $R_4$, $R_5$, $R_6$, $R_2$ and –OH at C-3)

| $R_2$ | $R_4$ | $R_5$ | $R_6$ | m.p. | Derivatives | Ref. |
|---|---|---|---|---|---|---|
| H | H | H | H | 123–129° | | 176 (i); 227 (g), (t);<br>317 (m.p.); 320 (m.p.);<br>342 (t); 404 (f);<br>478 (i), (r); 512 (u);<br>514 (n); 520 (n);<br>525 (h), (u); 527; 529;<br>648 (u); 659 (u);<br>899 (n); 900 (f);<br>901 (n); 902 (i);<br>904 (u); 905 (m.p.), (u);<br>906 (a); 909 (u);<br>910 (j); 911 (t);<br>912 (u); 915 (m);<br>918 (s); 925 (t);<br>1036 (m.p.); 1037 (m.p.);<br>1038 (m.p.); 1039 (m);<br>1040 (f); 1041 (e);<br>1042 (n); 1043 (g), (u);<br>1044 (u); |
| | | | | | picrate,<br>m.p. 205–206° | 227 (m.p.); |
| | | | | | hydrochloride,<br>m.p. 105–107° | 1036 (m.p.); |
| | | | | | deuterio | 1045 (i); |
| | | | | | sodium salt | 545 (i), (r); |
| | | | | | *O*-acetyl | 404 (f), (g); 414 (i),<br>(n), (g); |
| | | | $CH_2NO_2$ | 190° | | 193 (m.p.); |
| $CH_3$ | | | | 165–169° | | 176 (m.p.) (i); 178 (f);<br>179 (m.p.); 181 (m.p.);<br>210 (m.p.), (u); 369 (m.p.);<br>514 (n); 515 (n); 520 (n);<br>899 (n); 925 (t);<br>1036 (m.p.); 1038 (m.p.);<br>1039 (m); 1046 (m.p.); |

| | | | | | | |
|---|---|---|---|---|---|---|
| | | | | | m.p. 203 | 178 (m.p.); |
| | | | | | $NH_4Cl$ salt, m.p. 230–232° | 179 (m.p.); |
| | | | | | hydrochloride, m.p. 225–227° | 1036 (m.p.); |
| $CH_3$ | | | | | *O*-acetyl, b.p. 66°/4.5 mm | 367 (b.p.), (g), (i), (m); (n); 369 (b.p.), (g); 386 (g); |
| | | | | | *O*-acetyl picrate, m.p. 143° | 369 (m.p.); |
| | $CH_3$ | | | 117–119° | | 317 (m.p.); 369 (m.p.); 925 (t); |
| | | | | | picrate, m.p. 200° | 369 (m.p.); |
| | | | | | $^{18}O$ | 392 (m.p.); |
| | | | | | m.p. 119–121° | |
| | | | | | *O*-acetyl, b.p. 98°/4.5 mm | 369 (b.p.), (g); 387 (n); 392 (g); 414 (g), (i), (n); |
| | | | | | *O*-acetyl picrate, m.p. 173° | 369 (m.p.); |
| | | $CH_3$ | | 133–135° | | 227 (m.p.), (g), (t); 317 (m.p.); |
| | | | | | picrate, m.p. 189–190° | 227 (m.p.); |
| | | | $CH_3$ | 164–170° | | 176 (i); 183 (m.p.); 369 (m.p.); 514 (n); 515 (n); 520 (n); 1047 (m.p.); |
| | | | | | *O*-acetyl, b.p. 80°/4 mm | 367 (g), (i), (m), (n); 369 (b.p.), (g); 1047 (b.p.); |
| | | | | | picrate, m.p. 150–152° | 1047 (m.p.); |
| | | | | | *O*-acetyl picrate, m.p. 144–145° | 369 (m.p.); 918 (s); |
| $CH_2OH$ | $CH_2NH_2$ | | | | di-HCl m.p. 179° | 1155 (m.p.), (u); |
| | | | $CH_2NH_2$ | 157–160° | | 193 (m.p.), (n); |
| $COCH_3$ | | | | 55–56° | | 302 (m.p.); |
| $CH_3$ | $CH_3$ | | | 167° | | 420 (m.p.), (m), (n), (u); |
| $CH_3$ | | | $CH_3$ | 190–212° | | 182 (m.p.); 210 (m.p.); 300 (m.p.); 369 (m.p.); |

TABLE XII-49. Alkyl and Aryl 3-Pyridinols (Continued)

(Structure: 3-pyridinol ring with substituents $R_2$, $R_4$, $R_5$, $R_6$ and OH)

| $R_2$ | $R_4$ | $R_5$ | $R_6$ | m.p. | Derivatives | Ref. |
|---|---|---|---|---|---|---|
| | | | | | | 386 (m.p.), (i); 390 (m.p.);<br>420 (m.p.), (m), (n), (u);<br>515 (n); 520 (n);<br>721 (m.p.); 925 (t);<br>1039 (m); |
| | | | | | O-acetyl,<br>b.p. 80°/0.8 mm | 369 (b.p.);<br>386 (b.p.), (g), (i);<br>414 (g), (n); |
| $C_2H_5$ | $CH_3$ | | $CH_3$ | 135°<br>134° | | 420 (m.p.), (m), (n), (u);<br>176 (m.p.), (i); 386 (i); |
| | | | | | picrate,<br>m.p. 173–174° | 176 (m.p.); |
| | | | | | O-acetyl | 386 (g); |
| | $C_2H_5$ | | | 100° | | 283 (m.p.), (u); |
| | | | | | oxalate,<br>m.p. 231° | 283 (m.p.); |
| $CH_3$ | | $CH_2OH$ | $C_2H_5$ | | O-acetyl | 386 (g); |
| $CH_3$ | | | $CH_2OH$ | 153° | hydrochloride,<br>m.p. 169–170° | 211 (m.p.);<br>368 (m.p.); 369 (m.p.);<br>721 (m.p.), (n); |
| | | | | | picrate,<br>m.p. 190–191° | 386 (m.p.); |
| | | | | | hydrochloride,<br>m.p. 157–158° | 721 (m.p.); |
| | | | | | 3,6-diacetyl,<br>b.p. 112°/1 mm | 369 (g); 721 (b.p.), (n); |
| $CH_2OH$ | | | $CH_3$ | 157° | | 368 (m.p.); 369 (m.p.); |
| | | | | | picrate,<br>m.p. 174° | 368 (m.p.); |
| | | | | | 2,3-diacetyl | 369 (g);<br>918 (s); |
| $CH_2OH$ | $CH_2OH$ | $CH_2OH$ | $CH_2OH$ | | hydrochloride,<br>m.p. 125–126° | 204 (m.p.), (f); |

| | | | | | | |
|---|---|---|---|---|---|---|
| | | | | | m.p. 166–168 | 204 (m.p.), (f); |
| | $CH_2NH_2$ | $CH_2OPO_3H_2$ | | | di-$H_2O$ | 204 (f); |
| $CH_3$ | CHO | $CH_2OH$ | | | | 1048 (n); |
| $CH_3$ | $CH_2OD$ | $CH_2OD$ | | | | 1049 (n); |
| $-C=CH_2$ | | | | | hydrate, | |
| $CH_3$ | | | | | m.p. 52–53° | 302 (m.p.); |
| $CH_3$ | $CH_2Cl$ | $CH_2Cl$ | | | hydrochloride, | |
| | | | | | m.p. 175–190° | 1050 (m.p.); |
| $CH_3$ | $CH_2Br$ | $CH_2Br$ | | | hydrobromide, | |
| | | | | | m.p. 224–228° | 1050 (m.p.); |
| $CH_3$ | $CH_2I$ | $CH_2I$ | | | hydroiodide, | |
| | | | | | m.p. 120–160° | 1050 (m.p.); |
| $CH_3$ | | $COCH_3$ | | 253–254° | | 204 (m.p.), (n), (t), (u); |
| $CH_3$ | $CH_2Br$ | | $CH_3$ | | hydrobromide, | |
| | | | | | m.p. 170–171° | 723 (m.p.); |
| $CH_3$ | $CH_2OH$ | $CH_2Cl$ | | | | 1051 (e); |
| $CH_3$ | $CH_2Br$ | $CH_2OH$ | | | hydrobromide, | |
| | | | | | m.p. 159–160° | 1052 (m.p.); |
| $CH_3$ | CHO | $CH_2OPO_3H_2$ | | | | 1048 (n); |
| $CH_3$ | $CH_3$ | $CH_3$ | | 178° | | 1053 (m.p.); |
| | | | | | hydrochloride, | |
| | | | | | m.p. 216° | 1053 (m.p.); 1054 (m.p.); |
| $CH_3$ | $CH_3$ | | $CH_3$ | 134–138° | | 185 (m.p.); 300 (m.p.); 370 (m.p.), (i); 386 (m.p.), (i); 390 (m.p.); 514 (n); 520 (n); 721 (m.p.), (n); 723 (m.p.); 925 (t); 1039 (m); |
| | | | | | picrate, | |
| | | | | | m.p. 166° | 370 (m.p.); |
| | | | | | *O*-acetyl, | |
| | | | | | b.p. 88°/1.5 mm | 386 (b.p.), (g), (i); |
| $CH_3$ | | $C_2H_5$ | | | *O*-acetyl | 386 (g); |
| $C_2H_5$ | | $CH_3$ | | 134–136° | | 227 (m.p.), (g), (t), (u); |
| | | | | | picrate, | |
| | | | | | m.p. 180–182° | 227 (m.p.); |
| $C_2H_5$ | | | $CH_3$ | 171–172° | | 182 (m.p.); |
| | | | *n*-$C_3H_7$ | 90–91° | | 386 (m.p.), (i), (n); |
| | | | | | *O*-acetyl | 386 (g); |
| $CH_3$ | $CH_3$ | $CH_2OH$ | | 271° | | 1048 (n); 1053 (m.p.); |
| | | | | | hydrochloride, | |
| | | | | | m.p. 254° | 1054 (m.p.); |
| $CH_3$ | $CH_2OH$ | $CH_3$ | | | hydrochloride, | |
| | | | | | m.p. 140–144° | 1053 (m.p.); 1054 (m.p.); |

TABLE XII-49. Alkyl and Aryl 3-Pyridinols (Continued)

| $R_2$ | $R_4$ | $R_5$ | $R_6$ | m.p. | Derivatives | Ref. |
|---|---|---|---|---|---|---|
| $CH_3$ | $CH_2OH$ | | $CH_3$ | 160–162° | | 723 (m.p.); |
| | | | | | hydrochloride, m.p. 159–160° | 723 (m.p.); |
| | | | | | 3,4-di-*O*-acetyl, b.p. 128°/1 mm | 723 (b.p.), (n); |
| $CH_3$ | $CH_2OH$ | $CH_2OH$ | | 197–206° | | 204 (m.p.), (f), (n); 525 (h), (u); 527; 912 (u); 918 (s); 1048 (n); 1051 (e); 1055 (e); 1056 (m.p.); |
| | | | | | hydrochloride, m.p. 203–206° | 200 (m.p.); 1057 (m.p.); 1058 (m.p.); |
| $CH_3$ | $CH_2OH$ | $^*CH_2OH$ | | | $^{14}C$·HCl m.p. 208–209° | 1059 (m.p.); |
| $CH_3$ | $CH_2OH$ | | $CH_2OH$ | 166–167° | | 721 (m.p.), (n); |
| | | | | | hydrochloride, m.p. 130–131° | 721 (m.p.); |
| | | | | | 3,4,6-tri-*O*-acetyl, m.p. 50° b.p. 165°/0.01 mm | 721 (b.p.), (m.p.), (n); |
| $CH_3$ | $CH(OH)_2$ | $CH_2OH$ | | | | 525 (h), (u); |
| $CH_3$ | $CH_2NH_2$ | $CH_2Br$ | | | dihydrobromide, m.p. > 250° | 1060 (m.p.); |
| $CH_3$ | $CH_2OH$ | $CH{=}NNH_2$ | | 175–176° | | 1053 (m.p.); |
| $CH_3$ | $CH_3$ | $CH_2OPO_3H_2$ | | | | 1048 (n); 1061 (u); |
| $CH_3$ | $CH_2OH$ | $CH_2OPO_3H_2$ | | | | 527; |
| $CH_3$ | $CH_2NH_2$ | $CH_3$ | | | di-HCl m.p. 260–263° | 1054 (m.p.); |
| | | | | | diacetate, m.p. 176–177° | 1054 (m.p.); |
| | | | | | ditosylate·HCl m.p. 194–195° | 1054 (m.p.); |
| CH N(CH ) | | | | 59–60° | | 515 (n); 520 (n); |

| | | | | | | |
|---|---|---|---|---|---|---|
| | | | | | m.p. 226–227° | 1062 (m.p.), (f); |
| | | | | | picrate, | |
| | | | | | m.p. 198° | 1062 (m.p.); |
| | | | | | di-HCl | |
| $CH_3$ | $CH_2OH$ | $CH_2NH_2$ | | | | |
| | | | | | m.p. 178° | 1063 (m.p.); |
| | | | | | di-HCl, | |
| | $CH_2NH_2$ | $CH_2OH$ | $CH_3$ | | | |
| | | | | | m.p. 234–240° | 204 (m.p.), (f); |
| | | | | | | 525 (h), (u); |
| $CH_3$ | $CH(OH)_2$ | $CH_2OPO_3H_2$ | | | | 527; 1048 (n); |
| $CH_3$ | $CH_2NH_2$ | $CH_2OPO_3H_2$ | | | | |
| | $CH_2NH_2$ | $CH_2OPO_3H_2$ | $CH_3$ | | dihydrate | 204 (f); |
| | | | | | hydrochloride, | |
| $CH_3$ | $CH_2NH_2$ | $CH_2NH_2$ | | | m.p. 278–283° | 1064 (m.p.); |
| | | | | 179° | | 208 (m.p.); |
| $CH_3$ | $-CH_2OCH_2OCH_2-$ | | | | | |
| | | | | | hydrochloride, | |
| $CH_3$ | $CH_2Cl$ | $CH_2OCH_3$ | | | m.p. 139–141° | 1063 (m.p.); |
| $CH_3$ | $C_2H_5$ | | $CH_3$ | 120–122° | | 185 (m.p.); |
| $n$-$C_3H_7$ | | | $CH_3$ | 161–162° | | 182 (m.p.); |
| iso$C_3H_7$ | | | $CH_3$ | 189–190° | | 182 (m.p.); 520 (n); |
| | | | | | hydrochloride, | |
| $CH_3$ | $CH_2OH$ | $CH(OH)CH_3$ | | | m.p. 159–160° | 204 (m.p.), (n); |
| | | | | | hydrochloride, | |
| $CH_3$ | $CH_2OH$ | $CH_2OCH_3$ | | | m.p. 141–142° | 1063 (m.p.); |
| $CH_3$ | $CH_2OCH_3$ | $CH_2OH$ | | 173° | | 1048 (n); 1056 (m.p.); |
| $CH_3$ | $CH(OH)CH_3$ | $CH_2OH$ | | 177–178° | | 1053 (m.p.); 1065 (m.p.); |
| | | | | | hydrochloride, | |
| | | | | | m.p. 173–174° | 204 (m.p.), (f), (n); |
| $CH_3$ | $CH_2OH$ | $CH_2OH$ | $CH_3$ | 174–177° | | 204 (m.p.), (f); |
| $CH_3$ | | | $CH_2N(CH_3)_2$ | 159–160° | | 520 (n); 721 (m.p.), (n); |
| | | | | | di-HCl, | |
| | | | | | m.p. 210–211° | 721 (m.p.); |
| | | | $CH_3$ | 68–69° | | 520 (n); 720 (m.p.), (n); |
| $CH_2N(CH_3)_2$ | | | | | di-HCl, | |
| $CH_3$ | $CH_2NH_2$ | $CH(OH)CH_3$ | | | m.p. 188–191° | 204 (m.p.); |
| | | | $CH_3$ | | di-HCl | |
| $CH_3$ | $CH_2NH_2$ | $CH_2OH$ | | | m.p. 182–184° | 204 (m.p.), (f); |
| $CH_3$ | $CH_2NH_2$ | $CH_2OPO_3H_2$ | $CH_3$ | | dihydrate | 204 (f); |
| $CH_3$ | $CH_2OD$ | $CH_2CH_2CH_2OD$ | | | | 1049 (n); |
| $CH_3$ | $CH_2Br$ | $CH_2CH_2CH_2Br$ | | | hydrobromide | 1049; |
| $n$-$C_4H_9$ | | | $CH_3$ | 141–143° | | 182 (m.p.); |
| iso$C_4H_9$ | | | $CH_3$ | 179–180° | | 182 (m.p.); |
| iso$C_5H_{11}$ | | | | 149–150° | | 722 (m.p.), (n); |
| | | | $-CH_2C(CH_3)_3$ | 118–119° | | 386 (m.p.), (i), (n); |
| | | | | | *O*-acetyl | 386 (g); |
| | | | | | picrate, | |
| $CH_3$ | $CH_2OC_2H_5$ | $CH_3$ | | | m.p. 137° | 1054 (m.p.); |

TABLE XII-49. Alkyl and Aryl 3-Pyridinols (Continued)

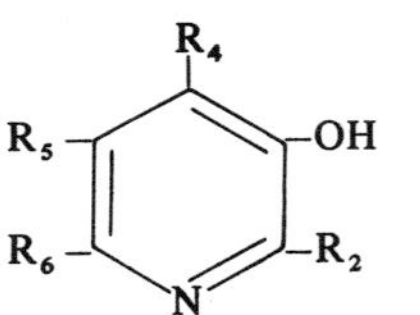


| R$_2$ | R$_4$ | R$_5$ | R$_6$ | m.p. | Derivatives | Ref. |
|---|---|---|---|---|---|---|
| $CH_3$ | $-CH(OC_2H_5)-OCH_2-$ | | | | | 1048 (n); |
| $CH_3$ | $CH_2OH$ | $CH_2CH_2CH_2OH$ | | | hydrochloride, m.p. 142–143° | 1049 (m.p.), (i), (u); |
| $CH_3$ | $CH_2OCH_3$ | $CH_2OCH_3$ | | | hydrochloride, m.p. 144-145° | 199 (m.p.), (u); 200 (m.p.); 1046 (m.p.); 1057 (m.p.); 1058 (m.p.); 1064 (m.p.); |
| $CH_3$ | $CH_2CH_2CH_2OH$ | $CH_2OH$ | | | picrate, m.p. 168–169° | 1046 (m.p.); |
| iso$C_3H_7$ | $CH_2OH$ | $CH_2OH$ | | | hydrochloride, m.p. 161–162° | 1065 (m.p.), (i), (n); |
| $CH_3$ | $CH_2OH$ | $CH(OCH_3)_2$ | | 167–168° | hydrochloride, m.p. 190–191° | 204 (m.p.), (f); 1066 (m.p.), (u); |
| $CH_3$ | $CH_2N(CH_3)_2$ | | $CH_3$ | b.p. 110°/2 mm | di-HCl m.p. 255–256°; *O*-acetyl, b.p. 100°/1 mm | 724 (b.p.), (n); 724 (m.p.); 723 (b.p.); |
| $CH_2N(C_2H_5)_2$ | | | | b.p. 85°/1 mm | di-HCl, m.p. 201–202° | 520 (n); 720 (b.p.), (n); 720 (m.p.); |
| iso$C_3H_7$ | $CH_2NH_2$ | $CH_2OH$ | | | di-HCl m.p. 169–170° | 204 (m.p.), (f); |
| $-C_6H_3(NO_2)Cl$ | | | | 325–326° | | 711 (m.p.), (n); |
| $C_6H_4NO_2$-*p* | | | | 275–276° | | 710 (m.p.), (n); |
| $-CH{=}CH-C_4H_2O-NO_2$ | | | | 238–258° | | 1067 (m.p.); |

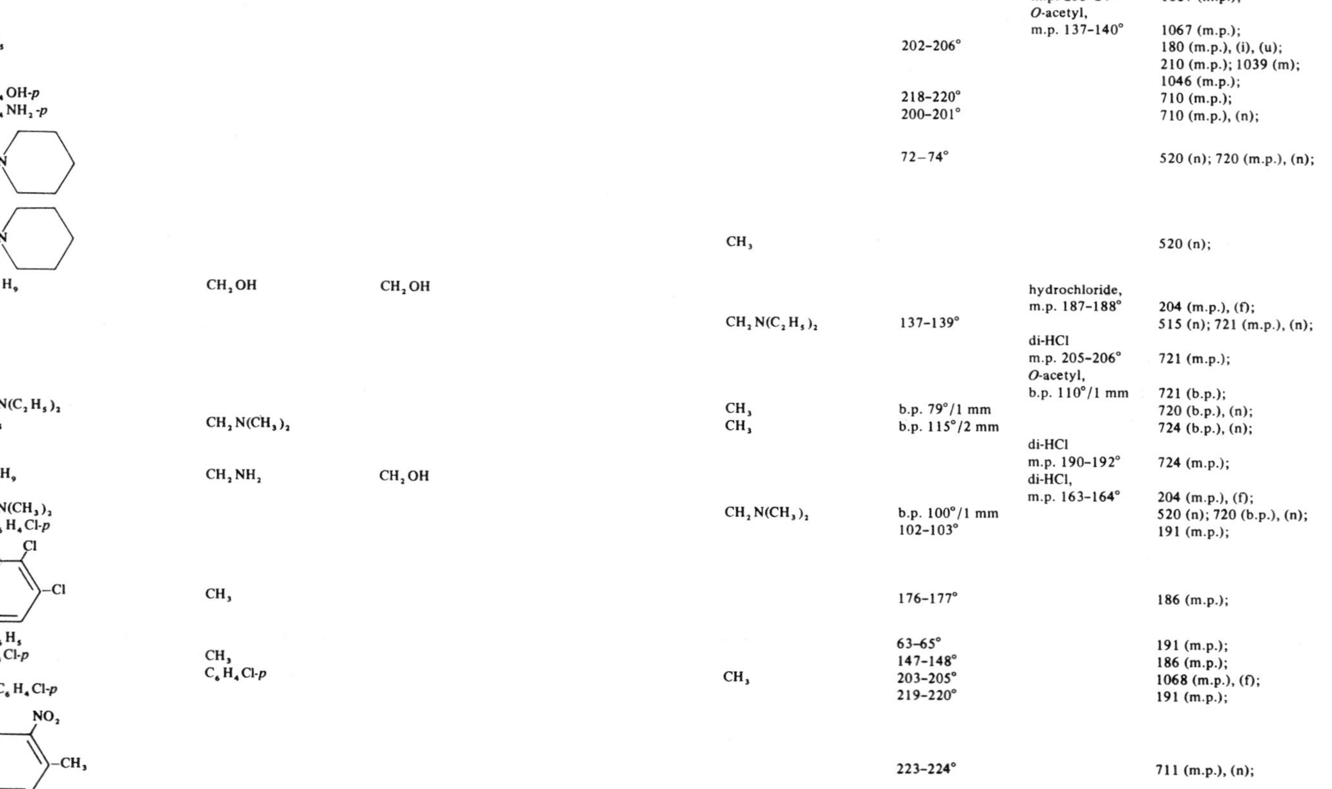

| | | | | | | |
|---|---|---|---|---|---|---|
| | | | | | m.p. 235–24?°<br>*O*-acetyl,<br>m.p. 137–140° | 1067 (m.p.);<br>1067 (m.p.); |
| $C_6H_5$ | | | | 202–206° | | 180 (m.p.), (i), (u);<br>210 (m.p.); 1039 (m);<br>1046 (m.p.); |
| $C_6H_4OH$-*p* | | | | 218–220° | | 710 (m.p.); |
| $C_6H_4NH_2$-*p* | | | | 200–201° | | 710 (m.p.), (n); |
| $CH_2N$ | | | | 72–74° | | 520 (n); 720 (m.p.), (n); |
| $CH_2N$ | | | $CH_3$ | | | 520 (n); |
| *n*-$C_4H_9$ | $CH_2OH$ | $CH_2OH$ | | | hydrochloride,<br>m.p. 187–188° | 204 (m.p.), (f); |
| $CH_3$ | | | $CH_2N(C_2H_5)_2$ | 137–139° | <br>di-HCl<br>m.p. 205–206°<br>*O*-acetyl,<br>b.p. 110°/1 mm | 515 (n); 721 (m.p.), (n);<br><br>721 (m.p.);<br><br>721 (b.p.); |
| $CH_2N(C_2H_5)_2$ | | | $CH_3$ | b.p. 79°/1 mm | | 720 (b.p.), (n); |
| $C_2H_5$ | $CH_2N(CH_3)_2$ | | $CH_3$ | b.p. 115°/2 mm | <br>di-HCl<br>m.p. 190–192° | 724 (b.p.), (n);<br><br>724 (m.p.); |
| *n*-$C_4H_9$ | $CH_2NH_2$ | $CH_2OH$ | | | di-HCl,<br>m.p. 163–164° | 204 (m.p.), (f); |
| $CH_2N(CH_3)_2$ | | | $CH_2N(CH_3)_2$ | b.p. 100°/1 mm | | 520 (n); 720 (b.p.), (n); |
| $COC_6H_4Cl$-*p* | | | | 102–103° | | 191 (m.p.); |
| Cl, –Cl | $CH_3$ | | | 176–177° | | 186 (m.p.); |
| $COC_6H_5$ | | | | 63–65° | | 191 (m.p.); |
| $C_6H_4Cl$-*p* | $CH_3$ | | | 147–148° | | 186 (m.p.); |
| | $C_6H_4Cl$-*p* | | $CH_3$ | 203–205° | | 1068 (m.p.), (f); |
| $CH_2C_6H_4Cl$-*p* | | | | 219–220° | | 191 (m.p.); |
| $NO_2$, –$CH_3$ | | | | 223–224° | | 711 (m.p.), (n); |

TABLE XII-49. Alkyl and Aryl 3-Pyridinols (Continued)

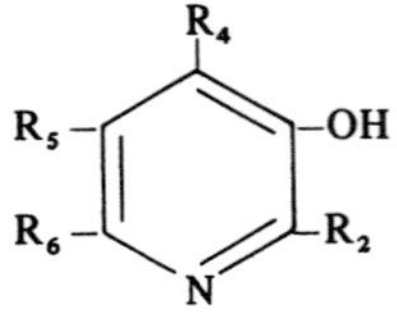

(Structure: pyridine ring bearing $R_4$, $R_5$, $R_6$, –OH at position 3, $R_2$, and N)

| $R_2$ | $R_4$ | $R_5$ | $R_6$ | m.p. | Derivatives | Ref. |
|---|---|---|---|---|---|---|
| 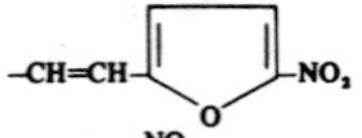 $-CH=CH-$[furan ring, O]$-NO_2$ | | | $CH_3$ | 250° | | 1047 (m.p.); 1069 (m.p.); |
| | | | | | *O*-acetyl, m.p. 167–168° | 1047 (m.p.); 1069 (m.p.), (g); |
| 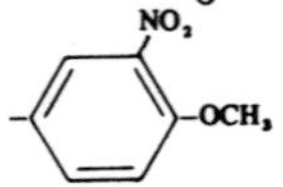 [benzene ring with $NO_2$ and $-OCH_3$] | | | | 133–134° | | 711 (m.p.), (n); |
| $C_6H_5$ | $CH_3$ | | | 115–116° | | 186 (m.p.); |
| $C_6H_4CH_3$-*p* | | | | 200° | | 180 (m.p.), (i), (u); |
| | $CH_3$ | $C_6H_5$ | | 197–198° | | 216 (m.p.); |
| | | | | | *O*-acetyl, oil | 218 (u); |
| | | | | | *p*-toluene-sulfonate, m.p. 71–72° | 218 (m.p.); |
| 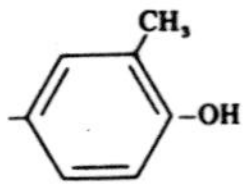 [benzene ring with $CH_3$ and $-OH$] | | | | 241–242° | | 180 (m.p.), (i), (u); |
| $-CH(OH)C_6H_5$ | | | | 117–119° | | 191 (m.p.); |
| | $CH_3$ | $C_6H_{11}$ | | 198–201° | | 218 (m.p.), (i), (u); |
| $CH_3$ | $-CH_2OCH(CH(CH_3)_2)OCH_2-$ | | | 163° | | 208 (m.p.); |
| $CH_3$ | | | 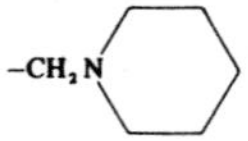 $-CH_2N$[piperidine ring] | 182–184° | | 721 (m.p.), (n); |
| | | | | | di-HCl m.p. 208–210° | 721 (m.p.); |

| | | | | | | |
|---|---|---|---|---|---|---|
|  $CH_2N$ (ring, cut off) | | | | | | |
| $CH_3$ | $CH_2OH$ | $CH_2N(CH_2)_4$ | | | hydrobromide, m.p. 200–203° | 1070 (m.p.); |
| $CH_3$ | $CH_2OCH_2OCH_3$ | $CH_2OCH_2OCH_3$ | | 63–65° | | 202 (m.p.), (u); |
| $CH_3$ | $-CH(OCH_3)_2$ | $-CH(OCH_3)_2$ | | | hydrochloride, m.p. 149–150° | 1066 (m.p.), (u); |
| $n$-$C_4H_9$ | | | $-CH_2N(CH_3)_2$ | 113–114° | | 722 (m.p.), (n); |
| iso$C_4H_9$ | | | $-CH_2N(CH_3)_2$ | 118–119° | | 722 (m.p.), (n); |
| $CH_3$ | $-CH_2N(CH_3)_2$ | | $-CH_2N(CH_3)_2$ | b.p. 125° 1 mm | | 721 (b.p.), (n); |
| | | | 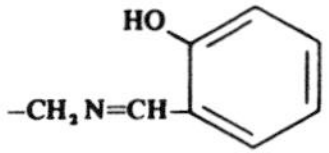 $-CH_2N{=}CH-$ (ring) HO | 173–175° | | 193 (m.p.); |
| 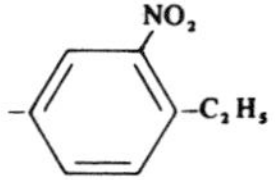 $-$(ring) $NO_2$, $-C_2H_5$ | | | | 125–126° | | 711 (m.p.), (n); |
| | | | $-CH_2CH_2C_6H_5$ | 146–147° | | 386 (m.p.), (i), (n); |
| | | | | | $O$-acetyl, b.p. 125°/0.4 mm | 386 (b.p.), (g), (i); |
| 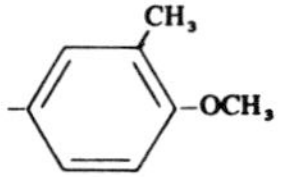 $-$(ring) $CH_3$, $-OCH_3$ | | | | 181–182° | | 180 (m.p.), (i), (u); |
| $C_6H_4OCH_3$-$p$ | $CH_3$ | | | 131–132° | | 186 (m.p.); |
| $CH_2NH_2$ | $CH_3$ | $C_6H_5$ | | | 2,3-diacetyl, m.p. 135–136° | 222 (m.p.), (i); |
| $C_2H_5$ | 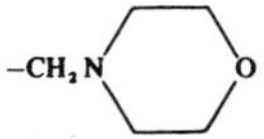 $-CH_2N$ (ring) O | | $CH_3$ | b.p. 160°/2 mm | | 724 (b.p.), (n); |
| | | | | | dihydrochloride, m.p. 227–229° | 724 (m.p.); |
| iso$C_5H_{11}$ | | | $-CH_2N(CH_3)_2$ | | dihydrochloride, m.p. 220–221° | 722 (m.p.), (n); |

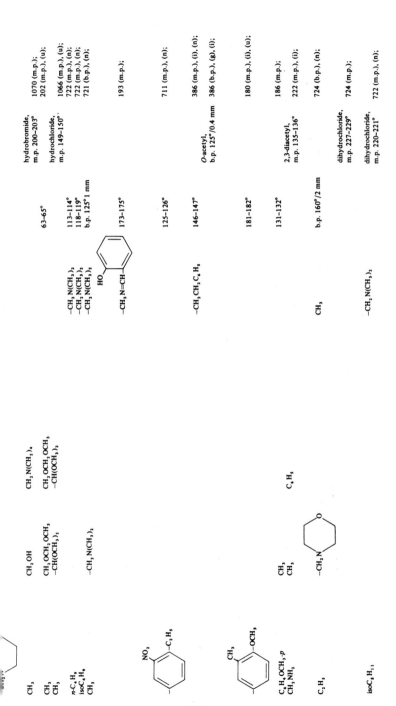

TABLE XII-49. Alkyl and Aryl 3-Pyridinols (Continued)

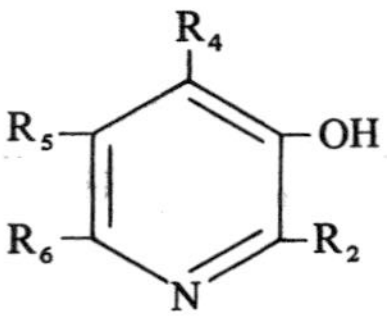

| $R_2$ | $R_4$ | $R_5$ | $R_6$ | m.p. | Derivatives | Ref. |
|---|---|---|---|---|---|---|
| 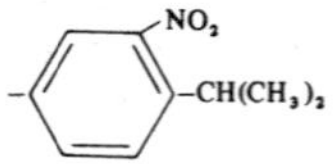 –(ring) NO$_2$, –CH(CH$_3$)$_2$ | | | | 190–191° | | 711 (m.p.), (n); |
| 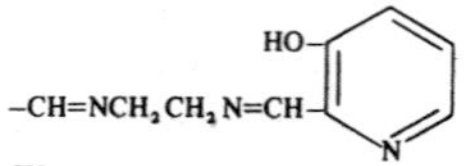 –CH=NCH$_2$CH$_2$N=CH–(ring, HO, N) | | | | 144° | | 519 (m.p.), 827 (u); |
| CH$_3$ | –CH$_2$C$_6$H$_5$ | CH$_2$OH | | 190–191° | | 1065 (m.p.), (n); |
| CH$_3$ | 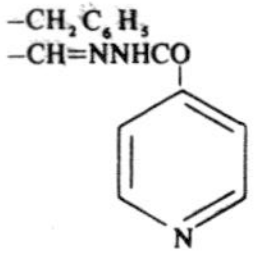 –CH=NNHCO–(ring, N) | CH$_2$OPO$_3$H$_2$ | | 260° | | 1061 (m.p.); |
| 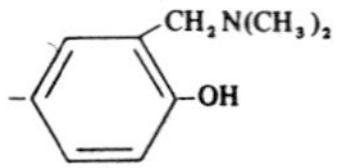 –(ring) CH$_2$N(CH$_3$)$_2$, OH | | | | | dihydrochloride, m.p. 227–228° | 727 (m.p.); |
| CH$_3$ | CH$_3$ | –CH$_2$N(CH$_2$)$_6$ | | 117–120° | | 1070 (m.p.); |
| C$_2$H$_5$ | 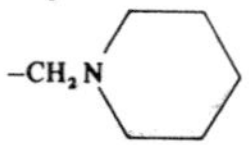 –CH$_2$N(ring) | | CH$_3$ | b.p. 155°/2 mm | | 724 (b.p.), (n); |
| | | | | | dihydrochloride, m.p. 208–209° | 724 (m.p.); |
| $n$-C$_3$H$_7$ | | |  –CH$_2$N(ring) [illegible] | [illegible] | | [illegible] |

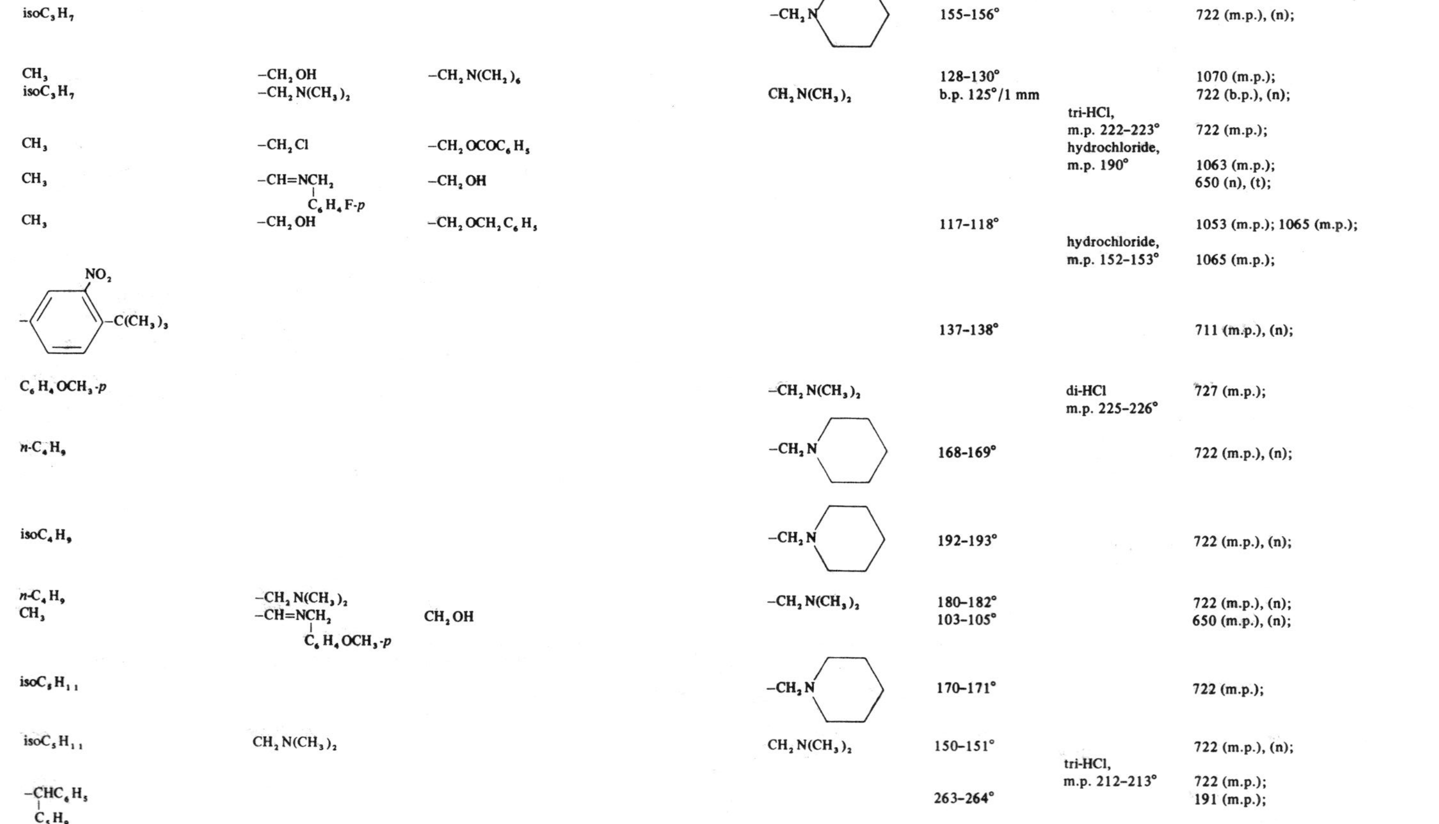

| | | | | | | |
|---|---|---|---|---|---|---|
| iso$C_3H_7$ | | | $-CH_2N$ | 155–156° | | 722 (m.p.), (n); |
| $CH_3$<br>iso$C_3H_7$ | $-CH_2OH$<br>$-CH_2N(CH_3)_2$ | $-CH_2N(CH_2)_6$ | <br>$CH_2N(CH_3)_2$ | 128–130°<br>b.p. 125°/1 mm | | 1070 (m.p.);<br>722 (b.p.), (n); |
| | | | | | tri-HCl,<br>m.p. 222–223° | 722 (m.p.); |
| $CH_3$ | $-CH_2Cl$ | $-CH_2OCOC_6H_5$ | | | hydrochloride,<br>m.p. 190° | 1063 (m.p.); |
| $CH_3$ | $-CH{=}NCH_2$<br>$C_6H_4F$-*p* | $-CH_2OH$ | | | | 650 (n), (t); |
| $CH_3$ | $-CH_2OH$ | $-CH_2OCH_2C_6H_5$ | | 117–118° | | 1053 (m.p.); 1065 (m.p.); |
| | | | | | hydrochloride,<br>m.p. 152–153° | 1065 (m.p.); |
| $-C_6H_3(NO_2)$–$C(CH_3)_3$ | | | | 137–138° | | 711 (m.p.), (n); |
| $C_6H_4OCH_3$-*p* | | | $-CH_2N(CH_3)_2$ | | di-HCl<br>m.p. 225–226° | 727 (m.p.); |
| *n*-$C_4H_9$ | | | $-CH_2N$ | 168–169° | | 722 (m.p.), (n); |
| iso$C_4H_9$ | | | $-CH_2N$ | 192–193° | | 722 (m.p.), (n); |
| *n*-$C_4H_9$<br>$CH_3$ | $-CH_2N(CH_3)_2$<br>$-CH{=}NCH_2$<br>$C_6H_4OCH_3$-*p* | <br>$CH_2OH$ | $-CH_2N(CH_3)_2$ | 180–182°<br>103–105° | | 722 (m.p.), (n);<br>650 (m.p.), (n); |
| iso$C_5H_{11}$ | | | $-CH_2N$ | 170–171° | | 722 (m.p.); |
| iso$C_5H_{11}$ | $CH_2N(CH_3)_2$ | | $CH_2N(CH_3)_2$ | 150–151° | | 722 (m.p.), (n); |
| | | | | | tri-HCl,<br>m.p. 212–213° | 722 (m.p.); |
| $-CHC_6H_5$<br>$C_5H_9$ | | | | 263–264° | | 191 (m.p.); |

## TABLE XII-49. Alkyl and Aryl 3-Pyridinols (Continued)

(General structure: pyridine ring with $R_4$, $R_5$, $R_6$, –OH (position 3) and $R_2$)

| $R_2$ | $R_4$ | $R_5$ | $R_6$ | m.p. | Derivatives | Ref. |
|---|---|---|---|---|---|---|
| [phenyl ring bearing $CH_2N$ (piperidine ring) and –OH] | | | | | di-HCl, m.p. 259–260° | 727 (m.p.); |
| $CH_3$ | $-CH_2NHCH(CH_3)CH_2C_6H_5$ | $CH_2OH$ | | | di-HCl, m.p. 229–230°<br>dipicrate m.p. 168–170° | 1052 (m.p.);<br>1052 (m.p.); |
| [phenyl ring bearing two $CH_2N(CH_3)_2$ groups and –OH] | | | | | tri-HCl, m.p. 230–231° | 727 (m.p.); |
| $-CH_2N$ (piperidine ring) | | | $-CH_2N$ (piperidine ring) | b.p. 125°/1 mm | | 520 (n); 720 (b.p.), (n); |
| 9-Xanthyl | | | | 267–270° | | 919 (m.p.); |
| $-CH(C_6H_5)C_6H_4Cl$-*o* | | | | 226–229° | | 191 (m.p.); |
| $-CH(C_6H_5)C_6H_4Cl$-*p* | | | | 193-195° | | 191 (m.p.); |
| $CH(C_6H_5)_2$ | | | | 221–222° | <br>methosulfate, m.p. 152–154°<br>*O*-acetyl, m.p. 133–135° | 191 (m.p.); 1071 (m.p.);<br>191 (m.p.);<br>1071 (m.p.); |
| $-C(C_6H_5)_2OH$ | | | | 168–170° | | 191 (m.p.); 1071 (m.p.); |

| | | | | | | |
|---|---|---|---|---|---|---|
| $-\underset{\mid}{C}HC_6H_{11}$ $C_6H_5$ | | | | 272–275° | | 191 (m.p.); |
| $C_6H_4OCH_3$-*p* | | | 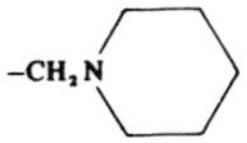 $-CH_2N$ | | di-HCl, m.p. 199–200° | 727 (m.p.); |
| $CH_3$ | $-CHO$ | 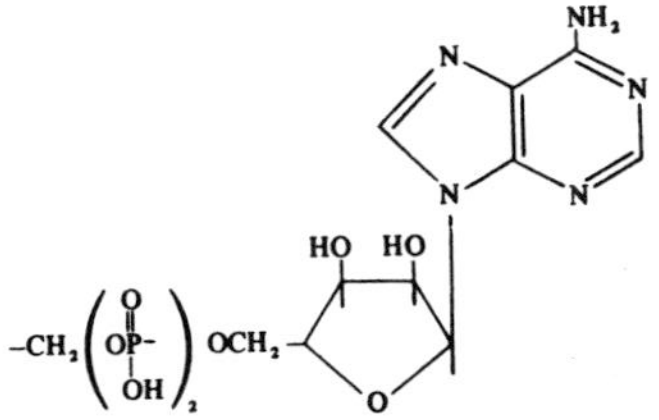 $-CH_2\left(\overset{O}{\overset{\|}{O\underset{\mid}{P}}}-\atop OH\right)_2OCH_2-$ | | | | 1072 (i), (u); |
| $CH_3$ | $-CH_2OH$ | 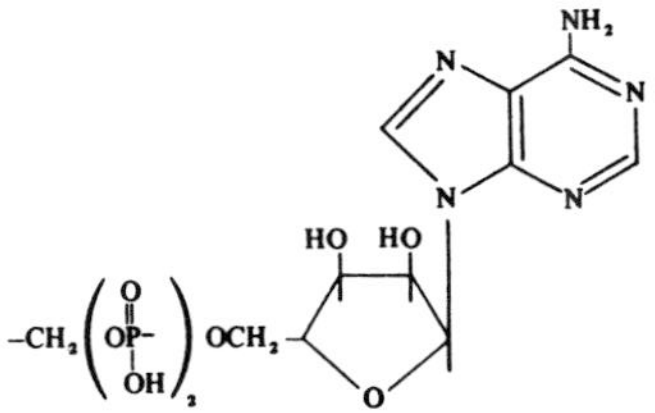 $-CH_2\left(\overset{O}{\overset{\|}{O\underset{\mid}{P}}}-\atop OH\right)_2OCH_2-$ | | | | 1072 (i), (u); |
| $CH_3$ | $-CH_2NH_2$ | 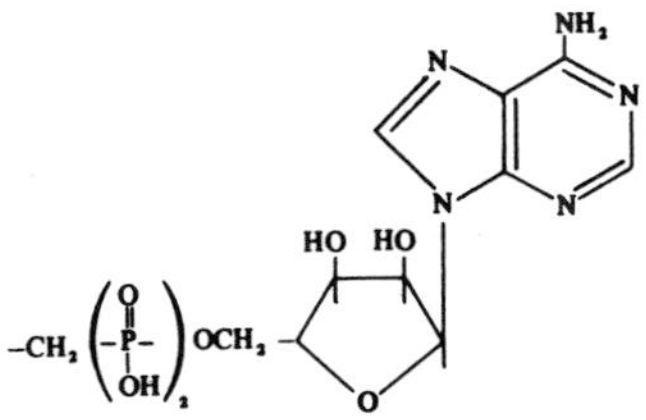 $-CH_2\left(-\overset{O}{\overset{\|}{\underset{\mid}{P}}}-\atop OH\right)_2OCH_2-$ | | | | 1072 (i), (u); |
| $C_6H_4OCH_3$-*p* | $-CH_2N(CH_3)_2$ | | $-CH_2N(CH_3)_3$ | | tri-HCl, m.p. 215–216° | 727 (m.p.); |

TABLE XII-49. Alkyl and Aryl 3-Pyridinols (Continued)

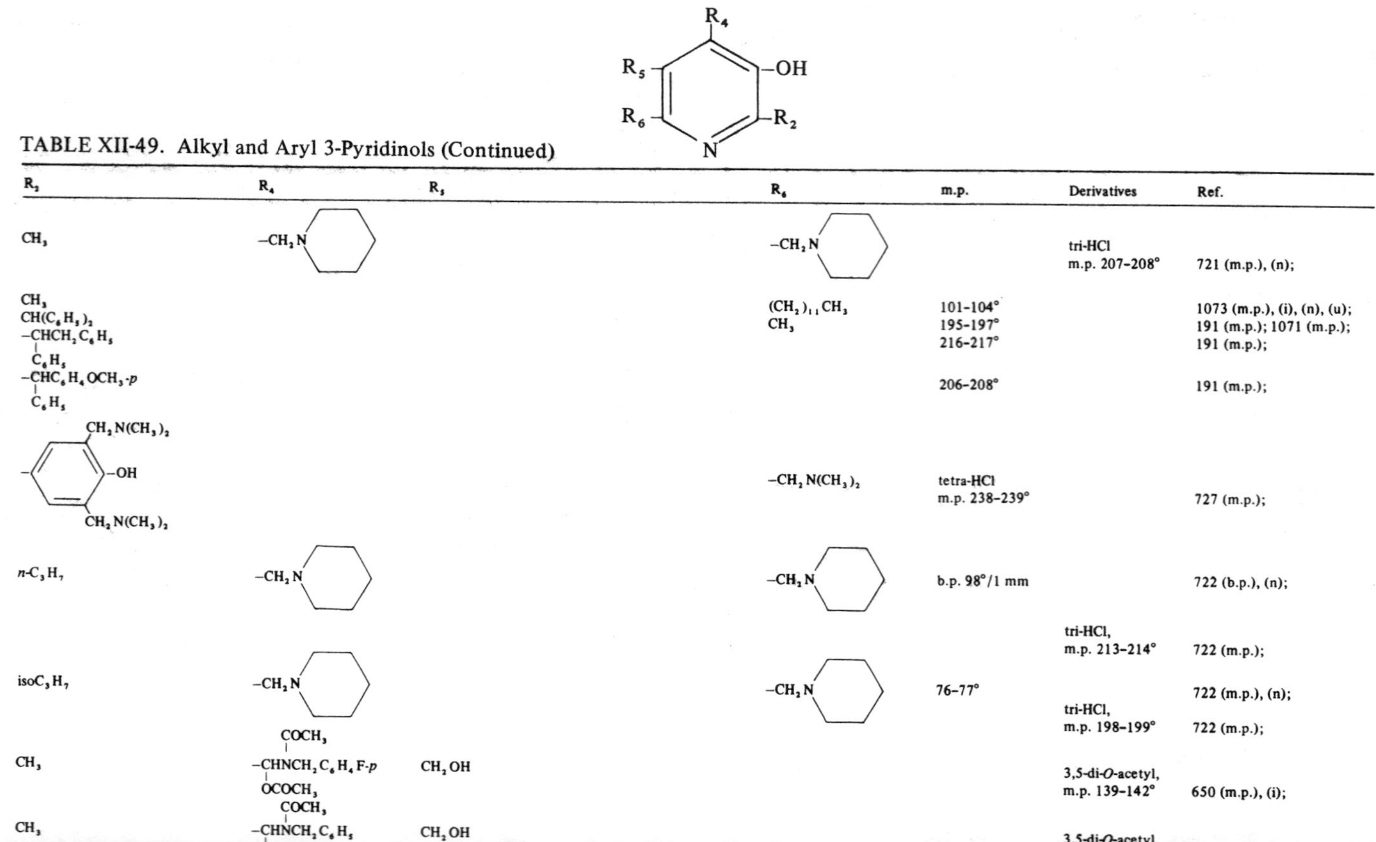

| $R_2$ | $R_4$ | $R_5$ | $R_6$ | m.p. | Derivatives | Ref. |
|---|---|---|---|---|---|---|
| $CH_3$ | $-CH_2N$ (ring) | | $-CH_2N$ (ring) | | tri-HCl m.p. 207–208° | 721 (m.p.), (n); |
| $CH_3$ | | | $(CH_2)_{11}CH_3$ | 101–104° | | 1073 (m.p.), (i), (n), (u); |
| $CH(C_6H_5)_2$ | | | $CH_3$ | 195–197° | | 191 (m.p.); 1071 (m.p.); |
| $-CH(C_6H_5)CH_2C_6H_5$ | | | | 216–217° | | 191 (m.p.); |
| $-CH(C_6H_5)C_6H_4OCH_3$-*p* | | | | 206–208° | | 191 (m.p.); |
| $-C_6H_2(OH)[CH_2N(CH_3)_2]_2$ | | | $-CH_2N(CH_3)_2$ | tetra-HCl m.p. 238–239° | | 727 (m.p.); |
| *n*-$C_3H_7$ | $-CH_2N$ (ring) | | $-CH_2N$ (ring) | b.p. 98°/1 mm | | 722 (b.p.), (n); |
| | | | | | tri-HCl, m.p. 213–214° | 722 (m.p.); |
| iso$C_3H_7$ | $-CH_2N$ (ring) | | $-CH_2N$ (ring) | 76–77° | | 722 (m.p.), (n); |
| | | | | | tri-HCl, m.p. 198–199° | 722 (m.p.); |
| $CH_3$ | $-CH(OCOCH_3)N(COCH_3)CH_2C_6H_4F$-*p* | $CH_2OH$ | | | 3,5-di-*O*-acetyl, m.p. 139–142° | 650 (m.p.), (i); |
| $CH_3$ | $-CH(\ldots)N(COCH_3)CH_2C_6H_5$ | $CH_2OH$ | | | 3,5-di-*O*-acetyl, | |

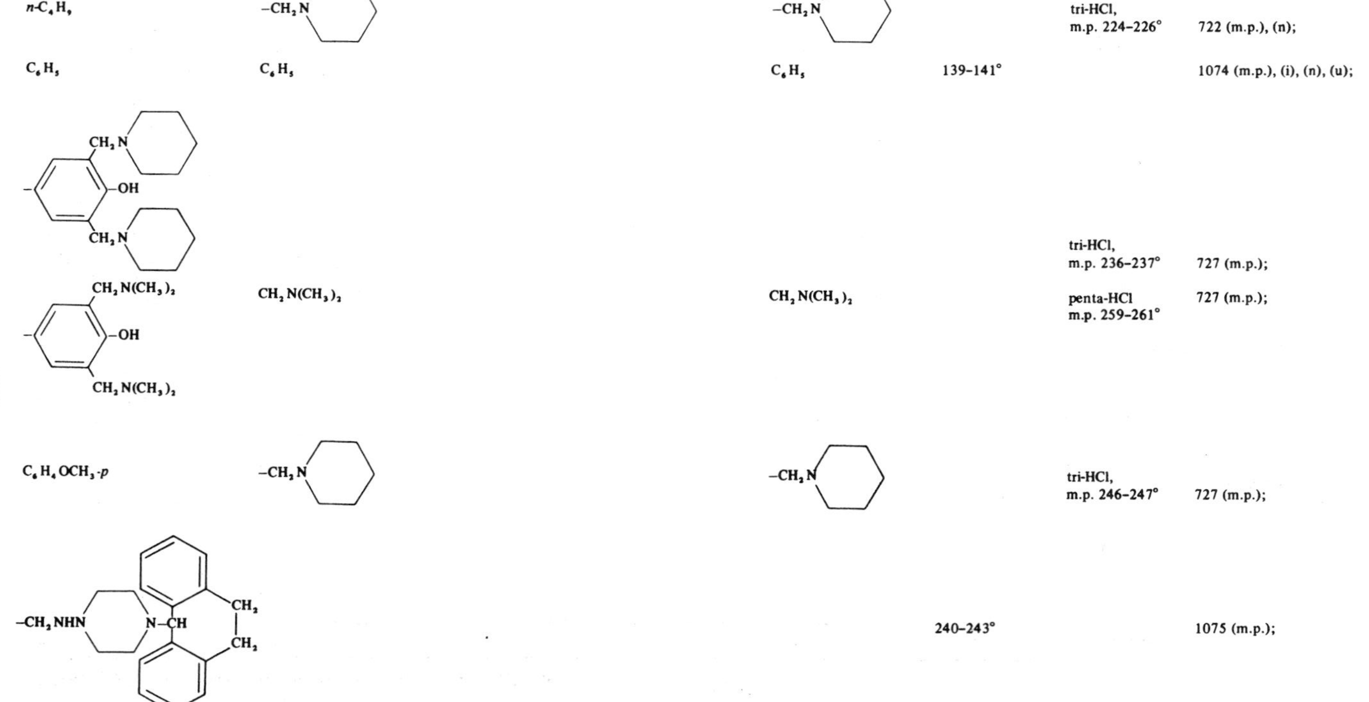

| | | | | | |
|---|---|---|---|---|---|
| $n$-$C_4H_9$ | $-CH_2N$ | $-CH_2N$ | | tri-HCl,<br>m.p. 224–226° | 722 (m.p.), (n); |
| $C_6H_5$ | $C_6H_5$ | $C_6H_5$ | 139–141° | | 1074 (m.p.), (i), (n), (u); |
| $CH_2N$, –OH, $CH_2N$ | | | | tri-HCl,<br>m.p. 236–237° | 727 (m.p.); |
| $CH_2N(CH_3)_2$, –OH, $CH_2N(CH_3)_2$ | $CH_2N(CH_3)_2$ | $CH_2N(CH_3)_2$ | | penta-HCl<br>m.p. 259–261° | 727 (m.p.); |
| $C_6H_4OCH_3$-$p$ | $-CH_2N$ | $-CH_2N$ | | tri-HCl,<br>m.p. 246–247° | 727 (m.p.); |
| $-CH_2NHN$, N–CH, $CH_2$, $CH_2$ | | | 240–243° | | 1075 (m.p.); |

TABLE XII-49. Alkyl and Aryl 3-Pyridinols (Continued)

| $R_2$ | $R_4$ | $R_5$ | $R_6$ | m.p. | Derivatives | Ref. |
|---|---|---|---|---|---|---|
| 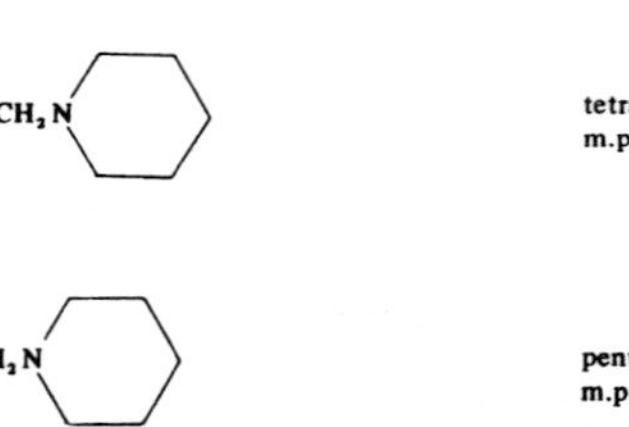 | | | | 222–223° | | 600 (m.p.), (f), (i), (k), (m), (u); |
| | | | | | $CH_3I$ m.p. 244–246° | 600 (m.p.); |
| | | | | | *O*-acetyl, m.p. 124–125° | 600 (m.p.), (i), (m), (t); |
| 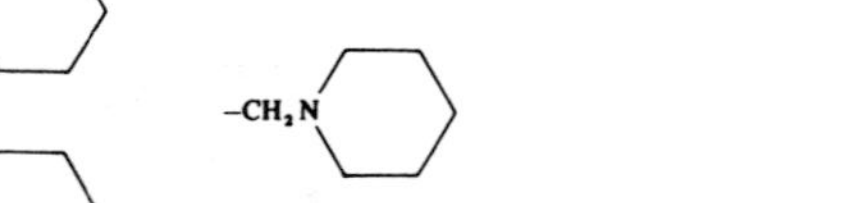 | | | $-CH_2N$(piperidine ring) | | tetra-HCl, m.p. 244–243° | 727 (m.p.); |
| (aryl ring: –OH, two $CH_2N$ piperidine groups; structure cut off at page edge) | $-CH_2N$(piperidine ring) | | $-CH_2N$(piperidine ring) | | penta-HCl m.p. 246–247° | 727 (m.p.); |

TABLE XII-50. Alkyloxy and Aryloxy 3-Pyridinols

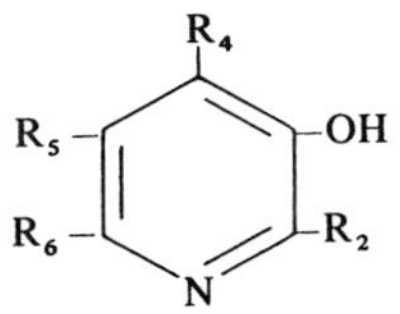

| $R_2$ | $R_4$ | $R_5$ | $R_6$ | m.p. | Derivatives | Ref. |
|---|---|---|---|---|---|---|
| $OCH_3$ | | | | | | 467 (u); |
| $CH_3$ | $OCH_3$ | | | 162–164° | | 125 (m.p.), (u); |
| | | $OC_2H_5$ | | 126–128° | | 317 (m.p.); 319 (m.p.); |

TABLE XII-51. Amino 3-Pyridinols

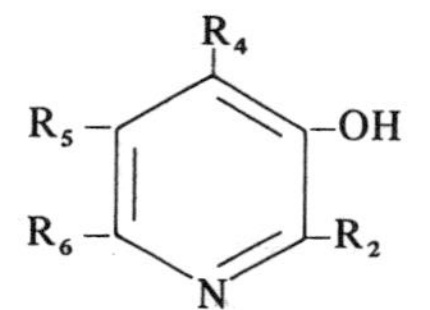


| $R_2$ | $R_4$ | $R_5$ | $R_6$ | m.p. | Derivatives | Ref. |
|---|---|---|---|---|---|---|
| $NH_2$ | | | | 172–174° | | 187 (m.p.); 188 (m.p.); 906 (a); |
| | | | | | sulfate, m.p. 124–125° | 187 (m.p.); |
| | | | | | benzamido, m.p. 95–96° | 292 (m.p.); |
| | | | | | benzamido picrate, m.p. 237–238° | 292 (m.p.); |
| | | | $NH_2$ | 116–117° | | 292 (m.p.); |
| | | | | | picrate, m.p. 225–227° | 292 (m.p.); |
| | | | | | hydrochloride, m.p. 125–126° | 292 (m.p.); |
| | | | | | benzamido, m.p. 181–182° | 292 (m.p.), (i), (u); |
| | | | | | benzamido·HCl, m.p. 215–220° | 292 (m.p.); |
| $NH_2$ | | | $NH_2$ | >300° | | 971 (m.p.); |
| $NH_2$ | | | $CH_3$ | 153–155° | | 1076 (m.p.); |
| $CH_3$ | | $NH_2$ | $CH_3$ | 259–260° | | 295 (m.p.); |
| $-N{=}NC_6H_4NO_2$-*p* | | | | 234–235° | | 292 (m.p.); |

| | | | | | | |
|---|---|---|---|---|---|---|
| | | | $-N{=}NC_6H_4NO_2$-*p* | 231–232° | | 292 (m.p.); |
| $-N{=}NC_6H_4NO_2$-*p* | | | $CH_3$ | 157–158° | | 741 (m.p.); |
| $CH_3$ | | | $-N{=}NC_6H_4NO_2$-*p* | 196–197° | | 741 (m.p.); |
| $-N{=}NC_6H_5$ | | | $CH_3$ | 166–168° | | 741 (m.p.); |
| $CH_3$ | | | $-N{=}NC_6H_5$ | 175–177° | | 741 (m.p.); |
| $NH_2$ | $CH_3$ | $C_6H_5$ | | 200–210° | | 222 (m.p.), (u); |
| | | | | | picrate, m.p. 260° | 292 (m.p.), (u); |
| | $CH_3$ | $C_6H_5$ | $NH_2$ | 190–195° | | 221 (m.p.); 222 (m.p.), (n); |
| | | | | | | 292 (m.p.), (u); |
| | | | | | *N'*-acetyl m.p. 110–112° | 221 (m.p.); |
| $NHCH_3$ | $CH_3$ | $C_6H_5$ | | 200° | | 222 (m.p.), (n), (u); |
| | $CH_3$ | $C_6H_5$ | $NHCH_3$ | 155–156° | | 225 (m.p.), (u); |
| $-N{=}NC_6H_4NO_2$-*p* | $CH_3$ | $C_6H_5$ | | 230–235° | | 292 (m.p.); |
| | $CH_3$ | $C_6H_5$ | $-N{=}NC_6H_4NO_2$-*p* | 261–269° | | 292 (m.p.); |
| $NHCOC_6H_5$ | $CH_3$ | $C_6H_5$ | | | picrate, m.p. 213° | 292 (m.p.); |
| | $CH_3$ | $C_6H_5$ | $NHCOC_6H_5$ | 216–217° | | 221 (m.p.), (u); 292 (m.p.); |
| | $CH_3$ | $C_6H_5$ | $-N(CH_3)COC_6H_5$ | 150–151° | | 225 (m.p.), (i), (u); |

TABLE XII-52. 3-Pyridinol Carboxylic Acids and Derivatives

| $R_2$ | $R_4$ | $R_5$ | $R_6$ | m.p. | Derivatives | Ref. |
|---|---|---|---|---|---|---|
| CHO | | | | 79–80° | | 513 (m.p.), (i);<br>524 (n);<br>1077 (i); |
| | CHO | | | 133–134° | | 513 (m.p.), (i);<br>524 (n); 1077 (i); |
| | | | CHO | | | 1078 (u); |
| | COOH | | | | ethyl ester,<br>m.p. 43–45° | 595 (m.p.), (n); |
| | | | COOH | | | 176 (i); |
| COOH | | | $CH_3$ | 235° | | 1047 (m.p.); |
| | CHO | $CH_2OH$ | | | | 204 (u); 1155 (u); |
| | | | | | hydrochloride,<br>m.p. 144–147° | 204 (m.p.), (f); |
| $CH_3$ | COOH | | | | ethyl ester,<br>m.p. 51–54° | 209 (m.p.), (i); |
| $CH_3$ | | COOH | | | ethyl ester,<br>m.p. 201–202° | 199 (m.p.), (n), (u);<br>211 (m.p.), (f), (i);<br>1079 (m.p.); |
| | | | | | amide,<br>m.p. 305° | 211 (m.p.); |
| | | | | | hydrazide,<br>m.p. >330° | 211 (m.p.); |
| $CH_3$ | | $COOC_2H_5$ | | | *O*-acetyl,<br>m.p. 201–202° | 199 (m.p.), (n); |

| | | | | | | |
|---|---|---|---|---|---|---|
| | CHO | $CH_2OPO_3H_2$ | | | hydrate | 204 (i); |
| $CH_3$ | CHO | CHO | | 150–160° | | 203 (m.p.), (n), (u); |
| | | | | | $HCl \cdot H_2O$, m.p. 158–161° | 203 (m.p.); 1066 (m.p.), (u); |
| $CH_3$ | $CH_2OH$ | CHO | | | lactone, m.p. 275–277° | 1059 (m.p.); |
| $CH_3$ | COOH | COOH | | 270–272° or 252–254° | | 199 (m.p.), (i), (u); 1064 (m.p.); |
| | | | | | dimethyl ester, m.p. 140–141° | 205 (m.p.); |
| | | | | | diethyl ester, b.p. 130°/ 0.5 mm | 200 (b.p.); 205 (b.p.); 1057 (b.p.); |
| | | | | | diethyl ester·HCl m.p. 143–144° | 199 (m.p.); |
| | | | | | | 200 (m.p.); 211 (m.p.), (f); 1057 (m.p.); 1058 (m.p.); 1064 (m.p.); |
| | COOH | COOH | $CH_3$ | 239° | | 197 (m.p.), (u); |
| $-CH{=}NCH_2COOH$ | | | | | potassium salt | 1077 (i); |
| | $-CH{=}NCH_2COOH$ | | | | potassium salt | 1077 (i); |
| $CH_3$ | CHO | $CH_3$ | | | | 1088 (m), (n); |
| $CH_3$ | CHO | $CH_2OH$ | | | | 204 (u); 524 (n); 527; 1081 (n); |
| | | | | | hydrochloride | 1045 (i); |
| | CHO | $CH_2OH$ | $CH_3$ | | | 204 (u); |
| | | | | | hydrochloride, m.p. > 170° | 204 (m.p.), (f); |
| $CH_3$ | $CH_3$ | COOH | | | ethyl ester, m.p. 146–148° | 212 (m.p.), (i); |
| $CH_3$ | COOH | $CH_2OH$ | | 256° | | 1053 (m.p.); |
| | | | | | cation | 525 (h), (u); |
| $CH_3$ | CHO | $CH_2OPO_3H_2$ | | | | 527; 1045 (i); 1061 (u); 1082 (n); |

TABLE XII-52. 3-Pyridinol Carboxylic Acids and Derivatives (Continued)

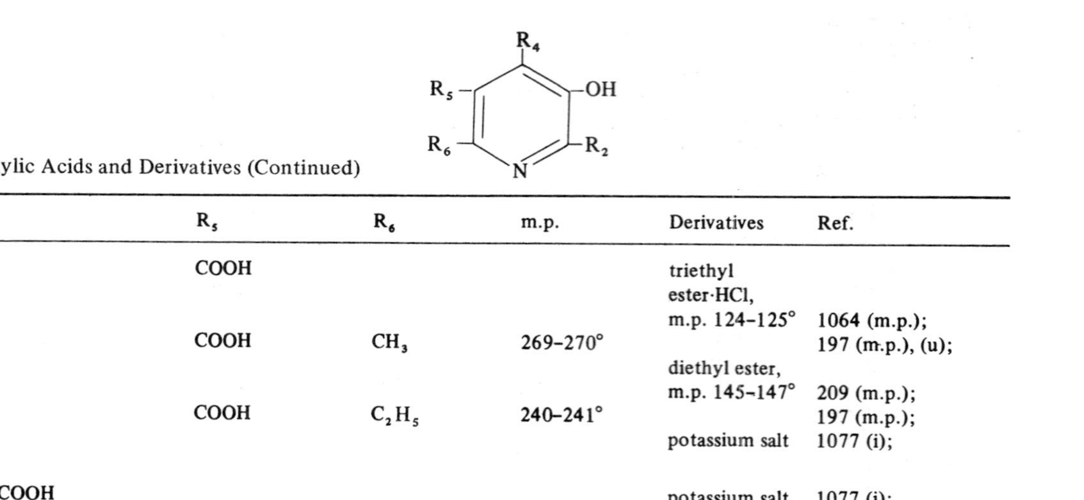


| $R_2$ | $R_4$ | $R_5$ | $R_6$ | m.p. | Derivatives | Ref. |
|---|---|---|---|---|---|---|
| $CH_2COOH$ | COOH | COOH | | | triethyl ester·HCl, m.p. 124–125° | 1064 (m.p.); |
| $CH_3$ | COOH | COOH | $CH_3$ | 269–270° | diethyl ester, m.p. 145–147° | 197 (m.p.), (u); 209 (m.p.); |
| | COOH | COOH | $C_2H_5$ | 240–241° | | 197 (m.p.); |
| $-CH{=}NCH(CH_3)COOH$ | | | | | potassium salt | 1077 (i); |
| | $-CH{=}NCH(CH_3)COOH$ | | | | potassium salt | 1077 (i); |
| $CH_3$ | CHO | $CH_2OH$ | $CH_3$ | >170° | | 204 (m.p.), (f); |
| $CH_3$ | CHO | $-CH(OH)CH_3$ | | | hydrochloride, m.p. 106–107° | 204 (m.p.); |
| $CH_3$ | CHO | $-CH_2OPO_3H_2$ | $CH_3$ | | | 204 (f); |
| $CH_3$ | CHO | $CH_2CH_2COOH$ | | 191–192° | oxime, m.p. 211–212° | 1080 (m.p.), (i), (n), (t); 1080 (m.p.); |
| | COOH | COOH | $CH_2CH_2CH_3$ | 216–217° | | 197 (m.p.); |
| $CH_3$ | COOH | COOH | $CH_2CH_3$ | 233–235° | | 197 (m.p.); |
| $isoC_3H_7$ | CHO | $CH_2OH$ | | | hydrochloride, m.p. 119–121° | 204 (m.p.), (f); |
| $CH_3$ | CHO | $-CH(OCH_3)_2$ | | 58–59° | | 1066 (m.p.), (u); |

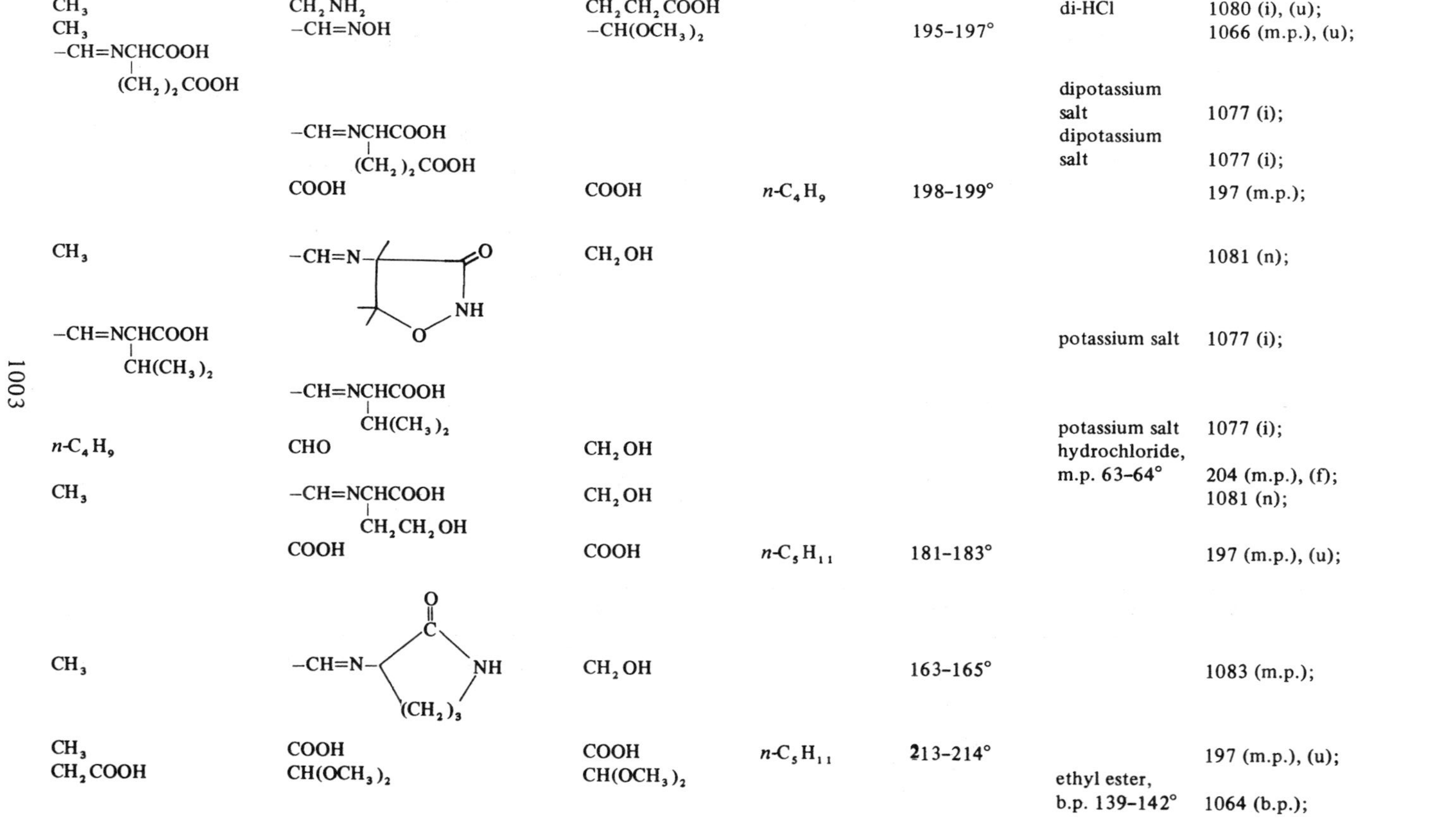

| | | | | | | |
|---|---|---|---|---|---|---|
| $CH_3$ | $CH_2NH_2$ | $CH_2CH_2COOH$ | | | di-HCl | 1080 (i), (u); |
| $CH_3$ | $-CH{=}NOH$ | $-CH(OCH_3)_2$ | | 195–197° | | 1066 (m.p.), (u); |
| $-CH{=}NCHCOOH$ with $(CH_2)_2COOH$ | | | | | dipotassium salt | 1077 (i); |
| | $-CH{=}NCHCOOH$ with $(CH_2)_2COOH$ | | | | dipotassium salt | 1077 (i); |
| | COOH | COOH | $n$-$C_4H_9$ | 198–199° | | 197 (m.p.); |
| $CH_3$ | $-CH{=}N-$ [ring structure] | $CH_2OH$ | | | | 1081 (n); |
| $-CH{=}NCHCOOH$ with $CH(CH_3)_2$ | | | | | potassium salt | 1077 (i); |
| | $-CH{=}NCHCOOH$ with $CH(CH_3)_2$ | | | | potassium salt | 1077 (i); |
| $n$-$C_4H_9$ | CHO | $CH_2OH$ | | | hydrochloride, m.p. 63–64° | 204 (m.p.), (f); |
| $CH_3$ | $-CH{=}NCHCOOH$ with $CH_2CH_2OH$ | $CH_2OH$ | | | | 1081 (n); |
| | COOH | COOH | $n$-$C_5H_{11}$ | 181–183° | | 197 (m.p.), (u); |
| $CH_3$ | $-CH{=}N-$ [ring structure] | $CH_2OH$ | | 163–165° | | 1083 (m.p.); |
| $CH_3$ | COOH | COOH | $n$-$C_5H_{11}$ | 213–214° | | 197 (m.p.), (u); |
| $CH_2COOH$ | $CH(OCH_3)_2$ | $CH(OCH_3)_2$ | | | ethyl ester, b.p. 139–142° | 1064 (b.p.); |

TABLE XII-52. 3-Pyridinol Carboxylic Acids and Derivatives (Continued)

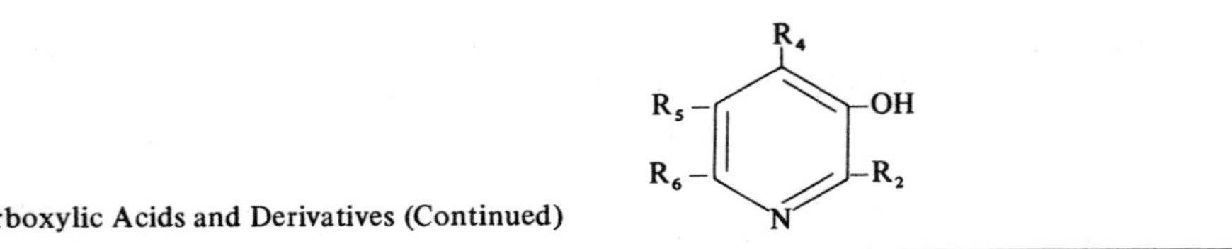


| $R_2$ | $R_4$ | $R_5$ | $R_6$ | m.p. | Derivatives | Ref. |
|---|---|---|---|---|---|---|
| $CH_3$ | O<br>‖<br>C<br>/ \ <br>–CH=N– NH<br>\ /<br>$(CH_2)_4$ | $CH_2OH$ | | 211–213° | | 1083 (m.p.); |
| $CH_3$ | –CONCO–<br>\|<br>$C_6H_5$ | | | 229–230° | | 209 (m.p.), (i); |
| $CH_3$ | H<br>\|<br>N<br>COOH<br>N NH | $CH_2OH$ | | | | 1081 (n); |
| $CH_3$ | $–CH_2NH(CH_2)_4CHCOOH$<br>\|<br>$NH_2$ | $CH_2OH$ | | 214–215° | | 1084 (m.p.); |
| –CH=NCHCOOH<br>\|<br>$CH_2C_6H_5$ | | | | | potassium salt<br>$\cdot H_2O$ | 1077 (i); |
| | –CH=NCHCOOH<br>\|<br>$CH_2C_6H_5$ | | | | potassium salt<br>$\cdot H_2O$ | 1077 (i); |

TABLE XII-53. 3-Pyridinol Nitriles

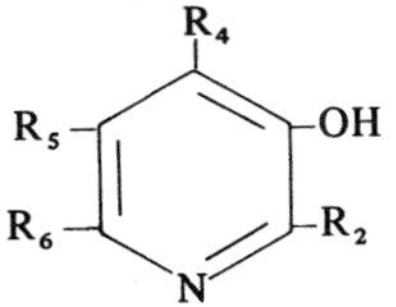

| $R_2$ | $R_4$ | $R_5$ | $R_6$ | m.p. | Derivatives | Ref. |
|---|---|---|---|---|---|---|
| $CH_3$ | | CN | | 246–247° | | 209 (m.p.), (i); 211 (m.p.), (i), (f); 1079 (m.p.); |
| $CH_3$ | CN | CN | | 187–191° or 120–125° | | 199 (m.p.); 201 (m.p.), (u); 205 (m.p.); 1064 (m.p.), (i), (u); |
| | | | | | hydrate, m.p. 89° | 205 (m.p.); |
| $CH_3$ | | CN | $CH_3$ | 249–251° | | 209 (m.p.), (i); |
| $CH_3$ | CN | $CH_2OH$ | | | hydrochloride, m.p. 186–190° | 205 (m.p.); 207 (m.p.); |
| $CH_2COOH$ | $-C(=NH)OC_2H_5$ | CN | | | ethyl ester, m.p. 164–166° | 1064 (m.p.); |

TABLE XII-54. Halo 3-Pyridinols

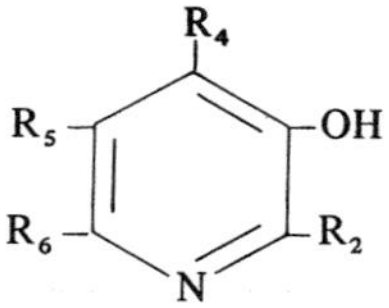

| $R_2$ | $R_4$ | $R_5$ | $R_6$ | m.p. | Derivatives | Ref. |
|---|---|---|---|---|---|---|
| Cl | | | | 169–170° | | 418 (m.p.); |
| | | Cl | | 158° | | 317 (m.p.); |
| Br | | | | 185–186° | | 583 (n), (u); 708 (m.p.); |
| | | Br | | 165–166° | | 319 (m.p.); |
| Br | | | Br | 168–169° | | 708 (m.p.); |
| I | | | I | 198–199° | | 298 (m.p.); |
| | I | | I | 230° | | 298 (m.p.), (n); |
| I | I | | I | 153–154° | | 599 (m.p.); |
| Br | Br | | $CH_3$ | 107–109° | | 583 (m.p.), (n), (u); |
| Br | | | $CH_3$ | 187–189° | | 583 (m.p.), (i), (n), (u); |
| I | | | $CH_3$ | 174° | | 583 (m.p.), (i), (n), (u); |
| $C_6H_4Cl$-*p* | | | I | 162–163° | | 709 (m.p.), (n); |
| | | | | | hydrochloride, m.p. 180–181° | 709 (m.p.); |

| | | | | |
|---|---|---|---|---|
| $C_6H_5$ | [illegible] | [illegible] | hydrochloride, m.p. 165–166° | [illegible]; 709 (m.p.); |
| $C_6H_4CH_3$-*p* | I | 167–168° | | 709 (m.p.), (n); |
| | | | hydrochloride, m.p. 175–176° | 709 (m.p.); |
| $C_6H_4OCH_3$-*p* | I | 171–172° | | 709 (m.p.); |
| | | | hydrochloride, m.p. 187–188° | 709 (m.p.); |
| $C_6H_4C_2H_5$-*p* | I | 147–148° | | 709 (m.p.), (n); |
| | | | hydrochloride, m.p. 168–169° | 709 (m.p.); |
| $C_6H_4CH(CH_3)_2$-*p* | I | 144–145° | | 709 (m.p.), (n); |
| | | | hydrochloride, m.p. 170–171° | 709 (m.p.); |
| $C_6H_4C(CH_3)_3$-*p* | I | 140–141° | | 709 (m.p.); |
| | | | hydrochloride, m.p. 185–186° | 709 (m.p.); |

TABLE XII-55. Nitro and Nitroso 3-Pyridinols

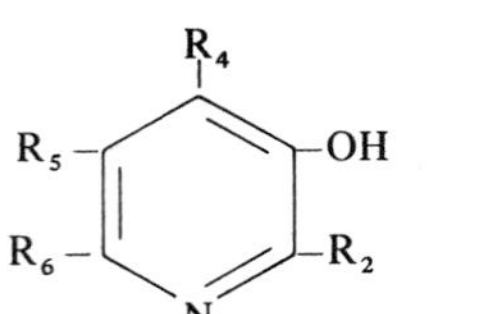


| $R_2$ | $R_4$ | $R_5$ | $R_6$ | m.p. | Derivatives | Ref. |
|---|---|---|---|---|---|---|
| $NO_2$ | | | | 67–69° | | 515 (n); 647 (m.p.), (n), (u); 648 (u); |
| | | | | | *O*-acetyl, m.p. 50–51° | 647 (m.p.); |
| | | | | | b.p. 101°/0.03 mm | 647 (b.p.); |
| | | | | | m.p. 76–77° | 1085 (m.p.), (b.p.); |
| | | | | | ammonium salt, m.p. 142–145° | 1086 (m.p.); |
| | | | | | magnesium salt, m.p. 330° | 1086 (m.p.); |
| | | | | | sodium salt, m.p. 300° | 1086 (m.p.); |
| | $NO_2$ | | | 132° | | 256a (m.p.); 261 (m.p.); |
| | | | $NO_2$ | 213–215° | | 647 (m.p.), (c), (n), (w); |
| | | | | | *O*-acetyl, m.p. 111–112° | 647 (m.p.); |
| Cl | $NO_2$ | | | 70–71° | | 647 (m.p.), (n), (u); |
| Cl | | | $NO_2$ | 197° | | 647 (m.p.), (n), (u); |
| $NO_2$ | | | $NO_2$ | | | 647 (n), (u); |
| $CH_3$ | $NO_2$ | | $NO_2$ | 90–91° | | 719 (m.p.), (n); |
| $NO_2$ | | $CH_3$ | | 133–135° | | 647 (m.p.), (n), (u); |
| $NO_2$ | | | $CH_3$ | 175–176° | | 719 (m.p.), (n); |

| | | | | | | |
|---|---|---|---|---|---|---|
| | | | | | | 714 (m.p.), (n); |
| | | | | | | 718 (m.p.), (n); |
| $CH_3$ | | | $NO_2$ | 262–263° | | 515 (n); 647 (m.p.), (n), (u); |
| | | | | | | 714 (m.p.); 718 (m.p.); 719 (m.p.), (n); |
| $CH_3$ | $CH_3$ | $NO_2$ | | 155° | | 379 (m.p.); |
| $CH_3$ | $NO_2$ | | $CH_3$ | 137–138° | | 295 (m.p.), (i); 719 (m.p.), (n); |
| $CH_3$ | | $NO_2$ | $CH_3$ | 208° | | 379 (m.p.); |
| | $CH_3$ | $NO_2$ | $CH_3$ | 200° | | 379 (m.p.), (i); |
| $n$-$C_3H_7$ | | | $NO_2$ | | ammonium salt, m.p. 116–117° | 719 (m.p.); |
| iso$C_3H_7$ | | | $NO_2$ | | ammonium salt, m.p. 164–166° | 719 (m.p.); |
| $CH_3$ | $CH_3$ | $NO_2$ | $CH_3$ | 170° | | 379 (m.p.), (i); |
| $C_2H_5$ | $NO_2$ | | $CH_3$ | 129–130° | | 719 (m.p.), (n); |
| $n$-$C_3H_7$ | $NO_2$ | | | 86–87° | | 714 (m.p.), (n); 718 (m.p.), (n); |
| | | | | | | 719 (m.p.); 726 (m.p.); |
| | | | | | ammonium salt, m.p. 116–117° | 719 (m.p.); |
| iso$C_3H_7$ | $NO_2$ | | | 66–67° | | 714 (m.p.), (n); 718 (m.p.), (n); |
| | | | | | | 719 (m.p.); 726 (m.p.); |
| | | | | | ammonium salt, m.p. 116–117° | 719 (m.p.); |
| $n$-$C_3H_7$ | | | $NO_2$ | 162–163° | | 714 (m.p.), (n); 718 (m.p.); |
| iso$C_3H_7$ | | | $NO_2$ | 160–162° | | 714 (m.p.), (n); 718 (m.p.); |
| $NO_2$ | | | $CH_2N(CH_3)_2$ | | ammonium salt, m.p. 181–182° | 725 (m.p.), (n); |
| $NO_2$ | | | $CH_2CH_2N(CH_3)_2$ | 88–89° | | 726 (m.p.); |
| $CH_2CH_2N(CH_3)_2$ | $NO_2$ | | | 143–144° | | 726 (m.p.), (n); |
| $n$-$C_4H_9$ | $NO_2$ | | $CH_3$ | 39–40° | | 719 (m.p.), (n); |
| iso$C_4H_9$ | $NO_2$ | | $CH_3$ | 63–65° | | 719 (m.p.), (n); |
| iso$C_5H_{11}$ | $NO_2$ | | | 72–73° | | 714 (m.p.), (n); 718 (m.p.) (n); |
| | | | | | | 719 (m.p.); |
| iso$C_5H_{11}$ | | | $NO_2$ | 200–202° | | 714 (m.p.), (n); 718 (m.p.); |

TABLE XII-55. Nitro and Nitroso 3-Pyridinols (Continued)

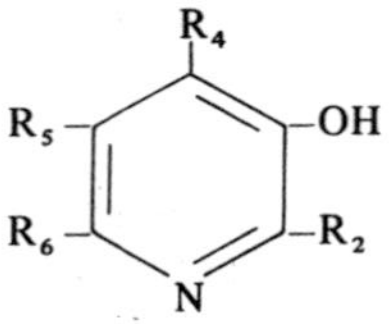

| $R_2$ | $R_4$ | $R_5$ | $R_6$ | m.p. | Derivatives | Ref. |
|---|---|---|---|---|---|---|
| $NO_2$ | | | $CH_2N(C_2H_5)_2$ | 128–129° | | 725 (m.p.), (n); |
| $C_6H_4NO_2$-*p* | | | $NO_2$ | 160–161° | | 710 (m.p.), (n); |
| NO | | | | | | |
| $NO_2$ | | | $CH_2N$(ring) 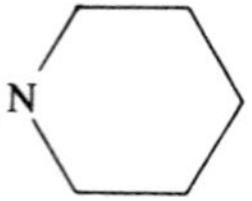 | | ammonium salt, m.p. 172–173° | 725 (m.p.), (n); |
| $CH_2CH_2N(C_2H_5)_2$ | $NO_2$ | | | 113–114° | | 726 (m.p.), (n); |
| 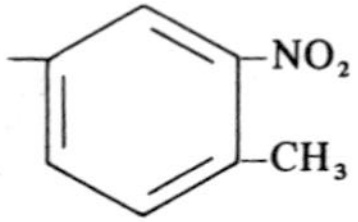 (ring)–$NO_2$, –$CH_3$ | | | $NO_2$ | 159–160° | | 711 (m.p.), (n); |
| NO | $CH_3$ | $C_6H_5$ | $NH_2$ | >280° | | 292 (m.p.); |
| $NO_2$ | | | $-CH_2CH_2N$(ring)  | 160° | | 726 (m.p.), (n); |
| $-CH_2CH_2N$(ring) 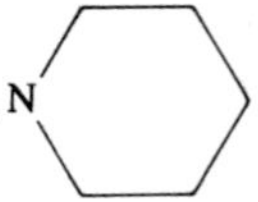 | $NO_2$ | | | 210° | dihydrochloride m.p. 125–127° | 726 (m.p.), (n); 726 (m.p.); 726 (n) |
| $-CH(CH_2N(CH_3)_2)_2$ | $NO_2$ | | | | | |

TABLE XII-56. Sulfur containing 3-Pyridinols

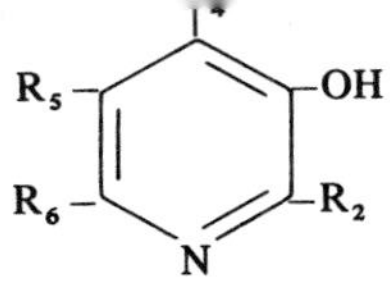

(Structure: pyridine ring bearing $R_5$, OH, $R_2$, $R_6$, N, with $R_4$ at the 4-position)

| $R_2$ | $R_4$ | $R_5$ | $R_6$ | m.p. | Derivatives | Ref. |
|---|---|---|---|---|---|---|
| SH | | | | 144–145° | | 583 (m.p.), (n), (u); |
| $SO_3H$ | | | | 300–302° | | 317 (m.p.); 320 (m.p.); |
| 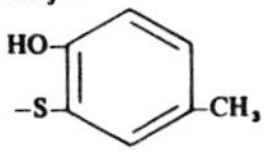 –S– (benzene ring with HO–, –$CH_3$) | | | $CH_3$ | 200–206° | | 416 (m.p.); |
| 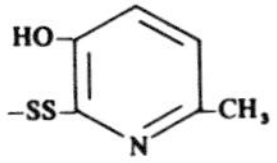 –SS– (pyridine ring with HO–, N, –$CH_3$) | | | $CH_3$ | 209–210° | | 583 (m.p.), (n), (u); |
| SH | | | $CH_3$ | 174–177° | | 583 (m.p.), (n), (u); |
| $CH_3$ | | $CSNH_2$ | | 259–260° | | 211 (m.p.); |
| $–CH{=}NNHCSNH_2$ | | | | | | 1078 (u); |
| | | | $–CH{=}NNHCSNH_2$ | | | 1078 (u); |
| $SC_2H_5$ | | | | 88–90° | | 416 (m.p.); |
| | $CH_2SH$ | $CH_3$ | | | hydrochloride, m.p. 194° | 1087 (m.p.); |
| $SCH_3$ | | | $CH_3$ | 127–129° | | 597 (m.p.), (u); |
| $CH_3$ | CHO | 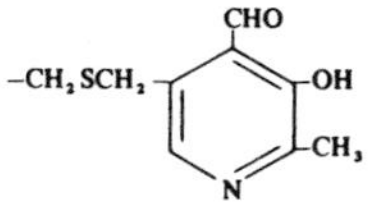 $–CH_2SCH_2–$ (pyridine ring with CHO, OH, $CH_3$, N) | | 169° | | 1060 (m.p.); |
| $CH_3$ | $–CH_2Cl$ | 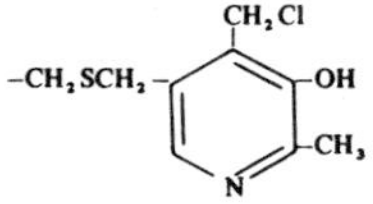 $–CH_2SCH_2–$ (pyridine ring with $CH_2Cl$, OH, $CH_3$, N) | | | di-HCl m.p. 205° | 1088 (m.p.); |
| $–SCH_2COOH$ | | | $CH_3$ | | hydrobromide, m.p. 168–170° | 597 (m.p.), (u); |

TABLE XII-56. Sulfur containing 3-Pyridinols (Continued)

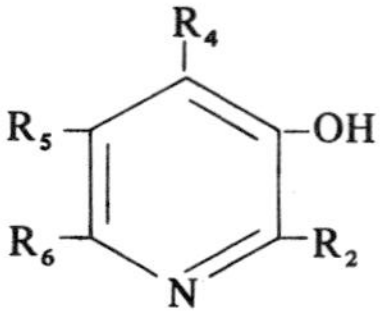

(Structure: pyridine ring with $R_4$, $R_5$, $R_6$, $R_2$ substituents and –OH at position 3)

| $R_2$ | $R_4$ | $R_5$ | $R_6$ | m.p. | Derivatives | Ref. |
|---|---|---|---|---|---|---|
| $CH_3$ | $CH_2OCOCH_3$ | 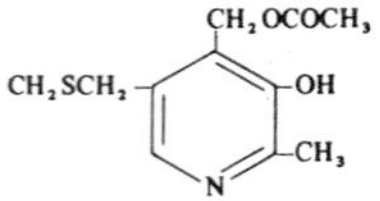 $CH_2SCH_2$– (pyridine ring with $CH_2OCOCH_3$, –OH, $CH_3$, N) | | | di-HCl, m.p. 196° | 1088 (m.p.); |
| $CH_3$ | $CH_2OH$ | 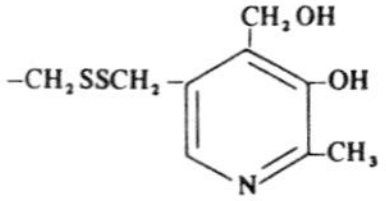 $-CH_2SSCH_2$– (pyridine ring with $CH_2OH$, –OH, $CH_3$, N) | | 222° | | 1089 (m.p.);<br>1090 (m.p.), (f), (i), (u); |
| $CH_3$ | 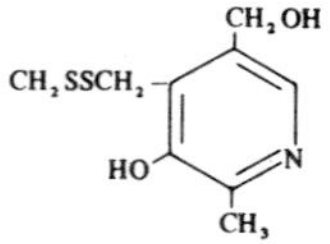 $CH_2SSCH_2$– (pyridine ring with $CH_2OH$, HO, $CH_3$, N) | $CH_2OH$ | | 207–208° | | 1091 (m.p.), (f), (u); |
| $CH_3$ | $CH_2OH$ | $CH_2SSO_3H$ | | | sodium salt·$3H_2O$ m.p. 192° | 1090 (m.p.), (f), (i); |
| $CH_3$ | $-CH{=}NOH$ | $CH_2SO_4H$ | | 207–209° | | 1092 (m.p.), (u); |
| $SC_2H_5$ | | | $CH_3$ | 47–49° | | 597 (m.p.), (u); |
| $CH_3$ | $CH_2SH$ | $CH_3$ | | | hydroiodide, m.p. 126–128° | 597 (m.p.); |
| $CH_3$ | $CH_2SH$ | $CH_2SH$ | | | hydrochloride, m.p. 194° | 1060 (m.p.);<br>1051 (e); |
| $SCH_2CH_2OH$ | | | $CH_3$ | 102–104° | | 583 (m.p.); |
| $CH_3$ | $CH_2SH$ | $CH_2OH$ | | | hydrochloride, m.p. 135–141°<br>hydrochloride, m.p. 174–180°<br>3,4,5-triacetate, | 583 (m.p.);<br>1051 (e); 1055 (e);<br>1063 (m.p.); 1087 (m.p.);<br>1091 (m.p.); |

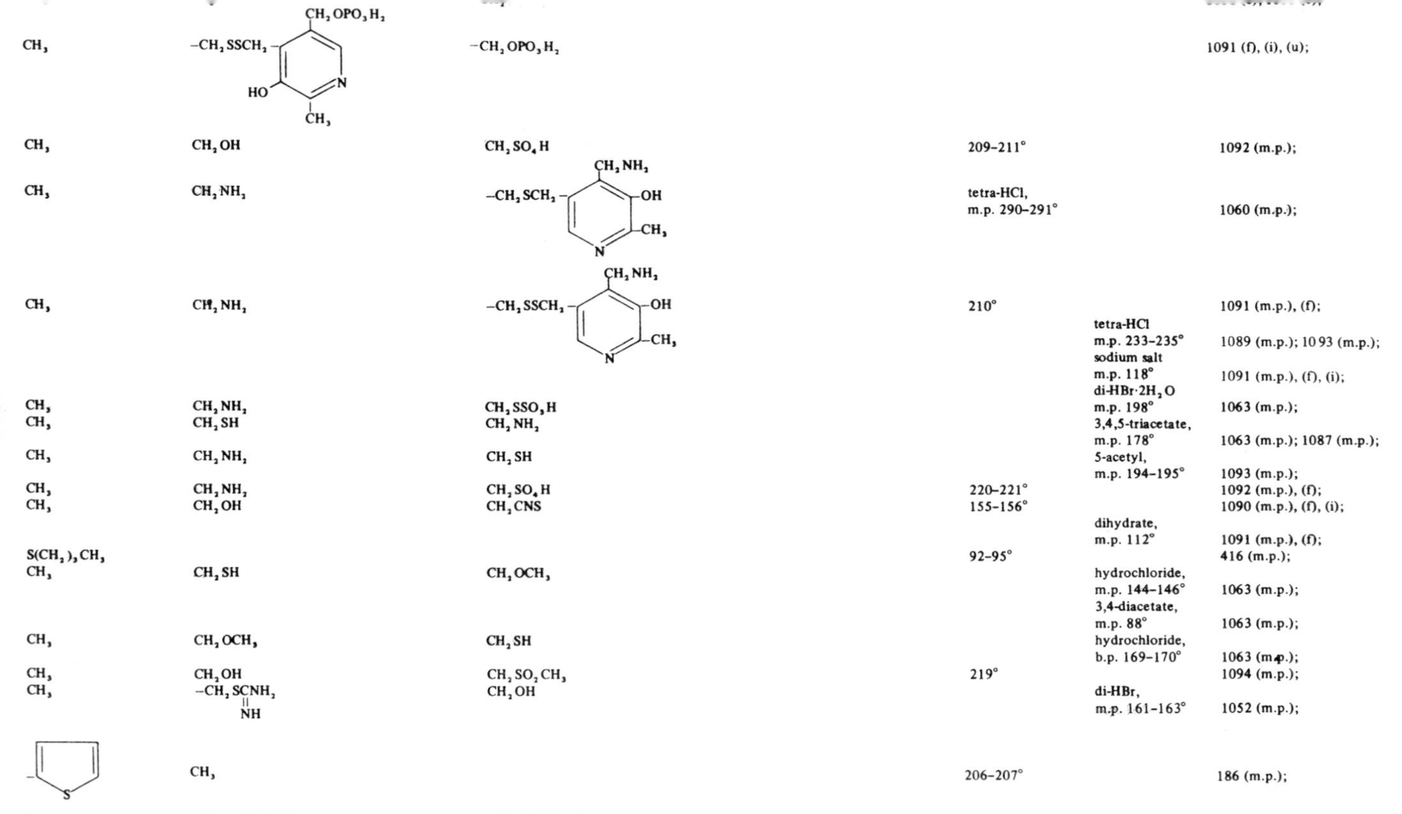

| | | | | | |
|---|---|---|---|---|---|
| $CH_3$ | $-CH_2SSCH_2-$ (pyridyl: $CH_2OPO_3H_2$, HO, $CH_3$) | $-CH_2OPO_3H_2$ | | | 1091 (f), (i), (u); |
| $CH_3$ | $CH_2OH$ | $CH_2SO_4H$ | 209–211° | | 1092 (m.p.); |
| $CH_3$ | $CH_2NH_2$ | $-CH_2SCH_2-$ (pyridyl: $CH_2NH_2$, OH, $CH_3$) | tetra-HCl, m.p. 290–291° | | 1060 (m.p.); |
| $CH_3$ | $CH_2NH_2$ | $-CH_2SSCH_2-$ (pyridyl: $CH_2NH_2$, OH, $CH_3$) | 210° | | 1091 (m.p.), (f); |
| | | | | tetra-HCl m.p. 233–235° | 1089 (m.p.); 1093 (m.p.); |
| | | | | sodium salt m.p. 118° | 1091 (m.p.), (f), (i); |
| $CH_3$ | $CH_2NH_2$ | $CH_2SSO_3H$ | | di-HBr·$2H_2O$ m.p. 198° | 1063 (m.p.); |
| $CH_3$ | $CH_2SH$ | $CH_2NH_2$ | | 3,4,5-triacetate, m.p. 178° | 1063 (m.p.); 1087 (m.p.); |
| $CH_3$ | $CH_2NH_2$ | $CH_2SH$ | | 5-acetyl, m.p. 194–195° | 1093 (m.p.); |
| $CH_3$ | $CH_2NH_2$ | $CH_2SO_4H$ | 220–221° | | 1092 (m.p.), (f); |
| $CH_3$ | $CH_2OH$ | $CH_2CNS$ | 155–156° | | 1090 (m.p.), (f), (i); |
| | | | | dihydrate, m.p. 112° | 1091 (m.p.), (f); |
| $S(CH_2)_3CH_3$ | | | 92–95° | | 416 (m.p.); |
| $CH_3$ | $CH_2SH$ | $CH_2OCH_3$ | | hydrochloride, m.p. 144–146° | 1063 (m.p.); |
| | | | | 3,4-diacetate, m.p. 88° | 1063 (m.p.); |
| $CH_3$ | $CH_2OCH_3$ | $CH_2SH$ | | hydrochloride, b.p. 169–170° | 1063 (m.p.); |
| $CH_3$ | $CH_2OH$ | $CH_2SO_2CH_3$ | 219° | | 1094 (m.p.); |
| $CH_3$ | $-CH_2SC(=NH)NH_2$ | $CH_2OH$ | | di-HBr, m.p. 161–163° | 1052 (m.p.); |
| 2-thienyl | $CH_3$ | | 206–207° | | 186 (m.p.); |
| $CH_3$ | $-CH=NNHCSNH_2$ | $-CH=NNHCSNH_2$ | 172–174° | | 1066 (m.p.); |

TABLE XII-56. Sulfur containing 3-Pyridinols (Continued)

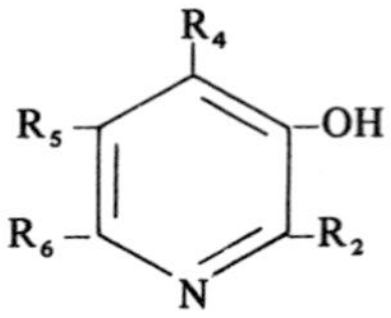

[structure: 3-pyridinol ring with substituents $R_2$, $R_4$, $R_5$, $R_6$, OH, N]

| $R_2$ | $R_4$ | $R_5$ | $R_6$ | m.p. | Derivatives | Ref. |
|---|---|---|---|---|---|---|
| $CH_3$ | $CH_3$ | $CH_2SCH_2CH_2Cl$ | | | hydrochloride, m.p. 150–152° | 1070 (m.p.); |
| $CH_3$ | $CH_3$ | $CH_2SCH_2CH_2OH$ | | 140–142° | | 1070 (m.p.); |
| $CH_3$ | $CH_2OH$ | $CH_2SCH_2CH_2OH$ | | | hydrochloride, m.p. 180–182° | 1070 (m.p.); |
| | | | | | hydrochloride, m.p. 146–148° | 1070 (m.p.); |
| $CH_3$ | $CH_2N(CH_3)_2$ | $CH_2SCH_2$– 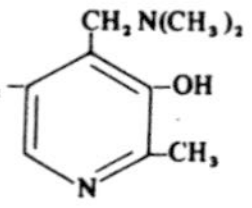 [structure] | | 150° | tetra-HCl·$4H_2O$, m.p. 140° | 1088 (m.p.); |
| $CH_3$ | 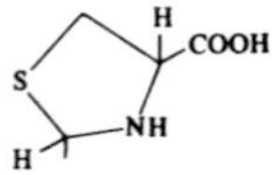 [structure] | $CH_2OH$ | | | | 1081 (n); |
| $CH_3$ | $CH_3$ | $-CH_2SCH(CH_3)CH_2Cl$ | | | hydrochloride, m.p. 128–132° | 1070 (m.p.); |
| $CH_3$ | $-CH_2NH_2$ | $-CH_2(C_2H_4OCS_2)$ | | | di-HBr m.p. > 230° | 1093 (m.p.); |
| $S(CH_2)_4CH_3$ | | | $CH_3$ | 34° b.p. 120°/1.5 mm | | 597 (m.p.), (b.p.), (u); |
| $CH_3$ | $CH_3$ | $-CH_2SCH(CH_3)CH_2OH$ | | 143–146° | | 1070 (m.p.); |
| $CH_3$ | | | $-N{=}NC_6H_4SO_3H$ | none | | 741; |
| $CH_3$ | 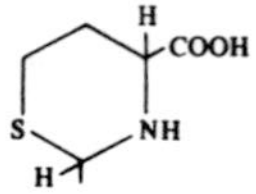 [structure] | $CH_2OH$ | | | | 1081 (n); |
| $CH_3$ | $-CH_2N$  [structure] | 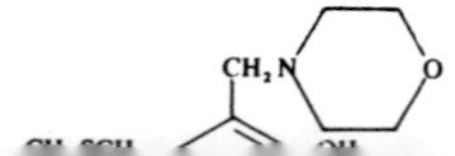 [structure] | | 200° | tetra-HCl m.p. 260° | 1060 (m.p.); 1088 (m.p.); |

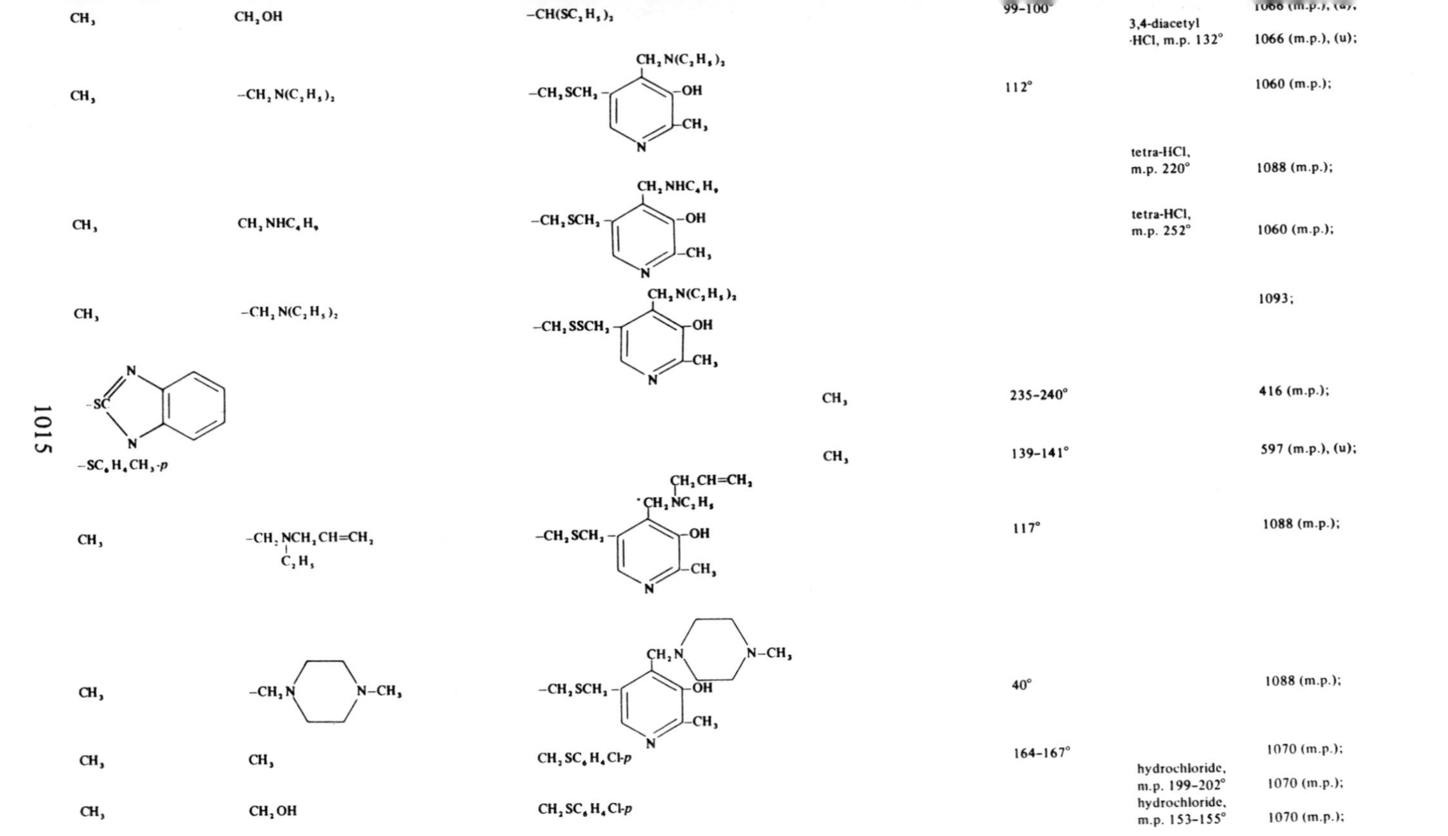

| | | | | | | |
|---|---|---|---|---|---|---|
| $CH_3$ | $CH_2OH$ | $-CH(SC_2H_5)_2$ | | 99–100° | | 1066 (m.p.), (u); |
| | | | | | 3,4-diacetyl ·HCl, m.p. 132° | 1066 (m.p.), (u); |
| $CH_3$ | $-CH_2N(C_2H_5)_2$ | $-CH_2SCH_2-$ (ring: $CH_2N(C_2H_5)_2$, OH, $CH_3$, N) | | 112° | | 1060 (m.p.); |
| | | | | | tetra-HCl, m.p. 220° | 1088 (m.p.); |
| $CH_3$ | $CH_2NHC_4H_9$ | $-CH_2SCH_2-$ (ring: $CH_2NHC_4H_9$, OH, $CH_3$, N) | | | tetra-HCl, m.p. 252° | 1060 (m.p.); |
| $CH_3$ | $-CH_2N(C_2H_5)_2$ | $-CH_2SSCH_2-$ (ring: $CH_2N(C_2H_5)_2$, OH, $CH_3$, N) | | | | 1093; |
| $-SC$ (ring: N, N) | | | $CH_3$ | 235–240° | | 416 (m.p.); |
| $-SC_6H_4CH_3$-*p* | | | $CH_3$ | 139–141° | | 597 (m.p.), (u); |
| $CH_3$ | $-CH_2NCH_2CH{=}CH_2$ ($C_2H_5$) | $-CH_2SCH_2-$ (ring: $CH_2NC_2H_5$ ($CH_2CH{=}CH_2$), OH, $CH_3$, N) | | 117° | | 1088 (m.p.); |
| $CH_3$ | $-CH_2N$ (ring) $N-CH_3$ | $-CH_2SCH_2-$ (ring: $CH_2N$ (ring) $N-CH_3$, OH, $CH_3$, N) | | 40° | | 1088 (m.p.); |
| $CH_3$ | $CH_3$ | $CH_2SC_6H_4Cl$-*p* | | 164–167° | | 1070 (m.p.); |
| | | | | | hydrochloride, m.p. 199–202° | 1070 (m.p.); |
| $CH_3$ | $CH_2OH$ | $CH_2SC_6H_4Cl$-*p* | | | hydrochloride, m.p. 153–155° | 1070 (m.p.); |

TABLE XII-56. Sulfur containing 3-Pyridinols (Continued)

R₄, R₅, OH, R₆, R₂ substituted pyridine (N)

| $R_2$ | $R_4$ | $R_5$ | $R_6$ | m.p. | Derivatives | Ref. |
|---|---|---|---|---|---|---|
| $CH_3$ | $CH_2SH$ | $CH_2OCOC_6H_5$ | | 182–184° | <br>3,4-diacetyl, m.p. 142–143° | 1063 (m.p.);<br>1063 (m.p.); |
| $CH_3$ | COOH<br>$-CH{=}NHCH(CH_2)_2S$<br>$H_2NCH(CH_2)_2S$<br>COOH | $CH_2OH$ | | | | 1081 (n); |
| $CH_3$ | $CH_2N(C_4H_9)_2$ | $CH_2N(C_4H_9)_2$<br>$-CH_2SCH_2-$ (pyridine ring) $-OH$<br>$-CH_3$<br>N | | | tetra-HCl, m.p. 220° | 1088 (m.p.); |
| $CH_3$ | $CH_2NH_2$ | $H_2NCH_2$ OH<br>$-CH_2SSCH_2-$ (pyridine ring) $-CH_3$<br>N | | | 233–255° | 1095 (m.p.); |
| $CH_3$ | $CH_2CH_3$ | $-CH_2SSCH_2$<br>(pyridine ring) $CH_2CH_3$<br>N $-OH$<br>$CH_3$ | | 220–221° | | 1095 (m.p.); |
| $CH_3$ | $-CH_2SSCCH_2CH_2OH$<br>$CH_3\overset{\parallel}{C}NCH_2-$ (pyrimidine ring) N<br>CHO<br>$H_2N$ N $-CH_3$ | $-CH_2OH$ | | | di-HCl·1/2$H_2O$, m.p. 178° | 1096 (m.p.); |
| $CH_3$ | $CH_2OH$ | $H_2N$ N $CH_3$<br>$CH_3$<br>$-CH_2SSC{=}CN(CHO)CH_2-$ (pyrimidine ring) N | | | di-HCl<br>m.p. 187–188° | 1096 (m.p.), (i), (u); |

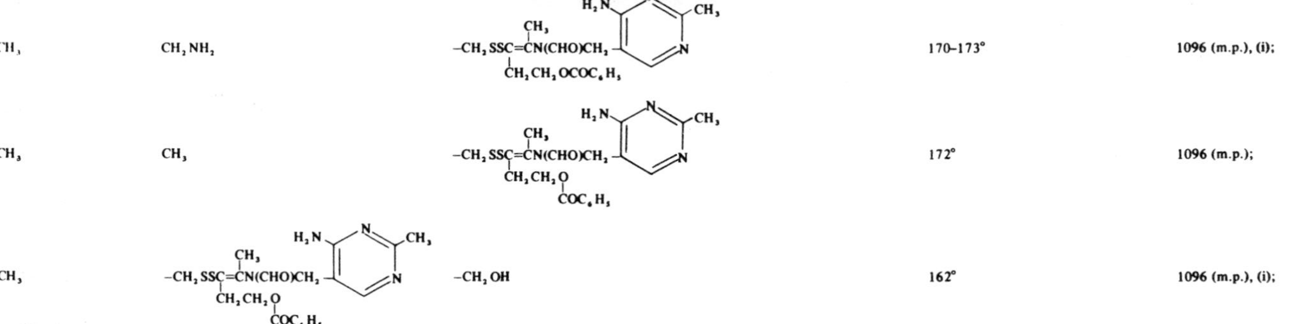

| | | | | |
|---|---|---|---|---|
| $CH_3$ | $CH_2NH_2$ | $-CH_2SSC(CH_3)=C(CH_2CH_2OCOC_6H_5)N(CHO)CH_2-$(4-amino-2-methylpyrimidin-5-yl) | 170–173° | 1096 (m.p.), (i); |
| $CH_3$ | $CH_3$ | $-CH_2SSC(CH_3)=C(CH_2CH_2OCOC_6H_5)N(CHO)CH_2-$(4-amino-2-methylpyrimidin-5-yl) | 172° | 1096 (m.p.); |
| $CH_3$ | $-CH_2SSC(CH_3)=C(CH_2CH_2OCOC_6H_5)N(CHO)CH_2-$(4-amino-2-methylpyrimidin-5-yl) | $-CH_2OH$ | 162° | 1096 (m.p.), (i); |

TABLE XII-57. Ethers and Esters of Alkyl and Aryl 3-Pyridinols

(Structure: pyridine ring with OR at position 3 and substituents $R_2$, $R_4$, $R_5$, $R_6$)

| R | $R_2$ | $R_4$ | $R_5$ | $R_6$ | m.p. | Derivatives | Ref. |
|---|---|---|---|---|---|---|---|
| $CH_3$ | | | | | b.p. 68°/12 mm | | 525 (h), (u); 594 (b.p.), (u); 964 (j), (w); 1042 (n); |
| | | | | | | hydrochloride, m.p. 160–161° | 594 (m.p.); |
| | | | | | | picrate, m.p. 136–139° | 594 (m.p.); |
| $CH_3$ | $CH_3$ | | | | | | 515 (n); |
| $-CH_2CH_2=CH_2$ | | | | | b.p. 90°/10 mm | | 595 (b.p.), (n); 598 (b.p.), (i); |
| | | | | | | picrate, m.p. 56–58° | 595 (m.p.); |
| $CH_3$ | $CH_3$ | $CH_2OH$ | $CH_2OH$ | | | | 1048 (n); |
| 2-Pyridyl | | | | | 30–31° | | 992 (m.p.); |
| | | | | | | picrate, m.p. 163–164° | 273 (m.p.); 992 (m.p.); |
| $COCH(CH_3)_2$ | | $CH_3$ | | | b.p. 97°/7 mm | | 387 (b.p.), (n); |
| | | | | | | picrate, m.p. 148–149° | 387 (m.p.); |
| $-(CH_2)_4CH_3$ | | | | | b.p. 116°/9 mm | | 596 (b.p.); |
| β-D-Glucosyl | | | | | 187–188° | | 603 (m.p.), (f); |
| | | | | | | tetraacetyl, m.p. 141–142° | 603 (m.p.); |
| $-COC_6H_5$ | | | | | 49–50° | | 342 (m.p.); |
| | | | | | b.p. 168°/12 mm | | 637 (b.p.), (m.p.); |
| | | | | | | picrate, m.p. 153–154° | 342 (m.p.); |
| | | | | | | $AgCl_2$ double salt, m.p. 175–176 | 342 (m.p.); |
| $CH_3$ | $-COC_6H_4Cl$-*p* | | | | 90–93° | | 191 (m.p.); |
| $CH_3$ | $-COC_6H_5$ | | | | b.p. 140°/1 mm | | 191 (b.p.); |
| | | | | | | hydrochloride, m.p. 148–151° | 191 (m.p.); 1071 (m.p.); |
| $CH_3$ | $-CH_2C_6H_4Cl$-*p* | | | | 74–76° | | 191 (m.p.); |
| $CH_3$ | $-CH_2C_6H_5$ | | | | b.p. 120°/3 mm | | 191 (b.p.), (m.p.); |
| | | | | | m.p. 46–47° | | 1071 (b.p.), (m.p.); |

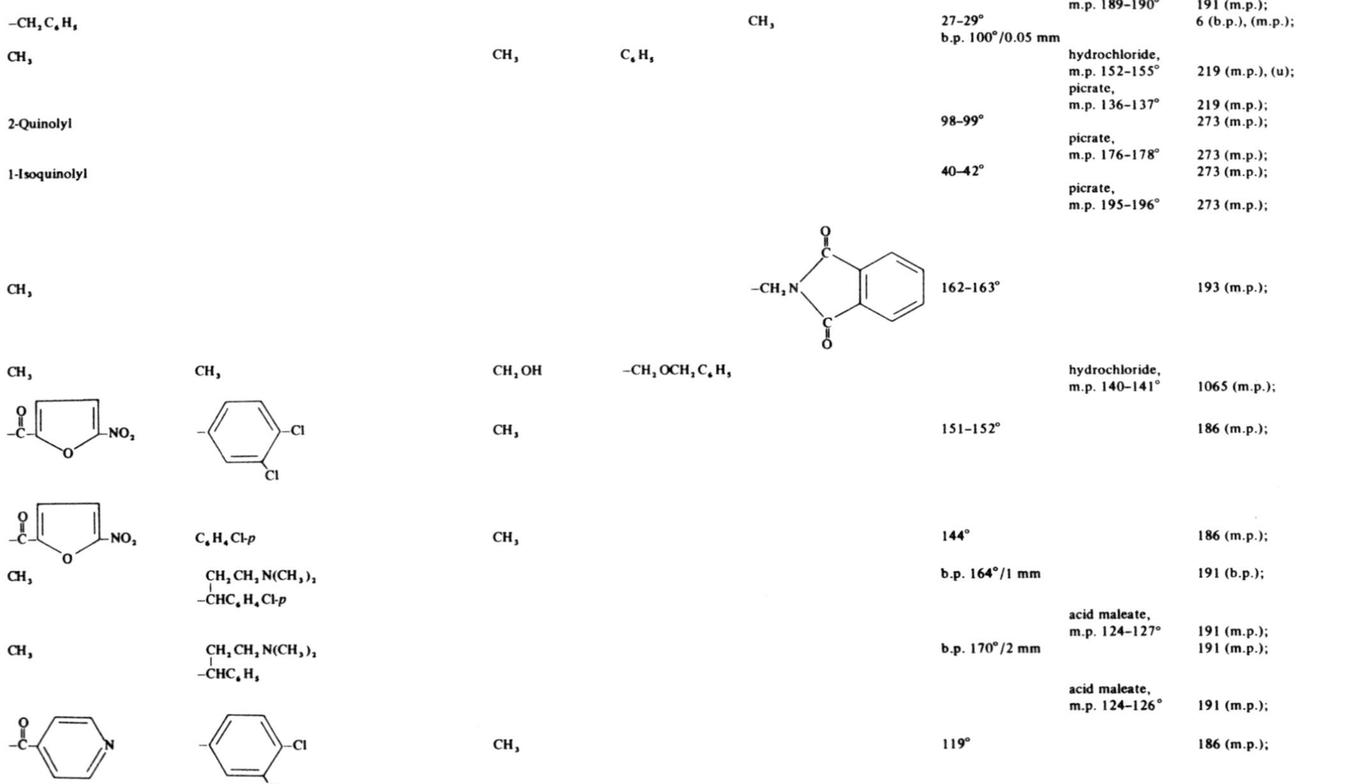

| | | | | | | | |
|---|---|---|---|---|---|---|---|
| | | | | | | hydrochloride, m.p. 189–190° | 191 (m.p.); |
| $-CH_2C_6H_5$ | | | | $CH_3$ | 27–29° b.p. 100°/0.05 mm | | 6 (b.p.), (m.p.); |
| $CH_3$ | | $CH_3$ | $C_6H_5$ | | | hydrochloride, m.p. 152–155° | 219 (m.p.), (u); |
| | | | | | | picrate, m.p. 136–137° | 219 (m.p.); |
| 2-Quinolyl | | | | | 98–99° | | 273 (m.p.); |
| | | | | | | picrate, m.p. 176–178° | 273 (m.p.); |
| 1-Isoquinolyl | | | | | 40–42° | | 273 (m.p.); |
| | | | | | | picrate, m.p. 195–196° | 273 (m.p.); |
| $CH_3$ | | | | | 162–163° | | 193 (m.p.); |
| $CH_3$ | $CH_3$ | $CH_2OH$ | $-CH_2OCH_2C_6H_5$ | | | hydrochloride, m.p. 140–141° | 1065 (m.p.); |
| | | $CH_3$ | | | 151–152° | | 186 (m.p.); |
| | $C_6H_4Cl$-*p* | $CH_3$ | | | 144° | | 186 (m.p.); |
| $CH_3$ | $-CH(CH_2CH_2N(CH_3)_2)C_6H_4Cl$-*p* | | | | b.p. 164°/1 mm | | 191 (b.p.); |
| | | | | | | acid maleate, m.p. 124–127° | 191 (m.p.); |
| $CH_3$ | $-CH(CH_2CH_2N(CH_3)_2)C_6H_5$ | | | | b.p. 170°/2 mm | | 191 (m.p.); |
| | | | | | | acid maleate, m.p. 124–126° | 191 (m.p.); |
| | | $CH_3$ | | | 119° | | 186 (m.p.); |

TABLE XII-57. Ethers and Esters of Alkyl and Aryl 3-Pyridinols (Continued)

| R | $R_2$ | $R_4$ | $R_5$ | $R_6$ | m.p. | Derivatives | Ref. |
|---|---|---|---|---|---|---|---|
| $-C(=O)C_6H_4N{=}NC_6H_4NO_2$-*p* | | | | | | | 1097 (g); |
| $-C(=O)$-(4-pyridyl) | $C_6H_4Cl$-*p* | $CH_3$ | | | 181° | | 186 (m.p.); |
| $CH_3$ | $-CH(C_6H_5)C_5H_9$ | | | | 103–106° | | 191 (m.p.); |
| $-CH_2$-(3,4-dichlorophenyl) | 3,4-dichlorophenyl | $CH_3$ | | | 90–91° | | 186 (m.p.); |
| $-CH_2$-(3,4-dichlorophenyl) | $C_6H_4Cl$-*p* | $CH_3$ | | | 74° | | 186 (m.p.); |
| $CH_3$ | $-CH(C_6H_5)C_6H_4Cl$-*p* | | | | 106–109° | | 191 (m.p.); |
| $CH_3$ | $-C(OH)(C_6H_4Cl$-*p*$)C_6H_5$ | | | | 133–134° | | 191 (m.p.); |
| $CH_3$ | $-CH(C_6H_5)_2$ | | | | 120–122° | | 191 (m.p.); |
| $CH_3$ | $-C(OH)(C_6H_5)_2$ | | | | 129–131° | | 191 (m.p.); |
| $CH_3$ | $-CH(C_6H_5)C_6H_{11}$ | | | | 142–144° | | 191 (m.p.); |
| $CH_3$ | $-CH(C_6H_5)CH_2C_6H_5$ | | | | 71–73° | | 191 (m.p.); |
| $C_2H_5$ | $-CH(C_6H_5)_2$ | | | | 96–98° | | 191 (m.p.); |

| | | | | | | | |
|---|---|---|---|---|---|---|---|
| 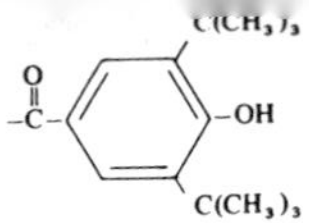 $-\overset{O}{\overset{\|}{C}}-$ (ring with –OH and two $C(CH_3)_3$) | | | | | 176–177° | | 1013 (m.p.); |
| $CH_2C_6H_5$ | $CH_3$ | $CH_2Cl$ | $-CH_2OCH_2C_6H_5$ | | | hydrochloride, m.p. 145–148°<br>picrate, m.p. 154–155° | 1065 (m.p.);<br>1065 (m.p.); |
| $CH_2C_6H_5$ | $CH_3$ | $CH_2OH$ | $-CH_2OCH_2C_6H_5$ | | 68–69° | | 1053 (m.p.); 1065 (m.p.); |
| $C_6H_5$ | $C_6H_5$ | | | $C_6H_5$ | 91–95° | | 1098 (m.p.); |
| $CH_2C_6H_5$ | $CH_3$ | $-\underset{OH}{\underset{\|}{C}}HCH_3$ | $-CH_2OCH_2C_6H_5$ | | 83-84° | | 1053 (m.p.);<br>1065 (m.p.), (n); |
| $CH_2C_6H_5$ | $CH_3$ | $-(CH_2)_3OH$ | $-CH_2OCH_2C_6H_5$ | | 64–65° | | 1065 (m.p.); |
| $CH_2C_6H_5$ | $CH_3$ | $-\underset{OH}{\underset{\|}{C}}HC_6H_5$ | $-CH_2OCH_2C_6H_5$ | | 102–104° | | 1065 (m.p.), (n); |
| $C_6H_4Cl$-*p* | $C_6H_5$ | $C_6H_5$ | | $C_6H_4Br$-*p* | 171–173° | | 74 (m.p.); |
| $C_6H_4Cl$-*p* | $C_6H_5$ | $C_6H_5$ | | $C_6H_5$ | 177–179° | | 74 (m.p.);<br>1098 (m.p.); |
| $CH_3$ | 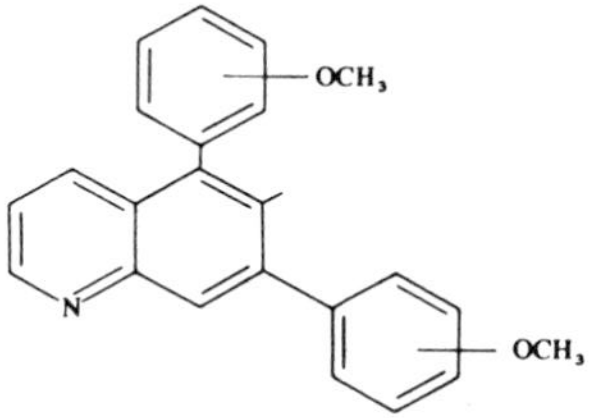 (structure: N, $OCH_3$, $OCH_3$) | | | | | | 600 (t); |
| $CH_3$ | 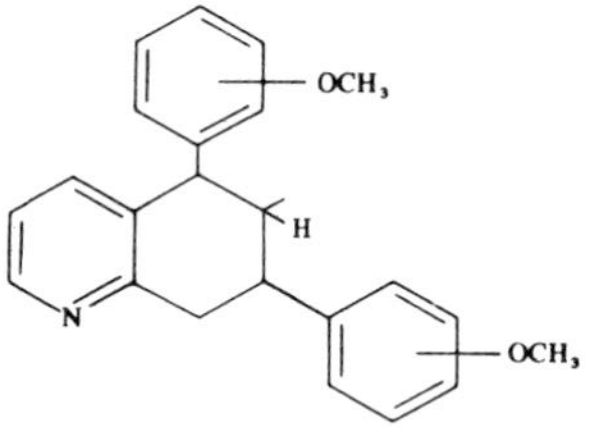 (structure: N, H, $OCH_3$, $OCH_3$) | | | | 179–180° | | 600 (m.p.); |

TABLE XII-58. Ethers and Esters of Amino 3-Pyridinols

| R | $R_2$ | $R_4$ | $R_5$ | $R_6$ | m.p. | Derivatives | Ref. |
|---|---|---|---|---|---|---|---|
| $C_2H_5$ | | | $-NHNO_2$ | | 160° | | 319 (m.p.); |
| $C_2H_5$ | | | $-NH_2$ | | 6° | | 319 (m.p.); |
| | | | | | | picrate, m.p. 189–191° | 319 (m.p.); |
| $C_2H_5$ | | | | $-NH_2$ | 47–48° | | 292 (m.p.); |
| | | | | | | hydrochloride, m.p. 143–145° | 292 (m.p.); |
| | | | | | | picrate, m.p. 240–241° | 292 (m.p.); |
| | | | | | | *N'*-acetyl, m.p. 108–109° | 292 (m.p.); |
| $-CH_2C{\equiv}CH$ | $-NH_2$ | | | | b.p. 91°/0.2 mm m.p. 58–59° | | 596 (b.p.), (m.p.); |
| $-CH_2CH{=}CH_2$ | $-NH_2$ | | | | b.p. 82–83° m.p. 42–44° | | 596 (b.p.), (m.p.); |
| $-CH_2CH(OCH_3)_2$ | $-NH_2$ | | | | b.p. 111°/0.3 mm | | 596 (b.p.); |
| $-(CH_2)_3N(CH_3)_2$ | $-NH_2$ | | | | b.p. 112°/0.2 mm m.p. 67–69° | | 596 (b.p.), (m.p.); |
| $-CH_2C_6H_5$ | $-NH_2$ | | | | b.p. 150°/0.05 mm m.p. 96–97° | | 596 (b.p.), (m.p.); |
| $-COC_6H_5$ | | | | $-N(COC_6H_5)_2$ | m.p. 182–183° | | 292 (m.p.), (i), (u); |
| $-COC_6H_5$ | | $CH_3$ | $C_6H_5$ | $-NHCOC_6H_5$ | m.p. 198–200° | | 222 (m.p.), (i); 292 (m.p.), (i); |
| $COC_6H_5$ | $NHCOC_6H_5$ | $CH_3$ | $C_6H_5$ | | 195–196° | | 292 (m.p.); |
| $COC_6H_5$ | $-N(COC_6H_5)_2$ | $CH_3$ | $C_6H_5$ | | 182° | | 292 (m.p.), (i); |

TABLE XII-59. Ethers and Esters of 3-Pyridinol Carboxylic Acids and Derivatives

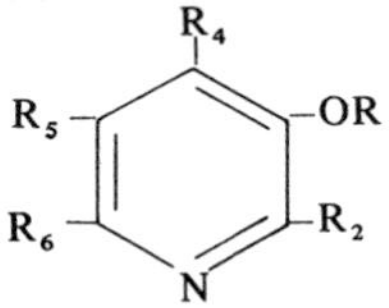

| R | $R_2$ | $R_4$ | $R_5$ | $R_6$ | m.p. | Derivatives | Ref. |
|---|---|---|---|---|---|---|---|
| COOH | | | | | | ethyl ester | 659 (u); |
| $CH_3$ | CHO | | | | | | 513 (i); 1077 (i); |
| $CH_3$ | | CHO | | | 36–37° | | 513 (m.p.), (i);<br>1077 (i); |
| $CH_3$ | COOH | | | | | amide,<br>m.p. 198–200°<br>methyl ester,<br>b.p. 85°/0.1 mm | <br>596 (m.p.);<br><br>596 (b.p.); |
| $-CH_2COOH$ | | | | | | ethyl ester,<br>b.p. 118°/1 mm<br>diethylamide,<br>b.p. 137°/0.4 mm<br>diethylamide·HCl<br>m.p. 140–142° | <br>595 (b.p.), (n);<br><br>596 (b.p.);<br><br>596 (m.p.); |
| $-CH_2COOH$ | $COOCH_3$ | | | | | amide,<br>m.p. 185–186°<br>methyl ester,<br>m.p. 78–79° | <br>596 (m.p.);<br><br>596 (m.p.); |
| $-CH_2COOH$ | $CONH_2$ | | | | | methyl ester,<br>m.p. 141–143°<br>diethylamide,<br>m.p. 134–135° | <br>596 (m.p.);<br><br>596 (m.p.); |
| $-CH_2COOH$ | $CONHC_6H_5$ | | | | | methyl ester,<br>m.p. 109–110° | <br>596 (m.p.); |
| $-CH_2COOH$ | | $COOC_2H_5$ | | | | ethyl ester,<br>oil | <br>595 (n); |

TABLE XII-59. Ethers and Esters of 3-Pyridinol Carboxylic Acids and Derivatives (Continued)

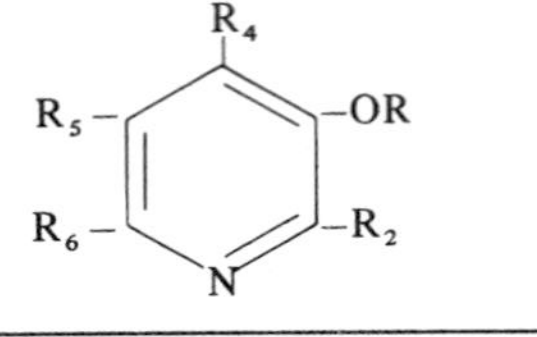


| R | $R_2$ | $R_4$ | $R_5$ | $R_6$ | m.p. | Derivatives | Ref. |
|---|---|---|---|---|---|---|---|
| $-CH(CH_3)COOH$ | | | | | | ethyl ester, b.p. 86°/0.2 mm | 595 (b.p.), (n); |
| $CH_2COOH$ | | | | $CH_3$ | | methyl ester, m.p. 50–51°<br>ethyl ester, b.p. 111°/0.1 mm, m.p. 10°<br>amide, m.p. 154–156° | 596 (m.p.);<br>596 (b.p.), (m.p.);<br>596 (m.p.); |
| $CH_2COOH$ | $CH_2OH$ | | | | | amide, m.p. 157–158° | 596 (m.p.); |
| $-CH(CH_3)COOH$ | | $COOC_2H_5$ | | | | ethyl ester, oil | 595 (n); |
| $CH_2CH{=}CH_2$ | COOH | | | | | amide, m.p. 125–126°<br>N′-benzylamide, m.p. 72–73°<br>N′-*p*-methoxy-benzylamide, m.p. 74–75° | 596 (m.p.);<br>596 (m.p.);<br>596 (m.p.); |
| $-CH(C_2H_5)COOH$ | $CONH_2$ | | | | | ethyl ester, m.p. 101–108° | 596 (m.p.); |
| $-COOH$ | $CH_3$ | $-CH_2NHCOOC_2H_5$ | $-CH_2OH$ | | | ethyl ester, m.p. 70–71° | 659 (m.p.); |
| $-(CH_2)_3CH_3$ | COOH | | | | | amide, m.p. 109–110° | 596 (m.p.); |

| | | | | | | |
|---|---|---|---|---|---|---|
| $-CH_2CH(CH_3)_2$ | COOH | | | | amide, m.p. 109–110° | 596 (m.p.); |
| $CH_3$ | $-CH{=}NCH(CH(CH_3)_2)COOH$ | | | | potassium salt | 1077 (i); |
| $CH_3$ | | $-CH{=}NCH(CH(CH_3)_2)COOH$ | | | potassium salt | 1077 (i); |
| $-CH_2C_6H_5$ | COOH | | | | amide, m.p. 106–107° | 596 (m.p.); |
| $CH_3$ | $-CH(C_6H_5)COOH$ | | | | ethyl ester, b.p. 175°/2 mm m.p. 105–107° | 191 (b.p.), (m.p.); |
| $-CH_2C_6H_5$ | $CH_3$ | CHO | $-CH_2OCH_2C_6H_5$ | 72° | | 1053 (m.p.); |
| | | | | | hydrate, m.p. 60–70° | 1053 (m.p.); |
| | | | | | hemihydrate, m.p. 72° | 1065 (m.p.), (n); |
| $-COCH_2NHCO-O-CH_2C_6H_5$ | 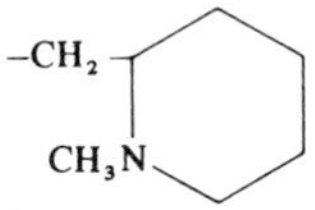 $-CH_2-$(2-piperidyl ring, $CH_3N$) | | | 102° | | 654 (m.p.); |
| $CH_2C_6H_5$ | $CH_3$ | $CH_2CH_2COOH$ | $CH_2OCH_2C_6H_5$ | | hemihydrate, m.p. 175–176° | 1065 (m.p.); |
| $CH_2C_6H_5$ | $CH_3$ | $CH_2CH(COOH)_2$ | $CH_2OCH_2C_6H_5$ | 173–174° | | 1065 (m.p.); |
| | | | | | diethyl ester·HCl m.p. 109–110° | 1065 (m.p.); |

TABLE XII-60. Ethers of 3-Pyridinol Nitriles

| R | $R_2$ | $R_4$ | $R_5$ | $R_6$ | m.p. | Derivatives | Ref. |
|---|---|---|---|---|---|---|---|
| $CH_3$ | CN | | | | | 111–113° | 596 (m.p.); |
| $CH_2C_6H_5$ | CN | | | | | 99–101° | 596 (m.p.); |

TABLE XII-61. Ethers of Halo 3-Pyridinols

| R | $R_2$ | $R_4$ | $R_5$ | $R_6$ | m.p. | Derivatives | Ref. |
|---|---|---|---|---|---|---|---|
| $CH_3$ | Br | | | | 45°<br>b.p. 123°/9 mm | | 596 (b.p.), (m.p.); |
| $CH_3$ | | | Br | | 33–34° | | 705 (m.p.); |
| | | | | | | hydrobromide,<br>m.p. 178–180° | 705 (m.p.); |
| $CH_2COOH$ | I | I | | I | 202–203° | | 599 (m.p.); |
| | | | | | | ethyl ester,<br>m.p. 137–138° | 599 (m.p.); |
| $CH_2COOH$ | Br | | | | | methyl ester,<br>m.p. 65–69° | 596 (m.p.); |

| | | | | | |
|---|---|---|---|---|---|
| $C_2H_5$ | | Br | Br | 67–69° | 712 (m.p.); |
| $CONHCH_3$ | | Cl | | 128° | 972 (m.p.); |
| $CH_3$ | Br | | $CH_3$ | 54° | 597 (m.p.), (u); |
| $C_2H_5$ | | Br | | 8.2–8.8° | |
| | | | | b.p. 111°/5 mm | 705 (b.p.), (m.p.); |
| $-CH_2CHCH_2$<br>O O<br>C<br>O | Cl | | | 124–126° | 596 (m.p.); |
| $-CH_2CHCH_2$<br>O O<br>C<br>O | Br | | | 143–144° | 596 (m.p.); |
| $-CH_2CHCH_2$<br>O NH<br>C<br>O | Cl | | | 186–188° | 596 (m.p.); |
| $-CH_2COC_6H_5$ | Br | | | 105–107° | 596 (m.p.); |
| $CH_3$ | | Cl | 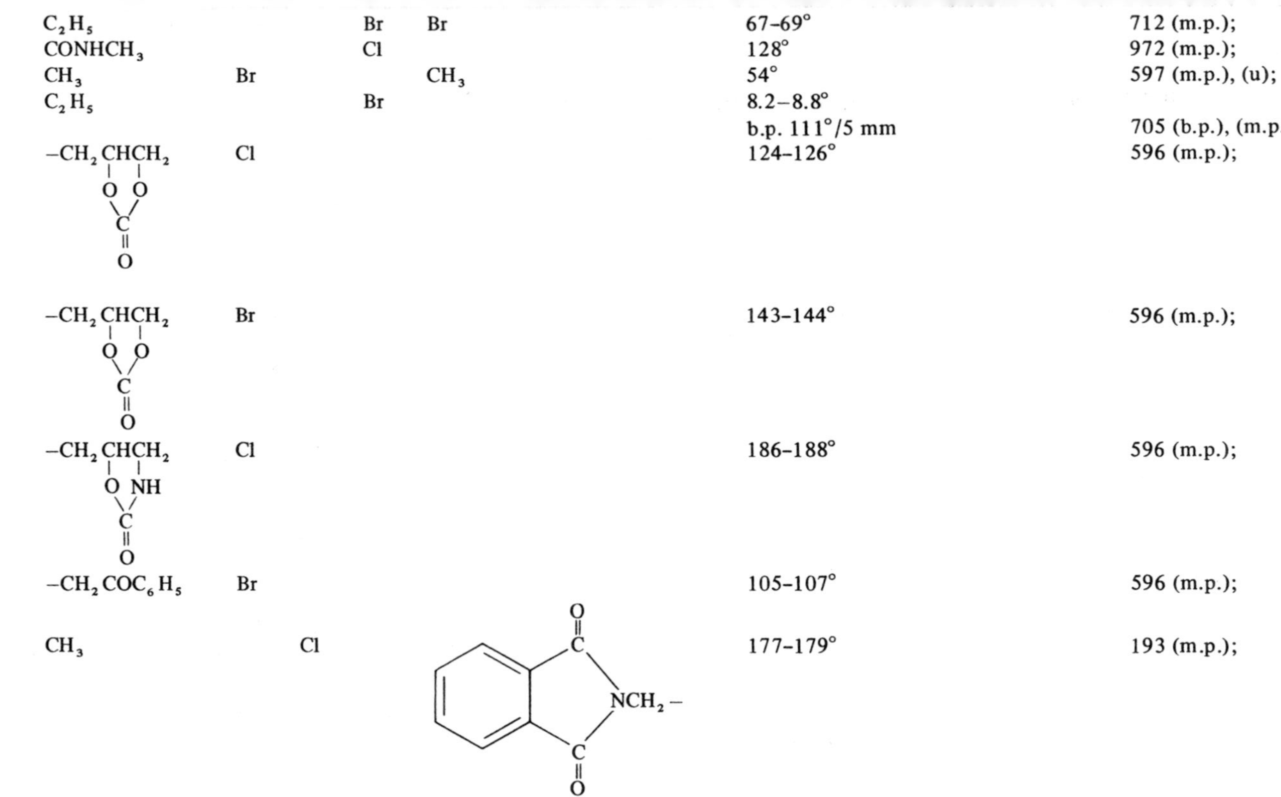 | 177–179° | 193 (m.p.); |

TABLE XII-62. Ethers and Esters of Nitro 3-Pyridinols

| R | $R_2$ | $R_4$ | $R_5$ | $R_6$ | m.p. | Derivatives | Ref. |
|---|---|---|---|---|---|---|---|
| $CH_3$ | $NO_2$ | | | | | | 515; |
| $CH_3$ | | $NO_2$ | | | 46° | | 256a (m.p.); 261 (m.p.); |
| $CONHCH_3$ | $NO_2$ | | | | 132–135° | | 1085 (m.p.); |
| $CH_3$ | $NO_2$ | | | $CH_3$ | 88–89° | | 596 (m.p.); |
| $CH_3$ | $CH_3$ | | | $NO_2$ | 101–103° | | 714 (m.p.), (n); |
| $C_2H_5$ | | $NO_2$ | | | 40° | | 256a; 261 (m.p.); |
| $-CH_2CHCH_2$ (cyclic carbonate: O–C(=O)–O) | $NO_2$ | | | | 162–164° | | 596 (m.p.); |
| $-COCH_2O$–(2,4-dichlorophenyl) | $NO_2$ | | | | 109–113° | | 1085 (m.p.); |
| $-COCH_2NH$–CO–O$CH_2C_6H_5$ | iso$C_4H_9$ | $NO_2$ | | $CH_3$ | 96–98° | | 654 (m.p.); |

TABLE XII-63. Ethers and Esters of Sulfur Containing 3-Pyridinols

| R | $R_2$ | $R_4$ | $R_5$ | $R_6$ | m.p. | Derivatives | Ref. |
|---|---|---|---|---|---|---|---|
| $SO_3H$ | $NO_2$ | | | | | potassium salt | 648 (i), (u); |
| $SO_3H$ | | | | | | | 648 (i), (u); |
| $SO_2CH_3$ | | | | | 60°<br>b.p. 103°/0.01 mm | | 649 (b.p.), (m.p.); |
| $COSCH_3$ | | | | | 47–48° | | 583 (m.p.); |
| $SO_2N(CH_3)_2$ | | | | | b.p. 99°/0.25 mm | | 1099 (b.p.); |
| $CH_2COOH$ | $SCH_3$ | | | $CH_3$ | | methyl ester, m.p. 74–75° | 597 (m.p.), (i), (n), (u); |
| $SO_2C_6H_4Cl$-*p* | $NO_2$ | | | | 75–77° | | 1085 (m.p.); |
| $COSC_6H_5$ | | | | | 60–62° | | 1100 (m.p.); |
| $S^{18}O_2C_6H_4CH_3$-*p* | | | | | b.p. 170°/3 mm<br>m.p. 78° | | 410 (b.p.), (m.p.); |
| 2-($SC_2H_5$)$C_6H_4$–C(=O)– | | | | | 53° | | 1101 (m.p.); |
| $-COCH(NHSO_2C_6H_4CH_3$-*p*$)CH(CH_3)_2$ | | | | | 153° | | 656 (m.p.), (k); |

TABLE XII-63. Ethers and Esters of Sulfur Containing 3-Pyridinols (Continued)

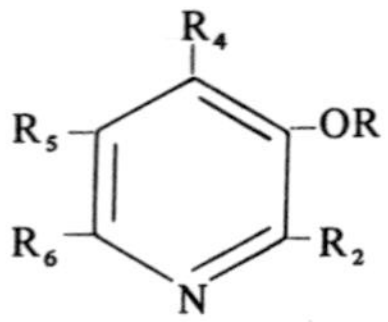

| R | $R_2$ | $R_4$ | $R_5$ | $R_6$ | m.p. | Derivatives | Ref. |
|---|---|---|---|---|---|---|---|
| 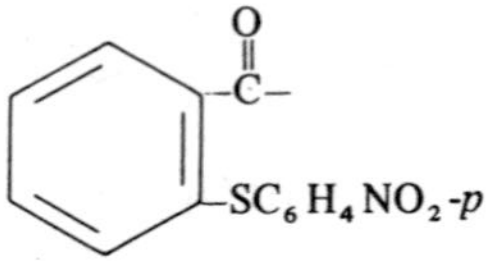 2-($SC_6H_4NO_2$-*p*)$C_6H_4C(=O)$– | | | | | | hydrochloride, m.p. 196° | 1101 (m.p.); |
| $CH_3$ | 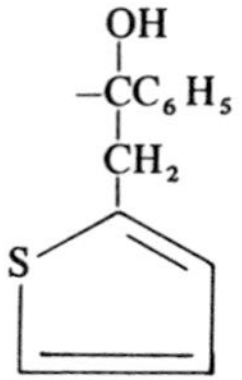 –C(OH)($C_6H_5$)$CH_2$-(2-thienyl) | | | | 127–129° | | 191 (m.p.); |
| –COCH($NHSO_2C_6H_4CH_3$-*p*)$(CH_2)_4NH_2$ | | | | | | hydrobromide, hygroscopic | 655; |
| 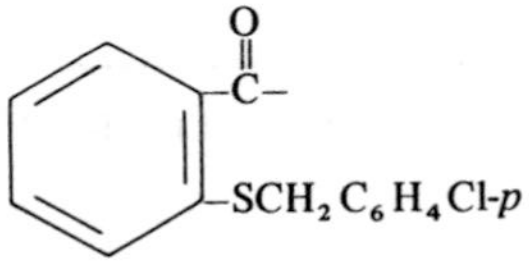 2-($SCH_2C_6H_4Cl$-*p*)$C_6H_4C(=O)$– | | | | | | hydrochloride, m.p. 195° | 1101 (m.p.); |

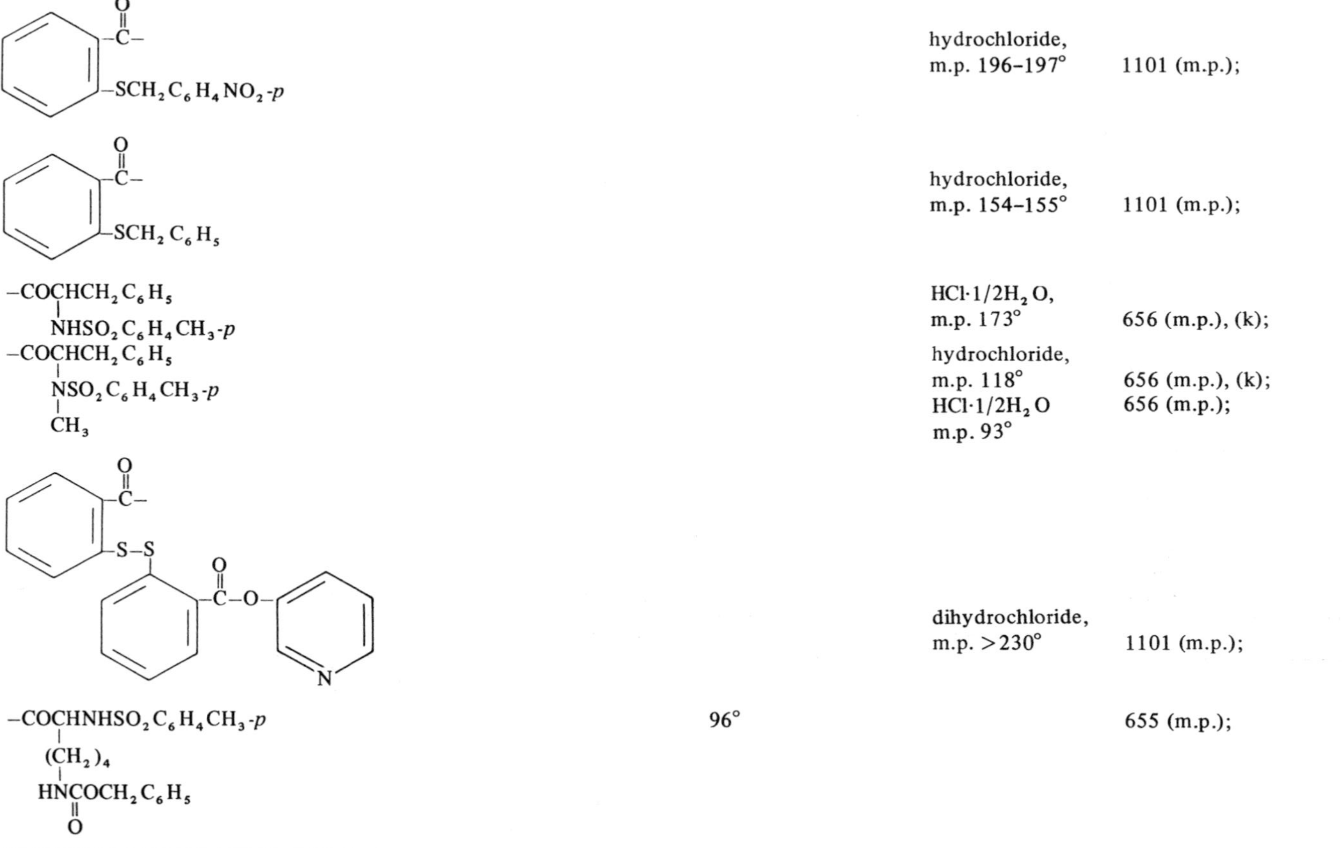

| Substituent | m.p. | Derivative | Reference |
|---|---|---|---|
| | | hydrochloride, m.p. 196–197° | 1101 (m.p.); |
| | | hydrochloride, m.p. 154–155° | 1101 (m.p.); |
| $-COCH(CH_2C_6H_5)NHSO_2C_6H_4CH_3$-*p* | | $HCl \cdot 1/2H_2O$, m.p. 173° | 656 (m.p.), (k); |
| $-COCH(CH_2C_6H_5)N(CH_3)SO_2C_6H_4CH_3$-*p* | | hydrochloride, m.p. 118° | 656 (m.p.), (k); |
| | | $HCl \cdot 1/2H_2O$ m.p. 93° | 656 (m.p.); |
| | | dihydrochloride, m.p. >230° | 1101 (m.p.); |
| $-COCH(NHSO_2C_6H_4CH_3$-*p*$)(CH_2)_4NHCOCH_2C_6H_5$ | 96° | | 655 (m.p.); |

TABLE XII-64. Alkyl and Aryl 4-Pyridinols

| $R_2$ | $R_3$ | $R_5$ | $R_6$ | m.p. | Derivatives | Ref. |
|---|---|---|---|---|---|---|
| H | H | H | H | 148–149° | | 404 (f); 460 (m.p.), (i); 462 (i); 466 (i); 478 (i), (r); 486 (n); 537 (n); 648 (u); 659 (u); 735 (n); 819 (i); 899 (n); 900 (f); 901 (n); 902 (i); 904 (u); 905 (m.p.), (u); 906 (a); 909 (u); 910 (j); 911 (t); 912 (u); 915 (m); 918 (s); 1040 (f); 1102 (u); 1103 (i), (n); 1104 (u); |
| | | | | | picrate | 1104 (u); |
| | | | | | $^{15}N$ | 462 (i); |
| | | | | | $^{18}O$ | 462 (i), (m); |
| | | | | | sodium salt | 545 (i), (r); 921 (i); |
| | | | | | *O*-acetyl, m.p. 130–145° | 404 (f); 414 (m.p.), (g), (i), (n); |
| | | | | | b.p. 120°/2 mm | 747 (b.p.), (m.p.), (i), (u); |
| | | | | | *O*-acetyl, m.p. 80° | 1105 (m.p.); |
| | | | | | $CoCl_2$ complex m.p. 219° | 824 (m.p.), (i), (v); |
| | | | | | $Co(NH_3)_5$ complex | 820 (i), (u); |
| | | | | | perchlorate | 821 (i), (u); |
| | | | | | hydrochloride | 536 (n); 1103 (i), (r); |
| | | | | | hydrate, m.p. 59–61° | 1106 (m.p.); |
| | | | | | hexachloro- | |

| | | | | | | |
|---|---|---|---|---|---|---|
| | | | | | hexachloro-stannate, m.p. 247–249° | 464 (m.p.), (i); |
| | | | | | *N*-deuterio | 460 (i), (r); 462 (i); |
| | | | | | *N*,*O*-$D_2$ chloride | 1045 (i); 1103 (i), (r); |
| D | D | D | D | | | 460 (i), (r); 1102 (u); 1103 (i); |
| | | | | | *N*,*O*-$D_2$ chloride | 1103 (i), (r); |
| | | | | | hydrochloride | 1103 (i), (r); |
| $CH_3$ | | | | | nitrate, m.p. 157–161° | 293 (m.p.); |
| | $CH_3$ | | | 92–98° | | 273 (m.p.); 278 (m.p.); 297 (m.p.); 789 (m.p.); |
| | | | | | picrate, m.p. 209–210° | 273 (m.p.); |
| $CH_3$ | | $CH_3$ | | 205–206° | | 297 (m.p.); |
| | | | | | picrate, m.p. 220–221° | 297 (m.p.); |
| $CH_3$ | | | $CH_3$ | 222–232° | | 25 (m.p.); 137 (m.p.), (n); 295 (m.p.); |
| | | | | | hydrochloride | 536 (n); |
| $CH_2OH$ | | | $CH_3$ | 225–227° | | 411 (m.p.); |
| $C_2H_5$ | | | $CH_3$ | 118-119° | | 25 (m.p.); |
| | $Si(CH_3)_3$ | | | 174° | | 377 (m.p.), (i); |
| $CH_3$ | $COCH_3$ | | $CH_3$ | 233–242° | | 69 (m.p.); 136 (m.p.), (u); 141 (m.p.); |
| | | | | | picrate, m.p. 172–174° | 141 (m.p.); |
| | 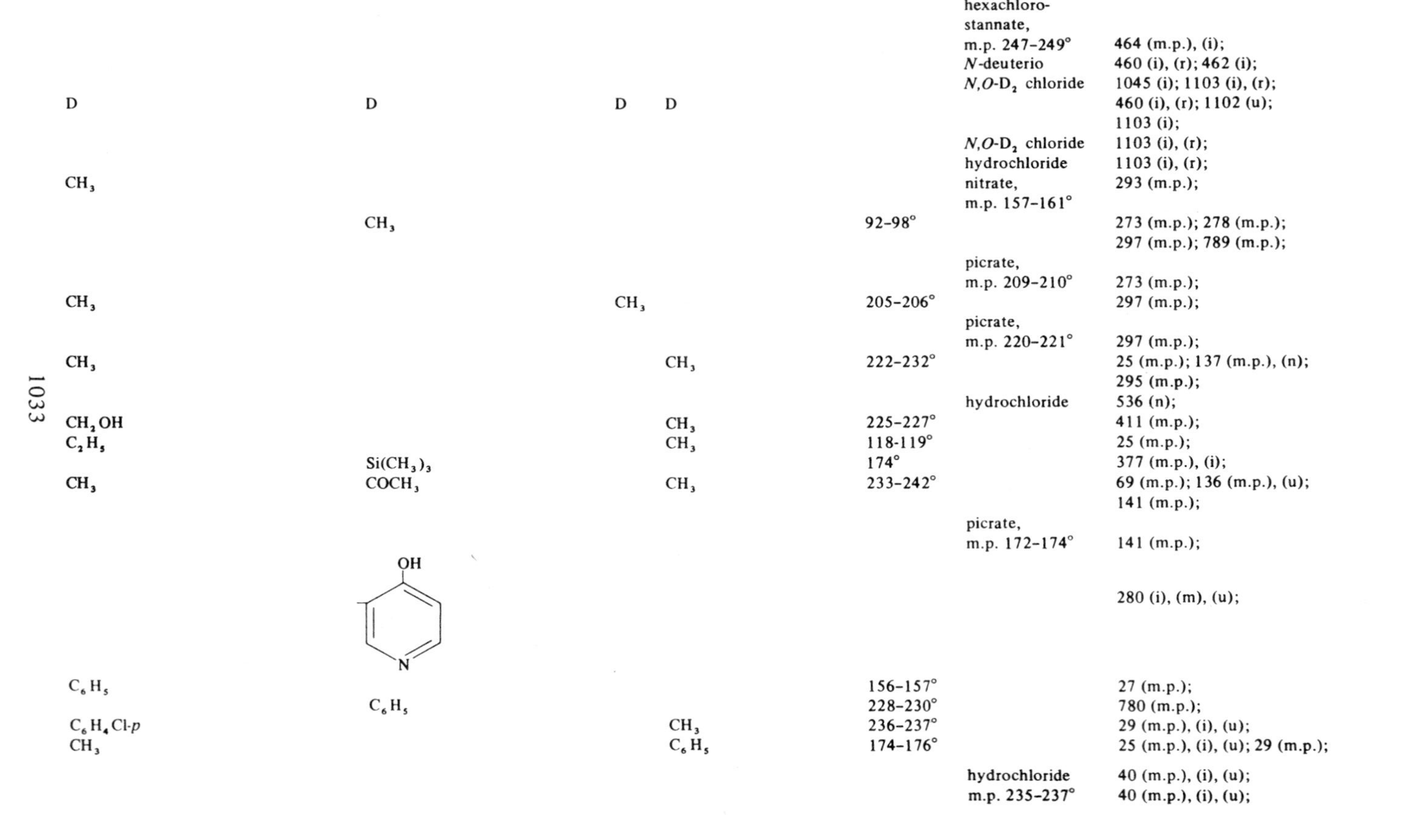 | | | | | 280 (i), (m), (u); |
| $C_6H_5$ | | | | 156–157° | | 27 (m.p.); |
| | $C_6H_5$ | | | 228–230° | | 780 (m.p.); |
| $C_6H_4Cl$-*p* | | | $CH_3$ | 236–237° | | 29 (m.p.), (i), (u); |
| $CH_3$ | | | $C_6H_5$ | 174–176° | | 25 (m.p.), (i), (u); 29 (m.p.); |
| | | | | | hydrochloride | 40 (m.p.), (i), (u); |
| | | | | | m.p. 235–237° | 40 (m.p.), (i), (u); |

TABLE XII-64. Alkyl and Aryl 4-Pyridinols (Continued)

(Structure: 4-pyridinol ring with OH at C-4 and substituents $R_2$, $R_3$, $R_5$, $R_6$)

| $R_2$ | $R_3$ | $R_5$ | $R_6$ | m.p. | Derivatives | Ref. |
|---|---|---|---|---|---|---|
| | $-CH_2CH_2-C_6H_3(NO_2)_2$ (2-$NO_2$, 4-$NO_2$) | | | 191–192° | | 310 (m.p.); |
| $C_2H_5$ | | | $C_6H_5$ | | | 25 (m.p.), (i), (u); |
| | $-CH_2CH_2C_6H_5$ | | | | hemihydrate, m.p. 162° | 310 (m.p.); |
| $t$-$C_4H_9$ | | | $t$-$C_4H_9$ | 170–172° | | 671 (m.p.); |
| | | | | | chloroaurate, m.p. 199–201° | 671 (m.p.); |
| $CH_3$ | $COCH_3$ | | $C_6H_4Br$-$p$ | 251° | | 136 (m.p.), (u); |
| $CH_3$ | $COC_6H_5$ | | $CH_3$ | 257–258° | | 69 (m.p.); |
| $CH_3$ | $COCH_3$ | | $C_6H_5$ | 214–215° | | 135 (m.p.), (i), (m), (u); |
| iso$C_3H_7$ | | | $C_6H_5$ | 117–118° | | 25 (m.p.), (i), (u); |
| $-CH_2$-(2-quinolyl) | | | $CH_3$ | 315° | | 138 (m.p.), (i); |
| $C_6H_4Cl$-$p$ | | | $C_6H_5$ | 208–212° | | 29 (m.p.), (i), (u); |
| $C_6H_5$ | | | $C_6H_5$ | 175–179° | | 29 (m.p.), (i), (u); 40 (m.p.), (i), (n), (u); 530 (m.p.); |
| $C_6H_5$ | $-CH_2C_6H_5$ | | | 247–248° | | 27 (m.p.), (i), (u); |
| $C_6H_4OCH_3$-$p$ | | | $C_6H_5$ | 219–222° | | 29 (m.p.), (i), (u); |
| $CH_3$ | $COC_6H_5$ | | $C_6H_5$ | 267–269° | | 135 (m.p.), (i), (m), (u); |
| $C_6H_5$ | | | $-CH_2CH_2C_6H_5$ | 172–173° | | 40 (m.p.), (u); |
| $C_6H_4OCH_3$-$p$ | | | $C_6H_4OCH_3$-$p$ | 274–275° | | 26 (m.p.), (i); |

| | | | |
|---|---|---|---|
| $C_6H_5$ | $-CH_2CH(OH)C_6H_4OCH_3$-*p* | 191–192° | 30 (m.p.), (u); |
| *n*-Nonyl | $C_6H_5$ | 88–89° | 25 (m.p.); |
| $C_6H_5$ | $-CH_2C(OH)(C_6H_5)_2$ | 205–206° | 30 (m.p.), (u); |
| $C_6H_4OCH_3$-*p* | $-CH_2C(OH)(C_6H_5)_2$ | 211–213° | 30 (m.p.), (u); |
| 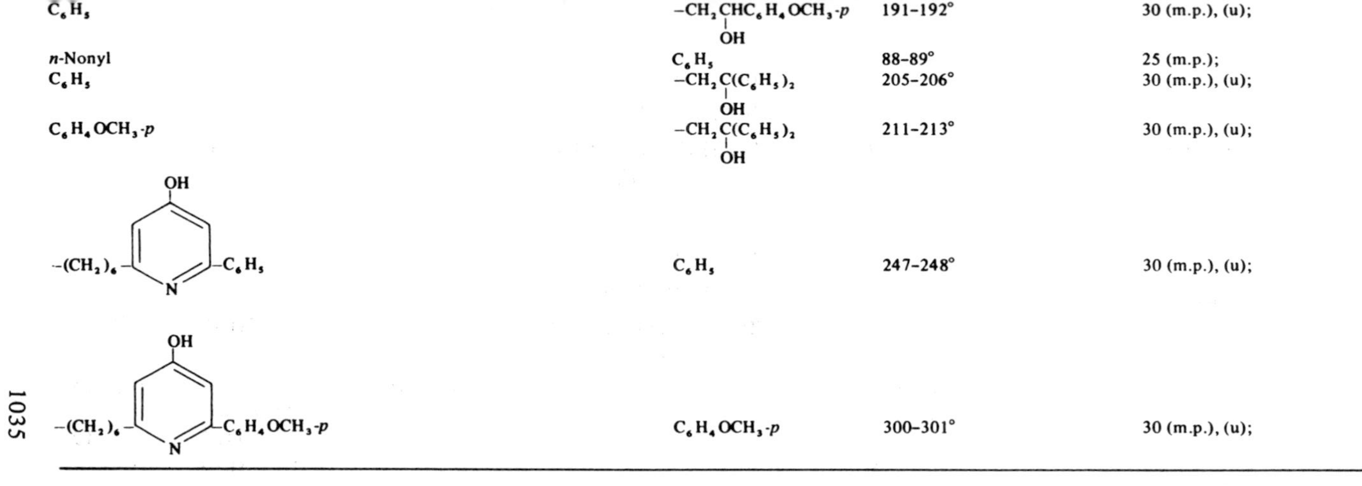 $-(CH_2)_6-$[2-$C_6H_5$-4-OH-pyridin-6-yl] | $C_6H_5$ | 247–248° | 30 (m.p.), (u); |
| $-(CH_2)_6-$[2-$C_6H_4OCH_3$-*p*-4-OH-pyridin-6-yl] | $C_6H_4OCH_3$-*p* | 300–301° | 30 (m.p.), (u); |

TABLE XII-65. Alkyloxy and Aryloxy 4-Pyridinols

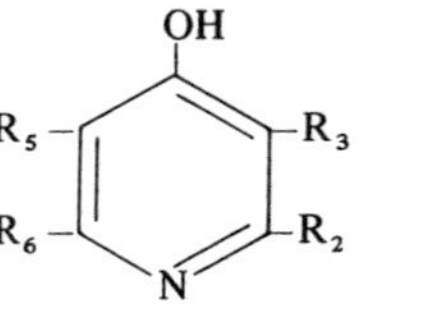

| $R_2$ | $R_3$ | $R_5$ | $R_6$ | m.p. | Derivatives | Ref. |
|---|---|---|---|---|---|---|
| $CH_3$ | $OCH_3$ | | | 155–156° | | 125 (m.p.), (u); |
| $CH_2OH$ | | $OCH_3$ | | 172° | | 794 (m.p.); |
| $OCH_3$ | $OCH_3$ | | $CH_2OH$ | 94–95° | | 160 (m.p.), (i), (u); |
| | | | | | 4,6-diacetyl, oil | 160 (c), (i), (u); |
| | $OCH_2C_6H_5$ | | | 195–197° | | 134 (m.p.); |
| $CH_3$ | β-D-Gluco-pyranosyloxy | | | 116–118° | | 131 (m.p.); 858 (m.p.); |
| $CH_2OH$ | | $OCH_2C_6H_5$ | | 224–226° | | 793 (m.p.); |
| $OCH_3$ | $OCH_3$ | $CH_3$ | $CH_3$<br>–$CH_2CH{=}CCH{=}CHCH_2$<br>$CH_3CCH(OH)CHCH{=}C$<br>$CH$   $CH_3$   $CH_3$<br>$CH_3$ | | | 591 (n), (u); 870 (n); |
| $OCH_3$ | $OCH_3$ | $CH_3$ | $CH_3$<br>–$CH_2CH_2CHCH_2CH_2CH_2$<br>$CH(CH_3)C{-}CHCH_2CH$<br>$CH_2$   $O$ $CH_3$   $CH_3$<br>$CH_3$ | *O*-acetyl | | 869 (m); |

| | | | | | |
|---|---|---|---|---|---|
| $OCH_3$ | $OCH_3$ | $CH_3$ | 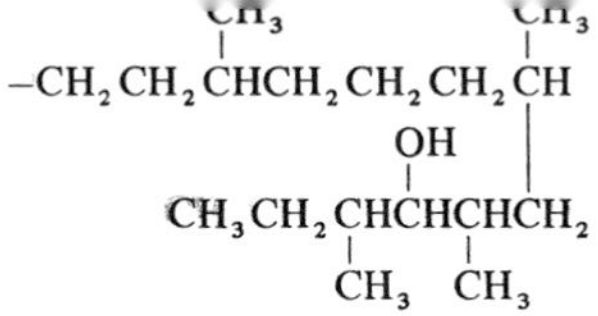 $-CH_2CH_2CH(CH_3)CH_2CH_2CH_2CH(CH_3)-CH_2CH(CH_3)CH(OH)CH(CH_3)CH_2CH_3$ | *O*-acetyl | 869 (m); |
| $OCH_3$ | $OCH_3$ | $CH_3$ | 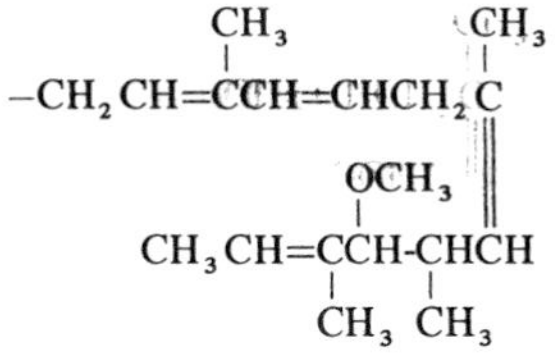 $-CH_2CH{=}C(CH_3)CH{=}CHCH_2C(CH_3){=}CHCH(CH_3)CH(OCH_3)C(CH_3){=}CHCH_3$ | | 871 (i), (n); 872 (i), (k), (n), (u); |
| | | | | *O*-acetyl | 872 (i), (k), (n); |
| $OCH_3$ | $OCH_3$ | $CH_3$ | 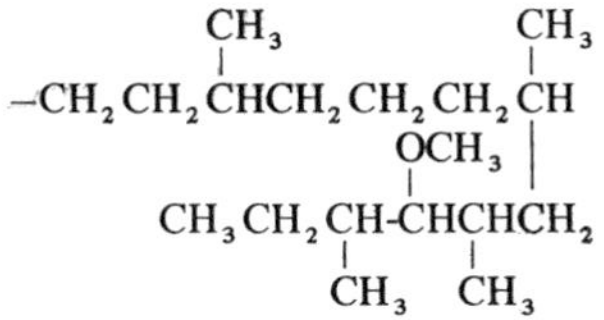 $-CH_2CH_2CH(CH_3)CH_2CH_2CH_2CH(CH_3)-CH_2CH(CH_3)CH(OCH_3)-CH(CH_3)CH_2CH_3$ | *O*-acetyl | 872 (m); |
| $OCH_3$ | $OCH_3$ | $CH_3$ | 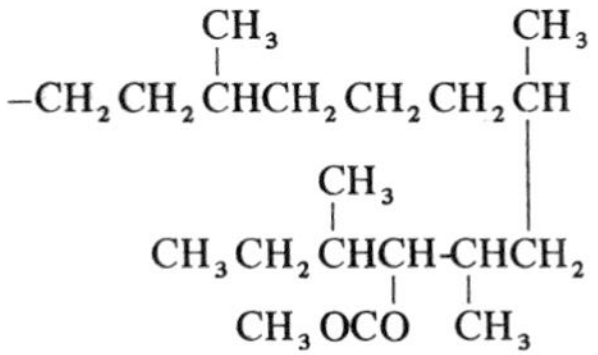 $-CH_2CH_2CH(CH_3)CH_2CH_2CH_2CH(CH_3)-CH_2CH(CH_3)-CH(OCOCH_3)CH(CH_3)CH_2CH_3$ | *O*-acetyl | 869 (m); |

TABLE XII-67. 4-Pyridinol Carboxylic Acids and Derivatives

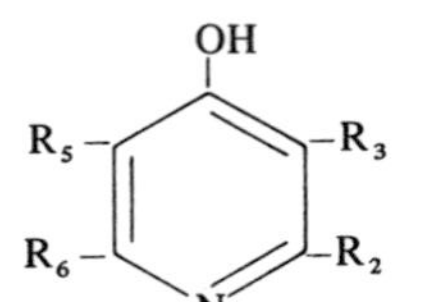


| $R_2$ | $R_3$ | $R_5$ | $R_6$ | m.p. | Derivatives | Ref. |
|---|---|---|---|---|---|---|
| COOH | | | | | ethyl ester·HCl m.p. 126–128° | 1107 (m.p.); |
| | COOH | | | 250–255° | | 101 (m.p.), (u); 789 (m.p.), (u); |
| | | | | | amide, m.p. 263° | 789 (m.p.), (i), (u); |
| | | | | | methyl ester | 101 (u); |
| | | | | | ethyl ester, m.p. 228–230° | 100 (m.p.); |
| | | | | | methyl ester-4-acetyl | 101 (u); |
| COOH | | | COOH | | diamide, m.p. 320–322° | 456 (m.p.); |
| | | | | | dimethyl ester, m.p. 170–171° | 456 (m.p.), (i), (n), (u); |
| | | | | | diethyl ester, m.p. 120–121° | 1107 (m.p.); |
| | COOH | COOH | | | dimethyl ester, m.p. 262–264° | 100 (m.p.); |
| CHO | | $OCH_3$ | | 176–178° | | 1108 (m.p.); |
| COOH | | $CH_3$ | | 225–226° | | 297 (m.p.); |
| | | | | | amide, m.p. 221–222° | 297 (m.p.); |

| | | | | | |
|---|---|---|---|---|---|
| $CH_3$ | COOH | COOH | | diethyl ester, m.p. 156–158° | 456 (m.p.), (i), (n), (u); |
| $CH_3$ | COOH | | $CH_3$ | hydrate, m.p. 257–258° | 147 (m.p.); |
| | | | | ethyl ester, m.p. 162–164° | 69 (m.p.), (i); |
| | | | | *p*-nitroanilide, m.p. 335–340° | 1030 (m.p.); |
| 2–Pyridyl | COOH | COOH | 2-Pyridyl | dimethyl ester | 500 (i); |
| | | | | diethyl ester | 500 (i), (u); |
| $C_6H_5$ | COOH | COOH | $C_6H_5$ | diethyl ester m.p. 199° | 448 (m.p.), (i); |
| 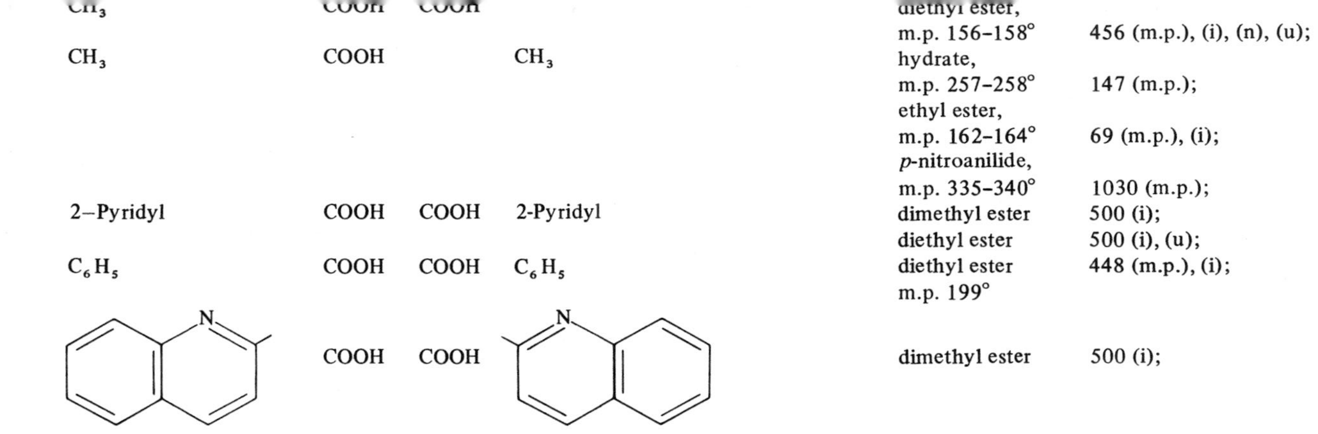 | COOH | COOH | | dimethyl ester | 500 (i); |
| | | | | diethyl ester | 500 (i); |

TABLE XII-68. 4-Pyridinol Nitriles

| $R_2$ | $R_3$ | $R_5$ | $R_6$ | m.p. | Derivatives | Ref. |
|---|---|---|---|---|---|---|
| | CN | | | 238–240° | | 789 (m.p.), (u); |

TABLE XII-69. Halo 4-pyridinols

| $R_2$ | $R_3$ | $R_5$ | $R_6$ | m.p. | Derivatives | Ref. |
|---|---|---|---|---|---|---|
| F | | | | 157° | | 266 (m.p.); |
| | F | | | 153° | | 266 (m.p.); |
| Cl | | | | 170° | | 241 (u); 253 (m.p.); |
| | Cl | | | 204–205° | | 254 (m.p.); |
| Br | | | | 173° | | 253 (m.p.); |
| | Br | | | 228–230° | | 254 (m.p.); 299 (m.p.); |
| [illegible] | | | | [illegible] | | [illegible] |

| | | | | | | |
|---|---|---|---|---|---|---|
| | I | | | 298–301° | | 254 (m.p.); 299 (m.p.); 704 (u); |
| Cl | | | Cl | 196° | | 241 (i), (u); 296 (m.p.); |
| | Cl | Cl | | | | 460 (i); 831 (i); |
| | Br | Br | | | | 460 (i), (u); |
| | I | I | | 300° | | 704 (u); 1109 (m.p.); |
| Cl | Cl | Cl | | 214–216° | | 456 (m.p.), (i), (n), (u); 1110 (m.p.); |
| Cl | Cl | | Cl | 193–195° | | 98 (m.p.), (i), (n), (u); |
| F | F | F | F | 73–75° or 95–97° | | 246 (m.p.), (i), (n); 950 (m.p.); 951 (m.p.); 1070 (m.p.); |
| | | | | | anilinium salt, m.p. 132° | 248 (m.p.); |
| F | Cl | F | F | 122–124° | | 246 (n); 951 (m.p.); |
| F | Cl | Cl | F | 101–102° | | 246 (m.p.), (i), (n); 950 (m.p.); 951 (m.p.); |
| | | | | | potassium salt, m.p. <300° | 951 (m.p.); |
| | | | | | *O*-acetyl, m.p. 55–56° | 951 (m.p.); |
| Cl | Cl | Cl | Cl | 232–233° | | 239 (m.p.); 241 (i), (u); |
| $CH_3$ | Br | Br | | 294–295° | | 293 (m.p.); |
| $CH_3$ | | | Cl | 156° | | 743 (m.p.); |
| $CH_3$ | Br | Br | $CH_3$ | 282° | | 141 (m.p.); |
| Br | Br | Br | $-N(CH_3)_2$ | oil | | 952; |
| | | | | | hydrochloride, m.p. 129–131° | 952 (m.p.); |
| $CH_3$ | $COCH_3$ | Br | $CH_3$ | 295–297° | | 141 (m.p.); |
| $C_4H_9$-*t* | Br | Br | $C_4H_9$-*t* | 146–147° | | 671 (m.p.); |

TABLE XII-70. Nitro 4-Pyridinols

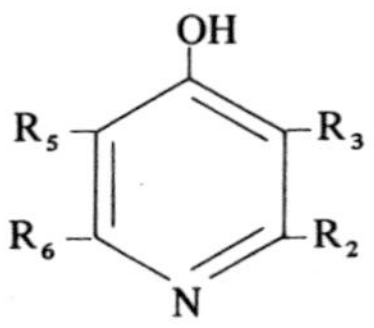

| $R_2$ | $R_3$ | $R_5$ | $R_6$ | m.p. | Derivatives | Ref. |
|---|---|---|---|---|---|---|
| | $NO_2$ | | | 270–279° | | 479 (u); 667 (m.p.); 700 (m.p.); 789 (m.p.); 955 (u); |
| | $NO_2$ | Cl | | 263° | | 664 (m.p.); |
| | $NO_2$ | $NO_2$ | | >300° | | 456 (i), (n), (u); 700 (m.p.); 716 (m.p.), (i); |
| | $NH_2$ | $NO_2$ | | 295° | | 556 (m.p.); |
| | | | | | hydrochloride, m.p. 270° | 556 (m.p.); |
| | | | | | *N'*-formyl, m.p. 298° | 556 (m.p.); |
| | | | | | *N'*-acetyl, m.p. 239–240° | 556 (m.p.); |
| | | | | | *N'*-carbethoxy, m.p. 165, 295° | 556 (m.p.); |
| $CH_3$ | $NO_2$ | Br | | 350° | | 293 (m.p.); |
| $CH_3$ | $NO_2$ | | | 273–275° | | 293 (m.p.); |
| $CH_3$ | | $NO_2$ | | 247–249° | | 293 (m.p.); |
| $CH_3$ | $NO_2$ | Cl | $CH_3$ | >320° | | 294 (m.p.); |
| $CH_3$ | $NO_2$ | Br | $CH_3$ | 340–345° | | 294 (m.p.); |
| $CH_3$ | $NO_2$ | I | $CH_3$ | 292–295° | | 294 (m.p.); |
| $CH_3$ | $NO_2$ | | $CH_3$ | 294° | | 294 (m.p.); 670 (m.p.); |
| $CH_3$ | | $NO_2$ | $CH_3$ | 296° | | 295 (m.p.); |

TABLE XII-71. Sulfur containing 4-Pyridinols

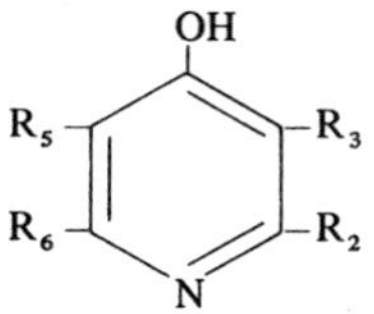

| $R_2$ | $R_3$ | $R_5$ | $R_6$ | m.p. | Derivatives | Ref. |
|---|---|---|---|---|---|---|
| | $SC_2H_5$ | $SC_2H_5$ | | 147–149° | | 962 (m.p.); |
| $C_4H_9$-*t* | $SO_3H$ | | $C_4H_9$-*t* | 256–258° | | 671 (m.p.); |

TABLE XII-72. Ethers and Esters of Alkyl and Aryl 4-Pyridinols

| R | $R_2$ | $R_3$ | $R_5$ | $R_6$ | m.p. | Derivatives | Ref. |
|---|---|---|---|---|---|---|---|
| $CH_3$ | D | D | D | D | | | 1102 (u); |
| $CH_3$ | | | | | 4°<br>b.p. 107°/50 mm<br>b.p. 185°/712 mm | | 275 (b.p.), (m.p.); 482 (z); 735 (n); 899 (n); 925 (t); 964 (j); 1102 (u); 1103 (n); 1106 (b.p.); |
| | | | | | | picrate, m.p. 170–172° | 1106 (m.p.); |
| | | | | | | hexachloro-antimonate, m.p. 135–138° | 464 (m.p.), (i); |
| $C_2H_5$ | | | | | | hydrochloride | 536 (n); |
| $CH_3$ | $CH_3$ | | | | | nitrate | 293; |
| $CH_2CH{=}CH_2$ | | | | | b.p. 104°/11 mm | | 598 (b.p.), (i), (n); |
| | | | | | | picrate, m.p. 109–110° | 598 (m.p.); |
| $N{=}C(CH_3)_2$ | | | | | 59–60°<br>b.p. 70°/4.5 mm | | 966 (b.p.), (m.p.); |
| | | | | | | hydrochloride, m.p. 192–193° | 966 (m.p.); |
| $CH_3$ | $CH_3$ | | | $CH_3$ | b.p. 70°/20 mm | | 137 (b.p.), (n); |
| $C_2H_5$ | $CH_3$ | | | | | nitrate, m.p. 84–86° | 293 (m.p.); |

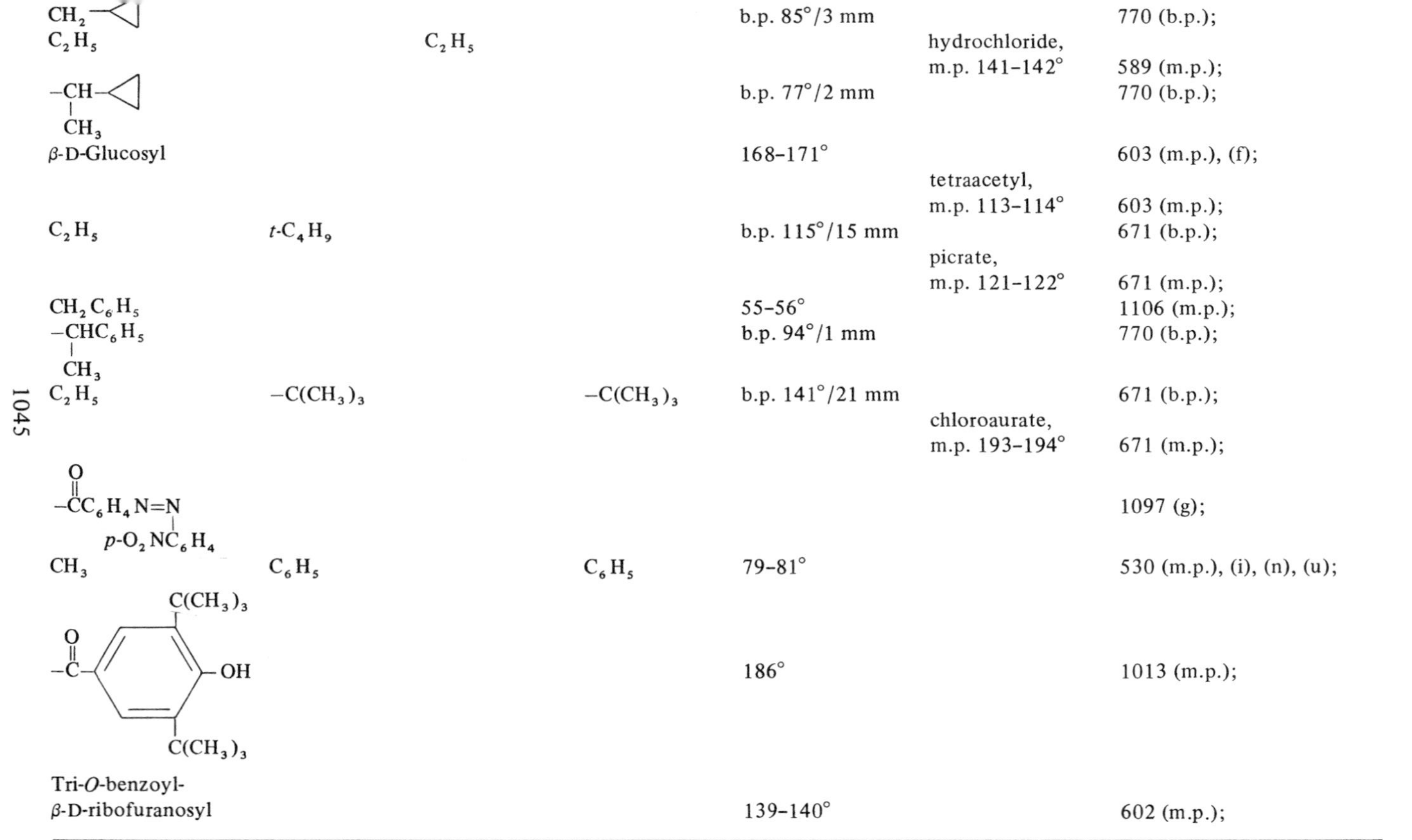

| | | | | | | |
|---|---|---|---|---|---|---|
| $CH_2$—(cyclopropyl) | | | | b.p. 85°/3 mm | | 770 (b.p.); |
| $C_2H_5$ | | $C_2H_5$ | | | hydrochloride, m.p. 141–142° | 589 (m.p.); |
| $-CH(CH_3)-$(cyclopropyl) | | | | b.p. 77°/2 mm | | 770 (b.p.); |
| β-D-Glucosyl | | | | 168–171° | | 603 (m.p.), (f); |
| | | | | | tetraacetyl, m.p. 113–114° | 603 (m.p.); |
| $C_2H_5$ | $t$-$C_4H_9$ | | | b.p. 115°/15 mm | | 671 (b.p.); |
| | | | | | picrate, m.p. 121–122° | 671 (m.p.); |
| $CH_2C_6H_5$ | | | | 55–56° | | 1106 (m.p.); |
| $-CH(CH_3)C_6H_5$ | | | | b.p. 94°/1 mm | | 770 (b.p.); |
| $C_2H_5$ | $-C(CH_3)_3$ | | $-C(CH_3)_3$ | b.p. 141°/21 mm | | 671 (b.p.); |
| | | | | | chloroaurate, m.p. 193–194° | 671 (m.p.); |
| $-C(=O)C_6H_4N{=}N-$ ($p$-$O_2NC_6H_4$) | | | | | | 1097 (g); |
| $CH_3$ | $C_6H_5$ | | $C_6H_5$ | 79–81° | | 530 (m.p.), (i), (n), (u); |
| $-C(=O)-$(3,5-$C(CH_3)_3$, 4-OH-phenyl) | | | | 186° | | 1013 (m.p.); |
| Tri-*O*-benzoyl-β-D-ribofuranosyl | | | | 139–140° | | 602 (m.p.); |

TABLE XII-73. Ethers of Amino 4-Pyridinols

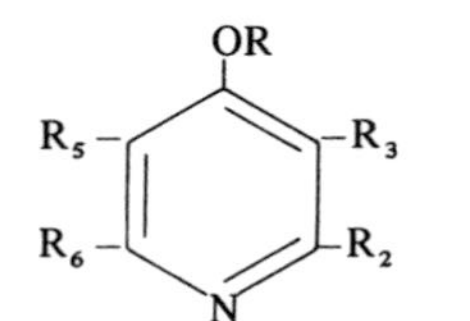


| R | $R_2$ | $R_3$ | $R_5$ | $R_6$ | m.p. | Derivatives | Ref. |
|---|---|---|---|---|---|---|---|
| $CH_3$ | | $NH_2$ | | | 83° | | 497 (u); 789 (m.p.); |
| $C_2H_5$ | $CH_3$ | $NH_2$ | | $CH_3$ | 62–63° | | 294 (m.p.); |

TABLE XII-74. Ethers of 4-Pyridinol Carboxylic Acids

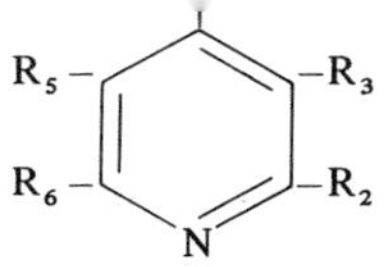

| R | $R_2$ | $R_3$ | $R_5$ | $R_6$ | m.p. | Derivatives | Ref. |
|---|---|---|---|---|---|---|---|
| COOH | | | | | | ethyl ester | 659 (u); |
| $CH_3$ | | COOH | | | | amide, m.p. 151–153° | 789 (m.p.), (u); |
| $CH_3$ | COOH | | | COOH | | dimethyl ester, m.p. 127–128° | 456 (m.p.), (i), (n), (u); 1107 (m.p.); |
| $C_2H_5$ | COOH | | | | | ethyl ester, b.p. 127°/1.3 mm | 1107 (b.p.); |
| | | | | | | ethyl ester picrate, m.p. 104–106° | 1107 (m.p.); |
| $C_2H_5$ | COOH | | | COOH | | diethyl ester, m.p. 85–87° | 1107 (m.p.); |

TABLE XII-75. Ethers of 4-Pyridinol Nitriles

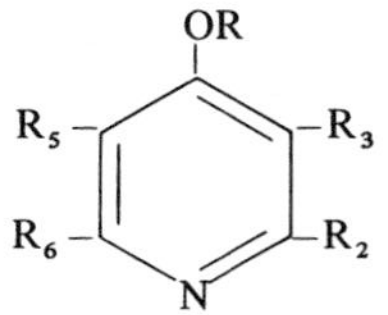

| R | $R_2$ | $R_3$ | $R_5$ | $R_6$ | m.p. | Derivatives | Ref. |
|---|---|---|---|---|---|---|---|
| $CH_3$ | | CN | | | 124° | | 789 (m.p.); |

TABLE XII-76. Ethers and Esters of Halo 4-Pyridinols

| R | $R_2$ | $R_3$ | $R_5$ | $R_6$ | m.p. | Derivatives | Ref. |
|---|---|---|---|---|---|---|---|
| $CH_3$ | Cl | | | | 229–230°<br>b.p. 106°/16 mm | picrate,<br>m.p. 168° | 241 (u); 252 (m.p.);<br>253 (m.p.); 768 (b.p.);<br>252 (m.p.); |
| $CH_3$ | | Cl | | | b.p. 105°/15 mm | picrate,<br>m.p. 159° | 255 (b.p.);<br>254 (m.p.); 255 (m.p.); |
| $CH_3$ | Br | | | | b.p. 120°/10 mm | picrate,<br>m.p. 115° | 252 (b.p.); 253 (b.p.);<br>252 (m.p.); |
| $CH_3$ | | Br | | | b.p. 114°/12 mm | picrate,<br>m.p. 160° | 255 (b.p.);<br>254 (m.p.); 255 (m.p.); |
| $CH_3$ | I | | | | 35° | picrate,<br>m.p. 154° | 252 (m.p.); 253 (m.p.);<br>252 (m.p.); |
| $CH_3$ | | I | | | 78° | picrate,<br>m.p. 162° | 254 (m.p.); 255 (m.p.);<br>255 (m.p.); |
| $CH_3$ | | Cl | Cl | | 225–227° | | 241 (m.p.), (u); |
| $CH_3$ | Cl | | | Cl | 100–102° | | 241 (m.p.), (u); |

| | | | | | | | |
|---|---|---|---|---|---|---|---|
| $CH_3$ | | Br | Br | | 85–86° | | 460 (i), (u); 705 (m.p.); |
| | | | | | | | 952 (m.p.), (n); |
| $CH_3$ | Cl | Cl | Cl | | 60–65° | | 236 (m.p.); |
| | | | | | | | 456 (m.p.), (n), (u); |
| $CH_3$ | Cl | Cl | | Cl | 136–138° | | 778 (m.p.), (n); |
| $CH_3$ | $NH_2$ | Cl | Cl | F | 125° | | 953 (m.p.); |
| $CH_3$ | F | Cl | Cl | F | b.p. 214° | | 246 (b.p.), (g), (i), (n); |
| | | | | | | | 950 (b.p.); |
| $CH_3$ | F | F | F | F | b.p. 161° | | 245 (b.p.), (i), (u); |
| | | | | | | | 248 (b.p.), (i), (u); |
| | | | | | | | 1111 (b.p.); |
| $CH_3$ | Cl | Cl | Cl | Cl | 107–114° | | 236 (m.p.); 239 (m.p.); |
| | | | | | | | 778 (m.p.), (n); |
| $CONHCH_3$ | Cl | Cl | Cl | Cl | 155° | | 972 (m.p.); |
| $C_2H_5$ | Cl | Cl | Cl | Cl | | | 241 (u); |
| $CH_2CH_2OH$ | F | Cl | Cl | F | 47–48° | | 953 (m.p.); |
| $CONHCH_3$ | Cl | Cl | Cl | | 140° | | 972 (m.p.); |
| $C_2H_5$ | Cl | | | | 55–57° | | 252 (m.p.); 253 (m.p.); |
| | | | | | | picrate, m.p. 133° | 252 (m.p.); |
| $C_2H_5$ | | Cl | | | b.p. 118°/20 mm | | 255 (b.p.); |
| | | | | | | picrate, m.p. 160° | 254 (m.p.); 255 (m.p.); |
| $C_2H_5$ | Br | | | | 38° | | 252 (m.p.); 253 (m.p.); |
| | | | | | | picrate, m.p. 115° | 252 (m.p.); |
| $C_2H_5$ | | Br | | | b.p. 116°/12 mm | | 255 (b.p.); |
| | | | | | | picrate, m.p. 161–162° | 254 (m.p.); 255 (m.p.); |
| $C_2H_5$ | I | | | | b.p. 145°/11 mm | | 252 (b.p.); |
| | | | | | | picrate, m.p. 132–133° | 252 (m.p.); 253 (m.p.); |
| $C_2H_5$ | | I | | | b.p. 143°/15 mm | | 255 (b.p.); |

TABLE XII-76. Ethers and Esters of Halo 4-Pyridinols (Continued)

(Structure: 4-OR pyridine ring with substituents $R_2$, $R_3$, $R_5$, $R_6$)

| R | $R_2$ | $R_3$ | $R_5$ | $R_6$ | m.p. | Derivatives | Ref. |
|---|---|---|---|---|---|---|---|
| | | | | | | picrate, m.p. 173° | 254 (m.p.); 255 (m.p.); |
| $C_2H_5$ | Br | Br | | | 80–81° | | 712 (m.p.); |
| $C_2H_5$ | $NH_2$ | Br | Br | | 100–101° | | 284 (m.p.); |
| $C_2H_5$ | $NH_2$ | Br | | | 147–148° | | 284 (m.p.); |
| $C_2H_5$ | $NH_2$ | | Br | | 149–150° | | 284 (m.p.), (g); |
| $CON(CH_3)_2$ | Cl | Cl | Cl | Cl | 156–159° | | 972 (m.p.); |
| $CONHC_2H_5$ | Cl | Cl | Cl | Cl | 150° | | 972 (m.p.); |
| $CH_2CH_2OCH_3$ | F | Cl | Cl | F | b.p. 75°/1 mm | | 953 (b.p.); |
| $CH_2CH_2OCH_3$ | Cl | Cl | Cl | Cl | 51–52° | | 236 (m.p.); |
| $CH_2CH_2OCH_3$ | Cl | Cl | Cl | | fluid | | 236; |
| $-CONHC_3H_7$-*n* | Cl | Cl | Cl | Cl | 130° | | 972 (m.p.); |
| $-CONHC_3H_7$-*n* | Cl | Cl | Cl | | 80–85° | | 972 (m.p.); |
| $C_2H_5$ | | $-CH=CH_2$ | | Cl | 70–72° | | 589 (m.p.); |
| $-COC(CH_3)_3$ | F | Cl | Cl | F | 63–64° | | 951 (m.p.); |
| –N (piperidine ring, drawn) | Cl | Cl | Cl | Cl | 135° | | 240 (m.p.); |
| $-CONHC_4H_9$-*n* | Cl | Cl | Cl | Cl | 113° | | 972 (m.p.); |
| $C_6H_4NO_2$-*p* | F | Cl | Cl | F | 85–87° | | 953 (m.p.); |
| $C_6H_5$ | F | Cl | Cl | F | 69–71° | | 953 (m.p.); |

| | | | | | | |
|---|---|---|---|---|---|---|
| $C_6H_5$ | Cl | Cl | Cl | Cl | 111° | 240 (m.p.); |
| $CH_3$ | Br | Br | Br | $-N$ | 81° | 952 (m.p.); |
| $CH_2CH_2OC_4H_9$ | Cl | Cl | Cl | | oil | 236; |
| $-Sn(C_2H_5)_3$ | | Cl | Cl | | 179–181° | 831 (m.p.), (i); |
| $-COC_6H_5$ | F | Cl | Cl | F | 96–97° | 951 (m.p.); |
| $-COC_6H_4OCH_3$-*p* | F | Cl | Cl | F | 69–70° | 951 (m.p.); |
| mesityl | F | Cl | Cl | F | b.p. 96°/0.08 mm | 953 (b.p.); |
| $-COC_{11}H_{23}$-*n* | F | Cl | Cl | F | 44–45° | 951 (m.p.); |
| $Pb(C_6H_5)_3$ | | Cl | Cl | | 241–242° | 831 (m.p.); |
| $Sn(C_6H_5)_3$ | | Cl | Cl | | 254–255° | 831 (m.p.), (i); |
| $CONHC_{18}H_{38}$ | Cl | Cl | Cl | Cl | 82–86° | 972 (m.p.); |

TABLE XII-77. Ethers of Nitro 4-Pyridinols.

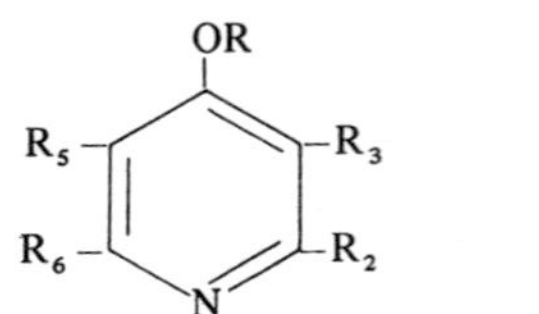


| R | $R_2$ | $R_3$ | $R_5$ | $R_6$ | m.p. | Derivatives | Ref. |
|---|---|---|---|---|---|---|---|
| $CH_3$ | | $NO_2$ | | | 72–74° | | 497 (u); 700 (m.p.); 705 (m.p.); 789 (m.p.); |
| $CH_3$ | | $NO_2$ | $NO_2$ | | 54–55° | 1:1 adduct with methoxide ion | 775 (n); 776 (m.p.); 776 (i), (n), (u), (v); |
| $C_2H_5$ | | $NO_2$ | | | 48–49° | | 705 (m.p.); 789 (m.p.); 955 (u); |
| $C_2H_5$ | $CH_3$ | $NO_2$ | | | 69–72° | | 293 (m.p.); |
| $C_2H_5$ | $CH_3$ | | $NO_2$ | | 61–63° | | 293 (m.p.); |

TABLE XII-78. Sulfur Containing Esters of 4-Pyridinols

| R | $R_2$ | $R_3$ | $R_5$ | $R_6$ | m.p. | Derivatives | Ref. |
|---|---|---|---|---|---|---|---|
| $SO_3H$ | | | | | | potassium salt | 648 (i), (u); |
| $-COSCH_3$ | Cl | | | Cl | 37–39° | | 976 (m.p.); |
| $-COSCH_3$ | Cl | Cl | Cl | | 70–72° | | 976 (m.p.); |
| $-COSCH_3$ | Cl | Cl | Cl | Cl | 99–101° | | 976 (m.p.); |
| $-COSC_6H_5$ | Cl | Cl | Cl | Cl | 90–91° | | 976 (m.p.); |
| $-SO_2C_6H_4CH_3$-*p* | F | Cl | Cl | F | 101–102° | | 951 (m.p.); |

TABLE XII-79. Alkyl and Aryl 4-Pyridones

(Structure: 4-pyridone ring, C=O at position 4, N-substituent $R_1$, ring substituents $R_2$, $R_3$, $R_5$, $R_6$)

| $R_1$ | $R_2$ | $R_3$ | $R_5$ | $R_6$ | m.p. | Derivatives | Ref. |
|---|---|---|---|---|---|---|---|
| $CD_3$ | | | | | | | 1112 (m); |
| $CH_3$ | D | | | D | 91–93° | | 139 (m.p.), (i), (n), (u); 530 (m); 731 (n); 1112 (m); |
| $CH_3$ | | D | D | | 92–94° | | 139 (m.p.), (i), (m), (n); (u); 1112 (m); |
| $CH_3$ | | | | | 91–99° b.p. 178°/4 mm | | 139 (m.p.); 460 (i); 464 (b.p.), (m.p.); 466 (i); 478 (i); 480 (z); 481 (z); 486 (n); 731 (m.p.); 735 (n); 819 (i); 980 (w); 1103 (n); 1112 (m); 1113 (i); |
| | | | | | | $^{15}N$ | 1112 (m); |
| | | | | | | hydrochloride | 536 (n); 1103 (i), (r); |
| | | | | | | hexachloro-antimonate, m.p. 155–159° | 464 (m.p.), (i); 542 (i); |
| | | | | | | hexachloro-stannate, m.p. 130–135° | 464 (m.p.), (i); |

| | | | | | | | |
|---|---|---|---|---|---|---|---|
| $COCH_3$ | | | | | 125–135° | | 633 (m.p.), (i), (n); |
| $CH_3$ | $CD_3$ | | | $CD_3$ | 247–249° | | 139 (m.p.), (i), (m), (n), (u); |
| $CH_3$ | D | $CH_3$ | $CH_3$ | D | 132–133° | | 139 (m.p.), (i), (m), (n), (u); |
| $CH_3$ | $CH_3$ | | | $CH_3$ | 240–250° | | 108 (m.p.), 137 (m.p.), (n); 139 (m.p.), (n), (u); 143 (m.p.); 147 (m.p.); 148 (i); 539 (i); 541 (i); 706 (m.p.); 819 (m.p.), (i); |
| | | | | | | trihydrate, m.p. 110° | 148 (m.p.); 706 (m.p.); 731 (n); 819 (m.p.), (i); |
| | | | | | | picrate, m.p. 197–199° | 148 (m.p.); 706 (m.p.); |
| | | | | | | hydrochloride, m.p. 260–270° | 706 (m.p.); 819 (m.p.), (i), |
| | | | | | | $CdCl_2$ complex, m.p. >300° | 819 (m.p.); |
| | | | | | | $HgCl_2$ complex, m.p. 191–193° | 819 (m.p.); |
| | | | | | | $ZnCl_2$ complex, m.p. >300° | 819 (m.p.); |
| | | | | | | $CaCl_2$ complex, m.p. >300° | 819 (m.p.); |
| | | | | | | $SnBr_4$ complex, m.p. 284–286° | 819 (m.p.); |
| | | | | | | hydrobromide, m.p. 272–273° | 539 (m.p.); |
| | | | | | | hydroiodide, m.p. 252–253° | 539 (m.p.); |

TABLE XII-79. Alkyl and Aryl 4-Pyridones (Continued)

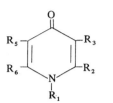

| $R_1$ | $R_2$ | $R_3$ | $R_5$ | $R_6$ | m.p. | Derivatives | Ref. |
|---|---|---|---|---|---|---|---|
| | | | | | | $HAsF_6 \cdot H_2O$, m.p. 124–126° | 539 (m.p.); |
| | | | | | | $HAsF_6$, m.p. 151° | 539 (m.p.); |
| | | | | | | $HSbCl_6$, m.p. 239–240° | 539 (m.p.); |
| | | | | | | $HPF_6 \cdot H_2O$, m.p. 118–119° | 539 (m.p.); |
| | | | | | | $HPF_6$, m.p. 150–155° | 539 (m.p.); |
| $CH_3$ | | $CH_3$ | $CH_3$ | | 131–133° | | 139 (m.p.), (i), (n), (u); 731 (m.p.), (n); |
| $-C(OC_2H_5)=CH_2$ | | | | | b.p. 140°/0.75 mm | | 746 (b.p.), (i), (n), (u); |
| $-(CH_2)_4Br$ | | | | | | hydrobromide, m.p. 132–135° | 588 (m.p.); |
| | | | | | | picrate, m.p. 112–115° | 588 (m.p.); |
| $-(CH_2)_4I$ | | | | | | hydroiodide, m.p. 177–179° | 588 (m.p.); |
| | | | | | | picrate, m.p. 102–104° | 588 (m.p.); |

|  |  |  |  |  |  |
| --- | --- | --- | --- | --- | --- |
|  |  |  |  | polymer·HI, m.p. 210–212° | 588 (m.p.), (u); |
| $C_2H_5$ | $CH_3$ | $CH_3$ | 158–164° |  | 108 (m.p.); 143 (m.p.); 148 (i); 539 (i); 819 (m.p.), (i); |
|  |  |  |  | hydrate, m.p. 59–74° | 41 (m.p.), (u); 143 (m.p.); 144 (m.p.); 148 (m.p), (i); 541 (i); 706 (m.p.); 819 (m.p.); |
|  |  |  |  | picrate, m.p. 191° | 148 (i); 706 (m.p.); |
|  |  |  |  | hydrochloride, m.p. 242–244° | 819 (m.p.); |
|  |  |  |  | $CoCl_2$ complex, m.p. 267–269° | 819 (m.p.); |
|  |  |  |  | $CdCl_2$ complex, m.p. 258–261° | 819 (m.p.); |
|  |  |  |  | $HgCl_2$ complex, m.p. 165–167° | 819 (m.p.); |
|  |  |  |  | $ZnCl_2$ complex, m.p. 257–262° | 819 (m.p.); |
|  |  |  |  | $SnBr_4$ complex, m.p. 110–114° | 819 (m.p.); |
|  |  |  |  | $SbCl_5$ complex, m.p. 160–165° | 819 (m.p.); |
|  |  |  |  | hydrobromide, m.p. 234–235° | 539 (m.p.); |
|  |  |  |  | hydroiodide, m.p. 208–210° | 539 (m.p.); |
|  |  |  |  | $H_2SiF_6$ complex, m.p. 150–151° | 539 (m.p.); |
|  |  |  |  | $HClO_4$ complex, m.p. 208–210° | 539 (m.p.); |

TABLE XII-79. Alkyl and Aryl 4-Pyridones (Continued)

| $R_1$ | $R_2$ | $R_3$ | $R_5$ | $R_6$ | m.p. | Derivatives | Ref. |
|---|---|---|---|---|---|---|---|
| | | | | | | $H_2SnBr_6$ complex, m.p. 202–204° | 539 (m.p.); |
| | | | | | | $H_2PtCl_6$ complex, m.p. 221–222° | 539 (m.p.); |
| | | | | | | $H_2SnCl_6$ complex, m.p. 256–259° | 539 (m.p.); |
| | | | | | | $HBF_4{\cdot}H_2O$, m.p. 68–70° | 539 (m.p.); |
| | | | | | | $HBF_4$ complex, m.p. 164–168° | 539 (m.p.); |
| | | | | | | $HAsF_6$ complex, m.p. 183–184° | 539 (m.p.); |
| | | | | | | $HPF_6$ complex, m.p. 165–168° | 539 (m.p.); |
| $CH_3$ | $CH_3$ | $CH_3$ | | $CH_3$ | b.p. 100°/10 mm | | 113 (b.p.); |
| | | | | | | hydrochloride, m.p. 252–254° | 113 (m.p.); |
| | | | | | | 2,4-dinitro-phenylhydrazone, m.p. 112–113° | 113 (m.p.); |
| $CH_2CH_2OH$ | $CH_3$ | | | $CH_3$ | 224–225° | | 147 (m.p.); |
| 4-Pyridyl | | | | | 177–178° | | 1104 (u); 1113 (m.p.); 1114 (m.p.); |

| | | | | | | |
|---|---|---|---|---|---|---|
| | | | | | picrate, m.p. 203–205° | 1114 (m.p.); |
| $n$-$C_3H_7$ | $CH_3$ | | $CH_3$ | | picrate, m.p. 175–176° | 143 (m.p.); |
| iso$C_3H_7$ | $CH_3$ | | $CH_3$ | 49–53° | | 41 (m.p.), (u); |
| $-(CH_2)_3OH$ | $CH_3$ | | $CH_3$ | 188–189° | | 145 (m.p.); |
| $C_6H_5Hg-$ | | | | 297–298° | | 831 (m.p.), (i); |
| β-D-Glucosyl | | | | amorphous | | 623 (f), (k); 1026 (k); |
| | | | | | tetraacetyl, amorphous | 623 (k); 1026 (k); |
| $n$-$C_4H_9$ | $CH_3$ | | $CH_3$ | 65° | | 118 (m.p.); |
| | | | | | hydrate, m.p. 88–89° | 143 (m.p.); |
| | | | | | picrate, m.p. 151–152° | 143 (m.p.); |
| $CH_2C_6H_4Br$-$o$ | | | | 121–122° | | 1017 (m.p.); |
| $CH_2C_6H_4NO_2$-$o$ | | | | 136–138° | | 564 (m.p.); |
| $CH_2C_6H_4NO_2$-$m$ | | | | 220–222° | | 564 (m.p.); |
| $CH_2C_6H_4NO_2$-$p$ | | | | 184–185° | | 564 (m.p.); |
| $CH_2C_6H_5$ | | | | 98–99° | | 279 (m.p.); |
| | | | | 109–111° | | 466 (m.p.), (i); |
| $CH_2C_6H_4NH_2$-$o$ | | | | 168–169° | | 564 (m.p.); |
| $CH_2C_6H_4NH_2$-$m$ | | | | 204–205° | | 564 (m.p.); |
| $CH_2C_6H_4NH_2$-$p$ | | | | 206–207° | | 564 (m.p.); |
| N; $CH_3$ | | $CH_3$ | | 197–198° | | 273 (m.p.); 278 (m.p.); |
| | | | | | picrate, m.p. 209–210° | 273 (m.p.); |

TABLE XII-79. Alkyl and Aryl 4-Pyridones (Continued)

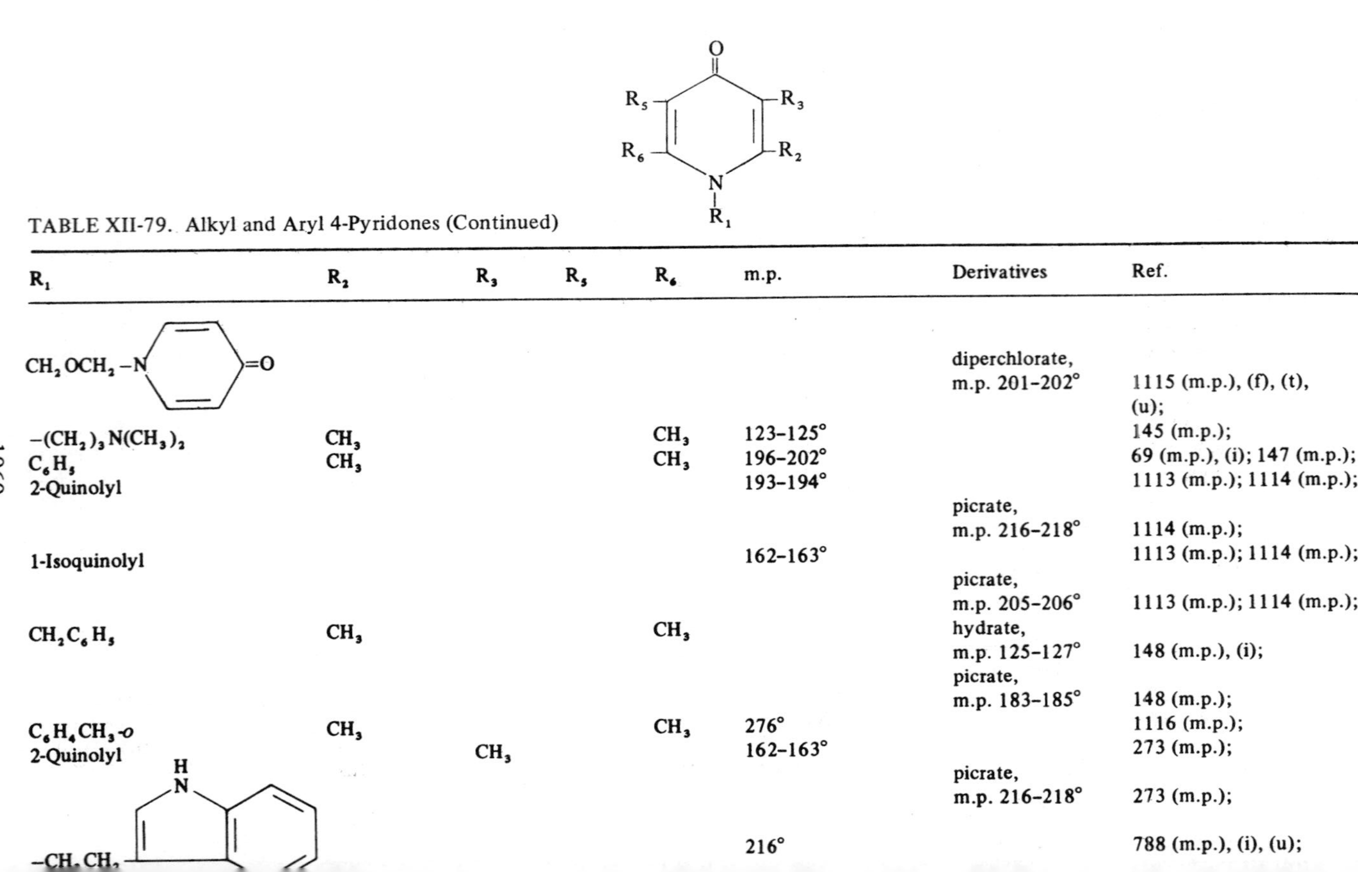

| $R_1$ | $R_2$ | $R_3$ | $R_5$ | $R_6$ | m.p. | Derivatives | Ref. |
|---|---|---|---|---|---|---|---|
| $CH_2OCH_2$–N[4-pyridone ring]=O | | | | | | diperchlorate, m.p. 201–202° | 1115 (m.p.), (f), (t), (u); |
| $-(CH_2)_3N(CH_3)_2$ | $CH_3$ | | | $CH_3$ | 123–125° | | 145 (m.p.); |
| $C_6H_5$ | $CH_3$ | | | $CH_3$ | 196–202° | | 69 (m.p.), (i); 147 (m.p.); |
| 2-Quinolyl | | | | | 193–194° | | 1113 (m.p.); 1114 (m.p.); |
| | | | | | | picrate, m.p. 216–218° | 1114 (m.p.); |
| 1-Isoquinolyl | | | | | 162–163° | | 1113 (m.p.); 1114 (m.p.); |
| | | | | | | picrate, m.p. 205–206° | 1113 (m.p.); 1114 (m.p.); |
| $CH_2C_6H_5$ | $CH_3$ | | | $CH_3$ | | hydrate, m.p. 125–127° | 148 (m.p.), (i); |
| | | | | | | picrate, m.p. 183–185° | 148 (m.p.); |
| $C_6H_4CH_3$-*o* | $CH_3$ | | | $CH_3$ | 276° | | 1116 (m.p.); |
| 2-Quinolyl | | $CH_3$ | | | 162–163° | | 273 (m.p.); |
| | | | | | | picrate, m.p. 216–218° | 273 (m.p.); |
| $-CH_2CH_2$–[3-indolyl] | | | | | 216° | | 788 (m.p.), (i), (u); |

| | | | | | |
|---|---|---|---|---|---|
| $CH_2CH_2C_6H_5$ | $CH_3$ | $CH_3$ | 164–168° | | 145 (m.p.), (i), (u); 1117 (m.p.), (t); |
| | | | | hydrate, m.p. 167–168° | 148 (m.p.), (i); |
| | | | | picrate, m.p. 218–220° | 148 (m.p.), (i); |
| $-(CH_2)_4OC_6H_5$ | | | 52–55° | | 588 (m.p.); |
| | | | | picrate, m.p. 139–141° | 588 (m.p.); |
| | | | | iodide, m.p. 149–151° | 588 (m.p.); |
| 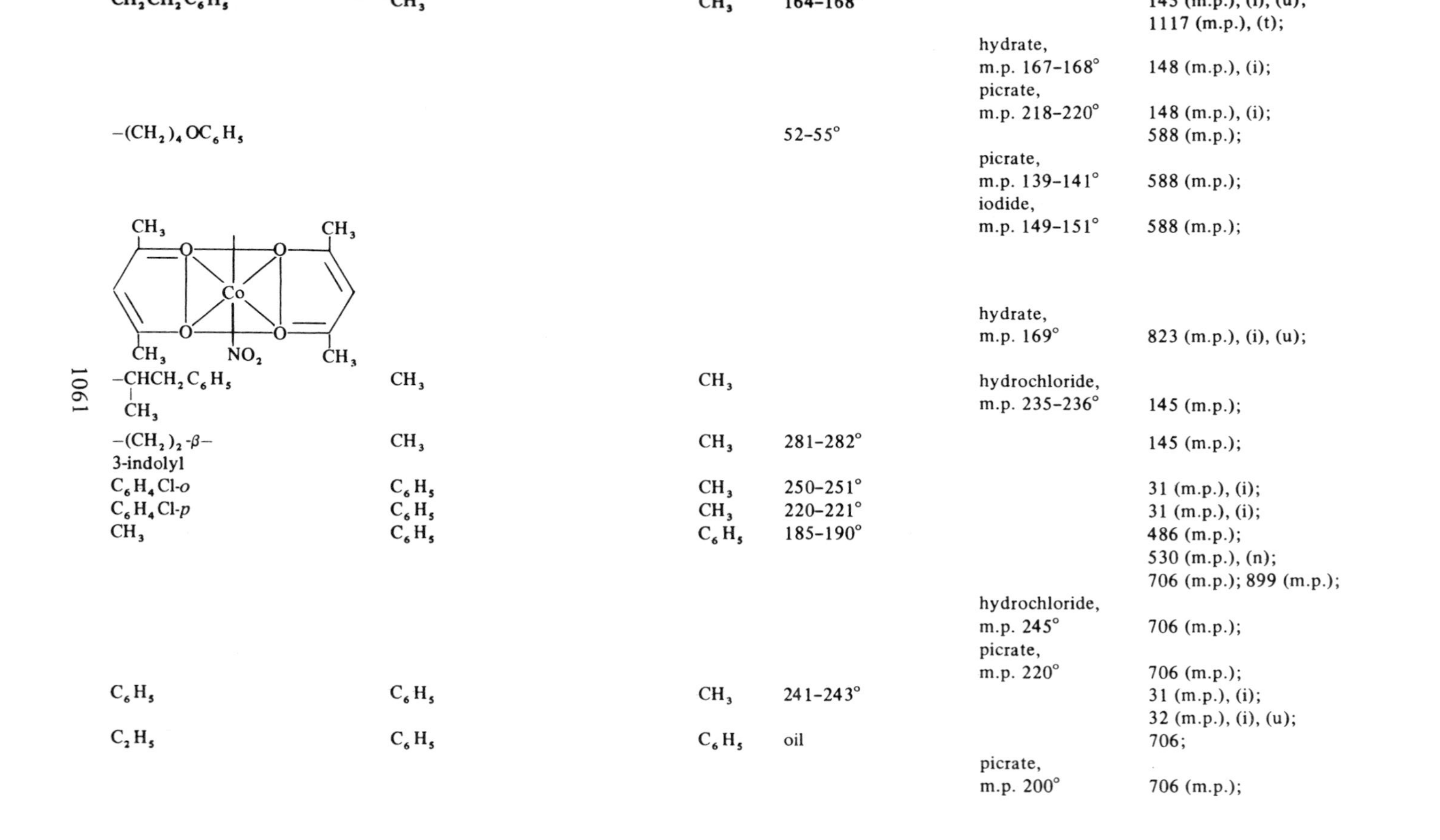 | | | | hydrate, m.p. 169° | 823 (m.p.), (i), (u); |
| $-CH(CH_3)CH_2C_6H_5$ | $CH_3$ | $CH_3$ | | hydrochloride, m.p. 235–236° | 145 (m.p.); |
| $-(CH_2)_2$-β-3-indolyl | $CH_3$ | $CH_3$ | 281–282° | | 145 (m.p.); |
| $C_6H_4Cl$-*o* | $C_6H_5$ | $CH_3$ | 250–251° | | 31 (m.p.), (i); |
| $C_6H_4Cl$-*p* | $C_6H_5$ | $CH_3$ | 220–221° | | 31 (m.p.), (i); |
| $CH_3$ | $C_6H_5$ | $C_6H_5$ | 185–190° | | 486 (m.p.); 530 (m.p.), (n); 706 (m.p.); 899 (m.p.); |
| | | | | hydrochloride, m.p. 245° | 706 (m.p.); |
| | | | | picrate, m.p. 220° | 706 (m.p.); |
| $C_6H_5$ | $C_6H_5$ | $CH_3$ | 241–243° | | 31 (m.p.), (i); 32 (m.p.), (i), (u); |
| $C_2H_5$ | $C_6H_5$ | $C_6H_5$ | oil | | 706; |
| | | | | picrate, m.p. 200° | 706 (m.p.); |

TABLE XII-79. Alkyl and Aryl 4-Pyridones (Continued)

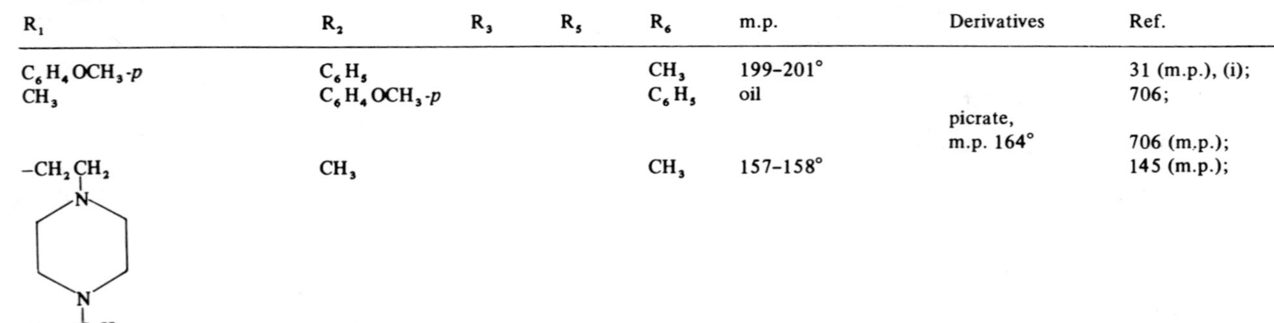


| $R_1$ | $R_2$ | $R_3$ | $R_5$ | $R_6$ | m.p. | Derivatives | Ref. |
|---|---|---|---|---|---|---|---|
| $C_6H_4OCH_3$-*p* | $C_6H_5$ | | | $CH_3$ | 199–201° | | 31 (m.p.), (i); |
| $CH_3$ | $C_6H_4OCH_3$-*p* | | | $C_6H_5$ | oil | | 706; |
| | | | | | | picrate, m.p. 164° | 706 (m.p.); |
| $-CH_2CH_2$-(4-$C_6H_5$-piperazin-1-yl) | $CH_3$ | | | $CH_3$ | 157–158° | | 145 (m.p.); |

| | | | | | |
|---|---|---|---|---|---|
| $C_2H_5$ | $C_6H_4OCH_3$-*p* | $C_6H_5$ | oil | picrate,<br>m.p. 199° | 706;<br>706 (m.p.); |
| $-CH_2CH_2CH_2-N(CH_2CH_2)_2N-C_6H_5$ | $CH_3$ | $CH_3$ | 144–145° | | 145 (m.p.); |
| $C_6H_4N(CH_3)_2$-*p* | *n*-$C_4H_9$ | *n*-$C_4H_9$ | 116–118° | | 1118 (m.p.), (i); |
| $CH_2CH_2C_6H_5$ | $C_6H_5$ | $C_6H_5$ | 247–252° | | 145 (m.p.); |
| $C_6H_4N(CH_3)_2$-*p* | $C_6H_5$ | $C_6H_5$ | 289–290° | | 1118 (m.p.), (i); |

TABLE XII-80. Alkyloxy, Aryloxy, and Hydroxy 4-Pyridones

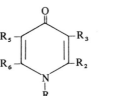

| $R_1$ | $R_2$ | $R_3$ | $R_5$ | $R_6$ | m.p. | Derivatives | Ref. |
|---|---|---|---|---|---|---|---|
| $OCH_3$ | | | | | | | 466 (i); 486 (n); |
| $CH_3$ | $CH_3$ | OH | | | 266–268° | | 131 (m.p.); |
| $C_2H_5$ | $CH_3$ | OH | | | 216–217° | | 131 (m.p.); |
| $CH_2CH_2OH$ | $CH_3$ | OH | | | 204–205° | | 131 (m.p.); |
| $CH_2CH_2Cl$ | $CH_2Cl$ | | $OCH_3$ | | 151–152° | | 1119 (m.p.); |
| | | | | | | hydrochloride, m.p. 195–196° | 1119 (m.p.); |
| $n$-$C_3H_7$ | $CH_3$ | OH | | | 164–165° | | 131 (m.p.); |
| iso$C_3H_7$ | $CH_3$ | OH | | | 254–256° | | 131 (m.p.); |
| –O–(4-pyridyl) | | | | | 130° | | 280 (m.p.), (e), (i), (m), (n), (u); |
| β-D-Arabinopentofuranosyl-2-hydroxy | | | | | | 2,2′-anhydro, m.p. 201°/218° | 617 (m.p.), (k), (u); |
| $C_3H_6Cl$ | $CH_2Cl$ | $OCH_3$ | | | 117–120° | | 1119 (m.p.); |
| | | | | | | hydrochloride, m.p. 169–177° | 1119 (m.p.); |
| $n$-$C_4H_9$ | $CH_3$ | OH | | | 126–127° | | 131 (m.p); |
| iso$C_4H_9$ | $CH_3$ | OH | | | 173–174° | | 131 (m.p.); |
| $C_6H_4Cl$-$p$ | | OH | | | 236–237° | | 129 (m.p.), (u); 1120 (u); |
| $C_6H_4Br$-$p$ | | OH | | | 247–248° | | 129 (m.p.), (u); 1120 (u); |
| $C_6H_5$ | | OH | | | 164° | hydrochloride, m.p. 197° picrate, | 128 (m.p.); |

| | | | | | | | |
|---|---|---|---|---|---|---|---|
| $C_6H_5$ | OH | | | OH | 84° | | 801 (m.p.); |
| | | | | | | hydrochloride, m.p. 205° | 801 (m.p.); |
| | | | | | | anil, m.p. 239° | 801 (m.p.); |
| | | | | | | phenyl hydrazone, m.p. > 250° | 801 (m.p.); |
| β-D-Glucosyloxy | | | | | 125–127° | | 623 (m.p.), (f); 1026 (m.p.), (k); |
| | | | | | | tetraacetyl, m.p. 134–136° | 623 (m.p.); 1026 (m.p.), (k); |
| $n$-$C_5H_{11}$ | $CH_3$ | OH | | | 127–128° | | 131 (m.p.); |
| $CH_2CH_2N(CH_3)_2$ | $CH_2OH$ | | $OCH_3$ | | 134–137° | | 130 (m.p.); |
| $OCH_2C_6H_5$ | | | | | | | 466 (i); |
| $C_6H_5$ | | $OCH_3$ | | | | hydrochloride, m.p. 180–181° | 128 (m.p.); |
| | | | | | | picrate, m.p. 150° | 128 (m.p.); |
| $C_6H_4CH_3$-$o$ | | OH | | | 206° | | 128 (m.p.); |
| | | | | | | hydrochloride, m.p. 250–252° | 128 (m.p.); |
| | | | | | | picrate, m.p. 146° | 128 (m.p.); |
| $C_6H_4CH_3$-$m$ | | OH | | | 158–160° | | 128 (m.p.); |
| | | | | | | $HCl \cdot H_2O$ m.p. 190–192° | 128 (m.p.); |
| | | | | | | picrate, m.p. 167–168° | 128 (m.p.); |
| $C_6H_4CH_3$-$p$ | | OH | | | 196° | | 128 (m.p.); |
| | | | | | | hydrochloride, m.p. 215–216° | 128 (m.p.); |
| | | | | | | picrate, m.p. 174° | 128 (m.p.); |

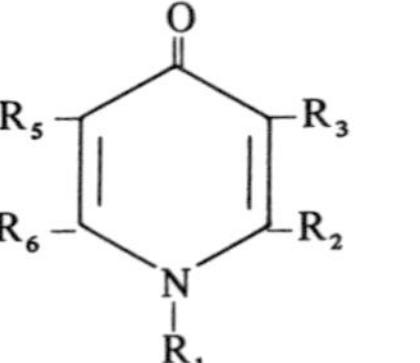

TABLE XII-80. Alkyloxy, Aryloxy, and Hydroxy 4-Pyridones (Continued)

| $R_1$ | $R_2$ | $R_3$ | $R_5$ | $R_6$ | m.p. | Derivatives | Ref. |
|---|---|---|---|---|---|---|---|
| $C_6H_{11}$ | $CH_3$ | OH | | | 206–207° | | 131 (m.p.); |
| $n$-$C_6H_{13}$ | $CH_3$ | OH | | | 125–126° | | 131 (m.p.); |
| $CH_2CH_2N(C_2H_5)_2$ | $CH_3$ | OH | | | 147–148° | | 131 (m.p.); |
| $CH_2C_6H_5$ | $CH_3$ | OH | | | 204–205° | | 131 (m.p.); |
| $C_6H_4CH_3$-$o$ | | $OCH_3$ | | | | hydrochloride, m.p. 196–197°<br>picrate, m.p. 187–188° | 128 (m.p.);<br>128 (m.p.); |
| $C_6H_4CH_3$-$m$ | | $OCH_3$ | | | | $HCl \cdot .5H_2O$, m.p. 186°<br>picrate, m.p. 163° | 128 (m.p.);<br>128 (m.p.); |
| $C_6H_4CH_3$-$p$ | | $OCH_3$ | | | | HCl, m.p. 204°<br>picrate, m.p. 164–165° | 128 (m.p.);<br>128 (m.p.); |
| $CH_3$ | $CH_3$ | β-D-Glucosyloxy | | | 228° | | 131 (m.p.); |
| $-CH_2CH_2-N$ | $CH_2OH$ | | $OCH_3$ | | 153–155° | | 130 (m.p.); |
| $n$-$C_7H_{15}$ | $CH_3$ | OH | | | 128–129° | | 131 (m.p.); |
| $-CH_2CH_2N(C_2H_5)_2$ | $CH_2OH$ | | $OCH_3$ | | 90–93° | <br>hydrochloride, m.p. 182–185° | 130 (m.p.); 1119 (m.p.);<br>130 (m.p.); |
| $C_2H_5$ | $CH_3$ | β-D-Glucosyloxy | | | 241–243° | | 131 (m.p.); |

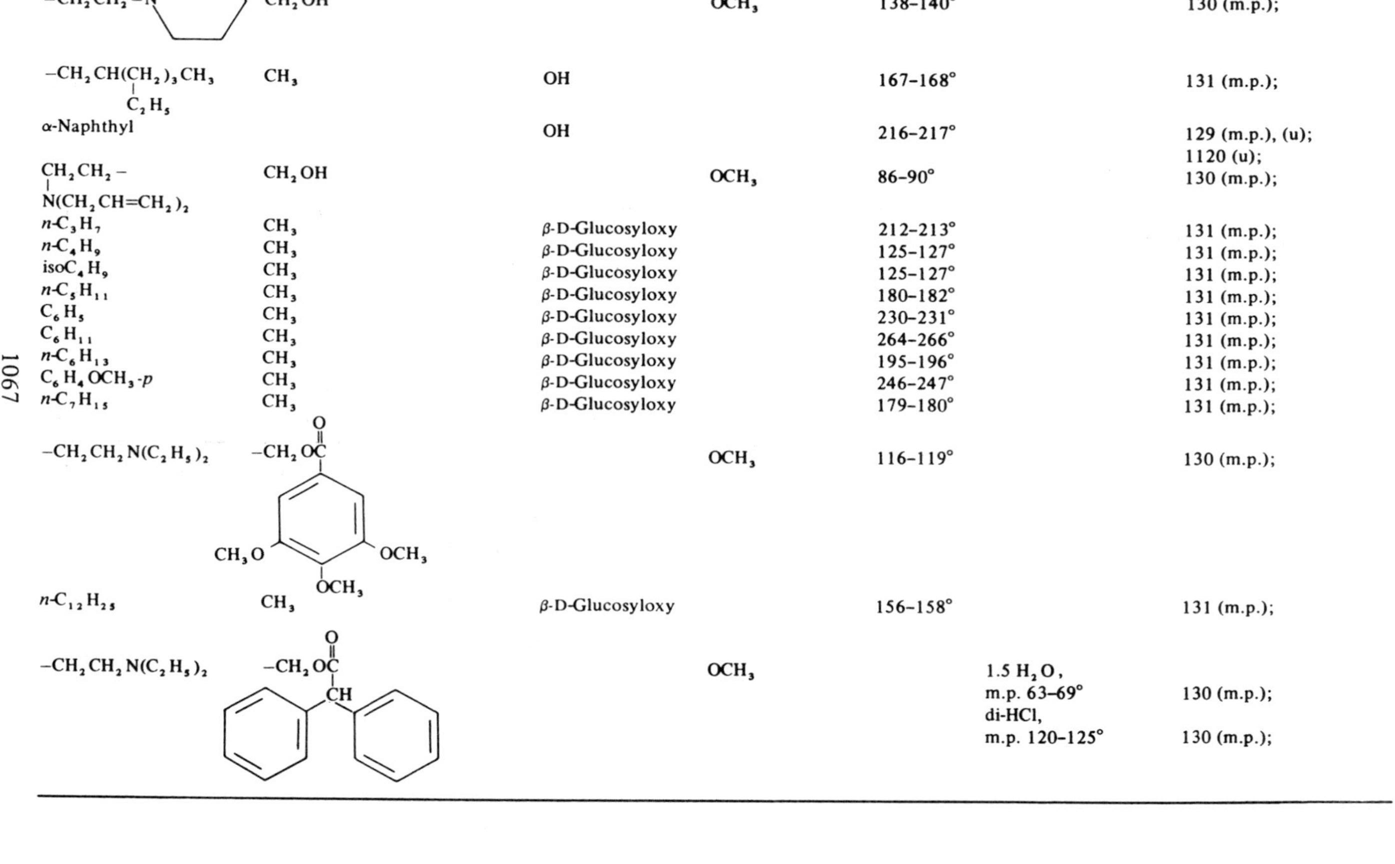

| | | | | | | |
|---|---|---|---|---|---|---|
| $-CH_2CH_2-N$(piperidino) | $CH_2OH$ | | $OCH_3$ | 138–140° | | 130 (m.p.); |
| $-CH_2CH(CH_2)_3CH_3$ with $C_2H_5$ branch | $CH_3$ | OH | | 167–168° | | 131 (m.p.); |
| α-Naphthyl | | OH | | 216–217° | | 129 (m.p.), (u); 1120 (u); |
| $CH_2CH_2-N(CH_2CH{=}CH_2)_2$ | $CH_2OH$ | | $OCH_3$ | 86–90° | | 130 (m.p.); |
| $n$-$C_3H_7$ | $CH_3$ | β-D-Glucosyloxy | | 212–213° | | 131 (m.p.); |
| $n$-$C_4H_9$ | $CH_3$ | β-D-Glucosyloxy | | 125–127° | | 131 (m.p.); |
| iso$C_4H_9$ | $CH_3$ | β-D-Glucosyloxy | | 125–127° | | 131 (m.p.); |
| $n$-$C_5H_{11}$ | $CH_3$ | β-D-Glucosyloxy | | 180–182° | | 131 (m.p.); |
| $C_6H_5$ | $CH_3$ | β-D-Glucosyloxy | | 230–231° | | 131 (m.p.); |
| $C_6H_{11}$ | $CH_3$ | β-D-Glucosyloxy | | 264–266° | | 131 (m.p.); |
| $n$-$C_6H_{13}$ | $CH_3$ | β-D-Glucosyloxy | | 195–196° | | 131 (m.p.); |
| $C_6H_4OCH_3$-$p$ | $CH_3$ | β-D-Glucosyloxy | | 246–247° | | 131 (m.p.); |
| $n$-$C_7H_{15}$ | $CH_3$ | β-D-Glucosyloxy | | 179–180° | | 131 (m.p.); |
| $-CH_2CH_2N(C_2H_5)_2$ | $-CH_2OC(=O)$-3,4,5-$(OCH_3)_3C_6H_2$ | | $OCH_3$ | 116–119° | | 130 (m.p.); |
| $n$-$C_{12}H_{25}$ | $CH_3$ | β-D-Glucosyloxy | | 156–158° | | 131 (m.p.); |
| $-CH_2CH_2N(C_2H_5)_2$ | $-CH_2OC(=O)CH(C_6H_5)_2$ | | $OCH_3$ | | 1.5 $H_2O$, m.p. 63–69° | 130 (m.p.); |
| | | | | | di-HCl, m.p. 120–125° | 130 (m.p.); |

TABLE XII-81. Amino 4-Pyridones

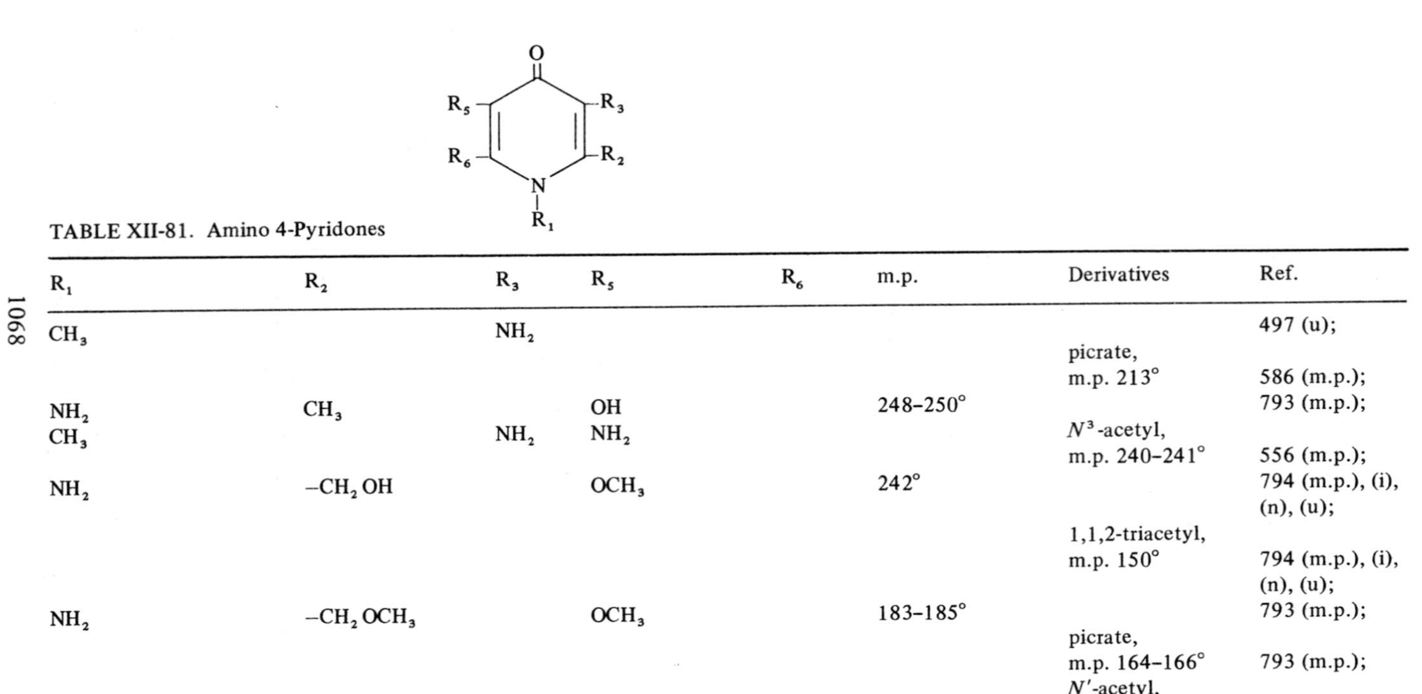

| $R_1$ | $R_2$ | $R_3$ | $R_5$ | $R_6$ | m.p. | Derivatives | Ref. |
|---|---|---|---|---|---|---|---|
| $CH_3$ | | $NH_2$ | | | | | 497 (u); |
| | | | | | | picrate, m.p. 213° | 586 (m.p.); |
| $NH_2$ | $CH_3$ | | OH | | 248–250° | | 793 (m.p.); |
| $CH_3$ | | $NH_2$ | $NH_2$ | | | $N^3$-acetyl, m.p. 240–241° | 556 (m.p.); |
| $NH_2$ | $-CH_2OH$ | | $OCH_3$ | | 242° | | 794 (m.p.), (i), (n), (u); |
| | | | | | | 1,1,2-triacetyl, m.p. 150° | 794 (m.p.), (i), (n), (u); |
| $NH_2$ | $-CH_2OCH_3$ | | $OCH_3$ | | 183–185° | | 793 (m.p.); |
| | | | | | | picrate, m.p. 164–166° | 793 (m.p.); |
| | | | | | | $N'$-acetyl, m.p. 207° | 793 (m.p.); |

| | | | | | | | |
|---|---|---|---|---|---|---|---|
| $-NHC_6H_4NO_2$-*p* | OH | | | OH | 152° | | 140 (m.p.); |
| | | | | | | hydrochloride, m.p. 185° | 140 (m.p.); |
| $-CH_2C_6H_5$ | | $NH_2$ | | | 161–162° | | 564 (m.p.); |
| $-N{=}CHC_6H_4NO_2$-*p* | $CH_3$ | | OH | | 259° | | 793 (m.p.); |
| $NH_2$ | $CH_2Cl$ | | $-OCH_2C_6H_5$ | | 157° | | 793 (m.p.); |
| | | | | | | hydrochloride, m.p. 209° | 793 (m.p.); |
| $-NHC_6H_4NO_2$-*p* | $CH_3$ | | | $CH_3$ | 178° | | 140 (m.p.); |
| | | | | | | hydrochloride, m.p. 203° | 140 (m.p.); |
| $-NHC_6H_4NO_2$-*p* | $-OCH_3$ | | | $OCH_3$ | 153° | | 140 (m.p.); |
| $NH_2$ | $CH_3$ | | $-OCH_2C_6H_5$ | | 195° | | 793 (m.p.); |
| | | | | | | acetate, m.p. 206° | 793 (m.p.); |
| $NH_2$ | $-CH_2OH$ | | $-OCH_2C_6H_5$ | | 174° | | 793 (m.p.); |
| $-N(CH_3)C_6H_5$ | $CH_3$ | | | $CH_3$ | 155–156° | | 1121 (m.p.); |
| $-N(C_6H_5)_2$ | $CH_3$ | | | $CH_3$ | 193–194° | | 1121 (m.p.); |
| $-NHCOC_2H_5$ | $-CH_2OCOC_2H_5$ | | $-OCH_2C_6H_5$ | | 154–155° | | 793 (m.p.); |
| $-N{=}CHC_6H_4NO_2$-*p* | $CH_2OH$ | | $-OCH_2C_6H_5$ | | 184–185° | | 793 (m.p.); |
| $-N{=}CHC_6H_5$ | $CH_2OH$ | | $-OCH_2C_6H_5$ | | 110°, 171° | | 793 (m.p.); |

TABLE XII-82. 4-Pyridone Carboxylic Acids

| $R_1$ | $R_2$ | $R_3$ | $R_5$ | $R_6$ | m.p. | Derivatives | Ref. |
|---|---|---|---|---|---|---|---|
| COOH | | $NH_2$ | | | | ethyl ester picrate·½$H_2O$, m.p. 217–218° | 586 (m.p.); |
| COOH | | NHCOOH | | | | diethyl ester, m.p. 85° | 586 (m.p.); |
| $CH_3$ | | COOH | | | | amide, 179–181° | 789 (m.p.), (i), (u); |
| $CH_3$ | COOH | | | COOH | | dimethyl ester, m.p. 97–98° | 456 (m.p.), (i), (n), (u); |
| $-CH_2CH(NH_2)COOH$ | | OH | | | 224–226° | | 134 (m.p.); |
| $-CH_2COOH$ | $CH_3$ | | | $CH_3$ | 297° | | 41 (m.p.), (u); |
| | | | | | | amide, m.p. > 320° | 41 (m.p.), (u); |
| $-CH_2CH_2COOH$ | $CH_3$ | | | $CH_3$ | | | 41 (u); |
| $-CH(CH_3)COOH$ | $CH_3$ | | | $CH_3$ | 284–300° | | 41 (m.p.), (u); |
| | | | | | | amide, m.p. > 320° | 41 (m.p.), (u); |
| $-CH_2CONH-CH_2$ | $CH_3$ | | | $CH_3$ | 300–310° | | 41 (m.p.), (u); |

| | | | | | | | |
|---|---|---|---|---|---|---|---|
| $C_6H_4Cl$-*p* | COOH | | OH | | | hydrate, m.p. 236° | 129 (m.p.); |
| | | | | | | ethyl ester, m.p. 203–204° | 129 (m.p.), (u); 1120 (u); |
| $C_6H_4Br$-*p* | COOH | | OH | | | hydrate, m.p. 192–193° | 129 (m.p.); |
| | | | | | | ethyl ester, m.p. 204–205° | 129 (m.p.), (u); 1120 (u); |
| $C_6H_4NO_2$-*m* | COOH | | OH | | 256–257° | | 129 (m.p.); |
| | | | | | | ethyl ester, m.p. 182–183° | 129 (m.p.), (u); 1120 (u); |
| $C_6H_5$ | COOH | | OH | | 200° | | 128 (m.p.); |
| | | | | | | methyl ester, m.p. 197–198° | 129 (m.p.), (u); 1120 (u); |
| | | | | | | ethyl ester, m.p. 177–178° | 129 (m.p.), (u); 1120 (u); |
| $-CH(CH_3)C(O)NHCH_2COOH$ | $CH_3$ | | | $CH_3$ | | ethyl ester, m.p. 139° | 41 (m.p.), (u); |
| $C_6H_4CH_3$-*o* | COOH | | OH | | 206° | | 128 (m.p.); |
| $C_6H_4CH_3$-*m* | COOH | | OH | | 190° | | 128 (m.p.); |
| $C_6H_4CH_3$-*p* | COOH | | OH | | 197–199° | | 128 (m.p.); |
| | | | | | | ethyl ester, m.p. 179–181° | 129 (m.p.), (u); 1120 (u); |
| $C_6H_5$ | COOH | | $OCH_3$ | | 176–177° | | 128 (m.p.); |
| $(CH_2)_5COOH$ | $CH_3$ | | | $CH_3$ | | hydrochloride, m.p. 180–184° | 41 (m.p.), (u); |
| $-CH(COOH)CH_2CH(CH_3)_2$ | $CH_3$ | | | $CH_3$ | 250–253° | | 41 (m.p.), (u); |
| | | | | | | amide, m.p. 286–290° | 41 (m.p.), (u); |
| $C_6H_5$ | $CH_3$ | COOH | | $CH_3$ | 267–270° | | 69 (m.p.); |

TABLE XII-82. 4-Pyridone Carboxylic Acids (Continued)

| $R_1$ | $R_2$ | $R_3$ | $R_5$ | $R_6$ | m.p. | Derivatives | Ref. |
|---|---|---|---|---|---|---|---|
| | | | | | | ethyl ester, m.p. 150–152° | 69 (m.p.); |
| $C_6H_4CH_3$-*o* | COOH | | $OCH_3$ | | 193° | | 128 (m.p.); |
| $C_6H_4CH_3$-*m* | COOH | | $OCH_3$ | | 177–178° | | 128 (m.p.); |
| $C_6H_4CH_3$-*p* | COOH | | $OCH_3$ | | 176–177° | | 128 (m.p.); |
| $-CH_2COOH$ | $CH_3$ | β-D-Glucosyloxy | | | 236–237° | | 131 (m.p.); |
| $-CH_2C_6H_5$ | $CH_3$ | COOH | | $CH_3$ | | ethyl ester, m.p. 178–180° | 69 (m.p.), (i); |

| | | | | | | | |
|---|---|---|---|---|---|---|---|
| $-CH_2CONHCH(CH_2CH(CH_3)_2)COOH$ | $CH_3$ | | | $CH_3$ | 305–310° | | 41 (m.p.), (u); |
| α-Naphthyl | COOH | | OH | | 203° | | 129 (m.p.); |
| | | | | | | ethyl ester, m.p. 182–183° | 129 (m.p.), (u); 1120 (u); |
| $-CH(COOH)CH_2C_6H_5$ | $CH_3$ | | | $CH_3$ | 229–231° | | 41 (m.p.), (u); |
| | | | | | | amide, m.p. > 320° | 41 (m.p.), (u); |
| $CH_3$ | $C_6H_5$ | COOH | COOH | $C_6H_5$ | | diethyl ester, m.p. 244° | 448 (m.p.), (i); |
| $CH_2CH{=}CH_2$ | $C_6H_5$ | COOH | COOH | $C_6H_5$ | | diethyl ester, m.p. 166° | 448 (m.p.), (i); |
| $CH_2C_6H_5$ | $C_6H_5$ | COOH | COOH | $C_6H_5$ | | diethyl ester, m.p. 204° | 448 (m.p.), (i); |

TABLE XII-83. 4-Pyridone Nitriles

| $R_1$ | $R_2$ | $R_3$ | $R_5$ | $R_6$ | m.p. | Derivatives | Ref. |
|---|---|---|---|---|---|---|---|
| CN | | | | | 164–167° | | 460 (m.p.), (n), (u); |
| CN | $CH_2Cl$ | | OH | | 164–165° | | 132 (m.p.), (i); |
| CN | $CH_2OH$ | | OH | | 155–156° | | 132 (m.p.), (i); |
| CN | $CH_3$ | | | $CH_3$ | 122–123° | | 132 (m.p.); |
| CN | $CH_2OH$ | | $OCH_3$ | | 169° | | 132 (m.p.), (i); |

TABLE XII-84. Halo 4-Pyridones

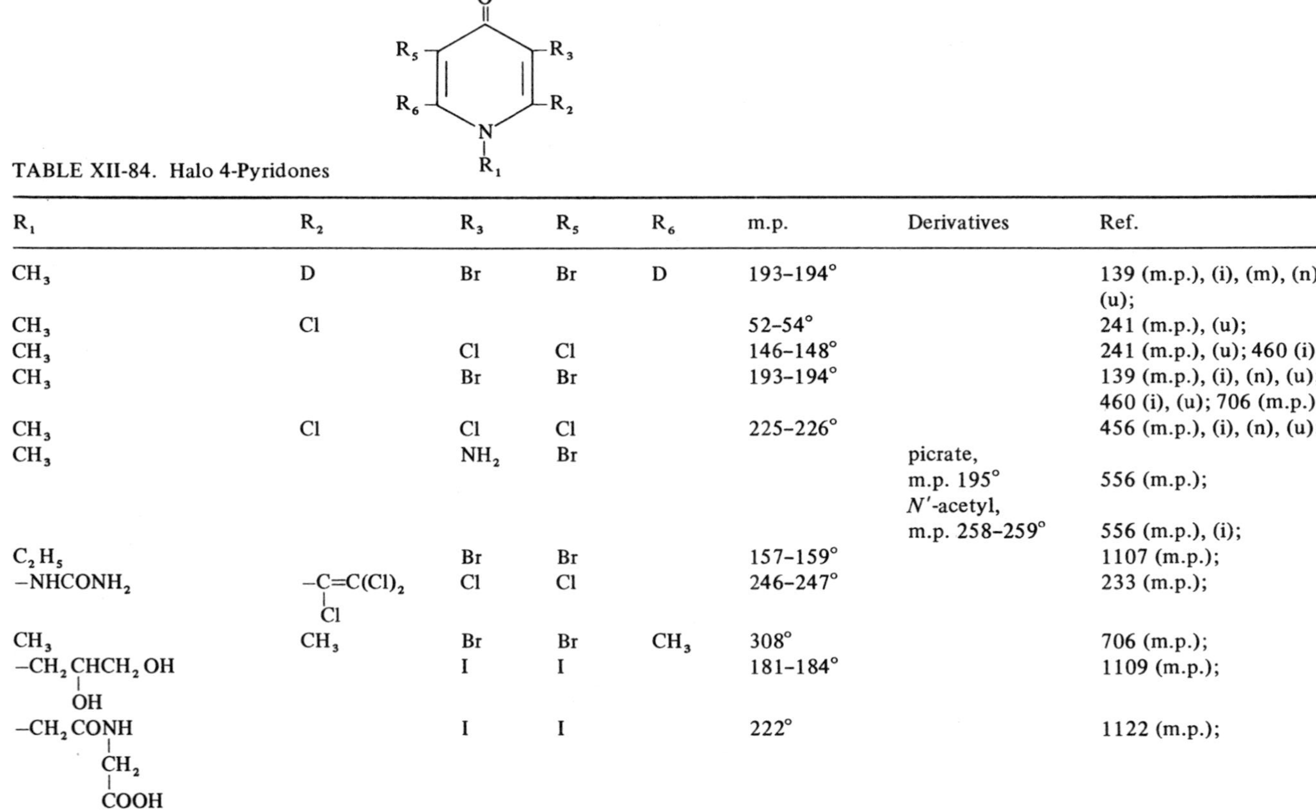

| $R_1$ | $R_2$ | $R_3$ | $R_5$ | $R_6$ | m.p. | Derivatives | Ref. |
|---|---|---|---|---|---|---|---|
| $CH_3$ | D | Br | Br | D | 193–194° | | 139 (m.p.), (i), (m), (n), (u); |
| $CH_3$ | Cl | | | | 52–54° | | 241 (m.p.), (u); |
| $CH_3$ | | Cl | Cl | | 146–148° | | 241 (m.p.), (u); 460 (i); |
| $CH_3$ | | Br | Br | | 193–194° | | 139 (m.p.), (i), (n), (u); 460 (i), (u); 706 (m.p.); |
| $CH_3$ | Cl | Cl | Cl | | 225–226° | | 456 (m.p.), (i), (n), (u); |
| $CH_3$ | | $NH_2$ | Br | | | picrate, m.p. 195° | 556 (m.p.); |
| | | | | | | *N′*-acetyl, m.p. 258–259° | 556 (m.p.), (i); |
| $C_2H_5$ | | Br | Br | | 157–159° | | 1107 (m.p.); |
| $-NHCONH_2$ | $-C(Cl)=C(Cl)_2$ | Cl | Cl | | 246–247° | | 233 (m.p.); |
| $CH_3$ | $CH_3$ | Br | Br | $CH_3$ | 308° | | 706 (m.p.); |
| $-CH_2CH(OH)CH_2OH$ | | I | I | | 181–184° | | 1109 (m.p.); |
| $-CH_2CONHCH_2COOH$ | | I | I | | 222° | | 1122 (m.p.); |

TABLE XII-84. Halo 4-Pyridones (Continued)

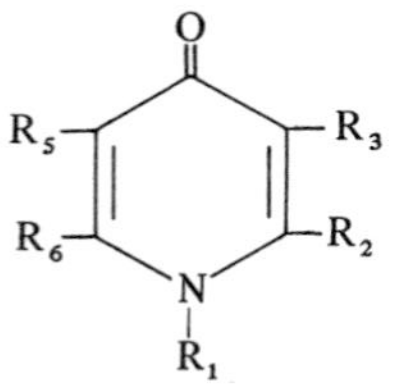

| $R_1$ | $R_2$ | $R_3$ | $R_5$ | $R_6$ | m.p. | Derivatives | Ref. |
|---|---|---|---|---|---|---|---|
| $CH_2CONHCH(COOH)CH_2$ |  | I | I |  |  | ethyl ester, m.p. 151–153° | 1122 (m.p.); |
| $C_2H_5$ | $CH_3$ | Br | Br | $CH_3$ | 248° |  | 706 (m.p.); |
| $(CH_2)_4COOH$ |  | I | I |  | 198–199° |  | 587 (m.p.); |
|  |  |  |  |  |  | amide, m.p. 210–211° | 587 (m.p.); |
|  |  |  |  |  |  | methyl ester, m.p. 172° | 587 (m.p.); |
|  |  |  |  |  |  | ethyl ester, m.p. 96° | 587 (m.p.); |
|  |  |  |  |  |  | *n*-butyl ester, m.p. 174° | 587 (m.p.); |
|  |  |  |  |  |  | *n*-amyl ester, m.p. 73° | 587 (m.p.); |
| $C_6H_5Hg-$ |  | Cl | Cl |  | 338–340° |  | 831 (m.p.), (i); |
| $C_6H_4NH_2$-*p* | Cl | Cl | Cl |  |  | hydrochloride, m.p. 260° | 232 (m.p.), (i); |
| $(CH_2)_5COOH$ |  | I | I |  | 190–191° |  | 587 (m.p.); |
|  |  |  |  |  |  | methyl ester, m.p. 81° | 587 (m.p.); |
| *n*-$C_4H_9$ | $CH_3$ | Br | Br | $CH_3$ | 180° |  | 118 (m.p.); |

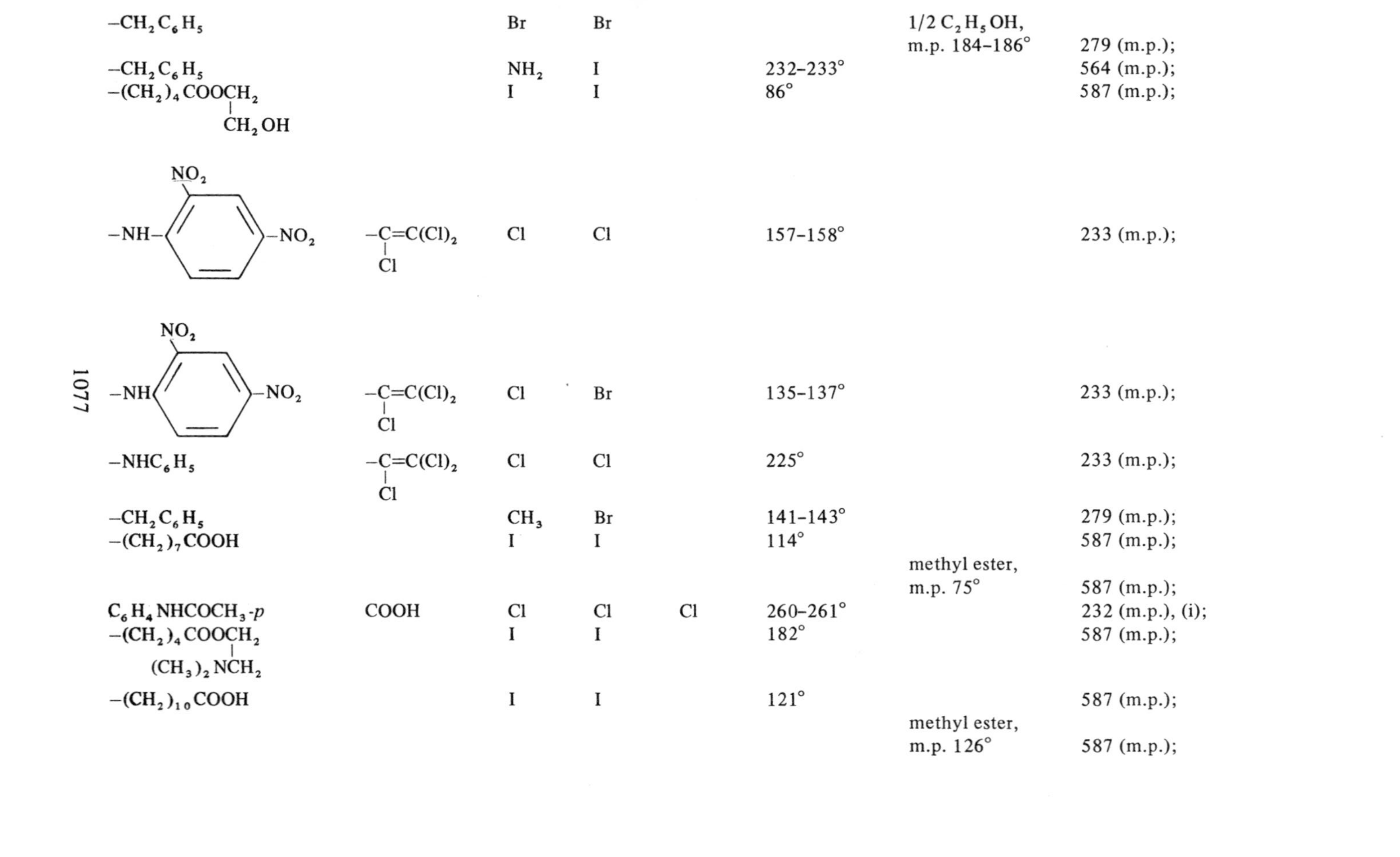

| | | | | | | | |
|---|---|---|---|---|---|---|---|
| $-CH_2C_6H_5$ | | Br | Br | | | 1/2 $C_2H_5OH$, m.p. 184–186° | 279 (m.p.); |
| $-CH_2C_6H_5$ | | $NH_2$ | I | | 232–233° | | 564 (m.p.); |
| $-(CH_2)_4COOCH_2CH_2OH$ | | I | I | | 86° | | 587 (m.p.); |
| $-NH-C_6H_3(NO_2)_2$ | $-C(Cl){=}C(Cl)_2$ | Cl | Cl | | 157–158° | | 233 (m.p.); |
| $-NH-C_6H_3(NO_2)_2$ | $-C(Cl){=}C(Cl)_2$ | Cl | Br | | 135–137° | | 233 (m.p.); |
| $-NHC_6H_5$ | $-C(Cl){=}C(Cl)_2$ | Cl | Cl | | 225° | | 233 (m.p.); |
| $-CH_2C_6H_5$ | | $CH_3$ | Br | | 141–143° | | 279 (m.p.); |
| $-(CH_2)_7COOH$ | | I | I | | 114° | | 587 (m.p.); |
| | | | | | | methyl ester, m.p. 75° | 587 (m.p.); |
| $C_6H_4NHCOCH_3$-*p* | COOH | Cl | Cl | Cl | 260–261° | | 232 (m.p.), (i); |
| $-(CH_2)_4COOCH_2CH_2N(CH_3)_2$ | | I | I | | 182° | | 587 (m.p.); |
| $-(CH_2)_{10}COOH$ | | I | I | | 121° | | 587 (m.p.); |
| | | | | | | methyl ester, m.p. 126° | 587 (m.p.); |

TABLE XII-84. Halo 4-Pyridones (Continued)

| $R_1$ | $R_2$ | $R_3$ | $R_5$ | $R_6$ | m.p. | Derivatives | Ref. |
|---|---|---|---|---|---|---|---|
| 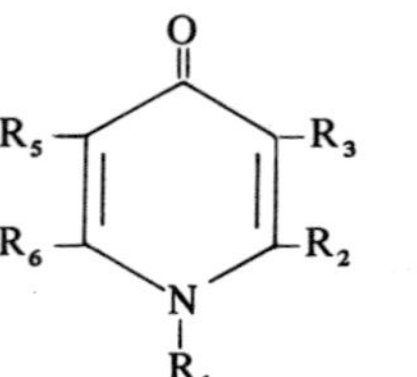 | $C_6H_5$ | Cl | Cl | | 271–273° | | 233 (m.p.); |
| $-(CH_2)_{11}COOH$ | | I | I | | 102° | | 587 (m.p.); |
| | | | | | | methyl ester, m.p. 92° | 587 (m.p.); |
| $CH_3$ | $C_6H_5$ | Br | Br | $C_6H_5$ | 313° | | 706 (m.p.); |
| $-CH_2CONHCH_2CH_2$–O–$(CH_2CH_2O)_4CH_3$ | | I | I | | 90–91° | | 1123 (m.p.); |
| $C_2H_5$ | $C_6H_5$ | Br | Br | $C_6H_5$ | 285° | | 706 (m.p.); |

TABLE XII-85. Nitro 4-Pyridones

| $R_1$ | $R_2$ | $R_3$ | $R_5$ | $R_6$ | m.p. | Derivatives | Ref. |
|---|---|---|---|---|---|---|---|
| COOH | | $NO_2$ | | | | ethyl ester, m.p. 114–115° | 586 (m.p.), (u); |
| $CH_3$ | | $NO_2$ | $NO_2$ | | 218° | | 456 (m.p.), (i), (n), (u); 775 (m.p.), (i), (n); |
| $CH_3$ | | $NO_2$ | | | 233° | | 497 (u); 586 (m.p.), (u); |
| $CH_3$ | | $NH_2$ | $NO_2$ | | 270° | | 556 (m.p.); |
| | | | | | | *N*′-acetyl, m.p. 323° | 556 (m.p.); |
| | | | | | | *N*′-benzoyl, m.p. 244° | 556 (m.p.); |
| $CH_2C_6H_5$ | | $NO_2$ | I | | 175–176° | | 564 (m.p.); |
| $CH_2C_6H_5$ | | $NO_2$ | | | 113–114° | | 564 (m.p.); |

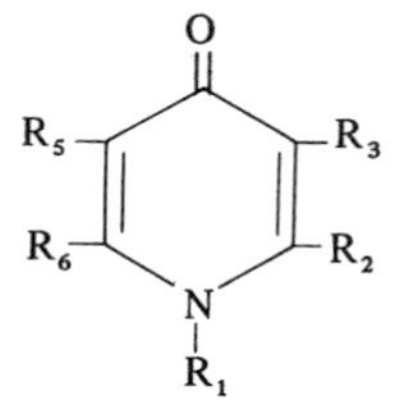

TABLE XII-86. Sulfur containing 4-Pyridones

| $R_1$ | $R_2$ | $R_3$ | $R_5$ | $R_6$ | m.p. | Derivatives | Ref. |
|---|---|---|---|---|---|---|---|
| $SO_3H$ | | | | | | potassium salt | 648 (i), (u); |

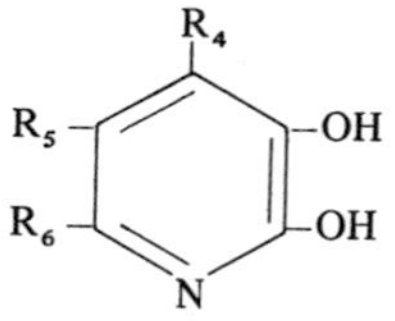

TABLE XII-87. 2,3-Pyridinediols

| $R_4$ | $R_5$ | $R_6$ | m.p. | Derivatives | Ref. |
|---|---|---|---|---|---|
| H | H | H | 246–255° | | 3 (m.p.); 176 (i); 187 (m.p.); 342 (m.p.), (t), (u); 361 (m.p.), (i); 467 (i), (u); |
| | | | | 1:1 complex with $UO_2$ | 825 (y); |
| | | | | 1:1 complex with Cu | 825 (y); |
| | | | | 1:1 complex with Pb | 825 (y); |
| | | | | 1:1 complex with Zn | 825 (y); |

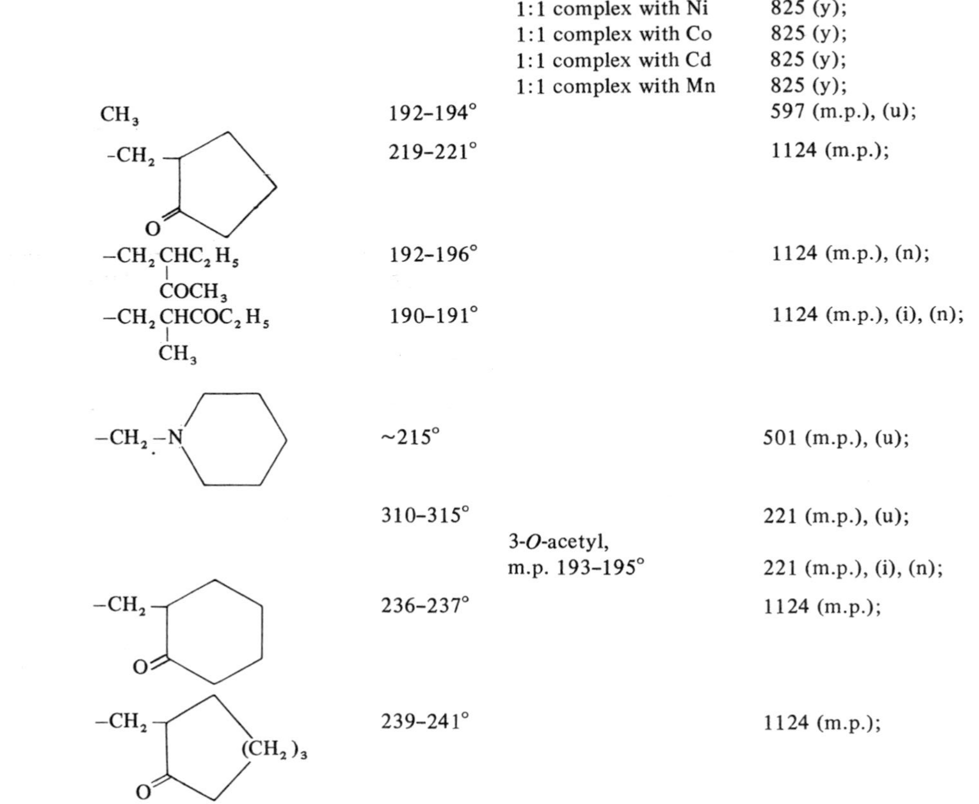

| | | | | | |
|---|---|---|---|---|---|
| | | | | 1:1 complex with Ni | 825 (y); |
| | | | | 1:1 complex with Co | 825 (y); |
| | | | | 1:1 complex with Cd | 825 (y); |
| | | | | 1:1 complex with Mn | 825 (y); |
| | | $CH_3$ | 192–194° | | 597 (m.p.), (u); |
| | | $-CH_2$–(2-oxocyclopentyl) | 219–221° | | 1124 (m.p.); |
| | | $-CH_2CH(COCH_3)C_2H_5$ | 192–196° | | 1124 (m.p.), (n); |
| | | $-CH_2CH(CH_3)COC_2H_5$ | 190–191° | | 1124 (m.p.), (i), (n); |
| | | $-CH_2-N$(piperidino) | ~215° | | 501 (m.p.), (u); |
| $CH_3$ | $C_6H_5$ | | 310–315° | | 221 (m.p.), (u); |
| | | | | 3-*O*-acetyl, m.p. 193–195° | 221 (m.p.), (i), (n); |
| | | $-CH_2$–(2-oxocyclohexyl) | 236–237° | | 1124 (m.p.); |
| | | $-CH_2$–(2-oxo ring with $(CH_2)_3$) | 239–241° | | 1124 (m.p.); |

TABLE XII-87. 2,3-Pyridinediols (Continued)

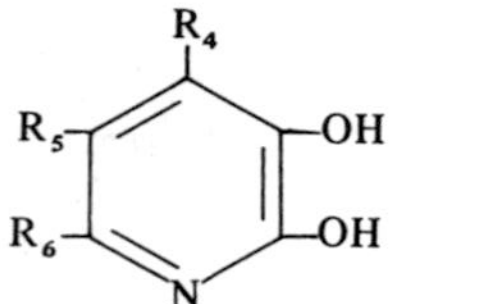


| $R_4$ | $R_5$ | $R_6$ | m.p. | Derivatives | Ref. |
|---|---|---|---|---|---|
| | | $-CH_2$– (ring with C=O and $(CH_2)_4$, i.e. 2-oxocyclohexyl) | 251–252° | | 1124 (m.p.), (i), (n); |
| $CH_2C_6H_5$ | $-COCH_3$ | $-CH_3$ | 210° | | 231 (m.p.); |
| $CH_2C_6H_5$ | $-CO_2C_2H_5$ | $C_2H_5$ | 163–164° | | 231 (m.p.); |
| $-CH_2N$ (piperidine ring) | | $-CH_2N$ (piperidine ring) | 184–185° | | 501 (m.p.), (u); |
| $CH_2C_6H_5$ | $-COC_6H_5$ | $CH_3$ | 227° | | 231 (m.p.); |

TABLE XII-88. 2,4-Pyridinediols

| $R_3$ | $R_5$ | $R_6$ | m.p. | Derivatives | Ref. |
|---|---|---|---|---|---|
| H | H | H | | | 906 (a); |
| | | $CH_3$ | 312–330° | | 50 (m.p.); 87 (m.p.); 109 (m.p.), (i), (u); 502 (m.p.), (n), (u); 644 (m.p.); 1034 (m); |
| | | | | 4-*O*-acetyl, m.p. 197° | 644 (m.p.), (i); 645 (i), (u); |
| | $CH_2CH_2Cl$ | | 154–155° | | 589 (m.p.); |
| $CH_3$ | | $CH_3$ | 268–270° | | 158 (m.p.), (u); 159 (i); (u); |
| | | $C_2H_5$ | 275° | | 50 (m.p.); |
| $COCH_3$ | | $CH_3$ | 256–261° | | 104 (m.p.); 106 (m.p.), (i), (u); 109 (m.p.), (u); 110 (n); 644 (m.p.), (i), (u); 645 (i), (u); 1005 (m.p.); |
| | | | | phenylhydrazone, m.p. 224–225° | 1005 (m.p.); |
| | | | | 2,4-dinitrophenylhydrazone, m.p. 305° | 106 (m.p.); |
| | $COCH_3$ | $CH_3$ | 282–296° | | 87 (m.p.); 90 (m.p.); 94 (m.p.), (u); |

TABLE XII-88. 2,4-Pyridinediols (Continued)

(Structure: pyridine ring with 4-OH, 3-$R_3$, 2-OH, 5-$R_5$, 6-$R_6$)

| $R_3$ | $R_5$ | $R_6$ | m.p. | Derivatives | Ref. |
|---|---|---|---|---|---|
| | | | | 2,4-dinitrophenylhydrazone, m.p. 240° | 87 (m.p.); |
| | | | | 4-*O*-acetyl, m.p. 173° | 87 (m.p.); |
| $-C(=NH)-CH_3$ | | $CH_3$ | 314–317° | | 104 (m.p.); 105 (m.p.); |
| $-C(=NOH)-CH_3$ | | $CH_3$ | 199° | | 151 (m.p.), (i), (n); |
| | | $CH_2CH_2CH_3$ | 279° | | 150 (m.p.); |
| $COC_2H_5$ | | $CH_3$ | 225–226° | | 644 (m.p.), (i), (u); 1005 (m.p.); |
| | | | | phenylhydrazone, m.p. 196–197° | 1005 (m.p.); |
| $-COCH_3$ | $COCH_3$ | $CH_3$ | 264° | | 157 (m.p.); |
| $C_2H_5$ | $COCH_3$ | $CH_3$ | 197° | | 89 (m.p.); |
| $-COC_3H_7$-*n* | | $CH_3$ | 183–184° | | 644 (m.p.), (i), (u); 1005 (m.p.); |
| | | | | phenylhydrazone, m.p. 218–220° | 1005 (m.p.); |
| | | $C_6H_4Cl$-*p* | 324° | | 50 (m.p.); |
| | | $C_6H_5$ | 315–322° | | 50 (m.p.); |
| $CH_3$ | $CH_3$ | (ring structure, C=O) | | 4-*O*-acetyl, m.p. 209–210° | 162 (m.p.), (u); |

| | | | | | |
|---|---|---|---|---|---|
| $-COC_4H_9$-*n* | | $CH_3$ | 188–189° | | 644 (m.p.), (i), (u); 1005 (m.p.); |
| | | $CH_3$ | | phenylhydrazone, m.p. 224–225° | 1005 (m.p.); |
| $-COC_4H_9$-iso | | $CH_3$ | 157–159° | | 644 (m.p.), (i), (u); 1005 (m.p.); |
| | | | | phenylhydrazone, m.p. 227–228° | 1005 (m.p.); |
| $C_6H_5$ | | $CH_3$ | 293° | | 163 (m.p.); |
| | | $CH_2C_6H_5$ | 261° | | 50 (m.p.); |
| $-COC_5H_{11}$-*n* | | $CH_3$ | 175–176° | | 644 (m.p.), (i), (u); 1005 (m.p.); |
| | | | | phenylhydrazone, m.p. 194–195° | 1005 (m.p.); |
| $-C(=N-C_4H_9\text{-}n)-CH_3$ | | $CH_3$ | 170–171° | | 106 (m.p.), (i); |
| | $COCH_3$ | $C_6H_5$ | 300–320° | | 87 (m.p.); 90 (m.p.); 94 (m.p.), (u); |
| $CH_2C_6H_5$ | | $CH_3$ | 247° | | 163 (m.p.); |
| | | 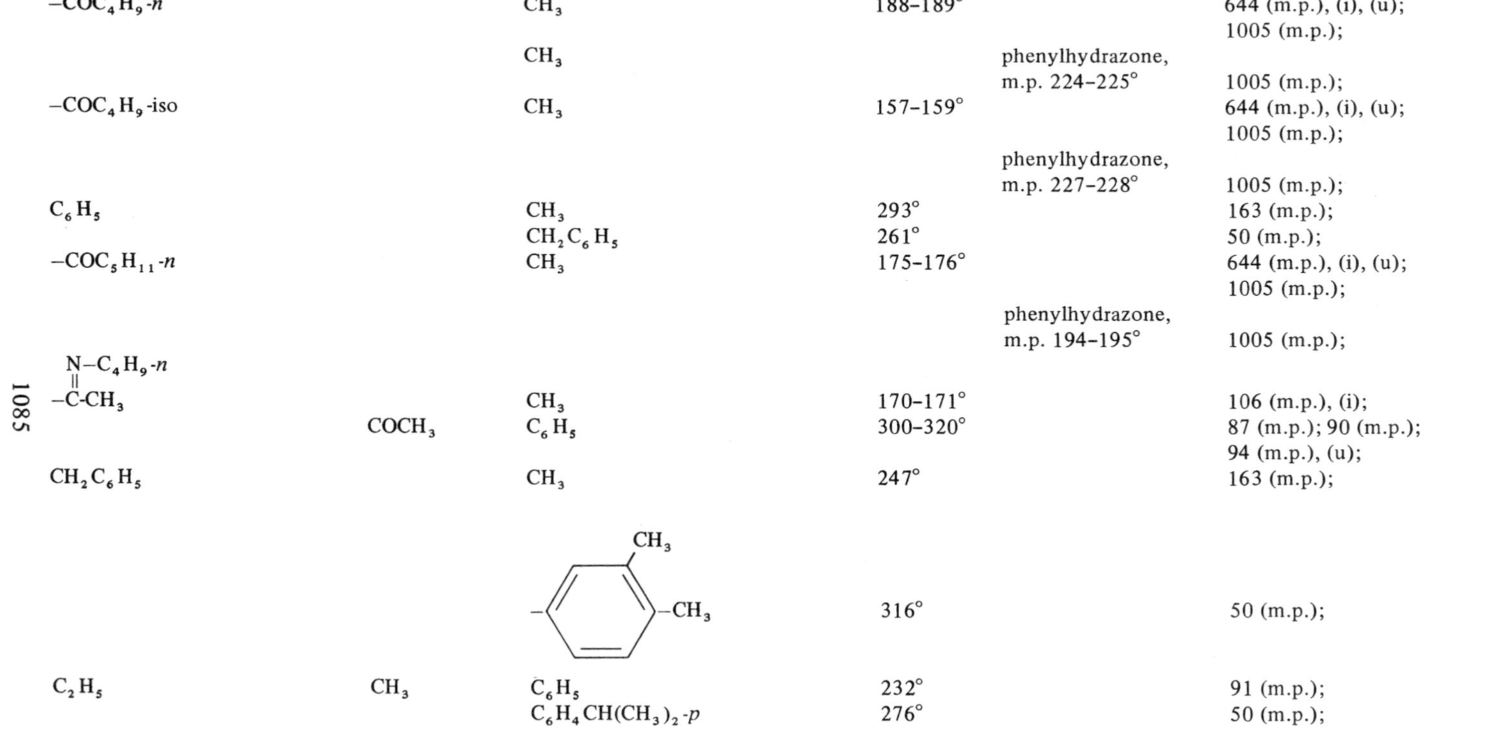 | 316° | | 50 (m.p.); |
| $C_2H_5$ | $CH_3$ | $C_6H_5$ | 232° | | 91 (m.p.); |
| | | $C_6H_4CH(CH_3)_2$-*p* | 276° | | 50 (m.p.); |

TABLE XII-88. 2,4-Pyridinediols (Continued)

| $R_3$ | $R_5$ | $R_6$ | m.p. | Derivatives | Ref. |
|---|---|---|---|---|---|
| | | 3,4,5-$(CH_3O)_3C_6H_2$– | 262° | | 50 (m.p.); |
| $-COCH{=}CHC_6H_4Cl$-*p* | | $CH_3$ | 313–314° | | 1031 (m.p.), (u); |
| $-COCH{=}CHC_6H_4NO_2$-*m* | | $CH_3$ | 277–278° | | 1031 (m.p.), (u); |
| $-COCH{=}CHC_6H_5$ | | $CH_3$ | 262–263° | | 1031 (m.p.), (u); |
| $-COCH{=}CHC_6H_4OH$-*p* | | $CH_3$ | 288–289° | | 1031 (m.p.), (u); |
| $-CH_2C_6H_5$ | $COCH_3$ | $CH_3$ | 224–226° | | 90 (m.p.); |
| $-CH_2C_6H_5$ | $CH_3$ | $C_2H_5$ | 210° | | 96 (m.p.); |
| $C_6H_5$ | | $C_6H_5$ | 300° | | 163 (m.p.); |
| $-COCH{=}CHC_6H_4N(CH_3)_2$-*p* | | $CH_3$ | 284–285° | | 1031 (m.p.), (u); |
| $-CH_2C_6H_5$ | $CH_3$ | $C_6H_5$ | 222–223° | | 91 (m.p.); 96 (m.p.); |
| $-CH_2C_6H_5$ | $COCH_3$ | $C_6H_5$ | 267° | | 90 (m.p.); |
| $CH_3$ | $CH_3$ | [oxygen ring]$=CHC(CH_3){=}CHC_6H_4NO_2$-*p* | 238–239° | | 162 (m.p.), (u); |
| | | | | acetate, m.p. 211–213° | 162 (m.p.); |

TABLE XII-89. 2,5-Pyridinediols

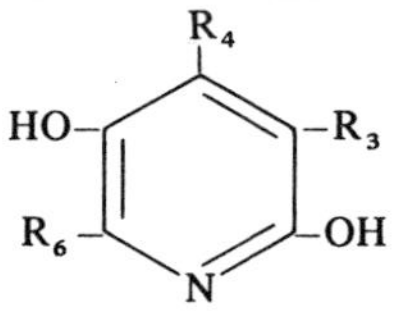

| $R_3$ | $R_4$ | $R_6$ | m.p. | Derivatives | Ref. |
|---|---|---|---|---|---|
| H | H | H | 245–250° | | 342 (m.p.), (t), (u); 699 (u); |
| | | | | 2,5-di-*O*-acetyl | 432 (n); |
| | | $CH_3$ / HO-(3-hydroxy-6-hydroxypyridin-2-yl)- [structure] | | | 176 (i); |
| | | | | 2,2′,5,5′-tetra-acetyl, m.p. 171° | 432 (m.p.), (n), (u); |
| $C_6H_5$ | $CH_3$ | | 250–260° | | 296 (m.p.); |

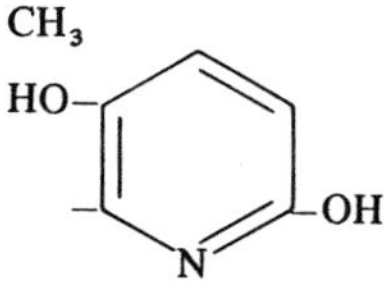

TABLE XII-90. 2,6-Pyridinediols

| $R_3$ | $R_4$ | $R_5$ | m.p. | Derivatives | Ref. |
|---|---|---|---|---|---|
| H | H | H | 198–200° | | 455 (m.p.), (n), (u); 467 (i), (u); 504 (i), (u); 899 (n); 906 (a); 918 (s); |
| | | | | deuterio | 504 (i); |
| | | | | hydrochloride | 543 (b); 546 (i), (u); |
| | $CH_3$ | | 192–194° | | 1000 (m.p.); |
| $CH_3$ | $CH_3$ | $CH_3$ | 179° | | 928 (m.p.), (i), (u); |
| | | | | $HCl{\cdot}H_2O$ m.p. 170° (unstable) | 928 (m.p.); |
| $C_2H_5$ | $CH_3$ | | 172–173° | | 510 (m.p.), (i), (n), (u); |
| $C_2H_5$ | $CH_3$ | $CH_3$ | 106° | | 928 (m.p.), (i), (u); |
| 2-Deoxy-D-ribosyl | | | 185° | | 618 (m.p.), (n), (u); |
| D-Ribosyl | | | 190° | | 618 (m.p.), (u); |
| $C_2H_5$ | $CH_3$ | $C_2H_5$ | 105° | | 928 (m.p.), (i), (u); |
| | | | | 2,6-di-*O*-acetyl, m.p. 99–100° | 928 (m.p.); |
| | $-(CH_2)_6CH_6$ | | 190–191° | | 6 (m.p.); |
| | $-(CH_2)_6-$(2,6-dihydroxy-4-pyridyl) | | 262–265° | | 6 (m.p.); |

TABLE XII-91. 3,4-Pyridinediols

| $R_2$ | $R_5$ | $R_6$ | m.p. | Derivatives | Ref. |
|---|---|---|---|---|---|
| H | H | H | 235–237° | | 134 (m.p.); |
| $CH_3$ | | | >250° | | 125 (m.p.), (u); |
| | | | 293–295° | | 131 (m.p.); 858 (m.p.); |
| | | | | hydrochloride, m.p. 184–186° | 858 (m.p.); |
| | | | | acetate, m.p. 205–207° | 858 (m.p.); |
| | | $CH_2OH$ | 246° | | 793 (m.p.); |

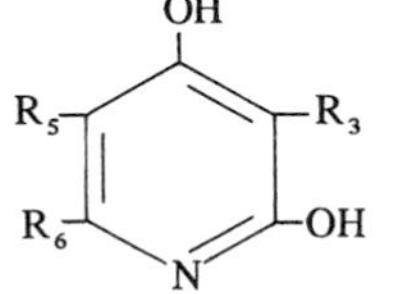


TABLE XII-92. Alkoxy 2,4-Pyridinediols

| $R_3$ | $R_5$ | $R_6$ | m.p. | Derivatives | Ref. |
|---|---|---|---|---|---|
| $OCH_3$ | | $CH_2OH$ | | $\cdot CH_3OH$ m.p. 169–171° | 160 (m.p.), (i), (u); |
| $OCH_3$ | | $CH_2OCOCH_3$ | | 4-*O*-acetyl, m.p. 143–146° | 160 (m.p.), (i), (u); |

TABLE XII-93. Alkoxy 2,6-Pyridinediols

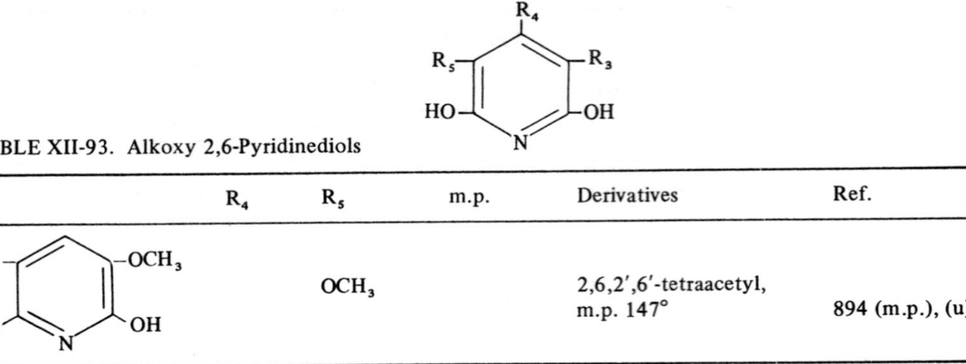


| $R_3$ | $R_4$ | $R_5$ | m.p. | Derivatives | Ref. |
|---|---|---|---|---|---|
| | | $OCH_3$ | | 2,6,2′,6′-tetraacetyl, m.p. 147° | 894 (m.p.), (u); |

TABLE XII-94. Amino 2,4-Pyridinediols

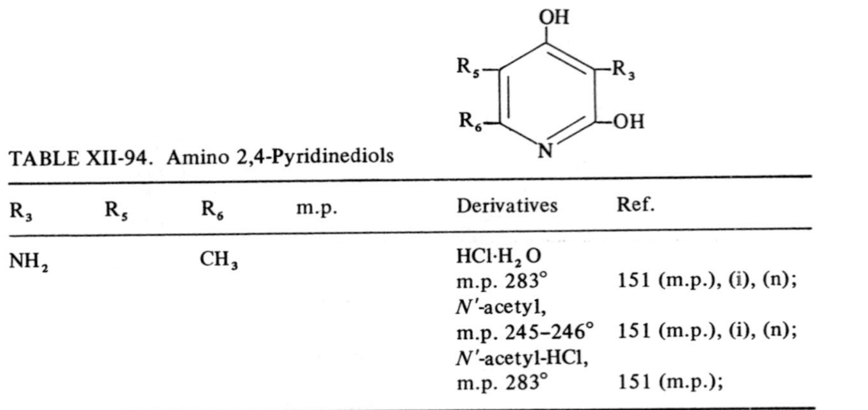


| $R_3$ | $R_5$ | $R_6$ | m.p. | Derivatives | Ref. |
|---|---|---|---|---|---|
| $NH_2$ | | $CH_3$ | | $HCl{\cdot}H_2O$ m.p. 283° | 151 (m.p.), (i), (n); |
| | | | | *N′*-acetyl, m.p. 245–246° | 151 (m.p.), (i), (n); |
| | | | | *N′*-acetyl-HCl, m.p. 283° | 151 (m.p.); |

TABLE XII-95. Amino 2,6-Pyridinediols

| $R_3$ | $R_4$ | $R_5$ | m.p. | Derivatives | Ref. |
|---|---|---|---|---|---|
| $NH_2$ | | | | hydrochloride | 846 (n), (u); |
| | $NH_2$ | | 300° | | 34 (m.p.); |
| | | | | $N'$-acetyl, m.p. 296–297° | 740 (m.p.); |
| $N(COCH_3)_2$ | | | | 2,6-di-*O*-acetyl, m.p. 89° | 846 (m.p.), (n), (u); |
| $-N{=}NC_6H_5$ | | | 218–220° | | 740 (m.p.), (i), (n); 846 (m.p.), (i), (u); |
| $-N{=}N-C_6H_3(OH)_2$ (2-OH, 4-OH) | | | 300° | | 939 (m.p.), (m), (u); |
| $-N{=}NC_6H_4Cl$-*p* | $NH_2$ | | 315° | | 34 (m.p.); |
| $-N{=}NC_6H_5$ | | $-N{=}NC_6H_5$ | 200–202° | | 740 (m.p.), (i), (n); |
| | $=NNHC_6H_4NO_2$-*p* | | | 1-*p*-nitroanilino, m.p. >210° | 800 (m.p.); |
| | $=NNHC_6H_5$ | | | 1-anilino, m.p. 188° | 801 (m.p.); |
| $-N{=}N-C_6H_3(OH)-OC(=O)-C_6H_5$ (2-OH, 4-OCOC$_6$H$_5$) | | | 271° | | 939 (m.p.), (m), (u); |

TABLE XII-96. 2,3-Pyridinediol Carboxylic Acids and Derivatives

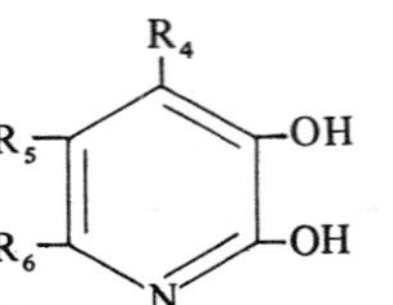


| $R_4$ | $R_5$ | $R_6$ | m.p. | Derivatives | Ref. |
|---|---|---|---|---|---|
| | | $-CH_2CH(CHO)C_5H_{11}$ | 172–173° | | 1124 (m.p.), (i), (n); |
| $CH_2C_6H_5$ | COOH | $CH_3$ | | ethyl ester, m.p. 148° | 231 (m.p.); |
| | | | | amide, m.p. 271–272° | 231 (m.p.); |
| | | $-CH_2CH(CHO)C_6H_{13}$ | 168–170° | | 1124 (m.p.), (i); |
| $CH_2C_6H_5$ | COOH | $C_6H_5$ | | ethyl ester, m.p. 174–175° | 231 (m.p.); |
| $CH_2C_6H_5$ | COOH | $CH_2C_6H_5$ | | methyl ester, m.p. 192° | 231 (m.p.); |

TABLE XII-97. 2,4-Pyridinediol Carboxylic Acids and Derivatives

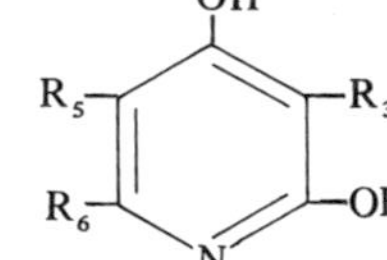


| $R_3$ | $R_5$ | $R_6$ | m.p. | Derivatives | Ref. |
|---|---|---|---|---|---|
| COOH | | $CH_3$ | | ethyl ester | 1034 (m); |
| | COOH | $CH_3$ | 280° | | 87 (m.p.); |
| | | | | ethyl ester, m.p. 228–229° | 87 (m.p.); 1034 (m); |
| | COOH | $C_2H_5$ | | ethyl ester, m.p. 182–183° | 87 (m.p.); 1034 (m); |
| | COOH | $CH_2CH_2CH_3$ | | ethyl ester, m.p. 160° | 87 (m.p.); 1034 (m); |

TABLE XII-98. 2,6-Pyridinediol Carboxylic Acids and Derivatives

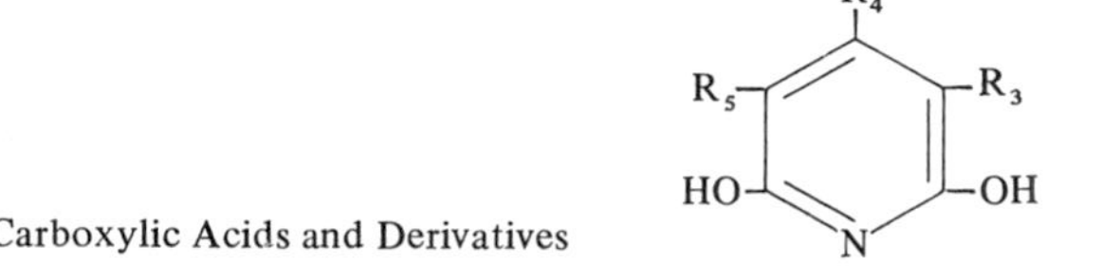


| $R_3$ | $R_4$ | $R_5$ | m.p. | Derivatives | Ref. |
|---|---|---|---|---|---|
| | $COOCH_3$ | | | 2,6-di-*O*-acetyl, m.p. 78–80° | 507 (m.p.); |
| $CONH_2$ | $NH_2$ | | 240° | | 9 (m.p.); |
| COOH | $NH_2$ | $-N{=}NC_6H_4Cl$-*p* | | ethyl ester, m.p. 259° | 34 (m.p.); |

TABLE XII-99. 2,3-Pyridinediol Nitriles

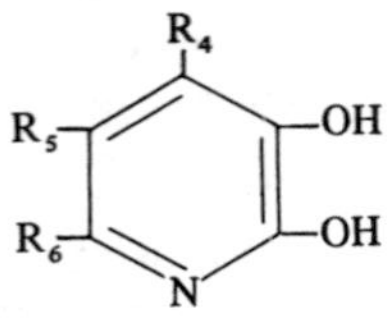

| $R_4$ | $R_5$ | $R_6$ | m.p. | Derivatives | Ref. |
|---|---|---|---|---|---|
| $CH_2C_6H_5$ | CN | $CH_3$ | 238° | | 231 (m.p.); |

TABLE XII-100. 2,6-Pyridinediol Nitriles

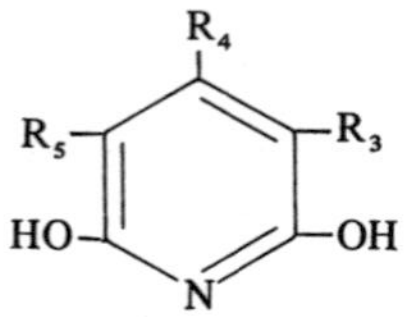

| $R_3$ | $R_4$ | $R_5$ | m.p. | Derivatives | Ref. |
|---|---|---|---|---|---|
| CN | | | | | 504 (i), (u); |
| | | | | 2,6-dideuterio | 504 (i); |
| CN | $NH_2$ | | 270° | | 9 (m.p.); |
| CN | $CF_3$ | | | 1.5 hydrate, m.p. 243–246° | 55 (m.p.), (u); |
| | | | | ammonium salt, m.p. 305–306° | 55 (m.p.); |
| CN | $CH_3$ | CN | | ammonium salt, m.p. 340° | 111 (m.p.), (i), (u); |
| | | | | piperidinium salt, m.p. 299° | 111 (m.p.), (u); |

| | | | | | |
|---|---|---|---|---|---|
| CN | $C_2H_5$ | CN | | hydrate, m.p. 216° | 111 (m.p.), (i), (u); |
| | | | | sodium salt·$H_2O$ m.p. 350° | 111 (m.p.), (i), (u); |
| | | | | potassium salt, m.p. 330° | 111 (m.p.), (i), (u); |
| | | | | ammonium salt, m.p. 312° | 111 (m.p.), (i), (u); |
| | | | | piperidinium salt, m.p. 298° | 111 (m.p.), (u); |
| CN | $n$-$C_3H_7$ | CN | | ammonium salt, m.p. 265° | 111 (m.p.), (i), (u); |
| CN | $n$-$C_6H_{13}$ | CN | | ammonium salt, m.p. 304° | 111 (m.p.), (i), (u); |
| | | | | piperidinium salt, m.p. 288° | 111 (m.p.), (u); |

TABLE XII-101. Halo 2,3-Pyridinediols

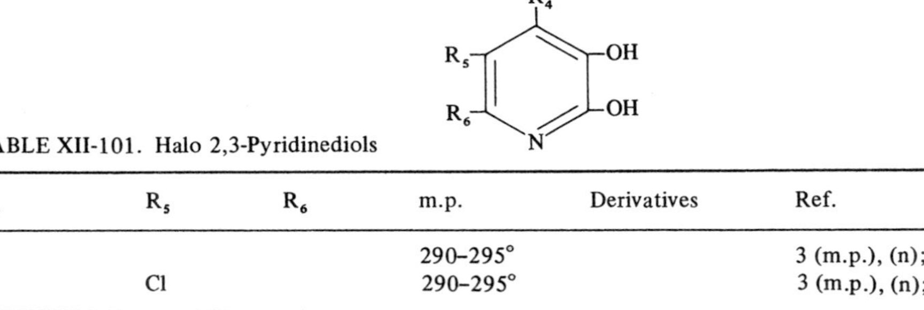


| $R_4$ | $R_5$ | $R_6$ | m.p. | Derivatives | Ref. |
|---|---|---|---|---|---|
| Cl | | | 290–295° | | 3 (m.p.), (n); |
| | Cl | | 290–295° | | 3 (m.p.), (n); |

TABLE XII-102. Halo 2,4-Pyridinediols

| $R_3$ | $R_5$ | $R_6$ | m.p. | Derivatives | Ref. |
|---|---|---|---|---|---|
| | Cl | Cl | 273–274° | | 98 (m.p.), (i), (m), (n), (u); |
| F | F | F | 188° | | 248 (m.p.), (i); 1111 (m.p.); |
| Cl | Cl | F | 219–221° | | 953 (m.p.); |
| Br | Cl | Cl | 230° | | 98 (m.p.), (u); |
| Br | COOH | Cl | | ethyl ester, m.p. 238° | 97 (m.p.); |
| CN | | Cl | 190° | | 98 (m.p.), (i); |
| | CN | Cl | 218–219° | | 98 (m.p.), (i), (m), (n), (u); |
| Cl | Cl | $CH_2Cl$ | 166–170° | | 502 (m.p.), (n); |
| | COOH | Cl | | ethyl ester, m.p. 219° | 97 (m.p.), (u); 98 (n); |
| Cl | Cl | $CH_3$ | 250–252° | | 502 (m.p.), (n); |
| | $CH_2Cl$ | Cl | 198° | | 98 (m.p.), (i), (n); |
| Br | $CH_3$ | Cl | 198° | | 97 (m.p.); |
| | $CH_2Br$ | Cl | 220° | | 98 (m.p.), (i), (n); |
| Cl | | $CH_3$ | >300° | | 502 (m.p.), (n); |
| | $CH_3$ | Cl | 302° | | 97 (m.p.), (i), (u); 98 (n); |
| Cl | COOH | $CH_3$ | | ethyl ester, m.p. 278–280° | 707 (m.p.), (i), (n); |
| | $CH_2CH_2Cl$ | Cl | 257° | | 97 (m.p.), (u); |
| | | | | 2,4-di-*O*-acetyl, b.p. 178°/4 mm | 589 (b.p.); |
| | $C_2H_5$ | Cl | 285° | | 98 (m.p.), (n); |

| | | | | |
|---|---|---|---|---|
| Br | $-COCH_2Br$ | $CH_3$ | 180° | 94 (m.p.), (n); |
| | $-CH_2CH_2CH_3$ | Cl | 305° | 97 (m.p.), (u); |
| | $-CH_2OC_2H_5$ | Cl | 210° | 98 (m.p.), (n); |
| | $-CH_2CH_2CH_2CN$ | Cl | 190° | 98 (m.p.), (i), (m), (n), (u); |
| Cl | Cl | $C_6H_5$ | 270–272° | 707 (m.p.); |
| | $C_6H_4Cl$-*p* | Cl | $>300°$ | 98 (m.p.), (i), (n), (u); |
| | $C_6H_4NO_2$-*p* | Cl | $>300°$ | 98 (m.p.), (i), (m), (n), (u); |
| | $C_6H_5$ | Cl | 310° | 97 (m.p.), (u); |
| | 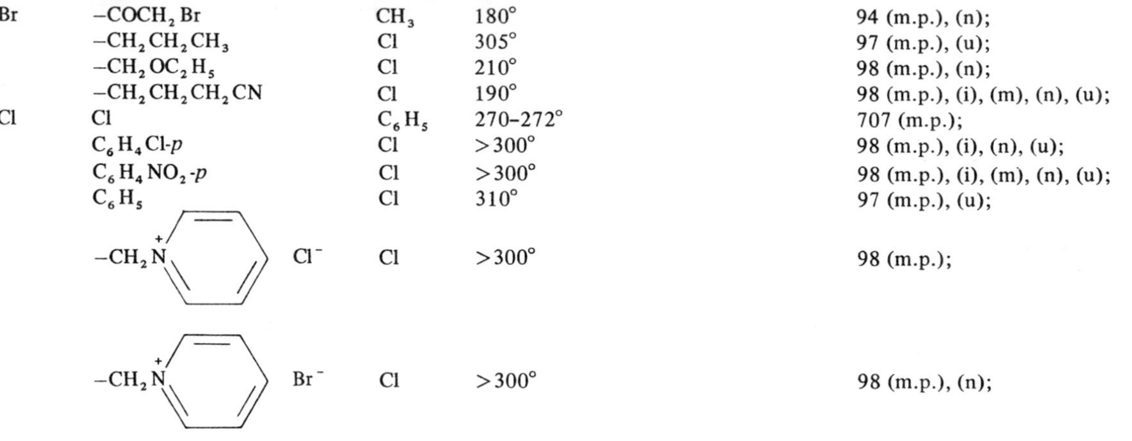 $-CH_2\overset{+}{N}$ (pyridinium) $Cl^-$ | Cl | $>300°$ | 98 (m.p.); |
| | $-CH_2\overset{+}{N}$ (pyridinium) $Br^-$ | Cl | $>300°$ | 98 (m.p.), (n); |
| | $-(CH_2)_{13}CN$ | Cl | 250° | 98 (m.p.), (i), (n), (u); |

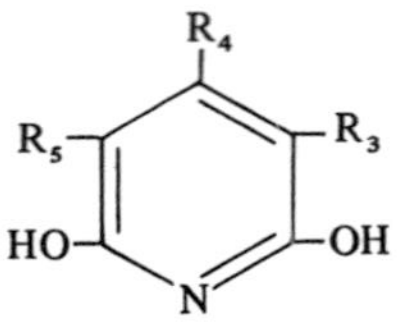

TABLE XII-103. Halo 2,6-Pyridinediols

| $R_3$ | $R_4$ | $R_5$ | m.p. | Derivatives | Ref. |
|---|---|---|---|---|---|
| Cl | Cl | Cl | 193–195° | | 18 (m.p.); |
| $C_4H_9$-*n* | Cl | Cl | 265–270° | | 7 (m.p.); |

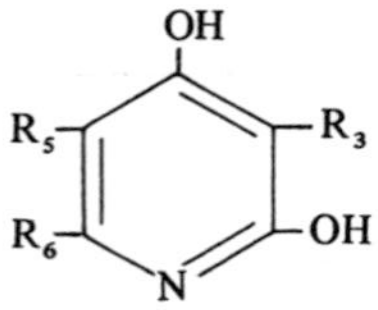

TABLE XII-104. Nitro 2,4-Pyridinediols

| $R_3$ | $R_5$ | $R_6$ | m.p. | Derivatives | Ref. |
|---|---|---|---|---|---|
| $NO_2$ | Cl | Cl | 215° | | 98 (m.p.), (u); |
| $NO_2$ | | | 264° | | 305 (m.p.); |
| $NO_2$ | $CH_3$ | Cl | 233° | | 97 (m.p.); |
| $NO_2$ | Br | $CH_3$ | 263–265° | | 502 (m.p.); |
| $NO_2$ | | $CH_3$ | 281–286° | | 86 (m.p.); |
| $NO_2$ | COOH | $CH_3$ | | ethyl ester, m.p. 250–251° | 86 (m.p.); |
| | | | | *n*-butyl ester, m.p. 183–184° | 86 (m.p.); |
| $NO_2$ | | $CH_2CH_2CH_3$ | 145° | | 150 (m.p.); |
| $NO_2$ | $COC_6H_5$ | $CH_3$ | 219° | | 86 (m.p.); |

TABLE XII-105. Nitroso 2,5-Pyridinediols

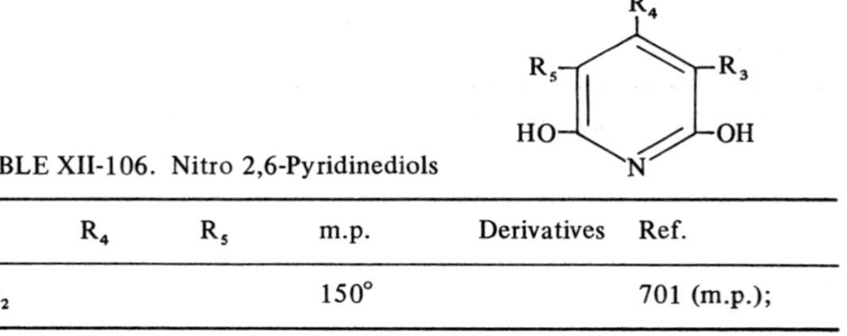


| $R_3$ | $R_4$ | $R_6$ | m.p. | Derivatives | Ref. |
|---|---|---|---|---|---|
| | | NO | 210° | | 292 (m.p.), (u); |
| $C_6H_5$ | $CH_3$ | NO | 250–253° | | 292 (m.p.), (u); |

TABLE XII-106. Nitro 2,6-Pyridinediols

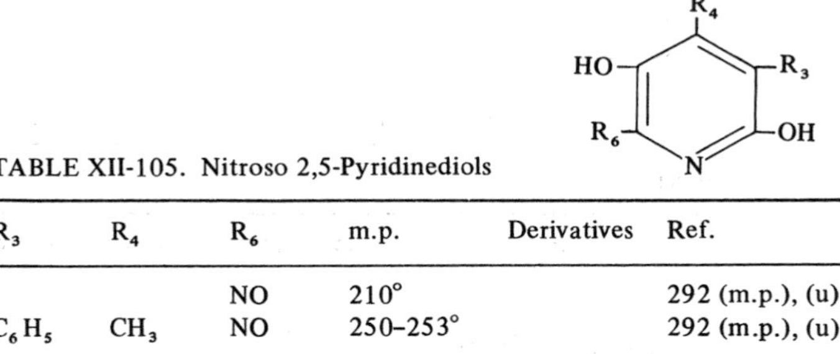


| $R_3$ | $R_4$ | $R_5$ | m.p. | Derivatives | Ref. |
|---|---|---|---|---|---|
| $NO_2$ | | | 150° | | 701 (m.p.); |

TABLE XII-107. Sulfur Containing 2,4-Pyridinediols

| $R_3$ | $R_5$ | m.p. | Derivatives | Ref. |
|---|---|---|---|---|
| COOH | COOH | | *N*-phenyl-3,5-diethyl ester, m.p. 156° | 171 (m.p.), (u); |

TABLE XII-108. Alkyl and Aryl 2,4-Pyridinediol Ethers

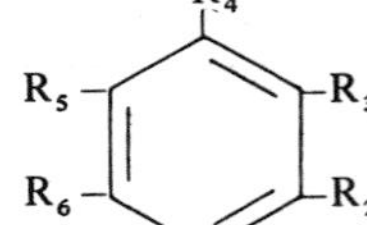


| $R_2$ | $R_3$ | $R_4$ | $R_5$ | $R_6$ | m.p. | Derivatives | Ref. |
|---|---|---|---|---|---|---|---|
| $OCH_3$ | | $OCH_3$ | | | b.p. 89°/12 Torr<br>b.p. 200–201° | | 252 (b.p.);<br>768 (b.p.); |
| | | | | | | picrate, m.p. 158–159° | 252 (m.p.); 768 (m.p.); |

TABLE XII-109. Alkyl and Aryl 2,6-Pyridinediol Ethers

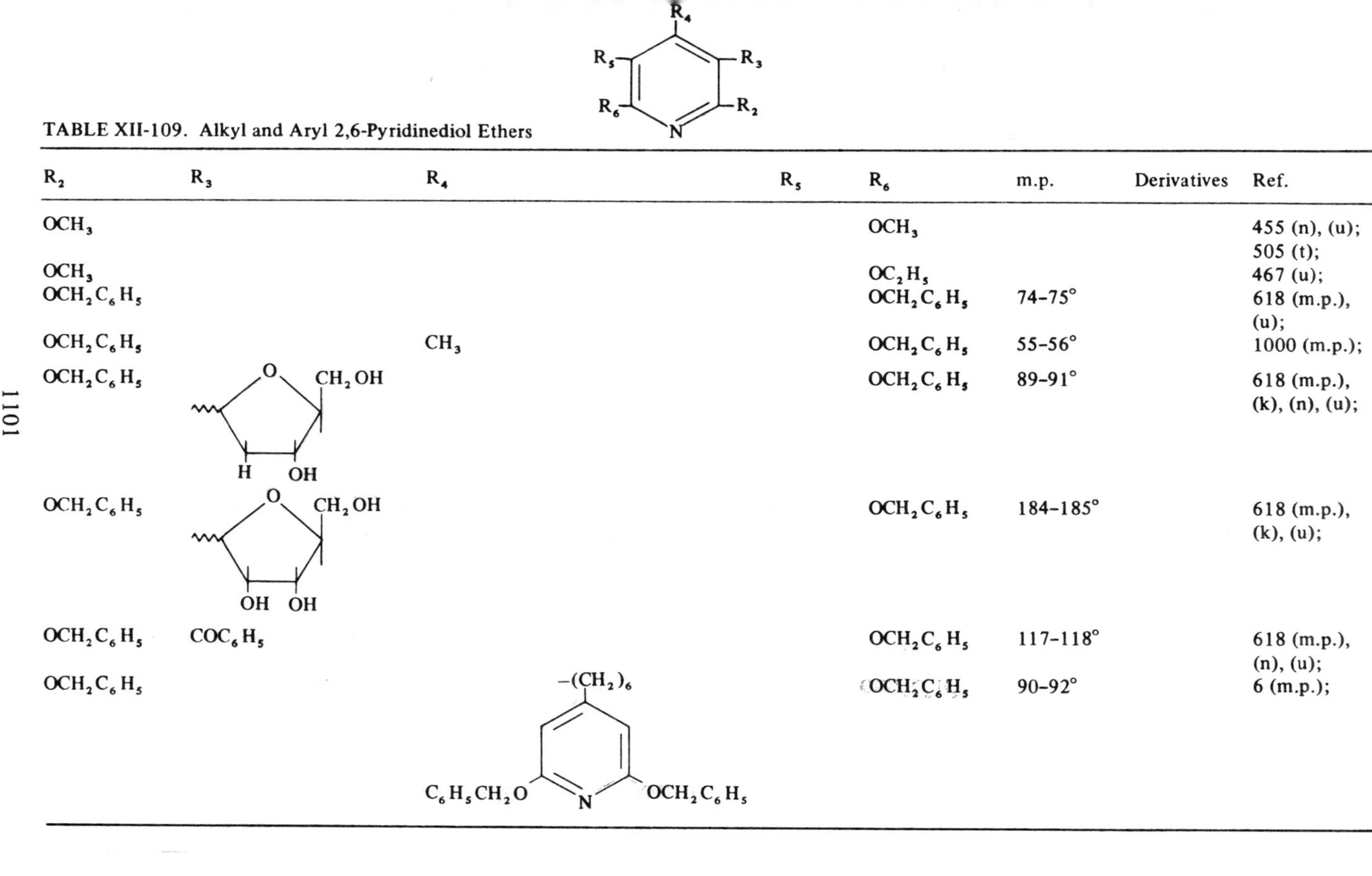

| $R_2$ | $R_3$ | $R_4$ | $R_5$ | $R_6$ | m.p. | Derivatives | Ref. |
|---|---|---|---|---|---|---|---|
| $OCH_3$ | | | | $OCH_3$ | | | 455 (n), (u); 505 (t); |
| $OCH_3$ | | | | $OC_2H_5$ | | | 467 (u); |
| $OCH_2C_6H_5$ | | | | $OCH_2C_6H_5$ | 74–75° | | 618 (m.p.), (u); |
| $OCH_2C_6H_5$ | | $CH_3$ | | $OCH_2C_6H_5$ | 55–56° | | 1000 (m.p.); |
| $OCH_2C_6H_5$ | (furanose ring: O, $CH_2OH$, H, OH) | | | $OCH_2C_6H_5$ | 89–91° | | 618 (m.p.), (k), (n), (u); |
| $OCH_2C_6H_5$ | (furanose ring: O, $CH_2OH$, OH, OH) | | | $OCH_2C_6H_5$ | 184–185° | | 618 (m.p.), (k), (u); |
| $OCH_2C_6H_5$ | $COC_6H_5$ | | | $OCH_2C_6H_5$ | 117–118° | | 618 (m.p.), (n), (u); |
| $OCH_2C_6H_5$ | | $-(CH_2)_6-$ (2,6-bis($C_6H_5CH_2O$)-4-pyridyl) | | $OCH_2C_6H_5$ | 90–92° | | 6 (m.p.); |

TABLE XII-110. Alkyl and Aryl 3,5-Pyridinediol Ethers

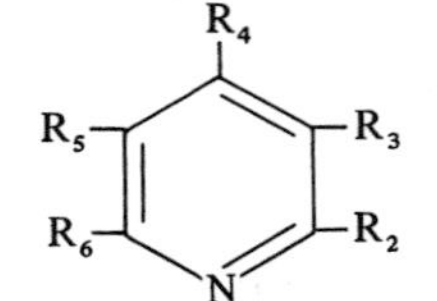


| $R_2$ | $R_3$ | $R_4$ | $R_5$ | $R_6$ | m.p. | Derivatives | Ref. |
|---|---|---|---|---|---|---|---|
|  | $OCH_3$ |  | $OCH_3$ |  |  | chloroplatinate, m.p. 212–213° | 705 (m.p.); |
| $C_6H_5$ | $OC_6H_5$ |  | $OC_6H_5$ | $C_6H_5$ | 162–164° |  | 1098 (m.p.); |

TABLE XII-111. Amino 2,4-Pyridinediol Ethers

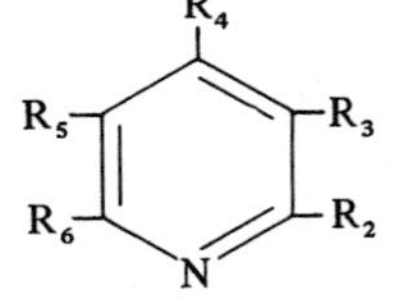


| $R_2$ | $R_3$ | $R_4$ | $R_5$ | $R_6$ | m.p. | Derivatives | Ref. |
|---|---|---|---|---|---|---|---|
| $OC_2H_5$ |  | $OC_2H_5$ |  | $NH_2$ | 35–36° |  | 284 (m.p.); |
|  |  |  |  |  |  | picrate, m.p. 176–177° | 284 (m.p.); |

TABLE XII-112. Amino 2,6-Pyridinediol Ethers

| $R_2$ | $R_3$ | $R_4$ | $R_5$ | $R_6$ | m.p. | Derivatives | Ref. |
|---|---|---|---|---|---|---|---|
| $OCH_3$ | | $NH_2$ | | $OCH_3$ | | | 504 (i), (u); |
| $OCH_2C_6H_5$ | | $NH_2$ | | $OCH_2C_6H_5$ | 74–75° | | 740 (m.p.), (i), (n); |
| | | | | | | *N'*-acetyl, m.p. 88–89° | 740 (m.p.), (i); |

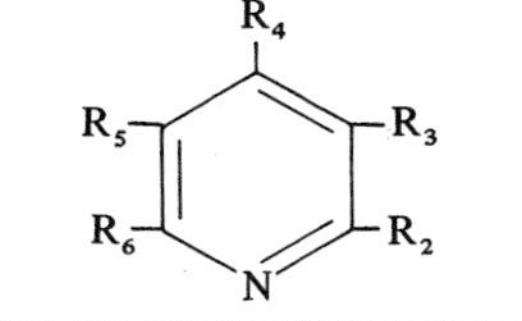

TABLE XII-113. Ethers of 2,6-Pyridinediol Carboxylic Acids

| $R_2$ | $R_3$ | $R_4$ | $R_5$ | $R_6$ | m.p. | Derivatives | Ref. |
|---|---|---|---|---|---|---|---|
| $OCH_3$ | COOH | | | $OCH_3$ | 136–137° | | 618 (m.p.), (u); |
| $OCH_3$ | | COOH | | $OCH_3$ | | methyl ester, m.p. 68–69° | 507 (m.p.); |
| | | | | | | diethylamide, m.p. 87–88° | 975 (m.p.); |
| $OCH_2C_6H_5$ | COOH | | | $OCH_2C_6H_5$ | 131–133° | | 618 (m.p.), (u); |

TABLE XII-114. Ethers of 2,6-Pyridinediol Nitriles

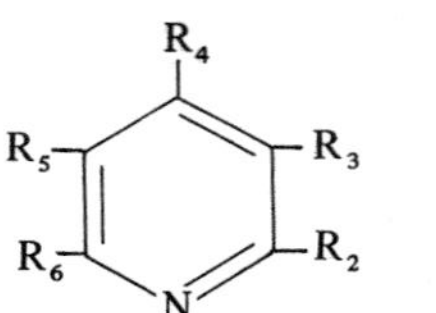


| $R_2$ | $R_3$ | $R_4$ | $R_5$ | $R_6$ | m.p. | Derivatives | Ref. |
|---|---|---|---|---|---|---|---|
| $OCH_3$ | CN | $CH_3$ | CN | $OCH_3$ | 129° | | 111 (m.p.), (i), (u); |
| $OCH_3$ | CN | $C_2H_5$ | CN | $OC_2H_5$ | 103° | | 111 (m.p.), (i), (u); |
| $OC_2H_5$ | CN | | CN | $OC_2H_5$ | 144–145° | | 100 (m.p.); |
| $OC_2H_5$ | CN | $C_2H_5$ | CN | $OC_2H_5$ | 103° | | 111 (m.p.); |

TABLE XII-115. Halo 2,4-Pyridinediol Ethers

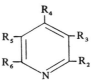

(Structure: pyridine ring with substituents $R_2$, $R_3$, $R_4$, $R_5$, $R_6$)

| $R_2$ | $R_3$ | $R_4$ | $R_5$ | $R_6$ | m.p. | Derivatives | Ref. |
|---|---|---|---|---|---|---|---|
| $OCH_3$ | F | $OCH_3$ | F | F | b.p. 87°/15 mm | | 245 (b.p.), (g), (i), (n), (u); 248 (b.p.), (i), (u); 1111 (b.p.); |
| $OCH_3$ | Cl | $OCH_3$ | Cl | F | 65–68° | | 246 (m.p.), (g), (i); 950 (m.p.); 953 (m.p.); |
| $OCH_3$ | Cl | $OCH_3$ | Cl | Cl | 116–125° | | 239 (m.p.), (u); 778 (m.p.), (n); |
| $OCH_3$ | Br | $OCH_3$ | Br | Br | 143° | | 952 (m.p.); |
| $OC_2H_5$ | Cl | $OCH_3$ | Cl | F | 50–51° | | 953 (m.p.); |
| $OCH(CH_3)_2$ | Cl | $OCH_3$ | Cl | Cl | b.p. 130°/3 mm | | 240 (b.p.); |
| $OC_2H_5$ | | $OC_2H_5$ | Br | | 71–72° | | 284 (m.p.); |
| $ON(CH_2CH_2)_2O$ | Cl | $OCH_3$ | Cl | Cl | 96° | | 240 (m.p.); |
| $ON(CH_2)_5$ | Cl | $OCH_3$ | Cl | Cl | 56° | | 240 (m.p.); |
| $OC_2H_4OCH_3$ | Cl | $OC_2H_4OCH_3$ | Cl | F | b.p. 135°/15 mm | | 953 (b.p.); |
| $OC_6H_4NO_2$-*p* | Cl | $OC_6H_4NO_2$-*p* | Cl | F | 235–236° | | 953 (m.p.); |
| *O*-Mesityl | Cl | *O*-Mesityl | Cl | F | 154–155° | | 953 (m.p.); |

TABLE XII-116. Halo 2,6-Pyridinediol Ethers

| $R_2$ | $R_3$ | $R_4$ | $R_5$ | $R_6$ | m.p. | Derivatives | Ref. |
|---|---|---|---|---|---|---|---|
| $OCH_3$ | Br | | | $OCH_3$ | b.p. 115°/29 mm | | 618 (b.p.), (u); |
| $OCH_3$ | F | | F | $OCH_3$ | 87° | | 949 (m.p.); |
| $OCH_3$ | Br | | Br | $OCH_3$ | 89–91° | | 618 (m.p.), (u); |
| $OCH_3$ | F | Br | F | $OCH_3$ | 120–122° | | 247 (m.p.), (n); 949 (n); |
| $OCH_3$ | Cl | Cl | Cl | $OCH_3$ | 140–141° | | 239 (m.p.), (u); |
| $OCH_3$ | Br | Br | Br | $OCH_3$ | 131–132° | | 952 (m.p.); |
| $OCH_3$ | Cl | $-NHNH_2$ | Cl | $OCH_3$ | 129–132° | | 239 (m.p.); |
| $OC_2H_5$ | Cl | Cl | Cl | $OC_2H_5$ | 105–106° | | 239 (m.p.); |
| $OCH_3$ | Cl | $-NHNHCOCH_3$ | Cl | $OCH_3$ | 201–203° | | 239 (m.p.); |
| $OC_6H_{11}$ | Cl | Cl | Cl | $OC_6H_{11}$ | 122–123° | | 239 (m.p.); |
| $OCH_2C_6H_5$ | Br | | | $OCH_2C_6H_5$ | 57–59° | | 618 (m.p.), (u); |
| $OCH_2C_6H_5$ | Br | | Br | $OCH_2C_6H_5$ | 74° | | 740 (m.p.), (i), (n); |
| $OCH_2C_6H_5$ | Cl | Cl | Cl | $OCH_2C_6H_5$ | 98–99° | | 239 (m.p.); |
| $OCH_2C_6H_5$ | Br | Br | Br | $OCH_2C_6H_5$ | 108–109° | | 740 (m.p.), (i), (n); |

TABLE XII-117. Nitro 2,4-Pyridinediol Ethers

| $R_2$ | $R_3$ | $R_4$ | $R_5$ | $R_6$ | m.p. | Derivatives | Ref. |
|---|---|---|---|---|---|---|---|
| $OCH_3$ | $NO_2$ | $OCH_3$ | | $CH_3$ | 108° | | 86 (m.p.); |
| $OCH_3$ | $NO_2$ | $OCH_3$ | $COOC_2H_5$ | $CH_3$ | 53–54° | | 86 (m.p.); |

TABLE XII-118. Nitro 2,6-Pyridinediol Ethers

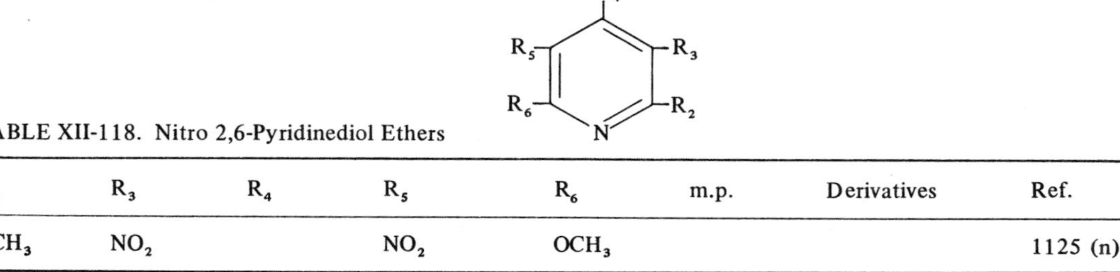


| $R_2$ | $R_3$ | $R_4$ | $R_5$ | $R_6$ | m.p. | Derivatives | Ref. |
|---|---|---|---|---|---|---|---|
| $OCH_3$ | $NO_2$ | | $NO_2$ | $OCH_3$ | | | 1125 (n); |

TABLE XII-119. Sulfur Containing Esters of 2,4-Pyridinediols

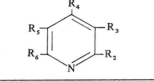

| $R_2$ | $R_3$ | $R_4$ | $R_5$ | $R_6$ | m.p. | Derivatives | Ref. |
|---|---|---|---|---|---|---|---|
| $OSO_2C_6H_5$ | Cl | $OSO_2C_6H_5$ | Cl | $CH_3$ | 97–98° | | 502 (m.p.); |
| $OSO_2C_6H_5$ | | $OSO_2C_6H_5$ | | $CH_3$ | 54–55° | | 502 (m.p.), (u); |

TABLE XII-120. Alkyl and Aryl 2,3,6-Pyridinetriols

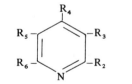

| $R_2$ | $R_3$ | $R_4$ | $R_5$ | $R_6$ | m.p. | Derivatives | Ref. |
|---|---|---|---|---|---|---|---|
| OH | OH | $CH_3$ | | OH | 209–210° | | 506 (m.p.), (i); |
| | | | | | | 2,3,6-triacetyl, m.p. 69° | 506 (m.p.), (i), (u); |
| OH | OH | | $CH_3$ | OH | 240° | | 506 (m.p.); |
| | | | | | | 2,3,6-triacetyl, m.p. 86° | 506 (m.p.); |
| OH | OH | $CH_3$ | $CH_3$ | OH | | 2,3,6-triacetyl, m.p. 102° | 506 (m.p.); |
| OH | OH | $CH_3$ | $C_6H_5$ | OH | | 2,3,6-triacetyl, m.p. 106–107° | 292 (m.p.); |

TABLE XII-121. Alkyl and Aryl 2,4,6-Pyridinetriols

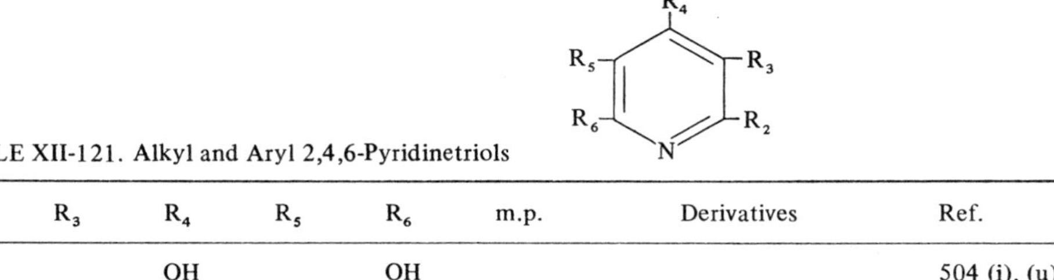


| $R_2$ | $R_3$ | $R_4$ | $R_5$ | $R_6$ | m.p. | Derivatives | Ref. |
|---|---|---|---|---|---|---|---|
| OH | | OH | | OH | | | 504 (i), (u); |

TABLE XII-122. Ethers of 2,3,4-Pyridinetriols

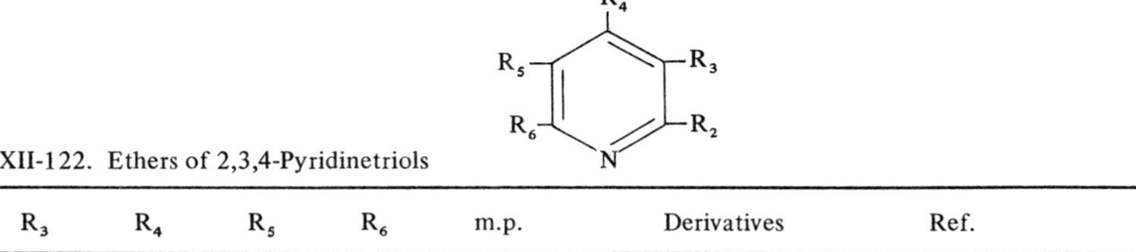


| $R_2$ | $R_3$ | $R_4$ | $R_5$ | $R_6$ | m.p. | Derivatives | Ref. |
|---|---|---|---|---|---|---|---|
| $OCH_3$ | $OCH_3$ | $OCH_3$ | | $CH_2OH$ | 48–49° | | 160 (m.p.), (c), (i), (u); |

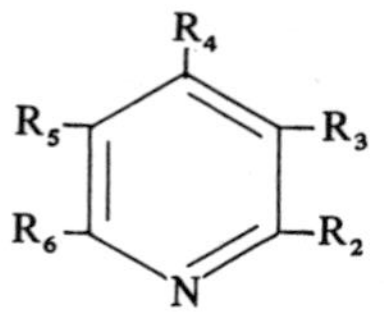

TABLE XII-123. Ethers of 2,4,6-Pyridinetriols

| $R_2$ | $R_3$ | $R_4$ | $R_5$ | $R_6$ | m.p. | Derivatives | Ref. |
|---|---|---|---|---|---|---|---|
| $OCH_3$ | F | $OCH_3$ | F | $OCH_3$ | 50–54° | | 248 (m.p.), (i), (u);<br>1111 (m.p.); |
| $OCH_3$ | Cl | $OCH_3$ | Cl | $OCH_3$ | 93–94° | | 16 (m.p.), 239 (m.p.); |

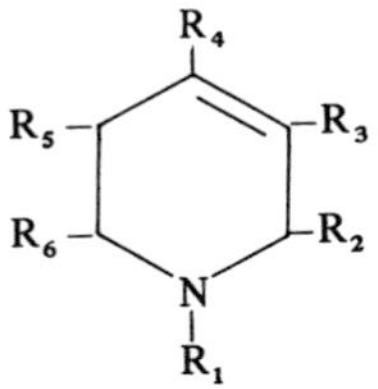

TABLE XII-124. Dihydro-2-pyridinols and Dihydro-2-pyridones

| $R_1$ | $R_2$ | $R_3$ | $R_4$ | $R_5$ | $R_6$ | m.p. | Derivatives | Ref. |
|---|---|---|---|---|---|---|---|---|
| | Keto | | | | | 120° or<br>65–67° | | 1126 (m.p.);<br>1127 (m.p.), (i), (n), (u); |
| $CH_3$ | Keto | | | | | b.p. 129°/0.25 mm | | 861 (b.p.), (i); |
| | Keto | | | | $CH_3$, H | 103–109° | | 441 (m.p.), (i), (u);<br>1128 (m.p.), (i), (u); |
| | Keto | | | | $CH_3$, H | b.p. 109°/13 mm | | 441 (b.p.), (u);<br>695 (b.p.); 1128 (i); |
| | Keto | $CH_3$ | $CH_3$ | | $CH_3$, H | 76–77° | | 1129 (m.p.), (i), (n), (u); |
| | Keto | | $CH_3$ | | $CH_3$, $CH_3$ | 115–118° | | 1130 (m.p.), (u);<br>1131 (m.p.), (i); |
| $C_2H_5$ | Keto | | | | $CH_3$, H | b.p. 111°/13 mm | | 1128 (b.p.), (i), (u); |
| | Keto | CN | $CH_3$ | | $CH_3$, $CH_3$ | 156–157° | | 1131 (m.p.), (i); |
| | Keto | COOH | $CH_3$ | | $CH_3$, $CH_3$ | 106–110° | | 1131 (m.p.), (i); |
| $CH_3$ | Keto | | $CH_3$ | | $CH_3$, $CH_3$ | 43–44° | | 1130 (m.p.), (i), (n), (m); |

| | | | | | | | | |
|---|---|---|---|---|---|---|---|---|
| | Keto | $COCH_3$ | $CH_3$ | | $CH_3, CH_3$ | 145–146° | | 1131 (m.p.), (i); 1132 (m.p.); 1149 (m.p.); |
| | | | | | | | semicarbazone, m.p. 245° | 1131 (m.p.); |
| $NH_2$ | Keto | | $C_6H_5$ | | | 60–62° | | 1133 (m.p.), (n), (u); |
| | | | | | | | hydrochloride, m.p. 197–198° | 1133 (m.p.); |
| | | | | | | | N′-acetyl, m.p. 205–206° | 1133 (m.p.), (i), (n), (u); |
| | | | | | | | N′-diacetyl, m.p. 131–132° | 1133 (m.p.), (i), (n), (u); |
| | Keto | $COCH_3$ | $CH_3$ | | $CH_3, C_2H_5$ | 121° | | 1132 (m.p.); 1149 (m.p.); |
| | Keto | $COCH_3$ | $C_2H_5$ | | $CH_3, CH_3$ | 120–122° | | 1149 (m.p.); |
| $-(CH_2)_3N(CH_3)_2$ | Keto | | | | $CH_3, H$ | b.p. 92°/0.25 mm | | 1128 (b.p.), (i); |
| $CH_3$ | Keto | | | | $C_6H_5, H$ | b.p. 186°/13 mm | | 441 (b.p.); |
| $C_6H_5$ | Keto | | | | $CH_3, H$ | b.p. 180°/10 mm | | 441 (b.p.); |
| | Keto | $COCH_3$ | $CH_3$ | | $-(CH_2)_4-$ | 134° | | 1132 (m.p.); 1149 (m.p.); |
| | Keto | $COC_3H_7$-n | $CH_3$ | | $CH_3, CH_3$ | 70–71° | | 1149 (m.p.); |
| | Keto | $COCH_3$ | $CH_3$ | | $C_2H_5, C_2H_5$ | 87° | | 1132 (m.p.); 1149 (m.p.); |
| | Keto | $COCH_3$ | $CH_3$ | | iso$C_3H_7$, $CH_3$ | 110–111° | | 1132 (m.p.); 1149 (m.p.); |
| | Keto | $COCH_3$ | n-$C_3H_7$ | | $CH_3, CH_3$ | 109–110° | | 1149 (m.p.); |
| | Keto | $COCH_3$ | iso$C_3H_7$ | | $CH_3, CH_3$ | 125–127° | | 1149 (m.p.); |
| | Keto | | $COCH_3$ | $-CH(OH)CH_2CH(CH_3)_2$ | | 189–190° | | 856 (m.p.), (m), (n); |
| | Keto | 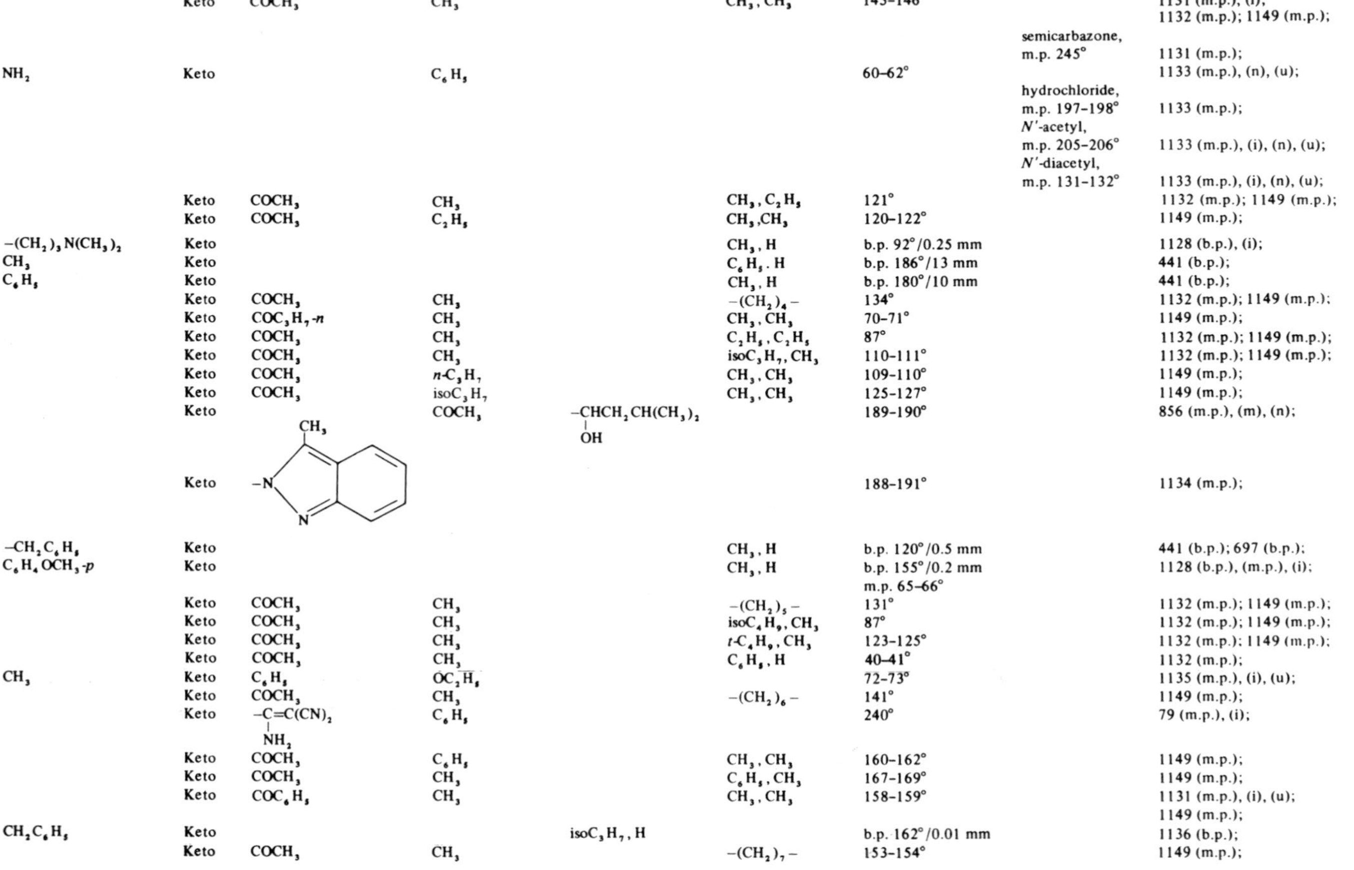 | | | | 188–191° | | 1134 (m.p.); |
| $-CH_2C_6H_5$ | Keto | | | | $CH_3, H$ | b.p. 120°/0.5 mm | | 441 (b.p.); 697 (b.p.); |
| $C_6H_4OCH_3$-p | Keto | | | | $CH_3, H$ | b.p. 155°/0.2 mm m.p. 65–66° | | 1128 (b.p.), (m.p.), (i); |
| | Keto | $COCH_3$ | $CH_3$ | | $-(CH_2)_5-$ | 131° | | 1132 (m.p.); 1149 (m.p.); |
| | Keto | $COCH_3$ | $CH_3$ | | iso$C_4H_9$, $CH_3$ | 87° | | 1132 (m.p.); 1149 (m.p.); |
| | Keto | $COCH_3$ | $CH_3$ | | t-$C_4H_9$, $CH_3$ | 123–125° | | 1132 (m.p.); 1149 (m.p.); |
| | Keto | $COCH_3$ | $CH_3$ | | $C_6H_5, H$ | 40–41° | | 1132 (m.p.); |
| $CH_3$ | Keto | $C_6H_5$ | $OC_2H_5$ | | | 72–73° | | 1135 (m.p.), (i), (u); |
| | Keto | $COCH_3$ | $CH_3$ | | $-(CH_2)_6-$ | 141° | | 1149 (m.p.); |
| | Keto | $-C(NH_2)=C(CN)_2$ | $C_6H_5$ | | | 240° | | 79 (m.p.), (i); |
| | Keto | $COCH_3$ | $C_6H_5$ | | $CH_3, CH_3$ | 160–162° | | 1149 (m.p.); |
| | Keto | $COCH_3$ | $CH_3$ | | $C_6H_5, CH_3$ | 167–169° | | 1149 (m.p.); |
| | Keto | $COC_6H_5$ | $CH_3$ | | $CH_3, CH_3$ | 158–159° | | 1131 (m.p.), (i), (u); 1149 (m.p.); |
| $CH_2C_6H_5$ | Keto | | | iso$C_3H_7$, H | | b.p. 162°/0.01 mm | | 1136 (b.p.); |
| | Keto | $COCH_3$ | $CH_3$ | | $-(CH_2)_7-$ | 153–154° | | 1149 (m.p.); |

TABLE XII-124. Dihydro-2-pyridinols and Dihydro-2-pyridones (Continued)

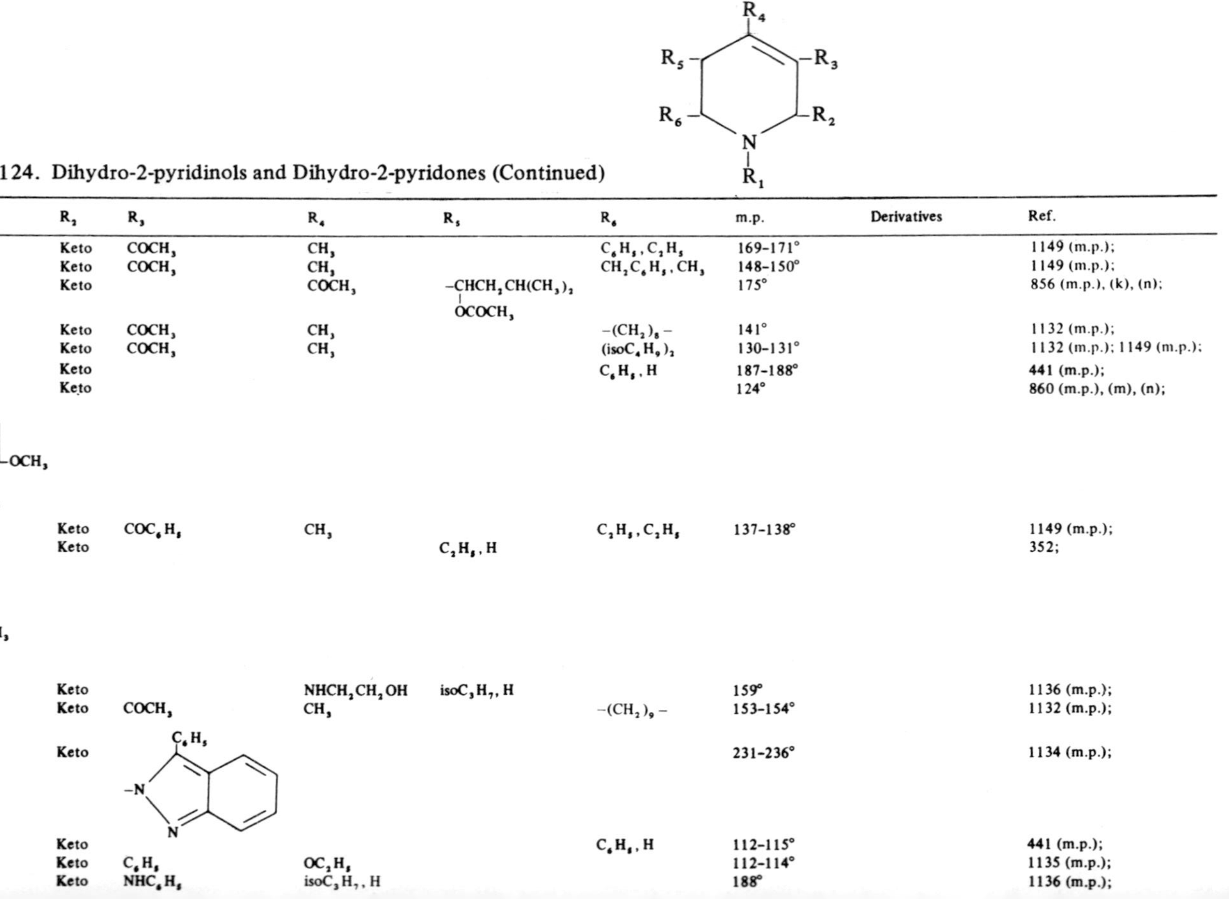


| $R_1$ | $R_2$ | $R_3$ | $R_4$ | $R_5$ | $R_6$ | m.p. | Derivatives | Ref. |
|---|---|---|---|---|---|---|---|---|
| | Keto | $COCH_3$ | $CH_3$ | | $C_6H_5, C_2H_5$ | 169–171° | | 1149 (m.p.); |
| | Keto | $COCH_3$ | $CH_3$ | | $CH_2C_6H_5, CH_3$ | 148–150° | | 1149 (m.p.); |
| $COCH_3$ | Keto | | $COCH_3$ | $-\underset{OCOCH_3}{CH}CH_2CH(CH_3)_2$ | | 175° | | 856 (m.p.), (k), (n); |
| | Keto | $COCH_3$ | $CH_3$ | | $-(CH_2)_8-$ | 141° | | 1132 (m.p.); |
| | Keto | $COCH_3$ | $CH_3$ | | $(isoC_4H_9)_2$ | 130–131° | | 1132 (m.p.); 1149 (m.p.); |
| $C_6H_5$ | Keto | | | | $C_6H_5, H$ | 187–188° | | 441 (m.p.); |
| $-COCH=CH$– (structure) $CH_3O$, $OCH_3$, $OCH_3$ | Keto | | | | | 124° | | 860 (m.p.), (m), (n); |
| | Keto | $COC_6H_5$ | $CH_3$ | | $C_2H_5, C_2H_5$ | 137–138° | | 1149 (m.p.); |
| $-CH_2CH_2$– (structure) $OCH_3$, $OCH_3$ | Keto | | | $C_2H_5, H$ | | | | 352; |
| $CH_2C_6H_5$ | Keto | | $NHCH_2CH_2OH$ | $isoC_3H_7, H$ | | 159° | | 1136 (m.p.); |
| | Keto | $COCH_3$ | $CH_3$ | | $-(CH_2)_9-$ | 153–154° | | 1132 (m.p.); |
| | Keto | (structure) $C_6H_5$, $-N$, N | | | | 231–236° | | 1134 (m.p.); |
| $CH_2C_6H_5$ | Keto | | | | $C_6H_5, H$ | 112–115° | | 441 (m.p.); |
| $CH_2C_6H_5$ | Keto | $C_6H_5$ | $OC_2H_5$ | | | 112–114° | | 1135 (m.p.); |
| $CH_2C_6H_5$ | Keto | $NHC_6H_5$ | $isoC_3H_7, H$ | | | 188° | | 1136 (m.p.); |

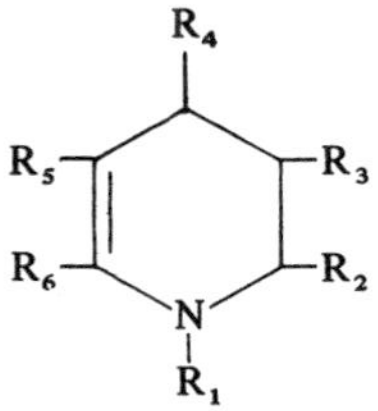

TABLE XII-124a. Dihydro-2-pyridinols and Dihydro-2-pyridones

| $R_1$ | $R_2$ | $R_3$ | $R_4$ | $R_5$ | $R_6$ | m.p. | Derivatives | Ref. |
|---|---|---|---|---|---|---|---|---|
|  | keto |  | $CH_3$ |  |  | 95–96° |  | 47 (m.p.), (n); |
|  | keto | OH, H |  |  | $CH_3$, H |  | 3-*O*-acetyl<br>m.p. 74° | 24 (m.p.); |
| $CH_3$ | keto |  |  |  | $CH_3$, H | b.p. 104°/13 mm |  | 695 (b.p.);<br>1128 (b.p.); (i); |
|  | keto |  |  | $C_2H_5$ |  | 79–80° |  | 1127 (m.p.), (i), (n), (u); |
| $C_2H_5$ | keto |  |  |  | $CH_3$, H | b.p. 106°/13 mm |  | 1128 (b.p.), (i), (u); |
| $COCH_3$ | OH, H | $SCH_3$, H |  |  | $SCH_3$, H |  | 2-*O*-acetyl<br>m.p. 115–117° | 1158 (m.p.), (c), (i), (m), (u); |
| $C_6H_4Cl$-*p* | keto | OH, H |  |  |  | 157–158° |  | 24 (m.p.); |
|  | keto | $C_2H_5$, $C_2H_5$ | $OCH_3$ | $CH_3$ |  | b.p. 74°/6 mm |  | 1146 (b.p.); |
| $C_6H_4Cl$-*p* | keto | OH, H |  |  | $CH_3$, H | 103° |  | 24 (m.p.); |
|  |  |  |  |  |  |  | 3-*O*-acetyl<br>m.p. 120–124° | 24 (m.p.);<br>1159 (m.p.); |
| $C_6H_5$ | keto | OH, H |  |  | $CH_3$, H | 102° |  | 24 (m.p.); |
|  |  |  |  |  |  |  | 3-*O*-acetyl<br>m.p. 117° | 24 (m.p.); |
| $COCH_3$ | OH, H | $-SC_4H_9$-*t*, $CH_3$ |  | $CH_3$ | OH, H |  | 2,6-di-*O*-acetyl<br>m.p. 125–127° | 1158 (m.p.), (c), (i), (m), (u); |
| $COCH_3$ | OH, D | $SC_4H_9$-*t* |  |  | $SC_4H_9$-*t*, D |  | 2-*O*-acetyl<br>m.p. 116–117° | 1158 (m.p.), (i), (m); |
| $COCH_3$ | OH, H | $SC_4H_9$-*t* |  |  | $SC_4H_9$-*t*, H |  | 2-*O*-acetyl<br>m.p. 116–117° | 1158 (m.p.), (i), (m), (u); |
| $COCH_3$ | OH, H | $SC_4H_9$-*t* | $CH_3$ |  | $SC_4H_9$-*t*, H |  | 2-*O*-acetyl<br>m.p. 105–107° | 1158 (m.p.), (c), (i), (m), (u); |
| $COCH_3$ | OH, H | $SC_4H_9$-*t* | $C_4H_9$-*t* |  | $SC_4H_9$-*t*, H |  | 2-*O*-acetyl<br>m.p. 98–100° | 1158 (m.p.), (i), (m), (u); |

TABLE XII-124a. Dihydro-2-pyridinols and Dihydro-2-pyridones (Continued)

| $R_1$ | $R_2$ | $R_3$ | $R_4$ | $R_5$ | $R_6$ | m.p. | Derivatives | Reference |
|---|---|---|---|---|---|---|---|---|
| $COCH_3$ | OH, H | $SC_4H_9$-*t* | $C_6H_5$ | | $SC_4H_9$-*t*, H | | 2-*O*-acetyl m.p. 140–142° | 1158 (m.p.), (i), (m), (u); |
| | keto | | | | $CH_3$ | 119–120° | | 1160 (m.p.); 1161 (m.p.), (i), (n); |
| | keto | | | COOH | $CH_3$ | | ethyl ester m.p. 156° | 70 (m.p.), (u); 84 (m.p.), (n); 1034 (m); |
| | | | | | | | *t*-butyl ester m.p. 127–128° | 84 (m.p.); |
| | keto | | | $CH_3$ | $CH_3$ | 130–131° | | 1160 (m.p.); 1162 (m.p.); |
| | keto | | | $COCH_3$ | $CH_3$ | 141° | | 84 (m.p.); 1034 (m.p.); |
| $CH_3$ | keto | | | COOH | $CH_3$ | | ethyl ester b.p. 105°/0.3 mm | 84 (b.p.); 1034 (m); |
| | keto | $CH_3$, H | | COOH | $CH_3$ | | ethyl ester m.p. 157–158° | 84 (m.p.); |
| | keto | | $CH_3$, H | COOH | $CH_3$ | | ethyl ester m.p. 143–147° | 84 (m.p.); 1034 (m); |
| | keto | | | $C_2H_5$ | $CH_3$ | 87–88° | | 1163 (m.p.); |
| $CH_3$ | keto | | | $COCH_3$ | $CH_3$ | b.p. 97°/0.1 mm | | 84 (b.p.); |
| | keto | | | *n*-$C_3H_7$ | $CH_3$ | 68–70° | | 1162 (m.p.); 1163 (m.p.); |
| | keto | | | *n*-$C_4H_9$ | $CH_3$ | 57–58° | | 1163 (m.p.); |

| | | | | | | | | |
|---|---|---|---|---|---|---|---|---|
| | keto | | | | $C_6H_5$ | 148–159° | | 938 (n); |
| | | | | | | | | 1160 (m.p.); |
| | | | | | | | | 1162 (m.p.); |
| | keto | | | $CH_3$ | $C_6H_4Cl$-*p* | 159–160° | | 1142 (m.p.); |
| | keto | | | $CH_3$ | $C_6H_5$ | 136–137° | | 1142 (m.p.); |
| | keto | | | $C_6H_5$ | $CH_3$ | 169–170° | | 787 (m.p.); |
| | keto | OH, H | $NH_2$, $CH_3$ | $C_6H_5$ | | 280° | | 221 (m.p.), (i), (n), (u); |
| | | | | | | | 3-*O*-acetyl, 4-*N*-acetyl m.p. 232–233° | 221 (m.p.), (i), (n); |
| | keto | | | COOH | $CH_2C_6H_5$ | | methyl ester m.p. 108–110° | 70 (m.p.), (u); |
| | keto | $C_6H_5$, H | | $CH_3$ | $CH_3$ | 132° | | 1164 (m.p.); |
| | keto | | | $CH_3$ | $C_6H_4CH_3$ *p* | 146–147° | | 1142 (m.p.); |
| | keto | CN, H | OH, H | | $CH{=}CHC_6H_5$ | 315° | | 57 (m.p.), (i), (u); |
| | | | | | | | 4-*O*-acetyl m.p. 289° | 57 (m.p.), (i), (u); |
| | keto | $C_6H_5$, H | $CH_3$, $CH_3$ | | $CH_3$ | 125° | | 1164 (m.p.); |
| | keto | $C_6H_5$, H | | $C_6H_5$ | Br | 190–192° | | 1165 (m.p.), (i); |
| | keto | | | $C_6H_5$ | $C_6H_5$ | 214–215° | | 35 (m.p.), (i); |
| | | | | | | | | 443 (m.p.), (i), (u); |
| 3,4,5-$(CH_3O)_3C_6H_2$–CH=CHCO– (—COCH=CH– with $CH_3O$, $OCH_3$, $OCH_3$ on ring) | keto | | | | | 124° | | 859 (m.p.), (i), (m), (n), (u); |
| | | | | | | | | 1143 (m.p.), (i), (n), (u); |
| | keto | $C_6H_5$, H | | $C_6H_5$ | $CH_3$ | 194° | | 1164 (m.p.); |
| | keto | | $C_6H_5$, H | $C_6H_5$ | $CH_3$ | 178° | | 444 (m.p.), (i); |
| | keto | $C_6H_5$, H | | $CH_3$ | $C_6H_5$ | 135° | | 1164 (m.p.) ; |
| | keto | | $C_6H_5$, H | $CH_3$ | $C_6H_5$ | 187° | | 444 (m.p.), (i); |
| | keto | COOH, H | $C_6H_5$, H | $CH_3$ | $C_6H_5$ | | | 444 (i); |
| | keto | $C_6H_5$, H | | $C_6H_5$ | $C_6H_5$ | 180–181° | | 1164 (m.p.); |

TABLE XII-125. Dihydro-3-pyridinols and Dihydro-3-pyridones

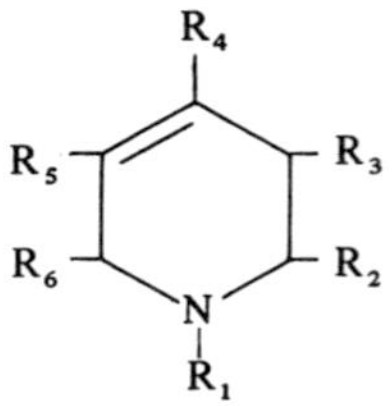

| $R_1$ | $R_2$ | $R_3$ | $R_4$ | $R_5$ | $R_6$ | m.p. | Derivatives | Ref. |
|---|---|---|---|---|---|---|---|---|
| $CH_2C_6H_5$ | | keto | | OH | | 87–89° | hydrate | 1166 (m.p.); |
| $CH_3$ | | OH, H | $C_6H_5$ | | | 104–106° | | 1151 (m.p.), (u); 1167 (m.p.); |
| | | | | | | | hydrochloride m.p. 211–214° | 1151 (m.p.); |
| | | | | | | | 3-*O*-acetyl ·HBr m.p. 230–232° | 1167 (m.p.), (u); |
| $C_6H_5$ | $NHC_6H_5$ | keto | | | | 140–142° | | 1153 (m.p.), (i), (u); |
| $C_6H_4CH_3$-*p* | $NHC_6H_4CH_3$-*p* | keto | | | | 164–165° | | 1153 (m.p.), (i), (u); |
| | 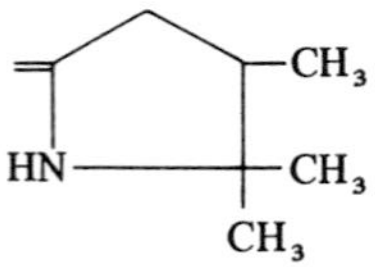 =C(ring: HN, $CH_3$, $CH_3$, $CH_3$) | keto | | $CH_3$, $CH_3$ | | 147° | | 1168 (m.p.), (i), (u); |
| | | | | | | | hydrochloride m.p. 227–229° | 1168 (m.p.); |
| | | | | | | | picrate m.p. 197° | 1168 (m.p.); |

# TABLE XII-126. Dihydro-4-pyridinols and Dihydro-4-pyridones

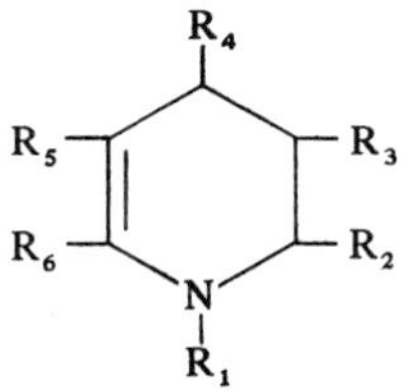

| $R_1$ | $R_2$ | $R_3$ | $R_4$ | $R_5$ | $R_6$ | m.p. | Derivatives | Ref. |
|---|---|---|---|---|---|---|---|---|
| | H, $C_6H_5$ | | keto | | $CH_3$ | 160–161° | | 48 (m.p.), (i), (n), (u); |
| $CH_3$ | H, $CH_2C_6H_5$ | | keto | | | 68–70° | | 767 (m.p.), (m), (n); |
| | H, $CH_2CH_2C_6H_5$ | | keto | | $CH_3$ | 143–145° | | 1138 (m.p.), (i), (n), (u); |
| 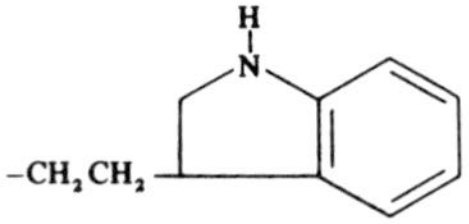 $-CH_2CH_2-$ (H, N) | | | keto | | | 164° | | 788 (m.p.), (i), (u); |
| 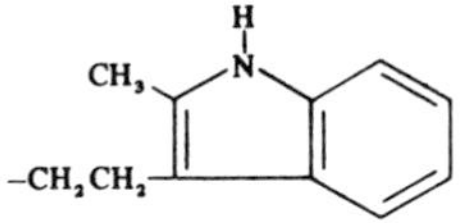 $-CH_2CH_2-$, $CH_3$ (H, N) | | | keto | | | 206° | | 788 (m.p.), (i), (u); |
| | H, $C_6H_5$ | | keto | | $C_6H_5$ | 149–151° | | 48 (m.p.), (i), (n), (u); |
| 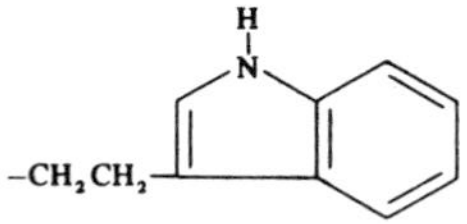 $-CH_2CH_2-$ (H, N) | | | keto | $COCH_3$ | | 198° | | 1169 (m.p.), (i), (n), (u); |
| $C_6H_5$ | H, $C_6H_5$ | | keto | I | $CH_3$ | 151° | | 449 (m.p.), (i), (n), (u); |
| $C_6H_5$ | H, $C_6H_5$ | | keto | | $CH_3$ | 84–85° | | 39 (m.p.), (i), (n), (u); 449 (m.p.), (i), (u); 765 (m.p.), (i), (n), (u); 1137 (n); |

TABLE XII-126. Dihydro-4-pyridinols and Dihydro-4-pyridones (Continued)

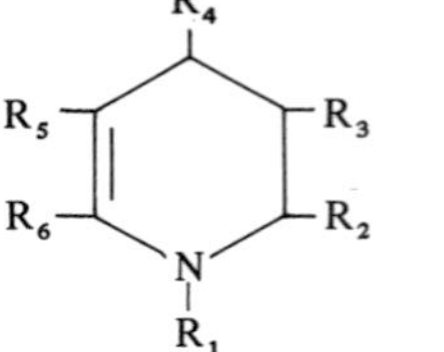


| $R_1$ | $R_2$ | $R_3$ | $R_4$ | $R_5$ | $R_6$ | m.p. | Derivatives | Ref. |
|---|---|---|---|---|---|---|---|---|
| | | | | | | | hydrobromide, m.p. 153° | 449 (m.p.); |
| | | | | | | | hydrochloride, m.p. 173° | 449 (m.p.); |
| | H, $C_6H_5$ | | keto | $CH_3$ | $C_6H_5$ | | | 765 (i), (n), (t), (u); |
| | H, $C_6H_5$ | | keto | $C_6H_5$ | $CH_3$ | 257° | | 765 (m.p.), (i), (n), (u); |
| $-N{=}CHC_6H_5$ | H, $C_6H_5$ | H, $CH_2C_6H_5$ | keto | | $CH_3$ | 176–177° | | 48 (m.p.), (i), (u); |
| | H, $C_6H_5$ | | keto | | $CH_3$ | 159–160° | | 48 (m.p.), (f), (n), (u); |
| | H, $C_6H_5$ | | keto | | $CH_2CH_2C_6H_5$ | 124–126° | | 48 (m.p.), (i), (n), (u); |
| $-NHCH_2C_6H_5$ | H, $C_6H_5$ | | keto | | $CH_3$ | | | 48 (f), (n), (u); |
| $C_6H_5$ | H, $CH{=}CHC_6H_5$ | | keto | | $CH_3$ | 62–63° | | 1138 (m.p.), (i), (n), (u); |
| $C_6H_5$ | H, $CH_2CH_2C_6H_5$ | | keto | | $CH_3$ | | | 1138 (t); |
| | H, $CH_2CH_2C_6H_5$ | H, $CH_2C_6H_5$ | keto | | $CH_3$ | | | 1138 (u); |
| | H, $CH_2CH_2C_6H_5$ | | keto | | $CH_2CH_2C_6H_5$ | 144–146° | | 1138 (m.p.), (i), (u); |
| $C_6H_5$ | H, $C_6H_5$ | | keto | | $C_6H_5$ | 149–151° | | 39 (m.p.), (i), (n), (u); |
| $-N{=}CHCH{=}CHC_6H_5$ | H, $CH{=}CHC_6H_5$ | | keto | | $CH_3$ | 185–186° | | 1138 (m.p.), (i), (n), (u); |
| $-N{=}CHCH{=}CHC_6H_5$ | H, $CH_2CH_2C_6H_5$ | | keto | | $CH_3$ | oil | | 1138 (n), (u); |
| $-NH(CH_2)_3C_6H_5$ | H, $CH_2CH_2C_6H_5$ | | keto | | $CH_3$ | oil | | 1138 (n); |
| $-N{=}CHC_6H_5$ | H, $C_6H_5$ | | keto | | $C_6H_5$ | 178–179° | | 48 (m.p.), (i), (n), (u); |
| $C_6H_5$ | H, $C_6H_5$ | | keto | | $CH{=}CHC_6H_5$ | 164–165° | | 1137 (m.p.), (c), (f), (i), (n), (t), (u); |
| $C_6H_5$ | H, $C_6H_5$ | H, $CH_2C_6H_5$ | keto | | $CH_3$ | 179–180° | | 39 (m.p.), (i), (n), (u); |

| | | | | | | | |
|---|---|---|---|---|---|---|---|
| $C_6H_5$ | H, $C_6H_5$ | | keto | $CH_2CH_2C_6H_5$ | 99–100° | | 39 (m.p.), (i), (n), (u); |
| $C_6H_5$ | H, $C_6H_5$ | H, $CHC_6H_5$ (OH) | keto | $CH_3$ | 122–123° | | 1137 (m.p.), (c), (i), (n), (u); |
| $-N=CHC_6H_5$ | H, $C_6H_5$ | H, $CH_2C_6H_5$ | keto | $CH_3$ | 129–131° | | 48 (m.p.), (i), (n), (u); |
| $-N=CHC_6H_5$ | H, $C_6H_5$ | | keto | $CH_2CH_2C_6H_5$ | 118–121° | | 48 (m.p.), (i), (n), (u); |
| $C_6H_5$ | H, $CH=CHC_6H_5$ | H, $CH_2C_6H_5$ | keto | $CH_3$ | 155–156° | | 1138 (m.p.), (i), (n), (t), (u); |
| $C_6H_5$ | H, $CH=CHC_6H_5$ | | keto | $CH_2CH_2C_6H_5$ | 174–176° | | 1138 (m.p.), (i), (n), (t), (u); |
| $-N=CHCH=CHC_6H_5$ | H, $CH=CHC_6H_5$ | | keto | $C_6H_5$ | | | 1138 (u); |
| $-N=CHCH=CHC_6H_5$ | H, $CH=CHC_6H_5$ | H, $CH_2C_6H_5$ | keto | $CH_3$ | | | 1138 (i), (t); |
| $-N=CHCH=CHC_6H_5$ | H, $CH=CHC_6H_5$ | | keto | $CH_2CH_2C_6H_5$ | | | 1138 (i), (n), (u); |
| $-NH(CH_2)_3C_6H_5$ | H, $CH_2CH_2C_6H_5$ | | keto | $CH_2CH_2C_6H_5$ | | | 1138 (i), (n), (u); |
| $-NH(CH_2)_3C_6H_5$ | H, $CH_2CH_2C_6H_5$ | H, $CH_2C_6H_5$ | keto | $CH_3$ | | | 1138 (i); |
| | $CH_3$ | COOH | keto | | | methyl ester m.p. 142–145° | 70 (m.p.), (u); |
| | $CH_2C_6H_5$ | COOH | keto | | | methyl ester m.p. 116–118° | 70 (m.p.), (u); |
| | | | | | | *t*-butyl ester m.p. 155–158° | 70 (m.p.), (u); |
| | $-CH_2C_6H_4OCH_3$-*p* | COOH | keto | | | methyl ester m.p. 124–125° | 70 (m.p.), (u); |
| | $-CH_2C_6H_3(OCH_3)_2$ | COOH | keto | | | methyl ester m.p. 129–131° | 70 (m.p.), (u); |

TABLE XII-127. Ethers of Dihydropyridinols

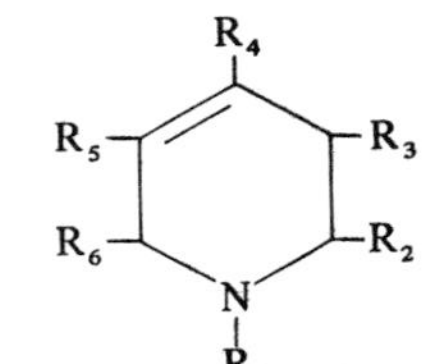


| $R_1$ | $R_2$ | $R_3$ | $R_4$ | $R_5$ | $R_6$ | m.p. | Derivatives | Ref. |
|---|---|---|---|---|---|---|---|---|
| $CH_3$ | | | $OCH_3$ | | | b.p. 168°/760 mm | | 1170 (b.p.), (i), (n); |
| | | | | | | | picrate m.p. 109–112° | 1170 (m.p.); |
| $CH_3$ | $CH_2C_6H_5$ | | $OCH_3$ | | | | picrate m.p. 116–118° | 767 (m.p.), (m), (n); |
| $CH_3$ | $CH_2C_6H_4OCH_3$-*p* | | $OCH_3$ | | | | picrate m.p. 126–127° | 767 (m.p.), (i), (n); |
| $CH_3$ | $CH_2C_6H_4OCH_3$-*p* | | $OCH_3$ | | | | picrate m.p. 162–163° | 1170 (m.p.), (i), (n); |

TABLE XII-128. 2,4-Diketotetrahydropyridines

| $R_1$ | $R_3$ | $R_{3'}$ | $R_5$ | $R_6$ | m.p. | Derivatives | Ref. |
|---|---|---|---|---|---|---|---|
| | Cl | Cl | COOH | $CHCl_2$ | | ethyl ester, m.p. 161–163° | 707 (m.p.), (i), (n); |
| | Cl | Cl | COOH | $CH_2Cl$ | | ethyl ester, m.p. 148° | 707 (m.p.), (i), (n); |
| | $CH_3$ | $CH_3$ | | | | | 1139 (n); |
| | $C_2H_5$ | $C_2H_5$ | | | | | 1139 (n); |
| | Cl | Cl | Cl | $C_6H_5$ | 160–161° | | 707 (m.p.); |
| [structure: 2,4-dichlorophenyl] | | | | $CH_3$ | 280–283° | | 149 (m.p.); |
| [structure: 2,5-dichlorophenyl] | | | | $CH_3$ | 237–239° | | 149 (m.p.); |
| $C_6H_4Cl$-*o* | | | | $CH_3$ | 263–265° | | 149 (m.p.); |
| $C_6H_4Br$-*o* | | | | $CH_3$ | 278–281° | | 149 (m.p.); |
| $C_6H_5$ | $C(=NH)CH_3$ | H | | $CH_3$ | 258–260° | | 1030 (m.p.), (i); |
| $C_6H_5$ | $CH_2C_6H_5$ | Cl | | $C_6H_5$ | 174° | | 92 (m.p.); |

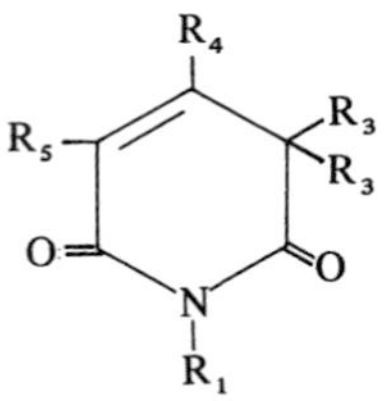

TABLE XII-129. 2,6-Diketotetrahydropyridines

| $R_1$ | $R_3$ | $R_3{}'$ | $R_4$ | $R_5$ | m.p. | Derivatives | Ref. |
|---|---|---|---|---|---|---|---|
| | $CH_3$ | H | Cl | Cl | 196–197° | | 7 (m.p.); |
| | | | COOH | | | cyclohexyl | |
| | | | | | | ester, m.p. 214° | 507 (m.p.); |
| $CH_3$ | | | | | | | 505 (n); |
| | $CH_3$ | H | | OH | oil | | 506; |
| | | | | | | triacetate, | |
| | | | | | | m.p. 86° | 506 (m.p.) (i), (u); |
| | $C_2H_5$ | H | Cl | Cl | 179–182° | | 7 (m.p.), (i), (u); |
| $CH_3$ | | | COOH | | | | 507 (i); |
| | | | | | | methyl ester, | |
| | | | | | | m.p. 198° | 507 (m.p.); |

| | | | | | | |
|---|---|---|---|---|---|---|
| | $CH_3$ | $CH_3$ | | | | 455 (n), (u); |
| | 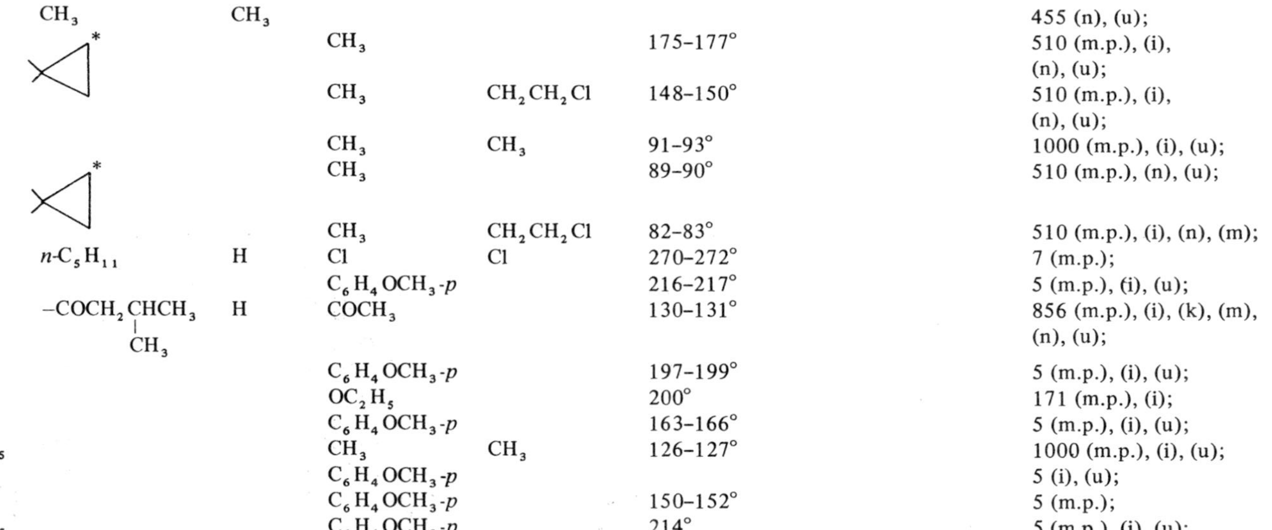 | | $CH_3$ | | 175–177° | 510 (m.p.), (i), (n), (u); |
| | | | $CH_3$ | $CH_2CH_2Cl$ | 148–150° | 510 (m.p.), (i), (n), (u); |
| $CH_3$ | | | $CH_3$ | $CH_3$ | 91–93° | 1000 (m.p.), (i), (u); |
| $CH_3$ | | | $CH_3$ | | 89–90° | 510 (m.p.), (n), (u); |
| $CH_3$ | | | $CH_3$ | $CH_2CH_2Cl$ | 82–83° | 510 (m.p.), (i), (n), (m); |
| | $n$-$C_5H_{11}$ | H | Cl | Cl | 270–272° | 7 (m.p.); |
| OH | | | $C_6H_4OCH_3$-$p$ | | 216–217° | 5 (m.p.), (i), (u); |
| | $-COCH_2CH(CH_3)CH_3$ | H | $COCH_3$ | | 130–131° | 856 (m.p.), (i), (k), (m), (n), (u); |
| $CH_3$ | | | $C_6H_4OCH_3$-$p$ | | 197–199° | 5 (m.p.), (i), (u); |
| $C_6H_5$ | | | $OC_2H_5$ | | 200° | 171 (m.p.), (i); |
| $-OCH_3$ | | | $C_6H_4OCH_3$-$p$ | | 163–166° | 5 (m.p.), (i), (u); |
| $-OCH_2C_6H_5$ | | | $CH_3$ | $CH_3$ | 126–127° | 1000 (m.p.), (i), (u); |
| $C_6H_5$ | | | $C_6H_4OCH_3$-$p$ | | | 5 (i), (u); |
| $-CH_2C_6H_5$ | | | $C_6H_4OCH_3$-$p$ | | 150–152° | 5 (m.p.); |
| $-OCH_2C_6H_5$ | | | $C_6H_4OCH_3$-$p$ | | 214° | 5 (m.p.), (i), (u); |

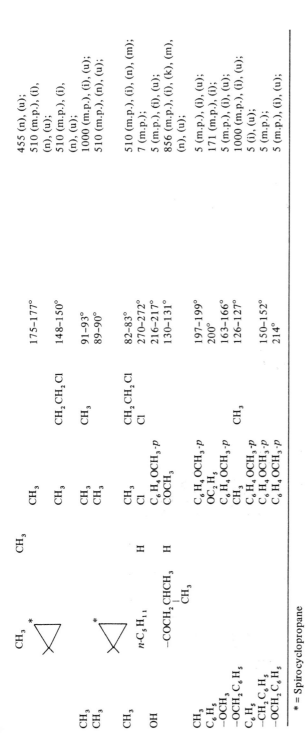

* = Spirocyclopropane

TABLE XII-130. 1,2,3,6-Tetrahydro-2,6-diketo-3-substituted hydrazonopyridines

$R_4$ / $R_5$ / =N–NHR / O= / =O / N / $R_1$

| R | $R_1$ | $R_4$ | $R_5$ | m.p. | Derivatives | Ref. |
|---|---|---|---|---|---|---|
| $CH_3$ | $CH_3$ | | | 163° | | 505 (m.p.), (i), (m), (n), (t), (u); |
| $C_6H_4Cl$-*o* | $C_6H_5$ | $C_6H_4OCH_3$-*p* | | 287–289° | | 739 (m.p.); |
| $C_6H_4Cl$-*m* | $C_6H_5$ | $C_6H_4OCH_3$-*p* | | 230–232° | | 739 (m.p.); |
| $C_6H_4Cl$-*p* | $C_6H_5$ | $C_6H_4OCH_3$-*p* | | 224–226° | | 739 (m.p.); |
| $C_6H_4Br$-*p* | $C_6H_5$ | $C_6H_4OCH_3$-*p* | | 233–234° | | 739 (m.p.); |
| $C_6H_4NO_2$-*o* | $C_6H_5$ | $C_6H_4OCH_3$-*p* | | 213–215° | | 739 (m.p.); |
| $C_6H_4NO_2$-*m* | $C_6H_5$ | $C_6H_4OCH_3$-*p* | | 210–212° | | 739 (m.p.); |
| $C_6H_4NO_2$-*p* | $C_6H_5$ | $C_6H_4OCH_3$-*p* | | 218–220° | | 739 (m.p.); |
| $C_6H_5$ | $C_6H_5$ | $C_6H_4OCH_3$-*p* | | 245–247° | | 739 (m.p.); |
| $C_6H_4CH_3$-*p* | $C_6H_5$ | $C_6H_4OCH_3$-*p* | | 230–231° | | 739 (m.p.); |
| $C_6H_4OCH_3$-*o* | $C_6H_5$ | $C_6H_4OCH_3$-*p* | | 286–288° | | 739 (m.p.); |
| $C_6H_4OCH_3$-*p* | $C_6H_5$ | $C_6H_4OCH_3$-*p* | | 222–224° | | 739 (m.p.); |

TABLE XII-131. 2-Piperidones

(Structure: 2-piperidone ring with substituents $R_1$ on N, $R_3$, $R_{3'}$, $R_4$, $R_{4'}$, $R_5$, $R_{5'}$, $R_6$, $R_{6'}$ on ring carbons, C=O at position 2)

| $R_1$ | $R_3R_{3'}$ | $R_4R_{4'}$ | $R_5R_{5'}$ | $R_6R_{6'}$ | m.p. | Derivatives | Ref. |
|---|---|---|---|---|---|---|---|
| | | COOH | | | 174–176° | | 1140 (m.p.), (i), (n); |
| | | | | | | methyl ester, m.p. 126–127° | 1140 (m.p.), (i), (n); |
| | | | | | | ethyl ester, m.p. 108° | 1140 (m.p.); |
| $CH_3$ | | | | | | | 861 (m.p.); |
| $CH_3$ | | $NH_2$ | | | b.p. 95°/0.01 mm | | 861 (b.p.), (i); |
| $CH_3$ | | COOH | | | | methyl ester, b.p. 120°/0.007 mm | 861 (b.p.), (i); |
| $CH_3$ | | $-CONH_2$ | | | 194–196° | | 861 (m.p.), (i); |
| $CH_3$ | | $-NHCONH_2$ | | | 203–204° | | 861 (m.p.), (i); |
| $CH_3$ | | | $-NHCONH_2$ | | > 230° | | 861 (m.p.), (i); |
| $C_2H_5$ | $CH_3$ | | | | b.p. 103°/15 mm | | 690 (b.p.); |
| | CN | $CH_3$ | | $CH_3$, $CH_3$ | 156–157° | | 1131 (m.p.); |
| $NH_2$ | | $C_6H_5$ | | | 100–102° | | 1133 (m.p.); |
| | | | | | | hydrochloride, m.p. 179–180° | 1133 (m.p.); |
| | | | | | | *N'*-diacetyl, m.p. 122–124° | 1133 (m.p.); |
| $CH_3$ | | | $-(CH_2)_4NHCH_3$ | | b.p. 145°/1.5 mm | | 691 (b.p.); |
| | | | | | | dihydrochloride, m.p. 166–168° | 691 (m.p.); |
| | | | | | | dipicrate, m.p. 166–167° | 691 (m.p.); |
| $-CH_2C_6H_4F$-*o* | | | | | b.p. 159°/3 mm | | 689 (b.p.); |
| $-CH_2C_6H_4Br$-*o* | | | | | 76–77° | | 558 (m.p.); |
| $CH_3$ | $C_6H_5$ | =NOH | | | 151–155° | | 1141 (m.p.); |
| $-CH_2C_6H_5$ | | | | | b.p. 162°/4 mm | | 689 (b.p.), (i); |
| $-CH_2-C_6H_3(Cl)-NH_2$ (ring structure with Cl and $NH_2$) | | | | | 139–140° | | 558 (m.p.); |
| $-CH_2C_6H_4NH_2$-*m* | | | | | b.p. 218°/2 mm | | 558 (b.p.); |
| $-CH_2C_6H_4NH_2$-*p* | | | | | b.p. 220°/3 mm | | 558 (b.p.); |
| $-CH_2C_6H_{11}$ | | | | | b.p. 153°/7 mm | | 689 (b.p.); |
| | | | | | | hydrochloride, m.p. 132–133° | 689 (m.p.); |

TABLE XII-131. 2-Piperidones (Continued)

(Structure: 2-piperidone ring, N-$R_1$, C=O at C-2, substituents $R_3,R_{3'}$; $R_4,R_{4'}$; $R_5,R_{5'}$; $R_6,R_{6'}$)

| $R_1$ | $R_3,R_{3'}$ | $R_4,R_{4'}$ | $R_5,R_{5'}$ | $R_6,R_{6'}$ | m.p. | Derivatives | Ref. |
|---|---|---|---|---|---|---|---|
| $CH_3$ | | | $-(CH_2)_4N(CH_3)_2$ | | b.p. 134°/2 mm | | 691 (b.p.); |
| | | | | | | dihydrochloride, m.p. 125–127° | 691 (m.p.); |
| | | | | | | dipicrate, m.p. 107–109° | 691 (m.p.); |
| $-CH_2-C_6H_3(Cl)-COOH$ | | | | | 225° | | 558 (m.p.); |
| $-CH_2C_6H_4COOH$-*p* | | | | | 167° | | 558 (m.p.); |
| | | | $CH_3, CH_3$ | OH, $C_6H_4Cl$-*p* | 175–176° | | 1142 (m.p.); |
| | | | $CH_3, CH_3$ | OH, $C_6H_4$ | 175–176° | | 1142 (m.p.); |
| $-CH_2(CH_2)_2-N(CH_2CH_2)_2NCH_3$ | | | | | 45° | | 558 (m.p.); |
| | | | $CH_3, CH_3$ | OH, $C_6H_4CH_3$-*p* | 172–174° | | 1142 (m.p.); |
| | | | $CH_3, C_2H_5$ | OH, $C_6H_5$ | 160–161° | | 1142 (m.p.); |
| | | | $CH_3, CH_3$ | OH, $C_6H_4OCH_3$-*p* | 188–190° | | 1142 (m.p.); |
| $-CH_2(CH_2)_2COC_6H_5$ | | | | | b.p. 108°/3 mm | | 558 (b.p.); |
| $-CH_2(CH_2)_2-C_6H_3(O_2CH_2)$ | | | | | b.p. 195°/1 mm | | 340 (b.p.), (i); |
| $-CH_2CH_2C_6H_5$ | | | $C_2H_5, H$ | | b.p. 130°/1 mm | | 352 (b.p.); |
| $-CH_2CH_2-C_6H_3(OCH_3)_2$ | | | $C_2H_5, H$ | | oil | | 350 (i); |
| $-COCH=CH-C_6H_3(OCH_3)_2$ | | | | | 116–117° | | 860 (m.p.), (n), (u); |

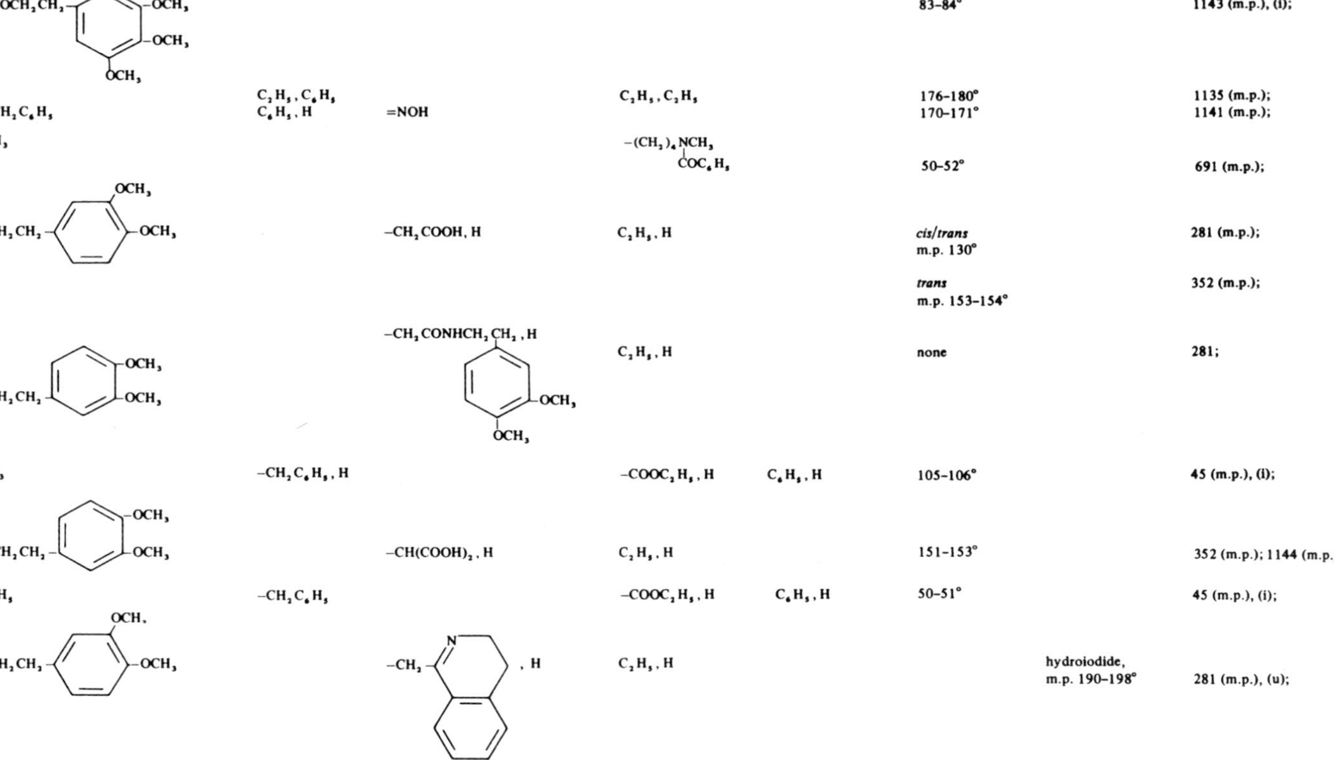

| | | | | | | | |
|---|---|---|---|---|---|---|---|
| $-COCH_2CH_2-$ (ring: $-OCH_3$, $-OCH_3$, $OCH_3$) | | | | | 83–84° | | 860 (m.p.), (n);<br>1143 (m.p.), (i); |
| $-CH_2C_6H_5$ | $C_2H_5, C_6H_5$<br>$C_6H_5, H$ | $=NOH$ | $C_2H_5, C_2H_5$ | | 176–180°<br>170–171° | | 1135 (m.p.);<br>1141 (m.p.); |
| $CH_3$ | | | $-(CH_2)_4N(CH_3)COC_6H_5$ | | 50–52° | | 691 (m.p.); |
| $-CH_2CH_2-$ (ring: $OCH_3$, $-OCH_3$) | | $-CH_2COOH, H$ | $C_2H_5, H$ | | *cis/trans*<br>m.p. 130° | | 281 (m.p.); |
| | | | | | *trans*<br>m.p. 153–154° | | 352 (m.p.); |
| $-CH_2CH_2-$ (ring: $-OCH_3$, $-OCH_3$) | | $-CH_2CONHCH_2CH_2-$ (ring: $-OCH_3$, $OCH_3$), H | $C_2H_5, H$ | | none | | 281; |
| $CH_3$ | $-CH_2C_6H_5, H$ | | $-COOC_2H_5, H$ | $C_6H_5, H$ | 105–106° | | 45 (m.p.), (i); |
| $-CH_2CH_2-$ (ring: $-OCH_3$, $-OCH_3$) | | $-CH(COOH)_2, H$ | $C_2H_5, H$ | | 151–153° | | 352 (m.p.); 1144 (m.p.); |
| $C_6H_5$ | $-CH_2C_6H_5$ | | $-COOC_2H_5, H$ | $C_6H_5, H$ | 50–51° | | 45 (m.p.), (i); |
| $-CH_2CH_2-$ (ring: $OCH_3$, $-OCH_3$) | | $-CH_2-$ (3,4-dihydroisoquinoline ring: N, $CH_3O$, $OCH_3$), H | $C_2H_5, H$ | | | hydroiodide,<br>m.p. 190–198° | 281 (m.p.), (u); |

TABLE XII-132. 2,4-Diketopiperidines

| $R_1$ | $R_3$ | $R_{3'}$ | $R_5$ | $R_6$ | m.p. | Derivatives | Ref. |
|---|---|---|---|---|---|---|---|
| | $C_2H_5$ | $C_2H_5$ | | | | | 1139 (n); |
| | $C_2H_5$ | $C_2H_5$ | $CH_3$ | | | | 1139 (n); |
| $CH_3$ | $C_6H_5$ | H | | | 136–137° | | 1135 (m.p.); |
| $CH_3$ | $C_6H_5$ | $C_2H_5$ | | | 82–84° | | 1135 (m.p.), (i), (u); |
| $-CH_2CH_2C_6H_3(OCH_3)_2$ | | | $C_2H_5$, H | | 38–40° | | 281 (m.p.); |
| $CH_2C_6H_5$ | $C_6H_5$ | H | | | 139–141° | | 1135 (m.p.); |
| $-CH_2CH_2C_6H_5$ | $C_6H_5$ | H | | | 133–135° | | 1135 (m.p.); |
| $-CH_2C_6H_5$ | $C_6H_5$ | $C_2H_5$ | | | oil | | 1135 (i); |
| $-CH_2C_6H_5$ | $C_6H_5$ | $C_2H_5$ | $C_2H_5$, $C_2H_5$ | | | | 1135 (i), (u); |

TABLE XII-133. 2,6-Diketopiperidines

| $R_1$ | $R_3$ | $R_{3'}$ | $R_4$ | $R_5$ | m.p. | Derivatives | Ref. |
|---|---|---|---|---|---|---|---|
| | CN | | $CF_3$, H | | 158–160° | | 1145 (m.p.), (i); |
| | | | | | | sodium salt, m.p. 313° | 1145 (m.p.); |
| | CN | | $CF_3$, $CH_3$ | | 195–198° | | 1145 (m.p.), (i); |
| | | | | | | sodium salt, m.p. 317° | 1145 (m.p.); |
| | Cl | $C_6H_5$ | $C_6H_5$, H | | 190–191° | | 242 (m.p.), (i), (n), (u); |
| | $C_6H_5$ | H | $C_6H_5$, H | | 225–227° | | 242 (m.p.), (n); |

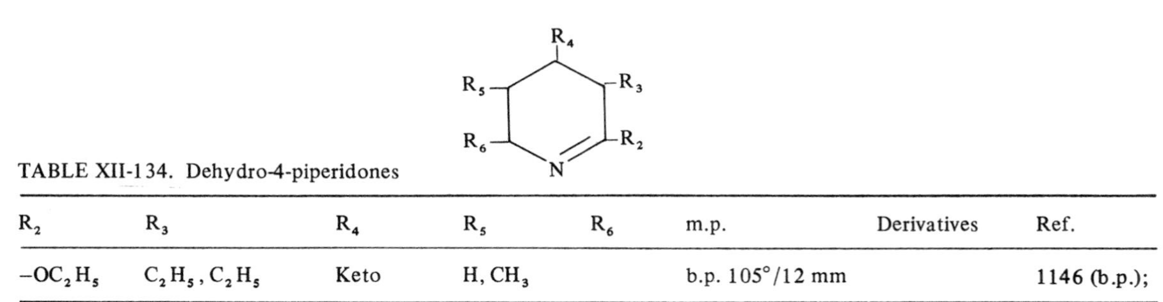

TABLE XII-134. Dehydro-4-piperidones

| $R_2$ | $R_3$ | $R_4$ | $R_5$ | $R_6$ | m.p. | Derivatives | Ref. |
|---|---|---|---|---|---|---|---|
| $-OC_2H_5$ | $C_2H_5$, $C_2H_5$ | Keto | H, $CH_3$ | | b.p. 105°/12 mm | | 1146 (b.p.); |

TABLE XII-135. Dehydro-2-piperidones

| $R_2$ | $R_3$ | $R_4$ | $R_5$ | $R_6$ | m.p. | Derivatives | Ref. |
|---|---|---|---|---|---|---|---|
| Keto | | | $CH_3$, $-N(C_2H_5)_2$ | $CH_3$ | b.p. 134°/5 mm | | 932 (b.p.); |
| | | | | | | picrate, m.p. 131–132° | 932 (m.p.); |
| Keto | | | $(-CH_2CH_2CN)_2$ | $CH_3$ | 222–223° | | 1147 (m.p.); |
| Keto | | | $n$-$C_3H_7$, $-N(C_2H_5)_2$ | $CH_3$ | b.p. 138°/4 mm | | 932 (b.p.); |
| | | | | | | picrate, m.p. 128–129° | 932 (m.p.); |
| Keto | | | $n$-$C_4H_9$, $-N(C_2H_5)_2$ | $CH_3$ | b.p. 138°/4 mm | | 932 (b.p.); |
| | | | | | | picrate, m.p. 138–139° | 932 (m.p.); |
| Keto | | | $n$-$C_3H_7$, – | $CH_3$ | | triamine, m.p. 219–220° | 932 (m.p.); |
| Keto | | | iso$C_3H_7$, – | $CH_3$ | | triamine, m.p. 229–230° | 932 (m.p.); |
| Keto | | | $n$-$C_4H_9$, – | $CH_3$ | | triamine, m.p. 198–200° | 932 (m.p.); |
| Keto | F, F | F, F | F, F | Cl | b.p. 111° | | 1148 (b.p.); |
| Keto | F, F, | F, F | F, F | $-OPCl_4$ | b.p. 101°/4 mm | | 1148 (b.p.); |
| Keto | | | | $CH_3$ | 117–118° | | 1147 (m.p.); |
| Keto | | | $CH_3$, Br | $CH_3$ | 91-93° | | 932 (m.p.); |
| Keto | | | $CH_3$, H | $CH_3$ | 130–131° | | 1147 (m.p.); |

| | | | | | | | |
|---|---|---|---|---|---|---|---|
| Keto | F, F | F, F | F, F | –N(ring)O | 127–129° | | 1148 (m.p.); |
| Keto | F, F | F, F | F, F | $N(C_2H_5)_2$ | b.p. 135°/11 mm | | 1148 (b.p.); |
| Keto | | | $-CH_2CH_2CN$, H | $CH_3$ | 166° | | 1147 (m.p.); |
| Keto | | | $n$-$C_3H_7$, Br | $CH_3$ | 113–114° | | 932 (m.p.); |
| Keto | | | iso$C_3H_7$, Br | $CH_3$ | 107–108° | | 932 (m.p.); |
| Keto | | | $n$-$C_3H_7$, H | $CH_3$ | b.p. 124°/11 mm | | 932 (b.p.); |
| | | | | | | picrate, m.p. 198° | 932 (m.p.); |
| Keto | F, F | F, F | F, F | –N(ring) | 68–69° | | 1148 (m.p.); |
| Keto | | | $-CH_2CH_2CN$, $CH_3$ | $CH_3$ | 157–158° | | 1147 (m.p.); |
| Keto | | | $n$-$C_4H_9$, Br | $CH_3$ | 108–109° | | 932 (m.p.); |
| Keto | | | Cl, H | $C_6H_5$ | 170–171° | | 447 (m.p.); |
| Keto | | | | $C_6H_5$ | 152–154° | | 1147 (m.p.); |

TABLE XII-136. 3-Piperidinols

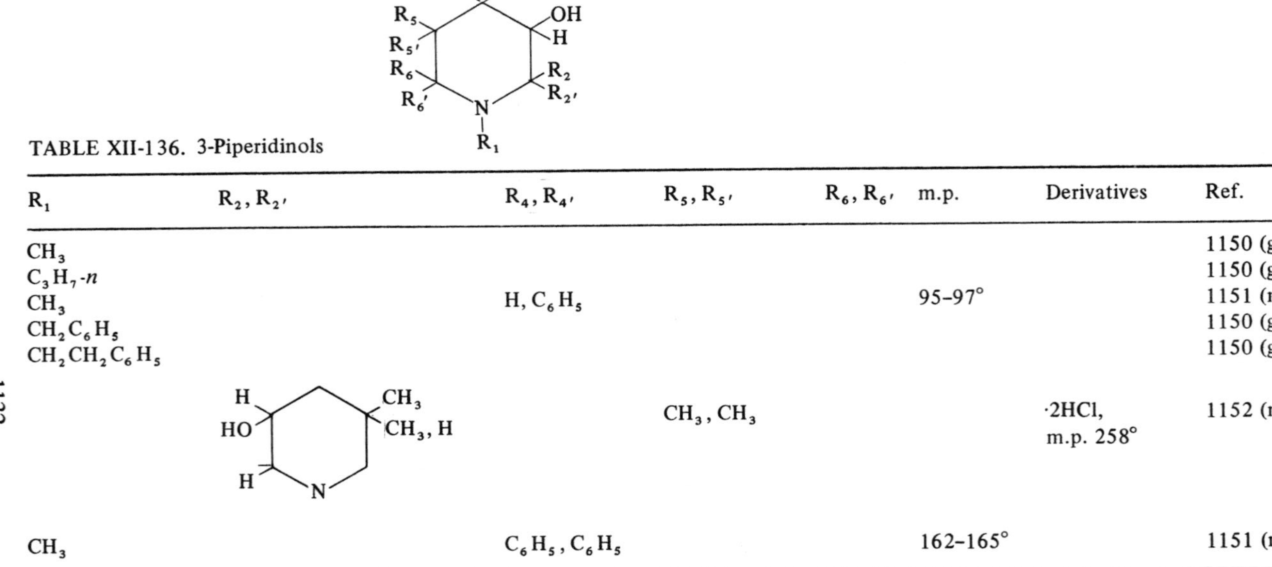

| $R_1$ | $R_2, R_{2'}$ | $R_4, R_{4'}$ | $R_5, R_{5'}$ | $R_6, R_{6'}$ | m.p. | Derivatives | Ref. |
|---|---|---|---|---|---|---|---|
| $CH_3$ | | | | | | | 1150 (g); |
| $C_3H_7$-*n* | | | | | | | 1150 (g); |
| $CH_3$ | | H, $C_6H_5$ | | | 95–97° | | 1151 (m.p.), (n); |
| $CH_2C_6H_5$ | | | | | | | 1150 (g); |
| $CH_2CH_2C_6H_5$ | | | | | | | 1150 (g) |
| | | | $CH_3$, $CH_3$ | | | ·2HCl, m.p. 258° | 1152 (m.p.), (i); |
| $CH_3$ | | $C_6H_5$, $C_6H_5$ | | | 162–165° | | 1151 (m.p.), (i); |

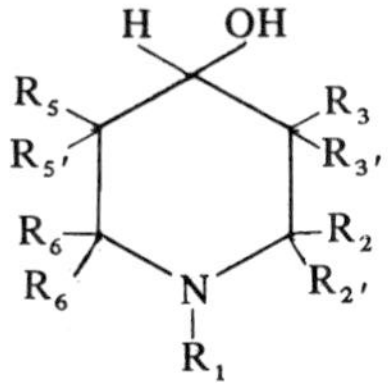

TABLE XII-137. 4-Piperidinols

| $R_1$ | $R_2, R_{2'}$ | $R_3, R_{3'}$ | $R_5, R_{5'}$ | $R_6, R_{6'}$ | m.p. | Derivatives | Ref. |
|---|---|---|---|---|---|---|---|
| $CH_3$ | | | | | | | 1150 (g); |
| $C_3H_7$-*n* | | | | | | | 1150 (g); |
| $CH_2C_6H_5$ | | | | | | | 1150 (g); |
| $CH_2CH_2C_6H_5$ | | | | | | | 1150 (g); |

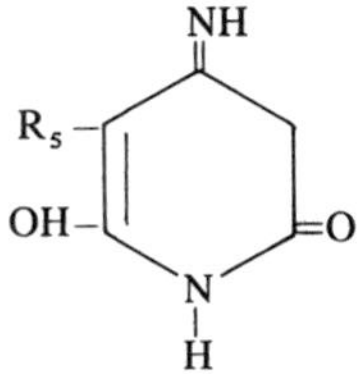

TABLE XII-138. Glutazines

| $R_5$ | m.p. | Derivatives | Ref. |
|---|---|---|---|
| H | | | 504 (i), (u); |
| CN | | | 504 (i), (u); |
| COOH | | ethyl ester | 504 (i), (u); |
| | | amide | 504 (i), (u); |

TABLE XII-139. 1-Azaquinones

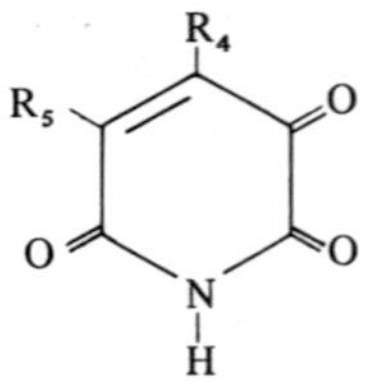

| $R_4$ | $R_5$ | m.p. | Derivatives | Ref. |
|---|---|---|---|---|
| | $CH_3$ | 158–160° | | 843 (m.p.), (i), (n), (u); |
| $CH_3$ | $CH_3$ | 126–127° | | 506 (m.p.), (u); |
| 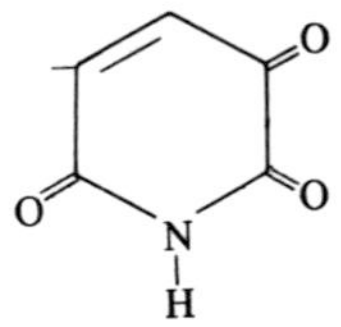 | | | | 894 (u); |
| $CH_3$ | $C_6H_5$ | 160–161° | | 292 (m.p.), (u); |

TABLE XII-140. 2-Alkoxypyridinium salts

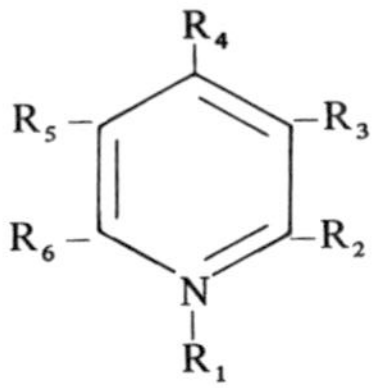

| $R_1$ | $R_2$ | $R_3$ | $R_4$ | $R_5$ | $R_6$ | m.p. | Anion | Ref. |
|---|---|---|---|---|---|---|---|---|
| $CH_3$ | $OCH_3$ | | | | D | 81–83° | tetrafluoroborate | 530 (m.p.); |
| $CH_3$ | $OCH_3$ | | | | | 81–83° | tetrafluoroborate | 530 (m.p.), (i), (n), (u); |
| | | | | | | 249–250° | hexachloro-antimonate | 464 (m.p.), (i); 542 (i); |
| $C_2H_5$ | $OCH_3$ | | | $NO_2$ | | 140–145° | tetrafluoroborate | 572 (m.p.); |
| $C_2H_5$ | $OC_2H_5$ | | | | | 47–49° | tetrafluoroborate | 571 (m.p.); |
| | | | | | | 153–154° | tetraphenylboron | 571 (m.p.); |
| $CH_2CH_2$– 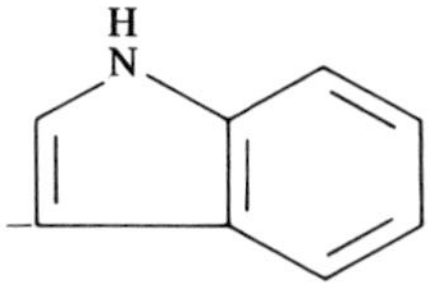 (indol-3-yl) | $OC_2H_5$ | | | | | 134–135° | bromide | 680 (m.p.), (u); |

TABLE XII-141. 3-Hydroxypyridinium salts

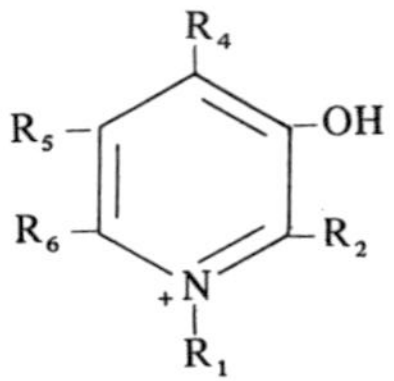

| $R_1$ | $R_2$ | $R_4$ | $R_5$ | $R_6$ | m.p. | Anion | Ref. |
|---|---|---|---|---|---|---|---|
| $CH_3$ | | | | | 179–181° | iodide | 593 (m.p.), (u); |
| | | | | | | picrate | 594 (u); |
| $CH_3$ | CN | | | | 209–213° | chloride | 189 (m.p.); |
| $CH_3$ | $CONH_2$ | | | | 269° | chloride | 189 (m.p.); |
| $CH_3$ | Br | | | $CH_3$ | 234–235° | iodide | 597 (m.p.), (u); |
| $C_2H_5$ | | | | | 108–111° | chloride | 189 (m.p.); |
| $C_2H_5$ | CN | | | | 220–223° | chloride | 189 (m.p.); |
| $C_2H_5$ | $CONH_2$ | | | | 249° | chloride | 189 (m.p.); |
| $C_2H_5$ | I | | | $CH_3$ | 187–192° | iodide | 597 (m.p.), (u); |
| $CH_3$ | $SCH_3$ | | | $CH_3$ | 190–193° | iodide | 597 (m.p.), (u); |
| $OC_2H_5$ | $CH_3$ | | CN | | 129–130° | ethosulfate | 212 (m.p.); |
| $CH_3$ | $SC_2H_5$ | | | $CH_3$ | 184–186° | iodide | 597 (m.p.), (u); |
| $C_6H_5$ | | | | | 171–176° | bromide | 1153 (m.p.), (i); |
| | | | | | 211–215° | chloride | 189 (m.p.); 1153 (m.p.), (i); |
| | | | | | 65–68° | nitrate | 1153 (m.p.), (i); |
| | | | | | 150° | perchlorate | 1153 (m.p.), (i); |
| | | | | | 330° | sulfate | 1153 (m.p.), (i); |
| $CH_2C_6H_5$ | | | | | 160–161° | chloride | 189 (m.p.); |
| $CH_2C_6H_5$ | CN | | | | 162–164° | chloride | 189 (m.p.); |
| $CH_2C_6H_5$ | $CONH_2$ | | | | 225° | chloride | 189 (m.p.); |

TABLE XII-142. 3-Alkoxypyridinium salts

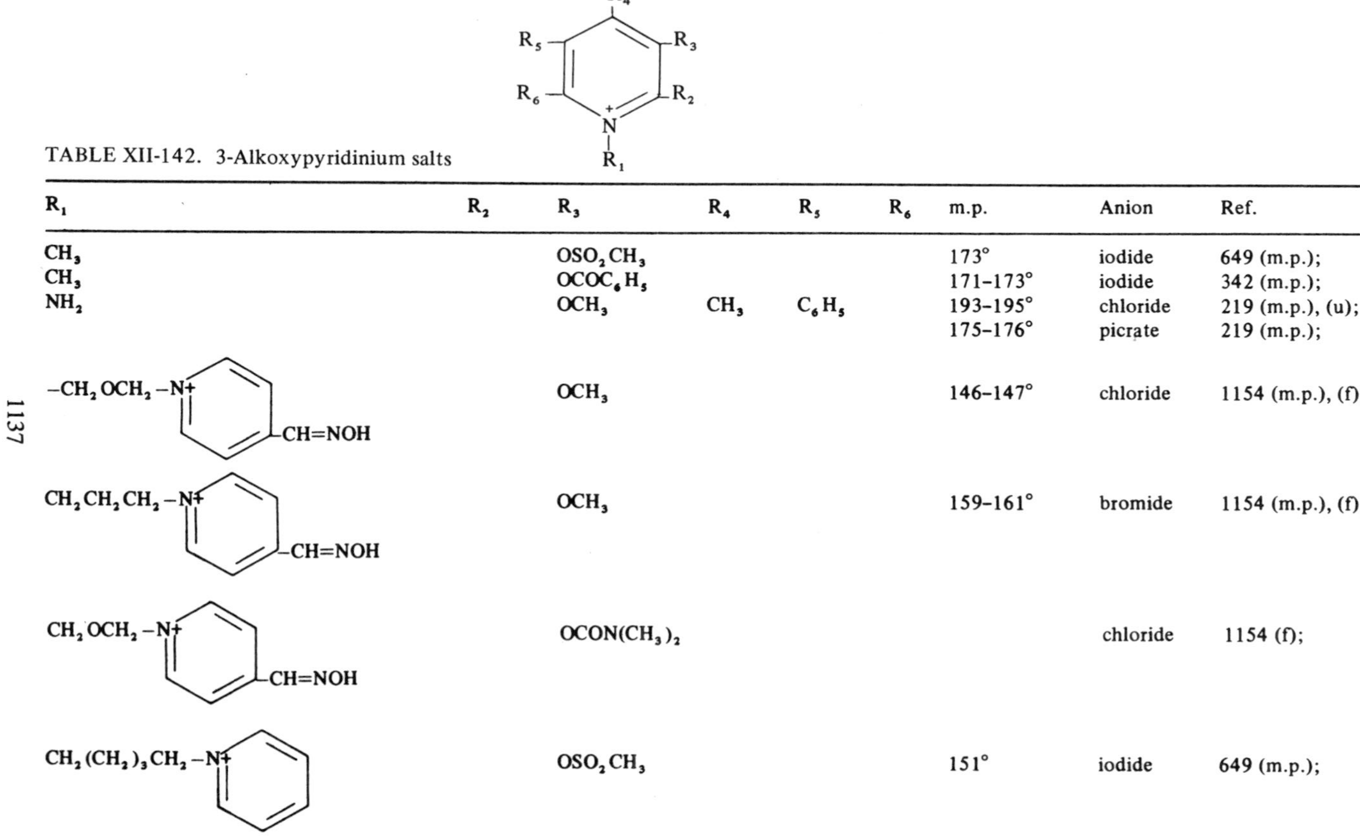

| $R_1$ | $R_2$ | $R_3$ | $R_4$ | $R_5$ | $R_6$ | m.p. | Anion | Ref. |
|---|---|---|---|---|---|---|---|---|
| $CH_3$ | | $OSO_2CH_3$ | | | | 173° | iodide | 649 (m.p.); |
| $CH_3$ | | $OCOC_6H_5$ | | | | 171–173° | iodide | 342 (m.p.); |
| $NH_2$ | | $OCH_3$ | $CH_3$ | $C_6H_5$ | | 193–195° | chloride | 219 (m.p.), (u); |
| | | | | | | 175–176° | picrate | 219 (m.p.); |
| $-CH_2OCH_2-N^+$ (pyridinium) –CH=NOH | | $OCH_3$ | | | | 146–147° | chloride | 1154 (m.p.), (f); |
| $CH_2CH_2CH_2-N^+$ (pyridinium) –CH=NOH | | $OCH_3$ | | | | 159–161° | bromide | 1154 (m.p.), (f); |
| $CH_2OCH_2-N^+$ (pyridinium) –CH=NOH | | $OCON(CH_3)_2$ | | | | | chloride | 1154 (f); |
| $CH_2(CH_2)_3CH_2-N^+$ (pyridinium) | | $OSO_2CH_3$ | | | | 151° | iodide | 649 (m.p.); |

TABLE XII-142. 3-Alkoxypyridinium salts (Continued)

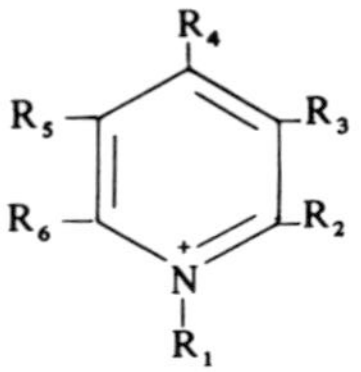

| $R_1$ | $R_2$ | $R_3$ | $R_4$ | $R_5$ | $R_6$ | m.p. | Anion | Ref. |
|---|---|---|---|---|---|---|---|---|
| $CH_2(CH_2)_3CH_2-N^+$ 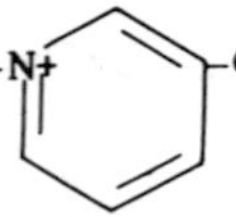 (3-$OSO_2CH_3$-pyridinium) | | $OSO_2CH_3$ | | | | 174° | iodide | 649 (m.p.); |
| $CH_2CH_2C_6H_5$ | | $OC_6H_5$ | | | | 155–156° | bromide | 211 (m.p.); |

TABLE XII-143. 4-Alkoxypyridinium salts

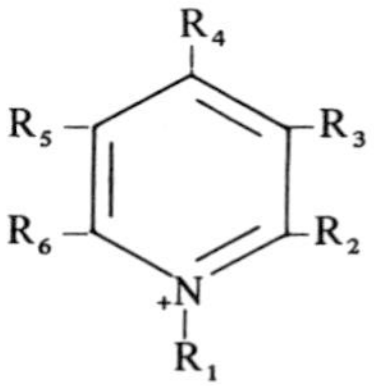

| $R_1$ | $R_2$ | $R_3$ | $R_4$ | $R_5$ | $R_6$ | m.p. | Anion | Ref. |
|---|---|---|---|---|---|---|---|---|
| $CH_3$ | | | $OCD_3$ | | | 52–55° | tetrafluoroborate | 530 (m.p.), (i), (n); |
| $CH_3$ | D | | $OCH_3$ | | D | 55–57° | tetrafluoroborate | 530 (m.p.), (i), (n); |
| $CH_3$ | | | $OCH_3$ | | | 56–58° | tetrafluoroborate | 139 (m.p.), (i), (n), (u); |
| | | | | | | 171–174° | hexachloroantimonate | 464 (m.p.), (i); |

| | | | | | | |
|---|---|---|---|---|---|---|
| $CH_3$ | | $OC_2H_5$ | | 125° | iodide | 588 (m.p.), (n), (u); |
| $CH_3$ | | $-OCH_2-$cyclopropyl | | 110–111° | iodide | 770 (m.p.); |
| | | | | 87–88° | perchlorate | 770 (m.p.); |
| $CH_3$ | | $-OCH(CH_3)-$cyclopropyl | | oil | iodide | 770; |
| $CH_3$ | | $OCH_2C_6H_5$ | | 149–151° | iodide | 770 (m.p.); |
| | | | | 141–142° | perchlorate | 770 (m.p.); |
| 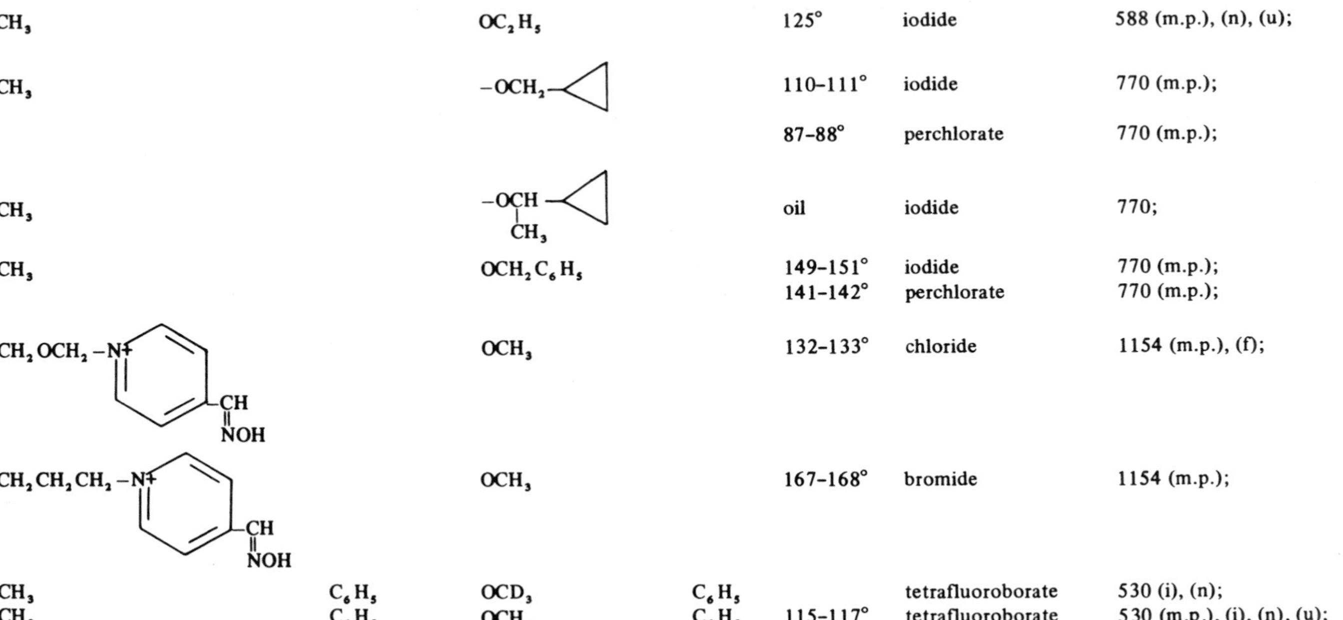 $CH_2OCH_2-N^+$ (4-CH=NOH pyridinium) | | $OCH_3$ | | 132–133° | chloride | 1154 (m.p.), (f); |
| $CH_2CH_2CH_2-N^+$ (4-CH=NOH pyridinium) | | $OCH_3$ | | 167–168° | bromide | 1154 (m.p.); |
| $CH_3$ | $C_6H_5$ | $OCD_3$ | $C_6H_5$ | | tetrafluoroborate | 530 (i), (n); |
| $CH_3$ | $C_6H_5$ | $OCH_3$ | $C_6H_5$ | 115–117° | tetrafluoroborate | 530 (m.p.), (i), (n), (u); |

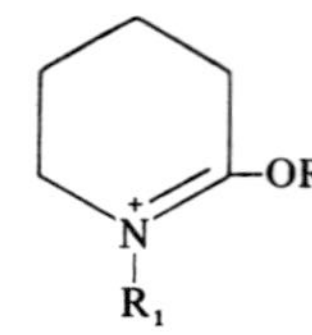


TABLE XII-144. 2-Alkoxy 3,4,5,6-tetrahydropyridinium salts

| R | $R_1$ | m.p. | Anion | Ref. |
|---|---|---|---|---|
| $CH_3$ | $CH_3$ | | tetrafluoroborate, m.p. 141–143° | 530 (m.p.), (i), (n); |

TABLE XII-145. 3-Alkoxypyridinium *N*-ylides

| R | $R_1$ | $R_4$ | $R_5$ | $R_6$ | m.p. | Derivatives | Ref. |
|---|---|---|---|---|---|---|---|
| H | $:\bar{N}COCF_3$ | $CH_3$ | $C_6H_5$ | | 179–180° | | 219 (m.p.), (u); |
| H | $:\bar{N}COCH_3$ | $CH_3$ | $C_6H_5$ | | 216–217° | | 219 (m.p.), (u); |
| $CH_3$ | $:\bar{N}COCF_3$ | $CH_3$ | $C_6H_5$ | | 134–135° | | 219 (m.p.), (u); |
| | | | | | | picrate, m.p. 171–172° | 219 (m.p.); |
| H | $:\bar{N}COCH_2CH_3$ | $CH_3$ | $C_6H_5$ | | 151°/175° | | 219 (m.p.); |
| $CH_3$ | $:\bar{N}COCH_3$ | $CH_3$ | $C_6H_5$ | | oil | | 219; |
| | | | | | | picrate, m.p. 152–153° | 219 (m.p.); |

TABLE XII-146. 3-Oxypyridyl betaines

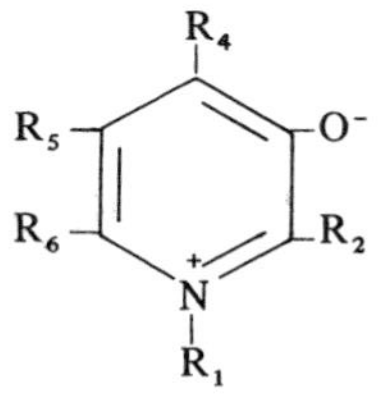

| $R_1$ | $R_2$ | $R_4$ | $R_5$ | $R_6$ | m.p. | Derivatives | Ref. |
|---|---|---|---|---|---|---|---|
| $CH_3$ | | | | | | hydroiodide, m.p. 109–111° | 592 (m.p.); |
| | | | | | | picrate, m.p. 201–202° | 592 (m.p.); |
| $C_2H_5$ | | | | | | hydrobromide, m.p. 99–102° | 592 (m.p.); |
| | | | | | | picrate, m.p. 153–155° | 592 (m.p.); |
| $-CH_2CH_2OH$ | | | | | | hydrochloride, m.p. 139–144° | 592 (m.p.); |
| | | | | | | hydrobromide, m.p. 122–123° | 592 (m.p.); |
| $-CH_2CH_2NH_2$ | | | | | | dihydrobromide, m.p. 218–222° | 592 (m.p.); |
| $-CH_2C(Cl)=CH_2$ | | | | | | hydrochloride, m.p. 153–155° | 592 (m.p.); |
| $-CH_2CH=CH_2$ | | | | | | hydrobromide, m.p. 97–99° | 592 (m.p.); |
| $-CH_2COCH_3$ | | | | | | hydrochloride, m.p. 134–136° | 592 (m.p.); |

TABLE XII-146. 3-Oxypyridyl betaines (Continued)

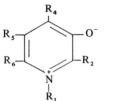

| $R_1$ | $R_2$ | $R_4$ | $R_5$ | $R_6$ | m.p. | Derivatives | Ref. |
|---|---|---|---|---|---|---|---|
| $-CH_2CH{-}CH_2$ (epoxide, $\backslash O /$) | | | | | | hydrochloride, m.p. 133–135 | 592 (m.p.); |
| $n$-$C_3H_7$ | | | | | | $p$-toluenesulfonate, m.p. 129–130° | 592 (m.p.); |
| | | | | | | picrate, m.p. 156–157° | 592 (m.p.); |
| $-CH_2CH{=}C(Cl)CH_3$ | | | | | | hydrochloride, m.p. 181–182° | 592 (m.p.); |
| $-CH_2C(CH_3){=}CH_2$ | | | | | | hydrochloride m.p. 139–144° | 592 (m.p.); |
| $n$-$C_4H_9$ | | | | | | picrate, m.p. 128–130° | 592 (m.p.); |
| $CH_3$ | $CH_3$ | $CH_2OH$ | $CH_2OH$ | | | | 1048 (n); |
| $-CH_2CH_2N(CH_3)_2$ | | | | | | dihydrochloride, m.p. 238–239° | 592 (m.p.); |
| | | | | | | di-$CH_3I$, m.p. 167–171° | 592 (m.p.); |
| $-(CH_2)_4CCl_3$ | | | | | | hydrochloride, m.p. 175–176° | 592 (m.p.); |
| $-(CH_2)_4CN$ | | | | | | hydrobromide, m.p. 191–194° | 592 (m.p.); |

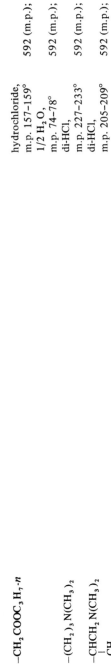

| | | |
|---|---|---|
| $-CH_2COOC_3H_7$-*n* | hydrochloride, m.p. 157–159° | 592 (m.p.); |
| | 1/2 $H_2O$, m.p. 74–78° | 592 (m.p.); |
| $-(CH_2)_3N(CH_3)_2$ | di-HCl, m.p. 227–233° | 592 (m.p.); |
| $-CH(CH_3)CH_2N(CH_3)_2$ | di-HCl, m.p. 205–209° | 592 (m.p.); |
| $-CH_2-N^+$ 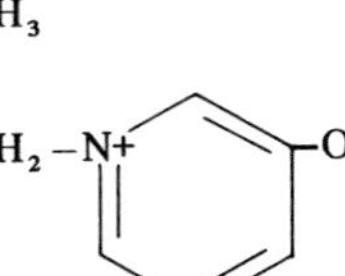 | di-HI, m.p. 205–208° | 592 (m.p.); |
| | picrate, m.p. 239° | 592 (m.p.); |
| $-(CH_2)_5CN$ | hydrochloride, m.p. 119–121° | 592 (m.p.); |
| $-(CH_2)_3CH_2OCOCH_3$ | hydrobromide, m.p. 94–96° | 592 (m.p.); |
| $-CH_2C_6H_4Cl$-*p* | hydrochloride, m.p. 184–185° | 592 (m.p.); |
| | picrate, m.p. 191–194° | 592 (m.p.); |
| | $\cdot CH_3I$, m.p. 125–127° | 592 (m.p.); |
| $-CH_2C_6H_5$ | hydrochloride, m.p. 154–157° | 592 (m.p.); |
| | hydrobromide, m.p. 125–128° | 592 (m.p.); |
| | hydrate, m.p. 107–111° | 592 (m.p.); |

TABLE XII-146. 3-Oxypyridyl betaines (Continued)

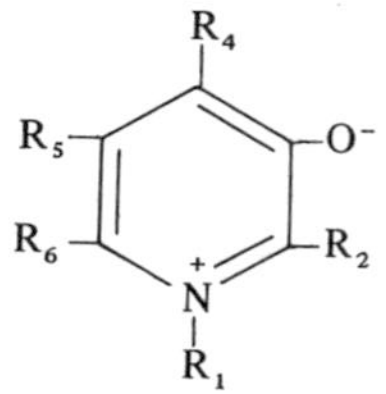

(Structure: pyridinium ring with $R_1$ on $\overset{+}{N}$, $R_2$, $O^-$, $R_4$, $R_5$, $R_6$)

| $R_1$ | $R_2$ | $R_4$ | $R_5$ | $R_6$ | m.p. | Derivatives | Ref. |
|---|---|---|---|---|---|---|---|
| $NH_2$ | | $CH_3$ | $C_6H_5$ | | | hydrochloride, m.p. 191–195° | 216 (m.p.), (u); 219 (m.p); |
| | | | | | | hemihydrate, m.p. 170°/200° | 219 (m.p.); |
| | | | | | | cyclopentanone hydrazone, m.p. 205–206° | 216 (m.p.); |
| $-CH_2-CH_2-N^+$ 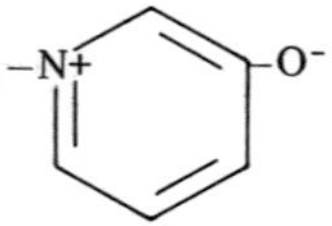 (3-oxidopyridinium, $-O^-$) | | | | | | dihydrobromide, m.p. 295–296° | 592 (m.p.); |
| | | | | | | *p*-toluenesulfate, m.p. 290–292° | 592 (m.p.); |
| $-CH_2OCH_2-N^+$ 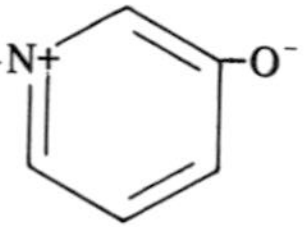 (3-oxidopyridinium, $-O^-$) | | | | | | dihydrochloride, m.p. 211–213° | 592 (m.p.); |
| | | | | | | picrate, m.p. 236° | 592 (m.p.); |
| $-CH_2COC_6H_4Cl$-*p* | | | | | | m.p. 231–233° | 592 (m.p.); |
| $-CH_2COC_6H_4Br$-*p* | | | | | | hydrobromide, m.p. 255–257° | 592 (m.p.); |

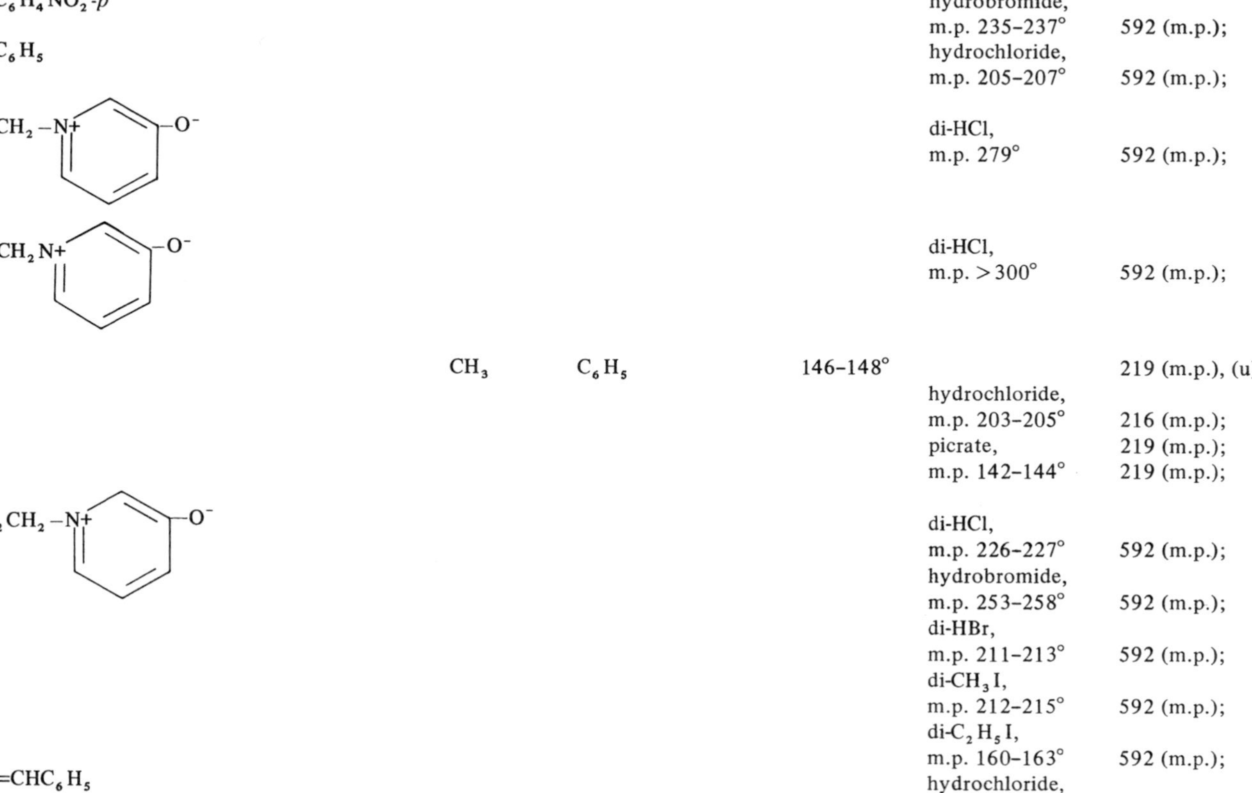

| | | | | | |
|---|---|---|---|---|---|
| $-CH_2COC_6H_4NO_2$-*p* | | | | hydrobromide, m.p. 235–237° | 592 (m.p.); |
| $-CH_2COC_6H_5$ | | | | hydrochloride, m.p. 205–207° | 592 (m.p.); |
| $CH_2=C-CH_2-N^{+}$ … $-O^{-}$ | | | | di-HCl, m.p. 279° | 592 (m.p.); |
| $-CH_2COCH_2N^{+}$ … $-O^{-}$ | | | | di-HCl, m.p. > 300° | 592 (m.p.); |
| $NHCH_3$ | $CH_3$ | $C_6H_5$ | 146–148° | | 219 (m.p.), (u); |
| | | | | hydrochloride, m.p. 203–205° | 216 (m.p.); |
| | | | | picrate, | 219 (m.p.); |
| | | | | m.p. 142–144° | 219 (m.p.); |
| $-CH_2CH_2CH_2-N^{+}$ … $-O^{-}$ | | | | di-HCl, m.p. 226–227° | 592 (m.p.); |
| | | | | hydrobromide, m.p. 253–258° | 592 (m.p.); |
| | | | | di-HBr, m.p. 211–213° | 592 (m.p.); |
| | | | | di-$CH_3I$, m.p. 212–215° | 592 (m.p.); |
| | | | | di-$C_2H_5I$, m.p. 160–163° | 592 (m.p.); |
| $-CH_2CH=CHC_6H_5$ | | | | hydrochloride, m.p. 137–140° | 592 (m.p.); |

TABLE XII-146. 3-Oxypyridyl betaines (Continued)

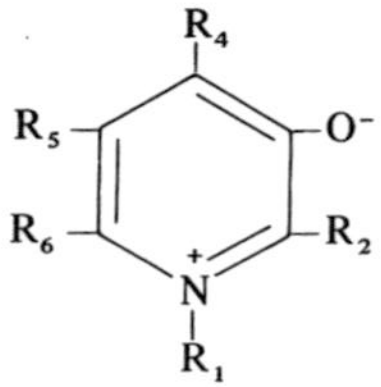

| $R_1$ | $R_2$ | $R_4$ | $R_5$ | $R_6$ | m.p. | Derivatives | Ref. |
|---|---|---|---|---|---|---|---|
| 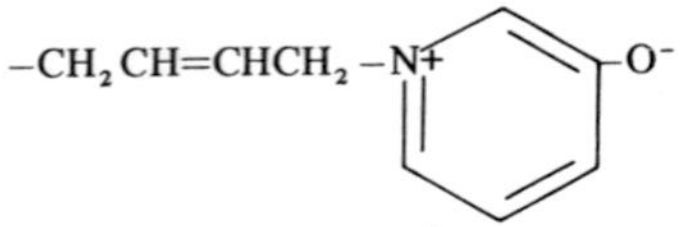 $-CH_2CH{=}CHCH_2-$(3-oxido-1-pyridinio, $N^+$, $O^-$) | | | | | | di-HBr, m.p. 220–222° | 592 (m.p.); |
| 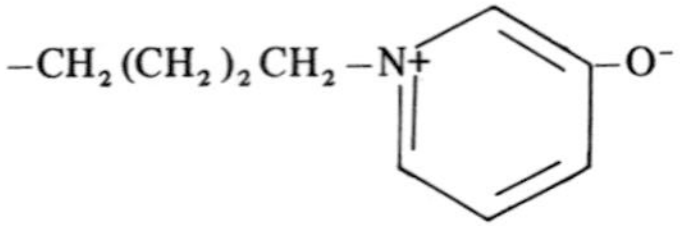 $-CH_2(CH_2)_2CH_2-$(3-oxido-1-pyridinio, $N^+$, $O^-$) | | | | | | di-HCl, m.p. 267–269° | 592 (m.p.); |
| | | | | | | di-$H_2O$, m.p. 195–198° | 592 (m.p.); |
| | | | | | | dipicrate, m.p. 224–226° | 592 (m.p.); |
| | | | | | | di-$CH_3I$, m.p. 194–198° | 592 (m.p.); |
| | | | | | | di-$C_2H_5I$, m.p. 166–167° | 592 (m.p.); |
| 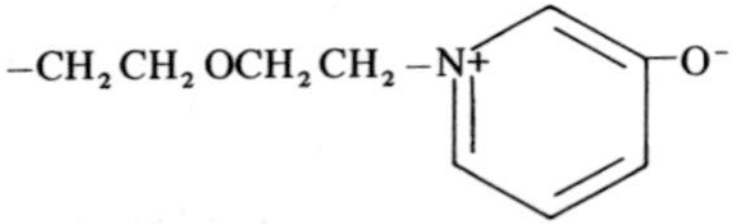 $-CH_2CH_2OCH_2CH_2-$(3-oxido-1-pyridinio, $N^+$, $O^-$) | | | | | | di-HCl, m.p. 169–172° | 592 (m.p.); |

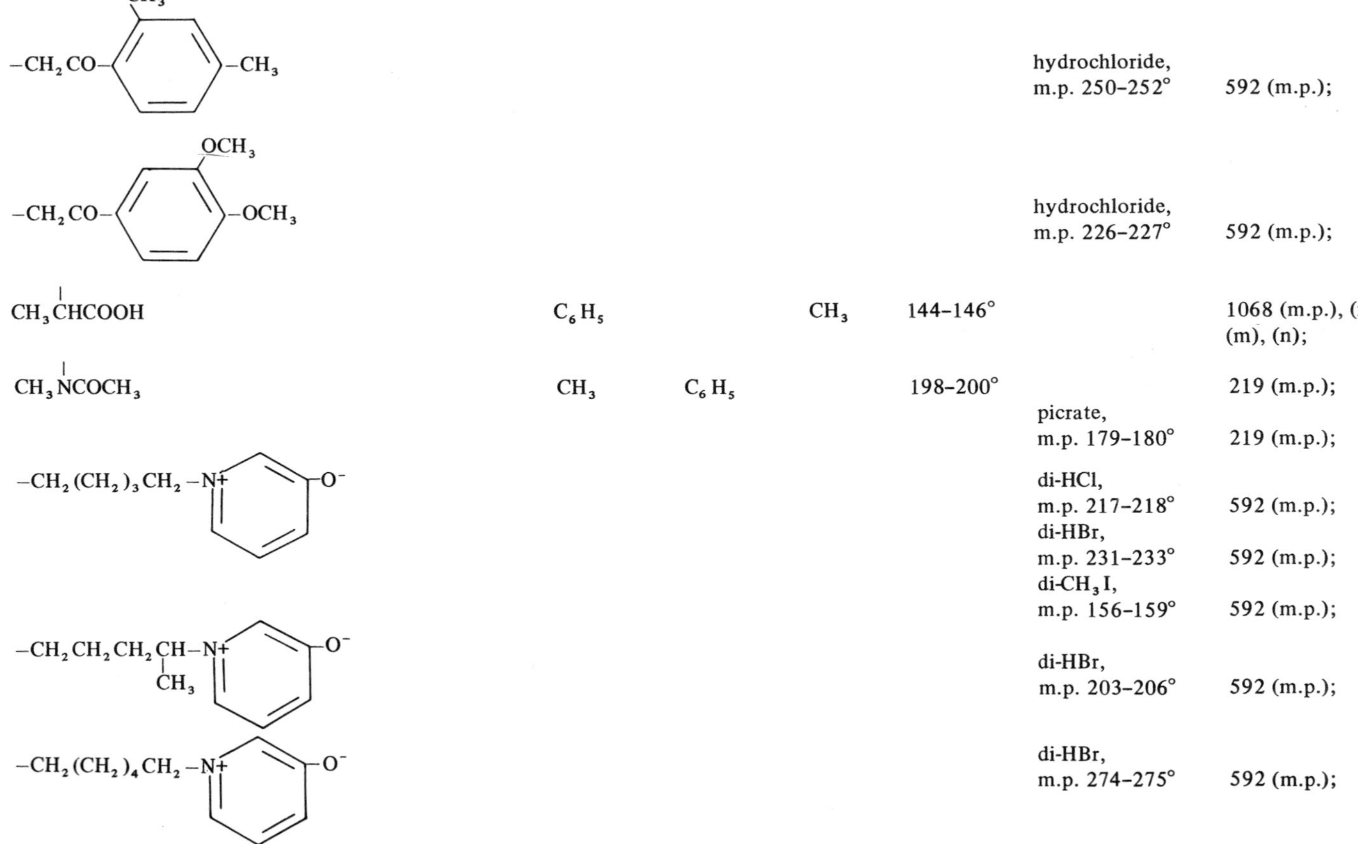

| | | | | | | |
|---|---|---|---|---|---|---|
| $-CH_2CO-$(2,4-dimethylphenyl) [$CH_3$, $-CH_3$] | | | | | hydrochloride, m.p. 250–252° | 592 (m.p.); |
| $-CH_2CO-$(3,4-dimethoxyphenyl) [$OCH_3$, $-OCH_3$] | | | | | hydrochloride, m.p. 226–227° | 592 (m.p.); |
| $CH_3CHCOOH$ | $C_6H_5$ | | $CH_3$ | 144–146° | | 1068 (m.p.), (i), (m), (n); |
| $CH_3NCOCH_3$ | $CH_3$ | $C_6H_5$ | | 198–200° | | 219 (m.p.); |
| | | | | | picrate, m.p. 179–180° | 219 (m.p.); |
| $-CH_2(CH_2)_3CH_2-N^+$ (pyridinium-$O^-$) | | | | | di-HCl, m.p. 217–218° | 592 (m.p.); |
| | | | | | di-HBr, m.p. 231–233° | 592 (m.p.); |
| | | | | | di-$CH_3I$, m.p. 156–159° | 592 (m.p.); |
| $-CH_2CH_2CH_2CH(CH_3)-N^+$ (pyridinium-$O^-$) | | | | | di-HBr, m.p. 203–206° | 592 (m.p.); |
| $-CH_2(CH_2)_4CH_2-N^+$ (pyridinium-$O^-$) | | | | | di-HBr, m.p. 274–275° | 592 (m.p.); |

TABLE XII-146. 3-Oxypyridyl betaines (Continued)

| $R_1$ | $R_2$ | $R_4$ | $R_5$ | $R_6$ | m.p. | Derivatives | Ref. |
|---|---|---|---|---|---|---|---|
| $CH_3CH(CH_2)_2CHCH_3$ (1-substituted 3-oxypyridinium) | | | | | | di-HBr, m.p. 247–249° | 592 (m.p.); |
| 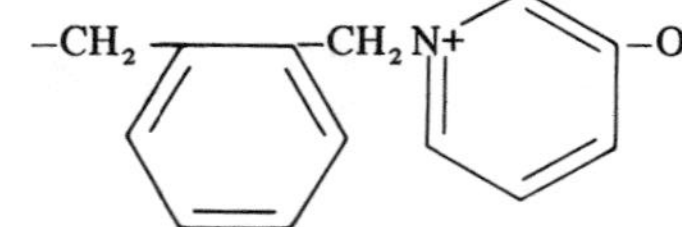 | | | | | | di-HBr, m.p. 252-255° | 592 (m.p.); |

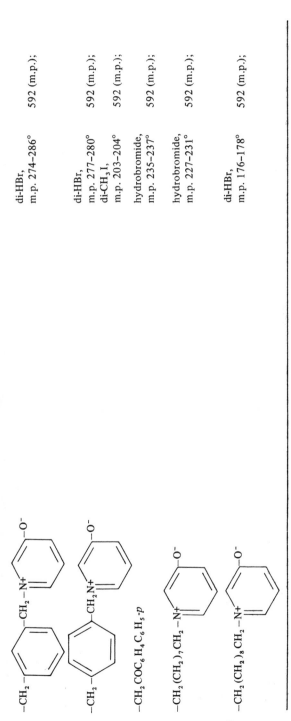

| Substituent | Derivative | Reference |
|---|---|---|
| $-CH_2-C_6H_4-CH_2-N^+C_5H_4-O^-$ | di-HBr, m.p. 274–286° | 592 (m.p.); |
| $-CH_2-C_6H_4-CH_2N^+C_5H_4-O^-$ | di-HBr, m.p. 277–280° | 592 (m.p.); |
| | di-$CH_3I$, m.p. 203–204° | 592 (m.p.); |
| $-CH_2COC_6H_4C_6H_5$-*p* | hydrobromide, m.p. 235–237° | 592 (m.p.); |
| $-CH_2(CH_2)_7CH_2-N^+C_5H_4-O^-$ | hydrobromide, m.p. 227–231° | 592 (m.p.); |
| $-CH_2(CH_2)_8CH_2-N^+C_5H_4-O^-$ | di-HBr, m.p. 176–178° | 592 (m.p.); |

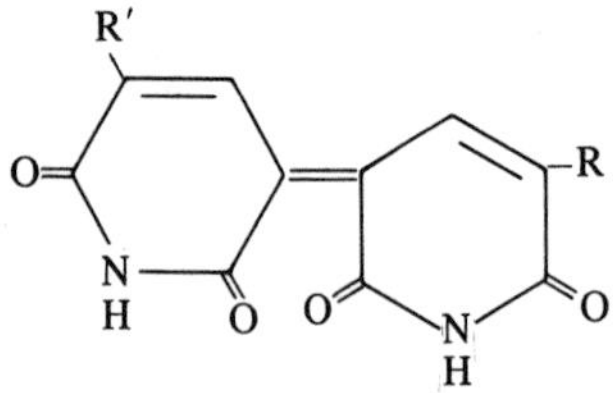

TABLE XII-147. Diazadibenzoquinones

| R | R′ | m.p. | Derivatives | Ref. |
|---|---|---|---|---|
| OH | OH | | | 894 (i), (u); |
| $CH_3$ | $CH_3$ | 240° | | 843 (m.p.), (i), (n), (u); |
| $OCH_3$ | $OCH_3$ | | | 894 (u); |

## References


1. J. Thesing and A. Müller, *Chem. Ber.*, **90**, 711 (1957).
2. H. Meislich, in "Pyridine and Its Derivatives," E. Klingsberg, Ed., Part 3, Interscience, New York, 1962, Chapter 12.
3. J. B. Petersen, J. Lei, N. Clauson-Kaas, and K. Norris, *Kgl. Danske Videnskab. Selskab. Mat.-Fys. Medd.*, **36**, 23 (1967); *Chem. Abstr.* **69**, 59063 (1968).
4. Ref. 2, pp. 511–516.
5. D. E. Ames and T. F. Grey, *J. Chem. Soc.*, 2310 (1959).
6. D. E. Ames and J. L. Archibald, *J. Chem. Soc.*, 1475 (1962).
7. A. Roedig, K. Grohe, and W. Mayer, *Tetrahedron*, **24**, 1851 (1968).
8. G. Simchen, *Chem. Ber.*, **103**, 389 (1970).
9. H. Junek and A. Schmidt, *Monatsh. Chem.*, **98**, 1097 (1967).
10. S. G. Cottis and H. Tieckelmann, *J. Org. Chem.*, **26**, 79 (1961).
11. A. P. Krapcho and P. S. Huyffer, *J. Org. Chem.*, **28**, 2461 (1963).
12. M. R. S. Weir, K. E. Helmer, and J. B. Hyne, *Can. J. Chem.*, **41**, 1042 (1963).
13. A. S. Alvarez-Insua, M. Lora-Tamayo, and J. L. Soto, *J. Heterocycl. Chem.*, **7**, 1305 (1970).
14. H. J. Prins, *Rec. Trav. Chim. Pays-Bas*, **65**, 455 (1946).
15. A. Roedig and G. Märkl, *Ann. Chem.*, **659**, 1 (1962).
16. A. Roedig and K. Grohe, *Chem. Ber.*, **98**, 923 (1965).
17. A. Roedig, R. Kohlhaupt, and G. Märkl, *Chem. Ber.*, **99**, 698 (1966).
18. A. Roedig and G. Märkl, *Ann. Chem.*, **636**, 1 (1960).
19. A. Roedig, G. Märkl, W. Ruch, H.-G. Kleppe, R. Kohlhaupt, and H. Schaller, *Ann. Chem.*, **692**, 83 (1966).
20. A. Roedig, G. Märkl, and V. Schaal, *Chem. Ber.*, **95**, 2844 (1962).
21. T. B. Windholz, L. H. Peterson, and G. J. Kent, *J. Org. Chem.*, **28**, 1443 (1963).
22. G. Kresze and O. Korpiun, *Tetrahedron*, **22**, 2493 (1966).
23. G. Kresze and J. Firl, *Tetrahedron*, **24**, 1043 (1968).
24. J. Firl and G. Kresze, *Chem. Ber.*, **99**, 3695 (1966).
25. S. D. Work and C. R. Hauser, *J. Org. Chem.*, **28**, 725 (1963).
26. M. L. Miles, T. M. Harris, and C. R. Hauser, *J. Org. Chem.*, **30**, 1007 (1965).
27. T. M. Harris, S. Boatman, and C. R. Hauser, *J. Amer. Chem. Soc.*, **87**, 3186 (1965).
28. R. P. Mariella, *Org. Synth. Coll. Vol.* **IV**, 210 (1963).

29. R. J. Light and C. R. Hauser, *J. Org. Chem.*, **25**, 538 (1960).
30. K. G. Hampton, T. M. Harris, C. M. Harris, and C. R. Hauser, *J. Org. Chem.*, **30**, 4263 (1965).
30a. K. G. Hampton and C. R. Hauser, *J. Org. Chem.*, **30**, 2934 (1965).
31. S. Boatman, R. E. Smith, G. F. Morris, W. G. Kofron, and C. R. Hauser, *J. Org. Chem.*, **32**, 3817 (1967).
32. S. Boatman and C. R. Hauser, *J. Org. Chem.*, **31**, 1785 (1966).
33. D. Fischer, G. Scheibe, P. Merkel, and R. Muller, *J. Prakt. Chem.*, **100**, 91 (1919).
34. H. N. Rydon and K. Undheim, *J. Chem. Soc.*, 4676 (1962).
35. A. D. Campbell and I. D. R. Stevens, *J. Chem. Soc.*, 959 (1956).
36. N. P. Shusherina, K. Khua-min', and R. Ya. Levina, *Zh. Obshch. Khim.*, **33**, 3613 (1963).
37. K. Grohe and A. Roedig, *Chem. Ber.*, **100**, 2953 (1967).
38. D. Münzner, H. Lettau, and H. Schubert, *Z. Chem.*, 7, 278 (1967).
39. N. Sugiyama, M. Yamamoto, and C. Kashima, *Bull. Chem. Soc. Jap.*, **42**, 1357 (1969).
40. C. Kashima, M. Yamamoto, S. Kobayashi, and N. Sugiyama, *Bull. Chem. Soc. Jap.*, **42**, 2389 (1969).
41. M. Broust-Bournazel, *Ann. Chim.* (Paris), **5**, 1409 (1960).
42. J. Chauvelier, M. Chauvin, and J. Aubouet, *Bull. Soc. Chim. Fr.*, 1721 (1966).
43. F. Hoffmann-LaRoche and Co. A. G., Brit. patent, 872,190 (1961); *Chem. Abstr.*, **56**, 462 (1962).
44. E. Gegner, *Tetrahedron Lett.* 287 (1969).
45. M. Shamma, R. W. Lagally, P. Miller, and E. F. Walker, Jr., *Tetrahedron*, **21**, 3255 (1965).
46. J. C. Powers and I. Ponticello, *J. Amer. Chem. Soc.*, **90**, 7102 (1968).
47. Th. Haug, F. Lohse, K. Metzger, and H. Batzer., *Helv. Chim. Acta*, **51**, 2069 (1968).
48. N. Sugiyama, M. Yamamoto, and C. Kashima, *Bull. Chem. Soc. Jap.*, **43**, 901 (1970).
49. H. Junek, A. Metallidis, and E. Ziegler, *Monatsh. Chem.*, **100**, 1937 (1969).
50. Th. Kappe, I. Maninger, and E. Ziegler, *Monatsh. Chem.*, **99**, 85 (1968).
51. Ref. 2, pp. 528–529.
52. U. Schmidt, *Ann. Chem.*, **657**, 156 (1962).
53. G. N. Walker and B. N. Weaver, *J. Org. Chem.*, **26**, 4441 (1961).
54. T. M. Harris, S. Boatman, and C. R. Hauser, *J. Amer. Chem. Soc.*, **85**, 3273 (1963).
55. S. Portnoy, *J. Org. Chem.*, **30**, 3377 (1965).
56. J. L. Greene, Jr., and J. A. Montgomery, *J. Med. Chem.*, **6**, 294 (1963).
57. L. Rateb, G. A. Mina, and G. Soliman, *J. Chem. Soc., C,* 2140 (1968).
58. T. V. Protopopova and A. P. Skoldinov, *J. Gen. Chem. USSR*, **27**, 1360 (1957).
59. G. F. Holland and J. N. Pereira, *J. Med. Chem.*, **10**, 149 (1967).
60. C. Bonsall and J. Hill, *J. Chem. Soc., C,* 1836 (1967).
61. Hoffman, LaRoche and Co., Swiss patent, 217,480 (1941); *Chem. Abstr.*, **42**, 6377 (1948).
62. Otto Schnider, U.S. patent, 2,384,136 (1945); *Chem. Abstr.*, **40**, 611 (1946).
63. F. Zymalkowski and P. Messinger, *Arch. Pharm.* (Weinheim), **300**, 168 (1967).
64. Ref. 2, pp. 526–527.
65. A. Sakurai and H. Midorikawa, *Bull. Chem. Soc. Jap.*, **41**, 165 (1968).
66. Ref. 2, p. 540.
67. T. Kametani, K. Ogasawara, and M. Shio, *Yakugaku Zasshi*, **86**, 809 (1966); *Chem. Abstr.*, **65**, 20092 (1966).
68. R. E. Smith, G. F. Morris, and C. R. Hauser, *J. Org. Chem.*, **33**, 2562 (1968).

69. E. Ziegler, I. Herbst, and Th. Kappe, *Monatsh. Chem.,* **100**, 132 (1969).
70. H. G. O. Becker, *J. Prakt. Chem.,* **12**, 294 (1961).
71. J. Thesing, Ger. patent, 1,092,016 (1960); *Chem. Abstr.,* **55**, 19957 (1961).
72. F. Kröhnke, K.-E. Schnalke, and W. Zecher, *Chem. Ber.,* **103**, 322 (1970).
73. M. A. Akhtar, W. G. Brouwer, J. A. D. Jeffreys, C. W. Gemenden, W. I. Taylor, R. N. Seelye, and D. W. Stanton, *J. Chem. Soc., C,* 859 (1967).
74. W. Zecher and F. Kröhnke, *Chem. Ber.,* **94**, 698 (1961).
75. W. Zecher and F. Kröhnke, *Chem. Ber.,* **94**, 690 (1961).
76. A. Sakurai and H. Midorikawa, *Bull. Chem. Soc. Jap.,* **40**, 1680 (1967).
77. T. Kametani, A. Kozuka, and S. Tanaka, *Yakugaku Zasshi,* **90**, 1574 (1970); *Chem. Abstr.,* **74**, 53461 (1971).
78. H. Junek, H. Sterk, and A. Schmidt, *Z. Naturforsch.,* **21B**, 1145 (1966).
79. H. Junek and A. Schmidt, *Monatsh. Chem.,* **99**, 635 (1968).
80. H. Junek, *Monatsh. Chem.,* **96**, 2046 (1965).
81. S. M. Kvitko and V. V. Perekalin, *Zh. Obshch. Khim.,* **32**, 144 (1962).
82. S. M. Kvitko, V. V. Perekalin, and N. V. Buival, *Zh. Organ. Khim.,* **2**, 2253 (1966).
83. F. Duran and L. Ghosez, *Tetrahedron Lett.,* 245 (1970).
84. G. Schroll, P. Klemmensen, and S.-O. Lawesson, *Ark. Kemi,* **26**, 317 (1967).
85. Ref. 2, pp. 536–538.
86. A. Dornow and H. vonPlessen, *Chem. Ber.,* **99**, 244 (1966).
87. E. Ziegler and F. Hradetzky, *Monatsh. Chem.,* **95**, 1247 (1964).
88. E. Ziegler, F. Hradetzky, and M. Eder, *Monatsh. Chem.,* **97**, 1394 (1966).
89. E. Ziegler, H. Wittmann, and V. Illi, *Monatsh. Chem.,* **100**, 1741 (1969).
90. E. Ziegler, F. Hradetzky, and K. Belegratis, *Monatsh. Chem.,* **96**, 1347 (1965).
91. E. Ziegler and K. Belegratis, *Monatsh. Chem.,* **98**, 219 (1967).
92. E. Ziegler and G. Kleineberg, *Monatsh. Chem.,* **96**, 1360 (1965).
93. E. Ziegler, G. Kleineberg, and K. Belegratis, *Monatsh. Chem.,* **98**, 77 (1967).
94. M. A. Butt and J. A. Elvidge, *J. Chem. Soc.,* 4483 (1963).
95. E. Ziegler and F. Hradetzky, *Monatsh. Chem.,* **97**, 710 (1966).
96. E. Ziegler and K. Belegratis, *Monatsh. Chem.,* **99**, 995 (1968).
97. S. J. Davis, J. A. Elvidge, and A. B. Foster, *J. Chem. Soc.,* 3638 (1962).
98. J. A. Elvidge and N. A. Zaidi, *J. Chem. Soc., C,* 2188 (1968).
99. S. J. Davis, and J. A. Elvidge, *J. Chem. Soc.,* 3553 (1962).
100. K. R. Huffman, F. C. Schaefer, and G. A. Peters, *J. Org. Chem.,* **27**, 551 (1962).
101. M. L. Peterson, *J. Org. Chem.,* **25**, 565 (1960).
102. R. Fuks, *Tetrahedron,* **26**, 2161 (1970).
103. L. Claisen and K. Meyer, *Chem. Ber.,* **35**, 583 (1902).
104. T. Kato, H. Yamanaka, and T. Shibata, *Chem. Pharm. Bull.* (Tokyo), **15**, 921 (1967).
105. T. Kato, H. Yamanaka, J. Kawamata, and T. Shibata, *Chem. Pharm. Bull.* (Tokyo), **16**, 1835 (1968).
106. V. I. Gunar, L. F. Ovechkina, and S. I. Zav'yalov, *Izv. Akad. Nauk. SSSR, Ser. Khim.,* 1885 (1965).
107. Z. Bukac and J. Sebenda, *Collect. Czech. Chem. Commun.,* **32**, 3537 (1967).
108. S. Garratt, *J. Org. Chem.,* **28**, 1886 (1963).
109. S. Seto, H. Sasaki, and K. Ogura, *Bull. Chem. Soc. Jap.,* **39**, 281 (1966).
110. H. Toda, *Yakugaku Zasshi,* **87**, 1351 (1967); *Chem. Abstr.,* **69**, 2687 (1968).
111. J. S. A. Brunskill, *J. Chem. Soc., C,* 960 (1968).
112. M. C. Seidel, G. C. VanTuyle, and W. D. Weir, *J. Org. Chem.,* **35**, 1475 (1970).
113. S. A. Vartanyan, A. S. Noravyan, and V. N. Zhamagortsyan, *Khim. Geterotsikl. Soedin.,* **2**, 670 (1966); *Chem. Abstr.,* **66**, 55339 (1967).

114. F. C. Uhle and H. Schröter, *J. Org. Chem.*, **26**, 4169 (1961).
115. A. R. Casola and F. E. Anderson, *J. Pharm. Sci.*, **54**, 1686 (1965).
116. I. Steffan and B. Prijs, *Helv. Chim. Acta*, **44**, 1429 (1961).
117. K. E. Schulte, J. Reisch, and O. Heine, *Arch. Pharm.* (Weinheim), **294**, 234 (1961).
118. M. A.-F. Elkaschef, M. H. Nosseir, and A. Abdel-Kader, *J. Chem. Soc.*, 4647 (1963).
119. N. P. Shusherina, E. A. Luk'yanets, and R. Ya. Levina, *Zh. Organ. Khim.*, **1**, 679 (1965); *Chem. Abstr.*, **64**, 3459 (1966).
120. I. E. El-Kholy, F. K. Rafla, and G. Soliman, *J. Chem. Soc.*, 2588 (1959).
121. I. E. El-Kholy, F. K. Rafla, and G. Soliman, *J. Chem. Soc.*, 4490 (1961).
122. I. E. El-Kholy, F. K. Rafla, and M. M. Mishrikey, *J. Chem. Soc.*, *C*, 1950 (1969).
123. Ref. 2, pp. 552–560.
124. J. E. Hodge and E. C. Nelson, *Cereal Chem.*, **38**, 207 (1961).
125. B. E. Fisher and J. E. Hodge, *J. Org. Chem.*, **29**, 776 (1964).
126. A. Peratoner and A. Tamburello, *Gazz. Chim. Ital.*, **36**, 33, 50 (1906).
127. L. L. Woods and H. C. Smitherman, *J. Org. Chem.*, **26**, 2987 (1961).
128. V. Hahn and S. Kukolja, *Croat. Chem. Acta*, **33**, 137 (1961).
129. K. Blazevic and V. Hahn, *Croat. Chem. Acta*, **38**, 113 (1966).
130. C. P. Krimmel, U.S. patent, 2,965,641 (1960); *Chem. Abstr.*, **55**, 13457 (1961).
131. M. Yasue, N. Kawamura, and J. Sakakibara, *Yakugaku Zasshi*, **90**, 1222 (1970); *Chem. Abstr.*, **74**, 23102 (1971).
132. L. L. Woods, *Chem. Ind.* (London), 1567 (1960).
133. I. D. Spenser and A. D. Notation, *Can. J. Chem.*, **40**, 1374 (1962).
134. A. D. Notation and I. D. Spenser, *Can. J. Biochem.*, **42**, 1803 (1964).
135. E. M. Kaiser, S. D. Work, J. F. Wolfe, and C. R. Hauser, *J. Org. Chem.*, **32**, 1483 (1967).
136. M. Simalty, H. Stryzelecka, and M. Dupre, *C. R. Acad. Sci. Ser. Paris, Ser. C.*, **265**, 1284 (1967).
137. P. Bellingham, C. D. Johnson, and A. R. Katritzky, *J. Chem. Soc.*, *B*, 866 (1968).
138. T. Kato and H. Yamanaka, *Chem. Pharm. Bull.* (Tokyo), **12**, 18 (1964).
139. P. Beak and J. Bonham, *J. Amer. Chem. Soc.*, **87**, 3365 (1965).
140. S. W. Nakhre and S. S. Deshapande, *Vikram J., Vikram. Univ.*, **4**, 153 (1960); *Chem. Abstr.*, **57**, 2185 (1962).
141. K. Hamamoto, T. Isoshima, and M. Yoshioka, *Nippon Kagaku Zasshi*, **79**, 840 (1958).
142. Ref. 2, p. 557.
143. J. D. Edwards, J. E. Page, and M. Pianka, *J. Chem. Soc.*, 5200 (1964).
144. D. Cook, *Can. J. Chem.*, **41**, 1435 (1963).
145. R. N. Schut, W. G. Strycker, and T. M. H. Liu, *J. Org. Chem.*, **28**, 3046 (1963).
146. S. Iguchi, A. Inoue, and C. Kurahashi, *Chem. Pharm. Bull.* (Tokyo), **9**, 1016 (1961).
147. S. Iguchi, A. Inoue, and C. Kurahashi, *Chem. Pharm. Bull.* (Tokyo), **11**, 385 (1963).
148. S. Iguchi and A. Inoue, *Chem. Pharm. Bull.* (Tokyo), **11**, 390 (1963).
149. A. K. Mallams, *J. Org. Chem.*, **29**, 3555 (1964).
150. C. A. Salemink, *Rec. Trav. Chim. Pays-Bas*, **80**, 545 (1961).
151. T. Kato, J. Kawamata, and T. Shibata, *Yakugaku Zasshi*, **88**, 106 (1968); *Chem. Abstr.*, **69**, 27302 (1968).
152. Ref. 2, p. 558.
153. C. S. Wang, Abst. Papers, 160th Meeting, American Chemical Society, 1970, ORGN 80.
154. S. Garratt and D. Shemin, *J. Org. Chem.*, **28**, 1372 (1963).
155. P. Petrenko-Kritschenko and J. Schöttle, *Chem. Ber.*, **44**, 2826 (1911).
156. F. Feist, *Chem. Ber.*, **23**, 3726 (1890).

157. F. Hradetzky and E. Ziegler, *Monatsh. Chem.*, **97**, 398 (1966).
158. P. E. Brenneisen, T. E. Acker, and S. W. Tanenbaum, *J. Amer. Chem. Soc.*, **86**, 1234 (1964).
159. T. E. Acker, P. E. Brenneisen, and S. W. Tanenbaum, *J. Amer. Chem. Soc.*, **88**, 834 (1966).
160. A. Suzuki, N. Takahashi, and S. Tamura, *Agr. Biol. Chem.* (Tokyo), **30**, 13 (1966).
161. H. Junek, H. Budschedl, and E. Ziegler, *Monatsh. Chem.*, **98**, 2252 (1967).
162. Y. Hirata, H. Nakata, K. Yamada, K. Okuhara, and T. Naito, *Tetrahedron*, **14**, 252 (1961).
163. F. K. Splinter and H. Arold, *J. Prakt. Chem.*, **311**, 869 (1969).
164. S. J. Davis and J. A. Elvidge, *J. Chem. Soc.*, 2251 (1953).
165. J. A. Elvidge, *J. Chem. Soc.*, 2606 (1962).
166. S. J. Davis and J. A. Elvidge, *J. Chem. Soc.*, 3550 (1962).
167. A. Butt, I. A. Akhtar, and M. Khurshid, *Pakistan J. Sci. Ind. Res.*, **9**, 335 (1966).
168. A. Butt, S. M. A. Hai, and I. A. Akhtar, *Tetrahedron*, **22**, 455 (1966).
169. A. Butt, I. A. Akhtar, S. A. Qureshi, and M. Akhtar, *Pakistan J. Sci. Ind. Res.*, **10**, 240 (1967).
170. A. Butt and I. A. Akhtar, *Tetrahedron*, **21**, 1917 (1965).
171. M. A. Butt, J. A. Elvidge, and A. B. Foster, *J. Chem. Soc.*, 3069 (1963).
172. M. A. Butt, I. A. Akhtar, and M. Akhtar, *Tetrahedron*, **23**, 1551 (1967).
173. A. Butt, M. Akhtar, and R. Parveen, *Pakistan J. Sci. Ind. Res.*, **10**, 243 (1967).
174. A. Butt, I. A. Akhtar, and M. Akhtar, *Tetrahedron*, **23**, 199 (1967).
175. Ref., pp. 560–567.
176. H. Sugisawa, and K. Aso, *Tohoku J. Agr. Res.*, **10**, 137 (1959); *Chem. Abstr.*, **54**, 11015 (1960).
177. H. Sugisawa and K. Aso, *Nippon Nogei Kagaku Kaishi*, **33**, 259 (1959); *Chem. Abstr.*, **59**, 8695 (1963).
178. H. Sugisawa, *Tohoku J. Agr. Res.*, **11**, 389 (1960); *Chem. Abstr.*, **55**, 19917 (1961).
179. H. Sugisawa, *Tohoku J. Agr. Res.*, **11**, 397 (1960); *Chem. Abstr.*, **55**, 19917–19918 (1961).
180. H. Sugisawa, H. Sugiyama, and K. Aso, *Tohoku J. Agr. Res.*, **12**, 245 (1961); *Chem. Abstr.*, **57**, 16535 (1962).
181. H. Sugisawa and K. Aso, *Chem. Ind.* (London), 781 (1961).
182. L. D. Smirnov, S. I. Sholina, K. E. Kruglyakova, and K. M. Dyumaev, *Izv. Akad. Nauk SSSR. Otd. Khim.*, 890 (1963).
183. N. Clauson-Kaas and M. Meister, *Acta Chem. Scand.*, **21**, 1104 (1967).
184. N. Elming and N. Clauson-Kaas, *Acta Chem. Scand.*, **10**, 1603 (1956).
185. L. D. Smirnov, K. M. Dyumaev, N. I. Shuikin, and I. F. Bel'skii, *Izv. Akad. Nauk. SSSR*, 2246 (1962).
186. H. Leditschke, *Arch. Pharm.* (Weinheim), **295**, 323 (1962).
187. J. R. Geigy A.-G., Fr. patent, 1,477,998 (1967); *Chem. Abstr.*, **68**, 29606 (1968).
188. J. R. Geigy A.-G., Br. patent, 1,108,975 (1968); *Chem. Abstr.*, **69**, 67226 (1968).
189. J. B. Petersen, K. Norris, N. Clauson-Kaas, and K. Svanholt, *Acta Chem. Scand.*, **23**, 1785 (1969).
190. Ref. 2, pp. 560–563.
191. L. A. Walter, C. K. Springer, J. Kenney, S. K. Gaien, and N. Sperber, *J. Med. Chem.*, **11**, 792 (1968).
192. H. Paulsen, *Angew. Chem. Intern. Ed.*, **1**, 454 (1962).
193. H. Paulsen, *Ann. Chem.*, **665**, 166 (1963).
194. S. J. Norton, C. G. Skinner, and W. Shive, *J. Org. Chem.*, **26**, 1495 (1961).
195. M. Ya. Karpeiskii and V. L. Florent'ev, *Russ. Chem. Rev.*, **38**, 540 (1969).

196. G. Ya. Kondrat'eva and C.-H. Huang, *Dokl. Akad. Nauk SSSR,* **141**, 628 (1961); *Chem. Abstr.,* **56**, 14229 (1962).
197. G. Ya. Kondrat'eva and C.-H. Huang, *Dokl. Akad. Nauk SSSR,* **141**, 861 (1961); *Chem. Abstr.,* **56**, 14229 (1962).
198. G. Ya. Kondrat'eva, *Izv. Akad. Nauk SSSR, Otd. Khim. Nauk,* 484 (1959); *Chem. Abstr.,* **53**, 21940 (1959).
199. R. A. Firestone, E. E. Harris, and W. Reuter, *Tetrahedron,* **23**, 943 (1967).
200. K. Pfister, III, E. E. Harris, and R. A. Firestone, U.S. patent, 3,227,724 (1966); *Chem. Abstr.,* **64**, 8149–8150 (1966).
201. E. E. Harris, R. A. Firestone, K. Pfister, III, R. R. Boettcher, F. J. Cross, R. B. Currie, M. Monaco, E. R. Peterson, and W. Reuter, *J. Org. Chem.,* **27**, 2705 (1962).
202. Merck and Co., Nether. patent, 6,614,801 (1967); *Chem. Abstr.,* **68**, 87190 (1968).
203. T. Naito, K. Ueno, M. Sano, Y. Omura, I. Itoh, and F. Ishikawa, *Tetrahedron Lett.,* 5767 (1968).
204. N. D. Doktorova, L. V. Ionova, M. Ya. Karpeisky, N. Sh. Padyukova, K. F. Turchin, and V. L. Florentiev, *Tetrahedron,* **25**, 3527 (1969).
205. T. Miki and Y. Matsuo, Jap. patent, 25,664 (1967); *Chem. Abstr.,* **69**, 43807 (1968).
206. W. Kimel and L. Leimgruber, Fr. patent, 1,384,099 (1965); *Chem. Abstr.,* **63**, 4263 (1965).
207. T. Maruyama, E. Araki, and N. Tokai, Jap. patent, 11,745 (1967); *Chem. Abstr.,* **68**, 87177 (1968).
208. T. Miki and T. Matsuo, *Yakugaku Zasshi,* **87**, 323 (1967); *Chem. Abstr.,* **67**, 32549 (1967).
209. T. Naito and T. Yoshikawa, *Chem. Pharm. Bull.* (Tokyo), **14**, 918 (1966).
210. T. Naito, T. Yoshikawa, F. Ishikawa, S. Isoda, Y. Omura, and I. Takamura, *Chem. Pharm. Bull.* (Tokyo), **13**, 869 (1965).
211. T. Yoshikawa, F. Ishikawa, Y. Omura, and T. Naito, *Chem. Pharm. Bull.* (Tokyo), **13**, 873 (1965).
212. T. Yoshikawa, F. Ishikawa, and T. Naito, *Chem. Pharm. Bull.* (Toyko), **13**, 878 (1965).
213. A. E. A. Porter and P. G. Sammes, *Chem. Commun.,* 1103 (1970).
214. E. Ziegler, K. Belegratis, and G. Brus, *Monatsh. Chem.,* **98**, 555 (1967).
215. J. A. Moore and R. W. Medeiros, *J. Amer. Chem. Soc.,* **81**, 6026 (1959).
216. J. A. Moore and J. Binkert, *J. Amer. Chem. Soc.,* **81**, 6029 (1959).
217. A. Nabeya, F. B. Culp, and J. A. Moore, *J. Org. Chem.,* **35**, 2015 (1970).
218. J. A. Moore and H. H. Püschner, *J. Amer. Chem. Soc.,* **81**, 6041 (1959).
219. J. A. Moore and J. Binkert, *J. Amer. Chem. Soc.,* **81**, 6045 (1959).
220. J. A. Moore, F. J. Marascia, R. W. Medeiros, and E. Wyss, *J. Amer. Chem. Soc.,* **84**, 3022 (1962).
221. J. A. Moore, R. L. Wineholt, F. J. Marascia, R. W. Medeiros, and F. J. Creegan, *J. Org. Chem.,* **32**, 1353 (1967).
222. J. A. Moore and E. C. Zoll, *J. Org. Chem.,***29**, 2124 (1964).
223. J. A. Moore, H. Kwart, G. Wheeler, and H. Bruner, *J. Org. Chem.,* **32**, 1342 (1967).
224. R. K. Bly, E. C. Zoll, and J. A. Moore, *J. Org. Chem.,* **29**, 2128 (1964).
225. J. A. Moore and W. J. Theuer, *J. Org. Chem.,* **30**, 1887 (1965).
226. Y. Inubushi, Y. Sasaki, Y. Tsuda, B. Yasui, T. Konita, J. Matsumoto, E. Katarao, and J. Nakano, *Tetrahedron,* **20**, 2007 (1964).
227. R. Kuhn and I. Löw, *Chem. Ber.,* **95**, 1748 (1962).
228. S. Masamune and K. Fukumoto, *Tetrahedron Lett.,* 4647 (1965).

229. N. C. Castellucci, M. Kato, H. Zenda, and S. Masamune, *Chem. Commun.*, 473 (1967).
230. A. S. Monahan and S. Tang, *J. Org. Chem.*, **33**, 1445 (1968).
231. W. Ried and F. Bätz, *Ann. Chem.*, **725**, 230 (1969).
232. H. Iida, I. Doi, H. Iida, *Kogyo Kagaku Zasshi,* **69**, 2164 (1966); *Chem. Abstr.*, **66**, 86580 (1967).
233. A. Roedig, G. Märkl, and H. Schaller, *Chem. Ber.*, **103**, 1022 (1970).
234. R. W. Meikle and E. A. Williams, *Nature,* **210**, 523 (1966).
235. W. W. Muelder and M. N. Wass, *J. Agr. Food Chem.*, **15**, 508 (1967).
236. Dow Chemical Co., Neth. patent, 6,402,443 (1965); *Chem. Abstr.*, **64**, 8152 (1966).
237. A. R. Sexton, U.S. patent, 3,355,456, (1967); *Chem. Abstr.*, **68**, 29602 (1968).
238. W. T. Flowers, R. N. Haszeldine, and S. A. Majid, *Tetrahedron Lett.*, 2503 (1967).
239. A. Roedig, K. Grohe, and D. Klatt, *Chem. Ber.*, **99**, 2818 (1966).
240. S. M. Roberts and H. Suschitzky, *J. Chem. Soc., C*, 1537 (1968).
241. A. R. Katritzky, J. D. Rowe, and S. K. Roy, *J. Chem. Soc., B,* 758 (1967).
242. P. I. Mortimer, *Aust. J. Chem.*, **21**, 467 (1968).
243. R. D. Chambers, J. Hutchinson, and W. K. R. Musgrave, *J. Chem. Soc.*, 3573 (1964).
244. R. E. Banks, R. N. Haszeldine, J. V. Latham, and I. M. Young, *J. Chem. Soc.*, 594 (1965).
245. R. D. Chambers, J. Hutchinson, and W. K. R. Musgrave, *J. Chem. Soc.*, 3736 (1964).
246. R. D. Chambers, J. Hutchinson, and W. K. R. Musgrave, *J. Chem. Soc.*, 5634 (1964).
247. R. D. Chambers, J. Hutchinson, and W. K. R. Musgrave, *J. Chem. Soc.*, 5040 (1965).
248. R. E. Banks, J. E. Burgess, W. M. Cheng, and R. N. Haszeldine, *J. Chem. Soc.*, 575 (1965).
249. T. Kato, H. Hayashi, and T. Anzai, *Chem. Pharm. Bull.* (Tokyo), **15**, 1343 (1967).
250. R. A. Abramovitch, F. Helmer, and M. Liveris, *J. Chem. Soc., B*, 492 (1968).
251. R. A. Abramovitch, F. Helmer, and M. Liveris, *J. Org. Chem.*, **34**, 1730 (1969).
252. Z. Talik, *Rocz. Chem.*, **35**, 475, 487 (1961).
253. Z. Talik, *Rocz. Chem.*, **36**, 1313 (1962).
254. T. Talik, *Rocz. Chem.*, **37**, 69 (1963).
255. T. Talik, *Rocz. Chem.*, **36**, 1465 (1962).
256. T. Talik and Z. Talik, *Rocz. Chem.*, **38**, 777 (1964).
256a. T. Talik and Z. Talik, *Rocz. Chem.*, **40**, 1187 (1966).
257. T. Talik and Z. Talik, *Rocz. Chem.*, **40**, 1457 (1966).
258. Z. Talik and B. Brekiesz, *Rocz. Chem.*, **41**, 279 (1967).
259. T. Talik and Z. Talik, *Rocz. Chem.*, **42**, 1647 (1968).
260. T. Talik and Z. Talik, *Rocz. Chem.*, **41**, 1507 (1967).
261. T. Talik and Z. Talik, *Bull. Acad. Pol. Sci.*, **16**, 7 (1968); *Chem. Abstr.*, **69**, 59052 (1968).
262. T. Talik and Z. Talik, *Rocz. Chem.*, **41**, 1721 (1967).
263. Z. Talik and B. Brekiesz-Lewandowska, *Rocz. Chem.*, **44**, 1325 (1970).
264. T. Talik and Z. Talik, *Rocz. Chem.*, **43**, 1961 (1969).
265. Z. Talik and E. Plazek, *Bull. Acad. Pol. Sci.*, **8**, 219 (1960); *Chem. Abstr.*, **60**, 9241 (1964).
266. T. Talik and Z. Talik, *Rocz. Chem.*, **42**, 1861 (1968).

267. J. D. Reinheimer, J. T. Gerig, R. Garst, and B. Schrier, *J. Amer. Chem. Soc.*, **84**, 2770 (1962).
268. J. A. Zoltewicz and A. A. Sale, *J. Org. Chem.*, **35**, 3462 (1970).
269. E. Koenigs, H. C. Gerdes, and A. Sirot, *Chem. Ber.*, **61**, 1022 (1928).
270. M. Liveris and J. Miller, *J. Chem. Soc.*, 3486 (1963).
271. S. Kajihara, *Nippon Kagaku Zasshi,* **86**, 839 (1965); *Chem. Abstr.*, **65**, 16935 (1966).
272. Ref. 2, pp. 610–611.
273. S. Kajihara, *Nippon Kagaku Zasshi,* **86**, 1060 (1965); *Chem. Abstr.*, **65**, 16936 (1966).
274. S. Kajihara, *Nippon Kagaku Zasshi,* **86**, 93 (1965); *Chem. Abstr.*, **63**, 578 (1965).
275. E. Spinner and J. C. B. White, *Chem. Ind.* (London), 1784 (1967).
276. P. Tomasik and E. Plazek, *Rocz. Chem.*, **39**, 1911 (1965).
277. Ref. 2, p. 583.
278. E. C. Taylor and J. S. Driscoll, *J. Org. Chem.*, **26**, 3001 (1961).
279. F. Kröhnke and H. Schäfer, *Chem. Ber.*, **95**, 1104 (1962).
280. T. Kosuge, H. Zenda, and Y. Suzuki, *Chem. Pharm. Bull.* (Tokyo), **18**, 1068 (1970).
281. M. Barash, J. M. Osbond, and J. C. Wickens, *J. Chem. Soc.*, 3530 (1959).
282. W. Herz and D. R. K. Murty, *J. Org. Chem.*, **26**, 122 (1961).
283. T. Takahashi and K. Kariyone, *Chem. Pharm. Bull.* (Tokyo), **8**, 1106 (1960).
284. M. J. Pieterse and H. J. den Hertog, *Rec. Trav. Chim. Pays-Bas,* **81**, 855 (1962).
285. J. Kozlowska and E. Plazek, *Rocz. Chem.*, **33**, 831 (1959).
286. T. Talik and Z. Talik, *Proc. Intern. Symp. Warsaw,* 81 (1963); *Chem. Abstr.*, **64**, 2046 (1966).
287. T. Batkowski, *Rocz. Chem.*, **43**, 1623 (1969).
288. T. Batkowski, *Rocz. Chem.*, **37**, 385 (1963).
289. T. Batkowski and M. Tuszynska, *Rocz. Chem.*, **38**, 585 (1964).
290. T. Batkowski, *Rocz. Chem.*, **41**, 729 (1967).
291. T. Batkowski, *Rocz. Chem.*, **42**, 2079 (1968).
292. J. A. Moore and F. J. Marascia, *J. Amer. Chem. Soc.*, **81**, 6049 (1959).
293. P. Nantka-Namirski, C. Kaczmarczyk, and L. Toba, *Acta Pol. Pharm.*, **24**, 228 (1967); *Chem. Abstr.*, **69**, 2827 (1968).
294. P. Nantka-Namirski, *Acta Pol. Pharm.*, **18**, 449 (1961); *Chem. Abstr.*, **58**, 3424 (1963).
295. H. Ban-Oganowska, Z. Skrowaczewska, and L. Syper, *Rocz. Chem.*, **40**, 1215 (1966).
296. T. Talik and E. Plazek, *Rocz. Chem.*, **33**, 387 (1959).
297. J. Koncewicz and Z. Skrowaczewska, *Rocz. Chem.*, **42**, 1873 (1968).
298. K. Lewicka and E. Plazek, *Rocz. Chem.*, **40**, 1875 (1966).
299. T. Talik, *Rocz. Chem.*, **36**, 1049 (1962).
300. T. Batkowski and E. Plazek, *Rocz. Chem.*, **36**, 51 (1962).
301. P. Tomasik and E. Plazek, *Rocz. Chem.*, **39**, 365 (1965).
302. C. M. Atkinson and B. N. Biddle, *J. Chem. Soc., C,* 2053 (1966).
303. C.-H. Huang and G. Ya. Kondrat'eva, *Izv. Akad. Nauk SSSR, Otd. Khim. Nauk,* 525 (1962).
304. A. Koshiro, *Yakugaku Zasshi,* **79**, 1129 (1959); *Chem. Abstr.*, **54**, 3418 (1960).
305. T. Talik and Z. Talik, *Rocz. Chem.*, **37**, 75 (1963).
306. A. Dornow, H. v. Plessen, and R. Huischen, *Chem. Ber.*, **99**, 254 (1966).
307. J. M. Cox, J. A. Elvidge, and D. E. H. Jones, *J. Chem. Soc.*, 1423 (1964).
308. M. Wahren, *Tetrahedron,* **24**, 441 (1968).
309. M. Wahren, *Tetrahedron,* **24**, 451 (1968).

310. W. Herz and D. R. K. Murty, *J. Org. Chem.*, **26**, 418 (1961).
311. R. Adams and T. R. Govindachari, *J. Amer. Chem. Soc.*, **69**, 1806 (1947).
312. H. J. den Hertog, J. P. Wibaut, F. R. Schepman, and A. A. van der Wal, *Rec. Trav. Chim. Pays-Bas*, **69**, 700 (1950).
313. E. J. Behrman and B. M. Pitt, *J. Amer. Chem. Soc.*, **80**, 3717 (1958).
314. Y. Ban and T. Wakamatsu, *Chem. Ind.* (London), 710 (1964).
315. Ref. 2, pp. 591–592.
316. Y. L. Gol'dfarb, L. V. Antik, and V. A. Petukhov, *Izv. Akad. Nauk, SSSR, Otd. Khim. Nauk*, 887 (1961); *Chem. Abstr.*, **55**, 22307 (1961).
317. W. Czuba, *Bull. Acad. Pol. Sci.*, **8**, 281 (1960); *Chem. Abstr.*, **60**, 2883 (1964).
318. W. Czuba, *Rocz. Chem.*, **34**, 905 (1960).
319. W. Czuba, *Rocz. Chem.*, **34**, 1639 (1960).
320. W. Czuba, *Rocz. Chem.*, **35**, 1347 (1961).
321. W. Czuba, *Rocz. Chem.*, **34**, 1647 (1960).
322. Ref. 2, p. 596.
323. J. Klosa, *J. Prakt. Chem.*, **8**, 168 (1959).
324. A. P. Gray, D. E. Heitmeier, and E. E. Spinner, *J. Amer. Chem. Soc.*, **81**, 4351 (1959).
325. K. Okamoto and M. Tetsuo, *Yakugaku Zasshi*, **82**, 769 (1962); *Chem. Abstr.*, **58**, 2433 (1963).
326. U. M. Teotino, L. PoloFriz, A. Gandini, and D. Della Bella, *Farmaco, Ed. Sci.*, **17**, 988 (1962).
327. D. J. Brown and J. S. Harper, *J. Chem. Soc.*, 5542 (1965).
328. E. A. Prill and S. M. McElvain, *Org. Synth., Coll. Vol.* **II**, 419 (1943).
329. J. Gripenberg, *Tetrahedron*, **10**, 135 (1960).
330. J. Gripenberg and T. Hase, *Acta Chem. Scand.*, **17**, 2250 (1963).
331. Ref. 2, p. 600.
332. H. Tomisawa, T. Agatsuma, and Y. Kamura, *Yakugaku Zasshi*, **81**, 947 (1961); *Chem. Abstr.*, **55**, 27301 (1961).
333. E. G. Podrebarac and W. E. McEwen, *J. Org. Chem.*, **26**, 1386 (1961).
334. R. A. Abramovitch and A. R. Vinutha, *J. Chem. Soc., B*, 131 (1971).
335. Ref. 2, pp. 596–603.
336. R. Lukes, A. A. Arojan, J. Kovar, and K. Blaha, *Collect. Czech. Chem. Commun.*, **27**, 751 (1962).
337. A. Raczka and H. Bojarska-Dahlig, *Acta Pol. Pharm.*, **22**, 307 (1965); *Chem. Abstr.*, **63**, 13201 (1965).
338. D. W. Brown, S. F. Dyke, W. G. D. Lugton, and A. Davis, *Tetrahedron*, **24**, 2517 (1968).
339. Y. Suzuta, *Yakugaku Zasshi*, **79**, 1314 (1959); *Chem. Abstr.*, **54**, 4580 (1960).
340. S. Akaboshi and S. Ikegami, *Chem. Pharm. Bull.* (Tokyo), **14**, 622 (1966).
341. L. A. Paquette and N. A. Nelson, *J. Org. Chem.*, **27**, 1085 (1962).
342. H. Möhrle and H. Weber, *Tetrahedron*, **26**, 3779 (1970).
343. E. M. Kosower and J. W. Patton, *Tetrahedron*, **22**, 2081 (1966).
344. S. Hünig and W. Kniese, *Ann. Chem.*, **708**, 198 (1967).
345. S. Hünig and W. Kniese, *Ann. Chem.*, **708**, 178 (1967).
346. S. Hünig and W. Kniese, *Ann. Chem.*, **708**, 170 (1967).
347. I. Enander, A. Sundwall, and B. Sörbo, *Biochem. Pharmacol.*, **11**, 377 (1962).
348. P. M. S. Miranda and J. L. Way, *Mol. Pharmacol.*, **2**, 117 (1966).
349. J. A. Berson and T. Cohen, *J. Amer. Chem. Soc.*, **78**, 416 (1956).
350. J. A. Berson and J. S. Walia, *J. Org. Chem.*, **24**, 756 (1959).
351. T. Kametani and Y. Nomura, *Chem. Pharm. Bull.* (Tokyo), **8**, 741 (1960).

352. A. R. Battersby and J. C. Turner, *J. Chem. Soc.*, 717 (1960).
353. E. Matsumura, T. Nashima, and F. Ishibashi, *Bull. Chem. Soc. Jap.*, **43**, 3540 (1970).
354. H.-J. Teuber, G. Wenzel, and U. Hochmuth, *Chem. Ber.*, **96**, 1119 (1963).
355. M. Stanulović and S. Chaykin, *Fed. Proc.*, **29**, A343 (1970).
356. H. J. Shine, "Aromatic Rearrangements," Elsevier, Amsterdam, 1967, p. 284.
357. E. Ochiai, "Aromatic Amine *N*-Oxides," Elsevier, Amsterdam, 1967, p. 290.
358. A. R. Katritzky, *Quart. Rev.*, **10**, 395 (1956).
359. V. J. Traynelis, in "Mechanisms of Molecular Migrations," B. S. Thyagarajan, Ed., Vol. 2, Interscience, New York, 1969, pp. 31–37.
360. J. H. Markgraf, H. B. Brown, Jr., S. C. Mohr, and R. G. Peterson, *J. Amer. Chem. Soc.*, **85**, 958 (1963).
361. B. M. Bain and J. E. Saxton, *J. Chem. Soc.*, 5216 (1961).
362. V. Boekelheide and W. L. Lehn, *J. Org. Chem.*, **26**, 428 (1961).
363. M. P. Cava and B. Weinstein, *J. Org. Chem.*, **23**, 1616 (1958).
364. E. C. Taylor and J. S. Driscoll, *J. Org. Chem.*, **25**, 1716 (1960).
365. Y. I. Chumakov and Z. M. Korsakova, *Metody Polucheniya Khim. Reak. Prep*, **No. 4–5**, 62 (1962). (*Gos. Kom. Sov., Min. SSSR po Khim*); *Chem. Abstr.*, **60**, 14468 (1964).
366. T. Cohen and G. L. Deets, *J. Amer. Chem. Soc.*, **89**, 3939 (1967).
367. P. W. Ford and J. M. Swan, *Aust. J. Chem.*, **18**, 867 (1965).
368. T. Kato, T. Kitagawa, T. Shibata, and K. Nakai, *Yakugaku Zasshi*, **82**, 1647 (1962); *Chem. Abstr.*, **59**, 559 (1963).
369. C. Kaneko, S. Yamada, and I. Yokoe, *Shika Zairyo Kenkyusho Hokoku*, **2**, 475 (1963).
370. L. Syper, Z. Skrowaczewska, and A. Bytnar, *Rocz. Chem.*, **41**, 1027 (1967).
371. W. Baker, K. M. Buggle, J. F. W. McOmie, and D. A. M. Watkins, *J. Chem. Soc.*, 3594 (1958).
372. Y. Murakami and J. Sunamoto, *Bull. Chem. Soc. Jap.*, **42**, 3350 (1969).
373. R. Adams and A. W. Schrecker, *J. Amer. Chem. Soc.*, **71**, 1186 (1949).
374. M. Hamana and M. Yamazaki, *Yakugaku Zasshi*, **81**, 574 (1961); *Chem. Abstr.*, **55**, 24743 (1961).
375. Y. Kobayashi and I. Kumadaki, *Chem. Pharm. Bull.* (Tokyo), **17**, 510 (1969).
376. G. F. van Rooyen, C. v. d. M. Brink, and P. A. deVilliers, *Tydskr. Natuurwetenskappe*, **4**, 182 (1964); *Chem. Abstr.*, **63**, 13202 (1965).
377. Y. Sakata, K. Adachi, Y. Akahori, and E. Hayashi, *Yakugaku Zasshi*, **87**, 1374 (1967); *Chem. Abstr.*, **68**, 59649 (1968).
378. W. Sliwa and Z. Skrowaczewska, *Rocz. Chem.*, **44**, 1941 (1970).
379. L. Achremowicz, T. Moroz-Banas, Z. Skrowaczewska, and L. Syper, *Rocz. Chem.*, **42**, 1499 (1968).
380. Ref. 2, p. 609.
381. S. Furukawa, *Yakugaku Zasshi*, 77, 11 (1957); *Chem. Abstr.*, **51**, 8745 (1957).
382. S. Oae, T. Kitao, and Y. Kitaoka, *Chem. Ind.* (London), 515 (1961).
383. S. Oae, T. Kitao, and Y. Kitaoka, *J. Amer. Chem. Soc.*, **84**, 3359 (1962).
384. V. J. Traynelis and Sr. A. I. Gallagher, *J. Org. Chem.*, **35**, 2792 (1970).
385. R. Bodalski and A. R. Katritzky, *Tetrahedron Lett.*, 257 (1968).
386. R. Bodalski and A. R. Katritzky, *J. Chem. Soc., B*, 831 (1968).
387. V. J. Traynelis and Sr. A. I. Gallagher, *J. Amer. Chem. Soc.*, **87**, 5710 (1965).
388. C. W. Muth and R. S. Darlak, *J. Org. Chem.*, **30**, 1909 (1965).
389. V. J. Traynelis and P. L. Pacini, *J. Amer. Chem. Soc.*, **86**, 4917 (1964).
390. S. Oae, Y. Kitaoka, and T. Kitao, *Tetrahedron*, **20**, 2677 (1964).

391. S. Oae, Y. Kitaoka, and T. Kitao, *Tetrahedron,* **20**, 2685 (1964).
392. S. Oae, T. Kitao, and Y. Kitaoka, *J. Amer. Chem. Soc.*, **84**, 3362 (1962).
393. S. Oae, T. Kitao, and Y. Kitaoka, *J. Amer. Chem. Soc.*, **84**, 3366 (1962).
394. S. Oae and S. Kozuka, *Tetrahedron,* **20**, 2691 (1964).
395. V. J. Traynelis, Sr. A. I. Gallagher, and R. F. Martello, *J. Org. Chem.*, **26**, 4365 (1961).
396. S. Oae, S. Tamagaki, T. Negoro, K. Ogino, and S. Kozuka, *Tetrahedron Lett.*, 917 (1968).
397. H. Iwamura, M. Iwamura, T. Nishida, and I. Miura, *Tetrahedron Lett.*, 3117 (1970).
398. T. Koenig, *J. Amer. Chem. Soc.*, **88**, 4045 (1966).
399. S. Oae, S. Tamagaki, T. Negoro, and S. Kozuka, *Tetrahedron,* **26**, 4051 (1970).
400. H. Iwamura, M. Iwamura, T. Nishida, and S. Sato, *J. Amer. Chem. Soc.*, **92**, 7474 (1970).
401. T. Cohen and J. H. Fager, *J. Amer. Chem. Soc.*, **87**, 5701 (1965).
402. T. Koenig, *Tetrahedron Lett.*, 2751 (1967).
403. D. M. Pretorius and P. A. deVilliers, *J. S. Afr. Chem. Inst.*, **18**, 48 (1965).
404. A. Klaebe and A. Lattes, *J. Chromatogr.*, **27**, 502 (1967).
405. T. Cohen, I. H. Song, and J. H. Fager, *Tetrahedron Lett.*, 237 (1965).
406. T. Cohen, I. H. Song, J. H. Fager, and G. L. Deets, *J. Amer. Chem. Soc.*, **89**, 4968 (1967).
407. C. Rüchardt, O. Krätz, and S. Eichler, *Chem. Ber.*, **102**, 3922 (1969).
408. J. F. Vozza, *J. Org. Chem.*, **27**, 3856 (1962).
409. H. J. den Hertog, D. J. Buurman, and P. A. deVilliers, *Rec. Trav. Chim. Pays-Bas,* **80**, 325 (1961).
410. S. Oae, T. Kitao, and Y. Kitaoka, *Tetrahedron,* **19**, 827 (1963).
411. T. Kato, Y. Goto, and Y. Yamamoto, *Yakugaku Zasshi,* **82**, 1649 (1962); *Chem. Abstr.*, **59**, 2765 (1963).
412. L. Bauer and T. E. Dickerhofe, *J. Org. Chem.*, **29**, 2183 (1964).
413. L. Bauer and A. L. Hirsch, *J. Org. Chem.*, **31**, 1210 (1966).
414. F. M. Hershenson and L. Bauer, *J. Org. Chem.*, **34**, 655 (1969).
415. R. A. Abramovitch and E. E. Knaus, *J. Heterocycl. Chem.*, **6**, 989 (1969).
416. K. Undheim, V. Nordal, and L. Borka, *Acta Chem. Scand.*, **23**, 2075 (1969).
417. Y. Nakai, N. Ohishi, S. Shimizu, and S. Fukui, *Bitamin,* **35**, 213 (1967).
418. K. Lewicka and E. Plazek, *Rec. Trav. Chim. Pays-Bas,* **78**, 644 (1959).
419. G. G. Spence, E. C. Taylor, and O. Buchardt, *Chem. Rev.*, **70**, 231 (1970).
420. J. Streith and C. Sigwalt, *Bull. Soc. Chim. Fr.*, 1157 (1970).
421. J. Streith, B. Danner, and C. Sigwalt, *Chem. Commun.*, 979 (1967).
422. J. Streith and C. Sigwalt, *Tetrahedron Lett.*, 1347 (1966).
423. F. J. Dinan and H. Tieckelmann, *J. Org. Chem.*, **29**, 1650 (1964).
424. R. Adams and S. Miyano, *J. Amer. Chem. Soc.*, **76**, 3168 (1954).
425. R. B. Greewald and C. L. Zirkle, *J. Org. Chem.*, **33**, 2118 (1968).
426. W. A. Lott and E. Shaw, *J. Amer. Chem. Soc.*, **71**, 70 (1949).
427. S. A. Safarov, *Kinet. Katal.*, **3**, 449 (1962); *Chem. Abstr.*, **58**, 2052 (1963).
428. B. Cercek and M. Ebert, *Trans. Faraday Soc.*, **63**, 1687 (1967).
429. L. G. Shevchuk, V. S. Zhikharev, and N. A. Vysotskaya, *Zh. Org. Khim.*, **5**, 1655 (1969); *Chem. Abstr.*, **72**, 2784 (1970).
430. E. K. Fields and S. Meyerson, *J. Org. Chem.*, **34**, 62 (1970).
431. Y. Koda, *Nagoya Kogyo Gijutsu Shikensho, Hokoku,* **15**, 155 (1966); *Chem. Abstr.*, **69**, 48876 (1968).
432. H. Loth and K. Eichner, *Arch. Pharm.* (Weinheim), **302**, 264 (1969).

433. K. Kanig, W. Koransky, G. Münch, and P. E. Schulze, *Arch. Exp. Pathol. Pharmakol.*, **249**, 43 (1964); *Chem. Abstr.*, **62**, 5752 (1965).
434. V. Neuhoff and F. Köhler, *Naturwissenschaften*, **52**, 475 (1965).
435. V. Neuhoff and F. Köhler, *Arch. Exp. Pathol. Pharmakol.*, **254**, 301 (1966); *Chem. Abstr.*, **65**, 11168 (1966).
436. T. Harris and V. Neuhoff, *Arch. Exp. Pathol. Pharmakol.*, **253**, 221 (1966); *Chem. Abstr.*, **64**, 18424 (1966).
437. V. Neuhoff and T. Harris, *Arch. Exp. Pathol. Pharmakol.*, **249**, 11 (1964); *Chem. Abstr.*, **62**, 8262 (1965).
438. V. Neuhoff and T. Harris, *Naturwissenschaften*, **51**, 290 (1964).
439. C. Houghton, K. A. Wright, and R. B. Cain, *Biochem. J.*, **106**, 51P (1968).
440. C. Houghton, G. K. Watson, and R. B. Cain, *Biochem. J.*, **114**, 75P (1969).
441. M. Shamma and P. D. Rosenstock, *J. Org. Chem.*, **26**, 718 (1961).
442. M. Shamma and P. D. Rosenstock, *J. Org. Chem.*, **26**, 2586 (1961).
443. O. Y. Magidson, *Zh. Obshch. Khim.*, **33**, 2173 (1963).
444. P. Cordier and E. Brändli, *C.R. Acad. Sci., Paris, Ser. C*, **258**, 4091 (1964).
445. A. I. Meyers and G. Garcia-Muñoz, *J. Org. Chem.*, **29**, 1435 (1964).
446. N. P. Shusherina, A. V. Golovin, and R. Ya. Levina, *Zh. Obshch. Khim.*, **30**, 1762 (1960); *Chem. Abstr.*, **55**, 7410 (1961).
447. N. P. Shusherina, K. Khua-min', and R. Ya. Levina, *Zh. Obshch. Khim.*, **33**, 2829 (1963); *Chem. Abstr.*, **60**, 4101 (1964).
448. E. Müller, R. Haller, and K. W. Merz, *Chem. Ber.*, **99**, 445 (1966).
449. N. Sugiyama, M. Yamamoto, and C. Kashima, *Bull. Chem. Soc. Jap.*, **42**, 2690 (1969).
450. Y. I. Chumakov and V. P. Sherstyuk, *Tetrahedron Lett.*, 771 (1967).
451. R. J. Highet, *J. Org. Chem.*, **29**, 471 (1964).
452. A. Albert, "Heterocyclic Chemistry," 2nd Ed., Athlone Press, London, 1968.
453. A. Albert and J. N. Phillips, *J. Chem. Soc.*, 1294 (1956).
454. A. Albert and G. B. Barlin, *J. Chem. Soc.*, 2384 (1959).
455. A. R. Katritzky, F. D. Popp, and J. D. Rowe, *J. Chem. Soc., B*, 562 (1966).
456. A. A. Gordon, A. R. Katritzky, and S. K. Roy, *J. Chem. Soc., B*, 556 (1968).
457. G. Simchen, *Chem. Ber.*, **103**, 398 (1970).
458. E. Spinner and G. B. Yeoh, *Tetrahedron Lett.*, 5691 (1968).
459. A. R. Katritzky and A. P. Ambler, "Physical Methods in Heterocyclic Chemistry," A. R. Katritzky, Ed., Vol. II, Academic Press, New York, 1963, p. 259.
460. B. D. Batts and E. Spinner, *Aust. J. Chem.*, **22**, 2581 (1969).
461. L. J. Bellamy and P. E. Rogasch, *Spectrochim. Acta*, **16**, 30 (1960).
462. R. A. Coburn and G. O. Dudek, *J. Phys. Chem.*, **72**, 3681 (1968).
463. G. H. Keller, L. Bauer, and C. L. Bell, *Can. J. Chem.*, **46**, 2475 (1968).
464. J. P. Shoffner, L. Bauer, and C. L. Bell, *J. Heterocycl. Chem.*, **7**, 479 (1970).
465. R. A. Coburn and G. O. Dudek, *J. Phys. Chem.*, **72**, 1177 (1968).
466. A. R. Katritzky and R. A. Jones, *J. Chem. Soc.*, 2947 (1960).
467. E. Spinner and J. C. B. White, *J. Chem. Soc., B*, 991 (1966).
468. H. G. Mautner, S. Chu, and C. M. Lee, *J. Org. Chem.*, **27**, 3671 (1962).
469. M. H. Krackov, C. M. Lee, and H. G. Mautner, *J. Amer. Chem. Soc.*, **87**, 892 (1965).
470. G. G. Hammes and H. O. Spivey, *J. Amer. Chem. Soc.*, **88**, 1621 (1966).
471. G. G. Hammes and A. C. Park, *J. Amer. Chem. Soc.*, **91**, 956 (1969).
472. R. H. Cox and A. A. Bothner-By, *J. Phys. Chem.*, **73**, 2465 (1969).
473. B. R. Penfold, *Acta Cryst.*, **6**, 591 (1953).
474. A. Kvick and I. Olovsson, *Ark. Kemi*, **30**, 71 (1969).

475. G. G. Hammes and P. J. Lillford, *J. Amer. Chem. Soc.*, **92**, 7578 (1970).
476. J. A. Elvidge and L. M. Jackman, *J. Chem. Soc.*, 859 (1961).
477. G. G. Hall, A. Hardisson, and L. M. Jackman, *Tetrahedron,* **19**, Suppl. 2, 101 (1963).
478. A. Albert and E. Spinner, *J. Chem. Soc.*, 1221 (1960).
479. H. H. Perkampus and U. Krüger, *Chem. Ber.*, **100**, 1165 (1967).
480. E. M. Evleth, Jr., J. A. Berson, and S. L. Manatt, *Tetrahedron Lett.*, 3087 (1964).
481. J. A. Berson, E. M. Evleth, Jr., and S. L. Manatt, *J. Amer. Chem. Soc.*, **87**, 2901 (1965).
482. H. Lumbroso and D. M. Bertin, *Bull. Soc. Chim. Fr.*, 1728 (1970).
483. R. A. Jones and A. R. Katritzky, *J. Chem. Soc.*, 3610 (1958).
484. S. F. Mason, *J. Chem. Soc.*, 5010 (1957).
485. W. E. Stewart and T. H. Siddall, III, *J. Phys. Chem.*, **74**, 2027 (1970).
486. R. A. Y. Jones, A. R. Katritzky, and L. M. Lagowski, *Chem. Ind.* (London), 870 (1960).
487. E. S. Levin and G. N. Rodionova, *Dokl. Akad. Nauk, SSSR,* **164**, 584 (1965); *Chem. Abstr.*, **64**, 4905 (1966).
488. E. S. Levin and G. N. Rodionova, *Dokl. Akad. Nauk, SSSR,* **172**, 607 (1967); *Chem. Abstr.*, **66**, 115110 (1967).
489. A. R. Katritzky, *Chimia,* **24**, 134 (1970).
490. A. Gordon and A. R. Katritzky, *Tetrahedron Lett.*, 2767 (1968).
491. L.N. Yakhontov, D. M. Krasnokutskaya, E. M. Peresleni, Yu. N. Sheinker, and M. V. Rubtsov, *Tetrahedron,* **22**, 3233 (1966).
492. L.N. Yakhontov, D. M. Krasnokutskaya, E. M. Peresleni, Yu. N. Sheinker, and M. V. Rubtsov, *Dokl. Akad. Nauk, SSSR,* **172**, 118 (1967); *Chem. Abstr.*, **66**, 104506 (1967).
493. L.N. Yakhontov, D. M. Krasnokutskaya, E. M. Peresleni, Yu. N. Sheinker, and M. V. Rubtsov, *Dokl. Akad. Nauk, SSSR,* **176**, 613 (1967); *Chem. Abstr.*, **68**, 78170 (1968).
494. E.M. Peresleni, L. N. Yakhontov, D. M. Krasnokutskaya, and Yu.N. Sheinker, *Dokl. Akad. Nauk, SSSR,* **177**, 592 (1967); *Chem. Abstr.*, **68**, 95634 (1968).
495. E. M. Peresleni, M. Ya. Utritskaya, V. A. Loginova, Yu. N. Sheinker, and L. N. Yakhontov, *Dokl. Akad. Nauk, SSSR,* **183**, 1102 (1968); *Chem. Abstr.*, **70**, 77831 (1969).
496. Yu. N. Sheinker, E. M. Peresleni, I. S. Rezchikova, and N. P. Zosimova, *Dokl. Akad. Nauk, SSSR,* **192**, 1295 (1970).
497. R. A. Jones and B. D. Roney, *J. Chem. Soc., B,* 84 (1967).
498. V. P. Zvolinskii, M. E. Perel'son, and Yu. N. Sheinker, *Dokl. Akad. Nauk, SSSR,* **179**, 1137 (1968); *Chem. Abstr.*, **69**, 51446 (1968).
499. E. M. Kosower, *J. Amer. Chem. Soc.*, **80**, 3253 (1958).
500. R. Haller, *Tetrahedron Lett.*, 3175 (1965).
501. A. Nakamura and S. Kamiya, *Chem. Pharm. Bull.* (Tokyo), **16**, 1466 (1968).
502. C. Wang, *J. Heterocycl. Chem.*, 7, 389 (1970).
503. H. J. den Hertog and D. J. Buurman, *Rec. Trav. Chim. Pays-Bas,* **75**, 257 (1956).
504. H. Sterk and H. Junek, *Monatsh. Chem.*, **98**, 1763 (1967).
505. S. Nesnow and R. Shapiro, *J. Org. Chem.*, **34**, 2011 (1969).
506. H.-J. Knackmuss, *Chem. Ber.*, **101**, 2679 (1968).
507. J. Pitha, *Collect. Czech. Chem. Commun.*, **28**, 1408 (1963).
508. J. R. Stevens, R. H. Beutel, and E. Chamberlin, *J. Amer. Chem. Soc.*, **64**, 1093 (1942).
509. E. Ritchie, *Aust. J. Chem.*, **9**, 244 (1956).

510. D. W. Jones, *J. Chem. Soc., C,* 1678 (1969).
511. Ref. 2, p. 617.
512. Y. V. Morozov, N. P. Bazhulina, V. I. Ivanov, M. Y. Karpeisky, and A. I. Kuklin, *Biofizika,* **10**, 595 (1965); *Chem. Abstr.,* **63**, 12527 (1965).
513. D. Heinert and A. E. Martell, *J. Amer. Chem. Soc.,* **81**, 3933 (1959).
514. V. P. Lezina, V. F. Bystrov, L. D. Smirnov, and K. M. Dyumaev, *Theor. Eksp. Khim.,* **4**, 379 (1968); *Chem. Abstr.,* **69**, 76370 (1968).
515. V. P. Lezina, L. D. Smirnov, K. M. Dyumaev, and V. F. Bystrov, *Izv. Akad. Nauk, SSSR, Ser. Khim.,* **25**, (1970); *Chem. Abstr.,* **72**, 116523 (1970).
516. J. S. Kwiatkowski, *Theor. Chim. Acta,* **16**, 243 (1970).
517. L. Paoloni, M. L. Tosato, and M. Cignitti, *Theor. Chim. Acta,* **14**, 221 (1969).
518. G. Berthier, B. Levy, and L. Paoloni, *Theor. Chim. Acta,* **16**, 316 (1970).
519. E. Suenaga, *Nippon Kagaku Zasshi,* **81**, 1710 (1960); *Chem. Abstr.,* **56**, 1059 (1962).
520. V. P. Lezina, V. F. Bystrov, L. D. Smirnov, and K. M. Dyumaev, *Theor. Eksp. Khim.,* **1**, 281 (1965); *Chem. Abstr.,* **63**, 13049 (1965).
521. D. Metzler and E. Snell, *J. Amer. Chem. Soc.,* 77, 2431 (1955).
522. H. C. Brown and X. Mihm, *J. Amer. Chem. Soc.,* 77, 1723 (1955).
523. I. I. Grandberg, G. K. Faizova, and A. N. Kost, *Khim. Geterotsikl. Soedin.,* **2**, 561 (1966); *Chem. Abstr.,* **66**, 10453 (1967).
524. O. A. Gansow and R. H. Holm, *Tetrahedron,* **24**, 4477 (1968).
525. J. W. Bridges, D. S. Davies, and R. T. Williams, *Biochem. J.,* **98**, 451 (1966).
526. Y. V. Morozov, N. P. Bazhulina, M. Y. Karpeisky, V. I. Ivanov, A. I. Kuklin, and Y. N. Breusov, *Bioenerg. Biol. Spectrofotom. Dokl. IV. Skets. Vses, Soveshch Upr. Biosin Biofiz. Pop. Krasnoyarsk. SSSR,* 179 (1965); *Chem. Abstr.,* **68**, 36204 (1968).
527. Y. V. Morozov, N. P. Bazhulina, and M. Y. Karpeisky. *Pyridoxal Catal: Enzymes, Model. Syst. Proc. Int. Symp.* 2nd Ed. 1966, p. 53; *Chem. Abstr.,* **70**, 101 (1969).
528. J. W. Bridges, P. J. Creaven, D. S. Davies, and R. T. Williams, *Biochem. J.,* **88**, 65P (1963).
529. D. D. Rosebrook and W. W. Brandt, *J. Phys. Chem.,* **70**, 3857 (1966).
530. P. Beak, J. Bonham, and J. T. Lee, Jr., *J. Amer. Chem. Soc.,* **90**, 1569 (1968).
531. P. Beak and J. Bonham, *Chem. Commun.,* 631 (1966).
532. P. Beak and J. T. Lee, Jr., *J. Org. Chem.,* **34**, 2125 (1969).
533. R. A. Abramovitch and G. N. Knaus, *J.C.S. Chem. Comm.,* 238 (1974).
534. E. Spinner, *J. Chem. Soc.,* 1226 (1960).
535. A. R. Katritzky and R. A. Y. Jones, *Chem. Ind.* (London), 722 (1961).
536. A. R. Katritzky and R. A. Y. Jones, *Proc. Chem. Soc.* (London), 313 (1960).
537. P. J. Van der Haak, and Th. J. deBoer, *Rec. Trav. Chim. Pays-Bas,* **83**, 186 (1964).
538. A. R. Katritzky and R. E. Reavill, *J. Chem. Soc.,* 753 (1963).
539. D. Cook, *Can. J. Chem.,* **41**, 2575 (1963).
540. D. Cook, *Can. J. Chem.,* **43**, 749 (1965).
541. D. Cook, *Can. J. Chem.,* **42**, 2292 (1964).
542. C. L. Bell, J. Shoffner, and L. Bauer, *Chem. Ind.* (London), 1435 (1963).
543. S. A. Mason, J. C. B. White, and A. Woodlock, *Tetrahedron Lett.,* 5219 (1969).
544. Yu. N. Sheinker and Y. I. Pomerantsev, *Russ. J. Phys. Chem.,* **33**, 174 (1959).
545. E. Spinner, *J. Chem. Soc.,* 1232 (1960).
546. E. Spinner and J. C. B. White, *J. Chem. Soc., B,* 996 (1966).
547. G. C. Hopkins, J. P. Jonak, H. J. Minnemeyer, and H. Tieckelmann, *J. Org. Chem.,* **32**, 4040 (1967).

548. G. C. Hopkins, J. P. Jonak, H. Tieckelmann, and H. J. Minnemeyer, *J. Org. Chem.*, **31**, 3969 (1966).
549. N. M. Chung and H. Tieckelmann, *J. Org. Chem.*, **35**, 2517 (1970).
550. K. R. Brower, R. L. Ernst, and J. S. Chen, *J. Phys. Chem.*, **68**, 3814 (1964).
551. N. P. Buu-Hoi, M. Gauthier, and N. D. Xuong, *Bull. Soc. Chim. Fr.*, 52 (1965).
552. Ref. 2, pp. 636–640.
553. K. Okon, G. Adamska, and W. Waclawek, *Rocz. Chem.*, **43**, 1653 (1969).
554. K. Winterfeld and M. Michael, *Chem. Ber.*, **93**, 61 (1960).
555. Ref. 2, p. 633.
556. T. Takahashi and A. Koshiro, *Chem. Pharm. Bull.* (Tokyo), **9**, 426 (1961).
557. B. I. Mikhant'ev, E. I. Fedorov, A. I. Kucherova, and V. P. Potapova, *Zh. Obshch. Khim.*, **29**, 1874 (1959); *Chem. Abstr.*, **54**, 8808 (1960).
558. J. Piechaczek and B. Gogolimska, *Acta Pol. Pharm.*, **25**, 254 (1968); *Chem. Abstr.*, **70**, 47245 (1969).
559. D. J. Cook, R. E. Bowen, P. Sorter, and E. Daniels, *J. Org. Chem.*, **26**, 4949 (1961).
560. S. Kajihara, *Nippon Kagaku Zasshi*, **86**, 933 (1965); *Chem. Abstr.*, **65**, 16936 (1966).
561. B. Gogolimska, *Acta Pol. Pharm.*, **21**, 363 (1964); *Chem. Abstr.*, **62**, 10404 (1965).
562. B. Gogolimska, *Acta Pol. Pharm.*, **25**, 391 (1968).
563. B. Gogolimska, *Acta Pol. Pharm.*, **25**, 397 (1968).
564. A. Raczka, A. Swirska, and H. Bojarska-Dahling, *Acta Pol. Pharm.*, **20**, 137 (1963); *Chem Abstr.*, **62**, 1630 (1965).
565. A. Raczka and F. Zahn, *Acta Pol. Pharm.*, **22**, 310 (1965); *Chem. Abstr.*, **63**, 13201 (1965).
566. A. Sacha, *Acta Pol. Pharm.*, **23**, 460 (1966); *Chem. Abstr.*, **66**, 104984 (1967).
567. A. Sacha, *Acta Pol. Pharm.*, **25**, 123 (1968).
568. N. A. Nelson and L. A. Paquette, *J. Org. Chem.*, **27**, 964 (1962).
569. Ref. 2, p. 641.
570. N. Kornblum and G. P. Coffey, *J. Org. Chem.*, **31**, 3447 (1966).
571. N. Kornblum and G. P. Coffey, *J. Org. Chem.*, **31**, 3449 (1966).
572. T. Severin, D. Bätz, and H. Lerche, *Chem. Ber.*, **103**, 1 (1970).
573. L. A. Paquette, *J. Org. Chem.*, **29**, 3545 (1964).
574. H. Kühmstedt and G. Wagner, *Arch. Pharm.* (Weinheim), **301**, 660 (1968).
575. S. Boatman, T. M. Harris, and C. R. Hauser, *J. Org. Chem.*, **30**, 3593 (1965).
576. S. Boatman, T. M. Harris, and C. R. Hauser, *J. Amer. Chem. Soc.*, **87**, 5198 (1965).
577. T. M. Harris and C. R. Hauser, *J. Org. Chem.*, **27**, 2967 (1962).
578. U. Basu, *J. Indian. Chem. Soc.*, **12**, 299 (1935).
579. R. E. Smith, S. Boatman, and C. R. Hauser, *J. Org. Chem.*, **33**, 2083 (1968).
580. R. L. Gay, S. Boatman, and C. R. Hauser, *Chem. Ind.* (London), 1789 (1965).
581. D. E. Ames and B. T. Warren, *J. Chem. Soc.*, Suppl. 1, 5518 (1964).
582. R. W. Balsiger, J. A. Montgomery, and T. P. Johnston, *J. Heterocycl. Chem.*, **2**, 97 (1965).
583. K. Undheim, V. Nordal, and K. Tjonneland, *Acta Chem. Scand.*, **23**, 1704 (1969).
584. P. Laland, J. O. Alvsaker, F. Haugli, J. Dedichen, S. Laland, and N. Thorsdalen, *Nature*, **210**, 917 (1966).
585. K. Undheim and L. Borka, *Acta Chem. Scand.*, **23**, 1715 (1969).
586. T. Takahashi and A. Koshiro, *Yakugaku Zasshi*, **79**, 1123 (1959); *Chem. Abstr.*, **54**, 3418 (1960).
587. J. Kejha, O. Radek, V. Jelinek, and O. Nemecek, *Cesk. Farm.*, **16**, 92 (1967); *Chem. Abstr.*, **67**, 108535 (1967).

588. J. P. Baker, A. R. Katritzky, and T. M. Moynehan, *J. Chem. Soc.*, Suppl., 6138 (1964).
589. L. N. Yakhontov, M. Y. Uritskaya, E. I. Lapan, and M. V. Rubtsov, *Khim. Geterotsikl. Soedin.*, 18 (1968); *Chem. Abstr.*, **69**, 106589 (1968).
590. L. N. Yakhontov, M. Y. Uritskaya, and M. V. Rubtsov, *Khim. Geterotsikl. Soedin.*, 918 (1965); *Chem. Abstr.*, **64**, 17563 (1966).
591. N. Takahashi, A. Suzuki, and S. Tamura, *Agr. Biol. Chem.* (Tokyo), **30**, 1 (1966); *Chem. Abstr.*, **64**, 12630 (1966).
592. S. L. Shapiro, K. Weinberg, and L. Freedman, *J. Amer. Chem. Soc.*, **81**, 5140 (1959).
593. K. Mecklenborg and M. Orchin, *J. Org. Chem.*, **23**, 1591 (1958).
594. D. A. Prins, *Rec. Trav. Chim. Pays-Bas,* **76**, 58 (1957).
595. M. P. Mertes, R. F. Borne, and L. E. Hare, *J. Heterocycl. Chem.*, **5**, 281 (1968).
596. P. Nedenskov, N. Clauson-Kaas, J. Lei, H. Heide, G. Olsen, and G. Jansen, *Acta Chem. Scand.*, **23**, 1791 (1969).
597. K. Undheim, P. O. Tveita, L. Borka, and V. Nordal, *Acta Chem. Scand.*, **23**, 2065 (1969).
598. R. B. Moffett, *J. Org. Chem.*, **28**, 2885 (1963).
599. D. Pitre and L. Fumagalli, *Farmaco, Ed. Sci.*, **18**, 495 (1963); *Chem. Abstr.*, **59**, 9971 (1963).
600. R. Tschesche, D. Klöden, and H. W. Fehlhaber, *Tetrahedron,* **20**, 2885 (1964).
601. H. Pischel and G. Wagner, *Z. Chem.*, **5**, 227 (1965).
602. H. Pischel and G. Wagner, *Arch. Pharm.* (Weinheim), **300**, 602 (1967).
603. G. Wagner and H. Pischel, *Arch. Pharm.* (Weinheim), **295**, 373 (1962).
604. G. Wagner and H. Frenzel, *Arch. Pharm.* (Weinheim), **299**, 536 (1966).
605. G. Wagner and E. Fickweiler, *Arch. Pharm.* (Weinheim), **298**, 62 (1965).
606. G. Wagner and H. Gentzsch, *Z. Chem.*, **7**, 310 (1967).
607. P. Nuhn, A. Zschunke, and G. Wagner, *Z. Chem.*, **9**, 335 (1969).
608. G. Wagner and E. Fickweiler, *Arch. Pharm.* (Weinheim), **298**, 297 (1965).
609. D. Thacker and T. L. V. Ulbricht, *Chem. Commun.*, 122 (1967).
610. G. Wagner and H. Gentzsch, *Arch. Pharm.* (Weinheim), **301**, 346 (1968).
611. D. Heller and G. Wagner, *Z. Chem.*, **8**, 415 (1968).
612. D. Heller, *Pharmazie,* **24**, 580 (1969).
613. D. Orth and T. Wieland, *Chem. Ber.*, **102**, 196 (1969).
614. H. Pischel and G. Wagner, *Z. Chem.*, **7**, 15 (1967).
615. H. Pischel and G. Wagner, *Arch. Pharm.* (Weinheim), **300**, 856 (1967).
616. L. Birkhofer, A. Ritter, and H.-P. Kühlthau, *Chem. Ber.*, **97**, 934 (1964).
617. M. J. Robins and B. L. Currie, *Chem. Commun.*, 1547 (1968).
618. M. P. Mertes, J. Zielinski, and C. Pillar, *J. Med. Chem.*, **10**, 320 (1967).
619. P. Nuhn, A. Zschunke, D. Heller, and G. Wagner, *Tetrahedron,* **25**, 2139 (1969).
620. G. Wagner and H. Pischel, *Naturwissenschaften,* **48**, 454 (1961).
621. H. Pischel and G. Wagner, *Pharmazie,* **25**, 45 (1970).
622. M. P. Mertes, *J. Med. Chem.*, **13**, 149 (1970).
623. G. Wagner and H. Pischel, *Arch. Pharm.* (Weinheim), **295**, 897 (1962).
624. G. Wagner and H. Gentzsch, *Arch. Pharm.* (Weinheim), **301**, 201 (1968).
625. T. Ukita, R. Funakoshi, and Y. Hirose, *Chem. Pharm. Bull.* (Tokyo), **12**, 828 (1964).
626. G. Wagner and H. Pischel, *Arch. Pharm.* (Weinheim), **296**, 699 (1963).
627. H. Pischel and A. Holy, *Collect. Czech. Chem. Commun.*, **34**, 89 (1969).
628. H. Pischel and A. Holy, *Collect. Czech. Chem. Commun.*, **33**, 2066 (1968).

629. Ref. 2, pp. 643–645.
630. D. Y. Curtin and L. L. Miller, *J. Amer. Chem. Soc.*, **89**, 637 (1967).
631. A. McKillop, M. J. Zelesko, and E. C. Taylor, *Tetrahedron Lett.*, 4945 (1968).
632. A. W. K. Chan, W. D. Crow, and I. Gosney, *Tetrahedron*, **26**, 2497 (1970).
633. I. Fleming and D. Philippides, *J. Chem. Soc., C*, 2426 (1970).
634. F. Arndt and A. Kalischek, *Chem. Ber.*, **63**, 587 (1930).
635. Ref. 2, pp. 645, 677.
636. J. I. G. Cadogan, *J. Chem. Soc.*, 2844 (1959).
637. Y. Ueno, T. Takaya, and E. Imoto, *Bull. Chem. Soc. Jap.*, **37**, 864 (1964).
638. Y. Ueno, S. Asakawa, and E. Imoto, *Nippon Kagaku Zasshi*, **89**, 101 (1968).
639. K. Yokota, M. Sasaki, and Y. Ishii, *J. Polym. Sci.*, **6**, 2935 (1968).
640. W. Kampe, *Angew. Chem. Intern. Edit.*, **2**, 479 (1963).
641. N. P. Buu-Hoï, M. Gauthier, and N. D. Xuong, *Bull. Soc. Chim. Fr.*, 52 (1965).
642. A. Koshiro, *Chem. Pharm. Bull.* (Tokyo), **7**, 725 (1959).
643. T. Takahashi and A. Koshiro, *Chem. Pharm. Bull.* (Tokyo), **7**, 720 (1959).
644. N. S. Vul'fson, G. M. Sukhotina, and V. N. Zheltova, *Izv. Akad. Nauk SSSR, Ser. Khim.*, 887 (1966); *Chem. Abstr.*, **65**, 10556 (1966).
645. N. S. Vul'fson, G. M. Sukhotina, and L. B. Senyavina, *Izv. Akad. Nauk SSSR, Ser. Khim.*, 1605 (1966).
646. Ref. 2, pp. 643–644.
647. R. C. DeSelms, *J. Org. Chem.*, **33**, 478 (1968).
648. A. Jerfy and A. B. Roy, *Aust. J. Chem.*, **23**, 847 (1970).
649. S. Ginsburg, *J. Med. Pharm. Chem.*, **5**, 1364 (1962).
650. W. Korytnyk, H. Ahrens, and N. Angelino, *Tetrahedron*, **26**, 5415 (1970).
651. B. Rzeszotarska and G. Palka, *Bull. Acad. Pol. Sci.*, **16**, 23 (1968); *Chem. Abstr.*, **69**, 59542 (1968).
652. E. Taschner, B. Rzeszotarska, and L. Lubiewska, *Angew. Chem.*, **77**, 619 (1965).
653. E. Taschner, B. Rzeszotarska, and L. Lubiewska, *Ann. Chem.*, **690**, 177 (1965).
654. V. A. Shibnev, K. M. Dyumaev, T. P. Chuvaeva, L. D. Smirnov, and K. T. Poroshin, *Izv. Akad. Nauk SSSR, Ser. Khim.*, 1634 (1967); *Chem. Abstr.*, **68**, 22235 (1968).
655. D. T. Elmore, D. V. Roberts, and J. J. Smyth, *Biochem. J.*, **102**, 728 (1967).
656. D. T. Elmore and J. J. Smyth, *Biochem. J.*, **107**, 103 (1968).
657. D. T. Elmore and J. J. Smyth, *Proc. Chem. Soc.* (London), 18 (1963).
658. W. Kampe, *Chem. Ber.*, **98**, 1031 (1965).
659. J. Larrouquere, *Bull. Soc. Chim. Fr.*, 329 (1968).
660. W. Kampe, *Chem. Ber.*, **98**, 1038 (1965).
661. W. Kampe, *Chem. Ber.*, **99**, 593 (1966).
662. K.-H. Scheit and W. Kampe, *Chem. Ber.*, **98**, 1045 (1965).
663. B. Miller, *J. Amer. Chem. Soc.*, **84**, 403 (1962).
664. P. Nantka-Namirski, *Acta Pol. Pharm.*, **18**, 391 (1961); *Chem. Abstr.*, **57**, 16553 (1962).
665. G. C. Finger and L. D. Starr, *J. Amer. Chem. Soc.*, **81**, 2674 (1959).
666. T. Kametani, K. Ogasawara, and T. Yamanaka, *J. Chem. Soc., C*, 1006 (1968).
667. A. Albert and G. B. Barlin, *J. Chem. Soc.*, 5156 (1963).
668. J. W. Clark-Lewis and R. P. Singh, *J. Chem. Soc.*, 2379 (1962).
669. A. Roedig, K. Grohe, D. Klatt, and H.-G. Kleppe, *Chem. Ber.*, **99**, 2813 (1966).
670. T. Kato, H. Hayashi, and T. Anzai, *Yakugaku Zasshi*, **87**, 387 (1967); *Chem. Abstr.*, **67**, 64211 (1967).
671. H. C. van der Plas and H. J. den Hertog, *Rec. Trav. Chim. Pays-Bas*, **81**, 841 (1962).
672. S. Portnoy, *J. Heterocycl. Chem.*, **6**, 223 (1969).

673. M. Satomi, M. Hasegawa, Y. Osawa, and K. Yamamoto, *Nippon Daigaku Yakugaku Kenkyu Hokoku,* **5–6**, 55 (1963); *Chem. Abstr.,* **61**, 13275 (1964).

674. J. L. Greene, Jr., and J. A. Montgomery, *J. Med. Chem.,* **7**, 17 (1964).

675. W. Czuba, *Rocz. Chem.,* **41**, 479 (1967).

676. A. Signor, E. Scoffone, L. Biondi, and S. Bezzi, *Gazz. Chim. Ital.,* **93**, 65 (1963); *Chem. Abstr.,* **59**, 2811 (1963).

677. Ref. 2, p. 649.

678. D. F. DeTar, *Org. React.,* **9**, 407 (1957).

679. R. A. Abramovitch, *Adv. Free Radical Chem.,* **2**, 87 (1967).

680. Y. Ban, R. Sakaguchi, and M. Nagai, *Chem. Pharm. Bull.* (Tokyo), **13**, 931 (1965).

681. Ref. 2, pp. 650, 674–675.

682. R. H. Wiley, N. R. Smith, and L. H. Knabeschuh, *J. Amer. Chem. Soc.,* **75**, 4482 (1953).

683. S. Sugasawa, S. Akaboshi, and Y. Ban, *Chem. Pharm. Bull.* (Tokyo), **7**, 263 (1959).

684. Y. Ban, O. Yonemitsu, T. Oishi, S. Yokoyama, and M. Nakagawa, *Chem. Pharm. Bull.* (Tokyo), **7**, 609 (1959).

685. R. F. Evans and H. C. Brown, *J. Org. Chem.,* **27**, 1329 (1962).

686. E. Ager, B. Iddon, and H. Suschitzky, *J. Chem. Soc., C,* 193 (1970).

687. M. S. Newman and H. A. Karnes, *J. Org. Chem.,* **31**, 3980 (1966).

688. W. A. Roelfsema and H. J. den Hertog, *Tetrahedron Lett.,* 5089 (1967).

689. B. Gogolimska, *Acta Pol. Pharm.,* **25**, 248 (1968); *Chem. Abstr.,* **70**, 37621 (1969).

690. M. Ferles and J. Caplovic, *Collect. Czech. Chem. Commun.,* **28**, 1434 (1963).

691. Y. L. Gol'dfarb and V. V. Kiseleva, *Izv. Akad. Nauk SSSR, Otd. Khim. Nauk,* 2208 (1960); *Chem. Abstr.,* **55**, 15481 (1961).

692. K. H. Büchel, A. K. Bocz, and F. Korte, *Chem. Ber.,* **99**, 724 (1966).

693. M. Ferles and M. Holik, *Collect. Czech. Chem. Commun.,* **31**, 2416 (1966).

694. M. Holik, A. Tesarova, and M. Ferles, *Collect. Czech. Chem. Commun.,* **32**, 1730 (1967).

695. M. Holik and M. Ferles, *Collect. Czech. Chem. Commun.,* **32**, 2288 (1967).

696. M. Ferles and H. Hruba, *Z. Chem.,* **9**, 450 (1969).

697. R. Lukes, Z. Koblicova, and K. Blaha, *Collect. Czech. Chem. Commun.,* **28**, 2182 (1963).

698. R. A. Abramovitch and J. G. Saha, *Adv. Heterocycl. Chem.,* **6**, 253 (1966).

699. A. Albert, *J. Chem. Soc.,* 1020 (1960).

700. P. J. Brignell, A. R. Katritzky, and H. O. Tarhan, *J. Chem. Soc., B,* 1477 (1968).

701. A. R. Katritzky, H. O. Tarhan, and S. Tarhan, *J. Chem. Soc., B,* 114 (1970).

702. Ref. 2, p. 659.

703. R. H. Rigterink and E. E. Kenaga, *J. Agr. and Food Chem.,* **14**, 304 (1966); *Chem. Abstr.,* **65**, 681 (1966).

704. F. W. Broekman and H. J. C. Tendeloo, *Rec. Trav. Chim. Pays-Bas,* **81**, 107 (1962).

705. K. Clarke and K. Rothwell, *J. Chem. Soc.,* 1885 (1960).

706. M. A.-F. Elkaschef and M. H. Nosseir, *J. Amer. Chem. Soc.,* **82**, 4344 (1960).

707. Th. Kappe, G. Baxevanidis, and E. Ziegler, *Monatsh. Chem.,* **100**, 1715 (1969).

708. K. Lewicka and E. Plazek, *Rocz. Chem.,* **40**, 405 (1966).

709. L. D. Smirnov, V. I. Kuz'min, V. P. Lezina, and K. M. Dyumaev, *Izv. Akad. Nauk SSSR, Ser. Khim,* 2400 (1970); *Chem. Abstr.,* **74**, 141464 (1971).

710. L. D. Smirnov, V. I. Kuz'min, V. P. Lezina, and K. M. Dyumaev, *Izv. Akad. Nauk SSSR, Ser. Khim.,* 1897 (1970); *Chem. Abstr.,* **74**, 64184 (1971).

711. L. D. Smirnov, V. I. Kuz'min, V. P. Lezina, and K. M. Dyumaev, *Izv. Akad. Nauk SSSR, Ser. Khim.,* 2385 (1970); *Chem. Abstr.,* **74**, 141466 (1971).

712. H. N. M. van der Lans and H. J. den Hertog, *Rec. Trav. Chim. Pays-Bas,* **87**, 549 (1968).
713. W. A. Waters, *J. Chem. Soc.*, 727 (1948).
714. L. D. Smirnov, R. E. Lokhov, V. P. Lezina, B. E. Zaitsev, and K. M. Dyumaev, *Izv. Akad. Nauk SSSR, Ser. Khim.*, 1567 (1969); *Chem. Abstr.*, **71**, 112765 (1969).
715. E. Plazek, *Rec. Trav. Chim. Pays-Bas,* **72**, 569 (1953).
716. C. O. Okafor, *J. Org. Chem.*, **32**, 2006 (1967).
717. W. Czuba and E. Plazek, *Rec. Trav. Chim. Pays-Bas,* **77**, 92 (1958).
718. L. D. Smirnov, V. P. Lezina, B. E. Zaitsev, and K. M. Dyumaev, *Izv. Akad. Nauk SSSR, Ser. Khim.*, 1652 (1968); *Chem. Abstr.*, **70**, 11516 (1969).
719. L. D. Smirnov, S. L. Orlova, V. P. Lezina, T. P. Kartasheva, and K. M. Dyumaev, *Izv. Akad. Nauk SSSR, Ser. Khim.*, 2752 (1967); *Chem. Abstr.*, **69**, 86782 (1968).
720. L. D. Smirnov, V. P. Lezina, V. F. Bystrov, and K. M. Dyumaev, *Izv. Akad. Nauk SSSR, Ser. Khim.*, 198 (1965); *Chem. Abstr.*, **62**, 11774 (1965).
721. L. D. Smirnov, V. P. Lezina, V. F. Bystrov, and K. M. Dyumaev, *Izv. Akad. Nauk SSSR, Ser. Khim.*, 1836 (1965); *Chem. Abstr.*, **64**, 5038 (1966).
722. L. D. Smirnov, S. L. Orlova, V. P. Lezina, and K. M. Dyumaev, *Izv. Akad. Nauk SSSR, Ser. Khim.*, 1816 (1967); *Chem. Abstr.*, **68**, 87111 (1968).
723. L. D. Smirnov, V. P. Lezina, V. F. Bystrov, and K. M. Dyumaev, *Izv. Akad. Nauk SSSR, Ser. Khim.*, 752 (1963); *Chem. Abstr.*, **59**, 7474 (1963).
724. K. M. Dyumaev, L. D. Smirnov, and V. F. Bystrov, *Izv. Akad. Nauk SSSR, Ser. Khim.*, 883 (1962); *Chem. Abstr.*, **57**, 12424 (1962).
725. L. D. Smirnov, T. P. Kartasheva, V. P. Lezina, and K. M. Dyumaev, *Izv. Akad. Nauk SSSR, Ser. Khim.*, 2742 (1967); *Chem. Abstr.*, **69**, 67191 (1968).
726. L. D. Smirnov, V. P. Lezina, T. P. Kartasheva, and K. M. Dyumaev, *Izv. Akad. Nauk SSSR, Ser. Khim.*, 198 (1968); *Chem. Abstr.*, **69**, 77086 (1968).
726a. L. D. Smirnov, Personal Communication (see ref. 726).
727. L. D. Smirnov, V. I. Kuz'min, V. P. Lezina, and K. M. Dyumaev, *Izv. Akad. Nauk SSSR, Ser. Khim.*, 2784 (1970); *Chem. Abstr.*, **74**, 111873 (1971).
728. K. M. Dyumaev, L. D. Smirnov, V. P. Lezina, and V. F. Bystrov, *Teor. Eksp. Khim.*, **1**, 290 (1965); *Chem. Abstr.*, **63**, 13049 (1965).
729. H. Tomisawa, Y. Kobayashi, H. Hongo, and R. Fujita, *Chem. Pharm. Bull.* (Tokyo), **18**, 932 (1970).
730. B. S. Thyagarajan and K. Rajagopalan, *Tetrahedron,* **19**, 1483 (1963).
731. P. Beak and J. Bonham, *Tetrahedron Lett.*, 3083 (1964).
732. P. Beak and E. M. Monroe, *J. Org. Chem.*, **34**, 589 (1969).
733. P. Bellingham, C. D. Johnson, and A. R. Katritzky, *Chem. Commun.*, 1047 (1967).
734. P. Bellingham, C. D. Johnson, and A. R. Katritzky, *Chem. Ind.* (London), 1384 (1964).
735. P. Bellingham, C. D. Johnson, and A. R. Katritzky, *J. Chem. Soc., B,* 1226 (1967).
736. Y. Kawazoe and Y. Yoshioka, *Chem. Pharm. Bull.* (Tokyo), **16**, 715 (1968).
737. N. S. Vul'fson and G. M. Sukhotina, *Izv. Akad. Nauk SSSR, Ser. Khim.*, 1785 (1966); *Chem. Abstr.*, **66**, 94944 (1967).
738. J. Barycki and E. Plazek, *Rocz. Chim.*, **38**, 553 (1964).
739. A. D. Palekar and R. A. Kulkarni, *Curr. Sci.*, **37**, 229 (1968).
740. B. C. Pathak, S. P. Dutta, and T. J. Bardos, *J. Heterocycl. Chem.*, **6**, 447 (1969).
741. K. M. Dyumaev, R. E. Lokhov, B. E. Zaitsev, and L. D. Smirnov, *Izv. Akad. Nauk SSSR, Ser. Khim.*, 2601 (1970); *Chem. Abstr.*, **74**, 113181 (1971).
742. Ref. 2, pp. 673–674.
743. H. Bojarska-Dahlig and I. Gruda, *Rocz. Chem.*, **33**, 505 (1959).
744. R. A. Abramovitch and A. D. Notation, *Can. J. Chem.*, **38**, 1445 (1960).

745. B. S. Thyagarajan, K. Rajagopalan, and P. V. Gopalakrishnan, *J. Chem. Soc., B,* 300 (1968).
746. L. A. Paquette, *J. Org. Chem.,* **30**, 2107 (1965).
747. B. Weinstein and D. N. Brattesani, *J. Org. Chem.,* **32**, 4107 (1967).
748. R. M. Acheson and P. A. Taskor, *J. Chem. Soc., C,* 1542 (1967).
749. N. P. Shusherina, O. V. Slavyanova and R. Ya. Levina, *Zh. Obshch. Khim.,* **39**, 1182 (1969); *Chem. Abstr.,* **71**, 70466 (1969).
750. L. Bauer, C. L. Bell, and G. E. Wright, *J. Heterocycl. Chem.,* **3**, 393 (1966).
751. E. B. Sheinin, G. E. Wright, C. L. Bell, and L. Bauer, *J. Heterocycl. Chem.,* **5**, 859 (1968).
752. H. Tomisawa and H. Hongo, *Tetrahedron Lett.,* 2465 (1969).
753. H. Tomisawa and H. Hongo, *Chem. Pharm. Bull.* (Tokyo), **18**, 925 (1970).
754. H. Tomisawa, R. Fujita, K. Noguchi, and H. Hongo, *Chem. Pharm. Bull.* (Tokyo), **18**, 941 (1970).
755. A. R. Katritzky and Y. Tacheuchi, *J. Amer. Chem. Soc.,* **92**, 4134 (1970).
756. E. C. Taylor and W. W. Paudler, *Tetrahedron Lett.,* 1 (1960).
757. W. A. Ayer, R. Hayatsu, P. de Mayo, S. T. Reid, and J. B. Stothers, *Tetrahedron Lett.,* 648 (1961).
758. G. Slomp, F. A. MacKeller, and L. A. Paquette, *J. Amer. Chem. Soc.,* **83**, 4472 (1961).
759. M. Laing, *Proc. Chem. Soc.* (London), 343 (1964).
760. E. C. Taylor, R. O. Kan, and W. W. Paudler, *J. Amer. Chem. Soc.,* **83**, 4484 (1961).
761. E. C. Taylor and R. O. Kan, *J. Amer. Chem. Soc.,* **85**, 776 (1963).
762. L. A. Paquette and G. Slomp, *J. Amer. Chem. Soc.,* **85**, 765 (1963).
763. A. Fozard and C. K. Bradsher, *J. Org. Chem.,* **32**, 2966 (1967).
764. E. J. Corey and J. Streith, *J. Amer. Chem. Soc.,* **86**, 950 (1964).
765. C. Kashima, M. Yamamoto, Y. Sato, and N. Sugiyama, *Bull. Chem. Soc. Jap.,* **42**, 3596 (1969).
766. A. Senning, *Acta Chem. Scand.,* **18**, 269 (1964).
767. M. Takeda, A. E. Jacobson, K. Kanematsu, and E. L. May, *J. Org. Chem.,* **34**, 4154 (1969).
768. M. Forchiassin, G. Illuminati, and G. Sleiter, *J. Heterocycl. Chem.,* **6**, 879 (1969).
769. C. Abbolito, C. Iavarone, G. Illuminati, F. Stegel, and A. Vazzoler, *J. Amer. Chem. Soc.,* **91**, 6746 (1969).
770. M. Vogel and J. D. Roberts, *J. Amer. Chem. Soc.,* **88**, 2262 (1966).
771. R. Foster and C. A. Fyfe, *Rev. Pure Appl. Chem.,* **16**, 61 (1966).
772. C. A. Fyfe, *Tetrahedron Lett.,* 659 (1968).
773. M. E. C. Biffin, J. Miller, A. G. Moritz, and D. B. Paul, *Aust. J. Chem.,* **23**, 963 (1970).
774. G. Illuminati and F. Stegel, *Tetrahedron Lett.,* 4169 (1968).
775. J. E. Dickeson, L. K. Dyall, and V. A. Pickles, *Aust. J. Chem.,* **21**, 1267 (1968).
776. P. Bemporad, G. Illuminati, and F. Stegel, *J. Amer. Chem. Soc.,* **91**, 6742 (1969).
777. M. E. C. Biffin, J. Miller, A. G. Moritz, and D. B. Paul, *Aust. J. Chem.,* **23**, 957 (1970).
778. R. A. Fernandez, H. Heaney, J. M. Jablonski, K. G. Mason, and T. J. Ward, *J. Chem. Soc., C,* 1908 (1969).
779. J. D. Cook and B. J. Wakefield, *J. Chem. Soc., C,* 2376 (1969).
780. W. G. Duncan and D. W. Henry, *J. Med. Chem.,* **11**, 909 (1968).
781. K. B. Wiberg, T. M. Shryne, and R. R. Kintner, *J. Amer. Chem. Soc.,* **79**, 3160 (1957).

782. L. A. Cohen and B. Witkop, in "Molecular Rearrangements", P. de Mayo, Ed., Vol. 2, Interscience, New York, 1964, p. 981.
783. D. J. Brown and R. V. Foster, *J. Chem. Soc.*, 4911 (1965).
784. D. J. Brown, "The Pyrimidines," Suppl. I, Interscience, New York, 1970, pp. 280–282.
785. R. G. Hiskey and J. Hollander, *J. Org. Chem.*, **29**, 3687 (1964).
786. C. K. Bradsher and M. F. Zinn, *J. Heterocycl. Chem.*, **1**, 219 (1964).
787. C. K. Bradsher and M. F. Zinn, *J. Heterocycl. Chem.*, **4**, 66 (1967).
788. E. Winterfeldt, *Chem. Ber.*, **97**, 2463 (1964).
789. T. Wieland, C. Fest, and G. Pfleiderer, *Ann. Chem.*, **642**, 163 (1961).
790. F. J. Dinan and H. Tieckelmann, *J. Org. Chem.*, **29**, 892 (1964).
791. H. F. Stewart and R. P. Seibert, *J. Org. Chem.*, **33**, 4560 (1968).
792. A. F. Thomas and A. Marxer, *Helv. Chim. Acta,* **41**, 1898 (1958).
793. A. F. Thomas and A. Marxer, *Helv. Chim. Acta,* **43**, 469 (1960).
794. I. Ichimoto, K. Fujii, and C. Tatsumi, *Agr. Biol. Chem.* (Tokyo), **31**, 979 (1967).
795. R. Kotani and C. Tatsumi, *Bull. Univ. Osaka Prefect., Ser. B,* **10,** 33 (1960); *Chem. Abstr.*, **58,** 11318 (1963).
796. R. G. Jones and M. J. Mann, *J. Amer. Chem. Soc.*, **75**, 4048 (1953).
797. C. Ainsworth and R. G. Jones, *J. Amer. Chem. Soc.*, **76**, 3172 (1954).
798. S. S. Deshapande, Y. V. Dingankar, and D. N. Kopil, *J. Indian Chem. Soc.*, **11**, 595 (1934).
799. D. N. Bedekar, R. P. Kaushal, and S. S. Deshapande, *J. Indian Chem. Soc.*, **12**, 465 (1935).
800. S. W. Nakhre and S. S. Deshapande, *Vikram J., Vikram Univ.*, **4**, 12 (1961).
801. S. W. Nakhre and S. S. Deshapande, *Vikram J., Vikram Univ.*, **5**, 16 (1961).
802. I. E. El-Kholy and F. K. Rafla, *J. Chem. Soc., C,* 974 (1969).
803. V. A. Volkova and M. I. Goryaev, *Vestn. Akad. Nauk Kaz. SSR,* **17**, 38 (1961); *Chem. Abstr.*, **55**, 24743 (1961).
804. J. A. Moore, F. J. Marascia, R. W. Medeiros, and R. L. Wineholt, *J. Org. Chem.*, **31**, 34 (1966).
805. M. A. Robinson and T. J. Hurley, *J. Inorg. Nucl. Chem.*, **28**, 1747 (1966).
806. P. R. Rony, *J. Amer. Chem. Soc.*, **90**, 2824 (1968).
807. P. R. Rony, *J. Amer. Chem. Soc.*, **91**, 6090 (1969).
808. P. R. Rony, W. E. McCormack, and S. W. Wunderly, *J. Amer. Chem. Soc.*, **91**, 4244 (1969).
809. E. Sacher and K. J. Laidler, *Can. J. Chem.*, **42**, 2404 (1964).
810. H. C. Beyerman and W. M. van den Brink, *Proc. Chem. Soc.* (London), 266 (1963).
811. H. C. Beyerman, W. M. van den Brink, F. Weygand, A. Prox, W. König, L. Schmidhammer, and E. Nintz, *Rec. Trav. Chim. Pays-Bas,* **84**, 213 (1965).
812. B. Liberek and A. Michalik, *Rocz. Chem.*, **40**, 781 (1966).
813. F. Pietra and D. Vitali, *Tetrahedron Lett.*, 5701 (1966).
814. W. C. Kuryla, *J. Appl. Polym. Sci.*, **9**, 1019 (1965).
815. T. J. Wallace and J. J. Mahon, *J. Org. Chem.*, **30**, 1502 (1965).
816. C. C. Houk and K. Emerson, *J. Inorg. Nucl. Chem.*, **30**, 1493 (1968).
817. I. Baxter and G. A. Swan, *J. Chem. Soc.*, 3011 (1965).
818. D. Cook, *Can. J. Chem.*, **43**, 741 (1965).
819. D. Cook, *Can. J. Chem.*, **41**, 515 (1963).
820. E. S. Gould, *J. Amer. Chem. Soc.*, **89**, 5792 (1967).
821. E. S. Gould, *J. Amer. Chem. Soc.*, **90**, 1740 (1968).
822. J. Reedijk, *Rec. Trav. Chim. Pays-Bas,* **88**, 1139 (1969).
823. L. J. Boucher and J. C. Bailar, Jr., *J. Inorg. Nucl. Chem.*, **27**, 1093 (1965).

824. J. R. Wasson, K. K. Ganguli, and L. Theriot, *J. Inorg. Nucl. Chem.*, **29**, 2807 (1967).
825. D. P. Goel, Y. A. G. Dutt, and R. P. Singh, *J. Inorg. Nucl. Chem.*, **34**, 3119 (1970).
826. G. Anderegg, *Helv. Chim. Acta,* **46**, 1011 (1963).
827. E. Suenaga, *Nippon Kagaku Zasshi,* **79**, 1551 (1958); *Chem. Abstr.*, **54**, 1515 (1960).
828. S. D. Paul, M. S. Krishnan, and C. Dass, *Indian J. Chem.*, **7**, 299 (1969).
829. B. W. Dale, R. J. P. Williams, P. R. Edwards, and C. E. Johnson, *Trans. Faraday Soc.*, **64**, 620 (1968).
830. R. M. Ismail, *J. Organometal. Chem.*, **6**, 663 (1966).
831. D. N. Kravtsov, E. M. Rokhlina, and A. N. Nesmeyanov, *Izv. Akad. Nauk SSSR, Ser. Khim,* 1035 (1968); *Chem. Abstr.*, **70**, 4254 (1969).
832. W. R. Rimm, D. O. Johnston, C. H. Oestreich, D. G. Lambert, and M. M. Jones, *J. Inorg. Nucl. Chem.*, **29**, 2401 (1967).
833. C. W. McDonald and J. H. Bedenbaugh, *Anal. Chem.*, **39**, 1476 (1967).
834. C. W. McDonald and R. Carter, *Anal. Chem.*, **41**, 1478 (1969).
835. C. W. McDonald and J. H. Bedenbaugh, *Mikrochim. Acta,* 474 (1970).
836. C. W. McDonald and J. H. Bedenbaugh, *Mikrochim. Acta,* 612 (1970).
837. M. J. Herak, M. Janko, and K. Blazevic, *Croat. Chem. Acta,* **41**, 85 (1969); *Chem. Abstr.*, **72**, 6659 (1970).
838. L. H. Toporcer, R. E. Dessy, and S. I. E. Green, *Inorg. Chem.*, **4**, 1649 (1965).
839. W. Selig, *Explosivstoffe,* **15**, 76 (1967); *Chem. Abstr.*, **67**, 108015 (1967).
840. E. J. Behrman and P. P. Walker, *J. Amer. Chem. Soc.*, **84**, 3454 (1962).
841. E. J. Behrman, *J. Amer. Chem. Soc.*, **85**, 3478 (1963).
842. J. H. Boyer and S. Kruger, *J. Amer. Chem. Soc.*, **79**, 3552 (1957).
843. H.-J. Knackmuss, *Chem. Ber.*, **101**, 1148 (1968).
844. N. L. Weinberg and E. A. Brown, *J. Org. Chem.*, **31**, 4054 (1966).
845. D. E. Ames, R. E. Bowman, and T. F. Grey, *J. Chem. Soc.*, 3008 (1953).
846. H.-J. Knackmuss, *J. Heterocycl. Chem.*, **7**, 733 (1970).
847. Ref. 2, pp. 715–719.
848. P. F. Juby and L. Marion, *Biochem. Biophys. Res. Commun.*, **5**, 461 (1961).
849. P. F. Juby and L. Marion, *Can. J. Chem.*, **41**, 117 (1963).
850. G. R. Waller and L. M. Henderson, *Biochem. Biophys. Res. Commun.*, **6**, 398 (1961).
851. G. R. Waller and L. M. Henderson, *J. Biol. Chem.*, **236**, 1186 (1961).
852. R. F. Dawson, D. R. Christman, A. F. D'Adamo, M. L. Solt, and A. P. Wolf, *J. Amer. Chem. Soc.*, **82**, 2628 (1960).
853. T. Griffith, K. P. Hellman, and R. U. Byerrun, *Biochemistry,* **1**, 336 (1962).
854. H. Tamir and D. Ginsburg, *J. Chem. Soc.*, 2921 (1959).
855. R. Mukherjee and A. Chatterjee, *Tetrahedron,* **22**, 1461 (1966).
856. C. G. Casinovi, G. Grandolini, R. Mercantini, N. Oddo, R. Olivieri, and A. Tonolo, *Tetrahedron Lett.*, 3175 (1968).
857. M. Yasue and N. Kawamura, *Chem. Pharm. Bull.* (Tokyo), **14**, 1443 (1966).
858. M. Yasue, N. Kawamura, and T. Kato, *Yakugaku Zasshi,* **87**, 732 (1967); *Chem. Abstr.*, **67**, 100382 (1967).
859. A. Chatterjee and C. P. Dutta, *Tetrahedron,* **23**, 1769 (1967).
860. B. S. Joshi, V. N. Kamat, and A. K. Saksena, *Tetrahedron Lett.*, 2395 (1968).
861. T. Endo, N. Otake, S. Takeuchi, and H. Yonehara, *J. Antibiot.* (Tokyo), *Ser. A,* **17**, 172 (1964).
862. H. Yonehara, S. Takeuchi, N. Otake, T. Endo, Y. Sakagami, and Y. Sumiki, *J. Antibiot.* (Tokyo), *Ser. A,* **16**, 195 (1963).

863. J. W. Hylin, *Phytochemistry,* **3**, 161 (1964).
864. E. Bianchi, C. Djerassi, H. Budzikiewicz, and Y. Sato, *J. Org. Chem.*, **30**, 754 (1965).
865. N. Takahashi, A. Suzuki, and S. Tamura, *J. Amer. Chem. Soc.*, 87, 2066 (1965).
866. M. Jeng, C. Hall, F. L. Crane, N. Takahashi, S. Tamura, and K. Folkers, *Biochemistry,* **7**, 1311 (1968).
867. N. Takahashi, A. Suzuki, S. Miyamoto, R. Mori, and S. Tamura, *Agr. Biol. Chem.* (Tokyo), **27**, 583 (1963).
868. N. Takahashi, A. Suzuki, and S. Tamura, *Agr. Biol. Chem.* (Tokyo), **27**, 798 (1963).
869. A. Suzuki, N. Takahashi, and S. Tamura, *Agr. Biol. Chem.* (Tokyo), **30**, 18 (1966).
870. N. Takahashi, S. Yoshida, A. Suzuki, and S. Tamura, *Agr. Biol. Chem.* (Tokyo), **32**, 1108 (1968).
871. N. Takahashi, A. Suzuki, Y. Kimura, S. Miyamoto, and S. Tamura, *Tetrahedron Lett.*, 1961 (1967).
872. N. Takahashi, A. Suzuki, Y. Kimura, S. Miyamoto, S. Tamura, T. Mitsui, and J. Fukami, *Agr. Biol. Chem.* (Tokyo), **32**, 1115 (1968).
873. N. Takahashi, Y. Kimura, and S. Tamura, *Tetrahedron Lett.*, 4659 (1968).
874. L. I. Hochstein and S. C. Rittenberg, *J. Biol. Chem.*, **234**, 151 (1959).
874a. L. I. Hochstein and S. C. Rittenberg, *J. Biol. Chem.*, **234**, 156 (1959).
875. L. I. Hochstein and S. C. Rittenberg, *J. Biol Chem.*, **235**, 795 (1960).
876. S. H. Richardson and S. C. Rittenberg, *J. Biol. Chem.*, **236**, 959 (1961).
877. S. H. Richardson and S. C. Rittenberg, *J. Biol. Chem.*, **236**, 964 (1961).
878. K. Deppe, *Zentr. Bakteriol. Parasitenk. Abt.*, II, **119**, 589 (1965); *Chem. Abstr.*, **64**, 14621 (1966).
879. K. Decker, H. Eberwein, F. A. Gries, and M. Brühmüller, *Z. Physiol. Chem.*, **319**, 279 (1960).
880. F. A. Gries, K. Decker, and M. Brühmüller, *Z. Physiol. Chem.*, **325**, 229 (1961).
881. G. D. Griffith, R. U. Byerrum, and W. A. Wood, *Proc. Soc. Exptl. Biol. Med.*, **108**, 162 (1961).
882. K. Decker, H. Eberwein, F. A. Gries, and M. Brühmüller, *Biochem. Z.*, **334**, 227 (1961).
883. K. Decker and H. Bleeg, *Biochim. Biophys. Acta,* **105**, 313 (1965).
884. K. Decker and V. D. Dai, *Eur. J. Biochem.*, **3**, 132 (1967); *Chem. Abstr.*, **68**, 46648 (1968).
885. M. L. Wu Chang and B. C. Johnson, *J. Nutr.*, **76**, 512 (1962).
886. J. C. Ensign, *Dissertation Abstr.*, **24**, 31 (1963); *Chem. Abstr.*, **59**, 15627 (1963).
887. J. C. Ensign and S. C. Rittenberg, *Arch. Mikrobiol.*, **47**, 137 (1963).
888. J. C. Ensign and S. C. Rittenberg, *Bact. Proc.*, 107 (1962).
889. R. L. Gherna and S. C. Rittenberg, *Bact. Proc.*, 107 (1962).
890. J. C. Ensign and S. C. Rittenberg, *Bact. Proc.*, 54 (1963).
891. R. L. Gherna, S. H. Richardson, and S. C. Rittenberg, *J. Biol. Chem.*, **240**, 3669 (1965).
892. J. C. Ensign and S. C. Rittenberg, *Arch. Mikrobiol.*, **51**, 384 (1965).
893. R. Kuhn, H. Bauer, H.-J. Knackmuss, D. A. Kuhn, and M. P. Starr, *Naturwissen.*, **51**, 409 (1964).
894. R. Kuhn, H. Bauer, and H.-J. Knackmuss, *Chem. Ber.*, **98**, 2139 (1965).
895. R. Kuhn, M. P. Starr, D. A. Kuhn, H. Bauer, and H.-J. Knackmuss, *Arch. Mikrobiol.*, **51**, 71 (1965).
896. L. J. Bellamy and P. E. Rogasch, *Proc. Roy. Soc., Ser. A,* **257**, 98 (1960).
897. D. A. Brent, J. D. Hribar, and D. C. DeJongh, *J. Org. Chem.*, **35**, 135 (1970).

898. N. A. Broisevich and N. N. Khovratovich, *Zh. Prikl. Spektrosk.*, 7, 538 (1967); *Chem. Abstr.*, **68**, 86651 (1968).
899. W. Brugel, *Z. Electrochem.*, **66**, 159 (1962); *Chem. Abstr.*, **57**, 307 (1962).
900. D. L. Gumprecht, *J. Chromatogr.*, **37**, 268 (1968).
901. D. Herbison-Evans and R. E. Richards, *Mol. Phys.*, **8**, 19 (1964).
902. R. Isaac, F. F. Bentley, H. Sternglanz, W. C. Coburn, Jr., C. V. Stephenson, and W. S. Wilcox, *Appl. Spectrosc.*, **17**, 90 (1963); *Chem. Abstr.*, **59**, 7077 (1963).
903. G. H. Keller, L. Bauer, and C. L. Bell, *J. Heterocycl. Chem.*, **5**, 647 (1968).
904. J. Kracmar, J. Kracmarova, and J. Zyka, *Cesk. Farm.*, **17**, 68 (1968); *Chem. Abstr.*, **69**, 5266 (1968).
905. J. Kracmar, J. Kracmarova, and J. Zyka, *Pharmazie*, **23**, 567 (1968); *Chem. Abstr.*, **70**, 31719 (1969).
906. J. S. Kwiatkowski and J. Olzacka, *Bull. Acad. Pol. Sci., Ser. Sci. Chim.*, **18**, 215 (1970).
907. R. Lawrence and E. S. Waight, *J. Chem. Soc., B*, 1 (1968).
908. A. Neszmelyi and J. Bayer, *Kem. Kozlem*, **32**, 221 (1969); *Chem. Abstr.*, **72**, 116593 (1970).
909. K. Nishimoto, *Bull. Chem. Soc. Jap.*, **39**, 645 (1966); *Chem. Abstr.*, **65**, 4833 (1966).
910. K. Nishimoto and L. S. Forster, *J. Phys. Chem.*, **71**, 409 (1967).
911. H.-J. Petrowitz, G. Pastuska, and S. Wagner, *Chem.-Ztg., Chem. App.*, **89**, 7 (1965); *Chem. Abstr.*, **62**, 11125 (1965).
912. R. M. Pinyazhko, *Farm. Zh.* (Kiev), **20**, 17 (1965); *Chem. Abstr.*, **64**, 9514 (1966).
913. G. N. Rodionova and E. S. Levin, *Dokl. Akad. Nauk SSSR*, **174**, 1132 (1967); *Chem. Abstr.*, **68**, 95190 (1968).
914. R. Soda, *Bull. Natl. Inst. Ind. Health*, **5**, 32 (1961); *Chem. Abstr.* **57**, 8864 (1962).
915. G. Spiteller and M. Spiteller-Friedmann, *Monatsh. Chem.*, **93**, 1395 (1962).
916. A. Veillard, *J. Chim. Phys.*, **59**, 1056 (1962).
917. V. R. Williams and J. G. Traynham, *J. Org. Chem.*, **28**, 2883 (1963).
918. H. W. Yurow, L. B. Morton, and S. Sass, *Microchem. J.*, **11**, 237 (1966); *Chem. Abstr.*, **65**, 7984 (1966).
919. B. E. Zaitsev and Y. N. Sheinker, *Izv. Akad. Nauk SSSR Otd., Khim. Nauk*, 2070 (1962); *Chem. Abstr.*, **58**, 6339 (1963).
920. R. M. Ismail, Fr. patent, 1,475,896 (1967); *Chem. Abstr.*, **68**, 13178 (1968).
921. Y. N. Sheinker and Y. I. Pomerantzev, *Zh. Fiz. Khim.*, **33**, 1819 (1959); *Chem. Abstr.*, **54**, 12156 (1960).
922. T. Koenig and J. Wieczorek, *J. Org. Chem.*, **35**, 508 (1970).
923. C. L. Bell, R. S. Egan, and L. Bauer, *J. Heterocycl. Chem.*, **2**, 420 (1965).
924. L. A. Paquette, U.S. patent, 3,318,899 (1967); *Chem. Abstr.*, **68**, 49579 (1968).
925. I. I. Grandberg, G. K. Faizova, and A. N. Kost, *Zh. Anal. Khim.*, **20**, 268 (1965); *Chem. Abstr.*, **62**, 15402 (1965).
926. W. Ried and E.-U. Kocher, *Ann. Chem.*, **647**, 116 (1961).
927. M. V. Lomonosov, K. Khua-min', and R. Y. Levina, *Zh. Obshch. Khim.*, **33**, 2829 (1963).
928. A. S. Bailey and J. S. A. Brunskill, *J. Chem. Soc.*, 2554 (1959).
929. E. Bullock, B. Gregory, and A. W. Johnson, *J. Chem. Soc.*, 1632 (1964).
930. R. F. Childs and A. W. Johnson, *Chem. Ind.* (London), 542 (1964).
931. B. Jaques and J. W. Hubbard, *J. Med. Chem.*, **11**, 178 (1968).
932. N. P. Shusherina, R. Y. Levina, and T. G. Rymareva, *Zh. Obshch. Khim.*, **32**, 89 (1962); *Chem. Abstr.*, **57**, 13716 (1962).
933. R. N. Seelye and D. W. Stanton, *Tetrahedron Lett.*, 2633 (1966).

934. Fr. Addn., 90, 356 (1967); *Chem. Abstr.*, **70**, 87584 (1969).
935. Fr. patent, 1,458,066 (1967); *Chem. Abstr.*, **70**, 68183 (1969).
936. E. Haack, H. Berger and W. Vomel, Br. patent, 1,014,050 (1965); *Chem. Abstr.*, **64**, 6624 (1966).
937. Nether. Appl. 6,613,362 (1967); *Chem. Abstr.*, **67**, 100,007 (1967).
938. C. F. H. Allen, *Can. J. Chem.*, **43**, 2486 (1965).
939. J. H. Snyders and M. J. Pieterse, *J. S. Afr. Chem. Inst.*, **22**, 37 (1969); *Chem. Abstr.*, **70**, 114,969 (1969).
940. T. Kametani, K. Ogasawara, M. Shio, and A. Kosuka, *Yakugaku Zasshi*, **87**, 260 (1967); *Chem. Abstr.*, **67**, 54317 (1967).
941. I. E. El-Kholy, F. K. Rafla, and M. M. Mishrikey, *J. Chem. Soc.*, *C*, 1578 (1970).
942. B. D. Sharma, *Acta Crystallogr.*, **20**, 921 (1966).
943. P. Tomasik and E. Plazek, *Rocz. Chem.*, **39**, 1671 (1965); *Chem. Abstr.*, **64**, 19549 (1966).
944. R. Adams and W. Reifschneider, *J. Amer. Chem. Soc.*, **81**, 2537 (1959).
945. F. Cuiban, S. Cilianu-Bibian, S. Popescu, and I. Rogozea, Fr. patent, 1,366,064 (1964); *Chem. Abstr.*, **61**, 14643 (1964).
946. J. C. Martin, U.S. patent, 3,141,880 (1964); *Chem. Abstr.*, **61**, 11977 (1964).
947. J. W. Streef and H. J. den Hertog, *Rec. Trav. Chim. Pays-Bas*, **85**, 803 (1966).
948. G. N. Smith, *J. Econ. Entomol.*, **61**, 793 (1968); *Chem. Abstr.*, **69**, 34964 (1968).
949. R. D. Chambers, J. Hutchinson, and W. K. R. Musgrave, Belg. patent, 660,873 (1965); *Chem. Abstr.*, **65**, 7152 (1966).
950. R. D. Chambers, J. Hutchinson, and W. K. R. Musgrave, Brit. patent, 1,107,882 (1968); *Chem. Abstr.*, **69**, 86821 (1968).
951. Imperial Chemical Industries, Ltd., Nether. Appl., 6,611,714 (1967); *Chem. Abstr.*, **68**, 21842 (1968).
952. I. Collins and H. Suschitzky, *J. Chem. Soc.*, *C*, 1523 (1970).
953. Imperial Chemical Industries, Ltd., Nether. Appl., 6,611,766 (1967); *Chem. Abstr.*, **68**, 59438 (1968).
954. W. J. Link, R. F. Borne, and F. L. Setliff, *J. Heterocycl. Chem.*, **4**, 641 (1967).
955. G. B. Barlin, *J. Chem. Soc.*, 2150 (1964).
956. P. Tomasik and Z. Skrowaczewska, *Rocz. Chem.*, **42**, 1427 (1968).
957. P. Tomasik and Z. Skrowaczewska, *Rocz. Chem.*, **42**, 795 (1968).
958. P. Tomasik, *Rocz. Chem.*, **44**, 341 (1970).
959. B. Glowiak, *Bull. Acad., Pol. Sci., Ser. Sci. Chim.*, **10**, 9 (1962); *Chem. Abstr.*, **58**, 501 (1963).
960. P. Tomasik and Z. Skrowaczewska, *Rocz. Chem.*, **42**, 1583 (1968).
961. B. Glowiak, *Zes. Nauk Politech. Wroclaw Chem.*, 7, 49 (1961); *Chem. Abstr.*, **58**, 498 (1963).
962. W. Reifschneider and J. S. Kelyman, U.S. patent, 3,335, 146 (1967); *Chem. Abstr.*, **68**, 87178 (1968).
963. J. K. Landquist, U.S. patent, 3,074,954 (1963); *Chem. Abstr.*, **59**, 635 (1963).
964. M. Berndt and J. S. Kwiatkowski, *Theor. Chim. Acta*, **17**, 35 (1970).
965. A. R. Katritzky and A. R. Hands, *J. Chem. Soc.*, 2202 (1958).
966. L. A. Paquette, U.S. patent, 3,218,329 (1965); *Chem. Abstr.*, **64**, 3495 (1966).
967. F. Cramer, Ger. patent, 1,203,777 (1965); *Chem. Abstr.*, **64**, 6625 (1966).
968. J. J. Eatough, L. S. Fuller, R. H. Good, and R. K. Smalley, *J. Chem. Soc.*, *C*, 1874 (1970).
969. J. W. Streef and H. J. den Hertog, *Tetrahedron Lett.*, 5945 (1968).
970. J. Büchi, S. Allisson, and W. Vetsch, *Helv. Chim. Acta*, **48**, 1216 (1965).
971. M. Melandri, V. Gerosa, and A. Buttini, *Ann. Chim.* (Rome), **50**, 125 (1960).

972. H. Johnston, U.S. patent, 3,249,619 (1966); *Chem. Abstr.*, **65**, 693 (1966).
973. R. H. Rigterink, Fr. patent, 1,360,901 (1964); *Chem. Abstr.*, **61**, 16052 (1964).
974. M. C. Ivey and H. V. Claborn, *J. Assoc. Offic. Anal. Chem.*, **51**, 1245 (1968).
975. T. A. Mikhailova, N. I. Kudryashova, and N. V. Khromov-Borisov, *Zh. Obshch. Khim.*, **39**, 26 (1960); *Chem. Abstr.*, **70**, 114460 (1969).
976. H. Johnston and M. S. Tomita, U.S. patent, 3,234,228 (1966); *Chem. Abstr.*, **64**, 14173 (1966).
977. H. V. Claborn, H. D. Mann, and D. D. Oehler, *J. Assoc. Offic. Anal. Chem.*, **51**, 1243 (1968).
978. R. H. Rigterink, U.S. patent, 3,385,859 (1968); *Chem. Abstr.*, **69**, 59105 (1968).
979. A. J. W. Brook and R. K. Robertson, *Chem. Ind.* (London), 2110 (1967).
980. E. M. Evleth, *Theor. Chim. Acta*, **11**, 145 (1968).
981. T. Gramstad and W. J. Fuglevik, *Acta Chem. Scand.*, **16**, 1369 (1962).
982. M. Holik, V. Skala, and J. Kuthan, *Collect. Czech. Chem. Commun.*, **33**, 394 (1968).
983. A. R. Katritzky, F. W. Maine, and S. Golding, *Tetrahedron*, **21**, 1693 (1965).
984. M. Ohtsuru, K. Tori, and H. Wantanabe, *Chem. Pharm. Bull.* (Toyko), **15**, 1015 (1967); *Chem. Abstr.*, **67**, 112621 (1967).
985. D. W. Turner, *J. Chem. Soc.*, 847 (1962).
986. D. Cook, *Chem. Ind.* (London), 1259 (1964).
987. B. I. Mikhant'ev and E. I. Fedorov, *Izv. Vyssh. Ucheb. Zaved., Khim. Khim. Tekhnol.*, **2**, 390 (1959); *Chem. Abstr.*, **54**, 4565 (1960).
988. A. Sacha, *Acta Pol. Pharm.*, **22**, 516 (1965); *Chem. Abstr.*, **64**, 9723 (1966).
989. A. Sacha, Polish patent, 51,367 (1966); *Chem. Abstr.*, **67**, 73526 (1967).
990. Upjohn Co., Nether. Appl., 6,505,770 (1965); *Chem. Abstr.*, **64**, 11185 (1966).
991. L. A. Paquette, U.S. patent, 3,432,507 (1969); *Chem. Abstr.*, **70**, 96834 (1969).
992. K. Hamamoto and T. Kajiwara, Jap. patent, 13,985 (1964); *Chem. Abstr.*, **65**, 20108 (1966).
993. W. Sauermilch, *Arch. Pharm.* (Weinheim), **293**, 452 (1960).
994. T. Kajiwara, Jap. patent, 20,295 (1967); *Chem. Abstr.*, **69**, 27265 (1968).
995. L. A. Paquette, *J. Amer. Chem. Soc.*, **87**, 5186 (1965).
996. E. C. Taylor, F. Kienzle, and A. McKillop, *J. Org. Chem.*, **35**, 1672 (1970).
997. L. A. Paquette, *J. Amer. Chem. Soc.*, **87**, 1407 (1965).
998. H. W. Chambers and J. E. Casida, *Toxicol. Appl. Pharmacol.*, **14**, 249 (1969).
999. B. B. Greene and K. G. Lewis, *Tetrahedron Lett.*, 4759 (1966).
1000. D. E. Ames and T. F. Grey, *J. Chem. Soc.*, 631 (1955).
1001. B. B. Greene and K. G. Lewis, *Aust. J. Chem.*, **21**, 1845 (1968).
1002. L. A. Paquette, U.S. patent, 3,213,101 (1965); *Chem. Abstr.*, **64**, 2063 (1966).
1003. L. A. Paquette, *Tetrahedron*, **22**, 25 (1966).
1004. P. H. Schroeder, S. Afr. patent, 07,133 (1967); *Chem. Abstr.*, **70**, 47,309 (1969).
1005. N. S. Vul'fson and G. M. Sukhotina, *Metody Poluch Khim., Reactiovov, Prep.*, 83 (1966); *Chem. Abstr.*, **67**, 32552 (1967).
1006. L. A. Paquette, U.S. patent, 3,213,100 (1965); *Chem. Abstr.*, **64**, 2069 (1966).
1007. S. Mizukami, E. Hirai, and M. Morimoto, *Shionogi Kenkyiesho Nempo (Shionogi Research Lab) Annual Report*, **16**, 29 (1966); *Chem. Abstr.*, **66**, 10827 (1967).
1008. H. Tomisawa, *Yakugaku Zasshi*, **79**, 1167 (1959); *Chem. Abstr.*, **54**, 3416 (1960).
1009. S. Grudzinski, *Rocz. Chem.*, **40**, 335 (1966); *Chem. Abstr.*, **65**, 681 (1966).
1010. J. Bernstein and K. A. Losee, U.S. patent, 3,356,683 (1967); *Chem. Abstr.*, **68**, 105013 (1968).
1011. H. Tomisawa, *Yakugaku Zasshi*, **79**, 1173 (1959); *Chem. Abstr.*, **54**, 3417 (1960).
1012. T. Kajiwara, Jap. patent, 20,294 (1967); *Chem. Abstr.*, **69**, 27264 (1968).

1013. G. M. Coppinger and E. R. Bell, *J. Phys. Chem.*, **70**, 3479 (1966).
1014. E. Bernasek, U.S. patent, 3,230,226 (1966); *Chem. Abstr.*, **64**, 8256 (1966).
1015. L. A. Paquette, U.S. patent, 3,321,482 (1967); *Chem. Abstr.*, **68**, 49577 (1968).
1016. American Cyanamid Co., Nether. Appl., 6,612,653 (1967); *Chem. Abstr.*, **68**, 21847 (1968).
1017. B. Gogolimska and I. Gruda, *Acta Pol. Pharm.*, **20**, 130 (1963); *Chem. Abstr.*, **62**, 1584 (1965).
1018. B. Gogolimska, H. Bojarska-Dahlig, I. Gruda, A. Raczka, and A. Swirska, Pol. patent, 48,708 (1964); *Chem. Abstr.*, **63**, 18,041 (1965).
1019. J. F. Willems, A. L. VandenBerghe, and G. F. VanVeelen, Belg. patent, 706,086 (1968); *Chem. Abstr.*, **70**, 47,303 (1969).
1020. A. Sacha, Pol. patent, 55,633 (1968); *Chem. Abstr.*, **70**, 106,557 (1969).
1021. K. Hamamoto and T. Kajiwara, Jap. patent, 20,556 (1965); *Chem. Abstr.*, **64**, 3505 (1966).
1022. H. Fugimura, K. Okamoto, and M. Tetsuo, Jap. patent, 19,649 (1964); *Chem. Abstr.*, **62**, 13,129 (1965).
1023. J. N. Ashley, R. F. Collins, M. Davis, and N. E. Sirett, *J. Chem. Soc.*, 3880 (1959).
1024. A. Sacha, Pol. patent, 52,123 (1966); *Chem. Abstr.*, **68**, 114,650 (1968).
1025. J. Faust, G. Speier, and R. Mayer, *J. Prakt. Chem.*, **311**, 61 (1969).
1026. G. Wagner and H. Pischel, *Z. Chem.*, **2**, 308 (1962).
1027. J. K. Sutherland and D. A. Widdowson, *J. Chem. Soc.*, 4651 (1964).
1028. V. Tortorella, F. Macioci, and G. Poma, *Farmaco, Ed. Sci.*, **23**, 236 (1968); *Chem. Abstr.*, **69**, 59,058 (1968).
1029. H. Tomisawa and H. Hongo, *Tohoku Yakka Daigaku Kenkyo Nempo*, 39 (1963); *Chem. Abstr.*, **61**, 4308 (1964).
1030. T. Kato and Y. Kubota, *Yakugaku Zasshi*, 87, 1212 (1967); *Chem. Abstr.*, **68**, 95,644 (1968).
1031. N. S. Vul'fson and G. M. Sukhotina, *Khim. Geterotsikl. Soedin.*, **3**, 682 (1967); *Chem. Abstr.*, **68**, 39,432 (1968).
1032. M. Eder, E. Ziegler, and E. Prewedourakis, *Monatsh. Chem.*, **99**, 1395 (1968).
1033. H. Wittmann, V. Illi, and E. Ziegler, *Monatsh. Chem.*, **98**, 1108 (1967).
1034. A. M. Duffield, C. Djerassi, G. Schroll, and S.-O. Lawesson, *Acta Chem. Scand.*, **20**, 361 (1966).
1035. M. Hamana and K. Funakoshi, *Yakugaku Zasshi*, **84**, 23 (1964); *Chem. Abstr.*, **61**, 3068 (1964).
1036. N. Clauson-Kaas and N. Elming, Ger. patent, 1,134,378 (1962); *Chem. Abstr.*, **58**, 1438 (1963).
1037. B. F. Duesel, and S. Emanuele, U.S. patent, 3,218,330 (1965); *Chem. Abstr.*, **64**, 3503 (1966).
1038. N. Elming, S. V. Carlsten, B. Lennart, and I. Ohlsson, Brit. patent, 862,581 (1961); *Chem. Abstr.*, **56**, 11574 (1962).
1039. A. D. Filyugina, K. M. Dyumaev, A. A. Dubrovin, and I. A. Rotermel, *Zh. Organ. Khim.*, **6**, 2131 (1970); *Chem. Abstr.*, **74**, 12306 (1971).
1040. L. Fishbein and M. A. Cavanaugh, *J. Chromatogr.*, **20**, 283 (1965).
1041. E. J. Poziomek, *Chemist-Analyst*, **55**, 78 (1966).
1042. Y. Sasaki, M. Hatanaka, I. Shiraishi, M. Suzuki, and K. Nishimoto, *Yakugaku Zasshi*, **89**, 21 (1969); *Chem. Abstr.*, **70**, 92,138 (1969).
1043. I. Schmeltz and R. L. Stedman, *Chem. Ind.* (London), 1244 (1962).
1044. P. W. Wigler and L. E. Wilson, *Anal. Biochem.*, **15**, 421 (1966).
1045. F. J. Anderson and A. E. Martell, *J. Amer. Chem. Soc.*, **86**, 715 (1964).

1046. T. Naito, T. Yoshikawa, F. Ishikawa, and H. Omura, Jap. patent, 22,740 (1965); *Chem. Abstr.*, **64**, 3496 (1966).
1047. A. Fujita, M. Nakata, S. Minami, and H. Takamatsu, *Yakugaku Zasshi*, **86**, 1014 (1966); *Chem. Abstr.*, **66**, 75885 (1967).
1048. W. Korytnyk and R. P. Singh, *J. Amer. Chem. Soc.*, **85**, 2813 (1963).
1049. W. Korytnyk, *J. Med. Chem.*, **8**, 112 (1965).
1050. G. E. McCasland, L. K. Gottwald, and A. Furst, *J. Org. Chem.*, **26**, 3541 (1961).
1051. H. Monig and R. Koch, *Biophysik.*, **3**, 11 (1966).
1052. L. A. Petrova and N. N. Bel'tsova, *Zh. Obshch. Khim.*, **34**, 2765 (1964); *Chem. Abstr.*, **61**, 14634 (1964).
1053. W. Korytnyk and M. Ikawa, *Methods in Enzymology*, **18A**, 524 (1970).
1054. T. Naito and K. Ueno, *Yakugaku Zasshi*, **79**, 1277 (1959); *Chem. Abstr.*, **54**, 4566 (1960).
1055. R. Koch, H. Lagendorff, and H. Monig, *Atomkernenergie*, **11**, 209 (1966); *Chem. Abstr.*, **65**, 9302 (1966).
1056. O. Kondo, Y. Miyazaki, T. Maruyama, and M. Ikariwa, Jap. patent, 2,356 (1968); *Chem. Abstr.*, **69**, 67232 (1968).
1057. K. Pfister, III, E. E. Harris, and R. A. Firestone, U.S. patent, 3,227,721 (1966); *Chem. Abstr.*, **64**, 9689 (1966).
1058. K. Pfister, III, E. E. Harris, and R. A. Firestone, U.S. patent, 3,227,722 (1966); *Chem. Abstr.*, **65**, 16949 (1966).
1059. C. J. Argoudelis and F. A. Kummerow, *J. Org. Chem.*, **29**, 2663 (1964).
1060. G. Schorre, Ger. patent, 1,238,473 (1967); *Chem. Abstr.*, **68**, 39479 (1968).
1061. T. Kuroda, *Bitamin*, **28**, 211 (1963); *Chem. Abstr.*, **62**, 515 (1965).
1062. K. Okumura, *Bitamin*, **23**, 236 (1961); *Chem. Abstr.*, **62**, 1630 (1965).
1063. U. Schmidt and G. Geisselmann, *Ann. Chem.*, **657**, 162 (1962).
1064. Tanabe Seiyaku Co., Ltd., Belg. patent, 671,385 (1966); *Chem. Abstr.*, **65**, 15348 (1966).
1065. W. Korytnyk and B. Paul., *J. Med. Chem.*, **13**, 187 (1970).
1066. P. D. Sattsangi and C. J. Argoudelis, *J. Org. Chem.*, **33**, 1337 (1968).
1067. C. F. Boehringer and Soehne, G. m.b.H., Nether. Appl., 6,402,960 (1964); *Chem. Abstr.*, **62**, 7732 (1965).
1068. K. Undheim and V. Nordal, *Acta Chem. Scand.*, **23**, 1975 (1969).
1069. K. Nakamura, Y. Utsui, and Y. Ninomija, *Yakugaku Zasshi*, **86**, 404 (1966); *Chem. Abstr.*, **65**, 18429 (1966).
1070. J. L. Greene, A. M. Williams, and J. A. Montgomery, *J. Med. Chem.*, **7**, 20 (1964).
1071. L. A. Walter and N. Sperber, U.S. patent, 2,997,478 (1958); *Chem. Abstr.*, **56**, 1434 (1962).
1072. T. Kuroda, *Bitamin*, **28**, 354 (1963); *Chem. Abstr.*, **62**, 515 (1965).
1073. R. J. Highet and P. F. Highet, *J. Org. Chem.*, **31**, 1275 (1966).
1074. P. L. Kumler and O. Buchardt, *Chem. Commun.*, 1321 (1968).
1075. J. W. Cusic, P. Yonan, U.S. patent, 3,377,344 (1968); *Chem. Abstr.*, **69**, 52172 (1968).
1076. N. Clauson-Kaas, F. Ostermayer, E. Renk, and R. Denss, Swiss patent, 452,528 (1968); *Chem. Abstr.*, **69**, 96746 (1968).
1077. D. Heinert and A. E. Martell, *J. Amer. Chem. Soc.*, **84**, 3257 (1962).
1078. R. Tomchick and J. A. R. Mead, *Biochemical Medicine*, **4**, 13 (1970).
1079. T. Naito, T. Yoshikawa, F. Ishikawa, and H. Omura, Jap. patent, 23908 (1965); *Chem. Abstr.*, **64**, 3495 (1966).
1080. C. Iwata and D. E. Metzler, *J. Heterocycl. Chem.*, **4**, 319 (1967).
1081. E. H. Abbott and A. E. Martell, *J. Amer. Chem. Soc.*, **92**, 1754 (1970).

1082. Y. Kobayashi and K. Makino, *Biochim. Biophys. Acta,* **208**, 137 (1970).
1083. K. Okumura, Jap. patent, 02,716 (1968); *Chem. Abstr.,* **69**, 59107 (1968).
1084. W. B. Dempsey and E. E. Snell, *Biochemistry,* **2**, 1414 (1963).
1085. R. C. DeSelms, U.S. patent, 3,409,624 (1968); *Chem. Abstr.,* **70**, 57665 (1969).
1086. R. C. DeSelms, U.S. patent, 3,409,630 (1968); *Chem. Abstr.,* **70**, 57663 (1969).
1087. U. Schmidt, U.S. patent, 3,075,987 (1963); *Chem. Abstr.,* **59**, 9999 (1963).
1088. G. Schorre, Ger. patent, 1,238,474 (1967); *Chem. Abstr.,* **68**, 39480 (1968).
1089. M. Iwanami, I. Osawa, and M. Nurakami, *Bitamin,* **36**, 122 (1967); *Chem. Abstr.,* **68**, 12817 (1968).
1090. T. Kuroda, *Bitamin,* **30**, 431 (1964); *Chem. Abstr.,* **62**, 10402 (1965).
1091. T. Kuroda and M. Masaki, *Bitamin,* **30**, 436 (1964); *Chem. Abstr.,* **62**, 10402 (1965).
1092. T. Kuroda, *Bitamin,* **28**, 21 (1963); *Chem. Abstr.,* **62**, 515 (1965).
1093. E. Merck A-G., Fr. patent, M2071 (1963); *Chem. Abstr.,* **60**, 10656 (1964).
1094. H. Hirano, K. Masuda, and T. Fushimi, Jap. patent, 02,712 (1968); *Chem. Abstr.,* **69**, 86827 (1968).
1095. M. Murakami and M. Iwanami, Jap. patent, 19,067 (1966); *Chem. Abstr.,* **66**, 37776 (1967).
1096. I. Utsumi, T. Watanabe, and G. Tsukamoto, *Bitamin,* **37**, 276 (1968); *Chem. Abstr.,* **69**, 36065 (1968).
1097. G. Neurath and W. Lüttich, *J. Chromatogr.,* **34**, 253 (1968).
1098. W. Zecher and F. Kröhnke, *Chem. Ber.,* **94**, 707 (1961).
1099. C. Corral and A. M. Municio, *Anales Real. Soc. Espan. Fis. Quim., Ser. B.,* **60**, 341 (1964); *Chem. Abstr.,* **62**, 7670 (1965).
1100. G. Wilbert and H. Wetstein, U.S. patent, 3,284,459 (1966); *Chem. Abstr.,* **68**, 49463 (1968).
1101. R. Ponci, A. Baruffini, and F. Gialdi, *Farmaco, Ed. Sci.,* **18**, 288 (1963); *Chem. Abstr.,* **59**, 7478 (1963).
1102. B. D. Batts and E. Spinner, *J. Chem. Soc., B,* 789 (1968).
1103. B. D. Batts and E. Spinner, *Aust. J. Chem.,* **22**, 2595 (1969).
1104. J. P. Wibaut and F. W. Broekman, *Rec. Trav. Chim. Pays-Bas,* **78**, 593 (1959).
1105. Y. Takahasi, T. Kato, and M. Kurihara, Jap. patent, 00,055 (1969); *Chem. Abstr.,* **70**, 114853 (1969).
1106. E. Hayashi and H. Yamanaka, Jap. patent, 15,616 (1961); *Chem. Abstr.,* **56**, 12863 (1962).
1107. D. G. Markees, *J. Org. Chem.,* **29**, 3120 (1964).
1108. H. D. Becker, *Acta Chem. Scand.,* **16**, 78 (1962).
1109. A. Sugii, Y. Kabasawa, Y. Niki, and Y. Yamazaki, *Yakugaku Zasshi,* **89**, 1066 (1969); *Chem. Abstr.,* **71**, 105251 (1969).
1110. M. J. Huraux and H. M. Lawson, *Symp. New Herbicides, 2nd, Paris,* 269 (1965); *Chem. Abstr.,* **65**, 12799 (1966).
1111. R. N. Haszeldine and R. E. Banks, U.S. patent, 3,317,542 (1967); *Chem. Abstr.,* **68**, 59442 (1968).
1112. J. Bonham, E. McLeister, and P. Beak, *J. Org. Chem.,* **32**, 639 (1967).
1113. K. Hamamoto and T. Kajiwara, Jap. patent, 23,180 (1965); *Chem. Abstr.,* **64**, 3504 (1966).
1114. S. Kajihawa, *Nippon Kagaku Zasshi,* **85**, 672 (1964); *Chem. Abstr.,* **62**, 14624 (1965).
1115. I. Christenson, *Acta Pharm. Suecica,* **5**, 23 (1968); *Chem. Abstr.,* **68**, 89904 (1968).
1116. K. H. Boltze, H. D. Dell, H. Lehwald, D. Lorenz, and M. Rueberg-Schweer, *Arzeimittel-Forsch.,* **13**, 688 (1963); *Chem. Abstr.,* **63**, 4289 (1965).

1117. S. Goto, A. Kono, and S. Iguchi, *J. Pharm. Sci.*, **57**, 791 (1968).
1118. T. Metler, A. Uchida, and S. I. Miller, *Tetrahedron*, **24**, 4285 (1968).
1119. C. P. Krimmel, U.S. patent, 2,999,860 (1961); *Chem. Abstr.*, **56**, 4780 (1962).
1120. K. Blazevic and N. Trinajstic, *Croatica Chem. Acta*, **39**, 25 (1967); *Chem. Abstr.*, **67**, 69116 (1967).
1121. K. H. Schuendehuette and K. Trautner, Fr. patent, 1,412,980 (1965); *Chem. Abstr.*, **64**, 8360 (1966).
1122. E. Habicht, U.S. patent, 2,995,561 (1959).
1123. E. Habicht and R. Zubiani, Belg. patent, 654,331 (1965); *Chem. Abstr.*, **64**, 17553 (1966).
1124. A. Nakamura and S. Kamiya, *Chem. Pharm. Bull.* (Tokyo), **17**, 425 (1969).
1125. A. P. Chatrousse, F. Terrier, and R. Schaal, *C. R. Acad. Sci., Paris, Ser. C.*, **271**, 1477 (1970).
1126. G. Stork and R. Matthews, *Chem. Commun.*, 445 (1970).
1127. R. J. Sundberg, P. A. Bukowick, and F. O. Holcombe, *J. Org. Chem.*, **32**, 2938 (1967).
1128. A. J. Verbiscar and K. N. Campbell, *J. Org. Chem.*, **29**, 2472 (1964).
1129. G. DiMaio and P. A. Tardella, *Gazz. Chim. Ital.*, **96**, 387 (1966).
1130. E. Cavalieri and D. Gravel, *Can. J. Chem.*, **48**, 2727 (1970).
1131. C. H. Eugster, L. Leichner, and E. Jenny, *Helv. Chim. Acta*, **46**, 543 (1963).
1132. H. Kuehnis, R. Denss, and C. H. Eugster, Swiss patent, 417,591 (1967); *Chem. Abstr.*, **68**, 39482 (1968).
1133. D. G. Cheesman, P. Garside, A. C. Ritchie, and J. M. Waring, *J. Chem. Soc., C*, 1134 (1967).
1134. J. C. Sheehan and M. M. Nagissu, *J. Org. Chem.*, **35**, 4246 (1970).
1135. K. Hohenlohe-Oehringen and G. Zimmer, *Monatsh. Chem.*, **94**, 1225 (1963).
1136. K. Kariyone, *Yakugaku Zasshi*, **83**, 398 (1963); *Chem. Abstr.*, **59**, 7481 (1963).
1137. N. Sugiyama, M. Yamamoto, and C. Kashima, *Bull. Chem. Soc. Jap.*, **43**, 3937 (1970).
1138. N. Sugiyama, M. Yamamoto, and C. Kashima, *Bull. Chem. Soc. Jap.*, **43**, 3556 (1970).
1139. P. Nuhn and W.-R. Bley, *Tetrahedron Lett.*, 611 (1970).
1140. H. K. Reimschuessel, K. P. Klein, and G. J. Schmitt, *Makrolmol. Chem.*, **2**, 567 (1969).
1141. K. Hohenlohe-Oehringen, *Monatsh. Chem.*, **96**, 262 (1965).
1142. P. A. Ratto and A. A. Liebman, *J. Pharm. Sci.*, **53**, 480 (1964).
1143. A. Chatterjee and C. P. Dutta, *Sci. Cult.* (Calcutta), **29**, 568 (1963).
1144. A. R. Battersby, Brit. patent, 895,910 (1962); *Chem. Abstr.*, **58**, 6878 (1963).
1145. S. Portnoy, *J. Heterocycl. Chem.*, **3**, 363 (1966).
1146. R. G. Glushkov and O. Y. Magidson, *Khim. Geterotsikl. Soedin. Akad. Nauk Latv. SSR*, 240 (1965); *Chem. Abstr.*, **63**, 13259 (1965).
1147. Dynamit-Nobel, A-G., Belg. patent, 635,427 (1963); *Chem. Abstr.*, **61**, 11977 (1964).
1148. H. Ulrich, E. Kober, H. Schroeder, R. Rätz, and G. Grundmann, *J. Org. Chem.*, **27**, 2585 (1962).
1149. H. H. Kuehnis, R. Denss, and C. H. Eugster, U.S. patent, 3,073,838 (1963); *Chem. Abstr.*, **58**, 12519 (1963).
1150. R. E. Lyle, K. R. Carle, C. R. Ellefson, and C. K. Spicer, *J. Org. Chem.*, **35**, 802 (1970).
1151. R. E. Lyle and W. E. Krueger, *J. Org. Chem.*, **32**, 2873 (1967).

1152. R. F. C. Brown, L. Subrahmanyan, and C. P. Whittle, *Aust. J. Chem.*, **20**, 339 (1967).
1153. C.-T. Chen, S.-J. Yan, and J.-C. Ho, *Bull. Inst. Chem. Acad. Sinica,* **13**, 49 (1967); *Chem. Abstr.*, **67**, 118073 (1967).
1154. E. Dirks, A. Scherer, M. Schmidt, and G. Zimmer, *Arzneimittel-Forsch.*, **20**, 55 (1970); *Chem. Abstr.*, **73**, 2338 (1970).
1155. J. E. Ayling, E. E. Snell, *Biochemistry,* **7**, 1626 (1968).
1156. C. Holstead, Belg. patent, 619,423 (1962); *Chem. Abstr.*, **59**, 12826 (1963).
1157. W. Seiffert and H. H. Mantsch, *Tetrahedron,* **25**, 4569 (1969).
1158. F. M. Hershenson and L. Bauer, *J. Org. Chem.*, **34**, 660 (1969).
1159. G. Kresze, G. Schulz, and F. Firl, *Angew. Chem.*, **75**, 375 (1963).
1160. H. Krimm, Ger. patent 1,092, 919 (1956).
1161. J. J. Vill, T. R. Steadman, and J. J. Godfrey, *J. Org. Chem.*, **29**, 2780 (1964).
1162. N. P. Shusherina, R. Ya. Levina, and K. Khua-min', *Zh. Obshch. Khim.*, **32**, 3599 (1962); *Chem. Abstr.*, **58**, 12507 (1963).
1163. N. P. Shusherina, R. Ya. Levina, and Z. S. Sidenko, *Zh. Obshch. Khim.*, **29**, 398 (1959); *Chem. Abstr.*, **54**, 519 (1960).
1164. A. Vigier and J. Dreux, *Bull. Soc. Chim. Fr.*, 2292 (1963); *Chem. Abstr.*, **60**, 5448 (1964).
1165. L. G. Duquette and F. Johnson, *Tetrahedron,* **23**, 4517 (1967).
1166. F. Zymalkowski and P. Messinger, *Arch. Pharm.* (Weinheim), **300**, 91 (1967).
1167. R. E. Lyle and W. E. Krueger, *J. Org. Chem.*, **32**, 3613 (1967).
1168. R. F. C. Brown, V. M. Clark, and L. Todd, *Tetrahedron, Supplement 8 Part 1,* 15 (1966).
1169. E. Winterfeldt, H. Radung, and P. Strehlke, *Chem. Ber.*, **99**, 3750 (1966).
1170. M. Takeda, A. E. Jacobson, and E. L. May, *J. Org. Chem.*, **34**, 4158 (1969).

# Index

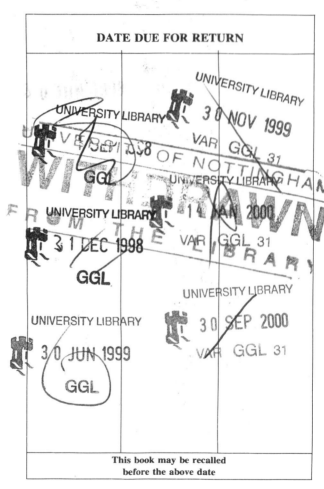